D0937233

OCEANOGRAPHIC INDEX

Subject Cumulation 1946-1971

Compiled by

Dr. Mary Sears

of the

Woods Hole Oceanographic Institution

Volume 2

Ci – Instrumentation H

G. K. HALL & CO., 70 LINCOLN STREET, BOSTON, MASSACHUSETTS

1972

This publication is printed on permanent/durable acid-free paper.

ISBN 0-8161-0932-X

Ciguatera

ciguatera
Bagnis, R., 1967.
Contribution a l'étude de l'ichtyotoxisme en Polynésie française.
Rev. intern. Océanogr. Med., 6/7:89-110.

Ciguatera
Hashimoto, Yoshiro, Shoji Konosu, Takeshi Yasumoto, 1969.
Ciguatera in the Ryukyu and Amami islands.
Bull. Jap. Soc. scient. Fish., 35 (3): 316-326

ciguatera
*Morelon, R., et P. Niaussat, 1967.
Ciguatera et ichtyosarcotoxisme.
Cah. Pacif. (10):7-41.

ciguatera
Morton, Rose Ann, and Mary Ann Burklew 1970.
Incidence of ciguatera in barracuda from the West Coast of Florida.
Toxicon 8: 317-318

ciguatoxin, effect of
Ehrhardt, J.-P. 1971.
État actuel des expérimentations en France sur les extraits ciguatoxiques.
Rev. int. Océanogr. Méd. 22-23: 151-164.

ciguatoxin, effect of
Ohshika, Hideyo 1971.
Marine toxins from the Pacific IX. Some effects of ciguatoxin on isolated mammalian atria.
Toxicon 9(4): 337-343

ciguatoxin
*Rayner, Martin D., Thomas I. Kosaki and Enid L. Fellmeth, 1968.
Ciguatoxin: more than an anticholinesterase.
Science, 160(3823):70-71.

ciguatoxin
Rayner, Martin D., Morris H. Baslow and Thomas I. Kosaki, 1969.
Marine toxins from the Pacific - Ciguatoxin: not an in vivo anticholinesterase.
J. Fish. Res. Bd, Can., 26 (8): 2208-2210

Cilia

cilia
Horridge, G.A., 1965.
Non-motile sensory cilia and neuromuscular junctions in a ctenophore independent effector organ.
Proc. R. Soc., London, (B), 162(988):333-350.

macrocilia
Horridge, G.A., 1965.
Macrocilia with numerous shafts from the lips of the ctenophore Beroe.
Proc. R. Soc., London, (B), 162(988):351-364.

Circadian

Circadian
Wurtman, Richard J., 1967.
Ambiguities in the use of the term circadian.
Science, 156(3771):104.

Circulation

A

circulation
Aagard, K., and L. K. Coachman, 1964.
Notes on the physical oceanography of the Chukchi Sea.
U.S.C.G. Oceanogr. Rept., No. 1(CG373-1):13-16.

circulation
Anonymous, 1951.
Circulation of water in the oceans. Nature 167: 543-545.

circulation
Arakawa, H., 1954.
On the maintenance of ocean circulation.
J. Ocean. Soc., Japan, 9(3/4):125-130.

circulation, oceanic
Arakawa, H. and S. Ooma, 1935
On the general circulation of the ocean. Geophys. Mag. 9:83-104, J. Met. Soc. Jap., 12:595-598. (In Japanese)

circulation
Asaoka, Osamu, and Shigeo Moriyasu 1966
On the circulation in the East China Sea and the Yellow Sea in winter (preliminary report).
Oceanogr. Mag., Jap. Met. Soc., 18(1/2): 73-81.

circulation
Arseniev, V.S., 1965.
Circulation in the Bering Sea. (In Russian; English abstract).
Okeanolog. Issled., Rezult. Issled. Programme Mezhd. Geofiz. Goda, Mezhd. Geofiz. Komitet, Presidiume, Akad. Nauk, SSSR, No.13:61-65.

B

circulation
Baczyk, Joseph, 1966.
La differenciation des masses d'eaux leurs mouvements et les influence atmospheriques dans la Baltique meridionale.
Cahiers Oceanogr., 18(9):775-799.

circulation
Baker, D. James, Jr., 1971.
A source-sink laboratory model of the ocean circulation.
Geophys. fluid Dyn. 2(1): 17-29

circulation
Baker, D. James, Jr. 1970
Models of oceanic circulation
Scient. Am. 222(1):114-121

circulation
Baker, D.J., Jr., and A.R. Robinson, 1969.
A laboratory model for the general ocean circulation.
Phil. Trans. R. Soc. (A) 265 (1168): 533-566

circulation
Balech, E., 1965.
Nuevas contribuciones a los esquemas de circulación oceánica frente a la Argentina.
Anais Acad. bras. Cienc., 37(Supl.):159-166.

circulation
Barnes, C.A., D.F. Bumpus, and J. Lyman, 1948
Ocean circulation in Marshall Islands area. Trans. Am. Geophys. Union, 29(6):871-876, 7 text figs.

circulation
Barnes, Clifford A., and R. G. Paquette, 1957
Circulation near the Washington coast. Proc. Eighth Sci. Congr. Oceanogr. 3: 585-608.

circulation
Berezkin, Vs. A., 1937.
[Warming in the Arctic and increased circulation of the waters of the polar basin.]
Morskoi sbornik 4:105-132.

Translated by David Kraus, A.M.S.

circulation
Berrill, N. J., 1949.
The spin of the sea. Nat. Hist., N.Y. 58(6): 270-274.

Circulation
Bessonov, N.M., and V.N. Kochikov 1967.
On some variable oceanographic conditions off the productive regions of Dakar and Takograd. (In Russian)
Atlant. nauchno-issled. Inst. rybn. Khoz. okeanogr. (AtlantNIRO). Materialy Konferentsii po Resul'tatam Okeanologischeskikh Issledovanii v Atlanticheskom Okeane, 94-100.

circulation
Bigelow, H.B., and M. Leslie, 1930
Reconnaissance of the waters and plankton of Monterey Bay, July 1928.
Bull. M.C.Z., 70(5):429-481, 43 text figs.

circulation
Birchfield, G.E., 1969.
Response of a circular model Great Lake to a suddenly imposed wind stress. J. geophys. Res., 74(23): 5547-5554.

circulation
Blumsack, Steven L., 1972.
The traverse circulation near a coast. J. phys. Oceanogr, 2(1): 34-40.

circulation
Bogdanov, M.A., 1965.
On the circulation changes in the atmosphere and hydrosphere. (In Russian).
Trudy, vses. nauchno-issled. Inst. morsk. ryb. Khoz. Okeanogr. (VNIRO), 57:43-52.

Circulation
Bogdanova, A. K., and L. N. Kropachev, 1959.
[Circulation and its role in the hydrological regime of the Black Sea.]
Meteorol. i. Gidiol. (4):26-33.

circulation
*Bogurtsev, B.N., 1968.
Surface and deep circulation of the Antarctic waters of the Pacific Ocean. (In Russian; English abstract).
Fisika Atmosfer. Okean., Izv. Akad. Nauk, SSSR, 4(10): 1070-1085.

circulation

Bolgurtsev, B.N., 1971.
On some specific features of water circulation in the south Pacific. (In Russian; English abstract). Okeanologiia 11(6): 1008-1015.

circulation

Bolshakov, V.S., D.M. Tolmazin and M. Sh. Rozengurt, 1964.
On the horizontal circulation in the Black Sea. (In Russian).
Izv., Akad. Nauk, SSSR, Ser. Geofiz., (6):924-929.

circulation

Bowden, K.F., 1960?
Circulation and mixing in the Mersey estuary. I.A.S.H., Commission of Surface Waters, No. 51: 352-360.

circulation

Bowden, K. F., 1959.
Quelque études recentes sur la circulation océanique. Cahiers Océanographiques (XI année) No. 1, p. 37-43
Translation by B. Saint-Guily from original article in J. du Cons. Int. pour l'Expl. de la Mer vol. 23 No. 3 1958.

circulation

Bowden, K.F., 1958.
Some recent studies of oceanic circulation. Essay Review. J. du Cons., 23(3):453-461.

circulation

Bowden, K.F., and S.H. Sharaf El Din, 1966.
Circulation, salinity and river discharge in the Mersey Estuary.
Geophys. J., R. astr. Soc., 10(4): 383-399.

circulation

Brodie, J.W., 1965.
Oceanography.
In: Antarctica, Trevor Hatherton, editor, Methuen & Co., Ltd., 101-127.

circulation

Bryan, Kirk, 1969.
Climate and the ocean circulation. III. The ocean model.
Mon. Wea. Rev. 97(11): 806-834

circulation

*Bryan, Kirk, and Michael D. Cox, 1968.
A nonlinear model of an ocean driven by wind and differential heating: I. Description of the three-dimensional velocity and density fields. 2. An analysis of the heat, vorticity and energy balance.
J. atmos. Sci., 25(6):945-967;968-983.

circulation

Bryan, Kirk, and Michael D. Cox, 1967.
Anumerical investigation of the oceanic general circulaton.
Tellus, 19(1):54-80.

Buchan, A. 1895. Circulation

Report on Oceanic Circulation, based on the Observations made on board H.M.S. "Challenger", and Other Observations. Rep.Sci.Res.Voy. "Challenger". A Summary of the Scientific Results. Appendix (Physics and Chemistry, Part VIII).

circulation

Bumpus, Dean F., 1960

Sources of water contributed to the Bay of Fundy by surface circulation. J. Fish. Res. Bd., Canada 17(2): 181-197.

Int. Passamaquoddy Fish Bd., 1956-1959 Sci. Rept. No. 11.

circulation

Bumpus, D.F., and Joseph Chase, 1965.
Changes in the hydrography observed along the east coast of the United States.
ICNAF Spec. Publ. No. 6:847-853.

circulation

Burkov, V.A., 1963.
Circulation of waters in the North Pacific. (In Russian).
Okeanologiia, Akad. Nauk, SSSR, 3(5):761-777.

circulation

Buscaglia, J.L., 1971.
On the circulation of the Intermediate Water in the southwestern Atlantic Ocean. J. mar. Res. 29(3): 245-255.

C

Circulation

Campbell, William J., 1965.
The wind-driven circulation of ice and water in a polar ocean.
J. Geophys. Res., 70(14):3279-3301.

circulation

Carruthers, James N., 1968.
Suggestions for increasing knowledge of the circulation of the North Sea waters to serve pollutant-drift studies.
Helgoländer wiss. Meeresunters. 17: 74-80

circulation

Chekotillo, C.A., 1961.
The investigation of horizontal and vertical water circulation in the North Pacific during the winter season. (Abstract).
Tenth Pacific Sci. Congr., Honolulu, 21 Aug.-6 Sept., 1961, Abstracts of Symposium Papers, 342-343.

circulation

Chew, F., 1955.
On the offshore circulation and a convergence mechanism in the red tide region of the west coast of Florida. Trans. Amer. Geophys. Un., 36(6):963-974.

Circulation

Clowes, A.J., 1933

Influence of the Pacific on the Circulation in the South-West Atlantic Ocean. Nature CXXXI:189-191.

circulation

Clowes, A.J., and G.E.R. Deacon, 1935.
The deep-water circulation of the Indian Ocean. Nature 136:936-938, 4 textfigs.

circulation

Coleman, J.S., 1950.
The sea and its mysteries. London, G. Bell & Sons Ltd., 285 pp., 1 fold-in.

circulation

Coachman, L.K., and C.A. Barnes, 1963
The movement of Atlantic water in the Arctic Ocean.
Arctic, J. Arctic Inst., N. America, 16(1): 9-16.

circulation

Cochrane, John D., 1969.
Water and circulation on Campeche Bank in May.
Bull. Jap. Soc. fish. Oceanogr. Spec. No. (Prof Uda Commem. Pap.): 123-128.

circulation

Collin, A.E., and M.J. Dunbar, 1964
Physical oceanography in Arctic Canada.
In: Oceanography and Marine Biology, Harold Barnes, Editor, George Allen & Unwin, Ltd. 2: 45-75.

circulation

Cooper, L.H.N., 1961.
Vertical and horizontal movements in the ocean.
Publ. Amer. Assoc. Adv. Sci., No. 67:599-621.

Abstr. in:
J. Mar. Biol. Assoc., U.K., 42(2):465.

Cooper, L. H. N., 1960 circulation
The effect of continental slopes upon vertical and horizontal circulation in the North Atlantic Ocean.
Proc. Verb., Assoc. Oceanogr. Phys., U.G.G.I., Totonto, 7:175

D

circulation

Dawson, E.Y., 1952.
Circulation within Bahia Vizcaino, Baja California and its effect on the marine vegetation.
Am. J. Bot. 39(7):425-432, figs.

circulation

Defant, A., 1950.
Die zwei Hauptprobleme der physikalischen Ozeanographie. Deutsche Hydro. Zeits. 3(1/2):13-20.

circulation

Defant, A., 1946.
Die Stromvorgänge in Meeresstrassen und ihre Bedeutung für die Tiefzirculation der Ozeans.
Naturwissenschaften 33:15-19.

circulation

Defant, A., 1938.
Aufbau und Zirculation des atlantischen Ozeans.
Sitzber. Preuss. Akad. Wiss., Physik.-math. Kl. 14:145-171, 8 pls.

circulation

Dietrich, G., 1957.
Stratification and circulation of the Irminger Sea. Ann. Biol., Cons. Perm. Int. Expl. Mer, 12: 36-37.

continents

Dietz, Robert S. 1972.
Geosynclines, mountains and continent-building.
Scient. Am. 226(3): 30-38

circulation

Dobrovolsky, A.D., Editor, 1968.
Hydrology of the Pacific Ocean. (In Russian).
P.P. Shirshov Inst. Okeanol., Akad. Nauk, Isdatel "Nauka", Moskva, 524 pp.

circulation
Dodimead, A.J., 1961.
The geostrophic circulation and its relation to the water characteristics in the Northeast Pacific Ocean during the period 1955-60. (Abstract).
Tenth Pacific Sci. Congr., Honolulu, 21 Aug.-6 Sept., 1961, Abstracts of Symposium Papers, 343.

circulation
Donguy, J.R., C. Oudot, and F. Rougerie, 1970.
Circulation superficielle et subsuperficielle en mer du Corail et a 170°E. Cah. O.R.S.T.O.M. sér. Océanogr., 8(1): 3-20.

circulation
Dunbar, M.J., 1951.
Eastern Arctic waters. Fish. Res. Bd., Canada, Bull. No. 88:131 pp., 32 textfigs.

circulation
Dunbar, M. J., 1950.
Feed-back systems and oceanography. Am. Sci. 38(4):599-603, 1 textfig.

circulation
*Dyer, K.R., and K. Ramamoorthy, 1969.
Salinity and water circulation in the Vellar estuary. Limnol. Oceanogr. 14(1): 4-15.

E

circulation
Eber, Laurence E., and Oscar E. Sette, 1959
Indices of mean monthly geostrophic wind over the north Pacific Ocean. USF&W. Sp. Sci. Rept. Fisheries No. 323: 108 pp.

circulation
Egorov, K.L., 1971.
Peculiarities of β-effect in water circulation of the near-polar region. Okeanologiia 11(4): 563-567. (In Russian; English abstract).

circulation
Emilsson I., 1964.
Dinâmica e natureza das águas adjacentes às praias do Rio de Janeiro (Brasil) tendo em vista o lançamento submarino de esgotos.
Contrib. Avulsas, Inst. Oceanogr., Univ. São Paulo, Oceanogr. Física, No. 6:16 pp.

circulation
Estoque, M.A., 1960
Dynamical prediction of arctic circulation. Tellus 12(1): 41-53.

F

circulation
Favorite, Felix, 1961.
Pacific subarctic circulation. (Abstract).
Tenth Pacific Sci. Congr., Honolulu, 21 Aug.-6 Sept., 1961, Abstracts of Symposium Papers, 343-344.

circulation
Fedenov, K.N., 1957.
(On yearly and half year fluctuations of the general circulation in oceans). DAN 116(3):393.
DAN = Dokl. Akad. Nauk SSSR

circulation
Felsenbaum, A.I., 1966.
Oceanic circulation and Ekman's problem. (In Russian).
Doklady, Acad. Nauk, SSSR, 167(4):807-810.

circulation
Filippov, D.M., 1961.
On horizontal circulation of deep waters in the Black Sea. Dynamics of the coastal zone of the Black Sea. (In Russian).
Trudy Inst. Okeanol., Akad. Nauk, SSSR, 53:112-122.

circulation
Fish, C.J., and M.W. Johnson, 1937
The biology of the zooplankton population in the Bay of Fundy and the Gulf of Maine with special reference to production and distribution. J. Biol. Bd., Canada 3(3):189-322, 29 tables, 45 text figs.

circulation
Fofonoff, N.P., 1967.
Variability of ocean circulation.
Trans. Am. geophys. Un., 48(2):575-578.

circulation
Fofonoff, N.P., 1953.
A theoretical example of wind-induced circulation in a semicircular canal. Trans. Amer. Geophys. Union 34(5):725-728, 3 textfigs.

circulation
Fomin, L.M., 1969.
V.B. Stockman's method of total flows for a variable depth ocean. (In Russian; English abstract).
Okeanologiia, 9(1):119-124.

circulation
Forsbergh, Eric D., 1969.
On the climatology, oceanography and fisheries of the Panama Bight. (In Spanish and English).
Bull. Inter-Am. Trop. Tuna Comm. 14(2): 385 pp.

culture, shrimp
Forster J.R.M., 1970.
Further studies on the culture of the prawn, Palaemon serratus Pennant, with emphasis on the post-larval stages.
Fish. Invest. Min. Agric. Fish. Food, London, (2) 26(6): 40 pp

circulation
Føyn, Ernst, 1965.
Biochemical and dynamic circulation of nutrients in the Oslofjord.
Helgoländer wiss. Meeresunters. 17: 489-495

circulation
Friedrich, Hans J., 1970.
Preliminary results from a numerical multilayer model for the circulation in the North Atlantic.
Dt. Hydrogr. Z. 23(4): 145-164.

circulation
Friedrich, Hans, 1966.
Numerische Berechnungen der allgemeinen Zirkulation im Meere nach einem Differenzenverfahren, vorhehmlich für den Atlantischen Ozean.
Dissertation zur Erlangung des Doktorgrades der Mathematisch-Naturwissenschaftlichen Fakultät der Universität Hamburg.
Mitt. Inst. Meereskunde, Univ. Hamburg, No. 3: 60 pp. (mimeographed).

circulation
Fukuoka, Jiro, 1962.
Abyssal circulation in the Atlantic near the poles and abyssal circulation in the Pacific and other oceans in relation to the former.
J. Oceanogr. Soc., Japan, 18(1):5-12.
JEDS Contrib. No. 27.

circulation
Fultz, D., and R.R. Long, 1951.
Two dimensional flow around a circular barrier in a rotating spherical shell. Tellus 3(2):61-68, 9 textfigs.

G

circulation
Ganapati, P.N., and D.V.Rama Sarma, 1965.
Mixing and circulation in Gautami-Godaveri Estuary.
Current Science, 34(22):631-632.

Circulation
Gates, W. Lawrence, 1970.
Effects of western coastal orientation on Rossby-wave reflection and the resulting large-scale oceanic circulation. J. geophys. Res., 75(21): 4105-4120.

circulation
Gill, A.E. and K. Bryan, 1971.
Effects of geometry on the circulation of a three-dimensional southern-hemisphere ocean model. Deep-Sea Res. 18(7): 685-721.

circulation
Gordienko P.A., and A.F. Laktionov, 1969.
Circulation and physics of the Arctic Basin waters.
Annls. int. geophys. Year. 46: 94-112.

circulation
*Gordon, Arnold L., 1967.
Circulation of the Caribbean Sea.
J. geophys. Res., 72(24):6207-6223.

circulation
Gordon, Arnold L., 1966.
Potential temperature, oxygen and circulation of bottom water in the Southern Ocean.
Deep-Sea Res., 13(6):1125-1138.

circulation
Groen, P. 1967.
The waters of the sea.
D. Van Nostrand Co., Ltd, 328 pp. $9.00

circulation
Grovel, Alain P. 1970.
Etude d'un estuaire dans son environment le Blavet maritime et la region de Lorient.
Trav. Lab. Geol. mar., Fac. Sci. Nantes (C.N.R.S. A.O 45-52): 122 pp. (multilith)

circulation
Gruzinov, V.M., 1965.
Hydrological front as a natural boundary of inhereent zones in the ocean. (In Russian).
Trudy, Gosudarst. Oceanogr. Inst., No. 84:252-262.

circulation
Gudkovich, Z.M., and E.G. Nikiforov, 1965.
The study of the nature of water circulation in the Arctic basin with the use of a model. (In Russian).
Okeanologiia, Akad. Nauk, SSSR, 5(1):73-83.

H

circulation
Hachey, H.B., 1961.
Oceanography and Canadian Atlantic waters.
Fish. Res. Bd., Canada, Bull. No. 134:120 pp.

circulation
Hachey, H.B., F. Hermann and W.B. Bailey, 1954.
The waters of the ICNAF Convention area.
Int. Comm. Northwest Atlantic Fish., Ann. Proc. 4:68-102, 29 textfigs.

circulation
Hachey, H.B., L. Lauzier and W.B. Bailey, 1956.
Oceanographic features of submarine topography.
Trans. R. Soc., Canada, (3)50:67-81.

ciculation
Halim, Youssef, Shoukry K. Guergues and Hamed H. Saleh, 1967.
Hydrographic conditions and plankton in the south east Mediterranean during the last normal Nile flood (1964).
Int. Rev. ges. Hydrobiol., 52(3):401-425.

circulation
*Hamon, B.V., 1968.
Spectrum of sea level at Lord Howe Island in relation to circulation.
J. geophys. Res., 73(22):6925-6927.

circulation
Haney, Robert L., 1971.
Surface thermal boundary condition for ocean circulation models. J. phys. Oceanogr. 1(4): 241-248.

circulation
Hansen, Donald V., and Maurice Rattray, Jr., 1965.
Gravitational circulation in straits and estuaries.
J. Mar. Res., 23(2):104-122.

circulation
Hansen, W., 1956.
Ein einfache Modell der Zirkulation im Meere.
Deutsche Hydrogr. Zeits., 9(2):102-106.

circulation
Hansen, W., 1950.
Bemerkungen zu neuzeitlichen Problemen der Ozeanographie. Deutsche Hydro. Zeits. 3(1/2):20-33.

circulation
Hansen, W., 1948.
Bewegungsvorgange im Meere. Geophys. Dieterich'sche Verlagsbuchhandlung, Wiesbaden, Pt. 2, Vol 18:108-131.

circulation
Hela, I., 1956.
A pattern of coastal circulation inferred from synoptic salinity.
Bull. Mar. Sci., Gulf and Caribbean, 6(1):74-83.

circulation
Herrera, Luis E., and John W. Snooks, 1969.
An investigation of the circulation pattern in the western tropical Atlantic Ocean during Equalant I and III.
Bol. Inst. oceanogr. Univ. Oriente 8(1/2): 35-45

circulation
Hidaka, K., 1957.
On the Pacific circulation.
Proc. UNESCO Symp., Phys. Ocean., Tokyo, 1955:137

circulation
Hidaka, K., 1950.
Circulation in a zonal ocean induced by a planetary wind system. Geophys. Notes, Tokyo, 3(10): 8 pp., 2 textfigs.

circulation
Hidaka, K., 1955.
A theoretical study on the general circulation of the Pacific Ocean. Pacific Sci. 9(2):183-220, 5 textfigs.

circulation
Hidaka, K., and M., Koizumi, 1949.
Vertical circulation due to winds as inferred from the buckling experiments of elastic plates.
Ocean. Mag., Tokyo, 1(2):89-98, 12 textfigs.

circulation
Hidaka, K., and T. Yamagiwa, 1949.
On the absolute velocity of the Subarctic Intermediate Current to the south of Japan. Geophys. Notes, Tokyo, 2(5):1-4, 3 textfigs.

circulation
Highley, E. 1967.
Oceanic circulation patterns off the east coast of Australia.
Comm. Sci. Ind. Res. Org. Australia Div. Fish. Oceanogr. Tech. Pap. 23:1-19.

circulation
Hirano, T., 1961
The oceanographic study on the subarctic region of the northwestern Pacific Ocean. IV. On the circulation of the subarctic water.
Bull. Tokai Reg. Fish. Res. Lab., No. 29: 11-39.

circulation
*Holopainer, E.O., 1967.
A determination of the wind-driven ocean circulation from the vorticity budget of the atmosphere.
Pure appl. Geophys. 67(2):156-165.

I

circulation
Ichiye, Takeshi, 1966.
Rotating model experiment on circulation in the Antarctic seas.
Antarctic J., United States, 1(5):223-224.

circulation
Ichiye, T., 1960
A theory of circulation in an ocean consisting of a homogeneous upper layer and a stratified lower layer.
J. Oceanogr. Soc., Japan, 16(2): 47-54.

circulation
Ichiye, T, 1958.
On convection circulation and density distribution in a zonally uniform ocean.
The Ocean. Mag. Vol. 10: No. 1, pp. 97-136.
Jap. Meteor Agency

circulation
Ichiye, T., 1955.
On the variation of oceanic circulation (V).
Geophys. Mag., Tokyo, 26(4):283-342, 23 textfigs.

circulation
Ichiye, T., 1954.
On the variation of oceanic circulation. VII.
Ocean. Mag., Tokyo, 6(1):1-14, 11 textfigs.

circulation
Ichijo, T., 1953.
On the variation of oceanic circulation. IV.
Ocean. Mag., Tokyo, 5(1):23-44, 9 textfigs.

circulation
Ilyin, A.M., V.M. Kamenkovich, T.G. Zhugrina, and M.M. Silkina, 1969.
On the calculation of the total flows in the World Ocean (Stationary problem). (In Russian; English abstract)
Fizika Atmosfer. Okean. Akad. Nauk SSSR, 5(11):1160-1171.

circulation
*Ingraham, W. James, Jr., 1968.
The geostrophic circulation and distribution of water properties off the coasts of Vancouver Island and Washington spring and fall 1963.
Fishery Bull. Fish Wildl. Serv. U.S. 66(2):223-250.

circulation
Iselin, C. O'D., 1956.
Recent advances in our understanding of the circulation problem, and their implications.
J. Mar. Res., 14(4):315-322.

circulation
Iselin, C. O'D., 1955.
Recent advances in our understanding of the circulation problem and their implications.
J. Mar. Res. 14(3):315-322.

circulation
Iselin, C. O'D., 1936
A study of the circulation of the western North Atlantic. PPOM, 4(4):

circulation
Ivanov, Yu. A., 1961
[Horizontal circulation of the Antarctic waters.]
Mezhd. Kom. Mezhd. Geofiz. Goda, Presidiume Akad. Nauk, SSSR, Okeanolog. Issled. (3): 5-29.

J

circulation
Jeffries, Harry P., 1962.
Environmental characteristics of Raritan Bay, a polluted estuary.
Limnol. and Oceanogr., 7(1):21-31.

circulation (Baltic)
Jurva, R., 1952.
Seas. Fennia 72:138-160.

K

Circulation

Kato, Kenji, 1961.
Oceanographical studies in the Gulf of Cariaco.
1. Chemical and hydrographical observations in January, 1961.
Bol. Inst. Oceanograf., Univ. Oriente, Cumana, Venezuela, 1(1):49-72.

circulation

Kawai, H., and M. Sasaki, 1961
An example of the short-period fluctuation of the oceanographic condition in the vicinity of the Kuroshio front.
Bull. Tohoku Reg. Fish. Res. Lab., (19): 119-134.

circulation

Kiilerich, A., 1945
On the Hydrography of the Greenland Sea.
Medd. om Grønland 144(2):63 pp., 22 text figs., 3 pls.

circulation

Klepikov, V.V., 1958(1960)
The origin and diffusion of Antarctic ocean-bed water. Problemy Severa, (1):
Translation in:
Problems of the North, 1: 321-333

circulation

Knauss, John A., 1962
On some aspects of the deep circulation of the Pacific.
J. Geophys. Res., 67(10):3943-3954.

circulation

Kochergin, V.P. 1970.
A numerical method for the solution of some problems on ocean circulation. (In Russian; English abstract)
Met. Gidrol. 1970 (5): 67-75.

circulation

Kollmeyer, Ronald C., Thomas C. Wolford, and Richard M. Morse 1966.
Oceanography of the Grand Banks region of Newfoundland in 1965.
U.S. Cst Gd Oceanogr. Rept. 11 (CG373-11): 157pp.

circulation

Koshliskov, M.N., 1966.
Physical foundation for the boxes method in studying deep sea circulation. (In Russian; English abstract).
Fizika Atmosferi i Okeana, Izv.Akad.Nauk,SSSR, 2(9):945-955.

circulation

Koshliakov, M.N., 1958.
Some problems concerning general circulation of ocean waters.
Izv., Akad. Nauk, SSSR, Ser. Geograf., (4):11-

circulation

Kowalik, Zygmunt 1970.
Wind-driven circulation in a shallow sea with application to the Baltic Sea. II. (In Polish; English abstract).
Przeglad Geofiz. polsk. Towarzyst. Geofiz. Warzawa 15 (23)(2): 145-162.

circulation

Kowalik, Zygmunt, and Andrzej Wróblewski 1971
Wind-driven circulation in the Bay of Gdańsk. (In Polish; English abstract)
Acta geophys. polonica 19(2):111-125.

circulation

Kozlov, V.F. 1971.
The results of approximate calculation of the integral circulation in the Japan Sea. (In Russian; English abstract).
Met. Gidrol. (4): 57-63.

circulation

Kozlov, V.F., 1971.
Some results of the approximated calculations of the integral circulation in the Pacific. (In Russian; English ab stract). Fizika Atmosfer. Okean. Izv. Akad. Nauk SSSR 7(4): 421-430.

circulation

Krauss, W., 1957.
Das Zirkulationstheorem von V. Bjerknes und sein Anwendung in der Ozeanographie.
Deutsche Hydrogr. Zeits. 10(1):13-19.

circulation

Kullenberg, Börje, 1968.
Studiet av den oceanografiska circulationen:en översikt.
Geophysica Gothoburgensia 1: 18 pp.

circulation

Kusunoki, Kou, 1962
Hydrography of the Arctic Ocean with special reference to the Beaufort Sea.
Contrib. Inst., Low Temp. Sci., Sec. A(17): 1-74.

L

circulation

Lacombe, H. 1951.
Application de la méthode dynamique à la circulation dans l'Océan Indien au printemps boréal et dans l'Océan antarctique pendant l'été austral.
Bull. D'Info. C.C.O.E.C 3(10): 459-473

circulation

Lacombe, H., 1950.
Aperçu de la circulation océanique dans la partie Sud-Est de l'Atlantique Nord. Bull. d'Info. C.O.E.C., 2:44-53.

circulation

Lacombe, H., P. Guibout et L. Gamberoni 1965.
Études de la circulation marine superficielle et profonde entre France continentale et Corse. (Abstract only).
Rapp. P.-v. Réun. Commn int. Explor. scient. Mer Méditerr. 18(3): 797.

Circulation

LaFond, E.C. and K.G. LaFond, 1969.
Studies of oceanic circulation in the Bay of Bengal. Bull. natn. Inst. Sci., India, 38(1): 164-183.

circulation

Laikhtman, D.L., B.A. Kagan, L.A. Oganesian, and R.V. Piaskovskii, 1971.
On global circulation in barotropic ocean of variable depth. (In Russian)
Dokl. Akad. Nauk, SSSR 198(2): 333-336

circulation

Lal, D., 1962
Cosmic ray produced radionuclides in the sea.
J. Oceanogr. Soc., Japan, 20th Ann. Vol., 600-614.

circulation

Lamb, H.H., 1965.
Climatic changes and variations in the atmosphere and ocean circulations.
Geol. Rundschau, Stuttgart, 54(1):486-504.

circulation

Leuzier, L.M., et A. Marcotte, 1965.
Comparison du climat marin de Grande-Rivière (baie des Chaleurs) avec celui d'autres stations de la côte atlantique.
J. Fish. Res. Bd., Canada, 22(6):1321-1334.

circulation

Lee, A.J., 1967.
Monitoring the ocean.
Marine Obs., 37(218):193-198.

circulation

LeFloch, J., 1951.
Hydrologie et circulation de quelques masses d'eaux de l'Océan Indien. Bull. d'Info., C.C.O.E.C., 3(10):433-458, Pls. 1-8.

circulation

Le Floch, Jean, et V. Romanovsky, 1966.
L'eau intermédiaire en Mer Tyrrhénienne en régime d'été. (no abstract).
Cahiers océanogr., 18(3):185-228.

circulation

Lida, Michitaka, 1964.
Dissolved oxygen in the Pacific Ocean and adjacent seas as an important element to study the circulation and structure in the changing marine environment.
In: Recent researches in the fields of hydrosphere, atmosphere and nuclear geochemistry, Ken Sugawara festival volume, Y. Miyake and T. Koyama, editors, Maruzen Co., Ltd., Tokyo, 349-356.

circulation

Lineykin, P.S., 1961.
Wind and thermal circulation in the oceans. (In Russian).
Doklady, Akad. Nauk, SSSR, 138(6):1341-1344.
Translation: Consultants Bureau for Amer. Geol. Inst., (Earth Sciences Section), 138(6):635-637. (1962).

circulation

Lineikin, P.S., 1961.
Determination of the wind elements in oceanic circulation. (In Russian).
Trudy Gosud. Okean. Inst., Leningrad, 61:4-21.

M

circulation

Mao, H., and K. Yoshida, 1955.
Physical oceanography in the Marshall Islands area. Bikini and nearby atolls, Marshall Islands Geol. Survey, Prof. Paper 260-R:645-684, Figs. 179-218.

circulation

Marchuk G. I. 1969.
Numerical solution of the Poincaré problem for oceanic circulation. (In Russian).
Dokl. Akad. Nauk SSSR 185(5):

circulation

Maximov, I.V. 1967.
On the gravitational nature of main peculiarities of the general circulation of the ocean and of the atmosphere in high latitudes of the earth. (In Russian; English abstract)
Trudy polyar. nauchno-issled. Inst. morsk. ryb. Khoz. Okeanogr. (PINRO) 20:336-342

circulation

Maksimov, I.V. 1966.
Statistical index of general circulation of waters in the northern Atlantic Ocean. (In Russian).
Trudy polyar. nauchno-issled. Proektn. Inst. Morsk. ribn. Khoz. Okeanogr. N.M. Knipovich, PINRO 17:279-285.

Mc

circulation

McManus, A.B., 1933.
Water circulation in Pacific regions.
Proc. Pacific Sci. Congr., 5th, 1:703-311.

circulation

Meincke, Jens 1971.
Der Einfluss der Grossen Meteorbank auf Schichtung und Zirkulation der ozeanischen Deckschicht.
Meteor Forsch.-Ergebn. (A) 9: 67-96

circulation

Merle, Jacques, Henri Rotschi and Bruno Voituriez, 1969.
Zonal circulation in the tropical western South Pacific at 170°E. Bull. Jap. Soc. fish. Oceanogr. Spec. No. (Prof Uda Commem. Pap.): 91-98.

circulation

Metallo, Antonio, 1961.
Il campo stabile meteoro-oceanografico del Mediterraneo.
Atti del XVIII Congresso Geografico Italiano, Trieste, 4-9 aprile 1961:3-12.

circulation
(atmospheric)

Meyer, Arno, 1964
Zusammenhang zwischen Eisdrift, atmosphärischer Zirkulation und Fischerei im Bereich der Fangplätze vor der südostgrönländischen Küste während der ersten Jahreshälfte.
Arch Fischereiwiss., 15(1):1-16.

circulation

Miyazaki, M., 1958.
Note on the zonal circulation in the Antarctic Ocean.
J. Oceanogr. Soc., Japan, 14(1):7-10.

circulation

Miyazaki, M., 1948.
Zonal circulation current caused by wind. Geophys Mag., Tokyo, 15:50-57.

circulation

Montgomery, R.B., 1938.
Circulation in upper layers of southern North Atlantic deduced with use of isentropic analysis. Papers in Phys. Oceanogr. and Meteorol., 6(2): 55 pp., 19 charts. 10 text figs.

Reviewed:
J. du Cons., 14(3):411-412. G. Dietrich.

circulation

Moore, D.E. 1969.
Construction of f/H maps and comparison of measured deep-sea motions with the f/H map for the North Atlantic Ocean.
Techn. Rept. Chesapeake Bay Inst. 51 (Ref. 69-4): 61 pp.

circulation

Moreira da Silva, Paulo de Castro, 1965.
Problemas da circulação oceânica nas águas brasileiras. (Abstract).
Anais Acad. bras. Cienc., 37 (Supl.):157-158.

circulation

Moroshkin, K.V., V.A. Bubnov, and R.P. Bulatov 1970.
Water circulation in the southeastern Atlantic. (In Russian; English abstract).
Okeanologiia 10(1): 38-47.

circulation

Moroshkin, K.V., V.A. Bubnov, and R.P. Bulatov, 1969.
New data on winter circulation in the south east Atlantic. (In Russian)
Dokl. Akad. Nauk SSSR 188(3): 681-684

circulation

Munk, W. H., 1950.
On the wind-driven circulation. J. Met. 7(2):79-93.

circulation

Munk, W.H., and G. F. Carrier, 1950.
The wind-driven circulation in ocean basins.
Tellus 2(3):158-167, 12 textfigs.

oceanic circulation

Munk, W.H., and R.L. Miller, 1950.
Variation in the earth's angular velocity resulting from fluctuations in atmospheric and oceanic circulation. Tellus 2(2):93-101, 2 textfigs.

circulation

Muromtsev, A.M., 1959
[The basis for the hydrology of the Indian Ocean.] Gidrometeoizdat, Leningrad, 437 pp.

circulation

Muromtsev, A. M., 1958.
Scheme of general circulation of the Pacific Ocean water.
Izv. Akad. Nauk, SSSR, Geogr. Ser. (4):24-32.

circulation

Muromtsev, A.M., 1958
[The basis for the hydrology of the Pacific Ocean.]
Gidrometeoizdat., Leningrad, 431 pp.

circulation

Murty, T.S., and J.D. Taylor, 1970.
A numerical calculation of the wind-driven circulation in the Gulf of St. Lawrence.
J. oceanogr. Soc., Japan, 26(4): 203-214.

N

circulation

*Neev, David, and K.O. Emery, 1967.
The Dead Sea: depositional processes and environments of evaporites.
Bull. Geol. Surv., Israel, 41: 146 pp.

circulation

Neumann, A. Conrad, and David A. McGill, 1961 (1962)
Circulation of the Red Sea in early summer.
Deep-Sea Res., 8(3/4):223-235.

circulation

Neumann, Gerhard, 1962
On the oceanic circulation in the equatorial region of the Atlantic.
J. Geophys. Res., 67(9):3583-3584.

circulation

Neumann, G., 1956.
Notes on the horizontal circulation of ocean currents. Bull. A.M.S., 37(3):96-100.

circulation

Neyman, V.G., 1963
The circulation of waters in the north-eastern area of the Indian Ocean during the period of the summer monsoon. (In Russian).
Okeanologiia, Akad. Nauk, SSSR, 3(3):418-423.

circulation

Neyman, V.G., B.N. Filjushkin, and A.D. Shcherbinin, 1966.
Structure and circulation of the water masses in the eastern Indian Ocean during the summer monsoon. (In Russian; English abstract).
Okeanol. Issled., Mezhd. Geofiz. Kom. Presid., Akad. Nauk, SSSR, No. 15:5-22.

Circulation

Niiler, P.P. and P.S. Dubbelday, 1970.
Circulation in a wind-swept and cooled ocean.
J. mar. Res., 28(2): 135-149.

circulation

Niiler, P.P., A.R. Robinson and S.L. Spiegel, 1965.
On thermally maintained circulation in a closed ocean basin.
J. Mar. Res., 23(3):222-230.

circulation

Nikiforov, E.G., E.I. Chaplygin, A.O. Shpaikher, 1969.
Pressure systems and dynamic processes in the Arctic seas. (In Russian; English abstract).
Okeanologiia, 9(5): 782-790.

circulation

Novitsky, V.P., 1965.
On the dynamics of Marmora sea water in the pre-Bosporus shelf area of the Black Sea. (In Russian).
Okeanologiia, Akad. Nauk, SSSR, 5(5):841-848.

circulation

Novitskii, V.P. 1964.
Vertical structure of the waters and the general characteristics of the water circulation in the Black Sea. (In Russian). Trudy azov.-chernomorsk. nauchno-issled. Inst. morsk. ryb. Khoz. Okeanogr. 23:3-22

circulation

Nowlin, Worth D., Jr., and Robert O. Reid, 1968.
Formulation of a three-dimensional, steady ocean circulation model using a similarity hypothesis.
J. Oceanogr. Soc., Japan 24(1):1-15.

circulation

Nowlin, Worth D. Jr., 1971.
Water masses and general circulation of the Gulf of Mexico.
Oceanol. int. 6(2):28-33

O

circulation

*O'Brien, James J., 1971.
A two dimensional model of the wind-driven north Pacific. Investigación pesq. 35(1): 331-349.

circulation

Oudot, C., P. Hisard et B. Voituriez 1969.
Nitrite et circulation méridienne à l'Equateur dans l'océan Pacifique occidental.
Cah. ORSTOM sér. Océanogr. 7(4): 67-82.

circulation

Olausson, Eric, 1969.
Le climat au Pleistocène et la circulation des océans. Revue Géogr. phys. Géol. dyn. 11(3): 251-264.

circulation

Ovchinnikov, I.M., 1966.
Circulation in the surface and intermediate layers of the Mediterranean Sea. (in Russian).
Okeanologiia, Akad. Nauk, SSSR, 6(1):62-75.

circulation

Ovchinnikov, I.M., 1961
[Circulation of waters in the northern Indian Ocean during the winter monsoon.] (In Russian; English abstract).
Okeanolog. Issledov., Mezhd. Komitet Proved. Mezhd. Geofiz. Goda, Prezidiume Akad. Nauk, SSSR, (4):18-24.

P

circulation

Parr, A. E., 1937.
A contribution to the hydrography of the Caribbean and Cayman Seas, based upon the observations made by the Research Ship "Atlantis", 1933-1934. Bull. Bingham Ocean. Coll., V(4):1-110, 82 textfigs.

circulation

Paskausky, D.F., 1971.
Numerically predicted changes in the circulation of the Gulf of Mexico accompanying a simulated hurricane passage. J. mar. Res. 29(3): 214-225.

circulation

Pavlova, Yu.V. and V.I. Byshev, 1971.
On time variability of the water circulation in the northeastern Pacific. (In Russian; English abstract). Okeanologiia 11(2): 217-222.

circulation

Pedlosky, Joseph, 1969.
Linear theory of circulation of a stratified ocean.
J. fluid Mech. 35(1):185-205

circulation

Pedlosky, Joseph 1967.
Fluctuating winds and the ocean circulation.
Tellus 19(2):250-257.

circulation

Pedlosky, Joseph, 1965.
A study of the time dependent ocean circulation.
J. Atmos. Sci., 22(3):267-272.

circulation

Pedlosky, J. And H.P. Greenspan, 1967.
A simple laboratory model for the oceanic circulation.
J.Fluid Mech., 27(2):291-304.

circulation

Pelletier, B.R. 1968.
Submarine physiography, bottom sediments, and modern sediment transport in Hudson Bay.
In: Earth Science Symposium on Hudson Bay, February 1968, Peter J. Hood, editor, GSC Pap., Geol. Surv. Can., Ottawa, 68-53: 100-136.

circulation

Peluchon, Georges et Jean Rene Donguy, 1962
Hydrologie en mer d'Alboran. Travaux oceanographiques de l'"Origny" dans le detroit de Gibraltar, Campagne internationale - 15 mai - 15 juin, 1961.
Cahiers Oceanogr., C.C.O.E.C., 14(8):573-578.

circulation

Peterson, Clifford L., 1960.
The physical oceanography of the Gulf of Nicoya, Costa Rica, a tropical estuary.
Inter-American Tropical Tuna Comm., Bull., 4(4): 139-216.

circulation

Pettersson, O., 1929
Changes in the oceanic circulation and their climatic consequences. Geogr. Rev. 19:121-131

circulation

Pettersson, O., 1923.
Aperçu d'Orientation vers la conception actuelle de la circulation océanique dans l'Atlantique.
Svenska Hydrogr.-Biol. Komm. Skr., n.s., Hydrogr 5:21 pp.

circulation

Pickard, G.L., 1962.
Oceanographic features of inlets in the British Columbia mainland coast.
J. Fish. Res. Bd., Canada, 18(6):907-999.

circulation

Pieterse, F., and D.C. van der Post, 1967.
The pilchard of South West Africa (Sardinops ocellata): oceanographic conditions associated with red-tides and fish mortalities in the Walvis Bay region.
Investl Rept., Mar. Res. Lab., SWest Africa, 14: 125 pp.

circulation

Pritchard, D.W., 1955.
Estuarine circulation patterns. Proc. Amer. Soc. Civil Eng., Separate 717, Vol. 81:717-1 - 717-11.

circulation

Pritchard, D.W., 1952.
Salinity distribution and circulation in the Chesapeake Bay estuarine system. J. Mar. Res. 11(2):106-123, 11 textfigs.

circulation

Pritchard, D.W., R.O. Reid, A. Okubo and H.H. Carter, 1971.
Physical processes of water movement and mixing. Radioactivity in the marine environment, U.S. Nat. Acad. Sci., 1971. 90-136.

Q

R

circulation

Rama, 1970.
Using natural radon for delineating monsoon circulation. J. geophys. Res., 75(12): 2227-2229.

circulation

Rao, T.S. Satyanarayana, 1963
On the pattern of surface circulation in the Indian Ocean as deducted from drift bottle recoveries.
Indian J. Meteorol. & Geophys., 14(1): 1-4.

circulation

Rasool, S.I. and J.S. Hogan, 1969.
Ocean circulation and climatic changes.
Bull. Am. met. Soc., 50(3): 130-134.

circulation

Rattray, Maurice, Jr., and Donald V. Hansen, 1962
A similarity solution for circulation in an estuary.
J. Mar. Res., 20(2):121-133.

Circulation

Rattray, M., Jr., and D. V. Hansen, 1962.
Similarity solutions for circulation in an estuary. (Abstract).
J. Geophys. Res., 67(4):1654.

circulation

Redfield, A.C., 1955.
The hydrography of the Gulf of Venezuela.
Pap. Mar. Biol. and Oceanogr., Deep-Sea Res., Suppl. to Vol. 3:115-133.

circulation
Redfield, A. C. and A. Beale, 1940.
Factors determining the distribution of populations of chaetognaths in the Gulf of Maine. Biol Bull., 79(3):459-487, 11 text figs.

circulation
Reznik, G.M. 1970.
On the problems of a three-dimensional circulation in a homogeneous ocean. (In Russian; English abstract)
Fisika Atmosfer. Okean. Izv. Akad. Nauk SSSR 6 (11): 1163-1177.

Circulation
Roa Morales, Pedro, and Francois Ottmann, 1961.
Primer estudo topografico y geologico del Golfo de Cariaco.
Bol. Inst. Oceanograf., Univ. Oriente, Cumana, Venezuela, 1(1):5-20.

circulation
Robinson, Allan R., 1962
On the theory of the wind-driven ocean circulation. (Abstract).
J. Geophys. Res., 67(9):3592.

circulation
Robinson, A.R., and Pierre Welander, 1963
Thermal circulation on a rotating sphere; with application to the oceanic thermocline.
J. Mar. Res., 21(1):25-38.

circulation
Rochford, D.J. 1969.
Origin and circulation of water types of the 25.00 sigma-t surface of the south-west Pacific.
Aust. J. mar. Freshwat. Res. 20(2): 105-114.

circulation
*Rochford, D.J., 1968.
Origin and circulation of water types on the 26.00 sigma-t surface of the south-west Pacific.
Aust.J.mar.Freshwat.Res., 19(2):107-127.

circulation
Rochford, D.J., 1951.
Studies in Australian estuarine hydrology.
Australian J. Mar. and Freshwater Res. 2(1):1-116, 1 pl., 7 textfigs.

circulation
Roden, Gunnar I., 1969
Winter circulation in the Gulf of Alaska.
J. geophys. Res., 74(18): 4523-4534.

circulation
Rodewald, Martin, 1958.
Beiträge zur Klimaschwankung im Meere. 10. Die Anomalie der Wassertemperatur und der Zirculation im Nordpazifischen Ozean und an der Küste Perus im Jahre 1955.
Deutsche Hydrogr. Zeits., 11(2):78-82.

circulation
Rossov, V.V., 1966.
On water circulation in the Gulf of Mexico and the Caribbean Sea. (In Russian).
Doklady, Akad. Nauk, SSSR, 166(3):705-708.

circulation
Rotschi, Henri, 1960
Sur la circulation et les masses d'eau dans le Nord-Est de la mer de Corail. C.R. Acad. Sci., Paris, 251(7): 965-967.

circulation
Rotschi, H., 1959
Remarques sur la circulation océanique entre la Nouvelle-Calédonie et l'Isle Norfolk.
Cahiers Oceanogr., C.C.O.E.C., 11(6):416-424.

circulation
Rotschi, H., and L. Lemasson, 1967.
Oceanography of the Coral and Tasman Seas.
Oceanogr. Mar.Biol. Ann.Rev., Harold Barnes, editor, George Allen and Unwin, Ltd., 5:49-97.

Circulation
Rumer, Ralph R., and Lance Robson 1968.
Circulation studies in a rotating model of Lake Erie.
Proc. 11th Conf. Great Lakes Res 1968: 487-495.

circulation
Russell, G.F., 1950.
Surface temperatures in the S.W. Pacific and Tasman Sea. N.Z. Dept. Sci. Ind. Res., Geophys. Conf., 1950, Paper No. 13:2 pp., 3 figs. (mimeographed).

circulation
*Rzheplinsky, D.G., and W.B. Stockmann (posthumous), 1968.
A study of some features of a circulation around oceanic islands. (In Russian; English abstract).
Fisika Atmosfer.Okean., Izv.Akad.Nauk,SSSR, 4(12): 1261-1274.

S

circulation
Sandoval, Eliseo, 1970.
Distribución de los atunes en el primer trimestre del año en relación con las condiciones oceanograficas generales frente a Chile y Perú.
Bol. cient. Inst. Fomento pesq. Chile 14: 86pp.

circulation
Sarkisjan, A.S. 1969.
On the defects of the barotropic models of oceanic circulation. (In Russian; English abstract)
Fizika Atmosfer. Okean. Izv. Akad. Nauk SSSR 5(8): 818-835.

circulation
Shannon, L.V., 1966.
Hydrology of the south and west coasts of South Africa.
Investl Rep. Div. Sea Fish. Un. S. Afr., 22 pp.

circulation
Shcherbinin, A.D. 1971.
On the relationship of circulation and structure of the Indian Ocean waters.
Dokl. Akad. Nauk SSSR 199(6): 1413-1416.

circulation, nearshore
Shepard, F.P., and D.L. Inman, 1950.
Nearshore water circulation related to bottom topography and wave refraction. Trans. Am. Geophys. Union 31(2):196-212, 12 textfigs.

circulation
Shtokman, V.B., 1947.
Theoretical model of the circulation on the surface of the ocean in the region of the Equatorial Counter Current. Dok. Akad. Nauk SSSR 57(7):4 pp.

circulation
Shtokman, V. B., 1947.
[New indicators of the significance of the irregularity of the wind as one of the reasons for circulation in the sea.] Dok. Akad. Nauk SSSR 58(1):4 pp.

circulation
*Siedler, G., 1969.
General circulation of water masses in the Red Sea.
In: Hot brines and Recent heavy metal deposits in the Red Sea, E.T. Degens and D.A. Ross, editors, Springer-Verlag, New York, Inc., 131-137.

circulation
Smith, Robert L., David B. Enfield, Thomas S. Hopkins and R. Dale Pillsbury 1971.
The circulation in an upwelling ecosystem: The Pisco cruise.
Investt. pesq. Barcelona 35(1): 9-24.

circulation
Soskin, I.M., and V.V. Denisov, 1957.
Calculation of the stationary horizontal circulation of the waters of the Baltic Sea for two types of wind fields. (In Russian).
Trudy Gosud. Okeanogr. Inst., 41:31-45.

circulation
Spangenberg, W.G., & W.R. Rowland, 1961
Convective circulation in water induced by evaporative cooling.
Physics of Fluids, 4(6): 743-750.

circulation
Spencer, R.S., 1956.
Studies in Australian estuarine hydrology.
Australian J. Mar. & Freshwater Res. 7(2):193-253.

circulation
Stepanov V.N. 1969.
Water circulation on the meridional plane of the oceans. (In Russian; English abstract)
Okeanologiia 9(3): 387-397.

circulation
Stepanov, V.N., 1960.
[The heat budget (balance) in the circulation of the world ocean.]
Trudy Okeanograf. Komissii, Akad. Nauk, SSSR, 10(1):79-81.

circulation
Stevenson, Merritt, 1970.
Circulation in the Panama Bight. J. geophys. Res., 75(3): 659-672.

circulation
Stevenson, Merritt R., Oscar Guillen G., and José Santoro de Ycaza 1970.
Marine atlas of the Pacific coastal waters of South America.
Univ. Calif. Press, 20 pp. 99 charts, $40.00.

circulation
Steward, R.W., 1957.
A note on the dynamic balance in estaurine circulation. J. Mar. Res., 16(1):34-39.

circulation
*Stockmann, W.B., 1967.
Development of the theory of oceanic and sea circulation in the USSR for 50 years. (In Russian; English abstract).
Okeanologiia, Akad. Nauk, SSSR, 7(5):761-773.

circulation
Stockmann, W.B., 1966.
A qualitative analysis of the causes of abnormal circulation around oceanic islands. (In Russian; English abstract).
Fizika Atmosferi i Okeana, Izv., Akad. Nauk, SSSR, 2(11):1175-1185.

circulation
Stockman, W.B., 1949.
[An investigation of the influence of the wind and bottom relief on the resulting circulation and the distribution of mass in a non-uniform (baroclinic) ocean or sea.] Trudy Inst. Okeanol., 3:3-65.

circulation
Stockmann, W.B., 1946
Characteristic features of the coastal circulation in the sea and their connection with the transverse non-uniformity of the wind. Comptes Rendus (Doklady) de l'Acad. des Sci. de l'URSS, LIV (3):227-230, 2 textfigs.

circulation
Stommel, Henry M., 1966.
The large-scale oceanic circulation.
In: Advances in earth science, P.M. Hurley, Editor, M.I.T. Press, 175-184.

circulation
Stommel, H., 1954.
Circulation in the North Atlantic Ocean. Nature 173(4411):886-888, 2 textfigs.

circulation
Stommel, H., 1950.
Note on the deep circulation of the Atlantic Ocean. J. Met. 7(3):245-246.

circulation
Stommel, Henry, & A. B. Arons, 1960.
On the abyssal circulation of the world ocean. 1. Stationary planetary flow patterns on a sphere.
Deep-Sea. Res. 6(2):140-154.

circulation
Stommel, Henry, and Claes Rooth 1968.
On the interaction of gravitational and dynamic forcing in simple circulation models.
Deep-Sea Res. 15(2): 165-170.

circulation
Sverdrup, H.U., 1939
Oceanic circulation. Fifth Intern. Congr. Applied Mech., Trans. 279-293.

circulation
Sverdrup, H.U., 1933
On vertical circulation in the ocean due to the action of the wind with application to conditions within the Antarctic Circumpolar Current. Discovery Rept., VII: 139-170

T

circulation
Taft, Bruce, 1971.
Ocean circulation in monsoon areas. In: Fertility of the Sea, John D. Costlow, editor, Gordon Breach, 2: 565-579.

circulation
Takano, Kengo 1966.
Effet de la sphéricité de la terre sur la circulation générale dans un océan.
J. oceanogr. Soc., Japan, 22(6): 255-263

circulation
Takano, Kengo 1965.
General circulation due to the horizontal variation of water density with the longitude maintained at the surface of an ocean.
Rec. Oceanogr. Wks, Japan 8(1):1-11 (not seen)

circulation
Takano, Kenzo, 1964.
Variation annuelle de la circulation générale dans les océans.
La Mer, 1(2):51-61.
Also in:
Geophys. Notes, Tokyo, 17(1)(Contrib. No. 4).

circulation
Takano, Kenzo, 1962
Circulation générale permanente dans les océans un calcul numérique complémentaire.
J. Oceanogr. Soc., Japan, 18(2):59-67.
English abstract.

circulation
Takano, Kenzo, 1962
Circulation Generale permanente dans un ocean.
Rec. Oceanogr. Wks., Japan, 6(2):59-155.
Also in:
Collected Oceanographical Papers, Geophys. Inst., Tokyo Univ., 6(Contrib. No. 21) (1962).

circulation
Takano, Kenzo, 1961.
Circulation générale permanente dans les océans. (Pts. 1 and 2). (In French).
J. Oceanogr. Soc., Japan, 17(3):121-131; 132-140
Also in:
Geophys. Notes, Tokyo Univ., 14(2).

circulation
Takano, Kenzo, 1959
Sur la circulation océanique.
Cahiers Océan., C.C.O.E.C., 11(9):644-660.

circulation
Takano, K., 1956.
Influence des échanges thermiques, de la précipitation et de l'évaporation sur le circulation dans les océans. C.R. Acad. Sci., Paris, 242(18): 2245-2247.
General

circulation
Tareev, B.A., 1960.
[Contribution to a theory of convectional circulation in oceanic deeps.]
Izv., Akad. Nauk, SSSR, Ser. Geofiz., (7):1022-1029.
Eng. Ed., Pergamon Press, (7):683-387.

circulation
Tareyev, B.A., 1958.
Stationary wind set-up and circulation in a shallow rectangular basin. (In Russian).
Izv., Akad. Nauk, SSSR, Ser. Geofiz., (9):1139-1144.
English Edit., (9):661-663.

circulation
Thoulet, J., 1927.
Étude densimétrique dans le Pacifique. Ann. Inst Océan., n.s., 4(7):261-273, 3 pls.

circulation
Tolbert, W.H., and G.G. Salsman, 1964
Surface circulation in the eastern Gulf of Mexico as determined by drift-bottle studies.
J. Geophys. Res., 69(2):223-230.

circulation
Tolmazin, D.M., and V.A. Shnaidman 1968.
Calculations of integral circulation and currents in the northwestern Black Sea. (In Russian; English abstract).
Fizika Atmosfer. Okean., Izv. Akad. Nauk SSSR 4(6): 634-646

circulation
Tomczak, G., 1961.
Ergebnisse der ozeanographischen Forschung im Internationalen Geophysikalischen Jahr. 1. Erforschung der Zirkulation im Atlantischen Ozean.
Umschau, (16):499-502.
Abstract in:
Ozeanogr., 1961, Deutsch. Hydrogr. Inst.

circulation
Toporkov, L.G. 1968
Water circulation in the southeastern part of the Pacific. (In Russian)
Inform. Biull. Sovetsk Antarkt. Exped. 71: 36-42.
Translation: Scripta Tecnica, Inc. for AGU.
Vol. 7(3):199-202

circulation
Trites, R.W., 1956.
The ocean floor and bottom movement.
Trans. R. Soc., Canada, (3)50:83-91.

circulation
Trites, R.W., 1956.
The oceanography of Chatham Sound, British Columbia. J. Fish. Res. Bd., Canada, 13(3):385-434.

circulation

Tsujita, T., 1954.
On the observed oceanographic structure of the Tsushima fishing grounds in winter and ecological relationship between the structure and the fishing conditions. J. Ocean. Soc., Japan, 10(3): 158-170, 12 textfigs.

circulation

Tully, J.P., and F.G. Barber, 1961
An estuarine model of the sub-Arctic Pacific Ocean.
In: Oceanography, Mary Sears, Edit., Amer. Assoc. Adv. Sci., Publ. No. 67:425-454.

Previously published under title "An estuarine analogy in the sub-Arctic Pacific Ocean" J. Fish. Res. Bd., Canada, 18:91-112(1960).

U

circulation

Uda, Michitaka, 1961.
Pulsation theory of the changing pattern in the North Pacific circulation. (Abstract).
Tenth Pacific Sci. Congr., Honolulu, 21 Aug.-6 Sept., 1961, Abstracts of Symposium Papers, 345-346.

circulation

Uda, M., 1955.
Rechearches on the fluctuation of the North Pacific circulation. 1. The fluctuation of Oyasiwo Current in relation to the atmospheric circulation and to the distribution of the dichothermal waters of the North Pacific Ocean. Rec. Ocean. Wks., Japan, 2(2):43-55, 17 textfigs.

circulation

Ueno, Takeo, 1967.
Numerical experiments of the wind-driven circulation in the Pacific Ocean.
Oceanogr. Mag., Jap. Met. Agency, 19(2): 119-156

V

circulation

Väisänen, A., 1961.
A study of the symmetrical general circulation by the aid of a rotating water tank experiment. Geophysica, Meteorology, 8(1):39-62.

circulation

Vallaux, C., 1927.
Expédition scientifique du "Meteor" au sud de l'Atlantique et dans l'Ocean Austral (1925-1926):premiers résultats. Ann. Inst. Océan., n.s., 4(1):1-24, 6 textfigs.

circulation

Varadachari, V.V.R., and G.S. Sharma, 1967.
Circulation of the surface waters in the North Indian Ocean.
J. India Geophys. Un., 4(2):61-73
Also in: Coll. Repr. Nat. Inst. Oceanogr. New Delhi, 1 (1963-1965)

circulation

Varlet, F. and M. Menache, 1947
L'Estuaire du Moros à Concarneau (Finistère) Etude du melange des eaux douces et salees. Bull. l'Inst. Ocean., Monaco, 917:26 pp., 2 text figs.

circulation

*Veronis, George, 1970.
Effect of fluctuating winds on ocean circulation. Deep-Sea Res., 17(3): 421-434.

circulation

Veronis, George, 1963
Wind-driven and thermal ocean circulation.
United States National Report, 1960-1963.
Thirteenth General Assembly, I.U.G.G.
In: Trans. Amer. Geophys. Union, 44(2):501-503.

circulation

Veronis, G., and G.W. Morgan, 1955.
A study of the time-dependent wind-driven circulation in a homogeneous, rectangular ocean. Tellus 7(2):232-242, 6 textfigs.

circulation

Vilks, G., 1970.
Circulation of surface waters in parts of the Canadian Arctic Archipelago based on foraminiferal evidence.
Arctic 23(2):100-111

circulation

Von Arx, W.S., 1957.
An experimental approach to problems in physical oceanography. Physics and chemistry of the earth, Pergamon Press:1-29.

circulation

Von Arx, W.S., (1953) 1956.
Some techniques for laboratory study of the primary ocean circulation. Fluid Models in Geophysics, Proc. First Symposium on Use of Models in Geophysical Fluid Dynamics, Johns Hopkins Univ., Sept. 1-4, 1953:89-99.

circulation

Von Arx, W.S., 1954.
A laboratory model of the wind driven ocean circulation. Weather 9:170-176.

circulation

von Arx, W.S., 1948.
The circulation systems of Bikini and Rongelap Lagoons. Trans. Am. Geophys. Union, 29(6):861-870, 8 textfigs.

circulation

Vorobjev, V.N., 1966.
The importance of long-period tides for studying the circulation of water and drift of ice in the outlying areas of the Arctic. (In Russian).
Materiali, Sess. Uchen. Soveta PINRO Rez. Issled., 1964, Minist. Ribn. Khoz., SSSR, Murmansk, 75-80.

W

circulation

Waldichuk, Michael, 1957.
Physical oceanography of the Strait of Georgia. J. Fish. Res. Bd., Canada, 14(3):321-486.

circulation

Waldichuk, M., 1955.
Effluent disposal from the proposed pulp mill at Crofton, B.C. Fish. Res. Bd., Canada, Pacific Coast Stas., Prog. Repts. No. 102:6-9, 3 textfigs.

circulation

Warren, Bruce A., 1970.
General circulation of the South Pacific.
In: Scientific exploration of the South Pacific, W.S. Wooster, editor, Nat. Acad. Sci., 33-49.

circulation

Warsh, K.L., E.Z. Stakhiv and M. Garstang, 1970.
On the relation between the surface temperatures of the Gulf of Mexico and its circulation.
Bull. mar. Sci., Miami, 20(4): 803-812.

circulation

Woodhead, P.M.J., 1970.
Sea-surface circulation in the southern region of the Great Barrier Reef, spring 1966.
Aust. J. Mar. Freshwat. Res. 21(2): 89-102.

circulation

Worthington, L.V., 1962
Evidence for a two gyre circulation system in the North Atlantic.
Deep-Sea Res., 9(1): 51-68.

circulation

Worthington, L.V., 1954.
A preliminary note on the time scale in North Atlantic circulation. Deep-Sea Res. 1(4):244-251, 5 textfigs.

circulation

Worthington, L.V., 1953.
Oceanographic results of Project SKIJUMP I and SKIJUMP II in the Polar Sea, 1951-1952. Trans. Am. Geophys. Union 34(4):543-551, 4 text figs.

circulation

Wüst, Georg, 1959
Remarks on the circulation of the intermediate and deep water masses in the Mediterranean Sea and the methods of their further exploration.
Annali Ist. Universitario Navale, Napoli, 28: 3-16.

circulation

Wüst, G., 1950.
Blockdiagramme der atlantischen Zirculation auf Grund der "Meteor"-Ergebnisse. Kieler Meeresf. 7(1):24-34, 3 textfigs., 1 fold-in.

circulation

Wüst, G., 1949.
Die Kreisläufe der atlantischen Wassermassen, ein neuer Versuch räumlicher Darstellung. Forsch. Fortschritte 23/24:287-290.

circulation

Wüst, Georg (with Arnold Gordon) 1964
Stratification and circulation in the Antillean-Caribbean basins. 1. Spreading and mixing of the water types with an oceanographic atlas.
Vema Research Series, Columbia Univ. Press, New York and London, No. 2: 201 pp.

circulation

*Wyrtki, Klaus, 1967
Circulation and water masses in the eastern equatorial Pacific Ocean.
Int. J. Oceanol. Limnol., 1(2): 117-147.

circulation

Wyrtki, Klaus, 1966.
Oceanography of the eastern equatorial Pacific Ocean.
In: Oceanography and marine biology, H. Barnes, editor, George Allen & Unwin Ltd., 4:33-68.

circulation

Wyrtki, Klaus, 1965.
The average annual heat balance of the North Pacific Ocean and its relation to ocean circulation.
J. geophys. Res., 70(18):4547-4559.

circulation

Wyrtki, Klaus, 1964.
Total integrated mass transports and actual circulation in the eastern South Pacific Ocean.
In: Studies on Oceanography dedicated to Professor Hidaka in commemoration of his sixtieth birthday, 47-52.

circulation

Wyrtki, K., 1962
Geopotential topographies and associated circulation in the Southeastern Indian Ocean.
Austral. J. Mar. & Freshwater Res., 13(1): 1-17.

circulation

Wyrtki, Klaus, 1961
The thermohaline circulation in relation to the general circulation in the oceans.
Deep-Sea Res., 8(1): 39-64.

circulation

Wyrtki, Klaus, 1960.
Circulation of surface water in the Coral and Tasman seas.
Fisheries Newsletter, Australia, 19(2):17.

circulation

Wyrtki, Klaus, 1960
The surface circulation in the Coral and Tasman seas.
C.S.I.R.O., Australia, Div. Fish. Oceanogr., Tech. Paper, (8): 1-44.

XYZ

circulation

*Yoshida, Kozo, 1967.
Circulation in the eastern tropical oceans with special reference to upwelling and undercurrents.
Jap. J. Geophys., 4(2):1-75.

circulation

Yoshida, K., 1955.
An example of variations in oceanic circulation in response to the variations in wind field.
J. Ocean. Soc., Japan, 11(3):103-108.

Also in: Geophys. Notes, Tokyo, 8(2):Contr. 33.

circulation

Yoshida, K., H.-L. Mao, and P.L. Horrer, 1953.
Circulation in the upper mixed layer of the Equatorial North Pacific. J. Mar. Res. 12(1): 99-120, 4 textfigs.

circulation

Zaitsev, G.N., 1961.
On the exchange of waters between the Arctic basin and the Pacific and Atlantic Oceans.
Okeanologiia, Akad. Nauk, SSSR, 1(4):743-744.

circulation

Zhukov, L.A., 1970.
On the annual cycle of variations in the circulation and thermohaline field in the North Atlantic.
Oceanologiia, 10(5): 770-773. (in Russian; English abstract)

circulation

Zore-Armanda, Mira, 1963.
Mixing of three water types in the south Adriatic
Rapp. Proc. Verb., Réunions, Comm. Int., Expl. Sci., Mer Méditerranée, Monaco, 17(3):879-885.

circulation

Zuta, Salvador, y Oscar Guillen, 1970.
Oceanografia de las aguas costeras del Peru.
Bol. Inst. Mar., Peru, 2(5):157-324.

Circulation, abyssal

circulation, abyssal

Arons, Arnold B., and Henry Stommel, 1967.
On the abyssal circulation of the World Oceans. III. An advection-lateral mixing model of the distribution of a tracer property in an ocean basin.
Deep-Sea Research, 14(4):441-457.

circulation, abyssal

Stommel, H., 1958.
The abyssal circulation. Letter to the Editors.
Deep-Sea Res., 5(1):80-82.

circulation, abyssal

Stommel, H., 1957.
The abyssal circulation under the ocean. Nature 180(4589):733-734.

circulation, abyssal

Stommel, Henry, and A.B. Arons, 1960.
On the abyssal circulation of the world ocean. II. An idealized model of the circulation pattern and amplitude in oceanic basins.
Deep-Sea Res., 6(3):217-233.

circulation, abyssal

Stommel, Henry, and A. B. Arons, 1960.
On the abyssal circulation of the world ocean. 1. Stationary planetary flow patterns on a sphere.
Deep-Sea Res., 6(2):140-154.

circulation, abyssal

Tolmazin, D.M., and M. Sh. Rosengurt, 1965.
On the abyssal horizontal and vertical movement of waters in the Black Sea. (In Russian).
Okeanologiia, Akad. Nauk, SSSR, 5(5):849-853.

circulation, abyssal

Uda, M., 1958.
On the abyssal circulation in the northwest Pacific area. Geophys. Mag., Tokyo, 28(3):411-416.

circulation, abyssal

Vladimirtsev, Y. A., 1964.
On the problem of the abyssal circulation in the Black Sea. (In Russian).
Okeanologiia, Akad. Nauk, SSSR, 4(6):1013-1019.

Circulation, annual variation

circulation, annual variations

Maksimov, I.V. 1966.
Index of annual changes in the circulation of the northern Atlantic Ocean. (In Russian).
Trudy Poliarn. Nauchno-Issled. Proektn. Inst. Morsk. Ribn. Khoz. Okeanogr. N.M. Kniporich (PINRO) 17:265-277.

circulation, annual variation

Takano, Kenzo, 1964.
Variation annuelle de la circulation générale dans les océans (suite et fin).
La Mer, Bull. Soc. Franco-Japonaise d'Océanogr. 2(1):1-21.

Also in:
Geophys. Notes Tokyo, 17(2): (Contr.16). 1964.

circulation, Antarctic

Gordon, Arnold L., 1970.
Antarctic circulation. EOS, Trans. Am. geophys. Un. 52(6): 320-232.

Circulation, atmospheric

circulation (atmosphere)

Bjerknes, J.A.B., 1959

North Atlantic sea surface temperatures and atmospheric circulation. Am. Geophys. U., Pac. Southwest Regional Meeting, Feb. 5-6, 1959.

Abstract in J. Geophys. Res., 64(6): 689

circulation, atmospheric

Hastenrath, Stefan L. 1971.
On meridional circulation and heat budget of the troposphere over the Equatorial Central Pacific.
Tellus 23(1): 60-73.

circulation, atmospheric

Kletter, L., 1959.
Charakterische Zirkulationstypen in mittleren Breiten der nördlichen Hemisphäre.
Arch. Met. Geophys., Bioklim., (A), 11(2):161-196.

circulation, atmospheric, effect of

Saha, Kshudiram, 1970.
Interhemispheric drift of radioactive debris and tropical circulation.
Tellus 22(6): 688-698

circulation (air) (monsoon)

Saha, Kshudiram 1970.
Zonal anomaly of sea surface temperature in equatorial Indian Ocean and its possible effect upon monsoon circulation.
Tellus 22(4): 403-409

circulation, atmospheric

Sorkina, A.J., 1965
Comparison of the atmospheric circulation in the northern parts of the Atlantic and Pacific oceans. (In Russian).
Okeanologiia, Akad. Nauk, SSSR, 5(3):426-430.

circulation, atmospheric

*Winston, Jay S., 1967.
Zonal and meridional analysis of 5-day averaged outgoing long-wave radiation data from TIROS IV over the Pacific sector in relation to the northern hemisphere circulation.
J. app. Met., 6(3):453-463.

Circulation, bottom

circulation, bottom

Bulatov, R.P., 1971.
On the structure and circulation of the bottom layer in the Atlantic Ocean. (In Russian; English abstract). Usloviia sediment. Atlantich. Okeana. Rez. Issled. Mezhd. Geofiz. Proekt. Mezhd. Geofiz. Kom. Prezid. Akad Nauk SSSR :43-59.

circulation, bottom

Burkov, V.A., 1969.
The bottom circulation in the Pacific. (In Russian; English abstract). Okeanologiia 9(2): 223-234.

circulation coastal

Ramanadham, R., B.S. Reddy and K.R.G.K. Murty 1971.
Coastal circulation near Kakinada Bay during the monsoon period.
Pure appl. Geophys. 89: 183-191

circulation cells

circulation cells
Faller, Alan J., 1962
The instability of Ekman boundary-layer flow and its application as a possible explanation of 'Langmuir' circulation cells in the ocean. (Abstract).
J. Geophys. Res., 67(9):3556.

circulation convective

circulation, convective
Takano, K., 1955.
Note on the convective circulation (II).
J. Ocean. Soc., Japan, 11(2):39-41.

circulation, convective
Takano, K., 1955.
Note on the convective circulation (1).
Rec. Ocean. Wks., Japan, 2(2):21-29.

circulation, convective
Takano, K., 1955.
Note on the convective circulation. III.
J. Ocean. Soc., Japan, 11(3):97-102.
Also in:
Geophys. Notes, Tokyo, 8(2):Contr. 35.

circulation, cyclonic

circulation, cyclonic (tropical)
Bulatov, R.P., and V.N. Stepanov 1968.
Tropical cyclonic circulation of ocean waters. (In Russian).
Dokl. Akad. Nauk SSSR 178(6):1403-1406.

circulation, deep

circulation, deep
Barbee, William D. 1966.
Deep circulation, central North Pacific Ocean, 1961, 1962, 1963.
Res. Pap. U.S. Cst Geodet. Surv. 1-104
Also in: Coll Repr. ESSA Inst. Oceanogr. 1966

circulation, deep
Bolin, Bert, and Henry Stommel, 1961
On the abyssal circulation of the World Ocean. IV. Origin and rate of circulation of deep ocean water as determined with the aid of tracers.
Deep-Sea Res., 8(2):95-110.

circulation, deep
Deacon, G.E.R., 1957.
The deep-water circulation in the Pacific Ocean.
Proc. UNESCO Symp., Phys. Ocean., Tokyo, 1955: 11~~0~~115.

deep circulation
Fukai, R., 1958.
On the deep circulation in the north-western North Pacific with reference to vertical distribution of dissolved oxygen.
Proc. UNESCO Symp. Phys. Oceanogr., Tokyo, 1955, 149-152.
Included in COLLECTED REPRINTS, 1958, Tokai Reg. Fish. Res. Lab., B-237.

circulation, deep
Fukai, R., 1957.
On the deep circulation in the north-western North Pacific with reference to the vertical distribution of dissolved oxygen.
Proc. UNESCO Symp., Phys. Ocean., Tokyo, 1955: 119-152.

circulation, deep.
Japan, National Committee for the International Geophysical Year 1960.
The results of the Japanese oceanographic project for the International Geophysical Year, 1957/58: 198 pp.

circulation, deep
Knauss, John A., 1961.
Water movements in the deep Pacific. 1. Evidence from direct current and temperature observation. (Abstract).
Tenth Pacific Sci. Congr., Honolulu, 21 Aug.-6 Sept., 1961, Abstracts of Symposium Papers, 336-337.

circulation, deep
Laird, N.P., 1971.
Panama Basin deep water - properties and circulation. J. mar. Res. 29(3): 226-234.

circulation, deep
LePichon, X., 1960
The deep water circulation in the southwest Indian Ocean. J. Geophys. Res., 65(12): 4061-4074.

circulation, deep
Le Pichon, Xavier, et Jean-Paul Troadec, 1963
La couche superficielle de la Méditerranée au large des côtes Provençales durant les mois d'été.
Cahiers Océanogr., C.C.O.E.C., 15(5):299-314

circulation, deep
Masuzawa, Jotaro, 1961.
Deep-water circulation in the western boundary of the North Pacific. (Abstract).
Tenth Pacific Sci. Congr., Honolulu, 21 Aug.-6 Sept., 1961, Abstracts of Symposium Papers, 337.

circulation, deep
Neumann, G., 1943.
Über den Aufbau und die Frage der Tiefenzirkulation des Schwarzen Meeres.
Ann. Hydrogr., usw., 71(1):1-20.

circulation, deep
Olivet, Jean-Louis, Xavier Le Pichon et Lucien Laubrier, 1970.
La fracture Gibbs et la circulation profonde en Atlantique Nord.
C.r. hebd. Séanc. Acad. Sci. Paris (D) 271(21): 1848-1851

circulation, deep
Reid, Joseph L., 1970.
Deep circulation. EOS. Trans. Am. geophys. Un. 52(6): 227-229.

circulation, deep
Reid, Joseph L., 1967.
Intermediate and deep circulation.
Trans. Am. geophys. Un., 48(2):572-575.

circulation, deep
Rochford, D.J., 1960
Some aspects of the deep circulation of the Tasman and Coral Seas. Austral. J. Mar. & Freshwater Res., 2(1): 166-181.

circulation, deep
Suess, Hans E., and John A. Knauss, 1961.
Water movements in the deep Pacific, II. Evidence from radiocarbon measurements. (Abstract).
Tenth Pacific Sci. Congr., Honolulu, 21 Aug.-6 Sept., 1961, Abstracts of Symposium Papers, 338.

circulation, deep
Takenouti, Y., 1960
X. Oceanography. Japanese Contrib. I.G.Y., 1957/8, 2: 124-145.

circulation, deep
Thoulet, J., 1925.
Contribution à l'étude de la circulation océanique. Annal. Inst. Océan., Monaco, 2:409-423.

circulation, deep
Thoulet, J., 1921.
Sur la circulation océanique profonde.
Bull. Inst. Océan., Monaco, No. 404:1-10.

circulation, deep
Vladimirtsev, Yu. A., 1966.
A deep-ocean circulation in the midland basins of the World Ocean. (In Russian; English abstract).
Okeanologiia, Akad. Nauk, SSSR, 6(4):569-579.

circulation, deep-sea
*Welander, Pierre, 1969.
Effects of planetary topography on the deep-sea circulation.
Deep-Sea Res., Suppl. 16: 369-391.

circulation, deep
Wooster, Warren S., and Gordon H. Volkmann, 1961.
Indications of deep Pacific circulation from the distribution of properties at five kilometers. (Abstract).
Tenth Pacific Sci. Congr., Honolulu, 21 Aug.-6 Sept., 1961, Abstracts of Symposium Papers, 339.

circulation, deep
Wooster, Warren S., and Gordon H. Volkmann, 1960.
Indications of deep Pacific circulation from the distribution of properties at five kilometers.
J. Geophys., Res. 65(4):1239-1250.

circulation, deep
*Wright, Redwood, 1969.
Deep water movement in the western Atlantic as determined by use of a box model.
Deep-Sea Res., Suppl. 16: 433-446.

circulation, deep

Wüst, Georg, 1960.
Die Tiefenzirkulation des Mittelländischen Meeres in den Kernschichten des Zwischen- und des Tiefenwassers.
Deutsche Hydrogr. Zeits., 13(3):105-131.

circulation, deep

Wüst, G., 1929.
Schichtung und Tiefenzirkulation des Pazifischen Ozeans. Veröff. Inst. Meeresk., Geogr. Reihe, 20: 1-64.

Transl. cited:
USFWS Spec. Sci. Rept., Fish., 227.

circulation, deep

Wyrtki, Klaus, 1961.
The flow of water into the deep sea basins of the western South Pacific Ocean.
Australian J. Mar. & Freshwater Res., 12(1):1-16

circulation, deep

*Yasuoka, Takeo, 1968.
Hydrography in the Okhotsk Sea. (2).
Oceanogrl Mag., 20(1):55-63.

circulation, deep

Zaklinsky, G.V., 1963
On the deep-circulation of the Indian Ocean waters. (In Russian).
Okeanologiia, Akad. Nauk, SSSR, 3(4):591-598.

CIRCULATION, DEEP, EFFECT OF

Berger, Wolfgang H., 1970.
Biogenous deep-sea sediments: fractionation by deep-sea circulation.
Bull. geol. Soc. Am., 81(5): 1385-1402.

circulation, drift

circulation, drift

Postnova, I.D., 1962.
Theoretical scheme of steady drift circulation in basins in the presence of cross-section profiles. (In Russian).
Izv., Akad. Nauk, SSSR, Ser. Geofiz., (11):1663-1670.

circulation, dynamic

Arseniev, V.S., 1965. circulation, dynamic
Circulation in Bering Sea. (In Russian; English abstract).
Okeanolog. Issled., Rez. Issled. po Programme Mezhd. Geofiz. Goda, Mezhd. Geofiz. Komitet, Prezidiume Akad. Nauk, SSSR, No.13:61-65.

circulation, effect of

Cromwell, T., 1953. circulation, effect of
Circulation in a meridional plane in the central equatorial Pacific. J. Mar. Res. 12(2):196-213, 9 textfigs.

circulation effect of

Dawson, E.Y., 1952.
Circulation within Bahia Vizcaino, Baja California and its effect on the marine vegetation.
Am. J. Bot. 39(7):425-432, figs.

circulation, effect of

Hulburt, Edward M., 1970.
Relation of heat budget to circulation in Casco Bay, Maine.
J. Fish. Res. Bd, Can., 27(12): 2255-2260

circulation, effect of

Ketchum, B.H., 1954.
Relation between circulation and planktonic populations in estuaries. Ecology 35(2):191-200, 7 textfigs.

circulation, effect of

McEwen, G.F., 1919.
Ocean temperatures, their relation to solar radiation and oceanic circulation.
Semicent. Publ. Univ. Calif., Misc. Stud., 334-421.

circulation, effect of

McEwen, G.F., 1914.
Peculiarities of the California climate on the basis of general principles of atmospheric and oceanic circulation.
Monthly Weather Review, 42:14-23.

circulation, effect of

Parin, N.V., and K.V. Beklemishev 1966.
On the possible effect of long-term variation in water circulation on the distribution of pelagic animals in the Pacific Ocean. (In Russian; English abstract)
Gidrobiol. Zhurn. 2(1): 3-9.

circulation, effect of

Selitskaia, E.S., 1959.
Variability of surface water temperature for separate regions of the North Sea. (In Russian).
Trudy Gosud. Okeanogr. Inst., 48:104-111.

circulation, effect of

*Voronina, N.M., 1968.
The distribution of zooplankton in the Southern Ocean and its dependence on the circulation of the water.
Sarsia, 34:277-284.

circulation, equatorial

circulation, equatorial

Knauss, John A., 1961.
Comparison of the equatorial circulation in the Atlantic, Pacific and Indian oceans. (Abstract).
Tenth Pacific Sci. Congr., Honolulu, 21 Aug.-6 Sept., 1961, Abstracts of Symposium Papers, 341-342.

circulation, equatorial

Tsuchiya, Mizuki, 1970.
Equatorial circulation of the South Pacific.
In: Scientific exploration of the South Pacific, W.S. Wooster, editor, Nat. Acad. Sci., 69-74

circulation equatorial

Yoshida, Kozo, 1961.
Some calculations on the equatorial circulation.
Rec. Oceanogr. Wks., Japan, 6(1):101-105.

Also in:
Geophys Notes, Tokyo Univ., 14(2)(24):reprint

circulation, estuaries

circulation (estuaries)

Bowden, K.F., 1967.
Circulation and diffusion.
In: Estuaries, G.H. Lauff, editor, Publs Am. Ass. Advmt Sci., 83:15-36.

circulation estuarine

Carriker, Melbourne R., 1967.
Ecology of estuarine benthic invertebrates: a perspective.
In: Estuaries, G.H. Lauff, editor, Publs Am. Ass. Advmt Sci., 83:442-487.

circulation (estuaries)

Hansen, Donald V., 1967.
Salt balance and circulation in partially mixed estuaries.
In: Estuaries, G.H. Lauff, editor, Publs Am. Ass. Advmt Sci., 83:45-51.

circulation, estuaries

Pritchard, Donald W., 1967.
Observations of circulation in coastal plain estuaries.
In: Estuaries, G.H. Lauff, editor, Publs Am. Ass. Advmt Sci., 83:37-44.

circulation (estuaries)

Riley, Gordon A., 1967.
The plankton of estuaries.
In: Estuaries, G.H. Lauff, editor, Publs Am. Ass. Advmt Sci., 83:316-326.

circulation, fjords

circulation (fjords)

Rattray, Maurice, Jr., 1967.
Some aspects of the dynamics of circulation in fjords.
In: Estuaries, J.H. Lauff, Editor, Publs Am. Ass. Advmt Sci., 83:52-62.

circulation (fjords)

Saelen, Odd H., 1967.
Some features of the hydrography of Norwegian fjords.
In: Estuaries, G.H. Lauff, editor, Publs Am. Ass. Advmt Sci., 83:63-70.

circulation, fluctuation

circulation, fluctuation of

Federov, K.N., 1957.
On yearly and half year fluctuations of the general circulation in the oceans. (In Russian).
Doklady, Akad. Nauk, SSSR, 116(3):393-

circulation, fluctuations

Uda, M., 1957.
Research on the fluctuation of the North Pacific circulation (1): Proc. UNESCO Symp., Phys. Ocean. Tokyo, 1955:112-113.

circulation, geostrpphic

Bulatov, R.P. 1971.
Circulation of waters of the Atlantic Ocean in different space-time scales. (In Russian).
Okeanol. Issled. Rezult. Issled. Meghd. Geofiz. Proekt. 22: 7-93.

circulation, geostrophic

Engel, I., 1967.
Currents in the eastern Mediterranean.
Int. hydrogr. Rev., 44(2):23-40.

circulation, geostrophic

Garner, D.M. 1970.
Hydrological studies in the New Zealand region 1966 and 1967: oceanic hydrology north-west of New Zealand, hydrology of the north-east Tasman Sea.
Bull. N.Z. Dept. scient. industr. Res. 202: 7-49.
Also: Mem. N.Z. Oceanogr. Inst. 58.

circulation, instability of

circulation, instability of

Robinson, A.R., 1962
The instability of a thermal ocean circulation
J. Mar. Res., 20(3):189-200.

circulation, island

Stockmann, W.B., and V.M. Kamenkovich, 1964.
On the effect of the wind and lateral friction on circulation around islands. (In Russian; English abstract).
In: Studies on Oceaography dedicated to Professor Hidaka in commemoration of his sixtieth birthday, 30-37.

circulation, intermediate (in depth)

Reid, Joseph L., 1967.
Intermediate and deep circulation.
Trans. Am. geophys. Un., 48(2):572-575.

circulation, longitudinal

*Wüst, Georg, 1968.
History of investigations of the longitudinal deep-sea circulation (1800-1922).
Bull. Inst. océanogr., Monaco, No. spécial 2: 109-120.

circulation, meridional

Hidaka, Koji 1968.
Non-linear computation of the meridional circulation in an equatorial section of the ocean.
Mitt. Inst. Meereok. Univ. Hamburg 10:140-160.

circulation, mid-ocean

Schulman, Elliott E. 1967.
The baroclinic instability of a mid-ocean circulation.
Tellus 19(2):292-305.

circulation models

Solomon, Harold, 1971.
On the representation of isentropic mixing in ocean circulation models. J. phys. Oceanogr. 1(3): 233-234.

circulation, monsoon-generated

Schmitz, H.P., 1967.
Eine numerische Approximation der monsunerzeugten Zirkulationen im Arabischen Meer
Dt. hydrogr. Z., 20 (5): 205-217.

circulation, nearshore

Harris, T.F.W., 1961.
The nearshore circulation of water.
Marine Studies off the Natal coast, C.S.I.R. Symposium, No. S2:18-30 (Multilithed).

circulation, seasonal

Bogdanov, K.I., 1961.
The water circulation in the Gulf of Alaska and its seasonal variation.
Okeanologiia, Akad. Nauk, SSSR, 1(5):815-824.

circulation, seasonal

Kutalo, A.A., 1971.
On seasonal variations of circulation in the North Atlantic. (In Russian; English abstract).
Fisika Atmosfer. Okean., Izv. Akad. Nauk SSSR, 7(3): 317-326.

circulation, surface

Burkov, V.A., and Iu. V. Pavlova, 1963
Geostrophic circulation on the surface of the northern part of the Pacific in summer. (In Russian).
Okeanolog. Issled., Rezhult. Issled. po Programme Mezhd. Geofiz. Goda, Mezhd. Geofiz. Kom., pri Presidiume Akad. Nauk, SSSR, No. 9: 21-31.

circulation, surface

Metallo, V., 1958.
Meccanismo della circolazione superficiale del Mediterraneo. Rivista Marittima, 92(1):

circulation, surface

Rotschi, Henri, 1960
Récents progrès des recherches océanographiques entreprises dans le Pacifique Sud-Ouest.
Cahiers Oceanogr., C.C.O.E.C., 12(4):248-267.

circulation, surface

Stepanov, V.N., 1960.
A general scheme for the circulation of the surface water in the World Ocean.
Trudy Okeanograf. Komissii, Akad. Nauk, SSSR, 10(1):69-78.

circulation, surface

Stevenson, Merritt R., Oscar Guillen G., and José Santoro de Ycaza, 1970
Marine atlas of the Pacific coastal waters of South America.
Univ. Calif. Press. 20pp., 99 charts, $40.00.

circulation, surface

Treshnikov, A.F., 1964.
Surface water circulation in the Antarctic Ocean. (In Russian).
Sovetsk. Antarkt. Eksped., Inform. Biull., 45:5-

Translation:
Scripta Tecnica, Inc., for AGU, 5(2):81-83. 1965.

circulation, surface

U.S.N. Hydrographic Office, 1957.
Oceanographic atlas of the Polar seas. 1. Antarctic. H.O. Pub., No. 705:70 pp.

circulation theorem (V. Bjerknes)

Krauss, W., 1957.
V. Bjerknes' circulation theorem and its application to oceanographic research.
Deutsche Hydrogr. Zeits., 10(1):13-19.

circulation, theoretical

Ichiye, T., 1960.
A theory of circulation in an ocean consisting of a homogeneous upper layer and a stratified lower layer.
J. Oceanogr. Soc., Japan, 16(2):47-54.

circulation (theoretical)

Moore, D.W., 1963.
Rossby waves in ocean circulation.
Deep-Sea Research, 10(6):735-747.

circulation (theoretical)

Takano, Kenzo, 1962
Circulation générale permanente dans un océan.
Rec. Oceanogr. Wks., Japan, N.s., 6(2):59-155.

circulation, thermohaline

Fukuoka, Jiro, 1967.
Masas de agua y dinámica de los océanos.
In: Ecología marina, Monogr. Fundación La Salle de Ciencias Naturales, Caracas, 14:130-183.

circulation thermohaline

Kozlov, V.F. 1968.
An application of one parameter density models to the investigation of thermohaline circulation in an ocean of finite depth (In Russian; English abstract).
Fizika Atmosfer. Okean. Izv. Akad. Nauk SSSR 4(6): 622-633.

circulation, thermohaline

Kozlov, V.F., 1967.
On the theory of the thermohaline circulation in the finite depth ocean. (In Russian; English abstract).
Fisika Atmosfer. Okean., Izv. Akad. Nauk, SSSR, 3(4):434-446.

circulation, thermohaline

*Needler, G.T., 1967.
A model for thermohaline circulation in an ocean of finite depth.
J. mar. Res., 25(3):329-342.

circulation, thermohaline

Stewart, R.W., 1969.
The atmosphere and the ocean.
Scient. Am., 221(3):76-86. (popular).

circulation, tidal

Owen, Wadsworth, 1969.
A study of the physical hydrography of the Patuxent River and its estuary.
Techn. Rept. Chesapeake Bay Inst., 53 (Ref. 69-6): 37 pp. (multilithed) (unpublished manuscript.)

circulation, tropical

Yoshida, K., 1968.
Dynamical oceanographic studies on upwelling and tropical circulation. (In Japanese).
J. oceanogr. Soc., Japan, 24(3):129-136.

circulation, variability

Ichiye, T., 1957.
On the variation of oceanic circulation in the adjacent waters of Japan.
Proc. UNESCO Symp., Phys. Ocean., Tokyo, 1955: 116-129.

circulation, variability of

Ichiye, T., 1957.
The change of various kinds of vortices in the sea east of Honshu. On the variation of oceanic circulation. Oceanogr. Mag., Tokyo, 8(2):123-142.

circulation, variation of

Johnson, J.A., C.B.Fandry and L.M.Leslie 1971.
On the variation of ocean circulation produced by bottom topography.
Tellus 23(2): 113-121.

circulation, variability of

Webster, Ferris, 1970.
Variability of ocean circulation. EOS, Trans. Am. geophys. Un. 52(6) : 225-227.

circulation, variability

Yoshida, K., 1955.
An example of variations in oceanic circulation in response to the variations in wind field.
J. Ocean. Soc., Japan, 11(3):103-108.

circulation, vertical

circulation, vertical

Adrov, M.M. 1968.
The influence of the winter vertical circulation upon the hydrochemical regime of near-bottom waters in the Barents Sea and their relation to geophysical processes. (In Russian; English abstract).
Trudy polyar. nauchno-issled. Inst. morsk. ryb. Khoz. Okeanogr. (PINRO) 23: 116-137.

circulation, vertical

Bogdanov, D.V., 1967.
Sobre la variabilidad de las condiciones oceanográficas en el Mar Caribe y el Golfo de Mexico. (In Russian; Spanish abstract).
Sovetsk.-Kubinsk. Ribokhoz. Issled. Vses. Nauchno-issled. Inst. Morsk. Ribn. Khoz. Okeanogr (VNIRO), Centr Ribokhoz. Issled. Natsion. Inst. Ribolov. Respibl. Kuba (TSRI), 2:21-38.

Cooper, L.H.N., 1958 circulation, vertical
Oceanographie experimentale en Méditerranée.
Trav. C.R.E.O., 3(2):17-25

circulation, vertical

Currie, Ronald I., 1965.
The oceanography of the southeast Atlantic.
Anais Acad. bras. Cienc., 37(Supl.):11-22.

vertical circulation

*Düing, Walter, 1966.
Die Vertikalzirkulation in den Küstennahen Gewässern des Arabischen Meeres während der Zeit des Nordostmonsuns.
"Meteor" Forschungsergebnisse, (A)(3):67-83.

circulation, vertical

Hidaka, K., and M. Koizumi, 1950.
Vertical circulation of water due to winds as inferred from the buckling experiments of elastic plates. Geophys. Notes, Tokyo Univ., 3(3):9 pp., 12 textfigs.

circulation, vertical

Hidaka, K., and M. Koizumi, 1950.
Vertical circulation of water due to winds as inferred from the buckling experiments of elastic plates. Geophys. Notes, Tokyo, 3(3):1-9, 12 textfigs.

circulation, vertical

Koshliakov, M.N., 1961.
The vertical circulation of water in the Kurosio area. Okeanologiia, Akad. Nauk, SSSR, 1(5):805-814.

circulation (vertical and horizontal)

LaFond, E.C., 1958
On the circulation of the surface layers of the East Coast of India.
Andhra Univ. Mem., Ocean., 2: 1-11.

circulation vertical

Nakano, M., 1957.
On a problem concerning the vertical circulation of sea water produced by winds with special reference to its bearing on submarine geology and submarine topography.
PROC. UNESCO Symp., Phys. Oceanogr., Tokyo, 1955 :107-111.

circulation, vertical

Nakano, M., 1955.
On a problem concerning the vertical circulation of sea water produced by winds, with special reference to its bearing on submarine geology and submarine topography.
Rec. Ocean. Wks., Japan, 2(2):68-81.

circulation, vertical

Olausson Eric, 1969.
On the Würm-Flandrian boundary in deep-sea cores.
Geologie Mijnb., 48(3): 349-361.

circulation, vertical

Privalova, I.V. 1971
Meridional and vertical circulation of waters in the northern part of the Atlantic Ocean. (In Russian)
Okeanol. Issled. Rezult. Issled. Meghd. Geofiz. Proekt. 22:154-219.

circulation, vertical

Rose, M., 1988.
La température et la circulation verticale des eaux méditerranéenes.
Bull. Sta. Aquic. et Pêche Castiglione 2:65-91.

circulation, vertical

Spidchenko, A.N., 1971.
On the possible use of Zubov's scheme of winter vertical circulation for machine data processing. (In Russian; English abstract).
Okeanologiia 11(5):913-916.

circulation, vertical

Stepanov, V.N., 1969.
Specific features in the vertical circulation of the Arctic basin waters. (In Russian).
Dokl. Akad. Nauk SSSR, 187(1): 187-190

circulation, vertical

Tcherniy, E.I. 1967.
Vertical circulation of waters in the Gulf of Alaska and seasonal variability. (In Russian).
Izv. tikhookean. nauchno-issled. Inst. ryb. Khoz. Okeanogr. 60: 163-172

circulation, vertical

Thomas, C. W., 1966.
Vertical circulation off the Ross Ice Shelf.
Pacific Science, 20(2):239-245.

circulation wind driven

circulation, wind-driven

Beardsley, R.C. 1969.
A laboratory model of the wind-driven ocean circulation.
J. fluid Mech. 38(2): 255-271.

circulation, wind-driven

Bryan, Kirk, 1963.
A numerical investigation of a nonlinear model of a wind-driven ocean.
J. Atmos. Sci., 20(6):594-606.

circulation, wind-driven

Bryan, Kirk, and Solomon Hellerman, 1966.
A test of convergence of a numerical calculation of the wind-driven ocean circulation.
J. Atmosph. Sci., 3(3):360-361.

circulation, wind-driven

Carrier, George F. and A.R. Robinson, 1962.
On the theory of wind-driven ocean circulation.
J. Fluid Mechanics, 12(1):49-80.

circulation, wind

Cooper, L.H.N., 1967.
The physical oceanography of the Celtic Sea.
Oceanogr. Mar.Biol.,Ann.Rev., H. Barnes, editor, George Allen and Unwin, Ltd., 5:99-110.

circulation, wind-driven

Csanady, G.T., 1968.
Wind-driven summer circulation in the Great Lakes.
J. geophys. Res., 73(8):2579-2589.

circulation, wind-driven

Csanady, G.T., 1968.
Simple analytical models of wind-driven circulation in the Great Lakes.
Proc. 11th Conf. Great Lakes Res., 1968: 371-354

circulation, wind.

Fel'senbaum, A.I., and N.B. Shapiro, 1966.
Steady wind circulation in a homogeneous ocean. (In Russian) Doklady, Akad. Nauk., S.S.S.R., 168(3):569-572.

circulation, wind-driven

Hassan, E.M., 1958.
On the winddriven ocean circulation.
Deep-Sea Res., 5(1):36-43.

circulation, wind-driven

Hendershot, Myrh C., 1968.
Physical and hydrodynamic factors. Ch. 7 in: Ocean engineering: goals, environment, technology, J.F. Brahtz, editor, John Wiley & Sons, 202-258.

circulation, wind-driven

*Holland, William R., 1967.
On the wind-driven circulation in an ocean with bottom topography.
Tellus, 19(4):582-600.

circulation, wind-driven

*Holopainen, E.O., 1967.
A determination of the wind-driven ocean circulation from the vorticity budget of the atmosphere.
Pure Appl. Geophys. 67:156-165.

Also: Finnish Met.Off.Contrib. 67.

circulation, wind driven

Johnson, J.A., 1971.
On the wind-driven circulation of a stratified ocean. J. mar. Res. 29(3): 197-213.

circulation, wind-driven

*Johnson, J.A., 1968.
A three-dimensional model of the wind-driven ocean circulation.
J. fluid Mech., 34(4):721-734.

circulation, wind-driven

Kowalik, Zygmunt, 1969.
Wind-driven circulation in a shallow sea with application to the Baltic Sea.
Acta geophys. pol. 17(1):13-35

circulation, wind-driven

Kowalik, Zygmunt, Andrzej Wróblewski, 1971
Wind-driven circulation in the Bay of Gdańsk. (In Polish; English and Russian abstracts).
Acta geophys. polon. 19 (2): 111-125.

circulation, wind-driven

Morgan, G.W., 1956.
On the wind driven ocean circulation. Tellus 8(3):301-320.

circulation, wind-driven

Niiler, Pearn P., 1966.
On the theory of wind-driven ocean circulation.
Deep-Sea Research, 13(4):597-606.

circulation, wind-driven

Robinson, A.R., 1964.
On the arbitrary suppression of vertical motion in wind-driven oceanic models.
J. Mar. Res., 22(2):168-174.

circulation, wind-driven

Robinson, Allan R., Editor, 1963.
Wind-driven ocean circulation. A collection of theoretical studies.
Blaisdell Publ. Corp., New York, Toronto and London, 161 pp.

circulation, wind-driven

Schulman, Elliott E., and Pearn P. Niiler, 1970
Topographic effects on the wind-driven ocean circulation.
Geophys. fluid Dyn., 1(4):439-462

circulation, wind-driven

Takano, Kenzo, 1965.
Periodic variation of the barotropic components of a wind-driven circulation in an ocean.
J. Oceanogr. Soc., Japan, 21(1):1-5.

circulation, wind-driven

Takano, Kenzo, 1962
Circulation générale permanente dans les océans - un calcul numérique complémentaire- (suite et fin).
J. Oceanogr. Soc., Japan, 20th Ann. Vol., 200-212.

English abstract

circulation, wind-driven

Ueno, Takeo 1967.
Numerical experiments of the wind-driven circulation in the Pacific.
Oceanogrl Mag. 19(2): 119-156.

circulation, wind-driven

Veronis, George, 1966.
Generation of mean ocean circulation by fluctuating winds.
Tellus, 18(1):67-76.

circulation, wind driven

Veronis, George, 1966.
Wind-driven ocean circulation. 1. Lenear theory and perturbations analysis. 2. Numerical solutions of the non-linear problem.
Deep-Sea Research. 13(1):17-29:31-55.

circulation, wind-driv

Veronis, George, 1965.
On parametric values and types of representation in wind-driven ocean circulation studies.
Tellus, 17(1):77-84.

circulation, wind-driven

Veronis, George, 1963.
An analysis of wind-drive ocean circulation with a limited number of Fourier components.
J. Atmos. Sci., 20(6):577-593.

circulation, wind-driven

von Arx, W.S., 1954.
A laboratory model of wind-driven ocean circulation. Weather 9(6):170-176, 4 textfigs.

CIRCULATION, WIND-DRIVEN

Welander, Pierre, 1958.
Wind-driven circulation in one- and two-layer oceans of variable depth.
Tellus 20(1): 1-16

circulation, wind-driven

Welander, Pierre, 1966.
A two-layer frictional model of wind-driven motion in a rectangular oceanic basin.
Tellus, 18(1):54-62.

civilization, effect of

civilization, effect of

Mann, Hans, 1965.
Die Beeinflussung der Fischerei in der Unterelbe durch zivilisatorische Massnahmen.
Helgolander wiss. Meeresunters. 17:168-181.

civilization, effect of

Mileikovsky, Simon A., 1968.
The influence of human activities on breeding and spawning of littoral marine bottom invertebrates
Helgoländer wiss. Meeresunters. 17: 200-205

clam industry

clam industry

Dow, Robert L., and Dana E. Wallace, 1961.
The soft-shell clam industry of Maine.
U.S.F.W.S. Circ., 110:1-36.

clapotis

clapotis

Suquet, F., and A. Wallet, 1953.
Basic experimental wave research.
Proc. Minnesota Int. Hydraulics Conv., Sept. 1-4, 1953:173-191, 18 figs.

classification

classification

Blackwelder, Richard E., 1963.
Classification of the animal kingdom.
Southern University Press, 94 pp. $7.00.

classification

Copeland, Herbert F., 1956.
The classification of lower organisms.
Pacific Books, Palo Alto, California, 302 pp.

classification

Mayr, Ernst, 1968.
Theory of biological classification.
Nature, Lond., 220(5167):545-548.

classification code (data)

Holm-Hansen, O., T.T. Packard and L.R. Pomeroy, 1970.
An improved code to classify the location of marine and terrestrial data. Limnol. Oceanogr., 15(5): 827-835.

clastics

clastics

*Neev, David, and K.O. Emery, 1967.
The Dead Sea: depositional processes and environments of evaporites.
Bull. Geol. Surv., Israel, 41: 146 pp.

clasts

clasts

Konzewitsch, Nicolas, 1961.
La forma de los clastos.
Rep. Argentina, Sec. Marina, Serv. Hidrogr. Naval, Publ. H. 626:113 pp.

English summary

clay

clay

Chamley, Hervé, et Françoise Picard, 1970.
L'héritage détritique des fleuves provençaux en milieu marin.
Téthys 2(1):211-226

clay

Conolly, J.R., and M. Ewing, 1967.
Sedimentation in the Puerto Rico Trench.
J. sedim. Petrol., 37(1):44-59.

clays

Lerman, A., 1966.
Boron in clays and estimation of paleosalinities.
Sedimentology, 6(4):267-286.

Clay

Neiheisel, James, 1965.
Source and distribution of sediments at Brunswick and vicinity, Georgia.
U.S. Army Coastal Eng. Res. Center, Techn. Memo., No. 12:49 pp.

clay

Pastouret, Léo, 1970
Étude sédimentologique et paléoclimatique de carottes prélevées en Méditerranée orientale.
Téthys 2(1):227-266.

clays

Seed, D.P. 1968.
The analysis of the clay content of some glauconitic oceanic sediments
J. sedim. Petrol. 38 (1): 229-231.

clay

Turekian, Karl K., 1967.
Estimates of the average Pacific deep-sea clay accumulation rate from material balance calculations.
Progress in Oceanography, 4:227-244.

clay, effect of

clay, effect of

Button, D.K. 1969.
Effect of clay on the availability of dilute organic nutrients to steady-state heterotrophic populations.
Limnol. Oceanogr. 14 (1): 95-100.

clay outcrops

Chamley, Hervé, 1969.
Relations entre la nature des minéraux argileux, leur origine pétrographique et leur environnement continental littoral ou marin. Cas de Nosy Bé (N.W. de Madagascar).
Rec. Trav. St. mar. Endoume, hors sér. Suppl. 9: 193-207.

clay, pelagic

clays, pelagic

Emiliani, Cesare, and John D. Milliman, 1966.
Deep-sea sediments and their ecological record.
Earth-Sci. Rev., Elsevier Publ. Co., 1:105-132.

clay, red

clay, red

Emiliani, Cesare, and John D. Milliman, 1966.
Deep-sea sediments and their ecological record.
Earth-Sci. Rev., Elsevier Publ. Co., 1:105-132.

clay suspensions

Pak, Hasong, J. Ronald V. Zaneveld and George F. Beardsley, Jr. 1971.
Mie scattering by suspended clay particles.
J. geophys. Res., 76(21): 5065-5069.

clay minerals
see also: minerals, clay

Clay minerals

See: minerals, clay

cleaning

Efford, Ian E. 1971.
The antennule cleaning setae in the sand crab, Emerita (Decapoda, Hippidae).
Crustaceana 21 (3): 316-318.

cleaning

Hobson, Edmund S. 1971.
Cleaning symbiosis among California inshore fishes.
Fish. Bull. nat. mar. fish. Serv. NOAA 69 (3): 491-523.

cleaning fishes

Holzberg, S., 1971.
Beobachtung einer Putzsymbiose zwischen der Garnele Leandrites cyrtorhynchus und Riffbarschen. Helgoländer wiss. Meeresunters 22(3/4): 362-365.

cliffs

Belderson, R.H. , N.H. Kenyon and A.H. Stride, 1970.
10-km wide views of Mediterranean deep-sea floor
Deep-Sea Res., 17(2): 267-270.

cliffs

Cotton, C.A. 1969.
Marine cliffing according to Darwin's theory.
Trans. R. Soc. N.Z., Geol. 6 (14): 187-208.

cliffs

Wright, L.W., 1970.
Variation in the level of the cliff shore platform junction along the south coast of Great Britain.
Marine Geol., 9 (5): 347-353.

Climatology

climatology

Ahlmann, H.W., 1949.
Climatic changes in the Arctic. Rapp. Proc. Verb. Cons. Perm. Int. Expl. Mer, 125:9-16.

climate

Arrhenius, G., 1950.
Late Cenozoic climatic changes as recorded by the Equatorial Current System. Tellus 2:83-88.

climate

Dunbar, M.J., 1954.
A note on climate change in the sea. Arctic 7(1):27-30.

climate

Friedman, Gerald M., and John E. Sanders 1970.
Coincidence of high sea-level with cold climate and low sea level with warm climate: evidence from carbonate rocks.
Bull. geol. Soc. Am. 81 (8): 2457-2458.

climate

Groissmayr, F.B., 1945.
Weltwetter und Klimaschwankung im Nordpolargebiet. Polarforsch., Jahrg. 18(1/2):5-9.

climatology

Lee, A.J., 1949.
The forecasting of climatic fluctuations and its importance to the Arctic fishery. Rapp. Proc. Verb Cons. Perm. Int. Expl. Mer, 125:40-41.

climatology

Lysgaard, L., 1949.
Recent climatic fluctuations. Rapp. Proc. Verb., Cons. Perm. Int. Expl. Mer, 125:17-20.

climatology

Ovey, C. D., 1949.
Note on the evidence for climatic changes from sub-oceanic cores. Weather 4(7):228-231, Pls. 3-4, 1 textfig.

Climatic changes

climatic change

*Hays, James D., and Neil D. Opdyke, 1967.
Antarctic Radiolaria, magnetic reversals and climatic change.
Science, 158(3804):1001-1011.

climate changes

Kukla, J., 1969.
The cause of the Holocene climatic change.
Geologie Mijnb., 48 (3): 307-334.

Climatic change

Mesolella, Kenneth J., R.K. Matthews, Wallace S. Broecker and David L. Thurber, 1969.
The astronomical theory of climatic change: Barbados data.
J. Geol. 77 (3): 250-274.

Climatic changes

Olausson, Eric, 1965.
Evidence of climatic changes in North Atlantic deep-sea cores with remarks on isotopic, paleotemperature analysis.
Progress in Oceanography, 3:221-252.

climate change, CO2 theory

Plass, G., 1956.
The carbon dioxide theory of climate change.
Tellus 8(2):140-154.

climatic change

Rasool, S.I., and J.S. Hogan, 1969.
Ocean circulation and climatic changes.
Bull. Am. met. Soc., 50(3): 130-134.

CLIMATE CHANGE

Vowinckel, E., and Svenn Orvig, 1969.
Climate change over the Polar Ocean. II. A method for calculating synoptic energy budgets. III. The energy budget of an Atlantic cyclone.
Arch. Met. Geophys. Biokl. (B) 17 (2/3): 121-146, 147-174.

climatic change

Weyl, Peter K., 1968.
The role of the oceans in climatic change: a theory of the ice ages. Met. Monogr. 8(30): 37-62. Also in: Coll. Reprints, Dept. Oceanogr. Univ. Oregon, 7(1968).

climatic changes

Wollin, Goesta, David B. Ericson and William B.F. Ryan 1971
Variations in magnetic intensity and climatic changes.
Nature, Lond., 232 (5312): 549-551

Climate, effects of

climate, effect of

Broecker, Wallace S., Yuan-Hui Li and Tsung-Hung Peng, 1971.
Carbon dioxide - man's unseen artifact. (11) In: Impingement of man on the oceans, D.W. Hood, editor, Wiley Interscience: 287-324.

climate, effect of

Crisp, D. J., 1959.
The influence of climatic changes on animals and plants.
Geogr. J. 125(1): 1-19.

climate, effect of

Jensen, Ad. S., 1939.
Concerning a change of climate during recent decades in the arctic and subarctic regions from Greenland in the west to Eurasia in the east and contemporary biological and geophysical changes.
Det. Kongl. Danske Vidensk. Selsk. Biol. Medd. 14(8):1-75.

Climate fluctuations in

climate, fluctuations in

Brown, P.R., 1953.
Climatic fluctuations in the Greenland and Norwegian Seas. Q.J.R. Met. Soc. 79(340):272-281.

climatic fluctuations

Cita, Maria Bianca, and Sara d'Onofrio, 1967.
Climatic fluctuations in submarine cores from the Adriatic Sea.
Progress in Oceanography, 4:161-178.

climatic fluctuations

Donahue, Jessie G. 1967.
Diatoms as indicators of Pleistocene climatic fluctuations in the Pacific sector of the Southern Ocean.
Progress in Oceanography 4:133-140.

climatic fluctuations

Hays, James D., 1967.
Quaternary sediments of the Antarctic Ocean.
Progress in Oceanography, 4:117-131.

climatic fluctuations

Wiles, William W., 1967.
Pleistocene changes in the pore concentration of a planktonic foraminiferal species from the Pacific Ocean.
Progress in Oceanography, 4:153-160.

Clouds

clouds

*Bunker, Andrew F., 1967.
Cloud formations leeward of India during the northeast monsoon.
J. atmos. Sci., 24(5):497-507.

clouds

Hayden, C.M., 1970.
An objective analysis of cloud cluster dimensions and spacing in the tropical North Pacific.
Mon. Wea. Rev., U.S. Dept. Comm. 98(7) 534-540.

clouds

Imai, Ichiro, Tetsu Ogata, Masami Akagawa, Akihito Kajihara and Ryoji Honda, 1969.
Band-shaped clouds over Funka Bay under winter monsoon situations. (In Japanese; English abstract).
Umi to Sora, 44(4):102-115.

Clouds

Magono, Choji, Katsuhiro Kikuchi and Toshiyuki Kasai 1969.
Comparison of aerial cloud pictures with satellite pictures - clouds over the Pacific Ocean. Part IV.
J. met. Soc. Japan 47(3): 227-234.

Clouds

Matsumoto, Seiichi, Kozo Ninomiya and Takako Akiyama 1968.
A statistical study on the relation between cloud amount and supply from the Japan Sea surface in January.
Pap. Met. Geophys. Tokyo 19(4): 551-558.

Clouds

Mossop, S.C., A. Ono, and E.R. Wishart, 1970.
Ice particles in maritime clouds near Tasmania.
Q. Jl R. Met. Soc. 96(409): 487-508.

clouds

*Ninimiya, K., 1966.
Some aspects of the cloud formation and the air-mass modification over the Japan Sea in winter revealed by TIROS observation. II.
Pap. Met. Geophys., Tokyo, 17(3):135-149.

cloud distribution

Saha, Kshudiram 1971.
Mean cloud distributions over tropical oceans.
Tellus 23(2): 183-195.

clouds

Sharp, F.A., 1958.
Cloud over the open sea.
Meteorol. Mag., 87(1033):202-204.

clouds

*Shenk, William E., 1970.
Meteorological satellite infrared views of cloud growth associated with the development of secondary cyclones.
Mon. Weath. Rev. U.S. Dept. Comm. 98(11):861-868.

clouds

*Srinivasan, V., 1968.
Some aspects of broad-scale cloud distribution over Indian Ocean during Indian southwest Monsoon season.
Indian J. Meteorol. Geophys. 19(1):39-54.

clouds

*Tsuchiya, Kiyoshi, and Tetsuya Fujita, 1967.
A satellite study of evaporation and cloud formation over the western Pacific under the influence of the winter monsoon.
J. Met. Soc., Japan, 45(3):232-250.

clouds

Turner, J.S., 1965.
Laboratory models of evaporation and condensation.
Weather, Roy. Meteorol. Soc., 20(4):124-128, Figs. on pp. 120-121.

Cloud cover

cloud cover

Tabata, Susumu, 1964.
Insolation in relation to cloud amount and sun's altitude.
In: Studies on Oceanography dedicated to Professor Hidaka in commemoration of his sixtieth birthday, 202-210.

Cloudiness

Bjerknes, Jacob, Lewis J. Allison, Earl R. Kreins, Frederic A. Godshall and Guenter Warnecke 1969.
Satellite mapping of the Pacific tropical cloudiness.
Bull. Am. met. Soc. 50(5):313-322.

clumps

Seki, H., 1971.
Microbial clumps in seawater in the euphotic zone of Saanich Inlet (British Columbia).
Mar. Biol. 9(1): 4-8.

Clouds, cumulus

cumulus convection

Bunker, Andrew F. 1971.
Energy transfer and tropical cell structure over the central Pacific.
J. atmos Sci. 28(7): 1101-1106.

Clouds, cumulus

Simpson, Joanne 1970.
Cumulus cloud modification: progress and aspects.
A century of meteorological progress, Am. Met. Soc., 143-155

clouds, cumulus

Malkus, Joanne S., and Robert H. Stimpson, 1964.
Modification experiments on tropical cumulus clouds. "Exploding" cumulus clouds by silver iodide seeding is used as a controlled experiment on their dynamics.
Science, 145(3632):541-548.

cumulus clouds

Novozhilov, N.I., 1962
Clouds over the sea. (In Russian).
Okeanologiia, Akad. Nauk, SSSR, 2(6):970-973.

clouds, cumulus

Oliver, Vincent J., 1968.
Cellular cumulus over the Pacific Ocean.
Mon. Weath. Rev. U.S. Dep. Comm. 96(7): 470-471

Clouds, cumulus

Simpson, Joanne 1967.
An experimental approach to cumulus clouds and hurricanes.
Weather 22(3): 95-114.

cumulus

*Simpson, Joanne, Glenn W. Brier and R.H. Simpson, 1967.
Storm fury cumulus seeding experiment 1965; statistical analysis and main results.
J. atmos. Sci., 24(5):508-521.

cumulus clouds (seeding)

Simpson, Joanne, and Victor Wiggert, 1971.
1968 Florida cumulus seeding experiment: numerical model results.
Mon. weath. Rev. U.S. Dept. Comm., 99(2):87-118.

clouds, cumulus towers

Simpson, Joanne and Victor Wiggert 1969.
Models of precipitating cumulus towers.
Mon. Weath. Rev. 97(7):471-489

Clouds effect of

clouds, effect of

Franceschini, Guy A., 1969.
The influence of clouds on solar radiation at sea.
Dt. hydrogr. Z. 21(4):162-168.

clouds, effect of

Steemann Nielsen, E., 1940.
Die Produktionsbedingungen des Phytoplanktons im Übergangsgebiet zwischen der Nord- und Ostsee.
Medd. Komm. Havundersøgelser, Ser. Plankton, 3(4):55 pp., 17 textfigs., 22 tables.

clouds, electrical properties

Takahashi, Tsutomu, Robert Uchida and Charles M. Fullerton, 1969
Surface measurements of the electrical properties of warm clouds over the sea.
Nature, Lond., 224(5223):113-114

cloud nuclei

Blanchard, Duncan C., 1971
The oceanic production of volatile cloud nuclei.
J. atmos. Sci. 28(5):811-812

Cloud nuclei

Dinger, J.E., H.B. Howell and T.A. Wojciechowski, 1970.
On the source and composition of cloud nuclei in a subsident air mass over the North Atlantic.
J. atmos. Sci., 27(5):791-797

Cloud, physics

cloud physics

Blanchard, D.C., 1953.
Raindrop size-distribution in Hawaiian rains.
J. Met. 10(6):457-473, 13 textfigs.

cloud water, chemistry

Lazrus, A.L., H.W. Baynton and J.P. Lodge, Jr. 1970.
Trace constituents in oceanic cloud water and their origin.
Tellus 22(1):106-114

clustering

clustering, intertidal

Moulton, James M., 1962
Intertidal clustering of an Australian gastropod.
Biol. Bull., 123(1):170-178.

coagulation

coagulation

Ichiye, T., 1955.
A note on the coagulation of small particles in the electrolytes. Ocean. Mag., Tokyo, 7(1):79-86.

coagulation

*Sakamoto, Wataru, 1968.
Study on the turbidity in an estuary. II. Observations on coagulation and settling processes of particles in the boundary of fresh and salt water. (In Japanese; English abstract).
Bull. Fac. Fish., Hokkaido Univ., 18(4):317-327.

Coal

coal

Anon. 1968.
Ocean-bottom minerals.
Ocean Industry 3(6):61-73

coal

Eden, R.A., Anne V.F. Carter and M.C. McKeown, 1969.
Submarine examination of lower Carboniferous strata on inshore regions of the continental shelf of southeast Scotland.
Marine Geol. 7(3):235-251.

Coastal bars

coastal bars

Nikiforov, L.G., 1964.
On the problem of the formation of coastal bars based on the study of the Ogurchinsky Island. (In Russian).
Okeanologiia, Akad. Nauk, SSSR, 4(4):654-659.

Coastal defense

coastal defense

DeRouville, A., 1950.
Renseignements et réflexions sur les ouvrages de défence des côtes. Conf. C.R.E.O. No. 2:40pp.

Extracted from ANN. Ponts Chausées, Sept.-Oct. 1950.

Coastal engineering

coastal engineering

Castanho, J., 1966.
Rebentação das ondas e transporte litoral.
Lab. Nac. Engenharia Civil, Memoria, Lisboa, No. 275-278 pp.
English synopsis (2 pp.)

coastal engineering

France, Laboratoire National d'Hydraulique, Chatou, 1957.
Port de Bone: étude sur modèle réduit. Ser. A: 25 pp.

coastal engineering

Hood, Donald W., editor 1971
Impingement of man on the oceans.
Wiley-Interscience, 738 pp.

coastal engineering

Johnson, J.W., ed., 1955.
Proceedings of the Fifth Conference on Coastal Engineering, Grenoble, Sept. 1954:669 pp.

coastal enineering

Reis de Carvalho, J.J., 1965.
Praia da Vitoria Harbour (Azores). Damages in the breakwater due to the storm of the 26th-27th December, 1962.
Lab. Nac. Engenharia Civil, Memoria, Lisboa, No. 271:20 pp.

coastal engineering

Terry, Richard D., editor, 1966.
Ocean Engineering, 1. Energy sources and energy conversion, waste conversion and disposal. 2. Undersea construction, habitation and vehicles; recreation.
Western Periodicals Co., North Hollywood, Calif. Vol. 3:431 pp. (multilithed).

Coastal geology
SEE ALSO GEOLOGY, COASTAL

coastal geology

Aibulatov, N. A., V. L. Boldyrev, V. P. Zenkovitch 1960.
Some new data on alongshore movements of the seashores of the USSR. Morsk. Geol., Doklady Sovetsk. Geol., 21 Sess., Int. Geol. Congress, 164-174.
English abstract

Coastal geology

Arbey, F., 1961.
Études littorales sur la côte des Maures. 1. Les croissants de plage (beach cusps).
Cahiers Océanogr., C.C.O.E.C., 13(6):380-396.

coastal geology

Auzel, Melle Marguerite, 1960.
Observations faites aux Îles de Glenan.
Cahiers Océanogr., C.C.O.E.C., 12(5):333-337.

Also in:
Trav. Lab. Géol. Sous-Marine, 10(1960).

coastal geology

Ballade, P., 1953.
Études des fonds sableux en Loire Maritime. Nature et evolution des ridens. Bull. d'Info., C.C.O.E.C. 5(4):163-176, Pls. 1-5.

coastal geology

Bascom, W., 1954.
The control of stream outlets by wave refraction.
J. Geol. 62(6):600-605, 4 textfigs.

coastal geology

Bremond, E., 1953.
Le talus continental marocain entre Mazagan et Fedala. Bull. d'Info., C.C.O.E.C. 5(6):242-244, 1 fig.

coastal geology

Caldwell, Joseph M., 1963
Shore processes and coastal engineering: 305-307.
Ocean wave spectra. Proceedings of a conference. Easton, Maryland, May 1-4, 1961.
Prentice-Hall, Inc., viii and 357 pp.

coastal geology

Charcot, J.B., 1934.
Rapport préliminaire sur la campagne du "Pourquoi Pas?" en 1933. Ann. Hydrogr., Paris, (3)13:1-85.

coastal geology

Charcot, J.B., 1933.
Rapport préliminaire sur la campagne du "Pourquoi Pas?" en 1932. Ann. Hydrogr., Paris, (3)12:1-29.

coastal geology

Charcot, J.B., 1925-1926.
Rapport préliminaire sur la campagne du "Pourquoi Pas?" en 1925. Ann. Hydrogr., Paris, (3)7:191-389, figs.

coastal geology

Charcot, J.B., 1925-1926.
Rapport préliminaire sur la campagne du "Pourquoi-Pas?" en 1924. Ann. Hydrogr., Paris, (3)7:1-96, figs.

coastal geology

Charcot, J.B., 1923-1924.
Rapport préliminaire sur la campagne du "Pourquoi-Pas?" en 1923. Ann. Hydrogr., Paris, (3) 6:1-89, figs.

coastal geology
Drach, P., and J. Monod, 1935-1936.
Rapport préliminaire sur les observations d'histoire naturelle et de géographie physique.
Ann. Hydrogr., Paris, Ser. 3, 14:29-37.

coastal geology
Geffrier, C.de, et J. Milliau, 1938-1939.
Mission hydrographique de la Guinée Française, de la Mauritanie et du Sénégal, 1936-1937.
Ann. Hydrogr., Paris, (3)16:119-183, figs.

Coastal geology
Guilcher, A., 1959
Coastal and submarine morphology
J. Wiley and Sons, Inc., N. Y., 274 pp.

coastal geology
Inman, D.L., 1953.
Areal and seasonal variations in beach and near-shore sediments at La Jolla, California.
B.E.B. Tech. Memo. No. 39:82 pp., 27 figs., numerous tables.

coastal geology
Ionin, A.S., P.A. Kaplin, and V.S. Medvedev, 1961
[Results of regional coastal investigations in the USSR.]
Trudy Inst. Okeanol., Akad. Nauk, SSSR,48: 3-33.

coastal geology
Kanaev, V. Ph., 1960
[Recent geomorphological observations on the Kurile Islands.] Trudy Inst. Okeanol., 32: 215-231.

coastal geology
King, L.C., and R.H. Belderson, 1961.
Origin and development of the Natal coast.
Marine Studies off the Natal Coast, C.S.I.R. Symposium, No. S2:1-9 (multilithed).

coastal geology
Lafon, M., 1953.
Recherches sur les sables cotiers de la Basse-Normandie et sur quelques conditions de leur peuplement zoologique. Ann. Inst. Océan., Monaco, 28(3):113-161.

coastal géology
Logatchev, L., 1956 Ref.
[Sur l'evaluation de l'etendu des dépôts dans les chenaux maritimes.] Inst. Nat. pour les Projets de Ports Marit., Trav., Moscow, 3:

coastal geology
Michailov, V.N., 1962.
River channel processes at the mouth of a one-branch rive discharging into a sea without tides. Marine geology and dynamics of coasts. (In Russian).
Trudy Okeanogr. Komissii, Akad. Nauk, SSSR, 10(3):123-134.

coastal geology
Navarro, F. de P., F. Lozano, J.M. Navaz, E. Otero J. Sáinz Pardo and others, 1943.
La pesca de Arrastre en los fondos del Cabo Blanco y del Banco Arguin (Africa Sahariana).
Trab. Inst. Español Ocean. No. 18:225 pp., 38 pls.

coastal geology
Seibold, Eugen, 1963
Geological investigation of near-shore sand-transport-examples of methods and problems from the Baltic and North seas.
Progress in Oceanography, M. Sears, Edit., Pergamon Press, London, 1:1-70.

coastal geology
Seshappa, G., 1953.
Phosphate content of mudbanks along the Malabar coast. Nature 171(4351):526-527.

coastal geology
Tromeur, J.Y., 1938-39.
Mission hydrographique du Saloum, 1930-1931.
Ann. Hydrogr., Paris, (3)16:5-33, figs.

coastal geology
Ulst, V. G., 1960.
[The general morphodynamic features of the Soviet coasts of the Baltic Sea.] Morsk.-Geol., Doklady Sovetsk. Geol., 21 Sess., Int. Geol. Congress, 197-203. (English Abstract)

coastal geology
Vercelli, F., 1923-24.
Ricerche di oceanografia fisica. Pt. 1. Correnti e maree. Ann. Idrogr. 11:13-208, charts, tables, diagrams.

coastal geology
Zenkovitch, V.P., 1962.
Basic state of the theory of the development of the accumulative forms in the coastal zones of the sea. Marine geology and dynamics of coasts.
Trudy Okeanogr. Komissii, Akad. Nauk, SSSR, 10(3):87-101.

coastal geology
Zenkovitch, V. P., O.K. Leontiev, and E. N. Nevessky 1960.
The influence of the eustatic post-glacial transgression on the development of the coastal zone of the USSR seas. Morsk. Geol., Doklady Sovetsk, Geo., 21 Sess., Int. Geol. Congress, 154-163.

Coastal hydrography

coastal hydrography
Geary, Edmund L., 1968.
Hidrografía costera.
Bol. Servicio Hidrograf. naval 5(3): 241-245
Translation of: Photogrammetric Engng 34(1):44-50.

COASTAL MARGIN
Gibson, Thomas G., 1970.
Late Mesozoic-Cenozoic tectonic aspects of the Atlantic coastal margin.
Bull. geol. Soc. Am., 81(6): 1713-1722.

Coastal processes

coastal processes
Johnson, J.W., 1956.
Nearshore sediment movement
Bull. Amer. Assoc. Petr. Geol., 40(9):2211-2232.

Coasts

A

coasts
Agron, S.L., 1963.
Oriented wave-splash fretwork in sandstone.
J. Sed. Petr., 33(2):469-471.

coasts
Ajbulatov, Nikolai, Helmut Griesseier, and Igor Sadrin, 1963.
Über den Sedimenttransport längs einer unregelmässig gegliederten Meeresküste.
Acta Hydrophysica, 8(1):5-21.

coasts
Ajbutalov, N., H. Griesseier and I Sadrin, 1962.
Küstendynamische Untersuchungen in der Uferzone der Anapa Nehrung.
Acta Hydrophysica, 7(2):105-150.

coasts
Aksenov, A.A., 1957.
[The effects of subdividing the administration on studies of sea coasts and reservoirs by the Oceanographic Commission of the Praesidium of the AN SSSR in 1952-1955.]
Trudy Okeanogr. Komissii, Akad. Nauk, SSSR, 2: 195-198.

coasts
Aksenov,A.A., A.S. Ionin, and F.A. Shcherbakov, 1965.
Some aspects of the coastal development and the near-shore sedimentation of the north coast of th Okhotsk Sea during post-glacial time. Sediment drift and the genesis of the heavy mineral fields in the coastal zone of the sea. (In Russian;English abstract).
Trudy Inst. Okeanol., Akad.Nauk,SSSR,76:76-102.

coasts
Aksenov,A.A., E.N. Nevessky, Yu. A. Pavlidis and F.A. Shcherbakov,1965.
The questions of the sea-coast fields formation. Sediment drift and the genesis of the heavy mineral fields in the coastal zone of the sea. (In Russian;English abstract).
Trudy Inst. Okeanol.,Akad. Nauk,SSSR,76:5-53.

coasts
Aksentjev, G.N., 1959
[Some peculiar processes of the land-slide s shore abrasion near Odessa (Black Sea). Questions of studies of marine coasts.]
Trudy Okeanogr. Komissii Akad. Nauk,SSSR, 4:118-121.

coasts
Allen, J.R.L., 1965.
Coastal geomorphology of eastern Nigeria: Beach-ridge barrier islands and vegetated tidal flats.
Geol. en Mijnbouw, 44(1):1-21.

Coasts
Anguenot, François, et André Monaco 1967.
Etude de transits sédimentaires, sur le littoral du Roussillon par la méthode des traceurs radioactifs.
Cah. océanogr. 19(7): 579-589.

coastline
Arber, M.A., 1948.
Factors controlling the Atlantic coastline of Europe. Nature 162:741-742.

coasts
Arbey, F., 1961.
Études littorales sur la côte des Maures. III. Les sinuosites rocheuses. IV. Les festons de plage.
Cahiers Oceanogr., C.C.O.E.C., 13(10):728-733.

coasts
Athearn, W.D., and C. Ronne, 1963.
Shoreline changes at Cape Hatteras: an aerial photographic study of a 17 year period.
Naval Research Reviews, 16(6):17-24.

coasts
Aybulatov, N.A., 1961.
[Effect of the wind on the dynamics of the accumulative shallow shore.]
Trudy Inst. Okeanol., Akad. Nauk, SSSR, 48:341-346.

coasts
Aybulatov, N.A., 1961.
Observations of the transport of sand along the shallow accumulative shore ("otmely"). Dynamics of the coastal zone of the Black Sea. (IN Russian
Trudy Inst. Okeanol., Akad. Nauk, SSSR, 53:3-18.

coasts
Aybulatov, N.A., & I.E. Schadrin, 1961
Role of discontinuous currents in the transport of sand-drifts near the coastal zone.
Trudy Inst. Okean., 53: 19-28.

coasts
Aybulatov, N.A., and I.E. Schadrin, 1961.
Some data on the transport of sand-drifts along the coastal zone near natural obstacles (capes, river branches). Dynamics of the coastal zone of the Black Sea. (In Russian).
Trudy Inst. Okeanol., Akad. Nauk, SSSR, 53:29-36.

B

Babin, C., J. Didier, A. Moign et Y. Plus-quellec, 1969.
Goulet et rade de Brest: essai de géologie sous-marine.
Revue Géogr. phys. Géol. dynam. (2) 11(1). 55-64

coasts
*Bacescu, M., E. Dumitrescu, M.T. Gomoiu, et A. Petran, 1967.
Elements pour la caracterisation de la zone sedimentaire medio-littorale de la Mer Noire.
Trav. Mus. Hist. nat. "Grigore Antipa", 7:1-14.

coasts
Bandy, Orville L., James C. Ingle, Jr., and Johanna M. Resig, 1964.
Facies trends, San Pedro Bay, California.
Geol. Soc., Amer., Bull., 75(5):403-424.

coasts
Bandy, Orville L., James C. Ingle, Jr., and Johanna M. Resig, 1964.
Facies trends, San Pedro Bay, California. (Abstract).
Geol. Soc., Amer., Special Paper, No. 76:7.

coasts
Barkovskaya, M.G., 1961
General laws governing the distribution of terrigenic material in the shoreline zone USSR of the Black Sea.
Trudy Inst. Okean., 53: 64-94.

coasts
Barsanova, N.G., 1961
[Some comparative characteristics of the rocky littoral fauna in the Barents and White seas.]
Trudy Inst. Okeanol., SSSR, 46:140-146.

coasts
Beall, Arthur O., 1968.
Sedimentary processes operative along the western Louisiana shoreline.
J. sedim. Petrol. 38(3): 869-877.

coasts
*Beauchesne, Pierre, et Guy Courtois, 1967.
Étude du mouvement des galets le long de la côte des Bas-Champs de la Somme utilisation de traceurs radioactifs.
Cah. océanogr., 19(8):613-625.

coasts
Beigbeder, Yvonne, 1967.
Problèmes géomorphologiques et sédimentologiques dans la partie orientale de la Baie de Saint-Brieuc.
Cah. Océanogr., 19(7):549-577.

Coasts
Bernard Francis 1967.
Réalisation d'une station automatique d'enregistrement des facteurs physico-chimiques dans la zone des marées.
Helgoländer wiss. Meeresunters. 15(1/4): 353-360.

coasts
Berthois, L., and A. Guilcher, 1959
Les bancs de Saint-March et du Moulin Blanc (Rade de Brest) et remarques sur la sédimentation du Maërl.
Cahiers Océanogr., C.C.O.E.C., 11(1):13-24.

Coasts
Bianchi, A., et R. Marquet 1965.
Étude de la pollution du Golfe de Marseille. II. La pollution des sables.
Rapp. P.-v. Réun. Commn int. Explor. scient. Mer Méditerr. 18(3): 599-606.

coasts
Bigarella, João Jose, 1949.
Contribuiçoã ao estudo da planicie sedimentar da parte norte da Ilha de Santa Catarina.
Arquivos de Biol. e Tecnologia, Inst. Biol. e Comércio, Estado do Parana, Brazil, 4(16):107-140.

coasts
*Blanc, Jean, Claude Froget et Gérard Guieu, 1968.
Géologie littorale et sous-marine dans la région de Marseille, Relations avec les structures de la Basse Provence.
Bull. Soc. géol. France, (7):9 (4):561-571.

coasts
Blanc-Vernhet, L., 1963.
Note préliminaire sur les Foraminifères des fonds détritiques côtiers et de la vase terrigène côtière dans la baie de Marseille.
Rec. Trav. Sta. Mar., Endoume, Bull., 30(45):83-93.

coasts
Bodere, Jean-Claude 1971.
Observations sur la côte de la baie d'Audierne entre Penhors et Porz-Carn (sud-ouest de la Bretagne).
Cah. océanogr. 23(6): 519-543.

coasts
Boekschoten, G.J., 1963
Some geological observations on the coasts of Crete.
Geologie en Mijnbouw, 42: 241-247.

coasts
Boldyrev, V.L., 1961.
[Underwater bars as indicators of sand shore drifting.]
Trudy Int. Okeanol., Akad. Nauk, SSSR, 48:193-201.

Coasts
Bondar, Constantin, and Sever Grăcium 1970.
Soundings of sea sediment movement at the Sulina branch outlet using labeled sands. (In Roomanian; English, French and Russian abstracts).
Studii Hidrol. Bucuresti 29:57-80.

coasts
Bourcart, Jacques, et Gilbert Boillot, 1960.
Étude des dépôts flandriens de l'anse Duguesclin près de Cancale (ille-et-Vilaine).
Bull. Soc. Géol., France, (7), 2:45-49.

Also in:
Trav. Lab. Géol. Sous-Marine, 10(1960).

coasts
Bowden, K.F., 1962.
Estuaries and coastal waters.
Proc. R. Soc., London, (A), 265(1322):320-325.

coasts
Bowen, A.J., and Douglas L. Inman, 1966.
Budget of littoral sands in the vicinity of Point Arguello, California.
Tech. Memo., Coast. Engng Res. Center, 19:41 pp.

coasts
Boynagryan, V.P., 1966.
Shore dynamics and morphology of the Sambian Peninsula. (In Russian; English abstract).
Okeanologiia, Akad. Nauk, SSSR, 6(3):458-465.

coasts (fauna)
Brattegard, Torleiv, 1966.
The natural history of the Hardanger fjord. 7. Horizontal distribution of the fauna of rocky shores.
Sarsia, 22:1-54.

coasts
Branckhoff K., 1970
Der morphologische Shorreaufbau der Flachuferküste Ostrügens und seine Gesetzmässigkeiten.
Acta Hydrophys 15(2). 93-103

Coasts

Brochet, L., and P. Giraudet, 1961.
Études hydrographiques recentes du Port d'Alger.
Annales des Ponts et Chaussees, 131(5):583-610.

coasts

Bruun, Per M. 1967.
By-passing and back-passing off Florida.
J. WatWays Harb. Div. Am. Soc. civ. Engrs 93(WW2): 101-128.

coasts

Bruun, Per, 1964.
Coastal protection procedures with special reference to conditions in Florida. 1. Withdrawn dikes and preservation lines. 2. Setback lines. 3. Coastal protection codes.
Eng. Prog., Univ. Florida, 18(12):14 pp.

coasts

Bruun, P., 1955.
Coastal development and coastal protection.
Eng. Prog., Univ. Fla., (Bull. Ser. 76), 9(11): 1-30.

Coast

Bruun, Per, 1954.
Coast erosion between beach profile No. 14A Lyngby, and beach profile No. 61, Dyba, with special consideration to the erosion of the lime inlet barriers. Pts. I & II
Atelier Elektra, København, 171 pp., figs., tables.

coasts

Bruun, Per, 1954.
Coast stability, Atelier Elektra, Kobenhavn, 7 pp (unnumbered).

coasts

Bruun, Per, 1954.
Forms of equilibrium of coasts with a littoral drift.
Atelier Elektra, Kobenhavn, 22 pp., 32 figs.

coasts

Bruun, Per, 1954.
Small scale experiments in plans for coastal protection.
Trans. A.G.U., 35(3):445-452.

coasts

Bruun, Per, 1953.
Breakwaters for coastal protection. Hydraulic principles in design.
XVIII Int. Navig. Congr., Rome, SII-Q1:5-39.

coasts

Bruun, Per, 1953.
Measures against erosion at groins and jetties.
Coastal Engineering, 3:137-164.
(Proc. third Conf. Coastal Eng., Cambridge)

coasts

Bruun, P., & F. Gerritsen, 1960
Stability of coastal inlets.
North Holland Publ. Co., Amsterdam: 123 pp.
Rev. in Science Progress, 49(194): 377

coasts

Bruun, P., and F. Gerritson, 1958.
Stability of coastal inlets.
J. Waterways Harb. Div., Proc. Amer. Soc. Civ. Eng., Pap., 1644 84:49 pp.

coasts

Bruun, Per, F. Gerritsen and W.H. Morgan, 1958
Florida Coastal Problems, Engineering Prog. at U. of Fla., Vol. 12(2) pp 33-79.
Selected paper from Proceedings of 6th Cong. on coastal engineering.

coasts

Bruun, Per, and Madhav Manohar, 1963.
Coastal protection for Florida. Development and design.
Engin. Progress, Univ. Florida, 18(8):1-56.

coasts

Budanev, V.I., 1951.
On understanding the shores of seas. (In Russian).
Trudy, Inst. Okeanol., Akad. Nauk, SSSR, 6:83-87.

coasts

Budanov, I.V., 1956.
[On the formation and development of spits of the "Azov" type. Studies of sea coasts and reservoirs.]
Trudy Okeanogr. Komissii, Akad. Nauk, SSSR, 1: 90-97.

coasts

Budanov, V.I., and A.S. Ionin, 1956.
[Contemporaneous vertical motion on the western shore of the ering Sea. Studies of sea coasts and reservoirs.]
Trudy Okeanogr. Komissii, Akad. Nauk, SSSR, 1: 65-72.

coasts

Budanov, V.I., A.S. Ionin, P.A. Kaplin, V.S. Medvedev, and A.T. Vladimirov, 1961
Present day vertical movements of Far-Eastern seacoasts of the USSR.
Proc. Ninth Pacific Sci. Congr., Pacific Sci. Assoc., 1957, 12(Geol.-Geophys.):42-48.

coasts

Budanov, V.I., and V.S. Medvedev, 1961
Microforms of nearshore bottom relief. New investigations of sea coasts and reservoirs. (In Russian).
Trudy Okeanogr. Komissii, Akad. Nauk, SSSR, 12: 73-77.

Coasts

Burton, A.J., S.R. Rogers and M.P.S. Berry, 1970.
A reassessment of the effect of effluent on the littoral fauna at Umkomaas, Natal
S Afr. J. Sci. 66 (5): 141-142

C

coasts

Caldwell, Joseph M. 1966.
Coastal processes and beach erosion.
J. Boston Soc. Civil Engrs. 53(2): 142-157.

coasts

Cameron, H.L., 1965.
Sequential air photo interpretation in coastal change studies.
Maritime Sediments, 1(2):8-13, (mimeographed)

coasts

Castanho, J.P., 1962
Metodos empregados na defesa contra a erosao costeira.
Lab. Nacional de Engenhara Civil, Memoria, Lisbon No. 196: 22 pp.

coasts

Chamard, Philippe, et Christian Barbey, 1970.
Contribution à l'Étude de la côte occidentale de la presqu'île du Cap-Vert. Indices morphométriques et morphoscopiques des galets de plage.
Bull. Inst. fond. Afr. Noire 32(4): 859-868. (A)

coasts

Chapman, Wilbert M., Francis T. Christy, Jr., Richard Baxter, Edward W. Allen and Giulio Pontecorvo, 1967.
A symposium on national interests in coastal waters.
In: The law of the sea, L.M. Alexander, editor, Ohio State Univ. Press, 125-138.

coasts

Chatel, G., 1958.
Essai d'interpretation des processes d'erosion littorale dans la region de Coutainville.
Bull. d'Info., C.C.O.E.C., 10(7):413-421.

coasts

Clifton, H. Edward, Ralph E. Hunter and R. Lawrence Phillips 1971.
Depositional structures and processes in the non-barred high-energy nearshore.
J. sedim. Petrol. 41(3): 651-670

coasts

Cloet, R.L., 1963.
Hydrographic analysis of the sandbanks in the approaches to Lowestoft Harbour.
Admiralty Mar. Sci. Publ., (6):1-15. (multilithed).

Coasts

Clos-Arceduc, A., 1961. ref.
Etude sur vues aeriennes des alluvions littorales d'allure periodique. Cordons litteraux et festons.
Societe Francaise de Photogrammetrie, Bull. (4):13-22.

coasts

Coch, Nicholas K., and David H Krinsley, 1971.
Comparison of stratigraphic and electron microscopic studies in Virginia Pleistocene coastal sediments.
J. Geol. 79 (4):426-437.

Coleman, James M., 1966.
Recent coastal sedimentation: central Louisiana coast.
Louisiana State Univ. Press, Coastal Studies 17: 73 pp (multilithed)

coasts

Collignon, J., 1965.
Les côtes et le plateau continental marocains.
Bull. Inst. Pêches Marit., Maroc, No. 13:21-37.

coasts
Colman, John S., and Anne Stephenson, 1966.
Aspects of the ecology of a "tideless" shore.
In: Some contemporary studies in marine biology, H. Barnes, editor, George Allen & Unwin, Ltd., 163-170.

coasts
Constantine, O., 1941.
Die aegyptische Mittelmeerküste zwischen Sollum und Alexandrien. Geogr. Anz. 42:345-346.

coasts
Cotecchia, V., G. Dai Pra e G. Magri, 1969.
Oscillazioni Tirreniane e oloceniche del livello mare nel Golfo di Taranto, corredate da datazioni col metodo del radiocarbonio.
Geol. appl. Idrogeol., Bari, 4:93-147.

coasts
Cotet, Petre V., 1967.
Le littoral de la mer Noire entre Eforie et Costinesti avec la région environnante (plus particulièrement le lac Tekirghiol).
(In Roumanian: French summary).
Hydrobiologie, Bucuresti, 7:264-281.

coastal drift
Courtois, G., and A. Monaco, 1969.
Radioactive methods for the quantitative determination of coastal drift rate.
Marine Geol. 7 (3): 183-206

coasts
Czekańska, Maria, 1971.
The Baltic Sea and coast in the papers by Stanisław Pawłowski. (In Polish). Zeszyty Naukowe UG, Geogr. 1: 223-229.

coasts
Czock, Hermann, und Peter Wieland, 1965.
Naturnaher Küstenschutz am Beispiel der Hörnum-Düne auf der Insel Sylt nach der Sturmflut vom 16/17 Februar 1962.
Die Küste, 13:61-72.

D

coasts
* da Franca, P., 1968.
Breves comentarios acerca da biogeografia marinha angolana.
Notas Centro Biol aquát. trop., Lisboa, 12: 1-22

coasts
Day, J.H., 1969.
A guide to marine life on South African shores.
Univ. Cape Town, A.A. Balkema, 300 pp.

coasts
DeGroot, A.J., 1964.
Origin and transport of mud (fraction <16 microns) in coastal waters from the western Scheldt to the Danish frontier.
In: Developments in Sedimentology, L.M.J.U. van Straaten, Editor, Elsevier Publishing Co., 1:93-100.

coasts
Derijard, Raoul, 1965.
Contribution à l'etude du peuplement des sédiments Sablo-vaseux et vaseux intertidaux, compactés ou fixés par la végétation de la région de Tuléar (Madagascar).
Recl. Trav. Stn. mar. Endoume (Trav. Stn. mar. Tulear (horserie) Suppl. 43:94 pp.

Coasts
Derijard, R., 1963.
Note préliminaire sur la localisation et le peuplement de certains atterrissements sablo-vaseux et vaseux intertidaux de la région de Tuléar (Madagascar).
Annls Malgaches 1:201-219.

coasts, protection of
Devaux, P., 1957.
La protection des digues a la mer.
La Nature (3267):265-267.

coasts
Devirts, A.L., N.G. Markova, and L.R. Serebriannyi, 1968.
Geological estimate of the age of old coastal structures of the Baltic Sea as checked by C14 data. (In Russian).
Dokl. Akad. Nauk, SSSR, 182(6):1387-1390.

coasts
Diaz-Marta, M., 1957.
Protection works on the Mexican coast.
Dock and Harbor Authority 37(435):306-309.

coasts
Dolan, Robert, 1971.
Coastal landforms: crescentic and rhythmic.
Bull. geol. Soc. Am. 82(1): 177-180.

coasts
Dolotov, J.S., 1962
On the development of coastal shoals in the course of the most recent transgression observed in the area of the Azov and Black Sea basin.
Okeanologiia, Akad. Nauk, SSSR, 2(2):278-283.

coasts
Dolotov, U.S., 1961.
[Formation and types of marine accumulative terraces on the rising shores.]
Trudy Inst. Okeanol., Akad. Nauk, SSSR, 48:172-192.

coasts
Dolotov, U.S., 1959
[Contributions to the evolution of embankments on the emerging sea shores. Questions of studies of marine coasts.]
Trudy Okeanogr., Komissii Akad. Nauk, SSSR, 4:66-80.

Coasts
Duane, David B., 1970.
Tracing sand movement in the littoral zone: progress in the radioisotope sand tracer (RIST) study, July 1968-February 1969.
Misc. Pap. Coast. Engng Res. Cent. U.S. Army Engrs 4-70: 46 pp.

coasts
Dumitrashko, N.V., 1966.
Recent tectonic movements of the Black Sea coast of the Caucasus. (In Russian; English abstract).
Stroenie Chermomorskoi Vladini, Rez. Issled., Mezhd. Geofiz. Proekt., Mezhd. Geofiz. Komitet, Presid., Akad. Nauk, SSSR, 79-84.

E

coasts
Easton, W.H., 1963.
New evidence for a 40-foot shore line on Oahu.
In: Essays in Marine Geology in Honor of K.O. Emery, Univ., Southern California Press, 51-68.

coasts
Egorov, E.N., and B.L. Kasyanov, 1961.
Intense recent transformation of the sea coast under the influence of the advance of river deltas and the installation of harbour piers.
Dynamics of the coastal zone of the Black Sea. (In Russian).
Trudy Inst. Okeanol., Akad. Nauk, SSSR, 53:42-51

coasts
Eisma, D., 1968.
Composition origin and distribution of Dutch coastal sands between Hoek van Holland and the island of Vlieland.
Netherlands J. Sea Res. 4(2): 123-267.

Coasts
El-Ashry, Mohamed T., and Harold R. Wanless 1968.
Photo interpretation of shoreline changes between Cape Hatteras and Cape Fear (North Carolina).
Marine Geology 6(5): 347-379.

coasts
El-Ashry, Mohamed T., and Harold R. Wanless, 1964
Photo interpretation of coast line changes due to hurricanes. (Abstract).
Geol. Soc., Amer., Special Paper, No. 76:53.

coasts
Ellis, C.W., 1962.
Marine sedimentary environments in the vicinity of the Norwalk islands, Connecticut.
State Geol. Nat. Hist. Survey, Connecticut, Bull. No. 94:1-89.

coasts
Emery, K.O., 1967.
Some stages in the development of knowledge of the Atlantic coast of the United States.
Progress in Oceanography, 4:307-322.

coasts
Emery, K.O., and Carl J. George, 1963
The shores of Lebanon.
Amer. Univ., Beirut, Misc. Papers in the Nat. Sci. No. 1:1-10, 4 figs.

coasts
Ertel, H. 1971.
Eine Betrachtung zur geomorphologisch wirksamen Arbeit der Brandungswellen an Flachküsten.
Acta hydrophys. 16(1): 5-10.

coast

Extel, H., 1970.
Der Rückgang der Strandlinie an Flachküsten.
Acta hydrophys. 14(4): 327-332

coasts

Extel, H., 1965.
Der Ausgleich von Störungen der Strandlinie bei zeitlich variablem Küstenstrom.
Acta Hydrophysica, 13(2) 85-90

coasts

*Etchichury, M.C., y J.R. Remiro, 1967.
Los sedimentos litorales de la Provincia de Sante Cruz entre Punta Dungeness y Punta Desengaño.
Revta Mus. argent. Cienc. nat. Bernardino Rivadavia Inst. nac. Invest., Geol., 6(8): 323-376.

coasts

Evans, G., C.G. St.C. Kendall and Patrick Skipwith, 1964.
Origin of the coastal flats, the Sabkha, of the Trucial coast, Persian Gulf.
Nature, 202 (4934): 759-761.

Also in: Collected Reprints, Int. Indian Ocean Exped., UNESCO, 3. 1966.

F

coasts

Fedorov, P.V., 1967.
Recent data on the correlations between the Old Euxinian and Uzunlar terraces of the Caucasian coast of the Black Sea. (In Russian)
Dokl. Akad. Nauk, SSSR, 174 (4): 924-926

coasts

Federov, P.V., D.A. Lilienberg and Vl. I. Popov, 1964.
New data on the Black Sea wave-cut platforms of Bulgaria. (In Russian).
Doklady, Akad. Nauk, SSSR, 144(2): 431-434.

Translation:
Earth Sci. Sect., Doklady, Akad. Nauk, SSSR, AGI, 144(1-6): 52-55. (1964).

coasts

Freira Motta, Victor, 1965
Resultados de algumas medicoes de transporte litoraneo em modelo costeiro esquematico.
Anais Acad. bras. Cienc., 37(Supl.) : 109-130.

coasts

Freistadt, Heinrich, 1962.
Die Sturmflut vom 16./17. Februar 1962. in Hamburg.
Die Küste, 10(1): 81-92.

coasts

Freykman, A.I., 1962
Certain problems relating to the stabilization of the gravel shores of the Black Sea coast of the Caucasus.
Okeanologiia, Akad. Nauk, SSSR, 2(1): 153-159.
Abstracted in: Soviet Bloc Res., Geophys. Astron. and Space, 1962(35): 27.
(OTS61-11147-35 JPRS13739)

coasts

Fridman, Melle Ruth, 1960.
Les sediments recents de l'Anse de l'Aiguillon et de ses limites marines à l'ouest.
Cahiers Océanogr., C.C.O.E.C., 12(4): 268-274.

coasts

Fujii, S. and A. Mogi, 1970.
On coasts and shelves in their mutual relations in Japan during the Quaternary.
Quaternaria 12: 155-164.

G

coasts

Gamazhenkov, V.S., 1962
The dynamics of coastal alluvium in the Gagra Sound.
Okeanologiia, Akad. Nauk, SSSR, 2(2): 284-292.

coasts

Gautier, Marcel, 1971
Le maërl sur le littoral de la Bretagne.
Cah. océanogr. 23(2): 171-191

coasts

Gautier, Françoise 1971.
Les processus de l'attaque des falaises sur le littoral continental de la baie de Bourgneuf.
Norois 18(70): 221-236

COASTS

Gavish, Eliezer, and Gerald M. Friedman, 1969.
Progressive diagenesis in Quaternary to late Tertiary carbonate sediments: sequence and time scale.
J. sedim. Petrol., 39(3): 980-1006.

coasts

Gaye, J., 1955.
Die deutsche Küstenforschung und der Seewasserbau
Die Küste, 3(1/2): 13-17.

coasts

Germany, Ministeriums für Ernährung, Landwirtschaft und Forsten-Landesamt für Wasserwirtschaft-Schleswig-Holstein, 1962.
Die Sturmflut vom 16/17 Februar 1962 an der Schleswig-Holsteinischen Westküste.
Die Küste, 10(1): 55-80.

coasts

Germany, Niedersächsischen Minister für Ernährung, Landwirtschaft und Forsten [eingesetzten Ingenieur-Kommission, 1962.
Die Sturmflut vom 16/17 Februar 1962 im niedersächsischen Küstengebiet.
Die Küste, 10(1): 17-53.

coasts

Gill, Edmund D., 1970.
Coast and continental shelf of Australia.
Quaternaria 12: 115-128.

coasts

*Gill, E.D., 1967.
The dynamics of the shore platform process, and its relation to changes in sea-level.
Proc. R. Soc., Victoria, N.S., 80(2): 183-192.

coasts

Gill, Edmund D., 1967.
Description of Quaternary shorelines with special reference to the tectonic factor.
J. Geosci., Osaka City Univ., 10: 131-134.

coasts

Golev, B.T., and A.A. Veselov, 1969.
On the problem involved in the age of analogues of the Kuberlin horizon of the Paleozene of the northern Black Sea littoral and the Crimea. (In Russian).
Dokl. Akad. Nauk. SSSR 187(3): 629-631.

coasts

Golikov, A.N., and O.G. Kussakin, 1962.
Fauna and ecology of Gastropoda, Prosobranchia, from the intertidal zone of the Kuril Islands. (In Russian).
Issled. Dalnevostochnich Moreia, SSSR, Zool. Inst., Akad. Nauk, SSSR, 8: 248-346.

coasts

Gomes, Nelson, 1964.
Protection des côtes contre l'érosion maritime et formation des plages de sable.
La Houille Blanche 19(6): 692-705.

coasts.

Govindankutty, A.G., and N.B. Nair, 1966.
Preliminary observations on the interstitial fauna of the south-west coast of India.
Hydrobiologia, 28(1): 101-122.

coasts

Grechishchev, E.K., 1957.
[On the determination of contemporary tectonic motion of the coasts of Lake Baikal. Introductory discussion. Material subdivisions for the study of sea coasts and reservoirs.]
Trudy Okeanogr. Komissii, Akad. Nauk, SSSR, 2: 129-146.

coasts

Greensmith, J.T., and E.V. Tucker, 1969.
The origin of Holocene shell deposits in the Chenier Plain facies of Essex (Great Britain)
Marine Geol. 7(5): 403-425

coasts

Grigoriev, N.F., 1963
The role of the cryogenic factors in the dynamics of the shore area of Yakutia. (In Russian).
Okeanologiia, Akad. Nauk, SSSR, 3(3): 477-481.

Coasts

Grovel, A.P., et R. Brosse 1968
Évolution actuelle du littoral de la région de Lorient (Morbihan)
Cah. océanogr. 20(5): 395-402.

Coasts

Gudelis, Vytautas 1969.
Die Küstenentwicklung der südöstlichen Ostsee während der Spät- und Nacheiszeit.
Beitr. Meereskunde 24-25: 10-14.

coasts

Gugnjaeff, J.E., 1961
[Wave action on the artificial shore slopes.]
Trudy Okeanogr. Komissii, Akad. Nauk, SSSR, 8: 104-108.

coasts

Gugnjaiev, J.E., 1959
[On the methodics of the shore-debris model experiments. Questions of studies of marine coasts.]
Trudy Okeanogr., Komissii Akad. Nauk, SSSR, 4: 149-155.

coasts
Guilcher, A., 1967.
Morphologie et sédimentologie littorales et sous-marines.
Rapp. Nat. Trav. français, 1963-1966, Com. Nat. Français Géodes.- Géophys., 244-255.

coasts
Guilcher, Andre, 1964.
Present-time trends in the study of Recent marine sediments and in marine physiography.
Marine Geology, 1(1):4-15.

coasts
Guilcher, A., and F. Joly, 1955.
Recherches sur la morphologie de la cote atlantique du Maroc. Inst. Sci. Chérifien, Ser. Geol.-Phys., No. 2:140 pp.

H

coasts
Hartke, D. 1970.
Ausgleich einer dreieckförmigen Strandlinienstörung.
Acta hydrophys. 15(1):15-21.

coasts
Hatai, Kotora, 1963.
Observations along the coast of the Muroto Peninsula, Kochi Prefecture, Shikoku, Japan.
Rec. Oceanogr. Wks., Japan, n.s., 7(1):89-94.

coasts
Hatai, K., Y. Funayama, and H. Mii, 1957.
A note on the development of certain marine pot-holes along the west coast of Wakayama Prefecture, Japan
Rec. Oceanogr. Wks., Japan, n.s., 4(1):45-48.

coasts
Hatai, Kotora, and Minoru Saito, 1963.
Observations on the sea coast of northern Shikoku.
Rec. Oceanogr. Wks., Japan, n.s., 7(1):101-106.

coasts
Hecht, Alan D., 1969.
Miocene distribution of molluscan provinces along the east coast of the United States.
Bull. geol. Soc. Am., 80(8): 1617-1620

coasts
Herm, Dietrich 1969
Marines Pliozän und Pleistozän in Nord- und Mittel-Chile unter besonderer Berücksichtigung der Entwicklung der Mollusken-Faunen.
Zitteliana 2: 159 pp.

coasts
Heydemann, Berndt, 1963.
Deiche der Nordseeküste als besonderer Lebensraum - Ökologische Untersuchungen über die Arthropoden-Besiedlung.
Die Küste, 11:90-130.

coasts
Hodgson, W.A., 1966.
Coastal processes around the Otago Peninsula.
New Zealand J. Geol. Geophys., 9(1/2):76-90.

coasts
Hom-ma, Masashi, Kiyoshi Horikawa and Choule Sonu, 1962
Field investigations at Tokai, Japan, conducted by combined procedure of macroscopic and microscopic approaches.
Coastal Engineering in Japan, 5:93-110.

coasts
Hommeril, P., and C. Larsonneur, 1963
Les effets des tempêtes du premier semestre 1962 sur les côtes Bas-Normandes.
Cahiers Océanogr., C.C.O.E.C., 15(5):320-334.

coasts
Horikawa, Kiyoshi, 1962
Brief review of coastal engineering problems. (In Japanese; English abstract.)
J. Oceanogr. Soc., Japan, 20th Ann. Vol., 351-367.

coasts
Horikawa, Kiyoshi, and Tsuguo Sunamura 1970.
A study on erosion of coastal cliffs and of submarine bedrocks.
Coastal Engng Japan 13:129-139.

coasts
Hoshino, Michihei, 1964
Rias coasts around Japanese islands. (In Japanese; English abstract).
In: Studies in Oceanography dedicated to Professor Hidaka in commemoration of his sixtieth birthday, 479-485.

coasts
*Hoyt, John H., 1967.
Occurrence of high-angle stratification in littoral and shallow neritic environments, central Georgia coast.
Sedimentology, 8(3):229-238.

coasts
Hoyt, John H., and John R. Hails, 1967.
Pleistocene shoreline sediments in coastal Georgia: deposition and modification.
Science, 155(3769):1541-1543.

coasts
Hoyt, John H., and Vernon J. Henry, Jr., 1964.
Formation of high-angle marine stratification, central Georgia coast. (Abstract).
Geol. Soc., Amer., Special Paper, No. 76:246.

coasts
*Hume, James D., and Marshall Schalk, 1967.
Shoreline processes near Barrow, Alaska: a comparison of the normal and of the catastrophic
Arctic, 20(2):86-103.

I

coasts
Inman, Douglas L., and Jeffrey D. Frautschy, 1965
Littoral processes and the development of shorelines.
Proc. Santa Barbara Speciality Conf., Coast. Engng, Oct. 1965, 511-536.

coasts
Inman, D.L., and C.E. Nordstrom, 1971.
On the tectonic and morphologic classification of coasts.
J. Geol. 79(1):1-21

Coasts
Ionin, A.S., 1967.
Relief of the near-shore zone of the island of Cuba. (In Russian; English abstract).
Okeanologiia, Akad. Nauk, SSSR, 7(2):287-301.

coasts
Ionin, A.S., 1961
[Some degrading shore-embayments in the Bering Sea.]
Trudy Okeanogr. Komissii, Akad. Nauk, SSSR, 8:85-97.

coasts
Ionin, A.S., 1961
[Some observations on the dynamics and morphology of the Kommandor islands shores.]
Trudy Okeanogr. Komissii, Akad. Nauk, SSSR, 8:206-210.

coasts
Ionin, A.S., 1959
[Shore dynamics and morphology researches of the Bering and Chucota sea shores. Questions of studies of marine coasts. (Information).]
Trudy Okeanogr., Komissii, Akad. Nauk, SSSR, 4:205-214.

coasts
Ionin, A.S., 1958.
Some peculiarities of the dynamics and morphology of the coasts of the Bering Sea. Oceanographic investigations in the northwest part of the Pacific Ocean. (In Russian)
Trudy Okeanogr. Komissii, Akad. Nauk, SSSR, 3:55-65.

coasts
Ionin, A.S., 1956.
[On the question of the evolution of bay shores.]
Trudy Okeanogr. Komissii, Akad. Nauk, SSSR, 1:82-89.

coasts
Ionin, A.S., and I.S. Dolotov, 1958.
[Characteristics of the dynamics and morphology of raising coasts illustrated on the example of Novaya Zemlya.] Trudy Inst. Okeanol., 28:71-84.

coasts
Ionin, A.S., P.A. Kaplin, and V.S. Medvedev, 1963.
A review of O.K. Leontiev's book "Fundamentals in the geomorphology of marine coast studies", Moscow State University Publishers, 1961, 417 pp. (In Russian).
Okeanologiia, Akad. Nauk, SSSR, 3(5):946-948.

coasts
Ionin, A.S., P.A. Kaplin and V.S. Medvedev, 1961
Classification of the world sea-coasts (according to the World Geographical Atlas) (A discussion). New investigations of sea coasts and reservoirs. (In Russian).
Trudy Okeanogr. Komissii, Akad. Nauk, SSSR, 12:94-108.

coasts
Ionin, A. S., P. A. Kaplin & V. S. Medvedev, 1961
On the methods of investigating the coastal zone of the USSR seas.
Okeanologiya, 1: 148-162.

coasts
Ionin, A.S., P.A. Kaplin, and V.S. Medvedev, 1961
[Results of regional coastal investigations in the USSR.]
Trudy Inst. Okeanol., Akad. Nauk, SSSR, 48:3-33.

coasts

Ippen, Arthur T., 1966.
Estuary and coastline hydrodynamics. McGraw Hill Book Co, Inc., 744 pp.

coasts

Ippen, Arthur T., 1966.
Waves and tides in coastal processes.
J. Boston Soc., Civil Eng., 53(2):158-181.

coasts

Ivanov, R.N., 1962.
Effect of wave action on the water pileup and driving away phenomena along the sea shore. (In Russian).
Izv., Akad. Nauk, SSSR, Ser. Geofiz., (7):955-964.

English Edit., (7):608-613.

J

coasts

Jacoby, Gustav, 1958.
Die Küste Dithmarschens im Kartenbild von 1559.
Deutsche Hydrogr. Zeits., 11(4):166-177.

coasts

Jakobsen, B., 1961.
Vadehavets sediment om saetning belyst ved kvantitative målinger.
Geografisk Tidsskr., 60:87-103.

coasts

Japan, Kobe Marine Observatory, 1951
Meteorology and oceanography in the Seto Inland Sea.
Bull. Kobe Mar. Obs., No. 161:211 pp.

coasts

Jipa, Dan, 1969.
Observations sur le modèle des ripple marks actuels sur le littoral de la mer Noire (secteur Navodari- Agigea)
Rapp. P.-v. Réun. Commn int. Explor. scient. Mer Mediterr. 19(4): 599-

coasts

Johnson, J.W., 1961.
Historical photographs and the coastal engineer.
Shore and Beach, April 1961:5 pp. (unnumbered on pages).

coasts

Jones, W. Eifion, and Andreas Demetropoulos 1968.
Exposure to wave action: measurements of an important ecological parameter on rocky shores on Anglesey.
J. exp. mar. Biol. Ecol. 2(1):46-63

coasts

Jussy, Maurice et Andre Guilcher, 1962
Les cordons littoraux entre la presquéile de Quiberon et l'estuaire de la Milaine (Golfe du Morbihan exclu).
Cahiers Oceanogr., C.C.O.E.C., 14(8):543-572.

K

coasts

Kachjugin, E.G., 1961
Main results of long time observations of the artificial lake coasts near Moscow and in the upper Volga region. (Information). New investigations of sea coasts and reservoirs. (In Russian).
Trudy Okeanogr. Komissii, Akad. Nauk, SSSR 12:109-119.

coasts

Kakinuma, Tadao, 1961
Size-frequency distribution of nearshore sediments at Ohsu beach along the Atsumi Bay coast.
Proc. 8th Conf. Coastal Eng., Japan, 156-160.
Also in:
Papers in Oceanogr. and Hydrol., Geophys. Inst., Kyoto Univ., (1949-1962).

coasts

Kaplin, P.A., 1959
[Shore-lines evolution of fiord regions. Questions of studies of marine coasts. Questions of studies of marine coasts.]
Trudy Okeanogr., Komissii Akad. Nauk, SSSR, 4: 54-65.

coasts

Kaplin, P.A., 1957.
[On some peculiarities of lagoons along the north-east coasts of the SSSR. Material subdivisions for the study of marine coasts and reservoirs.]
Trudy Oceanogr. Komissii, Akad. Nauk, SSSR, 2: 104-110.

coasts

Kaplin, P.A., and V.L. Boldyrev, 1961
[Joint Polish-Soviet investigations on the Baltic shores in 1958.]
Trudy Okeanogr.Komissii, Akad. Nauk, SSSR, 8:245-250.

coasts

Kaplina, T.N., 1959
[Some characteristics of the washing-away of everfrost consolidated shores. Questions of studies of marine coasts.]
Trudy Okeanogr., Komissii Akad. Nauk, SSSR, 4:113-117.

coasts

Kashin, Iu. S., 1956.
[Investigations of "galechnix" deposits along the "Kavkazhsk" coast of the Black Sea between Gelendzhik and Tuapse. Studies of sea coasts and reservoirs.]
Trudy Okeanogr. Komissii, Akad. Nauk, SSSR, 1: 73-76.

coasts

Kaye, C.A., 1959.
Shoreline features and Quarternary shoreline changes. Puerto Rico.
Geol. Survey Prof. Pap., 317-B:140.

coasts

Kempf, Marc, 1970.
Nota preliminar sôbre os fundos costeiros da região de Itamaracá (Norte do estado de Pernambuco, Brasil). Trabhs oceanogr. Univ. Fed. Pernambuco 9/11: 95-110.

Coast

Kendall, Christopher G. St.C., and Patrick A. d'E. Skipwith, 1969.
Geomorphology of a Recent shallow-water carbonate province: Khor Al Bazam, Trucial Coast Southwest Persian Gulf.
Bull. Geol. Soc. Am., 80(5): 865-892

coasts

Kenkovich, V.P., and V.V. Longinov, 1961
[On the organization of an International Commission devoted to the study of coast dynamics and near-shore marine zones.]
Okeanologiia, Akad. Nauk, SSSR, (2):343-347.

coasts

Kestner, A.P., 1956.
[Work with electrical devices in the coastal zone. Studies on sea coasts and reservoirs.]
Trudy Okeanogr. Komissii, Akad. Nauk, SSSR, 1: 134-140.

coasts

Khodyrev, N.A., 1962
Coastal protection in the German Democratic Republic. (In Russian).
Okeanologiia, Akad. Nauk, SSSR, 2(5):939-942.

coasts

King, C.A., 1961.
Beaches and coasts.
Edward Arnold Ltd., London, 403 pp., 65 shill.

coasts

*Klausewitz, Wolfgang, 1967.
Die physiographische Zonierung der Saumriffe von Sarso.
Meteor Forschungsergebn. (D)(2)(Biol.):44-68.

coasts

Klein, George deVries, 1967.
Paleocurrent analysis in relation to modern marine sediment dispersal patterns.
Bull.Am.Ass.PetrolGeol., 51(3):366-382.

Coasts

Kluyev, Ye.V., 1970
Thermal abrasion of the coast line of Arctic seas. (In Russian).
Izv. Vses. Geogr. Obshch. 102 (2): 129-135

coasts

*Knaps, R.J., 1968.
Computations of the power of the seashore sand drifts. (In Russian; English abstracts).
Okeanologiia, Akad.Nauk, SSSR, 8(5):848-857.

coasts

Knaps, R.J., 1959
[Some general laws of longshore bars and troughs development. Questions of studies of marine coasts.]
Trudy Oceanogr., Komissii Akad. Nauk, SSSR, 4:44-53.

coasts

Knaps, R. Ia., 1957.
[Investigations on the dynamics of marine coasts in the German Democratic Republic. (On the work ----- ----- Science K. Folibrechta). Material subdivisions for the study of sea coasts and reservoirs.]
Trudy Okeanogr., Komissii, Akad. Nauk, SSSR, 2: 176-188.

coasts

Knop, Friedrich, 1963.
Küsten- und Wattveränderungen Nordfrieslands - Methoden und Ergebnisse ihrer Überwachung.
Die Küste, 11(1):1-33.

coasts

Kofoed, J.W., 1963.
Coastal development in Volusia and Brevard counties, Florida.
Bull. Mar. Sci., Gulf and Caribbean, 13(1):1-10.

coasts
*Kranck, Kate, 1967.
Bedrock and sediments of Kouchibouguac Bay, New Brunswick.
J. Fish. Res. Bd., Can., 24(11):2241-2265.

coasts
Krejci-Graf, K., 1955.
Küstonzerstorung. Natur und Volk 85(8):252-261.

coasts
Krishnan, M.S., 1969.
The evolution of the coasts of India. Bull. natn. Inst. Sci., India, 38(1): 398-404.

coast
Krivonosova, N.M., 1962
On the method of compiling atlases reflecting the dynamics and the morphology of marine coasts. (In Russian).
Okeanologiia, Akad. Nauk, SSSR, 2(5):912-916.

coasts
Kuzminskaya, G.G., 1965.
On the problem of spectral structure of sea disturbances in the near-shore zone. (In Russian).
Okeanologiia, Akad. Nauk, SSSR, 5(3):404-407.

L

coasts
*Laborel, Jacques Louis, e Marc Kempf, 1965/1966.
Formações de vermetos e algas calcárias nas costas do Brasil.
Trabhs Inst. Oceanogr., Univ. Fed. Pernambuco, Recife, (7/8):33-50.

coasts
Ladochin, N.P., 1957.
[Some peculiarities of the dynamics and distribution of alluvium in connection with the character of the relief of coasts ---- of the south-eastern Baikal. Material subdivisions for the study of sea coasts and reservoirs.]
Trudy Okeanogr. Komissii, Akad. Nauk, SSSR, 2: 24-34.

coasts
Lang, Hans Dietrich, 1963.
Untersuchungen an Nordsee-Küstenschlicken im Hinblick auf ihre balneologische Verwendbarkeit.
Geol. Jahrb., 80:49-68.

coasts
Larsonneur Claude et Michel Rioult, 1969.
Le Trias et le Lias des côtes de la Manche au nord et à l'est du Cotentin.
C.r. hebd. Séanc. Acad. Sci. Paris. (D) 268(16): 2019-2022

Leontiev, O.K., also Lontiev, Leontjev, Leonyjev

coasts, coral
Leontiev, O.K., 1969.
Some structural features and dynamics of the coral shores. (In Russian; English abstract).
Okeanologiia, 9(2): 271-281.

coasts
Leclaire, Lucien, 1963
Facteurs d'évolution d'une côte sablonneuse rectiligne très ouverte. Etude préliminaire à l'implantation d'un port de pêche et de plaisance.
Cahiers Océan., C.C.O.E.C., 15(8):540-556.

coasts
Leontiev, O.K., 1965.
On the geomorphology of the Caspian sea coast. Sediment drift and the genesis of the heavy mineral fields in the coastal zone of the sea. (In Russian; English abstract).
Trudy, Inst. Okeanol., Akad. Nauk, SSSR, 76:225-249.

coasts
Leont'ev, O.K., 1949.
[Rebuilding of the profile of an accumulative shore during the lowering of sea level.] Dok. Akad Nauk, USSR, 66:377-380.

coasts
Leontiev, O.K., 1957.
[Some peculiarities of the dynamics and morphology of the coastal zone of the northwestern shores of the Caspian. Material subdivisions for the study of sea coasts and reservoirs.]
Trudy Okeanogr. Komissii, Akad. Nauk, SSSR, 2:35-50.

coasts
Leontiev, O.K., 1956.
[On the terminology in the science of marine coasts. Studies on sea coasts and reservoirs.]
Trudy Okeanogr. Komissii, Akad. Nauk, SSSR, 1: 141-149.

coasts
Leontiev, O.K., and A.I. Khalikov, 1965
Formation of the shores of the Caspian Sea. (In Russian).
Inst. Geograf., Akad. Nauk, Azerbaid. SSR, Baku, 205 pp.

coasts
Leontiev, O.K., and V.K. Leontiev, 1957.
[On the question of the genesis and regularity of development of coastal lagoons. Material subdivisions for the study of marine coasts and reservoirs.]
Trudy Okeanogr. Komissii., Akad. Nauk, SSSR, 2: 86-103.

coasts
Leontjev, O.K., 1961
[General features of the morphology and evolution of the North Azerbydgean Coast (Caspian Sea).]
Trudy Okeanogr. Komissii, Akad. Nauk, SSSR, 8:3-32.

coasts
Leontjev, O.K. 1959
[Some considerations of the height and age of "New Caspian" transgression. Questions of studies of marine coasts.]
Trudy Okeanogr., Komissii, Akad. Nauk, SSSR, 4:81-90.

coasts
Leonyjev, O.K., M.E. Bachtina and T.A. Dobrynina, 1959
[Nearshore sediments study of the N-W Caspian Sea. Questions of studies of marine coasts.]
Trudy Oceanogr., Komissii Akad. Nauk, SSSR, 4:18-30.

coasts
*Lewis, John R., 1968.
Water movements and their role in rocky shore ecology.
Sarsia, 34:13-36.

coasts (terraces)
Lilienberg, D.A., K.I. Mishev and V.I. Popov, 1964.
Recent data on Black Sea terraces of the Strandjin shore of Bulgaria. (In Russian).
Doklady, Akad. Nauk, SSSR, 159(3):552-555.

coasts
Limarev, V.I., 1957.
[The type of coast in the Aral Sea. Material subdivisions for the study of sea coast and reservoirs.]
Trudy Oceanogr. Komissii, Akad. Nauk, SSSR, 2: 59-68.

coasts
Longhurst, A.R., 1964.
The coastal oceanography of western Nigeria.
Bull. Inst. Français, Afrique Noire, (A), 26(2): 337-402.

coasts
Longinov, V. V., 1964.
On the possibility of calculating the discharge of sand drifts along a shoaled sea coast. (In Russian).
Okeanologiia, Akad, Nauk, SSSR, 4(6):1035-1043.

coasts
Longinov, V.V., 1962.
Recent ideas on the dynamics of coastal zones of the sea. Marine geology and dynamics of coasts. (In Russian).
Trudy Okeanogr. Komissii, Akad. Nauk, SSSR, 10(3):102-112.

coasts
Longinov, V.V., 1959
[Some historical aspects of the progress of the shore-dynamics investigations. Questions of studies of marine coasts.]
Trudy Okeanogr., Komissii Akad. Nauk, SSSR, 4:161-196.

coasts
Longuinov, V., 1958.
[On the dynamics of the surf zone and of the shore zone as a whole.]
Mokslinai Pranešimai, Lietvos TSR Mokslu Akad. Geol. Geogr. Inst., 7:131-181.

coasts
Lopatnikov, V.I., 1962
Analysis of the monsoon field over the sea in an area with a complex shore line. (In Russian).
Doklady, Akad. Nauk, SSSR, 147(2):353-356.

Abstr. in:
Soviet Bloc Res., Geophys., Astron. & Space, 52:25.

coasts
Lorenzen, J.M., 1955.
Hundert Jahre Küstenschutz an der Nordsee.
Die Küste 3(1/2):18-32.

coasts
Lyslov, I.A., 1961.
Model study of some submerged breakwater constructions. New investigations of sea coasts and reservoirs. (In Russian).
Trudy Okeanogr. Komissii, Akad. Nauk, SSSR, 12: 5-16.

M

coasts
Mabesoone, J.M., 1964.
Origin and age of the sandstone reefs of Pernambuco (northeastern Brazil).
J. Sed. Petr., 34(4):715-726.

coasts

Maloney, Neil J., 1965
Geomorphology of the central coast of Venezuela.
Bol. Inst. oceanogr., Univ. Oriente, 4(2): 241-265.

coasts

Mamaeva, R.B., 1956.
[Experiments on the determination of differential, tectonic motion of marine sea coasts by geomorphological methods. Studies of sea coasts and reservoirs.]
Trudy Okeanogr. Komissii, Akad. Nauk, SSSR, 1: 77-81.

coasts

Mamykina, V.A., 1961
[Shore--types of the north-eastern Azov Sea and peculiarities of their dynamics.]
Trudy Okeanogr. Komissii, Akad. Nauk, SSSR, 8:33-44.

coasts

Martin, P.L.C., 1966.
Contribucion de la sedimentologia a la hidrografia en el estudio de las transformaciones de las costas y estuarios.

Bol. Serv. Hidrogr. Naval, Buenos Aires (Publ. H. 106), 3(1):9-22.

coasts

Martin, P.L.C., 1965.
The contribution of sedimentology to hydrography in the study of coast and estuary changes.
Int. Hydrogr. Rev., 42(2):159-171.

coasts

Martin Salazar, Felipe, 1956

Oceanografia de la region central.
El Agriculator Venezolano, 21(190): 28-35; 68-69.

coasts

Mary, Guy, 1968.
Les formations marines actuelles et fossiles à l'embouchure du Rio Esba (Lucarca-Asturies-Espagne).
Cah. océanogr., 20(8):683-693.

coasts

Mashima, Yasuo, 1961
Stable configuration of coast line.
Coastal Eng., Japan, 4:47-59.

Mc

Coasts

McCann, S.B. and R.J. Chorley 1967.
Trend surface mapping of raised shorelines.
Nature, Lond. 215 (5101): 611-612.

Coasts

McMaster, Robert L., and Louis E. Garrison 1967.
A submerged Holocene shoreline near Block Island, Rhode Island.
J. Geol. 75(3):335-340.

coasts

McMurry, J.H., 1958.
Shoretype classification of the Gulf coast of Florida. Pap. Mich. Acad. Sci., Arts & Letters, 43:259-266.

coasts

Meade, Robert H. 1971.
The coastal environment of New England
New England River Basins Commission, 47 pp.

coasts

Medvedev, V.S., 1964
On the study of the role of tides in the dynamics of the coastal zone. (In Russian).
Okeanologiia, Akad. Nauk, SSSR, 4(2):277-283.

coasts

Medvedev, V.S., 1961
Influence of some physical-geographical factors upon coastal dynamics (based on the data of the Sea of Japan). New investigation of sea coasts and reservoirs. (In Russian).
Trudy Okeanogr. Komissii, Akad. Nauk, SSSR, 12: 42-53.

coasts

Medvedev, V.S., 1961.
Peculiarities of bench formation on tidal sea shores. (Abstract)
Tenth Pacific Sci. Congr., Honolulu, 21 Aug.-6 Sept., 1961, Abstracts of Symposium Papers, 378.

coasts

Medvedev, W.S., 1961
[Some morphological and dynamic features of the western Sakhalin coast.]
Trudy Okeanogr. Komissii, Akad. Nauk, SSSR, 8:65-84.

coasts

Medvedev, V.S., 1959

[Erosion remnants of loose deposits on the shores of the White Sea.] Izv. Akad. Nauk, SSSR, Ser. Geograf., (3): 96-

coasts

Medvedev, V.S., 1959
Some questions of western Sakhalin Island shore dynamics and morphology in connection with fishery ports construction. Questions of studies of marine coasts.
Trudy Okeanogr., Komissii Akad. Nauk, SSSR, 4: 3-12.

coasts

Medvedev, V.S., 1957.
A brief sketch of the dynamics and morphology of the western coasts of the White Sea. Material dubdivisions for the study of sea coasts and reservoirs.]
Trudy Okeanogr. Komissii, Akad. Nauk, SSSR, 2:69-85.

Coasts

Medvedev, V.S., E.A. Nevessky, Yu. A. Pavlidis and F.A. Shcherbakov 1968.
Relief and history of the south coast of the Kola Peninsula during the Holocene.
(In Russian, English abstract).
Okeanologiia, Akad. Nauk SSSR 8(2): 257-269.

coasts

Mekhtiyev, N.N., 1960.
[Some problems of the dynamics and morphology of the western shore of the southern Caspian.]
Izv. Akad. Nauk, Ser. Geol.-Geogr. Nauk. (5): 151-162.

OTS-61-11147-18 JPRS:8807:15-16.

coasts

Michlke, D. 1968.
Aufgaben und Perspektiven der Küstenforschung in der Deutschen Demokratischen Republik.
Acta hydrophysica 12(3): 115-132.

coasts

Mihailescu, Nicolae, 1969.
Étude sédimentologique des graviers se trouvant sur le littoral de la Mer Noire entre Constantza et Vama Vecle (Roumanie)

Rapp. P.-v. Reun. Commn int. Explor. scient. Mer Mediterr., 19(4): 623-626.

coasts

Mitt, K.L., 1964.
On the morphology and dynamics of the Anabaro-Olenek shore in the area of the Laptev Sea. (In Russian).
Okeanologiia, Akad. Nauk, SSSR, 4(4):660-668.

coasts

Mizoguchi, Yutaka, and Masamichi Ugi, 1963.
Size distribution of coastal sand.
Coastal Engineering in Japan, 6:95-101.

coasts

Moign, Yvon et Annik Moign, 1970.
Les Iles Heimaey et Surtsey de l'archipel volcanique Vestmannaejar (Islande): Étude du littoral.
Norois 17 (67):305-334.

coasts

Mokyevsky, O.B., 1969.
The biogeocoenotic system of the marine littoral zone. (In Russian; English abstract).
Okeanologiia 9(2): 211-222.

Coasts

Mokyevsky, O.B. 1967.
The changes of the intertidal biota from the Bering Sea to the Sunda Strait.
(In Russian; English abstract).
Okeanologiia 7(2): 321-324.

coasts

Molinier, Roger, et Marcelle Vidal, 1963.
Études écologiques et biocénotiques dans la baie du Brusc (Var). 4. Contribution à l'étude du peuplement infralittoral de substrat solide dans le port du Brusc.
Bull. Inst. Océanogr., Monaco, 61(1273):12 pp.

coasts

Møller, J.T., 1963.
Vadehavet mellem mandø og Ribe Å.
Folia Geograph. Danida, 8(4):40 pp.

coasts

Morgan, J.P., and P.B. Larimore, 1957
Changes in the Louisiana shoreline.
Trans. Gulf Coast Assoc. Geol. Soc., 7:303-310.

Coasts

Morra, R. H. J., H. M. Oudshoorn, J. N. Svasek, and F. J. de Vos, 1961.
De zandbeweging in het getijgebied van Zuidwest-Nederland.
Rapp. Deltacommissie, Bijdragen, (5):327-380.

coasts
Murray, Grover E., 1961.
Geology of the Atlantic and Gulf Coastal province of North America. Harper, New York, xvii 692 pp., illus. $24.

Reviewed by: Monroe, W.H., 1962.
Science, 135(3508):1057.

N

coasts
Nakano, M., 1947
An investigation on the effect of prevailing winds upon the depths of bays.
Geophys. Mag. 14(2/4):57-86, 7 figs.

coasts
Nevesskaya, L.A., 1959
[The Mollusca-complexes of the Upper Quarternary nearshore sediments of Anapa-region (Black Sea). Questions of studies of marine coasts.]
Trudy Oceanogr., Komissii Akad. Nauk, SSSR, 4: 132-145.

coasts
Neveskaya, L.A., and Ye. N. Nevessky, 1961.
The relationship between the Karangat and New Euxine beds in littoral regions of the Black Sea (In Russian).
Doklady, Akad. Nauk, SSSR, 137(4):934-937.

Translation:
Consultants Bureau for Amer. Geol. Inst., 137 (1-6):345-348. (1962).

coasts
Nevesskaia, L.A., and E.N. Nevesskii, 1961.
[On the correlation of Karangat and New Euxenian strata in the littoral areas of the Black Sea.]
Doklady Akad. Nauk, SSSR, 137(4):934-937.

coasts
Nevessky, E.N., 1959
[New data on the genesis of Markitant and Bakal bottom-embayments on the Black Sea. Questions of studies of marine coasts.]
Trudy Oceanogr., Komissii, Akad. Nauk, SSSR, 4: 122-126.

coasts
Nevessky, E.N., and Ph. A. Shcherbakov, 1959
[The history of development of the Tuapse-bay during the last Black Sea transgression. Question of studies of marine coasts.]
Trudy Oceanogr., Komissii Akad. Nauk, SSSR, 4:127-131.

coasts
Newman, Walter S., and Stanley March, 1968.
Littoral of the northeastern United States: Quaternary warping.
Science, 160(3832):1110-1112.

O

coasts
Oldale, R.N., J.D. Friedman and R.S. Williams Jr. 1971.
Changes in coastal morphology of Monomoy Island, Cape Cod, Massachusetts.
Prof. Pap. U.S. Geol. Surv. 750-B: B101-B107.

coasts
Olson, J.S., 1958.
Lake Michigan dune development. 3. Lake-level, beach and dune oscillations. J. Geol., 66(5):473-483.

coasts
Onofre de Morais, Jáder 1969
Aspectos correlativos de geologia litoral e submarina no nordeste do Brasil.
Arq. Ciên. Mar, Fortaleza, Ceará, Brasil, 9(2): 127-131.

coasts
Orviku, K., 1965.
On the accumulation of boulders along the sea coasts of Esthonia. (In Russian).
Okeanologiia, Akad. Nauk, SSSR, 5(2):316-321.

coasts
Osokin, S.D., 1962
The USSR and the length of its coastal line. (In Russian).
Okeanologiia, Akad. Nauk, SSSR, 2(5):943.

coasts
Ostrovskii, A.B., 1966.
On the occurrence of the Kuyalnitskii marine terrace and Upper-Pliocene red-brown clays on the Black Sea shore of the Caucasus, to the south of Anapa. (In Russian).
Dokl., Acad. Nauk, SSSR, 171(5):1160-1163.

P

coasts
*Panin N., and St. Panin, 1967.
Regressive sand waves on the Black Sea shore.
Mar. Geol., 5(3):221-226.

Coasts
Pascu, Mircea Radu, et Matei Mihai Ghiką 1969
Contributions à la connaissance de l'hydrogéologie karstique du littoral roumain de la mer Noire.

Rapp. P.-v. Reun. Commn int. Explor. scient. Mer Mediterr., 19(4): 615-616.

coasts
*Paskoff, Roland, 1967.
Recent state of investigations on Quarternary sea levels along the Chilean coast between Lat. 30°and 33°S.
J. Geosci., Osaka City Univ., 10:107-113.

coasts
Pavlidis, Y. A., 1964
Peculiar features of the lithology of the nearshore deposits in the Kuril Islands. (In Russian).
Okeanologiia, Akad, Nauk, SSSR, 4(6):1044-1051.

coasts
Pavlidis, Iu. A., 1961
Peculiarities of the post glacial transgression of the Baltic Sea and its relation to the modern transgression of other seas. New investigations of sea coasts and reservoirs. (In Russian).
Trudy Okeanogr. Komissii, Akad. Nauk, SSSR, 12:86-93.

coasts
Pavlidis, Iu. A., and V.L. Boldirev, 1961
Postglacial development of the southern coast of the Baltic sea (according to the investigations made in the Polish Peoples Republic). New investigations of sea coasts and reservoirs. (In Russian).
Trudy Okeanogr. Komissii, Akad. Nauk, SSSR, 12:30-42.

coasts
Penrith, M.-L., and B.F. Kensley, 1970
The constitution of the intertidal fauna of rocky shores of South West Africa. I. Lüderitzbucht.
Cimbebasia (A). (9). 191-239

coasts
Phylippova, I.V., 1959
[Some model studies of the unconsolidated shore outwash processes of oblique waves. Questions of studies of marine coasts.]
Trudy Okeanogr., Komissii Akad. Nauk, SSSR, 4: 156-160.

Coasts
Picard-Tarbouriech, Françoise 1969.
Contribution à l'étude des minéraux lourds dans les sables littoraux de Toulon au Cap Lardier (Provence).
Téthys 1(2): 539-560.

coasts
*Plafker, George, and Meyer Rubin, 1967.
Vertical tectonic displacements in south-central Alaska during and prior to the Great 1964 earthquake.
J. Geosci., Osaka City Univ., 10:53-66.

coasts
Pomerol, Ch., 1957.
Une excursion géologique et océanographique du Mont Saint-Michel au Finistère.
Rev. Géogr. Phys., Géol. Dynam., (2)1(4):249-251.

coasts
Popov, B.A., 1965.
Approximations for the form of accumulative coastal areas. Sediment drift and the genesis of the heavy mineral fields in the coastal zone of the sea. (In Russian;English abstract).
Trudy Inst. Okeanol., Akad. Nauk, SSSR, 76:167-188.

coasts
Popov, B.A., 1961.
[Influence of wave refraction on the formation of nearshore equilibrium profiles.]
Trudy Inst. Okeanol., Akad. Nauk, SSSR, 48:309-327.

coasts
Popov, E.A., 1959
[Some considerations on the "ice-crust" and "ice-fringe" (prepay effects on the seashore dynamics. Questions of studies of marine coasts.]
Trudy Okeanogr., Komissii Akad. Nauk, SSSR, 4: 109-112.

coasts
Price, J.B., and Gordon Gunter, 1964.
Studies of the chemistry of fresh and brackish waters in Mississippi and the boundary between fresh and brackish water.
Int. Revue Ges. Hydrobiol., 49(4):629-636.

coasts
Price, W. Armstrong, 1962
Tidal inlets and headland-flanking barriers.
Proc. First Nat. Coastal and Shallow Water Symposium, Oct. 1961: 47.

coasts
Provotorov, I.A., 1965.
On the total energy flows, that is, the coast formation factors. (In Russian).
Okeanologiia, Akad. Nauk, SSSR, 5(3):473-475.

Q

R

coasts
Raison, Jean-Pierre, 1961.
La falaise et l'estran rocheux du Cap d'Alprech (Pas de Calais).
Cahiers Oceanogr., C.C.O.E.C., 13(9):636-651.

Coasts
Rao, G. Prabhakar, 1969.
Age of the Warkalli (Varkala) formations and the emergence of the present Kerala Coast, India. Bull. natn. Inst. Sci., India, 38(1): 449-456.

coasts
Reid, W.J., 1958.
Coast experiments with radioactive tracers.
Dock & Harbour Auth., 39(453):84-88.

coasts
Richards, Horace G., 1967.
A summary of the marine Quaternary of the Pacific coast of North and South America.
J. Geosci., Osaka City Univ., 10:83-90.

coasts
Rikhter, V.G., Ye. A. Gofman and Ye. G. Mayev, 1960.
[A contribution to the study of the shorelines on the floor of the Caspian Sea.]
Doklady Akad. Nauk, SSSR, 135(6):1476-1479.

translation: (1961), 135(1-6):1226-1228.

Coasts
Robinson, A.H.W. 1966.
Residual currents in relation to shoreline evolution of the East Anglican Coast.
Marine Geology 4(1): 57-84.

coasts
Robinson, A.H.W., 1960
Ebb-flood channels systems in sandy bays and estuaries.
Geography, 45: 183-199.

coasts
Rosa, B. 1970
Einige Probleme der Geomorphologie, Paläogeographie und Neotektonik des Südbaltischen Küstenraumes.
Baltica, Lietuv. TSR Mokslu Akad. Geogr. Skyr. INQUA Tarpt. Sek., Vilnius, 4:197-210.

coasts
Ruhe, R.V., J.M. Williams and E.L. Hill, 1965.
Shorelines and submarine shelves, Oahu, Hawaii.
J. Geol., 73(3):485-497.

coasts
Russell, Richard J., 1967.
River plains and sea coasts.
Univ.Calif.Press, 173 pp.

coasts
Russell, R.J., 1963.
Recent recession of tropical cliffy coasts.
Elevated benches and other coastal forms give evidence of eustatic change in sea level.
Science, 139(3549):9-15.

coasts
Russell, R.C.H., 1961
The use of fluorescent tracers for the measurement of littoral drift. Ch. 24, in: Proc. Seventh Conf. on Coastal Engineering, 418-444.

S

coasts
Sabir, V.A., 1959.
[Fortune of the bays of Kuronova and Agrachansk in connection with land reclamation and with the sinking of the sea level. Problems of the Caspian]
Trudy Okeanogr. Komissii, Akad. Nauk, SSSR, 5: 378-380.

coasts
Safianov, G.A., 1965.
The erosional effect of debris in the coastal zone. (In Russian).
Okeanologiia, Akad. Nauk, SSSR, 5(2):304-310.

coasts
Safianov, G.A., 1962
Chemical leaching in the coastal area and the abrasive process. (In Russian).
Okeanologiia, Akad.Nauk, SSSR, 2(4):673-683.

Coasts
Saint-Requier, Anne, et André Guilcher, 1969.
Un grès coquillier de faciès littoral, immergé à -34 m en Baie d'Audierne (Finistère), est daté de 15,000 ans avant l'Actuel.
C. r. hebd. Séanc., Acad. Sci. Paris.

coasts
Sakagishi, Syokichi, 1963.
Investigation of the movement of mobile bed material in the estuary and on the coast by means of radioactive material.
J. Oceanogr. Soc., Japan, 19(1):27-36.

coasts
Sakou, Toshitsugu, 1965.
Surf on the coral-reefed coast.
Ocean Sci. and Ocean Eng., Mar. Techn. Soc., Amer. Soc. Limnol. Oceanogr., 2:700-705.

coasts
Sato, Seiichi, and Tsutomu Kishi, 1958.
Experimental study of wave run-up on sea wall and shore slope.
Coastal Engineering in Japan, 1:39-43.

coasts
Sato, Shoji, 1962
Sand movement at Fukue coast in Atsumi Bay, Japan and its observation by radioactive glass sand.
Coastal Engineering in Japan, 5:81-92.

Coasts
Sato, Shoji, Norio Tanaka, and Isao Irie, 1969.
Study on scouring at the foot of coastal structures. Coast. Engng. Japan, 12: 83-98.

coasts
Saudry, Yves, et Monique Juignet-Jardin, 1964.
Étude bionomique des milieux marins et maritimes de la région de Luc-sur-Mer. III. Le Roque-Mignon, rocher des hauts niveaux à Luc-sur-Mer.
Bull. Soc. Linnéenne, Normandie, (10), 5:155-178.

coasts
Savage, R.P., 1959
Laboratory data on wave run-up on roughened and permeable slopes.
Beach Erosion Bd., Tech. Memo., No. 109: 1-28.

coasts
Scherbakov, Ph. A., 1959
[Lithological study of nearshore sediments in Anadyr-gulf (Bering Sea). Questions of studies of marine coasts.]
Trudy Okeanogr., Komissii Akad. Nauk, SSSR, 4: 31-43.

coasts
Schipani, R., 1957.
Variazioni della linea di costa nel littorale P. ta S. Salvatore-C. Peloro (Messina).
Atti Soc. Peloritana, 3(3):221-242.

coasts
Schmitz, H.P., 1957.
Küstenschutz und wissenschaftliche Grundlagenforschung. Wasserwirtschaft-Wassertechnik, 7: 64-74.

coasts
Schofield, J.C., 1970.
Coastal sands of Northland and Auckland
N.Z. Jl Geol. Geophys. 13(3):767-824

coasts
*Schwartz, Maurice L., 1968.
The scale of shore erosion.
J. Geol., 76:508-517.

Coasts
Selley, R.C. 1969.
Near-shore marine and continental sediments of the Sirte basin, Libya.
Q. Jl geol. Soc. London 124 (496-4):419-452.

coasts
Sevon, W.D., 1966.
Sediment variation on Farewell Spit, New Zealand.
New Zealand J. Geol. Geophys., 9(1/2):60-75.

coasts
Shigley, C.M., 1951.
Engineering problems of coastal waters. Texas J. Sci. 3:21-29.

Shuisky, Yu.D.
Shuysky, Yu.D.

coasts
Shuisky, Yu.D., 1969.
On the influence of heavy gales on the shallow sand coasts of the eastern Baltic Sea. Okeanologiia 9(3): 475-478.
(In Russian; English abstract)

coasts
Shuyski, Yu., D., 1965.
Some relief forms of the sand shores in the northwestern part of the Black Sea. (In Russian)
Isv., Vses. Geograf. Obshch., 97(5):456-460.

coasts
Shvisxii, Iu. D., V. L. Boldruyev and B.V. Kochetkov 1970.
On conditions and peculiarities of the coastal marine placer formation in the eastern Baltic. (In Russian).
Dokl. Akad. Nauk SSSR 194(1): 187-190.

coasts
Shyshov, N.D., 1959
[Changing of the wind generated waves parameters around the isolated pier. Questions of studies of marine coasts.]
Trudy Okeanogr., Komissii Akad. Nauk, SSSR, 4:146-148.

coasts
Skriptunov, N.A., 1960.
[Basic peculiarities of the hydrology of shallow "predust'evogo" coastal waters with swell depth (based on the example of the coastal waters of the Volga).]
Trudy Okeanogr. Komissii, Akad. Nauk, SSSR, 10(1):107-118.

coasts
Slomianko, P.N., 1961
[An experimental study of applying luminescent sand for the investigation of the bore operation along the Polish coast of the Baltic Sea.]
Okeanologiia, (3):537-542.

Coasts
Southward, A. J., 1965.
Life on the Sea - Shore.
Harvard Univ. Press., 153 pp. $3.00

coasts
Spataru, A., 1965.
Some features of morphology and dynamics in the Rumanian coastal area of the Black Sea. (In Russian).
Okeanologiia, Akad. Nauk, SSSR, 5(2):311-315.

coasts
Stapor, F.W., 1971.
Sediment budgets on a compartmented low-to-moderate energy coast in northwest Florida.
Marine Geol. 10(2): M1-M7.

coasts
Stearns, Charles E., and David L. Thurber, 1967.
Th^{230}/U^{234} dates of Late Pleistocene marine fossils from the Mediterranean and Moroccan littorals.
Progress in Oceanography, 4:293-305.

coasts
Steers, J.A., 1960
Defense against the Sea.
Advancement of Science, 17(65):7-15.

coasts
Steers, J.A., 1957.
The coastline of Great Britain: a brief word picture.
Tijdschr. Konink. Nederl. Aardrijk. Gen., 74(3): 392-398.

coasts
Steers, J.A., 1951.
The sea coast. Collins, London, 276, $3.75.
Reviewed, J. Geol. 63(1):94-97.

coasts
Stephenson, W., 1961.
Experimental studies on the ecology of intertidal environmenta at Hebron Island. II. The effect of substratum.
Australian J. Mar. Freshwater Res., 12(2):164-176.

coasts
Stewart, Harris B., Jr., 1960
Coastal water temperature and sea level - California to Alaska.
California Coop. Ocean. Fish. Invest. Rept. 7:97-104.

coasts
Stewart, J. Q. 1945
Coast, Waves, and Weather for Navigators. vii and 348 pp., maps, diagr., ill., bibliogr., index. Ginn and Co. $3.75.

coasts (uplift of)
Stovas, M.V., 1963.
Recent tectonic uplift of the coast of the White and Barents seas. (In Russian).
Doklady, Akad. Nauk, SSSR, 153(1/6):1415-1417.

Translation:
AGI Earth Sci. Sect. 153(1/6):119-121.

coasts
Striggow, Klaus, 1966.
Strömungsmessung in der Brandungszone Aufgabenstellung und Voruntersuchungen zur Lösung eines Messtechnischen Problems der Küstenforschung.
Beitr. Meereskunde, 17-18:111-126.

coasts
Stuiver, Minze, and Joseph J. Daddario, 1963
Submergence of the New Jersey coast.
Science, 142(3594):951.

coasts
Suda, Kanji, Mitsuo Iwashita, Yukio Yamamoto, Nobuo Watanabe, Zinziro Nakai and Michihei Hoshino, 1963.
Outline of coastal oceanographic surveys off Miho Key. (In Japanese).
Bull. Coast. Oceanogr., 2(2): 1-13

Also in:
Coll. Repr. Fac. Oceanogr. Tokai Univ., 1965.

coasts
Sunamura, Tsuguo and Kiyoshi Horikawa, 1969.
A study of erosion of coastal cliffs by using aerial photographs. Coast. Engng. Japan, 12: 99-120.

coasts
Swift, Donald J.P., Charles E. Dill Jr. and John McHone 1971.
Hydraulic fractionation of heavy mineral suites on an unconsolidated retreating coast.
J. Sedim. Petrol. 41 (3): 683-690.

coasts
Swift, D.J.P., R.B. Sanford, C.E. Dill Jr. and N.F. Avignone 1971
Textural differentiation on the shore face during erosional retreat of an unconsolidated coast, Cape Henry to Cape Hatteras, western North Atlantic shelf.
Sedimentology 16(3/4): 221-250.

coasts
Syvakov, I.K., 1961
[Development of shore-face terrace outer edge caused by wave action.]
Trudy Okeanogr. Komissii, Akad. Nauk, SSSR, 8:129-135.

T

Coasts
Tanner, W. F., 1962.
Reorientation of convex shores.
Amer. J. Sci., 260(1):37-43.

coasts
Tanner, W.F., 1959.
Near-shore studies in sedimentology and morphology along the Florida panhandle coast.
J. Sed. Petr., 29(4):564-574.

coasts
Tanner, W.F., R.G. Evans and C.W. Holmes, 1963.
Low-energy coast near Cape Romano, Florida.
J. Sed. Petr., 33(3):713-722.

coasts
Taylor, J.C.M., and L.V. Illing, 1969.
Holocene intertidal calcium carbonate cementation, Qatar, Persian Gulf.
Sedimentology, 12(1/2): 69-107.

coasts
Tomczak, G., 1952.
Der Einfluss der Küstengestalt und des vorgelagerten Meeresboden auf den windbedingten Anstau des Wassers, betrachtet am Beispiel der Westküste Schleswig-Holsteins.
Deutsches Hydrogr. Zeits. 5(2/3):114-131, 4 textfigs.

coasts
Tommasi, Luiz Roberto, 1965.
Faunistic provinces of the western South Atlantic littoral region. (Abstract).
Anais.Acad. bras. cienc., 37(Supl.):261-262.

coasts
Townsley, S.J., Lamarr Trott and Edith Trott, 1962
A preliminary report of the rehabilitation of the littoral marine community on a new lava flow at Kapoho, Hawaii.
Ecology, 43(4):728-730.

coasts
Traeger, Günther, 1962.
Die Sturmflut vom 16./17.Februar 1962 im Lande Bremen.
Die Küste, 10(1):93-112.

coasts
Tricart, Jean, 1962.
Étude générale de la desserte portuaire de la "Sasca". Conditions morphologiques générales du littoral occidental de Côte d'Ivoire.
Cahiers Océanogr., C.C.O.E.C., 14(2):88-97.

coasts

Tricart, J., 1959
Problemes geomorphologiques du littoral oriental du Bresil.
Cahiers Ocean., C.C.O.E.C., 11(5): 276-308.

coasts

True, Merrill A., 1970.
Étude quantitative de quatre peuplements sciaphiles sur substrat rocheux dans la région marseillaise.
Bull. Inst. océanogr. Monaco 69 (1401): 48pp.

U

coasts

Uchio, Takayasu, 1962
Recent Foraminifera thanatocoenoses of beach and nearshore sediments along the coast of Wakayama-Ken, Japan.
Publ. Seto Mar. Biol. Lab., 10(1) (Article 8): 133-144.

coasts

Ulst, V., 1959
[Some general conditions of the eolian accumulationsoon sea-shores. Questions of studies of marine coasts.]
Trudy Okeanogr., Komissii Akad. Nauk,SSSR, 4:91-100.

coasts

USA, Army Coastal Engineering Research Center, 1964.
Land against the sea.
Misc. Paper, No. 4-64:43 pp.

V

coasts

Van Straaten, L.M.J.U., 1965.
Coastal barrier deposits in South- and North Holland in particular in the areas around Scheveningen and IJmuiden.
Mededelingen van de Geologische Stichting, n.s., No. 17:41-75.

coasts

Van Straaten, L.M.J.U., 1961
Directional effects of winds, waves and currents along the Dutch North Sea coast.
Geologie en Mijnbouw, 40:333-346.

Also:
Publikatie, Geol. Inst., Groningen, No. 144.

coasts

Van Straaten, L.M.J.U., 1959.
Minor structures of some littoral and neritic sediments. Geol. en Mijnbouw (nw. ser.) 21:197-216.

coasts

Varjo, Uuno, 1969.
Über Riffbildungen und ihre Entstehung an den Küsten des Sees Oulujärvi (Finnland).
Die Küste, 17:51-80.

coasts

Vasiliev, Iu. F. and S. M. Popov, 1960
[Temperature field in front of protruding capes]
Izv. Akad. Nauk, SSSR, Ser. Geofiz. (4): 557-565.

coasts

Veenstra, H.J., 1970.
Quaternary North Sea coasts. Quaternaria 12: 169-184.

coasts

Venzo, G.A., e A. Brambati, 1968.
Evoluzione e difesa delle coste dell'alto Adriatico da Venezia a Trieste.
Riv. ital. Geotecn., 1968(3): 3-19.
Reprinted: Comm. Studio Oceanograf. Limnol., (a) (11)

coasts

Vidal, Annie 1967.
Étude des fonds rocheux circalittoraux le long de la côte du Roussillon.
Vie Milieu (B) 18 (1B): 167-219.

coasts

Vladimyrov, A.T., 1959
[Quarternary evolutions of the west Kamchatka coast. Questions of studies of marine coasts.]
Trudy Okeanogr., Komissii Akad. Nauk SSSR, 4:101-108.

coasts

Vladimyrov, A.T., and V.S. Medvedev, 1959
[Shore dynamics and morphology researches of the Japan and Okhotsk sea shores. (Information). Questions of studies of marine coasts.]
Trudy Okeanogr., Komissii, Akad. Nauk,SSSR, 4:215-221.

coasts

Vollbrecht, Kurt, 1966
The relationships between wind records, energy of longshore drift, and energy balance off the coast of a restricted body of water, as applied to the Baltic.
Marine Geology, 4(2): 119-147.

coasts

Völpel, Fred, 1960.
Die Bildung charakteristischer Sedimente in den küstennahen Flachsee.
Deutsche Hydrogr. Zeits., 13(6):290-307.

coasts

Voskoboynikov, V.M., 1966.
Shore dynamics of the elongated aquateria at earlier and later stages of development. (In Russian; English abstract).
Okeanologiia, Akad. Nauk, SSSR, 6(4):659-665.

W

coasts

Wait, J.R., 1957.
Amplitude and phase of the low frequency ground wave near a coastline.
J. Res. Nat. Bur. Standards, 58(5):237-242.

coasts

Whittow, John B., 1970.
Shoreline evolution on the eastern coast of the Irish Sea. Quaternaria 12: 185-196.

coasts

Wiegel, Robert L., 1964.
Oceanographical engineering.
Prentice-Hall Series in Fluid Mechanics, 532 pp.

coasts

Wilson, Ronald F., 1963.
Organic carbon levels in some aquatic ecosystems.
Publ. Inst. Mar. Sci., Port Aransas, 9:64-76.

coasts

Wohlenberg, Erich, 1963.
Der Deichbruch des Ülvesbüller Kooges in der Februar-Sturmflut 1962 Versalzung-Übersandung-Rekultivierung.
Die Küste, 11:52-89.

XYZ

coasts

Zabelina, E.K., 1966.
On recent vertical movements of the northwestern coast of the Okhotsk Sea. (In Russian; English abstract).
Okeanologiia, Akad. Nauk, SSSR, 6(5):830-834.

coasts

Zabelina, E.K., 1961.
Migration of the Ochota-river mouth and its effect on coastal dynamics. New investigations of sea coasts and reservoirs. (In Russian).
Trudy Okeanogr. Komissii, Akad. Nauk, SSSR, 12:67-72.

coasts

Zarva, A.V., 1959
[Shingle beaches study of the Caucasus Black Sea shores between Tuapse and Adler. Questions of studies of marine coasts.]
Trudy Okeanogr., Komissii Akad. Nauk, SSSR, 4:13-17.

coasts

Zenkovich, V.P., 1971.
Problems of the sea shore dynamics. (In Russian; English abstract). Okeanologiia 11(5): 865-872.

coasts

Zenkovich, V.P., 1970.
Nature of the USSR marine shelves and coasts.
Quaternaria 12: 71-77.

coasts

Zenkovitch, V.P., 1965.
Buts et principaux axes de recherches des etudes sur les zones maritimes littorales.
Cahiers Oceanogr., C.C.O.E.C., 17(9):605-623.

coasts

Zenkovich, V.P., 1963
In the coastal area of Democratic Republic Vietnam. (In Russian).
Okeanologiia, Akad. Nauk, SSSR, 3(3):470-476.

coasts

Zenkovich, V.P., 1962
The basis of the concept on the development of sea coasts. Publishing House, Akademia Nauk, SSSR, 710 pp.

Reviewed by:
A.D. Dobrovolsky, 1962, Okeanologiia, 2(5): 950-953.

coasts

Zenkovich, V.P., 1961
Aufgaben und Methoden des Studiums der Meeresufer der USSR (in Russian).
Gerland's Beitr. z. Geoph., 70(3):168-183.

coasts

Zenkovich, V.P., 1959.
On the genesis of cuspate spits along lagoon shores.
J. Geology, 67(3):269-277.

coasts

Zenkovich, V.P., 1958.
Basic problems in the study of coasts of the Far Eastern Seas. Oceanographic investigations in the northwest part of the Pacific Ocean. (In Russian).
Trudy Okeanogr. Komissii, Akad. Nauk, SSSR, 3:52-54.
(Abridged text of a lecture).

coasts

Zenkovich, V.P., 1958.
Coast erosion and silting of harbors. (In Russian).
Priroda, 47(7):65-

coasts

Zenkovich, V.P., 1958
Morphology and dynamics of the Soviet coast of the Black Sea.
Akad. Nauk, SSSR, Vol. 1:68-78; 100-108; 126-146; 165-175.
USN HO TRANS-141
M. Slessers
MO 16104-c
PO 04297
1962

coasts

Zenkovich, V.P., 1958.
On the dynamics of the coast of the Poland's Baltic.
Izv. Vses. Geograph. Obshsh., 90(3):269-279.

coasts

Zenkovitch, V.P., 1958.
On the profiles of underwater shore slopes of the Crimea west coasts. Trudy Inst. Okeanol., 28:93-99.

coasts

Zenkovich, V.P., 1957.
On the formation of the coasts of the south-eastern Caspian. Material subdivisions for the study of sea coasts and reservoirs.
Trudy Okeanogr. Komissii, Akad. Nauk, SSSR, 2:51-58.

coasts

Zenkovich, V.P., 1957.
New works on the study of the dynamics of sea shores.
Meteorol. i Gidrol., (10):43-

coasts

Zenkovich, V.P., 1957.
Polish Baltic coasts. Material subdivisions for the study of sea coasts and reservoirs.
Trudy Okeanogr. Komissii, SSSR, 2:189-194.
Akad. Nauk,

coasts

Zenkovich, V.P., 1956.
Certain mechanisms for the development of the coasts of western Kamchatka. Studies of sea coasts and reservoirs.
Trudy Okeanogr. Komissii, Akad. Nauk, SSSR, 1:57-64.

coasts

Zenkovich, V.P., 1948.
Developing an abrasion design in the process of raising the sea level. Dok. Akad. Nauk, USSR, 63:183-186.

coasts

Zenkovich, V.P., y A.S. Ionin, 1969.
Breve resumen sobre las investigaciones de la estructura y dinamica de la zona litoral de la Isla de Cuba.
Serie Oceanol., Inst. Oceanol., Acad. Cienc., Cuba, 8:1-22.

coasts

Zenkovich, V.P., and A.S. Ionin, 1962
On the movement of the pebble material in the coastal zone. (In Russian).
Okeanologiia, Akad. Nauk, SSSR, 2(5):864-873.

coasts

Zenkovitch, V.P., A.S. Ionin and P.A. Kaplin,
Abrasion as the source of debris in the coastal zone. Sediment drift and the genesis of the heavy mineral fields in the coastal zone of the sea. (In Russian; English abstract).
Trudy Inst. Okeanol., Akad. Nauk, SSSR, 76:102-125.

coasts

Zenkovitch, Z.P., and P.A. Kaplin, 1965.
Submarine geomorphological investigations (Dalmatian coast). (In Russian).
Izv., Akad. Nauk, SSSR, Ser. Geogr., (3):18-34.

coasts

*Zenkovich, V.P., and B.A. Shulak, 1967.
Advances in the near-shore zone investigations of the seas. (In Russian; English abstract).
Okeanologiia, Akad. Nauk, SSSR, 7(5):811-827.

coasts

Zhdanov, A.M., 1962.
Methods for reinforcing marine coasts and on recent developments. Marine geology and dynamics of coasts. (In Russian).
Trudy Okeanogr. Komissii, Akad. Nauk, SSSR, 10(3):113-122.

coasts

Zhdanov, A.M., 1956.
Reinforcement of coast with "galechnimi" alluvium (deposits) dikes (dams) along the complete profile. Studies of sea coasts and reservoirs.
Trudy Okeanogr. Komissii, Akad. Nauk, SSSR, 1:18-36.

coasts

Zhiliaev, A.P., and N.V. Esin, 1965.
Methods for quantitative estimation of abrasion based on flysch coastal studies. (In Russian).
Okeanologiia, Akad. Nauk, SSSR, 5(6):1107-1109.

coasts

Zubenko, F.S., 1961.
Use of aerophotography for studies of artificial-lake shores. (Information). New investigations of sea coasts and reservoirs. (In Russian).
Trudy Okeanogr. Komissii, Akad. Nauk, SSSR, 12:120-124.

Coasts, biology of

coasts, biology of

Herberts, Charmaine, 1964.
Contribution à l'étude du Peuplement rocheux sessile dans la zone a Fucus serratus L. (1).
Bull. Lab. Marit., Dinard, 49/50:5-61.

coasts, lists of spp.

Plante, R., 1964.
Contribution à l'étude des peuplements de hauts niveaux sur substrats solides non récifaux dans la région de Tuléar, Madagascar.
Rec. Trav. Sta. Mar., Endoume, hors sér., Suppl., No. 2:205-315. (Trav. Sta. Mar., Tuléar).

coasts, biology of

Vermeij, Geerat J. and James W. Porter, 1971.
Some characteristics of the dominant intertidal molluscs from rocky shores in Pernambuco, Brazil. Bull. mar. Sci. Miami 21(2): 440-454.

"coastal "development" effect of

Taylor, John L., and Carl H. Saloman, 1968.
Some effects of hydraulic dredging and coastal development in Boca Ciega Bay, Florida.
Fish. Bull. U.S. Dept. Comm. 67(2): 213-241

Coast, effect of

coast, effect of

Båth, M., 1953.
Comparison of microseisms in Greenland, Iceland, and Scandinavia. Tellus 5(2):109-134, 10 textfigs.

coast, effect of

Blumsack, Steven L., 1972.
The traverse circulation near a coast. J. phys. Oceanogr. 2(1): 34-40.

coasts, effect of

Evanov, R.N., 1957.
The influence of the shore on the direction of wind surface currents. (In Russian).
Trudy, Morsk Gidrofiz. Inst., Akad. Nauk SSSR, 11:84-96.

Translation:
OTS, SLA, or ETC

coast, effect of

Hobbs, Carl H. III 1970.
Shoreline orientation and storm surge.
Marit. Sed. 6(3): 113-115. (Halifax)

coast, effect of

McPherson, Ronald D., 1970.
A numerical study of the effect of a coastal irregularity on the sea breeze.
J. appl. Met. 9 (5): 767-777.

coast, effect of

Piries de Carvalho, Maria da Gloria, 1970.
Componentes biogênicos dos sedimentos da plataforma dos estados do Espírito Santo e Rio de Janeiro.
Publ. Inst. Pesquisas Marinha, Brasil, 46: 14 pp.

coasts, effect of

Takahasi, R., 1964.
Coastal effects upon tsunamis and storm surges.
Bull. Earthquake Res. Inst., Tokyo, 42(1):175-180.

Coastline, fossil

coastline,fossil

*Friedman,Gerald M.,1965.
A fossil shoreline reef in the Gulf of Eilat (Aqaba).
Israel J.Earth Sci., 14(3/4):86-90.

coasts, lists of spp.

Kosler, Albert, 1969.
Zur Makrofauna des Sublittorals bei Hiddensee.
Beitr. Meereskunde 24-25: 56-80

Coastal margins

coasts, margins of

Worzel, J. Lamar 1965.
Deep structure of coastal margins and midoceanic ridges
In: Submarine geology and geophysics, Colston Papers W.F. Whittard and R. Bradshaw, editors, Butterworths, London 17: 335-359.

Coastal processes

coastal processes

Steers, J.A., 1967.
Geomorphology and coastal processes.
In: Estuaries, G.H. Lauff, editor, Publs Am. Ass. Advmt Sci., 83:100-107.

Coast protection

coast protection

Egorov, E.N., 1961.
On one of the ways of developing and protecting coastal installations. Dynamics of the coastal zone of the Black Sea. (In Russian).
Trudy Inst. Okeanol., Akad. Nauk, SSSR, 53:37-41.

Coasts, rising

coasts, rising

Cotton, Sir Charles, 1968.
Relation of the continental shelf to rising coasts.
Geogr. J. 134(3): 382-389.

coastal water, effect of

Prakash, A., 1971.
Terrigenous organic matter and coastal phytoplankton fertility. (English and Portuguese abstracts). In: Fertility of the Sea, John D. Costlow, editor, Gordon Breach, 2: 351-368.

Coating, epoxy

Coating, epoxy

Jorda, Robert M., 1965.
Applying protective epoxy coatings to submerged metal, wood and concrete structures.
Trav. Centre de Recherches et d'Etudes Oceanogr. n.s., 6(1/4):121-126.

cobble

Bowin,C.O., A.J. Nalwalk and J.B. Hersey,1966.
Serpentinized peridotite from the north wall of the Puerto Rico Trench.
Geol., Soc., Amer., Bull., 77(3):257-270.

coatings

Garoyan, John R., 1969.
Coatings and encapsulants — preservers in the sea.
Ocean Engng 1(6): 435-456

cobbles

cobbles

Laughton, A.S., 1963.
18. Microtopography.
In: The Sea, M.N. Hill, Editor, Interscience Publishers, 3:437-472.

Coelacanth

See: fish, coelacanth

coherence, temporal

Namias, Jerome and Robert M. Born, 1970.
Temporal coherence in North Pacific sea-surface temperature patterns. J. geophys. Res., 75(30): 5952-5955.

coherence, horizontal

SCHOTT FRIEDRICH 1971.
On horizontal coherence and internal wave propagation in the North Sea. Deep-Sea Res., 18(3): 291-307.

coiling

Bolli, H.M. 1971
The direction of coiling in planktonic Foraminifera.
In: Micropalaeontology of oceans, B.M. Funnell and W.R. Riedel, editors, Cambridge Univ. Press, 639-648.

cold, effect of

cold, effect of

Dahlberg, Michael D. and Frederick G. Smith, 1970.
Mortality of estuarine animals due to cold on the Georgia Coast. Ecology, 51(5): 931-933.

cold, effect of

Holme, N.A., 1967.
Changes in the bottom fauna of Weymouth Bay and Poole Bay following the severe winter of 1962-1963.
J. mar. biol. Ass., U.K., 4 (2):397-405.

cold, effect of

Kühlmorgen-Hille, Georg, 1965.
The effect of the severe winter 1962/63 on the bottom fauna of Kiel Bay.
Ann. Biol., Cons. Perm. Int. Expl. Mer, 1963, 20:98-99.

cold, effect of

Tokioka, Takasi, 1963
Supposed effects of the cold weather of the winter of 1962-63 upon the intertidal fauna in the vicinity of Seto.
Publ. Seto Mar. Biol. Lab., Kyoto Univ., 11(2): 415-424 (245-254).

collision at sea

collisions at sea

Wylie, F.J., 1963.
Collision at sea in fog: the common sense approach.
J. Inst. Navigation, 16(1):100-110.

Color

color

Arkhipova, E.G., and G.V. Rzheplinskiy, 1961.
Transparency and color of water in the Apsheronsk area of the Caspian Sea. (In Russian).
Trudy Gosudarst. Okeanogr. Inst., 61:153-158.

Translation:
USN Oceanogr. Off. TRANS-211(1964).

color

Arseniev, V.S., and V.I. Voitov, 1968.
Relative transparency and colour of the Bering Sea waters. (In Russian; English abstract).
Okeanologiia, Akad. Nauk, SSSR, 8(1):55-57.

color

*Brandhorst,Wilhelm, y José Raúl Cañón,1967.
Resultados de estudios oceanografico-pesqueros aeros en el norte de Chile.
Pub. Inst.Fomento Pesquero, Chile, 29: 44 pp. (mimeographed)

color

Brennecke, W., 1921.
Die ozeanographischen Arbeiten der Deutschen Antarktischen Expedition 1911-1912.
Arch. Deutschen Seewarte 39(1):1-216, 15 pls., 41 textfigs.

color

Brennecke, W., 1915.
Oceanographische Arbeiten S.M.S. "Planet" im westlichen Stillen Ozean 1912-1913. Ann. Hydr., usw., 43:145-158.

color (data)

Calmels, Augusto Pablo, y Hugo Alfredo Taffetani 1969.
Reconocimientos oceanograficos en la ria interior de la Bahía Blanca.
Contrib. Inst. Oceanogr. Univ. Nac. Sur, Bahía Blanca, Argentina 1969 (3): 1-21.

Color (Data)

Calmels, Augusto Pablo, y Hugo Alfredo Taffetani, 1969
Nuevos aportes al conocimiento oceanografico de la ria de Bahía Blanca: Puerto de Ingeniero White.
Contrib. Inst. Oceanogr. Univ. Nac. Sur, Bahía Blanca, Argentina, 4: 35pp.

color

Charcot, J.B., 1931-1932.
Rapport préliminaire sur la campagne du "Pourquoi Pas?" en 1931. Ann. Hydrogr., Paris, (3)11:57-139.

color

Chosen. Fishery Experiment Station, 1938.
Research in the neighboring sea of the Korean Gulf on board the R.M.S. Misago-maru in Feb-Mar 1933. Ann. Rept. Hydrogr. Obs., Chosen Fish. Exp. Sta., 8:7-22.

color

Chosen. Fishery Experiment Station, 1938.
Oceanographic investigations in the Japan Sea made during Aug. and Oct. 1933, on board the R.M.S. Misago-maru. (In Japanese; English headings for tables and charts.) Ann. Rept. Hydrogr. Obs., Chosen Fish. Exp. Sta. 8:77-113, tables, charts.

color

Chosen. Fishery Experiment Station, 1938.
Oceanographical investigations during Jun-Jul 1933, in the Japan Sea offshore along the east coast of Tyosen on board the R.M.S. Misago-maru. (IN Japanese). Ann. Rept. Hydrogr. Obs., Chosen Fish. Exp. Sta., 8:11, 23-29, 38-60, 62-65.

color
Chosen. Fishery Experiment Station 1936-1938. Monthly observations from Urusaki to Kawaziri misaki during the year 1932-1933. Ann. Rept. Hydrogr. Obs., Chosen Fish. Exp. Sta., 7:94-113; 8:139-153.

color
Chosen. Fishery Experiment Station, 1930. Report on the hydrographical investigations in the adjacent sea off the east coast of Tyosen, during May and July 1929. (In Japanese; English summary). Ann. Rept. Hydrogr. Obs., Chosen Fish. Exp. Sta. 4:17-44, 6 pls., tables, charts.

color
Czerpa, O., 1955. Eichwerte der Farbskala von Forel zur Bestimmung der Meerwasserfarbe. Acta Hydrophysica 2(4):145-147.

color
DeBuen, F., 1955. Notas sobre un viaje de estudio de oceanografia aplicada en el extremo norte de la costa chilena Bol. Cient., C.A.G., 2:25-39.

color
Dobrovolsky, A.D., V.G. Zavriev, and A.N. Kosarev, 1961. Colour and transparency as characteristic features of river water in the sea area. Okeanologiia, Akad. Nauk, SSSR, 1(4):626-629.

color
Elliott, F.E., W.H. Myers and W.L. Tressler, 1955. A comparison of the environmental characteristics of some shelf areas of eastern United States. J. Washington Acad. Sci. 45(8):248-259, 4 figs.

color
Ewing, Gifford, 1967. What we can learn from the color of the sea. Ocean Industry, 2(5):26-27.

color
Francis-Boeuf, C., 1941. Résultats des mesures physico-chimiques effectuées à bord du Chausseur 2 le long de la côte marocaine entre Port-Lyautey et Mazagan, au mois de janvier 1941. Bull. Inst. Océan., Monaco, 804:1-16.

color(data)
Fuse, Shin-ichiro, and Eiji Harada, 1960
A study on the productivity of Tanabe Bay.(3) Result of the survey in the summer of 1958. Rec. Oceanogr. Wks., Japan, Spec. No. 4: 13-28.
Oceanographic conditions of Tanabe Bay.

color
Fuse, S., I. Yamazi and E. Harada, 1958. A study on the productivity of the Tanabe Bay. (Part 1). 1. Oceanographic conditions of the Tanabe Bay. Results of the survey in the autumn of 1956. Rec. Oceanogr. Wks., Japan, Spec. No. 2:3-9.

colour
Ganesan, A.S., 1961. The colour of the sea. Current Science, India, 30(8):285-286.

color
Guarrera, S.A., 1950. Estudios hidrobiologicos en el Rio de la Plata. Rev. Inst. Nac., Invest. Cienc. Nat., Ciencias Botanicas 2(1):62 pp.

color
Gunther, E. R., 1936. A report on oceanographical investigations in the Peru coastal current. Discovery Rept., XIII: 107-276, Pls. XIV-XVI.

color
Hahn, Sangbok, 1968. The relationship between the water color and the transparency in the seas around Korea. J. Oceanogr. Soc., Korea, 3(2):55-62.

color
Hanzawa, M., and T. Tsuchida, 1954. A report on the oceanographical observations in the Antarctic carried out on board the Japanese whaling fleet during the years 1946 to 1952. J. Ocean. Soc., Japan, 10(3):99-111, 7 textfigs.

color
Hirasawa, K., 1940. On the Yaku-mizu. J. Oceanogr. 12(2):365-370.

color
Hopkins, T.L., 1961. Natural coloring matter as an indicator of inshore water masses. Limnol. & Oceanogr., 6(4):484-486.

color (data only)
Hoshino, Zen-ichiro 1970. Oceanographical conditions observed at definite station off Asamushi during 1969. Bull. mar. biol. Sta. Asamushi 14 (1): 63 pp. (fold out)

color
*Ishino Makoto, Keiji Nasu, Yoshimi Morita and Makoto Hamada, 1968. Oceanographic conditions in the west Pacific Southern Ocean in summer of 1964-1965. J. Tokyo Univ. Fish., 9(1): 115-208.

color (data only)
Japan, Defense Agency, Maritime Self Defense Force, Oceanographic Unit 1970 (1971?) JMSDF annual oceanographic report for 1970, 522 pp.

color
Japan, Hakodate Marine Observatory, 1956. Report of the oceanographical observations in the sea off Tohoku District made aboard the R.M.S. Yushio-maru in spring and summer 1952. 1. Bull. Hakodate Mar. Obs., (3):205-215. 1-11.

color
Japan, Maizuru Marine Observatory, 1963 Report of the oceanographic observations in the Wakasa Bay in August, 1961. (In Japanese) Bull. Maizuru Mar. Obs., No. 8:89-95.

color
Japan, Maizuru Marine Observatory, 1956. Report of the oceanographical observations north off Sanin and Hokuriku districts in summer 1954. Report of the oceanographical observations north off Sanin and Hokuriku districts in summer 1955. Bull. Maizuru Mar. Obs., (5):85-94; 49-56.

color
Japan, Maizuru Marine Observatory, 1956. Report of the oceanographic observations off Kayoga-misaki during the latter half of 1955. Bull. Maizuru Mar. Obs., (5):31-37.

color
Japan, Maizuru Marine Observatory, 1956. Report of the serial observations off Kyoyamisaki during the first half of 1955. Bull. Maizuru Mar. Obs., (5):27-32.

color
Japan, Maizuru Marine Observatory, 1955. Report of the oceanographical observations (taken) off San'in and Hokuriku (July-August, 1953). Bull. Maizuru Mar. Obs., (4):13-27. 217-231.

color
Japan, Maizuru Marine Observatory, 1955. Report of the Oceanographical observations off San'in and Hokuri-ku (July-August, 1952). Bull. Maizuru Mar. Obs., (4):49-63 (168-181).

color
Jerlov, N.G., 1964 Factors influencing the color of the oceans. In: Studies in Oceanography dedicated to Professor Hidaka in commemoration of his sixtieth birthday, 260-264.

color
Jerlov, N.G., 1963. Optical oceanography. In: Oceanography and Marine Biology, H. Barnes, Editor, George Allen and Unwin, 1:89-114.

color
Kalle, K., 1940. Ein neuer optischer Streuungeffekt an Meerwasser. Ann. Hydr., Jahrg. 68(10):358-360.

colors
Lythgoe, J.N. 1968. Red and yellow as conspicuous colours underwater. Rept. Underwat. Ass. 1968: 51-53

color
Kalle, K., 1939. Die Farbe des Meeres. Rapp. Proc. Verb., Cons. Perm. Int. Expl. Mer, 109:98-105, 7 textfigs.

color
Kalle, K., 1939 V. Die chemischen Arbeiten auf der "Meteor" Fahrt Januar bis Mai 1938. Bericht über die zweite Teilfahrt des Deutschen Nordatlantischen Expedition des Forschungs-und Vermessungsschiffes "Meteor", Januar bis Juli, 1938. Ann. Hydro. u. Mar. Meteorol. 1939: 23-30, 6 text figs.

color
Koizumi, M., 1955. Researches on the variations of oceanographic conditions in the region of the Ocean Weather Station "Extra" in the North Pacific Ocean. (1). "Normal" values and annual variations of oceanographic variations of oceanographic elements. Pap. Met. Geophys., Tokyo, 6(2):185-201.

color
Koizumi, M., 1953. On the annual variation in oceanographical elements at a fixed point (39°N., 153°E) in the Pacific Ocean. Rec. Ocean. Wks., Japan, n.s., 1(1):36-43, 6 textfigs.

color
Koizumi, M., 1952. The annual variation in and correlation between colour of the sea, transparency and plankton volume around the fixed point "Extra". J. Ocean. Soc., Japan, 8(2):79-83, Fig. 2.

color

Kokubo, S., 1952
Results of the observations on the plankton and oceanography of Mutsu Bay during 1950, reference being made also to the period 1946-1950. Bull.Mar.Biol.Sta., Asamushi 5(1/4): 1-54, 3 tables,(fold-in), 1 fold-in.

color

Krümmel, Otto, 1907
Die Farbe der Meere.
Handbuch der Ozeanographie, Vol. 1, Chapter III:266-279.

USN-HO-TRANS-106
M. Slessers
M.O. 16104
P.O. 13377. 1961

color

Lenoble, J., 1956.
Remarque sur le couleur de la mer.
C.R. Acad. Sci., Paris, 242(5):662-664.

colors

Lythgoe,J.N., 1968.
Red and yellow as conspicuous colours underwater.
Rept. Underwater Ass., 1968:51-53.

color

Mielke, H., 1928.
Wasserverfärbungen an der südamerikanischen Ostküste. Ann. Hydrogr., usw., Jahrg. 56(Heft(IX): 294-295.

color

Motoda, S., 1940.
Comparison of the conditions of water in the bay, lagoon and open sea in Palao.
Palao Trop. Biol. Sta. Studies 2(1):41-48, 2 textfigs.

color

Mazzarelli, G., 1949.
Ricerche sul colore del mare esequite tra la Sicilia e la Libia e lungo le coste della Puglia e della Calabria. Boll. Soc. Naturalisti, Napoli 57:115-119.

Nakamura, Nakaroku, 1958 color

On the seasonal variation of chlor content, transparency and water-colour of the eel-culture ponds in Tokai district in relation to the particle size of phytoplankton. Bull. Jap. Soc. Sci. Fish, 24:495-600.

color

Nan'niti, T., 1953.
An improvement on the case of Forel's sea colour scale. Meteorol. Res. Inst., Tokyo, 4(2):

color

Nan'niti, T., 1953.
An improvement on the case of Forel's sea color scale. Pap. Met. Geophys., Tokyo, 4(2):1 pp., 1 fig.

color

Nasu, K., 1960

Oceanographic investigation in the Chukchi Sea during the summer of 1958.
Sci. Repts., Whales Res. Inst., 15: 143-157.

color(data)

Nigeria, Federal Fisheries Service, 1963.
Quarterly research report - January-March 1963, 18 pp. (mimeographed).

color(data)

Nigeria, Federal Fisheries Service, 1963.
Research report, April to October 1963, 27 pp. (mimeographed).

color

Okitsu, T., T. Tokui and B. Tsubata, 1954.
Oceanographical observations off Asamushi during 1952. Bull. Mar. Biol. Sta., Asamushi, 8(1):21-25, 4 textfigs.

color

Oshite, Kei, 1968.
Suspended matter. J. Tokyo Univ. Fish., (2): 5-26.

color

Parenzan, Pietro, 1960
Il Mar Piccolo di Taranto.
Giovanni Semerano, Editore, Roma, 254 pp.

color

Parke, M.L.Jr., K.O. Emery, Raymond Szymankiewicz and L.M. Reynolds 1971.
Structural framework of continental framework in South China Sea.
Bull. Am. Ass. Petrol. Geol. 55(5): 723-751.

color

Parke, M.L.Jr., K.O. Emery, Raymond Szymankiewicz and L.M. Reynolds 1971.
Structural framework of the continental margin in the South China Sea.
Techn. Bull. Econ. Comm. Asia Far East, 4:103-142

Colour

Piatek, W., 1959.
Colour and transparency of sea water near Gdynia in 1954.
Prace Morsk. Inst. Ryback., Gdyni, 10A:175-192.

color

Ray,I.A., and K. Bonder,1963.
Turbidity,transparency and color of water in an estuary. (In Russian).
Gidrologiye Ust'yevoy Oblasti Dunaya, Chap.12: 326-332.

Translation: USN Oceanogr. Off.,TRANS. 258. (M. Slessers). 1967.

color

Schott, G., 1942
Geographie des Atlantischen Ozeans im auftrage der Deutschen Seewarte vollständy neu bearbeitete dritte auflage. 438 pp., 141 text figs., 27 plates. C. Boysen. Hamburg.

color

Schulz, B., and A. Wulff, 1929
Hydrographie und Oberflächen plankton des westlichen Barentsmeeres im Sommer 1927. Ber. deutschen wissensch. Komm. F. Meeresforsch. n.s. 4(5):232-372, 13 tables, 25 text figs.

color

Slaucitajs, L., 1947
Ozeanographie des Rigaischen Meerbusens. Teil 1, Statik. Contrib. Baltic Univ. No.45, Pinneberg, 110 pp., 69 text figs.

color

Snezhinskii, V.A., 1951.
[Observations on the transparency and color of the sea.] Prakticheskaia Okeanografia, Leningrad, 524-554.

RT-1494, Bibl. Transl. Rus. Sci. Tech. Lit. 10.

color

Somma, A., 1952.
Elementi di Meteorologia ed Oceanografia. Pt. II. Oceanografia. Casa Editrice Dott. Antonio Milano, Padova, Italia, xviii 758 pp., 322 textfigs., maps, tables. (4500 lire).

color

Somma, A., 1937.
Osservazioni sul colore e sulla temperatura delle acque marine del Golfi di Napoli. Ann. R. Ist. Sup. Navale 6:205-210.

water color

Steenstrup, K.J.V. 1893
Beretning om Undersøgelsesrejserne i Nord-Grønland i Aarene 1878-1880.
Medd. Grønland, 5:1-41.

color

Taguchi, K., 1955.
The movement of water masses in the seas of West Aleutian Is. and off the east coast of Kamchatka salmon fishing ground. I. Inference from the distribution of water temperature, water colour, and transparency. Bull. Japan. Soc., Sci. Fish., 20(9):774-779.

color

Tamura, T., and J. Saguira, 1950.
A report on oceanographical observations of the surface sea waters extending between Japan and Antarctic whaling grounds. Bull. Fac. Fish., Hokkaido, Univ., 1(1):12-17, 2 figs. (In Japanese; English summary).

color(data)

Tanaka, Otohiko, Haruhiko Irie, Shozi IIzuka and Fumihiro Koga, 1961
The fundamental investigation on the biological productivity in the north-west of Kyushu. 1. Rec. Oceanogr. Wks., Japan, Spec. No. 5: 1-58.

color

Tyler, John E., 1965.
Colour of 'pure' water.
Nature, 208(5010):549-550.

color

Tyler, J.E., 1964.
Colour of the ocean.
Nature, 202(4939):1262-1264.

color(data)

Uda, Michitaka, and Makoto Ishino, 1960
Researches on the currents of Kuroshio.
Rec. Oceanogr. Wks., Japan, Spec. No. 4: 59-72.

color

U.S.N. Hydrographic Office, 1958
Oceanographic atlas of the Polar seas. Pt. II. Arctic. H.O. Publ. No. 705: 139 pp., charts.

color (data)
*Usuki, Itaru, 1967.
A record of hydrographic conditions of inshore water AROUND the vicinity of the Sado Marine Biological Station during 1964 to 1966.
Sci. Rep. Niigata Univ. (D) 4:87-107.

color
Velokurova, N.I., and D.K. Starov, 1946.
[Hydrometeorological characteristics of the Black Sea: transparency and water color.]
Gidrometeorologicheskaia Kharakterrstika Chenoga Moria, Moscow:30-33.

RT-1474 Bibt. Transl. Rus. Sci Tech Lit. 10

color
Vercelli, F., 1950.
Trasparenza e colore delle acque della Laguna di Venezia. (Ist. Talassogr., Trieste, Pubbl. No. 266). Archiv. Ocean. e Limnol. 7(1):3-14.

color
Vercelli, F., 1940.
Colore e trasparenza delle acque nella Laguna di Venezia. Atti del Reale Ist. Venato di Sci. Lett. ed Art., Cl. di Sc. Mat. e Nat., XCIX(II):53-59.

color
Vercelli, F., 1923-24.
Ricerche di oceanografia fisica. ricerche di ottica marina. Pt. II. Ann. Idrogr. 11:209-266, tables, charts.

color
Visser, M.P. 1967.
Secchi disc and sea colour observations in the North Atlantic Ocean during the Navado III Cruise, 1964-65, aboard H. Neth. M.S. Snellius (Royal Netherlands Navy).
Neth. J. Sea Res., 3(4):553-563.

color
Wageman, John M., Thomas W.C. Hilde and K.O. Emery, 1970.
Structural framework of East China Sea and Yellow Sea.
Bull. Am. Ass. Petrol. Geol. 54(9):1611-1643.

color
Wattenberg, H., 1938.
Untersuchungen über Durchsichtigkeit und Farbe des Seewassers. I. Kieler Meeresf. 2:293-300.

color
Wilson, Robert C., Eugene L. Nakamura, and Howard O. Yoshida, 1958
Marquesas area fishery and environmental data, October 1957-June 1958.
USFWS Spec. Sci. Rept. Fish., No. 283: 105pp.

color
Yamazi, I., 1955.
Plankton investigations in inlet waters along the coast of Japan. VIII. The plankton of Miyazu Bay in relation to water movement.
Publ. Seto Mar. Biol. Lab., 4(2/3):269-284, 16 textfigs.

color
Yamazi, I., 1953.
Plankton investigations of inlet waters along the coast of Japan. X. The plankton of Kamaisi Bay on the eastern coast of Tohoku District.
Publ. Seto Mar. Biol. Lab. 3(2)(Article 18):189-204, 16 textfigs.

color
Yamazi, I., 1952?
Plankton investigations in inlet waters along the coast of Japan. VI. The plankton of Nanao Bay.
Seto Mar. Biol. Lab. Contr. No. 191:309-319, 11 textfigs.

color
Yamazi, I., 1952.
Plankton investigations in inlet waters along the coast of Japan. III. The plankton of Imari Bay in Kyusyu. IV. The plankton of Nagasaki Bay and Harbour in Kyusyu. V. The plankton of Hiroshima Bay in the Seto-Naikai (Inland Sea).
Publ. Seto Mar. Biol. Lab., Kyoto Univ., 2(2): 289-304; 305-318; 319-330, 8 and 8 and 7 textfigs.

color
Yamazi, I., 1951.
Plankton investigations in inlet waters along the coast of Japan. II. The plankton of Hakodate Harbour and Yoichi Inlet in Hokkaido.
Publ. Seto Mar. Biol. Sta., Kyoto Univ., 1(4): 185-194, 3 textfigs.

color
Yamazi, I., 1950.
Plankton investigations in inlet waters along the coast of Japan. I. Publ. Seto Mar. Biol. Lab. 1(3):93-113, 14 textfigs.

color
Yentsch, C.S., 1960
The influence of phytoplankton on the colour of sea-water.
Deep-Sea Res., 7(1): 1-9.
WHOI Contrib. No. 1058.

color adaptation

color adaptation
Chassard, C., 1960.
L'adaptation chromatique chez Athanas nitescens Leach (Crustacé décapode.
Cahiers Biol. Mar., Roscoff, 1(4):453-463.

color adaptation
Chassard-Bouchard, Colette, 1965.
L'adaptation chromatique chez les Natantia (Crustacées decopodes).
Cahiers Biol. Mar., 6:469-576.

color changes
Humbert, Chantal, 1965.
Etude experimentale du role de l'organe X (paro distalis) dans les changements de couleur et la mue de la crevette Palaemon serratus.
Trav. Inst. scient. cherif., Ser. Zool. No. 32, 96 pp.

color changes
Pautsch, Fryderyk 1967.
Pigmentation and colour changes in decapod larvae.
Proc. Symp. Crustacea, Ernakulam, Jan. 12-15, 1965, 3: 1108-1123.

colors of animals
Kondakov, N.N., 1962.
On the color of the inhabitants of the sea at different depths. (In Russian).
Methods and results of submarine investigations.
Trudy, Okeanogr. Komissii, Akad. Nauk, SSSR, 14: 39-42.

color (of organisms)
Kondakov, N.N., 1962.
The color of marine populations at various depths. (In Russian).
Trudy Okeanogr. Komissii, 14:39-42.

Translation:
USN Oceanogr. Off. TRANS -213(1964).

Color (data)

color (data only)
Anon., 1950.
[The results of the surface observations.]
J. Ocean., Kobe Obs., 2nd ser., 1(3):43-50.

color (data only)
Anon., 1950.
[Report of the oceanographical observations from Fisheries Experimental Stations (Sept. 1950).]
J. Ocean., Kobe Obs., 2nd ser., 1(2):5-11.

color (data only)
Anon., 1950.
Table 11. Marine meteorological observations made on board the "Fugen-maru" in the Iki-suido near Ikizaki Is. Res. Mar. Met., Ocean. Obs., Tokyo, July-Dec., 1947, No. 2:162-163.

color (data only)
Anon., 1950.
Table 3. Surface observations made on board S.S. "Chofuku-maru"; physical and chemical conditions of the surface water of the Goto-Nada between Nagasaki and Tomie. Res. Mar. Met., Ocean. Obs., Tokyo, July-Dec. 1947, No. 2:81-88.

color (data only)
Anon., 1950.
Table 1. Surface observations made on board R.M.S. "Syunpu-maru"; physical and chemical conditions of the surface water of the sea from Kobe to Maizuru. Res. Mar. Met., Ocean. Obs., Tokyo, July-Dec. 1947, No. 2:77-79.

color (data only)
Anon., 1950.
Table 14. Oceanographical observations taken in the Maizuru Bay; physical and chemical data for stations occupied by R.M.S. "Syunpu-maru". Res. Mar. Met., Ocean. Obs., Tokyo, July-Dec. 1947, No. 2:71-76.

color (data only)
Anon., 1950.
Table 13. Oceanographical observations taken in the Nagasaki Harbour; physical and chemical data for stations occupied by a cutter. Res. Mar. Met. Ocean. Obs., Tokyo, July-Dec. 1947, No. 2:65-70.

color (data only)
Anon., 1950.
Table 12. Oceanographical observations taken in the Omura Bay; physical and chemical data for stations occupied by the "Tsuru-maru". Res. Mar. Met., Ocean. Obs., Tokyo, July-Dec. 1947, No. 2: 61-64.

color (data only)
Anon., 1950.
Table 11. Oceanographical observations taken in the Omura Bay; physical and chemical data for stations occupied by a cutter. Res. Mar. Met., Ocean. Obs., Tokyo, July-Dec. 1947, No. 2:59-60.

color (data only)
Anon., 1950.
Table 10. Oceanographical observations taken in the sea northwest of Kyushu; physical and chemical data for stations occupied by the "Fugen-maru". Res. Mar. Met., Ocean. Obs., Tokyo, July-Dec. 1947, No. 2:57-58.

color(data only)
Anon., 1950.
Table 8. Oceanographical observations taken in the adjacent sea to the west of Kyushu; physical and chemical data for stations occupied by R.M.S. "Yushio-maru". Res. Mar. Met., Ocean. Obs., Tokyo, July-Dec. 1947, No. 2:51-54.

color(data only)
Anon., 1950.
Table 7. Oceanographical observations taken in the Sagami Bay; physical and chemical data for stations occupied by R.M.S. "Arashio-maru". Res. Mar. Met., Ocean. Obs., Tokyo, July-Dec. 1947, No. 2:48-50.

color(data only)
Anon., 1950.
Table 6. Oceanographical observations taken at the "fixed point" (39°N, 153°E) in the North Pacific Ocean; physical and chemical data for stations occupied by R.M.S. "Ryofu-maru". Res. Mar. Met., Ocean. Obs., Tokyo, July-Dec. 1947, No. 2:44-47.

color(data only)
Anon., 1950.
Table 5. Oceanographical observations taken in the North Pacific Ocean east off Tohoku District; physical and chemical data for stations occupied by R.M.S. "Ryofu-maru". Res. Mar. Met., Ocean. Obs., Tokyo, July-Dec. 1947, No. 2:22-43.

color(data only)
Anon., 1950.
Table 4. Oceanographical observations taken in the North Pacific Ocean along the coast of Sanriku; physical and chemical data for stations occupied by R.M.S. "Kuroshio-maru". Res. Mar. Met., Ocean. Obs., Tokyo, July-Dec. 1947, No. 2: 13-21.

color(data only)
Anon., 1950.
Table 3. Oceanographical observations taken in the North Pacific Ocean east off Sanriku; physical and chemical data for stations occupied by "Oshoro-maru". Res. Mar. Met., Ocean. Obs., Tokyo July-Dec. 1947, No. 2:8-12.

color(data only)
Anon., 1950.
Table 2. Oceanographical observations taken in the Uchiura Bay; physical and chemical data for stations occupied by R.M.S. "Kuroshio-maru". Res. Mar. Met., Ocean. Obs., Tokyo, July-Dec. 1947, No. 2:3-7.

color (data only)
Anon., 1950.
Table 1. Oceanographical observations taken in the Uchiura Bay; physical and chemical data for stations occupied by a fishing boat belonging to Mori Fishery Assoc. Res. Mar. Met., Ocean. Obs., Tokyo, July-Dec. 1947, No. 2:1-2.

color (data only)
Anon., 1950.
Surface observations made on board a steam-ship. Physical conditions of the surface water of the Goto Nada between Nagasaki and Tomi (continued). Res. Mar. Met., Ocean. Obs., Jan.-June 1947, No. 46-52, Table 4.

color (data only)
Anon., 1950.
Surface observations made on board R.M.S. "Syunpu-maru". Physical and chemical conditions of the surface water of the Seto Inland Sea between Kobe and Hiroshima. Res. Mar. Met., Ocean. Obs., Jan.-June 1947, No. 1:44-45, Table 3.

color (data only)
Anon., 1950.
Surface observations made on board the R.M.S. "Syunpu-maru"; physical and chemical conditions of the surface water of the Pacific coast of Japan between Kobe and Tokyo. Res. Mar. Met., Ocean. Obs., Jan.-June 1947, No. 1:42-43,

color (data only)
Anon., 1950.
Surface observations made on board the R.M.S. "Kuroshio-maru"; physical and chemical conditions of the surface of the sea near Hideshima Island. Res. Mar. Met., Ocean. Obs., Jan.-June 1947, No. 1:39-41, Table 1.

color (data only)
Anon., 1950.
Oceanographical observations taken in the Nagasaki Harbour; physical and chemical data for stations occupied by the cutter. Res. Mar. Met., Ocean. Obs., Jan.-June 1947, No. 1:38, Table 5.

color (data only)
Anon., 1950.
Oceanographical observations taken in the North Pacific Ocean (east of Sanriku); physical and chemical data for stations occupied by R.M.S. "Yushio-maru" (continued). Res. Mar. Met., Ocean. Obs., Jan.-June 1947, No. 1:28-37, Table 4.

color (data only)
Anon., 1950.
Oceanographic observations taken in the Miyako Bay; physical and chemcial data for stations occupied by R.M.S. "Kuroshio-maru". Res. Mar. Met., Ocean. Obs., Jan.-June 1947, No. 1:24-27, Table 3.

color (data only)
Anon., 1950.
Oceanographical observations taken in the North Pacific Ocean (east off Hideshima Is.); physical and chemical data for stations occupied by R.M.S. "Kuroshio-maru" (continued). Res. Mar. Met., Ocean., Obs., Jan.-June 1942, No. 1:12-23, Table 2.

color (data only)
Argentina, Secretaria de Marina, Servicio de Hidrografia Naval, 1961.
Trabajos oceanograficos realizados en la campana Antartica 1960/1961. Resultados preliminares.
Publico, H. 629:unnumbered pp.

Color (data only)
Argentina, Secretaria de Marina, Servicio de Hidrografia Naval, 1961.
Operacion oceanografica, Vema - Canepa 1, Resultados preliminares.
Publico, H. 628:30 pp.

color (data only)
Argentina, Secretaria de Marina, Servicio de Hidrografia Naval, 1961.
Trabajos oceanograficos realizados en la campana Antartica 1959/1960. Resultados preliminares
Publico, H. 623:unnumbered pp.

color (data only)
Berrit, G.R., R. Gerard and L. Vercesi, 1968.
Observations océanographiques exécutées en 1966. II. Stations côtières d'Abidjan, Lome et Cotonou - observations de surface et de fond.
ORSTOM Centre Rech. Océanogr., Côte d'Ivoire, Document scient. provis. 017: 71 pp.

color (data only)
Berrit, G.R., R. Gerard, L. Lemasson, J.P. Rebert et L. Vercesi, 1968.
Observations océanographiques exécutées en 1967.
1. Stations hydrologiques; observations de surface at de fond; stations côtières.
Doc. sci. provis., Centre Recherces océanogr., Côte d'Ivoire, 026:133pp (mimeographed)

color(data only)
Callaway, Richard J., 1957.
Oceanographic and meteorological observations in the northeast and central North Pacific, July-December 1956. USFWS Spec.Sci. Rept., Fish. No. 230:49 pp.

color (data only)
Canada, Canadian Oceanographic Data Center, 1967.
Ocean Weather Station 'P' North Pacific Ocean, May 27 to August 10, 1967.
1967 Data Record Ser., 5: 185 pp.

color (data only)
Canada, Canadian Oceanographic Data Centre, 1966.
Ocean Weather Station 'P' North Pacific Ocean, March 3 to June 2 1966.
1966 Data Record Series 11: 168 pp.

color (data only)
Canada, Canadian Oceanographic Data Centre, 1966.
Arctic Hudson Bay and Hudson Strait, August 5 to October 4, 1962.
1966 Data Record Series, No. 4:247 pp.

color (data only)
Canada, Canadian Oceanographic Data Centre, 1965.
Data Record, Grand Banks to the Azores, June 1 to July 15, 1964.
1965 Data Record Series, No. 13:221 pp. (mimeographed).

color (data only)
Canadian Oceanographic Data Centre, 1965.
Data record, Seguenay and Gulf of St. Lawrence, August 19 to August 30, 1963.
1965 Data Record Series, No. 12:123 pp. (multilithed).

color (data only)
Canada, Canadian Oceanographic Data Centre, 1965.
Data Record Gulf of St. Lawrence, July 23 to August 23, 1964.
1965 Data Record Ser., No. 9:262pp. (multilith)

color(data only)
Canada, Canadian Oceanographic Data Center, 1964
Ocean weather Station "P", North Pacific Ocean.
1964 Data Record Series, No. 15:95 pp. (multilithed)

color(data only)
Canada, Canadian Oceanographic data Center, 1964
ICNAF Norwestlant-2 Survey, Canada.
1964 Data Record Series, No. 14:185 pp. (multilithed)

color(data)
Capart, A., 1953.
Liste des stations. Rés. Sci., Expéd. Océan. Belge dans les Eaux Côtières Africaines de l'Atlantique Sud (1948-1949) 1: 65 pp., 2 pls.

color (data only)
Dragovich, Alexander, John H. Finucane, John A. Kelly, Jr., and Billie Z. May, 1963.
Counts of red-tide organisms, Gymnodinium breve, and associated oceanographic data from Florida west coast, 1960-61.
U.S. Fish and Wildlife Service, Spec. Sci. Repts., Fish., No. 455:1-40.

color (data only)
Gallardo, Y., A. Crosnier, Y. Gheno, J.M. Guillerm, J.C. Le Guen et J.P. Rebert, 1969.
Resultats hydrologiques des campagnes du Centre ORSTOM de Pointe-Noire (Congo-Brazza) devant l'Angola, de 1965 à 1967.
Cah. Océanogr., 21(4):387-400.

color (data only)
Graham, Joseph J., and William L. Craig, 1961
Oceanographic observations made during a cooperative survey of albacore (Thunnus germo) off the North American west coast in 1959.
U.S.F.W.S. Spec. Sci.Rept., Fish. No. 386: 31 pp.

color (data only)
Japan, Central Meteorological Observatory, 1953
The results of Marine meteorological and oceanographical observations. Jan.-June 1952, No. 11:362, 1 fig.

color (data only)
Japan, Central Meteorological Observatory, 1952
The results of Marine Meteorological and oceanographical observations, July - Dec. 1951, No. 10:310 pp., 1 fig.

color (data only)
Japan, Central Meteorological Observatory, 1952.
The results of Marine meteorological and oceanographical observations. July - Dec. 1950. No. 8: 299 pp.

color (data only)
Japan, Central Meteorological Observatory, 1952.
The Results of Marine Meteorological and oceanographical observations. Jan. - June 1950. No. 7: 220 pp.

color (data only)
Japan, Central Meteorological Observatory, 1951.
The results of marine meteorological and oceanographic observations. July - Dec. 1949. No. 6: 423 pp.

color (data only)
Japan, Central Meteorological Observatory, 1951.
Table 14. Oceanographical observations taken in Miyazu Bay: physical and chemical data for stations occupied by R.M.S. "Asanagi-maru".
Res. Mar. Met. Ocean. Obs., Jan.-June, 1949, No. 5:77-84.

color (data only)
Japan, Central Meteorological Observatory, 1951.
Table 12. Oceanographical observations taken in Sasebo Bay: physical and chemical data for stations occupied by R.S.S. "Asakaze-maru".
Res. Mar. Met. Ocean. Obs., Jan.-June, 1949, No. 5:74-75.

color (data only)
Japan, Central Meteorological Observatory, 1951.
Table 11. Oceanographical observations taken in Nagasaki Harbor: physical and chemical data for stations occupied by R.S.S. "Asakaze-maru".
Res. Mar. Met. Ocean. Obs., Jan.-June, 1949, No. 5:70-74.

color (data only)
Japan, Central Meteorological Observatory, 1951.
Table 9. Oceanographical observations taken in the Goto-nada and the Tsushima Straits. (A) Physical and chemical data for stations occupied by the fishing boat "Daikoku-maru". (B) ------- R.S.S. "Umikaze-maru".
Res. Mar. Met. Ocean. Obs., Jan.-June, 1949, No. 5:61-67.

color (data only)
Japan, Central Meteorological Observatory, 1951.
Table 8. Oceanographical observations taken in the Ariake Sea; physical and chemical data for stations occupied by R.S.S. "Umikaze-maru".
Res. Mar. Met. Ocean. Obs., Jan.-June, 1949, No. 5:50-60.

color (data only)
Japan, Central Meteorological Observatory, 1951.
Table 7. Oceanographical stations taken in Osaka Bay; physical and chemical data for stations occupied by R.M.S. "Shumpu-maru".
Res. Mar. Met. Ocean. Obs., Jan.-June, 1949, No. 5:48-49.

color (data only)
Japan, Central Meteorological Observatory, 1951.
Table 5. Oceanographical observations taken in the sea area to the south of the Kii Peninsula along "E"-line; physical and chemical data for stations occupied by R.M.S. "Chikubu-maru".
Res. Mar. Met. Ocean. Obs., Jan.-June, 1949, No. 5:37-40.

color (data only)
Japan, Central Meteorological Observatory, 1951.
Table 4. Oceanographical observations taken in Sagami Bay; physical and chemical data for stations occupied by R.M.S. "Asashio-maru".
Res. Mar. Met. Ocean. Obs., Jan.-June, 1949, No. 5:30-37.

color (data only)
Japan, Central Meteorological Observatory, 1951.
Table 3. Oceanographical observations taken in the North Pacific Ocean along "C" line; (A) Physical and chemical data for stations occupied by R.M.S. "Ukuru-maru"; (B) --- R.M.S. Ikuna-maru"; (C) --- R.M.S. "Ryofu-maru"; (D) --- "Shinnan-maru"; (E) --- "Kung-maru"; (F) --- R.M.S. Chikubu-maru"; (G) --- "Ryofu-maru"; (H) --- "Ikuna-maru"; (I) --- "Chikubu-maru"; (K) --- "Shinnan-maru"; (L) "Ukuna-maru"; (M) --- "Ryofu-maru".
Res. Mar. Met. Ocean. Obs., Jan.-June, 1949, No. 5:13-30.

color (data only)
Japan, Central Meteorological Observatory, 1951.
Table 2. Oceanographical observations taken in the sea area along the east coast of Tohoku District. (A). Physical and chemical data for stations occupied by R.M.S. "Oyashio-maru": (B) Physical and chemical data for stations occupied by R.M.S. "Kuroshio-maru". Res. Mar. Met. Ocean. Obs., Jan.-June 1949, No. 5:3-12.

color (data only)
Japan, Central Meteorological Observatory, 1951.
Table 1. Oceanographical observations taken in the sea area south of Hokkaido; physical and chemical data for stations occupied by R.M.S. "Yushio-maru". Res. Mar. Met. Ocean. Obs., Jan.-June 1949, No. 5:2-3.

color (data only)
Japan, Central Meteorological Observatory, Japan, 1951.
The results of marine meteorological and oceanographical observations, July-Dec. 1948, No. 4: vi+ 414 pp., of 57 tables, 1 map, 1 photo.

color (data only)
Japan, Central Meteorological Observatory, Japan, 1951.
The results of marine meteorological and oceanographical observations, Jan.-June 1948, No. 3: 256 pp.

color (data only)
Japan, Central Meteorological Observatory, Japan, 1950.
The results of marine meteorological and oceanographical observations, Jan.-June 1947, 1:113 pp

color (data only)
Japan, Central Meteorological Observatory, 1949
Report on sea and weather observation on Antarctic Whaling Ground (1947-48). Ocean. Mag., Japan, 1(1):49-88, 17 text figs.

color (data only)
Japan, Defense Agency, Oceanographic Unit (undated).
JMSDF annual oceanographic observation report for 1969: 555 pp.

Color (data only)
Japan, Defense Agency, Oceanography and Intelligence Service Unit (undated)
JMSDF annual oceanographical observation report for 1968: 507 pp.

color (data only)
Japan, Defense Agency, Oceanography and Intelligence Service Unit (undated)
JMSDF annual oceanographical observation report for 1967: 517 pp.

color (data only)
Japan, Defense Agency, Oceanography and Intelligence Service Unit (undated)
JMSDF annual oceanographical observation report for 1966: 527 pp.

color (data only)
Japan, Defense Agency, Maritime Intelligence Service Unit (undated)
JMSDF annual oceanographical observation report for 1965: 431 pp.

color (data only)
Japan, Defense Agency, Maritime Intelligence Service Unit (undated)
JMSDF annual oceanographical observation report for 1964: 363 pp.

color (data only)
Japan, Defense Agency, Maritime Intelligence Service Unit (undated)
JMSDF annual oceanographical observation report for 1963: 301 pp.

color (data only)
Japan, Defense Agency, Maritime Staff Office (undated)
JMSDF annual oceanographical observation report for 1961: 231 pp.

color (data only)
Japan, Hokkaido University, Faculty of Fisheries. 1965.
Data record of oceanographic observations and exploratory fishing, No. 9: 343 pp.

color (data only)
Japan, Hokkaido University, Faculty of Fisheries, 1963
Data Record of Oceanographic observations and exploratory fishing, No. 7:262 pp.

color(data only)
Japan, Japan Meteorological Agency, 1959.
The results of marine meteorological and oceanographical observations, July-December, 1958, No. 24:289 pp.

color (data only)
Japan, Japan Meteorology Agency, 1959.
The results of marine meteorological and oceanographical observations, January-June 1958, No. 23:240 pp.

color (data only)
Japan, Japanese Oceanographic Data Center, 1971. Shumpu Maru, Kobe Marine Observatory, Japan Meteorological Agency, 16 July - 8 August 1970, south of Japan. Data Rept. CSK (KDC Ref. 49K127) 279: 7 pp. (multilithed).

color (data only)
Japan, Japanese Oceanographic Data Center, 1971. Tenyo Maru, Shimonoseki University of Fisheries, Japan, 13-30 June 1969, Southwest of the Japan Sea. Data Rept. CSK (KDC Ref. 49K108) 243: 17 pp. (multilithed).

color (data only)
Japan, Japanese Oceanographic Data Center, 1971. Kofu Maru, Hadkodate Marine Observatory, Japan Meteorological Agency, 27 June - 9 August 1968, East of Japan & Okhotsk Sea. Data Rept. CSK (KDC Ref. 49K084) 178: 32 pp. (multilithed).

color (data only)
Japan,Japanese Oceanographic Data Center,1968. Kaiyo,Hydrographic Division,Maritime Safety Agency,May 10-29,1967,South of Japan. Prelim.Data Rept.CSK (KDC 49K047),96:15 pp. (multilithed).

color (data only)
Japan, Japanese Oceanographic Data Center,1967. Nagasaki Maru,Faculty of Fisheries, Nagasaki University, June 13-17,1967, South of Japan. Prelim.Data Rept.CSK (KDC Ref.No. 49K061)113:6pp.

color (data only)
Japan, Japanese Oceanographic Data Center, 1967. Chofu Maru (NMO), Jan.13-14,1967; Mar.19-20,1967, Apr.15-15,1967, May 11-12,1967,Japan Meteorological Agency, South-East of Yakushima. Prelim.Data Rept.CSK (KDC Ref.Nos.49K315,49K316, 49K317, 49K318) 104: 18 pp.

color (data only)
Japan, Japanese Oceanographic Data Center,1967. Chofu maru, Nagasaki Marine Observatory, Japan Meteorological Agency, Japan, May 17-18,1967, East China Sea.
Prelim. Data Rept. CSK (KDC Ref. 49K051) 100:5pp

color (data only)
Japan, Japanese Oceanographic Data Center,1967. Shumpu maru, Kobe Marine Observatory, Japan Meteorological Agency, Japan, May 13-14,1967, South of Japan.
Prelim.Data Rept. CSK (KDC Ref. 49K050)99: 5 pp.

color (data only)
Japan, Japanese Oceanographic Data Center,1967. Nagasaki maru, The Faculty of Fisheries,Nagasaki University, Japan, January 19-22,1967, East China Sea.
Prelim. Data Rept. CSK (KDC Ref.49K046) 88: 7 pp.

color (data only)
Japan, Japanese Oceanographic Data Center, 1967. Oshoro maru, the Faculty of Fisheries, Hokkaido University, Japan, January 15-February 1, 1967, South of Japan.
Prelim. Data Rept.CSK (KDC Ref. 49K045) 87:9 pp.

color (data only)
Japan, Japanese Oceanographic Data Center,1967. Chofu maru,Nagasaki Marine Observatory, Japan Meteorological Agency, Japan, January 20-February 22,1967, East China Sea.
Prelim.Data Rept.CSK (KDC Ref.49K043) 85:11 pp.

color (data only)
Japan, Japanese Oceanographic Data Center,1967. Shumpu maru, Kobe Marine Observatory, Japan Meteorological Agency, Japan,February 26-27, 1967, south of Japan.
Prelim. Data Rept. CSK (KDC Ref. 49K042) 84: 6 pp

color (data only)
Japan, Japanese, Oceanographic Data Center, 1967. Suro No. 1, Hydrographic Division, Korea, October 12-November 2, 1966, south of Korea.
Prelim. Data Rept., CSK (KDC Ref. 24K011) 70:12 pp.

color (data only)
Japan, Japanese Oceanographic Data Center, 1967. Bukhansan, Fisheries Research and Development Agency, Korea, July 16- August 9, 1966, East of the Yellow Sea.
Prelim. Data Rept., CSK (KDC Ref. 24K010) 69:17 pp.

color (data only)
Japan, Japanese Oceanographic Data Center,1967. Yang Ming, Chinese National Committee on Oceanic Research, Republic of China, September 10-October 14,1966, Adjacent Sea of Taiwan.
Prelim. Data Rept. CSK (KDC Ref. 21K003)67:17pp.

color (data only)
Japan, Japanese Oceanographic Data Center,1967. Chofu Maru (NMO) Jul. 2-3,1966; Chofu Maru (NMO), Aug. 27-29,1966; Shumpu Maru (KMO), Oct. 27-31, 1966, Chofu Maru,Dec. 7-8,1966, Japan Meteorological Agency of Japan,south-east of Yakushima.
Prelim. Data Rept., CSK (KDC Ref. 49K311,49K312, 49K313; 49K314) 61: 18 pp.

color data only)
Japan, Japanese Oceanographic Data Center,1967. Kagoshima maru, the Faculty of Fisheries, Kagoshima University, Japan, August 5-14,1966, East China Sea.
Prelim. Data Rept.CSK(KDC Ref. 49K031) 57: 9pp.

color (data only)
Japan, Japanese Oceanographic Data Center,1967. Seifu Maru,Maizuru Marine Observatory,Japan Meteorological Agency,Japan, August 21-September 19,1966,Japan Sea.
Prelim. Data Rept., CSK(KDC Ref. 49K029)55:15 pp.

color (data only)
Japan, Japanese Oceanographic Data Center, 1966. Preliminary data report of CSK, Takuyo, Japanese Hydrographic Division, February 23-March 15,1966, South eastern Sea of Japan. No. 27(October 1966). KDC Ref. No. 49K016:17 pp. (multilithed).

color (data only)
Japan, Japanese Oceanographic Data Center,1966. Preliminary data report of CSK. Kofu Maru, Hakodate Marine Observatory, Japan Meteorological Agency, July 22-28, 1965, Eastern Sea of Japan, No. 14 (October 1966). KDC Ref. No. 49K004:10 pp. (multilithed).

color (data only)
Japan,Japanese Oceanographic Data Center,1966. Preliminary data report of CSK. Ryofu Maru, Marine Division, Japan Meteorological Agency, July 7-August 3,1965,Eastern Sea of Japan. No. 10 (October 1966). KDC Ref. No. 49K003:31 pp. (multilithed).

color (data only)
Japan, Maritime Safety Agency, 1966.
The results of oceanographic observations in the northwestern Pacific Ocean (11), 1943.
Data Rept. Hydrogr. Obs., Ser. Oceanogr. Publ. 792(2):198 pp.

color (data only)
Japan, Maritime Safety Board, Tokyo, 1962.
The results of oceanographic observations in the north-western Pacific Ocean, No. 9 (1941). Hydrographic Bull., Tokyo, (Publ. No. 981), No. 71:180 pp.

color (data only)
Japan, Maritime Safety Board, 1956.
Tables of results from oceanographic observations in 1952 and 1953. Hydrogr. Bull. (Publ. 981)(51): 1-171.

color (data only)
Japan, Maritime Safety Board, 1955.
The results of oceanographic observation in the north-western Pacific Ocean, 1939-1941. No. 7. Pub. 981, Hydrogr. Bull., Spec. No. 16:1-152.

color (data only)
Japan, Maritime Safety Board, Tokyo, 1954.
Tables of results from oceanographic observations in 1951 Publ. No. 981, Hydrogr. Bull., Spec. No. 15:31-129.

color(data)
Japan, Maritime Safety Board, 1954.
Tables of results from oceanographic observations in 1950. Publ. No. 981, Hydrogr. Bull., Maritime Safety Bd., Spec. Number (Ocean. Repts.) No. 14:26-164, 5 textfigs.

color (data only)
Japan, Nagasaki University Research Party for CSK, 1966.
The results of the CSK - NU65S.
Bull. Fac. Fish. Nagasaki Univ., No. 21:273-292.

color (data only)
Japan, Shimonoseki University of Fisheries 1970.
Oceanographic surveys of the Kuroshio and its adjacent waters, 1967 and 1968.
Data. Oceanogr. Obs. Explor. Fish. 5: 117 pp.

color (data only)
Japan, Shimonoseki University of Fisheries, 1965.
Data of oceanographic observations and exploratory fishings, International Indian Ocean Expedition 1962-63 and 1963-64. No. 1: 453 pp.

color (data only)
Japan, Tokai Regional Fisheries Research Laboratory, 1959.
IGY Physical and chemical data by the R. V. Soyo-maru, 25 July-14 September 1958. 17 pp. (multilithed).

color (data only)
Japan, Tokai Regional Fisheries Research Laboratory, 1952.
Report oceanographical investigation (Jan-Dec. 1950, collected), No. 74:143 pp.

color(data only)
Japan, Tokai Regional Fisheries Research Laboratory, 1951.
Report oceanographical investigation (Jan. 1943-Dec. 1944 collected), No. 72:105 pp., 1 fig.

color (data only)
Korea, Republic of, Fisheries Research and Development Agency, 1971
Annual report of oceanographic observations, 19:717 pp.

color (data only)
Korea, Fisheries Research and Development Agency, 1968.
Annual report of oceanographic observations, 16: ~~964~~ 691 pp.

Color (data only)
Republic of Korea, Fisheries Research and Development Agency 1966.
Annual report of observations, 1965, 14:343pp.

color, data only
Republic of Korea, Fisheries Research and Development Agency. 1965.
Annual report of oceanographic observations. 1964. 13: 222pp.

color (data only)
Republic of Korea, Hydrographic Office 1964.
Technical reports.
H.O. Publ 1101:136pp.

color (data only)
Korea, Fisheries Research and Development Agency, 1967.
Annual report of oceanographic observations, 15 (1966):459 pp.

color (data only)
Korea, Fisheries Research and Development Agency, 1965.
Annual Report of oceanographic observations, 1963, 12:173 pp.

Color (data only)
Korea, Republic of, Fisheries Research and Development Agency, 1964.
Annual Report of Oceanographic Observations, 1962, 11:203 pp.

Color (data only)
Korea, Republic of, Fisheries Research and Development Agency, 1964.
Annual Report of Oceanographic Observations, 1960, 9:184 pp.

Color (data only)
Korea, Hydrographic Office 1970.
Technical reports for the year 1969.
H.O. Pub. 1101:214pp.

color (data only)
Korea (Republic of), Hydrographic Office, 1965.
Technical reports.
H.O. Pub., Korea, No.1101:179 pp.

color (data only)
Korea (South), R.O.K. Navy Hydrographic Office, 1962
Technical reports.
H.O. Publ. South Korea, No. 1101: 85 pp.

color(data only)
Longhurst, A.R., 1961.
Cruise report of oceanographic cruise, 6/61.
Federal Fisheries Service, Nigeria, 4 pp. (mimeographed)

color (data only)
Republic of Korea, Hydrographic Office, 1963.
Technical reports. (In Korean).
H. O. Pub., Korea, No. 1101:111 pp.

Color (data only)
Nishyia, Moritaka 1968.
Oceanographical conditions observed at definite station off Asamushi during 1967.
Bull. mar. biol. Sta. Asamushi 13(2): 151.

color(data only)
Picotti, Mario, 1960
Crociera Talassografica Adriatica 1955. III. Tabelle delle osservazioni fisiche, chimiche, biologiche e psammografiche.
Archivio di Oceanograf. e Limnol., 11(3): 371-377, plus tables.

Color (data only)
Saloman, Carl H., John H. Finucane and John A. Kelly Jr. 1964.
Hydrographic observations of Tampa Bay, Florida, and adjacent waters, August 1961 through December 1962.
U.S. Dept. Interior, Fish Wildl. Serv. Bur. Comm. Fish. Data Rept. 4:6cards (microfiche).

color(data)
Schott, G., and B. Schulz, 1914.
Die Forschungsreise: S.M.S. "Möwe" im Jahre 1911.
Arch. Deutschen Seewarte 37(1):1-80, 8 pls., 16 textfigs.

Color (data only)
South Africa, Division of Sea Fisheries, Department of Industry 1968.
Hydrological station list, 1961-1962, 1:342 pp.

Color (data only)
Suarez-Caabro, José 1965.
Datos meteorologicos, hidrographicos y planctonicos del litoral de Veracruz, Ver.
Anales Inst. Biol. Univ. Mexico 36(1/2): 25-46.

color(data only)
Tsubata, B., 1958.
Oceanographical conditions observed at definite station off Asamushi during 1956.
Bull. Mar. Biol. Sta., Asamushi, 9(1):43.

color (data)
Tsubata, B., 1955.
Oceanographical observations off Asamushi during 1954. Bull. Mar. Biol. Sta., 8(1):43-48.

color(data)
Tsuruta, A., T. Satow, K. Hayama and T. Chiba, 1957.
Oceanographical and planktonological studies of tuna-fishing ground in the eastern part of the Indian Ocean. J. Shimonoseki Coll. Fish., 7(1): 1-17.

color (data only)
Uda Michitaka, Yoshima Morita, and Makoto Ishino, 1957.
Results from the oceanographic observations in the North Pacific (1955-56) with Umitaka Maru and Shinyo-Maru. Rec. Ocean. Wks., Japan (Spec. No.):1-20

color(data)
U.S. Navy Hydrographic Office, 1957.
Operation Deep Freeze II, 1956-1957. Oceanographic survey results. H.O. Tech. Rept., 29: 155 pp. (multilithed).

color (data)
Van Goethem, C., 1951.
Étude physique et chimique du milieu marin.
Rés. Sci., Exped. Océan. Belge dans les Eaux Côtières Africaines de l'Atlantique Sud (1948-1949) 1:151, 1 pl.

color (data only)
Wilson, R.C., and M.O. Rinkel, 1957.
Marquesas area oceanographic and fishery data, January-March 1957.
USFWS Spec. Sci. Rept., Fish., No. 238:136 pp.

Color, effect of

effects of colors
Saito, T., 1931
Researches in fouling organisms of the Ships' bottom. Zosen Kiokai (T. Soc. Naval Arch.) 47pp:13-64, 51 figs., 9 graphs.

color preference
Gerberg, Gary and Frank J. Schwartz, 1970.
Color preferences of the xanthid mud crab Rhithropanopeus harrisii
Ohio J. Sci., 70(2): 115-118

color (profiles)

color(profiles)
Yamazi, I., 1953.
Plankton investigations in inlet waters along the coast of Japan. VII. The plankton collected during the cruises to the new Yamamoto Bank in the Sea of Japan. Publ. Seto Mar. Biol. Lab. 3(1):75-108, 19 textfigs.

Commensalism

commensals
Boss, Kenneth J., 1965.
A new mollusk (Bivalvia, Erycinidae) commensal on the stomatopod crustacean Lysiosquilla.
Amer. Mus. Novitates, No. 2215:11 pp.

commensals
Bouligand, Y., 1966.
Recherches recentes sur les copepodes associés aux Anthozoaires.
In: The Cnidaria and their evolution, W.J. Rees, editor, Zool.Soc.,London,Academic Press,267-306.

commensalism
Bruce, A.J., 1971.
On a new commensal shrimp Periclimenes hirsutus sp. nov. (Crustacea, Decapoda Natantia, Pontoniinae) from Fiji.
Pacific Sci. 25(1): 91-99

commensals
Bruce, A.J. 1970.
Report on some commensal pontoniinid shrimps (Crustacea: Palaemonidae) associated with an Indo-Pacific gorgonian host (Coelenterata: Gorgonacea).
J. Zool. Lond. 160 (4): 537-544.

commensalism
Dales, R.P., 1957.
Interrelations of organisms. A. Commensalism.
In: Ch. 15, Treatise on Marine Ecology and Paleo-ecology, Vol. 1. Mem., Geol. Soc., Amer., 67: 391-412.

commensals
*Davis, William P., and Daniel M. Cohen, 1968.
A gobiid fish and palaemonid shrimp living on an antipatharian sea whip in the Tropical Pacific.
Bull. mar. Sci., 18(4): 749-761.

Commensals
Fenchel, Tom., 1966.
On the ciliated Protozoa inhabiting the mantle cavity of lammelibranchs.
Malacologia, 5(1): 35-36.

Commensals
Gore, Robert H. and John B. Shoup 1968.
A new starfish host and an extension of range for the commensal crab, Minyocerus angustus (Dana, 1852) (Crustacea: Porcellanidae).
Bull. mar. Sci. Miami 18(1): 240-248

commensals
Hampson, George R., 1964.
Redescription of a commensal pelecypod Rochefortia cuneata, with notes on ecology.
Nautilus, 77(4): 125-128.

commensals
Hart, C.W. Jr., N. Balakrishnan Nair and Dabney G. Hart, 1967.
A new ostracod (Ostracoda: Entocytheridae) commensal on a wood-boring marine isopod from India.
Notulae Naturae, 409: 11 pp.

commensals
Herbst, H.V., 1962.
Marine Cyclopoida Gnathostoma (Copepoda) von der Bretagne-Küste als Kommensalen won Polychaeten.
Crustaceana, 4(3): 243.

commensalism
Hollowday, E.D., 1947.
On the commensal relationship between the amphipod, Hyperia galba (Mont.) and the Scyphomedusa, Rhizostoma pulmo Agassiz var. octopus Oken.
J. Quekett Microsc. Club, Ser. 4, 2(4): 187-190, Pls. 25-26.

Commensals
Jeffrey, S.W. 1968.
Pigment composition of Siphonales algae in the brain coral, Favia.
Biol. Bull. mar. biol. Lab. Woods Hole 135(1): 141-148.

commensals
*Johnson, D.S., 1967.
On some commensal decapod crustaceans from Singapore (Palaemonidae and Porcellanidae).
J. Zool., Lond., 153(4): 499-526.

commensals
Johnson, D.S., and Margaret Laing, 1966.
On the biology of the WATCHMAN prawn, Anchistus custos (Crustacea: Decapoda; Palaemonidae), an Indo-West Pacific commensal of the bivalve Pinna.
J. Zool., Lond., 150(4): 433-455.

COMMENSALS
Kirkegaard, J.B., and L.N. Santhakumaran, 1967
On a new species of annelid associate of marine wood borers, Cirriformia limnoricola n. sp. (Polychaeta)
Vidensk. Meddr dansk naturh. Foren. 130: 213-216

commensal larvae
Kramp, P.L., 1957.
Hydromedusae from the Discovery collections.
Discovery Repts., 29: 1-128.

commensalism
Kujawa, Stanisław M., 1971.
A particular case of commensalism between the crab Geryon quinquedens Smith and two barnacles Octolasmis geryonophile Pilsbry and O. lowei Lesson. (In Polish)
Przeglad Zool. 15 (3): 285-286.

Commensals
McCloskey, L. R. 1970.
A new species of Dulichia (Amphipoda, Podoceridae), commensal with a sea urchin.
Pacific Sci. 24 (1): 90-98.

commensals
Morris, Robert A., and Louis S. Mobray, 1966.
An unusual barnacle attachment on the teeth of the Hawaiian spinning dolphin.
Norsk Hvelfangst-Tidende, 55(1): 15-16.

commensals
Patton, Wendell K., 1968.
Feeding habits, behavior and host specificity of Caprella grahami, an amphipod commensal with the starfish Asterias forbesi.
Biol. Bull. mar. biol. Lab., Woods Hole, 134(1): 148-153.

Commensals
Patton, Wendell K. 1967.
Commensal Crustacea.
Proc. Symp. Crustacea, Ernakulam, Jan. 12-15 1965 3: 1228-1243.

commensals
Patton, Wendell K., 1967.
Studies on Domecia acanthophora, a commensal crab from Puerto Rico, with particular reference to modifications of the coral host and feeding habits.
Biol. Bull. mar. biol. Lab., Woods Hole, 132(1): 56-67.

commensals
Sastry, A.N., and R. Winston Menzel, 1962
Influence of hosts on the behavior of the commensal crab Pinnotheres maculatus Say.
Bioil Bull., 123(2): 388-395.

commensals
Serène, R., 1961.
A megalopa commensal in a squid.
Proc. Ninth Pacific Sci. Congr., Pacific Sci. Assoc., 1957, Fish, 10: 35-36.

Commensals
Stevčić Zdravko 1968.
Parasiten und Kommensalen des Bruthaumes der Seespinne (Maja squinado Herbst)
Thalassia Jugoslavica 4: 5-10.

commensalism
Strickland, David L., 1971.
Differentiation and commensalism in the hydroid Proboscidactyla flavicirrata.
Pacific Sci. 25(1): 88-90

commensals
Thorson, Gunnar, 1965.
A neotenous dwarf-form of Capulus ungaricus (L.) (Gastropoda, Prosobranchia) commensalistic on Turritella communis Risso.
Ophelia, 2(1): 175-210.

commensal [on baleen plates.]
Vervoort, W., and D. Tranter, 1961
Balaenophilus unisetus P.O.C. Aurivillius (Copepoda Harpacticoida) from the southern hemisphere.
Crustaceana, 3(1): 70-84.

commensals
Wigley, Roland L., and Paul Shave, 1966.
Caprella grahami, a new species of caprellid (Crustacea: Amphipod) commensal with starfishes.
Biol. Bull., 130(2): 289-296.

commensals, effect of
Ross, D.M., 1971.
Protection of hermit crabs (Dardanus spp.) from Octopus by commensal sea anemones (Calliactis sp.).
Nature, Lond., 230(5293): 401-402.

commercial netting and trawling, effect of
Woodburn, Kenneth D. 1965.
A discussion and selected annotated references of subjective or controversial marine matters.
Techn. Ser. Fla Bd Conserv. 46: 50 pp.

Commission

Commissions
Chapman, William McLeod 1968.
The theory and practice of international fisheries commissions and bodies.
Proc. Gulf Carib. Fish. Inst., 20th Ann. Sess. 77-105.

commissions
Peluchon, G., 1958.
Commission Internationale pour l'Exploration Scientifique de la Mer Méditerranée. Compte Rendu de la XVI e Assemblée Pleniere.
Bull. d'Info., C.C.O.E.C., 10(10): 637-641.

Inter-American Tropical Tuna Commission
Royce, William F., 1958.
Tuna Commission activities in Pacific Ocean.
Proc. Ninth Pacific Sci. Congr., Pacific Sci. Assoc., 1957, Oceanogr., 16: 42-43.

Inter-American Tropical Tuna Comm.
Schaefer, M.B., 1955.
Report on the investigations of the Inter American Tropical Tuna Commission for the year 1954.
Ann. Rept. Inter American Tropical Tuna Comm., 1954:24-59.

Committee

committees, Interagency Committee on Ocean.
Abel, Robert B., 1963.
The ICO and the long-range plan.
Naval Research Reviews, 16(9):2-3.

committees
Aksenov, A.A., 1960
[An enlarged assembly of the Oceanographic Committee. (January 5-12, 1959).]
Biull. Okeanograf. Komissii, Akad. Nauk, SSSR, (6): 5-8.

ICSU
Aksenov, A.A., D.V. Bogdanov, M.E. Vinogradov and G.E. Feldman, 1961
[The fourth session of SCOR in Helsinki held on July 23-24th, 1960 and the results of the XIIth Session of ICSU in regard to the SCOR activities.]
Okeanologiia, Akad. Nauk, SSSR, (2): 339-343

Harrison Brown Committee
Anon., 1959.
Oceanography - the science of the oceans.
Amer. Sci., 47(3):234-249.

committees
Anon., 1958.
Special Committee on Oceanic Research.
Deep-Sea Research 5(1):75-78.

committees
Deveze, L., 1963.
Rapport sur les travaux du Comite de Microbiologie et de Biochimie.
Rapp. Proc. Verb., Réunions, Comm. Int. Expl. Sci., Mer Méditerranée, Monaco, 17(3):671-674.

Long Core Committee (LOCO)
Ewing, Maurice, and John Ewing, 1963
Sediments at proposed LOCO drilling sites.
J. Geophys. Res., 68(1):251-256.

committees
France, Comité Central d'Océanographie et d'Etude des Côtes, 1955.
Comité consultatif provisoire des sciences de la mer. Bull. d'Info., C.C.O.E.C., 7(10):438-441.

committees
France, Comité National Français de Géodésie et Géophysique, 1960
Section d'Oceanographie Physique.
C.R., Com. Nat. Francais, Géod. et Géophys., Année 1959: 133-148.

committees
International Biological Programme, Sub-Committee C, 1964
Proposals for a programme on the productivity of marine communities.
IBP (Marine), Edinburgh 1/64:14 pp. (multilithed).

committees
Kort, V.G., 1965.
On the draft of a general scientific framework for World Ocean study. (In Russian).
Okeanologiia, Akad. Nauk, SSSR, 5(4):753-755.

Comite de Physique de la Mer
Lacombe, H., 1961.
Rapport sur l'activité du Comité de Physique de la Mer.
Rapp. Proc. Verb., Réunions, Comm. Int. Expl. Sci. Mer Méditerranée, Monaco, 16(3):567.

committees
Ovey, C.D., editor, 1954.
International Committee on the Nomenclature of Ocean Bottom Features. Minutes of meeting held at the International Hydrograhic Bureau at Monaco 22 September 1952. Int. Hydrogr. Rev. 31(1):93-99

Committees, Mexico
Porraz, Mauricio, 1971.
Notes on the Mexican Engineering Committee on Ocean Resources.
J. mar. techn. Soc. 5(5): 20-21.

committees
Tregouboff, G., 1963.
Rapport du président sur l'activité du Comité du Planeton pendant la XVIIIe Assemblée plenière.
Rapp. Proc. Verb., Réunions, Comm. Int. Expl. Sci. Mer Méditerranée, Monaco, 17(2):433-448.

committees
Myslovsky, M., 1963
The Committee for Marine Research of the Polish Academy of Sciences. (In Russian).
Okeanologiia, Akad. Nauk, SSSR, 3(3):558-559.

committees
Rochford, D.J., and A.W.B. Powell, 1958
Report of the chairman of the Standing Committee on Pacific Oceanography.
Proc. Ninth Pacific Sci. Congr., Pacific Sci. Assoc., 1957, 16(Oceanogr.): 1-5.

committees, SCOR
Saijo, Yatsuka, 1971.
Activity of SCOR working group 15. Phytosynthetic radiant energy, in reference to Discoverer Expedition. Bull. Plankt. Soc. Japan, 18(1): 99-102. (In Japanese and English).

IOC
Snodgrass, James M., 1962
An oceanographic data and communication system.
Proc. 2nd Interindustrial Oceanogr. Symposium Lockheed Aircraft Corp., 31-39.

committees
Sokolov, O.A., 1962
The recent activities of the Submarine Research Committee. The Xth Pacific Science Congress.
Okeanologiia, Akad. Nauk, SSSR, 2(3):527-529.

committees
Union Geodesique et Geophysique Internationale, 1964.
Scientific Commitee on Oceanic Research.
Chronique de l'U.G.G.I., No. 57:283-288.

committees (Joint Panel on Oceanographic Tables and Standards)
UNESCO, 1965.
First report of the Joint Panel on Oceanographic Tables and Standards held at Copenhagen 5-6 October 1964 sponsored by UNESCO, ICES, SCOR and IAPO.
UNESCO Techn. Papers in Mar. Sci., No. 1: 9 pp. + appendices (mimeographed).

committees
UNESCO, Comite Consultatif International des Sciences de la Mer, 1957.
Bull. d'Info., C.C.O.E.C., 9(1):4-8.

committees, etc.
1948
United States Committee on the Oceanography of the Pacific.
Pacific Science II(4):299.

TENOC
U.S.A. National Academy of Sciences-National Research Council, 1959.
American Oceanography - survey and proposals for a ten-year program.
Science, 129(3348):550-551.

committees
United States, President's Science Advisory Committee, Panel on Oceanography, 1966.
Effective Use of the Sea.
U.S. Government Printing Office - 60 cents - 144 pp.

CSAGI meeting in Moscow
USSR, Committee for the International Geophysical Year, 1959.
Information. Biull. No. 6:95 pp.

committees
Zenkevich, L.A., 1958.
Activities of the Oceanographic Commission in 1951-1956. (In Russian).
Biull., Okeanogr. Komissiia, Akad. Nauk, SSSR, (1):7-

committees
Zenkovich, V.P., 1964.
A commission for the preservation of sea shores. (In Russian).
Okeanologiia, Akad. Nauk, SSSR, 4(2):363.

Communications

communications
Clarke, S.W., and R. Mitchell, 1961.
Fifty years of maritime radio communications (A brief history of the Post Office coast stations).
Marine Observer 31(191):33-37.

communication
Evans, William E., and Jarvis Bastian, 1969
Marine mammal communication: social and ecological factors.
In: The biology of marine mammals, H.T. Andersen, editor, Academic Press, 425-475.

communication
Fay, W J., and D.R. Munoz, 1965.
Communication problems associated with data transmission from weather stations.
Ocean Sci. and Ocean Eng., Mar. Techn. Soc., Amer. Soc. Limnol. Oceanogr., 1:472-497.

Communications
Martin, W.R., and R.L. Adams, 1970.
Physiologic factors in the design of underwater communication.
J. mar. techn. Soc. 4(3): 55-58

Communications (acoustic)
Painter, Daniel W. II. 1968.
Underwater acoustic communications.
Oceanology int. 3(5): 43-45.

Communication
Ray, B. 1969.
Communications between divers.
Oceanol. int. 69(3): 4 pp., 4 figs.

Communications (verbal)
Sergeant, Russell L. 1968.
Limitations in voice communications during deep submergence-helium dives.
Mar. Sci. Instrument. 4: 405-411.

communications
Snodgrass, James M., 1968.
Instrumentation and communications. Ch. 12 in: Ocean engineering: goals, environment, technology, J.F. Brahtz, editor, John Wiley & Sons, 393-477.

communications
Sullivan, Gerard E., 1967.
International regulation of communications for oceanographic equipment.
In: The law of the sea, L.M. Alexander, editor, 195-203.

communication
Tang, T.G., and H.A.P. Smith, 1965.
Communication between dolphins in separate tanks by way of an electronic acoustic link.
Science, 150(3705): 1839-1843.

Communities

communities, invertebrate
*Bretsky, Peter W., 1968.
Evolution of Paleozoic marine invertebrate communities.
Science, 159(3820): 1231-1233.

communities
Fager, Edward W., 1971.
Pattern in the development of a marine community.
Limnol. Oceanogr. 16(2): 241-253.

communities
Fager, E.W., 1963
19. Communities of organisms. In: The Sea, M.N. Hill, Edit., Vol. 2 (IV) Biological Oceanography, Interscience Publishers, New York and London, 415-437.

communities
Margalef, R., 1967.
Some concepts relative to the organization of plankton.
Oceanogr. Mar. Biol., Ann. Rev., H. Barnes, editor, George Allen and Unwin, Ltd., 5: 257-289.

communities
Mills, Eric L., 1969.
The community concept in marine zoology with comments on continua and instability in some marine communities: a review.
J. Fish. Res. Bd., Can., 26 (6): 1415-1428

community concept
Mills, Eric L. 1971.
Views on the community concept with comments on continua and the role of instability in some marine benthic communities.
Vie Milieu Suppl. 22 (1): 145-153

compaction

compaction
Boswell, P.G.H., 1961
Muddy sediments. Geotechnical studies for geologists, engineers and soil scientists.
W. Heffer & Sons, Ltd., Cambridge, 140 pp.

compaction
Collette, B.J., and K.W. Rutten, 1970.
Differential compaction vs. dispersion in abyssal plains.
Mar. Geophys. Res. 1(1): 104-107

compaction
Dangeard, L., and M. Rioult, 1965.
Le domaine de la géologie marine et ses frontières: confromtation de l'océanographie et de géologie.
In: Submarine geology and geophysics, Colston Papers, W.F. Whittard and R. Bradshaw, editors, Butterworth's, London, 17: 93-105.

compaction
Law, J., 1959.
In place measurement of compaction. (Abstract).
J. Geophys. Res., 64(6): 691.

Compaction
Wangersky, Peter J. 1967.
Sodium chloride content and the compaction in deep-sea cores.
J. Geol. 75 (3): 332-335.

Companies

Companies, Aluminum Company of America
Anon., 1967.
Aluminum alloys keyed for deep ocean applications: profile - Aluminum Company of America oceanographic program.
Undersea Techn. 8(4): 20-21

companies - Straza Industries
Anon., 1967.
CTFM sonar goes deep: profile - Straza Industries.
Undersea Techn., 8(5): 49-51.

Companies, Braincon Corp
Anon., 1967.
Remote control launches make continental shelf surveys easier: profile - Braincon Corp.
Undersea Techn., 8(4): 18-19.

companies, Lockheed
Anon., 1965.
Quest into the deep- profile, Lockheed Aircraft Co.
Undersea Techn., 7(3): 39-41.

Companies Francis Associates, Inc.
Booda, Larry L., 1966.
From ocean products to space systems: profile- Sippican Corporation and Francis Associates, Inc.
Undersea Techn., 7(12): 33-35.

companies, Raytheon
Booda, Larry L., 1967.
Raytheon - pioneer in underwater acoustics.
Undersea Techn., 8(8): 14-15. (popular)

companies, Global Marine, Inc.
Booda, Larry L., 1967.
Oil drilling know-how, hardware, applied to sub-sea mining and engineering.
Undersea Techn., 8(2): 26-27.

Companies Sippican Corporation
Booda, Larry L., 1966.
From ocean products to speace systems: profile- Sippican Corporation and Francis Associates, Inc.
Undersea Techn., 7(12): 33-35.

companies
Covey, Charles W., 1965.
Edgerton, Germeshausen & Grier, Inc: company profile.
Undersea Technology, 6(6): 23-27.

companies
Lill, Gordon G., 1963
The role of the private company in oceanic development.
Presented before the American Chemical Society, Los Angeles, California, 3 April, 1963, 17 pp. (multilithed).

compensation point
Khalemsky, E.N., 1971.
The position of the compensation point in the Pacific. (In Russian)
Dokl. Akad. Nauk SSSR, 196(2): 445-447

compressibility
Lepple, F.K. and F.J. Millero, 1971.
The isothermal compressibility of seawater near one atmosphere. Deep-Sea Res. 18(12): 1233-1254.

compression

compressibility
Bulgakov, N.P., 1960
[On the phenomenon of compressibility during the mixing of waters] Izv. Aka. Nauk, USSR, Ser. Geophys. 2: 229-234. (Vol. and pagination of English Edition.)

compressibility
Cox, R.A., 1965.
The physical properties of sea water.
In: Chemical oceanography, J.P. Riley and G. Skirrow, editors, Academic Press, 1: 73-120.

compression
ZoBell, C. E., 1959.
Thermal changes accompanying the compression of aqueous solutions to deep-sea conditions
Limnol. & Ocean., 4(4): 463-471.

Computers

See chiefly: instrumentation, computers methods, computers

Computer processing

computer processing

Sweers, H.E., 1967.
Some Fortran II programs for computer processing of oceanographic observations.
NATO Subcomm. Oceanogr. Res., Tech. Rept. 37 (Irminger Sea Project):104 pp. (mimeographed).

Computer programs

computer programs

Akiba, Yoshio, 1971.
Program for oceanographic computation and data processing on the electronic digital computer AICOM-C3. (In Japanese; English abstract).
Bull. Hokkaido Univ. Fac. Fish. 22(1): 47-57.

computer programs

Akiba, Yoshio, 1971.
Note on the programing for oceanographic data processing. (In Japanese; English abstract).
Bull. Hokkaido Univ. Fac. Fish. 21(4): 299-304.

computer programmes

Allen, K. Radway, 1967.
Computer programmes available at St. Andrews Biological Station.
Tech. Rep. Fish. Res. Bd., Can., 20: 32 pp. (mimeographed).

computers

Gay, S.M., 1968.
Computer analysis and design of undersea cable systems.
UnderSea Techn., 9(10):43,48-49.

computer programs

Godin, Gabriel, S.E. Eldring and J.D. Taylor 1967
The analysis of nineteen years of observations on the high and low water with the aid of the German method.
Manuscr. Rep. Ser. Mar. Sci. Br., Dept. Energy, Mines, Resources, [Ottawa], 3: 115 pp. (multilithed)

Computer programs

Grim, Paul J. 1970.
Computer program for automatic plotting of bathymetric and magnetic anomaly profiles.
ESSA Techn. Memo. ERLTM-AOML8: 31 pp. (mimeographed).

computer programs

Grossling, Bernardo F., 1967.
The internal magnetization of seamounts and its computer calculation
Geol. Surv., Prof. Pap., U.S., 554-F: F1-F26.

computer programs

Keyte, F.K., 1967.
Computer vs. manual calculations of temperature and depths.
J. Fish. Res. Bd., Can., 24(11):2491.

computers, shipboard

Ross, D.I., 1969.
Experience with a shipboard PDP-8 computer.
Oceanol. int., 69, 2: 5 pp., 5 figs.

computer programs

Schle, John, and Jacqueline Webster, 1967
A computer program for grain-size data.
Sedimentology, 8(1): 45-53.

computer routines

Taylor, J., P. Richards and R. Halstead 1971.
Computer routines for surface generation and display.
Manuscript Rept. Ser. Mar. Sci. Br., Dept. Energy, Mines Resources, Ottawa 16: 47 pp.

computer programs

Thomas, Terence M., and Clifford L. Fishback, 1969.
A computer program for handling data on the abundance of shrimp and associated animals.
FAO Fish. Repts. 3(57) (FRm/57.3 (Trm)): 1041-1053 (mimeographed)

computer programs

Wilson, J.R., 1967.
A temperature-salinity plotting program.
Manuscr. Rep. Ser. Mar. Sci. Br. Dept. Energy, Mines Resources, Ottawa, 6: 5-14 (multilithed)

Concentration

Concentration of animals

*Bittel, R., et G. Lacourly, 1968.
Discussion sur le concept de facteur de concentration entre les organismes marins et l'eau en vue de l'interprétation des mesures.
Rev. intern. Océanogr. Med., 11:107-128.

Concentration of chemicals within organisms

See: Chemistry, chemicals concentrated within organisms

concrete

Hansen, F.J., 1969.
Concrete construction off-shore. Oceanol. int., 69, 2: 7 pp., 11 figs.

concrete

Sakoda Shigemi, Shoichi Ichikawa, Takashi Uchida, Shinsei Narisawa, and Ushio Hatae, 1970.
The influence of sea water on the strength and dynamical elastic properties of concrete. (In Japanese; English abstract).
J. Coll. mar. Sci. Technol., Tokai Univ. 4: 25-36

Concretions

Concretions

See also: Chemistry, manganese nodules, etc.

concretions, calcareous

*Kempf, Marc, et Jacques Laborel, 1968.
Formations de vermets et d'algues calcaires sur les côtes du Brésil.
Recl Trav. Stn mar. Endoume, 43(59):9-23.

concretions

Laborel, J., 1961.
Sur un cas particulier de concrétionnement animal. Contrétionnement à Cladocera cespitosa L. dans le golfe de Talante.
Rapp. Proc. Ver., Réunions, Comm. Int. Expl. Sci. Mer Méditerranée, Monaco, 16(2):429-432.

concretions, fossiliferous

*Stanley, Daniel J., Donald J.P. Swift, and Horace G. Richards, 1967.
Fossiliferous concretions on Georges Bank.
J. sedim. Petrol. 37(4):1070-1083.

concretions, coralline algae

Laborel, Jacques, 1961.
Le concretionnement algal "Coralligène" et son importance géomorphologique en Méditerranée.
Rec. Trav. Sta. Mar., Endoume, 23(37):37-60.

concretions

Volkov, I.I., 1957.
[Concretions of iron sulfide in Black Sea deposits]
Doklady, Akad. Nauk, SSSR, 116(4):654.

concretions, manganese

Zenkevich, N.L., and N.S. Skornyakova, 1961
[Iron and manganese on the ocean floor.]
Priroda, 1961(2):47-50.
OTS Soviet-bloc Res. Geophys. Astron. and Space (10): 7.

Condensation

condensation

*Gerard, Robert D., and J. Lamar Worzel, 1967.
Condensation of atmospheric moisture from tropical maritime air masses as a freshwater source.
Science, 157(3794):1300-1302.

condensation

Miyazaki, M., 1952.
The heat budget of the Japan Sea. Bull. Hokkaido Regional Fish. Res. Lab., Fish. Agency, No. 4: 1-54, 45 charts.

condensation

Turner, J.S., 1965.
Laboratory models of evaporation and condensation
Weather, Roy. Meteorol. Soc., 20(4):124-128, Figs on pp. 120-121.

condensation

Tsuji, M., 1950.
On the rate of evaporation and condensation of falling drops. Geophys. Mag., Japan, 22(1):11-14, 1 textfig.

condensation

Zubov, N.N., 1957
Condensation near the merging of marine waters based on temperature and salinity.
Gidrometerizdat, Leningrad, 1957: 39 pp.

Condensation nuclei

condensation nuclei

Blanchard, Duncan C., 1963.
Condensation nuclei and raindrop spectra.
J. Atmos. Sci., 20(6):624-625.

Griffith, Rosemay M., 1963.
Reply.
J. Atmos. Sci., 20(6):625-626.

condensation nuclei

Blanchard, Duncan C., and A. Theodore Spencer, 1964.
Condensation nuclei and the crystallization of saline drops.
J. Atmos. Sci., 21(2):182-186.

CONDENSATION NUCLEI

Graborskii, R.I., 1952.
[The ocean as a source of condensation nuclei in the atmosphere.] Izvest. Akad. Nauk. SSSR, Ser. Geofiz., 2: 56-74.
Listed in Bibl. Transl. Rus. Sci & Tech. Lit. 17: RT 2405.

condensation nuclei
Keith, C.H., and A.B. Arons, 1954.
The growth of sea-salt particles by condensation of atmospheric water vapor. J. Met. 11(3):173-184, 9 textfigs.

condensation nuclei
Köhler, H., and M. Bath, 1953.
Quantitative chemical analysis of condensation nuclei from sea water. Nov. Act. Reg. Soc. Sci., Upsal., (4) 15(7):24 pp.

condensation nuclei
Moore, D.J., 1954.
Measurement of condensation nuclei over the ocean. Weather (M.O. 579), 24(163):18-21.

condensation nuclei
Moore, D.J., 1952.
Measurement of condensation nuclei over the North Atlantic. Q.J.R. Met. Soc. 78(338):596-602, 4 textfigs.

condensation nuclei
Ogiwara, S., 1950.
On the solid condensation nuclei which is not soluable in water. J. Met. Soc., Japan, 28(11):391-397, 4 textfigs. (English summary)

condensation nuclei
Ohta, S., 1950.
On the contents of condensation nuclei and uncharged nuclei on the Pacific Ocean and Japan Sea. J. Met. Soc., Japan, 2nd ser., 28(10):357-367, 3 textfigs., (English summary).

condensation nuclei
Matui, H., 1948.
Chemical studies of condensation and sublimation nuclei. II. Chemical studies on the fogs at Hitoyosi, Kumamoto Prefecture. (In Japanese) J. Met. Soc., Japan, 2nd ser., 26(6):169-172, with English summary p. 14.

condensation nuclei
Woodcock, A.F., C.F. Kientzler, A.B. Arons, and D.C. Blanchard, 1953.
Giant condensation nuclei bursting from bubbles. Nature 172(4390):1144-1145, 2 textfigs.

condensation nuclei
Yamamoto, G., and T. Ohtake, 1955
Electron microscope study of cloud and fog nuclei II. Sci. Repts. Tohoku Univ., (5) Geophys., 7(1):10-16.

condensation nuclei chemistry of

Conductance electrical

conductance, electrical
Bradshaw, A., and K.E. Schleicher, 1965.
The effect of pressure on the electrical conductance of sea water.
Deep-Sea Research. 12(2):151-162.

conductance, electrolyte
Connors, Donald N., and Kilho Park, 1967.
The partial equivalent conductance of electrolytes in seawater: a revision.
Deep-Sea Research, 14(4):481-484.

conductance, electrical
Park, K., and H.C. Curl, Jr., 1965.
Effect of photosynthesis and respiration on the electrical conductance of sea-water.
Nature, 205(4968):274-275.

Conductance, heat

conductance (heat)
Osborne, M.F.M., 1964.
The interpretation of infrared radiation from the air in terms of its boundary layers
Deutsche Hydrogr. Zeits., 17(3):115-136.

conduction (heat)
Sverdrup, H.U., 1943
On the ratio between heat conduction from the sea surface and heat used for evaporation. Ann. N.Y. Acad. Sci. XLIV (1):81-88.

Conductivity

conductivity
Chanu, J. et Y. Le Grand 1965.
Influence de la composition de l'eau de mer sur sa conductibilité électrique. (Abstract only).
Rapp. P.-v. Réun. Comm. int. Explor. scient. Mer Méditerr. 18(3):793.

conductivity
Charnock, H., and J. Crease, 1966.
A salinity estimator based on measurement of conductivity ratio.
UNESCO, Techn. Papers, Mar. Sci., No. 4:App. B. (mimeographed)

conductivity
Cox, Roland A., 1963
The salinity problem.
Progress in Oceanography, M. Sears, Edit., Pergamon Press, London, 1:243-261.

conductivity
Cox, R.A., 1962.
Report on the scientific results of the programme for comparing physical properties of sea water.
Rapp. Proc. Verb., Cons. Perm. Int. Expl. Mer, 1961:94-98. (Reunion)

conductivity
Cox, R.A., F. Culkin, R. Greenhalgh and J.P. Riley, 1962.
Chlorinity, conductivity and density of sea water.
Nature, 193(4815):518-520.

conductivity
Holzkamm, F., G. Krause and G. Siedler, 1964.
On the processes of renewal of the North Atlantic deep water in the Irminger Sea.
Deep-Sea Res., 11(6):881-890.

conductivity
Kwiecinski, Bogdan, 1965.
The relation between the chlorinity and the conductivity in Baltic water.
Deep-Sea Res., 12(2):113-120.

conductivity
Lewis, Bernard L., 1966.
The effect of frequency on the conductivity of sea water.
IEEE Proc., 54(9):1210-1211.

conductivity
Mangelsdorf, Paul C., Jr., 1967.
Salinity measurements in estuaries.
In: Estuaries, G.H. Lauff, editor, Publs Am. Ass. Advmt Sci., 83:71-79.

conductivity
Morcos, Selim A., and J.P. Riley, 1966.
Chlorinity, salinity, density and conductivity of sea water from the Suez Canal region.
Deep-Sea Res., 13(4):741-749.

conductivity
Mosetti, F., 1966.
Una nuova formula per la conduttività dell' acqua di mare in funzione della temperatura e della salinità.
Boll. Geofis. teor. Appl., 8(31):213-217.

conductance
Park, Kilho, 1964
Partial equivalent conductance of electrolytes in sea water.
Deep-Sea Res., 11(5):729-736.

conductivity
Park, Kilho, 1964
Reliability of standard sea water as a conductivity standard.
Deep-Sea Research, 11(1):85-87.

conductivity
Ullyott, P., and O. Ilgaz, 1942.
Apparatus and methods for measuring the conductivity of natual waters in marine and semi-marine conditions. Rev. Fac. Sci., Univ. Istanbul, Ser. A, 8:190-227.

conductivity
Weyl, Peter K., 1964
On the change in electrical conductance of seawater with temperature.
Limnology and Oceanography, 9(1):75-78.

cones

cones, abyssal
Heezen, Bruce C., and H. W. Menard, 1963.
12. Topography of the deep-sea floor.
In: The Sea, M.N. Hill, Editor, Interscience Publishers, 3:233-280.

Conferences

See: meetings institutions, etc.

confused seas
*Ewing, J.A., 1969.
A note on wavelength and period in confused seas.
J. Geophys. Res., 74(6):1406-1408.

Congresses

congresses
Anon., 1966.
Report on The Second International Oceanographic Congress. (In Japanese).
J. oceanogr. Soc., Japan., 22(5):226-230.

Conservation

Conservation
Collier, J.A., 1964.
The regime of the seas: exploitation and conservation.
In: Report of Conference on Law and Science held at Niblett Hall, July 11-12, 1964, David Davies Memorial Institute of International Studies and British Institute of Comparative and International Law, 54-60.

Conservation

Crutchfield, James A. editor 1965.
The fisheries, problems in resource management.
Univ. Washington Press, Seattle 136pp.

Conservation

Currie, R.I., 1964.
Conservation and exploitation of the sea.
In: Report of Conference on Law and Science held at Niblett Hall, July 11-12, 1964, David Davies Memorial Institute of International Studies and British Institute of Comparative and International Law, 49-53.

conservation

Herrington, William C., 1967.
The convention of fisheries and conservation of living resources: accomplishments of the 1958 Geneva Conference.
In: The law of the sea, L.M.Alexander, editor, Ohio State Univ. Press, 26-35.

conservation

McKernan, Donald L., and Wilvan G. Van Campen, 1968.
The emerging pattern of territorial seas policies and international cooperation in fishery conservation and development.
Proc. Gulf Carib. Fish. Inst. 20th Ann. Sess., 106-111

conservation

U.S.A., University of California, Institute of Marine Resources, La Jolla, 1965.
California and use of the ocean: a planning study of marine resources.
IMR Ref. 65-21: app. 350 pp. (multilithed, paged by chapters)

CONSHELF II

Conshelf II

See also: man-in-the-sea

CONSHELF II

Booda, Larry, 1965.
Man-in-the-sea projects will go deeper, stay longer.
Undersea Techn., 6(3):15-16,18-19.

Conshelf

Cousteau, Jacques-Yves, Philippe Cousteau and Bates littleheles, 1966.
Working for weeks on the sea floor.
National Geographic. 129(4):498-537.

CONSHELF III

Cousteau, Jacques-Yves, 1966.
Exploring the sea.
Industrial Research, 8(3):42-50,52.

CONSHELF III

MacInnis, Joseph B., 1966.
Living under the sea. To learn more about the ocean and harvest its resources, men must be able to live and work as free divers on the continental shelf.
Scientific American, 214(3):24-33.

consolidation

Miller, Donald F., Jr., and Adrian F. Richards, 1969.
Consolidation and sedimentation-compression studies of a calcareous core, Exuma Sound, Bahamas.
Sedimentology, 12(3/4):301-306

consortisms

Norris, Richard E. 1967.
Algal consortisms in marine plankton.
In: Proc. Seminar, Sea, Salt and Plants, V. Krishnamurthy, editor, Bhavnagar, India, 178-189.

Construction

construction

Terry, Richard D., editor, 1966.
Ocean Engineering, 1. Energy sources and energy conversion, waste conversion and disposal. 2. Undersea construction, habitation and vehicles; recreation.
Western Periodicals Co., North Hollywood, Calif.
Vol. 3:431 pp. (multilithed).

Contamination

contamination, radioactive

Bernhard, M., 1971.
Studies on the radioactive contamination of the sea. Annual Report 1968-69.
Com. Naz. Energia Nucleare, CNEN Rept. RT/BIO (70)-11:109 pp.

contamination, radioactive

Bernhard, M., 1970.
Evaluation of the risks related to the discharge of radio-active isotopes in a marine ecosystem. 1. A comparison between the concentration factor approach and the specific activity approach.
Rev. int. Océanogr. Méd. 20: 125-131

contamination, radioactive

Bernhard, M., 1967.
Studies on the radioactive contamination of the sea. Annual Report 1966.
Com. Naz. Energia Nucleare EUR 3635e, RT/BIO(67)35:29 pp.

contamination

Braarud, T., and Adam Bursa, 1939
On the phytoplankton of the Oslo Fjord, 1933-1934. Hvaldrådets Skr. No.19:1-63; 9 text figs. Reviewed. J. du. Cons. 14(3): 418-420. A.C. Gardiner.

contamination (Nansen bottle)

Ivanenkov, V.N., 1964.
Changes of pH, alk, P. and O2 caused by the inner surface of Nansen bottles. Investigations in the Indian Ocean (33rd voyage of E/S "Vitiaz"). (In Russian).
Trudy Inst. Okeanol., Akad. Nauk, SSSR, 64:80-84.

contamination, heavy-metal

Merlini, Margaret, 1971.
Heavy-metal contamination. (12) In: Impingement of man on the oceans, D.W. Hood, editor, Wiley Interscience: 461-486.

contamination, radioactive

Miyake, Yasuo, 1971.
Radioactive models. (22) In: Impingement of man on the oceans, D.W. Hood, editor, Wiley Interscience: 565-588.

contamination (bacteria)

Orlob, G.T., 1956.
Evaluating bacterial contamination in sea water samples.
Pub. Health Rept., 71:1246-1252.

Abstr. in:
Publ. Health Eng. Abstr., 37(5):24.

contamination

Pritchard, D.W., Akira Okubo and Emanuel Mehr, 1962
A study of the movement and diffusion of an introduced contaminant in New York Harbor waters.
Chesapeake Bay Inst., Techn. Rept. (Ref. 62-21) 31: 89 pp. (multilithed)

contaminants

Smith, F.I.P., and Alex D.D. Craik 1971.
Wind-generated waves in thin liquid films with soluble contaminant.
J. fluid Mech. 45(3):527-544.

Continents

continents

*Beloussov, V.V., and I. P. Kosminokaya, 1968.
Structure and development of transition zones between the continents and oceans.
Can. J. Earth Sci., 5(4-2):1011-1026.

continents

Dietz, Robert S., 1967.
Passive continents, spreading sea floors and continental rises: a reply.
Am. J.Sci., 265(3):231-237.

continents

Dietz, Robert S., 1966.
Passive continents, spreading sea floors, and collapsing continental rises.
Am. J. Sci., 264(3):177-193.

contaminants

Foster, R.F., I.L. Ophel and A. Preston, 1971.
Evaluation of human radiation exposure. Radioactivity in the marine environment, U.S. Nat. Acad. Sci., 1971: 240-260.

continents

Gilvarry, J.J., 1961.
The origin of ocean basins and continents.
Nature, 190(4781):1048-1053.

continents

Wilson, J. Tuzo, 1966.
Patterns of growth of ocean basins and continents.
In: Continental margins and island arcs, W.H. Poole, editor, Geol.Surv.Pap.,Can.,66-15:388-391.

continents

Young, Grant, M., 1967.
Passive continents, spreading sea floors, and continental rises.
Am. J. Sci., 265(3):225-230.

Continents vs. oceans

continents vs. oceans

Dietz, Robert S., 1964
Commotion in the ocean: the growth of continents and ocean basins.
In: Studies in Oceanography dedicated to Professor Hidaka in commemoration of his sixtieth birthday, 465-478.

continents vs. oceans
Gainanov, A.G., and O.N. Soloviev, 1963.
On the nature of magnetic anomalies in the region of transition from the Asiatic continent to the Pacific. (In Russian).
Doklady, Akad. Nauk, SSSR, 151(6):1399-1401.

Continents vs. oceans
Harrison, Christopher G.A. 1966.
Antipodal location of continents and oceans.
Science 153 (3741): 1246-1248.

continents, role of
Vine, F.J. and H.H. Hess, 1970.
III Concepts. 1. Sea-floor spreading. In: The sea: ideas and observations on progress in the study of the seas, Arthur E. Maxwell, editor, Wiley-Interscience 4(2): 587-622.

continental borderland
Krause, Dale C., 1964.
Lithology and sedimentation in the southern continental borderland.
In: Papers in Marine Geology, R.L. Miller, Editor, Macmillan Co., N.Y., 274-318.

continents
Markov, K.K., 1968.
On the unity of nature of the ocean and continents.
Izv. Uses. Geograf. Obohch., 100(6):481-487.

continental borderland
Smith, Stephen V. 1971.
Budget of calcium carbonate, southern California continental borderland.
J. sedim. Petrol. 41 (3): 798-808.

continental break up
Sleep, Norman H. 1971.
Thermal effects of the formation of Atlantic continental margins by continental break up.
Geophys. J. R. astr. Soc. 24 (4): 325-350.

Continental drift

Continental drift
Ahmad, F. 1968.
Orogeny, geosyncline and continental drift.
Tectonophysics 5 (3): 177-189.

Continental drift
Ahmad, F. 1966.
An estimate of the rate of continental drift in the Permian period.
Nature, Lond. 210 (5031): 81-83.

continental drift
Allan, T.D. and C. Morelli, 1970.
II. Regional observations. 13. The Red Sea (May 1969). In: The sea: ideas and observations on progress in the study of the seas, Arthur E. Maxwell, editor, Wiley-Interscience 4(2): 493-542.

continental drift
Arur, M.G. and Ivan I. Mueller, 1971.
Latitude observations and the detection of Continental Drift. J. geophys. Res., 76(8): 2071-2076.

continental drift
*Baker, J., N.H. Gale and J. Simons, 1967.
Geochronology of the St. Helens Volcanoes.
Nature, 215(5109):1451-1456.

continental drift
Barnett, C.H., 1962
A suggested reconstruction of the land masses of the earth as a complete crust.
Nature, 195(4840):447-448.

Appended comments by Jeffries, H., 1962
Nature, 195(4840):448.

Continental drift
Behrendt, John C., and Cletus S. Wotorson, 1970.
Aeromagnetic and gravity investigations of the coastal area and continental shelf of Liberia, West Africa, and their relation to continental drift.
Bull. Geol. Soc. Am. 81 (12): 3563-3574.

continental drift
Berger, Jon, A.E. Cok, J.E. Blanchard and M.J. Keen, 1966.
Morphological and geophysical studies on the eastern seaboard of Canada: The Nova Scotian Shelf.
In: Continental drift, G.D. Garland, editor, Spec. Publs. R. Soc. Can., No. 9:102-113.

Continental drift
Bonatti, E., M. Ball, and Carl Schubert 1970.
Evaporites and continental drift.
Naturwissenschaften 57 (3): 107-108.

continental drift
*Briden, J.C., 1967.
Recurrent continental drift of Gondwanaland.
Nature 215(5108):1334-1339.

continental drift
Bullard, Edward, J.E. Everett and A. Gilbert Smith, 1965.
Continental reconstructions IV. The fit of the continents around the Atlantic.
Phil. Trans. R. Soc., A., 258(1088):41-51.

continental drift
Cameron, H.L., 1966.
The Cabot fault zone.
In: Continental drift, G.D. Gerland, editor, Spec. Publs. R. Soc. Can. No. 9:129-140.

continental drift
Collinson, D.W. and S.K. Runcorn, 1960.
Paleomagnetic observations in the United States new evidence for polar wandering and continental drift. Bull. Geol. Soc. Amer, 71(7): 915-958.

continental drift
Cooper, L.H.N., 1965.
Radiolarians as possible chronometers of continental drift.
Progress in Oceanography, 3:71-82.

continental drift
Deutsch, Ernst R., 1966.
The rock magnetic evidence for continental drift.
In: Continental drift, G.D. Garland, editor, Spec. Publs. R. Soc. Can., No. 9:28-52.

continental drift
Dietz, Robert S. 1971.
Those shifty continents
Sea Frontiers 17 (4): 204-212 (popular)

continental drift
Dietz, Robert S. 1967.
More about continental drift.
Sea Frontiers 13 (2): 66-82.

continental drift
Dietz, Robert S. and Walter P. Sproll, 1970.
Fit between Africa and Antarctica: a continental drift reconstruction. Science, 167(3925): 1612-1614.

continental drift
*Donnelly, Thomas W., 1967.
Some problems of island-arcs tectonics, with reference to the northeastern West Indies.
Studies Trop. Oceanogr., Miami, 5:74-87.

continental drift
Doumain, G.A., and W.E. Long, 1962.
The ancient life of the Antarctic.
Scientific American, 207(3):169-184.

continental drift
Elsasser, Walter M., 1968.
The mechanics of continental drift.
Proc. Am. phil. Soc., 112(5):344-353.

continental drift
Erimesco, P., 1963.
The expanding ocean floor.
Bull. Inst. Pêches Marit., Maroc, (9-10):3-31.

continental drift
Ewing, Maurice and John Ewing, 1964.
Continental drift and ocean-bottom sediments. (Abstract).
Geol. Soc., Amer., Special Paper, No. 76:55-56.

Continental drift
Facer, F.A. 1970.
Continental drift: implications of paleomagnetic studies, meteorology, physical oceanography and climatology: a discussion.
J. Geol. 78 (5): 630-633.

Continental drift
Friend, P.F. 1967.
The growth of the North Atlantic Ocean by the spreading of its floor.
Polar Rec. 13 (86): 579-588.

continental drift
Garland, G.D., editor, 1968.
Continental drift.
Univ. Toronto Press. $5.95 (not seen).

continental drift
Garland, G.D., editor, 1966.
Continental drift.
Spec. Publs. R. Soc. Can., No. 9:140 pp.

continental drift
*Girdler, R.W., 1969.
The Red Sea - a geophysical background.
In: Hot brines and Recent heavy metal deposits in the Red Sea, E.T. Degens and D.A. Ross, editors, Springer-Verlag, New York, Inc., 38-58.

continental drift
Gilluly, James, 1969.
Oceanic sediment volumes and continental drift.
Science, 166(3908): 992-993.

continental drift
*Girdler, R.W., 1968.
Drifting and rifting of Africa.
Nature, 217(5134):1102-1106.

continental drift
Godley, E.J., 1967.
Widely distributed species, land bridges and continental drift.
Nature 214(5083):74-75.

continental drift
Goguel, J., 1965.
Tectonics and continental drift.
Phil. Trans. R. Soc., A, 258(1088):194-198.

continental drift
Goodacre, A.K. and E. Nyland, 1966.
Underwater gravity measurements in the Gulf of St. Lawrence.
In: Continental drift, G.D. Garland, editor, Spec. Publs. R. Soc. Can., No. 9:114-128.

continental drift
Hales, A.L., 1969.
Gravitational sliding and continental drift.
Earth Planet. Sci. Lett. 6(1): 31-34.

continental drift
*Hallam, Anthony, 1967.
The bearing of certain Palaeogeographic data on continental drift.
Palaeogr. Palaeoclimatol. Palaeoecol., 3(2):201-24

continental drift
Harland, W.B., 1967.
Tectonic aspects of continental drift.
Sci. Prog. Lond., 55(217):1-14.

continental drift
Havemann, H., 1964.
The Pacific rotation and the circumpacific movement. (In German).
Geologie, Berlin, 13(5):505-523.

continental drift
Heezen, B.C., and Marie Tharp, 1965.
Tectonic fabric of the Atlantic and Indian oceans and continental drift.
Phil. Trans. R. Soc., A, 258(1088):90-106.

continental drift
*Heirtzler, J.R., G.O. Dickson, E.M. Herron, W.C. Pitman, III, and X. Le Pichon, 1968.
Marine magnetic anomalies, geomagnetic field reversals and motions of the ocean floor and continents.
J. geophys. Res., 73(6):2119-2136.

continental drift
*Hernes, I., 1968.
Continental drift and orogenesis.
Tectonophysics, 5(2):151-154.

~~continental drift~~
Herz, Norman, 1969.
Anorthosite belts, continental drift and the anorthosite event, 1969.
Science, 164(3882):944-947.

continental drift
Hess, H.H., 1965.
Mid-oceanic ridges and tectonics of the sea-floor.
In: Submarine geology and geophysics, Colston Papers, W.F. Whittard and R. Bradshaw, editors, Butterworth's, London, 17:317-332.

Continental drift
Hood, Peter 1966.
Flemish Cap, Galicia Bank and continental drift.
Earth Planet. Sci. Letts 1(4): 205-208.

continental drift
Howard, L.N., W.V.R. Malkus and J.A. Whitehead, 1970.
Self-convection of floating heat sources: a model for continental drift.
Geophys. fluid Dyn. 1(1/2): 123-142

continental drift
*Hurley, P.M., F.F.M. de Almeida, G.C. Melcher, U.G. Cordani, J.R. Rand, K. Kawashita, P. Vandoros, W.H. Pinson, Jr., and H.W. Fairbairn, 1967.
Test of continental drift by comparison of radiometric ages: a pre-drift reconstruction shows matching geologic age provinces in West Africa and northern Brazil.
Science, 157(3788):495-500.

~~continental drift~~
Hurley, Patrick M., and John R. Rand, 1969.
Pre-drift continental nuclei: two ancient nuclei appear to have a peripheral growth and no predrift fragmentation and dispersal.
Science, 164(3885):1229-1242.

Continental drift
Illies, H., 1966.
Kontinental drift - mit oder ohne Konvektionsstromungen.
Tectonophysics, 2(6):521-557.

continental drift
Jacobs, J.A., 1960.
Continental drift.
Nature, 185(4708):231-232.

continental drift
Jones, J.G., 1971.
Australia's Caenozoic drift.
Nature, Lond. 230(5291): 237-239

continental drift
Karp, Edwin, 1967.
Continental drift and spreading of ocean floors.
Science, 158(3803):947.

continental drift
Katz, H.R. 1971.
Continental margin in Chile - is tectonic style compressional or extensional?
Bull. Am. Ass. Petrol. Geol. 55(10): 1753-1758.

continental drift
Kay, Marshall, 1968.
North Atlantic continental drift.
Proc. Am. phil. Soc. 112(5):321-324.

Continental drift
Kay, Marshall 1967.
Stratigraphy and structure of northeastern Newfoundland bearing on drift in North Atlantic.
Bull. Am. Ass. Petrol. Geol. 51(4): 579-600.

continental drift
Keen, M.J., B.D. Loncarevic and G.N. Ewing, 1970.
II. Regional observations. 7. Continental Margin of Eastern Canada: Georges Bank to Kane Basin. In: The sea: ideas and observations on the progress in the study of the seas, Arthur E. Maxwell, editor, Wiley-Interscience 4(2): 251-291.

Continental drift
Knopoff, L. 1970.
Models of continental drift.
Phys. Earth Planet. Interiors 2(5): 386-392.

continental drift
Kraus, E.C. 1967.
Die Bodenstruktur des Indischen Ozeans und dessen Geschichte.
Geol. Rdsch. 56(2): 373-393

continental drift
Laughton, A.S., 1966.
The birth of an ocean.
New Scientist, 29(480):218-220.

Continental drift
LePichon, Xavier 1968.
Sea-floor spreading and continental drift
J. geophys. Res. 73(12): 3661-3697

continental drift
Le Pichon, Xavier, Guy Pautot, Jean-Marie Auzende et Jean-Louis Olivet 1971.
La Méditerranée occidentale depuis l'Oligocène: schéma d'évolution.
Earth planet. Sci. Letts 13(1): 145-152.

continental drift
Ma, Ting Ying H., 1957.
Continental drift and the present velocity of shift of the continental margin of eastern Asia.
Res. on the Past Climate and Continental Drift, 12:1-22.

continental drift
MacDonald, Gordon J.F., 1966.
Mantle properties and Continental drift.
In: Continental drift, G.D. Garland, editor, Spec. Publs. R. Soc. Can., No. 9:18-27.

continental drift
Macdonald, G.J. F., 1965.
Continental structure and drift.
Phil. Trans. R. Soc., A, 258(1088):215-227.

continental drift
May, Paul R., 1971.
Pattern of Triassic-Jurassic diabase dikes around the North Atlantic in the context of predrift position of the continents.
Bull. geol. Soc. Am. 82(5): 1285-1292.

Continental drift
McElhinny, M.W. and J.C. Briden 1971
Continental drift during the Palaeozoic.
Earth planet. Sci. Letts 10(4): 407-416.

continental drift
McElhinny, M.W., and P.R. Luck 1970.
Paleomagnetism and Gondwanaland.
Science 168 (3933):830-832.

Continental drift
Meyerhof, A.A. 1970.
Continental drift: implications of paleomagnetic studies, meteorology, physical oceanography and climatology: a reply.
J. Geol. 78(5): 633-634.

continental drift
Meyerhoff A.A., 1970.
Continental drift: implications of paleomagnetic studies, meteorology, physical oceanography and climatology.
J. Geol. 78(1): 1-51.

continental drift
Meyerhoff, A.A. 1970
Continental drift II: high latitude evaporite deposits and geologic history of Arctic and North Atlantic oceans.
J. Geol. 78(4): 406-444.

continental drift
Meyerhoff A.A., and James L. Harding 1971.
Some problems in current concepts of continental drift.
Tectonophysics 12(3): 235-260.

continental drift
Meyerhoff, A.A., and Howard A. Meyerhoff 1972.
Continental drift, IV: The Caribbean "Plate".
J. Geol. 80(1): 34-60.

continental drift
Meyerhoff, A.A. and Curt Teichert 1971
Continental drift, III: late Paleozoic glacial centers, and Devonian-Eocene coal distribution.
J. Geol. 79(3): 285-321.

continental drift
Miller, J.A., 1965.
Geochronology and continental drift - the North Atlantic.
Phil. Trans. R. Soc., A, 258(1088):180-191.

continental drift
*Myers, George Sprague, 1967.
Zoogeographical evidence of the age of the South Atlantic Ocean.
Stud. trop. Oceanogr., Miami, 5:614-621.

continental drift
North, F.K. 1971.
Alpine serpentinites, oceanic ridges and continental drift.
Geol. Mag. 108(2): 81-192.

continental drift
Osmaston, M.F., 1967.
Core convection, the earth's figure and continental drift.
Nature, Lond., 216(5120):1095-1096.

continental drift
Oppenheim, Victor, 1967.
Critique of hypothesis of continental drift.
Bull. Am. Ass. Petr. Geol. 51(7):1354-1367.

continental drift.

continental drift
Palmer, Harold, 1968.
East Pacific Rise and westward drift of North America.
Nature, Lond., 220(5165):341-345.

continental drift
Pelletier, B.R., 1966.
Development of submarine physiography in the Canadian Arctic and its relation to crustal movements.
In: Continental drift, G.D. Garland, editor, Spec. Publs. R. Soc. Can., No. 9:77-101.

continental drift
Radforth, N.W., 1966.
The ancient flora and continental drift.
In: Continental drift, G.D. Garland, editor, Spec. Publs. R. Soc. Can., No. 9:53-70.

continental drift
Ramberg, Hans, 1971.
Dynamic models simulating rift valleys and continental drift.
Lithos 4(3): 259-276.

Continental drift
Reyment, R.A. 1969.
Ammonite biostratigraphy, continental drift and oscillatory transgressions.
Nature Lond 224 (5215): 137-140.

continental drift
Ridd, M.F. 1971.
South-east Asia as a part of Gondwanaland
Nature, Lond. 234 (5331): 531-533.

continental drift
Runcorn, S.K., 1965.
Changes in the convection pattern in the Earth's mantle and continental drift: evidence for a cold origin of the Earth.
Phil. Trans. R. Soc., A, 258(1088):228-251.

continental drift
Runcorn, S.K., 1965.
Continental reconstructions. I. Palaeomagnetic comparisons between Europe and North America.
Phil. Trans. R. Soc., A, 258(1088):1-11.

continental drift
Runcorn, S.K., 1962.
Continental drift.
Research applied in industry, Butterworth's London, 15(3):103-108.

continental drift
Runnells, Donald D., 1970.
Continental drift and economic minerals in Antarctica.
Earth Planet Sci. Letters 8(6): 400-402.

continental drift
Rupke, N.A., 1970.
Continental drift before 1900.
Nature, Lond., 227(5256): 349-350

continental drift
Schenk, Paul E., 1971.
Southeastern Atlantic Canada, northwestern Africa, and continental drift.
Can. J. Earth Sci. 8(10): 1218-1251

continental drift theory
Sewell, R.B. Seymour 1956.
The continental drift theory and the distribution of the Copepoda. Proc. Linn. Soc., London, 166 (1/2):149-177.

continental drift
Shuiling, R. D., 1966.
Continental drift and oceanic heat-flow.
Nature, 210(5040):1027-1028.

continental drift
Smith, A. Gilbert and A. Hallam, 1970.
The fit of the southern continents.
Nature, Lond., 225(5228): 139-144

continental drift
Sneath, P.H.A., 1967.
Conifer distributions and continental drift.
Nature, 215 (5100): 467-470

continental drift
Stock, J.H., 1957.
The pycnogonid Family Austrodecidae.
Beaufortia, 6(68):1-81.

continental drift
Stoneley, R., 1966.
The Niger Delta region in the light of the theory of continental drift.
Geol. Mag., 103(5):385-397.

Continental drift
Strangway, D.W., and P. R. Vogt 1970.
Aeromagnetic tests for continental drift in Africa and South America.
Earth Planet. Sci. Letts 7(5): 429-435

continental drift
Stubbs, Peter, 1967.
Ocean floor conveyor belts.
Sea Frontiers, 14(1):41-49. (popular).

Continental drift
Takin, Manoochehr 1972
Iranian geology and continental drift in the Middle East.
Nature, Lond. 235 (5334): 147-150

continental drift
Tanner, R.W., 1966.
Astronomical evidence on the present rate of continental drift.
In: Continental drift, G.D. Garland, editor, Spec. Publs. R. Soc. Can., No. 9:71-74.

continental drift
Tarling, D.H., 1971.
Gondwanaland, Palaeomagnetism and continental drift.
Nature, Lond., 229(5279):17-21, 71.

continental drift
Tarling, D.H., and M.P. Tarling, 1970.
Continental drift.
Bell, 112 pp. 30s.

continental drift
Turcotte, D.L., and E.R. Oxburgh 1967.
Finite amplitude convective cells and continental drift.
J. fluid Mech. 28(1):29-42.

continental drift
Van Bemmelen, R.W., 1965.
Mega-undations as a cause of continental drift.
Geol. en Mijnbouw, 44(9):320-333.

continental drift
Van Hilten, D., 1964.
Interpretation of the wandering paths of ancient magnetic poles.
Geol. en Mijnbouw, 43(6):209-221.

continental drift
Vilas, J.F., and D.A. Valencio, 1970.
The recurrent Mesozoic drift of South America and Africa.
Earth Planet. Sci. Letts 7(5): 441-444.

continental drift
Vogt, Peter R., and Ned A. Ostenso, 1967.
Steady state crustal spreading.
Nature, Lond., 215(5103): 810-817.

continental drift
Watkins, N.D., and A. Richardson 1968.
Comments on the relationship between magnetic anomalies, crustal spreading and continental drift.
Earth Planet. Sci. Letts 4(3): 256-264.

continental drift
Westoll, T.S., 1965.
Continental reconstructions. II Geological evidence bearing upon continental drift.
Phil. Trans. R. Soc., A. 258(1088):12-26.

continental drift
Wilson, J. Tuzo, 1970.
III. Concepts. 2. Continental Drift, trans-current, and transform faulting. In: The sea: ideas and observations on progress in the study of the seas, Arthur E. Maxwell, editor, Wiley-Interscience 4(2): 623-644.

continental drift
Wilson, J. Tuzo 1969.
The current revolution in earth science.
Trans R. Soc. Can. (4) 6(3): 273.

continental drift
Wilson, J.T., 1966.
Some rules for continental drift.
In: Continental drift, G.D. Garland, editor, Spec. Publs. R. Soc. Can., No. 9:3-17.

continental drift
Wilson, J. Tuzo, 1965.
Convection currents and continental drift evidence from ocean islands suggesting movement in the earth.
Phil. Trans. R. Soc., A. 258(1088):145-167.

continental drift
Wilson, J.T., 1963.
Continental drift.
Scientific American, 208(4):86-100.

continental drift
Wilson, J.T., 1962.
Some further evidence in support of the Cabot fault, a great Paleozoic transcurrent fault zone in the Atlantic provinces and New England.
Trans. R. Soc., Canada, (3), 56(3)(1):31-36.

continental drift
Wright, J.B., 1971.
Comments on "Timing of break-up of the continents round the Atlantic as determined by palaeomagnetism" by E.E. Larson and L. La Fontain, Earth and Planetary Science Letters 8 (1970) 341-344.
Earth Planet. Sci. Letts, 10(2): 271-272

Continental drift
Wright, J. B., 1966.
Convection and continental drift.
Tectonophysics, 3(2):69-81.

continental drift
Yungul, S.H., 1971.
Magnetic anomalies and the possibilities of continental drifting in the Gulf of Mexico.
J. geophys. Res., 76(11): 2639-2642.

continental fitting
Al-Chalabi, M., 1971.
Reliability of the rotation pole in continental fitting.
Earth planet Sci. Letts 11(4): 257-262

continental fitting
Bullard, E.C. and D.P. McKenzie, 1971.
Remarks on uncertainties in poles of rotation in continental fitting.
Earth planet. Sci. Letts 11(4): 263-264

Continental margins

continental margin
Bamford, S.A.D. 1971.
An interpretation of first-arrival data from the continental margin refraction experiment.
Geophys. J. R. astr. Soc. 24(3): 213-229

Continental margin
Barker, P.H. 1967.
Bathymetry of the fiordland continental margin.
N.Z. Jl Sci. 10(1):128-137.

continental margin
Bellaiche, Gilbert, 1970.
Géologie sous-marine de la marge continentale au large du massif des Maures (Var, France), et de la plaine abyssale Ligure.
Revue Géogr. phys. Géol. dyn. (2) 12(5): 403-440.

Continental margin
Bird, John M., and John F. Dewey 1970
Lithospheric plate-continental margin tectonics and the evolution of the Appalachian orogen.
Bull. geol. Soc. Am. 81(4): 1031-1060.

continental margins
Bott, M.H.P. 1971.
Evolution of young continental margins and formation of shelf basins.
Tectonophysics 11(5): 319-327.

continental margin
Burk, C.A., 1966.
The Aleutian Arc and Alaska continental margin.
In: Continental margins and island arcs, W.H. Poole, editor, Geol. Surv. Pap., Can., 66-15: 206-214.

continental margin
Byrne, John V., Gerald A. Fowler and Neil J. Maloney, 1966.
Uplift of the continental margin and possible continental accretion off Oregon.
Science, 154(3757):1654-1656.

continental margin
*Cholet, J., B. Damotte, et G. Grau, J. Debyer et L. Montadert, 1968.
Recherches préliminaires sur la structure géologique de la marge continentale du Golf de Gascogne: commentaires sur quelques profiles de sismique réflexion "Flexotir".
Revue Inst. français Pétrole, 23(9):1029-1045.

continental margin
Clarke, R.H., R.J. Bailey and D. Taylor Smith 1970.
Seismic reflection profiles of the continental margin west of Ireland. In: The geology of the East Atlantic continental margin. 2. Europe, ICSU/SCOR Working Party 31 Symposium, Cambridge 1970, Rept. 70/14: 67-76.

continental margins
Dewey, J.F., 1969.
Continental margins: a model for conversion of Atlantic type to Andean type.
Earth Planet. Sci. Letters, 6(3): 189-197.

continental margins
Dickenson, William R., 1971.
Plate tectonic models for orogeny at continental margins.
Nature, Lond., 232(5305): 41-42.

continental margin
Drake, Charles L., 1966.
Recent investigations of the continental margin of eastern United States.
In: Continental margins and island arcs, W.H. Poole, editor, Geol. Surv. Pap., Can., 66-15:33-47.

continental margin
*Drake, Charles L., John I Ewing and Henry Stockard, 1968.
The continental margin of the eastern United States.
Can. J. Earth Sci., 5(4-2):993-1010.

continental margin

Dunham, K.C., 1970.
Gravel, sand, metallic placer and other mineral deposits on the East Atlantic continental margin. In: The geology of the East Atlantic continental margin, 1. General and economic papers, ICSU/SCOR Working Party 31 Symposium, Cambridge 1970, Rept. No. 70/13: 79-85.

continental margin

Eden, R.A., J.E. Wright and W. Bullerwell, 1970.
The solid geology of the East Atlantic continental margin adjacent to the British Isles. In: The geology of the East Atlantic continental margin. 2. Europe, ICSU/SCOR Working Party 31 Symposium, Cambridge 1970, Rept. 70/14: 111-128.

continental margins

Elmendorf, C.H., and B.C. Heezen, 1957.
Oceanographic information for engineering submarine cable systems. Bell Syst. Tech. J., 36(5): 1047-1094.

continental margin

Emery, K.O. 1970.
Continental margins of the world.
In: The geology of the East Atlantic Continental margin, 1. General and economic papers, ICSU/SCOR Working Party 31 Symposium, Cambridge 1970, Rept. No. 70/13: 7-29

continental margin

Fairbridge, Rhodes W., 1961
Continental margin of the southwest Pacific: advancing or retreating? (Abstract).
Proc. Ninth Pacific Sci. Congr., Pacific Sci. Assoc., 1957, 12(Geol-Geophys.): 69.

continental margin

Field, Michael E., and Orrin H. Pilkey, 1969.
Feldspar in Atlantic continental margin sands off the southeastern United States.
Bull. geol. Soc. Am. 80 (10): 2097-2102.

continental margins

Guilcher, A., 1963.
12. Continental shelf and slope (continental margin).
In: The Sea, M.N. Hill, Editor, Interscience Publishers, 3:281-311.

continental margin

Hawkins, L.V., J.F. Hennion, J.E. Nafe and H.A. Doyle, 1965.
Marine seismic refraction studies on the continental margin to the south of Australia.
Deep-Sea Res., 12(4):479--495.

continental margin

Heezen, B.C. 1966.
Physiography of the Indian Ocean.
Phil. Trans. R. Soc. London (A) 259 (1099): 137-149.

continental margins

Heezen, Bruce C., and H.W. Menard, 1963.
12. Topography of the deep-sea floor.
In: The Sea, M.N. Hill, Editor, Interscience Publishers, 3:233-280.

continental margin

Hyndman, R.D., and N.A. Cochrane 1971.
Electrical conductivity structure by geomagnetic induction at the continental margin of Atlantic Canada.
Geophys. J. R. astron. Soc. 25 (5): 425-446.

continental margin

JOINT OCEANOGRAPHIC INSTITUTIONS' DEEP EARTH SAMPLING PROGRAM (JOIDES), 1965.
Ocean drilling on the continental margin.
Science, 150(3697):709-716.

continental margin

Katz, H.R. 1971.
Continental margin in Chile - is tectonic style compressional or extensional?
Bull. Am. Ass. Petrol. Geol. 55 (10): 1753-1758.

continental margins

Kuno, Hisashi, 1966.
Lateral variation of basalt magma across continental margins and island arcs.
In: Continental margins and island arcs, W.H. Poole, editor, Geol. Surv. Pap., Can., 66-15:317-335.

continental margins

Lee, W.H.K., S. Uyeda and P.T. Taylor, 1966.
Geothermal studies of continental margins and island arcs.
In: Continental margins and island arcs, W.H. Poole, editor, Geol. Surv. Pap., Can., 66-15:398-414.

continental margin

*Ludwig, William J., John I. Ewing and Maurice Ewing, 1968.
Structure of Argentine continental margin.
Bull. Am. Ass. Petr. Geol., 52(12):2337-2368.

continental margin

Ludwig, W.J., J.E. Nafe, E.S.W. Simpson and S. Sacks 1968.
Seismic-refraction measurements on the Southeast African continental margin.
J. geophys. Res. 73 (12): 3707-3719.

continental margin

*Maloney, Neil J., 1967.
Geomorphology of the continental margin of Venezuela. 3. Bonaire Basin (66oW to 70o W longitude).
Bol. Inst. Oceanogr., Univ. Oriente, 6(2) 286-302.

continental margin

Maloney, Neil J. 1967.
Geomorphology of the continental margin off central Oregon U.S.A.
Bol. Inst. Oceanogr. Univ. Oriente 6 (1): 116-146

continental margin

*Manheim, F.T., and M.K. Horn, 1968.
Composition of deeper subsurface waters along the Atlantic continental margin.
SEast. Geol., Duke Univ., 9(4):215-236.

continental margins

Mitchell, Andrew H., and Harold G. Reading, 1969.
Continental margins, geosynclines, and ocean floor spreading.
J. Geol., 77 (6): 629-646.

continental margins

Nagata, Takesi, 1966.
A review of recent studies on conductivity anomalies along continental margins.
In: Continental margins and island arcs, W.H. Poole, editor, Geol. Surv. Pap., Can., 66-15:418-429.

continental margins

*Narain, Hari, K.L. Kaila and R.K. Verma, 1968.
Continental margins of India.
Can. J. Earth Sci., 5(4-2):1051-1065.

continental margin

Newton, John G., and Orrin H. Pilkey, 1969.
Topography of the continental margin off the Carolinas.
SEast. Geol., Duke Univ., 10(2): 87-92

continental margin

Page, Benjamin M., 1970
Sur-Nacimiento Fault Zone of California: continental margin tectonics.
Bull. geol. Soc. Am., 81 (3): 667-690.

continental margin

Rabinowitz Philip D., 1971.
Gravity anomalies across the East African Continental Margin. J. geophys. Res. 76(29): 7107-7117.

continental margin

Rona, Peter A. 1970.
Comparison of continental margins of eastern North America at Cape Hatteras and northwestern Africa at Cap Blanc
Bull. Am. Ass. Petrol. Geol. 54 (1): 129-157.

continental rise hills

Rona, Peter A. 1969.
Linear "lower continental rise hills" off Cape Hatteras.
J. sedim. Petrol. 39 (3): 1132-1141.

continental margin

Sheridan, R.E., R.W. Houtz, C.L. Drake and M. Ewing 1969.
Structure of continental margin off Sierra Leone, West Africa.
J. geophys. Res. 74 (10): 2512-2530.

continental margin

Silver, Eli A. 1971.
Transitional tectonics and late Cenozoic structure of the continental margin off northernmost California
Bull. geol. Soc. Am. 82 (1): 1-22

continental margin

*Simpson, E.S.W., and A. du Plessis, 1968.
Bathymetric, magnetic, and gravity data from the continental margin of southwestern Africa.
Can. J. Earth Sci., 5(4-2):1119-1123.

continental margin

Sleep, Norman H. 1971.
Thermal effects of the formation of Atlantic continental margins by continental break up.
Geophys. J. R. astr. Soc. 24 (4): 325-350.

continental margin

*Uchupi, Elazar, 1967.
Slumping on the continental margin southeast of Long Island, New York.
Deep-Sea Res., 14(5):635-639.

continental margin
Veevers, J.J., 1971.
Shallow stratigraphy and structure of the Australian continental margin beneath the Timor Sea.
Mar. Geol. 11 (4): 209-249.

continental margin
*Worzel, J. Lamar, 1968.
Advances in marine geophysical research of continental margins.
Can. J. Earth Sci., 5(4-2):963-983.

continental margins
Worzel, J. Lamar, 1966.
Structure of continental margin and development of ocean trenches.
In: Continental margins and island arcs, W.H. Poole, editor, Geol. Surv. Pap., Can., 66-15:357-375.

continental margins
Worzel, J. Lamar, and J.C. Harrison, 1963.
9. Gravity at sea.
In: The Sea, M.N. Hill, Editor, Interscience Publishers, 3:134-174.

continental margin
*Zarudzki, E.F.K., and Elazar Uchupi, 1968.
Organic reef alignments on the continental margin south of Cape Hatteras.
Bull. geol. Soc. Am., 79(12):1867-1870.

Continental remnant

continental remnant
Kerr, J. Wm, 1967.
A submerged continental remnant beneath the Labrador Sea.
Earth Planet. Sci. Letts 2 (4):283-289.

continental rise
See also: rises

continental rises

See also: rises

continental rise
Ballard, James A., 1966.
Structure of the lower continental rise hills of the western North Atlantic.
Geophysics, 31(3):506-523.

continental rise
Belderson, R.H., and A.S. Laughton, 1966.
Correlation of some Atlantic turbidites.
Sedimentotogy, 7(2):103-116.

continental rise
Clay, C.S., John Ess and Irving Weisman, 1964
Lateral echo sounding of the ocean bottom on the continental rise.
J. Geophys. Res., 69(18):3823-3835.

continental rise
*Conolly, John R., 1969.
Western Tasman sea floor.
N.Z. Jl Geol. Geophys. 12:310-343.

continental rise
*Conolly, John R., and C.C. von der Borch, 1967.
Sedimentation and physiography of the sea-floor south of Australia.
Sedimentary Geol., 1(2):181-220.

continental rises
Dietz, Robert S., 1967.
Passive continents, spreading sea floors and continental rises: a reply.
Am. J. Sci., 265(3):231-237.

continental rise
Emery, K.O., Bruce C. Heezen and T.D. Allan, 1966.
Bathymetry of the eastern Mediterranean Sea.
Deep-Sea Res., 13(2):173-192.

continent rise
Emery, K.O., Elazar Uchupi, J.D. Phillips, C.O. Bowin, E.T. Bunce and S.T. Knott, 1970
Continental rise off eastern North America.
Bull. Am. Ass. Petrol. Geol. 54(1): 44-108.

continental rise
Field, Michael E., and Orrin H. Pilkey, 1971.
Deposition of deep-sea sands: Comparison of two areas of the Carolina Continental rise.
J. sedim. Petrol. 41 (2): 526-536.

continental rise
Hadley, M.L., 1964
The continental margin southwest of the English Channel.
Deep-Sea Res., 11(5):767-779.

continental rise
Heezen, Bruce C., Charles D. Hollister and William F. Ruddiman, 1966.
Shaping of the continental rise by deep geostrophic contour currents.
Science, 152(3721):502-508.

continental rise
Heezen, Bruce C., and H.W. Menard, 1963.
12. Topography of the deep-sea floor.
In: The Sea, M.N. Hill, Editor, Interscience Publishers, 3:233-280.

continental rise
Hoskin, Hartley, 1967.
Seismic reflection observations on the Atlantic continental shelf, slope and rise southeast of New England.
J. Geol., 75(5):598-611.

continental rise
Ilyin, A.V., 1971.
Main features of geomorphology of Atlantic bottom. (In Russian; English abstract).
Okeanol. Issled. Rezult. Issled. Mezhd. Geofiz. Proekt. 21: 107-246.

continental rise
Newton, John G., and Orrin H. Pilkey, 1969.
Topography of the continental margin off the Carolinas.
SEast. Geol., Duke Univ., 10(2): 87-92

continental rise
Rona, Peter A., and C.S. Clay, 1967.
Stratigraphy and structure along a continuous seismic reflection profile from Cape Hatteras, North Carolina to the Bermuda Rise.
J. geophys. Res., 72(8):2107-2130.

continental rise
*Rona, Peter A., Eric D. Schneider and Bruce C. Heezen, 1967.
Bathymetry of the continental rise off Cape Hatteras.
Deep-Sea Res., 14(5):625-633.

continental rise
Stanley, Daniel J., Harrison Sheng and Carlos P. Pedraza 1971.
Lower continental rise east of the middle Atlantic states: predominant sediment dispersal perpendicular to isobaths.
Bull. geol. Soc. Am. 82 (7): 1831-1840.

continental rise
Uchupi, Elazar, and K.O. Emery, 1963
The continental slope between San Francisco, California and Cedros Island, Mexico.
Deep-Sea Research, 10(4):397-447.

continental rises
Young, Grant M., 1967.
Passive continents, spreading sea floors, and continental rises.
Am. J. Sci., 265(3):225-230.

Continental shelf

A

continental shelf
Allen, J.R.L., 1964.
The Nigerian continental margin: bottom sediments submarine morphology and geological evolution.
Submarine Geology, 1(4):289-332.

continental shelf
Alvares Perez, P., 1965.
Trabajos de investigacion y busqueda sobre la plataforma continental de Venezuela. (In Russian: Spanish abstract).
Sovetsk.-Cub. Ribokhoz. Issled., VNIRO: Tsentr Ribokhoz. Issled. Natsional. Inst. Ribolovsta Republ. Cuba, 285-288.

continental shelf

Amor, I. Asensio, 1970.
Remaniements de formations quaternaires fluviatiles par action marine dans la région littorale Galicio-Asturienne (NW de l'Espagne).
Quaternaria 12: 219-221.

continental shelf
Antoine, John, William Bryant and Bill Jones, 1967.
Structural features of continental shelf, slope, and scarp, northeastern Gulf of Mexico.
Bull. Am. Ass. Petr. geol., 51(2):257-262.

continental shelf
Antoine, John W., and James L. Harding, 1965.
Structure beneath continental shelf, northeastern Gulf of Mexico.
Bull. Amer. Assoc. Petr. Geol., 49(2):157-171.

continental shelf
Antoine, John W., and Vernon J. Henry, Jr., 1965.
Seismic refraction study of shallow part of the continental shelf off Georgia coast.
Bull. Amer. Assoc. Petr. Geol., 49(5):601-609.

continental shelf
Avilov, I.K., 1965.
Some data on the bottom topography and grounds of the West African shelf. (In Russian).
Trudy vses. nauchno-issled. Inst. morsk. ryb. Khoz. Okeanogr. (VNIRO), 57:235-259.

B

continental shelf

Bailey, R.J., J.S. Buckley and R.H. Clarke 1971.
A model for the early evolution of the Irish continental margin.
Earth planet. Sci. Letts. 13(1): 79-84

continental shelf

*Ballard, Robert D. and Elazar Uchupi, 1970.
Morphology and Quaternary history of the continental shelf of the Gulf Coast of the United States. Bull. mar. Sci., 20(3): 547-559.

continental shelf

Bellaiche, Gilbert, 1970.
Géologie sous-marine de la marge continentale au large du massif des Maures (Var, France), et de la plaine abyssale Ligure.
Revue Géogr. phys. Géol. dyn. (2) 12(5): 403-440.

continental shelf

Battistini, R., 1970.
Les relations entre rivages et plate-forme continentale à Madagascar, Océan Indien.
Quaternaria 12: 129-136.

continental shelf

Bellaiche, Gilbert, 1968.
Reconnaissance du socle sous-marin des Maures et de sa couverture sédimentaire par sismique continue ("Flexotir")
C.r. hebd. Séanc. Acad. Sci., Paris (D) 266 (10): 994-996.

Continental Shelf

Bellaiche, Gilbert, 1968.
Précisions apportées à la connaissance de la pente continentale et de la plaine abyssale à la suite de trois plongées en bathyscaphe Archimède.
Rev. Géogr. phys. Géol. dynam. (2) 10(2): 137-145.

continental shelf

Berger, J., J.E. Blanchard, M.J. Keen and C.F. Tsong, 1964.
Seismic crustal studies on the east coast of Canada: the edge of the continental shelf.
Trans. Amer. Geophys. Union, 45(1):72.

Continental Shelf

Berthois, Leopold, 1965.
Essai de correlation entre la sedimentation actuelle sur le bord externe des plateformes continentales et la dynamique fluviale.
Progress in Oceanography, 3:49-62.

abstract

continental shelf

Berthois, Léopold, René Battistini, et Alain Crosnier, 1964.
Recherche sur le relief et la sédimentologie du plateau continental de l'extrême sud de Madagascar.
Cahiers Océanogr., C.C.O.E.C., 16(7):511-527.

continental shelf

Boillot, Gilbert, et Laurent d'Ozouville, 1970.
Etude structurale du plateau continental nord-espagnol entre Aviles et Llanes.
C.r. hebd. Séanc. Acad. Sci., Paris (D) 270 (15): 1865-1868

continental shelf

Boillot, Gilbert, et André Rousseau, 1971.
Étude structurale du plateau continental nord-espagnol entre 2°20 et 3°30 de longitude Ouest.
C.r. hebd. Séanc. Acad. Sci., Paris (D) 272 (16): 2056-2059

continental shelf

Bourcart, Jacques, 1963
Le modelé du précontinent,
Bull., Assoc. Française Étude des Grandes Profondeurs Océaniques, (1):5-9.

continental shelf

Bourcart, J., 1959.
Le plateau continental de la Mediterranee occidentale.
Comptes Rendus, Acad. Sci., Paris, 249(15):1380-1382.

continental shelf

Bourcart, Jacques, 1959
Morphologie du precontinent des Pyrenées à la Sardaigne. Colloques Int. Centre Nat. de la Recherche Sci., 83 (La Topographie et la Geologie des Profondeurs Oceaniques): 33-52.

continental shelf

Bourcart, J., 1955.
Recherches sur le plateau continental de Banyuls-sur-Mer.
Vie et Milieu, Bull. Lab. Arago, 6(4):435-524.

continental shelf

Bourcart, J., 1955.
Résultats généraux d'une étude du socle continental de la France en Méditerranée.
Bull. d'Info., C.C.O.E.C. 7(8):336-340, 4 pls.

continental shelf

Bourcart, J., 1949.
Peut-on étudier directement la Géologie du Plateau Continental? C.R.S. Soc. Géol., France, No. 2, Séance du 24 Jan. 1949:12-14.

continental shelf

Bourcart, J., and L. Glangeaud, 1954.
Morphotectonique de la marge continentale Nord-Africaine. Bull. Soc. Géol., France (6) 4:751-772

continental shelf

Bourcart, J., and P. Marie, 1951.
Sur la nature du "rebord continental" à l'ouest de la Manche. C.R. Acad. Sci., Paris, 232:2346-2348.

continental shelf

Brodie, J.W., 1964.
The fjordland shelf and Milford Sound.
New Zealand Dept. Sci. Ind. Res. Bull., 157: New Zealand Oceanogr. Inst., Memoir, No. 17: 15-23.

continental shelf

Brodie, J. W , 1959.
A shallow shelf around Franklin Island in the Ross Sea, Antarctica.
N. Z. J. Geol. Geophys., 2(1):108-119.

continental shelf

Browning, David S.,
Who has what rights on the continental shelves.
Ocean Industry 3(2):52-54.

continental shelves

Brückner, W.D., and H.J. Morgan, 1964.
Heavy mineral distribution on the continental shelf off Accra, Ghana, West Africa.
In: Developments in Sedimentology, L.M.J.U. van Straaten, Editor, Elsevier Publishing Co., 1:54-61.

continental shelf

Bullard, E.C., 1961
Forces and processes in ocean basins. In: Oceanography, Mary Sears, Edit., Amer. Assoc Adv. Sci., 39-50.

continental shelf

Bullard, E.C., and R.G. Mason, 1963.
10. The magnetic field over the oceans.
In: The Sea, M.N. Hill, Editor, Interscience Publishers, 3:175-217.

continental shelf

Bullerwell, W., 1969.
Surveys on the continental shelf around Britain.
Oceanol. int., 69, 1: 8 pp.

continental shelf

Bunce, Elizabeth T., Stuart Crampin, J.B. Hersey and M.N. Hill, 1964
Seismic refraction observations on the continental boundary west of Britain.
J. Geophys. Res., 69(18):3853-3863.

continental shelf

Burke, William T., Northcutt Ely, Richard Young, Bernard E. Jacob, Bruce A. Harlow and Quincy Wright, 1967.
A symposium on limits and conflicting uses of the continental shelf.
In: The law of the sea, L.M. Alexander, editor, 172-187.

continental shelf

Byrne, J.V., 1962.
Here's a look at offshore Oregon.
The Oil and Gas Jour., 116-119.

Also in: COLLECTED REPRINTS, Dept. Oceanogr., Oregon State Univ., Vol. 1, 1956-1962.

C

Continental Shelf

Caicedo Castilla, Jose Joaquin, 1962.
Plataforma Continental y mar territorial.
Peces y Conservas, Colombia, No. 12:40-43.

continental shelf

Carbonnel, J.P., 1962
La soucoupe plongeante permet l'exploration du plateau continental.
La Nature, 3327: 294-299.

continental shelf

Carrigy, M.A., and R.W. Fairbridge, 1954.
Recent sedimentation, physiography and structure of the continental shelves of western Australia.
J.R. Soc. West. Australia, 38:65-95, 8 textfigs.

continental shelf

Cartwright, D., and A.H. Stride, 1958.
Large sand waves near the edge of the continental shelf. Nature 181(4601):41.

Continental Shelf

Chapman, W. McL., 1970.
Outer boundaries of the continental shelf.
J. mar. techn. Soc. 4 (3): 7-15

continental shelf

Collette, B.J., J.A. Schouten, K.W. Rutten, D.J. Doornbos and W.H. Staverman, 1971.
Geophysical investigations off the Surinam Coast. Hydrogr. Newsletter, R. Netherlands Navy, Spec. Publ. 6: 17-24.

continental shelf
Collignon, J., 1965.
Les côtes et le plateau continental marocains.
Bull. Inst. Pêches Marit., Maroc, No. 13:21-37.

continental shelf
Collin, A.E., 1961
Oceanographic activities of the Polar Continental Shelf Project.
J. Fish. Res. Bd. Can., 18(2): 253-258.

continental shelf
Cotton, Sir Charles, 1968.
Relation of the continental shelf to rising coasts.
Geogr. J. 134(3): 382-389.

Continental shelf
Cotton, C.A., 1966.
The continental shelf.
New Zealand J. Geol. Geophys. 9(1/2):105-110

continental shelf
*Conolly, John R., and C.C. von der Borch, 1967.
Sedimentation and physiography of the sea-floor south of Australia.
Sedimentary Geol., 1(2):181-220.

continental shelf
Creager, Joes S., and Dean A. McManus, 1961.
Marine geology of the continental shelf north of Bering Strait. (Abstract).
Tenth Pacific Sci. Congr., Honolulu, 21 Aug.-6 Sept., 1961, Abstracts of Symposium Papers, 369.

continental shelf
Crosnier, A., 1964.
Fonds de pêche le long des côtes de la République Fédérale du Cameroun.
Cahiers, O.R.S.T.R.O.M., No. special:133 pp.

continental shelf
Curray, Joseph R., 1970.
Quaternary influence, coast and continental shelf of western U.S.A. and Mexico. Quaternaria 12: 19-34.

continental shelves
Curray, Joseph R., 1964
Transgressions and regressions.
In: Papers in Marine geology, R.L. Miller, Editor, Macmillan Co., N.Y., 175-203.

continental shelf
Curray, Joseph R., and David G. Moore, 1964.
Holocene regressive littoral sand, Costa de Nayarit, Mexico.
In: Developments in Sedimentology, L.M.J.U. van Straaten, Editor, Elsevier Publishing Co., 1:76-82.

Continental shelf
Curry, D., D. Hamilton and A.J. Smith 1970.
Geological evolution of the western English Channel and its relation to the nearby continental margin.
In: The geology of the East Atlantic continental margin. 2. Europe, ICSU/SCOR Working Party 31 Symposium, Cambridge, 1970, Rept. 70/14: 129-142

D

continental shelf
D'Arrigo, A., 1954.
Caratterizzazione fisiografica della piattaforme litorale in Sicilia e suoi rapporti con la pesca e con i giacimenti profondi d'idrocarburi.
Atti Accad. Gioenia 6(9):25-36.

continental shelf
Day, Alan A., 1959
The continental margin between Brittany and Ireland. Deep-Sea Research 5(4): 249-265.

continental shelf
Dietrich, G., 1963.
Die Meere.
Die Grosse Illustrierte Länderkunde, C. Bertelsmann Verlag, 2:1523-1606.

continental shelf
Dietrich, Gunter, 1959
Zur Topographie und Morphologie des Meeresboden im nördlichen Nordatlantischen Ozean.
Deutschen Hydrogr. Zeits., Ergänzungsheft Reihe B (4): No. 3: 26-34. Also in: Deutsches Hydrogr. Inst., Ozeanogr., 1959 (1960), No. 8.

continental shelf
Dietz, R.S., 1952.
Geomorphic evolution of continental terrace (continental shelf and slope).
Bull. Amer. Assoc. Petr. Geol. 36(9):1802-1819, 8 textfigs.

continental shelves
Dietz, Robert S., A.J. Carsola, E.C. Buffington, and Carl J. Shipek, 1964
Sediments and topography of the Alaskan shelves.
In: Papers in Marine Geology, R.L. Miller, Editor, Macmillan Co., N.Y., 241-256.

continental shelf
*Drake, Charles L., John I. Ewing and Henry Stockard, 1968.
The continental margin of the eastern United States.
Can. J. Earth Sci., 5(4-2):993-1010.

Continental shelf
Dupeuple, Pierre-Alain, et Michel Lamboy 1969.
Le plateau continental au nord de la Galice et des Asturies: premières données sur la constitution géologique.
C.r. hebd. Séanc. Acad. Sci. Paris (D) 269 (5): 548-551.

continental shelf
Duran S., Luis G., 1964.
Interpretación geofísica de la plataforma continental del Caribe.
Caldasia, Bol., Inst. Ciencias Nat., Univ. Nac., Colombia, Bogota, 9(42):137-150.

E

Continental shelf
Edgerton, H. E., et O. Leenhardt, 1966.
Mesures d'épaisseur de la vase sur les fortes pentes du précontinent.
C. R. herb. séanc. Acad. Sci., Paris (D) 262 (19):2005-2007.

continental shelf
Eisma, D. and A.J. van Bennekom, 1971.
Oceanographic observations on the eastern Surinam Shelf. Hydrogr. Newsletter, R. Netherlands Navy, Spec. Publ. 6: 25-29.

continental shelves
Emery, K.O., 1969.
The continental shelves.
Scient. Am., 221(3):106-114,116,118,120-122. (popular).

Continental shelf
Emery, K.O. 1968?
The continental shelf and its mineral resources.
Selected Papers from the Governor's Conference on Oceanography, October 11 and 12, 1967, at the Rockefeller University, New York, N.Y., 36-51.

continental shelves
*Emery, K.O., 1968.
Relict sediments on continental shelves of the world.
Bull.Am.Ass.Petrol.Geol., 52(3):445-464.

Continental shelves
Emery, K.O. 1967.
Estuaries and lagoons in relation to continental shelves.
In: Estuaries, G.H. Lauff, editor, Publs Am. Ass. Advmt Sci., 83: 9-11.

continental shelf
Emery, K.O., 1966.
The Atlantic continental shelf and slope of the United States: geologic background.
Prof.Pap.U.S.geol.Surv., 529-A:A1-A23.

continental shelves
Emery, K.O., 1965.
Characteristics of continental shelves and slopes
Bull. Amer. Assoc. Petr. Geol., 49(9):1379-1384.

continental shelf
Emery, K.O., 1956(1953)
Some salient problems of continental shelves.
Proc. 8th Pacific Sci. Congr., Geol. Geophys. Meteorol., (Nat. Res. Counc. Philippines) 2A: 810-814.

continental shelf
Emery, K. O., and Y. K. Bentor, 1960
The continental shelf of Israel. Ministry of Agric., Div. Fish., Israel, Sea Fish Res. Sta. Bull., No. 28: 25-41. Also, on same cover: Ministry of Develop., Israel, Geol. Surv. Bull. No. 26: 25-41.

continental shelf
Emery, K.O., Bruce C. Heezen and T.D. Allen, 1966.
Bathymetry of the eastern Mediterranean Sea.
Deep-Sea Res., 13(2):173-192.

continental shelf
Emery, K.O. and John D. Milliman, 1970.
Quaternary sediments of the Atlantic Continental Shelf of the United States. Quaternaria 12: 3-18.

continental

Emery, K.O., and L.C. Noakes, 1965.
Economic placer deposits of the continental shelf.
Techn. Bull. Econ. Comm. Asia, Far East, Comm. Co-ordin. Joint Prospect. Mineral Res. Asian Offshore Areas 1:95-

continental shelf
*Emery, K.O., R.L. Wigley, Alexandra S. Bartlett, Meyer Rubin and E.S. Barghoorn, 1967.
Freshwater peat on the continental shelf.
Science, 158(3806):1301-1307.

continental shelf
Etchichury, Maria C., and Joaquin R. Remiro, 1960
Muestras de fondo de la plataforma continental comprendida entre los paralelos 34 y 36 30 de latitud sur y los meridianos 53 10 y 56 30 de longitud oeste.
Revista Argentino Ciencias Nat. "Bernardino Rivadavia", Ciencias Geol., 6(4):199-263.

continental shelf
Ewing, J., B. Luskin, A. Roberts and J. Hirshman 1960.
Sub-bottom reflection measurements on the continental shelf, Bermuda Banks, West Indies Arc, and in the West Atlantic basins.
J. Geophys. Res., 65(9):2849-2860.
Check spelling before citing

F

continental shelf
Fairbridge, Rhodes W. and Horace G. Richards, 1970.
Eastern coast and shelf of South America.
Quaternaria 12: 47-55.

continental shelf
Fisk, H. N., & B. McClelland, 1959.
Geology of continental shelf off Louisiana: Its influence on offshore foundation design.
Bull. Geol. Soc. Amer., 70(10):1369-1394.

continental shelf
Friedman, Gerald M., 1966.
Study of continental shelf and slope on the coasts of Long Island, N.Y., and New Jersey.
Marit. Sediments, 2(1):21-22. (mimeographed)

continental shelf
Fujii, S. and A. Mogi, 1970.
On coasts and shelves in their mutual relations in Japan during the Quaternary.
Quaternaria 12: 155-164.

continental shelf
Fujiwara, Kenzo, 1971.
Soundings and submarine topography of the glaciated continental shelf in Lützow-Holm Bay, East Antarctica. (In Japanese; English abstract). Antarct. Rec., Repts. Japan Antarctic Res. Exped. 41: 81-103.

G

continental shelf
Garde, S.C., B.K. Rastogi and C.P. Gupta, 1970.
Magnetic anomalies and the sub-shelf geologic structure off the Bombay coast (India).
Marine Geol. 9(5):355-363

continental shelf
Garrison, Louis E., 1970.
Development of the continental shelf south of New England.
Bull. Am. Ass. Petrol. Geol., 54(1): 109-124.

continental shelf
Geddes, Wilburt H., and Joel S. Watkins, 1966.
Atlantic shelf magnetic anomaly.
In: Continental margins and island arcs, W.H. Poole, editor, Geol. Surv. Pap., Can., 66-15:48-56.

continental shelf
Gennessaux, M., 1963.
Structure et morphologie de la pente continentale de la region niçoise.
Rapp. Proc. Verb., Réunions, Comm. Int., Expl. Sci., Mer Méditerranée, Monaco, 17(3):991-998.

continental shelves
Gershanovich, D.E., 1966.
On the principles in the classification of the shelf zone. (In Russian).
Trudy vses. nauchno-issled. Inst. morsk. ryb. Khoz. Okeanogr. (VNIRO), 60:79-88.

continental shelf
Giermann, Günter 1969.
Morphologie et tectonique du plateau continental entre le cap Cavallo et Saint-Florent (Corse).
Bull. Inst. océanogr. Monaco, 69 (1397): 6pp.

continental shelf
Gill, Edmund D., 1970.
Coast and continental shelf of Australia.
Quaternaria 12: 115-128.

continental shelf
Gilluly, James, 1964.
Atlantic sediments, erosion rates and the evolution of the continental shelf: some speculations.
Geol. Soc., Amer., Bull., 75(6):483-492.

continental shelf
Glangeaud, L., 1955.
Résultats acquis dans la connaissance de la marge continental nord-africaine.
Trav. Lab. Géol. Sous-Marine, 6:12-14, Pls. 5-6.

Continental shelf
Gorgy, Samy, 1966.
Les pêcheries et le milieu marin dans le Secteur Méditerranéen de la Republique Arabe linie.
Rev. Trav. Inst. Pêches Marit. 30(1):25-

continental shelf
*Got, Henri, et Andre Monaco, 1969.
Sédimentation et tectonique plio-quaternaire du pré continent méditerranéen au large du Roussillon.
C.r. hebd. Seanc. Acad. Sci., Paris, (1) 268(8): 1171-1174.

continental shelves
Guilcher, A., 1970
Quaternary events on the continental shelves of the world. In: The geology of the East Atlantic continental margin, 1. General and economic papers, ICSU/SCOR Working Party 31 Symposium, Cambridge 1970, Rept. No. 70/13: 31-46.

continental shelf
Guilcher, André, 1967.
Morpho-sédimentologie de la proche plate-forme continentale Atlantique entre Ouessant et les Sables d'Olonne.
Revue Géogr. phys. Géol. dyn., 9(3): 181-190

continental shelf
Guilcher, A., 1963.
24. Estuaries, deltas, shelf, slope.
In: The Sea, M.N. Hill, Editor, Interscience Publishers, 3:620-654.

continental shelf
Guilcher, A., 1963.
13. Continental shelf and slope (continental margin).
In: The Sea, M.N. Hill, Editor, Interscience Publishers, 3:281-311.

H

continental shelf
Hatori, Tokutaro, 1967.
The wave form of tsunami on the continental shelf.
Bull. Earthq. Res. Inst., Tokyo Univ., 45(1): 79-90.

continental shelf
Hayes, Miles O., 1967.
Relationship between coastal climate and bottom sediment type on the inner continental shelf.
Mar. Geol., 5(2):111-132.

continental shelf
Hayes, Miles O., 1964.
Lognormal distribution of inner continental shelf widths and slopes.
Deep-Sea Research, 11(1):53-78.

Continental shelf
Hedberg, Hollis D. 1970.
Continental margins from viewpoint of the petroleum geologist.
Bull. Am. Ass. Petrol. Geol. 54(1): 3-43.

continental shelf
Heezen, Bruce C., and H.W. Menard, 1963.
12. Topography of the deep-sea floor.
In: The Sea, M.N. Hill, Editor, Interscience Publishers, 3:233-280

continental shelf
*Henry, Vernon J., Jr., and John H. Hoyt, 1968.
Quaternary paralic and shelf sediments of Georgia.
SEast. Geol., Duke Univ., 9(4):195-214.

continental shelf
Hernandez-Pacheco, F., 1955.
Caracteristicas del zocalo continental del Africa occidental española.
Bol. Inst. Español Ocean., 70:3-24.

continental shelf
Hersey, J.B., E.T. Bunce, R.F. Wyrick and F.T. Dietz, 1959.
Geophysical investigations of the continental region between Cape Henry, Virginia, and Jacksonville Florida.
Bull. Geol. Soc., Amer., 70:437-466.

continental shelf
Hersey, J.B., and W.F. Whittard, 1966.
The geology of the western approaches of the English Channel. V. The continental margin and shelf under the South Celtic Sea.
In: Continental margins and island arcs, W.H. Poole, editor, Geol. Surv. Pap., Can., 66-15:80-105.

continental shelf
Holtedahl, H., 1958
Some remarks on geomorphology of continental shelves off Norway, Labrador and southeast Alaska. J. Geol., 66(4):461-471.

continental shelf
Holtedahl Olaf, 1970.
On the morphology of the West Greenland shelf with general remarks on the "marginal channel" problem.
Mar. Geol. 8(2):155-172.

continental shelf
Hoshino, Michichei, 1970.
Three flat plains on the Japanese insular shelf. Quaternaria 12: 165-168.

continental slope
Hoshino, M., and Y. Ichihara, 1960
[Submarine topography and bottom sediments off Kumanonada and Enshu-nada.]
J. Oceanogr. Soc., Japan, 16(2): 41-46.

continental shelf
Hood, Peter 1967.
Geophysical surveys of the continental shelf south of Nova Scotia.
Marit. Sed., Halifax-Fredericton, N.S. Can. 3(1):6-11 (multilithed).

shelf
*Hoskin, Hartley, 1967.
Seismic reflection observations on the Atlantic continental shelf, slope and rise southeast of New England.
J. Geol., 75(5):598-611.

continental shelf
Hülsemann, Jobst, 1967.
Der Kontinental-Schelf - geologische Grenze Land/Meer.
Umschau 67(4): 105-111

continental shelf
Hunkins, Kenneth, 1966.
The Arctic continental shelf north of Alaska.
In: Continental margins and island arcs, W.H. Poole, editor, Geol. Surv. Pap., Can., 66-15:197-205.

I

continental shelf
Ilyin, A.V., 1971.
Main features of geomorphology of Atlantic bottom. (In Russian; English abstract).
Okeanol. Issled. Rezult. Issled. Mezhd. Geofiz. Proekt. 21: 107-246.

continental shelf
Ilyin, A.V., 1962
[Geomorphology of the continental shelf of the North Atlantic.]
Trudy Inst. Okeanol. Akad. Nauk, SSSR, 56: 3-14.

continental shelf
Inman, D.L., and G.A. Rusnak, 1956.
Changes in sand level on the beach and shelf at La Jolla, California. B.E.B. Tech. Memo. 82: 39 pp.

continental shelf
Iwabuti, Y., 1962
[Continental slope and outer continental shelf off Zyoban.]
Hydrogr. Bull., Tokyo. (Publ. No. 981), 70: 33-38.

J

continental shelf
Jakobi, Hans, 1971.
On the interstitial fauna of the continental shelf of the state of Paranã (Preliminary report). In: Fertility of the Sea, John D. Costlow, editor, Gordon Breach, 1: 209-213.

continental shelves
Japan, Maritime Safety Board, 1955.
[To research continental shelves is a pressing need of the hour.] Publ. No. 981, Hydrogr. Bull., Spec. No. 17:1-4.

continental shelf
Jordan, G.F., 1962
Submarine physiography of the United States continental margins.
Techn. Bull. U.S Dept. Commerce, Coast & Geodetic Survey, 18: 28 pp.

K

continental shelf
Kaplin, P.A., E.N. Nevessky, Yu. A. Pavlidis, and F.A. Shcherbakov, 1968.
Peculiarities of structure and history of development of the upper shelf and the nearshore zone during the Holocene. (In Russian: English abstract).
Okeanologiia, Akad. Nauk, SSSR, 8(1):3-13.

Continental shelf
Keary R., 1970.
Coastal climate and shelf-bottom sediments: a comment.
Mar. Geol., 5(5): 363-365.

continental shelf
Keen, M.J., and J.E. Blanchard, 1966.
The continental margin of eastern Canada.
In: Continental margins and island arcs, W.H. Poole, editor, Geol. Surv. Pap., Can., 66-15:

continental shelf
King, L.C., and R.H. Belderson, 1961.
Origin and development of the Natal coast.
Marine Studies off the Natal Coast, C.S.I.R. Symposium, No. S2:1-9 (multilithed).

continental shelf
Klein, George deVries, 1967.
Paleocurrent analysis in relation to modern marine sediment sispersal patterns.
Bull. Am. Ass. Petrol. Geol., 51(3):366-382.

Continental shelf
Klingebiel, André, Francis Lapierre, Janine Larroude et Michel Vigneaux 1968.
Présence d'affleurements de roches d'âge Miocène sur le plateau continental du Golfe de Gascogne.
C.r. hebd. Séanc. Acad. Sci Paris (D) 266 (11): 1102-1104.

continental shelf
Klingebiel, André, Claude Pujol et Michel Vigneaux, 1970.
Sur la stabilité des positions relatives du plateau et du talus continental dans le Golfe de Gascogne depuis le Miocène moyen.
C.r. hebd. Séanc. Acad. Sci., Paris (D) 270 (26): 3175-3176

Continental Shelf
Knott, S.T., and H. Hoskins, 1968.
Evidence of Pleistocene events in the structure of the continental shelf off the northeastern United States.
Marine Geol., 6(1):5-43.

continental shelf
Krueger, Robert B., 1971.
International and national regulation of pollution from offshore oil production. (2) In: Impingement of man on the oceans, D.W. Hood, editor, Wiley Interscience: 630-634.

continental shelf
Kukkuteswera Rao, B., and E.C. LaFond, 1954.
The profile of the continental shelf off Visakapatnam coast. Andhra Univ. Ocean. Mem. 1:78-85, 3 textfigs.

L

continental shelf
Landisman, Mark and Stephan Mueller, 1966.
Seismic studies of the earth's crust in continents. II. Analysis of wave propagation in continents and adjacent shelf areas.
Geophys. J. R. astr. Soc., 10:539-548.

continental shelf
Lapierre, F., 1970.
Fleuves et rivage préflandriens sur le plateau continental du Golfe de Gascogne.
Quaternaria 12: 207-217.

continental shelf
LeClaire, Lucien, Jean-Pierre Caulet et Philippe Bouysse, 1965.
Prospection sédimentologique de la marge continentale nord-africaine.
Cahiers Océanogr., C.C.O.E.C., 17(7):467-479.

Continental shelf
Leenhardt, O., et G. Alla 1969.
Structure du plateau continental provençal.
Vie Milieu (B) 20(1): 1-12.

continental shelves
Lomnitz, C., 1963.
Tidal variation due to subsidence of a continental shelf.
Proc. Tsunami Meet., 10th Pacific Sci. Congr., Univ. Hawaii, Aug.-Sept., 1961.
Union Géodés. et Géophys. Int., Monogr., No. 24:

continental shelf
Lyman, J., 1958.
A review of present knowledge concerning the continental shelves of the Americas.
Geol. Survey Bull. No. 1067:27-41.

M

continental shelf
Ma, T.Y.H., 1955.
The geological significance of the continental shelf and the accumulation of continental slope deposits. Oceanographia Sinica 2(2):1-9, 1 fig.

continental shelf
MacIntyre, I.G., and J.D. Milliman, 1970
Physiographic features on the outer shelf and upper slope, Atlantic continental margin, southeastern United States.
Bull. geol. Soc. Am. 81 (9): 2577-2598.

continental shelf
MacIntyre, Ian G., and Orrin H. Pilkey, 1969.
Preliminary comments on linear sand-surface features, Onslow Bay, North Carolina, continental shelf: problems in making detailed sea-floor observations.
Marit. Sediments 5 (1): 26-29.

continental shelf
*Maloney, Neil J., 1967.
Geomorphology of continental margin of Venezuela, Cariaco Basin.
Boln Inst. Oceanogr. Univ. Oriente, 5(½):38-53.

continental shelf
Martin, L., 1969.
Introduction à l'étude géologique du plateau continental ivoirien. Premiers résultats.
Doc. Centre Rech. océanogr. Abidjan, 034, 163 pp. Also Doc. scient. Côte d'Ivoire 034.

Mc

continental shelf
McIlhenny, W.F., 1966.
The oceans: technology's new challenge.
Chem. Engng., (Nov. 7):247-254.

McIver, N.L. 1972 continental shelf
Cenozoic and Mesozoic stratigraphy of the Nova Scotia Shelf.
Can. J. Earth. Sci. 9 (1): 54-70.

continental shelf
*McMaster, Robert L., and Thomas P. Lachance, 1968.
Seismic reflectivity studies of northwestern African continental shelf: Strait of Gibraltar to Mauritania.
Bull. Am. Ass. Petr. Geol., 52(12):2387-2395.

continental shelf
McMaster, Robert L., Thomas P. Lachance and Asaf Ashraf, 1970.
Continental shelf geomorphic features off Portuguese Guinea, Guinea, and Sierra Leone (West Africa).
Marine Geol. 9 (3): 203-213.

continental shelf
McMaster, Robert L., John D. Milliman and Asaf Ashraf. 1971.
Continental shelf and upper slope sediments off Portuguese Guinea, Guinea, and Sierra Leone West Africa.
J. sedim. Petrol. 41 (1): 150-158.

continental shelf
Mero, John L., 1965.
The mineral resources of the sea.
Elsevier Oceanogr. Ser., 312 pp.

continental shelf
Mogi, A., 1953.
On the depth of the continental shelf margin along the north-west coast of Honsyu. Publ. 981, Hydrogr. Bull., Tokyo, Spec. Number, No. 12:54-57
3 textfigs.

continental shelf
Moodie, A.E., 1954.
Some territorial problems associated with the continental shelf. Adv. Sci., 11(41):42-48.

continental shelf
Moodie, A.E., 1954.
The continental shelf. Adv. Sci., 11(41):42-47.

continental shelf
Moore, Clyde H., Jr., 1969.
Factors controlling carbonate sand distribution in the shallow shelf environment: illustrated by the Texas Cretaceous. Trans. Gulf Coast Ass. geol. Socs. 19 : 507.

continental shelves
Moore, David G., 1964.
Acoustic-reflection reconnaissance of continental shelves: eastern Bering and Chukchi seas.
In: Papers in Marine Geology, R.L. Miller, editor Macmillan Co., N.Y., 319-362.

continental shelf
Moore, David G., 1963.
Geological observations from the bathyscaph "Trieste" near the edge of the continental shelf off San Diego, California.
Geol. Soc., Amer., Bull., 74(8):1057-1062.

continental shelf
Mouton, M.W., 1952.
The continental shelf. Martinus Nijhoff, The Hague, 367 pp.

"critical and comprehensive study of the juridical position of the continental shelf and of the questions concerning the utilization of the sea covering it and of its soil and subsoil beyond the limits of the territorial waters."

continental shelf
Muraour, P., et M. Genesseaux, 1965.
Quelques remarques à la suite d'une étude séismique par refraction sur la pente continentale.
Comptes Rendus, Acad. Sci., Paris, 260(1):227-230.

continental shelf
Muraour, Pierre, Jacques Merle et Jean Ducrot, 1962.
Observations sur le plateau continental à la suite d'une étude séismique par réfraction dans le golfe du Lion.
Comptes Rendus, Acad. Sci., Paris, 254(15):2801-2803.

N

continental shelf
Nanda, J.N., 1957.
Seismic exploration of the continental shelf off the west coast of India. J. Geophys. Res., 62(1):113-115.

continental shelves
Nasu, Noriyuki, 1964
The provenance of the coarse sediments on the continental shelves and the trench slopes off the Japanese Pacific coast.
In: Papers in Marine Geology, R.L. Miller, Editor, Macmillan Co., N.Y., 65-101.

continental shelf
Nasu, Noriyuki, and Takahiro Sato, 1962.
VII. Geological results in the Japanese Deep Sea Expedition in 1961 (JEDS-4).
Oceanogr. Mag., Japan Meteorol. Agency, 13(2): 166.

JEDS Contrib. No. 135.

continental shelf
Nesteroff, W.D., 1966.
Deux exemples de bordure continentale francaise: le Gouf de Cap Breton et le Cap Cartaya.
In: Continental margins and island arcs, W.H. Poole, editor, Geol. Surv. Pap., Can., 66-15:107-113.

continental shelf
Nevessky, E.N., 1970.
Holocene history of the coastal shelf zone of the USSR in relation with processes of sedimentation and condition of concentration of useful minerals. Quaternaria 12: 79-88.

Continental shelf
Newman, Walter S., and Stanley March 1968.
Littoral of the northeastern United States: late Quaternary warping.
Science 160 (3832): 1110-1112.

continental shelf
Niino, Hiroshi, 1965.
Geology and mineral resources of the Japanese continental shelf (abstract).
In: Submarine geology and geophysics, Colston Papers, W.F. Whittard, and R. Bradshaw, editors, Butterworth's, London, 17:177.

continental shelf
Niino, Hiroshi, 1956(1953).
Bottom characters of the continental shelf around the Japanese islands.
Proc. 8th Pacific Sci. Congr., Geol. Geophys., Meteorol., 2A:901-909.

continental shelf
Niino, Hiroshi and K.O. Emery, 1966.
Continental-shelf sediments off northeastern Asia.
J. Sed. Petr., 36(1):152-161.

O

continental shelf
Ottmann, Francois, 1965.
Introduction a la Geologie Marine et Littorale.
Masson et Cie, Paris, 259 pp. 47 F

P

continental shelf
Paskoff, Roland, 1970.
Sur les relations possibles entre un ensemble de terrasses marines émergées et les caractères de la marge continentale au Chili entre 30° et 33° sud. Quaternaria 12: 35-45.

continental shelf
Pautot, Guy, 1970.
La marge continentale au large de l'Estérel (France) et les mouvements verticaux pliocènes.
Mar. Geophys. Res. 1 (1): 61-84

continental shelf
Pepper, J.F., 1958.
Potential mineral resources of the continental shelves of the western hemisphere.
Geol. Survey Bull., 1067:43-65.

continental shelf
Pevear, David R., and Orrin H. Pilkey, 1966.
Phosphorite in Georgia continental shelf sediments.
Geol. Soc., Am., Bull., 77(8):849-858.

continental shelf
Phipps, Charles V.G., 1966.
Evidence of Pleistocene warping of the New South Wales continental shelf.
In: Continental margins and island arcs, W.H. Poole, editor, Geol. Surv. Pap., Can., 66-15:280-293.

continental shelf
Pilkey, Orrin H., 1963.
Heavy minerals of the U.S. South Atlantic continental shelf and slope.
Geol. Soc., Am., Bull., 74:641-648.
Also in:
Collected Reprints, Mar. Inst., Univ. Georgia, 1965.

continental shelf
Pilkey, Orrin H., and Dirk Frankenberg, 1964.
The relict-recent sediment boundary on the Georgia continental shelf.
Bull. Georgia Acad. Sci., 22(1):37-40.
Also In:
Collected Reprints, Mar. Inst., Univ. Georgia, 4 (1965).

continental shelf
Pilkey, Orrin H., and Robert T. Giles, 1965.
Bottom topography of the Georgia Continental Shelf.
Southeast, Geol., Duke liniv., 7(1):15-18.

continental shelf
Pilkey, Orrin, Ian G. Macintyre and Elazar Uchupi, 1971
Shallows structures: shelf edge of continental margin between Cape Hatteras and Cape Fear, North Carolina. Am. Ass. Petrol. Geol. Bull., 55(1): 110-115.

Continental shelf
Pilkey, Orrin H., Detmar Schnitker, and D. R. Pevear, 1966.
Oolites on the Georgia continental shelf.
J. Sed. Petr., 36(2):462-467.
abstract

continental shelf
Pomerancblum, Malvina, 1966.
The distribution of heavy minerals and their hydraulic equivalents in sediments of the Mediterranean continental shelf of Israel.
J. Sed. Petr., 36(1):162-174.

Continental Shelf
Posada, Jaime, 1962.
Las relaciones internacionales y la plataforma continental.
Peces y Conservas, Colombia, No. 12:43-44.

continental shelves
Pratt, Wallace E., 1965.
Petroleum on continental shelves.
Bull. Amer. Assoc. Petr. Geol., 49(10):1711-1712.

continental shelf
Price, W.A., 1954.
Dynamic environments: reconnaissance mapping, geologic and geomorphic, of continental shelf of Gulf of Mexico. Trans. Gulf Coast Assoc. Geol. Soc., 4:75-107., 16 figs.

Q

R

Redfield, A. C., 1958 — continental shelf, effect of
The influence of the continental shelf on the tides of the Atlantic coast of the United States J. Mar. Res., 17:432-448.

continental shelves
Reineck, H.E., 1967.
Layered sediments of the tidal flats, beaches, and shelf bottoms of the North Sea.
In: Estuaries, G.H. Lauff, editor, Publs Am. Ass. Advmt Sci., 83:191-206.

continental shelf
Rona, Peter A., and C.S. Clay, 1964.
Continuous seismic profiles and cores from the shelf and slope off Cape Hatteras. (Abstract).
Trans. Amer. Geophys. Union, 45(1):72.

continental shelf
Roots, E.F., 1962.
Canadian polar continental shelf project, 1959-62.
Polar Record, 11(72):270-276.

continental shelf
Rosfelder, A., 1956.
Sur les facteurs de répartition des sédiments sur le plateau continental algérien.
C.R. Acad. Sci., Paris, 242(9):1196-1199.

continental shelf
Rosfelder, A., 1955.
Carte provisoire au 1/500,000 de la marge continental algérienne. Note de présentation.
Publ. Service Carte Géol., Algérie, n.s., Bull. No. 5, Trav. des Coll., 1954:57-106, 6 figs., 1 pl., 1 map.

continental shelf
*Rvachev, V.D., 1968.
On some features of geomorphology of the shelf and continental slope of the Newfoundland area. (In Russian; English abstract).
Okeanologiia, Akad. Nauk, SSSR, 8(4):659-665.

continental shelf
Ryzhkov, Yu. G., 1966.
On the vortices with horizontal axes over an edge of the continental shelf of South America. (In Russian).
Fizika Atmosferi i Okeana, Izv., Akad. Nauk, SSSR, 2(1):88-91.

S

continental shelf
Scheidegger, K.F., L.D. Kulm and E.J. Runge 1971.
Sediment sources and dispersal patterns of Oregon continental shelf sands.
J. sedim. Petrol. 41 (4): 1112-1120.

continental shelf
Schlee, John, and Richard M. Pratt, 1970
Atlantic continental shelf and slope of the United States - gravels of the northeastern part.
Prof. Pap., U.S. Geol. Surv. 529-H: H-1 - H-39.

CONTINENTAL SHELF
Scholl, David W., and David M. Hopkins, 1969.
Newly discovered Cenozoic basins, Bering Sea shelf, Alaska.
Bull. Am. Ass. Petrol. Geol. 53 (10): 2067-2078.

continental shelf
Shalowitz, A.L., 1955.
Boundary problems associated with the continental shelf. Int. Hydrogr. Rev. 23(1):111-139, 17 text-figs.

continental shelf
Shepard, Francis P., 1964.
Criteria in modern sediments useful in recognizing ancient sedimentary environments.
In: Deltaic and shallow marine deposits, Developments in Sedimentology, L.M.J.U. van Straaten, editor, Elsevier Publishing Co., 1:1-25.

continental shelf
Shepard, F.P., 1964.
Criteria in modern sediments in recognizing ancient sedimentary environments.
In: Developments in Sedimentology, L.M.J.U. van Straaten, Editor, Elsevier Publishing Co., 1:1-25

continental shelf
Shepard, F.P., J.R. Curray, D.L. Inman, E.A. Murray, E.L. Winterer and R.F. Dill, 1964.
Submarine geology by diving saucer. Bottom currents and precipitous submarine canyon walls continue to a depth of at least 300 meters.
Science, 145(3636):1042-1046.

continental shelf
Stetson, H.C., 1955.
Patterns of deposition at the continental margin
Pap. Mar. Biol. and Oceanogr., Deep-Sea Res., Suppl. to Vol. 3:299-308.

continental shelves
Stride, A.H., 1963
The geology of some continental shelves. In: Oceanography and Marine Biology, H. Barnes, Edit. George Allen & Unwin, 1: 77-88.

continental shelf
Subba Rao, M., 1964.
Some aspects of continental shelf sediments off the east coast of India.
Marine Geology, 1(1):59-87.

continental shelf
Summerhayes, C.P., 1969.
Submarine geology and geomorphology off northern New Zealand.
N.Z. Jl Geol. Geophys. 12(2/3): 507-525

continental shelf
*Summerhayes, C.P., 1967.
Marine environments of economic mineral deposition around New Zealand: A review.
N.Z. Jl mar. Freshwat. Res., 1(3):267-282.

continental shelf
Sundvor, Eirik 1971.
Seismic refraction measurements on the Norwegian continental shelf between Andøya and Fugløybanken.
Mar. geophys. Researches 1(3): 303-313

continental shelf
Swift, Donald J.P., Daniel J. Stanley and Joseph R. Curray, 1971.
Relict sediments on continental shelves: a reconsideration.
J. Geol. 79 (5): 322-346

T

continental shelf
Terekhov, A.A., E.M. Khakhaler and J.P. Malovitskii, 1970.
New data on geological structure of Cis-Caucasian shelf of the Black Sea. (In Russian)
Dokl. Akad. Nauk SSSR 195(1):174-177.

continental shelf
Themasson, E.M., 1958.
Problems of petroleum development on the continental shelf of the Gulf of Mexico.
Geol. Survey Bull. 1067:67-90.

continental shelves
Thompson, Warren C., 1961.
A genetic classification of continental shelves.
Proc. Ninth Pacific Sci. Congr., Pacific Sci. Assoc., 1957 (Geol. & Geophys.), 12:30-39.

continental shelf
Trumbull, J., 1958.
Continents and ocean basins and their relation to continental shelves and continental slopes.
Geol. Survey Bull., 1067:1-26.

U

continental shelf
Uchupi, Elazar, and K.O.Emery, 1967.
Structure of continental margin off Atlantic coast of United States.
Bull.Am.Ass.Petr.Geol., 51(2):223-234.

continental shelf
Ulrich, J., 1963.
Der Formenschatz des Meeresbodens.
Geogr. Rundschau, 15(4):136-148.

Abstr.: Geomorph. Abstr., (16):6. (1964).

Continental shelf
United States, U.S. Army Engineering Research Center 1966.
Intraagency conference on continental shelf research.
Misc. Paper, No. 1-66: 37pp.

continental shelf
University of Southern California, Allan Hancock Foundation, 1965.
An Oceanographic and biological survey of the southern California mainland shelf.
State of California, Resources Agency, State Water Quality Control Board, Publ. No. 27:232 pp.
Appendix, 445 pp.

continental shelf
Urien, Carlos, 1970.
Les rivages et le plateau continental du Sud du Brésil, de l'Uruguay et de l'Argentine.
Quaternaria 12: 57-69.

V

continental shelves
van Andel, Tjeerd H., and John J. Veevers, 1965.
Submarine morphology of the Sahul Shelf, northwestern Australia.
Bull. Geol. Soc., Amer., 76(6):695-700.

continental shelf
van Hees, G.L. Strang, 1971.
Gravity measurements on the continental shelf of Surinam. Hydrogr. Newsletter, R. Netherlands Navy, Spec. Publ. 6: 11-12.

continental shelf
Veldkamp, J. and H.J.A. Vesseur, 1971.
Geomagnetic anomalies in the continental shelf of Surinam. Hydrogr. Newsletter, R. Netherlands Navy, Spec. Publ. 6: 13-15.

continental shelf
Verger, F., 1970.
Les rivages quaternaires de la plate-forme continentale de l'Ouest de la France.
Quaternaria 12: 197-206.

continental shelf
Vila, Fernando, 1965.
Conocimiento actual de la plataforma Continental Argentina.
Republica Argentina, Servicio Hidrograf. Naval Publ. H. 644: 25pp.

continental shelf
Virgili, Carmina, 1967.
El límite de los oceanos.
In: Ecología marina, Monogr. Fundación La Salle de Ciencas Naturales, Caracas, 14:1-34.

continental shelf
Vivier, 1949.
Le plateau continental marocain. Bull. Sci. Com. Océan et d'Etudes de Côtes du Maroc, No. 5:28-30, 2 fold-ins.

continental shelf
Vogt, P.R., 1970.
Magnetized basement outcrops on the southeast Greenland continental shelf.
Nature, Lond., 226 (5247): 743-744

continental shelf
Von Metzsch, E.H., 1967.
A survey of the continental shelf of Surinam.
Hydrogr. Newsl., Neth., 1(6): 343-348

continental shelf, effect of
von Trepka, Lothar, 1967.
Anwendung des hydrodynamisch-numerischen Verfahrens zur Ermittlung des Schelfeinflusses auf die Gezeiten in Modellkanälen und Modellozeanen.
Mitt. Inst.Meeresk. Univ. Hamburg, 9:73 pp. (multilithed).

W

continental shelf
Wang, Chao-Siang, 1961.
Sand fraction of the shelf sediments off the China coast. (study)
Proc. Geol. Soc., China, (4):33-49.

continental shelf
Watkins, Joel S., and Wilbur H. Geddes, 1965.
Magnetic anomaly and possible orogenic significance of geologic structure of the Atlantic shelf
J. Geophys. Res., 70(6):1357-1361.

continental shelves
Weaver, Paul, 1950.
Variations in history of continental shelves.
Bull. Amer. Assoc. Petr. Geol. 34(3):351-360, 3 textfigs.

continental shelf
Weber, Alban, 1966.
Our newest frontier: the seabottom, some legal aspects of the continental shelf status.
Exploiting the Ocean, Trans. 2nd Mar. Techn. Soc. Conf., June 27-29, 1966, 405-411.

continental shelf
Wolfle, Dael, 1965.
The continental shelf. [Editorial].
Science, 148(3666):25.

continental shelf
*Worzel, J. Lamar, 1968.
Advances in marine geophysical research of continental margins.
Can. J. Earth Sci., 5(4-2):963-983.

XYZ

continental shelf
Yung-shan, Ch'in, 1963.
Relief and bottom deposits of the continental shelf in the Sea of China. (In Chinese).
Oceanol., et Limnol. Sinica, 5(1):71-86.

Abstracted in:
Soviet Bloc Res., Geophys., Astron., Space, 95:48

continental shelf
Zenkovich, V.P., 1970.
Nature of the USSR marine shelves and coasts.
Quaternaria 12: 71-77.

continental shelf
Zhivago, A.V. and S.A. Evteev, 1970.
Shelf and marine terraces of Antarctica.
Quaternaria 12: 89-114.

Continental shelf, convention on

Continental Shelf, convention on, etc.
Tubman, William C., 1966.
The legal status of minerals located on or beneath the ocean floor beyond the continental shelf.
Exploiting the Ocean, Trans. 2nd Mar., Techn. Soc. Conf., June 27-29, 1966, 379-404.

continental shelf, effect of

continental shelf, effect of
*Aida, Isamu, Tokutaro Hatori, Morio Koyama and Kinjiro Kajuira, 1968.
A model experiment on long-period waves travelling along a continental shelf.
(In Japanese; English abstract).
Bull.Earthq.Res.Inst., Tokyo Univ., 46(3):707-739.

continental shelf, effect of
Ichiye, T., 1962.
Circulacion y distribucion de la masa de agua en el Golfo de Mexico. (In Spanish and English).
Geofisica Internacional, Rev. Union Geofis. Mexicana, Inst. Geofis., Univ. Nacional Autonoma de Mexico, 2(3):47-76, 22 figs.

Continental shelf, structure of

continental shelf, structure of
Antoine, John W., and James L. Harding, 1963.
Structure of the continental shelf, northeastern Gulf of Mexico.
Texas A. & M., Dept. of Oceanogr. and Meteorol., Ref. 63:13T:18 pp.

Continental slope

continental slope
Allen, J.R.L., 1964.
The Nigerian continental margin: bottom sediments submarine morphology and geological evolution.
Marine Geology, 1(4):289-332.

continental slope
Antoine, John, William Bryant and Bill Jones, 1967.
Structural features of continental shelf, slope, and scarp, northeastern Gulf of Mexico.
Bull.Am.Ass.Petr.Geol., 51(2):257-262.

continental slope
*Antoine, J.W., and B.R. Jones, 1967.
Geophysical studies of the continental slope, scarp, and basin, eastern Gulf of Mexico.
Trans. Gulf Coast Assoc. Geol. Soc., 17th Ann. Meet.: 268-277.

continental slope
Athearn, William D., 1968.
Upper slope bathymetry off Guayanilla and Ponce, Puerto Rico. Trans. Fifth Carib. Geol. Conf. St. Thomas, V.I., 1-5 July 1968, Peter H. Mattson; editor. Geol. Bull. Queens Coll., Flushing, 5:41-43.

continental slope
Aubert, M., M. Gennesseaux, E. Groubert, and P. Muraour, 1964.
Mission de seismique refraction sur la pente continentale nicoise.
Cahiers C.E.R.B.O.M., Nice, (1):13-28.

continental slope
Avilov, I.K., 1965.
Relief and bottom sediments of the shelf and continental slope of the Northwestern Atlantic. Investigations in line with the programme of the International Geophysical Year, 2. (In Russian).
Trudy, Vses. Nauchno-Issled. Inst. Morsk. Ribn. Choz. i Okeanogr. (VNIRO), 57:173-234.

continental slope
Bartlett, Grant A., 1968.
Mid-Tertiary stratigraphy of the continental slope off Nova Scotia.
Marit. Sed., 4(1):22-31.

continental slope
Bellaiche, Gilbert, 1970.
Geologie sous-marine de la marge continentale au large du massif des Maures (Var, France), et de la plaine abyssale Ligure.
Revue Geogr. phys. Geol. dyn. (2) 12 (5): 403-440.

continental slope
Berthois, L., and R. Brenot, 1960
La morphologie sous-marine du talus du plateau entre le sud de l'Irlande et le Cap Ortegal. J. du Cons., 25(2):111-114.

continental slope
Bourcart, Jacques, 1959.
Morphologie du precontinent des Pyrenées à la Sardaigne. Colloques Int. Centre Nat. de la Recherche Sci., 83 (La Topographie et la Geologie des Profondeurs Oceaniques): 33-52.

continental slope
Beynagryan, V.P., 1966.
Morphometric analysis of short-term relief changes. (In Russian; English abstract).
Okeanologiia, Akad. Nauk, SSSR, 6(4):651-658.

continental slope
Brodie, J.W., 1964.
The fjordland shelf and Milford Sound.
New Zealand Dept. Sci. Ind. Res. Bull., 157; New Zealand Oceanogr. Inst., Memoir, No. 17: 15-23.

continental slope
*Conolly, John R., 1969.
Western Tasman sea floor.
N.Z. Jl Geol. Geophys. 12:310-343.

continental slope
*Conolly, John R., and C.C. Von der Borch, 1967.
Sedimentation and physiography of the sea-floor south of Australia.
Sedimentary Geol., 1 (2):181-220.

continental slope
Curray, J.R., D.G. Moore, R.H. Beldersen and A.H. Stride, 1966.
Centrinental margin of western Europe: slope progradation and erosion.
Science, 154(3746):265-266.

continental slope
Curry, D., E. Martini, A.J. Smith and W.F. Whittard, 1962.
The geology of the western approaches of the English Channel. 1. Chalky rocks from the upper reaches of the continental slope.
Phil. Trans., R. Soc., London, (B), 245(724): 267-290.

continental slope
Dietz, Robert S., 1964.
Origin of continental slopes.
American Scientist, 52(1):50-69.

continental slope
Dietz, R.S., 1952.
Geomorphic evolution of continental terrace (continental shelf and slope.).
Bull. Amer. Assoc. Petr. Geol. 36(9):1902-1819, 8 textfigs.

continental slopes
Dietz, R.S., 1952.
Some Pacific and Antarctic sea-floor features discovered during the U.S. Navy Antarctic Expedition, 1946-1947. (Abstract).
Proc. Seventh Pacific Sci. Congr., Met. Ocean. 3:335-344, 4 textfigs.

continental slope
Dietz, R.S., and H.W. Menard, 1951.
Origin of abrupt change in slope at continental shelf margin. Bull. Amer. Assoc. Petr. Geol. 35(9):1994-2016, 12 textfigs.

continental slope
Emery, K.O., 1966.
The Atlantic continental shelf and slope of the United States: geologic background.
Prof. Pap. U.S. geol. Surv., 529-A:A1-A23.

continental slope
Emery, K.O., 1965.
Characteristics of continental shelves and slopes.
Bull. Amer. Assoc. Petr. Geol., 49(9):1379-1384.

continental slope
Emery, K.O., Bruce C. Heezen and T.D. Allan, 1966.
Bathymetry of the eastern Mediterranean Sea.
Deep-Sea Res., 13(2):173-192.

Continental slope
Emery, K.O. and David A. Ross 1968.
Topography and sediments of a small area of the Continental slope south of Martha's Vineyard.
Deep-Sea Res. 15(4): 415-422.

Continental slope
Emiliani, Cesare, 1965.
Preciptious continental slopes and considerations on the transitional crust.
Science, 147 (3654):145-148.

continental slope
Ewing, Maurice, and John Antoine, 1966.
New seismic data concerning sediments and diapiric structures in Sigsbee Deep and upper continental slope, Gulf of Mexico.
Bull. Amer. Assoc. Petr. geol., 50(3):479-504.

continental slope
Friedman, Gerald M.,
Study of continental shelf and slope on the coasts of Long Island, N.Y., and New Jersey.
Marit. Sediments, 2(1):21-22. (mimeographed).

continental slope
Gakkel, J.J., 1957.
The continental slope as a geographical zone of the Arctic Ocean. Izv. Vses. Geograf. Obshsh., 89(6):493-507.

continental slope
Glangeaud, L., 1955.
Résultats acquis dans la connaissance de la marge continentale nord-africaine.
Bull. d'Info., C.C.O.E.C. 7(8):342-344, Pls. 5-6.

continental slope
Gealy, B.J., 1955.
Topography of the continental slope of northwest Gulf of Mexico. Bull. G.S.A. 66(2):203-208, 7 textfigs., 1 fold-in.

continental slope
Gibson, Thomas G., 1965.
Eocene and Miocene rocks off the northeast coast of the United States.
Deep-Sea Res., 12(6):975-981.

continental slope
Guilcher, A., 1963.
24. Estuaries, deltas, shelf, slope.
In: The Sea, M.N. Hill, Editor, Interscience Publishers, 3:620-654.

continental slope
Guilcher, A., 1963.
13. Continental shelf and slope (continental margin).
In: The Sea, M.N. Hill, Editor, Interscience Publishers, 3:281-311.

continental slope
Hadley, M.L., 1964
The continental margin southwest of the English Channel.
Deep-Sea Res., 11(5):767-779.

continental slope
Hanna, G.D., 1952.
Geology of the continental slope off central California. Proc. Calif. Acad. Sci., 4th ser., 27(9):325-358, Pls. 7014, 1 textfig.

continental slope
shelf
Hayes, Miles O., 1964.
Lognormal distribution of inner continental/ wides and slopes.
Deep-Sea Research, 11(1):53-78.

CONTINENTAL SLOPE
Hoshino, Michihei, and Yoshio Iwabuchi 1966.
Topography of continental slopes around the Japanese islands. (In Japanese; English abstract).
J. Fac. Oceanogr., Tokai Univ. (1): 37-49

continental slope
*Hoskin, Hartley, 1967.
Seismic reflections observations on the Atlantic continental shelf, slope, and rise southeast of New England.
J. Geol., 75(5):598-611.

continental slope
Hülsemann, Jobst, 1968.
Morphology and origins of sedimentary structure on submarine slopes.
Science 161(3836):45-47.

continental slope
Iijima, Azuma, and Hideo Kagami, 1962
Cainozoic tectonic development of the continental slope, northeast of Japan. (In Japanese: English abstract).
Repts. JEDS, Deep-Sea Res. Comm., Japan Soc., Promotion of Science, 3:561-577.

JEDS Contrib. No. 26.

continental slope
Iijima, Azuma and Hideo Kagami, 1961.
Origin of the continental slope off northeastern Japan. (Abstract).
Tenth Pacific Sci. Congr., Honolulu, 21 Aug.-6 Sept., 1961, Abstracts of Symposium Papers, 373.

continental slope
Ilyin, A.V., 1971.
Main features of geomorphology of Atlantic bottom. (In Russian; English abstract).
Okeanol. Issled. Rezult. Issled. Mezhd. Geofiz. Proekt. 21: 107-246.

continental slope
Iwabuti, Y., 1962
[Continental slope and outer continental shelf off Zyoban.]
Hydrogr. Bull., Tokyo, (Publ. No. 981), 70: 33-38.

continental slope
Jordan, G.F., 1951.
Continental slope off Apalachicola River, Florida
Bull. Amer. Assoc. Petr. Geol. 35(9):1978-1993, 16 textfigs.

continental slope
Jordan, G.F., and H.B. Stewart, Jr., 1959.
Continental slope off southwest Florida.
Bull. Amer. Assoc. Petr. Geol., 43(5):974-99.

continental slope
Kagami, Hideo, and Azuma Iijima, 1960.
On the bottom sediments off Onagawa and Kushiro, the adjacent continental slope of Japan Trench. Repts. of JEDS, 1:233-242.

Also in Oceanogr. Mag. 11(2):233-242.

Continental slope
Kanenko, Shiro 1966.
Rising promontories associated with a subsiding coast and sea-floor in south-western Japan.
Trans. R. Soc. N.Z., Geol. 4 (11): 211-228

continental slope
Koblents, Ya. P., 1963.
Description of the continental slope and floor of the Antarctic region of the Southern Ocean. (In Russian).
Sovetsk. Antarkt. Eksped., Inform. Biull., 43: 31-

Translation:
Scripta Tecnica for AGU, 5(1):16-17. 1965.

continental slope
Krause, Dale C., Mark A. Chramiec, George M. Walsh and Serge Wisotsky, 1966.
Seismic profile showing Cenozoic development of the New England continental margin.
J. geophys. Res., 71(18):4327-4332.

Continental slope
Lagaay, R.A., and B. J. Collette 1967.
A continuous seismic section across the continental slope off Ireland.
Mar. Geol. 5 (2): 155-157.

CONTINENTAL SLOPE
Lebedev, L.I., 1962
The origin of the relief in the continental slope of the mid-Caspian Sea area. (In Russian)
Okeanologiia, Akad. Nauk, SSSR, 2(5):874-880.

continental slope
Lehner, Peter, 1969.
Salt tectonics and Pleistocene stratigraphy on continental slope of northern Gulf of Mexico.
Bull. Am. Ass. Petrol. Geol., 53(12): 2431-2479

continental slope
Lewis, K.B., 1971.
Slumping on a continental slope inclined at 1°-4°.
Sedimentology 16(1/2): 97-110.

continental slope
*Maloney, Neil J., 1967.
Geomorphology of continental margin of Venezuela, Cariaco Basin.
Boln Inst. Oceanogr. Univ. Oriente, 5(½):38-53.

continental slope
MacIntyre, I.G., and J.D. Milliman, 1970
Physiographic features on the outer shelf and upper slope, Atlantic continental margin, southeastern United States.
Bull. geol. Soc. Am. 81 (9): 2577-2598.

continental slope
*Manheim, Frank T., 1967.
Evidence for submarine discharge of water on the Atlantic continental slope of the southern United States, and suggestions for further search.
Trans. N.Y. Acad. Sci., (2)29(7):839-853.

continental slope
Mogi, Akio, and Takahiro Sato, 1964.
Topography and sediment in the southern part of the Japan Trench.
J. Oceanogr. Soc., Japan, 20(2):51-56.

continental slope
*Morelock, Jack, 1969.
Shear strength and stability of continental slope deposits, western Gulf of Mexico.
J. geophys. Res., 74(2):465-482.

continental slope
Newton, John G., and Orrin H. Pilkey, 1969.
Topography of the continental margin off the Carolinas.
SEast. Geol., Duke Univ., 10(2): 57-92

continental slope
Parker, Robert H., 1964.
Zoogeography and ecology of some macroinvertebrates, particularly mollusks, in the Gulf of California and the continental slope off Mexico.
Vidensk. Medd., Dansk Naturh. Foren., 126:1-178.

continental slope
Pilkey, Orrin H., 1963.
Heavy minerals of the U.S. South Atlantic continental shelf and slope.
Geol. Soc., Am., Bull., 74:641-648.
Also in:
Collected Reprints, Mar. Inst., Univ., Georgia, 1965.

continental slope
*Pratt, Richard M., 1968.
Atlantic continental shelf and slope of the United States- physiography and sediments of the deep-sea basin.
Prof. Pap. U.S. geol. Surv., 529B:B1-B44, fold in chart.

continental slope
Richards, Adrian F., 1964.
Local sediment shear strength and water content.
In: Papers in Marine Geology, R.L. Miller, Editor, Macmillan Co., N.Y., 474-487.

Continental slope
Rona, Peter A., 1969.
The Middle Atlantic continental slope of United States: deposition and erosion
Bull. Am. Ass. Petrol. Geol. 53(7): 1453-1465.

Rona, Peter A., and C.S. Clay, 1967.
Stratigraphy and structure along a continuous seismic reflection profile from Cape Hatteras, North Carolina to the Bermuda Rise.
J. geophys. Res., 72(8):2107-2130.

continental slope
Rona, Peter A., and C.S. Clay, 1964.
Continuous seismic profiles and cores from the shelf and slope off Cape Hatteras. (Abstract).
Trans. Amer. Geophys. Union, 45(1):72.

continental slope
*Rvachev, V.D., 1968.
On some features of geomorphology of the shelf and continental slope of the Newfoundland area. (In Russian; English abstract).
Okeanologiia, Akad. Nauk, SSSR, 8(4):659-665.

continental slope
Trumbull, J., 1958.
Continenta and ocean basins and their relation to continental shelves and continental slopes.
Geol. Survey Bull., 1067:1-26.

continental slope
Uchupi, Elazar, and K.O. Emery, 1963
The continental slope between San Francisco, California and Cedros Island, Mexico.
Deep-Sea Research, 10(4):397-447.

continental slope
Vinogradova, P. S., 1964.
Some results of the study of the continental slope along the western border of the Barents Sea. (In Russian).
Material. Sess. Uchen. Sov. PINRO, Rez. Issled., 1962-1963, Murmansk, 20-34.

continental slope
Virgili, Carmina, 1967.
El límite de los océanos.
In: Ecología marina, Monogr. Fundación La Salle de Ciencias Naturales, Caracas, 14:1-34.

continental slope
*Worzel, J. Lamar, 1968.
Advances in marine geophysical research of continental margins.
Can. J. Earth Sci., 5(4-2):963-983.

Continental terrace

contimental terrace
Byrne, J.V. 1962
Geomorphology of the continental terrace off the central coast of Oregon.
The Ore Bin, 24(5):65-74.

Also in:
COLLECTED REPRINTS, Dept. Oceanogr., Oregon State Univ., Vol. 1, 1956-1962.

continental terrace
Curray, Joseph R., and David G. Moore, 1964.
Pleistocene deltaic progradation of continental terrace, Costa de Nayarit, Mexico.
In: Marine geology of the Gulf of California, a symposium, Amer. Assoc. Petr. Geol., Memoir, T. van Andel and G.G. Shor, Jr., editors, 3:193-215.

continental terrace
Dietz, Robert S., 1956(1953)
Salient problems of the continental terrace. Proc. 8th Pacific Sci. Congr., Geol. Geophys. Meteorol., (Nat. Res. Counc., Philippines) 2A: 815-819.

continental terrace
Dietz, R.S., 1952.
Geomorphic evolution of continental terrace (continental shelf and slope).
Bull. Amer. Assoc. Petr. Geol. 36(9):1802-1819, 8 textfigs.

continental terraces
Dietz, R.S., 1951.
Proposed geomorphic evolution of continental terraces. (Abstract). Bull. G.S.A. 62:501.

Continental terrace
Piermann, Günter, 1969.
Le précontinent entre le Cap Ferrat et le Cap Martin (Alpes maritimes, France)
Bull. Inst. océanogr., Monaco 68 (1392): 20 pp.

continental terrace
Gill, Edmund D., 1970.
Coast and continental shelf of Australia.
Quaternaria 12: 115-128.

continental terrace
Holtedahl, H., 1956.
On the Norwegian continental terrace, primarily outside Møre-Romsdal: its geomorphology and sediments. With contributions on the Quarternary geology of the adjacent land and on the bottom deposits of the Norwegian Sea.
Arbok, Univ. Bergen, Natur. Rekke, No. 14:1-209.

continental terrace
Kuenen, Ph. H., 1950.
The formation of the continental terrace.
Adv. Sci. 7(25):76-80, 3 textfigs.

continental terrace
Uchupi, Elazar, 1970.
Atlantic continental shelf and slope of the United States - shallow structure. Prof. Pap. U.S. Geol. Surv., 529-I: I1-I44.

continental terrace
Uchupi, Elazar, 1968.
Sedimentary frame work of the continental terrace off the east coast of the United States.
SEast Geol., Duke Univ., 9(4):269-271.

continental terraces
Wang, Chao-Siang, 1968.
Continental terrace: its initiation, growth, and destruction.
Acta geol. Taiwanica, 12:59-74.

Continental terrace
Weeks, L. Austin, and Robert K Lattimore, 1971.
Continental terrace and deep plain offshore central California.
Mar. geophys. Res. 1(2): 145-161.

Continuity equation

continuity equation
O'Connor, Donald J., and Dominic M. DiToro 1965
The solution of the continuity equation in cylindrical coordinates with dispersion and advection for an instantaneous release.
Symposium: Diffusion in oceans and fresh waters, Lamont Geol. Obs. 31 Aug.-2 Sept 1964: 80-85.

Continuous plankton recorder

See also: instrumentation, biological (plankton recorder), etc.

Continuous plankton recorder
Glover, R.S., 1967.
The continuous plankton recorder survey of the North Atlantic. In: Aspects of Marine Zoology, N.B. Marshall, editor, Symp. Zool. Soc., Lond., 19: 189-210.

Controls

controls
Carter, Luther J., 1968.
Deep seabed: "Who should control it," U.N. asks.
Science, 159(3810):66-68.

Convection

convection
Arakawa, H., 1960
Effect of rotation on convective motion.
Pap. Meteor. & Geophys., 11(2/4):191-195.

convection
Bourkov, V.A., and Yu. V. Pavlova, 1963
Geostrophic circulation on the surface of the northern part of the Pacific in summer. (In Russian; English abstract).
Okeanol. Issled., Rezhult. Issled., Programme Mezhd. Geofiz. Goda, Mezhd. Geofiz. Komitet, Presidiume Akad. Nauk, SSSR, No. 9:21-31.

convection
Brekhovskikh, Academician L.M., K.N. Fedorov, L.M. Fomin, M.N. Koshlyakov and A.D. Yampolsky, 1971.
Large-scale multi-buoy experiment in the tropical Atlantic. Deep-Sea Res. 18(12): 1189-1206.

convection
Bulgakov, N.P., 1962.
The investigation of convection and the process of autumn cooling in the sea.
Trudy Inst. Okeanol., Akad. Nauk, SSSR, 60:3-95.
In Russian; English summary

convection
Defant, A., 1949.
Konvektion und Eisbereitschaft in Polaren Schelfmeeren. Geografiska Annaler, Årg. 31(1/4):25-35, 4 textfigs.

convection
Fedorov, K.N., 1971.
A case of convection with occurrence of temperature inversion in connection with local instability in oceanic thermal wedge. (In Russian)
Dokl. Akad. Nauk SSSR 198(4): 822-825

convection
Filippov, D.M., S.E. Navrotskaya and Z.N. Matveeva, 1968.
On the depth of the autumn convection. (In Russian; English abstract).
Okeanologiia, Akad. Nauk, SSSR, 8(1):26-37.

convective circulation
Fofonoff, N.P., 1962
Sect. III. Dynamics of Ocean Currents. In: The Sea, Interscience Publishers, Vol 1, Physical Oceanography, 323-395.

convection
Foster, Theodore D., 1971.
A convective model for the diurnal cycle in the upper ocean. J. geophys. Res., 76(3): 666-675.

convection
Gill, A.E., and C.C. Kirkham, 1970.
A note on the stability of convection in a vertical slot.
J. fluid Mech., 42(1):125-127.

convection
Harleman, Donald R.F., and Arthur T. Ippen, 1961 ?
The turbulent diffusion and convection of saline water in an idealized estuary.
I.A.S.H., Comm. of Surface Waters, Publ., No. 51:362-378.

convection
Ichiye, T, 1958.
On convection circulation and density distribution in a zonally uniform ocean.
The Ocean. Mag. Vol. 10: No. 1, pp 97-136.
Jap. Meteor. Agency

convection
Koshljakov, M.N., 1960
Investigation of wind and convective zonal currents with application to the tropical part of the Pacific.
Trudy Inst. Okeanol., 40: 142-151.

convection
Kraus, Eric B., 1967.
Organized convection in the ocean surface layer resulting from slick and wave radiation stress.
Physics Fluids, 10(9-2):S294-S297.

convection
Kuenen, Ph. H., 1968.
Settling convection and grain-size analysis.
J. sedim. Petrol. 38(3): 817-831

convection, bottom
*Leontieva, V.V., 1968.
Some properties of the lower convective layer in the Pacific trenches. (In Russian; English abstract).
Okeanologiia, Akad. Nauk, SSSR, 8(5):807-813.

convection
Mortimer, C.H., and F.J.H. Mackereth, 1958.
Convection and its consequences in ice-covered lakes. Verh. Int. Ver. Limnol., 13:923-932.

convection
Nakagawa, Y., 1957.
Apparatus for studying convection under the simultaneous action of a magnetic field and rotation. Rev. Sci. Instr. 28(8):603-609.

convection

Negliad, K.V., 1965.
About the convection depth calculation in the seas with high water temperature. (In Russian).
Meteorol. i Gidrol., (3):31-34.

convection

Natarov, V.V., 1963
On the water masses and currents of the Bering Sea (In Russian).
Sovetsk. Ribokh. Issled. B Severo-Vostokh. Chasti Tikhogo Okeana, VNIRO 48, TINRO 50(1): 111-134.

convection

Novozhilov, N.I., 1960

[On the role of dynamic turbulence in the development of convection] Meteorol. i Gidrol., 10: 27-28.

convection

*Palm, E., T.Ellingsen and B. Gjevik, 1967.
On the occurrence of cellular motion in Bénard convection.
J. fluid Mech., 30(4):651-661.

convection

Plakhin, E.A., 1971.
Formation of deep water properties in the Mediterranean Sea under convective mixing. (In Russian; English abstract). Okeanologiia 11(4): 622-628.

convection

Ribaud, G., 1957.
Convection laminaire et convection turbulente.
La Houille Blanche, 12(1):12-18.

convection

Ryzhkov, Iu. G., and L.A. Koveshnikov, 1963.
Upwelling zone of convection and divergence of currents in regions with sudden changes of sloping bottom. (In Russian).
Izv., Akad. Nauk, SSSR, Ser. Geofiz., (6):953-959.

convection

Saint-Guily, B., 1959.

Note sur l'action de la force de Coriolis dans la circulation convective.
Bull. de l'Inst. Ocean., 1148: 7 pp.

convection

Schiff, L.I., 1966.
Lateral boundary mixing in a simple model of ocean convection.
Deep-Sea Res., 13(4):621-626.

convection

*Somerville, Richard C.J., 1967.
A non-linear spectral model of convection in a fluid unevenly heated from below.
J. atmos. Sci., 24(6):665-676.

convection

Spangenberg, W.G., & W.R. Rowland, 1961

Convective circulation in water induced by evaporative cooling.
Physics of Fluids, 4(6): 743-750.

convection

*Stern, Melvin E. and J. Stewart Turner, 1969.
Salt fingers and convecting layers. Deep-Sea Res., 16(5): 497-511.

convection

Stewart, R.W., 1969.
The atmosphere and the ocean.
Scient. Am., 221(3):76-86. (popular).

convection

Stommel, H., and G. Veronis, 1957.
Steady convective motion in a horizontal layer of fluid heated uniformly from above and cooled non-uniformly from below. Tellus 9(3):401-407.

convection

Stommel, Henry, Arthur Voorhis and Douglas Webb 1971.
Submarine clouds in the deep ocean.
Am. Scient. 59(6): 716-722.

convection

Takano, K., 1958.
Sur la circulation générale convective dans les océans. 1. Partie.
Bull. Inst. Océan., Monaco, 55 (No. 1128):15 pp.

convection

Takano, K., 1957.
Note on the convective circulation.
Proc. UNESCO Symp., Phys. Ocean., Tokyo, 1955: 138.

convection

Tareev, B.A., 1964.
The vertical convection in some regions of the Indian Ocean. Investigations in the Indian Ocean (33rd voyage of the E/S "Vitiaz"). (In Russian).
Trudy Inst. Okeanol., Akad. Nauk, SSSR, 64:59-62.

convection

Tareev, B.A., 1960.
[Contribution to a theory of convectional circulation in oceanic deeps.]
Izv., Akad. Nauk, SSSR, Ser. Geofiz., (7):1022-1029.

translation:
Eng. Ed., Pergamon Press, (7):683-687.

convection

Tareyev, B.A., 1959.
[On free convection in the deep water troughs of the oceans.]
Doklady Akad. Nauk, SSSR, 127(5):1005-1008.

Listed in: Tech. Transl., 3(8):499.

convection

Thomas, D.B., and A.A. Townsend, 1957.
Turbulent convection over a heated horizontal surface.
J. Fluid. Mech., 2(5):473-492.

convection

Townsend, A.A., 1964.
Natural convection in water over an ice surface.
Q.J. R. Meteorol. Soc., 90(385):248-259.

convection

Tsikunov, V.A., 1958.
Simplified theory of convectional mixing in the upper layers of the sea. (In Russian)
Trudy Gosud. Okeanogr. Inst., 42:115-127.

convection

Turner, J.S., and Henry Stommel, 1964.
A new case of convection in the presence of combined vertical salinity and temperature gradients.
Proc. Nat. Acad. Sci., 52(1):49-53.

convection (bottom)

Vladimirtsev, Yu. A., 1964.
Study of bottom convection in the ocean (review of Soviet and foreign investigations). (In Russian).
Okeanologiia, 4(4):553-563.

Abstract in:
Soviet Bloc Res. Geophys., Astron., Space, No. 95:49.

convection

Vladimirtsev, Yu. A., 1962.
On near-bottom convection in the Black Sea. (In Russian).
Izv., Akad. Nauk, SSSR, Ser. Geofiz., (7):974-977.

Engl. Transl., (7):621-623.

convection

Vladimirtsev, Yu. A., and A.N. Kosarev, 1963
Some peculiar features of convective mixing in the Black and Caspian seas. (In Russian).
Okeanologiia, Akad. Nauk, SSSR, 3(6):979-985.

convection

Woodcock, Alfred H., 1965.
Melt patterns in ice over shallow waters.
Limnol. and Oceanogr., Redfield Vol., Suppl. to 10:R290-R297.

Convection

Woodcock, A.H., 1940.
Convection and soaring over the open sea. J. Mar. Res. 3(3):248-253, Textfig. 60.

convection (air)

Woodcock, A.H., and J. Wyman, 1947.
Convective motion in air over the sea. Ann. N.Y. Acad. Sci. 48(8):749-776, 7 textfigs., 9 pls.

convection

Wulfson, N. I., 1958.

Statistical methods for determining the actual parameters of convection-currents from the observed parameters.
Bull. (Izv.) Acad. Sci. USSR, Geoph. Ser. 7 (Eng. Ed.): 498-506.

convection

Zakharov, V.F., 1966.
The role of leads off the edge of shore ice in the hydrological and ice regime of the Laptev Sea. (In Russian; English abstract).
Okeanologiia, Akad. Nauk, SSSR, 6(6):1014-1022.

convection

Zalogin, B.S., 1963.
Fall convection under real conditions. (In Russian).
Vestnik Moskovsk. Univ., Moscow, Ser. 5, Geogr., (3):39-44.

Abstract:
Soviet Bloc Res. Geophys. Astron., Space, 1963, (64):14.

Convection, Bénard

convection Bénard

* Rossby, H.T., 1969.
A study of Bénard convection with and without rotation.
J. fluid. Mech. 36(2): 309-335.

Convection Bénard

Schneck Paul and George Veronis 1967.
Comparison of some recent experimental and numerical results in Bénard convection.
Physics of Fluids 10(5): 927-930.

Convection, Bénard

Somerville, Richard C.J., 1971.
Bénard convection in a rotating fluid.
Geophys. fluid Dynam. 2(3): 247-262

convection, Benard

*Veronis, George, 1968.
Large-amplitude Bénard convection in a rotating fluid.
J. fluid Mech. 31(1):113-139.

Convection, Bénard

Veronis, George 1966.
Large-amplitude Bénard convection.
J. fluid Mech. 26(1):49-68.

Convection cells

convection, cellular

*Busse, F.H., 1967.
The stability of finite amplitude cellular convection and its relation to an extremum principle.
J. fluid Mech. 30(4):625-649.

convection cells

*Chen, Michael M., and John A. Whitehead, 1968.
Evolution of two-dimensional periodic Ray Leigh convection cells of arbitrary wave-numbers.
J. fluid Mech., 31(1):1-15.

convection cells

Munk, W. H., 1948.
Effect of earth's rotation on convection cells.
Ann. N. Y. Acad. Sci. 48(8):815-820.

convection cells

Munk, W.H., 1947.
Effect of earth's rotation upon convection cells
Ann. N.Y. Acad. Sci. 48(8):815-820, 3 textfigs.

convection cells

Neumann, G., 1948.
Bemerkungen zur Zellularkonvektion im Meer und in der Atmosphäre und die Beurteilung des statischen Gelichgewichts. Ann. Met. Juli/Aug. 1948L235-244, 5 textfigs.

convection, cellular

Nikitin, D.P. and B.A. Taryeev 1971
Cellular convection on the rotating earth near the equator. (In Russian; English abstract).
Fizika Atmosfer. Okean. Izv. Akad Nauk, SSSR 7(5):534-541.

convection cells

Stommel, H., 1949.
Trajectories of small bodies sinking slowly through convection cells. J. Mar. Res. 8(1): 24-29, 4 textfigs.

convection cells

Stommel, H., 1947
A summary of the theory of convection cells.
Ann. N.Y. Acad. Sci. 48:715-726.

convective cells

Takano, K., 1955.
Sur la dimension des tourbillons cellulaires dans les expériences de Henri Benard.
Rec. Ocean. Wks., Japan, 2(1):80-86.

Convective cells

Turcotte, D.L. and E.R. Oxburgh 1967.
Finite amplitude convective cells and continental drift.
J. fluid Mech. 28(1):29-42.

convection cells, effect of

LaFond, E.C., 1954.
Environmental factors effecting the vertical temperature structure of the upper layers of the sea. Andhra Univ. Ocean. Mem. 1:94-101, 25 textfigs.

Convection currents

convection currents

Miyazaki, M., 1948
On the computation of the stationary currents. Geophys. Mag. 15(2/4):58-61, 2 tables.

convection, haline

Foster, Theodore D., 1969.
Experiments on haline convection induced by the freezing of sea water. J. geophys. Res., 74(28): 6967-6974.

convection (haline)

*Foster, Theodore D., 1968.
Haline convection induced by the freezing of sea water.
J. geophys. Res., 73(6):1933-1938.

convection, intermittent

Foster, Theodore D., 1971.
Intermittent convection.
Geophys. fluid Dynam. 2(3): 201-217

Convection layer

convection layer

Glazkov, V.V., 1970.
A volumetric statistical T, S-analysis of the Black Sea water masses. (In Russian; English abstract). Okeanologiia, 10(6): 958-962.

convection, layers

Montgomery, R. B., 1947.
Introduction: problems concerning convective layers. Ann. N.Y. Acad. Sci. 48:707.

"convection layer"

Okubo, A., 1953.
Note on the growth and temperature variation of the "convection layer". Ocean. Mag., Tokyo, 5(1): 15-22, 4 textfigs.

convection (lateral)

Fedorov, K.N., 1971.
The new evidence of the lateral convection in the ocean. (In Russian; English abstract).
Okeanologiia 11(6): 994-998.

Convection, mantle

Convection, mantle

Ewing, M., T. Saito and X. LePichon 1967.
Reply to 'Comments on mantle convection and mid ocean ridges' by Peter R. Vogt and Ned A. Ostenso.
J. geophys. Res. 72(8): 2085.

Convection, mantle

Oxburgh, E.R., and D.L. Turcotte, 1968.
Mid-ocean ridges and geotherm distribution during mantle convection.
J. geophys. Res., 73(8):2643-2661.

convection, mantle

Vogt, Peter R., and Ned A. Ostenso, 1967.
Comments on mantle convection and mid-ocean ridges.
J. geophys. Res., 72(8):2077-2084.

convection, micro-

Chernous'ko, Yu.L. and A.V. Shumilov 1971.
Evaporation and microconvection in the thin surface layer. (In Russian; English abstract).
Okeanologiia 11(6): 982-986.

Convection, near-bottom

convection, near-bottom

*Filippov, D.M., and V.D. Egorikhin, 1967.
Potential temperature as an indicator of the near-bottom convection. (In Russian; English abstract).
Okeanologiia Akad. Nauk SSSR, 7(4):571-576.

convection, near bottom

Phillipov, D.M., 1966.
Near-bottom convection in the deep ocean.
(In Russian).
Fisika Atmosferi i Okeana, Izv., Akad. Nauk, SSSR, 2(6):668-671.

Convection thermal

convection, thermal

Montgomery, R.B., 1954.
Convection of heat. Arch. Met., Geophys. u. Bioklim., A, 7:125-132.

convection, thermal

Niiler, P.P., and F.E. Bisshopp, 1965.
On the influence of Coriolis force on onset of thermal convection.
J. Fluid Mech., 22(4):753-761.

convection, thermal

Saint-Guily, B., 1963.
On vertical heat convection and diffusion in the South Atlantic.
Deutsche Hydrogr. Zeits., 16(6):263-268.

convection, thermal

Stommel, H., 1950.
An example of thermal convection. Trans. Am. Geophys. Union 31(4):553-554, 1 textfig.

convection, thermal

Veneziani, Giulio, 1969.
Effect of modulation on the onset of thermal convection.
J. fluid Mech., 35(2): 243-254.

Convection thermohaline

convection, thermohaline

Lindberg, William R., 1971.
An upper bound on transport processes in turbulent thermohaline convection. J. phys. Oceanogr. 1(3): 187-195.

convection, thermohaline

Stommel, H., 1961
Thermohaline convection with two stable regimes of flow.
Tellus, 13(2):224-230.

convection, thermohaline

Thorpe, S.A., P.K. Hutt and R. Soulsby, 1969.
The effect of horizontal gradients on thermohaline convection.
J. fluid Mech. 38(2): 375-400.

convection, thermohaline

Veronis, George, 1965.
On finite amplitude instability in thermohaline convection.
J. Mar. Res., 23(1):1-17.

convection, transient (deep)

Schwartzlose, Richard A. and John D. Isaacs 1969
Transient circulation event near the deep ocean floor.
Science 165 (3896): 889-891.

Convection turbulent

convection, turbulent

Baines, W.D., and J.S. Turner, 1969.
Turbulent buoyant convection from a source in a confined region.
J. fluid Mech. 37 (1): 51-80

convection, vertical

convection, vertical

Krug, Joachim, 1963
Erneuerung des Wassers in der Kieler Bucht im Verlaufe eines Jahres am Beispiel 1960/61.
Kieler Meeresforsch., 19(2):158-174.

Conventions

conventions

Dean, Arthur H., 1967.
The Law of the Sea Conference, 1958-60, and its aftermath.
In: The law of the sea, L.M. Alexander, editor, 244-264.

conventions

Reiff, Henry, Ralph Johnson, L.F.E. Goldie, John Mero and Alexander Melamid, 1967.
A symposium on the Geneva conventions and the need for future modifications.
In: The law of the sea, L.M. Alexander, editor, 265-298.

Convergences

convergences

Allain, C., 1964.
Les poissons et les courants.
Rev. Trav. Inst. Pêches Marit., 28(4):401-426.

convergences

Babkov, A.I., 1964
Convergences at sea in the geographical aspect and their study by aeromethods. (In Russian).
Izv. Vses. Geograf. Obshch., 96(4):329-331.

convergence

Bogdanov, D.V., 1965.
Algunos rasgos de la oceanografía del Golfo de México y del Mar Caribe. (In Russian; Spanish abstract).
Sovetsk.-Cub. Ribokhoz. Issled., VNIRO:Tsentr. Ribokhoz. Issled. Natsional. Inst. Ribolovsta Republ. Cuba, 23-45.

convergence

*Boltovskoy, Esteban, 1968.
Hidrologia de las aguas superficiales en la parte occidental del Atlnatico Sur.
Revta Mus. argent. Cienc. nat. Bernardino Rivadavia Inst. nac. Invest. Cienc. nat., Hidrobiol. 2(6):199-224.

Convergences

Boltovskoy, Esteban 1965
Datos nuevos con respecto a la ubicacion de la zona de convergencia subtropica/subantartica en base al estudio de los foraminiferos planctonicos.
Anais Acad. bras. Cienc. 37 (Suppl): 146-155.

convergence

Botnikov, V.N., 1964.
Seasonal and lon-term fluctuations of the Antarctic Convergence zone. (In Russian).
Sovetsk. Antarkt. Ekspad., Inform. Biull., 45:17-

Translation:
Scripta Tecnica, Inc., for AGU, 5(2):92-95. 1965.

convergence

Brodie, J.W., 1965.
Oceanography.
In: Antarctica, Trevor Hatherton, editor, Methuen & Co., Ltd., 101-127.

convergences

Brooks, E.M., 1939.
Transport and convergence of the North Atlantic drift current computed from the average January pressure distribution. J. Mar. Res. 2(2):163-167, Textfig. 52, 1 table.

convergence

Bryan, Kirk, and Solomon Hellerman, 1966.
A test of convergence of a numerical calculation of the wind-driven ocean circulation.
J. Atmosph. Sci., 3(3):360-361.

convergences (subtropical)

Burling, R.W., 1961.
Hydrology of circumpolar waters south of New Zealand.
New Zealand Dept. Sci. Industr. Res. Bull., 143: 66 pp.

convergence

Burling, R. W., and D. M. Garner, 1959.
A section of 14C activities of sea water between 9°S and 66°S in the south-west Pacific Ocean.
N.Z. J. Geol. Geophys., 2(4):799-824.

convergence

Casellos, Alberto O., and Vicente J. Zubillaga, 1960
Algunos resultados de las actividades oceanograficas Argentinas en el Atlantico.
Contribucion del Inst. Antartico Argentino. No. 46: 1-22.

convergence

Central Meteorological Observatory, 1949.
Report on sea and weather observations on Antarctic whaling ground (1948-1949). Ocean. Mag., Tokyo, 1(3):142-173, 11 textfigs.

convergence

Chew, F., 1955.
On the offshore circulation and a convergence mechanism in the red tide region of the west coast of Florida. Trans. Amer. Geophys. Union, 36(6):963-974.

convergence

Crease, J., 1964.
The Antarctic Circumpolar Current and convergence.
Proc. R. Soc., London, (A), 281(1384):14-21.

convergence of organisms

Crisp, D.J., 1962
Swarming of planktonic organisms.
Nature, 193(4815):597-598.

convergences

Darbyshire, Mollie, 1966.
The surface waters near the coasts of southern Africa.
Deep-Sea Research, 13(1):57-81.

Convergence

Deacon, G.E.R. 1965.
The Southern Ocean and the convergence.
Anais Acad. bras. Cienc. 37 (Suppl): 23-29

convergence

Deacon, G.E.R., 1960
The southern cold temperate zone. Proc. Roy. Soc., Ser. B., 152(949): 441-447.

convergence

Deacon, G.E.R., 1959
The Antarctic ocean.
Science Progress, 47(188): 647-660.

convergence

Deacon, G.E.R., 1952.
Surface boundaries in the Southern Ocean.
Proc. Seventh Pacific Sci. Congr., Met. Ocean., 3:305-306.

convergences

Deacon, G. E. R., 1933
A general account of the hydrology of the South Atlantic Ocean. Discovery Repts. 7:173-238, pls.8-10.

convergences

*Duncan, C.P., 1968.
An eddy in the subtropical convergence southwest of South Africa.
J. geophys. Res., 73(21):531-534.

convergence

Garner, D. M., 1959
The sub-tropical convergence in New Zealand surface waters.
J. Geol. & Geophysics, 2 (2): 315-337.

convergence

Garner, D.M., 1958.
The Antarctic convergence south of New Zealand.
N.Z.J. Geol. Geophys., 1(3):577-594.

convergence

Goedecke, Erich, 1965
Über die hydrographische Struktur der Deutschen Bucht im Hinblick auf die Verschmutzung in der Konvergenzzone.
Helgoländer wiss. Meeresunters. 7: 105-125.

convergence

Gorbenko, Yu. A., 1964.
On the possibility of revealing the areas of water rise and water lowering in the central part of the Pacific by applying the microbiological method. (In Russian).
Okeanologiia, Akad. Nauk, SSSR, 4(1):167-174.

convergence

Hasle, Grethe Rytter, 1969.
An analysis of the phytoplankton of the Pacific Southern Ocean: abundance, composition, and distribution during the Brategg Expedition, 1947-1948.
Hvalråd. Skr. 52: 168 pp.

convergence, subtropical

*Heath, R.A., 1968.
Geostropic currents derived from oceanic density measurements north and south of the subtropical convergence east of New Zealand.
N.Z. Jl. mar. Freshwat. Res., 2(4):659-677.

convergence

Hidaka, K., 1955.
3. Divergence of surface drift currents in terms of wind stresses, with special application to the location of upwelling and sinking.
Jap. J. Geophys. 1(2):47-56, textfig.

convergence
Hidaka, K., and T. Kusunoki, 1951.
On the mixing coefficient and the meridional component of velocity in the Equatorial Counter Current. J. Ocean. Soc., Japan, (Nippon Kaiyo Gakkaisi), 6(3):168-193, 4 tables, 5 textfigs.

convergence
Hikosaka, S., and R. Watanabe, 1957.
Areas of divergence and convergence of surface currents, in the north-western Pacific.
Proc. UNESCO Symp., Phys. Ocean., Tokyo, 1955: 101-103.

Convergence
Houtman, Th. J. 1967.
Water masses and fronts in the Southern Ocean south of New Zealand.
Bull. N.Z. Dept. Sci Ind. Res. 174: 1-40.

convergences
Houtman, Th. J., 1964.
Surface temperature gradients at the Antarctic Convergence.
N.Z. J. Geol. Geophys., 7(2):245-270.

convergence
India, Indian National Committee on Oceanic Research, Council of Scientific and Industrial Research, New Delhi, 1964.
International Indian Ocean Expedition, Newsletter India, 2(4):48 pp.

convergence
Istapoff, Feodor, 1962
The salinity distribution at 200 meters and the Antarctic frontal zones.
Deutsche Hydrographische Zeits., 15(4): 133-142.

convergences
Ivanenkov, V.N., and F.A. Gebin, 1960.
Water masses and hydrochemistry in the western and southern parts of the Indian Ocean. Physics of the Sea. Hydrology. (In Russian).
Trudy Morsk. Gidrofiz. Inst., 22:33-115.

Translation: Scripta Technica, Inc. for Amer. Geophys. Union, 27-99.

convergence, subarctic
Kawai, H., 1955.
On the polar frontal zone and its fluctuations in the waters to the northeast of Japan. 1.
Bull. Tohoku Reg. Fish. Res. Lab., No. 4:46 pp.

Convergence
Kawai, H. and Hisao Sakamoto 1969.
A study on convergence and divergence in surface layer of the Kuroshio - II. Direct measurement of convergence and divergence at the top and bottom of surface mixed layer. (In Japanese; English abstract). Bull. Nansei reg. Fish. Res. Lab., 2: 19-38.

convergences
Knox, G.A., 1960.
Littoral ecology and biogeography of the southern oceans.
Proc. R. Soc., London, (B), 152(949):577-624.

Convergence
Laevastu, T., 1962
The components of the surface currents in the sea and their forecasts.
Proc. Symposium on Mathematical-Hydrodynamical Methods of Phys. Oceanogr., Sept. 1961, Inst. Meeresk., Hamburg, 321-338.

convergence, effect of
LaFond, E.C., 1954.
Environmental factors affecting the vertical temperature structure of the upper layers of the sea. Andhra Univ. Ocean. Mem. 1:94-101, 25 textfigs.

convergence
Lane, R.K., 1962
A review of the temperature and salinity structures in the approaches to Vancouver Island, British Columbia.
J. Fish. Res. Bd., Canada, 19(2):45-91.

convergence
Lebedeva, M.N., 1962
Microbiological indication of the convergence and divergence zones in the Indian Ocean. (In Russian).
Okeanologiia, Akad. Nauk, SSSR, 2(6):1104-1109.

convergence
Lusquinos, Andres Jorge, 1966.
Determinación de la convergencia subtropical sobre la base de los datos de temperatura y salinidad obtenidos en las campañas Tridente I of II.
Bol. Servicio Hidrogr. Naval, Publ., Argentina, H. 106:79-97.

convergence
Mackintosh, N. A., 1946.
The Antarctic convergence and the distribution of surface temperature in Antarctic waters. Discovery Repts. 23:177-212.

convergence
*Mazlika, P.A., 1968.
Some features of the Gulf Stream off Chesapeake Bay in the spring of 1963.
Fishery Bull. Fish Wildl. Serv. U.S., 66(2):387-423.

convergence
Midttun, L., and J. Natvig, 1957.
Pacific Antarctic waters. Scientific results of the "Brategg" Expedition, 1947-48, No. 3.
Publikasjon, Komm. Chr. Christensens Hvalfangstmus., Sandefjord, No. 20:1-130, 39 figs.

convergence
Mratov, K.A., 1971.
Water upwelling and sinking zones of the Atlantic near West Africa. (In Russian; English abstract). Okeanol. Issled. Rezult. Issled. Mezhd. Geofiz. Proekt. 21: 97-106.

convergence
Namias, Jerome, 1963.
Large-scale air-sea interactions over the North Pacific from summer 1962 though a subsequent winter.
J. Geophys. Res., 68(22):6171-6186.

convergence
Neumann, Gerhard, 1965.
Oceanography of the tropical Atlantic.
Anais Acad. bras. Cienc., 37 (Supl.):63-82.

convergence
Oceanographic Section. C.M.O., 1949.
Report on sea and weather observations on Antarctic Whaling Ground (1948-1949). Ocean. Mag., Tokyo, 1(3):142-173, 11 textfigs.

convergence
Owen, Robert W., Jr.,
Small-scale, horizontal vortices in the surface layer of the sea.
J. Mar. Res., 24(1):56-66.

convergences
*Pizarro, Mariano Javier, 1967.
Distribución del oxígeno disuelto en la zona oeste de la convergencia subtropical del Atlantico sud.
Bolm. Inst. oceanogr. S. Paulo, 16(1):67-85.

convergence
Pomeroy, A.S., 1965.
Notes on the physical oceanographic environment of the Republic of South Africa.
Naval Oceanogr. Res. Unit, 41 pp. (Unpublished manuscript).

convergence
Pritchard, D.W., and E.C. LaFond, 1952.
Some recent temperature sections across the Antarctic Convergence. Proc. Seventh Pacific Sci. Congr., Met., Ocean., 3:290-297, 5 textfigs

convergence
Rotschi, Henri, 1963
Sur les flux d'eau en mer Corail entre Nouvelle-Caledonie et Norfolk.
C.R. Acad. Sci., Paris, 256:2461-2464.

convergence
Rotschi, H., and L. Lemasson, 1967.
Oceanography of the Coral and Tasman seas.
Oceanogr. Mar. Biol. Ann. Rev., Harold Barnes, editor, George Allen and Unwin, Ltd., 5:49-97.

convergence, subtropical
Russell, G.F., 1950.
Surface temperatures in the S.W. Pacific and Tasman Sea. N.Z. Dept. Sci. Ind. Res., Geophys. Conf., 1950, Paper No. 13:2 pp., 3 figs. (mimeographed).

convergences
Ryzhkov, Yu. G., and L.A. Koveshnikov, 1963.
Genesis of convergence and divergence zones above sharp changes in the slope of the ocean floor. (In Russian).
Izv., Akad. Nauk, SSSR, Ser. Geofiz. (6):953-959.

Translation (AGU), (6):585-588.

Convergence
Schooley, A. H. 1969.
Convergence and strain waves caused by a submerged turbulent disturbance in stratified fluids
Science 164 (3886): 1393-1394.

convergences
Schumacher, A., 1949.
Über das subtropische Konvergenzgebeit im Südatlantischen Ozean. Mecking-Festschrift, Hannover :41-49, 2 pls.

convergence
Schumacher, A., 1946.
Ueber das subtropische Konvergenzengebeit im sudatlantischen Ozean. G.H.I. Rept. No. 1.

convergence
Semonov, A.I., 1961.
Estimate of the influence of the convergence-divergence phenomenon on the height of the average-year level (according to the example of the Azov Sea).
Trudy Gosud. Okean. Inst., Leningrad, 61:116-132.

In Russian

convergences
Stepanov, V.N., 1960
[The main water convergences and divergences in oceans.]
Biull. Okeanograf. Komissii. Akad. Nauk, SSSR, (6): 15-22.

convergence
Sudo, Hideo, 1960
On the distribution of divergence and convergence of surface drift vectors in the western Pacific Oceans. Rec. Oceanogr. Wks., Japan, n.s., 5(2): 25-43.

convergences
Tchernia, P., 1951.
Compte-rendu preliminaire des observations océanographiques faites par le Batiment Polaire "Commandant Charcot" pendant la campagne 1949-1950. Bull. d'Info., C.O.E.C., 3(1):13-22, 11 figs.

convergence
Uda, M., 1955.
On the subtropical convergence and the currents in the northwestern Pacific.
Rec. Ocean. Wks., Japan, 2(1):141-150, 10 textfigs.

convergence
Uda, M., 1953.
On the convergence and divergence in the NW Pacific in relation to the fishing grounds and productivity. Bull. Japan. Soc. Sci. Fish. 19(4):435-438, 3 textfigs.

convergence
Uda, M., N. Watanabe, and M. Ishino, 1956.
General results of the oceanographic surveys (1952-1955) on the fishing grounds in relation to the scattering layer. J. Tokyo Univ. Fish., 42(2):169-207.

convergence
Vallaux, C., 1938.
Recherches du Discovery II sur la dynamique de l'Océan Austral. Bull. Inst. Océan., Monaco, 751:1-15.

convergence
Varadachari, V.V.R., and G.S. Sharma, 1964.
On the vergence field in the north Indian Ocean. Bull. Nat. Geophys. Res. Inst., New Delhi, 2(1):1-14.

Vazzoler, A. E. A. de M., 1963. convergences
Deslocamentos sazonais da corvina relacionados com as massas de água.
Contrib. Avulsas, Inst. Oceanogr., Univ. São Paulo, Oceanogr. Biol. No. 5:4 pp., 4 pls.

convergence
Vladimirov, O.A., and E.A. Nikulicheva, 1960.
Survey of some contemporary methods for computing "convergence-divergence" phenomenon and tests of its application in understanding of the Finnish Bays. (In Russian).
Trudy Gosud. Okean. Inst., Leningrad, 56:9-29.

convergence
Wüst, W., 1950.
Blockdiagramme der atlantischen Zirculation auf Grund der "Meteor"* Ergebnisse. Kieler Meeresf. 7(1):24-34, 3 textfigs., 1 fold-in.

convergence
Wyrtki, Klaus, 1961
The thermohaline circulation in relation to the general circulation in the oceans.
Deep-Sea Res., 8(1): 39-64.

convergence
Wyrtki, Klaus, 1960
The Antarctic Circumpolar Current and the Antarctic Polar Front. Deutsche Hydrogr. Zeits., 13(4): 153-174.

convergences
Wyrtki, Klaus, 1960
The Antarctic convergence and divergence.
Nature, 187(4737): 581-582.

convergences
Wyrtki, Klaus, 1960
The surface circulation in the Coral and Tasman seas.
C.S.I.R.O., Australia, Div. Fish. Oceanogr. Tech. Paper, (8): 1-44.

Convergence effect of

convergence, effect of
Osborne, M.F.M., 1965.
The effect of convergent and divergent flow patterns on infrared and optical radiation from the sea.
Deutsche Hydrogr. Zeits., 18(1):1-25.

convergence, effect of
*Semina, H.J., 1968.
Water movement and the size of phytoplankton cells.
Sarsia, 34:267-272.

Convergence, inter-tropical

convergence, intertropical
Johnson, H. McClure, 1965.
Intertropical convergence zone conditions as observed from research ships and aircraft and TIROS satellites during EQUALANT I. (Abstract).
Ocean Sci. and Ocean Eng., Mar. Techn. Soc., Amer. Soc. Limnol. Oceanogr., 1:504.

convergence, subtropical
Yoshida, Kozo, and Toshiko Kidokoro, 1967.
A subtropical counter-current in the North Pacific- an eastward flow near the subtropical convergence.
J. oceanogr. Soc., Japan, 23(2):88-91.

Convergence, sound

convergence, sound
Leroy, C.C., 1967.
Sound propagation in the Mediterranean Sea.
In: Underwater Acoustics, V.M. Albers, editor, Plenum Press, 2: 203-241.

convergence zones, sound
Urick, R.J., 1965.
Caustics and convergence zones in deep-water sound transmission.
J. Acoustic Soc., Amer., 38(2):348-358.

CONVERGENCE, sound
Urick, R.J., and G.R. Lund, 1968.
Coherence of convergence zone sound.
J. acoust. Soc. Am., 43(4): 723-729.

Convergences tropical

convergences, tropical
Stanton, B.R., 1969.
Hydrological observations across the Tropical Convergence north of New Zealand.
N.Z. Jl mar. Freshwat. Res. 3(1): 124-146.

Cooling

cooling
Bowden, K. F., 1948.
The process of heating and cooling in a section of the Irish Sea. Mon. Not. Roy. Astron. Soc., 5:270-281.

autumnal cooling
Bulgakov, N.P., 1962.
The investigation of convection and the process of autumn cooling in the sea.
Trudy Inst. Okeanol., Akad. Nauk, SSSR, 60:3-95.

cooling
Girdyuk, G.V., 1965.
On the role of the effective radiation during the processes of winter sea cooling. (In Russian)
Material Ribochoz. Issled. Severn. Bassin. Poliarn. Nauchno-Issled. i Proekt. Inst. Morsk. Rib. Choz. i Okeanogr., N.M. Knipovich, 5:123-127

cooling
Lineikin, P.S., 1951.
[The cooling of the surface layer of the sea.]
Dok. Akad. Nauk, SSSR, 80(2):205.

cooling
Nansen, F., 1912.
Das Bodenwasser und die Abkühlung des Meeres.
Int. Rev. Ges. Hydrobiol. u. Hydrogr., 5(1):1-42.

cooling (fresh water) (fr. lake)
Schmitz, H.P., 1953.
Über die Abkühlung von Binnenseen.
Acta Hydrophysica 1(1):7-30, 3 textfigs.

cooling
Tully, John P., 1964.
Oceanographic regions and assessment of temperature structure in the seasonal zone of the North Pacific.
J. Fish. Res. Bd., Canada, 21(5):941-970.

cooling
Tully, J.P., 1964.
Oceanographic regions and processes in the seasonal zone of the North Pacific Ocean.
In: Studies on Oceanography dedicated to Professor Hidaka in commemoration of his sixtieth birthday, 68-84.

Cooling, abnormal

cooling, abnormal
Kitani, Kozo, and Michitaka Uda, 1969
Variability of the deep cold water in the Japan Sea - particularly on the abnormal cooling in 1963.
J. oceanogr. Soc. Japan, 25(1): 10-20

cooling, effect of

cooling, effect of
Krüger, F., 1957.
Temperature-Regelung durch gesteuerte Kühlung.
Helgoland. Wiss. Meeresunters., 6(1):71-75.

Coprophagy

coprophagy
*Frankenberg, Dirk, and K.L. Smith, Jr., 1967.
Coprophagy in marine animals.
Limnol. Oceanogr., 12(3):443-450.

corals

corals
Boschma, H., 1962.
On milleporine corals from Brazil.
Konink. Nederl. Akad. Wetenschappen. Proc. (C) 65(4):302-312.

corals
Broecker, Wallace S., and David L. Thurber, 1965
Uranium-series dating of corals and oolites from Bahaman and Florida Key limestones.
Science, 149(3679):58-60.

corals
Burkholder, Paul R., Lillian M. Burkholder and Juan A. Rivero, 1959.
Chlorophyll A in some corals and marine plants.
Nature 183:1338-1339.

corals
Carpine, Christian, 1964.
Un Octocoralliaire nouveau pour la Méditerranée: Scleranthelia musiva Studer 1878.
Bull. Inst. Océanogr., Monaco, 64(1327):10 pp.

corals
Chevalier, J.-P., 1964
Compte Rendu des missions effectuées dans le Pacifique en 1960 et 1962 (Mission d'études des récifs coralliens de Nouvelle-Caledonie).
Cahiers du Pacifique, Mus. Nat. Hist. Nat., Paris (6):171-175.

corals
Cloud, P.E., jr., 1952.
12. Preliminary report on geology and marine environment of Onotoa Atoll, Gilbert Islands.
Atoll Res. Bull., N.R.C.:1-73 (mimeographed), 8 figs. (multilithed).

coral
Crooker, G. R., 1966.
Radiochemical data correlations and coral-surface bursts.
Nature, 210(5040):1028-1031.

corals
Eguchi, Motoki, 1964.
A study of Stylasterina from the Antarctic Sea.
Jap. Antarct. Res. Exped., 1956-62, Sci. Repts. (E), (JARE Sci. Repts., Biol.), No. 20:10 pp., 2 pls.

corals
George, V., 1955.
Quelques remarques concernant les côtes de Madagascar. Bull. d'Info., C.C.O.E.C. 7(10): 442-448, 4 pls.

corals, physiol.
Goreau, Thomas F., 1959
The physiology of skeleton formation in corals 1. A method for measuring the rate of calcium deposition by corals under different conditions
Biol. Bull., 116(1): 59-75.

corals
Goreau, T.F., and V.T. Bowen, 1955.
Calcium uptake by a coral. Science 122(3181): 1188-1189.

corals
Goreau, Thomas F., and Nora I. Goreau, 1960
The physiology of skeleton formation in corals IV. On isotopic equilibrium exchanges of calcium between corallum and environment in living and dead reef-building corals. Biol. Bull., 119(3): 416-427.

corals
Goreau, T. F., and N. I. Goreau, 1959.
The physiology of skeleton formation in corals. II. Calcium deposition by hermatypic corals under various conditions in the reef.
Biol. Bull., 117(2):239-250.

corals
Goreau, T. F. Goreau, N. I. Goreau and C.M. Yonge 1972
On the mode of boring in Fungiacava eilatensis (Bivalvia: Mytilidae).
J. Zool. Lond 166(1):55-60

corals
Hand, Cadet, 1956.
Are corals really herbivores? Ecology 37(2):384-385.

corals, anat.-physiol
Kanwisher, John W., and Stephen A. Wainwright, 1967.
Oxygen balance in some reef corals.
Biol.Bull.,mar.biol.Lab.,Woods Hole, 133(2):378-390.

corals
Kawaguti, S., and D. Sakumoto, 1948.
The effect of light on the calcium deposition of corals. Bull.Ocean. Inst., Taiwan, No. 4: 65-70.

corals
Kornicker, Louis S., and Donald F. Squires, 1962
Floating corals: a possible source of erroneous distribution data.
Limnol. and Oceanogr., 7(4):447-452.

corals
Lorch, J., 1965.
The history of theories on the nature of corals.
Colloque Internat., Hist. Biol. Mar., Banyuls-sur-Mer, 2-6 sept., 1963, Suppl., Vie et Milieu, No. 19:337-345.

corals
Ma, Tong Yong H., 1961.
Climate and the relative positions of continents during the Permian as deduced from the growth rate of corals.
Proc. Geol. Soc., China, (4):91-102.

corals
Ma, Ting Ying H., 1958
The effects of the warm and cold currents in the southwest Pacific on the growth rate of reef corals. (Abstract).
Proc. Ninth Pacific Sci. Congr., Pacific Sci. Assoc., 1957, 16(Oceanogr.):135.
Published in:
Oceanographica Sinica, 5(2), 1957.

corals
Ma, Ting Ying H., 1958.
The relation of growth rate of reef corals to surface temperature of sea water as basis for study of causes of diastrophisms instigating evolution of life.
Res. on Past Climate and Continental Drift 14: 60 pp.

Corals, deep
Montaggioni, Lucien, 1969.
Sur la présence de coraux profonds et de thanatocoenoses quaternaires dans l'archipel de Madère (Océan Atlantique).
C. r. hebd. Séanc. Acad. Sci., Paris (D), 268(26): 3160-3163.

corals
Nesteroff, Wladimir D., 1965.
Sur l'état de développement des madréporaires des récifs coralliens de Nouvelle Calédonie.
Comptes Rendus, Acad. Sci., Paris, 260(8):2278-2279.

corals
Odum, H.T., and E.P. Odum, 1956.
Corals as producers, herbivores, carnivores and possibly decomposers. Ecology 37(2):385.

corals
Osmond, J.K., J.R. Carpenter and H.L. Windom, 1965.
Th^{230}/U^{234} Age of the Pleistocene corals and oolites of Florida.
J. Geophys. Res., 70(8):1843-1847.

corals
Pasternak, F.A., 1959
On the finding of Bathypathes patula Brook in high latitudes in Antarctica.
Info. Biull. Sovetsk. Antarkt. Exped., (9): 40-42

corals
Pasternak, Th. A., 1961.
[Pennatularia (Octocorallia) and Antipatharia Hexacorallia) obtained by the Soviet Antarctic Expedition in 1955-1958.]
Trudy Inst. Okeanol., 56:217-230.

corals
Renson, Gilbert, 1966.
Biologie des coraux: IV Croissance des coraux.
Cah. Pacif. No. 9:29-45.

corals
Ranson, Gilbert, 1964.
Biologie des coraux: III. Rapports des coraux avec leur milieu.
Cahiers du Pacifique, Mus. Nat. Hist. Nat., Paris (6) 51-69.

corals
Ranson, Gilbert, 1961.
Biologie des coraux.
Cahiers du Pacifique, Foundation Singer-Polignac Mus. Nat. Hist. Nat., No. 3:75-94.

corals
Squires, Donald F., 1965.
Neoplasia in a coral?
Science, 148(3669):503-505.

corals
Squires, Donald F., 1965.
Deep-water coral structure on the Campbell Plateau, New Zealand.
Deep-Sea Res., 12(6):785-788.

corals
Squires, Donald F., 1964.
The Southern Ocean: a potential for coral studies.
Ann. Rept., Smithsonian Inst., 1963, Publ. 4530: 447-459.

Corals
Squires, D.E., 1962.
The fauna of the Ross Sea. 2. Scleractinian corals.
New Zealand Dept. Sci. Industr. Res., Bull., 147: 28 pp.
Also:
New Zealand Oceanogr. Inst. Mem., No. 19.

corals
Wells, J.W., 1967.
Corals as bathometers.
Marine Geol., 5(5/6):349-365.

corals
Wells, John W., 1963.
Coral growth and geochronometry.
Nature, 197(4871):948-950.

corals, precious

Nishijima, S., K. Yamazato and S. Kamura, 1969. Notes on the characteristics of precious coral grounds in the Ryukyu islands. (In Japanese; English abstract). Bull. Jap. Soc. fish. Oceanogr. Spec. No. (Prof. Uda Commem. Pap.): 291-297.

coral skeletons

Barnes, David J., 1970. Coral skeletons: an explanation of their growth and structure. Science 170(3964): 1305-1308.

coral atolls

coral atolls

Bellamy, D., E. Drew, D. Jones and J. Lythgoe, 1969. Aldabra: a preliminary report of the work of Phase VI of the Royal Society Expedition. Rept. Underwat. Ass. 4:100-104.

coral atolls

Berthois, Léopold, André Guilcher, François Doumenge et Alain Michel, 1963. Le renouvellement des eaux du lagon dans l'atoll de Maupihaa-Mopelia (Îles de la Societe). Comptes Rendus, Acad. Sci., Paris, 257(25):3992-3995.

coral atolls

Dana, Thomas F., 1971. On the reef corals of the world's most northern coral atoll (Kure: Hawaiian Archipelago). Pacific Sci. 25(1): 80-87.

coral atolls

Fosberg, F.R., 1962. A brief survey of the cays of Arrecife Alacran, a Mexican atoll. Atoll. Res. Bull., 93 of 88-94:25 pp.

coral atolls

Haas, H., 1962. Central subsidence. A new theory of atoll formation. Atoll. Res. Bull., 91 of 88-94:4 pp.

coral atolls

Labeyrie, Jacques, Claude Lalou and Georgette Delibrias, 1969. Étude des transgressions marines sur l'atoll de Mururoa par la datation des différents niveaux de corail. Cah. Pacifique 13: 59-68.

coral atolls

Ladd, H.S., and J.I. Tracy, Jr., 1957. Fossil land shells from deep drill holes in Western Pacific atolls. Deep-Sea Res., 4(3):218-219.

coral atolls

Plessis, Yves, 1969. Les atolls des Tuamotu en tant qu'écosystème marin. Bull. Mus. natn. Hist. Nat. Paris (2)40(6): 1232-1236.

coral "atolls"

Sakou, Toshitsugu, 1965. Surf on the coral-reefed coast. Ocean Sci. and Ocean Eng., Mar. Techn. Soc., Amer. Soc. Limnol. Oceanogr., 2:700-705.

coral atolls

Shor, G.G., Jr., and R.P. Phillips, 1964. Measurements of coral thickness at Midway Atoll. (Abstract). Trans. Amer. Geophys. Union, 45(1): 73.

coral atolls

Stoddart, David R., editor, 1966. Reef studies at Addu Atoll, Maldive Islands. Atoll Res. Bull., 116:122 pp. (mimeographied).

coral atolls

Stoddart, D.R., 1965. The shape of atolls. Marine Geology, 3(5):369-383.

coral atolls

Verstappen, H.T., 1954. The influence of climatic changes on the formation of coral islands. Amer. J. Sci. 252(7): 428-435.

coral atolls

Wells, J.W., 1954. Recent corals of the Marshall Islands. An ecologic and taxonomic analysis of living reef- and non-reef-building corals at Bikini and other Marshall Island atolls. Bikini and nearby atolls Pt. 2. Oceanography (Biologic). Geol. Surv. Prof. Pap. 260-I:385-486, 184 pls.

coral atolls

Wiens, H. J., 1962. Atoll environment and ecology. Yale University Press, 532 pp.

coral atolls

Wiens, Herold J., 1961. The evolution and destruction of atoll land. Proc. Ninth Pacific Sci. Congr., Pacific Sci. Assoc., 1957, 12(Geol.-Geophys.):367-376.

coral banks

Kühlmann, Dietrich H.H. 1971. Die Korallenriffe Kubas. II Zur Ökologie der Bankriffe und ihrer Korallen. Int. Revue ges. Hydrobiol. 56(2):145-199.

coral banks

Pratt, Richard M., 1971. Lithology of rocks dredged from the Blake Plateau. SEast. Geol. 13(1):19-38.

coral banks

Wong, H.K., W.D. Chesterman and J.D. Bromhall 1970. Comparative side-scan sonar and photographic survey of a coral bank. Int. Hydrogr. Rev. 47(2): 11-23.

coral islands

coral islands

*Schäfer, Wilhelm, 1969. Sarso, Modell der Biofazies-Sequenzen im Korallenriff-Bereich des Schelfs. Senckenberg. maritima 50:165-188.

corals, lists of spp.

corals, lists of spp.

Pichon, M., 1964. Contribution à l'étude de la répartition des Madréporaires sur le récif de Tuléar, Madagascar. Rec. Trav. Sta. Mar. Endoume, hors sér., Suppl., No. 2:79-203. (Trav. Sta. Mar., Tuléar).

corals, lists of spp.

Vasseur, P., 1964. Contribution à l'étude bionomique des peuplements sciaphiles infralittoraux de substrat dur dans les récifs de Tuléar, Madagascar. Rec. Trav. Sta. Mar., Endoume, hors sér., Suppl. No. 2:1-77 (Trav. Sta. Mar. Tuléar).

Coral predators

Weber, J.N. and P.M.J. Woodhead, 1970. Ecological studies of the coral predator Acanthaster planci in the South Pacific. Marine Biol., 6(1): 12-17.

coral reefs

coral reefs

Abe, N., M. Eguchi, and F. Hiro, 1937. Preliminary survey of the coral reef of Iwayama Bay, Palao. Palao Trop. Biol. Sta. Stud. 1(1):17-35, map, 2 pls., 1 fig.

coral reefs

Agassiz, A., 1903. The coral reefs of the tropical Pacific. Mem MCZ, 28:410 pp.

coral reefs

Asano, K., 1942. [Coral reefs in Oceania.] (In Japanese). Mem. Geol Palaeontol Inst., Fac. Sci., Tohoku Univ. 39: 27-45.

coral reefs

Asashina, H., 1931. [On the coral reefs in the South Sea Islands under Japanese mandate.] (In Japanese). Hydrogr. Bull. (Suiro Yoho), 10:113-127.

coral reefs

Bakus, Gerald J., 1966. Some relationships of fishes to benthic organisms on coral reefs. Nature, 210 (5033):280-284.

coral reefs

Barnes, J., D.J. Bellamy, D.J. Jones, B.A. Whitton, E. Drew and J. Lythgoe, 1970. Sublittoral reef phenomena of Aldabra. Nature Lond., 225(5229): 268-269.

coral reefs

Boschma, H., 1950. Notes on the coral reefs near Suva in the Fiji Islands. Proc. Kon. Neder. Akad. Wetensch. 53(3): 294-298, pls. 2 textfigs.

coral reefs

*Broecker, Wallace S., David L. Thurber, John Goddard, Teh-Lung Ku, R.K. Matthews and Kenneth J. Mesolella, 1968. Milankovitch hypothesis supported by precise dating of coral reefs and deep-sea sediments. Science, 159(3812):297-300.

coral reefs

Clausade, Mireille, Nicole Gravier, Jacques Picard, Michel Pichon, Marie-Louise Roman, Bernard Thomassin, Pierre Vasseur, Mireille Vivien et Pierre Weydert 1971. Morphologie des récifs coralliens de la région de Tuléar (Madagascar): éléments de terminologie récifale. Tethys Suppl. 2: 1-74.

coral reefs

Cloud, P.E., jr., 1951. Facies relationships of reefs. Bull. G.S.A. 62: 1499.

coral reefs
Coleman, J.S., 1950.
The sea and its mysteries. London, G. Bell & Sons, Ltd., 285 pp., 1 fold-in.

coral reefs
Cotton, C.A., 1948.
The present-day status of coral reef theories.
New Zealand Sci. Rev. 6(6):111-113.

coral reefs
Daly, R. A., 1948.
Coral reefs ---- a review. Amer. J. Sci., 246: 193-207.

coral reefs
Daly, R. A., 1916.
A new test of the subsidence theory of coral reefs. Proc. Nat. Acad. Sci. 2:664-670.

Funafuti

coral reefs
Davis, W.M., 1923.
The coral reef problem. Amer. Geogr. Soc., Spec. Publ. 9:596 pp.

coral reefs
Dawson, E. Yale, 1961.
The rim of the reef.
Natural History 70(6):8-17.

coral reefs
Dietz, R.S., 1954.
Marine geology of northwestern Pacific: description of Japanese Bathymetric Chart/ 6901.
Bull. G.S.A. 65(1):1199-1224, 6 textfigs.

coral reefs
DiSalvo, Louis H., 1971.
Regenerative functions and microbial ecology of coral reefs: labelled bacteria in a coral reef microcosm. J. exp. mar. Biol. Ecol. 7(2): 123-136.

coral reefs
DiSalvo, Louis H., 1971.
Regenerative functions and microbial ecology of coral reefs. II Oxygen metabolism in the regenerative system.
Can. J. Microbiol. 17(8): 1091-1100

coral reefs
Emery, K.O., 1962.
Marine geology of Guam: geology and hydrology of Guam, Mariana Islands.
U.S. Geol. Survey Prof. Paper, 403-B:76 pp.

coral reefs
Emery, K.O., J.I. Tracy jr., and H.S. Ladd, 1954.
Geology of Bikini and nearby atolls. Bikini and nearby atolls; Part 1, Geology.
Geol. Survey Prof Pap. No. 260-A:1-265, 64 pls., 11 charts.

coral reefs
Fairbridge, R.W., 1950.
Recent and Pleistocene coral reefs of Australia.
J. Geol. 58(4):330-401, 12 figs.

Coral reefs
Faure, Gérard, et Lucien Montaggioni 1971.
Les récifs coralliens Sous-Le-Vent de l'Île Maurice (Archipel des Mascareignes, Océan Indien): morphologie et bionomie de la pente externe.
C.R. hebd. Séanc. Acad. Sci. Paris (D) 273 (21): 1914-1916

coral reefs
Faure, Gérard, et Lucien Montaggioni, 1970.
Le récif corallien de Saint-Pierre de la Réunion (Océan Indien): géomorphologie et répartition des peuplements.
Rec. Trav. mar. Endoume, hors sér. Suppl. 10: 271-284.

coral reefs
Fischer, P.-H., 1961.
Coup d'oeil sur le Grande-Barrière d'Australie et en particulier sur un ~~relief~~ récif du groupe de Capricorne.
Cahiers du Pacifique, Fondation Singer-Polignac, Mus. Nat. Hist. Nat., No. 3:53-74.

coral atolls
Fosberg, F.R., 1961.
Qualitative description of the coral atoll ecosystem.
Atoll. Res. Bull., 81:11 pp.

coral atolls
Fosberg, F.R., 1957.
Some geological processes at work on coral atolls.
Trans. N.Y. Acad. Sci., (2), 19(5):411-422.

coral reefs
Gardiner, J.S., 1936.
XIII. The reefs of the western Indian Ocean. I. Chagos Archipelago. II. The Mascarene region.
Trans. Linn. Soc., London, Ser. 2, 9:393-436, Pl. 24, Textfigs. 1-10.

coral reefs
Gerlach, S.A., 1961.
The tropical coral reef as a biotope.
Atoll. Res. Bull., No. 80:6 pp.

coral reefs
Gordon, Malcolm S., and Hamilton M. Kelly, 1962
Primary productivity of an Hawaiian coral reef: A critique of flow respirometry in turbulent waters.
Ecology, 43(3):473-480.

coral reefs
Goreau, T.F., 1964
Mass expulsion of zooxanthellae from Jamaican reef communities after Hurricane Flora.
Science, 145(3630):383-386.

coral reefs
Goreau, T., and Kevin Burke, 1966.
Pleistocene and Holocene geology of the island shelf near Kingston Jamaica.
Marine geol., 4(3):207-225.

coral reefs
Goreau, T.F., and Goreau, N.I., 1960.
Distribution of labelled carbon in reef-building corals with and without zooxanthellae.
Science, 131(3401):668-669.

coral reefs
Goreau, Thomas F., Nora I. Goreau and C.M. Yonge, 1971.
Reef corals: autotrophs or heterotrophs?
Biol. Bull. mar. biol. Lab. Woods Hole 141(2): 247-260.

coral reefs
Goreau, Thomas F., and Willard D. Hartman, 1966.
Sponges: effect on the form of coral reefs.
Science, 151(3708):343-344.

coral reefs
Goreau, T.F., and J.W. Wells, 1967.
The shallow-water Schleractinia of Jamaica: revised list of species and their vertical distribution range.
Bull. mar. Sci., Miami, 17(2):442-453.

coral reefs
Gravier Nicole, Jean-Georges Harmelin, Michel Pichon, Bernard Thomassin, Pierre Vasseur et Pierre Weydert, 1970
Les récifs coralliens de Tuléar (Madagascar): morphologie et bionomie de la pente externe.
C.r. hebd. Séanc. Acad. Sci., Paris (D) 270(8): 1130-1133.

coral reefs
Guerin-Ancey, Odile, 1970.
Etude des intrusions terrigènes fluviatiles dans les complexes récifaux: délimitation et dynamique des peuplements des vases et des sables vaseux du chenal postrécifal de Tuléar (S.W. de Madagascar).
Rec. Trav. Sta. mar. Endoume, hors sér suppl. 10: 3-46.

coral reefs
Guilcher, André, 1965.
Coral reefs and lagoons of Mayotte Island, Comoro Archipelago, Indian Ocean, and of New Caledonia, Pacific Ocean.
In: Submarine Geology and geophysics, Colst Papers, W.F. Whittard and R. Bradshaw, edit Butterworth's, London, 17: 21-44.

coral reefs
Guilcher, Andre, 1964.
Present-time trends in the study of Recent marine sediments and in marine physiography.
Marine Geology, 1(1):4-15.

coral reefs
Guilcher, A., 1956.
Étude géomorphologique des récifs coralliens du Nord-ouest de Madagascar.
Ann. Inst. Océan., Monaco, 33(2):65-136.

coral reefs
Guilcher, A., 1955.
Géomorphologie de l'extremité septentrionale du Banc Farsan (Mer Rouge). Rés. Sci. Camp. Calypso. 1. Camp. 1951-1952 en Mer Rouge.
Ann. Inst. Océan., Monaco, 30:56-97, 9 textfigs., 20 pls.

coral reefs
Guilcher, A., 1952.
Morphologie sous-marins et récifs coralliens du Nord du Banc Farsan (Mer Rouge). Bull. Assn. Géogr. Francaise (224-225):52-63.

coral reefs
Guilcher, A., L. Berthois, R. Battistini and P. Fourmanoir, 1959.
Les récifs coralliens des îles Radama et de la baie Ramanetaka (cote Nord-ouest de Madagascar), étude géomorphologique et sédimentologique.
Mem. Inst. Sci. Madagascar, 1958, Ser. F:117-200.

coral reefs
Guilcher, André, Michel Denizot et Léopold Berthois, 1966.
Sur la constitution de la crête externe de l'atoll de Mopelia ou Mauphihaa (Îles de la Société), et de quelques autres récifs voisins.
Cah. océanogr., 18(10):851-856.

coral reefs
Haeberle, F.R., 1952.
Coral reefs of the Loyalty Islands. Am. J. Sci. 250(9):656-666, figs.

coral reefs
Hirata, K., 1956.
Ecological studies on the recent and raised coral reefs in Yoron Island. Sci. Rept., Kagoshima Univ 5:97-128.

coral reefs
Hoffmeister, J. Edward, and H. Gray Multer, 1964.
Growth-rate estimates of a Pleistocene coral reef of Florida.
Geol. Soc., Amer., Bull., 75(4):353-358.

coral reefs
Hoskin, Charles M., 1966.
Coral pinnacle sedimentation, Alacran reef lagoon, Mexico.
J. Sedim. Petrol., 36(4):1058-1074.

coral reefs
Jeffrey, S.W. 1968.
Photosynthetic pigments of the phytoplankton of some coral reef waters
Limnol. Oceanogr. 13(2): 350-355.

coral reefs
Jones, J.A., 1963.
Ecological studies of the southeastern Florida patch reefs. 1. Diurnal and seasonal changes in the environment.
Bull. Mar. Sci., Gulf and Caribbean, 13(2):282-307.

coral reefs
Kawaguchi, S., 1942.
[Investigation on reef-corals.] (In Japanese).
Kakaku Nabyo (Sci. South Sea) 5(1):95-106.

coral reefs
Kinsey, D.W., and Barbara E. Kinsey, 1967.
Diurnal changes in oxygen content of the water over the coral reef platform at Heron I.
Aust. J mar. Freshwat. Res., 18(1): 23-34

coral reefs
Kinsman, D.J.J., 1964.
Reef coral tolerance of high temperatures and salinities.
Nature, 202(4939):1280-1282.

Also in :
Collected Reprints, Int. Indian Ocean Exped., UNESCO, #.1964.

coral reefs
Kohn, A.J., and P. Helfrich, 1957.
Primary organic productivity of a Hawaiian coral reef. Limnol. & Oceanogr., 2(3):241-251.

coral reefs
Kornicker, Louis S., and Donald W. Boyd, 1962.
Shallow-water geology and environments of Alacran Reef complex, Campeche Bank, Mexico.
Bull. Amer. Assoc. Petr. Geol., 46(5):640-673.

coral reefs
Kuenen, Ph. H., 1954.
Eniwetok drilling results. Deep-Sea Res. 1(3): 187-189, 1 textfig.

coral reefs
Kuenen, Ph. H., 1951.
An argument in favor of glacial control of coral reefs. J. Geol. 59(5):503-507, 2 textfigs.

coral reefs
Kumpf, H.E., and H.A. Randall, 1961.
Charting the marine environments of St. John, U.S. Virgin Islands.
Bull. Mar. Sci., Gulf and Caribbean, 11(4):542-551.

coral reefs
Laborel, J., 1965.
On braxilian coral reefs. (abstract).
Anais Acad. bras. Cienc., 37 (Supl.):258.

coral reefs
Ladd, H.S., 1961
Reef building. The growth of living breakwaters has kept pace with subsidence and wave erosion for fifty million years.
Science, 134(3481):703-715.

coral reefs
Ladd, Harry S., 1953(1956).
Coral reef problems in the open Pacific.
Proc. 8th Pacific Sci. Congr., Geol. Geophys. Meteorol., 2A:833-849.

coral reefs
Ladd, H.S., 1950.
Recent reefs. Bull. Amer. Soc. Petr. Geol. 34(2):203-314, textfigs.

coral reefs
Ladd, H.S., and J.I. Tracy, jr., 1949.
The problem of coral reefs. Sci. Month. 69(5): 297-305, figs.

coral reefs
Lalou, Claude, Jacques Labeyrie et Georgette Delibrias, 1966.
Datation des calcaires coralliens de l'atoll de Mururoa (archipel des Taumotu) de l'époque actuelle jusqu' à -500,000 ans.
C.r.hebd. Seanc. Acad. Sci., Paris, (D)263 (25): 1946-1949.

coral reefs
LeDanois, Y., 1959.
Adaptations morphologiques et biologiques des poissons des massifs coralliens.
Bull. Inst. Francais, Afrique Noire, 21(4):1304-1325.

Coral Reefs
Légarde, E., 1965.
Recherches de microbiolgie marine en Nouvelle-Calédonie dans le cadre de la Mission d'Etudes des Récifs Coralliens.
Cahiers de Pacifique, No. 7:3-6.

coral reefs
Lewis, John B 1970.
Spatial distribution and pattern of some Atlantic reef corals.
Nature, Lond., 227 (5263): 1158-1159.

coral reefs
Lewis, J.B., 1960
The coral reefs and coral communities of Barbados, W.I., Canadian J. Zool., 38(6): 1133-1146.

coral reefs, fringing
Lewis, Michael Samuel, 1969.
Sedimentary environments and unconsolidated carbonate sediments of the fringing coral reefs of Mahé, Seychelles.
Marine Geol., 7(2): 95-127.

coral reefs
*Lewis, Michael Samuel, 1968.
The morphology of the fringing coral reefs along the east coast of Mahé, Seychelles.
J. Geol. 76(2):140-153.

coral reefs
Link, T.A., 1950.
Theory of transgressive and regressive reef (Bioherm) development and origin of oil. Bull. Amer. Assoc. Petr. Geol. 34(2):263-294, textfigs.

coral reefs
Ma, Ting-Ying H., 1964.
A comparison of the study of the upper Jurassic climate based on growth values of reef corals with that by the oxygen isotope method.
In: Studies on Oceanography dedicated to Professor Hidaka in commemoration of his sixtieth birthday, 496-514.

coral reefs
Ma, Ting Ying H., 1961.
Marine terraces in the western Pacific and the origin of coral reefs. (Abstract).
Proc. Ninth Pacific Sci. Congr., Pacific Sci. Assoc., 1957, 12(Geol.-Geophys.):377.
Published in Oceanographica Sinica, 5(2):1957.

coral reefs
Ma, Ting Ying H., 1959
History of the Pacific, Atlantic and Indian Ocean Basins as deduced from growth values of reef corals. Proc. Geol. Soc., China, No. 3: 67-82.

coral reefs
Ma, T.Y.H., 1957.
Reef corals used for proving the occurrence of shift in crustal masses and the equator and submarine features used to prove the sudden total displacement of the solid earth shell.
Proc. UNESCO Symp., Phys. Ocean., Tokyo, 1955: 220-224.

reef corals
Ma, Ting Ying H., 1957.
The effect of warm and cold currents in the south-western Pacific on the growth rate of reef corals.
Oceanographia Sinica 5(1):1-34.

coral reefs
Ma, Ting Ying H., 1957.
Marine terraces in the western Pacific and the origin of coral reefs.
Oceanographia Sinica, 5(2):1-5.

coral reefs
Ma, Ting Ying H., 1956.
Coral-reefs and the problem of sial in oceanic areas. Oceanografia Sinica 3:4 pp.

coral reefs
Ma, T.Y.H., I. Hayasaka, and S. Kawaguchi, 1948.
Observations on the coral reefs of Penghu Islands, Taiwan. Bull. Ocean. Inst., Taiwan, No. 1: 1-7, 1 textfigs.

coral reefs
Macintyre, I.G., 1968.
Some submerged coral reefs in the Caribbean.
Trans. Fifth Carib. Geol. Conf., St. Thomas, V.I., 1-5 July 1968, Peter H. Mattson, editor.
Geol. Bull. Queens Coll., Flushing, 5: 49-54.

coral reefs
Macintyre, Ian G., and Orrin H. Pilkey 1969.
Tropical reef corals: tolerance of low temperatures on the North Carolina continental shelf.
Science 166 (3903): 374-375.

coral reefs
MacNeil, F.S., 1950.
Planation of recent reef flats on Okinawa.
Bull. G.S.A. 61:1307-1308, 1 pl.

Coral reefs
Maiklem, W.R., 1968.
The Capricorn Reef Complex, Great Barrier Reef, Australia.
J. Sedim. Petrol., 38(3): 785-798.

coral reefs
Margalef, Ramón, 1959.
Pigmentos asimiladores extraídos de las colonias de celentéreos de los arrecifes de coral y su significado ecológico.
Investigaciones Pesquera, Barcelona, 15:81-101.

coral reefs
*Marsh, James A.,Jr., 1970.
Primary productivity of reef-building calcareous red algae. Ecology 51(2): 255-263.

coral reefs
*Marshall, N., 1968.
Observations on organic aggregates in the vicinity of coral reefs.
Marine Biol., 2(1):50-53.

Marshall, Nelson, 1965. coral reefs
Detritus over the reef and its potential contribution to adjacent waters of Eniwetok Atoll.
Ecology, 46(3):343-344.

coral reefs
Mesolella, Kenneth J., 1967.
Zonation of uplifted Pleistocene coral reefs on Barbados, West Indies.
Science, 156 (3775):638-640.

coral reefs
Milliman, John D., 1965.
An annotated bibliography of recent papers on corals and coral reefs.
Atoll. Res. Bull., No. 111:58 pp.

coral reefs
Montaggioni, Lucien 1970.
Essai de chronologie relative des épisodes récifaux à l'île de la Réunion (Océan Indien); leur incidence sur la morphologie récifale actuelle.
C.r. hebd. Séanc. Acad. Sci. Paris (D) 270 (15): 1869-1871.

coral reefs
*Morelock, Jack, and Karl J. Koenig, 1967.
Terrigenous sedimentation in a shallow water coral reef environment.
J. sedim. Petrol., 37(4):1001-1005.

coral reefs
*Motoda, Sigeru, 1969.
An assessment of primary productivity of a coral reef lagoon in Palau, western Caroline Islands, based on the data obtained during 1935-1937.
Rec. oceanogr. Wks. Japan, 10(1):65-74.

coral reefs
Motoda, S., 1940.
A study of growth rate in the massive reef coral, Goniastrea aspera Verrill. Palao Trop. Biol. Sta. Studies 2(1):1-6, 1 textfig.

coral reefs
Motoda, S., 1940. reef
The environment and the life of massive/coral, Goniastrea aspera Verrill, inhabiting the reef flat in Palao. Palao Trop. Biol. Sta. Studies 2(1):61-104, 9 textfigs.

coral atolls
Munk, W.H. and M.C. Sargent, 1948
Adjustment of Bikini Atoll to ocean waves.
Trans. Am. Geophys. Union 29(6):855-860, 3 text figs.

coral reefs
Muscatine, Leonard and Elsa Cernichiari, 1969.
Assimilation of photosynthetic products of zooxanthellae by a reef coral. Biol. Bull. mar. biol. Lab., Woods Hole, 137(3): 506-523.

coral reefs
Nair, P.V. Ramachandran, 1970
Primary productivity in the Indian seas.
Bull. cent. mar. Fish. Res. Inst. 22: 56pp. (mimeographed)

coral reefs
Nesteroff, W.D., 1955.
Les récifs coralliens du Banc Farsan Nord (Mer Rouge). Rés. Sci. Camp. Calypso. 1. Camp. 1951-1952 en Mer Rouge. Ann. Inst. Océan., Monaco, 30:8-53, 21 pls., 11 textfigs.

coral reefs
Nesteroff, W., 1952.
Coupes transversales de la Mer Rouge. Contribution aux théories de formation des recifs coralliens. C.R. Acad. Sci., Paris, 235:1503-1505, 1 textfig.

coral reefs
Nugent, L., 1946.
Coral reefs in the Gilbert, Marsahll, and Caroline islands. Bull. G.S.A. 57(8):735-779, Fig. 15.

coral reefs
Odum, H.T., and E.P. Odum, 1955.
Trophic structure and productivity of a windward coral reef community on Eniwetok Atoll.
Ecol. Monogr., 25:291-320.

coral reefs
Pichon, M., 1964.
Contribution à l'étude de l'écologie et des méthodes de pêches des Palinuridae dans la

la région de Nosy-Bé (Madagascar).
Cahiers, O.R.S.T.R.O.M., Océanographie, Paris, 11(3):71-101.

coral reefs
Pichon, M., 1963.
Note préliminaire sur la topographie et la géomorphologie des recifs coralliens de la région de Tuléar.
Annls malgaches, Univ. Madagascar, 1:153-168.

coral reefs
Pichon, 1962.
Note préliminaire sur la topographie et la géomorphologie des récifs coralliens de la région de Tuléar.
Ann. Faculté Sciences Techniques, Madagascar, 153-168.

Also:
Rec. Trav. Sta. Mar., Endoume, Fasc. hors series, Suppl. (1).
Trav. Sta. Mar. Tuléar (République Malgache), Int. Indian Ocean Exped.

coral reefs
Powers, Dennis, 1970.
A numerical taxonomic study of Hawaiian reef corals.
Pacific Sci., 24(2): 180-186

coral reefs
Pressick, Mary Lou, 1970
Zonation of stony coral of a fringe reef southeast of Icacos Island, Puerto Rico.
Carib. J. Sci., 10 (3-4): 137-139.

coral reefs
Quiguer, Jean-Paul, 1969.
Quelques données sur la répartition des poissons des récifs coralliens. Cah. Pacifique 13: 181-185.

coral reefs
Ranson, G., 1958.
Coraux et récifs coralliens (bibliographie).
Bull. Inst. Océan., Monaco, No. 1121:80 pp.

coral reefs
Ranson, Gilbert, 1958
Coraux et Recifs Coralliens.
Cahiers du Pacifique No. 1:15-36.

coral reefs
Roberts, Harry H. 1971.
Environments and organic communities of North Sound, Grand Cayman Island, B.W.I.
Carib. J. Sci. 11 (1/2): 67-79

coral reefs
Robinson, D.E. 1971.
Observations on Fijian coral reefs and the crown of thorns starfish.
J. Roy. Soc. N.Z. 1 (2): 99-112

coral reefs
Rosen, B.R. and J.D. Taylor 1969.
Reef coral from Aldabra: new mode of reproduction.
Science 166 (3901): 119-121.

coral reefs
Rotschi, H., 1953.
Expédition océanographique "Capricorn" de la "Scripps Institution of Oceanography" de l'Université de Californie, Novembre 1952-Février 1953. Bull. d'Info., C.C.O.E.C. 5(10):439-466, 2 pls.

coral reefs
Roy, K.J., and S.V. Smith, 1971.
Sedimentation and coral reef development in turbid water: Fanning lagoon.
Pacific Sci., 25(2): 234-248

coral atolls
Sachet, M.-H., 1962
Geography and land ecology of Clipperton Island.
Atoll Res. Bull., 86:115 pp.

coral reefs
Salvat, B., 1970
Études quantitatives (comptages et biomasses) sur les mollusques récifaux de l'atoll de Fangataufa (Tuamotu-Polynésie).
Cah. Pacifique 14:1-58.

coral reefs
#Sara, Michele, 1967.
Un coralligeno di piattaforma (coralligène de plateau) lungo el litorale pugese.
Arch. Oceanogr. Limnol. 15(Suppl.):139-150.

coral reefs
Schäfer, Wilhelm, 1969.
Sarso, Modell der Biofazies-Sequenzen im Korallenriff-Bereich des Schelfs. Senckenberg. marit. [1] 50: 165-188.

coral reefs
Schlanger, Seymour O., and Charles J. Johnson, 1969.
Algal banks near La Paz, Baja California—modern analogues of source areas of transported shallow-water fossils in pre-Alpine flysch deposits.
Palaeogr. Palaeoclimatol. Palaeoecol. 6(2):141-157

coral reefs (theory)
Schlee, Susan 1971.
The curious controversy over coral reefs.
Sea Frontiers 17(4):214-223 (popular)

coral reefs
Sheridan, R.E., J.D. Smith and J. Gardner, 1969
Rock dredges from Blake Escarpment near Great Abaco Canyon.
Bull. Am. Ass. Petrol. Geol., 53(12): 2551-2558.

coral reefs
Silvester, R., 1965.
Coral reefs, atolls and guyots.
Nature, 207(4998):681-688.

coral reefs
Stanley, Daniel J., and Donald J.P. Swift, 1967
Bermuda's southern aeolianite reef tract.
Science, 157(3789).

coral reefs
Stearns, T. H., 1946.
An integration of coral-reefs hypotheses. Am. J. Sci. 244:245-262, 6 textfigs., 1 pl.

coral reefs
Stephenson, T.H., 1946.
Coral reefs. Endeavour 5(19):83-95, 96-106, Fig. 8.

coral banks
Stetson, T.R., D.F. Squires and R.M. Pratt, 1962.
Coral banks occurring in deep water on the Blake Plateau.
Amer. Mus. Nat. Hist., Novitates, (2114):39 pp.

coral reefs
Storr, John F., 1964.
Ecology and oceanography of the coral-reef tract, Abaco Island, Bahamas.
Spec. Papers, Geol. Soc., Amer., 79:98 pp.

coral reefs
Sugiyama, T., 1942.
[Reef-building corals of Yap Islands and its fringing reef.] (In Japanese). Mem. Geol. Palaeontol. Inst., Fac. Sci. Tôhoku Imp. Univ., 39: 7-26.

coral reefs
Taisne, Capitan de Corvette, 1964.
Mission d'études des récifs coralliens de Nouvelle-Calédonie.
Cahiers du Pacifique, Mus. Nat. Hist. Nat., Paris (6):71-75.

coral reefs
Tayama, R., 1952.
Coral reefs of the South Seas. Publ. No. 941, Bull. Hydrogr. Off. 11:292 pp. (English resume, pp. 184-292) with Appendix 1 of 133 pp. of photos and appendix 2 of 18 charts (bathymetric).

coral reefs
Teichert, C., 1958.
Cold- and deep-water coral banks.
Bull. Amer. Assoc. Petr. Geol., 42(5):1064-1082.

coral reefs
Teichert, C., and R.W. Fairbridge, 1948
Some coral reefs of the Sahul Shelf.
Geogr. Rev. 38(2):222-249, 17 text figs.

coral atolls
Thurber, David L., Wallace S. Broecker, Richard L. Blanchard and Herbert A. Potratz, 1965.
Uranium-series ages of Pacific atoll coral.
Science, 149(3679):55-58.

coral reefs
Tracy, J.I., jr., H.S. Ladd and J.E. Hoffmeister, 1948
Reefs of Bikini, Marshall Islands. Bull. G.S.A. 59(9):861-878, 11 pls.

Coral reefs
Umbgrove, J.H.F., 1947
Coral reefs of the East Indies. Bull. G.S.A. 58(8):729-778.

coral reefs, chemistry
Veeh, H. Herbert, and Karl K. Turekian, 1968.
Cobalt, silver, and uranium concentrations of reef-building corals in the Pacific Ocean.
Limnol. Oceanogr., 13(2):304-308.

coral reefs
Wainwright, Stephen A., 1965.
Reef communities visited by the Israel South Sea Expedition, 1962.
Sea Fish. Res. Sta., Israel, Bull., No. 38:40-53.

coral reefs
Weber, Jon N., and Peter M.J. Woodhead 1970
Carbon and oxygen isotope fractionation in the skeletal carbonate of reef-building coral reefs.
Chem. Geol. 6(2):93-117

coral reefs
Wells, J.W., 1957.
Coral reefs. Ch. 20 in: Treatise on Marine Ecology and Paleoecology, Vol. 1.
Mem. Geol. Soc., Amer., 67:609-632.

coral reefs
Weydert, Pierre, 1971.
Étude sédimentologique et hydrodynamique d'une coupe de la partie médiane du Grand récifs de Tuléar (S.W. de Madagascar)
Téthys Suppl. 1: 237-290

corals
Woodhead, P.M.J. and J.N. Weber, 1969.
Coral genera of New Caledonia. Marine Biol., 4(3): 250-254.

coral reefs
Yabe, H., 1942.
[Coral reef problem.] (In Japanese). Mem. Geol. Palaeontol. Inst., Fac. Sci., Tôhoku Imp. Univ. 39:1-6.

coral reefs
Wiens, Herold J., 1961.
The role of mechanical abrasion in the erosion of coral reefs and land areas.
Proc. Ninth Pacific Sci. Congr., Pacific Sci. Assoc., 1957, 12(Geol.-Geophys.):361-366.

coral reefs
Yabe, H., and T. Sugiyama, 1941.
Recent reef-building corals from Japan and the South Sea Islands under the Japanese Mandate. II.
Sci. Repts., Tohoku Imp. Univ., Ser. 2, Geol., Spec. vol. 2:67-91, illus.

coral reefs
Yonge, C.M., 1963
The biology of coral reefs.
In: Advances in Marine Biology, F.S. Russell, Editor, Academic Press, London and New York, 1:209-260.

coral reefs
Yonge, C.M., 1951.
The form of coral reefs. Endeavour 10(39):136-144, figs.

corals, feeding of
Roushdi, H.M., and V. Kr. Hansen, 1961.
Filtration of phytoplankton by the octocoral Alcyonium digitatum L.
Nature, 190(4778):649-650.

coral reef communities
Taylor, J.D., 1968.
Coral reef and associated invertebrate communities (mainly molluscan) around Mahe Seychelles.
Phil. Trans. R. Soc., (B)255(793):129-206.

coral reefs, microbial population
DiSalvo, L., and K. Gundersen 1971.
Regenerative functions and microbial ecology of coral reefs. I. Assays for microbial population.
Can. J. Microbiol 17(8): 1081-1089.

corals, effect of
Thompson, G., and H.D. Livingston 1970.
Strontium and uranium concentrations in aragonite precipitated by some modern corals.
Earth Planet. Sci. Letts 8(6): 439-442.

corallines
Adey, W.H., 1966.
The genera Lithothamnion, Leptophytum (nov.gen.) and Phymatolithon in the Gulf of Maine.
Hydrobiologia, 28 (3/4):321-350.

Corals, zooxanthellae

corals, zooxanthellae in

Kawaguti, Siro, 1964.
An electron microscope proof for a path of nutritive substances from zooxanthellae to the reef coral tissue.
Proc. Japan Acad., 40(10):832-835.

coral reefs
(zooxanthellae pigments)

Margaleff, Ramón, 1959

Pigmentos asimiladores etraídos de las colonias de celentéreos de los arrecifes de coral y su significado ecológico. Invest. Pesquera, 15: 81-101.

Cores

A

cores

Alliata, E. di Napoli, 1963.
Conclusions de l'analyse de la carotte No. 11 croisière méditerranéenne du "Vema" en 1956.
Rapp. Proc. Verb., Réunions, Comm. Int., Expl. Sci., Mer Méditerranée, Monaco, 17(3):1073-1077.

cores

Akers, W.H., 1965.
Pliocene-Pleistocene boundary, northern Gulf of Mexico.
Science, 149(3685):741-742.

cores

Anon., 1965.
Pacific corers hit legal snag.
Oil and Gas Jour., 63(25):105.

cores

Arrhenius, G., 1952.
Sediment cores from the East Pacific.
Repts. Swedish Deep-sea Exped., 1947-1948, 5(1): 227 pp., textfigs. with app. of pls.

cores

Athearn, W.D. 1965?
Sediment cores from the Cariaco Trench, Venezuela.
Proc Fourth Geol. Conf., Trinidad 1965: 343-352.

cores

Ault, Wayne U., 1959.
Oxygen isotope measurements on Arctic cores.
Sci. Studies, Fletcher's Ice Island, T-3, Vol. 1, Air Force Cambridge Res. Center, Geophys. Res. Pap. No. 63:159-168.

cores

Australia, Commonwealth Scientific and Industrial Research Organization, Division of Fisheries and Oceanography, 1962.
Oceanographic observations in the Pacific Ocean in 1960, H.M.A.S. Gascoyne, Cruises G 1/60 and G 2/60.
Oceanographical Cruise Report No. 5:255 pp.

B

cores

Baranov, V.I., and Kuz'mina, L.A., 1954.
[An ionium method for determining the age of sea cores. The direct determination of ionium.]
Dokl. Akad. Nauk, SSSR, 92(3):483.

core

Behre, Karl-Ernst, und Burchard Menke, 1969.
Pollenanalytische Untersuchungen an einem Bohrkern der südlichen Doggerbank.
Beitr. Meereskunde 24-25: 122-129

cores

Belderson, R.H., and A.S. Laughton, 1966.
Correlation of some Atlantic turbidites.
Sedimentology, 7(2):103-116.

cores

Bellaiche, Gilbert, Colette Vergnaud-Grazzini et Louis Glangeaud 1969.
Les épisodes de la transgression flandrienne dans le golfe de Fréjus.
C.r. hebd. Séanc. Acad. Sci. Paris (D) 268: 2765-2770.

cores

Belshé, J.C., 1964.
The magnetization of sediment cores from the Indian Ocean. (Abstract).
Trans. Amer. Geophys. Union, 45(1):71.

cores

*Bender, Michael L. and Cynthia Schultz, 1969.
The distribution of trace metals in cores from a traverse across the Indian Ocean.
Geochim. cosmochim. Acta, 33(2):292-297.

cores

*Berggren, W.A., 1969.
Micropaleontologic investigations of Red Sea cores - summation of synthesis of results.
In: Hot brines and Recent heavy metal deposits in the Red Sea, E.T. Degens and D.A. Ross, editors, Springer-Verlag, New York, Inc., 329-335.

cores

*Berggren, W.A., 1968.
Micropaleontology and the Pliocene/Pleistocene boundary in a deep-sea core from the south-central North Atlantic.
G. Geol., (2)35(2):291-312.

cores

Berggren, W.A., J.D. Phillips, A. Bertels and D. Wall, 1967.
Late Pliocene-Pleistocene stratigraphy in deep sea cores from the south-central North Atlantic.
Nature 216(5112):253-255.

cores

Berrit, G.R., 1955.
Étude des teneurs en manganese et en carbonates de quelques carottes sedimentaires Atlantiques et Pacifiques. Göteborgs K. Vetenskaps- och Vitterhets-Samhalles Handl., Sjätte Följden, (B), 6(12):61 pp.

Also:
Medd. Oceanografiska Inst., Göteborg, No. 23.

cores

Berry, Richard W., and William D. Johns, 1966.
Mineralogy of the clay-sized fractions of some North Atlantic-Artic Ocean bottom sediments.
Geol. Soc., Amer., Bull., 77(2):183-196.

cores

*Berthois, Léopold, 1968.
Sur la présence d'affleurements calcaires organogènes mio-pliocènes au large de Gibraltar.
C.r. hebd. Séanc., Acad. Sci., Paris, (D)267(15):1186-1189.

cores

Bezroukov, P.L., 1961
[Investigation of sediments in the northern Indian Ocean.] (In Russian; English abstract).
Okeanolog. Issled., Mezhd. Komitet Proved. Mezhd. Geofiz. Goda, Prezidiume, Akad. Nauk, SSSR, (4):76-90.

cores

Besrukov, P. L., 1960.
Sedimentation in the northwestern Pacific Ocean.
Morsk. Geol., Doklady Sovetsk. Geol., 21 sess. Int. Geol. Kongress, 45-58.
(English abstract)

cores

Bezrukov, P.L., V.P. Petelin, 1951.
[Account of work with corers] Trudy Inst. Okeanol. 5:14-16.

cores

*Black, D.I., 1967.
Cosmic ray effects and faunal extinctions at geomagnetic field reversals.
Earth Planet. Sci. Letters., 3(3):225-236.

cores

Blackman, Abner, and B.L.K. Somayajulu, 1966.
Pacific Pleistocene cores: faunal analyses and geochronology.
Science, 154(3751):886-889.

cores

Blanc-Vernet, Laure, 1965.
Note préliminaire sur la microfaune de quelques carottes de Méditerranée occidentale.
Rec. Trav. Sta. Mar. Endoume, Bull., 39(55): 305-305.

cores

Blanc-Vernet, L., H. Chamley et C. Froget, 1969
Analyse paléoclimatique d'une carotte de Méditerranée nord-occidentale. Comparaison entre les résultats de trois études: foraminifères, ptéropodes, fraction sédimentaire issue du Continent.
Palaeogeogr. Palaeoclimatol. Palaeoecol., 6(3): 215-235.

cores

Blanc, Jean et Colerine Froget, 1967.
Recherches de géologie marine et sédimentologie. Campagne de la Calypso en Méditerranée orientale (quatrième mission, 1964).
Annls Inst. Océanogr., n.s. 45(2): 257-292.

cores

Bolli, H.M., J.E. Boudreaux, Cesare Emiliani, W.W. Hay, R.J. Hurley and J.I. Jones, 1968.
Biostratigraphy and paleotemperatures of a section cored on the Nicaragua Rise, Caribbean Sea.
Bull. geol. Soc., Am., 79(4): 459-470.

cores

Bonatti, E., K. Bostrum, B. Eyl and E. Rona, 1969.
Geochemistry, mineralogy and absolute ages of a Caribbean sediment core. Trans. Gulf Coast Ass. geol. Socs, 19 : 506.

cores

Bonner, Francis T., and Alzira Soares Lourenço, 1965.
Nickel content of Pacific Ocean cores.
Nature, 207(5000):933-935.

cores

Bonner, Francis T., and Alzira Soares, 1964.
Nickel content of ocean core samples. (Abstract).
Trans. Amer. Geophys. Union, 45(1):118.

cores
Boudreaux, Joseph E., and William W. Hay 1969.
Calcareous nannoplankton and biostratigraphy of the late Pliocene-Pleistocene-Recent sediments in submarine cores.
Revist. esp. Micropaleontol. 1(3): 249-292.

cores
Bouma, A.H., 1965.
Sedimentary characteristics of samples collected from some submarine canyons.
Marine Geology Elsevier Publ. Co., 3(4):291-320.

cores
Bouma, A.H., and J.A.K. Boerma 1968.
Vertical disturbances in piston cores.
Marine Geol. 6(3): 231-241.

cores
Bottema, S., and L.M.J.U. Van Straaten, 1966.
Malacology and palynology of two cores from the Adriatic sea floor.
Marine Geol., 4(6):553-564.

cores
Bourcart, Jacques, Maurice Gennesseaux et Éloi Klimek, 1961.
Sur le remplissage des canyons sous-marine de la Méditerranée française.
Comptes Rendus, Acad. Sci., Paris, 252:3693-3698.
Also in:
Travaux, Lab. Géol. Sous-Marine, Sta. Oceanogr., Villefranche, 11(1961-1962).

cores
Bourcart, Jacques, Maurice Gennesseaux et Éloi Klimek, 1960.
Écoulements profonds de sables et de galets dans la grande vallée sous-marine de Nice.
C.R. Acad. Sci., Paris, 250:3761-3765.
Also in:
Trav. Lab. Géol. Sous-Marine, 10(1960).

cores
Bradley, W.H., 1942.
Geology and Biology of the North Atlantic deep-sea cores between Newfoundland and Ireland.
U.S. Geol. Surv., Prof. Pap. No. 196:

CORES
Broecker, Wallace S., and Jan Van Donk, 1970.
Insolation changes, ice volumes, and the O[18] record in deep-sea cores.
Rev. Geophys. Space Phys. 8(1): 169-198.

cores
Bryant, William, Adrian F. Richards and George H. Keller, 1969.
Shear strength of sediments measured in place near the Mississippi Delta compared to measurements obtained from cored material. Trans. Gulf Coast Ass. geol. Socs, 19: 267.

cores
*Buchan, S., F.C.D. Dewes, D.M. McCann and D. Taylor Smith, 1967.
Measurements of the acoustic and geotechnical properties of marine sediment cores.
In: Marine geotechnology, A.F. Richards, editor, Univ. Illinois Press, 65-92.

cores
Burckle, Lloyd H., Jessie H. Donahue, James D. Hays and Bruce C. Heezen, 1966.
Radiolaria and diatoms in sediments of the southern oceans.
Antarctic J., United States, 1(5):204.

cores
Burckle, Lloyd H., Tsunemasa Saito and Maurice Ewing, 1967.
A Cretaceous (Turonian) core from the Naturaliste Plateau, southeast Indian Ocean.
Deep-Sea Research, 14(4):421-426.

cores
Burke, Kevin, 1967.
The Yallahs Basin: a sedimentary basin southeast of Kingston Jamaica.
Marine Geol., 5(1):45-60

C

core
Caralp, Michelle 1970.
Essai de stratigraphie du Pléistocène marin terminal d'après les paléoclimats observés dans des carottes du golfe de Gascogne.
Bull. Soc. géol. France (7) 12 (3): 403-412

cores
*Caralp, Michelle, 1967.
Les foraminifères planctoniques d'une carotte atlantique (golfe de Gascogne) dans la mise en évidence d'une glaciation.
C.r.hebd.Séanc.Acad.Sci., Paris, (D)265(21):1588-1591.

cores
Caralp, Michelle, et Michel Vigneaux 1970
Le Pléistocène marin sur un dôme sous-marin du Golfe de Gascogne.
C.r.hebd. Séanc. Acad. Sci., Paris, (D) 271 (22): 1949-1952.

cores
Chalmers, G.V., and A.K. Sparks, 1959.
An ecological survey of the Houston ship channel and adjacent bays.
Publ. Inst. Mar. Sci., 6:213-250.

cores
Chamley, Hervé, 1969.
Intérêt paléoclimatique de l'étude morphologique d'argiles méditerranéennes.
Téthys 1(3): 923-926.

cores
Chester, R., and M.J. Hughes, 1966.
The distribution of manganese, iron and nickel in a North Pacific deep-sea clay core.
Deep-Sea Res., 13(4)627-634.

cores
Chmelik, Frank B., Arnold H. Bouma and Richard Rezak, 1969.
Comparison of electrical logs and physical parameters of marine sediment cores. Trans. Gulf Coast Ass. geol. Socs, 19 :63-70.

cores
Chow, T.J., M. Tatsumoto and C.C. Patterson, 1962
Lead isotopes and uranium contents in experimental Mohole cores (Guadalupe site).
J. Sed. Petr., 32(4):866-869.

cores
Cita, M.B., and M.A. Chierici, 1963
Crociera Talassografica Adriatica 1955.
V. Ricerche sui Foraminiferi contenuti in 18 carote prelavate sul fondo del mare Adriatico.
Arch. Oceanogr. & Limnol., 12(3):297-360.

cores
Cita, Maria Bianca, and Sara d'Onofrio, 1967.
Climatic fluctuations in submarine cores from the Adriatic Sea.
Progress in Oceanography, 4:161-178.

cores
Conolly, J.R., and M. Ewing, 1967.
Sedimentation in the Puerto Rico Trench.
J. sedim. Petrol., 37(1):44-59.

cores
*Conolly, John R., and C.C. von der Borch, 1967.
Sedimentation and physiography of the sea-floor south of Australia.
Sedimentary Geol., 1(2):181-220.

cores, chemistry
Coquema, C., R. Coulomb et J. Ros, 1963.
Étude géochimique de deux carottes de sédiment du bassin occidental de la Méditerranée.
Rapp. Proc. Verb., Réunions, Comm. Int. Expl. Sci., Mer Méditerranée, Monaco, 17(3):1009-1019.

D

cores
Damiani, A., L. Favretto and G.L. Morelli, 1964.
Le argille della fossa mesoadriatica.
Archiv. Oceanogr. e Limnol., 13(2):187-196.

cores
Dergunov, I.D., 1958.
[Contemporary presentation on the thermal regime of sediment cores.] Izv. Akad. Nauk, SSSR, (1): 65-74.

cores
Dickson, G.O., and J.H. Foster, 1966.
The magnetic stratigraphy of a deep sea core from the North Pacific Ocean.
Earth Planet. Sci. Letters, 1(6):458-462.

cores
Di Napoli Alliata, Enrico, Gabriella Fierzmont et Sergio Stefanini, 1968
Etude de quelques carottes provenant de la plate-forme continentale de la Sardaigne méridionale.
Rapp. P.-v. Réun. Commn int. Explor. scient. Mer Mediterr., 19(4): 639-641.

cores
Donahue, Jack 1971.
Burrow morphologies in north-central Pacific sediments.
Mar. Geol 11(1):M1-M7.

E

cores
Emery, K.O., Bruce C. Heezen and T.D. Allan, 1966.
Bathymetry of the eastern Mediterranean Sea.
Deep-Sea Res., 13(2):173-192.

cores
Emery, K.O., and J. Hülsemann, 1964.
Shortening of sediment cores collected in open barrel gravity corers.
Sedimentology, 3(2):144-154.

Cores

Emery, K. O., and J. Hulsemann, 1961 (1962).
The relationships of sediments, life and water in a marine basin.
Deep-Sea Res., 8(3/4):165-180.

cores

Emiliani, Cesare, 1966.
Paleotemperature analysis of Caribbean cores P6304-8 and P6304-9 and a generalized temperature curve for the past 425,000 years.
J. Geol., 74(2):109-124.

cores

Emiliani, Cesare, 1964.
Paleotemperature analysis of the Caribbean cores A 254-Br-C and CP-28.
Geol. Soc., Amer., Bull., 75(2):129-144.

cores

Emiliani, Cesare, 1961.
Cenozoic climatic changes as indicated by stratigraphy and chronology of deep-sea cores of Globigerina-ooze facies. In: Solar variations climatic change and related geophysical problems.
Ann. N.Y. Acad. Sci., 95(1):521-536.

cores

Emiliani, C., 1957.
Temperature and age analysis of deep-sea cores.
Science 125(3244):383-387.

cores

Enbysk, Betty Joyce, 1961.
Foraminifera from northeast Pacific cores. (Abstract).
Tenth Pacific Sci. Congr., Honolulu, 21 Aug.-6 Sept., 1961, Abstracts of Symposium Papers, 369-370.

Ericson, David B.

cores

Ericson, D.B., 1963.
31. Cross-correlation of deep-sea sediment cores and determination of relative rates of sedimentation by micropaleontological techniques.
In: The Sea, M.N. Hill, Editor, Interscience Publishers, 3:832-842.

cores

Ericson, David B., 1961.
Pleistocene climatic record in some deep-sea sediment cores. In: Solar variations, climatic change and related geophysical problems.
Ann. N.Y., Acad. Sci., 95(1):537-541.

cores

Ericson, David B., Maurice Ewing and Goesta Wollin, 1963
Pliocene-Pleistocene boundary in deep-sea sediments. Extinctions and evolutionary changes in microfossils clearly define the abrupt onset of the Pleistocene.
Science, 139(3556):727-737.

cores

Ericson, David B., Maurice Ewing and Goesta Wollin, 1964
Sediment cores from the Arctic and Subarctic seas.
Science, 144(3623):1183-1192.

cores

Ericson, David B., Maurice Ewing and Gosta Wollin, 1964
The Pleistocene Epoch in deep-sea sediments. A complete time scale dates the beginning of the first ice age at about 1 1/2 million years ago.
Science, 146(3645):723-732.

cores

Ericson, D.B., Maurice Ewing, Goesta Wollin and Bruce C. Heezen, 1961.
Atlantic deep-sea sediment cores.
Bull. Geol. Soc., Amer., 72:193-286.

cores

Ericson, David B., and Gosta Wollin, 1959.
Micropaleontology and Lithology of Arctic sediment cores.
Sci. Studies, Fletcher's Ice Island, T-3, Vol. 1, Air Force Cambridge Res. Center, Geophys. Res. Pap., No. 63:50-58.

cores

Ericson, D.B., and G. Wollin, 1956.
Correlation of six cores from the equatorial Atlantic and the Caribbean. Deep-Sea Res. 3(2): 104-125.

cores

Ericson, D.B., and G. Wollin, 1956.
Micropaleontological and isotopic determinations of Pleistocene climates. Micropalenotology 2(3): 257-270.

cores

Emiliani, Cesare, 1966.
Isotopic paleotemperature: Urey's method of paleotemperature analysis has greatly contributed to our knowledge of past climates.
Science, 154(3751):851-857.

cores

Ericson, D.B., G. Wollin and J. Wollin, 1955.
Coiling direction of Globorotalia truncatulinoides in deep-sea cores. Deep-Sea Res. 2(2):152-158, 4 textfigs.

cores

Eriksson, K. Gösta, and Ingrid U. Olson, 1963.
Some problems in connection with C14 dating of tests of Foraminifera.
Bull. Geol. Inst., Univ. Uppsala, 42:1-13.

Also:
Publ. Inst. Quarternary Geol., Univ. Uppsala, Octavo Ser., No. 7.

F

cores

Fisher, Robert L., 1961.
Middle America Trench: topography and structure.
Bull. Geol. Soc., Amer., 72(5):703-720.

cores

Fox, Paul J., William F. Ruddiman, William B.F. Ryan and Bruce C. Heezen, 1970.
The geology of the Caribbean crust. I: Beata Ridge.
Tectonophysics 10(5/6): 495-513.

G

cores

Geitzenauer, Kurt R., Stanley V. Margolis and Dennis S. Edwards 1968.
Evidence consistent with Eocene glaciation in a South Pacific sediment-bry core
Earth Planet. Sci. Letts 4(2): 173-177.

cores

Gennesseaux, M., 1962.
Travaux du Laboratoire de Géologie Sous-Marine concernant les grands carottages effectués sur le précontinent de la region Nicoise.
In: Océanographie Géologique et Géophysique de la Méditerranée Occidentale, Centre National de la Recherche Scientifique, Villefranche sur Mer, 4 au 8 avril 1961, 177-181.

cores

Gevirtz, Joel L., and Gerald M. Friedman, 1966.
Deep-sea carbonate sediments of the Red Sea and implications on marine lithification.
J. Sed. Petr., 36(1):143-151.

Cores

Glass, Bill 1967.
Microtectites in deep-sea sediments.
Nature, Lond. 214 (5086): 372-374.

cores

Goodell, H.G., 1964.
Marine geology of the Drake Passage, Scotia Sea and South Sandwich Trench.
USNS Eltanin Mar. Geol. Cruise, 1-8, Florida State Univ., Sedimentol. Lab., 263 pp. (mimeographed).

cores

Goodell, H.G., and J.K. Osmund, 1967.
Marine geological investigations.
Antarctic Jl., U.S.A., 2(5):182-183.

cores

Goodell, H.G., and J.K. Osmond, 1966.
Marine geological investigations in the South Pacific Ocean.
Antarctic J., United States, 1(5):203.

Cores

Goodell, H.G., N.D. Watkins, T.T. Mather and S. Koster 1968.
The Antarctic glacial history recorded in sediments of the Southern Ocean.
Palaeogr. Palaeoclimatol. Palaeoecol. 5(1): 41-62.

cores

Gorshkova, T. I., 1960.
(Sediments of the Norwegian Sea.) Morsk. Geol., Doklady Sovetsk. Geol., 21 Sess., Int. Geol. Congress, 132-139.
English abstract

cores

Gorsline, D.S., and R.A. Stewart, 1962.
Benthic marine exploration of Bahia de San Quintin, Baja California, 1960-61 marine and Quarternary geology.
Pacific Naturalist, 3(8):283-319.

cores

Gouleau, Dominique, et Juan Soriano, 1969.
Étude sédimentologique et géotechnique d'une carotte de la Baie de Bourgneuf.
Cah. océanogr., 21(1):57-70.

cores

Gow, Anthony J., 1971.
Relaxation of ice in deep drill cores from Antarctica. J. geophys. Res., 76(11): 2533-2541.

cores

Gram, Ralph, 1969
Grain surface features in Eltanin cores and Antarctic glaciation.
Antarctic J., U.S., 4(5): 174-175

cores
Grazzini, Colette Vergnaud et Yvonne Herman Rosenberg 1969.
Étude paléoclimatique d'une carotte de Méditerranée orientale.
Revue Geogr. phys. Géol. dynam. (2) 11 (3): 279-292

cores
Great Britain, Intergovernmental Oceanographic Commission, 1963.
H.M.S. "Owen" oceanographic cruise (North and South Atlantic), 1961-1962.
Admiralty Mar. Sci. Publ., (1):17 pp. (multilithed).

cores
Groot, J.J., C.R. Groot, M. Ewing, L. Burckle and J.R. Conolly, 1967.
Spores, pollen, diatoms and provenance of the Argentine Basin sediments.
Progress in Oceanography, 4:179-217.

H

cores
*Habib, Daniel, 1969.
Middle Cretaceous palynomorphs in a deep-sea core from the seismic reflector Horizon A outcrop area.
Micropaleontology, 15(1):85-101.

cores
Hamaguchi, Hiroshi, Masumi Osawa and Naoki Onima, 1962.
The vertical distribution of elements in core samples from the Japan Trench. (In Japanese; English abstract).
Repts. JEDS, Deep-Sea Res. Comm., Japan Soc., Promotion of Science, 3:1-2.

Originally published:
Nippon Kagaku Zasshi, 82:691-693. (1961).

JEDS Contr. No. 21.

cores compaction of
Handy R.L. and R.A. Lohnes 1968.
Sodium chloride content and compaction in deep-sea cores: a discussion.
J. Geol. 76 (3): 365-368.

cores
Harris, W.F., 1964.
A note on pollen distribution in a core from Milford Sound.
New Zealand Dept. Sci. Ind. Res., Bull., No. 157; New Zealand Oceanogr. Inst., Memoir, No. 17:77-78

sediment cores
Hayes, F. R., and M. A. MacAulay, 1959.
Lake water and sediment. V. Oxygen consumed in water over sediment cores.
Limnol. & Oceanogr., 4(3):291-298.

cores
Heezen, Bruce C., 1964.
Discussion of "A note on some possible misinformation from cores obtained by piston-coring device".
J. Sed. Petr., 34(3):699.

Reply by Robert E. Burns: 699.

cores
Heezen, B.C., E.T. Bunce, J.B. Hersey and Marue Tharp, 1964.
Chain and Romanche Fracture zones.
Deep-Sea Research, 11(1):11-33.

cores
Heezen, Bruce C., Bill Glass and H.W. Menard 1966.
The Manihiki Plateau.
Deep-Sea Res. 13 (3): 445-458.

cores
Hendey, N.I., 1958.
Diatoms from equatorial Indian Ocean cores.
Nature 181(4616):953-954.

cores
*Herman, Yvonne and P.B. Rosenberg, 1969.
Mineralogy and micropaleontology of a goethite-bearing Red Sea core.
In: Hot brines and Recent heavy mineral deposits in the Red Sea, E.T. Degen an d D.A. Ross, editors, Springer-Verlag, New York, Inc., 448-459.

coring
Hersey, J. B., 1960.
Acoustically monitored bottom coring.
Deep-Sea Res., 6(2):170-171.

cores
Höhnk, W., 1959
Ein Beitrag zur ozeanischen Mykologie.
Deutsche Hydrogr. Zeits., Ergänzungsheft, Reihe B, 4(3):81-86.

cores
Holmes, Charles W., J.K. Osmond and H.G. Goodell, 1968.
The geochronology of foraminiferal ooze deposits in the Southern Ocean.
Earth Planet. Sci. Letters, 5(6): 367-374

cores
Holmes, C.W., J.K. Osmond and H.G. Goodell, 1966.
The geochronology of Eltanin cores from the South Pacific Ocean.
Antarctic J., United States, 1(5):203-204.

cores
Hough, J.L., 1953.
Pleistocene climatic record in a Pacific Ocean core sample. J. Geol. 61(3):252-262, 3 textfigs.

cores
Hülsemann, J., and K.O. Emery, 1961.
Stratification in Recent sediments of Santa Barbara Basin as controlled by organisms and water character.
J. Geol., 69(3):279-290.

cores
Hunt, John M., Earl E. Hays, Egon T. Degens and David A. Ross, 1967.
Red Sea: detailed survey of hot-brine areas.
Science, 156(3774):514-516.

I

cores
Isaac, N., and E. Picciotto, 1953.
Ionium determination in deep-sea sediments.
Nature 171(4356):742-743.

cores
Inderbitzen, A.L. 1968.
A study of the effects of various core samples on mass physical properties in marine sediments.
J. Sedim. Petrol. 38 (2): 473-489.

cores
Inderbitzen, Anton L., 1965.
An investigation of submarine slope stability
Ocean Sci. and Ocean Eng., Mar. Techn. Soc., Amer. Soc. Limnol. Oceanogr., 2:1309-1344.

J

cores
Joensuu, Oiva, and Eric Olausson, 1964.
Barium content in deep-sea cores and its relationship to organic matter. (Abstract).
Geol. Soc., Amer., Special Paper, No. 76:86-87.

cores
JOINT OCEANOGRAPHIC INSTITUTIONS' DEEP EARTH SAMPLING PROGRAM (JOIDES), 1965.
Ocean drilling on the continental margin.
Science, 150(3697):709-716.

K

cores
Keen, M.J., 1963
The magnetization of sediment cores from the eastern basin of the North Atlantic Ocean.
Deep-Sea Res., 10(5):607-622.

cores
Keen, M. J., 1960
Magnetization of sediment cores from the eastern Atlantic.
Nature, 187(4733): 220-222.

Cores
Kennett, James P., 1970.
Pleistocene paleoclimates and foraminiferal biostratigraphy in subantarctic deep-sea cores.
Deep-Sea Res., 17(1): 125-140.

cores
Kennett, J.P., and K.R. Geitzenauer, 1969.
Pliocene-Pleistocene boundary in a South Pacific deep-sea core.
Nature, Lond., 224 (5222): 899-

cores
Kennett, James P. and Kurt R. Geitzenauer, 1969
Relationships between Globorotalia truncatulinoides and G. tosaensis in a Pliocene-Pleistocene deep-sea core from the South Pacific. Trans. Gulf Coast Ass. geol. Socs, 19 : 613.

cores
Kharkar, D.P., K.K. Turekian and Martha Scott, 1969.
Comparison of sedimentation rates obtained by ^{32}Si and uranium decay series determinations in some siliceous Antarctic cores.
Earth Planet. Sci. Lett. 6(1): 61-68.

cores
Klingebiel, André, Alain Rechiniac et Michel Vigneaux 1967.
Étude radiographique de la structure des sédiments meubles.
Marine Geol. 5 (1): 71-76.

cores
Kobayashi, Kazuo, Kaoru Oinuma and Toshio Sudo, 1961.
20. Clay minerals in the core sample of deep sea sediment from Japan Trench. (English abstract).
Repts. JEDS, 2(16):

cores
Kobayashi, K., K. Oinuma and T. Sudo, 1960.
Clay mineralogical study on Recent marine sediments. (1) Samples JEDS-1-R', JEDS-1-R".
Oceanogr. Mag., Japan, 11(2):215-222.

cores

Kobayashi, Kazuo, Kazuhiro Kitazawa, Taro Kanaya and Toyosaburo Sakai, 1971.
Magnetic and micropaleontological study of deep-sea sediments from the west-central equatorial Pacific. Deep-Sea Res. 18(11): 1045-1062.

cores

Kolbe, R.W., 1957.
Diatoms from Equatorial Indian Ocean cores.
Rept. Swedish Deep-Sea Exped., 1947-48, 9(1): 3-60

cores

Kolbe, R.W., 1955.
Diatoms from Equatorial Atlantic cores.
Repts., Swedish Deep-Sea Exped., Sediment Cores from the N. Atlantic Ocean, 7(3):151-184, 2 pls.

cores

Kolbe, R.W., 1954.
Diatoms from Equatorial Pacific cores. Repts. Swedish Deep-Sea Exped., 1947-48, Sediment Cores from the W. Pacific, 6(1):3-49, 4 pls.

cores, chemistry

Koons, Charles B., 1970.
JOIDES cores: organic geochemical analyses of four Gulf of Mexico and western Atlantic sediment samples.
Geochim. cosmochim. Acta, 34(12):1353-1356

cores

Koreneva, K.B., 1961.
The study of two marine sediment cores from the Sea of Japan by applying the spore pollen method.
Okeanologiia, Akad. Nauk, SSSR, 1(4):651-657.

cores

Krause, Dale C., 1964.
Lithology and sedimentation in the southern continental borderland.
In: Papers in Marine Geology, R.L. Miller, Editor, Macmillan Co., N.Y., 274-318.

cores

Kröll, V.S., 1955.
The distribution of radium in deep-sea cores.
Rept. Swedish Deep-Sea Exped., 1947-1948, 10(1):1-32.

cores

Kulikov, N.N., 1962.
A bottom core from the area of Peter I Island. (In Russian).
Sovetsk. Antarktich. Expedits., Inform. Biull., (35):14-17.

Translation:
(Scripta Tecnica, Inc., for AGU), 4(3):138-140.

cores

Kutschale, Henry 1966.
Arctic Ocean geophysical studies: the southern half of the Siberia Basin.
Geophysics 31(4): 683-710.

L

cores

Lalou, Mlle. Claude, 1962
Formation des carbonates dans les vases prises a differents niveaux dans une carotte du fond de la baie de Villefranche-sur-Mer (Alpes-Maritimes, France).
Trav., Centre de Recherches et d'Etudes Oceanogr., 4(4): 11-18.

cores

Landergren, Sture, 1964.
On the geochemistry of deep-sea sediments.
Repts. Swedish Deep-Sea Exped., 10(Spec. Invest. 5):61-148.

cores

Landergren, S., 1954.
On the geochemistry of the North Atlantic sediment core No. 238. Repts. Swedish Deep-Sea Exped., 1947-48, Sediment cores from N. Atlantic Ocean, 7(2):125-148.

Cores

Landergren, Sture and Oiva Joensuu, 1965.
Studies on trace element distribution in a sediment core from the Pacific Ocean.
Progress in Oceanography, 3:179-189.

coring

Lankford, R.R., and F.P. Shepard, 1960.
Facies interpretations in Mississippi Delta borings.
J. Geology, 68(4):408-426.

cores

Laughton, A.S., 1960.
Some bottom cores from the Swedish Deep-Sea Expedition, 1947-48.
Nature 186(4725):609-610.

cores

Lavrov, V.M., 1964.
Deep-sea sediment cores of the North Atlantic basin of the Atlantic Ocean. Investigation of marine bottom sediments and suspended matter. (In Russia; English abstract).
Trudy Inst. Okeanol., Akad. Nauk, SSSR, 68:136-156.

cores

Linkova, T.I., and A.P. Lisitsyn 1971.
Paleomagnetic investigation of a long column from the western part of the Indian Ocean. (In Russian)
Dokl. Akad. Nauk SSSR 201(2):335-338

cores

Lisitzin, A. P., 1960.
Sedimentation in southern parts of the Indian and Pacific Oceans. Morsk. Geol., Doklady Sovetsk. Geol, 21 Sess, Int. Geol. Kongress, 69-87.
(English abstract)

cores

Lissitzyn, A.P., 1956.
Humidity variation in long samples taken from the Bering Sea. Doklady Akad. Nauk, SSSR, 109(2): 313-316.

M

cores

Margolis, Stanley V. and James P. Kennett, 1970
Antarctic glaciation during the Tertiary recorded in sub-antarctic deep-sea cores.
Science, 170(3962): 1085-1087.

cores

Martini, Erlend, 1965.
Mid-Tertiary calcareous nannoplankton from Pacific Deep-Sea cores.
In: Submarine geology and geophysics, Colston Papers, W.F. Whittard and R. Bradshaw, editors, Butterworth's, London, 17:393-410.

cores

Matsudaira, T., and T. Kawamoto, 1953.
On the subsidence of ground level at Amagasaki region. (II). Analytical data for boring cores and submarine muds. Umi to Sora 30(3): 31-39, 3 textfigs.

cores

Mayudu, Y. Rammohanroy, 1964.
Volcanic ash deposits in the Gulf of Alaska and problems of correlation of deep-sea ash deposits.
Marine Geology, 1(3):194-212.

Mc

cores

McDougall, J.C., 1961
Ironsand deposits offshore from the west coast, North Island, New Zealand.
New Zealand J. Geol. Geophys., 4(3):283-300.

cores

Michard, Gil, 1971.
Theoretical model for manganese distribution in calcareous sediment cores. J. geophys. Res., 76(9): 2179-2186.

cores

*Miyake, Yasuo, Yukio Sugimura and Eiji Matsumoto, 1968.
Ionium-thorium chronology of the Japan Sea cores.
Rec. oceanogr. Wks., Japan, n.s., 9(2):189-195.

cores

Mo, Tin, Barbara C. O'Brien and A.D. Suttle, Jr., 1971.
Uranium: further investigation of uranium content of Caribbean cores P6304-8 and P6304-9.
Earth Planet. Sci. Letts 10(2): 175-178.

cores

Moore, D.G., and A.F. Richards, 1962
Conversion of "relative shear strength" measurements by Arrhenius on east Pacific deep-sea cores to conventional units of shear strength.
Geotechnique, March, 1962, Inst. Civil Engineers, London 55-59.

cores

Morin, Ronald W. 1971.
Late Quaternary biostratigraphy of cores from beneath the California Current.
Micropaleontology 17(4):475-491.

cores

Muller, Jan, 1959.
Palynology of Recent Orinoco delta and shelf sediments. Report of the Orinoco Shelf Expedition, Volume 5.
Micropaleontology, 5(1):1-32.

cores

Murata, K.J., and R.C. Erd, 1964.
Composition of sediments in cores from the experimental Mohole Project (guadalupe site).
J. Sed. Petr., 34(3):633-655.

N

cores

*Nesteroff, Wladimir D., 1966.
Quelques résultats sédimentologiques des premiers forages du précontinent américain (JOIDES).
Bull. Soc. géol. Fr. (7)8(6):773-785.

cores

Nesteroff, W.D., 1961
La "sequence type" dans les turbidites terrigènes modernes.
Rev. Géogr. Phys. et Géol. Dyn., (2), 4(4): 263-268.

Also in:
Trav. Lab. Géol. Dyn. et Centres de Recherches Géodynamiques, Fac. Sic., Univ. Paris, No 2, 1964.

cores
Nesteroff, Wladimir D., Solange DuPlaix, Jacqueline Sauvage, Yves Lancelot, Frédéric Melières et Edith Vincent 1968.
Les dépôts récents du canyon de Cap Breton.
Bull. Soc. géol. France (7) 10:218-252.

cores
Ninkovich, D., and B.C. Heezen, 1967.
Physical and chemical properties of volcanic glass sherds from Pozzuolana ash, Thera Island, and from upper and lower ash layers in eastern Mediterranean deep sea sediments.
Nature, Lond., 213(5076):582-584.

cores
Ninkovich, Dragoslav, and Bruce C. Heezen, 1965.
Santorini tephra.
In: Submarine geology and geophysics, Colston Papers, W.F. Whittard and R. Bradshaw, editors, Butterworth's, London, 17:413-452.

O

cores
Olausson, Eric 1969.
On the Würm-Flandrian boundary in deep-sea cores.
Geologie Mijnb. 48(3):349-361.

cores
Olausson, Eric, 1961
Remarks on some Cenozoic core sequences from the Central Pacific, with a discussion of the role of coccoli hophorids and Foraminifera in carbonate deposition.
Medd. Oceanogr. Inst., Göteborg, 29:35 pp.
Also:
Göteborgs Kungl. Vetenskaps-och Vitterhets-Samhälles Handl., Sjätte Följden, (B), 8(10)

cores
Olausson, Eric, 1961.
Studies of deep sea cores. Sediment cores from the Mediterranean Sea and the Red Sea.
Repts. Swedish Deep-Sea Exped., 1947-1948, 8(4): 337-391.

cores
Olausson, E. and I.U. Olsson, 1969.
Varve stratigraphy in a core from the Gulf of Aden. Paleogrogr., Palaeoclimatol., Palaeoecol. 6: 87-103.

cores
Opdyke, N.D., B. Glass, J.D. Hays and J. Foster, 1966.
Paleomagnetic study of Antarctic deep-sea cores. Paleomagnetic study of sediments in a revolutionary method of dating events in Earth's History.
Science, 154(3747):349-357.

cores
Ovey, C.D., 1951.
International cooperation in the analysis of the deep-sea cores collected by the Swedish Deep Sea Expedition 1947-48. Nature 168(4265):148-149.

cores
Ovey, C.D., 1950.
On the interpretation of climatic variationsas revealed by a study of samples from an Atlantic deep sea core. Centenary Proc. R. Met. Soc.: 211-215.

cores
Ovey, C. D., 1949.
Note on the evidence for climatic changes from sub-oceanic cores. Weather 4(7):228-231, Pls. 3-4, 1 textfig.

P

cores
Pantin, H.M., 1964.
Sedimentation in Milford Sound.
New Zealand Dept. Sci. Ind. Res., Bull., No. 157:
New Zealand Oceanogr. Inst. Memoir, No. 17:35-47.

cores
Pastouret, Léo, 1970
Étude sédimentologique et paléoclimatique de carottes prélevées en Méditerranée orientale.
Téthys 2(1):227-266.

cores
Peronne, Dominique Jacques, 1968.
Contribution à l'étude sédimentologique de carottages sous-marins en Méditerranée.
Cah. océanogr., 20(5):369-394.

cores
Perry, P.B., 1961.
A study of the marine sediments of the Canadian Eastern Arctic Archipelago.
Fish. Res. Bd., Canada, Manuscript Rept. Ser., (Oceanogr. and Limnol.), No. 89:80 pp. (multilithed).

cores
Pettersson, H., 1960.
Poussière d'étoile.
Scientia, 95(12):367-369.

cores
Pettersson, H., 1949.
Exploring the bed of the ocean. Nature 164(4168) 468-470.

cores
Pettersson, H., 1948
Three sediment cores from the Tyrrhenian Sea. Medd. Ocean. Inst., Göteborg 15 (Göteborgs Kungl. Vetenskaps-och Vitterhets-Samhälles Handlingar. Sjätte Följden, Ser.B 5(13)):94 pp.

cores
Phillips, J.D., W.A. Berggren, A. Bertels and D. Wall, 1968.
Paleomagnetic stratigraphy and micropaleontology of three deep sea cores from the central North Atlantic Ocean.
Earth Planet. Sci. Letts 4(2):118-130.

core
Phleger, Fred B, jr., 1948.
Foraminifera of a submarine core from the Caribbean Sea. Medd. Ocean. Inst., Göteborg, 16 (Göteborgs Kungl. Vetenskaps- och Vitterhets-Samhälles Handlingar, Sjätte Följden, Ser. B, 5(14)):9 pp., 1 pl., 2 tables.

cores
Phleger, F. B, jr., 1947.
Foraminifera of three submarine cores from the Tyrhennian Sea. Göteborgs Kungl. Vetenskaps-och Vitterhets- samhälles Handl., Sjätte Följden Ser. B., 5(5):1-19, 1 textfig. (Medd. Från Oceanografiska Institutet i Göteborg, 13). With a foreword by H. Pettersson.

cores
Phleger, F.B, F.L. Parker and J.F. Peirson, 1953.
Sediment cores from the North Atlantic Ocean.
Repts. Swedish Deep-Sea Exped., 1947-1948, 7(1): 1-122, 12 pls., 26 textfigs.

cores
Pilkey, Orrin H., 1964.
Mineralogy of the fine fraction in certain carbonate cores.
Bull. Mar. Sci., Gulf and Caribbean, 14(1):126-139

cores
*Pyle, Thomas E., 1968.
Late Tertiary history of Gulf of Mexico based on a core from Sigshee Knolls.
Bull. Am. Ass. Petr. Geol., 52(11):2242-2246.

Q

R

cores
Ramsay, A.T.S. 1968.
A preliminary study of some Barbados Ridge cores.
Marit. Sed. 4(3): 108-112.

cores
Rees, Anthony I., Ulrich von Rad and Francis P. Shepard, 1968.
Magnetic fabric of sediments from the La Jolla Submarine Canyon and Fan, California.
Mar. Geol., 6(2): 145-178.

cores
Rezak, Richard, Arnold H. Bouma and Lela M. Jeffrey, 1969
Hydrocarbons cored from knolls in southwestern Gulf of Mexico. Trans. Gulf Coast Ass. geol. Socs, 19: 115-118.

cores
Richards, Adrian F., 1962
Investigations of deep-sea sediment cores. II. Mass physical properties.
U.S. Navy Hydrogr. Off. Techn. Rept., TR-106: 146 pp.

cores
Richards, A.F., 1961
Investigations of deep-sea sediment cores. 1. Shear strength, bearing capacity and consolidation.
U.S. Navy Hydrogr. Off., Techn. Rept., TR-63: 70 pp.

cores
Richards, A.F., and G.H. Keller, 1962
Water content variability in a silty clay core off Nova Scotia
Limnol. and Oceanogr., 7(3):426-427

cores
Riedel, W.R., M N. Bramlette and F.L. Parker, 1963
"Pliocene-Pleistocene" boundary in deep-sea sediments.
Science, 140(3572):1238-1240.

cores
Riedel, W.R., and B.M. Funnell, 1965.
Tertiary sediment cores and microfossils from the Pacific Ocean floor.
Proc. Geol. Soc., London, No.1615:48-49. (not seen)

core layer
*Rochford, D.J., 1968.
The continuity of water masses along the western boundary of the Tasman and Coral seas.
Aust. J. mar. Freshwat. Res., 19(2):77-90.

cores
Romankevich, E.A., and N.V. Petrov, 1961.
Oxidation-reduction potential Eh and pH of sediments in the northeastern Pacific.
Trudy Inst. Okeanol., Akad. Nauk, SSSR, 45:72-85.

cores
Rone, Peter A., and C.S. Clay, 1964.
Continuous seismic profiles and cores from the shelf and slope off Cape Hatteras. (Abstract).
Trans. Amer. Geophys. Union, 45(1):72.

cores
Rosenberg, E., 1964.
Purine and pyrimidines in sediments from the experimental Mohole.
Science, 146(3652):1680-1681.

cores
Rosholt, J.N., C. Emiliani, J. Geiss, F.F. Koczy and P.J. Wangersky, 1961.
Absolute dating of deep-sea cores by the Pa 231/Th 230 method.
J. Geology, 69(2):162-185.

cores
Round, F.E. 1968.
The phytoplankton of the Gulf of California. II. The distribution of phytoplanktonic diatoms in cores.
J. exp. mar. Biol. Ecol. 2(1): 64-86.

cores
Ruddiman, William F., 1971.
Pleistocene sedimentation in the equatorial Atlantic: stratigraphy and faunal paleoclimatology.
Bull. Geol. Soc. Am., 82(2): 283-302

cores
Rusnak, Gene A., 1963.
"Absolute dating of deep-sea cores by the Pa 231/Th 230 method" and accumulation rates: a discussion.
J. Geology, 71(6):809-810.

Rosholt, J.N., C. Emiliani, J. Geiss, F.F. Koczy, and P.J. Wangersky, 1963.
"Absolute dating of deep-sea cores by the Pa 231/Th 230 method" and accumulation rates: a reply.
J. Geology, 71(6):810.

cores
Rusnak, Gene A., Albert L. Bowman and H. Göte Östlund, 1964
Miami natural radiocarbon measurements. III.
Radiocarbon, 6:208-214.

cores
Rusnak, Gene A., and W.D. Nesteroff, 1964.
Modern turbidites: terrigenous abyssal plain versus bioclastic basin.
In: Papers in Marine Geology, R.L. Miller, Editor, Macmillan Co., N.Y., 488-507.

cores
Russell, K.L., K.S. Deffeyes, G.A. Fowler and R.M. Lloyd, 1967.
Marine dolomite of unusual isotopic composition.
Science, 155(3759):189-191.

cores
Ryan, William B.F., Fifield Workum Jr. and J.B. Hersey 1965
Sediments on the Tyrrhenian Abyssal Plain.
Bull. Geol. Soc. Am. 76(11): 1261-1282.

S

cores
Sackett, William L., 1964.
The depositional history and isotopic organic carbon composition of marine sediments.
Marine Geology, 2(3):173-185.

cores
Saito, Tsunemasa, Lloyd H. Burckle and Maurice Ewing, 1966.
Lithology and paleontology of the reflective layer Horizon A.
Science, 154(3753):1173-1176.

cores
Saito, T., L.H. Burckle and D.R. Horn, 1967.
Palaeocene core from the Norwegien Basin.
Nature, Lond., 216(5113):357-359.

cores
Sato, T., 1962
A scoriaceous bed in Tokyo Kaiwan. (In Japanese; English abstract).
Hydrogr. Bull. (Publ. No. 981), Tokyo, No. 72: 26-28.

cores
Scafe, D.W., and G.W. Kunze, 1971.
A clay mineral investigation of six cores from the Gulf of Mexico
Marine Geol. 10(1): 69-85.

cores
*Schreiber, B. Charlotte, 1968.
Sound velocity in deep sea sediments.
J. geophys. Res., 73(4):1259-1267.

cores
Sechkina, T.V., 1959.
Diatoms in long cores of bottom sediments in the Sea of Japan. (In Russian).
Doklady Akad. Nauk, SSSR, 126(1):171-174.

cores
Shepard, Francis P., 1964.
Criteria in modern sediments useful in recognizing ancient sedimentary environments.
In: Deltaic and shallow marine deposits, Developments in Sedimentology, L.M.J.U. van Straaten, editor, Elsevier Publishing Co., 1:1-25.

cores
Shepard, F.P., and G. Einsele, 1962.
Sedimentation in San Diego Trough and contributing submarine canyons.
Sedimentology, 1(1):81-133.

cores
Shepard, F.P., & R.R. Lankford, 1959
Sedimentary facies from shallow borings in lower Mississippi delta.
Bull. Amer. Assoc. Petrol. Geol., 43(9): 2051-2067.

cores
Siever, Raymond, Robert M. Garrels, John Kanwisher and Robert A. Berner, 1961
Interstitial waters of Recent marine muds off Cape Cod.
Science, 134(3485):1071-1072.

cores
*Smith, Jerry D. and John H. Foster, 1969.
Geomagnetic reversal in Brunhes normal polarity epoch.
Science 163(3867):565-567.

cores
Stearns, Harold T., and Theodore K. Chamberlain 1967.
Deep cores of Oahu, Hawaii and their bearing on the geologic history of the Central Pacific Basin.
Pacif. Sci., 21(2):153-165.

cores
St. Kröll, V., 1955.
The distribution of radium in deep-sea cores.
Repts. Swedish Deep-Sea Exped., 1947-48, Spec. Invest., 10(1):3-32.

cores
St. Kroll, V., 1953.
Vertical distribution of radium in deep-sea sediments. Nature 171(4356):742, 1 textfig.

T

cores
Thomas, Charles W., 1969.
Paleontological analyses of North Pacific Ocean-bottom cores
Pacific Sci., 23(4): 473-482

cores
Thomas, Charles W., 1959.
Lithology and zoology of an Antarctic Ocean bottom core.
Deep-Sea Res., 6(1):5-15.

cores
Todd, Ruth, 1970.
Maestrichtian (Late Cretaceous) Foraminifera from a deep-sea core off southwestern Africa,
Rev. esp. Micropaleontol. 2(2): 131-154

cores
*Tsuchi, Ryuichi, 1966.
Report of geologic survey by the Umitaka maru, International Indian Ocean Expedition in the winter of 1963-1964. (In Japanese; English abstract).
J. Tokyo Univ. Fish. (Spec. ed.) 8(2):71-81.

cores
Turekian, Karl K. 1968.
Deep-sea deposition of barium, cobalt and silver.
Geochim. cosmochim. Acta 32(6): 603-612

cores
Turekian, K.K., 1957.
The significance of variation in the strontium content of deep sea cores. Limnol. & Oceanogr., 2(4):309-314, 2 textfigs.

cores
Turner, Ralph R., 1971.
The significance of color banding in the upper layers of Kara Sea sediments.
Oceanogr. Rept. U.S. Cst Gd 36 (CG 373-36): 36pp.

U

V

cores
*Van Andel, Tjeerd H., G. Ross Heath, T.C. Moore and David F.R. McGeary, 1967.
Late Quaternary history, climate and oceanography of the Timor Sea, northwestern Australia.
Am. J. Sci., 265(9):737-758.

Core
Van Straaten, Lambertus M.J.U., 1969
Solution of aragonite in a core from the south-eastern Adriatic Sea.
Rapp. P.-v. Réun. Commn int. Explor. scient. Mer Mediterr., 19(4): 643-644.

cores
*Van Straaten, L.M.J.U.,1967.
Solution of aragonite in a core from the southeastern Adriatic Sea.
Marine Geol., 5(4):241-248.

cores
Van Straaten,L.M.J.U.,1966.
Micro-malacological investigation of cores from the southern Adriatic Sea.
Proc.K. ned.Akad. Wet., (B),69(3):429-445.

cores
Van Straaten, L.M.J.U., 1965.
Coastal barrier deposits in South- and Nort Holland in particular in the areas around Scheveningen and IJmuiden.
Mededelingen van de Geologische Stichting, No. 17:41-75.

cores
Van Straaten, L.M.J.U., 1964
Turbidite sediments in the southeastern Adriatic Sea.
In: Turbidites, A.H. Bouma and A. Brouwer, Editors. Developments in Sedimentology, Elsevier Publishing Co., 3:142-147.

cores
Vikhrenko, M.M., 1962
[Luminescent-bituminological characteristic of the organic matter in cores from the North Atlantic Ocean.]
Trudy Inst. Okeanol., Akad. Nauk, SSSR: 56: 32-58.

cores
Vronskiy, V.A., 1970.
Spore-and-pollen analysis of a sediment core from the Mediterranean Sea. (In Russian; Englis abstract). Okeanologiia, 10(6): 1028-1033.

W

cores
Wangersky, Peter J. 1967.
Sodium chloride content and the compaction in deep-sea cores.
J. Geol. 75 (3): 332-335.

cores
Wangersky, Peter J., 1958.
The sea as a clue to the earth's history.
Yale Scientific Magazine, 33(3):6 pp.

cores
Wangersky, P.J., 1962
Sedimentation in three carbonate cores.
J. Geology, 70(3):364-374.

cores
Wangersky, Peter J., and Oiva I. Joensuu 1967.
The fractionation of carbonate deep-sea cores.
J. Geol. 75(2): 148-177.

cores
Wangersky, Peter J., and Oiva Joensuu, 1964
Strontium, magnesium and manganese in fossil foraminiferal carbonates.
J.Geol., 72(4):477-483.

cores
Watkins, N.D. 1969
Continuing studies of Eltanin sedimentary cores and dredged rocks
Antarctic Jl, U.S. 4(5): 177.

cores
Watkins, N.D., and H.G. Goodell, 1966.
The stratigraphic use of paleomagnetism in sedimentary cores from the Southern Ocean. (Abstract only).
Trans. Am. Geophys. Un., 47 (3):478.

cores
Welte, Dietrich H., und Götz Ebhardt, 1965.
Die Verteilung höherer, geradkettiger Paraffine und Fettsäuren in einem Sedimentprofil aus dem Persischen Golf.
Meteor Forschungsergebn. (C) 1: 43-52.

cores
*Windisch, Charles C., R.J. Leyden, J.L. Worzel, T. Saito and J. Ewing, 1968.
Investigation of Horizon Beta.
Science, 162(3861):1473-1479.

core
Wiseman, J.D.H., 1959.
The relation between paleotemperature and carbonate in an equatorial Atlantic pilot core.
J. Geology, 67(6):685-690.

cores
Wiseman, J.D.H., 1958.
Secondary oscillations in an equatorial Atlantic deep-sea core.
Nature, 182(4648):1534-1535.

cores
Wiseman, J.D.H., 1954.
The determination and significance of past temperature changes in the upper layer of the equatorial Atlantic Ocean. Proc. R. Soc., London, A, 222:296-323, 3 textfigs.

cores
Wiseman, J.D.H., C. Emiliani, R. Yalkovsky, 1959
The relationship between paleotemperatures and carbonate content in a deep-sea core: A discussion.
J. Geol. 67(5):572-576.

XYZ

cores
Yalkovsky, Ralph 1967.
Time-series analysis of Caribbean deep-sea core A172-6.
J. Geol. 75(2): 225-

cores
Yalkovsky, Ralph, 1967.
Signs test applied to Caribbean Core A 176-6.
Science, 155(3768):1408-1409.

cores
Yalkovsky, Ralph, 1964
Time series analysis of Caribbean core A172-6. (Abstract).
Trans. Amer. Geophys. Union, 45(1): 71.

cores
Yalkovsky, R., 1957.
The relationship between the paleotemperature and cabonate content in a deep-sea core.
J. Geol., 65(5):480-496.

cores
Young, Edward J. 1965.
Spectrographic data on cores from the Pacific Ocean and the Gulf of Mexico.
Geochim. cosmochim. Acta. 32(4): 466-471

cores
Zagwijn, W.H., and H.J. Veenstra 1966
A pollen-analytical study of cores from the Outer Silver Pit, North Sea.
Marine Geol., 4(6): 539-551.

cores
Zhuze
Gzuse, A.P., 1963
The problems of stratigraphy and paleogeography in the northern part of the Pacific Ocean according to the data of diatom analysis. (In Russian).
Okeanologiia, Akad. Nauk, SSSR, 3(6):1017-1028.

Cores, acoustical properties

cores acoustics of
Horn, D.R., B.M. Horn and M.N. Delach, 1968.
Correlation between acoustical and other physical properties of deep-sea cores.
J. geophys. Res., 73(6):1939-1957.

cores, carbonate
Bartlet, Grant A., and Robert G. Greggs 1970.
A reinterpretation of stylolitic solution surfaces deduced from carbonate cores from San Pablo Seamount and the Mid-Atlantic Ridge.
Can. J. Earth Sci. 7(2): 274-279.

Cores, chemistry of

cores, chemistry
Bender, Michael, Wallace Broecker, Vivian Gornitz, Ursula Middel, Robert Kay, Shine-Soon Sun, and Pierre Biscaye 1971.
Geochemistry of three cores from the East Pacific Rise.
Earth planet. Sci. Letts. 12(4): 425-433.

cores, chemistry of
Chester, R. and M.J. Hughes, 1969.
The trace element geochemistry of a North Pacifi pelagic clay core. Deep-Sea Res., 16(6): 639-654.

cores
Thompson, Geoffrey, 1968.
Analyses of B, Ga, Rb and K in two deep-sea sediment cores, consideration of their use as paleoenvironmental indicators.
Marine Geol. 6(6): 463-477

cores, chemistry of
Turekian, Karl K., 1964.
The marine geochemistry of strontium.
Geochim. et Cosmochim. Acta, 28(9):28(9):1479-1496.

Cores, Compaction

Cores, Consolidation of

cores, consolidation of
*Almagor, Gideon, 1967.
Interpretation of strength and consolidation data from some bottom cores off Tel-Aviv-Palmakhim coast of Israel.
In: Marine geotechnique, A.F. Richards, editor, Univ. Illinois Press, 131-153.

cores, consolidation
*Richards, Adrian F., and Edwin L. Hamilton, 1967.
Investigations of deep-sea sediment cores. III. Consolidation.
In: Marine geotechnology, A.F. Richards, editor, Univ. Illinois Press, 93-117.

Cores, geomagnetism of

cores, geomagnitism of

Watkins, N.D. 1968.
Short period geomagnetic polarity events in deep-sea sedimentary cores.
Earth Planet Sci. Letters, 4(5): 341-349.

cores, polarity of

Opdyke, N.D., 1970.
1. General observations. 5. Paleomagnetism (June 1968). In: The sea: ideas and observations on progress in the study of the seas, Arthur E. Maxwell, editor, Wiley-Interscience, 4(1): 157-182.

Cores properties of

cores, properties of
*Einsele, Gerhard, 1967.
Sedimentary processes and physical properties of cores from the Red Sea, Gulf of Aden, and off the Nile Delta.
In: Marine geotechnique, A.F. Richards, editor, Univ. Illinois Press, 154-176.

Cores samples

cores samples
Lissitzyn, A.P., 1956.
[Humidity variation in long samples taken from the Bering Sea.] Dokl. Akad. Nauk, SSSR, 108(2):313-316.

Cores, strength of

cores, strength of
*Almagor, Gideon, 1967.
Interpretation of strength and consolidation data from some bottom cores off Tel-Aviv-Palmakhim coast of Israel.
In: Marine geotechnique, A.F. Richards, editor, Univ. Illinois Press, 131-153.

Coring

coring
Peterson, M.N.A., 1970.
Ocean sediment coring program. EOS, Trans. Am. geophys. Un. 52(6): 256-258.

coring
Taylor, Donald M., 1968.
The Challenger's adventure begins.
Ocean Indust. 3(10): 35-50.

Coriolis force

Coriolis force
Abbott, M.R., 1960
Coriolis effects on the tidal flow in an exponential estuary.
Houille Blanche, 15(5):616-624.

Abstr. in:
Appl. Mech. Rev., 14(9):745.

Coriolis force
Arakawa, H., 1960
Effect of rotation on convective motion.
Pap. Meteor. & Geophys., 11(2/4):191-195.

Coriolis force, effect of
Barcelon, Victor, 1965.
Stability of a non-divergent Ekman layer.
Tellus, 17(1):53-68.

Coriolis force
Bonnefille, R., 1957.
Formation experimentale d'un point amphidromique sous l'effet de la force de Coriolis.
Bull. d'Info., C.C.O.E.C., 9(5):265-269.

Coriolis force
Bonnefille, R., 1956.
Étude de la circulation des courants dans la Manche en tenant compte du frottement et de la force de Coriolis.
Bull. d'Info., C.C.O.E.C., 8(7):306-313, 9 pls.

Coriolis Force
Campbell, William J., 1965.
The wind-driven Circulation of ice and water in a polar ocean.
J. Geophys. Res., 70(14):3279-3301.

Coriolis force
Fell H. Barraclough 1967.
Resolution of Coriolis parameters for former epochs.
Nature, Lond., 214 (5094): 1192-1195.

Coriolis force
Felsenbaum, A.I., 1956.
[An extension of Ekman's theory to the case of a non-uniform wind and an arbitrary bottom relief in a closed sea.] Dokl. Akad. Nauk, SSSR, 109(2): 299-302.

Coriolis force
Hachey, H.B., 1961.
Oceanography and Canadian Atlantic waters.
Fish. Res. Bd., Canada, Bull. No. 134:120 pp.

Coriolis force
Haurwitz, B., 1950.
Internal waves of tidal character. Trans Am. Geophys. Union 31(1):47-52, 2 textfigs.

Coriolis force
Hayami, S., H. Kawai and M. Ouchi, 1955.
On the theorem of Helland-Hansen and Ekman and some of its applications. Rec. Ocean. Wks., Japan 2(2):56-67, 9 textfigs.

Coriolis force
Hela, I., 1946.
Coriolis-voiman vaikutuksesta suomenlahden hidrografisiin oloihin. Terra 58(2):52-59, Fig. 2.

Coriolis force
Hidaka, K., and H. Miyoshi, 1949.
On the neglect of the inertia terms in dynamical oceanography. Ocean. Mag., Tokyo, 1(4):185-193, 6 textfigs.

Coriolis force
Hidaka, K., and H. Miyoshi, 1949.
On the neglect of the inertia terms in dynamical oceanography. Gepphys. Notes, Tokyo, 2(22):1-9, 6 textfigs.

Coriolis force
Howell, B.F., Jr., 1970.
Coriolis force and the new global tectonics.
J. geophys. Res., 75(14): 2769-2772.

Coriolis' Force
Ichie, T., 1949.
On the diffusion in the field of Coriolis' Force. Ocean. Mag., Tokyo, 1(2):121-127, 4 textfigs.

Coriolis force
Jung, K., 1951.
Die Corioliskraft im Anfängerunterricht. Deutsch. Hydro. Zeits. 4(1/2):46-51.

Coriolis' force
Kuenen, Ph.H., 1948
Influence of the earth's rotation on ventilation currents of the Moluccan deep-sea basins. Proc. Koninklijke Nederlandsche Akademie van Wetenschappen, Ll(4):417-426, 4 text figs.

Coriolis force
Miyazaki, M., 1950.
On drift currents in the ocean surrounding the earth (an extension of Goldsbrough's theory).
Mem., Kobe Mar. Obs. 8:41-43.

Coriolis Force
Mortimer, C.H., 1955.
Some effects of the earth's rotation on water movements in stratified lakes.
Verh. Int. Ver. Limnol. 12:66-77, 7 textfigs.

Coriolis Force
Pomeroy, A.S., 1965.
Notes on the physical oceanographic environment of the Republic of South Africa.
Naval Oceanogr. Res. Unit, 41 pp. (Unpublished manuscript).

Coriolis force
Saint-Guily, B., 1962
On the general form of the Ekman problem.
Proc. Symposium on Mathematical-Hydrodynamical Methods of Phys. Oceanogr., Sept. 1961, Inst. Meeresk., Hamburg, 61-73.

Coriolis force
Saint-Guily, B., 1960.
Écoulement plan autour d'un cercle en présence d'une force de Coriolis de paramètre variant avec la latitude.
C.R. Acad. Sci., Paris, 250(17):2920-2921.

Coriolis force
Saint-Guily, B., 1959.
[The influence of Coriolis forces on sea currents] Houille Blanche 14(5):556-559.

Rev. in: Appl. Mech. Rev., 13(11):#6126.

Coriolis effect
Saint-Guily, B., 1959.
Essai en vue d'une théorie d'Ekman generalisée.
Cahiers Océanogr., C.C.O.E.C., 11(2):101-129.

Coriolis
Saint-Guily, B., 1959.
Note sur l'action de la force de Coriolis dans la circulation convective.
Bull. de l'Inst. Ocean., 1148: 7 pp.

Coriolis Force
Saint-Guily, B., 1958.
Mouvements radiaux de Hamel en presence d'une force de Coriolis. Bull. d'Info., C.C.O.E.C., 10(6):324-334.

Coriolis force
Saint-Guily, B., 1957.
Les mouvements radiaux de Hamel lorsqu'il existe unde force de Coriolis et la structure de certains courants océaniques.
C.R. Acad. Sci., Paris, 244(11):1528-1529.

Coriolis force
Saint-Guily, B., 1956.
Sur la théorie des courants marins induits par le vent. Ann. Inst. Océan., Monaco, n.s., 33(1):1-64.

Coriolis force
Sasaki, Yoshikazu, 1964.
Resonance phenomena of inertia-internal gravity waves.
In: Studies on Oceanography dedicated to Professor Hidaka in commemoration of his sixtieth birthday.
244-253.

Coriolis force
Shtokman, V.B., 1950.
[Determination of the flow rate and density distribution in the cross-section of an infinite channel in relation to the effect of wind and lateral friction in the field of Coriolis force.] Dok. Akad. USSR, 71:41-44.

Coriolis force
Stewart, R.W., 1969.
The atmosphere and the ocean.
Scient. Am., 221(3):76-86 (popular).

Coriolis force, etc.
Tanner, W.F., 1963.
Spiral flow in rivers, shallow seas, dust devils and models.
Science, 139(3549):41-42.

Coriolis force
Unoki, S., 1951.
[On the upheaving effect of stationary and non-stationary winds upon sea level (continued). An investigation on meteorological tides in the neighboring seas of Japan. 3rd paper.]
J. Met. Soc., Japan, 29(10):336-346, 4 textfigs.

Coriolis force
Waldichuk, Michael, 1957.
Physical oceanography of the Strait of Georgia.
J. Fish. Res. Bd., Canada, 14(3):321-486.

Coriolis force
Walker, M.J., 1958.
Graphic aids for teaching Coriolis force.
Amer. J. Phys., 26(6):392-395.

Coriolis force, effect of

Coriolis force, effect of
Al'tshuler, V.M., 1961.
Divergence due to force of rotation of the earth and its calculation in computing tidal currents. (In Russian).
Trudy Gosud. Okeanogr. Inst., 63:3-7.

Coriolis force, effect of
Arbot, M.R., 1960
L'influence de la force de Coriolis sur les courants de marée dans un estuarie exponential
La Houille Blanche (5): 616-624.

Coriolis Force, effect of
Bonnefille, R., 1957.
Étude expérimentale de l'influence de la force de Coriolis sur la propagation de la marée dans la Manche. La Houille Blanche, 12(Oct. 1957).

Coriolis force, effect of
Bonnefille, R., 1957.
Étude expérimentale de l'influence de la force de Coriolis sur la propagation de la marée dans la Manche. Mém. et Trav. Soc. Hydrotech., France, 2

Coriolis force, effect of
Bonnefille, R., 1957.
Formation experimentale d'un point amphidromique sous l'effet de la force de Coriolis.
Bull. d'Info., C.C.O.E.C., 9(5):265-269.

Coriolis force, effect of
Chisnell, R.F., 1966.
Outflow from a circular cylinder under the action of Coriolis force.
Tellus, 18(1):77-78.

Coriolis force, effect of
Defant, F., 1953.
Theorie der Seiches des Michigansees und ihre Abwandlung durch Wirkung der Corioliskraft.
Arch. Met., Vienna, A6:218-241.

Coriolis force, effect of
France, Institut Geographique National, 1964.
La photographie aérienne et l'étude des dépôts prélittoraux.
Études de Photo-Interprétation, Clos-Arceduc, Paris, No. 1:53 pp.

Coriolis Force, effect of
Hidaka, Koji, 1962
Non-linear theory of an equatorial flow with special application to the Cromwell Current.
J. Oceanogr. Soc., Japan, 20th Ann. Vol., 223-241.

Coriolis Force, effect of
Ichiye, Takashi, 1964
The effect of horizontal component of Coriolis Force on internal waves. (Abstract).
Trans. Amer. Geophys. Union, 45(1): 67.

Coriolis force, effect of
Kajiura, K., 1958
Effect of Coriolis force on edge waves (II) Specific examples of free and forced waves.
J. Mar. Res. 16(2): 145-157.

Coriolis force, effect of
Kosiba, A., 1956.
[On the deflecting forces action on the air and water currents (a contribution to the teaching methods in meteorology).] Przeglad Geofiz., Poland 1(9)(1):37-47.

Coriolis force, effect of
Krauss, Wolfgang, 1961.
Über den Einfluss der Erdrotation auf interne Wellen.
Kieler Meeresf., 17(1):8-16.

Coriolis force
Lacombe, Henri, 1965.
Cours d'oceanographie physique. (Theories de la circulation. Houles et vagues).
Gauthier-Villars, Paris, 392 pp.

Coriolis force, effect of
Moore, D.E. 1969.
Construction of f/H maps and comparison of measured deep-sea motions with the f/H map for the North Atlantic Ocean.
Techn. Rept. Chesapeake Bay Inst. 51 (Ref. 69-4): 61 pp.

Coriolis force, effect of
Nakamura, Kohei, 1961
Velocity of long gravity waves in the ocean.
Sci. Repts., Tohoku Univ., (5 Geophys.), 13 (3):164-173.

Coriolis force, effect of
Namikawa, Tomikazu, 1961.
The occurrence of over-stability of a layer of fluid heated from below and subject to the simultaneous action of magnetic field and rotation.
J. Inst. Polytechnics, Osaka City Univ., (G)5(1):7011.

Coriolis force, effect of
Niiler, P.P., and F.E. Bisshopp, 1965.
On the influence of Coriolis force on onset of thermal convection.
J. Fluid Mech., 22(4):753-761.

Coriolis force, effect of
Reid, R.O., 1958.
Effect of Coriolis force on edge waves (I). Investigation of the normal modes.
J. Mar. Res., 16(2):109-144.

Coriolis force, effect of
Röber, Klaus, 1970
Analytische und numerische Lösungen für Mitschwingungsgezeiten in einem Rechteckbecken konstanter Tiefe unter Berücksichtigung von Bodenreibung, Corioliskraft und horizontalem Austausch.
Mitt. Inst. Meeresk. Univ. Hamburg, 16: 119 pp. (multilithed)

Coriolis force, effect of
Rzheplinsky, D.G., 1970.
The effect of Coriolis force and bottom topography on wind currents around oceanic islands. (In Russian; English abstract).
Fisika Atmosfer. Okean., Izv. Akad. Nauk SSSR, 6(7): 715-727.
circulation

Coriolis force, effect of
Saint-Guily, Bernard, 1961
Influence de la variation avec la latitude du parametre de Coriolis sur les mouvements plans d'un fluide parfait.
Cahiers Oceanogr., C.C.O.E.C., 13(3): 167-175.

Coriolis force, effect o
Saint-Guily, B., 1958.
Remarque sur l'importance de la force de Coriolis dans les courants marins. Bull. d'Indo., C.C.O.E.C., 10(4):215-218.

Coriolis Force, effect of
Shtokman, 1950.
[Determination of current-velocities and density distribution on a cross-section of an infinitely long channel, as related to the action of the wind and of side-friction in the Coriolis-force field.]
Dok. Akad. Nauk, SSSR, 71(1):41-44.
T6OR

Coriolis force, effect of
Takano, K., 1955.
A complimentary note on the diffusion of the seaward river flow off the mouth.
J. Ocean. Soc., Japan, 11(4):147-149.

corrections

corrections
Kuwahara, S., 1938
Correction to the echo-depth for the density of water in the Pacific Ocean.
Japan. J. Astron. & Geophys., 14(2/3): 43-78.

correction, Eötvös
Sagitov, M.U., and G.D. Marchuk, 1962
[The Eötvös correction for currents during gravity determinations at sea.]
Geofiz. Biull., Mezhduved. Geofiz. Komitet, Prezid., Akad. Nauk, SSSR, No. 11:40-42.

correction
*Stern, Melvin E., 1969.
Salt fingers convection and the energetics of the general circulation.
Deep-Sea Res., Suppl. 16: 263-267.

Correlations

correlations
Moriyasu, Shigeo 1967.
On the anomaly of the sea surface temperature in the East China Sea.
Oceanogrl Mag. 19(2): 201-220.

corrosion

Corrosion

Acker, Robert F. 1967.
The why of marine microbiology.
Naval Res. Rev. (March): 10-17.

corrosion

Arbuzova, K.S., 1961.
The effect of fouling on steel corrosion in the Black Sea. (In Russian).
Trudy Inst. Okeanol., Akad. Nauk, SSSR, 49:266-273

USN-HO-TRANS 183
M. Slessers 1963
P.C. 32672

corrosion

Arbuzova, K.S., and V.V. Patrikeev, 1960.
The role of Balanus in the corrosion of spotless steel in the Black Sea.
Doklady Akad. Nauk, SSSR, 132(3):693-695.

corrosion

Arocena, A. A., 1958.

La fosfatacion "directa" de las estructuras marinas.
B. Inst. Esp. Ocean., 91:1-21.

corrosion

Arocena, Antonio Arevalo 1958.

La fosfatacion del hierro y su accion anticorrosiva frente al medio marino.
Inst. Espagnol de Oceanografia
Serie Trabajos No. 24- 5-5-58.

corrosion

Atkins, W.R.G., 1944.
Measurement of potential difference as a method for studying the action of water on lead pipes.
Nature 154:211.

corrosion

Barnard, K.N., G.L. Christie, and J.H. Greenblatt 1953.
Cathodi protection of active ships in sea water with graphite anodes. Corrosion 9(8):246-250, 8 textfigs.

corrosion

Barnes, H., 1956.
Corrosion and shell composition in barnacles.
Nature 177(4502):290.

corrosion

Barriety, L., J. Debyser and A. Hache, 1956.
La station d'essais de corrosion au Musée de la mer. Bull., CERS, Biarritz, 1(2):283-296.

corrosion

Bartha, S., and S. Henriksson 1971
The growth of sea-organisms and the effect on the corrosion-resistance of stainless steel and titanium.
Trav. Cent. Rech. Études océanogr. n.s. 10(4): 20pp.

Corrosion

Boom, J.W., 1965.
Paint tests in sea water on floating pontoons.
Trav. Centre de Recherches et d'Etudes Oceanogr.
n.s., 6(1/4):31-36.

Corrosion

Booth, G.H., Pamela M. Shinn and D.S. Wakerly, 1965.
The influence of various strains of actively growing sulfate-reducing bacteria on the anaerobic corrosion of mild steel.
Trav. Centre de Recherches et d'Etudes Oceanogr.
n.s., 6(1/4):363-371

corrosion

Borriety, L., J. Bebyser and A. Hache, 1956.
La station d'essais de corrosion au musée de la mer.
Bull., C.E.R.S., Biarritz, 1(2):283-296.

corrosion

Boyd, W.K., 1969.
Comments on "Coping with the problem of the stress-corrosion cracking of structural alloys in sea-water" by B.F. Brown.
Ocean Engng 1(3): 297

corrosion

Bradley, W.G., 1955.
Corrosion control on the hulls of harbor craft and small ships. Texas J. Sci. 7(4):396-401.

Corrosion

Brisou, J., et Y. de Rautlin de la Roy, 1965.
Le role des bacteries aerobies sulfhydrogenes dans la corrosion des metaux.
Trav. Centre de Recherches et d'Etudes Oceanogr.
n.s., 6(1/4):373-375.

corrosion

* Brown, B.F., 1969.
Coping with the problem of the stress-corrosion cracking of structural alloys in sea water.
Ocean Engng 1(3): 291-296.

corrosion

Callame, B., 1959.

Caractères physico-chimiques et biologiques de la station de corrosion de la Rochelle.
Trav. Centr. Recherches et d'Études Oceanogr., ns, 3(3): 19-28.

corrosion

Campbell, Hector S., 1969.
The compromise between mechanical properties and corrosion resistance in copper and aluminium alloys for marine applications.
Ocean Engng 1(4): 387-393

Corrosion

Cauchetier, Martinez et Orlowski, 1965.
Protection contre la corrosion par metallisation en mer et en climat marin.
Trav. Centre de Recherches et d'Etudes Oceanogr.
n.s., 6(1/4):107-111.

Corrosion

Cornet, I., T.W. Pross, Jr., and B.C. Bloom, 1965.
Current requirements for cathodic protection of disks rotatingin salt water - a mass transfer analysis.
Trav. Centre de Recherches et d'Etudes Oceanogr.
n.s.,,6(1/4):175-181.

Corrosion

Defrancux, J.M., 1965.
Sur la resistance a la corrosion des aciers inoxydables dans l'eau de mer.
Trav. Centre de Recherches er d'Etudes Oceanogr.
n.s., 6(1/4):57-64.

corrosion

Dellies, O.J., N.J. Levenson and C.L. Hikes, 1971
Underwater corrosion experience with towed vehicles instrument housings and tow cables.
J. mar. techn. Soc. 5(1): 47-56

Corrosion

Domanski, Andrezej, and Jersy Birn, 1965.
Problems of ship's propeller protection.
Trav. Centre de Recherches et d'Etudes Oceanogr.
n.s., 6(1/4):19-30.

Corrosion

Fink, F.W., R.G. Fuller, L.J. Nowacki, B.G. Brand, and W.K. Boyd, 1965.
Navigational buoy corrosion and deterioration.
Trav. Centre de Recherches et d'Etudes Oceanogr.
n.s.,6(1/4):77-82.

Corrosion

Gatzek, L.E., 1965.
Corrosion prevention techniques for a shipboard missile system.
Trav. Centre de Recherches et d'Etudes Oceanogr.
n.s., 6 (1/4):83-92.

corrosion

Geld, Isidore, and Samuel A. Davey, 1970.
Stress corrosion of a coated titanium alloy plate in sea water.
Ocean Engng. 1(6): 611-616

Corrosion

Gherardi, D., M. Troyli, L. Rivola, et G. Bombara, 1965.
Experiences relatives a la corrosion et a la protection des aciers dans l'eau de mer.
Trav. Centre de Recherches et d'Etudes Oceanogr.
n.s., 6 (1/4):93-101.

Corrosion

Godard, Hugh P., andF.F. Booth, 1965.
Corrosion behaviour of aluminium alloys in seawater.
Trav. Centre de Recherches et d'Etudes Oceanogr.
n.s., 6(1/4):37-52.

Corrosion

Guillen Rodrigo, Miguel A., 1965.
Influence de la temperature et de l'agitation sur la formation du depot magnesium-calcaire dans l'acier protege cathodiquement dans l'eau de mer.
Trav. Centre de Recherches et d'Etudes Oceanogr.
n.s.,6(1/4):197-203.

Corrosion

Hache, A. 1966.
Étude du comportement de quelques revêtements de zinc dans l'eau de mer.
Bull. Cent. Étud. Rech. scient. Biarritz 6(2): 329-339

corrosion

Hache, A., 1962.
Compte rendu résumé des Journées d'Études de la corrosion en atmosphère marine des matériaux metalliques.
Bull. Centre Études Recherches Scientifiques, Biarritz, 4(1):67-76.

salt water corrosion

Hache, A., 1960

Journées préparatoires d'études de la corrosion marine et de la protection cathodique de l'Acier, Biarritz 2-4 avril 1959. Bull. C.E.R.S. Biarritz 3(1): 109-126.

corrosion

Hache, A., 1959

La Corrosion de l'Acier en eau de mer. Trav. Centre Recherches et d'Etudes Oceanogr., n.s. 3(3): 29-49.

Corrosion

Hache, A., et L. Barriety, 1965.
La teneur en Chlorure d'une atmosphere marine et son influence sur la corrosion de l'acier.
Trav. Centre de Recherches et d'Etudes Oceanogr. n.s., 6(1/4):295-300.

corrosion

Hache, A., and P. Deschamps, 1954.
Etude de la corrosivité de l'eau de mer.
Corr. et Anticorr. 2:134-140.
Contr. C.R.E.O. Vol. 5.

corrosion

Heinemann, G., 1951.
Problems of industries using sea water. Texas J. Sci. 3(2):311-321, 4 textfigs.

Corrosion

Herzog, Eugene, Lazzlo Backer et Antonio Niguel Valero, 1965.
Mecanisme de la corrosion des acieres dans des solutions de chlorure de sodium et dans L'eau de mer.
Trav. Centre de Recherches et d'Etudes Oceanogr. n.s., 6(1/4):165-174.

corrosion

Hoar, T.P., 1961
Electrochemical principles of the corrosion and protection of metals.
J. App. Chem., 11(4): 121-129.

Corrosion

Hoffmann, Klaus, 1965.
Ueber die Korrosion einiger Kupfer-Zink-Legierungen im Meerwasser.
Trav. Centre de Recherches et d'Etudes Oceanogr. n.s. ,6(1/4):103-106.

corrosion

Ishchenko, N. I., and I. B. Ulanovskii, 1963.
Protective effect of aerobic bacteria against corrosion of carbon steel in sea water. (In Russian).— English abstract)
Mikrobiologiia, 32(3):521-525.

translation for NSF:
32(3):445-448.

corrosion

Ivanov, S.A., I.B. Ulanovskki, and E.S. Rit, 1950.
[A more rapid method of determining the required current density in cathodic protection against Corrosion] Zavodskaya Laboratoriya 7:833-835.

Translation: E. Hope, D.S.I. Service, D.R.B., Canada, T91R dated 15 Apr. 1953.

corrosion

LeGrand, R., and M. Lambert, 1962.
Mesures électrochimiques appliquées à l'étude de la protection cathodique des ouvrages de la Rance.
Mém. et Trav., Soc. Hydrotechn., France, Suppl. to Vol. 1, 1962:65-81.

Corrosion

La Que, F.L. 1968.
Materials selection for ocean engineering.
In: Ocean engineering: goals, environment, technology, John F. Brahtz, editor, John Wiley + Sons, pp. 588-632.

corrosion

Leidheiser, Henry, Jr., 1965.
Corrosion: sometimes good is mostly bad.
Chemical and Engineering News, 43(14):78-92.

corrosion

Licheron, S., 1962.
La lutte contre la corrosion du matériel des usines marémotrices.
Mem. et Trav., Soc. Hydrotechn., France, Suppl. au Vol. 1, 1962:52-64.

corrosion

Littauer, E.L., 1966.
Impressed current systems for corrosion protection.
GeoMarine Techn., 2(6):17-23.

corrosion

Marette, D., 1960
Comportement à la corrosion et à la protection cathodique de l'acier immergé dans une solution saline simple en mouvement.
Bull. du C.E.R.S., Biarritz, 3(2): 231-248.

corrosion

Marette, D., 1960
Réunion franco-espagnole d'étude de la corrosion marin, Biarritz-Madrid (mai 1960).
Bull. C.E.R.S., Biarritz, 3(1): 127-130.

corrosion

Marette, D., and M.-M. Cugnier, 1962.
Comportement à la corrosion et à la protection cathodique de l'acier immergé dans une eau de mer naturelle en mouvement.
Bull. Centre Études Recherches, Scientifiques, Biarritz, 4(1):77-91.

corrosion

Masseille, H., 1948.
Notes sur la protection des carènes métalliques.
Chim. Peint. Belg., 11:218-221.

Corrosion

May. T. P., and B.A. Weldon, 1965.
Copper-nickel alloys for service in sea water.
Trav. Centre de Recherches et d'Etudes Oceanogr. n.s.,6(1/4):141-156.

corrosion

Mine, S., 1953.
Cathodic protection research against steel sheet-piling corrosion in the Gdansk abd Gydnia-harbours. Prace Morsiego Inst. Tech., No. 2:3-30.

Corrosion

Mor, E., et G. Milanese, 1965.
Recherches sur la corrosion de cables en acier en miliur marin.
Trav. Centre de Recherches et d'Etudes Oceanogr. n.s., 6(1/4):495-502.

corrosion

Olson, F.C.W., 1953.
Tampa Bay studies.
Ocean. Inst., Florida State Univ., Rept. 1:27 pp. (mimeographed), 21 figs. (multilithed).
UNPUBLISHED

corrosion

Proctor, I.A., 1957.
Study of corrosion patterns on a steel immersed in sea water.
Bull. Centre Etudes Rech. Sci., 1(3):435-444.

corrosion

Raclot, B., 1960.
Les actions secondaires dans la protection cathodique des structures immergées par anodes de magnésium.
Trav. Centre Recherches et d'Etudes Océanogr., n.s., 3(3):11-17.

corrosion

Reinhart, Fred M., 1965.
First results - deep ocean corrosion.
GeoMar. Techn., 1(9):15-26.

corrosion

Richards, Francis A., 1968.
Chemical and biological factors in the marine environment. Ch. 8 in: Ocean engineering: goals, environment, technology, J.F. Brahtz, editor, John Wiley & Sons, 259-303.

Corrosion

Richaud, H., 1965.
Etude potentiostatique de la protection de l'aluminium par les chromates en presence de chlorures.
Trav. Centre de Recherches et d'Etudes Oceanogr. n.s.,6(1/4):217-220.

Corrosion

Richardson, J.I., 1965.
Zinc silicate coatings for protection of steel structures exposed to marine corrosion.
Trav. Centre de Recherches et d'Etudes Oceanogr. n.s., 6(1/4):221-223.

corrosion

Rogers, T. Howard, 1968.
Marine corrosion.
GeorgeNewnes Ltd., $7.50.

corrosion

Rogers, T.H., 1960
The marine corrosion handbook.
McGraw-Hill Co. of Canada Ltd., Toronto: 297 pp.

Corrosion

Romanovsky, V., Editor, 1965.
Congres International de la Corrosion Marine et des Salissures, 8-12 join 1964, Cannes.
Trav. Centre de Recherches et d'Etudes Oceanogr. n.s., 6(1/4):509 pp.

corrosion

Romanovsky, V., 1959.
Essai d'étalonnage des stations française de corrosion marine.
Trav. Centre Recherches et d'Études Océanogr., n.s., 3(3):53-61.

corrosion

Rosenberg, L.A., 1963.
Depolarizing role of some sulphate reducing and saprophyte bacteria in the electrochemical corrosion of stainless and carbon steel. (In Russian).
Trudy Inst. Okeanol., Akad. Nauk, SSSR, 70:231-245.

corrosion

Rozenberg, L.A., 1963.
Role of bacteria in the electrochemical corrosion of steel in sea water. (In Russian). (English summary)
Mikrobiologia, 32(4):689-694.

Translation for NSF (1964)
32(4):586-590.
Microbiology.

Corrosion

Ross, F. Fraser, 1965.
The control of mussels in sea water cooling systems.
Trav. Centre de Recherches et d'Etudes Oceanogr. n.s., 6(1/4):437-439.

Corrosion

Rozenfeld, I.L., O.I.Vashkov and K.A. Zhigalova, 1965.
Electrochemistry of metals in sea water.
Trav. Centre de Recherches et d'Etudes Oceanogr. n.s., 6(1/4):235-250.

corrosion
Sanhes, J., 1962.
Protection contre la corrosion marine de la station marémotrice expérimentale de Saint-Malo.
Mém. et Trac., Soc., Hydrotechn., France, Suppl. au Vol. 1:82-92.

corrosion
Schaufele, H.J., 1950.
Erosion and corrosion on marine structures. Inst. Coastal Eng., Univ. Ext., Univ. Calif., Long Beach, 11-13 Oct. 1950:6 pp., 5 pls. (multilithed).

corrosion
Schwerdtfeger, W.J., 1958.
Current and potential relations for the cathodic protection of steel in salt water.
J. Res. Nat. Bur. Standards, 60(3):153-160.

corrosion
Shal'nev, K.K., 1954.
Resistance of metals to cavitation corrosion in fresh water and sea water. Dokl. Akad. Nauk, SSSR, 95(2):229-232.

T152R

Shreir, L.L., Editor, 1963. Corrosion
Corrosion.
George Newnes Ltd., London.

Corrosion
Skulikidis, Th., et Th. Tassios, 1965.
Corrosion des armatures de beton et leur protection par un nouvel addatif.
Trav. Centre de Recherches et d'Etudes Oceanogr. n.s., 6(1/4):235-233.

corrosion
Snyder, Robert M. and Seabrook Hull, 1965.
Parametric variables affecting corrosion.
GeoMar. Techn., 1(7):17-24.

Corrosion
Sorel, R., et A. Maurin-Larcade, 1965.
Aperçu sur l'exploitation petroliere off shore.
Trav. Centre de Recherches et d'Etudes Oceanogr. n.s., 6(1/4):127-139.

corrosion
Spencer, K.A., 1958.
Cathodic protection of ships and marine structures. Dock & Harbour Auth. 38(447):307-312.

corrosion
Sysoev, N.N., 1940.
[About some ways of sheltering hydrological instruments from electochemical destruction.]
Meteorol. i. Gidrol. 6(12):77-78.

corrosion
Tarasov, N.I., 1949.
[Corrosion and fouling.] Priroda (11):32.

corrosion
Tarasov, N.I., and I.B. Ulanovskii, 1960.
[The effect of a corn barnacles on the corrosion of carbonaceous steel.]
Doklady Akad. Nauk, SSSR, 132(3):696-699.

corrosion
*Tuthill, A.H., and C.M. Schillmoller, 1967.
Guidelines for selection of marine materials.
J. Ocean Tech., 2(1):6-36.

Uhlig, H.H., Editor, 1948. corrosion
The corrosion handbook.
John Wiley & Sons, New York, New York.

corrosion
Ulanovsky, I.B., and A.D. Gerasimenka, 1963.
Influence of algae on the corrosion of carbon steel in sea water and the effect of ultrasonic vibrations on the intensity of photosynthesis. (In Russian; English summary).
Trudy Inst. Okeanol., Akad. Nauk, SSSR, 70:246-251.

Corrosion
Vlieger, J.H., 1965.
The performance in seawater of organic coatings on sprayed zinc steel.
Trav. Centre de Recherches et d'Etudes Oceanogr. n.s., 6(1/4):65-75.

corrosion
Watkins, Laverne L., 1969.
Corrosion and protection of steel piling in sea water.
Techn. Memo, Coast. Engng Res. Cent. 27: 98 pp.

Corrosion
White, E.E., 1965.
Maintenance paints for the protection of ships' hulls.
Trav. Centre de Recherches et d'Etudes Oceanogr. n.s., 6(1/4):113-120.

corrosion
Woodcock, A.H., 1950.
Impact deposition of atmospheric sea salts on a test plate. Proc. Amer. Soc., Testing Materials 50:1151-1160, 9 textfigs., with discussion 1161-1166, Textfigs. 10-12.

Corrosion, prevention of

corrosion, prevention of
Wisely, B., 1962.
Prevention of marine corrosion of steel plates by an epoxy resin coating.
Australian J. Sci., 25(1):24-26.

Corrosion protection

corrosion, protection against
Marette, D., A. Hache and M. -M. Cugnier, 1961.
Contribution à l'étude de la protection cathodique de l'acier immergé en eau de mer.
Bull. Centre Études Recherches Sci., Biarritz, 3(4):513-538.

corrosion protection
Milano, J., 1961
Cathodic protection of marine terminal facilities.
J. Waterways & Harb. Div., Proc. Amer. Soc. Civil Eng., 87(WW2): 27-44.

cosmic dust

cosmic dust
Hunter, W., and D.W. Parkin, 1960
Cosmic dust in recent deep-sea sediments.
Proc. Roy. Soc., London, Ser. A., 255(1282): 382-397.

cosmic dust
Pettersson, H., 1960.
Poussière d'étoile.
Scientia, 95(12):367-369.

cosmic dust
Pettersson, H., 1958.
Rate of accretion of cosmic dust on earth.
Nature 181(4605):330.

cosmic rays

cosmic rays
Allen, J.E., and A.J. Apostolakis, 1961.
Sea-level cosmic ray spectra at large zenith angles.
Proc. R. Soc., London, (A), 265(1320):117-132.

cosmic rays
Allkofer, O.C., R.D. Andresen und W.D. Dau, 1970.
Der Einfluss des Erdmagnetfeldes auf die kosmische Strahlung. II Untersuchungen der Myonenkomponente der kosmischen Strahlung während der Atlantischen Expedition IQSY 1965 auf dem Forschungsschiff Meteor.
Meteor Forsch.-Ergebnisse (B) 5:1-22
Atlantischen Exped. IQSY 1965

cosmic rays
Sandström, Arne Eld., Martina A. Pomerantz and Bengt-Olov Grönkvist, 1963.
Sea level cosmic ray intensity and threshold rigidity.
Tellus 15(2):184-193.

cosmic radiation
*Waddington, C.J., 1967.
Paleomagnetic field reversals and cosmic radiation.
Science, 158(3803):913-915.

cosmic rays, effect of
Somayajulu, B.L.K., 1969.
Cosmic ray produced silicon-32 in near-coastal waters.
Proc. Indian Acad. Sci. (A), 69(6): 338-346

Cotidal lines, charts

cotidal lines
Bogdanov, K.T., 1961.
[New maps of cotidal lines for diurnal waves (k_1 and O_1) in the Pacific.]
Doklady, Akad. Nauk, SSSR, 139(3):713-716

cotidal charts
Bogdanov, C.T., & V.F. Nefediev, 1962
New cotidal charts of diurnal tidal waves (K_1 and O_1) in Australo-Asiatic seas.
Doklady Akad. Nauk, SSSR, 144(5): 1034-1038.

Cotidal Lines
Bogdanov, K. T., 1962.
(Propagation of semi-diurnal tidal waves in the Pacific Ocean water area.)
Mezhd. Geofiz. Komitet, Prezidiume Akad. Nauk, SSSR, Rezult. Issled. Programme Mezhd. Geofiz. Goda, Okeanol. Issled. No. 5:5-18.

cotidal lines
Bogdanov, K.T., 1961.
[New charts of cotidal lines for semidiurnal constituents (M_2+ S_2) in the Pacific Ocean.]
Doklady Akad. Nauk, SSSR, 138(2):441-444.

Engl. Abstr: New charts of cotidal lines of semidiurnal tides in the Pacific Ocean.
OTS-61-11147-17 JPRS:8710:8

cotidal charts
Bogdanov, K.T., and V.P. Nefediev, 1961.
New cotidal charts of semidiurnal tidal waves (M2 and S 2) in Australo-Asian seas.
Doklady Akad. Nauk, SSSR, 141(5):1078-1082.

tides (cotidal lines)
Boris, L.I., 1964
Some methodical results of the computation of tidal phenomena. (In Russian).
Materiali Vtoroi Konferentsii, Vzaimod. Atmosfer. i Gidrosfer. v Severn. Atlant. Okean Mezhd. Geofiz. God., Leningrad. Gidrometeorol. Inst., 114-122.

cotidal lines
Brettschneider, Gottfried, 1967.
Anwendung des hydrodynamisch-numerischen Verfahres zur Ermittlung der M_2-Mitschwingungegezeit der Nordsee.
Mitt. Inst. Meeresk., Univ. Hamburg, 7:65 pp. (multilithed).

cotidal charts
Duvanin, A.I., 1964
On the practical use of cotidal charts. (In Russian).
Okeanologiia, Akad. Nauk, SSSR, 4(4):576-582.

cotidal lines
Gohin, F., 1961.
Calcul de la dénivellation et trace des lignes cotidales dans les océans.
Cahiers Océanogr., C.C.O.E.C., 13(6):363-375.

cotidal lines
Hikosaka, S., 1948.
The semidiurnal lunar tidal motion of Shimabara Kaiwan. Geophys. Notes, Tokyo, No. 39:14 pp., 2 figs. (mimeographed).

cotidal lines
Imbert, Bertrand, 1956.
Analyse et discussion des resultats.
Expeds. Polaires Francaises, Exped. Antarct., Res. Sci., n.s., 2(4):35-84.

Cotidal Lines
Lennon, G. W., 1961.
The deviation of the vertical at Bidston in response to the attraction of ocean tides.
Geophys. J., R. Astron. Soc., 6(1):64-84.

cotidal lines
Proudman, J., 1947.
The tides. Mem. & Proc. Manchester Lit. & Phil. Soc., 1946-1947, 88:77-84, 12 figs.

cotidal charts
Sergeev, Y.N., 1964.
The application of the method of marginal values for the calculation of the charts of tidal harmonic constants in the South China Sea. (In Russian).
Okeanologiia, Akad. Nauk, SSSR, 4(4):595-602.

cotidal charts
Sergeev, J.N., 1963.
About cotidal chart computation of the M2 tidal wave in the South China Sea. Questions of Physical Oceanography. (In Russian; English abstract).
Trudy Inst. Okeanol., Akad. Nauk, SSSR, 66:66-78.

cotidal lines
Ueno, Takeo, 1964.
Theoretical studies on tidal waves travelling over the rotating globe (1).
Oceanogr. Mag., Tokyo, 16(1/2):47-51.

cotidal lines
United States Navy Hydrographic Office, 1959.
Climatological and oceanographic atlas for mariners. Vol. 1.
North Atlantic Ocean. 182 charts.

cotidal lines
United States, Weather Bureau and Hydrographic Office, 1961
Climatological and oceanographic atlas for Mariners. Vol. II. North Pacific Ocean.
Unnumbered pages.

cotidal lines
Zahel, Wilfried, 1970.
Die Reproduktion Gezeitenbedingter Bewegungsvorgänge im Weltozean mittels des hydrodynamischen-numerischen Verfahrens.
Mitt. Inst. Meereskunde Univ. Hamburg, 17: 50 pp., 18 figs. (multilited)

Couette flow

Couette flow
Craik, Alex D.D. 1969.
The stability of plane Couette flow with viscosity stratification.
J. fluid Mech. 36(4): 685-693.

Couette flow
Snyder, H.A., 1969.
Change in wave-form and mean flow associated with wavelength variations in rotating Couette flow. 1.
J. fluid Mech. 35(2): 337-352

Couette flow
Snyder, H.A., 1969.
Wave-number selection at finite amplitude in rotating Couette flow.
J. fluid Mech. 35(2): 273-298.

Couette flow
Snyder, H.A. 1968.
Stability of rotating Couette flow. II. Comparison with numerical results.
Physics Fluids 11(8): 1599-1605.

Couette flow
Snyder, H.A. 1968.
Stability of rotating Couette flow. I. Asymmetric wave forms.
Physics Fluids 11(4): 728-734.

Countercurrent
Also check under currents.

countercurrent, warm-core
Blandford, Robert, 1965.
Inertial flow in the Gulf Stream.
Tellus, 17(1):69-76.

countercurrents
Ghosh, S.B., R.K. Sinha and P. Nandi, 1961.
An expression for the evaluation of partition coefficient in countercurrent distribution.
J. Indian Chem. Soc., 38(10):789-796.

counter currents, equatorial
Hantel, Michael 1971.
Zum Einfluss des Entrainmentprozesses auf die Dynamik der Oberflächenschicht in einem tropisch-subtropischen Ozean.
Dt. hydrogr. Z. 24(3): 120-137.

Countercurrents, equatorial
Hisard, Ph. et P. Rual, 1970.
Courant équatorial intermédiaire de l'océan Pacifique et contre-courants adjacents. Cah. O.R.S.T.O.M., sér. Océanogr., 8(1): 21-45.

countercurrents, equatorial
Khanaichenko, N.K., and N.F. Khlystov, 1966.
The southern branch of the equatorial countercurrent in the Atlantic. (In Russian).
Doklady, Akad. Nauk, SSSR, 166(3):709-712.

countercurrents, equatorial
Khanaichenko, N.K., N.Z. Khlystov, and V.G. Zhidkov, 1965.
On the system of equatorial countercurrents in the Atlantic Ocean. (In Russian).
Okeanologiia, Akad. Nauk, SSSR, 5(2):222-229.

countercurrent, deep
Ponomarenko, G.P., 1965.
Discovery of a deep countercurrent at the Equator in the Atlantic Ocean on the Research Vessel "Mikhail Lomonosov". (In Russian; English abstract).
Okeanolog. Issled., Rezult. Issled. Programme Mezhd. Geofiz. Goda, Mezhd. Geofiz. Komitet Presidiume, Akad. Nauk, SSSR, No. 13:77-81.

countercurrents
Korotaev, G.K. and N.B. Shapiro 1971.
On the problem of a coastal countercurrent near the Gulf Stream. (In Russian)
Fisika Atmosfer. Okean. Izv. Akad. Nauk SSSR 7(3): 359-362.

countercurrents
Masuzawa, Jotaro 1967.
An oceanographic section from Japan to New Guinea at 137°E in January 1967.
Oceanogrl Mag. 19(2): 95-118.

countercurrents
Reid, Joseph L., Jr., 1962
Measurements of the California Countercurrent at a depth of 250 meters.
J. Mar. Res., 20(2):134-137.

countercurrents
Reid, J.L., Jr., 1959.
Evidence of the existence of a South Equatorial Countercurrent in the Pacific.
American Geophysical Union, Pacific Southwest Regional Meeting, Feb. 5-6, 1959.
Abstracted in:
J. Geophys. Res., 64(6):693.

counter currents, subtropical
Robinson, Margaret K., 1969.
Theoretical predictions of subtropical countercurrent confirmed by bathythermograph (BT) data
Bull. Jap. Soc. fish. Oceanogr. Spec. No. (Prof Uda Commem. Pap.): 115-121.

countercurrent
Römer, E., 1939
Der Gegenstrom unter der süd und südostafrikanischen Küste. Seewart 7:175-178.

countercurrents
Uda, Michitaka, and Keiichi Hasunuma, 1969.
The Eastward Subtropical Countercurrent in the western North Pacific Ocean.
J. oceanogr. Soc. Japan 25(4): 201-210

counter currents
Wyrtki, Klaus, 1966.
Oceanography of the eastern equatorial Pacific Ocean.
In: Oceanography and marine biology, H. Barnes, editor, George Allen & Unwin, Ltd., 4:33-68.

counter currents, subtropical
Yoshida, Kozo, 1969.
Subtropical counter currents - a preliminary note on further observational evidences.
Rec. oceanogr. Wks. Japan, 10(1):123-125.

COUNTER CURRENTS
Yoshida, Kozo, and Toshiko Kidokoro, 1967.
A subtropical countercurrent (II) - A prediction of eastward flows at lower subtropical latitudes.
J. oceanogr. Soc., Japan, 23(5): 231-246.

counter currents
Yoshida, Kozo, and Toshiko Kidokoro, 1967.
A subtropical counter-current in the North Pacific - an eastward flow near the subtropical convergence.
J. oceanogr. Soc., Japan, 23(2):88-91.

counting fish
Craig, Robert E., and Sinclair T. Forbes, 1969.
Design of a sonar for fish counting.
FiskDir. Skr. Ser. HavUnders. 15(3): 210-219.

craters
Schwartz, M.L., J.A. Scrimger, W.H. Halliday and C. Kelly, 1971.
Union Seamount: site of a flank crater.
Nature, Lond., 230(1): 20-22. (Phys. Sci.)

Creeks, tidal

CREEKS, tidal
Schou, Axel, 1967.
Estuarine research in the Danish moraine archipelago.
In: Estuaries, G.H. Lauff, editor, Publs Am. Ass. Advmt Sci., 83:129-145.

Creep

creep
Kuenen, Ph. H., 1964
Deep-sea sands and ancient turbidites.
In: Turbidites, A.H. Bouma and A. Brouwer, Editors, Developments in Sedimentology, Elsevier Publishing Co., 3:3-33.

Crescentic landforms

crescentic landforms
Ball, M.M., A.C. Neumann, Dolan, Robert and John C. Ferm, 1968.
Crescentic landforms along the Atlantic coast of the United States.
Science, 161(3842):710.

critical depth

"critical depth"
Cushing, D.H., 1962.
An alternative method of estimating the critical depth.
J. du Cons., 27(2):131-140.

cross-lamination
Allen, J.R.L. 1970.
A quantitative model of climbing ripples and their cross-laminated deposits.
Sedimentology 14(1/2): 5-26

"crown of thorns"
Robinson, D.E. 1971.
Observations on Fijian coral reefs and the crown of thorns starfish.
J. Roy. Soc. N.Z. 1(2): 99-112

Crustal movements

crustal movements
Pelletier, B.R., 1966.
Development of submarine physiography in the Canadian Arctic and its relation to crustal movements.
In: Continental drift, G.D. Garland, editor, Spec. Publs. R. Soc. Can., No.9:77-101.

crust, see also: earth's crust

crust
Bacon, M., and F. Gray, 1971
Evidence for crust in the deep ocean derived from continental crust.
Nature, Lond. 229(5283): 331-332.

crust
Beloussov, V.V., 1968.
The earth's crust and upper mantle of the oceans. (In Russian; English abstract).
Rez. Issled. Mezhdunarod. Geofiz. Proekt., Mezhduvedomst. Geofiz. Kom., Akad. Nauk, SSSR, 253 pp.

crust
Cann, J.R. 1970.
New model for the structure of the ocean crust.
Nature, Lond., 226(5249): 928-930.

crust
Cann, J.R. 1968
Geological processes at mid-ocean ridge crests.
Geophys. J. R. astr. Soc. 15(3): 331-341

CRUST
Christensen, Nikolas I., 1970.
Possible greenschist facies metamorphism of the oceanic crust.
Bull. geol. Soc. Am., 81(3): 905-908

crust
Dietz, Robert S., and John C. Holden, 1971
Pre-Mesozoic oceanic crust in the eastern Indian Ocean (Wharton Basin)?
Nature, Lond., 229(5283): 309-312

crust
Hajnal, Z., 1968.
A two-layer model for the earth's crust under Hudson Bay.
In: Earth Science Symposium on Hudson Bay, Ottawa, February, 1968, Peter J. Hood, editor, GSC Pap., Geol. Surv. Can., 68-53:326-336.

crust
Hales, A.L., C.E. Helsley, J.J. Dowling and J.B. Nation, 1968.
The east coast onshore-offshore experiment. I. The first arrival phases.
Bull. seismol. Soc. Am., 58(3): 757-819.

crust
Hales, A.L., C.E. Helsley and J.B. Nation, 1970.
P travel times for an oceanic path. J. geophys. Res., 75(35): 7362-7381.

crust
Hall, Donald H. 1968.
A seismic-isostatic analysis of crustal data from Hudson Bay.
In: Earth Science Symposium on Hudson Bay, Ottawa, February 1968, Peter J. Hood, editor, GSC Pap. Geol. Surv. Can. 68-53: 337-364.

crust
Heezen, B.C., C. Gray, A.G. Segre and E.F.K. Zarudzki, 1971
Evidence of foundered continental crust beneath the central Tyrrhenian Sea.
Nature, Lond., 229(5283): 327-329

crust, oceanic
*Henningsen, Dierk, und Richard Weyl, 1967.
Ozeanische Kruste im Nicoya-Komplex von Costa Rica (Mittelamerika).
Geol. Rdsch, 57(1):33-47.

crust
Isaacs, John D., 1968.
General features of the ocean. Ch. 6 in; Ocean engineering: goals, environment, technology, J.F. Brahtzm editor, John Wiley & Sons, 157-201.

cruise lists
Joyce, Edwin A., Jr., and Jean Williams 1969.
Rationale and pertinent data.
Mem. Hourglass Cruises, Mar. Res. Lab. Fla. Dept. Nat. Res. 1(1): 1-50

crust
*Kosminskaya, I.P., and S.M. Zvev, 1968.
Abilities of explosion seismology and continental crust amd mantle studies.
Can. J. Earth Sci., 5(4-2):1091-1100.

crust
Macdonald, Gordon A., 1965.
The lithologic constitution of the crust and mantle in the Hawaiian area.
Pacific Science, 19(3):285-286.

crust
Malovitskii, J. P., Iu. P. Neprochnov, I.A. Garkalenko, E.A. Starshinova, K.G. Mileshina, M.J. Komornaia, L.N. Rykunov, B.V. Kholopov and V.V. Sedov, 1969.
Earth crust structure in the western part of the Black Sea. (In Russian).
Dokl. Akad. Nauk, SSSR, 186(4): 905-907.

crust
Melson, William G., and Tjeerd H. Van Andel, 1966.
Metamorphism in the Mid-Atlantic Ridge, 22°N. latitude.
Marine Geol., 4(3):165-186.

crust
Moores E.M. and F.J. Vine, 1971.
The Troodos Massif, Cyprus and other ophiolites as oceanic crust: evaluation and implications.
Phil. Trans. R. Soc. Lond. (A)268(1192): 443-466.

crust
*Nag, S.K., 1967.
Surface wave dispersion and crustal structure in the Indian Ocean.
Indian J. Met. Geophys., 18(1):119-122.

crust

*Neprochnov, Yu. P., 1968.
Structure of the earth's crust of epi-continental seas: Caspian, Black and Mediterranean.
Can. J. Earth Sci., 5(4-2):1037-1043.

crust

Neprochnov, Yu. P., I.P. Kosminskaya and Ya. P. Malovitsky, 1970
Structure of the crust and upper mantle of the Black and Caspian seas.
Tectonophysics, 10 (5/6): 517-533

crust

Nicholls, I.A. 1971.
Santorini Volcano, Greece - tectonic and petrochemical relationships with volcanics of the Aegean region.
Tectonophysics, 11 (5): 377-385.

crust

*Rikitake, T., S. Miyamura, I. Tsubokawa, S. Murauchi, S. Uyeda, H. Kuno and M. Gorai, 1968.
Geophysical and geological data in and around the Japan Arc.
Can. J. Earth Sci., 5(4-2):1101-1118.

crust

Peyve, A.V., 1969.
Oceanic crust of the geologic past (In Russian)
Geotekton. Akad. Nauk SSSR 1969 (4): 5-23
Translation: Geotectonics, Scripta Technica, for AGU 1970: 210-224

crust

Sclater, John G., Roger N. Anderson and M. Lee Bell, 1971.
Elevation of ridges and evolution of the central eastern Pacific. J. geophys. Res. 76(32): 7888-7915.

crust, oceanic

Shiraki, Keiichi 1971.
Metamorphic basement rocks of Yap Islands, western Pacific: possible oceanic crust beneath an island arc.
Earth planet. Sci. Letts. 12 (1): 117-126

crust

Shor, G.G., Jr., H.K. Kirk and H.W. Menard, 1971.
Crustal structure of the Melanesian Area.
J. geophys. Res., 76(11): 2562-2586.

crust, oceanic

Vine, F.J. and H.H. Hess, 1970.
III Concepts. 1. Sea-floor spreading. In: The sea: ideas and observations on progress in the study of the seas, Arthur E. Maxwell, editor, Wiley-Interscience 4(2): 587-622.

crust, oceanic

Zietz, Isidore, 1970.
II. Regional observations. 8. Eastern Continental Margin of the United States (July 1968). Part 1: A magnetic study. In: The sea: ideas and observations on progress in the study of the seas, Arthur E. Maxwell, editor, Wiley-Interscience 4(2): 293-310.

crustal blocks

Bonatti, Enrico and Jose Honnorez, 1971.
Nonspreading crustal blocks at the mid-Atlantic Ridge. Science 174(4016): 1329-1331.

crustal genesis

Krause, D.C., and N.D. Watkins, 1970.
North Atlantic crustal genesis in the vicinity of the Azores.
Geophys. J. R. astr. Soc. 19(3): 261-283.

crustal plate, interactions of

Malfait, Bruce T., and Menno G. Dinkelman 1972.
Circum-Caribbean tectonic and igneous activity and the evolution of the Caribbean plate.
Bull. geol. Soc. Am. 83(2): 251-272.

crustal plates

Van Andel, Tjeerd H., Richard P. Von Herzen and J.D. Phillips 1971.
The Vema Fracture Zone and the tectonics of transverse shear zones in oceanic crustal plates.
Mar. geophys. Researches 1 (3): 261-283.

Crustal structure

crustal structure

*Blundell, D.J., and R. Parks, 1969.
A study of the crustal structure beneath the Irish Sea.
Geophys. J. R. astr. Soc., 17(1):45-62.

crustal structure

Davies, D., and T. J. G. Francis, 1964.
The crustal structure of the Seychelles Bank.
Deep-Sea Res., 11(6):921-927.

crust, original

Dietz, R.S., 1965.
Earth's original crust - lost quest ?
Tectonophysics, 2(6):515-520.

crustal structure

Ewing, Maurice, Laric V. Hawkins and William J. Ludwig, 1970.
Crustal structure of the Coral Sea. J. geophys Res., 75(11): 1953-1962.

crustal structure

Francis, T.J.G., D. Davies and M.N. Hill 1966.
Crustal structure between Kenya and the Seychelles.
Phil. Trans. R. Soc. London (A) 259 (1099): 240-261.

crustal structure

Gainanov, A.G., Ye. N. Isaev, P.A. Stroev and S.A. Ushakov 1971.
Isostasy and crustal structure of the Okhotsk region. (In Russian; English abstract)
Geofiz. Biull. Mezhd. Geofiz. Kom. Presid. Akad. Nauk SSSR, 22: 37-43.

crust

Girdler, R.W., 1965.
The formation of new oceanic crust.
Phil. Trans. R. Soc., A, 258(1088):123-136.

crustal structure

Gough, D.I., 1967.
Magnetic anomalies and crustal structure in eastern Gulf of Mexico.
Bull. Am. Ass. Petrol. Geol., 51(2):200-211.

crust (oceanic)

Hess, Harry H., 1964.
The oceanic crust, the upper mantle and the Mayaguez serpentinitized peridotite.
Nat. Acad. Sci.-Nat. Res. Counc., Publ., No. 1188:169-175.

oceanic crust

Hess, H.H., 1955.
The oceanic crust. J. Mar. Res., 14(4):423-439.

crustal structure

*Hobson, George D., A. Overton, D.N. Clay and W. Thatcher, 1967.
Crustal structure under Hudson Bay.
Can. J. Earth Sci., 4(5):929-947.

crust

*Hunter, J.A., and R.F. Mereu, 1967.
The crust of the earth under Hudson Bay.
Can. J. Earth Sci., 4(5):949-960.

crustal structure

Payo, Gonzalo, 1969
Crustal structure of the Mediterranean Sea. II. Phase velocity and travel times.
Bull. seismol. Soc. Am. 59 (1): 23-42.

crustal structure, effect of

*Ripper, I.D. and R. Green, 1967.
Tasmanian examples of the influence of bathymetry and crustal structure upon seismic T-wave propagation.
N.Z. Jl Geol. Geophys. 10(5):1226-1230.

crust, thickness of

Hales, Anton L., 1970.
II. Regional observations. 9. Eastern Continental Margin of the United States (July 1968). Part 2: A review. In: The sea: ideas and observations on progress in the study of the seas, Arthur E. Maxwell, editor, Wiley-Interscience 4(2): 311-320.

Crust-mantle transition

crust-mantle transition

*Helmberger, Donald V., 1968.
The crust-mantle transition in the Bering Sea.
Bull. Seismol. Soc., Am., 58(1):179-214.

Crustal plates

Andrews, James E. 1971.
Gravitational subduction of a western Pacific crustal plate.
Nature, Lond. (Nat. phys. Sci.) 233 (39): 81-83

crustal shortening

Rabinowitz, Philip D., and William B.F. Ryan, 1970.
Gravity anomalies and crustal shortening in the eastern Mediterranean.
Tectonophysics, 10 (5/6): 585-608.

cultivation

cultivation
Yamamoto, Tadasu, Takemi Ichimura, Naoya Tajino and Yusuki Ishikaea, 1960
Maintenance of bottom conditions of a pond for culturing the Kuruma-ebi shrimp, Penaeus japonicus, by means of underdrainage bottom structure.
Aquiculture, 8:133-137.
Abstracted in: Rec. Res., Fac. Agricult., Univ. Tokyo, 11:51-52.

Culture.

A

cultures
Adshead, Patricia C., 1967.
Collection and laboratory maintenance of living planktonic Foraminifera.
Micropaleontology, 13(1):32-40.

cultures
Akinina, D.K., 1966.
Dependence of light saturation of two mass species of dinoflagellates on a number of factors. (In Russian; English abstract).
Okeanologiia, Akad. Nauk, SSSR, 6(5):861-868.

cultures
*Aldrich, David V., Sammy M. Ray and William B. Wilson, 1967.
Gonyaulax monilata: population growth and development of toxicity in cultures.
J. Protozool., 14(4):636-639.

cultures
Allen, E.J., 1914.
On the culture of the plankton diatom Thalassiosira gravida Cleve, in artificial sea water. J.M.B.A., n.s., 10:417-439.

culture.
Allen, E. J., and E. W. Nelson., 1910.
On the artificial culture of marine plankton organisms. JMBA, VIII:421-474.

cultures
Antia, N.J., and J.Y. Cheng 1970.
The survival of axenic cultures of marine planktonic algae from prolonged exposure to darkness at 20°C.
Phycologia 9(2):179-184.

cultures
*Arnold, Zach M., 1967.
II. Application à la technique des cultures des foraminifères utilisation des antibiotiques dans la realisation des cultures de foraminifères sous faible volume.
Vie Milieu (A) 18(1-A): 36-45.

B

cultures
Baker, A. de C., 1963
The problem of keeping planktonic animals alive in the laboratory.
J. Mar. Biol. Assoc., U.K., 43(2):291-294.

cultures
Baslavskaia, S.S., and R.F. Kulikova, 1956.
[The photometric method for measuring the growth of algal cultures.] Moskovskoe Obshchestvo Ispytatelei Prirody, Biull., 61(6):77-

cultures
Battaglia, B., 1970.
Cultivation of marine copepods for genetic and evolutionary research. Helgoländer Meeresunters, 20(1/4): 385-392.

cultures
Belyaeva O.I., 1969.
On the methods of Pontella mediterranea Claus keeping under laboratory conditions. (In Russian).
Gidrobiol. Zh. 5(5):133-135.

cultures
Bentley-Mowat J.A., and S.M. Reid, 1969.
Effect of gibberellins, kinetin and other factors on the growth of unicellular marine algae in cultures.
Botanica mar. 12(1/4):155-199

cultures
Berland, Brigitte, 1966.
Contribution à l'étude des cultures de diatomees marines.
Recl. Trav. Stn. mar., Endoume, 40(56):3-82.

cultures
Berland, B.R. and S.Y. Maestrini, 1969.
Study of bacteria associated with marine algae in culture. II. Action of antibiotic substances.
Marine Biol., 3(4): 334-335.

cultures
*Berland, Brigitte R., et Serge Y. Maestrini, 1969.
Action de quelques antibiotiques sur le développement de cinq diatomées en cultures.
J. mar. Biol. Ecol., 3(1):62-75.

cultures (copepods)
Bernard, M., 1961
Adaptation de quelques copépodes pélagiques méditerranéens à different milieux de survie en aquarium.
Rapp. Proc. Verb., Réunions. Comm. Int. Expl. Sci. Mer Méditerranée, Monaco, 16(2):165-176.

cultures
Bernhard, M., preparator, 1965.
Studies on the radioactive contamination of the sea, annual report 1964.
Com. Naz. Energ. Nucleare, La Spezia, Rept., No. RT/BIO (65) 18:35 pp.

cultures
Bernhard, M., L. Rampi and A. Zattera 1971
First trophic level of the food chain.
CNEN Rept. RT/BIO (70)-11, M. Bernhard, editor: 23-40

Cultures
Bernhard, M. and A. Zattera, 1970.
The importance of avoiding chemical contamination for a successful cultivation of marine organisms. Helgoländer wiss. Meeresunters, 20(1/4): 655-675.

cultures
Bonin, Daniel J., 1969.
Influence de différents facteurs écologiques sur la croissance de la diatomée marine Chaetoceros affinis Lauder en culture.
Tethys 1(1): 173-256

culture
Braarud, Trygve, 1958.
Observations on Peridinium trochoideum (Stein) Lemm. in culture.
Nytt Mag. Botan., 6:39-42.

cultures
Braarud, T., 1945.
Morphological observations on marine dinoflagellate cultures (Porella perforata, Goniaulax tamarensis, Protoceratium reticulatum). Avhandl. Norske Videns.-Akad., Oslo, Mat.-Natur. Kl., 1944, (11):1-18, 4 pls., 6 textfigs.

cultures
Behrend, H., 1950.
Notiz über die Wirkung intermittuerenden Lichtes auf das Wachstum der Diatomeen. Arch. Mikrobiol. 14:531-533.

cultures
Boalch, G.T., 1961
Studies on Ectocarpus in culture. I. Introduction and methods of obtaining uni-algal and bacteria-free culture.
J. Mar. Bio. Ass. U.K., 41(2): 287-304.

cultures
Braarud, Trygve, 1961.
Cultivation of marine organisms as a means of understanding environmental influences on populations.
Oceanography, M. Sears, Edit., Amer. Assoc. Adv. Sci. Publ., No. 67:271-298.

cultures
Braarud, T., 1948.
On variations in form of Sceletonema costatum and their bearing on the supply of silica in cultures of diatoms. Nytt Mag. f. Naturvidensk. 86:31-44.

cultures
Braarud, T., and Ellen Rossavik, 1952.
Observations on the marine dinoflagellate, Prorocentrum micans Ehrenb. in culture.
Avhandl. Norske Videnskaps-Akad., Oslo 1951, Mat.-Naturvid. Kl., No. 1:1-18, 8 textfigs.

cultures
Braarud, T., and E. Fagerland, 1946.
A Coccolithophoride in laboratory culture: Syracosphaera carterae, n.sp. Avhandl. Norske Videnskaps-Akademi, Oslo. I. Mat.-Naturv. Kl. 1946(2):1-10, 1 pl.

C

cultures
Canada, Nova Scotia Research Foundation, 1966.
Selected bibliography on algae, No. 7:105 pp.

cultures
Carpenter, Edward J., 1970.
Phosphorus requirements of two planktonic diatoms in a steady state culture.
J. Phycol. 6(1):28-30

cultures
Chapman, V.J., 1961.
An underwater growth chamber for large algae. (Abstract).
Tenth Pacific Sci., Congr., Honolulu, 21 Aug.-6 Sept., 1961, Abstracts of Symposium Papers, 158.

Cultures
Chakravarty, Dilip K., 1970.
Production of pure culture of Lagenisma coscinodisci Drebes parasitising the marine diatom Coscinodiscus.
Veröff. Inst. Meeresforsch., Bremerh. 12(3): 305-312.

cultures
Chin, T.G., C.F. Chen, S.C. Liu and S.S. Wu, 1965.
Influence of temperature and salinity on the growth of three plankton diatom species. (In Chinese; English abstract). (Not seen).
Oceanol. et Limnol. Sinica, 7(4):373-384.

cultures

Chu, S. P., 1949
Experimental studies on the environmental factors influencing the growth of phytoplankton. Sci. & Tech. in China 2(3):37-52.

cultures

Chu, S.P., 1947
Note on the technique of making bacteria-free cultures of marine diatoms. JMBA 26(3): 296-302.

cultures

Conover, R.J., 1970.
Cultivation of plankton populations. Convener's report on an informal session, held on September 10, during the International Helgoland Symposium 1969. Helgoländer wiss. Meeresunters. 21(4): 401-444.

cultures

Cook, H.L. 1969.
A method of rearing penaeid shrimp larvae for experimental studies. FAO Fish. Repts 3(57) (FRm) (R57.3 (Tam)): 709-715 (mimeographed)

cultures

Corkett, C.J. 1970.
Techniques for breeding and rearing marine calanoid copepods. Helgoländer wiss. Meeresunters. 20(1/4): 318-324.

cultures

Corkett, C.J., 1967.
Technique for rearing marine calanoid copepods in laboratory conditions. Nature, Lond., 216(5110):58-59.

CULTURES

*Corkett, C.J., and D.L. Urry, 1968.
Observations on the keeping of adult female Pseudocalanus elongatus under laboratory conditions.
J. mar. biol. Ass., U.K., 48(1):97-105.

cultures

Costlow, J.D., Jr., C.G. Bookhout and R. Monroe, 1962.
Salinity-temperature effects on the larval development of the crab Panopeus herbstii Milne-Edwards reared in the laboratory.
Physiol. Zool., 35(1):79-93.

cultures

Crisp, D.J., 1962.
The planktonic stages of the Cirripedia Balanus balanoides (L.) and Balanus balanus (L.) from north temperate waters.
Crustaceana, 3(3):207-221.

cultures

Culliney, John L., 1971.
Laboratory rearing of the larvae of the mahogany date mussel Lithophaga bisulcata. Bull. mar. Sci. Miami 21(2): 591-602.

D

cultures

D'Asaro, Charles N., 1970.
Egg capsules of prosobranch mollusks from South Florida and the Bahamas and notes on spawning in the laboratory. Bull. mar. Sci., 20(2): 414-440.

cultures

Davey, Earl W., John H. Gentile, Stanton J. Erickson and Peter Betzer, 1970.
Removal of trace metals from marine culture media. Limnol. Oceanogr., 15(3): 486-488.

cultures

Davies, Anthony G., 1970.
Iron, chelation and the growth of marine phytoplankton. 1. Growth kinetics and chlorophyll production in cultures of the euryhaline flagellate Dunalliela tertiolecta under iron-limiting conditions. J. mar. biol. Ass., U.K., 50(1): 65-86.

cultures

Davis, W.P., 1970.
Closed systems and the rearing of fish larvae. Helgoländer wiss. Meeresunters, 20(1/4): 691-696.

cultures

de Lépiney, L., 1962.
Sur l'élevage de copépodes au laboratoire. Hydrobiologia, 20(3):217-222.

cultures

Denffer, D.von, 1948.
Über einen Wachstumshemmstoff in alternden Diatomkulturen. Biol. Zentralb. 67:7-13.

cultures

*de Sousa e Silva, Estela, 1967.
Cochlodinium heterolobatum n. sp.: Structure and cytophysiological aspects.
J. Protozool., 14(4):745-754.

cultures

De Sousa e Silva, Estela, 1962.
Some observations on marine dinoflagellate cultures. II. Glenodinium foliaceum Stein and Goniaulax diacantha (Meunier) Schiller.
Botanica Marina, 3(3/4):75-100.

Also in: Notas e Estudos, Inst. Biol. Marit., Lisbon, No. 24: 75-100

cultures

Donnelly, Patricia V., M.A. Burklew and Rosa A. Overstreet, 1967.
Amino acids and organic nitrogen content in Florida Gulf Coast waters and in artificial cultures of marine algae. Prof. Pap. Ser., Florida Bd. Conserv., Mar. Lab. St. Petersburg, 9: 90-97.

cultures

Droop, M.R., 1970.
Vitamin B_{12} and marine ecology. V. Continuous culture as an approach to nutritional kinetics. Helgoländer wiss. Meeresunters, 20(1/4): 629-636.

cultures

*Droop, M.R., 1968.
Vitamin B_{12} and marine ecology. IV. The kinetics of uptake, growth and inhibition in Monochrysis lutheri.
J. mar. biol. Ass., 48(3):689-733.

cultures

Droop, M.R., 1961

Some chemical considerations in the design of synthetic culture media from marine algae. Botanica Marina, 2(3/4):231-246.

cultures

Droop, M.R., 1954.
A note on the isolation of small marine algae and flagellates for pure cultures. J.M.B.A. 33 (2):511-514.

culture

Dunstan, William M. and David W. Menzel, 1971.
Continuous cultures of natural populations of phytoplankton in dilute, treated sewage effluent. Limnol. Oceanogr. 16(4): 623-632.

E

cultures

Ehrhardt, J.P., R. Moncoulon et P. Niausset 1971.
Comportement in vitro de la Chlorophycée Dunaliella salina Dunal dans les milieux à salinité différente détermination d'un optimum de salinité.
Vie Milieu Suppl. 22(1): 203-217

cultures

Eppley, R.W., A.F. Carlucci, O. Holm-Hansen, D. Kiefer, J.J. McCarthy, Elizabeth Venrick and P.M. Williams, 1971.
Phytoplankton growth and composition in shipboard cultures supplied with nitrate, ammonium, or urea as the nitrogen source. Limnol. Oceanogr. 16(5): 741-751.

cultures

Eppley, Richard W., and James L. Coatsworth, 1966.
Culture of the marine phytoplankter, Dunaliella tertiolecta, with light-dark cycles.
Ark. Mikrobiol., 55:17-25.

cultures

Eppley, Richard W., Jane N. Rogers, James J. McCarthy and Alain Sournia, 1971.
Light/dark periodicity in nitrogen assimilation of the marine phytoplankters Skeletonema costatum and Coccolithus huxleyi in N-limited chemostat cultures.
J. Phycol. 7(2): 150-154

F

cultures

Fahy, William E., 1964
A temperature-controlled salt-water circulating apparatus for developing fish eggs and larvae.
Journal du Conseil, 28(3):364-384.

cultures,

*Finenko, Z.Z., and L.A. Lanskaya, 1968.
The amount of pigments in marine planktonic algae grown in the laboratory. (In Russian; English abstract).
Okeanologiia, Akad. Nauk, SSSR, 8(5):839-847.

cultures

Fogg, G.E., 1965.
Algal cultures and phytoplankton ecology.
The University of Wisconsin Press, Madison & Milwaukee, 126 pp., $5.50.

cultures, decapod

Forss, Carl Albert, and Harold G. Coffin, 1960
The use of the brine shrimp nauplii Artemia salina, as food for the laboratory culture of decapods. Walla Walla Coll. Publ. Dept. Biol. Sci. and Biol. Sta. No. 26: 17 pp.

cultures
Fudinami, M., and H. Kasahara, 1942.
[Rearing and metamorphosis of Balanus amphitrite hawaiiensis Bloch.] Zool. Mag., Tokyo, 54(3):108-118.

G

cultures
Gaertner, A., 1970.
Einiges zur Kultur mariner neiderer Pilze.
Helgoländer wiss. Meeresunters, 20(1/4): 29-38.

culture
Galtsoff, P.S., F.E. Lutz, P.S. Welch and J.G. Needham, 1937.
Culture methods for invertebrate animals.
Comstock Publ. Col., Ithaca, 590 pp., $4.00.

culture experiments
Gilson, H. C., 1937
Chemical and Physical Investigations. The nitrogen cycle. John Murray Exped., 1933-34, Sci. Repts., 2(2):21-81, 16 text figs.

cultures
Glooschenko, W.A. and H. Curl, 1968.
Obtaining synchronous cultures of algae. Nature, Lond. 218(5141): 573-574. Also in: Coll. Reprints, Dept. Oceanogr. Univ. Oregon, 7(1968).

CULTURES
Golterman, H.L., 1967.
Tetraethylsilicate as a "molybdate unreactive" silicon source for diatom cultures.
In: Chemical environment in the aquatic habitat Proc. I.B.P. Symp., Amsterdam, Oct. 1966:56-62.

cultures
Goryunova, S.V., and M.N. Ovsyannikova, 1961.
[Technique of cultivation of some marine forms of diatoms under laboratory conditions.]
Mikrobiologiya, 30(6):995-997.

English Edit., (1962):810-811.

cultures
Götting, Klaus-Jurgen, 1963.
Zur Reinecultur von Dunaliella.
Helgoländer Wiss. Meeresuntersuch., 8(4):404-424.

cultures
Gold, K., 1970.
Cultivation of marine ciliates (Tintinnida) and heterotrophic flagellates. Helgoländer wiss. Meeresunters, 20(1/4): 264-271.

cultures
Greve, W., 1970.
Cultivation experiments on North Sea ctenophore Helgoländer wiss. Meeresunters, 20(1/4): 304-317.

culture
Gross, F., 1937.
Notes on the culture of some marine plankton organisms. JMBA, XXI:753-768.

cultures
Guillard, Robert R.L., and John H. Ryther, 1962
Studies of marine planktonic diatoms. I. Cyclotella nana Hustedt and Detonula confervacea (Cleve).
Canadian J. Microbiol., 8(2):229-240.

H

Cultures
Hamilton, R. D., and A. F. Carlucci, 1966.
Use of ultra-violet irradiated sea water in the preparation of culture media.
Nature, 211 (5048):483-484.

Cultures
Hamilton, R.D. and Janet E. Preslan, 1970.
Observations on the continuous culture of a planktonic phagotrophic protozoan. J. exp. mar Biol. Ecol., 5(1): 94-104.

cultures
Hamilton, R.D. and Janet E. Preslan, 1969.
Cultural characteristics of a pelagic marine hymenostome ciliate, Uronema sp. J. exp. mar. Biol. Ecol., 4(1): 90-99.

cultures
*Hayward, J., 1968.
Studies on the growth of Phaeodactylum tricornutum. IV. Comparison of different isolates.
J. mar. biol. Ass., U.K., 48(3):657-666.

cultures
Hedley, R.H., and J. St. J. Wakefield, 1967.
Clone culture studies of a new rosalinid foraminifer from Plymouth, England and Wellington, New Zealand.
J. mar. biol. Ass. U.K., 47(1):121-128.

cultures
Hirano, Reijiro, 1966.
Plankton culture and aquatic animal's seedling production. (In Japanese; English abstract).
Inf. Bull. Planktol. Japan, No. 13:72-75.

cultures
Heinle, D.R., 1970.
Population dynamics of exploited cultures of calanoid copepods. Helgoländer wiss. Meeresunters, 20(1/4): 360-372.

cultures
*Heinle, Donald R., 1969.
Culture of calanoid copepods in synthetic sea water.
J. Fish. Res. Bd. Can., 26(1):150-153.

cultures
Hinegardner, Ralph T., 1969.
Growth and development of the laboratory cultured sea urchin. Biol. Bull. mar. biol. Lab., Woods Hole, 137(3): 465-475.

cultures
Hochachka, Peter W., and John M. Teal, 1964.
Respiratory metabolism in a marine dinoflagellate
Biol. Bull., 126(2):274-281.

cultures
Humphrey, G.F., 1963.
Chlorophyll a and c in cultures of marine algae.
Australian J. Mar. Freshwater Res., 14(2):148-154.

I

cultures, copepods
Inoue, Motoo and Mitsuyoshi Aoki, 1969.
Reproduction of Copepoda, Tisbe furcata, cultured with seawater-acclimatized Chlorella as a basic diet. Bull. Japan. Soc. scient. Fish. 35(9). Also in: Coll. Repr. Coll. mar. Sci. Techn. Tokai Univ. 4: 47-52. (In Japanese English abstract).

cultures
Inoue Motoo, and Mitsuyoshi Aoki, 1969
Reproduction of Copepoda, Tisbe furcata, cultured with sea-water acclimatized Chlorella as a basic diet. (In Japanese; English abstract)
Bull. Jap. Soc. scient. Fish. 35(9):862-867.

J

cultures
Jacobs, J., 1961.
Laboratory cultivation of the marine copepod Pseudodiaptomus coronatus Williams.
Limnol. & Oceanogr., 6(4):443-446.

cultures
Jannasch, H.H., 1967.
Enrichments of aquatic bacteria in continuous cutlure.
Archiv. für Mikrobiol., 59: 165-173.

cultures
Jannasch, Holger W., 1962
Die kontinuierliche Kultur in der experimentellen Ökologie mariner Mikroorganismen.
Kieler Meeresf., 18(3) (Sonderheft): 67-73.

cultures
Jatzke, P., 1970.
The trichterkreisel, an in situ device for cultivating marine animals in tidal currents.
Helgoländer wiss. Meeresunters, 20(1/4): 685-690.

cultures
Jitts, H.R., C.D. McAllister, K. Stephens and J.D.H. Strickland, 1964.
The cell division rates of some marine phytoplankters as a funtion of light and temperature.
J. Fish. Res. Bd., Canada, 21(1):139-157.

cultures
Johnson, T.W., Jr., and H.S. Gold, 1959.
A system of continual-flow sea-water cultures.
Mycologia, 51(1):89-94.

cultures
Johnston, R., 1963
Antimetabolites as an aid to the study of phytoplankton nutrition.
J. Mar. Biol. Assoc., U.K., 43(2):409-425.

cultures
Johnston, R., 1963
Effects of gibberellins on marine algae in mixed cultures.
Limnol. and Oceanogr., 8(2):270-275.

culture
Johnston, R., 1963
Sea water, the natural medium of phytoplankton. 1. General features.
J. Mar. Biol. Assoc., U.K., 43(2):427-456.

cultures
Jørgensen, Erik G., and E. Steemann Nielsen, 1961
Effect of filtrates from cultures of unicellular algae on the growth of Staphylococcus aureus.
Physiol. Plantarum, 14: 896-908.

K

cultures
Kabanova, J.G., 1961.
[On cultivation of diatoms and dinoflagellates under laboratory conditions.]
Trudy Inst. Okeanol., Akad. Nauk, SSSR, 47:203-216.

cultures
Kabanova, Iu. G., 1959.
[The influence of extract from "cystozir" and "phyllophora" on some microphytes.]
Trudy Inst. Okeanol. 30:156-

cultures
Kain, Joanna M., and G.E. Fogg, 1960.
Studies on the growth of marine plankton. III. Prorocentrum micans Ehrenberg.
J.M.B.A., U.K., 39:33-50.

cultures
Kalber, F.A. 1970.
Osmoregulation in decapod larvae as a consideration in culture techniques.
Helgoländer wiss. Meeresunters. 20(1/4): 697-706.

cultures
Karande, A.A., and M.K. Thomas, 1971.
Laboratory rearing of Balanus amphitrite communis (D.)
Current Sci 40(5):109-110

cultures
Kayama, M., Y. Tsuchiya and J.P. Mead, 1963.
A model experiment of aquatic food chain with special significance in fatty acid conversion.
Bull. Jap. Soc. Sci. Fish., 29(5):452-458.

cultures
Kayser, H. 1970
Experimental-ecological investigations on Phaeocystis pouchetii (Haptophyceae): cultivation and waste water test.
Helgoländer wiss. Meeresunters. 20(1/4): 195-212

cultures
Kayser, H., 1969.
Züchtungsexperimente an zwei Marinen Flagellaten (Dinophyta) und ihre Anwendung im toxikologischen Abwassertest.
Helgoländer wiss. Meeresunters. 19(1): 21-44

cultures
*Keller, Steven E., S.H. Hutner and Dolores E. Keller, 1968.
Rearing the colorless marine dinoflagellate Cryptothe codinium cohnii for use as a biochemical tool.
J. Protozool., 15(4):792-795.

cultures
Kensler, Craig B., 1967.
Notes on laboratory rearing of juvenile spiny lobsters, Jasus edwardsii (Hutton) (Crustacea: Decapoda: Polinuridae).
N.Z. Jl. mar. Freshwat. Res. 1(1): 71-75

cultures
Ketchum, B. H., 1939
The absorption of phosphate and nitrate by illuminated cultures of Nitschia closterium.
Am. J. Bot., 26:399-407.

cultures
Ketchum, B. H., L. Lillick, and A.C. Redfield, 1949.
The growth and optimum yields of unicellular algae in mass culture. J. Cell.&Comp. Physiol. 33(3):267-279, 3 textfigs.

cultures
Ketchum, B. H. and A.C. Redfield, 1949.
Some physical and chemical characteristics of algae grown in mass culture. J. Cell. & Comp. Physiol. 33(3):281-299, 2 textfigs.

cultures
Ketchum, B. H., and A. C. Redfield, 1938
A method for maintaining a continuous supply of marine diatoms by culture. Biol. Bull., 75: 165-169

cultures, diatoms
*Kim, Sung Ki, and Yong Kil Ro, 1967.
Experiment on the plankton culture for larvae fish. (In Korean; English abstract).
Bull. Fish. Res. Develop. Agency, 1:133-139.

cultures
Knaggs, F.W. 1965.
A simplified system for the controlled illumination of algal cultures.
Netherlands J. Sea Res. 4(1): 21-26

cultures
Knight, Margaret D., 1966.
The larval development of Polyonyx quadriungulatus Glassell and Pachycheles rudis Stimpson (Decapoda, Porcellanidae) cultured in the laboratory.
Crustaceana, 10(1):75-97.

cultures
Ko, Yatsuzuka, 1962
Studies on the artificial rearing of the larval Brachyura, especially of the larval blue-crab, Neptunus pelagicus Linnaeus.
Repts., USA Marine Biol. Sta., Kochi Univ., 9(1):88 pp. (In Japanese; English abstract).

cultures
Komaki, Yuze, 1966.
Technical notes on keeping euphausids alive in the laboratory, with a review of experimental studies on euphausids.
Inf. Bull. Planktol. Japan, No. 13:95-105.

cultures
Kornmann, P., 1970.
Advances in marine phycology on the basis of cultivation. Helgoländer wiss. Meeresunters. 20(1/4): 39-61.

cultures
Kornmann, Peter, 1963.
Der Lebenszyklus einer marinen Ulothrix-Art.
Helgolander Wiss. Meeresuntersuch., 8(4):357-360.

cultures
Kornmann, P., 1955.
Beobachtungen an Phaeocystis-Kulturen.
Helgoländer Wiss. Meeresunters., 5(2):218-233.

cultures
Krumbein, W.E., 1970.
On the behaviour of pure cultures of marine micro-organisms in sterilized and re-inoculated sediments. Helgoländer wiss. Meeresunters. 20 (1/4): 17-28.

cultures (aquaria)
Kühl, H., and H. Mann, 1960.
Vergleich des Stickstoffabbaus in See- und Süsswasseraquarien.
Vie et Milieu, 11(4):532-545.

L

cultures
Lagarde, E., 1967.
Étude de l'action des antibiotiques sur les microflores hétérotrophes marines. Utilisation des antibiotiques dans la réalisation des cultures de foraminifères sous faible volume.
Vie Milieu (A)18 (1-A):27-35.

cultures
Lanskaia, L.A., and T.I. Pshenina, 1963.
Comparison of the chemical composition of some species of diatoms in culture and in the sea. (In Russian).
Trudy Sevastopol Biol. Sta., 16:457-462.

cultures
Lanskaya, L.A., and S.I. Sivkov, 1950.
[The relation between the rates of development of cultures of marine diatoms and amounts of radiation.] Doklady Akad. Nauk, SSSR, 73(3):581-

cultures
Lasker, Reuben, and Gail H. Theilacker, 1965.
Maintenance of euphausiid shrimp in the laboratory.
Limnology and Oceanography, 10(2):287-288.

cultures
Laval, Philippe 1968.
Développement en élevage et systématique d'Hyperia schizogeneios Stebb. (amphipode hypéride).
Arch. Zool. exp. gén. 109(1): 25-67.

cultures
*Lee, Byung Don, and Taek Yuil Lee, 1968.
Experiments on the rearing of Metapenaeus joyneri (Miers).
Publ. Haewundae mar. Lab., 1:39-42.

cultures
Lefèvre, M., and H. Jakob, 1949.
Sur quelques propriétés des substances (actives) tirées des cultures d'algues d'eau douce.
C.R. Acad. Sci., Paris, 229:234-236.

cultures
Lefèvre, M., M. Nisbet, E. Jakob, 1949.
Action des substances excrétées en culture par certaines espèces d'algues, sur le métabolisme d'autres espèces d'algues. Verhandl. Int. Verein. f. Theoret. u. Angewandte Limnol. 10:259-264.

cultures
Levring, T., 1945
Some culture experiments with marine plankton diatoms. Medd. Oceanografiska Institutet i Göteborg, No.9 (Göteborgs Kungl. Vetenskaps-ochVitterhets-Samhälles Handlingar, Sjätte Följden. Ser.B. Vol.3(12) 1-17.

cultures
Lewin, Ralph A., Editor, 1962.
Physiology and biochemistry of algae.
Academic Press, New York and London, 929 pp.

cultures
Lewis, Alan G., 1967.
An enrichment solution for culturing the early developmental stages of the planktonic marine copepod Euchaeta japonica Marukawa.
Limnol. Oceanogr., 12(1):147-148.

cultures
Ling, S.W., and A.B.O. Merican, 1962.
Notes on the life and habits of the adults and larval stages of Macrobrachium rosenbergi (DeMan).
Indo-Pacific Fish. Council, FAO, Proc., 9th Sess., (2/3):55-61.

cultures
Loosanoff, V.L., J.E. Hanks and A.E. Ganaros, 1957.
Control of certain forms of zooplankton in mass algal cultures. Science 125(3257):1092-1093.

cultures
Lutze, G.F., B. Grabert und E. Seibold 1971.
Lebendbeobachtungen an Gross-Foraminiferen (Heterostegina) aus dem Persischen Golf.
Meteor-Forsch.-Ergebnisse (C) 6:21-40

M

cultures
Mackie, G.O., and D.A. Boag, 1964.
Fishing, feeding and digestion in siphonophores.
Pubbl. Staz. Zool., Napoli, 33(3):178-196.

cultures
*Mackintosh, N.A.,1967.
Maintenance of living Euphausia Superba and frequency of moults.
Norsk Hvalfangsttid., 56(5):97-102.

cultures
Maddux, William S., and Raymond F. Jones, 1964.
Some interactions of temperature, light intensity and nutrient concentration during the continuous culture of Nitzschia closterium and Tetraselmis sp.
Limnology and Oceanography, 9(1):79-86.

cultures
Malone, Ph. G., and K.M. Towe 1970.
Microbial carbonate and phosphate precipitates from sea water cultures.
Marine Geol. 9(5): 301-309.

cultures
Margalef, Ramon, 1967.
Laboratory analogues of estuarine plankton systems.
In: Estuaries, G.H. Lauff, editor, Publs Am. Ass. Advmt Sci., 83:515-521.

cultures
Margalef, Ramon, 1963
Algunas regularidades en la distribución a escala pequeña y media de las poblaciones marinas de fitoplancton y en sus características funcionales.
Invest. Pesquera, Barcelona, 23:169-230.

cultures
Margalef, Ramon, 1963
Desarrollo experimental de picnoclinas en pequeños volumenes de agua.
Invest. Pesquera, Barcelona, 23:3-10.

cultures
Margalef, Ramón, 1963
Modelos simplificados del ambiente marino para el estudio de la sucesión y distribución del fitoplancton y del valor indicador de sus pigmentos.
Invest. Pesquera, Barcelona, 23:11-52.

cultures (fresh water)
Margalef, Ramon, 1956.
Cultivos experimentales de algas unicelulares.
Inv. Pesq., Barcelona, 3:3-19.

cultures
Margalef, R., 1954.
Un aparato para el cultivo de algas en condiciones regulables. Publ. Inst. Biol. Aplic. 17:65-69, 3 textfigs.

cultures
Matsudaira Chikayoshi, 1957
[Culturing of Copepoda, Sinocalanus tenellus.]
Info. Bull. Plankton., Japan, (5):1-6.

cultures
Matsue, Y., 1954.
On the culture of the marine plankton diatom, Skeletonema costatum (Grev.) Cleve.
Rev. Fish. Sci., Japan, (Suisangaku no Gaikano): 1-4.

Mc

McAllister, C.D., T.R. Parsons and J.D.H. Strickland, 1960 — cultures
Primary productivity at station "P" in the north-east Pacific Ocean. J. du Cons., 35(3):240-259.

cultures
McAllister, C.D., N. Shah and J.D.H. Strickland, 1964.
Marine phytoplankton photosynthesis as a function of light intensity: a comparison of methods.
J. Fish. Res. Bd., Canada, 21(1):159-181.

cultures
McConnell, William J., 1962
Productivity relations in carboy microcosms.
Limnol. and Oceanogr., 7(3):335-343.

cultures
McDaniel, H.R., 1961.
Maintenance of sea urchins in laboratory aquaria
Texas J. Sci., 13(4):490-492.

cultures
McIntire, C. David, Robert L. Garrison, Harry K. Phinney and Charles E. Warren, 1964.
Primary production in laboratory streams.
Limnology and Oceanography, 9(1):92-102.

cultures
McLachlan, J., 1959.
The growth of unicellular algae in artificial and enriched sea water media.
Canadian J. Microbiology, 5(1):9-15.

cultures
McLachlan, J., and G.C. MacLeod, 1959.
The use of conversion factors for the determination of the concentration of nutrients in culture media.
Limnol. & Oceanogr., 4(2):218-219.

cultures
McLachlan, Jack, and Charles S. Yentsch, 1959
Observations on the growth of Dunaliella euchlora in culture. Biol. Bull., 116(3): 461-483.

cultures
McLaughlin, John J. A., and Paul A. Zahl, 1961.
In vitro culture of Pyrodinium.
Science, 134(3493):1878.
Also in:
Contrib., Inst. Mar. Sci., Univ. Puerto Rico, 3.

cultures
Meixner, R., 1966.
Eine Methode zur Aufzucht Von Crangon Crangon (L.) (Crust.Decap. Netantia).
Arch. Fischereiwiss., 17(1):1-4.

Cultures
Millar, R.H., and J. M.Scott, 1965.
The use of a direct counting technique for bacteria in marine culture experiments.
Journal du Conseil, 29(3):253-255.

cultures
*Miller, R.A., J.P. Shyluk, O.L. Gamborg, and J.W. Kirkpatrick, 1968.
Phytostat for continuous culture and automatic sampling of plant-cell suspensions.
Science, 159(3814):540-542.

cultures
Mjaaland, Gunnar, 1956
Laboratory experiments with Coccolithus huxleyi
Oikos, 7(2): 251-255.

cultures
Moyse, J., 1960.
Mass rearing of barnacle cyprids in the laboratory.
Nature 185(4706):120.

cultures
*Mullin, Michael M., and Elaine R. Brooks, 1967.
Laboratory culture, growth rate and feeding behavior of a planktonic marine copepod.
Limnol. Oceanogr., 12(4):657-666.

cultures
Munda, I., 1963.
Kulturversuche mit Ascophyllum nodosum (L) Le Jol und Fucus vesiculosus L., in Medien von verschiedenem Salzgehalt.
Botanica Marina, 5(2/3):84-96.

cultures
Murakami, Akio, 1966.
Rearing experiments of a chaetognath, Sagitta crassa. (In Japanese; English abstract).
Inf. Bull. Planktol. Japan, No. 13:62-65.

cultures
Murano, Masaaki, 1966.
Culture of Isaza-Ami, Neomysis intermedia. (In Japanese; English abstract).
Inf. Bull. Planktol. Japan, No. 13:65-68.

N

cultures, copepods
Nassogne, A., 1971
First heterotrophic level of the food chain.
CNEN Rept. RT/BIO (70)-11, M. Bernhard, editor: 41-60.

cultures
Nassogne, A., 1970.
Influence of food organisms on the development and culture of pelagic copepods. Helgoländer wiss. Meeresunters, 20(1/4): 333-345.

cultures, copepod
Nassogne Armand 1969.
La coltura dei Copepodi in laboratorio.
Pubbl. Staz. Zool. Napoli 37 (2 Suppl.): 203-218

cultures
Natarajan, K.V., and R.C. Dugdale, 1966.
Bioassay and distribution of thiamine in the sea.
Limnol. Oceanogr., 11(4):621-629.

cultures
Neunes, Jeinz W., and Gian-Franco Pongolini, 1965.
Breeding a pelagic copepod, Euterpina acutifrons (Dana) in the laboratory.
Nature, 208(5010):571-573.

cultures
Nixon, Scott W., 1969.
A synthetic microcosm. Limnol. Oceanogr., 14 (1): 142-145.

cultures
Nordli, E., 1957.
Algal flour extract as a stimulating agent for marine dinoflagellate cultures. Nytt Mag. Bot., 5:13-16.

cultures
Nordli, Erling, 1957.
Algal flour extract as a stimulating agent for marine dinoflagellate cultures.
Nytt Mag. Botan., 5:13-16.

cultures
Nordli, E., 1957.
Experimental studies on the ecology of Ceratia.
Oikos, 8(2):201-265.

culture
North, Wheeler J., and Carl L. Hubbs, editors, 1968.
Utulization of kelp-bed resources in southern California.
Fish Bull., Dept. Fish Game, Calif. 139:264.

O

cultures
Oppenheimer, Carl H., editor, 1966.
Phytoplankton, Marine Biology II, Proceedings of the Second International Interdisciplinary Conference.
New York Acad. Sci., 369pp. ($8.00).

P

cultures
Paasche, E. 1968.
Marine plankton algae grown with light-dark cycles. II. Ditylum brightwelli and Nitzschia turgidula.
Physiologia Pl. 21(1): 66-77.

cultures
Paffenhöfer, G.-A., 1970.
Cultivation of Calanus helgolandicus under controlled conditions. Helgoländer wiss. Meeresunters. 20(1/4): 346-359.

cultures
Parker, Richard A., 1960.
Competition between Simocephalus vetulus and Cyclops viridis.
Limnol. & Oceanogr., 5(2):180-189.

cultures
Parsons, T.R., 1961
On the pigment composition of eleven species of marine phytoplankters.
J. Fish. Res. Bd. Canada 18(6):1-17-1025.

cultures
Parsons T.R., K. Stephens and J.D.H. Strickland, 1961
On the chemical composition of eleven species of marine phytoplankters.
J. Fish. Res. Bd. Canada, 18(6):1001-1016.

cultures
Persoone, Guido, 1966.
Contributions à l'étude des bactéries marines du littoral belge. III Milieux de culture et ensemencements.
Bull. Inst. r. Sci. Nat. Belg., 42(6):1-14.

cultures
Pincemin, J.-M. 1971.
Télémédiateurs chimiques et équilibre biologique océanique. 3. Étude in vitro de relations entre populations phytoplanctoniques.
Rev. int. Océanogr. Méd. 22-23:165-196

cultures
Pintner, I.J., and L. Provosoli, 1958.
Artificial cultivation of red-primented marine blue-green alga, Phormidium persicinum.
J. Gen. Microbiol., 18(1):190-197.

cultures
Plessis, Y., 1956.
Note sur le control de la salinité en milieu marin artificiel.
Bull. Mus. Nat. Hist. Nat., 28:583-589.

cultures
Polikarpov, G.G., and L.A. Lanskaia, 1961.
Culture of unicellular algae Prorocentrum micans Ehr. in bulk in the presence of S35.
Trudy Sevastopol Biol. Sta. (14):329-333.

cultures
Prakash, A., 1967.
Growth and toxicity of a marine dinoflagellate, Gonyaulax tamarensis
J. Fish. Res. Bd. Can., 24 (7): 1589-1606

cultures
Pratt, David M., 1966.
Competition between Skeletonema costatum and Olisthodiscus luteus in Narragansett Bay and in culture.
Limnol. Oceanogr., 11(4):447-455.

cultures
Pringsheim, E.G., 1952.
Observations on some species of Trachelomonas grown in culture. The New Phytologist 52(2):93-113, 5 textfigs.

cultures
Provasoli, L., D.E. Conklin and A.S. D'Agostino 1970.
Factors inducing fertility in aseptic Crustacea
Helgoländer wiss. Meeresunters, 20(1/4): 443-454.

cultures
Provasoli, L., and J.F. Howell, 1952.
Culture of a marine Gyrodinium in a synthetic medium. Proc. Amer. Soc. Protozool. 3:6.

cultures
Provasoli, L., and I.J. Pintner, 1953.
Ecological implications of in vitro nutritional requirements of algal flagellates.
Ann. N.Y. Acad. Sci. 56(5):839-851.

Cultures
Provenzano, Anthony J., Jr., 1967.
Recent advances in the laboratory culture of decapod larvae.
Proc. Symp. Crustacea, Ernakulam, Jan. 12-15, 1965, 2: 940-945

cultures
*Provenzano. Anthony J., Jr., 1967.
The zoeal stages and glaucothoe of the tropical eastern Pacific hermit crab.
Trizopagurus magnificus (Bouvier, 1898) (Decapoda; Diogenidae) reared in the laboratory.
Pacif. Sci., 21(4):457-473.

Q

cultures
Quraishi, F.O. and C.P. Spencer, 1971.
Studies on the growth of some marine unicellular algae under different artificial light sources.
Marine Biol., 8(1): 60-65.

R

cultures
Ramamurthy, V.D., and K. Krishnamurthy, 1965.
On the culture of the diatom Melosira sulcata (Ehr.) Kutzing.
Proc. Indian Acad. Sci., B, 42 (1):25-31.
Also in; Collected Reprints, Mar. Biol. Sta., Porto-Novo, 1963/64.

cultures
Ramamurthy, V. D., and K. Krishnamoorthy, 1965.
On the culture of the diatom Melosira sulcata (Ehr.) Kutzing.
Proc. Indian Acad. Sci., (B), 62(1):25-31.

culture
Ranson, M.G., 1936
Essais de culture, dans la nature, de la Navicule bleue, cause du verdissement des Huîtres. Ostréiculture, Cultures marines, No.4

cultures
Rao, D.V. Subba and T. Platt, 1969.
Optimal extraction conditions of chlorophylls from cultures of five species of marine phytoplankton.
J. Fish. Res. Bd. Can. 26 (6): 1625-1630

cultures, fungi
Ray, Sammy M., 1964?
A review of the culture method for detecting Dermocystidium marinum with suggested modifications and precautions.
1963 Proc., Nat. Shellfish. Assoc., 55-69.

cultures
Ray, S.M., and W.B. Wilson, 1957.
The effects of unialgal and bacteria-free cultures of Gymnodinium brevis on fish and notes on related studies with bacteria.
U.S.F.W.S. Spec. Sci. Repts., Fish, No. 211.
Also: U.S.F.W.S. Fish. Bull. (in press).

cultures
Ray, S.M., and W.B. Wilson, 1957.
Effects of unialgal and bacteria-free cultures of Gymnodinium brevis on fish.
USFWS Fish. Bull., 123:469-496.

cultures
Raymont, J.E.G., and F. Gross, 1942.
On the feeding and breeding of Calanus finmarchicus under laboratory conditions. Proc. Roy. Soc., Edinburgh, Sec. B (Biol.), 61(3); No. 20: 267-287, 2 textfigs.

cultures
Raymont, J.E.G., and R.S. Miller, 1962.
Production of marine zooplankton with fertilization in an enclosed body of water.
Int. Revue Ges. Hydrobiol., 47(2):169-209.

cultures
Reed, Paul H. 1969.
Culture methods and effects of temperature and salinity on survival and growth of Dungeness crab (Cancer magister) larvae in the laboratory.
J. Fish. Res. Bd. Can. 26(2): 389-397.

cultures
Reeve, M.R., 1970.
Complete cycle of development of a pelagic chaetognath in culture.
Nature, Lond., 227(5256): 381.

cultures
Rice, A.L. and D.I. Williamson, 1970.
Methods for rearing larval decapod Crustacea.
Helgoländer wiss. Meeresunters, 20(1/4): 417-434.

cultures
Riley, Gordon A., Peter J. Wangersky and Denise Van Hemert, 1964
Organic aggregates in trppical and subtropical surface waters of the North Atlantic Ocean
Limnology and Oceanography, 9(4):546-550.

cultures
Riley, J.P. and Igal Roth, 1971.
The distribution of trace elements in some species of phytoplankton grown in culture.
J. mar. biol. Ass., U.K., 51(1): 63-72.

cultures
Rust, John D., and Frank Carlson, 1960.
Some observations on rearing blue crab larvae.
Chesapeake Science, 1(3/4):196-197.

cultures
Ryther, J. H., and R. R. L. Guillard, 1959.
Enrichment experiments as a means of studying nutrients limiting to phytoplankton production.
Deep-Sea Res., 6(1):65-69.

S

cultures
Sastry, A.N., 1970.
Culture of brachyuran crab larvae using a re-circulating sea water system in the laboratory. Helgoländer wiss. Meeresunters, 20(1/4): 406-416.

Cultures
Sastry, A.N., 1965.
The development and external morphology of pelagic larval and post-larval stages of the bay scallop, Aequipecten irradians concentricus Say, reared in the laboratory.
Bull. Mar. Sci.,15(2):417-435.

cultures
Satomi, Yoshihiro, 1968.
Über dem einfacheren Bestimmungsverfahren der Phytoplankton-und Bakteriendichte in einartigem Phytoplankton-Kulturmedium. (In Japanese; German abstract). Bull. Plankton Soc., Japan, 15(2): 10-13.

cultures
*Sazhina,L.I.,1968.
A method of cultivation of marine pelagic Copepods in the laboratory. (In Russian;English abstract).
Zool.Zh., 47(11):1713-1716.

cultures
Scagel, Robert F., and Janet R. Stein, 1961.
Marine nannoplankton from a British Columbia fjord.
Canadian J. Botany, 39:1205-1213.

cultures
Schumann, George O., and Herbert C. Perkins, 1966.
A jar for culturing small pelagic marine organisms in running water during shipboard or laboratory studies.
J. Cons. perm. int. Expl. Mer, 30(2):171-176.

cultures
Seki, Humitake,1966.
Studies on microbiol participation to food cycle in the sea. III. Trial cultivation of brine shrimp to adult in a chemostat.
J. Oceanogr.Soc.,Japan, 22(3):105-110.

cultures
Sguros, Peter L., and Jacqueline Simms, 1963.
Role of marine fungi in the biochemistry of the oceans. II. The effect of glucose, inorganic nitrogen, and tris(hydroxymethyl) aminomethane on growth and pH changes in synthetic media.
Mycologia, 55(6):728-741.

cultures
Shiraishi, Kagehide,1966.
Bacteria-free culture and nutrition of micro-crustaces. (In Japanese;English abstract).
Inf. Bull. Planktol. Japan, No. 13:49-54.

cultures
Sick, Lowell V., James W. Andrews and David B. White 1972.
Preliminary studies of selected environmental and nutritional requirements for the culture of penaeid shrimp.
Fish. Bull. U.S. Nat. Fish. Serv. 70(1): 101-109

cultures
Skulberg, O.M., 1970.
The importance of algal cultures for the assessment of the eutrophication of the Oslofjord.
Helgoländer wiss. Meeresunters, 20(1/4): 111-125.

cultures
Sliter, William V., 1970
Bolivina doniezi Cushman and Wickenden in clone culture.
Contrib. Cushman Fdn foramin Res. 21(3): 87-99

cultures
Smayda, T.J., 1970.
Growth potential bioassay of water masses using diatom cultures: Phosphorescent Bay (Puerto Rico) and Caribbean waters. Helgoländer wiss. Meeresunters, 20(1/4): 172-194.

cultures
Smayda, Theodore J., 1964.
Enrichment experiment using the marine centric diatom Cyclotella nana (clone 13-1) as an assay organism.
Narragansett Mar. Lab., Univ.Rhode Island, Occ. Publ., No. 2:25-32.

cultures
Soli, Giorgio, 1964
A system for isolating phytoplankton organisms in unialgal and bacteria-free culture.
Limnology and Oceanography, 9(2):265-267.

cultures
Spencer,C.P.,1966.
Theoretical aspects of the control of pH in natural sea water and synthetic culture media for marine algae.
Botanica mar., 9(3/4): 81-99.

cultures
Spencer, C.P., 1954.
Studies on the culture of a marine diatom.
J.M.B.A. 33(1):265-290, 16 textfigs.

cultures
Spencer, C.P., 1952.
On the use of antibiotics for isolating bacteria-free cultures of marine phytoplankton organisms.
J.M.B.A. 31(1):97-106, 1 textfig.

cultures
Steele, J.H., and I.E. Baird, 1962.
Carbon-chlorophyll relations in cultures.
Limnol. and Oceanogr., 7(1):101-102.

cultures
Stickney, Alden P., 1964
Salinity, temperature and food requirements of soft-shell clam larvae in laboratory culture.
Ecology, 45(2):283-291.

cultures
Strand,John A., Joseph T. Cummins and Burton E. Vaughn,1966.
Artificial culture of marine sea weeds in recirculation aquarium systems.
Biol. Bull., mar. biol. Lab., Woods Hole, 131(3): 487-500.

cultures, phytoplankton
Strickland, J.D.H., O.Holm-Hansen, R.W. Eppley, and R.J. Linn, 1969.
The use of a deep tank in plankton ecology. I. Studies of the growth and composition of phytoplankton crops at low nutrient levels.
Limnol. Oceanogr., 14(1): 23-34.

cultures
Subrahmanyan, R., 1952.
Notes on growing diatoms in cultures. Microscope (Jan.-Feb.):

cultures

Svoboda, A., 1970.
Simulation of oscillating water movement in the laboratory for cultivation of shallow water sedentary organisms. Helgoländer wiss. Meeresunters, 20(1/4): 676-684.

T

cultures

Takahashi, Masayuki, Sooji Shimura, Yukuya Yamaguchi and Yoshihiko Fujita 1971.
Photo-inhibition of phytoplankton photosynthesis as a function of exposure time.
J. oceanogr. Soc. Japan. 27(2): 43-50

cultures

Takano, Hideaki, 1971.
Notes on the raising of an estuarine copepod Gladioferens imparipes Thomson. Bull. Tokai reg. Fish. Res. Lab. 64: 81-88.

cultures

*Takano, Hideaki, 1967.
Some culture experiments of Cyclotella nana Hustedt. (In Japanese; English abstract).
Inf. Bull. Planktol. Japan, Comm. No. Dr. Y. Matuse, 231-237.

cultures

Takano, Hideaki, 1965.
Diatom culture in artificial sea water. III. Growth of diatoms in small flasks.
Bull. Tokai Reg. Fish. Res. Lab., No. 44: 17-24.

cultures

Takano, Hideaki, 1964
Diatom culture in artificial sea water. II. Cultures without using soil extract.
Bull. Tokai Reg. Fish. Res. Lab., No. 38: 45-56.

cultures

Takano, Hideaki, 1963.
Culture media for diatoms. (In Japanese; English abstract).
Info. Bull., Plankt., Japan, No. 10: 5-9.

cultures

Takano, Hideaki, 1963.
Diatom culture in artificial sea water. 1. Experiments on five pelagic species.
Bull. Tokai Reg. Fish. Res. Lab., No. 37: 17-25.

cultures

Takano, Hideaki, 1963.
Notes on marine littoral diatoms of Japan.
Bull. Tokai Reg. Fish. Res. Lab., No. 36: 1-8.

cultures

Takeda, Keiji, 1970.
Culture experiments on the relative growth of a marine centric diatom, Chaetoceros calcitrans Takano, in various concentrations of nitrate nitrogen. (In Japanese; English abstract).
Bull. Plankt. Soc., Japan, 17(1): 11-19.

cultures

Thomas, Pierre, et Raoul Dumas, 1970
Contribution à l'étude de Dunaliella salina en cultures bactériennes: nutrition et composition.
Téthys 2(1): 19-28

cultures

Thomas, William H., 1969.
Phytoplankton nutrient enrichment experiments off Baja California and in the eastern Equatorial Pacific Ocean.
J. Fish. Res. Bd., Can. 26(5): 1133-1145

cultures

Thomas, William H., 1964.
An experimental evaluation of the C14 method for measuring phytoplankton production, using cultures of Dunaliella primolecta Butcher.
U.S.F.W.S. Fish. Bull., 63(2): 273-292.

cultures

*Thomas, William H., and Anne N. Dodson, 1968.
Effects of phosphate concentration on cell division rates and yield of a tropical oceanic diatom.
Biol. Bull. mar. biol. Lab., Woods Hole, 134(1): 199-208.

cultures

Tighe-Ford, D.J., M.J.D. Power and D.C. Vaile, 1970.
Laboratory rearing of barnacle larvae for anti-fouling research. Helgoländer wiss. Meeresunters 20(1/4): 393-405.

cultures

*Tilton, R.C., G.J. Stewart, and G.E. Jones, 1967.
Marine thiobacilli. II. Culture and ultra structure.
Can. J. Microbiol., 13(11): 1529-1534.

cultures

Tokioka, Takasi, 1963.
Problems in maintaining an aquarium.
Bull. Mar. Biol. Sta., Asamushi, Tohoku Univ., 11(3): 117-120.

cultures

Toriumi, Saburo, 1968.
Cultivation of marine Ceratia using natural sea water culture media. I. Bull. Plankton Soc. Japan, 15(1): 1-6.
In Japanese; English

cultures

Toriumi, Saburo, 1966.
Cultivation of marine dinoflagellates.
(In Japanese; English abstract).
Inf. Bull. Planktol. Japan, No. 13: 41-49.

cultures

Trüper, H.G., 1970.
Culture and isolation of phototrophic sulfur bacteria from the marine environment. Helgoländer wiss. Meeresunters., 20(1/4): 6-16.

U

cultures

Umezu, Takeshi, and Masamichi Saiki, 1967.
Gross uptake of radionuclides ny marine micro-organisms in batch culture. (In Japanese; English abstract).
Bull. Naikai reg. Fish. Res. Lab., 24: 1-9

cultures

Uno, Shiroh, 1971.
Turbidometric continuous culture of phytoplankton constructions of the apparatus and experiments on the daily periodicity in photosynthetic activity of Phaeodactylum tricornutum and Skeletonema costatum. Bull. Plankt. Soc. Japan 18(1): 14-27.

cultures

Urry, D.L., 1965.
Observations on the relationship between the food and survival of Pseudocalanus elongatus in the laboratory.
Jour. Mar. Biol. Assoc., U.K., 45(1): 49-58.

V

cultures

Vaccaro, Ralph F., and John H. Ryther, 1960
Marine phytoplankton and the distribution of nitrite in the sea. J. du Cons., 35(3): 260-271.

cultures

von Stosch, H.A., 1962
Kulturexperiment und Oekologie bei Algen.
Kieler Meeresf., 18(3) (Sonderheft): 13-27.

cultures

Van Valkenburg, Shirley D., and Richard E. Norris, 1970
The growth and morphology of the silicoflagellate Dictyocha fibula Ehrenberg in culture.
J. Phycol. 6(1): 48-50

cultures

von Thun, Wolf, 1966.
Eine Methode zue Kultivierung der Mikrofauna.
Veröff. Inst. Meeresforsch., Bremerh., Sonderband II: 277-280.

cultures

Vilela, Marie Helena, 1969.
The life cycle of Tisbe sp. (Copepoda Harpacticoida) under laboratory conditions.
Notas Estud. Inst. Biol marit., Lisboa, (36): 16 pp.

cultures

* Vosjan, J.H., and R.J. Siezin, 1968.
Relation between primary production and salinity of algal cultures.
Netherlands J. Sea Res. 4(1): 11-20.

cultures

*Vuillemin, S., 1968.
Élevage de Serpulines (Annélides Polychètes).
Vie Milieu, 19(1-A): 195-199.

W

cultures

Wall, D., R.R.L. Guillard, and B. Dale 1967.
Marine dinoflagellate cultures from resting spores.
Phycologia 6(2/3): 83-86

cultures

Werner, Bernhard, 1963.
Effect of some environmental factors on differentiation and determination in marine Hydrozoa with a note on their evolutionary significance.
Ann. N.Y. Acad. Sci., 105(8): 461-488.

cultures

Werner, Dietrich, 1971.
Der Entwicklungscyclus mit Sexualphase bei der marinen Diatomee Coscinodiscus asteromphalus. I. Kultur und Synchronisation von Entwicklungsstadien.
Arch. Mikrobiol. 80(1): 43-49.

cultures

Werner, D., 1970.
Productivity studies on diatom cultures.
Helgoländer wiss. Meeresunters. 20(1/4): 97-103.

cultures

*Williams, Barbara G., 1968.
Laboratory rearing of the larval stages of Carcinus maenas (L.) (Crustacea:Decapoda).
J. nat.Hist., 2(1):121-126.

cultures

Wilson, William B., 1967.
Forms of the dinoflagellate Gymnodinium breve Davis in cultures.
Contrib. mar. Sci., Port Aransas 12: 120-134

cultures

Wilson, W.B., and A. Collier, 1955.
Preliminary notes on culturing of Gymnodinium brevis Davis. Science 121(3142):394-395

cultures

Wisely, B., 1960
Experiments on rearing the barnacle Elminius modestus Darwin to the settling stage in the laboratory.
Australian J. Marine and Freshwater Research 1(1): 42-54.

XYZ

cultures

Yasuie, S. and I. Wada, 1971.
Study on utilization of wild plankton as larval fish feed. (4) Seasonal abundance of marine sticking animals. (In Japanese). Bull. Fish. Exp. Sta. Okayama Pref. (1970): 207-208.

cultures

Yoo, Sung Kyoo, 1969
Culture conditions and growth of larvae of Mytilus coruscus Gould. (In Korean, English abstract).
J. Oceanogr. Soc., Korea 4(1): 36-45.

cultures

Yoo, Sung Kyoo, 1968.
Studies on the growth of algal food, Cyclotella nana, Chaetoceros calcitrans and Monochrysis lutheri. (In Korean; English abstract).
Bull. Fusan Fish. Coll. (Nat. Sci.), 8(2):123-126.

cultures

Zemtsova, E.V., and A.E. Kriss, 1962.
The survival of sea microorganisms (heterotrophs) grown under laboratory conditions.
Doklady Akad. Nauk, SSSR, 142(3):695-698.

cultures

Zillioux, E.J., 1969.
A continuous recirculating culture system for planktonic copepods. Mar. Biol., 4(3): 215-218.

cultures

Zillioux, E.J., and N.F. Lackie 1970.
Advances in the continuous culture of planktonic copepods.
Helgoländer wiss. Meeresunters. 20(1/4): 325-332.

cultures

Zillioux, Edward J., and Donald F. Wilson, 1966.
Culture of a planktonic calanoid copepod through multiple generations.
Science, 151(3713):996-997.

cultures

Zobell, C.E., and J.H. Long, 1938
Studies on the isolation of bacteria-free cultures of marine phytoplankton. J. Mar. Res.,1:328-334.

culture age, effect of

Pugh, P.R., 1971.
Changes in the fatty acid composition of Coscinodiscus eccentricus with culture-age and salinity. Mar. Biol. 11(2): 118-124.

cultures, algae

Burns, Richard L. and Arthus C. Mathieson, 1972.
Ecological studies of economic red algae. II. Culture studies of Chondrus crispus Stackhouse and Gigartina stellata (Stackhouse) Batters. J. exp. mar. Biol. Ecol. 8(1): 1-6.

cultures, decapod

Savage, Thomas 1971.
Effect of maintenance parameters on growth of the stone crab, Menippe mercenaria (Say)
Spec. scient. Rept. Mar. Res. Lab., St. Petersburg, Fla. Dept. Nat. Resources 28: 19 pp.

cultures, crabs

*Veragnolo, Sergio, 1967.
Pesca e cultura del granchio Carcinus maenas L. nella Laguna di Venezia.
Arch.Oceanogr.Limnol.15(Suppl.):83-96.

culture, fish

cultures, fish

Agalides, Eugene 1966.
Synthetic seawater for sharks
Geo Mar. Techn. 2(7): 11-15.

cultures, fish

Araga, Chuichi, 1963.
On fish rearing in the "oceanariums".
Bull. Mar. Biol. Sta., Asamushi, Tohoku Univ., 11(3):105-112.

cultures

Blaxter, J.H.S., 1968.
Rearing herring larvae to metamorphosis and beyond.
J. mar. biol. Ass., U.K., 48(1):17-28.

cultures, fish

Bowers, A.B., 1966.
Marine fish culture in Britain, VI., the effect of the acclimatization of adult plaice to pond conditions on the viability of eggs and larvae.
J. Cons. perm. int. Expl. Mer, 30(2):196-203.

culture, fish

Cooper, L.H.N., and G.A. Steven, 1948
An experiment in marine fish cultivation.
Nature 161:631.

cultures, fish

El-Zarka, Salah El-Din, 1963
Acclimatization of Solea vulgaris (Linn) in Lake Quarun, Egypt.
J. du Conseil, 28(1):126-136.

cultures, fish

Furukawa, Atsushi, Takeshi Umezu, Hiroko Tsukahara, Katsumi Funae and Isamu Iwata, 1966.
Studies on feed for fish. 5. Results of the small floating net culture test to establish the artificial diet as complete yellow-tail foods (1964).(In Japanese; English abstract).
Bull. Naikai reg. Fish. Res. Lab., 23:45-56.

cultures, fish

Gohar, H.A.F., and F.M. Mazhar, 1964.
Keeping elasmobranchs in vivaria.
Publ. Mar. Biol. Sta. Ghardaqa, No. 13:241-250.

Cultivation, fish

Gross, F., 1948
Marine Fish Cultivation. Nature 162(4114): 378, with reply by L.H.N. Cooper and G.A. Steven.

cultures, fish

Hepher, B., 1962
Primary production in fishponds and its application to fertilization experiments.
Limnology and Oceanography, 7(2):131-136.

cultures, fish

Houde, E.D. and B.J. Palko, 1970.
Laboratory rearing of the clupeid fish Harengula pensacolae from fertilized eggs. Marine Biol., 5(4): 354-358.

cultures, fish

Inoue, Motoo, 1970.
Possibility of tuna cultivation and propagation.
Bull. Japan Soc. Fish. Oceanogr. 16. Also in: Coll. Repr. Coll. mar. Sci. Techn. Tokai Univ. 4: 197-210. (In Japanese; English abstract).

cultures, fish

*Inoue, Motoo, Ryohei Amano, Yukinobu Iwasaki, and Mitsuyoshi Aoki, 1967.
Ecology of various tunas under captivity. I. Preliminary rearing experiments. (In Japanese; English abstract).
J. Coll.mar.Sci.Techn.,Tokai Univ., (2):197-209.

cultures (eel)

Ito, Takashi, and Tashio Iwai, 1960.
Studies on the "Mizukawari" in eel-culture ponds. 19. The effect of the sea water supply upon the plankton in mixohaline eel-culture pond.
Rept. Fac. Fish., Pref. Univ. of Mie, 3(3):649-655.

cultures (eel)

Ito, Takashi, Toshio Iwai and Yasihiko Akamine, 1960.
Studies on the "Mizukawari" in eel-culture ponds 20. The effect of winter drainage upon the plankton and formation of the plankton community after the water-renewal in eel-culture ponds.
(English resume.)
Rept. Fac. Fish., Pref. Univ. Mie, 3(3):656-679.

cultures, fish

Huet, Marcel, 1970.
Traité de pisciculture. 4th edit., Editions Ch. de Wyngaert, Bruxells XXIV + 718 pp.
$19.00

Cultures, fish

Lasker, R., H.M. Feder, G.H. Theilacker and R.C. May, 1970.
Feeding, growth, and survival of Engraulis mordax larvae reared in the laboratory. Marine Biol., 5(4): 345-353.

cultures, fish

LeMare, D.W., 1951.
Application of the principles of fish culture to estuarine conditions in Singapore. Proc. Indo-Pacific Fish. Counc., 17-28 Apr. 1950, Cronulla N.S.W., Australia, Sects. II-III:180-183.

cultures, fish

Lloyd, Robert E., 1965.
An approach to aquaculture -- United Kingdom White Fish Authority.
GeoMar. Techn.,1(7):14-16.

cultures, fish

Murakami, Akio 1969.
A balance sheet of the nitrogen as the plant nutrient in a laver culture ground. Bull. Nansei reg. Fish. Res. Lab. 2: 1-18.
(In Japanese; English abstract).

culture, fish

Matsui, I., 1952.
[Studies on the morphology, ecology and pond-culture of the Japanese eel (Anguilla japonica Temminck & Schlegel)] J. Simonoseki Coll. Fish. 2(2):1-245, 3 pls., 85 textfigs.
(English summary).

cultures, fish

Riley, John S., 1966.
Marine fish culture in Britain. VII. Plaice (Pleuronectes platessa L.) post-larval feeding on Artemia salina L. naupli and the effects of varying feeding levels.
J. Cons. perm. int. Expl. Mer, 30(2):204-221.

cultures, fish

Riley, John D., and Graham T. Thacker, 1963
Marine fish culture in Britain. III. Plaice (Pleuronectes platessa (L.)) rearing in closed circulation at Lowestoft, 1961.
J. du Conseil, 28(1):80-90.

cultures, fish

Rosenthal, H., 1970.
Anfütterung und Wachstum der Larven und Jungfische des Hornhechtes Belone belone.
Helgoländer wiss. Meeresunters, 21(3): 320-332.

cultures, fish

Schuurman, J.J., 1966.
A method for the determination of the suitability of coastal regions for the construction of brackish water ponds.
Indo-Pacific Fish. Counc. Proc. 11th Sess., 116-121.

"cultures", fish

Shelbourne, J.E., 1964.
The artificial propagation of marine fish.
In: Advances in Marine Biology, F.S. Russell, editor, Academic Press, 2:1-83.

cultures, fish

Shelbourne, J.E., 1963
Marine fish culture in Britain. II. A plaice rearing experiment at Port Erin, Isle of Man, during 1960, in open sea water circulation.
J. du Conseil, 28(1):70-79.

culture, fish

Shelbourne, J.E., 1963
Marine fish culture in Britain. IV. High survivals of metamorphosed plaice during salinity experiments in open circulation at Port Erin, Isle of Man, 1961.
J. du Conseil, 28(2):246-261.

cultures, fish

Shelbourne, J.E., J.D. Riley and G.T. Thacker, 1963
Marine fish culture in Britain. I. Plaice rearing in closed circulation at Lowestoft, 1957-1960.
J. du Conseil, 28(1):50-69.

cultures, fish

Steven, G.A., and L.H.N. Cooper, 1951.
A further experiment in marine fish culture.
Nature 167(4248):518.

Cultures mass

cultures, mass

Adams, M.N.E., and J.E.G. Raymont, 1958.
Studies on the mass culture of Phaeodactylum.
Ann. Rept. Challenger Soc., 1958, 3(10):

cultures, mass

Burlew, J.S., ed., 1953.
Algal culture: from laboratory to pilot plant.
Carnegie Inst., Washington, Publ. 600:357 pp.

cultures, mass

Ansell, Alan D., J. Coughlan, K.F. Lander and F.A. Loosmore, 1964.
Studies on the mass culture of Phaeodactylum. IV Production and nutrient utilization in outdoor mass culture.
Limnology and Oceanography, 9(3):334-342.

cultures, mass

Ansell, A.D., J.E.G. Raymont and K.F. Lander, 1963
Studies on the mass culture of Phaeodactylum. III. Small- scale experiments.
Limnol. and Oceanogr., 8(2):207-213.

cultures, mass

Ansell, A.D., J.E.G. Raymont, K.F. Lander, E. Crowley and P. Shackle, 1963
Studies on the mass culture of Phaeodactylum. II. The growth of Phaeodactylum and other species in outdoor tanks.
Limnol. and Oceanogr., 8(2):184-206.

cultures, mass

Davis, Harry C., and Ravenna Ukeles, 1961
Mass culture of phytoplankton as foods for metazoans.
Science, 134(3478):562-564.

cultures, mass

Denffer, D. von, 1948.
Die planktische Massenkultur pinnater Grunddiatomeen. Arch. Microbiol. 14:159-202.

cultures

Gold, K., 1971.
Growth characteristics of the mass-reared tintinnid Tintinnopsis beroidea. Marine Biol., 8(2): 105-108.

cultures, mass

Harder, R., and H. von Witsh, 1942.
Über Massenkultur von Diatomeen.
Ber. Deutsch. Bot. Ges. 60:146-152.

cultures, mass

Krauss, R., 1955.
Nutrient supply for large-scale algal cultures.
Sci. Mon. 80(1):21-28.

cultures, mass

Loosanoff, V.L., 1951.
Culturing phytoplankton on a large scale. Ecol. 32(4):748-750, 1 fig.

cultures, mass

McAllister, C.D., T.R. Parsons, K. Stephens and J.D.H. Strickland, 1961
Measurements of primary production in coastal sea water using a large volume plastic sphere.
Limnol. & Oceanogr., 6(3): 237-258.

cultures, mass

McLaughlin, John J. A., Paul A. Zahl, Andrew Nowak, John Machisotto and Jan Prager, 1960.
Mass cultivation of some phytoplankton.
Ann. N.Y. Acad. Sci., 90(3):856-865.

cultures, mass

Milner, H.W., 1955.
Some problems in large-scale culture of algae.
Sci. Mon. 80(1):15-20.

cultures, mass

Pettersson, H., F. Gross, and F. Koczy, 1939
Large scale plankton cultures. Medd. från Oceanografiska Institutet i Göteborg No.3 (Göteborgs Kungl. vetenskaps-och Vitterhets-Samhälles Handlingar Femte Följden. Ser. B) Vol.6(13):1-24.

1. The plankton shaft. H. Pettersson
2. Experiments with phytoplankton. F. Gross and F. Koczy
3. Experiments with zooplankton.

cultures, mass

Raymont, J.E.G., and M.N.E. Adams, 1958.
Studies on the mass culture of Phaeodactylum.
Limnol. & Oceanogr., 3(2):119-136.

cultures, mass

Sato, Tadao and Makoto Serikawa, 1968.
Mass culture of a marine diatom, Nitzschia closterium. (In Japanese; English abstract).
Bull. Plankton Soc., Japan, 15(1): 13-16.

cultures, mass

Spiktorova, L.V., 1970.
The sea flagellate, Platymonas viridis, Rouch sp. nov. as an object of mass culture. (In Russian)
Dokl. Akad. Nauk SSSR 192(3):662-664

cultures, mass

Theilacker, G.H. and M.F. McMaster, 1971.
Mass culture of the rotifer Brachionus plicatilis and its evaluation as a food for larval anchovies. Mar. Biol. 10(2): 183-188.

cultures, mass

Ukeles, Ravenna, 1965.
A simple method for the mass culture of marine algae.
Limnol., Oceanogr., 10(3):492-495.

cultures, mass

Vinberg, G.G., 1957.
[Mass culture of unicellular algae as a new source of food and industrial raw materials.]
Uspekhi Sovremenoi Biol., 43(2):332-

cultures mollusc

cultures mollusc

Ansell, Alan D., 1969.
Thermal releases and shellfish culture: possibilities and limitations.
Chesapeake Sci. 10(3/4): 256-257.

cultures, molluscs
Azouz, Abderrazak, 1966.
Étude des peuplements et des possibilités d'ostreiculture du Lac de Bizerte.
Inst. nat. Sci. tech. Océanogr. Pêche, Salammbo, Ann 15: 69 pp.

cultures, molluscs
Carriker, M.R., 1961.
Interrelation of functional morphology, behavior and autecology in early stages of the bivalve Mercenaria mercenaria.
J. Elisha Mitchell Sci. Soc., 77(2):168-241.

culture, oyster
Curtin, L., 1968.
Cultivated New Zealand rock oysters.
Fish. Techn. Rept., N.Z. mar. Dept., 25:47 pp. (multilithed).

cultures (oyster)
Deltreil, Jean-Pierre, 1969.
Observations sur les sols ostreicoles du Bassin d'Arcachon.
Revue Trav. Inst. Pêches marit., 33(3): 343-349

cultures, molluscs
Furukawa, Atsushi, Kazuhiko Nogami, Minoru Hisaoka, Yoshimitsu Ogasawara, Ryo Okamoto and Utao Kobayashi, 1961.
[A study of the suspended matter in the sea and its importance as a factor characterizing the environment of shellfish culture ground.]
Bull. Nakai Reg. Fish. Res. Lab., Fish. Agency, No. 14: (Contrib. No. 94:1-151.

cultures, oysters
Gabbott, P.A. and A.J.M. Walker, 1971.
Changes in the condition index and biochemical content of adult oysters (Ostrea edulis L.) maintained under hatchery conditions. J. Cons. int. Explor. Mer 34(1): 98-105.

cultures, molluscs
*Hrs-Brenko, Mirjana, 1967.
Index of conditions in cultured mussels on the Adriatic coast.
Thalassia jugosl., 3(1/6):173-181.

cultures, molluscs
Itami, K., Y. Izawa, S. Maeda and K. Nakai, 1963.
Notes on the laboratory culture of the octopus. (In Japanese; English abstract).
Bull. Jap. Soc. Sci. Fish., 29(6):514-520.

cultures, mollusc
Kirby-Smith, William W., 1972.
Growth of the bay scallop: the influence of experimental water currents. J. exp. mar. Biol. Ecol. 8(1): 7-18.

cultures, oysters
*Krakatitsa, T.F., 1968.
Experience of Ostrea taurica Kryn. breeding in Yagorlytsky Bay of the Black Sea. (In Russian; English abstract).
Gidrobiol. Zh., 4(5):34-38.

cultures, mollusc
Kuwatani, Yukimasa, 1964.
Rearing of the pearl oyster Pinctada martensii (Dunker) on the diet of Skeletonema costatum. (In Japanese; English abstract).
Bull. Jap. Soc. Sci. Fish., Tokyo, 30(2):104-113

cultures, oysters
LeDantec, Jean, 1968.
Ecologie et reproduction de l'huitre portugaise (Crassostrea angulata Lamarck) dans le bassin d'Arcachon et sur la rive gauche de la Gironde.
Revue Trav. Inst. (scient. Tech.) Pêches marit. 32(3):241-362.

cultures, mollusc
Leloup, E., Editor, with collaboration of L. Van Meel, Ph. Pold, R. Halewyck and A. Gryson, undated.
Recherches sur l'ostreiculture dans le Bassin de Chasse d'Ostende en 1962.
Ministere de l'Agriculture, Commission T.W.O.Z. Groupe de Travail - "Ostreiculture", 58 pp.

cultures
Longwell, A. Crosby and S.S. Stiles, 1970.
The genetic system and breeding potential of the commercial American oyster.
Endeavour 29 (107): 94-99.

cultures, mollusc
Loosanoff, Victor L., and Harry C. Davis, 1963
Rearing of bivalve mollusks. In: Advances in Marine Biology, F.S. Russell, Editor, Academic Press, London and New York, 1:1-136.

culture, oysters
Marin, Jean 1971.
Croissance, condition et mortalité des huîtres du Belon.
Rev. Trav. Inst. Pêches marit. 35(2): 201-212.

cultures, oyster
Marteil, Louis 1971
Environnement et mortalité des huîtres plates de la rivière de Belon (1961-1970)
Rev. Trav. Inst. Pêches marit. 35(2): 103-108.

cultures, oyster
May, Edwin B. 1969.
Feasibility of off bottom oyster culture in Alabama.
Bull. Alabama mar. Resources 3: 14 pp.

cultures, mollusc.
Meixner, R. 1971.
Wachstum und Ertrag von Mytilus edulis bei Flosskultur in der Flensburger Förde.
Arch. Fischereiwiss. 22 (1): 41-50.

cultures, mollusc
Menzel, R.W., 1962
Shellfish Mariculture.
Proc. Gulf and Caribbean Fish. Inst., Inst. Mar. Sci., Univ. Miami, 14th Ann. Session: 195-199.

cultures, mollusc
*Millar, R.H., and J.M. Scott, 1968.
An effect of water quality on the growth of cultured larvae of the oyster Ostrea edulis L.
J. Cons. perm. int. Explor. Mer, 32(1):123-130.

cultures, oyster
*Mon, Bae Kyung, 1967.
Study on spat collections of oyster (Crassostrea gigas Thunberg). (In Korean; English abstract).
Bull. Fish. Res. Develop. Agency, 1:109-115.

cultures, molluscs
Morović, Dinko, et Ante Šimunović, 1969
Contribution à la connaissance de la croissance de l'huitre (Ostrea edulis, L.) et de la moule (Mytilus galloprovincialis, Lk.) dans la baie de Mali Ston. (In Jugoslavian; French and Italian abstracts)
Thalassia Jugoslavica 5:237-247

cultures, shellfish
Nikolic, Miroslav, 1964.
Shellfish culture and the productivity of our rearing places. (In Jugoslavian; English resume)
Acta Adriatica, 11(1):239-242.

cultures (oyster)
Nikolic, Miroslav, 1964.
Causes of oyster (Ostrea edulis L.) mass mortality, Limski Kanal, Istra, 1960. (In Jugoslavian; English resume).
Acta Adriatica, 11(1):227-238.
resume, p. 237-238.

cultures, mollusc
*Padilla, Miguel, y Julio Orrego, 1967.
La fijación larval de ostras sobre colectores experimentales en Quetalmahue, 1966-67.
Bol. cient., Inst. Fomento Pesquero, Chile, 5:15pp.

culture of molluscs
Peruško, Gillian H. 1970.
The effects of several environmental factors on the spawning cycle of Ostrea edulis L. in the North Adriatic.
Thalassia Jugoslavica, 6:101-105

cultures, molluscs
Peruško, Gillian H., 1970.
Laboratory experiments on the effects of water temperature, water salinity and light intensity on the spawning and sexual development of mature oysters (Ostrea edulis L.) in Limski Kanal
Thalassia Jugoslavica, 6:91-99

cultures, molluscs
*Pierantoni, Angiolo, 1968.
La mitilicoltura nel golfo di Napoli.
Boll. Soc. Nat. Napoli, 76(1-1967):219-228.

oyster culture
Poisbeau, J., P. Moré, et M.-T. Moré, 1968.
Le croisic et l'ostréiculture.
Bull. Soc. Sci. nat. Ouest France, 65:23-37.

cultures
Pyen, Choong-Kyu, and Il-Mann Shong, 1970.
The culture of food organisms for the production of edible molluscan seedlings. (In Korean; English abstract).
Bull. Res. Dev. Agency, Pusan, 6:233-239.

cultures, oyster
Quayle, D.B., 1971.
Pacific Oyster culture in British Columbia.
Bull. Fish. Res. Bd. Can. 178: 34 pp.

cultures, oyster
Quayle, D.B., 1969.
Pacific oyster culture in British Columbia.
Bull. Fish. Res. Bd. Can. 169: 192 pp.

cultures (oysters)
Ranson, Gilbert et Mlle. Paretias, 1965.
Les huîtres biologie - culture: bibliographie
Bull. Inst. océanogr., Monaco, 67(1388): 51 pp.

cultures, mollusc
Saenz, Braulio A., 1965.
El ostion antillano: Crassostrea rhizophorae Guilding y su cultivo experimental en Cuba.
Centro Invest. Pesq., Cuba, Nota sobre Invest., No. 6:34 pp.

cultures, mollusc

Serventy, Vincent, 1967.
Pearl culture in Australia.
Pacific Discovery, Calif. Acad. Sci., 20(1):22-

cultures (oysters)

*Shaw, William N., 1967.
Advances in the off-bottom culture of oysters.
Proc. Gulf Caribb. Fish. Inst., 19th Sess:108-115.

cultures, molluscs

Šimunović, Ante, 1969.
Contribution à la connaissance de la faune bentique de la baie de Pirovac et la possibilité d'élevage de certains coquillages. (Communication préliminaire. (In Jugoslavian; French and Italian abstracts).
Thalassia Jugoslavica 5:309-314.

cultures, mollusc

Uyeno, Fukuzo, 1964.
Relationships between production of foods and oceanographical condition of sea water in pearl farms. II. On the seasonal changes of sea water constituents and of bottom condition, and the effect of bottom cultivation. (In Japanese; English abstract)
J. Fac. Fish., Pref. Univ. Mie, 6(2):145-169.

culture, oyster

Walne, P.R., 1970.
Studies on the food value of nineteen genera of algae to juvenile bivalves of the genera Ostrea, Crassostrea, Mercenaria and Mytilus.
Fish. Invest. Min. Agric. Fish. Food, London (2) 26(5): 62pp.

Cultures, mollusc

Walne, P.R., 1966.
Experiments in the large-scale culture of the larvae of Ostrea edulis L.
Minist. Agric., Fish., Food. Fish Invest., Great Britain, (2)25(4):1-53.

cultures, molluscs

Yoo, Sung Kyoo, 1969.
Food and growth of the larvae of certain important bivalves. (In Korean; English abstract)
Bull. Pusan Fish. Coll. 9(2): 65-87

cultures pearls

Serventy, Vincent, 1967.
Pearl culture in Australia.
Pacific Discovery, Calif. Acad. Sci., 20(1):22-28.

Cultures, pearl

Tranter, David J., 1957.
Pearl culture in Austalia.
Australian, J. Sci., 19(6):230-232.

Cultures, prawn

cultures prawns

Cook, Harry L., and M. Alice Murphy, 1969.
The culture of larval penaeid shrimp.
Trans. Am. Fish. Soc. 98(4): 751-754

cultures, prawns

Delmendo, M.N., and H.R. Rabanal, 1958.
Cultivation of suppo (Jumbo Tiger shrimp) Penaeus monodon Fabricius in the Philippines.
Proc. Indo-Pacific Fish. Counc., Tokyo, 30 Sept. -14 Oct., 1955, Sect. 2-3(6th Session):424-431.

cultures, prawns

Fujinaga (Hudinaga), Motosaku, 1969.
Kuruma shrimp (Penaeus japonicus) cultivation in Japan.
FAO Fish. Repts. 3(57) (FRm/57.3 (Tm)) 811-832. (mimeographed).

cultures, prawns

Gopinath, K., 1958.
Prawn culture in the rice fields of Travancore-Cochin India.
Proc. Indo-Pacific Fish. Counc., Tokyo, 30 Sept.-14 Oct., 1955, Sect. 2-3 (6th Sess.):419-424.

cultures, prawns

*Ishikawa, Y., 1967.
A disease of young cultured karuma-prawn, Penaeus japonicus Bate. (In Japanese; English abstract).
Bull. Fish. Exp. Stn, Okayama Pref., 1966:5-9.

cultures, prawn

India, Central Inland Fisheries Institute, 1963.
Information on prawns from Indian waters; synopsis of biological data.
Indo-Pacific Fish. Council, Proc., 10th Sess., Seoul, Korea, 10-25 Oct., 1962, FAO, Bangkok, Sect. II:124-133.

cultures

Lee, Byung Don, and Taek Yuil Lee, 1970.
Studies on the rearing of larvae and juveniles of Metapenaeus joyneri (Miers) under various feeding regimes.
Publ. mar. Lab. Pusan Fish. Coll. 3: 27-35
(In Korean; English abstract)

cultures

Lee, Byung Don, and Taek Yuil Lee, 1968
Experiments on the rearing of Metapenaeus joyneri (Miers).
Publ. mar. Lab. Pusan Fish. Coll. 1: 39-42.

cultures, prawn

Ling S.W., 1969.
Methods of rearing and culturing Macrobrachium rosenbergii (DeMan).
FAO Fish. Repts 3(57) (FRm/R57.3 (Tri)): 607-619 (mimeographed)

cultures, prawn

*Miyake, Yoshio, Yusuke Ishikawa and Noboru Hoshino, 1968.
Changes in body color of cultured kuruma prawn, Penaeus japonicus Bate, by different diets and bottom conditions. (In Japanese: English abstract)
Bull. Fish. Exp. Sta., Okayama Pref. 42:27-35.

cultures, prawn

Phillips, Graham, 1971.
Incubation of the English prawn Palaemon serratus. J. mar. biol. Ass., U.K., 51(1): 43-48.

cultures, prawn

Reeve, M.R. 1969.
The suitability of the English prawn, Palaemon serratus (Pennant) for cultivation - a preliminary assessment.
FAO Fish Repts. 3(57) (FRm) (57.3 (Tm)): 1067-1073 (mimeographed).

cultures, prawn

Reeve, M.R., 1969.
The laboratory culture of the prawn Palaemon serratus.
Fish. Invest. Min. Agric. Fish. Food. (2) 26(1): 1-38.

cultures, prawn

Subrahmanyam, M., and K. Janardhana Rao 1970.
Observations on the postlarval prawns (Penaeidae) in the Pulicat Lake with notes on their utilization in capture and culture fisheries.
Proc. Indo-Pacific Fish. Counc. FAO 13(2): 113-117.

cumulus: SEE Clouds

currents

currents

Aasen, O., and E. Akyuz, 1956.
Further observations on the hydrography and occurrence of fish in the Black Sea.
Repts. Fish. Res. Center, Meat & Fish Off., Ser. Mar. Res., 1(6):5-34.

current

Abdullah, A. J., 1949
Wave motion at the surface of a current which has an exponential distribution of vorticity. Ann. N. Y. Acad. Sci. 51(3):425-441, 2 text figs.

currents

Accerboni, E. and F. Mosetti, 1971.
Sulla possibilità di previsione del livello marino e dell'acqua alta nella laguna di Venezia mediante la misura continua della corrente marina nei canali di accesso col metodo elettromagnetico. Boll. Geofis. teor. appl. 13(49): 3-17.

currents

Adams, W., 1931
Ocean currents and life in Philippine seas.
Pub. Manila obs. 3:205-210.

currents

Ahlmann, H.W., 1955.
Forskning och händelser i Arktis efter Vega-Expeditionen. Ymer 75(2):81-97, 8 textfigs.

currents

Akagawa, M., 1956.
On the oceanographical conditions of the North Japan Sea (west of the Tsugaru Straits) in summer. (Part 2).
Bull. Hakodate Mar. Obs., No. 3:5-11.
1-7.

currents

Akamatsu, Hideo and Tsutomu Sawara 1967.
Cruise report of CSK survey by JMA in 1966.
Oceanogr. Mag., Jap. Met. Agency, 19(2): 157-162.

currents

Akiba, Yoshio, Shoichi Yamamoto and Motokazu Ueno, 1959.
On the oceanographical conditions of the Okhotsk Sea in summer of 1958. Bull. Fac. Fish., Hokkaido Univ., 10(1): 37-46.

currents

Akiyama, T., 1954
On the currents in the vicinity of Mikomoto Sima in summer. (In Japanese).
Hydrogr. Bull. (Suiro Yoho), 13:391-398.

currents

Aleem, A.A., 1967.
Concepts of currents, tides and winds among medie-val Arab geographers in the Indian Ocean.
Deep-Sea Research, 14(4):459-463.

currents

Alekseev, A.P., and B.V. Istoshin, 1960
Some results of oceanographic investigations in the Norwegian and Greenland Seas. Soviet Fish. Invest., North European Seas, VNIRO, PNIRO, Moscow, 1960, 23-

currents
Alekseev, A.P., and B.V. Istoshin, 1956.
[Scheme for the permanent currents of the Norwegian and Greenland Seas.] Trudy, Polar Sci. Inst., Sea Fish. Econ. & Oceanogr., No. 9:62-68.

currents
Alekseev, A.P., and B.V. Istoshin, 1956.
[Chart of constant currents in the Norwegian and Greenland Seas.] Polyarnyy nauchno-issledov. Inst. Morsk. Rybn. Khozyay. Okean., Murmansk, Vypusk, 9:

Pagination not given in citation, original not seen.
Translation: USFWS Spec. Sci. Rept., Fish. No. 327.

currents
Alekseev, A.P., B.V. Istoshin and L.R.Shmarina, 1964.
Results of the oceanographic investigations in the Norwegian and Greenland seas. (In Russian). Trudy, Poliarn. Nauchno-Issled. i Proektn. Inst. Morsk. Ribn. Choz. i Okeanogr. im N.M. Knipovicha, 16:133-149.

currents
Algeria, Centre de Geologie Marine et de Sedimentologie, 1962
Utilisation des radio-traceurs à courbet-marine. Etude de l'ensablement d'un port du littoral algerois. Methodes sedimentologiques utilisation de traceurs radio-actifs.
Cahiers Océanogr., C.C.O.E.C., 14(8):526-542.

currents
Allain, Charles, 1964.
L'hydrologie et les courants du Détroit de Gibraltar pendant l'été de 1959.
Rev. Trav. Inst. Pêches Marit., 28(1):1-102.

currents
Allard, P., 1948.
Pt. 1. Situation geographique de la rade d'Agadir et régime de la côte marocaine au sud du Cap Ghis. Pt. 2. Influence de la construction de la jetée sur le régime de la rade d'Agadir. Pt. 3. Synthese des perturbations observées. Bull. Sci. C.O.E.C., Maroc, 1:36-44.

currents
Allard, M.P., 1948
Compte Rendu des Travaux. Mission Hydrographique du Maroc (Mai 1947). No. 1018:18 pp., 1 pl., 1 chart.

currents
Allen, George P., 1971.
Relationship between grain size parameter distribution and current patterns in the Gironde Estuary (France)
J. sedim. Petrol. 41(1): 74-88.

currents
Altman, E.N., 1962
The sea research with the application of the new oceanography instruments. (In Russian).
Meteorol. i Gidrol., (12):33-38.

currents
Ananiadis, Const. I., 1962
Ocean currents. (In Greek; English summary).
Thalassina Epiztnmonika Fulla, 6:59-77.

currents
Andersen T., F. Beyer and E. Føyn 1970.
Hydrography of the Oslofjord: report on the study course in chemical oceanography arranged in 1969 by ICES with support of UNESCO.
Coop. Res. Rept (A), Int. Cons. Explor. Sea, 20: 62 pp. (multilithed)

currents
Andreev, V., 1938.
Kuril'skie Ostrova. (Kurile Islands) Morsk. sborn. 1938 (11):75-87, (12):87-99.

currents
Anon., 1969.
Currents - the underwater rivers.
Oceans, Trident Publishers, 1(1):50-51.(popular).

currents
Anon, 1946
Physalia alias Portuguese Man-of-War alias Blue-Bottle. Fisheries Newsletter, Australia, 5(5):12-13

currents
Anon. 1914
Ice observation, meteorology and oceanography in the North Atlantic Ocean. Rept. on the work carried out by the S.S. "Scotia" 1913, 139 pp.

currents
Antoine, M.P., 1949.
1. Explication sommaire de deux phenomenees. a) Les iles ridees flottantes. b) Explication de certains courants marins. Bull. d'Info., C.O.E.C 8(2):4-7, figs.

currents
Antonov, V.S., 1958(1960).
[The role of continental drainage in the current regime of the Arctic Ocean.]
Problemy Severa, (1):

Translation in:
Problems of the North, 1:55-69.

currents
Aota, Masaaki, 1968.
Study of the variation of oceanographic condition northeast off Hokkaido in the Sea of Okhotsk. (In Japanese; English abstract)
Low Temp. Sci. (A) 26: 351-361.

currents
Argentina, Secretaria de Marina, Servicio de Hidrografia Naval, 1961.
Dunal gigantes en el Golfo San Matias.
Publico, H. 622:unnumbered pp.

currents
Arsenyev, V.S., A.C. Shcherbinin, 1963
Investigation of currents in the Aleutian waters and in the Bering Sea. (In Russian).
Mezhd. Geofiz. Komitet. Prezidiume. Akad. Nauk SSSR. Rezult. Issled. Programme Mezhd. Geofiz. Goda. Okeanol. Issled., (8):58-66.

currents
Arthur, R.S., 1951.
The effect of islands on surface waves. Bull. S.I.O. 6(1):1-26, 1 pl., 13 textfigs.

currents
Asano, H., 1927.
[Results of hydrographical observations in the adjacent waters of Japan, Dec. 1925-Nov. 1926.] (In Japanese). Suisan Kai 530:48.

currents
Asano, H., 1913.
Hydrographical observations made on board the H.M.S. Matsue during December 1912. Rep. Imp. Fish. Inst. Sci. Invest. 2:8-17, photos, charts.

Muroto Zaki to Mikomoto Shima (Izu Shoto).

currents
Ashkin, A.A., 1969.
The depth variation of the intensity of currents induced in a model ocean.
Geophys. J. R. astron. Soc. 17(3): 321-325.

currents
Association d'Oceanographie Physique, 1957.
Bibliography on generation of currents and changes in surface level in oceans seas and lakes by wind and atmospheric pressure, 1726-1955.
Publ. Sci., 18:83 pp.

currents
Association D'Océanographie Physique, 1939
Bibliography on tides and certain kindred matters. Fifth Installment. Publ. Sci. No.6, 19 pp.

currents
Atkins, G.R., 1970.
Winds and current patterns in False Bay.
Trans. roy. Soc. S.Afr. 39(2): 139-148.

currents
Aubert, M., 1963.
Etude de la situation hydrologique de la Baie des Anges.
Rapp. Proc. Verb., Réunions, Comm. Int., Expl. Sci., Mer Méditerranée, Monaco, 17(3):895-905.

currents
Aubert, Maurice, Henri Lebout et Jacqueline Aubert, 1964.
Etude de la situation hydrologique et bacteriologique de la partie ouest de la Baie des Anges (face est et sud du Cap d'Antibes).
Cahiers Centre Etudes Recherches Biol. Océanogr. Medicale, Nice, 14(2):11-143.

currents
Aubert, M., H. Lebout, J. Aubert et M. Gauthier 1965.
Répartition et évolution des eaux de surface au large du Cap d'Antibes.
Rapp. P.-v. Réun. Commn int. Explor. scient. Mer Mediterr. 18(3): 783-789.

currents
Aubert, M., and P. Michel, 1961
Recherches sur la repartition des eaux d'apport du bassin de la Tour Rouge. Les Cahiers du C.E.R.B.O.M. (1): 5-11 (mimeographed).

currents
Audouin, Jacques, 1962.
Hydrologie de l'Étang de Thau.
Rev. Trav. Inst. Pêches Marit., 26(1):5-104.

currents
Audouin, J., 1961.
Contribution à l'étude des courants de l'étang de Thau. (Résumé).
Rapp. Proc. Verb., Réunions, Comm. Int. Expl. Sci., Mer Méditerranée, Monaco, 16(3):781-782.

Complete thesis appears in:
Rev. Trav. Inst. Pêches Marit., 1962, 26(1):

B

currents
Băcescu, M., G. Müller, H. Skolka, A. Petran, V. Elian, M.T. Gomoiu, N. Bodeanu şi S. Stănescu, 1965.
Cercetări de ecologie marină în sectorul predeltaic în condiţiile anilor 1960-1961.
In: Ecologie marină, M. Băcescu, redactor, Edit. Acad. Republ. Pop. Romañe, Bucureşti, 1: 185-344.

currents
Backhausen, F., 1955.
Stromangaben an de NO- und O-Küste Brasiliens.
Der Seewarte 16(3):92-97.

currents
Backhausen, F., 1953.
Über Stromversetzingen im Arabischen und Bengalischen Meer während des SW-Monsuns.
Der Seewart 14(5):27-30, 1 textfig.

currents
Baczyk, Joseph, 1963.
Influence des conditions météorologiques sur les courants marins du golfe de Gdansk.
Cahiers Océanogr., C.C.O.E.C., 15(9):601-616.

currents
Bailey, W.B., 1958.
On the dominant flow in the Strait of Belle Isle.
J. Fish. Res. Bd., Canada, 15(6):1165-1174.

currents
Bailey, W.B., and H.B. Hachey, 1951.
The vertical temperature structure of the Labrador Current. Proc. Nova Scotian Inst. Sci. 22 (4):34-48, 8 textfigs.

currents
Balay, Marciano A., 1961.
El Rio de la Plata entre la atmosfera y el mar.
Rep. Argentina, Sec. Marina, Serv. Hidrogr. Naval, Publ., H. 621:153 pp.

currents
Balech, E., 1949.
Estudio critico de las corrientes marinas del litoral Argentino. Physis 20(57):159-169, 1 textfig.

currents
Barandiarán P., J.F., 1954.
Investigaciones oceanográficas en la Costa Norte del Perú. 27 pp., 13 pls.

currents
Barnanov, A.N., 1965.
Nearshore currents in the Gulf of Riga. (In Russian).
Trudy, Gosudarst. Okeanogr. Inst., No. 87:51-57.

currents
Barber, F.G., 1965.
Current observations in Fury and Hecla Strait.
J. Fish. Res. Bd., Canada, 22(1):225-229.

currents
Barber, F.G., 1958.
Currents and water structure in Queen Charlotte Sound, British Columbia.
Proc. Ninth Pacific Sci. Congr., Pacific Sci. Assoc., 1957, Oceanogr., 16:196-199.

currents
Barber, F.G., 1957.
Observations of currents north of Triangle Island B.C. Prog. Repts., Pacific Coast Stas., Fish. Res. Bd., Canada, No. 108:15-18.

currents
Barber, F.G., and A.W. Groll, 1955.
Current observations in Hecate Strait.
Prog. Repts., Pacific Coast Stas., Fish. Res. Bd. Canada, No. 103:23-25, 4 figs.

currents
Barber, N., and M. S. Longuet-Higgins, 1948.
Water movements and earth currents: electrical and magnetic effects. Nature 161(4081):192-193.

currents
Barber, N. F., and M. S. Longuet-Higgins, 1948.
Electrical and magnetic effects of marine currents. The Observatory 68(843):55-59.

currents (data)
Bardin, I.P., 1958.
[Hydrological, hydrochemical, geological and biological studies, Research Ship "Ob", 1955-1957.] Trudy Kompleks. Antarkt. Exped., Akad. Nauk, SSSR, Mezh. Geofiz. God, Gidrometeorol. Izdatel., Leningrad, 217 pp.

currents
Barlow, E.W., 1956.
Currents of the western South Pacific.
Mar. Obs., 26(174)(M.O. 608):224-226.

currents
Barlow, E.W., 1939.
Currents of the Mediterranean Sea and the South-eastern portion of the North Atlantic Ocean (November to April). Mar. Obs. XVI:134, 63.

currents
Barlow, E.W., 1939
The charting of ocean currents. Mar. Observ 16:99-103.

currents
Barlow, E.W., 1938.
The 1910 to 1937 survey of the currents of the South Pacific Ocean. Mar. Obs. XV:132, 140-169, 8 figs., 3 tables.

currents
Barlow, E.W., 1938.
Currents in the South Pacific Ocean, western and central portions. Mar. Obs. 15:105-108, charts at the end of each number in Vol. 15

currents
Barlow, E.W., 1935.
Currents in the China Seas and East Indian Archipelago. 1. Meteorological conditions which effect the currents, with summary of current information previous to the present charting. Mar. Obs. 12:69-70, 114-116, charts at the end of each number in Vol. 12.

currents
Barnes, C.A., D.F. Bumpus, and J.Lyman, 1948
Ocean circulation in Marshall Islands area. Trans. Am. Geophys. Union, 29(6):871-876, 7 text figs.

currents
Baschin, O., 1927.
Die Polflucht des Meerwassers. Naturwiss. 15(27):559-561.

theoretical

currents
Baskokov, G. A., A. V. Kopteva, 1956.
[Expedition for the investigation of currents in the Kara Sea.]
Problemy Arktiki 3:122-124.

currents
Bassett, H., 1910.
The flow of water through the Irish Sea.
Proc. & Trans. Liverpool Biol. Soc. 24:210-219.

currents
Becher, A.B., 1864.
Navigation of the Pacific Ocean with an account of the winds, weather and currents found there throughout the year:128-131. London.

currents
Beliaev, L.I., 1960.
Methods for the computation of the distribution of marine particles in currents by fractions. Sea chemistry. Hydrology. Oceanic Geology. (In Russian).
Trudy Morsk. Gidrofiz. Inst., 19:42-56.

Beliaev, B.N.

currents
Beliaev, B.N., and V.S. Boldyrev, 1963
The application of the theory of accidental functions to the study of sea currents. (In Russian).
Okeanologiia, Akad. Nauk, SSSR, 3(6):953-961.

currents
Belyaev, B.N., and V.D. Pozdynin, 1966.
Multi-dimensional statistical characteristics of sea currents and their use. (In Russian; English abstract).
Okeanologiia, Akad. Nauk, SSSR, 6(6):1059-1069.

currents
Belinsky, N.A., and M.G. Glagoleva, 1960.
[Methods of study and calculation of non-periodic currents in the sea.]
Meteorol. i Gidrol., (3):18-25.

OTS$0.50.

currents
Belleville, M., 1993.
Sur le régime des courants et des matières alluvionnaires dans l'estuaire de la Seine.
Assoc. Française Avance, Sci., 12eme Session, Rouen:258-262.

currents
Belousova, L.E., 1964
On the problem of the influence of the earth's magnetic field upon ocean currents. (In Russian).
Okeanologiia, Akad. Nauk, SSSR, 4(4):574-575.

currents
Belyaeva, I. P., 1964.
On the calculation of currents in the Norwegian and Greenland seas. (In Russian).
Material, Sess. Uchen. Sov. PINRO, Rez. Issled., 1962-1963., Murmansk, 155-158.

currents
Benioff, H., and B. Gutenberg, 1939.
Waves and currents recorded by electromagnetic barographs. Bull. Amer. Met. Soc. 20:422-426.

currents
Bennett, Edward B., 1965.
Currents observed in Panama Bay during September-October 1958. (In Spanish and English).
Bol. Com. Interamericana Atun Tropical 10(7):399-457.

currents (data)
Bennett, Edward B., 1963.
An oceanographic atlas of the Eastern Tropical Pacific Ocean, based on data from EASTROPIC expedition. October-December 1955.
Bull. Inter-American Trop. Tuna Comm., 8(2):33-165.

currents
Bennett, E.B., 1959.
Some oceanographic features of the northeast Pacific Ocean during August 1955. J. Fish. Res. Bd., Canada, 16(5): 565-633.

currents
Berenbeym, D.J., 1961.
Long-term fluctuations of the Black Sea level and streamflow.
Rapp. Proc. Verb., Réunions, Comm. Int. Expl. Sci., Mer Méditerranée, Monaco, 16(3):643-646.

currents
Berlage, H. P., 1927.
Monsoon-currents in the Java Sea and its entrances. Verhand. Kong. Magn en Meteor. Observatorium, Batavia.

currents
Bernard, F., 1956.
Eaux atlantiques et méditerranéenes au large de l'Algérie. II. Courants et nannoplancton, de 1951 à 1953. Ann. Inst. Océan., 31(4):231-334.

currents
Bernard, F., 1953.
Le courant atlantique en Algérie.
Algérie maritime, No. 5:

currents
Bernhard, M., editor, 1971.
Studies on the radioactive contamination of the sea: annual report 1968-69.
CNEN Rept. RT/BIO (70)-11:109 pp.

currents
Berson, F. A., 1949.
Summary of a theoretical investigation into the factors controlling the instability of long waves in zonal currents. Tellus 1(4):44-52, 3 textfigs.

Berthois, L.

currents
Berthois, L., 1958.
Observation des variations de la salinité et de la température sur une transversale Saint-Nazaire-Mindin, le 8 juillet 1956.
Bull. d'Info., C.C.O.E.C., 10(8):484-497.

currents
Berthois, L., 1958.
Observation d'une tranche d'eau suivie dans ses déplacements entre Basse-Indre et Saint Nazaire.
Bull. d'Info., C.C.O.E.C., 10(4):192-208.

currents
Berthois, L., and M. Barbier, 1953.
La sédimentation en Loire maritime en période d'étiage. C.R. Acad. Sci., Paris, 236(20):1984-1986, 1 fig.

currents
Berthois, Léopold et Philippe Bois, 1969.
Le cours inférieur et l'estuaire de la rivière du Château en période d'Etiage, Îles de Kerguelen. Étude hydraulique, sédimentologique et chimique.
Cah. océanogr. 21(8): 727-771.

currents
Berthois, L., R. Brenot, C. Auffret et Marie-Henriette De Buit, 1969.
La sédimentation dans la région de la Bassurelle (Manche orientale): Étude hydraulique, bathymétrique, dynamique et granulométrique.
Trav. Cent. Rech. Etud. océanogr. ns. 8(3): 13-29

currents
Berthois, L. et A. Crosnier 1966.
Étude dynamique de la sédimentation au large de l'estuaire de la Betsiboka.
Cah. ORSTOM, Sér. Océanogr. 4(2): 49-130.

currents
Berthois, L., A. Crosnier et Y. Le Calvez, 1968.
Contribution a l'étude sédimentologique du plateau continental dans la Baie de Biafra.
Cah. ORSTOM, sér. Océanogr., 6(3/4): 55-86.

currents
Berthois, Léopold, et Yolande Le Calvez, 1961.
Étude dynamique de la sedimentation dans la baie de Sangarea (République de Guinée).
Cahiers Océan., C.C.O.E.C., 13(10):694-714.

currents
Bhattacharya, S.K. and S.T. Ghotankar, 1969.
The estuary of the Hooghly. Bull. natn. Inst. Sci., India, 38(1): 25-32.

currents
Biays, P., 1960.
Le courant du Labrador et quelques-unes de ses conséquences géographiques.
Cahiers Géogr., Quebec, 4(8):237-302.

currents
Biroulin, G.M., and T.T. Binokurova, 1957.
[On the investigation of currents in an open sea.]
Izv. Tichoozean. Nauch.-Issle. Inst. Rib. Choz. I Okean., 44:261-265.

currents
Biske, S.F., 1947.
[New hypothesis on currents in the Bosphorus.]
Izvest. Vses. Geogr. Obshch. 79:486-489.

currents
Bishop, S.E., 1904.
The cold-current system of the Pacific and the source of the Pacific Coast Current. Science 20:338-341.

currents
Bloom, G.L., 1956.
Current, temperature, tide and ice growth measurements, eastern Bering Strait-Cape Prince of Wales, 1953-55. U.S.N. Electronics Lab. Res. Rept 739:25 pp.

currents
Blumenstock, D.I., and D. F. Rex, 1959
Microclimatic observations at Eniwetok NAS-NRC, Pac. Sc. Bd., Atoll Res. Bull., No. 71: 158.

currents
Bogdanov, D.V., 1967.
Sobre la variabilidad de las condiciones oceanográficas en el Mar Caribe y el Golfo de México. (In Russian; Spanish abstract).
Sovetsk.-Kubinsk. Ribokhoz. Issled. Vses. Nauchno-Issled. Inst. Morsk. Ribn. Khoz. Okeanogr (VNIRO), Centr Ribokhoz. Issled. Natsion. Inst. Ribolov. Respibl. Kuba (TSRI), 2:21-38.

currents
*Bogdanov, M.A., G.N. Zaitsev and S.I. Potaichuk, 1967.
Water mass dynamics in the Iceland-Faroe Ridge area.
Rapp. P.-V. Réun. Cons. perm. int. Explor. Mer. 157:150-156.

currents
Bogdanov, M.A., 1965.
On the dynamics of the Faroe-Island Ridge waters. Investigations in line with the programme of the International Geophysical Year, 2. (In Russian).
Trudy, Vses. Nauchno-Issled. Inst. Morsk. Ribn. Choz. i Okeanogr. (VNIRO), 57:33-41.

currents
Böhl, D., 1961.
Untersuchungen zur Strömungsmessung mit Fix- und Schleppelektroden.
Beitr. Meeres., 1:48-55.

currents
Boisvert, William E., 1962.
A reinterpretation of surface ship drift observations.
H.O. Informal Mss. Repts., No. 0-49-62 (duplicated) (unpublished manuscript).

currents
Bolgurtsev, B.N., 1966.
Calculation of currents in the Antarctic Pacific. (In Russian; English abstract).
Fizika Atmosferi i Okeana, Izv., Akad. Nauk, SSSR, 2(11):1162-1174.

currents
Bolgurtsev, B.N., V.F. Kozlov and L.A. Molchanova 1969.
The results of calculation of the currents in the Antarctic Pacific. (In Russian; English abstract).
Fizika Atmosfer. Okean. Izv. Akad Nauk SSSR 5(8): 846-859.

Currents
Bondar, Constantin et Vasile Roventa, 1969.
Les courants formés le long du littoral roumain de la mer Noire et leur influence sur la stratification des masses d'eau.
Stud. Hidraul. 19:5-21
Abstract in: Rapp. P.-v. Reun. Commn int. Expl scient. Mer Mediterr., 19(4): 683.

currents
Bonnefille, R., 1956.
Étude de la circulation des courants dans la Manche en tenant compte du frottement sur le fond et de la force de Coriolis.
Bull. d'Info., C.C.O.E.C., 8(7):306-313, 9 pls.

currents
Bossolasco, M., 1957.
Le correnti marine e la determinazione delle loro caratteristiche. Geofis. e Meteorol., 5(3/4):37-42.

currents
Bossolasco, M., and I. Dagnino, 1959
La diffusione delle acque Ioniche nel Tirreno attraverso lo Stretto di Messina.
Geofis. Pura e Appl., Milano, 44(3):168-178.

currents
Bossolasco, M., and I. Dagnino, 1957.
Sulle turbolenza delle correnti marina nello Stretto di Messina. Geofis. Pura e Applic., 37: 318-324.

currents
Bougis, P., M. Ginat and M. Ruivo, 1956.
Recherches hydrologiques sur le Golfe du Lion.
Vie et Milieu, 7(1):1-18.

currents
Bougis, P., and M. Ruivo, 1955.
Sur l'utilisation des flotteurs en matière plastique (modèle siphonophore) pour l'étude des courants. Bull. d'Info., C.C.O.E.C. 7(4):159-171, 2 pls.

currents
Bourcart, J., 1957.
Géologie sous-marine de la Baie de Villefranche.
Ann. Inst. Océan., Monaco, 23(3):137-199.

currents
Bourcart, J., 1952.
Les frontières de l'océan. Albin Michel, Paris, 317 pp., 77 textfigs.

currents
Bourkov, V.A., 1963
Some results of oceanographic observations with express methods to the east and south of Japan.
Okeanol. Issled., Rezhult. Issled., Programme Mezhd. Geofiz. Goda, Mezhd. Geofiz. Komitet, Presidiume Akad. Nauk. SSSR, No. 9:32-41.

currents
Bousfield, E.L., 1951.
Pelagic amphipods of the Belle Isle Strait region. J. Fish. Res. Bd., Canada, 8(3):134-163, 14 textfigs.

currents
Bowden, K.F., 1960.
The dynamics of flow on a submarine ridge. Tellus, 12(4):418-425.

currents
Bowden, K.F., 1956.
Physics in oceanography. Brit. J. Appl. Phys., 7: 273-281.

currents
Bowden, K.F., 1955.
Deep-water movements in the ocean. Nature 174 (4444):24-26.

currents
Bowden, K.F., 1955.
Physical oceanography of the Irish Sea. Fish. Invest., Min. Agric., Fish. & Food (2) 18(8):1-67, 18 textfigs.

currents
Bowdem, K.F., 1950.
Processes affecting the salinity of the Irish Sea Monthly Notices, R. Astron. Soc., Geophys. Suppl. 6(2):63-90, 7 textfigs.

currents
Boyd, L. A., 1948.
The coast of northeast Greenland, with hydrographic studies in the Greenland Sea. Am. Geogr. Soc., Spec. Pub. 30:340 pp., illus., maps.

currents
*Bøyum,Gunnvald,1967.
Hydrology and currents in the area west of Gibraltar: results from the Helland-Hansen Expedition,May 1965.
Tech. Rep.,Mediterranean Outflow Project,NATO Subcomm. Oceanogr.Res.,36:24pp. numerous figs. (mimeographed).

currents
Bøyum, Gunnvald, 1963
Hydrology and currents in the area west of Gibraltar. Results from the "Helland-Hansen" Expedition, May-June 1961.
NATO Subcommittee on Oceanogr. Res., Geophy Inst., Bergen, 16 pp., 11 figs.(mimeographed).

currents
Braarud, T. and J. T. Ruud, 1932
The "Øst" Expedition to the Denmark Strait 1929. I. Hydrography. Hvalrådets Skr., No. 4:44 pp., 19 text figs.

currents
Brazil, Diretoria de Hidrografia e Navegacao, 1957
Ano Geofisico Internacional, Publicacao DG-06-VI : unnumbered mimeographed pages.

currents
Brekhovskikh, Academician L.M., K.N. Fedorov, L.M. Fomin, M.N. Koshlyakov and A.D. Yampolsky, 1971.
Large-scale multi-buoy experiment in the tropical Atlantic. Deep-Sea Res. 18(12): 1189-1206.

currents
Brennecke, W., 1921.
Die ozeanographischen Arbeiten der Deutschen Antarktischen Expedition 1911-1912.
Arch. Deutschen Seewarte 39(1):1-216, 15 pls., 41 textfigs.

currents
Briggs, Peter, 1969
Rivers in the sea.
Weybright and Talley, N.Y. 126 pp. $5.50. (not seen)

currents
Bright, K.M.F., 1938
The South African intertidal zone and its relation to ocean currents. II, III. Areas of the west coast. Trans. Roy. Soc. S. Africa, 26:49-88.

currents
Broc, J., 1953.
Mesure des courants marins par radar.
Bull. Inst. Océanogr., Monaco, No. 1029:12 pp., 4 pls., 7 textfigs.

currents
Brodie, J.W., 1965.
Oceanography.
In: Antarctica, Trevor Hatherton, editor, Methuen & Co., Ltd., 101-127.

current
Brooks, N. H., 1959.
Lateral diffusion of mass in a steady current in the ocean or atmosphere.
Am. Geophys. U., Pac. Southwest Regional Meeting, Feb. 5-6, 1959.
Abst. in J. Geophys. Res., 64(6):689.

currents
Brown, P.R., 1951.
Ice in the Newfoundland region during February 1950. M.O. 546, Mar. Obs. 21(152):111-114, 5 figs

currents
Brown, A.R., 1874.
Winds and currents in the vicinity of the Japanese Islands. Trans. Asiatic Soc., Japan, 2:159-173.

currents
Browne, Irene M., 1959.
Ice drift in the Arctic Ocean.
Trans. A.G.U., 40(2):195-200.

currents
Bruce,John G.,Jr., and Edward M. Thorndike,1967.
Photographic measurements of bottom currents.
In: Deep-sea photography,J.B.Hersey,editor, Johns Hopkins Oceanogr.,Studies, 3:107-111.

currents
Bryan, E.H., Jr., 1961.
Geography of the Pacific (Abstract).
Tenth Pacific Sci. Congr., Honolulu, 21 Aug.-6 Sept., 1961, Abstracts of Symposium Papers, 224.

currents
Buchanan, J.B., 1958.
The bottom fauna communities across the continental shelf off Accra, Ghana (Gold Coast).
Proc. Zool. Soc., London, 130(1):1-56.

currents
Bulson, P. S., 1961
Currents produced by an air curtain in deep water. Report on recent experiments at Southampton.
Dock & Harb. Autho., 42(487): 15-22.

Burkov, V.A

currents
Burkov, V.A., 1966.
The structure of the currents in the Pacific Ocean and their nomenclature. (in Russian).
Okeanologiia, Akad. Nauk, SSSR, 6(1):3-14.

currents
Burkov, V.A., 1961.
Investigation of currents in the western tropical part of the Pacific during the International Geophysical Year, 1957-1958. (Abstract).
Tenth Pacific Sci. Congr., Honolulu, 21 Aug.-6 Sept., 1961, Abstracts of Symposium Papers, 34C

currents
Burkov, V.A., 1960
[The studies of the Equatorial currents of the Pacific Ocean] Okean. Issle., IGY Com. SSSR: 117-126.

currents
Burkov, V.A., and K.M. Gobdanov, 1957
[Short-term free floating stations for observations upon the currents.] Met. i Gidrol. (10): 37.

current
Burkov, V. A., & M. N. Koshliakov, 1959.
Dynamic balance in the deep current field of the Pacific.
Doklady Akad. Nauk SSSR, 127(1): 70-73.

currents
Burling, R.W., 1961.
Hydrology of circumpolar waters south of New Zealand.
New Zealand Dept. Sci. Industr. Res. Bull., 143: 66 pp.

currents
Burt, W.V., and S. Borden,1966.
Ocean current observations from offshore drilling platforms.
Ore Bin, 28(3):61-64.
Also in: Coll. Repr. Dep. Oceanogr., Oregon State Univ., 5.

currents
Buzdalin, Yu. I., and A.A. Elisarov, 1962.
Hydrological conditions in Newfoundland and Labrador areas in 1960.
Sovetskie Riboch. Issledov. v Severo-Zapadnoi Atlant. Okeana, VINRO-PINRO, Moskva, 155-171.
In Russian; English summary

C

currents
Cabioch, Louis, 1968.
Contribution à la connaissance des peuplements benthiques de la Manche occidentale.
Cah. Biol. mar. 9 (5-Suppl.): 493-720

currents
California Academy of Sciences
California Division of Fish and Game
Scripps Institution of Oceanography
U. S. Fish and Wildlife Service 1950
California Cooperative Sardine Research Program.
Progress Rept. 1950:54 pp., 37 text figs.

currents
California, Humboldt State College, 1964.
An oceanographic study between the points of Trinidad Head and Eel River.
State Water Quality Control Bd., Resources Agency California, Sacramento, Publ., No. 25:136 pp.

currents
Canada, Fisheries Research Board of Canada, 1959.
Bathythermograms and meteorological data record.
Swiftsure Bank and Umatilla Reef Lightships 1958.
Mss. Rept. Ser. (Oceanogr. Limnol.) No. 37:121 pp.

currents
Cano Lucayo, Natalio 1968.
Contribución al conocimiento del mar de Alborán. I. Superficie de referencia.
Bol. Inst. esp. Oceanogr. 135:1-27.

currents
*Canò, M., e C. Stocchino, 1966.
Sulla circolazione delle correnti nel Golfo dell' Asinara e nelle Bocche di Bonifacio.
Atti XV Convegno ann.Ass.Geofis.ital., 523-530.

currents
Capurro, Luis R.A., 1967.
Medicion de corrientes superficiales y profundas en el Pasaje Drake.
Publ. Serv. Hidrogr. nav., Argentina H 641:5-58.
Also in: Contr. Oceanogr. Texas A.M Univ., 12 (377). (Contr)

Currents
Capurro, Luis R.A., 1965.
(Current measurements in the Drake Passage Southern Ocean). (Abstract).
Anais Acad. bras. Cienc., 37(Supl.) :94.

currents
Carapiperis, L.N., 1952.
On the surface temperature of Greek waters.
Geofis. Pura e Applic. 23:153-161, 7 textfigs.

Current
Carranza, Luis, 1891.
Contra-corriente maritima, observada en Paita y Pacasmayo. Bol. Soc. Geogr. Lima, 1:334-345.

currents
Carrillo, D.C.N., 1892
Estudics sobre las corrientes y especialmente de la corriente Humboldt. Bol. Soc. Geogr., Lima, II:72-110

Carruthers, J.N.

currents
Carruthers, J.N., 1952.
Some new work on current -observing in the seas around Great Britain.
Assoc. Océan. Phys., Proc. Verb., Brussels, No. 5(G1):153-154.

currents
Carruthers, J.N., 1951.
Observations continues sur les mouvements de l'eau de mer dans les zones de peches.
Bull. d'Info., C.O.E.C., 3(2):36-39.

currents
Carruthers, J.N., 1951.
Observations conitiués sur les mouvements de l'eau de mer dans les zones de pêche.
Conf., C.R.E.O. No. 1:1-5.
Is this the same article as that in Bull. d'Info., C.C.O.E.C. 3(2):36-39?

currents
Carruthers, J.N., 1951.
An attitude of "Fishery hydrography". J. Mar. Res. 10(1):101-118, 3 textfigs.

currents
Carruthers, J.N., 1950.
Studies of water movements and winds at various light vessels in 1938, 1939, and 1940. 2. Some unpublished records of overall waterflow past various light vessels off the Netherlands coast from 1939-1940. Ann. Biol. 6:120-121.

currents
Carruthers, J.N., 1939.
The investigation of surface current eddies in the sea by means of short-period drift bottles.
Sborn. Posvashch. Nauch. Deyatel. N.M. Knipoviche (1885-1939):39-46.

currents
Carruthers, J.N., 1939.
First annual report on vertical log observations in the southern North Sea and the eastern English Channel. Cons. Perm. Int. Expl. Mer, Rapp. Proc. Verb. 109(3):37-45.

currents
Carruthers, J.N., 1939.
The investigation of surface current eddies in the sea by means of short period drift bottles.
USSR Inst. Mar. Fish., Ocean., Sbornik Knipovich :39-46.

currents
Carruthers, J.N., 1939.
Water movements past certain light vessels in the southern North Sea and eastern English Channel, under different wind conditions. Experiences of winter, 1938-1939. Mar. Obs. 16:104-106.

currents
Carruthers, J.N., A.L. Lawford, and V.F.C. Veley, 1951.
Water movements at the North Goodwin Light Vessel
Mar. Obs. 21(151)(M.O.546):36-46, 5 textfigs.

currents
Carruthers, J.N., A.L. Lawford, and V.F.C. Veley, 1950.
Studies of water movements and winds at various light vessels in 1938, 1939, and 1940. 1. At the Varne Lightship and her successors. Ann. Biol. 6: 115-120, 4 textfigs.

currents
Carruthers, J.N., and A.L. Lawford, 1950.
Water movement and winds. Weather 5(8):278-283, 3 textfigs.

currents
Cartwright, D.E., 1961.
A study of the currents in the Strait of Dover.
J. Inst. Navigation, 14(2):130-151.

currents
Castens, G., 1934.
Das Bodenwasser und die Gliederung des Atlantischen Ozeans. Ann. Hydrogr., usw., Jahrg 62, Heft 5:185-191.

currents
Castens, G., 1931.
Strömung und Isolinenform. Ann. Hydrogr., usw., Jahrg. 59(Heft II):41-46.

currents
Caye, G., et B. Thomassin, 1967.
Note préliminaire à une étude écologique de la levée détritique et du platier friable du Grand Récif de Tuléar: morphologie et hydrodynamisme.
Recl. Trav. Stn.mar.Endoume, hors série, Suppl. 6: 25-35.

currents
Chalmers, G.V., and A.K. Sparks, 1959.
An ecological survey of the Houston ship channel and adjacent bays.
Publ. Inst. Mar. Sci., 6:213-250.

currents
Chang, Sun-duck, 1971.
Oceanographic studies in Chinju Bay. Bull. Pusan Fish. Coll. (nat. Sci.) 11(1): 1-43.
(In Korean; English abstract).

currents
Chang, Sun-duck, 1969.
The circulation in Chinju Bay. 1. Observations of the tidal currents by drifting current-drags.
(In Korean; English abstract).
Bull. Pusan Fish. Coll. 9(2): 95-103

currents
Chaplygin, Ye. I., and Yu. K. Alekseyev, 1957.
Guide to observations on currents. Chap. 6.
Study of marine currents and ice drifts with the aid of automatic drifting radio beacons.
(In Russian).
Rudovodstvo po Nablyudeniyam nad Techeniyami, Moscow, 152-176.
OTS, $0.75

currents
Charnock, H., 1960
Ocean currents.
Science Progress, 48: 257-270.

currents
Charnock, H., 1959.
Tidal friction from currents near the seabed.
Geoph. J. Roy. Astron. Soc., 2(3):215-221.

currents
Charcot, J.B., 1929.
Rapport préliminaire sur la campagne du "Pourquoi Pas?" en 1928. Ann. Hydrogr., Paris, (3)9:15-84.

currents
Charcot, J.B., 1927-1928.
Rapport preliminaire sur la Campagne du "Pourquoi Pas?" en 1927. Ann. Hydrogr., Paris, (3)8:99-154, figs.

currents
Charcot, J.B., 1925-1926.
Rapport préliminaire sur la campagne du "Pourquoi-Pas?" en 1924. Ann. Hydrogr., Paris, (3)7:1-96, figs.

currents
Charney, Jules G., 1960.
Non-linear theory of a wind-driven homogeneous layer near the equator. Deep-Sea Res., 6(4): 303-310.

currents
Chase, J., 1954.
A comparison of certain North Atlantic wind, tide gauge and current data. J. Mar. Res. 13(1): 22-31.

currents
Chekotillo, K.A., 1964
A study of sea currents affected by turbulent stresses. (In Russian).
Okeanologiia, Akad. Nauk, SSSR, 4(5):920-921.

currents
Chekotillo, K.A., 1961.
[Peculiarities of the circulation of intermediate waters in the northern Pacific.]
Trudy Inst. Okeanol., Akad. Nauk, SSSR, 45:113-122.

currents
Chekotillo, K.A., 1958
The dynamics of sea currents.
Trudy Okeanol. Ak. Nauk, SSSR. Spec. Issue 1: 11-14.
Abstr. in: Appl. Mech. Res., 15(6): #3699.

currents
Chernyshev, M. P., 1960
[Determination of direction and velocity of sea currents according to the slope or deviation of sparbuoy from the shore.] Meteorol. i Gidrol. (6): 35-37.

currents
Chevanier, M.C., 1946.
Mission hydrographique de l'Afrique occidental française. Ann. Hydrogr., Paris, (3)17:57-82.

currents
Chevey, P., and P. Carton, 1935.
Les courants de la Mer de Chine meridionale et leurs rapports avec le climat de l'Indochine.
Note Inst. Ocean., Indochine, 26:13 pp.

currents
Chidambaram, K., and A.D.I. Rajendran, 1951.
On the hydro-biological data collected on the Wadge Bank early in 1949. J. Bombay Nat. Hist. Soc. 49(4):738-748.

currents
Churkina, N.A., 1958(1960)
[Drift of buoys in the central Arctic and in the Arctic seas.]
Problemy Severa, (1):
Translation in:
Problems of the North, (1): 342-345.

currents
Gabatti, M., P. Colantoni ed F. Rabbi, 1968.
Ricerche oceanografiche nell'alto Adriatico antistante il delta del Po: crociera estiva 1966.
G. Geol., (2) 34: 479-430. Also in: Raccolta Dati Oceanogr., Comm. ital. Oceanogr. Consig. naz. Ricerche, (A) (17) (1969).

currénts (surface and subsurface)
Clowes, A.J., 1954.
The temperature, salinity and inorganic phosphate of the surface layer near St. Helena Bay, 1950-52. Union of S. Africa, Dept. Commerce & Industries, Div. Fish., Invest. Rept. No. 16: 1-47, 17 pls.

Reprinted from "Commerce and Industry" Aug. 1954.

currents
Clowes, A.J., 1950.
An introduction to the hydrology of South African waters. Fish. & Mar. Biol. Survey Div., Dept. Commerce & Industries, Union of S. Africa, Investigational Rept. No. 12:42 pp., 20 fold-in charts, 14 figs.

currents
Clowes, A.J., 1934.
Hydrology of the Bransfield Strait. Discovery Rept. 9:1-64, 68 textfigs.

currents
Coachman, L.K., and K. Aagaard, 1966.
On the water exchange through Bering Strait.
Limnol. and Oceanogr., 11(1):44-59.

currents
Coachman, L.K., and D.A. Rankin 1968.
Currents in Long Strait, Arctic Ocean
Arctic 21(1):27-38.

Currents
Coachman, L.K. and R.B. Tripp, 1970.
Currents north of Bering Strait in winter.
Limnol. Oceanogr., 15(4): 625-632.

currents
Codispoti, Louis A. 1968.
Some results of an oceanographic survey in the northern Greenland Sea summer 1964.
Techn. Rept. U.S.N. Oceanogr. Off. TR202: 49 pp.

currents
Coe, W.R., 1946.
The means of dispersal of bathypelagic animals in the North and South Atlantic Oceans. Amer. Nat. 80(794):453-469, 2 figs.

currents
Coleman, J.S., 1950.
The sea and its mysteries. London, G. Bell & Sons, Ltd., 285 pp., 1 fold-in.

currents
Collins, C.A., C.N.K. Mooers, M.R. Stevenson, R.L. Smith and J.G. Pattullo, 1969.
Direct current measurements in the frontal zone of a coastal upwelling region.
Jl oceanogr. Soc. Japan, 24(6): 295-306

currents
Comite Central d'Oceanographie et d'Etude des Cotes, 1950.
Remarks on wave forecasting. By. M. Celci.
Relations between microseismic activity, ~~Bx~~ and waves. By J. Debrach.
Wave diffraction. By H. Lacomb.
An explanation of marine currents. By. M. P. Antoine
Bull. B.E.B. 4(1):36-38.

currents
Comité local d'Oceanographie et d'Etudes des Cotes de l'Afrique Occidentale Francaise, 1948.
Procer verbal. 30 pp. numerous fold-ins.

currents
Conseil Permanent International pour l'Exploration de la Mer, 1909.
II. Summary of the results of investigations.
Rapp. Proc. Verb. 10:24-43, 11 textfigs.

currents
Cooper, L.H.N., 1967.
The physical oceanography of the Celtic Sea. Oceanogr. Mar. Biol. Ann. Rev., H. Barnes, editor, George Allen and Unwin, Ltd., 5:99-110.

Cooper, L.H.N., 1960 currents
Some theorems and procedures in shallow-water oceanography applied to the Celtic Sea. JMBA 39:155-171.

Cooper, L.H.N., 1960 currents
The water flow into the English Channel from the south-west. J.M.B.A., 39:173-208.

currents
Cooper, L.H.N., A.L. Lawford and V.F.C. Veley, 1960
On variations in the current of the Seven Stones Light Vessel. J.M.B.A., U.K., 39(3): 659-666.

currènts
Cooper, L.H.N., and D. Vaux, 1949.
Cascading over the continental slope of water from the Celtic Sea. J.M.B.A. 28(3):719-750, 14 textfigs.

currents
Copenhagen, W.J., 1953.
The periodic mortality of fish in the Walvis Region. A Phenomenon within the Benguela Current. Union of S. Africa, Div. Fish., Investig. Rept. No. 14:1-35, 9 pls.

currents
Cot, D., 1949.
Sur une particularité des courants de la Manche C.R. Acad. Sci., Paris, 228(14):1173-1176.

Abstr. Appl. Mech. Rev. 3(1):30

currents
Cot, D., 1948.
Les mouvements de la mer. Rev. Gen. Hydraul. 14: 239-244.

Currents
Countryman, Kenneth A., 1969.
Some summer oceanographic features of the Norwegian Sea summer 1963.
Techn. Rept. U.S.N. Oceanogr. Off. TR-216: 35 pp.

currents
Courtier, A., 1921.
Mission hydrographique de l'Indochine (1913-14).
Ann. Hydrogr., Paris, Ser. 3, 4:1-44.

currents

Cox, C.S., J.H. Filloux, and J.C. Larsen, 1970. 1. General observations. 17. Electromagnetic studies of ocean currents and electrical conductivity below the ocean-floor. In: The sea: ideas and observations on progress in the study of the seas, Arthur E. Maxwell, editor, Wiley-Interscience 4(1): 637-693.

currents

Crease, J., 1964. The Antarctic Circumpolar Current and convergence. Proc. R. Soc., London, (A), 281(1384):14-21.

currents

Crease, James, 1959. Ocean currents. The New Scientist, 6:492-404.

currents

Crease, J., 1952. The origin of ocean currents. J. Inst. Navigation 5(3):280-284.

currents

Crepon, Michel, 1967. Hydrodynamique marine en régime impulsionnel. Cah. Oceanogr. 19(8):627-655.

currents

Cromwell, Townsend, 1958. Thermocline topography, horizontal currents and "ridging" in the Eastern Tropical Pacific. Inter-Amer. Trop. Tuna Comm., 3(3):135-164.

currents

Cromwell, T., 1951. Mid-Pacific oceanography, January through March 1950. Spec. Sci. Rept., Fish., No. 54:9 pp., 17 textfigs., stationdata.

currents

Cromwell, T., R.B. Montgomery and E.D. Stroup, 1954. Equatorial undercurrent in Pacific Ocean revealed by new methods. Science 119(3097):648-649, 2 textfigs.

Csanady, G.T., 1967. currents

Large-scale motion in the Great Lakes. J. geophys. Res., 72(16):4151-4162.

currents

Csanady, G.T., 1966. Accelerated diffusion in the skewed shear flow of lake currents. J. geophys. Res., 71(2):411-420.

currents

Cunningham, C.M., 1909. Report on the drift of the Irish Sea. Fish., Ireland, Sci. Invest., 1907, 7:1-11.

D

currents

Dall, W.H., 1904. Currents of the North Pacific. Science 20:436-437

currents

Dall, W.H., 1882. Report on the currents and temperatures of Bering Sea and the adjacent waters. Ann. Rept. U.S.C.G. S., 1880, App., 16:297-340, Pl. 80. Amer. J. Sci. 21:104-111.

currents

Dall, W.H., 1881. Hydrologie der Beringmeeres und der benachbarten Gewässer. Petermanns Mitt. 27:361-380; 443-448.

currents

Daniel, R.J., and H.M. Lewis, 1930. Surface drift bottle experiments in the Irish Sea, July 1925-June 1927. Proc. & Trans. Liverpool Biol. Soc. 44:36-86.

currents

Darbyshire, J., 1964 A hydrological investigation of the Agulhas Current. Deep-Sea Res., 11(5):781-815.

currents

Dawson, W.B., 1909. Effect of the wind on currents and tidal streams. Trans. R. Soc., Canada, Ser. 3, 3(3):20 pp.

currents

Day, J.H., 1961. The Benguela Current. Nature, 190(4781):1069-1070.

Is a review of: Hart, T.J., and R.I. Currie, 1960. The Benguela Current. Discovery Rept., 31:123-298.

Deacon, G.E.R

Currents

Deacon, G.E.R., 1957. Charting the waves and currents. Many pressing problems for the oceanographers. International Geophysical Year. The Scotsman, July, 1, 1957.

currents

Deacon, G.E.R., 1952. Electrical and magnetic effects of ocean current. Proc. Seventh Pacific Sci. Congr., Met. Ocean., 3:220.

currents

Deacon, G.E.R., 1934. Die Nordgrenzen antarktischen und subantarktischen Wassers im Weltmeer. Ann. Hydrogr., usw., Jahrg. 62(Heft 3):129-136.

currents

Deacon, G. E. R., 1933 A general account of the hydrology of the South Atlantic Ocean. Discovery Repts. 7:173-238, pls.8-10.

currents

*Deacon, Margaret, 1968. Some early investigations of the currents in the Strait of Gibraltar. Bull. Inst. océanogr., Monaco, No. spécial 2: 63-75.

currents

de Buen, F., 1949. El Mar de Solis y su fauna de peces. (1 Partie). S.O.Y.P., Publ. Cient. No. 1:43 pp., 15 textfigs.

currents

DeBuen, R., 1927. Résultats des investigations espagnoles dans le détroit de Gibraltar. Cons. Perm. Int. Expl. Mer, Rapp. Proc. Verb. 44:60-91, Textfigs. 29-50.

currents

Defant, A., 1956. Il recente sviluppo della teoria delle correnti marina. Geofis. Met., Genova, 4:1-6.

currents

Defant, A., 1952. Neue Ansichten über die Theorie der Meeresströmung. Die Pyramide (8/9):142-147.

currents

Defant, A., 1941. Zur Dynamik der Äquatorialengegenströmes. Ann. Hydrogr., usw., Berlin, 69:249-260.

currents, theoretical

Defant, A., 1941. Die absolute Berechnung ozeanischer Ströme nach dem dynamischen Verfahren. Ann. Hydr., Jahrg. 69 (6):169-173.

currents

Defant, A., 1940. Die ozeanographischen Verhältnisse während der Ankerstation des "Altair" am Nordrand des Hauptstromstriches des Golfstroms nördlich der Azoren. Wiss. Ergeb. Internat. Golfstrom-Unternehmung 1938, Lief. 4, in November Beiheft, Ann. Hydr., usw., 68:35 pp., 20 textfigs.

currents

Defant, A., 1939. Das Druck- und Stromfeld in Stromsystemen und ihre Wechselbeziehungen zueinander. Ann. Hydrog. 67:234-242.

currents

Defant, A., 1935 Die Aquatoriale Gegenstrom. Sitzber. Preuss. Akad. Wis., Phys.-Math Kl. 28:1-25, 12 text figs.

currents

Defant, A., 1931 Bericht über die Ozeanographischen Untersuchungen des Vermessungsschiffes "Meteor" in der Dänemarkstrasse und in der Irmingersee. Zweiter Bericht. Sitzber. Preuss. Akad. Wiss., Phys. Math. Kl. 19: 17 pp., 5 text figs.

currents

Defant, A., 1929. Stabile Lagerung ozeanische Wasserkörper und dazu gehörige Stromsysteme. Veroff. Inst. Meereskunder, n.f., A. Geogr.-Naturwiss. Reihe, 19: 33 pp., 20 textfigs.

currents

Dell, R.K., 1952. Ocean currents affecting New Zealand. Deductions from the drift bottle records of H.C. Russell. N.Z.J. Sci. and Tech., Sect. B, 34(2):86-91.

currents

Della Croce, Norberto, 1959

Copepod: Pelagici Raccolti nelle Crociere Talassografiche del "Robusto" nel Mar Ligure ed alto Tirreno. Boll. Mus. e Ist. Biol., Univ Genova 29(176): 29-114.

currents

Dellà Croce, N., 1952. Lanci di galleggianti per lo studio delle correnti superficiali nel Bacino Tirrenico. Publ. Centro Talassografico Tireno, No. 13:3-7.

currents

d'Erceville, A., 1938-1939. Mission hydrographique de l'Indo-Chine, 1936-1937 Ann. Hydrogr., Paris, Ser. 3, 16:79-118, figs.

currents

de Saint-Bon, Marie Catherine, 1963. Les chaetognathes de la Côte d'Ivoire (espèces de surface). Rev. Trav. Inst. Pêches Marit., 27(3):301-346.

Dietrich, Günter

currents
Dietrich, Gunter, 1964.
Oceanic Polar Front Survey in the North Atlantic.
In: Research in Geophysics. Solid Earth and Interface Phenomena, 2:291-308.

currents
Dietrich, G., 1963.
Die Meere.
Die Grosse Illustrierte Länderkunde, C. Bertelsmann Verlag, 2:1523-1606.

currents
Dietrich, G., 1957.
Schichtung und Zirkulation der Irminger See in juni 1955. Ber. Deutsch. Wiss. Komm. Meeresf., n.f., 14(4):255-312.

currents
Dietrich, G., 1951.
Oberflächenströmungen im Kattegat, im Sund und in der Beltsee. Deutsch. Hydro. Zeits. 4(4/5/6): 129-150, 4 textfigs., Pls. 8-10.

currents
Dietrich, G., 1946.
Die Beziehung zwischen Quergefalle und Oberflächenstrom, ein Beitrag zur Synoptischen Ozeanographie. Rept. Sect.2, Ger. Hydro. Inst., No. 11 (manuscript).

currents (surface)
Dietrich, G., 1939
Das Amerikanische Mittelmeer; ein meereskundlicher über blick. Zeitschr. Gesellschaft für Erdkunde zu Berlin. Jahrgang 1939 (3/4):108-130, 38 textfigs.

currents
Dietrich, G., 1936.
Aufbau und Bewegung von Golfstrom und Agulhasstrom eine vergleichende Betractung. Naturwis. 24:225-230.

currents
Dietrich, G., and H. Weidemann, 1952.
Strömungsverhältnisse in der Lübecker Bucht. "Die Küste", Arch. f. Forsch. u. Technik an der Nord- u. Ostsee, 1(2):66-89, 12 textfigs.

currents
Dobrovolsky, A.D., and V.S. Arsenyev, 1961.
Hydrological characteristics of the Bering Sea.
Trudy Inst. Okeanol., Akad. Nauk, SSSR, 38:64-96.

currents
Dodimead, A.J. 1968.
Oceanographic conditions in the central subarctic Pacific region, winter 1966.
Techn. Rept. Fish. Res. Bd. Can., 75: 10pp., + 11 figs. (multilithed).

currents
Dodimead, A.J., F. Favorite and T. Hirano, 1964.
Review of oceanography of the subarctic Pacific Region. Salmon of the North Pacific Ocean. II. Collected Reprints, Tokai Reg. Fish. Res. Lab., No. 2:187 pp.
(In Japanese).

currents
*Dodimead, A.J., and R.H. Herlinveaux, 1968.
Some oceanographic features of the waters of the central British Columbia coast.
Techn. Rept. Fish. Res. Bd., Can., 70:26pp. 104 figs. (multilithed).

currents
Dodimead, A.J., and J.P. Tully, 1958.
Canadian oceanographic research in the northeast Pacific Ocean.
Proc. Ninth Pacific Sci. Congr., Pacific Sci. Assoc., 1957, Oceanogr., 16:180-195.

currents
Doe, L.A.E., 1955.
Offshore waters of the Canadian Pacific coast.
J. Fish. Res. Bd., Canada, 12(1):1-34, 19 text-figs.

Currents (data)
Dohler, G. C., 1961.
Current survey, St. Lawrence River, Montreal-Quebec, 1960. Canadian Hydrographic Survey, Surveys and Mapping Branch, Department of Mines and Technical Surveys, Ottawa, 58 pp. (multilithed).

currents
Donguy, Jean-Rene, Jean Hardiville et Jean-Claude, Le Guen, 1965.
Le parcours maritime des eaux du Congo.
Cahiers Oceanogr., C.C.O.E.C., 17(2):85-97.

currents
Dorrestein, R., 1960.
On the distribution of salinity and of some other properties of the water in the Ems-Estuary.
Verh. Kon. Ned. Geol. Mijnb. k. Gen., Geol. Ser., 19:43-74.

Currents
Douvanine, A. I., 1961.
(Calculations of the periodic fluctuations of sea level and of currents.)
Trudy Kom. Okeanogr., Akad. Nauk, SSSR, 11:7-12.

currents
Drummond, Kenneth H., and George B. Austin, Jr., 1958.
Some aspects of the physical oceanography of the Gulf of Mexico.
USFWS Spec. Sci. Rept., Fish., No. 249:5-13.

currents
Duboul-Razavet, C., 1958
Sur quelques lancers de cartes siphonophores et de bidons lestés au large de la baie de Marseille. Bull. l'Inst. Océan. Monaco, 55 (1132): 16

currents
Düing, Walter 1970.
The monsoon regime of the currents in the Indian Ocean.
Int. Indian Oceanogr. Monogr. East-West Center Press, Honolulu, 1: 68 pp. $7.50.

currents
Düing, Walter, 1965.
Strömungsverhaltnisse im Golf von Neapel.
Pubbl. Staz. Zool., Napoli, 34:256-316.

currents
*Düing, Walter, Klaus Grasshoff und Gunther Krause, 1967.
Hydrographische Beobachtungen auf einem Äquatorschnitt im Indischen Ozean.
"Meteor Forschungsergebnisse (A)(3):84-92.

Dunbar, M.J.

currents
Dunbar, M.J., 1966.
The sea waters surrounding the Quebec-Labrador peninsula.
Cahiers Geogr., Quebec, 10(19):13-35.

currents
Dunbar, M.J., 1955.
Arctic and subarctic marine ecology.
Arctic 7(3/4):213-228.

current
Dunbar, M.J., 1946
The state of the west Greenland current up to 1944. J. Fish. Res. Bd., Canada, 6(7):460-471, 6 figs.

currents
Duncan, C.P., 1967.
Current measurements of the Cape coast.
Fish. Bull., Misc. Contr. Oceanogr. Mar. Biol., S. Afr., 4:9-14.

currents
Duncan, C.P., 1966.
Modifications of the Pisa tube and some results from observations off Cape Point.
J. Mar. Res., 24(2):124-130.

currents
Duvanin, A.I., 1961.
Computations of periodic fluctuations of sea level and currents. Forecasting and computation of physical phenomena in the sea. (In Russian).
Trudy Okeanogr. Komissii, Akad. Nauk, SSSR, 11: 7-12.

currents
Dvorkin, E.N., 1961.
On the deviation of magnetic systems for current measurements. Methods for hydrological investigations for stations aboard ship. Methods for oceanological investigations, a collection of papers. (In Russian).
Trudy Arktich. i Antarktich. Nauchno-Issled. Inst., 210:85-90.

currents
Dybern, Bernt I. 1967.
Topography and hydrography of Kvitundvikpollen and Vågsböpollen on the west coast of Norway.
Sarsia 30:1-27.

E

currents
Edelman, M.S., 1965.
Some peculiarities in the hydrology of the waters adjacent to the Indian Peninsula. (In Russian).
Trudy vses. nauchno-issled. Inst. morsk. ryb. Khoz. Okeanogr. (VNIRO), 57:79-92.

currents
Egedal, J., 1949.
Abnorme Vandstandsforhold i de Danske Farvande. II. Resultater as Vandstandsobservationer foretaget under abnorme Vandstandsforhold i Aarene 1916-1940. Publ. Dansk Met. Inst. Medd. No. 11: 67 pp., 9 textfigs.

currents
Einarsson, H., and U. Stefansson, 1953.
Drift bottle experiments in the waters between Iceland, Greenland and Jan Mayen during the years 1947 and 1949. Rit Fiskideildar 1953(1):20 pp., 15 textfigs.

currents
Ekman, V.W., 1953.
Studies on ocean currents. Results of a cruise on board the "Armauer Hansen" in 1930 under the leadership of Bjørn Helland-Hansen. Geograf. Pub. 19(1):106 pp. (text), 122 pp. (92 pls.).

currents
Ekman, V.W., 1949.
Nagra ord angaende grundlaget för Kännedomen om havströmmarna. Svenska Geogr. Aarb. 25:130-139.

currents
Ekman, V.W., 1931.
Meereströmungen. Handb. Phys. Techn. Mech., F. Auerbach & W. Hort, 5:177-206, 10 textfigs., Leipzig.

current
Ekman, V.W., 1928
A survey of some theoretical investigations on ocean currents. J. Cons.,III:295-327.

currents
Ekman, V.W., 1906.
Beiträge zur Theorie der Meereströmungen. Ann. Hydrogr., usw., 34:423-430; 472-484; 527-540; 566-583; 38 textfigs.

current
Ekman, V.W., 1905
On the influence of the earth's rotation on ocean currents. Ark. Mat. Astr. Fys., II, 2, No.11:1-52.

currents
El Din, S.H. Sharaf, E.M. Hassan and R.W. Trites, 1970.
The physical oceanography of St. Margaret's Bay. Techn. Rept. Fish. Res. Bd Can. 219: 242 pp (multilith).

currents
Elizarov, A.A., 1960.
Oceanographic investigations in the Labrador and Newfoundland areas.
Ann. Proc. ICNAF' 1959-1960, 10:95-101.

currents
Elliott F.E., 1968.
On the mechanics of meanders of ocean currents.
J. Ocean Techn. 2(3):30-32

currents
Elliott, F.E., W.H. Myers and W.L. Tressler, 1955.
A comparison of the environmental characteristics of some shelf areas of eastern United States.
J. Washington Acad. Sci. 45(8):248-259, 4 figs.

currents
Ellison, J. J., D. E. Powell, and H. H. Hildebrand, 1950
Exploratory fishing expedition to the northern Bering Sea in June and July 1949. Fishery Leaflet 369: 56 pp., 23 figs. (multilith).

currents
Emery, K.O., W.S. Butcher, H.R. Gould, and F.P. Shepard, 1952.
Submarine geology off San Diego, California.
J. Geol. 60(6):511-548, 15 textfigs.

currents
Emilsson, I., 1959.
Alguns aspectos fisicos ó quimicos das águas marinhas brasileiras.
Ciencia e Cultura, 11(2):44-54.

currents
Engel, I., 1967.
Currents in the eastern Mediterranean.
Int. hydrogr. Rev., 44(2):23-40.

currents (data)
Engler, Robert G., 1963.
Data current sources for surface current observations.
U.S. Naval Oceanogr.Off., IMR No. 0-63-63:16 pp.

currents
Eremeiva, G.V., 1964
On the problem of current calculations in the North Atlantic. (In Russian).
Okeanologiia, Akad. Nauk, SSSR, 4(5):913-914.

currents
Eriksson, E., and F. Mosetti, 1963.
Quelques renseignements sur l'emploi du tritium comme traceur dans des mesures de courants. Exemple d'application dans un problème d'hydrologie souterraine pour le fleuve Timavo.
Rapp. Proc. Verb., Reunions, Comm. Int., Expl. Sci. Mer Méditerranée, Monaco, 17(3):947-951.

currents
Erman, G.A., 1847.
Ortsbetimmungen bei einer Ueberfahrt von Ochozk nach Kamtschatka und darauf begründete Untersuchung der Strömungen in der Nordhälfte des Ochozkischen Meeres. Arch. f. Wiss. Kunde, Russland, 5:530-560, tables.

currents
Ertel, H. 1966.
Corrientes marinas bajo la acción de un campo giratorio de la fuerza tangencial del viento.
Beitr. Geophys 75(6):465-468.

currents
Ertel,H., 1966.
Corrientes marinas con disipacion binaria.
Beitr. Geophys., 75(5):414-418.

currents
Evans, F., 1957.
Sea currents off the Northumberland coast.
J.M.B.A., 36(3):493-499.

currents
Evgenov, N.I., 1954.
[Sea currents] Leningrad, Gidrometeorologisheska Izdwo, 108 pp., 20 figs., 14 sketch maps.
Nauchna Populiarnaia Biblioteka.

currents
Evtushenko, V.A., 1963
On the water dynamics in the Riga Gulf.
(In Russian).
Okeanologiia, Akad. Nauk, SSSR, 3(2):235-242.

currents
Eyre, J., 1939
The South African intertidal zone and its relation to ocean currents. VII An area in False Bay. Ann. Natal Mus., 9:283-306.

currents
Eyre,J., Brock Ruysen, G.J., and M.I. Crichton 1938
The South African intertidal zone and its relation to ocean currents. VI The East London District. Ann. Natal Mus., 9:83-112.

currents
Eyre, J., and T.A. Stevenson, 1938
The South African intertidal zone and its relation to ocean currents. V A sub-tropical Indian shore. Ann. Natal Mus. 9:21-46

currents
Eyries, M., 1956.
Activité océanographique de l'Escoteur "L'Aventure" en 1955. Bull. d'Info., C.C.O.E.C., 8(10):527-545, 14 pls.

F

currents
Fairbridge, R.W. and C. Teichert, 1948
The Low Isles of the Great Barrier Reef: a new analysis. Geogr. Jour. CXI (1/3): 67-88, 6 photos; 1 fold-in chart, 1 map.

currents
Farquharson, W.I., 1962
Tides, tidal streams and currents in the Gulf of St. Lawrence
Marine Sciences Branch, Department of Mines and Technical Surveys, Ottawa, 76 pp.

currents
Favorite, Felix, 1967.
The Alaskan Stream.
Bull. Int. North Pac. Fish. Comm., 21:1-20.

currents
Fedorov, K. N., 1961
Density discontinuity layer in the variable currents.
Okeanologiya, 1: 25-29.

currents
Fedorov, K.N., 1960.
[Seasonal variations in ocean currents.]
Izv. Akad. Nauk, SSSR, Geophys., 2:185-190.
(English edition volume number and pagination)

currents
Federov, K.N., 1956.
[Sea level and currents during the catastrophic North Sea gale of 1953.]
Izv. Akad. Nauk, SSSR, Ser. Geophys., (4):437-451

currents
Fedorov, K. N., 1956.
Results of modelling full currents produced by wind action on the sea.
Trudy Inst. Okeanol. Akad. Nauk SSSR. 19: 83-97.
Rev. in Appl. Mech. Ref., 12(8): 580

currents
Felsenbaum, A.I., 1960
[Theoretical basis and methods for calculating currents in the sea.]
Inst. Okeanol., Akad. Nauk, SSSR, 123 pp.

currents
Felsenbaum, A.I., 1957.
[On extension of the theory of steady currents in a shallow sea to the case where the coefficient of vertical exchange is variable.]
Dokl. Akad. Nauk, SSSR, 113(1):86-89.

ocean currents

Fel'zenbaum, A. I., 1956.
Method of full flows in the classical theory of ocean currents.
Trudy Inst. Okeanol. Adad. Nauk SSSR, 19: 57-82.
Rev. in Appl. Mech. Rev., 12(8): 579.

currents

Felzenbaum, A.I., 1956.
[Methods for total currents in classical theories of ocean currents.]
Trudy Inst. Okeanol., 19:57-82.

currents

Fel'zenbaum, A.I., 1955.
Method of "complete currents" in the theory of flow of a shallow sea. (In Russian).
Meteorol. i Gidrol., (3):16-22.

Abstr. in:
Meteorol. and Geoastrophys. Abstr., 11(1):31.

currents

Ferrando, H.J., 1957.
Hipotesis sôbre productividad en el area biocean-ografica correspondiente a los litorales marit-imos de Argentina, Uruguay y sur del Brasil.
IV Reunion del Grupo de Tabajo del Ciencias del Mar, Montevideo, 22-24 de Mayo de 1957, Actas de Sesiones y Trabajos Presentados:71-94.

current

Fidman, B. A., 1948.
[Use of high speed motion pictures to study the speed of a turbulent current.] Izvest. Akad. Nauk SSSR, Ser. Geogr. Geofiz. 12:99-106.

currents

Fineisen, W., 1935.
Über Beobachtungen auffälliger Wellenbildung auf dünner Wasserschicht auf dem Neuwerker Watt.
Ann. Hydrogr., usw., Jahrg. 63(Heft 5):186-189.

currents

Fjeldstad, J.E., 1958.
Ocean current as an initial problem.
Geofys. Publikasjoner, Geofys. Norvegica, 20(7): 24 pp.

ocean current

Fjeldstad, J. E., 1958.

Ocean current as an initial problem.
Journees des 24 et 25 fevrier 1958: 67-89.
Publ. by Centre Belge d'Ocean. Recherches Sous-marines.

currents

Fjeldstad, J.E., 1955.
Eddy viscosity, current and wind.
Astrophysica Norvegica, 5(5):153-166.

currents

Fleming, R. H., 1942.
Oceanographic observations of the "E.E. Scripps" cruises of 1938. Results in physical oceanography. Rec. Observ., S.I.O., 1(1):13-14.

currents

Fleming, R.H., 1940-1941
Character of currents off southern California. Sixth Pac. Sci. Congr. 1939, Proc., 3:149-160.

currents

Fleming, R.H., 1938
Tides and currents in the Gulf of Panama. J. Mar. Res.1:192-206.

currents

Fofonoff, N.P., and F. Webster 1971.
Current measurements in the western Atlantic.
Phil. Trans. R. Soc. (A) 270: 423-436

currents

Fomichev, A.V., 1965.
Application of weather observations made on board to indirect studies of currents. Investigations in line with the programme of the International Geophysical Year. (In Russian).
Trudy, Vses. Nauchno-Issled. Inst. Morsk. Ribn. Choz. i Okeanogr., (VNIRO), 57:109-112.

currents

Forrester W.D. 1970.
Geostrophic approximation in the St. Lawrence estuary.
Tellus 22 (1):53-65

currents

France, Laboratoire National d'Hydraulique, 1958.
Estuaire de la Gironde. Port Autonome de Bordeaux Étude sur modèle réduit de la région du Bec d'Ambès et des Îles. Ser. A, Mars 1958:90 pp.

currents

France, Laboratoire National d'Hydraulique, 1956.
Île de la Réunion, Port de la Pointe des Galets. Études sur modèles réduits, Sér. A., 56 pp.

currents

Francis, J.R.D. 1967.
Wind generated waves on a water current.
Q. Jl. R. Met. Soc. 93 (397): 381.

currents

Francis-Boeuf, C., 1941.
Résultats des mesures physico-chimiques effectuées à bord du Chausseur 2 le long de la côte marocaine entre Port-Lyautey et Mazagan, au mois de janvier 1941. Bull. Inst. Océan., Monaco, 804:1-16.

currents

Francis-Boeuf, C., and V. Romanofsky, 1946.
L'envasement du port de Honfleur. Bull. Inst. Océan., Monaco, No. 901:8 pp., 5 figs.

currents

Fraser, J H., 1958
The drift of the planktonic stages of fish in the Northeast Atlantic and its possible significance to the stocks of commercial fish. Some problems for Biological Fishery Survey and Techniques for their Solution. Symposium held at Biarritz, March 1-10, 1956. Int. Comm., Northwest Atlantic Fish., Spec. Publ., No. 1:289-310.

currents

Frassetto, R. 1965.
A study of the turbulent flow and character of the water masses over the Sicilian Ridge in both summer and winter. Summary.
Rapp. P.-v. Réun. Commn int. Explor. scient. Mer Méditerr. 18 (3): 811-815.

curreant

Fuglister, F.C., and L.V. Worthington, 1951.
Some results of a multiple ship survey of the Gulf Stream. Tellus 3(1):1-14, 11 textfigs.

currents

Fukuoka, Jiro, 1965 (1967).
Condiciones meteorologicas e hidrograficas de los mares adyacentes a Venezuela 1962-1963.
Memoria Soc. Cienc. nat. La Salle, 25(70/71/72):11-38.

currents

Fukuoka, J., 1957.
A note on the westward instensification of ocean current. Rec. Oceanogr. Wks., Japan, n.s., 4(1): 7-13.

currents

Fukuoka, J., 1957.
On the Tsushima Current.
J. Oceanogr. Soc., Japan, 13(2):57-60.

currents

Fukushima, N., 1950.
Progressive changes in the current system of the bay disturbance. Geophys. Notes, Tokyo, 3(22): 10 pp.

currents

Funder, Th.P., 1916.
Hydrographic investigations from the Danish School Ship "Viking" in the southern Atlantic and Pacific in 1913-14. Medd. Komm. Havundersøgelser, Ser. Hydrogr., 2(6):28 pp.

currents

Furnestin, J., 1959.
Hydrologie du Maroc atlantique.
Rev. Trav. Inst. Pêches Marit., 23(1):5-77.

currents

Furnestin, J., 1948.
Hydrologie cotière du Maroc, étudiée d'après les températures de l'eau de mer. Bull. Sci., C.O.E.C., Maroc, 4:7-27.

G

currents

Gade, Herman G., 1963
Some hydrographic observations of the inner Oslofjord during 1959.
Hvalrådets Skrifter, No. 46:1-62.

currents

Gade, Herman G., 1961.
Further hydrographic observations in the Gulf of Cariaco, Venezuela. The circulation and water exchange.
Bol. Inst. Oceanogr., Univ. de Oriente, Venezuela 1(2):359-395.

currents

Gade, Herman G., 1961.
On some oceanographic observations in the southeastern Caribbean Sea and adjacent Atlantic Ocean with special reference to the influence of the Orinoco River.
Bol. Inst. Oceanogr., Univ. de Oriente, Venezuela, 1(2):287-342.

currents

Gallagher, B.S., K.M. Shimada, F.I Gonzalez, Jr., and E.D. Stroup 1971.
Tides and currents in Fanning Atoll lagoon.
Pacific Sci. 25(2): 191-205.

currents
Gallardo, Yves, 1970.
Contribution à l'étude du Golfe de Guinée hydrologie et courants dans la région de l'Île Annobon.
Cah. océanogr. 22(3): 277-288

currents
Gallardo, Y., A. Crosnier, J.C. Le Guen et J.P. Rebert, 1968.
Resultats d'observations hydrologiques et courantologiques effectuees autour de l'Ile Annobon (1°25'S - 5°37'E).
Cah. oceanogr., 20(8):711-726.

currents
Galtsoff, P.S., ed., 1954.
Gulf of Mexico, its origin, waters and marine life. Fish. Bull., Fish and Wildlife Service, 55:1-604, 74 textfigs.

currents
Gamutilov, A.E., 1959
[Hydrological features of the Kronotski Bay.] Trudy Inst. Okeanol., 36: 40-58.

currents
Ganapathy, P.N., and V.S.R. Sastry, 1954.
Salinity and temperature variations of the surface waters off the Visakapatnam coast.
Andhra Univ. Ocean. Mem. 1:125-141, 12 textfigs.

currents
Gandin, L.S., 1956.
[On the stability of waves of the surface of separation of currents directed at an angle to one another.] Izv. Akad. Nauk, SSSR, Ser. Geofiz., 1956(3):408-

currents
Gaughn, James L., and Bruce A. Taft, 1967.
Current measurements at the mouth of the Gulf of Thailand, November 1964.
J. geophys. Res., 72(6):1691-1695.

currents (data)
Gaul, Roy, 1961
The occurrence and velocity distribution of short-term internal temperature variations near Texas Tower No. 4.
U.S. Navy H.O. Tech. Rept. (ASWEPS Rept. No.1) Tr.0107:45 pp. (multilithed).

currents
Geffrier, C. de, et J. Milliau, 1938-1939.
Mission hydrographique de la Guinée Française, de la Mauritanie et du Sénégal, 1936-1937.
Ann. Hydrogr., Paris, (3);16:119-183, figs.

currents
Genovese, S., 1954.
Osservazioni oceanografiche eseguite sui campi di pesca dell'alalunga delle isola Eolia. Boll. Pesca, Piscicol. Idrobiol., n.s., 9(2):186-196.

currents
Genovese, S., 1952.
Osservazioni idrologiche esequite nella Tonnara del "Tono" (Milazzo) durante la Campagna di Pesca 1952. Bol. Pesca, Piscicult., Idrobiol., n.s., 7(2):196-200, 3 textfigs.

currents
Gerard, Robert, Robert Sexton and Paul Mazeika, 1965.
Parachute drogue measurements in the eastern Tropical Atlantic in September, 1964.
J. geophys. Res., 70(22): 5696-5698.

currents
Germany, Deutsche Seewarte, 1905.
Wind, Strom, Luft- und Wassertemperatur auf den wichtigsten Dampferwegen des Mittelmeeres.
Beilage zu Ann. Hydrogr., usw., 33:

currents
Germany, Institut für Meereskunde der Universität Hamburg, 1966.
Die Reproduktion der Bewegungsvorgänge im Meere mit Hilfe hydrodynamisch-numerischer Verfahren.
Mitt. Inst. Meeresk., Univ. Hamburg, 5:58 pp. (multilithed)

currents
Germany, Oberkommando der Kriegsmarine, 1940.
Die Naturverhaltnisse des sibirischen Seeweges.
169 pp.

currents
Germany Oberkommando der Kriegsmarine, 1940
Die Naturverhältnisse des Sibirischen Seeweges. Beilage zum Handbuch des Sibirischen Seeweges. 169 pp., 64 text figs., 2 bathymetric charts.

currents
Germany, Seehydrographische Dienst, Hydro-Meteorologische Institut, Berlin, 1952.
Atlas des Salzgehaltes, der Wassertemperatur und der Strömungen im Mittelmeer.

currents
Gershanovich, D.E., 1962.
New data on Recent sediments of the Bering Sea.
Trudy Vses. Nauchno-Issledov. Inst. Morsk. Ribn. Chos. i Okean., VNIRO, 46:128-164.

In Russian

currents (data)
Gesenzvey, A.N., 1963.
Orders of magnitude of oceanological properties.
Questions of Physical Oceanography. (In Russian; English abstract).
Trudy Inst. Okeanol., Akad. Nauk, SSSR, 66:91-124.

currents
Geyer, R.A., 1955.
Effect of the Gulf of Mexico and the Mississippi River on the hydrography of Redfish Bay and Blind Bay. Publ. Inst. Mar. Sci., Texas, 4(1):155-168, 8 textfigs.

currents
Giddings, J.L., jr., 1952.
Driftwood and problems of Arctic sea currents.
Proc. Am. Phil. Soc. 96(2):129-142, 6 textfigs.

currents
Gigoux, E.E., 1947.
Contribución a la Oceanografia chilena. Mar, Año XVIII(123):187-188.

currents
Gilmour, A.E., 1960.
Currents in Cook Strait.
New Zealand J. Geol. and Geophys., 3(3):410-431.

currents
Gilmour, A.E., W.J.P. Macdonald and F.G. van der Hoeven, 1962.
Winter measurements of sea currents in McMurdo Sound.
New Zealand J. Geol. Geophys. (Spec. Antarctic Issue), 5(5):778-789.

currents
Godin, Gabriel, 1968.
The 1965 current survey of the Bay of Fundy-a new analysis of the data and an interpretation of the results.
Manuscript Rept. Ser., Mar. Sci. Br. Dept. Energy, Mines, Resources, Ottawa, 8:97 pp.

currents
Goedecke, E., 1939.
Über unperiodische Wasserversetzungen in der Deutschen Bucht. Cons. Perm. Int. Expl. Mer, Rapp. Proc. Verb. 109:89-92, 1 textfig.

currents
Gogolev, V.M., 1957
[The approximate design of fluid current in a channel.] Vestnik, Leningrad Univ., Ser. Matem. Mekh. i Astron. 1(1): 197

currents
Golemis, A., 1936.
Contribution on the regime of water circulation in the Mediterranean, especially in the Greek seas. Greek Ocean. Dept., Bull. 16(1):

currents
Golubeva, O.V., 1948.
[On the simplification of the hydrodynamic equation in the course of investigation of surface currents in the ocean.] Dok. Akad. Nauk SSSR, 61:453-456.

currents
Gonella, Joseph 1971.
Sur la "polarisation" des courants marins par la rotation terrestre.
C.r. hebd. Séanc., Acad. Sci. Paris 273:162-166.

currents
Gonella, Joseph, 1970
Au sujet de l'applicabilité de la théorie d'Ekman sur les courants marins.
C.r. hebd. Séanc. Acad. Sci. Paris (B) 271: 530-532. Also in: Recl Trav. Lab. Océanogr. phys. Mus. natn. Hist. Nat. Paris 8 (157)

currents
Gonella, J., 1970.
Les courants marins
Connaissance de la houle du vent du courant pour le calcul des ouvrages pétroliers, Publ. Inst. français Pétrole, Coll. Colloques Séminaires 16: 67-83.

currents (data)
Gonella, Joseph, 1969.
Analyse des mesures de courant et de vent à la Bouée-Laboratoire (Position B), juillet 1968.
Cah. océanogr. 21(9): 855-862

currents
Gonella, Joseph, Michel Crepon et François Madelain, 1969.
Observations de courant, de vent et de température à la bouée-laboratoire (Position A), septembre-octobre 1966.
Cah. océanogr. 21(9): 845-854

currents
Gonella, Joseph, Gerard Eskenazi et Jean Fropo, 1967.
Résultats des mesures de vent et de courant à la bouée-Laboratoire au cours de l'année 1964.
Cah. océanogr., 19(3):195-218.

currents
Goodman, J.R., J.H. Lincoln, T.G. Thompson, and F.A. Zeusler 1942.
Physical and chemical investigations: Bering Sea, Bering Strait, Chukchi Sea during the summers of 1937 and 1938.
Univ. Washington Publ. Oceanogr. 3(4): 105-169.

currents
Goodman, J., and T.G. Thompson, 1940.
Characteristics of the waters in sections from Dutch Harbor, Alaska to the Strait of Juan de Fuca and from the Strait of Juan de Fuca to Hawaii. Univ. Washington Publ. Ocean. 3(3):81-103, app. 1-48, 17 figs.

currents
Gordienko, P.A., 1961
The Arctic Ocean.
Scientific American, 204(5): 88-102.

currents
Gordon, A.H., and E.W. Barloe, 1952.
Summary of weather and currents at the original ocean weather stations 'Item' and 'Jig'. Mar. Obs 22(156):91-98.

currents
Gorgy, Samy, 1966.
Les pêcheries et le milieu marin dans le Secteur Méditerranéen de la Republique Arabe linie.
Rev. Trav. Inst. Pêches Marit, 30(1):25-

currents
Gormatjuk, Yu. K., and A.S. Sarkisjan, 1965.
The results of calculation of currents in the North Atlantic by means of a four-level model. (In Russian; English abstract).
Fisika Atmos. i Okeana, 1(3):313-326.

currents
Gorshkova, T.I., 1965.
Carbonates in the sediments of the Norwegian-Greenland Sea basin as indicators of the distribution of water masses. (In Russian).
Trudy vses. nauchno-issled. Inst. morsk.ryb. Khoz. Okeanogr. (VNIRO), 57:297-312.

currents
Gorshkova, T.I., 1965.
Carbonates in the sediments of the Norwegian-Greenland Sea basin as indicators of the distribution of water masses. Investigations in line with the programme of the International Geophysical Year, 2. (In Russian).
Trudy, Vses. Nauchno-Issled. Inst. Morsk. Ribn. Choz. i Okeanogr. (VNIRO), 57:297-312.

currents
Gorsline, Donn S., 1963.
Oceanography of Apalachichola Bay, Florida.
In: Essays in Marine Geology in honor of K.O. Emery, Thomas Clements, Editor, Univ. Southern California Press, 69-96.

currents
Görtler, H., 1941.
Einfluss der Bodentopographie auf Strömungen über der Rotierenden Erde. Z. angew. Math Mech. 21: 279-303.

currents
Gouleau, Dominique, 1971.
Le régime hydrodynamique de la Baie de Bourgneuf et ses conséquences sur la sédimentation.
Cah. océanogr. 23(7):629-647.

currents
Gouleau, Dominique 1968.
Etude hydrologique et sedimentologique de la Baie de Bourgneuf.
Trav. Lab. Geol. mar. Fac. Sci. Nantes, 187pp. (multilithed).

currents
Graham, Joseph J., 1967.
Ocean currents.
Am. Biol. Teacher, 29(6):453-459.

currents
*Grant, D.A., 1968.
Current, temperature, and depth data from a moored oceanographic buoy.
Can. J. earth Sci., 5(5):1261-

currents
Gray, I.E., and M.J. Cerame-Vivas, 1963
The circulation of surface waters in Raleigh Bay, North Carolina.
Limnology and Oceanography, 8(3):330-337.

currents
Green, Charles K., 1960.
Physical hydrography and temperature. Limnological survey of eastern and central Lake Erie, 1928-1929.
USFWS Spec. Sci. Rept., Fish., No. 334:11-70.

currents
Greslou, L., 1964.
Méthodes d'investigation sur modèles réduits des differents problèmes d'hydraulique côtière.
La Houille Blanche, 19(6):706-717.

currents
Griesseier, H., 1969.
Über die äquatoriale Unstetigkeit des Ablenkungswinkels im geostrophisch-antitriptischen Windfeld.
Acta Hydrophysica, 14(1/2): 95-106

Groen, P., 1967.
The waters of the sea.
D. Van Nostrand Co., Ltd. 328 pp. $9.00

currents
*Grossman, Stuart, 1967.
Ecology of Rhizopoda and Ostracoda of southern Pamlico Sound region, North Carolina. I. Living and subfossil rhizopod and ostracode populations.
Paleontol. Contrib. Univ. Kansas, 44(1):7-82.

currents
Grousson, R., and A. Comolet - Tirmen, 1957.
Hydrologie dans la partie orientale de l'Atlantique Nord. Bull. d'Info., C.C.O.E.C., 9(7):359-369.

currents
Grovel, Alain P. 1970.
Etude d'un estuaire dans son environnement le Blavet maritime et la region de Lorient.
Trav. Lab. Geol. mar., Fac. Sci. Nantes (C.N.R.S. AO 45-52): 122 pp. (multilithed)

currents
Guelke, R. W. & C. A. Schoute-Vanneck, 1947.
The measurement of sea-water velocities by electromagnetic induction.
J. Inst. Elect. Engineers, 94(37): 71-74.

currents
Guibout, P., and J. C. Liseray, 1959.
Mesures effectuees dans l'ocean Indien a l'aide du courantometre a electrodes remorquees.
Cahiers Oceanogr., C.C.O.E.C., 11(3):155-157.

currents
Guiler, Eric R., 1960.
The intertidal zone-formaing species on rocky shores of the east Australian coast.
J. Ecology, 48(1):1-28.

currents.
Gunther, E. R., 1936.
A report on oceanographical investigations in the Peru coastal current. Discovery Rept., XIII; 107-276, Pls. XIV XVI.

current
Gunther, E.R., 1936
Variations in the behaviour of the Peru coastal current - with a historical introduction. J.Roy Geogr. Soc., LXXXVIII (1):37-65.

currents
Gustofson, T., and B. Kullenberg, 1930.
Untersuchungen von Tragheitsstromungen in der Ostsee.
Svenska Hydrogr.-Biol. Komm. Skr., n.s., Hydrogr.

currents
Gustafson, T., and B. Ottestedt, 1932.
Observations de courants dans le Baltique, 1931.
Svenska Hydrogr.-Biol. Komm. Skr., n.s., Hydrogr. 11:28 pp.

currents
Gustafson, T., and B. Otterstadt, 1931.
Svenska Strömmätningar i Kattegat 1930.
Svenska Hydrogr.-Biol. Komm. Skr., n.s., Hydrogr. 10:43 pp.

currents
Guyot, M.A., 1951.
L'hydrologie du Canal de Sicile. Bull. d'Info., C.C.O.E.C., 3(7):269-280, 8 pls.

H

currents
Hachey, H.B., 1961.
Oceanography and Canadian Atlantic waters.
Fish. Res. Bd., Canada, Bull. No. 134:120 pp.

currents
Hachey, H.B., L. Lauzier and W.B. Bailey, 1956.
Oceanographic features of submarine topography.
Trans. R. Soc., Canada, (3)50:67-81.

currents
Hada, Yoshine, 1942.
Oceanic currents and plankton in the sea adjacent to the Palau Islands. (In Japanese).
Kagaku Nanyo (Sci. South Seas) 5(1):78-84.

currents
Hansen, Walter 1966.
Motion in the North Sea, the influence of coastal engineering structures on tides, and currents and wind effects.
Dock and Harbour Auth. 47(549): 85-89

currents
Hansen, W., 1956.
Theorie zur Errechnung des Wasserstandes und der Strömungen in Randmeeren nebst Anwendungen.
Tellus 8(3):287-300.

currents
Hachey, H.B., 1947
Water transports and current patterns for the Scotian shelf. J. Fish. Res. Bd., Canada 7(1):1-16, 7 text figs.

currents
Hamon, B.V., 1961
The structure of the East Australian Current.
C.S.I.R.O., Div. Fish. & Ocean. Tech. Pap. 11: 11 pp.

currents
Hansen, Donald V., 1965.
Currents and mixiing in the Columbia River, estuary.
Ocean Sci. and Ocean Eng., Mar. Techn. Soc., Amer. Soc., Limnol. Oceanogr., 2:943-955.

currents
Hansen, P. M., 1949
Studies on the biology of the cod in Greenland waters. Rapp. Proc. Verb., Int. Cons. 123:1-77, 34 text figs.

currents
Hansen, W., 1954.
Neuere Untersuchungen über Meeresströmungen.
Die Naturwissenschaften 41(9):202-205, 5 textfigs.

currents
Hansen, W., 1936.
Die Strömungen im Barents-Meer im Sommer 1927 auf Grund der Dichteverteilung.
Arch. Deutschen Seewarte 55(5):1-41, 4 pls.

currents
Hanzawa, M., 1949.
Oceanographical conditions on the bad rice crop in the Tohoku District of Japan. (Synthetic review). Ocean. Mag., Tokyo, 1(4):194-203, 2 textfigs.

current
Hanzawa, M., and T. Tomada, 1954.
A report on the oceanographical observations in the Antarctic carried out on board the Japanese whaling fleet during the years 1946 to 1952.
J. Ocean. Soc., Japan, 10(3):99-111, 7 textfigs.

currents
Hart, T. John, and Ronald I. Currie, 1960
The Benguela Current.
Discovery Repts. 31: 123-298.

currents
Harvey, H.W., 1925.
Hydrography of the English Channel.
Cons. Perm. Int. Expl. Mer, Rapp. Proc. Verb. 37:59-89, Figs. 17-29, 1 pl.

currents
Haurwitz, B., 1948.
The effect of ocean currents on internal waves.
J. Mar. Res. 7(3):217-228, Text-figs. 1-3.

Currents
Hayami, Shoitiro, Nobuo Watanabe and Sanae Unoki, 1967.
On the distribution and fluctuation of current and salinity in the estuary of Kiso-Sansen (In Japanese; English abstract)
Rep. Rept. Kisosansen Surv. Team, 4 and 5.
Also in: Coll. Repr. Coll. mar. Sci. Techn., Tokai Univ., 1967-68, 3: 129-185; 159-216.

currents
Hebard, J. F., 1959.
Currents in southeastern Bering Sea and possible effects upon king crab larvae.
USFWS Spec. Sc. Rept., Fish. No. 293 11 pp.

currents
Hedgpeth, J.W., 1953.
An introduction to the zoogeography of the northwestern Gulf of Mexico with reference to the invertebrate fauna. Publ. Inst. Mar. Sci. 3(1): 111-224, 46 textfigs.

currents
Hela, I., 1958
A hydrographical survey of the waters in the Aland Sea.
Geophysica, Helsinki, 6(3/4):219-242.

currents
Hela, Ilmo, and Wolfgang Kraus, 1959.
Zum Problem der starken Veränderlichkeit der Schichtungsverhältnisse im Arkona-Becken.
Kieler Meeresf., 15(2):125-144.

currents
Hela, I., and E. Laurila, 1941.
Hydrographische Beobachtungen an Bord vom finnischen Forschungsdampfer "Aranda" während der internationalen Ostseeexpedition in Juli und August 1939. Strombeobachtungen. Merent Julk. 131:1-14.

currents
Hela, I., H.B. Moore and H. Harding, 1953.
Seasonal changes in the surface water masses and in their plankton in the Bermuda area.
Bull. Mar. Sci., Gulf and Caribbean, 3(3), 157-167, 10 textfigs.

currents
*Helm, Roland, 1968.
Zum Problem der Strömungen im Fehmarnbelt.
Beiträge Meeresk., 22:25-40.

currents
Hentsch, Jean-Marc, 1962
Etude des courants dans la baie de Villefranche
Trav., Centre de Recherches et d'Etudes Océanogr. 4(4):19-44.

currents
*Herlinveaux, R.H., 1968.
Features of water movement over the "sill" of Saanich Inlet June-July 1966.
Techn.Rept.Fish.Res.Bd.,Can., 99: 34 pp. (multilithed).

currents
Herlinveaux, R.H., 1962
Oceanography of Saanich Inlet in Vancouver Island, British Columbia.
J. Fish. Res. Bd., Canada, 19(1):1-37.

currents
Hermann, F., 1949
Hydrographic conditions in the Faroe-Shetland Channel in July 1947. Ann. Biol., Int. Cons. 4:8-9, 3 figs.

currents
Hermann, F., and H. Thomsen, 1946.
Drift-bottle experiments in the northern North Atlantic. Medd. Komm. Danmarks Fisk. of Havundersøgelser 3(4):87 pp., 4 pls.

currents
Herschdörfer, S., and G. Kuipers, 1961.
Hydrographic survey for oil exploration and exploitation.
Int. Hydrogr. Rev., 38(1):49-62.

Hidaka, Koji

currents
Hidaka, Koji, 1961
A contribution to the computation of three dimensional ocean currents by high-speed computers.
Rec. Oceanogr. Wks., Japan, 6(1):16-28.

currents
Hidaka, K., 1958.
Dynamics of ocean currents parallel to a long straight coast. Geophys. Mag., Tokyo, 28(3):357-366.

currents
Hidaka, Koji, 1958
Boundaries in the theory of ocean currents. J. Mar. Res., 17:235-246.

currents, drift
Hidaka, K., 1952.
Drift currents in an enclosed ocean. Pt. III.
J. Ocean. Soc., Japan, 8(2):51-66, 5 textfigs.

currents
Hidaka, K., 1950.
Mass transport in ocean currents and the lateral mixing. J. Ocean. Soc., Tokyo, 6(1):48-52.

currents, convection
Hidaka, K., 1950.
Mass transport in convection currents and lateral mixing. Geophys. Notes, Tokyo, 3(2):1-11, 7 textfigs.

currents
Hidaka, K., 1949.
Mass transport in ocean currents and lateral mixing. J. Mar. Res. 8(2):132-136.

currents
Hidaka, K., 1949.
Mass transport in ocean currents and the lateral mixing. Geophys. Notes, Tokyo, 2(3); 4 pp.

currents
Hidaka, K., and H. Miyoshi, 1949.
On the neglect of the inertia terms in dynamical oceanography. Ocean. Mag., Tokyo, 1(4):185-193, 6 textfigs.

currents
Hidaka, K., and H. Miyoshi, 1949.
On the neglect of the inertia terms in dynamical oceanography. Geophys. Notes, Tokyo, 2(22):1-9, 6 textfigs.

current

Hidaka, K., & Y Nagata, 1959.
Dynamical computation of the equatorial current system of the Pacific, with special application to the Equatorial Undercurrent.
Geophys. Notes. 12(1): Contr. 9.

currents, dynamic

Hidaka, Koji, and Yutaka Nagata, 1958
Dynamical computation of the Equatorial Current System of the Pacific, with special application to the Equatorial Undercurrent.
Geophys. J.R. Astron. Soc., 1(3): 198-207.

currents

Hidaka, K., and K. Ogawa, 1958.
On the seasonal variations of surface divergence of the ocean currents in terms of wind stresses in the oceans. Rec. Oceanogr. Wks., Japan, n.s., 4(2):124-169.

currents

Hidaka, K., and K. Ogawa, 1958.
On the seasonal variations of surface divergence of the ocean currents in terms of wind stresses over the oceans. Geophys. Notes, Tokyo, 11(1): No. 10.

currents

Hidaka, K., and T. Susuki, 1949
On the secular variation of Tusima Current.
Ocean. Mag., Japan, 1(1):39-42, 3 text figs.

currents

Hidaka, K., and M. Tsuchiya, 1953.
On the Antarctic Circumpolar Current.
J. Mar. Res. 12(2):214-222, 2 textfigs.

currents (velocity)

Hidaka, K., and T. Yamagiwa, 1949.
On the absolute velocity of the Subarctic Intermediate Current to the south of Japan. Geophys. Notes, Tokyo, 2(5):1-4, 3 textfigs.

Hikosaka, S.

currents

Hikosaka, S., 1953.
[On the ocean-currents (non-tidal currents) in the Tugaru Strait.] Publ. 981, Hydrogr. Bull., Tokyo, No. 39:279-285, 5 textfigs.

currents

Hishida, Kozo, 1958.
[On the current in Sakai Suido (San-in district)].
Umi to Sora, 33(6)45-48.
34(3)

currents

Hjort, J., 1928.
The first cruise with the steamship "Michael Sars". Cons. Perm. Int. Expl. Mer, Rapp. Proc. Verb. 47:188-195, 6 textfigs.

Currents

Hollan, Eckard, 1970.
Eine physikalische Analyse kleinräumiger Änderungen chemischer Parameter in den tiefen Wasserschichten der Gotlandsee.
Kieler Meeresforsch. 25(2): 255-267

currents

Hollan, Eckard, 1969.
Die Veränderlichkeit der Strömungsverteilung im Gotland-Becken am Beispiel von Strömungsmessungen im Gotland-Tief.
Kieler Meeresforsch. 25(1): 19-70.

currents

Holmboe, J., 1948.
On dynamic stability of zonal currents. J. Mar. Res. 7(3):163-174, Figs. 1-7.

currents (data)

Horrer, Paul L., 1962
Oceanographic studies during Operation "Wigwam". Limnol. and Oceanogr., Suppl. to Vol. 7: vii-xxvi.

currents

Horn, W., W. Hussels und J. Meincke 1971.
Schichtungs- und Strömungsmessungen im Bereich der Grossen Meteorbank.
Meteor Forsch.-Ergebn. (A) 9: 31-46.

currents

Houtman, Th. J., 1966.
A note on the hydrological regime in Foveaux Strait.
N.Z. J. Sci., 9 (2):472-483.

currents

Hsueh, Y., 1967.
On the effect of bottom topography and variable wind stress on ocean water movements.
J. geophys. Res., 72(16):4101-4107.

currents

*Hulburt, Edward M., 1968.
Stratification and mixing in coastal waters of the western Gulf of Maine during summer.
J. Fish. Res. Bd., Can., 25(12)2609-2621.

currents

Hutton, R.F., B. Eldred, K.D. Woodburn and R.M. Ingle, 1956.
The ecology of Boca Ciega Bay with special reference to dredging and filling operations.
Florida State Bd., Conserv., Mar. Lab., St. Petersburg, Tech. Ser., 17(1):87 pp.

I

currents

Ianes, A.V., 1959
[On the calculation of the heat currents in the ice cover.] Problemi Arktiki i Antarktiki. (1): 49-58.

currents

Iapogov, B.A., 1960.
[Formation of the East Islandic Current.]
Trudy Okeanograf. Komissii, Akad. Nauk, SSSR, 10(1):96-99.

Ichiye, Takashi

currents

Ichiye, Takashi, 1960
A note on determination of the depth of no meridional motion.
J. Oceanogr. Soc., Japan, 16(1):177-179.

currents

Ichiye, T., 1960
A note on ocean currents produced by source and sinks.
J. Oceanogr. Soc., Japan, 16(3): 111-116.

currents

Ichiye, Takashi, 1960
On critical regimes and horizontal concentration of momentum in ocean currents with two-layered systems.
Tellus, 12(2): 149-158.

currents

Ichiye, Takashi, 1959.
On the response of western boundary currents to variable wind stresses.
J. Geophys. Res., 64(2):175-189.

currents

Ichiye, T., 1957.
A note on the horizontal eddy viscosity in the Kuroshio. Rec. Oceanogr. Wks., Japan, n.s., 3(1): 16-25.

currents

Ichiye, T., 1952.
On the variation of oceanic circulation (III).
Ocean. Mag. 4(2):37-47, 2 textfigs.

currents

Ichiye, T., 1952.
A short note on energy transfer from wind to waves and currents. Ocean. Mag., Tokyo, 4(3): 89-93, 1 textfig.

currents

Ichiye, T., 1951.
On the variation of the oceanic circulation. II.
Ocean. Mag. 3(3):89-96, 7 textfigs.

currents

Ichiye, T., 1951.
On the variation of oceanic circulation. (1st paper). Ocean. Mag. 3(2):79-82, 4 textfigs.

Abstr. Appl. Mech.Rev. 6:45

currents

Ichiye, T., 1951.
The theory of ocean currents (1)a. Mem., Kobe Mar. Obs., 9:28-29, 1 textfig.

currents

Ichiye, T., 1951.
On the surface wave in a current. Ocean. Mag. 3(1):23-26, 2 textfigs.

currents

Ichie, T., 1950.
[On the mutual effect of the coastal and offshore water.] J. Ocean. Soc., Japan (Nippon Kaiyo Gakkaisi) 6(2):1-7. (In Japanese; English summary

currents

Ichie, T., 1950.
On the theory of lateral mixing in the ocean current. Ocean. Mag., Tokyo, 2(3):105-111, 4 textfigs.

ocean currents

Ichie, T., 1950.
A note on the friction terms in the equation of ocean currents. Ocean. Mag., Tokyo, 2(2):49-52, 2 textfigs.

currents

Ichie, T., 1950.
The theory of oceancurrents (1). Mem., Kobe Mar. Obs. 8:31-34., 5 textfigs.

Iizuka, Syoji
currents

[Iizuka, Syoji, and Haruhiko Irie, 1967.
Studies on the oceanographic characteristics of Haiki Channel and the adjacent waters and of the effects of closing of the channel on pearl farms. II Movement of sea water and amount of dissolved oxygen in the farm. (In Japanese; English abstract).
Bull. Fac. Fish. Nagasaki Univ., 22:1-14.

currents
Il'in, A.M., and V.M. Kamenkovich, 1963.
The influence of friction on ocean currents. (In Russian).
Doklady, Akad. Nauk, SSSR, 150(6):1274-1277.

Currents
Ingraham, W. James, Jr., and F. Favorite 1968
The Alaskan Stream south of Adak Island.
Deep-Sea Res., 15(4): 493-496.

currents
Isaacs, J.D., 1948
Discussion of "Refraction of Surface Waves by Currents" by J.W. Johnson, Trans.28:867-874, 1947
Trans. Am. Geophys. Un. 29(5):739-742.

currents
Iselin, C., 1930
A report on the coastal waters of Labrador based on explorations of the "Chance" during the summer of 1926. Proc. Am. Acad. Arts Sci., 66(1):1-37, 14 text figs.

currents
Iselin, C. O'D., and F. C. Fuglister, 1948.
Some recent developments in the study of the Gulf Stream. J. Mar. Res. 7(3):317-329, 6 textfigs.

currents
Ishtoshin, Ju. V., and G.N. Kuklin, 1962
The flows near the equatorial zone of the Pacific. (In Russian).
Meteorol. i Gidrol., (11):28-32.

currents
Istomin, Yu. V., and G.N. Kuklin, 1962
Currents in the equatorial zone of the Pacific Ocean. (In Russian).
Meteorol. i Gidrol., (11):28-32.

Abstract in:
Soviet Bloc Res. Geophys., Astron. & Space, 52: 19.

currents
Ivanov, U.A., 1959
[Concerning the seasonal variability of the Antarctic circumpolar current] Doklady Akad. Nauk, SSSR, 127(1): 74-78.
Translation NIOT/40

currents
Ivanov, Iu. A., and V.M. Kamenkovich, 1959.
Bottom relief as the main factor responsible for the non-zonal course of the Antarctic Circumpolar Current.
Doklady Akad. Nauk, SSSR, 128(6):1167-1170.

currents
Ivanov, R. N., 1958.
On the mechanism of the transfer of wind-energy to currents.
Bull. (Izv.) Acad. Sci., SSSR. Geoph. Ser., 5 (Eng. Ed.): 386-388.

currents
Ivanov, V.M., 1961.
On the relationship of hydrometeorological processes in the Atlantic part of the Arctic and the thermal and dynamic state of the Gulf Stream. (In Russian).
Problemy Severa, 4:27-45.

Translation, 1962:
Problems of the North, 4:25-43.

currents
Izvekov, M.N., 1959
[Observational results on currents in the area of the West Shelf ice.] Inform. Buill. Sovets. Antarct. Exped. (13): 25-28.

J

currents(data)
Jacobsen, J.P., 1918.
Hydrographische Untersuchungen im Randersfjord (Jylland). Medd. Komm. Havundersøgelser, Ser. Hydrogr., 2(7):46 pp., 15 textfigs.

currents
Jacobsen, J.P., 1915
Hydrographical investigations in Faeroe Waters cruise of the M/S "Margrethe" in 1913. Medd. fra Komm. for Havundersøgelser, serie Hydrografi II(4): 47 pp., 15 textfigs.

currents(data)
Jacobsen, J.P., 1913.
Beitrag zur Hydrographie der Dänischer Gewässer. Medd. Komm. Havundersøgelser, Ser. Hydrogr., 2(2):94 pp., 47 tables, 17 textfigs., 14 pls.

currents
Jacobsen, J.P., 1910.
Gezeitenstroeme und resultierende Stroeme im Grosse Belt in verschiedenen Tiefer im Monat Juni 1909. Medd. Komm. Havundersøgelser, Ser. Hydrogr., 1(14):19 pp., 7 textfigs.

currents
Jacobsen, J.P., and A.J.C. Jensen, 1925.
Examination of hydrographical measurements from the Research Vessels "Explorer" and "Dana" during the summer of 1924. Cons. Perm. Int. Expl. Mer, Rapp. Proc. Verb. 39:31-84, Figs. 8-52.

currents
Jakhelln, A., 1936
Oceanographic investigations in East Greenland waters in the summers of 1930-1932. Skr. om Svalbard og Ishavet No. 67:79 pp., Pl. 2, 28 text figs.

currents
Jakhelin, A., 1936
The water transport of gradient currents. Geofysiske Publ. XI (11): 14 pp., 6 text figs.

currents
Janke, J., 1920.
Strömungen und Oberflächtemperaturen im Golfe von Guinea. Arch. Deutschen Seewarte 38(6):1-68, 5 textfigs., 7 pls.

Japan, Anon.

currents
Japan, Anon., 1941.
[Hydrographic data on Southern Seas areas - currents] Japan H.O. Publ. No. 8100:31-39.

Transl. cited: USFWS Spec. Sci. Rept., Fish:227

currents
Japan, Anon., 1934.
[Currents off the Pescadores Islands area.]
Japan, Hydrogr. Off. Publ., No. 616:1-17.

Transl. cited USFWS Spec. Sci. Rept., Fish., No. 227.

currents
Japan, Hakodate Marine Observatory, 1970.
Report of the oceanographic observations in the sea east of Honshu and Hokkaido, and in the Tsugaru Straits from April to May, 1966. (In Japanese)
Bull. Hakodate mar. Obs. 15:11-16

currents
Japan, Hakodate Marine Observatory, 1961.
[Report of the oceanographic observations in the sea east of Tohoku District and in the Okhotsk Sea from August to September, 1959.]
Bull. Hakodate Mar. Obs., (8):6-14.

currents
Japan, Hakodate Marine Observatory, 1961
[Report of the oceanographic observations in the sea east of Tohoku District from February to March 1959.]
Bull. Hakodate Mar. Obs., (8):3-7.

currents
Japan, Maizuru Marine Observatory, 1967.
Report of the oceanographic observations in the Japan Sea from October to November, 1964. (In Japanese).
Bull. Maizuru mar. Obs., 10:87-94.

currents
Japan, Maizuru Marine Observatory, 1967.
Report of the oceanographical observations in the Japan Sea from August to September, 1964. (In Japanese).
Bull. Maizuru mar. Obs., 10:74-86.

currents
Japan, Maizuru Marine Observatory, 1967.
Report of the oceanographic observations in the Japan Sea from May to June, 1964.
Bull. Maizuru mar. Obs., 10:65-76.
(In Japanese)

currents
Japan, Maizuru Marine Observatory, 1963
Report of the oceanographic observations in the Japan Sea from May to June 1960. (In Japanese).
Bull. Maizuru Mar. Obs., No. 8:56-67.

Japan, Marine Safety Agency

currents
Japan, Marine Safety Agency, Hydrographic Division, undated
State of the adjacent seas of Japan, 1960-1964.
2:numerous charts.

currents
Japan, Maritime Safety Board, 1957.
Currents in Sagami-wan. Hydrogr. Bull. (Publ. 981) 53:44-47.

currents
Japan, Maritime Safety Board, 1955.
[Oceanographic conditions in the district of the Indian Ocean. Pt. 1.] Publ. 981, Hydrogr. Bull. No. 49:186-188, 9 charts.

currents
Japan, Maritime Safety Board, Tokyo, 1954.
[On the transformation of the equations of oceanic currents.] Publ. No. 981, Hydrogr. Bull., Spec. No. 15:24-26.

currents
Japan, Maritime Safety Board, 1954.
The ocean current in the southwestern sea area of Kyusyu in Jan., 1954. Publ. 981, Hydrogr. Bull., Maritime Safety Bd., No. 42:138, 1 fold-in.

currents
Japan, Maritime Safety Agency, 1951.
[Distribution of currents in Mikuni Kō] Hydro. Bull. (Publ. No. 981) No. 27:359-375, figs.

currents
Japan, Maritime Safety Agency, 1951.
[Oceanographic result in the southern area off Honsyu.] Publ. No. 981, Hydro. Bull. No. 27:376, 1 fold-in.

currents
Japan, Maizuru Marine Observatory 1969.
Report of the oceanographic observations in the western part of the Japan Sea in November 1965. (In Japanese).
Bull. Maizuru mar. Obs. 11: 98-103

Japan, Nagasaki Marine Observatory

currents
Japan, Nagasaki Marine Observatory 1971.
Report of the oceanographic observations in the sea southward of Kyushu from October to November, 1965. (In Japanese).
Oceanogr. Met. Nagasaki mar Obs. 18: 78-82

currents
Japan, Nagasaki Marine Observatory, 1971.
Report of the oceanographic observations in the sea southeast of Kyushu in September, 1965. (In Japanese)
Oceanogr. Met. Nagasaki mar. Obs. 18: 75-77

currents
Japan, Nagasaki Marine Observatory 1971.
Report of the oceanographic observations in the sea west of Japan from June to August, 1965. (In Japanese)
Oceanogr. Met. Nagasaki Mar. Obs. 18: 59-74

currents
Japan, Nagasaki Marine Observatory 1971
Report of the oceanographic observations in the sea south of Kyushu in April, 1966 (In Japanese).
Oceanogr. Met. Nagasaki mar. Obs. 18: 53-57.

currents
Japan, Nagasaki Marine Observatory 1971
Report of the oceanographic observations in the sea south of Kyushu in March 1966.
Oceanogr. Met. Nagasaki Mar. Obs. 18: 50-52
(In Japanese)

currents
Japan, Nagasaki Marine Observatory 1971.
Report of the oceanographic observations in the sea west of Japan from January to February 1966. (In Japanese)
Oceanogr. Met. Nagasaki Mar. Obs. 18: 34-49.

Currents
Japan, Nagasaki Marine Observatory, Oceanographical Section 1960.
Report of the oceanographic observation in the sea west of Japan from January to February, 1960. Report of the Oceanographic observation in the sea north-west of Kyushu from April to May, 1960.
Results Mar. Meteorol. & Oceanogr., J.M.A., 27:42-50; 51-67.
Also in:
Oceanogr. & Meteorol., Nagasaki Mar. Obs., (1961) 11 (202).

currents
Japan, Nagasaki Marine Observatory, Oceanographical Section, 1960
[Report of the oceanographic observation in the sea west of Japan from January to February 1959.]
Res. Mar. Meteorol. & Ocean., J.M.A., 25: 48-56.
Also in:
Oceanogr. & Meteorol., Nagasaki Mar. Obs., (1961) 11(199).

currents
Japan, National Committee for the International Geophysical Coordination, Science Council of Japan, 1961
The results of the Japanese oceanographic project for the International Geophysical Cooperation 1959:1-65.

currents
Japan, National Committee for the International Geophysical Year 1960.
The results of the Japanese oceanographic project for the International Geophysical Year, 1957/8: 198 pp.

currents
Japan, Science Council, National Committee for IIOE, 1966.
General report of the participation of Japan in the International Indian Ocean Expedition.
Rec. Oceanogr. Wks., Japan, n.s. 8(2): 133 pp.

currents
Japan, Tokai Regional Fisheries Research Laboratory, 1959.
IGY Physical and chemical data by the R. V. Soyo-maru, 25 July-14 September 1958.
17 pp. (multilithed).

currents
Jeffries, Harry P., 1962
The atypical phosphate cycle of estuaries in relation to benthic metabolism.
The Environmental Chemistry of Marine Sediments, Proc. Symp., Univ. R.I., Jan. 13, 1962, Occ. Papers, Narragansett Mar. Lab., No. 1: 58-67.

currents
Jenness, J.L., 1953.
The physical geography of the waters of the western Canadian Arctic. Geogr. Bull., Canada, No. 4: 33-64, 4 text figs.

currents
Jennings, F.D., & R.A. Schwartzlose, 1960
Measurements of the California Current in March 1958.
Deep-Sea Res., 7(1): 42-47.

current
Jensen, A.J.C., 1940
The influence of the currents in the Danish waters on the surface temperature in winter, and on the winter temperature of the air. Medd. Komm. Danmarks Fiskeri-og Havundersø., Ser. Hydrografi, 3(2):52 pp., 14 text figs.

currents
Jensen, A.J.C., 1937.
Fluctuations in the hydrography of the Transition Area during 50 years. Cons. Perm. Int. Expl. Mer, Rapp. Proc. Verb. 102:3-49, 14 textfigs.

currents
Jerlov, N.G., 1956.
The Equatorial currents in the Pacific Ocean. Repts. Swedish Deep-Sea Exped., 3(Phys. Chem. 5): :129-154.

currents
Johannessen, Ola M. 1968.
Some current measurements in the Drøbak Sound, the narrow entrance to the Oslofjord.
Hvalråd. Skr. 50: 38 pp.

currents
Johannessen, Ola M. 1967.
Note on some vertical current profiles below ice floes in the Gulf of St. Lawrence and near the North Pole.
J. geophys. Res. 75(15): 2857-2861.

currents
Johnson, James H., 1958.
Surface-current studies of Saginaw Bay and Lake Huron, 1956.
U.S.F.W.S. Spec. Sci. Rept., Fish., No. 267:84 pp

currents
Johnson, J.W., 1947
The refraction of surface waves by currents
Trans. Am. Geophys. Union, 28(6):867-874, 7 text figs.

currents
Johnson, M. W., 1949.
Relation of plankton to hydrographic conditions in Sweetwater Lake. J. Am. Water Works Assoc. 41(4):347-356, 12 textfigs.

currents
Johnstone, J., 1903.
On some experiments with "drift bottles".
Proc. & Trans. Liverpool Biol. Soc. 17:154-164.

currents
Jones, W.M., 1950.
Progress in hydrography and physical oceanography. N.Z. Dept. Sci. Ind. Res., Geophys. Conf., 1950, Paper No. 11:2 pp. (mimeographed).

currents
Joseph, Edwin B., William H. Massmann and John J. Norcross, 1960.
Investigations of inner continental shelf waters off lower Chesapeake Bay. 1. General introduction and hydrography.
Chesapeake Science, 1(3/4):155-167.

currents
*Joseph, J., 1967.
Current measurements during the International Iceland-Faroe Ridge Expedition, 30 May to 18 June, 1960.
Rapp. P.-V. Réun. Cons. perm. int. Explor. Mer, 157:157-172.

K

currents
Kajiura, J., 1953.
On the influence of bottom topography on ocean currents. J. Ocean. Soc., Japan, 9(1):1-14, 3 textfigs.

currents

Kalinin, G.P. and Milinkov, P.I., 1957

On the calculation of the non-stationary movements of water in open channels. Met. i Gidrol. (10): 10

currents

Kalle, K., 1948
Eine neuartige Methode der Oberflächen-strommessung im Meere vom verankertin Schiff aus. German Hydro. Jour. 1(5/6):164-168; 4 text figs.

currents

Kamenkovich, V.M., 1961.
The integration of the marine current theory equations in multiply connected regions. (In Russian).
Doklady, Akad. Nauk, SSSR, 138(5):1076-1079.

Translation: Consultants Bureau for Amer. Geol. Inst. (Earth Sciences Section only), 138(1-6): 629-631. 1962.

currents

Kamenkovich, V.M., 1960

Antarctic circumpolar current as influenced by the bottom relief. Doklady Akad. Nauk, SSSR, 134 (5): 1076-1078.

currents

Kan, S.I., 1961.
On the computation and forecasting of currents in the Kerchensk sound. Forecasting and computation of physical phenomena in the sea. (In Russian).
Trudy Okeanogr. Komissii, Akad. Nauk, SSSR, 11: 130-141.

currents

Kan, S.I., 1960

Improvement of methods for current forecasting in the Kerchensky Strait. Meteorol. i Gidrol., (12): 25-27.

currents

Kandler, R., 1949
Jahreszeitliches Vorkommen und unperiodisches Auftreten von Fischbrut, Medusen und Decapodenlarven in Fehmarnbelt in den Jahren 1934-1943. Ber. Deutschen Wiss. Komm. f. Meeresf. nf. 12(1):49-85, 6 figs.

currents

Kao, S.-K., and G.R. Farr, 1966.
Turbulent kinetic energy in relation to jet streams, cyclone tracks and ocean currents. J. geophys. Res., 71(18):4289-4296.

currents

#Karlovac, Jozica, 1967.
Etudé de l'écologie de la Sardine, Sardina pilchardus Walb., dans la phase planctonique de sa vie en Adriatique moyenne.
Acta adriat., 13(2):1-109.

currents

Kashiwamura, M., 1960.

The coastal current in the Volcano Bay. J. Oceanogr. Soc., Japan, 16(1): 1-5.

currents

Kawai, H., and M. Sasaki, 1961.
An example of the short-period fluctuation of the oceanographic condition in the vicinity of the Kuroshio front.
Bull. Tohoku Reg. Fish. Res. Lab., (19):119-134.

currents

Kel'der, E.G., 1953.
Seasonal fluctuations in the thermal regime of currents. Trudy Morsk. Gidrofiz. Inst., 3:82-91.

currents

Keller, George H., and Adrian F. Richards, 1967.
Sediments of the Malacca Strait, Southeast Asia.
J. sedim. Petrol., 37(1):102-127.

currents

Kenney, B.C. and Ian S.F. Jones, 1971.
Relative diffusion as related to quasi-periodic current structures. J. phys. Oceanogr. 1(3): 224-232.

currents

Kiilerich, A., 1945
On the Hydrography of the Greenland Sea. Medd. om Grønland 144(2):63 pp., 22 text figs., 3 pls.

currents

Kiilerich, A., 1943.
The hydrography of the west Greenland fishing banks. Medd. Komm. Dammarks Fisk- og Havundersøgelser, Ser. Hydrogr., 3(3):45 pp., 7 pls.

currents

Kimura, K. (undated)
Investigation of ocean current by drift bottle experiment. 1. Current in Suruga Bay with special reference to its anticlockwise circulation within the Bay. J. Ocean. Soc., Tokyo, (Nippon Kaiyo Gakkaisi) 5(2/4):70-84, 6 figs. (In Japanese; with English summary).

currents

King, Joseph E., and Thomas S. Hida, 1957
Zooplankton abundance in the Central Pacific. II.
Fish. Bull., U.S. Fish and Wildlife Service (Fish. Bull., 118), 57:365-395.

currents

Kislyakov, A.G., 1964.

Horizontal circulation of waters on the watershed of the Norwegian and Barents seas. (In Russian).
Trudy, Poliarn. Nauchno-Issled. i Proektn. Inst. Morsk. Ribn. Choz. i Okeanogr. im N.M. Knipovicha, 16:183-194.

currents

Kisliakov, A.G., 1960.
The northern branch of the North Cape Current. Nauchno-Techn. Biull., PINRO, 12(2):12-14.

currents

Klepikov, V.V., 1958(1960)

The origin and diffusion of Antarctic ocean-bed water. Problemy Severa, (1):

Translation in:
Problems of the North, 1: 321-333

currents

Klepikov, V.V., and N.P. Shesterijov, 1959.
Currents on three diurnal stations in coastal waters of East Antarctic.
Info. Biull. Sovetsk. Antarkt. Exped., (8):16-20

currents (data)

Klepikova, V.V., Edit., 1961
Third Marine Expedition on the D/E "Ob", 1957-1958. Data.
Trudy Sovetskoi Antarktich. Exped., Arktich. i Antarktich. Nauchno-Issled. Inst., Mezhd. Geofiz. God, 22:1-234 pp.

currents

Klimenkov, A.I., and V.I. Pakhorukov, 1964.
Hydrological observations in the western Atlantic. (In Russian).
Trudy, Poliarn. Nauchno-Issled. i Proektn. Inst. Morsk. Ribn. Choz. i Okeanogr. im. N.M. Knipovicha, 16:15-23.

currents

Klimenkov, A.I., and V.I. Pakhorukov, 1962
Hydrological observations in the northwest Atlantic in spring-summer 1960. (In Russian; English summary).
Sovetskie Ribochoz. Issledov. Severo-Zapad. Atlant. Okeana, VNIRO-PINRO, 189-200.

currents

Knauss, John A., 1961

The structure of the Pacific Equatorial Counter current. J. Geophys. Res., 66(1): 143-156.

currents

Knox, G.A., 1960.
Littoral ecology and biogeography of the southern oceans.
Proc. R. Soc., London, (B), 152(949):577-624.

currents

Koblentz-Mishke, O.I., 1958.
The distribution of certain varieties of phytoplankton in relation to the principal currents in the western part of the Pacific Ocean.
Doklady Akad. Nauk, SSSR, 121(6):1012-1014.

Translation NIOT/83

currents

Koczy, F.F., M.O. Rinkel and S.J. Niskin, 1960.
The current patterns on the Tortugas shrimp grounds.
Proc. Gulf and Caribbean Fish. Inst., 12th Ann. Session, Nov. 1959:112-125.

currents

Koh, Ryuji, 1966.
Littoral drift along Iwafune Port.
Coast. Engng Japan 9: 127-136

currents

Koh, Kwan Soh, and Kyu Hwan Cho, 1962
Current conditions of Pusan Harbor. (In Korean)
Bull. Fish. Coll. Pusan Nat. Univ., 4(1/2): 49-58.

currents

Kolesnikov, A.G., S.G. Boguslavskii, G.P. Ponomarenko, and V.G. Zhidkov, 1971.
General diagram of currents of Central Atlantic in the zone of trade winds.
Dokl. Akad. Nauk, SSSR, 196(3): 689-692.

currents

Kol'man, O.V., 1961.
Steady currents calculated by the dynamic method from instrumental observations. Some results of investigations of oscillations of sea level and currents. Methods of oceanological investigation a collection of papers. (In Russian).
Trudy Arktich. i Antarktich. Nauchno-Issled. Inst., 210:164-167.

currents

Koninklijk Nederlands Meteorologisch Instituut, 1949.
Rode Zee en Golf van Aden. Oceanographic and meteorological data. No. 129:26 pp. (charts).

currents
Koninklijk Nederlands Meteorologisch Instituut, 1949.
Sea areas round Australia. Oceanographic and meteorological data. No. 124 (Atlas):46 charts.

currents
Konovalova, I.Z., 1969.
Space discreteness of current observation in the near-shore zone of sea. (In Russian; English abstract). Okeanologiia, 9(6): 944-952.

currents
Konovalova, I.Z., 1961.
Computation of tables of steady currents.
Trudy Gosud. Okean. Inst., Leningrad, 61:133-141.
In Russian.

currents
Korneva, L.A., 1951.
[Anomalous geomagnetic field and its equivalent in a system of currents in the world ocean.]
Dok. Akad. Nauk, SSSR, 76:49-52.

currents
Kort, V.G., 1971.
Ocean currents from recent data. (In Russian; English abstract). Okeanologiia 11(5): 811-818.

currents
Kort, V.G., 1968.
The role of heat advection by sea currents in large-scale interaction between the ocean and the atmosphere. (In Russian).
Dokl. Akad. Nauk, SSSR, 182(5):1059-1062.

currents
Kort, V.G., 1962.
The Antarctic Ocean.
Scientific American, 207(3):113-128.

currents
Kosiba, A., 1956.
[On the deflecting forces action on the air and water currents (a contribution to the teaching methods in meteorology).] Przeglad Geofizyczny, Poland, 1(9)(1):37-47.

currents
Koshljakov, M.N., 1960
[Investigation of wind and convective zonal currents with application to the tropical part of the Pacific.]
Trudy Inst. Okeanol., 40: 142-151.

currents
Koshliakov, M.N., and V.G. Neiman, 1965.
Some measurements and calculations of zonal currents in the equatorial part of the Pacific Ocean. (In Russian).
Okeanologiia, Akad. Nauk, SSSR, 5(2):235-249.

currents
Koslowski, G., 1960.
Über die Strömungsverhältnisse und den Volumtransport im Nordatlantischen Ozean zwischen Kap Farvel und der Flämischen Kappe im Spätwinter und Spätsommer 1958.
Deutsche Hydrogr. Zeits., 13(6):269-281.

currents
Koto, Hideto, and Takeji Fujii, 1958.
Structure of the waters in the Bering Sea and the Aleutian region.
Bull. Fac. Fish., Hokkaido Univ., 9(3):149-170.

currents
Kozlyaninov, M.V., 1961.
[Relation between transparency and currents in the northeastern Pacific.]
Trudy Inst. Okeanol., Akad. Nauk, SSSR, 45:102-112.

currents
Kramp, P.L., 1963.
Summary of the zoological results of the "Godthaab" Expedition, 1928.
Medd. om Grønland, 81(7):115 pp.

currents
Kramp, P. L., 1927
The Hydromedusae of the Danish waters.
Det. Kgl. Vidensk Selsk. Skrifter Naturv. - Math. Afdl., 8RK XII(1):292 pp., 24 charts, 3 text figs.

currents
Krauss, Wolfgang, 1966.
Die Spektrum der Temperaturschwankingen und der Strömung im Gebiet nordwestlich von Fehmarn.
Kieler Meeresforsch., 22(1):35-38.

currents
Krauss, W., 1955.
Zum System der Meeresströmungen in der höhen Breiten. Deutsche Hydrogr. Zeits. 8(3):102-111.

currents
Krauss, Wolfgang, und Walter Düing, 1964.
Schichtung und Strom im Golf von Neapel.
Pubbl. Staz. Zool., Napoli, 33(3):243-263.

currents
Krey, J., 1954.
Beziehungen zwischen Phytoplankton, Temperatursprungschicht und Trübungsschirm in der Nordsee im August 1952. Kieler Meeresf. 10(1):3-18, 11 textfigs.

currents
Krügler, Fredrich, 1966.
Über eine optisch markante Stromgrenze an der Polarfront in der Dänemark Strasse.
Dt. hydrogr. Z. 19(4):159-170.

currents
Kruglov, L.A., 1964
Preliminary data on current behavior in fishery areas of Conakry-Takoradi (western coast of Africa). (In Russian).
Okeanologiia, Akad. Nauk, SSSR, 4(5):922-923.

currents
Kubota, T., and K. Iwasa, 1961
[On the currents in Tugaru Strait.]
Hydrogr. Bull., Maritime Safety Bd., Tokyo, (Publ. No. 981), No. 65:19-26.

current
Kudersky, S.K., 1962.
Some peculiarities of the currents in the area of South West Africa (17-24°S). (In Russian).
Baltisk. Nauchno-Issled. Inst. Morsk. Ribn. Khoz. i Okeanogr. (BALTNIRO), Trudy, 9: 39-45

currents
Kuksa, V.I., 1959
[Hydrological features of the North Kuril waters.]
Trudy Inst. Okeanol., 36: 191-214.

currents
Kullenberg, B., and I. Hela, 1942.
Om Tröghetssvängningar i Östersjön.
Svenska Hydrogr.-Biol. Komm. Skr., n.s., Hydrogr. 16:14 pp.

currents
*Kurashina, Syoji, Koji Nishida and Syuji Nakabayashi, 1967.
On the open water in the southeastern part of the frozen Okhotsk Sea and the currents through the Kurile islands. (In Japanese; English abstract).
J. oceanogr. Soc. Japan, 23(2):57-62.

currents
Kusunoki, K., and T. Kashima, 1951.
[Ocean current variations off the western coast of Hokkaido in the Sea of Japan.] J. Ocean. Soc., Japan, (Nippon Kaiyo Gakkaisi) 6(3):133-142, 5 textfigs.

currents
*Kvinze, Thor, 1967.
On the special current and water level variations in the channel of São Sebastião.
Bolm Inst. oceanogr. S. Paulo, 16(1):23-38.

L

currents
Labaye, G., 1948.
Note sur le debit solide des cours d'eau.
Houille Blanche 3, No. spec. "A":600-627, 19 textfigs.

currents
Labeish, V.G., 1959.
On the dynamics of coastal currents.
Vestnik, Ser. Geol. Geogr., 14(6)91):139-143.
Transl. listed in: Techn. Transl., 1962, 7(4):205.

Lacombe, Henri

currents
Lacombe, H., 1970.
Courants marins
Annuaire du Bureau des Longitudes, 1970:221-248. Also in: Recl Trav. Lab. Océanogr. phys. Mus. natn. Hist. nat. 8 (109).

currents
Lacombe, Henri, 1965.
Cours d'oceanographie physique. (Theories de la circulation. Houles et vagues).
Gauthier-Villars, Paris, 392 pp.

currents
Lacombe, Henri, 1961
Année Geophysique Internationale 1957-1958, participation française. Contribution à l'étude du regime du detroit de Gibraltar. 1 Etude dynamique. Cahier Océanogr., C.C.O.E.C., 13 (2): 73-106.

currents
Lacombe, H., 1961.
Note sur le régime du détroit de Gibraltar. Résumé).
Rapp. Proc. Verb., Réunions, Comm. Int. Expl. Sci., Mer Méditerranée, Monaco, 16(3):597.
Cahiers Océanogr., C.C.O.E.C., 1961, 13(2):108 } for complete
13(5):276 } articles

currents
Lacombe, H., 1954.
Contribution à l'étude hydrologique de la Méditerranee occidentale.
Bull. d'Info., C.C.O.E.C. 6(1):31-35, 2 pls.

currents
Lacombe, H., and J.-C. Lizeray, 1959.
Sur le régime des courants dans le détroit de Gibraltar.
C.R. Acad. Sci., Paris, 248(17):2502-2504.

Lacombe, Henri, and Paul Tchernia, 1960 currents
Quelques traits generaux de l'hydrologie Méditerranéenne d'après diverses campagnes hydrologiques recentes en Méditerranée, dans le proche Atlantique et dans le détroit de Gibraltar. Cahiers Océanogr., C.C.O.E.C., 12(8) 527-547.

Leivastu, Taivo

currents
Laevastu, T., 1960
Factors affecting the temperature of the surface layer of the sea.
Merent. Julk. (Havsforskningsinst. Skr.), No. 195:136 pp.

LaFond, E.C.

currents
LaFond, E.C., 1968.
Detailed temperature and current data sections in and near the Kuroshio Current.
Oceanogr. Data Rept. CSK, 22 pp. (multilithed).

currents
*LaFond, E.C., 1967.
Movements of benthic organisms and bottom currents as measured from the bathyscaph Trieste.
In: Deep-sea photography, J.B. Hersey, editor, Johns Hopkins Oceanogr. Studies, 3:295-302.

currents
LaFond, E.C., 1962
Internal waves. Part 1, Ch. 21, Sect. V. Waves. In: The Sea, Vol. 1, Physical Oceanography, Interscience Publishers, 731-751.

currents
LaFond, E.C., 1954.
Physical oceanography and submarine geology of the seas to the west and north of Alaska. Arctic 7(2):93-101, 11 textfigs.

currents
LaFond, E. C., 1949.
Oceanographic research at the U.S. Navy Electronics Laboratory. Trans. Am. Geophys. Union 30(6):894-896.

currents
LaFond, E. C., 1949
The use of bathythermograms to determine ocean currents. Trans. Am. Geophys. Union, 30(2):231-237, 6 text figs.

Laktonov, A.F., V.A. Shamontyev and A.V. Yanes, 1960 currents
Oceanographic characteristics of the North Greenland Sea. Soviet Fish. Invest., North European Seas, VNIRO, PNIRO, Moscow, 1960:51-65.

currents
Lalou, Claude, et Maurice Gennesseaux, 1959
Travaux de la Station Océanographique de Villefranche. Rev. Géogr. Phys. et Géol. Dyn. (2) 2(4): 231-252.

currents
Lamb, H.H., 1948
Topography and weather in the Antarctic. Geogr. Jour. CXI (1/3):48-66.

currents
Lamb, H.H., and A.I. Johnson, 1959.
Climatic variations and observed changes in the general circulation.
Geografiska Annaler, 41(2/3):94-134.

currents
Lane, R.K., 1962
A review of the temperature and salinity structures in the approaches to Vancouver Island, British Columbia.
J. Fish. Res. Bd., Canada, 19(2):45-91.

currents
Lappo, S.S., 1966.
The dynamic effect of atmospheric pressure perturbation on sea currents. (In Russian).
Dokl., Akad. Nauk, SSSR, 171(5):1088-1091.

current
Lastres, D.E.C. y 1935
La Corriente del Peru. Rev. del Cons. Oceanogr. Obero-Americano. Año VI(1): 45.

currents
Latun, V.S., 1960
On the type of dependence of the horizontal turbulent mixing coefficient in the sea on the period of averaging the velocity current oscillations. Meteorol. i Gidrol., No. 7: 35-36.

currents
de Laubenfels, M. W., 1950.
Ocean currents in the Marshall Islands. Geogr. Rev. 40(2):254-259, 1 textfig.

currents
La Violette, Paul E., and Paul Chabot, 1967.
A satellite photograph of Pacific Ocean currents.
Deep-Sea Research, 14(4):485-486.

currents
Lavoinne, M., 1883.
Observations présentées au sujet du régime des courants et des alluvions dans l'estuaire de la Seine. Assoc. Francaise Avance Sci., 12ème Session, Rouen:262-264.

currents
Lavalle, J.A., yGarcia, 1917
La contra corriente ecuatorial como causa determinante del fenomeno marino conocido con el nombre de "Aguaje".
Biol. Soc. Geogr., Lima, XXXIII:313-330.

currents
Lawford, A.L., 1954.
Currents in the North Sea during the 1953 gale.
Weather 9(3):67-72, 1 textfig.

currents
Lazarenko, N.N., 1962.
Experiments in the application of aerophotographs for study of currents in the Baltic Sea. (In Russian).
Trudy Gosud. Okeanogr. Inst., 70:71-87.

currents
Ledenev, V.G., 1964.
Currents in the Lazarev Ice Shelf region. (In Russian).
Sovetsk. Antarkt. Eksped., Inform. Biull., 45:23-
Translation:
Scripta Tecnica Inc., for AGU, 5(2):95-98. 1965.

currents
Ledentsev, V.G., 1963.
Measurement of currents and their characteristics (In Russian).
Trudy Sovetsk. Antarktich. Exped., 39:142-164.
Abstract in:
Soviet Bloc Res., Geophys., Astron., Space, 95:46

currents
Ledniev, A., 1956.
Le régime thermique des courants de l'Atlantique
Inst. Nauchnoj Informacii, Akad. Nauk, SSSR, Trav. Inst. Océan d'Etat, No. 32(44):
Bull. d'Info., 8(10):

currents
Lee, A.J. 1969.
Maff hydrographic buoy study. On the use of moored current meter networks in the shelf seas around Britain.
Oceanol. int. 69 (2): 5pp.

currents
Lee, Arthur, 1963.
The hydrography of the European Arctic and subarctic seas.
In: Oceanography and Marine Biology, H. Barnes, Editor, Geroge Allen and Unwin, 1:47-76.

currents
Lee, Arthur, 1962
The effect of water movements in the Norwegian and Greenland seas.
Proc. Symposium on Mathematical-Hydrodynamical Methods of Phys. Oceanogr., Sept. 1961, Inst. Meeresk., Hamburg, 353-373.

currents
Lee, Chang-Ki and Jong-Hon Bong, 1968.
On the current of the Korean Eastern Sea (West of the Japan Sea). (In Korean; English abstract).
Bull. Fish. Res. Develop Agency, Korea, 3: 7-26

currents
Lee, J., 1950.
Sea surface temperature and floods and droughts in China. J. Chinese Geophys. Soc., Acad. Sinica, 2(2):169-189, 17 textfigs.

currents
Le Floch, Jean, 1970
Mesures différentielles de courants au large de la côte d'Ivoire.
Cah. océanogr. 22(8): 781-799

currents
Le Floch, Jean, 1963
Regimes de courants non permanents à evolution rapide dans le Canal de Corse. Etude de cette evolution pendant une semaine.
Cahiers Océanogr. C.C.O.E.C., 15(7):456-469.

currents (data)
Le Floch, J., 1962
Rapport sur les recherches océanographiques faites en Nord Tyrrhénienne en fevrier 1960.
Trav., Centre de Recherches et d'Etudes Océanogr. 4(4):45-63.

currents
Le Floch, J., 1951.
Caracteristiques hydrologiques et transport des masses d'eaux dans le canal de Sicile en Mars et Juin 1950. Bull. d'Info, C.C.O.E.C. 3(7):281-300, 4 pls.

currents
Le Floch, J., 1951.
Observations hydrologiques au voisinage du Groenland et des Banos de Terre-Neuve en 1948, 1949, et 1950. Bull. d'Info., C.O.E.C., 3(6): 191-205, 19 figs.

currents
Leipper, Dale F., 1970.
A sequence of current patterns in the Gulf of Mexico. J. geophys. Res., 75(3): 637-657.

currents
Leipper, D.F., 1953.
Computed ocean currents in the Gulf of Mexico. (Abstract). Proc. Gulf Caribbean Fish. Inst., 5th Ann. Session:146.

currents
Leipper, D.F., 1951.
Nature of ocean currents in the Gulf of Mexico. Texas J. Sci. 3:41-44.

Lemasson, L. and J.P. Rebert, 1968.
Observations de courants sur le plateau continental ivoirien: mise en évidence d'un sous-courant. Doc. Centre Rech. Océanogr. Abidjan, 022: 67 pp.

currents
Lendenfeld, R. von, 1884.
The geographical distribution of the Australian Scyphomedusae. Proc. Linn. Soc., N.S. Wales, 9: 421-433.

currents
Leontyeva, V.V., 1961.
[Waters of Kurosio in the northwestern Pacific in summer 1953 and 1954.] Trudy Inst. Okeanol., Akad. Nauk, SSSR, 38:3-30.

currents
Le Pichon, Xavier, et Jean-Paul Troadec, 1963
La couche superifcielle de la Méditerranée au large des côtes Provençales durant les mois d'été.
Cahiers Océanogr., C.C.O.E.C., 15(5):299-314

currents
Lesser, R.M., 1951.
Some observations of the velocity profile near the sea floor. Trans. Am. Geophys. Union 32(2): 207-211.

currents
Levin, L., and M. Voyinovia, 1953.
Rôle de la distorsion sur certains modèles d'hydraulique maritime à fond mobile.
Proc. Minnesota Hydraulics Conv., Sept. 1-4, 1953:235-240, 8 textfigs.
Int.

currents
Lilly, S.J., J.F. Sloane, R. Bassindale, F.J. Ebling and J.A. Kitching, 1953.
The ecology of the Lough Ine rapids with special reference to water currents. IV The sedentary fauna of sublittoral boulders. J. Animal Ecol. 22(1):87-122, 22 textfigs.

Lineikin P.S

Currents
Lineykin, P.S., 1969.
On the theory of currents in an ocean of finite depth. (In Russian; English abstract). Okeanologiia, 9(1):58-62.

currents
Lineikin, P.S., 1962
On some recent work on the hydrodynamics of ocean currents. (In Russian).
J. Oceanogr. Soc., Japan, 20th Ann. Vol., 448-457.

currents
Lineikin, P.S., 1960.
[On the question of the development of investigations of ocean currents.]
Trudy Okeanogr. Komissii, Akad. Nauk, SSSR, 10 (1):16-21.

currents
Lineikin, P.S., 1960.
[A simplified method for the determination of currents in the shallow and deep layers of the sea far out from shore.]
Trudy Gosudarst. Okeanogr. Inst., 50:5-26.

currents
Lineikin, P.S., 1959.
Currents and stratification of the waters in the sea. (In Russian).
Trudy Gosud. Okeanogr. Inst., 47:13-29.

currents
Lineikin, P.S., 1956.
[Dynamics of the established currents in a heterogeneous sea.] Dokl. Akad. Nauk, SSSR, 105(6):1215.

currents
Lineikin, P.S., 1956.
[On winter currents in a deep sea.]
Met. i Gidr., No. 12:26-33.

currents
Lipps, J.H., and J.E. Warme, 1966.
Planktonic foraminiferal biofacies in the Okhotsk.
Contrib. Cushman Found. Foram. Res., 17(4):125-134.

currents
Lisitzin, A.P., 1966.
Processes of Recent sedimentation in the Bering Sea. (In Russian).
Inst. Okeanol., Kom. Osad. Otdel Nauk o Zemle, Isdatel, Nauka, Moskva, 574 pp.

Lisitzin, Eugénie

currents
Lisitzin, Eugénie, 1964.
La pression atmosphérique comme cause primaire des processus dynamique dans les océans.
Cahiers Océanogr., C.C.O.E.C., 16(1):17-22.

Currents
Lisitzin, Eugenie, 1962.
Some characteristics of the variation in the water volume in the Baltic as a function of air pressure gradient changes
Soc. Sci. Fennica, Comment. Phys.-Math., 26(9):15 pp.

currents
Lisitzin, E., 1955.
Contribution à la connaissance des courants dans la Mer Ligure et la Mer Tyrrhénienne.
Bull. Inst. Océan., Monaco, No. 1060, 1-11, 2 fig

currents
Lisitzin, E., 1951.
A brief report on the scientific results of the hydrological expedition to the Archipelago and Åland Sea in the year 1927. Fennia 73(4):21 pp., 8 textfigs.

currents
Lisitzin, E., 1950/51.
A brief report on the scientific results of the hydrological expedition to the Archipelago and Åland Sea in the year 1922. Fennia 73(4):3-21.

currents
Lisitzin, E., 1949.
Observations on currents and winds made on board Finnish Light-ships during the years 1943-1947. Merent. Julk., Havsforskningsinst. Skrift. No. 143:66 pp., 1 textfig., tables.

currents
Lisitzin, E., 1949.
On current velocity in the northern Baltic. Fennia 71(3):1-14, 1 textfig.

currents
Lisitzen, E., 1947.
Observations on currents and winds made on board Finnish light-ships during the years 1940, 1941, and 1942. Havsforskningsinst. Skrift. No. 139, 36 pp.

currents
Littlepage, Jack L., 1965.
Oceanographic investigations in McMurdo Sound, Antarctica.
In: Biology of Antarctic seas, II.
Antarctic Res. Ser., Am. Geophys. Union, 5:1-38.

currents
Lizeray, J.-C., 1957.
Travaux océanographiques récents de l'Elie Monnier en Méditerranée occidentale. Bull. d'Info., C.C.O.E.C., 9(8):413-415.

currents
Ljøen, R., 1962.
The waters of the western and northern coasts of Norway in July-August 1957.
Fiskeri Skrifter, Ser. Havundersøgelser, 13(2): 39 pp.

currents
Longinov, V.V., 1951.
[Role of compensatory currents near the bottom in moving materials on a bottom slope.] Izvest Akad. Nauk, Ser. Geogr. Geofiz., 15(2):78-81.

currents
Longinov, V. V., 1948.
[Relation between the course of waves and the corresponding maximum rate of displacement of alluvium from sea-bottom slopes along the shore.] Izvest. Akad. Nauk. SSSR, Ser. Geogr. Geofiz. 12:362-368.

Longuet-Higgins

currents
Longuet-Higgins, M.S., 1965.
Symons Memorial Lecture for 1965. Some dynamical aspects of ocean currents.
Q.J. Roy. Meteorol Soc., 91(390):425-451.

currents
Longuet-Higgins, M.S., & R.W. Stewart, 1961
The changes in amplitude of short gravity waves on steady non-uniform currents.
J. Fluid Mech., 10(4): 529-549.

currents
Loosanoff, V. L., and P.B. Smith, 1950.
Apparatus for maintaining several streams of water of different constant salinities. Ecology 31(3): 473-474, 1 textfig.

M

currents
MacGregor, O.G., 1956.
Currents and transport in Cabot Strait.
J. Fish. Res. Bd., Canada, 13:435-448.

currents
Machado, L. de B., 1951.
Resultados cientificos do cruzeiro do "Baependi" e do "Vega" a Ilha de Trindade. Oceanografia fisica. Contribuição para o conhecimento das caracteristicas fisicas e quimicas das aguas.
Bol. Inst. Paulista Ocean. 2(2):95-110, 5 pls.

currents
Maksimov, I.V., 1962.
Nature of the Great East Drift. (In Russian).
Sovetsk. Antarktich. Exped. Inform. Biull., (32):(32):5-9.

Translation:
Scripta Tecnica, Inc., for AGU, 4(1):33-35.

currents
Maksimov, I.V., 1960.
[Currents in the Bellingshausen Sea area.]
Inform. Biull. Sovets. Antarkt. Exped., (14):19-23.

currents
Maksimov, I.V., 1958.
A study of the western coastal Antarctic current (In Russian).
Sovetskaya Antark. Eksped. Inform. Byull., (2):31-35.

Transl. cited:
Techn. Transl., 1963, 10(2):188.
OTS-SLA, $1.10 63-13866

Transl. M. Slessers, 1962, H.O. Transl. No. 168: 7 pp.

currents
Maksimov, I.V., 1945.
[On the determination of the relative volume of flow of Pacific waters through Bering Strait.]
Problemy Arktiki 1944(2):51-58, tables, diagrams, sketch maps.

currents
Mallory, J.K., 1961.
Bathymetric and hydrographic aspects of marine studies off the Natal coast.
Marine Studies off the Natal Coast, C.S.I.R. Symposium, S2:31-39 (multilithed).

currents
Malmberg, Svend-Aage, 1962
Schichtung und Zirkulation in den südisländischen Gewassern.
Kieler Meeresf., 18(1):3-28.

currents
Manegold, W., 1935.
Die Wetterabhängigkeit der Oberflächenströmungen in den Pforten der Ostsee.
Arch. Deutschen Seewarte 54(4):1-40, 2 pls.

currents
Mann, G., F., 1954.
Delimitacion de los sectores Pacifico y Atlantico en el Mar de Drake. Invest. Zool. Chilenas 2(6):97-100, 3 textfigs.

currents
Marchi, L. de, 1920.
Le correnti dell'Adriatico secondo la distribuzione superficiale della salsedine et della temperatura. (Appendice). R. Comm. Talassogr. Ital. Mem. 55:95-109, 14 textfigs.

currents
Margineanu, Carmen, 1968.
Quelques observations sur les changements qualitatifs et quantitatifs du zooplancton marin dans une station de 24 heures. (In Roumanian; French summary).
In: Lucrările Sesiunii Ştiinţifice a Staţiunii de Cercetări Marine "Prof. Ioan Borcea", Agigea, (1-2 Noiembre 1966), Volum Festiv, Iaşi, 1968: 273-280.

currents
Marlowe, J.I., 1966.
Mineralogy as an indicator of long-term current fluctuations in Baffin Bay.
Canadian J. Earth Sci., 3(2):191-202.

currents
Marmer, H.A., 1951.
The Peru - and Nino - Currents. Geogr. Rev. 41 337-338.

currents
Marmer, H.A., 1926
Coastal currents along the Pacific Coast of the United States. U.S.C. and G.S. Ser. No.330, Spec. Publ. No.121, 80 pp., 23 textfigs.

currents
Martin, C., 1948.
Étude des mouvements des sables. Bull. Sci., C.O.E.C., No. 3:5-9.
Mars

currents, G.E.K.
Martin, Jean, 1964.
Expérience de G.E.K. vertical dans le Détroit de Gibraltar.
Cahiers Océanogr., C.C.O.E.C., 16(5):377-392.

currents
Martin, J.H.A., 1966.
The bottom waters of the Faroe-Shetland Channel.
In: Some contemporary studies in marine science, H. Barnes, editor, George Allen & Unwin, Ltd., 469-478.

currents
Martin Salazar, Felipe, 1956
Oceanografia de la region central.
El Agriculator Venezolano, 21(190): 28-35; 68-69.

currents
Marukawa, H., and T. Kamiya, 1926.
Outline of the hydrographical features of the Japan Sea.
Annot. Oceanogr. Res., 1(1):1-7, tables and charts

currents
Maslennikov, V.V., 1971.
Oceanographic surveys in The Andaman Sea and the northeastern Bay of Bengal. (In Russian)
Vses. nauchno-issled. Inst. morsk. Ribn. Khoz. Okeanogr. VNIRO, Trudy 72: 46-55

Currents
Masuzawa, Jotaro 1967.
An oceanographic section from Japan to New Guinea at 137° E. in January 1967.
Oceanogr. Mag. 19(2): 95-118.

currents
Masuzawa, J., 1954.
On the Kuroshio south of Shiono-Misaki of Japan. (Currents and water masses of the Kuroshio System. I.). Ocean. Mag., Tokyo, 6(1):25-33, 7 textfigs.

currents
Masuzawa, J., 1950.
On the intermediate water in the southern Sea of Japan. Ocean. Mag., Tokyo, 2(4):137-144, 6 textfigs.

currents
Masuzawa, J., and T. Nakai, 1955.
Notes on the cross-current structure of the Kuroshio (Currents and water masses of the Kuroshio system. V).
Rec. Ocean. Wks., Japan, 2(2):96-101, 8 textfigs.

currents
Matida, Y., 1953.
The cycles of phosphorus and nitrogen in Tokyo Bay. Bull. Japan. Soc. Sci. Fish. 19(4):429-434, 7 textfigs.

currents
Matsudeira, Yasuo, 1965.
An opinion on the forecasting of the fishing grounds and on the fundamental sea states from the water temperature distribution in the neighboring Sea of Japan. (In Japanese: English summary).
J. Fac. Fish., Animal Husbandry, Hiroshima Univ. 6(1):313-321.

currents (data)
Matthews, D.J., 1926.
VII. Physical oceanography. Trans. Linn. Soc., London, Ser. 2, 19:169-205, Pls. 10-13, textfigs. 1-8.

currents
Matzenauer, L., 1933
Die Dinoflagellaten des indischen Ozeans (mit Ausnahme der Gattung Ceratium.) Bot. Arch. 35:437-510, 77 text figs., 2 charts.

currents
Maury, M.F., 1844
Remarks on the Gulf Stream and Currents of the Sea. Amer. J. Sci. and Arts. XLVII:161-181.

currents
Mauvais, J.L., 1971.
Calcul des vitesses moyennes instantanées en Loire maritime.
Cah. océanogr. 23(3):251-266

currents
Maximov, I.V., 1961
[The centenary cycle of solar activity and the North Atlantic current.]
Okeanologia, 2: 206-212.

currents
Maximov, I.V., 1958
[On the study of the west coastal Antarctic current.]
Inform. Biull. Sovetsk. Antarkt. Exped., (4): 31-36.

currents
Maximov, I.V., and N.P. Smirnov, 1965.
On the origin of the semiannual period of ocean currents. (In Russian; English abstract).
Fizika Atmosferi i Okeana, Izv. Akad. Nauk, SSSR, 1(10):1079-1087.

currents
Mazzarelli, G., 1938.
I vortici i tagli altrifenomeni delle correnti delle Stretto di Messine. Alti Accad. Peloritana 40:

currents
McEwen, G.F., 1915.
Oceanic circulation and temperature off the Pacific coast.
Nature and Science on the Pacific Coast:133-140.

currents
McKenzie, Kenneth G., 1964.
The ecologic association of an ostracode fauna from Oyster Harbour, a marginal marine environment near Albany, Western Australia.
Pubbl., Staz. Zool., Napoli, 33(Suppl.):421-461.

currents
*McLain, Douglas R., 1968.
Oceanographic surveys of Traitors Cove, Revillagigedo Island, Alaska.
Spec. scient. Rep. U.S. Fish Wildl. Serv. Fish, 576: 15 pp. (multilithed).

currents
McLellan, J.H., 1957.
On the distinctness and origin of the slope water off the Scotian shelf and its easterly flow south of the Grand Banks.
J. Fish. Res. Bd., Canada, 14(2):213-239.

currents
Menaché, Maurice, 1963
Première campagne océanographique du "Commandant Robert Giraud" dans le Canal de Mozambique (11 octobre-28 novembre 1957).
Cahiers Océanogr., C.C.O.E.C., 15(4):224-235, 27 figs.

currents
Menendez, N., 1960
Les courants de gradient dans la mer d'Alboran.
Comm. Int. Sci., Mer Méditerranée, Rapp. Proc. Verb., Monaco, 15(3): 271.

currents
Mesteroff, J., 1960.
German hydrographic observations in the ICNAF area (subareas 2 and 3) in September and November 1959.
Ann. Proc., Int. Comm. North Atlantic Fish., 10:50-56.

currents
Metcalf, W.G., and M.C. Stalcup, 1967.
Origin of the Atlantic Equatorial Undercurrent.
J. geophys. Res., 72(20):4959-4975.

currents
Mikhailov, Iu. D., 1962.
Expedition investigations of the level of oscillations currents in the Finnish Gulf. (In Russian).
Trudy Gosud. Okeanogr. Inst., 69:73-86.

currents
Mikhailov, Iu. D., and M.A. Babanina, 1965.
Study on the connection between the atmospheric pressure field and currents in the Gulf of Finland. (In Russian).
Trudy, Gosudarst. Okeanogr. Inst., No. 87:58-63.

currents
Mikhailova, E.N., A.I. Felsenbaum and N.B. Shapiro, 1967.
On the calculation of unsteady oceanic currents and tides. (In Russian).
Dokl. Akad. Nauk, SSSR, 175(5):1041-1044.

currents
Mikhailova, E.N., A.I. Felsenbaum and N.B. Shapiro, 1967.
On the calculation of ice-field drifts and currents in the Arctic basin. (In Russian).
Dokl., Akad. Nauk, SSSR, 175(6):1273-1276.

currents
Milburn, K., and E. W. Barlow, 1949.
Ocean currents. Mar. Observ. 19(146):225-232.

Currents
Miller, Gerald S., 1968.
Currents at Toledo Harbor.
Proc. 11th Conf. Great Lakes Res. 1968: 437-453.

currents
Miyake, Y., Y. Sugiura and K. Kameda, 1955.
On the distribution of radioactivity in the sea around Bikini Atoll in June 1954.
Pap. Met. Geophys., Tokyo, 5(3/4):253-262.

currents
Miyazaki, N., 1950.
Lateral mixing effect on the distribution of ocean currents. Ocean. Mag., Tokyo, 2(3):117-121, 6 textfigs.

currents
Miyazaki, M., 1950.
A theory of the transversal velocity distribution in ocean currents. Ocean. Mag., Tokyo, 2(2):53-57, 4 textfigs.

currents
Miyazaki, M., 1948.
On the computation of the stationary currents.
Geophys. Mag., Tokyo, 15:58-62.

current
Miyazaki, M., 1948
Zonal circulating current caused by winds.
Geophys. Mag. 15(2/4):50-57, 4 figs.

currents
Miyazaki, M., and S. Abe, 1960
[On the water masses in the Tsushima Current.]
J. Oceanogr. Soc., Japan, 16(2): 59-68.

currents
Model, Fr., 1958
Strombeobachtungen in der südwestlichen Nordsee.
Deutsche Hydrogr. Zeits., 11(5): 217-218.

currents
Model, F., 1950.
Pillsburys Strommessungen und der Wasserhaushalt des Amerikanischen Mittelmeeres. Deutsche Hydro. Zeits. 3(1/2):57-61, 1 textfig.

currents
Model, F., 1948.
Wasserstand, Strom und Singularitäten in der Ostsee. Ann. Meteorol., Juli-Aug., 1948:245-247.

currents
Møller, J.T., 1956.
Map of the tidal area of Juvre Dyb in relation to some topographic and hydrographical conditions. Geografisk Tidsskrift 55:88-105.

currents
Montgomery, R.B., and E.D. Stroup, 1962.
Equatorial waters and currents at 150°W in July-August 1952.
The Johns Hopkins Oceanogr. Studies, No. 1:68 pp.

currents
Moreau-Defarges, Alain, 1964.
Étude par analogie rhéoélectrique des courants induits par les vents sur l'Atlantique.
Cahiers Océanogr., C.C.O.E.C., 16(9):755-780.

currents
*Morcos, Selim A., 1967.
The chemical composition of sea water from the Suez Canal region. I. The major anions.
Kieler Meeresforsch., 23(2):80-91.

currents
*Morgan, Charles W., 1969.
Oceanography of the Grand Banks region of Newfoundland in 1967.
Oceanogr. Rept. U.S.C.G. 19(CG 373-19):1-209.

currents
Mortimer, C.H., 1963.
Frontiers in physical limnology with particular reference to long waves in rotating basins.
Great Lakes Res. Div., Univ. Michigan, Publ., No. 10:9-42.

Moroshkin, K.V

currents
Moroshkin, K.V., 1960
[Characteristics of the Circumpolar current in the Pacific Ocean.]
Okean. Issle., IGY Com., SSSR, 86-90.

currents
Moroshkin, K.V., 1959.
[Hydrological investigations.]
Arktich. i Antarkt. Nauchno-Issled. Inst., Mezhd. Geofiz. God, Sovetsk. Antarkt. Eksped., 5:107-124.

currents
Moroshkin, K.V., 1959
[Measurements of currents in the South-west Indian Ocean by the use of electromagnetic method.]
Info. Biull. Sovetsk. Antarkt. Exped., (7): 22-25.

currents
Morozov, A.P., 1957.
Currents near Litavsk lightship. (In Russian).
Trudy Gosud. Okeanogr. Inst., 41:46-53.

currents
Mosby, H., 1947.
Experiments on turbulence and friction near the bottom of the sea. Bergens Museums Aarbok 1946/47, Naturvitenskap. rekke, 6 pp., 3 textfigs.

currents
Mosby, H., 1944
On en teori for havstrømmene. Naturen, No. 9: 274-280, 4 textfigs.

currents
Mosby, H., 1938
Svalbard waters. Geofysiske Publ. 12(4): 85 pp., 34 textfigs.

currents
*Mosetti, F., 1967.
Considerazioni preliminari sulla dinamica dell' Adriatico settentrionale.
Arch. Oceanogr. Limnol. 15(Suppl.):237-244.

currents
Munk, W.H., and E. Palmen, 1951.
Notes on the dynamics of the Antarctic Circumpolar Current. Tellus 3(1):53-55.

currents
Munk, W.H., G.C. Ewing, and R.R. Revelle, 1949
Diffusion in Bikini Lagoon. Trans. Am. Geophys. Un., 30(1):59-66; 9 text figs.

currents

Muñoz, F., y J.M. San Feliu 1969.
Hidrografía y fitoplancton de las costas de Castellón de febrero a junio de 1967.
Inv. pesq. 33(1):313-334.

currents

Murray, J.W., 1966.
A study of the seasonal changes of the water mass of Christchurch Harbour, England.
Jour. mar. biol. Assoc., U.K., 46(3):561-578.

N

currents

Nair, K.V.K., and P.M.A. Bhattathiri 1969.
A note on current measurements at Angria Bank in the Arabian Sea.
Bull. natn. Inst. Sci. India 38(1):289-293.

currents

Nair, K.V.K., V.R. Neralla and A.K. Ganguly, 1969.

Current measurements off Mormugao. Bull. natn. Inst. Sci., India, 38(1): 254-262.

current

Nakai, Zinziro, Shigemasa Hattori, Koji Honjo, Taisuke Watanabe, Takashi Kidachi, Takashi Okutani, Hideya Suzuki, Shigeichi Hayashi, Masao Hayaishi, Keiichi Kondo and Shuzo Hayaishi, 1964.
Preliminary report of marine biological anomalies on the Pacific coast of Japan in early months of 1963, with reference to oceanographic conditions.
Bull. Tokai Reg. Fish. Res. Lab., No. 38:57-75.

currents

Nakamiya, T., and K. Suda, 1950.
On the variation of "Kuroshio". Hydrogr. Div., Maritime Safety Agency, Tokyo, 14 pp., 14 figs.

currents

Nakano, Masito 1968.
Interchanging flows through straits and canals. (In Japanese; English abstract)
Bull. coast. Oceanogr. 6(2): 113-115.
Also in: Coll. Repr. Coll. mar. Sci. Techn., Tokai Univ. 1967-68, 3.

Nan-niti, Tosio

currents

Nan'niti, Tosio, 1962
Deep-sea current measurements.
J. Oceanogr. Soc., Japan, 18(2):73-77.

currents

Nan'niti, T., 1960
Long-period fluctuations in the Kuroshio.
Pap. Meteorol. & Geophys., 11(2-4):339-347.

currents

Nan'niti, T., 1959.
On the structure of ocean currents. III. A new Nan'niti-Iwamiya current meter and current profile observed by it. Pap. Meteorol. Geophys., Tokyo, 10(2):124-134.

Also in: Coll. Reprints, Oceanogr. Lab., Meteorol. Res. Inst. Tokyo, 6

currents

Nanniti, T., 1951.
On the fluctuation of the Kuroshio and/the Oyashio.
Pap. Met. Geophys. 2(1):102-111, 6 textfigs.

currents

Nan-niti, Toshio, Hideo Akamatsu and Takeo Yasuoka 1966.
A deep current measurement in the Japan Sea.
Oceanogrl Mag. 18(1/2): 63-71.

currents

Nansen, F., 1905.
Die Ursachen der Meeresströmungen. Petermanns Geogr. Mitt., 51(1):1-4; (2):25-31.

currents

Natarov, V.V., 1963
On the water masses and currents of the Bering Sea (In Russian).
Sovetsk. Ribokh. Issled. B Severo-Vostokh. Chasti Tikhogo Okeana, VNIRO 48, TINRO 50(1): 111-134.

currents

Navrotsky, V.V., 1964.
The study of the interaction between oceanic currents and the atmospheric processes in the northern Atlantic. (In Russian).
Okeanologiia, Akad. Nauk, SSSR, 4(3):396-407.

currents

*Neev, David, and K.O. Emery, 1967.
The Dead Sea: depositional processes and environments of evaporites.
Bull. Geol. Surv., Israel, 41: 146 pp.

currents (data)

Nehring, D. and H.J. Brosin 1968
Ozeanographische Beobachtungen im äquatorialen Atlantik und auf dem Patagonischen Schelf während der 1. Südatlantik Expedition mit dem Fischereiforschungsschiff Ernst Haeckel von August bis Dezember 1966.
Geod. Geoph. Veröff. 4(3):93pp.

currents

Nelson, Raymond M., 1967.
Sensing ocean currents from space.
Ocean Industry, 2(5):40-42

currents,

Netchaeff, A., and S. Tcherneff, 1939.
Premiers résultats de l'étude des courants près de la côte d'ouest de la Mer Noire. Courant du Diable.
Trav. (Trudy) Sta. Ichthyol. Sozopol, Bulgaria, 1937, 6:7-75, maps and diagrams.

currents

Netherlands Hydrographer, 1965.
Some oceanographic and meteorological data of the southern North Sea.
Hydrographic Newsletter, Spec. Issue No. to Vol. 1. numerous pp. not sequentially numbered.

Neumann, Gerhard

currents

Neumann, G., 1968.
Ocean currents.
Elsevier Publishing Co., 352 pp. $20.00.

currents

Neumann, Gerhard, 1965.
Oceanography of the tropical Atlantic.
Anais Acad bras. Cienc., 37 (Supl.):63-82.

currents

Neumann, Gerhard, 1960.
On the effect of bottom topography on ocean currents.
Deutsche Hydrogr. Zeits., 13(3):132-141.

currents

Neumann, G., 1956.
Notes on the horizontal circulation of ocean currents. Bull. Amer. Met. Soc., 37(3):96-100.

currents

Neumann, G., 1943.
Periodische Strömungen im Finnischen Meerbusen im Zusammenhang mit den Eigenschwingungen der Ostsee. Gerlands Beitr. Geophys. 59(1):1-

currents

Neumann, G., 1942.
Die absolute Topographie des physikalischen Meeresniveaus und die Oberflächenströmungen des Schwarzen Meeres. Ann. Hydr., usw., 70:265-282, 3 textfigs.

currents

Neumann, G., 1940.
Die ozeanographischen Verhältnisse an der Meeres-oberfläche im Golfstromsektor nördlich und nordwestlich der Azoren. Wiss. Ergeb. Internat. Golfstrom-Unternehmung 1938, Lief. 1, in Juni Beiheft, Ann. Hydr., usw., 68:87 pp., 36 textfig 3 pls.

currents

Neumann, G., and A. Schumacher, 1944.
Strömungen und dichte der Meeresoberfläche vor der Ostküste Nordamerikas. Ann. Hydr., usw., 72: 277-279, Pl. 12.

currents

Newell, B.S., 1959

The hydrography of the British East African coastal waters. II. Colonial Off., Fish. Publ. London, No. 12: 18 pp.

currents

Neyman, V.G., 1960

[Some results of the hydrological observations carried out on board of the d/s "Ob" in the Tasman Sea.]
Okean. Issle., IGY Com., SSSR:96-99.

currents

Nielsen, J.N., 1908.
Contribution to the understanding of the currents in the northern part of the Atlantic Ocean.
Medd. Komm. Havundersøgelser, Ser. Hydrogr., 1(11):15 pp., 1 pl.

currents

Nikiforov, E.G., 1961
[Certain hydrodynamic effects in unsteady currents caused solely by the wind.]
Doklady, Akad. Nauk, SSSR, 140(2):358-360.

currents

Nikiforov, E.G., 1960.
[On the relationship between wind waves and currents.]
Trudy Okeanograf. Komissii, Akad. Nauk, SSSR, 10 (1):22-30.

currents

Nikitin, I.S., 1957
The radar method of studying sea currents.
Meteorol. i Gidrol., SSSR, 4:47-50.

currents
Nishizawa, Tanzo and T. Muraki, 1940
Chemical studies in the sea adjacent to Palau. I. A survey crossing the sea from Palau to New Guinea. Kagaku Nanyo (Sci. of the South Sea) 2(3):1-7.

currents
Nitani, H., K. Iwasa and W. Inada, 1959.
On the oceanic and tidal current observation in the channel by making use of induced electric potential. Hydrogr. Bull., Tokyo, (Publ. 981) (61):14-24.

currents
Noble, Vincent E., and Robert F. Anderson, 1968.
Temperature and current in the Grand Haven Michigan vicinity during thermal bar conditions.
Proc. 11th Conf. Great Lakes Res. 1968: 470-479.

currents
Novaro, L., 1923-24.
Campagna idrografica della "R.N. Ammiraglio Magnaghi" in Mar Rosso, 1923-24. Ann. Idrogr. 11:1-11.

currents
Novitskii, V.P., 1961.
Constant currents of the northern part of the Barents Sea. (In Russian).
Trudy Gosud. Okeanogr. Inst., 64:3-32.

currents
Nutt, D.C., and L.K. Coachman, 1956.
The oceanography of Hebron Fjord, Labrador.
J. Fish. Res. Bd., Canada, 13(5):709-758.

O

currents
O'Brien, M.P., 1952.
Salinity currents in estuaries. Trans. Amer. Geophys. Union 33(4):520-522, 1 textfig.

currents
O'Brien, M.P. and R.G. Folsom, 1948
Notes on the design of current meters. Trans. Am. Geophys. Union 29(2):243-250, 5 text figs.

currents
Oceanographic Section, C.M.O., 1949.
Report on sea and weather observations on Antarctic Whaling Ground (1948-1949). Ocean. Mag., Tokyo, 1(3):142-173, 11 textfigs.

currents
Okada, M., 1939
Two dimensional ocean currents. Bull. Japan Soc. Sci. Fish.8:1-14.

currents
Okubo, Akira 1970.
Oceanic mixing.
Chesapeake Bay Inst., Ref. 70-1: 140 pp. (multilithed).

currents
Okuda, Taizo, 1960
Metabolic circulation of phosphorus and nitrogen in Matsushima Bay (Japan) with special reference to exchange of these elements between sea water and sediments. (Portuguese, French resumés)
Trabalhos, Inst. Biol. Maritima e Oceanogr., Universidade do Recife, Brasil, 2(1):7-153.

currents
Olson, F.C.W., 1953.
Tampa Bay studies. Ocean. Inst., Florida State Univ., Rept. 1:27 pp. (mimeographed), 21 figs. (multilithed). UNPUBLISHED

currents
Orr, A. P., 1933
Physical and chemical conditions in the sea in the neighborhood of the Great Barrier Reef. Brit. Mus. (N.H.) Great Barrier Reef Exped., 1928-29, Sci. Repts. 2(3):37-86, 7 text figs.

currents
Orren, M.J., 1963.
Hydrological observations in the south west Indian Ocean.
Rept., S. Africa, Dept. Comm. Industr., Div. Sea Fish., Invest. Rept., No. 45:61 pp.
Reprinted from: Commerce & Industry, May 1963

currents
Osiecimski, Roman, 1961
[Characteristic of the eroding phenomena in the Swinoujscie-Szczecin fairway.] (In Polish; English summary).
Prace Inst. Morsk., Gdansk, (1) Hydrotech., II. Sesja Naukowa Inst. Morsk., 20-21 wrzesnia 1960:63-105; English summary, 146-148. (mimeographed).

currents
Ovchinnikov, I.M., 1961
[Circulation of waters in the northern Indian Ocean during the winter monsoon.] (In Russian; English abstract).
Okeanolog. Issledov., Mezhd. Komitet Proved. Mezhd. Geofiz. Goda, Prezidiume Akad. Nauk, SSSR, (4):18-24.

currents
Ozmidov, R.V., 1965.
On the energy distribution between oceanic motions of different scales. (In Russian; English abstract).
Fisika Atmosferi i Okeana, 1 (4):439-448.

currents
Ozmidov, R.V., 1960(1961)
[On the rate of dissipation of turbulent energy in sea currents and on the dimensionless universal constant in the "4/3" power law.]
Izv. Akad. Nauk, SSSR, Ser. Geofiz., (8): 1234-1237.
English translation:
(8): 821-823

P

currents
Paasche, E., 1960
Phytoplankton distribution in the Norwegian Sea in June, 1954, related to hydrography and compared with primary production data.
Fiskeridirektoratets Skr., Ser. Havundersøgelser 12(11): 77 pp.

currents
Pagliari, Marcello, 1966.
First experimental results of the diffusion of fresh water in a shallow bay.
Air and Water Pollution, 10(8):537-548.

currents
Palausi, G., 1968.
Étude du mode de dispersion de flotteurs colorés dans la baie de Cannes.
Rev. int. Océanogr. Méd. 9:191-205.

currents
Palmen, E., 1920
Untersuchungen über die Strömungen in der Finnland umgebenden Meeren. Soc. Scient. Fennica, Comm. V:12

currents
Palmieri, Sabino, e Carlo Finizio, 1971.
Un metodo di calcolo della vorticità nei bassi strati atmosferici sul Golfo di Taranto.
Riv. Met. aeronaut. 31(1):45-48.

currents
Parenzan, Pietro, 1960
Il Mar Piccolo di Taranto.
Giovanni Semerano, Editore, Roma, 254 pp.

currents
Parr, A. E., 1937
Report on hydrographic observations at a series of anchor stations across the Straits of Florida. Bull. Bingham Ocean. Coll. VI (3):1-62, 36 text figs.

currents
Paulmier, Gérard, 1965.
Le microplancton de la Rivière d'Auray.
Rev. Trav. Inst. Pêches Marit., 29(2):211-224.

currents
Pedlosky, Joseph, 1964.
The stability of currents in the atmosphere and the oceans. II.
J. Atmos. Sci., 21(4):342-353.

currents
Peluchon, Georges et Jean Rene Donguy, 1962
Hydrologie en mer d'Alboran. Travaux oceanographiques de l'"Origny" dans le detroit de Gibraltar, Campagne internationale - 15 mai - 15 juin, 1961.
Cahiers Oceanogr., C.C.O.E.C., 14(8):573-578.

currents
Ponin, V.V., 1966.
Hydrological conditions in the Norwegian Sea in the period of summer feeding of herring from 1958 to 1963. (In Russian).
Trudy, Poliarn. Nauchno-Issled. Proektn. Inst. Morsk. Ribn. Khoz. Okeanogr., N.M. Knipovich, PINRO, 17:55-66.

currents
Perlroth, Irving, 1969
The distribution of water type structure in the first 300 feet of the equatorial Atlantic
Actes Symp. Oceanogr. Ressources halieut. Atlant. trop., Abidjan, 20-28 Oct. 1966, UNESCO, 185-191.

currents
Pettersson, O., 1908.
Stromstudier vid Ostersjons portat.
Svenska Hydrogr.-Biol. Komm. Skr., 3(5):13 plus 9 pp.

current
Pezet, F.A., 1896
La Contra-corriente "El Nino", en la costa Norte del Perú. Bol. Soc. Geogr. Lima, V:457-461.

currents
Phleger, Fred B, 1965.
Sedimentology of Guerrero Negro Lagoon, Baja California, Mexico.
In: Submarine geology and geophysics, Colston Papers, W.F. Whittard and R. Bradshaw, editors, Butterworth's, London, 17:205-235.

currents
Pickard, G.L., 1953.
Oceanography of British Columbia mainland inlets. II. Currents. Prog. Repts., Pacific Coast Stas., Fish. Res. Bd., Canada, No. 97:12-13, 1 textfig.

currents
Pickard, G. L., and Keith Rodgers, 1959.
Current measurements in Knight Inlet, British Columbia.
J. Fish. Res. Bd., Canada, 16(5):635-684.

currents
Pierce, E.L., 1953.
The Chaetognatha over the continental shelf of North Carolina with attention to their relation to the hydrography of the area. J. Mar. Res. 12(1):75-92, 4 textfigs.

currents
Pirie, Robert Gordon, 1965.
Petrology and physical-chemical environment of bottom sediments of the Rivière Bonaventure-Chaleur Bay area, Quebec, Canada.
Rept. B.I.O. 65-10:182 pp. (multilithed).

currents
Plakhotnik, A.F., 1964.
Hydrological characteristics of the Gulf of Alaska. (In Russian).
Gosudarst. Kom. Sov. Ministr., SSSR, Ribn. Choz., Trudy VNIRO, 49, Izv., TINRO, 51:17-50.

currents
Pochapsky, T.E., 1962.
Measurement of currents below the surface at a location in the western equatorial Atlantic.
Nature, 195(4843):767-768.

currents
Polli, Silvio, 1961
Sul fenomeno dell'acqua alta nell'Adriatico settentrionale.
Rapp. Prelim., Comm. di Studio per la Conservazione della Laguna e della Laguna e della Citta di Venezia, Ist. Veneto di Scienze, Lettere ed Arti, Venezia, 1:1-15.

Also: Ist Sperimental Talassografico, Pubbl., No. 384.

currents
Poloukarov, G.V., 1957.
[Les methodes numeriques de definition du niveau de la maree et la vitesse des courants de flot et de jusant.] Trudy Gosud. Inst. Okeanol., 38:

currents
Polushkin, V.A., 1962
The study of currents in the Caspian Sea.
Okeanologiia, Akad. Nauk. SSSR, 2(1):143-145.

Abstracted in: Soviet Bloc Res., Geophys. Astron. and Space, 1962(35):25
(OTS61-11147-35 JPRS13739)

currents
Pomeroy, A.S., 1965.
Notes on the physical oceanographic environment of the Republic of South Africa.
Naval Oceanogr. Res. Unit, 41pp. (Unpublished manuscript).

currents
Ponomarenko, G.P., 1963.
Investigation of the North Atlantic currents based on the data of the IGY. Physics of the sea. (In Russian).
Trudy, Morsk. Gidrofiz. Inst., Akad. Nauk, Ukrain. SSR, 28:112-123.

currents
Ponomarenko, G.P., 1960
[A study of the currents in the north-eastern Atlantic during the third cruise of R/S "M. Lomonosov".]
Biull. Okeanograf. Komissii, Akad. Nauk, SSSR, (6): 41-42.

currents
Ponomarienko, G.P., 1962
[The study of currents in the Atlantic during the sixth cruise of the "M. Lomonossov".]
Trudy Inst. Gidrofiz. Morsk, 25:17-47.

currents
Popovici, Z., 1936.
Einige Beobachtungen über die Strömungen an der Westküste des Schwarzen Meeres.
Bull. Acad. Roumanie 18(3/5):1-2.

currents
Postma, H., 1961.
Transport and accumulation of suspended matter in the Dutch Wadden Sea.
Netherlands J. Sea Res., 1(1/2):148-190.

currents
Postma, H., 1958.
Chemical results and a survey of water masses and currents.
Snellius Exped., Eastern Part of the East Indian Archipelago, 1929-1930, Oceanogr. Res., 2 (8):116 pp.

currents
Pova, E.A., 1946.
Problèmes de physiologie animale dans la Mer Noire. Bull. Inst. Océan., Monaco, No. 903:43 pp. 21 textfigs.

currents
Prahm, Gertrud, 1964.
Strömungen, Temperatur und Salzgehalt im Seegebiet zwischen Grönland und Labrador-Neufundland.
Handbuch für die Fischereigebiet des Nordwest Atlantischen Ozeans, 134-157.

Also in:
Oceanographie 1964, Deutsches Hydrogr. Inst.

currents
Princeton University, Department of Civil Engineering, River and Harbor Program, 1958 (21 July).
Bibliography: Currents along the Eastern Seaboard of the United States and Coast Erosion. 6 pp. (mimeographed).

currents
Pritchard, D., 1948.
Streamlines from a discrete vector field with application to ocean currents. J. Mar. Res., 7(3):296-303, 1 fold-in.

currents
Pritchard, D.W., and W.V. Burt, 1951.
An inexpensive and rapid technique for obtaining current profiles in estuarine waters.
J. Mar. Res. 10(2):180-189, 4 textfigs.

currents
Pritchard, D.W., and R.E. Kent, 1953.
The reduction and analysis data from the James River, Operation Oyster Spat. Chesapeake Bay Inst. Tech. Rept. VI, Ref. 53-12:92 pp., 11 tables.
UNPUBLISHED MANUSCRIPT.

currents
Proudman, J., 1939.
Currents in the North Channel of the Irish Sea.
Mon. Not. R. Astron. Soc., Geophys. Suppl. 4: 387-403.

currents, longshore
Putnam, J. A., W. H. Munk, and M. A. Traylor, 1949.
The prediction of longshore currents. Trans. Am. Geophys. Union, 30(3):337-345, 8 textfigs.

current
Putnam, J.A., K.J. Bermel, and J.W. Johnson, 1947
Suspended matter sampling and current observations in the vicinity of Hunters Point, San Francisco Bay. Trans. Am. Geophys. Un. 28(5):742-746, 4 text figs.

currents
Proudman, J., 1946
On the salinity of the surface waters of the Irish Sea. Phil. Trans. Roy. Soc., London, Ser.A. Math. and Phys. Sci. 239 (812):579-592.

currents
Puri, Harbans S., Gioacchino Bonaduce and John Malloy, 1964.
Ecology of the Gulf of Naples.
Pubbl. Staz. Zool., Napoli, 33(Suppl.):87-199.

R

currents (data)
Ramalho, A. de M., 1942
Observacoes oceanograficas. "Albacora", 1940. Trav. St. Biol. Mar., Lisbonne, No. 45: 96 XVI pp., 2 figs. (fold-in). (Bumpus reprint)

currents (data)
Ramalho, A. de M., L. S. Dentinho, C. A. M. de Sousa, Fronteira, F. L. Mamede, and H. Vilela, 1935.
Observacoes oceanograficas. "Albacora" 1934. Trav. St. Biol. Mar., Lisbonne, No. 35: 35 pp., 1 fig. (Bumpus reprint).

currents
Rama Sastry, A.A., 1961
Exploration of the Indian Ocean - Physical Oceanography.
Indian J. Meteorol. & Geophys., 12(2):182-188.

currents
Ramasastry, A.A., and C. Balaramamurty, 1957.
Thermal field and oceanic circulation along the east coast of India. Proc. Indian Acad. Sci., B, 46(5):293-323.

currents
Ramster, J.W., and H.W. Hill, 1969.
Current system in the northern Irish Sea.
Nature, Lond., 224(5214):59-61

currents
Rao, C. P., 1956.
Ocean currents off Visakhapatnam.
Indian J. Meteorol. Ocean., 7(4):377-379.

currents
*Rebaudi, Roberto S., 1967.
Sistema de boyas para medir corrientes sobre la plataforma submarina.
Bol. Servicio Hidrografia, Naval, Armada Argentina, 4(2):225.

currents
Redfield, A. C., 1941
The effect of the circulation of water on the distribution of the calanoid community in the Gulf of Maine. Biol. Bull., LXXX(1):86-110.

currents
Redfield, A. C., 1939
The history of a population of Limacina retroversa during its drift across the Gulf of Maine. Biol. Bull., 76:26-47.

currents
Reed, R.K., 1971.
Nontidal flow in the Aleutian Island passes. Deep-Sea Res., 18(3): 379-380.

currents
Regula, H., 1937.
Bodenwindbeobachtungen un Hohenwindmessungen auf M.S. "Schwabenland". Ann. Hydrogr., usw., Jahrg. 65(Heft 7):307-311.

currents
Reid, G.K., jr., 1955.
A summer study of the biology and ecology of East Bay, Texas.
Texas J. Sci., 7(3):316-343, 6 textfigs.

currents
Reid, R. O., 1948.
A model of the vertical structure of mass in equatorial wind-driven currents of a baroclinic ocean. J. Mar. Res. 7(3):304-312, 4 textfigs.

currents
Reid, R.O., 1948
The equatorial currents of the eastern Pacific as maintained by the stress of the wind. J. Mar. Res. 7(2):74-99, figs.6-14.

currents
Reid, Joseph L., Jr., 1962
On circulation, phosphate-phosphorus content and zooplankton volumes in the upper part of the Pacific Ocean.
Limnol. and Oceanogr., 7(3):287-306.

currents
Renner, O., 1946.
Ein neues Gerät zum Messen der Meeresströmung. G.H.I. Rept. No. 25.

currents
Revel, François 1971
Marée et courants dans les atolls polynésiens
Cah. océanogr. 23(7):593-601

currents
Revelle, R., 1950.
E.W. Scripps cruise to the Gulf of California. 5. Sedimentation and oceanography: survey of field observations. Mem. G.S.A., 43:1-6.
1940

currents
Reynolds, O., 1881-1900.
Report of the committee appointed to investigate the action of waves and currents on the beds and foreshores of estuaries by means of models. Sci. Papers, Cambridge, 2:380+ pp., 48+ figs

currents
Riccardi, R., 1946.
Lezioni di oceanografia. Ed. Perrele, 147 pp., 16 figs., 7 pls.

currents
Richter, C. M., 1887.
Ocean currents contiguous to the coast of California. Bull. Calif. Acad. Sci. 2(7):337-350
8 pls.

currents
Riley, G.A., 1952.
Hydrography of the Long Island and Block Island Sounds. Bull. Bingham Ocean. Coll. 13(3):5-39, 11 textfigs.

currents system
Riley, G.A., 1951.
Oxygen, phosphate, and nitrate in the Atlantic Ocean. Bull. Bingham Ocean. Coll. 8(1):126 pp., 33 textfigs.

currents
Riley, G. A., H. Stommel, and D. F. Bumpus, 1949
Quantitative ecology of the plankton of the western North Atlantic. Bull. Bingham Ocean. Coll. 12(3):169 pp., 39 text figs.

currents
Robinson R.A., H. Tong, and Tham Ah Kow, 1953.
A study of drift in the Malacca and Singapore Straits from salinity determinations.
Proc. Indo-Pacific Fish. Counc., 4th Meet.,:105-110.

currents
Robson, R.A., and L.V. Cox, 1956.
Radar and ocean current studies.
Proc. Inst. Radio Eng., Australia, 17:419-423.

currents
Roche, A., and A. Roubault, 1947.
Observations sur les courants superficiels de la Mer de Monaco. Bull. Inst. Océan., Monaco, No. 909:11 pp., 2 textfigs.

currents
Rochford, D.J., 1967.
The phosphate levels of the major currents of the Indian Ocean.
Aust. J. mar. Freshwat. Res., 18(1):1-22.

currents
Rochford, D.J., 1951
Studies in Australian estuarine hydrology. Australian J. Mar. and Freshwater Res. 2(1):1-116, 1 pl., 7 textfigs.

currents
Roden, Gunnar I., 1964.
Oceanographic aspects of Gulf of California.
In: Marine geology of the Gulf of California, a symposium, Amer. Assoc. Petr. Geol., Memoir, T. van Andel and G.G. Shor, Jr., editors, 3:30-58

currents
Roelofs, E.W., and D.F. Bumpus, 1953.
The hydrography of Pamlico Sound.
Bull. Mar. Sci., Gulf and Caribbean, 3(3):181-205, 11 textfigs.

currents
Romanovsky, V., 1964.
Coastal effects of the Cape Sicié sewer outfall (French Mediterranean coast west of Toulon). Air and Pollution, 8(10):557-589.
Water

currents
Romanovsky, V., 1964.
Mesures de courants dans la portion septentrionale de la mer Tyrrhénienne. Res. Sci. Camp. "Calypso".
Ann. Inst. Océanogr., Monaco, n.s., 41:289-300.

currents
Romanovsky, V., 1961.
Les courants dans la région septentrionale de la Mer Tyrrhenienne. (Resume).
Rapp. Proc. Verb., Reunions, Comm. Int. Expl. Sci., Mer Mediterranee, Monaco, 16(3): 589.

currents
Romanovsky, V., 1960
Les courants dans le détroit de Bonifacio.
Trav. C.R.E.O. 3(4): 31-37.

currents
Romanovsky, V., 1955.
Étude de la circulation littorale dans le Golfe Juan. Bull. d'Info., C.C.O.E.C., 7(3):119-126, 2 pls.

currents
Romanovsky, V., 1952.
Étude de la circulation littorale dans la Baie de Nice. Bull. d'Info., C.C.O.E.C. 4(5):181-193, 8 figs.

currents
Romanovsky, M. V., 1950.
Direct measurement of ocean currents. Hydr. Rev. 27(1):102.

Extr. Publ. 1357, Navy Hydr. Serv., Paris, 1949.

currents
Romanovsky, V., 1949.
Les marins directes des courants marins.
Ann. hydrographiques, No. 1357:34 pp., 15 textfigs.

currents
Romanovskt, V., 1949.
Les mesures directes des courants marins. Ann. Hydro. No. 1357:34 pp.

currents
Romanovsky, V., 1948.
Quelques données sur la Météorologie et l'Océanographie de l'anse de Ponteau dans le Golfe de Fos. Bull. Inst. Océan., Monaco, No. 932:16 pp., 7 textfigs.

currents
Romanovsky, V., 1948
Quelques données sur la météorologie et l'océanographie de l'anse de Ponteau dans le golfe de Fos. Bull. Inst. Océan., Monaco, No. 932, 16 pp., 7 text figs.

currents
Romanovsky, V., 1948
Quelques donnees sur la Meteorologie et l'Oceanographie de l'anse du Ponteau dans le golfe de For. Bull. l'Inst. Ocean., Monaco, No.932:16 pp., 7 text figs.

currents
Romanovsky, V., 1946
La mer à l'assaut des côtes. Peut-on empêcher la mer d'eroder les côtes et d'envaser les ports? Problèmes Éditions Elzevir 11, 63 pp., 18 text figs. 73 bis. Quai D'Orsay, Paris VII

currents
Romanovsky, Vsevolod, and Jean Le Floch, 1963
Hydrologie et courantométrie dans le Détroit de Gibraltar.
Cahiers Océanogr., C.C.O.E.C., 15(5):315-319.

currents
Rosfelder, A., 1955.
Carte provisoire au 1/500, 000 de la marge continental algérienne. Note de présentation. Publ. Service Carte Géol., Algérie, n.s., Bull. No. 5, Trav. des Coll., 1954:57-106, 6 figs., 1 pl., 1 map.

currents
Rossby, C.-G., 1951.
On the vertical and horizontal concentration of momentum in air and ocean currents. 1. Introductory comments and basic principles with particular reference to the vertical concentration of momentum in ocean currents. Tellus 3(1):15-27, 1 textfig.

currents
Rossby, C.-G., 1951?
2. A comparison of current patterns in the atmosphere and in the ocean basins.
C.R. Symposium sur la circulation générale des océans et de l'atmosphère, Assoc. Météorol., U.G.G.I., 1951:9-30, 14 textfigs.

currents
Rossby, C. -G., 1937, 1938.
On the mutual adjustment of pressure and velocity distributions in certain simple current systems. I and II. J. Mar. Res. 1:15-28, textfigs. 5-9, 1(3):239-263, figs. 84-87.

Reviewed: J. du Cons. 14(3):408-409, A. Defant.

currents
Rossby, C.G., 1936
Dynamics of steady ocean currents in the light of experimental fluid mechanics.
P.P.O.M. 5(1):1-43, 2 charts, 26 textfigs.

currents
Rossby, H. Thomas, 1969.
A vertical profile of currents near Plantagenet Bank. Deep-Sea Res., 16(4): 377-385.

currents
*Rossov, V.V., 1967.
Sobre el sistema de corrientes del Mediterráneo americano.
Estudios Inst. Oceanol., Acad. Ciencias, Cuba, 2(1): 31-49.

Rossov, V.V., 1964. currents
Observations on currents from a drifting vessel. (In Russian).
Materiali, Ribochoz. Issled. Severn. Basseina, Poliarn. Nauchno-Issled. i Proekt. Inst. Morsk. Rib. Choz. i Okeanogr. in. N.M. Knipovicha, PINRO, Gosud. Proizvodst. Komm. Ribn. Choz. SSSR 4:99-100.

currents
Rotschi, Henri, 1963.
Sur les flux d'eau en mer de Corail entre la Nouvelle-Calédonie et l(archipel des Salomon.
Comptes Rendus, Acad. Sci., Paris, 256(9):2009-2012.

currents
Rotschi, H., and L. Lemasson, 1967.
Oceanography of the Coral and Tasman seas.
Oceanogr. Mar. Biol. Ann. Rev., Harold Barnes, editor, George Allen and Unwin, Ltd., 5:49-97.

currents
Rouch, J., 1954.
Les iles flottantes. Arch. Met., Geophys., Bioklim., A, 7:528-532.

currents
Rouch, J., 1948.
Traité d'océanographie physique. Les mouvements de la mer. Payot, Paris, 413 pp., 66 textfigs.

currents
Rounsefell, George A., 1964.
Preconstruction study of the fisheries of the estuarine areas traversed by the Mississippi River-Gulf outlet project.
U.S.F.W.S. Fish. Bull., 63(2):373-393.

current
Roux, G., 1943
Les Eaux d'origine Méditeranéene dans la Région Nord-Africaine de l'Atlantique
Ann. de l'Inst. Océan., Monaco, 21(4): 171-228, 27 text figs.

currents
Roventa, Vasile, 1969
Contribution à l'étude des Courants totaux de la mer Noire.
Stud. Hidraul. 19:21-35
Abstract in: Rapp. P.-v. Reun. Commn int. Explor. scient. Mer Méditerr., 19(4): 685-686.

currents
Ruud, J. T., 1932.
On the biology of southern Euphausiidae.
Hvalrådets Skrifter No. 2:1-105, 37 text figs.

S

currents
Saelen, O.H., 1961
Preliminary report on current measurements in 1958 on the Galicia Bank west of Cape Finisterre. Contribution to Special IGY Meeting, 1959.
Cons. Perm. Int. Expl. Mer, Rapp. Proc. Verb. 149: 130-132.

currents
Saelen, O.H., 1950.
The hydrography of some fjords in northern Norway. Balsfjord, Ulfsfjord, Grøtsund, Vengsøyfjord, and Malangen. Tromsø Mus. Årshefter, Naturhist. Afd. 38, 70(1):102 pp., 10 pls., 42 textfigs.

currents
Saint-Guily, B., 1960
La dynamique des courants marins. Separatum Experentia 16(8): 6 pp.

currents
Saint-Guily, Bernard, 1959
Mesures de courant à l'ouvert de la Baie de Villefranche.
Cahiers Océan., C.C.O.E.C., 11(8):604.

currents
Saint-Guily, B., 1959.
The influence of Coriolis forces on sea currents.
Houille Blanche 14(5):556-559.

Rev. in: Appl. Mech. Rev., 13(11):#6126.

currents
Saint-Guily, B., 1958.
Remarque sur l'importance de la force de Coriolis dans les courants marins.
Bull. d'Info., C.C.O.E.C., 10(4):215-218.

currents
Saint-Guily, B., 1957.
Quelques compléments sur la théorie des courants de derive non-stationnaires. Bull. d'Info., C.C.O.E.C., 9(4):213-239.

currents
Saint-Guily, B., 1957.
Sur le coefficient vertical de turbulence dans les courants de vent et de pente.
Bull. Inst. Océan., Monaco, No. 1090:8 pp.

currents
Saint-Guily, B., 1957.
Les mouvements radiaux de Hamel lorsqu'il existe une force de Coriolis et la structure de certains courants océaniques.
C.R. Acad. Sci., Paris, 244(11):1528-1529.

currents
Saint-Guily, B., 1956.
Sur la théorie des courants marins induits par le vent. Bull. d'Info., C.C.O.E.C., 8(3):111-123, 3 pls.

currents
Saito, Y., 1952.
On the diffusion of salinity in a horizontal current field. J. Tokyo Univ. Fish. 38(3):351-361-, 1 textfig.

currents
Saito, Y., and S. Kanari, 1960
On the relations between the distribution of sea water temperature, wind-stress-curl, upwelling and sinking.
J. Tokyo Univ., Fish., 46(1/2): 1-19.

currents
Samnov, V.P. 1970.
Calculation of currents for a real ocean. (In Russian; English abstract).
Met. Gidrol. 1970(1):69-79.

currents
Sandström, J.W., 1904.
Einfluss des Windes auf die Dichte und die Bewegung des Meereswassers.
Publ. de Circ., Cons. Perm. Int. Expl. Mer, No. 18:1-6.

currents
Sandström, J.W., and B. Helland-Hansen, 1905.
On the mathematical investigation of ocean currents. Fish. Bd., Scotland, North Sea Fish. Invest. Comm., 1902-1903 (Rept. No. 1):135-163, 5 textfigs.

currents
San Feliu Lozano, 1962
Consideraciones sobre la hidrografía y el zooplancton del puerto de Castellón.
Inv. Pesq., Barcelona, 21:3-27.

currents
San Feliu, J.M., y F. Muñoz, 1967
Hidrografía y fitoplancton de las costas de Castellón, de mayo de 1965 a julio de 1966.
Investigación pesq., 31(3):419-461

currents
Santibañez, E.J., 1945.
La corriente de Humboldt en las costas de Chile.
Mar, Valparaiso, 16: 59-63; 153-160; 200-224.

Sarkisyan, A.S.

currents
Sarkisyan, A.S., 1969
On dynamics of currents in the equatorial Atlantic.
Actes Symp. Oceanogr. Ressources halieut. Atlant. trop., Abidjan, 20-28 oct. 1966, UNESCO, 23-31

currents
*Sarkisyan, A.S., 1967.
Dynamics of the non-periodic sea currents. (In Russian; English abstract).
Fisika Atmosfer. Okean., Izv. Akad. Nauk, SSSR, 3 (9):915-927.

currents
Satow, T., 1955.
Oceanographical conditions on the coastal waters west of the Sunda Archipelago. J. Shimonoseki Coll., Fish., 4(1): 1-20, 34 textfigs.

currents
Sauskan, E.M., 1966.
On the problem of current studies in the Atlantic Ocean according to data produced by multi-diurnal buoy stations. (In Russian)
Okeanologiia, Akad. Nauk, SSSR, 6(1):53-61.

currents
Scaccini, Cicatelli, Marta, 1959.
Caratteri idrodinamici e batimetrici del Golfo dell'Asinara.
Note Lab. Biol. mar. Fano, 1(13):109-120.

currents
Schachter, D., 1954.
Contribution à l'étude hydrographique et hydrologique de l'étang de Berre (Bouches-du-Rhone).
Bull. Inst. Océan., Monaco, No. 1048:20 pp.

currents
Schaefer, Milner B., 1958.
Report on the investigations of the Inter-American Tropical Tuna Commission for the Year 1957, 31.

Ref.
currents
Schakourov, P., 1959
[Measurement of the direction and speeds of currents with a ball and a submerged float.]
Trudy, Morsk Hydrophys. Inst., 15: 80-85.

currents
Schmidt-Ries, H., 1949.
Kurzgefasste Hydrographie Griechenlands.
Arch. Hydrobiol. 43(1):95-141.

currents
Schmitz, H.P., 1962.
A relation between the vectors of stress, wind and current at water surfaces and between the shearing stress and velocities at solid boundari-es.
Deutsche Hydrogr. Zeits., 15(1):23-36.

currents
Schott, G., 1942
Geographie des Atlantischen Ozeans im auftrage der Deutschen Seewarte vollständy neu bearbeitete dritte auflage. 438 pp., 141 text figs., 27 plates. C. Boysen. Hamburg.

currents, streamline
Schott, G., 1942.
Die Grundlagen einer Weltkarte der Meeresströmungen (Deutsche Admiralitätskarte Nr. 1942, 2 Blätter). Ann. Hydrogr. 70(11):329-340.

currents
Schott, G., 1935
Geographie des indischen und stillen Ozeans. Hamburg 413 pp., 37 pls., 114 textfigs.

currents
Schott, G., 1932.
Zur Ozeanographie der Hudsonbai und Hudsonstrasse. Ann. Hydrogr., usw., Jahrg. 60(Heft XI) 453-455.

current
Schott, G., 1932
The Humboldt Current in relation to Land and Sea Conditions on the Peruvian Coast. Geography, XVII (96):87-98.

currents
Schott, G., 1891
Die Meereströmungen und Temperatureverhältnisse in den Ostasiatischen Gewassern. Petermanns Geogr. Mitteilungen, XXXVII:215.

currents
Schou, A., 1945(1950).
Det marine Forland. Geografiske Studien over Danske Fladkystlandskabers dannelse og Formudvikling samt traek af disse omraaders Kulturgeografi. Folia Geogr. Danica IV (Medd. Skalling Lab. 9):236 pp., 85 textfigs., 1 fold-in.

currents
Schubert, O. von, 1944.
Ergebnisse der Strommessungen und der ozeanographischen Serienmessungen auf den beiden Ankerstationen der zweiten Teilfahrt. Wiss. Ergeb. Deutschen Nordatlantischen Exped., 1937 u. 1938, Lief. 1, Januar Beiheft, Ann. Hydr., usw., 72: 74 pp., 30 textfigs., 1 pl.

currents
Schumacher, A., 1950.
Zur kartographischen Darstellung der Oberflächenströmungen des Meeres. Mitt. Geogr. Gesellschaft, Hamburg, 49:116-133, 2 fold-ins.

currents
Schumacher, A., 1943.
Monatskarten der Oberflächenströmungen im äquatorischen und südlichen Atlantischen Ozean. Ann. Hydrog. u. Mar. Met., Apr./Juni 1943, 209-219, 4 pls.

Reviewed: Gerlands Beitr. Geophys. 60:235-236, by A. Defant.

currents
Schumacher, A., 1940.
Monatskarten der Oberflächenströmungen im Nordatlantischen Ozean (5°S bis 50°N.). Ann. Hydr 68:109-123 and 435.

currents
Schumacher, A., 1935.
Neuere Arbeiten über den Golfstrom im westlichen Atlantischen Ozean. Ann. Hydrogr., usw., Jahrg. 63(Heft 2):53-58.

currents
Schwiegger, E., 1943
Pesqueria y Oceanografia del Peru y Proposiciones para su Desarrollo Futuro: Informe Elevada a la Compañia Administradora del Guano. 1. Los Peces y la Vida en el Mer. 2. Oceanografia. 3 Proposiciones para el Futuro. 356 pp. 67 charts in text, 8 graphs, 4 tables in appendix.

current
Schweigger, E.H., 1931
Observaciones oceanográficas solre la corriente de Humboldt. Bol. Compañia Admin. Guano, VII:3-39

currents
Scruton, P.C., 1956.
Oceanography of Mississippi delta sedimentary environments. Bull. Amer. Assoc. Petr. Geol., 40(12):2864-2952.

currents
Seguin, G., 1966.
Contribution à l'etude de la biologie du plancton de surface de la baie de Dakar (Sénégal). Etude quantitative, qualitative et observations écologiques au cours d'un cycle annual.
Bull. Inst. Francais Afrique Noire., 28(1):1-90.

currents
Selvarajah, V., 1962
A study of drift in the North Malacca Strait from salinity determinations.
Indo-Pacific Fish. Council. FAO. Proc., 9th Sess. (2/3):1-6.

currents (data)
Serpoianu, G.H., 1967.
Considérations sur la pénétration des eaux méditerranéennes dans le bassin de la Mer Noire.
Hydrobiologie, Bucuresti, 8: 239-251

currents
Servant, Jean 1966.
La radioactivite de l'eau de mer.
Cah. océanogr. 18(4): 277-318.

currents
Shannon, L.V., 1966.
Hydrology of the south and west coasts of South Africa.
Investl Rep. Div. Sea Fish. Un. S. Afr., 58: 22 pp.

currents
Shapiro, N.B., 1968.
Development of westerly intensification in oceanic currents (a numerical experiment). (In Russian).
Dokl. Akad. Nauk, SSSR, 183(5):1064-1067.

currents
Sharukov, P., 1959
[Determining the velocity and direction of currents by anchored spar buoys.]
Trudy, Morskoi Gidrofiz. Inst., Akad. Nauk, SSSR, 15: 80-

currents
Sheerey, V.A., 1961.
Hydrographic features in the northeastern Pacific in the autumn-winter of 1958-1959.
Trudy Inst. Okeanol., Akad. Nauk, SSSR, 45:86-97.

currents
Shepard, Francis P., and Neil F. Marshall, 1969.
Currents in La Jolla and Scripps submarine canyons.
Science 165(3889): 177-178

currents
Shepard, F.P., R. Revelle, and R.S. Dietz, 1939
Ocean-bottom currents off the California Coast. Science 89:488-489

currents
Shesterikov, N. P., 1959.
[On the currents in the coastal area of the Davis Sea.]
Info. Biull., Sovetsk. Antarkt. Exped., No. 10:24-28.

currents
Shimomura, T., and K. Miyata, 1957.
[The oceanographical conditions of the Japan Sea and its water systems, laying stress on the summer of 1955.]
Bull. Japan Sea Reg. Fish. Res. Lab., No. 6 (General survey of the warm Tsushima Current, 1) 123-120.

currents
Shkudova, G. Ya., 1963
The calculation of oceanic currents in the Northern Atlantic. (In Russian).
Okeanologiia, Akad. Nauk, SSSR, 3(3):405-417.

currents
Shonting, David H., 1969.
Rhode Island Sound square kilometer study 1967: flow patterns and kinetic energy distribution.
J. geophys. Res., 74(13): 3386-3395.

Shtokman V.B. - Stockman W.B.

currents
Shtokman, V.B., 1954.
On the cause of circular currents near islands and contrary currents on the shores of channels.
Izvest. Akad. Nauk, SSSR, Ser. Geogr., 4:29.

currents
Shtokman, V.B., 1953.
Several questions on the dynamics of ocean currents.
Izvest. Akad. Nauk, Ser. Geofiz., SSSR, (1):69 ff.

Translated by Leslie Brown, WHOI Ref. 54-90.

ocean currents
Shtokman, V.B., 1949.
Influence of the relief of the bottom on the direction of ocean currents. Priroda 38(1):19-23.

currents
Shtokman, V.B., 1949.
Effect of bottom topography on the direction of currents/ in the sea. Priroda 38(11):10-23.

RT-2108 Bibl. Transl. Rus. Sci. Tech. Lit. 15.

currents
Shtokman, V.B., 1948.
Relationships between the field of the wind, the field of "total currents" and the mean field of masses in a non-homogeneous ocean. Dokl. Akad. Nauk, SSSR, 59(4):675-678.

currents
Shtokman, V.B., 1947.
Effect of bottom topography on the direction of currents in the sea. Priroda 11:10-23.

T57R

currents
Shtokman, V.B., 1946.
Equations for a field of flow induced by the wind in a non-homogeneous sea. Comptes Rendus (Dok.) Akad. Sci. U.S.S.R. 54:403.

current
Stockmann, W.B., 1946
Characteristic features of the coastal circulation in the sea and their connection with the transverse non-uniformity of the wind. Comptes Rendus (Doklady) de l'Acad. des Sci de L'URSS, LIV (3):227-230, 2 textfigs.

currents
Stockmann, W.B., 1946
Equations for a field of total flow induced by the wind in a non-homogeneous sea. Comptes Rendus (Doklady), Acad. des Sci de l'URSS, LIV (5):403-406.

currents
Shtokman, V.B., 1946.
On the dissipation of energy in permanent ocean currents. Trudy Inst. Okean. 1:16 pp. (p. 62).

currents
Stockman, W., 1941.
Durch wind erzeugte Zirculation und Staueffekt in einem homogenen geschlossenen nicht tiefen Meer. Bull Acad. Sci. U.S.S.R., ser. Geogr. 1: 67-87. (In Russian with German summary, 87-88)

Shuleikin, V.V.

currents
Shuleykin, V.V., 1965.
Analysis of complicated thermal conditions in the region of closed cyclic Atlantic currents. (In Russian, English abstract).
Fisika Atmosferi i Okeana, 1(4):413-425.

currents
Shuleikin, V.V., 1962.
A single characteristic of turbulent viscosity for sea waves and currents.
Doklady Akad. Nauk, SSSR, 144(4):781-784.

currents
Shuleikin, V.V., 1960.
Hydrodynamic resonance in currents of the summer monsoons.
Izv. Akad. Nauk, SSSR, Ser. Geofiz., (6):828-838.

currents
Shuleikin, V.V., 1958.
Telluric currents in the ocean and the magnetic declination.
Doklady Akad. Nauk, SSSR, 119(2):257-

currents
Shuliak, B.A., 1962
Vitesse de migration de rides et courant de particules au-dessus d'un fond à structure periodique.
Cahiers Océanogr., C.C.O.E.C., 14(7):452-466.

currents
Shuliak, B.A., 1961.
The problem of the simulation of processes in currents with solid particles and bodies.
Okeanologiia, Akad. Nauk, SSSR, 1(1):78-85.

Translation in:
Deep-Sea Res., 9(3):240-245.

currents
Skogsberg, T., 1936.
Hydrography of Monterey Bay, California. Thermal conditions, 1929-1933. Trans. Am. Phil. Soc. 29(1):1-152, 45 textfigs.

currents
Siedler, G., 1966.
Zum Mechanismus des Wasseraustausches zwischen dem Roten Meer und dem Golf Von Aden.
Z. Geophys., 32 (sonderheft):335-339.

current
Simpson, A., 1948.
Note on current determination in 60°N, 20°W. (Station "Item"). Mar. Obs., M.O. 493, 18(142): 223-224.

currents
Sisoev, N.N., 1959.
On currents in the oceans. (In Russian).
Doklady, Akad. Nauk, SSSR, 125(5):1123-1125.

currents
Smayda, Theodore J., 1966.
A quantitative analysis of the phytoplankton of the Gulf of Panama. III General ecological conditions and the phytoplankton dynamics at 8o 45'N, 79o 23'W from November 1954 to May 1957.
Inter-Amer. Trop. Tuna Comm., Bull., 11(5): 355-612.

currents
Smidt, E.L.B., 1944
Biological Studies of the Invertebrate Fauna of the Harbor of Copehagen. Vidensk. Medd. fra Dansk naturh. Foren. 107:235-316, 23 text figs.

currents
Smith, G. L., 1940
The Great Bahama Bank. 1. General hydrographical and chemical features. J. Mar. Res., 3 (2):147-170, 8 figs.

currents
Smith, E.H., F.M. Soule, and O. Mosby, 1937
The Marion and General Greene Expeditions to Davis Strait and Labrador Sea under Direction of the United States Coast Guard 1928-1931-1933-1934-1935. Scientific Results. Part 2. Physical Oceanography. C.G. Bull. 19: 259 pp., 155 figs., tables.

currents
Smith, F.G. Walton, 1946
Effect of water currents upon the attachment and growth of barnacles. Biol. Bull. 90(1): 51-70, 6 textfigs.

currents
Smith, W.E., 1946
Some observations on water-level and other phenomena along the Bosphorus. Trans. Am. Geophys. Union, 27(1):61-68, 15 textfigs.

currents
Soeriaatmadja, Rd. E., 1956.
Surface salinities in the Strait of Malacca.
Mar. Res., Indonesia, No. 2:27-48.

currents
Soika, A. Giordani, 1955
Ricerche sull'ecologia e sul popolamento della zona intercotidale delle Spiagge di Sabbia fina.
Boll. Mus. Civico de Storia Nat., Venezia, 8: 7-151.

currents
Soliankin, E.V., 1962.
Some peculiarities of the hydrology of the Danish Strait based on material from the cruise of the E.S. "Sevastopol" in the summer of 1958.
Trudy Vses. Nauchno-Issledov. Inst. Morsk. Ribn. Khos. i Okean., VNIRO, 46:74-92.

In Russian

currents
Solovjov, J., 1968.
Correlation between the atmospheric pressure field and the current in the open part of the Baltic Sea in the Klaipeda area. (In Russian)
Statii Gidrometeorol., Vilnius, 1: 179-191

current
Somerville, B.T. 1923
Ocean Passages of the World-Winds and Currents.

currents
Somma, A., 1952.
Elementi di Meteorologia ed Oceanografia. Pt. II. Oceanografia. Casa Editrice Dott. Antonio Milani, Padova, Italia, xviii 758 pp., 322 textfigs., maps, tables. (4500 lire).

currents
Soskin, I.M., I. A. Vavilov, and B.M. Rossyisky, 1960.
Use of radio-navigation system "coordinator" for the watching of currents.
Meteorol. i Gidrol., (11):35-36.

currents
Soule, F.M., 1941.
Physical oceanography, the Grand Banks region and the Labrador Sea. U.S.C.G. Bull. 30:36-56.

currents
Soule, F.M., 1938
Oceanography. Excerpt from International Ice Observation and Ice Patrol Service in the North Atlantic Ocean. Season of 1936. C.G. Bull. No.26:33-82, 58 text figs., tables of data.

currents
Soule, Floyd M., and R.M. Morse, 1960
Physical oceanography of the Grand Banks Region and the Labrador Sea in 1958. International Ice Observation and Ice Patrol Service in the North Atlantic.
U.S. Coast Guard Bull., No. 44:29-99.

currents
Sourie, R., 1954.
Contribution à l'étude écologique des côtes rocheuses du Sénégal.
Mém. Inst. Francais d'Afrique Noire No. 38: 342 pp., 23 pls.

currents
South Africa, Division of Fisheries, 1954.
Twenty-fourth annual report:1-199.

currents
Spilhaus, A. F., 1937.
Note on the flow of streams in a rotating system.
J. Mar. Res. 1(1):29-33, textfigs. 10-11.

currents
Stalcup, Marvel C. and William G. Metcalf, 1972.
Current measurements in the passages of the Lesser Antilles. J. geophys. Res. 77(6): 1032-1049.

currents
Steele, John H., 1967.
Current measurements on the Iceland-Faroe Ridge.
Deep-Sea Research, 14(4):469-473.

currents
Stefansson, Unnstein, 1962.
North Icelandic waters.
Rit Fiskideildar, Reykjavik, 3:269 pp.

ocean currents
Stephenson, T.A., A. Stephenson, and C.A. du Toit, 1937
The South African intertidal zone and its relation to ocean currents. I. A temperate Indian Ocean shore. Trans. Roy. Soc., S. Africa, 24:341-382.

currents
Stetson, H. C., 1937.
Current-measurement in the Georges Bank Canyons.
Trans. Am. Geophys. Un., 18th Meeting, 216-219, 2 textfigs.

currents
Stetson, H. C., and J. F., Smith, 1937.
Behavior of suspension currents and mud slides on the continental slope. Am. J. Sci., 5th ser., 35(205):1-13, 2 textfigs.

currents
Stewart, R.W., 1969.
The atmosphere and the ocean.
Scient.Am., 221(3):76-86. (popular).

currents
Stocchino, C., ed A. Testoni, 1969.
Le correnti nel canale di Corsica e nell' Arcipelago Toscano.
Raccolta Dati Oceanogr. Comm. ital Oceanogr. Consig. naz. Ricerche (A) (19): 1-15.

Stommel, Henry

currents
Stommel, Henry, 1960
Wind-drift near the Equator. Deep-Sea Res., 6(4):298-302.

currents
Stommel, H., 1957.
A survey of ocean current theory. Deep-Sea Res., 4(3):149-184.

currents
Stommel, H., 1951.
An elementary explanation of why ocean currents are stongest in the west. Bull. Amer. Met. Soc. 32:21-23.

currents
Stommel, H., 1951.
An elementary explanation of why ocean currents are strongest in the west. Bull. A.M.S. 32:21-23.

currents
Stommel, H., 1948
The westward intensification of wind-driven ocean currents. Trans. Am. Geophys. Union 29(2):202-206, 6 text figs.

currents
Storr, John F., 1964.
Ecology and oceanography of the coral-reef tract, Abaco Island, Bahamas.
Spec. Papers, Geol. Soc., Amer., 79:98 pp.

currents
Strazzeri, G., 1915-22.
Regime delle acque delle Yang-Tze-Kiang nel 2° semestre, 1914. Ann. Idrogr. 10:70-75.

currents
Subrahmanyan, R., 1959.
Studies on the phytoplankton of the west coast of India. 1. Quantitative and qualitative fluctuation of the total phytoplankton crop, the zooplankton crop and their interrelationship, with remarks on the magnitude of the standing crop and production of matter and their relationship to fish landings. 2. Physical and chemical factors influencing the production of phytoplankton, with remarks on the cycle of nutrients and on the relationship of the phosphate content to fish landings.
Proc. Indian Acad. Sci., 1:113-252.

currents
Sugiura, J., 1961.
On the currents south off Hokkaido in the western North Pacific. III.
Oceanogr. Mag., 12(2):123-156.

Reprinted in: Bull. Hakodate Mar. Obs., (8):123-156.

currents
Sugiura, J., 1960.
On the currents south off Hokkaido in the western North Pacific II.
Bull. Hakodate Mar. Obs., 7(4):79-97.

Reprinted from:

Oceanogr. Mag., 1960, 11(2):79-97.

currents
Suguira, J., 1959
On the currents south off Hokkaido in the western North Pacific. Oceanogr. Mag., Tokyo 11(1): 1-12.

currents
Sugiura, J., 1958.
On the Tsugaru warn current. Geophys. Mag., Tokyo, 28(3):399-410.

currents
Sündermann, Jürgen, 1966.
Ein Vergleich zwischen der analytischen und numerischen Berechnung winderzeugter Strömungen und Wasserstände in einem Modellmeer mit Anwendungen auf die Nordsee.
Mitt.Inst.Meeresk.Univ.Hamb., 4:77 pp. (mimeographed).

currents
Svansson, A., 1959
Some computations of water heights and currents in the Baltic.
Tellus, 11(2): 231-238.

currents
Svendsen, Harald 1971.
Investigation of the Norwegian coast current off Egersund September 1968.
Rept. Geophys. Inst. (A) Phys. Oceanogr. Bergen, 28: 15 pp, 13 figs (multilithed).

Sverdrup, H.U.

currents
Sverdrup, H.U., 1956.
Transport of heat by currents of the North Atlantic and North Pacific Oceans. Festskrift til Professor Bjørn Helland-Hansen fra Venner of Kolleger ved Chr. Michelsens Institutt, Bergen, 1956:226-236.

currents
Sverdrup, H.U., 1953/54.
The currents off the coast of Queen Maud Land.
Norsk Geogr. Tidskr. 14:239-249.

currents
Sverdrup, H.U., 1953.
The current of the coast of Queen Maud Land.
Medd. Norsk Polarinst. 75:

currents.
Sverdrup, H.U., 1953.
The currents off the coast of Queen Maud Land.
Norsk Geografisk Tidsskrift 14(1/4):239-249, 5 textfigs.

currents
Sverdrup. H.U. 1950.
Physical oceanography of the North Polar Sea.
Arctic 3(3):178-186, textfigs.

currents
Sverdrup, H. U., 1949.
Det ekvatoriale strømsystem. Årbok 1948. Norske Videnskaps-Akademi i Oslo:22-33.

currents
Sverdrup, H.U., 1941
Water masses and currents of the North Pacific Ocean. Science 93:436

currents
Sverdrup, H.U., 1941
The influence of bottom topography on ocean currents. pp. 66-75 in Contrib. Appl. Mechanics and Related Subjects, Theodore von Karman Anniv. Vol., Pasadena, California Inst. Tech. 337 pp.

currents
Sverdrup, H.U., 1940
The currents of the Pacific Ocean and their bearing on the climates of the coasts.
Sci. 91:273-282.

currents
Sverdrup, H.U., 1933.
Als Meeresforscher mit dem Unterseeboot "Nautilus" im Nordpolargebiet. Polarbuch:1-22.

current
Sverdrup, H.U., 1933
On vertical circulation in the ocean due to the action of the wind with application to conditions within the Antarctic Circumpolar Current. Discovery Repts., VII: 139-170

currents
Sverdrup, H. U., and R. H. Fleming, 1941.
The waters of the coast of southern California, March to July, 1937. Bull. S.I.O. 4(10):261-375, 66 textfigs.

Swallow, John C.

currents
Swallow, J.C., 1962.
Ocean circulation.
Proc. R. Soc., London, (A), 265(1322):326-328.

currents
Swallow, J.C., and J.G. Bruce, 1966.
Current measurements off the Somali coast during the southwest monsoon of 1964.
Deep-Sea Res., 13(5):861-888.

current
Sǿmme, J. D., 1933.
A possible relation between the production of animal plankton and the current-system of the sea. Am. Nat., 67:30-52.

currents
Sǿmme, J. D., 1934.
Animal plankton of the Norwegian coast waters and the open sea. 1. Production of Calanus finmarchicus (Gunner) and Calanus hyperboreus (Krǿyer) in the Lofoten area. Report on Norwegian Fishery and Marine Investigations, IV(9):3-163.

currents
Sysoev, N.N., 1953.
Procedure for the measurement of currents while drifting. Trudy Inst. Oceanolog. 7:32-326.
RT 2456 in Bibl. Transl. Rus. Sci. & Tech. 18 Lit.

T

currents
Taft, Bruce A., and John A. Knauss, 1967.
The equatorial undercurrent of the Indian Ocean as observed by the Lusiad Expedition.
Bull. Scripps Inst. Oceanogr. 9:1-163.

currents
Taguchi, K., and Y. Shoji, 1955.
The movement of water masses in the seas of West Aleutian Is. and off the east coast of Kamchatka salmon fishing ground. II. Inference after observation by the current meters in two layers.
Bull. Japan. Soc., Sci. Fish., 20(9):780-782.

currents
Tait, J.B., 1952.
Hydrography. Northern North Sea and approaches. Cons. Perm. Int. Expl. Mer, Ann. Biol. 8:94-98.

currents
Tait, J. B., 1934.
Surface drift bottle results in relation to temperature, salinity and density distributions in the northern North Sea. Rapp. Proc. Verb. 89(3):69-79, 9 textfigs.

currents
Tait, J.B., 1928.
Hydrographical survey of the Moray Firth. Rapp. Proc. Verb., Cons. Perm. Int. Expl. Mer, 52:48-57, 3 figs.

currents
*Takahashi, Tadao, and Masaaki Chaen, 1967.
Oceanic conditions near the Ryukyu islands in summer of 1965.
Mem.Fac.Fish., Kagoshima Univ., 16:63-75.

currents
Takano, Kenzo, 1965.
Courants marins induits par le vent et la non-uniformité de la densité de l'eau superficielle dans un océan.
La Mer: Bull. Soc. franco-japon. Océanogr., 2(2): 81-86.
Also in:
Geophys. Notes, 18(1)(2). 1965. Tokyo.

currents
Takano, K., 1954.
On the velocity distribution off the mouth of a river. J. Ocean. Soc., Japan, 10(2):60-64, 7 textfigs.

currents
Taning, A.V., 1931.
Drift-bottle experiments in Icelandic waters. Rapp. Proc. Verb., Cons. Perm. Int. Expl. Mer 72:3-20, 3 textfigs.

currents
*Tanioka, Katsumi, 1968.
On the East Korean Warm Current (Tosen Warm Current).
Oceanogrl Mag., 20(1):31-38.

currents
Tannehill, I.R., 1945.
Weather around the world. Princeton Univ. Press, 200 pp., 55 textfigs.

currents
Tantsura, A. I., 1959.
About the currents in the Barents Sea.
Trudy Pol. Nauch. Issle. Inst. Moskogo, 11:35-53.

currents
Tapager, J.R.D., 1955.
Local environmental factors affecting ice formation in Sǿndre Strǿmfjord, Greenland.
U.S.H.O. Tech. Rept., TR-22:26 pp.

currents
Tapanes, Juan J., 1963
Afloramiento y corrientes cercanas a Cuba. 1. Contrib., Centro Invest. Pesqueras, Cons. Nac. Pesca, No. 17: 1-29.

currents
Tareev, B.A., 1965.
Unstable Rossby waves and nonstationarity of the oceanic currents. (In Russian, English abstract).
Fisika Atmosferi i Okeana, 1(4):426-438.

Currents
Tchaplyguine, E. I., 1961. ref.
[The importance of the heat of the currents in the Kara Sea.]
Probl. Ark. i Antarkt., (8):19-28.

currents
Templeman, Wilfred, 1966.
Marine resources of Newfoundland.
Bull. Fish. Res. Bd., Can., 154: 170 pp.

currents
Templeman, W., and S. N. Tibbo, 1945.
Lobster investigations in Newfoundland 1938 to 1941. Res. Bull. (Fisheries), No. 16:98 pp., 20 textfigs.

currents
Theisen, E. 1946
Tanafjorden. Enfinmarksfjords oceanografi. Fiskeridirektoratets skrifter Ser. Havundersǿkelser. (Repts. on Norwegian Fishery and Marine Investigations) VIII (8):1-77, 23 text figs., 8 pls.

currents
Thiel, B., 1943.
Einiges über die Ergebnisse von Strombeobachtungen in der westlichen Ostsee. Ann. Hydrogr. 71(4/6), 226-231, Fig. 2, Tab. 2.

currents
Thiel, R., 1942.
Die Bennung der Wind und Meeresströmungen nach ihren Richtungen. Geogr. Anz. 43:29.

currents
Thiel, G., 1938.
Strombeobachtungen in der westlichen Ostsee im Juli 1936. Arch. Deutschen Seewarte 58(7):1-28, 5 textfigs., 7 pls.

currents
Thiel, G., 1937.
Über Stromkonvergenslinien und Kabbelungen. Ann. Hydrogr., usw., Jahrg. 65(Heft 3):109-110, f: figs.

currents
*Thomas, Robert W., 1968.
Oceanography survey results off Point Arguello, California, January and November-December 1964.
Tech.Rept., U.S.Naval Oceanogr.Off., TR-201:53 pp. (multilithed).

currents.
Thomsen, H., 1954.
Danish waters.
Ann. Biol., Cons. Perm. Int. Expl. Mer, 10:121.

currents
Thomsen, H., 1953.
Hydrography. Danish waters.
Ann. Biol., Cons. Perm. Int. Expl. Mer, 9:145-146.

currents
Thomsen, H., 1952.
Hydrography. Danish waters. Cons. Perm. Int. Expl. Mer, Ann. Biol. 8:118-119.

currents
Thomsen, H., 1950.
Hydrography. 1. Kattegat and Belts. Ann. Biol. 6:143.

currents
Thomsen, H., 1949
Hydrography. Kattegat and Belts. Ann. Biol., Int. Cons., 4:125-126.

currents (data)
Thomsen, H., 1947.
Hydrografiske undersǿgelser i Limfjorden 1943. 1. Observationsmaterialet. Publ. Danske Met. Inst Medd. No. 9:104 pp., 1 fold-in, 6 textfigs.

current
Thomsen, H., 1939.
Note on the secular variation of the current at the Light Vessel "Horns Rev" and "Vyl". J. du Cons. 14(3):400a-400d, 1 textfig.

currents
Thomsen, H., 1938.
Hydrography of Icelandic waters. Zool. Iceland 1(4):36 pp., 1 fold-in, 22 textfigs.

currents

Thomsen, H., 1934.
Danish hydrographical investigations in the Denmark Strait and the Irminger Sea during the years 1931, 1932 and 1933. Rapp. Proc. Verb., Cons. Perm. Int. Expl. Mer 86:1-14, 15 textfigs.

currents

Thorade, H., 1941.
Der Äquatoriale Gegenstrom im Atlantischen Ozean und seine Entstehung. Ann. Hydr. 69:210-209.

currents

Thorade, H., 1909.
Über die Kalifornische Meeresströmung. Oberflachentemperaturen und Strömungen an der Westküste Nordamerikas. Inaug. Diss. 31 pp., 3 pls., Göttingen.

currents

Thoulet, J., 1927.
Densimétrie en mer Tyrrheniene. Bull. Inst. Océan., Monaco, No. 492:8 pp.

currents

Thoulet, J., 1927.
Densimétrie et volcanicité abyssale dans le Pacifique. Ann. Inst. Océan., n.s., 4(2):25-45.

currents

Thoulet, J., 1926.
Essai d'une densimétrie des océans. Ann. Inst. Océan., n.s., 3(3):137-160, figs.

currents

Tickner, E.G., 1961

Transient wind tides in shallow water.
Beach Erosion Bd., Tech. Memo. No. 123: 1-48.

currents

Timonov, V.V., 1961
A study of the variations of the hydrometeorological conditions in the Northern Atlantic in relation to the IGY and IGC.
Okeanologiia, Akad. Nauk, SSSR (2):220-225.

currents

Timonov, V.V., 1960.
Resultant and secondary currents in seas with tides.
Trudy Okeanograf. Komissii, Akad. Nauk, SSSR, 10(1):43-44.

currents

Timonov V.V., and I.I. Soskin, Ed., 1955.
Collected works on methods of studying ocean currents and tidal phenomena.
Trudy Gosud. Okean. Inst., 30(42):1-290.

currents

Todeschini, B., 1957.
Sulle correnti piane poco rotazionali.
Atti Accad. Naz. Lincei, 22(2):167-171.

currents

Tokioka, T., 1940.
The chaetognath fauna of the waters of western Japan. Rec. Oceanogr. Wks., Japan, 12(1):1-22, 3 charts, 4 tables. (Contrib. 87, Seto M.B.L.).

Tolmasin
Tolmazin D.M.

currents

Tolmazin, D.M., 1962
The current field and water exchange in the Bosporus.
Okeanologiia, Akad. Nauk, SSSR, 2(1):44-50.
Abstracted in Soviet Bloc Res., Geophys., Astron. and Space, 1962(35):16.
(OTS61-11147-35 JPRS13739)

currents

Tolmazin, D.M., and V.A. Shnaidman, 1968.
Calculations of integral circulation and currents in the northwestern Black Sea. (In Russian; English abstract).
Fizika Atmosfer. Okean., Izv. Akad. Nauk, USSR, 4(6):634-646.

Tomczak, Gerhard

currents

Tomczak Gerhard, 1968.
Die Wassermassenverteilung und Strömungsverhältnisse am Westausgang des Skagerraks während der internationalen Skagerrak-Expedition im Sommer 1966.
Dt. hydrogr. Z. 21(3): 97-105

currents

Tomczak, G., 1961
Ergebnisse der ozeanographischen Forschung im Internationalen Geophysikalischen Jahr.
II. Topographie des Meeresbodens - Die grossen Meereströmungen im Pazifischen Ozean Weiteres Forschungsziel: Indischer Ozean.
Umschau, (18):570-572.

currents

*Tooma, Samuel G., Jr., and Harry Iredale II, 1968.
Oceanography in the Channel Islands area off southern California, September and October 1965.
Tech.Rept., U.S.Naval Oceanogr.Off., TR-203:50 pp. (multilithed).

currents

Trabin, V.I., 1961.
Fishing in the northwest Atlantic.
Murmanskoe Knizhnoe Izdatel, 1961, 93 pp.

currents

Tregouboff, G., 1961
Rapport sur les travaux intéressant la planctonologie méditerranéene publiés entre juillet 1958 et octobre 1960.
Rapp. Proc. Verb., Réunions, Comm. Int. Expl. Sci. Mer Méditerranée, Monaco, 16(2):33-68.

currents

Tregouboff, G., 1956.

Rapport sur les travaux concernant le plancton Méditerranéen publiés entre Novembre 1952 et Novembre 1954.
Rapp. Proc. Verb., Comm. Int. Expl. Sci., Mer Méditerranée, 13:65-100

currents

Tressler, Willis L., 1960.
Oceanographic observations at IGY Wilkes station Antarctica.
Trans., Amer. Geophys. Union, 41(1):98-104.

currents

Tressler, Willis L., 1960
Oceanographic and hydrographic observations at Wilkes IGY Station, Antarctica.
J. Washington Acad. Sci., 50(5):1-13.

currents

Trites, R.W., 1956.
The oceanography of Chatham Sound, British Columbia. J. Fish. Res. Bd., Canada, 13(3):385-434.

currents

Tromeur, J.Y., 1938-39.
Mission hydrographique du Saloum, 1930-1931.
Ann. Hydrogr., Paris, (3)16:5-33, figs.

currents

Tsujita, T., 1954.
On the observed oceanographic structure and the development of stratified water currents in the Straits of Tsushima and the Sea of Toto-Amakusa in the western Japan.
Bull. Seikai Reg. Fish. Res. Lab., Nagasaki, 1:

currents

Tsujita, T., 1953.
A marine ecological study on the Bay of Omura.
J. Ocean. Soc., Japan, 9(1):23--32, 6 textfigs.

currents

Tully, J.P., 1964.
Oceanographic regions and processes in the seasonal zone of the North Pacific Ocean.
In: Studies on Oceanography dedicated to Professor Hidaka in commemoration of his sixtieth birthday, 68-84.

currents

Tully, J.P., and L.F. Giovando, 1963
Seasonal temperature structure in the eastern Subarctic Pacific Ocean.
In: Marine distributions, M.J. Dunbar, Editor.
R. Soc., Canada, Spec. Publ., No. 5:10-36.

U

currents

Uda, Michitake, 1963.
Oceanography of the subarctic Pacific Ocean.
J. Fish. Res. Bd., Canada, 20(1):119-179.

currents

Uda, Michitaka, 1962
Subarctic oceanography in relation to whaling and salmon fisheries.
Sci. Repts., Whales Res. Inst., No. 16: 105-119.

currents

Uda, M., 1955.
On the subtropical convergence and the currents in the northwestern Pacific.
Rec. Ocean. Wks., Japan, 2(1):141-150, 10 textfigs.

currents

Uda, M., 1952.
On the fluctuation of oceanic current. (Second report). Drift current in Japan Sea, Yellow Sea, and East China Sea. Bull. Tokai Regional Fish. Res. Lab., Fish. Agency, (Contrib. B) No. 3: 55-69, 4 textfigs.

currents

Uda, M., 1949
On the correlated fluctuation of the Kuroshio Current and the cold water mass.
Ocean. Mag., Japan 1(1):1-12, 8 text figs.

currents

Uda, M., ?
On the fluctuation of oceanic current. J. Ocean. Soc., Japan, (Nippon Kaiyo Gakkaisi) 5(2/4): 55-69, 4 figs.

currents

Uda, Michitaka, and Makoto Ishino, 1960
Researches on the currents of Kuroshio.
Rec. Oceanogr. Wks., Japan, Spec. No. 4: 59-72.

currents
U.S. Coast and Geodetic Survey, 1950.
Manual of current observations. Spec. Publ. No. 215 (Revised(1950) edition):87 pp., 34 textfigs.

currents
USN Hydrographic Office, 1957.
Oceanographic atlas of the Polar seas. 1. Antarctic. H.O. Pub., No. 705:70 pp.

currents
U.S.N. Hydrographic Office, 1958
Oceanographic atlas of the Polar seas. Pt. II. Arctic. H.O. Publ. No. 705: 139 pp., charts.

currents
*Urien, Carlos Maria, 1967.
Los sedimentos modernos del Rio de la Plata exterior.
Bol. Servicio Hidrografia Naval, Armada Argentina, 4(2):113-213.

V

currents
Vallaux, C., 1938.
Recherches du Discovery II sur la dynamique de l'Océan Austral. Bull. Inst. Océan., Monaco, 751:1-15.

currents
Van Dijk, W., 1956.
An investigation of the vergence field of the wind and ocean currents in the Indian Ocean. Arch. Met. Geophys. Bioklim., A, 9(1):158-177.

currents
Van Goethem, C., 1951.
Étude physique et chimique du milieu marin. Rés. Sci., Expéd. Océan. Belge dans les Eaux Côtières Africaines de l'Atlantique Sud (1948-1949) 1:1-151, 1 pl.

current
van Riel, P.M., 1939
Current measurements in the southern North Sea. Cons. Perm. Internat. p. l'Explor. de la Mer. Rapp. et Proc. - Verb. 109(3):120.

currents
Van Roosendaal, A.M., and C.H. Wind, 1905.
Pruefung von Strommessern und Strommessungsversuche in der Nordsee.
Publ. de Circ., Cons. Perm. Int. Expl. Mer, No. 26:1-10, 2 pls.

currents
Van Straaten, L.M.J.H., 1950.
Giant ripples in tidal channels. Waddensymposium, Tijdschr. Kon. Nederl. Aardrijkskundig Genootschap 78-81, 3 figs., 2 photos.

currents
VanVeen, J., 1947.
Analogie entre marees et courants alternatifs. Houille Blanche 2:401-416.

currents
Varlet, F., 1958
Les traits essentials du régime côtise de l'Atlantique près d'Abidjan (Côte d'Ivoire). Bull. d'I.F.A.N., (A), 20(4):1089-1102.

currents
Varlet, F. and M. Menaché, 1947
L'Estuaire du Moros a Concarneau (Finistère). Etude du Mélange des eaux douces et salées. Bull. Inst. Ocean., No.917, 26 pp., 21 text figs.

currents
Vasiliev, G.D., y Iu. A. Torin, 1965.
Carastics oceanografica pesquera y biologica del Golfo de México y del Mar Caribe. (In Russian; Spanish abstract).
Sovetsk.-Cub. Ribokhoz. Issled., VNIRO:Tsentr. Ribokhoz. Issled. Natsional. Inst. Ribolovsta Republ. Cuba, 241-266.

Vercelli, F.

currents
Vercelli, F., 1927.
Ricerche di oceanografia fisica. 1. Correnti e marea. Ann. Idrogr., 11:13-208.

Transl. cited: Fish.
USFWS Spec. Sci. Rept., 227.

currents
Vercelli, F., 1925.
Pt. 1. Il regime delle correnti e delle maree nello Stretto di Messina. Crociere per lo studio dei fenomeni nello Stretto di Messina. (Campagne delle R. Nave "Marsigli" negli anni 1922 e 1923) Comm. Int. Medit., Delag. Ital., 209 pp., 56 fig

currents
Vercelli, F., 1923-24.
Ricerche di oceanografia fisica. Pt. 1. Correnti e maree. Ann. Idrogr. 11:13-208, charts, tables, diagrams.

currents
Vercelli, F., and M. Picotti, 1926.
Pt. 2. Il regime fisico-chemico della acque nello Stretto di Messina. Crociere per lo studio dei fenomeni nello Stretto di Messina. (Campagne della R. Nave "Marsigli" negli anni 1922 e 1923). Comm. Int. Medit., Delag. Ital., 161 pp., 40 figs

Currents
Verploegh, G., 1960.
On the annual variation of climatic elements of the Indian Ocean.
K. Nederlands Meteorol. Inst., Mededelingen en Verhandl., 77 (1):1-64, (2) 28 charts.

currents
Vila, Fernando, 1966.
Tecnicas de medicion de corrientes aplicadas en le campena Drake III
Bol. Serv. Hidrogr. Naval, Buenas Aires (Publ. H. 106) 3(1):5-8.

Currents
Visser, M. P., 1966.
Note on the estimation of eddy diffusivity from salinity and current observations.
Netherlands J. Sea Res., 3(1):21-22.

currents
Volokhonsky, L.V., 1964
A theory for current studies in shallow seas. (In Russian).
Okeanologiia, Akad. Nauk. SSSR, 4(5):921-922.

W

currents
Waldichuk, Michael, 1965.
Water exchange in Port Moody, British Columbia, and its effect on waste disposal.
J. Fish. Res. Bd., Canada, 22(3):801-822.

currents
Waldichuk, Michael, 1964.
Dispersion of kraft-mill effluent from submarine diffuser in Stuart Channel, British Columbia. J. Fish. Res. Bd., Canada, 21(5):1289-1316.

currents
Waldichuk, M., and J.P. Tully, 1953.
Pollution study in Nainaimo Harbour. Prog. Rept., Pacific Coast Stas., Fish. Res. Bd., Canada, No. 97:14-17, 4 textfigs.

currents
Walford, L. A., 1938
Effect of currents on distribution and survival of the eggs and larvae of the haddock (Melanogrammus aeglefinus) on Georges Bank.
Bull. U.S.Bur. Fish., 49:1-73

currents
Walton, William R., 1964.
Ecology of benthonic Foraminifera in the Tampa-Sarasota Bay area, Florida.
In: Papers in Marine Geology, R.L. Miller, Editor, Macmillan Co., N.Y., 429-454.

Warren, Bruce A.

currents
*Warren, Bruce A., 1967.
Notes on translatory movement of rings of current with application to Gulf Stream eddies. Deep-Sea Res., 14(5):505-524.

currents
Warren, Bruce, Henry Stommel and J.C. Swallow, 1966.
Water masses and patterns of flow in the Somali Basin during the southwest monsoon of 1964. Deep-Sea Res., 13(5):825-860.

currents
Watanabe, Kantaro, 1963.
On the reinforcement of the East Sakhalin Current preceding to the sea ice season off the coast of Hokkaido. Study on sea ice in the Okhotsk Sea.
Oceanogr. Mag., Tokyo, 14(2):117-130.

Also in:
Bull. Hakodate Mar. Obs., (10).

currents (data)
*Watanabe, Nobuo, 1967.
Estimation of chlorinity change due to effluent from a hydraulic power plant. (In Japanese; English abstract).
J.Coll. mar.Sci.Techn., Tokai Univ., (2):1-19.

currents
Watanabe, N., 1954.
A report on oceanographical investigations in the salmon fishing grounds of the North Pacific, 1952 and 1953.
Tokai Regional Fish. Res. Lab., Spec. Publ., No. 3:1-5, 1 fig.

currents
Watson, E.E., 1936.
Mixing and residual currents in tidal waters as illustrated in the Bay of Fundy. J. Biol. Bd., Canada, 2(2):141-208, tables, 26 textfigs.

CURRENTS
Wattenberg, H., 1949. (died 1944)
Die Salzgehaltsverteilung in der Kieler Bucht und ihre Abhängigkeit von Strom- und Wetterlage. Kieler Meeresforschungen 6:17-30, 10 textfigs.

currents
Webster, Ferris, 1967.
A scheme for sampling deep-sea currents from moored buoys.
Trans. 2nd Int. Buoy Techn. Symp., 419-431.

currents
Webster, Ferris, 1963.
A preliminary analysis of some Richardson current meter records.
Deep-Sea Res., 10(4):389-396.

currents
Webster, Ferris, 1961
The effect of meanders on the kinetic energy balance of the Gulf Stream.
Tellus, 13(3):392-401.

currents
Weenink, M.P.H., and P. Groen, 1952.
On the computation of ocean surface current velocities in the equatorial regions from wind data.
Proc. K. Ned. Akad. Wet., B, 55:239-246.

Abstr. Appl. Mech. Rev. 6:166 (1953).

currents
Weidemann, H., 1961
Results of current measurements by towed electrodes in the northern North Atlantic during the IGY. Contribution to Special IGY Meeting, 1959.
Cons. Perm. Int. Expl. Mer, Rapp. Proc. Verb. 149:115-117.

currents
Weidemann, Hartwig, 1959
Strommessungen vom fahrenden Schiff auf F.F.S. "Anton Dohrn" und V.F.S. "Gauss" während der Fahrten im Internationalen Geophysikalischen Jahr 1958. Deutschen Hydrogr. Zeits., Ergänzungsheft Reihe B (4°), No. 3: 59-66. Also in: Deutsches Hydrogr. Inst., Ozeanogr., 1959(1960)

currents
Weidemann, H., 1955.
Strömungsuntersuchungen in Fehmarnsund (Ostsee).
Deutsche Hydrogr. Zeits. 8(3):89-102.

Currents
Weiler, H.S., 1965.
Current measurements in Lake Ontario in 1967.
Proc. 11th Conf. Great Lakes Res. 1968:500-511

currents
Werenskiold, W., 1935.
Coastal currents. Geofys. Publ. 10(13):14 pp.

currents
Wiborg, K.F., 1948
Investigations on cod larvae in the coastal waters of northern Norway. Occurrence of cod larvae, and occurrence of food organisms in the stomach contents and in the sea. Repts. Norwegian Fishery and Marine Invest., 9(3):26 pp., 7 text figs.

currents
Wicker, C.F., and O. Rosenzweig, 1950.
Theories of tidal hydraulics. Evaluation of present state of knowledge of factors affecting tidal hydraulics and related phenomena. Comm. on Hydraulics, Corps Eng., U.S.A., Rept. No. 1:

currents
Wiegel, Robert L., 1964.
Oceanographical engineering.
Prentice-Hall Series in Fluid Mechanics, 532 pp.

currents
Wiegel, R.L., 1953.
Waves, tides, currents and beaches: glossary of terms and list of standard symbols.
Council on Wave Research, Berkeley, California, 113 pp.

currents
Williams, H.F., 1951.
The Gulf of Mexico adjacent to Texas. Texas J. Sci. 3(2):237-250, 11 textfigs.

currents
Williamson, Gordon R., 1970.
Hydrography and weather of the Hong Kong fishing grounds.
Hong Kong Fish. Bull. 1:43-49, 71 figs.

currents
Wilson, A.M., and H. Thompson, 1934.
III. Hydrographic and biological investigations. A. Notes on conditions in 1933. Ann. Rept., Newfoundland Fish. Res. Comm., 2(2):40-47, Text-figs. 4-5.
physical

currents
Wilson, A.M., and H. Thompson, 1933.
III. Hydrographic and biological investigations. A. Notes on the physical conditions in 1932. Ann. Rept., Newfoundland Fish. Res. Comm. 2(1): 49-58, Textfigs. 12-15.

currents
Wilson, B.W., 1963.
Strong currents of the deep ocean.
Dock & Harbour Authority, 44(511):16-20.

currents
Worthington, L. Valentine, 1964.
Anomalous conditions in the slope water area in 1959.
J. Fish. Res. Bd., Canada, 21(2):327-333.

currents
Wulfson, N. I., 1958.
Statistical methods for determining the actual parameters of convection-currents from the observed parameters.
Bull. (Izv.) Acad. Sci. USSR, Geoph. Ser. 7(Eng. Ed.): 498-506.

currents
Wüst, Georg, 1963
On the stratification and the circulation in the cold water sphere of the Antillean-Caribbean basins.
Deep-Sea Res., 10(3):165-187.

currents
Wüst, Georg, 1961.
On the vertical circulation of the Mediterranean Sea.
J. Geophys. Res., 66(10):3261-3272.

currents
Wüst, G., 1938.
Bodentemperatur und Bodenstrom in der atlantischen, indischen, und pazifischen Tiefsee.
Gerlands Beiträge zur Geophysik 54(1)1-8, 1 map.

currents
Wüst, G., 1937
Bodentemperatur und Bodenstrom in der Pazifischen Tiefsee. Veröffentlichen des Instituts für Meereskunde, n.f., A. Geographisch-naturwiss. Reihe, 35:56 pp., 19 textfigs., 3 pls.

currents
Wyrtki, K., 1962.
Geopotential topographies and associated circulation in the western South Pacific Ocean.
Australian J. Mar. Freshwater Res., 13(2):89-105.

Wyrtki, K., 1960 currents
On the presentation of surface currents. Int. Hydrogr. Rev., 37(1):111-128.

currents
Wyrtki, Klaus, 1960
The Antarctic Circumpolar Current and the Antarctic Polar Front. Deutsche Hydrogr. Zeits., 13(4): 153-174.

currents
Wyrtki, K., 1954.
Die Dynamik der Wasserbewegungen im Fehmarnbelt. II. Kieler Meeresf. 10(2):162-181, Figs. 11-17.

currents
Wyrtki, K., 1954.
Der gross Salzeinbruch in der Ostsee im November und Dezember 1951. Kieler Meeresf. 10(1):19-25, Textfigs. 12-15.

currents
Wyrtki, K., 1952.
Der Einfluss des Windes auf die Wasserbewegung durch die Strasse von Dover.
Deutsch Hydrogr. Zeits. 5(1):21-27, 8 textfigs.

Reviewed: J. du Cons. 19(1):95-96 by David Vaux
Translation: Fish Lab., Aberdeen

currents
*Yada, Shigeaki, Shigeo Abe, Shoroku Inoue, and Yusho Akishige, 1967.
Survey on summer tuna long-line fishing grounds in the southern water of Java. (In Japanese; English abstract).
Bull. Fac. Fish., Nagasaki Univ., 23: 217-222.

currents
Yamagiwa, T., 1951.
Variation of the flow in the intermediate water at the fixed point 153°E., 39°N. J. Ocean. Soc., Japan, (Nippon Kaiyo Gakkaisi), 6(3):160-164, 4 textfigs.

currents
Yamazi, I., 1955.
Plankton investigations in inlet waters along the coast of Japan. VIII. The plankton of Miyazu Bay in relation to water movement.
Publ. Seto Mar. Biol. Lab., 4(2/3):269-284, 16 textfigs.

currents
Yamazi, I., 1955.
Plankton investigations in inlet waters along the coast of Japan. XVI. The plankton of Tokyo Bay in relation to the water movement.
Publ. Seto Mar. Biol. Lab., 4(2/3):285-309, Pls. 19-20, 22 textfigs.

currents
Yaragov, B. A., 1959.
East-Icelandic current (problems on the formation and transformation of water masses).
Trudy Pol. Nauch. Issle. Inst. Moskogo, 11:94-105.

currents
Yasui, Z., and K. Hata, 1960.
On the seasonal variations of the sea conditions in the Tsugaru Warm Current region.
Bull. Hakodate Mar. Obs., 7(1):1-10.
Mem. Kobe Mar. Obs., 14:3-12.

currents
Yasui, Z., and M. Matuno, 1953.
On the circulation of currents in China Eastern Sea. Rec. Ocean. Wks., Japan, n.s., 1(1):28-35, 3 textfigs.

currents
Yeskin, L.I., 1964.
"Permanent" currents in Leningrad Bay (according to instrumental data).(In Russian).
Sovetsk. Antarkt. Eksped., Inform. Biull., 49:19-
Translation:
Scripta Tecnica, for AGU, 5(4):232-234. 1965.

currents
Yih, C.S., 1957.
Fonctions de courant dans les écoulements à trois dimensions. La Houille Blanche (3):

currents
Yoshida, S., 1953.
Oceanographic observations on the eastward area off Honsyu in July and August 1952. Publ. No. 981 Hydrogr. Bull. No. 35:56-59, 1 fold-in of 3 figs.

currents
Yoshida, K., 1951.
Dynamics on the co-existent system of waves and currents in the ocean. J. Ocean. Soc., Japan, 7(3/4):97-104.
Pt. 1.
Geophys. Notes 5(1), Contrib. 11, 1952.

currents
Zenkovich, V.P., 1948.
[Currents of coastal deposition of the Caucasus littoral of the Black Sea.] Dok. Akad. Nauk SSSR, 60:263-266.

currents
Zenkovich, V.P., 1946
The process of bottom drifting. CR (Doklady) Acad. Sci. URSS LIII (5):421-423.

currents
Zhukov, L.A., 1961
[On the advection of heat by the currents in the upper water layer of the Atlantic Ocean.] Issled. Severnoi Chasti Atlanticheskogo Okeana, Mezhd. Geofiz. God, Leningradskii Gidrometeorol Inst., 1:38-42.

currents
Zoppritz, K., 1878.
Hydrodynamic problems in reference to the theory of ocean currents. Phil. Mag., ser. 5, 6:192-211.

currents
Zore-Armanda, Mira 1966.
The system of currents found at a control station in middle Adriatic.
Acta adriat. 10(11):1-19.

currents
Zore-Armanda, Mira, 1964.
Results of direct current measurements in the Adriatic. (In Jguoslavian; English resume).
Acta Adriatica, 11(1): 293-308.

currents
Zore, Mlle. M., 1961.
Deep water movements in the Adriatic (a preliminary account).
Rapp. Proc. Verb., Réunions, Comm. Int. Expl. Sci. Mer Méditerranée, Monaco, 16(3):625-630.

currents
Zore, M. (Mlle), 1960
Preliminary results of direct current measurements in Adriatic. Comm. Int. Expl. Sci. Mer Méditerranée, Rapp. Proc. Verb. Reunions, Monaco, 15(3): 241-246.

currents, abyssal

currents, abyssal
Bennett, E.B., 1959.
Some oceanographic features of the northeast Pacific Ocean during August 1955. J. Fish. Res. Bd., Canada, 16(5): 565-633.

currents, abyssal
Latun, V.S., 1964
On the dynamics of abyssal currents beneath the layer of no motion. (In Russian).
Okeanologiia, Akad. Nauk, SSSR, 4(5):913.

currents, abyssal
Pyzkin, Iu. G., A.A. Pivovarov and G.G. Khundzhua 1968.
Abyssal bottom currents in the Black Sea. (In Russian).
Dokl. Akad. Nauk SSSR 179 (3): 585-588.

currents, alongshore (effect of)
Hsueh, Y. and James J. O'Brien, 1971.
Steady coastal upwelling induced by an alongshore current. J. phys. Oceanogr. 1(3): 180-186

currents, ancient

currents, ancient
Stehli, F.G., 1965.
Paleontologic technique for defining ancient ocean currents.
Science, 148(3672):943-946.

current anomalies

current anomalies
Nakai, Zinziro, Shigemasa Hattori, Koji Honjo, Taisuke Watanabe, Takashi Kidachi, Takashi Okutani, Hideya Suzuki, Shigeichi Hayashi, Masao Hayaishi, Keiichi Kondo, and Shuzo Usami, 1964.
Preliminary report on marine biological anomalies on the Pacific coast of Japan in early months of 1963, with reference to oceanographic conditions.
Bull. Tokai Reg. Fish. Res. Lab., No. 38:57-75.

currents, aperiodic

currents, aperiodic
Kudryavtsev, N.F., 1964.
Experience in compiling aperiodic currents in the sea from data on coastal level observations. (In Russian).
Problemy Arktiki i Antarktiki, 16:83-87.

currents, assisting

currents, assisting
Francis, J.R.D., and C.R. Dudgeon, 1967.
An experimental study of wind-generated waves on a water current.
Q. Jl. R. Met. Soc., 93(396):247-

current axis

current axis
Endo, H., 1961
[On the correlation between the surface water temperature and current axis in the Kurosio region.]
Hydrogr. Bull. Maritime Safety Bd., Tokyo, (Publ. 981, No. 65: 42-47.

currents, baroclinic boundary
Csanady, G.T., 1971.
Baroclinic boundary currents and long edge-waves in basins with sloping shores. J. phys. Oceanogr., 1(2): 92-104.

currents, bottom

currents, bottom
Bartolini, C., and C.E. Gehin, 1970.
Evidence of sedimentation by gravity-assisted bottom currents in the Mediterranean Sea.
Marine Geol. 9 (2): M1-M5.

currents, bottom
Bourcier, M., 1968.
Étude du benthos du plateau continental de la baie de Cassis.
Rec. Trav. Sta. mar. Endoume, 44(60): 63-108

currents, bottom
Carruthers, J.N., 1963.
History, sand waves and near-bed currents of La Chapelle Bank.
Nature, 197(4871):942-946.

currents, bottom (residual)
Dietrich, G., 1956.
Ergebnisse synoptischen ozeanographischen Arbeiten in der Nordsee. Deutsche Hydrogr. Inst., Ozeanogr., 1956(23):376-383.

residual bottom current
Dietrich, G., 1955.
Ergebnisse synoptischer ozeanographischer Arbeiten in der Nordsee. Deutscher Geographentag, Hamburg, 1-5 August 1955, Tangungsbericht und Wissenschaftliche Abhandlungen:376-383.

currents, bottom
Emery, K.O. and David A. Ross 1968.
Topography and sediments of a small area of the continental slope south of Martha's Vineyard
Deep-Sea Res. 15(4): 415-422.

currents, bottom
Fenner, Peter, Gilbert Kelling and Daniel J. Stanley, 1971
Bottom currents in Wilmington Submarine Canyon.
Nature, Lond. 229(2): 52-54.

currents, bottom
Fliegel, Myron and Ali A. Nowroozi, 1970.
Tides and bottom currents off the coast of northern California. Limnol. Oceanogr., 15(4): 615-624.

currents, bottom
Garner, D.M. 1972.
Flow through the Charlie-Gibbs Fracture Zone, Mid-Atlantic Ridge.
Can. J. Earth Sci. 9 (1): 116-121

currents, bottom
Hadley, M.L., 1964.
Wave-induced bottom currents in the Celtic Sea.
Marine Geology, 2(1/2):164-167.

currents, bottom
Hartley, Robert P., 1968.
Bottom currents in Lake Erie.
Proc. 11th Conf. Great Lakes Res. 1968: 398-405.

currents, bottom
Heaps, N.S., 1966.
Wind effects on the water in a narrow two-layered lake. 1. Theoretical analysis.
Phil. Trans. R. Soc., (A), 259: 393-416.
(1102)

currents, bottom
Heezen, Bruce C., and Charles D. Hollister, 1967.
Physiography and bottom currents in the Bellingshausen Sea.
Antarctic Jl. U.S.A., 2(5):184-185.

currents, bottom
Heezen, Bruce C., and A.S. Laughton, 1963.
14. Abyssal plains.
In: The Sea, M.N. Hill, Editor, Interscience Publishers, 3:312-364.

currents, bottom
Hunkins, Kenneth, Edward M. Thorndike and Guy Mathieu, 1969.
Nepheloid layers and bottom currents in the Arctic Ocean. J. geophys. Res., 74(28): 6995-7001.

currents, bottom
Ichie, Takashi, 1951
[On the hydrography of the Kii-Suido (1951).]
Bull. Kobe Mar. Obs., No. 164:253-278(top of page); 35-60(bottom of page).

currents
Inter-American Tropical Tuna Commission, 1961
Annual report for the year 1960: 183 pp.

Currents bottom
Kawai, Hideo and Hisao Sakamoto, 1969.
On bottom currents in the East China Sea measured with sea-bed drifters - I. (In Japanese; English abstract). Bull. Nansei reg. Fish. Res. Lab. 2: 39-48.

currents, bottom
Laird, N.P. and T.V. Ryan, 1969.
Bottom current measurements in the Tasman Sea.
J. geophys. Res., 74(23): 5433-5438.

currents, bottom
LeFloch, J., et J.L. Mauvais, 1968.
Mesures de courant au voisimage de fond dans le golfe de Gascogne.
Cah. océanogr. 20(10):885-892.

currents, bottom
Leloup, Eugène, 1966.
Observations sur la derive des courants au large de la côte belge au moyen de flotteurs de fond.
Bull. Inst. r. Sci. nat. Belg., 42(20):1-20.

currents, bottom
Mauvais, J.L., et J. le Floch, 1969.
Sur quelques mesures du gradient de vitesse près de fond dans le Golfe de Gascogne.
Cah. océanogr., 21(6): 571-580

currents, bottom
Mauvais, J.L., et J. Le Floch, 1969.
Traitement statistique de mesures de courant effectuées près du fond dans le Golfe de Gascogne.
Cah. océanogr., 21(4):379-386.

Currents, bottom
McCoy, F.W., Jr., 1969.
Bottom currents in the western Atlantic Ocean between the Lesser Antilles and the Mid-Atlantic Ridge. Deep-Sea Res., 16(2): 179-184.

currents, bottom
*Morse, Betty-Ann, M. Grant Gross and Clifford A. Barnes, 1968.
Movement of seabed drifters near the Columbia River.
J. WatWays Harb. Div. Am. Soc. civ. Engrs., 94(WW1): 93-103.

Currents bottom
Murray, Stephen P., 1970.
Bottom currents near the coast during hurricane Camille. J. geophys. Res., 75(24): 4579-4582.

Currents, bottom
Neumann, A. Conrad, and Mahlon M. Bell 1970.
Submersible observations in the Straits of Florida: geology and bottom currents.
Bull. geol. Soc. Am. 81(10): 2861-2874.

currents, bottom
Oulianoff, N., 1961.
Rides (ripple marks) sur les fonds océaniques et courants sous-marins.
C.R. Acad. Sci., Paris, 253(3):507-509.

currents, bottom
Park Won Cheon, and Sang Yong Kim, 1968.
Submarine topography and grain size distribution of sediments in the southern part of the East Sea of Korea. (In Korean; English abstract).
Bull. Fish. Res. Develop. Agency, Korea, 3:105-118

currents, bottom
Pettersson, H., 1952.
Current observations close to the bottom.
Medd. Ocean. Inst., Göteborg [Göteborgs K. Vetenskaps- och Vitterhets Samhälles Handl., Sjätte Följden B 6(3)] :19-21.

currents, bottom
Phillips, Ada, 1969.
Sea-bed water movements in Morecambe Bay.
Dock Harb. Auth. 49(580): 379-382.

currents, bottom
Pratt, Richard M., 1963
Bottom currents on the Blake Plateau.
Deep-Sea Res., 10(3):245-249.

currents, bottom
Rattig, H., 1956.
Beitrag zum Problem der Wasserbewegung im Boden.
Meteorol. Rundschau, 9(9/10):182-184.

currents, bottom
Reimnitz, Erk, 1971
Surf-beat origin for pulsating bottom currents in the Rio Balsas Submarine Canyon, Mexico.
Bull. geol. Soc. Am. 82(1): 81-90.

currents, bottom (measurement)
Revelle, R., 1940.
Current measurements near the sea bottom.
Proc. Verb., Assoc. Océan. Phys. 3:114-115.

currents, bottom
Romanovsky, V., and J. Le Floch, 1960
Mesure de courants au voisinage du fond.
C.R. Acad. Sci., Paris, 251(19): 2059-2060.

currents, bottom
Rowe, Gilbert T., and Robert J. Menzies, 1968.
Deep bottom currents off the coast of North Carolina.
Deep-Sea Res., 15(6):711-719.

currents, bottom
Shepard, F.P., J.R. Curray, D.L. Inman, E.A. Murray, E.L. Winterer and R.F. Dill, 1964.
Submarine geology by diving saucer. Bottom currents and precipitous submarine canyon walls continue to a depth of at least 300 meters.
Science, 145(3636):1042-1046.

currents, bottom
Sleath, J.F.A., 1970.
Velocity measurements close to the bed in a wave tank.
J. fluid Mech., 42(1): 111-123.

currents, bottom
Spataru, Arcadie 1969.
Considérations sur la fluctuation des vitesses de fond.
Rapp. P.-v. Réun. Commn int. Explor. scient. Mer Méditerr. 19(4): 713-715

currents, bottom
Stanley, Daniel J., and Gilbert Kelling, 1965.
Photographic investigation of sediment texture, bottom current activity, and benthonic organisms in the Wilmington Submarine Canyon.
U.S.C.G. Oceanogr. Rept. 22 (CG 373-22): 95 pp.

currents, bottom
Stewart, H.B., jr., 1958.
Upstream bottom currents in New York Harbor.
Science, 127(3306):1113-1115.

currents, bottom
Stride, A.H., 1963
Current-swept sea floors near the southern half of Great Britain.
Quart. Jour. Geol. Soc., London, 119:175-199.

Also in:
Collected Reprints, N.I.O., Vol. 11(458).1963.

bottom currents
Wattenberg, H., 1938.
Kalkauflösung und Wasserbewegung am Meeresboden.
Ann. Hydr., usw., 63:387-391.

currents (bottom), effect of

Johnson, David A. and Thomas C. Johnson, 1970
Sediment redistribution by bottom currents in the central Pacific. Deep-Sea Res., 17(1): 157-169.

currents, bottom, effect of

Hubert, John F., 1964.
Textural evidence for deposition of many western North Atlantic deep-sea sands by ocean-bottom currents rather than by turbidity currents.
J. Geol., 72(6):757-785.

currents (bottom), effect of

Rowe, Gilbert T., 1971.
Observations on bottom currents and epibenthic populations in Hatteras Submarine Canyon. Deep-Sea Res, 18(6): 569-581.

currents, bottom, effect of

Weser, Oscar E., 1970.
Lithologic summary. Initial Repts. Deep Sea Drilling Project, Glomar Challenger 5: 569-620.

currents, bottom

Zimmerman, Herman B., 1971.
Bottom currents on the New England Continental Rise. J. geophys. Res, 76(24): 5865-5876.

currents, bottom (velocity)

currents, bottom (velocity of)
Pyrkin, J.G., 1966.
The measurement of bottom current velocities in the Atlantic Ocean. (In Russian).
Fisika Atmosfer Okean., Izv.Akad.Nauk.SSSR, 2(12):1316-1317.

currents, bottom + surface

currents, boundary

Arthur, Robert S., 1965. currents, boundary
On the calculation of vertical motion in eastern boundary currents from determinations of horizontal motion.
J. Geophys. Res., 70(12):2799-2803.

current boundaries (visible)
Brosin, Hans-Jürgen und Dietwart Nehring, 1967.
Ozeanologische Beobachtungen an einer Stromgrenze auf dem patagonischen Schelf.
Beitr. Meereskunde, 21:76-78.

currents, boundary
Kort, V.G., Iu. A. Ivanov, K.A. Chekotillo and V.G. Neiman, 1969.
New data concerning the system of western boundary currents in the tropical Atlantic Ocean. (In Russian).
Dokl. Akad. Nauk SSSR 188(3): 677-680

currents, boundary

Masuzawa, J., 1960

Western boundary currents and vertical motions in the subarctic North Pacific Ocean.
J. Oceanogr. Soc., Japan, 16(2): 69-73.

boundary current, deep
Reid, Joseph, Jr., Henry Stommel, E. Dixon Stroup and Bruce A. Warren, 1968.
Detection of a deep boundary current in the western South Pacific.
Nature, Lond., 217(5132):237.

currents (boundary)
Spiegel, S.L., and A.R. Robinson 1968.
On the existence and structure of inertial boundary currents in a stratified ocean.
J. fluid Mech. 32(3): 569-607.

currents, boundary
Stewart, R.W., 1969.
The atmosphere and the ocean.
Scient. Am., 221(3):76-86. (popular).

currents, boundary

Warren, B.A., 1971
Evidence for a deep western boundary current in the South Indian Ocean.
Nature, Lond., 229(1): 18-19.

currents, boundary (eastern)
Wooster, W.S., and J.L. Reid, Jr., 1963
11. Eastern boundary currents. In: The Sea, M.N. Hill, Edit., Vol. 2. (III) Currents, Interscience publishers, New York and London, 253-280.

currents, boundary

Wyrtki, Klaus, 1961
The thermohaline circulation in relation to the general circulation in the oceans.
Deep-Sea Res., 8(1): 39-64.

current boundaries, effect of

current boundaries, effect of
Suzuki, Tsuneyoshi, 1963
Studies on the relationship between current boundary zones in waters to the southeast of Hokkaido and migration of the squid, Omnastrephes sloani pacificus (Steenstrup).
Mem., Fac. Fish., Hokkaido Univ., 11(2):153 pp

current, branching

current branching
*Warren, Bruce A., 1969.
Divergence of isobaths as a cause of current branching.
Deep-Sea Res., Suppl. 16: 339-355.

currents, coastal

currents, coastal
Arlman, J.J., J.N., Svasek and B. Verkerk, 1960.
The use of radioactive isotopes for the study of littoral drift. Importance of accurate investigation of coastal currents.
Dock and Harbour Authority, 41(476):57-64.

currents, coastal
Avery, Don E., Doak C. Cox and Taivo Laevastu, 1963.
Currents around the Hawaiian Islands (A study of coastal currents in respect to sewage disposal).
Hawaii Inst. Geophys., Rept., No. 26:22 pp.

currents, coastal
Bossolasco, M., and I. Dagnino, 1957.
Sulle correnti costiere nel Golfo di Genova.
Geofis. Pura e Applic., Milano, 38:3-20.

currents, coastal'

Dmitriev, A. A., & T. V. Bonchkovskaya, 1954.
[Observations on models of the motions appearing as the result of a frontal impact of a wave against a slope, and some ideas on the circulation caused by the oblique approach of a wave to a smooth sloping shore]
Trudy Mor. Gidrofiz. In-ta Akad. Nauk. SSSR, 4: 31-71.
Rev. in Appl. Mech. Rev., 12(8): 580.

currents, coastal
Ertel, H., 1965.
Der Ausgleich von Störungen der Strandlinie bei zeitlich variablem Küstenstrom.
Acta Hydrophysica, 13(2) 85-90

currents, coastal
Filimonov, A.I., 1965.
Longshore and normal-to-the-shore movements of water under shallow seashore conditions. (In Russian; English abstract).
Okeanologiia, Akad. Nauk, SSSR, 6(4):645-650.

currents, coastal
Graham, Joseph J. 1970.
Coastal currents of the western Gulf of Maine.
Res. Bull. Int. Comm. NW Atlantic Fish. 7: 19-31.

currents, coastal
Harrison, W., Morris L. Brehmer and Richard B. Stone, 1964.
Nearshore tidal and nontidal currents, Virginia Beach, Virginia.
U.S. Army, Coastal Engineering Res. Center, Techn. Memo., No. 5:20 pp.

currents, coastal
Hidaka, K., 1954.
A contribution to the theory of upwelling and coastal currents. Trans. Amer. Geophys. Union 35(3):431-444, 7 textfigs.

currents, coastal
Johnson, J.W., and R.L. Wiegel, 1958.
Investigation of current measurement in estuarine and coastal waters.
State Water Pollution Control Bd., California, 19:233 pp.

Abstr. in: Publ. Health Eng. Abstr., 39(9):32.

currents, coastal

Labeish, V. G., 1959.
[Dynamics of coastal currents.]
Vestnik, Leningrad Univ., Ser. Geol. i Geograf. 1) 14(6):139-142.

currents, coastal
Martiel, L., 1956.
Études des courants du littoral de la Bretagne.
Rev. Trav. Inst. Pêches Marit., 20(3):263-280.

currents, coastal
Mikhailov, J.D., 1960.
[Formation of residual currents near the capes in the tidal seas.] Izv. Vses. Geograf. Obshch., 92(6):525-526.

currents, coastal

Murthy, C.R., 1972.
Complex diffusion processes in coastal currents of a lake. J. phys. Oceanogr. 2(1): 80-90.

currents, coastal

Poncet, Jacques, 1961

Existence de "formes festonnées" prelittorales en certains points de la côte algerienne, mise en evidence de "courants d'arrachement", Cahiers Oceanogr., C.C.O.E.C. 13(1): 32-37.

currents, coastal
Popov, E.A., 1956.
On "away from current" "driven away" water in the coastal zone. Studies on sea coasts and reservoirs. Trudy Okeanogr. Komissii, Akad. Nauk, SSSR, 1: 98-104.

currents, coastal
Romanovsky, V., 1963
Effet du vent sur le mouvement des eaux littorales.
Trav. Centre de Recherches et d'Études Océanogr. 31-36.

currents, coastal
Saito, Y., 1956.
The theory of the transient state concerning upwelling and coastal current.
Trans. Amer. Geophys. Union 37(1):38-42.

currents, coastal
Samilov, I.B., V.N. Michailov, A.I. Simonov and N.A. Skriptunov, 1960.
[Circulation of water in "predust'ebom" coastal waters and accompanying processes.]
Trudy Okeanogr. Komissii, Akad. Nauk, SSSR, 10(1):100-106.

currents, coastal
Selleck, R.E., and E.C. Peixoto, 1963.
Nearshore winds and currents of Guanabara State, Brazil.
Int. J. Air Water Poll., 7(4/5):445-458.

currents, coastal (forecasting)
Shadrin, I.F., 1961.
[Possibility of forecasting of coastal currents in nontidal seas.]
Trudy Inst. Okeanol., Akad. Nauk, SSSR, 48:328-340.

currents, coastal
Shtokman, V.B., 1954.
[Ueber die Ursache von Kreisströmungen in der Nähe von Inseln und Gegenströmungen an den Küsten von Meerengen.]
Nachrichten Akad. Wiss., USSR, Ser. Geogr., 1954(4):29-37.

currents, coastal
Striggow, Klaus, 1966.
Strömungsmessung in der Brandungszone Aufgabenstellung und Voruntersuchungen zur Lösung eines Messtechnischen Problems der Küstenforschung.
Beitr. Meereskunde, 17-18:111-126.

currents, coastal
Volkov, P.A., 1961.
[On the values of maximum bottom speeds in the coastal zone under disturbance conditions.]
Okeanologiia, (3):432-438.

currents, coastal
Wiegel, R.L., and J.W. Johnson, 1960
Ocean currents, measurement of currents and analysis of data.
Waste disposal in the marine environment, Pergamon Press: 175-245.

currents, coastal
Wolf, Stephen C., 1970.
Coastal currents and mass transport of surface sediments over the shelf regions of Monterey Bay, California.
Mar. Geol. 8(5): 321-336

currents, coastal
Yoshida, K., 1958.
Coastal upwelling, coastal currents and their variations. Rec. Oceanogr. Wks., Japan, Spec. No. 2:85-89.

currents, coastal
Zenkovitch, V., 1956.
[Le transport d'alluvions le long des rives soviétiques de la mer Noire.] Inst. Nat. pour les Projets de Ports Marit., Trav., Moscow, 3:

coastal currents
Zhdanov, A.M., 1951.
[Determination of flow power in shore deposits by direct observations.] Izvest. Akad. Nauk, Ser. Geogr. Geofiz. 15(2):81-90.

currents (coastal)
Zeigler, J.M. and H.J. Tasha, 1969.
Measurement of coastal currents. Proc. Eleventh Conf. on Coastal Engin., London: 436-445.
Also in: Contrib. Univ. Puerto Rico, Dept. mar. Sci. 8 (1968-1969).

currents, coastal
Zoubkova, L.A., 1959
[Currents in a coastal area above a gently sloping bottom according to the direction and speed of the wind.]
Trudy Morsk. Hydrophys. Inst., 15: 43-55.

currents, (coastal), effect of
Higuchi, Haruo and Yuichi Iwagaki, 1969.
Hydraulic model experiment on the diffusion due to the coastal current. Coast. Engng, Japan, 12: 129-138.

currents, coherence of
Webster, Ferris, 1972.
Estimates of the coherence of ocean currents over vertical distances. Deep-Sea Res. 19(1): 35-44.

Currents, Computation of

current computation
Kozlov, V.F., 1971.
The experience of computing currents in the sub-Arctic front zone in the northwestern Pacific. (In Russian; English abstract). Okeanologiia 11(4): 568-577.

currents, computation of
Sarkisyan, A.S., 1961.
Computations of oceanic currents under the influence of air pressure. (In Russian).
Papers of the Conference on the Problem Inter-action between the Atmosphere and the Hydrosphere in the North Atlantic, Leningrad (3/4):13-19.
Translation:
National Lending Library for Science and Technology
Boston Spa, Yorkshire, England

Currents, Contour

currents, contour
Hollister, Charles D., and Robert B. Elder, 1969.
Contour currents in the Weddell Sea.
Deep-Sea Res., 16(1):99-101.

currents, contour
Schneider, Eric D., Paul J. Fox, Charles D. Hollister, H. David Needham and Bruce C. Heezen, 1967.
Further evidence of contour currents in the western North Atlantic.
Earth Planet. Sci. Letters, 2(4): 251-259.

Currents, convection

currents, convection
Byzova, N.L., 1951.
[Spontaneous fluctuations of convection currents (in the sea).] Izv. Akad. Nauk, SSSR, Ser. Geofiz. (5):84-

currents, convection
Byzova, N.L., 1950.
Self-excited oscillations of a thermal convection current. Dok. Akad. Nauk, USSR, 72(4):675-678.

currents, convection
Defant, A., 1947.
Der Einfluss des Windes auf das Massenfeld und die Konvektionsströme im Meer. Forsch. u. Fortschr. 21-23.

convection currents, effect of
Gould, H.R., and T.F. Budinger, 1958
Control of sedimentation and bottom configuration by convection currents, Lake Washington, Washington.
J. Mar. Res., 17:183-198.

convection currents
Hidaka, K., 1950.
Mass transport in convection currents and the lateral mixing. J. Ocean. Soc., Tokyo, 6(1):57-67, 7 textfigs.

currents, convection
Hidaka, K., 1949.
Mass transport in convection currents and lateral mixing. Ocean. Mag., Tokyo, 1(4):175-184, 7 textfigs.

currents, convection
Morgan, W. Jason, 1965.
Gravity anomalies and convection currents.
2. The Puerto Rico Trench and the Mid-Atlantic Rise.
J. geophys. Res., 70(24):6189-6204.

currents, convection
Morgan, W. Jason, 1965.
Gravity anomalies and convection currents.
1. A sphere and cylinder sinking beneath the surface of a viscous fluid.
J. geophys. Res., 70(24):6175-6187.

currents, convection
Phillips, O.M., 1966.
On turbulent convection currents and the circulation of the Red Sea.
Deep-Sea Res., 13(6):1149-1160.

currents, convection
Selin, E.A., 1964
Ionized layers and convection currents in sea waters. (In Russian).
Okeanologiia, Akad. Nauk, SSSR, 4(5):774-777.

currents, longshore
Shepard, F. P., 1950.
Longshore current observations in southern California. B.E.B. Tech. Memo. No. 13:54 pp., 22 figs. (multilithed).

currents
Shepard, F. P., 1949.
Dangerous currents in the surf. Physics Today 2: 20-29.

currents, convection
Shuleikin, V.V., 1952.
[Oscillations of the regime in a convection current.] Dok. Akad. Nauk, SSSR 82(2):241.

currents, convection
Takano, Kenzo, 1961
Circulation générale permanente dans les océans. Deuxième partie (suite et fin).
J. Oceanogr. Soc., Japan, 17(4):179-189.

currents, convection
Takano, K., 1960.
Sur la circulation générale convective dans les océans.
Bull. Inst. Océanogr., Monaco, 1178:23 pp.

currents, convective
Takano, K., 1955.
Sur la différence critique de température dans les expériences de Douchan Avsec au sujet des courants thermoconvectifs.
Rec. Ocean. Wks., Japan, 2(1):93-96, 1 textfig.

currents, convective
Takano, K., 1955.
Sur les courants thermoconvectifs dans la nappe fluide horizontale chauffee par en dessous.
Rec. Ocean. Wks., Japan, 2(1):97-105, 1 textfig.

currents, thermal convecti
Takano, K., 1955.
An example of thermal convective current.
Rec. Ocean. Wks., Japan, 2(1):76-79.

currents, convection
Tareev, B. A., 1960.
[On the theory of convectional circulation in deep water depressions of the ocean].
Izv. Akad. Nauk, SSSR (7):1022-1029.

convection currents
Tozer, D.C., 1965.
Heat transfer and convection currents.
Phil. Trans. R. Soc., A. 258(1088):252-271.

currents, convection
Yampol'kiy, A.D., 1948.
[Convection currents provided by thermal processes in the sea.] Met. Gidrol. 5:63-66.

Currents, data interpretation

currents, data interpretation
Burkov, V.A., and Yu. V. Pavlova, 1963
On the use of statistical characteristics for investigation of the Northern trade-wind current in the western tropical part of the Pacific. (In Russian).
Mezhd. Geofiz. Komitet, Prezidiume, Akad. Nauk SSSR, Rezult, Issled. Programme Mezhd. Geofiz. Goda, Okeanol. Issled., (8):52-57.

current, data only

A

current (data only)
Anon., 1950.
[Report of the oceanographical observations from Fisheries Experimental Stations (Sept. 1950).]
J. Ocean., Kobe Obs., 2nd ser., 1(2):5-11.

currents (data only)
Anon., 1939 (?)
Synopsis of the Hydrographical investigations (July-December 1939). Semiannual Rept. Oceanogr. Invest. (July-Dec. 1939).
Imp. Fish. Expt. Sta., Tokyo, No.65, 147 pp.

B

currents (data only)
Berrit, G.R., R. Gerard, L. Lemasson, J.P. Rebert et L. Vercesi, 1968.
Observations océanographiques exécutées en 1967. 1. Stations hydrologiques; observations de surface at de fond; stations côtières.
Doc. sci. provis., Centre Recherces océanogr., Côte d'Ivoire, 026:133pp (mimeographed)

C

currents (data only)
Chesapeake Bay Institute, 1954.
Data Report 24. Choptank River winter cruise, 7 December to 10 December 1952. Ref. 54-11:1-37, 1 fig.

currents, (data only)
Chesapeake Bay Institute, 1954.
Data Report 23 Patuxent River Winter Cruise, 3 December - 7 December 1952. Ref. 54-10:1-44, 1 textfig.

currents (data only)
Chesapeake Bay Institute, 1954.
Choptank River autumn cruise, 23 Sept-27 Sept 1952. Data Rept. 22, Ref. 54-9:37 pp., 1 fig.

current (data only
Chesapeake Bay Institute, 1954.
Choptank River spring cruise, 28 April-1 May 1952
Ref. 54-1:37 pp.

currents (data only)
Chosen. Fishery Experiment Station, 1938.
Oceanographical investigations during Jun-Jul 1933, in the Japan Sea offshore along the east coast of Tyosen on board the R.M.S. Misago-maru.
Ann. Rept. Hydrogr. Obs., Chosen Fish. Exp. Sta., 8:11, 23-29, 38-60, 62-65.

currents (data only)
Chosen, Fishery Experiment Station, 1936.
Number of the drift-bottles picked up at each locality, which had been liberated in the neighboring sea of Tyosen during the years 1929-32.
Ann. Rept. Hydogr. Obs., Chosen Fishery Exp. Sta., 8:114-116, tables.

currents (data only)
Chosen. Fishery Experiment Station, 1936.
Current measurements at Sts. A and O, along the west coast of Tyosen in Nov. 1932.
(English headings to tables). Ann. Rept. Hydrogr. Obs., Chosen Fish. Exp. Sta., 7:75-85.

currents (data only)
Chosen. Fishery Experiment Station, 1936.
Current measurements at station (M) 10 mi. south off the shore of Husan, on Jan. 20-21, 1932.
Ann. Rept. Hydrogr. Obs., Chosen Fish. Exp. Sta. 7:70-74.

currents (data only)
Chosen. Fishery Experiment Station, 1937.
Results of the current measurements in the Southern Sea of Tyosen and the western part of the Japan Sea, on board the R.M.S. Misago-maru in June -Sept. 1932. (English headings for tables). Ann. Rept. Hydrogr. Obs., Chosen Fish. Exp. Sta., 7:15-62, incl. tables.

currents (data only)
Chosen, Fishery Experiment Station, 1930.
Report on the hydrographical investigations in the adjacent sea off the east coast of Tyosen, during May and July 1929. (English summary) Ann. Rept. Hydrogr. Obs., Chosen Fish. Exp. Sta. 4:17-44, 6 pls., tablos, charts.

currents, (data only)
Conseil International pour l'Exploration de la Mer, 1968.
Cooperative synoptic investigations of the Baltic 1964. 2. Germany.
ICES Oceanogr., Data Lists. 176 pp. (multilithed).

currents (data only)
Conseil International pour l'Exploration de la Mer, 1968.
Cooperative synoptic investigation of the Baltic 1964. 3. Poland.
ICES Oceanogr. Data Lists: 266 pp. (multilithed).

currents (data only)
Conseil International pour l'Exploration de la Mer, 1968.
Cooperative synoptic investigation of the Baltic 1964. 5. U.S.S.R.
ICES Oceanogr. Data Lists: 173 pp. (multilithed).

D

currents (data only)
Denmark, Det Danske Meteorologiske Institut, 1971.
Oceanografiske observationer fra danske fyrskibe og kyststationer 1970.
Publ. danske met. Inst. Arbøg. 147 pp.

currents (data only)
Denmark, Det Danske Meteorologiske Institut, 1969.
Oceanografiske observationer fra Danske fyrskibe og kyststationer 1968.
Arboger: 171 pp.

currents (data only)
Denmark, Det Danske Meteorologiske Institut, 1968.
Oceanografiske observationer fra danske fyrskibe og kyststationer 1967.
Publner danske met. Inst. Arbog: 171 pp.

currents (data only)
Denmark, Danske Meteorologiske Institut 1967.
Oceanografiske observationer fra Danske fyrskibe og kyststationer.
Publner danske met. Inst. 1966: 171 pp.

currents (data only)
Danmark, Danske Meteorologiske Institut, 1965.
Oceanografiske Observationer fra Danske fyrskibe og kuststationer. 1964.
Publ. Danske Meteorol. Inst., Arbøger, 171 pp.

currents (data only)
Denmark, Det Danske Meteorologiske Institut, 1964.
Oceanografiske observationer fra Danske fyrskibe og kyststationer, 1963.
Publikationer, Danske Meteorol. Inst., Arboger, 171 pp.

currents (data only)
Denmark, Det Danske Meteorologiske Institut, 1964
Oceanografiske observationer fra Danske fyrskibe og kyststationer.
Publ., Danske Meteorologiske Inst., Årbøger, 1962 169 pp.

currents (data only)
Denmark, Danske Meteorologiske Institut, 1964.
Oceanografiske observationer fra Danske fyrskibe og kyststationer, 1962.
Publ. Sanske Meteorol. Inst., Arbøger, 169 pp.

currents (data only)
Denmark, Danske Meteorologiske Institut, 1962.
Nautisk-meteorologisk årbog, 1961, 168 pp.

currents(data only)
Denmark, Det Danske Meteorologiske Institut, 1960
Nautisk-Meteorologisk Arbog, 1959:171 pp., charts.

currents (data only)
Denmark, Det Danske Meteorologiske Institut, 1959.
Nautisk-Meteorologisk Arbog, 1958: 169 pp.

currents (data only)
Denmark, Danske Meteorologiske Institut, 1958.
Nautisk-meteorologisk årbog, 1957:170 pp.

currents (data only)
Denmark, Danske Meteorologiske Institut, 1957.
Nautisk-meteorologisk aarbog, 1956:170 pp., charts.

currents(data only)
Danske Meteorologiske Institut, 1954.
Nautisk-Meteorologisk Årbog, 1953:171 pp., 12 charts.

currents(data only)
Denmark, Danske Meteorologiske Institut, 1953.
Nautisk-Meteorologisk Aarbog, 1952: 171 pp., 12 charts.

currents (data only)
Dansk Meteorologisk Institut, 1951.
Nautisk-Meteorologisk Årbog, 1950, 165 pp., 12 charts.

currents (data only)
Danske Meteorologiske Institut, 1950.
Nautisk-Meteorologisk Årbog, 1935:72 pp.

currents (data only)
Danske Meteorologiske Institut, 1948
Nautisk-Meteorologisk Aarbog 1946, 159 pp., 13 charts.

currents, data only
Denmark, Dansk Meteorologiske Institut, 1956.
Nautisk-Meteorologisk Årbog, 1955: 164 pp., 12 charts

currents (data only)
Denmark, Kommission for Havundersøgelser, 1923.
Current measurements from Danish Lightships.
Medd. Komm. Havundersøgelser, Ser. Hydrogr., 2(8):78 pp.

E

F

currents (data only)
Fjeldstad, Jonas Ekman, 1964.
Internal waves of tidal origin. 1. Theory and analysis of observations. 11. Tables.
Geofys. Publikasjoner, Geophysica Norvegica, 25(5):1-73; 1-155.

currents, data only
France, Comité Central d'Océanographie et d'Études des Côtes, 1955.
Observations hydrologiques des bâtiments de la marine nationale. Avisogarde-peche "Ailette", Mars à Octubre, 1954. Bull. d'Info., C.C.O.E.C. 7(7):314- , 5 pls.

currents (data only)
France, Service Hydrographique de la Marine, 1968.
Marée à Rikitea, Iles Gambier-Archipel des Tuamotu. Résultats d'observations courantologiques effectuées autout de l'Ile Annobon. Résultats des observations de température et de salinité des eaux de surface à 12h Tu-Ile Annobon.
Cah. océanogr., 20(9):813-829.

currents(data only)
France, Service Hydrographique de la Marine, 1963
Résultats d'observations.
Cahiers Océanogr., C.C.O.E.C., 15(5):344-355.

G

currents (data only)
Germany, Deutsches Hydrographisches Institut, 1970.
Strombeobachtungen 1963, Deutsche Bucht und mittlere Nordsee.
Meeresk. Beobacht. Ergebn. 29 (2149/1): unnumbered pp.

currents (data only)
Germany, Deutsches Hydrographisches Institut, 1970.
Beobachtungen auf den deutschen Feuerschiffen der Nord- und Ostsee im Jahre 1968 sowie Monatsmittelwerte von Temperatur und Salzgehalt des Jahres 1968
Meeresk. Beobacht. Ergebn. (2149/2) 30: unnumbered pp.

currents (data only)
Germany, Deutsches Hydrographisches Institut 1969
Strombeobachtungen 1962, Nordsee und Fehmarnbelt.
Meeresk. Beobacht. Ergeb. 27 (2148): numerous unnumbered pp.

currents (data only)
Germany, Deutsches Hydrographisches Institut 1969
Strombeobachtungen 1961, Nordsee und Flensburger Förde.
Meeresk. Beobacht. Ergeb. 26 (2147): numerous unnumbered pp.

currents (data only)
Germany, Deutsches Hydrographisches Institut, 1968.
Beobachtungen auf den deutschen Feuerschiffen der Nord-und Ostsee im Jahre 1967 soure Monatsmittelwerte von Temperatur und Salzgehalt.
Meeresk.Beobacht.Ergebn.25(2146):unnumbered pp.

currents (data only)
Germany, Deutsches Hydrographisches Institut, 1968.
Beobachtungen auf den deutschen Feuerschiffen der Nord- und Ostsee im Jahre 1966 soure Monatsmittelwerte von Temperatur und Salzgehalt.
Meeresk.Beobacht.Ergebn. 24(2145):unnumbered pp.

currents (data only)
Germany, Deutsches Hydrographisches Institut, 1967.
Beobachtungen auf den deutschen Feuerschiffen der Nord- Und Ostsee im Jahre 1965 sowie Monatsmittelwerte von Temperatur und Salzgehalt des Jahres 1965.
Meeresk Beobacht. Ergeb. 23(2144) unnumbered pp.

currents (data only)
Germany, Deutsches Hydrographisches Institut., 1965.
Beobachtungen auf den deutschen Fluerschiffen der Nord- und Ostsee im Jahre 1964 sowie Monatsmittelwerte von Temperatur und Salzgehalt.
Meeresk. Biobacht. N. Ergebn. No. 22(Nr. 2143): Unnumbered pp(Multilithed)

currents (data only)
Germany, Deutsches Hydrographisches Institut, 1964.
Beobachtungen auf den deutschen Feuerschiffen der Nord- und Ostsee im Jahre 1963 sowie Monatsmittelwerte von Temperatur und Salzgehalte des Jahres 1963.
Meereskund. Beobacht. und Ergeb., 21(2142): unnumbered pp. (multilithed).

currents, (data only)
Germany, Deutsches Hydrographisches Institut, 1963.
Beobachtungen auf den deutschen Feuerschiffen der Nord- und Ostsee im Jahre 1962 sowie Monatsmittelwerte von Temperatur und Salzgehalt des Jahres 1962.
Meereskund. Beobacht. u. Ergebn., Nr. 20(2141): unnumbered pp.

currents,(data only)
Germany, Deutsches Hydrographisches Institut, 1963.
Strombeobachtungen in der Nordsee in den Jahren 1959 und 1960.
Meereskundliche Beobachtungen und Ergebnisse (19): :unnumbered pp.

currents (data only)
Germany, Deutsches Hydrographisches Institut, 1962.
Beobachtungen aud die deutschen Feuerschiffen der Nord- und Ostsee im Jahre 1961 sowie Monatsmittelwerte von Temperatur und Salzgehalt des Jahres 1961. Nr. 2139.
Meereskund. Beobacht. und Ergebn., No. 18: unnumbered pp.

currents (data only)
Germany Deutsches Hydrographisches Institut, 1956.
Beobachtungen auf den deutschen Feuerschiffen der Nord- und Ostsee in Jahre 1955.
Meereskund. Beobacht. u. Ergeb. (2122/8):No. 8: unnumbered pp.

currents (data only)
Germany, Deutsches Hydrographisches Institut, 1955.
Beobachtungen auf den deutschen Feuerschiffen der Nord- und Ostsee im Jahre 1954.
Meereskundliche Beobacht. u. Ergebn., No. 7(2122/7):numerous pp.

currents (data only)
Germany, Deutsches Hydrographisches Institut, 1954.
Beobachtungen auf den deutschen Feuerschiffen der Nord- und Ostsee in Jahre 1953.
Meereskundliche Beobacht. u. Ergeb. No. 6(2122/6): numerous pp. (unnumbered).

currents(data)
Germany Deutsches Hydrographisches Institut, 1954.
Beobachtungen auf den deutschen Feuerschiffen der Nord- und Ostsee im Jahre 1951.
Meeresk. Beobacht. u. Ergeb. No. 4(2122/4): numerous pp. (unnumbered).

currents(data only)
Germany Deutsches Hydrographisches Institut, 1953.
Beobachtungen auf den deutschen Feuerschiffen der Nord- und Ostsee im Jahre 1950.
Meeresk. Beobacht. u. Ergeb. No. 3(2122/3): numerous pp. (unnumbered).

Germany currents (data only)
Deutsches Hydrographisches Institut, 1953.
Beobachtungen auf den deutschen Feuerschiffen der Nord- und Ostsee im Jahre 1949.
Meeresk. Beobacht. u. Ergeb. No. 2(2122/2): numerous pp. (unnumbered).

currents (data only)
Germany, Deutsches Hydrographisches Institut, 1953.
Beobachtungen auf den deutschen Feuerschiffen der Nord- und Ostsee in Jahre 1948. Meereskundliche Beobachtungen und Ergebnisse No. 1(2122/1):

Germany currents (data only)
Deutsche Seewarte, 1928, 1929, 1930, 1931.
Meereskundliche Beobachtungen auf deutschen Feuerschiffen der Nord- und Ostsee. Jahr 1924 und 1925:84 pp., 2 charts; Jahr 1926 und 1927: 84 pp., 2 charts; Jahr 1928, 44 pp., 2 charts; Jarh 1929:61 pp., 2 charts.

Currents (data only)
Great Britain, Ministry of Defence Hydrographic Department 1967.
Equatorial current investigations, Gilbert Islands, H.M.S. Cook 1963.
Publ. Adm. Mar. Sci. (H.D. 550): 40 pp. (multilithed)

H

currents, (data only)
Hollister, H.J., 1961
Bathythermograms and meteorological data record, Swiftsure Bank and Umatilla Reef Lightships, January 1, 1960 to June 30, 1961.
Fish. Res. Bd., Canada, Manuscript Rept. Ser. (Oceanography & Limnology), No. 99: 89 pp. (multilithed).

I

J

currents, (data only)
Japan, Central Meteorological Observatory, 1955.
The results of marine meteorological and oceanographical observations, January-June 1951. No. 9:177 pp.

currents (data only)
Japan, Central Meteorological Observatory, 1954.
The results of marine meteorological and oceanograhical observationr. Part 1, Oceanography, July-December, 1952, No. 12:138 pp.

currents (data only)
Japan, Central Meteorological Observatory, 1952
The results of Marine Meteorological and oceanographical observations, July - Dec. 1951, No. 10:310 pp., 1 fig.

currents (data only)
Japan, Central Meteorological Observatory, 1952.
The results of Marine meteorological and oceanographical observations. July - Dec. 1950. No. 8: 299 pp.

currents (data only)
Japan, Central Meteorological Observatory, 1952.
The Results of Marine Meteorological and oceanographical observations. Jan. - June 1950. No. 7: 220 pp.

currents (data only)
Japan, Central Meteorological Observatory, 1951.
The results of marine meteorological and oceanographic observations. July - Dec. 1949. No. 6: 423 pp.

currents (data only)
Japan, Central Meteorological Observatory, 1951.
Tables 37-38. Direct current measurements for stations occupied by R.M.S. "Shumpu-maru" in the Akashi-seto and Kitan-seto; --- in Osaka Bay.
Res. Mar. Met., Ocean. Obs., Jan.-June 1949, No. 5:332-337.

Japan currents (data only)
Central Meteorological Observatory, Japan, 1951.
The results of marine meteorological and oceanographical observations, July-Dec. 1948, No. 4: vi+414 pp., (of 57 tables,) 1 map, 1 photo.

Japan currents (data only)
Central Meteorological Observatory, Japan, 1951.
The results of marine meteorological and oceanographical observations, Jan.-June 1948, No. 3: 256 pp.

Japan currents (data only)
Central Meteorological Observatory, 1949.
Report on sea and weather observations on Antarctic whaling ground (1948-1949). Ocean. Mag., Tokyo, 1(3):142-173, 11 textfigs.

currents (data only)
Japan, Maritime Safety Agency, 1952.
The results of oceanographic observation in the northwestern Pacific Ocean, 1938-1941, No. 5. Publ. 981, Hydrogr. Bull. (Ocean. Repts.), Special Number No. 9:194 pp., figs.

currents, (data only)
Japan, Fisheries Agency, 1967.
The results of Fisheries Oceanographical Observation, January-December 1964, 1643 pp.

currents (data only)
Japan, Japan Meteorological Agency 1971.
The results of marine meteorological and oceanographical observations,
July - December 1969, 46:270 pp.

Currents (data only)
Japan, Japan Meteorological Agency, 1970
The results of marine meteorological and oceanographical observations. (The results of the Japaneses Expedition of Deep Sea (JEDS-11); January-June 1967 41: 332 pp.

currents (data only)
Japan, Japan Meteorological Agency 1970.
The results of marine meteorological and oceanographical observations, July-December 1966, 40:336 pp.

currents (data only)
Japan Meteorological Agency 1969
The results of marine meteorological and oceanographical observations, Jan. - June 1966, 39:349 pp. (multilithed)

currents (data only)
Jepan, Japan Meteorological Agency, 1966.
The results of the Japanese Expedition of Deep Sea (JEDS-8).
Results mar.met.oceanogr.Obsns.Tokyo, 35:328 pp.

currents, surface
Japan, Japen Meteorological Agency, 1966.
The results of the Japanese Expedition of Deep Sea (JEDS-8).
Results mar.met.oceanogr.Obsns.Tokyo, 35:328 pp.

currents (data only)
Japan, Japan Meteorological Agency. 1965.
The results of marine meteorological and oceanographical observations. July-December 1963, No. 34: 360 pp.

currents (data only)
Japan, Japan Meteorological Agency, 1964.
The results of marine meteorological and oceanographical observations, January-June 1963, No. 33:289 pp.

currents (data only)
Japan, Japan Meteorological Agency, 1964.
The results of the Japanese Expedition of Deep Sea (JEDS-5).
Res. Mar. Meteorol. and Oceanogr. Obs., July-Dec. 1962, No. 32:328 pp.

currents (data only)
Japan Meteorological Agency, 1964.
Oceanographic observations.
Res. Mar. Meteorol. Oceanogr. Observ., (31):220 pp

currents (data only)
Japan, Japan Meteorological Agency, 1962
The results of marine meteorological and oceanographical observations, January-June 1961, No. 29:284 pp.

currents (data only)
Japan, Japan Meteorological Agency, 1961
The results of marine meteorological and oceanographical observations, January-June 1960. The results of the Japanese Expedition of Deep-Sea (JEDS-2, JEDS-3), No. 27: 257 pp.

currents (data only)
Japan, Japan Meteorological Agency, 1960.
The results of marine meteorological and oceanographical observations, July-December, 1959, No. 26:256 pp.

currents (data only)
Japan, Japan Meteorological Agency, 1960.
The results of marine meteorological and oceanographical observations, Jan-June 1959, No. 25: 258 pp.

currents (data only)
Japan, Japan Meteorological Agency, 1959.
The results of marine meteorological and oceanographical observations, July-December, 1958, No. 24:289 pp.

currents (data only)
Japan, Japan Meteorological Agency 1959
The results of marine meteorological and oceanographical observations, January-June 1958, No. 23: 240 pp.

currents (data only)
Japan, Japan Meteorological Agency, 1958
The results of marine meteorological and oceanographical observations, July-December 1957, No. 22: 183 pp.

currents (data only)
Japan, Kobe Marine Observatory, 1961
Data of the oceanographic observations in the sea south of Honshu from February to March and in May, 1959.
Bull. Kobe Mar. Obs., No. 167(27):99-108; 127-130; 149-152; 161-164; 205-218.

currents, (data only)
Japan, Maritime Safety Agency, 1963. Tokyo
The results of oceanographic observations in the north-western Pacific Ocean, No. 10. 1942.
Hydrogr. Bull., (Publ. No. 981), No. 74:212 pp.

Currents, data only (parachute drogue)
Japan, Maritime Safety Board 1961.
Tables of results from oceanographic observations in 1958.
Hydrogr. Bull. Tokyo No 66 (Publ. No. 981): 153 pp.

currents (data only)
Japan, Maritime Safety Agency, 1952.
Currents and other phenomena in the southern area of Honsyu in February through May 1952.
Publ. No. 981, Hydrogr. Bull. No. 32:*

*Pages cannot be determined as everything in Japanese.

currents (data only)
Japan, Maritime Safety Board, Tokyo, 1962.
The results of oceanographic observations in the north-western Pacific Ocean, No. 9 (1941).
Hydrographic Bull., Tokyo (Publ. No. 981), No. 71:180 pp.

Currents (data only)
Japan, Maritime Safety Board, 1962
The results of oceanographic observations in the north-western Pacific Ocean. 8. 1931-1940.
Hydrogr. Bull., Tokyo, (Publ. No. 981), No. 69: 245 pp.

currents (data only)
Japan, Maritime Safety Board, 1961
Tables of results from oceanographic observation in 1959.
Hydrogr. Bull. (Publ. No. 981), No. 68: 112 pp.

currents (data only)
Japan, Nagasaki Marine Observatory
Tables 9, 10, 11, 22, 23, 24, 34, 35 and 44
Oceanogr. Met. Nagasaki mar. Obs. 15: 92-120, 176-180, 191-192, 213-

currents (data only)
Japan, Nagasaki Marine Observatory, 1967.
Report of the Oceanographic observations in the sea west of Japan from January to February, 1969 (In Japanese)
Oceanogr. Met. Nagasaki Mar. Obs. 17 (242): 26 pp

currents (data only)
Japan, Nagasaki Marine Observatory, 1967
Report of the oceanographic observations in the sea west of Japan from July to September 1965. (In Japanese)
Oceanogr. Met. Nagasaki mar. Obs. 17 (241): 26 pp.

currents (data only)
Japan, Tokai Regional Fisheries Research Laboratory, 1952.
Report oceanographical investigation (Jan.-Dec. 1950), No. 74:143 pp.

K

Currents (data only)
Kusunoki, K., T. Minoda, K. Fujino and A. Kawamura 1967.
Data from oceanographic observations at Drift Station ARLIS II in 1964-1965.
Arctic Inst. N. Am. unnumbered pp. (duplicated)

L

currents (data only)
Lacombe, Henri, 1964
Mésures de courant, d'hydrologie et de météorologie effectuees à bord de la "Calypso" Campagne Internationale d'observations dans le détroit de Gibraltar (15 mai-15 juin 1961).
Cahiers Océanogr., C.C.O.E.C., 16(1):23-94.

M

Mc

currents (data only)
Mosby, Haakon, 1963
Current measurements in the Faroe-Shetland Channel 1962. Tables.
NATO Subcommittee on Oceanographic Research, Bergen, 59 pp. (multilithed).

currents (data only)
Mosby, Hakon, 1963
Current measurements in the Norwegian Sea and in the North Sea, 1923, 1924, 1928, 1929 Tables.
NATO Subcommittee on Oceanography, Bergen, 67 pp. (multilithed).

currents (data only)
Mosby, Hakon, 1962
Current measurements in the Faroe-Shetland Channel. 1960 and 1961. Tables.
NATO Sbucommittee on Oceanogr. Res., 173 pp. (mimeographed).

N

currents (data only)
Nato Subcommittee on Oceanographic Research, 1962
Current measurements, meteorological observations and soundings of the M/S "Helland-Hansen" near the Strait of Gibraltar, May-June 1961. Tables. 29 pp. (multilithed).

currents (data only)
Netherlands
Koninklijk Nederlandsch Meteorlogsch Instituut, 1921.
Oceanographsche en Meteorologsch Waarnemingen in den Atlantischen Oceaan. Maart, April, Mai, (1856-1920). Tabellen. Kon. Neder. Meteorol. Inst. No. 110:186 pp.

currents (data only)
Nikitin, M.M., 1955.
Observations of currents. Material. Nabluid. Nauch.-Issledov. Dreifuius. 1950/51, Morskoi Transport, 1:171-178; 180-403.
David Knauss, translator, A.M.S.-Astia Doc. 117134.

currents (data only)
Nordstrom, Svante, 1958
Finnish light-ship observations on currents and winds in 1955, 1956 and 1957.
Merent. Julk., No. 185: 70 pp.

O

P

Q

R

currents (data only)
Romanovsky, V., 1958.
Résultats de quelques mesures de courants prodonds. Trav. C.R.E.O., n.s., 3(1):3-7.

S

currents, data only
Saelen, O.H., 1963.
Studies in the Norwegian Atlantic Current. II. Investigations during the years 1954-59 in an area west of Stad.
Geofys. Publik., Geoph. Norv., 23(6):1-82.

currents (data only)
Scripps Institution of Oceanography, 1949.
Marine life research program. Progress report, 1 May to 31 July 1949. 24 pp. (mimeographed), 16 figs. (ozalid).

currents (data only)
Scripps Institution of Oceanography, 1949
Marine Life Research Program. Progress Report to 30 April 1949, 25 pp. (mimeographed), numerous figs. (photo.+ozalid). 1 May 1949.

currents (data only)
Svansson, Artur, compiler, 1971.
Hydrographical observations on Swedish lightships and fjord stations in 1969.
Fish. Bd Sweden, Ser. Hydrogr. 25: 67 pp.

Currents (data only)
Svansson, Artur 1970.
Hydrographical observations on Swedish lightships and fjord stations in 1968
Rept. Fish. Bd, Sweden, Ser. Hydrogr. 24: 83 pp.

currents (data only)
Svansson, Artur, 1969.
Hydrographical observations on Swedish lightships and fjord stations in 1967.
Rept. Fish. Bd Sweden, Ser. Hydrogr., 22: 94 pp.

currents (data only)
Svansson, Artur, 1968.
Hydrographical observations on Swedish lightships and fjord stations in 1966.
Rept. Fish. Bd., Sweden, Ser. Hydrogr. 21: 91 pp.

currents (data only)
Svansson, Artur, editor, 1967.
Hydrographical observations on Swedish lightships and fjord stations in 1965.
Rep. Fish. Bd., Sweden, Ser. Hydrogr., 19:98 pp.

current (data only)
Svansson, Artur, 1966.
Hydrographical observations on Swedish Lightships and fjord stations.
Fish. Bd., Sweden, Ser. Hydrogr., Rept. No. 18: 95 pp.

currents, (data only)
Svansson, Artur, Compiler, 1965.
Hydrographical observation on Swedish lightships in 1963.
Fish. Bd., Sweden, Ser-Hydrogr., Rept. No. 17: 83 pp.

currents (data only)
Sweden, Fishery Board of Sweden, 1964.
Hydrographical observations on Swedish lightships in 1962.
Fish. Bd., Sweden, Ser. Hydrogr., Rept., No. 16: 96 pp.

currents (data only)
Sweden, Fishery Board of Sweden, 1963
Hydrographical observations on Swedish lightships in 1961. (Artur Svansson, Editor).
Fish. Bd., Sweden. Ser. Hydrogr., Rept., No. 15
112 pp.

currents (data only)
Sweden, Fishery Board of Sweden, 1962
Hydrographical observations on Swedish lightships in 1960.
Fish. Bd., Sweden. Ser. Hydrogr. Rept. No. 14:
117 pp.

currents (data only)
Sweden, Fishery Board of Sweden, 1961
Hydrographical observations on Swedish Lightships in 1959.
Fish. Bd. Sweden. Ser. Hydrography. Rept. No. 12: 130 pp.

currents (data only)
Sweden, Fishery Board of Sweden, 1959
Hydrographical observations on Swedish lightships in 1958.
Fish. Bd., Sweden. Ser. Hydrogr., Rept.
No. 11: 124 pp.

currents (data only)
Sweden, Fishery Board of Sweden, 1958
Hydrographical observations on Swedish lightships in 1957.
Rept. Hydrogr. Ser., No. 10: 129 pp.

currents (data only)
Sweden, Fishery Board, 1957.
Hydrographical observations on Swedish lightships in 1956. Fish. Bd., Sweden, Ser. Hydrogr., Rept.
No. 9:110 pp.

currents (data only)
Sweden, Fishery Board of Sweden, 1956.
Hydrographical observations on Swedish lighships in 1955. Fish. Bd., Sweden, Ser. Hydrogr., Rept.
No. 8:129 pp.

currents (data only)
Sweden, Fishery Board of Sweden, 1955.
Hydrographical observations on Swedish lightships in 1954. Ser. Hydrogr., Rept. No. 6:130 pp.

currents (data)
Sweden, Fishery Board, 1954.
Hydrographical observations on Swedish lightships in 1953.
Fish. Bd., Sweden, Hydrogr. Ser., Rept. No. 4: 1-130, 1 fig.

currents (data
Sweden, Fishery Board of Sweden, 1953.
Hydrographic observations on Swedish lightships in 1952. Fish Bd., Sweden, Ser. Hydrogr., Rept.
No. 3:133 pp.

currents (data only)
Sweden, Fishery Board of Sweden, 1952.
Hydrographical observations on Swedish lightships in 1951. Series Hydrography, Rept. 1:133 pp., 1 chart(multilithed).

currents (data only)
Sweden, K. Fiskeristyrelsen, 1951.
Fyrskeppsundersökning, År 1950; numerous pp., (unnumbered).

currents (data only)
Sweden, Kungl. Fiskeristyrelsen, Hydrografiska Avd., 1950.
Fyrskeppsundersökning. År 1948-1949. 104 pp.

currents (data only)
Svenska Hydrografisk-Biologiska Kommission, 1950
Fyrskeppsundersökning. År 1946-1947:78 pp., 2 fig

T

Currents (data only)
Tabata, S., L.F. Giovando, J.A. Stickland and J. Wong, 1970
Current velocity measurements in the Strait of Georgia-1967.
Techn. Rept Fish. Res. Bd, Can. 169: 245 pp. (multilithed)

currents (data only)
Tabata, S., L.F. Giovando, J.A. Stickland and J. Wong, 1970
Current velocity measurements in the Strait of Georgia-1969.
Techn. Rept, Fish. Res. Bd, Can. 181: 72 pp.

currents, data only
Tabata, S., J.A. Stickland and B.R. de Lange Boom, 1971.
The program of current velocity and water temperature observations from moored instruments in the Strait of Georgia-1968-1970 and examples of records obtained.
Techn. Rept. Fish. Res. Bd Can. 253: 222 pp. (multilithed)

U

Current (data only)
Uda, Michitaka, Yoshima Morita, and Makoto Ishino, 1957
Results from the oceanographic observations in the North Pacific (1955-56) with Umitaka Maru and Shinyo-Maru. Rec. Ocean. Wks., Japan (Spec.No1):1-20.

currents, (data only)
U.S. Navy Hydrographic Office, 1960.
Oceanographic observations, Arctic waters, Task Force Five and Six, summer-autumn 1956, USS Requisite (AGS-18), USS Eldorado (AGC-11), USS Atka (AGB-3) and USCGC Eastwind (WAGB-279).
U.S. Navy Hydrographic Off., Techn. Rept., TR-58:89 pp.

currents (data only)
U.S. Navy Hydrographic Office, 1957.
Operation Deep Freeze II, 1956-1957. Oceanographic survey results. H.O. Tech. Rept., 29: 155 pp. (multilithed).

V

W

XYZ

currents (data only)
Zvereva, A.A., Edit., 1959.
Data, 2nd Marine Expedition, "Ob", 1956-1957.
Arktich. i Antarkt. Nauchno-Issled. Inst., Mezhd. Geofiz. God, Sovetsk. Antarkt. Exped., 6: 1-387.

Currents, deep

currents, deep
Akamatsu, Hideo, 1969.
On the results of deep current measurements. (In Japanese; English abstract). Bull. Jap. Soc. fish. Oceanogr. Spec. No. (Prof. Uda Commem. Pap.): 135-138.

currents, deep
Aliverti, Giuseppina, 1961.
Considerazione sulle misure dirette di correnti marine con il metodo di Swallow e sopra un dispositivo di recupero dei "Pinger".
Consiglio Nazionale delle Ricerche, Comm., Naz. Italiana, per lo Cooperazione Geofisica Int., Pubbl., No. 1:3-11.

currents, deep
Allain, Charles, 1960.
Topographie dynamique et courants generaux dans le bassin occidental de la Mediterranee (Golfe du Lion, Mer Catalane, Mer d'Alboran et ses abords, secteur a l'est de la Corse.
Rev. Trav. Inst. Peches Marit., 24(1):121-145.

currents, deep-sea
Anon., 1956.
Measurement of deep sea currents.
Dock and Harbour Authority 37(427):32.

currents, deep
Belevich, R.R., 1964.
Experience in determining sea currents at great depths from a drifting vessel. (In Russian).
Trudy, Dal'nevostochnogo Nauchno-Issled. Gidrometeorol. Inst., No. 17:95-98.

Abstracted in:
Soviet Bloc Res., Geophys., Astron., Space, 95: 57.

currents, deep
Brown, C.H., 1909.
Report on the deep currents of the North Sea as ascertained by experiments with drift bottles.
Fish. Bd., Scotland, North Sea Fish. Invest. Comm., Northern Area, Rept. No. 4:125-142.

currents, deep
Bubnov, V.A., 1971.
On some peculiar features of the Lusitanian deep-ocean current. (In Russian; English abstract). Okeanologiia 11(2): 203-210.

currents, deep
Capurro, Luis R.A., 1966.
Surface and deep current measurements in the Drake Passage.
Antarctic J., United States, 1(5):223.

currents, deep
Deacon, G.E.R., 1957.
Deep ocean currents. Discovery, 18:386-387.

currents, deep
Deacon, G.E.R., 1931.
Velocity of deep currents in the South Atlantic.
Nature 128:267, 2 textfigs.

Currents, deep
Derjugin, K.K., 1961
On the study of deep currents of the active sea layer in the North Atlantic.
Issled. Severnoi Chasti Atlanticheskogo Okeana, Mezhd. Geofiz. God, Leningradskii Gidrometeorol Inst., 1: 32-37.

currents, deep
Doebler, Harold J., 1967.
Savonius current meter measurements of deep ocean currents on the slope of Plantagenet Bank, Bermuda.
J. geophys. Res., 72(2):511-519.

currents, deep

Fedorov, C.N., 1960
[Estimation of deep currents in the region under investigation.]
Trudy Inst. Okeanol., 40:162-166.

currents, deep

Fofonoff, Nicholas,P., 1960

Description of the northeastern Pacific oceanography. Cal. Coop. Ocean. Fish. Invest. Rept. Vol. 7: 91-95.

currents, deep

Fukuoka, Jiro, 1964.
Hydrography of the adjacent sea. (1). The circulation in the Japan Sea.
J. Oceanogr., Soc., Japan, 21(3):95.

currents, deep

Gurikova, Z.F., 1966.
Computations of the surface and deep ocean currents in the North Pacific Ocean in summer. (In Russian; English abstract).
Okeanologiia, Akad. Nauk, SSSR, 6(4):615-631.

currents, deep

Heezen, Bruce C., and Charles Hollister, 1964.
Deep-sea current evidence from abyssal sediments
Marine Geology, 1(2):141-174.

currents,deep

Heezen,Bruce C., Charles D.Hollister and William F.Ruddiman,1966.
Shaping of the continental rise by deep geostrophic contour currents.
Science,152(3721):502-508.

CURRENTS DEEP

Hidaka, Koji, 1969.
Deep-sea currents research in Japan. Okeanologiia 9(3): 430-434. (In Russian; English abstract)

currents, deep

Knauss,John A.,1968.
Measurements of currents close to the bottom in the deep ocean.
Sarsia, 34:217-226.

currents, deep

Kolesnikov, A.G., G.P. Ponomarenko and S.G. Boguslavsky, 1966.
Deep current in the Atlantic. (In Russian; English abstract).
Okeanologiia, Akad. Nauk, SSSR, 6(2): 234-239.

currents, deep

Koshiakov, M.N., 1964
Indirect methods for the calculation of deep currents in the ocean. (In Russian).
Okeanologiia, Akad. Nauk, SSSR, 4(5):910.

currents, deep

Koshlyakov, M.N., 1961.
An example of the computation of the deep currents in the northeastern Pacific. (Abstract).
Tenth Pacific Sci. Congr., Honolulu, 21 Aug.-6 Sept., 1961, Abstracts of Symposium Papers, 344-345.

currents, deep

Koshlyakov, M. N., 1961

[Study of dynamics and kinematics of deep meridional currents applied to the north-eastern Pacific]
Trudy Inst. Okean., 52: 133-154.

currents, deep

Krivelevich, L.M. 1967.
Formation of the temperature field by the drift currents in the equatorial region. (In Russian).
Fisika Atmosfer. Okean. Izv. Akad. Nauk SSSR, 3(3): 344-347.

currents, deep

Lacombe, Henri, 1961

Mesures de courant à 1000 metres de profondeur à l'ouest de la côte espagnol (Cap Finisterre).
Cahiers Océanogr., C.C.O.E.C. 13(1): 9-13.

currents, deep

LaFond, E.C., 1962
Deep Current measurements with the Bathyscaph Trieste.
Deep-Sea Res., 9(2):115-116.

currents, deep

Lappo, S.S., 1963
On the deep currents in the northern part of the Atlantic Ocean. (In Russian).
Okeanologiia, Akad. Nauk, SSSR, 3(5):808-813.

currents, deep

Latun, V.S., 1963.
On the north-south deep water current below the zero-flow surface [i.e., layer of no motion]. (In Russian).
Izv., Akad. Nauk, SSSr, Ser. Geofiz., (8):1251-1258.

Translation:
Amer. Geophys. Union, (8):761-765.

currents, deep

Laughton, A.S., 1959.
The sea floor.
Science Progress, 47(186):230-249.

currents, deep

Lineikin, P.S., 1962.
On the zero surface and deep water currents of the northern part of the Atlantic Ocean.
Izv., Akad. Nauk, SSSR, Ser. Geofiz., (6):776-794.

In Russian

currents, deep

Lineikin, P.S., 1962.
The zero surface and the deep currents of the North Atlantic. (In Russian).
Izv., Akad. Nauk, SSSR, Ser. Geofiz., (6):776-794.

English Edit., (6):498-507.

currents, deep

Lineykin, P.S., 1959.
Certain problems of dynamics of deep-sea currents.
Meteorol. i Gidrol., (10):45-49.

Listed in Techn. Transl., 3(10):647.

(In Russian).

currents, deep

Maksimov, I.V., and V.N. Vorob'yev, 1962.
Contribution to the study of deep currents in the Antarctic Ocean. (In Russian).
Sovetsk. Antarktich. Expedits. Inform. Biull., (31):35-40.

Translation:
Scripta Tecnica, Inc., for AGU, 4(1):17-19.

currents, deep

Maksimov, I.V., and V.N. Vorob'yev, 1961.
The study of deep currents in the Antarctic Ocean (In Russian).
Biull. Sov. Antarkt. Eksped., 31:35-39.

Abstracted in:
Soviet-Bloc Res. Geophys., Astron., and Space, 1962(42):7.

JPRS 15, 068 OTS 61-11147-42.

currents, deep

McAllister, Raymond F., 1962.
Deep-current measurements near Bermuda.
Marine Sciences Instrumentation, Instr. Soc., Amer., Plenum Press, N.Y., 1:210-222.

currents, deep

Meshchersvii, V.I., 1959
[Measurement of deep currents.]
Trudy Gosud. Okeanogr. Inst., Leningrad,37: 79-84.
Translator: M. Slessers
M.O. 15047-62
P.O.:08836
USN-HO-TRANS-98-1960

currents, deep

Nan'niti, Tosio, and Hideo Akamatsu, 1966.
Deep current observations in the Pacific Ocean near the Japan Trench.
J. Oceanogr. Soc., Japan, 22(4):154-160.

currents, deep (southward)

Nan-niti, Tosio, Hideo Akamatsu and Toshisuke Nakai, 1965.
A deep current measurement in the Honshu Nankai, the sea south of Honshu, Japan.
Oceanogr. Mag., Tokyo, 17(1/2):77-86.

currents, deep

Nan'niti, T., H. Akamatsu, T. Nakai and K. Fuji, 1964.
An observation of a deep current in the south-eastern east sea of Tori shima.
Oceanogr. Mag., Tokyo, 15(2):113-122.

currents, deep

Nan'niti, T., A. Watanabe, H. Akamatsu and T. Nakai, 1963.
Deep current measurements in the cold water mass and vicinity, the Enshu Nada Sea.
Oceanogr. Mag., Japan Meteorol. Agency, 14(2):135-139.

JEDS Contrib. No. 43.

currents, deep

Nikitin, M.M., and N.I. Demianov, 1965.
On the deep-sea currents in the Arctic basin. (In Russian).
Okeanologiia, Akad. Nauk, SSSR, 5(2):261-263.

currents, deep

Nitani, Hideo, 1963.
On the analysis of deep sea in the region of the Kurile-Kamchatka, Japanese and Izu-Bonin Trench.
J. Oceanogr. Soc., Japan, 19(2):82-92.

currents, deep
*Nowroozi, Ali A., M. Ewing, J.E. Nafe and M. Fleigel, 1968.
Deep ocean current and its correlation with the ocean tide off the coast of northern California.
J. geophys. Res., 73(6):1921-1932.

currents, deep
Pochapsky, T.E., 1963.
Measurement of small-scale oceanic motions with neutrally-buoyant floats.
Tellus, 15(4):352-362.

currents, deep
Ponomarenko, G.P., 1963.
Study of deep currents of the equatorial region of the Atlantic Ocean on the tenth voyage of the 'Mikhail Lomonosov'. (In Russian).
Voprosy Geogr., No. 62:35-53.

Not seen

Abstract: Soviet Bloc Res., Geophys., Astron., Space, No. 95:50-51.

currents, deep
Pochapsky, T.E., 1966.
Measurements of deep water movements with instrumented neutrally buoyant floats.
J. geophys. Res., 71(10):2491-2504.

currents, deep
Romanovsky, V., 1962
Les méthodes de mesures de courants près du fond par grande profondeur.
In: Océanographie Géologique et Géophysique de la Méditerranee Occidentale, Centre National de la Recherche Scientifique, Villefranche sur Mer, 4 au 8 avril, 1961, 37-40.

currents, deep
Ryzhkov, Yu. G., 1961.
[Measurement of deep currents in the Black Sea by neutrally buoyant ultrasonic floats.]
Doklady Akad. Nauk, SSSR, 141(1):74-75.

currents, deep
Ryzhkov, Iu. G., and N.N. Karnaushenko, 1961.
[Measurement of deep currents in the Black Sea with the aid of an ultrasonic buoy of neutral buoyancy.]
Doklady Akad. Nauk, SSSR, 141(1):74-76.

currents, deep
Shkudova, G. Ya., 1964
The calculation of deep currents in the ocean. (In Russian).
Okeanologiia, Akad. Nauk, SSSR, 4(5):912.

currents, deep
Steele, J.H., J.R Barrett and L.V. Worthington, 1962
Deep currents south of Iceland.
Deep-Sea Res., 9(5):465-474.

~~currents, deep~~
Swallow, J.C., 1969.
Some features of deep currents measured by means of neutrally buoyant floats. (In Russian; English abstract).
Okeanologiia, 9(1):92-96.

currents, deep
Swallow, J.C., 1955.
A neutral-buoyancy float for measuring deep currents. Deep-Sea Res. 3(1):74-81.

currents, deep
Swallow, J.C., 1957.
Some further deep current measurements using neutrally buoyant floats. Deep-Sea Res., 4(2):93-104.

currents, deep
Swallow, J. C., and B. V. Hamon, 1960.
Some measurements of deep currents in the Eastern North Atlantic.
Deep-Sea Res., 6(2):155-168.

currents, deep
*Swallow, J.C. and L.V. Worthington, 1969.
Deep currents in the Labrador Sea.
Deep-Sea Res., 16(1):77-84.

currents, deep
Swallow, J.C., and L.V. Worthington, 1961
An observation of a deep countercurrent in the western North Atlantic.
Deep-Sea Res., 8(1): 1-19.

currents, deep
Swallow, J.C., and L.V. Worthington, 1957.
Measurement of deep currents in the western North Atlantic. Nature 179(4571):1183-1184.

currents, deep
Swallow, Mary, 1962
Deep currents in the oceans.
Discovery, 23:17-22.

Also in:
Collected Reprints, Nat. Inst. Oceanogr., Wormley, 10.

currents, deep
Sysoyev, N.N., 1961.
[The study of deep currents in the Pacific during the IGY period.]
Informats. Biull., Mezhd. Geofiz. God, Akad. Nauk, SSSR, No. 9:17-21.

currents, deep
Takenouti, Yositada, Tosio Nan'niti and Masashi Yasui, 1962.
The deep-current in the sea east of Japan.
Oceanogr. Mag., Japan Meteorol. Agency, 13(2): 89-101.

currents, deep
Treshnikov, A.F., 1960.
[The Arctic discloses its secrets: new data on the bottom topography and the waters of the Arctic Basin.]
Priroda, 1960(2):25-32.

Translation by E.R. Hope T357R

currents, deep(data)
Volkmann, G., 1962
Deep current observations in the Western North Atlantic.
Deep-Sea Res., 9(5):493-500.

currents, deep
Wüst, Georg, 1958.
Über Stromgeschwindigkeiten und Strommengen in der Atlantischen Tiefsee.
Geol. Rundschau, 47(1):187-195.

currents, deep
Wüst, Georg, 1957.
Strongeschwindigkeiten und Strommengen in den Tiefen des Atlantischen Oceans.
Wiss. Ergeb. Deutschen Atlantischen Exped., --- Meteor. 1925 - 1927, 6(2) (6):261-420.

Reviewed by H. Stommel in:
Trans. A. G. U., 39 (6):1171-1172.

currents, deep
Wyrtki, Klaus, 1961
The thermohaline circulation in relation to the general circulation in the oceans.
Deep-Sea Res., 8(1): 39-64.

currents, deep
Yasui, Masashi, and Tosio Nan'niti, 1962.
III. Deep-sea current measurements during the JEDS-4.
Oceanogr. Mag., Japan Meteorol. Agency, 13(2): 133-136.

JEDS Contrib. No. 31.

Current deflection

current deflection
Oi, M., 1956.
Deflection of an ocean current due to a submarine ridge. Pap. Oceanogr. Inst., Fla. Univ. Studies, 22(2):7-12.

Currents, density

density currents
Allen, F. H., and W. A. Price, 1959.
Density currents and siltation in docks and tidal basins.
Dock & Harbour Auth., July, 1959: 5 pp.

Density currents
Allen, F. H., and W. A. Price, 1959.
Density currents and siltation in docks and tidal basins.
Dock & Harb. Auth., 40(465):72-76.

currents, density
Balaramamurty, C., and A.A. Ramasastry, 1957.
Distribution of density and the associated currents at the sea surface in the Bay of Bengal. Indian J. Meteorol., & Geophys., 8(1): 88-92.

density currents
Bell, H.S., 1947
The effect of entrance mixing on the size of density currents in Shaver Lake. Trans. Am. Geophys. 28(5):780-791, 5 text figs.

density currents
Bell, H.S., 1942.
Stratified flow in reservoirs and its use in prevention of silting. U.S. Dept. Agri. Misc. Publ. 491:46 pp., 40 textfigs.

currents, density
Berrit, G.R., 1961
Etude des conditions hydrologiques en fin de saison chaude entre Pointe-Noire et Loanda.
Cahiers Oceanogr., C.C.O.E.C., 13(7): 456-461.

currents, density
Bradley, W.H. 1969.
Vertical density currents - II.
Limnol. Oceanogr. 14 (1): 1-3

currents, density (vertical)
Bradley, W.H., 1965.
Vertical density currents.
Science, 150(3702):1423-1428.

currents
Craya, A., 1951.
Critical regimes of flow with density stratification. Tellus 3:28-42.

currents, density
Crosnier, A., 1964.
Fonds de peche le long des cotes de la Republique Federale du Cameroun.
Cahiers, O.R.S.T.R.O.M., No. special:133 pp.

currents, density
*Daly, Bart J., and William E. Pracht, 1968.
Numerical study of density-current surges.
J.fluid Mech., 11(1):15-30.

density currents
Fry, A.S., M.A. Churchill and R.A. Elder, 1953.
Significant effects of density currents in TVA's integrated reservoir and river system.
Proc. Minnesota Int. Hydraulics Conv., Sept. 1-4, 1953:335-354, 10 textfigs.

density currents
Geza, B., and K. Bogich, 1953.
Some observations on density currents in the laboratory and in the field.
Proc. Minnesota Int. Hydraulics Conv., Sept. 1-4, 1953:387-400, 5 textfigs.

density currents
Hamada, T., 1953.
Density current problems in an estuary.
Proc. Minnesota Int. Hydraulics Conv., Sept. 1-4, 1953:313-320, 2 textfigs.

density currents
Howard, C.S., 1953.
Density currents in Lake Meade. Proc. Minnesota Int. Hydraulics Conv., Sept. 1-4, 1953:355-368, 6 textfigs.

currents, density
Ippen, A.T., and D.R.F. Harleman, 1952.
Steady state characteristics of subsurface flow.
Proc. N.B.S. Semicent. Symp., Gravity Waves, June 18-20, 1952, Nat. Bur. Stand. Circ. 521: 79-93, 9 textfigs.

currents, density
Ito, Takeshi, Sei-ichi Sato, Tsutomu Kishi and Masateru Tominaga, 1960
On the density currents in the estuary.
Coastal Eng., Japan, 3: 21-31.

currents, density
Kiilerich, 1943.
The hydrography of the west Greenland fishing banks. Medd. Komm. Danmarks Fisk- og Havundersøgelser, Ser. Hydrogr., 3(3):45 pp., 7 pls.

currents, density
Lacombe, H., 1965.
Courants de densité dans le détroit de Gibraltar.
La Houille Blanche, 20(1):38-43.

currents, density
Lebedev, V.L., 1957.
Use of hydraulic functions in calculating density currents in ocean straits.
Vestnik, Mosk. Univ., 2:229-232.

currents, density
Mason, David T., 1966.
Density-current plumes.
Science, 152 (3720):354-355.

currents, density
Michon, X., J. Goddet and R. Bonnefille, 1955.
Etude theorique et experimentale des courants de densite. B500/53, 2 vols., 186 pp.

currents, density
Middleton, Gerard V., 1966.
Experiments on density and turbidity currents.
II. Uniform flow of density currents.
Canadian J. Earth Sci., 3(5):627-637.

currents, density
Moreira da Silva, P., 1956.
Um nuvo metodo para e determinacao das correntes de densidadeadaptado a Tabua. H.O. 614. Anais Hidr., 15:173-177.

currents, density
Nanniti, T., 1951.
On the variation of the oceanographical condition along the so-called "C" line (38 N., from 141 E. to 153E) from August 1948 to December 1949.
Ocean. Mag., Tokyo, 3(1):27-41.

density currents
Nizery, A., and J. Bonnin, 1953.
Observations systematiques de courants de densité dans une retenue hydroelectrique.
Proc. Minnesota Int. Hydraulics Conv., Sept. 1-4, 1953:369-386, 28 textfigs.

currents, density
Price, W.A., and M.P. Kendrick, 1962.
Density currents in estuary models.
La Houille Blanche, 17(5):611-628.

currents, density
* Riddell, John F., 1969.
A laboratory study of suspension-effect density currents.
Can. J. Earth Sci., 6(2): 231-246.

density currents
Smith, W.E., 1946.
Some observations on water-level and other phenomena along the Bosphorus. Trans. Am. Geophys. Union, 27(1):61-68, 15 textfigs.

density currents
Stommel, H., 1953.
The role of density currents in estuaries.
Proc. Minnesota Int. Hydraulics Conv., Sept. 1-4, 1953:305-312, 7 textfigs.

currents, density
Sysoev, N.N., 1961.
Thermal streams from the bottom of the Black Sea.
Doklady Akad. Nauk, SSSR, 139(4):974-975.

currents, density
Valembois, J., 1965.
Travaux recents sur les courants de densité.
La Houille Blanche, 20(1):15-20.

currents, density
Wiegel, R.L., and J.W. Johnson, 1960
Ocean currents, measurement of currents and analysis of data.
Waste disposal in the marine environment. Pergamon Press: 175-245.

current (direction)
Chidambaram, K., A.D.I. Rajandran, and A.P. Valsan, 1951.
Certain observations on the hydrography and Biology of the Pearl Bank, Tholayviam Paar off Tuticorin in the Gulf of Manaar in April 1949.
J. Madras Univ., Sect. B, 21(1):48-74, 2 figs.

dip currents
Van Straaten, L.M.J.U., 1952.
Current rips and dip currents in the Dutch Wadden Sea. K. Nederl. Akad. Wetensch., Proc., B, 55(3):228-238, 2 photos, 10 textfigs.

currents, drift
Belyaev, V.S., and A.G. Kolensnikov, 1966.
On the cause of formation of intertial oscillation in the drift currents. (In Russian).
Fisika Atmosferi i Okeana, Izv., Akad. Nauk, SSSR, 2(10):1104.

currents, drift
Belyaeva, I.P., 1966.
Surface drift currents in the Norwegian and Greenland seas. (In Russian).
Material. Sess. Uchen. Soveta PINRO Result. Issledovan. (6):130-141.

currents, drift
Belyaeva, I.P. and S. Boganov, 1966.
Surface drift currents in the Norwegian and Greenland seas. (In Russian).
Materiali, Sess. Uchen. Soveta PINRO Rez. Issled., 1964, Minist. Ribn. Khoz., SSSR, Murmansk, 141-148.

currents, drift
Belyaeva, I.P., and S.I. Sorochenko, 1967.
Drift currents in the eastern Norwegian Sea in the spring of 1965. (In Russian).
Mater. Sess. Uchen. Sovet. PINRO Rezult. Issled. 1965, Poliarn. Nauchno-Issled. Proektn. Inst. Morsk. Ribn. Khoz. Okeanogr. (PINRO), 8:133-141.

currents drift
Belyakov, L.N., 1969.
Summer desalination of the sea surface and dynamics of the drift current and ice movement in the Arctic Basin.
Okeanologiia 9(3): 424-429. (In Russian; English abstract)

currents, drift
Dobroklonsky, S.V., 1969.
Drift currents in the sea with an exponentially decaying viscosity coefficient. (In Russian; English abstract).
Okeanologiia, 9(1):26-33.

currents, drift
Ekman, V.W., 1953.
Studies on ocean currents. Results of a cruise on board the "Armauer Hansen" in 1930 under the leadership of Bjørn Helland-Hansen. Geograf. Pub. 19(1):106 pp. (text), 122 pp. (92 pls.).

CURRENTS, drift
Fomin, L.M., 1968.
A frequency analysis of the drift current in the ocean forced oscillations in the velocity field.
Mitt. Inst. Meeresk. Univ. Hamburg, 10:81-95.

currents, drift
Gezentsvei, A.N., 1954.
Divergence of drift currents and heat transmission by the currents in the North Pacific and North Atlantic. Trudy Okeanol. Inst., 9:54-118.

currents, drift
Gonella, J., 1971.
The drift current from observations made on the Bouée Laboratoire.
Cah. océanogr. 23(1): 19-33.

currents, drift
Hansen, W., 1951.
Beobachtungen des Windstaus und Triftstroms in Modellkanal. Deutsche Hydrogr. Zeits. 4(3):81-91, 8 textfigs.

currents
Hansen, W., 1950.
Triftstromund Windstau. Deutsch. Hydro. Zeits. 3(5/6):303-313.

currents, drift
Hela, I., 1952.
Drift currents and permanent flow. Soc. Sci. Fenn., Comm. Phys.-Math. 16(14):27 pp.

drift currents
Hidaka, K., 1955.
3. Divergence of surface drift currents in terms of wind stresses with special application to the location of upwelling and sinking. Jap. J. Geophys. 1(2):47-56. textfig.

currents, drift
Hidaka, K., 1951?
10. Drift currents in an enclosed ocean. III.
11. Circulation in a zonal ocean induced by a planetary wind system. C.R. Symposium sur la circulation générale des océans et de l'atmosphere, Assoc. Météorol., U.G.G.I., 1951:70, 71.

currents
Hidaka, K., 1951.
Drift currents in an enclosed ocean. Pt. III. Geophys. Notes 4(3):19 pp., 5 figs.

currents, drift
Hidaka, K., 1950.
Drift currents in an enclosed ocean. Pt. 1. Geophys. Notes, Tokyo Univ., 3(23):23 pp., 15 fig

currents, drift
Hidaka, K., 1950.
Drift currents in an enclosed ocean. Pt. II. Geophys Notes, Tokyo Univ., 3(38):10 pp., 1 text-fig.

currents, drift
Hidaka, K., 1950.
Drift current in an enclosed ocean. Pt. II. J. Ocean. Soc., Japan, (Nippon Kaiyo Gakkaisi) 6(2):39-48.

currents, surface drift
Hidaka, K., 1950.
The relation between surface drift current and wind. J. Ocean. Soc., Japan, (Nippon Kaiyo Gakkaisi) 6(2):8-10. (In Japanese; English summary).

currents
Hidaka, K., 1947
Drift currents in an enclosed sea and the rotation of the earth. Trans. Am. Geophys. Union, 28(4):549-558, 2 textfigs.

currents
Hidaka, K., 1946.
A method of integration of differential equations of drift currents in the ocean. Geophys. Mag., Tokyo, 14:47-55.

currents drift (Ekman)
Hunkins, Kenneth, 1966.
Ekman drift currents in the Arctic Ocean.
Deep-Sea Res., 13(4):607-620.

currents, drift
Ichie, T., 1950.
A note on the effect of inertia terms upon the drift currents. Ocean. Mag., Tokyo, 2(2):41-44, 2 textfigs.

drift current
Ichie, T., 1949.
On the theory of drift current in an enclosed sea. Ocean. Mag., Tokyo, 1(2):128-132, 1 textfig

currents, drift
Ivanov, Iu. A., and B.A. Tareev, 1960.
On the computation of the vertical component of the velocity of drift currents. Physics of the sea. Hydrology. (In Russian).
Trudy Morsk. Gidrofiz. Inst., 22:3-4.
USN-HO TRANS-157
M. SLESSERS
HO 16782
PO 45201
1962

currents, drift
Johnson, J.W., 1960
The effect of wind and wave action on the mixing and dispersion of wastes.
Waste Disposal in the Marine Environment, Pergamon Press: 328-343.

currents, drift
Kagan, B.A., 1968.
On the relationship of drift currents and turbulence in the upper layer of the sea with a changing wind stress. (In Russian).
Fisika Atmosfer. Okean., Izv. Akad. Nauk, SSSR, 4(9):1004-1007.

currents, drift
Kagan, B.A., 1965.
The use of the quasistationary model of the interaction between atmospheric and oceanic boundary layers for the computation of the water temperature and current in the North Atlantic. (In Russian; English abstract).
Fizika Atmosferi i Okeana, Izv., Akad. Nauk, SSSR 1(8):845-852.

currents, drift
Kagan, B.A., 1964.
An estimation of energy dissipation of turbulence in purely drift currents. (In Russian).
Okeanologiia, Akad. Nauk, SSSR, 4(1):3-8.

currents, drift
Kagan, B.A., 1962
On the established drift currents in the deep sea areas. (In Russian).
Okeanologiia, Akad. Nauk, SSSR, 2(6):974-980.

currents, surface drift
Kagan, B.A., 1961.
Concerning the computation of surface drift currents.
Issled. Severnoi Chasti Atlanticheskogo Okeana, Mezhd. Geofiz. God, Leningradskii Gidrometeorol. Inst., 1:98-106.

currents, drift
Kamenkovich, V.M., 1960.
The quasistationary property of drift currents in the ocean. (In Russian).
Izv., Akad. Nauk, SSSR, Ser. Geofiz., (1):74-82.

English transl., (1):45-49.

currents, drift
Kamenokovich, V.M., 1956.
On the influence of the relief of the bottom on a pure drift current in a homogeneous shoreless sea
Izv. Akad. Nauk, SSSR, Ser. Geofiz., 1956(10): 11820

drift current (Ekman)
Katz, B., R. Gerard and M. Costin, 1965.
Response of dye tracers to sea surface conditions
J. geophys. Res., 70(22):5505-5513.

currents, drift
Kozlov, B.F., 1963.
On the influence of the variations of the coefficients of vertical exchange on drift currents. (In Russian).
Izv., Akad. Nauk, SSSR, Ser. Geofiz., (7):1100-1107.

currents, drift
Krauss, W., 1965
Theorie des Triftstromes und der virtuellen Reibung im Meer.
Deutsche Hydrop Zeits. 18(5):193-210

currents, drift
Krivelevich, L. M., 1964.
A hydrodynamic model for drift currents at the equator. (In Russian).
Okeanologiia, Akad. Nauk, SSSR, 4(6):962-967.

currents, drift
Krivelevich, L.M., 1964
A hydrodynamic model for the study of drift currents in the equatorial area. (In Russian)
Okeanologiia, Akad. Nauk, SSSR, 4(5):919-920.

CURRENTS, DRIFT
Miyazaki, M., 1950.
On the distribution of the drift current and the suspending sands in a channel. J. Ocean. Soc. Tokyo, 6(1):15-17, 4 textfigs. (In Japanese with English abstract).

currents, drift
Ogura, S., 1930.
On the tidal and drift currents at Urkt Roads, North Sakhalin. Hydrogr. Rev. 7:57-65, 3 textfigs

currents, drift
Ozmidov, R.V., 1961.
Pure drift current generated by arbitrary conditions.
Trudy Inst. Okeanol., Akad. Nauk, SSSR, 38:110-120.

currents, drift
Ozmidov, R.V., 1959.
Extension of Ekman's theory of unsteady purely drift currents to the case of arbitrary wind.
Doklady Akad. Nauk, SSSR, 128(5):913-916.

drift currents
Saito, Y., 1950.
One problem of the stationary drift-circulative current in a polytropic ocean. J. Met. Soc., Japan, 28(2):61-66, textfigs. (In Japanese with English summary).

drift currents
Saito, Y., (undated in English).
One problem of the stationary drift-circulative current in a polytropic ocean. J. Ocean. Soc., Tokyo, (Nippon Kaiyo Gakkaisi) 5(2/4):130-142.

currents, drift
Stommel, H., 1954.
Serial observations of drift currents in the Central North Atlantic Ocean. Tellus 6(3):203-214, 6 textfigs.

currents, drift
Tareev, B.A., 1958.
Drift currents in the sea under the action of a wind varying with time. (In Russian).
Izv., Akad. Nauk, SSSR, Ser. Geofiz., (5):55-

currents, drift
Tareyev, B.A., A.V. Fomitchev, 1963
On surface currents of the South Ocean. (In Russian).
Mezhd. Geofiz. Komitet, Prezidiume, Akad. Nauk SSSR, Rezult. Issled. Programme Mezhd. Geofiz. Goda, Okeanol. Issled., (8):24-33.

currents, drift
Yampolsky, A.D., 1966.
On the relation of drift current and wind stress spectra. (In Russian;English abstract).
Fisika Atmosferi i Okeana,Izv.Akad.Nauk,SSSR, 2(11):1186-1192.

currents, drift (theoretical)
Yampolsky, A.D., 1963.
Computing methods for the inertial fluctuations of hydrological elements. Questions of physical oceanology. (In Russian; English abstract).
Trudy Inst. Okeanol., Akad. Nauk, SSSR, 66:142-149.

currents, drift
Yampol'skiy, A.D., 1961.
[On variations of hydrological elements with an inertial period.]
Izv. Akad. Nauk, SSSR, Ser. Geofiz., (3):445-452.

Engl. Abstr. in:
Soviet-Bloc Research on Geophysics, Astronomy and Space,(10):8. OTS

currents, drift
Zubkova, I.D., 1959.
Scheme for drift currents in the coastal zone on a sloping inclined bottom in relation to the direction and velocity of the wind. Swell and currents in the sea. Loss of salt from the sea into the atmosphere. (In Russian).
Trudy Morsk. Gidrofiz. Inst., 15:43-55.

currents (drogues)

currents (drogues)
Griffiths, Raymond C., 1965.
A study of ocean fronts off Cape San Lucas, Lower California.
U.S.F.W.S. Spec. Sci. Rept., Fish., No. 499:54 pp

currents due to swell

currents due to swell
Laurent, J., 1949.
Essais sur modèle réduit en liaison avec des observations en nature. Leur importance pour l'étude des courants de houle.
C.R. Acad. Sci., Paris, 229(23):12 3-1204.

Vennin, J., 1949. currents, caused by swell
Études poursuivies par la Société d'Études Hydrographiques et Océanographiques au Maroc sur la région de la côte Atlantique du Maroc en liaison avec les essais sur modèle réduit executé au Laboratoire Central Hydraulique.
Rev. Gen. Hydraul. 15(54):285-294.

currents, dynamic

Alekseev, A.P., 1959. dynamic currents
[Polar front in the Norwegian and Greenland Seas.]
Trudy Poliarnozo-Nauchno-Issledovat. Inst. Morsk. Ribn. Chozia i Okean. imeni N.M. Knipovich (PINRO) 11:60-73.

dynamic currents
Allain, Ch., J. Dardignac et A. Vincent,1967.
L'hydrologie et les courants généreux du Détroit de Danemark et du nord de la Mer d'Irminger du 20 mars au 8 mai 1963.
Revue Trav. Inst. (scient. tech.) Pêches marit. 31 (3):275-305.

currents, dynamic
Arseniev, V.S., 1965.
Circulation in the Bering Sea. (In Russian; English abstract).
Okeanolog. Issled., Rezult. Issled. Programme Mezhd. Geofiz. Goda, Mezhd. Geofiz. Komitet Presidiume, Adad. Nauk, SSSR, No. 13:61-65.

currents, dynamic
Astok, V.K., 1965.
Currents inthe Gulf of Finland studied on the basis of treatment of hydrographical factors with the aid of the dynamic method.(In Russian)
Okeanologiia, Akad. Nauk, SSSR, 5(5):825-833.

currents, dynamic
Astok, V.K., 1964
Currents in the Gulf of Finland based on the treatment of hydrological observations and the application of the dynamic method. (In Russian).
Okeanologiia, Akad. Nauk, SSSR, 4(5):922.

currents, dynamic
Austin, G.B., jr., 1955.
Some recent oceanographic surveys of the Gulf of Mexico. Trans. Amer. Geophys. Union 36(5):885-898.

dynamic currents
Austin, T.S., 1954.
Mid-Pacific oceanography III. Transequatorial waters, August-October 1951.
U.S.F.W.S. Spec. Sci. Rept. - Fish. No. 131: numerous pp. (unnumbered), 13 figs. (multilithed)

currents, dynamic (lakes)
Ayers, J.C., 1956.
A dynamic height method for the determination of currents in deep lakes. Limnol. Oceanogr., 1(3): 150-161.

dynamic currents
Ayers, John C., David C. Chandler, George H. Lauf, Charles F. Powers and E. Bennett Henson, 1958
Currents and water masses of Lake Michigan.
Great Lakes Res. Inst., Publ. No. 3:169 pp.

currents, dynamic
Bailey, W.B., 1957.
Oceanographic features of the Canadian Archipelago. J. Fish. Res. Bd., Canada, 14(5):731-769.

Dynamic Currents
Bennett, E. B., 1959.
Some oceanographic features of the Northeast Pacific Ocean during August 1955. J. Fish. Res. Bd., Canada 16(5):565-633.

currents, dynamic
Bennett, Edward B., and Milner B. Schaefer, 1960
Studies of physical, chemical and biological oceanography in the vicinity of the Revilla Gigedo Islands during "Island Current Survey" of 1957.
Bull. Inter-American Tropical Tuna Comm., 4(5): 219-317. (Also in Spanish).

currents, dynamic
Berrit, Georges Roger et Jean-Rene Donguy,1964
La petite saison chaude en 1959 dans la region orientale du Golfe de Guinee.
Cahiers Océanogr., C.C.O.E.C., 16(8):657-672.

dynamic currents
Bjerknes,Jacob,1966.
Survey of El Niño 1957-58 in its relation to tropical Pacific Meteorology.
Bull.inter-Am.trop.Tuna Commn., 12(2):25-86.

currents, dynamic
Burkov, V.A., and I.M. Ovchinnikov, 1960
[Structure of zonal streams and meridional circulation in the central Pacific during the Northern Hemisphere winter.]
Trudy Inst. Okeanol., 40: 93-107.

currents, dynamic
Burkov, V.A., S.G. Panfilova L.K. Moisseev, and A.B. Rubin 1971.
Circulation and water masses of the southeastern Pacific. (In Russian; English abstract).
Trudy Inst. Okeanol. P.P. Shirshova, Akad. Nauk SSSR 89:9-32

currents, dynamic
Bourkov, V.A., and Yu. V. Pavlova, 1963
Geostrophic circulation on the surface of the northern part of the Pacific in summer. (In Russian; English abstract).
Okeanol. Issled., Rezhult. Issled. Programme Mezhd. Geofiz. Goda, Mezhd. Geofiz. Komitet, Presidiume Akad. Nauk, SSSR, No. 9:21-31.

dynamic currents
Buzdalin, Yu. I., and A.A. Elisarov, 1962.
Hydrological conditions in Newfoundland and Labrador areas in 1960.
Sovetskie Riboch. Issledov. v Severo-Zapadnoi Atlant. Okeana, VINRO-PINRO, Moskva, 155-171.

In Russian; English summary

dynamic currents
Callaway, E.B., 1951.
Graphical determination of specific-volume anomaly and current. Trans. Amer. Geophys. Union 32(5):719-728, 6 textfigs.

dynamic currents
Central Meteorological Observatory, 1949.
Report on sea and weather observations on Antarctic whalingground (1948-1949). Ocean. Mag., Tokyo, 1(3):142-173, 11 textfigs.

dynamic currents
Cheney, L.A., and F.M. Soule, 1951.
International ice observations and Ice Patrol Service in the North Atlantic Ocean.
U.S.C.G. Bull. 35:116 pp., 33 textfigs.

dynamic currents
Clowes, A.J., 1934.
Hydrology of the Bransfield Strait. Discovery Repts. 9:1-64, 68 textfigs.

dynamic currents
Coachman, L.K., and C.A. Barnes, 1961
The contribution of Bering Sea water to the Arctic Ocean.
Arctic 14(3):146-161.

dynamic currents
Defant, A., 1941.
Die absolute Berechnung ozeanischen nach dem dynamischen Verfahren. Ann. Hydrogr., usw., 69: 169-173.

dynamic currents
Defant, A., 1938.
Aufbau und Zirculation des atlantischen Ozeans.
Sitzber. Preuss. Akad. Wiss., Physik.-Math. Kl. 14:145-171, 7 pls.

currents, dynamic
Deryugin, K.K., 1964
The first voyage of the complex training and research oceanological expedition on the "Bataisk" ship designed for practical and research work. (In Russian).
Materiali Vtoroi Konferentsii, Vzaimod. Atmosfer. i Gidrosfer. v Severn. Atlant. Okean., Mezhd. Geofiz. God, Leningrad. Gidrometeorol. Inst., 269-278.

dynamic currents (data)
Dietrich, G., 1957.
Schichtung und Zirkulation der Irminger-See im Juni 1955. Ber. Dtsch. Komm. Meeresf.14(4): 255-312.

dynamic currents
Dietrich, G., 1937.
Über Bewegung und Herkunft des Golfstromwassers. Veröff. Inst. Meereskunde, n.f., A. Geogr.-Naturwiss. Reihe, 33:53-91, 26 textfigs.

dynamic currents
Dietrich, G., 1937.
Die Lage der Meeresoberfläche im Bruckfeld von Ozean und Atmosphäre mit besonderer Berücksichtigung des westlichen Nordatlantischen Ozeans und des Golfs von Mexico. Veröff. Inst. Meereskunde, n.f., A. Geogr.-Naturwiss. Reihe, 33:1-52, 19 textfigs.

currents, dynamic
Dobrovolski, A.D., 1949.
[On the equation of the position of the surface of no motion for dynamical calculation in the northern Pacific Ocean.] Trudy Inst. Okeanol., 4:3-26.

currents dynamic
Dodimead, A.J., F. Favorite and T. Hirano, 1964.
Review of oceanography of the subarctic Pacific Region. Salmon of the North Pacific Ocean. II. Collected Reprints, Tokai Reg. Fish. Res. Lab., No. 2:187 pp.
(In Japanese).

dynamic currents
Doe, L.A.E., 1955.
Offshore waters of the Canadian Pacific Coast. J. Fish. Res. Bd., Canada, 12(1):1-34, 29 textfigs.

dynamic currents
Doe, L.A.E., 1952.
Currents and net transport in Loudoun Channel, April 1950. J. Fish. Res. Bd., Canada, 9(1): 42-64, 17 textfigs.

dynamic currents
Eggvin, J., 1933.
A Norwegian fat-herring fjord. An oceanographical study of the Eidsfjord. Repts. Norwegian Fish. Mar. Invest. 4(6):22 pp., 11 textfigs.

currents, dynamic
Elizarov, A.A., 1967.
Some results of oceanographic studies in the fishery areas off South-West Africa. (In Russian; English abstract).
Okeanologiia, Akad. Nauk, SSSR, 7(3):445-449.

dynamics
Erimesco, P., 1959
Quelques observations sur la dynamique de l'Océan Atlantique sur la côte Marocaine. Bul. l'Inst. Pêches Mar. Maroc. (3): 3-10.

dynamic currents
Eyries, M., and M. Menache, 1953.
Contribution à la connaissance hydrologique de l'Océan Indien entre Madagascar et La Réunion. Bull. d'Info., C.C.O.E.C. 5(10):433-438, 9 pls.

currents, dynamic
Favorite, Felix, 1967.
The Alaskan Stream.
Bull. Int. North Pac. Fish. Comm., 21:1-20.

currents, dynamic
Fhomin, L.M., 1959
[Possibilities of calculations of sea currents with dynamic method.]
Biull Okeanograf.Komissii, Akad. Nauk,SSSR, (3): 82-900

currents, dynamic
Fomichev, A.V., 1968.
Studies on the currents in the northeastern Indian Ocean. (In Russian).
Trudy, Vses. Nauchno-Issled. Inst. Morsk. Ribn. Okeanogr (VNIRO) 64. Trudy Azovo-Chernomorsk. Nauchno-Issled. Inst. Morsk. Ribn. Khoz. Okeanogr. (AscherNIRO), 28:65-93.

currents, dynamic
Fomitchev, A.V., 1964.
The investigation of the north Indian Ocean currents. Investigations in the Indian Ocean (33rd voyage of the E/S "Vitiaz"). (In Russian).
Trudy Inst. Okeanol., Akad. Nauk, SSSR, 64:43-50.

dynamic currents
Fukuoka, Jiro, 1965(1967).
Condiciones meteorologicas e hidrograficas de los mares adyacentes a Venezuela 1962-1963.
Memoria Soc.Cienc.nat.La Salle, 25(70/71/72):11-38.

dynamic currents
Ginés, Hno., y R. Margalef, editores, 1967.
Ecología marina.
Fundación La Salle de Ciencias Naturales, Caracas Monografía 14: 711 pp.

Dynamic currents
Grancini, Gianfranco, Antonino Lavenica
Ferruccio Mosetti, 1969
Ricerche oceanografiche nel Golfo di Taranto (indaginifisiche del luglio 1968)
Atti Ipt. Veneto Sci. 127: 309-326.
Also Conv. Osserv. Geofio, Sper. Trieste 192 (bio)

currents, (geostrophic)
dynamic
Gruzinov, V.M., 1964
Geostrophic currents in the subpolar front zone in the northern part of the Atlantic Ocean. (In Russian).
Okeanologiia, Akad. Nauk, SSSR, 4(2):243-248.

dynamic currents
Guyot, M.A., 1951.
L'hydrologie du canal de Sicile. Bull. d'Info., C.C.O.E.C., 3(7):269-280, 8 pls.

dynamic currents
Hachey, H.B., 1947
Water transports and current patterns for the Scotian shelf. J. Fish. Res. Bd., Canada 7(1):1-16, 7 text figs.

currents, dynamic
Hata, Katsumi, 1963.
The report of drift bottles released in the North Pacific Ocean. (In Japanese; English abstract).
J. Oceanogr. Soc., Japan, 19(1):6-15.

currents, dynamic
Hayami, S., H. Kawai and M. Ouchi, 1955.
On the theorem of Helland-Hansen and Ekman and some of its applications.
Rec. Ocean. Wks., Japan, 2(2):56-67, 9 textfigs.

dynamic currents (figs.)
Hermann, F., 1952.
Hydrographic conditions in the southwestern part of the Norwegian Sea, 1951. Cons. Perm. Int. Expl. Mer, Ann. Biol. 8:23-26, Textfigs. 4-12.

currents, dynamic
Hidaka, K., 1958
Surface dynamical topography over 1000-db surface derived from the result of the "Operation EQUAPAC" in the summer 1956. (Abstract).
Proc. Ninth Pacific Sci. Congr., 16: 226.

currents, dynamic
Hidaka, K., 1957.
Influence of friction on geostrophic flow: sources of errors in dynamic computation of ocean current. J. Oceaogr. Soc., Japan, 13(2): 37-49.

currents, dynamic
Hidaka, K., 1955.
Dynamical computation of ocean currents in a vertical section occupied across the Equator.
Jap. J. Geophys., 1(2):57-60.

currents, dynamic
Highley, E. 1967.
Oceanic circulation patterns off the east coast of Australia.
Comm. Sci. Ind. Res. Org. Australia, Div. Fish. Oceanogr. Tech. Pap. 23:1-19.

dynamic currents
Ichiye, T., 1962.
Circulacion y distribucion de la masa de agua en el Golfo de Mexico. (In Spanish and English).
Geofisica Internacional, Rev. Union Geofis. Mexicana, Inst. Geofis., Univ. Nacional Autonoma de Mexico, 2(3):47-76, 22 figs.

dynamic currents
Ichie, T., K. Tanioka, and T. Kawamoto, 1950.
Reports of the oceanographical observations on board the R.M.S. "Yushio Maru" off Shionomisaki (Aug. 1949). Papers and Repts., Ocean., Kobe Mar. Obs., Ocean. Dept., No. 5:15 pp., 33 figs. (Odd atlas-sized pages - mimeographed).

currents, dynamic
IIda, H., 1960
On the dynamical structure of the Tsugaru warm current. Mem. Kobe Mar. Obs., 14: 13-18.

currents, dynamic
Ivanov, J.A., and N.S. Smetanina, 1960
[Geostrophic currents in the Indian sector of the Antarctic waters] Okean. Issle., IGY Com., SSSR: 100-103.

dynamic currents
Ivanov, Yu. A., 1961
[Horizontal circulation of the Antarctic waters.]
Mezhd. Kom. Mezhd. Geofiz. Goda, Presidiume Akad. Nauk, SSSR, Okeanolog. Issled., (3): 5-29.

dynamic currents
Jacobsen, J.P., 1943.
The Atlantic Current through the Faroe-Shetland Channel and its influence on the hydrographical conditions in the northern part of the North Sea the Norwegian Sea and the Barents Sea.
Cons. Perm. Int. Expl. Mer, Rapp. Proc. Verb. 112:5-47, 29 textfigs.

dynamic currents
Jacobsen, J.P., and A.J.C. Jensen, 1925.
Examination of hydrographical measurements from the Research Vessels "Explorer" and "Dana" during the summer of 1924. Cons. Perm. Int. Expl. Mer, Rapp. Proc. Verb. 37:31-84, Figs. 8-52.

dynamic currents
Jakleln, A., 1936
Oceanographic investigations in East Greenland waters in the summers of 1930-1932. Skr. om Svalbard og Ishavet No. 67:79 pp., Pl. 2, 28 text figs.

dynamic currents
Japan, Hakodate Marine Observatory, 1957.
[Report of the oceanographic observations in the sea east of Tohoku District from February to March 1956.]
Bull. Hakodate Mar. Obs., No. 4:49-57.
1-9.

dynamic currents
Japan, Hakodate Marine Observatory, 1957.
[Report of the oceanographic observations in the sea east of Tohoku District from May to June 1956]
Bull. Hakodate Mar. Obs., No. 4:113-119.
9-15.

currents, dynamic
Japan, Hakodate Marine Observatory, 1956.
[Report on the oceanographical observation of the sea off Tohoku District made on board the R.M.S. Yoshio-maru in summer 1953 (1st paper).]
Bull. Hakodate Mar. Obs., No. 2:1-3.

currents, dynamic
Japan, Kobe Marine Observatory 1967.
Report of the oceanographic observations in the sea south of Honshu from October to November 1963. (In Japanese)
Bull. Kobe mar. Obs. No. 178: 41-49.

dynamic currents
Japan, Kobe Marine Observatory, 1962
Report of the oceanographic observations in the cold water region south of Enshu Nada from October to November, 1960. (In Japanese).
Res. Mar. Meteorol. and Oceanogr., July-Dec., 1960, Japan Meteorol. Agency, No. 28: 43-51.

dynamic currents
Japan, Kobe Marine Observatory, Oceanographical Section, 1962
Report of the oceanographical observations in the sea south of Honshu from July to August 1961. (In Japanese).
Res. Mar. Meteorol. & Oceanogr., 30:39-48.

Also in:
Bull. Kobe Mar. Obs., No. 173(3). 1964

dynamic currents
Japan, Kobe Marine Observatory 1962
Report of the oceanographic observations in the sea south of Honshu from July to August. 1960. (In Japanese).
Res. Mar. Meteorol. and Oceanogr., July-Dec., 1960, Japan Meteorol. Agency, No. 28: 36-42.

dynamic currents
Japan, Kobe Marine Observatory, 1958.
[Report of the oceanographic observations in the sea south of Honshu from November to December, 1957.]
J. Oceanogr., Kobe Mar. Obs., (2), 10(1):21-28.

in May 1957.]
Ibid., 9(2):69-78.

in August 1957.]
Ibid., 79-86.

dynamic currents
Japan, Kobe Marine Observatory, 1956.
[Report of the oceanographical observations in the sea south off Honshu in May 1955.]
J. Ocean., Kobe (2)7(2):17-25.

pagination repeats that of previous article.

dynamic currents
Japan, Kobe Marine Observatory, 1956.
[Report of the oceanographical observations in the sea south off Honshu in March 1955.]
J. Ocean., Kobe, (2)7(2):17-24.

currents, dynamic
Japan, Kobe Marine Observatory, 1956?
[Report on the oceanographical observations south off Honshu in 1954.] J. Ocean., Kobe (2)7(2):46-69.

dynamic currents
Japan, Kobe Marine Observatory, 1954.
[The outline of the oceanographical observations in the southern area of Honshu on board the R.M.S. "Syunpu maru".]
Aug.-Sept. 1954 - J. Ocean. (2)5(9):1-44.

dynamic currents
Japan, Kobe Marine Observatory, 1954.
[Results of oceanographic observations off Enshu-Nada (June 1954).] J. Ocean. (2)5(6):1-6, 4 figs.

currents, dynamic
Japan, Maizuru Marine Observatory, 1963
Report of the oceanographic observations in the central part of the Japan Sea in September, 1960. (In Japanese).
Bull. Maizuru Mar. Obs., No. 8:60-68.

currents dynamic
Japan, Maizuru Marine Observatory, 1963
Report of the oceanographic observations in the Japan Sea in June 1961. (In Japanese).
Bull. Maizuru Mar. Obs., No. 8:59-79.

dynamic currents
Japan, Maritime Safety Board, 1955.
[Oceanographical state in the adjacent waters of Mar hall Islands between May and June in 1954.]
Publ. No. 981, Hydrogr. Bull., Spec. No. 17:70-79.

dynamic topography
Jerlov, N.G., 1953.
Studies of the Equatorial Currents in the Pacific
Tellus 5(3):308-314, 5 textfigs.

dynamic currents
Johnson, M. W., 1939.
The correlation of water movements and dispersal of pelagic larval stages of certain littoral animals, especially the sand crab, Emerita.
J. Mar. Res. 2(3):236-245, figs. 68-71.

currents, dynamic
Kean, D.J., and W. Chimiak, 1955.
Caribbean current survey, spring 1953.
H.O. Tech. Rept., TR-12:14pp., 14 figs.

currents, dynamic
Kharchenko, A.M., 1968.
Currents and water masses of the West China Sea. (In Russian; English abstract).
Okeanologiia, Akad. Nauk, SSSR, 8(1):38-48.

currents, dynamic
Kitano, Kiyomitsu, 1958.
Oceanographic structure of the Bering Sea and the Aleutian waters. 1. Based on oceanographic observations by R.V. "Tenyo-maru" of 1957.
Bull. Hokkaido Reg. Fish. Res. Lab., No. 19:1-9.

dynamic currents
Klimenkov, A.I., and V.I. Pakhorukov, 1962.
Hydrological observations in the northwest Atlantic in spring-summer 1960. (In Russian; English summary)
Sovetskie Riboch. Issledov. v Severo-Zapadnoi Atlant. Okean., VINRO-PINRO, Moskva, 189-200.

dynamic currents
*Konaga, Shunji, 1968.
Variations of the oceanographic condition south of Japan in relation to mean sea level at Kushimoto and Uragami tidal stations, Japan No. 2- oceanographic condition and the difference of the mean sea level at both stations. (In Japanese; English abstract).
Umi to Sora, 44(1):1-12.

currents, dynamic
Kozlov, V.F. 1971.
The results of approximate calculation of the integral circulation in the Japan Sea. (In Russian; English abstract).
Met. Gidrol. (4): 57-63.

dynamic currents
Kudlo, B.P., and T.A. ershtadt., 1965.
On the adaptability of the dynamic method of calculating elements of sea currents in the Barents Sea. (In Russian).
Trudy, Gosudarst. Okeanogr. Inst., No. 86:100-111.

dynamic currents
Lacombe, Henri, 1965.
Cours d'océanographie physique. (Theories de la circulation. Houles et vagues).
Gauthier-Villars, Paris, 392 pp.

currents, dynamic
Lacombe, H., 1954.
Contribution à l'étude hydrologique de la Méditerranée occidentale.
Bull. d'Info., C.C.O.E.C., 6(1):31-35, 2 pls.

dynamic currents
LaFond, E. C., 1949
The use of bathythermograms to determine ocean currents. Trans. Am. Geophys. Union, 30(2):231-237, 6 text figs.

dynamic currents
LaFond, E.C., and D.W. Pritchard, 1952.
Physical oceanographic investigations in the eastern Bering and Chukchi Seas during the summer of 1947. J. Mar. Res. 11(1):69-86, 17 text-figs.

currents, dynamic
LeFloch, Jean 1970.
Évolution rapide de régimes de circulation non permanents des couches d'eaux superficielles dans le secteur sud-est du Golfe de Gascogne.
Cah. océanogr. 22(3): 269-276.

currents, dynamic
Le Floch, J., 1962
Rapport sur les recherches océanographiques faites en Nord Tyrrhénienne en fevrier 1960.
Trav., Centre de Recherches et d'Etudes Océanogr. 4(4):45-63.

currents, dynamic
LeFloch, Jean, 1961
Mesures de courants par électrodes remorquées dans le Canal de Corse. Relation avec le relief dynamique. Application au choix d'une surface de référence pour le calcul des vitesses en profondeur et des debits.
Cahiers Océanogr., C.C.O.E.C., 13(9):619-626.

currents, dynamic
Leonov, A.K., 1958.
[Thermics and currents of the Japan Sea.]
Izv. Vses. Geograph. Obshsh., 90(3):244-264.

currents, dynamic
Madelain, Francois, 1967.
Calculs dynamiques au large de la peninsule Iberique.
Cah. oceanogr., 19(3):181-193.

currents, dynamic
Mao, H., and K. Yoshida, 1955.
Physical oceanography in the Marshall Islands area. Bikini and nearby atolls, Marshall Islands.
Geoll Survey Prof. Paper 260-R:645-684, Figs. 179-218.

currents, dynamic
Martsinkevich, L.M., 1964.
Geostrophic currents of the northwest part of the Sargasso Sea and adjoining part of the Gulf Stream. Hydrophysical Incestigations. (Results of the investigations of the seventh cruise of the Scientific Research Ship "Mikhail Lomonosov") (In Russian).
Trudy Morsk. Gidrofiz. Inst., Akad. Nauk Ukrain., SSR, 29:43-53.

dynamic currents
Masuzawa, J., 1955.
An outline of the Kuroshio in the Eastern Sea of Japan. (Currents and waters masses of the Kuroshio system 4.). Ocean. Mag., Tokyo, 7(1):29-48.

dynamic currents
Masuzawa, J., 1955.
Preliminary report on the Kuroshio in the Eastern Sea of Japan (currents and water masses of the Kuroshio system III).
Rec. Ocean. Wks., Japan, 2(1):132-140, 5 textfigs

dynamic currents
McGary, Jame W., 1955.
Mid-Pacific oceanography, Pt. 6. Hawaiian offshore waters, December 1949-November 1951.
Spec. Fish. Rept.:Fish. No. 152:1-138, 33 figs.

currents, dynamic
McGary, J.W., and E.D. Stroup, 1956.
Mid-Pacific oceanography. VIII. Middle latitude waters, January-March, 1954.
Spec. Sci. Publ., Fish., No. 180:173 pp.

currents, dynamic
Menaché, M., 1955.
Étude hydrologique rapide de la region d'Anjouan (8-15 octobre 1953). Bull. d'Info., C.C.O.E.C., (9):419-420, 2 pls.

currents, dynamic
Montgomery, R.B., and E.D. Stroup, 1962.
Equatorial waters and currents at 150° W in July-August 1952.
The Johns Hopkins Oceanogr. Studies, No. 1:68 pp.

dynamic currents
Mosby, H., 1938
Svalbard Waters. Geofysiske Publ. 12(4): 85 pp., 34 textfigs.

currents, dynamic
Mosetti, Ferruccio 1967
Caratteristiche idrologiche dell'Adriatico settentrionale: situazione estiva.
Atti Ist. veneto Sci. 125: 147-175.

currents, dynamic
Natarov, V.V., and V.N. Pashkin, 1968.
Influence of some oceanographic factors on the formation of fish-productive areas off the Australian coast. (In Russian).
Trudy. Vses. Nauchno-Issled. Inst. Morsk. Ribn. Okeanogr (VNIRO) 64, Trudy Azovo -Chernomorski Nauchno-Issled. Inst. Morsk. Ribn. Khoz. Okeanogr. (AscherNIRO), 28: 130-141.

dynamic currents
Navrotsky, V.V., 1964.
Some results of research studies on the interaction of the ocean and atmosphere in the Gulf Stream during the IGY. (In Russian).
Okeanologiia, Akad. Nauk, SSSR, 4(4):603-611.

dynamics currents
Neyman, V.G., 1961
[Dynamic map of Antarctica.]
Mezhd. Kom. Mezhd. Geofiz. Goda, Presidiume Akad. Nauk, SSSR, Okeanol. Issled., 3:117-123.

currents, dynamic
Novitsky, V.P., 1963.
Hydrological features of the Aegean Sea in October 1959. (In Russian; English abstract).
Rez. Issled. Programme Mezhd. Geofiz. Goda, Okeanolog. Issled., Akad. Nauk, SSSR, No. 8:78-89

dynamic currents
Owen, R.W., Jr., 1967.
Atlas of July oceanographic conditions in the northeast Pacific Ocean, 1961-64.
Spec. scient. Rep. U.S. Fish. Wildl. Serv., Fish., 549:85 pp.

currents, dynamic
Penin, V.V., 1966.
Hydrological conditions in the Norwegian Sea in the period of summer feeding of herring from 1958 to 1963. (In Russian).
Trudy, Poliarn. Nauchno-Issled. Proektn. Inst. Morsk. Ribn. Khoz. Okeanogr., N.M. Knipovich, PINRO, 17:55-68.

currents, dynamic
Penin, V.V., 1965.
On the dynamics of currents in the Norwegian and Greenland seas. (In Russian).
Materiel. Ribochoz. Issled. Severn. Bassin. Poliarn. Nauchno-Issled. i Proekt. Inst. Morsk. Rib. Choz. i Okeanogr., N.M. Knipovich, 5:80-90.

dynamic currents
Redfield, A. C. and A. Beale, 1940.
Factors determining the distribution of populations of chaetognaths in the Gulf of Maine. Biol Bull., 79(3):459-487, 11 text figs.

dynamic currents
Riis-Carstensen, E., 1938
Fremsaettelse af et dynamisk-topografisk Kort over Østgrønlandsstrømmen mellem 74° og 79°N Br. paa Grundlag af hidtidig gjorte Undersøgelser i disse Egne. Geografisk Tidsskrift 41(1): 25-51, 5 figs.

dynamic currents
Romanov, Yu. A., 1961
[Dynamic method as applied to the Equatorial Indian Ocean.] (In Russian; English abstract).
Okeanolog. Issled., Mezhd. Komitet Proved. Mezhd. Geofiz. Goda, Prezidiume, Akad. Nauk, SSSR (4):25-30.

dynamic currents
Rotschi, Henri, 1960
Orsom III, resultats de la Croisière "Dillon" 1ère Partie. Océanographie Physique.
Off. Recherche Sci. et Techn., Outre-Mer, Inst. Français d'Océanie, Centre d'Océanogr., Rapp. Sci. Noumea, No. 18: 58 pp.

currents, dynamic
Rotschi, Henri, 1959.
Resultats des observations scientifiques du "Tiare" croisiere "Dauphin" (8-14 Aout 1959) sous le commandement du Lieutenant de Vaisseau Cerbelaud.
Rapp. Sci., Inst. Francais d'Oceanie, Centre d'Oceanogr., No. 14:18 pp.

dynamic currents
Rouch, J., 1949.
Type de calcul pour la determination des courants d'apres la methode de Bjerknes. Bull. d'Info., C.O.E.C., No. 10:5-9 (mimeographed).

currents, dynamic
Saelen, Odd H., 1963
Studies in the Norwegian Atlantic Current. II. Investigations during the years 1954-1959 in an area west of Stad.
Geofysiske Publikasjoner, 23(6):1-82.

currents, dynamic
Sandström, J.W., 1921.
Deux theoremes fondamentaux de la dynamique de la mer. Traite elementaire experimental.
Svenska Hydrogr.-Biol. Komm Skr., 7:1-6.

dynamic currents
Scripps Institution of Oceanography, 1949
Marine Life Research Program, Progress Report to 30 April 1949, 25 pp. (mimeographed), numbrous figs. (photo.+ozalid). 1 May 1949.

dynamic currents
Semina, H.J., 1961
[The phytoplankton of the mixing zone between Oyashio and Kuroshio in spring 1955.]
Trudy Inst. Okeanol., 51: 3-15.

dynamic currents
Seriy, V.V., 1968.
Peculiarities of seasonal variations in hydrological conditions in the Gulf of Aden. (In Russian)
Trudy, Vses. Nauchno-Issled. Inst. Morsk. Ribn. Okeanogr. (VNIRO) 64, Trudy Azovo-Chernomorsk. Nauchno-Issled. Inst. Morsk. Ribn. Khoz. Okeanogr. (AscherNIRO), 28: 117-129.

dynamic currents
Seryi, V.V., and V.A. Khimitza, 1963
On the hydrology and chemistry of the Gulf of Aden and the Arabian Sea. (In Russian).
Okeanologiia, Akad. Nauk, SSSR, 3(6):994-1003.

currents, dynamic
Shannon, L.V., and M. van Rijswijck 1969.
Physical oceanography of the Walvis Ridge region.
Invest'l Rept. Div. Sea Fish. SAfr. 70:19 pp.

dynamic currents
Soule, F.M., 1953.
Physical oceanography of the Grand Banks region and the Labrador Sea in 1952. U.S.C.G. Bull. No. 38:31-100, 17 textfigs.

dynamic currents
Soule, F. M., 1942
Oceanography: excerpts from International Ice Observations and Ice Patrol Service in the North Atlantic Ocean. Season of 1940. USCG Bull., 36-89; 33 figs; tables of data

currents - dynamic
Soule, F. M., 1940.
Oceanography: excerpt from International Ice Observation and Ice Patrol Service in the North Atlantic Ocean. Season of 1939. U.S.C.G. Bull., pp. 79-133;32 textfigs; Tables of data.

currents, dynamic
Soule F.M. 1940.
Oceanography. Season of 1938.
Bull. Int. Ice Obs. Ice Patrol Serv. N. Atlantic Ocean 28:113-175.

dynamic currents
Soule, F. M., 1939.
Consideration of the depth of the motionless surface near the Grand Banks of Newfoundland. J. Mar. Res. 2(3):170-180, textfigs. 53-58.

dynamic currents
Soule, F. M., 1938.
Oceanography. Season of 1937. International Ice Observation and Ice Patrol Service, U.S.C.G. Bull. No. 27:71-126, 45 textfigs., tables of data.

dynamic currents
Soule, F.M., 1938
Oceanography. Excerpt from International Ice Observation and Ice Patrol Service in the North Atlantic Ocean. Season of 1936. C.G. Bull No.26:33-82, 58 text figs., tables of data.

dynamic currents
Soule, F.M., and C.A. Barnes, 1950.
Physical oceanography of the Ice Patrol area in 1941. U.S.C.G. Bull. No. 31:1-62, 23 textfigs., fold-in charts.

dynamic currents
Soule, F.M., P.S. Branson, and R.P. Dinsmore, 1952.
Physical oceanography of the Grand Banks region and the Labrador Sea in 1951. U.S.C.G Bull. No. 37:17-85, 21 textfigs.

currents, dynamic
Soule, F.M., A.J. Bush and J.E. Murray, 1955.
International ice observations and Ice Patrol Service in the North Atlantic Ocean. Season of 1953. U.S.C.G. Bull. No. 39:45-138, 43 textfigs.

dynamic currents
Soule, F.M., H.H. Carter and L.A. Cheney, 1950.
Oceanography of the Grand Banks region and Labrador Sea, 1948. U.S.C.G. Bull. No. 34:67-118. Figs. 27-37, fold-in charts.

currents, dynamic
Soule, F.M., and J.E. Murray, 1955.
Physical oceanography of the Grand Banks region and the Labrador Sea.
U.S.C.G. Bull. No. 40:79-168.

currents, dynamic
Sverdrup, H. U., and R.H. Fleming, 1941.
The waters of the coast of southern California, March to July, 1937. Bull. S.I.O. 4(10):261-375, 66 textfigs.

dynamic currents
Takahashi, Tadao, 1959.
Hydrographical researches in the western equatorial Pacific.
Mem. Fac. Fish., Kagoshima Univ., 7:141-147.

currents, dynamic
Takahashi, Tadao, and Masaaki Chaen 1969.
Oceanic conditions near the Ryukyu islands. II. Oceanic conditions on 125°E in spring and summer 1966.
Mem. Fac. Fish. Kagoshima Univ. 18:99-114

currents, dynamic
Tareyev, B.A., and A.V. Fomichev, 1960
[Dynamic map of the Pacific sector of the Antarctic waters.]
Okean. Issle., IGY Com., SSSR:104-107.

currents, dynamic
Timonov, V.V., and L. A. Zhukov, 1956.
The dynamic method as a research tool in the study of ocean currents. (In Russian).
Meteorol. i Gidrol., (5):50-55.

Rev. in:
Appl. Mech. Rev., (5):50-55.

currents, dynamic
Trotti, L., 1954.
Report on the oceanographic investigations in the Ligurian and North Tirrenian Seas. Hydrography.
Centro Talassografico Tirreno, Pubbl. 16:1-39, 5 pls.

currents, dynamic
Tully, J.P., A.J. Dodimead and S. Tabata, 1960.
An anomalous increase of temperature in the ocean off the Pacific coast of Canada through 1957 and 1958.
J. Fish. Res. Bd., Canada, 17(1) ;61-80.

currents, dynamic
Uda, M., 1955.
On the subtropical convergence and the currents in the northwestern Pacific.
Rec. Ocean. Wks., Japan, 2(1):141-150, 10 textfigs.

dynamic currents
U.S.N. Hydrographic Office, 1958
Oceanographic atlas of the Polar seas. Pt. II. Arctic. H.O. Publ. No. 705: 139 pp., charts.

currents, dynamic, computation of
Weenink, M.P.H., 1951.
Een methode van bepalding van absolute stroomsnelheden in de zee uit relatieve topografieen van isobarische vlakken met ~~xxx~~ behulp van de diffusievergelijking.
K. Nederl. Met. Inst., Afd. Ocean. Marit. Met., Rapp. 2:5 pp. (mimeographed).

currents, dynamic
Vapniar, D.U., 1967.
Some results of computing the gradient currents in the western equatorial Atlantic. (In Russian; English abstract).
Okeanologiia, Akad. Nauk, SSSR, 7(3):437-444.

dynamic currents
Varadachari, V.V.R., C.S. Murty and Piyush Kanti Das, 1969.
On the level of least motion and the circulation in the upper layers of the Bay of Bengal.
Bull. natn. Inst. Sci., India, 38(1): 301-307.

dynamic currents
Visser, G.A. and M.M. Van Niekerk, 1965.
Ocean currents and water masses at 1,000, 1,500, and 3,000 metres in the south-west Indian Ocean.
S. Africa, Dept. Commerce and Industries, Div. Sea Fish., Invest. Rept. No. 52: 46 pp.

dynamic currents
Wyrtki, Klaus, 1964.
Upwelling in the Costa Rica dome.
U.S.F.W.S. Fish. Bull., 63(2):355-372.

dynamic currents
Yasui, M., 1955.
On the rapid determination of the dynamic depth anomaly in the Kuroshio area.
Rec. Ocean. Wks., Japan, 2(2):90-95, 5 textfigs.

currents, dynamic
Zore-Amanda, Mira, 1963.
Les masses d'eau de la mer Adriatique. (In French ;jugoslavian resume).
Acta Adriatica, 10(3):1-89.

currents, dynamic
Zore-Armanda, Mira, 1964.
Results of direct current measurements in the Adriatic. (In Jugoslavian; English resume).
Acta Adriatica, 11(1):293-308.

Currents dynamics 2

currents, dynamics of
Longuet-Higgins, M.S., 1965.
Some dynamical aspects of ocean currents.
Q. Jl. R. Met. Soc., 91:425-451.
Also in:
Collected Reprints, Nat. Inst. Oceanogr., 13.

currents, east coast
Durance J.A. and J.A. Johnson 1970.
J. fluid Mech. 44 (1): 161-172

current, eastward
Thompson, Rory O.R.Y., 1971.
Why there is an intense eastward current in the North Atlantic but not in the South Atlantic.
J. phys. Oceanogr. 1(3): 235-237.

currents, eastward
White, Warren B., 1971.
A Rossby wake due to an island in an eastward current. J. phys. Oceanogr. 1(3): 161-168.

currents, eastern boundary
Wooster, Warren S., 1970.
Eastern boundary currents in the South Pacific.
In: Scientific exploration of the South Pacific, W.S. Wooster, editor, Nat. Acad. Sci., 60-68.

currents, eddy

currents, eddy
Shuleikin, V.V., 1960
[More on eddy sea currents] Doklady Akad. Nauk, SSSR, 134(6): 1343-1346.

currents, effect of

currents, effect of
Adem, Julian 1970.
Incorporation of advection of heat by mean winds and by ocean currents in a thermodynamic model for long-range weather prediction.
Mon. Weath. Rev. U.S. Dept. Comm. 78(10): 776-786.

currents, effect of
Allain, C., 1964.
Les poissons et les courants.
Rev. Trav. Inst. Pêches Marit., 28(4):401-426.

currents, effect of
Arlman, J., P. Santema and J.N. Svasek, 1957.
Movement of bottom sediment in coastal waters by current and waves; measurements with the aid of radioactive tracers in the Netherlands.
Rijkswaterstaat, The Hague:63 pp.

currents, effect of
Aubert, M., J.P. Gamberotta et F. Laumond 1967.
Étude de la répartition du fer au large des côtes de Provence et de Corse: étude de la dispersion des apports terrigènes.
Rev. intern. Océanogr. Méd. 5:23-61.

currents, effect of
Aybulatov, N.A., & I.E. Schadrin, 1961
Role of discontinuous currents in the transport of sand-drifts near the coastal zone.
Trudy Inst. Okean., 53: 19-28.

currents, effect of
Batalin, R.I., 1961.
Application of transverse circulation methods to problems of erosion and siltation.
Dock & Harbour Authority, 41(486):407-408.

currents, effect of
Bernard, Francis, 1958
Le courant Atlantique en Méditerranée.
Rapp. Proc. Verb., Comm. Int. Expl. Sci., Mer Medit., n.s., 14:97-100.

currents, effect of
Bernard, F., 1954.
Vents, courants et fertilité marine au large de l'Algerie.
Bull. Soc. Hist. Nat., Afrique du Nord, 45(3/4): 79-88, 2 textfigs.

Bishai, H.M., 1960 currents, effect of
The effect of water currents on the survival and distribution of fish larvae. J. du Cons., 25(2):134-146.

currents, effect of
Braarud, T., and B. Hope, 1952.
The annual phytoplankton cycle of a landlocked fjord near Bergen (Nordåsvatn). Rep. Norwegian Fish. Mar. Invest. 9(16):26 pp., 4 textfigs.

currents, effect of
Bryan, Kirk, 1962
Measurements of meridional heat transport by ocean currents.
J. Geophys. Res., 67(9):3403-3414.

currents, effect of
Bryan, Kirk, 1962
Measurements of meridional heat transport by ocean currents. (Abstract).
J. Geophys. Res., 67(9):3546.

currents, effect of
Cameron, F.E., 1957.
Some factors influencing the distribution of pelagic copepods in the Queen Charlotte Islands area. J. Fish. Res. Bd., Canada, 14(2):165-202.

currents, effect of
Carbonel, Pierre, Jean Moyes et Michel Vigneaux 1971.
La répartition des Thanatocoenoses d'ostracodes dans l'estuaire de la Gironde et ses relations avec les courants. (D)
C. r. hebd. Séanc. Acad. Sci. Paris 273 (19): 1679-1682.

currents, effect of
Carruthers, J.N., A.L. Lawford, and V.F.C. Veley, 1951.
Fishery hydrography: brood-strength fluctuations in various North Sea fish, with suggested methods of prediction. Kieler Meeresf. 8(1):5-15, 8 figs.

currents, effect of
Collins, J.I., 1964
The effect of currents on the mass transport of progressive waves.
J. Geophys. Res., 69(6):1051-1056.

currents, effect of
Corlett, John, 1965.
Winds, currents, plankton and the year-class strength of cod in the western Barents Sea.
ICNAF Spec. Publ., No. 6:373-378.

currents, effect of
Crisp, D.J., 1953.
Changes in the orientation of barnacles of certain species in relation to water currents.
J. Animal Ecology 22(2):331-343, 6 pls., 6 text-figs.

currents, effect of
Crisp, D.J., and H.G. Stubbings, 1957.
The orientation of barnacles to water currents.
J. Animal Ecology, 26(1):179-196.

currents, effect of
Demyanov, N.I., 1959.
Effect of pulsations of sea current on log readings (In Russian).
Problemy Arktiki, (3):23-27.

currents, effect of
Dmitriev, A.A., and T.V. Bonchkovskaya, 1954.
Effects of currents on the wave motion of a fluid. Izv. Akad. Nauk, SSSR, Ser. Geofiz., 4: 360-374.

currents, effect of
Doochin, H., and F.G.W. Smith, 1951.
Marine boring and fouling in relation to velocity of water currents. Bull. Mar. Sci., Gulf and Caribbean 1(3):196-208, 4 textfigs.

currents, effect of
*Edwards, C., 1968.
Water movements and the distribution of Hydromedusae in British and adjacent waters.
Sarsia, 34:331-346.

currents, effect of
Emig, Marlies, 1967.
Heat transport by ocean currents.
J. geophys. Res., 72(10):2519-2529.

Currents, effect of
Emig, C.C. et F. Becherini, 1970.
Influence des courants sur l'éthologie alimentaire des phoronidiens. Étude par séries de photographies cycliques. Mar. Biol., 5(3): 239-244.

currents, effect of
Forest, G., and P. Jaffry, 1958
Emploi de traceurs radioactifs dans l'étude des mouvements de sediments sous l'effet de la houle et des courants. Communication, Assoc Recherches Hydraul., Lisbonne, Juillet, 1957 (D16): 1-10.

currents, effect of
France, Institut Geographique National, 1964.
La photographie aérienne et l'étude des dépôts prélittoraux.
Études de Photo-Interprétation, Clos-Arceduc, Paris, No. 1:53 pp.

currents, effect of
*French, Robert R., and W. Bruce McAlister, 1970.
Winter distribution of salmon in relation to currents and water masses in the northeastern Pacific Ocean and migrations of sockeye salmon.
Trans. Am. Fish. Soc., 99(4):649-663.

currents, effect of
Furnestin, J., and R. Coupé, R. Gain, Ch. Maurin, and M. Rossignol, 1953.
Ultra-sons et pêche à la sardine au Maroc.
Bull. Inst. Pêches Marit., Maroc, No. 1:1-57, 8 textfigs., 39 echograms.

currents, effect of
*Gamulin-Brida, Helena, 1968.
Mouvements des masses d'eau et distribution des organismes marins et des diocoenoses benthiques en Adriatique.
Sarsia, 34:149-162.

currents, effect of
*Golikov, A.N., 1968.
Distribution and variablility of long-lived benthic animals as indicators of currents and hydrological conditions.
Sarsia, 34:199-208.

currents, effect of
*Golikov, A.N., and O.A. Scarlato, 1968.
Vertical and horizontal distribution of biocoenoses in the upper zones of the Japan and Okhotsk seas and their dependence on the hydrological system.
Sarsia, 34:109-116.

currents, effect of
Gomes, A. L., 1955.
Possibilidades da pesca maritima no Brasil.
Rev. Biol. Mar., Valparaiso, 6(1/3):6-20.

currents, effect of
*Gomoiu, M.-T., 1968.
On the effects of water motion on marine organisms in the mesolittoral and infralittoral zones of the Rumanian shore of the Black Sea.
Sarsia, 34:95-108.

currents, effect of
Gorodenskii, N.B., N.F. Kudriavtsev, and V.G. Labeish, 1961.
Investigations by means of models of the effect of currents and swell at autonomous stations in observing currents. Methods for investigating oscillations of sea level and currents with the help of autonomous stations of long duration. Methods of oceanological investigation, a collection of papers. (In Russian).
Trudy Arktich. i Antarktich. Nauchno-Issled. Inst., 210:13-22.

currents, effect of
*Grandperrin, R., et M. Legand, 1967.
Influence possible du système des courants équatoriaux du Pacifique sur la répartition et la biologie de deux poissons bathypélagiques.
Cah. ORSTOM, Sér. Océanogr., 5(2):69-77.

currents, effect of
Griffin, G.M., 1962
Regional clay-mineral facies--products of weathering intensity and current distributio in the northeastern Gulf of Mexico.
Geol. Soc. Amer. Bull., 73(6):737-768.

currents, effect of
*Gurjanova, E.F., 1968.
The influecne of water movements upon the species composition and distribution of the marine fauna and flora throughout the Arctic and North Pacific intertidal zones.
Sarsia, 34:83-94.

currents, effect pf
Hamon, B.V. 1969.
"Current effect" on sea level.
Rept. CSIRO Div. Fish. Oceanogr. 46:1-12. Austr.

Currents, effect of
Heezen, Bruce C., Bill Glass and N.W. Menard 1966.
The Manihiki Plateau.
Deep-Sea Res. 13(3): 445-458.

currents, effect of
Hermann, Frede, Paul M. Hansen and Sv. Aa. Horsted, 1965.
The effect of temperature and currents on the distribution and survival of cod larvae at West Greenland.
ICNAF Spec. Publ., No. 6:389-395.

currents, effect of
Ichiye, T., 1957.
On the relationship between the plankton distribution and hydrographic conditions in the adjacent seas of Japan. Rec. Ocean. Wks., Japan, (Spec. No.):34-41.

currents, effect of
Ingle, Robert M., Bonnie Eldred, Harold W. Sims, and Eric A Eldred, 1963
On the possible Caribbean origin of Florida' spiny lobster populations.
State of Florida, Bd., Conserv., Div. Salt Water Fish., Techn. Ser., No. 40:12 pp.

currents, evvect of
Inglis, C.C.L., and F.H. Allen, 1957.
The regimen of the Thames estuary as affected by currents, salinities and river flow.
Proc. Inst. Civ. Eng., 7:827-878.

currents, effect of
Inman, D.L., 1953.
Areal and seasonal variations in beach and nearshore sediments at La Jolla, California.
B.E.B. Tech. Memo. No. 39:82 pp., 27 figs., numerous tables.

currents, effect of
Iselin, C.O'D., 1955.
Coastal currents and fisheries.
Pap. Mar. Biol. and Oceanogr., Deep-Sea Res., Suppl. to Vol. 3:474-478.

currents, effect of
Iwata, Noriyuki, 1962
Effekt der Meeresströmung auf interne Wellen im offenen Meer.
J. Oceanogr. Soc., Japan, 18(2):69-72.

currents, effect of
Jacquette, R., 1962.
Étude des fonds de maerl de Méditerranée.
Rec. Trav. Sta. Mar., Endoume, Bull., 26(41):141-

currents, effect of
Kenyon, Kern E., 1971.
Wave refraction in ocean currents. Deep-Sea Res. 18(10): 1023-1034.

currents, effect of
Kirby-Smith, William W., 1972.
Growth of the bay scallop: the influence of experimental water currents. J. exp. mar. Biol. Ecol. 8(1): 7-18.

currents, effect of
Knox, G.A., 1963.
The biogeography and intertidal ecology of the Australasian coasts.
In: Oceanography and Marine Biology, H. Barnes, Editor, George Allen and Unwin, 1:341-404.

currents, effect of
Kojima, S., 1954.
[On the relationship between the ocean currents, and fluctuations in the sardine catch off Shimane Prefecture.] Bull. Jap. Soc. Sci. Fish. 20(5):372-374, 1 textfig.

currents, effect of
Kudrjavtsev, N.F., 1959.
[On the current influence on the immersion depth of automatic station recorders.]
Probl. Arktiki i Antarktiki, (1):11-24.

currents, effect of
LaFond, E.C., 1954.
Environmental factors effecting the vertical temperature structure of the upper layers of the sea. Andhra Univ. Ocean. Mem. 1:94-101, 25 textfigs.

currents, effect of
LaFond, E.C., and R. Prasada Rao, 1954.
Beach erosion cycles near Waltair on Bay of Bengal.
Andhra Univ. Ocean. Mem. 1:63-77, 10 textfigs.

currents, effect of
Laughton, A.S., 1963.
18. Microtopography.
In: The Sea, M.N. Hill, Editor, Interscience Publishers, 3:437-472.

currents, effect of
Lazarenko, A.A., and E.V. Semenov 1967.
The mechanism of pebble orientation in current deposits. (In Russian).
Dokl. Akad. Nauk SSSR 172(6): 1397-1400.

currents, effect of
Lee, A.J., 1952.
Influence of currents on northern fisheries.
World Fishing 1(3):81-87.

currents, effect of
*Lindquist, Armin, 1968.
On fish eggs and larvae in the Skagerak.
Sarsia, 34:347-354.

currents, effect of
*Lowrie, Allen, Jr., and Bruce C. Heezen, 1967.
Knoll and sediment drift near Hudson Canyon.
Science, 157(3796):1552-1553.

currents, effect of
Ma, Ting Ying H., 1957.
The effect of warm and cold currents in the southwestern Pacific on the growth rate of reef corals.
Oceanographia Sinica 5(1):1-34.

currents, effect of
MacNae, W., 1962.
The fauna and flora of the eastern coasts of southern Africa in relation to ocean currents.
S. African J. Sci., 58(7):208-212.

currents, effect of
McManus, Dean A., and Joe S. Creager, 1961.
Offshore current control of sedimentary environments associated with large underwater shoals.
Tenth Pacific Sci. Congr., Honolulu, 21 Aug.-6 Sept., 1961, Abstracts of Symposium Papers, 377-378. (Abstract)

currents, effect of
Meade, Robert H. 1971.
The coastal environment of New England
New England River Basins Commission, 47 pp.

currents, effect of
*Michel, A., 1968.
Dérive des larves de stomatopodes del'est de l'Océan Indien.
Cah. O.R.S.T.O.M., ser. Océanogr., VI(1):13-41.

currents, effect of
*Mileikovsky, Simon A., 1968.
Some common features in the drift of pelagic larvae and juvenile stages of bottom invertebrates with marine currents in temperature regions.
Sarsia, 34:209-216.

currents, effect of
Mileikovskii, S.A., 1960
[About the range of dispersal of pelagic larvae of bottom invertebrates with marine currents. (On the example of Limapontia capitata Müll. (Gastropoda Opisthobranchia) of Norwegian and Barents seas.]
Doklady Akad. Nauk, SSSR, 135(4): 965-967.

effects of currents
Moore, H. B. and J. A. Kitching, 1939
The biology of Chthamalus stellatus (Poli).
JMBA 23:521-541.

currents, effect of
Munk, W., and G. Groves, 1952.
The effect of winds and ocean currents on the annual variation in latitude. J. Met. 9(6):385-396, 3 textfigs.

currents, effect of
Mysak, Lawrence A., 1968.
Effects of deep-sea stratification and current on edgewaves.
J. mar. Res., 26(1):34-42.

currents, effect of
*Nagle, J. Stewart, 1967.
Wave and current orientation of shells.
J. sedim. Petr., 37(4):1124-1138.

currents, effect of

Nakano, Masito, and Tomosaburo Abe, 1959.
Standing oscillations of bay water induced by currents.
Rec. Oceanogr. Wks.,Japan, Spec. No. 3:75-96.

currents, effect of
Nakano, M., and T. Abe, 1958.
Standing oscillation of bay water induced by currents. Geophys. Mag., Tokyo, 28(3):375-398.

currents, effect of
Nozharov,Peter B., 1968.
Comments on the influence of fluid motion upon the magnetic orientation of sediments.
Pure Appl. Geophys., 70:81-87.

currents, effect of
Osterberg, Charles, June Pattullo and William Pearcy, 1964
Zinc-65 in euphausiids as related to Columbia River water off the Oregon Coast.
Limnology and Oceanography, 9(2):249-257.

currents, effect of
Otto, L. 1971.
The frequency distribution of the current speed at the Netherlands lightvessels and its possible influence on the composition of sediments in the southern North Sea.
Geol. Mijnb. 50(3):475-478.

currents, effect of
Otto, L., 1963
Note on the influence of currents upon wind estimates.
J. Appl. Meteorol., 2(1):186-190.

currents, effect of
Oulianoff, N., 1961.
Rides (ripple marks) sur les fonds océaniques et courants sous-marins.
Comptes Rendus, Acad. Sci., Paris, 253(1):507-509.

currents, effect of
Pavshtics, E.A., 1969
Effect of currents on seasonal changes of zooplankton in Davis Strait. (In Russian)
Gidrobiol. Zh. 5(5): 75-92

currents, effect of
*Pavshtiks,E.A.,1968.
The influence of currents upon seasonal fluctuations in the plankton of Davis Strait.
Sarsia, 34:383-392.

currents, effect of
Pavshtiks, E.A., and M... Gogoleva, 1964.
Plankton distribution in the area of Georges Bank and Browns Bank in 1962. (In Russian).
Trudy, Poliarn. Nauchno-Issled. i Proktn. Inst. Morsk. Ribn. Choz. i Okeanogr. im N.M. Knipovicha 16:15-48.

currents, effect of
Peters, Nicolaus, 1934
Die Bevölkerung des Sudatlantischen Ozeans mit Ceratien. Biol. Sonderuntersuchungen 1.
Wiss Ergeb. Deutschen Atlantischen Exped.---"Meteor" 1925-1927, 12(1): 1-69, 28 text figs.

currents, effect of
Phillips, O.M., W.K. George and R.P. Mied, 1968
A note on the interaction between internal gravity waves and currents.
Deep-Sea Research, 15(3):267-273.

currents, effect of
Plate, Erich and Michael Trawle, 1970.
A note on the celerity of wind waves on a water current. J. geophys. Res., 75(18): 3537-3544.

currents, effect of
Priymachenko, A.D., 1961.
The current as a factor determining the development of phytoplankton in water basins. (In Russian).
Pervichnaya Produkt. Morey i Vnutrennikh Vod, 314-318.

USN-HO-TRANS 202
M. Slessers 1963
P.O. 39010.

currents, effect of
Ragotzkie, R.A., and R.A. Bryson, 1953.
Correlation of currents with the distribution of adult Daphnia in Lake Mendota. J. Mar. Res. 12(2):157-172, 12 textfigs.

currents, effect of
Rance, Peter John, 1966.
Investigation of wind-induced currents and the effect on the performance of sea outfalls.
Inst. Civil Eng., London, Proc., 33:231-260.

currents, effect of
Rao, T.S. Satyanarayana, 1958
Studies on Chaetognatha in the Indian Seas. IV. Distribution in relation to currents.
Andhra Univ. Mem., Oceanogr. 2:164-167.

water currents, effect of
Rees, A.I., 1961
The effect of water currents on the magnetic remanence and anistropy of susceptibility of some sediments.
Geophys. J., 5(3):235-251.

currents, effect of
Reyss, Daniel, 1964.
Observations faites en soucoupe plongeante dans deux vallées sous-marines de la mer Catalane: le rech du Cap et le rech Lacaze-Duthiers.
Bull. Inst. Océanogr., 63(1308):8 pp.

currents, effect of
Sagitov, M.U., and G.D. Marchuk, 1962
[The Eötvös correction for currents during gravity determinations at sea.]
Geofiz. Biull., Mezhduved. Geofiz. Komitet, Prezid., Akad. Nauk, SSSR, No. 11:40-42.

Currents, effect of
Saint-Guily, Bernard 1970.
On internal waves. Effects of the horizontal component of the earth's rotation and of a uniform current.
Dt. hydrogr. Z. 23(1): 16-23

currents, effect of
Sanderson, R.M. 1971.
Ice-edge movements in the Greenland Sea.
Mar. Obs. 41 (234) (M.O.839): 173-183.

Seryakov, E. I., 1960 currents, effect of
On the advection of heat by currents in the Barents Sea.
Soviet Fish. Invest., North European Seas, VNIRO, PNIRO, Moscow, 1960: 80-88.

currents, effect of
Sitarz, Jean A., 1960
Côtes Africaines. Étude des profils d'equilibre de plage. Trav. C.R.E.O., 3(4): 43-62.

currents, effect of
Smirnova, A.I., 1970.
Heat transfer by currents in the North Atlantic (In Russian; English Abstract). Okeanologiia, 10(1): 30-37.

currents, effect of
Solyankin, E.V., 1966.
On the relation of hydrometeorologic processes in the northern and southern hemispheres. (In Russian).
Trudy vses. Nauchno-issled. Inst. morsk. ryb. Khoz. Okeanogr. (VNIRO), 60:49-58.

currents, effect of
Stapor, F.W., 1971.
Sediment budgets on a compartmented low-to-moderate energy coast in northwest Florida.
Marine Geol. 10(2): M1-M7.

currents, effect of
*Takeuchi, Shoichi,1968.
Relation between the direction of a current and the catch in a trap net with two bags. (In Japanese;English abstract).
J. Tokyo Univ.Fish., 54(2):123-127.

currents, effect of
Terwindt, J.H.J. 1971.
Litho-facies of inshore estuarine and tidal-inlet deposits.
Geol. Mijnb. 50(3): 515-526.

currents, effect of
Thorson, Gunnar, 1961.
Length of pelagic larval life in marine bottom invertebrates as related to larval transport by ocean currents. In: Oceanography, Mary Sears, Editor, Amer. Assoc., Adv. Sci., 455-474.

currents, effect of
Tominaga, Masahide, 1962
On the waves generated over the steady uniform current (Part 1).
J. Oceanogr. Soc., Japan, 20th Ann. Vol., 189-199.

currents, effect of
Tsuruta, Arao, 1963
Distribution of plankton and its characteristics in the oceanic fishing grounds, with special reference to their relation to fishery.
J. Shimonoseki Univ., Fish., 12(1):13-214.

currents, effect of
Uda, M., 1953.
The Kuroshio and its branch currents in the seas adjacent to Hachijo Island in relation to fisheries. (Report 1). Rec. Ocean. Wks., Japan, n.s., 1(1):1-10, 8 textfigs.

currents, effect of
Uralov, N.S., 1959
[The influence of the Nordkapp (North Cape) Current on the iciness of the Barents Sea.]
Trudy Gosud. Okeanogr. Inst., Leningrad, 37: 14-33.

USN-HO-TRANS-102(NO. 2)
M. Slessers
M.O. 16104
P.O. 13377. 1961

currents, effect of
Uyeno, Fukuzo, 1961
Oceanographical and ecological studies on primary production of the sea, with special references to relationship between diatom production and temperature and chlorinity of water.
Rept., Fac. Fish., Pref. Univ., Mie, 4(1): 1-64.

currents, effect of
Van Straaten, L.M.J.U., 1961
Directional effects of winds, waves and currents along the Dutch North Sea coast.
Geologie en Mijnbouw, 40:333-346.
Also:
Publikatie, Geol. Inst., Groningen, No. 144.

currents, effect of
Vigare, Andro, 1965.
Les modalités du remblaiement alluvial dans l'estuaire de la Seine.
Cahiers Océanogr., C.C.O.E.C., 17(5):301-330.

currents, effect of
Visser, M.P., 1966.
Note on the estimation of eddy diffusivity from salinity and current observations.
Netherlands J. Sea Res., 3(1):21-27.

currents, effect of
Wells, Harry W., 1965.
Maryland records of the gastropod, Littorina littorea, with a discussion of factors controlling its southern distribution.
Chesapeake Science, 6(1):38-42.

currents, effect of
Werner, Friedrich, 1964.
Sedimentkerne aus den Rinnen der Kieler Bucht.
Meyniana, 14:52-65.

currents, effect of
Yu, Yi-Yuan, 1952.
Breaking of waves by an opposing current.
Trans. Amer. Geophys. Union 33(1):39-41, 2 textfigs.

currents, effect of
Zhukov, L.A., 1961.
Approximate calculation of the temperature and salinity fluctuation in the surface layers of the sea and their relationship to currents. Forecasting and computation of physical phenomena in the oceans. (In Russian).
Trudy Okeanogr. Komissii, Akad. Nauk, SSSR, 11: 158-163.

Currents, electromagnetic

currents, electromagnetic
Cagniard, L., 1957.
Sur la théorie de la mesure électromagnétique des courants marins. Annales de Geophys., 13(2): 155-157.

currents, electrical
LeGrand, Y., 1956.
Courants electriques et differences de potential dans la mer. Bull. d'Info., C.C.O.E.C., 8(1):11-20.

currents, electromagnetic
Mironov, A.T., 1950.
[Electrodes for measuring electrical currents in the sea.] Dokl. Akad. Nauk, SSSR, 70(5):825-827.
T44R

electric current
Ryzhkov, Yu. G., 1957.
[Measurement of electric current in the ocean.]
Doklady Akad. Nauk, SSSR, 113(4):787-790.
OTS translation $0.50

Currents, energy of

currents, energy of (utilization)
Stas', I.I., 1960.
Problem concerning the utilization of the energy of currents. (In Russian).
Trudy, Morsk. Gidrofiz. Inst., Akad. Nauk, SSSR, 20:58-67.
Translation:
Scripta Tecnica, for AGU, 46-53.

Currents, equatorial

currents, equatorial
Burkov, V.A., and I.M. Ovchinnikov, 1960
[Investigations of equatorial currents to the north of New Guinea.]
Trudy Inst. Okeanol., 40: 121-134.

currents, equatorial
Charney, Jule G. and Stanley L. Spiegel, 1971.
Structure of wind-driven equatorial currents in homogeneous oceans. J. phys. Oceanogr. 1(3): 149-160.

currents, equatorial
Federov, K. N., 1963
Some specific features of currents and ocean level at the equator. (In Russian).
Okeanologiia, Akad Nauk, SSSR, 3(1):3-12.

currents, equatorial
Fedorov, K. N., 1961
[Seasonal variations in oceanic equatorial currents.]
Trudy Inst. Okean., 52: 155-159.

currents, equatorial
Gill, A.E., 1971.
The equatorial current in a homogeneous ocean.
Deep-Sea Res., 18(4): 421-431.

currents, equatorial (theoretical)
Hidaka, Koji, 1963.
A hydrodynamical computation of an equatorial flow. n.s.
Rec. Oceanogr. Wks., Japan, 7(1):1-7.
Also in:
Geophysical Notes, Tokyo, 16 (1)(Contrib. No. 7)

currents, equatorial
Ichiye, Takashi, 1964
An essay on the equatorial current system.
In: Studies in Oceanography dedicated to Professor Hidaka in commemoration of his sixtieth birthday, 38-46.

equatorial currents
Knauss, John A., 1963
Equatorial currents.
United States National Report, 1960-1963, Thirteenth General Assembly, I.U.G.G.
In: Trans. Amer. Geophys. Union, 44(2):477-478.

currents, equatorial
Knauss, John A., 1960.
Equatorial currents.
Trans. Amer. Geophys. Union, 41(2):255-257.

currents, equatorial
Knauss, John A., 1960.
Measurements of the Cromwell Current.
Deep-Sea Res., 6(4):265-286.

currents, equatorial
Knauss, John A., and R. Pepin, 1959.
Measurements of the Pacific equatorial current.
Trans. Amer. Geophys. Union, 40(1):78-79.
Might also be referred to as:
IGY Bull. No. 21:78-79.

currents, equatorial
Knauss, John A., and Bruce A. Taft, 1963.
Measurments of currents along the equator in the Indian Ocean.
Nature, 198(4878):376-377.

currents, equatorial
Kort, V.G., V.A. Burkov, and K.A. Chekotillo, 1966.
Recent data on the equatorial currents in the western part of the Pacific. (In Russian).
Dokl. Akad. Nauk, SSSR, 171(2):337-339.

currents, equatorial
Kozlov, V.F., 1965.
On the meridional structure of currents at the Equator. (In Russian; English abstract).
Fishika Atmosferi i Okeans, Izv. Akad. Nauk SSSR 1(2):214-223.

currents, equatorial
Kozlov, V.F., 1964
On the meridional structure of currents in the equatorial area. (In Russian).
Okeanologiia, Akad. Nauk, SSSR, 4(5):919.

currents, equatorial
Krivelevich, L.M., 1970.
On the effect of water temperature and salinity on the formation of the ocean currents at the equator. Okeanologiia, 10(3): 406-412.
(In Russian; English abstract)

currents, equatorial
Krivelevich, L.M., 1967.
On the role of gradient pressure in the overall balance of energies in the formation of currents at the equator. (In Russian).
Atlant. nauchno-issled. Inst. ribn. Khoz. okeanogr. (AtlantNIRO). Materialy Konferentsii po Resul'tatam Okeanologischeskikh Issledovanii v Atlanticheskom Okeane, 15-21

Currents equatorial
Moiseev, L.K., 1970.
Thermal structure of equatorial currents in the western Pacific. Okeanologiia, 10(3): 413-425.
(In Russian; English abstract)

equatorial currents
Neiman, V.G., 1964
On the structure of zonation currents in the equatorial area of the Indian Ocean. (In Russian).
Okeanologiia, Akad. Nauk, SSSR, 4(5):920-

equatorial currents (Pacific)
Noel Jacques, et Jacques Merle, 1969
Analyse des courants superficiels et subsuperficiels equatoriaux durant une periode de six jours à 170°est. Courant Equatorial Pacifique et courant de Cromwell.
Cah. océanogr. 21(7): 663-671

currents (equatorial counter currents)
Ponomarenko, G.P., 1963.
Lomonosov's abyssal equatorial counterflow in the Atlantic. (In Russian).
Doklady, Akad. Nauk, SSSR, 149(5):1178-1181.

equatorial countercurrent
Reid, Joseph L., Jr., 1964
Evidence of a South Equatorial Countercurrent in the Atlantic Ocean in July 1963.
Nature, 203(4941):182.

currents, equatorial
Rinkel, M.O., P. Sund and G. Neumann, 1966.
The location of the termination area of the Equatorial Current in the Gulf of Guinea based on observations during Equalant III.
J. geophys. Res., 71(6):3893-3901.

currents, equatorial
Robinson, Allan R., 1960
The general thermal circulation in equatorial regions.
Deep-Sea Res., 6(4):311-317.

currents, equatorial
Rotschi, Henri, 1970.
Variations of equatorial currents.
In: Scientific exploration of the South Pacific, U.S. Wooster, editor, Nat. Acad. Sci., 75-

currents, equatorial (deep)
Rual, P., 1969.
Courants équatoriaux profonds. Deep-Sea Res., 16(4): 387-391.

currents, equatorial
Sarkisyan, A.S., and A.A. Serebryakov, 1969.
A non-stationary model of equatorial currents. (In Russian; English abstract).
Okeanologiia, 9(1):87-91.

currents, equatorial
Shapiro, N.B., 1965.
Analytical investigation of the relations between wind and current in the equatorial zone of the ocean. (In Russian).
Doklady, Akad. Nauk, SSSR, 164(2):319-322.

currents, equatorial
Stommel, Henry, 1963
Varieties of oceanographic experience. The ocean can be investigated as a hydrodynamical phenomenon as well as explored geographically.
Science, 139(3555):572-576.

currents, equatorial
Stone, Peter H., 1971.
The symmetric baroclinic instability of an equatorial current.
Geophys. fluid Dynam. 2(2): 147-164

equatorial currents
Stroup, E.D., and F.W. Hunt, 1963.
Measurements of equatorial currents in the Gilbert Islands area, July-August 1963.
Nature, 200(4910):1001-1002.

Also in:
Collected Reprints, The Johns Hopkins Univ., Chesapeake Bay Inst., and Dept. Oceanogr., 6. 1965.

currents, equatorial
Tomczak, Gerhard, 1963
Unbekannte Meeresströme am Äquator.
Atlantis (7) (Juli 1963).

In:
Oceanogr., Deutsches Inst., 1963(3):(1964)

equatorial currents
Vinogradov, M.E., and N.M. Voronina, 1965.
Some peculiarities of plankton distribution in the Pacific and Indian equatorial current areas. (In Russian; English abstract).
Okeanolog. Issled., Rez. Issled. po Programme Mezhd. Geofiz. Goda, Mezhd. Geofiz. Komitet, Prezidiume Akad. Nauk, SSSR, No.13:128-136.

equatorial countercurrent
Voit, S.S., and S.S. Strekalov, 1964
Some features of the equatorial subsurface Lominosov current in the Atlantic Ocean. (In Russian).
Okeanologiia, Akad. Nauk, SSSR, 4(5):809-812.

currents, equatorial
Yamanaka, Hajime, 1969.
Relation between the fishing grounds of tunas and the equatorial current system. (In Japanese English abstract). Bull. Jap. Soc. fish. Oceanogr. Spec. No. (Prof. Uda Commem. Pap.): 227-230.

Currents, Equatorial
Yoshida, Kozo, 1961.
Some calculations on the equatorial circulation.
Rec. Oceanogr. Wks., Japan, 6(1):101-105.

currents, equatorial, effect of

currents, equatorial, effect of
Hubbs, Carl L., and Richard H. Rosenblatt, 1961.
Effects of the equatorial currents of the Pacific on the distribution of fishes and other marine animals. (Abstract).
Tenth Pacific Sci. Congr., Honolulu, 21 Aug.-6 Sept., 1961, Abstracts of Symposium Papers, 340-341.

currents, equatorial, effect of
Scheltema, Rudolf S., 1968.
Dispersal of larvae by equatorial ocean currents and its importance to the zoogeography of shoal-water tropical species.
Nature, 217(5134):1159-1162.

currents, equatorial undercurrent

currents, undercurrents, equatorial
Arthur, R.S., 1960.
A review of the calculation of ocean currents at the equator. Deep-Sea Res., 6(4):287-297.

currents, undercurrent, equatorial
*Brosin, Hans-Jürgen, und Dietwart Nehring, 1968.
Der Äquatoriale Unterstrom im Atlantischen Ozean auf 29°30'W im September und Dezember 1966.
Beiträge Meeresk., 22:5-17.

currents, equatorial undercurrent
Crease, J., and A. Pogson, 1964.
Observations of the equatorial undercurrent by submarine.
Deep-Sea Res., 11(3):391-393.

currents, undercurrents, equatorial
Istoshin, Yu. V., and A.A. Kalashnikov, 1965.
The Cromwell Current in the western Pacific.
Okeanologiia, Akad. Nauk, SSSR, 5(6):954-958.

currents, undercurrent, equatorial
Knauss, J.A., 1966.
Further measurements and observations on the Cromwell Current.
J. Mar. Res., 24(2):205-240.

equatorial undercurrent
Longuet-Higgins, M.S., 1965.
Some dynamical aspects of ocean currents.
Q. Jl. R. met. Soc., 91:425-451.
Also in:
Collected Reprints, Nat. Inst. Oceanogr., 13.

currents, equatorial undercurrent
Montgomery, R.B., 1961.
The equatorial undercurrent from geostrophic calculation. (Abstract).
Tenth Pacific Sci. Congr., Honolulu, 21 Aug.-6 Sept., 1961, Abstracts of Symposium Papers, 342.

equatorial undercurrent
Munk, Walter and Dennis Moore, 1968.
Is the Cromwell Current driven by equatorial Rossby waves?
J. fluid Mech., 33(2): 241-259

currents, undercurrents, equatorial
Neumann, Gerhard 1969.
The equatorial undercurrent in the Atlantic Ocean.
Actes Symp. Oceanogr. Ressources halieut. Atlant. trop., Abidjan, 20-28 Oct. 1966, UNESCO, 33-44.

currents, undercurrents, equatorial
Neumann, Gerhard, 1960
Evidence for an equatorial undercurrent in the Atlantic Ocean. Deep-Sea Res., 6(4): 328-334.

currents undercurrent, equatorial
Robinson, A.R., 1966.
An investigation into the wind as a cause of equatorial undercurrent.
J. Mar. Res., 24(2):179-204.

currents equatorial undercurrent
Schemainda, R., M. Sturm and K. Voigt, 1964.
Vorläufige Resultate der Untersuchungen im Bereich des Äquatorialen Unterstroms im Golf von Guinea mit MS "Professor Albrecht Penck" in der Zeit von April bis Juli 1964.
Beiträge zur Meereskunde, (15):13 pp.

currents undercurrents, equatorial
Stalcup, M.C., and W.G. Metcalf, 1966.
Direct measurements of the Atlantic Equatorial Undercurrent.
J. Mar. Res., 24(1):54-55.

currents, undercurrent, equatorial
*Swallow, John C., 1967.
The equatorial undercurrent in the western Indian Ocean in 1964.
Studies Trop. Oceanogr., Miami, 5:15-36.

equatorial undercurrent
Swallow, J.C., 1965.
Some observations of the equatorial undercurrent in the western Indian Ocean during 1964.
Trans. Am. geophys. Union, 46:100.

currents – undercurrents, equatorial
*Taft, Bruce A., 1967.
Equatorial undercurrent of the Indian Ocean, 1963.
Studies Trop. Oceanogr., Miami, 5:3-14.

currents, equatorial undercurrent
Taft, Bruce A., and John A. Knauss 1967.
The equatorial undercurrent of the Indian Ocean as observed by the Lusiad Expedition.
Bull. Scripps Inst. Oceanogr. 9:1-163

currents undercurrents, equatorial
Tsuchiya, M., 1961
An oceanographic description of the Equatorial Current System of the Western Pacific.
Oceanogr. Mag., Tokyo, 13(1):1-30.

currents
undercurrents, equatorial

Veronis, George, 1960
An approximate theoretical analysis of the equatorial uncercurrent.
Deep-Sea Res., 6(4):318-327.

currents
undercurrents, equatorial

Wooster, Warren S., 1960
Introduction - Investigations of equatorial undercurrents.
Deep-Sea Res., 6(4):263-264.

currents, undercurrents, equatorial

Wyrtki, Klaus, 1966.
Oceanography of the eastern equatorial Pacific Ocean.
In: Oceanography and marine biology, H. Barnes, editor, George Allen & Unwin, Ltd., 4:33-68.

Currents, estuarine

currents (estuaries)

Horrer, Paul L., 1967.
Methods and deviced for measuring currents.
In: Estuaries, G.H. Lauff, editor, Publs Am. Ass. Advmt Sci., 83:80-89.

currents, estuarine

Kulm, L.D., and John V. Byrne, 1967.
Sediments of Yaquina Bay, Oregon.
In: Estuaries, G.H. Lauff, editor, Publs Am. Ass. Advmt Sci., 83:226-238.

currents, estuarine

Wiegel, R.L., and J.W. Johnson, 1960
Ocean currents, measurement of currents and analysis of data.
Waste disposal in the marine environment, Pergamon Press: 175-245.

CURRENTS, fluctuations

Currents, fluctuations

See also: Currents, variations, etc.

currents, annual fluctuations

Nasukawa, Ikuo, 1960
[On the oceanographic conditions and the annual fluctuation in the sea adjacent to the Kurile Islands.]
Bull. Hokkaido Reg. Fish. Res. Lab., No. 21: 15-30. (English summary)

currents, short period fluctuations

Yoshida, S., 1961
[On the short period variation of the Kurosio in the adjacent sea of Izu Islands.]
Hydrogr. Bull., Maritime Safety Bd., Tokyo, (Publ. No. 981), No. 65:1-18.

Current forecasting

current forecasting

Laevastu, T., 1962
The components of the surface currents in the sea and their forecasts.
Proc. Symposium on Mathematical-Hydrodynamical Methods of Phys. Oceanogr., Sept. 1961, Inst. Meeresk., Hamburg, 321-338.

currents (surface), prediction

Laevastu, T., 1962.
The causes and predictions of surface currents in sea and lake.
Hawaii Inst. Geophys., Rept., No. 21:55 pp. (mimeographed).

current forecasting

Nikiforov, E.G., 1964
Nonstationary phenomena in baroclinic seas and the problem of current prognosis. (In Russian).
Okeanologiia, Akad. Nauk, SSSR, 4(5):914.

currents, prediction

Post, L.A., 1954.
A practical method for the prediction of the surface currents of the ocean. Tellus 6(1):59-62, 3 textfigs.

Currents freshwater

currents, freshwater

Ayers, J.C., D.C. Chandler and G.H. Lauff, 1955.
Currents and water masses in Lake Huron.
Bull. Amer. Phys. Soc., 30(4):13.

Abstr.
Bull. An. CNRS (1) 16:5836.

currents, freshwater

Wedderburn, E.M., 1909.
Current measurements in Loch Garry.
Proc. R. Soc., Edinburgh, 30:312.

currents, freshwater

Wedderburn, E.M., 1908.
Temperature observations in Loch Garry with notes on currents and seiches.
Proc. R. Soc., Edinburgh, 29:98.

currents fronts, oceanic

current fronts, oceanic

Wooster, Warren S., 1959
Oceanographic observations in the Panama Bight, "Askoy" Expedition, 1941.
Bull. Amer. Mus., N.H., 118(3):117-151.

currents, GEK

currents, GEK

Kitamura, Hiroyuki, and Takesni Sagi, 1964.
On the chemical elements in the sea south of Honshu, Japan. (In Japanese; English abstract).
Bull. Kobe Mar. Obs., No. 172:6-54.

currents, GEK

MacGregor, D.G., and H. McLellan, 1952.
Current measurements in the Grand Manan Channel.
J. Fish. Res. Bd., Canada, 9(5):213-222, 4 textfigs.

currents, GEK

McGary, J.W., and E.D.S. Stroup, 1956.
Mid-Pacific oceanography VIII. Middle latitude waters, January-March 1954. Spec. Sci. Rept., Fish., 180:173 pp.

currents - GEK

Novysh, V.V., 1957.
Approximate solution of the problem on the influence of electromotive forces, generated in a deep current, on the results of EMIT measurements. (In Russian).
Trudy Gos. Okeanogr. Inst., No. 40:24-34.

currents (GEK)

Tanioka, Katsumi, and Tsuneo Kamei, 1963
The oceanographical conditions of the Japan Sea. III. On the relation between the oceanographical condition and the sea level. (In Japanese); English abstract).
Umi to Sora, 39(3):91-96.

GEK measurements

Vaux, David, 1965.
Current measuring by towed electrodes; observations in the Arctic and North Seas, 1953-59.
Ministry of Agriculture and Fisheries, Fishery Invest., (2), 23(8):154 pp.

Currents, geostrophic

currents, geostrophic

Bakaev, V.G., editor, 1966.
Atlas Antarktiki, Sovetskaia Antarktich-eskaia Ekspeditsiia. 1.
Glabnoe Upravlenie Geodesii i Kartografii MG SSSR, Moskva-Leningrad, 225 charts.

Currents, geostrophic

Berrit, G.R., 1965.
Les conditions de saison chaude dans la region orientale du Golfe de Guinee.
Progress in Oceanography, 3:31-47.

currents, geostrophic

Bogdanov, M.A., F.N. Zaitsev (deceased), and S.I. Potaichuk 1967.
Dynamics of the water masses in the vicinity of the Faroe-Icelandic Channel. (In Russian).
Atlant. nauchno-issled. Inst. ribn. Khoz. Okeanogr. (ATLANTNIRO). Materialy Konferentsii po Resul'tatam Okeanologischeskikh Issledovanii v Atlantischeskom Okeane, 43-62.

currents, geostrophic

Bulatov, R.P. 1971.
Circulation of waters of the Atlantic Ocean in different space-time scales. (In Russian).
Okeanol. Issled. Rezult. Issled. Mezhd. Geofiz. Proekt. 22: 7-93.

currents, geostrophic

Burkov, V.A., 1967.
Determination of the absolute values of geostrophic current velocities in the oceans. (In Russian; English abstract).
Okeanologiia, Akad. Nauk, SSSR, 7(1):41-50.

currents, geostrophic

Burman I. and O.H. Oren 1970.
Water outflow close to the bottom from the Aegean.
Cah. Océanogr. 22(8): 775-780.

currents, geostrophic

Cooper, L.H.N., 1963.
Essay review: Georg Wust, "Quantitative Untersuchungen zur Statik und Dynamik des atlantischen Ozeans; sechste Lieferung: Stromgeschwindigkeiten und Strommengen in den Tiefen des atlantischen Ozeans". Wiss. Ergebn. deutsch. atlant. Exped. "Meteor" 1925-1927, 6(2):260-420. 1927

Journal du Conseil, 28(2):316-320.

currents, geostrophic

Dodimead, A.J. 1965.
Oceanographic conditions in the central subarctic Pacific region, winter, 1962.
Techn. Rept. Fish. Res. Bd. Can., 75: 10 pp., + 11 figs. (multilithed).

currents, geostrophic

Donguy, Jean Rene, and Michel Prive, 1964.
Les conditions de l'Atlantique entre Abidjan et l'Equateur.
Cahiers Oceanogr., C.C.O.E.C., 16(5):393-398.

currents, geostrophic

Favorite, Feliz, Betty-Ann Morse, Alan H. Haselwoods and Robert A. Preston, Jr., 1964.
North Pacific oceanography, February-April 1962.
U.S.F.W.S., Spec. Sci. Rept., Fish., No. 477:66 pp

currents, geostrpphic

Fedoseev, A., 1970.
Geostrophic circulation of surface waters on the shelf of north-west Africa.
Rapp. P.-v. Réun. Cons. int. Explor. Mer 159: 32-37.

currents, geostrophic
Fofonoff, N.P., 1962
Sect. III. Dynamics of Ocean Currents. In: The Sea, Interscience Publishers, Vol. 1, Physical Oceanography, pp. 323-395.
(Mss received November 1960)

currents, geostrophic
Forrester, W.D. 1970.
Geostrophic approximation in the St. Lawrence estuary.
Tellus 22 (1):53-65

currents, geostrophic
Gostan, Jacques, 1967.
Remarques sur les minimums de salinité observés dans les eaux littorales du golfe de Gênes.
Cah. océanogr., 19(6):469-476.

currents, geostrophic
Gruzinov, V.M., 1965.
Hydrological front as a natural boundary of inherent zones in the ocean. (In Russian).
Trudy, Gosudarst. Oceanogr. Inst., No. 84:252-262.

currents, geostrophic
Hamon, B.V., 1965.
The East Australian Current, 1960-1964.
Deep-Sea Res., 12(6):899-921.

currents, geostrophic
Hamon, B.V., 1965.
Geostrophic currents in the south-eastern Indian Ocean.
Australian J. Mar. Freshwater Res., 16(3):255-271.

current, geostrophic
*Healey, Davis, and P.H. LeBlond, 1969.
Internal wave propagation normal to a geostrophic current.
J. mar. Res., 27(1):85-98.

currents, geostrophic
*Heath, R.A., 1968.
Geostrophic currents derived from oceanic density measurements north and south of the subtropical convergence east of New Zealand.
N.Z. Jl. mar. Freshwat. Res., 2(4):659-677.

currents, geostrophic
Hidaka, K., 1956.
Geostrophic currents and mixing.
J. Ocean. Soc., Japan, 12(4):109-110.

currents, geostrophic
Ingraham, W. James, Jr. 1968.
The geostrophic circulation and distribution of water properties off the coasts of Vancouver Island and Washington, spring and fall 1963.
Fishery Bull. Fish. Wildl. Serv. U.S. 66(2): 223-250.

currents, geostrophic
Japan, Meteorological Agency, 1962
Report of the Oceanographic Observations in the sea south and west of Japan from October to November, 1960. (In Japanese).
Res. Mar. Meteorol. and Oceanogr., July-Dec., 1960, Japan Meteorol. Agency, No. 28:30-35.

current, geostrophic (eastward)
Jarrige, Francois, 1968.
On the eastward flow of water in the western Pacific south of the equator.
J. mar. Res., 26(3):286-289.

currents, geostrophic
Kitano, Kiyomitsu, 1967.
Oceanographic structure near the western terminus of the Alaskan Stream.
Bull. Hokkaido reg. Fish. Res. Lab. No.32: 23-40

currents, geostrophic
Knauss, John A., 1963
10. Equatorial current systems. In: The Sea, M.N. Hill, Edit., Vol. 2. (III) Currents, Interscience Publishers, New York and London, 235-252.

currents, geostrophic
Koshlyakov, M.N., L.I. Galerkin, Chu'Ong Dinh Hiên, 1970.
On mesostructure of geostrophic currents in the open ocean. Oceanologiia, 10(5): 805-814.
(In Russian; English abstract)

currents, geostrophic
Kozlov, V.F., 1971.
On Geostrophic currents in the north Pacific. (In Russian; English abstract). Okeanologiia 11(2): 211-216.

currents, geostrophic
Kozlov V.F., 1969.
The influence of bottom relief on geostrophic currents in the Pacific Ocean. Okeanologiia, 9(4): 608-615.
(In Russian; English abstract)

currents, geostrophic
Kozlov, V.F., 1966.
On geostrophic currents. (In Russian; English abstract).
Okeanologiia, Akad. Nauk, SSSR, 6(2):208-216.

currents, geostrophic
Kuznetsova, L.N., 1966.
On the vertical circulation of water masses in the northwestern North Atlantic. (In Russian).
Mater. Ribokhoz. Issled. severn. Basseina, Poliarn. Nauchno-Issled. Proektn. Inst. Morsk. Ribn. Khoz. Okeanogr. (PINRO), 7:129-136.

currents, geostrophic
Kvinge, Tor, 1963.
The "Conrad Holmboe" Expedition to East Greenland waters in 1923.
Årbok, Univ. i Bergen, Mat.-Naturv. Ser., (15): 44 pp.

currents, geostrophic
Magaard, L. 1968
Ein Beitrag zur Theorie der internen Wellen als Störungen geostrophischer Strömungen.
Dt. hydrogr. Z. 21(6) :241-278

currents, geostrophic
Masuzawa, Jotaro, 1964.
A typical hydrographic section of the Kuroshio extension.
Oceanogr. Mag., Tokyo, 16(1/2):21-30.

currents, geostrophic
Masuzawa, Jotaro, 1962
The deep water in the western boundary of the North Pacific.
J. Oceanogr. Soc., Japan, 20th Ann. Vol., 279-285.

Also JEDS Contrib. No. 39.

currents, geostrophic
Masuzawa, Jotaro, 1961
Preliminary report of the Japanese Expedition of Deep-Sea, the Third Cruise (JEDS-3).
Oceanogr. Mag., Tokyo, 12(2):207-218.

currents, geostrophic
Masuzawa, Jotaro, and Hideo Akamatsu, 1962.
1. Hydrographic observations during the JEDS-4.
Oceanogr. Mag., Japan Meteorol. Agency, 13(2): 122-130.

currents, geostrophic (data)
McGary, James W., Edward D. Stroup, 1958
Oceanographic observations in the central North Pacific, September 1954-August 1955.
U.S.F.W.S. Spec. Sci. Rept., Fish., No.252: 250 pp.

currents, geostrophic
Metcalf, W.G., A.D. Voorhis and M.C. Stalcup, 1962.
The Atlantic Equatorial Undercurrent.
J. Geophys. Res., 67(6):2499-2508.

currents, geostrophic
Mihaljan, John M., 1963.
The exact solution of the Rossby adjustment problem.
Tellus, 15(2):150-154.

currents, geostrophic
Nan-niti, Tosio, Hideo Akamatsu and Toshisuke Nakai, 1964.
A further observation of a deep current in the east-north-east sea of Torishima.
Oceanogr. Mag., Tokyo, 16(1/2):11-19.

currents, geostrophic
Neiman, V.G. 1970.
New maps of Indian Ocean currents.
(In Russian)
Dokl. Akad. Nauk SSSR, 195(4): 948-952.

currents, geostrophic
Ostapoff, Feodor, 1961
A contribution to the problem of the Drake Passage circulation.
Deep-Sea Res., 8(2): 111-120.

currents, geostrophic
Pashkin, V.N., 1968.
Some hydrological features of the shelf waters of the western and southern coasts of Australia. (In Russian).
Trudy, Vses. Nauchno-Issled. Inst. Morsk. Ribn. Okeanogr (VNIRO) 64, Trudy Azovo-Chernomorsk. Nauchno-Issled. Inst. Morsk. Ribn. Khoz. Okeanogr. (AscherNIRO), 28: 142-151.

currents, geostrophic
Ramam, K.V. Sundara and K.V. Sreerama Murthy, 1969.
Geostrophic currents in the Eastern Arabian Sea during the monsoon season. Bull. natn. Inst. Sci., India, 38(1): 221-235.

currents, geostrophic
Reid, R.O., 1959
Influence of some errors in the equation of state or in observations on geostrophic currents.
Nat. Acad. Sci.-Nat. Res. Counc., Publ. No. 600: 19-29.
Also in: Contrib. Oceanogr. Meteorol., A.&M. Coll. Texas, Vol. 5, Contrib. 150.

currents, geostrophic
Reid, Joseph L., Jr., and Richard A. Schwartzlose, 1962
Direct measurements of the Davidson Current off central California.
J. Geophys. Res., 67(6):2491-2497.

geostrophic currents
Rotschi, Henri, 1959
Resultats des observations scientifiques du "Tiare", Croisière DAUPHIN (8-14 août 1959) sous le commandemant du Lieutenant de Vaisseau Cerbelaud.
C.L.O.E.C., Inst. Français d'Océanie, Rapp. Sci., No. 14: 18 pp.

currents, geostrophic
Sdubbundhit, Char Erb, and A.E. Gilmour, 1964
Geostrophic currents derived from oceanic density over the Hikurangi Trench.
New Zealand J. Geol. & Geophys., 7(2):271-278.

Currents, gradient

current gradients
Pochapsky, T.E., 1968.
Oceanic current and temperature gradients at 12°N. 27°W.
J. geophys. Res., 73(4):1221-1237.

currents, gradient
Tully, J.P., 1952.
Notes on the behavious of fresh water entering the sea. Proc. Seventh Pacific Sci. Congr., Met. Ocean., 3:267-289, 10 textfigs.

currents, gradient
Zore, M., 1956.
On gradient currents in the Adriatic Sea.
Acta Adriatica, 8(6):1-38.

currents, gradient
Zore, M., 1956.
[On gradient currents in the Adriatic Sea.]
Hidrografiski Godisnjak, Split, 1955:

currents, gradient
Zore, M., 1960
[Variations of the sea level along the eastern Adriatic coast and the system of gradient currents in the Adriatic.]
Hidrografski Godisnjak, Split, 1959(30):59-65.

Currents, gravity

currents, gravity
*Benjamin, T. Brooke, 1968.
Gravity currents and related phenomena.
J. fluid Mech., 31(2):209-248.

currents, horizontal

currents, horizontal
Cromwell, Townsend, 1958.
Thermocline topography, horizontal currents and "ridging" in the Eastern Tropical Pacific.
(Abstract).
Proc. Ninth Pacific Sci. Congr., Pacific Sci. Assoc., 1957, Oceanogr., 16:72.

Published in:

Bull. Inter-American Tropical Tuna Comm., 3(3), (1958).

Currents, hurricane

currents, hurricane
*Ball, Nahlon M., Eugene A. Shinn and Kenneth W. Stockman, 1967.
The geologic effects of Hurricane Donna in South Florida.
J. Geol., 75(5):583-597.

Current indicators

current indicator (geological)
Berthel, K. Warner, 1966.
Concentric marks: current indicators.
J. Sedim. Petrol., 36(4):1156-1162.

current indicators
Hoyt, John H., and Vernon J. Henry, Jr., 1963.
Rhomboid ripple mark, indicator of current direction and environment.
J. Sed. Petr., 33(3):604-608.
Also in:
Collected Reprints, Mar. Inst., Univ. Georgia, 4 (1965).

current, inertial

current, inertial
Bretschneider, Charles L. 1967.
Calculating storm surge criteria for the continental shelf.
Ocean Industry, 2(12):42-47.

currents, inertial
Greenspan, H.P., 1965.
A note concerning topography and inertial currents.
J. Mar. Res., 21(3):147-154.

current, inertial
Hoshi, T., 1958.
On the inertial current in the ocean.
Geophys. Mag., Tokyo, 28(3):367-374.

currents, inertia
Johns, B., 1965.
Inertia-currents.
Deep-Sea Res., 12(6):825-830.

currents, inertial
Kozlov, V.F., 1970.
The model of the meandering of the inertial currents in the baroclinic ocean. Fizika Atmosfer. Okean., Izv. Akad. Nauk SSSR, 6(9): 922-933.
(In Russian; English abstract)

currents, inertial
Laevastu, T., 1962
The components of the surface currents in the sea and their forecasts.
Proc. Symposium on Mathematical-Hydrodynamical Methods of Phys. Oceanogr., Sept. 1961, Inst. Meeresk., Hamburg, 321-338.

currents, inertial
*Niiler, P.P., and S.L. Spiegel, 1968.
Formation of an inertial current on a continental shelf.
J. mar. Res., 26(1):13-23.

currents, inertial
Robinson, A.R., and P.P. Niiler 1967.
The theory of free inertial currents. I.
Path and structure.
Tellus 19(2): 269-291.

currents, inertial
Titov, V.B., 1968.
Simple techniques for evaluating the spectrum of the oscillations of tidal and inertial currents in the sea. (In Russian; English abstract).
Okeanologiia, Akad. Nauk, SSSR, 8(3):514-521.

currents, instability

currents, instability of
Tareev, B.A., 1965.
Quasigeostrophic instability of oceanic currents.
(In Russian).
Doklady, Akad. Nauk, SSSR, 162(1):74-77.

currents, kinetic energy of
Webster, Ferris, 1971.
On the intensity of horizontal ocean currents.
Deep-Sea Res. 18(9): 885-893.

currents, large scale
Gutman, L.N., 1970.
On large-scale currents in a baroclinic ocean.
(In Russian; English abstract). Fizika Atmosfer. Okean., Izv. Akad. Nauk SSSR, 6(9): 908-921.

Currents, littoral

currents
Allard, P., 1949.
Relations hydrodynamiques dans les mers littorales. Application au bassin oriental de la Manche. Ann. Geophys. 5:25:60.

currents, littoral
Martiel, L., 1956.
Études des courants du littoral sud de la Bretagne. Rev. Trav. Inst. Pêches Marit., 20(3): 263-280.

currents, littoral (data only)
Moreira da Silva, Paulo de Castro, 1968.
Correntes litorais produzidas pelo vento.
Publicões Inst. Pesq. marin., Brasil, 020:21 pp.

currents, littoral
Niaussat, P.M., and R. Bourcart, 1963.
Importance des courants locaux, au Nord de l'embouchure de la Gironde, comme cause eventuelle d'accumulation du phytoplancton.
Cahiers Océanogr., C.C.O.E.C., 15(8):521-526.

currents, littoral, effect of
Palausi, Guy, 1968.
Influence des courants Côtiers sur sédimentation littorale dans la region de Cannes (A.-M.).
Cah. océanogr., 20(9):769-774.

current, littoral
Sato, Shoji, and Norio Tanaka, 1967.
Field investigations on sand drift at Port Kashima facing the Pacific Ocean.
Proc. 10th Conf. Coast. Engng. Tokyo, 1966, 1:595-614.

currents, littoral
Tanner, W.F., 1961
Offshore shoals in area of energy deficit.
J. Sed. Petr., 31(1): 87-95.

Currents, local

currents, local
Wiegel, R.L., and J.W. Johnson, 1960
Ocean currents, measurement of currents and analysis of data.
Waste disposal in the marine environment, Pergamon Press: 175-245.

Currents, longshore

currents, longshore
Aiboulatov, N.A., 1961.
Quelques données sur le transfert des sediments sableux le long d'un littoral, obtenues à l'aide de luminophores.
Cahiers Océanogr., C.C.O.E.C., 13(5):292-300.

longshore currents
Aibulatov, N.A., 1958.
New investigations of marine sand drifts along seashores. (In Russian).
Biull., Okeanogr. Komissiia, Akad. Nauk, SSSR, (1):72-

currents, longshore
Boldyrev, V.L., and E.N. Nevesskiy, 1961
[Alongshore sand-stream to the west of Temrjuk (Azov Sea).]
Trudy Okeanogr. Komissii, Akad. Nauk, SSSR, 8:45-59.

currents, longshore
Bowen, A.J., 1969.
The generation of longshore currents on a plane beach. J. mar. Res., 27(2): 206-215.

currents, longshore
Bretschneider, Charles L. 1968.
On wind tides and longshore currents over the continental shelf due to winds blowing at an angle to the coast.
Mitt. Inst. Meeresk. Univ. Hamburg, 10:96-128.

currents, longshore
Bruun, Per, 1963
Longshore currents and longshore troughs.
J. Geophys. Res., 68(4):1065-1078.

currents, longshore
Caulet, Jean-Claude, 1963.
Étude des plages entre Arzew et Port aux Poules.
Cahiers Océanogr., C.C.O.E.C., 15(9):617-637.

Currents, longshore
Dronkers, J. J., 1961.
De invloed van de deltawerken op de getijbeweging en de stormvloedstanden langs de kust van Zuidwest-Nederland.
Rapp. Deltacommissie, Bijdragen (5):35-84.

currents, longshore
Filimenev, A.I., 1965.
Longshore and normal-to-the-shore movements of water under shallow seashore conditions. (In Russian; English abstract).
Okeanologiia, Akad. Nauk, SSSR, 6(4):645-650.

longshore currents
*Galvin, Cyril J. Jr., 1967.
Longshore current velocity: a review of theory and data.
Review of Geophysics, 5(3):287-304.

currents, longshore.
Galvin, Cyril J., Jr., 1964.
Longshore currents on a laboratory beach. (Abstract).
Geol. Soc., Amer., Special Paper, No. 76:65.

currents, longshore
Galvin, C.J., Jr., and P.S. Eagleson, 1965.
Experimental study of longshore currents on a plane beach.
U.S. Army Engin. Res. Center, Techn. Memo., No. 10:80 pp.
Coastal

currents, longshore
Galvin, Cyril J., and Richard A. Nelson 1967.
Computation of longshore current data.
Misc. Pap. U.S. Army Coast. Engng Res. Center 2-67: 1-19 (multilithed).

currents, longshore
Galvin, C.J., Jr., and R.P. Savage, 1966.
Longshore currents at Nags Head, North Carolina.
CERC Bull. and Summary Rept. of Res. Prog., 1965-1966:11-29.

currents, longshore
Harris, T.F.W., 1961.
The nearshore circulation of water.
Marine Studies off the Natal Coast, C.S.I.R. Symposium, No. S2:18-30. (Multilithed).

currents, longshore (velocity of)
*Harrison, W., 1968.
Empirical equation for longshore current velocity
J. geophys. Res., 73(22):6929-6936.

longshore currents
Hishida, K., 1950.
Hydrography of the mouth of Kumihama Bay.
Ocean. Mag., Tokyo, 2(2):67-68, 2 textfigs.

currents, longshore (generation of)
Hogg, Nelson G., 1971.
Longshore current generation by obliquely incident internal waves.
Geophys. fluid Dynam. 2(4): 361-376

currents, longshore
Hom-ma, Masashi, Kiyoshi Horikawa, and Choule Sonu, 1960
A study on beach erosion at the sheltered beaches of Katase and Kamakura, Japan.
Coastal Eng., Japan, 3:101-122.

longshore currents
Inman, D. L., 1950.
Report on beach study in the vicinity of Mugu Lagoon, California. B.E.B. Tech. Memo. 14:47 pp., 27 figs. (multilithed).

currents, longshore
Inman, D.L., and R.A. Bagnold, 1963.
21. Beach and nearshore processes. 2. Littoral processes.
In: The Sea, M.N. Hill, Editor, Interscience Publishers, 3:529-553.

currents, longshore
Inman, D.L., and W.H. Quinn, 1952?
Currents in the surf zone. Coastal Engineering, Ch. 3:24-36, Textfigs. 5-9 (multilithed).

longshore currents
Iwata, Noriyuki, 1970.
A note on the wave set-up, longshore currents and undertows.
J. oceanogr. Soc. Japan, 26(4): 233-236.

currents, littoral
Johnson, J.W., 1953.
Sand transport by littoral currents. Proc. Fifth Hydraulics Conf., June 9-11, 1952, State Univ. Iowa Studies in Eng., Bull. 34(426):89-109, 9 textfigs.

currents, longshore
Jordaan, J.M., Jr., 1961.
Basic model studies of nearshore wave action.
Marine Studies off the Natal Coast, C.S.I.R. Symposium, No. S2:118-134 (multilithed).

currents, longshore
Klein, George deVries, 1967.
Paleocurrent analysis in relation to modern marine sediment dispersal patterns.
Bull. Am. Ass. Petrol. Geol., 51(3):366-382.

currents, longshore
Kühn, H., 1962.
Der hydrodynamische Kennwert des Bodens.
Zeits. Deutschen Geol. Gesell., 114(1):153-156.

currents, longshore
Larras, Jean 1969.
Vitesse des courants de transport dans les brisants (longshore currents).
Cah. Océanogr. 21(3): 283-285.

Currents, longshore
Longuet-Higgins, M.S., 1970.
Longshore currents generated by obliquely incident sea waves, 2. J. geophys. Res., 75(33): 6790-6801.

currents, longshore
Longuet-Higgins, M.S., 1970.
Longshore currents generated by obliquely incident sea waves, 1. J. geophys. Res., 75(33): 6778-6789.

currents, longshore
Mashima, Yasuo, 1958.
Study on littoral drift and longshore current.
Coastal Engineering, Japan, 1:85-96.

currents, longshore
Mashimo, Yasuo, 1958.
Study on the typhoon characteristics in respect of wave development and the distribution of longshore current.
Coastal Engineering, Japan, 1:1-20.

Currents, longshore
Morra, R. H. J., H. M. Oudshoorn, J. N. Svasek, and F. J. de Vos, 1961.
De zandbeweging in het getijgebied van Zuidwest-Nederland.
Rapp. Deltacommissie, Bijdragen, (5):327-380.

currents, longshore
Nasu, Noriyuki, 1964
The provenance of the coarse sediments on the continental shelves and the trench slopes off the Japanese Pacific coast.
In: Papers in Marine Geology, R.L. Miller, Editor, Macmillan Co., N.Y., 65-101.

currents, longshore
O'Rourke, J.C. and P.H. LeBlond, 1972.
Longshore currents in a semicircular bay.
J. geophys. Res. 77(3): 444-452.

currents, littoral
Saville, T., jr., 1950.
Model study of sand transport along an infinitely long, straight beach. Trans. Am. Geophys Union 31(4):555-565, 6 textfigs.

longshore currents
Schou, A., 1945(1950).
Det marine Forland. Geografiske Studien over danske Fladkystlandskabers dannelse of Formudvikling samt traek af disse omraaders Kulturgeografi. Folia Geogr. Danica IV(Medd. Skalling Lab. 9):236 pp., 85 textfigs., 1 fold-in.

currents, longshore
Shadrin, I.F., 1961
[Alongshore and compensative currents near the smooth-slope accumulative shore.]
Trudy Okeanogr. Komissii, Akad. Nauk, SSSR, 8:158-169.

currents, longshore

Shepard, F.P., 1950.
Longshore-bars and longshore-troughs. B.E.B. Tech. Memo. 15:32 pp., 19 figs. (multilithed).

currents, longshore

Shepard, F.P., and D.B. Sayner, 1953.
Longshore and coastal currents at the Scripps Institution Pier. Bull. B.E.B. 7(1):1-9, 3 figs.

longshore currents

Shepard, F.P., and D.L. Inman, 1950.
Nearshore circulation. Inst. Coastal Eng., Univ. Ext., Univ. Calif., Long Beach, 11-13 Oct. 1950: 12 pp. (mimeographed), 9 figs. (ozalid).

currents, longshore

*Sonu, Choule J., James M. McCloy, and David S. McArthur, 1967.
Longshore currents and nearshore topographies. Proc.10th Conf.Coast.Engng.Tokyo,1966,1:525-549.

Longshore Currents

Todd Thomas W., 1968.
Dynamic diversion: influence of longshore current-tidal flow interaction on Chenier and barrier island plains.
J. Sedim. Petrol., 38(3): 734-746.

currents, longshore

Vera-Cruz, D., 1960
Transporte sólido em costas arenosas. Laboratorio Nacional de Engenharia Civil, Ministerio das Obras Publicas, Lisboa, Memoria No. 145: 1-12.

currents, longshore

Wiegel, R.L., D.A. Patrick and H.L. Kimberley, 1954.
Wave, longshore current and beach profile records for Santa Margarita River, Oceanside, California, 1949. Trans. Amer. Geophys. Union 35(6):887-896.

longshore currents, effect of

Wunderlich, Friedrich, 1971.
Der Golf von Gaeta (Tyrrhenisches Meer). II. Strandaufbau und Stranddynamik. Senckenbergiana maritima 3(1): 135-183.

currents, longshore

Wyrtki, K., 1953.
Die Bilanz des Längstransportes in der Brandungszone. Deutsche Hydrogr. Zeits. 6(2):65-76, 5 text figs.

Currents, longshore, effects of

Currents, longshore, effects of

Aubert, M., and H. Lebout, 1961. ref.
Étude de la pollution des plages de Nice par les résidus ou déchets amenés par les courants ou les fleuves côtiers. Cahiers C.E.R.B.O., Nice, (4):74 pp.

currents, longshore, effect of

Hartke, D. 1970.
Ausgleich eines dreieckförmigen Strandlinienstörung.
Acta hydrophys. 15(1): 15-21.

Current markings

current markings

Owen, David M., and K.O. Emery, 1967.
Current markings on the continental shelf. In: Deep-sea photography, J.B.Hersey, editor, Johns Hopkins Oceanogr.Studies, 3:167-172.

currents (mean)

currents (mean) ocean

Barlow, E.W., 1956.
Five-year current means at Weather Stations I and J. Mar. Obs., 26(172):108-109.

current meanders

currents, meanders of

Hikosaka, Shigeo, 1958
Possible role of horizontal divergence to the meanders of currents in a stratified ocean. Hydrogr. Bull., Tokyo (Publ. No. 981): 1-3.

current meanders

Japan, Maritime Safety Board, 1958.
[Possible role of horizontal divergence to the meanders of currents in a stratified ocean] Hydrogr. Bull. (Publ. 981), No. 55:1-3.

Current measurements

current measurement

Anderson, B., 1958.
On measurement of velocity by Pitot tube. Arkiv f. Math., 3(5):391-394.

currents, measurements

Anon., 1950.
Table 1. Direct current measurement in the Sagami Bay for several stations occupied by a cutter. Res. Mar. Met., Ocean. Obs., Tokyo, July-Dec. 1947, No. 2:243.

currents, measurements

Bagnold, R.A., 1951.
Measurement of very low velocities of water flow. Nature 167:1025.

Current measurements

Berthois, Léopold, et Gérard Auffret, 1970.
Contribution à l'étude des conditions de sédimentation dans la rade de Brest. 3. Hydrologie et Courantométrie.
Cah. océanogr. 22 (7): 701-726

current measurement

Böhnecke, G., 1955.
The principles of measuring currents. Assoc. d'Océan. Phys., UGGI, Publ. Sci., 14:28 pp.

current measurements

Böhnecke, G., 1937
III. Bericht über die Strommessungen auf der Ankerstation 369. Bericht über die erste Teilfahrt der Deutschen Nordatlantischen Expedition des Forschungs-und Vermessungsschiffes "Meteor" Februar bis Mai 1937. Ann Hydro u. mar. Meteorol., 1937, 14-16, 3 text figs.

current measurement

Bowden, K.F., 1954.
The direct measurement of subsurface currents in the oceans. Deep-Sea Res. 2(1):33-47, 1 textfig.

current measurements

Bruce, J.G. and G.H. Volkmann, 1969.
Some measurements of current off the Somali coast during the northeast monsoon. J. geophys. Res., 74(8): 1958-1967.

currents, measurements

Cameron, H.L., 1952.
The measurements of water current velocities by parallax methods. Photogram. Eng. 18:99-104.

currents, measurements

Carruthers, J.N., 1947
Realism in current-measuring in the upper layers of the sea. Int. Hydr. Rev. 24:3-13, 4 tables, 5 text figs.

current, measurements

Carruthers, J.N., 1947
Realism in current-measuring in the upper layers of the sea. Int. Hydr. Rev. 24:54-65, 5 figs., 3 tables.

currents, measurements

Carruthers, J.N., 1947
Practical proposals for a continuous programme of thick-layer current measuring in all weather, with remarks on relevant wind observations and other related matters. J. du Cons. 15(1):13-26.

currents, measurements

Carruthers, J.N., 1938.
Continuous current measurements from light vessels. Review of progress, with results for a third winter, 1937/38. Cons. Perm. Int. Expl. Mer Rapp. Proc. Verb. 107:16-20, 1 textfig.

currents, measurements

Carruthers, J.N., 1937.
Continuous current observations for fishery research application. Cons. Perm. Int. Expl. Mer Rapp. Proc. Verb. 105:16-27, 3 textfigs.

currents, measurements

Carruthers, J.N., A.L. Lawford, and V.F.C. Veley, 1950.
Continuous observations from anchored vessels on water movements in the opensea. Experiences at the "Royal Sovereign" (50°43'N., 0°26'E.) and "Bromer Knoll" (53°16' N., 1°18'E.) light vessels. Deutsch. Hydro. Zeits. 3(5/6):277-286, 7 textfigs.

current measurements

Coston, J. Michael 1968.
Direct current measurements in the Antilles Current.
J. geophys. Res. 73(10): 3341-3344.

Day, C. Godfrey, and Ferris Webster, 1965.
Some current measurements in the Sargasso Sea. Deep-Sea Res., 12(6):805-814.

current measurements.

Deacon, G.E.R., 1964
Sea current measurements. The Royal Society IGY Expedition, Halley Bay, 1955-59, 4:348-352.

current measurements

Eisma, D. and A.J. van Bennekom, 1971.
Oceanographic observations on the eastern Surinam Shelf. Hydrogr. Newsletter, R. Netherlands Navy, Spec. Publ. 6: 25-29.

current measurements

Fofonoff, N. P. 1967.
Current measurements from moored buoys.
Trans. 2nd int. Buoy Techn. Symp. 409-418.

current measurements

Gallagher, James J., 1968.
Discussion of paper by J.Michael Coston, 'Direct current measurements in the Antilles Current'. J.geophys.Res.,73(22):7148.

current measurements
Günzpp, Hans, und Gerhard Tomczak 1965
Strömungsmessungen in der Deutschen Bucht bei Sturmfluten.
Helgoländer wiss. Meeresunters. 17: 94-107.

current measurements
Gould, W.J., 1971
Observations of an event in some current measurements in the Bay of Biscay. Deep-Sea Res., 18(1): 35-49.

current measurements (data only)
Granqvist, G., 1955.
The summer cruise with M/S Aranda in the northern Baltic, 1954. Merent. Julk., No. 166;56 pp.

current measurements
Grulich, W., 1952.
Strömmessungen im Tidegebiet und zur Frage ihrer Umrechnung auf mittlere Tide. Wasserwirtschaft, Hamburg Conf., Sept. 19-20, 1951:19-22.

currents, measurements
Guelke, R.W., and C.A. Schoute-Vanneck, 1947.
The measurement of sea-water velocities by electromagnetic induction. J. Inst. Eng. 1(94): 71-74.

currents, measurements
Helland-Hansen, B., 1907.
Current measurements in Norwegian fjords, the Norwegian Sea in 1906. Bergens Mus. Aarb., 1907, (15):61 pp., 2 pls., 13 textfigs., tables, charts

current measurements
*Hidaka, Koji, 1968.
Computation of upwelling and sinking from direct measurements of horizontal currents.
J.oceanogr.Soc.,Japan,24(4):167-172.

currents, measurement of
Karnaushenko, N.N., 1965.
Optimal parameters of depth for the measurement of currents. (In Russian).
Trudy Morsk. Gidrofiz. Inst., Akad. Nauk, SSSR, 31:96-104.

current measurements
Kvinge, Thor, 1968.
Technical report on project to measure currents related to the formation of Antarctic bottom water in the Weddell Sea: United States Antarctic research program International Weddell Sea oceanographic expedition, 1968, conducted on board USCGC Glacier WAGB-4).
Geofys.Inst.,Univ.Bergen, 19 pp. (mimeographed).

current measurements
*Malone, Frank D., 1968.
An analysis of current measurements in Lake Michigan.
J.geophys.Res., 73(22):7065-7081.

current measurements
Okubo, Akira, 1969.
A note on the effect of dispersion on mean current measurements
Techn. Rept. Chesapeake Bay Inst. 55 (Ref. 69-8): 1-20pp. (unpublished manuscript).

currents, measurements of
Popchapsky, T.W., 1962.
Measurement of currents below the surface at a location in the Western Equatorial Atlantic. Nature, 195(4843):767-768.

current, measurements
Reed, R.K., and N.E. Taylor, 1965.
Some measurements of the Alaska Stream with parachute drogues.
Deep-Sea Res., 777-784.

current measurements
Revelle, R., and F.P. Shepard, 1943.
Current measurements off the California coast in 1938 and 1939. Rec. Observ., S.I.O. 1(2):85-87.

currents, measurements
Romanovsky, V., 1949.
La probleme de la mesure des courants marins. La Houille Blanche 4(2):150-162.

currents, measurement of
Stommel, H., 1955.
Direct measurement of subsurface currents. Deep-Sea Res. 2(4):284-285.

current measurements
Vaux, D., 1955.
Current measuring in shallow waters by towed electrodes. J. Mar. Res. 14(2):187-194, 4 textfigs.

current measurements
Witting, R.J., 1905.
Ftliches ueber Strommessung.
Publ. de Circ., Cons. Perm. Int. Expl. Mer, No. 31:1-11.

Currents, monsoon

currents, monsoon
Berlage, H.P., 1927.
Monsoon-currents in the Java Sea and its entrances. Verh. Magn. Met. Obs., Batavia, 19:28 pp. tables, charts.

currents, monsoon
Desai, B.N. 1968
Interaction of the summer monsoon current with water surface over the Arabian Sea.
Indian J. Met. Geophys. 19 (2): 159-166

currents, monsoon
Graham, H.W., 1952.
A contribution to the oceanography of the Sulu Sea. Proc. Seventh Pacific Sci. Congr., Met. Ocean., 3:225-266, 29 textfigs.

Currents, near-bottom

currents, near-bottom
Harlett, John C. and L.D. Kulm, 1972.
Some observations of near-bottom currents in deep-sea channels. J. geophys. Res. 77(3): 499-504.

currents, near-bottom
Isaacs, John D., Joseph L. Reid, Jr., George B. Schick and Richard A. Schwartzlose, 1966.
Near-bottom currents measured in 4 kilometers depth off the Baja California coast.
J. geophys. Res., 71(18):4297-4303.

Currents, nearshore

currents, nearshore
Deane, R.E., 1961.
Preliminary investigations of near-shore currents in Lake Ontario. (Abstract)
Proc. Fourth Conf., Great Lakes Res., Univ. Michigan, 115.

currents, nearshore
Düing, Walter. 1964.
Deutung der Besonderheiten im Masstransport der küstennahen Strömung im Golf von Neapel.
Kieler Meeresf., 20(2):101-108.

currents, nearshore
Filimonov, A.I., 1965.
On the movement of waters in the near-bottom layer of the near-shore sea zone during the off-and-on shore storms. (In Russian).
Okeanologiia, Akad. Nauk, SSSR, 5(3):397-403.

currents, nearshore
Gaul, R.D. & H.B. Stewart, Jr., 1960
Nearshore ocean currents off San Diego, California. J. Geophy. Res. 65(5): 1543-1556.

currents, nearshore
Shimano, Teizo, Masashi Hom-ma and Kiyoshi Horikawa, 1958
Effect of a jetty on nearshore currents. Model experiment.
Coastal Engineering in Japan, 1:45-58.

currents, nearshore
Watson, Richard L., and E. William Behrens 1970.
Nearshore surface currents, southeastern Texas Gulf coast.
Contrib. mar. Sci., Port Aransas 15: 133-143

currents, near surface
Warsh, K.L., K.L. Echternacht and M. Garstang, 1971.
Structure of near-surface currents east of Barbados. J. phys. Oceanogr., 1(2): 123-129.

currents, non-stationary

currents, non-stationary
Gurikova, Z.F., 1964
The formation of the density field and the calculation of non-stationary currents in the Pacific Ocean. (In Russian).
Okeanologiia, Akad. Nauk, SSSR, 4(5):911-912.

Currents, non-tidal

currents, non-tidal
Bowden K.F., 1953.
Physical oceanography of the Irish Sea.
Brit. Assoc. Sci. Survey Merseyside:69-80.

In: Collected Reprints, National Institute of Oceanography.

Currents, oceanic

currents, open-sea
Bernikov, R. G., 1962.
First experiment on the observation of currents with a "dori" in the open sea. (In Russian).
Trudy, Baltiisk, Nauchno-Issled. Inst., Morsk. Ribn. Khoz. i Okeanogr., (BALTNIRO), 8:245-246.

currents, oceanic
University of Southern California, Allan Hancock Foundation, 1965.
An oceanographic and biological survey of the southern California mainland shelf.
State of California, Resources Agency, State Water Quality Control Board, Publ. No. 27:232 pp. Appendix, 445 pp.

Currents, opposing

currents, opposing

Francis, J.R.D., and C.R. Dudgeon, 1967.
An experimental study of wind-generated waves on a water current.
Q. Jl. R. Met. Soc., 93(396):247-

current, oscillatory

currents, oscillatory

Draper, L., 1967.
Wave activity at the sea bed around northwestern Europe.
Mar. Geol., 5(2):133-146

current, oscillatory

Kajiura, K., 1964.
On the bottom friction in an oscillatory current.
Bull. Earthquake Res. Inst., Tokyo, 42(1):147-174.

current, oscillatory

*Okubo, Akira, 1967.
The effect of shear in an oscillatory current on horizontal diffusion from an instantaneous source.
Int. J. Oceanol. Limnol., 1(3):194-204.

Currents, palaeo-

Currents, palaeo-
(effect of)

Allen, J.R.L. 1966.
On bed forms and palaeocurrents.
Sedimentology 6(3): 153-190

currents (parachute drogue) data only

current profile

current profile

Japan, Kobe Marine Observatory, Oceanographical Observatory, 1962
Report of the oceanographic observations in the cold water region off Enshu Nada in May, 1961. (In Japanese).
Res. Mar. Meteorol. and Oceanogr. Obs., Jan.-June, 1961, No. 29: 25-35.

Currents, random

currents, random

Keller, Joseph B. and George Veronis, 1969.
Rossby Waves in the presence of random currents.
J. geophys. Res., 74(8): 1941-1951.

Currents, residual

currents, residual

Anderson, F.P., 1961.
The use of drift-cards to deduce currents along the Natal coast.
Marine Studies off the Natal Coast, C.S.I.R. Sympos., No. S2:40-45.

currents, residual

Bowden, K.F., and P. Hughes, 1961
The flow of water through the Irish Sea and its relation to wind.
Geophys. J., R. Astron. Soc., 5(4):265-291.

currents, residual

Bowden, K.F., and S.H. Sharaf El Din, 1966.
Circulation and mixing processes in the Liverpool Bay area of the Irish Sea.
Geophys. J., R. Astr. Soc., 11(3):279-292.

currents, residual

Brekhovskikh, L.M., G.N. Ivanov-Frantskevich, M.N. Koshlyakov, K.N. Fedorov, L.M. Fomin and A.D. Yampol'sky 1971.
Some results of the hydrophysical experiment on the "polygon" in the tropical Atlantic.
Fisika atmosfer. okean., Izv. Akad. Nauk SSSR 7(5): 511-527

(In Russian; English abstract

currents, residual

Dietrich, G., 1957.
Hydrographic conditions in the southern North Sea in March 1955, based on a multiple ship survey. Ann. Biol., Cons. Perm. Int. Expl. Mer, 12: 74, Figs. 21-23.

residual currents

Dietrich, G., 1953.
Verteilung, Ausbreitung und Vermischung der Wasserkörper in der südwestlichen Nordsee auf Grund der Ergebnisse der "Gauss"-Fahrt in Februar /März 1952. Ber. Deutsch. Wiss. Komm. Meeresf., Stuttgart, 13(2):104-129, 12 textfigs.

currents, residual

Johns, B. 1970.
On the determination of the tidal structure and residual current system in a narrow channel.
Geophys. Jl. R. astron. Soc. 20(2): 159-175

currents, residual

*Joseph, J., 1960.
Current measurements during the International Iceland-Faroe Ridge Expedition, 30 May to 18 June, 1960.
Rapp. P.-V. Réun. Cons. perm. int. Explor. Mer, 157:157-172.

currents, residual (bottom)

Lee, Arthur, and John Ramster, 1968.
The hydrography of the North Sea. A review of our knowledge in relation to pollution problems.
Helgoländer wiss. Meeresunters. 17: 44-63

currents, residual

*Mandelbaum, Hugo, 1968.
Tidal currents and residual currents at Norderrey, Elbe 1, and Aussen-Eider lightships and their variation under the influence of prevailing winds.
Pure appl. Geophys. 71(3):66-117.

currents, residual

Mandelbaum, H., 1963.
On the variability of computed residual current vectors.
J. Geophys. Res., 68(2):597-598.

currents, residual

Mandelbaum, H., 1934.
Gezeitenströme und Reströme bei Borkum-riff Feuerschiff auf Grund von Beobachtungen der Jahre 1924/28. Arch. Deutschen Seewarte 53(4):1-80, 29 figs.

currents, residual

Mikhaylov, Yu. D., 1960.
[On the formation of residual currents at capes in seas with tides.]
Izv. Vses. Geograph. Obsh., 92(6):521-524.

Engl. Abstr.: The behavior of residual currents near capes.
OTS-61-11148 18. JPRS:8807:15.

currents, residual

Netherlands Hydrographer, 1965.
Some oceanographic and meteorological data of the southern North Sea.
Hydrographic Newsletter, Spec. Issue No. to Vol. 1: numerous pp. not sequentially numbered.

currents, residual

Njun'ko, V.G., 1960.
[Estimation of permissible errors under determining the residual currents.]
Meteorol. i Gidrol., (11):32-34.

currents, residual

Otto, L., 1964.
Results of current observations at the Netherlands lightvessels over the period 1910-1939
1. Tidal analysis and the mean residual currents
K. Nederl. Meteorol. Inst., Mededel. Verhandel. No. 85:56 pp.

currents, residual

Welsh, J.G., 1964
Measurements of currents on the Agulhas Bank with an Ekman current meter.
Deep-Sea Research, 11(1):43-52.

currents, resultant

Zore-Armanda, Mira, 1964.
Results of direct current measurements in the Adriatic. (In Jugoslavian; English resume).
Acta Adriatica, 11(1):293-308.

currents, residual
(effect of)

Robinson, A.H.W. 1966.
Residual currents in relation to shoreline evolution of the East Anglian coast.
Mar. Geol. 4(1): 57-84

Currents, resultant

currents, resultant

Gibson, Blair W. 1965.
Isotachs of resultant current.
Dt. Hydrogr. Z. 18(4): 160-172.

Currents, reversing

currents, reversing of

Warren, Bruce A., 1966.
Medieval Arab references to the seasonally reversing currents of the North Indian Ocean.
Deep-Sea Research, 13(2):167-171.

Currents, rip

currents, rip

Bowen, Anthony J. and Douglas L. Inman, 1969.
Rip currents, 2, laboratory and field observations. J. geophys. Res., 74(23): 5479-5490.

currents, rip

Bowen, Anthony J., 1969.
Rip currents, 1. Theoretical investigations.
J. geophys. Res., 74(23): 5467-5478.

rip currents

Bruun, Per, 1963
Longshore currents and longshore troughs.
J. Geophys. Res., 68(4):1065-1078.

current rip

Great Britain, Meteorological Office, 1961

The marine observers' log - Jan. Feb. March
Marine Observer, 31(191): 6-23.

current rips

Imai, Y., 1960

[Oceanographical studies on the behaviour of chemical elements. 1. On iron and manganese in the sediments around "Shiome" (current rips) in Urado Bay.]
J. Oceanogr. Soc., Japan, 16(3): 134-138.

rip currents
McKenzie, P., 1958.
Rip-current systems. J. Geol., 66:103-113.

currents, rip
Shepard, F.P., K.O. Emery, and E.C. LaFond, 1941
RIp currents: a process of geological importance. J. Geol. 49:337-369

rip currents
Shepard, F.P., and D.L. Inman, 1950.
Nearshore circulation. Inst. Coastal Eng., Univ. Ext., Univ. Calif., Long Beach, 11-13 Oct. 1950: 12 pp. (mimeographed), 9 figs. (ozalid).

currents rips
Van Straaten, L.M.J.U., 1952.
Current rips and dip currents in the Dutch Wadden Sea. K. Nederl. Akad. Wetensch., Proc., B, 55(3):228-238, 2 photos and 10 textfigs.

current rips
Watanabe, Kantaro 1971.
On an understanding of the multiple structure in radiative temperature patterns over current-rips and of the low skin temperature of slicks.
Umi to Sora 46(2): 57-65

current ripples

current ripples
Allen, John R.L., 1968.
Current ripples. North-Holland, 433 pp.
Hfl. 108 or £12.12s or $30.00 (not seen)

current ripples
Dzulynski, Stanislaw, and John E. Sanders, 1962.
Current marks on firm mud bottoms.
Trans., Connecticut Acad. Arts Sci., 42:57-96.

(current) ripples
Shepard, F.P., 1964.
Criteria in modern sediments in recognising ancient sedimentary environments.
In: Developments in Sedimentology, L.M.J.U. van Straaten, Editor, Elsevier Publishing Co., 1: 1-25.

currents (rivers)

currents (river)
Mikhailov, V.N., 1960.
The equations of the juntion of river currents and reservoirs. Physics of the Sea. Hydrology. (In Russian).
Trudy Morsk. Gidrofiz. Inst., 22:5-14.

currents (rivers)
Mikhailov, V.N., 1959.
Dynamics of river streams flowing into streams. (In Russian).
Trudy Gosud. Okeanogr. Inst., 45:73-90.

currents roses

currents roses!
Germany, Deutsches Hydrographisches Institut 1958
[Westindien-Handbuch, 1. Die Nordküste Süd- und Mittelamerikas.] 3rd Edit., No. 2048: 118-153.
Translation:
USN-HO TRANS-66
translator: M. Slessers.
M.O. 16086
P.O.: 20412

currents, rotary

currents, rotary
LaFond, E.C., 1958
On the circulation of the surface layers of the East Coast of India.
Andhra Univ. Mem., Ocean., 2: 1-11.

CURRENTS, ROTARY
LaFond, E.C., and C. Borreswara Rao, 1954.
Rotary currents in the Bay of Bengal.
Andhra Univ. Ocean. Mem. 1:102-106, 7 textfigs.

currents, shallow

currents, shallow
*Metcalf, William G., 1968.
Shallow currents along the northeastern coast of South America.
J. mar. Res., 26(3):232-243.

current shear

current, shearing
Bowden, K.F., 1965.
Horizontal mixing in the sea due to a shearing current.
J. Fluid Mech., 21(1):83-95.

current shear
Lovett, J.R., 1968.
Vertical temperature gradient variations related to current shear and turbulence.
Limnol. Oceanogr., 13(1):127-142.

current shear
Metcalf, W.G., A.D. Voorhis and M.C. Stalcup, 1962.
The Atlantic Equatorial Undercurrent.
J. Geophys. Res., 67(6):2499-2508.

currents, shift in

current shifts
Hanzawa, Masao, 1964
Preliminary report on the abnormal oceanic conditions in the seas adjacent to Japan in the winter of 1963.
In: Studies in Oceanography dedicated to Prof. Hidaka in commemoration of his sixtieth birthday, 59-67.

currents, short period fluctuations

currents, size of

currents, size of
Kuenen, Ph. H., 1964
Deep-Sea sands and ancient turbidites.
In: Turbidites, A.H. Bouma and A. Brouwer, Editors, Developments in Sedimentology, Elsevier Publishing Co., 3:3-33.

currents, small-scale

currents, small-scale
Crisp, D.J., and A.J. Southward, 1956.
Demonstration of small-scale water currents by means of milk. Nature 178(4541):1076.

current spectra
Germany, Deutsches Hydrographisches Institut, 1969.
Spektren von Strömungsmessungen in der Deutschen Bucht im Periodenbereich von 3 bis 110 Stunden während des Winters 1965/66.
Meeresk. Beobacht. Ergebn. 28 (2149): 22 pp.

current speeds
Webster, Ferris, 1971.
On the intensity of horizontal ocean currents.
Deep-Sea Res. 18(9): 885-893.

current speed
Esin, N.V. 1971.
Periodical fluctuations in the field of ocean current speed. (In Russian)
Met. Gidrol. (8): 87-90.

currents, stationary

currents, stationary
Kagan, B.A., S.P. Rebenck, 1961.
[On the method of calculating stationary currents under conditions of instable stratification of the atmosphere.]
Okeanologiia, Akad. Nauk, SSSR, 1(6):1003-1006.

currents, stationary
Koshlyakov, M.N., 1969.
Calculation of stationary ocean currents. (In Russian; English abstract).
Okeanologiia, 9(1):52-57.

currents, stationary
Sarkisyan, A.S., and A.F. Pastukhov, 1970
Density field as a basic indicator of stationary sea currents. (In Russian; English abstract).
Fizika Atmosfer. Okean. Izv. Akad. Nauk SSSR, 6(1): 64-75.

currents, stationary
Tolmazin, D.M. and V.A. Shnaidman, 1971.
On the dynamics of stationary currents in the sea of Azov. (In Russian; English abstract).
Okeanologiia 11(6): 1016-1024.

currents, steady

currents, steady
Felsenbaum, A.I., 1956.
[The relation of the wind to the water level and to steady currents in a shallow sea.]
Dokl. Akad. Nauk, SSSR, 109(1):80-84.

currents, steady
Longuet-Higgins, M.S. 1970.
Steady currents induced by oscillations round islands.
J. fluid Mech. 42(4): 701-720.

currents, steady
Mikhailova, E. N., A. I. Felsenbaum and N. B. Shapiro, 1966.
On the computation of steady currents in seas and oceans. (In Russian).
Doklady, Akad. Nauk, SSSR, 168(4):788-791.

currents (air and sea)
Pedlosky, Joseph, 1964.
The stability of currents in the atmosphere and the ocean. Part 1.
J. Atmospheric Science, 21(2):201-219.

currents, stratified

currents, stratified
Beyer, F., E. Føyn, J.T. Ruud and E. Totland, 1967.
Stratified currents measured in the Oslofjord by means of a new, continuous depth-current recorder, the bathyrheograph.
J. Cons. perm. int. Explor. Mer, 31(1):5-26.

currents, subsurface

currents, subsurface
Barrett, Joseph R., Jr., 1965.
Subsurface currents off Cape Hatteras.
Deep-Sea Research, 12(2):173-184.

currents, subsurface
Fuglister, F.C., 1963.
Gulf Stream '60.
Progress in Oceanography, 1:263-373.

currents, subsurface
Stevenson, Merritt R., June G. Pattullo and Bruce Wyatt, 1969.
Subsurface currents off the Oregon coast as measured by parachute drogues. Deep-Sea Res., 16(5): 449-461.

currents, subsurface
Takenouti, Y., 1958
Measurements of subsurface currents in the cold-belt along the northern boundary of the Kuroshio.
Oceanogr. Mag., Tokyo, 10(1):13-18.

currents, subsurface
Von Arx, William S., 1963
Measurement of subsurface currents by submarine.
Deep-Sea Research, 10(3):189-194.

Currents, surf zone

currents, surf zone
Inman, D.L., and R.A. Bagnold, 1963.
21. Beach and nearshore processes. 2. Littoral processes.
In: The Sea, M.N. Hill, Editor, Interscience Publishers, 3:529-553.

Currents, surface

currents, surface
Akamatsu, Hideo, and Tsutomu Sawara 1967.
Cruise report of CSK survey by JMA in 1966.
Oceanogr. Mag. 19(2): 157-162.

currents, surface
Allain, Charles, 1960.
Topographie dynamique et courants generaux dans le bassin occidental de la Mediterranee (Golfe du Lion, Mer Catalane, Mer d'Alboran et ses abords, secteur a l'est de la Corse).
Rev. Trav. Inst. Peches Marit., 24(1):121-145.

currents, surface
Australia, C.S.I.R.O., Division of Fisheries and Oceanography, Marine Biological Laboratory, Cronulla, 1959.
F.R.V. "Derwent Hunter", Scientific report of Cruise DH9/57, Aug. 19-25, 1957; Cruise DH10/57, Sept. 4-11, 1957; Cruise DH11/57, Sept. 18-21, 1957; Cruise DH12/57, Sept. 26-Oct. 11, 1957.
CSIRO Div. Fish. & Ocean., Rept., No. 20:20 pp. (mimeographed).

currents, surface
Ayers, John C., David C. Chandler, George H. Lauf, Charles F. Powers and E. Bennett Henson, 1958
Currents and water masses of Lake Michigan.
Great Lakes Res. Inst., Publ. No. 3:169 pp.

currents, surface
Balaramamurty, C., and A.A. Ramasastry, 1957.
Distribution of density and the associated currents at the sea surface in the Bay of Bengal.
Indian J. Meteorol. and Geophys., 8(1):88-92.

currents, surface
Barlow, E.W., 1954.
Surface currents of the oceans. II.
J. Inst. Navig. 7(4):347-355, 5 textfigs.

currents, surface
Belevich, R.R., 1960.
Use of bifilar suspension (GR-6) for the determination of the current direction in the surface layer.
Meteorol. i Gidrol., (11):36-37.

currents, surface
Bent, S., 1869.
An address ---- upon the thermometric gateways to the pole, surface currents of the ocean, and the influence of the latter upon the climate of the world:22-24. London.

currents, surface
Bin-Sian, Guan, 1962.
The connection between wind and surface currents in the coastal waters of China. (In Russian).
Sbornik Dokl. II Plen., Komissii Ribochoz. Issledov. Zapadnoi Chasti Tichogo Okeana, 31-40.

currents, surface
Blackburn, M., 1962.
An oceanographic study of the Gulf of Tehuantepec
U.S.F.W.S. Spec. Sci. Rept., No. 404:28 pp.
Fish.

currents, surface
Bogdanov, C.T., and B.G. Popov, 1960
Currents of the surface layer of the Western Pacific.
Trudy Inst. Okeanol., 40: 135-141.

currents, surface (coastal)
Bossolasco, Mario, and Ignazio Dagnino, 1961
Ricerche di fisica marina.
Contr. Ist. Geofis. e Geodet., Univ. Genova, all'Anno Geofis. Internaz., 1957-58 & 1959, Memoria, No. 1: 62 pp.
Reprinted from:
Geofisica e Meteorologia, (Boll. Soc. Ital. di Geofis. e Meteorol., Genova), Vol. 7(1959): 99-112: Vol. 8(1960):22-32; 87-96; 142-155: 9(1961):55-64.

currents, surface
Bougis, P., 1958
Contribution à la connaissance des courants superficiels dans le Nord-Ouest de la Mediterananée occidentale. Trav. St. Zool. Villefranche-sur-Mer. Fasc. 17(8): 67-84. (From Rapp. et Proc. Verb. Reunions 14 n.s. Sept. 1958)

currents, surface
Brodie, J.W., 1960
Coastal surface currents around New Zealand.
N.Z.J. Geol. and Geophys. 3(2): 235-252.

currents, surface
Bruns, E., 1961.
Meereskundliche Expeditionen der DDR auf dem Forschungsschiff "M. Lomonosov" im Atlantischen Ozean.
Beitr. Meeres., 1:7-18.

currents
surface
Capart, Jean-Jacques, and Marc Steyaert, 1963
Mission OTAN en mer d'Alboran, juillet-août 1962. Rapport préliminaire. Températures et courants de surface enregistrés à bord du navire Belge "Euren".
Documents de Travail, Inst. Roy. Sci. Nat. Belg. No. 1: 16 pp. numerous charts. (mimeographed and multilithed).

currents, surface
Clowes, A.J., 1954.
Inshore surface currents on the west coast of the Union of South Africa. Nature 173(4412): 1003-1004, 1 textfig.

currents, surface
Cross, Ford A., and Lawrence F. Small, 1967.
Copepod indicators of surface water movements off the Oregon coast.
Limnol. Oceanogr., 12(1):60-72.

currents, surface
Darbyshire, Mollie, 1963
Computed surface currents off the Cape of Good Hope.
Deep-Sea Res., 10(5):623-632.

currents, surface
Della Croce, U., 1954.
Lanci de galleggianti per lo studio delle correnti superficiali nei bacini ligure e tirrenico.
An. Geofis. 7:241-279.

currents, surface
Della Crose, N., 1952.
Lanci di galleggianti per le studio delle correnti superficiali nel bacino tirrenico.
Cent. Talassografico Tirreno, Genova, No. 18:7 pp.

currents, surface
Dietrich, Günter, 1960
Die Überströmung des Island-Faröer-Rückens, eine Voruntersuchung zum internationalen "Overflow-Program" im Juni 1960. Kieler Meeresf., 16(1): 9-12.

currents, surface
Dobrovolski, A.D., 1949.
Charts of surface currents of the northern part of the Pacific Ocean. Trudy Inst. Okeanol., 3:66-74.

currents, surface
Dobrovol'skii, A.D., and V.S. Arsenev, 1959(1961).
The Bering Sea currents.
Problemy Severa, 3:3-9.

Translation in:
Problems of the North, 3:1-7.

currents, surface
Donguy, J.R., 1962.
Etude du régime des courants superficiels dans le détroit de Gibraltar en relations avec la température de surface. 4ème partie: Trav. Océan. de l'"Origny" dans le détroit de Gibraltar, campagne internationale, 15 mai-15 juin, 1961, par G. Peluchon.
Cahiers Océanogr., C.C.O.E.C., 14(9):626-632.

Currents, surface
Entel, H. 1967.
Influencia de una fuerza perturbadora oscilante del viento sobre las corrientes superficiales del mar.
Beitr. Geophys. 76(1): 64-68.

currents, surface (wind)
Evanov, R.N., 1957.
The influence of the shore on the direction of wind surface currents. (In Russian).
Trudy, Morsk Gidrofiz. Inst., Akad. Nauk, SSSR, 11:84-96.

Translation:
OTS, SLA or ETC

currents, surface
Felber, O.H., 1934.
Oberflächenströmungen des Nordatlantischen Ozeans zwischen 15° und 50° N.B. Arch. Deutschen Seewarte 53(1):1-18, 6 pls.

currents, surface
Fell, H. Barraclough, 1967.
Cretaceous and Tertiary surface currents in the oceans.
Oceanogr. Mar. Biol., Ann. Rev., H. Barnes, editor, George Allen and Unwin, Ltd., 5:317-341.

currents, surface
Frankcom, C.E.N., 1954.
Surface currents of the ocean. I. J. Inst. Navig 7(4):343-347.

currents, surface
Ganapati, P.N., and D. Venkata Rama Sarma, 1958.
Hydrography in relation to the production of plankton off Waltair coast.
Andhra Univ. Mem. Oceanogr., 2:168-192.

Currents, surface
Garner, D.M., 1955.
Some hydrological features of the tropical south-west Pacific Ocean.
New Zealand J. Sci. & Tech., B, 37(1):39-46, 4 textfigs.

currents, surface
Germany, Deutsches Hydrographisches Institut, 1958
[Westindien-Handbuch. 1. Die Nordküste Süd- und Mittelamerikas.] 3rd Edit., No. 2048: 118-153.
Translation:
USN-HO Trans-66
translator: M. Slesse s.
M.O. 16086
P.O.:20412

currents, surface
Gibson, B.W., 1962.
The nature of the sea surface as deduced from composite temperature analysis.
Deutsche Hydrogr. Zeits., 15(2):72-77.

currents, surface
Giddings, J.L., 1952.
Driftwood and problems of Arctic sea currents.
Proc. Amer. Phil. Soc. 96(2):129-142, figs.

currents, surface
Giovando, L.F., and Susumu Tabata 1970.
Measurements of surface flow in the Strait of Georgia by means of free-floating current followers.
Techn. Rept. Fish. Res. Bd Can. 163: 69 pp. (multilithed).

currents, surface
Gopovachev, V. S. and B. A. Iarogov, 1962.
Preliminary results surface current observations in the southwestern Norwegian Sea. (In Russian).
Baltiisk. Nauchno-Issled. Inst. Morsk. Ribn. Khoz. i Okeanogr., 51-55.

currents, surface
Graham, H.W., 1952.
A contribution to the oceanography of the Sulu Sea. Proc. Seventh Pacific Sci. Congr., Met. Ocean., 3:225-266, 29 textfigs.

currents, surface
Grousson, Roger, and Jean Faroux, 1963
Mesure de courants de surface en Mer d'Alboran.
Cahiers Ocean., C.C.O.E.C., 15(10):716-721.

currents, surface
Gurikova, Z.F., 1966.
Computations of the surface and deep ocean currents in the orth Pacific Ocean in summer. (In Russian; English abstract).
Okeanologiia, Akad. Nauk, SSSR, 6(4):615-631.

currents, surface
Gurikova, Z.F., 1962
On the calculation of the currents in the surface layer in the northern sector of the Pacific.
Izv. Akad. Nauk, SSSR, Ser. Geophys, 9:1240-1250. Eng. Ed., 9:776-781.

currents, surface
Harris, T.F.W., 1970.
Features of the surface currents in the south west Indian Ocean.
Symp. Oceanogr. SAfr. C.S.I.R.: 12 pp.
Also in: Coll. Repr. Oceanogr. Inst. Univ. Cape Town 9.

currents, surface
Hart, T. John, and Ronald I. Currie, 1960
The Benguela Current.
Discovery Repts., 31: 123-298.

currents, surface
Hela, Ilmo, 1963.
Surface currents of the Ligurian Sea.
Bull. Inst. Oceanogr., Monaco, 60(1268):15 pp.
Also:
Int. Atomic Energy Agency, Radioactivity in the Sea, Publ. No. 4.

currents, surface
Hela, I., 1954.
The surface current field in the western part of the North Atlantic. Bull. Mar. Sci., Gulf and Caribbean, 3(4):241-272, 22 textfigs.

currents, surface
Hikosaka, S., and R. Watanabe, 1957.
Areas of divergence and convergence of surface currents in the north-western Pacific.
Proc. UNESCO Symp., Phys. Oceanogr., Tokyo, 1955 :101-103.

currents, surface
Hubert, W.E., 1965.
Computer produced synoptic analyses of surface currents and their application for navigation.
Navigation, J. Inst. Navigation, 12(2):101-107.

currents, surface
Iarogov, B.A., and V.S. Golovachev, 1967
Results of the study of the constant surface currents in the western part of the Norwegian Sea with the electromagnetic current meter (EMIT). (In Russian).
Atlant. nauchno-issled. Inst. ribn. Khoz. okeanogr. (AtlantNIRO). Materialy Konferentsii po Resul'tatam Okeanologicheskikh Issledovanii v Atlanticheskom Okeane, 64-72.

currents, surface
Ichiye, Takashi, and Meredith L. Jones, 1961
On the hydrography of the St. Andrew Bay system, Florida.
Limnol. & Oceanogr., 6(3): 302-311.

currents, surface
Ishino, M., 1955.
Hydrographic survey in the Equatorial Pacific Ocean, the South China Sea and the Formosa-Satsunan-Kuroshio region.
Rec. Ocean. Wks., Japan, 2(1):125-131, 4 textfigs

currents, surface (data)
Japan, Fisheries Agency, Research Division, 1956.
Radiological survey of western area of the dangerous zone around the Bikini-Eniwetok Atolls, investigated by the "Shunkotsu maru" in 1956, Part 1:143 pp.

currents, surface
Japan, Hakodate Marine Observatory, 1969.
Report of the oceanographic observations in the Tsugaru Straits from December 7-9, 1965. (In Japanese)
Bull. Hakodate mar. Obs. 14:22-23

currents, surface
Japan, Hakodate Marine Observatory, 1969.
Report of the oceanographic observations in the sea south of Hokkaido and in the Sea of Okhotsk from October to November 1965 (In Japanese)
Bull. Hakodate mar. Obs. 14:16-21

currents, surface
Japan, Hakodate Marine Observatory, 1969.
Report of the oceanographic observations in the sea east of Hokkaido and the Kurile islands from May to June 1965. (In Japanese)
Bull. Hakodate mar. Obs. 14: 12-17.

currents, surface
Japan, Hakodate Marine Observatory 1969.
Report of the oceanographic observations in the Tsugaru Straits in April 1965. (In Japanese)
Bull. Hakodate mar. Obs. 14:10-11.

currents, surface
Japan, Hakodate Marine Observatory, 1969.
Report of the oceanographic observations in the sea east of Hokkaido and the Kurile islands, and in the Okhotsk Sea from July to September, 1965.
Bull. Hakodate mar. Obs. 14: 3-15. (In Japanese)

currents, surface
Japan, Hakodate Marine Observatory, 1969.
Report of the oceanographic observations in the sea east of Hokkaido and Tohoku District from February to March, 1965. (In Japanese)
Bull. Hakodate mar. Obs. 14:3-9

currents, surface
Japan, Hakodate Marine Observatory, 1961
Report of the oceanographic observations in the Okhotsk Sea and in the sea east of Tohoku District from May to June 1959.
Bull. Hakodate Mar. Obs., (8):8-16.

currents, surface
Japan, Hakodate Marine Observatory, 1961
[Report of the oceanographic observations in the sea east of Tohoku District from November to December 1969.]
Bull. Hakodate Mar. Obs., (8):15-19.

currents, surface
Japan, Hakodate Marine Observatory, 1961
[Report of the oceanographic observations in the sea south of Hokkaido from July to August 1959.]
Bull. Hakodate Mar. Obs., (8):3-5.

currents, surface
Japan, Japan Meteorological Agency. 1965.
The results of marine meteorological and oceanographical observations, July-December 1963. No. 34: 360 pp.

currents, surface (GEK)
Japan, Kobe Marine Observatory, Oceanographical Section, 1962
Report of the oceanographic observations in the sea south of Honshu in May, 1960. (In Japanese).
Bull. Kobe Mar. Obs., No. 169(12):27-33.

currents, surface (GEK)
Japan, Kobe Marine Observatory, Oceanographical Section, 1962
Report of the oceanographic observations in the sea south of Honshu in March 1960. (In Japanese).
Bull. Kobe Mar. Obs., No. 169(12):22-33.

currents, surface (GEK)
Japan, Kobe Marine Observatory, Oceanographical Section, 1962
Report of the oceanographic observations in the sea south of Honshu from October to November, 1959. (In Japanese).
Bull. Kobe Mar. Obs., No. 169(11):44-50.

currents, surface
Japan, Kobe Marine Observatory, Oceanographical Section, 1962
Report on the oceanographic observations in the sea south of Honshu from July to August, 1959.
Bull. Kobe Mar. Obs., No. 169(11):37-43.
(In Japanese)

currents, surface
Japan, Maizuru Marine Observatory, 1963
Report of the oceanographic observations in the Japan Sea in June, 1961. (In Japanese).
Bull. Maizuru Mar. Obs., No.8:59-79.

currents, surface
Johnson, James H., 1960.
Surface currents in Lake Michigan.
U.S.F.W.S. Spec. Sci. Rept., Fish., No. 338:120 pp.

currents, surface
Johnson, Sven I., and James L. Squire, 1970.
Surface currents as determined by drift card releases on the continental shelf off the northwestern United States.
Techn. Pap. USFWS Bur. Sport Fish. Wildl., (45):1-12.

currents, surface
Jones, Ian S.F., 1968.
Surface layer currents in Lake Huron.
Proc. 11 Conf. Great Lakes Res. 1968: 406-4[illegible]

surface currents
Kasahara, S., 1958.
A study on the surface currents in the adjacent waters of Noto Peninsula in spring of 1956, with a consideration upon the drift of sardine eggs and larvae.
Ann. Rept. Japan Sea Reg. Fish. Res. Lab., 1(4):77-86.

surface currents, GEK
Keen, D.J., and W. Chimiak, 1955.
Caribbean current survey, spring 1953.
H.O. Tech. Rept., TR-12:14 pp., 14 figs.

currents, surface
Klevtsova, N.D., 1966.
Surface currents in the mean and southern parts of the Caspian Sean under conditions of different winf fields. (In Russian).
Okeanologiia, Akad. Nauk, SSSR, 6(1):82-88.

currents, surface
Koopmann, G., 1953.
Entstehung und Verbreitung von Divergenzen in der oberflächennalen Wasserbewegung der antarktischen Gewässer. Deutsche Hydrogr. Zeits. Ergänzungsheft 2:1-38, 2 pls., 18 textfigs.

Currents, surface
Korea, Republic of, Fisheries Research and Devlopment Agency, 1964.
Oceanographic Handbook of the Neighboring Seas of Korea, 214 pp.

currents, surface
Kramp, P.L., 1963
Summary of the zoological results of the "Godthaab" Expedition 1928.
Medd. om Gronland, 81(7): 115 pp.

currents, surface
Labesh, V.G., 1955.
[Measurement of the direction of ~~fixw~~ the surface flow in the open sea.] Meteorol. i Gidrol., (6): 55-56.
Transl. cited USFWS Spec. Sci. Rept., Fish. 227

currents, surface
Laevastu, T., 1962
The components of the surface currents in the sea and their forecasts.
Proc. Symposium on Mathematical-Hydrodynamical Methods of Phys. Oceanogr., Sept. 1961, Inst. Meeresk., Hamburg, 321-338.

currents, surface
Lanfredi, Néstor W., y Cesar D. Vara 1970.
Mediciones directas de corrientes superficiales en Golfo Nuevo.
Bol. Serv. Hidrograf. Nav. Argentina 7(2): 121-151

currents, surface
Ledenev, V.G., 1964.
Surface currents to the north of Enderby Land. (In Russian).
Sovetsk. Antarkt. Eksped., Inform. Biull., 47:40-
Translation:
Scripta Tecnica, Inc. for AGU, 5(3):172-173. 1965

currents, surface
Ledenev, V.G., 1964.
Direction of surface currents in the west wind drift zone. (In Russian).
Sovetsk. Antarkt. Eksped., Inform. Biull., 48:12-
Translation:
Scripta Tecnica, Inc., for AGU, 5(3):198-200. 1965.

currents, surface
Le Floch, Jean 1970.
Evolution rapide de régimes de circulation non permanents des couches d'eaux superficielles dans le secteur sud-est du Golfe de Gascogne.
Cah. océanogr. 22(3): 269-276.

currents, surface
Le Floch, Jean, 1963.
Sur les variations saisonnières de la circulation superficielle dans le secteur nord-est de la Méditerranée occidentale.
Trav. Centre de Recherches et d'Etudes Océanogr., n.s., 5(1):5-10.

currents, surface
LeFloch, J., and V. Romanovsky, 1954.
Circulation superficielle des eaux dans la partie orientale du bassin occidental de la Mediterranée. Trav. C.R.E.O. 1(1):1-17, 16 textfigs.

Currents, Surface
Leontyeva, V. V., 1961.
[Current and water masses in the western part of the Pacific Ocean in summer 1957.]
Mezhd. Kom. Mezhd. Geofiz. Goda, Presidiume Akad. Nauk, SSSR, Okeanol. Issled., (3):137-150.

surface currents (data)
Lisitzin, E., 1954.
Observations on currents and winds made on board the Finnish Light-Ships during the years 1950 and 1951. Merent. Julk. No. 162:1-45.

currents, surface
Lukyanov, V.V., N.P. Nefedyev, Yu. A. Romano 1962
[On the scheme of surface currents in the Indian Ocean during the winter monsoon. Mezhd. Geofiz. Komitet, Prezidiume Akad.] Nauk, SSSR, Rezult. Issled. Programme Mezhd. Geofiz. Goda, Okeanol. Issled., No. 5:19-24.

currents, surface
Marinelli, G., 1932.
Esperimenti e rilievi sulle corrente superficialá del Tirreno.
Ann. R. Inst. Sup. Navale, Napoli, 1(1):49-81.

currents, surface
Marini, L., 1927.
Risultati dei lanci di gallegzianti per lo studio delle correnti superficiali del mar ligure eseguiti negli anni 1914, 1920-22.
Atti Soc. Ligustica di Sci. e Lett., Pavia 6(3):173-230.

currents, surface
Marr, J.W.S., 1956.
Euphausia superba og de antarktiske overflatestrømmer. En foreløpig meddelelse om hvalfødens itbredølse. Norsk Hvalfanst-Tid. 45(3):127-134.

currents, surface
Meseck, M., 1954.
Meereströmungen auf der Reise zum Persischen Gulf. Der Seewart 15(5):189-192, 1 textfig.

currents, surface
Metallo, Antonio, 1962.
L'onda portante meteoro-oceanografica del Mediterraneo.
Rivista Marittima, Feb., 59-67.

currents, surface
Minevich, A., 1963
Investigations of the surface currents in the Chukchi and Barents seas. (In Russian).
Okeanologiia, Akad. Nauk, SSSR, 3(5):940-942.
from Cahiers Oceanogr., No. 9, 1962

currents, surface
Moroshkin, K.V., 1964.
New map of the surface currents of the Sea of Okhotsk.
Okeanologiia 4(4):641-643.

Abstract in:
Soviet-Bloc Res., Geophys. Astron., Space. 94:47.

currents, surface
Moroshkin, K.V., 1964.
A new pattern of surface currents in the Okhotsk Sea. (In Russian).
Okeanologiia, Akad. Nauk, SSSR, 4(4):641-643.

currents, surface
Mosby, H., 1954.
Oberflächenströmungen in der Meerenge bei Tromsø.
Arch. Met., Geophys., Bioklim., A, 7:378-384, 4 textfigs.

currents, surface
Mratov, K.A., 1971.
Water upwelling and sinking zones of the Atlantic near West Africa. (In Russian; English abstract). Okeanol. Issled. Rezult. Issled. Mezhd. Geofiz. Proekt. 21: 97-106.

currents, surface
Neumann, A. Conrad, 1963.
Processes of recent carbonate sedimentation in Harrington Sound, Bermuda.
Mar. Sci. Center, Lehigh Univ., 130 pp. (Unpublished manuscript).

currents, surface
Niwa, M., and T. Senta, 1962.
Ecological and oceanographical studies on the purse seine fishing frounds around Oki Isls. I. The surface current in autumn 1961. (In Japanese; English summary).
Bull. Japan. Soc. Sci. Fish., 28(9):862-869.

currents, surface
NORPAC Committee, 1960.
The NORPAC Atlas. Oceanic observations of the Pacific, 1955.

surgace currents(direction and velocity)
Netherlands, Koninklijk Nederlands Meteorologisch Instituut, 1952.
Indian Ocean oceanographic and meteorological data. 2nd edition. 31 pp., 24 charts.

Reviewed by P.R. Brown in Met. Mag. (M.O. 581), 83(378-379. (990):

currents, surface
Paasche, E., 1960.
Phytoplankton distribution in the Norwegian Sea in June, 1954, related to hydrography and compared with primary production data.
Fiskeridirektoratets Skr., Ser. Havundersøgelser, 12(11):1-77.

currents, surface
Peluchon, Georges, and Jean Rene Donguy, 1962.
Travaux océanographiques de l'"Origny" dans le détroit de Gibraltar - 15 mai, 15 juin 1961. 2. Courants de surface dans le détroit de Gibraltar.
Cahiers Océanogr., C.C.O.E.C., 14(7):474-483.

currents, surface
Popov, B.G., 1963
The surface currents in the Solomon Sea. (In Russian).
Okeanologiia, Akad. Nauk, SSSR, 3(4):599-605.

currents, surface
Rama Raju, V.S., 1963.
Note on the sea surface currents of the western part of the Indian Ocean.
Bull. Nat. Geophys. Res. Inst., Hyderabad, 1(3):175-178.

currents, surface
Reid, Joseph L., Jr., Richard A. Schwartzlose, Daniel M. Brown, 1963
Direct measurements of a small surface eddy off northern Baja California.
J. Mar. Res., 21(3):205-218.

currents, surface
Ridgway, N.M., 1962.
Nearshore surface currents in southern Hawke Bay, New Zealand.
New Zealand J. Geol. and Geophys., 5(4):545-566.

currents, surface
Ridgway, N.M., 1960
Surface water movements in Hawke Bay, New Zealand.
New Zealand J. Geol. Geophys., 3(2):253-261.

currents, surface
Rochford, D.J. 1969.
Seasonal variation in the Indian Ocean along 110°E. 1. Hydrological structure of the upper 500m.
Austr. J. mar. Freshwat. Res. 20(1):1-50.

Currents, surface
Rodin, G.I., 1958.
Oceanographic and meteorological aspects of the Gulf of California. Pacific Science 12(1):21-45.

currents, surface
Romanovsky, V., 1955.
Résultats de la détermination dans le bassin occidental de la Méditerranée des courants superficiels par la méthode des flotteurs derivants. Trav. C.R.E.O., 2(1/2):1-16.

currents, surface
Saur, J.F.T., J.P. Tully and E.C. LaFond, 1954.
Oceanographic cruise to the Bering and Chukchi seas, summer 1949. Pt. 4. Physical oceanographic studies. Vol. 1. Descriptive report. USNEL Rept. 416(Vol. 1):29 pp., 13 figs.

currents, surface
Schaefer, Milner B., 1958.
Report on the investigations of the Inter-American Tropical Tuna Commission for the Year 1957, 31-

currents, surface (data)
Schott, G., and B. Schulz, 1914.
Die Forschungsreise: S.M.S. "Möwe" im Jahre 1911.
Arch. Deutschen Seewarte 37(1):1-80, 8 pls., 16 textfigs.

currents, surface
Schumacher, A., 1923.
Die Oberflächenströmungen in der Nordsee nach G. Böhnecke. Fischerbote 15:185-188.

Translation by Fish. Lab., Aberdeen

currents, surface
Shannon, L.V. 1970.
Oceanic circulation off South Africa.
Fish. Bull. SAfr. 6:27-33.

currents, surface
Sharikov, Ju. D., 1961
On the problem of fluorescene application for determination of surface currents in the sea.
Meteorol. i Gidrol., (6):51-52.

surface currents (data)
Sitarz, J., 1955.
Resultats de la determination dans le Golfe de Gascogne et La Manche des courants superficiels par la methode des flotteurs derivants.
Trav. C.R.E.O., 2(8/9):1-15.

currents, surface
Stalcup, M. C., and C.E. Parker, 1965.
Drogue measurements of shallow currents on the equator in the western Atlantic Ocean.
Deep-Sea Res., 12(4):535-536.

Currents, surface
Stevenson, Merritt R. 1970.
On the physical and biological oceanography near the entrance of the Gulf of California, October 1966-August 1967. (In English and Spanish).
Bull. int. Am trop. tuna Commn 14(3):389-504.

currents, surface
Stomianko, Pawel, 1960
Analysis of the littoral transport by using marked luminescent sands as tracers. Sea-coast investigations on Hal Peninsula. Prace Instytutu Morskiego 1. Hydrotechnika, No. 4: 146 pp. multilithed or mimeographed.

currents, surface
Suda, K., and D. Syôzi, 1954.
Geomagnetic Electro Kinetograph. Publ. 981, Hydrogr. Bull., Maritime Safety Tokyo, Spec. Number (Ocean. Repts.) No. 4:1-18, 16 figs.

currents, surface
Taguchi, K., 1959
On the surface currents in the waters fished by Japanese salmon motherships from the results of drift float experiments.
Bull. Jap. Soc. Sci. Fish., 25(2):117-121.

currents, surface
Tareyev, B.A., A.V. Fomitchev, 1963
On surface currents of the South Ocean. (In Russian).
Mezhd. Geofiz. Komitet, Prezidiume, Akad. Nauk SSSR, Rezult. Issled. Programme Mezhd. Geofiz. Goda, Okeanol. Issled., (8):24-33.

currents, surface
Thomsen, H., 1951.
1. Kattegat and Belts.
Ann. Biol., Cons. Perm. Int. Expl. Mer, 7:100-101

currents, surface
Tomczak, G., 1964.
Investigations with drift cards to determine the influence of the wind on surface currents.
In: Studies on Oceanography dedicated to Professor Hidaka in commemoration of his sixtieth birthday, 129-139.

currents, surface
Tomczak, G., 1963.
Der Einfluss des Windes auf Oberflächenströmungen im Meer.
Die Umschau in Wiss. und Technik, (13):401-402; (18):568-569.

Also in: Ozeanographie, Deutsches Hydrographisches Inst., 1964.

currents, surface

Tully, J.P., 1942.
Surface non-tidal currents in the approaches to Juan de Fuca Strait. J. Fish. Res. Bd., Canada, 5(4):398-409, 7 figs.

currents, surface

United States Navy Hydrographic Office, 1959.
Climatological and oceanographic atlas for mariners. Vol. 1.
North Atlantic Ocean. 182 charts.

currents, surface

United States, Weather Bureau and Hydrographic Office, 1961
Climatological and oceanographic atlas for Mariners. Vol. II. North Pacific Ocean.
Unnumbered pages.

currents, surface

Vize, V.I., 1945.
[The drift of buoys in arctic seas.]
Problemy Arktiki 1944(2):122, tables.

currents, surface

Vize, V. Iu., 1924.
[Surface current in the Kara Sea.] (German summary)
pp. 1-16.
reprint in MBL library

Currents, surface

Voorhis, A.D., and J.B. Hersey, 1964.
Oceanic thermal fronts in the Sargasso Sea.
J. Geophys. Res., 69(18):3809-3814.

currents, surface

Wüst, Georg (with Arnold Gordon), 1964
Stratification and circulation in the Antillean-Caribbean basins. 1. Spreading and mixing of the water types with an oceanographic atlas.
Vema Research Series, Columbia Univ. Press, New York and London, No. 2:201 pp.

currents, surface

Wyrtkin, Klaus, 1965.
Surface currents of the eastern tropical Pacific Ocean. (In Spanish and English).
Inter-American Tropical Tuna Comm., Bull., 9(5): 271-304.

currents, surface

Wyrtki, K., 1957.
Die Zirkulation an der Oberfläche der südostasiatischen Gewässer. Deutsche Hydrogr. Zeits., 10(1):1-13.

currents, surface

Yamanaka, Hajime, and Noboru Anraku, 1965.
Surface currents in the Indian Ocean as seen from the drift of tuna longline gear.
Rept. Nankai Res. Fish. Res. Lab., No. 22:21-33.

currents, surface

Zore-Armanda, Mira, 1969.
Oceanographic conditions in the middle Adriatic area — II. System of currents in the surface layer and their effect on the temperature distribution. (Jugoslavian and Italian abstracts)
Thalassia Jugoslavica 5: 465-475

currents, surface and bottom

Pickard, G.L., 1956.
Surface and bottom currents in the Strait of Georgia. Bull. Fish. Res. Bd., Canada, 13(4):581-590.

currents, surface (data only)

CURRENTS, SURFACE (data only)

Blackburn, M., R.C. Griffiths, R.W. Holmes and W.H. Thomas, 1962.
Physical, chemical and biological observations in the eastern tropical Pacific Ocean: three cruises to the Gulf of Tehuantepec, 1958-1959.
U.S.F.W.S. Spec. Sci. Rept., Fish., No. 420:170pp

currents (surface) data only

FRANCE, Service Hydrographique de la Marine, 1964
Résultats d'observations. Campagne internationale d'observations dane le détroit de Gibraltar (15 mai-15 juin 1961). Mesures de courant d'hydrologie et de météorologie effectuées à bord de la "Calypso".
Cahiers Océanogr., C.C.O.E.C., 16(1):23-94.

currents, surface (data only)

India, Naval Headquarters, Office of Scientific Research & Development, New Delhi, 1960.
Indian oceanographic stations list, Ser. No. 6 (Jan.): 11 pp. (mimeographed).

currents, surface (data only)

Japan, Nagasaki Marine Observatory, 1967.
Report of the oceanographic observations in the sea southeast of Yakushima Island from April to May 1965. (In Japanese)
Oceanogr. Met. Nagasaki mar. Obs., 17(240): 26 pp.

surface currents (data only)

Lisitzin, E., 1955.
Observations on currents and winds made on board Finnish light-ships during the years 1952, 1953, and 1954. Merent. Julk., 167:76 pp.

currents, surface data

Matsudaira, Yasuo, Haruyuki Koyama and Takuro Endo, 1961
[Hydrographic conditions of Fukuyama Harbor.]
J. Fac. Fish. and Animal Husbandry, Hiroshima Univ., 3(2):247-296.

currents, surface effect of

currents, surface, effect of

Frankcom, C.E.N., and E.W. Barloe, 1954.
Surface currents of the ocean and their effects on navigation. J. Inst. Navig. 7:343-361.

currents, surface, effect of

Taylor, G., 1955.
The action of a surface current used as a breakwater. Proc. R. Soc., London, A, 231(1187):466-478.

Currents, surface monthly

currents (surface monthly)

Wooster, Warren S., Milner B. Schaefer and Margaret K. Robinson, 1967.
Atlas of the Arabian Sea for fishery oceanography.
Inst. Mar. Resources, Univ. Calif., San Diego, IMR Ref. 67-12: numerous pp. (unnumbered) (multilithed).

Currents, surface (offshore)

currents, surface (offshore)

Bainbridge, V., 1960
The plankton of inshore waters off Freetown, Sierra Leone. Colonial Off., Fish. Publ., London, No. 13: 48 pp.

current system

current system

Day, J.H., 1951.
The ecology of South African estuaries. 1. A review of estuarine conditions in general. Trans. R. Soc., S. Africa, 33(1):53-91, 2 textfigs.

current system

Masuzawa, J., 1950.
On the intermediate water in the Southern Sea of Japan. Ocean. Mag., Tokyo, 2(4):137-144, 6 textfigs.

currents, telluric

currents, telluric

Fonarev, G.A., 1961
Some data on telluric currents in the Barents Sea.
Geomag. i Aeronomiya, 1(4): 599-605.
Listed in: Techn. Transl., 7(12): 1024.

current, telluric

Swift, Daniel W., and Victor P. Hessler, 1964.
A comparison of telluric current and magnetic field observations in the Arctic Ocean.
J. Geophys. Res., 69(9):1883-1893.

telluric current variations

Westcott, Eugene M., 1967.
Coastal effects in magnetic and telluric current variations near a complex land, shelving seawater boundary.
J. geophys. Res., 72(7):1959-1969.

currents, temperature

currents, theoretical

currents, theoretical

Bolochonskii, L. Sh., 1963.
Theory of currents in shallow seas. (In Russian).
Trudy Gosud. Okeanogr. Inst., 74:3-32.

currents, theoretical

Chekotillo, K.A., 1966.
Determination of the velocity field of a quasistationary flow in an ocean. (In Russian).
Doklady, Akad. Nauk, SSSR, 169 (5):1071-1074.

currents, theoretical

Duxbury, Alyn C., 1963
An investigation of stable waves along a velocity shear boundary in a two-layer sea with a geostrophic flow regime.
J. Mar. Res., 21(3):246-283.

currents, theory

Fel-zenbaum, A.I., 1971.
Discussion on the theory of ocean currents.
(In Russian). Okeanologiia 11(2): 337.

currents, theoretical

Gormatjuk, Ya. K., and A.S. Sarkisyan, 1964
The results of current calculations in the North Atlantic area in accordance with the four-plane model of the ocean. (In Russian).
Okeanologiia, Akad. Nauk, SSSR, 4(5):910.

currents, theoretical

Ichiye, T., 1960.
A note on ocean currents produced by sources and sinks.
J. Oceanogr. Soc., Japan, 16(3):111-116.

currents, theory

Kamenkovich, V.M., 1971.
Discussion on the theory of ocean currents.
(In Russian). Okeanologiia 11(2): 335-336.

currents, theoretical

Kamenkovich, V.M., 1964
A method of total flows and its use for the theory of oceanic currents. (In Russian).
Okeanologiia, Akad. Nauk, SSSR, 4(5):908-909.

currents (theoretical study)
Kamenkovich, V.M., 1958.
[Certain simplifications of dynamics equations for steady currents in a baroclinic sea.]
Dokl. Akad. Nauk, SSSR, 119(6):1134-1137.

currents (theoretical)
Katkov, V.L., 1968.
One exact solution in the theory of ocean currents. (In Russian).
Fisika Atmosfer. Okean., Izv. Akad. Nauk, SSSR, 4(1): 97-101.

currents, theoretical
Koshliakov, M.N., 1961.
[The calculation of oceanic circulation at depth.]
Okeanologiia, Akad. Nauk, SSSR, 1(6):997-1002.

currents, theoretical
Lebedkina, L.G., 1963.
Determination of currents within a spherical rotating layer of viscous fluid from given surface currents. Theories of waves and currents. (In Russian).
Trudy, Morsk. Gidrofiz. Inst., Akad. Nauk Ukrain., SSR, 27:128-150.

Translation:
Soviet Oceanogr., Trans., Mar. Hydrophys. Inst., SSSR, 27:107-125.

(Agu - Issue 1 - 1964).

currents, theory
Lineykin, P.S., 1970.
On the solution of a boundary value problem of oceanic currents theory. (In Russian; English abstract).
Met. Gidrol. (12): 34-51

currents, theoretical
Lineikin, P.S., 1961.
Review for new foreign researches on theory of sea currents. (In Russian).
Meteorol. i Gidrol., (9):51-56.

currents, theory
Monin, A.S., 1971.
Discussion on the theory of ocean currents. (In Russian). Okeanologiia 11(2): 334-335.

currents, theoretical
Postnova, I.D., 1962.
A theoretical calculation of a stable drifting circulation in a basin with an indented bottom profile. (In Russian).
Izv., Akad. Nauk, SSSR, (11):1663-1670.

Translation:
Amer. Geophys. Union, (11):1036-1040. (1963).

currents (theoretical)
Saito, Y., 1954.
On the Oyashio Current. II. Theoretical consideration on the currents near the Polar Front.
J. Inst. Polytech., Osaka City Univ., G, 2:1-11.

currents, theory
Sarkisian, A.S., 1971.
Discussion on the theory of ocean currents. (In Russian). Okeanologiia 11(2): 337-338.

currents, theoretical
Sarkisyan, A.S., 1964
Formulation of the problem and the calculation layout of nonstationary currents and a temperature field in baroclinic ocean. (In Russian).
Okeanologiia, Akad. Nauk, SSSR, 4(5):909.

currents, theory
Tolmasin, D. M., 1969.
Sur la Théorie des courants dans les détroits des mers du sud
Rapp. P.-v. Réun. Commn int. Explor. scient. Mer Méditerr., 19(4): 701-704.

currents (theoretical)
Tolmazin, D.M., 1964.
A contribution to the theory of currents in straits. (In Russian).
Doklady, Akad. Nauk, SSSR, 159(1):77-80.

Translation:
Earth Sci. Sect. AGI, 159(1/6):1-3. 1965.

currents, theory
Welander, P., 1971.
Discussion on the theory of ocean currents. (In Russian). Okeanologiia 11(2): 336-337.

currents, theory
Zarubin, A.G. 1968.
Boundary layer equation in the two-dimensional theory of ocean currents.
Dokl. Akad. Nauk SSSR 179(4): 798-891.

currents, thermal
*Garrett, W.D., 1967.
Damping of capillary waves at the air-sea interface by oceanic surface-active material.
J. mar. Res., 25(3):279-291.

currents, thermal
Huang, Joseph Chi Kan, 1971.
The thermal current in Lake Michigan. J. phys. Oceanogr., 1(2): 105-122.

currents, temperature
Kloosterman, J., 1956.
Recent investigations of ocean bottoms, especially of their warmth current. Tijd. K. Nederl. Aardrijk. Genoot. (R) 73(2):157-170.

currents, thermohaline
Tjuriakov, B.I. and L.N. Kuznetsova, 1970.
On variability of wind-driven and thermohaline currents in the North Atlantic as related to the changes of the macrosynoptic processes.
Okeanologiia, 10(5): 757-769.

currents through straits
currents through straits
Defant, A., 1955.
Die Strömungen in Meerestrassen.
Deutsche Hydrogr. Zeits. 8(1):1-15, 11 textfigs.

currents, time-varying
Longuet-Higgins, M.S., 1969.
On the transport of mass by time-varying ocean currents. Deep-Sea Res., 16(5): 431-447.

currents, trade-wind
Khanaichenko, N.K., 1968.
Some problems of seasonal precalence of trade-wind currents in the Atlantic. (In Russian).
Meteorologiya Gidrol. (6):93-96.

currents, transient (deep)
Schwartzlose, Richard A., and John D. Isaacs, 1969.
Transient circulation event near the deep ocean floor.
Science, 165(3896):889-891.

current transport
Broida, Saul 1969.
Geostrophy and direct measurements in the Straits of Florida.
J. mar. Res. 27 (3): 278-292.

currents, tidal
SEE ALSO: Tidal currents

currents, tidal
See: tidal currents

currents, turbidity
currents, turbidity
Bagnold, R.A., 1963.
21. Beach and nearshore processes. 1. Mechanics of marine sedimentation.
In: The Sea, M.N. Hill, Editor, Interscience Publishers, 3:507-528.

currents, turbidity
Biricenmater, K., 1957.
[Turbidity currents in the marine environment.]
Przegl. Geofiz., Poland, 2(10):165-178.

currents, turbidity
Chichov, N., 1956.
[Méthode de calcul du volume et de l'extension du courant d'alluvions dans les mers et des grands lacs.] Inst. Nat. pour les Projets de Ports Marit., Trav., Moscow, 3:

currents, turbidity
Ericson, D.B., M. Ewing, and B.C. Heezen, 1952.
Turbidity currents and sediments in the North Atlantic. Bull. Amer. Assoc. Petr. Geol. 36(3): 489-511, 4 textfigs.

currents, turbidity
Heezen, Bruce C., 1963.
27. Turbidity currents.
In: The Sea, M.N. Hill, Editor, Interscience Publishers, 3:742-775.

currents, turbidity
Heezen, B.C., 1956.
Corrientes de Turbidez del Rio Magdalena, Colombia. Bol. Soc. Geogr. Columbia, Bogota, 51-52; 135-143.

currents, turbidity
Heezen, Bruce C., and A.S. Laughton, 1963.
14. Abyssal plains.
In: The Sea, M.N. Hill, Editor, Interscience Publishers, 3:312-364.

currents, turbidity
Hubert, John F., 1964.
Textural evidence for deposition of many western North Atlantic deep-sea sands by ocean-bottom currents rather than turbidity currents.
J. Geol., 72(6):757-785.

currents, turbidity
Johnson, M.A., 1964
Turbidity currents.
In: Oceanography and Marine Biology, Harold Barnes, Editor, George Allen and Unwin, Ltd. 2:31-43.

currents, turbidity
Johnson, M.A., 1962.
Turbidity currents.
Science Progress, 50(198):257-273.

currents,turbidity
Klein,George deVries,1967.
Paleocurrent analysis in relation to modern marine sediment dispersal patterns.
Bull.Am.Ass.Petrol.Geol.,51(3):366-382.

turbidity currents
Kuenen, Ph. H., 1964
Deep-sea sands and ancient turbidites.
In: Turbidites, A.H. Bouma and A. Brouwer, Editors, Developments in Sedimentology, Elsevier Publishing Co., 3:3-33.

turbidity currents
Kuenen, Ph. H., 1953.
Features of graded bedding.
Bull. Amer. Assoc. Petr. Geol. 37(5):1044-1066, 14 textfigs.

currents, turbidity
Kuenen, Ph. H., 1952.
Estimated size of the Grand Banks turbidity current. Am. J. Sci. 250:874-884.

currents, turbidity
Kuenen, Ph. H., 1952.
Transportation and deposition by turbidity currents. (Abstract). Proc. Seventh Pacific Sci. Congr., Met. Ocean., 3:216-218.

currents, turbidity
Kuenen, Ph. H., 1951.
Turbidity flow, a significant geological phenomena. Proc. 3rd Int. Congress Sedimentol., Netherlands, 1951:11-13.

currents, turbidity
Kuenen, Ph. H., 1948.
Turbidity currents of high density. Int. Geol. Congress, Rept. 18th Session, 1948:44-52, 3 textfigs.

Geol. Inst., Groningen Publ. No. 56.

currents, turbidity
Kuenen, Ph. H., and A. Carozzi, 1953.
Turbidity currents and sliding in geosynclinal basins of the Alps. J. Geol. 61(4):363-373, 3 textfigs.

currents
Kuenen, Ph. H., and C.I. Migliorini, 1950.
Turbidity currents as a cause of graded bedding. J. Geol. 58(2):91-127, 7 figs.

currents, turbidity
Laughton, A.S., 1959.
The sea floor.
Science Progress, 47(186):230-249.

currents, turbidity
Longinov, V.V., 1971.
The problem of turbidity currents in the lithodynamics of the ocean. Okeanologiia 11(3): 363-373. (In Russian; English abstract).

currents, turbidity
Nesteroff, Wladimir D., 1962.
Essai d'interprétation du mécanisme des courants de turbidité.
Bull. Soc. Géol., France, (7): 4:849-855.

Also in:
Trab. Lab. Géol. Dyn. et Centres de Recherches Geodynamiques, Fac. Sci., Univ. Paris, No. 2: (1964).

currents, turbidity
Nesteroff, W.D., and B.C. Heezen, 1963.
Essais de comparaison entre les turbidites modernes et le flysch.
Rev. Géogr. Phys. et Géol. Dyn., (2), 5(2):115-127.

Also in:
Trav. Lab. Géol. Dyn. et Centres de Recherches Géodynamiques, Fac. Sci., Univ. Paris, No. 2.

currents, turbidity
Pettijohn, F. J., 1950.
Turbidity currents and greywackes -- a discussion
J. Geol. 58:169-171.

turbidity currents
Plapp, John E., and James P. Mitchell, 1960.
A hydrodynamic theory of turbidity currents.
J. Geophys., 65(3):983-992.

currents, turbidity, effect of
Conolly, John R. and Maurice Ewing, 1969.
Redeposition of pelagic sediment by turbidity currents; a common process for building abyssal plains. Trans. Gulf Coast Ass. geol. Socs, 19 : 506.

currents, turbidity (effect of)
Emery, K.O., Jobst Hülsemann and K.S. Rodolfo, 1962
Influence of turbidity currents upon basin waters.
Limnol. & Oceanogr., 7(4):439-446.

currents, turbulent

Currents, turbulent
Ozmidov, R. V., 1960
Translation:
[On the speed of dissipation of turbulent energy by marine currents and on irregular universal extraneous (effects) by the "4/3Law"]
Izv. Akad. Nauk, SSSR, Ser. Geophys.,(8): 1234-1237

currents, two-layered
*Tareev, B.A., 1968.
Nongeostrophic distribances and baroclinic instability of the two-layer current in the ocean. (In Russian; English abstract).
Fisika Atmosfer.Okean.,Izv.Akad.Nauk,SSSR,4(12): 1275-1284.

currents, undercurrents

undercurrents
Carruthers, J.N., 1962
The Bosporus undercurrent: some bed measurements.
Nature, 201(1917): 363-365.

undercurrents
Fukuoka, Jiro, 1971.
Coastal upwelling and sinking in the Caribbean Sea - especially about the existence of the under current. (Abstracts in English and Portuguese). In: Fertility of the Sea, John D. Costlow, editor, Gordon Breach, 1: 123-142.

currents - undercurrents
Halim, Youssef, Shoukry K. Guergues and Hamed H. Saleh, 1967.
Hydrographic conditions and plankton in the south east Mediterranean during the last normal Nile Flood (1964).
Int.Rev.ges.Hydrobiol., 52(3):401-425.

currents undercurrents
Heezen, Bruce C., and G. Leonard Johnson, 1969
Mediterranean undercurrent and microphysiography west of Gibraltar.
Bull. Inst. océanogr., Monaco, 67 (1380): 95 pp.

currents undercurrents
Masuzawa, Jotaro, 1967.
An oceanographic section from Japan to New Guinea at 137°E in January 1967.
Oceanogr. Mag., Jap. Met. Agency, 19 (2): 95-118

currents undercurrents
Neumann, Gerhard, 1965.
Oceanography of the tropical Atlantic.
Anais Acad. bras. Cienc., 37 (Supl.):63-82.

currents undercurrents
Wooster, W.S., and Malvern Gilmartin, 1961
The Peru-Chile Undercurrent.
J. Mar. Res., 19(3):97-122.

currents undercurrents
Yoshida, Kozo, 1967.
Circulation in the eastern tropical oceans with special reference to the upwelling and undercurrents.
Jap. J. Geophys., 4(2):1-75.

currents, undercurrents, equatorial

Currents, undercurrents (equatorial)
See: Currents, equatorial undercurrents

currents, under ice

currents(under ice)
Lomniewski, Kazimierz, 1958
[The Firth of Vistula.] Polska Akad. Nauk, Inst Geografii, Prace Geograficzne, No. 15: 106 pp.

current variations

Currents, variations, etc.
See also: Current fluctuations

currents, variation
Gudkovich, Z.M., 1961.
[On the problem of the nature of the Pacific current in the Bering Strait and the causative factors for the change of its intensity.]
Okeanologiia, Akad. Nauk, SSSR, 1(4):608-612.

currents, variability of
Kuklin, G.N., 1964
On the variability of currents in the ocean in connection with atmospheric processes. (In Russian).
Okeanologiia, Akad. Nauk, SSSR, 4(5):914.

currents, variability of
Pavlychev, V.P., 1971.
To the question of Kuroshio changeability at the south-western coast of Honshu. (In Russian)
Izv. Tikhookean. nauchno-issled. Inst. ribn. Khoz. Okeanogr, 75:85-93.

current variations

Pettersson, O., 1905.
On the probable occurrence in the Atlantic current of variations, periodical and otherwise, and their bearing on meteorological and biological phenomena, with an introduction.
Rapp. Proc. Verb., Cons. Perm. Int. Expl. Mer, 3(A):26 pp., 16 textfigs.

current variations, coherence of

Siedler, G., 1971.
Vertical coherence of short-periodic current variations. Deep-Sea Res., 18(2): 179-191.

currents, surface, monthly variations

Wyrtki, Klaus, 1960.
The surface circulation in the Coral and Tasman seas.
C.S.I.R.O., Australia, Div. Fish. Oceanogr., Tech. Paper, (8):1-44.

Current velocity

currents, low velocity

Beliakov, L.N., 1966.
Peculiar features involved with the measurement of low velocity currents with the aid of BVP. (in Russian).
Okeanologiia, Akad. Nauk, SSSR, 6(1):159-161.

current velocity

Biays, Pierre, 1960.
Le courant Labrador et quelques-unes de ses consequences geographique. Cahiers Geogr. Quebec, 4(8)237-302.

current velocity

Castens, G., 1932.
Tiefenstrom-Geschwindigkeiten im Nordatlantischen Ozean. Ann. Hydrogr., usw., Jahrg. 60(Heft I): 39-40.

current, velocities

Crease, J., 1962.
Velocity measurements in the deep water of the western North Atlantic, summary.
J. Geophys. Res., 67(8):3173-3176.

currents, mean velocity(data only)

Denmark, Danske Meteorologiske Institut, 1955.
Nautisk-meteorologisk Årbog, 1954:171 pp.

current velocity

Dietrich, G., 1953.
Verteilung, Ausbreitung und Vermischung der Wasserkörper in der südwestlichen Nordsee auf Grund der Ergebnisse der "Gauss"-Fahrt in Februar/März 1953. Ber. Deutsch. Wiss. Komm. Meeresf., Stuttgart, 13(2):104-129, 12 textfigs.

current velocities

Dietrich, Günter, H. Aurich, and A. Kotthaus, 1961
On the relationship between the distribution of redfish and redfish larvae and the hydrographical conditions in the Irminger Sea.
Rapp. Proc. Verb., Cons. Perm. Int. Expl. Mer, 150:124-139.

current velocity

Dyer, K.R., 1970.
Current velocity profiles in a tidal channel.
Geophys. J.R. astron. Soc. 22 (2): 153-161.

current velocity

*Fedorov, K.N., 1967.
Current velocity in the Dover Strait as derived from salinity obervations. (In Russian; English abstract).
Okeanologiia, Akad. Nauk, SSSR, 7(4):601-606.

current velocity

Ford, W.L., and A.R. Miller, 1952.
The surface layer of the Gulf Stream and adjacent waters. J. Mar. Res. 11(3):267-280, 7 textfigs.

current, velocities (data)

Germany, Deutsches Hydrographisches Institut, 14 1960.
Strombeobachtungen in der Deutschen Bucht in den Jahren 1956-1958.
Meereskundliche Beobachtungen und Ergebnisse, No. 15:2122/15): numerous unnumbered pp.

current velocities

Gesenzwei, A.N., 1959
On the dependence of the coefficient of the horizontal macroexchange in sea upon the average period of the current's velocity pulsations. Izvestia Acad. Sci. Geophys. Series 105-107, English Edition, A.G.U. 1960: p. 63-64.

current velocities

Griswold, W.R., 1952.
Loran survey of the Gulf Stream.
Int. Hydrogr. Rev. 29(2):93-104, 10 textfigs.

current velocities

Harvey, J.G., 1968.
The flow of water through the Menai Straits.
Geophys. J.R. astr. Soc., 15(6): 517-528.

Current, velocity

Hidaka, Koji, 1964.
Non-linear computation of the Equatorial Current System of the Pacific. 2. Computation of the velocity on both sides of the equator.
Jour. Oceanogr. Soc., Japan, 20(5):203-208.

currents, velocity of

Hidaka, Koji, 1962
Non-linear theory of an equatorial flow, with special application to the Cromwell Current.
J. Oceanogr. Soc., Japan, 20th Ann. Vol., 223-241.

current velocities

Hidaka, K., 1941.
Absolute evaluation of velocity of ocean currents in dynamical calculations. Mem. Imp. Mar. Obs. 7 (3):379-389.

current velocities

Hidaka, Koji, and Takao Momoi, 1961
Determination of the vertical eddy viscosity in sea water from wind stresses and surface current velocities.
Rec. Oceanogr. Wks., Japan, 6(1): 1-10.

current velocity

Hidaka, K., and T. Suzuki, 1950.
Secular variation of the Tsushima Current.
J. Ocean. Soc., Tokyo, 6(1):28-31, 3 textfigs. (In Japanese with English summary).

current velocity

Hidaka, K., and T. Yamagiwa, 1949.
On the absolute velocity of the subarctic Intermediate Current to the south of Japan. Ocean. Mag., Tokyo, 1(2):99-102, 3 textfigs.

current velocities

Hirano, T., 1953.
Horizontal mixing and flow in a boundary area of the Kuroshio off Nojima-Zaki.
J. Ocean. Soc., Japan, 8(3/4):105-112, 6 textfigs

Also in: Bull. Tokai Regional Fish. Res. Lab., Fish. Agency, 6(Contrib. B) 1953

current velocity (data)

Hong, Sung-Myong, 1966.
A change in salinity to tidal cycle at Kunsen Hang in spring. (In Korean).
Tech. Rep. H.O. Pub. No. 1101:156-163.

current velocity (profile)

Ichiye, T., 1953.
On the variation of circulation. IV.
Ocean. Mag., Tokyo, 5(1):23-44, 9 textfigs.

current velocity

Ichiye, Takashi, 1952?
On the hydrographical condition in the Kuroshio region (1952). 1. Southern area of Honshu.
Bull. Kobe Mar. Obs., No. 163: 1-30.

current velocity

Ichiye, Takashi, 1951
On the hydrography off Shionomisaki and Enshu-Nada (1951).
Bull. Kobe Mar. Obs., No. 164:231 (top of page -240; 13 (bottom of page)-22.

current velocity

Ichie, T., 1949.
Report on the oceanographical observations on board R.M.S. "Syunpu Maru" in the Akashi Seto and the Yura Seto in March 1949. Papers and Repts., Ocean., Kobe Mar. Obs., Ocean. Dept., No. 4:10 pp., numerous figs. and tables. (Atlas-sized pages - mimeographed).

current velocity

Ichie, T., and K. Tanioka, 1949.
On the report of the observations of the hydrographical conditions in the Osaka-wan after the heavy rain caused by the typhoon "Della". Papers and Repts., Ocean., Kobe Mar. Obs., Ocean. Dept., No. 3:21 pp., 19 figs. (mimeographed).

current velocities

Ichie, T., K. Tanioka, and T. Kawamoto, 1950.
Reports of the oceanographical observations on board the R.M.S. "Yushio Maru" off Shionomisaki (Aug. 1949). Papers and Repts., Ocean., Kobe Mar. Obs., Ocean. Dept., No. 5:15 pp., 33 figs. (Odd atlas-sized pages - mimeographed).

current velocities

Japan, Hydrographic Office, Maritime Safety Board, Undated
State of the adjacent seas of Japan, 1955-1959.
Vol. 1: numerous charts.

current velocity (computed)

Japan, Kobe Marine Observatory, Oceanographical Section, 1962
Report of the oceanographic observations in the sea south of Honshu from February to March, 1961. (In Japanese).
Res. Mar. Meteorol. and Oceanogr. Obs., Jan.-June, 1961, No. 29:22-27.

current velocity

Japan, Kobe Marine Observatory, 1956?
Report on the oceanographical observations south off Honshu in 1954. J. Ocean., Kobe (2)7(2):46-69

current velocity
Japan, Kobe Marine Observatory, 1956.
[Report of the oceanographic observations in the sea south of Honshu in August 1955.]
J. Ocean., Kobe (2)7(3):23-32.

current velocity
Japan, Kobe Marine Observatory, 1955.
[Report of the oceanographical observations in the sea south off Honshu in May 1955.]
J. Ocean., Kobe, (2)7(2):17-25.

pagination repeats that of previous article.

current velocity
Japan, Kobe Marine Observatory, 1954.
[The outline of the oceanographical observations off Shionomisaki on board the R.M.S. "Syunpu-maru" (May 1954).] J. Ocean. (2)5(5):1-11, 14 figs.

current velocity
Japan, Kobe Marine Observatory, 1954.
[Results of oceanographical observations off Enshu-Nada (June 1954).] J. Ocean. (2)5(6):1-6, 4 figs.

current velocity
Japan, Kobe Marine Observatory, 1954.
[The outline of the oceanographical observations in the southern area of Honshy on board the R.M.S. "Syunpu-maru".]
Aug.-Sept. 1954 - J. Ocean. (2)5(9):1-44.
Oct. 1954 - J. Ocean. (2)5(10):1-44.

current velocity
Japan, Maizuru Marine Observatory, 1965.
Report of the oceanographic observations in the central part of the Japan Sea from February to March, 1962.---in the Japan Sea from June to July 1962.---in the western part of Wakasa Bay from Janurey to April, 1962.---in the central part of the Japan Sea from September to October, 1962.--in the western part of Wakasa Bay from May to November, 1962.---in the central part of the Japan Sea in March, 1963.---in the Japan Sea in June, 1963.---in Wakasa Bay in July, 1963.---in the central part of the Japan Sea in October, 1963. (In Japanese)
Bull. Maizuru Mar. Obs., No. 9:67-73; 74-88-89-95; 71-80; 81-87; 59-65; 66-77; 80-84; 85-91.

current velocities
Japan, Maizuru Marine Observatory, 1963
Report of the oceanographic observations in the Japan Sea in June 1961. (In Japanese).
Bull. Maizuru Mar. Obs., No. 8:59-79.

current velocity
Japan, Maizuru Marine Observatory and Hakodate Marine Observatory, Oceanographical Sections, 1962
Report of the oceanographic observations in the Japan Sea in June, 1961. (In Japanese).
Res. Mar. Meteorol. and Oceanogr. Obs., Jan.-June, 1961, No. 29:59-79.

current velocity
Japan, Maizuru Marine Observatory, Oceanographical Section, 1961
[Report of the oceanographic observations in the Japan Sea from June to July, 1959.]
Bull. Maizuru Mar. Obs., No. 7:57-64.

current velocity
Japan, Maizuru Marine Observatory, Oceanographical Section, 1961
[Report of the oceanographic observations in the sea north of Kyoga-misaki from August to December 1958.]
Bull. Maizuru Mar. Obs., No. 7:60-63.

current velocity
Japan, Maizuru Marine Observatory, 1958
[Report of the oceanographic observations in the sea north of Sanin and Hokuriku districts in August 1957.]
Bull. Maizuru Mar. Obs., No. 6: 245-253. [107-115.]

current velocity
Japan, Maizuru Marine Observatory, 1956.
[Report of the oceanographical observations north off Sanin and Hokuriku districts in summer 1954.]
[Report of the oceanographical observations north off Sanin and Hokuriku districts in summer 1955.]
Bull. Maizuru Mar. Obs., (5):85-94; 49-56.

current velocity
Japan, Maizuru Marine Observatory, 1955.
[Report of the oceanographical observations (taken) off San'in and Hokuriku (July-August, 1953).] Bull. Maizuru Mar. Obs., (4):13-27.

current velocities
Japan, Maizuru Marine Observatory, 1956.
[Report of the serial observations of Kyoyamisaki during the first half of 1955.]
Bull. Maizuru Mar. Obs., (5):27-32.

current velocities
Japan, Maizuru Marine Observatory, 1955.
[Report of the oceanographical observations off San'in and Hokuri-ku (July-August, 1952).]
Bull. Maizuru Mar. Obs., (4):49-63 (168-181).

current velocities
Japan, Maritime Safety Board, 1955.
[Oceanographical state in the adjacent waters of Marshall Islands between May and June in 1954.]
Publ. No. 981, Hydrogr. Bull., Spec. No. 17:70-79.

current velocity
Japan, Nagasaki Marine Observatory, Oceanographical Section, 1962
Report of the oceanographic observations in the sea west of Japan from April to May, 1961 (In Japanese).
Res. Mar. Meteorol. and Oceanogr. Obs., Jan.-June, 1961, No. 29: 45-53.

current velocity
Japan, Nagasaki Marine Observatory, Oceanographical Section, 1962
Report of the oceanographic observations in the sea west of Japan from February to March, 1961. (In Japanese).
Res. Mar. Meteorol. and Oceanogr. Obs., Jan.-June, 1961, No. 29:36-44.

current velocity
Japan, Nagasaki Marine Observatory, 1962
Report of the oceanographic observations in the sea west of Japan from October to November, 1960. (In Japanese).
Res. Mar. Meteorol. and Oceanogr., July-Dec., 1960, Japan Meteorol. Agency, No. 28:52-59.

current velocity
Japan, Nagasaki Marine Observatory, Oceanographical Section, 1960
[Report of the oceanographic observations in the sea west of Japan from June to July, 1959.]
Res. Mar. Meteorol. & Oceanogr., J.M.A., 26: 51-57.
Also in:
Oceanogr. & Meteorol., Nagasaki Mar. Obs., (1961), 11(200).

current velocity
Japan, Nagasaki Marine Observatory, Oceanographical Section, 1960
[Report of the oceanographic observation in the sea west of Japan from January to February 1959.]
Res. Mar. Meteorol. & Ocean., J.M.A., 25: 48-56.
Also in:
Oceanogr. & Meteorol., Nagasaki Mar. Obs., (1961) 11(199).

current velocity
Japan, Nagasaki Marine Observatory, 1959.
Report of the oceanographic observations in the sea west of Japan from June to July 1958 and in the sea north west of Kyushu in October 1958.
Results Mar. Meteorol. & Oceanogr., J.M.A., 24: 47-60.

Also in:
Oceanogr. & Meteorol., Nagasaki Mar. Obs., 10(195). 1960.

current velocity
Jerlov, N.G., 1953.
Studies of the Equatorial Currents in the Pacific. Tellus 5(3):308-314, 9 textfigs.

current velocity (horizontal, cm/sec).
Jerlov, N.G., 1953.
Studies of the Equatorial Currents in the Pacific
Tellus 5(3):308-314, 5 textfigs.

current velocities (horizontal, cm/sec)
Jerlov, N.G., 1953.
The Equatorial Currents in the Indian Ocean.
Repts. Swedish Deep-sea Exped., 1947-1948, Phys. Chem., 3(5):115-125, 14 textfigs.

currents, velocity
Kajiura, K., 1952.
The velocity distribution of wind currents in the eastern part of the equatorial Pacific.
J. Ocean. Soc., Japan, 8(1):15-22, 3 textfigs.

current velocities
Keen, D.J., and W. Chimiak, 1955.
Caribbean current survey, spring 1953.
H.O. Tech. Rept. TR-12:14 pp., 14 figs.

current velocity
Koslowski, Gerhard, 1960.
Über die Stromungsverhältnisse und den Volumentransport im Nordatlantischen Ozean zwischen Kap Farvel und der Flämischen Kappe im Spätwinter und Spätsommer, 1958.
Deutsche Hydrogr. Zeits., 13(6):269-282.

current velocity
Korgen, Ben J., Gunnar Bodvarsson and L.D. Kulm, 1970.
Current speeds near the ocean floor west of Oregon. Deep-Sea Res., 17(2): 353-357.

current velocity (gradient)
Korobova, V.A., 1967.
Vertical component of the gradient current velocity in the North Atlantic in autumn. (In Russian)
Mater. Sess. Uchen. Sovet. PINRO Rezult. Issled. 1965, Poliarn. Nauchno-Issled. Proektn. Inst. Morsk. Ribn. Khoz. Okeanogr. (PINRO), 8:178-184.

currents, velocity of
Lacombe, H., 1965.
Courants de densité dans le détroit de Gibraltar.
La Houile Blanche, 20(?):38-43.

current velocities
Lamoen, J., 1949.
Tides and current velocities in a sea level canal. Engineering 168(4357):97-99, 3 figs.

Current velocity
Lee, Chang-Ki and Jong-Hon Bong, 1968.
On the current of the Korean Eastern Sea (West of the Japan Sea). (In Korean; English abstract).
Bull. Fish. Res. Develop Agency, Korea, 3: 7-26

current velocity
Le Floch, Jean 1970.
Sur quelques observations de fluctuations de température et de vitesse de courant associées à des ondes internes à courts période ou à la turbulence.
Cah. océanogr. 22(7): 687-699.

currents, velocity of
Le Floch, J., 1963.
Rapport sur les mesures effectuées dans le canal de Corse en août 1960.
Trav. Centre de Recherches et d'Études Océanogr., n.s., 5(1):11-26.

currents, velocity of
Le Floch, J., 1962
Hydrologie d'hiver dans le canal de Capraia, le canal de Piombino et les régions adjacentes. Confrontation avec les résultats de mesure directe de courant.
In: Océanographie Géologique et Géophysique de la Mediterranée Occidentale, Centre National de la Recherche Scientifique, Villefranche sur Mer, 4 au 8 avril 1961, 41-49.

current velocities
LeFloch, Jean, 1961
Mesures de courants par électrodes remorquées dans le Canal de Corse. Relation avec le relief dynamique. Application au choix d'une surface de référence pour le calcul des vitesses en profondeur et des débits.
Cahiers Océanogr., C.C.O.E.C., 13(9):619-626.

current velocity
Lisitzin, E., 1958
Determination of the current velocity in the Kvark on the basis of sea-level records.
Geophysica, Helsinki, 6(3/4):299-307.

current velocity(data).
Littlepage, Jack L., 1965.
Oceanographic investigations in McMurdo Sound, Antarctica.
In: Biology of Antarctic seas, II.
Antarctic Res. Ser., Am. Geophys. Union, 5:1-37.

current velocities
MacGregor, D.G., 1956.
Currents and transport in Cabot Strait.
J. Fish. Res. Bd., Canada, 13(3):435-448.

current velocity
Masuzawa, J., 1960
Statistical characteristics of the Kuroshio Current. Oceanogr. Mag., Tokyo, 12(1): 7-16.

current velocities
Masuzawa, J., 1955.
Preliminary report on the Kuroshio in the Eastern Sea of Japan (Currents and water masses of the Kuroshio system III).
Rec. Ocean. Wks., Japan, 2(1):132-140, 5 textfigs

current velocity
Masuzawa, J., 1954.
On the Kuroshio south of Shiono-Misaki of Japan (Currents and water masses of the Kuroshio System. I.). Ocean. Mag., Tokyo, 6(1):25-33, 7 textfigs.

current velocity
*Matsuda, Yoshihiro, 1967.
Difference in current velocity in a tidal channel with a constant flux. (In Japanese; English abstract).
J.Coll.mar.Sci.,Techn., Tokai Univ., (2):21-27.

current velocity
Mosby, Håkon, 1961.
Veines et artères de la mer.
Bull. Inst. Océanogr., Monaco, No. 1195:27 pp.

current velocity
Nan'niti, Tosio, 1959.
Relation between the velocity and the location of the front of the Kuroshio off the Tohoku district. Oceanogr. Mag., Tokyo, 10(2):185-192. Also in: Collected Reprints, Oceanogr. Lab., Meteorol. Res. Inst., Tokyo, June 1960.
(In English)

current velocity(data)
*Nasu, Keiji, and Tsugio Shimano, 1966.
The physical results of oceanographic survey in the south east Indian Ocean in 1963/64.
J. Tokyo Univ. Fish., (Spec.ed.)8(2):133-164.

current velocity
Okuda, Taizo, and Ramon Nobrega, 1960
Estudoda Barra das Jangadas. 1. Distribuicão e movimento da clorinidade - quantidade de corrente. (In Portuguese; English and French resumés).
Trabalhos, Inst. Biol. Marit. e Oceanogr., Universidade do Recife, Brasil, 2(1):175-191.

current velocity
Pavlova, Yu. V., 1966.
Seasonal variability of the California Current. (In Russian; English abstract).
Okeanologiia, Akad. Nauk, SSSR, 6(6):1003-1013.

currents velocity(data)
Pieterse, F., and D.C. van der Post, 1967.
The pilchard of South West Africa (Sardinops ocellata): oceanographic conditions associated with red-tides and fish mortalities in the Walvis Bay region.
Investl Rept., Mar. Res. Lab., SWest Africa, 14: 1s5 pp.

current velocities
Popov, B.A., 1958.
[A calculation of velocities and acceleration of wave currents as measured by pressure gauges.]
Trudy Inst. Okeanol., 28:195-200.

currant velocity
Postma, H., 1967.
Sediment transport and sedimentation in the estuarine environment.
In: Estuaries, G.H. Lauff, editor, Publs Am. Ass. Advmt Sci., 83:158-179.

current velocity
Postma, H., and J. Verwey, 1950.
Resultaten van hydrografisch Onderzoek in de Waddenzee. Tijdjschrift K. Nederl. Aardrijkskundig Genootschap 67(3):14-33, 14 textfigs.

current velocity
Pritchard, D.W., 1954.
A study of the salt balance in a coastal plain estuary. J. Mar. Res. 13(1):133-144, 4 textfigs.

current, velocity
Reid, Joseph L. and Worth D. Nowlin, Jr., 1971.
Transport of water through the Drake Passage.
Deep-Sea Res., 18(1): 51-64.

current velocities
Rhodes, R. F., 1950.
Effect of salinity on current velocities. Evaluation of present state of knowledge of factors affecting tidal hydraulics andrelated phenomena.
Comm. on Hydraulics, Corps Eng., U.S.A., Rept. No. 1:41-100, 19 figs.

current velocities
Romanov, Yu. A., 1961
[Dynamic method as applied to the equatorial Indian Ocean.]
Okeanol. Issledov., Mezhd. Komit., Mezhd. Geofiz. God, Presidiume Akad. Nauk, SSSR: 25-30.

current velocities
Saelen, Odd H., 1963
Studies in the Norwegian Atlantic Current. II. Investigations during the years 1954-1959 in an area west of Stad.
Geofysiske Publikasjoner, 23(6):1-82.

current velocities
Sarkisyan, A.S., and V.V. Knysh 1969.
The experience in calculation of a level surface and current speed in the Caribbean Sea. (In Russian, English abstract).
Met. Gidrol. 1969(3): 87-93.

current velocities
Shakurov, P., 1959.
Determination of the velocity and direction of currents around a submerged buoy with a pole. Swell and currents in the sea. Loss of a salt from the sea into the atmosphere. (In Russian).
Trudy Morsk. Gidrofiz. Inst., 15:80-85.

current velocity
Shoji, D., R. Watanabe, N. Suzuki and K. Hasuike, 1958.
On the "shiome" at the boundary zone of the Kuroshio and the coastal waters off Shionomisaki.
Rec. Oceanogr. Wks., Japan, (Spec. No. 2):78-84.

current velocities
Shtokman, V.B., 1950.
[Determination of current-velocities and density distribution on a cross-section of an infinitely long channel, as related to the action of the wind and of side friction in the Coriolis-force field.] Dok. Akad. Nauk SSSR 71(1):41-44.
T60R

current velocity
*Siedler, Gerold, 1968.
Schichtungs- und Bewegungsverhälnisse am Südausgang des Roten Meeres.
Meteor Forschungsergeb. (A):4:1-76.

current velocity
Soeriaatmadja, Rd. E., 1957.
The coastal current south of Java.
Penjelidikan Laut, Indonesia (Mar. Res.) No. 3: 41-55.

current velocities
Soule, F.M., and J.E. Murray, 1957.
Physical oceanography of the Grand Banks region and the Labrador Sea in 1956.
U.S.C.G. Bull., No. 42:35-100.

current velocity
Stefansson, Unnstein, 1962.
North Icelandic waters.
Rit Fiskideildar, Reykjavik, 3:269 pp.

current velocity, mean
Tanioka, K., 1966.
Oceanographical conditions of the Japan Sea (IV). (In Japanese; English abstract)
Umi to Sora, 41(1/2): 50-57.

current velocity
Tanioka, Katsumi, and Tsuneo Kamei, 1965.
The oceanographical conditions of the Japan Sea. III. On the relation between the oceanographical condition and the sea level (In Japanese; English abstract).
Bull. Maizuru Mar. Obs., No. 9:91-96.

current velocity(data)
Timofeev, V.T., 1961.
The effect of the deep layers of Atlantic waters on the hydrological regime of the Kara Sea. (In Russian).
Problemy Severa, 4:46-58.
Translation:
Problems of the North, 1962, 4:45-56.

current velocities
Troadec, Jean-Paul, 1963
Mesures directes de courant au large de Saint-Tropez (Var.).
Cahiers Oceanogr., C.C.O.E.C., 15(3):170-182.

current velocity
Tsuchiya, M., 1955.
On a simple method of estimating the current velocity at the Equator. J. Ocean. Soc., Japan, 11(1):1-4.
Also in: Geophys. Notes, Tokyo, 8(2):Contr. 36.

current velocity
Tsuchiya, M., 1955.
On a simple method for estimating the current velocity at the equator. Rec. Ocean. Wks., Japan, 2(2):37-42.

current velocities (subsurface)(data)
Tsujita, T., 1953.
Studies on naturally occurring suspended organic matter in waters adjacent to Japan II.
Rec. Ocean. Wks., Japan, n.s., 1(2):94-100, 2 textfigs., 1 pl. of 4 figs.

current velocity
Von Arx, W.S., 1952.
Notes on the surface velocity profile and horizontal shear across the width of the Gulf Stream. Tellus 4(3):211-214, 3 textfigs.

current velocity
von Arx, W.S., 1950.
Some surface measurements of ocean current velocities obtained from a ship under way by means of the geomagnetic electrokinetograph. (Abstract)
Trans. Am. Geophys. Union, 31(2):331.

current velocity
Warren, Bruce A., and Arthur D. Voorhis, 1970.
Velocity measurements in the deep western boundary current of the South Pacific.
Nature, Lond., 228 (5274): 849-850

current velocity
Watanabe, Nobuo, Toshiyuki Hirano, Rinnosuke Fukai, Kozi Matsumoto and Fumiko Shiokawa, 1957.
A preliminary report on the oceanographic survey in the "Kuroshio" area, south of Honshu, June-July 1955.
Rec. Oceanogr. Wks., Japan (Suppl.):197-

current velocities
Wüst, G., 1957.
Stromgeschwindigkeiten and Strommengen in den Tiefen des Atlantischen Ozeans unter besonderer Berücksichtigung des Tiefen- und Bodenwassers. Quantitative Untersuchungen zur Statik und Dynamik des Atlantischen Ozeans.
Wiss. Ergebn. Deutschen Atlant. Exped., "Meteor", 1925-1927, 6(2):262-420, 36 figs.

current velocities
Wüst, Georg, 1957.
Die Stromgeschwindigkeiten, besonders im Tiefenund Bodenwasser.
Wiss. Ergebn. dt. atlant. Exped. 'Meteor', 6(2): 321-351.
Translation: USN Oceanogr. Off. TRANS 348. (M. Slessers). 1967.

currents, velocity of
Wüst, G., 1955.
Stromgeschwindigkeiten im Tiefen- und Bodenwasser des Atlantischen Ozeans auf Grund dynamischer der Meteor-Profile der Deutschen Atlantischen Expedition 1925/27.
Pap. Mar. Biol. and Oceanogr., Deep-Sea Res., Suppl. to Vol. 3:373-397.

current velocity
Wyrtki, K., 1953.
Die Dynamik der Wasserbewegungen im Fehmarnbelt. I. Kieler Meeresf. 9(2):155-170, 4 pls. (13 figs.).

current velocity
Yamanaka, Hajime, Noboru Anraku and Jiro Morita, 1965.
Seasonal and long-term variations in oceanographic conditions in the western North Pacific Ocean.
Rept. Nankai Reg. Fish. Res. Lab., No. 22:35-70.

current velocity(data)
Yasui, Z., and M. Matuno, 1953.
On the circulation of currents in China Eastern Sea. Rec. Ocean. Wks., Japan, n.s., 1(1):28-35, 3 textfigs.

current velocity
Yoshida, S., H. Nitani and N. Suzuki, 1959.
Report of multiple ship survey in the equatorial region (I.G.Y.), Jan.-Feb., 1958.
Publ. 981, Hydrogr. Bull. 59:1-30.

Currents, velocity-bottom

currents, bottom velocities
Wilde, Pat, 1965.
Estimates of bottom current velocities from grain size measurements for sediments from the Monterey deep-sea fan.
Ocean Sci. and Ocean Eng., Mar. Techn. Soc., Amer. Soc. Limnol. Oceanogr., 2:718-727.

current velocity components
Belyaev, V.S., and R.V. Ozmidov 1971.
Probability distribution laws of the current velocity components in the ocean. (In Russian; English abstract).
Fizika Atmosfer. Okean, Izv. Akad. Nauk SSSR 7(5): 528-533.

Current velocity data only

current velocity (data only)
Japan, Japan Meteorological Agency 1971.
The results of marine meteorological and oceanographical observations, July - December 1969, 46:270 pp.

current velocity (data only)
Japan Meteorological Agency, 1968.
The results of marine meteorological and oceanographical observations, July-December, 1965, 38: 404 pp. (multilithed)

current velocity (data only)
Lemasson, L., et J.P. Rebert, 1968.
Observations des courants sur le plateau continental Ivoirien: mise en evidence d'un sous-courant. Doc. Sci. Provisoire, Min. Product. Animale Cote d'Ivoire, (022):66 pp. (mimeographed).

Current velocity (data only)
Tabata S., L.V. Giovando, J.A. Strickland and J. Wong, 1970.
Current velocity measurements in the Strait of Georgia - 1968.
Techn. Rept. Fish. Res. Bd., Can., 178: 112 pp. (multilithed)

Currents, velocity profile

velocity profile
Ellison, T.H., 1960
A note on the velocity profile and longitudinal mixing in a broad open channel. J. Fluid Mech. 8(1): 33-40.

Currents, vertical

currents, vertical
Glowinska, A., 1954.
Vertical currents in the Baltic Sea.
Rept. Sea Fish. Inst., Gydnia, 7:153-158.

currents, vertical
Ichiye, Takashi, 1966
Vertical currents in the equatorial Pacific Ocean.
J. oceanogr. Soc., Japan, 22(6): 274-284

currents, vertical
Ichiye, Takashi, 1966.
Vertical currents in the equatorial Pacific Ocean. (Abstract only).
Trans. Am. Geophys. Un., 47(3):478.

currents, vertical velocity
Ivanov, Yu. A., 1961
On frontal zones in the Antarctic waters.
Mezhd. Kom. Mezhd. Geofiz. Goda, Presidiume Akad. Nauk, SSSR, Okeanol. Issled., (3):30-51.

currents, vertical structure
Marchuk, G.I., and V.P. Kochergin 1968.
On the vertical current structure in a baroclinic ocean. (In Russian; English abstract)
Meteorologiya i Gidrol. (1): 3-10.

currents, vertical
Shuleikin V.V., 1967.
Calculation of vertical currents in a sea with a complex bottom relief. (In Russian)
Dokl. Akad. Nauk, SSSR, 175(3): 575-551
РАСЧЕТ ВИХРЕВЫХ ТОКОВ В МОРЕ СО СЛОЖНЫМ РЕЛЬЕФОМ ДНА.

currents, vertical component
*Sudo, Hideo, 1969.
An attempt to estimate the vertical component of the current velocity in the south off the main island of Japan on the basis of heat conservation.
Rec. oceanogr. Wks. Japan, 10(1):1-11.

Currents, vertical
Voorhis, A.D., and D.C. Webb, 1970.
Large vertical currents observed in a winter sinking region of the northwestern Mediterranean.
Cah. océanogr., 22(6): 571-580

currents, vertical profiles
*Webster, Ferris, 1969.
Vertical profiles of horizontal ocean currents.
Deep-Sea Res., 16(1):35-98.

currents (vertical structure)

Titov, V.B. and L.M. Fomin, 1971.
Vertical structure of currents based on the measurements in the Indian Ocean. (In Russian; English abstract). Okeanologiia 11(4):578-587.

currents, vortical

current vortices (helical)

Katz, B., R. Gerard and M. Costin, 1965.
Response of dye tracers to sea surface conditions.
J. geophys. Res., 70(22):5505-5513.

Currents, vortical

Takahashi, Tadeo, 1960.
Existence of a contra solen vortical motion in the Coral Sea.
Rec. Oceanogr. Wks., Japan, 5(2):52-54.

Currents, warm

currents, warm

Sustanov, Y.V., 1965.
On the specific features of marine warm currents. (In Russian).
Okeanologiia, Akad. Nauk, SSSR, 5(1):56-62.

Currents, wave-drift

currents, wave-drift

Bye, John A.T., 1967.
The wave-drift current.
J. mar. Res., 25(1):95-102.

currents, wave induced

currents, wave induced

Wiegel, R.L., and J.W. Johnson, 1960
Ocean currents, measurement of currents and analysis of data.
Waste disposal in the marine environment, Pergamon Press: 175-245.

currents, weak

Alais, Pierre, et Bernard Saint-Guilly, 1971.
Procede ultrasonore de mesure des courants marins faibles ou à fluctuations rapides.
Cah. océanogr. 23 (2): 131-134

Currents westward intensification

currents, western intensification of

Felzenbaum, A.I., and G.S. Dvoryaninov, 1969.
On western intensification of currents in a two-layer ocean. (In Russian; English abstract).
Okeanologiia, 9(1):111-118.

currents, western boundary

Hamon, Bruce V., 1970.
Western boundary currents in the South Pacific.
In: Scientific exploration of the South Pacific, W.S. Wooster, editor, Nat. Acad. Sci., 50-59.

currents, westward intensification

Longuet-Higgins, M.S., 1965.
Some dynamical aspects of ocean currents.
Q.Jl. R. met. Soc., 91:425-451.
Also in:
Collected Reprints, Nat. Inst., Oceanogr., 13.

currents, western intensification of

Saint-Guily, B., 1961

Quelques solutions simple du problème d'Ekman illustrant l'intensification ouest des courants océaniques.
C.R. Acad. Sci., Paris, 252(7): 1051-1053.

Currents western

Verber, J.L., 1955.
Rotational water movements in Lake Erie.
Verh. Int. Ver. Limnol. 12:97-104, 8 textfigs.

currents, wind

currents, (wind) (velocity)

Bashkirov, G.S., 1966.
The limiting wind current velocity in the surface layer and phenomena in the contact layer. (In Russian: English abstract).
Okeanologiia, Akad. Nauk, SSSR, 6(2):217-223.

currents, wind

*Belyaev, V.S., 1967.
On the dependence of spectra of the wind-driven current velocity components on the spectrum of tangential wind stress. (In Russian; English Abstract).
Fisika Atmosfer. Okean., Izv. Akad. Nauk, SSSR, 3(11): 1217-1226.

currents, wind

Belyaev, V.S. 1966.
On periodical wind-driven currents in a homogeneous ocean. (In Russian; English abstract).
Fizika Atmosferi i Okeana, Izv. Akad Nauk SSSR 2(4): 414-422.

currents, wind

Birchfield, G.E. 1967.
Wind-driven currents in a long rotating channel.
Tellus, 19(2): 243-249.

currents

Blink, H., 1887.
Wind- und Meereströmungen in Gebiet der Kleinen Sund-inseln. Beitr. Geophys. 1:1-58.

currents, wind

Bowden, K.F., 1956.
Physics in oceanography. Brit. J. Appl. Phys., 7:273-281.

currents, wind

Bowden, K.F., 1953.
Note on wind drift in a channel in the presence of tidal currents. Proc. R. Soc., London, A, 219(1139):426-446, 9 textfigs.

currents, wind

Bowden, K.F., 1953.
Measurement of wind currents in the sea by the method of towed electrodes. Nature 171(4356): 735-737, 1 textfig.

currents, wind-driven

Bretschneider, Charles L. 1967.
Estimating wind-driven currents over the continental shelf.
Ocean Industry 2(6): 45-48.

currents, wind induced

Cooper, L.H.N., 1960.

Some theorems and procedures in shallow-water oceanography applied to the Celtic Sea. J. Mar Biol. Ass'n. U.K., 39(2): 155-171.

currents, wind

Dooley, H.D., and J.H. Steele 1969.
Wind driven currents near a coast.
Dt. hydrogr. Z., 22(5): 213-223

currents, wind

Drogaytsev, D.A., 1958

Wind currents in the Arctic Ocean. Problemy Severa, USSR, 2: 1-16. Rev. in Tech. Transl. 4(9): 511.

currents, wind

Ekman, V.W., 1923.
Über Horizontalzirkulationen bei winderzeugten Meereströmmungen. Arkiv. Mat. Astr. Fysik, 7(26): 1-74.

currents, wind

Federov, K.N., 1956.
Results of operations with models of total current induced by the wind in the sea. Trudy Inst. Okean. 19:83-97.

currents, wind

Federov, K.N., 1955.
Wind currents in a sea of variable depth.
Izvestia Akad. Nauk, SSSR, Ser. Geofiz., 3:223-233.

Rev. in:
Met. Abstr. & Bibl. 7(7):823.

Currents, wind

Fjeldstad, Jonas Ekman, 1962.
A circulation theorem for stationary wind currents.
Geofys. Publik., Geophysica Norvegica, 24:241-242.

current velocity (wind-driven)

*Fomin, L.M., 1968.
The spectrum of the wind-driven current velocity in the ocean. (In Russian; English abstract).
Fisika Atmosfer. Okean., Izv. Akad. Nauk, SSSR, 4(12): 1285-1290.

currents, wind

Fomin, L.M., 1963.
The steady wind-driven currents in the non-homogeneous ocean of variable depth. Questions of Physical Oceanography. (In Russian: English abstract).
Trudy Inst. Okeanol., Akad. Nauk, SSSR, 66:29-45.

currents, wind

Fomin, L.M., 1956.
The depth of penetration of wind current into the sea. Izvestia Akad. Nauk, SSSR, Ser. Geofiz., 2:172-181.

currents, winddriven

Francis, J.R.D., 1953.
A note on the velocity distribution and bottom stress in a wind-driven water current system.
J. Mar. Res. 12(1):93-98, 4 textfigs.

currents, wind

Fukuoka, Jiro, 1967.
Masas de agua y dinámica de los océanos.
In: Ecología marina, Monogr. Fundación La Salle de Ciencias Naturales, Caracas, 14:130-183.

currents, wind-driven

Gurikova, Z.F., 1962
On the calculation of the currents in the surface layer in the northern sector of the Pacific.
Izv. Akad. Nauk. SSSR. Ser. Geophys. 9:1240-1250, Eng. Ed., 9:776-781.

currents, wind

Gurikova, Z.F., Vinokurova, T.G., and V.V. Natarov, 1964
The scheme of the wind circulation of the Bering Sea current in August 1959 and 1960. (In Russian).
Gosudarst. Kom. Sov. Ministr. SSSR, Ribn. Choz., Trudy VNIRO, 49; Izv., TINRO, 51: 51-76.

currents, wind

Gushchin, O.A., and L.M. Krivelech, 1970.
Three-dimensional hydrodynamical model of non-stationary wind-driven currents in a region of low latitudes. (In Russian; English abstract).
Fizika Atmosfer. Okean. Izv. Akad. Nauk, SSSR 6(10):1035-1042.

currents, wind

Hachey, H.B., and N.O. Fothergill, 1954.
The wind currents and dominant surface flow at Sambro Lightship. Joint Comm. Ocean. 14 pp.

currents, wind

Hachey, H.B., and N.O. Gothergill, 1953.
Wind currents and storm effects on water movement at Sambro Lightship. Trans. R.Soc., Canada, 3rd ser., 47:1-13, 6 textfigs.

currents, wind-driven

Hansen, W., 1951.
Winderzeugte Strömungen im Ozean. Deutsche Hydro Zeits. 4(4/5/6):161-172, Textfigs. 1-11, Pls. 12-13.

currents, wind

Hidaka, K., 1952.
Circulations in a zonal ocean induced by a planetary wind system. J. Ocean. Soc., Japan, 8(1):7-13, 2 textfigs.

currents, wind

Hidaka, K., 1952.
Wind-driven current and lateral mixing in a zonal ocean. Geophys. Mag. 23(4):487-495, 3 textfigs.

Geophys. Mag. 5(1):Contrib. No. 7.

currents

Hidaka, K., 1950.
Effect of land barriers upon the wind-driven currents in the sea. Geophys. Notes, Tokyo, 3(1):1-4, 2 textfigs.

currents

Hidaka, K., 1950.
Wind-driven currents and surface contours in an enclosed ocean and the lateral mixing. Geophys. Notes, Tokyo, 3(12):16 pp., 5 textfigs.

currents, wind-driven

Hidaka, K., 1950.
Lateral mixing and wind-driven currents in an enclosed ocean. J. Mar. Res. 9(2):55-64, 4 textfigs.

currents, wind-driven

Hidaka, K., 1950.
Effect of the land barrier upon the wind-driven currents in the ocean. J. Ocean. Soc., Japan, 6(1):53-56, 2 textfigs.

currents, wind-driven

Hidaka, K., 1950.
Effect of land barriers upon the wind-driven currents in the ocean. Ocean. Mag., Tokyo, 2(1):27-29, 3 textfigs.

currents, wind-driven

Hidaka, K., 1950.
Surface contours due to wind-driven currents in an enclosed ocean. Ocean. Mag., Tokyo, 2(1):17-26, 3 textfigs.

Currents, wind driven

Hidaka, Koji, and Takao Momoi, 1961.
Determination of the vertical eddy viscosity in sea water from wind stresses and surface current velocities.
Rec. Oceanogr. Wks., Japan, 6(1):1-10.

currents, wind-driven

Hidaka, K., and K. Takano, 1952.
Wind-driven current and lateral mixing in a zonal ocean. Geophys. Mag. 23:487-495.

currents, wind-wave

Hunt, J.N., and B. Johns, 1963.
Currents induced by tides and gravity waves. Tellus, 15(4):343-351.

currents, wind-driven

Ichie, T., 1950.
A note on wind-driven currents under a circular wind stress. Ocean. Mag., Tokyo, 2(2):45-48, 3 textfigs.

currents, wind

Ivanov, R.N., 1965.
On the development of wind-driven currents. Investigations of sea roughness. (In Russian).
Trudy Morsk. Gidrofiz. Inst., Akad. Nauk, SSSR, 31:69-74.

currents, wind

Ivanov, R.N., 1962.
Dependence of tangential wind stress over a distrubed water surface on wind speed. (In Russian).
Trudy, Morsk. Gidrofiz. Inst., 20:20-32.

Translation:
Scripta Tecnica, for AGU (4):14-24.1964.

current, wind

Ivanov, R.N., 1961
Use of a storm basin to study wind currents. Physics of the Sea. (In Russian).
Trudy Morsk. Gidrofiz. Inst., 23: 94-121.

Translation: Scripta Technica, Inc. for Amer. Geophys. Union, pp. 76-96.

currents, wind

Kajiura, K., 1952.
The velocity distribution of wind currents in the eastern part of the equatorial Pacific.
J. Ocean. Soc., Japan, 8(1):15-22, 3 textfigs.

currents, wind

Kajiura, K., 1952.
The velocity distribution of wind currents in the the eastern part of the equatorial Pacific.
J. Ocean. Soc., Japan 8(1):16-22, 3 textfigs.

Also:
Geophys. Notes 5(2); No. 25.

currents, wind

Kowalik, Z., 1965.
Survey of computation of stationary wind currents (In Polish).
Przeglad Geofis., Warszawa, 10(18)(3/4):311-321.

currents, wind

Krasyuk V.S. and E.M. Sauskan, 1970
On the computation of wind streams velocity in the ocean. (In Russian)
Met. Gidrol. 1970 (9): 68-74

Currents, wind-driven

Krivelevich, L.M. 1967.
Numerical calculations of non-stationary wind-driven currents at the equator. (In Russian; English abstract).
Okeanologiia 7(2): 232-237.

currents, wind

Laevastu, T., 1962
The components of the surface currents in the sea and their forecasts.
Proc. Symposium on Mathematical-Hydrodynamical Methods of Phys. Oceanogr., Sept. 1961, Inst. Meeresk., Hamburg, 321-338.

currents, wind-driven

Lighthill, M.J., 1969.
Unsteady wind-driven ocean currents.
Q.Jl R. met. Soc 95(406): 675-688.

currents, wind

Lighthill, M.J., 1969.
Dynamic response of the Indian Ocean to onset of the southwest monsoon.
Phil. Trans. R. Soc., 265 (1159): 45-92

currents, wind

Lineikin, P.S., 1958.
On the influence of an interface on the wind currents of the deep sea. (In Russian)
Trudy Gosud. Okeanogr. Inst., 42:89-104.

currents, wind

Lineikin, P.S., 1956.
On the theory of unsteady wind currents in a deep sea. Dokl. Akad. Nauk, SSSR, 106(1):47-50.

currents, wind

Lineikin, P.S., 1955.
On wind currents and the baroclinic layer over the sea.
Trudy Gos. Okeanogr., No. 29:34-64.

Abstr in:
Appl. Mech. Rev., 11:401.

currents, wind (theory)

Lineikin, P.S., 1955.
Determination of thickness of the baroclinic layer of the sea. Doklady, Akad. Nauk, SSSR, 101(3):461-464.

Currents, wind

Mandelbaum, H., 1925.
Wind-generated ocean currents at Amrum Bank Lightship. Trans. Amer. Geophys. Union 36(1): 72-86.

currents, wind

Miyazaki, M., 1952.
Notes on the theory of the wind-driven oceanic circulation. Ocean. Mag. 4(2):31-36, 3 textfigs.

currents, wind

Moroskin, K.V., 1951.
On the laws of the change of dissipation in the stationary wind-driven currents of the Baltic Sea. (In Russian).
Trudy Inst. Okeanol., Akad. Nauk, SSSR, 6:49-58.

wind-driven currents

Neumann, Gerhard, 1958.
On the mass transport of wind-driven currents in a baroclinic owean with application to the North Atlantic.
Zeits. Meteorol., 12(4):138-147.

currents, wind

Neumann, G., 1955.
On the dynamics of wind-driven currents.
Met. Pap., N.Y.U., 2(4):1-33.

currents, wind

Nikiforov, E.G., 1961.
On the theory of transient wind currents under conditions of a highly stratified sea. Some results of investigation of oscillations of sea level and currents. Methods of oceanological investigations, a collection of papers. (In Russian).
Trudy Arktich. i Antarktich. Nauchno-Issled. Inst., 210:141-163.

Currents, wind
Nikiforov, E.G., 1956.
On the connection between currents and waves produced by wind. Izv. Akad. Nauk, SSSR, Ser. Geofiz, 1956(12):1450-

currents, wind
Popov, B.A., 1958.
A calculation of velocities and acceleration of wave currents as measured by pressure gauges. Trudy Inst. Okeanol., 28:195-207.

currents, wind
Revillon, de V., 1953.
Houle, état de la mer, courants dus au vent. La Météorol. (4)30:103-135.

currents, wind
Rzheplinsky, D.G., 1970.
The effect of Coriolis force and bottom topography on wind currents around oceanic islands. (In Russian, English abstract). Fisika Atmosfer. Okean, Izv. Akad. Nauk. SSSR, 6(7): 715 -727
circulation

currents, wind
Saint-Guily, Bernard, 1962
Le problème d'Ekman pour un océan formé par deux couches d'eaux superposées.
Deep- Sea Res., 9(3):199-207.

currents, wind
Saint-Guily, B., 1957.
Sur le coefficient vertical de turbulence dans les courants de vent et de pente.
Bull. Inst. Océan., Monaco, No. 1090:8 pp.

currents, wind
Saint-Guily, B., 1956.
Sur la théorie des courants marins induits par le vent. Ann. Inst. Océan., Monaco, ns 33(1):1-64.

currents, wind
Saint-Guily, B., 1956.
Sur la théorie des courants marins induits par le vent. C.R. Acad. Sci., Paris, 242(3):403-405.

currents, wind-driven
Saito, Y., 1951.
On the theory of ocean currents driven by winds in an anisotropic ocean. J. Tokyo Univ. Fish. 38 (2):87-179, 7 textfigs.

currents, wind
Saito, Y., 1951.
On the theory of ocean currents driven by winds in an anisotropic ocean. Pt. 1. J. Tokyo Univ. Fish. 39(2):87-189, 7 textfigs.

currents, wind
Sarkisjan, A.S., 1962
On the dynamics of the origin of wind currents in the baroclinic ocean.
Okeanologiia, Akad. Nauk, SSSR, 2(3):393-409.

currents, wind
Sarkisan, A.S., 1960
Density advection and the intensification of wind currents on the west coast of the oceans. Doklady Akad. Nauk, SSSR, 134(6): 1339-1342.

currents, wind
Sarkissian, A.S., 1958.
Unsteady wind currents in a baroclinic sea. Doklady Akad. Nauk, USSR, 119(4):698-701.

currents, wind
Sarkisian, A.S., 1957.
Determination of stationary wind currents and the slopes of the levels in a homgeneous ocean.
Izv. Akad. Nauk, SSSR, Ser. Geophys., 1957(5): 616-627.

currents, wind
Sarkissiane, A.S., 1957.
Determination des courants permanents produits par le vent et des courants de gradient dans un océan homogène (Application a l'Océan Pacifique) Izv. Akad. Nauk, SSSR, Ser. Geofiz., (5):616-627

currents, wind
Sarkisian, A.S., 1957.
On non-stationary wind currents.
Izv. Akad. Nauk, SSSR, Ser. Geofiz, No. 8:1008-1019.

currents, wind
Sarkisian, A.S., 1957.
On the theory of unsteady wind currents in a homogeneous ocean. Izv. Akad. Nauk, SSSR, Ser. Geofiz., (10):1232-1237.

currents, wind
Sarkisian, A.S., 1957.
On unsteady wind currents in a baroclinic sea. Doklady Akad. Nauk, SSSR, 117(6):975-978.

currents, wind
Sarkisyan, A.S., 1956.
Contribution to the question concerning the determination of fixed wind flow in a density layer of the ocean. (In Russian).
Trudy, Geofiz. Inst., Akad. Nauk, SSSR, 37(164): 50-

currents, wind
Sarkisyan, A.S., 1954.
Calculation of stationary wind currents in the ocean. Izv. Akad. Nauk, SSSR, Ser. Geofis., 6: 554-

currents, wind
Soskin, I.M., 1962.
Empirical dependence for the calculation of wind currents. (In Russian).
Trudy Gosud. Okeanogr. Inst., 70:3-27.

currents, wind
Stockmann, W.B., 1957.
Influence of wind on currents in Bering Strait, causes of their large velocity and predominantly northward direction. (In Russian)
Trudy Inst. Okeanol., Akad. Nauk, SSSR, 25:177-197.

Abstr. in:
Techn. Transl., 3(12):783.

currents, wind-driven
Shtokman, V.B., 1948.
Effect of bottom topography on the direction of the transport of water set up by the wind or the mass-field in a non-homogeneous ocean.
Dok. Akad. Nauk, SSSR 59(5):889-892.

Translation: E.R. Hope - Canada Defence Sci. Info., DRB - 25 Oct. 1952. T57R

currents, wind
Stommel, H., 1948.
The westward intensification of wind-driven ocean currents. Trans. A.G.U., 29(2):202-206.

currents
Sverdrup, H.U., 1947
Wind-Driven Currents in a Baroclinic Ocean; with application to the equatorial currents of the Eastern Pacific. Proc. Nat. Acad. Sci., 33(11):318-326, 2 textfigs.

currents, wind
Takano, Kenzo, 1961
Circulation générale permanente dans les océans. Deuxième partie (suite et fin).
J. Oceanogr. Soc., Japan, 17(4):179-189.

currents, wind
Takano, Kenzo, 1965.
Un exemple numérique des courants marins induits par le vent et la non-uniformité de la densité de l'eau superficielle dans un océan.
La Mer, Bull. Soc. franco-japon. Océanogr., 3:57-65. (not seen).

Currents, wind
Tjurjakov, B.I., 1961
On the correlation of the wind-induced current components in the North Atlantic.
Issled. Severnoi Chasti Atlanticheskogo Okeana, Me.hd. Geofiz. God, Leningradskii Gidrometeorol Inst., 1: 20-31.

Currents, wind
Tjuriakov, B.I. and L.N. Kuznetsova, 1970.
On variability of wind-driven and thermohaline currents in the North Atlantic as related to the changes of the macrosynoptic processes.
Okeanologiia, 10(5): 757-769.

currents, wind-driven
Von Arx, W.S., 1952.
A laboratory study of the wind-driven ocean circulation. Tellus 4(4):311-318, 4 textfigs.

currents, wind
Weenink, M.P.H., and P. Groen, 1952.
On the computation of ocean surface current velocities in the equatorial regions from wind data. Proc. K. Nederl. Akad. Wetensch., A, 55(3) :239-246, 2 textfigs.

currents, wind-driven
Welander, Pierre, 1962
Application of a two-layer Ekman model to the problem of the wind-driven currents.
Proc. Symposium on Mathematical-Hydrodynamical Methods of Phys. Oceanogr., Sept. 1961, Inst. Meeresk., Hamburg, 51-59.

currents, wind
Wiegel, R.L., and J.W. Johnson, 1960
Ocean currents, measurement of currents and analysis of data.
Waste disposal in the marine environment, Pergamon Press:175-245.

currents, wind
Zaitsev, G.N., 1967.
On the wind currents in the Norwegian and Greenland seas. (In Russian).
Trudy vses. nauchno-issled. Inst. morsk. ryb. Khoz. Okeanogr. (VNIRO), 57:21-31.

currents, wind
Zaitsev, G.N., 1965.
On the wind currents in the Norwegian and Greenland seas. Investigations in line with the programme of the International Geophysical Year, 2. (In Russian).
Trudy, Vses. Nauchno-Issled. Inst. Morsk. Ribn. Choz. i Okeanogr., (VNIRO), 57:21-31.

Currents, wind decay of

currents, wind, decay of
Noble, Vincent E., 1965.
On the decay of wind-driven currents.
Ocean Sci. and Ocean Eng., Mar. Techn. Soc., Amer. Soc. Limnol. Oceanogr., 1:544-554.

abstract

Currents, wind mixing

currents, wind mixing

Freeman, J.C., Jr., 1954.
Wind mixing currents. J. Mar. Res. 13(2):157-165, 4 textfigs.

CURV

CURV

Smith,H.D.,1968.
Useful underwater work with CURV.
J. Ocean Techn. 2(4):19-23.

curricula

curricula

U.S.A., Federal Council for Science and Technology, Interagency Committee on Oceanography, 1963.
University curricula in oceanography, Academic Year, 1963-64.
ICO Pamphlet, No. 14:162 pp.

cusps

cusps

*Dolan,Robert, and John C. Ferm,1968.
Crescentic land forms along the Atlantic coast of the United States.
Science, 159(3815):627-629.

cusps

Komar, Paul D., 1971.
Nearshore cell circulation and the formation of giant cusps.
Bull. geol. Soc. Am. 82(9):2643-2650

cusps

*Panin,N.,1967.
Structure des dépôts de plage sur la côte de la Mer Noire.
Mar. Geol. 5(3):207-219.

cusps

Worrall G.A., 1969.
Present-day and subfossil beach cusps on the West African coast.
J.Geol. 77(4): 484-457.

cycles

cycles

Stephens, G.C., 1957.
Twenty-four hour cycles in marine organisms.
Amer. Nat., 91(858):135-152.

cyclogenesis

cyclogenesis

Mowla, K.G. 1970.
Cyclogenesis in the Bay of Bengal and the Arabian Sea.
Tellus 22(6):716-718.

CYCLOGENESIS

Mowla, K.G. 1965.
Cyclogenesis in the Bay of Bengal and Arabian Sea.
Tellus, 20(1): 151-162

cyclogenesis

Rao, Y.P., and B.N. Desai 1970.
Cyclogenesis in the Bay of Bengal and Arabian Sea.
Tellus 22(4):466-469

Cyclogenèsis,baroclinic

Tareev, B.A., 1969.
Baroclinic cyclogenesis and large-scale horizontal turbulence in the ocean. (In Russian; English abstract).
Okeanologiia, 9(1):103-105.

Cyclones
See also: tropical cyclones

cyclones

Algue, J., 1904
Cyclones of the Far East. 2nd Ed. Manita, Bureau of Public Printing, 283 pp., illus.

cyclones

Algue, J., 1900
Cyclones aux Philippines et dans le mer de Chine. (Transl. from the Spanish).
Paris. 253 pp.

Cyclones, tropical

Brand Samson 1970.
Interaction of binary tropical cyclones of the western North Pacific Ocean.
J. appl. Met. 9(3): 433-441

cyclones

Chaussard, Albert, et Laurent LaPlace, 1964.
Les cyclones du sud-ouest de l'océan Indien.
Mém. Météorol. Nat., Min. Trav. Publics et des Transports, Paris, No. 49:1-155.

cyclones, tropical

Chin, P.C., 1959.
A climatological study of tropical cyclones over the China seas.
Proc. Ninth Pacific Sci. Congr., Pacific Sci. Assoc., 1957, Meteorol., 13:92-98.

cyclones

Colon, José A., C.R.V. Raman and V. Srinivasan, 1970.
On some aspects of the tropical cyclone of 20-29 May 1963 over the Arabian Sea.
Indian J. Met. Geophys. 21(1): 1-22.

cyclones

Davy, E.G., 1971.
The cyclones of January, February and March 1970 in the S.W. Indian Ocean
Mar. Obs. (Met. O.839) 41(231): 33-36

cyclones

*Desai, B.N., 1967.
On the formation, direction of movement and structure of the Ara bian Sea cyclone of 20-29 May 1963.
Indian J. Met. Geophys., 18(1):61-68.

cyclones

Desai, B.N., and Y.P. Rao, 1955.
Some aspects of depressions and cyclones in the Indian seas.
Proc. UNESCO Symp., Typhoons, Nov. 1954:175-198, 16 textfigs.

cyclones, effect of

Grimaldi, M., 1955.
The disastrous cyclone damage at Genoa. Causes of breakwater failures examined.
Dock and Harbour Authority 36(418):117-120.

cyclones, tropical

Jayaraman, S., T.R. Srinivasan and N.C. Rai Sircar, 1966.
Persistence of the movement of the tropical cyclones/depressions in the Bay of Bengal during the premonsoon and post-monsoon periods.
Indian J. Met. Geophys., 17(3):395-398.

cyclone tracks

Kao, S.-K., and G.R. Farr, 1966.
Turbulent kinetic energy in relation to jet streams, cyclone tracks and ocean currents.
J. geophys. Res., 71(18):4289-4296.

cyclones, extratropical

Kruger, Paul, Charles L. Hosler and Larry G. Davis 1966.
Radiochemical study of two extratropical cyclones containing nuclear debris of different ages.
J. geophys. Res., 71(18):4257-4266.

cyclonic gyres,effect of

*Legendre,L. and W.D. Watt,1970.
The distribution of primary production relative to a cyclonic gyre in Baie des Chaleurs.
Marine Biol., 7(2):167-170.

cyclones

Leigh, R.M., 1969.
A meteorological satellite study of a double vortex system over the western Pacific Ocean.
Aust. Met. Mag. 17(1): 45-62.

cyclones

Malkus, Joanne S., 1962
Large-scale interactions. Ch. 4, Sect. II. Interchange of properties between sea and air. In: The Sea, Interscience Publishers, Vol. 1, Physical Oceanography, 88-294.

cyclones

Mashkova, T.B., 1959.
The formation of Mediterranean cyclones passing across the Black Sea. Swell and currents in the sea. Loss of salt from the sea into the atmosphere. (In Russian).
Trudy Morsk. Gidrofiz. Inst., 15:146-166.

cyclones

Mori, Goro and Ken Akedo, 1970.
On the cyclone in forming stage in the vicinity of Formosa (so-called Taiwan Bozu), in its center Chofu-maru ran into.
(In Japanese; English abstract).
Umi to Sora 45(4): 145-153

cyclones

Morskoi, G.I., ;064.
Mexican cyclone. Hydrophysical investigations. (Results of the investigations of the seventh cruise of the Scientific Research Ship "Mikhail Lomonosov"). (In Russian).
Trudy Morsk. Gidrofiz. Inst., Akad. Nauk Ukrain., SSR, 29:22-28.

tropical cyclones

Ooyama Katsuyuki, 1969.
Numerical simulation of the life cycle of tropical cyclones.
J. atmos. Sci., 26(1): 3-40.

cyclone

Ota, M., 1961
[Statistical investigation on the acceleration of tropical cyclone's movement.]
Bull. Kobe Mar. Obs., No. 167(1): 13-22.

A reprint of a 1960 article.

cyclones

Pedgley, D.E., 1969.
Cyclones along the Arabian Coast.
Weather, 24(11): 456-468.

cyclones

Pyke, Charles B., 1965.
On the role of air-sea interaction in the development of cyclones.
Bull. Amer. Meteorol. Soc., 46(1):4-15.

cyclones
Ramage, C.S., 1961.
The subtropical cyclone. (Abstract).
Tenth Pacific Sci. Congr., Honolulu, 21 Aug.-6 Sept., 1961, Abstracts of Symposium Papers, 329.

cyclones
Renard, Robert J. 1968.
Forecasting the motion of tropical cyclones using a numerically derived steering current and its bias.
Mon. Weath. Rev. U.S. Dep. Comm. 96(7): 453-469

cyclones
Sen, S.N., 1959.
Influence of upper level troughs and ridges on the formation of post-monsoon cyclones in the Bay of Bengal.
Indian J. Meteorol. & Geophys., 10(1):7-24.

cyclones
*Shenk, William E., 1970.
Meteorological satellite infrared views of cloud growth associated with the development of secondary cyclones.
Mon.Weath.Rev.U.S.Dept.Comm. 98(11):861-868.

Cyclones
Spar, Jerome, Howard A. Friedman and Fred L. Zuckerberg 1969.
Aircraft reconnaissance of winter coastal cyclones in the northeastern United States.
Bull. Am. met. Soc. 50(11): 857-867.

cyclones, effect of
Unoki, Sanae, 1959.
On the ocean waves due to tropical cyclones.
Voastal Eng., Japan, 2:1-8.

cyclones
Van Loon, H., 1962.
On the movement of lows in the Ross and Weddell Sea sectors in summer.
Notos, 11(1/4):47-50.

Cyclones
Vowinckel, E. and Svenn Orvig 1969.
Climate change over the Polar Ocean. II. A method for calculating synoptic energy budgets. III. The energy budget of an Atlantic cyclone.
Arch. Met. Geophys. Bioklim. (B) 17 (2/3): 121-146.

Dacite
Karig, Daniel E., and William E. Glassley, 1970.
Dacite and related sediment from the West Mariana Ridge, Philippine Sea.
Bull. geol. Soc. Am. 81(7): 2143-2146.

'Daly gap'

"Daly gap"
Baker, Ian 1968.
Intermediate oceanic volcanic rocks and the "Daly gap".
Earth Planet. Sci. Letts 4(2):103-106.

dams

dams (Bering Strait)
Poland, Fred 1968.
Melting the Arctic ice.
New Scientist 39 (608): 244.

dams, effect of
Pople, W., and the late M. Rogoyski
The effect of the Volta River hydro-electric project on the salinity of the lower Volta River.
Ghana J. Sci. 9 (1): 9-20.

Dams, effect of

dam, effect of
Chang, Sun-duck, 1971.
Influence of dam discharge on the oceanography and fisheries. J. oceanogr. Soc. Korea 6(1): 49-55. (In Korean; English abstract).

dams, effect of
Norris, Robert M., 1964.
Dams and beach-sand supply in southern California.
In: Papers in Marine Geology, R.L. Miller, Editor, Macmillan Co., N.Y., 154-171.

dams, effect of
Peelen, R., 1970.
Changes in salinity in the delta area of the rivers Rhine and Meuse resulting from the construction of a numbers of enclosing dams.
Netherl. J. Sea Res. 5(1):1-19.

Dams, effect of
Santema, P., and J. N. Svasek, 1961.
De invloed van de afdamming van de zeegaten op het ijsbezwaar op de Zeeuwse en Zuidhollandse stromen.
Rapp. Deltacommissie, Bijdragen, (5):295-326.

dams, effect of
Trites, R.W., 1961
Probable effects of proposed Passamaquoddy Power Project on oceanographic conditions.
J. Fish. Res. Bd. Can., 18(2):163-201.

damping see also: waves, damping of

Damping.
See also: waves, damping of

damping
*Garrett, W.D., 1967.
Damping of capillary waves at the air-sea interface by oceanic surface-active material.
J. mar. Res., 25(3):279-291.

damping
Hayashi, Taizo, 1965.
Virtual mass and the damping factor of the breakwater during rocking, and the modification of their effect of the expression of the thrusts exerted upon breakwaters by the action of breaking waves.
Coastal Engin., Japan, 8:105-117.

damping, waves
*Iwagaki, Yuichi, and Yoshito Tsuchiya, 1967.
Linear damping of oscillatory waves due to bottom friction.
Proc. 10th Conf. Coast. Engng, Tokyo, 1966, 1:149-174.

damping of waves
Jarvis, N.L., W.D. Garrett, M.A. Scheiman and C.O. Timmons, 1967.
Surface chemical characterization of surface-active material in seawater.
Limnol. Oceanogr., 12(1):88-96.

damping
Kato, Juichi, Toshifumi Noma, Yukio Uekita and Seiya Hagino, 1969.
Damping effect of floating breakwater.
J. WatWays Harb. Div. Am. Soc. civ. Engrs. 95 (WW3): 337-344

damping
Manabe, D., and I. Ishida, 1967.
On the types of the mean damping curves of rolling ships and their relations to the sea state. (In Japanese; English abstract)
Umi to Sora 42(3/4): 95-114

damping
*Warren, F.W.G., 1968.
Gravity wave damping of hydrostatic oscillations for a buoyant disk.
J. fluid Mech., 31(2):309-319.

damping, effect of
Katō, Jūichi, and Toshifumi Noma, 1969.
On the wave damping effect of double curtain wall breakwater. Coast. Engng., Japan, 12: 41-45.

dangerous animals

dangerous animals
Halstead, Bruce W., 1959
Dangerous marine animals. Cornell Maritime Press, Cambridge, Maryland. 146 pp. Reviewed by John P. Wise, 1960. Trans. Amer. Fish. Soc. 89(3): 318.

Data

DATA
After a time this section became unwieldy and it was time-consuming to retrieve wanted references. Therefore sources of data may be found for more recent years under headings for the parameter concerned.
These cards are for references which may be all data or merely contain some data.

data
Abramson, H.N., and C.L. Bretschneider, 1954.
Some observations concerning the analsis of surface waves when the bottom is non rigid.
B.E.B. Tech. Memo. No. 46(App. 1.):1-11, 3 figs.

With App. 2.- Summary of wave data, Pure Oil Structure A.

data
Anon., 1951.
[Regular offshore observations.] Semi-Ann. Rept., Ocean. Invest., Tokyo, Jan. 1943-Dec. 1944, No. 72:103-105.

data
Anon., 1951.
[Coastal observations. 1. Table of water temperature. 2. Table of salinity.] Semi-Ann. Rept., Ocean. Invest., Tokyo, Jan. 1943-Dec. 1944, No. 72:79-102.

data
Anon., 1951.
[Sectional observations. 1. List of sectional observations carried out. 2. Table of sectional observations.] Semi-Ann. Rept., Ocean. Invest., Tokyo, Jan. 1943-Dec. 1944, No. 72:4-78.

data
Anon., 1951.
["eport of the coastal observations (Dec. 1950)]
J. Ocean., Kobe Obs., 2nd ser., 1(6):30-31.

data

Anon., 1951.
Report on the surface observations of the coastal weather stations (Dec. 1950). J. Ocean., Kobe Obs., 2nd ser., 1(6):22, 2 figs.

data

Anon., 1951.
Report of the surface observations of the coastal weather stations. J. Ocean., Kobe Obs., 2nd ser., 1(5):34-36.

data

Anon., 1950.
Report of the coastal observations (Oct. 1950). J. Ocean., Kobe Obs., 2nd ser., 1(4):21-22.

data

Anon., 1950.
The results of the surface observations. J. Ocean., Kobe Obs., 2nd ser., 1(3):43-50.

data

Anon., 1950.
The results of the coastal observations (Aug. 1950). J. Ocean., Kobe Obs., 2nd ser., 1(3):41-42.

data

Anon., 1950.
The results of the oceanographical observations reported by the Fisheries Experimental Stations (Aug. 1950). J. Ocean., Kobe Obs., 2nd ser., 1(3):37.

data

Anon., 1950.
Record of ships weather observations (1948). J. Ocean., Kobe Obs., 2nd ser., 1(2):15-22.

data

Anon., 1950.
Report of ships weather observations (Sept. 1950). J. Ocean., Kobe Obs., 2nd ser., 1(2):12-14.

data

Anon., 1950.
Report of the oceanographical observations from Fisheries Experimental Stations (Sept. 1950). J. Ocean., Kobe Obs., 2nd ser., 1(2):5-11.

data

Anon., 1950.
Report of the surface observations on board the R.M.S. "Shyunpu Maru" in Harima Nada (Sept.-Oct. 1950) J. Ocean., Kobe Obs., 2nd ser., 1(2):1-4, 1 fig.

data

Anon., 1950.
Report of the surface observations along the route between Kobe and Sumoto. J. Ocean., Kobe Obs., 2nd ser., 1(2):1, 1 fig.

data

Anon., 1950.
The results of harmonic analysis of tidal currents, water temperature and chemical components in the Kii-Suido. J. Ocean., Kobe Obs., 2nd ser., 2(1):28-31, 2 figs.

data

Anon., 1950.
Table 1. Direct current measurement in the Omura Bay for several stations occupied by a cutter. Res. Mar. Met., Ocean. Obs., Tokyo, July-Dec. 1947, No. 2:243.

data

Anon., 1950.
Table 19. Marine meteorological observations made on board R.M.S. "Shinnan-maru" in the North Pacific Ocean, between Tokyo and the fixed point, 39 N., 153 E. Res. Mar. Met., Ocean. Obs., Tokyo, July-Dec. 1947, No. 2:224-239.

data

Anon., 1950.
Table 18. Marine meteorological observations made on board R.M.S. "Ryofu-maru" in the North Pacific Ocean, between Tokyo and the fixed point, 39° E., 153° E. Res. Mar. Met., Ocean. Obs., Tokyo, July-Dec. 1947, No. 2:212-223.

data

Anon., 1950.
Table 17. Marine meteorological observations made on board R.M.S. Chikubu-maru" in the North Pacific Ocean between Tokyo and the fixed point, 39N. 153° E. Res. Mar. Met., Ocean. Obs., Tokyo, July-Dec., 1947, No. 2:210-211.

data

Anon., 1950.
Table 16. Marine meteorological observations made on board R.M.S. "Shinnan-maru" in the North Pacific Ocean, between Tokyo and the fixed point, 39°N., 153° E. Res. Mar. Met., Ocean. Obs., Tokyo, July-Dec., 1947, No. 2:192-209.

data

Anon., 1950.
Table 15. Marine meteorological observations made on board R.M.S. "Ryofu-maru" in the North Pacific Ocean, between Tokyo and the fixed point, 39° N., 153° E. Res. Mar. Met., Ocean. Obs., Tokyo, July-Dec., 1947, No. 2:178-191.

data

Anon., 1950.
Table 14. Marine meteorological observations made on board the "Kyoei-maru" in the Tsugaru Straits and the Sea of Japan between Aomori and Esashi. Res. Mar. Met., Ocean. Obs., Tokyo, July-Dec., 1947, No. 2:176-177.

data

Anon., 1950.
Table 13. Marine meteorological observations made on board the "Oki-maru" in the Japan Sea between Oki Is. and Sakai. Res. Mar. Met., Ocean. Obs., Tokyo, July-Dec. 1947, No. 2:170-175.

data

Anon., 1950.
Table 12. Marine meteorological observations made on board R.M.S. "Syunpu-maru" in the Seto Inland Sea and the Japan Sea between Kobe and Maizuru. Res. Mar. Met., Ocean. Obs., Tokyo, July-Dec., 1947, No. 2:164-169.

data

Anon., 1950.
Table 11. Marine meteorological observations made on board the S.S. "Fugen-maru" in the Iki-suido near Ikizaki Is. Res. Mar. Met., Ocean. Obs., Tokyo, July-Dec., 1947, No. 2:162-163.

data

Anon., 1950.
Table 10. Marine meteorological observations made on board S.S. "Tsuru-maru" in the Goto-Nada, the Iki-suido, and the Higashi-suido. Res. Mar. Met., Ocean. Obs., Tokyo, July-Dec., 1947, No. 2:158-161.

data

Anon., 1950.
Table 9. Marine meteorological observations made on board S.S. "Chofuku-maru" in the Goto-Nada. Res. Mar. Met., Ocean. Obs., Tokyo, July-Dec. 1947, No. 2:132-157.

data

Anon., 1950.
Table 8. Marine meteorological observations made on board S.S. "Tsuru-maru" in the Omura Bay. Res. Mar. Met., Ocean. Obs., Tokyo, July-Dec. 1947, No. 2:130-131.

data

Anon., 1950.
Table 7. Marine meteorological observations made on board a cutter in the Omura Bay. Res. Mar. Met., Ocean. Obs., Tokyo, July-Dec. 1947, No. 2:128-129.

data

Anon., 1950.
Table 6. Marine meteorological observations made on board R.M.S. "Yoshio-maru" in the Seto Inland Sea between Kobe and Nagasaki. Res. Mar. Met., Ocean. Obs., Tokyo, July-Dec. 1947, No. 2:126-127.

data

Anon., 1950.
Table 5. Marine meteorological observations made on board R.M.S. "Yushio-maru" in the adjacent sea to the west of Kyushu. Res. Mar. Met., Ocean. Obs., Tokyo, July-Dec. 1947, No. 2:124-125.

data

Anon., 1950.
Table 4. Marine meteorological observations made on board R.M.S. "Syunpu-maru" in the Seto Inland Sea between Kobe and Takamatsu. Res. Mar. Met., Ocean. Obs., Italy, July-Dec., 1947, No. 2:122-123.

data

Anon., 1950.
Table 3. Marine meteorological observations made on board R.M.S. "Ryufu-maru" in the North Pacific Ocean east off Tohoku District. Res. Mar. Met., Ocean. Obs., Tokyo, July-Dec. 1947, No. 2:108-121.

data

Anon., 1950.
Table 2. Marine meteorological observations made on board the "Oshoro-maru" in the North Pacific ocean east off Tohoku District. Res. Mar. Met., Ocean. Obs., Tokyo, July-Dec., 1947, No. 2:100-107.

data

Anon., 1950.
Table 1. Marine meteorological observations made on board the ferry boat "Iki-maru" in the Tsugaru Straits between Aomori and Hakodate. Res. Mar. Met., Ocean. Obs., Tokyo, July-Dec. 1947, No. 2:90-99.

data

Anon., 1950.
Table 3. Surface observations made on board S.S. "Chofuku-maru"; physical and chemical conditions of the surface water of the Goto-Nada between Nagasaki and Tomie. Res. Mar. Met., Ocean. Obs., Tokyo, July-Dec. 1947, No. 2:81-88.

data

Anon., 1950.
Table 1. Surface observations made on board R.M.S. "Syunpu-maru"; physical and chemical conditions of the surface water of the sea from Kobe to Maizuru. Res. Mar. Met., Ocean. Obs., Tokyo, July-Dec. 1947, No. 2: 77-79.

data

Anon., 1950.
Table 9. Oceanographical observations taken in the Goto Nada; physical data for stations occupied by R.M.S. "Yushio-maru". Res. Mar. Met., Ocean. Obs., Tokyo, July-Dec. 1947, No. 2:55.

data

Anon., 1950.
Table 14. Oceanographical observations taken in the Maizuru Bay; physical and chemical data for stations occupied by R.M.S. "Syunpu-maru". Res. Mar. Met., Ocean. Obs., Tokyo, July-Dec. 1947, No. 2:71-76.

data

Anon., 1950.
Table 13. Oceanographical observations taken in the Nagasaki Harbour; physical and chemical data for stations occupied by a cutter. Res. Mar. Met. Ocean. Obs., Tokyo, July-Dec., 1947, No. 2:65-70.

data
Anon., 1950.
Table 12. Oceanographical observations taken in the Omura Bay; physical and chemical data for stations occupied by the "Tsuru-maru". Res. Mar. Met., Ocean. Obs., Tokyo, July-Dec. 1947, No. 2: 61-64.

data
Anon., 1950.
Table 11. Oceanographical observations taken in the Omura Bay; physical and chemical data for stations occupied by a cutter. Res. Mar. Met., Ocean. Obs., Tokyo, July-Dec. 1947, No. 2:59-60.

data
Anon., 1950.
Table 10. Oceanographical observations taken in the sea northwest of Kyushu; physical and chemical data for stations occupied by the Fugen-maru" Res. Mar. Met., Ocean. Obs., Tokyo, July-Dec. 1947, No. 2:57-58.

data
Anon., 1950.
Table 8. Oceanographical observations taken in the adjacent sea to the west of Kyushu; physical and chemical data for stations occupied by R.M.S. "Yushio-maru". Res. Mar. Met., Ocean. Obs., Tokyo, July-Dec. 1947, No. 2:51-54.

data
Anon., 1950.
Table 7. Oceanographical observations taken in the Sagami Bay; physical and chemical data for stations occupied by R.M.S. "Arashio-maru". Res. Mar. Met., Ocean. Obs., Tokyo, July-Dec. 1947, No. 2:48-50.

data
Anon., 1950.
Table 6. Oceanographical observations taken at the "fixed point" (39°N, 153°E) in the North Pacific Ocean; physical and chemical data for stations occupied by R.M.S. "Ryofu-maru". Res. Mar. Met., Ocean. Obs., Tokyo, July-Dec. 1947, No. 2:44-47.

data
Anon., 1950.
Table 5. Oceanographical observations taken in the North Pacific Ocean east off Tohoku District Physical and chemical data for stations occupied by R.M.S. "Ryofu-maru". Res. Mar. Met., Ocean. Obs., Tokyo, July-Dec. 1947, No. 2:~~44-47.~~ 22-43

data
Anon., 1950.
Table 4. Oceanographical observations taken in the North Pacific Ocean along the coast of Sanriku; physical and chemical data for stations occupied by R.M.S. "Kuroshio-maru". Res. Mar. Met., Ocean. Obs., Tokyo, July-Dec. 1947, No. 2: 13-21.

data
Anon., 1950.
Table 3. Oceanographical observations taken in the North Pacific Ocean east off Sanriku; physical and chemical data for stations occupied by "Oshoro-maru". Res. Mar. Met., Ocean. Obs., Tokyo July-Dec. 1947, No. 2:8-12.

data
Anon., 1950.
Table 2. Oceanographical observations taken in the Uchiura Bay; physical and chemical data for stations occupied by R.M.S. "Kuroshio-maru". Res. Mar. Met., Ocean. Obs., Tokyo, July-Dec. 1947, No. 2:3-7.

data
Anon., 1950.
Table 1. Oceanographical observations taken in the Uchiura Bay; physical and chemical data for stations occupied by a fishing boat belonging to Mori Fishery Assoc. Res. Mar. Met., Ocean. Obs., Tokyo, July-Dec. 1947, No. 2:1-2.

data
Anon., 1950.
Table 1, Marine meteorological observations made on board the ferry boart "Soya-maru" in the Tsugaru Straits between Aomori and Hakodate.
Table 2, Marine meteorological observations on board R.M.S. "Oyashio-maru" in the North Pacific Ocean along the coast of Sanriku.
Table 3, Marine meteorological observations made on board R.M.S. "Kuroshio-maru" in the North Pacific Ocean along the coast of Sanriku.
Table 4, Marine meteorological observations made on board R.M.S. "Syunpu-maru" along the Pacific coast of Japan between Kobe and Tokyo.
Table 5, Marine meteorological observations made on board R.M.S. "Toyotomi-maru" in the Mikawa Bay.

data
Anon., 1950.
Surface observations made on board a steamship. Physical conditions of the surface water of the Goto Nada between Nagasaki and Tomi (continued). Res. Mar. Met., Ocean. Obs., Jan.-June 1947, No. 1:46-52, Table 4.

data
Anon., 1950.
Surface observations made on board R.M.S. "Syunpu-maru". Physical and chemical conditions of the surface water of the Seto Inland Sea between Kobe and Hiroshima. Res. Mar. Met., Ocean. Obs., Jan.-Juma 1947, No. 1:44-45, Table 3.

data
Anon., 1950.
Surface observations made on board the R.M.S. "Syunpu-maru"; physical and chemical conditions of the surface water of the Pacific coast of Japan between Kobe and Tokyo. Res. Mar. Met., Ocean. Obs., Jan.-June 1947, No. 1:42-43, Table 2.

data
Anon., 1950.
Surface observations made on board the R.M.S. "Kuroshio-maru"; physical and chemical conditions of the surface of the sea near Hideshima Island. Res. Mar. Met., Ocean. Obs., Jan.-June 1947, No. 1:39-41, Table 1.

data
Anon., 1950.
Oceanographical observations taken in the Nagasaki Harbour; physical and chemical data for stations occupied by the cutter. Res. Mar. Met., Ocean. Obs., Jan.-June 1947, No. 1:38, Table 5.

data
Anon., 1950.
Oceanographical observations taken in the North Pacific Ocean (east of Sanriku); physical and chemical data for stations occupied by R.M.S. "Yushio-maru" (continued). Res. Mar. Met., Ocean Obs., Jan.-June 1947, No. 1:28-37, Table 4.

data
Anon., 1950.
Oceanographic observations taken from the Miyako Bay; physical and chemical data for stations occupied by R.M.S. "Kuroshio-maru". Res. Mar. Met., Ocean. Obs., Jan.-June 1947, No. 1:24-27, Table 3.

data
Anon., 1950.
Oceanographical observations taken in the North Pacific Ocean (east off Hideshima Is.); physical and chemical data for stations occupied by R.M.S. "Kuroshio-maru" (continued). Res. Mar. Met., Ocean. Obs., Jan.-June 1947, No. 1:12-23, Table 2

data
Anon., 1950.
Oceanographical observations taken off the Sanriku coast; physical and chemical data for stations occupied by R.M.S. "Oyashio-maru". Res. Mar. Met., Ocean. Obs., Jan.-June 1947, No. 1: 1-11, Table 1.

data
Anon., 1950.
Exploracion oceanográfica del Africa Occidental. Campaña del "Malaspina" en enero de 1950 en aguas del Sahara desde Punta Durnford a Cabo Barbas. Bol. Inst. Español Ocean., No. 38:12 pp., 2 figs.

data
Anon., 1949-1950.
[The results of the regular monthly oceanographical observations on board the R.M.S. "Syunpu Maru" and "Takatori Maru" in the Osaka Wan.] J. Ocean., Kobe Obs., 2nd ser., 2(1):1-28.

data
Anonymous, 1949.
Tables of temperatures and salinities obtained at various stations during the years 1947-1949. Prak. Hellen. Hidrobiol. Inst., 1949, 3(1):77-111, 1 fold-in.

data
Anon., 1949.
Exploracion oceanografica del Africa Occidental. Campañas del "Malaspina" en 1947 y 1948 en aguas del Sahara desde Cabo Juby a Punta Durnford. Registro de operaciones. Bol. Inst. Español Ocean. No. 23: 28 pp., 2 textfigs.

data
Anonymous, 1947
Discovery investigations Station List 1937-1939. Discovery Report, XXIV: 197-422, pls. IV-VI.

data
Anon., 1939 (?)
Synopsis of the Hydrographical investigations (July-December 1939). Semiannual Rept. Oceanogr. Invest. (July-Dec. 1939). Imp. Fish. Expt. Sta., Tokyo, No.65, 147 pp.

data
Anon., 1934.
Appendix B. Temperatures and salinities at depth m (Meters). Ann. Rept., Newfoundland Fish. Res. Comm. 2(2):86-99.

data
Anon., 1933.
Appendix B. Hydrographic data. Ann. Rept., Newfoundland Fish. Res. Comm. 2(1):107-119.

data
Anon., 1932.
Appendix B. Hydrographic data. Ann. Rept., Newfoundland Fish. Res. Comm. 1(4):92-98.

data
Asano, H., 1913.
Hydrographic observations made on board the H.M.S. Matsue during December 1912. Rep. Imp. Fish. Inst. Sci. Invest. 2:8-17, plates, charts.

Muroto Zaki to Mikomoto Shima (Izu Shoto).

data
Austin, T.S., 1957.
Summary, oceanographic and fishery data, Marquesas Islands area, August-September 1956, (EQUAPAC). USFWS Spec. Sci. Rept., Fish., No. 217:186 pp.

zooplankton volume
salinity
oxygen
temperature
density
phosphate

data
Austin, T.S., 1954.
Mid-Pacific oceanography. V. Transequatorial waters, May-June 1952, August 1952. Spec. Sci. Rept., Fish., No. 136:1-85, 31 figs.

phosphate
density
salinity
temperature

data
Austin, T.S., 1954.
Mid-Pacific oceanography III. Transequatorial waters, August-October 1950. U.S.F.W.S.Spec. Sci. Rept. - Fish. No. 131: numerous pp (unnumbered), 13 figs. (multilithed)

temperature
salinity
density
oxygen
PO4-P

data
Bardin, I.P., 1958.
[Hydrological, hydrochemical, geological and biological studies, Research Ship "Ob", 1955-1957.] Trudy Kompleks. Antarkt. Exped., Akad. Nauk, SSSR, Mezh. Geofis. God, Gidrometeorol. Izdatel, Leningrad, 217 pp.

data
Beach Erosion Board, 1948.
Proof test of water transparency method of depth determination. Eng. Notes No. 29: 29 pp., 34 figs., data sheets (multilith).

data
Bernard, F., 1956.
Contribution à la connaissance de détroit de Gibraltar (Hydrographie et nannoplancton en juin 1954). Bull. Inst. Océan., Monaco, No. 1074:1-22.

salinity
temperature
phosphate
nitrate
nannoplancton

data
Bernard, F., 1952.
Eaux atlantiques et mediterraneennes au large de l'Algerie. 1. Hydrographie, sals nutritifs et phytoplancton en 1950. Ann. Hydr., Monaco, n.s., 27(1):48 pp., 15 textfigs.

data
Birkenes, E., and T. Braarud, 1954.
Phytoplankton in the Oslo Fjord during a "Coccolithus huxleyi-summer".
Avhandl. Norske Videnskaps.-Akad., Oslo, 1. Mat.-Naturvid. Kl., 1952(2):1-23, 1 textfig.

Coccolithus huxleyi
temperature
salinity
oxygen

data
Bougis, P., M. Ginat and M. Ruivo, 1956.
Recherches hydrologiques sur le Golfe du Lion. Vie et Milieu, 7(1):1-18.

temperature density
salinity

data
Bougis, P., and M. Ruivo, 1954.
Sur un descente d'eaux superficielles en profondeur (cascading) dans le sud du Golfe du Lion. Bull. d'Info., C.C.O.E.C. 6(4):147-154, 2 pls.

data
Braarud, T., K.R. Gaarder, and J. Grøntved, 1953.
The phytoplankton of the North Sea and adjacent waters in May 1948. Rapp. Proc. Verb., Cons. Perm Int. Expl. Mer, 133:1-87, 29 tables, Pls. A-B, Textfigs. 1-18.

data
Braarud, T., and A. Klem, 1931.
Hydrographical and chemical investigations in the coastal waters off Møre and in the Romsdalfjord. Hvalrådets Skrifter, 1931(1):88 pp., 19 figs.

data(of sorts)
Brennecke, W., 1921.
Die ozeanographischen Arbeiten der Deutschen Antarktischen Expedition, 1911-1912.
Arch. Deutschen Seewarte 39(1):1-216, 15 pls., 41 textfigs.

data
Brennecke, W., 1915.
Oceanographische Arbeiten S.M.S. "Planet" im westlichen Stillen Ozean 1912-1913. Ann. Hydr., usw., 43:145-158.

data
Brennecke, W., 1910.
Lotungen des Kabeldampfers "Grossherzod von Oldenburg" im östlichen Teil des Nordatlantischen Ozeans. Arch. Deutschen Seewarte 33(3):1-20, 1 textfig., 2 pls.

bathymetry
bottom sediments

data
Bruewicz, S.W., 1957.
On certain chemical features of waters and sediments in north-west Pacific.
Proc. UNESCO Symp., Phys. Oceanogr., Tokyo, 1955: 277-292.

oxygen pH
salinity T-S
phosphate silicate

data
Bruneau, L., N.G., Jerlov and F.F. Koczy, 1953.
Physical and chemical methods.
Repts. Swedish Deep-sea Exped., 1947-1948, Phys. Chem. 3(4):101-112, i-iv, 7 textfigs., 2 tables.

temperature oxygen
salinity phosphate
density silicate
pH particle content
alkalinity
carbon dioxide

data, magnetism
Bullard, E.C., 6. Freedman, H. Gellman, and J. Nixon, 1950.
The westward drift of the earth's magnetic field. Proc. Roy. Soc., London, Ser. A, 243(859):67-92, 11 textfigs., 13 tables.

data
Burling, R.W., 1961.
Hydrology of circumpolar waters south of New Zealand.
New Zealand Dept. Sci. Industr. Res. Bull., 143: 66 pp.

data
Callaway, R.J., 1957.
Oceanographic and meteorological observations in the northeast and central North Pacific, July-December 1956. USFWS Spec. Sci. Rept., Fish., No. 230:49 pp.

salinity(surface) temperature(surface)
phosphate(surface) waves
color transparency
sea

data
Capurro, L.R.A., 1955.
Expedicion Argentina al Mar de Weddell (diciembre 1954 a enero de 1955). Ministerio de Marina, Argentina, Direccion Gen. de Navegacion e Hidrografia, 184 pp.

bathymetry
temperature
salinity
density
oxygen

data
Carlsberg Foundation, 1937.
Hydrographical observations made furing the "Dana" Expedition 1928-30 with an introduction by Helge Thomsen. Dana Rept. No. 12: 46 pp.

data
Capart, A., 1951.
Liste des stations. Rés. Sci., Expéd. Océan. Belge dans les Eaux Côtières Africaines de l'Atlantique Sud (1948-1949) 1: 65 pp.

bottom sediments
color
transparency
oxygen
density
temperatures
sea

data
Carruthers, J.N., 1937.
Continuous current observations for fishery research application. Con. Perm. Int. Expl. Mer, Rapp. Proc. Verb. 105:16-27, 3 textfigs.

data
Caspers, H., 1951.
Quantitative Untersuchungen über die Bodentierwelt des Schwarzen Meeres im bulgarischen Küstenbereich. Arch. f. Hydrobiol. 45(1/2):192 pp., 66 textfigs.

data
Chauvet, R., 1926.
Observations marégraphiques faites au port de Monaco (1902-1921). Bull. Inst. Océan., Monaco, No. 481: 5 pp.

data
Cheney, L.A., and F.M. Soule, 1951.
International ice observation and Ice Patrol Service in the North Atlantic Ocean.
U.S.C.G. Bull. 35:116 pp., 33 textfigs.

data
Chew, F., 1953.
Results of hydrographic and chemical investigations in the region of the "red tide" bloom on the west coast of Florida in November 1952.
Bull. Mar. Sci., Gulf and Caribbean 2(4):610-625, 10 textfigs.

temperature
salinity
oxygen
total phosphorus

data
Chidambaram, K., and A.D.I. Rajendran, 1951.
On the hydro-biological data collected on the Wadge Bank early in 1949. J. Bombay Nat. Hist. Soc. 49(4):738-748.

data
Chosen. Fishery Experiment Station, 1938.
Oceanographical investigations during Jun-Jul 1933, in the Japan Sea offshore along the east coast of Tyosen on board the R.M.S. Misago-maru. (In Japanese). Ann. Rept. Hydrogr. Obs., Chosen Fish. Exp. Sta. 8:11, 23-29, 38-60, 62-65.

data
Chosen. Fishery Experiment Station, 1938.
Oceanographic investigations in the Japan Sea made during Aug. and Oct. 1933, on board the R.M.S. Misago-maru. (In Japanese; English headings for tables and charts). Ann. Rept. Hydrogr. Obs., Chosen Fish. Exp. Sta. 8:77-113, tables, charts.

data
Chosen. Fishery Experiment Station, 1938.
Hydrographic observations during Aug. 1933, on board the R.M.S. Misago-maru in the southwest part of the Japan Sea. (In Japanese; English abstract). Ann. Rept. Hydrogr. Obs., Chosen Fish. Exp. Sta. 8:iii-iv; 67-76.

data
Chosen. Fishery Experiment Station, 1936-1938.
Monthly observations from Urusaki to Kawaziri Misaki, during the year 1932-1933. Ann. Rept. Hydrogr. Obs., Chosen Fish. Exp. Sta., 7:94-113; 8:139-153.

data
Chosen. Fishery Experiment Station, 1937.
Results of the current measurements in the Southern Sea of Tyosen and the western part of the Japan Sea, on board the R.M.S. Misago-maru in Jun.-Sept. 1932. (In Japanese; English headings for tables). Ann. Rept. Hydrogr. Obs., Chosen Fish. Exp. Sta., 7:15-62, tables.

data
Chosen, Fishery Experiment Station, 1936.
Hydrographical investigations off the shore along the east coast of Tyosen, on board the Misago-maru in Oct. 1932. (In Japanese; English title and headings for tables.) Ann. Rept. Hydrogr. Obs., Chosen Fish. Exp. Sta. 7:86-93.

data
Chosen. Fishery Experiment Station, 1936.
Current measurements at station (M) 10 mi. south off the shore of Husan, on Jan. 20-31, 1932. (In Japanese). Ann. Rept. Hydrogr. Obs., Chosen Fish. Exp. Sta., 7:70-74.

data
Chosen. Fishery Experiment Station, 1936.
Current measurements at Sta. A and O, along the west coast of Tyosen in Nov. 1932. (In Japanese; English headings to tables). Ann. Rept. Hydrogr. Obs., Chosen Fish. Exp. Sta., 7:75-85.

data
Chosen. Fishery Experiment Station, 1930.
Report on the hydrographical investigations in the adjacent sea off the east coast of Tyosen, during May and July 1929. (In Japanese; English summary). Ann. Rept. Hydrogr. Obs., Chosen Fish. Exp. Sta. 4:17-44, 6 pls., tables, charts.

data
Collier, A., and J. W. Hedgpeth, 1950.
An introduction to the hydrography of tidal waters of Texas. Publ., Inst. Mar. Sci., 1(2):120-194, 32 textfigs.

data
Comité Central d'Océanographie et d'Études des Côtes, 1951.
Liste des stations hydrologiques profondes éxecutées par le bâtiment polaire "Commandant Charcot" de la Marine Nationale dans l'Océan Indien et dans l'Océan Antarctique (Campagnes 1948-1949 et 1949-50). Bull. d'Info., C.C.O.E.C. 3(10):473-479.

data
Comité local d'Oceanographie et d'études des Côtes de l'Afrique Occidentale Francaise, 1949.
Travaux d'océanographie physique effectués en 1949 par la Section Technique des Pêches Maritimes de Dakar. A. Dakar et environs B. Côtes du Sénégal. C.L.O.E.C. de l'A.O.F., Annee. 1949:27-39, 4 fold-in charts.

data
Cordini, I.R., 1955.
Contribucion al conocimiento del sector Antartico Argentino. Inst. Antartico Argentino, Publ., 1: 277 pp., 82 textfigs., 56 pls.

ice
density
chlorinity
temperature
salinity
sea state

data
Conseil Permanent International pour l'Exploration de la Mer, 1933.
Croisières du Navire "Muirchu", 1931 Rapp. Proc. Verb. 84:11-13.

temperature
salinity

data(surface)
Conseil Permanent International pour l'Exploration de la Mer, 1933.
Croisières du Navire H.M.S. "Salpa" (Janvier-Novembre, 1931). Rapp. Proc. Verb. 84:9-10.

temperature
salinity

data
Conseil Permanent International pour l'Exploration de la Mer, 1933.
Croisière du Garde Pêche "Estafette".
Rapp. Proc. Verb. 84:6-8.

temperature
salinity

data
Conseil Permanent International pour l'Exploration de la Mer, 1932.
Croisière du Navire "Sentinelle" de la Marine National. Rapp. Proc. Verb. 47:8.

temperature
salinity

data (surface)
Conseil Permanent International pour l'Exploration de la Mer, 1932.
Croisières du Navire H.M.S. "Salpa" (Fevrier 1930 -Decembre 1930). Rapp. Proc. Verb. 67:9-10.

temperature
salinity

data
Conseil Permanent International pour l'Exploration de la Mer, 1932.
Croisières du Navire "Muirchu", 1930.
Rapp. Proc. Verb. 67:11-14.

temperature
salinity

data
Conseil Permanent International pour l'Exploration de la Mer, 1931.
Croisières du Navire "Muirchu", 1929.

Rapp. Proc. Verb. 70:13-16.

temperature
salinity

data
Conseil Permanent International pour l'Exploration de la Mer, 1931.
Croisières du Navire H.M.S. "Salpa". Rapp. Proc. Verb. 70:10-12.

temperature
salinity

data
Conseil Permanent International pour l'Exploration de la Mer, 1930.
Croisières du Navire "Sentinelle" de la Marine National.
Croisières du Navire "Pourquoi-Pas?" de la Marine National.
Croisières du Garde-Peche "Petrel".
Rapp. Proc. Verb. 70:7-9.

salinity
temperature

data
Conseil Permanent International pour l'Exploration de la Mer, 1927.
Croisières periodiques des Canonnières de la Marine Nationale "Engageante" et "Vaillante".
Rapp. Proc. Verb. 44:10-11.

temperature
salinity

data
Conseil Permanent International pour l'Exploration de la Mer, 1927.
Liste des stations océanographiques, "Albacora" 1926, République Portugaise. Rapp. Proc. Verb. 44:30-31.

data
Conseil Permanent International pour l'Exploration de la Mer, 1927.
Liste des Stations, Royaume d'Espagne.
Cons. Perm. Int. Expl. Mer, Rapp. Proc. Verb. 44: 27-29.

data
Conseil Permanent International pour l'Exploration de la Mer, 1927.
Croisieres du Navire "Muirchu", 1926.
Rapp. Proc. Verb. 44:24-26.

data
Conseil Permanent International pour l'Exploration de la Mer, 1927.
Croisiere du Navire "H.M.S. Salpa". Cons. Perm. Int. Expl. Mer., Rapp. Proc. Verb. 44:20-23.

data(temperature, salinity)
Conseil Permanent International pour l'Exploration de la Mer, 1927.
Croisière du "Pourquoi Pas?" Rapp. Proc. Verb. 44:13.

data (temperature)
Conseil Permanent International pour l'Exploration de la Mer, 1927.
Croisière du Navire "Estafette" de la Marine National (France). Rapp. Proc. Verb. 44:12.

data
Conseil permanent internations pour l'Exploration de la Mer, 1927.
Bulletin Hydrographique Trimestrial, No. 3, 1927 (Juillet-Septembre):16 pp. (multilith), charts.

data
Conseil Permanent International pour l'Exploration de la Mer, 1926.
Republique portugaise. Rapp. Proc. Verb. 40:32-35.

salinity
temperature

data
Conseil Permanent International pour l'Exploration de la Mer, 1926.
Royaume d'Espagne.
Rapp. Proc. Verb. 40:27-31.

salinity
temperature

data
Conseil Permanent International pour l'Exploration de la Mer, 1926.
Croisières du Navire "Muirchu", 1925.
Rapp. Proc. Verb. 40:23-26.

salinity
temperature

data
Conseil Permanent International pour l'Exploration de la Mer, 1926.
Grande Bretagne. Rapp. Proc. Verb. 40:19-22.

salinity
temperature

data
Conseil Permanent International pour l'Exploration de la Mer, 1926.
Les croisières françaises en 1925.
Rapp. Proc. Verb. 40:13-18.

temperature
salinity

data
Conseil Permanent International pour l'Exploration de la Mer, 1925.
Croisières du Navire H.M.S. "Salpa".
Rapp. Proc. Verb. 37:12-16.

data
Conseil Permanent International pour l'Exploration de la Mer, 1925.
Croisières du Navire "H.M.S. Salpa".
Rapp. Proc. Verb. 35:19-22.

data
Conseil Permanent International pour l'Exploration de la Mer, 1925.
Republique portugaise. Rapp. Proc. Verb. 35:30-34, 11 pls.

data
Conseil Permanent International pour l'Exploration de la Mer, 1925.
Les croisières françaises en 1923.
Rapp. Proc. Verb. 35:12-18.

data
Conseil Permanent International pour l'Exploration de la Mer, 1925.
Les croisières françaises en 1924.
Rapp. Proc. Verb. 37:9-12.

data
Conseil Permanent International pour l'Exploration de la Mer, 1925.
Croisières du Navire "Muirchu", 1924.
Rapp. Proc. Verb. 37:17-19.

temperature
salinity

data
Conseil Permanent International pour l'Exploration de la Mer, 1923.
Croisières du Navire "H.M.S. Salpa".
Rapp. Proc. Verb. 29:15-18.

salinity
temperature

data
Conseil Permanent International pour l'Exploration de la Mer, 1923.
Croisières du Navire "H.M.S. Salpa".
Rapp. Proc. Verb. 31:13-18.

salinity
temperature

data
Conseil Permanent International pour l'Exploration de la Mer, 1923.
Croisières du Navire "Helga", 1922.
Rapp. Proc. Verb. 31:19-21.

salinity
temperature

data
Conseil Permanent International pour l'Exploration de la Mer, 1923.
Croisières du Navire "Helga" en 1920-21.
Rapp. Proc. Verb. 29:19-23.

temperature
salinity
Atlantic, eastern

data
Copenhagen W.J., 1953.
The periodic mortality of fish in the Walvis Region. A phenomenon within the Benguela Current.
Union of S. Africa, Div. Fish., Investig. Rept. No. 14:1-35, 9 pls.

pH
oxygen
hydrogen sulfide
temperature

data
Cromwell, T., 1951.
Mid-Pacific oceanography, January through March 1950. Spec. Sci. Rept., Fish., No. 54:9 pp., 18 figs., station data.

data
Conseil Permanent International pour l'Exploration de la Mer, 1923.
Les croisières françaises en 1922.
Rapp. Proc. Verb. 31:3-13, figs.

salinity
temperature

data
Conseil Permanent International pour l'Exploration de la Mer, 1923.
Les croisières françaises en 1921.
Rapp. Proc. Verb. 29:6-15, 19 pls.

salinity
temperature

data
Crehuet, Ramón Fernández y Maria Jesús del Val Cordón, 1960
Observaciones oceanograficas en la Bahia de Malaga (Marzo a marzo 1957). Bol. Inst. Esp. Ocean., 98: 1-29.

data
Cromwell, T., 1954.
Mid-Pacific oceanography II. Transequatorial waters, June-August 1950, January-March 1951.
U.S.F.W.S., Spec. Sci. Rept. - Fish. No. 131: numerous pp. (unnumbered), 32 figs. (multilithed)

temperature
salinity
density
oxygen
PO4-P

data only
Danske Meteorologiske Institut, 1954.
Nautisk-Meteorologisk Årbog, 1953:171 pp., 12 charts

currents
temperature
salinity
sea

data
Danske Meteorologiske Institut, 1951.
Nautisk-Meteorologisk Årbog, 1950:165 pp., 12 charts.
currents
temperature
salinity

data
Danske Meteorologiske Institut, 1950.
Nautisk-Meteorologisk Årbog, 1935:72 pp.

data
DeBuen, R., 1937.
Déterminations physico-chimiques dans les eaux de la Baie de Santander pendant l'été de 1935.
Cons. Perm. Int. Expl. Mer, Rapp. Proc. Verb. 104:26-31, 6 textfigs.

data
DeBuen, F., 1937.
Sur quelques résultats de la IIIème croisière du Garde-Côte "Xauen" dans les eaux de Guipuzea (Espagne) en 1934. Rapp. Proc. Verb., Cons. Perm. Int. Expl. Mer, 104:18-25, 3 textfig

data
deBuen, F., 1953.
La oceanografía frente a las costas del Uruguay. Resultado de dos viajes del pesquero "Antares" a la planacie continental en julio de los años 1949 y 1950, y de una corta campaña oceanografica frent al Departamenti de Rocha en Mayo de 1951.
An. Mus. Hist. Nat., Montevideo, 6(1):1-37, charts

temperature
salinity
oxygen
turbidity

data
DeBuen, O., 1925.
Croisière océanographique du Transport "Almirante Lobo". Cons. Perm. Int. Expl. Mer, Rapp. Proc. Verb. 37:33-57.

data
de Buen, O., 1916.
Première campagne de l'Institut espagnol d'Océanographie dans la Mediterranée. Liste des stations et des operations. Bull. Inst. Océan., Monaco, No. 314:23 pp.

data
DeBuen, R., and F. de P. Navarro, 1935.
Condiciones oceanográficas de la costa catalana entre la frontera francesa y el Golfo de San Jorge. (Campañas del "Xauen" en Mayo de 1933 y de 1934). Trab. Inst. Español Ocean. No. 14:47 pp., 1 pl., 20 textfigs.

temperature
density
salinity

data
Defant, A., 1940.
Die Lage des Forschungsschiffes "Altair" auf der Ankerstation 16 bis 20 Juni 1938 und das auf ihr gewonnene Beobachtungsmaterial. Wiss. Ergeb. Internat. Golfstrom-Unternehmung 1938, Lief. 3, in Oktober Beiheft, Ann. Hydr., usw., 68:35 pp., 6 textfigs.

data
Denamur, J., 1955.
Étude physico-chimique de Bonifacio.
Trav. C.R.E.O. 2(10-11):14 pp.

salinity
pH
temperature
oxygen
des eaux du detroit

data
Deutschen Seewarte, 1930.
Temperatur-und Salzgehaltbeobachtungen an der Oberfläche der Weltmeere, 1914-1929.
Arch. Deutschen Seewarte 49(2):1-33.

data
Deutsche Seewarte, 1928, 1929, 1930, 1931.
Meereskundliche Beobachtungen auf deutschen Feuerschiffen der Nord- und Ostsee. Jahr 1924 und 1925: 84 pp., 2 charts; Jahr 1926 und 1927: 84 pp., 2 charts; Jahr 1928: 44 pp., 2 charts: Jahr 1929: 61 pp., 2 charts.

data
Dietrich, G., 1957.
Schichtung un Zirkulation der Irminger See in juni 1955. Ber. Deutsch. Wiss. Komm. Meeresf., n.f., 14(4):255-312.

temperature
currents
salinity

data, hydrographic
Dinsmore, R.P., R.M. Morse, Floyd M. Soule, 1960
International ice observation and ice patrol service in the north Atlantic Ocean-season of 1958. U.S. Treasury Dept. Coast Guard: Bull. No. 44: 1-99.

data
Dontas, S., 1947. proceedings of
[Yearly report of the Greek Hydrobiological Institute of the Athens Academy.] Praktica toi Hellinikoi Hidrobiologikoi Institutoi, 1947, 1(1):66 pp.

data
Dunbar, M.J., 1958.
Physical oceanographic results of the "Calanus" Expeditions in Ungava Bay, Frobisher Bay, Cumber-land Sound, Hudson Strait and Northern Hudson Bay, 1949-1955.
J. Fish. Res. Bd., Canada, 15(2):155-201.

data
Eggvin, J., 1933.
A Norwegian fat-herring fjord. An oceanographical study of the Eidsfjord. Repts. Norwegian Fish. Mar. Invest. 4(6):22 pp., 11 textfigs.

data (chiefly)
Emilsson, I., 1956.
Relatorio e resultados fisico-quimicos de tres cruzeiros oceanograficos em 1956.
Contrib. Avulsas, Inst. Ocean., Univ. Sao Paulo, Ocean. Fisica, No. 1:1-70.

temperature
oxygen
salinity
density

data
Eyries, M., 1956.
Activité océanographique de l'Escoteur "l'Aventure" en 1955.
Bull. d'Info., C.C.O.E.C., 8(10):527-545.

temperature
salinity
density

data
Eyries, M., 1954.
La campagne hydrologique de l'escoteur "L'Emporte" entre Madère et la côte marocaine (Avril-Mai 1953). Bull. d'Info., C.C.O.E.C. 6(4):163-168, 5 pls.

data
Eyries, M., and M. Menache, 1953.
Contribution à la connaissance hydrologique de l'Océan Indien entre Madagascar et La Réunion.
Bull. d'Info., C.C.O.E.C., 5(10):433-438, 9 pls.

temperature
density

data only
Favorite, Felix, John W. Schantz and Charles R. Hebard, 1961
Oceanographic observations in Bristol Bay and the Bering Sea 1939-41 (USCGT Redwing).
U.S.F.W.S. Spec. Sci. Rept., Fish. No. 381: 323 pp.

data
Filarski, J., 1955.
[Hydrographical conditions of the southern Baltic for the period from April 1952 to May 1953.]
Prace Morskiego Inst. Ryback., Gydni, No. 8:255-282

temperature
salinity
density
oxygen

data
Fisheries Research Board of Canada, 1947
Sea water temperature, salinity, and density on the Pacific coast of Canada. Vol.II including data from 1935-1937. 109 pp., 1 fig. (mimeographed)

data
Fisheries Board of Canada, 1947.
Observations of sea water temperature, salinity, and density on the Pacific coast of Canada. Vol. 1, including data from 1914 to 1934, 113 pp., 1 chart (mimeographed).

data
France, Service Central Hydrographique, 1957.
A. Observations hydrologiques des bâtiments de la Marine Nationale. B. Flotteurs-temoins de courant
Bull. d'Info., C.C.O.E.C., 9(2):147-153.

temperature, surface
salinity, surface
drift bottles

data
France, Service Central Hydrographique, 1956.
A. Observations hydrologiques des bâtiments de la Marine Nationale. B. Flotteurs-temoins du courant
Bull. d'Info., C.C.O.E.C., 8(8():408-411.

temperature
salinity

data
France, Comité Central d'Océanographie et d'Études des Côtes, 1955.
Observations hydrologiques des bâtiments de la marine nationale. Avisogarde-peche "Ailette", Mars à Octubre 1954. Bull. d'Info., C.C.O.E.C., 7(7):314- , 5 pls.

temperature, salinity, density currents

data
France, Comité Central d'Océanographie et d'Étude des Côtes, 1955.
Observations hydrologiques des bâtiments de la Marine Nationale. A. Travaux de l'escorteur l'"Aventure" (Avril-Octobre). B. Observations hydrologiques au large de l'entrée de la Manche. Bull. d'Info., C.C.O.E.C. 7(8):378-386.
temperature
salinity
density
T-S

data
France, Comite Central d'Oceanographie et d'Etude des Cotes, 1954.
Observations hydrologiques effectuees par les batiments de la Marine Nationale. Pbservations en Mer du Labrador par le Frigate "L'Aventure" (1953). Bull. d'Info., C.C.O.E.C. 6(9):419-430.

temperature
salinity
density

data only
France, Service Hydrographique de la Marine, 1964
Resultats d'observations.
Cahiers Océanogr., C.C.O.E.C., 16(5):399-409.

data only
France, Service Hydrographique de la Marine, 1962
Resultats d'Observations.
Cahiers Océanogr., C.C.O.E.C., 14(3):187-199.

data
Funder, Th. P., 1916.
Hydrographic investigations from the Danish School Ship "Viking" in the southern Atlantic and Pacific in 1913-14. Medd. Komm. Havundersøgelser, Ser. Hydrogr., 2(6):28 pp.

data
Gaarder, T., 1927.
Die Sauerstoffverhältnisse im östlichen Teil des Nord atlantischen Ozeans. Cruises of the "Armauer Hansen", No. 2. Geofys. Publ. 4:72 pp., 2 figs.

oxygen

data
Gaarder, T., and R. Spärck, 1932.
Hydrographisch-biochemische Untersuchungen in norwegischen Austern-Pollen. Bergens Mus. Aarbok, Naturvidensk.-rekke, No. 1:5-144, 75 textfigs.

temperature
salinity
density
pH
oxygen

data
Ganapati, P.N., E.C. La Fond and P.V. Bhavanarayana, 1956.
On the vertical distribution of chemical constituents in the shelf waters off Waltair.
Proc. Indian Acad. Sci., 44:68-71.

salinity
phosphate
silicate
oxygen

data
Gardiner, J.S., 1936.
XVI. Concluding remarks on the distribution of the land and marine flora with a list of water temperature observations. Trans. Linn. Soc., London, Ser. 2, 19:447-464.

data
Garner, D.M., 1958.
The Antarctic convergence south of New Zealand.
N.Z.J. Geol. Geophys., 1(3):577-594.

data
Genovese, S., 1954.
Osservazioni oceanografiche eseguite sui campi di pesca dell'alalunga delle isola Eolia.
Bol. Pesca, Piscicol., Idrobiol., n.s., 9(2): 186-196.

temperature,
salinity,
density

data
Germany, Deutsches Hydrographisches Institut, 1955.
Beobachtungen auf den deutschen Feuerschiffen der Nord- und Ostsee im Jahre 1954.
Meereskundl. Beobacht. u. Ergebn., No. 7(2122/7): numerous pp.

wind
sea
currents
surface temperature (some deep)
surface salinities (some deep.)

data
Germany, Dautsches Hydrographisches Institut, 1953.
Beobachtungen aud den deutschen Feuerschiffen der Nord- und Ostsee in Jahre 1948. Meereskundliche Beobachtungen und Ergebnisse No. 1(2122/1): numerous pp.

sea
temperature
currents

data
Giral, J., and O. Gomez Ibanez, 1930.
Determination des composés azotés dans l'eau de mer. Cons. Perm. Int. Expl. Mer, Rapp. Proc. Verb 67:93-99.

data
Glowinska, A., 1954.
[Hydrologic research in the southern Baltic in 1951.] Rept. Sea Fish. Inst., Gdynia, No. 7:159-190, 27 textfigs.

temperature
salinity
oxygen
density
phosphate

data
Glowinska, A., 1951.
[Hydrographical conditions in the southern Baltic in the time from August 1949 to May 1951.]
Trudy Morskogo Rybnogo Inst., Gdyna (Rept. Sea Fish. Inst.) No. 6:119-130, 7 textfigs.

temperature

salinity
density

data
Glowinska, A., 1948.
[Hydrographical conditions in the Gulf of Gdansk in the second part of 1946.] (In Polish, with English summary). Bull. Lab. Mar., Gdynia, No. 4: 171-185, 9 textfigs.

data
Gomez, Ibanez, O., 1929.
Determinacion del nitrogeno en sus formes, amoniacal, nitroso y nitrico en el agua, de mar.
Inst. Espagnol Oceanogr., Notas y Res., 2nd ser., No. 36:24 pp., figs.

data
Gomez Ibáñez, O., 1928.
Contribucion a la determinación de la materia orgánica contenida en el agua de mar.
Notas y Res., Inst. Espanol Ocean., 2nd sér., No.

organic matter
salinity

data
Graham, H. W. and N. Bronikovsky, 1944
The genus Ceratium in the Pacific and North Atlantic Oceans. Sci. Res. Cruise VII of the Carnegie, 1928-1929 ----- Biol. V (565):209 pp., 54 charts, 27 figs., 54 tables.

data
Gran, H.H., 1900.
Hydrographic-biological studies of the North Atlantic Ocean and the coast of Nordland.
Rept. Norwegian Fish. and Mar. Invest. 1(5): 1-89, 2 textfigs., 39 hydro. tables, 13 plankton tables.

temperature
salinity
phytoplankton

data (chiefly)
Granqvist, G., 1955.
Vattnets temperatur och salthalt i Nagu Skärgård. 1. Observationsmaterialet, 1952-54.
Merent. Julk, No. 168:26 pp.

temperature
salinity
density

data
Granqvist, G., 1954.
Surface temperature and salinity records along the coast of Finland, July 1940-June 1952.
Merent. Julk. No. 155:1-105, 1 textfig.

data
Granqvist, G., 1951.
Regular observations of temperature and salinity in the seas round Finland, July 1946-June 1950.
Merent. Julk. No. 150:1-35.

temperature
salinity

data
Granqvist, G., 1948.
Regular observations of temperature and salinity in the seas around Finland, July 1940-June 1946.
Merent. Julk., Havsforskningsinst. Skrift., No. 142:39 pp., tables.

data
Grøntved, J., 1952.
Investigations on the phytoplankton in the south -ern North Sea in May 1947. Medd. Komm. Danmarks Fisk. og Havundersøgelser, Plankton ser., 5(5): 1-49, 1 pl., 21 tables, 24 textfigs.

temperature
salinity
phosphate
nitrate
oxygen
density
phytoplankton

data

Grousson, R., and A. Comolet-Tirmen, 1957.
Hydrologie dans la partie orientale de l'Atlantique Nord. Bull. d'Info., C.C.O.E.C., 9(7):359-369.

density
temperature
salinity

data

Gudkovich, Z.M., 1955.
[Results of a preliminary analysis of the deep-water hydrological observations.]
Material. Nabluid. Nauch. Issledov. Dreifuiuš. 1950/51. Morskoi Transport, 1:41-46, 48-170.

David Krauss translator in AMS-Astia Doc. 117133

temperature
salinity
chlorinity
oxygen
pH
density

data

Halldal, P., 1953.
Phytoplankton investigations from Weather Ship M in the Norwegian Sea, 1948-49 (including observations during the "Armauer Hansen" cruise, July 1949). Hvalrådets Skrifter No. 38:91 pp., 20 tables, 21 textfigs.

temperature
salinity
phytoplankton (numbers)
nutrient salts

data

Hanaoka, T., and A. Murakami, 1954.
[On the submarine illumination of bay water.]
Nakai Regional Fish. Res. Lab. 6:7-14, 6 text-fig
(Bull.)

temperature
chlorinity
transparency

data

Hasegawa, Y., M. Yokoseki, E. Fukuhara, and K. Terai, 1952.
On the environmental conditions for the culture of laver in su0Bay, *buri Prov., Hokkaido.
Bull. Hokkaido Regional Fish. Res. Lab. 6:1-24, 8 textfigs., 9 tables.

phosphate
silicate
salinity
chlorinity
oxygen
temperature

data

Hela, I., 1953.
[Vattenståndensvärden 1947 och 1948.] (English summary.). Merent. Julk. No. 157:1-78.

data

Helland,-Hansen, B., 1907.
Current measurements in Norwegian fjords, the Norwegian Sea in 1906. Bergens Mus. Aarb., 1907 (15):61 pp., 2 pls., 13 textfigs., tables, charts

data

Henjes, F., 1910.
Ein Beitrag zur Morphographie des Meeresbodens im südwestlichen Pazifischen Ozean.
Arch. Deutschen Seewarte 32(3):1-42, 5 pls., 6 textfigs.

bathymetry
uneveness of bottom

data

Hidaka, K., 1941.
A stereophotogrammetric survey of waves and swells in the ocean. Mem. Imp. Mar. Obs., Tokyo, 7(3):231-368, 104 textfigs.

data

Hishida, K., 1953.
[Physical studies on the turbidity in the sea water with special reference to the relation of the radiant energy.] J. Ocean. Soc., Japan, 9(3/4):143-180, 31 textfigs.

data

Hydrografiska Avd., Kungl. Fiskeristyrelsen, 1950.
Fyrskeppsundersökning. År 1948-1949. 104 pp.

data

Ichie, T., K. Tanioka, and H. Yoshzawa, 1949.
On the tidal currents and other hydrological conditions at Matsuyama Port. Papers and Repts., Ocean., Kobe Mar. Obs., Ocean. Dept., No. 1:15 pp, 15 figs., followed by tables of data. (Odd atlas-sized pages - mimeographed).

data

Jacobsen, J.P., 1925.
Die Wasserumsetzung durch den Öresund, den Grossen und Kleinen Belt. Medd. Komm. Havundersøgelser, Ser. Hydrogr., 2(9):72 pp.

temperature
salinity
density

data

Jacobsen, J.P., 1918.
Hydrographische Untersuchungen im Randersfjord (Jylland). Medd. Komm. Havundersøgelser, Ser. Hy Hydrogr. 2(7):46 pp., 15 textfigs.

currents
salinity

data

Jacobsen, J.P., 1913.
Beitrag zur Hydrographie der Dänischer Gewässer. Medd. Komm. Havundersøgelser, Ser. Hydrogr. 2(2):94 pp., 47 tables, 17 textfigs., 14 pls.

data

Jacobsen, J.P., 1908.
Mittelwerte von Temperatur und Salzgehalt bearbeitet nach hydrographischen Beobachtungen in ischen Gewässern, 1880-1907. Medd. Komm. Havundersøgelser, Ser. Hydrogr., 1(10):26 pp., 11 pls.

salinity
density
temperature

data

Jacobsen, J.P., 1908.
Der Sauerstoffgehalt des Meereswassers in den daenischen Gewässern innerhalb Skagens.
Medd. Komm. Havundersøgelser, Ser. Hydrogr., 1(12):23 pp., 5 pls.

oxygen
temperature
salinity
density

data

Jakleln, A., 1936
Oceanographic investigations in East Greenland waters in the summers of 1930-1932. Skr. om Svalbard og Ishavet No. 67:79 pp., Pl. 2, 28 text figs.

data

Japan, Central Meteorological Observatory, 1953
The results of Marine meteorological and oceanographical observations. Jan.-June 1952, No. 11:362, 1 fig.

temperature
density
silicate
swell
phosphate
color
chlorinity
oxygen
sea
transparency

data

Japan, Central Meteorological Observatory, 1952
The results of Marine Meteorological and oceanographical observations, July - Dec. 1951, No. 10:310 pp., 1 fig.

temperature
chlorinity
density
oxygen
silicate
sea
swell
color
transparency
phosphate
pH
nitrite
phytoplankton, lists of species
lists of species quantitative
zooplankton

data

Japan, Central Meteorological Observatory, 1952.
The results of Marine meteorological and oceanographical observations. July - Dec. 1950. No. 8: 299 pp.

data

Japan, Central Meteorological Observatory, 1951.
The results of marine meteorological and oceanographic observations. July - Dec. 1949. No. 6: 423 pp.

data

Japan, Central Meteorological Observatory, 1951.
Tables 37-38. Direct current measurements for stations occupied by R.M.S. "Shumpu-maru" in the Akashi-seto and Kitan-seto; ---- in Osaka Bay. Res. Mar. Met., Ocean. Obs., Jan.-June 1949, No. 5:332-337.

data

Japan, Central Meteorological Observatory, 1951.
Tables 23-36. Marine meteorological observations. Res. Mar. Met., Ocean. Obs., Jan.-June 1949, No. 5:118-329.

data

Japan, Central Meteorological Observatory, 1951.
Table 21. Surface observations made on board R.S.S. "Umikaze-maru"; physical and chemical conditions of the surface waters of the Tsushima Straits and Goto-nada. Res. Mar. Met., Ocean. Obs. Jan.-June 1949, No. 5:115-116.

data

Japan, Central Meteorological Observatory, 1951.
Table 20. Surface observations made on board the "Kuroshio-maru"; physical and chemical conditions of the surface water of the sea between Tokyo and Hachijo Is. Res. Mar. Met. Ocean. Obs., Jan.-June, 1949, No. 5:108-115.

temperature
salinity
phosphate
silicate
sea
swell

data

Japan, Central Meteorological Observatory, 1951.
Table 19. Surface observations made on board the "Tokitsu-maru"; physical and chemical conditions of the surface water of the sea between Tokyo and Hakodate. Res. Mar. Met. Ocean. Obs., Jan.-June, 1949, No. 5:106-108.

temperature
salinity
phosphate
silicate
sea
swell

data(surface)

Japan, Central Meteorological Observatory, 1951.
Surface observations made on board R.M.S. "Oyashio-maru"; physical and chemical conditions of the surface water of the sea between Hakodate and Hachinoe. Res. Mar. Met. Ocean. Obs., Jan.-June, 1949, No. 5:106.

temperature
salinity
density

data

Japan, Central Meteorological Observatory, 1951.
Table 14. Oceanographical observations taken in Miyazu Bay: physical and chemical data for stations occupied by R.M.S. "Asanagi-maru". Res. Mar. Met. Ocean. Obs., Jan.-June 1949, No. 5:77-84.

temperature
salinity
density
oxygen
P
Si
pH
color
transparency
sea
swell

data

Japan, Central Meteorological Observatory, 1951.
Table 13. Oceanographical observations taken in Yosa-naikai: physical and chemical data for stations occupied by a boat. Res. Mar. Met. Ocean Obs., Jan.-June, 1949, No. 5:76.

data

Japan, Central Meteorological Observatory, 1951.
Table 12. Oceanographical observations taken in Sasebo Bay: physical and chemical data for stations occupied by R.S.S. "Asakaze-maru". Res. Mar. Met. Ocean. Obs., Jan.-June, 1949, No. 5:74-75.

temperature
salinity
density
transparency
color
sea
swell

data
Japan, Central Meteorological Observatory, 1951. Table 11. Oceanographical observations taken in Nagasaki Harbor:physical and chemical data for stations occupied by R.S.S. "Asakaze-maru". Res. Mar. Met. Ocean. Obs., Jan.-June, 1949, No. 5:70-74.
temperature
salinity
density
sea
swell
color
transparency
bottom sediments

data
Japan, Central Meteorological Observatory, 1951. Table 10. Oceanographical observations taken in the East China Sea. (A) Physical and chemical data for stations occupied by the "Akebono-maru No. 9"; (B) --------- by the "Hatsutaka-maru". Res. Mar. Met. Ocean. Obs., Jan.-June, 1949, No. 5:68-70.
sea
swell
temperature
salinity
density
P
Si

data
Japan, Central Meteorological Observatory, 1951. Table 9. Oceanographical observations taken in the Goto-Nada and the Tsushima Straits. (A) Physical and chemical data for stations occupied by the fishing boat "Daikoku-maru": (B) ------- R.S.S. "Umikaze-maru". Res. Mar. Met. Ocean. Obs., Jan.-June, 1949, No. 5:61-67.
temperature
salinity
density
transparency
oxygen
sea
swell
bottom deposits
color
phosphate
silicate

data
Japan, Central Meteorological Observatory, 1951. Table 8. Oceanographical observations taken in the Ariake Sea; physical and chemical data for stations occupied by R.S.S. "Umikaze-maru". Res. Mar. Met. Ocean. Obs., Jan.-June, 1949, No. 5:50-60.
temperature
salinity
density
oxygen
bottom sediments
transparency
sea
swell
phosphate
silicate
color

data
Japan, Central Meteorological Observatory, 1951. Table 7. Oceanographical stations taken in Osaka Bay; physical and chemical data for stations occupied by R.M.S. "Shumpu-maru". Res. Mar. Met. Ocean. Obs., Jan.-June, 1949, No. 5:48-49.
temperature
salinity
sea
color
transparency
density
swell

data
Japan, Central Meteorological Observatory, 1951. Table 6. Oceanographical observations taken in the Akashi-seto, the Yura-seto and the Kii Suido physical and chemical data for stations occupied by R.M.S. "Shumpu-maru". Res. Mar. Met. Ocean. Obs., Jan.-June 1949, No. 5:40-47.
temperature
salinity
density
oxygen
phosphate
silicate
nitrite
ammonia
iron
sea
swell
transparency

data
Japan, Central Meteorological Observatory, 1951. Table 5. Oceanographical observations taken in the sea area to the south of the Kii Peninsula along "E"-line; physical and chemical data for stations occupied by R.M.S. "Chikubu-maru". Res. Mar. Met. Ocean. Obs., Jan.-June, 1949, No. 5:37-40.
temperature
chlorinity
oxygen
silicate
transparency
color

data
Japan, Central Meteorological Observatory, 1951. Table 4. Oceanographical observations taken in Sagami Bay; physical and chemical data for stations occupied by R.M.S. "Asashio-maru". Res. Mar. Met. Ocean. Obs., Jan.-June, 1949, No. 5:30-37.
temperature
chlorinity
density
oxygen
silicate
transparency
color
sea
swell

data
Japan, Central Meteorological Observatory, 1951. Table 3. Oceanographical observations taken in the North Pacific Ocean along "C" line; (A) Physical and chemical data for stations occupied by R.M.S. "Ukuru-maru"; (B) --- R.M.S. "Ikuna-maru" (C) --- R.M.S. "Ryofu-maru"; (D) --- R.M.S. "Shinnan-maru"; (E) --- R.M.S. "Kung-maru"; (F) --- R.M.S. "Chikubu-maru"; (G) --- "Ryofu-maru"; (H) --- "Ikuna-maru"; (I) --- "Chikubu-maru"; (J) --- "Ukuru-maru"; (K) --- "Shinnan-maru"; (L) --- "Ukune-maru"; (M) --- "Ryofu-maru". Res. Mar. Met. Ocean. Obs., Jan.-June 1949, No. 5:13-30.
temperature- salinity density
oxygen silicate phosphate
sea swell- color transparency

data
Japan, Central Meteorological Observatory, 1951. Table 2. Oceanographical observations taken in the sea area along the east coast of Tohoku District: (A) Physical and chemical data for stations occupied by R.M.S. "Oyashio-maru": (B) Physical and chemical data for stations occupied by R.M.S. "Kuroshio-maru". Res. Mar. Met. Ocean. Obs., Jan.-June 1949, No. 5:3-12.
temperature
salinity
density
oxygen
transparency
sea
swell
color

data
Japan, Central Meteorological Observatory, 1951. Table 1. Oceanographical observations taken in the sea area south of Hokkaido; physical and chemical data for stations occupied by R.M.S. "Yushio-maru". Res. Mar. Met. Ocean. Obs., Jan.-June 1949, No. 5:2-3.
temperature
chlorinity
density
silicon
transparency
color
sea
swell

data
Japan, Fisheries Agency, Research Division, 1956. Radiological survey of western area of the danger-ous zone around Bikini-Eniwetok Atolls, investigated by the "Shunkotsu maru" in 1956, Part 1:143 pp.
radioactivity
currents, surface
chlorinity
oxygen
phosphate
dynamic depths
temperature
pH
silicate
density

data
Japan, Maritime Safety Agency, 1952. The results of oceanographic observation in the northwestern Pacific Ocean, 1938-1941, No. 5. Hydrogr. Bull., Publ. No. 981, (Ocean. Repts.) Special Number No. 9:194 pp., figs.

data only
Japan, Maritime Safety Board, 1962
The results of oceanographic observations in the north-western Pacific Ocean. 8. 1931-1940.
Hydrogr. Bull., Tokyo, (Publ. No. 981), No. 69: 245 pp.

data
Japan, Maritime Safety Bd., 1954.
Tables of results from oceanographic observations in 1950. Publ. 981, Hydrogr. Bull., Maritime Safety Bd., Spec. Number (Ocean. Repts.), No. 14:26-164, 5 textfigs.
temperature
chlorinity
pH
oxygen
silicate
density
sea
swell
color
transparency
ship's roll

data
Japan, National Committee for the International Geophysical Year, 1960
The results of the Japanese oceanographic project for the International Geophysical Year, 1957/58:198 pp.

data
Japan, Tokai Regional Fisheries Research Laboratory, 1951.
Report, Oceanographical investigation (Jan. 1943-Dec. 1944 collected), No. 72:105 pp., 1 fig.
bottom sediments
temperature
salinity
transparency
color

data
Jensen, A.J.C., 1944.
The hydrography of the Praestø Fjord.
Folia Geogr. Danica 3(2):47-55, 2 textfigs., 4 tables.
salinity
temperature
oxygen
pH

data
Kalle, K., 1956.
Chemisch-hydrographische Untersuchungen in der innered Deutschen Bucht. Deutsche Hydrogr. Zeits. 9(2):55-65.
salinity
fluorescence
phosphate
"Eiweiss"
yellow stuff
silicate
chlorophyll
temperature

density
Kalle, K., and H. Thorade, 1940.
Tabellen und Tafeln für die Dichte des Seewassers (σt). Arch. Deutschen Seewarte 60(2):1-12, 3 textfigs., 7 pls.

data
Karlovac, O., 1956.
Station list of the M.V. "Hvar" fishery-biological cruises, 1948-1949. Inst. Ocean. i Ribarstvo, Split, Repts., 1(3):177 pp.
temperature,
salinity
sea
chlorinity
density
swell

data
King, J.E., T.S. Austin and M.S. Doty, 1957.
Preliminary report on Expedition EASTROPIC
U.S.F.W.S. Spec. Sci. Rept., Fish., 201:155 pp.
temperature
oxygen
quantitative
salinity
phosphate
density

data
Knox, G.A., 1957.
General account of the Chatham Islands 1954 Expedition.
N. Zealand D.S.I.R. Bull. (Mem. N.Z. Ocean. Inst. 2) 122:1-17.
sea
temperature
bottom sediments
swell
salinity

data
Knudsen, M., 1911.
Danish hydrographical investigations at the Faroe Islands in the spring of 1910. Medd. Komm. Havundersøgelser, Ser. Hydrogr., 2(1):17 pp., 5 textfigs., 2 pls.
salinity
temperature
oxygen

data
Knudsen, M., 1905.
Contribution to the hydrography of the North Atlantic Ocean. Medd. Komm. Havundersøgelser, Ser. Hydrogr., 1(6):13 pp., 21 pls.
temperature
salinity

data
Koczy, F., 1954.
Swedish observations.
Ann. Biol., Cons. Perm. Int. Expl. Mer, 10:134-136.
temperature
salinity
oxygen

data
Koczy, F.F., 1952.
Hydrography. Swedish observations. Cons. Perm. Int. Expl. Mer, Ann. Biol. 8:130-131.
temperature
salinity
phosphate
pH

data
Koninklijk Nederlandsch Meteorlogisch Instituut, 1921.
Oceanographische en Meteorlogische Waarnemingen in den Atlantischen Oceaan. Maart, April, Mai, (1856-1920). Tabellen. Kon. Neder. Meteorl. Inst. No. 110:186 pp.

data
Lacombe, H., 1956.
Quelques elements de l'extension des eaux méditerraneennes dans l'Océan Atlantique.
Bull. d'Info., C.C.O.E.C., 8(5):210-224, 7 pls.

temperature
salinity
density

data
LaFond, E.C., 1958
On the circulation of the surface layers of the East Coast of India.
Andhra Univ. Mem., Ocean., 2: 1-11.

data
LeGall, J., 1931.
Travaux à la mer du Laboratoire de l'Office des Pêches Maritimes à Boulogne sur Mer pendant l'année 1929.
Cons. Perm. Int. Expl. Mer, Rapp. Proc. Verb. 70: 33-46, 19 textfigs.

salinity
temperature

data
LeGall, J., 1927.
Croisière de la "Tanche", 25 juillet-13 septembre 1926. Cons. Perm. Int. Expl. Mer, Rapp. Proc. Verb. 44:13-17.

data
LeFloch, J., 1955.
Esquisse de la structure hydrologique de l'Atlantique Equatorial au large de la Guyane et de l'embouchure de l'Amazone.
Bull. d'Info., C.C.O.E.C., 7(10):449-467, 13 pls.

temperature
salinity
density

data
Lenoble, J., 1955.
Sur quelques nouvelles mesures de la pénétration du rayonnement ultraviolet dans la Mediterranée et leur interprétation théorique.
C.R. Acad. Sci., Paris, 241(20):1407-1409.

ultraviolet light.

data
Lisitzin, E., 1949.
Observations on currents and winds made on board Finnish Light-ships during the years 1943-1947.
Merent. Julk., Havsforskningsinst. Skrift. No. 143:66 pp., 1 textfig., tables.

data
Lizeray, J.-C., 1957.
Travaux océanographique recents de l'"Elie Monnier" en Méditerranée occidentale.
Bull. d'Info., C.C.O.E.C., 9(8):413-415.

currents (GEK)
temperature
salinity
density

data
Lüneburg, H., 1939.
Hydrochemische Untersuchungen in der Elbmündung mittels Elektrokolorimeter.
Arch. Deutschen Seewarte 59(1):1-27, 8 pls.

temperature
chlorinity
gelbstoff
phosphate
silicate
nitrite
turbidity

data (not much!)
Lütgens, R., 1911.
Ergebnisse einer ozeanographischen Forschungsreise in dem Atlantischen und dem südöstlichen Stillen Ozean. Arch. Deutschen Seewarte 34(1):1-74, 4 textfigs., 4 pls.

temperature(surface)
salinity (surface)
waves

data
Machado, L. de B., 1952.
Pesquisas fisicas e quimicas do sistema hidrografico da regeiao Lagunar de Cananeia. Bol. Inst. Ocean., Univ. Sao Paulo, 3(1/2):55-75, 4 graphs, 1 chart.

data
Maeda, H., 1953.
Studies on Yosa-Naikai. 3. Analytical investigations on the influence of the River Noda and the benthonic communities. J. Shimonoseki Coll. Fish 3(2):141-149, 3 textfigs.

temperature
chlorinity
oxygen
pH
silicate
phosphate
bottom sediments

data
Maeda, H., 1953.
Studies on Yosa-Naikai. 2, Considerations upon the range of the stagnation and the influences by the River Noda and the open sea.
J. Shimonoseki Coll. Fish. 3(2):133-140, 2 textfigs.

transparency
temperature
chlorinity
oxygen
pH
silicate
phosphate
bottom sediments
plankton (quant.)

data
Mankowski, W., 1955.
[Plankton investigations in southern Baltic in 1951.] Prace Morskiego Inst. Ryback., Gydni, No. 8:197-233.

Temperature, surface and bottom
salinity, surface and bottom

data
Mankowski, W., 1951.
[Macroplankton of the southern Baltic, in 1949.] Trudy Morskogo Rybnogo Inst., Gdyna, (Rept. Sea Fish. Inst.) No. 6:83-94.

temperature
salinity

data
Marchand, J.M., 1957.
Twenty-seventh annual report for the period 1st April 1955 to 31st March 1956.
Comm. & Ind., S. Africa, July, 1957:159 pp.

temperature
density
salinity
phosphate.

data(chiefly)
Marchand, J.M., 1956.
Twenty-sixth annual report for the period 1st April 1954 to 31st March 1955. Union of S. Africa Dept. Comm. & Industr., Div. Fish.:183 pp.

Reprinted from:"Commerce & Industry", July 1956.

temperature
sea
density
salinity
swell
phosphate

data
Marchand, J.M., 1953.
Pilchard-research programme. First Progress Report. Annexure "A", 23rd Ann. Rept., Div. Fish., Union S. Africa:17-181, 5 charts

temperature
salinity
density
sea
Phosphorus

data
Marija, Marinkovic, 1959.
Observations hydrographiques terminées a Rovinj en 1955-1957.
Thalassia Jugoslavica, 1(6-10): 41-68.
(French summary)

data
Marvin, K.T., 1955.
Oceanographic observations in west coast Florida waters, 1949-52. U.S. Fish & Wildlife Service, Spec. Sci. Rept., No. 149:6 pp., 2 figs., numerous pp tables.

temperature
oxygen
total phosphate
inorganic phosphate
pH
salinity
transparency
nitrate

data
Matthews, D.J., 1926.
VII. Physical oceanography. Trans. Linn. Soc., London, Ser. 2, 19:169-205, Pls. 10-13, textfigs. 1-8.

data
Mazzarelli, G., 1949.
Ricerche sul colore del mare esequite tra la Sicilia e la Libia e lungo le coste della Puglia e della Calabria. Boll. Soc. Naturalisti, Napoli, 57:115-119.

data
McDougall, J.C., 1961
Ironsand deposits offshore from the west coast, North Island, New Zealand.
New Zealand J. Geol. Geophys., 4(3):283-300.

data
McGary, J.W., 1955.
Mid-Pacific oceanography, Pt. 6. Hawaiian offshore waters, December 1949-November 1951.
Spec. Fish. Rept:Fish.No. 152:1-138, 33 figs.

salinity
temperature
density
PO4-P
oxygen

data(chiefly)
McGary, James W., Edward D. Stroup, 1958
Oceanographic observations in the central North Pacific, September 1954-August 1955.
U.S.F.W.S. Spec. Sci. Rept., Fish., No. 252: 250 pp.

data
McGary, J.W., and E.D. Stroup, 1956.
Mid-water oceanography. VIII. Middle latitude waters, January-March 1954.
Spec. Sci. Rept., Fish., No. 180:173 pp.

temperature
density
phosphate
salinity
oxygen

data
Menaché, M., 1954.
Étude hydrologique sommaire de la région d'Anjouan en rapport avec la pêche des coelacanthes. Mém. Inst. Sci., Madagascar, A, 9:151-185, 34 textfigs.

temperature
chlorinity
salinity
density
sea
swell

data
Metcalf, W.G., 1955.
On the formation of bottom water in the Norwegian Basin. Trans. Amer. Geophys. Union, 36(4):596-600, 2 textfigs.

temperature
salinity
oxygen

no station positions given

data
Midttun, L., and J. Natvig, 1957.
Pacific Antarctic waters. Scientific results of the Brategg" Expedition, 1947-48, No. 3.
Publikasjon, Komm. Chr. Christensens Hvalfangstmus., Sandefjord, No. 20:1-130, 39 figs.

temperature
density
salinity
oxygen

data
Mishima, S., and S. Nishigawa, 1955.
Report on hydrographic investigations in Aleutian waters and the southern Bering Sea in the early summers of 1953 and 1954.
Bull. Fac. Fish., Hokkaido Univ., 6(2):85-124, 29 textfigs.

temperature
salinity

data

Miyake, Y., 1952.
A table of the saturated vapour pressure of sea water. Ocean. Mag., Tokyo, 4(3):95-118.

data (for several stations)

Miyazaki, M., 1953.
[On the water masses of the Japan Sea.] Bull. Hokkaido Regional Fish. Res. Lab., Fish. Agency, No. 7:1-65.

data

Moreira da Silva, Paulo, 1957
Oceanografia do trinagulo Cabo-Frio-Trinidade-Salvador.
Anais Hidrograficos, Brasil, 16:213-308.

data (8 stas.)

Moroshkin, K.V., 1959.
[Hydrological investigations.]
Arktich. i Antarkt. Nauchno-Issled. Inst., Mezhd. Geofiz. God, Sovetsk.Antarkt. Eksped., 5:107-124.

data

Mosby, H., 1938
Svalbard waters. Geophys. Publ. 12(4): 85 pp., 34 text figs.

data

Motoda, S., 1940.
Comparison of the conditions of water in the bay, lagoon and open sea in Palao.
Palao Trop. Biol. Sta. Studies 2(1):41-48, 2 figs

data

Nagaya, Y., 1959
On the distribution of the nutrient salts in the Equatorial region of the Western Pacific, Jan.-Feb., 1958.
Publ. 981, Hydrogr. Bull., No. 59:31-40.

data

Navarro, F. de P., 1947.
Exploración oceanografía del Africa occidental desde el Cabo Ghir al Cabo Judy: resultados de las campañas del "Melaspina" y del "Xauen" en Mayo 1946. Trab. Inst. Espagñol Ocean. No. 20: 40 pp., 8 figs.

temperature
salinity
density
phosphate
bathymetry
bottom sediments

data

Navarro, F. de P., F. Lozano, J.M. Navaz, E. Otero J. Sáinz Pardo and others, 1943.
La pesca de Arrastre en los fondos del Cabo Blanco y del Banco Arguín (Africa Shaariana).
Trab. Inst. Español Ocean. No. 18:225 pp., 38 pls.

pH
oxygen
density
temperature
chlorinity
salinity

data

Netherlands, Koninklijk Nederlands Meterologisch Instituut, 1956.
Meteorologische en oceanografische Waarnemingen verricht aan boord van Nederlands Lichtschepen in de Noordzee, Jaargand 6, 1954: unnumbered pages.

tidal currents — wave heights
temperature(surface)

data

Newell, B.S., 1959
The hydrography of the British East African coastal waters. II. Colonial Off., Fish. Publ. London, No. 12: 18 pp.

data

Neumann, G., 1940.
Die ozeanographischen Verhältnisse an der Meeresoberfläche im Golfstromsektor nördlich und nordwestlich der Azoren. Wiss. Ergeb. Internat. Golfstrom-Unternehmung 1938, in Juni Beiheft, Ann. Hydr., usw., 68:87 pp., 36 textfigs., 3 pls.

data

Nielsen, J.N., 1907.
Contribution to the hydrography of the north-eastern part of the Atlantic Ocean.
Medd. Komm. Havundersøgelser, Ser. Hydrogr., 1(9):25 pp., 3 pls.

temperature
salinity
density

data

Nielsen, J.N., 1905.
Contribution to the hydrography of the waters north of Iceland. Medd. Komm. Havundersøgelser, Ser. Hydrogr., 1(7):28 pp., 2 pls.

temperature
salinity
density

data

Nielsen, J.N., 1904.
Hydrography of the waters by the Faroe Islands and Iceland during the cruises of the Danish Research Steamer "Thor" in the summer 1903.
Medd. Komm. Havundersøgelser, Ser. Hydrogr., 1(4):29 pp., 8 pls.

temperature
salinity

data

Nutt, D.C., and L.K. Coachman, 1956.
The oceanography of Hebron Fjord, Labrador.
J. Fish. Res. Bd., Canada, 13(5):709-758.

temperature — salinity
oxygen — inorganic phosphate

data

Picotti, M., 1923-24.
Richerche di oceanografia chimica. Tabelle generali della analisi chlorometriche e dei dati di temperatura, salinita e densita. Ann. Idrogr. 11 bis:69-116, charts, tables.

temperature
salinity
density

data

Picotti, M., and A. Vatova, 1942.
Osservazioni fisiche e chemiche periodiche nell' Alto Adriatico (1920-1938). Thalassia 5(1):1-157.

sea temperature
sea
chlorinity
salinity
density
oxygen
pH

data

Polli, S., 1949.
100 anni di osservazioni meteorologiche eseguite a Trieste, 1841-1940. Pt. 4. La velocità del vento. Ist. Talass., Trieste, Publ. 234:42-87.

data

Rakestraw, N.W., P.L. Horrer and W.S. Wooster, 1957.
Oceanographic observations of the Pacific, 1949. University of California, Press, 363 pp.

data

Ramalho, A. de M., 1942
Observacoes oceanograficas. "Albacora", 1940. Trav. St. Biol. Mar., Lisbonne, No. 45: 96 XVI pp., 2 figs. (fold-in). (Bumpus reprint)

data

Ratmanoff, G.E., 1937.
[Explorations of the seas of Russia.] Hydrological Inst., Leningrad, Publ. No. 25:1-175.

chemical
physical

data

Reichelt, W., 1941.
Die ozeanographischen Verhältnisse bis zur warmen Zwischenschicht an der antarktischen Eisgrenze in Südsommer 1936/37. Nach Beobachtungen auf dem Walfang-Mutterschiff "Jan Wellem" im Weddell-Meer Arch. Deutschen Seewarte 61(5):1-54, 11 pls.

temperature
salinity
density
phosphate
oxygen

data

Reish, D.J., and H.A. Winter, 1954.
The ecology of Alamitos Bay, California, with special reference to pollution.
Calif. Fish. and Game 40(2):105-221, 1 textfig.

transparency
temperature
dissolved oxygen
pH
chlorinity

data

Rochford, D.J., 1953.
Estuarine hydrological investigations in eastern and south-western Australia, 1951. Oceanographical station list of investigations made by the Division of Fisheries, Commonwealth Scientific and Industrial Research Organization, Australia, 12:111 pp.

temperature
chlorinity
oxygen
pH
phosphate
nitrate

data

Rochford, D.J., 1953.
Analysis of bottom deposits in eastern and south-western Australia, 1951 and records of twenty-four hourly hydrological observations at selected stations in eastern Australian estuarine systems 1951. Oceanographical station list of investigations made by the Division of Fisheries, Commonwealth Scientific and Industrial Research Organization, Australia, 13:68 pp.

chemistry of bottom sediments
temperature — phosphate
chlorinity — nitrate
oxygen

data

Roll, H.U., 1953.
Höhe, Länge und Steilheit der Meereswellen in Nordatlantik. (Statistik der Wellenbeobachtungen der Ozean-Wetterschiffe). Deutscher Wetterdienst, Seewetterampt, Hamburg, Einzelveröff. No. 1: 58 pp., figs.

data, waves

Roll, H.U., 1952.
Statistik von Wellenbeobachtungen, Nordatlantischen Wetterschiffe (Nov. 1950-Okt. 1951) und deutscher Feuerschiffe (1949/50). Abt. Marit. Met., Met. Amt f. Nordwestdeutschland, 25 pp. (mimeographed), 5 charts (ozalid)

Wave height
wave period.

data

Rotschi, Henri, 1959.
Orsom III. Resultats de la croisiere "Boussole" 1. Oceanographie physique.
O.R.S.T.O.M., Inst. Francais d'Oceanie, Rapp. Sci., No. 12:67 pp.

data

Rotschi, Henri, Michel Angot, et Michel Legrand, 1959
Orsom III. Resultats de la croisiere "Astrolabe" 2. Chimie, productivite et zooplankton.
Rapp. Sci., Inst. Francais d'Oceanie, No. 9: 97 pp. (mimeographed).

data

Rotschi, Henri, Michel Angot, Michel LeGand, and H.R. Jitt, 1959
Chimie, productivité et zooplancton. "Orsom III". Resultats de la Croisiere "Boussole". Resultats "production primaire" de la croisière 56-5.
Rapp. Sci. Inst. Francais d'Oceanie, Centre d'Oceanogr., No. 13:

data

Rouch, J. and J. Vernet, 1949.
La température de la mer comparée à la température de l'air à Monaco. Bull. Inst. Océan., Monaco, No. 960: 8 pp.

data

Saiz, Fernando, Manuel Lopes-Benito and Emilio Anadon, 1957
Estudio hidrográfico de la ría Vigo.
Inv. Perq., Barcelona, 8:29-87.

data

Satow, T., 1955.
[Oceanographical conditions on the coastal waters west of the Sunda Archipelago.] J. Shimonoseki Coll., Fish. 4(1): 1-20, 34 textfigs.

temperature
salinity

data

Schmidt, J., 1929.
Introduction to the oceanographical reports including list of stations and hydrographical observations.
Danish "Dana" Exped., 1920-1922, 1:1-87, 6 pls.
Rept.

data

Schott, G., and P. Perlewitz, 1906.
Lotungen I.N.M.S. "Edi" und des Kabeldampfers "Stephen" im westlichen Stillen Ozeans.
Arch. Deutschen Seewarte 29(2):1-38, 4 pls.

bottom sediments
bathymetry

data

Schott, G., and B. Schulz, 1914.
Die Forschungsreise:S.M.S. "Möwe" im Jahre 1911.
Arch. Deutschen Seewarte 37(1):1-80, 8 pls., 16 textfigs.

surface currents
temperature
salinity
density
color
bottom sediments

data

Schubert, O. von, 1944.
Ergebnisse der Strommessungen und der ozeanographischen Serienmessungen auf den beiden Ankerstationen der zweiten Teilfahrt. Wiss. Ergeb. Deutschen Nordatlantischen Exped., 1938 u. 1938, Lief. 1, Januar Beiheft, Ann. Hydr., usw., 72: 74 pp., 30 textfigs., 1 pl.

data

Schulz, B., 1923.
Hydrographische Untersuchungen besonders über den Durchlüftungszustand in der Ostsee in Jahre 1922. (Forschungsschiffe "Skagerak" und "Nautilus").
Arch. Deutschen Seewarte 41(1):1-64, 2 textfigs., 5 pls.

temperature
salinity
density
pH
oxygen
Carbon dioxide

data

Schulz, B., 1922.
Hydrographischen Beobachtungen inbesondere über die Kohlensäures in der Nord- und Ostsee im Sommer 1921. (Forschungsschiffe "Poseidon" und "Skagerak"). Arch. Deutschen Seewarte 40(2):1-44 2 textfigs., 4 pls.

salinity
density
oxygen
alkalinity
pH
carbon dioxide
temperature

data

Schulz, B., and A. Wulff, 1929
Hydrographie und Oberflächen plankton des westlichen Barentsmeeres im Sommer 1927.
Ber. deutschen wissensch. Komm. F. Meeresforsch. n.s. 4(5):232-372, 13 tables, 25 text figs.

data

Sears, Mary, 1954.
Notes on the Peruvian coastal current. 1. An introduction to the ecology of Pisco Bay.
Deep-Sea Res. 1(3):141-169, 4 textfigs.

temperature
salinity

data (chiefly)

Seckel, G.R., 1955.
MidPacific oceanography. VII. Hawaiian offshore waters, Sept. 1952-August 1953.
Spec. Sci. Repts., Fish. No. 164:250 pp., 38 figs

temperature
salinity
density
phosphate
oxygen

data

Seiwell, H. R., 1939.
Daily temperature variations in the North Atlantic. J. du Cons. 14(3):357-369, 6 textfigs., 7 tables.

data

Servicio meteorologico del Ecuador, 1945
Resumens de las observaciones meteorologicas, correspondientes a los anos de 1935-1943, inclusive. Boletin Meteorologico No.2, 191 pp

data

Shimomura, T., and K. Miyata, 1957.
[The oceanographical conditions of the Japan Sea and its water systems, laying stress on the summer of 1955.]
Bull. Japan Sea Reg. Fish. Res. Lab., No. 6 (General survey of the warm Tsushima Current 1): 23-120.

temperature — salinity
chlorinity — oxygen
pH — density
silicate — phosphate
nitrite — nitrate

data

Sjarif, Sjarmilah, 1959
Seasonal fluctuations in the surface salinity along the coast of the southern part of Kalimantan (Borneo). Mar. Res., Indonesia, (Penjelidikan Laut di Indonesia), No. 4:29 pp.

data

Sjöstrand, J., 1921.
De hydrografiska förhållanden i Norre Ishavet mellan norska kusten och Spetsbergen samt väster om Spetsbergen ävenom i Isfjorden och Van Mijans-fjord år 1920. Svenska Hydrogr.-Biol. Komm. Skr 7:1-8.

temperature — salinity
density

data

Soule, F.M., 1953.
Physical oceanography of the Grand Banks region and the Labrador Sea in 1952. U.S.C.G. Bull. No. 38:31-100, 17 textfigs.

temperature
salinity
density

data

Soule, F. M., 1940.
Oceanography. Season of 1938. Bull. No. 28, Int. Ice Obs. and Ice Patr. Service. N. Atlantic Ocean:113-173, 56 textfigs.

data

Soule, F. M., 1938.
Oceanography. Season of 1937. International Ice Observations and Ice Patrol Service, U.S.C.G. Bull. No. 27:71-126, 45 textfigs., tables of data.

data

Soule, F.M., 1938
Oceanography. Excerpt from International Ice Observation and Ice Patrol Service in the North Atlantic Ocean. Season of 1936. C.G. Bull. No.26:33-82, 58 text figs., tables of data.

data

Soule, F.M., P.S. Branson, and R.P. Dinsmore, 1952.
Physical oceanography of the Grand Banks region and the Labrador Sea in 1951. U.S.C.G. Bull. No. 37:17-85, 21 textfigs.

temperature
salinity
density

data

Soule, F.M., A.J. Bush and J.E. Murray, 1955.
International ice observation and Ice Patrol Service in the North Atlantic Ocean. Season of 1953. U.S.C.G. Bull. No. 39:45-168, 43 textfigs.

Total P/µg /L
density
temperature
salinity

data

Soule, F.M., A.P. Franceschetti and R.M. O'Hagen, 1963.
Physical oceanography of the Grand Banks region and the Labrador Sea in 1961.
U.S. Coast Guard Bull., No. 47:19-82.

data

Soule, F.M., and J.E. Murray, 1957.
Physical oceanography of the Grand Banks region and the Labrador Sea in 1956. U.S.C.G. Bull., No. 42:35-100.

temperature
salinity
density

data

Soule, F.M., and J.E. Murray, 1956.
Physical oceanography of the Grand Banks and Labrador Sea in 1955. U.S.C.G. Bull., No. 41:59-114.

temperature — phosphate
salinity — density

data

Soule, F.M., and J.E. Murray, 1955.
Physical oceanography of the Grand Banks region and the Labrador Sea.
U.S.C.G. Bull. No. 40:79-168.

temperature
salinity
density

data

Soot-Ryen, T., 1956.
Report on hydrographical conditions in West-Finmark, March-April, 1953.
Acta Borealia, Tromsø Mus., A. Scientia, No. 10: 1-37

salinity — temperature
oxygen — pH
density — P mg/m3

data

Soot-Ryen, T., 1938.
Hydrographical investigations in the Tromsø district in 1931. Tromsø Mus. Aarsheft., Naturhist. Avd. No. 10, 54(2):1-6, plus 41 pp. of tables.

temperature — phosphorus
salinity — nitrate
density — nitrite
pH — ammonia
oxygen

data

Spain, 1932.
Data. Rapp. Proc. Verb., Cons. Perm. Int. Expl. Mer, 77:15-20.

temperature
salinity
density

data
Spain, 1931.
Liste des stations. Rapp. Proc. Verb., Cons. Perm Int. Expl. Mer, 70:17-19

density
temperature
salinity

data
Soot-Ryen, T., 1947.
Hydrographical investigations in the Tromsø-district 1934-1938 (Tables).
Tromsø Mus. Aarsheften, Naturhist. Avd. No. 33, 66(1943)(1): numerous pp. (unnumbered).

temperature
salinity
density
pH
oxygen
phosphates
nitrate
nitrite
ammonia

data
Soot-Ryen, T., 1932.
The Folden Fjord. Hydrography. Tromsø Mus. Skr. 1(3):7 pp.

temperature
salinity
density

data
Soule, F.M., and C.A. Barnes, 1950.
Physical oceanography of the Ice Patrol area in 1941.
Bull. U.S.C.G. No. 31:1-62, 23 textfigs., fold-in charts.

temperature
salinity
density

data
Soule, F.M., H.H. Carter and L.A. Cheney, 1950.
Oceanography of the Grand Banks region and Labrador Sea, 1948. U.S.C.G. Bull, No. 34:67-118, Figs. 27-37, fold-in charts.

temperature
salinity
density.

data
South Africa, Fisheries and Marine Biological Survey Division, 1952.
Twenty-first annual report for the year ended December 1941. Station list R.S. "Africana". Commerce and Industry:51 pp., 6 charts.

data
South Africa, Division of Fisheries, 1950.
Station list - R.S. "Africana II", R.V. "Sbipa" and P.B. "Palinurus". 22nd Ann. Rept. 21-169, 2 charts.
bottom sediments
temperature
salinity
density
phosphorus
sea

data
Strickland, J., 1959.
Assessment of the accuracy and precision of data. Nat. Acad. Sci.-Nat. Res. Counc. Publ. 600:98-100.

data
Stroup, E.D., 1954.
Mid-Pacific oceanography IV. Transequatorial waters, January-March 1952.
U.S.F.W.S. Spec. Sci. Rept. - Fish. - No. 135: 1-52, 18 figs. (multilithed).

temperature
salinity
density
oxygen
PO4-P

data
Svenska Hydrografisk-Biologiska Kommission, 1950?
Fyrskeppsundersökning År 1946-1947: 78 pp., 2 figs

data
Sverdrup, H.U., and Staff, 1947.
Oceanographic onservations on the E.W.Scripps cruises of 1941. Obs. Rec. SIO 1:249-407.

data
Sverdrup, H.U., and Staff, 1943
Oceanographic observations of the Scripps Institution in 1939. SIO Rec. of Obs. 1(2): 65-159.

data
Sverdrup, H.U., and Staff, 1942
Oceanographic observations on the E.W. Scripps cruises of 1938. SIO Records of observation 1(1):63 pp.

data only
Sweden, Fishery Board, 1954.
Hydrographical observations on Swedish light-ships in 1953.
Fishery Bd., Sweden, Ser. Hydrogr., Rept. No. 4: 1-130, 1 fig.

temperature
salinity
currents

data
Sweden, Fishery Board of Sweden, 1953.
Hydrographic observations on Swedish lightships in 1952.
Fish. Bd., Sweden, Hydrogr. Ser., Rept. No. 3: 133 pp.

temperature
salinity
currents

data
Sweden, Fishery Board of Sweden, 1952.
Hydrographical observations on Swedish lightships in 1951. Series Hydrography, Rept. No. 1:133 pp., 1 chart (multilithed).

currents
temperature
salinity

data
Sweden, K. Fiskeristyrelsen, 1951.
Fyrskeppsundersökning, År 1950: numerous pp. (unnumbered).

currents
temperature
salinity

data
Syazuki, K., 1953.
[Studies on the foul-water drained from factories 2. On the harmful components and the water pollution of the foul-water drained from the metal plants.] J. Shimonoseki Coll. Fish. 3(2):181-185, 2 textfigs.

ferrous sulfate
pH
temp.

data
Szarejko-Lukaszewicz, D., 1957.
[Qualitative investigations of phytoplankton of Firth of Vistula in 1953.]
Prace Morsk. Inst. Rybacki, Gdyni, No. 9:439-451

temperature
salinity
oxygen

data
Tamura, T., and J. Saguira, 1950.
A report on oceanographical observations of the surface sea waters extending between Japan and Antarctic whaling grounds. Bull. Fac. Fish., Hokkaido Univ., 1(1):12-17, 2 figs. (In Japanese; English summary).

data
Tchernia, P., 1956.
Contribution à l'étude hydrologique de la Méditerranée occidentale. Hydrologie.
Bull. d'Info., C.C.O.E.C., 8(9):427-454, 11 pls.

temperature
density
salinity

data
Tchernia, P., 1954.
Contributions à l'étude hydrologique de la Méditerranée occidentale. Premier rapport préliminaire. Bull. d'Info., C.C.O.E.C 6(1):7-30, 11 pls.

salinity
temperature

data
Tchernia, P., H. Lacombe and P. Guibout, 1958.
Sur quelques nouvelles observations hydrologiques relatives à la région équatoriale de l'océan Indien. Bull. d'Info., C.C.O.E.C., 10(3):115-143.

temperature
salinity
density

data
Tchernia, Paul, and Bernard Saint-Guily, 1959.
Nouvelles observations hydrologiques d'hiver en Mediterranee occidentale.
Cahiers Ocean., C.C.O.E.C., 11(8):499-542.

data
Thomsen, H., 1947.
Hydrografiske undersøgelser i Limfjorden 1943. 1. Observationsmaterialet. Publ. Danske Met. Inst. Medd. No. 9:104 pp., 1 fold-in, 6 text-figs.

currents
temperature
salinity
oxygen

data("Nautilus" 1931; "Stalin", 1940)
Timofeyev, V.T., 1957
[Formation of the bottom waters of the central part of the Arctic Basin.] Problemy Arktiki, 1: 29-33.
T 349R-Jan. 1961-E.R. Hope

data
Townsend, C.H., 1901.
Dredging and other records of the United States Fish Commission Steamer "Albatross" with bibliography relative to the work of the vessel.
U.S. Comm. Fish. Fisheries Rept., 1900:387-562, 7 pls.

data
Trotti, L., 1953.
Risultati delle crociere talassografiche nel Mar Ligure e nell'alto Tirreno. Introduzione, Osservazioni meteorologiche e idrografiche.
Centro Talassografico Tirreno, Pubbl. No. 14: 7 pp., photos, 12 fold-in tables.

density
salinity
temperature
oxygen
wind
sea

data
Trotti, L., 1953.
Risultati delle crociere talassografiche nel Mar Ligure e nell'alto Tirreno. Introduzione, osservazione meteorologiche e idrografiche.
Centro Talassograf. Tirreno, Pubbl. No. 14:12 pp

oxygen
density
temperature
salinity

data
Trotti, L., 1951.
Ricerche idrografiche sulle acque costiere Ligustiche comprese tra l'isola Palmaria e Capo Melo. Pt. 1. Introduzione, osservazioni meteorologiche e fisico-chemiche. Centro Talassografico, Tirreno, Pubbl. No. 8:29 pp.

temperature
salinity
density

data
Tsubata, B., 1956.
Oceanographical observations off Asamushi during 1954.
Bull. Mar. Biol. Sta., 8(1):43-48.

TRANSPARENCY
temperature
chlorinity
oxygen
pH
color

data
Tsuruta, A., T. Satow, K. Hayama and T. Chiba, 1957.
[Oceanographical and planktonological studies of tuna-fishing ground, in the eastern part of the Indian Ocean.] J. Shimonoseki Coll. Fish., 7(1):1.

color
temperature
transparency
salinity

data
Uda, M., M. Watanabe and M. Ishino, 1956.
General results of the oceanographic surveys (1952-1955) on the fishing grounds in relation to the scattering layer. J. Tokyo Univ. Fish., 42(2):169-207.

temperature
density
sea
chlorinity
oxygen
swell

data
USA, National Academy of Sciences-National Research Council, Committee on Oceanography, 1960.
Oceanography, 1960-1970. 9. Ocean-wide surveys. 22 pp.

data
Uriarte, L.B., 1932.
Premiers travaux du Laboratoire Oceanographique des Iles Canaries.
Rapp. Proc. Verb., Cons. Perm. Int. Expl. Mer, 67:65-88, 8 textfigs.

temperature
density
salinity
oxygen

data
U.S. Navy Hydrographic Office, 1957.
Operation Deep-Freeze, II, 1956-1957. Oceanographic survey results. H.O. Tech. Rept., TR-29: 155 pp. (multilithed).

temperature
salinity
density
oxygen
ice
bottom sediments
sea
swell
color
transparency
deep-scattering layer
currents

data
Van Goethem, C., 1951.
Étude physique et chimique du milieu marin. Rés. Sci., Expéd. Océan Belge dans les Eaux Côtières Africaines de l'Atlantique Sud (1948-1949) 1:1-151, 1 pl.

color
transparency
temperature
salinity
oxygen
pH
silicate
nitrate
bottom sediments

data
Van Riel, P. M., 1943.
The bottom water. Introductory remarks and oxygen content. Snellius Exped. Vol. 2, Pt. 5, Ch. 1:77 pp., 34 textfigs.

data
Van Riel, P.m., H.C. Hamaker, and L. van Eyck, 1950.
Tables. Serial and bottom observations. Temperature, salinity, and density. Ocean. Res., Snellius Exped., 1929-1930:2(6):44 pp.

data
Vatova, A., and P.M. di Villagrazia, 1950.
Sulle condizioni idrografiche de Canal di Leme in Istria. Nova Thalassia 1(8):24 pp., graphs.

temperature
salinity
transparency
oxygen
pH

data
Vercelli, F., and M. Picotti, 1926.
Pt. 2. Il regime fisico-chimico della acque nello Stretto di Messina. Crociere per lo studio dei fenomeni nello Stretto di Messina. (Campagne della R. Nave "Marsigli" negli anni 1922 e 1923).
Comm. Int. Medit., Delag. Ital., 160 pp., 40 figs

data
von Bonde, C., 1950.
Twentieth annual report for the year ended December 1948. Commerce and Industry, April 1950 412 pp., 12 charts.

data
Watanabe, Nobuo, Toshiyuki Hirano, Rinnosuke Fukai, Kozi Matsumoto and Fumiko Shiokawa, 1957.
A preliminary report on the oceanographic survey in the "Kuroshio" area, south of Honshu, June-July 1955.
Rec. Oceanogr. Wks., Japan, (Suppl.):197-

data
Wojnicz, B., 1957.
Deversement d'eaux salées dans la Baltique observé au mois de novembre 1951. Przeglad Geofiz., Rocznik 2(10)(Zeszyt 1/2):53-58.

temperature
salinity
oxygen

data
Worthington, L. V., 1959.
Oceanographic observations.
Sci. Studies at Fletcher's Ice Island, T-3, Vol. 1, Air-Force Cambridge Res. Center, Geophys. Res. Pap., No. 31-35.
This is WHOI Ref. 53-92. It does not bear a contribution number. Data for 1954. and 1955 added.

data
Worthington, L.V., 1953.
Oceanographic results of Project SKIJUMP I and SKIJUMP II in the Polar Sea, 1951-1952.
Trans. Am. Geophys. Union 34(4):543-551, 4 textfigs.

data
Wüst, G., 1957.
1. Die Verteilung von Temperatur, Salzgehalt und Dichte. Ergebnisse einer hydrographisch-Produktions-biologischen Längsschnitts durch die Ostsee im Sommer 1956. Kieler Meeresf., 13(2):163-185.

temperature
salinity
density

data
Yamazi, I., 1952?
Plankton investigations in inlet waters along the coast of Japan. VI. The plankton of Nanao Bay.
Seto Mar. Biol. Lab. Contr. 191:309-319, 11 textfigs.

data
Yasui, Z., and M. Matuno, 1953.
On the circulation of currents in China Eastern Sea. Rec. Ocean. Wks., Japan, n.s., 1(1):28-35, 3 textfigs.

current velocity

data
Zmudzinski, L., and D. Szarejko, 1955.
Hydrographical-biological investigation on the Firth of Vistula.
Prace Morskiego Inst. Ryback., Gydni, No. 8:283-312.

wind
temperature
salinity

data
Zorell, F., 1935.
Beiträge zur Hydrographie der Deutschen Bucht auf Grund der Beobachtungen von 1920 bis 1932 ----
Arch. Deutschen Seewarte 54(1):1-69, 24 pls.

Data only

A

data only
Anderson, W.W., and J.W. Gehringer, 1959.
Physical oceanographic, biological and chemical data, South Atlantic coast of the United States, M/V Theodore N. Gill Cruise 9.
USFWS Spec. Sci. Rept., Fish., No. 313:225 pp.

data only
Anderson, W.W., and J.W. Gehringer, 1959.
Physical oceanographic, biological and chemical data South Atlantic coast of the United States, M/V Theodore N. Gill Cruise 8.
USFWS Spec. Sci. Rept., Fish., No. 303:227 pp.

data only
Anderson, William W., and Jack W. Gehringer, 1959.
Physical oceanography, biological and chemical data, South Atlantic coast of the United States, M/V Theodore N. Gill, Cruise 7.
USFWS Spec. Sci. Rept., Fish. No. 278:277 pp.

data only
Anderson, William W., and Jack W. Gehringer, 1958.
Physical oceanographic, biological and chemical data, south Atlantic coast of the United States, M/V Theodore N. Gill Cruise 6.
USFWS SPEC. SCI Rept., Fish., No. 265 :99 pp.

data only
Anderson, W.W., and J.W. Gehringer, 1958.
Physical oceanographic, biological and chemical data, South Atlantic coast of the United States, M/V Theodore N. Gill Cruise 5.
USFWS Spec. Sci. Rept., Fish., 248:220 pp.

zooplankton, quantitative
phosphate
arabinose
tyrosinose
density
nitrate
temperature
salinity
oxygen

data only
Anderson, W.W., and J.W. Gehringer, 1957.
Physical oceanographic, biological and chemical data, South Atlantic coast of the United States, M/V Theodore N. Gill Cruise 4.
U.S.F.W.S. Spec. Sci. Rept., Fish., 234:192 pp.

Total P
phosphate
nitrate
arabinose
tyrosine
temperature
salinity
density
oxygen
zooplankton, nos./c.c.
zooplankton, volumes

data only
Anderson, W.W., and J.W. Gehringer, 1957.
Physical oceanographic, biological and chemical data, South Atlantic coast of the United States, Theodore N. Gill Cruise.
USFWS Spec. Sci. Rept., Fish., No. 210:208 pp.

phosphate
nitrate
arabinose
tyrosine
temperature
salinity
density
oxygen

data only
Anderson, W.W., J.W. Gehringer and E. Cohen, 1956
Physical oceanographic, biological and chemical data, South Atlantic coast of the United States, Theodore N. Gill Cruise 2.
Spec. Sci. Rept., Fish., 198:270 pp.
zooplankton (quantitative
temperature
salinity
density
oxygen
phosphate
nitrate
tyrosine
arabinose

data only
Anderson, W.W., J.W. Gehringer and E. Cohen, 1956.
Physical oceanographic, biological and chemical data, south Atlantic coast of the United States, M/V Theodore N. Gill, Cruise 1.
USFWS Spec. Sci. Rept., Fish., No. 178:160 pp.

temperature
salinity
density
nitrate
tyrosine
oxygen
zooplankton
phosphate
arabinose

data only
Anderson, William W., Joseph E. Moore and Herbert R. Gordy, 1961.
Water temperatures off the South Atlantic coast of the United States, Theodore N. Gill cruises 1-9, 1953-1954.
U. S. Fish and Wildlife Service, Spec. Sci. Rept., Fish., No. 380:206 pp.

data only
Australia, Commonwealth Scientific and Industrial research Organization, 1964.
Oceanographical observations in the Indian Ocean in 1962, H.M.A.S. Diamantina, Cruise Dm 2/62.
Oceanogr. Cruise Rept., Div. Fish. and Oceanogr. No. 15:117 pp.

data only
Australia, Commonwealth Scientific and Industrial Research Organization, 1963.
Coastal hydrological investigations in the New South Wales tuna fishing area, 1963.
Div. Fish. and Oceanogr., Oceanogr. Sta. List, 53:81 pp.
A.D. Crooks, compiler

data only
Australia, Commonwealth Scientific and Industrial Research Organization, 1963.
Coastal investigations at Port Hacking, New South Wales, 1960.
Div. Fish. and Oceanogr., Oceanogr. Sta. List, No 52:135 pp.
A.D. Crooks, Compiler.

data only
Australia, Commonwealth Scientific and Industrial Research Organization, 1963
Oceanographical observations in the Indian Ocean in 1961, H.M.A.S. Diamantina Cruise Dm 2/61.
Oceanogr. Cruise Rept., Div. Fish. and Oceanogr., No. 9:155 pp., 14 figs.

data only
Australia, Commonwealth Scientific and Industrial Research Organization, 1963.
Oceanographical observations in the Pacific Ocean in 1961, H.M.A.S. Gascoyne, Cruise G 1/61.
Oceanogr. Cruise Rept., Div. Fish. and Oceanogr., No. 8:130 pp., 12 figs.

data only
Australia, Commonwealth Scientific and Industrial Research Organization, Division of Fisheries and Oceanography, 1963.
Oceanographical observations in the Pacific Ocean in 1960, H.M.A.S. Gascoyne, Cruise G 3/60.
Oceanographical Cruise Report, No. 6: 115 pp.

data
Australia, Commonwealth Scientific and Industrial Research Organization, Division of Fisheries and Oceanography, 1963.
Oceanographical observations in the Indian Ocean in 1960, H.M.A.S. Diamantina, Cruise Dm 2/60.
Oceanographical Cruise Report No. 3: 347 pp.

data only
Australia, Commonwealth Scientific and Industrial Research Organization, Division of Fisheries and Oceanography, 1962.
Oceanographic observations in the Pacific Ocean in 1960, H.M.A.S. Gascoyne, Cruises G 1/60 and G 2/60.
Oceanographical Cruise Report No. 5:255 pp.

data only
Australia, Commonwealth Scientific and Industrial Organization, 1962.
Oceanographical observations in the Indian Ocean in 1960, H.M.A.S. Diamantina, Cruise Dm 1/60.
Oceanogr. Cruise Rept., Div. Fish. and Oceanogr. No. 2:128 pp.

data only
Australia, Commonwealth Scientific and Industrial Organization, 1962.
Oceanographical observations in the Indian Ocean in 1960, H.M.A.S. DIAMANTINA Cruise Dm 3/60.
Oceanogr. Cruise Rept., Div. Fish. and Oceanogr., No. 4:39 pp.

data only
Australia, Commonwealth Scientific and Industrial Organization, 1962.
Oceanographical observations in the Indian Ocean in 1959, H.M.A.S. Diamantina, Cruises Dm 1/59 and Dm 2/59.
Oceanogr. Cruise Rept., Div. Fish. and Oceanogr., No. 1:134 pp.

data only
Australia, Commonwealth Scientific and Industrial Organization, 1962.
Surface sampling in the Coral and Tasman seas, 1960.
Oceanogr. Sta. List, Div. Fish. and Oceanogr., 50:172 pp.

data only
Australia, Commonwealth Scientific and Industrial Research Organization, 1961
Coastal hydrological sampling at Rottnest Island, W.A., and Port Moresby, Papua, during the I.G.Y. (1957-58), and surface sampling in the Tasman and Coral seas, 1959.
Div. Fish. & Oceanogr., Oceanogr. Sta. List, 48:239 pp.

data only
Australia, Commonwealth Scientific and Industrial Research Organization, 1961
Coastal hydrological investigations in the New South Wales tuna fishing area, 1959.
Div. Fish. & Oceanogr., Oceanogr. Sta. List, 46: 132 pp.

data only
Australia, Commonwealth Scientific and Industrial Research Organization, 1961
Coastal investigations at Port Hacking, New South Wales, 1959.
Div. Fish. & Oceanogr., Oceanogr. Sta. List, No. 47: 135 pp.

data only
Australia, Commonwealth Scientific and Industrial Research Organization, 1961.
Oceanic investigations in Eastern Australian waters, F.R.V. "Derwent Hunter", 1959.
Div. Fish. and Oceanogr., Oceanogr. Sta. List, 48:84 pp.

data only
Australia, Commonwealth Scientific and Industrial Research Organization, 1960
Coastal hydrological investigations in eastern Australia, 1959.
Div. Fish. and Oceanogr., Oceanogr. Sta. List, 45: 24 pp.

data only
Australia, Commonwealth Scientific and Industrial Organization, 1960
Coastal hydrological investigations in south-eastern Australia, 1958. Oceanogr. Sta. List., Div. Fish. and Oceanogr., 60 pp.

data only
Australia, Commonwealth Scientific and Industrial Research Organization, 1960.
Coastal investigations at Port Hacking, New South Wales, 1958.
Oceanogr. Sta. List, Div. Fish. & Oceanogr., 42: 99 pp.

data only
Australia, Commonwealth Scientific and Industrial Research Organization, 1960
Oceanic observations in Antarctic waters, M.V. Magga Dan, 1959.
Div. Fish. and Oceanogr., Oceanogr. Sta. List, 44: 78 pp.

data only
Australia, Commonwealth Scientific and Industrial Research Organization, 1960.
Oceanic investigations in eastern Australia, H.M.A. Ships Queenborough, Quickmatch and Warrego, 1958.
Div. Fish. and Oceanogr., Oceanogr. Sta. List, No. 43:57 pp.

data only
Australia, Commonwealth Scientific and Industrial Research Organization, 1960
Oceanic investigations in eastern Australian waters, F.R.V. Derwent Hunter, 1958. Oceanogr. Sta. List., Div. Fish. and Oceanogr. No. 41: 232 pp.

data only
Australia, Commonwealth Scientific and Industrial Research Organization, 1960
Surface sampling in the Coral and Tasman Sea, 1958. Oceanogr. Sta. List, Div. Fish. and Oceanogr., No. 39: 276 pp.

data only
Australia, C.S.I.R.O., Division of Fisheries, 1959
Coastal hydrological investigations at Eden, New South Wales, 1957.
Oceanogr. Sta. List, 35: 36 pp.

data only
Australia, C.S.I.R.O., Division of Fisheries, 1959
Coastal hydrological investigations at Port Hacking, New South Wales, 1957.
Oceanogr. Sta. List, 34:72 pp.

data only
Australia, Division of Fisheries and Oceanography, C.S.I.R.O., 1959.
Coastal hydrological investigations in the New South Wales tuna fishing area, 1958.
Oceanogr. Sta. List, 38:96 pp.

data only
Australia, C.S.I.R.O., Division of Fisheries and Oceanography, 1959.
Hydrological investigations from F.R.V. Derwent Hunter, 1957.
Oceanogr. Sta. List, No. 37:96 pp.
(compiled by D. J. Rochford)

data only
Australia, C.S.I.R.O., Division of Fisheries, 1959.
Surface sampling in the Tasman and Coral Seas, 1957.
Oceanogr. Sta. List, 36:175 pp.

data only
Australia, Commonwealth Scientific and Industrial Research Organization, 1957.
Estuarine hydrological investigations in eastern and south-western Australia, 1956.
Oceanogr. Sta. List, Div. Fish. and Oceanogr., 32:170 pp.

temperature, chlorinity, oxygen, pH, phosphate, nitrate

data only
Australia, Commonwealth Scientific and Industrial Research Organization, 1957.
Estuarine hydrological investigations in eastern and south-western Australia, 1955.
Ocean. Station List, 29:93 pp.

temperature, chlorinity, oxygen, nitrate, pH, phosphate

data only

Australia, Commonwealth Scientific and Industrial-1 and Research Organization, 1957.
Onshore and oceanic hydrological investigations in eastern and southwestern Australia, 1955.
Oceanographical Station List, 27:145 pp.

temperature, chlorinity, density, oxygen, phosphate, nitrate, pH

data only

Australia, Commonwealth Scientific and Industrial Research Organization, 1957.
Surface sampling in the Tasman and Coral Seas, 1955. Oceanogr. Sta. List, 28:88 pp.

surface temperature, surface chlorinity, surface salinity

data only

Australia, Commonwealth Scientific and Industrial Research Organization, 1956.
Onshore hydrological investigations in eastern and south-western Australia, 1954.
D.R. Rochford, Compiler.
Ocean. Sta. List., Invest. Div. Fish., 24:119 pp.

temperature, chlorinity, density, oxygen, pH, phosphate, nitrate

data only

Australia, Commonwealth Scientific and Industrial Research Organization, 1956.
Surface sampling in the Tasman Sea, 1954.
Compiled by D.R. Rochford. Ocean. Sta. List., Invest Div. Fish., 25:79 pp.

temperature, surface; salinity, surface; chlorinity, surface

data only

Australia, Commonwealth Scientific and Industrial Research Organization, 1954.
Onshore hydrological investigations in eastern and south-western Australia, 1953.
Ocean. Sta. List., Invest. Div. Fish., 18:64 pp.

temperature, chlorinity, density, oxygen, phosphate, nitrate

data only

Australia, Commonwealth Scientific and Industrial Research Organization, 1953.
Onshore hydrological investigations in eastern and south-western Australia, 1951. Vol. 14:64 pp.

temperature, chlorinity, density, oxygen, pH, Phosphate, nitrate

data only

Australia, Commonwealth Scientific and Industrial Research Organization, 1952.
Records of twenty-four hourly hydrological observations at Shell Point, Georges River, New South Wales, 1942-50. Compiled by D.R. Rochford.
Ocean. Sta. List, Invest. Fish. Div., 10:134 pp.

temperature, chlorinity, oxygen, phosphate, nitrate

data only

Australia, Commonwealth Scientific and Industrial Research Organization, Melbourne, 1952.
Oceanographical station list of investigations made by the Division of Fisheries, C.S.I.R.O.
Vol. 5. Estuarine hydrological investigations in eastern Australia 1940-50:150 pp.
Vol. 6. Ibid.:138 pp.
Vol. 7: Ibid. 139 pp.
Vol. 8. Hydrological investigations in south-western Australia, 1944-50:152 pp.
Vol. 10. Records of twenty-fourly hourly hydrological observations at Shell Point, Geogges River, New South Wales, 1942-50:134 pp.

data only

Australia, Commonwealth Scientific and Industrial Research Organization, 1951.
Oceanographic station list of investigations made by the Division of Fisheries, C.S.I.R.O.
Vol. 1. Hydrological and planktological observations by F.R.V. Warreen in south-eastern Australian waters, 1938-39;109 pp.
Vol. 2. Ibid., 1940-42:133 pp.
Vol. 3. Hydrological and planktological observations by F.R.V. Warreen in south-western Australian waters, 1947-50:63 pp.
Vol. 4. Onshore hydrological investigations in eastern Australia, 1942-50:114 pp.

temperature, density, volumetric salinity, oxygen (Vol. 4), pH (Vol. 4)

B

data only

Berrit, G.R., 1962.
Resultats d'observations. Campagne No. 11, Campagne JONAS.
Cahiers Oceanogr., C.C.O.E.C., 14(1):54-76.

data only

Berrit, G.R., M. Rossignol and J.P. Troadec, 1961.
Résultats d'observations, Année 1959, Centre d'Océanographie de Pointe-Noire, Campagnes Nos. 7,8,9, et 10 de l'"Ombango".
Cahiers Océanogr., C.C.O.E.C., 13(5):319-354.

data only

Blackburn, M., R.C. Griffiths, R.W. Homes and W.H. Thomas, 1962.
Physical, chemical and biological observations in the eastern tropical Pacific Ocean: three cruises to the Gulf of Tehuantepec, 1958-1959.
U.S.F.W.S. Spec. Sci. Rept., Fish., No. 420:170 pp.

data only

Brasil, Marinha do Brasil, Diretoria de Hidrografia e Navegação, 1963
Operação "TRIDENTE I", Estudo das condições oceanográficas entre o Rio de Janeiro e o Rio da Prata, durante o inverno (Agosto-Septembro), ano de 1962.
DG-06-XV: unnumbered pp. (mimeographed).

data only

Brasil, Diretoria de Hidrografia e Navegação, Marinha do Brasil, 1962
Estudo das condições oceanográficas entre o Rio de Janeiro e o Rio da Prata, durante o outono (Maio de 1962).
DG-06-XIV:unnumbered pp., charts, (mimeographed

data only

Brasil, Diretoria de Hidrografia e Navegacao, 1961
Estudo das condições oceanográficas nas proximidades do Rio de Janeiro durante o mês de Dezembro.
DG-06-XIII:mimeographed sheets

data only

Brasil, Marinha do Brasil, Diretoria de Hidrografia e Navegacao, 1960.
Estudo das condicoes oceanograficas na regiao profunda a NorNordeste de Natal, Estado do Rio Grande do Norte. DG-06-XI(Sept. 1960): unnumbered pp., (Mimeographed).

data only

Brasil, Marinha do Brasil, Diretoria de Hidrografia e Navegação, 1960.
Estudo das condições oceanograficas sobre a plataforma continental, entre Cabo-Frio e Vitoria, durante o outono (abril-maio).
DG-06-X(Junho):unnumbered pp. (mimeographed).

data only

Brasil, Marinha do Brasil, Diretoria de Hidrografia e Navegacao, 1959.
Levantamento oceanografico da Costa nordeste.
DG-06-IX:unnumbered pp. (mimeographed).

data only

Brazil, Diretoria de Hidrografia e Navegacao, 1958.
Ano Geofisico Internacional, Publ. DG-06-VII: mimeographed sheets.

dynamic heights, temperature, salinity, density, pH, oxygen, phosphate, nitrate

data only

Brazil, Diretoria de Hidgrografia e Navegacao, 1957.
Ano Geofisico Internacional, Publicacao DG-06-VI: unnumbered mimeographed pages.

temperature, salinity, density, oxygen, phosphate, pH

data only

Brasil, Diretoria de Hidrografia e Navegacao, 1957.
Ano Geofisico Internacional. Publicacao DG-06-V: 3 pp., 18 figs., 3 pp. data tables.

temperature, salinity, density, pH, oxygen, phosphate

data only

Brazil, Diretoria de Hidrografia e Navegacao, 1957.
Ano Geofisico Internacional. Publ. DG-06-IIIV (mimeographed)

dynamic anomalies, temperature, T-S, salinity, oxygen-density, density, phosphate-density, oxygen, pH, phosphate

data only

Brazil, Diretoria de Hidrografia e Navigacão, 1957.
Ano Geofisico Internacional. Publicacão DG-06-II mimeographed.
reference room

temperature, salinity, density, oxygen, phosphate, nitrate

data only

Buljan, M., and Marinkovic, M., 1956.
Some data on hydrography of the Adriatic (1946-1951). Acta Adriatica 7(12):1-55.

temperature, chlorinity, salinity, oxygen, phosphate, silicate

data only

Bumpus, Dean F., 1961
Drift bottle records for the Gulf of Maine, Georges Bank and the Bay of Fundy, 1956-1958.
U.S.F.W.S. Spec. Sci. Repts., Fish., No. 378: 127 pp.

C

data only

Callaway, Richard J., and James W. McGary, 1959.
Northeastern Pacific albacore survey. 2. Oceanographic and meteorological observations.
USFWS Spec. Sci. Rept., Fish., No. 315:133 pp.

data only

Canada, Canadian Oceanographic Data Center, 1964
ICNAF Norwestlant-2 Survey, Canada.
1964 Data Record Series, No. 14:185 pp. (multilithed)

data only

Canada, Canadian Oceanographic Data Center, 1964
Ocean weather Station, "P", North Pacific Ocean.
1964 Data Record Series, No. 15:95 pp. (multilithed).

data only

Canada, Department of Mines and Technical surveys, 1960.
Tidal and oceanographic survey of Hudson Strait, August and September, 1959. Data Record, 55 pp. (multilithed.)

data only

Canada, Fisheries Research Board of Canada, 1959.
Bathythermograms and meteorological data record. Swiftsure Bank and Umatilla Reef Lightships 1958.
Mss. Rept. Ser. (Oceanogr. Limnol.) No. 37:121 pp.

data only
Canada, Fisheries Research Board, 1959.
Data record, Ocean Weather Station "P" (Latitude 50 00' N., Longitude 145 00'W) January 22 - July 11, 1958.
MSS Rept. Ser., (Oceanogr. Limnol.), No. 31: 112 pp.

data only
Canada, Fisheries Research Board, 1959.
Physical and chemical data record, coastal seaways project, November 12 to December 5, 1958.
MSS Rept. Ser. (Oceanogr. & Limnol.), No. 36: 120 pp.

data only
Canada, Pacific Oceanographic Group, Nanaimo, 1961
Data record of oceanographic observations and bathythermograms observed in the northeast Pacific Ocean by Fisheries Research Board chartered fishing vessels, May to July, 1960.
Fish. Res. Bd., Canada, Manuscript Rept. Ser. (Oceanogr. and Limnol.), No. 87: 30 pp.

data only
Canada, University of British Columbia, Institute of Oceanography, 1965.
British Columbia and Alaska inlet cruises, 1964.
Data Rept., No. 24:34 pp. (multilithed).

data only
Canada, University of British Columbia, Institute of Oceanography, 1964
Data Report 23, British Columbia Inlet Cruises, 1963:102 pp. (mimeographed).

data only
Canada, University of British Columbia, Institute of Oceanography, 1963.
Inlet cruises, 1962.
Data Report, 21:90 pp. (multilithed).

data only
Canada, University of British Columbia, Institute of Oceanography, 1956.
British Columbia Inlet Cruise, 1956. Data Rept. No. 8:33 pp., (Mimeographed)

temperature - oxygen
salinity density

data only
Canada, University of British Columbia, Institute of Oceanography, 1956.
Queen Charlotte Strait, 1956. Data Rept. No. 9: 20 pp. (mimeographed).

temperature oxygen
salinity phosphate

data only
Capurro, Luis R.A., compiler, 1961
Oceanographic observations in the intertropical region of the world during the IGY and IGC. Part IIa: Pacific Ocean (France, Japan and Union of Soviet Socialist Republics).
IGY World Data Center A: Oceanography, IGY Oceanogr. Rept., No. 3: 225 pp.

data only
Capurro, Luis R.A., compiler, 1961.
Oceanographic observations in the intertropical region of the World Ocean during IGY and IGC. Part IIb: Pacific Ocean (United States of America).
IGY World Data Center A, Oceanography, IGY Oceanogr. Rept., No. 3:298 pp.

Data Only
Chesapeake Bay Institute, 1963.
Bay Cruise XXIX, March 11-19, 1963. Bay Cruise XXX, July 30-September 11, 1963. Bay Cruise XXXI, October 21-24, 1963.
Data Rept. 50, Ref. 63-3:31 pp. (unpublished manuscript).

data only)
Chesapeake Bay Institute, 1962.
Bay Cruise 28, July 24-August 7, 1962.
Data Rept. 48, Ref. 62-16:17 pp. (unpublished manuscript).

data only
Chesapeake Bay Institute, Johns Hopkins University, 1959.
Data Report 31. Baltimore Harbor Cruises 1-16. July 1958-Nov. 1959. 211 pp. Ref. No. 59-4.

data only
Chesapeake Bay Institute, 1954.
Data report 24. Choptank River winter cruise, 7 December to 10 December 1952. Ref. 54-11:1-37, 1 fig.

oxygen
pH
chlorophyll
turbidity
currents
temperature
salinity

data (only)
Chesapeake Bay Institute, 1954.
Data Report 23, Patuxent River Winter Cruise, 3 December-7 December 1952. Ref. 54-10:1-44, 1 textfig.

oxygen
pH
chlorophyll
turbidity
temperature
salinity
currents

data only
Chesapeake Bay Institute, 1954.
Choptank River autumn cruise, 23 Sept-27 Sept 1952. Data Rept. 22, Ref. 54-9:37 pp., 1 fig.

temperature
salinity
current
oxygen
pH
chlorophyll
phosphate
turbidity

data only
Chesapeake Bay Institute, 1954.
Choptank River Spring cruise, 28 April-1 May 1952. Ref. 54-1:38 pp.

oxygen
pH
chlorophyll
phosphate
alkalinity
temperature
salinity
currents
turbidity

data only
Chesapeake Bay Institute, 1952.
Data Report, St. Mary's River cruise, June 19-July 18, 1951. Rept. 11, Ref. 52019:115 pp.

data only
Chesapeake Bay Institute, 1951.
Data report, Cruise VII, 14 Oct. 1950-2 Nov. 1950. Rept. No. 5:41 pp. (duplicated).

data only
Chesapeake Bay Institute, 1951.
Data report, Cruise VIII, Jan. 10, 1951-Jan. 23, 1951. Rept. No. 6:29 pp. (duplicated).

data only
Chesapeake Bay Institute, 1951.
Data report. Cruise VII. October 14, 1950 -November 2, 1950. Rept. No. 2:41 pp. (duplicated).

data only
Chesapeake Bay Institute, 1950.
Data Report, Cruises V and VI, May 20, 1950-May 25, 1950, July 14, 1950-July 19, 1950. Rept. No. 4:51 pp. (mimeographed).

data only
Chesapeake Bay Institute, 1950.
Data Report. Cruise IV, March 25, 1950-April 25, 1950. Rept. No. 3: 49 pp. (mimeographed).

data only
Chesapeake Bay Institute, 1949.
Data Report, Cruise III. October 10, 1949-October 25, 1949. Rept. No. 2:39 pp. (mimeographed).

data only
Chesapeake Bay Institute, 1949.
Quarterly report, July 1, 1949-October 1, 1949. Rept. 1:121 pp. (multilith).

data only
Collier, Albert, 1958.
Gulf of Mexico physical and chemical data from ALASKA Cruises.
USFWS Spec. Sci. Rept., Fish., No. 249:417 pp.

data only
Conseil Permanent International pour l'Exploration de la Mer, Service Hydrographique, 1963.
ICES oceanographic data lists, 1958(1):1-259

data only
Conseil Permanent International pour l'Exploration de la Mer, 1957.
Bulletin Hydrographique pour l'Année 1953:167 pp.

temperature
oxygen
alkalinity
Total P
silicate
salinity
pH
phosphate

data only
Conseil Permanent International pour l'Exploration de la Mer, 1956.
Bulletin Hydrographique pour l'année 1952:164 pp.

temperature
oxygen
nitrate
nitrite
salinity
phosphate
silicate

data only
Conseil Permanent International pour l'Exploration de la Mer, 1955.
Bulletin hydrographique pour l'annèe 1951, 131 pp.
phosphate
oxygen
temperature
salinity

data only
Conseil Permanent International pour l'Exploration de la Mer, 1954.
Bulletin Hydrographique pour l'Année 1950: 114 pp., 5 charts.

salinity
temperature
oxygen
pH
Total-P
Phosphate-P
Organic P
Nitrate-N
silicate-Si.

data only
Conseil Permanent International pour l'Exploration de la Mer, 1954.
Bulletin hydrographique pour l'Annee 1949, 85 pp., 4 charts.

temperature
salinity
oxygen
phosphate
silicate

data only
Conseil Permanent International pour l'Exploration de la Mer, 1952.
Bulletin Hydrographique, 1948:87 pp.

temperature
salinity
phosphate
oxygen
nitrate
nitrite
silicate

data only
Conseil Permanent International pour l'Exploration de la Mer, 1950.
Bulletin hydrographique pour les années 1940-1946 avec un appendix pour les années 1936-1939. 190pp.

data only
Conseil Permanent international pour l'Exploration de la Mer, 1950.
Bulletin Hydrographiqué, 1947:49 pp., 4 charts.

data only
Conseil Permanent International pour l'Exploration de la Mer, 1939
Bulletin Hydrographique pour l'annee 1937: 106 pp., 4 maps., 4 figs.

data only
Conseil Permanent International pour l'Exploration de la Mer, 1936.
Bulletin Hydrographique pour l'annee 1935: 105 pp., 4 maps, 4 figs.

data only
Bulletin Hydrographique pour l'année 1933 1944
Cons. perm. int. l'explor. de la mer. 1934 1935

data only
Crean, P.B., W.R. Harling, R.B. Tripp, F.W. Dobson, J.H. Meikle, and H.J. Hollister, 1962
Oceanographic data record, MONITOR project, July 24 to November 16, 1961.
Fish. Res. Bd., Canada, Mss. Rept. Ser. (Oceanogr. & Limnol.), No. 111: 409 pp. (multilithed).

data only
Crean, P.B., R.B. Tripp and H.J. Hollister, 1962.
Oceanographic data record, Monitor project, March 12 to April 5, 1962.
Fish. Res. Bd., Canada, Mss. Rept. Ser. (Ocean. and Limnol.), No. 129:210 pp. (multilithed).

data only
Crean, P.B., R.B. Tripp and H.J. Hollister, 1962
Oceanographic data record, MONITOR project, January 15 to February 5, 1962.
Fish. Res. Bd., Canada, MSS Rept. Ser. (Oceanogr. and Limnol.), No. 113: 169 pp. (multilithed).

D

data only
Day, C. Godfrey, 1960
Oceanographic observations, 1959, east coast of the United States. USFWS Spec. Sci. Rept., Fish. No. 359:114 pp.

data only
Day, C. Godfrey, 1959.
Oceanographic observations, 1958, east coast of the United States.
USFWS Spec. Sci. Rept., Fish. No. 318:119 pp.

data only
Day, C.G., 1959
Oceanographic observations, 1957, East coast of the United States. U.S. Fish Wild Serv.-Sp. Sci. Rept. Fish. No. 282, 1-123.

data only
Denmark, Det Danske Meteorologiske Institut, 1964.
Oceanografiske observationer fra Danske fyrskibe og kyststationer, 1963.
Publikationer, Danske Meteorol. Inst., Arbøger, 171. pp.

data only
Denmark, Danske Meteorologiske Institut, 1962.
Nautisk-meteorologisk årbog. Nautical-meteorological annual, 1961, 168 pp.

Data only
Denmark, Danske Meteorologiske Institut, 1961.
Nautisk-Meteorologisk Arbog, 1960:171 pp.

data only
Denmark, Det Danske Meteorologiske Institut, 1960
Nautisk-Meteorologisk Arbog, 1959:171 pp., charts.

data only
Denmark, Det Danske Meteorologiske Institut, 1959.
Nautisk-Meteorologisk Arbog, 1958: 169 pp.

data only
Denmark, Danske Meteorologiske Institut, 1958.
Nautisk-meteorologisk årbog, 1957:170 pp.

data only
Denmark, Danske Meteorologiske Institut, 1957.
Nautisk Meteorologisk Aarbog, 1956:170 pp., chart

temperature
salinity
currents

data only
Denmark, Denmarks Meteorologiske Institut, 1956.
Nautisk-Meteorologisk Aarbog, 1955:164 pp., 12 charts

currents
salinity
temperature

data only
Denmark, Danske Meteorologiske Institut, 1955.
Nautisk-meteorologisk Årbog, 1954:171 pp.

temperature
salinity
currents(mean velocity)

data only
Denmark, Danske Meteorologiske Institut, 1953.
Nautisk-Meteorologisk Aarbog, 1952:171 pp., 12 charts.

currents
salinity
temperature
oxygen
charts of surface temperature for N. Atlantic

data(currents)
Denmark, Kommission for Havundersøgelser, 1923.
Current measurements from Danish Lightships.
Medd. Komm. Havundersøgelser, Ser. Hydrogr., 2(8):78 pp.

data only
Deutsches Hydrographisches Institut, 1958
Beobachtungen auf den deutschen Feuerschiffen den Nord u. Ostsee im jahre 1956. So wie Monatsmittelwerte u. Temperatur u. Saltzgehalt der Jahre 1948-1956. Meereskundliche Beobachtungen u. Ergebnisse No. 10.

data only
Deutsches Hydrographisches Institut, 1953.
Beobachtungen auf den deutschen Feuerschiffen der Nord- und Ostsee im Jahre 1949.
Meeresk. Beobacht. u. Ergeb. No. 2(2122/2): numerous pp. (unnumbered).

currents
sea
temperature
salinity

data only
Deutsches Hydrographisches Institut, 1954.
Beobachtungen auf den deutschen Feuerschiffen der Nord- und Ostsee im Jahre 1952.
Meereskundl. Beobacht. u. Ergeb. No. 5(2122/5): numerous unnumbered pages, 1 fig.

sea
currents, tidal
temperature
salinity

data only
Deutsches Hydrographisches Institut, 1954.
Beobachtungen auf den deutschen Feuerschiffen der Nord- und Ostsee im Jahre 1951.
Meeresk. Beobacht. u. Ergeb. No. 4(2122/4): numerous pp. (unnumbered).

currents
sea
temperature
salinity

data only
Deutsches Hydrographisches Institut, 1953.
Beobachtungen auf den deutschen Feuerschiffen der Nord- und Ostsee im Jahre 1950.
Meeresk. Beobacht. u. Ergeb. No. 3(2122/3): numerous pp. (unnumbered).

currents
sea
temperature
salinity

data only
Discovery Committee, 1953.
Station list, R.R.S. "William Scoresby", 1950.
Discovery Repts. 26:211-258, Pls. 11-12.

temperature
salinity
phosphorus (inorganic & total)
oxygen
sea
swell

data only
Dodimead, A.J., F.W. Dobson, N.K. Chippindale, and H.J. Hollister, 1962
Oceanographic data record, North Pacific Survey, May 23 to July 5, 1962.
Fish. Res. Bd., Canada, Mss. Rept. Ser. (Ocean and Limnol.), No. 138: 383 pp.

data only
Dragovich, Alexander, John H. Finucane and Billie Z. May, 1961.
Counts of red tide organisms, Gymnodinium breve, and associated oceanographic data from Florida west coast, 1957-1959.
USFWS Spec. Sci. Rept., Fish., No. 369:175 pp.

E

data only
Einarsson, H., and J. Jonsson, 1952.
Investigations at sea during the years 1947 and 1948 with a list of stations (Station 1-1012).
Fjölrit Fiskideildar, No. 1:1-69, 12 charts.

temperature
salinity

F

data only
Favorite Felix Richard J. Callaway and James F. Hebard 1961
North Pacific and Bering Sea oceanography 1959.
U.S.F.W.S. Spec. Sci. Rept., Fish., No. 377: 212 pp.

data only
Favorite, Felix, and Glenn Pedersen, 1959.
North Pacific and Bering Sea oceanography, 1958.
USFWS Spec. Sci. Rept., Fish., No. 312:230 pp.

data only
Favorite, Felix, and Glenn M. Pedersen, 1959.
North Pacific and Bering Sea oceanography, 1957.
USFWS Spec. Sci. Rept., Fish., No. 292:106 pp.

data only
Finland, Havforskningsinstitutet, 1962
Temperature and salinity at the fixed Finnish stations, 1957-1959.
Havsforskningsinst. Skrift (Merent. Julk.), No. 201: 135 pp.

data only
Finucane, John H., and Alexander Dragovich 1959.
Counts of red tide organisms, Gymnodinium brevis and associated oceanographic data from Florida west coast 1954-1957.
USFWS Spec. Sci. Rept. Fish. No. 289: 200 pp.

data only
France, Centre de Recherches et d'Études Océanographiques, 1960.
Stations hydrologiques effectuées par le "Passeur du Printemps" dans le cadre de travaux de l'Année Géophysique Internationale.
Travaux, C.R.E.O., 3(4):17-22.

data only
France, Service Hydrographique de la Marine, 1959.
Resultats d'observations.
Cahiers Océanogr., C.C.O.E.C., 11(7):555-566.

temperature
salinity
density

data only
France, Service Central Hydrographique 1959.
Resultats d'observations océanographiques.
1. Observations hydrologiques de la station AGI de Madagascar.
Cahiers Océanogr., C.C.O.E.C., 11(5):323-331.

data only
France, Service Hydrographique de la Marine 1959.
Resultats d'observations hydrologiques.
Cah. Océanogr. 11(3):182-199.

data only
France, Service Hydrographique de la Marine, 1959
Resultats d'observations hydrologiques. 1. Batiments de la Marine Nationale. 2. Observations du "Pyrrhus". 3. Observations du niveau marin.
Cahiers Oceanogr., 11(1):54-73.

data only
France, Centre de Recherches et d'Études Oceanographiques, 1958
Observations oceanographiques des navires stationnaires meteorologiques. Trav. C.R.E.O. n.s. 3(2): 27-40.

data only)
France, Service Central Hydrographique, 1957.
Observations hydrologiques. A. Batiments de la Marine Nationale. B. Navires stationnaires météorologiques. Bull. d'Info., C.C.O.E.C., 9(9): 521-530.

salinity
density
temperature

data only
France, Service Central Hydrographique, 1957.
A. Observations hydrologiques. B. Flotteurs témoins de courant. Bull. d'Info., C.C.O.E.C., 9(6):342-348.

drift bottles
temperature, surface
salinity, surface

data only
France, Service Central Hydrographique, 1957.
Observations hydrologiques des bâtiments de la Marine. Bull. D'Info., C.C.O.E.C., 9(5):291-298.

temperature, surface
salinity, surface

data only
France, Comité Central d'Océanographie et d'Étude des Côtes, 1960.
Stations hydrologiques effectuées par le "Leon Coursin" batiment de la Mission Hydrographique des Travaux Publics d'A.O.F.
Cahier Océanogr., 12(7):493-507. (C.C.O.E.C.)

data only
France, Comité Central d'Oceanographie et d'Etude des Cotes, 1960
Stations hydrologiques effectivées par le "Léon Coursin", bâtiment de la mission hydrographique des Travaux Publics d'A.O.F.
Cahiers Oceanographiques 12(5): 345-356.

data only
France, Comité Central d'Océanographie et d'Étude des Côtes, 1956.
Observations hydrologiques des bâtiments de la Marine Nationale. A. Croiseur-École "Jeanne d'Arc". B. Transport-Petrolier "Var".
Bull. d'Info., C.C.O.E.C. 8(1):51-57.

temperature
salinity

data only
France, Comité Central d'Océanographie et d'Étude des Cotes, 1955.
A. Observations hydrologiques du Petrolier "Ashtarak". B. Observations hydrologiques des bâtiments de la Marine Nationale. C. Flotteurs temoins de courant. Bull. d'Info., C.C.O.E.C., 7(9):430-436.

temperature(surface)
salinity(surface)

data only
France, Comité Central d'Océanographie et d'Étude des Côtes, 1955.
Observations hydrologiques effectuées par les navires stationnaires météorologiques au point K en 1953-1954. Bull. d'Info., C.C.O.E.C. 7(4):188-190.

temperature
salinity
density

data only
France, Institut Francais d'Oceanie, Noumea, New Caledonie, 1958
Orsom III, Croisiere "Astrolabe", Océanographique Physique. Rapp. Sci. No. 8:79 pp. (mimeograph).

data only
France, Service Central Hydrographique, 1959.
1. Observations des bâtiments de la Marine Nationale.
Bull. d'Info., C.C.O.E.C., 11(8):509-514.

data only
France, Service Central Hydrographique, 1958
Observations hydrologiques des batiments de la Marine Nationale.
Bull. d'Info., C.C.O.E.C., 10(10):751-771.

data only
France, Service Central Hydrographique, 1958.
1. Observations hydrologiques des bâtiments de la Marine Nationale.
Bull. d'Informa., C.C.O.E.C., 10(8):509-514.

data only
France, Service Central Hydrographique, 1958.
Observations hydrographiques.
Bull. d'Info., C.C.O.E.C., 10(6):367-373.

salinity
density
temperature

data only
France, Service Central Hydrographique, 1958.
Observations hydrologiques des bâtiments. A. Marine Nationale. B. Navire Meteorologiques Stationnaires. C. Marine Marchande.
Bull. d'Info., C.C.O.E.C., 10(5):293-300.

salinity
temperature
density

data only
France, Service Central Hydrographique, 1958.
Observations hydrologiques des bâtiments de la Marine nationale.
Bull. d'Info., C.C.O.E.C., 10(2):100-108.

temperature
density
salinity

data only
France, Service Central Hydrographique, 1958.
Observations hydrologiques des batiments de la Marine Nationale. Bull. d'Info., C.C.O.E.C., 10(1):29-41.

temperature
density
salinity

data only
France, Service Central Hydrographique, 1958.
Observations océanographiques de surface.
Bull. d'Info., C.C.O.E.C., 10(3):165-177.

temperature, surface
salinity, surface

data only
France, Service Central Hydrographique, 1957.
Observations hydrologiques des bâtiments de la Marine Nationale. Bull. d'Info., C.C.O.E.C., 9(10):579-588.

temperature
salinity
density

data only
France, Service Central Hydrographique, 1957.
Observations hydrologiques des bâtiments de la Marine Nationale. Drageur Océanique "Ailètte" (mars 1955 à décembre 1955).
Bull. d'Info., C.C.O.E.C., 9(1):61-72.

temperature
density
salinity

data only
France, Service Central Hydrographique, 1956.
A. Observations hydrologiques du Bâtiment Hydrographe "Ingenieur Hydrographe Nicolas". B. Flotteurs temoins de courant.
Bull. d'Info., C.C.O.E.C. 8(3):140-144.

temperature
salinity
drift bottles.

data only
France, Service Central Hydrographique, 1956.
A. Observations hydrologiques des bâtiments de la Marine Nationale. B. Observations hydrologiques du Petrolier "Ashtarak".
Bull. d'Info., C.C.O.E.C., 8(4):195-197.

temperature, surface
salinity, surface

data only
France, Service Central Hydrographique, 1956. Observations hydrographiques du Petrolier "Ashtarak" en 1955. Bull. d'Info., C.C.O.E.C., 8(2):97-100.

temperature
salinity

data only
France, Service Hydrographique de la Marine, 1963
Resultats d'observations.
Cahiers Océanogr., C.C.O.E.C., 15(3):418-428.

data only
France, Service Hydrographique de la Marine, 1963.
Resultats d'observations.
Cahiers Oceanogr., C.C.O.E.C., 15(2):135-142.

data only
France, Service Hydrographique de la Marine, 1962
Resultats d'Observations. 1. Navire meteorologique stationnaire "France II". II. Flotteurs-temoins de courant. III. Observations du Niveau marin à la station littorale de l.A.G.I. à Matavai (Tahiti).
Cahiers Oceanogr., C.C.O.E.C., 14(8):604-609.

data only
France, Service Hydrographique de la Marine, 1962
Résultats d'observations.
Cahiers Océanogr., C.C.O.E.C., 14(7):498-512.

data only
France, Service Hydrographique de la Marine, 1962
Observations hydrologiques du bâtiment oceanographe "Origny", Campagne Internationale à Gibraltar (15 mai - 15 juin 1961).
Cahiers Océanogr., C.C.O.E.C., 14(5):340-375.

data only
France, Service Hydrographique de la Marine, 1962
Résultats d'observations.
Cahiers Oceanogr., C.C.O.E.C., 14(4):274-283.

data only
France, Service Hydrographique de la Marine, 1961
Resultats d'Observations.
Cahiers Ocean., C.C.O.E.C., 13(10):747-767.

data only
France, Service Hydrographique de la Marine, 1961
Resultats d'observations. Bâtiments de la Marine Nationale.
Cahiers Océanogr., C.C.O.E.C., 13(9):667-679

data only
France, Service Hydrographique de la Marine, 1961
Resultats d'observations. 1. Observations hydrologiques des bâtiments de la Marine Nationale. 2. Flotteurs-temoins de courant.
Cahiers Océanogr., C.C.O.E.C., 13(7): 498-515.

data only
France, Service Hydrographique de la Marine, 1961
Résultats d'observations. Observations hydrologiques du Navire Hydrologique "Lapérouse".
Cahiers Océanogr., C.C.O.E.C., 13(6):409-422.

data only
France, Service Hydrographique de la Marine, 1961
Resultats d'observations.
Cahiers Oceanogr., C.C.O.E.C., 13(4):250-263.

data only
France, Service Hydrographique de la Marine, 1961
Résultats d'observations. 1. Observations hydrologiques des bâtiments de la Marine National.
Cahiers Océanogr., C.C.O.E.C., 13(4):594-605.

data only
France, Service Hydrographique de la Marine, 1961
Resultats d'observations.
Cahiers Océanogr., C.C.O.E.C., 13(3): 187-201.

data only
France, Service Hydrographique de la Marine, 1961.
Resultats d'observations.
Cahiers Oceanogr., C.C.O.E.C., 13(2):116-132.

data only
France, Service Hydrographique de la Marine, 1960.
I. Observations hydrologiques des bâtiments de la Marine Nationale. II. Observations du niveau marin à la station littorale de AGI à Nosy-Be (Madagascar).
Cahiers Océanogr., C.C.O.E.C., 12(4):285-294.

data only
France, Service Hydrographique de la Marine, 1960.
Resultats d'Observations.
Cahiers Oceanogr., C.C.O.E.C., 12(1):68-71.

data only
France, Service Hydrographique de la Marine, 1960.
Resultats d'Observations hydrologiques.
Cahiers Oceanogr., 12(2):128-154.

data only
France, Service Hydrographique de la Marine, 1959.
Observations du "Pyrrhus" bâtiment de recherche de Travaux Publics d'Afrique Occidentale Française.
Cahiers Océanogr., C.C.O.E.C., 11(1):63-73.

data only
France, Service Hydrographique de la Marine, 1959
Resultats d'observations
Cahiers Ocean., C.C.O.E.C., 11(9):689-699

data only
France, Service Hydrographique de la Marine, 1959.
Resultats d'observations.
Cahiers Océanogr., C.C.O.E.C., 11(9):613-617.

data only
France, Service Hydrographique de la Marine, 1959.
Resultats d'observations océanographiques.
Cahiers Océanogr., C.C.O.E.C., 11(5):323-370.

data only
France, Service Hydrographique de la Marine, 1959
Resultats d'observations. Cahiers Ocean., C.C.O.E.C., 11(4): 253-264.

data only
France, Service Hydrographique de la Marine, 1959.
Stations hydrologiques du croiseur-école "Jeanne d'Arc" de la Marine Nationale.
Cahiers Océanogr., C.C.O.E.C., 11(1):54-62.

data only
Fuglister, F.C., Edit., 1960.
Atlantic Ocean atlas of temperature and salinity profiles and data from the International Geophysical Year of 1957-1958, Vol. 1:209 pp.

G

data only
Germany, Deutsches Hydrographisches Institut, 1956.
Beobachtungen auf den deutschen Feuerschiffen der Nord- und Ostsee in Jahre 1955.
Meereskund. Beobacht. u. Ergeb. (2122/8) No. 8: unnumbered pp.

currents
sea
temperature, surface
salinity, surface

data only
Germany, Deutsches Hydrographisches Institut, 1954.
Beobachtungen auf den deutschen Feuerschiffen der Nord- und Ostsee in Jahre 1953.
Meereskundliche Beobacht. u.Ergeb. No. 6(2122/6): numerous pp. (unnumbered).

wind
sea
currents
temperature(surface)
salinity(surface)

data only
Graham, Joseph J., and William L. Craig, 1961
Oceanographic observations made during a cooperative survey of albacore (Thunnus germo) off the North American west coast in 1959.
U.S.F.W.S. Spec. Sci. Rept., Fish., No. 386: 31 pp.

data only
Granqvist, G., 1955.
Regular observations of temperature and salinity in the seas around Finland, July 1952-June 1954.
Merent. Julk., No. 165:43 pp.

temperature
salinity

data only
Granqvist, G., 1955.
The summer cruise with M/S Aranda in the northern Baltic, 1954. Merent. Julk., No. 166:56 pp.

temperature
salinity
oxygen
pH
alkalinity
current measurements
bottom fauna
phosphates
silicates
Ca
Mg
transparency
plankton (nos./m3)

data only
Granqvist, G., 1954.
Depth temperature and salinity records along the coast of Finland, July 1950-June 1952.
Merent. Julk. No. 163:18 pp.

temperature
salinity

Density (data only)

Great Britain, Discovery Committee, 1957.
Station List 1950-51. Discovery Repts. 28: 299-398.

data only
Great Britain, Discovery Committee, 1955.
Station list, R.R.S. "William Scoresby".
Discovery Repts., 26:212-258, Pls. 11-12.

sea
temperature
density
phosphate
swell
salinity
oxygen

data only
Great Britain, Discovery Committee, 1949.
Station list, R.R.S. "William Scoresby".
Discovery Repts., 25:144-280, Pls. 34-37.

sea
temperature
density
oxygen
swell
salinity
phosphate

data only
Great Britain, Discovery Committee, 1947.
Station list, 1937-1939. Discovery Repts., 24: 198-422, Pls. 4-6.

sea
swell
temperature
density
phosphate
silicate
oxygen
salinity
pH
nitrite

data only
Great Britain, Discovery Committee, 1945.
Station list, 1935-1937. Discovery Repts. 24: 3-196, Pls. 1-3.

sea-
swell
temperature
density
phosphate
silicate
oxygen
salinity
pH
nitrite

data only
Great Britain, Discovery Committee, 1942.
Station list, 1933-1935. Discovery Rept. 22:3-196, Pls. 1-4.

temperature
salinity
density
phosphate
nitrite
nitrate
silicate
oxygen
sea

data only
Great Britain, Discovery Committee, 1949.
Station list, R.R.S. William Scoreby, 1931-1938.
Discovery Repts., 25:143-280, Pls. 34-37.

sea
temperature
density
oxygen
swell
salinity
phosphate

data only
Great Britain, 1932.
Station list. Discovery Repts. 4:3-230, Pls. 1-5.

temperature
density
pH
phosphate
nitrite.
salinity
swell
oxygen
nitrate

data only
Great Britain, Discovery Committee, 1931.
Station list. Discovery Repts. 3:3-132, Pls. 1-1[illegible].

temperature
sea
density
phosphate
nitrite
salinity
swell
pH
oxygen
nitrate

data only
Great Britain, Discovery Committee, 1929.
Station list. Discovery Repts. 1:3-140.

temperature
density
swell
phosphate
nitrate
salinity
sea
pH
oxygen
nitrite

data only
Great Britain, Fisheries Laboratory, Lowestoft, 1957.
Hydrographical observations, equipment, methods, etc., 60 pp.

data only
Great Britain, Fisheries Laboratory, Lowestoft, 1957.
Research Vessel 'Ernest Holt', hydrographical observations, 1955:88 pp.

temperature-
density
sea
salinity
oxygen
swell

data only
Great Britain, Fisheries Laboratory, Lowestoft, 1957.
Research Vessel 'Ernest Holt', hydrographical observations, 1951:unnumbered pages.

temperature
density
sea
salinity
oxygen
swell

data only
Great Britain, Fisheries Laboratory, Lowestoft, 1956.
Research vessel 'Ernest Holt'. Hydrographical observations, 192 pp. (multilithed).

temperature
density
sea
salinity
oxygen
swell

data only
Great Britain, Fisheries Laboratory, Lowestoft, 1956.
Research Vessel "Ernest Holt", Hydrographical observations: 130 pp.

temperature
salinity
oxygen
density
sea
swell

data only
Great Britain, Fisheries Laboratory, Lowestoft, 1954.
Research Vessel "Ernest Holt", hydrographical observations, 130 pp.

data only
Great Britain, Fisheries Laboratory, Lowestoft, 1950.
Research Vessel 'Ernest Holt', Hydrographical observations, 1950. 150 pp.

temperature
salinity
density
oxygen

data only
Great Britain, Ministry of Africulture, Fisheries and Food, 1953.
Fisheries Laboratory, Lowestoft, Research Vessel "Ernest Holt"; hydrographical observations.
Unnumbered pages.

temperature
density
oxygen
salinity
phosphate

data only
Great Britain, Ministry of Agriculture, Fisheries and Food, 1949.
Fisheries Laboratory, Lowestoft, Research Vessel "Ernest Holt"; hydrographical observations.
Unnumbered pages.

temperature-
density
oxygen
salinity
phosphate

H

data only
Hapgood, William, 1959.
Hydrographic observations in the Bay of Naples (Golfo di Napoli), January 1957-January 1958 (station lists).
Pubbl. Sta. Zool., Napoli, 31(2):337-371.

data only
Hela, Ilmo, and Folke Koroleff, 1958
Hydrographical and chemical data collected in 1957 on board the R/V Aranda in the Barents Sea.
Merent. Julk. No. 179: 67 pp.

data only
Hela, Ilmo, and F. Koroleff, 1958.
Hydrographical and chemical data collected in 1956 on board the R/V "Aranda" in the Baltic Sea
Merent. Julk., No. 183:1-52.

data only
Herlinveaux, R.H., 1963
Data record of oceanographic observations made in Pacific Naval Laboratory underwater sound studies, November 1961 to November 1962.
Fish. Res. Bd., Canada, Mss. Rept. Ser. (Ocean. and Limnol.), No. 146: 101 pp. (multilithed).

data only
Herlinveaux, R.H., 1963
Oceanographic observations in the Canadian Arctic Basin, Arctic Ocean, April-May, 1962.
Fish. Res. Bd., Canada, Mss. Rept. Ser. (Ocean. and Limnol.), No. 144: 25 pp. (multilithed).

data only
Herlinveaux, R.H., 1961.
Data record of oceanographic observations in Pacific Naval Laboratory underwater sound studies.
Fish. Res. Bd., Canada, Mss. Rept. Ser., (Ocean.& Limnol.), No. 108:85 pp.

data only
Hirota, Reiichiro, 1963.
Some oceanographical and meteorological conditions observed at a definite station of Mukaishima Marine Biological Station (1954-1962).
Contr. Mukashima Mar. Biol. Sta., Hiroshima Univ. 1962-1963, No. 73:107 pp.

data only
Hollister, H.J., 1962.
Observations of seawater temperature and salinity on the Pacific coast of Canada.
Fish. Res. Bd., Canada, Mss. Rept. Ser., (Ocean. and Limnol.), No. 131:85 pp.

data only
Holmes, Robert W., and Maurice Blackburn, 1960
Physical, chemical and biological observations in the eastern tropical Pacific Ocean, Scot Expedition, April-June 1958.
USFWS Spec. Sci. Rept. Fish., No. 345:106 pp/

I

data only
India, Naval Headquarters, Office of Scientific Research and Development, New Delhi, 1960.
Indian oceanographic stations list, Ser. No. 6 (Jan.): 11 pp. (mimeographed).

data only
India, Naval Headquarters, Office of Scientific Research & Development, 1958-1959.
Indian Oceanographic Stations List, Serial Nos. 3-5 (June 1958; Jan. & June 1959): unnumbered pp. (mimeographed).

data only
India, Naval Headquarters, New Delhi, 1958.
Indian oceanographic station list, Ser, No. 2: numerous pp., (mimeographed)

waves
temperature
salinity
phosphate
nitrite
oxygen
plankton (remarks only)
turbidity

data only
India, Naval Headquarters, Office of Scientific Research & Development, 1957.
Indian Oceanographic Stations List, No. 1/57: unnumbered pp. (mimeographed).

data only
Indonesia, Lembaga Penjelidikan Laut, 1957.
Oceanographic station list 1957.
Penjel. Laut, Indonesia (Mar. Res.) No. 3:57-69.

temperature
salinity
oxygen

data only
Indonesia, Institute of Marine Research, 1956.
Oceanographic station list, 1956.
Mar. Res., Indonesia, No. 2:49-55.

temperature
salinity
oxygen

data only
Ivanov, A.A., and V.K. Agenorov, Editors, 1962.
Part 1. Tables of observations on the first and second voyages of the "Mikhail Lomonosov". Results, "M. Lomonosov" in the Atlantic, 1957-1958. (In Russian).
Trudy Morsk. Gidrofiz. Inst., 21:13-448.
Scientific Research Vessel

J

data only
Japan, Central Meteorological Observatory, 1956.
The results of meteorological and oceanographic observations (NORPAC Expedition, Spec. No.) 17: 131 pp.
Marine

temperature
salinity
density
silicate
BT data
chlorinity
oxygen
phosphate
lists of species

data only
Japan, Central Meteorological Observatory, 1955.
The results of marine meteorological and oceanographical observations. 1. Oceanography, January-June 1954, No. 14:91 pp.
1. Oceanography, July-December 1954, No. 15:134 p

temperature
chlorinity
oxygen
phytoplankton (lists of species and numbers)

data only
Japan, Central Meteorological Observatory, 1955.
The results of marine meteorological and oceanographical observations, January-June 1951.
No. 9:177 pp.

temperature
oxygen
currents
chlorinity
lists of species

data only
Japan, Central Meteorological Observatory, 1954.
The results of marine meteorological and oceanographical observations. 13(1):1-210.

temperature
chlorinity
oxygen
phytoplankton, lists of species
zooplankton, lists of species

data only
Japan, Central Meteorological Observatory, 1954.
The results of marine meteorological and oceanographical observations. Part 1, Oceanography, July-December, 1952, No. 12:138 pp.

temperature
chlorinity
oxygen
phytoplankton, lists of species
zooplankton, lists of species
quantitative
currents

data only
Japan, Central Meteorological Observatory, 1952.
The Results of Marine Meteorological and oceanographical observations. Jan. - June 1950. No. 7: 220 pp.

data only
Japan, Central, Meteorological Observatory, Japan, 1950
The results of marine meteorological and oceanographical observations, Jan.-June 1947, No. 1:113 pp., 19 tables.

data only
Japan, Hokkaido University, Faculty of Fisheries, 1963
Data Record of Oceanographic observations and exploratory fishing, No. 7:262 pp.

data only
Japan, Hokkaido University, Faculty of Fisheries 1961.
Data record of oceanographic observations and exploratory fishing. No. 5: 391 pp.

data only
Japan, Hokkaido University, Faculty of Fisheries, 1960.
Data record of oceanographic observations and exploratory fishing, No. 4:221 pp.

data only
Japan, Hokkaido University, Faculty of Fisheries, 1959.
Data record of oceanographic observations and exploratory fishing, No. 3:296 pp.

data only
Japan, Hokkaido University, Faculty of Fisheries 1957.
Data record of oceanographic observations and exploratory fishing, No. 1:247 pp.

temperature
oxygen
whales
turbidity
stomach contents
pH
plankton volume
density
radioactivity
salinity
birds
dynamic depth
light penetration
phosphate
chlorophyll
silicate

data only
Hapan, Hokkaido University, Faculty of Fisheries 1955.
Correction of data presented in "Hydrographic data obtained principally in the Bering Sea by Training Ship 'Oshoro Maru' in the summer of 1955" published in January 1956:8 pp.

oxygen
phosphate
silicate

data only
Japan, Japan Meteorological Agency, 1964.
Coastal observations.
Res. Mar. Meteorol. and Oceanogr. Obs., (31):253-271.

data (pp.90-284) only
Japan, Japan Meteorological Agency, 1962
The results of marine meteorological and oceanographical observations, January-June 1961, No. 29: 284 pp.

data only
Japan, Meteorological Agency, 1962
The results of marine meteorological and oceanographical observations, July-December, 1960, No. 28: 304 pp.

data only
Japan, Japan Meteorological Agency, 1961
The results of marine meteorological and oceanographical observations, January-June 1960. The results of the Japanese Expedition of Deep-Sea (JEDS-2, JEDS-3), No. 27: 257 pp.

data only
Japan, Japan Meteorological Agency, 1960.
The results of marine meteorological and oceanographical observations. Jan.-June. 1959, No.: 25: 258 pp.

data only
Japan, Japan Meteorological Agency, 1960.
The results of marine meteorological and oceanographical observations, July-December, 1959, No. 26:256 pp.

data only
Japan, Japan Meteorological Agency, 1960.
The results of marine meteorological and oceanographical observations. Supplement, 149 pp.

data only
Japan, Japan Meteorological Agency, 1959.
The results of marine meteorological and oceanographical observations, July-December 1958, No. 24:289 pp.

data only
Japan, Japan Meteorological Agency 1959.
The results of marine meteorological and oceanographical observations, January-June 1958, No. 23:240 pp.

data only
Japan, Japan Meteorological Agency, 1958.
The results of marine meteorological and oceanographical observations, January-June, 1957, No. 21:168 pp.

data only
Japan, Japan Meteorological Agency, 1958
The results of marine meteorological and oceanographical observations, July-December 1957, No. 22: 183 pp.

data only
Japan, Japanese Meteorological Agency, 1957
The results of marine meteorological and oceanographical observations, Jan.-June, 1956: 184 pp.
July-December, No. 20: 191 pp.

data only
Japan, Kagoshima University, Faculty of Fisheries 1957.
Oceanographical observation made during the International Cooperation Expedition, EQUAPAC, in July-August 1956, by M.S. Kagoshima-maru and M.S. Keitan-maru, 68 pp. (mimeographed).

temperature
salinity
density
chlorinity
oxygen

data only
Japan, Kobe Marine Observatory, 1961.
Data of the oceanographic observations in the sea south of Honshu from February to March and in May, 1959.
Bull. Kobe Mar. Obs., No. 167(27):99-108; 127-130; 149-152; 161-164; 205-218.

data only
Japan, Maizuru Marine Observatory, 1963
Data of the oceanographic observations (1960-1961) (35-36):115-272.

data only
Japan, Maritime Safety Board, Tokyo, 1962.
The results of oceanographic observations in the north-western Pacific Ocean, No. 9 (1941).
Hydrographic Bull., Tokyo, (Publ. No. 981), No. 71:180 pp.

data only
Japan, Maritime Safety Board, 1961
Tables of results from oceanographic observation in 1959.
Hydrogr. Bull. (Publ. No. 981), No. 68: 112 pp.

data only
Japan, Maritime Safety Board, 1960.
Tables of results from oceanographic observation in 1957.
Hydrogr. Bull. (Publ. No. 981), No. 64:103 pp.

data only
Japan, Maritime Safety Board, 1955.
The results of oceanographic observation in the north-western Pacific Ocean, 1939-1941, No. 7.
Pub. 981, Hydrogr. Bull. Spec. No. 16:1-152.

temperature
salinity
color
transparency

data only
Japan, Maritime Safety Board, 1956.
Tables of results from oceanographic observations in 1952 and 1953. Hydrogr. Bull., (Publ. 981)(51):1-171.

temperature, pH, silicate, sea, color, density, chlorinity, oxygen, phosphate, swell, transparency

data only
Japan, Maritime Safety Board, Tokyo, 1954.
Tables of results from oceanographic observations in 1951. Publ. No. 981, Hydrogr. Bull., Spec. No. 15:31-129.

temperature, salinity, pH, oxygen, silicate, phosphate, sea, swell, transparency, color

data only
Japan, Tokai Regional Fisheries Research Laboratory, 1959.
IGY Physical and chemical data by the R. V. Soyo-maru, 25 July-14 September 1958.
17 pp. (multilithed).

data only
Joseph, J. (compiler), 1961.
Trübungs- und Temperature-Verteilung auf den Stationen und Schnitten von V.F.S. "Gauss" sowie Bathythermogramme von F.F.S. "Anton Dohrn" und V.F.S. "Gauss" im Internationalen Geophysikalischen Jahr 1957/1958.
Deutschen Hydrogr. Zeits., Erganz. B(4)(5):1-131.

K

data only
Kollmeyer, R.C., J.W. McGary and R.M. Morse, 1964.
Oceanographic observations, Tropical Atlantic Ocean, EQUALANT II, August 1963.
U.S.C.G. Oceanogr. Rept., No. 4(373-4):96 pp.
CG

data only
Korea (South), R.O.K., Navy Hydrographic Office 1962
Technical reports.
H.O. Publ. South Korea, No. 1101: 85 pp.

data only
Koroleff, Folke, 1959.
The Baltic cruise with R/V Aranda 1958, hydrographic data.
Merent. Julk. (Havforskningsinstitutets Skr.), No. 193:25 pp.

data only
Koroleff, Folke, 1959
Temperature and salinity at the fixed Finnish stations, July 1954-Dec. 1956.
Merent. Julk. (Havsforskningsinstitutets Skr.) No. 192: 147 pp.

data only
Koroleff, Folke, and Aarno Voipio, 1963
The Finnish Baltic cruise 1960. Hydrographical data.
Havsforskningsinst. Skrift (Merent. Julk.), No. 204: 27 pp.

Data only
Koroleff, Folke, and Aarno Voipio, 1961.
The Baltic cruise with R/V Aranda, 1959. Hydrographical data.
Merent. Julk. (Havforskningsinst. Skrift), No. 197:26 pp.

L

data only
Lacombe, Henri, et Jean-Claude Lizeray, 1960.
Liste des stations M.O.P. "Winnaretta-Singer". Campagne d'aout 1958 dans le detroit de Gibraltar. Cahiers Oceanogr., C.C.O.E.C., 12(9): 673-682.

data only
Lacombe, Henri, and Paul Tchernia, 1960
Liste des stations M.O.P. CALYPSO 176à 234 (Campagne 1957) pour servir à l'étude des échanges entre la mer Méditerranée et l'océan Atlantique. Cahiers Oceanogr., C.C.O.E.C., 12(3): 204-235.

data only
Lane, R.K., J. Butters, W. Atkinson and H.J. Hollister, 1961.
Oceanographic data record, coastal and seaways projects, February 6 to March 2, 1961.
Fish. Res. Bd., Canada, Manuscript Rept. Ser., (Oceanogr. & Limnol.), No. 91:128 pp.

data only
Lane, R.K., R.H. Herlinveaux, W.R. Harling and H.J. Hollister, 1960.
Oceanographic data record, Coastal Seaways Projects, October 3 to 26, 1960.
Fish. Res. Bd., Canada, Mss Rept. Ser., Oceanogr. & Limnol.), No. 83:142 pp. (multilithed).

data only
Lane, R.K., A.M. Holler, J.H. Meikle and H.J. Hollister, 1961
Oceanographic data record, Monitor and Coastal Projects, March 20 to April 14, 1961.
Fisheries Res. Bd., Canada, Manuscript Rept. Ser. (Oceanogr. & Limnol.), No. 94: 188 pp. (multilithed).

Data only
Lisitzin, Eugenie, 1961.
Vedenskorkeusarvoja 1960. Vattenstandsvarden 1960
English summary: Sea level records for the year 1960.
Merent. Julk. (Havsforskningsinst. Skrift), No. 198-59 pp.

data only
Lisitzin, E., 1954.
Observations on currents and winds made on board Finnish Light-Ships during the years 1950 and 1951.
Merent. Julk. No. 162:1-85.

wind
surface currents

data only
Longhurst, A.R., 1961.
Cruise report of oceanographic cruise 6/61.
Federal Fisheries Service, Nigeria, 4 pp. (mimeographed).

M

Mc

data only
McGary, James W., and Joseph J. Graham, 1960.
Biological and oceanographic observations in the central north Pacific, July-September 1958.
U.S.F.W.S., Spec. Sci. Rept., Fish., No. 358: 107 pp.

data only
McGary, J.W., E.C. Jones, and T.S. Austin, 1956.
Mid-Pacific oceanography. IX. Operation NORPAC.
Spec. Sci. Repts.: Fish. 168:127 pp.

temperature, waves, salinity, density, oxygen, bird counts, mammal counts, fish troll counts, zooplankton {volumetric, nos. per cu m}

data only
McGary, J.W., and E.D. Stroup, 1958.
Oceanographic observations in the central North Pacific, Sept. 1954-Aug. 1955.
U.S.F.W.S. Spec. Sci. Rept., Fish., No. 252:250 pp.

data only
Mosby, Hakon, 1963
Current measurements in the Norwegian Sea and in the North Sea, 1923, 1924, 1928, 1929. Tables. NATO Subcommittee on Oceanography, Bergen, 67 pp. (multilithed).

data only
Mosby, Haakon, 1963
Current measurements in the Faroe-Shetland Channel 1962. Tables.
NATO Subcommittee on Oceanographic Research, Bergen, 59 pp. (multilithed).

data only
Mosby, Hakon, 1963?
Oceanographical tables from Weather Ship Station M (66°N, 2°E).
Univ. i Bergen, Geofysisk Inst., 114 pp.
Also in: Collected Papers, Weather Station M 66°N, 2°E, Univ. Bergen, Geophys. Inst., 1963

data only
Motoda, S., H. Koto, K. Kato, and T. Fugii, 1956.
Hydrographic data obtained principally in the Bering Sea by Training Ship "Oshoro-maru" in the summer of 1955, 59 pp.

temperature
salinity
phosphate
silicate
pH

N

data only
Nato Subcommittee on Oceanographic Research, 1962
Current measurements, meteorological observations and soundings of the M/S "Helland-Hansen" near the Strait of Gibraltar, May-June 1961. Tables. 29 pp. (multilithed).

Data only
Netherlands, Koninklijk Nederlands Meteorologisch Instituut, 1960.
Meteorologische en oceanographische Waarnemingen verricht aan boord van Nederlandse Lichtschepen in de Noordzee, 11 (1959):307 pp.

data only
Netherlands, Koninklijk Nederlands Meteorologisch Instituut, 1959.
Meteorologische en oceanografische Waarnemingen verricht aan boord van Nederlandse Lichtschepen in de Noordzee. 9:283 pp.

data only
Netherlands, Koninklijk Nederlands Meteorologisch Instituut, DeBilt, 1957.
Meteorologische en Oceanografische Waarnemingen verricht aan boord van Nederlandse Lichtschep in de Noordzee, Jaargang 8, 1956:283 pp.

temperature, surface
meteorological
waves
tidal currents

data only
Netherlands, Koninklijk, Nederlands Meteorologisch Instituut, 1957.
Meteorologische en oceanografische waarnemingen verricht aan boord van Nederlandse Lichtschepen in de Noordzee. Jahrgang 5, 1953:283 pp.

tidal currents
waves
temperature, surface

data only
Netherlands, Koninklijk Nederlands Meteorologisch Instituut, 1956.
Meteorologische en Oceanografische Waarnemingen verricht aan boord van Nederlandse Lichtschepen in de Noordzee. Jaargang 7 (1955):283 pp.
tidal currents
temperature, surface
waves
tides

data only
Nordstrom, Svante, 1958
Finnish light-ship observations on currents and winds in 1955, 1956 and 1957.
Merent. Julk., No. 185: 70 pp.

data only
NORPAC Committee, 1955.
Oceanic observations of the Pacific, 1955.
The NORPAC data, 532 pp.

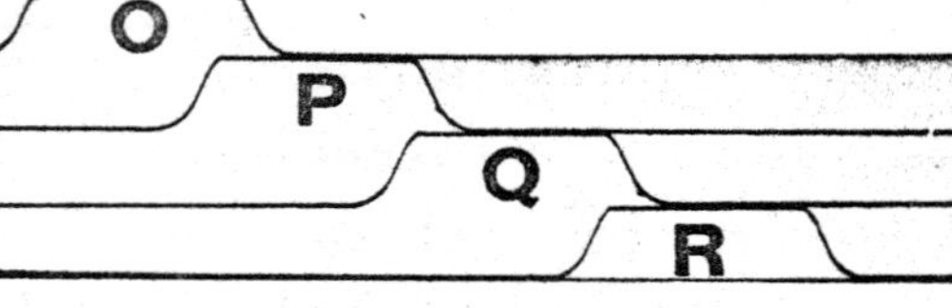

R

data only
Ramalho, A. de M., 1941
Observacoes oceanograficas. "Albacora", 1930, 1938 e 1939. Trav. St. Biol. Mar., Lisbonne, No. 44:52 pp., 1 fold-in (with 8 figs.)

data only
Ramalho, A. de M., L. S. Dentinho, C. A. M. de Sousa, Fronteira, F. L. Mamede, and H. Vilela, 1935.
Observacoes oceanograficas. "Albacora" 1934. Trav. St. Biol. Mar., Lisbonne, No. 35: 35 pp., 1 fig.

data only
Ramalho, A. de M., R. G. Boto, B. C. Goncalves, H. Vilela, 1936.
Observacoes oceanograficas. "Albacora", 1935-1936. Trav. St. Biol. Mar. Lisbonne, No. 39: 25 pp., 2 figs.

data only
Ramalho, A. de M., B. C. Goncalves, R. G. Boto, Z. Vilela, 1938.
Observacoes oceanograficas. "Albacora", 1937. Trav. St. Biol. Mar., Lisbonne, No. 43: 30 pp., 1 fig.

data only
Reid, Joseph L., Jr., Editor, 1960.
Oceanographic observations of the Pacific: 1955, the NORPAC atlas prepared by the NORPAC Committee University of California Press and Tokyo University Press.
$10, including data volume.
Reviewed by John D. Cochrane, Limnology and Oceanography, 8(3):366-367.

data only
Republic of Korea, Central Fisheries Experimental Station, 1962.
Annual Report of oceanographic observations, 1957, 7: 214 pp. (In Japanese).

data only
Republic of Korea, Central Fisheries Experimental Station, 1961.
Annual report of oceanographic observations, 1956-1957. (In Japanese, except Section III). 5 and 6:212 pp.

data only
Robertson, D.G., F.W. Dobson and H.J. Hollister, 1963
Oceanographic data record, Ocean Weather Station "P", August 1, 1962 to January 18, 196[3]
Fish. Res. Bd., Canada, Mss Rept. Ser., (Oceanogr. and Limnol.), No. 154:155 pp.

data only
Romanovsky, R., 1958.
Observations océanographiques des navires stationnaires météorologiques. Trav. C.R.E.O., n.s., 3(1):51-68.

temperature
salinity
Because of heavy expense this service now turned over to Service Central Hydrographique.

data only
Rotschi, Henri, 1961
Resultats des observations scientifiques du "Tiare", Croisière "Entrecasteaux" (11-24 août 1960) sous le commandement du Lieutenant de Vaisseau Carbelaud.
Comité Local d'Océanogr. et d'Études des Côtes de Nouvelle Caledonie, Rapp. Sci., Noumea, No. 21: 41 pp (mimeographed).

data only
Rotschi, Henri, 1960
Orsom III, resultats de la Croisière "Dillon" 1ère Partie. Océanographie Physique.
Off. Recherche Sci. et Techn., Outre-Mer, Inst. Francais d'Océanie, Centre d'Océanogr., Rapp. Sci. Noumea, No. 18: 58 pp.

data
Rotschi, Henri, 1959.
Resultats des observations scientifiques du "Tiare" croisiere "Dauphin" (8-14) Aout 1959) sous le commandement du Lieutenant de Vaisseau Cerbelaud.
Rapp. Sci., Inst. Francais d'Oceanie, Centre d'Oceanogr., No. 14:18 pp.

data (only)
Rotschi, Henri, 1958
"Orsom III", Océanographie physique.
Rapp. Tech. de la Croisière 56-5, Inst. Français d'Océanie, Rapp. Sci., No. 5 : 35 pp. (Mimeographed).

data only
Rotschi, Henri, 1958
Résultats des observations scientifiques du "Tiare", Cruisière "Bounty" sous le commandement du Lieutenant de Vaisseau Morlanne, 20-29 Juin, 1958.
Comité Local d'Océanographie et d'Etude des Côtes de Nouvelle-Caledonie, Rapp. Sci, I.F.O. No. 7: 20 pp.

data only
Rotschi, Henri, 1957
"Orsom III". Oceanographie physique.
Rapp. Tech. de l'Expedition EQUAPAC (Croisiere 56-4), Inst. Francais d'Oceanie, Rapp. Sci., 3: 52 pp. (mimeographed).

data only
Rotschi, Henri, Michel Angot, Michel Legand and Roget Desrosieres, 1961
Orsom III, Resultats de la Croisiere "Dillon", 2eme Partie. Chimie et Biologie.
ORSTOM, Inst. Francais d'Oceanie, Centre d'Oceanogr., Rapp. Sci., No. 19: 105 pp. (mimeographed).

data only
Rotschi, Henri, Michel Legand and Roger Desrosieres, 1961
Orsom III, Croisières diverses de 1960, physique chimie et biologie. ORSTOM, Inst. Francais d'Océanie, Centre d'Océanogr., Noumea, Rapp. Sci., No. 20: 59 pp. (mimeographed).

S

data only
Saelen, O.H., 1962
The natural history of the Hardangerfjord. 3. The hydrographical observations 1955-1956. Sarsia, 6:1-25.

data only
Schmidt, J., 1912.
Hydrographical observations.
Rept. Danish Oceanogr. Exped., 1908-1910, 1:51-75.
temperature
chlorinity
salinity
density
oxygen
pH

data only
Scripps Institution of Oceanography, 1949
Physical and chemical data, Cruise 1, February 28 to March 16, 1949. Marine Life Research Program. Physical and Chemical Data Report No.1:10 figs. (ozalid), tables of data (mimeographed).

data only
Scripps Institution of Oceanography, 1949
Physical and chemical data, Cruise 2, March 28 to April 12, 1949. Physical and Chemical Data Report No.2:10 figs. (ozalid) tables of data (mimeographed).

data only
Shipley, A.M., and P. Zoutendyk, 1964.
Hydrographic and plankton data collected in the south west Indian Ocean during the SCOR International Indian Ocean Expedition, 1962-1963.
Univ. Cape Town, Inst. Oceanogr., Data Rept., No 2:210 pp.

data only
Spain, Instituto Español de Oceanografía, 1961.
Campañas biologicas del "Xauen" en las costas del Mediterraneo Marroqui, Mar de Alboran, Baleares y Noroeste y Cantabrico Españoles en los años 1952, 1953 y 1954.
Bol. Inst. Español Oceanogr., 103:1-130.

data only
Spain, Instituto Español de Oceanografía, 1955.
Campañas del "Xauen" en la costa noroeste de España en 1949 y 1950. Registro de operaciones.
Bol. Inst. Español Ocean., 71:72 pp.

temperature, salinity, density, phosphate, oxygen

data only
Sweden, Fishery Board of Sweden, 1963
Hydrographical observations on Swedish lightships in 1961. (Arthr Svansson, Editor).
Fish. Bd., Sweden, Ser. Hydrogr., Rept. No. 15: 112 pp.

data only
Sweden, Fishery Board of Sweden, 1962
Hydrographical observations on Swedish lightships in 1960.
Fish. Bd., Sweden, Ser. Hydrogr. Rept., No. 14 117 pp.

data only
Sweden, Fishery Board of Sweden, 1961
Hydrographical observations on Swedish Lightships in 1959.
Fish. Bd. Sweden, Ser. Hydrography, Rept. No. 12: 130 pp.

data only
Sweden, Fishery Board of Sweden, 1959
Hydrographical observations on Swedish lightships in 1958.
Fish. Bd., Sweden, Ser. Hydrogr., Rept. No. 11: 124pp.

data only
Sweden, Fishery Board of Sweden, 1958
Hydrographical observations on Swedish lightships in 1957.
Rept. Hydrogr. Ser., No. 10: 129 pp.

data only
Sweden, Fishery Board, 1957.
Hydrographical observations on Swedish lightships in 1956. Fish. Bd., Sweden, Ser. Hydrogr., Rept. No. 9:110 pp.

temperature
salinity
currents

data only
Sweden, Fishery Board of Sweden, 1955.
Hydrographical observations on Swedish lightships in 1954. Ser. Hydrogr., Rept. No. 6:130 pp.

temperature
salinity
currents

T

data only
Tabata, S., C.D. McAllister, R.L. Johnston, D.G. Robertson, J.H. Meikle, and H.J. Hollister, 1961
Data record. Ocean Weather station "P" (Latitude 50 00'N, Longitude 145 00'W), December 9, 1959 to January 19, 1961.
Fish. Res. Bd., Canada, MSS. Rept. Ser. (Ocean & Limnol.), No. 98:296 pp. (multilithed).

data only
Tabata, S., D.G. Robertson, W. Atkinson, and H.J. Hollister, 1962
Oceanographic data record, Ocean Weather Station "P", September 12, 1961 to January, 21, 1962.
Fish. Res. Bd., Canada, Mss. Rept. Ser. (Oceanogr. & Limnol.), No. 125:187 pp.

data only
Tsubata, B., 1958.
Oceanographical conditions observed at definite station off Asamushi during 1956.
Bull. Mar. Biol. Sta. Asamushi, 9(1):43.

temperature, chlorinity, oxygen, pH, color, transparency

U

data only
United States, Agricultural and Mechanical College of Texas, 1961.
Oceanographic observations in the intertropical region of the World Ocean during IGY and IGC. Pt. 1. Atlantic and Indian Oceans.
IGY World Data Center A, Oceanography, IGY Oceanogr. Rept. No. 3:264 pp.

data only
U.S. Naval Oceanographic Office, 1963.
Oceanographic data report, Arctic, 1959.
IMR No. 0-43-63:245 pp.

data only
U.S. Navy Hydrographic Office, 1960.
Oceanographic observations, Arctic waters, Task Force Five and Six, summer-autumn 1956, USS Requisite (AGS-18), USS Eldorado (AGC-11), USS Atka (AGB-3) and USCGC Eastwind (WAGB-279).
U.S. Navy Hydrogr. Off., Techn. Rept., TR-58: 89 pp.

data only
University of California, Scripps Institution of Oceanography, 1960
Oceanic observations of the Pacific, 1950, 508 pp.

V

data only
Van Riel, P.N., P. Groen, M.P.H. Weenink, 1957
Quantitative data concerning the status of the East-Indonesian waters. Depths of standard pressure and stability values. Snellius Exped., 2(Ocean. Rec.)(7):45 pp.

temperature, salinity, density

W

DATA only
Wilson, Robert C., Eugene L. Nakamura, and Howard O. Yoshida, 1958
Marquesas area fishery and environmental data, October 1957-June 1958.
USFWS Spec. Sci. Rept. Fish., No. 283:105pp.

data only
Wilson, R.C., and M.O. Rinkel, 1957.
Marquesas area oceanographic and fishery data, January-March 1957.
USFWS Spec. Sci. Rept., Fish., 238:136 pp.

temperature, salinity, density, oxygen, phosphate, BT, zooplankton volume, waves, transparency, color

XYZ

data only
Zorè, M., & A. Zupan, 1960
[(Hydrogrographical data for the Kastela Bay 1953-1954)]
Acta Adriatica 9 (1): 32 pp.

data only
Zvereva, A.A., Edit., 1959.
Data, 2nd Marine Expedition, "Ob", 1956-1957.
Arktich. i Antarkt. Nauchno-Issled. Inst., Mezhd. Geofiz. God, Sovetsk. Antarkt. Exped., 6: 1-387.

Data acquisition

data-collecting systems
Gilbert, R.L.G., 1966.
Data-collecting systems in oceanography.
Boll. Geofis. teor. appl., 8(32):286-293.

data acquisition
O'Hagan, Robert M., and Paul Ferris Smith, 1965.
The use of a small computer with real time techniques for oceanographic data acquisition, immediate analysis and presentation. (abstract)
Ocean Sci. and Ocean Eng., Mar. Techn. Soc.,-Amer. Soc. Limnol. Oceanogr., 1:459-471.

data acquisition
Snodgrass, James M., 1962
An oceanographic data and communications system.
Proc. 2nd Interindustrial Oceanogr. Symposium Lockheed Aircraft Corp., 31-39.

Data classification

data catalogue

data catalogue
U.S.A., World Data Center A, Oceanography, 1966.
Catalogue of data in World Data Center A, Oceanography. Data received during the period 1 July - 31 December 1965.
Suppl. No. 4:99 pp. (loose leaf)

data (catalogues for)
United States, World Data Center, Oceanography, 1964.
Catalogue of data in World Data Center A, Oceanography. Data received during the period 1 July 1957 - 31 December 1963, numerous pp., loose-leaf multilithed.

data centers

data centers
Conella, Joseph, et Jean Martin, 1966.
Centrale de mesures océanographiques.
Cahiers Océanogr., C.C.O.E.C., 18(5): 381-422.

data centers
Saltykova, T. I., 1966.
Characteristics of the WDC B materials on oceanography. (In Russian).
Geofiz. Biull., Mezhd. Geofiz. Kom., Prezid., Akad. Nauk., SSSR, No. 17:116-128.

data classification
Tabata, S., 1957.
Classification of daily sea-water data.
Trans. A.G.U., 38(2):191-197.

data collection

data collection
Anon., 1969.
Big brother (scientific type) is up there!
Ocean Industry, 4(7):30-31. (secondary source)

data collection
Wolff, Paul M., 1968.
Oceanographic data collection.
Bull.Am.Met.Soc., 49(2):96-107.

data exchange

data files

Moiseev, L.K., 1971.
Volume increase of punched-card files of many-year averages of oceanographic parameters by current observational data. (In Russian; English abstract). Okeanologiia 11(2): 319-326.

data exchange

UNESCO, Intergovernmental Oceanographic Commission1965.
Manual on international oceanographic data exchange.
Int. Oceanogr. Comm., Techn. Ser., 1:28 pp.

data gathering

data gathering, systematic

Burkholder, Edward K., 1965.
A systematic approach to oceanographic data gathering. (Abstract).
Ocean. Sci. and Ocean Eng., Mar. Techn. Soc.,-Amer. Soc. Limnol. Oceanogr., 1:440-458.

data inventory

data inventory

U.S.A., National Oceanographic Data Center, 1966.
Inventory of Archived Data. Catalog series. Publ. C-3: unnumbered pp. ($2.00 -paper; $3.00 hard cover).

data processing

data processing

Barlow, G.E., 1967.
Instrument data processing systems.
In: The collection and processing of field data. E.F. Bradley and O.T. Denmead, editors. Interscience Publishers, 407-436.

data processing

Bradley, E.F., and O.T. Denmead, editors, 1967.
The collection and processing of field data: a CSIRO symposium.
Interscience Publishers, 597 pp.

data processing

Fofonoff, N.P., and C. Froese, 1960.
Programs for oceanographic computations and data processing on the electronic digital computer ALWAC III-E, M-1 miscellaneous programs.
Fish. Res. Bd., Canada, Mss Rept. Ser. (Oceanogr. & Limnol.), No. 72:35 pp.
Abstr. in: Meteorol. Geoastr. Abstr., 12(10):1995

data processing

Neidell, Norman, S., 1968.
Data processing for controlled energy acoustic sources.
Under Sea Techn., 9(10):28-29, 44-46.

data processing

O'Hagan, Robert M., 1965.
Data processing at sea.
Geomarine Techn., 1(2):18-21.

data, interpretation of

Pritchard, Donald W., 1964.
Interpretations and conclusions.
Symposium, Environmental Measurements, U.S. Dept. Health, Education and Welfare, Public Health Service, Environmental Health Series, Publ. No. 999-AP-15 or 999-WP-15, 235-243.
Also in:
Collected Reprints, The Johns Hopkins Univ., Chesapeake Bay Inst., and Dept. Oceanogr., 6. 1965.

Data Processing

Sachkova, A.I., 1964.
Mechanical processing of surface observation data on ships, (In Russian).
Trudy Inst. Okeanol., Akad., SSSR, 75:62-68.

data processing

Van Bavel, C.H.M., 1967.
Use and abuse of information processing by machine.
In: The collection and processing of field data. E.F. Bradley and O.T. Denmead, editors, Interscience Publishers, 437-446.

data retrieval

data retrieval

O'Connor, Joel S., and Saul B. Sailla, 1962.
A developing aquatic sciences information retrieval system.
Trans. Amer. Fish. Soc., 91(2):151-154.

data, satellite

data, satellite

Greaves, James R., Raymond Wexler and Clinton J. Bowley, 1966.
The feasibility of sea surface temperature determination using satellite infrared data.
Allied Res. Assoc., Inc., NASA Contr. Rept., CR-474:47 pp. (multilithed). ($3.00).

Data sources

data sources

Defant, A., 1928.
Die systematische Erforschung des Weltmeeres.
Zeits. Gesellschaft f. Erdk., Berlin, Jubiläumssonderband 459-505, Pl. 32, figs. 18-31.
Locations of stations 1000 m. or more.

data sources

Vaughn, T.W., 1937.
International aspects of oceanography, oceanographic data and provisions for oceanographic research. Nat. Acad. Sci.,

data inventory

U. S. Fish and Wildlife Service, Bureau of Commercial Fisheries, 1963.
Inventory of oceanographic data for the western North Atlantic Ocean and the Gulf of Mexico. (Oceanographic station data, bathythermograph and sea-surface temperature observations).
Circular, 176:39 sheets.

data transmission

data transmission

Fay, W. J., and D.R. Monoz, 1965.
Communication problems associated with data transmission from weather stations.
Ocean Sci. and Ocean Eng., Mar. Techn., Soc., Amer. Soc. Limnol. Oceanogr., 1:472-497.

dating

dating

See: age determination

day length

day length

Mintz, Y., M.H. Munk, 1951.
The effect of winds and tides on the length of day. Tellus. 3(3):117-121, 1 textfig.

day length

Rudloff, Willy, 1963
Geophysikalische Einflüsse auf die Länge des Sterntages.
Deutsche Hydrogr. Zeits., 16(2):76-85.

dead matter

Dead Matter

Krey, Johannes, 1961.
The balance between living and dead matter in the oceans.
Oceanography, Amer. Assoc. Adv. Sci., Publ. 67:539-548.

dead water

"dead water"

Ekman, V.W., 1904.
On dead water. Norwegian North Polar Exped., 1893-96, Sci. Res. 5(15):152 pp.

debris

debris, reef-derived

Gevirtz, Joel L., and Gerald M. Friedman, 1966.
Deep-sea carbonate sediments of the Red Sea and implications on marine lithification.
J. Sed. Petr. 36(1):143-151.

debris, woody

*Uchupi, Elazar, and Gilbert F. Jones, 1967.
Woody debris on the mainland shelf off Ventura, southern California.
Sedimentology, 8(2):147-151.

decimeter

decimeters

Menache, Maurice, 1965.
Litre et decimetre cube - densité et masse volumique: conséquences de l'abandon du litre comme unité de volume pour les mesures de précision.
Cahiers Océanogr., C.C.O.E.C., 17(9):625-646.

decomposition

Decomposition,

See also: chemistry

decomposition

Anderson, D.Q., 1939.
Distribution of organic matter in marine sediments and its availability for further decomposition. J. Mar. Res. 2(3):225-235, Textfig. 67, 2 tables.

chitin decomposition

Campbell, L.L., jr., and O.B. Williams, 1951.
A study of chitin-decomposing micro-organisms of marine origin. J. Gen. Microbiol. 5(5):894-905.

decomposition

Gorshkova, T.I., 1955.
[Rate of decomposition of organic matter of phytoplankton in Taganrog Bay.]
Doklady Akad. Nauk, SSSR, 104(1):112-

decomposition

Ikubo, A., 1954.
A note on the decomposition of sinking remains of plankton organisms and its relationship to nutrient liberation. J. Ocean. Soc., Japan, 10(3):121-131, 4 textfigs.

decomposition

Renn, C.E., 1940
Effects of marine mud upon the aerobic decomposition of plankton materials.
Biol.Bull. 78(3):454-462, 3 textfigs.

decomposition of plankton

Skopinzev, B.A., and E.S. Bruck, 1940.
[Contribution to the study of regeneration of nitrogen and phosphorus compounds in the course of decomposition od dead plankton.]
C. R. Acad Sci., Leningrad, 26:807-810.

decompression sickness
Hemplemann, H.V., 1968.
Bubble formation and decompression sickness.
Rept. Underwater Ass., 1968:63-65.

decomposition
von Brand, T., and N.W. Rakestraw, 1940.
Decomposition and regeneration of nitrogenous organic matter in sea water. III. Influence of temperature and source and condition of water.
Biol. Bull. 79(2):231-236, 2 textfigs.

Reviewed by H. Barnes in J. du Cons. 16(1):113-116.

decomposition
von Brand, Th., N.W. Rakestraw, and C.E. Renn, 1939.
Further experiments on the decomposition and regeneration of nitrogenous organic matter in sea water. Biol. Bull. 77(2):285-296, 3 textfigs.

Reviewed: J. du Cons. 16(1):113-116 by H. Barnes

decomposition
von Brand, Th., N.W. Rakestraw, and C.E. Renn, 1937.
The experimental decomposition and regeneration of nitrogenous organic matter in sea water.
Biol. Bull. 72(2):165-175, 2 textfigs.

Reviewed: J. du Cons. 16(1):113-116 by H. Barnes

decomposition
Waksman, S.A., 1936.
Decomposition of organic matter in sea water by bacteria. III. Factors influencing the rate of decomposition. Biol. Bull. 70(3):472-483, 2 text figs.

decomposition
Waksman, S.A., and C.L. Carey, 1935.
Decomposition of organic matter in sea water by bacteria. II. Influence of addition of organic substances upon bacterial activities. J. Bact. 29(5):545-561, 2 textfigs.

decomposition
Waksman, S.A., and C.L. Carey, 1935.
Decomposition of organic matter in sea water by bacteria. 1. Bacterial multiplication in stored sea water. J. Bact. 29(5):531-543, 4 textfigs.

decomposition
Waksman, S.A., and C.L. Carey, 1933.
Role of bacteria in decomposition of plant and animal residues in the ocean. Proc. Soc. Exp. Biol. and Med. 30:526-527.

decomposition
Waksman, S.A., M. Hotchkiss, C.L. Carey, and Y. Hardman, 1938.
Decomposition of nitrogenous substances in sea water by bacteria. J. Bact. 35(5):477-486.

deep currents

See: Currents, deep

deeps
Dietrich, G., 1963.
Die Meere.
Die Grosse Illustrierte Länderkunde, C. Bertelsmann Verlag, 2:1523-1606.

deeps
Dietz, R.S., and H.W. Menard, 1953.
Hawaiian swell, deep, and arch, and subsidence of the Hawaiian islands. J. Geol. 61(2):99-113, 8 textfigs.

deeps
Guilcher, A., 1956.
Quelques faits et problems des parties profondes de la mer. Scientia V. XC, N D XXXII:235-240.

"deeps"
Hunt, John M., Earl E. Hays, Egon T. Degens and David A. Ross, 1967.
Red Sea: detailed survey of hot-brine areas.
Science, 156(3774):514-516.

deeps
Matthews, D.H., A.S. Laughton, D.T. Pugh, E.J.W. Jones, J. Sunderland, M. Takin and M. Bacon, 1969
Crustal structure and origin of Peake and Freen deeps NE Atlantic.
Geophys. J. R. astr. Soc., 18(5):517-542

deeps
Nasu, Noriyuki, Azuma Iijima and Hideo Kagami, 1960.
Geological results in the Japanese Deep Sea Expedition in 1959. Exploration of the Ramapo Deep in the Japan Trench and the southernmost part of the Kurile Trench by the Ryofu Maru. Repts. of JEDS, 1:201-214.

Also in Oceanogr. Mag. 11(2):201-214

deeps
Selga, M., 1931.
The deeps of the Philippines. Pub. Manila Obs. 8:185-195.

deeps
Stocks, T., 1951.
Die grösste heute bekante Tiefe des Weltmeeres.
Deutsches Hydro. Zeits. 4(4/5/6):182-184, 1 textfig.

deeps
Tareev, B.A., 1960.
[Contribution to a theory of convectional circulation in oceanic deeps.]
Izv., Akad. Nauk, SSSR, Ser. Geofiz., (7):1022-1029.
translation
Eng. Ed., Pergamon Press, (7):683-687.

deeps
Udincev, G.B., 1955.
Neue Angaben über das Relief des Kurilen-Kamcatka-Grabens.
Petermanns Geogr. Mitt. 99(1):78-80, 2 textfigs.

Vorträge vom 6X1953 der Akad. Wiss. UdSSR 1954, 94(2):315-318.

deeps
Wüst, G., 1951.
Die grössten Tiefen des Weltmeeres in kritischer Betrachtung. Kieler Meeresf. 7:203-214.

Deeps
Yasui, Masashi, Jotaro Masuzawa, Tsutomu Sawara and Toshisuke Nakai, 1961.
Soundings of the Ramapo Deep.
Oceanogr. Mag., Tokyo, 13(1):41-49.

deeps
Zenkevich, L.A., 1954.
[Achievements of Soviet science in the study of ocean deeps.] Vsesoiuznoe Obshchestvo po Rasprostranenenilu Politicheskikh i Nauchnykh Zpanii Ser. III(3):1-31.

RT-4275 in Bibl. Transl. Russian Sci. Tech. Lit. No. 38.

deeps, Pacific
Wood, E.J.F., 1956.
Diatoms in the ocean deeps. Pacific Science 10(4):377-381.

deeps
Zenkevich, L.A., 1959.
[Some questions connected with the study of the ocean depths. Some problems and results of the oceanographic investigations.]
Comm. Int. Geophys Year, SSSR, 7-11.

Deep-Sea Fan
Piper, D.J.W., 1970.
Transport and deposition of Holocene sediment on La Jolla Deep Sea Fan California.
Mar. Geol. 8(3/4):211-227.

deep-drilling
Beliayevsky, N.A., and V.V. Fedynsky, 1964.
Deep drilling for crust investigations in the U.S.S.R.
Tectonophysics, 1(4):353-355.

deep drilling
JOINT OCEANOGRAPHIC INSTITUTIONS' DEEP EARTH SAMPLING PROGRAM (JOIDES), 1965.
Ocean drilling on the continental margin.
Science, 150(3697):709-716.

deep-sea organisms, role in nutrition
Andrievskaya, L.D., and B.M. Mednikov, 1956.
[The role of deep-sea organisms in the nutrition of Oncorhyncus.] Dokl. Akad. Nauk, SSSR, 109(2):387-388.

deep-sea problems
Bruun, A. Fr., 1953.
Problems of life in the deepest deep sea.
Geogr. Mag. 26(5):

DEEPSTAR
Booda, Larry L., 1966.
Diversification into the productive sea.
Undersea Techn., 7(2):30-32.

deep-submergence vehicles
Mackenzie, Kenneth V. 1969.
Acoustic behavior of near-bottom sources utilized for navigation of manned deep submergence vehicles.
J. mar. techn. Soc 3(2):45-56

deep submergence vehicles
Mackenzie, Kenneth V. 1970.
A decade of experience with velocimeters.
J. acoust. Soc. Am. 50(5-2):1321-1333

deep water
Aagaard, K., 1968.
Temperature variations in the Greenland Sea deep-water.
Deep-Sea Research, 15(3):281-296.

deep water
Bailey, W.B., 1956.
On the origin of deep Baffin Bay water.
J. Fish. Res. Bd., Canada, 13(3):303-308.

deep water, effect of

Beklemishev, C.W., 1960
The role of deep waters of polar and subpolar origin in enriching of the deep-sea plankton. Bull. Soc. Nat., Moscow, Ser. Biol. 65(3): 46-52.
Bull. M. O-VA ISP. Prirodi, otd. Biol.

deep water

Gruzinov, V.M., 1965.
Hydrological front as a natural boundary of inherent zones in the ocean. (In Russian). Trudy, Gosuderst. Oceanogr. Inst., No. 84:252-262.

deep water

Ichiye, T., 1960
On the deep water in the western North Pacific. Oceanogr. Mag., Tokyo, 11(2): 99-110.

deepwater

Laird, N.P., 1971.
Panama Basin deep water - properties and circulation. J. mar. Res. 29(3): 226-234.

deep water

Lynn, Ronald J., and Joseph L. Reid, 1968.
Characteristics and circulation of deep and abyssal waters.
Deep-Sea Res., 15(5):577-598.

deep water

Reed, R.K., 1970.
On the anomalous deep water south of the Aleutian Islands. J. mar. Res., 28(3): 371-372.

deep water

Saint-Guily, B., 1963.
Remarques sur le mécanisme de formation des eaux profondes en Méditerranée occidentale.
Rapp. Proc. Verb., Réunions, Comm. Int., Expl. Sci., Mer Méditerranée, Monaco, 17(3):929-932.

deep water

Shcherbinin, A.D., 1969.
Deep waters of the Indian Ocean. (In Russian; English abstract). Okeanologiia, 9(6): 967-987.

deep water

Wüst, Georg, 1957.
Die Stromgeschwindigkeiten, besonders im Tiefenund Bodenwasser.
Wiss. Ergebn. dt. atlant. Exped. 'Meteor', 6(2): 321-351.
Translation: USN Oceanogr. Off. TRANS 348. (M. Slessers). 1967.

deep water

Wüst, G., 1957.
Stromgeschwindigkeiten und Strommengen in den Tiefen des Atlantischen Ozeans unter besonderer Berücksichtigung des Tiefen- und Bodenwassers. Quantitative Untersuchungen zur Statik und Dynamik des Atlantischen Ozeans.
Wiss. Ergebn. Deutschen Atlant. Exped., "Meteor", 1925-1927, 6(2):262-420, 36 figs.

deep water

Wyrtki, Klaus, 1961
The thermohaline circulation in relation to the general circulation in the oceans.
Deep-Sea Res., 8(1): 39-64.

Deep water formation

Anati, David and Henry Stommel, 1970.
The initial phase of deep water formation in the northwest Mediterranean, during Medoc '69 on the basis of observations made by Atlantis II January 25 - February 12, 1969. Cah. océanogr., 22(4): 343-351.

deep water formation

Medoc Group, 1970
Observations of formation of deep water in the Mediterranean Sea, 1969
Nature, Lond., 227(5262): 1037-1040

deep-water formation

Plakhin, E.A., 1971.
Formation of deep water properties in the Mediterranean Sea under convective mixing. (In Russian; English abstract). Okeanologiia 11(4): 622-628.

Deep water, renewal of

Krause, Gunther 1970.
Ein Beitrag zum Problem der Erneuerung des Tiefenwassers im Arkona-Becken
Kieler Meeresforsch 25(2): 268-271

deep-water renewal

Sturges, Wilton, 1970.
Observations of deep-water renewal in the Caribbean Sea. J. geophys. Res., 75(36): 7602-7610.

defense

defense

Martell, Charles B., 1966.
Defending the sea.
Industrial Research, 8(3):95-97-100.

defense mechanisms

*Bakus, G.J., 1968.
Defensive mechanisms and ecology of some tropical holothurians.
Marine Biol., 2(1):23-32.

definitions

definitions

Shepard, F.P., 1954.
Nomenclature based on sand-silt-clay ratios.
J. Sed. Petrol. 24:151-158.

Defined coastal influence

Defined coastal influence

Craig, R.E., 1954.
Defined coastal influence. A measure for the presentation of salinity observations in coastal areas.
Ann. Biol. Cons. Perm. Int. Expl. Mer, 10:86-89, Figs. 10-11.

deflection

DEFLECTION, surface

Carlson, Herbert, Karl Richter und Hans Walden, 1967.
Messungen der statistischen Verteilung der Auslenkung der Meeresoberfläche im Seegang.
Dt. hydrogr. Z., 20(2): 59-64.

deflocculation

deflocculation

Postma, H., 1967.
Sediment transport and sedimentation in the estuarine environment.
In: Estuaries, G.H. Lauff, editor, Publs Am. Ass. Advmt Sci., 83:158-179.

deformation

deformation

Rance, Hugh, 1969.
Lineaments and torsional deformation of the earth: Indian Ocean. J. geophys. Res., 74(12): 3271-3272.

deformities

deformities

Barber, H.G., 1961.
A note on unusual diatom deformities.
J. Quekett Microsc. Club, (4) 5(13):365.

deforms

Behrens, E. William, 1966.
Surface salinities for Baffin Bay and Laguna Madre, Texas April 1964-March 1966.
Publs. Inst. Mar. Sci., Port Aransas, Univ. Texas, 11:168-173.

deformities

Heerebout, G.R. 1969.
Enige teratologische afwijkingen bij de strandkrab, Carcinus maenas (L.) (Crustacea Decapoda).
Zool. Bijdr. Leiden 11: 29-31.

deliquescence

Pueschel, R.F., R.J. Charlson and N.C. Ahlquist, 1969.
On the anomalous deliquescence of sea-spray aerosols
J. appl. Met., 8: 995-998.

deltas

deltas

Allen, J.R.L., 1964.
Sedimentation in the modern delta of the River Niger, West Africa.
In: Developments in Sedimentology, L.M.J.U. van Straaten, Editor, Elsevier Publishing Co., 1:26-34.

deltas

Axelson, V. 1967.
The Laitaure Delta.
Geografisker Annaler (A) 49(1):1-127.

deltas

Băcescu, M., G. Müller, H. Skolka, A. Petran, V. Elian, M.T. Gomoiu, N. Bodeanu și S. Stănescu, 1965.
Cercetări de ecologie marină în sectorul predeltaic în condițiile anilor 1960-1961.
In: Ecologie marină, M. Băcescu, redactor, Edit. Acad. Republ. Pop. Romîne, București, 1:185-344.

deltas

Baidin, S.S., 1959.
Approximate method for the calculation of the distribution of water flowing in river deltas. (In Russian).
Trudy Gosud. Okeanogr. Inst., 45:63-72.

deltas

Baidin, S.S., 1959.
Processes of delta formation and the hydrographic system of the Volga delta. (In Russian).
Trudy Gosud. Okeanogr. Inst., 45:5-50.

deltas
Baidin, S.S., 1959.
Water balance of small streams in the "ilmeni" of the western plain of the Volga delta (according to a model of "ilmenia" of fresh and salt lakes of oozes. (In Russian).
Trudy Gosud. Okeanogr. Inst., 45:91-108.

deltas
Baidin, S.S., 1958
Hydrological regime of the western near-plain "ilmeni" of the Volga delta. (In Russian)
Trudy Gosud. Okeanogr. Inst., 43:101-116.

deltas
Bates, C.C., 1953.
Rational theory of delta formation.
Bull. Amer. Assoc. Petr. Geol. 37(9):2119-2162, 21 textfigs.

deltas
Belevich, E.F., 1969.
On sedimentation at the outer edge of the delta (avant-delta) of the Volga River. (In Russian; English abstract). Okeanologiia, 9(6): 1007-1017.

deltas
Belevich, E.F., 1956.
On the history of the Volga delta. Studies on sea coasts and reservoirs.
Trudy Okeanogr. Komissii, Akad. Nauk, SSSR, 1: 37-56.

deltas
Berthois, Léopold 1967.
La sédimentation en période de crue dans le fleuve Konkouré (République de Guinée).
Revue Géogr. phys. Géol. dyn., 9(3): 253-262

definitions
Cox, Roland A., 1964.
Proposed redefinition of the litre and the effect on oceanographic data.
Nature, 202(4937):1098.

deltas
Curray, Joseph R., 1964
Transgressions and regressions.
In: Papers in Marine geology, R.L. Miller, Editor, Macmillan Co., N.Y., 175-203.

deltas
Curray, Joseph R., and David G. Moore, 1964.
Pleistocene deltaic progradation of continental terrace, Costa de Nayarit, Mexico.
In: Marine geology of the Gulf of California, a symposium, Amer. Assoc. Petr. Geol., Memoir, T. van Andel and G.G. Shor, Jr., editors, 3:193-215.

deltas
Den Hartog, C., 1964.
The amphipods of the deltaic region of the rivers, Rhine, Meuse and Scheldt in relation to the hydrography of the area.
Netherlands J. Sea. Res., 2(3):404-457.

deltas
Den Hartog, C., 1963.
The amphipods of the deltaic region of the rivers Rhine, Meuse and Scheldt in relation to the hydrography of the sea. I. Introduction and hydrography.
Netherlands J. Sea Res., 2(1):29-39.

Deltas
Dott, R. H., Jr., 1966.
Eocene deltaic sedimentation at Coos Bay, Oregon.
J. Geol., 74(4):373-420.

Deltas
Dronkers, J. J., 1961.
De invloed van de deltawerken op de getijbeweging en de stormvloedstanden langs de kust van Zuidwest-Nederland.
Rapp. Deltacommissie, Bijdragen (5):35-84.

Deltas
Dronkers, J. J., and H. J. Stroband, 1961.
De invloed van de deltawerken op de waterbeweging en de veiligheid tegen overstroingen in het getijgebied van Zuidwest-Nederland.
Rapp Deltacommissie, Bijdragen (5):85-190.

deltas
Egorov, V.V., 1960.
Formation of the Amu-Dari delta and the possible fluctuation of its hydrographic system. In: Study of river estuaries.
Trudy Okeanograf. Komissii, Akad. Nauk, SSSR, 6: 20-24.

deltas
El-Ashry, M.T., and H.R. Wanless, 1965.
Birth and early growth of a tidal delta.
J. Geol., 73(2):404-406.

deltas
Forkin, M.I., 1959.
Growth of the Volga delta. Trudy VNIRO, 38:106-133.

deltas
*Frazier, David E., 1967.
Recent deltaic deposits of the Mississippi River: their development and chronology.
Trans.Gulf Coast Assoc.Geol.Soc., 17th Ann.Meet.: 287-318.

Deltas
Giresse, Pierre, 1970.
Les limites de la province sedimentaires littorale du delta du Llobregat.
Cah. océanogr. 22(6): 581-612.

deltas
Giresse, Pierre 1969.
Carte sedimentologique des fonds sous-marins du delta de l'Ogoove.
Cah. océanogr. 21(10): 965-994.

deltas
Guilcher, A., 1963.
24. Estuaries, deltas, shelf, slope.
In: The Sea, M.N. Hill, Editor, Interscience Publishers, 3:620-654.

deltas
Jones, Bennett G., and William Shofnos, 1961
Mapping the low-water line of the Mississippi delta.
Int. Hydrogr. Rev., 38(1): 63-76.

deltas
Klein, George deVries, 1967.
Paleocurrent analysis in relation to modern marine sediment dispersal patterns.
Bull.Am.Ass.PetrolGeol., 51(3):366-382.

deltas
Lagaay, R., and F.P.H.W. Kopstein, 1964.
Typical features of a fluviomarine offlap sequence.
In: Developments in Sedimentology, L.M.J.U. van Straaten, Editor, Elsevier Publishing Co., 1: 216-226.

deltas
Lopatin, G.B., 1960.
The characteristics of the formation of deltas in Amu-Dari district. In: Study of River Estuaries.
Trudy Okeanograf. Komissii, Akad. Nauk, SSSR, 6: 9-19.

deltas
Mathews, W.H., and F.P. Shepard, 1962
Sedimentation of Fraser River Delta, British Columbia.
Bull. Amer. Assoc. Petr. Geol., 46(8):1416-1443.

deltas
McAllister, Raymond F., Jr., 1964.
Clay minerals from west Mississippi delta marine sediments.
In: Papers in Marine Geology, R.L. Miller, editor, Macmillan Co., N.Y., 457-473.

deltas
McMaster, Robert L., Thomas P. Lachance and Asaf Ashraf, 1970.
Continental shelf geomorphic features off Portuguese Guinea, Guinea, and Sierra Leone (West Africa).
Marine Geol. 9(3): 203-213.

deltas
Mikhailov, V.N., 1962.
Drainage distribution in the Dina River delta. (In Russian).
Trudy Gosud. Okeanogr. Inst., 66:5-25.

deltas
Moore, David G., and Joseph R. Curray, 1964.
Sedimentary framework of the drowned Pleistocene delta of Rio Grande de Santiago, Nayarit, Mexico.
In: Developments in Sedimentology, L.M.J.U. van Straaten, Editor, Elsevier Publishing Co., 1: 275-281.

deltas, (defined)
Moore, Derek, 1966.
Deltaic sedimentation.
Earth-Sci. Rev., 1(2/3):87-104.

deltas
Morgan, James P. 1967.
Ephemeral estuaries of the deltaic environment.
In: Estuaries, G.H. Lauff, editor, Publs Am. Ass. Advmt Sci., 83:115-120.

deltas
Morgan, James P., James M. Coleman and Sherwood M. Gagliano, 1963.
Mudlumps at the mouth of South Pass, Mississippi River: sedimentology, paleontology, structure, origin and relation to deltaic processes.
Louisiana State Univ., Studies, Coast Stud. Ser., (10):190 pp. (multilithed).

deltas

Muller, Jan, 1959.
Palynology of Recent Orinoco delta and shelf sediments. Reports of the Orinoco Shelf Expedition, Volume 5.
Micropaleontology, 5(1):1-32.

deltas

Nasu, N., and Y. Sato, 1957.
Particle size distribution of the Obitsu delta (The occurrence of the steep marginal slope of a small scales delta).
J. Fac. Sci., Univ. Tokyo, (II) 11(1):37-55.

deltas

Netherlands, Delta Committee, 1962
Final report delivered by the Advisory Committee to provide an answer to the question of what waterways-technical provisions must be made for the area devastated by the storm flood of February 1, 1953, (Delta Committee) instituted by decree of the Minister of Transport and Waterways of February 18, 1953. 100pp

deltas

Nienaber, James H., 1963.
Shallow marine sediments offshore from the Brazos River, Texas.
Publ. Inst. Mar. Sci., Port Aransas, 9:311-372.

deltas

Ottmann, Francois, 1965.
Introduction a la Geologie Marine et Littorale. Masson et Cie, Paris, 259 pp. 47 F

deltas

Pichazov, G.I., 1960.
[Age of the Tereka deltas In: Study of river estuaries.]
Trudy Okeanograf. Komissii, Akad. Nauk, SSSR, 6: 86-88.

deltas

Pomashin, V.V., 1960.
Some general rules for the formation of the Don delta. (In Russian).
Trudy Gosud. Okeanogr. Inst., 49:111-117.

deltas

Reimnitz, Erk, Neil F. Marshall, 1965.
Effects of the Alaska earthquake and tsunami on recent deltaic sediments.
Jour. Geophys. Res., 70(10):2363-2376.

deltas

Ruck, Klaus-Wolfgang, 1969.
Ein geologisches Profil durch die Eider-mündung.
Meyniana 19:113-118.

deltas

Russell, Richard J., 1967.
Origin of estuaries.
In: Estuaries, G.H. Lauff, editor, Publs Am. Ass. Advmt Sci., 83:93-99.

deltas

Sakamoto, Ichitaro and Jun Yamada, 1969
Sedimentological study of submarine delta at the mouth of the Kiso River (mechanical analyses of bottom sediment). (In Japanese; English Abstract.
J. Fac. Fish. Pref. Univ. Mie, 8(1):17-140

Deltas

Santema, P., and J. N. Svasek, 1961.
De invloed van de afdamming van de zeegaten op het ijsbezwaar op de Zeeuwse en Zuidhollandse stromen.
Rapp. Deltacommissie, Bijdragen, (5):295-326.

deltas

Scruton, P.C., 1960.
Delta building and the deltaic sequence.
In: Recent sediments, northwest Gulf of Mexico, 1951-1958. Amer. Assoc. Petr. Geol., Tulsa, 82-102, with consolidated bibliography, pp. 368-381

deltas

Shaposhnikov, A.S., 1960.
[Tests of the economic-geographical character of deltas (on the example of the Volga delta). In: Study of river estuaries.]
Trudy Okeanogr. Komissii, Akad. Nauk, SSSR, 6: 108-143.

deltas

Shepard, Francis J., 1964.
Criteria in modern sediments useful in recognizing ancient sedimentary environments.
In: Deltaic and shallow marine deposits: Developments in Sedimentology, L.M.J.U. van Straaten, editor, Elsevier Publishing Co., 1:1-25.

deltas

Shepard, F.P., 1964.
Criteria in modern sediments in recognising ancientsedimentary environments.
In: Developments in Sedimentology, L.M.J.U. van Straaten, Editor, Elsevier Publishing Co., 1:1-25.

deltas

Shepard, F.P., and P.C. Scruton, 1954.
Delta building in the northern Gulf of Mexico. (Abstract). Bull. G.S.A. 65(2):1304.

deltas

*Shuisky, Yu.D., 1968.
The dynamics of the sea margin of the Danube's Kilya Delta. (In Russian; English abstract).
Okeanologiia, Akad. Nauk, SSSR, 8(5):858-864.

deltas

Speden, I.G., 1960
Post-glacial terraces near Cape Chocolate, McMurdo Sound, Antarctica.
New Zealand J. Geol. Geophys., 3(2):203-217

deltas

Stoneley, R., 1966.
The Niger Delta region in the light of the theory of continental drift.
Geol. Mag., 103 (5):385-397.

deep water

Sukhovey, V.F. and A.P. Metalnikov 1968.
On deep-sea water exchange between the Caribbean Sea and the Atlantic Ocean. (In Russian; English abstract).
Okeanologiia 8(2): 203-209.

deltas

Swift, D.J.P., R.M. McMullen and A.K. Lyall, 1967.
A tidal delta with an ebb-flood channel system in the Minas Basin, Bay of Fundy: preliminary report.
Marit. Sed., Halifax-Fredericton, N.S., Can, 3(1): 12-16. (multilithed)

deltas

Tempier, Claude, 1964.
Sédimentologie et paléocéanographie des formations marneuses du Callovien et de l'Oxfordien inférieur dans la région de Vauvenargues (Bouches du Rhône).
Rec. Trav. Sta. Mar., Endoume, 32(48):203-219.

Deltas

Thijsse, J. Th., 1961.
Het Deltamodel in het Waterloopkundig Laboratorium te Delft.
Rapp. Deltacommissie, Bijdragen (5):F-34.

Deltas

Valken, K. F., 1961.
De waterstaatkundige aspecten van de waterhuishouding in het gebied van de benedenrivieren en de zeearmen na de uitvoering van de deltawerken.
Rapp. Deltacommissie, Bijdragen, (5):191-294.

deltas

Van Andel, Tj. H. 1967.
The Orinoco Delta.
J. Sed. Petrol. 37(2): 297-310.

deltas

Van Straaten, L.M.J.U., 1960.
Some recent advances in the study of deltaic sedimentations.
Liverpool and Manchester Geol. Jour., 2(3): 411-442.

demagnetization

Vogt, Peter R. 1969.
Can demagnetization explain seamount drift?
Nature, Lond. 224 (5219): 574-576

density

A

density
(sigma-t)

Acara, A., 1961
Poor water masses in the North Pacific Ocean. The distribution of the water mass. Pt.2.
Hidrobiologi, Istanbul, (B), 5(3/4): 113-128.

density

Akagawa, M., 1956.
[On the oceanographical conditions of the North Japanese Sea (west off Tsugaru-Straits) in summer (Part 1.)]
Bull. Hakodate Mar. Obs., No. 3:1-11
190-199.

density

Akagawa, M., 1954.
On the oceanographical conditions of the north Japan Sea (west off the Tsugaru-Straits) in summer.] J. Ocean. Soc., Japan, 10(4):189-199, 5 textfigs.

density

Akagawa, M., 1956.
On the oceanographical conditions of the North Japan Sea (west of the Tsugaru Straits) in summer (Part 2).
Bull. Hakodate Mar. Obs., No. 3: 5-11,
1-7

density

Allain, Charles, 1964.
L'hydrologie et les courants du Détroit de Gibraltar pendant l'été de 1959.
Rev. Trav. Inst. Pêches Marit., 28(1):1-102.

density
Anderson, William W., and Jack W. Gehringer, 1959.
Physical oceanography, biological and chemical data, South Atlantic coast of the United States, M/V Theodore N. Gill, Cruise 7.
USFWS Spec. Sci. Rept., Fish. No. 278:277 pp.

density (data)
Anderson, W.W., J.W. Gehringer and E. Cohen, 1956.
Physical oceanographic, biological and chemical data, south Atlantic coast of the United States, M/V Theodore N. Gill, Cruise 1.
USFWS Spec. Sci. Rept., Fish., No. 178:160 pp.

Angot, Michel

density
Angot, Michel, et Robert Gerard, 1966.
Hydrologie et phytoplancton de l'eau de surface en avril 1965 a Nosy Be.
Cah. ORSTOM, Ser. Oceanogr., 4(1):95-136.

density
Angot, Michel, 1965.
Le phytoplancton de surface pendant l'année 1964 dans la Baie d'Ambaro près de Nosy Bé. conclusion p. 13.
Océanographie, Cahiers, O.R.S.T.R.O.M., 3(4):5-18.

density (data)
Angot, Michel, et Robert Gerard, 1966.
Caractères hydrologiques de l'eau de surface au Centre ORSTOM de Nosy Bé de 1962 à 1965.
Cah. ORSTOM, Océanogr., 4 (3): 37-53.

density (data)
Angot, M., et R. Gerard, 1965.
Hydrologie de la région de Nosy-B e: decembre 1963 à mars 1964.
Cahiers, O.R.S.T.R.O.M. - Océanogr., 3(1):31-53.

density (data)
Angot, M. et R. Gerard, 1965.
Hydrologie de la région de Nosy-Bé: juillet à novembre 1963.
Cahiers, O.R.S.T.R.O.M. - Océanogr., 3(1):3-29.

density(data)
Angot, M., et R. Gerard, 1963.
Hydrologie de la région de Nosy-Bé: mars-avril-mai-juin 1963.
Trav. Centre Océanogr., Nosy-Bé, Cahiers, O.R.S.T.R.O.M., Océanogr., Paris, No. 6:255-283.

density
Anon., 1951.
[Report of the coastal observations.] J. Ocean., Kobe Obs., 2nd ser., 2(1):45-46.

density
Anon., 1951.
[Report of the coastal observations (Dec. 1950).] J. Ocean., Kobe Obs., 2nd ser., 1(6):30-31.

density
Anon., 1951.
[Report of the surface observations of the coastal weather stations (Dec. 1950).] J. Ocean., Kobe Obs., 2nd ser., 1(6):22, 2 figs.

density
Anon., 1951.
[Report of the surface observations of the coastal weather stations.] J. Ocean., Kobe Obs., 2nd ser., 1(5):34-36.

density
Anon., 1951.
[On the water temperature and density of coastal stations (1948).] J. Ocean., Kobe Obs., 2nd ser., 1(5):24-26.

density
Anon., 1950.
[The results of the coastal observations (Aug. 1950).] J. Ocean., Kobe Obs., 2nd ser., 1(3):41-42.

density
Anon., 1951.
[The results of the oceanographical observations reported by the Fisheries Experimental Stations (Aug. 1950).] J. Ocean., Kobe Obs., 2nd ser., 1(3):37.

density
Anon., 1950.
[The report of the oceanographical observations in the Bungo Suido and off Ashizuri-Misaki in August 1950.] J. Ocean., Kobe Obs., 2nd ser., 1(1):15-44, 7 figs., tables.

density
Anon., 1949-1950.
[The results of the regular monthly oceanographical observations on board the R.M.S. "Syunpu Maru" and "Takatori Maru" in the Osaka Wan.] J. Ocean., Kobe Obs., 2nd ser., 2(1):1-28.

density
Anon. 1914
Ice observation, meteorology and oceanography in the North Atlantic Ocean. Rept. on the work carried out by the S.S. "Scotia", 1913, 139 pp.

density
Aratskaya, V.V., 1964
On the annual variations of the density of water in the layer of reduced salinity in the Pacific Ocean. (In Russian).
Okeanologiia, Akad. Nauk, SSSR, 4(4):621-624.

density (data)
Argentina, Servicio de Hidrografia Naval, 1960.
Operacion Oceanografia Malvinas (Resultados preliminares).
Servicio de Hidrografia Naval, Argentina, Publ., H. 606: numerous unnumbered pp.

density (data)
Argentina, Servicio de Hidrografia Naval, 1960.
Operacion Oceanografica Drake II. Resultados preliminares.
Servicio de Hidrografia Naval, Argentina, Publ., H. 14: numerous unnumbered pp.

density(data)
Argentina, Servicio de Hidrografia Naval, 1959.
Trabajos oceanograficos realizados en la campana Antartica, 1957-1958. Resultados preliminares.
Servicio de Hidrografia Naval, Argentina, Publ., H. 651: numerous unnumbered pp.

density(data)
Argentina, Servicio de Hidrografia Naval, 1959.
Trabajos oceanograficos realizados en la campana Antartica, 1955-1956.
Servicio de Hidrografia Naval, Argentina, Publ., H. 620: numerous unnumbered pp.

density(data)
Argentina, Servicio de Hidrografia Naval, 1959.
Operacion Oceanografia Meridiano. Resultados preliminares.
Servicio de Hidrografia Naval, Argentina, Publ., H. 617: numerous unnumbered pp.

density(data)
Argentina, Servicio de Hidrografia Naval, 1959.
Trabajos oceanograficos realizados en la campana Antartica 1958/1959. Resultados preliminares.
Servicio de Hidrografia Naval, Argentina, Publ., H. 616:127 pp.

density(data)
Argentina, Servicio de Hidrografia Naval, 1959.
Operacion Oceanografica Drake. 1. Resultados preliminares.
Servicio de Hidrografia Naval, Argentina, Publ., H. 613: numerous unnumbered pp.

density(data)
Argentina, Servicio de Hidrografia Naval, 1959.
Operacion Oceanografica Atlantico Sur. Resultados preliminares.
Servicio de Hidrografia, Naval, Argentina, Publ., H. 608: numerous unnumbered pp.

density(data)
Argentina, Servicio de Hidrografia Naval, 1959.
Operacion Oceanografica Cuenca. Resultados preliminares.
Servicio Oceanografica Naval, Argentina, Publ., H. 607: numerous unnumbered pp.

density
Armstrong, F.A.J., and E.I. Butler, 1962.
Hydrographic surveys off Plymouth in 1959 and 1960.
J. Mar. Biol. Assoc., U.K., 42(2):445-463.

density
Asano, H., 1913.
Hydrographic observations made on board the H.M.S. Matsue during December 1912. Rep. Imp. Fish. Inst. Sci. Invest. 2:8-17, plates, charts.

MurotoZaki to Mikomoto Shima (Izu Shoto).

specific gravity
Asano, H., 1913
Temperature and specific gravity of sea water and floating organisms around the Kurile Islands. (In Japanese). Rep. Imp. Fish. Inst. Sci. Invest. 3:81-91.

density
Austin, G.B., jr., 1955.
Some recent oceanographic surveys of the Gulf of Mexico. Trans. Amer. Geophys. Union 36(5):885-892.

density
Austin, Thomas S., 1960
Oceanography of the east central equatorial Pacific as observed during expedition "Eastropic". U.S.F.& W.S., Fish. Bull, 168, Vol. 60: 257-282.

density
Austin, Thomas S., 1958.
Variations with depth of oceanographic properties along the equator in the Pacific.
Trans. A.G.U., 39(6):1055-1063.

density (data)
Austin, T.S., 1957.
Summary, oceanographic and fishery data, Marquesas Islands area, August-September 1956 (EQUAPAC). USFWS Spec. Sci. Rept., Fish., No. 217:186 pp.

density (data)
Austin, T.S., 1954.
Mid-Pacific oceanography III. Transequatorial waters, August-October 1951.
U.S.F.W.S. Spec. Sci. Rept. - Fish. No. 131: numerous pp. (unnumbered), 13 figs. (multilithed)

density (data)
Austin, T.S., 1954.
Mid-Pacific oceanography. V. Transequatorial waters, May-June 1952, August 1952.
Spec. Sci. Rept., Fish., No. 136:1-85, 31 figs.

B

density (data)
Babudieri, B., e R. Favento, 1969.
Ricerche di microbiologia marina nel Mediterraneo orientale.
Boll. Pesca, Piscic. Idrobiol 24(2): 151-158

density
Băcescu, M., 1961
Cercetări fiziço-chimice şi biologice romînesti la Marea Neagra, efectuate în perioda 1954-1959.
Hidrobiologia, Acad. Repub. Pop. Rom., (3): 17-46.

density
Bacgyk, Josef 1968.
Les masses d'eaux de la mer Baltique meridionale et l'influence de leurs mouvements sur la zone littorale polonaise. (In Polish; French résumé)
Prace Geograf. Warsaw 65:1-120.

density
Baker, A. de C., 1965.
The latitudinal distribution of Euphausia species in the surface waters of the Indian Ocean.
Discovery Repts., 33:309-334.

density
Balamaramamurty, C., and A.A. Ramasastry, 1957.
Distribution of density and the associated currents at the sea surface in the Bay of Bengal. Indian J. Meteorol. and Geophys., 8(1): 88-92.

density (data)
Ballestor, A., 1965.
Hidrografia y nutrientes de la Fosa de Cariaco.
Informe de Progresso del Estudio Hidrografico de la Fosa de Cariaco, Fundación, La Salle de Ciencias Naturales Estación de Investigacion -es Marinas de Margarita, Caracas, Sept. 1965. (mimeographed):3- 12.

density (data)
Ballester Antonio Enrique Arias, Antonio Cruzado, Dolores Blasco y José Maria Camps, 1967.
Estudio hidrográfico de la costa catalana de junio de 1965 a mayo de 1967.
Investigación, pesq. 31(3): 621-662

density
Barber, F.G., 1967.
A contribution to the oceanography of Hudson Bay.
Manuscript Rep.Ser. Dept.Energy,Mines,Resources, Can., No. 4:69 pp. (multilithed).

density (data)
Bardin, I.P., 1958.
Hydrological, hydrochemical, geological and biological studies, Research Ship "Ob", 1955-1957. Trudy Kompleks. Antarkt. Exped., Akad. Nauk, SSSR, Mezh. Geofiz. God, Gidrometeorol. Izdatel., Leningrad, 217 pp.

density
Barkley, Richard A., 1968.
Oceanographic atlas of the Pacific Ocean.
University of Hawaii Press. 156 figures.

density
Bennett, Edward B., 1963.
An oceanographic atlas of the Eastern Tropical Pacific Ocean, based on data from EASTROPIC expedition. October-December 1955.
Bull. Inter-American Trop. Tuna Comm., 8(2):33-165.

Density
Bennett, E. B., 1959.
Some oceanographic features of the Northeast Pacific Ocean during August 1955. J. Fish. Res. Bd., Canada 16(5):565-633.

Density
Berrit, G.R., 1965.
Les conditions de saison chaude dans la region orientale du Golfe de Guinee.
Progress in Oceanography, 3:31-47.

density (data)
Berrit, G.R., 1964.
Résultats d'observations. Centre d'Océanographie et des Pêches de Pointe-Noire (ORSTROM).
Campagnes 12 et 13 de l'"Ombango". Hydrologie.
Cahiers Océanogr., C.C.O.E.C., 16(2):151-155.

density
Berrit, G.R., 1961
Etude des conditions hydrologiques en fin de saison chaude entre Pointe-Noire et Loanda.
Cahiers Oceanogr., C.C.O.E.C., 13(7): 456-461.

densities
Bigelow, H. B., 1922
Exploration of the coastal water off the northeastern United States in 1916 by the U.S. Fisheries Schooner Grampus. Bull. M.C.Z. 65 (5):85-188, 53 text figs.

density
Bjornberg, T.K.S., 1964.
On the free-living copepods off Brazil.
Bol. Inst. Oceanogr., Sao Paulo, 13(1):3-142.

density
Blackburn, M., 1962.
An oceanographic study of the Gulf of Tehuantepec.
U.S.F.W.S. Spec. Sci. Rept., Fish., No. 404:28 pp.

buoyancy (density)
Böhnecke, G., 1943.
Auftriebwassergebiete im Atlantischen Ozean.
Ann. Hydr., usw., 71(4/6):114-117.

density
Böhnecke, G., 1936
Atlas zu Temperatur, Salzgehalt und Dichte an der Oberfläche des Atlantischen Ozeans. Deutsche Atlantische Expedition Meteor 1925-1927, Wiss. Erg., Bd. V

density
Boto, R. G., 1945
Contribuicao para os estudios de oceanografia ao longo da costa de Portugal. Fosfatos e nitratos. Trav. St. Biol. Mar., Lisbonne, No. 49:102 pp., 57 figs. (Bumpus reprint).

density(data)
Bougis, Paul, and Claude Carre, 1960.
Conditions hydrologiques à Villefranche-sur-Mer pendant les années 1957 et 1958. Cahiers Océanogr., 12(6): 392-408.

density (data)
Bougis, P., et M.C. Carre 1971.
Conditions hydrologiques à Villefranche-sur-Mer pendant les années 1964, 1965, 1966 et 1967.
Cah. océanogr. 23(8): 733-754.

density(data)
Bougis, Paul, and Robert Fenaux, 1961
Conditions hydrologiques à Villefranche-sur-Mer pendant les années 1959 et 1960.
Cahiers Océanogr., C.C.O.E.C., 13(9):627-635.

density (data)
Bougis, Paul, Lucienne Fernaux et Madeleine Degilière 1965
Conditions hydrologiques à Villefranche-sur-Mer pendant les années 1961, 1962 et 1963.
Cah. océanogr. 17(10): 685-701.

density(data)
Bougis, P., M. Ginat and M. Ruivo, 1957.
Contribution a l'hydrologie de la Mer Catalane.
Vie et Milieu Suppl., No. 6:123-164.

density(data)
Bougis, P., M. Ginat and M. Ruivo, 1956.
Recherches hydrologiques sur le Golfe du Lion.
Vie et Milieu, 7(1):1-18.

density
Bougis, P., and M. Ruivo, 1954.
Sur un descente d'eaux superficielles en profondeur (cascading) dans le sud du Golfe du Lion. Bull. d'Info., C.C.O.E.C. 6(4):147-154, 2 pls.

density
Braarud, T. 1945
A phytoplankton survey of the polluted waters of inner Oslo Fjord. Hvalradets Skrifter, No. 28, 142 pp.

density
Braarud, T., and Bjørg Føyn, 1958.
Phytoplankton observations in a brackish water locality of south-east Norway.
Nytt Mag. Botan., 6:47-73.

density
Braarud, T., and A. Klem, 1931.
Hydrographical and chemical investigations in the waters off Møre and in the Romsdalfjord.
Hvalrådets Skrifter, 1931(1):88 pp., 19 figs.
coastal

density(data)
Braarud, T., K.R. Ringdal and J. Grøntved, 1953.
The phytoplankton of the North Sea and adjacent waters in May 1948. Rapp. Proc. Verb., Cons. Perm. Int. Expl. Mer, 133:1-87, 29 tables Pls. A-B, 18 textfigs.

density
Braarud, T. and J. T. Ruud, 1937
The Hydrographic conditions and aeration of the Oslo Fjord, 1933-1934. Hvalrådets Skr. No. 15:56 pp., 24 figs.

Reviewed: J. du Cons. XIV(3):406-408. J. N. Carruthers.

density
Braarud, T. and J. T. Ruud, 1932
The "Øst" Expedition to the Denmark Strait 1929. I. Hydrography. Hvalrådets Skr., No. 4:44 pp., 19 text figs.

Braconnet, Jean-Claude

density (data)
Braconnot, Jean-Claude, Claude Carre, Jacqueline Goy, Philippe Laval et Eveline Sentz-Braconnot, 1966.
Conditions hydrologiques pendant les années 1963 et 1964 en un point au large de Villefranche-sur-mer (A.M.). Particularités dues à l'hiver froid de 1963.
Cahiers Océanogr., C.C.O.E.C., 18(5):423-437.

density(data)
Braconnot, Jean-Claude, Claude Carre, Jacqueline Goy, et Eveline Sentz-Braconnot, 1965.
Campagnes planctoniques en Mer de Ligure (1963-1964).
Cahiers Oceanogr., C.C.O.E.C.,17(3):185-206.

density(surface; bottom, averaged)
Brennecke, W., 1921.
Die ozeanographischen Arbeiten der Deutschen Antarktischen Expedition, 1911-1912.
Arch. Deutschen Seewarte 39(1):1-216, 15 pls., 41 textfigs.

density (data)
*Brewer,P.G.,C.D.Densmore,R.Munns and R.J. Stanley,1969.
Hydrography of the Red Sea brines.
In: Hot brines and Recent heavy metal deposits in the Red Sea,E.T.Degens and D.A.Ross,editors, Springer-Verlag,New York,Inc., 138-147.

density
*Brosin,Hans-Jürgen,und Dietwart Nehring,1968.
Der Aquatoriale Unterstrom im Atlantischen Ozean auf 29o30'W im September und Dezember 1966.
Beiträge Meeresk.,22:5-17.

density
Bruevich, S. V., 1948.
[The computed chlorosity and density of water from the Caspian Sea.] Trudy Inst. Okean.,

density(data)
Bruneau, L., N.G. Jerlov, F.F. Koczy, 1953.
Physical and chemical methods.
Repts. Swedish Deep-sea Exped., 1947-1948, Phys. Chem., 3(4):101-112, i-iv, 7 textfigs., 2 tables.

density
Brunel, A., 1938-1939.
Leve des bancs des Esquerquis.
Ann. Hydrogr., Paris, (3)16:61-77, figs.

density
Buchanan, J.Y., 1884.
Report on the specific gravity of ocean water, as observed on board H.M.S. Challenger, 1873-1876.
Challenger Rept., Phys. Chem., 1(2):46 pp., chart

density
Bulgakov, N.P., 1961.
[On determining the depth of density mixing.]
Izv. Akad. Nauk, SSSR, Ser. Geofiz., (2):319-323.

Engl. Edit., (2):204-207.

density
Bumpus, D.F., 1955.
The circulation over the continental shelf south of Cape Hatteras. Trans. Amer. Geophys. Union, 36(4):601-611, 18 textfigs.

density
Bumpus, D.F., and E.L. Pierce, 1955.
Hydrography and the distribution of chaetognaths over the continental shelf of North Carolina.
Pap. Mar. Biol. and Oceanogr., Deep-Sea Res., Suppl. to Vol. 3:92-109.

C

density
Cahierre, L., 1954.
Relevés de la densité de l'eau de mer à la station médimarémétrique de Marseille.
C.R. Comité Nat. Français, Géodésie et Géophys., 1953:135-136.

density(data)
Capart, A., 1951.
Liste des stations. Rés. Sci., Exped. Océan. Belge dans les Eaux Côtières Africaines de l'Atlantique Sud (1948-1949) 1: 65 pp.

density(data)
Capurro, L.R.A., 1955.
Expedicion Argentina al Mar de Weddell (diciembre 1954 a enero de 1955). Ministerio de Marina, Argentina, Direccion Gen. de Navegacion e Hidrografia, 184 pp.

density
Carritt, D.E., 1963
5. Chemical instrumentation. In: The Sea, M.N. Hill, Edit., Vol. 2. The composition of sea water, Interscience Publishers, New York and London, 109-123.

density
Carritt, Dayton E., and James H. Carpenter, 1959.

The composition of sea water and the salinity-chlorinity-density problems. Nat. Acad. Sci.-Nat. Res. Counc., Publ. No. 600: 67-86.

density
Cavaliere, Antonio, 1963.
Studi sulla biologia e pesca di Xiphias gladius. II.
Boll. Pesca, Piscicolt. e Idrobiol., 8(2):143-170.

density
Chacko, P.I., 1950.
Marine plankton from waters around Krusadai Island. Proc. Indian Acad. Sci., Sect. B, 31(3): 162-174, 1 textfig.

density
Chew, F., 1953.
Results of hydrographic and chemical investigations in the region of the "red tide" bloom on the west coast of Florida in November 1952.
Bull. Mar. Sci., Gulf and Caribbean, 2(4):610-625, 10 textfigs.

density
Chidambaram, K., and M.D. Menon, 1945.
The co-relation of the west coast (Malabar and South Kanara) fisheries with plankton and certain oceanographical factors. Proc. Indian Acad. Sci., Sect. B, 22(6):355-367, 5 textfigs.

density
Chidambaram, K., and A.D.I. Rajendran, 1951.
On the hydro-biological data collected on the Wadge Bank early in 1949. J. Bombay Nat. Hist. Soc. 49(4):738-748.

density
Chidambaram, K., A.D.I. Rajandran, and A.P. Valsan, 1951.
Certain observations on the hydrography and biology of the pearl bank, Tholayviam Paar off Tuticorin in the Gulf of Manaar in April 1949.
J. Madras Univ., Sect. B, 21(1):48-74, 2 figs.

density
Clowes, A.J., 1954.
The temperature, salinity and inorganic phosphate of the surface layer near St. Helena Bay, 1950-52. Union of S. Africa, Dept. Commerce & Industries, Div. Fish., Invest. Rept. No. 16: 1-47, 17 pls.

Reprinted from "Commerce & Industries" Aug. 1954.

density
Clowes, A.J., 1950.
An introduction to the hydrology of South African waters. Fish. & Mar. Biol. Surv. Div., Dept. Commerce & Industries, Union of South Africa, Investigational Rept. No. 12:42 pp., 20 fold-ins, 14 figs.

density
Clowes, A.J., 1934.
Hydrology of the Bransfield Strait. Discovery Rept. 9:1-64, 68 textfigs.

Coachman, L.K.

density
Coachman, L.K., and K. Aagaard, 1966.
On the water exchange through Bering Strait.
Limnol. and Oceanogr., 11(1):44-59.

density
Coachman, L.C., and C.A. Barnes, 1961
The contribution of Bering Sea water to the Arctic Ocean.
Arctic, 14(3): 147-161.

density
Collignon, J., 1955.
Observations hydrologiques sur les eaux superficielles de la region de Pointe-Noire.
Rapp. Proc. Verb., Cons. Perm. Int. Expl. Mer, 137:7-9.

density(data)
Comité Central d'Océanographie et d'Études des Côtes, 1951.
Liste des stations hydrologiques profondes éxécutées par le bâtiment polaire "Commandant Charcot" de la Marine Nationale dans l'Océan Indien et dans l'Océan Antartique (Campagnes 1948-1949 et 1949-50). Bull. d'Info., C.C.O.E.C. 3(10):473-479.

density
Comité local d'Oceanographie et d'études des Côtes de l'Afrique Occidentale Francaise, 1949.
Travaux d'océanographie physique effectués en 1949 par la Section Technique des Pêches Maritimes de Dakar. A. Dakar et environs B. Côtes du Sénégal. C.L.O.E.C. de l'A.O.F., Annee. 1949:27-39, 4 fold-in charts.

density
Comité local d'Oceanographie et d'Etudes des Cotes de l'Afrique Occidentale Francaise, 1948.
Procer verbal. 30 pp. numerous fold-ins.

density(data)
Conseil Permanent International pour L'Exploration de la Mer, 1927.
Liste des Stations, Royaume d'Espagne.
Rapp. Proc. Verb. 44:22-29.

Cooper, L.H.N., 1960 density
The water flow into the English Channel from the south-west. J.M.B.A., 30:173-208.

density
Cooper, L.H.N., 1952.
The physical and chemical oceanography of the waters bathing the continental slope of the Celtic Sea. J.M.B.A. 30:465-510, 15 textfigs.

density
Cooper, L.H.N., and D. Vaux, 1949.
Cascading over the continental slope of water from the Celtic Sea. J.M.B.A. 28(3):719-750, 14 textfigs.

density(data)
Cordini, I.R., 1955.
Contribucion al conocimiento del sector antartico Argentino. Inst. Antartico Argentino, Publ., 1: 277 pp., 82 textfigs., 56 pls.

density
*Coste, Bernard, 1970
Résultats de la mission Hydralante 1 du Jean Charcot. Sels nutritifs.
Téthys 2 (1): 3-18.

density
Cox, R.A., 1965.
The physical properties of sea water.
In: Chemical oceanography, J.P. Riley and G. Skirrow, editors, Academic Press, 1: 73-120.

density
Cox, Roland A., 1963
The salinity problem.
Progress in Oceanography, M. Sears, Edit., Pergamon Press, London, 1:243-261.

density
Cox, R.A., 1962.
Report on the scientific results of the programme for comparing physical properties of sea water.
Rapp. Proc. Verb., Reunion, Cons. Perm. Int. Expl. Mer, 1961:94-98.

density
Cox, R.A., F. Culkin, R. Greenhalgh and J.P. Riley, 1962.
Chlorinity, conductivity and density of sea water.
Nature, 193(4815):518-520.

density
Craya, A., 1951.
Critical regimes of flows with density stratification. Tellus 3:24-42.

density
Crean P.B. 1967.
Physical oceanography of Dixon Entrance British Columbia.
Bull. Fish. Res. Bd Can. 156:1-66.

density
Crehuet, Ramón Fernández y Maria Jesús del Val Cordón, 1960
Observaciones oceanograficas en la Bahia de Malaga (Marzo a marzo 1957). Bol. Inst. Esp. Ocean., 98: 1-29.

density (data)
Cromwell, T., 1954.
Mid-Pacific Oceanography II. Transequatorial waters, June-August 1950, January-March 1951.
U.S.F.W.S. Spec. Sci. Rept. - Fish. No. 131:

density
Cromwell, T., 1953.
Circulation in a meridional plane in the central equatorial Pacific. J. Mar. Res. 12(2):196-213, 9 textfigs.

density(data)
Cromwell, T., 1951.
Mid-Pacific oceanography, January through March 1950. Spec. Sci. Rept., Fish., No. 54:9 pp., 17 figs., station data.

D

density
Das, P.K., and N.C. Dhar, 1954.
Meteorological and oceanographic observations on I.N.S. Investigator (March-April 1952). Indian J. Met. Geophys. 5(1):16-28, 6 textfigs.

density
Deacon, G. E. R., 1933
A general account of the hydrology of the South Atlantic Ocean. Discovery Repts. 7:173-238, pls.8-10.

density
de Barros Machado, L., 1950.
Pequisas físicas e químicas do sistema hidrografico da região lagunar de Cananéia. 1. Cursos de águas. Bol. Inst. Paulista Oceanogr. 1(1):45-67, 3 textfigs.

density
DeBuen, F., 1937.
Sur quelques résultats de la IIIème croisière du Garde-côte "Xauen" dans les eaux de Guipuzea (Espagne) en 1934. Cons. Perm. Int. Expl. Mer, Rapp. Proc. Verb. 104:18-25, 3 textfigs.

density(data)
DeBuen, R., 1937.
Déterminations physico-chimiques dans les eaux de la Baie de Santander pendant l'été 1935.
Cons. Perm. Int. Expl. Mer, Rapp. Proc. Verb. 104:26-31, 6 textfigs.

density(fig.)
DeBuen, R., 1937.
Hydrographie du détroit de Gibraltar en 1934 (Croisière du Navire "Xauen"). Cons. Perm. Int. Expl. Mer, Rapp. Proc. Verb. 104:12-17, 8 textfigs.

density (data)
DeBuen, R., and F. de P. Navarro, 1935.
Condiciones oceanográficas de la costa catalana entre la frontera francesa y el Golfo de San Jorge. (Campañas del "Xauen" en mayo de 1933 y de 1934). Trab. Inst. Español Ocean. No. 14:47 pp., 1 pl., 20 textfigs.

density
DeBuen, O., 1925.
Croisière océanographique du Transport "Almirante Lobo". Cons. Perm. Int. Expl. Mer, Rapp. Proc. Verb. 37:33-57.

density(data)
Defant, A., 1940.
Die Lage des Forschungsschiffes "Altair" auf der Ankerstation 16 bis 20 Juni 1938 und das auf ihr gewonnene Beobachtungsmaterial. Wiss. Ergeb. Internat. Golfstrom-Unternehmung 1938, Lief. 3, In Oktober Beiheft, Ann. Hydr., usw., 68:35 pp., 6 textfigs.

density
Defant, A., 1935
Die Aquatoriale Gegenstrom. Sitzber. Preuss. Akad. Wis., Phys.-Math Kl. 28:1-25, 12 text figs.

density (σ_t)
Defant, A., 1931
Bericht über die Ozeanographischen Untersuchungen des Vermessungsschiffes "Meteor" in der Dänemarkstrasse und in der Irmingersee. Zweiter Bericht. Sitzber. Preuss. Akad. Wiss., Phys. Math. Kl.19: 17 pp., 5 text figs.

density
Defant, A., G. Böhnecke, H. Wattenberg, 1936.
I. Plan und Reiseberichte die Tiefenkarte das Beobachtungsmaterial. Die Ozeanographischen Arbeiten des Vermessungsschiffes "Meteor" in der Dänemarkstrasse und Irmingersee während der Fischereischutzfahrten 1929, 1930, 1933 und 1935, Veroffentlichungen des Instituts für Meereskunde, n.f., A. Geogr.-naturwiss. Reihe, 32:1-152 pp., 7 text figs., 1 plate.

density
de Quay, Alain, 1966.
Calcul théorique de la densité de l'eau de mer.
Cahiers Océanogr., C.C.O.E.C.,19(1):43-61.

density
Derbek, F.A., 1910-1912.
Iz otcheta po estestuenno- istoricheskim rabotam na opisnom parokhode "Okhotsk" vo vremia plavaniia Gidrograficheskoi ekspeditsii Vostochago Okeana v 1909 i 1910 gg. (From the report on the natural history of the Okhotsk trips during the Hydrographical Expeditions to the Eastern Ocean during 1909-1910). Ezhegodan. Zool. Muz., Akad. Nauk, 15(4):xx-xxx, 18(3/4):xxiii-lv.

density
Dietrich, G., 1953.
Verteilung, Ausbreitung und Vermischung der Wasserkörper in der südwestlichen Nordsee auf Grund der Ergebnisse der "Gauss"-Fahrt in Februar /März 1952. Ber. Deutsch. Wiss. Komm. Meeresf., Stuttgart, 13(2):104-129, 12 textfigs.

density
Dietrich, G., 1950.
Kontinentale Einflüsse auf Temperatur und Salzgehalt des Ozeanwassers. Deutsche Hydro. Zeits. 3(1/2):33-39, 3 textfigs., Pls. 1-2.

density
Di Maio, A., and L. Trotti, 1961
Sur la formation d'eau de fond et d'eau profonde dans la Mer Ligurienne.
Cahiers Oceanogr., C.C.O.E.C., 13(4): 227-233.

density(data)
Dinsmore, R.P., R.M. Morse and Floyd M. Soule, 1960.
International ice observation and Ice Patrol Service in the North Atlantic Ocean. Season of 1959.
U.S. Coast Guard Bull. No. 44:1-99.

density
Dobrobolsky, A.D., Editor, 1968.
Hydrology of the Pacific Ocean. (In Russian).
P.P. Shirshov Inst.Okeanol.,Akad.Nauk,Isdatel "Nauka",Moskva, 524 pp.

density
Dodimead, A.J. 1968.
Oceanographic conditions in the central subarctic Pacific region, winter 1966
Techn. Rept. Fish. Res. Bd. Can., 75: 10 pp., + 11 figs. (multilithed).

density
Dodimead, A.J., and F. Favorite, 1961.
Oceanographic atlas of Pacific Subarctic Region, Summer of 1958.
Fisheries Res. Bd., Canada, Manuscript Rept. Ser. (Oceanogr. and Limnol.), No. 92: 6 pp., 40 figs. (Multilithed).

Doe, L.A. Earlston

density
Doe L. A. E., 1965.
Physical conditions on the shelf near Karachi during the post-monsoonal calm, 1964. Ocean Sci. and Ocean Eng., Mar. Sci. Techn. Soc. - Amer. Soc. Limnol. Oceanogr., 1: 278-292.

density
Doe, L.A.E., 1955.
Offshore waters of the Canadian Pacific coast. J. Fish. Res. Bd., Canada, 12(1):1-34, 19 textfigs.

density (data)
*Drainville, Gerard, 1968.
Le fjord du Saguenay: 1. Contribution à l'océanographie.
Le Naturaliste canadien, 95(4):809-855.

density
*Düing, Walter, 1967.
Die Vertikalzirkulation in den Küstennahen Gewässern des Arabischen Meeres während der Zeit des Nordostmonsuns.
"Meteor" Forschungsergebnisse, (A)(3):67-83.

density
Düing, Walter, 1965.
Strömungsverhältnisse im Golf von Neapel.
Pubbl. Staz. Zool., Napoli, 34:256-316.

density
*Düing, Walter, Klaus Grasshoff und Gunther Krause, 1967.
Hydrographische Beobachtungen auf einem Äquatorschnitt im Indischen Ozean.
"Meteor Forschungsergebnisse (A)(3):84-92.

density
*Düing, Walter and Peter H. Koske, 1967.
Hydrographische Beobachtungen im Arabischen Meer während der Zeit des Nordostmonsuns 1964/65.
"Meteor" Forschungsergebnisse (A)(3):1-43.

density
Dunbar, M.J., 1958.
Physical oceanographic results of the "Calanus" Expeditions in Ungava Bay, Frobisher Bay, Cumberland Sound, Hudson Strait and Northern Hudson Bay, 1949-1955.
J. Fish. Res. Bd., Canada, 15(2):155-201.

density
Dunbar, M.J., 1955.
Density inversions in Canadian eastern Arctic waters. Nature 176:703.

E

density
Eggvin, J., 1933.
A Norwegian fat-herring fjord. An oceanographical study of the Eidsfjord. Repts. Norwegian Fish. Mar. Invest. 4(6):22 pp., 11 textfigs.

density(data)
El-Sayed, Sayed Z., and Enrique F. Mandelli, 1965
Primary production and standing crop of phytoplankton in the Weddell Sea and Drake Passage. In: Biology of Antarctic seas, II.
Antarctic Res. Ser., Am. Geophys. Union, 5:87-106

density
Emilsson, Ingvar, 1961
The shelf and coastal waters off southern Brazil.
Bol. Inst. Oceanogr., Univ. Sao Paulo, 11(2): 101-112.

density
Emilsson, I., 1959.
Alguns aspectos fisicos é quimicos das águas marinhas brasileiras.
Ciancia e Cultura, 11(2):44-54.

density(chiefly data)
Emilsson, I., 1956.
Relatorio e resultados fisico-quimicos de tres cruzeiros oceanograficos em 1956.
Contrib. Avulsas, Inst. Ocean., Univ. Sao Paulo, Ocean. Fisica, No. 1:1-70.

density(data)
Eyries, M., 1956.
Activité océanographique de l'Escoteur "l'Aventure" en 1955.
Bull. d'Info., C.C.O.E.C., 8(10):527-545, 14 pls.

density
Eyries, M., 1954.
La campagne hydrologique de l'escoteur "L'Emporte" entre Madère et la côte marocaine (Avril-Mai 1953). Bull. d'Info., C.C.O.E.C. 6(4):163-168, 5 pls.

density (data)
Eyries, M., and M. Menache, 1953.
Contribution à la connaissance hydrologique de l'Océan Indien entre Madagascar et La Réunion.
Bull. d'Info., C.C.O.E.C. 5(10):433-438, 9 pls.

F

density
Federov, K.N., 1956.
On the calculation of the field of density in the ocean. Izv. Akad. Nauk, SSSR, Ser. Geofiz., 1956 (9):1044-

density (data)
Fieux, Michèle 1971.
Observations hydrologiques hivernales dans le Rech Lacaze-Duthiers (Golfe du Lion).
Cah. océanogr. 23 (8): 677-686.

density(data)
Filarski, J., 1955.
Hydrographical conditions of the southern Baltic for the period from April 1952 to May 1953.
Prace Morskiego Inst. Ryback., Gydni, No. 8:255-282.

density (data)
Filippov, B.A., 1965.
On the question of the hydrology of coastal waters in the Antarctic. (In Russian).
Trudy, Gosudarst. Okeanogr. Inst., No. 87:64-76.

Fisheries Research Board of Canada

density
Fisheries Research Board of Canada, 1949.
Observations of the water temperature, salinity and density on the Pacific coast of Canada. 1948. Vol. 8:55 pp., 1 fig. (mimeographed).

density
Fisheries Research Board of Canada, 1949
Observations of sea water temperature, salinity, and density on the Pacific Coast of Canada. Vol. 7 incl. data from 1946 to 1947, 105 pp. (mimeographed), 1 fig.

density
Fisheries Research Board of Canada, 1948
Observations of sea water temperature, salinity, and density on the Pacific coast of Canada. Including data from 1944 to 1945. Vol.6:105 pp., 1 fig. (mimeographed).

density
Fisheries Board of Canada, 1948
Observations of sea water temperature, salinity, and density on the Pacific Coast of Canada. Vol.III. Including data from 1938 to 1939, 93 pp., 1 chart (mimeographed).

density
Fisheries Research Board of Canada, 1948
Observations of sea water temperature, salinity, and density on the Pacific Coast of Canada. Vol.V (including data from 1942-1943): 116 pp., 1 fig. (mimeographed). (Pacific Oceanographic Group, Nanaimo).

density
Fisheries Board of Canada, 1947.
Observations of sea water temperature, salinity and density on the Pacific coast of Canada. Vol. 1, including data from 1914 to 1934, 113 pp., 1 chart (mimeographed).

density
Fisheries Research Board of Canada, 1947
Sea water temperature, salinity and density on the Pacific coast of Canada. Vol.II including data from 1935-1937. 109 pp., 1 fig. (mimeographed)

Fleming, Jno. A.

density
Fleming, J.A., C.C. Ennis, H.U. Sverdrup, S.L. Seaton, W.C. Hendrix, 1945
Observations and results in physical oceanography. Graphical and tabular summaries. Ocean. 1-B, Sci. Res. of Cruise VII of the Carnegie during 1928-1929 under command of Capt. J.P. Ault, Carnegie Inst., Washington, Publ. 545:315 pp., 5 tables, 254 figs.

Sigma-T.
Fleming, R.H., 1939
Tables for Sigma-T. J. Mar. Res.2:9-11.

density
Flores P., Luis Alberto 1967.
Informe preliminar del crucero 6611 de la primavera de 1966 (Cabo Blanco-Punta Coles).
Informe, Inst. Mar Peru 17: 16pp.

density
Flores, Luis, Oscar Guillén y Rogelio Villanieva, 1966.
Informe preliminar del Crucero de invierno 1965 (Máncora-Morro Sama).
Inst. Mar. Peru, Informe, No.11:1-34 (multilithed)

density
Flores, Luis, Oscar Guillén y Rogelle Villanueva, 1966.
Informe preliminer del crucere de invierne 1965 (Mancera-Morra Sama).
Inst. Mar. Peru, Informe, No. 11:3-34. (multilithed)

density
Fofonoff, N. P. and S. Tabata, 1966.
Variability of oceanographic conditions between Ocean Station P and Swiftsure Bank off the Pacific coast of Canada.
J. Fish. Res. Bd., Canada, 23(6):825-868.

density
Fomin, L.M. 1968.
The relationship between the spectra of the velocity or of the pressure or density distributions in the ocean. (In Russian; English abstract)
Fisika Atmosfer. Okean. Izv. Akad. Nauk SSSR 4 (3): 335-340.

density
Fonselius, Stig H., 1960.
Hydrography of the Baltic deep basins. III.
Rept., Fish. Bd., Sweden, Ser. Hydrogr., 23: 1-97.

Density (data)
Forsbergh, Eric D., William W. Broenkow, 1965.
Observaciones oceanograficas del oceano Pacifico oriental recolectadas por el barco Shoyo Maru, octubre 1963-marzo 1964.
Comision Interamericana del Atun Tropical, Bol. 10(2):85-237.

density (data)
Franco, Paolo 1970.
Oceanography of northern Adriatic Sea. 1. Hydrologic features: cruises July-August and October-November 1965.
Archo Oceanogr. Limnol. 16 (Suppl.): 1-93.

density

Franco, Paolo, 1967.
Condizioni idrologiche e produttività primaria nel Golfo di Venezia - nota preliminare.
Arch. Oceanogr. Limnol. 15(1): 69-83.

density

Friedrich, H., 1962
About the surface structure as derived from the density distribution in the oceans.
Proc. Symposium on Mathematical-Hydrodynamical Methods of Phys. Oceanogr., Sept. 1961, Inst. Meeresk., Hamburg: 411-420.

density

Fukuo, Y., 1964.
On the deviation from Knudsen's formula of the density of sea water and its bearing on the productivity of the sea.
Mem. Coll. Sci., Univ. Kyoto, (A), 30(3):273-321.

density

*Fukuoka, Jiro, 1967.
Coastal upwelling near Venezuela. 2. Certain periodicities of hydrographical conditions.
Boln Inst. Oceanogr., Univ. Oriente, 5(½):84-95.

density

Fukuoka, Jiro, 1965.
Hydrography of the adjacent sea. (1). The circulation in the Japan Sea.
J. Oceanogr., Soc., Japan, 21(3):95-

density

Fukuoka, Jiro, 1962.
Abyssal circulation in the Atlantic near the poles and abyssal circulation in the Pacific and other oceans in relation to the former.
J. Oceanogr. Soc., Japan, 18(1):5-12.

JEDS Contrib. No. 27.

density

Fukuoka, Jiro, 1962
Characteristics of hydrography of the Japan Sea - in comparison with hydrography of the North Pacific -. (In Japanese; English abstract).
J. Oceanogr. Soc., Japan, 20th Ann. Vol., 180-188.

density

Fukuoka, J., 1955.
Oceanic condition in the north-western Pacific Ocean (preliminary report).
Rec. Ocean. Wks., Japan, 2(2):102-107, 4 figs.

density

Fukuo, Yoshiaki, 1962
On the deviation from Knudsen's formula of the density of sea water and its bearing on the productivity of the sea. (In Japanese: English abstract).
J. Oceanogr. Soc., Japan, 20th Ann. Vol., 386-408.

density

Funder, Th. P., 1916.
Hydrographic investigations from the Danish School Ship "Viking" in the southern Atlantic and Pacific in 1913-14. Medd. Komm. Havundersøgelser, Ser. Hydrogr., 2(6):28 pp.

density (data)

Furnestin, Jean, 1960.
Hydrologie de la Méditerranée occidentale (Golfe du Lion, Mer Catalane, Mer d'Alboran, Corse orientale) 14 juin-20 juillet, 1957.
Rev. Trav. Inst. Pêches Marit., 24(1):5-120.

G

density

Gaarder, Karen Ringdal, 1938.
Phytoplankton studies from the Tromsø district, 1930-31. Tromsø Mus. Årshefter, Naturhist. Avd., 11, 55(1):159 pp., 4 fold-in pls., 12 textfigs.

density (data)

Gaarder, T., and R. Spärck, 1932.
Hydrographisch-biochemische Untersuchungen in norwegischen Austern-Pollen. Bergens Mus. Aarbok, Naturvidensk.-rekke, No. 1:5-144, 75 textfigs.

density(data)

Gade, Herman G., 1963
Some hydrographic observations of the inner Oslofjord during 1959.
Hvalrådets Skrifter, No. 46:1-62.

density

Gade, Herman G., 1961.
On some oceanographic observations in the south-eastern Caribbean Sea and adjacent Atlantic Ocean with special reference to the influence of the Orinoco River.
Bol. Inst. Oceanogr., Univ. de Oriente, Venezuela, 1(2):287-342.

density

Gámez, J.C., 1942.
El indice de refraccion en el estudio de la densidad del agua del mar.
Anales Fis. y Quim., Madrid, 38:148-174.

Garner, D.M.

density(data)

Garner, D.M., 1958.
The Antarctic convergence south of New Zealand.
N.Z.J. GEOL. GEOPHYS., 1(3):577-594.

density (data)

Garner, D.M., and N.M. Ridgeway, 1965.
Hydrology of New Zealand offshore waters.
N.Z. Dept. Sci. Ind. Res. Bull., 162 (N.Z. Oceanogr. Inst. Mem. No. 12): 62 pp., 1 fold chart.

density(data)

Gennesseaux, Maurice, 1960.
L'oxygène, les phosphates et les nitrates dissous dans les eaux du bassin de la Méditerranée occidentale (entre Baléares & Gibraltar).
Trav. du Centre de Recherches et d'Études Océan. 3(4):5-22.

Also in:
Trav. Lab. Géol. Sous-Marine, 10(1960).

density (some data)

Genovese, S., 1954.
Osservazioni oceanografiche eseguite sui campi di pesca dell'alalungo delle isola Eolia.
Boll. Pesca, Piscicol., Idrobiol., n.s., 9(2): 186-196.

density

Genovese, S., 1953.
Osservazioni idrologiche esequite nella tonnara Capo San Marco (Sciacca) durante la campagna di pesca 1953. Boll. Pesca., Piscicolt., Idrobiol n.s., 8(2):241-251, 4 textfigs.

density (data)

Gesenzvey, A.N., 1963.
Orders of magnitude of oceanological properties. Questions of Physical Oceanography. (In Russian; English abstract).
Trudy Inst. Okeanol., Akad. Nauk, SSSR, 66:91-124.

density

Gilmartin, M., 1962.
Annual cyclic changes in the physical oceanography of a British Columbia fjord.
J. Fish. Res. Bd., Canada, 19(5):921-974.

density

Gilmour, A.E., 1958
The suspended plummet method for density measurements. N.Z.J. Sci. 1(3): 391-401.

density

Glowinska, A., 1954.
[Hydrologic research in the southern Baltic in 1951.] Rept. Sea Fish Inst., Gdynia, No. 7:159-190, 27 textfigs.

density (data)

Glowinska, A., 1951.
[Hydrographical conditions in the southern Baltic in the time from August 1949 to May 1951.]
Trudy Morskogo Rybnogo Inst., Gdyna (Rept. Sea Fish. Inst.) No. 6:119-120, 7 textfigs.

density

Goedecke, Erich, 1958.
Über Höhe und Eintrittszeit der Extreme sowie deren Schwankungen im mittleren Jahresgang von Temperatur, Salzgehalt und Dichte des Wassers in der Deutschen Bucht.
Deutschen Hydrogr. Zeits., 11(4):137-165.

density

Goedecke, E., 1955.
Über die Intensität der Temperatur-, Salzgehalts, und Dichteschichtung der Deutschen Bucht.
Deutsche Hydrogr. Zeits. 8(1):15-28, 9 textfigs.

density

Goedecke, E., 1939.
Beitrage zur Hydrographie der Helgoland umgebenden Gewasser. I. Die Oberflächenverhältnisse bei Helgoland-Reede. Ann. Hydr., Jahrg. 67, Heft. IV: 161-176, folded charts, diagrams, tables.

density (data)

Gorgy, Samy, 1966.
Les pêcheries et le milieu marin dans le Secteur Méditerranéen de la Republique Arabe linie.
Rev. Trav. Inst. Pêches Marit, 30(1):25-

density (data)

Gorgy, O.S., 1966.
Contribution a l'étude du milieu marin et de la pêche en Mer Rouge (secteur de la Republique Arabe linie).
Rev. Trav. Inst. Pêches Marit., 30(1):93-112.

density

Gostan, Jacques, 1967.
Comparaison entre les conditions hydrologiques et climatiques observées dans le golfe de Gênes pendant les hivers 1962-1963 et 1963-1964.
Ann. océanogr., 19(5):391-416.

density

Gostan, J., 1962.
Observations hydrologiques en mer Ligure pendant l'hiver 1961.
Bull. Inst. Océanogr., Monaco, 59(1250):19 pp., annexe.

density

Gran, H.H., 1927.
The production of plankton in the coastal waters off Bergen, March-April, 1922. Rept. Norwegian Fish. Mar. Invest. 3(8):74 pp., 8 textfigs.

density

Gran, H.H., and T. Braarud, 1935
A quantitative study of the phytoplankton in the Bay of Fundy and the Gulf of Maine (including observations on hydrography, chemistry, and turbidity).
J. Biol. Bd., Canada, 1(5):279-467, 69 text figs.

density

Groen, P., 1948.
Internal waves in certain types of density distribution. Nature 161(4081):92.

density (data)

Grøntved, J., 1952.
Investigations on the phytoplankton of the southern North Sea in May 1947. Medd. Komm. Danmarks Fisk.- og Havundersøgelser, Plankton ser., 5(5): 1-49, 1 pl., 21 tables, 24 textfigs.

density

Guarrera, S.A., 1950.
Estudios hidrobiologicos en el Rio de la Plata. Rev. Inst. Nac., Invest. Cienc. Nat., Ciencias Botanicas 2(1):62pp.

density (data)

Gudkovich, Z.M., 1955.
Results of a preliminary analysis of the deep-water hydrological observations.
Material. Nabluid. Nauch.-Issledov. Dreifuius. 1950/51, Morskoi Transport, 1:40-46, 48-170.

David Krauss, translator in AMS-Astia Doc. 11713

density

Guillén, Oscar, 1967.
Anomalies in the waters off the Peruvian Coast during March and April 1965.
Stud. trop. Oceanogr., Miami, 5:452-465.

density

Guillen G., Oscar, y Luis Alberto Flores P. 1968.
Informe preliminar del crucero 6702 del verano de 1967 (Cabo Blanco - Arica).
Informe Inst. Mar Peru 18: 12pp.

density

Gurikova, Z.F., 1964.
Computation of the surface field of water density in the Pacific Ocean. (In Russian).
Izv., Akad. Nauk, SSSR, Ser. Geofiz. (7):1070-1083

H

density

Hachey, H.B., 1947
Water transports and current patterns for the Scotian shelf. J. Fish. Res. Bd., Canada 7(1):1-16, 7 text figs.

density

Haefner, Paul A., Jr. 1967.
Hydrography of the Penobscot River (Maine) estuary.
J. Fish. Res. Bd., Can., 24(7):1553-1571

density (data)

Halim, Youssef, 1960.
Étude quantitative et qualitative du cycle écologique des dinoflagellés dans les eaux de Villefranche-sur-Mer. (1953-1955).
Ann. Inst. Océanogr. Monaco 38: 123-232

density

Hampel, C.A., 1950.
Densities and boiling points of sea water concentrates. Ind. Chem. Eng., 42:383-386.

density

Hansen, W., 1946.
Die Strömungen im Barents-Meer im Sommer 1927 auf Grund der Dichteverteilung.
Arch. Deutschen Seewarte 55(5):1-41, 4 pls.

density

Hela, I., 1958
A hydrographical survey of the waters in the Aland Sea.
Geophysica, Helsinki, 6(3/4):219-242.

density

Hela, I., 1955.
Ecological observations on a locally limited red tide bloom. Bull. Mar. Sci., Gulf and Caribbean, 5(4):269-291, 16 textfigs.

density

Helland, Palmar, 1963.
Temperature and salinity variations in the upper layers at Ocean Weather Station M (66°N, 2°E).
Acta Univ. Bergensis, Ser. Mathematica Rerumque Naturalium, Arbok, Univ. Bergen, Mat.-Naturv. Ser., 1963, No. 16:26 pp.

Also in:

Collected Papers, Weather Ship Station M, 66° N, Geophys. Inst., Univ. Bergen, 1963.

density

Helland-Hansen, B., 1907.
Current measurements in Norwegian fjords, the Norwegian Sea in 1906. Bergens Mus. Aarb., 1907, (15):61 pp., 2 pls., 13 textfigs., tables, charts

density (includes data)

Helland-Hansen, B., 1905.
On hydrographic investigations in the Faroe-Shetland Channel and the northern part of the North Sea in the year 1902. Fish. Bd., Scotland, North Sea Fish. Invest. Comm., Rept. No. 1:1-49.

density

Herlinveaux, R.H., 1962
Oceanography of Saanich Inlet in Vancouver Island, British Columbia.
J. Fish. Res. Bd., Canada, 19(1):1-37.

density

Herlinveaux, R.H., and J.P. Tully, 1961
Some oceanographic features of Juan de Fuca Strait.
J. Fish. Res. Bd., Canada, 18(6):1027-1071.

density(data)

Herrera, Juan, y Ramón Margalef, 1963
Hidrografía y fitoplancton de la costa comprendida entre Castellón y la desembocadura del Ebro, de julio de 1960 a junio de 1961.
Inv. Pesq., Barcelona, 24:33-112.

density(data)

Herrara, Juan, and Ramon Margalef, 1961
Hidrografía y fitoplancton de las costas de Castellón de julio de 1958 a junio de 1959.
Inv. Pesq., Bacelona, 20:17-63.

density (data)

Hidaka, K., and T. Yamagiwa, 1949.
On the absolute velocity of the Subarctic Intermediate Current to the south of Japan. Geophys. Notes, Tokyo, 2(5):1-4, 3 textfigs.

density

Hirasawa, K., 1940.
On the Yaku-mizu. J. Oceanogr. 12(2):365-370.

density

Hishida, K., 1950.
Studies on the material for the oceanographical observations along the coast. (1). Ocean. Mag., Tokyo, 2(3):123-128, 2 textfigs.

density

Høiland, E., 1953.
On the dynamic effect of variation in density on two dimensional perturbations of flow with constant sheer. Geofys. Publ. 18(10):1-12.

density

Horrer, Paul L., 1962
Oceanographic studies during Operation "Wigwam". Physical oceanography of the test area.
Limnol. and Oceanogr., Suppl. to Vol. 7: vii-xxvi.

density (data)

Houtman, Th. J., 1967.
Water masses and fronts in the Southern Ocean south of New Zealand.
Bull. N.Z. Dept. Sci. Ind. Res., 174: 1-40
Also: N.Z. Oceanogr. Inst. Mem. 36.

density (data)

Houtman, Th. J., 1965.
Winter hydrological conditions of coastal waters south of Kaikoura Peninsula.
N.Z.J. geol. geophys. 8(5):807-819.

density

Hulburt, Edward M., 1966.
The distribution of phytoplankton, and its relationship to hydrography, between southern New England and Venezuela.
J. Mar. Res., 24(1):67-81.

density

Huntsman, A.G., W.B. Bailey and H.B. Hachey, 1954
The general oceanography of the Strait of Belle Isle. J. Fish. Res. Bd., Canada, 11(3):198-260, 35 textfigs.

density (data)

Hureau, J.C., 1962.
Observations hydrologiques en Terre Adélie de janvier 1961 à janvier 1962 (Expéditions Polaires Francaises, (Missions P.E. Victor).
Bull. Mus. Nat. Hist. Nat., (2), 34(5):412-426.

I

density

Ichiye, T., 1962.
Circulacion y distribucion de la masa de agua en el Golfo de Mexico. (In Spanish and English).
Geofisica Internacional, Rev. Union Geofis. Mexicana, Inst. Geofis., Univ. Nacional Autonoma de Mexico, 2(3):47-76, 22 figs.

density

Ichiye, Takashi, 1962
On formation of the intermediate water in the northern Pacific Ocean.
Geofisica Pura e Applicata, 51(1):108-119.

density

Ichiye, T., 1960
On the hydrography near Mississippi Delta.
Oceanogr. Mag., Tokyo, 11(2): 65-78.

density

Ichiye, T., 1958.
On convection circulation and density distribution in a zonally uniform ocean.
The Ocean. Mag. Vol. 10 (1) :97-136.
Jap. Meteor. Agency

density

Ichiye, T., 1958.

A theory of vertical structure of density in the ocean.
J. Oceanogr. Soc., Japan, 14(2):35-46.

density

Ichiye, T., 1954.
[Change of density of coastal water due to precipitation.] Oceanogr. Rept. Centr. Meteorol. Obs., 3(3):73-79.

Transl. cited FWS Spec. Sci. Rept., Fish., 227.

density

Ichiye, Takashi, 1952?
[On the hydrographical condition in the Kuroshio region (1952). 1. Southern area of Honshu.]
Bull. Kobe Mar. Obs., No. 163: 1-30.

density

Ichie, T., K. Tanioka, and T. Kawamoto, 1950.
Reports of the oceanographical observations on board the R.M.S. "Yushio Maru" off Shionomisaki (Aug. 1949). Paper and Repts., Ocean., Kobe Mar. Obs., Ocean. Dept., No. 5:15 pp., 33 figs. (Odd atlas-sized pages - mimeographed).

density

Ichie, T., and K. Tanioka, 1949.
On the report of the observations of the hydrographical conditions in the Osaka-wan after the heavy rain caused by the typhoon "Della". Papers and Repts., Ocean., Kobe Mar. Obs., Ocean. Dept., No. 3:21 pp., 19 figs., (mimeographed).

density

Iselin, C., 1930
A report on the coastal waters of Labrador based on explorations of the "Chance" during the summer of 1926. Proc. Am. Acad. Arts Sci., 66(1):1-37, 14 text figs.

density

Ishino, Makoto, 1963.
Studies on the oceanography of the Antarctic Circumpolar waters.
J. Tokyo Univ., Fish., 49(2):73-181.

density, data

Istoshin, Yu. V., 1961
[The range of distribution of the "eighteen degree" waters of the Sargasso Sea.]
Okeanologiia, Akad. Nauk, SSSR, 1(4):600-607.

density (data)

Jacobs, Stanley S., Peter M. Bruchausen and Edward B. Bauer, 1970
Cruises 32-36, 1968, hydrographic stations, bottom photographs, current measurements. Eltanin Repts. Lamont-Doherty Geol. Obs., Nat. Sci. Found. U.S. Antarctic Res. Program, 460 pp. (multilithed)

density

Jacobsen, J.P., 1943.
The Atlantic Current through the Faroe-Shetland Channel and its influence on the hydrographical conditions in the northern part of the North Sea, the Norwegian Sea and the Barents Sea.
Cons. Perm. Int. Expl. Mer, Rapp. Proc. Verb, 112:5-47, 29 textfigs.

density(data)

Jacobsen, J.P., 1925.
Die Wasserumsetzung durch den Öresund, den Grossen und Kleinen Belt. Medd. Komm. Havundersøgelser, Ser. Hydrogr., 2(9):72 pp.

density

Jacobsen, J.P., 1916
Contribution to the Hydrography of the Atlantic. Researches from the M/S "Margrethe" 1913. Medd. fra Komm. for Havundersøgelser, Serie; Hydrografi, 2(5): 23 pp., 7 figs.

DENSITY

Jacobsen, J.P., 1915
Hydrographical investigations in Faeroe Waters cruise of the M/S "Margrethe" in 1913. Medd. fra Komm. for Havundersøgelser, serie Hydrografi II(4): 47 pp., 15 textfigs.

density(data)

Jacobsen, J.P., 1908.
Der Sauerstoffgehalt des Meereswassers in den daenischen, innerhalb Skagens. Medd. Komm. Havundersøgelser, Ser. Hydrogr., 1(12):23 pp., 5 pls.

Gewässern

density(data)

Jacobsen, J.P., 1908.
Mittelwerte von Temperatur und Salzgehalt bearbeitet nach hydrographischen Beobachtungen in Dänischen Gewässern 1880-1907. Medd. Komm. Havundersøgelser, Ser. Hydrogr., 1(10):26 pp., 11 pls.

density (data)

Jakleln, A., 1936
Oceanographic investigations in East Greenland waters in the summers of 1930-1932. Skr. om Svalbard og Ishavet No. 67:79 pp., Pl. 2, 28 text figs.

density (data only)

Japan, Japanese Oceanographic Data Center 1970.
Suro No.3, Hydrographic Office, Republic of Korea, July 22-August 9, 1968, South of Korea.
Data Rept. CSK (HDC Ref. 24KD23) 197: 17 pp. (multilithed).

DENSITY

Japan, Kobe Marine Observator., 1955.
[Report of the coastal stations.]
J. Ocean., Kobe Mar. Obs. (2)6(1):36.

density

Japan, Kobe Marine Observatory, 1954.
[The outline of the oceanographical observations off Shionomisaki on board the R.M.S. "Syunpumaru" (May 1954).] J. Ocean. (2)5(5):1-11, 14 figs

density

Japan, Kobe Marine Observatory, 1954.
[The reports of the coastal observations.]

Jan. 1954 - J. Ocean. (2)5(1):22-23.
Feb. 1954 - J. Ocean. (2)5(2):20-21.
Mar. 1954 - J. Ocean. (2)5(3):12-13.
Apr. 1954 - J. Ocean. (2)5(4):8-9.
May 1954 - J. Ocean. (2)5(5):37-39.
June 1954 - J. Ocean. (2)5(6):7-9.
July 1954 - J. Ocean. (2)5(7):21.
J. Ocean. (2)5(8):38-39.
J. Ocean. (2)5(11):7-9.

density

Japan, Kobe Marine Observatory, 1953.
[Report on the coastal stations.]
Jan. 1953 - J. Ocean. (2)4(1):33-34.
Feb. 1953 - J. Ocean. (2)4(2):38-39.
Mar. 1953 - J. Ocean. (2)4(3):22-23.
Apr. 1953 - J. Ocean. (2)4(4):36-37.
May, 1953 - J. Ocean. (2)4(5):20-21.
June 1953 - J. Ocean. (2)4(6):42-43.
July 1953 - J. Ocean. (2)4(7):21-22.
Aug. 1953 - J. Ocean. (2)4(8):29-30.
Sept. 1953 - J. Ocean. (2)4(9):40-41.
Oct. 1953 - J. Ocean. (2)4(10):25-26.
Nov. 1953 - J. Ocean. (2)4(11):27-28.
Dec. 1953 - J. Ocean. (2)4(12):18-19.

density

Japan, Kobe Marine Observatory, 1953?
[The report of the coastal observations (Nov. 1953).] J. Ocean., Kobe Mar. Obs., 2nd ser., 4(11):27-29.

There were similar reports in all other numbers of Vol. 4, but these have not been indexed.

density

Japan, Maizuru Marine Observatory, 1956.
[Report of the oceanographical observations north off San'in and Hokuriku districts in summer 1954.]
[Report of the oceanographical observations north off San'in and Hokuriku districts in summer 1955.]
Bull. Maizuru Mar. Obs., (5):85-94; 49-56.

density

Japan, Maizuru Marine Observatory, 1955.
[Report of the oceanographic observations off San'in and Hokuri-ku (July-August, 1952).]
Bull. Maizuru Mar. Obs., (4):49-63 (168-181).

density

Japan, Maritime Safety Board, 1955.
[Observation of Kuro-Sio region off Sio-no-misaki and Ensyunada from 1942-1943.] Publ. No. 981, Hydrogr. Bull. Spec. No. 17:63-69.

density(data)

Japan, Maritime Safety Board, 1954.
[Tables of results from oceanographic observations in 1950.] Publ. No. 981, Hydrogr. Bull., Maritime Safety Bd., Spec. Number (Ocean. Repts.) No. 14:26-164, 5 textfigs.

density

Japan, Maizuru Marine Observatory, 1955.
[Report of the oceanographical observations (taken) off San'in and Hokuriku (July-August, 1953).]
Bull. Maizuru Mar. Obs., (4):13-27. 217-231.

density

Japan, Nagasaki Marine Observatory, 1959.
Report of the oceanographic observations in the sea west of Japan from January to February 1958. Results Mar. Meteorol. & Oceanogr., J.M.A., 23: 43-49.

Also in:
Oceanogr. & Meteorol., Nagasaki Mar. Obs., 10(194). 1960.

density

Japan, National Committee for the International Geophysical Year, 1960
The results of the Japanese oceanographic project for the International Geophysical Year 1957/58: 198 pp.

density

Jebsen-Marwedel, H., 1962
Dynaktivität des Wassers Oberhalb des Gefrierpunktes.
Proc. Symposium on Mathematical-Hydrodynamical Methods of Phys. Oceanogr., Sept. 1961, Inst. Meeresk., Hamburg, 421-425.

density

Jerlov, N.G., 1956.
The Equatorial currents in the Pacific Ocean. Repts. Swedish Deep-Sea Exped., 3(Phys. Chem. 6): 129-154.

density

Jerlov, N.G., 1953.
The Equatorial Currents in the Indian Ocean. Repts. Swedish Deep-se Exped., 1947-1948, Phys. Chem., 3(5):115-125, 14 textfigs.

density
Jerlov, N.G., 1953.
Studies of the Equatorial Currents in the Pacific
Tellus 5(3):308-314, 5 textfigs.

density
Jerlov, N.G., 1953.
Studies of the Equatorial Currents in the Pacific
Tellus 5(3):308-314, 9 textfigs.

Density (data)
Jillett, J.B. 1969.
Seasonal hydrology of waters off the Otago Peninsula, south-eastern New Zealand.
N.Z. Jl mar. freshw. Res. 3(3):349-375.

density
Jones, Robert S., B.J. Copeland and H.D. Hoese, 1965.
A study of the hydrography of inshore waters in the western Gulf of Mexico off Port Aransas, Texas.
Publ. Inst. Mar. Sci., Port Aransas, 10:23-32.

K

density
Kado, Y., 1955.
The seasonal change of plankton and hydrography of the neighboring Sea of Mukaishima.
J. Sci., Hiroshima Univ., B, 15:193-204, 5 textfigs.

density(data)
Karlovac, O., 1956.
Station list of the M.V. "Hvar" fishery-biological cruises, 1948-1949. Inst. Ocean. i Ribarstvo Split, Repts., 1(3):177 pp.

Density
Kato, Kenji, 1961.
Oceanographical studies in the Gulf of Carisco.
1. Chemical and hydrographical observations in Januar y, 1961.
Bol. Inst. Oceanograf., Univ. Oriente, Cumana, Venezuela, 1(1):49-72.

density(data)
Kato, Kenji, 1961
Some aspects on biochemical characteristics of sea water and sediments in Mochima Bay, Venezuela.
Bol. Inst. Oceanogr., Univ. de Oriente, Venezuela 1(2):343-358.

density
Kay, H., 1954.
Untersuchungen zur Menge und Verteilung der organischen Substanz im Meerwasser.
Kieler Meeresf. 10(2):202-214, Figs. 18-25.

density
Ketchum, Bostwick H., and Nathaniel Corwin, 1964
The persistence of "winter" water on the continental shelf south of Long Island, New York.
Limnology and Oceanography, 9(4):467-475.

density
Ketchum, B.H., A.C. Redfield, and J.C. Ayers, 1951.
The oceanography of the New York Bight. P.P.O.M. 12(1):1-46, 20 textfigs.

Reviewed: J. du Cons. 18(2):246-247, by D. Vaux.

density(data)
Kiilerich, A., 1957.
Galathea-Ekspeditionens arbejde i Philippiner-graven. Ymer, 1957(3):200-222.

density
Kindiushev, V.I., 1965.
On the increase of water density in connection with mixing in the transport of Mediterranean water masses into the Atlantic Ocean. (In Russian).
Okeanologiia, Akad. Nauk, SSSR, 5(4):617-624.

density(data)
King, J.E., T.S. Austin and M.S. Doty, 1957.
Preliminary report on Expedition EASTROPIC.
USFWS Spec. Sci. Rept., Fish., 201:155 pp.

density
Kinzer, J., 1969.
Quantitative distribution of zooplankton in surface waters of the Gulf of Guinea during August and September 1963.
Actes Symp. Oceanogr. Ressources halieut. Atlant. trop., Abidjan, 20-28 Oct. 1966, UNESCO 231-240.

density(data)
Klepikova, V.V., Edit., 1961
Third Marine Expedition on the D/E "Ob", 1957-1958. Data.
Trudy Sovetskoi Antarktich. Exped., Arktich. i Antarktich. Nauchno-Issled. Inst., Mezhd. Geofiz. God, 22:1-234 pp.

density
Kolesnikov, A.G., 1960.
[Vertical turbulent exchange in a stably stratified sea.]
Izv., Akad. Nauk, SSSR, Ser. Geofiz., 11:1614-1623.

English Ed., 11:1079-1084.

density
Koopman, G., 1953.
Entstehung und Verbreitung von Divergenzen in der oberflächennalen Wasserbewegung der antarktischen Gewässer. Deutsche Hydrogr. Zeits., Erganzungsheft 2:1-38, 2 pls., 18 textfigs.

density(data)
Kollmeyer, Ronald C., Robert M. O'Hagan and Richard M. Morse, 1965.
Oceanography of the Grand Banks region and the Labrador Sea in 1964.
U.S.C.G. Oceanogr. Rept., No. 10(373-10):1-24; 34-285.

density(data)
Krauss, Wolfgang, und Walter Düing, 1964.
Schichtung und Strom im Golf von Neapel.
Pubbl. Staz. Zool., Napoli, 33(3):243-263.

Density
Kuhn, R., Compiler, 1967.
Western North Atlantic Ocean, water masses and density stratification.
U.S. Nat. Oceanogr. Data Center, Publ (Atlas 1) G-9: unnumbered sheets.

density (data)
Kullenberg, G.E.B. 1971.
Results of diffusion experiments in the upper region of the sea.
Rept. Inst. fys. Oceanogr. Univ. København 12: 42pp., 23 figs. (multilithed).

density
Kusunoki, Kou, 1962
Hydrography of the Arctic Ocean with special reference to the Beaufort Sea.
Contrib. Inst., Low Temp. Sci., Sec.A, (17): 1-74.

density(data)
Kvinge, Tor, 1963.
The "Conrad Holmboe" Expedition to East Greenland waters in 1923.
Årbok, Univ. i Bergen, Mat.-Naturv. Ser., (15): 44 pp.

L

density
Lacombe, Henri, 1965.
Cours d'océanographie physique. (Theories de la circulation. Houles et vagues).
Gauthier-Villars, Paris, 392 pp.

density (icluding data)
Lacombe, H., 1956.
Quelques elements de l'extension des eaux méditerranennes dans l'Océan Atlantique.
Bull. d'Info., C.C.O.E.C., 8(5):210-224, 7 pls.

density
Lacombe, Henri, and P. Tchernia, 1959
Stations hydrologiques effectuées en Méditerranée à bord de la "Calypso" en 1955 et 1956. 2ème Partie. Campagne de septembre-octobre 1956.
Cahiers Océanogr., C.C.O.E.C., 11(6): 433-460.

density(data)
Lacombe, H., and P. Tchernia, 1959.
Stations hydrologiques effectuées à bord de la "Calypso" en 1955 et 1956 (par l'équipe du Laboratoire d'Océanographie Physique du Muséum Nationale d'Histoire Naturelle).
Cahiers Océanogr., C.C.O.E.C., 11(5):332-370.

density
LaFond, E.C., 1958
On the circulation of the surface layers of the East Coast of India.
Andhra Univ. Mem., Ocean., 2: 1-11.

density
LaFond, E.C., 1957.
Seawater density at four stations on the east coast of India. Indian J. Meteorol. & Geophys. 8(2):213-217.

density
LaFond, E. C., 1949
The use of bathythermograms to determine ocean currents. Trans. Am. Geophys. Union, 30(2):231-237, 6 text figs.

density
LaFond, E.C., and D.W. Pritchard, 1952.
Physical oceanographic investigations in the eastern Bering and Chukchi Seas during the summer of 1947. J. Mar. Res. 11(1):69-86, 17 textfigs.

density(data)
Le Floch, J., 1963.
Rapport sur les mesures effectuées dans le canal de Corse en août 1960.
Trav. Centre de Recherches et d'Études Océanogr., n.s., 5(1):11-26.

density(data)
Le Floch, J., 1962
Rapport sur les recherches océanographiques faites en Nord Tyrrhénienne en février 1960.
Trav., Centre de Recherches et d'Etudes Océanogr. 4(4):45-63.

density(data)
LeFloch, J., 1955.
Esquisse de la structure hydrologique de l'Atlantique Equatorial au large de la Guyane et de l'embouchure de l'Amazone.
Bull. d'Info., C.C.O.E.C. 7(10):449-467, 13 pls.

density (data)
Lemasson, L., et Y. Magnier, 1966.
Résultats des observations scientifiques de La Dunkerquoise sous le commendement du Capitaine de Corvette Brosset. Croisière Hunter.
Cah. ORSTOM, Sér. Océanogr., 4(1):3-78.

Density (data)
Leontyeva, V. V., 1962.
[Hydrological studies of oceanic trenches of the Pacific Ocean and some problems of further research.]
Mezhd. Geofiz. Komitet, Prezidiume Akad. Nauk, SSSR, Rezult. Issled. Programme Mezhd. Geofiz. Goda, Okeanol. Issled., No. 5:31-42.

density(data)
Leontyeva, V.V., 1960
[Some data on the hydrology of the Tonga and Kermadec Trenches.]
Trudy Inst. Okeanol., 40: 72-82.

density
Le Pichon, Xavier, et Jean-Paul Troadec, 1963
La couche superficielle de la Méditerranée au large des côtes Provençales durant les mois d'été.
Cahiers Océanogr., C.C.O.E.C., 15(5):299-316

density
Leprevost, Alsedo, e Joao Jose Bigarella, 1949.
Notas preliminares sobre a composicao quimica de algumas aguas do litoral paranaense.
Arquivos de Biologia e Tecnologia, Inst. Biol. e Pesquisas Tecnolog., Secret. Agricult., Indust. e Comercio, Estado do Parana, Brazil, 4 (13):73-86.

density
Lisitzin, E., 1959.
Les variations saisonnières de niveau de la mer et de la densité de l'eau Méditerranée occidentale.
Cahiers Océanogr., 11(1):7-13.
C.C.O.E.C.

density
Lisitzin, Eugenie, 1959.
The influence of water density variations on sea level in the northern Baltic.
Int. Hydrogr. Rev., 36(1):153-159.

density(data)
Lizeray, J.-C., 1957.
Travaux océanographiques recents de l'"Elie Monnier" en Méditerranée occidentale.
Bull. d'Info., C.C.O.E.C., 9(8):413-415.

density (data)
Ljøen, R., 1962.
The waters of the western and northern coasts of Norway in July-August 1957.
Fiskeri Skrift., Ser. Havundersøgelser, 13(2): 39 pp.

density
Ljøen, Rikard, and Luis E. Herrera, 1965.
Some oceanographic conditions of the coastal waters of eastern Venezuela.
Bol. Inst. Oceanogr., Univ. Oriente, 4(1):7-50.

density
Lynn, Ronald J., and Joseph L. Reid, 1968.
Characteristics and circulation of deep and abyssal waters.
Deep-Sea Res., 15(5):577-598.

M

density
Machado, L. de B., 1952.
Pesquisas fisicas e quimicas do sistema hidrografico da regiao Lagunar de Cananeia. Bol. Inst. Ocean., Univ. Sao Paulo, 3(1/2)55-75, 4 graphs, 1 chart.

density
Machado, L. de B., 1950.
Pesquisas fisicas e quimicas do sistema hifrográfico da região lagunar de Cananeia. Curso de água. Bol. Inst. Paulista Ocean. 1(1):45-68.

data
Machado, L. de B., 1951.
Resultados científicos do cruzeiro do "Baependi" e do "Vega" à Ilha de Trindade. Oceanografia física. Contribuição para o conhecimento das caracteristicas fisicas e químicas das águas.
Bol. Inst. Paulista Ocean. 2(2):95-110, 5 pls.

density
Machado, L. de B., 1951.
Resultados científicos do cruzeiro do "Baependi" e do "Vega" à Ilha de Trindade. Oceanografia física. Contribuição para o conhecimento das caracteristicas físicas e químicas das águas.
Bol. Inst. Paulista Ocean. 2(2):95-110, 5 pls.

density
Maeda, Sonosuke 1968.
On the cold water belt along the northern coast of Hokkaido in the Okhotsk Sea. 1. Hydrography of the Okhotsk Sea.
Umi to Sora 43(3):71-90.

density (data)
Magazzù, Giuseppe, e Guglielmo Cavallaro 1969.
Rapporto sulle crociere di studio lungo le coste meridionali Calabresi e della Sicilia orientale (1967-1968). 1. Idrografia.
Programma Ricerca Risorse Mar. Fondo mar. (B) 45:5-70.

density
Mamaev, O.I. 1970.
On water density increase from mixing of four water masses of the ocean. (In Russian; English abstract)
Okeanologiia 10(3):396-405.

density
Mamaev, O.I., 1964.
A simplified relationship between density, temperature and salinity of sea water. (In Russian)
Izv., Akad. Nauk, SSSR, Ser. Geofiz., (2):309-311
Translation:
(AGU) (2):180-181.

density (data)
Marchand, J.M., 1957.
Twenty-seventh annual report for the period 1st April 1955 to 31st March 1956.
Comm. & Ind., S. Africa, July 1957:159 pp.

density(chiefly data)
Marchand, J.M., 1956.
Twenty-sixth annual report for the period 1st April 1954 to 31st March 1955. Union of S. Africa Dept. Comm. & Industr., Div. Fish.:183 pp.
Reprinted from:"Commerce & Industry", 1956(July)

density(data)
Marchand, J.M., 1955.
Twenty-fifth annual report for the period 1st April 1953 to 31st March 1954.
Commerce & Industry, S. Africa, July 1955:162 pp.

density(data)
Marchand, J.M., 1953.
Pilchard-research programme. First Progress Report Annexure "A", 23rd Ann. Rept., Fish. Div., Union S. Africa, :17-181, 5 charts

density
Margalef, Ramón, 1961
Hidrografía y fitoplancton de un área marina de la costa meridional de Puerto Rico.
Inv. Pesq., Barcelona, 18:38-96.

density(data)
Margalef, R., F. Cervignon and G. Yepez T., 1960
Exploracion preliminar de las caracteristicas hidrograficas y de la distribucion del fitoplancton en el area da la Isla Margarita (Venezuela).
Mem. Soc., Ciencias Nat. de la Salle, 22(57): 210-221.
Contribucion No. 2, Estacion de Investigaciones Marinas de Margarita, Fundacion La Salle de Ciencias Naturales.

density
Margalef, Ramón, y Juan Herrera, 1964.
Hidrografía y fitoplancton de la costa comprendido entre Castellón y la desembocadura del Ebro, de julio de 1961 a julio de 1962.
Inv. Pesq., Barcelona, 26:49-90.

density(data)
Margalef, Ramón, and Juan Herrera, 1963
Hidrografía y fitoplancton de las costas de Castellón, de julio de 1959 a junio de 1960.
Inv. Pesq., Barcelona, 22: 49-109.

density
Marija, Marinkovic, 1959.
Observations hydrographiques terminées à Rovinj en 1955-1957.
Thalassia Jugoslavica, 1(6-10): 41-68.
(French summary)

density(data)
Marinkovic-Roje, Marija, and Miroslav Nicolic, 1961
[Oceanographic researches in the areas of Rovinj and Limski Kanal from 1959-1961.] In Jugoslavian: English summary.
Hidrografskog Godisnjaka 1960, Split: 61-67.

density(data)
Matsudaira, Yasuo, Haruyuki Koyama and Takuro Endo, 1961
[Hydrographic conditions of Fukuyama Harbor.]
J. Fac. Fish. and Animal Husbandry, Hiroshima Univ., 3(2):247-296.

density(data)
McGary, James W., Edward D. Stroup, 1958
Oceanographic observations in the central North Pacific, September 1954-August 1955.
U.S.F.W.S. Spec. Sci. Rept., Fish., No. 252: 250 pp.

density(data)
McGary, J.W., and E.D. Stroup, 1956.
Mid-Pacific oceanography. VIII. Middle-latitude waters, January-March 1954.
USFWS Spec. Sci. Rept., Fish. 180:173pp.

density
Mejía, Jorge, y Luis Alberto Poma E., 1966.
Informe preliminar del Crucero de otoño 1966. (Cabo Blanco-Ilo).
Inst. Mar, Peru, Informe No.13:31pp (multilithed).

density
*Menaché, M., 1967.
Du problème de la masse volumique de l'eau.
Metrologia, 3(3):58-63.

density
Menache, Maurice, 1965.
Litre et decimetre cube - densité et masse volumique: conséquences de l'Abandon du litre comme unité de volume pour les mesures de précision.
Cahiers Océanogr., C.C.O.E.C., 17(9):625-646.

density
Menaché, Maurice, 1963
Première campagne océanographique du "Commandant Robert Giraud" dans le Canal de Mozambique (11 octobre-28 novembre 1957).
Cahiers Océanogr., C.C.O.E.C., 15(4):224-235, 27 figs.

density(data)
Menaché, M., 1954.
Étude hydrologique sommaire de la region d'Anjouan en rapport avec la peche des coelacanthes.
Mem. Inst. Sci., Madagascar, A, 9:151-185, 34 textfigs.

density (data)
Memendez, Nicanor, 1964.
Campagne dans le Détroit de Gibraltar. Résultats hydrologiques obtenus à bord du navire océanographique espanol "Xauen" en juin 1961.
Cahiers Océanogr., C.C.O.E.C., 16(7):565-590.

density(data)
Midttun, L., and J. Natvig, 1957.
Pacific Antarctic waters. Scientific results of the "Brategg" Expedition, 1947-48, No. 3.
Publikasjon, Komm. Chr. Christensens Hvalfangstmus., Sandefjord, No. 20:1-130, 39 figs.

density
Mif, N.P. 1970
Water density control in the deep-sea measurements. (In Russian; English abstract)
Okeanologiia 10 (4): 714-717.

density (data)
*Minas, Hans Joachim, 1968.
A propos d'une remontée d'eaux "profondes" dans les parages du Golfe de Marseille (octobre 1964): conséquences biologiques.
Cah. océanogr., 20(8):647-674.

density
Montgomery, R. B., 1941
Transport of the Florida Current off Habana, J. Mar. Res., III(3):199-220; textfigs. 42-49. Tables of data. (Also, Contrib. No. 303, WHOI in "Collected Reprints" for 1941)

density
Montgomery, R.B., and R.B. Sykes 1941.
Table for density of sea water at 0°C and atmospheric pressure for values of chlorinity between 18.00 and 20.99 according to Knudsen's formula.
J. mar. Res. 4 (1): 28-31.

Morcos, Selim A.

density
Morcos, Selim A., and J.P. Riley, 1966.
Chlorinity, salinity, density and conductivity of sea water from the Suez Canal region.
Deep-Sea Res., 13(4):741-749.

density(data)
Moreira da Silva, Paulo, 1957
Oceanografia do trinagulo Cabo-Frio-Trinidade-Salvador.
Anais Hidrograficos, Brasil, 16:213-308.

density (data)
*Morgan, Charles W., 1969.
Oceanography of the Grand Banks region of Newfoundland in 1967.
Oceanogr. Rept.U.S.C.G. 19(CG 373-19):1-209.

density
Mori, Isamu, and Haruhiko Irie, 1966.
The hydrographic conditions and the fisheries damages by the red water occurred in Omura Bay in summer 1965 - III. The oceanographic conditions in the offing region of Omura Bay during the term of the red water. (In Japanese; English abstract).
Bull. Fac. Fish. Nagasaki Univ., No. 21:103-113.

density
Moriyasu, Shigeo, 1960
The thickness of the upper homogeneous layer.
Rec. Oceanogr. Wks., Japan, n.s., 5(2): 44-51.

density(data)
Moroshkin, K.V., 1959.
[Hydrological investigations.]
Arktich. i Antarkt. Nauchno-Issled. Inst., Mezhd. Geofiz. God., Sovetsk. Antarkt. Eksped., 5:107-124. (2nd Marine Exped. "Ob", 1956-1957)

density
Mosby, H., 1950.
Recherches océanographiques de la Mer de Norvège à la station météorologique M. Cahiers du Centre de Recherches et d'Études Océanographiques, Ch. No. 1:7 pp., 5 textfigs.

density (data)
Mosby, H., 1938
Svalbard waters. Geophys. Publ. 12(4): 85 pp., 34 text figs.

density
Mosby, H., 1936.
Zur Hydrographie des nordlichen Barentsmeeres.
Ann. Hydrogr., usw., Jahrg. 64(Heft 9):407-408.

density
*Mosetti, F., 1967.
Considerazioni preliminari sulla dinamica dell' Adriatico settentrionale.
Arch.Oceanogr.Limnol.15(Suppl.):237-244.

density (data)
Mosetti, Ferruccio, 1967.
Caratteristiche idrologiche dell'Adriatico settentrionale: situazione estiva.
Atti Ist. veneto Sci., 125:147-175.

density
Motoda, S., 1940.
Comparison of the conditions of water in the bay, lagoon and open sea in Palao.
Palao Trop. Biol. Sta. Studies 2(1):41-48, 2 figs.

density
Motoda, S., 1940.
The environment and the life of massive reef coral, Goniastrea aspera Verrill, inhabiting the reef flat in Palao.
Palao Trop. Biol. Sta. Studies 2(1):61-104, 9 textfigs.

Density
Muñoz, Felipe, y Jose M. San Feliu, 1965.
Hidrografía y fitoplancton de las costas de Castellón de agosto de 1962 a julio de 1963.
Inv. Pesq., Barcelona, 28:173-209

density (data)
Muromtsev, A.M., 1959
[The basis for the hydrology of the Indian Ocean.] Gidrometeoizdat, Leningrad, 437 pp.

density
Muromtsev, A.M., 1959
[Basic data for the hydrology of the Indian Ocean. Appendix II. Atlas of vertical section and charts of temperature, salinity, density and oxygen content.]
Gosud. Okeanogr. Inst. Glabnoe Upravlenie Gidrometeorol. Sluzhbi pri Sovete Ministrov SSSR, Gidrometoizdat, Leningrad, 112 pp.

density
Muromtsev, A.M., 1958
[Basic data for the hydrology of the Pacific Ocean. Appendix II. Atlas of vertical sections and charts of temperature, salinity, density and oxygen content.]
Gosud. Okeanogr. Inst., Glabnoe Upravlenie Gidrometeorol. Sluzhbi pri Sovete Ministrov, SSSR, Gidrometeoizdat, Leningrad, 124 pp.

density(data)
Muromtsev, A.M., 1958
[The basis for the hydrology of the Pacific Ocean.]
Gidrometeoizdat., Leningrad, 431 pp.

density
Muromtsev, A.M., 1958.
Water density on the surface of the Indian Ocean. (In Russian).
Doklady, Akad. Nauk, SSSR, 123(6):1014-1017.

density
Muromtsev, A.M., 1957.
New charts of the distribution of the salinity and density on the surface of the Pacific Ocean. (In Russian).
Meteorol. i Gidrol., (4):15-19.

N

density
Nasu, K., 1960
Oceanographic investigation in the Chukchi Sea during the summer of 1958.
Sci. Repts., Whales Res. Inst., 15: 143-157.

density (data)
Navarro, F. de P., 1947.
Exploración oceanografía del Africa occidental desde El Cabo Ghir al Cabo Judy; resultados de la campañas del "Melaspina" y del "Xauen" en Mayo 1946, Trab. Inst. Español Ocean. No. 20: 40 pp., 8 figs.

density (data)
Navarro, F. de P., F. Lozano, J.M. Navaz, E.Otero J. Sáinz Pardo and others, 1943.
La pesca de Arrastre en los fondos del Cabo Blanco y del Banco Arguin (Africa Shariana).
Trab. Inst. Español Ocean. No. 18:225 pp., 38 pls

density
Navarro, F. de P., and M. Massuti, 1929.
Oceanografía, plancton y pesca en la Bahia de Palma de Mallorca en 1928. Notas y Res., Inst. Español Ocean., 2nd ser., No. 33:61 pp., 20 textfigs.

density
*Neev, David, and K.O. Emery, 1967.
The Dead Sea: depositional processes and environments of evaporites.
Bull. Geol. Surv., Israel, 41: 146 pp.

density
Nehring, Dietwart, Sigurd Schulz und Karl-Heinz Rohde 1969.
Untersuchungen über die Produktivität der Ostsee. 1. Chemisch-biologische Untersuchungen in der mittleren Ostsee und in der Bottensee im April/Mai 1967.
Beitr. Meeresk. 23: 5-36.

density
Nekrasova, V.A., and V.N. Stepanov, 1963.
Meridional hydrological profiles of the oceans from IGY data. (In Russian).
Mezhd. Geofiz. Kom., Prezidiume, Akad. Nauk, SSSR, Rezult. Issled. Programme Mezhd. Geofiz. Goda, Okeanol. Issled., (8):34-51.

density
Neumann, A. Conrad, and David A. McGill, 1961 (1962).
Circulation of the Red Sea in early summer.
Deep-Sea Res., 8(3/4):223-235.

density
Neumann, G., 1944.
Das Schwarze Meer. Zeitsch. Gesellschaft f. Erdkunde zu Berlin. 1944 (3/4):92-114, 25 text figs.

density(data)
Neumann, G., 1940.
Die ozeanographischen Verhältnisse an der Meeresoberfläche im Golfstromsektor nördlich und nordwestlich der Azoren. Wiss. Ergeb. Internat. Golfstrom-Unternehmung 1938, Lief. 1, in Juni Beiheft, Ann. Hydr., usw., 68:87 pp., 36 textfigs 3 pls.

density
Neumann, G., and A. Schumacher, 1944.
Strömungen und Dichte der Meeresoberfläche vor der Ostküste Nordamerikas. Ann. Hydr., usw., 72: 277-279, Pl. 12.

density(data)
Nichols, M.M., and R.C. Barnes, 1964.
Shelf observations - hydrography: cruise of August 21-26, 1962.
Virginin Inst. Mar. Sci., Spec. Sci. Rept., (41): 23 pp. (multilithed).

density(data)
Nichols, M.M., and M.P. Lynch, 1964.
Shelf observations - hydrography: cruises of January 22-25, July 15-19, 1963.
Virginia Inst. Mar. Sci., Spec. Sci. Rept., (48) :33 pp. (multilithed).

density(data)
Nielsen, J.N., 1907.
Contribution to the hydrography of the north-eastern part of the Atlantic Ocean.
Medd. Komm. Havundersøgelser, Ser. Hydrogr., 1(9):25 pp., 3 pls.

density (data)
Nielsen, J.N., 1905.
Contribution to the hydrography of the waters north of Iceland. Medd. Komm. Havundersøgelser, Ser. Hydrogr. 1(7):28 pp., 2 pls.

O

density
Okubo, A., 1954.
A note on the decomposition of sinking remains of plankton organisms and its relationship to nutrient liberation. J. Ocean. Soc., Japan, 10(3):121-131, 4 textfigs.

density
*Okuda, Taizo, 1967.
Vertical distribution of inorganic nitrogen in the equatorial Atlantic Ocean.
Boln Inst. Oceanogr., Univ. Oriente, 5(½):67-83.

density
Okuda, Taizo, 1962
Physical and chemical oceanography over continental shelf between Cabo Frio and Vitorio (Central Brazil).
J. Oceanogr. Soc., Japan, 20th Ann. Vol., 514-540.

density
Okuda, Taizo, and Lourinaldo Barreto Cavalcanti, 1963.
Algumas condicoes oceanograficas na area nordeste de Natal. (Septembro 1960).
Trab. Inst. Oceanogr., Univ. do Recife, 3 (1): 3-25.

density
Olson, F.C.W., 1956.
Nomograms for hydrometer salinity and sea water density. Pap. Oceanogr. Inst., Fla. State Univ., Studies, 22(2):13-18.

density
Oort, Abraham H., 1964.
Computations of the eddy heat and density transport across the Gulf Stream.
Tellus, 16(1):55-63.

density (data)
Oren, O.H., 1964.
Hydrography of Dahlak Archipelago (Red Sea).
Sea Fish. Res.Sta., Israel, Bull. No. 35:3-22.

density
Orr, A.P., 1947
An experiment in marine fish cultivation: II. Some physical and chemical conditions in a fertilized sea-loch (Loch Craiglin, Argyll).
Proc. Roy. Soc., Edinburgh, Sect.B, 63, Pt.1 (2):3-20, 19 text figs.

density
Orr, A. P., 1933
Physical and chemical conditions in the sea in the neighborhood of the Great Barrier Reef, Brit. Mus. (N.H.) Great Barrier Reef Exped., 1928-29, Sci. Repts. 2(3):37-86, 7 text figs.

density(data)
Orren, M.J., 1963.
Hydrological observations in the south west Indian Ocean.
Rept., S. Africa, Dept. Comm. Indust., Div. Sea Fish., Invest. Rept., No. 45:61 pp.

density
Oshima, Kazuo, 1963.
Ecological study of Usu Bay, Hokkaido, Japan, 1. Bottom materials and benthic fauna.
(In Japanese; English abstract).
Bull. Hokkaido Reg. Fish. Res. Lab., Fish. Ag., No. 27:32-51.

density
Owen, Robert W., Jr., 1968.
Oceanographic conditions in the northeast Pacific Ocean and their relation to the albacore fishery.
Fish. Bull. U.S. Fish. Wildl. Serv., 66(3): 503-526.

density
Owen, R.W., Jr., 1967.
Atlas of July oceanographic conditions in the northeast Pacific Ocean, 1961-64.
Spec. scient. Rep. U.S. Fish. Wildl. Serv., Fish., 549:85 pp.

P

density
Paccagnini, Ruben N., and Alberto O. Casellas, 1961
Estudios y resultados preliminares sobre trabajos oceanograficos en el area del Mar de Weddell. (In Spanish; Spanish, English, French, German and Italian resumes)
Contrib., Inst. Antartico Argentino, No. 64: 12 pp.

density
Palosuo, Erkki, 1964
A description of the seasonal variations of water exchange between the Baltic proper and the Gulf of Bothnia.
Merent. Julk. (Havsforskningsinst. Skrift), No. 215: 32 pp.

density
Parr, A. E., 1937
Report on hydrographic observations at a series of anchor stations across the Straits of Florida. Bull. Bingham Ocean. Coll. VI (3):1-62, 36 text figs.

density
Patil, M.R. and C.P. Ramamirtham 1963
Hydrography of the Laccadives offshore waters - a study of the winter conditions.
J. mar. biol. Ass., India 5(2):159-169.

density
Patil, M.R., C.P. Ramamirtham, P. Udaya Varma, C.P. Aravindakshan Nair and Per Myrland, 1964.
Hydrography of the west coast of India during the pre-monsoon period of the year 1962. 1. Shelf waters of Maharashtra and southwest Sourashtra coasts.
J. Mar. biol. Ass., India, 6(1):151-166.

Density (data)
Perin-Luca, Lilia, 1960.
Osservazioni di temperatura e salinita nel Golfo di Trieste
Boll. della Soc. Adriatica di Sci., 51:1-19.

Also:
Ist. Sperimentale Talassogr., Trieste, Pubbl., No. 374 (1961)

density(data)
Peterson, Clifford L., 1960.
The physical oceanography of the Gulf of Nicoya, Costa Rica, a tropical estuary.
Inter-American Tropical Tuna Comm., Bull., 4(4): 139-216.

density
Phifer, L. D. and T. G. Thompson, 1937.
Seasonal variations in the surface waters of San Juan channel during the five year period, January 1931 to December 30, 1935. J. Mar. Res., 1(1):34-59, text figs. 12-18, 12 tables.

density
Piatek, W., 1957.
[Changes in the density of the southern Baltic waters depending upon the temperature and salinity in 1949-1954.]
Prace Morsk. Inst. Ryback., Gdyni, No. 9:381-425.

density
Pickard, G.L., 1961.
Oceanographic features of inlets in the British Columbia mainland coast.
J. Fish. Res. Bd., Canada, 18(6):907-999.

density
Pickard, G.L., 1961
Oceanographic fdatures of inlets in the British Columbia mainland coast.
J. Fish. Res. Bd., Canada, 18(6):907-999.

density

Picotti,Mario,1965.
La crociera idrografico-talassografica di Capo Matapan.
Pubbl.Commissione Ital.Comitato int.Geofis., Ser.IGC, 42:63 pp.

density(data)

Picotti, Mario, 1960

Crociera Talassografica Adriatica 1955. III. Tabelle delle osservazioni fisiche, chimiche, biologiche e psammografiche.
Archivio di Oceanograf. e Limnol., 11(3): 371-377, plus tables.

density

Picotti, M., 1954.
Physikalisch-chemische Verhältnisse in der Nord-Adria längs des 45 Breitenkreises.
Arch. Met., Geophys., Bioklim., A, 7:466-476, 4 textfigs.

density

Picotti, M., 1930.
Ricerche di oceanografia chimica. 1. Tabelle generali delle analisi clorometrische e dei temperature, salinita e densita. Ann. Idrogr., 11(2):69-116.

translation cited:
USFWS Spec. Sci. Rept., Fish., 227.

density(data)

Picotti, M., 1923-24.
Richerche di oceanografia chimica. Tabelle generali della analisi chlorometriche e dei dati di temperatura, salinita e densita. Ann. Idrogr. 11 bis:69-116, charts, tables.

density(data)

Picotti, M., and A. Vatova, 1942.
Osservazioni fisiche e chemiche periodiche nell' Alto Adriatico (1920-1938). Thalassia 5(1):1-157

density

Picotti, M. and A. Vatova, 1942.
Osservazioni fisiche e chimiche periodiche nell' Alto Adriatico (1920-1938). Thalassia V (1):157 pp., 8 tables, 11 figs.

density (data)

Piejus,Pierre,1965.
Hydrologie dans l'océan antarctique.
Annls hydrogr., Paris (4)12:333-342.

density

Plakhotnik, A.F., 1964.
Hydrological characteristics of the Gulf of Alaska. (In Russian).
Gosudarst. Kom. Sov. Ministr., SSSR, Ribn. Choz., Trudy VNIRO, 49, IZV., TINRO, 51:17-50.

density

Pomeroy, A.S., 1965.
Notes on the physical oceanographic environment of the Republic of South Africa.
Naval Oceanogr. Res. Unit, 41 pp.(Unpublished manuscript).

density

Portugal, 1931.
"Albacora": croisière de février 1929 (17 février au 14 mars).
Rapp. Proc. Verb., Cons. Perm. Int. Expl. Mer, 70:20-23.

density

Postma, H., 195-.
Hydrography of the Dutch Waddensea. A study of the relations between water movement, the transport of suspended materials and the production of organic matter. Arch. Neerl. Zool., 10(4):405-511 55 textfigs.

density

Postma, H., and J. Vervey, 1950.
Resultaten van hydrografisch onderzoek in der Waddenzee. Waddensymposium, Tijdschr. Kon. Neder. Aardrijkskundig Genootschap:4-33, 9 textfigs., 2 photos. (English summary).

density

Pucher-Petković, Tereza 1966.
Vegetation des diatomées pélagiques de l'Adriatique moyenne.
Acta adriat. 13(1): 1-97.

Q

R

density (data)

Ramalho, A. de M., 1942
Observacoes oceanograficas. "Albacora", 1940. Trav. St. Biol. Mar., Lisbonne, No. 45: 96 XVI pp., 2 figs. (fold-in).

density (data)

Ramalho, A. de M., R. G. Boto, B. C. Goncalves, H. Vilela, 1936.
Observacoes oceanograficas. "Albacora", 1935-1936. Trav. St. Biol. Mar. Lisbonne, No.39: 25 pp., 2 figs.

density (data)

Ramalho, A. de M., L. S. Dentinho, C. A. M. de Sousa, Fronteira, F. L. Mamede, and H. Vilela, 1935.
Observacoes oceanograficas. "Albacora" 1934. Trav. St. Biol. Mar., Lisbonne, No. 35: 35 pp., 1 fig.

density (data)

Ramalho, A. de M., B. C. Goncalves, R. G. Boto, Z. Vilela, 1938.
Observacoes oceanograficas. "Albacora", 1937. Trav. St. Biol. Mar., Lisbonne, No. 43: 30 pp., 1 fig.

density

Ramamirtham, C.P., and M.R. Patil, 1965.
Hydrography of the west coast of India during the pre-monsoon period of the year 1962. 2. In and offshore waters of the Konkan and Malabar coasts.
J. Mar. biol. Ass., India, 7(1):150-168.

density

Ramasastry, A.A., and P. Myrland, 1959.
Distribution of temperature, salinity and density in the Arabian Sea along the South Malabar coast (South India) during the post-monsoon season.
India J. Fish., 6 (2):223-255.

density

Reichelt, W., 1941.
Die ozeanographischen Verhältnisse bis zur warmen Zwischenschicht an der antarktischen Eisgrenze im Südsommer 1936/37. Nach Beobachtungen auf dem Walfangmutterschiff "Jan Wellem" im Weddell-Meer.
Archiv. Deutschen Seewarte 61(5):54 pp., 11 pls.

density

Reid, Joseph L., Jr., 1964.
A transequatorial Atlantic oceanographic section ki July 1963 compared with other Atlantic and Pacific sections.
J. Geophys. Res., 69(24)5205-5215.

DENSITY

Rial, J.R. Besada, and L.R. Molins, 1962.
Determinación complexométrica de los iones calcio y magnesio en el agua de mar y estudio de las variaciones de su concentración en las aguas de la Ría de Vigo.
Bol. Inst. Español Oceanogr., 111:11 pp.

density (data)

Ridgway, N. M. 1970.
Hydrology of the southern Kermadec Trench region.
Bull. N.Z. scient. industr. Res. 205:7-28.
Also: Mem. N.Z. Oceanogr. Res. 56.

density

Riley, G.A., 1952.
Hydrography of the Long Island and Block Island Sounds. Bull. Bingham Ocean. Coll., 13(3): 5-39, 11 textfigs.

density

Riley, G.A., 1951.
Oxygen, phosphate, and nitrate in the Atlantic Ocean. Bull. Bingham Ocean. Coll. 8(1):126 pp., 33 textfigs.

density

Riley, G. A. 1942.
The relationship of vertical turbulence and spring diatom flowerings. J. Mar. Res., 5(1): 76-87.

density (data)

Robertson, A.J., 1913.
On hydrographical investigations in the North Sea and the Faroe Channel during the years 1909-1910. Fish. Bd., Scotland, North Sea Fish. Invest. Comm., Northern Area, Rept. No. 5:337-404, 10 pls.

density(data)

Robertson, A.J., 1909.
On hydrographical investigations in the North Sea and Faroe-Shetland Channel during the years 1907-1909. Fish. Bd., Scotland, North Sea Fish. Invest. Comm., Northern Area, Rept. No. 4:143-196 9 pls.

density(data)

Robertson, A.J., 1909.
On hydrographical investigations in the North Sea and Faroe-Shetland Channel during the year 1906. Fish. Bd., Scotland, North Sea Fish. Invest. Comm., Northern Area, Rept. No. 4:1-60, 14 pls.

density(some data)

Robertson, A.J., 1907.
On hydrographical investigations in the Faroe-Shetland Channel and the northern part of the North Sea during the years 1904-1905. Fish. Bd., Scotland, North Sea Fish. Invest. Comm., Northern Area, Rept. No. 2:1-140, 3 textfigs., 15 pls.

density (data)

Robertson, A.J., 1905.
Report on hydrographical investigations in the Faroe-Shetland Channel and the northern part of the North Sea in 1903. Fish. Bd., Scotland, North Sea Fish. Invest. Comm., 1902-1903 (Rept. No. 1):51-113, 6 figs.

density

Rochford, D.J., 1960

The intermediate depth waters of the Tasman and Coral Seas. I The 27.20 sigma-t surface. II The 26.80 sigma-t surface. Austral. J. Mar. & Freshwater Res., 2(2): 127-165.

density

Rochford, D.J., 1952.
A comparison of the hydrological conditions of the eastern and western coasts of Australia. Indo-Pacific Fish. Counc., Proc., 3rd meeting, 1-16 Feb. 1951. Sects. 2/3:61-68, 7 textfigs.

density

Rogalla, E.H., 1965.
On the results of North Sea research; hydrographic investigations on board FRV Anton Dohrn in summer 1959/60 and winter 1962/63.
Int. Hydrogr. Rev. 42(1):135-167.

density(data)
Romanovsky, V., 1963
Effet du vent sur le mouvement des eaux littorales.
Trav. Centre de Recherches et d'Études Océanogr. 31-36.

density
Romanovsky, V., 1948
Quelques données sur la météorologie et l'océanographie de l'anse de Ponteau dans le golfe de Fos. Bull. Inst. Océan., Monaco, No. 932, 16 pp., 7 text figs.

density
Romanovsky, V., 1948
Quelques donnees sur la Meteorologie et l'Oceanographie de l'anse du Ponteau dans le golfe de For. Bull. l'Inst. Ocean., Monaco, No.932:16 pp., 7 text figs.

density
Rotschi, H., 1961
Orsom III. Resultats de la croisière "Epi" Ière Partie. Océanographie Physique.
O.R.S.T.O.M., I.F.O., Rapp. Sci. 22: 64 pp.

density (data)
Rotschi, Henri, 1960
Orsom III, Resultats de la croisière "Choiseul" ler Partie. Océanographie Physique.
Centre d'Océanogr., Inst. Français d'Océanie Rapp. Sci., No. 15: 58 pp. (mimeographed).

density(data)
Rotschi, Henri, 1960
Orsom III, Resultats des Croisières diverses de 1959. Oceanographie physique.
Centre d'Océanogr., Inst. Français d'Océanie, Rapp. Sci., No. 17: 59 pp.

density (data)
Rotschi, Henri, 1959.
Resultats des observations scientifiques du "Tiaré croisière "Dauphin" (8-14 Août 1959) sous le commandement du Lieutenant de Vaisseau Cerbelaud.
Rapp. Sci., Inst. Français d'Océanie, Centre d'Océanogr. No. 14:18 pp.

density (data)
Rotschi, Henri, 1959.
Orsom III. Resultats de la croisiere "Boussole" 1. Oceanographie physique.
O.R.S.T.O.M., Inst. Francais d'Oceanie, Rapp. Sci., No. 12:67 pp.

density
Rotschi, Henri, 1959
Resultats des observations scientifiques du "Tiare", Croisière DAUPHIN (8-14 août 1959) sous le commandemant du Lieutenant de Vaisseau Cerbelaud.
C.L.O.E.C.P Inst. Francais d'Océanie, Rapp. Sci., No. 14: 18 pp.

density(data)
Rotschi, Henri, et Yves Magnier, 1963
Resultats des observations scientifiques de La DUNKERQUOISE-------------.
Inst. Français d'Océanie, Rapp. Sci., O.R.S.T.R.O.M., Noumea, No. 24: 67 pp.

density
Rouch, J., 1950.
Le Canal de Panama. Bull. Inst. Océan, Monaco, No. 975:20 pp., 4 textfigs.

density
Rouch, J., 1950
La densite de l'eau de mer et les courants au voisinage de Monaco. Bull. Inst. Ocean., Monaco, No. 968: 12 pp.

density
Rouch, J., 1948.
Stations hydrologiques des campagnes scientifiques du Prince Albert ler. Res. Camp. Sci., Monaco, 108:26 pp.

density
Rouch, J., 1946
Traité d'Océanographie physique. L'eau de mer. Payot, Paris, 349 pp., 150 text figs.

density
Rouch, J., 1940.
Observations océanographiques de surface dans l'océan Atlantique et dans la Mediterranee. Ann. Inst. Océan. 20(2):51-73, 11 textfigs.

density
Rouch, J., 1940.
La température et la densité de l'eau de mer à Marseille. C. R. Acad. Sci., Paris, 211:654-657.

density
Rouch, J., 1939.
Observations océanographiques de surface dans l'océan Atlantique et dans l'océan Pacifique.
Bull. Inst. Océan., Monaco, No. 781:10 pp., 4 figs.

density
Ryther, John H., and John R. Hall, Allan K. Pease, Andrew Bakun and Mark M. Jones, 1966.
Primary organic production in rélation to the chemistry and hydrography of the western Indian Ocean.
Limnol. Oceanogr. 11(3):371-380.

density
Ryther, John H., D.W. Menzel and Nathaniel Corwin, 1967.
Influence of the Amazon River outflow on the ecology of the western tropical Atlantic. I. Hydrography and nutrient chemistry.
J.mar.Res., 25(1):69-83.

S

density
Sabinin, K.D., 1966.
On the relationship between short period internal waves and the vertical density gradient in the sc (In Russian; English abstract).
Fisika Atmosferi i Okeana, Izv., Akad. Nauk, SSSR 2(8):872-882.

density
Saelen, O.H., 1950.
The hydrography of some fjords in northern Norway. Balsfjord, Ulfsfjord, Grøtsund, Vengsøyfjord, and Malangen. Tromsø Mus. Årshefter, Naturhist. Afd., 38, 70(1):192 pp., 10 pls., 42 textfigs.

density
Sandström, J.W., 1904.
Einfluss des Windes auf die Dichte und die Bewegung des Meereswassers.
Publ. de Circ., Cons. Perm. Int. Expl. Mer, No. 18:1-6.

density (data)
San Feliu, J.M., y F. Muñoz 1967.
Hidrografía y fitoplancton de las costas de Castellón de mayo de 1965 a julio de 1966.
Investigación pesq. 31 (3): 419-461.

density(data)
San Feliu, J.M., y F. Muñoz, 1965.
Hidrografía y planoton del puerto de Castellón de junio de 1961 a enero de 1963.
Inv. Pesq., Barcelona, 28:3-48.

density
Scaccini, A., 1963.
Quelques notes oecologiques sur les thons génétiques de la Sardaigne.
Rapp. Proc. Verb., Réunions, Comm. Int. Expl. Sci., Mer Méditerranée, Monaco, 17(2):367-369.

density
Schott, G., 1942
Geographie des Atlantischen Ozeans im auftrage der Deutschen Seewarte vollständy neu bearbeitete dritte auflage. 438 pp., 141 text figs., 27 plates. C. Boysen. Hamburg.

density
Schott, G., 1932.
Zur Ozeanographie der Hudsonbai und Hudsonstrasse. Ann. Hydrogr., usw., Jahrg. 60(Heft XI): 453-455.

density (data)
Schott, G., and B. Schulz, 1914.
Die Forschungsreise: S.M.S. "Möwe" im Jahre 1911.
Arch. Deutschen Seewarte 37(1):1-80, 8 pls., 16 textfigs.

density(data)
Schulz, B., 1923.
Hydrographische Untersuchungen besonders über den Durchlüftungszustand in der Ostsee in Jahre 1922. (Forschungsschiffe "Skagerak" und "Nautilus"). Arch. Deutschen Seewarte 41(1):1-64, 8 textfigs., 5 pls.

density(data)
Schulz, B., 1922.
Hydrographischen Beobachtungen inbesondere über die Kohlensäure in der Nord- und Ostsee im Sommer 1921. (Forschungsschiffe "Poseidon" und "Skagerak"). Arch. Deutschen Seewarte 40(2):1-48, 2 textfigs., 4 pls.

density
Schulz, B., and A. Wulff, 1929
Hydrographie und Oberflächen plankton des westlichen Barentsmeeres im Sommer 1927.
Ber. deutschen wissensch. Komm. F. Meeresforsch. n.s. 4(5):232-372, 13 tables, 25 text figs.

density
Seiwell, H. R., 1937.
Relationship of minimum oxygen concentration to density of the water column in the Western North Atlantic. Gerlands Beiträge z. Geophysik, 50: 302-306, 1 textfig.

density, data
Serpoianu, G.H., 1967.
Considérations sur la pénétration des eaux méditerranéennes dans le bassin de la Mer Noire.
Hydrobiologie, Bucureşti, 8: 239-251

density (data)
Serpoianu, Gheorghe, et Viorel Chiriţa 1969
Observations sur les particularités hydrologiques des eaux de la mer Noire dans la Couche où la vie commence à disparaître.
Rapp. P.-v. Réun. Commn int. Explor. scient. Mer Mediterr., 19(4): 689-692.

density
Shannon, L.V., 1966.
Hydrology of the south and west coast of South Africa.
Investl Rep. Div. Sea Fish. Un. S. Afr., 58: 22 pp.

density
Shannon, L.V. and M. van Rijswijck 1969
Physical oceanography of the Walvis Ridge Region.
Investl Rept. Div. Sea Fish SAfr. 70: 19 pp.

density (data)
Shimomura, T., and K. Miyata, 1957.
The oceanographical conditions of the Japan Sea and its water systems, laying stress on the summer of 1955.
Bull. Japan Sea Reg. Fish. Res. Lab., No. 6 (General survey of the warm Tsushima Current 1): 23-120.

density
Shtokman, V.B., 1950.
Determination of current-velocities and density distribution on a cross-section of an infinitely long channel, as related to the action of the wind and of side-friction in the Coriolis-force field. Dok. Akad. Nauk, SSSR, 71(1):41-44.
T60R

density
Simpson, John G., y Raymond C. Griffiths, 1967.
La distribucion de densidad en el Golfo de Cariaco, Venezuela oriental.
Serie Recursos Expl pesquer., (Venezuela) 1(8):305-325.

density (data)
Sjöstrand, J., 1921.
De hydrografiska förhållanden i Norra Ishavet mellan norska kuster och Spetsbergen samt väster om Spetsbergen ävensom i Isfjorden och Van Mijensfjord år 1920.
Svenska Hydrogr.-Biol. Komm. Skr., 7:8 pp.

density
Slaucitajs, L., 1947
Ozeanographie des Rigaischen Meerbusens. Teil 1, Statik. Contrib. Baltic Univ. No.45, Pinneberg, 110 pp., 69 text figs.

density
Smayda, Theodore J., 1966.
A quantitative analysis of the phytoplankton of the Gulf of Panama. III. General ecological conditions and the phytoplankton dynamics at 8o 45'N, 79o 23'W from November 1954 to May 1957.
Inter-Amer. Trop. Tuna Comm., Bull., 11(5): 355-612.

density
Somma, A., 1952.
Elementi di Meteorologia ed Oceanografia. Pt. II. Oceanografia. Casa Editrice Dott. Antonio Milani, Padova, Italia, xviii 758 pp., 322 textfigs., maps tables. (4500 lire)

density (data)
Soot-Ryen, T., 1956.
Report on the hydrographical conditions in West-Finnmark, March-April, 1935.
Acta Borealia, Tromsø Mus., A. Scientia, No. 10: 1-37.

density (data)
Soot-Ryen, T., 1947.
Hydrographical investigations in the Tromsø-district 1934-1938 (Tables). Tromsø Mus. Aarsheft., Naturhist. Avd. No. 33, 66(1943)(1): numerous pp. (unnumbered).

density (data)
Soot-Ryen, T., 1938.
Hydrographical investigations in the Tromsø district in 1931.
Tromsø Mus. Aarsheft., Naturhist. Avd. No. 10, 54(2):1-6 plus 41 pp. tables.

density (data)
Soot-Ryen, T., 1932.
The Folden Fjord. Hydrography. Tromsø Mus. Skr. 1(3):7 pp.

density (data)
Soule, F.M., 1953.
Physical oceanography of the Grand Banks region and the Labrador Sea in 1952. U.S.C.G. Bull. No. 38:31-100, 17 textfigs.

density
Soule, F. M., 1940.
Oceanography. Season of 1938. Bull. International Ice Observation and Ice Patrol Service in the North Atlantic Ocean, 28:113-173, textfigs. 45-56.

density
Soule, F. M., 1940.
Oceanography: excerpt from International Ice Observation and Ice Patrol Service in the North Atlantic Ocean. Season of 1939. U.S.C.G., Bull., pp. 79-133; 32 textfigs; Tables of data.

density
Soule, F. M., 1938.
Oceanography. Season of 1937. International Ice Observations and Ice Patrol Service, U.S.C.G. Bull. No. 27:71-126, 45 textfigs., tables of data.

density
Soule, F.M., 1938
Oceanography. Excerpt from International Ice Observation and Ice Patrol Service in the North Atlantic Ocean. Season of 1936. C.G. Bull. No.26:33-82, 58 text figs., tables of data.

density (data)
Soule, F.M., and C.A. Barnes, 1950.
Physical oceanography of the Ice Patrol area in 1941. U.S.C.G. Bull. No. 31:1-62, 23 textfigs., fold-in charts.

density (data)
Soule, F.M., P.S. Branson, and R.P. Dinsmore, 1952.
Physical oceanography of the Grand Banks region and the Labrador Sea in 1951. U.S.C.G. Bull. No. 37:17-85, 21 textfigs.

density (data)
Soule, F.M., M.J. Bush and J.E. Murray, 1955.
International ice observation and Ice Patrol Service in the North Atlantic Ocean. Season of 1953. U.S.C.G. Bull. No. 39:45-168, 43 textfigs.

density (data)
Soule, F.M., H.H. Carter and L.A. Cheney, 1950.
Oceanography of the Grand Bankd region and Labrador Sea, 1948. U.S.C.G. Bull. No. 34:67-118, Figs. 27-37, fold-in charts.

density (data)
Soule, F.M., A.P. Franceschetti and R.M. O'Hagen, 1963.
Physical oceanography of the Grand Banks region and the Labrador Sea in 1961.
U.S. Coast Guard Bull., No. 47:19-82.

density (data)
Soule, Floyd M., Alfred P. Franceschetti, R.M. O'Hagan and V.W. Driggers, 1963.
Physical Oceanography of the Grand Banks region, the Labrador Sea and Davis Strait in 1962.
U.S.C.G. Bull., No. 48:29-78;95-153.

density (data)
Soule, Floyd M., and R.M. Morse, 1960
Physical oceanography of the Grand Banks Region and the Labrador Sea in 1958. International Ice Observation and Ice Patrol Service in the North Atlantic.
U.S. Coast Guard Bull., No. 44: 29-99.

density (data)
Soule, F.M., and J.E. Murray, 1957.
Physical oceanography of the Grand Banks region and the Labrador Sea in 1956. U.S.C.G. Bull., No. 42 35-100.

density (data)
Soule, F.M., and J.E. Murray, 1956.
Physical oceanography of the Grand Banks and the Labrador Sea in 1955. U.S.C.G. Bull., 41:59-114.

density (data)
Soule, F.M., and J.E. Murray, 1955.
Physical oceanography of the Grand Banks region and the Labrador Sea.
U.S.C.G. Bull. No. 40:79-168.

density (data)
South Africa, Department of Commerce and Industries, Division of Fisheries, 1961
Thirtieth Annual Report, 1 April 1958 to 31 March 1959: 160 pp.

density (data)
South Africa, Division of Fisheries, 1954.
Twenty-fourth annual report:1-199.

density (data)
South Africa, Division of Fisheries, 1950.
Station list - R.S. "Africana II", R.V. "Shipa" and P.B. "Palinurus". 22nd Ann. Rept.:21-169, 2 charts.

density (data)
Spain, 1931.
Liste des stations.
Rapp. Proc. Verb., Cons. Perm. Int. Expl. Mer, 70:17-19.

density
Steele, J.H., J.R. Barrett and L.V Worthington 1962
Deep currents south of Iceland.
Deep-Sea Res., 9(5):465-474.

density
Steemann Nielsen, E., 1940.
Die Produktionsbedingungen des Phytoplanktons im Übergangsgebiet zwischen der Nord- und Ostsee.
Medd. Komm. Havundersøgelser, Ser. Plankton, 3(4): 55 pp., 17 textfigs., 22 tables.

density

Stefánsson, Unnsteinn and Larry P. Atkinson, 1971.
Nutrient-density relationships in the western North Atlantic between Cape Lookout and Bermuda. Limnol. Oceanogr. 16(1): 51-59.

density

Stefánsson, Unnsteinn, Larry P. Atkinson and Dean F. Bumpus, 1971.
Hydrographic properties and circulation of the North Carolina Shelf and slope waters. Deep-Sea Res., 18(4): 383-420.

density

Stevenson, Merritt R., Oscar Guillén G., and José Santoro de Ycaza, 1970.
Marine atlas of the Pacific coastal waters of South America.
Univ. Calif. Press, 20 pp., 99 charts. $40.00

density (data)

Steyaert, Marc, 1966.
Campagne internationale d'étude du régime des eaux dans le détroit de Gibraltar, mai-juin 1961. Résultats des observations hydrologiques effectuées à bord du Navire Belge "Eupen". Cah. océanogr., 18(Suppl.):19-94.

density (data)

Stimpson, John G., y Raymond C. Griffiths, 1967.
La distribución de densidad en el golfo de Cariaco, Venezuela oriental.
Invest. Pesqueras, Ser. Recursos Explotación pesqueros, 1(8):305-325.

density

Stirling, P.H., & H. Ho., 1961
Determining density.
Ind. & Eng. Chem., 53(10):48A-50A.

density

Stommel, H., 1961
Thermohaline convection with two stable regimes of flow.
Tellus, 13(2):224-230.

density(data)

Stroup, E.D., 1954.
Mid-Pacific oceanography IV. Transequatorial waters, January-March 1952.
U.S.F.W.S. Spec. Sci. Rept. - Fish. No. 135: 1-52, 18 figs. (multilithed).

density

Suda, K., 1930
On the seawater density with Akanuma-type hydrometer. Kobe Kaiyo Kishodia Iho, 24: 18-26.

density

Suzuki, T., and N. Sano, 1960.
On the current rip in Ishikari Bay caused by the Ishikari River flowing into that bay.
Bull. Fac. Fish., Hokkaido Univ., 11(3):132-161.

density

Sverdrup, H. U., 1933
On vertical circulation in the ocean due to the action of the wind with application to conditions within the Antarctic Circumpolar Current. Discovery Repts. 7:139-169, 23 text figs.

density

Sverdrup, H. U., and R. H. Fleming, 1941.
The waters of the coast of southern California, March to July, 1937. Bull. S.I.O. 4(10):261-375, 66 textfigs.

density

Tabata, S., and G.L. Pickard, 1957.
The physical oceanography of Bute Inlet, British Columbia. J. Fish. Res. Bd., Canada, 14(4):487-520.

density

Tait, J. B., 1934.
Surface drift bottle results in relation to temperature, salinity and density distributions in the northern North Sea. Rapp. Proc. Verb. 89(3):69-79, 9 textfigs.

density

Takano, Kenzo, 1965.
Courants marins induits par le vent et la non-uniformité de la densité de l'eau superficielle dans un océan.
La Mer: Bull. Soc. franco-japon. Océanogr., 2(2): 81-86.

Also in:
Geophys. Notes, Tokyo, 18(1)(2). 1965.

density

Takano, Kenzo, 1965.
Un exemple numérique des courants marins induits par le vent et la non-uniformité de la densité de l'eau superficielle dans un océan.
La Mer, Bull. Soc. franco-japon. Océanogr., 3:57-65. (not seen).

density

Takano, Kenzo, 1961
Distribution de densité à la surface d'un océan de longeur indéfinite en fonction de latitude.
J. Oceanogr. Soc., Japan, 17(4):190-196.

density(data)

Tanaka, Otohiko, Fumihiro Koga, Haruhiko Irie, Shozi Iizuka, Yosie Dotu, Keitaro Uchida, Satoshi Mito, Seiro Kimura, Osamu Tabeta and Sadahiko Imai, 1962.
The fundamental investigation of the biological productivity in the north-western sea area of Kyushu. II. Study on plankton productivity in the waters of Genkai-Nada region.
Rec. Oceanogr. Wks., Japan, Spec. No. 6:1-20.

Also in:
Contrib., Dept. Fish., and Fish. Res. Lab., Kyushu Univ., No. 8.

density

Tareyev, B.A., A.V. Fomitchev, 1963
On surface currents of the South Ocean.
(In Russian).
Mezhd. Geofiz. Komitet, Prezidiume, Akad. Nauk SSSR, Rezult. Issled. Programme Mezhd. Geofiz. Goda, Okeanol. Issled., (8):24-33.

density

Taylor, C.B., 1951.
Sea water density at different temperatures.
Jour. C.& G. S. 4:88-90.

density (data)

Tchernia, P., 1956.
Contribution à l'étude hydrologique de la Méditerranée occidentale. Hydrologie.
Bull. d'Info., C.C.O.E.C., 8(9):427-454, 11 pls.

density

Tchernia, P., 1951.
Compte rendu preliminaire des observations océanographiques faites par le Batiment Polaire "Commandant Charcot" pendant la campagne 1949-1950. Bull. d'Info., C.O.E.C., 3(2):40-56, 6 pls.

density

Tchernia, P., 1951.
Compte-rendu preliminaire des observations océan-ographiques faites par le Batiment Polaire "Commandant Charcot" pendant la campagne 1949-1950. Bull. d'Info., C.O.E.C., 3(1):13-22, 11 figs.

density (data)

Tchernia, P., H. Lacombe and P. Guibout, 1958.
Sur quelques nouvelles observations hydrologiques relatives à la région équatoriale de l'océan Indien. Bull. d'Info., C.C.O.E.C., 10(3):115-143.

density(data)

Tchernia, Paul, and Jean-Claude Lizeray, 1960
Ocean Indien, observations relatives à l'hydrologie du Bassin Nord-Australien.
Cahiers Oceanogr., 12(6):371-388.

density

Tchernia, Paul, and Bernard Saint-Guily, 1959.
Nouvelles observations hydrologiques d'hiver en Mediterranee occidentale.
Cahiers Ocean., C.C.O.E.C., 11(8):499-542.

density

Theisen, E. 1946
Tanafjorden. Enfinmarksfjords oceanografi. Fiskeridirektoratets skrifter Ser. Havundersøkelser. (Repts. on Norwegian Fishery and Marine Investigations) VIII (8):1-77, 23 text figs., 8 pls.

density

Thompson, E. F., 1939.
Chemical and physical investigations. The exchange of water between the Red Sea and the Gulf of Aden over the "sill". John Murray Exped., 1933-34, Sci. Repts. 2(4):105-119, 10 textfigs.

density

Thoulet, J., 1927.
Densimétrie en mer Tyrrhenienne. Bull. Inst. Océan., Monaco, No. 492:8 pp.

density

Thoulet, J., 1927.
Étude densimétrique dans le Pacifique. Ann. Inst. Océan., n.s., 4(7):261-273, 3 pls.

density

Thoulet, J., 1927.
Densimétrie et volcanicité abyssale dans le Pacifique. Ann. Inst. Océan., n.s., 4(2):25-45.

density

Thoulet, J., 1926.
Essai d'une densimétrie des océans. Ann. Inst. Océan., n.s., 3(3):137-160, figs.

density(data)

Trotti, L., 1953.
Risultati delle crociere talassografiche nel Mar Ligure e nell'alto Tirreno. Introduzione, Osservazioni meteorologiche e idrografiche.
Centro Talassografico Tirreno, Pubbl. No. 14: 7 pp., photos, 12 folding tables.

density(data)

Trotti, L., 1951.
Ricerche idrografiche sulle acque costiere Ligustiche comprese tra l'isola Palmarica e Capo Melo. Pt. 1. Introduzione, osservazioni meteorologiche e fisico-chemiche.
Centro Talassografico Tirreno, Pubbl. No. 8:29 pp.

density
Trotti, L., 1950.
Prima crociera talassografica del "Robusto" nel mare Ligure ed alto Tirreno.
Centro Talassografico Tirreno, Pubbl. No. 7:6 pp.

density
Tsujita, T., 1954.
(On the observed oceanographic structure of the Tsushima fishing grounds in winter and the ecolog-ical relationship between the structure and the fishing conditions.) J. Ocean. Soc., Japan, 10(3): 158-170, 12 textfigs.

density
Tyukov, I. Ya., 1964.
The problem of vertical distribution of the density field in a baroclinic sea. (In Russian). Izv., Akad. Nauk, SSSR, Ser. Geofiz., (3):422-425.

Translation:
(AGU) (3):255-257.

U

density
Uda, Michitake, 1963.
Oceanography of the subarctic Pacific Ocean.
J. Fish. Res. Bd., Canada, 20(1):119-179.

density
Uda, Michitaka, 1962
Subarctic oceanography in relation to whaling and salmon fisheries.
Sci. Repts., Whales Res. Inst., No. 16:105-119.

density
Uda, M., 1955.
On the subtropical convergence and the currents in the northwestern Pacific.
Rec. Ocean. Wks., Japan, 2(1):141-150, 10 textfigs.

density(data)
Uda, M., N. Watanabe and M. Ishino, 1956.
General results of the oceanographic surveys (1952-1955) on the fishing grounds in relation to the scattering layer. J. Tokyo Univ. Fish., 42(2):169-207.

U.S. Coast and Geodetic Survey

density
U.S. Coast & Geodetic Survey, 1957.
Density of sea water at tide stations, Atlantic coast, North and South America. USCGS Publ., 31 (2)(5th Edit.):72 pp.

density
U.S. Coast and Geodetic Survey, 1953.
Sea water temperature and density reduction tables Spec. Publ. 298:21 pp., 4 figs., 5 tables.

density
U. S. Coast and Deodetic Survey, 1950.
Density of sea water at Coast and Geodetic Survey tide stations, Pacific Ocean. Publ. No. D41-2 (revised 1950 edition):37 pp.

density
U.S. Coast and Geodetic Survey, 1949.
Density of sea water at Coast and Geodetic Survey tide stations, Atlantic and Gulf coasts. DW1(revised edition):31 pp.

density
U. S. Coast and Geodetic Survey, 1946.
Density of Sea Water at Coast and Geodetic Survey Tide Stations, Pacific Ocean. DW-2 (revised). 29 pp.

density
U. S. Coast and Geodetic Survey, 1945.
Density of Sea Water at Coast and Geodetic Survey Tide Stations, Atlantic and Gulf Coasts. DW-1 (revised). 23 pp.

Density (data)
United States, Department of Commerce, Environmental Sciences Services Administration, 1965
International Indian Ocean Expedition, USC&GS Ship Pioneer - 1964.
Vol. 1. Cruise Narrative and scientific results 139 pp.
Vol. 2. Data report: oceanographic stations, BT observations, and bottom samples, 183 pp.

U.S. Navy Hydrographic Office

density
U.S. Navy Hydrographic Office, 1960
Summary of oceanographic conditions in the Indian Ocean. Spec. Publ. SP-53: 142 pp.

density
United States Navy Hydrographic Office, 1959.
Climatological and oceanographic atlas for mariners. Vol. 1.
North Atlantic Ocean., 182 charts.

density(data)
U.S. Navy Hydrographic Office, 1957.
Operation Deep Freeze II, 1956-1957. Oceanographic survey results. H.O. Tech. Rept., 29: 155 pp., (multilithed)

density
U.S. Navy Hydrographic Office, 1952.
Tables of sea water density. H.O. Pub. 615.

density
United States, Weather Bureau and Hydrographic Office, 1961
Climatological and oceanographic atlas for Mariners. Vol. II. North Pacific Ocean.
Unnumbered pages.

density (data)
University of Southern California, Allan Hancock Foundation, 1965.
An oceanographic and biological survey of the southern California mainland shelf.
State of California, Resources Agency, State Water Quality Control Board, Publ. No. 27:232 pp.
Appendix, 445 pp.

density(data)
Uriarte, L.B., 1932.
Premiers travaux du Laboratoire Oceanographique des Iles Canaries.
Rapp. Proc. Verb., Cons. Perm. Int. Expl. Mer, 67:65-88, 8 textfigs.

V

density
Valdivia, Julio E., y Oscar Guillen, 1966.
Informe preliminer del Crucero de primavera 1965 (Cabo Blanco-Morro Sama).
Inst. Mer, Peru, Informe, No. 11:35-70.

density
Veldivia, Julie, y Oscar Guillén, 1966.
Informe preliminar del crucero de primavera 1965 (Cabe Blance-Morro Sama).
Inst. Mar. Peru, Informe, No. 11:35-70.

density
Van Riel, P.M., H.C. Hamaker, and L. Van Eyck, 1950.
Tables. Serial and bottom observations. Temperature, salinity and density. Ocean. Res., Snellius Exped., 1929-1930, 2(6):44 pp.

density
Van Weel, K. M., 1923.
Meteorological and hydrographical observations made in the western part of the Netherlands East Indian Archipelago. Treubia, 4: 559 pp.

density(data)
Varlet, F., 1960
Sur l'hydrologie du plateau continental africain du Cap des Palmes au Cap des Trois Pointes
Bull. Inst. Pêches Maritimes, Maroc, No. 5: 20 pp.

density
Vatova, A. 1965.
Les conditions hydrographiques de la Mar Piccolo de Tarante pendant l'année 1963.
Rapp. P.-v. Reun. Comm. int. Explr. scient. Mer. Mediterr. 18(3): 653-655.

density
Vatova, A., 1948.
Osservazioni idrografiche periodiche nell'alto Adriatico(1937-1944). Boll. Pesca, Pescicotura e Idrobiologia, Anno 24, n.s., 3(2):245-277.

density
Vatova, A., 1934.
L'anormale regime fisico-chimico dell' Alto Adriatico nel 1929 e le sue ripercussioni sulla fauna. Thalassia 1(8):49 pp., 3 fold-ins, 16 figs., tables.

density
Vatova, A., and P. Milo di Villagrazia, 1948.
Sulle condizioni chimicofisiche del Canale di Lema presso Rovigno d'Istria.
Bol. Pesca, Piscicol. Idrobiol., n.s., 3(1):5-27, 2 textfigs., 5 pls.

density (data)
Vega Rodriguez, Filiberto, y Virgilio Arenas Fuentes 1965.
Resultados preliminares sobre la distribución del plancton y datos hidrograficos del Arrecife "La Blanquilla", Veracruz, Ver.
Anales Inst. Biol. Univ. Mexico 36(1/2): 53-59.

density
Vercelli, F., and M. Picotti, 1926.
Pt. 2. Il regime fisico-chemico della acque nello Stretto di Messina. Crociere per lo studio dei fenomeni nello Stretto di Messina. (Campagne della R. Nave "Marsigli" negli anni 1922 e 1923).
Comm. Int. Medit., Delag. Ital., 161 pp., 40 figs

density
Visser, G.A., and M.M van Niekerk 1965.
Ocean currents and water masses at 1,000, 1,500 and 3000 meters in the south-west Indian Ocean.
Investl Rept S.Afr. Dept. Commerce Industries Div. Sea Fish. 52:46pp.

density
Visser, S. W., 1939(1940).
The "Snellius" expedition in the eastern part of the Netherlands Indies 1929-1930: Surface observations, temperature, salinity, density.
Proc. Sixth Pacific Sci. Congr., 3:143-145.

densities
Visser, S.W., 1938.
Surface observations; temperatures, salinities, densities. Snellius Exped. II, Oceanogr. Res. Pt. 4:1-60, 20 figs., 4 charts.

Reviewed: J. du Cons. 14(3):401-403, G. Schott.

density
Voipio, Aarno, and Erkki Häsänen, 1962
Relationships between chlorinity, density and specific conductivity in Baltic waters.
Annales Acad. Sci. Fennicae, (A) (Chemica) (111): 18 pp.

W

density
Waldichuk, Michael, 1965.
Water exchange in Port Moody, British Columbia, and its effect on waste disposal.
J. Fish. Res. Bd., Canada, 22(3):801-822.

density
Watanabe, Nobuo, Toshiyuki Hirano, Rinnosuke Fukai, Kozi Matsumoto and Fumiko Shiokawa, 1957.
A preliminary report on the oceanographic survey in the "Kuroshio" area, south of Honshu, June-July 1955.
Rec. Oceanogr. Wks., Japan (Suppl.):197-

density
Watson, E.E., 1936.
Mixing and residual currents in tidal waters as illustrated in the Bay of Fundy. J. Biol. Bd., Canada, 2(2):141-208, tables, 26 textfigs.

density
Weidemann, H., 1950.
Untersuchungen über unperiodische und periodische hydrographische Vorgänge in der Beltsee. Kieler Meeresf. 7(2):70-86, 10 textfigs.

density
Wennekens, M. P., 1965.
Diffusion processes vs. oceanic micro-structure. Symposium, Diffision in oceans and fresh waters, Lamont Geol. Obs., 31 Aug.-2 Sept., 1964, 108-113.

specific gravity
Wentworth, C.K. 1939.
Specific gravity of sea-water and the Ghyben-Herzberg ratio in Hawaii. Trans. Am. Geophys. Un. Pt. 4:690-692.

density
Wiborg, K.F., 1944
The production of zooplankton in a landlocked fjord, the Nordåsvatn near Bergen, in 1941-42, with special reference to the copepods... (Repts. Norwegian Fish. and Mar. Invest.) 7(7):83 pp., 40 text figs.

density
Wirth, H. E., 1940.
The problem of the density of sea water.
J. Mar. Res., 3(3):230-247, 3 figs.

density
Wüst, Georg, 1961
Das Bodenwasser und die Vertikalzirkulation des Mittelländischen Meeres. 3. Beitrag zum mittelmeerischen Zirkulationsproblem.
Deutsche Hydrogr. Zeits., 14(3):81-92.

density(data)
Wüst, G., 1957.
1. Die Verteilung von Temperatur, Salzgehalt und Dichte. Erhebnisse einer hydrographisch-produktions-biologischen Längsschnitts durch die Ostsee im Sommer 1956. Kieler Meeresf., 13(2):163-185.

density
Wüst, G., 1939.
Das submarine Relief bei den Azoren.
Abhandl. Preussischen Akad. Wissenschaften, Jahrgang 1939, Phys.-math. Kl. No.5:46-58, 7 text figs.

density(including data)
Wüst, G., and W. Brogmus, 1955.
Oceanographische Ergebnisse einer Untersuchungsfahrt mit Forschungskutter "Südfall" durch die Ostsee, Juni-Juli 1954 (anlässlich der totalen Sonnenfinsternis auf Öland).
Kieler Meeresf. 11(1):3-21, 8 textfigs.

density
Wüst, G., W. Brogmus and E. Noodt, 1954.
Die zonale Verteilung von Salzgehalt, Niederschlag, Verdunstung, Temperatur und Dichte an der Oberfläche der Ozeans. Kieler Meeresf. 10(2):137-161, 10 textfigs.

density
Wyrtki, K., 1950.
Über die Beziehungen zwischen Trübung und ozeanographischen Aufbau. Kieler Meeresf. 7(2):87-107, 10 textfigs.

XYZ

density
Yasui, M., 1957.
On the rapid estimation of the dynamic topography in the seas adjacent to Japan.
Rec. Oceanogr. Wks., Japan, n.s., 3(1):8-15.

density (data)
#Yasuoka, Takeo, 1967.
Hydrography of the Okhotsk Sea - (1).
Oceanogrl Mag., 19(1):61-72.

density
Yoshida, Kozo, 1965.
A theoretical model on wind-induced density field in the oceans. I.
J. Oceanogr. Soc., Japan, 21(4):154-173.

density
Zetler, B.D., 1953.
Some effects of the diversion of the Santee River on the waters of Charleston Harbor.
Trans. Amer. Geophys. Union 34(5):729-732, 2 text-figs.

density, annual variations

density(annual variations
Koizumi, M., 1955.
Researches on the variations of oceanographic conditions in the region of the Ocean Weather Station "Extra" in the North Pacific Ocean. 1. "Normal values and annual variations of oceanographic elements. Pap. Met., Geophys., Tokyo, 6(2):185-201.

density anomalies

density anomalies, effect of
Gudkovich, Z.M., and E.G. Nikiforov, 1965.
On some significant features related to the formation of water density anomalies and their effect upon ice and hydrological conditions in the Arctic basin and bordering seas. (In Russian).
Okeanologiia, Akad. Nauk, SSSR, 5(2):250-260.

density, calculations
Gascard, Jean-Claude, 1970.
Calcul de la salinité et de la densité de l'eau de mer à partir de mesures in situ de température, conductivité électrique et pression.
Cah. océanogr. 22(3): 239-257

density charts
Bock, Karl-Heinz 1971.
Monatskarten der Dichte des Wassers in der Ostsee dargestellt für verschiedene Tiefenhorizonte.
Ergänzhft dt hydrogr. Z. B(4°) 13:126pp.

density/conductance

density/conductance
Connors, Donald N., and Peter K. Weyl, 1968.
The partial equivalent conductances of salts in sea water and the density/conductance relationship.
Limnol. Oceanogr., 13(1):39-50.

Density currents see currents, density

Density currents

See: currents, density

Density – data only

A

density (data only)
Andersen, Henry S., and Martin J. Moynihan, 1971.
Oceanography of the Grand Banks region and the Labrador Sea, April-June 1968.
Oceanogr. Rept. U.S. Cst Gd 39 (CG 373-39). 308pp

density (data only)
Anderson, William W., and Jack W. Gehringer, 1959.
Physical oceanographic, biological and chemical data, South Atlantic coast of the United States, M/V Theodore N. Gill cruise 9.
USFWS Spec. Sci. Rept., Fish. No. 313:226 pp.

density(data only)
Anderson, W.W., and J.W. Gehringer, 1959.
Physical oceanographic, biological and chemical data South Atlantic coast of the United States, M/V Theodore N. Gill Cruise 8.
USFWS Spec. Sci. Rept., Fish., No. 303:227 pp.

density(data only)
Anderson, William W., and Jack W. Gehringer, 1958.
Physical oceanographic, biological and chemical data, south Atlantic coast of the United States, M/V Theodore N. Gill Cruise 6.
USFWS Spec. Sci. Rept., Fish., No. 265:99 pp.

density(data only)
Anderson, W.W., and J.W. Gehringer, 1958.
Physical oceanographic, biological and chemical data, South Atlantic coast of the United States, M/V Theodore N. Gill cruise 5.
USFWS Spec. Sci. Rept., Fish. 248:220 pp.

density (data only)
Anderson, W.W., and J.W. Gehringer, 1957.
Physical oceanographic, biological and chemical data, South Atlantic coast of the United States, M/V Theodore N. Gill cruise 4.
USFWS Spec. Sci. Rept., Fish., 234:192 pp.

density (data only)
Anderson, W.W., and J.W. Gehringer, 1957
Physical, oceanographic, biological and chemical data, South Atlantic coast of the United States, Theodore N.Gill Cruise
USFWS Spec. Sci. Rept., Fish., No. 210:208 pp.

density (data only)
Anderson, W.W., J.W. Gehringer, and E. Cohen, 1956.
Physical oceanographic, biological and chemical data, south Atlantic coast of the United States, M/V Theodore N. Gill, Cruise 1.
USFWS Spec. Sci. Rept., Fish., No. 178:160 pp.

density(data only)
Anderson, W.W., J.W. Gehringer and E. Cohen, 1956
Physical oceanographic, biological and chemical data, South Atlantic coast of the United States, Theodore N. Gill, Cruise 2.
Spec. Sci. Rept., Fish., No. 198:270 pp.

density (data only)
Angot, M., et R. Gerard, 1965.
Hydrologie de la region de Nosy-Bé: décembre 1963 à mars 1964.
Cahiers O.R.S.T.R.O.M., Océanogr., 4(1):31-53.

density (data only)
Angot, M., et R. Gerard, 1965.
Hydrologie de la région de Nosy-Bé: juillet à novembre 1963.
Cahiers, O.R.S.T.R.O.M., Océanogr., 3(1):3-29.

Anon., 1950. density (data only)
Table 14. Oceanographical observations taken in the Maizuru Bay; physical and chemical data for stations occupied by R.M.S. "Syunpu-maru". Res. Mar. Met., Ocean. Obs., Tokyo, July-Dec. 1947, No. 2:71-76.

density(data only)
Anon., 1950.
Table 13. Oceanographical observations taken in the Nagasaki Harbour; physical and chemical data for stations occupied by a cutter. Res. Mar. Met. Ocean. Obs., Tokyo, July-Dec. 1947, No. 2:65-70.

density (data only)
Anon., 1950.
Table 12. Oceanographical observations taken in the Omura Bay; physical and chemical data for stations occupied by the "Tsuru-maru". Res. Mar. Met., Ocean. Obs., Tokyo, July-Dec. 1947, No. 2: 61-64.

density(data only)
Anon., 1950.
Table 11. Oceanographical observations taken in the Omura Bay; physical and chemical data for stations occupied by a cutter. Res. Mar. Met., Ocean. Obs., Tokyo, July-Dec. 1947, No. 2:59-60.

density(data only)
Anon., 1950.
Table 10. Oceanographical observations taken in the sea northwest of Kyushu; physical and chemical data for stations occupied by the Fugen-maru". Res. Mar. Met., Ocean. Obs., Tokyo, July-Dec. 1947, No. 2:57-58.

density(data only)
Anon., 1950.
Table 8. Oceanographical observations taken in the adjacent sea to the west of Kyushu; physical and chemical data for stations occupied by R.M.S. "Yushio-maru". Res. Mar. Met., Ocean. Obs., Tokyo, July-Dec. 1947, No. 2:51-54.

density(data only)
Anon., 1950.
Table 7. Oceanographical observations taken in the Sagami Bay. physical and chemical data for stations occupied by R.M.S. "Arashio-Bay. Res. Mar. Met., Ocean. Obs., Tokyo, July-Dec. 1947, No. 2:48-50.

density(data only)
Anon., 1950.
Table 6. Oceanographical observations taken at the "fixed point" (39°N, 153°E) in the North Pacific Ocean; physical and chemical data for stations occupied by R.M.S. "Ryofu-maru". Res. Mar. Met., Ocean. Obs, Tokyo, July-Dec. 1947, No. 2:44-47.

density(data only)
Anon., 1950.
Table 5. Oceanographical observations taken in the North Pacific Ocean east off Tohoku District; physcial and chemical data for stations occupied by R.M.S. "Ryofu-maru". Res. Mar. Met., Ocean. Obs., Tokyo, July-Dec. 1947, No. 2:22-43.

density(data only)
Anon., 1950.
Table 4. Oceanographical observations taken in the North Pacific Ocean along the coast of Sanriku; physical and chemical data for stations occupied by R.M.S. "Kuroshio-maru". Res. Mar. Met., Ocean. Obs., Tokyo, July-Dec. 1947, No. 2: 13-21.

density (data only)
Anon., 1950.
Table 3. Oceanographical observations taken in the North Pacific Ocean east off Sanriku; physical and chemical data for stations occupied by "Oshoro-maru". Res. Mar. Met., Ocean. Obs., Tokyo July-Dec. 1947, No. 2:8-12.

density (data only)
Anon., 1950.
Table 2. Oceanographical observations taken in the Uchiura Bay; physical and chemical data for stations occupied by R.M.S. "Kuroshio-maru". Res. Mar. Met., Ocean. Obs., Tokyo, July-Dec. 1947, No. 2:3-7.

density (data only)
Anon., 1950.
Table 1. Oceanographical observations taken in the Uchiura Bay; physical and chemical data for stations occupied by a fishing boat belonging to Mori Fishery Assoc. Res. Mar. Met., Ocean. Obs., Tokyo, July-Dec. 1947, No. 2:1-2.

density (data only)
Anon., 1950.
Oceanographical observations taken in the Nagasaki Harbour; physical and chemical data for stations occupied by the cutter. Res. Mar. Met., Ocean. Obs., Jan.-June 1948, No. 1:38, Table 5.

density (data only)
Anon., 1950.
Oceanographic observations taken in the Miyako Bay; physical and chemical data for stations occupied by R.M.S. "Kuroshio-maru". Res. Mar. Met Ocean. Obs., Jan.-June 1947, No. 1:24-27, Table 3

density (data only)
Anon., 1950.
[Report of the coastal observations (Oct. 1950)]
J. Ocean., Kobe Obs., 2nd ser., 1(4):21-22.

density (data only)
Anon., 1950.
Oceanographical observations taken in the North Pacific Ocean (east off Hideshima Is.); physical and chemical data for stations occupied by R.M.S. "Kuroshio-maru" (continued). Res. Mar. Met., Ocean. Obs., Jan.-June 1947, No. 1:12-23, Table 2

density (data only)
Anon., 1950.
[Report of the oceanographical observations from Fisheries Experimental Stations (Sept. 1950)]
J. Ocean., Kobe Obs., 2nd ser., 1(2):5-11.

density (data only)
Anon., 1950.
Oceanographical observations taken off the Sanriku coast; physical and chemical data for stations occupied by R.M.S. "Oyashio-maru". Res. Mar. Met. Ocean. Obs., Jan.-June 1947, No. 1:1-11, Table 1.

density (data only)
Anonymous, 1947
Discovery investigations Station List 1937-1939. Discovery Report, XXIV: 197-422. plsIV-VI.

Density (Data only)
Aragno, Federico, Alberto Gomez, Aldo Orlando, y Andres J. Lusquiños, 1968.
Datos y resultados de las campañas pesqueria "Pesqueria II" (9 de noviembre al 12 diciembre de 1966)
Publ. (Ser. Informes tecn) Mar del Plata, Argentina 10(2): 1-129

density (data only)
Aragno Federico, Alberto Gomez, Aldo Orlando y Andres J. Lusquiños 1968.
Datos y resultados preliminares de las campañas pesquerias "Pesqueria I" (12 de agosto al 8 de setiembre de 1966).
Publ. (Ser. Informes tecn) Mar del Plata, Argentina 10(1): 1-159.

density (data only)
Aragno Federico J., Alberto Gomez, Aldo Orlando y Andres J. Lusquiños 1968.
Datos y resultados de las campañas pesquerias "Pesqueria III" (20 de febrero al 20 de marzo de 1967).
Publ. (Ser. Informes tecn.) Mar del Plata, Argentina 10(3): 1-162.

density (data only)
Argentina, Secretaria de Marina, Servicio de Hidrografia Naval, 1962.
Datos preliminares como contribucion de la influencia de los cambios volumetricos en la determinacion del nivel medio del mar.
Publico, H. 612:133 pp., foldins.

density (data only)
Argentina, Secretaria de Marina, Servicio de Hidrografia Naval, 1961.
Trabajos oceanograficos realizados en la campana Antartica 1959/1960. Resultados preliminares
Publico, H., 623:unnumbered pp.

density (data only)
Argentina, Secretaria de Marina, Servicio de Hidrografia Naval, 1961.
Operacion oceanografica, Vema - Canepa 1. Resultados preliminares.
Publico, H. 628:30 pp.

density (data only)
Argentina, Secretaria de Marina, Servicio de Hidrografia Naval, 1961.
Trabajos oceanograficos realizados en la campana Antartica 1960/1961. Resultados preliminares.
Publico, H. 629:unnumbered pp.

density (data only)
Artüz M. İlham, 1970.
Some observations on the hydrography of the Turkish Aegean waters during 4-25 September (1963).
Hidrobiol. Arastirma Enst. Yayinlarindan, Istanbul Univ. Fen Fakült., 6(3/4):1-9

density (data only)
Artüz, M. Ilham 1969.
Daily observations on the hydrographic conditions of the Bosphorus during the period of 1962-1966.
Istanbul Univ. Facült. Mecmuasi (B), 34 (3/4): 207-244.

Australia Commonwealth Scientific and Industrial Organization

density (data only)
Australia Commonwealth Scientific and Industrial Research Organization, Division of Fisheries and Oceanography, 1970.
Coastal investigations off Port Hacking, New South Wales, in 1965.
Oceanogr. Sta. List 85: 124 pp.

density (data only)
Australia, Commonwealth Scientific and Industrial Research Organization, 1969.
Oceanographic observations in the Indian Ocean in 1966 H.M.A.S. Diamantina Cruise Dm2/66.
Div. Fish. Oceanogr. Oceanogr. Cruise Rept. 54: 64 pp. (multilithed).

Density (Data only)
Australia, Commonwealth Scientific and Industrial Research Organization, 1969.
Oceanographical observations in the Indian Ocean in 1965 H.M.A.S. Diamantina Cruise Dm2/65. Div. Fish. Oceanogr. Oceanogr. Cruise Rept. 49: 57 pp. (multilithed)

density (data only)
Australia, Commonwealth Scientific and Industrial Research Organization, 1969.
Oceanographical observations in the Pacific Ocean in 1965 H.M.A.S. Gascoyne Cruise G3/65.
Div. Fish. Oceanogr. Oceanogr. Cruise Rept. 44: 24 pp. (multilithed).

density (data only)
Australia, Commonwealth Scientific and Industrial Organization, 1968.
Coastal investigations off Port Hadking, New South Wales, in 1964.
Oceanogr. Sta. List, 84: 49 pp. (multilithed)

density (data only)
Australia, Commonwealth Scientific and Industrial Organization, 1968.
Coastal investigations off Port Hacking, New South Wales, in 1961.
Oceanogr. Sta. List, 81: 55 pp. (multilithed)

density, surface (data only)
Australia, Commonwealth Scientific and Industrial Research Organization, 1968.
Investigations by F.R.V. Lancelin in western Australian waters in 1963.
Div. Fish. Oceanogr., Oceanogr. Sta. Lists, 80: 38 pp. (multilithed).

density (data only)
Australia, Commonwealth Scientific and Industrial Research Organization, 1968.
Investigations by F.V. Degei in New South Wales, south, and western Australian waters in 1966.
Div. Fish. Oceanogr., Oceanogr. Sta. Lists 75: 72 pp. (multilithed).

density (data only)
Australia, Commonwealth Scientific and Industrial Research Organization, 1968.
Investigations by F.V. Degei in Queensland waters in 1965.
Div. Fish. Oceanogr., Oceanogr. Sta. Lists, 73: 49 pp. (multilithed).

density (data only)
Australia, Commonwealth Scientific and Industrial Research Organization, 1968.
Investigations by F.R.V. Marelda on the eastern Australian tuna ground in 1965.
Div. Fish. Oceanogr., Oceanogr. Sta. List, 72: 58 pp. (multilithed).

density (data only)
Australia, Commonwealth Scientific and Industrial Research Organization, 1968.
Investigations by F.R.V. Investigator on the South Australian Tuna grounds in 1965.
Div. Fish. Oceanogr., Oceanogr. Sta. List, 70: 21 pp. (multilithed)

density (data only)
Australia, Commonwealth Scientific and Industrial Research Organization, 1968.
Investigations by F.R.V. Marelda on the eastern Australian tuna grounds in 1964.
Oceanogr. Sta. List, Div. Fish. Oceanogr., 68: 91 pp.

density (data only)
Australia, Commonwealth Scientific and Industrial Research Organization, 1968.
Investigations by F.R.V. Investigator on the South Australian tuna grounds in 1964.
Oceanogr. Sta. List, Div. Fish. Oceanogr., 67: 78 pp. (multilithed).

Australia Commonwealth Scientific and Industrial Organization, 1968.
Investigations by F.R.V. Marelda on the eastern Australian tuna grounds in 1963.
Div. Fish. Oceanogr., Oceanogr. Sta. List, 65: 144 pp.

density (data only)
Australia, Commonwealth Scientific and Industrial Research Organization, 1968.
Investigations by F.R.V. Investigator on the south Australian tuna ground in 1963.
Oceanogr. Data List, Div. Fish. Oceanogr., 64: 95 pp. (multilithed).

Australia, Commonwealth Scientific and Industrial Organization, 1968.
Investigations by F.R.V. Marelda on the eastern Australian grounds in 1962.
Div. Fish. Oceanogr., Oceanogr. List 61: 135 pp.

density (data only)
Australia Commonwealth and Scientific (& Industrial) Research Organization, 1968.
Investigations by F.R.V. Investigator on the South Australian tuna grounds in 1962.
Oceanogr. Sta. List, Div. Fish. Oceanogr. 60: 23 pp.

density (data only)
Australia, Commonwealth Scientific and Industrial Organization, 1968.
Investigations by F.R.V. Derwent Hunter on the eastern Australia tuna grounds in 1962.
Div. Fish. Oceanogr., Oceanogr. List 59: 74 pp.

density (data only)
Australia Commonwealth Scientific and Industrial Research Organization 1968.
Investigations by F.R.V. Marelda on the eastern Australian tuna grounds in 1961
Oceanogr. Sta. List, Div. Fish. Oceanogr. 56: 130 pp.

density (data only)
Australia, Commonwealth Scientific and Industrial Research Organization, 1968.
Investigations by F.R.V. Weerutta on the south Australian tuna ground in 1961.
Oceanogr. Sta. List, Div. Fish. Oceanogr., 55: 37 pp. (multilithed).

density (data only)
Australia, Commonwealth Scientific and Industrial Organization, 1968
Investigations by F.R.V. Derwent Hunter on the eastern Australian tuna grounds in 1961.
Div. Fish. Oceanogr., Oceanogr. List, 54: 235 pp.

density (data only)
Australia Commonwealth Scientific and Industrial Research Organization 1968.
Oceanographical observations in the Indian Ocean in 1965, H.M.A.S. Gascoyne, Cruise G 2/65.
Oceanogr. Cruise Rept. 43: 1-58.

density (data only)
Australia, Commonwealth Scientific and Industrial Research Organization 1968.
Oceanographical observations in the Pacific Ocean in 1964, H.M.A.S. Gascoyne, Cruise G6/64.
Oceanogr. Cruise Rept. 42: 53 pp.

density (data only)
Australia Commonwealth Scientific and Industrial Research Organization, 1968.
Oceanographical observations in the Indian Ocean in 1964, H.M.A.S. Gascoyne, Cruise G5/64.
Oceanogr. Cruise Rept. Div. Fish. Oceanogr. 41: 52 pp.

density (data only)
Australia Commonwealth Scientific and Industrial Research Organization 1968.
Oceanographical observations in the Indian Ocean in 1964 H.M.A.S. Diamantina Cruise Dm 5/64.
Oceanogr. Cruise Rept. Div. Fish. Oceanogr. 40: 48 pp.

density (data only)
Australia, Commonwealth Scientific and Industrial Research Organization, 1968.
Oceanographical observations in the Pacific Ocean in 1963, H.M.A.S. Gascoyne, Cruise G3/63.
Oceanogr. Cruise Rept., Div. Fish. Oceanogr., 26: 134 pp.

density (data only)
Australia Commonwealth Scientific and Industrial Research Organization 1967.
Oceanographical observations in the Indian Ocean in 1965, H.M.A.S. Gascoyne, Cruise G 5/65.
Div. Fish. Oceanogr. Oceanogr. Cruise Rep. 46: 62 pp.

density (data only)
Australia, Commonwealth Scientific and Industrial Research Organization, 1967.
Oceanographical observations in the Pacific Ocean in 1964, H.M.A.S. Gascoyne Cruise G4/64.
Oceanogr. Cruise Rept., Div. Fish. Oceanogr. 39: 39 pp.

density (data only)
Australia, Commonwealth Scientific and Industrial Research Organization, 1967.
Oceanographical observations in the Indian Ocean in 1964, H.M.A.S. Gascoyne, Cruise G 2/64.
Div. Fish. Oceanogr., Oceanogr. Cruise Rep.
34:46 pp. (multilithed)

density (data only)
Australia, Commonwealth Scientific and Industrial Research Organization, 1967.
Oceanographical observations in the Pacific Ocean in 1964, H.M.A.S. Gascoyne, Cruise G 1/64.
Div. Fish. Oceanogr., Oceanogr. Cruise Rep.
32:66pp (multilithed).

density (data only)
Australia, Commonwealth Scientific and Industrial Research Organisation, 1967.
Oceanographical observations in the Pacific Ocean in 1963, H.M.A.S. Gascoyne, Cruise G 4/63.
Div. Fish. Oceanogr., Oceanogr. Cruise Rep. 29:64 pp. (multilithed).

density (data only)
Australia, Commonwealth Scientific and Industrial Organization, 1967.
Oceanographic observations in the Pacific Ocean in 1963, H.M.A.S. Gascoyne, Cruise G 5/63.
Div. Fish. Oceanogr. Cruise Rep. 31:57 pp.

density (data only)
Australia, Commonwealth Scientific and Industrial Organization, 1967.
Oceanographical observations in the Indian and Pacific oceans in 1964, H.M.A.S. Gascoyne, Cruise G3/64.
Oceanogr. Cruise Rep. Div. Fish. Oceanogr., 35: 40 pp.

density (data only)
Australia, Commonwealth Scientific and Industrial Organization, 1967.
Oceanographical observations in the Indian Ocean in 1964, H.M.A.S. Diamantina, Cruise Dm 2/64.
Oceanogr. Cruise Rep. Div. Fish. Oceanogr. 36:53 pp.

density (data only)
Australia, Commonwealth Scientific and Industrial Research Organization, 1967.
Oceanographic observations in the Indian Ocean in 1963, H.M.A.S. Gascoyne, Cruise G 2/63.
Div. Fish. Oceanogr., Oceanogr. Cruise Rep. 22: 51 pp.

density (data only)
Australia, Commonwealth Scientific and Industrial Research Organization, 1967.
Oceanographical observations in the Indian Ocean in 1962, H.M.A.S. Diamantina Cruise Dm 4/62.
Div. Fish. Oceanogr., Oceanogr. Cruise Rep. 20: 138 pp.

Australia, Commonwealth Scientific and Industrial Research Organization, 1967.
Oceanographical observations in the Pacific Ocean in 1962 H.M.A.S. Gascoyne Cruise G 5/62.
Oceanogr. Cruise Rep, Div. Fish. Oceanogr., 14: 71 pp.

density (data only)
Australia, Commonwealth Scientific and Industrial Research Organization, 1966.
Oceanographical observations in the Indian Ocean in 1962 H.M.A.S. Gascoyne Cruise G 4/62.
Div. Fish. Oceanogr., Oceanogr. Cruise Rep. No. 17: 151 pp.

density (data only)
Australia, Commonwealth Scientific and Industrial Research Organization 1967.
Oceanographical observations in the Pacific and Indian oceans in 1962, H.M.A.S. Gascoyne, cruises G 2/62 and G 3/62.
Div. Fish. Oceanogr. Oceanogr. Cruise Rep. 16: 90 pp.

density (data only)
Australia, Commonwealth Scientific and Industrial Research Organization, 1967.
Oceanographical observations in the Pacific Ocean in 1962, H.M.A.S. Gascoyne, Cruise G 1/62.
Div. Fish. Oceanogr., Oceanogr. Cruise Rep. 13: 180 pp.

density (data only)
Australia, Commonwealth Scientific and Industrial Research Organization, 1967.
Oceanographical observations in the Pacific Ocean in 1961, H.M.A.S. Gascoyne Cruise G 3/61.
Div. Fish. Oceanogr., Oceanogr. Cruise Rep., 12: 126 pp.

density (data only)
Australia, Commonwealth Scientific and Industrial Research Organization, 1964.
Oceanographic observations in the Indian Ocean in 1962, H.M.A.S. Diamantina, Cruise Dm 1/62.
Oceanogr. Cruise Rept., Div. Fish Oceanogr., No. 10: 128 pp.

Density (data only)
Australia, Commonwealth Scientific and Industrial Research Organiz-tion.
Oceanographic observations in the Indian Ocean in 1963, H.M.A.S. DIAMANTINA Cruise DM 3/63.
Div. Fish. Oceanogr., Oceanogr. Cruise Rept. No. 25: 147 pp.

density (data only)
Australia, Commonwealth Scientific and Industrial Research Organization 1965.
Oceanographical observations in the Indian Ocean in 1963, H.M.A.S. Diamantina, Cruise Dm 2/63.
Div. Fish. Oceanogr. Oceanogr. Cruise Rept. 24: 153 pp.

density (data only)
Australia Commonwealth Scientific and Industrial Research Organization 1965.
Oceanographical observations in the Indian Ocean in 1963 H.M.A.S. Diamantina Cruise Dm 1/63.
Div. Fish. Oceanogr. Oceanogr. Cruise Rept. 23: 175 pp.

density (data only)
Australia, Commonwealth Scientific and Industrial Research Organization, 1965
Oceanographic observations in the Indian Ocean in 1963, H.M.A.S. Gascoyne, Cruise G 1/63.
Div. Fish. and Oceanogr., Oceanogr. Cruise Rept. 21: 135 pp.

density (data only)
Australia, Commonwealth Scientific and Industrial Research Organization, 1964.
Oceanographical observations in the Indian Ocean in 1962, H.M.A.S. Diamantina, Cruise Dm 2/62.
Oceanogr. Cruise Rept., Div. Fish. and Oceanogr. No. 15:117 pp.

density(data only)
Australia, Commonwealth Scientific and Industrial Research Organization, 1964.
Oceanographical observations in the Indian Ocean in 1961, H.M.A.S. Diamantina, Cruise Dm 3/61.
Div. Fish. and Oceanogr., Oceanogr. Cruise Rept., No. 11:215 pp.

density (data only)
Australia, Commonwealth Scientific and Industrial Research Organization, 1963.
Coastal investigations at Port Hacking, New South Wales, 1960.
Div. Fish. and Oceanogr., Oceanogr. Sta. List No. 52:135 pp.

A.D. Crooks, compiler

density(data only)
Australia, Commonwealth Scientific and Industrial Research Organization 1963
Coastal hydrological investigations in eastern Australia, 1960.
Div. Fish. and Oceanogr., Oceanogr. Sta. List 51: 46 pp.

density (data only)
Australia, Commonwealth Scientific and Industrial Research Organization, 1963.
Coastal hydrological investigations in the New South Wales tuna fishing area, 1963.
Div. Fish. and Oceanogr., Oceanogr. Sta. List, 53:81 pp.

A.D. Crooks, compiler.

density(data only)
Australia, Commonwealth Scientific and Industrial Research Organization, Division of Fisheries and Oceanography, 1963
Oceanographical observations in the Indian Ocean in 1960, H.M.A.S. Diamantina, Cruise Dm2/60.
Oceanographical Cruise Report No. 3:347 pp.

density(data only)
Australia, Commonwealth Scientific and Industrial Research Organization, 1963
Oceanographical observations in the Indian Ocean in 1961, H.M.A.S. Diamantina Cruise Dm 2/61.
Oceanogr. Cruise Rept., Div. Fish and Oceanogr. No. 9: 155 pp., 14 figs.

density (data only)
Australia, Commonwealth Scientific and Industrial Research Organization, 1963.
Oceanographical observations in the Pacific Ocean in 1961, H.M.A.S. Gascoyne, Cruise G 1/61.
Oceanogr. Cruise Rept., Div. Fish. and Oceanogr., No. 8:130 pp., 12 figs.

density(data only)
Australia, Commonwealth Scientific and Industrial Research Organization, Division of Fisheries and Oceanography, 1963.
Oceanographical observations in the Pacific Ocean in 1960, H.M.A.S. Gascoyne, Cruise G 3/60.
Oceanographical Cruise Report, No. 6:115 pp.

density(data only)
Australia, Commonwealth Scientific and Industrial Research Organization, Division of Fisheries and Oceanography, 1962.
Oceanographic observations in the Pacific Ocean in 1960, H.M.A.S. Gascoyne, Cruises G 1/60 and G 2/60.
Oceanographical Cruise Report No. 5:255 pp.

density(data only)
Australia, Commonwealth Scientific and Industrial Organization, 1962.
Oceanographical observations in the Indian Ocean in 1960, H.M.A.S. DIAMANTINA Cruise, Dm 3/60.
Oceanogr. Cruise Rept., Div. Fish. and Oceanogr., No. 4:39 pp.

density (data only)
Australia, Commonwealth Scientific and Industrial Organization, 1962.
Oceanographical observations in the Indian Ocean in 1960, H.M.A.S. Diamantina, Cruise Dm 1/60.
Oceanogr. Cruise Rept., Div. Fish. and Oceanogr., No. 2:128 pp.

density (data only)
Australia, Commonwealth Scientific and Industrial Organization, 1962.
Oceanographical observations in the Indian Ocean in 1959, H.M.A.S. Diamantina, Cruises Dm 1/59 and Dm 2/59.
Oceanogr. Cruise Rept., Div. Fish. and Oceanogr. No. 1:134 pp.

density(data only)
Australia, Commonwealth Scientific and Industrial Research Organization, 1961
Oceanic investigations in Eastern Australian waters, F.R.V."Derwent Hunter", 1959.
Div. Fish. and Oceanogr., Oceanogr. Sta. List, 48:84 pp.

density(data only)
Australia, Commonwealth Scientific and Industrial Research Organization, 1961
Coastal hydrological sampling at Rottnest Island, W.A., and Port Moresby, Papua, during the I.G.Y. (1957-58), and surface sampling in the Tasman and Coral seas, 1959.
Div. Fish. & Oceanogr., Oceanogr. Sta. List, 48:239 pp.

density(data only)
Australia, Commonwealth Scientific and Industrial Research Organization, 1961
Coastal investigations at Port Hacking, New South Wales, 1959.
Div. Fish. & Oceanogr., Oceanogr. Sta. List, No. 47: 135 pp.

density(data only)
Australia, Commonwealth Scientific and Industrial Research Organization, 1961
Coastal hydrological investigations in the New S outh Wales tuna fishing area, 1959.
Div. Fish. & Oceanogr., Oceanogr. Sta. List, 46: 132 pp.

density (data only)
Australia, Commonwealth Scientific and Industrial Organization, 1961.
Oceanographic observations in the Indian Ocean in 1961. H.M.A.S. Diamantina.
Div. Fish. Oceanogr., Cruise DM 1/61:88 pp.

density(data only)
Australia, Commonwealth Scientific and Industrial Research Organization, 1960
Coastal hydrological investigations in eastern Australia, 1959.
Div. Fish. and Oceanogr., Oceanogr. Sta. List, 45: 24 pp.

density(data only)
Australia, Commonwealth Scientific and Industrial Research Organization, 1960.
Coastal investigations at Port Hacking, New South Wales, 1958.
Oceanogr. Sta. List, Div. Fish. & Oceanogr., 42: 99 pp.

density (data only)
Australia, Commonwealth Scientific and Industrial Research Organization, 1960.
Oceanic investigations in eastern Australia, H.M.A. Ships Queenborough, Quickmatch and Warrego, 1958.
Div. Fish. and Oceanogr., Oceanogr. Sta. List, 43:57 pp.

density (data only)
Australia, Commonwealth Scientific and Industrial Organization, 1960
Oceanic investigations in eastern Australian waters, F.R.V. Derwent Hunter, 1958. Oceanogr. Sta. List., Div. Fish. and Oceanogr. No. 41: 232 pp.

density(data only)
Australia, Commonwealth Scientific and Industrial Research Organization, 1960
Coastal hydrological investigations in south-eastern Australia, 1958. Oceanogr. Sta. List. Div. Fish. and Oceanogr., 60 pp.

density(data only)
Australia, Division of Fisheries and Oceanography C.S.I.R.O., 1959.
Coastal hydrological investigations in the New South Wales tuna fishing area, 1958.
Oceanogr. Sta. Lists, 38:96 pp.

density (data only)
Australia, C.S.I.R.O., Division of Fisheries and Oceanography, 1959.
Hydrological investigations from F.R.V. Derwent Hunter, 1957.
Oceanogr. Sta. List, No. 37:96 pp.
(compiled by D. J. Rochford)

density (data only)
Australia, C.S.I.R.O., Division of Fisheries, 1959
Coastal hydrological investigations at Eden, New South Wales, 1957.
Oceanogr. Sta. List, 35: 36 pp.

density (data only)
Australia, C.S.I.R.O., Division of Fisheries, 1959
Coastal hydrological investigations at Port Hacking, New South Wales, 1957.
Oceanogr. Sta. List, 34:72 pp.

density (data only)
Australia, C.S.I.R.O., Division of Fisheries and Oceanography, 1959.
Scientific reports of a cruise on H.M.A. Ships "Queenborough" and "Quickmatch", March 24-April 26, 1958. Rept. No. 24: 24 pp. (mimeographed).

density (data only)
Australia, C.S.I.R.O., Division of Fisheries and Oceanography, Marine Biological Laboratory, Cronulla, 1959.
F.R.V."Derwent Hunter", Scientific report of Cruise DH9/57, Aug. 19-25, 1957; Cruise DH10/57, Sept. 4-11, 1957; Cruise DH11/57, Sept. 18-21, 1957; Cruise DH12/57, Sept. 26-Oct. 11, 1957.
CSIRO, Div. Fish. & Ocean., Rept. No. 20:20 pp. (mimeographed)

density (data only)
Australia, C.S.I.R.O., Division of Fisheries and Oceanography, 1958/1959.
FVR "Derwent Hunter". Rept. No. 19:16 pp.
No. 21:16 pp.
(mimeographed)

Density (data only)
Australia Commonwealth Scientific and Industrial Research Organization, 1957.
Onshore and oceanic hydrological investigations in eastern and southwestern Australia, 1956. Oceanogr. Sta. List, Div. Fish. & Oceanogr. 30:79pp.

density(data only)
Australia, Commonwealth Scientific Industrial Research Organization, 1957.
Onshore and oceanic hydrological investigations in eastern and southwestern Australia, 1955.
Oceanogr. Sta. List, 27:145 pp.

density (data only)
Australia, Marine Biological Laboratory, Cronulla, Sydney, 1957.
F.R.V. "Derwent Hunter", scientific report of Cruise 3/56, September 19-October 5, 1956; Cruise 4/56, October 9-November 6, 1956; Cruise 5/56, November 8-December 3, 1956.
C.S.I.R.O., Australia, Div. Fish. and Oceanogr., Rept., No. 5:16 pp. (mimeographed).

density (data only)
Australia, Commonwealth Scientific and Industrial Research Organization, 1956.
Onshore hydrological investigations in eastern and south-western Australia, 1954. Compiled by D.R. Rochford.
Ocean. Sta. List., Invest. Div. Fish., 24:119 pp.

density(data only)
Australia, Commonwealth Scientific and Industrial Research Organization, 1954.
Onshore hydrological investigations in eastern and south-western Australia, 1953.
Ocean. Sta. List, Invest. Fish. Div. 18:64 pp.

density(data only)
Australia, Commonwealth Scientific and Industrial Research Organization, 1953.
Onshore hydrological investigations in eastern and south-western Australia, 1951. Vol. 14:64 pp.

density (data only)
Australia, Commonwealth Scientific and Industrial Research Organization, 1951.
Oceanographic station list of investigations made by the Fisheries Division, C.S.I.R.O.
Vol. 1. Hydrological and Planktological observations by F.R.V. Warreen in south-eastern Australian waters, 1938-39:109 pp.
Vol. 2. Ibid., 1940-42:133 pp.
Vol. 3. Hydrological and planktological observations by F.R.V. Warreen in south-western Australian waters, 1947-50:63 pp.
Vol. 4. Onshore hydrological investigations in eastern Australia, 1942-50:114 pp.

density (data only)
Australia, Marine Biological Laboratory, Cronulla, 1960.
F.R.V. "Derwent Hunter", scientific report of cruises 10-20/58.
C.S.I.R.O., Div. Fish. & Oceanogr., Rept., 30: 53 pp., numerous figs. (mimeographed).
For complete "title", see author card

B

density (data only)
Bang, N.D., and F.C. Pearse 1970.
Hydrological data Agulhas Current project, March 1969, R.V. Thomas B. Davie.
Data Rept. Inst. Oceanogr. Univ. Cape Town, 4: 26 pp. Also in: Collected Repr. Oceanogr. Inst. Univ Cape Town 9.

density(data only)
Berrit, G.R., R. Gerard and L. Vercesi, 1968.
Observations océanographiques executées en 1966. II. Stations côtières d'Abidjan, Lome et Cotonou - observations de surface et de fond.
ORSTOM Centre Rech. Océanogr., Côte d'Ivoire, Document scient. provis. 017: 71 pp.

density (data only)
Berrit, G.R., R. Gerard, L. Lemasson, J.P. Rebert et L. Vercesi, 1968.
Observations océanographiques exécutées en 1967. 1. Stations hydrologiques; observations de surface at de fond; stations côtières.
Doc. sci. provis., Centre Recherces océanogr., Côte d'Ivoire, 026:133pp (mimeographed)

density(data only)
Berrit, G.R., 1962
Resultats d'observations. Campagne No. 11, Campagne JONAS.
Cahiers Oceanogr., C.C.O.E.C., 14(1):54-76.

density(data only)
Berrit, G.R., and J.R. Donguy, 1964
Radiale de Pointe-Noire.
Cahiers Océanogr., C.C.O.E.C., 16(3):231-247.

density(data only)
Berrit, G.R., M. Rossignol and J.P. Troadec, 1961.
Résultats d'observations, Année 1959, Centre d'Océanographie de Pointe-Noire, Campagnes Nos. 7,8,9, et 10 de l'"Ombango".
Cahiers Océanogr., C.C.O.E.C., 13(5):319-354.

density(data only)
Blackburn, M., R.C. Griffiths, R.W. Holmes and W.H. Thomas, 1962.
Physical, chemical and biological observations in the eastern tropical Pacific Ocean: three cruises to the Gulf of Tehuantepec, 1958-1959.
U.S.F.W.S. Spec. Sci. Rept., Fish., No. 420:170 pp.

density (data only)
Boudreault, F. Robert, 1967.
Observations d'océanographie physique à la station-pilote 112, 1952-1961.
Cah. inf. Stn. biol. Mar., Grande Rivière, 43: unnumbered pp.

Brasil, Diretoria

density (data only)
Brasil, Diretoria de Hidrografia e Navegação, 1969
III Comissão oceanográfica no Almirante Saldanha (20/3 a 16/4/1957), DG 20-III: 73 pp.

density (data only)
Brasil, Diretoria de Hidrografia e Navegação, 1969.
II Comissão Oceanográfica no Almirante Saldanha (15 A 28/2/1957).
DG20-II:60 pp.

density (data only)
*Brazil, Diretoria de Hidrografia e Navegacao, 1968.
XXXI Comissao oceanografica noc Almirante Saldanha (14/11 a 16/12/66), DG26-249 pp.

density (data only)
Brasil, Diretoria de Hidrografia e Navegação 1967.
XXXII Comissão Oceanográfia noc Almirante Saldanha (14/3 a 3/5/67), DG 26-X: 411 pp.

density(data only)
Brasil, Marinha do Brasil, Diretoria de Hidrografia e Navegação, 1963
Operação, "TRIDENTE I", Estudo das condições oceanográficas entre o Rio de Janeiro e o Rio da Prate, durante o inverno (Agosto-Setembro), ano de 1962.
DG-06-XV:unnumbered pp. (mimeographed).

density(data only)
Brasil, Diretoria de Hidrografia e Navegação, Marinha do Brasil, 1962
Estudo das condições oceanográficas entre o Rio de Janeiro e o Rio da Prata, durante o outono (Maio de 1962).
DG-06-XIV:unnumbered pp.,charts,(mimeographed)

density(data only)
Brasil, Diretoria de Hidrografia e Navegacao, Marinha do Brasil, 1961
Estudo das condições oceanográficas sobre a plataforma continental, entre Cabo-Frio e a Ponta do Boi, durante o mes de setembro (transição inverno-primervera).
DG-06-XII: unnumbered pp. (mimeographed).

density (data only)
Brasil, Marinha do Brasil, Diretoria de Hidrografia e Navegacao, 1960.
Estudo das condicoes oceanograficas na regiao profunda a Nornordeste de Natal, Estado do Rio Grande do Norte.
DG-06-XI (Sept. 1960):unnumbered pp. (mimeographed).

density (data only)
Brasil, Marinha do Brasil, Diretoria de Hidrografia e Navegação, 1960.
Estudo das condições oceanograficas sobre a plataforma continental, entre Cabo-Frio e Vitoria, durante o outono (abril-maio).
DG-06-X(junho):unnumbered pp. (mimeographed).

density(data only)
Brasil, Marinha do Brasil, Diretoria de Hidrografia e Navegação, 1959.
Levantamento oceanografico da costa nordeste.
DG-06-IX:unnumbered pp. (mimeographed).

density(data only)
Brazil, Diretoria de Hidrografia e Navegacao, 1958.
Ano Geofisico Internacional, Publ. DG-06-VII: mimeographed sheets.

density(data only)
Brazil, Diretoria de Hidrografia e Navegacao, 1957.
Ano Geofisico Internacional. Publ. DG-06-III+IV (mimeographed)

density(data only)
Brazil, Diretoria de Hidrografia e Navegacao, 1957.
Ano Geofisico Internacional, Publicacao DG-06-VI: unnumbered mimeographed pages.

density(data only)
Brasil, Diretoria de Hidrografia e Navegacao, 1957.
Ano Geofisico Internacional. Publicacao DG-06-V: 3 pp., 18 figs., 3 pp. of data.

density (data only)
Brazil, Diretoria de Hidrografia e Navegacão, 1957.
Ano Geofisico Internacional. Publicacão DG-06-II mimeographed.
reference room

density (data only)
Brettschneider, Gottfried, Klaus Grasshoff, Peter H. Koske und Lothar von Trepka, 1970.
Physikalische und chemische Daten nach Beobachtungen des Forschungsschiffes Meteor im Persischen Golf 1965.
Meteor Forschungsergebn. (A) 8: 43-90

density (data only)
Buljan, Miljenko, and Mira Zore-Armanda 1966.
Hydrographic data on the Adriatic Sea collected in the period from 1952 through 1962.
Acta adriat. 12: 438 pp.

C

density(data only)
Callaway, Richard J., and James W. McGary, 1959.
Northeastern Pacific albacore survey. 2. Oceanographic and meteorological observations.
USFWS Spec. Sci. Rept., Fish., No. 315:133 pp.

density (data only)
Canada, Canadian Committee on Oceanography, 1965.
Data Record, Gulf of St. Lawrence and Halifax section, August 26 to September 3, 1963.
Canadian Oceanogr. Data Centre, 1965 Data Rec. Ser. No. 2:98 pp. (unpublished manuscript).

density (data only)
Canada, Canadian Committee on Oceanography, 1965.
Ocean Weather Station "P", North Pacific Ocean, May 16 to August 12, 1964.
Canadian Oceanographic Data Centre, 1965 Data Record Series, No. 3:112 pp. (Un published manuscript).

Canadian Oceanographic Data Centre

density (data only)
Canada, Canadian Oceanographic Data Centre 1971.
Gulf of St. Lawrence, September 17 to October 7, 1969.
1971 Data Record Ser. 2: 276 pp.

density (data only)
Canada, Canadian Oceanographic Data Center 1971.
Gulf of St. Lawrence, two surveys, June 18 to June 30, 1968; June 16 to July 1, 1968.
1971 Data Record Series, 1: 300 pp. (multilithed).

density (data only)
Canada, Canadian Oceanographic Data Center, 1969.
Operation Tanquary, Ellesmere Island, N.W.T., 1963-1966.
1969 Data Record Ser. 13: 152 pp.

density (data only)
Canada, Canadian Oceanographic Data Center 1969.
Ocean Weather Station P, North Pacific Ocean, May 17 to August 15, 1968.
1969 Data Record Ser. 12: 163 pp.

density (data only)
Canada, Canadian Oceanographic Data Center 1969.
Gulf of St. Lawrence, November 14 to November 24, 1968.
1969 Data Record Ser. 11: 72 pp.

density (data only)
Canada, Canadian Oceanographic Data Center, 1969
Scotian Shelf, two surveys, June 3 to June 6, 1968, October 16 to October 29, 1968.
1969 Data Record Series No. 9: 71 pp. (multilithed)

density (data only)
Canada, Canadian Oceanographic Data Centre, 1969
Davis Strait and Northern Labrador Sea, August 27 to October 22, 1965.
1969 Data Record Series 7: 203 pp (multilithed)

density (data only)
Canada, Canadian Oceanographic Data Center 1969.
Ocean Weather Station 'P' North Pacific Ocean, December 3, 1967, to February 25, 1968.
1969 Data Record Series 6: 116 pp. (multilithed)

density (data only)
Canada, Canadian Oceanographic Data Center, 1969.
Cabot Strait, August 8 to October 8, 1966.
1969 Data Record Ser. 5: 79 pp. (mimeographed)

density (data only)
Canada, Canadian Oceanographic Data Center, 1969.
Labrador and Irminger seas, March 12 to May 12, 1966.
1969 Data Record Ser. 1: 152 pp. (multilithed)

density (data only)
Canada, Canadian Oceanographic Data Centre 1969.
East Greenland, Denmark Strait and Irminger Sea January 16 to April 5 1967.
1969 Data Rec. Ser. 4: 158 pp. (multilithed).

density (data only)
Canada, Canadian Oceanographic Data Centre,1968.
Scotian shelf, January 20 to January 27,1968.
1968 Data Record Series, 7: 49 pp. (multilithed).

density (data only)
Canada, Canadian Oceanographic Data Center, 1968
Ocean Weather Station 'P', North Pacific Ocean, April 7 to July 6, 1967.
1968 Data Record Ser. 5:140 pp. (multilithed)

density (data only)
Canada,Canadian Oceanographic Data Centre,1968.
Hudson Bay,Hudson Strait and Arctic, July 21 to September 9,1967.
1968 Data Record Ser.4: 123 pp. (multilithed).

density (data only)
Canada, Canadian Oceanographic Data Centre 1968.
Gulf of St. Lawrence, November 16 to November 27, 1967.
1968 Data Record Ser. 3: 81 pp. (multilithed).

Canada, Canadian Oceanographic Data Center, 1967.
Ocean Weather Station 'P', North Pacific Ocean, October 28, 1966 to January 9, 1967.
1967 Data Record Ser., 8: 111 pp.

density (data only)
Canada,Canadian Oceanographic Data Centre,1967.
Ocean Weather Station "P",August 5 to October 31, 1966.
1967 Data Record Ser., 6:164 pp.(mimeographed).

density (data only)
Canada, Canadian Oceanographic Data Center,1967.
Ocean Weather Station 'P' North Pacific Ocean, May 27 to August 10,1966.
1967 Data Record Ser., 5: 185 pp.

density (data only)
Canada, Canadian Oceanographic Data Centre, 1967.
Cabot Strait, August 16 to August 28, 1966.
1967 Data Record Ser., 3:89 pp.

density (data only)
Canada, Canadian Oceanographic Data Centre, 1967.
Sable Island to Grand Banks, September 1 to September 12, 1965.
1967 Data Record Ser., 2:89 pp.

density (data only
Canada, Canadian Oceanographic Data Centre, 1967.
Gulf of St. Lawrence, September 16, to October 16, 1965
1967 Data Record Series No. 1: 169 pp. (multilithed)

density (data only)
Canada, Canadian Oceanographic Data Centre, 1966.
Ocean Weather Station 'P' North Pacific Ocean, March 3 to June 2 1966.
1966 Data Record Series 11: 168 pp.

density (data only)
Canada, Canadian Oceanographic Data Center, 1966.
Gulf of St. Lawrence, Halifax section and Scotian Shelf to Grand Banks, three surveys, November 20 to December 22, 1965.
1966 Data Record Series, No. 10:113 pp. (multilithed).

density (data only)
Canadian Oceanographic Data Centre, 1966.
Gulf of St. Lawrence and Halifax Section, November 16 to November 25, 1964.
1966 Data Record Ser., 9: 86 pp. (multilithed)

density (data only)
Canada, Canadian Oceanographic Data Centre 1966.
Ocean Weather Station 'P', North Pacific Ocean, December 11, 1965, to March 9, 1966.
1966 Data Record Ser. 8: 144 pp. (multilithed)

density (data only)
Canada, Canadian Oceanographic Data Center, 1966.
Gulf Stream between Cape Cod and Bermuda, November 16 to December 15, 1964.
1966 Data Record Series, No. 7:59 pp. (multilithed).

Canada, Canadian Oceanographic Data Centre 1966.
Ocean Weather Station 'P', North Pacific Ocean, September 17 to December 15, 1965.
1966 Data Record Series No. 6: 170 pp.

density (data only)
Canada,Canadian Oceanographic Data Centre,1966.
Ocean Weather Station "p", January 23,1965 to April 19,1965.
1966 Data Redord Ser., No. 1:122 pp. (multilithed

density (data only)
Canada, Canadian Oceanographic Data Centre, 1966
Arctic 1961, August 2 to October 12, 1961.
1966 Data Record Series, 322 pp.

density (data only)
Canada,Canadian Oceanographic Data Centre, 1966.
Western North Atlantic and Caribbean Sea, February 1 to February 27,1965.
1966 Data Record Ser., No. 2:78 pp.(multilithed).

density (data only)
Canada,Canadian Oceanographic Data Centre,1966.
Arctic,Hudson Bay and Hudson Strait,August 5 to October 4,1962.
1966 Data Record Series, No. 4:247 pp.

density (data only)
Canada,Canadian Oceanographic Data Centre,1965.
Data Record,Grand Banks to the Azores,June 1 to July 15,1964.
1965 Data Record Series, No. 13:221 pp. (mimeographed),

density (data only)
Canada, Canadian Oceanographic Data Centre,1965
North Atlantic, east of Nova Scotia, south of the Grand Banks, March 5 to August 10, 1962.
1965 Data Record Series, No. 10:226 pp.

Sigma-t (data only)
Canada, Canadian Oceanographic Data Centre, 1965.
Data Record Gulf of St. Lawrence, July 23 to August 23, 1964.
1965 Data Record Ser., No. 9:262 pp.(multilith)

density,(data only)
Canadian Oceanographic Data Centre, 1966.
Ocean Weather Station "P" North Pacific Ocean, April 17 to June 3,1965.
1966 Data Record Ser. No. 3:150pp. (multilithed)

density (data only)
Canada, Canadian Oceanographic Data Centre, 1965.
Baffin Bay, Smith Sound to Strait of Belle Isle September 4 to October 24, 1964.
1965 Data Record Ser., No. 11:165 pp.

density(data only)
Canada, Canadian Oceanographic Data Centre,1965
Data record, St. Lawrence Estuary, June 10 to July 24, 1963. 1965 Data Record Series, No.1: 127 pp. (multilithed).

density (data only)
Canada, Department of Mines and Technical Surveys, 1960.
Tidal and oceanographic survey of Hudson Strait, August and September, 1959. Data Record, 55 pp. (multilithed.)

density(data only)
Canada, Fisheries Research Board, 1959.
Data record, Ocean Weather Station "P" (Latitude 50 00'N, Longitude 145 00'W) January 22 - July 11, 1958.
MSS Rept. Ser. (Oceanogr. & Limnol.), No. 31: 112 pp.

density(data only)
Canada, Fisheries Research Board, 1959.
Physical and chemical data record, coastal seaways project, November 12 to December 5, 1958.
MSS Rept. Ser. (Oceanogr. & Limnol.), No. 36: 120 pp.

density(data only)
Canada, Pacific Oceanographic Group, Nanaimo, 1961
Data record of oceanographic observations and bathythermograms observed in the northeast Pacific Ocean by Fisheries Research Board chartered fishing vessels, May to July, 1960.
Fish. Res. Bd., Canada, Manuscript Rept. Ser. (Oceanogr. and Limnol.), No. 87: 30 pp.

Canada University of British Columbia

Density (data only)
Canada, University of British Columbia, Institute of Oceanography 1970
British Columbia inlets and Pacific cruises, 1969.
Data Rept. 30: 65 pp. (mimeographed)

density (data only)
Canada, University of British Columbia, 1969.
British Columbia inlet cruises, 1968.
Data Rept., 28: 59 pp. (multilithed)

density (data only)
Canada, University of British Columbia, Institute of Oceanography, 1968.
British Columbia Inlet Cruises, 1967.
Data Rept. 27: 36 pp (mimeographed).

density (data only)
Canada,University of British Columbia,Institute of Oceanography,1967.
British Columbia and Alaska inlets and Pacific Cruises,1966.
Data Rep. 26:40pp. (mimeographed).

density (data only)
Canada, University of British Columbia, Institute of Oceanography, 1966.
British Columbia and Alaska Inlet Cruises, 1965.
Date Report 25: 39 pp. (mimeographed).

Canada, density(data only)
University of British Columbia, Institute of Oceanography, 1965.
British Columbia and Alaska inlet cruises, 1964.
Data Rept., No. 24:34 pp. (multilithed).

density (data only)
Canada, University of British Columbia, Institute of Oceanography, 1964.
British Columbia Inlet Cruises, 1963.
Data Report, No. 23:102 pp. (mimeographed).

density (data only)
Canada, University of British Columbia, Institute of Oceanography, 1963.
Inlet cruises, 1962.
Data Report 21:90 pp. (multilithed).

sigma-t, (data only
Canada, University of British Columbia, Institute of Oceanography, 1960.
Data Report No. 16, Indian Arm Cruises, 1959.

density(data only)
Canada, University of British Columbia, Institute of Oceanography, 1956.
British Columbia Inlet cruise, 1956. Data Rept. No. 9:33 pp. (mimeographed).

density(data only)
Capurro, Luis R.A., compiler, 1961
Oceanographic observations in the intertropical region of the world during the IGY and IGC. Part IIa: Pacific Ocean (France, Japan and Union of Soviet Socialist Republics).
IGY World Data Center A: Oceanography, IGY Oceanogr. Rept., No. 3:225 pp.

density(data only)
Capurro, Luis R.A., compiler, 1961.
Oceanographic observations in the intertropical region of the World Ocean during IGY and IGC. Part IIb: Pacific Ocean (United States of America).
IGY World Data Center A, Oceanography, IGY Oceanogr. Rept., No. 3:298 pp.

density (data only)
Champagnat, C., T. Boely, E. de Bondy, F. Conand, et J.L. Cremoux, 1969.
Observations océanographiques exécutées en 1968.
Centre Rech. océanogr. Dakar-Thiaroye ORSTOM (DSP 19): 169+ pp. (mimeographed)

Density (data only
Champagnat, C., F. Conand, J.L. Cremoux et J.P. Rebert 1969.
Campagne océanographique du Jean Charcot (Dakar-Cap Blanc-Iles du Cap Vert) du 29-7 au 5-8-68.
ORSTOM, Sénégal, DSP 17: 87 pp. (mimeographed)

density (data only)
*Charnell, Robert L., David W.K. Au and Gunter R. Seckel, 1967.
The Trade Wind Zone Oceanography Pilot Study 6. Townsend Cromwell cruises 16, 17 and 21.
Spec. Scient. Rep. U.S. Fish. Wildl. Serv. Fish. 557: 59 pp.

density (data only)
*Charnell, Robert L., David W.K. Au and Gunter R. Seckel, 1967.
The Trade Wind Zone Oceanography Pilot Study 5. Townsend Cromwell cruises 14 and 15, March and April 1965.
Spec. scient. Rep. U.S. Fish. Wildl. Serv., Fish. 556:54 pp.

density (data only)
Charnell, Robert L., David W.K. Au, and Gunter R. Seckel, 1967.
The Trade Wind Zone Oceanography Pilot Study IV: Townsend Cromwell cruises 11, 12, and 13, December 1964 to February 1965.
Spec. Scient Rep. U.S. Fish. Wildl. Serv. Fish., 555: 78 pp. (multilithed).

density (data only
Charnell, Robert L., David W.K. Au and Gunter R. Seckel, 1967.
The Trade Wind Zone Oceanography Pilot Study III: Townsend Cromwell cruises 8, 9, and 10, September to November 1964.
Spec. Scient. Rep. U.S. Fish. Wildl. Serv. Fish. 554: 78 pp. (multilithed).

density (data only)
Charnell, Robert L., David W.K. Au and Gunter R. Seckel, 1967.
The Trade Wind Zone Oceanography Pilot Study II: Townsend Cromwell cruises 4, 5, and 6, May to July 1964.
Spec. scient. Rep. U.S. Fish Wildl. Serv., Fish., 553: 78 pp. (multilithed).

density (data only)
Charnell, Robert L., David W.K. Au and Gunter R. Seckel, 1967.
The Trade Wind Zone Oceanography Pilot Study I. Townsend Cromwell cruises 1, 2, and 3, February to April, 1964.
Spec. scient. Rep. U.S. Fish. Wildl. Serv., Fish. 552. 75 pp. (multilithed)

density (data only)
Republica de Chile, Instituto Hidrografico de la Armada, 1966.
Operación Oceanográfica Marchile III: datos fisico-quimicos. Unnumbered pp.

density(data only)
China (Taiwan), Academia Sinica, Chinese National Committee on Oceanic Research, 1964
Oceanographic observations in the adjacent seas of Taiwan Research Vessel Yang Ming.
Oceanographic Cruise Report, No. 1: 55 pp.

density (data only)
Collier, Albert, 1958.
Gulf of Mexico physical and chemical data from ALSAKA cruises.
USFWS Spec. Sci. Rept., Fish., No. 249:417 pp.

density (data only)
Collins, C.A., C. DeJong, A. Huyer and L. Boilard, 1969
Oceanic observations at Ocean Station P (50°N, 145°W), 16 May - 3 July 1969.
Techn. Rept. Fish. Res. Bd., Can. 154: 302 pp. (multilithed).

Conseil Permanent International pour l'Exploration de la Mer

density (data only)
Conseil International pour l'Exploration de la Mer, 1969
ICES Oceanographic Data Lists, 1962, 9: 209 pp. (multilithed)

density (data only)
Conseil Permanent International pour l'Exploration de la Mer, 1968.
ICES Oceanographic data lists, 1962 (4): 171 pp.

density (data only)
Conseil, Permanent International pour l'Exploration de la Mer, 1968.
ICES Oceanographic Data Lists 1962(3): 27 pp.

density (data only)
Conseil Permanent International pour l'Exploration de la Mer, 1968.
ICES Oceanographic Data Lists, 1962, 1:153 pp., 2:85 pp.

density (data only)
Conseil International pour l'Exploration de la Mer, 1968.
ICES Oceanographic Data Lists, 1961, 10: 217 pp. (multilithed).

density (data only)
Conseil International pour l'Exploration de la Mer, 1968.
ICES oceanographic data lists, 1961, 9:187 pp. (mimeographed).

density (data only)
Conseil International pour l'Exploration de la Mer, 1968.
Cooperative synoptic investigation of the Baltic 1964. 1. Finland.
ICES Oceanogr. Data lists: 82 pp. (multilithed).

density (data only)
Conseil International pour l'Exploration de la Mer, 1968.
ICES Oceanographic Data Lists 1961, 8:101 pp. (multilithed).

density (data only)
Conseil International pour l'Exploration de la Mer. 1968.
Cooperative synoptic investigation of the Baltic 1964. Sweden.
ICES Oceanogr. Data Lists 4:181 pp. (mimeographed)

density (data only)
Conseil International pour l'Exploration de la Mer, 1968.
ICES Oceanographic Data Lists, 1962, 7:235 pp.

density (data only)
Conseil Permanent International pour l'Exploration de la Mer, 1968.
ICES oceanographic data lists, 1962, 6: 88pp.

density (data only)
Conseil Permanent International pour l'Exploration de la Mer, 1968.
ICES oceanographic data lists 1962. (5):107 pp. (mimeographed).

density (data only)
Conseil Permanent International pour l'Exploration de la Mer, 1968.
ICES Oceanographic data lists 1961, No. 5: 247 pp.

density (data only)
Conseil Permanent International pour l'Exploration de la Mer, 1968.
ICES Oceanographic Data Lists, 1960, 11: 178 pp.

density (data only)
Conseil Permanent International pour l'Exploration de la Mer, 1967.
ICES Oceanographic data lists, 1960, No. 8: 227 pp. (multilithed).

density (data only)
Conseil Permanent International pour l'Exploration de la Mer, 1967.
ICES Oceanographic data lists, 1961, 4:124 pp. (multilithed).

density (data only)
*Conseil Permanent International pour l'Exploration de la Mer, 1967.
ICES oceanographic data lists, 1960, No. 7:270 pp (multilithed).

density (data only)
Conseil Permanent International pour l'Exploration de la Mer, 1967.
ICES Oceanographic data lists. 1960, No. 6: 295 pp. (multilithed).

density (data only)
Conseil Permanent International pour l'Exploration de la Mer 1967.
ICES oceanographic data lists 1959, No. 9: 235 pp. (multilithed).

density (data only)
Conseil Permanent International pour l'Exploration de la Mer, 1967.
ICES Oceanographic data lists, 1958, 12:111 pp. (multilithed).

density (data only)
Conseil Permanent International pour l'Exploration de la Mer 1967.
ICES oceanographic data lists 1957, No. 9: 199 pp.

density (data only)
Conseil International pour l'Exploration de la Joint Skagerak Expedition 1966. 1. Oceanographic stations, temperature-salinity-oxygen content. 2. Oceanographic stations, chemical observations.
ICES Oceanogr. Data lists: 250 pp; 209 pp.

density (data only
Conseil Permanent International pour l'Exploration de la Mer, 1966.
ICES Oceanographic Data Lists, 1961 (3): 166 pp (multilithed)

density (data only
Conseil Permanent International pour l'Exploration de la Mer, 1966.
ICES Oceanographic data lists, 1960, No. 3:118 pp.
ICES Oceanographic data lists, 1960, No. 4:190 pp.

density (data only)
Conseil Permanent International pour l'Exploration de la Mer, 1966.
ICES Oceanographic data lists, 1960, No. 2:166 pp.

density (data only)
Conseil Permanent International pour l'Exploration de la Mer, 1966.
ICES oceanographic data lists, 1958 No. 9: 175 pp

density (data only)
Conseil Permanent International pour l'Exploration de la Mer 1965
ICES oceanographic data lists 1958
No. 7: 192 pp.
No. 8: 286 pp.

density (data only)
Conseil Permanent International pour l'Exploration de la Mer, Service Hydrographique, 1966.
ICES Oceanographic data lists, 1957(8):75 pp.
ICES Oceanographic data lists, 1958(10):72 pp.
ICES Oceanographic data lists, 1960(1):198 pp.

density, (data only)
Conseil Permanent International pour l'Exploration de la Mer. 1966.
ICES Oceanographic data lists, 1959, No. 6:202 pp.

density (data only)
Conseil Permanent International pour l'Exploration de la Mer, 1965.
ICES oceanographic data lists, 1958, No. 6:199 pp.

density (data only)
Conseil Permanent International pour l'Exploration de la Mer, 1965.
ICES oceanographic data lists, 1958, No. 5:284 pp.

density (data only)
Conseil Permanent International pour l'Exploration de la Mer, 1965.
ICES Oceanographic data lists, 1952, No. 3: 214 pp.

density (data only)
Conseil Permanent International pour l'Exploration de la Mer, 1964.
ICES oceanographic data lists, 1957, No. 3, 167 pp.
No. 4, 178 pp.
No. 5, 255 pp.
No. 6, 160 pp.
1958, No. 2, 157 pp.
No. 3, 184 pp.
No. 4, 241 pp.
1959, No. 1, 201 pp.

density (data only)
Conseil Permanent International pour l'Exploration de la Mer, Service Hydrographique, 1963.
ICES oceanographic data lists, 1958(1):1-259.

density (data only)
Conseil Permanent International pour l'Exploration de la Mer, 1963.
ICES Oceanogr. Data Lists, 1957(2):353 pp. (Multilithed).

density (data only)
Conseil Permanent International pour l'Exploration de la Mer, 1963.
ICES oceanographic data lists, 1957 (1):277 pp.

density (data only)
Contreras, Agustin 1970.
Resumen climatologico y mareografico 1968, Punta de Piedras, Estado Nueva Esparta, Venezuela
Mem. Soc. Cienc. nat. La Salle, Venezuela 30 (86): 75-101

density (data only)
Côte d'Ivoire, Centre de Recherches Océanographiques, 1963.
Résultats hydrologiques effectués au large de la Côte d'Ivoire de 1956-1963. Première partie: observations, 1956-1963.
Trav. Centre de Recherches Océanogr., Ministere de la Production Animale, Côte d'Ivoire, unnumbered pp. (multilithed).

density (data only)
Crean, P.B., R.B. Tripp and H.J. Hollister, 1962.
Oceanographic data record, Monitor project, March 12 to April 5, 1962.
Fish. Res. Bd., Canada, Mss. Rept. Ser. (Ocean. and Limnol.), No. 129:210 pp. (multilithed).

density (data only)
*Countryman, Kenneth A., and William L. Gsell, 1966.
Operations Deep Freeze 63 and 64, summer oceanographic features of the Ross Sea.
Tech. Rept., U.S. Naval Oceanogr. Off., Tr-190:193 pp. (multilithed).

density (data only)
Crean, P.B., R.B. Tripp, and H.J. Hollister, 1962
Oceanographic data record, Monitor Project, January 16 to February 5, 1962.
Fish. Res. Bd. Canada, Mss. Rept. Ser. (Oceanogr. & Limnol.), No. 113:169 pp. (multilithed).

density (data only)
Cremoux, J.L., 1970.
Observations océanographiques effectuées en 1969. Stations hydrologiques. Centre Rech. océanogr., Dakar-Thiaroye, ORSTOM, DSP 24: 217 pp.
1. Stations hydrologiques

Density, suraface-bottom (data only)
Cremoux, J.L., 1970.
Observations océanographiques effectuées en 1969. II Observations de surface et de fond — bathythermogrammes.
Centre Rech. océanogr. Dakar-Thiaroye ORSTOM, DSP 25: 23 + unnumbered pp.

density (data only)
Cremoux, J.L., with J. Diarra 1971.
Observations océanographiques effectuées en 1970.
Centre Rech. océanogr. Dakar-Thiaroye, Sénégal, DSP 33: unnumbered pp.

D

density (data only)
DeMaio, A., D. Bregant and E. Sanone, 1968.
Oceanographic data of the R.V. Bannock collected during the International NATO cruise in the Tyrrhenian Sea (16 September-24 October 1963).
Cah. océanogr., 20(Suppl.1):1-64.

density (data only)
Dodimead, A.J., F.W. Dobson, N.K. Chippindale, and H.J. Hollister, 1962
Oceanographic data record, North Pacific Survey, May 23 to July 5, 1962.
Fish. Res. Bd. Canada, Mss. Rept. Ser. (Ocean. and Limnol.), No. 138:383 pp.

density (data only)
Dragovich, Alexander, and James E. Sykes 1967.
Oceanographic atlas for Tampa, Florida, and adjacent waters of the Gulf of Mexico, 1958-61.
Circular, Fish Wildl. Serv. Bur. Comm. Fish. 255: 466 pp. (quarto).

E

density (data only)
Elder, Robert B., 1970
Oceanographic observations between Iceland and Scotland, July-Nov. 1965.
Oceanogr. Rept. U.S. Coast Guard 28 (CG 373-28): 118 pp. (multilithed).

F

density, (data only)
Favorite, Felix, Richard J. Callaway and James F. Hebard, 1961
North Pacific and Bering Sea oceanography, 1959.
U.S.F.W.S. Spec. Sci. Rept., Fish., No. 377: 212 pp.

density (data only)
Favorite, Felix, Betty-AnnMorse, Alan H. Haselwood and Robert A. Preston, Jr., 1964.
North Pacific oceanography, February-April 1962.
U.S.F.W.S. Spec. Sci. Rept., Fish., No. 477:66 pp

density(data only)
Favorite, Felix, and Glenn Pedersen, 1959.
North Pacific and Bering Sea oceanography, 1958.
U.S.F.W.S. Spec. Sci. Rept., Fish., No. 312:230pp

density(data only)
Favorite, Felix, and Glenn M. Pedersen, 1959.
North Pacific and Bering Sea oceanography, 1957.
USFWS Spec. Sci. Rept., Fish., No. 292:106 pp.

density(data only)
Favorite, Felix, John W. Schantz and Charles R. Hebard, 1961
Oceanographic observations in Bristol Bay and the Bering Sea 1939-41 (USCGT Redwing).
U.S.F.W.S. Spec. Sci. Rept., Fish. No. 381: 323 pp.

density(data only)
France, Centre de Recherches et d'Études Oceanographiques, 1958
Observations oceanographiques des navires stationnaires meteorologiques. Trav. C.R.E.O. n.s. 3(2): 27-40.

density (data only)
France, Service Hydrographique de la Marine, 1965.
Résultats d'Observations.
Cahiers Océanogr., C.C.O.E.C., 17(2):127-133.

density (data only)
France, Service Hydrographique de la Marine, 1965. Resultats d'observations.
Cahiers Oceanogr., C.C.O.E.C., 17(1):55-65.

density (data only)
France, Service Hydrographique de la Marine, 1964.
Résultats d'observations.
Cahiers Océanogr., C.C.O.E.C., 16(9):799-812.

density(data only)
France, Service Hydrographique de la Marine, 1964.
Resultats d'observations.
Cahiers Oceanogr., C.C.O.E.C., 16(6):475-482.

density(data only)
FRANCE, Service Hydrographique de la Marine, 1964
Résultats d'obsefvations. Campagnes internationale d'observations dans le détroit de Gibraltar (15 mai-15 juin 1961). Mesures de courant d'hydrologie et de météorologie effectuées à bord de la "Calypso".
Cahiers Océanogr., C.C.O.E.C., 16(1):23-94.

density(data only)
France, Service Hydrographique de la Marine, 1963
Resultats d'observations.
Cahiers Océan., 15(10):738-758.

density(data only)
France, Service Hydrographique de la Marine, 1963.
Resultats d'observations.
Cahiers Oceanogr., C.C.O.E.C., 15(9):666-684.

density (data only)
France, Service Hydrographique de la Marine, 1963.
Resultats d'observations.
Cahiers Oceanogr., C.C.O.E.C., 15(8):576-588.

density(data only)
France, Service Hydrographique de la Marine, 1963.
Resultats d'observations.
Cahiers Oceanogr., C.C.O.E.C., 15(7):492-507.

density(data only)
France, Service Hydrographique de la Marine, 1963
Résultats d'observations.
Cahiers Océanogr., C.C.O.E.C., 15(5):344-355.

density(data only)
France, Service Hydrographique de la Marine, 1963
Résultats d'observations. Première campagne océanographique du "Commandant Robert Giraud" en Canal de Mozambique.
Cahiers Océanogr., C.C.O.E.C., 15(4):260-285.

density(data only)
France, Service Hydrographique de la Marine, 1963
Résultats d'observations.
Cahiers Océanogr., C.C.O.E.C., 15(1):70-76.

density (data only)
France, Service Hydrographiques de la Marine, 1962.
Résultats d'observations. 1. Observations hydrologiques des bâtiments de la Marine Nationale et des navires météorologiques stationnaires. 2. Flotteurs-témoins du courant.
Cahiers Océanogr., C.C.O.E.C., 14(6): 424-435.

density(data only)
France, Service Hydrographique de la Marine 1962
Observations hydrologiques du bâtiment oceanographe "Origny", Campagne Internationa le à Gibraltar (15 mai - 15 juin 1961).
Cahiers Océanogr., C.C.O.E.C., 14(5):340-375

density(data only)
France, Service Hydrographique de la Marine, 1962
Résultats d'observations.
Cahiers Oceanogr., C.C.O.E.C., 14(4):274-283.

density(data only)
France, Service Hydrographique de la Marine 1962
Resultats d'Observations.
Cahiers Océanogr., C.C.O.E.C., 14(3):187-199.

density(data only)
France, Service Hydrographique de la Marine, 1961
Resultats d'observations. 1. Observations hydrologiques des bâtiments de la Marine Nationale. 2. Flotteurs-temoins de courant.
Cahiers Océanogr., C.C.O.E.C., 13(7):498-515.

density(data only)
France, Service Hydrographique de la Marine, 1961.
Résultats d'observations. Observations hydrologiques du Navire Hydrologique "Lapérouse".
Cahiers Océanogr., C.C.O.E.C., 13(6):409-422.

density(data only)
France, Service Hydrographique de la Marine, 1961
Résultats d'observations. 1. Observations hydrologiques des bâtiments de la Marine National.
Cahiers Océanogr., C.C.O.E.C., 13(4):594-605.

density(data only)
France, Service Hydrographique de la Marine, 1961
Resultats d'observations.
Cahiers Oceanogr., C.C.O.E.C., 13(4):250-263.

density(data only)
France, Service Hydrographique de la Marine, 1961
Resultats d'observations.
Cahiers Océanogr., C.C.O.E.C., 13(3): 187-201.

density (data only)
France, Service Hydrographique de la Marine, 1961.
Resultats d'observations.
Cahier Oceanogr., C.C.O.E.C., 13(2):116-132.

density (data only)
France, Comité Central d'Océanographie et d'Étude des Côtes, 1960.
Stations hydrologiques effectuées par le "Leon Coursin" bâtiment de la Mission Hydrographique des travaux publics d'A.O.F.
Cahiers Océanogr., C.C.O.E.C., 12(7):493-507.

density (data only)
France, Comité Central d'Oceanographie et d'Etude des Cotes, 1960
Stations hydrologiques effectivées par le "Léon Coursin", bâtiment de la mission hydrographique des Travaux Publics d'A.O.F.
Cahiers Océanographiques 12(5): 345-356.

density (data only)
France, Service Hydrographique de la Marine, 1960.
I. Observations hydrologiques des bâtiments de la Marine Nationale. II. Observations du niveau marin à la station littorale de AGI a Nosy-Be (Madagascar).
Cahiers Océanogr., C.C.O.E.C., 12(4):285-294.

density (data only)
France, Service Hydrographique de la Marine, 1960.
Resultats d'Observations hydrologiques.
Cahiers Oceanogr., 12(2):128-154.

density(data only)
France, Service Hydrographique de la Marine, 1959.
Resultats d'observations.
Cahiers Océanogr., C.C.O.E.C., 11(8):555-566.

density(data only)
France, Service Central Hydrographique, 1959.
1. Observations des bâtiments de la Marine Nationale.
Bull. d'Info., C.C.O.E.C., 11(8):509-514.

density(data only)
France, Service Hydrographique de la Marine, 1959.
Resultats d'observations océanographiques.
Cahiers Océanogr., C.C.O.E.C., 11(5):323-370.

density (data only)
France, Service Hydrographique de la Marine 1959.
Resultats d'observations hydrologiques.
Cah. océanogr. 11(3): 182-199.

density(data only)
France, Service Central Hydrographique, 1958
Observations hydrologiques des bâtiments de la Marine Nationale.
Bull. d'Info., C.C.O.E.C., 10(10):751-771.

density(data only)
France, Service Central Hydrographiques, 1958.
1. Observations hydrologiques des bâtiments de la Marine Nationale.
Bull. d'Info., C.C.O.E.C., 10(8):509-514.

density(data only)
France, Service Central Hydrographique, 1958.
Observations hydrographiques.
Bull. d'Info., C.C.O.E.C., 10(6):367-373.

density(data only)
France, Service Central Hydrographique, 1958.
Observations hydrologiques des bâtiments.
Bull. d'Info., C.C.O.E.C., 10(5):293-300.

density (data only)
France, Service Central Hydrographique, 1958.
Observations hydrologiques des bâtiments de la Marine nationale. Bull. d'Info., C.C.O.E.C., 10(2):100-108.

density (data only)
France, Service Central Hydrographique, 1957.
Observations hydrologiques des bâtiments de la Marine nationale. Bull. d'Info., C.C.O.E.C., 9(10):579-588.

density(data only)
France, Service Central Hydrographique, 1957.
Observations hydrologiques. A. Bâtiments de la Marine Nationale. 3. Navires stationnaires météorologiques. Bull. d'Info., C.C.O.E.C., 9(9):521-530.

density(data only)
France, Service Central Hydrographique, 1957.
Observations hydrologiques des bâtiments de la Marine Nationale. Drageur Océanique "Ailette" (mars 1955 à décembre 1955).
Bull. d'Info., C.C.O.E.C., 9(1):61-72.

density(data only)
France, Comité Central d'Océanographie et d'Étude des Côtes, 1955.
Observations hydrologiques des bâtiments de la Marine Nationale. A. Travaux de l'escorteur l'"Aventure" (Avril-Octobre). B. Observations hydrologiques au large de l'entrée de la Manche.
Bull. d'Info., C.C.O.E.C. 7(8):378-386.

density, (data only)
France, Comité Central d'Océanographie et d'Étude des Côtes, 1955.
Observations hydrologiques des bâtiments de la marine national. Avisogarde-peche "Ailette", Mars à Octubre, 1954. Bull. d'Info., C.C.O.E.C. 7(7): 314- , 5 pls.

density (data only)
France, Comite Central d'Oceanographie et d'Etude des Cotes, 1954.
Observations hydrologiques effectuees par les batiments de la Marine Nationale. Observations en Mer du Labrador par le Frigate "L'Aventure" (1953). Bull. d'Info., C.C.O.E.C. 6(9):419-430.

density (data only)
France, Comité Centrale d'Océanographie et d'Étude des Côtes, 1954.
Observations hydrologiques effectuées par les bâtiments de la Marine Nationale.
Bull. d'Info., C.C.O.E.C. 6(7):308-309.

density(data only)
France, Centre de Recherches et d'Études Océanographiques, 1960
Stations hydrologiques effectuées par le "Passeur du Printemps" dans le cadre des travaux de l'Année Géophysique Internationale.
Travaux, C.R.E.O., 3(4):17-22.

density(data only)
France Institut Français d'Océanie, Noumea, New Caledonie, 1958
Orsom III, Croisière "Astrolabe", Océanographique Physique. Rapp. Sci. No. 8:79 pp. (mimeograph).

density (data only)
France, Sous-Comité Océanographique de l'Organisation du Traité de l'Atlantique Nord, 1969.
Projet Mer Tyrrhénienne (1963). Résultats des mesures faites à bord du navire océanographique italien Bannock (16 septembre-24 octobre 1963) et a bord du navire océanographique français Origny (17 septembre - 15 octobre 1963).
Rapp. Techn OTAN, Serv. Hydrogr. de la Marine, 44:125 pp. (multilithed)

G

density (data only)
Gallardo, Y., A. Crosnier, Y. Gheno, J.M. Guillen, J. C. Le Guen et J.P. Rebert 1969.
Résultats hydrologiques des campagnes du Centre ORSTOM de Pointe-Noire (Congo-Brazza) devant l'Angola de 1965 à 1967.
Cah. océanogr. 21(6): 564-595.

density (data only)
Gallardo, Y., A. Crosnier, Y. Gheno, J.M. Guillerm, J.C. Le Guen et J.P. Rebert 1969.
Resultats hydrologiques des campagnes du Centre ORSTOM de Pointe-Noire (Congo-Brazza) devant l'Angola, de 1965 à 1967.
Cah. Océanogr., 21(4):387-400.

density (data only)
Gallardo, Y., A. Crosnier, J.C. Le Guen et J.P. Hebert, 1968.
Resultats d'observations hydrologiques et courantologiques effectuees autour de l'Ile Annobon (1°25' S - 5°37'E).
Cah. oceanogr., 20(8):711-726.

density (data only)
Gantzer, K.A., D.H. Joergensen and D.A. Healey 1969.
Oceanographic observations at Ocean Station P (50°N 145°W) 27 October to 26 February 1969.
Techn. Rept. Fish. Res. Bd Can. 143: 49 pp.

density (data only)
Garcia, Andrew W., 1969.
Oceanographic observations in the Kara and eastern Barents Seas.
Oceanogr. Rept. U.S.C.G. (CG373-25) 25: 99 pp. multilithed

density (data only)
Gostan, Jacques, 1967.
Résultats des observations hydrologiques effectuées entre les côtes de Provence et de Corse (6 août 1962-30 juillet 1964).
Cah. océanogr., 19(Suppl.1):1-69.

density (data only)
Gostan, Jacques, 1967.
Résultats des obsertations hydrologiques effectuees entre les côtes de Provence et de Corse (6 août 1962-30 juillet 1964).
Cah. océanogr., 19(Suppl.1):1-69.

density(chiefly data)
Granqvist G., 1955.
Vattnets temperatur och salthalt i Nagu Skärgård.
1. Observationsmaterialet, 1952-54.
Merent. Julk., No. 168:26 pp.

Great Britain Discovery Committee

density (data only)
Great Britain, Discovery Committee, 1957.
Station list, 1950-1951.
Discovery Rept., 28:300-398.

density(data only)
Great Britain, Discovery Committee, 1957.
Station list, 1950-1951. Discovery Repts., 28: 299-398.

density(data only)
Great Britain, Discovery Committee, 1955.
Station list, R.R.S. "William Scoresby".
Discovery Rept., 26:212-258, Pls. 11-12.

density(data only)
Great Britain, Discovery Committee, 1949.
Station list, R.R.S. "William Scoresby".
Discovery Rept., 25:144-280, Pls. 34-37.

density(data only)
Great Britain, Discovery Committee, 1947.
Station list, 1937-1939. Discovery Repts., 24: 198-422, Pls. 4-6.

density (data only)
Great Britain, Discovery Committee, 1945.
Station list, 1935-1937.
Discovery Rept., 24:3-196.

density(data only)
Great Briatin, Discovery Committee, 1942.
Station list. 1933-1935. Discovery Repts. 22:3-196, Pls. 1-4.

density(data only)
Great Britain, Discovery Committee, 1932.
Station list. Discovery Repts., 4:3-230, Pls. 1-

density (data only)
Great Britain, Discovery Committee, 1931.
Station list. Discovery Repts. 3:3-132, Pls. 1-10.

density(data only)
Great Britain, Discovery Committee, 1929.
Station list. Discovery Repts. 1:3-140.

density(data only)
Great Britain, Fisheries Laboratory, Lowestoft, 1957.
Hydrographical observations, equipment, methods, etc., 60 pp.

density(data only)
Great Britain, Fisheries Laboratory, Lowestoft, 1957.
Research Vessel 'Ernest Holt', hydrographical observations, 1955:88 pp.

density(data only)
Great Britain, Fisheries Laboratory, Lowestoft, 1957?
Research Vessel 'Ernest Holt', hydrographical observations, 1951:unnumbered pages.

density(data only)
Great Britain, Fisheries Laboratory, Lowestoft, 1956.
Research Vessel, 'Ernest Holt', hydrographical observations, 1950:150 pp.

density(data only)
Great Britain, Fisheries Laboratory, Lowestoft, 1956.
Research Vessel 'Ernest Holt'. Hydrographical observations, 192 pp. (multilithed)

density(data only)
Great Britain, Fisheries Laboratory, Lowestoft, 1954.
Research Vessel 'Ernest Holt', hydrographical observations:130 pp.

Density(data only)
Great Britain, Fisheries Laboratory, Lowestoft, 1952.
Research vessel, "Ernest Holt", Hydrographical observations, 1952. 98 pp.

density (data only)
Great Britain, Ministry of Agriculture, Fisheries and Food, 1953.
Fisheries Laboratory, Lowestoft, Research Vessel "Ernest Holt"; hydrographical observations.
Unnumbered pages.

density (data only)
Great Britain, Ministry of Agriculture, Fisheries and Food, 1949.
Fisheries Laboratory, Lowestoft, Research Vessel "Ernest Holt"; hydrographical observations.
Unnumbered pages

H

density(data only)
Hapgood, William, 1959.
Hydrographic observations in the Bay of Naples (Golfo di Napoli), January 1957-January 1958 (Station lists).
Pubbl. Sta. Zool., Napoli, 31(2):337-371.

density (data only)
Healey, D.A., and R.L.K. Tripe 1969.
Oceanographic observations at Ocean Station P (50°N, 145°W) 21 February - 9 April 1969.
Techn. Rept. Fish. Res. Bd. Can. 145:116pp.

density(data only)
Hela, Ilmo, and F. Koroleff, 1958.
Hydrographical and chemical data collected in 1956 on board the R/V"Aranda" in the Baltic Sea.
Merent. Julk., No. 183:1-52.

density (data only)
Henrotte-Bois, Maurice, 1969.
Résultats de mesures hydrologiques faites à bord de l'Origny: campagne "Mediterranean outflow (septembre-octobre 1965).
Cah. oceanogr. 21 (Suppl. 1): 49-192.

density (data only)
Henrotte-Bois, Maurice, 1968.
Résultats de mesures faites à bord de l'Origny en mer Tyrrhenienne. Campagne internationale de l'OTAN (16 septembre-15 octobre 1963).
Présentation de résults.
Cah. océanogr. 20(Suppl.1):65-125.

sigma T(data only
Herlinveaux, R.H., 1970.
Sapack 8/68 - oceanographic and biological observations.
Techn. Rept. Fish. Res. Bd. Can. 159: 60pp.

density(data only)
Herlinveaux, R.H., 1963
Data record of oceanographic observations made in Pacific Naval Laboratory underwater sound studies, November 1961 to November 1962.
Fish. Res. Bd., Canada, Mss. Rept. Ser. (Ocean. and Limnol.), No. 146: 101 pp. (multilithed).

density(data only)
Herlinveaux, R.H., 1963
Oceanographic observations in the Canadian Arctic Basin, Arctic Ocean, April-May, 1962.
Fish. Res. Bd., Canada, Mss. Rept. Ser. (Ocean. and Limnol.), No. 144: 25 pp. (multilithed).

density(data only)
Herlinveaux, R.H., 1961.
Data record of oceanographic observations made in Pacific Naval Laboratory underwater sound studies.
Fish. Res. Bd., Canada, Mss. Rept. Ser., (Ocean. & Limnol.), No. 108:85 pp.

specific gravity(data only)
Hirota, Reiichiro, 1964.
Some oceanographical and meteorological conditions observed at a definite station off Mukaishima Marine Biological Station in 1963.
Contrib. Mukaishima Mar. Biol. Sta., Hiroshima Univ., No. 76:13 pp.

density(data only)
Hirota, Reiichiro, 1963.
Some oceanographical and meteorological conditions observed at a definite station off Mukashima Marine Biological Station (1954-1962).
Contr. Mukashima Mar. Biol. Sta., Hiroshima Univ. 1962-1963, No. 73;107 pp.

density (data only)
Holmes, Robert W., and Maurice Blackburn, 1960
Physical, chemical and biological observations in the eastern tropical Pacific Ocean, Scot Expedition, April-June 1958.
USFWS Spec. Sci. Rept. Fish., No. 345:106 pp.

density (data only)
Hufford, Gary L., and James M. Seabrooke 1970.
Oceanography of the Weddell Sea in 1969 (1°W50°E).
Oceanogr. Rept. U.S. Cst Gd 31 (CG 373-31): 32 pp.

density (data only)
Husby, David M., 1971.
Oceanographic investigations in the northern Bering Sea and Bering Strait, June - July 1968.
Oceanogr. Rept. U.S. Cst Gd 40 (CG 373-40): 49 pp.

density (data only)
Husby, David M., 1969.
Oceanographic observations, North Pacific Ocean Station November, 30°00'N., 140°00'W., March 1967-March 1968.
Oceanogr. Rept. U.S. Coast Guard 26 (CG 373-26): 217 pp.

density (data only)
Husby, David M. 1969.
Oceanographic observations North Atlantic Ocean Station Delta, 44°N, 41°W, July 1966 - August 1967.
Oceanogr. U.S.C.G. 23 (CG 373-23): 161 pp.

density (data only)
#Husby, David M., 1967.
Oceanographic observations, North Atlantic Ocean Station BRAVO, 56o30'N., 51o00'W.
U.S.C.G. Oceanogr. Rep. 9 (C G 373-9):118pp.

density (data only)
#Husby, David M., 1967.
Oceanographic observations, North Atlantic Ocean Station BRAVO, 56o30'N., 51o00'W.
U.S.C.G., Oceanogr. Rep. 14 (C G 373-14):113 pp.

density (data)
Husby, David M., and Gary L. Hufford 1971.
Oceanographic investigation of the northern Bering Sea and Bering Strait 8-21 June 1969.
Oceanogr. Rept. U.S Cst Gd 42 (CG 373-42): 55 pp.

I

density (data only)
Ingham, Merton C., and Robert B. Elder, 1970.
Oceanic conditions off northeastern Brazil, February to March and October-November 1966.
Oceanogr. Rept. U.S. Cst Guard, 34 (CG 373-34): 155 pp.

density (data only)
Ingraham, W. James, Jr., 1967.
Distribution of physical-chemical properties and tabulations of station data, Washington and British Columbia coasts, October-November 1963.
Data Rep. U.S. Fish Wildl. Serv., Bur. Comm. Fish. 2 cards (microfiche).

density (data only
Ingraham, W. James, Jr., 1964.
North Pacific oceanography, February-March 1963.
U.S. Dept. Interior, Fish Wildl. Serv., Bur. Comm. Fish., Data Rept., 3:1 card (microfiche).

Density (Data only)

Ivanoff, A., A. Morel, J. Boutler, G. Copin-Montegut, F. Varlet, P. Courau, P. Geistdoerfer, F. Nyffeler, et J.P. Bethoux, 1969.
Resultats des observations effectuées en Mer Méditerranée principalement dans le Détroit de Sicile à bord du navire océanographique Calypso en mai 1965 et juillet 1966. Cah. océanogr. 21 (Suppl.):903-214.

J

density(data only)
Japan, Central Meteorological Observatory, 1956.
The results of marine meteorological and oceanographic observations (NORPAC Expedition, Spec. No) 17:131 pp.

density(data only)
Japan, Central Meteorological Observatory, 1953
The results of Marine meteorological and oceanographical observations. Jan.-June 1952, No. 11:362, 1 fig.

density(data only)
Japan, Central Meteorological Observatory, 1952
The results of Marine Meteorological and oceanographical observations, July - Dec. 1951, No. 10:310 pp., 1 fig.

density(data only)
Japan, Central Meteorological Observatory, 1952.
The results of Marine meteorological and oceanographical observations. July - Dec. 1950. No. 8: 299 pp.

density(data only)
Japan, Central Meteorological Observatory, 1952.
The Results of Marine Meteorological and oceanographical observations. Jan. - June 1950. No. 7: 220 pp.

density(data only)
Japan, Central Meteorological Observatory, 1951.
The results of marine meteorological and oceanographic observations. July - Dec. 1949. No. 6: 423 pp.

density(data, surface)
Japan, Central Meteorological Observatory, 1951.
Surface observations made on board R.M.S. "Oyashio-maru"; physical and chemical conditions of the surface water of the sea between Hakodate and Hachinoe. Res. Mar. Met. Ocean. Obs., Jan.-June 1949, No. 5:106.

density(data only)
Japan, Central Meteorological Observatory, 1951.
Table 14. Oceanographical observations taken in Miyazu Bay:physical and chemical data for stations occupied by R.M.S. "Asanagi-maru". Res. Mar. Met. Ocean. Obs., Jan.-June, 1949, No. 5:77-84.

density(data only)
Japan, Central Meteorological Observatory, 1951.
Table 12. Oceanographical observations taken in Sasebo Bay: physical and chemical data for stations occupied by R.S.S. "Asakaze-maru". Res. Mar. Met. Ocean. Obs., Jan.-June, 1949, No. 5:74-75.

density(data only)
Japan, Central Meteorological Observatory, 1951.
Table 11. Oceanographical observations taken in Nagasaki Harbor: physical and chemical data for stations occupied by R.S.S. "Asakaze-maru". Res. Mar. Met. Ocean. Obs., Jan.-June, 1949, No. 5:70-74.

density (data only)
Japan, Central Meteorological Observatory, 1951.
Table 10. Oceanographical observations taken in the East China Sea. (A) Physical and chemical data for stations occupied by the "Akebono-maru No. 9": (B) --------- by the Hatsutaka-maru". Res. Mar. Met. Ocean. Obs., Jan.-June, 1949, No. 5:68-70.

density(data only)
Japan, Central Meteorological Observatory, 1951.
Table 9. Oceanographical observations taken in the Goto-Nada and the Tsushima Straits. (A) Physical and chemical data for stations occupied by the fishing boat "Daikoku-maru": (B) ------- R.S.S. "Umikaze-maru". Res. Mar. Met. Ocean. Obs., Jan.-June, 1949, No. 5:61-67.

density(data only)
Japan, Central Meteorological Observatory, 1951.
Table 8. Oceanographical observations taken in the Ariake Sea; physical and chemical data for stations occupied by the R.S.S. "Umikaze-maru". Res. Mar. Met. Ocean. Obs., Jan.-June, 1949, No. 5:50-60.

density(data only)
Japan, Central Meteorological Observatory, 1951.
Table 7. Oceanographical stations taken in Osaka Bay; physical and chemical data for stations occupied by R.M.S. "Shumpu-maru". Res. Mar. Met. Ocean. Obs., Jan.-June, 1949, No. 5:48-49.

density(data only)
Japan, Central Meteorological Observatory, 1951.
Table 6. Oceanographical observations taken in the Akashi-seto, the Yura-seto, and the Kii Suido; physical and chemical data for stations occupied by R.M.S. "Shumpu-maru". Res. Mar. Met. Ocean. Obs., Jan.- June 1949, No. 5:40-47.

density(data only)
Japan, Central Meteorological Observatory, 1951.
Table 4. Oceanographical observations taken in Sagami Bay; physical and chemical data for stations occupied by R.M.S. "Asashio-maru". Res. Mar. Met. Ocean. Obs., Jan.-June, 1949, No. 5:30-37.

density (data only)
Japan, Central Meteorological Observatory, 1951.
Table 3. Oceanographical observations taken in the North Pacific Ocean along "C" line; (A) Phys-ical and chemical data for stations occupied by R.M.S. "Ukuru-maru";(B) --- R.M.S. "Ikuna-maru"; (C) --- R.M.S. "Ryofu-maru";(D) --- R.M.S. "Shinnan-maru";(E) --- "R.M.S. Kung-maru"; (F) --- R.M.S."Chikubu-maru";(G) --- R.M.S. "Ryofu-maru";(H) --- "Ikuna-maru";(I) --- "Chikubu-maru" (J) --- "Ukuru-maru";(K) --- "Shinnan-maru";(L) --- ("Ukuna-maru";(M) --- "Ryofu-maru". Res. Mar. Met. Ocean. Obs., Jan.-June 1949, No. 5:13-30.

density(data only)
Japan, Central Meteorological Observatory, 1951.
Table 2. Oceanographical observations taken in the sea area along the east coast of Tohoku District. (A). Physical and chemical data for stations occupied by R.M.S. "Oyashio-maru": (B) Physical and chemical data for stations occupied by R.M.S. "Kuroshio-maru". Res. Mar. Met. Ocean. Obs., Jan.-June 1949, No. 5:3-12.

density(data only)
Japan, Central Meteorological Observatory, 1951.
Table 1. Oceanographical observations taken in the sea area south of Hokkaido; physical and chemical data for stations occupied by R.M.S. "Yushio-maru". Res. Mar. Met. Ocean. Obs., Jan.-June 1949, No. 5:2-3.

density(data only)
Japan, Central Meteorological Observatory, Japan, 1951.
The results of marine meteorological and oceanographical observations, July-Dec. 1948, No. 4: vi+ 414 pp., of 57 tables, 1 map, 1 photo.

density(data only)
Japan, Central Meteorological Observatory, Japan, 1951.
The results of marine meteorological and oceanographical observations, Jan.-June 1948, No. 3: 256 pp.

density (data only)
Japan, Central Meteorological Observatory, Japan, 1950.
The results of marine meteorological and oceanographical observations, Jan.-June 1947, 1:113 pp

density(data only)
Japan, Fisheries Agency, Research Division, 1956.
Radiological survey of western area of the dangerous zone around the Bikini-Eniwetok Atolls, investigated by the "Shunkotsu maru" in 1956, Part 1:143 pp.

Japan, Hokkaido Univ., Faculty of Fisheries

Density (data only)
Japan, Hokkaido University, Faculty of Fisheries, 1970.
Data record of oceanographic observations and exploratory fishing, 13:406 pp.

density (data only)
Japan, Hokkaido University, Faculty of Fisheries, 1968.
Data record of oceanographic observations and exploratory fishing. No. 12:420 pp.

density (data only)
Japan, Hokkaido University, Faculty of Fisheries, 1968.
The Oshoro Maru cruise 23 to the east of Cape Erimo, Hokkaido, April 1967. Data Record Oceanogr. Obs. Expl. Fish., 12: 115-169.

density (data only)
Japan, Hokkaido University, Faculty of Fisheries, 1968.
The Oshoro Maru cruise 21 to the Southern Sea of Japan, January 1967. Data Record Oceanogr. Obs. Explor. Fish. 12: 1-97; 113-119.

density (data only)
Japan, Hokkaido University, Faculty of Fisheries, 1967.
The Oshoro Maru cruise 18 to the east of Cape Erimo, Hokkaido, April 1966. Data Record Oceanogr. Obs. Explor. Fish. 11: 121-164.

density (data only)
Japan, Hokkaido University, Faculty of Fisheries, 1967.
The Oshoro Maru cruise 16 to the Great Australian Bight November 1965-February 1966. Data Record Oceanogr. Obs. Explor. Fish., Fac. Fish., Hokkaido Univ. 11: 1-97; 113-119.

density (data only)
Japan, Hokkaido University, Faculty of Fisheries, 1967.
Data record of oceanographic observations and exploratory fishing, 11:383 pp.

Density (data only)
Japan Hokkaido University, Faculty of Fisheries, 1966.
Data record of oceanographic observations and exploratory fishery, No 10:388 pp.

density (data only)
Japan, Hokkaido University, Faculty of Fisheries. 1965.
Data record of oceanographic observations and exploratory fishing, No. 9: 343 pp.

density(data only)
Japan, Hokkaido University, Faculty of Fisheries, 1964.
Data record of oceanographic observations and exploratory fishing, No. 8:303 pp.

density(data only)
Japan, Hokkaido University, Faculty of Fisheries, 1963
Data Record of Oceanographic observations and exploratory fishing, No. 7:262 pp.

density (data only)
Japan Faculty of Fisheries, Hokkaido University 1961.
Data record of oceanographic observations and exploratory fishing, No.5: 391 pp

density (data only)
Japan, Hokkaido University, Faculty of Fisheries, 1960.
Data record of oceanographic observations and exploratory fishing, No. 4 :221 pp.

density (data only)
Japan, Hokkaido University, Faculty of Fisheries, 1959.
Data record of oceanographic observations and exploratory fishing, No. 3:296 pp.

density(data only)
Japan, Hokkaido University, Faculty of Fisheries, 1957.
Data record of oceanographic observations and exploratory fishing, No. 1:247 pp.

Japan Meteorological Agency

density (data only)
Japan, Japan Meteorological Agency. 1965.
The results of marine meteorological and oceanographical observations, July-December 1963, No. 34: 360 pp.

density (data only)
Japan, Japan Meteorological Agency, 1964.
The results of marine meteorological and oceanographical observations, January-June 1963, No. 33:289 pp.

density(data only)
Japan, Japan Meteorological Agency, 1964.
The results of the Japanese Expedition of Deep Sea (JEDS-5).
Res. Mar. Meteorol. and Oceanogr. Obs., July-Dec. 1962, No. 32:328 pp.

density(data only)
Japan, Japan Meteorological Agency, 1964.
Coastal observations.
Res. Mar. Meteorol. and Oceanogr. Obs., (31):253-271.

density (data only)
Japan, Japan Meteorological Agency, Oceanographical Section, 1962
Report of the oceanographic observations in the sea east of Honshu from February to March 1961. (In Japanese).
Res. Mar. Meteorol. and Oceanogr. Obs., Jan.-June, 1961, No. 29:13-21.

density (data only)
Japan, Japan Meteorological Agency, 1962
Report of the Oceanographic Observations in the sea south and west of Japan from October to November, 1960. (In Japanese).
Res. Mar. Meteorol. and Oceanogr., July-Dec., 1960, Japan Meteorol. Agency, No. 28: 30-35.

density(data only)
Japan, Japan Meteorological Agency, 1959.
The results of marine meteorological and oceanographical observations, July-December, 1958, No. 24:289 pp.

Japanese Oceanographic Data Center

density (data only)
Japan, Japanese Oceanographic Data Center, 1971.
Shumpu Maru, Kobe Marine Observatory, Japan Meteorological Agency, 16 July - 8 August 1970, south of Japan. Data Rept. CSK (KDC Ref. 49K127) 279: 7 pp. (multilithed).

density (data only)
Japan, Japanese Oceanographic Data Center, 1971.
Takuyo, Hydrographic Department, Maritime Safety Agency, Japan, 10-30 July 1970, south & east of Japan. Data Rept. CSK (KDC Ref. 49K126) 278: 19 pp. (multilithed).

density (data only)
Japan, Japanese Oceanographic Data Center, 1971.
Cape St. Mary, Fisheries Research Station, Hong Kong, 3-6 March 1970, South China Sea. Data Rept. CSK (KDC Ref. 74K015) 275: 9 pp. (multilithed).

density (data only)
Japan, Japanese Oceanographic Data Center, 1971.
Keiten Maru, Faculty of Fisheries, Kagoshima University, Japan, 23 April - 12 May 1969, East China Sea & Southeast of Taiwan. CSK Data Rept. (KDC Ref. 49K109) 247: 9 pp. (multilithed).

density (data only)
Japan, Japanese Oceanographic Data Center, 1971.
Tenyo Maru, Shimonoseki University of Fisheries, Japan, 13-30 June 1969, Southwest of the Japan Sea. Data Rept. CSK (KDC Ref. 49K108) 243: 17 pp. (multilithed).

density (data only)
Japan, Japanese Oceanographic Data Center, 1971.
Kofu Maru, Hadkodate Marine Observatory, Japan Meteorological Agency, 27 June - 9 August 1968, East of Japan & Okhotsk Sea. Data Rept. CSK (KDC Ref. 49K084) 178: 32 pp. (multilithed).

density (data only)
Japan, Japanese Oceanographic Data Center 1971.
Tensei Maru, Ocean Research Institute, University of Tokyo, Japan, 24-28 July 1967, (KDC Ref No. 49K433), 29 July - 9 August 1967 (KDC Ref. No. 49K434) east of Japan.
Data Rept. CSK 146: 5 pp. (multilithed).

DENSITY (DATA ONLY)
Japan, Japanese Oceanographic Data Center, 1970.
Suro No. 3, Hydrographic Office, Republic of Korea, July 23 - August 19, 1969, South of Korea Data Rept. CSK (KDC Ref. 24K035) 240: 19 pp. (multilithed).

density (data only)
Japan, Japanese Oceanographic Data Center 1970
Ji Ri San, Fisheries Research and Development Agency, Republic of Korea, August 7-29, 1969, South of Korea and East China Sea. Data Rept. CSK (KDC Ref. 24K034) 239: 15 pp.

density (data only)
Japan, Japanese Oceanographic Data Center 1970.
Han Ra San, Fisheries Research and Development Agency, Republic of Korea, August 9-21 1969, east of the Yellow Sea. Data Rept. CSK (KDC Ref 24K033) 238: 17 pp (multilithed).

DENSITY (DATA ONLY)
Japan, Japanese Oceanographic Data Center, 1970.
Takuyo, Hydrographic Department, Maritime Safety Agency, Japan, August 11 - September 2, 1969, South and East of Japan. Data Rept. CSK (KDC Ref. 49K112) 229: 21 pp. (multilithed).

density (data only)
Japan, Japanese Oceanographic Data Center, 197[illegible]
Shumpu Maru, Kobe Marine Observatory, Japan Meteorological Agency, July 18 - August 14, 19[illegible] South of Japan. Data Rept. CSK (KDC Ref. 49K[illegible] 228: 7 pp. (multilithed).

density (data only)
Japan, Japanese Oceanographic Data Center, 1969.
Ryofu Maru, Marine Division, Japan Meteorological Agency, Japan. October 7 - November 9, 1968. East China Sea. Data Rept. CSK (KDC Ref. No. 49K449) 217: 39 pp. (multilithed).

density (data only)
Japan, Japanese Oceanographic Data Center 1970.
Oshoro Maru, Faculty of Fisheries, Hokkaido University, Japan, November 5, 1968 - January 20, 1969, southwest of the North Pacific Ocean. Data Rept. CSK (KDC Ref. 49K104) 216: 13 pp.

density (data only)
Japan Japanese Oceanographic Data Center 1969.
Seifu Maru, Maizuru Marine Observatory, Japan Meteorological Agency, Japan, February 8-26, 1969, Japan Sea.
Data Rept. CSK (KDC Ref. 49K103) 215: 7 pp. (multilithed).

density (data only)
Japan, Japanese Oceanographic Data Center, 1969.
Takuyo, Hydrographic Department, Maritime Safety Agency, Japan. October 14 - November 6, 1968. South of Kyushu. Data Rept. CSK (KDC Ref. No. 49K454) 206:17 pp. (multilithed)

density (data only)
Japan, Japanese Oceanographic Data Center, 1969.
Researcher I, Bureau of Coast & Geodetic Survey, Manila, Philippines. April 20 - July 19, 1968. North & east of Philippine Region. Data Rept. CSK (KDC Ref. No. 66K001) 202: 49 pp. (Multilithed)

density (data only)
Japan, Japanese Oceanographic Data Center, 1970.
SURO No. 3, Hydrographic Office, Republic of Korea. October 12 - November 8, 1968, South of Korea. Data Report CSK (KDC Ref. 24K027) 201: 15 pp. (multilithed).

Density (Data only)
Japan, Japanese Oceanographic Data Center 1969.
Ostick, USSR, July 21 - August 20, 1968, northwest North Pacific Ocean.
Data Rept. CSK (KDC Ref. 90K020) 191: 32 pp (multilithed)

density (data only)
Japan, Japanese Oceanographic Data Center 1969. Iskatel', USSR January 26 - March 11, 1968, northwest North Pacific Ocean. Data Rept. CSK (KDC Ref. 90K019) 187: 32 pp. (multilithed).

density (data only)
Japan, Japanese Oceanographic Data Center, 1969. Kagoshima Maru, Faculty of Fisheries, Kagoshima University, Japan. August 15-26, 1968. East China Sea & southeast of Taiwan. Data Rept. CSK (KDC Ref. No. 49K089) 183:7 pp. (multilithed).

density (data only)
Japan, Japanese Oceanographic Data Center, 1969. Shumpu Maru, Kobe Marine Observatory, Japan Meteorological Agency, July 19-31, 1968, South of Japan. Data Rept. CSK (KDC Ref. 49K085) 179: 9 pp.

density (data only)
Japan, Japanese Oceanographic Data Center, 1969. Keiten Maru, Faculty of Fisheries, Kagoshima University, Japan. April 23 - May 12, 1968, East China Sea & southeast of Taiwan. Data Rept. CSK (KDC Ref. No. 49K079) 173: 10 pp.

density (data only)
Japan, Japanese Oceanographic Data Center 1969. Hakuho Maru, Ocean Research Institute, University of Tokyo, Japan, May 15-June 8, 1968, southwest of Japan and the East China Sea. Data Rept. CSK (KDC Ref. 49K078) 172: 7 pp. (multilithed).

density (data only)
Japan, Japanese Oceanographic Data Center, 1969. Takuyo, Hydrographic Division, Maritime Safety Agency, May 14-25, 1968, South of Japan. Data Rept. CSK (KDC Ref. 49K077) 171: 16 pp.

density (data only)
Japan, Japanese Oceanographic Data Center, 1968. Tae Baek San, Fisheries Research and Development Agency, Republic of Korea, February 16- March 12, 1968, South of Korea. Data Rept. CSK (KDC Ref. 24K022) 163: 12 pp. (multilithed)

DENSITY (data only)
Japan, Japanese Oceanographic Data Center, 1968. Han ra San, Fisheries Research and Development Agency, Republic of Korea, February 15-March 12, 1968, East of the Yellow Sea. Data Rept. CSK (KDC Ref. 24K021) 162: 18 pp. (multilithed)

density (data only)
Japan, Japanese Oceanographic Data Center, 1968. Baek du San, Fisheries Research and Development Agency, Republic of Korea, February 17-March 12, 1968, West of the Japan Sea. Data Rept. CSK (KDC Ref. 24K020) 161: 23 pp. (multilithed)

density (data only)
Japan, Japanese Oceanographic Data Center, 1969. Seifu Maru, Maizuru Marine Observatory, Japan Meteorological Agency, February 17-24, 1968, Japan Sea. Data Rept. CSK (KDC Ref. 49K073) 159:7 pp.

density (data only)
Japan, Japanese Oceanographic Data Center,1968. Chofu Maru, Nagasaki Marine Observatory, Japan Meteorological Agency, Japan, February 2-4, 1968, East China Sea. Data Rept. CSK (KDC Ref. 49K072) 158: 5 pp. (multilithed).

density (data only)
Japan, Japanese Oceanographic Data Center,1968. Shumpu Maru, Kobe Marine Observatory, Japan Meteorological Agency, Japan, February 17-28, 1968, South of Japan. Data Rept. CSK (KDC Ref. 49K071) 157: 6 pp. (multilithed).

density (data only)
Japan, Japanese Oceanographic Data Center, 1969. Kofu Maru, Hakodate Marine Observatory, Japan Meteorological Agency, February 7-26, 1968, east of Japan. Data Rept. CSK (KDC Ref. 49K070) 156:9 pp.

density (data only)
Japan, Japanese Oceanographic Data Center,1968. Takuyo, Hydrographic Division, Maritime Safety Agency, Japan, February 19-March 10, 1968, south and east of Japan. Data Rept. CSK (KDC Ref. 49K068) 154: 14 pp. (multilithed).

density (data only)
Japan, Japanese Oceanographic Data Center,1968. Takuyo, Hydrographic Divison, Maritime Safety Agency, Japan, September 25-October 18, 1967, Vicinity of Izu Islands. Data Rept. CSK (KDC Ref. 49L428) 142:12 pp. (multilithed).

density (data only)
Japan, Japanese Oceanggraphic Data Center, 1968. Fisheries Research Vessel No. 2, Department of Fisheries, Thailand, November 1- December 10, 1967, Gulf of Thailand and South China Sea. Data Rept. CSK (KDC Ref. 86K005) 139:29 pp. (multilithed)

density (data only)
Japan, Japanese Oceanographic Data Center,1968. Cape St. Mary, Fisheries Research Station, Hong Kong, September 19-October 11, 1967, South China Sea. Prelim. Data Rept. CSK (KDC Ref. 74K008) 138:16 pp. (multilithed).

density (data only)
Japan, Japanese Oceanographic Data Center,1968. Jalanidhi, Naval Hydrographic Office, Indonesia, October 4-19, 1967, South China Sea. Data Rept. CSK (KDC Ref. 42K002) 137: 15 pp. (multilithed).

density (data only)
Japan, Japanese Oceanographic Data Center,1968. Suro No. 1, Hydrographic Office, Korea, September 23-October 21, 1967, South of Korea. Data Rept. CSK (KDC Ref. 24K019) 136: 13 pp. (multilithed).

density (data only)
Japan, Japanese Oceanographic Data Center,1968. Chofu Maru, Nagasaki Marine Observatory, Japan Meteorological Agency, Japan, July 3-4, 1967, Aug. 28-29, 1967, Oct. 6-7, 1967, Nov. 9-10, 1967. Data Rept. CSK (KDC Refs. 49K319, 49K320, 49K321, 49K322) 134:18 pp. (multilithed).

density (data only)
Japan, Japanese Oceanographic Data Center,1968. Nagasaki Maru, Faculty of Fisheries, Nagasaki University of Fisheries, Japan, November 7-29, 1967, East China Sea. Data Rept. CSK (KDC Ref. 49K076) 133: 5 pp. (multilithed).

density (data only)
Japan, Japanese Oceanographic Data Center,1968. Chofu Maru, Nagasaki Marine Observatory, Japan Meteorological Agency, Japan, October 5-6, 1967, East China Sea. Data Rept. CSK (KCD Ref. 49K065) 130: 5 pp. (multilithed).

density (data only)
Japan, Japanese Oceanographic Data Center,1968. Kaiyo, Hydrographic Division, Maritime Safety Agency, Japan, November 13-December 1, 1967, South of Japan. Data Rept. CSK (KCD Ref. 49K063) 128: 13 pp. (multilithed).

density (data only)
Japan, Japanese Oceanographic Data Center, 1968. Oceanographic Vessel, No. 1, Hydrographic Department, Royal Thai Navy, Thailand, August 2-11, 1967, Gulf of Thailand. Data Rept. CSK (KDC Ref. 86K004) 127: 9 pp. (multilithed)

density (data only)
Japan, Japanese Oceanographic Data Center, 1968. Oceanographic Vessel, No. 1, Hydrographic Department, Royal Thai Navy, Thailand, June 7-12 1967, Gulf of Thailand. Data Rept. CSK (KDC Ref. 86K003) 126:9 pp. (multilithed)

density (data only)
Japan, Japenses Oceanographic Data Center, 1968. Oceanographic Vessel No. 1, Hydrographic Department, Royal Thai Navy, Thailand, January 18-24, 1967, Gulf of Thailand. Data Rept. CSK (KDC Ref. 86K001) 124:9 pp. (Multilithed)

density (data only)
Japan, Japanese Oceanographic Data Center,1968. Burudjulasad, Naval Hydrographic Office, Indonesia, August 25-September 5, 1967, South China Sea. Data Rept. CSK (KDC Ref. 42K001) 119: 11 pp. (multilithed).

density (data only)
Japan, Japanese Oceanographic Data Center,1968. Han Ra San, Fisheries Research and Development Agency, Korea, August 12-September 4, 1967, North of East China Sea. Prelim. Data Rept. CSK (KDC Ref. 24K018) 117:16 pp. (multilithed).

density (data only)
Japan, Japanese Oceanographic Data Center,1968. Chun Ma San, Fisheries Research and Development Agency, Korea, August 13-21, 1967, East of Yellow Sea. Prelim. Data Rept. CSK (KDC Ref. 24K017) 116: 16 pp. (multilithed).

density (data only)
Japan, Japanese Oceanographic Data Center,1968. Baek Du San, Fisheries Research and Development Agency, Korea, August 12-25, 1967, West of Japan Sea. Data Rept. CSK (KDC Ref. 24K016) 115:26 pp. (multilithed).

density (data only)
Japan, Japanese Oceanographic Data Center,1968. Hakuho Maru, Ocean Research Institute, University of Tokyo, Japan, September 6-20, 1967, South of Japan. Data Rept. CSK (KDC Ref. 49K059) 111: 9 pp. (multilithed).

density (data only)
Japan, Japanese Oceanographic Data Center,1968. Seifu maru, Maizuru Marine Observatory, Japan Meteorological Agency, Japan, August 2-24, 1967, Japan Sea. Data Rept. CSK (KDC Ref. 49K058) 110: 9 pp. (multilithed).

density (data only)
Japan, Japanese Oceanographic Data Center,1968. Takuyo, Hydrographic Division, Maritime Safety Agency, July 12-August 30, 1967, Central Part of the North Pacific Ocean. Prelim. Data Rept. CSK (KDC Ref. 49K053) 105:25 pp. (multilithed).

density (data only)
Japan, Japanese Oceanographic Data Center,1968. Yang Ming, Chinese National Committee on Ocean Research, Republic of China, April 1-May 14, 1967, Surrounding waters of Taiwan. Data Rept. CSK (KDC Ref. 21K004) 102: 15 pp. (multilithed).

density (data only)
Japan, Japanese Oceanographic Data Center, 1968.
Kaiyo, Hydrographic Division, Maritime Safety Agency, May 10-29, 1967, South of Japan.
Prelim. Data Rept. CSK(KDC 49K047), 96:15 pp. (multilithed).

density (data only)
Japan, Japanese Oceanographic Data Center, 1968.
Orlick, USSR, February 5-March 11, 1967, Northwest of North Pacific Ocean.
Prelim. Data Rept. CSK(KDC Ref. 90K012) 94:29pp.

density (data only)
Japan, Japanese Oceanographic Data Center, 1968.
Bering Strait, U.S. Coast Guard, January 14-18, 1967, East of Japan.
Data Rept. CSK (KDC Ref. 31K007) 91: 7 pp. (mutlilithed).

density (data only)
Japan, Japanese Oceanographic Data Center, 1968.
Seifu Maru, Maizuru Marine Observatory, Japan Meteorological Agency, February 10-March 2, 1967, Japan Sea.
Prelim. Data Rept. CSK(KDC 49K044) 86:14 pp. (multilithed).

density (data only)
Japan, Japanese Oceanographic Data Center, 1968.
Ryofu Maru, Marine Division, Japan Meteorological Agency, January 11-February 24, 1967, West of the North Pacific Ocean.
Prelim. Data Rept. CSK (KDC 49K040) 82:52 pp.

density (data only)
Japan, Japanese Oceanographic Data Center, 1968.
U.M. Schokalsky, USSR, July 20-August 23, 1966, North-west of North Pacific Ocean.
Prelim. Data Rept. CSK(KDC Ref. 90K010) 75:41 pp. (multilithed).

density (data only)
Japan, Japanese Oceanographic Data Center, 1968.
G. Nevelskoy, USSR, July 13-September 17, 1966,
Prelim. Data Rept. CSK (KDC Ref. 90K008) 73:32 pp.

density (data only)
Japan, Japanese Oceanographic Data Center, 1967.
Nagasaki Maru, Faculty of Fisheries, Nagasaki University, June 13-17, 1967, South of Japan.
Prelim. Data Rept. CSK(KDC Ref. No. 49K061) 113:6pp.

density (data only)
Japan, Japanese Oceanographic Data Center, 1967.
Chofu Maru (NMO), Jan. 13-14, 1967; Mar. 19-20, 1967, Apr. 15-15, 1967, May 11-12, 1967, Japan Meteorological Agency, South-East of Yakushima.
Prelim. Data Rept. CSK (KDC Ref. Nos. 49K315, 49K316, 49K317, 49K318) 104: 18 pp.

density (data only)
Japan, Japanese Oceanographic Data Center, 1967.
Chofu maru, Nagasaki Marine Observatory, Japan Meteorological Agency, Japan, May 17-18, 1967, East China Sea.
Prelim. Data Rept. CSK (KDC Ref. 49K051) 100:5 pp.

density (data only)
Japan, Japanese Oceanographic Data Center, 1967.
Shumpu maru, Kobe Marine Observatory, Japan Meteorological Agency, Japan, May 13-14, 1967, South of Japan.
Prelim. Data Rept. CSK (KDC Ref. 49K050) 99:5pp.

density (data only)
Japan, Japanese Oceanographic Data Center, 1967.
Nagasaki maru, The Faculty of Fisheries, Nagasaki University, Japan, January 19-22, 1967, East China Sea.
Prelim. Data Rept. CSK (KDC Ref. 49K046) 88: 7pp.

density (data only)
Japan, Japanese Oceanographic Data Center, 1967.
Oshoro maru, the Faculty of Fisheries, Hokkaido University, Japan, January 15-February 1, 1967, South of Japan.
Prelim. Data Rept. CSK (KDC Ref. (49045) 87: 9pp.

density (data only)
Japan, Japanese Oceanographic Data Center, 1967.
Chofu maru, Nagasaki Marine Observatory, Japan Meteorological Agency, Japan, January 20-February 22, 1967, East China Sea.
Prelim. Data Rept. CSK (KDC Ref. 49K043) 85:11 pp.

density (data only)
Japan, Japanese Oceanographic Data Center, 1967.
Shumpu maru, Kobe Marine Observatory, Japan Meteorological Agency, Japan, February 26-27, 1967, south of Japan.
Prelim. Data Rept. CSK(KDC Ref. 49K042) 84: 6pp.

density (data only)
Japan, Japanese Oceanographic Data Center, 1967.
Kofu Maru, Hakodate Marine Observatory, Japan Meteorological Agency, February 4-March 7, 1967, East of Japan.
Prelim. Data Rept. CSK (KDC Ref. No. 49K041) 83: 15 pp.

density (data only)
Japan, Japanese Oceanographic Data Center, 1967.
Takuyo, Hydrographic Division, Maritime Safety Agency, Japan, February 23-March 16, 1967, south and east of Japan.
Prelim. Data Rept., CSK (KDC Ref. 49K039) 81:17pp

density (data only)
Japan, Japanese Oceanographic Data Center, 1967.
Cape St. Mary, Fisheries Research Station, Hong Kong, November 24-December 7, 1966, South China Sea.
Prelim. Data Rept. CSK (KDC Ref. 74K005) 80:11 pp.

density (data only)
Japan, Japanese Oceanographic Data Center, 1967.
Tansei Maru, Ocean Research Institute, University of Tokyo, Japan, July 30-August 6, 1966, vicinity of Izu islands.
Prelim. Data Rept., CSK (KDC Ref. No. 49K419): 79 : 10 pp.

density (data only)
Japan, Japanese Oceanographic Data Center, 1967.
Cape St. Mary, Fisheries Research Station, Hong Kong, August 27-September 7, 1966, South China Sea.
Prelim. Data Rept. CSK (KDC Ref. 74K004) 78: 17pp.

density (data only)
Japan, Japanese Oceanographic Data Center, 1967.
Cape St. Mary, Fisheries Research Station, Hong Kong, June 1-8, 1966, South China Sea.
Prelim. Data Rept. CSK (KDC Ref. 74K003) 77:12 pp.

density (data only)
Japan, Japanese Oceanographic Data Center, 1967
Takuyo, Hydrographic Division, Maritime Safety Agency, October 14-24, 1966, Off Kii Peninsula.
Prelim. Data Rept., CSK(KDC Ref. 49K413) 76:6pp

density (data only)
Japan, Japanese Oceanographic Data Center, 1967.
Orlick, USSR, July 9-September 12, 1966, North-West of North Pacific Ocean.
Prelim. Data Rept. CSK (KDC Ref. 90K009) 74:34 pp.

density (data only)
Japan, Japanese Oceanographic Data Center, 1967.
Bering Strait, U.S. Coast Guard, July 8-14, 1966 South-east of Japan.
Prelim. Data Rept., CSK(KDC Ref. 31K005), 71:11 pp.

density (data only)
Japan, Japanese Oceanographic Data Center, 1967.
Bukhansan, Fisheries Research and Development Agency, Korea, July 16- August 9, 1966, East of the Yellow Sea.
Prelim. Data Rept., CSK (KDC Ref. 24K010) 69:17 pp.

density (data only)
Japan, Japanese, Oceanographic Data Center, 1967
Baekdusan, Fisheries Research and Development Agency, Korea, July 14-28, 1966, West of the Japan Sea.
Prelim. Data Rept. CSK (KDC Ref. 24K009) 65:22 pp.

density (data only)
Japan, Japanese Oceanographic Data Center, 1967.
Yang Ming, Chinese National Committee on Oceanic Research, Republic of China, September 10-October 14, 1966, Adjacent Sea of Taiwan.
Prelim. Data Rept. CSK (KDC Ref. 21K003) 67:17pp.

density (data only)
Japan, Japanese Oceanographic Data Center, 1966.
Koyo Maru, Shimonoseki University of Fisheries, October 26-29, south of Japan.
Prelim. Data Rept. CSK (KDC Ref. No. 49K038) 66: 7 pp. (multilithed).

density (data only)
Japan, Japanese Oceanographic Data Center, 1967.
Chofu Maru (NMO) Jul. 2-3, 1966; Chofu Maru (NMO), Aug. 27-29, 1966; Shumpu Maru (KMO), Oct. 27-31, 1966, Chofu Maru, Dec. 7-8, 1966, Japan Meteorological Agency of Japan, south-east of Yakushima.
Prelim. Data Rept., CSK (KDC Ref. 49K311, 49K312, 49K313, 49K314) 61: 18 pp.

density (data only)
Japan, Japanese Oceanographic Data Center, 1967.
Ryofu Maru, Marine Division, Japan Meteorological Agency, Japan, September 13-17, 1966, Eastern Sea of Japan.
Prelim. Data Rept., CSK, (KDC Ref. No. 49K033) 59:11 pp. (multilithed)

density (data only)
Japan, Japanese Oceanographic Data Center, 1967.
Nagasaki Maru, Faculty of Fisheries, Nagasaki University, Japan, June 16-21, 1966, southern Sea of Japan.
Prelim. Data Rept., CSK, (KDC Ref. No. 49K032) 58:8 pp. (multilithed).

density (data only)
Japan, Japanese Oceanographic Data Center, 1967.
Kagoshima maru, the Faculty of Fisheries, Kagoshima University, Japan, August 5-14, 1966, East China Sea.
Prelim. Data Rept. CSK (KDC Ref. 49K031) 57:9 pp.

density (data only)
Japan, Japanese Oceanographic Data Center, 1967.
Seifu Maru, Maizuru Marine Observatory, Japan Meteorological Agency, Japan, August 21-September 19, 1966, Japan Sea.
Prelim. Data Rept., CSK (KDC Ref. 49K029) 55:15 pp

density (data only)
Japan, Japanese Oceanographic Data Center, 1967.
Kofu Maru, Hakodate Marine Observatory, Japan Meteorological Agency, Japan, June 30-July 10, 1966, east of Japan.
Prelim. Data Rept., CSK(KDC Ref. 49K026) 52:14pp.

density (data only)
Japan, Japanese Oceanographic Data Center 1967.
Kaiyo, Hydrographic Division M.S.A. Japan, August 10-30, 1966, S.E. of Japan.
Prelim. Data Rept. CSK (KDC Ref 49K025) 51:15pp. (multilithed)

density (data only)
Japan, Japanese Oceanographic Data Center, 1967.
Yoko Maru, Seikai Regional Fisheries Research Laboratory, February 24-March 10, 1966, East China Sea.
Prelim. Data Rept., CSK, (KDC Ref. No. 49K526) 49:6 pp. (multilithed).

density (data only)
Japan, Japanese Oceanographic Data Center, 1967.
Shunyo Maru, Nankai Regional Fisheries Research Laboratory, February 23-March 13, 1966, S.W. of Kyushu, E. of Formosa.
Prelim. Data Rept., CSK,)KDC Ref. No. 49K525) 48:17 pp. (multilithed).

density (data only)
Japan, Japanese Oceanographic Data Center, 1967.
Vitjaz, USSR, December 17, 1965-April 15, 1966, west of north Pacific Ocean.
Prelim. Data Rept., CSK(KDC Ref. 90K007) 43:79 pp

density (data only)
Japan,Japanese Oceanographic Data Center,1966.
U.M. Shokalski, USSR,July 16-August 18,1965,
E. & S. of Japan.
Prelim. Data Rept., CSK,(KDC Ref. 90K001),23:
41 pp. (multilithed).

density (data only)
Japan, Japanese Oceanographic Data Center, 1966
Kerin Ho, Fisheries Research and Development
Agency, Korea, September 12-October 13, 1965,
East of Yellow Sea.
Prelim. Data Rept., CSK, (KDC Ref. 24K005), 38:
11 pp. (multilithed).

density (data only)
Japan, Japanese Oceanographic Data Center, 1966.
Bukhansan Ho, Fisheries Research and Development
Agency, Korea, December 2-12, 1965.
Prelim. Data Rept., CSK,(KDC 24K004),37: 15 pp.
(multilithed).

density (data only)
Japan,Japanese Oceanographic Data Center,1966.
Oshoro Maru,The Faculty of Fisheries,Hokkaido
University,Japan,November 30,1965-January 25,
1966,South of Japan.
Prelim.Data Rept., CSK (KDC Ref.(49K022),33:
11pp. (multilithed).

density (data only)
Japan,Japanese Oceanographic Data Center,1966.
Seifu Maru,Maizuru Marine Observatory,Japan
Meteorological Agency,Japan,February 12-
February 28,1966,Japan Sea.
Prelim. Data Rept., CSK (KDC Ref.48K021),32:
14 pp. (multilithed).

density (data only)
Japan,Japanese Oceanographic Data Center,1966.
Chofu Maru,Nagasaki Marine Observatory,Japan
Meteorological Agency,Japan,January 26-
February 25,1966,East China Sea.
Prelim.Data Rept., CSK,(KDC Ref. 49K020),31:
13 pp. (multilithed).

density (data only)
Japan, Japanese Data Center, 1966.
Preliminary data report of CSK, Shumpu Maru,
Kobe Marine Observatory, Japan Meteorological
Agency, February 19-24, 1966, Southern Sea of
Japan, No. 30 (October 1966).
KDC Ref. No. 49K019:6 pp (multilithed).

density (data only)
Japan,Japanese Oceanographic Data Center,1966.
Ryofu Maru,Japan Meteorological Agency,Japan,
February 4-February 28,1966,East of Japan.
Prelim. Data Rept., CSK (KDC Ref. 49K017),28:
17 pp. (multilithed).

density (data only)
Japan, Japanese Oceanographic Data Center,1966.
Preliminary data report of CSK, Takuyo,Japanese
Hydrographic Division, February 23-March 15,1966,
South eastern Sea of Japan, No. 27(October 1966).
KDC Ref. No. 49K016:17 pp. (multilithed).

density (data only)
Japan,Japanese Oceanographic Data Center,1966.
Shumpu Maru (KMO) May 11-12,1965 49K301
Chofu Maru (NMO) Jul. 1-2,1965 19K302
Chofu Maru (NMO) Aug. 2-13,1965 19K303
Chofu Maru (NMO) Sep. 25-26,1965 49K304
Chofu Maru (NMO) Nov. 1-2,1965 49K305
Shumpu Maru (KMO) Nov. 11-12,1965 49K306
Japan Meteorological Agency,Japan, South-East
of Yakushima.
Prelim. Data Rept., CSK (49K301-49K306),26:
26 pp. (multilithed).

density (data only)
Japan, Japanese Oceanographic Data Center,1966.
Zhyemchug,USSR,July 28-September 19,1965,
E. & S. of Japan.
Prelim.Data Rept.,CSK,(KDC Ref. 90K002),24:
32 pp. (multilithed).

density (data Only)
Japan,Japanese Oceanographic Data Center,1966.
Yang Ming,Chinese National Committee on Oceanic
Research,Republic of China,August 10-October
13,1965,Adjacent Sea of Taiwan.
Prelim.Data Rept., CSK (KDC Ref. 21K001),22:
18 pp. multilithed).

density (data only)
Japan, Japanese Oceanographic Data Center,1966.
Preliminary data report of CSK, Atlantis II,
Woods Hole Oceanographic Institution, USA,
August 4-September 23, 1965,Southern Sea of
Japan, No. 20,(October 1966).
KDC Ref. No. 31K001:68 pp. (multilithed).

density (data only)
Japan,Japanese Oceanographic Data Center,1966.
Cape St. Mary,Fisheries Research Station,Hong
Kong,October 2-October 10,1965,South China Sea.
Prelim. Data Rept., CSK,(KDC Ref.(74K001)20:12pp.
(multilithed)

density (data only)
Japan, Japanese Oceanographic Data Center,1966.
Preliminary data report of CSK. Kofu Maru,
Hakodate Marine Observatory, Japan Meteoro-
logical Agency, July 22-28, 1965, Eastern Sea
of Japan, No. 14 (October 1966.)
KDC Ref. No. 49K004:10 pp. (multilithed).

density (data only)
Japan, Japanese Oceanographic Data Center,1966.
Preliminary data report of CSK. Ryofu Maru,
Marine Division, Japan Meteorological Agency,
July 7-August 3,1965,Eastern Sea of Japan.
No.10(October)1966).
KDC Ref. No. 49K003:31 pp. (Multilithed).

density(data only)
Japan, Kagoshima University, Faculty of Fisheries
1957.
Oceanographical observation made during the Int-
ernational Cooperation Expedition, EQUAPAC, in
July-August 1956 by M.S. Kagoshima-maru and
M.S. Keitan-maru, 68 pp. (mimeographed)

density (data only)
Japan, Maritime Safety Board, 1956.
Tables of results from oceanographic observa-
tions in 1952 and 1953.
Hydrogr. Bull. (Publ. 981)(51):1-171.

density (data only)
Japan, Shimonoseki University of Fisheries
1970.
Oceanographic surveys of the Kuroshio and
its adjacent waters, 1967 and 1968.
Data. Oceanogr. Obs. Explor. Fish. 5: 117 pp.

density (data only)
Japan, Shimonoseki University of
Fisheries, 1970.
Data of oceanographic observations
and exploratory fishings. 5. Oceanographic
surveys of the Kuroshio and its adjacent
waters, 1967 and 1968: 117 pp.

K

density (data only)
Kollmeyer, Ronald C., Thomas C.
Wolford and Richard M. Morse
1966
Oceanography of the Grand Banks
region of Newfoundland in 1965.
U.S. Cst Gd Oceanogr. Rept. 11 (CG 373-11):
157 pp

density(data only)
Kollmeyer, R.C., J.W. McGary and R.M. Morse, 1964
Oceanographic observations, Tropical Atlantic
Ocean, EQUALANT II, Aigist 1963.
U.S.C.G. Oceanogr. Rept., No. 4(CG373-4):96 pp.

density (data only)
Korea,Fisheries College,Pusan,1968.
Baek-Kyung-Ho cruise to the central Pacific
Ocean,1967.
Data Rept.Oceanogr.Obs.Expl.Fish.,1:30 pp.

density (data only)
Republic of Korea, Hydrographic Office, 1966.
Coastal oceanographic observation (Inchon,
Kunsan, Mokop, Yosu, Cheju, Chinhae, Pusan, Mukho
Huksan Do, Ulneung Do).
Tech. Rep. H.O. Pub. No. 1101: 21-40.

density mean monthly (data only)
Korea(Republic of),Hydrographic Office,1965.
Technical reports.
H.O.Pub.,Korea,No.1101:179 pp.

density (data only)
Korea, Fisheries Research and Development Agency,
1968.
Annual report of oceanographic observations, 16:
~~961~~ 691 pp.

density (data only)
Koroleff, Folke, 1959.
The Baltic cruise with R/V Aranda 1958, hydro-
graphic data.
Merent. Julk. (Havforskningsinstitutets Skr.),
No. 193:25 pp.

density (data only)
Koroleff, Folke, and Aarno Voipio, 1963
The Finnish Baltic cruise 1960. Hydrographi-
cal data.
Havsforskningsinst. Skrift (Merent. Julk.),
No. 204: 27 pp.

Density (data only)
Koroleff, Folke, and Aarno Voipio, 1961.
The Baltic cruise with R/V Aranda, 1959. Hydrographical
data.
Merent. Julk. (Havforskningsinst. Skrift), No. 197:26 pp.

density (data only)
Krauel, David P., 1969/
Bedford Basin data report, 1967.
Techn. Rept. Fish. Res. Bd., Can., 120:84 pp
(multilithed).

L

density (data only)
Lacombe, Henri, 1969.
Résultats des mesures d'hydrologie et
de courants effectuées à bord de la
Calypso. Projet "Mediterranean outflow"
du Sous-Comité de Recherches Océanographiques
du Conseil Scientifique de l'OTAN (septembre-
octobre 1965).
Cah. océanogr. 21 (Suppl. 1):1-48.

density(data only)
Lacombe, Henri, 1964
Mesures de courant, d'hydrologie et de météor-
ologie effectuées à bord de la "Calypso".
Campagne Internationale d'observations dans
le détroit de Gibraltar (15 mai-15 juin 1961)
Cahiers Océanogr., C.C.O.E.C., 16(1):23-94.

density (data only)
Lacombe, Henri, et Jean-Claude Lizeray, 1960.
Liste des stations M.O.P. "Winnaretta-Singer".
Campagne d'aout 1958 dans le detroit de Gibralt-
ar. Cahiers Océanogr., C.C.O.E.C., 12(9): 673-
682.

density(data only)
Lacombe, Henri, and Paul Tchernia, 1960
Liste des stations M.O.P. CALYPSO 176à 234
(Campagne 1957) pour servir à l'étude des
échanges entre la mer Méditerranée et l'océan
Atlantique. Cahiers Oceanogr., C.C.O.E.C.,
12(3): 204-235.

density (data only)
Lacombe, H., and P. Tchernia, 1960.
Resultats d'observations, Année Géophysique
International, 1957-1958. Participation
française. II. Liste des stations M.O.P.
Calypso 241 à 297 (Campagne 1958) pour servir
à l'étude des échanges entre la Mer Méditerranée
et l'Ocean Atlantique.
Cahiers Océanogr., C.C.O.E.C., 12(6):417-439.

density(data only)
Lane, R.K., J. Butters, W. Atkinson and H.J. Hollister, 1961.
Oceanographic data record, coastal and seaways projects, February 6 to March 2, 1961.
Fish. Res. Bd., Canada, Manuscript Rept. Ser., (Oceanogr. & Limnol.), No. 91:128pp.

density(data only)
Lane, R.K., R.H. Herlinveaux, W.R. Harling and H.J. Hollister, 1960.
Oceanographic data record, Coastal Seaways Projects, October 3 to 26, 1960.
Fish. Res. Bd., Canada, Mss Rept. Ser. (Oceanogr. & Limnol.), No. 83:142 pp. (multilithed)

density (data only)
Lane, R.K., A.M. Holler, J.H. Meikle and H.J. Hollister, 1961
Oceanographic data record, Monitor and Coastal Projects, March 20 to April 14, 1961.
Fisheries Res. Bd., Canada, Manuscript Rept. Ser. (Oceanogr. & Limnol.), No. 94: 188 pp. (multilithed).

density (data only)
Le Floch, Jean et V. Romanovsky, 1966.
Résultats des mesures océanographiques effectuées en août 1954 en Mer Tyrrhénienne et on Mer Ligure.
Cahiers Océanogr., 18(3):229-242.

density (data only)
Lemasson, L., et J.P. Rebert, 1968.
Observations de courants sur le plateau continental Ivoirien: mise en evidence d'un sous-courant. Doc. Sci. Provisoire, Min. Product. Animale Cote d'Ivoire, (022):66 pp. (mimeographed).

density(data only)
Longhurst, A.R., 1961.
Cruise report of oceanographic cruise, 6/61.
Federal Fisheries Service, Nigeria, 4 pp.

density (data only)
Love, C.M., 1966.
Physical, chemical, and biological data from the northeast Pacific Ocean: Columbia river effluent area, January-June 1963. 6. Brown Bear Cruise 326:13-23 June:CNAV Oshawa Cruise Oshawa-3:17-30 June.
Univ. Washington, Dept. Oceanogr., Tech. Rep., No. 134:230 pp. (Unpublished manuscript).

M

density (data)
Marumo, Ryuzo, Editor, 1970.
Preliminary report of the Hakuho Maru Cruise KH-69-4, August 12-November 13, 1969, The North and Equatorial Pacific Ocean.
Ocean Res. Inst. Univ. Tokyo, 68pp.

Mc

density (data only)
McGary, James W., 1965.
Oceanographic observations, North Atlantic Ocean Station BRAVO, January-April 1964, 56° 30'N, 50° 00'W.
U.S. Coast Guard Oceanogr. Rept., (373-7), No.7: 57 pp.

density (data only)
McGary, J.W., 1955.
Mid-Pacific oceanography, Pt. 6. Hawaiian offshore waters, December 1949-November 1951.
Spec. Fish. Rept:Fish No. 152:1-138, 33 figs.

density(data only)
McGary, James W., and Joseph J. Graham, 1960.
Biological and oceanographic observations in the central north Pacific, July-September 1958.
U.S.F.W.S. Spec. Sci. Rept., Fish., No. 358: 107 pp.

density (data only)
McGary, J.W., E.C. Jones, and T.S. Austin, 1956.
Mid-Pacific oceanography. IX. Operation NORPAC.
Spec. Sci. Repts.: Fish. 168:127 pp.

density(data only)
McGary, J.W., and R.M. Morse, 1964.
Oceanographic observations, North Atlantic Ocean Station Delta 44° N, 41° W.
U.S.C.G. OCEANOGR. Rept., No. 3(CG373-3):30pp

density (data only)
Minas, H.J. 1971.
Résultats de la campagne "Mediprod I" du Jean Charcot.
Cah. océanogr. 23 (Suppl. 1):93-144.

density (data only)
Minkley, B.G., K.A. Gantzer, and D.A. Healey, 1970.
Oceanographic observations at Ocean Station P (50°N, 145°W) 19 September to 6 November 1969 and 5 December to 15 January 1970.
Techn. Rept. Fish. Res. Bd. Can. 211:97 pp.

Density (data only)
Minkley, B.G., R.L.K. Tripp and D.A. Healey, 1970.
Oceanographic observations at Ocean Station P (50°N, 145°W) 27 June to 25 September 1969.
Techn. Rept. Fish. Res. Bd Can. 184: 222 pp.

density (data only)
Mork, Martin, 1968.
Hydrographic and meteorological data from the Bear Island cruise, 1966.
Geophys. Inst., Univ. Bergen, 50 pp. (mimeographed).

density (data only)
Morse, Betty Ann, 1964.
North Pacific and Boring Sea oceanography, 1960 and 1961.
U.S. Dept. Interior, Fish Wildl. Serv., Bur. Comm. Fish., Data Rept., 7:8 cards (microfiche).

density(data only)
Morse, R.M., and J.W. McGary, 1964.
Oceanographic observations, North Atlantic Ocean Station Echo, 35°N, 48°W, January-February 1963.
U.S.C.G. Oceanogr. Rept., No. 2(CG373-2):31 pp.

density (data only)
Mosby, Håkon, 1969
Norwegian Atlantic Current, March and December 1965.
Rep. Geophys. Inst. Div. Phys. Oceanogr. Univ. Bergen, 17:53 pp.

density(data only)
Mosby, Hakon, 1961
Hydrological observations of the M/S "Helland Hansen" near the Strait of Gibraltar, May - June 1961. 67 pp. (Mimeographed).

density (data only)
Moynihan, Martin J., and Robin D. Muench 1971.
Oceanographic observations in Kane Basin and Baffin Bay, May and August-October 1969.
Oceanogr. Rept. U.S. Cst Gd (CG 373-44): 44:143 pp.

density (data only)
Muench, Robin D., Martin D. Moynihan, Edward J. Tennyson, Jr., W. Gordon Tidmarsh and Roger B. Theroux, 1971.
Oceanographic observations in Baffin Bay during July-September 1968.
Oceanogr. Rept., U.S. C.Guard 37 (373-37): 97 pp.

N

density (data only)
(NATO)
Organisation du Traité de l'Atlantique Nord, Sous-Comité Océanographique, 1969
Résultats des mesures d'hydrologie et de courants faites à bord du navire océanographique français Calypso (1er septembre-1er octobre 1965). Résultats des mesures d'hydrologie faites à bord du navire océanographique français Origny. Projet Mediterranean Outflow (septembre-octobre 1965)
Rapp. Techn. OTAN HS: 192 pp. (multilithed)

density (data)
Nehring, D. and H.-J. Brosin 1968
Ozeanographische Beobachtungen im äquatorialen Atlantik und auf dem Patagonischen Schelf während der I. Südatlantik Expedition mit dem Fischereiforschungsschiff Ernst Haeckel von August bis Dezember 1966.
Geod. Geoph. Veröff. 4 (3): 93 pp.

density (data only)
NORPAC Committee, 1955.
Oceanic observation of the Pacific, 1955.
The NORPAC data, 532 pp.

O

density (data only)
Oren, Oton Haim, 1967.
Croisière Chypre-04 en Méditerranée orientale, février-mars, 1965: résultats des observations hydrologiques.
Cah. océanogr., 19(9):783-798.

density (data only)
Oren, O.H., 1967.
The fifth cruise in the eastern Mediterranean, Cyprus-05, May 1967. Hydrographic data.
Bull. Sea Fish Res. Stn Israel, 47:55-63.

density (data only)
Oren, O.H., 1967.
Croisière Chypre-04 dans la Méditerranée orientale, février-mars 1965. Resultats des observations hydrographiques.
Bull. Sea Fish. Res. Stn Israel, 47:37-54.

density (data only)
Oren, O.H., 1967.
Oceanographic cruise of Cyprus-03 in the eastern Mediterranean, July 30-August 15, 1964. Results of oceanographic observations.
Bull. Sea Fish. Res. Stn Israel, 47:3-36.

density (data only)
Oren, Oton Haim, 1966.
Croisière "Chypre-02" en Méditerranée orientale, juillet-août 1963. Résultats des observations hydrologiques.
Cah. océanogr., 18(Suppl.):1-17.

P

density (data only)
Peluchon, Georges 1966.
Campagne Alboran I. Hydrologie en mer d'Alboran. I. Résultats des mesures faites à bord des navires Eupen (Belgique) et Origny (France) en juillet-août 1962. Présentation des résultats.
Cah. océanogr. 17 (Suppl. 1): 1-88.

density (data only)
Piton, B., et Y. Magnier 1971.
Observations physico-chimiques faites par le Vauban le long de la côte nord-ouest de Madagascar de janvier à septembre 1970.
Doc. scient. Cent. Nosy-Bé, Off. Rech. scient. techn. Outre-Mer 21: unnumbered pp. (mimeographed)

density (data only)
Poirier, Marcel, et Robert Boudreault, 1965.
Observations d'océanographie physique dans la Baie-des-Chaleurs, juin-octobre 1958.
Cah. Inf. Stn biol. mar. Grande-Rivière, 32: unnumbered pp. (multilithed).

density (data only)
Poirier, Marcel, et Robert Boudreault, 1965.
Observations d'océanographie physique dans la Baie-des-Chaleurs, juin-octobre 1957.
Cah. inf. Stn. biol. mar. Grande-Rivière, No. 31: unnumbered pp. (multilithed).

density (data only)
Poirier, Marcel, et Robert Boudreault 1965.
Observations d'océanographie physique dans la Baie-des-Chaleurs, juin-octobre 1956.
Cah. Inf. Stn biol. mar. Grande-Rivière 30: unnumbered pp. (multilithed).

density (data only)
Poirier, Marcel et Robert Boudreault, 1965
Observations d'océanographie physique dans la Baie-des-Chaleurs, mai-septembre 1955.
Cah. inf. Stn. biol. mar. Grande-Rivière, No. 25: unnumbered pp. (multilithed)

density (data only)
Portugal, Instituto Hidrografico, 1965.
Resultados das observacões oceanográficas no Canal de Mocambique, Cruzeiro al 1/64: abril-maio 1964.
Servico de Oceanografia, Publ.1:73 pp., 46 figs.

density (data only)
Portugal, Instituto Hidrográfico, Servico de Oceanografio, 1965.
Resultados das observacões oceanográficas no Canal de Moçambique, Cruzeiro al 1/64: Abril-Maio 1964, 73 pp., 46 figs.

Q

R

density (data only)
Ramalho, A. de M., 1941
Observacoes oceanograficas. "Albacora", 1930, 1938 e 1939. Trav. St. Biol. Mar., Lisbonne, No. 44:52 pp., 1 fold-in (with 8 figs.)

density (data only)
Raymond, Robert, et Robert Boudreault 1965.
Observations d'océanographie physique dans le nord du Golfe Saint-Laurent 1952-1954.
Cah. Inf. Stn biol. mar. Grande Rivière 27: unnumbered pp. (multilithed)

density (data only)
Raymond, Robert, et Robert Boudreault 1965.
Observations d'océanographie physique dans la Baie-des-Chaleurs, mai-septembre 1953.
Cah. Inf. Stn biol. mar. Grande-Rivière 25: unnumbered pp. (multilithed)

density (data only)
Raymond, Robert, et Robert Boudreault 1964.
Observations d'océanographie physique dans la Baie-des-Chaleurs, mai-octobre 1950.
Cah. Inf. Stn biol. mar. Grande-Rivière 24: unnumbered pp. (multilithed).

density (data only)
Republic of China, Chinese National Committee on Oceanic Research, Academia Sinica, 1968.
Oceanographic Data Report of CSK, 2: 126 pp.

density (data only)
Republic of China, Chinese National Committee on Oceanic Research, Academia Sinica 1966.
Oceanographic data report of CSK 1: 123 pp.

density (data only)
Robe, Robert Quincy 1971.
Oceanographic observations, North Atlantic standard monitoring sections A1, A2, A3, and A4.
Oceanogr. Rept. U.S. Cst Gd 43 (CG 373-43): 227 pp.

density
Rochford, D. J. 1969.
Origin and circulation of water types of the 25.00 sigma-t surface of the south-west Pacific.
Aust. J. mar. Freshwat. Res. 20(2): 105-114.

density (data only)
Rotschi, Henri, 1961
Resultats des observations scientifiques du "Tiare", Croisière "Entrecasteaux" (11-24 août 1960) sous le commandement du Lieutenant de Vaisseau Carbelaud.
Comité Local d'Océanogr. et d'Études des Côtes de Nouvelle Caledonie, Rapp. Sci., Noumea, No. 21: 41 pp. (mimeographed).

density (data only)
Rotschi, Henri, 1960
Orsom III, resultats de la Croisière "Dillon" lère Partie. Océanographie Physique.
Off. Recherche Sci. et Techn., Outre-Mer, Inst Francais d'Océanie, Centre d'Océanogr., Rapp. Sci. Noumea, No. 18: 58 pp.

density (data only)
Rotschi, Henri, 1958
"Orsom III", Océanographie physique.
Rapp. Tech. de la Croisière 56-5, Inst. Français d'Océanie, Rapp. Sci., No. 5: 35 pp (Mimeographed).

density (data only)
Rotschi, Henri, 1957
"Orsom III", Oceanographie physique.
Rapp. Tech. de l'Expedition EQUAPAC (Croisiere 56-4), Inst. Francais d'Oceanie, Rapp. Sci., 3: 52 pp. (mimeographed).

density (data only)
Rotschi, Henri, 1958
Résultats des observations scientifiques du "Tiare", Croisière "Bounty" sous le commandement du Lieutenant de Vaisseau Morlanne, 20-29 Juin, 1958.
Comité Local d'Océanographie et d'Etude des Côtes de Nouvelle-Caledonie, Rapp. Sci., I.F.O., No. 7: 20 pp.

density (data only)
Rotschi, Henri, Michel Legand and Roger Desrosieres, 1961
Orsom III, Croisieres diverses de 1960, physique chimie et biologie. ORSTOM, Inst. Francais d'Océanie, Centre d'Océanogr., Noumea, Rapp. Sci., No. 20: 59 pp. (mimeographed).

density (data only)
Rotschi, Henri, Ives Magnier, Maryse Tirelli, et Jean Garbe, 1964.
Résultats des observations scientifiques de "La Dunkerquoise " (Croisière "Guadalcanal").
Océanographie, Cahiers O.R.S.T.R.O.M., 11(1):49-154.

S

density(data only)
Schmidt, J., 1912.
Hydrographical observations.
Rept. Danish Oceanogr. Exped., 1908-1910, 1:51-75.

density (data only)
Scripps Institution of Oceanography, 1949
Physical and chemical data, Cruise 2, March 28 to April 12, 1949. Physical and Chemical Data Report No.2:10 figs. (ozalid) tables of data (mimeographed).

density (data only)
Scripps Institution of Oceanography, 1949
Physical and chemical data, Cruise 1, February 28 to March 16, 1949. Marine Life Research Program. Physical and Chemical Data Report No.1:10 figs. (ozalid), tables of data (mimeographed).

density(data chiefly)
Seckel, G.R., 1955.
MidPacific oceanography. VII. Hawaiian offshore waters, Sept. 1952-August 1953.
Spec. Sci. Repts., Fish. No. 164:250 pp., 38 figs.

density(data only)
Shipley, A.M., and P. Zoutendyk, 1964.
Hydrographic and plankton data collected in the south west Indian Ocean during the SCOR International Indian Ocean Expedition, 1962-1963.
Univ. Cape Town, Inst. Oceanogr., Data Rept., No. 2:210 pp.

density (data only)
South Africa Department of Industry, Division of Sea Fisheries, 1968.
Hydrological station list 1961-1962, 1:342 pp.

density(data only)
South Africa, Department of Commerce and Industries, Division of Sea Fisheries, 1964
Thirty-second annual report for the period 1st April 1960 to 31st March, 1961:267 pp.

density(data only)
Spain, Instituto Español de Oceanografia, 1961.
Campañas biologicas del "Xauen" en las costas del Mediterraneo Marroqui, Mar de Alboran, Baleares, y Noroeste y Cantabrico Españoles en los años 1952, 1953 y 1954.
Bol. Inst. Español Oceanogr., 102:1-130.

density (data only)
Spain, Instituto Español de Oceanografia, 1955.
Campaña del "Xauen" en la costa noroeste de España en 1949 y 1950. Registro de operaciones.
Bol. Inst. Español Ocean., 71:72 pp.

density (data only)
Stevenson, M.R. and F.R. Miller 1971.
Oceanographic and meteorological observations for project Little Window: March 1970. (In English and Spanish)
Data Rept. Inter-Am. Trop. Tuna Comm. 324 pp.

density (data only)
Sweden, Havsfiske laboratoriet, lysekil, 1969.
Hydrographica data, January-June 1968.
R.V. Skagerak, R.V. Thetis.
Meddn Havsfiskelab., Lysekil, hydrog. Avdeln., Goteborg 63: numerous unnumbered pp. (multilithed).

density (data only)
Sweden, Havsfiskelaboratoriet, Lysekil 1971.
Hydrographical data January-June 1970 R.V. Skagerak, R.V. Thetis and TV252, 1970.
Meddn. Hydrogr. avd. Göteborg, 101: unnumbered pp. (multilithed).

density (data only)
Sweden, Havsfiskelaboratoriet, Lysekil 1970.
Hydrographical data 1966, R.V. Skagerak, R.V. Thetis.
Meddn Hydrogr. avd. Göteborg 25:255pp.

density (data only)
Sweden, Havsfiskelaboratoriet, Lysekil, 1968.
Hydrographical data, July-December 1967 R.V. Skagerak.
Meddn Havsfiskebal., Lysekil, 52: unnumbered pp. (mimeographed).

density (data only)
Sweden, Havsfiskelaboratoriet, Lysekil, 1968.
Hydrographical data, July-December 1967, R.V. Skagerak.
Meddn Havsfiskelab., Lysekil, Hydrogr. Avd., Göteborg, 52: unnumbered pp. (multilithed).

density (data only)
Sweden, Havsfiskelaboratoriet, Lysekil, Hydrografiska Avd., Göteborg, 1967.
Hydrographical data, January-June 1967, R.V. Skagerak.
Meddn. Havsfiskelab., Lysekil, Hydrogr. Avd., Göteborg, 38: numerous pp. mimeographed).

density (data only)
Sweden, Havsfiskelaboratoriet, Lysekil, Hydrografiska Avd., Göteborg, 1967.
Hydrographical data, January-June 1967, R.V. Thetis.
Meddn, Havsfiskelab., Lysekil, Hydrogr. Avd., Göteborg, 41: numerous pp. (mimeographed).

T

density(data only)
Tabata, S., C.D. McAllister, R.L. Johnston, D.G. Robertson, J.H. Meikle, and H.J. Hollister, 1961.
Data record. Ocean Weather station "P" (Latitude 50 00'N, Longitude 145 00"W), December 9, 1959 to January 19, 1961.
Fish. Res. Bd., Canada, MSS. Rept. Ser. (Ocean & Limnol.), No. 98:296 pp. (Multilithed).

density(data only)
Tabata, S., C.D. McAllister, J.H. Meikle and H.J. Hollister, 1962
Oceanographic data record, Ocean Weather Station "P", January 17 to August 5, 1962.
Fish. Res. Bd., Canada, Mss. Rept. Ser. Ocean. and Limnol.), No. 139: 113 pp. (multilithed).

density(data only)
Tabata, S., D.G. Robertson, W. Atkinson, and H.J. Hollister, 1962
Oceanographic data record, Ocean Weather Station "P", September 12, 1961 to January, 21, 1962.
Fish. Res. Bd., Canada, Mss. Rept. Ser. (Oceanogr. & Limnol.), No. 125:187 pp.

density (data only)
Tchernia, Paul, et Michèle Fieux 1971.
Résultats des observations hydrologiques exécutées à bord du N/O Jean Charcot pendant la campagne MEDOC 1969 (30 janvier - 28 février) (18 mars - 31 mars).
Cah. océanogr. 23 (Suppl. 1): 1-91.

density (data only)
Tiphane, Marcel, 1962.
Observations océanographiques dans la Baie-des-Chaleurs juin-septembre 1962.
Sta. Biol. Mar., Grande Rivière, Québec, Cahiers d'Inform., No. 16: unnumbered pp. (multilithed).

density (data only)
Tiphane, Marcel, 1962.
Observations océanographique dans la Baie-des-Chaleurs juin-septembre 1961.
Sta. Biol. Mar., Grande-Rivière, Québec, Cahiers d'Inform., No. 15: unnumbered pp. (multilithed).

density (data only)
Tiphane, Marcel, et Robert Boudreault 1964.
Observations océanographiques dans la Baie-des-Chaleurs, été 1963.
Cah. Inf. Stn biol. mar. Grande-Rivière 21: unnumbered pp. (multilithed)

density (data only)
Trotti, Leopoldo, 1968.
Stazione mareografica di Trieste; osservazioni idrologiche, gennaio-dicembre 1967.
Atti Accad. ligure, 25: 3-19.
Also: Pubbl. Ist. speriment. talassogr., Trieste 450 (1968).

density (data only)
Trotti, Leopoldo, 1967.
Stazione Mareografica di taranto; osservazioni idrologiche, luglio 1966-guigno 1967.
Atti Accad. ligure, 34: 1-19.
Also: Pubbl. Ist. speriment. talassogr., Trieste 446. (1968).

density (data only)
Trotti, L., 1967.
Crociera Golfo Palmas e Canale di Sardegna, 1965: dati oceanografici
Pubbl. Ist. Speriment. Talassogr., Trieste, 439: unnumbered pp. (quarto).

density (data only)
Trotti, L., 1967.
Crociera Bocche di Bonifacio 1964.
Pubbl. Ist. Speriment. Talassogr., Trieste, 438: unnumbered pp. (quarto)

density (data only)
Trotti, L. 1967.
Dati oceanografici raccolti durante l'A.G.I. 1957-1958 dal Centro Talassografico Tirreno.
Pubbl. Ist. Speriment. Talassogr. Trieste, 437: unnumbered pp. (quarto).

density (data only)
Trotti, Leopoldo, 1967.
Stazione mareografica di Trieste: osservazioni idrologiche gennaio-dicembre 1965. (e 1966)
Atti Accad. ligure, 23(1): 52-65; 327-343.

density (data only)
Trotti, Leopoldo, 1966.
Stazione mareografica di Trieste: osservazioni idrologiche, gennaio-dicembre 1964
Atti Accad. ligure 22(1): 204-220.

density (data only)
Trotti, Leopoldo, 1966.
Stazione mareografica di Taranto: osservazioni idrologiche, giugno 1964-giugno 1965
Atti Accad. ligure, 22(1): 151-170

density (data only)
Trotti, Leopoldo, 1966.
Stazione Mareografica di Tarante, osservazioni idrologiche, Luglio 1965-Giugno 1966.
Atti Acad. ligure, 23:1-19.

density (data only)
Trotti, Leopoldo, 1966.
Stazione Mareografica di Trieste, osservazioni idrologiche, Gennaio-Dicembre 1965.
Atti Acad. ligure, 23:1-19.

density, (data only)
Trotti, Leopoldo, 1966.
Stazione mareografica di Trieste, Osservazioni idrologiche, gennaio-dicembre 1964.
Atti Accad. ligure, 22:3-19.

density (data only)
Trotti, Leopoldo, 1966.
Stazione mareografica di Taranto, osservazioni idrologiche, giugno 1964-giugno 1965.
Atti Accad. ligure, 22:3-22.

U

Density (data only)
Uda, Michitaka, Yoshimi Morita, and Makoto Ishino, 1957.
Results from the oceanographic observations in the North Pacific (1955-56) with Umitaka Maru and Shinyo-Maru. Rec. Ocean. Wks., Japan (Spec. No.):1-20

density (data only)
Union of South Africa, Department of Commerce and Industry, 1965.
Station list.
Div. Sea Fish., Invest. Rept., No. 51:50-67.

density(data only)
United States, Agricultural and Mechanical College of Texas, 1961.
Oceanographic observations in the intertropical region of the World Ocean during IGY and IGC. Pt. 1. Atlantic and Indian Oceans.
IGY World Data Center A, Oceanography, IGY Oceanogr. Rept. No. 3:264 pp.

density (data only)
U.S. Department of Commerce, Coast and Geodetic Survey, 1958.
Density of sea water at tide stations, Pacific coast, North and South America and Pacific Ocean Islands.
U.S. C.& G.S. Publ., 31-4(5th Ed.):79 pp.

density(data only)
United States, U.S. Coast Guard, 1965.
Oceanographic cruise USCGC Northwind: Chukchi, East Siberian and Laptev seas, August-September, 1963.
U.S.C.G. Oceanogr. Rept., No. 6(CG373-6):69 pp.

United States, National Oceanographic Data Center, 1965.
Data report EQUALANT III.
Nat. Oceanogr. Data Cent., Gen. Ser., G-7:339pp. $5.00
(Publ.)

density (data only)
United States, National Oceanographic Data Center, 1964.
Data report EQUALANT II.
NODC Gen. Ser., Publ. G-5: numerous pp. (not serially numbered; loose leaf - $5.00)

density (data)
United States, National Oceanographic Data Center, 1964.
A summary of temperature-salinity characteristics of the Persian Gulf.
Gen. Ser., Publ. 4:223 pp. (loose leaf). ($2.50).

merely a compilation of old data.

density(data only)
U.S.A., U.S. Coast Guard, 1964.
Oceanographic Cruise, USCGC NORTHWIND, Bering and Chukchi Sea, July-Sept. 1962.
U.S.C.G. Oceanogr. Rept., No. 1(CG373-1):104 pp.

density (data only)

University of California, Scripps Institution of Oceanography, 1960.
Oceanic observations of the Pacific, 1950, 508 pp.

V

density (data only)

Valdez, Alberto J., Alberto Gomez, Aldo Orlando y Andres J. Lusquiños 1969.
Datos y resultados de las campañas pesqueras "Pesqueria V" (28 de agosto al 7 de setiembre de 1967).
Publ. (Ser. Informes técn.) Mar del Plata, Argentina 10(5): unnumbered pp.

density (data only)

Van Riel, P.M., P. Groen and M.P.H. Weenink, 1957
Quantitative data concerning the status of the East-Indonesian waters. Depths of standard pressures and stability values. Snellius Exped., 2(Ocean. Res.)(7):45 pp.

density (data only)

Vicariot, Jean, 1967.
Résultats des mesures faites à bord du navire Origny en Méditerranée occidentale sur le méridien 6° est (12 septembre 1962 - 7 mai 1963).
Cah. océanogr., 19(Suppl. 1):71-155.

W

density (data only)

Wilson, R.C., and M.O. Rinkel, 1957.
Marquesas area oceanographic and fishery data, January-March 1957.
USFWS Spec. Sci. Rept., Fish., No. 238:136 pp.

XYZ

density (data only)

Zorè, M., & A. Zupan, 1960.
(Hydrogrographical data for the Kastela Bay. 1953-1954)
Acta Adriatica 9 (1):32 pp

density (data only)

Zoutendyk, Peter, 1963.
Hydrographic and plankton data collected in the Agulhas Current by RV John D. Gilchrist during July, 1959.
Univ. Cape Town, Oceanogr. Dept., Data Rept., 27 pp.

density (data only)

Zoutendyk, P., and D. Sacks, 1969
Hydrographic and plankton data, 1960-1965.
Data Rept., Inst. Oceanogr., Univ Cape Town, (3): 52 pp.

density (data only)

Zvereva, A.A., Edit., 1959.
(Data, 2nd Marine Expedition, "Ob", 1956-1957.)
Arktich. i Antarkt. Nauchno-Issled. Inst., Mezhd. Geofiz. God, Sovetsk. Antarkt. Exped., 6: 1-387.

density, determination of

Crease, J., 1971
Determination of the density of seawater.
Nature, Lond. 233(5318): 329

density, effect of

density, effect of

Barr, D.I.H., 1961.
Some aspects of densimetric exchange flow.
Dock & Harbour Authority, 42(494):253-258.

density, effect of

Hanaichenko, N.K., 1960
(On the influence of water density on the sea level deviation from geoidal surface) Meteorol i Gidrol. (3): 35-37.

density, effect of

Khanaychenko, N.K., 1960.
(The influence of the density of water on deviations of the sea level from the surface of the geoid.)
Meteorol. i Gidrol., (3):35-37.

density, effect of

Kuwahara, S., 1938
Correction to the echo-depth for the density of water in the Pacific Ocean.
Japan. J. Astron. & Geophys., 14(2/3): 43-78.

density, effect of

Peters, Nicolaus 1934.
Die Bevolkung des Sudatlantischen Ozeans mit Ceratien. Biol. Sonderuntersuchungen 1.
Wiss. Ergeb. dt. Atlant. Exped. Meteor 1925-1927, 12(1):1-69

density, effect of

Rose, D., 1962
The influence of the closed sea-arms on the water level and current off the coast of the Netherlands and in the new waterway.
Proc. Symposium on Mathematical-Hydrodynamical Methods of Phys. Oceanogr., Sept. 1961, Inst. Meeresk., Hamburg: 227-232.

density, effect of

Takano, Kenzo, 1965.
General circulation due to the horizontal variation of water density with the longitude, maintained at the surface of an ocean.
Rec. Oceanogr. Wks., Japan, 8(1):1-11.

density, effect of

Tolstoy, I., 1965.
Effect of density stratification on sound waves.
J. geophys. Res., 70(24):6009-6015.

density, effect of

Wyrtki, Klaus, 1961
The thermohaline circulation in relation to the general circulation in the oceans.
Deep-Sea Res., 8(1): 39-64.

density field

density field

Gurikova, Z.F., 1964
The formation of the density field and the calculation of non-stationary currents in the Pacific Ocean. (In Russian).
Okeanologiia, Akad. Nauk, SSSR, 4(5):911-912.

Density, gradients

density gradients

Dietrich, G., 1954.
Einfluss der Gezeitenstromturbulenz auf die hydrographische Schichtung der Nordsee.
Arch. Met., Geophys., Bioklim., A, 7:391-405, 6 textfigs.

density gradient, vertical

Mandelbrot, L., 1965.
Le nombre de Richardson et les criteres de stabilite des ecoulements stratifies.
La Houille Blanche, 20(1):24-28.

density interface, effect of

density interface, effect of

Turner, J.S., 1965.
The coupled turbulent transports of salt and heat across a sharp density interface.
Int. J. Heat Mass Transfer, 8:759-767.

density interface

Turner, J.S., T.G.L. Shirtcliffe and P.G. Brewer, 1970.
Elemental variations of transport coefficients across density interfaces in multiple-diffusive systems.
Nature, Lond. 228 (5276): 1083-1084.

Density, inversions

density inversions

Cooper, L.H.N., 1967.
Stratification in the deep ocean.
Sci. Prog., Lond., 55(217):73-90.

density inversions

Dunbar, M.J., 1958.
Physical oceanographic results of the "Calanus" Expeditions in Ungava Bay, Frobisher Bay, Cumberland Sound, Hudson Strait and Northern Hudson Bay, 1949-1955.
J. Fish. Res. Bd., Canada, 15(2):155-201.

density inversions

Dunbar, M.J., 1955.
Density inversions in the sea in regions of high tidal activity. (Abstract).
Proc. R. Soc., Canada, 1955:63-64.

density inversions

Ehrlich, A., 1954.
Subsurface density inversions off Nantucket Island. Trans. Amer. Geophys. Union, 35(4):573-584.

density inversions

Tait, J.B., 1957.
XVIII. Recent oceanographic investigations in the Faroe-Shetland Channel. Proc. R. Soc., Edinburgh, 64(3):239-289.

Density, mean

density, mean

Dannevig, A., 1956.
Hydrography of the Norwegian Skagerak coast.
Ann. Biol., Cons. Perm. Int. Expl. Mer, 11:52.

Density, mean surface (data only)

Korea, Republic of, Fisheries Research and Development Agency, 1964.
Oceanographic Handbook of the Neighboring Seas of Korea, 214 pp.

density of organisms

density of organisms/m3

Marr, J.W.S., 1962.
The natural history and geography of the Antarctic krill (Euphausia superba Dana).
Discovery Repts., 32:33-464, Pl. 3.

Density-O_2 correlation

density-oxygen ratio

Kawamoto, T., 1957.
(On the distribution of dissolved oxygen in the Pacific Ocean. Pt. 3. On the σ_t-O_2 diagram in the South Pacific Ocean and the general summary.)
Umi to Sora, 33(1/2):28-33.

density-oxygen diagrams

Kawamoto, T., 1956.
(On the distribution of the dissolved oxygen in the Pacific Ocean. 2. On the σ_t-O_2 diagram in the equatorial region and the eastern region of the North Pacific Ocean.) Umi to Sora, 32(5/6):92-98.

density-oxygen correlation
Kawamoto, T., 1954.
[On the dissolved oxygen in the sea. On the σ_t ~ O2 diagram in the Japan Sea.]
Spec. Publ. Japan Sea Reg. Fish. Res. Lab., 3rd Ann., 125-133.

density-oxygen diagram
Kawamoto, T., 1955.
On the distribution of the dissolved oxygen in the Pacific Ocean. Pt. 1. On the σ_t-O_2 diagram in the western North Pacific. Umi to Sora 32(2):23-37, 14 textfigs.

density-O2
Kitamura, H., 1955
[Sigma t-dissolved oxygen diagram. The study of horizontal σ_t-O2 relation to water mass separation.] Umi-to-Sora 32(3):51-56.

sigma -t - oxygen
Rochford, D. J., 1966.
Source regions of oxygen maxima in intermediate depths of the Arabian Sea.
Australian J. Mar. freshw. Res., 17(1):1-30.

density-oxygen
Rochford, D.J., 1958.
Characteristics and flow paths of the intermediate depth waters of the southeast Indian Ocean.
J. Mar. Res., 17:483-504.

density-phosphorus

density-phosphorus
Australia, C.S.I.R.O., Division of Fisheries and Oceanography, Marine Biological Laboratory, Cronulla, 1959.
F.R.V. "Derwent Hunter", Scientific report of Cruise DH9/57, Aug. 19-25, 1957; Cruise DH10/57, Sept. 4-11, 1957; Cruise DH11/57, Sept. 18-21 1957; DH12/57, Sept. 26-Oct. 11, 1957.
CSIRO, Division of Fisheries and Oceanography, Rept. No. 20:20 pp. (mimeographed).

density, potential

density, potential
Cooper, L.H.N., 1963.
Essay review: Georg Wust, "Quantitative Untersuchungen zur Statik und Dynamik des atlantischen Ozeans; sechste Lieferung: Stromgeschwindigkeiten und Strommengen in den Tiefen des atlantischen Ozeans". Wiss. Ergebn. deutsch. atlant. Exped. "Meteor" 1925-1927, 6(2):260-420. 1927.
Journal du Conseil, 28(2):316-320.

density, potential
Lynn, R.J., 1971.
On potential density in the deep South Atlantic Ocean. J. mar. Res. 29(2): 171-177.

density profiles

density profiles
Clark, Charles B., Philip J. Stockhausen and John F. Kennedy, 1967.
A method for generating linear density profiles in laboratory tanks.
J. geophys. Res., 72(4):1393-1395.

density stratification
*Kenari, Seiichi, 1966.
Some experiments on the sedimentation in estuaries with density stratification.
Spec. Contr. geophys. Inst. Kyoto Univ., 6:127-133.

Density spectra

density spectra
Fomin, L.M., and A.D. Yampolsky, 1966.
The relationship between velocity and pressure (or density) spectra in the ocean. (In Russian; English abstract).
Fizika Atmosferi i Okeana, Izv., Akad. Nauk, SSSR, 2(6):656-663.

density standards
Menaché, Maurice, 1971.
Vérification, par analyse isotopique, de la validité de la méthode de Cox, McCartney et Culkin tendant à l'obtention d'un étalon de masse volumique. Deep-Sea Res. 18(5): 449-456.

density stratification
Monin, A.S., V.G. Neiman and B.N. Filiushkin 1970.
Density stratification in the ocean. (In Russian).
Dokl. Akad. Nauk SSSR, 191 (6): 1277-1279.

density stratification
Ramberg, Hans, 1972
Theoretical models of density stratification and diapirism in the earth. J. geophys. Res. 77(5): 877-889.

density stratification
SIMPSON J.H. 1971.
Density stratification and microstructure in the western Irish Sea. Deep-Sea Res., 18(3): 309-319.

density, surface

density surfaces
Berrit, G.R., 1961
Etude des conditions hydrologiques en fin de saison chaude entre Pointe-Noire et Loanda.
Cahiers Oceanogr., C.C.O.E.C., 13(7): 456-461.

density, surface
Collignon, J., 1955.
Observations hydrologiques sur les eaux superficielles de la région de Pointe-Noire.
Cons. Perm. Inst. Expl. Mer, Rapp. Proc. Verb., 137:7-10.

density, surface
Khan, M.A. and D.I. Williamson, 1970.
Seasonal changes in the distribution of chaetognatha and other plankton in the eastern Irish Sea. J. exp. mar. Biol. Ecol., 5(3): 285-303.

density, surface
Muromtsev, A.M., 1957.
[New maps of the distribution of salinity and density of the surface of the Pacific Ocean.]
Meteorol. i Gidrol., (3):15-

density, surface
Murty, A.V.S., and P. Udaya Varma, 1964.
The hydrographical features of the waters of Palk Bay during March 1963.
J. Mar. Biol. Ass., India, 6(2):207-216.

density, surface
Reid, Joseph L., Jr., 1961
On the temperature, salinity and density differences between the Atlantic and Pacific oceans in the upper kilometre.
Deep-Sea Res., 7(4): 265-275.

density, surface
Stevenson, Merritt R., Oscar Guillen G., and José Santoro de Ycaza, 1970
Marine atlas of the Pacific coastal waters of South America.
Univ. Calif. Press. 20pp., 99 charts, $40.00.

density surfaces
*Turner, J.S., 1968.
The influence of molecular diffusivity on turbulent entrainment across a density surface.
J. fluid Mech. 33(4):639-656.

density, surface
USN Hydrographic Office, 1957.
Oceanographic atlas of the polar seas. 1. Antarctic. H.O. Pub. No. 705:70 pp.

density tables

density tables
Morcos, Selim A., 1967.
Sigma-t (σ_t) tables for sea water of high salinity (from 38°/oo to 48°/oo).
Beitr. Meereskunde, 21:5-14.

density tables
Oren, O.H., 1952.
Tables for computation of sigma-t, for salinities from 38.00 o/oo to 41.00 o/oo and for temperatures from 13.00°C. to 31.00° C.
Proc. Tech. Pap. Gen. Fish. Council, Medit.:50-65

density tables
U.S. Hydrographic Office, 1956.
Tables for rapid computation of density and electrical conductivity of sea water.
H.O. Pub., No. 619:24 pp.

density variation, effect of

density variations (annual)
*Meincke, Jens, 1967.
Die Tiefe der jahreszeitlichen Dichreschwankungen im Nordatlantischen Ozean.
Kieler Meeresforsch., 23(1):1-15.

density variation, effect of
Uusitalo, S., 1960.
The numerical calculation of wind effect on sea level elevations.
Tellus, 12(4):427-435.

deposit feeders
Hargrave, Barry T. 1970.
The effect of a deposit-feeding amphipod on the metabolism of benthic microflora
Limnol. Oceanogr. 15(1):21-30.

deposit feeders
Howell, B.R. and R.G.J. Shelton 1970.
The effect of China clay on the bottom fauna of St. Austell and Mevagissey bays.
J. mar. biol. Ass. U.K. 50(3): 593-607.

deposit feeders
Nicol, David, 1969.
Deposit-feeding pelecypods in Recent marine faunas. Trans. Gulf Coast Ass. geol. Socs. 19 423-424.

deposition

deposition, salt
Borchert, Hermann 1969.
Principles of oceanic salt deposition and metamorphism.
Bull. Geol. Soc. Am., 80(5): 821-864

deposition
Doyle, Larry J., William J. Cleary and O.H. Pilkey, 1968.
Mica: its use in determining shelf-depositional regimes.
Marine Geol., 6(5): 381-389

deposition, sand
Haruta, Tadao, 1961
Recent coastal processes in Niigata Prefecture.
Coastal Eng., Japan, 4:73-83.

deposition, carbonate

Olausson, Eric, 1967.
Climatological geoeconomical and paleooceanographical aspects on carbonate deposition.
Progress in Oceanography, 4:245-265.

deposition

Schmalz, Robert F. 1969.
Deep-water evaporite deposition: a genetic model.
Bull. Am. Ass. Petr. Geol., 53(4): 798-823.

deposition

Scruton, P.C., 1953.
Deposition of evaporites.
Bull. Amer. Soc. Petr. Geol., 37(11):2498-2512.

deposition of salts

Thompson, T.G., and K.H. Nelson, 1956.
Concentration of brines and deposition of salts from sea water under frigid conditions.
Am. J. Sci., 254(4):227-238.

depressions

Cox, David O., 1971.
Depressions in shallow marine sediment made by benthic fish.
J. Sedim. Petrol. 41 (2): 577-578.

depth

Cloet, R.L., 1970.
How deep is the Sea?
Navigation, R. geogr. Soc. 23(4):416-425

depth, baroclinic

depth, baroclinic

Shkudova, G. Ya., 1963.
The calculation of baroclinic depth. (In Russian)
Meteorol. i Gidrol., (6):35-40.

Depth contours

depth correction

depth correction

Dorrestein, R., 1951.
Berekening van dieptocorrecties bij oceanographische seriewaarnemingen uit de gemeten draadhoek. K. Nederl. Met. Inst., Afd. Ocean. Marit. Met., Rapp. 1:1-7 (mimeographed).

depth corrections

Koczy, F.F., 1956.
Korrektion av Djupbestämning med ekolod.
Fish. Bd., Sweden, Ser. Hydrogr., Rept. 7:9 pp.

depth determination

depth determination

Beach Erosion Board, 1948.
Proof test of water transparency method of depth determination. Eng. Notes No. 29:29 pp., 34 figs. data sheets (multilith).

depth determination, errors in

Bialek, Eugene L., 1966.
Errors in the determination of depth by pressure gauges utilizing a linear pressure-depth relationship.
Int. Hydrogr. Rev., 43(1):69-74.

depth

Katz, Eli Joel, 1963
A statistical model of the oceans' variable depth.
Deep-Sea Res., 10(1/2):11-16.

depth determination

Krause, Gunther, and Gerold Siedler, 1962
Zur kontinuierlichen Bestimmung der Tiefenlage von Schleppgeräten im Meere.
Kieler Meeresf., 18(1):29-33.
German and English summary.

depth determination

LeFur, A., 1952.
Note sur la détermination du gradient des plages par examen de la houle sur les photographies aériennes. Ann. Hydrogr., Paris, 4th ser., 3:167-176, Pls. 35-38.

depth determination

Moore, J. G., 1947.
The determination of the depths and extinction coefficients of shallow water by air photograph using colour filters. Phil. Trans. Roy. Soc., London, Ser. A., Math. Phys. Sci., 240(816)B 163-217.

depth determination

Nakai, J., 1935

Determination of the unprotected thermometer depth. Umi to Sora, 15(2): 53-55.

depth determination

Nakai, J., 1935

Two types of correction diagrams of an unprotected reversing thermometer. Umi to Sora 15(2): 56-60.

depth determination

Starshinova, E.A., and I.N. Galkin, 1960.
[Tests on the utilization of explosive sources of sound for the determination of the depths of the ocean.]
Trudy Okeanograf. Komissii, Akad. Nauk, SSSR, 10(1):152-169.

depth determinations

Suda, K., 1932
On the exact determinations of the depth of the observed layers by means of the non-protected reversing thermometer. J. Oceanogr., Kobe Mar. Obs., 3(3):708-714.

depth determination

Suda, K., 1931
On the exact determination of the depth of the observed layers by means of non-protected reversing thermometers. J. Oceanogr., Kobe Mar. Obs., 2(4):687-692.

depth determination

Teramoto, T., 1958.
Depth determination in oceanographic observations
Geophys. Notes, Tokyo, 11(1):No. 12.

depth determination

Teramoto, T., 1958.
Depth determination in oceanographic observations
Rec. Oceanogr. Wks., Japan, n.s., 4(2):170-185.

depth determination

Whitney, G.G., 1957.
Factors affecting the accuracy of thermometric depth determinations. J. du Cons. 22(2):167-173.

depth, effect of

depth, effect of

Fooks, V.R., and K.T. Bogdanov, 1965.
On the causes of variations of tidal current characteristics with depth. (In Russian).
Okeanologiia, Akad. Nauk, SSSR, 5(1):63-72.

depth, effect of

Ivanov, Y.A., 1965.
The role of boundary conditions and advection in the formation and distribution of extreme values of oceanological characteristics according to depth.
(In Russian).
Okeanologiia, Akad. Nauk, SSSR, 5(1):40-44.

depth, effect of

Menzies, Robert J., Robert Y. George and Gilbert Rowe, 1968.
Vision index for isopod Crustacea related to latitude and depth.
Nature, Lond., 217(5123):93-95.

depth, effect of

Perrone, Anthony J., 1970.
Ambient-noise-spectrum levels as a function of water depth.
J. acoust. Soc. Am., 48(1-2): 362-370

depth, effect of

Schinke, Hans, 1961
[Model-tank investigations on the mutual influence of wave and sea-bottom.] (In Polish; English summary).
Prace Inst. Morsk., Gdansk. (1) Hydrotech. II. Sesja Naukowa Inst. Morsk., 20-21 wrzesnia 1960:105-128; English summary 149.

depth, effect of

Števčić, Zdravko, 1969.
Spiny spider crab in relation to the depth.
(In Jugoslavian; Italian and English abstracts).
Thalassia Jugoslavica, 5: 353-360.

depth, effect of

Tsyplukhin, V.F., 1963.
The results of instrumental study of wave attenuation with depth in the sea. (In Russian).
Okeanologiia, Akad. Nauk, SSSR, 3(5):833-839.

depth, effect of

Varjo, Uuno, 1969.
Über Riffbildungen und ihre Entstehung an den Küsten des Sees Oulujärvi (Finnland).
Die Küste, 17:51-80.

depth indicators

depth indicators

*Allen, J.R.L., 1967.
Depth indicators of clastic sequences.
Marine Geol., 5(5/6):429-446.

depth indicators

*Bathurst, R.G.C., 1967.
Depth indicators in sedimentary carbonates.
Marine Geol., 5(5/6):447-471.

depth indicators

*Bromley, R.G., 1967.
Marine phosphorites as depth indicators.
Marine Geol., 5(5/6):503-509.

depth indicators

*Funnell, B.M., 1967.
Foraminifera and Radiolaria as depth indicators in the marine environment.
Marine Geol., 5(5/6): 333-347.

depth indicators

*McAlester, A-Lee, and Donald C. Rhoads, 1967.
Bivalves as bathymetric indicators.
Marine Geol., 5(5/6):383-388.

depth indicators

*Nichols, G.D., 1967.
Trace elements in sediments: an assessment of their possible utility as depth indicators.
Marine Geol., 5(5/6):539-555.

depth indicators
*Porrenza,D.H., 1967.
Glauconite and chemosite as depth indicators in the marine environment.
Marine Geol., 5 (5/6): 495-501.

depth indicators
*Price,N.B., 1967.
Some geochemical observations on manganese-iron oxide nodules from different depth environments.
Marine Geol., 5(5/6):511-538.

depth indicators
*Wells, J.W.,1967.
Corals as bathometers.
Marine Geol., 5(5/6):349-365.

depth indicators
*Williams,David B., and William A.S.Sarjeant, 1967.
Organic-walled microfossils as depth and shore-line indicators.
Marine Geol., 5(5/6):389-412.

depths, maximal

depths (maximal)
Hanson(Ganson), P.P., G.B. Udintsev, et al., 1958, 1959.
Maximal depths of the Pacific Ocean and discovery of a deep-sea trough in the western part of the Pacific Ocean. (In Russian).
Priroda, 47(7):85-88 and 48(6):84-87.

Translation:
OTS or SLA, $1.60.

depths, maximal
Hanson, P.P., N.L. Zenkevich and I.V. Sergeev, 1959.
Maximal depths of the World Ocean. (In Russian).
Priroda, (6):84-88.

depth of no meridional motion

depth of no motion
Ichiye, Takashi, 1960
A note on determination of the depth of no meridional motion.
J. Oceanogr. Soc., Japan, 16(1): 177-179.

depth of no motion
Ichiye, T., 1959.
A note on determination of the depth of no meridional motion. J. Oceanogr. Soc., Japan, 15(4): 177-179.

depth of no motion
Latun, V.S., 1963
Vertical movements at the depth of a zero surface in the Atlantic Ocean. (In Russian).
Okeanologiia, Akad. Nauk, SSSR, 3(2):206-212.

depth of no motion
Masuzawa, Jotaro, 1961
Preliminary report of the Japanese Expedition of Deep-Sea, the Third Cruise (JEDS-3).
Oceanogr. Mag., Tokyo, 12(2):207-218.

depth of no meridional motion
Stommel, H., 1956.
On the determination of the depth of no meridional motion. Deep-Sea Res. 3(4):273-278.

Desalting. see also "fresh water from sea"

desalination
Abelson, Philip H., 1964.
Desalination of water. (Editorial).
Science, 146(3651):1533.

desalination
American Chemical Society, 1964
Saline water conversion, II.
American Chemical Society, 199 pp. ($6.00).

desalination
Anon.,1966.
Saline-water conversion.
J. A. Wat.Wks. Ass., 58(10):1231-1237.

desalination
Anon., 1964.
Dual-purpose plant to desalt water, generate electricity.
Chem. & Eng. News, 42(29):50.

desalination
Anon., 1964.
Nuclear power-water desalting combinations possible by 1975.
Chemical & Engineering News, 42(15):86-88.

desalination
Barduhn,Allen J.,1967.
Fresh water from the sea.
Oceanology Int.,2(2):28-31.

desalination
Belyakov, L.N., 1969.
Summer desalination of the sea surface and dynamics of the drift current and ice movement in the Arctic Basin.
Okeanologiia 9(3): 424-429.
(In Russian: English abstract)

DESALINATION
Blanc, F. et M. Leveau, 1970.
Effets de l'eutrophie et de la dessalure sur les populations phytoplanctoniques. Marine Biol., 5(4): 283-293.

desalination
Brian, P.L.T. 1967.
Water desalination by freezing.
Ocean Indust. 2(8): 34-35 (popular).

desalinization
Brice, Donat B., 1964.
Saline water conversion by flash evaporation utilizing solar energy.
In: Saline water conversion, II, Advances in Chemistry Series, 38:99-116.

desalinization
Briggs, Frederick A., and Allen J. Barduhn, 1963.
Properties of the hydrates of fluorocarbons 142b and 12B1. Comparison of six agents for use in the hydrate process.
In: Saline water conversion, II. Advances in Chemistry Series, 38:190-199.

desalination
Cain, Stanley, 1967.
Desalination and water pollution control.
Undersea Techn., 8(1):36-39,62.

desalination
Calmon,C., and A.W. Kingsbury,1966.
Preparation of ultrapure water.
In: Principles of desalination,K.S. Spiegler, editor,Academic Press, 441-495.

desalination
Clawson, Marion, Hans H. Landsberg and Lyle T. Alexander, 1969.
Desalted seawater for agriculture: is it economic?
Science, 164(3884):1141-1148.

desalinization
Dodge, Barnett F., 1963.
Review of distillation processes for the recovery of fresh water from saline waters.
In: Saline Water Conversion II, Advances in Chemistry Series, 38:1-26.

desalination
Dodson, Roy E., and Stewart E. Mulford, 1965.
Use of distilled sea water at San Diego.
J. Am. Water Wks. Assoc., 57(9):1106-1112.

desalination
Durante, Raymond W. 1967.
U.S. desalting program moving into high gear.
Ocean Industry 2(4):64-66.

desalination
Emilsson, Ingvar 1968.
Investigaciones sobre la hidrologia en la ensenada de la Broa con vista a su posible transformación en un embalse de agua dulce.
Ser. Transformación Naturaleza, Acad. Cienc. Cuba, Inst. Oceanogr. 5:1-45.

desalination
Evans,Robert B., Gary L. Crellin and Myron Tribus,1966.
Thermoeconomic considerations of sea water demineralization.
In: Principles of desalination,K.S.Spiegler, editor,Academic Press,21-76.

desalination
Geller, S. Yu., 1964.
The role of desalting for the water balance of dry territories (compiled for the needs of economy). (In Russian).
Izv., Akad. Nauk, SSSR, Ser. Geogr., (1):24-35.

desalination
Gilliam,W.S., 1968.
Materials problems in desalination.
Ocean Engng., 1(2):137-142.

desalination
Gillam,W.S., and W.H. McCoy,1966.
Desalination research and water resources.
In: Principles of desalination, K.S. Spiegler, editor,Academic Press,1-20.

desalination
Girelli, Alburto, editor, 1965.
Fresh water from the sea. Proceedings of the International Symposium held in Milan by Federazione delle Associazioni Scientifiche e Techniche and Ente Autonomo Fiera di Milano.
Pergamon Press and Tamburini Editore, 179 pp.

desalination
Glueckauf, E. 1966.
Sea water desalination - in perspective.
Nature, 211 (5055): 1227-1230.

desalinization
Gould, Robert F., 1963.
Saline water conversion, II.
Advances in Chemistry Series, Amer. Chem. Soc., Appl. Publ. 38:199 pp.

desalination
Hassan, El Sayed Mohamed, 1963
Oceanic hydrostatic pressure as a possible source of energy for desalting water.
Deep-Sea Res., 10(1/2): 33.

desalinization
Herbert, L.S., and U.J. Sterns, 1963.
Saline water distillation-scaling experiments using the spray evaporator technique.
In: Saline water conversion, II, Advances in Chemistry Series, 38:52-64.

desalination
Hood, Donald W., and Richard R., 1960.
The place of solvent extraction in saline water conversion.
Advances in Chemistry Series, No. 28:40-49.

desalination
Howe, E.D., 1952.
Fresh water from salt water. Trans. Amer. Geophys. Union 33(3):417-422, 5 textfigs.

desalination
Hupfer, Peter, 1966.
Ozeanologische Aspekte der industriellen Meerwasserentsalzung.
Beitr. Meereskunde, 19:71-77.

desalination
Johnson, James S., Jr., Lawrence Dresner, and Kurt A. Kraus, 1966.
Hyperfiltration (reverse osmosis).
In: Principles of desalination, K.S. Spiegler, editor, Academic Press, 345-439.

desalinization
Katz, William E., 1963.
Design of medium-to-large electrodialytic water demineralizers. Several hundred thousand to several million gallons per day.
In: Saline water conversion, II, Advances in Chemistry Series, 38:158-167.

desalination
Keilin, Bertram, 1967.
Water desalination by reverse osmosis.
Ocean Indust. 2(8): 25-33 (popular)

desalinization
King, C., 1961.
Desalinization of the soil of tidal land in Taiwan. (Abstract).
Tenth Pacific Sci. Congr., Honolulu, 21 Aug.-6 Sept., 1961, Abstracts of Symposium Papers, 44

desalination
Koenig, Louis, 1966.
The cost of conventional water supply.
In: Principles of desalination, K.S. Spielger, editor, Academic Press, 515-550.

desalinization
Lacey, R.E., E.W. Lang and E.L. Huffman, 1963.
Economics of demineralization by electrodialysis.
In: Saline water conversion, II, Advances in Chemistry Series, 38:168-178.

desalination
Levine, Summer N., 1967.
Desalination and ocean technology.
Dover Publ. $4.00 (not seen)

desalination
Löf, George O.G., 1966.
Solar distillation.
In: Principles of desalination, K.S. Spiegler, editor, Academic Press, 151-198.

desalinization
Loeb, Sidnet, 1963.
Sea water demineralization by means of an osmotic membrane.
In: Saline water conversion, II, Advances in Chemistry Series, 38:117-132.

desalinization
Lotz, Charles W., 1963,
Saline water conversion using wiped thin-film distillation.
In: Saline water conversion, II, Advances in Chemistry Series, 38:78-85.

desalinization
Lustenader, E. L., 1963.
Saline water conversion by the diffusion still.
In: Saline water conversion, II, Advances in Chemistry Series, 38:86-98.

desalinization
McIlhenny, W.F., 1963.
Minimizing scale formation in saline water evaporators.
In: Saline water conversion, II, Advances in Chemistry Series, 38:40-51.

desalination
Messing, T. and E. Lassar, 1968.
Production of potable water and sea salt.
Chemical and Process Engineering, 49(8):59-70. (not seen).

desalination
Munro, C.H., and J.B. Clampett, 1962
The economics of de-salination for stock watering.
Water Res. Found., Australia, Rept. No. 5: 44 pp., appendices, figs.

desalination
Palmork, Karsten H., 1963.
Studies of the dissolved organic compounds in the sea.
Fiskeridirekt. Skrift., Ser. Havundersøgelser, 13(6):120-125.

desalination
Porter, J. Winston, 1967.
Water desalination by distillation.
Ocean Indust. 2(8): 39-45 (popular)

desalinization
Raben, Irwin A., George Commerford and Gale E. Nevill, Jr., 1963.
Effect of vibration on heat transfer and scaling in saline water conversion.
In: Saline water conversion, II, Advances in Chemistry Series, 38:65-77.

desalination
Rideal, Eric K., 1965.
Desalination of sea water.
Nature, 207(5002):1115.

desalinization
Salutsky, Murrell L., and Maria G. Dunseth, 1963
Recovery of minerals from sea water by phosphate precipitation.
In: Saline water conversion II, Advances in Chemistry Series, 38:27-39.

desalination
Schaefer, Heinz, 1964.
Beiträge zur Entsaalzung mit Retardion 11A8.
Helgoländer Wiss. Meeresuntersuchungen, 11(3/4): 301-322.

desalination
Schuessler, Raymond 1971.
Desalting the seas: a step toward world peace.
Oceans Mag. Menlo Park, Calif. 4(5): 64-69. (popular)

desalination
Scroggs, Schiller, 1968.
Desalination effluent.
Sea Frontiers, 14(1):11-16. (popular).

desalination
Sheffer, L.H., and M.S. Mintz, 1966.
Electrodialysis.
In: Principles of desalination, K.S. Spiegler, editor, Academic Press, 199-289.

desalination
Silver, R.S., 1966.
Distillation.
In: Principles of desalination, K.S. Spiegler, editor, Academic Press, 77-115.

desalination
Snyder, A.E., 1966.
Freezing methods.
In: Principles of desalination, K.S. Spiegler, editor, Academic Press, 291-343.

desalination (fresh water from salt)
Snyder, A.E., 1962.
Desalting water by freezing.
Scientific American, 207(6):41-47.

desalination
Spiegler, K.S., editor, 1966.
Principles of desalination.
Academic Press, 566 pp.

desalinization
Spiegler, K.S., 1963.
Saline water research in Isreal.
In: Saline water conversion, II, Advances in Chemistry Series, 38:179-189.

desalination
Sporn, Philip, 1966.
Fresh water from saline waters: the political, social, engineering and economic aspects of desalination.
Pergamon Press, 34 pp.

desalinization
Thompson, Lewis, and Carl N. Hodges, 1963
Solar radiation, water demand and desalination.
Solar Energy, 7(2):79-80.

desalting
Thompson, T.G., and K.H. Nelseon, 1954.
Desalting sea water by freezing.
Refrig. Engin., July 1954.

desalinization
USA, American Chemical Society, 1964.
Improved process to convert saline water
Chemical and Engineering News, 42(25):50-52.

desalination
United States, Department of the Interior, 1967.
Proceedings of the First International Symposium on Water Desalination, Washington, D.C. 3-9 October 1965, 1: 632 pp: 2:830 pp: 3:792 pp/

desalination
United States, Department of the Interior, 1964.
1963 saline water conversion report. Office of Saline Water, U.S. Gov't. Printing Office, 187 pp $1.50.

desalination
*Untersteiner, N., 1967.
Natural desalination and equilibrium salinity profile of old sea ice.
In: Physics of snow and ice, H. Oura, editor, Inst. Low Temp.Sci., Hokkaido Univ., 569-577.

desalination
*Untersteiner, Norbert, 1968.
Natural desalination and equilibrium salinity profile of perennial sea ice.
J. geophys.Res., 73(4):1251-1257.

desalinization
Volckman, O.B., 1963.
Operating experience on a large scale electrodialysis water demineralization plant.
In: Saline water conversion, II, Advances in Chemistry Series, 38:133-157.

desalination
Vulfson, V.I., 1967.
The problem of fresh water and methods of obtaining it from sea water. (In Russian; English summary).
Okeanologiia, Akad. Nauk, SSSR, 7(1):3-16.

desalination
Walsh, John, 1965.
Desalination: emphasis is on dual purpose nuclear power and desalting plants.
Science, 147(3662):1117-1119.

desalination
Woodward, Teynham, 1966.
Vapor Reheat Distillation.
In: Principles of desalination, K.S. Spiegler, editor, Academic Press, 117-150.

desalination
Woolrich, W.R., 1962
Ten years progress in desalinization of sea water by freezing in the laboratories of the United States.
Texas J. Sci., 14(4):480-487.

desalination
York, J. Louis and Bernard J. Schorle, 1966.
Scale formation and prevention.
In: Principles of desalination, K.S. Spiegler, editor, Academic Press, 497-514.

desiccation

desiccation.
Bajard, Jacques 1966.
Figures et structures sédimentaires dans la zone intertidale de la partie orientale de la Baie du Mont-Saint-Michel.
Rev. Géogr. phys. et Géol. dyn. (2) 8(1):39-111.

dessication, effect of
Foster, B.A., 1971
Desiccation as a factor in the intertidal zonation of barnacles. Marine Biol., 8(1): 12-29.

detection

detection
Ahrens, E., 1957.
Use of horizontal sounding for wreck detection.
Int. Hydrogr. Rev., 34(2):73-81.

detection underwater, sound
Albers, Vernon M., 1965.
Underwater acoustics handbook II.
The Pennsylvania State University Press, 356 pp. $12.50.

detection
Anon., 1957.
Asdic's future in fish finding. World Fish., London, 6(4):40-41.

detection
Backus, R.H., and J.B. Hersey, 1956.
Echo-sounder observations of mid-water nets and their towing cables. Deep-Sea Res. 3(4):237-241.

detection
Born, N. and B.W. Conoly, 1970.
Zero-crossing shift as a detection method.
J. acoust. Soc. Am., 47(5-2):1408-1411

detection
Chigusa, M., M. Takase and M. Hirose, 1955.
[Detection of the withered-tree below the surface by the ultrasonic echo-sounder.]
J. Shimonoseki Coll. Fish. 4(2):275-278, 5 text-figs.

detection (sound)
Dietz, R.S., and M.F. Sheehy, 1953.
TransPacific detection by underwater sound of Myojin volcanic explosions.
J. Ocean. Soc., Japan, 9(2):53-83, 6 textfigs.

Also published:
Bull. G.S.A. 65(10):941-956, 4 pls. 4 textfigs. in 1954

detection
Genka, T., 1955.
[On the detective effect of the radar upon the location of the tunny-long-line (1).]
Mem. Fac. Fish., Kagoshima Univ., 4:25-30.

detection
Good, C.M., 1956
Asdic in the fishing indurtry. World Fishing, (1) 5(3):26-27; (2) 26(4): 2 pp.

detection
Haines, G., 1957.
Getting the best from an echosounder. 1.
World Fish. News, London, 5(12):30-33.

detection
Hashimoto, Tomiju, Yoshinobu Maniwa, Osamu Omoto and Hidekuni Noda, 1963
Echo sounding of frozen lake from surface of ice.
Int. Hydrog. Rev., 40(2):31-40.

detection
Hazlett, R.W.G., 1962.
Measurement of the dimensions of fish to facilitate calculations of echo-strength in acoustic fish detection.
Journal du Conseil, 27(3):261-269.

detection (by radar and sonar)
Ishida, M., T. Suzuki, N. Sano, I Saito and S. Michima, 1960
On the detection of the boundary zone. Bull. Fac. Fish. Hokkaido, U., 10(4): 291-302. (In Japanese-figures with English headings).

detection
Jagodzinski, Z., 1957.
[Multiple echoes in echosounders and the probability of detection of small targets.]
Acad. Polon. Sci., Com. Geod., 9 pp.

detection
Kawada, S., and C. Yoshimuta, 1959.
Fundamental study of the detection of fish by supersonic wave. II. Supersonic reflection coefficient of fishing net.
Bull. Tokai Reg. Fish. Res. Lab., No. 23:25-38.

detection
Kibblewhite, A. C., 1966.
The acoustic detection and location of an underwater volcano.
New Zealand J. Sci., 9(1):178-199.

detection
Kuroki, T., and M. Chuman, 1958.
[An example of three-dimensional records of fish-school attracted by underwater lamps.]
Mem. Fac. Fish., Kagoshima Univ., 6:77-81.

detection
Kuroki, T., and M. Chuman, 1958.
[Studies on the horizontal finding of fish school. III. About errors of "reading" on the records of general fish finder.] Mem. Fac. Fish., Kagoshima Univ., 3(2):25-28.

detection
Middleton, D., 1962.
Acoustic signal detection by simple correlators in the presence of non-Gaussian noise. I. Signal-to-noise ratios and canonical forms.
J. Acoust. Soc., Amer., 34(10):1598-1609.

detection
Owatari, A., K. Furumo and S. Matsumoto, 1953.
Plankton detected by fish-finder. (In Japanese)
Rept. Fish. Sect. Nagasaki Pref. Office, No. 6: 1-8.

detection
Schärfe, J., 1953.
Messung der Öffnung von Schleppnetzen mit Echolot
Fischwirtschaft, 5(12):282-284.

Translation: Fish. Lab., Aberdeen

detection
Stewart, J.L., E.C. Westerfield and M.K. Brandon, 1961
Optimum frequencies for active sonar detection.
J. Acoust. Soc., Amer., 33(9):1216-1222.

detection
Tavolga, William N., 1965.
Review of marine bio-acoustics: state of the art: 1964.
Techn. Rept. NAVTRADEVCEN 1212-1:100 pp.

detections, eruptions (submarine)
Norris, Roger A. and David N. Hart, 1970.
Confirmation of sofar-hydrophone detection of submarine eruptions. J. geophys. Res., 75(11): 2144-2147.

detection, fish

detection, fish
Ahnens, E., 1955.
Horizontaltotung im Fischfang.
Arch. f. Fischereiwissenschaft 5(3/4):229-240.

detection, fish
Alander, H., 1950.
Ekolodningens principer. Svenska Västkustfiskeren 20(1):4-6.

detection
Anon., 1957.
Detection of Fish, Echo Sounding
Trade News Dec. 1957.

(World Fisheries Abstracts Vol. 9, No. 5. Sept. - Oct. 1958 p. 1)

detection, fish
Anon., 1952.
Shows fish on cathode ray tube. Fish. Newsletter 11(11):15, 1 fig.

detection, fish
Anon., 1951.
Echo sounding gear in tuna tenders cuts time required for making bait. Pacific Fisherman 49(2): 23.

detection, fish

Anon., 1951.
Echo sounders identify species. Pacific Fisherman 49(2):47.

detection, fish

Anon., 1950.
Fischlotungen im Ausland. Fischereiwelt 2(6): 89-90, 5 figs.

detection, fish

Anon., 1950.
Echolood in dienst van de visserij.
De Visseriwereld 9(42):978-979.

detection, fish

Anon., 1950.
El "ojo electronico". Espana Pesquera 1(4):14

Translation of article originally appearing in Readers Digest!

detection, fish

Anon., 1950.
Echo sounders pinpoint individual fish; tell trollers depth at which to run gear.
Pacific Fisherman 48(12):25.

detection, fish

Anon., 1950.
La electronica actuando como localizadora de pesca. Industria Conservera 16(134):182-184.

Translation of article in Food Industries, 1950.

detection, fish

Anon., 1950.
Electronics takes "guess" out of fishing.
Food Industries 22(6):79-81.

detection (fish)

Arata, G., jr., 1955.
The use of a portable depth recorder for locating fish. State of Florida, Bd., Cons., Tech. Ser. 15:1-17.

detection, fish

Balls, R., 1952.
Echo sounders in commercial fisheries. World Fishing 1(3):76-80, figs.

detection, fish

Balls, R., 1948.
Herring fishing with the echometer. J. du Cons. 15(2):193-206, 5 figs.

detection, fish

Bass, George A., and Mark Rascovich, 1965.
A device for sonic tracking of large fishes.
Zoologica., N.Y. Zool. Soc., 50(2):75-81.

detection, fish

Batenburg, 1951.
De onderwater-acoustiek in haar toepassing op het echolood. Visserij-Nieuws 3(9):105-108, 3 figs.

detection, fish

Bolla, R.N.B., 1950.
Lo scandaglio ultrasonoro come ausilio per la pesca. Boll. Pesca 25(7/8):2-6.

detection, fish

Boudreault, Yves, 1967.
Essais de petits sondburs à ultra-sons.
Rapp.Stn Biol.Mar. Grande-Riviere, 1966:137-140.

detection, fish

Boudreault, Yves, 1965.
Recherches sur l'utilisation des sondeurs à ultra-sons pour la détection des poissons.
Rapp. Ann., Sta. Biol. Mar., Grande Rivière, 1964:113-118.

detection

Busnel, R. G., 1959.

Etude d'un appeau acoustique pour la peche, utilise au Senegal et au Niger.
Bull. de l'Inst. Franc. d'Afr. Noire. Ser. A 21(1):346-360.

detection (fish)

Chatoba, O.I., and M.N. Shcherbino, 1964.
On the use of fish finders. (In Russian).
Materialy, Ribochoz. Issled. Severn. Basseina, Gosudarst. Kom. Rib. Choz., SNCH, SSSR, Poliarn. Nauchno-Issled. i Proektn. Inst. Morsk. Rib. Choz i Okeanogr., N.M. Knipovich, (PINRO), 2:145-148.

detection, fish

Chuman, M., 1953.
Studies on the horizontal finding of fish school. 2. About the test of the apparatus by the contrived gain echo.
Mem. Fac. Fish., Kagoshima Univ., 3(1):65-70.

detection, fish

Curnock, D., 1948.
British test sub detector on herring fishing banks. Inter. Fisherman and Allied Workers 8(6): 10.

detection, fish

*Cushing, D.H., 1968.
Direct estimation of a fish population acoustically.
J.Fish.Res.Bd.,Can.,25(11):2349-2364.

detection, fish

Cushing, D.H., 1967.
The acoustic estimation of fish abundance.
In: Marine bio-acoustics, W.N.Tavolga, editor, Pergamon Press, 2:75-90.

detection

Cushing, D.H., Finn Devold, J.C. Marr, and H. Kristjonsson, 1952.
Some modern methods of fish detection: echo sounding, echo ranging and aerial scouting.
F.A.O. Fish. Bull. 5(3/4):95-119, 9 textfigs.

detection, fish

de Boer, P.A., 1952.
Echoloodproeven aan boord van het onderzoekingsvaartuig "Antoni van Leeuwenhoek". Visserij-Nieuws 5(8):91-94, 5 figs.

detection, fish

de Boer, P.A., 1952.
Het belang van een goed echolood de visserij.
Visserij-Nieuws 5(8):94-96.

de Boer, P.A., 1951. detection, fish
Het echolood in het brandpunt van de Internationale Belangstelling. Visserij-Nieuws 4(5): 54-56.

detection, fish

deBoer, P.A., 1950.
Ervaringen met het echolood tijdens ijle haringteelt op de Belgische kust. Visserij Nieuws 3(4):42-45, figs.

detection, fish

deBoer, P.A., 1949.
Opsporen van visscholen. Visserij Nieuws 2(6): 66-71, figs.

detection, fish

Degterev, A. A., 1964.
The results of exploiting the hydroacoustic technique in the purse-seine fishing for herring in the Norwegian Sea. (In Russian).
Material. Sess. Uchen. Sov. PINRO, Rez. Issled., 1962-1963, Murmansk, 165-168.

detection, fish

Dowd, R.G., E. Bakken and O. Nakken, 1970
A comparison between two sonic measuring systems for demersal fish.
J. Fish. Res. Bd., Can., 27(4):737-742

detection, fish

Eddy, D., 1949.
Electronic fish-finder. Sci. Newsletter 56(5): 74-76.

detection, fish

Eggvin, J., 1950.
Ekkoskreimelding unter Lofotfisket. Aarsberet. ved Norges Fisherier, 1949, 7:47-50.

detection (fish)

Furnestin, J., and R. Coupé, R. Gail, Ch. Maurin, and M. Rossignol, 1953.
Ultra-sons et pêche à la sardine au Maroc.
Bull. Inst. Pêches Marit., Maroc, No. 1:1-57, 8 textfigs., 39 echograms.

also anchovies.

detection, fish

Gabler, H., 1951.
Zur Echolotung von Fischschwärmen. Jahresheft Deutsch. Fischwirtschaft 1951:

detection (fish)

Ganjkov, A.A., and O.N. Kiselev, 1964.
The use of echo sounder when fishing under ice. (In Russian).
Materialy, Ribochoz. Issled. Severn. Basseina, Gosudarst. Kom. Rib. Choz., SNCH, SSSR, Poliarn. Nauchno-Issled. i Proektn. Inst. Morsk. Rib. Choz i Okeanogr., N.M. Knipovich, (PINRO), 2:148-151.

detection, fish

Golenchenko, A.P., 1959

[Use of aviation in the sea fish detection and fishery investigations.]
Biull. Okeanograf. Komissii. Akad. Nauk. SSSR., (3): 91-98.

detection, fish

Golenchenko, A.P., 1958.
Acoustic soundings for fish from a helicopter. (In Russian).
Priroda, (4):124-

detection, fish

Hashimoto, T., 1952.
[Studies on the ultrasonic echo-sounder fish-finder and its applications.] Fish. Agency, Tokyo, 150 pp. (English abstract, pp. 135-150.)

detection, fish.

Haslett, R.W.G., 1962.
Measurement of the dimensions of fish to facilitate calculations of echo-strength in acoustic fish detection.
J. du Conseil, 27(3):261-269.

detection, fish
Hodgson, W.C., 1950.
Opsporen en identificieren van visscholen.
De Visserijwereld 9(42):979.

detection, fish
Hodgson, W.C., 1950.
Locating and identifying fish shoals by sound echoes. S. African Shipping News and Fish. Ind. Rev. 5(7):61-62.

detection, fish
Hodgson, W.C., and I.D. Richardson, 1949.
The experiments on the Cornish pilchard fishery in 1947-48. Min. Agric. Fish., Fish. Invest., Ser. 2, 17(2):1-21, 7 figs., 10 pls.

detection, fish
*Inoue, Motoo, 1965.
Movement of fishing grounds for the albacore and their migration with oceanographic conditions in the northwestern Pacific Ocean. (In Japanese; English).
Rep.Fish.Res.Lab., Tokai Univ., 2(1):1-98.

detection, fish
Japan, Japanese Fisheries Agency, 1951.
[Study of the ultrasonic echo sounder fish finder and its application.] Fish. Boat Ser. II: 1-150. (English summary).

detection, fish
Johns, W., 1934.
Das Echolot in der Fischerei. Der Fischmarkt (10):256-258, 8 textfigs.

detection, fish
Jones, F.R. Harden, and B.S. McCartney, 1962.
The use of electronic sector-scanning sonar for following the movements of fish shoals: sea trials on R.R.S. "Discovery II".
J. du Cons., 27(2):141-149.

detection, fish
Judanov, K.I., 1966.
Forms of distribution of fish and their records with acoustical devices. (In Russian).
Trudy vses. nauchno-issled. Inst. morsk. ryb. Khoz. Okeanogr. (VNIRO), 60:173-188.

detection
Kimura, K., 1929.
On the detection of fish-groups by an acoustic method. J. Tokyo Univ., Fish., 24(2):41-45, 2 pls.

detection, fish
Koyama, T., 1962
Telemetric estimation of set net catches by fish finder. (In Japanese; English summary).
Bull. Tokai Reg. Fish. Res. Lab., Tokyo, No. 32: 141-148.

detection, fish
Krefft, G., and F. Schüler, 1951.
Beobachtungen über die Tiefenverteilung von Heringsschwärmen der nördlichen und mittleren Nordsee im August 1950. Fischereiwelt 3(6):93-95, 8 figs.

detection, fish
Laevastu, Taivo, 1969.
Effects of ocean thermal structure on fish finding with sonar.
FiskDir. Skr. Ser. HavUnders. 15(3): 202-209

detection, fish
Lawrence, E., jr., 1951.
Use of echo sounder in fisheries. Comm. Fish. Rev. 13(3):1-5, 1 fig.

detection, fish
Lea, E., 1947.
L'asdic au service de la peche. Peche Marit. 30: 826.

detection, fish
Luling, K.H., 1952.
Thunfisch Beobachtungen und Thunfischgang.
Fischereiwelt 4(2):10-12, figs.

detection, fish
Marchena, J.A.F., 1951.
A utilizacao das ultra-sonoras na pesca da sardinha. Bol. da Pesca (30):40-66.

detection, fish
Maughan, Paul M., 1969.
Remote-sensor applications in fishery research
J. mar. techn. Soc., 3(2):11-20

detection, fish
McCartney, Brian, 1967.
Underwater sound in oceanography.
In: Underwater acoustics, V.M. Albers, editor, Plenum Press, 2: 185-201.

detection, fish
McCartney, B.S., A.R. Stubbs, and M.J. Tucker, 1965.
Low-frequency target strengths of pilchard shoals and the hypothesis of swimbladder resonance.
Nature, Lond., 207:39-40.
Also in:
Collected reprints, Nat. Inst. Oceanogr., 13.

detection, fish
Meyer, A., 1953.
Fischvorkommen und Echoaufzeichnungen.
Fischwirtschaft 5(2):45-47.

Translation, Pacific Biol. Sta., Fish. Res. Bd., Canada

detection, fish
Midttun, Lars, und Odd Nakken, 1971.
On acoustic identification, sizing and abundance estimation of fish.
FiskDir. Skr. Ser. HavUnders. 16(1): 36-48.

detection, squid
Mishima, S., 1951.
[Detection of squid school by fish detector.]
Bull. Fac. Fish., Hokkaido Univ., 1(2):97-101, 4 figs.

detection, fish
Mitson, R.B., and R. J. Wood, 1961
On automatic method of counting fish echoes.
J. du Conseil, 26(3):281-291.

detection, fish
Mohr, Hermann, 1969.
Echolotbeobachtungen über den Einfluss einiger Verhaltensweisen der Fische auf den Fang mit pelagischen Schleppnetzen.
Ber. dt. wiss. Komm. Meeresforsch. 20(3/4): 256-277.

detection, fish
Moltbahn, K., 1950.
Das Schwimmtrawl und sein Einsatz vor Island und Skagen. Fischereiwelt 2(2):20-21, 5 figs.

detection, fish
Odegaard, J.A., S. Abad Carpio y F. Malave 1971.
Prospecciones hidroacusticas en el oriente de Venezuela desde mayo hasta agosto de 1971.
Informe tecn. Proyecto Invest. Desarrollo Pesq. MAC-PNUD-FAO, Caracas 33: 31 pp.

diseases, fish
Oppenheimer, C.H., 1953.
Why study fish diseases? J. du Cons. 19(1):39-43.

detection, fish
Ortalda, F., 1927.
Sull'impiego degli scandagli acustici a bordo del naviglio pesca. Rivista Marit., May:525-526.

detection, fish
Radakov, D.V., 1960
[Herring observations in the cours of the research cruise of the submarine "Severyanka".]
Biull. Okeanograf. Komissii, Akad. Nauk, SSSR, (6): 39-40.

detection, fish
Renou, J., and P. Tchernia, 1947.
Detection des bancs des poissons par ultra-sons. Ministere de la Marine, Paris, :1-29, 20 pls.

detection, fish
Richardson, I.D., 1951.
Echo sounder surveys for sprat in the 1950-51 season. Cons. Perm. Int. Expl. Mer., Ann. Biol. 7:97.

detection, fish
Richardson, I.D., 1950.
The use of the echo sounder to chart sprat concentrations. Ann. Biol., Cons. Perm. Int. Expl. Mer, 6:136-139, 2 textfigs.

detection(fish)
Richardson, I.D., D.H. Cushing, F.R. Harden Jones, R.J.H. Beverton, R.W. Blacker, 1959
Echo sounding experiments in the Barents Sea.
Min. Agric. Fish. & Food, Fish. Invest., (2), 22(9): 57 pp.

detection, fish
Royce, R.F., 1950.
Echo sounding developments in the European fisheries as reported to the International Council for the Exploration of the Sea. Comm. Fish. Rev. 12(7):54-56, figs.

detection, fish
*Sano, Noritatsu, 1968.
On some techniques of detecting salmon by the echo sounding method. I. Estimation of the fishing grounds of salmon according to swimming speed and swimming depth calculated from echo traces on recording papers. II. On the trial of horizontal fish finder of salmon.
Bull.Jap.Soc.Sci.Fish. 34(8):660-669, 670-680.

detection, fish
Saunois, 1949.
Poursuite des experiences de detection des bancs de poissons a l'ultra-son sur les bancs de Terre Neuve (été 1948). Bull. d'Info., C.C.O.E.C. 1(2):11-14.

detection, fish.
Schärfe, J., 1956.
Echolotungen über das Verhalten eines Herringsschwarmes gegenüber einem Grundschleppnetz.
Fischwirtschaft 8(2):24.

detection, fish
Schärfe, J., 1952.
Besonderheiten bei Echolotungen über unebenem Grund. Fischereiwelt 4(7):99-101, figs.

Translations available
Fisheries Lab., Aberdeen
Pacific Biol. Sta., Nanaimo.

detection, fish
Schärfe, J., 1952.
Über Form und Grösse des Wirkbereiches bei Fischlotungen. Fischereiwelt 4(2):6-8, 6 figs.

detection
Schärfe, J., 1951.
Fischwanderungen im grossen Plöner See während einer Tagesperiode dargestellt an Echogrammen. Arch. f. Fischerei-wiss. 3(3/4):135-146, 22 textfigs.
Translation available
Fisheries Lab., Aberdeen

detection, fish
Schärfe, J., 1951.
Zur Frage der Fischscheuchung durch Lotschall. Fischereiwelt 2(2):30-31, 5 figs.
Translations available
Atlantic Biol. Sta., St. Andrews
Fisheries Lab., Aberdeen
Pacific Biol. Sta., Nanaimo

detection, fish
Schnackenbeck, W., 1950.
Ergebnisse der Versuche mit dem Echolot zur Feststellung von Fischschwärmen. Fischereiwelt 2(1):3 pp., 7 figs.
Translation available
Atlantic Biol. Sta., St. Andrews.

detection, fish
Schnakenbeck, W., 1950.
Fischlotungen im Ausland. Fischereiwelt 2(6):68-69, figs.

detection, fish
Schnakenbeck, W., 1934.
Versuche zur Feststellung von Fischschwärmen durch Echolot. Der Fischmarkt (8):204-207, 1 fig.

detection, echo sounding
Schüler, F., 1954.
Über die echographische Aufzeichnung des Verhaltens von Meeresfischen während der Sonnenfinsternis von 30 Juni 1954. Deutsche Hydrogr. Zeits. 7(3/4):140-143, 2 text figs.

detection, fish
Shibata, Keishi, 1965.
Analysis of fish-finder records. VI. On the fish-finder for biomeasurement. (In Japanese; English abstract).
Bull. Fac. Fish., Nagasaki Univ., No. 19:37-46.

detection, fish
Schubert, K., 1950.
Fischlotungen. Fischereiwelt 2(10):151-152.
Translation available
Fisheries Lab., Aberdeen

detection, fish
Schüler, F., 1954.
Versuche zur Erhöhung der Ergassungstiefe von Meeresfischen durch den Echolotschreiber. Die Fischwirtsschaft 6(12):277.

detection, fish
Schüler, Fr., 1952.
Über das Lesen Echogrammen. Fischereiwelt, 4(1):7-9, 22 figs.
Translation available
Fish. Lab., Aberdeen

detection, fish
Schüler, Fr., 1951.
Der Gegenwärtige Stand und die Grenzen Deutung von Fischzeichen im Echogram. Fischereiwelt 3(7):110-112, 5 figs; 128-129, 11 figs.
Translations available
Fish. Lab., Aberdeen
Pacific Biol. Sta., Nanaimo

detection, fish
Schüler, Fr., + G. Krefft 1951.
Zur Frage der Verwendung des Echographen in der Loggerfischerei. Fischereiwelt 3(4):63-66, 6 fig
Translations available
Fish. Lab. Aberdeen
Pacific Biol. Sta., Nanaimo.

detection, fish
Shatova, O.E., 1961.
[On recording fish density with self-writing and electronic marks of hydroacoustic instruments.]
Nauchno-Techn. Biull., PNIRO, 1(15):61-64.

detection, fish
Shibata, Keishi, 1970.
Analysis of echo-sounder records: acoustic information of fish size. (In Japanese; English abstract).
Bull. Jap. Soc. scient. Fish. 36(5):462-468.

detection fish
Shibata, Keishi, 1970.
Study on details of ultrasonic reflection from individual fish.
Bull. Fac. Fish. Nagasaki Univ. 29:1-82

detection, fish
Shibata, Keishi, 1969.
The use of echo-sounder for estimating fish abundance. (In Japanese; English abstract).
Bull. Jap. Soc. fish. Oceanogr. Spec. No. (Prof Uda Commem. Pap.): 319-322.

detection
Shibata, Keishi, 1964.
Analysis of fish-finder records V. on winter shrimp trawl in the Yellow Sea. (In Japanese; English abstract).
Bull., Fac., Fish. Nagasaki Univ., No. 17:25-43.

detection, fish
Shibata, Keishi and Minoru Nishimura, 1969.
Analysis of fish-finder records. VIII. Classification and interpretation of echo traces on the tuna fishing ground.
Bull. Fac. Fish. Nagasaki Univ. 28:43-67.

detection, fish
Smith, O.R., 1947.
The location of sardine schools by super-sonic echo ranging. Comm. Fish. Rev. 9(1):1-6.

detection, fish
Smith, O.R., and E.H. Ahlstrom, 1948.
Echo ranging for fish schools and observations on temperature and plankton in waters off central California in the spring of 1946. U.S. Fish and Wildlife Service, Spec. Rept. No. 44:1-30, 13 figs.

detection, fish
Sund, O., 1943.
The fat and small herring on the coast of Norway. Cons. Perm. Int. Expl. Mer., Ann. Biol., 1:58-72, 18 figs.

detection, fish
Sund, O., 1935.
Echo sounding in fishery research. Nature 135: 953, 1 fig.

detection, fish
Takayama, S., and C. Yoshimuta, 1958
Fundamental study of the detection of fish by supersonic wave. 1. Preliminary tests on the supersonic reflection of fish and fishing net with a 50 kc supersonic fish finder.
Bull. Tokai Reg. Fish. Res. Lab., No. 22: 47-52.

detection, fish
Tawara, Satoru, Goro Sakurai, Akio Fujiishi and Kazuyuki Omura, 1968.
Echo-survey of tuna fishing ground in the Western Pacific Ocean. 1. Analysis of echo-sounder records of the D.S.L., S. L. and tuna traces. (In Japanese; English abstract).
J. Shimonoseki Univ. Fish., 16(2/3):71-80.

detection
Tashima, I., 1953.
The usefulness of the echo-sounder as the detector for sardine schools from the statistical point of view. J. Shimonoseki Coll. Fish. 3(2):197-200, 2 textfigs.

detection
Tauti, M., 1953.
How to read images of fish schools.
Bull. Japan. Soc. Sci. Fish. 19(4):372-375, 6 textfigs.

detection
Trefethen, P.S., 1956.
Sonic equipment for tracking individual fish.
Spec. Sci. Rept., Fish. No. 179: 11 pp.

detection, fish
Trefethen, P.S., J.W. Dudley and M.R. Smith, 1957
Ultrasonic tracer follows tagged fish.
Electronics, 30(4):156-160.

detection, fish
Truskanov, M.D., and M.H. Shcherbino, 1960.
[Identification of the density of the fish stock with the help of hydroacoustic apparatus.]
Nauchno-Techn. Biull., PNIRO, 2(12):37-40.

detection, fish
Tsujita, T., 1957.
[Studies on the shoal behaviour of demersal fish in the East China Sea by means of echo-sounder. (1) An analysis of echograms taken by the bull-trawlers.]
Bull. Seikai Reg. Fish. Res. Lab., Nagasaki, No. 14:1-47.

detection, fish
Tucker, D.G., 1967.
Sonar in fisheries - a forward look.
Fishing News (Books) Ltd., 136 pp.

detection, fish
Valdez, V., and D.H. Cushing, 1966.
The diurnal variations in depth and quantity of echo traces and their distribution in area in the southern bight of the North Sea.
J. Cons. perm. int. Expl. Mer, 30(2):237-254.

detection, fish
Vestnes G., y G. Saetersdal, 1966.
El ecosonda y su aprovechamiento por los pescadores.
Publ. Inst. Fomento Perquero, (2):16 pp.

detection
Yokata, T., T. Kitagawa, and T. Asami, 1953.
Basic study of fish school research by fish finders. Bull. Japan. Soc. Sci. Fish. 19(4):341-371, 13 textfigs.

detection (fish)
Yudanov, K.I., 1958 (1960).
Use of hydroacoustic fish-locating techniques in the North Atlantic. (In Russian).
Soveschanie Biol. Osnovam Okeanich. Rybolovstva.
Also in:
Trudy Soveshch. Ikhtiolog. Komissii, Akad. Nauk, SSSR, 10:230-234.
Translation:
OTS 63-11126 (1963).

detection, fish

Yuen, Heeny S.H. 1970.
Behavior of skipjack tuna, Katsuwonus pelamis, as determined by tracking with ultrasonic devices.
J. Fish. Res. Bd, Can., 27(11): 2071-2079.

detection, kelp

detection, kelp
Carstens, R.W., 1950.
Bathogram interpretations of side echoes and kelp. J. USC&GS(3):50-52, 5 figs.

detection lost objects

detection, lost objects
Grice, C. Fitzhugh, 1968.
Finding underwater objects. 2. New developments. 3. Case histories: underwater search techniques. Ocean Industry, 3(2):28-33;33-39.

detection particles

Parsons, T.R. and H. Seki, 1969.
A short review of some automated techniques for the detection and characterization of particles in sea water. Bull. Jap. Soc. fish. Oceanogr. Spec. No. (Prof. Uda Commem. Pap.): 173-177.

detection, pollution
Bsch, W.K., M. Ricard et R. Cantin, 1970.
Utilisation des diatomées benthiques comme indicateurs de pollutions minières dans le bassin de la Miramichi N.W.
Techn. Rept. Fish. Res. Bd. Can. 202: 72pp. (multilithed)

detection, precipitation

detection, precipitation
Singer, S.F., and G.F. Williams, Jr. 1968.
Microwave detection of precipitation over the surface of the ocean.
J. geophys. Res. 73(10): 3324-3329.

detection, prey

detection, prey
*Horridge, G.A., and P.S. Boulton, 1967.
Prey detection by Chaetognatha via a vibration sense.
Proc. R. Soc., (B)168(1013):413-419.

detection, scatterers

detection, scatterers
Johnson, H.R., R.H. Backus, J.B. Hersey and D.M. Owen, 1956.
Suspended echo-sounder and camera studies of midwater sound scatterers. Deep-Sea Res. 3(4):266-272.

detection, shellfish

detection, shellfish
Chestnut, A.F., 1950.
The use of the fathometer for surveying shellfish areas. Science 111(2894):677, 1 fig.

detection, sonic

detection, sonic
Edgerton, Harold E., 1966.
Sonic detection of a fresh water-salt water interface.
Science, 154(3756):1555.

detection, storms

detection
Genka, Tomoyuki, 1957.
On the detective effect of radar upon the location of the tuna-long-line. (2). (In Japanese).
Mem. Fac. Fish., Kagoshima Univ., 5:53-59.

detection, radar, storms
Perry, R.E., 1953.
The radar detection of dangerous storms.
J. Inst. Navig. 6(3):238-239, 1 fig.

Detergents

See: chemistry, detergents

detection, visual
Sails, S.B. and J.M. Flowers, 1968
Some aspects of the visual detection of targets from research submarines
Trans. Nat. Symp. Ocean Sci. Engng Atlantic Shelf, Mar. Techn. Soc., March 19-20, 1968: 257-263

detritus, organic

detritus
Chamley, H., 1963.
Contribution à l'étude minéralogique et sédimentologique de vases méditerranéenes.
Rec. Trav. Sta. Mar., Endoume, 29(44):91-195.

detritus
Chesselet, Roger, et Claude Lalou, 1965.
Rôle du "détritus" dans la fixation de radio-éléments dans le milieu marin.
C.R., Acad. Sci., Paris, 260(4):1225-1227.

detritus
Conolly, John R., and Maurice Ewing, 1965.
Ice-rafted detritus as a climatic indicator in Antarctic deep-sea cores.
Science, 150(3705):1822-1824.

detritus (organic)
Darnell, Regnear M. 1967.
Organic detritus in relation to the estuarine ecosystem.
In: Estuaries, G.H. Lauff, editor, Publs Am. Ass. Advmt Sci. 83:376-382.

detritus
Darnell, Regnear M. 1967.
The organic detritus problem.
In: Estuaries, G.H. Lauff, editor, Publs Am. Ass. Advmt Sci. 83:374-375.

detritus
El Wardani, S.A., 1957.
On the biogeochemistry of igneous detritus.
Deep-Sea Res., 4(3):219-220.

detritus, organic
Fox, D.L., 1955.
Organic detritus in the metabolism of the sea.
Sci. Mon. 80(4):256-259.

Detritus
Kenchington, R.A., 1970.
An investigation of the detritus in Menai Straits plankton samples. J. mar. biol. Ass., U.K., 50(2): 489-498.

detritus
*Khailov, K.M., and Z.Z. Finenko, 1968.
The interaction of detritus with high-molecular components of dissolved organic matter in sea water. (In Russian; English abstract).
Okeanologiia, Akad. Nauk, SSSR, 8(6):980-991.

detritus
Körte, Friedrich (posthumous) 1966.
Plankton- und Detritusuntersuchungen zwischen Island und den Faröer im Juni 1960.
Kieler Meeresforsch., 22(1):1-27.

detritus
Krey, Johannes 1967.
Detritus in the ocean and adjacent sea.
In: Estuaries, G.H. Lauff, editor, Publs Am. Ass. Advmt Sci. 83: 389-394.

detritus
Krey, Johannes, 1964
Die mittlere Tiefenverteilung von Seston, Mikrobiomasse und Detritus im nördlichen Nordatlantik.
Kieler Meeresforsch., 20(1):18-29.

Detritus
Krey, Johannes, 1961
Beobachtungen über den Gehalt an Mikrobiomasse und Detritus in der Kieler Bucht 1958-1960.
Kieler Meeresf., 17(2):163-175.

detritus, quantitative and qualitative.
Krey, Johanne, 1961.
Der detritus im Meere.
J. Conseil, 26(3):263-280.

detritus
Kun, M.S., 1959.
[Some information on the distribution of detritus in the northern Caspian.]
Trudy VNIRO, 38:292-303.

detritus
*Lenz, Jürgen, 1968.
Die Teilchengrossenanalyse und Merrgenbestimmung des Detritus in Seewasserproben.
Kieler Meeresforsch. 24(2):85-94.

detritus, turbidite
MacGillavry, H.J., 1968.
Turbidite detritus and geosyncline history.
Trans. Fifth Carib. Geol. Conf., St. Thomas, V.I., 1-5 July 1968, Peter H. Mattson, editor. Geol. Bull. Queens Coll., Flushing 5: 39. (Abstract only).

detritus
Marshall, Nelson, 1965.
Detritus over the reef and its potential contribution to adjacent waters of Eniwetok Atoll.
Ecology, 46(3):343-344.

detritus, organic
Menzel, David W., and John J. Goering, 1966.
The distribution of organic detritus in the ocean.
Limnol. Oceanogr., 11(3):333-337.

detritus
*Nellen, Walter, 1967.
Horizontale und vertikale Vertrilung der Planktonproduktion im Golfe von Guinea und in angrenzenden Meeresgebieten während der Monate Februar bis Mai 1964.
Kieler Meeresforsch., 23(1):48-67.

detritus
Nishizawa, Satoshi, 1966.
Suspended material in the sea: from detritus to symbiotic microcosmos (review). (In Japanese: English abstract).
Inf. Bull. Planktol. Japan, No. 13:1-33.

detritus (organic)
Odum, Eugene P., and Armando A. de la Cruz 1967.
Particulate organic detritus in a Georgia salt marsh-estuarine ecosystem.
In: Estuaries, G.H. Lauff, editor, Publs Am. Ass. Advmt Sci. 83: 383-388.

detritus (source of food)
Parsons, T.R , and J.D.H. Strickland, 1962
Oceanic detritus.
Science, 136(3513):313-314.

detritus, effect of
Polikarpov, G.G., 1961.
The role of detritus formation in the migration of strontium-90, cesium-137 and cerium-144.
Experiments with the sea alga Cystoseira barbata
Doklady Akad. Nauk, SSSR, 136(4):921-923.
English Edit., 1962, 136(1-6):11-13.

detritus, chemistry of
Sato, Tadao, Yoshishige Horiguchi, and Rokuro Adachi, 1965.
On the chemical composition of coarse (mainly plankton) and the fine (mainly detritus) suspended particles in the water of Matoyo Bay.
(In Japanese; English abstract).
Inf. Bull. Planktol., Japan, No. 12:66-71.

detritus
Starr, T.J., 1956.
Relative amounts of vitamin B12 in detritus from oceanic and estuarine environments near Sapelo Island, Georgia. Ecology, 37(4):658-664.

detritus
*Trevallon, Ann, 1967.
An investigation of detritus in Southampton Water.
J. mar. biol. Ass., U.K., 47(3);523-532.

detritus
Wellershaus, Stefan, 1964.
Die Schichtungsverhältnisse im Pelagial des Bornholmbeckens. Über den Jahresgang einiger biotischer Faktoren.
Kieler Meeresf., 20(2):148-156.

detritus
Wheeler, E.H., Jr., 1967.
Copepod detritus in the deep sea.
Limnol. Oceanogr., 12(4):697-702.

detritus (carbonate)
Zeigler, John M. and Graham S. Giese, 1968.
Geology and hydrodynamics of Punta Arenas shoal. Trans. Fifth Carib. Geol. Conf., St. Thomas, V.I., 1-5 July 1968, Peter H. Mattson, editor. Geol. Bull. Queens Coll., Flushing, 5: 33-38.

detritus, effect of

detritus, effect of
Newell, Richard, 1965.
The role of detritus in the nutrition of two marine deposit feeders, the prosobranch Hydrobia ulvae and the bivalve Macoma baltica.

detritus feeders
Doty, Maxwell S., and Gertrudes Aguilar-Santos 1970.
Transfer of toxic algal substances in marine food chains.
Pacific Sci. 24(3):351-356.

Detritus, radioactivity of

detritus, radioactivity of
Chesselet, R., et Ul. Lalou, 1965.
Recherches recentes de la radioactivité du plancton et du detritus organique. 1. Étude de la radioactivité.
Cahiers du C.E.R.B.O.M., Nice, 2:67-85.

devices

diagenesis

diagenesis
Antal, Paul S., 1966.
Diagenesis of thorium isotopes in deep-sea sediments.
Limnol. Oceanogr., 11(2):278-292.

diagenesis
Arrhenius, Gustaf, 1967.
Deep-sea sedimentation: a critical review of U. S. work.
Trans. Am. geophys. Un., 48(2):604-631.

diagenesis
Berner, Robert A., 1967.
Diagenesis of iron sulfide in Recent marine sediments.
In: Estuaries, G.H. Lauff, editor, Publs Am. Ass. Advmt Sci., 83:268-272.

diagenesis
Berner, Robert A., 1966
Diagenesis of carbonate sediments: interaction of magnesium in sea water with mineral grains.
Science, 153(3732): 188-191.

Diagenesis
Choquette, Philip W., 1968.
Marine diagenesis of shallow marine lime-mud sediments: insights from δO^{18} and δC^{13} data.
Science, 161(3846):1130-1132.

diagenesis
Degens, Egon T., 1964.
Über biogeochemische Umsetzungen im Frühstadium der Diagenese.
In: Developments in Sedimentology, L.M.J.U. van Straaten, Editor, Elsevier Publishing Co., 1:81-92

diagenesis
Harris, William H., and R.K. Matthews, 1968.
Subaerial diagenesis of carbonate sediments: efficiency of the solution-reprecipitation process.
Science, 160(3823):77-79.

diagenesis
Irwin, M.L., 1965.
General theory of epeiric clear water sedimentation.
Bull. Amer. Assoc. Petr. Geol., 49(4):445-459.

diagenesis
Kaplan, I.R., and S.C. Rittenberg, 1963.
23. Basin sedimentation and diagenesis.
In: The Sea, M.N. Hill, Editor, Interscience Publishers, 3:583-619.

diagenesis
Lynn, D.C., and E. Bonatti 1965.
Mobility of manganese in diagenesis of deep-sea sediments.
Marine Geol., 3(6):457-474.

diagenesis
MacKenzie, Fred T., and Robert M. Garrels, 1966.
Silica-bicarbonate balance in the ocean and early diagenesis.
J. Sedim. Petrol., 36(4):1075-1084.

diagenesis
Marlowe, James I. 1971
Dolomite, phosphorite, and carbonate diagenesis on a Caribbean seamount.
J. Sedim. Petrol. 41(3): 809-827.

diagenesis
Siever, Raymond 1968
Sedimentological consequences of a steady-state ocean-atmosphere.
Sedimentology 11 (1/2):5-29.

diagenesis
Shishkina, O.V., 1960
Changes in the salinity of interstitial water in the process of diagenesis.
Trudy Comm. Oceanogr., Akad. Nauk, SSSR, 10(2): 13-20.

diagenesis, effect of

diagenesis, effect of
Bajor, M., et B.M. Van der Weide, 1967.
Effets de la diagenèse sur la distribution des amino-acides dans les sédiments.
Bull. Centre Rech. PAR-SNPA, 1(1):173-186.

diagrams

diagrams
Burt, W.V., 1956.
A light scattering diagram. J. Mar. Res. 15(1):76-80.

diagrams
Margalef, Ramón, 1965(1967).
Composición y distribución del fitoplancton.
Memoria Soc. Cienc. nat. La Salle, 25(70/71/72):141-205, numerous tables.

diamantkus

diamantkus
Wilson, Thomas A., 1965.
Offshore mining paves the way to ocean mineral wealth.
Engineering and Mining Jour., 166(6): 124-132.

diamonds

diamonds
Anon., 1968.
Ocean-bottom minerals.
Ocean Industry 3(6):61-73.

diamonds
Anon, 1957.
Diamond dredging produces new [from] of wealth from the sea.
Ocean Indust. 2(8): 67-72

diamonds
McIlhenny, W.F., 1966.
The oceans technology's new challenge.
Chem. Engng., (Nov. 7):247-254.

diamonds
Mero, John L., 1965.
The mineral resources of the sea.
Elsevier Oceanogr. Ser., 312 pp.

diamonds
Murray, L.G., R.H. Joynt, D.O'C. O'Shea, R.W. Foster and L. Kleinjan 1970.
The geological environment of some diamond deposits of South West Africa.
In: The geology of the East Atlantic continental margin. 1. General and economic papers. ICSU/SCOR Working Party 31 Symposium, Cambridge 1970, Rept. No. 70/13: 119-141.

diamonds
Webb, Bill, 1965.
Technology of sea diamond mining.
Ocean Sci. and Ocean Eng., Mar. Techn. Soc., Amer. Soc., Limnol. Oceanogr., 1:8-23.

diamonds
Wilson, Thomas A., 1965.
Offshore mining paves the way to ocean mineral wealth.
Engineering and Mining Jour., 166(6): 124-132.

diapir

diapirs
Alla, Geneviève, et Olivier Leenhardt 1971.
Découverte d'un affleurement de cap-rock sur le sommet d'un dôme de sel (dôme T, Sud-Toulon) avec le bathyscaphe.
C. r. hebd. Séanc. Acad. Sci. Paris (D) 272(10): 1347-1349.

diapirs
Ballard, J. Alan, and Robert H. Feden, 1970.
Diapiric structures on the Campeche shelf and slope, western Gulf of Mexico.
Bull. geol. Soc. Am., 81(2): 505-512.

diapirs
Baumgartner, Timothy R., and Tjeerd H. van Andel, 1971.
Diapirs of the continental margin of Angola, Africa.
Bull. geol. Soc. Am. 82(3): 793-802

diapirs
Collette, B.J., and K.W. Rutten, 1970.
Differential compaction vs. diapirism in abyssal plains.
Mar. geophys. Res. 1(1): 104-107

diapirs
*Emery, K.O., 1968.
Shallow structure of continental shelves and slopes.
SEast. Geol., Duke Univ., 9(4):173-194.

diapirs
Ensminger, H. Robert 1970
Seismic profiling in the Gulf of Mexico.
UnderSea Techn., 11(2): 22-25.

diapir
*Glangeaud, Louis, Jean Alinat, Jean Polveche, André Guillaume et Olivier Leenhardt, 1966.
Grandes structures de la mer Ligure, leur évolution et leurs relations avec les chaînes continentales.
Bull. Soc. géol. France, (7)8(7):921-937.

diapirs
Keen, M. J., 1970.
A possible diapir in the Laurentian Channel.
Can. J. Earth Sci., 7(6): 1561-1564

diapirs
King, Lewis H., and Brian MacLean, 1970.
A diapiric structure near Sable Island - Scotian Shelf.
Marit. Sed., 6(1): 1-4

diapirs
Montadert, Lucien, Jean Sancho, Jean-Pierre Fail, Jacques Debyser et Etienne Winnock, 1970.
De l'âge tertiaire de la série salifère responsable des structures diapiriques en Méditerranée Occidentale (Nord-Est des Baléares).
C. r. hebd. Séanc. Acad. Sci., Paris, (D) 271(10): 812-815

diapirs
Pautot, Guy, Jean-Marie Auzende and Xavier Le Pichon, 1970.
Continuous deep sea salt layers along North Atlantic margins related to early phase of rifting.
Nature, Lond., 227(5256): 351-354

diapirs
Ryan, William B.F., Daniel J. Stanley, J.B. Hersey, Davis A. Fahlquist and Thomas D. Allan, 1970.
II. Regional observations. 12. The tectonics and geology of the Mediterranean Sea (June 1969)
In: The sea: ideas and observations on progress in the study of the seas, Arthur E. Maxwell, editor, Wiley-Interscience 4(2): 387-492.

diapirs
Schneider, E.D., and G.L. Johnson, 1970
Deep ocean diapiric structures. In: The geology of the East Atlantic continental margin, 1. General and economic papers, ICSU/SCOR Working Party 31 Symposium, Cambridge 1970, Rept. No. 70/13: 153-175.

diapirs
Scholl, David W., and Michael S. Marlow, 1970.
Diapirlike structures in southeastern Bering Sea.
Bull. Am. Ass. Petrol. Geol. 54(9): 1644-1650.

diapirs
Summerhayes, C.P., A.H. Nutter and J.S. Tooms, 1971.
Geological structure and development of the continental margin of northwest Africa.
Mar. Geol. 11(1): 1-25.

diapirs
Uchupi, Elazar, and K.O. Emery, 1968.
Structure of continental margin off Gulf coast of United States.
Bull. Am. Ass. Petr. Geol. 52(7): 1162-1193.

diapirs
Watson, Jerry A., and G. Leonard Johnson, 1970.
Seismic studies in the region adjacent to the Grand Banks of Newfoundland.
Can. J. Earth Sci. 7(2-1): 306-316

diapirs
Wilson, J. Tuzo 1969.
Aspects of the different mechanics of ocean floors and continents.
Tectonophysics 8(4/6): 281-284.

diapirs
Worzel, J. Lamar, Robert Leyden and Maurice Ewing, 1968.
Newly discovered diapirs in Gulf of Mexico.
Bull. Am. Ass. Petr. Geol., 52(7): 1194-1203.

diapiric domes
*Shepard, Francis P., Robert F. Dill and Bruce C. Heezen, 1968.
Diapiric intrusions in foreset slope sediments of Magdalena Delta Colombia.
Bull. Am. Ass. Petr. Geol., 52(11):2197-2207.

diapiric domes
*Watson, Jerry A., and G. Leonard Johnson, 1968.
Mediterranean diapiric structures.
Bull. Am. Ass. Petr. Geol., 52(11);2247-2249.

diapirs
Wilhelm, Oscar, and Maurice Ewing, 1972.
Geology and history of the Gulf of Mexico.
Bull. geol. Soc. Am. 83(3): 575-600.

diapiric structures
Bryant, William R., A.A. Meyerhoff, Noel K. Brown, Jr., Max A. Furrer, Thomas E. Pyle and John W. Antoine, 1969.
Escarpments, reef trends, and diapiric structures, eastern Gulf of Mexico.
Bull. Am. Ass. Petrol. Geol., 53(12): 2506-2542.

diapirism
Auzende, Jean-Marie, Jean Bonnin, Jean-Louis Olivet, Guy Pautot and Alain Mauffret, 1971.
Upper Miocene salt layer in the western Mediterranean Basin.
Nature, Lond., 230(12): 82-84

diapirism
Ramberg, Hans, 1972
Theoretical models of density stratification and diapirism in the earth. J. geophys. Res. 77(5): 877-889.

diatom blooms

diatom blooms
See chiefly under diatoms in ORGANISMAL INDEX

diatom blooms
Pratt, David M., 1965.
The winter-spring diatom flowering in Narragansett Bay.
Limnology and Oceanography, 10(2):173-184.

diatom flowering
Riley, G. A., 1943
Physiological aspects of spring diatom flowerings. Bull. Bingham Oceanogr. Coll., VIII(4):1-53

diatomaceous earth

diatomaceous earth
Bens, Evertt M., and Charles M. Drew, 1967.
Diatomaceous earth: scanning electron microscope of "chromosorb P".
Nature, Lond., 216(5119):1046-1048.

dictionaries

dictionaries
Gorsky, N.N., and V.I. Goskaya, 1957.
English-Russian dictionary of oceanographic terms. State Publ. Off., Moscow, 292 pp.

dictionaries, French-English
Legendre, Vianney, W.B. Scott et Julien Bergeron 1964.
Noms français et anglais des poissons de l'Atlantique canadien.
Cah. inf. Stn biol. mar. Grande-Rivière 23: 178 pp. (multilithed).

dictionaries
Segditsas, P.E., 1965.
Elsevier's Nautical Dictionary. Vol. 1. Maritime Terminology in five languages, English-French-Italian-Spanish and German. Elsevier Publishing Co., 577 pp.

diffraction, internal waves
Manton, M.J., L.A. Mysak and R.E. McGorman 1970.
The diffraction of internal waves by a semi-infinite barrier.
J. fluid Mech. 43(1): 165-176

Diffraction, light

diffraction, light
Neuimin, G.G., and N.A. Sorokina, 1964.
On the optical dispersing layers in the sea. (In Russian).
Okeanologiia, Akad. Nauk, SSSR, 4(1):51-54.

diffraction, scalar

Diffraction, scalar
Poincelot, Paul, 1966.
Le Theorème de Babinet relatif à la diffraction scalaire.
Cahiers océanogr., 18(7):621-627.

diffraction, sound

diffraction, sound
Murphy, E.L., 1968.
Ray representation of diffraction effects in the split-beam sound field.
J. acoust. Soc. Am., 43(3):610-618.

diffraction, sound
Noble, W.J., 1956.
Theory of the shadow zone diffraction of under-water sound. J. Acoust. Soc., Amer., 28(6):1247-1252.

diffraction, sound
Zhitkovsky Yu. Yu. and Yu. P. Lysanov 1969.
On some peculiarities of the Fresnel diffraction of sound upon rough sea surface and bottom. (In Russian).
Fizika Atmosfer. Okean. Izv. Akad. Nauk, SSSR, 5(9): 982-985.

diffraction, swell

diffraction, swell
*Baraillier, L., et P. Gaillard, 1967.
Evolution recente des modeles mathématique d'agitation due à la houle. Calcul de la diffraction en profondeur non uniforme.
Houille blanche, 22(8):861-869.

diffraction, swell
Daubert, A., et J.C. Lebreton, 1965.
Diffraction de la houle sur des obstacles à parois verticales.
La Houille Blanche 20(4):337-344.

diffraction, swell
Larras, Jean, 1966.
Diffraction de la houle par les obstacles rectilignes semi-indefinis sous incidence oblique.
Cahiers Océanogr., C.C.O.E.C., 18(8):661-667.

diffraction, swell
Poincelot, Paul, 1964.
Diffraction de la houle par une jetée.
Cahiers Océanogr., C.C.O.E.C., 16(10):845-868.

diffraction, swell
Robin, Louis, 1966.
Diffraction de la houle par une île elliptique, par une bande plane et par une fente dans une jetée.
Cahiers Océanogr., C.C.O.E.C., 18(2):123-138.

diffraction, tides
Buchwald, V.T., 1971.
The diffraction of tides by a narrow channel.
J. fluid Mech. 46(3): 501-511.

diffraction, tsunami

diffraction, tsunami
Momoi, Takeo, 1966.
A further study of diffraction of tsunami invading a semi-circular peninsula.
Bull. Earthquake Res. Inst., 44(2):473-480.

diffraction, waves

diffraction, waves
Allard, M.P., 1950.
Le regime de la cote marocaine entre Safi et Mogador. Bull. d'Info., C.O.E.C., 2(10):369-378, 6 pls.

diffraction, waves
Baines, P.G. 1971.
The reflexion of internal/inertial waves from bumpy surfaces. 2. Split reflexion and diffraction.
J. fluid Mech. 49(1): 113-131

diffraction, waves
Brekhovskikh, L.M., 1952.
Diffraction of waves at an uneven surface. 1. General theory. II. Application of the general theory. Zh. Eksp. i Teor. Fiz., 23, 3(9):257-288, 289-304.
Translation T114R

diffraction, waves
*Buchwald, V.T., 1968.
The diffraction of Kelvin waves at a corner.
J. fluid Mech., 31(1):193-205.

diffraction, waves
Carr, J.H., and M.E. Stelzriede, 1951.
Diffraction of water waves by breakwaters.
Heat Transfer, Fluid Mechanics Inst., Stanford Univ.:5-16.
Abstr. Appl. Mech. Rev. 5:542(1952).

diffraction (waves)
Carr, J.H., and M.E. Stelzriede, 1952.
Diffraction of water waves by breakwaters.
Proc. N.B.S. Semicent. Symp., Gravity Waves, June 18-20, 1951, Nat. Bur. Stand. Circ. 521:109-125, 9 textfigs.

diffraction (wave)
Carry, C., and E. Chapus, 1951.
Calculation of diffracted wave height behind a semi-infinite jetty. Bull. B.E.B. 5(3):15-24, 4 figs.
Apparently translated from La Houille Blanche.

diffraction (waves)
Comite Central d'Oceanographie et d'Etudes des Cotes, 1950.
Remarks on wave forecasting. By. M. Celci.
Relations between microseismic activity and waves. By. J. Debrach.
Wave diffraction. By. H. Lacomb.
An explanation of marine currents. By. M. P. Antoine.
Bull. B.E.B. 4(1):36-38.

diffraction, waves
Darbyshire, J., 1963.
Wave diffraction, liquid, 687-688.
wave dissipation, liquid, 688-690.
wave energy, liquid, 690-691.
wave refraction, liquid, 722-724.
wave in liquids, 7250727.
wave spectra, liquid, 729-731.
Wave, water, long, 739-740.
In: Encyclopaedic Dictionary of Physics, Pergamon Press, Ltd., Vol. 7.

diffraction, waves
Dunham, J.W., 1950.
Refraction and diffraction diagrams. Inst. Coastal Eng., Univ. Ext., Univ. Calif., Long Beach, 11-13 Oct. 1950:30 pp. (mimeographed), 6 pls., (multilithed).

diffraction, waves
Ippen, Arthur T., 1966.
Estuary and coastline hydrodynamics. McGraw Hill Book Co, Inc., 744 pp.

diffraction, waves
Iwagaki, Yuichi, and Toru Sawaragi, 1962
A new method for estimation of the rate of littoral sand drift.
Coastal Engineering in Japan, 5:67-79.

diffraction (waves)
Johnson, J.W., 1952.
Engineering aspects of diffraction and refraction
Proc. Amer. Soc. Civil Eng. 78(Sep. No. 122): 32 pp.

diffraction, waves
* Kelly, R.E., 1969.
Wave diffraction in a two-fluid system.
J. fluid Mech. 36(1): 65-73.

diffraction, wave
Lacombe, H., 1953.
Diffraction de la houle en incidence oblique.
Amer. Hydr. (4); 3:205-243.

diffraction, waves
Lacombe, H., 1952.
Diffraction de la houle en incidence oblique.
Ann. Hydrogr., Paris, 4th ser., 3:205-243, 8 textfigs.

diffraction (wave)
Lacombe, H., 1952.
The diffraction of a swell. A practical approximate solution and its justification. Proc. N.B.S. Semicent. Symp., Gravity Waves, June 18-20, 1951, Nat. Bur. Stand. Circ. 521:129-140, 8 textfigs.

diffraction, swell
Lacombe, H., 1952.
Diffraction de la houle par une breche.
Bull. d'Info., C.C.O.E.C., 4(2):54-63, 4 pls.

wave diffraction
Lacombe, H., 1951.
Wave diffraction for oblique incidence. Bull. B.E.B. 5(2):13-18, 2 figs.
Translation from: Bull. d'Info., C.O.E.C. (exact reference not given).

diffraction, waves
Lacombe, H., 1950.
Note on diffraction of swell in perpendicular incidence. (Ann. Hydr., Paris, No. 1363). Int. Hydrogr. Rev. 27(2):122-137, 11 textfigs., 11 pls

diffraction (swell)
Lacombe, H., 1950.
La diffraction de la houle en incidence oblique.
Bull. d'Info., C.O.E.C., 2(10):387-394, 2 textfig

diffraction, waves
Lacombe, H., 1949.
Note sur la diffusion de la houle en incidence normale. Ann. Hydro., Paris, No. 1363: 49 pp., 11 pls., 11 textfigs.

diffraction, waves
MacCamy, R.C., and R.A. Fuchs, 1954.
Wave forces on piles: a diffraction theory.
Tech. Memo., B.E.B. No. 69:1-17, 7 figs.

diffraction, waves
*Mobarek, Ismail, and Robert L. Wiegel, 1967.
Diffraction of wind generated water waves.
Proc.10th Conf.Coast.Engng.Tokyo,1966,1:183-206.

diffraction, waves
Mogi, K., 1956.
Experimental study of diffraction of water surface waves. Bull. Earthquake Res. Inst., Tokyo, 34 (3):267-278.

diffraction, waves
*Momoi, Takao, 1968.
A long wave in the vicinity of an estuary. IV.
Bull.Earthq.Res.Inst.,Tokyo Univ., 46(3):631-650.

diffraction, swell
Montaz, J.P., 1964.
Étude expérimentale systématique en vue de l'utilisation des modèles mathématiques pour l'étude de la diffraction de la houle.
La Houille Blanche, (7):784-791.

diffraction, waves
Oberhettinger, F., 1954.
Diffraction of waves by a wedge.
Pure Appl. Math. 7(3):551-563, 3 textfigs.

diffration, waves
*Packham, B.A., and W.E. Williams, 1968.
Diffraction of Kelvin waves at a sharp bend.
J. fluid Mech., 34(3):517-529.

diffraction, waves
Pinsent, H.G., 1971.
The effect of a depth discontinuity on Kelvin wave diffraction.
J. fluid Mech., 45(4): 747-758.

diffraction
Putnam, J.A., and R.S. Arthur, 1948
Diffraction of water waves by breakwaters.
Trans. Am. Geophys. Union, 29(4):481-490, 11 text figs.

diffraction, waves
*Sebekin, B.I., 1967.
Diffraction of surface waves by a wedge.
(In Russian; English abstract).
Fisika Atmosfer.Okean.,Izv. Akad. Nauk,SSSR, 3(8):890-902.

diffraction, waves
*Sekerzh-Zen'kovich, S.Ya, 1968.
The diffraction of plane waves by a circular island. (In Russian; English abstract).
Fisika Atmosfer.Okean.,Izv.Akad.Nauk,SSSR, 4(1):69-79.

diffraction, waves
*Shem, M.C., R.E. Meyer and J.B. Keller, 1968.
Spectra of water waves in channels and around islands.
Phys.fluids., 11(11):2289-2304.

diffraction, waves
Sretenski, L.N., 1959
[The diffraction of waves in the CAUCHY-POISSON problem.]
Doklady Akad. Nauk, SSSR, 129(1):59-60.

diffraction, waves
Voit, S.S., and B.I. Sebekin, 1969.
Diffraction of unsteady surface and internal waves. (In Russian; English abstract).
Fisika Atmosfer. Okean., Izv. Akad. Nauk, SSSR, 5(2):180-187.

diffraction, waves
Wada, Akira, 1965.
On a method of solution of diffraction problems.
Coastal Engin., Japan, 8:1-19.

diffraction, waves
Williams, W.E., 1965.
Diffraction of surface waves on an incompressible fluid.
J. Fluid Mech., 22(2):253-256.

diffraction, waves
Zhudovets, A.M., 1963
The effect of the wind on defracted waves on the water. (In Russian).
Okeanologiia, Akad. Nauk, SSSR, 3(6):970-978.

diffusion

diffusion, chemical
*Aubert, M., et N. Desirotte, 1969.
Théorie formalisée de la diffusion des prodeuits ghimiques en mer.
Rev. intern.Océanogr., Méd., 13-14:125-156.

diffusion
Batchelor, G.K., 1957.
Diffusion in freee turbulent shear flows.
J. Fluid Mech., 3(1):67-80.

diffusion
*Bonham-Carter, G.F., and Alex J. Sutherland, 1967.
Diffusion and settling of sediments at river mouths: a computer simulation model.
Trans.Gulf Coast Assoc.Geol.Soc.,17th Ann.Meet.: 326-338.

Diffusion
Brush, Lucien M., Jr., 1962.
Exploratory study of sediment diffusion.
J. Geophys. Res., 67(4):1427-1433.

diffusion
Callame, B., 1962.
Observations sur les échanges par diffusion entre les sédiments et l'eau de mer qui les recouvre.
In: Océanographie Géologique et Géophysique de la Méditerranée Occidentale, Centre National de la Recherche Scientifique, Villefranche sur Mer, 4 au 8 avril 1961, 83-87.

diffusion
*Church, Michael, 1967.
Observations of turbulent diffusion in a natural channel.
Can. J. Earth Sci., 4(5):855-872.

Diffusion
Corrsin, S., 1962.
Discussion of paper by L. M. Brush, Jr., 'Exploratory study of sediment diffusion'.
J. Geophys. Res., 67(4):1435.

diffusion
Costin, J. Michael, 1965.
Dye tracer studies on the Bahama Banks. (summary)
Symposium, Diffusion in oceans and fresh waters, Lamont Geol. Obs., 31 Aug.- 2 Sept., 1964, 68-69

diffusion
Csanady, G.T., 1966.
Accelerated diffusion in the skewed shear flow of lake currents.
J. geophys. Res., 71(2):411-420.

diffusion
Daubert, A., et J.-C. Lebreton, 1965.
Effet de la houle sinusoidale sur la diffusion entre liquides de salinités différentes.
La Houille Blanche, 20(1):45-47.

diffusion
Duursma, E.K., 1966.
Molecular diffusion of radioisotopes in interstitial water of sediments.
Disposal of Radioactive Wastes into Seas, Oceans and Surface Water, IAEA, 355-371. (SM-72/20)

Diffusion
Duursma, E.K., and C.J. Bosch, 1970.
Theoretical, experimental and field studies concerning diffusion of radioisotopes in sediments and suspended particles of the sea. B. Methods and experiments.
Netherl. J. Sea Res. 4(4): 395-469.

diffusion
Ford, W.L., 1949
Radiological and Salinity relationships in the water at Bikini Atoll. Trans. Am. Geophys. Un., 30(1):46-54, 22 text figs.

diffusion
Gade, Herman G., 1965.
Horizontal and vertical exchanges and diffusion in the water masses of the Oslo Fjord.
Helgoländer wiss. Meeresunters., 17: 412-475.

diffusion
Gunnerson, Charles G., 1965.
Limitations of Rhodamine B and Pontacyl Brilliant Pink B as tracers in estuarine waters. (abstract)
Symposium, Diffusion in oceans and fresh waters, Lamont Geol. Obs., 31 Aug.-2 Sept., 1964, 53.

diffusion
Han Young Ho, and Gap Dong Yoon, 1970.
On diffusion experiments in the sea off the coast of Ko-Ri.
Bull. Pusan Fish. Coll. 10(1): 17-25

diffusion
Harleman, Donald R.F., and Arthur T. Ippen, 1961?
The turbulent diffusion and convection of saline water in an idealized estuary.
I.A.S.H., Comm. of Surface Waters, Publ., No. 51:362-378.

diffusion
Hela, Ilmo, and Aarno Voipio, 1960.
Tracer dyes as a means of studying turbulent diffusion in the sea.
Annales Acad. Sci. Fennicae, (A)(VI Physics), 69:1-9.

diffusion
*Higuchi, Haruo and Yuichi Iwagaki, 1969.
Hydraulic model experiment on the diffusion due to the coastal current. Coast. Engng, Japan, 12: 129-138.

diffusion
*Higuchi, Haruo and Takashige Sugimoto, 1966.
Hydraulic model experiment on the diffusion due to the tidal current.
Spec. Contr., geophys. Inst., Kyoto Univ., 6:113-125.

diffusion
*Ichiye, Takashi, 1967.
Upper ocean boundary-layer flow determined by dye diffusion.
Physics Fluids 10 (9-2):S270-S277.

diffusion
Ichiye, Takashi, 1965.
Diffusion experiments in coastal waters using dye techniques. (abstract).
Symposium, Diffusion in oceans and freshwaters, Lamont Geol. Obs., 31 Aug.-2 Sept., 1964, 54-67.

diffusion
Ichiye, Takashi, 1962.
Studies of turbulent diffusion of dye patches in the ocean.
J. Geophys. Res., 67(8):3213-3216.

diffusion of vorticity
Ichiye, T., 1956.
On the behaviour of the vortex in the Polar Front Region. (Hydrography of the Polar Front Region I). Ocean. Mag., Tokyo, 7(2):115-132.

diffusion
Ichiye, T., 1950.
Some remarks on Richardson's neighbour diffusion equation. Ocean. Mag., Tokyo, 2(3):101-104.

diffusion
Ichie, T., 1949.
On the diffusion in the field of Coriolis' Force. Ocean. Mag., Tokyo, 1(2):121-127, 4 textfigs.

diffusivity
Ichiye, T., and F.C.W. Olson, 1960
Über die neighboring diffusivity im Ozean.
Deutsche Hydrogr. Zeits., 13(1): 13-23.

diffusion
Ichiye, Takashi, and Noel B. Plutchak, 1966.
Photodensitometric measurement of dye concentration in the ocean.
Limnol. Oceanogr., 11(3):364-370.

diffusion
Ippen, Arthur T., 1966.
Estuary and coastline hydrodynamics. McGraw Hill Book Co, Inc., 744 pp.

diffusion
Ivanoff, A., 1959.
Introduction à une étude des propriétés diffusantes des eaux de la Baie de Naples.
Pubbl. Sta. Zool., Napoli, 31(1):xxxiii-xliii

diffusion
Japan, Maritime Safety Board, 1957.
On the diffusions of floatages.
Hydrogr. Bull. (Bull. 981) 53:27-29.

Diffusion
Jones, R.H., and R.E. Stewart, 1970.
Diffusion of sewage effluent from an ocean outfall.
Rev. intern. Oceanogr. méd. 17: 99-108

diffusion
Joseph, J., and H. Sender, 1962.
On the spectrum of the mean diffusion velocities in the ocean.
J. Geophys. Res., 67(8):3201-3206.

diffusion
Joseph, J., and H. Sendner, 1958.
Über die horizontale Diffusion im Meere.
Deutschen Hydrogr. Zeits., 11(2):49-77.

diffusion
Kanwisher, John W., 1963
On the exchange of gases between the atmosphere and the sea.
Deep-Sea Research, 10(3):195-207.

diffusion
Katayama, Katsusuke, Teruyuki Sugiyama, Kazuo Ukida, Motoyuki Shinohara and Kuniyasu Fujisawa, 1971.
The effects of effluent of night-soil treatment on the coastal fisheries grounds - I.
On diffusion of effluent and quality of bottom mud in the surrounding area. (In Japanese).
Bull. Fish. Exp. Sta. Okayama Pref. (1970): 42-68.

diffusion
Kenney, B.C. and Ian S.F. Jones, 1971.
Relative diffusion as related to quasi-periodic current structures. J. phys. Oceanogr. 1(3): 224-232.

Kilezhenko, V.P., 1965. diffusion, turbulent
Experimental studies of horizontal turbulent diffusion of a patch of admixture at sea. (In Russian).
Materal, Ribochoz. Issled. Severn. Basseina, Poliarn. Nauchno-Issled. i Proektn. Inst. Morsk. Rib. Choz. i Okeanogr. im. N.M. Knipovicha, PINRO, Gosud. Proizvodst. Komm. Ribn. Choz. SSSR 4:101-105.

diffusion
Kirwan, A.D., Jr., 1965.
On the use of Rayleigh-Ritz method of calculating the eddy diffusivity.
Symposium, Diffusion in ocean and fresh waters, Lamont Geol. Obs., 31 Aug.-2 Sept. 1964., 86-92.

diffusion
Kochina, N.N. 1968
A periodical solution to the diffusion equation with a non-linear boundary condition. (In Russian)
Dokl. Akad. Nauk SSSR 179(6): 1297-1300.

diffusion
Kullenberg, G.E.B., 1971.
Results of diffusion experiments in the upper region of the sea.
Rept. Inst. fys. Oceanogr. Univ. København, 12: 42 pp., 23 figs. (multilithed)

diffusion
Kullenberg, Gunnar, 1968.
Measurements of horizontal and vertical diffusion in coastal waters.
Rept. Københavns Univ. Inst. fys. Oceanogr., 3:50 pp. (multilithed).

diffusion
McEwen, G.F., 1950.
A statistical model of instantaneous point and disk sources with applications to oceanographic observations. Trans. Am. Geophys. Union 31 (1):35-46, 3 textfigs.

diffusion
Miyoshi, H., S. Hori and S. Yoshida, 1955.
Drift and diffusion of radiologically contaminated water in the ocean. Rec. Ocean. Wks., Japan, 2(2):30-36.

diffusion
Monin, A.S., 1969.
On the interaction between the vertical and horizontal diffusion of a mixture in the sea.
(In Russian; English abstract).
Okeanologiia, 9(1):76-81.

diffusion
Morinaga, Toyoko, Yoshimasa Toyota and Shiro Okabe, 1970.
Pollution in coastal waters of Japan (1) Inside and outside of Tagonoura Port, Shizuoka Prefecture. Bull. Coast Oceanogr. 8(2). Also in: Coll. Repr. Coll. mar. Sci. Techn. Tokai Univ. 4: 313-318.

diffusion
Munk, Walter H., 1966.
Abyssal recipes.
Deep-Sea Res., 13(4):707-730.

diffusion
Munk, W.H., G.C. Ewing, R.R. Revelle, 1950.
Diffusion in Bikini Lagoon. Trans. Am. Geophys. Union 30(1):59-66, 9 textfigs.

a
diffusion
Murthy, C.R. and C.T. Csanady 1971
Experimental studies of relative diffusion in Lake Huron.
J. phys. Oceanogr. 1(1): 17-24

diffusion
Nan'niti, T., and A. Okubo, 1957.
[An example of the diffusion of a floating dye patch in the sea] J. Ocean. Soc., Japan, 13(1):1-4.

diffusion
Noble, Vincent E., 1961.
Measurement of horizontal diffusion in the Great Lakes.
Proceedings Fourth Conf., Great Lakes Research, Great Lakes Res. Div., Inst. Sci. & Tech., Univ. Michigan Publ., (7):85-95.

diffusion
Novikov, E.A., 1966.
The relative diffusion in the turbulent shear flow, (In Russian).
Fisika Atmosferi i Okeana, Izv., Akad. Nauk, SSSR, 2(11):1198-1199.

diffusion
Okubo, Akira, 1971
Application of the telegraph equation to oceanic diffusion: another mathematical model.
Techn Rept. Chesapeake Bay Inst. 69 (Ref. 71-3): 35 pp. (multilithed)

diffusion
Okubo, Akira, 1971.
4. Horizontal and vertical mixing in the sea.
In: Impingement of man on the oceans, D.W. Hood, editor, Wiley Interscience: 89-168.

diffusion
Okubo, Akira 1970.
Oceanic mixing.
Chesapeake Bay Inst. Ref. 70-1: 140 pp.
(multilithed).

diffusion

Okubo, Akira, 1971.
Oceanic diffusion diagrams. Deep-Sea Res., 18(8): 789-802.

diffusion

*Okubo, Akira, 1968.
A new set of oceanic diffusion diagrams.
Techn. Rept. Chesapeake Bay Inst., 38:45 pp. (mimeographed).

diffusion

Okubo, Akira 1966.
A note on 'horizontal diffusion from an instantaneous source in a nonuniform flow.'
J. oceanogr. Soc., Japan, 22(2): 35-40

diffusion

Okubo, Akira, 1965.
A theoretical model of diffusion of dye patches. (summary).
Symposium, Diffusion in oceans and freshwaters, Lamont Geol. Obs., 31 Aug.-2 Sept., 1964, 74-78.

diffusion

Okubo, Akira, 1964.
Equations describing the diffusion of an introduced pollutant in a one-dimensional estuary.
In: Studies on Oceanography dedicated to Professor Hidaka in commemoration of his sixtieth birthday, 216-226.

diffusion

Okubo, Akira, 1964
Diffusion from an instantaneous source in a nonuniform flow. (Abstract).
Trans. Amer. Geophys. Union, 45(1):66.

diffusion

Okubo, Akira, 1962
A review of theoretical models for turbulent diffusion in the sea.
J. Oceanogr. Soc., Japan, 20th Ann. Vol., 286-320.

diffusion

Okubo, Akira, 1962
Horizontal diffusion from an instantaneous point-source due to oceanic turbulence.
Chesapeake Bay Inst., Techn. Rept. (Ref. 62-22) 32: 124 pp.

diffusion

*Okubo, Akira, and Michael J. Karweit, 1969.
Diffusion from a continuous source in a uniform shear flow.
Limnol. Oceanogr. 14(4):514-520.

diffusion

Olson, F.C.W., 1964
A simple derivation of the Noble equation for the diffusion of a circular dye patch.
J. Geophys. Res., 69(2):366.

diffusion

Olson, F.C.W., 1963.
Die horizontale Diffusion eines unendlich langen Streifens.
Deutsche Hydrogr. Zeits., 16(4):185-188.

diffusion

Olson, F. C. W., and T. Ichiye, 1959.
Horizontal diffusion.
Science, 130(3384):1255.

diffusion

Ozmidov, R.V., 1958.
[The calculation of a turbulent horizontal diffusion of admixture spots in a sea.]
Dokl. Akad. Nauk, SSSR, 120(4):761-763.

diffusion

Ozmidov, R.V., 1958.
[On the role of turbulence in the turbulent inequality rate in the process of diffusion.]
Izv. Akad. Nauk, SSSR, Ser. Geofiz., (2):272-273.

diffusion, turbulent

Ozmidov, P.V., 1957.
[Experimental investigation of horizontal turbulent diffusion in the sea and in an artificial reservoir of shallow depth.]
Izv. Akad. Nauk, SSSR, Ser. Geofiz., (6):756-764.

diffusion

Ozmidov, R.V., V.K. Astok., A.N. Gesentswei and M.K. Yukhat, 1971.
Statistical properties of concentration fields of a passive pollutant artificially introduced in the sea. Fiz. Atmosfer. Okean. Izv. Akad. Nauk SSSR, 7(9): 963-973. (In Russian; English abstract)..

diffusion

Ozmidov, R.V., A.N. Gezentswei and G.S. Karabashev, 1969.
New data on the admixture diffusion in the sea. (In Russian; English abstract).
Fizika Atmosfer. Okean. Akad. Nauk SSSR, 5(11): 1191-1204

diffusion

Ozmidov, R.V., A.N. Gezentsvey and G.S. Karabashev, 1969.
Some results of the mixture diffusion in the sea. (In Russian; English abstract).
Okeanologiia, 9(1):82-86.

diffusion

Pagliari, Marcello, 1966.
First experimental results of the diffusion of fresh water in a shallow bay.
Air and Water Pollution, 10(8):537-548.

diffusion

Palm, E., 1958.
On Reynolds stress, turbulent diffusion and velocity profile in a stratified fluid.
Geofys. Publikaj., Geophysica Norvegica, 20(2): 1-13.

diffusion

Philip, J.R., 1956.
An application of the diffusion equation to viscous motion with a free surface.
Australian J. Physics, 9(4):570-573.

diffusion

Pochapsky, T.E., 1965.
Diffusing particles in the deep ocean. (abstract).
Symposium, Diffusion in oceans and freshwaters, Lamont Geol. Obs., 31 Aug. - 2 Sept., 1964, 114-115.

diffusion

Pritchard, D.W., Akira Okubo and Emanuel Mehr, 1962
A study of the movement and diffusion of an introduced contaminant in New York Harbor waters.
Chesapeake Bay Inst., Techn. Rept. (Ref. 62-21) 31: 89 pp. (multilithed)

diffusion

Reinert, Richard ., 1965.
Near-surface oceanic diffusion from a continuous point source.
Symposium, Diffusion in oceans and fresh waters., Lamont Geol. Obs., 31 Aug.-2 Sept. 1964, 10-27.

diffusion, dyes

Reinert, Richard L., and Gerard J. McNally, 1965.
On the use of a tape recording system in dye diffusion studies. (abstract).
Ocean Sci. and Ocean Eng., Mar. Techn. Soc.,-Amer. Soc. Limnol. Oceanogr., 2:1095-1102.

diffusion

Rosset, François, 1962.
Expérience de dispersion d'un effluent dans la mer.
Cahiers Océanogr., C.C.O.E.C., 14(2):103-119.

diffusion

Rouse, H., and J. Dodu, 1955.
Diffusion turbulente à travers une discontinuité de densité. La Houille Blanche 10(4):522-529.

diffusion

Saint-Guily Bernard, 1969.
Sur la diffusion turbulente verticale dans le mer.
Rapp. P.-v. Réun. Commn int. Explor. scient. Mer Méditerr.
19(5):927

diffusion

Saint-Guily, B., 1963.
On vertical heat convection and diffusion in the South Atlantic.
Deutsche Hydrogr. Zeits., 16(6):263-268.

diffusion

Schönfeld, J.C., 1962.
Integral diffusivity.
J. Geophys. Res., 67(8):3187-3200.

diffusion

Schooley, Allen H., 1971.
Diffusion sublayer thickness over wind-disturbed water surfaces. J. phys. Oceanogr. 1(3): 221-223.

diffusion

Shtokman, V.B., 1944.
[The stationary distribution of tongue-shaped isotherms in the sea in turbulent diffusion and speed of flow.] Izvest. Akad. Nauk, USSR, Geogr. Geofiz., 8:176-182.

diffusion

Spencer, Derek W. and Peter G. Brewer, 1971.
Vertical advection, diffusion and redox potentials as controls on the distribution of manganese and other trace metals dissolved in waters of the Black Sea. J. geophys. Res, 76(24): 5877-5892.

diffusion

Stommel, H., 1949.
Horizontal diffusion due to oceanic turbulence.
J. Mar. Res. 8(3):199-225, 3 textfigs.

diffusion

Turbovich, L.T., 1947.
[Turbulent diffusion and its function in the evaporation of water from the ocean surface.] Zh. Eks Teor. Fiz. 17(4):6 pp.

diffusion

Uusitalo, S., 1961
On the diffusivity of turbulent flow
Geophysica, 7(3): 191-193.

diffusion

Waldichuk, Michael, 1963.
Vertical diffusion in the sea from a radioactive point source.
Fish. Res. Bd., Canada, Mss Rept. Ser., (Ocean. & Limnol.), No. 155:27 pp.

diffusion

Watanabe, Nobuo, 1963.
Diffusion of industrial drainage in a tidal river. An example in the lower reach of the Edo River with small natural discharge. (In Japanese; English abstract).
J. Oceanogr. Soc., Japan, 18(4):172-184.

diffusion

Welander, Pierre, 1966.
Note on the effect of rotation on diffusion processes.
Tellus, 18(1):63-66.

diffusion

Wennekens, M. P.
Diffusion processes vs. oceanic microstructure.
Symposium, Diffision in oceans and fresh waters, Lamont Geol. Obs. 31 Aug. - 2 Sept., 1964, 108-113.

diffusion

*Yoshido, Kozo, 1967.
Time-dependent responses of stratified oceans.
Rec. oceanogr. Wks. Japan, 9(1):7-22.

diffusion

Zatz, V.I., 1964.
On the problem of horizontal turbulence of diffusion in the near-coastal zone of the Black Sea. (In Russian).
Okeanologiia, Akad. Nauk, SSSR, 4(2):249-257.

diffusion, effect of

Phillips, O.M., 1970.
On flows induced by diffusion in a stably stratified fluid. Deep-Sea Res., 17(3): 435-443.

diffusion, estuarine

diffusion (estuarie

Bowden, K.F., 1967.
Circulation and diffusion.
In: Estuaries, G.H. Lauff, editor, Publs Am. Ass. Advmt Sci., 83:15-36.

diffusion, gases

diffusion, sediments

Calleme, Bernard, 1967.
Sur le diffusion de l'oxygène à l'intérieur des sédiments marine.
Trav. Cent. Rech. Etud. océanogr., 7(2):25-29.

diffusion heat, effect of

Welander, Pierre with Kjell Holmåker, 1971.
Instability due to heat diffusion in a stably stratified fluid.
J. fluid Mech. 47(1): 51-64

diffusion, horizontal

diffusion, horizontal

Ichiye, Takashi, 1962
On formation of the intermediate water in the northern Pacific Ocean.
Geofisica Pura e Applicata, 51(1):108-119.

diffusion, horizontal

Ichiye, T., 1959.
A note on horizontal diffusion of dye in the ocean. J. Oceanogr. Soc., Japan, 15(4): 171-176

diffusion, horizontal

Ito, Naoji, 1964
On the small-scale horizontal diffusion near the coast.
J. Oceanogr. Soc., Japan, 19(4):182-189.

diffusion, horizontal

Joseph, J., H. Sender and H. Weidemann, 1964.
Untersuchungen über die horizontale Diffusion in der Nordsee.
Deutsche Hydrogr. Zeits., 17(2):57-75.

diffusion, horizontal

*Okubo, Akira, 1968.
Some remarks on the importance of the "shear effect" on horizontal diffusion.
J. Oceanogr. Soc., Japan, 24(2):60-69.

diffusion, horizontal

*Okubo, Akira, 1967.
The effect of shear in an oscillatory current on horizontal diffusion from an instantaneous source.
Int. J. Oceanol. Limnol., 1(3):194-204.

diffusion (ions)

Manheim, F.T., 1970.
The diffusion of ions in unconsolidated sediments.
Earth planet. Sci. Letts. 9(4): 307-309.

diffusion, lateral

diffusion, lateral

Alsaffar, Adnan M., 1966.
Lateral diffusion in a tidal estuary.
J. geophys. Res., 71(24):5837-5841.

diffusion, lateral

Ichie, T., 1950.
On the mutual effect of the coastal and offshore waters. J. Ocean. Soc., Japan (Nippon Kaiyo Gakkaisi) 6(2):1-7 (In Japanese; English summary)

diffusion, lateral

Miyazaki, M., 1950.
A theory of the transversal velocity distribution in ocean currents. Ocean. Mag., Tokyo, 2(2): 53-57, 4 textfigs.

diffusion, light

diffusion, light

Bauer, D. 1965
Mesure dans la mer du coefficient de diffusion de la lumière pour des angles compris entre 3° et 14°. (abstract only)
Rapp. P.-v. Réun., Comm. int. Explor. scient., Mer Mediterr., 18(3):819-820.

diffusion, optical

Bauer, Daniel, et Alexandre Ivanoff 1971.
Description d'un diffusiomètre "intégrateur": quelques résultats concernant le coefficient de diffusion optique des eaux de mer.
Cah. océanogr. 23(9): 827-839.

diffusion, light

Bauer, Daniel, et Alexandre Ivanoff, 1965.
Au sujet de la mesure du coefficient de diffusion de la lumière par les eaux de mer pour des angles compris entre 14° et 1° 30'.
Compte Rendus, Acad. Sci., Paris, 260(2):631-634.

diffusion, light

Ivanoff, A., 1961
Quelques resultats concernant les proprietes diffusantes des eaux de mer. Symposium on radiant energy in the sea, Helsinki, 4-5 Aug. 1960.
U.G.G.I., Monogr., No. 10:45-51.

diffusion, light

LeGrand, Y., 1961
Au sujet de la definition et de la mesure de la diffusion. Symposium on radiant energy in the sea, Helsinki, 4-5 Aug., 1960.
U.G.G.I. Monogr., No. 10:9-11.

diffusion, light

Morel, André 1968.
Relations entre coefficients angulaires et coefficient total de diffusion de la lumière pour les eaux de mer.
Cah. océanogr. 20(4): 291-301.

diffusion, light

*Morel, Andre, 1968.
Note au sujet des constantes de diffusion de la lumière pour l'eau et l'eau de mer optiquement pure.
Cah. océanogr., 20(2):157-162.

diffusion, light

Morel, André 1965
Détermination de l'indicatrice de diffusion de la lumière de quelques eaux méditerranéennes (abstract only)
Rapp. P.-v. Réun., Comm. int. Explor. scient., Mer Mediterr., 18(3):817.

diffusion, light

Morel, André, 1965.
Résultat des mésures de diffusion de la lumière effectuées lors de la croisère No. 25 du batiment oceanographique "Chain".
Cahiers Oceanogr. C.C.).E.C., 17(2):107-121.

diffusion, thermal

Alexander, R.C. 1971.
On the advective and diffusive heat balance in the interior of a subtropical ocean.
Tellus. 23 (4/5): 393-403

diffusion, turbulent

diffusion, turbulent

Carstens, T., 1970.
Turbulent diffusion and entrainment in two-layer flow.
J. WatWays Harb. Div. Am. Soc. civ. Engrs. 96 (WW1): 97-104.

turbulent diffusion

*Church, Michael, 1967.
Observations of turbulent diffusion in a natural channel.
Can. J. Earth Sci., 4(5):855-872.

diffusion, turbulent

Hecht, A. Artur, 1964.
On the turbulent diffusion of the water of the Nile floods in the Mediterranean Sea.
Sea Fish. Res. Sta., Israel, Bull., No. 36:1-24.

diffusion (turbulent)

*Herrera, Luis E., 1967.
Un experiments sobre difusion turbulenta.
Bol. Inst. Oceanogr., Univ. Oriente, 6(2):163-185.

diffusion, turbulent

Ichiye, Takashi, 1966.
Turbulent diffusion of suspended particles near the ocean bottom.
Deep-Sea Res., 13(4):679-685.

diffusion, turbulent

Kochergin, V.P., 1970.
A numerical method for the solution of some problems on ocean circulation. (In Russian, English abstract)
Met. Gidrol. 1970 (5): 67-75

diffusivity, turbulent

Ozmidov, R.V., 1968.
On the dependence of the lateral turbulent diffusivity K_l in the ocean of the phenomenon seale. (In Russian).
Fisika Atmosfer. Okean., Izv. Akad. Nauk, SSSR, 4(11): 1224-1225.

diffusion, turbulent

Pritchard, D.W., and J.H. Carpenter, 1960
Measurements of turbulent diffusion in estuarine and inshore waters.
Bull. Int. Assoc., Sci. Hydrol., (20):37-50.

Also in: Chesapeake Bay Inst., Collected Reprints, 5(1963)

diffusion, turbulent
*Saint-Guily, B., 1968.
Effect de la thermocline sur la diffusion d'une substance à partir d'une source.
Vie Milieu 19(1-B): 1-8.

diffusion, turbulent
Schuert, Edward A., 1970.
Turbulent diffusion in the intermediate waters of the North Pacific Ocean. J. geophys. Res., 75(3): 673-682.

diffusion vertical

diffusion, vertical
Miyake, Yasuo, Katsuko Saruhashi, Yukio Katsuragi and Teruko Kanazawa, 1962
Penetration of 90 Sr and 137 Cs in deep layers of the Pacific and vertical diffusion rate of deep water.
Repts. JEDS, Deep-Sea Res. Comm., Japan Soc., Promotion of Science, 3:141-147.

Originally published (1962):
J. Radiation Res., 3(3):141-147.

JEDS Contrib. No. 25

diffusion, vertical
Murray, Stephen P., 1970.
Settling velocities and vertical diffusion of particles in turbulent water. J. geophys. Res., 75(9): 1647-1654.

diffusion, vertical
Saint-Guily B. 1965.
Diffusion verticale dans les eaux superficielles de l'ouest de la Mer Méditerranée et du sud de la Mer Rouge.
Vie Milieu 19(2-B): 225-231

diffusion, vertical
Saint-Guily, Bernard, 1966.
Diffusion verticale au niveau de l'eau intermédiaire.,
Bull. Inst. Océanogr., Monaco, 66(1367):12pp.

diffusion, vertical
Weichart, Günter, 1967.
Berechnung der Vertikaldifussion von natürlichen Stoffen und Abfallstoffen in der Iberischen Tiefsee-Elbene aus der vertikalen Konzentrationsverteilung der natürlichen Stoffe über dem Meeresboden.
Dt. hydrogr. Z., 19(6):266-284.

diffusive interface
Huppert, Herbert E., 1971.
On the stability of a series of double-diffusive layers. Deep-Sea Res, 18(10): 1005-1021.

diffusivity.

diffusivity
Brekhovskikh, Academician L.M., K.N. Fedorov, L.M. Fomin, M.N. Koshlyakov and A.D. Yampolsky, 1971.
Large-scale multi-buoy experiment in the tropical Atlantic. Deep-Sea Res. 18(12): 1189-1206.

diffusivity, neighbor
Ichiye, T., and Olson, F.C.W., 1960
Uber die neighbor diffusivity im Ozean.
Deutschen Hydrogr. Zeits., 13(1): 13-23.

diffusivity
Mongelli, F., 1961
Il coefficiente di diffusivita termica del suolo.
Bol. Geofis. Teorica ed Applicata, Trieste 3(9):1-12.

diffusivity (vertical)
Sweers, H.E. 1970.
Vertical diffusivity coefficient in a thermocline.
Limnol. Oceanogr. 15(2): 273-280.

diffusivity
*Turner, J.S., 1968.
The influence of molecular diffusivity on turbulent entrainment across a density surface.
J. fluid Mech., 3(4):639-656.

Diffusivity
Uusitalo, S., 1961.
On the diffusivity of turbulent flow.
Geophysica, Helsinki, 7(3):191-193.

dikes

dikes, diabase
May, Paul R., 1971.
Pattern of Triassic-Jurassic diabase dikes around the North Atlantic in the context of predrift position of the continents.
Bull. geol. Soc. Am. 82(5): 1285-1292.

dikes
Watt, W. Stuart 1969.
The coast-parallel dike swarm of southwest Greenland in relation to the opening of the Labrador Sea.
Can. J. Earth Sci., 6(5): 1320-1321

dikes
Zhdanov, A.M., 1956.
[Reinforcement of coast with "galechnimi" alluvium (deposits) dikes (dams) along the complete profile. Studies of sea coasts and reservoirs.]
Trudy Okeanogr. Komissii, Akad. Nauk, SSSR, 1:18-36.

diorites

dipole hypothesis

dipole hypothesis
Opdyke, N.D., and K.W. Henry, 1969.
A test of the dipole hypothesis.
Earth planet. Sci. Letters 6(2): 139-151

disasters

disasters
Halmos, E.E., Jr., 1963.
Disaster under water.
Undersea Techn., 4(1):36-37.

disasters at sea
Sekizuka, R., and N. Tamada, 1963.
Relations between marine disasters and tidal streams in Obatake Seto. (In Japanese; English abstract).
Hydrogr. Bull., Tokyo, No. 73 (Publ. No. 981): 29-45.

discoasters

discoasters, zonation of
Bandy, Orville L., and Mary E. Wade, 1967.
Miocene-Pliocene boundaries in deep-water environments.
Progress in Oceanography, 4:51-66.

discoasters
Boganov, Iu. A., and M.G. Vshakova, 1966.
Coccoliths of the discoaster group Tan Syn Hok in aqueous suspensions from the Pacific. (In Russian).
Dokl., Akad. Nauk, SSSR, 171(2):465-467.

discoasters
Bukry, David, 1971
Discoaster evolutionary trends.
Micropaleontology, Vol. 17 (1): 43-52

discoasters
Cohen, C.L.D., 1965.
Coccoliths and discoasters.
Geol. en Mijnbouw, 44110):337-344.

discoasters
Cohen, C.L.D., 1965.
Coccoliths and discoasters from the bottom sediments of the Adriatic.
Rapp. P.-v. Réun., Comm. int. Explor. scient., Mer Méditerr. 18(3):957.

discoasters
*Gartner, Stefan, Jr., 1967.
Calcareous mannofossils from Neogene of Trinidad, Jamaica, and Gulf of Mexico.
Paleontol. Contrib., Univ. Kansas, (29):1-6.

discoasters
*Hay, William W., Hans P. Mohler, Peter H. Roth, Ronald R. Schmidt and Joseph E. Boudreaux, 1967.
Calcareous nannoplankton zonation of the Cenozoic of the Gulf Coast and Caribbean-Antillean area, and transoceanic correlation.
Trans. Gulf Coast Assoc. Geol. Soc., 17th Ann. Meet.: 428-480.

discolored water
See also: Red water

discolored water
See also: red water

discolored water
Brodie, J.W., 1958
A note on tidal circulation in Port Nicholson, New Zealand. N.Z.J. Geol. Geophys. 1(4): 684-702.

discolored water
Hanks, Robert W., 1966.
Observations on "milky water" in Chesepeake Bay.
Chesapeake Sci., 7(3):175-176.

discoloration (water0
*Pieterse, F., and D.C. van der Post, 1967.
The pilchard of South West Africa (Sardinops ocellata): oceanographic conditions associated with red-tides and fish mortalities in the Walvis Bay region.
Investl Rept., Mar. Res. Lab., SWest Africa, 14: 1s5 pp.

discolored water
*Taga, Nobuo, 1967.
Microbial coloring of sea water in tidal pool, with special reference to massive development of photosynthetic bacteria.
Inf. Bull. Planktol. Japan, Comm. No. Dr. Y. Matsue, 219-229.

discontinuity, density

discontinuities
Ishida, Masami, Tsuneyoshi Suzuki, Noritatsu Sano, Ichiro Saito and Seikichi Mishima, 1960.
[On the detection of the boundary zone.] Bull. Fac. Fish., Hokkaido Univ., 10(4):291-302.

discontinuities (chlorophyll)
Lorenzen, Carl J., 1971.
Continuity in the distribution of surface chlorophyll. J. Cons. int. Explor. Mer, 34 (1): 18-23.

Discontinuities

Lumby, J.R., and D.J. Ellett, 1965.
Some considerations on oceanographic observations.
Journal du Conseil, 29(3):237-248.

discontinuity, horizontal
Pandolfe, Joseph P., and Philip S. Brown Jr., 1967.
Inertial oscillations in an Ekman layer containing a horizontal discontinuity surface.
J.mar.Res., 25(1):1-28.

discontinuity, density
Rouse, H., and J. Dodu, 1955.
Diffusion turbulente à travers und discontinuité de densité. La Houille Blanche 10(4):522-529.

discontinuity layer

discontinuity layers
Anderson, George C., and Karl Banse, 1963
Hydrography and phytoplankton production.
Proc. Conf., Primary Productivity Measurements, Marine and Freshwater, Univ. Hawaii, Aug. 21-Sept. 6, 1961, U.S. Atomic Energy Comm., Div. Techn. Info. TID-7633:61-90.

discontinuities
Bulgakov, N.P., 1966.
On the influence of initial conditions on the transformation of some characteristic surfaces by the vertical mixing of water masses. (In Russian; English abstract).
Okeanologiia, Akad. Nauk, SSSR, 6(5)L760-769.

discontinuity layer (density)
Fedorov, K. N., 1961
Density discontinuity layer in the variable currents.
Okeanologiya, 1: 25-29.

discontinuity layer, effect of
Gillbricht, M., 1954.
Das Verhalten von Zooplankton - Vorzugsweise von Tintinnopsis beroidea Entz - gegenüber Thermohalinen Sprungschichten. Kurze Mitt., Inst. Fischereibiol., Univ. Hamburg, No. 5:32-44, 5 textfigs.

discontinuity layer
Goering, J.J., D.D. Wallen and R.M. Nauman, 1970.
Nitrogen uptake by phytoplankton in the discontinuity layer of the eastern subtropical Pacific Ocean. Limnol. Oceanogr., 15(5): 789-796.

discontinuity layer
Hansen, K.V., 1951.
On the diurnal migration of zooplankton in relation to the discontinuity layer. J. du Cons. 17(3):231-241, 9 textfigs.

discontinuity layers, effect of
Harder, W., 1957.
Reaction of plankton-organisms to discontinuity layers. L'Année Biol., 33(5-6):227-232.

discontinuity layer
Kalle, K., 1953.
Zur Frage der inneren thermischen Unruhe des Meeres. Deutsche Hydrogr. Zeits. 6(4/5/6):145-170, Pls. 14-21, 30 textfigs.

discontinuity layer
Miyazaki, M., and S. Abe, 1960
[On the water masses in the Tsushima Current.]
J. Oceanogr. Soc., Japan 16(2): 59-68.

discontinuity layer
Yamamoto, H., 1958.
Oceanochemical studies on Japanese inlets (1).
Discontinuous layer in Urado Bay.]
J. Oceanogr. Soc., Japan, 14(1):25-30.

disease

diseases
Andrews, J.D., and W. Hewatt, 1957.
Oyster mortality studies in Virginia. II. The fungous disease caused by Dermocystidium marinum in oysters in Chesapeake Bay.
Ecological Monographs, 27(1):1-26.

diseases of oysters
Bonami, Jean Robert, Henri Grizel, Constantin Vago, et Jean-Louis Duthoit 1971.
Recherches sur une maladie épizootique de l'huitre plate, Ostrea edulis Linné.
Revue Trav. Inst. Pêches marit. 35(4): 415-418.

disease
Bruun, A. Fr., and B. Heiberg, 1932
The "Red Disease" of the eel in Danish Waters. Medd. fra Komm for Danmarks Fiskeri og Havundersøgelser, serie: Fiskeri, IX (6): 19 pp.

diseases
Dawe, Clyde J., and John C. Harshbarger, 1969.
A symposium on neoplasms and related disorders of invertebrate and lower vertebrate animals, Washington D.C., June 19-21, 1968
Nat. Cancer Inst. (U.S.A.) Monogr. 31: 772 pp.

diseases
Egusa, Syuzo, and Tomoko Nishikawa, 1965.
Studies of a primary infectious disease in the so-called fungus disease of eels.
Bull. Jap. Soc. Sci. Fish., 31(10):804-813.

diseases (fish)
Hall, David L., and E.S. Iversen, 1967.
Henneguya lagodon, a new species of myxosporidian, parasitizing the pinfish, Lagodon rhomboides.
Bull. mar. Sci., Miami, 17(2):274-279.

diseases
*Ishikawa, Y., 1967.
A disease of young cultured karuma-prawn, Penaeus japonicus Bate. (In Japanese; English abstract).
Bull.Fish.Exp.Stn, Okayama Pref., 1966:5-9.

diseases (fish)
Iversen, E.S., and N.N. van Meter, 1967.
a new myxosporidian (Sporozoa) infecting the Spanish mackerel.
Bull. mar. Sci., Miami, 17(2):268-273.

diseases
Moore, B., 1970.
The present status of diseases connected with marine pollution.
Rev. int. Océanogr. méd. 18-19: 193-223

diseases, fish
Nigrelli, Ross F., K.S. Ketchen and G.D. Ruggieri, 1965.
Studies on virus diseases of fishes. Epizootiology of epithelial tumors in the skin of flatfishes of the Pacific coast with special reference to the sand sole (Psettichthys Melanostictus) from northern Hecate Strait, British Columbia, Canada, Zoologica, N.Y. 50(3):115-122.

diseases, fish
Nigrelli, Ross, and George D. Ruggieri, 1965.
Studies on virus diseases of fishes, spontaneous and experimentally induced cellular hypertrophy (lymphocystis disease) in fishes of the New York aquarium, with a report of new cases and an annotated bibliography (1874-1965).
Zoologica, N.Y. Zool. Soc., 50(2):83-95.

diseases
Numachi, Ken-ichi, Juichi Oizumi, Shigeru Sato, and Takeo I mai, 1965.
Studies on the mass mortality of the oyster in Matsushima Bay. III. The pathological changes of the oyster caused by gram-postive bacteria and the frequency of their injection. (In Japanese: English abstract).
Bull. Tohoku Rep. Fish Res. Lab., No. 25:39-44.

disease
Oppenheimer, C., and C.L. Kesteven, 1953.
Disease as a factor in natural mortality of marine fish. FAO Fish. Bull. 6(6):215-222.

disease
Sato, Shigekatsu, Maryse Nogueira Paranagua and Enide Eskinazi (1963-1964) 1966.
On the mechanism of red tide of Trichodesmium in Recife, northeastern Brazil with some considerations of the relation to the human disease "Tamandaré" fever.
Trabhs Inst. Oceanogr. Univ. Recife 5(5/6): 7-49.

diseases
Sindermann, Carl J. 1968.
Bibliography of oyster parasites and diseases.
Spec. scient. Rep. U.S. Fish Wildl. Serv., Fish. 563: 13 pp.

diseases
Sprague, Victor, and Robert L. Beckett, 1966.
A disease of blue crabs (Callinectes sapidus) in Maryland and Virginia.
J. invert. Pathol., 8(2):287-289.

invertebrates, diseases of
Squires, Donald F., 1965.
Neoplasia in a coral?
Science, 148(3669):503-505.

diseases
Stewart, James E., John W. Cornick and B.M. Zwicker, 1969.
Influence of temperature on gaffkemia, a bacterial disease of the lobster Homarus americanus.
J. Fish. Res. Bd., Can., 26 (9): 2503-2510.

diseases
Thorshaug, K., and A. Fr. Rosted, 1956.
Researches into the prevalence of trichinosis in animals in Arctic and Antarctic waters.
Norsk Polarinst. Medd., 80:115-129.

disease (Minamata neurosis)
*Yoshida, Tamao, Toshiharu Kawabata and Yoshiyuki Matsue, 1967.
Transference mechanism of mercury in marine environment.
J. Tokyo Univ. Fish., 53(1/2):73-84.

dispersal

dispersal
Briggs, John C., 1967.
Dispersal of tropical marine shore animals: Coriolis parameters or competition?
Nature, Lond., 216(5113):350.

dispersal
Coe, W.R., 1946.
The means of dispersal of bathypelagic animals in the North and South Atlantic Oceans. Amer. Nat. 80(794):453-469, 2 figs.

dispersion
Dobrin, M.B., 1951.
Dispersion in seismiv surface waves. Geophys. 16(1):63-80, 17 textfigs.

dispersal, mechanism for
Marr, J.W.S., 1962.
The natural history and geography of the Antarctic krill (Euphausia superba Dana).
Discovery Repts., 32:33-464, Pl. 3.

dispersion

dispersion
Bailey, Thomas E., Charles A. McCullough and Charles G. Gunnerson, 1966.
Mixing and dispersion studies in San Francisco Bay.
J. sanit. Engng Div. Soc. civ. Engrs, 92(SA5):23-45.

dispersion
Carter, H.H., and A. Okubo 1970.
Longitudinal dispersion in non-uniform flow.
Techn. Rept. Chesapeake Bay Inst. 68 (Ref. 70-8): 46 pp. (multilithed)

dispersion
Geffriaud, Jean-Paul, 1968.
Étude sur la dispersion et sur le renouvellement de l'eau dans les rades du port de Cherbourg, mars 1967.
Cah. océanogr., 20(2):109-132.

dispersion
Horrer, Paul L., 1962
Oceanographic studies during Operation "Wigwam". Physical oceanography of the test area.
Limnol. and Oceanogr., Suppl. to Vol. 7:vii-xxvi.

dispersion
Ketchum, B.H., and W.L. Ford, 1952.
Rate of dispersion in the wake of a barge at sea.
Trans. Amer. Geophys. Union 33(5):680-684, 3 textfigs.

dispersion, sound
Kibblewhite, A.C., and R.N. Denham, 1965.
Experiment on propagation in surface sound channels.
J. Acoust. Soc., Amer., 38(1):63-71

dispersion
LeGoff, P., 1961.
Les dispersions dluide-dans fluide. Princeps physico-chimiques et utilisations en génie chimique.
Mem. Trav. Soc. Hydrotechn., France, 1(Suppl.): 2-14.

dispersion
O'Connor, Donald J., and Dominic M. DiToro, 1965.
The solution of the continuity equation in cylindrical coordinates with dispersion and advection for an instantaneous release.
Symposium, Diffusion in oceans and fresh waters Lamont Geol. Obs., 31 Aug.-2 Sept., 1964, 80-85.

dispersion
Okubo, Akira, 1969.
A note on the effect of dispersion on mean current measurements
Techn. Rept. Chesapeake Bay Inst. 55 (Ref. 69-8): 1-20pp. (unpublished manuscript)

dispersion
Palmer, Merv D. and J. Bryan Izatt, 1971.
Lake hourly dispersion estimates from a recording current meter. J. geophys. Res., 76(3) 688-693.

displacement, ocean bed
Le, Saty a Sankar 1969.
The effect of a sudden displacement of the ocean bed.
Beitr. Geophys. 78(5): 399-413 (German)

dissolved organic matter

distance measurements

distances, measurement of
Bigelow, Henry W., 1963.
Measurement of distances over water.
Naval Oceanogr. Off., IMR, No. N-2-63:9 pp.

distance measurements
Zverev, S.M., 1959.
The use of sound records for distance determination during operations of deep seismic sounding at sea.
Izv., Akad. Nauk, SSSR, Ser. Geofiz., (4):560-569.
English Edit. (1960), (4):385-392.

distribution of animals

distribution of animals
Ekman, S., 1947.
Ueber die Festigkeit der marinen Sedimente als Factor der Tierverbreitung, ein Beitrag zur Associationsanalyse. Zool. Bidrag., Uppsala, 25:1-20.

distribution, animals
Hutchins, L.W., 1947.
The bases for temperature zonation in geographical distribution. Ecol. Mon. 17:325-335, 8 text figs.

distribution by currents

distribution by currents
Bright, K.M.F., 1938
The South African intertidal zone and its relation to ocean currents. II,III. Areas of the west coast. Trans. Roy. Soc. S. Africa, 26:49-88.

distribution by currents
Eyre, J., 1939
The South African intertidal zone and its relation to ocean currents. VII An area in False Bay. Ann. Natal Mus., 9:283-306.

distribution by currents
Eyre, J., Brock Ruysen, G.J., and M.I. Crichton 1938
The South African intertidal zone and its relation to ocean currents. VI The East London District. Ann. Natal Mus., 9:83-112

distribution by currents
Eyre, J., and T.A. Stevenson, 1938
The South African intertidal zone and its relation to ocean currents. V A subtropical Indian shore. Ann. Natal Mus. 9:21-46

distribution, horizontal

horizontal distribution of properties
Austin, Thomas S., 1960
Oceanography of the east central equatorial Pacific as observed during expedition "Eastropic". U.S.F.&W.S., Fish. Bull. 168, Vol. 60: 257-282.

Diurnal migrations: See Vertical migration

diurnal migrations
See: vertical migrations
See also under animal groups in file of organisms

diurnal fluctuations
Honjo, Tsuneo, and Tasuku Hanaoka, 1969
Diurnal fluctuations of photosynthetic rate and pigment contents in marine phytoplankton.
J. Oceanogr. Soc. Japan, 25(4): 182-190
(In Japanese; English abstract)

diurnal migrations
Kawaguchi, Kouichi, 1969.
Diurnal vertical migration of micronektonic fish in the western North Pacific. Bull. Plankt. Soc Japan, 16(1): 63-66.
(In Japanese; English abstract)

Diurnal Migration
Zalkina, A.V., 1970.
Vertical distribution and diurnal migration of some Cyclopoida (copepoda) in the tropical region of the Pacific Ocean. Marine Biol., 5 (4): 275-282.

diurnal rhythm

diurnal rhythm
Hastings, J. Woodland, and Beatrice M. Sweeney, 1958.
A persistent diurnal rhythm of luminescence in Gonyaulax polyedra.
Biol. Bull., 115(3):440-458.

diurnal rhythms
Lewin, Ralph A., Editor, 1962.
Physiology and biochemistry of algae.
Academic Press, New York and London, 929 pp.

Diurnal rhythms
Mangum, C.P. and D.M. Miyamoto, 1970.
The relation between spontaneous activity cycles and diurnal rhythms of metabolism in the polychaetous annelid Glycera dibranchiata.
Marine Biol., 7(1): 7-10.

diurnal rhythms
Nozawa, K., 1940.
Problem in the diurnal rhythm in the cell division of the dinoflagellate, Oxyrrhis marina.
Annot. Zool. Japonensis, 19(3):170-174.

diurnal rhythms
Ohata, C.A., D.T. Matsuura, G.C. Whittow and S.W. Tinker 1972
Diurnal rhythm of body temperature in the Hawaiian monk seal (Monachus schauinslandi).
Pacific Sci. 26(1): 117-120.

diurnal variation

diurnal variations
Kolesnikov, A.G., and A.A. Pivovarov, 1955.
Computation of the diurnal variation of ocean temperatures by the total radiation and the temperature of the air. Doklady Akad. Nauk, SSSR, 102(2):261-264.

diurnal variation
Miyazaki, M., 1954.
On the diurnal variation of oceanographic elements in the Miyazu Bay.
Spec. Publ. Japan Sea Fish. Res. Lab., 3rd Anniv., 203-208.

diurnal variation
Sugiura, Y., 1954.
On the diurnal variation of dissolved oxygen. (1).
J. Ocean. Soc., Japan, 10(1):22-28, 2 textfigs.

diurnal variation
Sugiura, Y., 1954.
On the diurnal variation of dissolved oxygen. (II). J. Ocean. Soc., Japan, 10(2):65-70, 5 textfigs.

diurnal variation
Yentsch, C.S., and R.F. Scagel, 1958
Diurnal study of phytoplankton pigments. An in situ study in East Sound, Washington. J. Mar. Res., 17:567-583.

divergence

divergences
Allain, C., 1964.
Les poissons et les courants.
Rev. Trav. Inst. Pêches Marit., 28(4):401-426.

divergence
Al'tshuler, V.M., 1961.
Divergence due to force of rotation of the earth and its calculation in computing tidal currents. (In Russian).
Trudy Gosud. Okeanogr. Inst., 63:3-7.

divergence
Bang, N.D., 1971.
The southern Benguela Current region in February 1966: Part II. Bathythermography and air-sea interactions. Deep-Sea Res., 209-224.

divergences(convergence?)
Beklemishev, K.V., 1959
The Antarctic divergence and whaling grounds.
Izv. Akad. Nauk, SSSR, Ser. Geograf. (6): 90-93.

divergence
Bogdanov, D.V., 1965.
Algunos rasgos de la oceanografía del Golfo de México y del Mar. Caribe. (In Russian; Spanish abstract).
Sovetsk.-Cub. Ribokhoz. Issled., VNIRO:Tsentr. Ribokhoz. Issled. Natsional. Inst. Ribolovsta Republ. Cuba, 23-45.

divergences
Bogorov, B.G., and M. Ye. Vinogradov, 1961
Certain esculiaritipe (sic!) of the plankton mass distribution in the surface waters of the Indian Ocean in winter 1959/60.
Okeanol. Issledov., Mezhd. Komit., Mezhd. Geofiz. God, Presidiume Akad. Nauk, SSSR:(4)66-71.

divergence
Brodie, J.W., 1965
Oceanography.
In: Antarctica, Trevor Hatherton, editor
Methuen & Co., Ltd., 101-127.

divergences
Burling, R.W., 1961.
Hydrology of circumpolar waters south of New Zealand.
New Zealand Dept. Sci. Industr. Res. Bull., 143: 66 pp.

divergence
Burling, R. W., and D. M. Garner, 1959.
A section of 14C activities of sea water between 9°S and 66°S in the south-west Pacific Ocean.
NZ. J. Geol. Geophys., 2(4):799-824.

divergence
Chew, Frank, and G.A. Berberian 1971.
A determination of horizontal divergence in the Gulf Stream off Cape Lookout.
J. phys. Oceanogr. 1(1): 39-44.

divergences
Derbyshire, Mollie, 1966.
The surface waters near the coasts of southern Africa.
Deep-Sea Research, 13(1):57-81.

divergence
Gezentsvei, A.N., 1954.
Divergence of drift currents and heat transmission by the currents in the North Pacific and North Atlantic. Trudy Okeanol. Inst. 9:54-118.

divergence
Gorbenko, Yu. A., 1964.
On the possibility of revealing the areas of water rise and water lowering in the central part of the Pacific by applying the microbiological method. (In Russian).
Okeanologiia, Akad. Nauk, SSSR, (4):167-174.

divergence
Hidaka, K., 1955.
3. Divergence of surface drift currents in terms of wind stresses, with special application to the location of upwelling and sinking.
Jap. J. Geophys. 1(2):47-56, textfig.

divergence
Hidaka, K., and T. Kusunoki, 1951.
On the mixing coefficient and the meridional component of velocity in the Equatorial Counter Current. J. Ocean. Soc., Japan (Nippon Kaiyo Gakkaisi) 6(3):168-193, 4 tables, 5 textfigs.

divergence, currents
Hidaka, K., and K. Ogawa, 1958.
On the seasonal variations of surface divergence of the ocean currents in terms of wind stresses over the oceans. Geophys. Notes, Tokyo, 11(1): No. 10.

divergence
Hikosaka, Shigeo, 1958
Possible role of horizontal divergence to the meanders of currents in a stratified ocean.
Hydrogr. Bull., Tokyo (Publ. No. 981): 1-3.

divergence
Hikosaka, S., and R. Watanabe, 1957.
Areas of divergence and convergence of surface currents in the north-western Pacific.
Proc. UNESCO Symp., Phys. Ocean., Tokyo, 1955: 101-103.

divergence
Ichiye, Takashi, 1951
On the hydrography off Shionomisaki and Enshu-Nada(1951).
Bull. Kobe Mar. Obs., No. 164:231(top of page) -240; 13(bottom of page)-22.

divergence
India, Indian National Committee on Oceanic Research, Council of Scientific and Industrial Research, New Delhi, 1964.
International Indian Ocean Expedition, Newsletter India, 2(4):48 pp.

divergence
Istapoff, Feodor, 1962
The salinity distribution at 200 meters and the Antarctic frontal zones.
Deutsche Hydrographische Zeits., 15(4):133-142.

divergence
Ivanov, Yu. A., 1961
On frontal zones in the Antarctic waters.
Mezhd. Kom. Mezhd. Geofiz. Goda, Presidiume Akad. Nauk, SSSR, Okeanol. Issled., (3):30-51.

divergence (convergence?)
Ivanov, Yu. A., and B.A. Tarayev, 1959
On the problem of the structure of the Antarctic Divergence Zone. Izv. Akad. Nauk, USSR, Ser. Geograf. (6): 82-89.

divergence
Japan, Maritime Safety Board, 1958.
Possible roll of divergence to the meanders of currents in a stratified ocean.
Hydrogr. Bull. (Publ. 981), No. 55:1-3.
horizontal

divergence
Kawai, Hideo, and Hisao Sakamoto 1969.
A study on convergence and divergence in surface layer of the Kuroshio-II. Direct measurement of convergence and divergence at the top and bottom of surface mixed layer. (In Japanese; English abstract).
Bull. Nansei reg. Fish. Res. Lab. 2:19-38.

divergence
Koopmann, G., 1953.
Entstehung und Verbreitung von Divergenzen in der eberflachennalen Wasserbewegung der antarktischen Gewässer. Deutsche Hydrogr. Zeits. Ergänzungsheft 2:1-38, 2 pls., 18 textfigs.

divergence
Laevastu, T., 1962
The components of the surface currents in the sea and their forecasts.
Proc. Symposium on Mathematical-Hydrodynamical Methods of Phys. Oceanogr., Sept. 1961, Inst. Meeresk., Hamburg, 321-338.

divergence, effect of
LaFond, E.C., 1954.
Environmental factors effecting the vertical temperature structure of the upper layers of the sea. Andhra Univ. Ocean. Mem. 1:94-101, 25 textfigs.

divergence
Lane, R.K., 1962
A review of the temperature and salinity structures in the approaches to Vancouver Island, British Columbia.
J. Fish. Res. Bd.,Canada, 19(2):45-91.

divergence
Lebedeva, M.N., 1962
Microbiological indication of the convergence and divergence zones in the Indian Ocean. (In Russian).
Okeanologiia, Akad. Nauk, SSSR, 2(6):1104-1109.

divergence
Maeda, Sonosuke, and Ricardo Kishimoto 1970.
Upwelling off the coast of Peru.
J. oceanogr. Soc. Japan, 26(5): 300-309.

divergence
*Mazlika, P.A., 1968.
Some features of the Gulf Stream off Chesapeake Bay in the spring of 1963.
Fishery Bull. Fish Wildl. Serv. U.S., 66(2):387-423.

divergence
*Minas, H.J., 1971.
Résultats préliminaires de la campagne Médiprod I du Jean Carcot (1-15 Mars et 4-17 Avril 1969). (In French; Spanish abstract).
Investigación pesq. 35(1): 137-146.

divergence
Mratov, K.A., 1971.
Water upwelling and sinking zones of the Atlantic near West Africa. (In Russian; English abstract). Usloviia sediment. Atlantich. Okeana, Rez. Issled. Mezhd. Geofiz. Proekt. Mezhd. Geofiz. Kom. Prezid. Akad Nauk SSSR: 97-106.

divergence

Neumann, Gerhard, 1965.
Oceanography of the tropical Atlantic.
Anais Acad. bras. Cienc., 37 (Supl.):63-82.

divergence

Reid, Joseph L., Jr., 1962
On circulation, phospate-phosphorus content and zooplankton volumes in the upper part of the Pacific Ocean.
Limnol. and Oceanogr., 7(3):287-306.

divergence

Roden, Gunnar I., 1971.
Aspects of the transition zone in the northeastern Pacific. J. geophys. Res. 76(15): 3462-3475.

divergence

Rotschi, Henri, 1962.
Sur la divergence des Salomon.
Comptes Rendus, Acad. Sci., Paris, 255(21):2795-2797.

divergences (Solomons)

Rotschi, Henri, 1961
Contribution française en 1960 à la connaissance de la Mer de Coral: Océanographique physique.
Cahiers Océanogr., C.C.O.E.C., 13(7):434-455.

divergence

Rotschi, Henri, 1961
Influence de la divergence des Salomon sur la répartition de certaines propriétés des eaux.
C.R. Acad. Sci., Paris, 253:2559-2561.

divergence

Rotschi, H., and L. Lemasson, 1967.
Oceanography of the Coral and Tasman seas.
Oceanogr. Mar.Biol. Ann.Rev., Harold Barnes, editor, George Allen and Unwin, Ltd., 5:49-97.

divergences

Ryzhkov, Yu. G., and L.A. Koveshnikov, 1963.
Genesis of convergence and divergence zones above sharp changes in the slope of the ocean floor. (In Russian).
Izv., Akad. Nauk, SSSR, Ser. Geofiz., (6):953-959

Translation (AGU): (6)585-588.

divergence

Ryzhkov, Iu. G., and L.A. Koveshnikov, 1963.
Upwelling zone of convection and divergence of currents in regions with sudden changes of sloping bottom. (In Russian).
Izv. Akad. Nauk, SSSR, Ser. Geofiz., (6):953-959.

divergence

Semonov, A.I., 1961.
Estimate of the influence of the convergence-divergence phenomenon on the height of the average-year level (according to the example of the Azov Sea).
Trudy Gosud. Okean. Inst., Leningrad, 61:116-132.

In Russian

divergences

Stepanov, V.N., 1960
[The main water convergences and divergences in oceans.]
Biull. Okeanograf. Komissii. Akad. Nauk, SSSR, (6): 15-22.

divergence

Sudo, Hideo, 1960

On the distribution of divergence and convergence of surface drift vectors in the western Pacific Oceans. Rec. Oceanogr. Wks., Japan, n.s., 5(2): 25-43.

divergence

Uda, M., 1953.
On the convergence and divergence in the NW Pacific in relation to the fishing grounds and productivity. Bull. Japan. Soc. Sci. Fish. 19(4): 435-438, 3 textfigs.

divergence

Varadachari, V.V.R., and G.S. Sharma, 1964.
On the vergence field in the north Indian Ocean.
Bull. Nat. Geophys. Res. Inst., New Delhi, 2(1): 1-14.

divergence

Vladimirov, O.A., and E.A. Nikulicheva, 1960.
Survey of some contemporary methods for computing the "convergence-divergence" phenomenon and tests of its application in understanding of the Finnish Bays. (In Russian).
Trudy Gosud. Okean. Inst., Leningrad, 56:9-29.

divergence

Wickett, W. Percy, and J. Arthur Thompson, 1969.
Computation of divergence of Ekman transport for the North Pacific Ocean, 1946-1968.
Techn. Rept., Fish. Res. Bd., Can., 128: 27 pp.
(mimeographed)

divergence

Wüst., G., 1950.
Blockdiagramme der atlantischen Zirculation auf Grund der "Meteor"-Ergebnisse. Kieler Meeresf. 7(1):24-34. 3 textfigs., 1 fold-in.

divergence

Wyrtki, Klaus, 1961
The thermohaline circulation in relation to the general circulation in the oceans.
Deep-Sea Res., 8(1): 39-64.

divergences

Wyrtki, Klaus, 1960

The Antarctic convergence and divergence.
Nature, 187(4737): 581-582.

divergence

Wyrtki, Klaus, 1960

The Antarctic Circumpolar Current and the Antarctic Polar Front. Deutsche Hydrogr. Zeits., 13(4): 153-174.

divergence

Zernova, V.V., 1970.
Phytoplankton in the waters of the Gulf of Mexico and the Caribbean Sea. (In Russian; English and Spanish abstracts). Okeanol. Issled. Rez. Issled. Mezhd. Geofiz. Proekt. Mezhd. Geofiz. Kom. Prezid. Akad. Nauk SSSR 20: 69-104.

divergence

Zernova, V.V., 1970.
On water discoloring in the Gulf of Mexico caused by development of plankton algae.
(In Russian; English and Spanish abstracts).
Okeanol. Issled. Rez. Issled. Mezhd. Geofiz. Proekt. Mezhd. Geofiz. Kom. Prezid. Akad. Nauk SSSR 20: 105-109.

divergence, effect of

divergence, effect of

Minas, H.J., et F. Blanc, 1970.
Production organique primaire au large et près des côtes méditerranéennes françaises (juin-juillet 1965), influence de la zone de divergence.
Téthys 2(2):299-316

divergence, effect of

Osborne, M.F.M., 1965.
The effect of convergent and divergent flow patterns on infrared and optical radiation from the sea.
Deutsche Hydrogr. Zeits., 18(1):1-25.

divergence, effect of

Smith, Robert L., Christopher N.K. Mooers, and David B. Enfield, 1971.
Mesoscale studies of the physical oceanography in two coastal upwelling regions: Oregon and Peru. In: Fertility of the Sea, John D. Costlow, editor, Gordon Breach, 2: 513-535.

diversity

diversity

See also: faunal diversity

diversity

Arthur H. Clarke, 1969.
Diversity and composition of abyssal benthos.
Science, 166(3908): 1033-1034.

diversity

Deevey, Georgiana B., 1971.
The annual cycle in quantity and composition of the zooplankton of the Sargasso Sea off Bermuda. I. The upper 500 m. Limnol. Oceanogr. 16(2): 219-240.

diversity

Edden, Ann C., 1971.
A measure of species diversity related to the lognormal distribution of individuals among species. J. exp. mar. Biol. Ecol., 6(3): 199-209.

diversity

Kafescioglu, Ismail A. 1971.
Specific diversity of planktonic foraminifera on the continental shelves as a paleobathymetric tool.
Micropaleontology 17(4):455-470.

diversity

Kohn, Alan J., 1971.
Diversity, utilization of resources, and adaptive radiation in shallow-water marine invertebrates of tropical oceanic islands.
Limnol. Oceanogr. 16(2): 332-348.

diversity

Krylov, V. V. 1969.
On the affinity criterion of a pair of species and the identification of recurrent groups. (In Russian; English abstract).
Okeanologiia 9(1): 172-174.

diversity

Margalef, Ramon, 1969.
A practical proposal and a model of interdependence.
Diversity and stability in ecological systems, Brookhaven Symposia in Biology 22:25-37.
Diversity and stability.

diversity

Sanders, Howard L., 1969.
Benthic marine diversity and stability-time hypothesis.
Diversity and stability in ecological systems, Brookhaven Symposia in Biology, 22: 71-80

diversity

Sanders, Howard L. and Robert R. Hessler, 1969.
Diversity and composition of abyssal benthos.
Science, 166(3908): 1034.

diversity

Sanders, H.L. and R.R. Hessler, 1969.
Ecology of the deep-sea benthos. Science, 163(3874): 1419-1424.

diversity, taxonomic

Schopf, Thomas J.M., 1970.
Taxonomic diversity gradients of ectoprocts and bivalves and their geologic implications.
Bull. Geol. Soc. Am. 81(12):3765-3768.

diversity

Slobodkin, Lawrence B., and Howard L. Sanders, 1969.
On the contribution of environmental predictability to species diversity.
Diversity and stability in ecological systems, Brookhaven Symposia in Biology, 22:82-93

diversity

Valentine, James W. 1971.
Plate tectonics and shallow marine diversity and endemism, an actualistic model.
System. Zool. 20(3):253-264

divers

divers

Baddeley, A.D., 1965.
The relative efficiency at depth of divers breathing air and oxy-helium.
Malta '65: 13-18.

divers

Clarke, John W., 1965.
Methods and techniques for sea-floor tasks.
Ocean Sci. and Ocean Eng., Mar. Techn. Soc., Amer. Soc. Limnol. Oceanogr., 1:267-277.

divers

Nichols, A.K. 1969.
The personality of divers (and other sportsmen).
Rept. Underwat. Ass. 4:62-66

divers

Ross, Helen E., 1968.
Personality of student divers.
Rept. Underwat. Ass. 1968: 59-62

divers

Ross, Helen E., S.S. Franklin and G. Weltman, 1969.
Adaptation of divers to size-distance distortion underwater.
Rept. Underwat. Ass. 4: 56-57.

divers

Ross, Helen E., 1965.
Size and distance judgements under water and on land.
Malta '65: 19-22.

diver

Smith, F.G.Walton, 1965.
Space age and the diver.
Sea Frontiers, 11(5):258-267.

divers

Woodley, J.D., and Helen E. Ross, 1969.
Distance estimates of familiar objects underwater.
Rept. Underwat. Ass. 4: 58-61

diver's personalities

Ross, Helen E., 1968.
Personality of student divers.
Rept. Underwater Ass., 1968:59-62.

diving

diving

Achurch, I.C. and M.A. Garnett, 1969.
A comparative study of the economics of four diving systems. Oceanol. int. 69, 4: 10 pp., 8 figs.

diving observations

Banner, A.H., 1955.
Note on a visible thermocline. Science 121(3142): 402-403.

diving

Bascom, W.N., and R.R. Revelle, 1953.
Free diving: a new exploratory tool. Am. Sci. 41(4):624-627, 2 figs.

diving

Carr, J.R., 1969.
Diving research and the transition to operational practice. Oceanol. int. 69, 4: 7 pp., 3 fig

diving

Ceccaldi, H.J., 1962.
Sur une méthode de récolte du macroplancton.
Rec. Trav. Sta. Mar. Endoume, Bull., 26(41):3-6.

diving

Ciszewski, P., K. Demel, Z. Ringer and M. Szatybelko, 1962.
Resources of Furcellaria fastigiata in Puck Bay estimated by the diving method.
Prace Morsk. Inst., Ryback. Oceanogr.-Ichtiol., Gdyni, 11(A):9-36.

In Polish with Polish, English and Russian summaries

diving

Dawson, James 1972.
Safety in the sixth continent.
J. Mar. techn. Soc. 6(1): 28-31

diving

DeVries, Arthur L., and Donald E. Wohlschlag, 1964
Diving depths of the Weddell seal.
Science, 145(3629):292.

diving

Dill, R.F., and G. Shumway, 1954.
Geologic use of self-contained diving apparatus.
Bull. Amer. Assoc. Petr. Geol. 38(1):148-157, 9 textfigs.

diving

Elsner, Robert, 1969.
Cardiovascular adjustments to diving.
In: The biology of marine mammals, H.T. Andersen, editor, Academic Press, 117-145

diving

Everson, I., and M.G. White, 1969.
Antarctic marine biological research methods involving diving.
Rept. Underwat. Ass. 4: 91-95.

diving

Fager, E.W., A.O. Flechsig, R.F. Ford, R.I. Clutter and R.J. Ghelardi, 1966.
Equipment for use in ecological studies using Scuba.
Limnol. Oceanogr., 11(4):503-509.

diving

Flechsig, Arthur O., 1966.
Value of extended diving to the scientific community.
Man's extension into the sea, Trans. Symp., 11-12 Jan. 1966, Mar. Techn. Soc., 170-175.

diving

Forster, G.R., 1955.
Underwater observations on rocks off Stoke Point and Dartmouth. J.M.B.A. 34(2):197-199.

diving

Forster, G.R., 1954.
Preliminary note on a survey of Stoke Point rocks with self-contained diving apparatus.
J.M.B.A. 33:341-344.

diving

Fructus, P., 1969.
[Deep diving.] Oceanol. int. 69, 4: 14 pp.

diving

Goff, L.G., 1957.
Self-contained diving and underwater swimming.
Research Reviews, Sept., 13-20.

diving

Golikov, A.N. and O.A. Skarlato, 1971.
Some results of diving hydrobiological investigations of the Posiet Bay (The Sea of Japan). (In Russian; English abstract). Gidrobiol. Zh. 7(5): 32-37.

diving

Great Britain, Divcon International, 1968.
Diving: the contractor's view.
Hydrospace, 1(4):30-31:3435.

diving, skin
Guzman Peredo, Miguel, 1966.
Skin diving at high altitudes.
Undersea Techn., 7(2):33-37.

diving
Hamilton, Edwin L., 1961
Under-water mapping by diving geologists.
Proc. Ninth Pacific Sci. Congr., Pacific Sci. Assoc., 1957, 12(Geol.-Geophys.):351.

scuba diving
High, William L., and Larry D. Lusz, 1966.
Underwater observations on fish in an off bottom trawl.
J. Fish. Res. Bd., Canada, 23(1):153-154.

diving
Hughes, D. Michael 1972.
The real meaning of safety regulation to the diver, customer, and contractor.
J. mar. techn. Soc. 6(1):34-40

diving
Kaplin, P.A., 1961.
[Diving investigations of heads of underwater canyons.]
Okeanologiia, 1(6):1034-1038.

Cited in Techn. Transl., 7(10):780.

JPRS-R-2315-D 62-23797
OTS or SLA $1.60.

diving
Keays, Keatinge, and J. Vincent Herrington, 1966.
Diving and salvage operations on the continental shelf.
Man's extension into the sea, Trans. Symp., 11-12 Jan. 1966, Mar. Techn. Soc., 176-206.

diving
Kenny, John E., 1968.
Scientific diver. 1. The geo-diver.
Ocean Industry, 3(12):44-45.

diving
Kireeva, M.S., 1962.
The role of diving in the study of the distribution of the seaweed resources of the seas of the USSR. Methods and results of submarine investigations. (In Russian).
Trudy, Okeanogr. Komissii, Akad. Nauk, SSSR, 14: 69-72.

divers
Kolman, O.V., 1963
"Man at sea". (In Russian).
Okeanologiia, Akad. Nauk, SSSR, 3(6):1126-1127.

diving, deep
Kooyman, G.L., and H.T. Andersen, 1969.
Deep diving.
In: The biology of marine mammals, H.T. Andersen, editor, Academic Press, 65-94

diving
Kooyman, G.L., and W.B. Campbell, Jr., 1969.
Biology of deep diving in Antarctic birds and mammals.
Antarctic J., U.S.A., 4(4):115.

diving
LaRocca, Joseph J., 1972.
Federal regulation of diving
J. mar. techn. Soc. 6(1): 32-33.

underwater diving
Larson, Howard E., 1959.
A history of self-contained diving and swimming.
Nat. Sea. Sci.-Nat. Res. Council, Publ., No. 469: 50 pp.

diving
Lenz, Jürgen, und Hjalmar Thiel 1967.
Tauchbeobachtungen an Plankton und an Echostreuschichten.
Helgoländer wiss. Meeresunters. 15(1/4): 534-546.

diving observations
Limbaugh, C., and A.B. Rechnitzer, 1955.
Visual detection of temperature-density discontinuities in water by diving. Science 121(3142): 395-395, 1 textfig.

Diving
Martin, W.R., and R.L. Adams, 1970.
Physiologic factors in the design of underwater communication.
J. mar. techn. Soc. 4(3), 55-58

diving
Marx, Robert F. 1971.
The early history of diving.
Oceans Mag., Menlo Park, Calif. 4(5): 24-34. (popular)

diving
Menard, H.W., R.F. Dill, E.L. Hamilton, D.G. Moore, G. Shumway, M. Silverman and H.B. Stewart, 1954.
Underwater mapping by diving geologists.
Bull. Amer. Assoc. Petr. Geol. 38(1):129-147, 15 textfigs.

diving
Mundey, G.R., 1968.
Will man ever dive beyond 5000 ft. without armour?
Hydrospace, 1(5):36-38.

diving
Munroe, Frederick F., 1966.
Scuba diving to investigate in-situ behavier of mobile suspended-sediment sampler.
CERC Bull. and Summary Rept. of Res. Prog., 1965-1966:60-61.

diving
Neushul, M., 1965.
Diving observations of sub-tidal Antarctic marine vegetation.
Botanica Marina, 8(2/4):234-243.

diving
Neushul, M., 1961
Diving in Antarctic waters.
Polar Record, 10(67): 353-358.

diving
Nikolic, M., 1960.
[Diving as a method of sea research.]
Hidrografski Godisnjak, Split, 1959, (30):129-150.

diving
Propp, M.V., 1962.
Underwater observations of sublittoral of the Barents Sea. (In Russian).
Trudy Okeanogr. Komissii, 14:73-75.

Translation:
USN Oceanogr. Off. TRANS-215(1964).

diving
Ray, B., 1969.
Communications between divers. Oceanol. int. 69, 3: 4 pp., 4 figs.

diving
Saltzer, Ben 1970.
A deep submergence diver's navigation system.
Navigation, U.S.A. 17(1): 76-82.

diving
Shepard, F.P., 1947.
Diving operations in California submarine canyons. Bull. G.S.A. 58(12):1227.

diving
Shumway, George, David G. Moore and George B. Dowling, 1964.
Fairway Rock in Bering Strait.
In: Papers in Marine Geology, R.L. Miller, Editor, Macmillan, Co., N.Y., 401-407.

diving
Stewart, James R., 1970.
A touchy situation.
J. mar. techn. Soc. 4(3):53-54

diving
Taggart, Robert, 1966.
Propulsive efficiency of man in the sea.
Man's extension into the sea, Trans. Symp., 11-12 Jan. 1966, Mar. Techn. Soc., 63-73.

diving
Treherne, P.C., 1963
Diving as an aid to surveying.
Int. Hydrogr. Rev., 40(1):15-21.

diving
Woods, J.D., and J.N. Lythgoe, editors, 1971.
Underwater science: an introduction to experiments by divers.
Oxford Univ. Press 330 pp.

diving accidents

diving accidents
*Harter, John V., and Robert C. Bornmann, 1968.
Diving accidents not involving decompression.
J. Ocean Techn., 2(4):139-143.

diving depth

diving depth, maximum
Kooyman, Gerald L., 1966.
Maximum diving capacities of the Weddell seal, Leptonychotes weddelli.
Science, 151 (3717):1553-1554.

diving hazards

diving hazards
Anon., 1960.
Hazards and medical problems of skin and scuba diving.
Spectrum, Charles Pfizer Mag., 8(7):156-165.

diving operations

diving operations
Cox, T.L., 1968.
Safety in the cachalot saturation diving system operations.
Journal of Hydronautics 2(4):187-191. (not seen).

diving physiol of

diving (physiology of)
Brauer, Ralph W., 1968.
Seeking man's depth level.
Ocean Industry, 3(12):28-33.

diving, physiology
Chouteau, J., and J.H. Corriol, 1971.
Physiological aspects of deep sea diving.
Endeavour 30 (110): 70-76

diving, physiology of
Eliassen, Einar, 1962
Notes on the mechanism of brady-and tachycardia in full diving birds.
Acta Univ. Bergensis, Ser. Mat. Rerumque Natura Arbok for Univ. i Bergen, Mat.-Naturv. Ser., No. 14: 9 pp.

diving, physiology
Froeb, H.F., 1961
Ventilatory response of SCUBA divers to CO_2 inhalations.
J. Appl. Physiol., 16(1): 8-10.

diving, physiology
Murdaugh, H. Victor, Jr., 1966.
Adaptations to diving in the harbor seal; cardiac output during diving.
Am. J. Physiol., 210(1):176-180.

diving, physiol. of
Olsen, C.R., O.D. Fanestil and P. F. Scholander, 1962.
Some effects of apneic underwater diving on blood gases, lactate, and pressure in man.
J. Appl. Physiol., 17(6):938-942.

diving, physiology
Olsen, C.R., D.D. Fanestil, & P.F. Schôlander, 1962
Some effects of breath holding and apneic underwater diving on cardiac rhythm in man.
J. Appl. Physiol., 17(3):461-466.

diving, physiology
Schafer, Karl E., and Charles R. Carey, 1962
Alveolar pathways during 90-foot, breath-hold dives.
Science, 137(3535): 1051-1052.

diving, physiology
Schenck, H., jr., 1955.
Emergency ascent of an undersea diver from great depths. Amer. J. Phys. 23(1):58-60.

diving, physiology
Scholander, P.F., 1962.
Physiological adaptation to diving in animals and man.
In: The Harvey Lectures, Academic Press, N.Y., and London, pp. 93-110.

diving saucers

diving saucer
Barham, Eric G., 1966.
Deep scattering layer migration and composition: observations from a diving saucer.
Science, 151(3716):1399-1403.

diving saucers
Dangeard, Louis, Michel Lamboy, Yves Lemosquet et Claude Froget, 1969.
Observations géologiques dans le canyon de Planier et dans le "Petit Canyon".
Téthys 1(3): 915-922.

diving saucers
Dietz, R.S., 1963
23. Bathscaphs and other deep submersibles for oceanographic research. In: The Sea, M.N. Hill, Edit., Vol. 2(V) Oceanographic Miscellanea, Interscience Publishers, New York and London, 497-515.

DIVING SAUCER
Franqueville, C., 1970.
Étude comparative de macroplancton en Méditerranée nord-occidentale par plongées en soucoupe SP 350, et pêches au chalut pélagique.
Mar. Biol., 5(3): 172-179.

diving saucers
Giermann, G., 1968.
Geological survey from the diving saucer.
Rept. Underwater Ass., 1968: 3-7.

diving saucer
Giermann, Günter, 1966.
Tauchkugel Soucoupe Plongeante und Photoschlitten Troika, zwei neve Werkzeuge für die geologische Unterwasserkartierung.
Dt. hydrogr. Z., 19(4):170-177.

diving saucer
Giermann, Günter, 1966.
Phénomènes géologiques dans la baie d'ASPRA SPITIA (golfe de Corinthe, Grèce) e'tudies à l'aide de la soucoupe plongeante et de la troïka.
Bull. Inst. Océanogr., Monaco, 66(1364: 19 pp.

diving saucer
Guille, Alain, 1965.
Exploration en soucoupe plongeanate cousteau de l'entre nord-est de la baie de Rosas (Espagne).
Bull. Inst. Océanogr., Monaco, 65 (1357). 9pp.

diving saucer (internal stresses)
Lévy-Soussan, Guy, et Albano Trombetta, 1965.
Etude de la coque d'une soucoupe plongeante de grande profondeur Determination des contreintes localisées de part et d'eutre de l'ouverture (coque et parte). Recherche du profil de renfort.
Bull. Inst. Oceanogr., Monaco, 62(1300: 124 pp.

diving saucers
Reyss, Daniel, et Jacques Soyer, 1965.
Étude de deux Vallées sous-marines de la mer Catalane. (Compte rendu de plongeés en soucoupe plongeante SP300).
Bull. Inst. Océanogr. Monaco, 65 (1356):27pp.

DNA
Sutcliffe, W.H., Jr., R.W. Sheldon and A. Prakash, 1970.
Certain aspects of production and standing stock of particulate matter in the surface waters of the northwest Atlantic Ocean.
J. Fish. Res. Bd, Can., 27(11): 1917-1926.

docks

docks
Naylor, E., 1965.
Biological effects of a heated effluent in docks at Swansea, S. Wales.
Proc. Zool. Soc., London, 144(2):253-268.

domes
See also: salt domes

domes
Leenhardt, Olivier, 1968.
Le problème des dômes de la Méditerranée occidentale: étude géophysique d'une colline abyssale la structure A.
Bull. Soc. géol. France (7), 10(4): 497-509

dome of spray

dome of spray
Pekeris, C. L., 1949.
Determination of the depth of an underwater explosion from measurement of the dome of spray.
Ann. N.Y. Acad. Sci. 51(3):442-452, 6 textfigs.

doppler

Doppler effect
Bukhteev, V. G., 1964.
On the necessity of taking into account the Doppler effect for the analysis of internal wave observations. (In Russian).
Okeanologiia, Akad. Nauk, SSSR, 4(6):994-996.

doppler
Crombie, D.D., 1955.
Doppler spectrum of sea echo at 13.56 Mc/s.
Nature 175(4459):681-682.

doppler
Durst, C.S., 1958.
Doppler navigation. V. The sea surface and Doppler.
J. Inst. Navig., 11(2):143-149.

Doppler
Grocott, D.F.H., 1963.
Doppler correction for surface movement.
J. Inst. Navigation, 16(1):57-63.

doppler, shifts
Hester, Frank J., 1967.
Identification of biological sonar targets from body-motion doppler shifts.
In: Marine bio-acoustics, W.N. Tavolga, editor, Pergamon Press, 2:59-73.

doppler
*Kramer, S.A., 1967.
Doppler and acceleration tolerances of high-gain wideband linear FM correlation sonars.
Proc. Inst. Electr. Electronics Engrs. 55(5): 627-636.

doppler
*Pidgeon, V.W., 1968.
Doppler dependence of radar sea return.
J. geophys. Res., 73(4):1333-1341.

doppler
Valenzuela, G.R. and M.B. Laing, 1970.
Study of doppler spectra of radar sea echo.
J. geophys. Res., 75(3): 551-563.

downwarping

downwarping
Newman, Walter S., and Stanley March 1968.
Littoral of the northeastern United States: late Quaternary warping.
Science 160 (3832): 1110-1112.

downwarping slab

Toksöz, M. Nafi, John W. Minear and Bruce R. Julian, 1971.
Temperature field and geophysical effects of a downgoing slab. J. geophys. Res., 76(5): 1113-1138.

drag coefficient

drag

Landweber, L., and Jin Wu., 1963
The determination of the viscous drag of submerged and floating bodies by wake surveys.
Journal of Ship Research, June 1963: 1-6

drag coefficient

Deacon, E.L., and E.K. Webb, 1962
Samll-scale interactions. Ch. 3, Sect. II.
Interchange of properties between sea and air, In: The Sea, Interscience Publishers, Vol. 1, Physical Oceanography, pp. 43-87.

drainage

drainage patterns

Creager, Joe S., and Dean A. McManus, 1965.
Pleistocene drainage patterns on the floor of the Chukchi Sea.
Marine Geology, Elsevier Publ. Co., 3(4):279-290.

drainage, effect of

Gunter, Gordon, and Judith Clark Edwards, 1969.
The relation of rainfall and fresh-water drainage to the production of penaeid shrimps (Penaeus fluviatilis Say and Penaeus aztecus Ives) in Texas and Louisiana waters.
FAO Fish. Repts. 3(57)(FR10/57.3 (Trim)): 875-892. (mimeographed)

dredging

Herbich, John B., 1969.
Dredging industry problems that need solving.
Ocean Industry, 4(3):59-63.

dredging-filling, effect of

Woodburn, Kenneth D. 1965.
A discussion and selected annotated references of subjective or controversial marine matters.
Techn. Ser. Fla. Bd Conserv. 46: 50 pp.

drift

drift

Bashkirov, G.S., 1961
[Construction of model for the study of suspended load and bottom drifting executed together.]
Trudy Okeanogr. Komissii, Akad. Nauk SSSR, 8:125-128.

drift (of ice camp)

Herlinveaux, R.H., 1963
Oceanographic observations in the Canadian Arctic Basin, Arctic Ocean, April-May, 1962.
Fish. Res. Bd., Canada, Mss. Rept. Ser. (Ocean. and Limnol.), No. 144:25 pp. (multilithed).

drift

Lee, Arthur, and John Ramster, 1965.
The hydrography of the North Sea. A review of our knowledge in relation to pollution problems.
Helgoländer wiss. Meeresunters. 17: 44-63

drifts

Miyoshi, H., S. Hori and S. Yoshida, 1955.
Drift and diffusion of radiologically contaminated water in the ocean. Rec. Ocean. Wks., Japan, 2(2):30-36.

drifts

Petrichenko, A.N., 1940.
[Preliminary data on the drift of the Sedov for the years 1936 and 1938.] Problemy Arktiki, 2:69-85.

R-1835 in Transl. Mon.

drift

Reed, Richard J., and William J. Campbell, 1962.
The equilibrium drift of Ice Station Alpha.
J. Geophys. Res., 67(1):281-298.

drift

Rochford, D.J. 1969.
Seasonal variation in the Indian Ocean along 110°E. 1. Hydrological structure of the upper 500 m.
Austr. J. mar. Freshwat. Res. 20(1): 1-50.

drift

Shuleykin, V.V., 1941.
[The analysis of the drift of the "North Pole" station.] Dokl. Akad. Sci., SSSR, 31(9):886-891.

drift

Sychev, K.A., 1959.
Three years of drift of the floating ice island North Pole-6.
Morskoy Flot, USSR, 19(4):21-23.

drifts

Thompson, J.P., 1954.
Drif of the "San Ernesto", 1943-49.
Mar. Obs. 24(164):82-83.

drift

Wooster, Warren S. and Hellmuth A. Sievers, 1970.
Seasonal variations of temperature, drift, and heat exchange in surface waters off the west coast of South America. Limnol. Oceanogr. 15(4): 595-605.

drift

Vize, V.Y., 1944.
Dreff buyev v arkticheskikh moryakh. [Drift of buoys in Arctic seas.] Probl. Arkt. 2:122.

drift

Vize, V.Y., 1943.
Opyt primeneniya dreyfuyushchikh buyev dlya izucheniya arkticheskikh morey. [An experiment in using drifting buoys for the study of Arctic seas.] Probl. Arkt. 1:72-106.

drift, data only

drift (data only)

Great Britain, Ministry of Defence, Hydrographic Department 1967.
Equatorial current investigations Gilbert Islands H.M.S. Cook 1963.
Publ. Adm. Mar. Sci. (HD 580): 40 pp. (multilithed).

drift bottles

A

drift bottles

Akagawa, M., 1956.
On the oceanographical conditions of the North Japan Sea (west of the Tsugaru Straits) in summer (Part 2).
Bull. Hakodate Mar. Obs., No. 3:5-11.
1-7.

drift bottles

Akegawa, M., 1955.
[On the oceanographical conditions of the North Japan Sea (west of the Tsugaru-Straits)) in summer (Pt. 2.)] J. Ocean. Soc., Japan, 11(1):5-11.

drift bottles

Anapovskii, S.I., 1933.
[Results of the study of surface currents of the Caspian Sea by means of freely drifting bottles.]
Zapiska po Gidrogr. (3):87-88.

Transl. cited: USFWS Spec. Sci. Rept., Fish. 227

drift bottles

Australia, Commonwealth Scientific and Industrial Research Organisation, Hydrographic Investigations, 1961
Use of drift bottles in fisheries research.
Fisheries Newsletter, 20(1): 17-20

Also:
Australian Fisheries Leaflet, No. 3:

drift bottles

Ayers, John C., David C. Chandler, George H. Lauf, Charles F. Powers and E. Bennett Henson, 1958
Currents and water masses of Lake Michigan.
Great Lakes Res. Inst., Publ. No. 3:169 pp.

B

drift bottles

Bane, Gilbert W., 1965.
Results of drift bottle studies near Puerto Rico.
Caribb. J. Sci., 5(3/4):173-174.

drift bottles (data)

Barkley, Richard A., Bernard M. Ito and Robert P. Brown, 1964.
Releases and recoveries of drift bottles and cards in the Central Pacific.
U.S.F.W.S., Spec. Sci. Rept., Fish., No. 492:31 p

drift bottles

Barnes, H., and the late E.F.W. Goodley, 1961
The general hydrography of the Clyde Sea area, Scotland. 1. Description of the area; drift bottle and surface salinity data.
Bulletins, Mar. Ecol., 5(43):112-150.

drift bottles

Berner, Leo D., and Joseph L. Reid, Jr., 1961

On the response to changing temperature of the temperature-limited plankter Doliolum denticulatum Quoy and Gaimard 1835.
Limnol. & Oceanogr., 6(2): 205-215.

drift bottles

Bowden, K.W., 1955.
Physical oceanography of the Irish Sea.
Fish. Invest., Min. Agric., Fish. & Food (2) 18(8):1-67, 18 textfigs.

drift bottles

Brown, C.H., 1909.
Report on the deep currents of the North Sea as ascertained by experiments with drift bottles.
Fish. Bd., Scotland, North Sea Fish. Invest. Comm., Northern Area, Rept. No. 4:125-142.

drift bottles
Bumpus, D.F., 1955.
The circulation over the continental shelf south of Cape Hatteras. Trans. Amer. Geophys. Union, 36(4):601-611, 18 textfigs.

drift bottles
Bumpus, D.F., and Joseph Chase, 1965.
Changes in the hydrography observed along the east coast of the United States.
ICNAF Spec. Publ. No. 6:847-853.

drift bottles
Bumpus, D.F., and C.G. Day, 1957.
Drift bottle records for the Gulf of Maine and Georges Bank, 1931-1956
USFWS Spec. Sci. Rept., Fish., No. 242:61 pp.

drift bottles
Burt, Wayne V., and Bruce Wyatt, 1964
Drift bottle observations of the Davidson Current off Oregon.
In: Studies in Oceanography dedicated to Professor Hidaka in commemoration of his sixtieth birthday, 156-165.

C

drift bottles
Carruthers, J.N. 1969.
Floating messages - bottle post and other drifts.
PLA Monthly, Port of London Authority 44 (525): 203-213.
Also in: Coll. Repr. Nat. Inst. Oceanogr. 17 (1969).

drift bottles
Carruthers, J.N., 1956.
'Bottle post' and other drifts.
J. Inst. Navig., 9(3):261-281.

drift bottles
Carruthers, J.N., 1939.
The investigation of surface current eddies in the sea by means of surface-period drift bottles.
Sborn. Posvyashch. Nauch. Deyatel. N.M. Knipoviche (1885-1939):39-46.

drift bottles
Carruthers, J.N., 1939.
The investigation of surface current eddies in the sea by means of short period drift bottles.
USSR Inst. Mar. Fish., Ocean., Sbornik Knipovich: 39-46.

drift bottles
Chevrier, J.R. and R.W. Trites, 1960
Drift bottle experiments in the Quoddy region, Bay of Fundy. J. Fish. Res. Bd., Canada, 17(6): 743-762.

drift bottles
Chew, Frank, K.L. Drennan and W.J. Demoran, 1962
Some results of drift bottle studies off the Mississippi delta.
Limnology and Oceanography, 7(2):252-257.

drift bottles
Chosen. Fishery Experiment Station, 1936.
Number of the drift-bottles picked up at each locality, which had been liberated in the neighboring sea of Tyosen during the years 1929-32. (In Japanese). Ann. Rept. Hydrogr. Obs., Chosen Fish. Exp. Sta., 7:114-116, tables.

drift bottles
Colton, John B., Jr., and Robert F. Temple, 1961
The enigma of Georges Bank spawning.
Limnol. & Oceanogr., 6(3): 280-291.

drift bottles
Craig, R.E., 1956.
Hydrography, near northern seas and approaches.
Ann. Biol., Cons. Perm. Int. Expl. Mer, 11:33-36, Figs. 1-2.

D

drift bottles
Daniel, R.J., and H.M. Lewis, 1930.
Surface drift bottle experiments in the Irish Sea
Proc. & Trans. Liverpool Biol. Soc. 44:36-86.
July 1925 - June 1927.

drift bottles
Day, C. Godfrey, 1958.
Surface circulation in the Gulf of Maine as deduced from drift bottles.
U. S. F. W. S., Fishery Bull. 141, Vol. 58, p. 443-472.

drift bottles
Dell, R.K., 1952.
Ocean currents affecting New Zealand. Deductions from the drift bottle records of H.C. Russell.
N.Z.J. Sci. and Tech., Sect. B, 34(2):86-91.

drift bottles
Dodimead, A.J., F. Favorite and T. Hirano, 1964.
Review of oceanography of the subarctic Pacific Region. Salmon of the North Pacific Ocean. II. Collected Reprints, Tokai Reg. Fish. Res. Lab., No. 2:187 pp.
(In Japanese).

drift bottles (data).
Dodimead, A.J., and H.J. Hollister, 1962.
Canadian drift bottle releases and recoveries in the North Pacific Ocean.
Fish. Res. Bd., Canada, Mss. Rept. Ser., (Ocean. and Limnol.), No. 141:64 pp., 44 charts.

drift bottles
Dodimead, A.J., and H.J. Hollister, 1958.
Progress report of drift bottle releases in the northeast Pacific Ocean.
J. Fish. Res. Bd., Canada, 15(5):851-865.

drift bottles
Duboul-Razavet, C., 1958
Sur quelques lancers de cartes siphonophores et de bidons lestés au large de la baie de Marseille. Bull. l'Inst. Océan. Monaco, 55 (1132): 16

E

drift bottles
Einarsson, H., and U. Stefansson, 1953.
Drift bottle experiments in the waters between Iceland, Greenland and Jan Mayen during the years 1947 and 1949. Rit Fiskideildar 1953 (1):20 pp., 15 textfigs.

F

drift bottles
Favorite, Felix, 1964.
Drift bottle experiment in the northern North Pacific Ocean.
Jour. Oceanogr. Soc., Japan, 20(4):160-167.

bottle drifts
Fish, C.J., and M.W. Johnson, 1937
The biology of the zooplankton population in the Bay of Fundy and the Gulf of Maine with special reference to production and distribution. J. Biol. Bd., Canada 3(3):189-322, 29 tables, 45 text figs.

owned by MS

drift bottles
Fish, C.J., and M.W. Johnson, 1937.
The biology of the zooplankton population in the Bay of Fundy and Gulf of Maine with special reference to production and distribution.
J. Biol. Bd., Canada, 3(3):189-322.

drift bottles
Fisk, Donald M., 1971.
Recoveries from 1964 through 1968 of drift bottles released from a merchant vessel, S.S. Java Mail, en route Seattle to Yokohama, October 1964.
Pacific Sci., 25(2): 171-177

drift bottles
Fofonoff, Nicholas P., 1960
Description of the northeastern Pacific oceanography. Cal. Coop. Ocean. Fish. Invest. Rept. Vol. 7: 91-95.

drift bottles
France, Comité Central d'Océanographie et d'Étude des Côtes, 1955.
A. Observations hydrologiques du Petrolier "Ashtarak". B. Observations hydrologiques des bâtiments de la Marine Nationale. C. Flotteurs temoins de courant. Bull. d'Info., C.C.O.E.C., 7(9):430-436.

drift bottles
France, Service Central Hydrologique, 1957.
A. Observations hydrologiques des bâtiments de la Marine Nationale. B. Flotteurs-temoins de courant
Bull. d'Info., C.C.O.E.C., 9(2):147-153.

drift bottles
France, Service Hydrographique de la Marine, 1963
Resultats d'observations.
Cahiers Océanogr., C.C.O.E.C., 15(3):418-428.

drift bottles
France, Service Hydrographique de la Marine, 1962
Resultats d'Observations. 1. Navire meteorologique stationnaire "France II". II. Flotteurs-temoins de courant. III. Observations du Niveau marin à la station littorale de l. A.G.I. à Matavai (Tahiti).
Cahiers Oceanogr., C.C.O.E.C., 14(8):604-609.

drift bottles
France, Service Hydrographique de la Marine, 1962
Resultats d'Observations.
Cahiers Océanogr., C.C.O.E.C., 14(3):187-199.

drift bottles
France, Service Hydrographique de la Marine, 1961
Resultats d'observations. 1. Observations hydrologiques des bâtiments de la Marine Nationale. 2. Flotteurs-temoins de courant.
Cahiers Océanogr., C.C.O.E.C., 13(7):498-515.

drift bottles

France, Service Hydrographique de la Marine, 1961

Resultats d'observations.
Cahiers Oceanogr., C.C.O.E.C., 13(4):250-263.

drift bottles

France, Service Hydrographique de la Marine, 1961

Resultats d'observations.
Cahiers Océanogr., C.C.O.E.C., 13(3): 187-201.

drift bottles

France, Service Hydrographique de la Marine, 1959

Resultats d'observations.
Cahiers Ocean., C.C.O.E.C., 11(9):689-699

drift bottles

France, Service Hydrographique de la Marine 1959.
Resultats d'observations hydrologiques.
Cah. océanogr. 11 (3): 182-199.

drift bottles

France, Service Hydrographique de la Marine, 1959.
Flotteurs-temoins de courant.
Cahiers Oceanogr., C.C.O.E.C., 11(2):135-138.

drift bottles

France, Service Central Hydrographique, 1958.
Flotteurs-temoins de courant.
Bull. d'Info., C.C.O.E.C., 10(4):224-234.

drift bottles

France, Service Central Hydrographique, 1957.
Flotteurs-temoins de courant.
Bull. d'Info., C.C.O.E.C., 9(8):464-467.

drift bottles

France, Service Central Hydrographique, 1956.
A. Observations hydrologiques des bâtiments de la Marine National. B. Flotteurs-temoins du courant. Bull. d'Info., C.C.O.E.C., 8(10):563-578

G

drift bottles

Gade, Herman G., 1961.
Further hydrographic observations in the Gulf of Cariaco, Venezuela. The circulation and water exchange.
Bol. Inst. Oceanogr., Univ. de Oriente, Venezuela 1(2):359-395.

drift bottles

Gast, James A., 1966.
A drift bottle survey in northern California.
Limnol. Oceanogr., 11(3):415-417.

drift bottles

Gibbs, Pearl. 1969.
Drift bottle messages.
Frontiers 33(4): 12-15.

drift bottles

Gray, I.E., and M.J. Cerame-Vivas, 1963
The circulation of surface waters in Raleigh Bay, North Carolina.
Limnology and Oceanography, 8(3):330-337.

H

drift bottles

Harrison, W., Morris L. Brehmer and Richard B. Stone, 1964.
Nearshore tidal and nontidal currents, Virginia Beach, Virginia.
U.S. Army, Coastal Engineering Res. Center, Techn Memo., No. 5:20 pp. (multilithed).

drift bottles

Hata, Katsumi, 1963.
12. The report of drift bottles released in the North Pacific Ocean. (In Japanese; English abstract).
J. Oceanogr. Soc., Japan, 19(1):6-15.

Also in:
Bull. Hakodate Mar. Obs., (10).

drift bottles

Hattori, Shigemasa, and Hidetsugu Katori, 1966.
Surface current in the southern waters off Japan viewed from drift bottle experiment, with special reference to translocation of fish eggs and larvae.
Bull. Tokai Reg. Fish Res. Lab., No. 45:1-30.

drift bottles

Hermann, F., 1952.
Hydrographic conditions in the southwestern part of the Norwegain Sea, 1951. Cons. Perm. Int. Expl Mer, Ann. Biol. 8:23-26, Textfigs. 4-12.

drift bottles

Hermann, F., and H. Thomsen, 1946.
Drift-bottle experiments in the northern North Atlantic. Medd. Komm. Danmarks Fisk. og Havundersøgelser 3(4):87 pp., 4 pls.

drift bottles

Huntsman, A.G., W.B. Bailey and H.B. Hachey, 1954.
The general oceanography of the Strait of Belle Isle. J. Fish. Res. Bd., Canada, 11(3):198-260, 35 textfigs.

Drift Bottles

Huzii, M., and M. Kimura, 1961.
Drifting of 20,000 current bottles, released in the sea south-west of Kyusyu, July, 1960.
Hydrogr. Bull., Tokyo, Spec. No., 90th Anniv., (Publ. #981), No. 67:58-62.

I

drift bottles

Ingle, Robert M., Bonnie Eldred, Harold W. Sims, and Eric A. Eldred, 1963
On the possible Caribbean origin of Florida's spiny lobster populations.
State of Florida, Bd., Conserv., Div. Salt Water Fish., Techn. Ser., No. 40:12 pp.

J

drift bottles

Japan, Hakodate Marine Observatory, 1957
Report of the oceanographic observations in the Tsugaru Straits in July 1956.
Bull. Hakodate Mar. Obs., No. 4: 13-21.

drift bottles

Japan, Maritime Safety Board, 1955.
Observations of the current in Japan Sea with drift bottles. Pub. 981, Hydrogr. Bull., Spec. No. 17:13-21.

drift bottles

Japan, Science Council, National Committee for IIOE, 1966.
General report of the participation of Japan in the International Indian Ocean Expedition.
Rec. Oceanogr. Wks., Japan, n.s. 8(2): 133 pp.

drift bottles

Johnson, James H., 1960.
Surface currents in Lake Michigan.
U.S.F.W.S., Spec. Sci. Rept., Fish., No. 338: 120 pp.

drift bottles

Johnstone, J., 1903.
On some experiments with "drift bottles".
Proc. & Trans. Liverpool Biol. Soc. 17:154-164.

K

drift bottles

Karwowski, J., 1963
Measurement of sea currents by means of drift bottles.
Int. Hydrogr. Rev., 40(2):119-123.

drift bottles

Kasahara, S., 1957.
A study on the surface flow in the northern Japan Sea during the spring, 1955, with special reference to its bearing upon the drift of sardine eggs and larvae.
Ann. Rept., Japan Sea Reg. Fish. Res. Lab., No. 3:137-154.

drift bottles

Kasahara, S., 1957.
Some considerations regarding the use of drift envelopes for the investigation of surface ocean currents, with special reference to their quality as compared with that of drift bottles.
Ann. Rept., Japan Sea Reg. Fish. Res. Lab., No. 3 155-166.

drift-bottle

Kawakami, K., 1959.

The report of the drift-bottle experiments at the west entrance of Tugaru Strait, Hokkaido, Japan. 2.
J. Oceanogr. Soc., Japan, 15(1):5-10.

drift bottles

Kawakami, K., 1957.
The report of the drift-bottle experiments at the west entrance of Tsugaru Strait, Hokkaido, Japan. 1.
J. Oceanogr. Soc., Japan, 13(4):131-138.

drift bottles

Kimura, K., (undated in English).
Investigation of ocean current by drift bottle experiment. 1. Current in Surugo Bay with special reference to its anticlockwise circulation within the bay. J. Ocean. Soc. (Nippon Kaiyo Gakkaisi) 5(2/4):70-84, 6 figs. Tokyo, (In Japanese; English summary).

drift bottles

Kojima, S., 1954.
The fluctuation in fishing off Shimane Prefecture viewed from the recovery of current bottle experiments. Bull. Jap. Soc. Sci. Fish. 20(5): 368-371, 2 textfigs.

drift bottles

Kupetsky, V.N., 1960

Some news about the old method of drift bottle releases. Meteorol. i Gidrol., 10: 37.

drift bottles
Kuro, G., 1956.
Observations sur la dérive des flotteurs lancés par le "Président Théodore Tissier" pendant les campagnes de 1951 a 1954. Rev. Trav. Inst. Pêches Marit. 20(3):225-261.

L

drift bottles
Larkin Richard R., and Gordon A. Riley 1967.
A drift bottle study in Long Island Sound.
Bull. Bingham oceanogr. Coll. 19(2): 62-71.

drift bottles
Lauzier, L.M., 1965.
Drift bottle observations in Northumberland Strait, Gulf of St. Lawrence.
J. Fish. Res. Bd., Canada, 22(2):353-368.

drift bottles
Leclaire, L., 1963.
Études littorales en Baie d'Alger, zone de Fort-de-l'Eau - Ben Mered.
Cahiers Océanogr., C.C.O.E.C., 15(2):109-124.

drift bottles
Lee, Chang Ki, 1968.
The drift bottle experiment in the east of the Yellow Sea (the west coast of Korea) during the years of 1962-1966 (In Korean; English abstract).
Bull. Fish. Res. Develop. Agency, Korea, 3: 29-41.

Drift bottles
Lee, Chang-Ki and Jong-Hon Bong, 1968.
On the current of the Korean Eastern Sea (West of the Japan Sea). (In Korean; English abstract).
Bull. Fish. Res. Develop Agency, Korea, 3: 7-26

drift bottles
LeFloch, J., and V. Romanovsky, 1954.
Circulation superficielle des eaux dans la partie orientale du bassin occidental de la Méditerranée. Trav. C.R.E.O. 1(1):1-17, 16 textfigs.

Drift Bottles
Leontyeva, V. V., 1961.
[Current and water masses in the western part of the Pacific Ocean in summer 1957]
Mezhd. Kom. Mezhd. Geofiz. Goda, Presidiume Akad. Nauk, SSSR, Okeanol. Issled., (3):137-150.

drift bottles
*Luedeman, Ellen F., 1967.
Preliminary results of drift-bottle releases and recoveries in the western tropical Atlantic.
Bolm Inst.oceanogr.S. Paulo 16(1):13-22.

drift bottles
Luedemann, Ellen Fortlage and Norman John Rock, 1971.
Studies with drift bottles in the region off Cabo Frio. (Abstracts in English and Portuguese)
In: Fertility of the Sea, John D. Costlow, editor, Gordon Breach, 1: 267-283.

M

drift bottles
Martiel, L., 1956.
Études des courants du littoral sud de la Bretagne. Rev. Trav. Inst. Pêches Marit., 20(3): 263-280.

drift bottles
Mazelle, E., 1915.
Flaschenposten in der Adria zur Bestimmung der Oberflächenstromungen.
Denkskr. Kaiserlichen Akad. Wiss. 91:362-377.

2418 in Transl. Monthly 3(4) 1957.

Mc

drift bottles (theoretical)
McEwen, G. F., 1950.
A statistical model of instantaneous point and disk sources with applications to oceanographic observations. Trans. Am. Geophys. Union 31(1): 35-46, 3 textfigs.

drift bottles
Miyata, Kazuo, and K. Naganuma, 1959.
[On the drift-bottle experiments made from August 1955 to October 1958 in the Sea of Japan. (The primary report).]
Ann. Rept., Japan Sea Res. Fish. Res. Lab., 34(5):133-147.

"drift bottles"
Mosby, H., 1954.
Oberflächstremungen in der Meerenge bei Tromsö.
Arch. Met. Geophys., Bioklim., A, 7:378-384, 4 textfigs.

N

drift bottles
Nakamura, Akikazu, 1959.
A study on the productivity of Tanabe Bay.
II. 1. Tidal currents in Tanabe Bay (2nd rept.)
Rec. Oceanogr. Wks., Japan. Spec. No. 3:1-22.

drift bottles
Nakamura, Akikazu, 1958.
A study on the productivity of Tanabe Bay (1). II. Tidal currents in the Tanabe Bay. Observations of tidal currents by the method of drift bottles (introductory report).
Rec. Oceanogr. Wks., Japan, (Spec. No. 2):10-14.

drift bottles
Nielsen, J.N., 1908.
Contribution to the understanding of the currents in the northern part of the Atlantic Ocean.
Medd. Komm. Havundersøgelser, Ser. Hydrogr., 1(11):15 pp., 1 pl.

drift bottles
Niwa, M., and T. Senta, 1962.
Ecological and oceanographical studies on the purse seine fishing grounds around Oki Isls. I. The surface current in autumn 1961. (In Japanese; English summary).
Bull. Jap. Soc., Sci. Fish., 28(9):862-869.

drift bottles
Norcross, J.J., W.H. Massman and E.B. Joseph, 1962.
Data on coastal currents off Chesapeake Bay.
Virginia Inst., Mar. Sci., Gloucester Pt., Spec. Sci. Rept., No. 31: 2 pp., 15 tables. (mimeographed).

O

drift bottle (substitute)
Olson, F.C.W., 1951.
A plastic envelope substitute for drift bottles. J. Mar. Res. 10(2):190-193.

P

"drift bottles"
Palausi, G. 1968.
Étude du mode de dispersion de flotteurs colorés dans la baie de Cannes.
Rev. int. Océanogr. Méd. 9:191-205.

drift bottles
Payne, R., and R.E. Craig, 1966.
Temperature and salinity in Scottish waters 1964. Monthly releases of current indicators from four stations in the North Sea.
Annls biol., Copenh., 21:26-29.

drift bottles
Platania, G., 1923.
Experiments with Drift bottles. (Second Report).
Rept. Danish Oceanogr. Exped., 1908-1910, 3(5): 1-18.

Q

R

drift bottles
Rao, T.S. Satyanarayana, 1963
On the pattern of surface circulation in the Indian Ocean as deducted from drift bottle recoveries.
Indian J. Meteorol. & Geophys., 14(1):1-4.

drift bottles
Reid, Joseph L., Jr., 1960
Oceanography of the northwestern Pacific Ocean during the last ten years.
California Coop. Ocean. Fish. Invest., Rept. 7: 77-90.

"drift bottles"
Richards, A.F., 1958.
Transpacific distribution of floating pumice from Isla San Benedicto, Mexico. Deep-Sea Res., 5(1): 29-35.

drift bottles
Romanovsky, V., 1955.
Résultats de la détermination dans le bassin occidental de la Méditerranée des courants superficiels par la méthode des flotteurs derivants. Trav. C.R.E.O. 2(1-2):1-16.

drift bottles
Rouch, J., 1956.
Botellas en los mares.
Puntal 3(32):8-11.

drift bottles
Rouch, J., 1956.
Bouteilles à la mer. Scientia 91:199-205.

drift bottles
Rouch, J., 1954.
Bottle papers. Bull. Inst. Océan., Monaco, No. 1046:5 pp., 2 figs.

S

drift bottles
Schmidt, J., 1913.
Experiment with drift bottles.
Rept. Danish Oceanogr. Exped., 1908-1910, 3(1): 1-13, 6 maps.

drift bottles (data)
Sitarz, J., 1955.
Resultats de la determination dans le Golfe de Gascogne et La Manche des courants superficiels par la methode des flotteurs derivants.
Trav. C.R.E.O., 2(8/9):1-15.

drift bottles
Stefansson, Unnstein, 1962.
North Icelandic waters.
Rit Fiskideildar, Reykjavik, 3:269 pp.

drift bottles
Sverdrup, H. U., and R. H. Fleming, 1941.
The waters of the coast of southern California, March to July, 1937. Bull. S.I.O. 4(10):261-375, 66 textfigs.

T

drift bottles
Taguchi, K., 1956.
A report of drift bottle surveys off Kamchatka USSR in 1940 and 1941. Bull. Jap. Soc. Sci. Fish. 22(7):393-399.

drift bottles
Tait, J.B., 1957.
Near northern seas and approaches.
Ann. Biol., Cons. Perm. Int. Expl. Mer, 12:58-59.

drift bottles
Tait, J.B., 1954.
Northern North Sea and approaches.
Ann. Biol., Cons. Perm. Int. Expl. Mer, 10:82-84

drift bottles
Tait, J.B., 1953.
Northern North Sea and approaches.
Ann. Biol., Cons. Perm. Int. Expl. Mer, 9:98-99.

drift bottles
Tait, J.B., 1952.
Hydrography. Northern North Sea and approaches.
Cons. Perm. Int. Expl. Mer, Ann. Biol. 8:94-98.

drift bottles
Tait, J.B., 1951.
Hydrography. (N. North Sea). Ann. Biol. 7:69-72.

drift bottles
Tait, J. B., 1949
Scottish Hydrographical Investigations.
Ann. Biol., Int. Cons., 4:60-63.

drift bottles
Tait, J. B., 1934.
Surface drift-bottle results in relation to temperature, salinity and density distributions in the northern North Sea. Rapp. Proc. Verb. 89(3):69-79, 9 textfigs.

drift bottles
Talbot, Gerald B., 1964.
Drift bottle modification for air drops.
Trans. Amer. Fish. Soc., 93(2):203-204.

drift bottles
Tåning, Å.V., 1931.
Drift-bottle experiments in Icelandic waters.
Rapp. Proc. Verb., Cons. Perm. Int. Expl. Mer 72:3-20, 3 textfigs.

drift bottles
Tapanes, Juan J., 1963
Afloramiento y corrientes cercanas a Cuba. 1.
Contrib., Centro Invest. Pesqueras, Cons. Nac. Pesca, No. 17:1-29.

drift bottles
Tolbert, W.H., and G.G. Salsman, 1964
Surface circulation in the eastern Gulf of Mexico as determined by drift-bottle studies.
J. Geophys. Res., 69(2):223-230.

drift bottles
Tregouboff, G., 1956.
Rapport sur les travaux concernant le plancton Méditerranéen publiés entre Novembre 1952 et Novembre 1954.
Rapp. Proc. Verb., Comm. Int. Expl. Sci., Mer Méditerranée, 13:65-100

drift bottles
Trites, R.W., and R.E. Banks, 1958.
Circulation on the Scotian Shelf as indicated by drift bottles. J. Fish. Res. Bd., Canada, 15(1): 79-89.

drift bottles ref.
Trites, R.W., and R.E. Banks, 1956.
Circulation on the Scotian Shelf as indicated by using drift bottles.
Nav. Res. Est., Defence Bd., Canada, TM56/6.

U

drift bottles
Uda, M., 1955.
On the subtropical convergence and the currents in the northwestern Pacific.
Rec. Ocean. Wks., Japan, 2(1):141-150, 10 textfigs.

V

W

drift bottles
Waldichuk, M., 1958.
Drift bottle observations in the Strait of Georgia. J. Fish. Res. Bd., Canada, 15(5):1065-1102.

drift bottles
Waldichuk, Michael, 1957.
Physical oceanography of the Strait of Georgia.
J. Fish. Res. Bd., Canada, 14(3):321-486.

drift bottles
Webster, J. R., and R. J. Buller, 1950.
Drift bottle releases off New Jersey. A preliminary report on experiments begun in 1948.
Spec. Sci. epts., Fisheries, No. 10:21 pp., 4 textfigs. (multilithed).

drift bottles
Whitaker, T.W., and C.F. Carter, 1954.
Oceanic drift of gourds, experimental observations
Amer. J. Bot. 41(9):697-700.

Abstr. Biol. Abstr. 29(5):10381.

drift bottles
Wilson, A.M., and H. Thompson, 1934.
III. Hydrographic and biological investigations.
A. Notes on the physical conditions in 1933.
Ann. Repts., Newfoundland Fish. Res. Comm. 2(2): 40-47, Textfigs. 4-5.

drift bottles
Wilson, A.M., and H Thompson, 1933.
III. Hydrographic and biological investigations.
A. Notes on the physical conditions in 1932.
Ann. Rept., Newfoundland Fish. Res. Comm., 2(1):49-58, Textfigs. 12-15.

XYZ

drift bottles data only

drift bottles (data only)
Australia, Commonwealth Scientific and Industrial Research Organization, 1968.
Drift bottle releases and recoveries in Bass Strait and adjacent waters, 1958-1962.
Oceanogr. Sta. List, Div. Fish. Oceanogr., 78:237 pp. (multilithed).

drift bottles (data only
Bumpus, Dean F., 1961
Drift bottle records for the Gulf of Maine, Georges Bank and the Bay of Fundy, 1956-1958.
U.S.F.W.S. Spec. Sci. Repts., Fish. No. 378: 127 pp.

drift bottles (data)
Bumpus, Dean F., and Louis M. Lauzier, 1965.
Surface circulation on the continental shelf off eastern North America between Newfoundland and Florida.
Ser. Atlas, Mar. Enviroment, Folio 7:8 pls.

drift bottles (data only)
Favorite, Felix, and Glenn M. Pedersen, 1959.
North Pacific and Bering Sea oceanography, 1957.
USFWS Spec. Sci. Rept., Fish., No. 292:106 pp.

drift bottles (data only)
France, Service Central Hydrographique, 1957.
A. Observations hydrologiques. B. Flotteurs témoins de courant. Bull. d'Info., C.C.O.E.C., 9(6):342-348.

drift bottles (data only)
France, Service Central Hydrographique, 1956.
A. Observations hydrologiques des bâtiments de la Marine Nationale. B. Fottteurs-temoins du courant
Bull. d'Info., C.C.O.E.C., 8(8):408-411.

drift bottles, data only
France, Service Central Hydrographique, 1956.
A. Observations hydrologiques du Bâtiment Hydrographe "Ingenieur Hydrographe Nicolas". B. Flotteurs temoins de courant.
Bull. d'Info., C.C.O.E.C., 8(3):140-144.

drift bottles (data only)
France, Service Hydrographique de la Marine, 1963.
Resultats d'observations.
Cahiers Oceanogr., C.C.O.E.C., 15(9):666-684.

drift bottles (data only)
France, Service Hydrographique de la Marine, 1963
Resultats d'observations.
Cahiers Oceanogr., C.C.O.E.C., 15(7):492-507.

drift bottles (data only)
France, Service Hydrographique de la Marine, 1961
Resultats d'Observations.
Cahiers Ocean., C.C.O.E.C., 13(10):747-767.

drift bottles (data only)
France,
Service Hydrographique de la Marine, 1961.
Resultats d'observations.
Cahiers Oceanogr., C.C.O.E.C., 13(2):116-132.

drift bottom

drift, bottom
*Rehrer, R., A.C. Jones and M.A. Roessler, 1967.
Bottom water drift on the Tortugas grounds.
Bull. mar. Sci., Miami, 17(3):562-575.

drift, bottom
Squire, James L., Jr., 1969.
Observations on cumulative bottom drift in Monterey Bay using seabed drifters. Limnol. Oceanogr., 14(1): 163-167.

drift buoys

drift buoys
Churkina, N.A., 1958(1960).
[Drift of buoys in the central Arctic and in the Arctic seas.]
Problemy Severa, (1):
Translation in:
Problems of the North, (1):342-345.

drift buoys
Colton, John B., Jr., and Robert F. Temple, 1961
The enigma of Georges Bank spawning.
Limnol. & Oceanogr., 6(3):280-291.

drift buoys
Vize, V.I.U., 1945.
Drift buoys in the Arctic seas. Problemy Arktiki 1944(2):122.
RT4049 in Bibl. Transl. Russ. Sci. Tech. Lit. 38

drift cards

drift cards
Anderson, F.P., 1961.
The use of drift cards to deduce currents along the Natal coast.
Marine Studies off the Natal Coast, C.S.I.R. Symposium, No. S2:40-45. (Multilithed).

drift cards
Assaf, G., R. Gerard and A.L. Gordon, 1971.
Some mechanisms of oceanic mixing revealed in aerial photographs. J. geophys. Res. 76(27): 6550-6572.

drift cards
Brodie, J.W., 1960
Coastal surface currents around New Zealand.
N.Z.J. Geol. and Geophys. 3(2): 235-252.

drift card
Brodie, J.W., 1958
A note on tidal circulation in Port Nicholson, New Zealand. N.Z.J. Geol. Geophys. 1(4): 684-702.

drift cards (Pl. III)
Clowes, A.J., 1954.
The temperature, salinity and inorganic phosphate of the surface layer near St. Helena Bay, 1950-52. Union of S. Africa, Dept. Commerce & Industries, Div. Fish., Invest. Rept. No. 16: 1-47, 17 pls.
Reprinted from "Commerce and Industry" Aug. 1954

drift cards
Clowes, A.J., 1954.
Inshore surface currents on the west coast of the Union of South Africa. Nature 173(4412): 1003-1004, 1 textfig.

drift cards
Craig, R.E., 1956.
Hydrography, near northern seas and approaches.
Ann. Biol., Cons. Perm. Int. Expl. Mer, 11:33-36, Figs., 1-2.

drift cards
Duboul-Razavet, C., 1958.
Le régime des courants superficiels aux abords des côtes du delta de l'Ebre.
Bull. d'Info., C.C.O.E.C., 10(7):392-406.

drift cards
Duncan, C.P., 1967.
Current measurements of the Cape coast.
Fish. Bull., Misc. Contr. Oceanogr. Mar. Biol. S. Afr., 4:9-14.

drift cards
Duncan, C.P., 1965.
Disadvantages of the Olson drift card, and description of a newly designed card.
J. Mar. Res., 23(3):233-236.

drift cards
Duncan, C.P., and J.H. Nell, 1969.
Surface currents off the Cape coast.
Invest. Rept. Div. Sea Fish. S.Afr. 76:1-19.

drift cards
France, Service Hydrographique de la Marine, 1962.
Résultats d'observations. 1. Observations hydrologiques des bâtiments de la Marine Nationale et des navires météorologiques stationnaires. 2. Flotteurs-témoins de courant.
Cahiers Océanogr., C.C.O.E.C., 14(6):424-435.

drift cards
Fry, F.E.J., 1956.
Movement of drift cards in Georgian Bay in 1953.
J. Fish. Res. Bd., Canada, 13(1):1-5.

drift cards
Gautier, Y.V., 1957.
Résultats de trois lachers de cartes derivantes a l'est des Iles d'Hyères.
Bull. d'Info., C.C.O.E.C., 9(7):380-384.

drift cards
Gautier, Y., 1956.
Sur quelques lachers de cartes du type "siphonohore" en vue de l'etude des courants de surface devant le delta du Rhone. Bull. d'Info., C.C.O.E.C. 8(6):274-283.
Reprint from
Recueil Trav Sta. Mar. Endoume, 18, 1956.

drift cards
Heath, R.A., 1969.
Drift card observations of currents in the central New Zealand region.
N.Z. Jl mar. Freshwat. Res. 3(1):3-12

drift cards
Herdman, H.F.P., 1954.
Operation Driftcard. The Trident 16:196-197.

drift cards
Hirano, T., 1955.
[On the water exchange and the current in Hamano Lake in relation to occurrences of toxic shellfish.] Bull. Jap. Soc. Sci. Fish. 20(9):783-792.

driftcards
Ito, S., and S. Kasahara, 1959
On the teeth-marks found impressed on the recovered polyethylene drift envelopes. Ann. Rept., Japan Sea Reg. Fish. Res. Lab., (5): 39-46.

drift envelope
Ito, S., and S. Kasahara, 1958.
Drift of sardine eggs and larvae in the surrounding waters of Japan as discussed by the results of drift envelope release experiments. 1. Discussion of the 1957 season.
Ann. Rept., Japan Sea Reg. Fish. Res. Lab., 1(4):65-76.

drift cards
Johnson, James H., 1960.
Surface currents in Lake Michigan.
U.S.F.W.S. Spec. Sci. Rept., Fish., No. 338:120 pp.

drift cards
Johnson, Sven I., and James L. Squire, 1970.
Surface currents as determined by drift card releases on the continental shelf off the northwestern United States.
Techn. Pap. USFWS Bur. Sport Fish. Wildl., (45):1-12.

drift cards
Kasahara, S., 1957.
[Some considerations regarding the use of drift envelopes for the investigation of surface currents, with special reference to their quality as compared with that of drift bottles.]
Anon. Rept., Japan Sea Reg. Fish. Res. Lab., No. 3:155-166.

drift cards
Lawford, A.L., 1956.
Postscript to Operation Post Card. Trident 18 (208):350-351.

drift cards
Martin, John W., 1967.
New plastic drift cards.
Limnol. Oceanogr., 12(4):706-707.

drift cards
*Neumann, Heinrich, 1967.
Die Beziehung zwischen Wind und Oberflächenströmung auf Grund von Triftkartenuntersuchungen.
Dt. hydrogr. Z., 19(6):253-266.

drift cards
Olson, F.C.W., 1953.
Tampa Bay studies.
Ocean. Inst., Florida State Univ., Rept. 1:27 pp. (mimeographed), 21 figs. (multilithed).
UNPUBLISHED.

drift cards
Ridgway, N.M., 1960.
Surface water movements in Hawke Bay, New Zealand.
New Zealand J. Geol. Geophys., 3(2):253-261.

drift cards
Segawa, Sokichi, Takeo Sawada, Masahiro Higak Tadao Yoshida, Hajime Ohshiro and Fumio Hayashida, 1962.
Some comments on the movement of the floating seaweeds.
Rec. Oceanogr. Wks., Japan, Spec. No. 6, 153-159.

Also in: Contrib. Dept. Fish and Fish Res. Lab., Kyushu Univ., No. 8.

drift cards
Segawa, S., T. Sawada, M. Higaki and T. Yoshida, 1961
Studies on the floating seaweeds. VIII. The drifting movement of the floating seaweeds off northern coast of Kyushu.
Sci. Bull. Fac. Agric. Kyushu Univ., 19(1): 135-
In Japanese; English summary.
Also in: Contr. Dept. Fish., Fish. Res. Lab., Kyushu Univ., No. 7.

drift cards
Segawa, Sokichi, Takeo Sawada, Masahiro Higaki, Tadeo Yoshida, Hajime Ohshiro and Fumio Hayashida, 1962
Some comments on the movement of the floating seaweeds.
Rec. Oceanogr. Wks., Japan, Spec. No. 6:153-159.

drift cards
South Africa, Division of Fisheries, 1954.
Twenty-fourth annual report:1-199.

drift cards
Stander, G.H., L.V. Shannon, and J.A. Campbell, 1969.
Average velocities of some ocean currents as deduced from the recovery of plastic drift cards.
J. mar. Res., 27(3): 293-300.

drift cards
Tomczak, G., 1964.
Investigations with drift cards to determine the influence of the wind on surface currents.
In: Studies on Oceanography dedicated to Professor Hidaka in commemoration of his sixtieth birthday, 129-139.

drift envelopes

drift envelopes
Lawford, A. L., 1962.
The drift envelope experiment in the north-east Atlantic Ocean, 1954.
N.I.O. Internal Rept., No. A19:8 pp., tables, figs., (Unpublished manuscript).

drift, depositional

depositional drifts
Laughton, A.S., 1963.
18. Microtopography.
In: The Sea, M.N. Hill, Editor, Interscience Publishers, 3:437-472.

drift, littoral

drift, littoral
Anon., 1956.
The measurement of littoral drift by radio-isotopes. Field experiments to aid design of new Japanese port.
Dock and Harbour Authority 36(423):284-288.

drift, littoral
*Aramaki, Makoto, and Shigemi Takayama, 1967.
A petrographic study on littoral drift along the Ishikawa Coast, Japan.
Proc. 10th Conf. Coast. Engng. Tokyo, 1966, 1:615-631.

drift, littoral
Battjes, J.A., 1967.
Quantitative research on littoral drift and tidal inlets.
In: Estuaries, G.H. Lauff, editor, Publs Am. Ass. Advmt Sci., 83:185-190.

littoral drift
Beach Erosion Board, 1953.
Shore protection planning and design (Preliminary subject to revision). B.E.B. Bull., Special Issue No. 2:230 pp., 149 figs., plus app.

littoral (longshore) drift
Byerly, John Robert, 1963.
The relationship between watershed geology and beach radioactivity.
Beach Erosion Bd., Techn. Memo. (135):32 pp.

littoral drift
Fukushima, Hisao, and Yutaka Mizoguchi, 1958
Field investigation of suspended littoral drift. Coastal Engineering in Japan, 1: 131-134.

littoral drift
Great Britain, Department of Scientific and Industrial Research, 1962.
Hydraulics Research, 1961: the report of the Hydraulics Research Board, with the report of the Director of Hydraulics Research (Wallingford), 96 pp.

littoral drift
Great Britain, Department of Scientific and Industrial Research, 1961.
Waves and sea defences.
Hydraulics Research, 1960:62-69.

drifts, longshore
Hodgson, W.A., 1966.
Coastal processes around the Otago Peninsula.
New Zealand J. Geol. Geophys., 9 (1/2):76-90.

drift, littoral
Jolliffe, I. P , (undated reprint)
The use of tracers to study beach movements; and the measurement of littoral drift by a fluorescent technique.
Revue de Géomorphologie Dynamique, unnumbered pp.

drift, longshore
Jurkevich, M.G., and L.E. Jurkevich, 1967.
Quantitative determinations of longshore drifts. (In Russian).
Okeanologiia, Akad. Nauk, SSSR, 7(6):1125-1127.

drift, littoral
Kinmont, A., 1961.
The nearshore movement of sand at Durban.
Marine Studies off the Natal Coast, C.S.I.R. Symposium, No. S2:46-58. (Multilithed).

longshore drift
Klein, George deVries, 1967.
Paleocurrent analysis in relation to modern marine sediment dispersal patterns.
Bull. Am. Ass. Petrol. Geol., 51(3):366-382.

drift, littoral
Koh, Ryuji 1966.
Littoral drift along Iwafune Port.
Coast. Engng Japan 9: 127-136.

drift, littoral
Kulm, L.D., and John V. Byrne, 1967.
Sediments of Yaquina Bay, Oregon.
In: Estuaries, G.H. Lauff, editor, Publs Am. Ass. Advmt Sci., 83:226-238.

littoral drift
Munch-Peterson, V.P., 1938.
Munch-Peterson's littoral drift formula. Bull. B.E.B. 4(4), 1950; 31 pp., 25 figs.

Translation of a speech prepared by the late Prof. Munch-Peterson of the Academy for Technical Sciences, 5 April 1938, and delivered by Mr. Sv. Svendsen at meeting of Association of Governmental and Harbor Engineers, 27 August 1938 at Helsingfors.

littoral drift
O'Brien, M. P., 1950(1947).
Wave refraction at Long Beach and Santa Barbara, California. Bull. B.E.B. 4(1)11-12, 8 figs.

HE-116-246.

littoral drift
Orlova, G.A., 1961
[Model experiments on the unconsolidated slope wave abrasion and the littoral drift.]
Trudy Okeanogr. Komissii, Akad. Nauk, SSSR, 8:235-239.

littoral drift
Russell, R.C.H., 1961
The use of fluorescent tracers for the measurement of littoral drift. Ch. 24, in: Proc. Seventh Conf. on Coastal Engineering, 418-444.

littoral drift
Slomianko, Pawel, 1961
[Some coastal problems in the Maritime Institute's investigations.] (In Polish; English summary).
Prace Inst. Morsk., Gdansk, (1) Hydrotech. II. Sesja Naukowa Inst. Morsk., 20-21 wrzesnia 1960: 129-134; English summary, 149-150.

littoral drift
Smith, D.B., and J.D. Eakins, 1957.
Radioactive methods for labelling and tracing sand and pebbles in investigation of littoral drift. Int. Conf., Radioisotopes in Sci. Res., Paris, Sept., 1957, UNESCO/NS/RIC/63:1-12.

littoral drift
Silvester, Richard, 1956.
The use of cyclonicity charts in the study of littoral drift. Trans. Amer. Geophys. Union, 37(6):694-696.

longshore drift
Vollbrecht, Kurt, 1966
The relationships between wind records, energy of longshore drift, and energy balance off the coast of a restricted body of water, as applied to the Baltic.
Marine Geology, 4(2): 119-147.

littoral drift
Yurkevich, M.G., 1968.
The use of sand-transporting wave properties for littoral drift study. (In Russian; English abstract).
Okeanologiia, Akad. Nauk, SSSR, 8(3):442-451.

drift, non-tidal

drift, non-tidal
Bumpus, D.F., J. Chase, C.G. Day, D.H. Frantz Jr. D.D. Ketchum and R.G. Walden, 1957.
A new technique for studying non-tidal drift with results of experiments off Gay Head, Mass., and in the Bay of Fundy. J. Fish. Res. Bd., Canada, 14(6):931-944.

drift, non-tidal
Howe, Malcolm R., 1962
Some direct measurements of the non-tidal drift on the continental shelf between Cape Cod and Cape Hatteras.
Deep-Sea Res., 9(5):445-455.

drift, residual

drifts, residual
Bumpus, Dean F., 1965.
Residual drift along the bottom on the continental shelf in the Middle Atlantic Bight area.
Limnol. and Oceanogr., Redfield Vol., Suppl. to 10:R50-R53.

drift, residual
*Lauzier, L.M., 1967.
Bottom residual drift on the continental shelf of the Canadian Atlantic coast.
J. Fish. Res. Bd. Can., 24(9):1845-1859.

drift, sand
See also: transport, sand

drift, sand
See also: transport, sand

drift, sand
Sato, Shoji, and Norio Tanaka, 1967.
Field investigations on sand drift at Port Kashima facing the Pacific Ocean.
Proc. 10th Conf. Coast. Engng. Tokyo, 1966, 1:595-614.

drift, sea-surface

drift, surface
*Bumpus, Dean F., 1969.
Reversals in the surface drift in the Middle Atlantic Bight Area.
Deep-Sea Res., Suppl. 16: 17-23.

drift, sea-surface
Hughes, P., 1957.
A determination of the relation between wind and sea-surface drift. Q.J.R. Meteorol. Soc., 83 (356):276-277.

drift, sea-surface
Hughes, P., 1956.
A determination of the relation between wind and sea-surface drift. Q.J.R. Meteorol. Soc., 82 (354):494-502.

drift stations

drift stations
Brewer, Max C., 1967.
Soviet Drifting Ice Station North-67.
Arctic 20(4):263-267.

drift stations
Cottell, I.B., 1960
U.S. research at drifting stations in the Arctic Ocean.
Polar Record, 10(66): 269-274.

drifting stations (Ice stations)
Timofeyev, V.T., 1958.
Scientific inspection of drift stations and observatories in 1956. (In Russian).
Problemy Arktiki, (4):109-110.

drift stations
Zaklinskii, A.B., and L.N. Nazaretskii, 1960.
On work of powered ships in the drifting oceanographic stations. (In Russian).
Trudy Morsk. Gidrofiz. Inst., 19:103-111.

drift, Stokes
Huang, Norden E., 1971.
Derivation of Stokes drift for a deep-water random gravity wave field. Deep-Sea Res., 18(2) 255-259.

drift - surface to bottom

drift, surface to bottom
*Harrison, W., J.J. Norcross, N.A. Pore and E.M. Stanley, 1967.
Circulation of shelf waters off the Chesapeake Bight: surface and bottom drift of continental shelf waters between Cape Henlopen, Delaware and Cape Hatteras, North Carolina, June 1963-December 1964.
ESSA Prof. Pap., U.S. Dept. Comm. 3:1-82.

drift, Swallow float
Rossby, T. and D. Webb, 1971.
The four month drift of a Swallow float.
Deep-Sea Res. 18(10): 1035-1039.

drift: vessels

drift, ships
Hiraiwa, T., T. Fujii and S. Saito 1969.
An experimental study of drift and leeway.
J. Inst. Navig. 20(2): 131-145.

drifts of vessels in ice
U.S.N. Hydrographic Office, 1958
Oceanographic atlas of the Polar seas. Pt. II. Arctic. H.O. Publ. No. 705: 139 pp., charts.

drift, wind
SEE ALSO: wind drift

drift, wind
Cooper, L.H.N., 1961
The oceanography of the Celtic Sea. I. Wind drift. II. Conditions in the spring of 1950.
J. Mar. Bio. Ass. U.K., 41(2): 223-270.

driftwood
Inoue, Motoo, Ryohei Amano, Yukinobu Iwasaki and Minoru Yamauti, 1968.
Studies on environments alluring skipjack and tunas. IV. Tagging experiments on the experimental driftwoods as part of ecological study of tunas.
Bull. Jap. Soc. scient. Fish. 34(4).
Also in: Coll. Repr. Coll. mar. Sci. Techn. Tokai Univ., 1967-68, 3:99-105

drifters

drifters
*Morse, Betty-Ann, M. Grant Gross and Clifford A. Barnes, 1968.
Movement of seabed drifters near the Columbia River.
J. WatWays. Harb. Div. Am. Soc. civ. Engrs, 94(WW1): 93-103.

drifting

drifting, epiplanktonic
Fell, H. Barraclough, 1967.
Cretaceous and Tertiary surface currents in the oceans.
Oceanogr. Mar. Biol., Ann. Rev., H. Barnes, editor, George Allen and Unwin, Ltd., 5:317-341.

drifting object, effect of

drifting object, effect of
*Gooding, Reginald M., and John J. Magnuson, 1967.
Ecological significance of a drifting object to pelagic fishes.
Pacif. Sci., 21(4):486-497.

drilling

drilling
Anon., 1967.
Scientists launch coring program to explore deep ocean sediments.
Ocean Industr., 2(11):32-37.

drilling
Bascom, Willard, 1969.
Technology and the ocean.
Scient. Am., 221(3):198-204, 206, 208, 210, 213, 214, 216-217. (popular).

drilling
Coudray, Jean, 1971.
Nouvelles données sur la nature et l'origine du complexe récifal côtier de la Nouvelle-Calédonie: Étude sédimentologique et paléoécologique préliminaire d'un forage réalisé dans le récif-barrière de la côte sud-ouest.
Quatern. Res. 1(2): 236-246.

drilling
Findley, D.C., and C.H. Smith, editors, 1966.
Drilling for scientific purposes. Report of the International Upper Mantle Symposium, Ottawa, 2-3 September, 1965.
Geol. Surv. Pap. Can., 66-13:264 pp.

drilling
Laborde, Alden J. 1967.
Problems of a drilling contractor.
Ocean Industry 2(4): 22-25.

drilling
Lisitzyn, A.P. 1971.
Drilling of the ocean bottom. (In Russian).
Geofiz. Bull. Mezhd. Geofiz. Kom. Presid. Akad. Nauk, SSSR, 23: 60-67.

drilling, off shore
Mollenhauer, Frank H., 1969.
Offshore drilling.
Oceans, 2(2): 25-31. (popular)

drilling, deep ocean
Peterson, M.N.A. and N.T. Edgar, 1969.
Deep ocean drilling with Glomar Challenger.
Oceans Mag. 1(6): 17-32. (popular)

drilling
Saito, Tsunemasa, 1966.
Ocean drilling off the coast of eastern Florida. (Summary).
Geol. Surv. Pap. Can., 66-13:52-53.

drilling, deep-sea, effect of
Scheibnerová, Viera 1971.
Implications of deep sea drilling in the Atlantic for studies in Australia and New Zealand – some new views on Cretaceous and Cainozoic palaeogeography and biostratigraphy.
Search ANZAAS 2(7):251-254

drilling
Van Andel, Tjeerd H. 1968.
Deep-sea drilling for scientific purposes: a decade of dreams.
Science 160 (3835): 1419-1424

drilling rigs
Watts, J.S.Jr., and R.E. Faulkner, 1968.
Designing a drilling rig for severe seas.
Ocean Industry, 3(11):28-37.

DROGUES

drogues
Bowden, K.F., and P. Hughes, 1961
The flow of water through the Irish Sea and its relation to wind.
Geophys. J., R. Astron. Soc., 5(4):265-291

drogues
Chew, Frank, and George A. Berberian 1970.
Some measurements of current by shallow drogues in the Florida Current.
Limnol. Oceanogr. 15(1): 88-99.

drogues
Reid, Joseph L., Jr., and Richard A. Schwartlose, 1962
Direct measurements of the Davidson Current off central California.
J. Geophys. Res., 67(6):2491-2497.

Droplets

drops
See also: bubbles

drop size
Colgate, Stirling A. and John M. Romero, 1970.
Charge versus drop size in an electrified cloud.
J. geophys. Res., 75(30): 5873-5881.

droplets
Garabedian, P.R., 1965.
On the shape of electrified droplets.
Comm. Pure and Appl. Math., 18(1/2):31-34.

drops
Georgii, Hans-Walter and Dieter Wötzel, 1970.
On the relation between drop size and concentration of trace elements in rainwater. J. geophys. Res., 75(9): 1727-173 .

droplets
Golovin, A.M., 1963.
The solution of the coagulation equation for cloud droplets in a rising air current. (In Russian).
Izv., Akad. Nauk, SSSR, Ser. Geofiz., (5):785-791.

Translation:
Amer. Geophys. Union, (5):482-485.

drops
Jayaratne, O.W., and B.J. Mason, 1964.
The coalescence and bouncing of water drops at an air/water interface.
Proc. R. Soc., London, (A), 280(1383):545-565.

drops
Komabayashi, M., 1964.
Primary fractionation of chemical components in the formation of submicron spray drops from seasalt solution.
J. Meteorol. Japan, (2), 42(5):309-316.

Also in:
Collected Papers on Sciences of Atmosphere and Hydrosphere. Water Res. Lab., Nagoya Univ., 1964, 2.

drops
Komabayashi, M., T. Gonda and K. Isono, 1964.
Life time of water drops before breaking and size distribution of fragment droplets.
J. Meteorol. Japan, (2), 42(5):330-340.

Also in:
Collected Papers on Sciences of Atmosphere and Hydrosphere, Water Res. Lab., Nagoya Univ., 1964, 2.

drops
Paluch, Ilga R. 1970.
Theoretical collision efficiencies of charged cloud droplets.
J. geophys. Res. 75(9): 1633-1640.

water drops
Woods, J.D., 1965.
The effect of electric charges upon collisions between equal-size water drops in air.
Q.J.R. Meteorol. Soc., 91(389):353-355.

DROUGHTS

droughts
Priestley, C.H.B., and A.J. Troup, 1966.
Droughts and wet periods and their association with sea surface temperature.
Aust. J.Sci., 29(2):56-57.

droughts
Schindler, G., 1957.
Trockenheit auf dem Meere.
Der Seewart, 18(3):105-112.

drowned islands

drugs

drugs
Arehart, Joan Lynn, 1969.
Oceanic drug chest.
Sea Frontiers, 15(2): 98-107 (popular)

drugs
Burkholder, Paul R. 1963.
Drugs from the sea.
Armed Forces chem. J. March 1963: 6-16.

drugs
Der Marderosian, Ara Harold 1969.
Biodynamic agents from marine sources as potential drugs.
J. mar. techn. Soc. 3(4): 6-84.

drugs
Freudenthal, Hugo D., editor, 1968.
Drugs from the sea.
Marine Techn. Soc., 297 pp. $12.00 (paperbound). (not seen).

systemic drugs
Gruber, Michael, 1968.
The healing sea. (popular).
Sea Frontiers, 14(2):74-86.

drugs
Hillman, Robert E., 1967.
Drugs from the sea. (Popular)
Oceanol. int., 2(6):33-37.

drugs
Hood, Donald W. and C. Peter McRoy, 1971.
Uses of the ocean. (4) In: Impingement of man on the oceans, D.W. Hood, editor, Wiley Interscience: 667-698.

drugs
Jones, Albert C., 1968.
Food and drugs from the sea: sleeping giant or deceptive illusion. Transactions, Conference on Industry's Future in the Ocean...the Challenge and the Reality (March 1968), Miami, Fla., Florida Commission on Marine Sciences and Technology,: 114-121, 1968. Also in: Collected Repr. Div. Biol. Res., Bur. Comm. Fish., U.S. Fish Wildl. Serv., 1968, 1.

drugs
Miloy, L.F. 1968.
New life-saving drugs are on the way.
Ocean Industry 3(6): 74-76

drugs
Youngken, Heber W., Jr., 1967.
Drugs from the sea.
Maritimes, Univ. R.I. 11(4):5-6. (popular).

drying, effect of

drying
Jaworski, G. and J.W.G. Lund 1970.
Drought resistance and dispersal of Asterionella formosa Hass.
Beihefte Nova Hedwigia 31: 37-45

drying, effect of
Ogata, Eizi, and Toshio Matsui, 1965.
Photosynthesis in several marine plants of Japan as affected by salinity, drying and pH, with attention to their growth habitats.
Botanica Marina, 8(2/4):199-217.

dry weight

dry weight
Bsharah, L., 1957.
Plankton of the Florida Current 5. Environmental conditions, standing crop, seasonal and diurnal changes at a station forty miles east of Miami.
Bull. Mar. Sci., Gulf & Caribbean, 7(3):201-251.

NO IDENTIFICATIONS !

dry weight (data)
Corlett, J., 1961.
Dry weight of plankton in the western Barents Sea 1957-1959.
Ann.Biol., Cons. Perm. Int. Expl. Mer, 1959, 16:68-69.

dry weight

Corlett, J., 1953.
Dry weight and fat content of plankton near Bear Island, 1949-1952.
Ann. Biol., Cons. Perm. Int. Expl. Mer, 9:8-9, 2 textfigs.

dry weight-wet weight

Hagmeier, E., 1961.
Plankton-Äquivalente (Auswertung von chemischen und mikroskopischen Analysen).
Kieler Meeresf., 17(1):32-47.

dry weight (gms)

McIntyre, A.D., and J.H. Steele, 1956.
Hydro-biological conditions in the Denmark Strait, May 1954.
Ann. Biol., Cons. Perm. Int. Expl. Mer, 11:20-25, Figs. 25-28.

zooplankton, dry wgt.

Steele, J.H., 1956.
Plant production on the Fladen Ground. J.M.B.A., 35(1):1-33.

dry weight of plankton

Wimpenny, R.S., 1956.
The dry weight of Hansen-net plankton.
Ann. Biol., Cons. Perm. Int. Expl. Mer, 11:58-59

mg/m³

dry weight

Wimpenny, R.S., 1953.
The dry-weight and fat content of plankton with estimates from flagellate counts.
Ann. Biol., Cons. Perm. Int. Expl. Mer, 9:119-122, Textfigs., 25-27.

dry weight

Wimpenny, R.S., 1950.
The dry weight and fat content of plankton.
Ann. Biol. 6:124-126, Textfigs. 6-8.

dry weight

Wimpenny, R. S., 1949.
The dry weight and fat content of plankton.
Ann. Biol. 5:89.

Coscinodiscus concinnus = 30-50%

dumping

Saila, S.B., T.T. Polgar and B.A. Rogers, 1968.
Results of studies related to dredged sediment dumping in Rhode Island Sound.
Proc. Ann. NEast. Rlg. Antipollution Conf., Univ. R.I., July 22-24, 1968.

dunes

dunes

Argentina, Secretaria de Marina, Servicio de Hidrografia Naval, 1961.
Dunas gigantes en el Golfo San Matias.
Publico, H. 622:unnumbered pp.

dunes

Bigarella, João José, Rosemarie Doria Becker and Gerusa M. Duarte 1969.
Coastal dune structures from Paraná (Brazil).
Mar. Geol. 7(1): 5-55.

dunes

Inman, D.L., G.C. Ewing and J.B. Corliss, 1966.
Coastal sand dunes of Guerrero Negro, Baja California, Mexico.
Geol. Soc., Am., Bull., 77 (8):787-802.

dunes

Milling, Marcus E., and E. William Behrens, 1966.
Sedimentary structures of beach and dune deposits: Mustang Island, Texas.
Publs., Inst. mar.Sci., Univ.Texas.Port Aransas, 11:135-148.

dunes

Morton, J.K., 1957.
Sand-dune formation on a tropical shore.
J. Ecology, 45(2):495-498.

dunes

Shepard, Francis P., 1964.
Criteria in modern sediments useful in recognizing ancient sedimentary environments.
In: Deltaic and shallow marine deposits, Developments in Sedimentology, L.M.J.U. van Straaten, Editor, Elsevier Publishing Co., 1:1-25

dunes

Yalin, M. Selim, 1964.
Geometrical properties of sand waves.
J. Hudraulics Div., Proc. Amer. Soc. Civil Eng. 90 (HY5) (Proc. Paper 4055):105-119.

dune sands

*Rex, Robert W., Stanley V. Margolis and Betty Murray, 1970.
Possible interglacial dune sands from 300 meters water depth in the Weddell Sea, Antarctica.
Bull.geol.Soc.Am., 81(11):3465-3472.

dunites

dunites

Komarov, A.G., 1965.
Oceanic ridge and rift structure: geological nature of the magnetic and gravity anomalies over the rift valley. (In Russian).
Priroda, (7):95-98.

Translation:
E.R. Hope, T 437 R - Nov. 1965.

dust

atmospheric dust

Abramov, R.V., 1971.
Atmospheric dust over the Atlantic Ocean. (In Russian; English abstract). Okeanol. Issled. Rezult. Issled. Mezhd. Geofiz. Proekt. 21: 7-30.

dust, aeolian

Ashton, S., R. Chester and L. R. Johnson 1972.
Uptake of cobalt from sea water by aeolian dust.
Nature, Lond., 235 (5338): 380-381

dust, aeolian

Chester, R., and L.R. Johnson, 1971
Trace element geochemistry of North Atlantic aeolian dusts.
Nature, Lond., 231(5299): 176-178

dust

Ferguson, William S., John J. Griffin and Edward D. Goldberg, 1970.
Atmospheric dusts from the North Pacific - a short note on a long-range eolian transport.
J. geophys. Res., 75(6): 1137-1139.

dust fall

*Folger, D.W., L.H. Burckle and B.C. Heezen, 1967.
Opal phytoliths in a North Atlantic dust fall.
Science, 155(3767):1243-1244.

dustfall

Game, P.M., 1964.
Observations on a dustfall in the eastern Atlantic, February 1962.
J. Sed. Petr., 34(2):355-359.

dust, air-borne

Glasby, G.P. 1971.
The influence of aeolian transport of dust particles on marine sedimentation in the south-west Pacific.
J. R. Soc. N.Z. 1 (3/4): 285-300.

dust, atmospheric

Goldberg, Edward D., 1971.
3. Atmospheric transport. In: Impingement of man on the oceans, D.W. Hood, editor, Wiley Interscience: 75-88.

dust, atmospheric (radioactive)

Gopalakrishnan, S. and C. Rangarajan, 1972.
The role of the Indian monsoons in the inter-hemispheric transport of radioactive debris from nuclear tests. J. geophys. Res, 77(6): 1012-1016.

dust particles (Ni/Fe)

Hodge, Paul W., Frances W. Wright and Chester C. Langway, Jr., 1964.
Studies of particles for extraterrestrial origin. 3. Analysis of particles for extraterrestrial origin. 3. Analysis of dust particles from polar ice deposits.
J. Geophys. Res., 69(14):2919-2931.

dust

*Idso, Sherwood B. and Paul C. Kangieser, 1970.
Seasonal changes in the vertical distribution of dust in the lower troposphere. J. geophys. Res., 75(12): 2179-2184.

dust, Sahara

Jaenicke, Ruprecht, Christian Junge, und Hans Joachim Kanter 1971.
Messungen der Aerosolgrössenverteilung über dem Atlantik.
Meteor-Forsch.-Ergebnisse (B) 7:1-54.

dust

Jarke, Joachim, 1960.
Staubfall auf dem Schwarzen Meer.
Deutsche Hydrogr. Zeits., 13(5):225-229.

atmospheric dust

Kool, L.V., 1971.
Jet streams over the Atlantic Ocean. (In Russian; English abstract). Okeanol. Issled. Rezult. Issled. Mezhd. Geofiz. Proekt. 21: 31-42.

dust (atmospheric)

Mullen, Ruth E., Dennis A. Darby and David L. Clark 1971.
Significance of atmospheric dust and ice rafting for Arctic Ocean sediment.
Bull. Geol. Soc. Am. 83(1): 205-212

dust

Parkin, D.W., D.R. Phillips, R.A.L. Sullivan and L. Johnson 1970.
Airborne dust collections over the North Atlantic.
J. geophys. Res 75 (9): 1782-1793

dust

Prospero, Joseph M., Enrico Bonatti, Carl Schubert and Toby N. Carlson, 1970.
Dust in the Caribbean atmosphere traced to an African dust storm.
Earth Planet. Sci. Letts. 9(3): 287-293

dust

Prospero, Joseph M. and Toby N. Carlson, 1970.
Radon-222 in the North Atlantic trade winds: its relationship to dust transport from Africa.
Science, 167(3920): 974-977.

dust

Volz, F.E., 1970.
On dust in the tropical and midlatitude stratosphere from recent twilight measurements.
J. geophys. Res., 75(9): 1641-1646.

dust, atmospheric

Chester R. and R.J. Johnson, 1971.
Atmospheric dusts collected off the West African coast.
Nature, Lond., 229 (5280): 105-107

dust (atmospheric)

Goldberg, Edward D. and John J. Griffin, 1970.
The sediments of the northern Indian Ocean.
Deep-Sea Res., 17(3): 513-537.

dust, atmospheric

Hamilton, Wayne L. 1970.
Atmospheric dust records in permanent snowfields: implications to marine sedimentation: a discussion.
Bull. geol. Soc. Am. 81 (10): 3175-3176.

dust, Atmospheric

Sparrow, J.G., 1971
Stratospheric properties and Bali dust.
Nature, Lond., 229 (5280): 107.

dust (atmospheric)

Windom, Herbert L. 1970.
Atmospheric dust records in permanent snowfields: implications to marine sedimentation: a reply.
Bull. geol. Soc. Am. 81 (10): 3177-3178.

dust, atmospheric, effect of

Windom, Herbert L., 1969.
Atmospheric dust records in permanent snowfields: implications to marine sedimentation.
Bull. geol. Soc. Am., 80 (5): 761-782

dust (Sahara)

Volz, F.E., 1970.
Spectral skylight and solar radiance measurements in the Caribbean: maritime aerosols and Sahara dust.
J. atmos. Sci., 27 (7): 1041-1047.

dye patches

dye injections

Assaf, G., R. Gerard and A.L. Gordon, 1971.
Some mechanisms of oceanic mixing revealed in aerial photographs. J. geophys. Res. 76(27): 6550-6572.

dye patches

Brodie, J.W., 1958
A note on tidal circulation in Port Nicholson New Zealand. N.Z. J. Geol. Geophys. 1(4): 684-702.

dyes

*Ichiye, Takashi, 1967.
Upper ocean boundary-layer flow determined by dye diffusion.
Physics Fluids 10 (9-2): S270-S277.

dyes

Ichiye, Takashi, 1965.
Diffusion experiments in coastal waters using dye techniques. (abstract).
Symposium, Diffusion in oceans and freshwaters, Lamont Geol. Obs., 31 Aug.-2 Sept., 1964. 54-67.

dye experiment

Ozmidov, R.V., V.K. Astok., A.N. Gesentswei and M.K. Yukhat, 1971.
Statistical properties of concentration fields of a passive pollutant artificially introduced in the sea. Fiz. Atmosfer. Okean. Izv. Akad. Nauk SSSR, 7(9): 963-973. (In Russian; English abstract).

dyes

Reinert, Richard L., and Gerard J. McNally, 1965
On the use of a type recording system in dye diffusion studies. (abstract).
Ocean Sci. and Ocean Eng., Mar. Techn. Soc.,-Amer. Soc. Limnol. Oceanogr., 2:1095-1102.

dye markers

Hale, Alan M., 1971.
The feasibility of using continuous dye injection for underwater flow visualization. Limnol. Oceanogr. 16(1): 124-129.

dye marks

Woods, J.D., and G.G. Fosberry 1965.
Observations of the behaviour of the thermocline and transient stratifications in the sea made visible by dye markers.
Malta '65: 31-36.

dye tracers

Costin, J. Michael, 1965.
Dye tracer studies on the Bahama Banks, (summary)
Symposium, Diffusion in oceans and fresh waters, Lamont Geol. Obs., 31 Aug.-2 Sept., 1964, 68-69.

dye tracers

Foxworthy, James E., Richard B. Tibby and George Barsom, 1965.
Multidimensional aspects of eddy diffusion determined by dye diffusion experiments in coastal waters. (summary).
Symposium, Diffusion in oceans and fresh waters, Lamont Geol. Obs., 31 Aug.-2 Sept., 1964, 71-73.

dye tracers

Okubo, Akira, 1965.
A theoretical model of diffusion of dye patches. (summary).
Symposium, Diffusion in oceans and fresh waters, Lamont Geol. Obs., 31 Aug.-2 Sept., 1964, 74-78.

dykes

dykes

*Harrison, C.G.A., 1968.
Formation of magnetic anomaly patterns by dyke injection.
J. geophys. Res., 73(6): 2137-2142.

dynamics

dynamics

*Gougenheim, Andre, 1968.
Deux ingénieurs hydrography du XIXth Siècle precurseurs en marière de dynamique des mers.
Bull. Inst. océanogr., Monaco, No. spécial 2: 87-97.

dynamic anomalies
see: dynamic heights

dynamic anomalies

Lemasson, L. et B. Piton, 1968.
Anomalie dynamique de la surface de la mer le long de d'équateur dans l'océan Pacifique.
Cah. ORSTOM, Océanogr., 6 (3/4): 39-45.

dynamic computations

dynamic computations

Bruneau, L., N.G. Jerlov, and F.F. Koczy, 1953.
Physical and chemical methods.
Repts. Swedish Deep-sea Exped., 1947-1948, Phys. Chem., 3(4): 101-112, i-iv, 7 textfigs., 2 tables.

dynamic computations, accuracy of

Bulgakov, N.P., and I.F. Moroz, 1965.
On the effect of interpolation of sea water specific volume to standard horizons in regard to the accuracy of the dynamic method. (In Russian).
Okeanologiia, Akad. Nauk, SSSR, 5(3): 541-548.

Dynamic currents: SEE Currents, dynamic

dynamic circulation

Lemasson, L., et Y. Magnier, 1966.
Résultats des observations scientifiques de La Dunkerquoise sous le commandement du Capitaine de Corvette Brosset. Croisière Hunter.
Cah. ORSTOM, Sér. Océanogr., 4(1): 3-78.

dynamics, equatorial

Philander, S.G.H., 1971.
The equatorial dynamics of a shallow, homogeneous ocean.
Geophys. fluid Dynam. 2 (3): 219-245.

Dynamic Heights
Dynamic topography

dynamic height - dynamic topography
have been used essentially interchangeably and have been mixed

A

dynamic topography

Akagawa, M., 1956.
On the oceanographical conditions of the North Japan Sea (west of the Tsugaru Straits) in summer (Part 2).
Bull. Hakodate Mar. Obs., No. 3: 5-11. 1-7.

dynamic topography
Aliverti, Guiseppina, 1963
Sur les caractéristiques hydrologiques de la Mer Tyrrheniènne.
Cahiers Océan., C.C.O.E.C., 15(8):557-558.

dynamic topography
Allain, Charles, 1963.
Topographie dynamique et courants généraux dans le bassin occidental de la Méditerranée au nord du 42 parallèle (supplément à l'etude hydrologique de septembre-octobre 1958).
Rev. Trav. Inst. Pêches Marit., 27(2):127-135.

dynamic heights
Australia, Commonwealth Scientific and Industrial Research Organization, 1961.
F.R.V. "Derwent Hunter".
C.S.I.R.O., Div. Fish. and Oceanogr., Rept., No. 32:56 pp.

dynamic heights
Australia, C.S.I.R.O., Division of Fisheries and Oceanography, Marine Biological Laboratory, Cronulla, 1959.
F.R.V. "Derwent Hunter", Scientific report of Cruise DH9/57, Aug. 19-25, 1957; Cruise DH10/57, Sept. 4-11, 1957; Cruise DH11/57, Sept. 18-21, 1957; Cruise DH12/57, Sept. 26-Oct. 11, 1957.
CSIRO, Div. Fish. & Ocean., Rept., No. 20:20 pp. (mimeographed)

dynamic topography
Australia, Marine Biological Laboratory, Cronulla, 1960.
F.R.V. "Derwent Hunter", scientific report of cruises 10-20/58 -----.
C.S.I.R.O., Div. Fish. & Oceanogr., Rept., 30: 53 pp., numerous figs. (mimeographed).
For complete "title", see author card.

dynamic heights
Australia, Marine Biological Laboratory, Cronulla, Sydney, 1957.
F.R.V. "Derwent Hunter", scientific report of Cruise 3/56. September 19-October 5, 1956; ----- Cruise 4/56, October 9-November 6, 1956; ------- Cruise 5/56, November 8-December 3, 1956.
C.S.I.R.O.Australia, Div. Fish and Oceanogr., Rept., No. 5:16 pp. (mimeographed)

dynamic heights
Ayers, J.C., and R. Bachman, 1957.
Simplified computations for the dynamic height method of current determination in lakes.
Limnol. & Oceanogr., 2(2):155-157.

B

dynamic topography
Baker, D.J., Jr., 1968.
A demonstration of magnification of dynamic topography at the thermocline.
J. mar. Res., 26(3):283-285.

dynamic topography
Barbee, William D., 1965.
Deep circulation, central North Pacific Ocean: 1961, 1962, 1963.
U.S.Coast and Geodetic Survey, E.S.S.A., Res. Paper:104 pp.
(paper has no number).

dynamic topography
Barnes, C.A., D.F. Bumpus, and J.Lyman, 1948
Ocean circulation in Marshall Islands area. Trans. Am. Geophys. Union, 29(6):871-876, 7 text figs.

dynamic topography
Bennett, Edward B., 1963.
An oceanographic atlas of the Eastern Tropical Pacific Ocean, based on data from EASTROPIC expedition. October-December 1955.
Bull. Inter-American Trop. Tuna Comm., 8(2):33-165.

dynamic heights (data)
Blackburn, M., 1962.
An oceanographic study of the Gulf of Tehuantepec.
U.S.F.W.S. Spec. Sci. Rept., Fish., No. 404:28 pp.

dynamic height anomaly (data only)
Blackburn, M., R.C. Griffiths, R.W. Holmes and W.H. Thomas, 1962.
Physical, chemical and biological observations in the eastern tropical Pacific Ocean: three cruises to the Gulf of Tehuantepec, 1958-1959.
U.S.F.W.S. Spec. Sci. Rept., Fish., No. 420: 170 pp.

dynamic topography
Brandhorst, W., J.G. Simpson, M. Carreño and O. Rojas, 1965.
Anchoveta resources in northern Chile in relation to environmental conditions from January to February, 1965.
Inst. Fomento Pesquero. 14(2/3): 167-235

dynamic heights
Brazil, Diretoria de Hidrografia e Navegacao, 1958.
Ano Geofisico Internacional, Publ. DG-06-VII: mimeographed sheets.

dynamic topography
Brazil, Diretoria de Geofisico e Navegacao, 1957.
Ano Geofisico Internacional, Publicacao DG-06-VI: unnumbered mimeographed pages.

dynamic anomalies
Brasil, Diretoria de Hidrografia e Navegacao, 1957.
Ano Geofisico Internacional. Publ. DG-06-III (mimeographed).

dynamic anomalies
Brazil, Diretoria de Hidrografia e Navegacão, 1957
Ano Geofisico Internacional. Publicacão DG-06-II mimeographed

dynamic topography
Broenkow, William W., 1965.
The distributionof nutrients in the Costa Rica dome in the eastern tropical Pacific Ocean.
Limnology and Oceanography, 10(1):40-52.

dynamic topography
*Bruce, J.G., 1968.
Comparison of near surface dynamic topography during the two monsoons in the western Indian Ocean.
Deep-Sea Res., 15(6):665-677.

dynamic topography
Burkov, V.A., 1969.
A model for stationary water transport in a non-homogeneous ocean. (In Russian; English abstract
Okeanologiia, 9(1):15-17.

dynamic topography
Burkov, V.A., V.S. Arsentyev, and I.M. Ovchinnikov, 1960
On the notion of northern and southern tropical fronts in the ocean.
Trudy Inst. Okeanol., 40: 108-120.

dynamic heights
Burkov, V.A., and K.V. Maroshkin, 1965.
The reduction of dynamic heights to a uniform level. (In Russian).
Okeanologiia, Akad. Nauk, SSSR, 5(3):548-553.

dynamic topography
Burkov, V.A., and I.M. Ovchinnikov, 1960
Investigations of equatorial currents to the north of New Guinea.
Trudy Inst. Okeanol., 40: 121-134.

dynamic topography
Buzdalin, Yu. I., and A.A Elisarov, 1962
Hydrological conditions in Newfoundland and Labrador areas in 1960. (In Russian; English summary).
Sovetskie Ribochoz. Issledov. Severo-Zapad. Atlant. Okeana, VNIRO-PINRO, 155-171.

C

dynamic height anomalies
Coachman, L.C., and C.A. Barnes, 1961
The contribution of Bering Sea water to the Arctic Ocean.
Arctic, 14(3):147-161.

dynamic topography
Coachman, L.C., and C.A. Barnes, 1961
The contribution of Bering Sea water to the Arctic Ocean.
Arctic, 14(3):147-161.

dynamic topography
Codispoti, Louis A., 1968.
Some results of an oceanographic survey in the northern Greenland Sea, summer 1964.
Techn. Rept., U.S. Nav. Oceanogr. Off., TR202: 49 pp.

Dynamic topography
Countryman, Kenneth A., 1969.
Some summer oceanographic features of the Norwegian Sea summer 1963.
Techn. Rept. U.S.N. Oceanogr. Off. TR-216: 35 pp.

dynamic topography
*Countryman, Kenneth A., and William L. Gsell, 1966.
Operations Deep Freeze 63 and 64, summer oceanographic features of the Ross Sea.
Tech. Rept., U.S. Naval Oceanogr. Off., Tr-190:193 pp (multilithed).

D

dynamic heights
Darbyshire, Mollie, 1963
Computed surface currents off the Cape of Good Hope.
Deep-Sea Res., 10(5):623-632.

dynamic topography
Dietrich, Gunter, 1965.
New hydrographical aspects of the northwest Atlantic.
ICNAF Spec. Publ. No. 6:29-51.

dynamic topography
Dietrich, Gunter, 1964.
Oceanic Polar Front Survey in the North Atlantic.
In: Research in Geophysics. Solid Earth and Interface Phenomena, 2:291-308.

dynamic topography
Dietrich, Günter, H. Aurich, and A. Kotthaus, 1961
On the relationship between the distribution of redfish and redfish larvae and the hydrographical conditions in the Irminger Sea.
Rapp. Proc. Verb., Cons. Perm. Int. Expl. Mer, 150:124-139.

dynamic height anomaly
Dodimead, A.J., and F. Favorite, 1961.
Oceanographic atlas of Pacific Subarctic Region, Summer of 1958.
Fisheries Res. Bd., Canada, Manuscript Rept. Ser. (Oceanogr. and Limnol.), No. 92: 6 pp., 40 figs. (multilithed)

dynamic topography
Donguy, J.R. et M. Prive, 1968.
Esquisse du régime hydrologique au large de la côte des Graines et de la côte d'Ivoire entre Abidjan et Monrovia. Cah. ORSTOM sér. Océanogr. 6(2): 47-51.

dynamic topography
Duxbury Alyn C., 1962
Averaged dynamic topographies of the Gulf of Mexico.
Limnol. and Oceanogr., 7(3):428-430.

E

dynamic topography
Edelman,M.S.,1965.
Brief charecteristics of the water masses in the Gulf of Aden and the northern Arabian Sea (Data from the 2nd Indian Ocean Expedition of the Azov-Black Sea Research Institute of Marine Fisheries and Oceanography). (In Russian).
Trudy vses. nauchno-issled. Inst. morsk.ryb. Khoz. Okeanogr. (VNIRO),57:93-107.

dynamic topography
Edelman, M.S., 1965.
Brief characteristics of the waters masses in the Gulf of Aden and the northern Arabian Sea (data of the 2nd Indian Ocean Expedition of the Azov-Black Sea Research Institute of Marine Fisheries and Oceanography). Investigations in line with the programme of the International Geophysical Year. (In Russian).
Trudy, Vses. Nauchno-Issled. Inst. Morsk. Ribn. Choz. I Oceanogr. (VNIRO), 57:93-107.

F

dynamics heights
Favorite, Felix, John W. Schantz and Charles R. Hebard, 1961
Oceanographic observations in Bristol Bay and the Bering Sea 1939-41 (USCGT Redwing).
U.S.F.W.S. Spec. Sci. Rept., Fish. No. 381: 323 pp.

dynamic topography
Fernández, F. 1970.
Cálculos dinámicos en las aguas al oeste de la Península Ibérica.
Bol. Inst. esp. Oceanogr. 144: 21 pp.

dynamic heights
Fleming, J.A., C.C. Ennis, H.U. Sverdrup, S.L. Seaton, W.C. Hendrix, 1945
Observations and results in physical oceanography. Graphical and tabular summaries. Ocean. 1-B, Sci. Res. of Cruise VII of the Carnegie during 1928-1929 under command of Capt. J.P. Ault, Carnegie Inst., Washington, Publ. 545:315 pp., 5 tables, 254 figs.

dynamic topography
Fomichev, A.V., 1965.
Water masses and vertical structure of the Antarctic waters. (In Russian).
Trudy vses. nauchno-issled. Inst. morsk. ryb. Khoz. Okeanogr. (VNIRO), 57:53-77.

G

Dynamic Topography
Galerkin, L. I., 1962.
[On definition of the reading table for calculations of density fluctuations of the sea level.]
Mezhd. Geofiz. Komitet, Prezidiume Akad. Nauk, SSSR, Rezult. Issled. Programme Mezhd. Geofiz. Goda, Okeanol. Issled., No. 5:25-30.

dynamic topography
Garner,D.M., 1969.
The geo potential topography of the ocean surface around New Zealand.
N.Z. Jl mar.Freshwat.Res.3(2):209-219.

dynamic topography
*Garner,D.M.,1967.
Hydrology of the south-east Tasman Sea.
Bull.N.Z. Dept Scient.indust.Res. 181:1-40. (Mem. N.Z. oceanogr. Inst. 48).

dynamic heights
Gurikova, Z.F., 1962
On the calculation of the currents in the surface layer in the northern sector of the Pacific.
Izv. Akad. Nauk. SSSR, Ser. Geophys. 9:1240-1250, Eng. Ed., 9:776-781.

H

dynamic topography
Hamon, B.V., 1965.
The East Australian Current, 1960-1964.
Deep-Sea Res., 12(6):899-921.

dynamic heights
Hart, T. John, and Ronald I. Currie, 1960
The Benguela Current.
Discovery Repts., 31: 123-298.

dynamic heights
Hermann, F., 1949.
Hydrographic conditions in the south-western part of the Norwegian Sea. Ann. Biol. 5:19-21, Figs. 14-21.

dynamic topography
Hidaka, Koji, 1961.
Non-linear dynamic computation of the equatorial flows. (Abstract).
Tenth Pacific Sci. Congr., Honolulu, 21 Aug.-6 Sept., 1961, Abstracts of Symposium Papers, 340.

dynamic topography
Hidaka, K., 1960
Surface dynamical topography over 1000-db surface derived from the result of the "Operation EQUAPAC" in the summer 1956. Geophys. Notes, 13(1): No. 1 from Proc. North Pac. Cong. 1957, 1958 Vol. 16: 226

dynamic topography
Hidaka, Koji, 1958.
Surface dynamical topography over 1000-db surface derived from the result of the "Operation EQUAPAC" in the summer of 1956. (Abstract).
Proc. Ninth Pacific Sci. Congr., Pacific Sci. Assoc., 1957, Oceanogr., 16:226.

I

dynamic topography
Ichiye, T., 1955.
On the variation of oceanic circulation (V).
Geophys. Mag., Tokyo, 26(4):283-342, 23 t extfigs.

J

dynamic heights (data)
Jacobs, Stanley S., Peter M. Bruchausen and Edward B. Bauer, 1970
Cruises 32-36, 1968, hydrographic stations, bottom photographs, current measurements.
Eltanin Repts Lamont-Doherty Geol. Obs., Nat. Sci. Found. U.S. Antarctic Res. Program, 460 pp. (multilithed)

dynamic depth(data)
Japan, Fisheries Agency, Research Division, 1956.
Radiological survey of western area of the dangerous zone around the Bikini-Eniwetok Atolls, investigated by the "Shunkotsu maru" in 1956, Part 1:143 pp.

dynamic topography
Japan, Hakodate Marine Observatory, 1957.
[Report of the oceanographic observations of the Okhotsk Sea from April to May 1956.]
Bull. Hakodate Mar. Obs., No. 4:105-112. 1-8.

dynamic topography
Japan, Hakodate Marine Observatory, 1956.
[Report of the oceanographical observations east off Tohoku District from August to September 1955.]
Bull. Hakodate Mar. Obs., (3):13-21.

dynamic topography
Japan, Hakodate Marine Observatory, 1956.
[Report of the oceanograhic observations east of Tohoku District in November 1955.]
Bull. Hakodate Mar. Obs., (3):11-12.

dynamic topography
Japan, Hakodate Marine Observatory, 1956.
[Report of the oceanographical observations in the sea east off Tohoku District in May-June 1955.]
Bull. Hakodate Mar. Obs., (3):11-16.

dynamic depth
Japan, Hokkaido University, Faculty of Fisheries, 1962
II. The "Oshoro Maru" cruise 48 to the Bering Sea and northwestern North Pacific in June-July 1961.
Data Record Oceanogr. Obs., Expl. Fish., 6:22 149.

dynamic depth (data only)
Japan, Hokkaido University, Faculty of Fisheries, 1962.
III. The "Hokusei Maru" cruise 10(1-3) to the northwestern North Pacific in June-August 1961.
Data Record Oceanogr. Obs., Expl. Fish., No. 6: 152-283.

dynamic topography (data only)
Japan, Hokkaido University, Faculty of Fisheries, 1957.
Data record of oceanographic observations and exploratory fishing, No. 1:247 pp.

dynamic topography
Japan, Kobe Marine Observatory, 1967.
Report of the oceanographic observations in the sea south of Honshu from July to August 1965. (In Japanese).
Bull. Kobe Mar. Obs. No. 178: 31-40

dynamic topography
Japan, Kobe Marine Observatory, 1967.
Report of the oceanographic observations in the sea south of Honshu from February to March 1964. (In Japanese)
Bull. Kobe Mar. Obs. No. 178: 27-

dynamic topography
Japan, Kobe Marine Observatory, Oceanographical Section, 1964.
Report of the oceanographic observations in the Kuroshio and region east of Kyushu from October to November 1962. (In Japanese).
Res. Mar. Meteorol. and Oceanogr., Japan. Meteorol. Agency, 32: 41-50.
Also in: Bull. Kobe Mar. Obs., 175. 1965.

dynamic topography

Japan, Kobe Marine Observatory, Oceanographical Section, 1964.
Report of the oceanographic observations in the sea south of Honshu from February to March, 1963. Res. Mar. Meteorol. and Oceanogr., Japan Meteorol. Agency, 33: 27-32.

Also in: Bull. Kobe Mar. Obs., 175. 1965.

dynamic topography

Japan, Kobe Marine Observatory, Oceanographical Section, 1964.
Report of the oceanographic observations in the sea south of Honshu from July to August, 1962. Res. Mar. Meteorol. and Oceanogr., Japan. Meteorol. Agency, 32: 32-40. (In Japanese).

Also in: Bull. Kobe Mar. Obs. 175.

dynamic topography

Japan, Kobe Marine Observatory, 1963.
Report of the oceanographic observations in the sea south of Honshu from July to August, and from the cold water region south of Enshu Nada, October to November. (In Japanese).
Bull. Kobe Mar. Obs., 171(3):36-52.

dynamic topography

Japan, Kobe Marine Observatory, 1963.
Report of the oceanographical observations in the sea south of Honshu from February to March 1961.
Report of the oceanographical observations in the cold water region off Enshu Nada in May 1961.
(In Japanese.).
Bull. Kobe Mar. Obs., 171(4):22-35.

dynamic topography

Japan, Kobe Marine Observatory, Oceanographical Observatory, 1962
Report of the oceanographic observations in the cold water region off Enshu Nada in May, 1961. (In Japanese).
Res. Mar. Meteorol. and Oceanogr. Obs., Jan. June, 1961, No. 29: 28-35.

dynamic topography

Japan, Kobe Marine Observatory, Oceanographical Section, 1962
Report of the oceanographic observations in the sea south of Honshu from February to March, 1961. (In Japanese).
Res. Mar. Meteorol. and Oceanogr. Obs., Jan.-June 1961. No. 29:22-27.

dynamic topography

Japan, Kobe Marine Observatory, Oceanographical Section, 1962
Report of the oceanographic observations in the sea south of Honshu in May, 1960. (In Japanese).
Bull. Kobe Mar. Obs., No. 169(12):27-33.

dynamic topography

Japan, Kobe Marine Observatory, Oceanographical Section, 1962
Report of the oceanographic observations in the sea south of Honshu from October to November, 1959. (In Japanese).
Bull. Kobe Mar. Obs., No. 169(11):44-50.

dynamic potential

Japan, Kobe Marine Observatory, Oceanographical Section, 1962
Report on the oceanographic observations in the sea south of Honshy from July to August, 1959.
Bull. Kobe Mar. Obs., No. 169(11):31-43.
(In Japanese)

dynamic topography

Japan, Kobe Marine Observatory, 1961
[Report of the oceanographic observations in the sea south of Honshu in March 1958.]
Bull. Kobe Mar. Obs., No. 167(21-22):30-36.
--from May to June, 1958(21-22):37-42
--from July to September,1958(23-24):34-40.
--from October to December, 1958(23-24):41-47
--from February to March,1959(25-26):33-47.

dynamic topography

Japan, Kobe Marine Observatory, 1956.
[Report of the oceanographic observations in the sea south of Honshu in August 1955.]
J. Ocean., Kobe(2)7(3):23-32.

dynamic topography

Japan, Maizuru Marine Observatory, 1967.
Report of the oceanographic observations in the Japan Sea from May to June, 1964.
Bull. Maizuru mar. Obs., 10:65-76.
(In Japanese)

dynamic topography

Japan, Maizuru Marine Observatory and Hakodate Marine Observatory, Oceanographical Sections, 1962
Report of the oceanographic observations in the Japan Sea in June, 1961. (In Japanese).
Res. Mar. Meteorol. and Oceanogr. Obs., Jan.-June, 1961, No. 29:59-79.

dynamic topography

Japan, Maizuru Marine Observatory, 1962
Report of the oceanographic observations in the central part of the Japan Sea in September, 1960. (In Japanese).
Res. Mar. Meteorol. and Oceanogr., July-Dec., 1960, Japan Meteorol. Agency, No. 28:60-68.

dynamic topography

Japan, Maizuru Marine Observatory, Oceanographical Section, 1961
Report of the oceanographic observations in the Japan Sea from June to July, 1958.
Bull. Maizuru Mar. Obs., No. 7:60-67.

dynamic topography

Japan, Maizuru Marine Observatory, Oceanographical Section, 1961
[Report of the oceanographic observations off Kyoga-misaki in Japan Sea from October to December, 1957.]
Bull. Maizuru Mar. Obs., No. 7:29-36.

dynamic topography

Japan, Maizuru Marine Observatory, 1955.
[Report of the oceanographical observations off San'in and Hokuri-ku (July-August, 1952).]
Bull. Maizuru Mar. Obs., (4):49-63 (168-181).

dynamic depth (data only)

Japan, Maritime Safety Board 1961.
Tables of results from oceanographic observations in 1958.
Hydrogr. Bull. Tokyo No. 66 (Publ. No. 981): 153pp.

dynamic currents

Japan Meteorological Agency, 1968.
The results of marine meteorological and oceanographical observations, July-December, 1965, 38: 404 pp.
(multilithed)

dynamic anomalies

Japan, National Committee for the International Geophysical Year 1960.
The results of the Japanese oceanographic project for the International Geophysical Year 1957/58: 198pp.

dynamic depths

Japan, Japanese Oceanographic Data Center 1970.
CSK Atlas. 4. Winter 1967: 32 pp.

K

dynamic depths

Kawai, H., 1957.
On the natural coordinate system and its application to the Kuroshio system.
Bull. Tohoku Reg. Fish. Res. Lab., (10):141-171.

dynamic heights

Kitamura, Hiroyuki, and Takeshi Saga, 1964.
On the chemical elements in the sea south of Honshu, Japan. (In Japanese; English abstract).
Bull. Kobe Mar. Obs., No. 172:6-54.

dynamic heights

Kollmeyer, Ronald C., Thomas C. Wolford, and Richard M. Morse 1966.
Oceanography of the Grand Banks region of Newfoundland in 1965.
U.S. Cst Gd. Oceanogr. Rept. 11 (CG 373-11): 157 pp.

dynamic topography

Kollmeyer, Ronald C., Robert M. O'Hagan and Richard M. Morse, 1965.
Oceanography of the Grand Banks region and the Labrador Sea in 1964.
U.S.C.G. Rept., No. 10(373-10):1-24;34-285.
Oceanogr.

dynamic heights

Koshlyakov, M.N., 1961.
[Problems of the water dynamics of the northwestern Pacific.]
Trudy Inst. Okeanol., Akad. Nauk, SSSR, 38:31-55.

dynamic topography

Kozlov, V.F., 1966.
On determining the depth of the zero surface. (In Russian).
Doklady Akad. Nauk SSSR, 170(5):1068-1069.

dynamic topography

Kuksa, V.I., 1962
[On the formation and the distribution of the intermediate water layer of reduced salinity in the northern section of the Pacific Ocean.]
(In Russian).
Okeanologiia, Akad. Nauk, SSSR, 2(5):769-782.

dynamic topography

Kusunoki, Kou, 1962
Hydrography of the Arctic Ocean with special reference to the Beaufort Sea.
Contrib. Inst., Low Temp. Sci., Sec. A (17): 1-74.

L

dynamic topography

Lacombe, H., 1956.
Quelques elements de l'extension des eaux méditerraneennes dans l'Ocean Atlantique.
Bull. d'Info., C.C.O.E.C., 8(5):210-224, 7 pls.

dynamic topography

Lacombe, Henri, P. Tchernia and G. Benoist, 1958

Contribution à l'étude hydrologique de la Mer Egée en période d'été.
Bull. d'Info., C.C.O.E.C., 10(8): 454-468.
19 pls.

dynamic typography

*Lemasson, L. et B. Piton, 1968.
Anomalie dynamique de la surface de la mer le long de l'équateur dans l'océan Pacifique.
Cah. ORSTOM, sér. Océanogr., 6(3/4): 39-45.

dynamic topography

Le Pichon, Xavier, et Jean-Paul Troadec, 1963
La couche superficielle de la Méditerranée au large des côtes Provençales durant les mois d'été.
Cahiers Océanogr., C.C.O.E.C., 15(5):299-314.

M

dynamic topography

Mamaev, O. I., 1957.
On the problem of the zero dynamic surface and its topography in the Southern Ocean.
Doklady Akad. Nauk, SSSR, 117(5):808-810.

Mc

dynamic topography

Menaché, Maurice, 1963
Première campagne océanographique du "Commandant Robert Giraud" dans le Canal de Mozambique (11 octobre-28 novembre 1957).
Cahiers Océanogr., C.C.O.E.C., 15(4):224-235, 27 figs.

dynamic topography

*Morgan, Charles W., 1969.
Oceanography of the Grand Banks region of Newfoundland in 1967.
Oceanogr., Rept.U.S.C.G.19(CG 373-19):1-209.

dynamic depth anomaly

Moriyasu, Shigeo, 1960

The thickness of the upper homogeneous layer.
Rec. Oceanogr. Wks., Japan, n.s., 5(2): 44-51.

dynamic depth anomaly

Moriyasu, Shigeo, 1958.
An attempt to estimate the dynamic depth anomaly.
Umi to Sora, 33(6)
34(3):40-44.

dynamic topography

Murray, J.E., 1969.
Report of the International Ice Patrol Service in the North Atlantic Ocean.
U.S.C.G. Bull. 54(CG-188-23): 105 pp

dynamic topography

Murray, J.E., 1966.
Report of the International Ice Patrol Service in the North Atlantic Ocean. Season of 1966.
Bull. U.S. Cst Guard, 52: 1-27.

N

dynamic topography

Natarov, V.V., 1963
On the water masses and currents of the Bering Sea (In Russian).
Sovetsk. Ribokh. Issled. B Severo-Vostokh. Chasti Tikhogo Okeana, VNIRO 48, TINRO 50 (1):111-134.

dynamic topography

Novozhilov, V.N., 1961.
Hydrological conditions in the regions of the Komandorski Islands and Kamchatka Peninsula in the Pacific Ocean in summer 1956.
Trudy Inst. Okeanol., Akad. Nauk, SSSR, 38:56-60.

dynamic topography

Nowlin, Worth D. Jr., 1971.
Water masses and general circulation of the Gulf of Mexico.
Oceanol. int. 6(2):28-33

dynamic topography

Nowlin, W.D., Jr., and H.J. McLellan, 1967.
A characterization of the Gulf of Mexico waters in winter.
J.mar.Res., 25(1):29-59.

O

dynamic topography

Ohwada, Mamoru, and Hisanori Kon, 1963
A microplankton survey as a contribution to the hydrography of the North Pacific and adjacent seas. (II). Distribution of the microplankton and their relation to the character of water masses in the Bering Sea and northern North Pacific Ocean in the summer of 1960.
Oceanogr. Magazine, Japan Meteorol. Agency, 14(2):87-99.

dynamic heighths

Okhlopkova, A.P., 1961.
An experiment in applying the dynamic method to the study of water circulation in the Ladoga Lake.
Okeanologiia, Akad. Nauk, SSSR, 1(6):1025-1033.

dynamic topography

Ovchinnikov, I.M., 1963.
Some peculiarities of water circulation in the Alaska Bay. (In Russian).
Mezhd. Geofiz. Kom. Prezidiume, Akad. Nauk, SSSR Rezult. Issled. Programme Mezhd. Geofiz. Goda, Okeanol. Issled., (8):67-75.

dynamic topography

*Owen, Robert W., Jr., 1968.
Oceanographic conditions in the northeast Pacific Ocean and their relation to the albacore fishery.
Fish. Bull. U.S. Fish. Wildl. Serv., 66(3): 503-526.

P

dynamic height

Parr, A. E., 1937
Report on hydrographic observations at a series of anchor stations across the Straits of Florida. Bull. Bingham Ocean. Coll. VI (3):1-62, 36 text figs.

Q

R

dynamic heights (data)

Roden, G.I., 1963.
On sea level, temperature and salinity variations in the Central Tropical Pacific and Pacific Ocean islands.
J. Geophys. Res., 68(2):455-472.

dynamic topography

Rotschi, Henri, 1961
Contribution française en 1960 à la connaissance de la Mer de Coral: Océanographique physique.
Cahiers Océanogr., C.C.O.E.C., 13(7):434-455

dynamic topography

Rotschi, Henri, 1961

Resultats des observations scientifiques du "Tiare", Croisière "Entrecasteaux" (11-24 août 1960) sous le commandement du Lieutenant de Vaisseau Carbelaud.
Comité Local d'Océanogr. et d'Études des Côtes de Nouvelle Caledonie, Rapp. Sci., Noumea, No. 21: 41 pp. (mimeographed).

dynamic topography

Rotschi, Henri, 1960

Sur la circulation et les masses d'eau dans le Nord-Est de la mer de Corail.
C.R. Acad. Sci., Paris, 251(7): 965-967.

dynamic topography

Rotschi, H., 1959
Remarques sur la circulation océanique entre la Nouvelle-Calédonie et l'Isle Norfolk.
Cahiers Oceanogr., C.C.O.E.C., 11(6):416-424

dynamic topographic (data only)

Rotschi, Henri, Yves Magnier, Maryse Tirelli, et Jean Garbe, 1964.
Résultats des observations scientifiques de "La Dunkerquoise" (Croisière "Guadalcanal").
Oceanographie, Cahiers O.R.S.T.R.O.M., 11(1):49-154.

S

dynamic height anomalies

Scripps Institution of Oceanography, 1949
Physical and chemical data, Cruise 1, February 28 to March 16, 1949. Marine Life Research Program. Physical and Chemical Data Report No.1:10 figs. (ozalid), tables of data (mimeographed).

dynamic heights

Scripps Institution of Oceanography, 1949.
Marine life research program. Progress report, 1 May to 31 July 1949. 24 pp. (mimeographed), 16 figs. (ozalid).

dynamic height anomalies

Scripps Institution of Oceanography, 1949
Physical and chemical data, Cruise 2, March 28 to April 12, 1949. Physical and Chemical Data Report No.2:10 figs. (ozalid) tables of data (mimeographed).

dynamic height anomalies (data)

Sdubbundhit, Cha Erb, and A.E. Gilmour, 1964.
Geostrophic currents derived from oceanic density over the Hikurangi Trench.
New Zealand J. Geol. & Geophys., 7(2):271-278.

dynamic topography

Seckel, G.R., 1962.
Atlas of the oceanographic climate of the Hawaiian Islands region.
U.S.F.W.S., Fish. Bull., 61(193):371-427.

dynamic topography

Seco Serrano, Edmundo, 1963.
Distribución de presiones dinámicas en una bahía. Aplicacion a la bahía de Cádiz.
Bol. Inst. Español de Oceanogr., No. 117:26 pp.

dynamic topography
Shannon, L.V., and M. van Rijswijk, 1969.
Physical oceanography of the Walvis Ridge region.
Investl Rep. Div. Sea Fish. S.Afr. 70: 1-19.

Dynamic topography
Shoji, Daitaro, 1965.
Description of the Kuroshio (Physical aspect).
Proc. Symp., Kuroshio, Tokyo, Oct. 29, 1963, Oceanogr. Soc., Japan and UNESCO, 1-10.

dynamic topography
Shtokman, V.B., 1948.
[Significance of wind and relief of the bottom on the formation of observed features of the dynamic topography of the South Atlantic Ocean.]
Dok. Akad. Nauk SSSR, 60:985-988.

dynamic topography
Smetanin, D.A., 1959
[Hydrochemistry of the Kuril-Kamchatka deep-sea trench.] Trudy Inst. Okeanol., 33: 43-86.

dynamic topography
Soule, F.M., A.P. Franceschetti and R.M. O'Hagen 1963.
Physical oceanography of the Grand Banks region and the Labrador Sea in 1961.
U.S. Coast Guard Bull., No. 47:19-82.

dynamic topography
Soule, Floyd M., Alfred P. Franceschetti, R.M. O'Hagan and V.W. Driggers, 1963.
Physical oceanography of the Grand Banks region, the Labrador Sea and Davis Strait in 1962.
U.S.C.G. Bull., No. 48:29-78;95-153.

dynamic topography
Soule, Floyd M., and R.M. Morse, 1960
Physical oceanography of the Grand Banks Region and the Labrador Sea in 1958. International Ice Observation and Ice Patrol Service in the North Atlantic.
U.S. Coast Guard Bull., No. 44: 29-99.

dynamic topography
Soule, F.M., and J.E. Murray, 1957.
Physical oceanography of the Grand Banks region and the Labrador Sea in 1956.
U.S.C.G. Bull., 42:35-100.

dynamic topography
Soule, F.M., and J.E. Murray, 1956.
Physical oceanography of the Grand Banks and the Labrador Sea in 1955.
U.S.C.G. Bull., 41:59-114.

dynamic topography
Stefansson, Unnstein, 1962.
North Icelandic waters.
Rit Fiskideildar, Reykjavik, 3:269 pp.

dynamic topography
Stommel, Henry, 1964.
Summary charts of the mean dynamic topography and current field at the surface of the ocean, and related functions of the mean wind-stress.
In: Studies on Oceanography dedicated to Professor Hidaka in commemoration of his sixtieth birthday 53-58.

dynamic heights
Stommel, H., 1947
Note on the use of the T-s correlation for dynamic height anomaly computations.
J. Mar. Res. 6(2):85-92.

dynamic topography
Sturges, Wilton, 1968.
Sea-surface topography near the Gulf Stream.
Deep-Sea Res., 15(2):149-156.

T

dynamic topography
Takahashi, Tadao, 1960
Existence of a contra solem vertical motion in the Coral Sea. Rec. Oceanogr. Wks., Japan, n.s 5(2): 52-54.

dynamic topography
Takahashi, Tadao, Masaaki Chaen and Soichi Ueda, 1960
Report of the Kagoshima-maru IGY cruise, 1958.
Mem. Fac. Fish., Kagoshima Univ., 8: 82-86.

dynamic topography
Tsuchiya, M., 1961
An oceanographic description of the Equatorial Current System of the Western Pacific.
Oceanogr. Mag., Tokyo, 13(1):1-30.

dynamic topography
Tully, J.P., and F.G. Barber, 1961.
An estuarine model of the sub-Arctic Pacific Ocean.
In: Oceanography, Mary Sears Edit., Amer. Assoc. Adv. Sci., Washington, 425-454.

Previously published under title:
"An estuarine analogy in the sub-Arctic Pacific Ocean", J. Fish. Res. Bd., Canada, 17(1):91-112. (1960).

U

V

dynamic topography
Volokhonsky, L. Sh., 1965.
Nonlinear dynamic problems of shallow seas. (In Russian).
Trudy, Gosudarst. Okeanogr. Inst., No. 87:3-31.

dynamic topography
von Arx, William S., 1965.
Absolute dynamic topography.
Limnol. and Oceanogr., Redfield Vol., Suppl. to 10:R265-R273.

W

dynamic height anomaly
Worthington, L. V., 1955.
Oceanographic observations.
Sci. Studies at Fletcher's Ice Island, T-3, Vol. 1, Air Force Cambridge Res. Center, Geophys. Res. Pap., No. 31-35.
This is WHOI Ref. 53-92. It does not bear a contribution number. Data for 1954 and 1955 added.

XYZ

dynamic topography
Yasui, M., 1957.
On the rapid estimation of the dynamic topography in the seas adjacent to Japan.
Rec. Oceanogr. Wks., Japan, n.s., 3(1):8-15.

dynamic topography
*Yoshida, Kozo, 1967.
Circulation in the eastern tropical oceans with special reference to upwelling and undercurrents.
Jap. J. Geophys., 4(2):1-75.

dynamic heights
Yoshida, S., 1961
[On the short period variation of the Kuroshio in the adjacent sea of Izu Islands.]
Hydrogr. Bull., Maritime Safety Bd., Tokyo, (Publ. No. 981), No. 65:1-18.

dynamic heights
Yoshida, K., and M. Tsuchiya, 1957.
Note on the thermosteric anomaly.
J. Oceanogr. Soc., Japan, 13(4):127-130.

dynamic heights (data only)

dynamic topography (data only)
Charnell, Robert L., David W.K. Au, and Gunter R. Seckel, 1967.
The Trade Wind Zone Oceanography Pilot Study. I. Townsend Cromwell cruises 1, 2, and 3, February to April, 1964.
Spec. scient. Rep. U.S. Fish. Wildl. Serv., Fish. 552. 75 pp. (multilithed)

dynamic height anomaly

dynamic height anomaly
Fofonoff, N. P. and S. Tabata, 1966.
Variability of oceanographic conditions between Ocean Station P and Swiftsure Bank off the Pacific coast of Canada.
J. Fish. Res. Bd., Canada, 23(6):825-868.

dynamical reference surface

dynamical reference surface
Neumann, G., 1956.
Zum Problem der "dynamischen Bezugsfläche" insbesondere im Golfstromgebiet.
Deutsche Hydrogr. Zeits., 9(2):66-78.

dynamic theory

dynamic theory.
Lineikin, P.S., 1957.
Fundamental questions on the dynamic theory of the baroclinic layers of the ocean. (In Russian).
Gosud. Okeanogr. Inst., Glabnoe Upravlenie Gidrometeorol. Sluzhbi pri Sovete Ministrov, SSSR, Gidrometeorol, Isdatel., Leningrad, 139 pp.

dynamic topography
See: dynamic heights

earth-atmosphere-ocean system
Faegre, Aron, 1972.
An intransitive model of the earth-atmosphere-ocean system.
J. appl. Met. 11(1): 4-6

earth's chronology

earth's chronology
Pettersson, H., 1951.
Radium and deep-sea chronology. Nature 167:942.

earth's core

earth's core
Bullard, E.C., 1955.
Introduction to a discussion on "movements in the earth's cores and electrical conductivity".
(Ann. Géophys., 11(1)), I.A.G.A. Bull. 15a:4 pp.

Earth's core
Bullen, K.E., 1955.
Physical properties of the earth's core.
(Ann. Geophys., 11(1)), I.A.G.A. Bull., 15a:12 p

earth's core
Runcorn, S.K., 1955.
Core motions and reversals of the geomagnetic field. (Ann. Geophys. 11(1)), I.A.G.A. Bull. 15a: 7 pp.

earth's core
Urey, H.C., 1955.
Distribution of elements in the meteorites and the earth and the origin of heat in the earth's core. (Ann. Geophys., 11(1)), I.A.G.A. Bull. 15a18 pp.

earth's crust
See also: crust

earths crust

See also: crust

earth's crust
Adams, R.D., 1964.
Thickness of the earth's crust beneath the Pacific-Antarctic Ridge.
New Zealand J. Geol. Geophys., 7(3):529-542.

earth's crust
Adams, R.D., 1962
Thickness of the earth's crust beneath the Campbell Plateau.
New Zealand J. Geol. Geophys., 5(1):74-85.

earth's crust
Bath, M., and V. Karnik, 1963.
Investigation of the earth's crust.
I.U.U.G. Monograph, (22):50 pp.

Reviewed:
P.N.S. O'Brien, 1963.
Geophysical Jour., R. Astron. Soc., 8(2):280-283

earth's crust
Boosman, Jaap W., 1964.
Seismic crustal studies in the Arctic Ocean basin.
Naval Research Reviews, 17(2):7-12.

earths crust
Christman, Nikolas I., 1970.
Composition and evolution of the oceanic crust.
Mar. Geol., 8(2):139-154

earth's crust
Cohen, Lewis H., Kent C. Condie, Louis J. Kuest, Jr., Glenn S. MacKenzie, Fred H. Meister, Paul Pushkar and Alan M. Stueber, 1963.
Geology of the San Benito Islands, Baja California, Mexico.
Geol. Soc., Amer., Bull., 74(11):1355-1370.

earth's crust, structure of
Demenitskaya, R.M., and V.D. Dibner, 1966.
Morphological structure and the earth's crust of the North Atlantic region.
In: Continental margins and island arcs, W.H. Poole, editor, Geol. Surv. Pap. Can., 66-15:63-79.

earth's crust
Evison, F.F., 1963.
Thickness of the earth's crust in Antarctica and the surrounding oceans: a reply.
Geophysical J., London, 7(4):469-476.

earth's crust
Evison, F.F., C.E. Ingham, R.H. Orr and J.H. Le Fort, 1960
Thickness of the earth's crust in Antarctica and the surrounding oceans. Geophys. J., R. Astron. Soc., 3(3): 289-306.

crust structure
Hersey, J.B., 1966.
Marine geophysical investigations in the West Indies.
In: Continental margins and island arcs, W.H. Poole, editor, Geol. Surv. Pap. Can., 66-15:151-164.

earth's crust
Kovylin, V.M., B.J. Karp and R.B. Shaiakhmetov, 1966
The structure of the earth-crust and of the sedimentary rock-mass of the Sea of Japan from seismic data. (In Russian).
Doklady, Akad. Nauk, SSSR, 165(5):1048-1051

earth's crust
MacDonald, G.A., and H. Kuno, 1962.
The crust of the Pacific Basin.
Geophys. Monogr., (6):1-195.

earths crust
Maynard, G.L., 1970.
Crustal layer of seismic velocity 6.9 to 7.6 kilometers per second under the deep oceans.
Science 168(3927): 120-121.

earth's crust
Menard, H.W., 1967.
Transitional types of crust under small ocean basins.
J. geophys. Res., 72(12):3061-3073.

earths' crust
Neprochnov, Iu. P., I.N. Elnikov and B.V. Khlopov, 1967.
The structure of the earth crust in the Indian Ocean, according to the results of seismic investigations carried out during the 36th voyage of the E/V Vitiaz. (In Russian)
Dokl. Akad. Nauk, SSSR, 174(2): 429-...

earth's crust
Neprochnov, Yu. P., A.F. Neprochnov, G.N. Lunarsky, M.F. Mikhno, G. Yu. Murusidze, and V.K. Chichinadze, 1966.
The structure of the earth's crust in the eastern part of the Black Sea according to seismic depth sounding data (in Russian).
Oceanologia, Akad. Nauk, SSSR, 6(1):98-108.

earth's crust
Orlenok, V.V., 1968.
On the structure of the earth's crust in the North Atlantic from seismic data. (In Russian; English abstract).
Okeanologiia, Akad. Nauk, SSSR, 8(2):245-256.

earth's crust
Panov, D.G., 1961.
Type of structure of the oceanic part of the terrestrial crust. (In Russian).
Doklady, Akad. Nauk, Belorussia SSR, 5(3):118-121.

Translation:
T359R
Library, National Research Council, Canada
Sussex St., Ottawa

earth's crust
Raitt, R.W., 1963.
6. The crustal rocks.
In: The Sea, M.N. Hill, Editor, Interscience Publishers, 3:85-102.

earth's crust
Riznichenko, Ou. V., 1957.
On a study of the formation of the earth's crust in the period of the Third International Geophysical Year.
Izv. Akad. Nauk, Ser. Geofiz., 1957(2):129-140.

earth's crust
Stroev, P.A., and A.G. Gainanov, 1965.
On the structure of the earth's crust of the Indian Ocean area in accordance with the data of geophysical investigations. (In Russian).
Okeanologiia, Akad. Nauk, SSSR, 5(4):684-691.

earth's crust, structure of
Ud intsev, G.B., Yu. P. Neprochnov, and V.M. Kovylin, 1965.
Thickness of the sedimentary layer and structure of the earth's crust in seas and oceans based on seismo-acoustic measurements. (In Russian; English abstract).
Okeanolog. Issled., Rezult. Issled. Programme Mezhd. Geofiz. Goda, Mezhd. Geofiz. Komitet Presidiume, Akad. Nauk, SSSR, No. 13:181-188.

earth's crust
Ushakov, S.A., 1966.
The dynamics of the earth's crust within the transitional zones from continents to the oceans of the Atlantic-type. (In Russian).
Dokl. Akad. Nauk, SSSR, 171(1):91-94.

earth currents

earth currents
Barber, N.F., 1948
The magnetic field produced by earth currents flowing in an estuary or sea channel.
Mon. Notices, Roy. Astron. Soc., Geophys. Suppl. 5(7):258-269, 7 text figs.

earth currents
Bowden, K.F., and P. Hughes, 1961
The flow of water through the Irish Sea and its relation to wind.
Geophys. J., R. Astron. Soc., 5(4):265-291.

earth's electrical current
Diakonov, B.P., 1957.
On the nature of the earth's electrical current and its investigation on the bottom of the ocean.
Izv. Akad. Nauk, SSSR, Ser. Geophys., (6):800-802.

earth currents
Kato, Y., S. Utashiro, R. Shojii, J. Ossaka, M. Hayashi, and F. Inaba, 1950.
On the changes of the earth-current and earth's magnetic field accompanying the Fukui earthquake.
Sci. Repts., Tôhoku Univ., 5th ser., Geophys., 2(1):53-57, 4 textfigs.

earth currents
Petrachi, G., 1954.
Le correnti telluriche. Geofis. e Met. 2(5/6): 77-79.

earth currents, electrical
Stommel, H., 1956.
Electrical data from cable may aid hurricane prediction. Western Union Tech. Rev. 10(1):15-19.

earth currents
Sysoev, N.N., 1956
[On the calculation and elimination of the influence of the horizontal component of the magnetic pole of the earth currents by the electromagnetic method on a moving ship.]
Trudy Inst. Okeanol., 19:107-111.

earth currents
Troyickaya, V.A., 1955.
[Earth currents.] Priroda 1955(5):81-85.
T190R

earth's curvature, effect of

earth's curvature, effect of
Nakamura, Kohei, 1961
Velocity of long gravity waves in the ocean.
Sci. Repts., Tohoku Univ., (5 Geophys.), 13 (3):164-173.

earth's mantle

earth's mantle
Birch, F., 1961
Composition of the earth's mantle.
Geophys. J., 4: 295-311.

earth's mantle
Clark, Sydney P., Jr., and A.E. Ringwood, 1964.
Density distribution and constitution of the mantle.
Reviews of Geophysics, 2(1):35-88.

earth's mantle (upper)
Engel, A.E.J., Celeste Engel and R.G. Havens, 1965.
Chemical charistics of oceanic basalts and the upper mantle.
Geol. Soc., Amer., Bull., 76(7):719-734.

earth's mantle
Ewing, John I., 1963.
7. The mantle rocks.
In: The Sea, M.N. Hill, Editor, Interscience Publishers, 3:103-109.

earth's mantle
Gurarii, G.Z., and I.A. Solovieva, 1962.
Preliminary data on the density of the Earth's mantle. (In Russian).
Doklady Akad. Nauk, SSSR, 146(4):877-880.

earth's mantle
U.S.A., National Academy of Sciences, 1962.
The upper mantle project - proposed program of the United States.
Internat. Geophys. Bull., 66:1-18.

earth's obliquity, effect of
Sellers, William D., 1970.
The effect of changes in the earth's obliquity on the distribution of mean annual sea-level temperatures.
J. appl. Met. 9(6): 960-961.

earthquakes

earthquakes
Adams, R.D., and J.H. LeFort, 1963.
The Westport earthquakes, May, 1962.
New Zealand J. Geol. Geophys., 6(4):487-509.

earthquakes
Andreev, B.A., 1963.
Geophysical and geological characters of the zone of deep-seated earthquakes in the northwestern part of the Pacific. (In Russian).
Doklady, Akad. Nauk, SSSR, 150(1):140-142.

earthquakes
Balakina, L.M., 1962.
General regularities in the direction of the principal stresses effective in the earthquake foci of the seismic bolt of the Pacific Ocean. (In Russian).
Izv., Akad. Nauk, SSSR, Ser. Geofiz., (11):1471-1483.

English translation: AGU, (11):918-926. (1963).

earthquakes
*Banghar, A.R., and Lynn R. Sykes, 1969.
Focal mechanism of earthquakes in the Indian Ocean and adjacent regions.
J. geophys. Res., 74(2):632-649.

earthquakes
Berg, Edward, 1966.
Triggering of the Alaskan earthquake of March 27, 1964, and major aftershocks by low ocean tide loads.
Nature, 210 (5039): 893-896

earthquakes
Birch, F.S., 1966.
An earthquake recorded at sea.
Bull. Seismol. Soc. Amer., 56(2):361-366.

earthquakes
Bossolasco, Mario, e Claudio Eva, 1965.
Il terremoto del 19 luglio 1963 epicentro mel Mar Ligure.
Geofis. e Meteorol., Genova, 14 (1/2):6-18.

earthquakes
Brotherton, Minier K. and Bernard S. Sadowski, 1969
Seismic velocity determinations of Pn for near earthquakes in the vicinity of Vancouver Island.
Acta Geophysica polonica, 17 (3): 215-220
geophys. pol.

earthquakes
Bullard, Sir Edward, 1969.
The origin of the oceans.
Scient. Am., 221(3):66-75. (popular).

earthquakes
Couch, Richard W. and Leonard J. Pietrafesca, 1968.
Earthquakes off the Oregon coast: January 1968 to September 1968. Ore Bin, 30(10): 205-212. Also in: Coll. Reprints, Dept. Oceanogr. Univ. Oregon, 7(1968).

Earthquakes
Davidson, C., 1924.
Distortion of the sea-bed in the Japanese earthquake of 1 September 1923. Geogr. Jour. 63:241-243.

earthquakes
Denham, D., 1969.
Distribution of earthquakes in the New Guinea-Solomon Islands region. J. geophys. Res., 74(17): 4290-4299.

earthquakes
Evernden, J.F., 1970.
T-phase data on Kamchatka/Kurils earthquakes.
Bull. Seismol. Soc. Am. 60(4): 1061-1076.

earthquakes
Fairhead, J.D., and R.W. Girdler, 1970.
The seismicity of the Red Sea, Gulf of Aden and Afar triangle.
Phil. Trans. R. Soc. (A) 267 (1181): 49-74

earthquakes
Gupta, Supriya Sen, 1964.
Grand Banks earthquake of 1929 and the 'instantaneous' cable failure.
Nature, 204(4959):674-675.

earthquakes
Hanks, Thomas C. 1971.
The Kuril Trench-Hokkaido Rise System: large shallow earthquakes and simple models of deformation.
Geophys. J. R. astron. Soc. 23 (2): 173-189.

earthquakes
*Hédervári, P., 1967.
Investigations regarding the earth's seismicity. 5. On the earthquake geography of the Pacific Basin and the Seismotectonical importance of the Andesite Line.
(Gerlands) Beitr. Geophys. 76(5):393-405.

earthquakes
Ichikawa, Masaji, 1969.
Redeterminations of hypocenters of earthquakes occurring in northern Japan by use of ocean-bottom seismographic data and some related problems.
Geophys. Mag. 34 (3): 333-344

earthquakes
Iida, Kumizi, 1963.
Magnitude, energy and generation mechanisms of tsunamis and a catalogue of earthquakes associated with tsunamis.
Proc. Tsu. Mtgs., Tenth Pacific Sci. Congr. UGGI Monogr., No. 24:167-174.

earthquakes
*Isacks, Bryan, Jack Oliver and Lynn R. Sykes, 1968
Seismology and the new global tectonics.
J. geophys. Res., 73(18):5855-5899.

earthquakes
Kanamori, H. 1971.
Great earthquakes at island arcs and the lithosphere.
Tectonophysics 12 (3): 187-198

earthquakes
Krause, Dale C., 1966.
Geology and geomagnitism of the Bounty region east of the South Island, New Zealand.
New Zealand Dept. Sci. Ind. Res., Bull., 170:32pp.
New Zealand Oceanogr. Inst. Mem. No. 30:32pp.

Earthquakes
Mitronovas, Walter, Bryan Isacks and Leonard Seeber, 1969.
Earthquake locations and seismic propagation in the upper 250 km of the Tonga Island Arc.
Bull. seismol. Soc. Am., 59(3): 1115-1135.

earthquakes
Northrup, J., and R.W. Raitt, 1963.
Seaquakes in the Flores Sea.
Bull. Seismol. Soc., Amer., 53(3):577-583.

earthquakes
Northrop, John, and R.W. Raitt, 1962
Seaquakes in the Flores Sea. (Abstract).
J. Geophys. Res., 67(9):3584.

Earthquakes
Oike, Kazuo, 1969
The deep earthquake of June 22, 1966 in Banda Sea: a multiple shock.
Bull. Disas. Prev. Res. Inst., Kyoto Univ. 19(2)(155): 55-65

earthquakes, effect of
Plafker, George, 1972.
Alaskan earthquake of 1964 and Chilean earthquake of 1960: implications for arc tectonics.
J. geophys. Res. 77(5): 901-925.

earthquakes, submarine
Press, F., M. Ewing, and I. Tolstoy, 1950.
The airy phase of shallow-focus submarine earthquakes. Bull. Seis. Soc., America, 40(2):111-148, 17 textfigs.

earthquakes
Savage, J.C. and W.R.H. White, 1969.
A map of Rayleigh-wave dispersion in the Pacific.
Can. J. Earth Sci., 6(6): 1289-1300

earthquakes
*Smith, Warwick D., 1971.
Earthquakes at shallow and intermediate depths in Fiordland, New Zealand. J. geophys. Res. 76(20): 4901-4907.

earthquakes
Stauder, William, 1968.
Mechanism of the Rat Island earthquake sequence of February 4, 1965, with relation to island arcs and sea-floor spreading.
J. geophys. Res., 73(12):3847-3858.

earthquakes
Stauder, W., 1962.
S-wave studies of earthquakes of the North Pacific Part 1: Kamchatka.
Bull. Seismol. Soc., Amer., 52(3):527-550.

earthquakes
Stauder, William, and G.A. Bollinger, 1966.
The focal mechanism of the Alaska earthquakes of March 28, 1964, and of its aftershock sequence.
J. geophys. Res., 71(22):5283-5296.

earthquakes
Stauder, W., and A. Udias, 1963.
S-wave studies of earthquakes of the North Pacific. Part. II. Aleutian Islands.
Bull. Seismol. Soc., Amer., 53(1):59-77.

Earthquakes
Stefánsson, R., 1966
Methods of focal mechanism studies with application to two Atlantic earthquakes.
Tectonophysics, 3(3):209-243.

earthquakes
Sugimura, Arata, 1966.
Composition of primary magmas and seismicity of the earth's mantle in the island arcs (a preliminary note).
In: Continental margins and island arcs, W.H. Poole, editor, Geol. Surv. Pap., Can., 66-15:337-346.

Earthquakes
Sykes, Lynn R. 1970.
Seismicity of the Indian Ocean and a possible nascent island arc between Ceylon and Australia.
J. geophys. Res., 75(26): 5041-5055.

earthquakes
Sykes, Lynn R., 1966.
The seismicity and deep structure of island arcs.
J. geophys. Res., 71(12):2981-3006.

earthquakes
Sykes, Lynn R., 1964.
Deep-focus earthquakes in the New Hebrides region
J. Geophys. Res., 69(24):5353-5355.

earthquakes
Sykes, Lynn R., and Maurice Ewing, 1965.
The seismicity of the Caribbean region.
J. Geophys. Res., 70(2):5065-5074.

earthquakes
Sykes, Lynn R., Bryan L. Isacks and Jack Oliver, 1969.
Spatial distribution of deep and shallow earthquakes of small magnitudes in the Fiji-Tonga Region.
Bull. seismol. Soc. Am., 59(3):1093-1113

earthquakes
Sykes, Lynn R., and Mark Landisman, 1964.
The seismicity of East Africa, the Gulf of Aden and the Arabian and Red seas.
Bull. Seismol. Soc., Amer., 54(6A):1927-1940.

earthquakes
Sykes, Lynn R., Jack Oliver and Bryan Isacks 1970.
1. General observations. 10. Earthquakes and tectonics.
In: The sea: ideas and observations on progress in the study of the seas, Arthur E. Maxwell, editor, Wiley Interscience 4(1): 353-420.

earthquakes
Sylvester, Arthur G., Stewart W. Smith and C.H. Scholz, 1970.
Earthquake swarm in the Santa Barbara Channel, California, 1968.
Bull. seismol. Soc. Am., 60(4):1047-1060.

earthquakes
Tamrazyan, G.P., 1970
Some characteristic features of seismic energy release (in time) on the south-western margin of the Pacific Ocean.
N.Z. Jl Geol. Geophys. 13(2): 400-407.

earthquakes
Teng, Ta-Liang, and Ari Ben-Menahem, 1965.
Mechanism of deep earthquakes from spectrums of isolated body-wave signals. 1. The Banda Sea earthquakes of March 21, 1964.
J. geophys. Res., 70(20):5157-5170.

earthquakes, deep
Suyehiro, S., 1962
Deep earthquakes in the Fiji region.
Papers Meteor. Geophys., 13(3-4):216-238.

earthquake distribution
*Ripper, I.D. and R. Green, 1967.
Tasmanian examples of the influence of bathymetry and crustal structure upon seismic T-wave propagation.
N.Z. Jl Geol. Geophys. 19(5):1226-1230.

earthquakes, effect of
Barrett, Peter J., 1966.
Effects of the 1964 Alaskan earthquake on some shallow-water sediments in Prince William Sound, southeast Alaska.
J. Sedim. Petrol., 36(4):992-1006.

earthquakes, effect of
*Dill, Robert F., 1969.
Earthquake effects on fill of Scripps Submarine Canyon.
Bull. Geol. Soc. Am., 80(2):321-328.

earthquakes, effect of
Eckel, Edwin B., 1970.
The Alaska earthquake, March 27, 1964: lessons and conclusions.
Prof. Pap., U.S. geol. Surv., 546: 57pp.

Earthquakes, effect of
Francis, T.J.G. 1971.
Effect of earthquakes on deep-sea sediments.
Nature, Lond. 233 (5315): 98-102.

earthquakes, effect of
Hai, Nguyen, 1963.
Le Noyau terrestre d'après les séismes profonds du su de l'océan Pacifique.
C.R. Acad. Sci., Paris, 257(4):948-951.

earthquakes, effect of
Houtz, R.E., and H.W. Wellman, 1962.
Turbidity current at Kadavu Passage, Fiji.
Geol. Mag., 99(1):57-62.

earthquakes, effect of
Japan, Kobe Marine Observatory, 1953.
Abnormal tides caused by the Kamchatka earthquake and typhoon Agnes. J. Ocean. (2)4(4):13-19, figs.

earthquakes, effect of
Johnson, Rockne H., John Northrop and
Robert Eppley, 1963
Sources of Pacific T phases.
J. Geophys. Res., 68(14):4251-4260.

earthquakes, effect of
Julien, A., 1955.
Influence du seisme d'Orleansville sur les cables
de la Mediterranee (9 et 10 septembre 1954).
Bull. d'Info., C.C.O.E.C. 7(5):196-200, 1 fig.

Earthquakes, effect of
Kalle, Kurt, 1960
Die rätselhafte und "unheimliche" Naturerscheinungdes
"explodierenden" und des "rotierenden" Meeresleuchtens-
eine Folge lokaler Seeleben? Deutsche Hydrogr. Zeits., 13
(2): 49-77.

earthquakes, effect of
Kishinouye, Fujuhiko, and Heihachiro Kobayashi,
1965.
A submarine fault line found near Awashima after
the Japan Sea earthquake on June 16, 1964.
Bull. Earthquake Res. Inst., 43(1):205-208.

earthquakes, effect of
Mogi, A., 1959
On the depth change at the time of Kanto-
Earthquake in Sagami Bay.
Publ. 981, Hydrogr. Bull. No. 60:52-60.

earthquakes, effect of
Momoi, Takeo, 1965.
A motion of water excited by an earthquake. 1.
The case of a rectangular basin (one-dimensional)
Bull. Earthquake Res. Inst., 43(1):111-127.

earthquakes, effect of
Nagumo, Shozaburo, Shiji Hasegawa, Sadayuki
Koresawa and Heihachiro Kobayashi,
1970.
Ocean-bottom seismographic observation off
Sanriku - aftershock activity of the 1968 Tokachi-
Oki Earthquake and its relation to the ocean-
continent boundary fault.
Bull. Earthq. Res. Inst. 48(5): 793-809.

earthquakes, effect of
*Plafker, George, and Meyer Rubin, 1967.
Vertical tectonic displacements in south-central
Alaska during and prior to the Great 1964-
Earthquake.
J. Geosci., Osaka City Univ., 10:53-66.

earthquakes, effect of
Reimnitz, Erk, Neil F. Marshall, 1965.
Effects of the Alaska earthquake and tsunami
on recent deltaic sediments.
Jour. Geophys. Res., 70(10):2363-2376.

earthquakes, effect of
Suyehiro, Y., 1934.
Some observations on the unusual behavious of
fishes prior to an earthquake.
Bull. Earthquake Res. Inst., Suppl. Vol. 1:228-
231.

earthquakes, effect of
Sykes, Lynn R., 1963
Seismicity of the South Pacific Ocean.
J. Geophys. Res., 68(21):5999-6006.

earthquakes, effect of
Tamrazyan, G.P., 1970
The earthquakes of the Arctic and the
tide-generating forces.
Arbok Norsk Polarinst. 1968: 136-138

earthquakes, effect of
Teisseyre, R., 1961.
The dislocation processes in the Pacific Ocean.
Annals, Int. Geophys. Year, 11:411-413.

earthquakes, effect of
United States Naval Oceanographic Office, 1965.
Oceanographic Atlas of the North Atlantic Ocean.
Section V.
U.S. Oceanogr. Off., Publ. 700: 71 pp (quarto)

earthquakes, effect of
Yamaguti, Seiti, 1965.
On the changes in the heights of mean sea-levels
before and after the great Niigata earthquake on
June 16, 1964.
Bull. Earthquake Res. Inst., 43(1):167-172.

earthquakes, effect of
Yamaguti, S., 1955.
On the changes of the heights of mean sea-
levels before and after the great earthquakes.
Bull. Earthquake Res. Inst., Tokyo Univ. 33(1):
27-31.

earthquake epicenters

epicenter
*Hoshino, Michihei, 1969.
On the relationship between the distribution of
epicenter and the submarine topography and
geology. (In Japanese; English abstract).
J. Coll. mar. Sci. Technol., Tokai Univ., (3):1-10.

epicenters
Meyerhoff, A.A., and Howard A.
Meyerhoff 1972.
Continental drift, IV: The Caribbean
"Plate".
J. Geol. 80(1): 34-60.

earthquake epicenters
Stone, D.B., 1968.
Geophysics in the Bering Sea and
surroundings: a review.
Tectonophysics 6(6):433-460

earthquake epicenters
Stauder, William, 1968.
Mechanism of the Rat Island earthquake sequence
of February 4, 1965, with relation to island
arcs and sea-floor spreading.
J. Geophys. Res., 73(12):3847-3858.

earthquake epicenters
Stover, Carl W., 1968.
Seismicity of the South Atlantic Ocean.
J. geophys. Res., 73(12):3807-3820.

earthquake epicenters
Stover, Carl W., 1966.
Seismicity of the Indian Ocean.
J. geophys. Res., 71(10):2575-2581.

earthquake epicenters
Sykes, Lynn R., 1964.
Large oceanic fracture zones delineated by
earthquake epicenters. (Abstract).
Geol. Soc., Amer., Special Paper, No. 76:162.

earthquake epicenters
Tobin, Don G., and Lynn R. Sykes 1968.
Seismicity and tectonics of the northeast
Pacific Ocean.
J. geophys. Res. 73(12): 3821-3845.

earthquake epicenters
Vrana, Ralph S., 1971.
Seismic activity near the eastern end of the
Murray Fracture Zone.
Bull. geol. Soc. Am. 82(3): 789-792.

earthquake foci

earthquake foci
Andreyev, B.A., 1963.
Geophysical and geological features of the north-
west Pacific zone of deep-focus earthquakes.
(In Russian).
Doklady, Akad. Nauk, SSSR, 150(1):140-142.

Translation:
E.R. Hope, T397R:13-17.

earthquake foci
Balakina, L.M., 1961.
The character of stresses and ruptures in
earthquake foci of the Pacific seismic zone.
(Abstract).
Tenth Pacific Sci. Congr., Honolulu, 21 Aug.-
6 Sept., 1961, Abstracts of Symposium Papers,
354.

earthquakes, foci
* Mino, Kazuo, Toshiyuki Onoguchi and Takeshi
Mikumo, 1968.
Focal mechanism of earthquakes on island arcs in
the southwest Pacific region.
Bull. Disas. Prev. Res. Inst. Kyoto Univ. 18
(139-2): 7-8-96.

earthquake foci
Ruditch, E.M., 1961.
Relations between deep-focus earthquakes in the
eastern margin of Asia and the large structures
of the earth's crust. (Abstract).
Tenth Pacific Sci. Congr., Honolulu, 21 Aug.-
6 Sept., 1961, Abstracts of Symposium Papers,
363-364.

earthquake foci
*Stauder, William, 1968.
Tensional character of earthquake foci beneath
the Aleutian Trench with relation to sea-floor
spreading.
J. geophys. Res., 73(24):7693-7701.

earthquakes mechanism of

earthquakes, focal
mechanisms
Isacks, Bryan, Lynn R. Sykes and
Jack Oliver 1969.
Focal mechanisms of deep and shallow
earthquakes in the Tonga-Kermadec
region and the tectonics of island arcs.
Bull. geol. Soc. Am. 80(8): 1443-1470.

earthquakes, mechanism of
Sykes, Lynn R., 1967.
Mechanism of earthquakes and nature of faulting
on the mid-oceanic ridges.
J. geophys. Res., 72(8):2131-2153.

earthquakes submarine

earthquakes, sea bottom
Richards, Paul G., 1971.
A theory for pressure radiation
from ocean-bottom earthquakes.
Bull. seismol. Soc. Am. 61(3): 707-721.

earthquakes, submarine
Tocher, D., 1956.
Earthquakes off the North Pacific coast of the United States.
Bull. Seismol. Soc., Amer., 46(3):165-174.

earthquake swarms
Iwata, Takayuki, 1970.
On the earthquake swarm in the Galapagos Islands region in June and July 1968. (In Japanese; English abstract)
Bull. Earthq. Res. Inst. Tokyo Univ. 48(5): 935-953.

earthquakes swarms
*Sykes, Lynn R., 1970.
Earthquake swarms and sea-floor spreading.
J. geophys. Res., 75(32):6598-6611.

earthquake swarm
Thatcher, Wayne, and James N. Brune 1971.
Seismic study of an oceanic ridge earthquake swarm in the Gulf of California.
Geophys. J. R. astron. Soc. 22(5): 473-489.

earthquake swarm
Wetmiller, Robert J., 1971.
An earthquake swarm on the Queen Charlotte Islands Fracture Zone.
Bull. seismol. Soc. Am. 61 (6): 1489-1505

earth's rotation

earth's rotation
Jones, Sir H.S., 1961.
Variations of the earth's rotation.
Physics and Chemistry of the Earth, Pergamon Press, 4(4):186-210.

earth's rotation
MacDonald, G.J.F., 1967.
Implications for geophysics of the precise measurement of the earth's rotation.
Science, 157(3786):304-305.

earth's rotation
Munk, W., and R. Revelle, 1952.
On the geophysical interpretation of irregularities in the rotation of the earth.
Mon. Not., R. Astron. Soc., Geophys. Suppl., 6(6):331-347, 1 textfig.

earth's rotation, effect of
Backus, G.E., 1962
The effect of the earth's rotation on the propagation of ocean waves over long distances
Deep-Sea Res., 9(3):185-197.

earth's rotation
Görtler, H., 1941.
Einfluss der Bodentopographie auf Strömungen über der rotierenden Erde. Z. angew. Math. Mech. 21:279-303.

earths rotation, effect of
Hidaka, K., and H. Iida, 1956.
On the influence of the earth's rotation and the dimension of the oceans. Geophys. Notes 11(1): No. 8.

earth's rotation, effect of
Hidaka, K. and H. Iida, 1957
On the influence of the earth's rotation and the dimension of the oceans. J. Ocean. Soc., Japan, 13(4): 123-126.

earth's rotation, effect of
Khizanashvili, G.G., 1963.
The origin of the flooded sea terrace in the light of the hypothesis on the dynamics of the earth's rotation axis. (In Russian).
Okeanologiia, Akad. Nauk, SSSR, 3(5):930-935.

earth's rotation, effect of
Maximov, I.V., and N.P. Smirnov, 1964
Variation of earth rotation velocity and the mean level of the world ocean. (In Russian).
Okeanologiia, Akad. Nauk, SSSR, 4(1):9-18.

earth's rotation
Munk, W. H., 1947.
Effect of earth's rotation on convection cells.
Ann. N. Y. Acad. Sci. 43(8):815-820.

earths rotation
Munk, Walter H., and Gordon J.F. MacDonald, 1961.
The rotation of the earth: a geophysical discussion. Cambridge University Press, xix+323 pp., $13.50.

Reviewed by Edgar W. Woolard in:
Trans. Amer. Geophys. Union, 42(3):307.

earth rotation, effect of
Revelle, R., and W. Munk, 1955.
Evidence from the rotation of the earth.
(Ann. Géophys., 11(1)), I.A.G.A. Bull. 15a:1-5.

earth's rotation, effect of
Runcorn, S.K., 1970.
1. General observations. 20. Marine life and the rotation rates of the earth and moon (Nov. 1968). In: The sea: ideas and observations on progress in the study of the seas, Arthur E. Maxwell, editor, Wiley-Interscience 4(1): 759-763.

Earth's rotation, effect of
Saint-Guily, Bernard, 1970.
On internal waves. Effects of the horizontal component of the earth's rotation and of a uniform current.
Dt. hydrogr. Z. 23(1): 16-23

earth's rotation
Schatzman, E., 1960.
Sur le nouveau régime de rotation de la terre établi en juillet 1959.
Ann. Géophysique, 16(4):495-506.

Earth's rotation
Schlichter, Louis B., 1963
Secular effects of tidal friction upon the Earth's rotation.
J. Geophys. Res., 68(14):4281-4288.

EARTH'S ROTATION, effect of
Smirnov, N.P., 1967.
On a relation between changes of the Earth's rotation and variations of hydrological conditions in the North Atlantic. (In Russian).
Kater. Rybokhoz. Issled. Severn. Basseina, (PINRO), 10:90-94.

earth's rotation, effect of
Van den Dungen, F.H., J.F. Cox, and J. van Mieghem, 1953.
Sur les variations du niveau des mers et de la vitesse de rotation de la terre.
Acad. R. Belg., Bull., Cl. Sci., 39:29-34.

earth's rotation
Vestine, E.H., 1955.
Relations between fluctuations in the earths rotation, the variation of latitude, and geomagnetism. (Ann. Géophys. 11(1)), I.A.G.A. Bull. 15a Bl.

earths sphericity

earths sphericity, effect of
Takano, Kenzo, 1966.
Effet de la sphéricité de la terre sur la circulation générale dans un océan.
J. oceanogr. Soc., Japan, 22(6): 255-263

ecdysis

ecdysis
See also: moulting

ecdysis
Ito, Katsuchiyo, and Toshio Kobayashi, 1967.
A female specimen of the edible crab, Chionectes opilio O. Fabricius, with the unusual symptom of additional ecdysis. (In Japanese).
Bull. Japan Sea reg. Fish. Res. Lab., 18:127-128.

ecdysis
Lasker, Reuben 1966.
Feeding, growth, respiration and carbon utilization of a euphausiid crustacean.
J. Fish. Res. Bd. Can. 23 (9): 1291-1317.

ecdysis
*Stewart, James E., and H.J. Squires, 1968.
Adverse conditions as inhibitors of ecdysis in the lobster Homarus americanus.
J. Fish Res. Bd. Can. 25(9)1763-1774.

ecdysis
Watson, J., 1971.
Ecdysis of the snow crab, Chionoecetes opilio.
Can. J. Zool. 4

echinoderms

Echinoderms
Achituv, Y., 1969.
Studies on the reproduction and distribution of Asterina burtoni Gray and A. wega Perrier (Asteroidea) in the Red Sea and the eastern Mediterranean.
Israel J. Zool. 18(4): 329-342.

echinoderms
Adithiya, L.A., 1969.
Beche-de-mer in Ceylon. Spolia zeylanica, 31 (2): 405-411.

echinoderms
Alton, Miles S., 1966.
A new sea-water from the northeastern Pacific Ocean, Asthenactis fisheri n. sp., with a review of the family Myxasteridae.
Deep-Sea Res., 13(4):687-697.

Echinoderms
Ancona Lopes, Ana Amélia, 1965.
Contribution to the ecology of the Holothureidea of the coast of São Paulo state.
Anais Acad. bras. Cienc., 37(Supl.):171-174.

echinoderms
Astrahantteff, Sergei, and Miles S. Alton, 1965.
Bathymetric distribution of brittlestars (Ophiuroidea) collected off the northern Oregon coast.
J. Fish. Res. Bd., Canada, 22(6):1407-1424.

echinoderms
Belyaev, G.M., 1971.
Deep water holothurians of the genus Elpidia. (In Russian; English abstract). Trudy Inst. Okeanol. P.P. Shirshov 92: 326-367.

~~echinoderms holothurians~~ (pel gic)
Belyaev, G.M., and M.E. Vinogradov, 1969.
A new pelagic holothurian (Elasipoda; Psychropotidae) from the abyssal of the Kurilo-Kamchatka Trench. (In Russian; English abstract).
Zool. Zh., 48(5): 709-716

echinoderms
Bernasconi, Irene, 1966.
Los equinideos y asteroideos colectados por el buque oceanografico R/V Vema frente a las costas argentinas, uruguayas y sur de Chile.
Revta Mus. argent. Cienc. nat. Bernardino Rivadavia Inst. nac. Invest. Cienc. nat., Zool., 9(7):147-175.

echinoderms
Bernasconi, Irene, 1965.
Ophiuroidea de Puerto Deseado (Santa Cruz, Argentina).
Physis, B. Aires, 25(69):143-152.
Also in:
Centro Invest. Biol. Mar., Estación Puerto Deseado y Estación Austral, Contrib. Cient., No. 14. 1965.

echinoderms
Bernasconi, Irene, 1965.
Astrotoma agassizii Lyman, especie vivipara del Atlantico Sur (Ophiuroidea, Gorgonocephalidae) Physis, B. Aires, 25(69):1-5.
Also in:
Centro Invest. Biol. Mar., Estación Puerto Deseado y Estación Austral, Contrib. Cient., No. 15. 1965.

echinoderms
Bernasconi, Irene, 1964.
Asteroideos argentinos: claves para los órdenes familias, subfamilias y géneros.
Physis, B. Aires, 24(68):241-277.
Also in:
Centro Invest. Biol. Mar., Estación Puerto Deseado y Estación Austral, Contrib. Cient., No. 10(1964).

echinoderms
Bernasconi, Irene, 1964.
Distribucion geografica de los equinoideos y asteroideos de la extremidad austral de Sudamerica.
Bol., Inst. Biol. Mar., Mar del Plata, Argentina. No. 7:43-50.

echinoderms
Boolootion, Richard A., editor, 1966.
The physiology of Echinodermata.
John Wiley & Sons, Inc., 772 pp. $45.00.

echinoderms
Boolootian, Richard A., and David Leighton, 1966.
A key to the species of Ophiuroidea (brittle stars) of the Santa Monica Bay and adjacent areas.
Los Angeles County Mus., Contrib. Sci., No. 93: 20 pp.

echinoderms
Brun, Einar, 1969.
Aggregation of Ophiothrix fragilis (Abildgaard) (Echinodermata: Ophiuroidea).
Nytt Mag. Zool. 17: 153-160

starfish
Burkenroad, Martin D., 1957.
Intensity of settling of starfish in Long Island Sound in relation to fluctuations of the stock of adult starfish and in the setting of oysters.
Ecology, 38(1):164-165.

echinoderms-starfish
Chaet, Alfred B., 1966.
Neurochemical control of gamete release in starfish.
Biol. Bull., 130(1):43-58.

holothurians
Cherbonnier, Gustave, 1967.
Deuxième contribution à l'étude des holothuries de la mer Rouge collectées par des Israéliens.
Rep. Israel South Red Sea Exped., Sea Fish. Res. Stn. 26:55-68.
(Bull. Sea Fish. Res. Stn Israel 43).

echinoderms
Cherbonnier, Gustave, 1965.
Holothurides.
Exped. Océanogr. Belge, Eaux Cotières Africaines Atlant. Sud (1848-1949), Res. Sci., 3(11):1-23. 11 pls.

echinoderms
Cherbonnier, G., 1964.
Holothuries récoltées par A. Crosnier dans le Golfe de Guinée.
Bull. Mus. Nat. Hist. Nat., 36(5):647-676.

echinoderms, ecology of
Chia, Fu-Shiang, 1969.
Some observations on the locomotion and feeding of the sand dollar, Dendraster excentricus (Eschscholtz). J. exp. mar. Biol. Ecol., 3(2): 162-170.

echinoderms, physio.
Chia, Fu-Shiang, 1969.
Histology of the pyloric caeca and its changes during brooding and starvation in a starfish, Leptasterias hexactis. Biol. Bull., mar. biol. Lab., Woods Hole, 136(2): 185-192.

Echinoderms
Clark, Ailsa M., 1966.
Echinoderms from the Red Sea. 2. (crinoids, ophiuroids, echinoids and more asteroids).
Rep. Israel South Red Sea Exped., Sea Fish. Res. Stn, 21:26-58. (Bull. Sea Fish. Res. Stn. Israel 41).

echinoderms, crinoids
Clark, Ailsa M., 1966.
Some crinoids from New Zealand waters.
New Zealand J. Sci., 9(3):684-705.

echinoderms (starfish)
Clark, Ailsa M., 1962.
Starfishes and their relations.
British Museum (Natural History), 119 pp., illus. 11 shillings.

echinoderms
Clark, Helen E. Shearburn, 1970
A new species of Paralophaster (Asteroidea) from New Zealand.
Trans. R. Soc. N.Z., Biol. Sci., 12 (15): 177-179.

echinoderms
Domantay, Jose S., Sr., and Cecelia R. Domantay, 1966.
Studies on the classification and distribution of Philippine littoral Ophiuroidea (brittle stars).
Philippine J. Sci., 95(1):1-76.

Echinoderms
Downey, Maureen E., 1970.
Marsipaster acicula, new species (Asteroidea: Echinodermata), from the Caribbean and Gulf of Mexico
Proc. Biol. Soc. Wash., 83(28): 309-312.

echinoderms
Durham, J.W., and K.E. Caster, 1963
Helicoplacoidea: a new class of echinoderms.
Science, 140(3568):820-822.

echinoderms
D'yakonov, A.M., 1923.
Echinodermata. 1. Echinoidea.
Fauna Rossii i sopredel'nykh stran', Petrograd.
Translation: Y. Salkind, Israel Program for Scientific Translations, 265 pp. (multilithed).

echinoderms
Ebert, Thomas A., 1971.
A preliminary quantitative survey of the echinoid fauna of Kealakekua and Honaunau Bays, Hawaii.
Pacific Sci., 25(1): 112-131

echinoderms
Endean, R., 1965.
Queensland faunistic records. 7. Further records of Echinodermata (excluding Crinoidea) from Southern Queensland.
Univ. Queensland Pap. (Zool.) 2(11):229-235.

echinoderms
Evamy, B.D., and D.J. Shearman, 1965.
The development of over growths from echin derm fragments.
Sedimentology, 5(3):211-233.

echinoderms
Fell, H. Barraclough, 1966.
Ancient echinoderms in modern seas.
In: Oceanography and marine biology, H. Barnes, editor, George Allen & Unwin, Ltd., 4:233-245.

FELL echinoderms
Fell, H. Barraclough, 1964.
A list of Echinodermata collected by N.Z. O.I. from Milford Sound.
New Zealand Dept. Sci. Ind. Res. Bull., No. 157: New Zealand Oceanogr. Inst., Memoir, No. 17:95.

echinoderms
Fricke, H.W. 1970
Beobachtungen über Verhalten und Lebensweise des im Sand lebenden Schlangensternes Amphioplus sp.
Helgolander wiss. Meeresunters. 21(1/2): 124-133.

echinoderms
Goreau, Thomas F., 1964.
On the predation of coral by the spiny starfish Acanthoster planci (L.) in the Southern Red Sea.
Sea Fish. Res.Inst., Israel, Bull., No. 35:23-26.

echinoderms
Grace, R.V., 1967.
An underwater survey of two starfish species in the entrance to the Whangateau Harbour.
Tane, J. Auckland Univ. Fld Club 13:13-19

echinoderms
Gray, I.E., Maureen E. Downey and M.J. Cerame-Vivas, 1968.
Sea-stars of North Carolina.
Fish. Bull., Bur. Comm. Fish., U.S.F.W.S. 67(1): 127-163.

echinoderms
Halpern, Jerald A., 1970.
Biological investigations of the deep sea. 51. Goniasteridae (Echinodermata: Asteriodea) of the Straits of Florida. Bull. mar. Sci., Miami, 20(1): 193-286.

Echinoderms
Halpern, Jerald A., 1970.
Growth rate of the tropical sea star Luidia senegalensis (Lamarck). Bull. mar. Sci., 20(3): 626-633.

starfish
Hancock, D.A., 1958.
Notes on starfish on an Essex oyster bed.
J.M.B.A., U.K., 37(3):565-590.

echinoderms (holothurians)
*Hansen, Bent, 1967.
The taxonomy and zoogeography of the deep-sea holothurians in their evolutionary aspects.
Stud. trop. Oceanogr., Miami, 5:480-501.

echinoderms
Hinegardner, Ralph T., 1969.
Growth and development of the laboratory cultured sea urchin. Biol. Bull. mar. biol. Lab., Woods Hole, 137(3): 465-475.

echinoderms
Kyte, Michael A., 1969.
A synopsis and key to the recent Ophiuroidea of Washington State and southern British Columbia.
J. Fish. Res. Bd. Can., 26(7): 1727-1741.

echinoderms
Litvinova, N.M., 1971.
The brittle-stars of the genus Amphiophiura of the Pacific and Indian Oceans collected by Soviet expeditions on the R/V Vityaz and Academician Kurchatov. (In Russian; English abstract). Trudy Inst. Okeanol. P.P. Shirshov 92:298-316.

(echinoderm, ecology
(starfish
Loosanoff, Victor L., 1964
Variations in time and intensity of setting of the starfish, Asterias forbesi, in Long Island Sound during a twenty-five-year period.
Biol. Bull., 126(3):423-439.

echinoderms
Madsen, F. Jesenius, 1970.
West African ophiuroids.
Atlantide Rept. 11: 151-241

echinoderms
McCauley, James E., and Andrew G. Carey, Jr., 1967
Echinoidea of Oregon
J. Fish. Res. Bd, Can., 24(6): 1385-1401

echinoderms
McKnight, D. G., 1969.
An outline distribution of the New Zealand shelf fauna: benthos survey, station list, and distribution of Echinoidea.
Bull. N.Z. Dept. scient. industr. Res. 195: 1-91. Also: Mem. N.Z. oceanogr. Inst. 47.

echinoderms
*McKnight, D.G., 1967.
Echinoderms from Cape Hallett, Ross Sea.
N.Z. Jl mar. Freshwat. Res., 1(3):314-323.

echinoderms
*McKnight, D.G., 1967.
Additions to the echinoderm fauna of the Chatham Rise.
N.Z. Jl mar. Freshwat. Res., 1(3):291-313.

echinoderms
Menker, D., 1970.
Lebenszyklus, Jugendentwicklung und Geschlechtsorgane von Rhabdomolgus ruber (Holothuroidea: Apoda). Marine Biol., 6(2): 167-186.

echinoderms
Millott, N., editor, 1967.
Echinoderm biology.
Symp. Zool. Soc., Lond., Academic Press, 20: 240 pp. $11.00

echinoderms
Mironov, A.N., 1971.
Soft sea urchins of the family Echinothuriidae collected by the R/V Vityaz and the Academician Kurchatov in the Pacific and Indian oceans. (In Russian; English abstract). Trudy Inst. Okeanol. P.P. Shirshov 92: 317-325.

echinoderms
Nichols, David, 1964
Echinoderms: experimental and ecological.
In: Oceanography and Marine Biology, Harold Barnes, Editor, George Allen & Unwin, Ltd., 2 393-423.

echinoderms (A. forbesi)
*Oviatt, Candace A., 1969.
Light influenced movement of the starfish Asterias forbesi (Desor).
Behaviour 33(1/2):52-

echinoderms
Patent, D.H., 1970.
The early embryology of the basket star Gorgonocephalus caryi (Echinodermata, Ophiuroidea). Mar. Biol., 6(3): 262-267.

echinoderms
Pawson, David L., 1970.
The marine fauna of New Zealand: sea cucumbers (Echinodermata: Holothuroidea)
Bull. N.Z. Dept. scient. industr. Res. 201: 7-69

echinoderms
Pawson, David L., 1969.
Holothuroidea from Chile. Report No. 46 of the Lund University Chile Expedition 1948-1949.
Sarsia 38: 121-145.

echinoderms
*Pawson, David L., 1968.
Some holothurians from Macquarie Island.
Trans. R. Soc. N.Z., Zool., 10(15):141-150.

echinoderms
Pawson, David L., 1965.
The bathyal holothurians of the New Zealand region.
Zool. Publ., Victoria Univ., Wellington, No. 39: 33 pp.

echinoderms, holothurians
Pawson, D. L., 1964.
The Holothuroidea collected by the Royal Society Expedition to Southern Chile, 1958-1959.
Pacific Science, 18(4):453-470.

echinoderms
Pearse, J.S., 1969.
Slow developing demersal embryos and larvae of the antarctic sea star Odontaster validus.
Marine Biol. 3(2): 110-116.

echinoderms
Pearse, J.S., 1969.
Reproductive periodicities of Indo-Pacific invertebrates in the Gulf of Suez. II. The echinoid Echinometra mathaei (de Blainville). Bull. mar. Sci., 19(3): 580-613.

echinoderms
Pearse, J.S., 1969.
Reproductive periodicities of Indo-Pacific invertebrates in the Gulf of Suez. I. The echinoids Prionocidaris baculosa (Lamarck) and Lovenia elongata (Gray). Bull. mar. Sci., 19(2) 323-350.

echinoderms
Pearse, J.S., 1965.
Reproductive periodicities in several contrasting populations of Odontaster validus Koehler, a common Antarctic asteroid.
In: Biology of Antarctic seas. II.
Antarctic Res. Ser., Amer. Geophys. Union, 5:39-85.

echinoderms
Petersen, J.A., P. Sawaya y Liu Pin-Yi, 1965.
Alguns aspectos da ecologia de Echinodermata.
Anais Acad. bras. Cienc., 37(Supl.):167-170.

echinoderms,anat.
Raup,David M., and Emery F. Swan,1967.
Crystal orinetation in the apical plates of aberrant echinoids.
Biol.Bull.,mar.biol.Lab.,Woods Hole,133(3):618-629.

echinoderms
Robinson, D.E. 1971.
Observations on Fijian coral reefs and the crown of thorns starfish.
J. Roy. Soc. N.Z. 1(2): 99-112

echinoderms
Robison, Richard A. and James Sprinkle, 1969.
Ctenocystoidea: new class of primitive echinoderms. Science, 166(3912): 1512-1514.

echinoderms
Rowe, F.W.E., 1970.
A note on the British species of cucumarians, involving the erection of two new nominal genera. J. mar. biol. Ass., U.K., 50(3): 683-687.

echinoderms
Rowe, F.W.E., and D.L. Pawson, 1967.
New genus in the holothurian family Synaptidae.
Papers and Proc., R. Soc., Tasmania, 101:31-35

echinoderms
Salsman, G. G., and W.H. Tolbert, 1965.
Observations on the sand dollar, Mellita quinquiesperforata.
Limnology and Oceanography, 10(1):152-155.

echinoderms
Serafy, D. Keith, 1971.
Biological results of the University of Miami Deep-Sea Expeditions. 83. A redescription of Clypeaster pallidus H.L. Clark, 1914, and a description of juveniles of C. rosaceus (Linnaeus, 1758) (Echinodermata; Echinoidea). Bull. mar. Sci. Coral Gables, 21(3): 779-786.

echinoderms
Serafy, Donald Keith, 1971.
A new species of Clypeaster (Echinodermata, Echinoidea) from San Felix Island, with a key to the recent species of the eastern Pacific Ocean.
Pacific Sci., 25(2): 165-170

Echinoderms
Serafy, Donald Keith, 1970.
A new species of Clypeaster from the Gulf and Caribbean and a key to the species in the tropical northwestern Atlantic (Echinodermata: Echinoidea). Bull. mar. Sci., 20(3): 662-677.

starfish
Smith, L.S., 1961
Clam-digging behavior in the starfish.
Behaviour, 18(1/2): 148-153.

echinoderms
Stevenson, Robert A., 1965.
Differences in trace elements composition in the sea urchins Tripneustes esculentus (Leske) and Echio otra lucuntra (L.). (Abstract).
Ocean Sci. and Ocean Eng., Mar. Techn. Soc.,-Amer. Soc. Limnol. Oceanogr., 1:300.

echinoderms
Thomas, L.P., 1962.
The shallow water amphiurid brittle stars (Echinodermata, Ophiuroidea) of Florida.
Bull. Mar. Sci., Gulf and Caribbean, 12(4):623-694.

echinoderms
Tommasi, Luiz Roberto, 1971.
The echinoderms of the Ilha Grande region (RJ Brasil). Distribution and abundance of six species up to the isobath of 50 m. (Portuguese abstract). In: Fertility of the Sea, John D. Costlow, editor, Gordon Breach, 2: 581-592.

echinoderms
Tommasi, Luiz Roberto 1970.
Os ofiuróides recentes do Brasil e de regiões vizinhas.
Contrções Inst. oceanogr. São Paulo, sér. Oceanogr. Biol. 20: 1-146

echinoderms
Tommasi, Luiz Roberto, 1970.
Lista dos Asteróides recentes do Brasil.
Contrções Inst. oceanogr. Univ. S. Paulo, Ser. Ocean. Biol. 18: 1-61.

echinoderms
Tommasi, Luiz Roberto 1966.
Sobre algumas equinodermas da região do Golfo de Mexico e do mar das Antilhas.
Anales Inst. Biol., México 37 (1/2): 155-166.

echinoderms
Tommasi, L.R., 1965.
Lista dos crinoides recentes do Brasil.
Controes Inst. oceanogr., Univ. S Paulo, ser. Ocean. biol., No. 9:1-33.

echinoderms
Tommasi, L.R., 1965.
Alguns Amphiuridae (Ophiuroidea) do litoral de São Paulo e Santa Catarina.
Contrcões Inst. oceanogr. Univ. S Paulo, sér. Ocean. biol., No. 8:1-9.

echinoderms
Warner, G.F., 1971.
On the ecology of a dense bed of the brittle-star Ophiothrix fragilis. J. mar. biol. Ass. U.K., 51(2): 267-282.

echinoderms
Warner, G.F., 1969.
Brittle-star beds in Torbay, Devon.
Rept. Underwat. Ass. 4: 81-85.

Echinoderms
Weber, J.N. and P.M.J. Woodhead, 1970.
Ecological studies of the coral predator Acanthaster planci in the South Pacific.
Marine Biol., 6(1): 12-17.

echinoderms
Wolff, W.J., 1968.
The Echinodermata of the estuarine region of the rivers Rhine, Meuse and Scheldt, with a list of species occurring in the coastal waters of the Netherlands.
Netherlands J. Sea Res., 4(1): 59-85.

echinoderms, age determination
Jensen Margit, 1969.
Age determination of echinoids.
Sarsia, 37: 40-44.

echinoderms anat.-

echinoderms
Anderson,John Maxwell,1965.
Studies on visceral regeneration in sea stars. II. Regeneration of pyloric caeca in Asteriidae, with notes on the source of cells in regenerating organs.
Biol. Bull., 128(1):1-23.

echinoderms, anatomy of
Boolootian, R.A., and J.L. Campbell, 1964.
A primitive heart in the echinoid Strongylocentrotus purpuratus.
Science, 145(3628):173-175.

echinoderms, anat physiol.
*Chia,Fu-Shiang,1968.
The embryology of a brooding starfish, Leptasteria hexactis (Stimpson).
Acta Zoologica, 49(3):321-364.

echinoderms, anat-physiol
Chia, Fu-shiang, 1966.
Brooding behavior of a six-rayed starfish, Leptasterias hexactis.
Biol. Bull., 130(3):304-315.

echinoderms,anat. physiol.
Donnay, G. abrielle and D.L. Pawson, 1969.
X-ray diffraction studies of echinoderm plates.
Science, 166(3909): 1147-1160.

echinoderms, anat.-physiol.
Fenchel, Tom 1965.
Feeding biology of the sea star Luidia sarsi Düben + Koren.
Ophelia 2(2): 223-236.

echinoderms, anat.-physiol
Ferguson, John Carruthers, 1967.
Utilization of dissolved exogenous nutrients by the starfishes, Asterias forbesi and Henricia sanguinolenta.
Biol. Bull., mar. biol. Lab., Woods Hole, 132(2):161-173.

echinoderm physiology
Ferguson, John Carruthers, 1964.
Nutrient transport in starfish. II. Uptake of nutrients by isolated organs.
Biol. Bull., 126(3):391-406.

echinoderms
Ferguson, John Carruthers, 1964.
Nutrient transport in starfish. 1. Properties of the coelomic fluid.
Biological Bulletin, 126(1):33-53.

echinoderms, anat.-physiol
Fish, John D., 1967.
The digestive system of the holothurian Cucumaria elongata. 1. Structure of the gut and haemal system. 2. Distribution of the digestive enzymes.
Biol. Bull., mar. biol. Lab., Woods Hole, 132(3): 337-353; 354-361.

Echinoderms, anat.-physiol.
Fontaine, A.R., 1965.
The feeding mechanism of the ophiuroid Ophiocomina nigra.
Jour. Mar. Biol. Assoc., U.K., 45(2):373-385.

echinoderms, anat.-physiol
Fontaine, A.R., and Fu-Shiang Chia, 1968.
Echinoderms: an autoradiographic study of assimilation of dissolved organic molecules.
Science, 161(3846):1153-1155.

echinoderms, anat. physiol.
Giese, Arthur C. 1969.
General physiology of the echinoderm body wall with special reference to asteroids and echinoids.
Antarctic Jl U.S. 4(5): 192-193.

echinoderms, anat.
Holland, Nicholas D. 1969.
An electron microscope study of the papillae of crinoid tube feet.
Pubbl. Staz. Zool. Napoli. 37(4): 575-580.

echinoderms, anat.-physiol
*Holland, Nicholas D., 1967.
Some observations on the saccules of Antedon mediterranea (Echinodermata, Crinoides).
Pubbl. Staz. Zool., Napoli., 35(3):257-262.

echinoderms, anat.-physiol.
Holland, Nicholas D. and Michael T. Ghiselin, 1970.
A comparative study of gut mucous cells in thirty-seven species of the class Echinoidea (Echinodermata). Biol. Bull., mar. biol. Lab. Woods Hole, 138(3): 286-305.

echinoderms, physioloy of
Ishihara, Tadashi, and Masato Yasuda, 1965.
A study of the chonchiolin-decomposing enzyme present in the digestive organs of the starfish. (In Japanese; English abstract).
Bull. Fac. Fish., Nagasaki Univ., No. 19:85-90.

echinoderms, anat.-physiol
Kanatani, Haruo, and Miwako Ohguri, 1966.
Mechanism of starfish spawning. 1. distribution of active substance responsible for maturation of oocytes and shedding of gametes.
Biol. Bull., 131(1):104-114.

echinoderms, anat. physiol.
Leclerc, M. et J. Augarde, 1970.
Effets de l'ablation de l'organe axial sur l'activité respiratoire chez Asterina gibbosa (échinodermes, astérides). Marine Biol., 6(1): 77-80.

echinoderms, anat. -physiol
Lewis, John B., 1967.
Nitrogenous excretion in the tropical sea urchin Diadema antillarum Philippi.
Biol. Bull. mar. biol. Lab., Woods Hole, 132(1):34-37.

echinoderms, breeding of
Mileikovsky, S.A., 1968.
Breeding of Asterias rubens L. in the White, Barents, Norwegian and other European seas. (In Russian; English abstract).
Okeanologiia, Akad. Nauk, SSSR, 8(4):693-707.

echinoderms, anat.-hysiol.
Newell, R.C., and W.A.M., Courtney, 1965.
Respiratory movements in Holothuria forskali Delle Chiaje.
J. Exp. Biol., 42(1):45-57.

echinoderms
Nichols, David, 1964.
Echinoderms: experimental and ecological.
In: Oceanography and Marine Biology, Harold Barnes, Editor, George Allen & Unwin, Ltd., 2:393-423.

echinoderms, anat. physiol.
Nissen, H.-U., 1969.
Crystal orientation and plate structure in echinoid skeletal units. Science, 166(3909): 1150-1152.

echinoderms, anat. physiol.
Pentreath, R.J., 1971.
Respiratory surfaces and respiration in three New Zealand intertidal ophiuroids
J. Zool. Lond., 163(3): 397-412

echinoderms, anat. physiol.
Pentreath, R.J., 1970.
Feeding mechanisms and the functional morphology of podia and spines in some New Zealand ophiuroids (Echinodermata).
J. Zool. Lond., 161(3): 395-429.

echinoderms, anat. physiol.
Pentreath, V.W. and G.A. Cottrell, 1971.
'Giant' neurons and neurosecretion in the hypo-neural tissue of Ophiothrix fragilis Abildgaard.
J. exp. mar. Biol. Ecol., 6(3): 249-264.

echinoderms, anat. -physiol
Yoshida, M., and H. Ohtsuki, 1966.
Compound ocellus of starfish; its function.
Science, 153(3732):197.

Echinoderms chemistry

echinoderms, chemistry of
*Krishnan, S., 1968.
Histochemical studies on reproductive and nutritional cycles of the holothurian, Holothuria scabra.
Marine Biol., 2(1):54-65.

echinoderms, chemistry
Lawrence, John M. 1972
Carbohydrate and lipid levels in the intestine of Holothuria atra (Echinodermata, Holothuroidea)
Pacific Sci. 26(1): 114-116

echinoderms, chemistry of
Riley, J.P. and D.A. Segar, 1970.
The distribution of the major and some minor elements in marine animals. 1. Echinoderms and coelenterates. J. mar. biol. Ass., U.K., 50(3): 721-730.

echinoderms, chemistry of
Schroeder, Johannes H., Edward J. Dwornik and J.J. Papike 1969.
Primary protodolomite in echinoid skeletons.
Bull. geol. Soc. Am. 80(8): 1613-1616.

echinoderms, chemistry
*Weber, Jon N., 1968.
Fractionation of the stable isotopes of carbon and oxygen in carbon and oxygen in calcareous marine invertebrates - The Asteroidea, Ophiuroidea and Crinoidea.
Geochim. Cosmochim. Acta. 32(1):33-70.

echinoderms- coral-eating
Branham, J.M., S.A. Reed, and Julie H. Bailey, 1971.
Coral-eating sea stars Acanthaster planci in Hawaii. Science 172(3988): 1155-1157.

Echinoderms ecology

echinoderms, ecology of
*Bakus, G.J., 1968.
Defensive mechanism and ecology of some tropical holothurians.
Marine Biol., 2(1):23-32.

echinoderms, ecology of
Birkeland, Charles and Fu-Shiang Chia, 1971.
Recruitment risk, growth, age and predation in two populations of sand dollars, Dendraster excentricus (Eschscholtz). J. exp. mar. Biol. Ecol., 6(3): 265-278.

echinoderms, ecology of
*Mauzey, Karl P., Charles Birkeland and Paul K. Dayton, 1968.
Feeding behavior of asteroids and escape responses of their prey in the Puget Sound region.
Ecology, 49(4):603-619.

echinoderms, effect of
Chesher, Richard H. 1969.
Destruction of Pacific corals by the sea star Acanthaster planci.
Science 165(3890): 280-283.

echinoderms, effect of
Rhoads, D.C. and D.K. Young, 1971.
Animal-sediment relations in Cape Cod Bay, Massachusetts. II. Reworking by Molpadia oolitica (Holothuroidea). Mar. Biol. 11(3): 255-261.

echinoderms, feeding habits
Brun, Einar, 1972.
Food and feeding habits of Luidia ciliaris Echinodermata: Asteroidea. J. mar. biol. Ass. U.K. 52(1): 225-236.

Echinoderms, growth of
Hines, Judith, and Ron Kenny, 1967.
The growth of Arachnoides placenta (L.) (Echinoidea).
Pacif. Sci., 21(2):230-235.

echinoderms, growth of
Regis, Marie-Berthe, 1970.
Premières données sur la croissance de Paracentrotus lividus LmK.
Téthys 1(4): 1049-1056.

echinoderms, intraspecific variation

Serafy, D. Keith, 1971.
Intraspecific variation in the brittle-star Ophiopholis aculeata (Linnaeus) in the north-western Atlantic (Echinodermata; Ophiuroidea). Biol. Bull. Mar. Biol. Lab. Woods Hole, 140(2): 323-330.

echinoderms, learning of

Wells, , M. J., 1965.
Learning by marine invertebrates.
In: Advances in Marine Biology, Sir Frederick S. Russell, editor, Academic Press, 8:1-62.

echinoderm physiology

Campbell, A.C., and M.S. Laverack, 1968.
The responses of pedicellariae from Echinus esculentus (L.).
J. exp. mar. Biol. Ecol., 2(3):191-214.

echinoderms, lists of spp.

Santiago Lima-Verde, José, 1969
Primeira contribuição ao inventário dos equinodermas do nordeste brasileiro.
Arq. Ciên. Mar, Fortaleza, Ceará, Brasil 9(1): 9-13.

echinoderms, lists of spp.

Sendô, Jirô and Seiichi Irimura, 1968.
Ophiuroidea collected from around the Ross Sea in 1964 with descriptions of a new species.
J. Tokyo Univ. Fish., 9 (2): 147-154.

echinoderms, lists of spp.

Tommasi, L.R., 1966.
Lista dos equinóides recentes do Brasil.
Contrcões Inst. oceanogr. S. Paulo, 11:1-50.

echinoderm, periodicity of reproduction

Schoener, Amy, 1968.
Evidence for reproductive periodicity in the deep sea.
Ecology, 49(1):81-87.

echinoderms, physiol

Krishnan, S. and S. Krishnaswamy, 1970.
Studies on the transport of sugars in the holothurian Holothuria scabra. Marine Biol., 5(4): 303-306.

Echinoderm population

Tokioka, Takasi, 1969.
On the stability of population composition in a fixed echinid colony on the rocky shore of Hatakezima Island. Publ. Seto mar. biol. Lab., 17(3): 187-191.

echinoderm(resources)

Fukuda, Tomio, 1971.
Progress report on the exploitation of the sea urchin Anthocidaris crassispina (A. Agassiz) in the eastern region of Okayama Prefecture, 1970. (In Japanese). Bull. Fish. Exp. Sta. Okayama Pref. (1970): 104-116.

echinoderms, thermal tolerance

Singletary, Robert L., 1971.
Thermal tolerance of ten shallow-water ophiuroids in Biscayne Bay, Florida. Bull. mar. Sci., Coral Gables, 21(4): 938-943.

Echinoids

Chesher, Richard H., 1970.
Evolution in the genus Meoma (Echinoidea: Spatangoida) and a description of a new species from Panama. Bull. mar. Sci., 20(3): 731-761.

echinoids

Pearse, J.S., 1970.
Reproductive periodicities of Indo-Pacific invertebrates in the Gulf of Suez. III. The echinoid Diadema setosum (Leske). Bull. mar. Sci., 20(3): 697-720.

echiurids

Amor, Analía, 1965.
Una nueva localidad para Pinuca chilensis (Max Müller) en el Atlántico Sur (Echiurida). Aclaración sobre su sinonimia: Pinucidae Pinucidae Nom. Nov. para Urechidae Fisher & MacGinitie. Physis, B. Aires, 25(69):165-168.

Also in:
Centro Invest. Biol. Mar., Estación Puerto Deseado y Estación Austral, Contrib. Cient., No. 13. 1965.

echiuroids

Knox, G.A., 1957.
Urechis novae-zealandiae (Dendy): a New Zealand echiuroid.
Trans. R. Soc., New Zealand, 85(1):141-148.

Also in:
Collected Mar. Reprints., Edward Percival Mar Lab., 1951-1964.

echiuroids

Stephen, A.C., 1967.
Sipunculiens et echiuriens de Nouvelle-Calédonie.
Cah. Pacif., (10):44-50.

echiuroids

Zenkevitch, L.A., 1966.
Shallow water Echiuroidea from the Galathea Expedition.
Vidensk. Meddr dansk naturh. Foren. 129:275-277.

echiuroids

Zenkevitch, L.A., 1966.
The systematics and distribution of abyssal and hadal (ultra-abyssal) Echiuroidea.
Galathea Rep., 8:175-183.

echiurids

Zenkevitch, L.A., 1964.
New representatives of deep water echiurids (Alomasoma belyaevi Zenk. sp. n. and Choanostoma filatovae sp. n.) in the Pacific. (In Russian).
Zool. Zhurn., Akad. Nauk, SSSR, 43(12):1863-1864.

echoes, hyperbolic

Laughton, A.S., 1962
Discrete hyperbolic echoes from an otherwise smooth deep-sea floor.
Deep-Sea Res., 9(3):218.

echogram

Bezroukov, P.L., 1961
[Investigation of sediments in the northern Indian Ocean.] (In Russian; English abstract).
Okeanolog. Issled., Mezhd. Komitet Proved. Mezhd. Geofiz. Goda, Prezidiume, Akad. Nauk, SSSR, (4):76-90.

echograms

Dietrich, G., 1957.
Schichtung und Zirkulation der Irminger See in Juni 1955. Ber. Deutsch. Wiss. Komm. Meeresf., n.f., 14(4):255-312.

echograms

Udintsev, G.B., 1956.
[On the interpretation of echograms.]
Trudy Inst. Oceanol., 19:169-194.

echolocation

Altes, Richard A., 1971.
Computer derivation of some dolphin echolocation signals. Science, 173(4000): 912-914.

ECHOLOCATION

Belkovich, V.M., V.I. Borisov, V.S. Gurevich and N.L. Krushinskaya, 1962.
The ability of echolocation in Delphinus delphis. (In Russian; English abstract).
Zool. Zh., 41(6): 876-884

echolocation

Busnel, René-Guy, Albin Dziedzic et Sören Andersen, 1963
Sur certaines caractéristiques des signaux acoustiques du Marsouin, Phocoena phocoena L.
C.R. Acad. Sci., Paris, 257(17):2545-2548.

echolocation

McBride, A.F., (1947)1956.
Evidence for echolocation by cetaceans. Ltr. to the Editors. Deep-Sea Res. 3(2):153-154.

echo location

Norris, Kenneth S. 1969.
Echolocation of marine mammals.
In: The biology of marine mammals.
H.T. Andersen, editor, Academic Press, 391-423.

echolocation

Norris, K.S., 1964.
Some problems of echolocation in cetaceans.
In: Marine bio-acoustics, W.N. Tavolga, Editor, Macmillan Co., 317-336.

echolocation

Norris, Kenneth S., and William E. Evans, 1967.
Directionality of echolocation clicks in the rough-tooth porpoise Steno bredanensis (Lesson).
In: Marine bio-acoustics, W.N. Tavolga, editor, Pergamon Press, 2:305-314.

echo-location

Norris, K.S., J.H. Prescott, P.V. Asa-Dorian, and P. Perkins, 1961.
An experimental demonstration of echo-location behavior in the porpoise, Tursiops truncatus (Montagu).
Biol. Bull., 120(2):163-176.

echolocation

Pye, J.D., 1961.
Echolocation in bats.
Endeavour, 20(78):101-111.

echo-location

Romanenko, E.V., 1964.
The underwater echo-location (Sonar) capability of dolphins (review). (In Russian).
Akust. Zhurn., Akad. Nauk, SSSR, 10(4):385-397.

Translation:
Soviet Physics Acoustics, Amer. Inst. Phys., 10(4):331-342. 1965.

echolocation
Schevill, W.E., and B. Lawrence 1956
Food finding by a captive porpoise (Tursiops truncatus).
Breviora 53: 15pp.

Echo-ranging

echo ranging
Cushing, D.H., Finn Devold, J.C. Marr and H. Kristjonsson, 1952.
Some modern methods of fish detection:echo sounding, echo ranging and aerial scouting.
F.A.O. Fish. Bull. 5(3/4):95-119, 9 textfigs.

echo ranging
Gabler, H., 1950.
Quantitative Untersuchungen über Echolotungen mit Magnetostriktionsschwungern. Deutsch. Hydro. Zeits. 3(5/6):341-353, 12 textfigs.

echoranging
Griffiths, J.W.R., and A.W. Pryor, 1958
Underwater acoustic echoranging.
Electronic & Radio Engineering, Jan.:

echo-ranging
Vigoureux, P., and J.B. Hersey, 1962
Sound in the sea. Ch. 12, Sect. IV. Transmission of energy within the sea. In: The Sea, Vol. 1, Physical Oceanography, Interscience Publishers, 476-497.

"echo-ranging"
Japan, Maritime Safety Board, 1954.
Discrimination of bottom characteristics by listening sounds. Publ. 981, Hydrogr. Bull., Maritime Safety Bd., Tokyo, No.42:139-141, 7 textfigs.

echo-ranging
Kellogg, W.N., 1958.
Echo ranging in the porpoise.
Science, 128(3330):982-988.

echo-ranging
Smith, O.R. and E.H. Ahlstrom, 1948
Echo-ranging for fish schools and observations on temperature and plankton in waters off Central California in the spring of 1946. U.S. Fish and Wildlife Ser., Spec. Sci. Rept. No.44, 30 pp. plus 13 figs.

echo detection
Sund, O., 1941.
The fat and small herring on the coast of Norway in 1940. Ann. Biol. 1:58-74.

echo-ranging
Weston, D.E., 1965.
Correlation loss in echoranging.
J. Acoust. Soc., Amer., 37(1):119-124.

Echo sounder, echo sounder corrections, see INSTRUMENTATION

echosounding

echo - sounding
Agarate, Christian, 1965.
Etude de la sedimentation sous-marine superficielle par sondage sonore. C.R. hebd. seanc., Acad. Sci., Paris (D), 262(13):1427-1430.

echosounding
Ahrens, E., 1955.
Horizontallotung im Fischfang.
Arch. Fischereiwiss. 6(3/4):229-240.

echo sounding
Andreeva, I.B., 1965.
Acoustic properties of sound scattering layers in the ocean and data of echo-sounding and direct catch observations. (In Russian).
Okeanologiia, Akad. Nauk, SSSR, 5(6):1028-1037.

echosounding
Anon., 1957.
Echosounding corrections. Int. Hydrogr. Rev., 34(2):35-39.

echosounding
Anon., 1953.
Echo sounding and the fisherman. Fishing News 2126:7

echo-sounding
Balls, R., 1951.
Environmental changes in herring behaviour: a theory of light avoidance, as suggested by echo-sounding observations in the North Sea.
J. du Cons. 17(2):274-298, 4 textfigs.

echo-sounding, fish detection
Balls, R., 1948.
Herring fishing with the echometer. J. du Cons. 15(2):193-206.

echograms
Bezrukov, P.L., 1961
Investigation of sediments in the northern Indian Ocean.
Okeanol. Issledov., Mezhd. Komit., Mezhd. Geofiz. God, Presidium Akad. Nauk, SSSR: 76-90.

echosounding (fish)
Boer, P.A., 1954.
Het Asdic-echolood. Visserij-Nieuws 7(3):36.

echo-sounding (fish)
Boer, P.A., 1955.
Een Asdic-echolood voor visserij is gereed gekomen. Visserij-Nieuws 7(9):100.

echo sounding
Bogorov, B.G., S.M. Brujewicz, M.V. Fedosov, and G.B. Udintzev, 1961
Methods of investigation of oceans in the USSR.
Ann. Int. Geophys. Year, 11:311-316.

echo sounding
Bunce, Elizabeth T., 1966.
The Puerto Rico Trench.
In: Continental margins and island arcs, W.H. Poole, editor, Geol. Surv. Pap., Can., 66-15:165-175.

echo sounding
Carpart, A., 1955.
Quelques échosondages des fonds et des poissons aux environs de Monaco.
Bull. Inst. Océan., Monaco, No. 1068:1-10.

echo-sounding
Chesterman, W. D., P. R. Clynick and A. H. Stride, 1958.
An acoustic aid to sea bed survey.
Acustica, 8:285-290.

echo sounding
Clay, C.S., John Ess and Irving Weisman, 1964
Lateral echo sounding of the ocean bottom on the continental rise.
J. Geophys. Res., 69(18):3823-3835.

echo sounding
Cohen, Philip M., 1959.
Directional echo sounding on hydrographic surveys.
Int. Hydrogr. Rev., 36(1):29-42.

echosounding
Craig, R.E., 1953.
The future of echo detection. World Fishing 2(8):303-307.

echo sounding
Crease, J., A.S. Laughton and J.C. Swallow, 1964.
The significance of precision echo sounding in the deep oceans.
Int. Hydrogr. Rev., 41(2):63-72.

echo sounding
Crombie, D.D., 1955.
Doppler spectrum of sea echo at 13.56 Mc/s.
Nature 175(4459):681-682.

echosounding
Cushing, D.H., 1955.
Production and a pelagic fishery.
Fish. Invest., Min. Agric., Fish., & Food, (2), 18(7):1-104.

echo sounding
Cushing, D.H., 1955.
Some echo-sounding experiments on fish.
J. du Cons. 20(3):266-275, 6 textfigs.

echo-sounding
Cushing, D.H., 1952.
Echo-surveys of fish. J. du Cons. 18:45-60, 11 textfigs.

echo sounding
Cushing, D.H., Finn Devold, J.C. Marr, and H. Kristjonsson, 1952.
Some modern methods of fish detection:echo sounding, echo ranging, and aerial scouting.
F.A.O. Fish. Bull. 5(3/4):95-119, 9 textfigs.

echo-sounding
Cushing, D.H., and I.D. Richardson, 1955.
Echo sounding experiments on fish.
Min. Agric. Fish., Fish. Invest. (2)18(4):1-34.

echo sounding
Eaton, R.M., 1963
Airborne hydrographic surveys in the Canadian Arctic.
Int. Hydr. Rev., 40(2):45-51.

echo-sounding
Ennis, C.C., 1928.
Use of regional constant correction factors for reduction of echo-soundings. Proc. Third Pan-Pacific Sci. Congr., Tokyo, 1926(1):202-204, 2 textfigs.

echosounding

Evans, S., 1967.
Progress report on radio echo sounding.
Polar Rec., 13(85):413-420.

echo-sounding

Fisher, R.L., 1954.
On the sounding of trenches. Deep-Sea Res. 2(1): 48-58, 3 figs., 2 pls.

Echo-sounding

Furnestin, J., 1953.
Ultra-sons et pêches à la sardine au Maroc. Les essais du bateau-pilote-de-pêche, "Jean François".
Inst. Pêche Marit., Maroc. Bull. No. 1:1-57, 45 textfigs.

echo sounding, limitations of

Gabler, H.M., 1961
Limits of accuracy of echo soundings in ocean regions.
Int. Hydrogr. Rev., 38(2): 7-24.
shortened version of report appearing in Deutsche Hydrogr. Zeits., 1959(6): 229-243.

echo sounding, limitations of

Gabler, Heinz, 1959
Uber die Grenzen der Genauigkeit bei Echolotungen in ozeanischen Seeräumen. Deutsche Hydrogr. Zeits. 12(6): 229-243.

echo sounding

Gabler, H.M., 1951.
On the efficiency of acoustic waves in echo sounding of fish shoals. Int. Hydro. Rev. 28(1): 122-131, 9 textfigs.

echo-sounding

Gankov, A.A., and O.N. Kiselyov, 1962.
Use of the echo-sounder in the fishing under ice. (In Russian).
Nauchno-Technich. Biull., PINRO, Murmansk, No. 4(22):33-36.

echo-sounding

Goncharov, V.P., and O.V. Mikhailov, 1966.
On depth corrections for sound velocity change in water while echo-sounding in the Black and Mediterranean seas. (In Russian; English abstract
Okeanologiia, Akad. Nauk, SSSR, 6(4):707-71 .

echo-sounding

Greve, Sv., 1938.
Echo-soundings an analysis of results.
Dana Rept., No. 14:25 pp., 4 figs., 19 pls.

echosounding

Harris, V.E., 1954.
Specialist explains electro-fishing. Fish. Newsletter 13(1):8-9.

echo sounding

Hashimoto, Tomiju, Yoshinobu Maniwa, Osamu Omoto and Hidekuni Noda, 1963
Echo sounding of frozen lake from surface of ice.
Int. Hydrog. Rev., 40(2):31-40.

echo-sounding

Hashimoto Tomiju, and Minoru Nishimura, 1959.
Reliability of bottom topography obtained by ultrasonic echo-sounding.
Int. Hydrogr. Rev., 36(1):43-50.

echo sounding

Herdman, H.F.P., 1955.
Directional echo sounding. Deep-Sea Res. 2(4): 264-268, 6 textfigs.

echo-sounding

Hodgson, W.C., 1950.
Echo-sounding and the pelagic fisheries. Fish. Inv., Min. Agric. Fish., Gt. Brit., ser. 2, 17(4):24 pp., 24 figs.

echosounding

Hoffman, J., 1957.
Hyperbolic curves applied to echosounding.
Int. Hydrogr. Rev., 34(2):45-55.

echo sounding

Hughes, A. J., 1950.
Influence of echo sounding on hydrography. Hydr. Rev. 27(1):29-39.

echosounding

Hughes, A.J., 1949.
The influence of echo sounding. J. Inst. Navig. 2(3):243-257, 5 figs.

echosounding

Ishida, M., T. Suzuki and N. Sano, 1955.
An experimental study for estimating the approximate amount of fish shoals by echo sounding.
Bull. Fac. Fish., Hokkaido Univ., 5(4):362-367, 8 textfigs.

echo sounding

Iudanov, K. I., 1965.
Peculiarity of echosound records with a sudden drop in depth. (in Russian).
Rybnoe Khoz., (10):35-38.

echo sounding

Jagodzinski, Zenon, 1960
Multiple echoes in echo sounders and the probability of detection of small targets. Int. Hydrogr. Rev., 37(1): 63-68.

echo-location

Jonsson, J., and U. Stefansson, 1955.
Herring investigations with the Research Vessel "Aegir" in the summer 1954.
Fjölrit Fiskideildar, No. 5:1-34, 14 figs.
(In Icelandic: English summary)

echosounding

Kiilerich, A., 1957.
Galathea-Ekspeditionens arbejde i Phillipinergraven. Ymer, 1957(3):200-222.

echo sounding

Knott, S.T., E.T. Bunce and R.L. Chase, 1966.
Red Sea seismic reflection studies.
In: The world rift system, T.N. Irvine, editor, Dept. Mines Techn. Surveys, Geol. Survey, Can., Paper, 66-14:33-61.

echo-sounding

Knott, S.T., and J.B. Hersey, 1956.
Interpretation of hig-resolution echo-sounding techniques and their use in bathymetry, marine geophysics and biology. Deep-Sea Res., 4(1):36-44.

echo sounding

*Kranck, Kate, 1967.
Bedrock and sediments of Kouchibouguac Bay, New Brunswick.
J. Fish. Res. Bd., Can., 24(11):2241-2265.

echosounding

Kunze, W., 1957.
General aspects of application of horizontal echosounding method to shipping.
Int. Hydrogr. Rev., 34(2):63-72.

echo-sounding

Kuwahara, S., 1938
Correction to the echo-depth for the density of water in the Pacific Ocean.
Japan. J. Astron. & Geophys., 14(2/3): 43-78.

echosounding (data)

Larson, N., 1956.
Narrative of the expedition. Scientific results of the "Brategg" Expedition, 1947-48, No. 1.
Publikasjon, Komm. Chr. Christensens Hvalfangstmus., Sandefjord, No. 17:10-18.

effluent

Lawrance, C.H., 1958
Sewage effluent dilution in sea water.
Water & Sewage Works, 105: 116-122.
Abs. in Pub. Health Eng. Abs., 39(7): 30

echosounding

Lawrence, E., jr., 1951.
The use of echo sounders in fisheries. Proc. Gulf and Caribbean Fish. Inst., 3rd session:88-95.

echo sounding

*Leenhardt, Olivier, 1969.
Developments in mud probing.
Int. hydrogr. Rev. 46(2):99-114.

echo sounding, detection

Le Gall, J., 1952.
La detection des bancs de poissons. Rapp. Proc. Verb., Cons. Perm. Int. Expl. Mer, 132: 65-71, 4 textfigs.

echo-sounding

Lisitzin, A.P., and A.V. Zhivago, 1958.
VII. Submarine geology.
Opisanie Exped. D/E "Ob" 1955-1956, Mezhd. Geofiz. God, Trudy Kompaeksnoi Antarkt. Exped., Akad. Nauk, SSSR:103-144.

echo sounding

Maul George A., 1970.
Precise echosounding in deep water.
Int. Hydrogr. Rev. 47(2): 93-106.

echo-sounding

McCartney, B.S., and B. McK. Bary, 1965.
Echo-sounding on probable gas bubbles from the bottom of Saanich Inlet, British Columbia.
Deep-Sea Research, 12(3):285-294.

echo sounding (errors in)

Miyajima, Jiro, 1962
Analysis of errors of echosounder and of topographical analogy between recorded figures and natural features. (In Japanese; English abstract).
Hydrogr. Bull., (Publ. No. 981), Tokyo, No. 72:67-80.

echo sounding
Potts, C.B., 1949.
Precision echosounding in hydrography. Canad. Surveyor 10(2):1-10.

echosounding
Ridgway, N.M., 1962.
Echo sounding through sea ice.
Polar Record, 11(72):298-299.

echo-sounding
Rust. H.H., and H. Drubba, 1953.
Practische Anwendung des Unterwasserfunkens als Impuls-Schallgeber für die Echolotung.
Zeits. Angew. Phys. 5:251-252.

echosounding
Schuler, 1957.
Echogram profiles and how they can be established
Int. Hydrogr. Rev., 34(2):41-44.

echo-sounding
Schüler, F., 1952.
Deutung von Echogrammen bei Seitenechos.
Deutschen Hydrogr. Zeits. 5(2/3):144-147, 4 textfigs.

echo sounding
Schüler, Fre, 1952.
On the accuracy of configuration of sea bottom profiles with high frequency echo sounders.
Hydrogr. Rev. 29(1):126-135, 27 figs.

Shortened and amended version of report in Deutsche Hydrogr. Zeits. 1951(3):97-105.

echo-sounding
Schüler, F., 1951.
Die Abbildungstreue von Meeresboden-Profilen in Echogrammen. Deutsche Hydrogr. Zeits. 4(3):97-105 17 textfigs.

echo-sounding
Schüler, F., 1951.
Über ein neues Anzeigverfahren für Echolote.
Deutsch. Hydro. Zeits. 4(1/2):52-57, 3 textfigs.

echosounding
Schumacher, A., 1958.
Die Lotungen der "Schwabenland".
Deutsch. Antark. Exped., 1938/39, Wiss. Ergeb., 2(2):41-62.

echo sounding
Sergeev, L.A., 1958.
Ultrasonic echo soundings for geophysical purposes. (In Russian).
Prikladnaia Geofizika, 20:141-154.

Translation:
U.S. Naval Oceanogr. Off., by M. Slessers, TRANS-115:20 pp.

echo-sounding
Smith, D. Taylor, and W.N. Li, 1966.
Echo-sounding and sea floor sediments.
Marine Geol., 4(5):353-364.

echo sounding
Stubbs, A.R., 1963
Identification of patterns on Asdic records.
Int. Hydrogr. Rev., 40(2):53-63.

echo-sounding
Suau, P., and M.G. Larreneta, 1961.
Influencia de los ecosondas en una pesquería de sardina.
Inv. Pesq., Barcelona, 20:73-78.

echo location of cetaceans
Tomilin, A.G., 1955.
[On the behavior and sonic signalling of cetaceans]
Trudy Inst. Oceanol., 18:28-47.

echo sounding
Trout, G.C., A.J. Lee, I.D. Richardson, and F.R. Harden Jones, 1952.
Recent echo sounder studies. Nature 170:71-72.

echo sounding
Tucker, M.J., 1961
Beam identification in multiple-beam echo-sounders.
Int. Hydrogr. Rev., 38(2): 25-32.

echo-sounding
Tungate, D.S., 1958.
Echo-sounder surveys in the autumn of 1956.
Ministry Agric., Fish. & Food, Fish. Inv., (2), 22(2):17 pp.

echo-sounding
Udintsev, G.B., 1962.
Studies of the submarine relief. (In Russian; English abstract).
Mezhd. Geofiz. Komitet, Presidiume, Akad. Nauk, SSSR, Rezult. Issled. Programme Mezhd. Geofiz. Goda, Okeanol. Issled., No. 7:33-48.

echo-sounding
Udintsev, G.B., 1956.
[On deciphering echograms.] Trudy Inst. Okeanol., 19:169-194.

echo sounding
Udintsev, G.B., 1951.
[On the method of echo-sounder records in marine geological investigations.] Trudy Inst. Okeanol., 5:17-34.

echosounding
Ulrich, J., 1962
Echolotprofile der Forschungsfahrten von F.F.S Anton Dohrn und V.F.S Gauss im Internationalen Geophysikalischen Jahr, 1957/1958.
Deutsch. Hydrogr. Zeits., Erganzungsheft, 4(6): 15 pp.

echo sounding
Van Brandt, A.V., and J. Schärfe, 1950.
Zur Auswertung der Echolotungen. Fischereiwelt 2(11):164-166, 12 figs.

1951 (in Dutch) in Visscherij-Nieuws 3(11):128-9 figs., 3(12):140-144, 7 figs.

echosounding
Van Drimmelin, D.E.,
Electrovisserij. Visserij-Niews 6(2):18-20.

echo-sounding
Vigoureux, P., and J.B. Hersey, 1962
Sound in the sea. Ch. 12, Sect. IV. Transmission of energy within the sea. In: The Sea, Vol. 1, Physical Oceanography, Interscience Publishers, 476-497.

echo-sounding
Wood, H., and B.B. Parrish, 1950.
Echo-sounding experiments on fishing gear in action. J. du Cons. 17(1):25-36, 7 textfigs.

echo sounding
Zhivago, A.V., 1962
[Investigation methods of floor topography in the Antarctic.]
Mezhd. Geofiz. Komitet, Prezidiume Akad. Nauk SSSR, Rezult. Issled. Programme Mezhd. Geofiz Goda, Okeanol., Issled., No. 5:60-66.

eclipse, effect of

eclipse, effect of
Schüler, F., 1954.
Über die echographische Aufzeichnung des Verhaltens von Meeresfischen während der Sonnenfinsternis von 30 juni 1954. Deutsche Hydrogr. Zeits. 7(3/4):140-143, 2 textfigs.

ecology

ecology
Barnes, H. 1967.
Ecology and experimental biology.
Helgoländer wiss. Meeresunters. 15(1/4): 6-26.

ecology (marine)
Băcescu, M., redactor responsabil, 1965.
Ecologie marină.
Edit. Acad. Republ. Pop. Române, Bucuresti, 1: 344 pp.; 2:293 pp.

ecology
Battaglia Bruno, 1967.
Genetic aspects of benthic ecology in brackish waters.
In: Estuaries, G.H. Lauff, editor, Publs Am. Ass. Advmt. Sci., 83:574-577.

ecology
Herrington, W.C., 1947
The role of intraspecific competition and other factors in determining the population level of a major marine species. Ecol. Mon. 17(3):317-323, 5 textfigs.

ecology
*Hoppe, Brigitte, 1968.
Influence de la biologie marine sur l'évolution de la pensée écologique au XIX Siècle.
Bull. Inst. océanogr., Monaco, No. spécial 2: 407-416.

ecology
Hutchins, L.W., 1947
The basis for temperature zonation in Geographical distribution. Ecol. Mon. 17(3):325-335, 8 textfigs.

ecology
Ketchum, B.H., 1947
Biochemical relations between marine organisms and their environment. Ecol. Mon. 17(3):309-315, 5 textfigs.

ecology
Ladd, H.S., Chairman, 1946
Report of the Committee on Marine Ecology as related to Palaeontology, 1945-1946, 101 pp.

ecosystems

Margalef, Ramon, 1967
Fluctiones en abundancia y asequibilidad causadas por factores bioticos.
Methodological Paper, No. 1:1265-1285.

ecology

Nelson, T.C., 1947
Some contributions from the land in determining conditions of life in the sea. Ecol. Mon. 17(3):337-346, 7 textfigs.

ecology
Arctic

Thorson, G., 1944.
Technique and future work in/animal ecology. App. 4. De Danske Expeditioner til Øst Grønland 1926-1939. Medd. om Grønland 144(4):40 pp., 9 textfigs.

ecological barriers

*Khlebovich, V.V., 1968.
Some peculiar features of the hydrochemical regime and the fauna of mesohaline waters.
Marine Biol., 2(1):47-49.

economics

economics

Moore J. Jamison, 1970.
The ocean- an economic perspective.
J. mar. techn. Soc., 4(6):33-37.

economic considerations

U.S.A., University of California, Institute of Marine Resources, La Jolla, 1965.
California and use of the ocean: a planning study of marine resources.
IMR Ref. 65-21: app. 350 pp. (multilithed, paged by chapters)

ecosystems

ecosystems

Aron, William I. and Stanford H. Smith, 1971.
Ship canals and aquatic ecosystems. Science 174(4004): 13-30.

ecosystem

Copeland, B.J., 1965.
Evidence for regulation of community metabolism in a marine ecosystem.
Ecology, 46(4):563-564.

ecosystems

Dunbar, M.J. 1968.
Ecological development in polar regions: a study in evolution.
Prentice-Hall, Inc. 119 pp. $4.95.

ecosystems

Ginés, Hno., y R. Margelef, editores, 1967.
Ecología marina.
Fundación La Salle de Ciencias Naturales, Caracas
Monografía 14: 711 pp.

ecosystem, pelagic

Margalef, Ramon, 1969
El ecosistema pelagico del mar Caribe.
Memoria, Soc. Cienc. nat. La Salle, 29(82):5-36

ecosystems

Margalef, Ramón, 1967.
El ecosistema.
In: Ecología marina, Monogr. Fundación La Salle de Ciencias Naturales, Caracas, 14:377-453.

ecosystems

Margalef, D. Ramon, 1963.
El ecosistema pelagico de un area costera del Mediterraneo occidental.
Memorias, Real Acad., Ciencias y Artes de Barcelona, 35(1):3-48.

ecosystems

Margalef, R., 1953.
On certain unifying principles in ecology.
American Naturalist, 97(897):357-374.

ecosystems

Margalef, Ramon, 1962
Comunidades naturales.
Publicacion Especial, Inst. Biol. Marina, Univ. de Puerto Rico, Mayaquez, 469 pp.

ecosystems

Odum, Howard T., Walter L. Silver, Robert J. Byers and Neal Armstrong, 1963.
Experiments with engineering of marine ecosystems
Publ. Inst. Mar. Sci., Port Aransas, 9:373-403.

ecosystems

*Olivier, Santiago R. Ricardo Bastida y Maria Rosa Torti, 1968.
Sobre el ecosistema de las aguas litorales de Mar del Plata: niveles troficos y cadenas alimentarias pelagico - demersales y bentonico-demersales.
Publ. Serv. Hidrograf. Naval, Argentina, H 1025: 45 pp.

ecosystems

*Olivier, Santiago Raul. Ricardo Bastida y Maria Rosa Torti, 1968.
Resultados de las campañas oceanográficas, Mar del Plata I-V. Contribución al trazado de una carta bionómica del área de Mar del Plata. Las asociaciones del sistema litoral entre 12 y 70 m de profundidad.
Bol., Inst. Biol. mar, Mar del Plata, 16:85 pp.

ecosystems

Winberg, G.G., 1962. (sometimes spelled Vinberg)
Energetic principle in investigation of trophic relations and productivity of ecological systems. (In Russian; English abstract).
Zool. Zhurn., Akad. Nauk, SSSR, 41(11):1618-1630.

ectocrines

ectocrines

Fogg, G.E. and W.D. Watt, 1965.
The kinetics of release of extracellular products of photosynthesis by phytoplankton.
Mem. Ist. Ital. Idrobiol., 18(Suppl.):165-174.

ectocrines

Johnston, R., 1955.
Biologicall active compounds in the sea.
J.M.B.A. 34(2):185-195.

ectocrines

Lucas, C.E., 1955.
External metabolites in the sea.
Pap. Mar. Biol. and Oceanogr., Deep-Sea Res., Suppl. to Vol. 3:139-148.

ectocrenes

Provasoli, L., 1963
8. Organic regulation of phytoplankton fertility. In: The Sea, M.N. Hill, Edit., Vol. 2. (II) Fertility of the oceans, Interscience Publishers, New York and London, 165-219.

eddies

eddies

Chen, T.C., 1961.
Experimental study on the solitary wave reflection along a straight sloped wall at oblique angle of incidence.
Beach Erosion Bd., Tech. Mem., No. 124:24 pp.

eddies

Dobrovol'skii, A.D., and V.S. Arsenev, 1959 (1961)
The Bering Sea currents.
Problemy Severa, 3:3-9.

Translation in:
Problems of the North, 3:1-7.

eddies

*Duncan, C.P., 1968.
An eddy in the subtropical convergence southwest of South Africa.
J. geophys. Res., 73(2)531-534.

eddies

Fuglister, F.C., and L.V. Worthington, 1951.
Some results of a multiple ship survey of the Gulf Stream. Tellus 3(1):1-14, 11 textfigs.

eddies

Grant, H.L., 1958.
The large eddies of turbulent motion.
J. Fluid. Mech., 4(2):149-190.

eddies

Hamon, B.V. 1969
Oceanic eddies
Rept. CSIRO Div. Fish. Oceanogr., Austr., 46:13-21

eddies

Hata Katsumi, 1969.
Some problems relating to fluctuation of hydrographic conditions in the sea northeast of Japan. 2. Fluctuation of the warm eddy cut off northward from the Kuroshio.
Oceanogrl Mag. 21(1): 13-29.

eddies

*Health, R.A., 1968.
Geostrophic currents derived from oceanic density measurements north and south of the subtropical convergence east of New Zealand.
N.Z. Jl. mar. Freshwat. Res., 2(4):659-677.

eddies

Iselin, C. O'D., and F. C. Fuglister, 1948.
Some recent developments in the study of the Gulf Stream. J. Mar. Res. 7(3):317-329, 6 textfigs.

eddies

Iwata, K., 1955.
Decay of the large horizontal eddies in the ocean.
Rec. Ocean. Wks., Japan, 2(1):23-28, 1 textfig.

eddies
Kitano, K., 1956.
Note on the eddies generated around the tubulent jet flow. J. Ocean. Soc., Japan, 12(4):121-124.

eddies
McEwen, G. F., 1948.
The dynamics of large horizontal eddies (axes vertical) in the ocean off southern California. J. Mar. Res., 7(3):188-216, Textfigs. 1-33.

eddies
Moriyasu, Shigeo, 1961
An example of the conditions at the occurrence of the cold water region.
Oceanogr. Mag., Tokyo, 12(2): 67-76.
Also in:
Bull. Kobe Mar. Obs., No. 167(20):

eddies
Nakano, M., 1957.
On the eddies of Naruto Strait. Pap. Met. & Geophys., 7(4):415-424.

eddies
Nakano, M., 1957.
On the eddies of Naruto Strait. Pap. Meteorol. & Geophys., Tokyo, 7(4):425-434.

eddies
Nowlin, W.D., Jr., J.M. Hubertz and R.O. Reid 1968.
A detached eddy in the Gulf of Mexico.
J. mar. Res. 26(2): 185-186

eddies
Parker, Charles E., 1971.
Gulf Stream rings in the Sargasso Sea. Deep-Sea Res. 18(10): 981-993.

eddies (large-scale)
Phillips, N., 1966.
Large-scale eddy motion in the western Atlantic. J. geophys. Res., 71(6):3883-3891.

eddies
Reid, Joseph L., Jr., Richard A. Schwartzlose, Daniel M. Brown, 1963
Direct measurements of a small surface eddy off northern Baja California.
J. Mar. Res., 21(3):205-218.

eddies
Rouse, Hunter, 1963.
On the role of eddies in fluid motion.
American Scientist, 51(3):285-314.

eddies
Ryzhkov Yu. G. 1966.
Eddies with horizontal axes above the edge of the continental shelf of South America. (In Russian)
Fizika Atmosfer. Okean. Izv. Akad. Nauk, SSSR 2(1): 88-91.

eddies
Ryzhkov, Yu. G., and L.A. Koveshnikov, 1963.
Genesis of convergence and divergence zones above sharp changes in the slope of the ocean floor. (In Russian).
Izv., Akad. Nauk, SSSR, Ser. Geofiz., (6):953-959
Translation (AGU): (6):585-588.

eddies
Saelen, Odd H., 1963
Studies in the Norwegian Atlantic Current. II. Investigations during the years 1954-1959 in an area west of Stad.
Geofysiske Publikasjoner, 23(6):1-82.

eddies
Sargent, M. S. and T. J. Walker, 1948
Diatom populations associated with eddies off southern California in 1941. J. Mar. Res. 7(3):490-505, 15 text figs.

eddies, deep
*Swallow, J.C., 1969.
A deep eddy off Cape St. Vincent.
Deep-Sea Res., Suppl. 16: 285-295.

eddies
Szekielda, Karl-Heinz 1971.
Anticyclonic and cyclonic eddies near the Somali coast.
Dt. hydrogr. Z. 24(1): ~~28~~ 26-29

eddies, effect of
Uda, Michitaka, 1958.
Enrichment patterns resulting from eddy systems. Proc. Ninth Pacific Sci. Congr., Pacific Sci. Assoc., 1957, Oceanogr., 16:91-93.

eddies
*Warren, Bruce A., 1967.
Notes on translatory movement of rings of current with application to Gulf Stream eddies. Deep-Sea Res., 14(5):505-524.

eddies
Welsh J. G. and G.A. Visser, 1970
Hydrological observations in the south-east Atlantic Ocean. 2. The Cape Basin
Invest. Rep. Div. Sea Fish. S.Afr. 83: 24pp.

eddies, anticyclonic
Saunders, P.M., 1971.
Anticyclonic eddies formed from shoreward meanders of the Gulf Stream. Deep-Sea Res. 18(12): 1207-1219.

eddy conductivity

eddy conductivity
Gade, Herman G., 1963
Some hydrographic observations of the inner Oslofjord during 1959.
Hvalrådets Skrifter, No. 46:1-62.

eddy conductivity
Koizumi, M., 1955.
Researches on the variations of oceanographic conditions in the region of the Ocean Weather Station "Extra" in the North Pacific Ocean (1). "Normal" values and annual variations of oceanographic elements. Pap. Met. Geophys., Tokyo, 6(2):185-201.

eddy conductivity
Overstreet, Roy and Maurice Rattray, Jr., 1969.
On the roles of vertical velocity and eddy conductivity in maintaining a thermocline.
J. mar. Res., 27(2): 172-190.

eddy conductivity (vertical and horizontal)
Riley, Gordon A. 1967.
Transport and mixing processes in Long Island Sound.
Bull. Bingham oceanogr. Coll. 19(2): 35-61.

eddy conductivity (vertical)
Riley, G.A., 1956.
Oceanography of Long Island Sound, 1952-1954. 1. Introduction. 2. Physical Oceanography.
Bull. Bingham Oceanogr. Coll., 15:8-46.

eddy conductivity
Tabata, S., and G.L. Pickard, 1957.
The physical oceanography of Bute Inlet, British Columbia. J. Fish. Res. Bd., Canada, 14(4):487-520.

eddy conductivity (vertical)
Tsujita, T., 1963.
The fishery oceanography of the East China Sea and Tsushima Strait. 2. On the characteristics and fluctuations of the cold water masses in view point of the production oceanography.
Bull. Seikai Reg. Fish. Res. Lab., 28:57-68.

eddy conductivity
Watanabe, N., 1955.
Hydrographic conditions of the north-western Pacific. 1. On the temperature change in the upper layer in summer. J. Ocean. Soc., Japan, 11(3):111-122.
Also: Bull. Tokai Reg. Fish. Res. Lab., 11(B-224)

eddy correlation

eddy correlation
Deacon, E.L., and E.K. Webb, 1962
Small-scale interactions. Ch. 3, Sect. II. Interchange of properties between sea and air. In: The Sea, Interscience Publishers, Vol. 1, Physical Oceanography. pp. 43-87.

eddy diffusion

eddy diffusivity
Akagawa Masaomi, 1970
On the drift ice east of South Sakhalin in the Okhotsk Sea in June 1969 and some oceanographic phenomena related to the sea ice situation. (In Japanese; English abstract).
Umi to Sora 45(4): 133-144.

eddy diffusion
Ben-Yaakov, Sam, 1971.
A multivariable regression analysis of the vertical distribution of TCO_2 in the eastern Pacific. J. geophys. Res. 76(30): 7417-7431.

~~eddy diffusivity~~
Bodvarsson, G., J.W. Berg, Jr., and R.S. Mesecar 1967.
Vertical temperature gradient and eddy diffusivity above the ocean floor in an area west of the coast of Oregon.
J. geophys. Res., 72(10):2693-2694.

eddy diffusion

Bossolasco, M., and I. Dagnino, 1959
La diffusione delle acque Ioniche nel Tirreno attraverso lo Stretto di Messina.
Geofis. Pura e Appl., Milano, 44(3):168-178.

eddy diffusion (vertical)

Bowden, K.F., and S.H. Sharaf El Din, 1966.
Circulation and mixing processes in the Liverpool Bay area of the Irish Sea.
Geophys. J., R. Astr. Soc., 11(3):279-292.

eddy diffusion

Bowden, K.F., and S.H. Sharaf El Din, 1966.
Circulation, salinity and river discharge in the Mersey Estuary.
Geophys J., R. astr. Soc., 10(4):383-399.

eddy diffusion, vertical

Broecker, W.S., and T.-H. Peng, 1971.
The vertical distribution of radon in the Bomex area.
Earth planet. Sci. Letts 11(2): 99-108.

eddy diffusivity

Foxworthy, J.E., G.M. Barsom and R.B. Tibby, 1966.
On eddy diffusivity and the four-thirds law in highly stratified near-shore coastal water (Abstract only).
Trans. Am. Geophys. Un., 47(3):478.

eddy diffusion

Foxworthy, James E., Richard B. Tibby and George Barsom, 1965.
Multidimensional aspects of eddy diffusion determined by dye diffusion experiments in coastal waters. (summary).
Symposium, Diffusion in oceans and freshwaters, Lamont Geol. Obs., 31 Aug.-2 Sept., 1964, 71-73.

eddy diffusivity

Green, P., I. Claessen, A.J. Meerburg and L. Otto, 1971.
Measurements of deformation by turbulent motions in the sea and eddy diffusivities derived therefrom.
Netherlands J. Sea Res. 5(2): 267-274.

eddy diffusion

Hanzawa, M., 1953.
On the eddy diffusion of pumice ejected from Mijojin-Reef in the Southern Sea of Japan.
Ocean. Mag., Tokyo, 4(4):143-148, 2 textfigs.

eddy diffusion

Hanzawa, M., 1953.
On the eddy diffusion of pumices ejected from Myojin reef in the Southern Sea of Japan.
Rec. Ocean. Wks., Japan, n.s., 1(1):18-22, 2 text figs.

Eddy diffusivity

Hela, Ilmo, 1966.
Vertical eddy diffusivity of waters in the Baltic Sea.
Geophysica, Helsinki, 9(3):219-234.

eddy diffusivity (vertical)

Isaeva, L.S., and I. L. Isaev, 1963.
Determination of vertical eddy diffusivity in the upper layer of the Black Sea by a direct method. (In Russian).
Trudy Morsk. Gidrofiz. Inst., Akad. Nauk Ukrain., SSSR, 28:32-35.

Translation: Soviet Oceanogr., Trans. Mar. Phys. 28:22-24. (AGU - Issue No. 2 - 1965).

eddy diffusivity

Joseph, J., 1954.
Die Sinkstofführung von Gezeitenströmungen als Austauschproblem. Arch. Met., Geophys., Bioklim. A, 7:482-501, 12 textfigs.

eddy diffusion

Kirwan, A.D., Jr., 1965.
On the use of Rayleigh-Ritz method for calculating the eddy diffusion.
Symposium, Diffusion in oceans and fresh waters, Lamont Geol. Obs., 31 Aug.-2 Sept. 1964, 86-92.

eddy diffusion, vertical

Koczy, F.F., 1956.
Vertical eddy diffusion in deep water. Nature, 178(4533):585-586.

eddy diffusivity, vertical

Koczy, F., and B. Szabo, 1962
Renewal time of bottom water in the Pacific and Indian oceans.
J. Oceanogr. Soc., Japan, 20th Ann. Vol., 590-599.

eddy diffusion

Moon, F.A., Jr., C.L. Bretschneider and D.W. Hood, 1957.
A method for measuring eddy diffusion in coastal embayments. Publ. Inst. Mar. Sci., Port Aransas, 4(2):14-21.

eddy diffusivity

Nan'niti, T., 1960
Long-period fluctuations in the Kuroshio.
Pap. Meteorol. & Geophys., 11(2-4):339-347.

eddy, diffusivity

Okubo, A., and R.V. Ozmidov, 1970.
An empirical relationship between the horizontal eddy diffusivity in the ocean and the scale of the phenomenon. (In Russian)
Fizika Atmosfer. Okean., Izv. Akad. Nauk SSSR 6(5): 534-536

eddy diffusion

Parr, A. E., 1939
Analysis of evidence relating to eddy diffusion processes at mid-depth in the Eastern Caribbean. J. du Cons. 14 (3): 347-356, 3 text figs.

eddy diffusivity (vertical)

Pritchard, Donald W., 1967.
Observations of circulation in coastal plain estuaries.
In: Estuaries, G.H. Lauff, editor, Publs Am. Ass. Advmt Sci., 83:37-44.

eddy diffusion

Richardson, L.F. and H. Stommel, 1948
Note on eddy diffusion in the sea. J. Meteorol. 5(5):238-240.

eddy diffusion

Seiwell, H.R., 1938.
Eddy diffusion at mid-depths in the Caribbean Sea region. J. du Cons. 13(2):155-162, 2 textfigs.

eddy diffusion

Seiwell, H. R., 1938.
Use of non-conservative properties of sea water in physical oceanographical problems. Nature 142:164.

eddy diffusion

Sholkovitz, Edward R. and Joris M. Gieskes, 1971.
A physical-chemical study of the flushing of the Santa Barbara Basin. Limnol. Oceanogr. 16(3): 479-489.

eddy diffusion

Steele, J.H., 1957
The role of lateral eddy diffusion in the northern North Sea. J. du Cons., 22(2):152-162.

eddy diffusivity

Stommel, H., 1951.
Determination of the lateral eddy diffusivity in the climatological mean Gulf Stream. Tellus 3(1):43.

eddy diffusivity

Visser, M.P., 1966.
Note on the estimation of eddy diffusivity from salinity and current observations.
Netherlands J. Sea Res., 3(1):21-27.

eddy diffusivity (vertical)

Hela, Ilmo, 1965.
Vertical eddy diffusivity of waters in the Baltic Sea.
Geophysica, 9(3):219-234.

eddy fluxes

eddy fluxes

Hicks, B.B., and A.J. Dyer, 1970.
Measurements of eddy-fluxes over the sea from an off-shore oil rig.
Q. Jl R. Met. Soc. 96(409): 523-527

eddy flux (vertical)

Pritchard, Donald W., 1967.
Observations of circulation in coastal plain estuaries.
In: Estuaries, G.H. Lauff, editor, Publs Am. Ass. Advmt Sci., 83:37-44.

eddy fluxes

Webster, Ferris, 1965.
Measurements of eddy fluxes of momentum in the surface layer of the Gulf Stream.
Tellus, 17(2):239-245.

eddy pressure

eddy pressure

*Faller, Alan J., 1969.
The generation of Langmuir circulations by the eddy pressure of surface waves.
Limnol. Oceanogr. 14(4):504-513.

eddy velocities

Neumann, J., 1964.
On the use of moving averages of eddy velocities in turbulent diffusion.
In: Studies on Oceanography dedicated to Professor Hidaka in commemoration of his sixtieth birthday, 227-231.

eddy viscosity

eddy viscosity
Bowden, K.F., 1950.
The effect of eddy viscosity on ocean waves.
Phil. Mag., ser. 7, 41:907-917.

eddy viscosity
Corkan, R. H. 1950
The levels in the North Sea associated with the storm disturbance of 8 January 1949. Phil. Trans. Roy. Soc., London, Ser. A. 242 (853): 493-525, 10 textfigs.

eddy viscosity
Fjeldstad, J.E., 1955.
Eddy viscosity, current and wind.
Astrophysica Norvegica, 5(5):153-166.

eddy viscosity
Hidaka, Koji, and Takao Momoi, 1961
Determination of the vertical eddy viscosity in sea water from wind stresses and surface current velocities.
Rec. Oceanogr. Wks., Japan, 6(1):1-10.

eddy viscosity
Ichiye, T., 1957.
A note on the horizontal eddy viscosity in the Kuroshio. Rec. Oceanogr. Wks., Japan, n.s., 3(1):16-25.

eddy viscosity
Ichiye, Takashi, and Shigeo Moriyasu, 1951
[On the tidal currents in the Mihara Seto.]
Bull. Kobe Mar. Obs., No. 164:371 (top of page)-385; 77 (bottom of page)-91.

eddy viscosity
*Johns, B., 1969.
On the representation of the Reynolds stress in a tidal estuary.
Geophys.J.R.astr.Soc.17(1):39-44.

eddy viscosity, effect of
Nakamura, Kohei, 1961
Velocity of long gravity waves in the ocean.
Sci. Repts. Tohoku Univ., (5 Geophys.), 13 (3):164-173.

eddy viscosity
Olson, F.C.W., 1959.
The Ekman spiral for a linearly varying eddy viscosity. J. Oceanogr. Soc., Japan, 15(2): 49-51.

eddy viscosity
Ostapoff, Feodor, 1962.
On the frictionally induced transverse circulation of the Antarctic Circumpolar Current.
Deutsche Hydrogr. Zeits., 15(3):103-113.

eddy viscosity
Rimnitz, H.P., 1965.
Ein Differenzengleichungssystem zur Ermittlung instationärer Bewegung in einem Meer mit geringer Turbulenzreibung. Numerische Beiträge zur Meteoro-Hydrographie, III.
Deutsche Hydrogr. Zeits., 18(3):97-113.

eddy viscosity
Schulkin, M., 1963.
Eddy viscosity as a possible acoustic absorption mechanism in the ocean.
J. Acoustic. Soc., Amer., 35(2):253-254.

eddy viscosity
Wada, Akira, 1966.
Effect of winds on a two-layered bay.
Coast. Engng Japan, 9:137-156.

eddy viscosity, vertical
Wyrtki, Klaus, and Edward B. Bennett, 1963.
Vertical eddy viscosity in the Pacific Equatorial Undercurrent.
Deep-Sea Res., 10(4):449-455.

edge waves

edge waves
See: waves, edge

education

education
Charlier, Roger H., 1966.
Growth of oceanographic education in the United States.
Limnol. Oceanogr., 11(4):636-640.

education
Chave, Keith E., 1963.
Oceanography at an inland university.
J. Geol. Education, 11(1):10-16.

education
Kellogg-Smith, Ogden, 1968.
Biological oceanography, junior grade.
Bio Science, 18(10):975-976.

education
Munske, Richard E., Editor, 1964.
Grade-school oceanography.
Undersea Technology, 5(7):18-19.

education
Munskie, R.E., 1962.
New ideas in marine education.
Undersea Techn., 3(4):27.

education
Nierenberg, William A. 1968.
What's behind our manpower shortage.
Ocean Industry 3(5): 45-47.

education
Rakestraw, Norris W. 1968.
Training for oceanography should begin at the graduate level.
Ocean Industry 3(5): 47-48.

education
Ray, Dixie Ray, 1963.
Needs for research and education.
(Special Issue on Marine Biology), AIBS Bull., 13(5):41-44.

education
Redfield, A.C., et al., 1960
Education and recruitment of oceanographers in the United States. Limnol. & Oceanogr., Suppl. to Vol. 6: 1-xxiii.

education
United States of American, The Committee on Education and Recruitment, ASLO, 1960.
Education and recruitment of oceanographers in the United States.
Limnol. & Oceanogr., 6(Suppl.):1-22.

eel grass

eel grass
Anon., 1956.
Nova Scotia plant makes insulation from eel grass
Maine Coast Fish., 11(2):5.

eel grass (Zostera nana)
Arasaki, S., 1950.
[Studies on the ecology of Zostera Marina and Z. nana.] Bull. Japan. Soc. Sci. Fish. 16(2):70-78, 5 figs. (In Japanese)

eel grass
Arasaki, S., 1950.
[Studies on the ecology of Zostera marina and Z. nana.] Bull. Japan. Soc. Sci. Fish. 16(2): 70-86, 5 figs. (In Japanese).

eel grass (Zostera marina)
Arasaki, S., 1950.
[Studies on the ecology of Zostera marina and Zostera nana.] Bull. Japan. Soc. Sci. Fish. 15(10):567-572, 2 figs. (In Japanese).

eel grass
Armiger, Lois C., 1964.
An occurrence of Labyrinthula in New Zealand Zostera.
New Zealand J. Botany, 2(1):3-9.

(Zostera marina) eel grass
Atkins, W.R.G., 1947.
Disappearance of Zostera marina. Nature 159(4040):477.

Zostera, effect of
Azuma, M., S. Matsumura, H. Hattori and T. Fukuda, 1971.
Ecological studies on the significance of Zostera region for the biological production of fishes. (IV). Seasonal fluctuations of some environmental conditions and biota in the Zostera belt and surrounding regions in the coastal waters of the eastern part of Okayama Prefecture. (In Japanese). Bull. Fish. Exp. Sta. Okayama Pref. (1970): 194-195.

eel grass
den Hartog, Cornelis 1967.
The structural aspect in the ecology of sea grass communities.
Helgoländer wiss. Meeresunters 15(1/4): 648-658.

eel grass (Zostera)
Dexter, R. W., 1950.
Restoration of the Zostera faciation at Cape Ann, Massachusetts. Ecol. 31(2):286-288, 1 textfig.

eel grass Zostera
Harmelin, J.G., et R. Schlenz, 1964.
Contribution préliminaire à l'étude des peuplements du sédiment des herbiers de phanérogames marines de la Méditerranée.
Rec. Trav. Sta. Mar. d'Endoume, Bull., 31(47): 149-151.

(Zostera marina) eel grass
Kikuchi, Taiji, 1968.
Faunal list of the Zostera marina belt in Tomioke Bay, Amakusa, Kyushu.
Publ. Amakusa mar. biol. Lab., 1(2):163-192.

(Zostera marina) eel grass
Kikuchi, Taiji, 1966.
An ecological study on animal communities of the Zostera marina belt in Tomieksa Bay, Amakusa, Kyushu.
Publ. Amakusa Mar. biol. Lab., 1(1):1-106.

(Zostera) eel grass
Kitamori, Ryonosuke, Kizo Nagata and Shin-ichi Kobayashi, 1959.
The ecological study on "Moba" (Zone of Zostera marina, L.). II. Seasonal changes.
Bull. Nakai Reg. Fish. Res. Lab., Fish. Agency, No. 12:187-199.

(Zostera marina) eel grass
Ledoyer, Michel, 1964.
Les migrations nycthemerales de la faune vagile au sein des herbiers de Zostera marina de la zone intertidale en Manche et comparaison avec les migrations en Méditerranée.
Rec. Trav. Sta. Mar., Endoume. 34(50):241-247.

eel grass
Le Gall, Jean-Yves, 1969.
Étude de l'endofaune des pelouses de Zostéracées superficielles de la Baie de Castiglione (Côtes d'Algérie).
Téthys 1(2): 395-420.

(Zostera) eel grass
Masamune, Genkei, 1965.
Icones plantarum marinarum notoensis. II. Zostera nana. (In Japanese; English abstract).
Ann. Rept., Noto Mar. Lab., Univ. Kanazawa, 5:1-2

eelgrass
McRoy, C. Peter, 1970.
Standing stocks and other features of eelgrass (Zostera marina) populations on the coast of Alaska.
J. Fish. Res. Bd., Can. 27(10): 1511-1821

eelgrass
*McRoy, C. Peter, 1968.
The distribution and biogeography of Zostera Marina (eelgrass) in Alaska.
Pacif. Sci., 22(4):507-513.

eel grass
*McRoy, C. Peter and Robert J. Barsdate, 1970.
Phophate absorption in eelgrass. Limnol. Oceanogr., 15(1): 6-13.

Zostera marine
(eel grass)
Okuda, Taizo, 1960
Metabolic circulation of phosphorus and nitrogen in Matsushima Bay (Japan) with special reference to exchange of these elements between sea water and sediments. (Portuguese, French resumes).
Trabalhos. Inst. Biol. Maritima e Oc eanogr., Universidade do Recife, Brasil, 2(1):7-153.

(Zostera) eel grass
Porsild, A.E., 1932.
Notes on the occurrence of Zostera and Zannichellia in arctic North America. Rhodora 34:90-94.

eel-grass,
Renn, C. E., 1937.
The eel-grass situation along the middle Atlantic coast. Ecol. 18(2):323-325.

eel-grass
Renn, C. E., 1936.
Persistence of the eel-grass disease and parasite on the American Atlantic coast. Nature 138: 507.

eel-grass
Renn, C. E., 1936.
The wasting disease of Zostera marina. 1. A phytological investigation of the diseased plant. Biol. Bull. 70(1):148-158, 7 textfigs.

eel-grass
Renn, C.E., 1935.
A mycetozoan parasite of Zostera marina. Nature, 135:544.

eel-grass
Renn, C.E., 1934.
Wasting disease of Zostera in American waters. Nature, 134:416, 1 textfig.

eel grass
Russak, M.L., 1957.
A study of eel grass (Zostera marina L.).
Biol. Rev., City Coll., N.Y., 19(1):32-34.

eel grass
Sando, Hitoshi, 1964.
Faunal list of the Zostera marina region at Kugurizaka coastal waters, Aomori Bay.
Bull. Mar. Biol. Sta., Asamushi, 12(1):27-35.

eel grass
Zostera
Segawa, Sokichi, Takeo Sawada, Masahiro Higaki, Tadao Yoshida and Shintoku Kamura, 1961
The floating seaweeds of the sea to the west of Kyushu.
Rec. Oceanogr. Wks., Japan, Spec. No. 5:179-186.

eel grass
*Simonetti, Gualtiero, 1967.
Variazioni nei popolamenti di Foster acee nel Golfo di Trieste durante gli ultimi decenni.
Arch. Oceanogr. Limnol. 15(Suppl.):107-114.

eel grass
Stauffer, R. C., 1937.
Changes in the invertebrate community of a lagoon after disappearance of the eel grass. Ecol. 18(3) 427-431.

(Zostera) eel grass
Tamai, Naoto, and Kojiro Nishida, 1962.
On the photosynthetic carbon dioxide fixation in Zostera caulescens and Halophila ovalis. (In Japanese; English abstract).
Ann. Rept., Noto Mar. Lab., Univ. Kanazawa, 2:15-21.

eel grass
Taylor, A.R.A., 1957.
Studies of the development of Zostera marina L. II. Germination and seedling development.
Canadian J. Bot., 35(5):681-695.

eel grass
Taylor, A.R.A., 1954.
Control of eel-grass in oyster culture areas.
Fish. Res. Bd., Canada, Atlantic Biol. Sta., General Ser. No. 23:3 pp., 2 textfigs.

(Zostera marina) eel grass
Tremblay, J.L., and R. Gaudry, 1936.
Décimation des Zostères (Herbe a bernaches) dans la region de l'ile Verte. Publ. Sta. Biol. Saint-Laurent, Contr. No. 7:8 pp.

eel grass (Zostera)
True-Schlenz, Renat, 1965.
Données sur les peuplements des sédiments à petites phanérogames marines (Zostera nana Roth et Cymodocea nodosa Ascherson) comparés à ceux des habitats voisins dépourvus de végétation. (Côte de Provence).
Rec. Trav. Sta. Mar., Endoume, Bull. 39(55): 95-125

(Zostera marina) eel grass
Van Den Ende, G., and Pauli Hauge, 1963.
Beobachtungen über den Epiphytenbewuchs von Zostera marina L. an der bretonischen Küste.
Botanica Marina, 5(4):105-110.

eel grass, disease of
Watson, S.W., and E.J. Ordal, 1951.
Studies on Labyrinthula. Univ. Wash. Ocean. Labs. Tech. Rept. No. 3, Ref. 51-1:37 pp. (duplicated), 12 figs. (multilithed).

eel grass
Wilson, D. P., 1949.
The decline of Zostera marina L. at Salcombe and its effects on the shore. J.M.B.A. 28(2):395-412, 4 pls.

eelgrass infauna
Ollivier, Marie-Thérèse 1970
Étude des peuplements de Zostères, Lanice et Sabelles de la région Dinardaise.
Téthys 1(4): 1097-1138.

eelgrass infauna
English Channel

eels—see FISH
for Leptocephalus see 'LARVAE-fish'
FISHERIES

eels

See: fish, eels, Anguilla, etc.

larvae, fish for the Leptocephalus larvae, in the organismal index.

effectors, independent

effectors, independent
Horridge, G.A., 1965.
Non-motile sensory cilia and neuromuscular junctions in a ctenophore independent effector organ.
Proc. R. Soc., London, (B), 162(988):333-350.

Effluents: See river discharge

effluents

See also: runoff
river drainage, etc.

effluent
Katayama, Katsusuke, Teruyuki Sugiyama, Kazuo Ukida, Motoyuki Shinohara and Kuniyasu Fujisawa, 1971.
The effects of effluent of night-soil treatment on the coastal fisheries grounds - I. On diffusion of effluent and quality of bottom mud in the surrounding area. (In Japanese).
Bull. Fish. Exp. Sta. Okayama Pref. (1970): 42-68.

effluents

* Poli Molinas, Marinella, e Maria Vittoria Olmo, 1966.
L'apporto in sali nutritivi di alcuni corsi d'acqua sficianti nell'Adriatico.
Note Lab. Biol. mar.Pesca, Fano, Univ. Bologna, 2(6): 85-116.

effluents—chemistry of

Berthois, Léopold et Philippe Bois, 1969.
Le cours inférieur et l'estuaire de la rivière du Château en période d'étiage, Îles de Kerguelen. Étude hydraulique, sédimentologique et chimique.
Cah. océanogr. 21 (8): 727-771.

effluents, effect of

Aubert, M., J.P. Gambarotta et F. Laumond 1968.
Rôle des apports terrigènes dans la multiplication du phytoplancton marin: cas particulier du fer.
Revue int. Océanogr. Méd. 12: 75-121.

effluents, effect of

Boltovskoy, Esteban, 1968.
Hidrología de las aguas superficiales en la parte occidental del Atlántico Sur.
Revta Mus. argent. Cienc. nat. Bernardino Rivadavia, Hidrobiol., 2(6):199-224.

effluents, effect of

Goldberg, Edward D., 1971.
River-ocean interactions. (Abstracts in English and Portuguese) In: Fertility of the Sea, John D. Costlow, editor, Gordon Breach, 1: 143-156.

Effluents, effect of

Mashtakova, G.P., 1964.
The influence of the land drainage on the development of the phytoplankton in the northwestern part of the Black Sea. (In Russian).
Trudy azov. chernomorsk nauchno-issled. Inst. morsk. ryb. khoz. okeanogr. 23: 55-67.

effluents, heated

Mileikovsky, Simon A., 1968.
The influence of human activities on breeding and spawning of littoral marine bottom invertebrates
Helgoländer wiss. Meeresunters. 17: 200-208

effluent, effect of

Uyeno, Fukuzo, Kyoichi Kawaguchi, Nagao Terada and Tadashi Okada 1970.
Decomposition, effluent and deposition of phytoplankton in an estuarine pearl oyster area.
Rept. Fac. Fish. Prefect. Univ. Mie 7(1): 7-41

effluents

Wright, L.D. and J.M. Coleman, 1971.
Effluent expansion and interfacial mixing in the Presence of a salt wedge, Mississippi River delta. J. geophys. Res. 76(36): 8649-8661.

effluent
freshwater

Dubra, J., 1970.
The spread of fresh water along the coast of the Baltic sea in Lithuania SSR. (In Lithuanian; Russian and English abstracts). Hidromet. Straipsn. LietuvosTSR, Vilnius, 3: 73-82.

egg white

egg white

Banse, K., 1956.
Produktionsbiologische Serienbestimmungen im südlichen Teil der Nordsee im März 1955.
Kieler Meeresf., 12(2):166-179.

egg white

*Hickel, Wolfgang, 1967.
Untersuchungen über die Phytoplanktonblüte in der westlichen Ostsee.
Helgoländer wiss. Meeresunters., 16(1-3):1-66.

egg white (data)

Kalle, K., 1956.
Chemisch-hydrographische Untersuchungen in der inneren Deutschen Bucht. Deutsche Hydrogr. Zeits. 9(2):55-65.

egg white

Krey, J., 1956.
Die Trophie küstennaher Meeresgebiete.
Kieler Meeresf., 12(1):46-64.

egg white

Krey, J., 1952.
Die Untersuchung des Eiweissgehaltes in kleinen Planktonproben. (Methodische Bemerkungen und Beispiele zu der Arbeit: Quantitative Bestimmung von Eiweiss im Plankton mittels der Biuretmethode). Kieler Meeresf. 8(2):164-172, Pl. 5.

egg white

Krey, J., 1951.
Quantitative Bestimmung von Eiweiss im Plankton mittels der Biuretreaktion. Kieler Meeresf. 8(1): 16-29, Pls. 9-10.

egg white

Krey, J., K. Banse and E. Hagmeier, 1957.
Über die Bestimmung von Eiweiss im Plankton mittels der Biureaktion. Kieler Meeresf. 13(1): 35-40.

"egg white"

Lenz, Jürgen, 1970.
4. Planktologie. 4.1. Seston, Chlorophyll und Eiweissgehalt. 4.3. Zooplankton.
Chemische, mikrobiologische und planktologische Untersuchungen in der Schlei im Hinblick auf deren Abwasserbelastung. Kieler Meeresforsch 26(2): 180; 180-129; 203-213.

egg white

*Nellen, Walter, 1967.
Horizontale und vertikale Verteilung der Planktonproduktion im Golfe von Guinea und in angrenzenden Meeresgebieten während der Monate Februar bis Mai 1964.
Kieler Meeresforsch., 23(1):48-67.

Eighteen-degree water
See under "temperature" with sub-heading

Eiweiss

See: egg white

Ekman layer

Ekman boundary layer

*Benton, Edward R., 1968.
A composite Ekman boundary layer problem.
Tellus, 20(4):667-672.

Ekman boundary layer

Caldwell, D.R., and C.W. van Atta 1970.
Characteristics of Ekman boundary layer instabilities.
J. fluid Mech. 44(1): 79-95.

Ekman layer

*Csanady, G.T., 1967.
On the "resistance law" of a turbulent Ekman layer.
J. atmos. Sci., 24(5):467-471.

Ekman layer

Endoh, Masahiro and Takashi Nitta, 1971.
A theory of non-stationary oceanic Ekman Layer.
J. met. Soc. Japan, (2)49(4): 261-266.

Ekman boundary layers

*Faller, Alan J. and Robert Kaylor, 1969.
Oscillatory and transitory Ekman boundary layers.
Deep-Sea Res., Suppl.16: 45-58.

Ekman boundary layer

*Geisler, J.E. and E.B. Kraus, 1969.
The well-mixed Ekman boundary layer.
Deep-Sea Res., Suppl. 16: 73-84.

Ekman layer

Griessier, H., 1969.
Über die äquatoriale Unstetigkeit des Ablenkungswinkels im geostrophisch-antitriptischen Windfeld.
Acta Hydrophysica, 14 (1/2): 85-106

Ekmanlayer

Hsueh, Y., 1969.
Buoyant Ekman layer.
Phys. Fluids 12(9): 1757-1762

Ekman layer

Pandolfo, Joseph P., and Philip S. Brown Jr., 1967.
Inertial oscillations in an Ekman layer containing a horizontal discontinuity surface.
J. mar. Res., 25(1):1-28.

Ekman layer

Tatro, P.R., and E.L. Mollo-Christensen 1967.
Experiments on Ekman layer instability.
J. fluid Mech. 28(3): 531-544.

Ekman's spiral
theory
problem

Ekman Spiral

*Faller, Allan J., and Robert Kaylor, 1967.
Instability of the Ekman Spiral with applications to the planetary boundary-layers.
Physics Fluids, 10 (9-2): S212- S219.

Ekman Spiral

Gonella, J., 1968.
Observation de la spirale d'Ekman en Méditerranée occidentale. C.R. Acad. Sci. Paris, 266(B): 205-208.
Hebd. Séanc.

Ekmans spiral
Hay, R.F.M., 1954.
A verification of Ekman's theory relating wind and ocean current directions using ocean weather ships' data. Mar. Obs. M.O. 579, 24(166):226-230.

Ekman spiral
Hesselberg, Th., 1954.
The Ekman spirals. Arch. Met., Geophys., Bioklim. A, 7:329-343, 5 textfigs.

Ekman's spiral
Hesselberg, Th., 1953.
Om Ekman-spiralen i luft og hav.
Norsk Geografisk Tidsskrift 14(1/4):100-108, 5 textfigs.

Ekman spiral
Mosby, Håkon, 1961.
Veines et artères de la mer.
Bull. Inst. Oceanogr., Monaco, No. 1195;27 pp.

Ekman spiral

Olson, F .C.W., 1959.
The polar forms of Sinh $t^{\frac{1}{2}x}$ etc. and their application to the Ekman spiral. J. Oceanogr. Soc., Japan, 15(2): 53-55.

Ekman spiral

Olson, F.C.W., 1959.
The Ekman spiral for a linearly varying eddy viscosity. J. Oceanogr. Soc., Japan, 15(2): 49-51.

"Ekman problem"
Saint-Guily, B., 1962
On the general form of the Ekman problem.
Proc. Symposium on Mathematical-Hydrodynamical Methods of Phys. Oceanogr., Sept. 1961, Inst. Meeresk., Hamburg, 61-73.

Ekman spiral
Saint-Guily, B., 1959.
Essai en vue d'une théorie d'Ekman généralisée.
Cahiers Océanogr., C.C.O.E.C., 11(2):101-130.

Ekman spiral
Saint-Guily, B., 1959.
Sur la solution du problème d'Ekman.
Deutsche Hydrogr. Zeits., 12(6):262-270.

Ekman flow, instability of
Stern, M.E., 1960.
Instability of Ekman flow at large Taylor number.
Tellus, 12(4):399-417.

"Ekman problem"
Welander, Pierre, 1962
Application of a two-layer Ekman model to the problem of the wind-driven currents.
Proc. Symposium on Mathematical-Hydrodynamical Methods of Phys. Oceanogr., Sept., 1961, Inst. Meeresk., Hamburg, 51-59.

Ekman's theory
Felsenbaum, A.I., 1956.
[An extension of Ekman's theory to the case of a non-uniform wind and an arbitrary bottom relief in a closed sea.] Dokl. Akad. Nauk, SSSR, 109(2): 299-302.

Ekman transport

Ekman vertical velocity
*Gates, W.L., 1969.
The Ekman vertical velocity in an enclosed B-plane ocean.
J. mar. Res., 27(1):99-120.

electrical analogues

Makarov, V.A. and A.B. Menzin, 1970.
On the use of electrical analogues to study the dynamics of the sea. Oceanologiia, 10(5): 815-819.
(In Russian; English abstract)

Ekman transport
Wickett, W. Percy 1967.
Ekman transport and zooplancton concentration in the North Pacific Ocean.
J. Fish. Res. Bd Can. 24(3): 581-594

electric charge

electric charge, earth's
Blanchard, Duncan C., 1963.
The electrification of the atmosphere by particles from bubbles in the sea.
Progress in Oceanography, 1:71-202.

electrical charge

Bradley, W.E. and R.G. Semonin, 1969.
Effect of space charge on atmospheric electrification, cloud charging, and precipitation.
J. geophys. Res., 74(8): 1930-1940.

electrical charge

Colgate, Stirling A. and John M. Romero, 1970.
Charge versus drop size in an electrified cloud.
J. geophys. Res., 75(30): 5873-5881.

electric earth currents
D'Yakonov, B.P., 1957.
[Nature of terrestrial electric currents and their study at the ocean bed.]
Izv. Akad. Nauk, SSSR, Ser. Geofiz., (6):129-133.

electric charge

Latham, J. and V. Myers, 1970.
Loss of charge and mass from raindrops falling in intense electric fields. J. geophys. Res., 75(3): 515-520.

electrical charges
Paluch, Ilga R. 1970.
Theoretical collision efficiencies of charged cloud droplets.
J. geophys. Res. 75(9): 1633-1640.

electric charge
Takahashi, Tsutomu, and Tatsuo Harumi. 1970.
Vertical distribution of electric charge on precipitation elements in the cloud obtained by radiosonde
J. met. Soc., Japan, (2)48(2):85-90

electric charge, effect of
Woods, J.D., 1965.
The effect of electric charge upon collisions between equal-size water drops in air.
Q.J.R. Meteorol. Soc., 91(389):353-355.

electric currents

electrical currents (vertical)
Fonarev, G.A., 1963.
Vertical electrical currents in the sea.
(In Russian).
Geomagnet. i Aeron., 3(4):784-785.

English abstract in
Soviet Bloc Res., Geophys., Astron., Space, 1963 (68):25.

electrical currents
Jones, W.M., 1950.
Progress in hydrography and physical oceanography. N.Z. Dept. Sci. Ind. Res., Geophys. Conf., 1950, Paper No. 11:2 pp. (mimeographed).

electric currents
LeGrand, Y., 1957.
L'électricité océanique. Scientia (6)51(10):249-254.

electric currents (earth currents)
Meunier, Jean, 1965.
Enregistrements telluriques en mer.
Bull. Inst. Océanogr., Monaco, 65(1339):28 pp.

electric current
Ryzhkov, Yu. G., 1957.
[Measurement of electric current in the ocean.]
Dokl. Akad. Nauk, SSSR, 113(4):787-790.

Electric currents
Vestine, E.H., 1950.
Electric currents in the oceans. (Abstract).
Trans. Am. Geophys. Union 31(2):331.

electric currents
Yoshio, K., and T. Kikuchi, 1950.
On the phase difference of earth current induced by the changes of the earth's magnetic field. Pts I and II. Sci. Repts., Tohoku Univ., 5th ser., Geophys., 2(2):139-145, 13 textfigs.

electric field

electric field

*Bogorov, V.(B.)G., R.M. Demenitskaya, A.M. Gorodnitsky, M.M. Kazansky, V.M. Kontorovich, E.M. Litvinov, N.N. Trubyatchinsky, V.D. Fedorov, 1969.
On the character and causes of the vertical change of the electric field in the ocean.
(In Russian; English abstract). Okeanologiia, 9(5): 767-772.

electric field
Cox, Charles, Toshihiko Teramoto and Jean Filloux, 1964
On coherent electric and magnetic fluctuations in the sea.
In: Studies in Oceanography dedicated to Professor Hidaka in commemoration of his sixtieth birthday, 449-457.

electric field
Deacon, G.E.R., 1955.
Information from electric currents in the sea.
J. Inst. Navig. 8(2):117-120, 1 fig.

electric field
Enenshtein, B.S., 1948.
[Method for studying the establishment of an electric field in the earth.] Dokl. Akad. Nauk, SSSR, 59(2):239-242.

Bill Trans. Russ. Sci Tech Lit 24:8.

electric field
Frenkel, Y.I., and G.P.Vager, 1948.
[Effect of an electric field upon a liquid stream.] Izvest. Akad. Nauk, SSSR, Ser. Geogr. Geofiz., 12:3-6.

electric fields
*Larsen, J.C., 1968.
Electric and magnetic fields induced by deep sea tides.
Geophys. J.R.astr.Soc., 16(1):47-70.

electrical field
Longuet-Higgins, M.S., M.E. Stern and H. Stommel, 1954.
The electrical field induced by ocean currents and waves, with applications to the method of towed electrodes. P.P.O.M. 13(1):1-37, 28 text-figs.

electric field
Mironov, A.T., 1948.
[A survey of electric current in the Black Sea near the southern coast of the Crimea from May 1946 to March 1947.] Akad. Nauk, SSSR, Izv., Ser. Geogr. Geifiz. 12(2):89-97.

electric field, air
Mühleisen, R., und H. Riekert, 1970.
Luftelektrische Messungen auf dem Meer: Ergebnisse von der Atlantischen Expedition 1969. III. Untersuchungen zum Elektrodeneffekt beim luftelektrischen Feld über dem Meer und die Konsequenzen für den globalen luftelectrischen Stromkreis.
Meteor Forsch.-Ergebnisse (B) 5: 46-51.

ELECTRIC FIELD, AIR
Mühleisen, R., und H. Riekert, 1970.
Luftelektrische Messungen auf dem Meer: Ergebnisse von den Atlantischen Expeditionen 1965 und 1969. II Das luftelektrische Feld in Troposphäre und Stratosphäre über dem Atlantischen Ozean.
Meteor Forsch.-Ergebnisse (B) 5: 23-45.

electric fields
Sanford, Thomas B., 1971.
Motionally induced electric and magnetic fields in the sea. J. geophys. Res. 76(15): 3476-3492.

electrical field
Smirnov, R.V., 1962.
Short-period wave train type oscillations of the natural electric field in the sea. (In Russian).
Doklady, Akad. Nauk, SSSR, 145(6):1271-1274.

Translation:
Earth Sci. Sect. (Amer. Geol. Inst.), 145(1-6): 8-11. 1964.

electric fields (GEK)
Solov'yev, L.G., 1961.
[On measurement of electric fields in the sea.]
Doklady Akad. Nauk, SSSR, 138(2):445-447.

Engl. Abstr., Factors which can introduce errors in measurement of electric fields in the sea.
OTS-61-11147:17 JPRS:8710:7.

electric field
Stommel, H., 1948.
The theory of the electric field induced in deep ocean currents. J. Mar. Res. 7(3):3860392.

electric organs

electric organs
Grundfest, Harry, 1967.
Comparative physiology of electric organs of elasmobranch fishes. In: Sharks, skates, and rays, Perry W. Gilbert, Robert F. Mathewson and David P. Rall, editors, Johns Hopkins Univ. Press, 399-432.

electric potentials

electric potentials
Barber, N.F., 1948
The magnetic field produced by earth currents flowing in an estuary or sea channel.
Mon. Notices, Roy. Astron. Soc., Geophys. Suppl., 5(7):258-269, 7 text figs.

electrical potential
Margalef, Ramón, 1967.
Significado de las diferencias verticales de potencial eléctrico en el mar.
Investigación pesq., 31(2):259-263.

electrical potential differences
Stommel, H., 1954.
Exploratory measurements of electrical potential differences between widely spaced points in the North Atlantic Ocean. Arch. Met. Geophys. u. Bioklim., A, 7:292-304, 2 textfigs.

electric potential
Takahashi, Tsutomu, 1969.
Electric potential of a rubbed ice surface.
J. atmos. Sci., 26(6): 1259-1265. Also in: Coll. Pap. Sci. Atmos. Hydrosph., Water Res. Lab., Nagoya Univ., 7(1969).

electric potential
Takahashi, Tsutomu, 1969.
Electric potential of liquid water on an ice surface. J. atmos. Sci., 26(6): 1253-1258.
Also in: Coll. Pap. Sci. Atmos. Hydrosph., Water Res. Lab., Nagoya Univ., 7(1969).

electrical potential
Wertheim, G.K., 1954.
Studies of the electrical potential between Key West, Florida, and Havana, Cuba.
Trans. Amer. Geophys. Union, 35(6):872-882, 9 textfigs.

electrical conductivity

electrical conductivity
Accerboni, E., e F. Mosetti, 1967.
A physical relationship among salinity, temperature and electrical conductivity of sea water.
Boll. Geofis. teor. appl., 9(34):87-96.

electrical conductivity
*Accerboni, E., e F. Mosetti, 1967.
Localizzazione dei deflussi d' acqua dolce in mare mediante un conduttometro elettrico superficiale a registrazione continua.
Boll. Geofis teor. appl., 9(36):255-268.

electrical conductivity
Atkins, W.R.G., 1947.
The electrical conductivity of river, rain and snow water. Nature 159:674.

electrical conductivity
Bowden, K.F., and P. Hughes, 1961
The flow of water through the Irish Sea and its relation to wind.
Geophys. J., R. Astron. Soc., 5(4):265-291.

electrical conductivity
Carritt, D.E., 1963
5. Chemical instrumentation. In: The Sea, M.N. Hill, Edit., Vol. 2. The composition of sea water, Interscience Publishers, New York and London, 109-123.

electrical conductivity
Chanu, Jacques, et Yves Le Grand, 1965.
Influence de la composition de l'eau de mer sur la conductibilité électrique.
Cahiers Oceanogr., C.C.O.E.C., 17(4):249-253.

Electrical conductivity
Cobb, William E., and Howard J. Wells 1970.
The electrical conductivity of oceanic air and its correlation to global atmospheric pollution.
J. atmos. Sci., 27(5): 814-819

electrical conductivity
Cox, R.A., 1965.
The physical properties of sea water.
In: Chemical oceanography, J.P. Riley and G. Skirrow, editors, Academic Press, 1: 73-120.

electrical conductivity
Cox, Roland A., 1963.
The s alinity problem.
Progress in Oceanography, 1:241-261.

electrical conductivity
Cox, C.S., J.H. Filloux, and J.C. Larsen, 1970.
1. General observations. 17. Electromagnetic studies of ocean currents and electrical conductivity below the ocean-floor. In: The sea: ideas and observations on progress in the study of the seas, Arthur E. Maxwell, editor, Wiley-Interscience 4(1): 637-693.

electrical conductivity
Gheorgiu, V.G., and N. Calinicenco, 1940.
Étude sur la variation de la conductibilité électrique de l'eau de la Mer Noire avec la température et son emploi à la mesure de la température de l'eau marine. Ann. Sci., Univ. Jassy, Sec. 2, 26:3-18.

electrical conductivity
Hamon, B. V., 1958.
The effect of pressure on the electrical conductivity of sea-water.
J. Mar. Res., 16(2):83-89.

electrical conductivity
Hara, Akihiro, 1971.
Distribution of salinity in an estuary.
Sci. Repts, Tokyo Kyoiku Daigaku (C) 10: 277-352.

electrical conductivity

Hinkelmann, Hans, 1960
Über eine Darstellung der elektrischen Leitfähigkeit als Produkt zweier Funktionen zum Zwecke einer direkten Messung des Salzgehaltes mit elektrischen Sonden. Kieler Meeresf., 16 (1): 3-8.

electrical conductance

Horne, R.A., and R.A. Courant, 1964
Application of Walden's rule to the electrical conduction of sea water.
J. Geophys. Res., 69(10):1971-1977.

electrical conductivity

Horne, R.A., and R.A. Courant, 1964
The temperature dependence of the activation energy of the electrical conductivity of sea water in the temperature range 0°C to 10°C.
J. Geophys. Res., 69(6):1152-1154.

electrical conductivity

Horne, R.A., and G.R. Frysinger, 1963
The effect of pressure on the electrical conductivity of sea water.
J. Geophys. Res., 68(7):1967-1973.

electrical conductivity

Hyndman, R.D., and N.A. Cochrane 1971.
Electrical conductivity structure by geomagnetic induction at the continental margin of Atlantic Canada.
Geophys. J.R. astron. Soc. 25 (5): 425-446.

electrical conductivity

Koske, Peter H., 1963
Uber EMK-Messungen zur Bestimmung von Ionenkonzentrationen in Meerwasser.
Kieler Meeresforsch., 19(2):182-188.

electrical conductivity

*Krause, Gunther, 1968.
Atruktur und verteilung des Wassers aus dem Roten Meer im Nordwesten des Indischen Ozeans.
Meteor Forschungsergeb. (A)4:77-100.

electrical conductivity (air)

Misaki, M. and T. Takeuti, 1970.
The extention of air pollution from land over ocean as revealed in the variation of atmospheric electric conductivity. J. met. Soc. Japan, (2)48(4): 263-269.

electrical conductivity

Morita, Yasuhiro, 1971.
The diurnal and latitudinal variation of electric field and electric conductivity in the atmosphere over the Pacific Ocean.
J. met. Soc. Japan. 49 (1): 56-58

Electrolytic conductance

Park, Kilho, 1964.
Electrolytic conductance of sea water: effect of calcium carbonate dissolution.
Science, 146(3640):56-57.

electrolytic conductance

Park, P. Kilho, Alvin L. Bradshaw, David W. Menzel, Karl E. Schleicher and Herbert C. Curl, Jr. 1969
Changes in electrolytic conductance of seawater during photosynthesis and respiration.
J. oceanogr. Soc. Japan, 25 (3): 119-122.

electrolytic conductance

Park, K. and W.V. Burt, 1966.
Electrolytic conductance of sea water and the salinometer - an addendum to the review.
J. oceanogr. Soc. Japan, 22(1): 25-28.

electrical conductance of sea water

Park, Kilho, P.K. Weyl and Alvin Bradshaw, 1964.
Effect of carbon dioxide on the electrical conductance of sea-water.
Nature, 201(4926):1283-1284.

electrical conductivity

Pollak, M.J., 1954.
The use of electrical conductivity measurements for chlorinity determinations. J. Mar. Res. 13(2):228-231.

electrical conductivity

Reeburgh, W.S. 1965.
Measurements of the electrical conductivity of sea water.
J. mar. Res. 23: 187-199

electrical conductance

Siedler, Gerold, 1966.
Die Bestimmung der Zunahme der elektrischen Leitfähigkeit von Seewasser bei wachsendem Druck mit Hilfe eines Nomogrammes.
Kieler Meeresforsch., 22(1):39-41.

electrical conductivity

Siedler, Gerold, 1963
On the insitu measurement of temperature and electrical conductivity of sea water.
Deep-Sea Res., 10(3):269-277.

Electrical Conductivity

Siedler, Gerold, 1961.
Uber die kurzfristige Veranderlichkeit von Temperatur - und Salzgehaltsschichtung in der ostlichen und mittleren Ostsee im Sommer 1960.
Kieler Meeresf., 17(2):148-153.

electrical conductivity

Sopach, E.D., 1958.
Electric conductivity as a means of measuring the salinity of sea water. (In Russian)
Gidrometeorol. Izd-vo, Moscow, 138 pp.

electrical conductivity

Thomas, B.D., T.G. Thompson and C.L. Utterback, 1934.
The electrical conductivity of sea water.
J. du Conseil, 9:28-35.

electrical conductivity

Tiphane, Marcel, and Jacques St-Pierre, 1962.
Tables de détermination de la salinité de l'eau de mer par conductivité électrique.
Faculté des Sciences, Université de Montréal, Canada, 20 pp. (multilithed)(Unpublished manuscript).

electrical conductivity

Voigt, Klaus, 1963
Untersuchungen in der Deckschicht des Atlantischen Ozeans mit einem digital registrierenden Temperatur-Leitfähigkeit-Druck-Messgerät Aus den Ergebnissen der deutschen Expeditionsgruppe während der sowjetischen Forschungsreisen mit der "Michail Lomonossow" im IGJ/IGC
Beitrage zur Meereskunde, Deutsche Akad. Wiss. Berlin, (7/8):2-151.

electrical conductivity tables

U.S. Hydrographic Office, 1956.
Tables for rapid computation of density and electrical conductivity of sea water. H.O. Pub., 619: 24 pp.

electrical conductivity/chlorinity

electrical conductivity/chlorinity

Cox, R.A., F. Culkin and J.P. Riley 1967.
The electrical conductivity/chlorinity relationship in natural sea water.
Deep-Sea Res. 14(2): 203-220.

electrical effects

electrical effect

Barber, N. F., and M. S. Longuet-Higgins, 1948
Electrical and magnetic effects of marine currents. The Observatory, 68(843):55-59.

electrical effects

Barber, N., and M. S. Longuet-Higgins, 1948.
Water movements and earth currents: electrical and magnetic effects. Nature 161(4081):192-193.

electric current, effect of

Bodrova, N.V., and B.V. Krayukhin, 1960.
Role of the receptors of the body surface in the mechanism of the reaction of fish to electric currents. (In Russian).
Trudy Inst. Biol. Vodokhranilishch, 3(6):266-272.

Translation:
OTS 63 - 11111 (1963).

electrical effect

Longuet-Higgins, M. S., 1949.
The electrical and magnetic effects of tidal streams. Mon. Not. R. Astron. Soc. 5:285-307.

electrical effects

*Pruppacher, H.R., E.H. Steinberger and T.L. Wang, 1968.
On the electrical effects that accompany the spontaneous growth of ice in supercooled aqueous solutions.
J. geophys. Res., 73(2):571-584.

electricity, effect of

Schwartz, Frank J., 1961

A bibliography: effects of external forces on aquatic organisms. Chesapeake Biol. Lab., Solomons, Maryland, Contrib. No. 168: 85 pp. (multilithed).

electrical effects

Shuleikin, V.V., 1962.
Magnetic and electric marine phenomena. (In Russian).
Ocherka po Fizike Morya (Studies on Marine Physics), Moscow, 410-466.

Translations:
OTS or SLA, $1.10.

electrical power, tidal

D'Arrigo, A., 1954. electrical power, tidal
Comunicazioni, energia mareoelettrica e pesca nello Stretto di Messina. Atti Accad. Gioenia, 6(9):159-185.

electricity, volcanic

electricity, volcanic
*Blanchard, Duncan C., and Sveinbjörn Björnsson, 1967.
Water and the generation of volcanic electricity. Mon. Wea. Rev. 95(12):895-898.

electrification
Gathman, Stuart G. and William A. Hoppel, 1970
Electrification processes over Lake Superior. J. geophys. Res., 75(6): 1041-1048.

electrification
Gathman, Stuart G. and William A. Hoppel, 1970.
Surf electrification. J. geophys. Res., 75 (24): 4525-4529.

electrocardiogram

electrocardiograms
Carricaburu, P., et M.-H. Filliol, 1962.
Electrocardiogramme de la langouste Pallinurus regius. Electrocardiogramme intracellulaire de la langouste Palinurus regius.
Comptes Rendus, Soc. Biol., Paris, 156:150-153; 716-718.

electrochemistry

Electrochemistry
Rozenfeld, I.L., O.I. Vashkov and K.A. Zhigalova, 1965.
Electrochemistry of metals in sea water.
Trav. Centre de Recherches et d'Etudes Oceanogr. n.s., 6(1/4):235-250.

electro-kinetic potential
Pravdić, V., 1970.
Surface charge characterization of sea sedimen[ts]
Limnol. Oceanogr., 15(2): 230-233.

electrolytic conductance

electrolytic conductance
Park, Kilho, and Wayne Burt, 1965.
Electrolytic conductance of sea water and the salinometer (2).
J. Oceanogr. Soc., Japan, 21(3):30-38.

electrolytic conductance
Park, Kilho, and Wayne V. Burt, 1965.
Electrolytic conductance of sea water and the salinometer. 1.
J. oceanogr. Soc., Japan, 21(2):69-80.

electromagnetic field

electromagnetism
Coggon, J.H., and H.F. Morrison, 1970.
Electromagnetic investigation of the sea floor.
Geophysics 35(3): 476-489.

electromagnetism
Kostecki, Andrzej 1971.
Amplitude characteristics of marine wave zone frequency sounding for media with an insulating substrate. (In Polish: English and Russian abstracts)
Acta geophys. polon. 19(2):168-179

electromagnetism
LeGrand, Y., J. Broc, B. Saint-Guily, and J. Canu, 1952.
Introduction à l'électromagnétisme des mers.
Ann. Inst. Océan., n.s., 27(4):235-329, 29 text-figs.

electromagnetic field
*Solovyev, V.S., 1967.
Studies of the natural electric field oscillations in the ocean. (In Russian; English abstract).
Okeanologiia, Akad. Nauk, SSSR, 7(4):696-703.

electromagnetic field
Troitkaya, V.A., 1959.
[Some peculiarities of the short-period oscillation generation of the earth's electromagnetic field in the Antarctic.]
Inform. Biull. Sovets. Antarkt. Exped., (12):33-36.

electromagnetism
USSR, Akademia Nauk Ukrainckoi SSR, 1968.
Electromagnetic effects in the Sea.
Morskoi Gidrofizicheskii Institat. Naukova Dymka Kiev, 40: 149 pp.

Electromagnetic induction currents

electromagnetic induction
Bullard, E.C. and R.L. Parker, 1970.
1. General observations. 18. Electromagnetic induction in the oceans. In: The sea: ideas and observations on progress in the study of the seas, Arthur E. Maxwell, editor, Wiley-Interscience 4(1): 695-730.

electromagnetic induction
Guelke, R.W., and C.A. Schoute-Vanneck, 1947.
The measurement of sea-water velocities by electromagnetic induction. J. Inst. Elect. Eng. 94:71-74.

electromagnetic currents
Cagniard, L., 1957.
Sur la théorie de la mesure électromagnétique des courants marins. Ann. Géophys., 13(2):

electromagnetic currents
Kalashnikov, A.G., 1961.
On some types of pulsations of the geomagnetic field and the earth's currents occurring simultaneously on the USSR territory.
Annals, Int. Geophys. Year, 11:96-

electromagnetic propagation
Liebmann, L.N., 1962
Other electromagnetic radiation. Ch. 11, Sect. IV. Transmission of energy within the sea. In: The Sea, Interscience Publishers, Vol. 1: Physical Oceanography, 469-475.

electromagnetic effects
Malkus, W.V.R., and M.E. Stern, 1952.
Determination of ocean transports and velocities by electromagnetic effects. J. Mar. Res. 11(2): 97-105, 2 textfigs.

electromagnetic studies
Cox, C.S., J.H. Filloux, and J.C. Larsen, 1970.
1. General observations. 17. Electromagnetic studies of ocean currents and electrical conductivity below the ocean-floor. In: The sea: ideas and observations on progress in the study of the seas, Arthur E. Maxwell, editor, Wiley-Interscience 4(1): 637-693.

electromagnetic wave
Michaelov, V. I, 1960.
[On the theory of the dispersion of the electromagnetic wave in shallow seas.]
Izv. Akad. Nauk, SSSR, Ser. Geofiz., (8):1229-1233.

electronics

ELECTRONICS
Ferrara, Angelo A., 1968.
Electronic age in oceanography.
Navigation, U.S.A, 15(1): 29-33.

electronics
Hersey, J.B., 1957
Electronics in oceanography. Advances in Electronics and Electronic Physics, 9:239-296.

electronics
Neshyba, Steve, 1968.
Oceanography and electronics. Proc. Region SIX IEEE Conf., Oceanogr. Session 1-C-1, 5 pp. Also in: Coll. Reprints, Dept. Oceanogr. Univ. Oregon, 7: 1968.

electronics, effect of
Schwartz, Frank J., 1961
A bibliography of external forces on aquatic organisms.
Chesapeake Biol. Lab., Solomons, Maryland, Contrib. No. 168: 85 pp. (multilithed).

electronic position

Electronic Position
Burmister, C.A., 1948
Electronics in Hydrographic Surveying. Part I. Shoran, Part II. Electronic Position Indicator. J. Coast and Geodetic Survey 1:3-29, 22 figs.

electron microscope

electron microscope
*Bartlett, Grant A., 1967.
Scanning electron microscope: potentials in the morphology of microorganisms.
Science, 158(3806):1318-1319.

electron microscope
Bartlett, Grant A., 1967.
Planctonic Foraminifera - new dimensions with the scanning electron microscope
Can. J. Earth Sci., 5(2): 231-233

electron microscope
Bé, Allan W.H., Andrew McIntyre and Dee L. Breger, 1966.
Shell microstructure of a planktonic foraminife[r]
Globorotalia menardii (d'Orbigny).
Eclog. geol. Helv., 59(2):885-896.

electron microscope
Bens, Everett M., and Charles M. Drew, 1967.
Diatomaceous earth: scanning electron microscope of Chromosorb P".
Nature, Lond., 216(5119):1046-1045.

electron microscope
Braarud, T., 1955.
Electron microscopy in oceanographic processes.
Pap. Mar. Biol. and Oceanogr., Deep-Sea Res., Suppl. to Vol. 3:479-481.

electronmicroscop
Braarud, T., 1954.
Studiet av planktonalger i elektronmikreskep.
Blyttia 2:102-108, 4 pls.

electron microscope
Braarud, T., J. Markali and E. Nordli, 1958.
A note on the thecal structure of Exuviaella baltica Lohm.
Nytt Mag. Botan., 6:43-45.

electron microscope
Bradley, D.E., 1959
Electron-microscopic study of finback whale myoglobin crystals. Nature, 183(4666): 941-943.

electron microscope
Bresciani, José, and Tom Fenchel, 1967.
Studies on dicyemid Mesozoa. II
The fine structure of the infusoriform larva.
Ophelia, 4(1): 1-17.

electron microscope
Gaarder, Karen Ringdal, 1962.
Electron microscope studies on Holococcolithophorids.
Nytt Mag. for Botanikk, 10:35-51.

electron microscope
Gaarder, K.R., J. Markali and E. Ramsfjell, 1954.
Further observations on the coccolithophorid, Calciopappus caudatus. Norske Videnskaps- Akad., Oslo, 1954(1):1-9, 2 textfigs., 4 pls.

electron microscope
Halldal, P., and J. Markali, 1955.
Electron microscope studies on coccolithophor-ids from the Norwegian Sea, the Gulf Stream and Mediterranean.
Avhandl. Norske Videnskaps-Akad., Oslo, 1. Mat.-Naturvidensk. Kl., 1955(1):1-30, 27 pls.

electron microscope
Halldal, P., and J. Markali, 1954.
Morphology and microstructure of coccoliths studied in the electron microscope. Observations on Anthosphaera robusta and Calyptrosphaera papillifera. Nytt Mag. Bot. 2:117-119, 2 pls.

electron microscope
Halldal, P., and J. Markeli, 1954.
Observations on coccoliths of Syracosphaera mediterranea Lohm., S. pulchra Lohm., and S. molischi Schill. in the electron microscope.
J. du Cons. 19(3):329-336, 6 textfigs.

electron microscope
Halldal, P., J. Markali and T. Naeso, 1954.
A method for transferring objects from a light microscope to marked areas on electron microscope. Mikroskopie 9(5/6):197-200, 8 textfigs.

electron microscope
Hasle, G.R., and B.R. Heimdal 1970.
Some species of the centric diatom genus Thalassiosira studied in the light and electron microscope.
Beihefte Nova Hedwigia 31: 559-589.

electron microscope
Hay, W.W., and Kenneth M. Towe, 1962.
Electron-microscope studies of Braarudosphaera bigelowi and some related coccolithophorids.
Science, 137(3528):426-428.

electron microscope
Hendey, N.I., 1971.
Electron microscope studies and the classification of diatoms. In: Micropaleontology of oceans, B.M. Funnell and W.R. Riedel, editors, Cambridge Univ. Press, 625-631.

electron microscope
Hendey, N.Ingram, 1959.
The structure of the diatom cell wall as revealed by the electron microscope.
J. Quekett Micros. Club, (4)5(6):147-175.

electron microscope
Hendy, N.I., D.H. Cushing and G.W. Ripley, 1954.
Electron microscope study of diatoms. J. Microsc. Soc. 74(3):22-34.
Abstr. in: Kodak Abstr. Bull. 41(3):163.

electron microscope
Holland, Nicholas D. 1969.
An electron microscope study of the papillae of crinoid tube feet.
Pubbl. Staz. Zool. Napoli. 37(4): 575-580.

electronmicroscope photographs
Ikzuka, Shoji, and Haruhiko Irie, 1964
Electron micrographic study on the marine diatoms especially Skeletonema costatum (Grev.) Cleve.
Bull. Fac. Fish., Nagasaki Univ., 15:92-99.

electronmicroscope
Lecal, J., 1965.
À propos des modalités d'élaboration des formations épineuses des Coccolithophoridés.
Protistologica, 1(2): 63-70

electron microscope
Manton, I., and G.F. Leedale, 1961.
Further observations on the fine structure of Chrysochromulina ericina Parke & Manton.
J.M.B.A., U.K., 41(1):145-155.

electron microscope
*Moll, Georg, Renate Ahrens und Gerhard Rheinheimer, 1967.
Elektronenoptische Untersuchungen über sternbildende Bakterien aus der Ostsee.
Kieler Meeresforsch., 23(2):137-147.

electron microscope
Okuno, H., 1957.
Electron-microscopical study on fine structures of diatom frustules. 15. Observation on the genus Rhizosolenia. Bot. Mag., Tokyo, 71(826):101-107.

electron microscope
Okuno, H., 1954.
Electron-microscope fine structure of some marine diatoms. Rev. Cytol. et Biol. Veget. 15(3):237-246.

electron microscope
Okuno, H., 1951.
Electron microscope study on Antarctic diatoms. (1). (In Japanese).
J. Jap. Bot., 26(10):305-310.

electron microscope
Oliveira, Lejeune P.H. de, and H. Muth, 1960.
Microscopia electronica de sies diatomaceas Pleurosigma com uma critica do genero (Naviculaceae, Bacillariophyceae).
Mem. Inst. Oswaldo Cruz, Brasil, 58(1):1-38.

electron microscope
Ross, R., and Patricia A. Sims, 1970.
Studies of Aulacodiscus with the scanning electron microscope.
Beihefte Nova Hedwigia 31: 49-88

Electron microscope
Round, F.E., 1970.
The delineation of the genera Cyclotella and Stephanodiscus by light microscopy, transmission and reflecting electron microscopy.
Beihefte Nova Hedwigia 31: 591-604

electron microscope
Swift, Elijah, and Charles C. Remsen, 1970.
The cell wall of Pyrocystis spp. (Dinococcales)
J. Phycol. 6(1): 79-86.

electron microscope
Thomassin B.A. et C. Picard, 1972.
Etude de la microstructure des soies de polychètes Capitellidae et Oweniidae au microscope électronique à balyage: un critère systématique précis. Mar. Biol. 12(3): 229-236.

electro-perception

electro-perception
Kalmijn, A.J., 1966.
Electro-perception in sharks and rays.
Nature, Lond., 212(5067):1232-1233.

electrophoresis, effect of
Gabbrielli, G.S., and E. La Pergola, 1970.
First experiments of mussels purification with electrophoresis.
Rev. int. Océanogr. méd. 18-19: 261-270

electro-seismic effect

electro-seismic effect
Antsyferov, M.S., 1962
[The electro-seismic effect.]
Doklady Akad. Nauk, SSSR, 144(6):1295-1297.

elimination
Fowler, S.W., L.F. Small and J.M. Dean, 1971.
Experimental studies on elimination of zinc-65, cesium-137 and cerium-144 by euphausiids. Marine Biol., 8(3): 224-231.

embryology

embryology
Kumé, Matazo, and Katsuma Dan, editors, 1968.
Invertebrate embryology. Translated from Japanese by Jean C. Dan, NOLIT Publ. House, Belgrade, 605 pp.

embryos

Kuzin, A.E., 1970.
On weight features of northern fur seals embryons. (In Russian). Izv. Tichookean. nauchno-issled. Inst. Ribn. choz. Okeanogr. 70: 52-59.

embryos

embryos

Kuzin, A.E., 1970.
On lobe structure and lung asymmetry in fur seal embryons. (In Russian). Izv. Tichookean. nauchno-issled. Inst. Ribn. choz. Okeanogr. 70: 238-240

embryos

Kuzmin, A.A. and V.I. Privalikhin, 1970.
Some data on composition of embryons sperm whale and fin whale blood. (In Russian). Izv. Tichookean. nauchno-issled. Inst. Ribn. choz. Okeanogr. 70: 180-186.

embryos

Kuzin, A.E., A.S. Sokolov, 1970.
Features of red blood of northern fur seals (Callorhinus ursinus L.) embrions and new-born pups. (In Russian). Izv. Tichookean. nauchno-issled. Inst. Ribn. choz. Okeanogr. 70: 44-51.

embryology

May, Raoul-Michel, 1965.
La contribution de l'enbryologie des animaux marins à la théorie de l'évolution au XIXe siècle. Colloque Internat., Hist. Biol. Mar., Banyuls-sur-Mer, 2-8 sept., 1963, Suppl., Vie et Milieu, No. 19:233-257.

encapsulants

encapsulants

Garoyan, John R., 1969.
Coatings and encapsulants— preservers in the sea.
Ocean Engng 1(4): 435-456

encrustation

encrustations

Glasby, G.P., J.S. Tooms and J.R. Cann, 1971.
The geochemistry of manganese encrustations from the Gulf of Aden. Deep-Sea Res. 18(12): 1179-1187.

encrustation

Milliman, John D., and Frank T. Manheim, 1968.
Submarine encrustation of a Byzantine nail.
J. Sedim. Petrol., 38(3): 950-953

encyclopedias

encyclopedias

Fairbridge, Rhodes W., 1966.
The encyclopedia of Oceanography.
Reinhold Publ. Corp., 1021 pp.

encyclopedias

Wadati, K., K. Terada, 1960

Encyclopedia of the Oceans. Tokyo-dou, Tokyo 671 pp.

Review in: Trans. Amer. Geophys. Un., 41(4): 659. By F.C.W. Olson

endemism

endemism

Balech, E. 1970
The distribution and endemism of some antarctic microplankters
Antarctic Ecol. 1:143-146

endemism

Valentine, James W. 1971.
Plate tectonics and shallow marine diversity and endemism, an actualistic model.
Systm. Zool. 20(3): 253-264

endemism

Woodring, W.P., 1965.
Endemism in middle Miocene Caribbean molluscan faunas.
Science, 148(3672):961-963.

endocrines

Hughes, D.A., 1969.
Evidence for the endogenous control of swimming in pink shrimp, Penaeus duorarum. Biol. Bull. mar. biol. Lab., Woods Hole, 136(3): 398-404.

endocrines, effect of

endocrines, effect of

*Nagabhushanum, R., 1967.
The endocrine control of white chromatophores of the crab, Uca annulipes (H.Milne Edwards).
Crustaceana, 13(3):292-298.

endocrinology

Kleinholz, L.H., 1967.
Problems in crustacean endocrinology.
Proc. Symp. Crustacea, Ernakulam, Jan. 12-15, 1965, 3. 1029-1037.

enemies

enemies

Moore, H. B. and J. A. Kitching, 1939
The biology of Chthamalus stellatus (Poli).
JMBA 23:521-541.

energy

energy, tidal

Gibrat, R., 1962
Source de l'énergie des marées: énergie cinétique de la Terre ou énergie thermique du Soleil?
Mém. Trav. de la S.H.F. Suppl. au 1: 141-152.

Energy

Kielmann, Jürgen, Wolfgang Krauss und Lorenz Magaard, 1970
Über die Verteilung der kinetischen Energie im Bereich der Trägheits- und Seichesfrequenzen der Ostsee im August 1964 (Internationales Ostseeprogramm).
Kieler Meeresforsch. 25(2): 245-254

energy

Lacombe, H., 1957.
L'énergie des mers. La Météorol., 1957:371-391.

energy

Legendre, R., 1949.
Les ressources énergétiques de la mer. Bull. Inst. Océan., Monaco, No. 947:16 pp.

energy dissipation

Moroshkin, K.V., 1948.
Energy dissipation in the Baltic. Met. Gidrol. 3:41-51.

energy balance

Neumann, J., and N. Resenan, 1954.
The Black Sea: Energy balance and evaporation.
Trans. Amer. Geophys. Union, 35(5):767-774, 1 textfig.

energy, thermic, utilization

Nizery, A., 1947.
Project de construction d'une prise d'eau sous-marine profonde. Ann. des Ponts et Chaussées de Nov.-Dec. 1947:827-867, 44 figs.

energy, thermal

Nizery, A., and L. Nisolle, undated.
Etudes sur l'energie thermique des mers.
Soc. Ing. Civ., France, 50 pp., 24 textfigs.

Recd. 7 Dec. 1949.

energy distribution

Ozmidov, R.V., 1965.
On the energy distribution between oceanic motions of different scales. (In Russian; English abstract).
Fisika Atmosferi i Okeana, 1(4):439-448.

energy distribution

Rouse, Hunter, 1960
Répartition de l'énergie dans les zones de décollement.
Distribution of energy in regions of separation.
La Houille Blanche (3) (4) :1-26; 27-45.

energy problems

Ruppert, L.L., 1964.
Some problems of the energy of the atmosphere and the hydrosphere. (In Russian).
Materiali Vtoroi Konferentsii, Vzaimod. Atmosfer. i Gidrosfer. v Severn Atlant. Okean., Mezhd. Geofiz. God, Leningrad. Gidrometeorol. Inst. 49-76.

energy (currents)

Terry, Richard D., editor, 1966.
Ocean Engineering, 1. Energy sources and energy conversion, waste conversion and disposal. 2. Undersea construction, habitation and vehicles; recreation.
Western Periodicals Co., North Hollywood, Calif.
Vol. 3:431pp. (multilithed).

energy, temperature diffrential

Terry, Richard D., editor, 1966.
Ocean Engineering, 1. Energy sources and energy conversion, waster conversion and disposal. 2. Undersea construction, habitation and vehicles; recreation.
Western Periodicals Co., North Hollywood, Calif.
Vol. 3:431pp. (multilithed).

Energy

Vlymen, William J., 1970.
Energy expenditure of swimming copepods. Limnol. Oceanogr., 15(3): 348-356.

energy budget

energy budget

Bryan, Kirk, 1969.
Climate and the ocean circulation. III. The ocean model.
Mon. Wea. Rev. 97(11): 806-834

energy budget

Konaga, Shunji 1967.
Water temperature at the sea surface. VII A calculation of energy budget near the sea surface.
Bull. Kobe Mar. Obs. 145-154
No. 175
(In Japanese, English abstract)

energy budget

Konaga, Shunji, 1966.
Water temperature at the sea surface. VII. A calculation of the energy budget near the sea surface. (In Japanese; English abstract).
Umi to Sora, 41(3/4):148-154.

ENERGY BUDGET

Vowinckel, E., and Svenn Orvig, 1969.
Climate change over the Polar Ocean. II. A method for calculating synoptic energy budgets. III.
Arch. Met. Geophys. Biokl. (B) 17 (2/3). 121-146, 147-174.
The energy budget of an Atlantic cyclone.

energy considerations

energy considerations

McLellan, H.J., 1958.
Energy considerations in the Bay of Fundy system
J. Fish. Res. Bd., Canada, 15(2):115-134.

energy conversion

energy conversion

Terry, Richard D., editor, 1966.
Ocean Engineering, 1. Energy sources and energy conversion, waste conversion and disposal. 2. Undersea construction, habitation and vehicles; recreation.
Western Periodicals Co., North Hollywood, Calif.
Vol. 3:431pp. (multilithed).

energy dissipation

energy dissipation

Kagan, B.A., 1964.
An estimation of energy dissipation of turbulence in purely drift currents. (In Russian).
Okeanologiia, Akad. Nauk, SSSR, 4(1):3-8.

energy dissipation

Korvin-Kroukovsky, B.V., 1969.
The dissipation of energy in wind-generated waves: a hypothesis.
Dt. hydrogr. Z. 22(3): 97-118

energy dissipation

Selivanov, L.V., 1966.
Determination of friction and energy dissipation in waves using sea wind wave measurements.
(In Russian).
Fisika Atmosferi i Okeana, 2(5):545-547.

energy dissipation

Shtokman, V.B., 1946.
[On the dissipation of energy in permanent ocean currents.] Trudy Inst. Okean. 1:16 pp. (p. 69).

energy exchange

energy exchange

Barber, F.G., 1967.
A contribution to the oceanography of Hudson Bay
Manuscript Rep.Ser., Dept.Energy,Mines,Resources, Can., No. 4:69 pp. (multilithed).

energy exchange

Bøyum, Gunnvald, 1966.
The energy exchange between sea and atmosphere at Ocean Weather Stations M, I and A.
Geofys. Publr, 26(7):19 pp.

energy exchange

Deacon, G.E.R., 1950.
Energy exchange between the oceans and the atmosphere. Nature 165(4188):173-174.

energy exchange

Jacobs, W.C., 1951.
The energy exchange between sea and atmosphere and some of its consequences. Bull. S.I.O. 6(2): 27-122, 67 textfigs.

ocean-air energy exchange

Manabe, S., 1958

On the estimation of energy exchange between the Japan Sea and the atmosphere during winter based upon the energy budget of both the atmosphere and the sea. J. Met. Soc., Japan, 36(4):123-134.

energy exchange

McHugh, J.L., 1967.
Estuarine nekton.
In: Estuaries, G.H. Lauff, editor, Publs Am. Ass. Advmt. Sci., 83:581-620.

energy exchange

*Miller, Banner I., 1966.
Energy echange between the atmosphere and the ocean.
Hurricane Symposium, Oct.10-11, 1966, Houston
Publ.Am.Soc.Oceanogr., 1:134-157.

energy exchange

Terada, K., and K. Oosaway, 1952.
On the energy exchange between sea and atmosphere in the adjacent seas of Japan. Ocean., Meteorol., Nagasaki Mar. Obs. 6(1):3-18, 6 textfigs.

energy exchange

Wyrtki, K., 1957.
Precipitation, evaporation and energy exchange at the surface of the southeast Asian waters.
Penjelidikan Laut, Indonesia (Mar. Res.), no. 3: 7-40.

energy flow

Energy flow

Platt, Trevor, D.V. Subba Rao 1970.
Energy flow and species diversity in a marine phytoplankton bloom.
Nature, Lond., 227 (5262): 1059-1060.

enery Flow

Small, Lawrence F. 1967.
Energy flow in Euphausia pacifica
Nature, Lond., 215 (5100): 515-516

Energy, kinetic

energy, kinetic

Kao, S.-K., and G.R. Farr, 1966.
Turbulent kinetic energy in relation to jet streams, cyclone tracks and ocean currents.
J. geophys. Res., 71(18):4289-4296.

ENERGY (LIBERATED)

Szekielda, Karl-Heinz, 1970.
The liberated energy potentially available from oxidation processes in the Arabian Sea. Deep-Sea Res., 17(3): 641-646.

energy loss

energy loss

Faure, M., 1953.
Calcul des pertes d'énergie dans un estuaire à marée (Gironde). Principe et exécution du calcul a l'aide d'une machine mathématique.
La Houille Blanche, No. Spécial B/ 1953:157-169.

energy spectrum

Nan-niti, Tosio, 1970.
A hypothesis of turbulent energy spectrum.
J. oceanogr. Soc. Japan 26(5): 296-299.

energy transfer

energy transfer

Charnock, H., 1964.
Energy transfer by the atmosphere and the Southern Ocean.
Proc. R. Soc., London, (A), 281(1384):6-14.

energy transfer

Charnock, H., 1955.
Energy transfer between air and water.
Verh. Int. Ver. Limnol. 12:105.

energy transfer

Hino, Mikio, 1966.
A theory on the fetch graph, the roughness of the sea and the energy transfer between wind and wave
Coast. Engng Japan, 9:12-25.

energy transfer (wind)

Ichiye, T., 1952.
A short note on energy transfer from wind to waves and currents. Ocean. Mag., Tokyo, 4(3):89-93, 1 fig.

energy transfer, surface, effect of

LaFond, E.C., 1954.
Environmental factors effecting the vertical temperature structure of the upper layers of the sea. Andhra Univ. Ocean. Mem. 1:94-101, 25 text-figs.

energy transfer

Stoneley, R., (1952)1954.
The communication of energy from ocean waves to the ocean floor.
Pontif. Acad. Sci., Scripta Varia, 12:389.

energy transfer

Webster, Ferris, 1961
The effect of meanders on the kinetic energy balance of the Gulf Stream.
Tellus, 13(3):392-401.

energy transport

energy transport
Ellingsen, Torbjørn and Enok Palm, 1966.
The energy transfer from submarine seismic waves to the ocean.
Geofys. Publr, 26(8):22 pp.

energy transport
Jung, G.H., 1952.
Note on the meridional transport of energy by the oceans. J. Mar. Res. 11(2):139-146, 2 textfigs.

engineers

engineers
Holm, Carl 1967.
Ocean engineers --- the new need.
Ocean Industry 2(4):26-28, 63.

engineering
Suda, Kanji 1965.
The present status of oceanographic engineering and its problems. (In Japanese; English abstract).
J. Jap. Soc. Civ. Engrs 50(2):40-47.
Also in: Coll. Repr. Fac. Oceanogr. Tokai Univ. 1965.

engineering oceanography

engineering
Anon., 1965.
Navy begins deep-ocean engineering.
Undersea Techn., 6(9):40-44.

engineering, fisheries
Bullis, Harvey R., Jr., 1966.
Engineering needs for fishery development.
Exploiting the Ocean, Trans. 2nd Mar. Techn. Soc. Conf., June 27-29, 1966, 342-347.

engineering
Craven, John P., and Willard F. Searle, 1966.
The engineering of sea systems.
Exploiting the Ocean, Trans. 2nd Mar. Techn. Soc. Conf., June 27-29, 1966, 412-423.

engineering
Davis, B. W., 1966.
Exploration engineering and instrumentation problems in the marine environment.
Exploiting the Ocean, Trans. 2nd Ann. Mar. Techn. Soc. Conf., June 27-29, 1966, 134-146.

engineering
Deacon, G.E.R., 1969.
Ocean engineering. Proceedings of the International Marine & Shipping Conference, 10-20 June 1969, Section 10, 6 pp. London: Institute of Marine Engineers.

engineering
Hromadik, J.J., 1966.
Ocean engineering for human exploration.
Man's extension into the sea, Trans. Symp., 11-12 Jan. 1966, Mar. Techn. Soc., 74-88.

engineering
Santi, G.G., 1969.
Underwater engineering applications. Oceanol. int. 69, 2: 6 pp.

engineering oceanography
Stevenson, R.E., 1957.
Engineering oceanography in Santa Monica Bay, California. Géogr. Phys. et de Géol. Dynamique, (2)1(1):58-59.

engineering
Wenk, Jr., Edward, 1965.
Engineering for marine exploration.
Undersea Techn., 6(3):29, 31-32, 34-35.

engineering
Wiegel, Robert L., 1964.
Oceanographical engineering.
Prentice-Hall Series in Fluid Mechanics, 532 pp.

enrichment

enrichment
Ketchum, Bostwick H., 1967.
Phytoplankton nutrients in estuaries.
In: Estuaries, G.H. Lauff, editor, Publs Am. Ass Advmt Sci., 83:329-335.

enrichment
Menzel, D.W., E.M. Hulburt and J.H. Ryther, 1963
The effects of enriching Sargasso Sea water on the production and species composition of phytoplankton.
Deep-Sea Res., 10(3):209-219.

enrichment experiments
Smayda, Theodore J., 1971.
Further enrichment experiments using the marine centric diatom Cyclotella nana (clone 13-1) as an assay organism. (Portuguese abstract). In: Fertility of the Sea, John D. Costlow, editor, Gordon Breach, 2: 493-509.

enrichment
Tranter, D.J., and B.S. Newell, 1963
Enrichment experiments in the Indian Ocean.
Deep-Sea Res., 10(1/2): 1-9.

enrichment
Uda, Michitaka, 1958.
Enrichment patterns resulting from eddy systems.
Proc. Ninth Pacific Sci. Congr., Pacific Sci. Assoc., 1957, Oceanogr., 16:91-93.

enrichment, effect of
Vaccaro, Ralph F., 1969.
The response of natural microbial populations in seawater to organic enrichment. Limnol. Oceanogr., 14(5): 726-735.

enteropneust

enteropneusts
Bourne, Donald W., and Bruce C. Heezen, 1965.
A wandering enteropneust from the abyssal Pacific, and the distribution of spiral tracks on the sea floor.
Science, 150(3692):60-63.

enthalpy
Connors, Donald N., 1970.
On the enthalpy of seawater. Limnol. Oceanogr., 15(4): 587-594.

entrainment
Hantel, Michael, 1971.
The entrainment influence on the ocean surface layer in tropical latitudes. J. phys. Oceanogr. 1(2): 130-138.

environments

environments
Hedgpeth, J.W., 1957.
Classification of marine environments. Ch. 6: Treatise on marine ecology and palaeoecology, Vol. 2, Paleoecology. G.S.A. Mem., 67:93-100.

environments, abyssal
Menzies, Robert J., 1965.
Conditions for the existence of life on the abyssal sea floor.
Oceanogr. Mar. Biol., Ann. Rev., 3:195-210.

environments, bottom
Lackey, James B., 1961
Bottom sampling and environmental niches.
Limnol. & Oceanogr., 6(3):271-279.

environmental contamination
Rice, Theodore R. and Douglas A. Wolfe, 1971.
Radioactivity - chemical and biological aspects.
In: Impingement of man on the oceans, D.W. Hood, editor, Wiley Interscience: 325-379.

environment, effect of

environment, effect of
Amoureux, Louis, 1966.
Etude bionomique et écologique de quelques annelides polychites des sables intertidaux des cotes ouest de la France.
Arch. Zool. exp. et gen., 107(1):218 pp.

environmental monitoring
Caston, V.N.D. 1970.
Oil company environmental monitoring in the North Sea, 1968-69.
Electronic Engineering in Ocean Technology, Swansea, 21-24 Sept. 1970, I.E.R.E. Conf. Proc. 19: 281-287

enzymes
see also under chemistry

enzymes
See: chemistry, enzymes

epibiota

epibiota (Sargassum)
Conover, John T., and John McN. Sieburth, 1964.
Effect of Sargassum distribution on its epibiota
Botanica Marina, 6(½):147-157.

epibionts

Eggleston, D., 1971.
Synchronization between moulting in Calocaris macandreae (Decapoda) and reproduction in its epibiont Tricella koreni (Polyzoa Ectoprocta).
J. mar. biol. Ass. U.K. 51(2): 409-410.

epibionts, list of spp.

Monniot, Claude, 1965.
Les "blocs à microcosmus" des fonds chalutables de la région de Banyuls-sur-Mer.
Vie et Milieu, Bull. Lab. Arago, (B)16(2-B):819-849

epicenters

See: earthquake epicenter

epichthon

Lyman, John, 1969.
Naviface, oxyty, and epichthon: words versus terms. J. mar. Res., 27(3): 367-368.

epineuston

David, Peter M., 1965.
The surface fauna of the ocean.
Endeavour, 24(92):95-100.

epiphytes

Ferreira-Correia, M.M., 1969.
Epífitas de Digenia simplex (Wulfen) C. Agardh no estado do Ceará (Rhodophyta: Rhodomelaceae).
Arq. Ciên. Mar, Fortaleza, Ceará, Brasil, 9(1): 63-69

Epiphytes

Glynn, Peter W., 1970.
Growth of algal epiphytes on a tropical marine isopod. J. exp. mar. Biol. Ecol., 5(1): 88-93.

equations

equation of continuity

Takano, K., 1963.
Equation of continuity and equations of motion of sea water in spherical co-ordinates.
J. oceanogr. Soc., Japan, 20:93-94.

equations of motion (sea water)

Takano, K., 1963.
Equation of continuity and equations of motion of sea water in spherical co-ordinates.
J. oceanogr. Soc., Japan, 20:93-94.

equator

Equator

Hidaka, Koji, 1969.
Relationship between the meridional and vertical flows at the equator
J. oceanogr. Soc., Japan 25 :273-280.

equator, effect of

*Kozlov, V.F., 1967.
On the theory of a baroclinic layer at the equator. (In Russian; English abstract).
Okeanologiia, Akad. Nauk, SSSR, 7(4):577-585.

Equator

Romanov, Y. A., 1964.
On the connection of the surface wind with pressure in the area of the equator. (In Russian)
Okeanologiia, Akad. Nauk, SSSR., 4(6):954-961.

Equatorial currents
See currents equatorial

equatorial currents

See: currents, equatorial or under appropriate name in geographical index

equatorial flow

equatorial flow

See also: flow

equatorial flow

Hidaka, Koji, 1963
A hydrodynamical computation of an equatorial flow.
Rec. Oceanogr. Wks., Japan, 7(1): 1-7.

equatorial flow

Hidaka, Koji, 1962
A computation of non-linear equatorial flow.
Rec. Oceanogr. Wks., Japan, 6(2):1-8.
Also in:
Collected Oceanographical Papers, Geophys. Inst. Tokyo Univ., 6(Contrib. No. 20) (1962).

equatorial flow, non-linear

Hidaka, Koji, 1962
A computation of non-linear equatorial flow.
Rec. Oceanogr. Wks., Japan, n.s., 6(2):1-8.

equatorial region

equatorial region

Krivelevich, L.M., 1967.
Formation of the temperature field by the drift currents in the equatorial region. (In Russian)
Fisika Atmosfer. Okean., Izv. Akad. Nauk SSSR, 3(3):344-347.

equatorial regions

Krivelevich, L.M., 1967.
Numerical calculation of non-stationary wind-driven currents at the equator. (In Russian; English abstract).
Okeanologiia, Akad. Nauk, SSSR, 7(2):232-237.

equilibrium constants

Pytkowicz, Ricardo M., 1969
Use of apparent equilibrium constants in chemical oceanography, geochemistry, and biochemistry.
Geochem. J. 3 (2/3): 181-184

equivalent conductance

equivalent conductance

Connors, Donald N., and Peter K. Weyl, 1968.
The partial equivalent conductances of salts in sea water and the density/conductance relationship.
Limnol. Oceanogr., 13(1):39-50.

erosion

erosion

Aksenov, A.A., 1957.
Quelques particularités de l'abrasion littorale de la mer d'Azov. Trav. Inst. Ocean., Leningrad., 34:

erosion

Astre, G., 1965.
Modélites de l'érosion océanique sur les roches de l'Atalaye.
Bull. Cent. Etud. Rech. Sci., Biarritz, 5(3): 347-364.

erosion

Auzel, Melle Marguerite, 1960.
Observations faites aux Îles de Glenan.
Cahiers Océanogr., C.C.O.E.C., 12(5):333-337.

Also in:
Trav. Lab. Géol. Sous-Marine, 10(1960).

erosion

Batalin, R.I., 1961.
Application of transverse circulation methods to problems of erosion and siltation.
Dock & Harbour Authority, 41(486):407-408.

erosion

Berthois, Léopold, 1960
Etude expérimentale de l'érosion des vases d'estuaires. Comptes Rendus, l'Acad. Sci. Paris, 250 (24): 4020-4022.

erosion

Blanc, J.J., 1954.
Erosion et sédimentation littorale actuelle dans le Détroit de Messine. Bull. Inst. Océan., Monaco No. 1051:12 pp.

shore erosion

Bourcart, J., 1953.
Note sur l'erosion marine de la côte entre La Tronche (Vendée) et l'Ile Madame (Charente-Maritime). Bull. d'Info., C.C.O.E.C. 5(9):396-401, 1 fig.

erosion

Bradley, W.C., 1958.
Submarine abrasion and wave-cut platforms.
Bull. G.S.A., 69(8):967-974.

erosion

Bruun, Per M., 1967.
By-passing and back-passing off Florida
J. Waterways Harb. Div. Am. Soc. civ. Engrs. 93 (WW2): 101-128.

erosion

Bruun, Per, 1954.
Coast erosion between beach profile No. 14A Lyngby, and beach profile No. 61, Dybs, with special consideration to the erosion of the lime inlet barriers. Pts. I & II.
Atelier Elektra, København, 171 pp., figs., tables.

erosion

Bruun, Per, 1953.
Measures against erosion at groins and jetties.
Coastal Engineering, 3:137-164.
(Proc. third Conf. Coastal Eng., Cambridge)

erosion

Bruun, Per, and Madhav Manohar, 1963.
Coastal protection for Florida. Development and design.
Engineer. Progress, Univ. Florida, 17(8):1-56.

erosion

Cailleux, Andre, 1958.

Érosion et défense de la côte en Loire-Antlantique.
C. R. Acad. Sci., Paris, 247:1211-1214.

Also reprinted in:
Trav. Lab. Geol. Sous-Marine, 8, 1958.

erosion

Caldwell, Joseph M. 1966.
Coastal processes and beach erosion.
J. Boston Soc. Civil Engrs 53(2): 142-157

erosion (protection against)

Castanho, J.P., 1962
Metodos empregados na defesa contra a erosao costeira.
Lab. Nacional de Engenhar Civil, Memoria, Lisbon No. 196: 22 pp.

erosion

Caston, Vivian N.D., 1965.
Localised sediment transport and submarine erosion in Tremadoc Bay, northern Wales.
Marine Geol., 3(6): 401-40.

erosion

Cotton, C.A., 1952.
Cyclic resection of headlands by marine erosion.
Geol. Mag., 89(3):221-225.

erosion

Cotton, C.A., 1951.
Accidents and interruptions in the cycle of marine erosion. Geogr. J., 117(3):343-349.

erosion

Das, P.K., V. Hariharan and V.V.R. Varadachari 1966.
Some studies on wave refraction in relation to beach erosion along the Kerala coast.
Proc. Indian Acad. Sci. (A) 64(3): 192-202
Also in: Coll. Repr. Nat. Inst. Oceanogr. New Delhi 1 (1963-1968).

erosion

Dill, Robert F., 1964.
Sedimentation and erosion in Scripps Submarine Canyon Head.
In: Papers in Marine Geology, R.L. Miller, Editor, Macmillan Co., N.Y., 23-41.

erosion

Dillon, William P., and Herman B. Zimmerman, 1970.
Erosion by biological activity in two New England submarine canyons.
J. sedim. Petrol. 40(2): 542-547.

erosion

Dionne, Jean-Claude, 1969.
Tidal flat erosion by ice at La Pocatière, St. Lawrence estuary.
J. sedim. Petrol., 39(3): 1174-1181.

erosion

Durand de Saint-Front, Y., 1955.
Considerations sur l'erosion du littoral à Tahiti. Bull. d'Info., C.C.O.E.C., 7(3):127-136, 2 figs.

Erosion

Dyer, K.R., 1970
Linear erosional furrows in Southampton Water.
Nature, Lond. 225(5227): 56-58.

erosion

Egorov, E.N., 1962
[Stopping erosion in medidional bays]
Priroda (3):54-56.

erosion

Esin, N.V., 1964
On the problem of kinematics of formation of marine abrasive terraces. (In Russian).
Okeanologiia, Akad. Nauk, SSSR, 4(2): 284-289.

erosion

Evans, John W., 1970.
A method for measurement of the rate of intertidal erosion. Bull. mar. Sci., 20(2): 305-314.

erosion

Ewing, John, Maurice Ewing and Robert Leyden, 1966.
Seismic profiles survey of the Blake Plateau.
Bull. Am. Assoc. Petr. Geol., 50(9):1948-1971.

erosion, marine

Fairbridge, R.W., 1952.
Marine erosion. Proc. Seventh Pacific Sci. Cong. Met. Ocean., 3:347-359, 1 textfig.

Fairbridge, R.W., 1952.
Marine erosion. Seventh Pacific Sci. Congr. 3:1-11.

erosion

Forward, C.N., 1960

Shoreline changes in Egmont Bay and Bedeque Bay, Prince Edward Island. Canada, Geographical Branch, Department of Mines and Technical Surveys. Geographical Paper No. 26: 1-15.

erosion

*Gill, E.D., 1967.
The dynamics of the shore platform process, and its relation to changes in sea-level.
Proc. R. Soc., Victoria, N.S., 80(2):183-192.

erosion

Gostev, A., 1928.
On the scouring velocities. Vestnik Irrigatsii 6(1):65-69.

RT 1128 Bibl. Transl. Rus. Sci. Tech. Lit. 7.

erosion

Great Britain, Department of Scientific and Industrial Research, 1956.
Hydraulics Research, 1955:56 pp.

erosion

Guilcher, A., 1953.
Mesures de la vitesse de sédimentation et d'érosion dans les estuaires bretons.
C.R. Acad. Sci., Paris, 237(21):1345-1347.

Erosion

Guilcher, Andre, Leopold Berthois, and Rene Battistini, 1962.
Formes de corrosion littorale dans les roches volcaniques, particulierement a Madagascar et au Cap Vert (Senegal).
Cahiers Oceanogr., C.C.O.E.C., 14(4):208-240.

erosion

Gilluly, James, 1964.
Atlantic sediments, erosion rates and the evolution of the continental shelf.
Geol. Soc., Amer., Bull., 75(6):483-492.

erosion

Hamada, T., 1951.
Breakers and beach erosions. Rept. Transport. Tech. Res. Inst., Rept. No. 1:165 pp., 20 photos., 39 textfigs.

erosion

Haruta, Tadao, 1961
Recent coastal processes in Niigata Prefecture.
Coastal Eng., Japan, 4:73-83.

erosion, deep

Heezen, B.C., 1959.
Deep-sea erosion and unconformities.
J. Geology, 67(6):713-714.

erosion

Heezen, B.C., 1959.
Note on progress in geophysics: Dynamic processes of abyssal sedimentation: Erosion, transportation and redeposition on the deep-sea floor.
Geophys. J., 2(2):142-163.

erosion

Hom-ma, Masashi, Kiyoshi Horikawa, and Choule Sonu, 1960

A study on beach erosion at the sheltered beaches of Katase and Kamakura, Japan.
Coastal Eng., Japan, 3:101-122.

erosion

Hommeril, Pierre, 1958.
Erosion et sédimentation à Saint-Aubin sur Mer (Calvados). 2.
Bull. d'Info., C.C.O.E.C., 10(10):691-740.

erosion

Hommeril, Pierre, 1958.
Erosion et sedimentation à Saint-Aubin sur Mer (Calvados).
Bull. d'Info., C.C.O.E.C., 10(9):559-610.

erosion

Hommeril, P., and C. Larsonneur, 1963
Les effets des tempêtes du premier semestre 1962 sur les côtes Bas-Normandes.
Cahiers Océanogr., C.C.O.E.C., 15(5):320-334

erosion

Hommeril, Pierre, et Claude Larsonneur, 1963.
Quelques effets morphologiques du gel intense de l'hiver 1963 sur le littoral Bas-Normand.
Cahiers Océanogr., C.C.O.E.C., 15(9):638-650.

erosion

Horikawa, Kiyoshi and Tsuguo Sunamura, 1970.
A study on erosion of coastal cliffs and of submarine bedrocks. Coastal Engng Japan 13: 127-139.

erosion

Ishihara, Tojiro, Yuichi Iwagaki and Masashi Murakami, 1958
On the investigation of beach erosion along the north coast of Akashi Strait. Coastal Engineering in Japan, 1: 97-109.

erosion

Jarlan, G.E., 1961.
A perforated vertical wall break water. An examination of mass-transport effects in gravitational waves.
Dock & Harbour Authority, 41(486):394-398.

erosion

Jordan, J.F., 1961.
Erosion and sedimentation, Eastern Chesapeake Bay at the Choptank River.
U.S.C.&G.S., Techn. Bull., No. 16:8 pp., 1 chart

erosion

Kestner, F.J.T., and C.C. Ingles, 1956?
A study of erosion and accretion during cyclic changes in an estuary and their effect on reclamation of marginal land. J. Agric. Eng. Res 1(1):1-8.

erosion

Kinmont, A., 1961.
The nearshore movement of sand at Durban.
Marine Studies off the Natal Coast, C.S.I.R. Symposium, No. S2:46-58. (Multilithed).

erosions

Koh, Ryuji, 1969.
Beach erosion and Quaternary sea level.
Coast. Engng. Japan, 12: 121-128.

erosion

Komukai, Ryoshichi, 1959.
On the marine geology of beach erosion in Omori-hama vicinity, Hakodate City.
H. O. Pub., Japan, 943:582 pp.

erosion

Kuenen, Ph. H., 1964.
Experimental abrasion. 6. Surf action.
Sedimentology, 3(1):29-43.

erosion

Kuenen, Ph. H., 1959.
Experimental abrasion. 3. Fluviatile action on sand.
Amer. J. Sci., 257:172-190.

erosion

Kuenen, Ph. H., 1958.
Some experiments on fluviatile rounding.
Proc. K. Nederl. Akad. Wetenschappen, Amsterdam, (B), 61:47-53.
Geol. Inst., Groningen, Publ., 110.

erosion

Labzovski, N.A., 1961
Hydrological basement for shore-abrasion computations on the artificial lakes.
Trudy Okeanogr. Komissii, Akad. Nauk, SSSR, 8:98-103.

erosion

Ladd, H.S., 1961
Reef building. The growth of living breakwaters has kept pace with subsidence and wave erosion for fifty million years.
Science, 134(3481):703-715.

erosion

Langfelder, Leonard J., Donald B. Stafford and Michael Amein, 1970
Coastal erosion in North Carolina.
J. Waterways Harb. Div. Am. Soc. civ. Engrs 96(WW2):531-545.

erosion

Larras, J., 1957.
Recherches sur l'érosion des sables par la houle et le clapotis. Ann. Ponts et Chaussées, 127(5):

Erosion

Lates, M., and A. Spătaru, 1970
Consequences of using the seaside beaches as sand source. (In Romanian)
Hidrotehnica, Romania, 15(2):87-91

erosion

*Laughton, A.S., 1968.
New evidence of erosion on the deep ocean floor.
Deep-Sea Res., 15(1):21-29.

erosion

Lüneburg, Hans, 1961.
Über die Erosion und den Sedimenttransport am Knechtsand und Eversand (nordliche Wesermündung).
Veroff. Inst. Meeresf., Bremerhaven, 7(2):277-294.

Erosion

Martin, R. Torrence, 1962.
Discussion of paper by Walter L. Moore and Frank D. Masch, Jr., "Experiments on the scour resistance of cohesive sediments".
J. Geophys. Res., 67(4):1447-1449.

erosion

*McLean, Roger F., 1967.
Measurements of beachrock erosion by some tropical marine gastropods.
Bull. mar. Sci. Miami, 17(3):551-561.

erosion

Medvedev, V.S., 1959
Erosion remnants of loose deposits on the shores of the White Sea. Izv. Akad. Nauk, SSSR, Ser. Geograf., (3): 96-

erosion

Mierzynski, Stanislaw, 1961
Survey of research work of the Maritime Institute at Gdansk in the Swinoujscie-Szczecin fairway.
(In Polish; English summary).
Prace Inst. Morsk., Gdansk, (1) Hydrotech., II. Sesja Naukowa Inst. Morsk., 20-21 wrzesnia 1960:53-61, English summary, 144-146.
(mimeographed).

Erosion

Moore, Walter L., and Frank D. Masch, Jr., 1962.
Experiments on the scour resistance of cohesive sediments.
J. Geophys. Res., 67(4):1437-1446.

erosion

Neumann, A. Conrad, 1966.
Observations on coastal erosion in Bermuda and measurements of the boring rate of the sponge, Cliona lampa.
Limnol. and Oceanogr., 11(1):92-108.

erosion

Orlova, G.A., 1961
Model experiments on the unconsolidated slope wave abrasion and the littoral drift.
Trudy Okeanogr. Komissii, Akad. Nauk, SSSR, 8:235-239.

erosion

Osiecimski, Roman, 1961
Characteristic of the eroding phenomena in the Swinoujscie-Szczecin fairway. (In Polish; English summary).
Prace Inst. Morsk., Gdansk, (1) Hydrotech., II. Sesja Naukowa Inst. Morsk., 20-21 wrzesnia 1960: 63-105; English summary, 146-148.
(mimeographed).

erosion

Ottmann, Francois, 1965.
Introduction a la Geologie Marine et Littorale.
Masson et Cie, Paris, 259 pp. 47 F

erosion

Postma, H., 1967.
Sediment transport and sedimentation in the estuarine environment.
In: Estuaries, G.H. Lauff, editor, Publs Am. Ass. Advmt Sci., 83:158-179.

erosion

Princeton University, Department of Civil Engineering, River and Harbor Program, 1958 (21 July).
Bibliography: Currents along the Eastern Seaboard of the United States and Coast Erosion. 6 pp. (mimeographed).

erosion

Purdy, E.G., and L.S. Kornicker, 1958.
Algal disintegration of Bahamian limestone coasts. J. Geol., 66(1):97-99.

erosion

Revelle, R., and K.O. Emery, 1957.
Chemical erosion of beach rock and exposed reef rock. Geol. Survey Prof. Paper 260-T:699-709.

erosion

Richards, Adrian F., 1966.
Geology of the Islas Revillagigedo. 3. Effects of erosion on Isla San Benedicto, 1952-61 following the birth of Volcán Bárcena.
Bull. volcan., 28: 1-22.

erosion

Rona, Peter A., 1969.
The Middle Atlantic Continental slope of United States: deposition and erosion
Bull. Am. Ass. Petrol. Geol. 53(7): 1453-1465.

erosion

Safianov, G.A., 1965.
The erosional effect of debris in the coastal zone. (In Russian).
Okeanologiia, Akad. Nauk, SSSR, 5(2):304-310.

erosion

Sager, Günther, 1958.
Zur Beurteilung der zerstörenden Wirkungen von Hochwassern und Sturmfluten an Meeresküsten.
Zeits. Angewandte Geol., 4(12):576-578.

erosion

Sanders N.K., 1968.
Wave tank experiments on the erosion of rocky coasts.
Pap. Proc. R. Soc. Tasmania, 102:11-16

erosion

Schaufele, H.J., 1950.
Erosion and corrosion on marine structures. Inst Coastal Eng., Univ. Ext., Univ. Calif., Long Beach, 11-13 Oct. 1950:6 pp., 5 pls. (multilithed).

erosion

Schou, Axel, 1967.
Estuarine research in the Danish moraine archipelago.
In: Estuaries, G.H. Lauff, editor, Publs Am. Ass. Advmt Sci., 83:129-145.

erosion

*Schultz, Leonard P., and Wallace Ashby, 1967.
An analysis of an attempt to control beach erosion in Chesapeake Bay, at Scientist Cliffs, Calvert County, Maryland.
Chesapeake Science 8(4):237-252.

erosion, shore

*Schwartz, Maurice L., 1968.
The scale of shore erosion.
J. Geol., 76:508-517.

erosion

Schwartz, Maurice L., 1967.
The BRUUN theory of sea-level rise as a cause of shore erosion.
J. geol., 75(1):76-92.

erosion

Schwartz, Maurice, 1965.
Laboratory study of sea-level rise as a cause of shore erosion.
J. Geol., 73(3):528-534.

erosion

Shepard, F.P., 1951.
Submarine erosion, a discussion of recent papers.
Bull. G.S.A. 62(12-1):1413-1417.

erosion

Siever, Raymond 1968.
Sedimentological consequences of a steady state ocean-atmosphere.
Sedimentology, 11 (1/2):5-29

erosion

Siever, Raymond and Miriam Kastner, 1967.
Mineralogy and petrology of some Mid-Atlantic Ridge sediments.
J. mar. Res., 25(3):263-278.

erosion

Slavianov, V. N., 1946.
[Graphical comparison of abrasion action of the Black Sea in various localities on the south Crimean coast.] Dok. Akad. Nauk SSSR 61:1083-1086.

erosion

Southard, J.B., R.A. Young and C.D. Hollister, 1971.
Experimental erosion of calcareous ooze. J. geophys. Res. 76(24): 5903-5909.

erosion

Sunamura, Tsuguo and Kiyoshi Horikawa, 1969.
A study of erosion of coastal cliffs by using aerial photographs. Coast. Engng. Japan, 12: 99-120.

erosion

Tanner, W.F., 1961
Offshore shoals in area of energy deficit.
J. Sed. Petr., 31(1): 87-95.

erosion

Uchupi, Elazar, 1967.
The continental margin south of Cape Hatteras, North Carolina.
SEast Geol., Duke Univ., 8(4):155-177.

erosion

Warme, J.E., T.B. Scanland, and N.F. Marshall, 1971.
Submarine canyon erosion: contribution of marine rock burrowers. Science 173(4002): 1127-1129.

coastal erosion

Zenkovich, V.P., 1962
The sea coast exploratory works in Holland and measures of eroding control. (In Russian).
Okeanologiia, Akad. Nauk, SSSR, 2(4):684-698.

erosion

Zenkovich, V.P., 1947.
[Erosion phenomena of the Caucasus shore of the Black Sea.] Voprosy Geogr. Akad. Nauk, USSR, 63: 183-186.

EROSION, coastal

coastal erosion

Byrne, John V., 1963.
Coastal erosion, northern Oregon.
In: Essays in Marine Geology in honor of K.O. Emery, Thomas Clements, Editor, Univ. Southern California Press, 11-33.

erosion, effect of

erosion, effect of

Flemming, N.C., 1965
Derivation of Pleistocene marine chronology from morphometry of erosion profiles.
J. Geol. 76: 280-296
Also in: Coll. Repr. Nat. Inst. Oceanogr. Wormley, 16 (656)

erosion, effect of

Lavrov, V.M., 1965.
O the influence of water erosion of old rock shelves upon the processes of recent marine sediment formation. (In Russian).
Okeanologiia, Akad. Nauk, SSSR, 5(1):94-98.

errors

Errata

Japan, Japanese Oceanographic Data Center, 1969
List of Data Report of CSK issued by March 1969 and Errata. 9 pp. (multilithed).

erratics

Pratt, Richard M., 1971.
Lithology of rocks dredged from the Blake Plateau.
SEast. Geol. 13(1):19-38.

errors

Uda, M., 1934
On the error of sounding which is introduced by the inclination of the sounding wire.
Umi to Sora 14(4): 147-153.

eruptions

eruptions (Krakatoa)

Ewing, M., and F. Press, 1951.
Tide-guage disturbances from the great eruption of Krakatoa (Abstract). Bull. G.S.A. 62:1527.

escape reaction

Thomas, G.E. and Ll.D. Gruffydd, 1971.
The types of escape reactions elicited in the scallop Pecten maximus by selected sea-star species. Mar. Biol. 10(1): 87-93.

escarpments

escarpments

Bryant, William R., A.A. Meyerhoff, Noel K. Brown, Jr., Max A. Furrer, Thomas E. Pyle and John W. Antoine, 1969.
Escarpments, reef trends, and diapiric structures, eastern Gulf of Mexico.
Bull. Am. Ass. Petrol. Geol., 53(12): 2506-2542.

escarpments

Dietz, R.S., 1952.
Some Pacific and Antarctic sea-floor features discovered during the U.S. Navy Antarctic Expedition, 1946-1947. (Abstract).
Proc. Seventh Pacific Sci. Congr., Met. Ocean., 3:335-344, 4 textfigs.

escarpment

Fox, P.J., B.C. Heezen, W.F. Ruddiman and W.B. Ryan, 1968.
Igneous rocks from the Beata Ridge. Trans. Fifth Carib. Geol. Conf., St. Thomas, V.I., 1-5 July 1968, Peter H. Mattson, editor. Geol. Bull. Queens Coll., Flushing 5: 65.

escarpments

Heezen, Bruce C., Bill Glass and H. W. Menard, 1966.
The Manihiki Plateau.
Deep-Sea Research, 13(3):445-458.

escarpments

Henderson, Garry C., 1963.
Preliminary study of the crustal structure across the Campeche escarpment from gravity data.
Geophysics, 28(5)(2):736-744.

escarpment

Menard, H.W., and R.S. Dietz, 1951.
Mendocino submarine escarpment. Bull. G.S.A. 62:1507.

escarpments

Worzel, J. Lamar, and J.C. Harrison, 1963.
9. Gravity at sea.
In: The Sea, M.N. Hill, Editor, Interscience Publishers, 3:134-174.

Estuaries

A

estuaries

Abbott, M.R., 1960

L'influence de la force de Coriolis sur les courants de marée dans un estuaire exponential. La Houille Blanche (5) (Sept.-Oct.): 616-624.

estuaries

Abbott, M.R., 1960

Tides in estuaries. The Dock and Harbour Authority, 41(482):259-260

estuaries

Abbott, M.R., 1960

Boundary layer effects in estuaries. J. Mar. Res., 18(2): 83-100.

estuaries

Abbott, M.R., 1960

Salinity effects in estuaries. J. Mar. Res., 18(2): 101-111.

estuaries

Abbott, M. R., 1959.

The downstream effect of closing a barrier across an estuary with particular reference to the Thames. Proc. R. Soc., London, A, 251:426-439.

estuaries

Abecasis, D., 1949.
Amélioration des embouchures des voies d'eaux naturelles et artificielles. XVII Congr. Int. Navig., Sect. 1, Question 1:191-217, 8 figs.

estuaries

Abecasis, F., M.F Matias, J.J. Reis de Cavalho and D. Vera-Cruz, 1962
Methods of determining sand-and-silt movement along the coast, in estuaries and in maritime rivers.
Lab. Nacional de Engenhara Civil. Tech. Paper Lisbon, No. 186: 25 pp.

estuaries

Acara, A., and C. Erol, 1960

On the pollution of Golden Horn Estuary. Rapp. Proc. Verb. Reun. Monaco, 15(3): 27-32. Comm. Int. Expl. Sci. Mer. Med.

Ahnert, Frank, 1960. estuaries
Estuarine meanders in the Chesapeake Bay area. Geogr. Rev., 50(3):391-401.

estuaries

Ahrens, Renate, 1970.
3.3. Über das Vorkommen von sternbildenden Bakterien Chemische, mikrobiologische und planktologische Untersuchungen in der Schlei im Hinblick auf deren Abwasserbelastung. Kieler Meeresforsch 26(2): 159-161.

estuaries

Allard, P., 1952.
Note sur l'integration des equations aux derivées partielles des marées par le méthode des caractéristiques. (Analyse de l'ouvrage de J.P. Schönfeld - Propagation of tides and similar waves). Bull. d'Info., C.C.O.E.C., 4(2):45-53.

estuaries

Allen, F.H., 1954.
The Thames model investigation. A study of siltation problems. J. Inst. Water Eng. 8(3): 232-242, 6 textfigs.

estuaries

Allen, F.H., and J. Grindley, 1957.
Radioactive tracers in the Thames Estuary, report on an experiment carried out in 1955. Dock & Harbour Authority, 37(435):302-306.

estuaries

Allen, F. H., and W. A. Price, 1959.

Density currents and siltation in docks and tidal basins.
Dock & Harbour Auth., July, 1959: 5 pp.

estuaries

Allen, F.H., W.A. Price and C.C. Inglis, 1955.
Model experiments on the storm surge of 1953 in the Thames Estuary and the reduction of future surges.
Proc. Inst. Civil Eng., III, Hydraulic Paper No. 5; Apr., Aug., 1955:48-82; 557-574.

estuaries

Allen, F.H., W.A. Price and C.C. Inglis, 1955.
Model experiments on the storm surge of 1953 in the Thames estuary and the reduction of future surges. Proc. Inst. Civil Eng. (3)4:48-72.

estuaries

Allen, George P., 1971.
Relationship between grain size parameter distribution and current patterns in the Gironde Estuary (France)
J. sedim. Petrol. 41(1): 74-88.

estuaries

Allen, G.P., A. Klingebiel et A. de Resseguier 1970
Evolution et signification dynamique de quelques indices granulometriques des sédiments de l'embouchure de la Gironde.
Cah. oceanogr. 22(5): 801-813

estuaries

*Allen, George W., 1967.
A biologist's viewpoint of man-made changes in estuaries.
Proc. Gulf Caribb. Fish. Inst., 19th. Sess.:69-74.

estuaries

Allen, J., 1938.
Experiments on water waves of translation in small channels. Phil. Mag., 7th ser., 25:754-767

estuaries

Allersma, E., A.J. Hoekstra and E.W. Bijker, 1967.
Transport patterns in the Chao Phya estuary.
Proc. 10th Conf. Coast. Engng. Tokyo, 1966, 1:632-650.

estuaries

Almazov A.M., 1956.
On the proportions between ion concentrations in waters of open estuaries.
Dokl. Akad. Nauk, SSSR, 108(5):833-836.

estuaries

Alsaffar, Adnan M., 1966.
Lateral diffusion in a tidal estuary.
J. geophys. Res., 71(24):5837-5841.

echinoderms

Alton, Miles S. 1966.
Bathymetric distribution of sea stars (Asteroidea) off the northern Oregon Coast. J. Fish. Res. Bd., Can., 23(11):1673-1714.

estuaries

Alvarez Valderrama, E., 1949.
Amélioration des embouchures des voies d'eau naturelles et artificielles. Deux estuaires de petites dimensions: l'estuaire de Noya; l'estuaire de Betanzos.
XVIIe Congr. Int. Navigation, Lisbonne, Sect. 2, Question 1:13-27, 11 figs.

estuaries

Anadón, E., F. Saiz, and M. López-Bonite, 1961.
Estudio hidrografico de la Ría de Vigo.
Inv. Pesq., Barcelona, 20:83-130.

estuaries

Anderson, Franz E. 1970.
The periodic cycle of particulate matter in a shallow, temperate estuary.
J. sedim. Petrol. 40(4): 1128-1135.

estuaries

* Anderson, Franz E., 1968.
Seaward terminus of the Vashon continental glacier in the Strait of Juan de Fuca.
Marine Geol. 6(6): 419-438.

estuaries

Anon., 1954.
Survey of the Thames estuary.
British Waterworks Assoc. 36(272):36-37.

estuaries

Anon., 1951.
Les problèmes de marée posés par le "Central Valley Project". Construction - La Technique Moderne 6(11):380-381.

estuaries

Anon., 1951.
Exposé d'une enquête américaine en cours sur la mesure des transports de sédiments dans les fleuves à marée. La Houille Blanche A:223-224.

estuaries

Anon., 1948.
Tidal model of the Firth of Forth. The Engineer 185(4816):478-480.

estuaries

Anon., 1946.
Tidal rivers. J. Central Board of Irrigation 3(4):275-280.

estuaries

Anon., 1925.
The Soulina mouth of the River Danube.
Dock and Harbour Authority No. 60:366-369, 5 figs

estuaries

Anon., 1925.
River Danube entrance channel.
Dock and Harbour Authority No. 53:139-145, 10 figs.

estuaries
Anon., 1921.
The pory of Sulina and the mouth of the Danube.
Dock and Harbour Authority No. 4:106-108.

estuaries
Antonov, V.S., 1960.
[The Lena River delta. (Brief hydrological sketch). Study of river estuaries.]
Trudy Okeanograf Komissii, Akad. Nauk, SSSR, 6: 25-34.

estuaries
Apollov, B.A., 1960.
[The distribution of currents in the preexisting extent of the Volga delta. Study of river estuaries.]
Trudy Okeanograf. Komissii, Akad. Nauk, SSSR, 6 :35-38.

estuaries
Arbot, M.R., 1960
L'influence de la force de Coriolis sur les courants de marée dans un estuarie exponential
La Houille Blanche (5): 616-624.

estuaries
Armstrong, F.A.J., and G.T. Boalch, 1961
The ultra-violet absorption of sea water.
J. Mar. Biol. Assoc., U.K., 41(3):591-597.

estuaries
Arnaud, J., G. Brun, 1961.
Note préliminaire sur l'hydrologie et l'hydrographie de l'estuaire du Grand Rhone.
Rapp. Proc. Verb., Reunions, Comm. Int. Expl. Sci., Mer Méditerranée, Monaco, 16(3):783-786.

estuaries
Arthur, Robert S., 1964.
The equations of continuity for seawater and river water in estuaries.
J. Mar. Res., 22(2):197-202.

estuaries
Arthur, Robert S., 1964.
The equations of continuity for sea water and river water in estuaries. (Abstract).
Trans. Amer. Geophys. Union, 45(1):66.

estuaries
Astu, G., 1960
La baie de Chingoudy et ses formations d'estuaire.
Bull. du C.E.R.S., Biarritz, 3(2):201-212.

estuaries
Auninsh, E.A., 1965.
The hydrochemical regime of the estuary and preestuary offing of the Zapad Dvina (Daugav) River. (In Russian).
Trudy, Gosudarst. Okeanogr. Inst., 83:101-157.

estuaries
Australia, Commonwealth Scientific and Industrial Research Organization, 1957.
Estuarine hydrological investigations in eastern and south-western Australia, 1956.
Oceanogr. Sta. List, Div. Fish. and Oceanogr., 32:170 pp.

estuaries
Australia, Commonwealth Scientific and Industrial Research Organization, 1951.
Oceanographic station list of investigations made by the Fisheries Division, C.S.I.R.O. Vol. 4. Onshore hydrological investigations in eastern Australia, 1942-50. 114 pp.

estuaries
Avila, Quinto, y Harold Loesch, 1965.
Identificacion de los camarones (Peneidae) juveniles de los esteros del Ecuador.
Bol. Cient. Y. Techico, Inst. Nacional, Pesca, Ecuador, 1(3):24 pp.

estuaries
Ayers, John C., 1959.
The hydrography of Barnstable Harbor, Massachusetts. Limnol. & Oceanogr. 4(4): 448-462.

estuaries
Ayers, J.C., 1956.
Population dynamics of the marine clam, Mya arenaria. Limnol., Oceanogr., 1(1):26-34.

B

estuaries
Baas Becking, L. G. M., 1956.
Biological processes in the estuarine environment. IX. Observations on total base.
Proc. K. Nederl. Akad. Wetensk., Amsterdam, B, 59(5):408-420.

estuaries
Baas Becking, L.G.M., and I. R. Caplan, 1956.
Biological processes in the estuarine environment. III. Electrochemical considerations regarding the sulphur cycle. K. Nederl. Akad. Wetens., Amsterdam, Proc., B, 59(2):85-96.

estuaries
Baas Becking, L.G.M., and I.R. Kaplan, 1956.
Biological processes in the estuarine environment. IV. Attempts at interpretation of observed Eh-pH relations of various members of the sulphur cycle. Proc. K. Nederl. Akad. Wetensk., Amsterdam, B, 59(2):97-108.

estuaries
Baas Becking, L.G.M., and E. J. Ferguson Wood, 1955.
Biological processes in the estuarine environment. 1. Ecology of the sulphur cycle. Proc. K. Nederl. Akad. Wetensk., Amsterdam, B, 58(3):161-181.

estuaries
Baas Becking, L.G.M., and E.J. Ferguson Wood, 1953.
Microbial ecology of the estuarine sulphuretum.
Atti VI Congreso Int. Microbiol., Roma, 6-12 settembre 1953, 7(22):379-388.

estuaries
Bacon, Peter R. 1971
Plankton studies in a Caribbean estuarine environment.
Carib. J. Sci. 11(1/2): 81-89

estuaries
Baidin, S.S., 1960.
[On the question of the distribution of the water flow in the Volga delta. Study of river estuaries.]
Trudy Okeanograf. Komissii, Akad. Nauk, SSSR, 6: 39-54.

estuaries
Bailey, Thomas E., Charles A. McCullough and Charles G. Gunnerson, 1966.
Mixing and dispersion studies in San Francisco Bay.
J. sanit. Engng Div. Soc. civ. Engrs, 92(SA5):23-45.

estuaries
Bainbridge, V., 1960.
The plankton of the inshore waters off Freetown Sierra Leone.
Col. Off. Fish. Publ., (13):1-48.

estuaries
Balasubrahmanyan, K., 1962
Studies in the ecology of the Vellar Estuary. 2. Phosphates in the bottom sediments.
J. Zool. Soc., India, Calcutta, 13(2):166-169.

Also in: Annamalai Univ. Mar. Biol. Sta. Porto Novo, S. India, Publ., 1961-1962.

estuaries
Balay, Marciano A., 1961.
El Rio de la Plata entre la atmosfera y el mar.
Rep. Argentina, Sec. Marina, Serv. Hidrogr. Naval, Publ., H. 621:153. pp.

estuaries
Balay, Marciano A., 1958.
Causas y periodicidad de las grandes crecidas en el Rio de la Plata (Crecida del 27 y 28 de julio de 1958).
Servicio de Hidrografia Naval, Argentina, Publ., H. 611:1-34.

crecidas = freshets

estuaries
Balbi, R., 1951.
[Physical and chemical characteristics of sea water at the Lido of Venice as compared with the mineral water of Salsomaggiore.] Arch. Opped. al Mare 3:127-132.

Abstr. Chem. Abstr., 1952:2355f.

estuaries
Banal, M., 1954.
Observations et études hydrologiques en Seine maritime et dans l'estuaire. Travaux No. 234: 278-288, 16 figs.

Incomplete reference?

estuaries
Banal, M., 1950.
Travaux d'amélioration de l'estuaire de la Seine.
Travaux No. 188:417-428, 20 figs.

estuaries
Banal, 1950.
Le mascaret en Seine; historique, description et mesure du phénomène. Théorie. Comm. à la Soc. Hydrotechnique de France, Session du 15 Juin 1950 (Mém. et Trav. S.H.F. HB B/1950:658 - resume).

Banks, R.E

estuaries
Banks, R.E., 1966.
The cold layer in the Gulf of St. Lawrence.
J. Geophys. Res., 71(6):1603-1610.

estuaries
Banks, R.E., 1964.
The cold layer of the Gulf of St. Lawrence. (Abstract).
Trans. Amer. Geophys. Union, 45(1):65-66.

estuaries

Bansemir, Klaus, und Gerhard Rheinheimer 1970.
3.6. Bakterielle Sulfatreduktion und Schwefeloxydation. Chemische, mikrobiologische und planktologische Untersuchungen in der Schlei im Hinblick auf deren Abwasserbelastung. Kieler Meeresforsch 26(2): 170-173.

Barlow, J.P., 1958 estuaries
Spring changes in the phytoplankton abundance in a deep estuary, Hood Canal, Washington. J. Mar. Res., 17:53-67.

estuaries
Barlow, J.P., 1957.
Effect of wind on salinity distribution in an estuary. J. Mar. Res., 15(3):193-204.

estuaries
Barlow, J.P., 1955.
Physical and biological processes determining the distribution of zooplankton in a tidal estuary. Biol. Bull. 109(2):211-225.

estuaries
Barlow, John P., Carl J. Lorenzen and Richard T. Myren, 1963
Eutrophication of a tidal estuary. Limnol. and Oceanogr.,8(2):251-262.

estuaries
Barnes, H., 1953.
Considerazioni statistiche sulla distribuzione spaziale di alcuni organismi pelagici raccolti su un lungo percorso nel Golfo della Clyde. Mem. Ist. Ital. Idrobiol., "Dott. Marco De Marchi" Pallanza, 7:109-127, 1 textfig.
(planctonici)

estuaries
Baron, G., 1938.
Étude du plancton dans le bassin de Marennes. Rev. Trav. Off. Pêches Marit. 11(2):167-188, 2 textfigs.

estuaries
Bartlett,G.A., 1966.
The significance of foraminiferal distribution in waters influenced by the Gulf of St. Lawrence. (Abstract Only).
Second Int. Oceanogr. Congr.,30 May-9 June 1966, Abstracts,Moscow:22-23.

estuaries
Bartsch, A. F., R.J. Callaway, R.A. Wagner and C.E. Woelke, 1967.
Technical approaches toward evaluating estuarine pollution problems.
In: Estuaries, G.H. Lauff, editor, Publs Am. Ass Advmt Sci., 83:693-700.

estuaries
Bassindale, R., 1955.
Ch. 5. Fauna. In: Bristol and its adjoining counties, J.W. Arrowsmith, Ltd., Bristol.

estuaries
Bassindale, R., 1938.
The intertidal fauna of the Mersey Estuary. J.M.B.A. 23(1):83-98, 1 pl.

Reviewed: J. du Cons. 14(3):421-433:HCR.

estuaries
Battjes, J.A., 1967.
Quantitative research on littoral drift and tidal inlets.
In: Estuaries, G.H. Lauff, editor, Publs Am. Ass Advmt Sci., 83:185-190.

estuaries
Bayly, I.A.E., 1965.
Ecological studies on the planktonic Copepode of the Brisbane River Estuary with special reference to Gladioferens pectinatus (Bredy) (Calanoida).
Australian J. Mar. Freshwater Res., 16(3):315-350.

estuaries
Beardmore, Nathaniel, 1862.
Manual of hydrology - Division III. Tides of the sea, estuaries and tidal rivers. London, Waterlow and Sons:203-275.

estuaries
Beauge, L., 1956.
Le Golfe du Saint-Laurent.
Rev. Trav. Inst. Pêches Marit., 20:5-39.

estuaries
Beauvais, , 1948.
Effet des obstacles dans les estuaires à marée. Rev. Gén. Hydraulique No. 48:307-310.

estuaries
Beaven, G. Francis, 1960.
Temperature and salinity of surface water at Solomons, Maryland.
Chesapeake Science, 1(1):2-11.

estuaries
Beaven, G.F., 1946.
Effect of Susquehanna River stream flow on Chesapeake Bay salinities and history of past oyster mortalities on upper bay bars. Chesapeake Biol. Sta. Contrib. No. 68:11 pp., 7 textfigs.

estuaries
Belevich, E.F., 1960.
On the vertical height of the islands in the lower zone of the Volga delta. Study of river estuaries.
Trudy Okeanograf. Komissii, Akad. Nauk, SSSR, 6: 55-65.

estuaries
Beliaev, I.P., 1960.
The basic characteristics of the hydrology of the Tereka estuary. Study of river estuaries.
Trudy Okeanograf. Komissii, Akad. Nauk, SSSR, 6: 75-85.

estuaries
Bell, W.H., 1963.
Reproduction of estuarine structure and current observation techniques in the Hecate Model.
Fish. Res. Bd., Canada, Mss. Rept. Ser., (Ocean. & Limnol.), No. 158.

estuaries
Belleville, M., 1883.
Sur le régime des courants et des matières alluvionnaires dans l'estuaire de la Seine.
Assoc. Française Avance.Sci., 12eme Session, Rouen:258-262.

estuaries
Benson, Richard H., and Rosalie F. Maddocks, 1964.
Recent ostracods of Knysna Estuary, Cape Province, Union of South Africa.
Univ. Kansas, Paleontol. Contrib., Article 5: 1-39.

estuaries

Bergeron, Julien 1970.
Travaux sur l'anguille.
Rapp. ann. 1969, Serv. Biol. Québec, 129-142.

estuaries
Berman, L. Zh., 1960.
Types of estuaries of small rivers on the west coast of the Caspian Sea. Study of river estuaries.
Trudy Okeanograf. Komissii, Akad. Nauk, SSSR, 6: 102-107.

estuaries
Bennett, H.H., 1951.
Relation of soil erosion to coastal waters.
Texas J. Sci. 3(2):147-161, numerous photos.

estuaries
Bergeron, J., and G. Lacroix, 1963.
Prélèvements de larve de poisson dans le sud-ouest du golfe Saint-Laurent en 1962.
Rapp. Ann., 1962, Sta. Biol. Mar. Grande-Rivière, 69-79. Canada

Berthois, Léopold

Estuaries
Berthois, Leopold, 1965.
Essai de correlation entre la sedimentation actuelle sur le bord externe des plateformes continentales et la dynamique fluviale.
Progress in Oceanography, 3:49-62.

estuaries
Berthois, Leopold, 1963
Contribution à l'étude de la sédimentation dnas l'estuaire du Konkoure (en période d'étiage), Republique de Guinée.
Cahiers Océanogr., C.C.O.E.C., 15(1):16-52.

estuaries
Berthois, Leopold, 1960.
Etude dynamique de la sedimentation dans la Loire. Cahiers Oceanogr., C.C.O.E.C., 12(9): 631-657.

estuaries
Berthois, Léopold, 1960
Etude expérimentale de l'érosion des vases d'estuaires. Comptes Rendus, l'Acad. Sci., Paris. 250(24): 4020-4022.

estuaries
Berthois, L., 1959
Essai de correlation entre le transport en suspension des sediments grossiers et la dynamique de l'estuaire de la Loire pendant la crue de Mars 1957. Cahiers Ocean. C.C.O. E.C., 11(6): 407-415.

estuaries
Berthois, L., 1958.
Observation d'une tranche dEau suivie dans ses déplacements entre Basse-Indre et Saint Nazaire.
Bull. d'Info., C.C.O.E.C., 10(4):192-208.

estuaries
Berthois, L., 1956.
Déplacement des aires d'envasement dans l'estuaire de la Loire. C.R. Acad. Sci., Paris, 243(18):1343-1345.

estuaries
Berthois, L., 1956.
Turbidité des eaux à l'entrée de l'estuaire de la Loire. C.R. Acad. Sci., Paris, 243(25):2113-2115.

estuaries
Berthois, L., 1956.
Variations de la salinité et de la temperature sur une coupe transversale de la Loire entre Saint-Nazaire et Mindin. Bull. d'Info., C.C.O.E.C 8(10):546-554, 3 pls.

estuaries
Berthois, L., 1955.
Evaluation du tonnage des sables apportés en suspension pendant les crues de la Loire en 195
C.R. Acad. Sci., Paris, 241(22):1605-1606.

estuaries
Berthois, L., 1955.
Sedimentation dans l'estuaire de la Loire en periode hivernale. C.R. Acad. Sci., Paris, 240(17):1691-1693.

estuaries
Berthois, L., 1954.
Sur les déplacements transverseaux des eaux très turbides dans l'estuaire de la Loire en période d'étiage. C.R. Acad. Sci., Paris, 239(14):820-822

estuaries
Berthois, Léopold, et Gérard Auffret, 1966.
Dynamique de la sédimentation dans les rias et les estuaires des petits cours d'eau tributaires de la Manche.
Cahiers océanogr., 18(9):761-774.

estuaries
Berthois, L., and M. Barbier, 1954.
Recherches sur la sedimentations en Loire prefils intantanes du fleuve en period d'etiage.
Bull. d'Info., C.C.O.E.C., 6(9):387-397, 6 pls.

estuaries
Berthois, L., and M. Barbier, 1953.
La sédimentation en Loire maritime en période d'étiage. C.R. Acad. Sci., Paris, 236(20):1984-1986, 1 textfig.

estuaries
Berthois, L.and C., 1954.
Étude de la sedimentation dans l'estuaire de la Rance. Bull. Lab. Marit., Dinard, 40:4-15, 4 textfigs.

estuaries
Berthois, Léopold et Philippe Bois, 1969.
Le cours inférieur et l'estuaire de la rivière du Château en période d'étiage, Iles de Kerguelen. Étude hydraulique, sédimentologique et chimique.
Cah. océanogr. 21 (8): 727-771.

estuaries
Blanc, François, Hervé Chamley et Michel Leveau, 1969.
Les minéraux en suspension, témoins du mélange des eaux fluviatiles en milieu marin. Exemple du Rhône.
C.r. hebd. Séanc. Acad Sci., Paris (D) 269 (25): 2509-2512

estuaries
Berthois, L., P. Chatelin, and A. Margon, 1953.
Influence de la salinité et de la température sur la vitesse de sedimentation dans les eaux de l'estuaire de la Loire. C.R.Acad. Sci., Paris, 237(7):465-467, 1 textfig.

Berthois, L., et A. Crosnier, 1966.
Étude dynamique de la sédimentation au large de l'estuaire de la Betsiboka.
Cah. ORSTOM, Sér. Océanogr., 4(2):49-130.

estuaries
Berthois, Leopold, et Alain Crosnier, 1965.
La sédimentation dans l'estuaire de la Betsiboka (côte ouest de Madagascar) et sur le plateau continental au large de l'estuarire.
Comptes Rendus hebdom. seanc. Acad. Sci., 261(18)

estuaries
Berthois, Léopold, et Yolande Le Calvez, 1961
Étude dynamique de la sedimentation dans la baie de Sangarea (République de Guinée).
Cahiers Océan., C.C.O.E.C., 13(10):694-714.

estuaries
Berthois, L., and Y. Le Guilly, 1963.
La stratification des eaux dans l'estuaire de la Loire.
Comptes Rendus, Acad. Sci., Paris, 256(19):4060-4063.

estuaries
Berthois, L., and P. Stertz, 1954.
Mesure de la vitesse de dépôt des sédiments en suspension dans les eaux de la Loire.
C.R. Acad. Sci., Paris, 239(13):891-893, 1 textfig.

estuaries
Besnard, W., 1950.
Considerações gerais em torno do região lagunar de Cananéia-Iguapel. Bol. Inst. Paulista Ocean. 1(1):9-26.

estuaries
Bhattacharya, S.K. and S.T. Ghotankar, 1969.
The estuary of the Hooghly. Bull. natn. Inst. Sci., India, 38(1): 25-32.

estuaries
Biber, V.A., and N.S. Bogolyubova, 1952.
The humic acids of estuary mud and their biological activity. (In Russian).
Doklady, Akad, Nauk, SSSR, 82(6):939.

estuaries
Biggs, Robert B., 1967.
The sediments of Chesapeake Bay.
In: Estuaries, G.H. Lauff, editor, Publs Am. Ass. Advmt Sci., 83:239-260.

estuaries
Biggs, Robert Bruce, 1963.
Deposition and early diagenesis of modern Chesapeake Bay muds.
Mar. Sci. Center, Lehigh Univ., 119 pp. (Unpublished manuscript).

estuaries
Biggs, R.B. and D.A. Flemer, 1972
The flux of particulate carbon in an estuary.
Mar. Biol. 12(1): 11-17.

estuaries
Biglane, K.E., and Robert A. Lafleur, 1967.
Notes on estuarine pollution with emphasis on the Louisiana Gulf coast.
In: Estuaries, G.H. Lauff, editor, Publs Am. Ass. Advmt Sci., 83:690-692.

estuaries
Bishev, L. L., 1955
[Results of hydrochemical investigations the Kubansk delta estuaries.] Trudy VNIRO 31: 145-150.

estuaries
Black, W.A., 1962
Geographical Branch program of ice surveys of the Gulf of St. Lawrence, 1956 to 1962.
Cahiers de Geographie de Quebec, 6(11):65-74.

estuaries
Black, W.A., 1962.
Gulf of St. Lawrence ice survey, winter 1961.
Dept. Mines and Techn. Surveys, Ottawa, Canada, Geogr. Paper, No. 32:52 pp., 46 photos.

estuaries
Blanchard, Richard L., 1965.
U 234/U 238 ratios in coastal marine waters and calcium carbonates.
J. Geophys. Res., 70(16):4055-4061.

estuaries
Blanton, Jackson, 1969.
Energy dissipation in a tidal estuary. J. geophys. Res., 74(23): 5460-5466.

estuaries
Blench, T., 1953.
Regime theory equations applied to a tidal estuar
Proc. Minnesota Int. Hydraulics Conv., Sept. 1-4, 1953:77-83, 1 textfig.

estuaries
Blosset, Marcel, 1951.
Théorie et pratique des travaux à la mer.
Editions Eyrolles, Paris, 647 pp., 513 figs.

estuaries
Bochkov, N.M., 1960.
[Estuarine processes for the average amount of water storage. Study of river estuaries.]
Trudy Okeanograf. Komissii, Akad. Nauk, SSSR, 6: 89-93.

estuaries
Boltovskoy, E., y A. Boltovskoy, 1968.
Foraminiferos y tecamebas de la parte inferior del Rio Quequen Grande.
Revta Mus. argent. Cienc. nat. Bernardino Rivadavia Inst. nac. Hidrobiol., 2(4):127-164.

estuaries
Bond, Gerard C., and Robert H. Meade, 1966.
Size distributions of mineral grains suspended in Chesapeake Bay and nearby coastal waters.
Chesapeake Sci., 7(4):208-212.

estuaries
*Bonnefille, René, Marcel Heuzel et Léopold Pernecker, 1967.
Emploi des traceurs radioactifs pour etudier l'evolution d'un nuage de Vase dans un estuaire.
Proc.10th Conf.Coast.Engng.Tokyo,1966,1:730-745.

estuaries
Bonnet, L., and J. Lamoen, 1946.
Étude théorique et expérimentale des cours d'eau à marée. Ann. Trav. Publ. de Belgique, Bruxelles, 3:271-277, 3 figs.

estuaries
Bonnin, R., 1903.
Les travaux actuels dans l'estuaire de la Seine.
La Nature:97-99, 2 textfigs.

estuaries
Boon, John D. III, and William G. MacIntyre 1968.
The boron-salinity relationships in estuarine sediments of the Rappahannock River, Virginia.
Chesapeake Science 9(1): 21-26.

estuaries
Borde, J., 1938.
Étude du plancton au bassin d'Arcachon, des rivières et du golfe de Morbihan.
Rev. Trav. Off. Pêches Marit. 11(4):523-542.

estuaries
Borishanskii, L.S., 1958.
Distribution of salinity in near-estuarine areas of the sea. (In Russian)
Trudy Gosud. Okeanogr. Inst., 42:128-141.

estuaries
Borut, S.Y., and T.W. Johnson, Jr., 1962.
Some biological observations on fungi in estuarine sediments.
Mycologia, 54(2):181-193.

estuaries
Bose, B.B., 1956.
Observations on the hydrology of the Hooghly Estuary.
Indian J. Fish., 1(3):101-118.

estuaries
Boudreault, F.-Robert, 1970.
Recherches en océanographie physique à Grande Rivière, en 1969.
Rapp. ann., 1969, Serv. Biol., Québec, 19-23

estuaries
Boudreault, F.-Robert, 1970.
Observations sur la température des eaux superficielles à Grande Rivière en 1969.
Rapp. ann. 1969, Serv. Biol., Québec, 25-27.

estuaries
Boudreault, F.-Robert, 1969.
Régime des vents à Grande-Rivière (Baie des Chaleurs).
Naturaliste can. 96: 667-670.
Also in: Trav. Pêch. Québec 29.

estuaries
Boudreault, F.-Robert, 1968.
Revue des travaux d'océanographie physique effectués dans la Baie-des-Chaleurs (1924-1967).
Cah. Info., Sta. Biol. mar. Grande Rivière, 47: 24 pp. (multilithed).

estuaries
*Boudreault, F. Robert, 1967.
Régime thermique saisonnier d'une station-pilote à l'entrée de la baie des Chaleurs.
Naturaliste Can., 94:695-698.

Also: Trav. Pêch., Quebec, 18

estuaries
Boudreault, F. Robert, 1967.
Océanographie physique en 1966.
Rapp. Stn Biol. Mar. Grande-Rivière, 1966:15-19.

estuaries
Boudreault, F. Robert, 1967.
Température des eaux superficielles de Grande-Rivière en 1966.
Rapp. Stn Biol. Mar. Grande-Rivière, 1966: 21-23.

estuaries
Boudreault, F. Robert, 1966.
Océanographie physique de la Baie-des-Chaleurs, 1965.
Rapp. Stn. Biol. Mar., Grande-Rivière, 1965:17-30.

estuaries
Boudreault, F. Robert, 1965.
Hydrographie de la Baie-des-Chaleurs, été 1964.
Rapp. Ann., Sta. Biol. Mar., Grande Rivière, 1964:17-29.

estuaries
Bouniceau, M., 1845.
Étude sur la navigation des rivières à marées et la conquête des lais et relais de leur embouchure. Paris, M. Mathias Augustin, 204 pp.

estuaries
Bouquet de la Grye, 1864-66.
Recherches hydrographiques sur le régime des côtes (Troisième cahier). Rapp. LXVI. Rapport sur le régime de la Loire. Service Hydrographique de la Marine No. 13-603:64-118.

Bourcart, Jacques

estuaries
Bourcart, J., 1952.
Les frontières de l'océan. Albin Michel, Paris, 317 pp., 77 textfigs.

estuaries
Bourcart, J., 1939.
Sur les vasières des estuaires de la Manche. C.R.S. Soc. Geol., France, 73-74.

estuaries
Bourcart, J., 1939.
Essai d'une définition de la vase des estuaires.
C.R. Acad. Sci., Paris, 209:542-543.

estuaries
Bourcart, J., C. Francis-Boeuf, and B. Bajcevic, 1941.
Sur le méchanisme de la sédimentation des vases dans les estuaires. C.R. Acad. Sci., Paris, 213: 1025-1027.

estuaries
Bousfield, E.L., 1955.
Some physical features of the Miramichi estuary.
J. Fish. Res. Bd., Canada, 12(3):342-361, 11 textfigs.

Bowden, K.F.

estuaries
Bowden, K.F., 1967.
Circulation and diffusion.
In: Estuaries, G.H. Lauff, editor, Publs Am. Ass. Advmt Sci., 83:15-36.

estuaries
Bowden, K.F., 1963.
The mixing process in a tidal estuary.
Int. J. Air Water Pollution, 7(4/5):343-356.

Paper presented at International Conference on Water Pollution Research, London, 3-7 September 1962.

estuaries
Bowden, K.F., 1962.
Estuaries and coastal waters.
Proc. R. Soc., London, (A), 265(1322):320-325.

estuaries
Bowden, K.F., 1960?
Circulation and mixing in the Mersey estuary.
I.A.S.H., Commission of Surface Waters, No. 51: 352-360.

estuaries
Bowden, K.F., and S.H. Sharaf El Din, 1966.
Circulation, salinity and river discharge in the Mersey Estuary.
Geophys. J., R. Astr. Soc., 10(4):383-399.

estuaries
Bowden, K.F., and M.R. Howe, 1963.
Observations of turbulence in a tidal current.
J. Fluid Mechanics, 17(2):271-284.

estuaries
Bowden, K. F., and J. Proudman, 1949.
Observations on the turbulent fluctuations of a tidal current. Proc. Roy. Soc., London, A, 199: 311-327, 6 textfigs.

Estuaries
Bowman, Thomas E., 1961.
The copepod genus Acartia in Chesapeake Bay.
Chesapeake Science, 2(3/4):206-207.

estuaries
Bozic, B., 1965.
Copepodes de quelques petits estuaires méditerranéens.
Bull. Mus. Nat. Hist. Nat., (2), 37(2):351-356.

estuaries
Braarud, Trygve, Bjørg Føyn and Grethe Rytter Hasle, 1958
The marine and fresh-water phytoplankton of the Dramsfjord and the adjacent part of the Oslo-fjord, March - December 1951.
Hvalradets Skr., 43:102 pp.

estuaries
Braarud, T., and B. Hope, 1952.
The annual phytoplankton cycle of a landlocked fjord near Bergen (Nordåsvatn). Rep. Norwegian Fish. Mar. Invest. 9(16):26 pp., 4 textfigs.

estuaries
Brandsma, W., 1908.
Guiding of tidal currents at the mouth of rivers and the prevention of the sanding-up of channels. De Ingenieur, The Hague, 23:665-676.

Resume in Minutes of Proc. Inst. Civ. Eng. 1908/1909(3):331.

estuaries
Brattström, H., and E. Dahl, 1951.
Reports of the Lund University Chile Expedition, 1948-49. 1. General account, lists of stations, hydrography. Lunds Universitets Årsskrift, [K. Fysiograf. Sällskapets Handl. n.f., 61(8):] n.f., Avd. 2, 46(8):86 pp.

estuaries
Brisou, J., Menantaud and M. Doublet, 1958
Contribution à l'étude de la flore microbienne normale des estuaires. Les Pseudomonadaceae des boues de la Charente. Bull. l'Inst. Ocean. Monaco, 55(1129): 8.

estuaries
Broekhuysen, G.J., and H. Taylor, 1959
The ecology of South African estuaries. Part VIII. Kosi Bay estuary system. Ann. S. African Mus., 44(7): 279-296.

estuaries
Brooks, Ralph H., Jr., Patrick L. Brezonik, Hugh D. Putnam and Michael A. Keirn, 1971.
Nitrogen fixation in an estuarine environment: the Waccasassa on the Florida Gulf Coast. Limnol. Oceanogr. 16(5): 701-710.

estuaries
Brown, A.C., 1959.
The ecology of South African estuaries. IX. Notes on the estuary of the Orange River. Trans. R. Soc., S. Africa, 35(5):463-473.

estuaries
Brun, Guy, 1967.
Étude écologique de l'estuaire du "Grand Rhone". Bull. Inst. océanogr., Monaco, 66(1371):46 pp.

estuaries
Brun, G., 1962.
Contribution à l'étude écologique de l'estuaire du "Grand Rhône". Pubbl. Staz. Zool., Napoli, 32 (Suppl.):236-254.

estuaries
Brunel, Jules, 1962
Le phytoplancton de la Baie des Chaleurs. Contrib. Ministère de la Chasse et des Pêcheries, Province de Québec, No. 91: 365 pp.

estuaries
Brunel, Pierre, 1970.
Catalogue d'invertébrés benthiques du Golfe Saint-Laurent recueillis de 1951 à 1966 par la Station de Biologie Marine de Grande-Rivière.
Trav. Pêch., Québec 32: 54 pp.

estuaries
Brunel, Pierre, 1966.
Inventaire taxonomique des invertébrés benthiques marins du Golfe Saint-Laurent. Rapp. Stn. Biol. Mar., Grande-Rivière, 1965:83-85.

estuaries
Brunel, Pierre, 1965.
Inventaire taxonomique des invertébrés benthiques marins du Golfe Saint-Laurent. Rapp. Ann., Sta. Biol. Mar., Grande Rivière, 1964:63-64.

estuaries
Brunel, Pierre, 1963.
Inventaire taxonomique des invertébrés marins du Golfe Saint-Laurent. Ann. Rept., 1962, Sta. Biol. Mar., Grande-Rivière, Canada, 81-89.

estuaries
Brunel, P., 1963.
Variation journalières et saisonnières de l'alimentation de la morue au large de Grande Rivière. Rapp. Ann., 1962, Sta. Biol. Mar., Grande Rivière 101-117.

estuaries
Burbanck, W.D., 1963
Some observations on the isopod, Cyathura polita in Chesapeake Bay. Chesapeake Science, 4(2):104-105.

estuaries
Burbanck, W.D., M.E. Pierce and G.C. Whiteley, Jr 1956.
A study of the bottom fauna of Rand's Harbor, Massachusetts, an application of the ecotone concept. Ecol. Monogr., 26:213-243.

estuaries
Burt, W.V., and L.D. Marriage, 1957.
Computation of pollution in the Yaquina River estuary. Sewage & Ind. Wastes, 29(12):1385-1389.

estuaries
Burt, Wayne V., and W. Bruce McAlister, 1959.
Recent studies in the hydrography of Oregon estuaries. Research Briefs, Fish Comm., Oregon, 7(1):14-27.

estuaries
Burt, W.V., and J. Queen, 1957.
Tidal overmixing in estuaries. Science, 126(3280):973-974.

estuaries
Burt, W.V., and J. Queen, 1957.
Tidal overmixing in estuaries. Science 126(3280):973-974.

estuaries
Burton, A.J., 1961.
A quantitative assessment of the effect of factory effluent upon littoral and estuarine faunas at Umkomaas, Natal. Marine Studies off the Natal coast, C.S.I.R., Symposium, No. S2:73-80 (Multilithed).
Reprinted from: S. Africa J. Sci., 56(7):163-166. (1960).

Burton, J.D., 1970.
The behaviour of dissolved silicon during estuarine mixing. II. Preliminary investigations in the Vellar Estuary, Southern India. J. Cons. int. Explor. Mer, 33(2): 141-148.

estuaries
Burton, J.D. and P.C. Head, 1970.
Observations on the analysis of iron in seawater with particular reference to estuarine waters. Limnol. Oceanogr., 15(1): 164-167.

estuaries
Butler, Philip A., 1968.
Pesticide residues in estuarine mollusks. Proceedings of the National Symposium on Estuarine Pollution (August 1967), Stanford University, Stanford, Calif., p. 107-121, 1968. Also in: Collected Repr. Div. Biol. Bur. Comm. Fish., U.S. Fish Wildl. Serv., 1968, 1.

estuaries
Butler, Philip A., 1968.
Pesticides in the estuary. Proceedings of the Marsh and Estuary Management Symposium (July 1967), Baton Rouge, La.; 120-124. Also in: Collected Repr. Div. Biol. Res., Bur. Comm. Fish., U.S. Fish Wildl. Serv., 1968, 1.

estuaries
Butler, Philip A., 1966.
Fixation of DDT in estuaries. Trans. 31st N. Amer. Wildl. Nat. Resources Conf., Pittsburgh 184-189.

estuaries
Butler, Philip A., 1966.
The problem of pesticides in estuaries. Am. Fish. Soc. Spec. Publ., (3):110-115.

estuaries
Buzas, Martin A., 1969.
Foraminiferal species densities and environmental variables in an estuary. Limnol. Oceanogr., 14(3): 411-422.

estuaries
Byrne, John V., and L.D. Kulm, 1967.
Indicators of estuarine sediment movement. J. WatWays Harb. Div. Am. Soc. civ. Engrs, 93(WW2): 181-194.

C

estuaries
Cameron, W.M., 1951.
On the transverse forces in a British Columbia inlet. Trans. R. Soc., Canada, Sect. 5, 45:1-8.

estuaries
Cameron, W.M., and D.W. Pritchard, 1963
15. Estuaries. In: The Sea, M.N. Hill, Edit. Vol. 2 (III) Currents, Interscience Publisher New York and London, 306-324.

Canadian Oceanographic Data Center

estuaries
Canada, Canadian Oceanographic Data Centre 1971.
Gulf of St. Lawrence, September 17 to October 7, 1969.
1971 Data Record Ser. 2: 276pp.

estuaries
Canada, Canadian Oceanographic Data Center 1971.
Gulf of St. Lawrence, two surveys, June 18 to June 30, 1968; June 16 to July 1, 1968.
1971 Data Record Series, 1: 300 pp. (multilithed).

Estuaries
Canada Canadian Oceanographic Data Centre, 1969
Gulf of St. Lawrence, November 14 to November 24, 1965.
1969 Data Record Series 11: 72pp. (multilithed)

estuaries
Canada, Canadian Oceanographic Data Centre, 1967.
Gulf of St. Lawrence, September 16, to October 16, 1965
1967 Data Record Series No. 1: 169pp. (multilithed)

Canada, Canadian Oceanographic Data Center, 1966.
Gulf of St. Lawrence, Halifax section and Scotian Shelf to Grand Banks, three surveys, November 20 to December 22, 1965.
1966 Data Record Series, No. 10:113 pp. (multilithed).

estuaries
Canadian Oceanographic Data Centre, 1965.
Data record, Saguenay and Gulf of St. Lawrence, August 19 to August 30, 1963.
1965 Data Record Series, No. 12:123 pp. (multilithed).

Estuaries
Canada, Canadian Oceanographic Data Centre, 1965.
Data Record Gulf of St. Lawrence, July 23 to August 23, 1964.
1965 Data Record Ser., No. 9:262 pp. (multilith

estuaries
Canada, Canadian Oceanographic Data Centre, 1965
Data record, St. Lawrence Estuary, June 10 to July 24, 1963. 1965 Data Record Series, No.1: 127 pp. (multilithed).

estuaries
Cannon, Glenn A., 1971.
Statistical characteristics of velocity fluctuations at intermediate scales in a coastal plain estuary. J. geophys. Res, 76(24): 5852-5858.

estuaries
Carbonel, Pierre, Jean Moyes et Michel Vigneaux 1971.
La répartition des Thanatocoenoses d'ostracodes dans l'estuaire de la Gironde et ses relations avec les courants. (D)
C.r. hebd. Séanc. Acad. Sci. Paris 273 (19): 1679-1682.

estuaries
Carneiro, Olímpio, and Petrônio A. Coêlho, 1960
Estuto ecológico do Barra das Jangadas. Nota Prévia. (In Portuguese; English and French resumes).
Trabalhos, Inst. Biol. Marit. e Oceanogr., Universidade do Recife, Brasil, 2(1):237-248.

estuaries
Carpenter, J.H., 1960
The Chesapeake Bay Institute study of the Baltimore Harbor.
Proc. 33rd Ann. Conf., Maryland-Delaware Water and Sewage Assoc., June 9-10, 1960: 62-7
Also in:
Chesapeake Bay Inst., Collected Reprints, 5 (1963).

estuaries
Carpenter, Edward J., 1971.
Annual phytoplankton cycle of the Cape Fear River Estuary, North Carolina.
Chesapeake Sci., 12(2): 95-100

estuaries
Carpenter, Edward J., 1971.
Effects of phosphorus mining wastes on the growth of phytoplankton in the Pamlico River Estuary.
Chesapeake Sci., 12 (2): 85-94

estuaries
Carpenter, J.H., D.W. Pritchard and R.C. Whaley, 1969.
Observations of eutrophication and nutrient cycles in some coastal plain estuaries.
In: Eutrophication: causes, consequences, correctives: proceedings of a symposium, U.S. Nat. Acad. Sci., 210-221

estuaries
Carriker, Melbourne R., 1967.
Ecology of estuarine benthic invertebrates: a perspective.
In: Estuaries, G.H. Lauff, editor, Publs Am. Ass. Advmt Sci., 83:442-487.

estuaries
Carritt, D.E., and S. Goodgal, 1954.
Sorption reactions and some ecological implications. Deep-Sea Res. 1(4):224-243, 16 textfigs.

estuaries
Carruthers, J.N., H.G. Stubbings and A.L. Lawford 1950.
Water sampling in estuarial waters. Dock and Harbour Authority 31(362):253-259, 8 figs.

estuaries
Carruthers, J.N., and A.L. Lawford, 1950.
Water movements and winds at the Mouse Light Vessel, Thames Estuary. Weather 5:8

estuaries
Carter, H.H., 1968.
The distribution of excess temperature from a heated discharge in an estuary.
Techn. Rept. Chesapeake Bay Inst. 44: (Ref. 38-14) 39 pp. (multilithed).

Estuaries
Caspers, Hubert 1967.
Estuaries: analysis of definitions and biological considerations.
In: Estuaries, G.H. Lauff, editor, Publs Am. Ass. Advmt Sci., No. 83: 6-8.

estuaries
Caspers, H., 1959.
Die Einteilung der Brackwasser-Regionen in einem Aestuar.
Arch. Ocean. e Limnol., 11(Supplement):153-170.

estuaries
Chalmers, G.V., and A.K. Sparks, 1959.
An ecological survey of the Houston ship channel and adjacent bays.
Publ. Inst. Mar. Sci., 6:213-250.

estuaries
Chapman, Charles, 1968.
Channelization and spoiling in Gulf Coast & South Atlantic estuaries. Proceedings of the Marsh and Estuary Management Symposium (July 1967), Baton Rouge, La., : 93-106, 1968. Also in: Collected Repr. Div. Biol. Res., Bur. Comm. Fish., U.S. Fish Wildl. Serv., 1968, 1.

estuaries
Chapon, J.P., 1964.
L'ouverture du nouveau chenal d'acces au port de Rouen dans l'estuaire de la Seine.
La Houille Blanche, (4):509-520.

estuaries
Chatley, H., 1946.
The use and valifity of models in harbour and estuary investigation. Dock and Harbour Authority No. 311:115.

estuaries
Chatley, H., 1940.
Le probleme des ports dans les estuaires a maree.
Dock and Harbour Authority 20:113-114, 2 textfigs

estuaries
Chatley, H., 1938.
The Yantse estuary. Dock and Harbour Auth. 18:275-280, 5 textfigs.

estuaries
Chau, Y.K., 1958
Some hydrological features of the surface waters of the Pearl River estuary between Hong Kong and Macau. Hong Kong Univ. Fish. J. (2): 37-41.

estuaries
Chesapeake Bay Institute, 1963.
Bay Cruise XXIX, March 11-19, 1963. Bay Cruise XXX, July 30-September 11, 1963. Bay Cruise XXXI, October 21-24, 1963.
Data Rept. 50, Ref. 63-3:31 pp. (unpublished manuscript).

estuaries
Chesapeake Bay Institute, 1962.
Bay Cruise 28, July 24-August 7, 1962.
Data Rept. 48, Ref. 62-16:17 pp. (unpublished manuscript).

estuaries
Chesapeake Bay Institute, 1962.
Temperature and salinity data in the Chesapeake Bay and tributary estuaries, during the period 1-9 Feb. 1956.
Data Rept. 49, Ref. 62-23:22 pp. (unpublished manuscript).

estuaries
Chesapeake Bay Institute, Johns Hopkins University, 1959
Data Report 31. Baltimore Harbor Cruises 1-16, July 1958-Nov. 1959. 211 pp. Ref. No. 59-4.

estuaries
Chesapeake Bay Institute, 1954.
Data Report 23, Patuxent River Winter Cruise, 3 December-7 December 1952. Ref. 54-10:1-44, 1 textfig.

estuaries
Chesapeake Bay Institute, 1954.
Choptank River autumn cruise, 23 Sept-27 Sept 1952. Data Rept. 22, Ref. 54-9:37 pp., 1 fig.

estuaries
Chesapeake Bay Institute, 1954.
Choptank River spring cruise, 28 April-1 May 1952
Ref. 54-1:37 pp.

estuaries
Chevanier, M.C., 1946.
Mission hydrographique de l'Afrique occidental française. Ann. Hydrogr., Paris, (3)17:57-82.

estuaries
Chew, F., 1955.
Red tide and the fluctuations of conservative concentrations at an estuary mouth.
Bull. Mar. Sci., Gulf and Caribbean, 5(4):321-330

estuaries
*Choe, Sang, Tai Wha Chung and Hi-Sang Kwak, 1968.
Seasonal variations in primary productivity and pigments of downstream water of the Han River. (In Korean: English abstract).
J. oceanogr. Soc., Korea, 3(1):16-25.

estuaries
Chrimes, P.A.T., 1964.
Tidal flow in the Thames estuary.
Dock & Harbour Auth., 45(527):162.

estuaries
Codde, R.E.L., J. Lamoen, and J.E.L. Verschave, 1953.
La chlorinité de l'Escaut maritime.
XVIII Congr. Int. Navigation, Rome 1953, Sect.II. Navigation Maritime, Communication, 3:43-60, 2 figs., 4 tables.

estuaries
Cohen, B., and L. T. McCarthy, Jr., 1962
Salinity of the Delaware Estuary.
Geol. Survey Water-Supply Paper, 1586-B: 47 pp.

estuaries
Collier, A., and J.W. Hedgpeth, 1950.
An introduction to the hydrography of tidal waters of Texas. Publ., Inst. Mar. Sci., 1(2):120-194, 32 textfigs.

estuaries
Colman, J.S., and L.H.N. Cooper, 1954.
The "Rosaura" Expedition, 1937-1938. II. Underwater illumination and ecology in tropical waters
Brit. Mus. (N.H.)Zool., 6(2):131-137.

Estuaries
Conaghan, P. J., 1966.
Sediments and sedimentary processes in Gladstone Harbor, Queensland.
Univ. - Queensland Papers, Dept. Geol., 6 (1): 7-52.

estuaries
Cooke, C.W., 1950.
Carolina bays, traces of tidal eddies. (Abstract)
Bull. G.S.A. 61:1452.

estuaries
Cooper, L.H.N., and A. Milne, 1939.
The ecology of the Tamar Estuary. V. Underwater illumination. Revision of data for red light.
J.M.B.A. 23:391-396.

estuaries
Cooper, L.H.N., and A. Milne, 1938.
The ecology of the Tamar Estuary. II. Underwater illumination. J.M.B.A. 22:509-528.

estuaries
Cory, Robert L., 1967.
Epifauna of the Patuxent River estuary, Maryland, for 1963 and 1964.
Chesapeake Sci., 8(2):71-89.

estuaries
Cory, Robert L., 1965.
Installation and operation of a water quality data collection system in the Patuxent River Estuary, Maryland.
Ocean Sci. and Ocean Eng., Mar. Techn. Soc., Amer. Soc. Limnol. Oceanogr., 2:728-736.

estuaries
Cotton, C.A., 1951.
Atlantic gulfs, estuaries and cliffs.
Geol. Mag., 88:113-128.

estuaries
* Cotton de Bennetot, Michelle, 1969.
Etude sédimentologique et morphologique de l'estuaire du Goayen (Finistère).
Cah. océanogr., 21(4): 355-377.

estuaries
Cotton de Bennetot, Michelle, Andre Guilcher et Anne Saint-Requier, 1965.
Morphologie et sédimentologie de l'Aber Benoît (Finistère).
Cahiers Océanogr., C.C.O.E.C., 17(6): 377-387.

estuaries
Couillault, M., 1948.
Mission hydrographique d'Indochine (Octubre 1937-Septembre 1938). Ann. Hydrogr., Paris, (3)17:115-169, Fig. 8.

estuaries
Couture, Richard, 1971.
Les décapodes du Plateau Madelinien.
Cah. d'Info. Serv. Biol. Din. gén. Pêches Québec 55: 19pp. (multilithed).

estuaries
Couture, Richard, 1970.
Inventaire quantitatif des crevettes de la baie des Chaleurs, 1969.
Rapp. ann. 1969, Serv. Biol. Québec, 65-85.

estuaries
Couture, Richard, 1970.
Étude comparative des rendements de la pêche aux crevettes pour 1968 et 1969.
Rapp. ann. 1969, Serv. Biol. Québec, 99-106

ESTUARIES
Cox, R.A., 1956.
Measuring the tidal flow in the River Humber.
Dock & Harbour Authority, 37:96-97.

estuaries
Craig, R.E., 1952.
Hydrography of the Firth of Forth in the spring.
Cons. Perm. Int. Expl. Mer, Ann. Biol. 8:98-99, Textfig. 1.

estuaries
Crance, Johnie H. 1971.
Description of Alabama estuarine areas-Cooperative Gulf of Mexico Estuarine Inventory.
Bull. Alabama mar. Resources 6: 85pp.

estuaries
Crance, Johnie H. 1969.
A selected bibliography of Alabama estuaries.
Bull. Alabama mar. Resources 2: 21pp.

estuaries
Crean, P.B. 1967.
Physical oceanography of Dixon Entrance British Columbia.
Bull. Fish. Res. Bd Can. 156: 1-66.

estuaries
Cronin, L. Eugene, 1968.
The protection of Maryland's estuarine areas.
Proc. Gulf Carib. Fish. Inst. 20th Ann. Sess. 40-54

estuaries
Cronin, L. Eugene 1967.
The role of man in estuarine processes.
In: Estuaries, G.H. Lauf editor, Publs Am. Ass. Advmt Sci. 83: 376-382.

estuaries
Cronin, L.E., 1954.
Hydrography of the Delaware estuary.
2nd Ann. Rept., Mar. Lab., Univ. Delaware, 6-15.

estuaries
Cronin, L. Eugene, Joanne C. Daiber and Edward M. Hulburt, 1962
Quantitative seasonal aspects of zooplankton in the Delaware River estuary.
Chesapeake Science, 3(2):63-93.

estuaries
Cronin, L.E., J.C. Daiber and E.M. Hulburt, 1962
37. Quantitative seasonal aspects of zooplankton in the Delaware River estuary. (Abstract).
Contributions to symposium on zooplankton production, 1961.
Rapp. Proc. Verb., Cons. Perm. Int. Expl. Mer, 153:216.

estuaries

Cronin, William B. 1971.
Volumetric, areal, and tidal statistics of the Chesapeake Bay estuary and its tributaries.
Spec. Rept. Chesapeake Bay Inst. 20 (Ref. 71-2): 135pp. (unpublished mss.)

estuaries

Cunningham, B., 1938.
Estuary channels and embankments. Dock and Harbour Authority 18:202-207, 14 figs.

Also in Revue Hydrographique 16(31):115-119.

estuaries

Cunningham, B., 1938.
Estuary channels and embankments. J. Inst. Civil Eng. No. 8:361-364.

estuaries

Cunningham, B., 1938.
Estuary channels and embankments. Dock and Harbor Authority 18:202-207, 14 textfigs.

D

estuaries

Daiber, F.C., 1963.
Tidal creeks and fish eggs - role of tidal flow, salinity and light in distribution of fish eggs and larvae.
Estuarine Bull., 7(2/3):6-14.

estuaries

Daiber, Franklin C., 1960.

Mangroves, the tidal marshes of the tropics.
Estuarine Bull. 5(2): 10-15.

estuaries

D'Ancona, U., 1959.

The classification of brackish waters with reference to the north Adriatic lagoons.
Arch. Oceanogr. e Limnol., 11(Supplement):93-110.

estuaries

D'Ancona, U., 1954.
Il trofismo della Laguna Veneta e la vivificazione marina. Arch. Oceanogr. Limnol. 9(1/2):11-15.

estuaries

Danel, P., 1953.
Penetration des eaux salees dans les fleuves. Effets inconvenients, remedes.
XVIIIe Congr. Int. Navig., Rome 1953, Sect. 11, Navig. Marit., Comm. 3:109-141, 25 textfigs.

estuaries

d'Anglejan, B.F., 1969.
Preliminary observations on suspended matter in the Gulf of St. Lawrence.
Marit. Sediments 5(1): 15-18

estuaries

Darnell, Regnest M., 1967.
Organic detritus in relation to the estuarine echosystem.
In: Estuaries, G.H. Lauff, editor, Publs Am. Ass. Advmt Sci. 83: 376-382.

estuaries

Darnell, R.M., 1954.
An outline for the study of estuarine ecology.
Proc. Louisiana Acad. Sci. 17:52-59.

estuaries

D'Arrigo, A., 1949.
The migration of a Sicilian river's mouth over eight centuries. Dock Harb. Auth. 30:183-185.

estuaries

Daubert, A., et O. Graffe, 1967.
Quelques aspects des écoulements presque horizontaux à deux dimensions en plan et non permanents application aux estuaires.
Houille blanche 22(8):847-860.

estuaries

Day, J.H. 1967.
The biology of Knysna Estuary, South Africa.
In: Estuaries, G.H. Lauff, editor, Publs Am. Ass. Advmt Sci. 83: 397-407.

estuaries

Day, J.H., 1961.
The intertidal and estuarine fauna of the Natal coast.
Marine Studies off the Natal Coast, C.S.I.R. Symposium, No. S2:68-72. (Multilithed).

estuaries

Day, J.H., 1951.
The ecology of South African estuaries. 1. A review of estuarine conditions in general. Trans. R. Soc., S. Africa, 33(1):53-91, 2 textfigs.

estuaries

Day, J.H., N.A.H. Millard and A.D. Harrison, 1952
The ecology of South African estuaries. III. Knysna: a clear open estuary.
Trans. R. Soc., S. Africa, 33(3):367-414.

estuaries

Dean, David and Harold H. Haskin, 1964
Benthic repopulation of the Raritan River estuary following pollution abatement.
Limnology and Oceanography, 9(4):551-563.

estuaries

De Angelis, Costanzo M., 1961
Osservazioni sulle condizioni fisico-chimiche e sul fitoplancton del Canale Calambrone e del Fosso dei Navicelli (Livorno).
Boll. Pesca, Piscicolt. e Idrobiol., n.s., 16(2):307-332.

estuaries

de Barros Machado, L., 1950.
Pesquisas físicas e químicas do sistema hidrográfico da região lagunar de Cananéia. 1. Cursos de águas. Bol. Inst. Paulista Oceanogr. 1(2):45-67, 3 textfigs.

estuaries

de Buen, F., 1949.
El Mar de Solis y su fauna de peces. (1 partie).
S.O.Y.P., Publ. Cient., No. 1:43 pp., 15 textfigs.

estuaries

Debyser, J., 1957.
La sédimentation dans le bassin d'Arcachon.
Bull., E.E.R.S., Biarritz, 1(3):405-418.

estuaries

de Castro Barros, Aldemir, & Skapti Jonsson 1967.
Prospecção de camarões na região estuarina do Rio São Francisco.
Bol. Estud. Pesca, SUDENE, Recife 7(2):9-29

estuaries

DeFalco, Paul, Jr., 1967.
The estuary - septic tank of the megalopolis.
In: Estuaries, G.H. Lauff, editor, Publs Am. As Advmt. Sci., 83:701-703.

estuaries

DeGroot, A.J., 1964.
Origin and transport of mud (fraction < 16 microns) in coastal waters from the western Scheldt to the Danish frontier.
In: Developments in Sedimentology, L.M.J.U. van Straaten, Editor, Elsevier Publishing Co., 1:93-100.

estuaries

De Groot, A.J., J.J.M. De Goeij and C. Zegers, 1971.
Contents and behaviour of mercury as compared with other heavy metals in sediments from the rivers Rhine and Ems.
Geol. Mijnb. 50 (3): 393-398.

estuaries

de Joly, Georges, 1923.
Travaux maritimes. La mer et les côtes. Ch. 7. Propagation de la marée dans les fleuves.
J.B. Bailliere, Paris, pp. 148-173.

estuaries

De Medeiros Tinoco, Ivan, 1959.
Classificao sistematica dos foraminferos dos testemunhos de sondagens submarinas efetuadas pelo Navio Escola "Almirante Saldanha" na embocadura do Rio Amazonas (1).
Trabalhos. Inst. Ocean. e Oceanogr. 1(1):107-112.

estuaries

DeMendonca, A.Z., 1934.
Estudo da salidade das aguas do Tojo.
Ministerio das Obras Publicas e Comunicaves Anuario dos Servicos Hydraulicos, 1933, Portugal, :53-71.

Estuaries

den Hartog, C. 1971.
The border environment between the sea and the fresh water, with special reference to the estuary.
Vie et Milieu Suppl. 22 (2): 739-751.

estuaries

d'Erceville, I., 1938-1939.
Mission hydrographique de l'Indo-Chine, 1936-1937
Ann. Hydrogr., Paris, Ser. 3, 16:79-118, figs.

estuaries

*DeSousa e Silva, Estela, 1968.
Plancton da Lagoa de Óbidos (III). Abundância, variacões sazonais e grandes "blooms".
Notas Estudos, Inst.Biol. marit. Lisboa 34: 79 pp.

estuaries
de Sousa e Silva, Estela Maria Etelvina Assis and Maria Antonia M. Sampayo, 1969.
Primary productivity in the Tagus and Sado estuaries from May 1967 to May 1968.
Notas Estud. Inst. Biol. marit., Lisboa) 37: 46 pp.

estuaries
Destriau, G., and M. Destriau, 1951.
Le mascaret dans les rivières de Gironde.
Ann. Ponts et Chaussées No. 5:609-635, 32 text-figs.

estuaries
Diachischyn, A.N., S.G. Hess, and W.T. Ingram, 1953.
Sewage disposal in tidal estuaries. Proc. Amer. Soc. Civil Eng. 79(167):1-14, 6 figs.

estuaries
Digby, P.S.B., 1953.
Plankton production in Scoresby Sound, East Greenland. J. Animal Ecol. 22(2):289-323, figs.

estuaries
Dionne, Jean-Claude 1970.
Exotic pebbles in Quaternary deposits from the South coast of the St. Lawrence estuary, Quebec.
Marit. Sed., 6(3): 110-112.
Halifax

estuaries
Dionne Jean-Claude, 1969.
Tidal flat erosion by ice at La Pocatière, St. Lawrence estuary.
J. Sedim. Petrol., 39(3): 1174-1181.

estuaries
Dionne, Jean-Claude, 1968.
Action of shore ice on the tidal flats of the St. Lawrence Estuary.
Marit. Sed., 4(3): 113-115.

estuaries
Dobson, Max R., 1965.
Black mud in the outer Thames estuary.
Dock Harb. Auth., 46(535):18-21.

estuaries
D'Olier, B., and R.J. Maddrell, 1970
Buried channels of the Thames Estuary.
Nature, Lond., 226(5243): 347-348

estuaries
Donguy, Jean-Rene, Jean Hardiville et Jean-Claude, Le Guen, 1965.
Le parcours maritime des eaux du Congo.
Cahiers Oceanogr., C.C.O.E.C., 17(2):85-97.

estuaries
Dorrestein, R., 1960.
A method of computing the spreading of matter in the water of an estuary.
Disposal of Radioactive Wastes, Int. Atomic Energy Agency, Vienna, 163-166.

estuaries
Dorrestein, R., 1960.
Einige Klimatologische und Hydrologische Daten für das Ems-Estuarium.
Verh. Kon. Ned. Geol. Mijbn. k. Gen., Geol. Ser., 19:29-42.

estuary
Dorrestein, R., 1960.
On the distribution of salinity and of some other properties of the water in the Ems-Estuary.
Verh. Kon. Ned. Geol. Mijnb. k. Gen., Geol. Ser., 19:43-74.

estuaries
Dorrestein, R., and L. Otto, 1960.
On the mixing and flushing of the water in the Ems-Estuary.
Verh. Kon. Ned. Geol. Mijnb. k. Gen., Geol. Ser. 19:83-102.

estuaries
Dragovich, A., and B.Z. May, 1962
Hydrological characteristics of TampaBay tributaries.
U.S.F.W.S. Fish. Bull. (205) 62: 163-176.

estuaries
*Drainville, Gérard, 1968.
Le fjord du Saguenay: 1. Contribution à l'océanographie.
Le Naturaliste canadien, 95(4):809-855.

estuaries
Drainville, Gerard, Marcel Tiphane et Pierre Brunel, 1963.
Croisière océanographique dans le fjord du Saguenay, 14-22 juin 1962.
Rapp. Ann., 1962, Sta. Biol. Mar., Grande-Rivière, Canada, 133-146. (mimeographed).

Estuaries
Dronkers, J. J., 1961.
De invloed van de deltawerken op de getijbeweging en de stormvloedstanden langs de kust van Zuidwest-Nederland.
Rapp. Deltacommissie, Bijdragen (5):35-84.

estuaries
Dronkers, J.J., 1953.
Theories on which a method of calculation for th movement of salt water and of fresh water in a river is based in the case of tidal currents.
XVIIIe Congr., Int. Navig., Rome 1953, Sect. 11; Navig. Marit., Comm. 3:200-203, 1 textfig.

estuaries
Dronkers, J.J., 1951.
Berekeningen over de afsluiting van de Brielsche Maas en Botlek met praktische beschouwingen over getijberekeningen in het algemeen.
De Ingenieur, The Hague, No. 40:137-145, 15 figs

estuaries
Duboul-Razavet, Chr., 1951.
Delta du Rhone promontoire deltaique du Grand Rhone. Proc. 3rd Int. Congress, Sedimentol., Netherlands, 1951:99-108, 3 textfigs.

estuaries
Duke, Thomas W., John P. Baptist and Donald E. Hoss, 1966
Bioaccumulation of radioactive gold used as a sedimenttracer in the estuarine environment.
Fishery Bull., Fish Wildl. Serv., U.S., 65(2):427-436.

estuaries
*Duke, T.W., and T.R. Rice, 1967.
Cycling of nutrients in estuaries.
Proc. Gulf Caribb. Fish. Inst., 19th. Sess.:59-67.

estuaries
Duke, Thomas W., James N. Willis and Thomas J. Price, 1966.
Cycling of trace elements in the estuarine environment. 1. Movement and distribution of zinc 65 and stable zinc in experimental ponds.
Chesapeake Science, 7(1):1-10.

estuaries
Dulemba, Jean L., 1964.
Apercu sur la chloruration de l'eau de mer dans le Golfe de Saint Florent et de l'eau dans l'estuaire de l'Alisio (Corse).
Cahiers Oceanogr., C.C.O.E.C., 16(7):557-563.

estuaries
Dulk, A. den, 1951.
Zeldzame Ostracoden in de Biesbos.
Levende Natuur 54(9):176-178.

estuaries
Dupont, P., 1922.
L'aménagement de l'estuaire de la Seine.
Soc. Ingénieurs Civils de France:360-774.

estuaries
Dutta, N., J.C. Malhotra and B.B. Bose, 1954.
Hydrology and seasonal fluctuations of the plankton in the Hooghly estuary.
Symp. Mar. Fresh-water Plankton, Indo-Pacific, Bangkok, Jan. 25-26, 1954, FAO-UNESCO:35-47.

Estuaries
Dyer, K.R. 1970
Linear erosional furrows in Southampton Water.
Nature, Lond. 225(5227): 56-58.

estuaries
Dyer, K.R., and K. Ramamoorthy, 1969.
Salinity and water circulation in the Vellar estuary. Limnol. Oceanogr. 14(1): 4-15.

estuaries
Dyer, K.R., N. Hamilton and R.D. Pingree, 1969.
A seismic refraction line across the Solent.
Geol. Mag., 106(1): 92-95.

E

estuaries
Ehrhardt, Manfred, 1970
2.8. Partikulärer organischer Kohlenstoff und Stickstoff sowie gelöster organischer Kohlenstoff. Chemische, mikrobiologische und planktologische Untersuchungen in der Schlei im Hinblick auf deren Abwasserbelastung. Kieler Meeresforsch 26(2): 138-144.

estuaries
Einstein, H.A., 1950.
The bed-load function for sediment transportation in open flow channels. Dept. Agric., Tech. Bull., No. 1026:71 pp., 21 textfigs., 5 fold-ins.

estuaries
Einstein, H.A., & R.B. Krone, 1961
Estuarial sediment transport patterns.
J. Hydraulics Div., Proc. Amer. Soc. Civil Eng., 87(HY2): 51-60.

estuaries
Elbrächter, Malte 1970.
4.2. Phytoplankton und Ciliaten
Chemische, mikrobiologische und planktologische Untersuchungen in der Schlei im Hinblick auf deren Abwasserbelastung. Kieler Meeresforsch 26(2): 193-203.

estuaries
Eldridge, E.F., and G.T. Orlob, 1951.
Pollution of Port Gardiner Bay and Snohomish River Estuary. Sewage and Ind. Wastes 23:782-795.

Abstr. in Chem Abstr. 1194c, 1952.

estuaries
Elliott, H.A., R. Ellison and M.M. Nichols, 1966.
Distribution of Recent Ostracoda in the Rappahannock Estuary, Virginia.
Chesapeake Sci., 7(4):203-207.

estuaries
Ellison, R., M. Nichols and J. Hughes, 1965.
Distribution of Recent Foraminifera in the Rappahannock River Estuary.
Virginia Inst. Mar. Sci., Spec. Sci. Rept., No. 47:35 pp. (Unpublished manuscript).

estuaries
Elsden, Oscar, 1949.
Regularisation of natural and artificial estuaries, including estuaries of small dimensions, tideless and otherwise. XVIIe Congres Inst. Navi Lisbonne, Sect. II, Question 1:93-109, 5 figs.

estuaries
Emery, K. O., 1967.
Estuaries and lagoons in relation to continental shelves.
In: Estuaries, G.H. Lauff, editor, Publs Am. Ass Advmt Sci., 83:9-11.

estuaries
Emery, K.O., R.E. Stevenson and J.W. Hedgpeth, 1957.
Estuaries and lagoons. Ch. 23 in: Treatise on Marine Ecology and Paleoecology, Vol. 1.
Mem., Geol. Soc., Amer., 67:673-750.

estuaries
*Eskinazi, Enide, 1965/1966.
Estudio da Barra das Jangadas. VI. Distribuição das diatomaceas.
Trabh Inst. Oceanogr., Univ. Fed. Pernambuco, Recife, (7/8):17-32.

F

estuaries
Fass, Richard W., 1966.
Paleoecology of an Artic estuary.
Arctic, 19(4):343-348.

estuaries
Faganelli, A., 1951.
Influence of tides on lagoon waters. Atti Ist. Veneto, Cl. Sci. Mat.-Nat., 109:327-331.

Abst. Chem. Abstr., 1952:9235h.

estuaries
Farleigh, D.R.P., and Sir Claude Inglis, 1962
The behaviour and control of the Karnafuli Estuary, East Pakistan.
Proc. Conf. on Civil Engineering Problems Overseas, Inst. Civil Eng., London, 317-337.

estuaries
Farmer, D.G. 1971.
A computer simulation model of sedimentation in a salt wedge estuary.
Marine Geol. 10(2):133-143.

estuaries
Farmer, H.G., and G.W. Morgan, 1953.
Ch. 5: The salt wedge. Proc. Third Conf. Coastal Eng., Cambridge, Mass., Oct. 1952:54-64, 4 text-figs.

estuaries
Farquharson, W.I., 1962
Tides, tidal streams and currents in the Gulf of St. Lawrence.
Marine Sciences Branch, Department of Mines and Technical Sruveys, Ottawa, 76 pp.

estuaries
Farquharson, W.I., 1959
Report on tidal survey, 1958. Causeway investigation Northumberland Strait.
Surveys and Mapping Branch, Dept. Mines and Technical Surveys, Ottawa, 137 pp. (multilithed).

estuaries
Fassenko, E.A., and M.S. Sheinin, 1956.
Quantitative fluctuations of zooplankton of the lower reach of the Don River and the eastern part of the Taganrog Bay. Dokl. Akad. Nauk SSSR 111(1):202-205.

estuaries
Faure, M., 1953.
Calcul des pertes d'énergie dans un estuaire à marée. Soc. Hydrotech., France, 2:157-169.

estuaries
Faure, M., 1953.
Calcul des pertes d'énergie dans un estuaire à marée (Gironde). Principe et exécution du calcul a l'aide d'une machine mathématique.
La Houille Blanche, No. Spécial B/1953:157-169.

estuaries
Febres, Germán, 1966.
Ch. 1. Hidrografía.
Estudios Hidrobiologicos en el Estuario de Maracaibo, Inst. Venezolano de Investig. Cient., 1-20. (Unpublished manuscript).

estuaries
Febres, German, y Gilberto Rodriguez, 1966.
Ch. 3. Sedimentos.
Estudios hidrobiologicos en el Estuario de Maracaibo, Inst. Venezolano de Invest. Cient., 66-82. (Unpublished manuscript).

estuaries
Febres, Germán, Gilberto Rodríguez y Andrés Eloy Esteves, 1966
Ch. 2. Química del agua.
Estudios Hidrobiologicos en el Estuario de Maracaibo, Inst. Venezolano de Invest. Cient., 21-65. (Unpublished manuscript).

estuaries
Feral, Alain, Claude Caratini, André Klingebiel et Paul-Charles Levêque, 1971.
Concordance entre les résultats obtenus à l'aide de la sédimentologie, de la palynologie et de la datation au ^{14}C dans l'étude d'un faciès flandrien de l'estuaire de la Gironde.
C.r. hebd. Séanc. Acad. Sci., Paris, (D) 272(9): 1201-1203

estuaries
Ferguson Wood, E.J., 1951.
Phytoplankton studies in eastern Australia. Proc. Indo-Pacific Fish. Counc., 17-28 Apr. 1950, Cronulla, N.S.W., Australia, Sects. II-III:60-63.

estuaries
Ferrero, Letizia, 1961
Ricerche fisico-chimiche e biologiche sui La Laghi Salmastri Pontini in relazione alla produttivita. II. Il Lago di Paola (Sabaudia). Ricerche quantitative sulla fauna bentonica.
Boll. Pesca, Piscicolt. e Idrobiol., n.s., 16(2):173-203.

estuaries
Fischer-Piette, Edouard, et Jean Seoane-Camba, 1963.
Examen écologique de la Ria de Camarinas.
Bull. Inst. Océanogr., Monaco, 61(1277):1-38.

estuaries
Fisher, E., 1929.
Recherches de Bionomie et d'Océanographie littorales sur la Rance et le littoral de la Manche.
Ann. Inst. Océan., Monaco, 5(3):205-429, 1 chart, numerous figs.

estuaries
Fiske, John D., Clinton E. Watson and Philip G. Coates, 1966.
A study of the marine resources of the North River.
Comm. Mass., Div. Mar. Fish., Monogr. Ser. 3:53 pp.

estuaries
Flemer, David A., and Janet Olmon. 1971.
Daylight incubator estimates of primary production in the mouth of the Patuxent River, Maryland.
Chesapeake Sci. 12(2): 105-110.

Fleming, Georges. 1970.
Sediment balance of Clyde estuary.
J. Hydraul. Div. Am. Soc. civ. Engrs, 96 (HY11) 2219-2230

estuaries
Fraga, F., 1967.
Hidrografía de la ría de Vigo, 1962, con especial referencia a los compuestos nitrogens.
Investigación pesq., 31(1):145-149.

Estuaries
Fraga, F., 1960.
Variación estacional de la materia orgánica suspendida y disuelta en la Ria de Vigo.
Influencia de la luz y la temperatura.
Inv. Pesq., Barcelona, 17:127-140.

estuaries
Fraga, F., and F. Vives, 1961.
La descomposicion de la materia organica nitrogenada en el mar.
Inv. Pesq., Barcelona, 19:65-79.

estuaries
Fraga, F., and F. Vives, 1961.
Variación estacional de la materia orgánica en la Ría de Vigo.
Inv. Pesq., Barcelona, 20:65-71.

estuaries
France, Chambre de Commerce de Rouen, 1844.
Enquête sur les travaux à faire pour l'amélioration de la navigation dans la Basse-Seine. 128 pp

estuaries
France, Laboratoire National d'Hydraulique, 1958.
Etude sur modele reduit de la Loire Maritime.
154 pp.

estuaries
France, Laboratoire National Hydraulique, 1952.
Port de Donges. Étude sur modèle réduit. 56 pp., figs. (multilithed).

estuaries
Francis-Boeuf, C., 1946.
La sedimentation dans les estuaires. C.R., Sess. Extr., Soc. Belg. Geol., Sept. 1946:174-185, 4 textfigs.

estuaries
Francis-Boeuf, C., 1943.
Physico-chimie du milieu fluvio-marin.
C.R. Somm. Soc. Biogrogr. 169:

estuaries
Francis-Boeuf, C., 1941.
Observations sur les variations de quelques facteurs physicochimiques des eaux de la Penzé Maritime (Finistère). C.R. Acad. Sci., Paris, 212:805-806, 3 textfigs.

estuaries
Francis-Boeuf, C., 1939.
Remarques sur quelques mesures de salinité des eaux de l'Orne, entre Caen et l'embouchure (Franceville). C. R. Acad. Sci., Paris, 208(12): 916-918.

estuaries
Francis-Boeuf, C., 1938.
Premières résultats d'une étude des eaux de l'Aulne maritime. Revue Géographie Physique et de Géologie Dynamique 11(4):399-437, 18 textfigs.

estuaries
Francis-Boeuf, C., 1938.
Premières résultats d'une étude des eaux de l'Aulne Maritime (Brest). Rev. Géogr. Phys. et Geol. Dynamique 11(4):399-438, 18 textfigs.

estuaries
Francis-Boeuf, C., and V. Romanovsky, 1946.
L'envasement du port de Honfleur. Bull. Inst. Océan., Monaco, No. 901:8 pp., 5 figs.

estuaries
Freidzom, A.I., 1962.
On the computation of the height of lifting of the water level in the Neva estuary. (In Russian) Trudy Gosud. Okeanogr. Inst., 69:92-94.

estuaries
Frainville, Gérard, Marcel Tiphane et Pierre Brunel, 1963.
Croisière océanographique dans le fjord du Saguenay, 14-22 juin 1962.
Sta. Biol. Mar., Grande-Rivière, Cahiers Info., (17):133-144. (Multilithed).
Also in Rapp. Ann., 1962, Sta. Biol. Mar., Grande-Rivière, 133-144.

estuaries
Freidzon, A.I., N.I. Belskii, and A.A. Popov, 1960.
Procedures for calculating swinging oscillations of water in the Neva estuary. (In Russian).
Trudy Gosud. Okean. Inst., Leningrad, 56:65-79.

estuaries
Fuse, Shin-ichiro, 1959.
A study on the productivity of Tanabe Bay III.
Oceanographic conditions of Tanabe Bay (2) Stratification and fluctuation of hydrological conditions on two sectional survey lines.
Rec. Oceanogr. Wks., Japan. Spec. No. 3:31-45.

G

estuaries
Gaarder, T., and R. Spärck, 1932.
Hydrographisch-biochemische Untersuchungen in norwegischen Austern-Pollen. Bergens Mus. Aarbok, Naturvidensk.-rekke, No. 1:5-144, 75 textfigs.

estuaries
Gameson, A.L.H., 1954.
Self-purification in estuaries.
Bull. Centre Belge et Doc., Eaux, 24:71-77.
Abstre. Publ. Health Eng. Abstr. 35(6):24.

estuary
Ganapati, P. N., & M. V. Lakshmana Rao, 1959.
Incidence of marine borers in the mangroves of the Godavari estuary.
Current Science, 38(8): 332

estuaries
Ganapati, P.N., and D.V.Rama Sarma, 1965.
Mixing and circulation in Gautami-Godavari Estuary.
Current Science, 34(22):631-632.

estuaries
Garland, C.F., 1952.
A study of water quality in Baltimore Harbor.
Chesapeake Biol. Lab. Publ. 96:1-132, figs.

estuaries
Garner, D.M., 1964.
The hydrology of Milford Sound.
New Zealand Dept. Sci. Ind. Res., Bull., No. 157; New Zealand Oceanogr. Inst., Memoir, No. 17:25-33

estuaries
Gautier, Monique, 1967.
Salinité des eaux de surface de Grande-Rivière, août 1965-décembre 1966.
Rapp. Stn Biol. Mar. Grande-Rivière, 1966:25-28.

estuaries
Germaneau, J., 1969.
Étude de la sédimentation dans l'estuaire de la Seine. 2. Origine, déplacement et dépôt des suspensions.
Trav. Centre Rech. Études océanogr. n.s. 9 (1-4): 100 pp.

estuaries
Germaneau, J., 1968.
Caractères de la sédimentation dans l'estuaire de la Seine.
Bull. Inst. géol. Bassin d'Aquitaine 5: 140-167.

estuaries
Gibert, M., 1949.
The historic role and changes of the Loire estuary. Dock and Harbor Authority No. 318:298-300, 6 figs.

estuaries
Gibert, R., and P. Durepatre, 1949.
Amélioration des embouchures des voies d'eau naturelles et artificielles (Seine, Loire, Gironde Adour). XVII Congr. Int. Navigation, Lisbonne, Sect. II, Question 1:55-79, 13 figs.

estuaries
Gibson, Blair, W., 1959
A method for estimating the flushing time of estuaries and embayments. HO Tech. Rept. TR-62: 19 pp.

estuaries
#Giere, Olav, 1968.
Die Fluctuationen des marins Zooplanktons im Elbe-Aestuar: Beziehungen Zurischen Populationsschwankungen und hydrographischen Faktoren im Brackwasser.
Arch. Hydrobiol./Suppl. 31(3/4): 379-546.

estuaries
Gieskes, Joris M.T.M., 1968.
Some hydrographical observations on salt brine pollution in the Kiel Fjord.
Helgoländer wiss. Meeresunters. 17:411-421

Estuaries
Giffen, Malcolm H., 1970.
Contributions to the diatom flora of South Africa. IV. The marine littoral diatoms of the estuary of the Kowie River, Port Alfred, Cape Province.
Beiheft Nova Hedwigia 31:259-312

Gilmartin, Malvern, 1964. estuaries
The primary production of a British Columbia fjord.
J. Fish. Res. Bd., Canada, 21(3):505-538.

estuaries
Gilmartin, M., 1962.
Annual cyclic changes in the physical oceanography of a British Columbia Fjord.
J. Fish. Res. Bd., Canada, 19(5):921-974.

estuaries
Giresse, Pierre, 1966.
Contribution des traceurs radioactifs à l'étude de la dynamique des dépôts de l'estuaire de la Sienne.
Cahiers Océanogr., C.C.O.E.C., 18(1):35-42.

estuaries
Giresse, Pierre, 1965.
Modalites de la sedimentation dans l'estuaire de la Sienne: principales zones de depot.
Cahiers Oceanogr., C.C.O.E.C., 17(1):45-52.

estuaries
Glangeaud, L., 1951.
Mécanism des dépôts dans les estuaires.
Congres de Géologie du Quaternnaire, 1949, III:

estuaries
Glangeard, Louis, 1941.
Evolution morphologique et dynamique des estuaires
Assoc. Geogr. Français, No. 140:95-102.

estuaries
Glangeard, L., 1939.
Rôle de la suspension tourbillonnaire et du roulement sur le fond dans le formation des sédiments actuels de l'estuaire girondin entre Bourdeaux et la Pointe de Grave. C.R. Acad. Sci. Paris, 208:1595-1597, 2 textfigs.

estuaries
Glangeaud, L., 1941.
Le formation et la répartition des facies vaseux dans les estuaires. C.R. Acad. Sci., Paris, 213: 1022.

estuaries
Glover, R.E., 1955.
A new method for predicting transient states of salinity intrusion into the Sacremento-San Joaquin Delta. Trans. Amer. Geophys. Union, 36: 641-648.

estuaries
Goedecke, E., 1939.
Beitrag zur Hydrographie der Helgoland umgebenden Gewasser. I. Die Oberflächenverhältnisse bei bei Helgoland-Reede. Ann. Hydr. Jahr. 67, Heft IV: 161-176, folded charts, tables, diagrams.

estuaries
Gold, Harvey S., 1959.
Distribution of some lignicolous Ascomycetes and Fungi Imperfecti in an estuary.
J. Elisha Mitchell Sci. Soc., 75:25-26.

estuaries
Granger, Daniel, 1970
Distribution bathymétrique benthique et migrations verticales journalières des cumacés à l'entrée de la baie des Chaleurs en 1968 et 1969.
Rapp. ann. 1969, Serv. Biol. Québec, 53-64

estuaries
Grant, George C., 1963
Chaetognatha from inshore coastal waters off Delaware, and a northward extension of the known range of Sagitta tenuis.
Chesapeake Science, 4(1):38-42.

estuaries
Great Britain, Department of Scientific and Industrial Research, 1962.
Hydraulics Research, 1961: report of the Hydraulics Research Board, with the report of the Director of Hydraulics Research (Wallingford), 96 pp.

estuaries
Great Britain, Department of Scientific and Industrial "esearch, 1961.
Estuaries.
Hydraulics Research, 1960:37-61.

estuaries
Great Britain, Department of Scientific and Industrial Research, 1956.
Hydraulics Research, 1955:56 pp.

estuaries
Great Britain, Hydraulics Research Station, Wallingford, 1963
Hydraulics research, 1962: The report of the Hydraulics Research Board with the report of the Director of Hydraulics Research, 90 pp.

estuaries
Great Britain, Hydraulics Research Station, Wallingford, 1963
Karnafuli River, East Pakistan.
Notes, No. 5: 2 pp.

estuaries
Great Britain, Hydraulics Research Board, 1957.
Hydraulics research, 1956, 54 pp.

estuaries
Great Britain, Hydraulics Research Station, 1956.
Radioactive tracers in the Thames estuary. Report of an experiment carried out in 1955.
DSIR, HRS/PLA, Paper 20:24 pp.

estuaries
Greco, L., A. Bongiorno and G. Ferro, 1949.
Amélioration des embouchoures des voies d'eaux naturelles et artificielles. 1. Les embouchoures des lagunes de la haute mer Adriatique. 2. Les ports-canaux de la mer Adriatique. 3. Les ports-canaux de la mer Tyrrhénienne.
XVIIe Congr. Int. Navig., Lisbonne 1949, sect. 2, Quast,:137-156, 2 pls.

estuaries
Gresswell, R.K., 1964.
The origin of the Mersey and Dee estuaries.
Geolog. J., 4(1):77-86.

estuaries
Greze, V.N., 1956.
[The drainage of the Enisey River as connected with its production of plankton.] Dokl. Akad. Nauk, SSSR, 110(1):1108-1110.

Grindley, John, 1960? estuaries
The determination of the salinity of water in estuaries.
I.A.S.H., Comm. Surf. Waters, Publ. No. 51:379-386

estuaries
Grindley, J., and A.B. Wheatland, 1956.
Salinity and the biochemical oxygen demand of estuary water. Wat. Sanit. Eng., 6:10-14.

Abstr. in Anal. Abstr., 4(10):#3475.

estuaries
Grommelin, R.D., 1951.
Quelques aspects granulométriques et minéralogiques de la sédimentation le long de l'estuaire de l'Escaut.
Congr. Geol. du Quaternaire, 1949,:63-72, 1 fig.

estuaries
Groot, Johan J., 1966
Some observations on pollen grains in suspension in the estuary of the Delaware River.
Marine Geol., 4(6):409-416

estuaries
Gross, M. Grant, Sevket M. Gucluer, Joe S. Creager and William A. Dawson, 1963.
Varved marine sediments in a stagnant fjord.
Science, 141(3584):918-919.

estuaries
Grovel, Alain P. 1970.
Étude d'un estuaire dans son environment le Blavet maritime et la région de Lorient.
Trav. Lab. Geol. mar., Fac. Sci. Nantes (C.N.R.S. A.O 45-52): 122 pp. (multilithed)

estuaries
Grovel, Alain, 1965.
Dynamique de la sédimentation dans l'estuaire du Blavet.
C.R. Acad. Sci., Paris, 260(9):2553-2555.

estuaries
Guarrera, S.A., 1950.
Estudios hidrobiologicos en el Rio de la Plata.
Rev. Inst. Nac., Invest. Cienc. Nat., Ciencias Botanicas, 2(1):62 pp.

estuaries
Gueritaud, J., 1940-1945.
Mission hydrographique de l'A.O.F., 1938.
Ann. Hydrogr., Paris, (3):17:219-234.

estuaries
Guilcher, André, 1967.
Origin of sediments in estuaries.
In: Estuaries, G.H. Lauff, editor, Publs Am. Ass. Advmt Sci., 83:149-157.

estuaries
Guilcher, A., 1963.
25. Estuaries, deltas, shelf, slope.
In: The Sea, M.N. Hill, Editor, Interscience Publishers, 3:620-654.

estuaries
Guilcher, A., 1955.
La sédimentation vaseuse dans les estuaires de Bretagne occidentale. Geol. Rundschau, 43:398-408

estuaries
Guilcher, A., 1953.
Mesures de la vitesse de sédimentation et d'érosion dans les estuaires bretons.
C.R. Acad. Sci., Paris, 237(21):1345-1347.

estuaries
Gunnerson, Charles G., 1966.
Hydrologic data collection in tidal estuaries. (Abstract only).
Trans. Am. Geophys. Un., 47(3):477.

estuaries
Gunnerson, Charles G., 1965.
Limitations of Rhodamine B and Pontacyl Brilliant Pink B as tracers in estuarine waters. (abstract) Symposium, Diffusion in oceans and fresh waters. Lamont Geol. Obs., 31 Aug.-2 Sept., 1964, 53.

estuaries
Gunter, Gordon, 1967.
Some relationships of estuaries to the fisheries of the Gulf of Mexico.
In: Estuaries, G.H. Lauff, editor, Publs Am. Ass. Advmt Sci. 83: 621-638.

estuaries
Gunter, Gordon, 1961
Some relations of estuarine organisms to salinity.
Limnol. & Oceanogr., 6(2): 182-190.

estuaries
Gunter, G., 1952.
Historical changes in the Mississippi River and the adjacent marine environment. Publ. Inst. Mar. Sci. 2(2):121-139.

estuaries
Gunter, Gordon, J.Y. Christmas and R. Killebrew, 1964.
Some relations of salinity to population distributions of motile estuarime organisms with special reference to penaeid shrimp.
Ecology, 45(1):181-185.

estuaries
Gunter, Gordon, and Gordon E. Hall, 1965.
A biological investigation of the Caloosahatchee Estuary of Florida.
Gulf Res. Repts., Ocean Springs, Mississippi, 2(1):71 pp.

estuaries
Gunter, Gordon, and Gordon E. Hall, 1963.
Biological investigations of the St. Lucie Estuary (Florida) in connection with Lake Okeechobee discharges through St. Lucie Canal.
Gulf Res. Repts., Ocean Springs, Miss., 1(5):189-307.

H

estuaries
Haefner, Paul A., Jr., 1967.
Hydrography of the Penobscot River (Maine) estuary.
J. Fish. Res. Bd., Can., 24(7):1553-1571

estuaries
Haertel, Lois, and Charles Osterberg, 1967.
Ecology of zooplankton, benthos and fishes in the Columbia River estuary.
Ecology, 48(3):459-472.

estuaries
Hall, K.J., W.C. Weimer and G.Fredlee 1970.
Amino acids in an estuarine environment.
Limnol. Oceanogr. 15(1): 162-164

estuaries
*Halliwell, A.R., and B.A. O'Connor, 1967.
Suspended sediment in a tidal estuary.
Proc.10th Conf.Coast.Engng.Tokyo,1966,1:687-706.

estuaries
Hamada, T., 1953.
Density current problems in an estuary.
Proc. Minnesota Int. Hydraulics Conv., Sept. 1-4, 1953:313-320, 2 textfigs.

estuaries
Hamada, T., and K. Okubo, 1952.
An observations of velocity fluctuation of medium period in a tidal estuary. Rept. Transport. Tech. Res. Inst. No. 3:22-30, 5 textfigs.

estuaries
Hanks, Robert W., 1964.
A benthic community in the Sheepscott River Estuary, Maine.
U.S.F.W.S. Fish. Bull., 63(2):343-353.

Hansen, Donald V.

estuaries
Hansen, Donald V., 1967.
Salt balance and circulation in partially mixed estuaries.
In: Estuaries, G.H. Lauff, editor, Publs Am. Ass. Advmt Sci., 83:45-51.

estuaries
Hansen, Donald V., 1965.
Currents and mixing in the Columbia River estuary.
Ocean Sci. and Ocean Eng., Mar. Techn. Soc. Amer. Soc. Limnol. Oceanogr., 2:943-955.

estuaries
Hansen, Donald V., and Maurice Rattray, Jr., 1966.
New dimensions in estuary classification.
Limnol. Oceanogr., 11(3):319-326.

estuaries
Hansen, Donald V., and Maurice Rattray, Jr., 1965.
Gravitational circulation in straits and estuaries.
J. Mar. Res., 23(2):104-122.

estuaries
Hansen, Kaj., 1944.
1. Introduction and the bottom deposits. Investigations of the geography and natural history of the Praestø Fjord, Zealand. Folia Geogr. Danica 3(1):46 pp., 15 textfigs.

estuaries
Hansen, Walter, 1962
Hydrodynamical methods applied to oceanographic problems.
Proc. Symposium on Mathematical-Hydrodynamical Methods of Phys. Oceanogr., Sept. 1961, Inst. Meeresk., Hamburg, 25-34.

estuaries
Hansen, W., 1953.
Das Eindringen von Salzwasser in die Gezeitenflüsse und ihre Nebenflusse in Seekanale und Hafen. Franzius-Inst. Tech. Hochschule, Hannover No. 3:20-50, 28 figs.

estuaries
Hansen, W., 1953.
Pénétration des eaux salées dans les eaux fleuves à marée (Embouchures de l'Elbe, l'Ems, le Weser).
XVIII Congr. Int. Navigation, Rome 1953, Sect. II Navigation Maritime, Communication 3:5-42, 28 figs.

estuaries
Hansen, W., 1952.
Kleine Studien aud dem Tidegebiet, No. 1. Ueber die Fortschrittsgeschwindigkeit der Tidewelle in einem Flusse. Franzius-Inst. Tech. Hochschule, Hannover, No. 2:98-106, 12 figs.

estuaries
Hansen, Walter, 1939.
L'influence de la rotation de la terre sur les rivières à marée dans la nature et sur modele résuit. Bautechnik 21:285-

estuaries
Hara, Akihiro, 1971.
Distribution of salinity in an estuary.
Sci. Repts, Tokyo Kyoiku Daigaku (C) 10: 277-352.

estuaries
Harleman, Donald R.F., and Arthur T. Ippen, 1961?
The turbulent diffusion and convection of saline water in an idealized estuary.
I.A.S.H., Comm. of Surface Waters, Publ., No. 51:362-378.

estuaries
Hatcher, J.R., 1956.
On the boundary between fresh water and salt water in an estuary. Bull. Amer. Phys. Soc., 1: 265-266.

estuaries
Hardenberg, J.D.F., 1951.
Estuarine problems in South East Asia. Proc. Indo-Pacific Fish. Counc., 17-28 Apr. 1950, Cronulla N.S.W., Australia, Sects. II-III:185-180.

estuaries
Harrison, W., M.P. Lynch and A.G. Altschaeffl, 1964.
Sediments of lower Chesapeake Bay, with emphasis on mass properties.
J. Sed. Petr., 34(4):727-755.

estuaries
Harrison, W., and Marvin L. Wass, 1965.
Frequencies of infaunal invertebrates related to water content of Chesapeake Bay sediments.
Southeast Geol., 6(4):177-187.

estuaries
Harvey, J.G., 1968.
The flow of water through the Menai Straits.
Geophys. J.R. astr. Soc., 15(5): 517-528.

estuaries
Harvey, H.W., 1925.
Hydrography of the English Channel.
Cons. Perm. Int. Expl. Mer, Rapp. Proc. Verb. 37:59-89, Figs. 17-29, 1 pl.

estuaries
*Haven, Dexter S., and Reinaldo Morales-Alamo, 1968.
Occurrence and transport of faecal pellets in suspension in a tidal estuary.
Sediment.Geol., 2(2):141-151.

estuaries
Havinga, B., 1959.
Artificial transformation of salt and brackish water into fresh water lakes in the Netherlands and possibilities for biological investigations.
Arch. Oceanogr. e Limnol., 11(Supplement):47-52.

estuaries
Hayami, Shoitiro, and Yoshiaki Fukuo, 1959.
A study on the productivity of Tanabe Bay. VI. On the exchange of water and the productivity of a bay with special reference to Tanabe Bay.
Rec. Oceanogr. Wks., Japan. Spec. No. 3:61-68.

Estuaries
Hayami, Shoitiro, Nobuo Watanabe and Sanae Unoki, 1967.
On the distribution and fluctuation of current and chlorinity in the estuary of Kiso-Sansen (In Japanese; English abstract).
Rep. Rept. Kisosansen Surv. Team, 4 and 5.
Also in: Coll. Repr. Coll. mar. Sci. Techn., Tokai Univ., 1967-68, 3: 129-185; 189-216.

estuaries
Hebara, Toshiyuki, 1965.
Role of minerals constituents in mechanical properties of the estuarine sediments along the Hiroshima Bay.
J. Sci., Hiroshima Univ., (C) 4(4):429-454.

estuaries
Hedgpeth, Joel H. 1967.
The sense of the meeting.
In: Estuaries, G.H. Lauff, editor, Publs Am. Ass. Advmt Sci. 83: 707-710.

estuaries
Hedgpeth, Joel W., 1967.
Ecological aspects of the Laguna Madre, a hyper-saline estuary.
In: Estuaries, G.H. Lauff, editor, Publs Am. Ass. Advmt. Sci., 83:408-419.

estuaries
Hedgpeth, J.W., 1953.
An introduction to the zoogeography of the northwestern Gulf of Mexico with reference to the invertebrate fauna. Publ. Inst. Mar. Sci. 3(1): 111-224, 46 textfigs.

estuaries
Hela, I., C.A. Carpenter, Jr., and J.K. McNulty, 1957.
Hydrography of a positive, shallow, tidal, bar-built estuary (report on the hydrography of the polluted area of Biscayne Bay).
Bull. Mar. Sci., Gulf and Caribbean, 7(1):47-99.

estuaries
Hendey, N.I., 1951
Littoral diatoms of Chicester Harbour with special reference to fouling. J.Roy. Microscop. Soc. 71(1): 1-86, 18 pls.

estuaries
Herlinveaux, R.H., 1963
Data record of oceanographic observations made in Pacific Naval Laboratory underwater sound studies, November 1961 to November 1962. Fish. Res. Bd., Canada, Mss. Rept. Ser. (Ocean. and Limnol.), No. 146: 101 pp. (multilithed)

estuaries
Herlinveaux, R.H., 1962
Oceanography of Saanich Inlet in Vancouver Island, British Columbia.
J. Fish. Res. Bd., Canada, 19(1):1-37.

estuaries
Herlinveaux, R.H., and J.P. Tully, 1961
Some oceanographic features of Juan de Fuca Strait.
J. Fish. Res. Bd., Canada, 18(6):1027-1071.

estuaries
Hernández-Pacheco, Francisco, e Isidoro Asensio Amor, 1966
Estudio fisiográfico-sedimentológico de la ría de Guernica.
Bolm. Inst. esp. Oceanogr. 125: 1-31

estuaries
Hickling, C.F. 1970.
Estuarine fish farming.
Advmt mar. Biol., F.S. Russell and Maurice Yonge, editors, Academic Press 8: 119-213.

estuaries
Hicks, Steacy D., 1964
Tidal wave characteristics of Chesapeake Bay.
Chesapeake Science, 5(3):103-113.

estuaries
Hicks, Steacy D., 1959.
The physical oceanography of Narragansett Bay.
Limnol. & Oceanogr. 4(3):316-327.

estuaries
Hinwood, J.B., 1964.
Estuarine salt wedges - determining their shape and size.
Dock & Harbour Authority, 45(525):79-83.

estuaries
*Hirano, Reijiro, 1967.
Mechanism of development of red tide in estuarine waters. (In Japanese; English abstract).
Inf. Bull. Planktol. Japan, Comm. No. Dr. Y. Matsue, 25-29.

estuaries
Hishida, Kozo, Katsumi Tanioka, Takeo Yasuoka, Toshio Wakabayashi, 1961
[On the blocking of the mouth of the River Yura]
Bull. Maizuru Mar. Obs., No. 7:31-44, (51-64).

estuaries
*Hobbie, John E., Claude C. Crawford and Kenneth L. Webb, 1968.
Amino acid flux in an estuary.
Science, 159(3822):1463-1464.

estuaries
Hodgkin, E.P. and R.J. Rippingale, 1971.
Interspecies conflict in estuarine copepods.
Limnol. Oceanogr. 16(3): 573-576.

estuaries
Hoffman, G.R., 1954.
Tidal calculations applied to the estuary of the river Great Ouse. Proc. Inst. Civ. Eng. 3(3): 809-829.

estuaries
Hood, Donald W., editor 1971
Impingement of man on the oceans.
Wiley-Interscience, 738 pp.

estuaries
Hopkins, J.T., 1964
A study of the diatoms of the Ouse Estuary, Sussex. III. The seasonal variation in the littoral epiphyte flora and the shore plankton.
Jour. Mar. Biol. Assoc., U.K., 44(3):613-644.

estuaries
Hopkins, J.T., 1964.
A study of the diatoms of the Ouse Estuary, Sussex. II. The ecology of the mud-flat diatom flora.
J. Mar. Biol. Assoc., U.K., 44(2):333-341.

estuaries
Hopkins, J. Trevor, 1963
A study of the diatoms of the Ouse Estuary, Sussex. 1. The movement of the mud-flat diatoms in response to some chemical and physical changes.
J. Mar. Biol. Assoc., U.K., 43(3):653-663.

estuaries
Hopkins, Thomas L., 1965.
Mysid shrimp abundance in surface waters of Indian River Inlet, Delaware.
Chesapeake Science, 6(2):86-91.

estuaries
Hopkins, T.L., 1963
The variation in the catch of plankton nets in a system of estuaries.
J. Mar. Res., 21(1):39-47.

estuaries
Hopkins, T.L., 1961
Natural coloring matter as an indicator of inshore water masses.
Limnol. & Oceanogr., 6(4):484-486.

estuaries
Horrer, Paul L. 1967.
Methods and devices for measuring currents.
In: Estuaries, G.H. Lauff, editor, Publs Am. Ass. Advmt Sci. 83: 80-89.

estuaries
Hoshiai, T., 1961.
Synecological study on intertidal communities. IV. An ecological investigation on the zonation in Matsushima Bay concerning the so-called covering phenomenon.
Bull. Mar. Biol. Sta., Asamushi, Tohoku Univ., 10(3):203-211.

Estuaries
Hosokawa, Iwao, Fumio Ohshima and Norihiko Kondo, 1970.
On the concentrations of the dissolved chemical elements in the estuary water of the Chikugogawa River. (In Japanese; English abstract).
J. oceanogr. Soc., Japan, 26(1): 1-5

estuaries
Huang, Ter-Chien, and H.G. Goodell 1967.
Sediments of Charlotte Harbor, southwestern Florida.
J. sedim. Petrol. 37(2): 449-474.

estuaries
Hughes, Peter, 1958.
Tidal mixing in the narrows of the Mersey Estuary.
Geophys. J., R. Astron. Soc., 1(4):271-283.

estuaries
Huggett, Robert J., Michael E. Bender and Harold D. Slone 1971.
Mercury in sediments from three Virginia estuaries.
Chesapeake Sci. 12(4): 280-282.

estuaries
Hulburt, E.M., 1957.
Distribution of phosphorus in Great Pond, Massachusetts. J. Mar. Res., 16(3):181-192.

estuaries
Hull, C.H.J., 1963.
Photosynthetic oxygenation of a polluted estuary. Int. J. Air Water Pollution, 7(6/7):669-696.

estuaries
Hunt, J.N., 1964.
Tidal flow in the Thames estuary.
Dock & Harbour Auth., 45(527):160-161.

estuaries
Hunt, J.N., 1964.
Tidal oscillations in estuaries.
Geophys. J., R. Astron. Soc., 8(4):440-455.

I

estuaries
Ichiye, T., 1955.
A note on the stationary currents in an estuary.
J. Ocean. Soc., Japan, 11(4):141-145.

estuaries
Ichiye, T., 1953
On the abnormal high water of rivers.
Ocean. Mag., Tokyo, 5(1):45-60, 9 textfigs.

estuaries
Ichiye, T., 1953.
On the variation of oceanographical elements due to tidal currents. Ocean. Mag., Tokyo, 4(3): 81-88, 3 textfigs.

estuaries
Ichiye, T., M.L. Jones, N.C. Hulings and F.C.W. Olson, 1961
Salinity change in Alligator Harbor and Ochlockonee Bay, Florida.
J. Oceanogr. Soc., Japan, 17(1): 1-9.

estuaries
Ichiye, Takashi, Sigeo Moriyasu and Hiroyuki Kitamura, 1951
On the hydrography near the estuaries.
Bull. Kobe Mar. Obs., No. 164:349 (top of page)-369; 53 (bottom of page)-75.

estuaries
Ingerson, I.M., 1955.
Lunar-cycle measurements of estuarine flows.
Proc. Amer. Soc. Civil Engl 81(836):18 pp.

estuaries
Inglis, C.C.L., and F.H. Allen, 1957.
The regimen of the Thames estuary as affected by currents, salinities and river flow.
Proc. Inst. Civ. Eng., 7:827-878.

estuaries
Ippen, Arthur T., 1966.
Estuary and coastline hydrodynamics. McGraw Hill Book Co, Inc., 744 pp.

estuaries
Ishino, M., 1960
[On the anomalous phenomenon caused by the dry spell in the early summer of 1958 in the tidal estuary of the River Tone.]
J. Oceanogr. Soc., Japan, 16(4): 157-162.

estuaries
Ito, Takeshi, Sei-ichi Sato, Tsutomu Kishi and Masateru Tominaga, 1960
On the density currents in the estuary.
Coastal Eng., Japan, 3:21-31.

estuaries
Iwata, K., 1953.
[A example of potamic tide.] Publ. 981, Hydrogr. Bull. No. 34:1-6, 5 textfigs.

estuaries
Iyengar, M.O.P. and G.Venkataraman, 1951.
The ecology and seasonal succession of the algae flora of the River Cooum at Madras with special reference to the Diatomaceae. J. Madras Univ. 21, Sect. B(1): 140-192, 1 pl of 4 figs., 11 text figs.

J

estuaries
Jackson, W.H., 1964.
Effect of tidal range, temperature and fresh water on the amount of silt in suspension in an estuary.
Nature, 201(4923):1017.

estuaries
Jacob, J., and K. Rangarajan, 1962
Seasonal cycles of hydrological events in Vellar Estuary.
Proc. First All-India, Congr. Zool., 1959, Sci. Pap., 2:329-350.
Also in:
Annamalai Univ. Mar. Biol. Sta., Porto Novo, S. India, Publ. 1961-1962.

estuaries
Jacobsen, J.P., 1918.
Hydrographische Untersuchungen im Randersfjord (Jylland). Medd. Komm. Havundersøgelser, Ser. Hydrogr. 2(7):46 pp., 15 textfigs.

estuaries
Jacobsen, J.P., 1913.
Beitrag zur Hydrographie der Dänischer Gewässer. Medd. Komm. Havundersøgelser, Ser. Hydrogr. 2(2):94 pp., 47 tables, 17 textfigs., 14 pls.

estuaries
Jaffe, G., and J.M. Hughes, 1953.
The radioactivity of bottom sediments in Chesapeake Bay. Trans. Amer. Geophys. Union 34(4): 539-542, 2 textfigs.

estuaries
Jednoral, Tadeusz, 1963
Probability of occurrence of the two-way water flow in the Swinoujscie-Szczecin Channel. In Polish; English abstract.
Prace Inst. Morsk., Ser. 1. Hydrotechnika, Gdansk No. 18: 48 pp.

estuaries
Jeffries, H. Perry, 1967.
Saturation of estuarine zooplankton by congeneric associates.
In: Estuaries, G.H. Lauff, editor, Publs Am. Ass. Advmt Sci., 83:500-508.

estuaries
Jeffries, H. Perry, 1966.
Partitioning of the estuarine environment by two species of Cancer.
Ecology, 47(3):477-481.

estuaries
Jeffries, Harry P., 1964.
Comparative studies on estuarine zooplankton.
Limnology and Oceanography, 9(3):348-358.

estuaries (Raritan Bay)
Jeffries, Harry P., 1962
Copepod indicator species in estuaries.
Ecology, 43(4):730-733.

estuaries
Jeffries, Harry P., 1962.
Environmental characteristics of Raritan Bay, a polluted estuary.
Limnol. and Oceanogr., 7(1):21-31.

estuaries
Jeffries, H.P., 1962.
Salinity-space distribution of the estuarine copepod genus Eurytemora.
Int. Rev. Ges. Hydrobiol., 47(2):291-300.

estuaries
Jeffries, Harry P., 1962
Succession of two Acartia in estuaries.
Limnol. and Oceanogr., 7(3):354-364.

estuaries
Jeffries, Harry P., 1962
The atypical phosphate cycle of estuaries in relation to benthic metabolism.
The Environmental Chemistry of Marine Sediments, Proc. Symp., Univ. R.I., Jan. 13, 1962, Occ. Papers, Narragansett Mar. Lab., No. 1: 58-67.

estuaries
Jennings, J.N., and E.C.F. Bird, 1967.
Regional geomorphological characteristics of some Australian estuaries.
In: Estuaries, G.H. Lauff, editor, Publs Am. Ass. Advmt Sci., 83:121-128.

estuaries
Jensen, A.J.C., 1944.
The hydrography of the Praestø Fjord.
Folia Geogr. Danica 3(2):47-55, 2 textfigs., 4 tables.

estuaries
Jensen, Eugene T., 1968.
National estuarine pollution study.
Proc. Gulf Carib. Fish. Inst., 20th Ann. Sess. 69-74

estuaries
Jitts, H.R., 1959.
The adsorption of phosphate by estuarine bottom deposits. Mar. & Australian J./Freshwater Res., 10(1):7-21.

estuaries
Johns, B. 1970.
On the determination of the tidal structure and residual current system in a narrow channel.
Geophys. J. R. Astron. Soc. 20(2): 159-175.

estuaries
Johns, B. 1969.
Some consequences of an inertia turbulence in a tidal estuary.
Geophys. J. R. astron. Soc. 18(1):65-72

estuaries
*Johns, B., 1969.
On the representation of the Reynolds stress in a tidal estuary.
Geophys. J. R. astr. Soc., 17(1):39-44.

estuaries
Johns, B. 1968.
Some effects of topography on the tidal flow in a river estuary.
Geophys. J. R. astr. Soc. 15(5): 501-567

estuaries
Johns, B., and N. Odd, 1966.
On the vertical structure of tidal flow in river estuaries.
Geophys. J., R. Astr. Soc., 12(1):103-

estuaries
Johnson, J.W., and R.L. Wiegal, 1958.
Investigation of current measurement in estuarine and coastal waters.
State Water Pollution Control Bd., California, 19:233 pp.

Abstr. in: Publ. Health Eng. Abstr., 39(9):32.

estuaries
Johnson, T.W., Jr., 1967.
The estuarine mycoflora.
In: Estuaries, G.H. Lauff, editor, Publs Am. Ass. Advmt Sci., 83:303-305.

estuaries
Johnson, T.W., Jr., & F.K. Sparrow, Jr., 1961
Fungi in oceans and estuaries.
Hafner Publ. Co., New York: 668 pp.

estuaries
Jones, D., and M.S. Wills, 1956.
The attenuation of light in sea and estuarine waters in relation to the concentration of suspended solid matter. J.M.B.A., 35(2):431-444.

estuaries
Jones, H. G., 1969.
Spider crabs of the genus Mithrax from Barbados.
Zool. Anz., 182(5/6): 379-383

estuaries
Jordan, G.F., 1961.
Erosion and sedimentation, Eastern Chesapeake Bay at Choptank River.
U.S.C.&G.S., Techn. Bull., No. 16:8 pp., 1 char

estuaries
Joseph, E.B., W.H. Massmann, & J.J. Norcross, 1960
Investigations of inner continental shelf water off lower Chesapeake Bay. Part 1 - General introduction and hydrography.
Chesapeake Science, 1(3-4): 155-167.

K

estuaries
Kalber, Frederick A., 1959.
The role of tidemarshes in estuarine productivity.
Estuarine Bull., Univ. Delaware, 4(1):3, 14-15.

estuaries
Kamps, L. F., R. Dorrestein and L. Otto, 1960.
Note on the annual variation of salinity, temperature and oxygen content in the Ems Estuary.
Verh. Kon. Ned. Geol. Mijnb. k. Gen., Geol. Ser. 19:75-81.

estuaries
*Kanari, Seiichi, 1966.
Some experiments on the sedimentation in estuaries with density stratification.
Spec. Contr. geophys. Inst. Kyoto Univ., 6:127-133.

Kändler
estuaries
Kändler, R., 1961.
Über das Vorkommen von Fischbrut, Decapodenlarven und Medusen in der Kieler Förde.
Kieler Meeresf., 17(1):48-64.

estuaries
Kändler, R., 1959
Hydrographische Beobachtungen in der Kieler Förde 1952-57. Kieler Meeresforsch., 15: 145-157.

estuaries
Kandler, R., 1953.
Hydrographische Untersuchungen zum Abwasserproblem in den Buchten und Förden der Ostseeküste Schleswig-Holstein. Kieler Meeresf. 9(2): 176-200, Pls. 9-14(18figs.).

Estuaries
Kane, Henry E., 1966
Sediments of Sabine Lake, the Gulf of Mexico and adjacent water bodies, Texas - Louisiana.
J. Sed. Petr., 36(2):608-619.

estuaries
Kashiwamura, Masakazu, 1963.
Variation of surface velocity in a tidal river.
J. Oceanogr. Soc., Japan, 19(1):1-5.

estuaries
Kawamura, Teruyoshi, 1966.
Distribution of phytoplankton populations in Sandy Hook Bay and adjacent areas in relation to hydrographic conditions in June 1962.
Tech. Pap. Bur. Sport Fish. Wildl., U.S., (1):1-37. (multilithed).

estuaries
Keighton, W.B., 1954.
The investigation of chemical quality of water in tidal rivers. U.S. Geol. Surv., 54 pp.

estuaries
Kennedy, J.F., 1959.
Anti-dunes and standing waves in alluvial channels. (Abstract).
J. Geophys. Res., 64(6):690-691.

estuaries
Kent, Richard E., and D.W. Pritchard, 1959.
A test of mixing length theories in a coastal plain estuary.
J. Mar. Res., 18(1):62-72.

estuaries
Kestner, Friedrich Julius Theodor, 1961
Short-term changes in the distribution of fine sediments in estuaries.
Proc. Inst. Civil Eng., 19:185-208.

estuary
Kestner, F.J.T., and C.C. Ingles, 1956?
A study of erosion and accretion during cyclic changes in an estuary and their effect on reclamation of marginal land. J. Agric. Eng. Res. 1(1):1-8.

Ketchum B.H.
estuaries
Ketchum, Bostwick H., 1969.
Eutrophication of estuaries.
In: Eutrophication: causes, consequences, correctives, proceedings of a symposium, 197-209.
U.S. Nat. Acad. Sci.,

estuaries
Ketchum, Bostwick H., 1967.
Phytoplankton nutrients in estuaries.
In: Estuaries, G.H. Lauff, editor, Publs Am. Ass. Advmt Sci., 83:329-335.

estuaries
Ketchum, B.H., 1955.
Distribution of coliform bacteria and other pollutants in tidal estuaries.
Sewage and Ind. Wastes 27(11):1288-1296.

estuaries
Ketchum, B.H., 1954.
Relation between circulation and planktonic populations in estuaries. Ecology 35(2):191-200, 7 textfigs.

estuaries
Ketchum, B.H., 1953.
Ch. 6: Circulation in estuaries. Proc. Third Conf., Coastal Eng., Cambridge, Mass, Oct. 1952: 65-76, 8 textfigs.

estuaries
Ketchum, B.H., 1951.
The exchange of fresh and salt waters in tidal estuaries. J. Mar. Res. 10(1):18-38, 5 textfigs.

estuaries
Ketchum, B.H., 1951.
The flushing of tidal estuaries. Sewage Ind. Wastes, 23:198-208.

estuaries
Ketchum, B. H., 1950.
Hydrographic factors involved in the dispersion of pollutants introduced into tidal waters.
J. Boston Soc., Civil Engineers, 37(3):296-314, 8 textfigs.

estuaries
Ketchum, B.H., A.C. Ayers, and R.F. Vaccaro, 1952.
Processes contributing to the decrease of coliform bacteria in a tidal estuary. Ecology 33(2): 247-258, 4 textfigs.

estuaries
Ketchum, B.H., and D.J. Keen, 1953.
The exchanges of fresh and salt waters in the Bay of Fundy and in Passamoquoddy Bay.
J. Fish. Res. Bd., Canada, 10(3):97-124, 11 textfigs.

estuaries
Keuligan, G. H., and J. V. Hall, jr., 1950.
A formula for the calculation of the tidal discharges through an inlet. Bull. B.E.B. 4(1):15-29, 11 figs.

estuaries
Kimata, M., H. Kadota, Y. Hata and H. Miyoshi, 1958.
The formation of sulfide by sulphate-reducing bacteria in the estuarine zone of the river receiving a large quantity of organic drainage.
Rec. Oceanogr. Wks., Japan, Spec. No., 2:187-199.

estuaries
Kimata, M., H. Kadota, Y. Hana and H. Miyoshi, 1957.
Studies on the marine sulfate-reducing bacteria. 4. Production of sulfides in the estuarine region receiving a large amount of organic drainage (1). Bull. Jap. Soc. Sci. Fish., 22(11):701-707.

estuaries
Kinne, Otto, 1967.
Physiology of estuarine organisms with special reference to salinity and temperature: general aspects.
In: Estuaries, G.H. Lauff, editor, Publs Am. Ass. Advmt Sci., 83:525-540.

estuaries
Kinne, Otto, 1966.
Physiological aspects of animal life in estuaries with special reference to salinity.
Neth. J. Sea Res., 3(2):222-244.

estuaries
Kinne, Otto, 1964.
Physiologische und ökologische Aspekte des Lebens in Ästuarien.
Helgol. Wiss. Meeresuntersuch., 11(3/4):131-156.

estuaries
Kishi, Tsutomu, and Masahiro Taniguchi, 1961
Improvement of the mouth of the Musa River.
Coastal Eng., Japan, 4:115-122.

estuaries
Kishi, Tsutomu, Masateru Tominaga and Ichiro Oeda, 1960.
Studies on meteorological tides at the mouth of the Tone River.
Coastal Eng., Japan, 3:1-8.

estuaries
Kitching, J.A., S.J. Lilly, S.M. Lodge, J.F. Sloane, R. Bassindale and F.J. Ebling, 1952.
The ecology of the Lough Ine Rapids with special reference to water currents. J. Ecology 40(1):179-201.

estuaries
Klein, George deVries, 1967.
Comparison of Recent and ancient tidal flat and estuarine sediments.
In: Estuaries, G.H. Lauff, editor, Publs Am. Ass. Advmt Sci., 83:207-218.

estuaries
Klein, George deVries, 1967.
Paleocurrent analysis in relation to modern marine sediment dispersal patterns.
Bull.Am.Ass.Petrol.Geol., 51(3):366-382.

estuaries
Kohout, F.A., and M.C. Kolipinski, 1967.
Biological zonation related to groundwater discharge along the shore of Biscayne Bay, Miami, Florida.
In: Estuaries, G.H. Lauff, editor, Publs Am. Ass. Advmt Sci., 83:488-499.

estuaries
Konietzko, B., 1954.
Recherches sur les fossés lateraux de l'Elbe soumis à l'influence des marées.
K. Belg. Inst. Nat. Verhandl. 2(23):65 pp.

estuaries
Korringa, P., 1967.
Estuarine fisheries in Europe as affected by man's multiple activities.
In: Estuaries, G.H. Lauff, editor, Publs Am. Ass. Advmt Sci., 83:658-663.

estuaries
Korringa, P., 1956.
Hydrographical, biological and ostreological observations in the Knysna Lagoon with notes on conditions in other South African waters.
S. African Dept. Comm. & Fish., Div. Fish., Invest. Rept., No. 20:63 pp., 23 pls.

estuaries
Koske, P.H., H. Krumm, G. Rheinheimer und K.-H. Szielda, 1966.
Untersuchungen über die Einwirkung der Tide auf Salzgehalt, Schwebstoffgehalt, Sedimentation und Bakteriengehalt in der Unterelbe.
Kieler Meeresforsch., 22(1):47-63.

estuaries
Kostaanitsin, M.N., 1962.
The piling up-retreating oscillations in level of the water in the Bugsk estuary. (In Russian).
Trudy Gosud. Okeanogr. Inst., 66:55-79.

estuaries
Kostianitsin, M.N., 1960.
[On the observations on alluvium movements and sedimentation in the coastal zone of the sea and the estuaries.]
Meteorol. i Gidrol., No. 7:33-35.

estuaries
Kranel, David P. 1969.
A physical oceanographic study of the Margaree and Cheticamp River estuaries.
Techn. Rept. Fish. Res. Bd., Can., 115: 152pp (multilithed)

estuaries
Krey, H., 1926.
Die Flutwelle in Flussmündungen und Meeresbuchten
Comm. Versuchsanstalt f. Wasserbau und Schiffbau, Berlin, No. 3:59 pp., 31 figs.

estuaries
Kruit, C., 1951.
Aperçu de l'histoire recente du delta du Rhone.
Proc. 3rd Int. Congress, Sedimentol., Netherlands 1951:181-191, 3 figs.

estuaries
Kubo, Tadashi, Yoshikazu Sato and Shigeshi Komaki 1967.
On the backflow of sea water into Lake Abashiri. 1. An estimation of charged volume and the condition of sea water backflowed into Lake Abashiri. (In Japanese; English abstract)
Bull. Hokkaido reg. Fish. Res. Lab. No. 32:49-61

estuary
Kubo, Tamotsu, 1958.
Consideration by fundamental test of jetty in river mouth.
Coastal Eng., Japan, 1:135-147.

estuaries
Kühl, Heinrich 1971.
Die Hydromedusen der Wesermündung.
Vie et Milieu Suppl. 22(2):803-810.

estuaries
Kühl, Heinrich, 1967.
Observations on the ecology of barnacles in the Elbe-Estuary.
Proc. Symp. Crustacea, Ernakulam, Jan. 12-15, 1965, 3: 965-975

estuaries
Kühl, Heinrich und Hans Mann, 1969.
Über das Zooplankton der Unterweser und Wesermündung.
Veröff. Inst. Meeresforsch. Bremerh. 12(2): 43-64

estuaries
Kühl, Heinrich und Hans Mann, 1965.
Vergleichende Untersuchungen über Hydrochemie und Plankton deutscher Flussmündungen.
Helgoländer wiss. Meeresunters. 17:435-444.

estuaries
Kühl, Heinrich, and Hans Mann, 1962
Über das Zooplankton der Unterelbe.
Veröffentlichungen Inst. Meeresf., Bremerhaven 8(1):53-69.

estuaries
Kullenberg, B., 1955.
Restriction of the underflow in a transition.
Tellus 7(2):215-217, 3 textfigs.

estuaries
Kulm, L.D., and John V. Byrne, 1967.
Sediments of Yaquina Bay, Oregon.
In: Estuaries, G.H. Lauff, editor, Publs Am. Ass. Advmt Sci., 83:226-238.

estuaries
Kulm, L.D., and John V. Byrne 1966.
Sedimentary response to hydrography in an Oregon estuary.
Mar. Geol. 4(2):85-118.

estuaries
* Kunze, G.W., L.I. Knowles and Y. Kitano. 1968.
The distribution and mineralogy of clay minerals in the Taku Estuary of southeastern Alaska.
Marine Geol. 6(6): 439-448.

L

estuaries
Labat, M., 1890.
De l'influence de la surface de l'estuaire d'amont dans l'approfondissement des passes des pleuves a marees. C.R. Assoc. Francaise Adv. Sci. 19eme Session, Limoges, :222-225.

estuaries
Labetoulle, J., 1955.
Principaux problèmes soulevés par les modèles réduits d'estuaire à marée. Communication présentée au 6e Congrès Internationale de Recherches Hydrauliques, La Haye, 31 Août-6 Septembre 1955(A13):1-17, 21 figs.

estuaries
Lackey, James B., 1967.
The microbiota of estuaries and their roles.
In: Estuaries, G.H. Lauff, editor, Publs Am. Ass. Advmt Sci., 83:291-302.

estuaries
*Lacroix, Guy, 1967.
Recherches sur le zooplancton de la Baie-des-Chaleurs en 1966.
Rapp. Stn Biol. Mar. Grande-Rivière, 1966:37-46.

estuaries
Lacroix, Guy, 1965.
Production de zooplancton dans la Baie-des-Chaleurs en 1964.
Rapp. Ann. Sta. Biol. Mar., Grande Rivière, 1964:53-58.

estuaries
Lacroix, Guy, 1963.
Production de zooplancton dans la Baie-des-Chaleurs en 1962.
Rapp. Ann., 1962, Sta. Biol. Mar., Grande-Rivière, Canada, 39-52. (multilithed).

estuaries
Lacroix, Guy, et Julien Bergeron, 1963.
Liste préliminaire des invertébrés du Banc de Bradelle, 1962.
Rapp. Ann., 1962, Sta. Biol. Mar., Grande-Rivière Canada, 59-67. (Multilithed).

estuaries
Lacroix, Guy, et Gabriel Filteau, 1969.
Les fluctuations quantitatives du zooplancton de la Baie-des-Chaleurs (Golfe Saint-Laurent). 1. Conditions hydroclimatiques et analyse volumétrique.
Naturaliste can. 96 (3): 359-397.

estuaries
Lacroix, Guy, et Louis Legendre, 1964.
La zooplancton de l'estuaire de la Riviere Restigouche (Baie des Chaleurs): quantites et composition en aout 1962.
Le Naturaliste Canadien, 91(1):21-40.

Also: Trav. Pecheries du Quebec, No. 2.

estuaries
Lacroix, Guy, et Louis Legendre, 1964.
Le zooplancton de l'Estuaire de la Rivière Restigouche (Baie des Chaleurs): quantités et composition en août 1962.
Le Naturaliste Canadien, 91(1):21-40.

Also: (3), 35(1):21-40.

estuaries
Lacroix, G., and L. Legendre, 1963.
Étude préliminaire du zooplancton de l'estuaire de la Rivière Restigauche.
Rapp. Ann., 1962, Sta. Biol. Mar., Grande Rivière, Canada, 53-58. (multilithed)

estuaries
Lacroix, Guy, et Laurentio Méthot, 1967.
Deuxième Catalogue des échantillons de zooplancton du Musée de la Station de Biologie Marine (1962-1964), additions au premier Catalogue (1951-1961) et sommaire des observations pour la période 1951-1964.
Cah. Info. Stn Biol. mar., Grande Rivière, 41: 70pp. (multilithed)

estuaries
Lafon, M., M. Durchon and Y. Saudray, 1955.
Recherches sur les cycles saisonnières du plankton
Ann. Inst. Ocean., 31(3):125-230.

estuaries
Lamoen, J., 1936.
Sur l'hydraulique des fleuves à marée.
Rev. Gén. Hydraulique, Juill'et-Aout:545.
Sept.-Oct.:595.
Nov.-Dec.:643.

estuaries
Lance, Joan, 1964.
The salinity tolerances of some estuarine planktonic crustaceans.
Biol. Bull., 127(1):108-118.

estuaries
Lance, Joan, 1963.
The salinity tolerance of some estuarine plankton copepods.
Limnology and Oceanography, 8(4):440-449.

estuaries
Lance, Joan, 1962.
Effects of water of reduced salinity on the vertical migration of zooplankton.
J. Mar. Biol. Assoc., U.K., 42(2):131-154.

estuaries
Land, Lynton S., and John H. Hoyt, 1966.
Sedimentation in a meandering estuary.
Sedimentology, 6:191-207.

estuaries
Lane, R.K., A.M. Holler, J.H. Meikle and H.J. Hollister, 1961
Oceanographic data record, Monitor and Coastal Projects, March 20 to April 14, 1961.
Fisheries Res. Bd., Canada, Manuscript Rept. Ser. (Oceanogr. & Limnol.), No. 94: 188 pp. (multilithed).

estuaries
LaPorte, 1904.
Recherches sur le régime des côtes. Projet d'établissement d'une chambre d'epanouissement et d'un brise-lames à la fosse du Lazaret a l'embouchure de l'Adour. Service Hydrographique de la Marine 13-617:157-165.

estuaries
Lauff, George H., editor, 1967.
Estuaries.
Publs Am. Ass. Advmt Sci., 83:757 pp.

estuaries
Laurent, J., 1951.
Étude de l'estuaire de l'Oued-Sebou. Observations en nature: essai sur un modèle réduit.
Assemblée Générale de Bruxelles, 3 Publ. de l'Assoc. Int. d'Hydrol. No. 34:154-161.

estuaries
Lauzier, L., 1953.
The St. Lawrence spring run-off and summer salinities in the Magdalen Shallows.
J. Fish. Res. Bd., Canada, 10(3):146-147, 1 text-fig.

estuaries
Lavoinne, M., 1883.
Observations présentées au sujet du régime des courants et des alluvions dans l'estuaire de la Seine. Assoc. Francaise Avance. Sci., 12ème Sess. Rouen:262-264.

estuaries
Leach, J.H., 1971.
Hydrology of the Ythan Estuary with reference to distribution of major nutrients and detritus
J. mar. biol. Ass., U.K., 51(1): 137-157.

estuaries
Ledoyer, Michel, 1970.
Additions à la liste des invertébrés benthiques recueillis dans le Golfe Saint-Laurent (Baie de Chaleurs).
Rapp. ann. 1969, Serv. Biol. Québec, 37-43

estuaries
Le Fèvre-Lehoërff, Geneviève 1971.
Étude d'un cycle nycthéméral dans l'estuaire de la rivière de Morlaix: hydrologie et zooplancton.
Revue Trav. Inst. Pêches marit. 35(3):347-366.

estuaries
Legendre, Louis, 1970.
Étude des associations planctoniques en 1969.
Rapp. ann. 1969, Serv. Biol. Québec, 29-35

estuaries
LeMare, D.W., 1951.
Applications of the principles of fish culture to estuarine conditions in Singapore. Proc. Indo-Pacific Fish. Counc., 17-28 Apr. 1950, Cronulla N.S.W., Australia, Sects. II-III:180-183.

estuaries
Lenz, Jürgen, 1970.
4. Planktologie. 4.1. Seston, Chlorophyll- und Eiweissgehalt. 4.3. Zooplankton.
Chemische, mikrobiologische und planktologische Untersuchungen in der Schlei im Hinblick auf deren Abwasserbelastung. Kieler Meeresforsch 26(2): 180; 180-129; 203-213.

estuaries
Leppik, Egon, 1949.
Verbesserung von Flussmündungen an Meeren ohne Tidebewegung unter besondere Berücksichtigung der Ost- und Südküste der Ostsee. Studien zu Bau und Verkehrsproblemen der Wasserstrassen, Offenbach am Main:239-254, 2 figs.

estuaries
Liaw, Wen Kuang, 1967.
On the occurrence of chaetognaths in the Tanshui River estuary of northern Taiwan (Formosa).
Publs. Seto mar. biol. Lab., 15(1):5-18.

estuaries
Lighthill, M.J., and G.B. Whitham, 1955.
On kinematic waves. 1. Flood movement in long rivers. Proc. R. Soc., London, A, 229:281-316.

estuaries
Lilly, S.J., J.F. Sloane, R. Bassindale, F.J. Ebling and J.A. Kitching, 1953.
The ecology of the Lough Ine rapids with special reference to water currents. IV. The sedentary fauna of sublittoral boulders. J. Animal Ecol. 22(1):87-122, 22 textfigs.

estuaries
Linder, C.P., 1953.
Intrusion of sea water in tidal sections of fresh water streams. Amer. Soc. Civ. Eng. 79(358):35 pp.

estuaries
*Liss, P.S., and C.P. Spencer, 1970.
Abiological processes in the removal of silicate from sea water.
Geochim. cosmochim. Acta, 34(10):1073-1088.

estuaries
Livingstone, Robert, Jr., 1965.
A preliminary bibliography with KWIC index on the ecology of estuaries and coastal areas of the eastern United States.
Spec. Scient. Rep. U.S.Fish. Wildl.Serv.Fish. No. 509:352 pp. (multilithed)

estuaries
Lockwood, Mason G., and Henry P. Carothers, 1967.
Preservation of estuaries by tidal inlets.
J. Watways Harb.Div.Am.Soc.civ. Engrs.93 (WW4): 133-152.

estuaries
Lomniewski, Kazimierz, 1958
[The Firth of Vistula] Polska Akad. Nauk, Inst. Geografii, Prace Geograficzne, No. 15: 106 pp.

estuaries
López-Benito, Manuel, 1970.
Silicatos en el agua de mar.
Inv. pesq. Barcelona, 34(2):385-397.

estuaries
Lorenzen, Carl J., 1963
Diurnal variation in photosynthetic activity of natural phytoplankton populations.
Limnology and Oceanography, 8(1):56-62.

estuaries
Loring, Douglas, 1965.
Resume of marine geological investigations carried out by the Atlantic Oceanographic Group in the Gulf of St. Lawrence, 1961-1964.
Maritime Sediments, 1(1):8-9 (mimeographed).

estuaries
*Loring, D.H., D.J.G. Nota, 1968.
Occurrence and significance of iron, manganese, and titanium in glacial marine sediments from the estuary of the St. Lawrence River.
J.Fish.Res.Can., 25(11):2327-2347.

estuaries
Loring, D.H., D.J.G. Nota, W.D. Chesterman and H.K. Wong, 1970.
Sedimentary environments on the Magdalen Shelf, southern Gulf of St. Lawrence.
Mar. Geol. 8(5):337-354

estuaries
Lucht, F., 1953.
Die Sandwanderung im unteren Tidegebiet der Elbe.
Deutsche Hydrogr. Zeits. 6(4/5/6):186-207, 16 textfigs.

estuaries
Lucht, F., 1953.
Hydrographische Untersuchungen in der Brackwasser-zone der Elbe. Deutsche Hydrogr. Zeits. 6(1): 18-32, 12 textfigs.

estuaries
Lucht, F., 1952.
Hydrographische Messungen in der Aussenelbe.
Wasserwirtschaft, Hamburg Conf., Sept. 19-20, 22-26.

estuaries
Lüneburg, H., 1964.
Beiträge zur Sedimentologie der Weser- und Elbästuare.
Helgoländer Wiss. Meeresuntersuch., 10(1/4):217-230.

estuaries
Lüneburg, H., 1964.
Origin and significance of iron-oolitic sand-grains in the sediments of the Weser estuary.
Marine Geology, 1(1):106-110.

estuaries
Lüneburg, Hans, 1961
Über die Erosion und den Sedimenttransport am Knechtsand und Eversand (nordliche Wesermündung).
Veroff. Inst. Meeresf., Bremerhaven, 7(2):277-294.

estuaries
Lüneberg, H., 1955.
Beiträge zur Hydrographie der Wesermündung. IV. Verteilung der Sinkstoffe in der Seitenräumen der Wesermündung.
Veröff. Inst. Meeresf., Bremerhaven, 3(2):228-265, 7 pls. 1 fig.

estuaries
Lüneberg, H., 1953.
Beitrage zur Hydrographie der Wesermündung. 2. Die Probleme der Sinkstoffvertung in der Wesermündung. Veröff. Inst. Meeresf., Bremerhaven, 2(1):15-51, 13 pls., 2 textfigs.

estuaries
Lüneberg, H., 1939.
Hydrochemische Untersuchungen in der Elbmündung mittels Elektrokolorimeter.
Arch. Deutschen Seewarte 59(1):1-27, 8 pls.

M

estuaries
Mabesoone, J.M., 1966.
Depositional environment and provenance of the sediments in the Guadelete Estuary (Spain).
Geol. en Mijnbouw, 45(2):25-32.

estuaries
Machado, L. de B., 1952.
Pesquisas fisicas e quimicas do sistema hidrografico da regiao Lagunar de Cananeia.
Bol. Inst. Ocean., Univ. Sao Paulo, 3(1/2):55-75, 4 graphs, 1 chart.

estuaries
Machado, L. de B., 1950.
Pesquisas fisicase quimicas do sistema hidrográfico da regiäo lagunar de Cananéia. 1. Cursès de ágna. Bol. Inst. Paulista Ocean. 1(1):45-68.

estuaries
MacMillan, D.H.T., 1952.
The approach channels to Southampton.
J. Inst. Navig. 5:178-194.

estuaries
Maeda, H., 1955.
Studies on Yosa-Naikai. 4. Classification of phytoplankton communities and relation between communities and water masses.
J. Shimonoseki Coll. Fish. 4(2):301-310, 5 textfigs.

estuaries
Maeda, H., 1953.
The relation between chlorinity and silicate concentration of water observed in some estuaries.
J. Shimonoseki Coll. Fish. 3(2):167-180, 16 textfigs.

estuaries
Maeda, H., 1953.
Studies on Yosa-Naikai. 3. Analytical investigations on the influence of the River Noda and the benthonic communities. J. Shimonoseki Coll. Fish. 3(2):141-149, 3 textfigs.

estuaries
Maeda, H., 1953.
Studies on Yoda-Naikai. 2. Considerations upon the range of the stagnation and the influences by the River Noda and the open sea.
J. Shimonoseki Coll. Fish. 3(2):133-140, 2 textfigs.

estuaries
Maeda, H., 1952.
The relation between chlorinity and silicate concentration of the water observed in some estuaries. Publ. Seto Mar. Biol. Lab., Kyoto Univ., 2(2):249-255, 5 textfigs.

estuaries
Maeda, Hiroshi, and Kaoru Takesue, 1961
The relation between chlorinity and silicate concentration of waters observed in some estuaries. V. Consideration upon the fitness of [Cl] - [SiO2] relation of the river water for the relation formula computed from the estuarine waters and upon the seasonal variation of the estimated constants.
Rec. Oceanogr. Wks., Japan, 6(1):112-119.

estuaries
*Majewski, Aleksander, 1966.
Range of sea influences in Vistula and Odra estuaries. (In Polish; English abstract).
Przezlad Geofiz., 11(19-4):217-224.

estuaries
Makimoto, H., H. Maeda and S. Era, 1955.
The relation between chlorinity and silicate concentration of water observed in some estuaries.
Rec. Ocean. Wks., Japan, 2(1):106-112, 4 textfigs

estuaries
Makkaveeva, N.S., 1965.
The hydrochemical investigations of the Korovinskaya Inlet. (In Russian).
Materiel. Ribochoz. Issled. Severn. Bassin. Poliern. Nauchno-Issled. i Proekt. Inst. Morsk. Rib. Choz. i Okeanogr., N.M. Knipovich, 5:117-122.

estuaries
Mandelbrot, L., 1965.
Le nombre de Richardson et les criteres de stabilite des ecoulements stratifies.
La Houille Blanche, 20(1):24-28.

estuaries
Mangelsdorf, Paul C., Jr., 1967.
Salinity measurements in estuaries.
In: Estuaries, G.H. Lauff, editor, Publs Am. Ass. Advmt Sci., 83:71-79.

estuaries
Manguin, E., 1956.
Les diatomées de l'estuaire de la Rance.
Bull. Lab. Marit. Dinard, (42):62-76.

estuaries
Mann, I.J., 1881.
River bars. London, Crosby Lockwood & Co., 77 pp. 23 figs.

estuaries
Mansuett, Romeo J., 1962
Callico crab, Ovalipes o. ocellatus in mid-Chesapeake Bay, Maryland.
Chesapeake Science, 3(2):129-130.

estuaries
Marcello, A., 1961.
Fitofenologia nella Laguna di Venezia (Resume).
Rapp. Proc. Verb., Reunions, Comm. Int. Expl. Sci., Mer Mediterranée, Monaco, 16(3):793.

estuaries
Marchal, , 1854-1857.
Sediment at the mouth of streams. Ann. Ponts et Chaussées, 1st semestre, 1854:137.
2nd " , 1857:114-115.

estuaries
Marchesoni, V., 1954.
Il trofismo della Laguna Veneta e la vivificazione marina. III. Ricerche sulle variazioni quantitative del fitoplancton.
Arch. Oceanogr. e Limnol. 9(3):153-284.

estuaries
Marcotte, Alexandre, 1964.
Observations quotidiennes sur la température superficielle de l'eau de mer à Grande-Rivière (Baie-des-Chaleurs) 1951-1962.
Cah. Inf. Biol. mar., Grande-Riviere, No. 20. 4pp., 29 tables. (multilithed)

estuaries
Margalef, Ramon, 1967.
Laboratory analogues of estuarine plankton systems.
In: Estuaries, G.H. Lauff, editor, Publs Am. Ass. Advmt Sci., 83:515-521.

estuaries
Margalef, Ramón, 1961
Hidrografía y fitoplancton de un área marina de la costa meridional de Puerto Rico.
Inv. Pesq., Barcelona, 18:38-96.

estuaries
Marin, Jean 1971.
Croissance, condition et mortalité des huîtres du Belon.
Rev. Trav. Inst. Pêches marit. 35(2): 201-212.

estuaries
Marin, Jean 1971.
Etude physico-chimique de l'estuaire du Belon.
Rev. Trav. Inst. Pêches marit. 35(2): 109-156

estuaries
Marshall, Harold G., 1967.
Plankton in James River estuary, Virginia. 1. Phytoplankton in Willoughby Bay and Hampton Road
Chesapeake Sci., 8(2):90-101.

estuaries
Marshall, Nelson, 1960.
Studies of the Niantic River, Connecticut with special reference to the bay scallop, Aequipecten irradians. Limnol. & Oceanogr. 5(1):86-105.

Martin, P.L.C.

estuaries
Martin, P.L.C., 1966.
Contribucion de la sedimentologia a la hidrografia en el estudio de las transformaciones de las costas y estuarios.
Bol. Serv. Hidrogr. Naval, Buenos Aires (Publ. H. 106), 3(1):9-22.

estuaries
Martin, P.L.C., 1965.
The contribution of sedimentology to hydrography in the study of coast and estuary changes.
Int. Hydrogr. Rev., 42(2):159-171.

estuaries
Martin, P.L.C., 1961.
Determination of bottom shifts in Rio Nunez estuary.
Int. Hydrogr. Rev., 38(1):111-130.

estuaries
Massman, W.H., 1962.
Water temperatures, salinities and fishes collected during trawl surveys of Chesapeake Bay and York and Pamunkey rivers, 1956-1959.
Virginia Inst. Mar. Sci., Gloucester Pt., Spec. Sci. Rept., No. 27:27 pp. (mimeographed).

estuaries
Mauvais, J.L., 1971.
Calcul des vitesses moyennes instantanées en Loire maritime.
Cah. océanogr. 23(3): 251-266

estuaries
Maximon, L.C., and G.W. Morgan, 1955.
A theory of tidal mixing in a "vertically homogeneous" estuary.
J. Mar. Res. 14(2):157-175, 3 textfigs.

estuaries
Maxwell, B.E., 1956.
Hydrobiological observations for Wellington Harbour. Trans. R. Soc., New Zealand, 83(3):493-503.

estuaries
*McAlice, B.J., 1970.
Observations on the small-scale distribution of estuarine phytoplankton.
Marine Biol., 7(2):101-111.

estuaries
McCone, Alistair W. 1967
The Hudson River Estuary: sedimentary and geochemical properties between Kingston and Haverstraw, New York
J. sedim. Petrol. 37 (2): 475-486.

estuaries
McCone, Alisteir W., 1966.
The Hudson River Estuary; hydrology, sediments and pollution.
Geogr. Rev., 56(2):175-189.

estuaries
McCrone, Alistair W., and Charles Schafer, 1966.
Geochemical and sedimentary environments of Foraminifera in the Hudson River estuary, New York.
Micropaleontology, 12(4):505-509.

estuaries
McCowan, J., 1892.
On the theory of long waves and its application to the tidal phenomena of rivers and estuaries.
Phil. Mag. 33(5):250-265.

estuaries
McHugh, J.L., 1967.
Estuarine nekton.
In: Estuaries, G.H. Lauff, editor, Publs Am. Ass. Advmt. Sci., 83:581-620.

estuaries
McIntire, D. David and W. Scott Overton, 1971.
Distributional patterns in assemblages of attached diatoms from Yaquina Estuary, Oregon.
Ecology 52(5): 758-777.

estuaries
*McLain, Douglas R., 1968.
Oceanographic surveys of Traitors Cove, Revillagigedo Island, Alaska.
Spec. scient. Rep. U.S. Fish Wildl. Serv. Fish, 576: 15 pp. (multilithed).

estuaries
McMaster, Robert L., 1967.
Compactness variability of estuarine sediments: an in situ study.
In: Esturries, G.H. Lauff, editor, Publs Am. Ass. Advmt Sci., 83:261-267.

estuaries
* Meade, Robert H., 1969.
Landward transport of bottom sediments in estuaries of the Atlantic coastal plain.
J. sed. Petr. 39(1): 222-234.

estuaries
Menaché, M., 1959.
Étude hydrologique de l'estuaire de Fascène à Nosy Bé (Madagascar) durant la saison des pluies 1956.
Mem. Inst. Sci. Madagascar, 1958, Ser. F: 201-284.

estuaries
Michailov, V.N., 1960.
[Relief of the water surface in the region of inflow to the channel of the river in a reservoir (laboratory tests). Study of River Estuaries.]
Trudy Okeanograf. Komissii, Akad. Nauk, SSSR, 6: 94-101.

estuaries
Michon, X., and J. Goddet, 1958
Amenagement de l'estuaire de la Rance. Modèle reduit de la station d'essair de Saint-Malo-Saint-Servan. Rapp. Quatrième Journ. Hydraul., Paris, 13-15, juin, 1956: 336-343.
Edite: La Houille Blanche

estuaries
Millard, N., 1950.
On a collection of sessile barnacles from Knysna Estuary, South Africa. Trans. R. Soc., S. Africa, 32(3):265-273, figs.

estuaries
1970
Millard, N.A.H., and G. J. Broekhuysen,
The ecology of South African estuaries. X.
St. Lucia: a second report.
Zool. Africana, 5(2): 277-307.

estuaries
Millard, N.A.H., and K.M.F. Scott, 1954.
The ecology of South African estuaries. VI.
Milnerton Estuary and the Diep River, Cape.
Trans. R. Soc., S. Africa, 34(2):279-324,
8 textfigs.

estuaries
Miller, E.G., 1962
Observations of tidal flow in the Delaware
River.
Geol. Survey Water-Supply Paper, 1586-C:26 pp.

estuaries
*Mock, Cornelius R., 1967.
Natural and altered estuarine habitats of
penaeid shrimp.
Proc. Gulf Caribb. Fish. Inst. 19th. Sess., 86-98.

estuaries
Mommaerts, J.P., 1969.
Données sur l'écologie de l'estuaire du
Tamar (Plymouth).
Bull. K. Belg. Inst. Nat. Wet. 45(22): 1-26

estuaries
Momoi, Takao, 1968.
A long wave in the vicinity of an
estuary [V].
Bull. Earthq. Res. Inst. (A) 46(6): 1237-1268

estuaries
*Momoi, Takao, 1968.
A long wave in the vicinity of an estuary. IV.
Bull. Earthq. Res. Inst., Tokyo Univ., 46(3):631-650.

estuaries
Momoi, Takao, 1966.
A long wave in the vicinity of an estuary. III.
Bull. Earthq. Res. Inst., Tokyo Univ., 44(3):
1009-1040.

estuaries
Moncure, Richard, and Maynard Nichols, 1968.
Characteristics of sediments in the James
River estuary, Virginia.
Spec. scient. Rept. Virginia Inst. mar. Sci. 53:
40 pp (multilithed)

estuaries
Moore, H.B., 1931.
The muds of the Clyde Sea area. III. Chemical
and physical conditions; rate and nature of
sedimentation; and fauna.
J. Mar. Biol. Assoc., U.K., 17(2):325-358.

estuaries
Moore, H.B., 1930.
The muds of the Clyde Sea area. I. Phosphates and
nitrogen contents.
J. Mar. Biol. Assoc., U.K., 16(2):595-607.

ESTUARIES
Moore, Joseph G., Jr., 1968.
Bays and estuaries and the Texas Water
Plan.
Proc. Gulf Carib. Fish. Inst. 20th Ann. Sess.
60-65

estuaries
Morgan, James P. 1967.
Ephemeral estuaries of the deltaic environment.
In: Estuaries, G.H. Lauff, editor, Publs Am. Ass.
Advmt Sci., 83:115-120.

estuaries
*Mulkana, Mohammed Saeed, 1968.
Winter standing plankton biomass in Barataria
Bay, Louisiana, and its adjacent estuarine
systems.
Proc. Louisiana Acad. Sci., 31:65-69.

estuaries
Muller, Jan, 1959.
Palynology of Recent Orinoco delta and shelf
sediments. Reports of the Orinoco Shelf Expedition, Volume 5.
Micropaleontology, 5(1):1-32.

estuaries
Murakami, Akio, 1961
Plankton studies related to pollution in
estuaries and inshore water.
Info. Bull., Plankton., Japan, No. 7:38-39.

estuaries
Murakami, A., 1954.
[Oceanography of Kasaoka Bay in Seto Inland Sea.]
Bull. Nakai Regional Fish. Res. Lab. 6:15-57,
42 textfigs.

estuaries
Murray, J.W., 1966.
A study of the seasonal changes of the water
mass of Christchurch Harbour, England.
Jour. mar. biol. Assoc., U.K., 46(3):561-578.

estuaries
Murty, T.S., and J.D. Taylor, 1970.
A numerical calculation of the wind-driven
circulation in the Gulf of St. Lawrence.
J. oceanogr. Soc., Japan, 26(4): 203-214.

estuaries
Myrick, Robert M., and Luna B. Leopold, 1963.
Hydraulic geometry of a small tidal estuary.
U.S. Dept. Int., Geol. Survey Prof. Pap., 422-B:
1-18.

N

estuaries
Nadezhin, B.M., 1965.
The hydrological regime in the Pechora estuary
and its possible changes in the case of the
exclusion of a portion of the Pechora river
discharge. (In Russian).
Okeanologiia, Akad. Nauk, SSSR, 5(3):448-457.

estuaries
Nair, P.V. Ramachandran, 1970
Primary productivity in the Indian seas.
Bull. cent. mar. Fish. Res. Inst. 22: 56pp.
(mimeographed)

estuaries
Nash, C. B., 1947.
Environmental characteristics of a river estuary.
J. Mar. Res. 6(3):147-176.

estuaries
*Neal, Victor T., 1966.
Predicted flushing times and pollution
distribution in the Columbia River estuary.
Proc. 10th Conf. Coast Engng, 2: 1463-1480.
Also in: Coll. Repr., Dep. Oceanogr., Oregon
State Univ., 5.

estuaries
Nece, Ronald E., and J. Dungan Smith
1970.
Boundary shear stress in rivers and
estuaries.
J. Watways Harb. Div. Am. Soc. civ. Engrs
96 (WW2): 335-358.

estuaries
Nellen, Walter, 1970.
2.9 Sediment. Chemische, mikrobiologische und
planktologische Untersuchungen in der Schlei im
Hinblick auf deren Abwasserbelastung. Kieler
Meeresforsch 26(2): 144-149.

estuaries
Nellen, Walter 1970.
2.7. Phosphorkreislauf. Chemische, mikrobiologische und planktologische Untersuchungen
in der Schlei im Hinblick auf deren Abwasserbelastung. Kieler Meeresforsch 26(2): 132-138.

estuaries
Nellen, Walter 1970.
2.5. Wasserstoffionenkonzentration. Chemische
mikrobiologische und planktologische Untersuchungen in der Schlei im Hinblick auf deren
Abwasserbelastung. Kieler Meeresforsch 26(2):
128-130.

estuaries
Nellen, Walter 1970.
2. Hydrographie und Chemie. Chemische,
mikrobiologische und planktologische
Untersuchungen in der Schlei im Hinblick
auf deren Abwasserbelastung.
Kieler Meeresforsch. 26(2): 110-111.

estuaries
Nellen, Walter, und Gerhard Rheinheimer
1970
2.2. Wassertemperatur. 2.3. Salzgehalt. Chemische,
mikrobiologische und planktologische Untersuchungen in der Schlei im Hinblick auf deren
Abwasserbelastung. Kieler Meeresforsch 26(2):
119-122; 123-126.

estuaries
Nellen, Walter, und Gerhard Rheinheimer
1970
Chemische mikrobiologische und planktologische
Untersuchungen in der Schlei im Hinblick
auf deren Abwasserbelastung. 1. Einleitung
und Literaturzusammenstellung früherer
Arbeiten über die Schlei.
Kieler Meeresforsch. 26(2): 105-109.

estuaries
Nelson, Bruce W., 1962
Important aspects of estuarine sediment
chemistry for benthic ecology.
The Environmental Chemistry of Marine Sediments, Proc. Symp., Univ. R.I., Jan. 13, 1962.
Occ. Papers, Narragansett Mar. Lab., No. 1:
27-41.

estuaries
Netherlands, Delta Committee, 1962
Final report delivered by the Advisory Committee to provide an answer to the question of what waterways-technical provisions must be made for the area devastated by the storm flood of February 1, 1953, (Delta Committee) instituted by decree of the Minister of Transport and Waterways of February 18, 1953. 100 pp.

estuaries
Nevessky, E.N., 1960
[Evolution of the Kalamit Bay estuary.]
Biull. Oceanogr. Komm., (5): 54-68.

estuaries
Newsom, John D., editor, 1968.
Proceedings of the Marsh and Estuary Management Symposium, Louisiana State University, July 19-20, 1967: 250 pp.

estuaries
Newman, William A., 1967.
On physiology and behaviour of estuarine barnacles.
Proc. Symp. Crustacea, Ernakulam, Jan 12-15, 1965, 3, 1038-1066

estuaries
Niaussal, Pierre-Marie, and Roland Bourcart, 1963.
Contribution à l'etude du plancton dans les eaux de l'embouchure de la Gironde. Prédominance du dino-flagellé "Noctula miliaris".
Cahiers Océanogr., C.C.O.E.C., 15(10):722-725.

estuaries
Nichols, Maynard M., 1965.
Transport of suspended sediment in the James Estuary. (Abstract).
Ocean Sci. and Ocean Eng., Mar. Techn. Soc.,-Amer. Soc. Limnol. Oceanogr., 2:1112.

estuaries
*Nichols, Maynard M., and George Poor, 1967.
Sediment transport in a coastal estuary.
J. Waterways Harb. Div. Am. Soc. civ. Engrs 93 (WW4):83-96.

ESTUARIES
Nizery, A., 1950.
Le Laboratoire d'Hydraulique au service de l'océanographie. Conf. C.R.E.O. No. 9:9 pp.

estuaries
Nolte, Willy, 1965.
Die Küstenfischerei in der Unter und Aussenweser und die Abwasserbedrohung.
Helgoländer wiss. Meeresunters. 17: 151-167.

estuaries
Nonn, H., 1953.
Morphologie du littoral breton entre Saint-Cast et Dinard. Bull. d'Info. C.C.O.E.C. 5(1):30-46, Pls. 5-8 (mimeographed).

estuaries
Nordgaard, O., 1900.
Some hydrographical results from an expedition to the North of Norway. Bergens Mus. Aarb., 1899(8): 26 pp.

estuaries
Nota, D.J.G., 1968.
Geomorphology and sedimentary petrology in the southern Gulf of St. Lawrence.
Geologie Mijnb., 47 (1): 49-52.

estuaries
Nota, D.J.G., and D.H. Loring, 1964.
Recent depositional conditions in the St. Lawrence River and Gulf - a reconnaissance survey.
Marine Geology, 2(3):198-235.

estuaries
Notenboom, W., and J.C. Schönfeld, 1953.
The penetration of sea water into inland water situated beyond shipping-locks.
XVIIIe Congr. Int. Navig., Rome 1953, Sect. Navig. Marit., Comm. 3:204-223.

estuaries
Nusbaum, I., and H.E. Miller, 1952.
The oxygen resources of San Diego Bay.
Sewage Ind Wastes 24:1512-1527.

Abstr.: Chem. Abstr. 2405a, 1953.

estuaries
Nutt, D.C., and L.K. Coachman, 1956.
The oceanography of Hebron Fjord, Labrador.
J. Fish. Res. Bd., Canada, 13(5):709-758.

O

estuaries
O'Brien, M.P., 1952.
Salinity currents in estuaries. Trans. Amer. Geophys. Union 33(4):520-522, 1 textfig.

estuaries
O'Brien, M., 1935.
Models of estuaries. Trans. Amer. Geophys. Union, 16:485-492, 5 figs.

estuaries
Odum, Eugene P., 1961.
The role of tidal marshes in estuarine production.
The N.Y. State Conservationist, Information Leaflet, June-July 1961:L 60 - 4 pp.

estuaries
Odum, Eugene P., and Armando A. de la Cruz, 1967.
Particulate organic detritus in a Georgia salt marsh-estuarine ecosystem.
In: Estuaries, G.H. Lauff, editor, Publs Am Ass. Advmt Sci., 83:383-388.

estuaries
Ohl, Hans, 1959.
Temperatur- und Salzgehaltsmessungen an der Oberflache des Kieler Hafens in den Jahren 1952 bis 1957.
Kieler Meeresf., 15:157-160.

Estuaries
Okabe, Shiro, and Toyoko Morinaga, 1968.
Vanadium and molybdenum in the river and estuary water which pour into the Suruga Bay, Japan. (In Japanese; English abstract).
Nippon Kagaku Zasshi, 89 (3).
Also in: Coll. Repr. Coll. mar. Sci. Techn., Tokai Univ. 1967-1968, 3:329-332.

estuaries
Okabe, Shiro, Yoshimasa Toyota and Toyoko Morinaga, 1970.
Pollution in coastwise waters of Japan. Kagaku-no-Ryoiki 24(1). Also in: Coll. Repr. Coll. mar. Sci. Techn. Tokai Univ. 4: 303-309.

estuaries
Okubo, Akira, 1964.
Equations describing the diffusion of an introduced pollutant in a one-dimensional estuary. In: Studies on Oceanography dedicated to Professor Hidaka in Commemoration of his sixtieth birthday, 216-226.

estuaries
*Okuda, Setsuo, 1968.
On the change in salinity distribution and bottom topography after the closing of the mouth of Kojima Bay.
Bull. Disas. Prev. Inst. Kyoto Univ., 18(1):35-48.

estuaries
Okuda, S., 1960
[The exchange of fresh and salt waters in the bay after closing the outlet.] J. Oceanogr. Soc., Japan, 16(1): 7-14.

Okuda, Taizo

estuaries
Okuda, Taizo, 1965.
Consideraciones generales sobre las condiciones ambientales de la laguna y el rio Unare.
Bol. Inst. Oceanogr., Univ. Oriente, 4(1): 136-154.

estuaries
Okuda, Taizo, José Benítez Alvarez, José Rafael Gomez, 1965.
Características químicas de los sedimentos de la laguna y rio Unare.
Bol. Inst. Oceanogr., Univ Oriente, 4(1): 108-122.

estuaries
Okuda, Taizo, Lourinaldo Cavalcanti, and Manoel Pereira Borba, 1960
Estudo da Barra das Jangadas. 3. Variação de nitrogenio e fosfato durante o ano.
(In Portuguese; English and French abstracts).
Trabalhos, Inst. Biol. Marit. e Oceanogr., Universidade do Recife, Brasil, 2(1):207-218.

estuaries
Okuda, Taizo, Lourinaldo Cavalcanti and Manoel Pereira Borba, 1960
Estudo da Barra das Jangadas. 2. Variacao do pH, oxigenio dissolvido e consumo de permaganato durante o ano. (In Portuguese; English and French resumés).
Trabalhos, Inst. Biol. Marit. e Oceanogr., Universidade do Recife, Brasil, 2(1):193-205.

estuaries
Okuda, Taizo, Angel José García, José Benítez Alvarez, 1965.
Variacion estacional de los elementos nutritivos en el agua de la laguna y el rio Unare.
Bol. Inst. Oceanogr., Univ. Oriente, 4(1): 123-135.

estuaries
Okuda, Taizo, José Rafael Gómez, José Bepitez Alvarez and Angel José Garcia, 1965
Condiciones hidrograficas de la laguna y rio Unere.
Bol. Inst. Oceanogr., Univ. Oriente, 4(1):60-107.

estuaries
Okuda, Taizo, and Ramon Nobrega, 1960
Estudoda Barra das Jangadas. 1. Distribuicão e movimento da clorinidade - quantidade de corrente. (In Portuguese; English and French resumés).
Trabalhos, Inst. Biol. Marit. e Oceanogr., Universidade do Recife, Brasil, 2(1):175-191.

estuaries
Olson, F.C.W., 1953.
Tampa Bay studies.
Ocean. Inst. Florida State Univ. Rept. 1:27 pp. (mimeographed), 21 figs. (multilithed).
UNPUBLISHED

estuaries
Öltjen, J., 1919.
Über die Berechnung von Flutwellen linien in einem Tidefluss. Zentralblatt der Bauverwaltung No. 27:137-139.

estuaries
Ono, Yuiti, 1965.
On the ecological distribution of ocypoid crabs in the estuary.
Mem. Fac. Sci., Kyushu Univ., (E), 4(1):60 pp.

estuaries
Ono, Y., 1959.
The ecological studies on Brachyura in the estuary.
Bull. Mar. Biol. Sta., Asamushi, Tohoku Univ., 9(4):145-148.

Ottmann, François

estuaries
*Ottmann, François, 1968.
L'étude des problèmes estuariens.
Revue Géorgr.phys.Géol.dynam. (2)10(4):329-353.

estuaries
Ottmann, Francois, 1965.
Introduction a la Geologie Marine et Littorale.
Masson et Cie, Paris, 259 pp. 47 F

estuaries
Ottmann, Francois, 1959.
Estudo das Amostras do fundo recolhidas pelo N. E. "Almirante Saldanha" na regiao da embroucadura do Rio Amazonas. Missao de Diretoria de Hidrografia e Navegacao por ocasiao do Ano Geofisico International (Dezembro de 1958). Trabalhos, Inst. Biol. Marit. e Oceanogr., Recife 1 (1): 77-106.

Estuaries
Ottmann, F., L. Barbaroux, A. Mahe et G. Moulin, 1970.
Étude sédimentologique, géochimique et géotechnique des vases estuariennes (Vasière de Méan, près St-Nazaire, estuaire de la Loire).
C.R. Trav. 1967, 1968, Lab. Géol. Univ Nantes, 111 pp. (multilitted)

estuaries
*Ottmann, François, Taizo Okuda, Lourinaldo Cavaltanti, Olimpio C.Da Solva, Júlio V.A. De Araújo, Petrônio A Coelho, Maryse N. Paranaguá, e Enide Eskinazi, 1965/1966.
Estudo da Barra das Jangadas. V. Efeitos da poluicão sôbre a ecologia do estuário.
Trabhs Inst.Oceanogr.,Univ.Fed.Pernambuco,Recife, (7/8):7-16.

estuaries
Ottmann, Francois, and Jeanne-Marie Ottmann, 1960
Estudo da Barra das Jangadas. 4. Estudo dos sedimentos. (In Portuguese; English and French resumes).
Trabalhos, Inst. Biol. Marit. e Oceanogr., Universidade do Recife, Brasil, 2(1):219-233.

estuaries
Ottmann, François, et Jeanne-Marie Ottmann, 1959
La Marée de salinité dans le Capibaribe, Recife-Bresil.
Trabalhos Inst. Biol. Marit. e Oceanogr., Recife, 1(1) :39-49.

estuaries
Ottmann, François, et Jeanne-Marie Ottmann, 1959.
Les sediments de l'embouchure du Capibaribe, Recife Bresil. Trabalhos Inst. Biol. Marit. e Oceanogr., Recife, 1 (1) :51-69.

estuaries
Ottmann, François, y Carlos M. Urien, 1965.
Observaciones preliminares sobre la distribución de los sedimentos en la zona externa del Rio de la Plata.
Anais Acad. bras. Cienc., 37(Supl.):283-288.

Estuaries
Ottmann, Francois, Carlos Maria Urien, 1965.
Trabajos sobre la sedimentologia en el Rio de la Plata realizados por el Laboratorio de Geologia del SHN.
Bol. Servicio Hidrogr. Naval, Argentina. 11(1): 1-8.

estuaries
Ottman, François, et Carlos M. Urien 1965.
Le melange des eaux douces et marins dans le Rio de la Plata.
Cah. Océanogr. 17(10): 703-713.

estuaries
Owen, Wadsworth, 1969.
A study of the physical hydrography of the Patuxent River and its estuary.
Techn. Rept. Chesapeake Bay Inst., 53 (Ref. 69-1): 37 pp. (multilitted) (unpublished manuscript.)

P

estuaries
*Pallares, Rosa E., 1968.
Copépods marinos de la Ria Deseado (Santa Cruz, Argentina). Contribución sistemática-ecológica. 1.
Publ., Serv.Hidrograf., Armada Argentina, H:1024: 125 pp.

estuaries
Palmer, John D., and Frank E. Round, 1965.
Persistent, vertical-migration rhythms in benthic microflora. 1. The effect of light and temperatur on the rhythmic behaviour of Euglena obtusa.
J. mar. biol. Ass., U.K., 45(3):567-582.

estuaries
Panikkar, N.K., 1951.
Physiological aspects of adaptation to estuarine conditions. Proc. Indo-Pacific Fish. Cound., 17-28 Apr. 1950, Cronulla N.S.W., Australia, Sects. II-III:168-175.

estuaries
Parker, Patrick L., Ann Gibbs and Robert Lawler, 1963.
Cobalt, iron and manganese in a Texas bay.
Publ. Mar. Sci., Port Aransas, 9:28-32.

estuaries
Parry, W.K., and W.E. Adeney, 1902.
The discharge of sewage into a tidal estuary.
Inst. Civil Eng., London, 147(Paper 3279):70-121, 3 figs.

estuaries
Partiot, H.L., 1892-1894.
Étude sur les rivières à marée et sur les estuaires. Congrès Int. de Navigation 118:127 70 pp., 24 9 pls.

estuaries
Partiot, M., 1871.
Mémoire sur les marées fluviales. C.R. Acad. Sci. Paris, 73:91-95.

estuaries
Partiot, H.L., 1886,
Projet des travaux à faire à l'embouchure de la Seine. Tome 1, Tome 2, texte, planches, 35 pp., 3 figs., 21 pls.
Paris, Librairie Polytechnique, Baudry et Cie,

estuaries
Patrick, Ruth, 1967.
Diatom communities in estuaries.
In: Estuaries, G.H. Lauff, editor, Publs Am. Ass Advmt Sc., 83:311-315.

Patten, Bernard C.

estuaries
Patten, B.C., 1966.
The biocoenetic process in an estuarine phytoplankton community.
Oak Ridge Nat. Lab., ORNL-3946(UC-48-Biol.Med.) 97 pp. (Unpublished manuscript). $4.00.

estuaries
Patten, B.C., 1962
Species diversity in net phytoplankton of Raritan Bay.
J. Mar. Res., 20(1):57-75.

estuaries
Patten, B., 1961.
Plankton energetics of Raritan Bay.
Limnol. & Oceanogr., 6(4):369-387.

estuaries
Patten, Bernard C., and Brian F. Chabot, 1966.
Factorial productivity experiments in a shallow estuary: characteristics of response surfaces.
Chesapeake Sci., 7(3):117-136.

estuaries
Patten, Bernard C., Richard C. Mulford and J. Ernest Warinner, 1963.
An annual phytoplankton cycle in the lower Chesapeake Bay.
Chesapeake Science, 4(1):1-20.

estuaries
Patten, Bernard C., and George M. van Dyne, 1968.
Factorial productivity experiments in a shallow estuary: energetics of individual plankton species in mixed populations.
Limnol. Oceanogr., 13(2):309-314.

estuaries
Patten, Bernard C., David K. Young and Morris Roberts, 1963.
Suspended particulate material in the lower York River, Virginia, June 1961-July 1962.
Virginia Inst. Mar. Sci., Spec. Sci. Rept. (44): unnumbered pp. (Unpublished manuscript).

estuaries
Paulmier, Gérard, 1971.
Cycle des matières organiques dissoutes, du plancton et du micro-phytoplancton dans l'estuaire du Belon: leur importance dans l'alimentation des huîtres.
Rev. Trav. Inst. Pêches marit. 35(2): 157-200.

estuaries
Paulmier, Gérard, 1969.
Le microplancton des rivières de Morlaix et de la Penzé.
Revue Trav. Inst. Pêches marit. 33(3): 311-332.

estuaries
Peelen, R., 1970.
Changes in salinity in the delta area of the rivers Rhine and Meuse resulting from the construction of a number of enclosing dams.
Netherl. J. Sea Res. 5(1):1-19.

estuaries
Peelen, R. 1967.
Isohalines in the delta area of the rivers, of waters in the delta area according to their chlorinity and the changes in these waters caused by hydro-technical constructions.
Netherland J. sea Res. 3(4): 575-597.

estuaries
Peer, D.L., 1963
A preliminary study of the composition of benthic communities in the Gulf of St. Lawrence.
Fish. Res. Bd., Canada, Mss Rept. Ser. (Ocean. and Limnol.), No. 145: 24 pp. (multilithed).

estuaries
Perdriau, Jacques, 1964.
Pollution marine par les hydrocarbures cancérigènes - type benzo-3,4 pyrène - incidences biologiques. (Suite et fin).
Cahiers Océanogr., C.C.O.E.C., 16(3):205-229.

estuaries
Persoone, Guido 1971.
Ecology of fouling on submerged surfaces in a polluted harbour.
Vie Milieu Suppl. 22(2): 613-636

estuaries
Peterson, Clifford L., 1960.
The physical oceanography of the Gulf of Nicoya, Costa Rica, a tropical estuary.
Inter-American Tropical Tuna Comm., Bull., 4(4): 139-216.

estuaries
Petit, G., and P. Schachter, 1961.
Rapport sur les travaux concernant les étangs et salés et les lagunes.
Rapp. Proc. Verb., Réunions, Comm. Int. Expl. Sci., Mer Méditerranée, Monaco, 16(3):757-771.

estuaries
Pfitzenmeyer, H.T., and K.G. Drobeck, 1963.
Benthic survey for populations of soft-shelled clams, Mya arenaria, in the lower Potomac River, Maryland.
Chesapeake Science, 4(2):67-74.

estuaries
Pickard, G.L., 1967.
Some oceanographic characteristics of the larger inlets of southeast Alaska.
J. Fish. Res Bd. Can., 24(7) 1475-1506

estuaries
Pickard, G.L., 1961
Oceanographic features of inlets in the British Columbia mainland coast.
J. Fish. Res. Bd., Canada, 18(6):907-999.

estuaries
Pickard, G.L., 1953.
Oceanography of British Columbia mainland inlets. II. Currents. Prog. Repts., Pacific Coast Stas., Fish. Res. Bd., Canada, No. 87:12-13, 1 textfig.

estuaries
Pickard, G.L., and L.F. Giovando, 1960
Some observations of turbidity in British Columbia inlets. Limnol. & Oceanogr., 5(2): 162-170.

estuaries
Pickard, G.L., and Keith Rodgers, 1959.
Current measurements in Knight Inlet, British Columbia.
J. Fish. Res. Bd., Canada, 16(5):635-678.

estuaries
Pickard, G. L., and R. W. Trites, 1957.
Fresh water transport determination from the heat budget with application to British Columbia Inlets.
J. Fish. Res. Bd., Canada, 14(4):605-616.

estuaries
Pierce, J.W., 1970.
Tidal inlets and washover fans.
J. Geol., 78(2): 230-234

estuaries
Pillay, T.V.R., 1967.
Estuarine fisheries of the Indian Ocean coastal zone.
In: Estuaries, G.H. Lauff, editor, Publs Am. Ass. Advmt Sci., 83:647-657.

estuaries
Pillay, T.V.R., 1967.
Estuarine fisheries of West Africa.
In: Estuaries, G.H. Lauff, editor, Publs Am. Ass. Advmt Sci., 83:639-646.

estuaries
Pinard, J., 1965.
La Charente, ses débits, ses crues et les marées de l'estuaire. Norois, 12(47):357-363.

estuaries
Pirie, Robert Gordon, 1965.
Petrology and physical-chemical environment of bottom sediments of the Rivière Bonaventure-Chaleur Bay area, Quebec, Canada.
Rept. B.I.O. 65-10:182 pp. (multilithed).

estuaries
Plate, L., 1949.
Verbesserung von Mündungen natürlicher Wasserlaufe in ein Meer mit Tidebewegung mit besonderer Berücksicktingung der Nordsee.
Studien au Bau- und Verkehrsproblemen der Wasserstrassen:225-237.

estuaries
Poirier, Marcel, et Robert Boudreault, 1965.
Observations d'océanographie physique dans la Baie-des-Chaleurs, juin-octobre 1958.
Cah. Inf. Stn. biol. mar., Grande-Rivière, No. 32: unnumbered pp. (multilithed)

estuaries
Poirier, Marcel, et Robert Boudreault, 1965.
Observations d'océanographie physique dans la Baie-des-Chaleurs, juin-octobre 1957.
Cah. inf. Stn. biol. mar., Grande-Rivière, No. 31: unnumbered pp. (multilithed).

estuaries
Poirier, Marcel, et Robert Boudreault 1965.
Observations d'océanographie physique dans la Baie-des-Châleurs, juin-octobre 1956.
Cah. Inf. Stn biol. mar. Grande-Rivière No. 30: unnumbered pp. (multilithed).

estuaries
Poirier, Marcel et Robert Boudreault, 1965.
Observations d'océanographie physique dans la Baie-des-Chaleurs, mai-septembre 1955.
Cah. inf. Stn. biol. mar. Grande-Rivière, No. 25 unnumbered pp. (multilithed)

estuaries
Pomeranets, K.S., 1959
Level fluctuations of the Kamchatka estuary in connection with the eruption of the Bezymyannyy volcano on March 30, 1956.
Vestnik, Leningrad Univ., Ser. Geol. i Geogr., 14(6):143-146.

estuaries
Pomeroy, L. R., 1959.
Algal productivity in salt marshes of Georgia
Limnol. & Ocean., 4(4):386-398.

estuaries
Pomeroy, Lawrence, E.E. Smith, and Carol M. Grant, 1965.
The exchange of phosphates between estuarine water and sediments.
Limnology and Oceanography, 10(2):167-172.

estuaries
Pople, W., and the late M. Rogoyske
The effect of the Volta River hydroelectric project on the salinity of the lower Volta River.
Ghana J. Sci., 9(1): 9-20.

estuaries
Pore, Arthur, 1960.
Chesapeake Bay hurricane surges.
Chesapeake Science, 1(3/4):178-186.

estuaries
Postma, H., 1967.
Sediment transport and sedimentation in the estuarine environment.
In: Estuaries, G.H. Lauff, editor, Publs Am. Ass. Advmt Sci., 83:158-179.

estuaries
Postma, H., 1954.
Hydrography of the Dutch Waddensea. A study of the relations between water movement, the transport of suspended materials and the production of organic matter. Arch. Néerl. Zool. 10(4):405-511, 55 textfigs.

estuaries
Postma, H., and K. Kalle, 1955.
Die Entstehung von Trübungszonen im Unterlauf der Flüsse, speziell im Hinblick auf die Verhältnisse in der Unterelbe. Deutsche Hydrogr. Zeits. 8(4): 137-144.

estuaries
Powers, Charles F., 1963.
Some aspects of the oceanography of Little Port Walter Estuary, Baranof Island, Alaska.
U.S.F.W.S., Fish. Bull., 63(1):143-164.

estuaries
Preddy, W.S., 1954.
The mixing and movement of water in the estuary of the Thames. J.M.B.A. 33:645-662, 9 textfigs.

estuaries
Préfontaine, Georges, and Pierre Brunel, 1962
Liste d'invertébrés marins recueillis dans l'estuaire du Saint-Laurent de 1929 à 1934.
Naturaliste Canadien, Québec, 89(8/9):237-264.

Also:
Contrib., Ministère de la Chasse et des Pêcheries, Québec, No. 86.

estuaries
Price, Kent S., Jr., 1962
Biology of the sand shrimp, Crangon septemspinosa in the shore zone of the Delaware Bay region.
Chesapeake Science, 3(4):244-255.

estuaries
Price, W.A., 1963.
Patterns of flow and channelling in tidal inlets.
J. Sed. Petrol., 33(2):279-290.

estuaries
Price, W.A., and Mary P. Kendrick, 1964
Field and model investigation into the reasons for siltation in the Mersey estuary.
Proc. Inst. Civil Engineers, 27:613-647.

Discussion on Paper No. 6669, Proc. April 1963

estuaries
Price, William Alan, and Mary Patricia Kendrick, 1963
Field and model investigation into the reasons for siltation in the Mersey Estuary.
Proc. Instn. Civ. Eng., 24:473-518.

estuaries
Price, W.A., and M.P. Kendrick, 1962.
Density currents in estuary models.
La Houille Blanche, 17(5):611-628.

Pritchard, D.W.

estuaries
Pritchard, Donald W., 1967.
Observations of circulation in coastal plain estuaries.
In: Estuaries, G.H. Lauff, editor, Publs Am. Ass. Advmt Sci., 83:37-44.

estuaries
Pritchard, Donald W., 1967.
What is an estuary: physical viewpoint.
In: Estuaries, G.H. Lauff, editor, Publs Am. Ass. Advmt Sci., 83:3-5.

estuaries
Pritchard, D.W., 1960
Salt balance and exchange rate for Chincoteague Bay. Chesapeake Sci., 1(1):48-57.

estuaries
Pritchard, D.W., 1960
The movement and mixing of contaminants in tidal estuaries.
In: Proc. First International Conf. on Waste Disposal in the Marine Environment, Berkeley California, July 22-25, 1959, E.A. Pearson, Edit., Pearson, Edit., Pergamon Press, 512-525
Also in: Chesapeake Bay Inst., Collected Reprints, 5(1963)

estuaries
Pritchard, D.W., 1959
Factors affecting the dispersal of fission products in estuarine and inshore environments.
Proc. 2nd Int. Conf., Peaceful Uses of Atomic Energy, Sept. 1-13, 1959(Session D-19):410-413.

estuaries
Pritchard, D.W., 1959
Problems related to disposal of radioactive wastes in estuarine and coastal waters.
Trans. 2nd Seminar on Biol. Problems in Water Pollution, April 20-24, 1959: 11 pp.(unnumbered in COLLECTED REPRINTS, Chesapeake Bay Inst.).

Pritchard, D. W., 1958 estuaries
The equations of mass continuity and salt continuity in estuaries. J. Mar. Res., 17:412-423.

estuary
Pritchard, D.W., 1956.
The dynamic structure of a coastal plain estuary.
J. Mar. Res. 15(1):33-42.

estuaries
Pritchard, D.W., 1955.
Estuarine circulation patterns. Proc. Amer. Soc. Civil Eng., Separate 717, Vol. 81:717-1 - 717-11.

estuaries
Pritchard, D.W., 1954.
A study of the salt balance in a coastal plain estuary. J. Mar. Res. 13(1):133-144, 4 textfigs.

estuaries
Pritchard, D.W., 1952.
Estuarine hydrography. Advances in Geophysics, Academic Press, Inc., 243-280.

estuaries
Pritchard, D.W., 1952.
Salinity distribution and circulation in the Chesapeake Bay estuarine system. J. Mar. Res. 11(2):106-123, 11 textfigs.

estuaries
Pritchard, D.W., 1951.
The physical hydrography of estuaries and some applications to biological problems. Trans. 16th North American Wildlife Congr., Mar. 5-7., 1951: 368-376.

estuaries
Pritchard, D.W., and R.E. Bunce, 1959.
Physical and chemical hydrography of the Magothy River.
Chesapeake Bay Inst., Tech Rept., (Ref. 59-2), 17:1-22. (Unpublished manuscript).

estuaries
Pritchard, D.W., and W.V. Burt, 1951.
An inexpensive and rapid technique for obtaining current profiles in estuarine waters.
J. Mar. Res. 10(3): 180-189, 4 textfigs.

estuaries
Pritchard, D.W , and J.H. Carpenter, 1960
Measurements of turbulent diffusion in estuarine and inshore waters.
Bull. Int. Assoc., Sci Hydrol , (20):37-50.

Also in: Chesapeake Bay Inst., Collected Reprints, 5(1963).

estuaries
Pritchard, D.W., and R.E. Kent, 1956.
A method for determining mean longitudinal velocities in a coastal plain estuary. J. Mar. Res., 15(1):81-91.

estuaries
Pritchard, D.W., and R.E. Kent, 1953.
The reduction and analysis data from the James River Operation Oyster Spat. Chesapeake Bay Inst Tech. Rept. VI, Ref. 53-12:92 pp., 11 tables, UNPUBLISHED MANUSCRIPT.

estuaries
Pritchard, D.W., Akira Okubo and Emanuel Mehr, 1962
A study of the movement and diffusion of an introduced contaminant in New York Harbor waters.
Chesapeake Bay Inst., Techn. Rept. (Ref. 62-21) 31: 89 pp. (multilithed)

estuaries
Pritchard, D.W., D.K. Todd and L. Lau, 1957.
On estimating streamflow into tidal estuary - a discussion.
Trans. Amer. Geophys Union, 38(4):581-583.

estuaries
Proudman, J., 1958.
On the series that represent tides and surges in an estuary. J. Fluid Mech., 3(4):411-417.

estuaries
Proudman, J., 1955.
The propagation of tide and surge in an estuary.
Proc. R. Soc., London, A, 231:8-24.

estuaries
Prych, Edmund A., David W. Hubbell, and Jeryl L. Glenn, 1967.
Estuarine measuring equipment and techniques.
J. WatWays Harb. Div., Am. Soc. civ. Engrs. 93(WW2): 41-58

estuaries
Qasim, S.Z., P.M.A. Bhattathiri and S.A.H. Abidi, 1968.
Solar radiation and its penetration in a tropical estuary.
J. exp. mar. Biol. Ecol., 2(1):87-103.

estuaries
Qasim, S.Z., S. Wellershaus, P.M.A. Bhattathiri and S.A.H. Abidi, 1969.
Organic production in a tropical estuary.
Proc. Indian Acad. Sci. (B) 49(2): 51-94

R

estuaries
Ragotzkie, Robert A., 1959.
Plankton productivity in estuarine waters of Georgia.
Publ. Inst. Mar. Sci., 6:146-158.

estuaries
Ragotzkie, R.A., and R.A. Bryson, 1955.
Hydrography of the Duplin River, Sapelo Island, Georgia. Bull. Mar. Sci., Gulf and Caribbean, 5(4):297-314.

estuaries
Rajcevic, B.M., 1957.
Études des conditions de sédimentation dans l'estuaire de la Seine.
Suppl. Ann. Inst. Techn. Bâtiment et Trav. Publ., 10e annee (117):744-775.

estuaries
Rajyalakshmi, T., 1961.
Studies on maturation and breeding in some estuarine palaemonid prawns.
Proc. Nat. Inst. Sci., India, 27(4):179-188.

estuaries
Rajyalakshmi, K., 1960.
Observations on the embryonic and larval development of some estuarine palaemonid prawns.
Proc. N.I. Sci., India, B, 26(6):395-408.

estuaries
Ramamurthy, V.D., K. Krishnamurthy and R. Seshadri, 1965.
Comparative hydrographical studies of the near shore and estuarine waters of Porto-Novo, S. India.
J. Annamalai Univ., 26:154-164.

Also in: Collected Reprints, Mar. Biol. Sta., Porto Novo, 1963/64.

estuaries
Ramming, Hans-Gerhard, 1968.
Ermittlung von Bewegungsvorgängen im Meere und in Flussmündungen zur Untersuchung des Transportes von Verunreinigungen.
Helgoländer wiss. Meeresunters. 17: 64-73

estuaries
Ramming, H.G., 1962
Gezeiten und Gezeitenströme in der Eider.
Proc. Symposium on Mathematical-Hydrodynamical Methods of Phys. Oceanogr., Sept. 1961, Inst. Meeresk., Hamburg, 233-237.

exchange between lagoon and open sea
Ranzoli, F., 1954.
Il trofismo della Laguna Veneta e la vivificazione marine. II. Ricerche sulle variazioni quantitative dello zooplancton.
Arch. Oceanogr., Limnol. 9(1/2):115-146, 5 textfigs.

Rattray, Maurice Jr.

estuaries
Rattray, Maurice, Jr., 1967.
Some aspects of the dynamics of circulation in fjords.
In: Estuaries, J.H. Lauff, Editor, Publs Am. Ass. Advmt Sci., 83:52-62.

estuaries
Rattray, Maurice, Jr., and Donald V. Hansen, 1965.
The physical classification of estuaries. (Abstract).
Ocean Sci. and Ocean Eng., Mar. Techn. Soc., Amer. Soc. Limnol. Oceanogr., 1:543.

estuaries
Rattray, Maurice, Jr., and Donald V. Hansen, 1962
A similarity solution for circulation in an estuary.
J. Mar. Res., 20(2):121-133.

Estuaries
Rattray, M., Jr., and D. V. Hansen, 1962.
Similarity solutions for circulation in an estuary. (Abstract).
J. Geophys. Res., 67(4):1654.

estuaries
Rebelo Pinto, J.F., 1953.
Pénétration des eaux salées dans la Tage.
XVIIIe Congr. Int. Navig., Rome 1953, Sect. 11, Navig. Marit. Comm. 3:225-231.

estuaries
Ranzoli, F., 1954.
Il trofismo della Laguna Veneta e la vivificazione marina. II. Ricerche sulle variazioni quantitative dello zooplancton. Arch. Oceanogr. Limnol. 9(1/2):115-146, 5 textfigs.

estuaries
Ray, I.A., and K. Bondar, 1963.
Turbidity, transparency and color of water in an estuary. (In Russian).
Gidrologiya Ust'yevoy Oblasti Dunaya, Chap. 12: 326-332.

Translation: USN Oceanogr. Off., TRANS. 258. (M. Slessers). 1967.

estuaries
Raymond, Robert, et Robert Boudreault, 1965.
Observations d'océanographie physique dans le nord du Golfe Saint-Laurent 1952-1954.
Cah. Inf. Stn biol. mar., Grande Rivière No. 27: unnumbered pp. (multilithed)

estuaries
Raymond, Robert, et Robert Boudreault 1965.
Observations d'océanographie physique dans la Baie-des-Chaleurs, mai-septembre 1953.
Cah. Inf. Stn biol. mar. Grande-Rivière No. 25: unnumbered pp. (multilithed)

estuaries
Raymond, Robert, et Robert Boudreault, 1964.
Observations d'océanographie physique dans la Baie-des-Chaleurs, mai-octobre 1952.
Cah. Inf. Stn biol. mar., Grande-Rivière, No. 24: unnumbered pp. (multilithed).

estuaries
Raymont, J.E.G., and B.G.A. Carrie, 1964.
The production of zooplankton in Southampton Water.
Int. Rev. Ges. Hydrobiol., 49(2):185-232.

estuaries
Redfield Alfred C., 1967.
The ontogeny of a salt marsh estuary.
In: Estuaries, G.H. Lauff, editor, Publs Am. Ass. Advmt Sci., 83:108-114.

Also published in somewhat abbreviated form: Science, 147:50-55 (1965).

estuaries
Redfield, A.C., 1951.
The flushing of harbors and other hydrodynamic problems in coastal waters. In: Hydrodynamics in Modern Technology, M.I.T., 127-135, 7 textfigs.

estuaries
Redfield, A.C., 1950.
The analysis of tidal phenomena in narrow embayments. P.P.O.M. 11(4):36 pp., 16 textfigs.

estuaries
Redfield, A.C., B.H. Ketchum and F.A. Richards 1963
2. The influence of organisms on the composition of sea-water. In: The Sea, M.N. Hill Edit., Vol. 2. The composition of sea water, Interscience Publishers, New York and London, 26-77.

estuaries
Reid, G.K., 1961
Ecology of inland waters and estuaries.
Reinhold Publ. Corp., N.Y. : 375 pp.

estuaries
Reid, G.K., 1957.
Biologic and hydrographic adjustment in a disturbed Gulf Coast estuary. Limnol. & Oceanogr., 2(3): 198-212.

estuaries
Reid, G.K., 1956.
Ecological investigations in a disturbed Texas coastal estuary. Texas J. Sci., 8(3):296-327.

estuaries
Reid, K.K., Jr., 1955.
A summer study of the biology and ecology of East Bay, Texas.
Texas J. Sci. 7(3):216-343, 6 textfigs.

estuaries
Reid, Wallace J., 1958.
Tidal model of Eyemouth Harbour
Engineering (Great Britain), Dec. 19, 1958.
2 pp.

estuaries
Reimold, Robert J., and Franklin C. Daiber, 1967
Eutrophication of estuarine areas by rainwater.
Chesapeake Sci., 8(2):132-133.

estuaries
Reineke, H. 1921.
Die Berechnung der Tidewelle im Tideflusse.
Jahrb. f. Gewässerkund Nord-Deutschlands Mitt. 3(4):1-22, 13 figs.

estuaries
Reish, D.J., and H.A. Winter, 1954.
The ecology of Alamitos Bay, California, with special reference to pollution.
Calif. Fish. and Game 40(2):105-121, 1 textfig.

estuaries
Renaud, 1895-1900.
Projet d'amelioration de la Charente maritime.
Rapp. CCCXXI. Service Hydrographique de la Marine 13-616:68-75.

estuaries
Renaud, 1887-1895.
Projets d'ouverture du Grau de Rousten aux embouchures du Rhône Rapp. 271. Service Hydrogr. de la Marine No. 13-615, Recherches hydrographiques sur le régime des côtes, 15 cahier, Paris: 22-25.

estuaries
Renaud, 1887-1894.
Projet de dérasement des seuils de la Charente et du balisage du nouveau chenal. Rapp. CCLXXIII, Service Hydrographique de la Marine No. 13+615: 29-32.

estuaries
Renaud, 1879.
Rapport de la Commission des Études hydrographiques et des Travaux d'Amelioration de la Baie de Seine. Rapp. CLXXIV. Service Hydrographique de la Marine No. 13-612:131-155.

estuaries
Reyes Vasqyez, Gregorio, 1966.
Ch. 6. Fitoplancton.
Estudios hidrobiologicos en el Estuario de Maracibo, Inst. Venezolano de Invest. Cient., 122-145.

estuaries
Reynolds, O., 1901.
On certain laws relating to the regime of rivers and estuaries and on the possibility of experiments on a small scale. Sci. Papers, Cambridge 2:326-335.

estuaries
Reynolds, O., 1881-1900.
Report of the committee appointed to investigate the action of waves and currents on the beds and foreshores of estuaries by means of working models. Sci. Papers, Cambridge, 2:380+ pp., 48 figs.+ figs.

estuaries
Rhead, M.M., G. Eglinton, G.H. Draffan and P.J. England 1971.
Conversion of oleic acid to saturated fatty acids in Severn Estuary sediments.
Nature, Lond., 232 (5309): 327-330

estuaries
Rheinheimer, Gerhard 1970.
3.4 Einfluss verschiedener Faktoren auf die Bakterienflora. 3.5. Bakterien und Stickstoffkreislauf. Chemische, mikrobiologische und planktologische Untersuchungen in der Schlei im Hinblick auf deren Abwasserbelastung. Kieler Meeresforsch 26(2): 161-168; 168-170.

estuaries
Rheinheimer, Gerhard 1970.
3. Mikrobiologie. 3.1 Bakterienverteilung. 3.2. Coliztiken. Chemische, mikrobiologische und planktologische Untersuchungen in der Schlei im Hinblick auf deren Abwasserbelastung. Kieler Meeresforsch 26(2): 150; 150-156; 156-159.

estuaries
Rheinheimer, Gerhard, 1970.
2.6 Ammoniak-, Nitrit-, Nitrat- und Phosphatgehalt. Chemische, mikrobiologische und planktologische Untersuchungen in der Schlei im Hinblick auf deren Abwasserbelastung. Kieler Meeresforsch 26(2): 130-132.

estuaries
Rheinheimer, Gerhard, 1970.
2.4. Sauerstoffhaushalt. Chemische, mikrobiologische und planktologische Untersuchungen in der Schlei im Hinblick auf deren Abwasserbelastung. Kieler Meeresforsch 26(2): 126-128.

estuaries
Ribes and Bascou, 1955.
Étude sur la salinité des estuaires à marée.
Ann. Ponts et Chaussees (Mar.-Apr.):

estuaries
Riemen, Franz, 1966.
Die interstitielle Fauna im Elbe-Aestuar: Verbreitung und Systematik.
Arch. Hydrobiol. 3(1/2) (Suppl. 31):1-279.

estuaries
Riley, Gordon A., 1967.
The plankton of estuaries.
In: Estuaries, G.H. Lauff, editor, Publs Am. Ass. Advmt Sci., 83:316-326.

estuaries
Riviere, A., 1951.
Sur l'évolution estuarienne et littorale de la côte Vendéenne dans la région de l'Aiguillon.
Congres de Geologie du Quaternnaire, 1949:123-124.

estuaries
Robinson, A.H.W., 1960
Ebb-flood channels systems in sandy bays and estuaries.
Geography, 45: 183-199.

Rochford, D.J.

estuaries
Rochford, D. J., 1959.
Classification of Australian estuarine systems.
Arch. Oceanogr. e Limnol., 11(Supplement):171-178.

estuaries
Rochford, D.J., 1953.
Analysis of bottom deposits in eastern and south-western Australia, 1951 and records of twenty-four hourly hydrological observations at selected stations in eastern Australian estuarine systems 1951. Oceanographical station list of investigations made by the Division of Fisheries, Commonwealth Scientific and Industrial Research Organization, Australia, 13:68 pp.

estuaries
Rochford, D.J., 1953.
Estuarine hydrological investigations in eastern and south-western Australia, 1951. Oceanographical station list of investigations made by the Division of Fisheries, Commonwealth Scientific and Industrial Research Organization, Austalia, 12:111 pp.

estuaries
Rochford, D.J., 1951.
Hydrology of the estuarine development. Proc. Indo-Pacific Fish. Counc., 17-28 Apr. 1950, Cronulla N.S.W., Australia, Sects. II-III:157-168, 10 textfigs., Table 6.

estuaries
Rochford, D.J., 1951.
Studies in Australian estuarine hydrology.
Australian J. Mar. and Freshwater Res. 2(1):1-116, 1 pl., 7 textfigs.

estuaries
Rochford, D.J., 1951.
Summary to date of the hydrological work of the Fisheries Division C.S.I.R.O. Proc. Indo-Pacific Fish. Comm., 17-28 Apr. 1950, Cronulla N.S.W., Australia, Sects. II-III:51-59.

Estuaries
Rodewald, Martin 1969.
Das Kuriosum eines Scheinbaren "Tidewindes" im Revier Aussenelbe.
Dt. hydrogr. Z. 22(6): 241-255

estuaries
Rodriguez Perez, F., and A. Trueba Gomez, 1953.
Penetration des eaux salées dans le Guadalquivir XVIII Congr. Int. de Navigation, Rome 1953, Sect. II, Navigation Maritime, Communication 3:61-79, 6 figs.

Rodríguez, Gilberto

estuaries
Rodriguez, G., 1963.
The intertidal estuarine communities of Lake Maracaibo, Venezuela.
Bull. Mar. Sci. Gulf and Caribbean, 13(2):197-218.

estuaries
Rodríguez, Gilberto y Andrés Eloy Esteves, 1966.
Ch. 4. Organismos bentonicos.
Estudios hidrobiologicos en el Estuario de Maracaibo, Inst. Venezolanus de Invest. Cient., 83-92.

estuaries
Rodriguez, Gilberto, y Eduardo Ormeno, 1966.
Ch. 5. Zoolplancton.
Estudios hidrobiologicos en el Estuario de Maracaibo, Inst. Venezolano de Invest. Cient., 93-121.

estuaries
Roelofs, E.W., and D.F. Bumpus, 1953.
The hydrography of Pamlico Sound,
Bull. Mar. Sci., Gulf and Caribbean 3(3):181-205, 11 textfigs.

estuaries
Rogers, H.M., 1940.
Occurrence and retention of plankton within the estuary. J. Fish. Res. Bd., Canada, 5(2):164-171.

estuaries
Romanovsky, V., 1950.
Les fluctuations du rivage au voisinage de l'embouchure du Var. Contr. C.R.E.O. No. 5:173-179, 4 charts.

estuaries
Romashin, V.V., 1962.
Fundamental lines of contemporary dynamics of estuary river strip of the west Dvina (Daugav).
Trudy Gosud. Okeanogr. Inst., 66:96-115.
(In Russian).

estuaries
Romashin, V.V., 1962.
Formation of the estuary bar in shallow water. (In Russian).
Trudy Gosud. Okeanogr. Inst., 66:116-120.

estuaries
Rossiter, J.R., 1963.
Tides.
In: Oceanography and Marine Biology, H. Barnes Editor, George Allen and Unwin, 1:11-25.

estuaries
Rossiter, J.R., 1961
Interaction between tide and surge in the Thames.
Geophys. J., R. Astron. Soc., 6(1):29-33.

estuaries
Rossiter, John Reginald, and Geoffrey William Lennon, 1965.
Computation of tidal conditions in the Thames Estuary by the initial value method.
Proc. Instn. Civ. Engrs., 31:25-26.

estuaries
Rounsefell, George A., 1964.
Preconstruction study of the fisheries of the estuarine areas traversed by the Mississippi River-Gulf outlet project.
U.S.F.W.S. Fish. Bull., 63(2):373-393.

estuaries
Rouvillois, Armelle, 1967.
Observations morphologiques sedimentologiques et ecologiques sur la plage de la Ville Ger, dans l'estuaire de la Rance.
Cah. océanogr., 19(5): 375-389

estuaries
Ruhstaller, Roberto E., y Arturo R. Maillo, 1965.
Contribución al conocimiento del material inorgánico en suspensión del Rio de la Plata.
Comun. Mus. argent. Cienc. nat. Bernardino Rivadavia, Geol. 2(3): 31-37

estuaries
Rusnak, Gene A., 1967.
Rates of sediment accumulation in modern estuaries.
In: Estuaries, G.H. Lauff, editor, Publs Am. Ass Advmt Sci., 83:180-184.

estuaries
Russell, Richard J., 1967.
Origin of estuaries.
In: Estuaries, G.H. Lauff, editor, Publs Am. Ass. Advmt Sci., 83:93-99.

estuaries
Ryan, J.D., 1953.
The sediments of Chesapeake Bay. Maryland Bd. Nat Res., Dept. Geol., Mines and Water Res. Bull. 12: 120 pp.

Abstr: Chem. Abstr. 48:38681.

estuaries
Ryther, John H., 1969.
The potential of the estuary for shellfish production.
Proc. nat. shellfish. Ass., 59:15-22

S

estuaries
Said, R., 1953.
Foraminifera of Great Pond, Falmouth, Massachusetts. Contr. Cushman Found. Foram. Res. 4(1): 7-14, 1 table, 3 textfigs.

Estuaries
Saiz, F., M. Lopez-Benito, y E. Anadon, 1961.
Estudio hidrografico de la Ria de Vigo. II.
Inv. Pesq., Barcelona, 18:97-133.

estuaries
Saiz, Fernando, Manuel López-Benito y Emilio Anadón, 1957
Estudio hidrográfico de la Ria de Vigo.
Inv. Pesq., Barcelona, 8:29-88.

estuaries
Sakagishi, Syokichi, 1963.
Investigation of the movement of mobile bed material in the estuary and on the coast by means of radioactive material.
J. Oceanogr. Soc., Japan, 19(1):27-36.

estuaries
*Sakamoto, Wataru, 1968.
Study on the turbidity in an estuary. II. Observations on coagulation and settling processes of particles in the boundary of fresh and salt water. (In Japanese; English abstract).
Bull. Fac. Fish., Hokkaido Univ., 18(4):317-327.

estuaries
Sakamoto, Wataru, 1966.
Study on the turbidity in estuary. 1. Settling patterns of particles through the boundary between two layers. (In Japanese; English abstract)
Bull. Fac. Fish., Hokkaido Univ., 17(2):71-82.

estuaries
Salah, Mostafa, and Gizella Tamas, 1970.
General preliminary contribution to the plankton of Egypt. Bull. Inst. Oceanogr. Fish., Cairo, 1: 307-337.

estuaries
Samoilov, I.V., 1960.
[Theoretical problems and practical tasks in investigations of river estuaries of the SSSR. In: Study of River Estuaries].
Trudy Okeanograf. Komissii, Akad. Nauk, SSSR, 6: 3-8.

estuaries
Sananman, Michael, and Donald W. Lear, 1961.
Iron in Chesapeake Bay waters.
Chesapeake Science, 2(3/4):207-209.

estuaries
Sanders, H.L., P.C. Mangelsdorf, Jr., G.R. Hampson, 1965.
Salinity and faunal distribution in the Pocasset River, Massachusetts.
Limnol. and Oceanogr., Redfield Vol., Suppl. to 10:R216-R229.

Estuaries
Santema, P., and J. N. Svasek, 1961.
De invloed van de afdamming van de zeegaten op het ijsbezwaar op de Zeeuwse en Zuidhollandse stromen.
Rapp. Deltacommissie, Bijdragen, (5):295-326.

estuaries
Sasaki, K., 1952.
On the oceanographical condition at the estuary of the River Kitakami. Bull. Met. Soc. for Res., of Tohoku Distr. 7(1):62-63.

estuaries
Saville, Thorndike, 1966.
A study of estuarine pollution problems on a small unpolluted estuary and a small polluted estuary.
Eng. Prog. Univ. Florida, 20(8):1-202.

estuaries
Schachter, D., 1954.
Contribution à l'étude hydrographique et hydrologique de l'étang de Berre (Bouches-du-Rhone).
Bull. Inst. Océan., Monaco, No. 1048:20 pp.

estuaries
Schafer, C.T. 1970.
Studies of benthonic foraminifera in the Restigouche Estuary: 1. Faunal distribution patterns near pollution sources.
Marit. Sed., 6 (3): 121-134.
Halifax

estuaries
Schell, R., 1953.
Zur Frage der Sandwanderung im Küstebgebiet und im Mündungsgebiet eines Tideflusses.
Die Wasserwirtschaft No. 11:292-302, 23 figs.

estuaries
Schelskie, Claire L., and Eugene P. Odum, 1962.
Mechanisms maintaining high productivity in Georgia estuaries.
Proc., Gulf and Caribbean Fish. Inst., Inst. Mar Sci., Univ. Miami, 14th Ann. Sess., :75-80.

estuaries
Schijf, J.B., and J.C. Schönfeld, 1953.
Theoretical consideration on the motion of salt and fresh water. Proc. Minnesota Int. Hydraulics Conv., Sept. 1-4, 1953:321-333, 10 textfigs.

estuaries
Schlieper, Carl, 1964.
Ionale und osmotische Regulation bei Astuarlebenden Tieren.
Kieler Meeresf., 20(2):169-178.

estuaries
Schneider, Joachim 1970.
3.7 Niedere Pilze. Chemische, mikrobiologische und planktologische Untersuchungen in der Schlei im Hinblick auf deren Abwasserbelastung.
Kieler Meeresforsch 26(2): 173-178.

estuaries
Schodduyn, P., 1927.
Observations biologiques marines faites dans un parc à huitres. Bull. Inst. Océan., Monaco, No. 498:44 pp.

estuaries
Schonfeld, J.C., 1954.
Salt-wedge in a river mouth without tides. (Abstr
Ass. Int. Hydr. Sci., Assem. Gen., 3:99.

estuaries
Schou, Axel, 1967.
Estuarine research in the Danish moraine archipelago.
In: Estuaries, G.H. Lauff, editor, Publs Am. Ass. Advmt Sci., 83:129-145.

estuaries
Schubel, J.R. 1970.
Tidal variation of the suspended sediment size distribution at a station in upper Chesapeake Bay.
Chesapeake Bay Inst. Ref. 70-2: 28pp. (multilithed)

estuaries
Schultze, M., 1941.
Le calcul des hauteurs de marées dans les embouchures des fleuves. Bautechnik 25/26:274 (18 lines).

estuaries
Schuster, O., 1952.
Die Vareler Rinne im Jadebusen. Die Bestandteile und das Gefüge einer Rinne im Watt. Abhandl. Senckenbergischen Naturforsch. Gesellsheft 486: 1-38, 14 textfigs.

estuaries
Schütz, Liselotte, 1964.
Die tierische Besiedlung der Hartböden in der Schwentinemündung.
Kieler Meeresf., 20(2):198-217.

estuaries
Schwerdtfeger, P., 1960.
Observations on estuary ice.
Canadian J. Phys., 38(10):1391-1394.

estuaries
Scott, K.M.F., A.D. Harrison and W. Macnae, 1952.
The ecology of South African estuaries. II. The Klien River estuary, Harmanus, Cape.
Trans. R. Soc., S. Africa, 33(3):283-332.

estuaries
Segal, Earl 1967.
Physiological response of estuarine animals from different latitudes.
In: Estuaries, G.H. Lauff, editor, Publs Am. Ass. Advmt Sci. 83: 548-553.

estuaries
Segerstrale, Sven, G., 1959.
Brackishwater classification, a historical survey.
Arch. Ocean. e Limnol., 11 (supplement) 7-34.

estuaries
Seitz, R.C., 1971.
Measurement of a three-dimensional field of water velocities at a depth of one meter in an estuary. J. mar. Res. 29(2): 140-150.

estuaries
Sekutowicz, L., 1902.
La Seine maritime. Le Génie Civil 41:181-187, 23 figs.

estuaries
Seshadri, B., 1958
Seasonal variations in the total biomass and total organic matter of the plankton in the marine zone of the Vellar Estuary.
J. Zool. Soc., India, Calcutta, 9(2):183-191

estuaries
Shankland, E.C., 1941.
Yanklet Dredget Channel in the Thames Estuary.
Dock & Harbour Authority 21(246):117-122, 4 figs

estuaries
Shaw, Jack T., and G. R. Garrison, 1959.
Formation of thermal microstructure in a narrow embayment during flushing.
J. Geophys. Res., 64(5):533-540.

estuaries
Shepard, F.P., 1953.
Sedimentation rates in Texas estuaries and lagoons. Bull. Am. Assoc. Petr. Geol. 37(8):1919-1934, 5 textfigs.

estuaries
Seshadri, R., K. Krishnamurthy and V.D. Ramamurthy, 1966.
Bacteria and yeasts in marine and estuarine water of Portonovo (S. India).
Bull. Dept. Mar. Biol. Oceanogr. Univ. Kerala, 2: 5-11.

estuaries
Sheldon, R.W. 1968.
Sedimentation in the estuary of the River Crouch, Essex, England
Limnol. Oceanogr. 13(1): 72-83

estuaries
Shetty, H.P.C., S.B. Saha and B.B. Gosh, 1961.
Observations on the distribution and fluctuations of plankton in the Hoogly-Matlah estuarine system with notes on their relation to commercial fish landings.
Indian J. Fish., 8(2):326-363.

estuaries
Siddiquie, H.N., and P.C. Shrivastava, 1970.
Study of sediment movement by fluorescent tracers at Haldia anchorage.
Current Sci. 39(20): 457-454

estuaries
*Sieburth, John McN., 1967.
Seasonal selection of estuarine bacteria by water temperature.
J. exp. mar. Biol. Ecol., 1(1):98-121.

estuaries
Simmons, H.B., 1953.
Sedimentation in estuaries. XVIIIe Congr. Int. Na Navig., Rome 1953, Sect. Navig. Interieure, Question 3:75-85.

estuaries
Simmons, H.B., 1951.
Ch. 7. Salinity problems. Proc. 2nd Conf. Coastal Eng. 58-85, 5 figs.

estuaries
Simonsen, Reimer 1969.
Diatoms as indicators in estuarine environments.
Veröff. Inst. Meeresforsch. Bremerhaven, 11(2): 287-291

estuaries
Siyazuki, K., 1951.
[Studies on the foul-water drained from factories. 1. On the influence of foul-waters drained from factories by the coast on the water of Mitaziri Bay.] Contr. Simonoseki Coll. Fish. Pt. 1:155-158, 4 textfigs. (English summary).

estuaries
Slomianko, P., 1958.
Research of shoaling in the Vistula estuary near Swibno. Trans. Marit. Inst., Gdansk, 10:85-102.

estuaries
Smayda, T.J., 1957.
Phytoplankton studies in lower Narragansett Bay.
Limnol. & Oceanogr., 2(4):342-359.

estuaries
Smith, J. David, R.A. Nicholson and P.J. Moore 1971
Mercury in water of the tidal Thames.
Nature, Lond. 232 (5310): 393-394

estuaries
Smith Ralph I., 1959.
Physiological and ecological problems of brackish water.
Twentieth Ann. Biol. Coll., Apr. 3-4, 1959, Mar. Biol., Oregon State Coll: 59-69.

estuaries
Smith, R.I., 1956.
The ecology of the Tamar Estuary. VII. Observations on the interstitial salinity of intertidal muds in the estuarine habitat of Nereis diversicolor. J.M.B.A., 35(1):81-104.

estuaries
*Smith, Stuart D., 1967.
Thrust-anemometer measurements of wind-velocity spectra and of Reynolds stress over a coastal inlet.
J. mar. Res., 25(3):239-262.

estuaries
Spencer, R.S., 1956.
Studies in Australian estuarine hydrology.
Australian J. Mar. & Freshwater Res. 7(2):193-253.

estuaries
Sterr, T.J., 1956.
Relative amounts of Vitamin B12 in detritus from oceanic and estuarine environments near Sapelo Island, Georgia. Ecology 37(4):658--664.

estuaries
Stauber, L. E., 1943.
Graphic representation of salinity in a tidal estuary. J. Mar. Res. 5(2):165-167, fig. 1.

estuaries
Steers, J.R., 1967.
Geomorphology and coastal processes.
In: Estuaries, G.H. Lauff, editor, Publs Am. Ass. Advmt Sci., 83:100-107.

estuaries
Stelzenmuller, William B., 1965.
Tidal characteristics of two estuaries in Florida J. Waterways and Harbors Div., Amer. Soc., Civil Engineers, 91(WW3):25-36. Proc.

estuaries
Stephens, K., J.D. Fulton and O.D. Kennedy, 19[illegible]
Summary of biological oceanographic observations in the Strait of Georgia, 1965-1968
Techn. Rept. Fish. Res. Bd Can., 110:

estuaries
Stephensen, W., 1951.
Preliminary observations upon the release of phosphate from estuarine mud. Proc. Indo-Pacific Fish Cound., 17-28 Apr. 1950, Cronulla N.S.W., Australia, Sects. II-III:184-189, 1 textfig.

estuaries
Stewart, H.B., jr., 1958.
Upstream bottom currents in New York harbor. Science, 127(3306):1113-1115.

estuaries
Stewart, R.W., 1957.
A note on the dynamic balance in estuarine circulation. J. Mar. Res., 16(1):34-39.

estuaries
Stock, J.H., H. Nijssen and P. Kant 1966
La répartition écologique des amphipodes de la famille des Gammaridae dans la Slack et son estuaire.
Bull. zool. Mus. Univ. Amsterdam 1(3): 19-30.

estuaries
Stommel, H., 1953.
Computation of pollution in a vertically mixed estuary. Sewage and Ind. Wastes 25(9):1065-1071, 3 textfigs.

estuaries
Stommel, H., 1953.
The role of density currents in estuaries. Proc. Minnesota Int. Hydraulics Conv., Sept. 1-4, 305-312, 7 textfigs.

estuaries
Stommel, H., and H.G. Farmer, 1953.
Control of salinity in an estuary by a transition J. Mar. Res. 12(1):13-20, 4 textfigs.

estuaries
Stone, Alfred N., and Donald Reish, 1965.
The effect of fresh-water run-off on a population of estuarine polychaetous annelids. Bull. S. Calif. Acad. Sci., 64(3):111-119.

estuaries
Stormont, D.H., 1966.
Refiners join attack on water pollution. Oil and Gas Jour., 64(13):130-131.

estuaries
Stretta Etienne, 1959.
Position du Rio Capibaribe dans l'ensemble hydrogeologique du Bassin de Recife (Bresil). Trabalhos, Inst. Biol. Marit. e Geogr., Recife, 71-76.

estuaries
Stuart, Mary, and Hubert Fuller, 1968.
Saprolegnia parasitica Coker in estuaries. Nature, 217 (5134):1157-1158.

estuaries
Suryanarayana Rao, S.V., and P.G. George, 1959.
Hydrology of the Korapuzha Estuary Malabar, Kerala Strait.
J. Mar. Biol. Assoc., India, 1(2):212-223.

estuaries
Swain, A., and O.F. Newman, 1962.
Hydrographical survey of the Tyne River.
Min. Agric., Fish and Food, Fish. Inv., (1), 6(3):45 pp.

estuaries
Szarejko-Lukaszewicz, D., 1959.
Hydrographic investigations on the Firth of Vistula in 1953-1954.
Prace Morsk. Inst. Ryb. Gdyni, 10(A):215-228.

T

estuaries
Tabata, S., 1954.
Oceanography of British Columbia mainland inlets. V. Salinity and temperatures of Bute Inlet.
Prog. Repts., Pacific Coastal Stas., Fish. Res. Bd., Canada, No. 100:8-11, 3 textfigs.

estuaries
Tabata, S., and R.J. LeBrasseur, 1958.
Sea water intrusion into the Fraser River and it relation to the incidence of shipworms in Steveston Cannery Basin.
J. Fish. Res. Bd., Canada, 15(1):91-113.

estuaries
Tabb, D.C., D.L. Dubrow and R.B. Manning, 1962.
The ecology of northern Florida Bay and adjacent estuaries.
Mar. Lab., Inst. Mar. Sci., Univ. Miami, Techn. Ser., 39:82 pp.

ESTUARIES
Takano, K., 1954.
Note on the influence of an inlet and an intake upon the velocity distribution in the interior. J. Ocean. Soc., Japan, 10(3):87-91.

estuaries
*Talbot, J.W., 1967.
The hydrography of the estuary of the River Blackwater.
Fish. Invest. Min. Agric., Fish. Food, (2)25(6):92pp.

estuaries
Tambe, R.Y., and N.N. Sathaye, 1961.
The Houghly River survey.
Int. Hydrogr. Rev., 38(1):33-37.

estuaries
Tepley, Sandra, 1969.
Foraminiferal analysis of the Miramichi Estuary.
Marit. Sediments 5(1):30-39.

estuaries
Taylor, W. Rowland, and John E. Hughes 1967.
Primary productivity in the Chesapeake Bay during the summer of 1964.
Chesapeake Bay Inst., Johns Hopkins Univ., Ref. 67-1: 31 pp. (multilithed).

estuaries
Taysi, I., and N. van Uden, 1964.
Occurrence and population densities of yeast species in an estuarine-marine area.
Limnology and Oceanography, 9(1):42-45.

estuaries
Terwindt, J.H.J. 1971.
Litho-facies of inshore estuarine and tidal-inlet deposits.
Geol. Mijnb. 50(3): 515-526.

estuaries
Tett, P.B., 1971.
The relation between dinoflagellates and the bioluminescence of sea water. J. mar. biol. Ass., U.K., 51(1): 183-206.

estuaries
Thayer, Gordon W. 1971.
Phytoplankton production and the distribution of nutrients in a shallow unstratified estuarine system near Beaufort, N.C.
Chesapeake Sci. 12(4):240-253.

estuaries
Thiemann, K., 1934.
Das Plankton der Flussmundungen.
Wiss. Ergeb. Deutschen Atlantischen Exped. "Meteor", 1925-1927, 12(1):199-273.

estuaries
Thoma, A.S., 1949.
Investigation of tidal phenomena in the Clyde Estuary using a scale model. J. Inst. Civ. Eng. 33:100-125.

estuaries
Thomann, Robert V., 1965.
Recent results from a mathematical model of water pollution control in the Delaware Estuary.
Water Resources Research, Am. Geophys. Union, 1(3):349-359.

estuaries
Thomoson, Seton H., 1968.
Estuaries: an action to save them.
Proc. Gulf Carib. Fish. Inst. 20th Ann. Sess. 55-59.

estuaries
Thomsen, H., 1947.
Hydrografiske undersøgelser i Limfjorden 1943. 1. Observationsmaterialet. Publ. Danske Met. Inst. Medd. No. 9:104 pp., 1 fold-in, 6 textfigs.

estuaries
Thoulet, J., 1910.
Étude des fonds marins de la Baie de la Seine.
Ann. Hydrogr., Paris: 53-78.

estuaries
Tiphane, Marcel, 1965.
Topographie de la Baie-des-Chaleurs
Cah. Inf. Stn Biol. mar. Grande-Rivière, No. 29: 3pp., fold-in chart (multilithed)

estuaries
Tiphane, Marcel, 1963.
Etude des températures des eaux de la région de la Baie-des-Chaleurs, 1961-1962.
Rapp. Ann., 1962, Sta. Biol. Mar., Grande-Rivière, Canada, 19-38. (multilithed).

Tiphane, Marcel, et Robert Boudreault, 1964.
Observations océanographiques dans la Baie-des-Chaleurs, été 1963.
Cah. Inf. Stn Biol. mar., Grande-Rivière, No. 21 unnumbered pp. (multilithed).

estuaries
Todd, D.K., and L.-k Lau, 1956.
On estimating streamflow into a tidal estuary.
Trans. Amer. Geophys. Union, 37(4):468-473.

estuaries
Tommasi, Luiz Roberto, 1970.
Observações sôbre a fauna bêntica do complexo estuarino-lagunar de Cananéia. Bolm Inst. oceanogr. São Paulo, 19: 43-56.

estuaries
*Tommasi, Luiz Roberto, 1967.
Observações preliminares a fauna bêntica de sedimentos moles da Baie de Santos e regiões vizinhas.
Bolm Inst.oceanogr.S. Paulo, 16(1):43-65.

estuaries
Tossini, L., 1959
Sistema hidrografico y cuenca del Rio de la Plata.
Anales Soc. Cient. Argentina, 167(3&4): 41-64.

estuaries
Tricart, Jean, 1962
Étude générale de la desserte portuaire de la "Sasca". 11. Les sites portuaires, leurs caracteristiques morphodynamiques et leurs possibilites d'amenagement.
Cahiers Oceanogr., C.C.O.E.C., 14(3):146-161.

estuaries
Train, Russell E., 1969.
The challenge of the estuary.
Proc. nat. Shellfish. Ass., 59: 14-17

estuaries
Tranter, D.J. and S. Abraham, 1971.
Coexistence of species of Acartiidae (Copepoda) in the Cochin Backwater, a monsoonal estuarine lagoon. Mar. Biol. 11(3): 222-241.

estuaries
Tromeur, J.Y., 1938-1939.
Mission hydrographique du Saloum, 1930-1931.
Ann. Hydrogr., Paris, (3)16:5-33, figs.

Tully, J.P., 1958 estuaries
On structure, entrainment and transport in estuarine embayments. J. Mar. Res., 17:523-535.

estuaries
Tully, J.P., 1952.
Notes on the behaviour of fresh water entering the sea. Proc. Seventh Pacific Sci. Congr., Met. Ocean., 3:267-289, 10 textfigs.

estuaries
Tully, J.P., and F.G. Barber, 1961.
An estuarine model of the sub-Arctic Pacific Ocean. In:
Oceanography, Mary Sears, Edit., Amer. Assoc. Adv Sci., Washington, 425-454.

Published previously under title:
"An estuarine analogy in the sub-Arctic Pacific Ocean."
J. Fish. Res. Bd., Canada, 17:91-112 (1960).

estuaries
Tully, J.P., and F.G. Barber, 1960
An estuarine analogy in the Sub-Arctic Pacific Ocean.
J. Fish. Res. Bd., Canada, 17(1):91-112.

Tundisi, José G., 1970.
O plâncton estuarino.
Contrções Inst. Oceanogr. Univ. S. Paulo, Ser. Ocean. Biol. 19: 1-22.

estuaries
Turazza, Hyacinthe, 1905.
General report. Improving the mouths of rivers discharging into tideless seas. Navig. Congres X Milan, Section II, Question 1:1-20.

estuaries
Turekian, Karl K., 1971
2. Rivers, tributaries and estuaries. In: Impingement of man on the oceans, D.W. Hood, editor, Wiley Interscience, 9-73.

estuaries
Turmel, J.M., 1957.
Formation des mares et des ruisseaux dans les prés-salés des estuaires de l'ouest du Contention.
Bull. Lab. Mar., Dinard, 43:79-91.

estuaries
Twinberrow, Wulstan, 1924.
The Severn Barrage. Dock and Harbour Authority 4(47):2½ pp.

U

estuaries
*Ulken, Annemarie, 1967.
Einigi Beobachtungen über das Vorkommen von Phycomyceten aus der Reihe der Chytridiales im brackigen und marinen Wasser.
Veröff. Inst. Meeresforsch. Bremerh. 10(3):167-172.

estuaries
United States, Johns Hopkins University, Chesapeake Bay Institute, 1962.
Cruise 28, July 24-August 7, 1962, Data Rept. 48, Ref. 62-16: 17 pp. (mimeographed).

estuaries
U.S.A. Johns Hopkins University, Chesapeake Bay Institute, 1962.
Temperature and salinity data collected in the Chesapeake Bay and tributary estuaries during the period 1 February 1956 to 9 February 1956.
Data Rept., 49 (Ref. 62-23):22 pp. (unpublished mss.)

estuaries
Unoki, Sanae, 1969.
Investigation of tides in estuaries (2).
Proc. 16th Conf. Coast Engng. Also in: Coll. Repr. Coll. mar. Sci. Techn. Tokai Univ. 4: 93-100. (In Japanese).

estuaries
Unoki, Sanae 1968.
Investigation of tides in estuaries.
(In Japanese; English abstract)
Proc. 15th Conf. Coastal Engng Japan.
Also in: Coll. Repr. Coll. mar. Sci. Techn., Tokai Univ. 1967-68, 3: 69-78.

estuaries
Unoki, Sanae 1967.
Flow of estuarine waters in three rivers of Kiso. (In Japanese; English abstract).
Bull. coast. Oceanogr. 6(1):
Also in: Coll. Repr. Coll. mar. Sci. Techn., Tokai Univ. 1967-68: 41-54.

Estuaries
Uyeno, Fukuzo, 1969.
Relationship between sea water and bottom condition and mechanism of fisheries production in estuarine areas,.....with special reference to the deterioration of pearl oyster farm. (In Japanese; English abstract). Bull. Jap. Soc. fish. Oceanogr. Spec. No. (Prof. Uda Commem. Pap.): 193-196.

estuaries
Uyeno, Fukuzo, 1966.
Nutrient and energy cycles in an estuarine oyster area.
J. Fish. Res. Bd., Can., 23(11):1635-1652.

V

estuaries
Valembois, J., 1950.
Études d'un appareil permettant la repartition de l'onde marée dans un modèle d'estuaire.
Mém. et Trav. S.H.F. HB B 1950:674-679, 4 figs

estuaries
Valembois, J., 1950.
Étude d'un appareil permettant la reproduction de l'onde marée dans un modèle d'estuaire. Houille Blanche 5(2):150-154.

estuaries
Valembois, J., 1950.
Étude d'un appareil permettant la reproduction de l'onde marée dans un modèle d'estuaire.
La Houille Blanche, Nov. 1950, No. Spec. B:3-7, 4 textfigs.

estuaries
Valembois, J., 1949.
Étude d'un appareil permettant la reproduction de l'onde marée dans un modèle estuaire. Int. Assoc. Hydraulic Structures Research (IAHSR)-Assoc. Int. Recherches pour Trav. Hydrauliques (AIRTH), 3rd Meeting, Grenoble, 5-7 Sept., 1949: 11 pp., 5 textfigs.

(Reference may be incomplete as this was taken from a reprint).

Estuaries
Valken, K. F., 1961.
De waterstaatkundige aspecten van de waterhuishouding in het gebied van de benedenrivieren en de zeearmen na de uitvoering van de deltawerken.
Rapp. Deltacommissie, Bijdragen, (5):191-294.

estuaries
*Van de Kreeke, Jacobus, 1967.
Water-level flucuations and flow in tidal inlets.
J. watways Harb.Div.Am.Soc.civ.Engrs. 93 (WW4): 97-106.

estuaries (Congo)
Van Goethem, C., 1951.
Étude physique et chimique du milieu marin. Rés. Sci. Exped. Océan. Belge dans les Eaux Côtières Africaines de l'Atlantique Sud (1948-1949) 1: 1-151, 1 pl.

estuaries
Van Straaten, L.M.J.U., 1960.
Transport and composition of sediments.
Verhandel. Kon. Nederl. Geol. Mijnb. k. Gen., Geol. Ser. (Symposium Ems-Estuarium (Nordsee), 19:279-292.

estuaries
Van Straaten, L.M.J.U., 1952.
Current rips and dip currents in the Dutch Wadden Sea. K. Nederl. Akad. Wetensch., Proc B, 55(3):228-238, 2 photos, 10 textfigs.

estuaries
Van Straaten, L.M.J.U., 1952.
Biogene textures and the formation of shell beds in the Dutch Wadden Sea, K. Nederl. Akad. Wetensch., Proc., B, 55(5):500-516, 6 textfigs.

estuaries
Van Uden, N. 1967.
Occurrence and origin of yeasts in estuaries.
In: Estuaries, G.H. Lauff, editor, Publs. Am. Ass. Advmt Sci. 83:306-310.

estuaries
Van Veen, J., 1954.
Development of marine plains (Penetration of sea water in Dutch tidal rivers and inland waters).
Ass. Int. Hydr. Sci., Assem. Gen., Rome, 3:101-114.

estuaries
Van Veen, J., 1953.
The penetration of sea water into Dutch river mouths and estuaries. II
XVIII Congr. Int. Navig., Sect. Navig. Marit., Comm. 3, Rome 1953:185-200, 15 figs.

estuaries
Van Veen, J., 1950.
Zoutmetingen op en bij de benedenricieren in de droge jaren 1947 en 1949. Commissie Woor Hydrologisch Onderzoek T.N.O., Holland.

estuaries
Varlet, F. and M. Menaché, 1947
L'Estuaire du Moros à Concarneau (Finistère Étude du melange des eaux douces et salées.
Bull. l'Inst. Océan., Monaco, 917:26 pp., 2 text figs.

estuaries
Varney, Jean-René, 1964.
Morphologie sous-marine du Mor Bras.
Cahiers Océanogr., C.C.O.E.C., 16(7):529-546.

estuaries
Vatova, A., and P.M. di Villagrazia, 1950.
Sulle condizione idrographiche de Canal di Lema in Istria. Nova Thalassia 1(8):24 pp., graphs.

estuaries
Vatova, A., and P. Milo di Villagrazia, 1948.
Sulle condizioni chimicofisiche del Canale di Lema presso Rovigno d'Istria.
Bol. Pesca, Piscicol. Idrobiol., n.s., 3(1): 5-27, 2 textfigs., 5 pls.

estuaries
Verger, F., 1955.
Observations sur l'Anse de l'Aiguillon (Vendée et Charante Maritime). Bull. d'Info., C.C.O.E.C. 7(3):103-118, 12 pls.

estuaries
Verger, F., 1954.
Les travaux de conquete sur la mer, de l'embouchure de l'Elbe a la frontiere Germano-Danoise.
Bull. d'Info., C.C.O.E.C. 7(4):172-179, 5 pls.

estuaries
Vernberg, F. John 1967.
Some future problems in the physiological ecology of estuarine animals.
In: Estuaries, G.H. Lauff, editor, Publs Am. Ass. Advmt Sci. 83: 554-557.

estuaries
Vernon-Harcourt, L.F., 1888.
The principles of training rivers through tidal estuaries as illustrated by investigations into the methods of improving navigation channels of the estuary of the Seine. Proc. R. Soc., London, 45:504-524.

estuaries
Vernon-Harcourt, L.F., 1882.
Harbors and estuaries on sandy coasts. Proc. Inst Civil Eng., Paper No. 1772:1-30.

estuaries
Verploegh, G., and P. Groen, 1955.
De uitwerking van de wind over de Groningse Waddenzee op de hoogwaterstanden van Delfzijl.
K. Nederl. Met. Inst., Watens. Rapp., W.R. 55-009(IV-010):9pp.

estuaries
Vezzani, R., 1953.
Pénétration des eaux salées dans les fleuves à marée, procedés généraux des recherches.
XVIIIe Congr. Int. Navig., Rome 1953, Sect. 11, Navig. Marit. Comm. 3:143-153, 6 textfigs.

estuaries
Vigare, André, 1965.
Les modalités du remblaiement alluvial dans l'estuaire de la Seine.
Cahiers Océanogr., C.C.O.E.C., 17(5):301-330.

estuaries
Villagrezia, P. Milo de, 1950.
Ulteriore contributo alle conoscenza delle condizioni idrografiche e biologiche del Canali di Lema (Istria). Bol. Pesca, Piscicolt. Idrobiol. 5(2):225-247, 3 textfigs.

estuaries
Visentini, M., 1953.
Pénétration des eaux salées dans les fleuves à marée. Résultats particuliers pour le Po.
XVIIIe Congr. Int. Navig., Rome 1953, Sect. 11, Navig. Marit. Comm. 3:153-160, 7 figs.

estuaries
Vives, F., 1960.
Nota sobre el zooplancton superficial de la Ria de Pontevedra.
Bol. Real Soc. Esp. Hist. Nat., Sec. Biol., 58 (2):389-402.

estuaries
Vives, F., and F. Fraga, 1961
Floristica y sucesión del fitoplancton en la Ria de Vigo.
Inv. Pesq., Barcelona, 19:17-36.

estuaries
Vives, F., and F. Fraga, 1961.
Producción basica en la Ría de Vigo (NW de España).
Inv. Pesq., Barcelona, 19:129-137.

estuaries
Vuletic, A., 1953.
Structure geologique du fond du Malo et du Veliko Jesero, sur l'Ile de Mjlet.
Acta Adriatica 6(1):3-63, 16 textfigs.

W

estuaries
Waldichuk, Michael, 1965.
Water exchange in Port Moody, British Columbia, and its effect on waste disposal.
J. Fish. Res. Bd., Canada, 22(3):801-822.

estuaries
Waldichuk, Michael, 1964.
Dispersion of kraft-mill effluent from a submarine diffuser in Stuart Channel, British Columbia.
J. Fish. Res. Bd., Canada, 21(5):1289-1316.

Waldichuck, Michael, 1958 estuaries
Some oceanographic characteristics of a polluted inlet in British Columbia. J. Mar. Res., 17:536-551.

estuaries
Waldichuk, Michael, 1957.
Physical oceanography of the Strait of Georgia.
J. Fish. Res. Bd., Canada, 14(3):321-486.

estuaries

Waldichuk, M., 1955.
Effluent disposal from the proposed pulp mill at Crofton, B.C. Fish. Res. Bd., Canada, Pacific Coast Stas., Prog. Repts., No. 102:6-9, 3 text-FIGS.

estuaries

Waldichuk, M., and J.P. Tully, 1953.
Pollution study of Nainaimo Harbour. Prog. Repts. Pacific Coast Stas., Fish. Res. Bd., Canada, No. 97:14-17, 4 textfigs.

estuaries

Wallace, G.M., and L.E. Newman, 1954.
Bacteriological survey of Auckland Harbours. III. Tamaki River and estuary.(N.Z. Ocean. Comm. Publ. 10), N.Z.J.Sci. Tech., Sect. B, 36(2):129-135, 3 textfigs.

estuaries

Walther, F., 1954.
Veränderungen der Wasserstände und Gezeiten in der Unterweser als Folge des Ausbaues. Hansa 91(21/22):8 pp.

estuaries

Ward, Ronald W., Valerie Vreeland, Charles H. Southwick and Anthony J. Reading, 1965.
Ecological studies related to plankton productivity in two Chesapeake Bay estuaries.
Chesapeake Science, 6(4):214-225.

estuaries

Warren, P.J., and R.W. Sheldon 1967.
Feeding and migration patterns of the pink shrimp Pandalus montagui in the estuary of the River Crouch, Essex, England.
J. Fish. Res. Bd., Can., 24(3):569-580

Wastler, T.A., 1969.
Measuring estuarine pollution.
Oceanology intl, 4(3):43-45. (popular).

estuaries

Wastler, Thaddeus A., 1968.
Management of the national estuarine resources of the United States.
Helgolander wiss. Meeresuntersuch., 17:392-397.

estuaries

Watanabe, Nobuo, 1969.
On a trial of forecasting of the chlorinity var tion during short period in the Kiso-Sansen estuary. (In Japanese; English abstract). Bull. Jap. Soc. fish. Oceanogr. Spec. No. (Prof. Uda Commem. Pap.): 157-171.

estuaries

Watanabe, Nobuo, 1963.
Diffusion of industrial drainage in a tidal river An example in the lower reach of the Edo River with small natural discharge. (In Japanese; English abstract).
J. Oceanogr. Soc., Japan, 18(4):172-184.

Estuaries

Watts, J. C. D., 1961.
Seasonal fluctuations and distribution of nitrite in a tropical West African estuary.
Nature, 191(4791):929.

estuaries

Watts, J.C. D., 1960.
A summary of the meteorological and hydrological observations made in the Sierra Leone River Estuary area.
Bull. I.F.A.N., 22 (4):1159-1164.

estuaries

Watts, J.C.D., 1960

Sea-water as the primary source of sulfate in tidal swamp soils from Sierra Leone. Nature 186(721): 308-309.

estuaries

Watts, J.C.D., 1960

Sulfate absorption by muds of relatively high organic content from the Sierra Leone River Estuary. Bull. de l'I.F.A.N. (2) 22(4): 1153-1158.

estuaries

Watts, J.C.D., 1958.
The hydrology of a tropical West African estuary.
Bull., I.F.A.N., 20(A):697-752.

estuaries

Watts, J.C.D., 1957.
The chemical composition of the bottom deposits from the Sierra Leone River estuary.
Bull. Inst. Francais d'Afrique Noire(A) 19(3): 1020-1029.

estuaries

Weber, R.E., and D.H. Spaangaren, 1970.
On the influence of temperature on the osmoregulation of Crangon crangon and its significance under estuarine conditions
Netherl. J. Sea Res. 5(1): 108-120.

Estuaries

Wellershaus, Stefan 1970.
On the taxonomy of some Copepoda in Cochin Backwaters (a South Indian estuary).
Veröff. Inst. Meeresforsch. Bremerh. 12(3): 463-490.

estuaries

Wells, A.L., 1938.
Some notes on the plankton of the Thames Estuary.
J. Anim. Ecol. 7:105-124.

estuaries

Wells, H.W., 1961
The fauna of oyster beds with special reference to the salinity factor.
Ecolog. Mon., 31(3):239-266.

estuaries

Wells, J.B.J., 1963.
Copepoda from the littoral region of the estuary of the river Exe (Devon, England).
Crustaceana, 5(1):10-26.

estuaries

*Weyland, Horst, 1967.
Über die Verbreitung von Abwasserbakterien im Sediment des Weserästuars.
Veröff. Inst. Meeresforsch. Bremerh. 10 (3):173-182.

estuaries

Wheatland, A.B., 1955.
Some factors affecting the presence of sulphide in a polluted estuary. Verh. Int. Ver. Limnol. 12:772-782, 7 textfigs.

estuaries

Wheatland, A.B., and J. Wheatland, 1956.
Salinity and the biochemical oxygen demand of estuary water. Water & Sanit. Eng., 6:10-14.

Abstr. Publ. Health Eng. Abstr. 36(12):27.

estuaries

*Whitfield, M., 1969.
Eh as an operational parameter in estuarine studies.
Limnol. Oceanogr., 14(4):547-558.

estuaries

Wicker, C.F., 1953.
The transportation and deposition of sediments in estuaries. Proc. Fifth Hydraulics Conf., June 9-11, 1952, State Univ. Iowa, Studies in Eng., Bull 34(426):227-239.

estuaries

Wiktor, Josef, 1962.
Quantitative and qualitative investigations of the Szczecin Firth bottom fauna. II.
Prace Morsk. Inst., Ryback., Oceanol.-Ichtiol., Gdynia, 11(A):81-112.
In Polish with Polish, English and Russian summary

estuaries

Wiktorowie, Jozef i Krystyna, 1962
Some hydrological properties of the Pomeranian Bay water.
Prace Morsk. Inst., Ryback., Gdyni, Oceanogr. -Ichtiol., 11(A):113-136.

estuaries

Wiktor, K., 1959.
The Szczecin Firth zooplankton. Part II.
Prace Morsk Inst. Ryb., Gdyni, 10(A):229-258.

estuaries

Wiktor, K., and D. Zembrzuska, D., 1959.
Materials for hydrography of Firth of Szczecin.
Prace Morsk. Inst. Ryback., Gdyni, 10A:259-272.

estuaries

Wilkinson, D., 1963.
Nitrogen transformations in a polluted estuary.
Int. J. Air, Water Pollution, 7(6/7):737-752.

estuaries

Williams, Austin B., 1971
A ten-year study of meroplankton in North Carolina estuaries: annual occurrence of some brachyuran developmental stages.
Chesapeake Sci., 12(2): 53-61

estuaries

Williams, Austin B. and Earl E. Deubler 1968.
A ten-year study of meroplankton in North Carolina estuaries: assessment of environmental factors and sampling success among bothid flounders and penaeid shrimps.
Chesapeake Sci., 9(1): 27-41.

estuaries

Williams, Hulen B., Vincent Farrugia, Anthony Ekker and Elard L. Haden, 1964.
Chlorinity determination in estuarine waters by physical methods.
J. Mar. Res., 22(2):190-196.

estuaries

Williams, Richard B., 1966.
Annual phytoplankton production in a system of shallow temperature estuaries.
In: Some contemporary studies in marine science H. Barnes, editor, George Allen & Unwin, Ltd., 698-716.

estuaries

Williams, Richard B., 1965.
Annual phytoplankton production in a system of shallow, temperate estuaries. (Abstract).
Ocean Sci. and Ocean Eng., Mar. Techn. Soc.-Amer. Soc. Limnol. Oceanogr., 1:95.

estuaries

Williams, Richard B., Marianne B. Murdoch and Leon K. Thomas, 1965.
Standing crop and importance of zooplankton in a system of shallow estuaries.
Chesapeake Sci., 9(1): 42-51.

estuaries

Wilson, G.A., F.H. Allen and N.D.E. Stephens, 1957
Siltation in coastal waters, in estuaries, in channels, in tidal basins, in enclosed docks and in maritime canals.
19th Int. Navig. Congr., London, Sect. 2. Ocean Navig., Comm. 3:1-27.

estuaries

Windom, Herbert L., 1971.
Fluoride concentration in coastal and estuarine waters of Georgia. Limnol. Oceanogr. 16(5): 806-810.

estuaries

Windom, H.L., K.C. Beck, and R. Smith 1971
Transport of trace metals to the Atlantic Ocean by three southeastern rivers
SEast. Geol. 12(3):169-181.

estuaries

Windom, Herbert L., William J. Neal, and Kevin C. Beck, 1971.
Mineralogy of sediments in three Georgia estuaries
J. Sedim. Petrol. 41(2): 497-504.

estuaries

Winkel, R., 1923.
Études sur mode du mouvement des marées (ports maritimes, embouchure de rivieres). Bautechnik No 21.

estuaries

Winston, Judith E., and Frank E. Anderson, 1971.
Bioturbation of sediments in a northern temperate estuary.
Marine Geol. 10(1):39-49.

estuaries

Wiseman, W.J. 1969.
On the structure of high-frequency turbulence in a tidal estuary.
Techn. Rept. Chesapeake Bay Inst., 59 (Ref. 69-12): 76 pp. (multilithed)

estuaries

Wolfe, Douglas A., 1971.
Fallout cesium-137 in clams (Rangia cuneata) from the Neuse River Estuary, North Carolina.
Limnol. Oceanogr. 16(5): 797-805.

estuaries

Wolff, W.J. 1968.
The Echinodermata of the estuarine region of the rivers Rhine, Meuse and Scheldt, with a list of species occurring in the coastal waters of the Netherlands
Netherlands J. Sea Res., 4(1): 59-85.

estuaries

Wolff, W.J., and A.J.J. Sandee 1971.
Distribution and ecology of the Decapoda Reptantia of the estuarine area of the rivers Rhine, Meuse and Scheldt.
Netherlands J. Sea Res. 5(2): 197-226.

estuaries

Wollast, R., and F. De Broeu, 1971.
Study of the behavior of dissolved silica in the estuary of the Scheldt.
Geochim. cosmochim. Acta 35(6): 613-620.

estuaries

Won, Chong Hun, 1964.
Tidal variations of chemical constituents of the estuary water at the Lava bed in the Nack-Dong River from Nov. 1962 to Oct. 1963. (In Korean; English abstract).
Bull. Pusan, Fish. Coll., 6(1): 21-32.

estuaries

Won, Chong Hun, 1963.
Distribution of chemical constituents of the estuary water in Gwang-Yang Inlet. (In Korean; English abstract).
Bull. Fish. Coll., Pusan Nat. Univ., 5(1):1-10.

estuaries

Won, Chong Hun, and Kil Soon Park, 1968.
Tidal variations of chlorinity and pH at the Yong-Ho Basin from Mar. 1 to Mar. 20, 1968. (In Korean; English abstract).
Bull. Pusan Fish. Coll. (Nat. Sci.), 8(2):103-111.

estuaries

Wood, E.J. Ferguson, 1962
The microbiology of estuaries.
The Environmental Chemistry of Marine Sediments, Proc. Symp., Univ. R.I., Jan. 13, 1962 Occ. Papers, Narragansett Mar. Lab., No. 1: 20-26.

estuaries

Woodwell, George M., Charles F. Wurster, Jr., and Peter A. Isaacson, 1967.
DDT residues in an east coast estuary: a case of biological concentration of a persistent insecticide.
Science, 156(3776):821-823.

estuaries

Wright, Ramil, 1968.
Miliolidae (foraminiferos) recientes del estuario del Rio Quequeh Grande.
Revta Mus. argent. Cienc. nat. Bernardino Rivadavia, Hidrobiol., 2(7):225-256.

estuaries

Wulle, K., and A. Faehndrich, 1935.
Les embouchures de l'estuaire de Stettin.
XVI Congr. Int. Navigation, Bruxelles, 2 Sect., le question No. 64:4-13, 9 figs.

estuaries

Yamamoto, H., 1958
Oceanographic studies on Japan inlets. 2. Diurnal variation of hydrographical condition in Urado Bay. (In Japanese).
J. Oceanogr. Soc., Japan, 14(3):93-97.

estuaries

Yamamoto, H., 1958.
Oceanographical studies on Japanese inlets. 3. Relations between variation of chemical compositions and hydrographical conditions in Urado Bay. 1. (In Japanese).
J. Oceanogr. Soc., Japan, 14(4):155-158.

estuaries

Yamamoto, H., 1958.
Oceanographical studies on Japanese inlets. 4. Relations between variation of chemical compositions and hydrographical conditions in Susaki Bay. 2.
J. Oceanogr. Soc., Japan, 14(4):159-161.

estuaries

Yamamoto, H., 1959
Oceanographical studies on Japanese inlets. 5. Relations between variation of chemical compositions and hydrographic conditions in Uranouchi Bay. 6. On the consumption of dissolved oxygen in Urado Bay.
J. Oceanogr. Soc., Japan, 15(1):15-18, 19-22.

estuaries

Yamazi, Isamu, 1966.
Zooplankton communities of the Navesink and Shrewsbury rivers and Sandy Hook Bay, New Jersey.
Tech. Pap. Bur. Sport Fish. Wildl., U.S., (2):1-44. (multilithed).

estuaries

Yamazi, Isamu, 1959.
A study on the productivity of Tanabe Bay. IV. On the voluminal change of plankton communities caused by tidal current.
Rec. Oceanogr. Wks., Japan, Spec. No. 3:47-60.

estuaries

Yamazi, Isamu, 1959.
A study on the productivity of Tanabe Bay. II. On some plankton indicating the water exchange in Tanabe Bay in August, 1957.
Rec. Oceanogr. Wks., Japan, Spec. No. 3:23-30.

estuaries

Yamazi, I., 1953.
Plankton investigations of inlet waters along the coast of Japan. X. The plankton of Kamaisi Bay on the eastern coast of Tohoku District.
Publ. Seto Mar. Biol. Lab. 3(2)(Article 18):189-204, 16 textfigs.

estuaries

Yamazi, I., 1953.
Plankton investigations in inlet waters along the coast of Japan. IX. The plankton of Onagawa Bay on the eastern coast of Tohoku District.
Publ. Seto Mar. Biol. Lab. 3(2)(Article 17):173-187, 10 textfigs.

estuaries

Yamazi, I., 1952.
Plankton investigations in inlet waters along the coast of Japan. III. The plankton of Imari Bay in Kyusyu. IV. The plankton of Nagasaki Bay and Harbour in Kyusyu. V. The plankton of Hiroshima Bay in the Seto-Nakai (Inland Sea).
Publ. Seto Mar. Biol. Sta., Kyoto Univ., 2(2): 289-304, 305-318; 319-330, 8 and 8 and 7 textfigs.

estuaries
Yamazi, I., 1952?
Plankton investigations in inlet waters along the coast of Japan. VI. The plankton of Nanao Bay.
Seto Mar. Biol. Lab. Contr. 191:309-319, 11 textfigs.

estuaries
Yamazi, I., 1951.
Plankton investigations in inlet waters along the coast of Japan. II. The plankton of Hakodate Harbour and Yoichi Inlet in Hokkaido.
Publ. Seto Mar. Biol. Sta., Kyoto Univ., 1(4): 185-194, 3 textfigs.

estuaries
Yih, C.-S., 1953.
On tides in estuaries and around small islands.
Trans. A.G.U. 34(3):389-393.

estuaries
Zembrzuska, Donatylla, 1962
Szczecin Firth phytoplankton.
Prace Morsk. Inst. Ryback. Gdyni Oceanolog. Ichtiol. 11(A): 137-158.

estuaries
Zetler, B.D., 1953.
Some effects of the diversion of the Santee River on the waters of Charleston Harbor.
Trans. Amer. Geophys. Union 34(5):729-732, 2 textfigs.

estuaries
Zhuravleva, L.A., 1970.
Effect of atmospheric circulation on salinity distribution in the Dnieper-bug Estuary.
Gidrobiol. Zh., 6(6): 71-77. (In Russian; English abstract).

estuaries
Zmudzinski, L., and D. Szarejks, 1955.
Hydrographical-biological investigation on the Firth of Vistula. Prace Morskiego Inst. Ryback., Gydni, No. 8:283-312.

estuaries
ZoBell, Claude E., and Joseph F. Prokop, 1966.
Microbial oxidation of mineral oils in Barataria Bay bottom deposits.
Z. allg. Mikrobiol., 6(3):143-162.

estuaries
Zorell, F., 1933.
Beiträge zur Kenntniss der Alkalinität des Meerwassers. Ann. Hydrogr., usw., Jahrg. 61(Heft I/II):18-22.

estuaries
Zotin, M.I., 1962.
Peculiarity of opening of the lower reaches of the river and the river mouth (with an example of the River Enisei). (In Russian).
Trudy Gosud. Okeanogr. Inst., 66:26-54.

estuaries, biology of
Swingle, Hugh A. 1971.
Biology of Alabama estuarine areas - Cooperative Gulf of Mexico Estuarine Inventory.
Bull. Alabama mar. Resources 5: 123 pp.

estuaries circulation

estuaries, circulation in
Bowden, K.F. and R.M. Gilligan, 1971.
Characteristic features of estuarine circulation as represented in the Mersey Estuary. Limnol. Oceanogr. 16(3): 490-502.

estuaries, circulation in
Ekman, F.L., 1875.
Om de strömningår som uppsta i närheten af flodmynningar: ett bidrag till kännedomen af hafsströmmarnes natur. K. Vetenskapen Akad. Förhandl. 1875(7):43-135.

estuarine circulation
Vreugdenhil, C.B. 1970.
Two-layer model of stratified flow in an estuary.
Houille blanche 25 (1-70):35-40.

estuaries, ecology
MacNae, William, 1967.
Zonation within mangroves associated with estuaries in North Queensland.
In: Estuaries, G.H. Lauff, editor, Publs Am. Ass Advmt Sci., 83:432-441.

estuaries, ecology of
Vernberg, F. John 1967.
Some future problems in the physiological ecology of estuarine animals.
In: Estuaries, G.H. Lauff, editor, Publs Am. Ass. Advmt Sci. 83:554-557.

estuaries, effect of

estuaries, effect of
Rheinheimer, Gerhard, 1968.
Die Bedeutung des Elbe-Ästuars für die Abwasserbelastung der südlichen Nordsee in bakteriologischer Sicht.
Helgoländer wiss. Meeresunters. 17:445-454

estuaries, pollution of
Jones, D.J., 1971.
Ecological studies on macroinvertebrate populations associated with polluted kelp forests in the North Sea. Helgoländer wiss. Meeresunters 22(3/4): 417-441.

ethology

ethology
Reese, Ernst S., 1964.
Ethology and carine zoology.
In: Oceanography and Marine Biology, Harold Barnes, Editor, George Allen & Unwin, Ltd., 2: 455-488.

euphotic layer

euphotic layer
Bernard, Francis, 1967.
Research on phytoplankton and pelagic Protozoa in the Mediterranean Sea from 1953-1966.
Oceanogr. Mar. Biol., Ann.Rev., H. Barnes, editor, George Allen and Unwin, Ltd.,5:205-229.

euphotic zone
Dera, Jerzy 1970.
On two layers of different light conditions in the euphotic zone of the sea.
Acta geophys. polon. 18(3/4):287-294

euryhalinity

Euryhalinity
Berger, V. Ya., 1970.
Total loss of salts in distilled water and degree of euryhalinity of some water organisms.
Gidrobiol. Zh., 6(3): 67-73.
(In Russian; English abstract)

euryhalinity, evolutionary
Hutchinson, G.E., 1960
On evolutionary euryhalinity. Am. J. Sc., Bradley Vol., 258-A: 98-103.

eutrophication

eutrophication
Beyer, Fredrik 1972
Om Vannutvekslingen i Oslofjorden og dens Betydning for faunaen.
Rapp. Inst. mar. Biol. Avd. A+C, 3: 25 pp. (multilithed)

EUTROPHICATION
Blanc, F. et M. Leveau, 1970.
Effets de l'eutrophie et de la dessalure sur les populations phytoplanctoniques. Marine Biol., 5(4): 283-293.

eutrophication
Caperon, John, S. Allen Cattell, and George Krasnick, 1971.
Phytoplankton kinetics in a subtropical estuary: eutrophication. Limnol. Oceanogr. 16(4): 599-607.

eutrophication
Carpenter, J.H., D.W. Pritchard and R.C. Whaley, 1969.
Observations of eutrophication and nutrient cycles in some coastal plain estuaries.
In: Eutrophication; Causes, Consequences, Correctives; proceedings of a symposium, U.S. Nat. Acad. Sci., 210-221

eutrophication
Frontier, S., 1971.
Présentation de l'étude d'une baie eutrophique tropicale: la Baie D'Ambaro (côte nord-ouest de Madagascar). Cah ORSTOM, sér. Océanogr. 9(2) 147-148.

eutrophication
Ketchum, Bostwick H., 1969.
Eutrophication of estuaries.
In: Eutrophication: Causes, Consequences, Correctives, proceedings of a symposium, U.S. Nat. Acad. Sci., 197-209.

eutrophication
Piton, B. et Y. Magnier, 1971.
Les régimes hydrologiques de la baie d'Ambaro (nord-ouest de Madagascar). Contribution à l'étude d'une baie eutrophique tropicale.
Cah. ORSTOM, sér. Océanogr. 9(2): 149-166.

eutrophication
Reimold, Robert J., and Franklin C. Daiber, 1967
Eutrophication of estuarine areas by rainwater.
Chesapeake Sci., 8(2):132-133.

eutrophication
Skulberg, O.M., 1970.
The importance of algal cultures for the assessment of the eutrophication of the Oslofjord.
Helgolander wiss. Meeresunters, 20(1/4): 111-125.

eutrophication
Stirn, Joze 1971.
Ecological Consequences of marine pollution.
Rev. int. Océanogr. Biol. med. 24:13-46.

eutrophication
United States, National Academy of Sciences, 1969.
Eutrophication: Causes, Consequences, correctives; proceedings of a symposium, 661 pp.

Evaporation

A

evaporation
Aliverti, Giuseppina, Arturo de Maio and Mario Picorri, 1959.
Sulla evaporazione annua dal Tirreno Meridionale Consiglio Nazionale Delle Ricerche, Commissione Nazionale Italiana per l'Anno Geofisico Internazionale, 1957-58: 3-6.

evaporation
Anadón, E., F. Saiz and M. López-Benito, 1961.
Estudio hidrografico de la Ria de Vigo.
Inv. Pesq., Barcelona, 20:83-130.

evaporation
Arsen'eva, N. Ia., 1962.
Comparison of the results of the computation of evaporation of shallow sea with instrumental measurements. (In Russian).
Trudy Gosud. Okeanogr. Inst., 70:34-38.

evaporation
Audouin, Jacques, 1962.
Hydrologie de l'Étang de Thau.
Rev. Trav. Inst. Pêches Marit., 26(1):5-104.

B

evaporation
Barber, F.G., 1967.
A contribution to the oceanography of Hudson Bay.
Manuscript Rep.Ser. Dept.Energy,Mines,Resources, Can., No. 4:69 pp. (multilithed).

evaporation
Bogorodsky, M.M., 1964.
A study of tangential friction, vertical turbulent heat exchange and evaporation in the open ocean. (In Russian).
Okeanologiia, Akad. Nauk, SSSR, 4(1):19-26.

evaporation
Boguslavskii, S.G., 1961.
Latitudinal change in the heat balance of the Atlantic Ocean. Physics of the sea. (In Russian)
Trudy Morsk. Gidrofiz. Inst., 23:139-147.
Translation:
Scripta Technica, Inc. for Amer. Geophys. Union, 111-117.

evaporation
Bossolasco, M., 1951.
Zur Frage der Verdunstung auf dem Meere.
Arch. Met. Geophys. u. Bioklim., Ser. B, 3: 40-46.

evaporation
Bowden, K.F., 1950.
Processes affecting the salinity of the Irish Sea
Monthly Notices, R. Astron. Soc., Geophys. Suppl. 6(2):63-90, 7 textfigs.

evaporation
Bøyum, Gunnvald, 1966.
The energy exchange between sea and atmosphere at Ocean Weather Stations M, I and A.
Geofys. Publr, 26(7):19 pp.

evaporation
Bøyum, Gunnvald, 1962
A study of evaporation and heat exchange between the sea surface and the atmosphere.
Geofys. Pbblikasjoner, Oslo, 22(7):15 pp.

evaporation
Brocks, K., 1955.
Wasserdampferschichtung über Meer und "Rauhigkeit" der Meeresoberfläche.
Arch. Met., Geophys., Bioklim., A, 8(4):354-383, 8 textfigs.

evaporation
Brouquet-Leglaire, 1950.
Note sur les possibilités d'extraction du sal marin sur littorql de la Cote d'Ivoire. C.R. Première Conf. Int. des Africanistes de l'Ouest, 1:68-78.

evaporation
Budiko, M.I., editor, 1963.
Atlas of heat balance of the world. (In Russian).
Mezhd. Geofiz. Komitet, Prezidiume Akad. Nauk SSSR, Glabnaia Geofiz. Observ. im. I.A. Borikova, 69 colored charts.

C

evaporation
California Academy of Sciences
California Division of Fish and Game
Scripps Institution of Oceanography
U. S. Fish and Wildlife Service } 1950
California Cooperative Sardine Research Program.
Progress Rept. 1950:54 pp., 37 text figs.

evaporation
Chernous'ko, Yu.L. and A.V. Shumilov 1971.
Evaporation and microconvection in the thin surface layer. (In Russian; English abstract).
Okeanologiia 11(6): 982-986.

evaporation
Copeland, B.J., 1967
Environmental characteristics of hypersaline lagoons.
Contr. mar. Sci., Port Aransas, 12: 207-218

evaporation
Craig, R. A., and R. B. Montgomery, 1949.
Evaporation (measured) from ocean into hydrostatically stable air. J. Met. 6(6):426-427.

D

evaporation
Deardorff, James W., 1968.
Dependence of air-sea transfer coefficients on bulk stability.
J. geophys. Res., 73(8):2549-2557.

evaporation
Deardorff, James W., 1961
Local evaporation from a smooth water surface.
J. Geophys. Res., 66(2): 529-534.

evaporation
Deuser, Werner G., and Egon T. Degens, 1969.
O^{18}/O^{16} and C^{13}/C^{12} ratios of fossils from the hot-brine deep area of the central Red Sea.
In: Hot brines and Recent heavy metal deposits in the Red Sea, E.T. Degens and D.A. Ross, editors, Springer-Verlag, New York, Inc., 336-347.

evaporation
Dieulafait, 1883.
Evaporation des eaux marins et des eaux douces dans la delta du Rhône et a Constantine.
C.R. Acad. Sci., Paris, 97:500-502.

evaporation
Dodimead, A.J., F. Favorite and T. Hirano, 1964.
Review of oceanography of the subarctic Pacific Region. Salmon of the North Pacific Ocean. II. Collected Reprints, Tokai Reg. Fish. Res. Lab., No. 2:187 pp. (In Japanese).
Pacific, northeast
Pacific, northern north
bathymetry
evaporation
temperature, surface
currents
salinity
heat budget
oxygen
thermocline (interesting diagrams)
halocline
Gulf of Alaska
Bering Sea
currents, dynamic
drift bottles
volume transport

evaporation
Dodimead, A.J., and J.P. Tully, 1958.
Canadian Oceanographic Research in the northeast Pacific Ocean.
Proc. Ninth Pacific Sci. Congr., Pacific Sci. Assoc., 1957, Oceanogr., 16:180-195.

E

evaporation
Emilsson, Ingvar 1968.
Investigaciones sobre la hidrologia en la ensenada de la Broa con vista a su posible transformación en un embalse de agua dulce.
Ser. Transformación Naturaleza, Acad. Cienc. Cuba, Inst. Oceanogr. 5:1-45.

evaporation
Emon, J., 1957.
L'évaporation au Maroc. Ann. Serv. Phys. Globe, Météorol., 17:87-107.

F

evaporation
Finkelstein, J., 1961
Estimation of open water evaporation in New Zealand.
New Z.J. Science, 4(3): 506-522.

evaporation
Fontes, Jean-Charles, 1966.
Intérêt en géologie d'une étude isotopique de l'evaporation. Cas de l'eau de mer.
C.r.hebd. Séanc., Acad.Sci., Paris (D)263(25): 1950-1953.

evaporation
*Forsbergh, Eric D., 1969.
On the climatology, oceanography and fisheries of the Panama Bight. (In Spanish and English).
Bull. Inter-Am. Trop. Tuna Comm. 14(2): 385 pp.

G

evaporation
Garrett, William D., 1971.
A novel approach to evaporation control with monomolecular films. J. geophys. Res. 76(21): 5122-5123.

H

evaporation
Hanzawa, M., 1950.
On the annual variation of evaporation from the sea-surface in the North Pacific Ocean. Ocean. Mag., Tokyo, 2(2):77-82, 3 textfigs.

evaporation
Harbeck, G.E., jr., 1955.
The effect of salinity on evaporation.
Geol. Survey Prof. Paper, 272-A:1-6, 2 figs.

evaporation
Hasse, Lutz, 1964.
On the cooling of the sea surface by evaporation and heat exchange.
Tellus, 15(4):363-366.

evaporation
Hatcher, R.W., and J.S. Sawyer, 1948
Sea breeze structure with particular reference to temperature and water vapour gradients and associated radio-ducts. Q.J. Roy. Met. Soc. 74(319):117-118.

evaporation
Hishida, Kozo, and Katsunobu Nishiyama, 1969.
On the variation of heat exchange and evaporation at the sea surface in the western North Pacific Ocean.
J. oceanogr. Soc. Japan, 25 (1):1-9.

I

evaporation
Ivanov, Yu. A., 1967.
On the connection between the processes of heat exchange on the ocean surface and the precipitation-evaporation balance. (In Russian; English abstract).
Fisika Atmosfer. Okean., Izv. Akad. Nauk, SSSR, 3(7):757-763.

evaporation
Izotova, A.F., 1962.
Evaporation from the open surface of Ladoga Lake.
Doklady Akad. Nauk, SSSR, 146(5):1143-1146.

In Russian

J

evaporation
Jakovlev, G.N., 1959
[A thermal stream of evaporation from the surface of ice cover in Central Arctic] Problemi Arktiki i Antarktiki (1): 59-64.

K

evaporation
Kaczmarek, Zdzislaw, 1960
[The diurnal course of evaporation from a water surface.]
Przeglad Geofiz., Polskie Towarzystwo Meteorol. i Hydrol., 13(4): 265-273.

Evaporation
Kagan, B. A., 1961. ref.
[The turbulent thermal exchange between the sea surface and the atmosphere and the loss of heat by evaporation in the Arctic seas.]
Probl. Arkt. i Antarkt., (8):78-84.

evaporation
Kitaigorodsky, S.A., and Yu. A. Volkov, 1965.
On the calculation of turbulent heat and humidity fluxes in the near-water layer of the atmosphere. (In Russian;English abstract).
Fizika Atmosferi i Okeana, Izv., Akad. Nauk, SSSR 1 (12):1319-1336.

evaporation
Kochikov, N.N., 1961.
Evaporation from the surface of the Black Sea.
Rapp. Proc. Verb., Réunions, Comm. Int. Expl. Sci., Mer Méditerranée, Monaco, 16(3):639-642.

evaporation
Koizumi, M., 1956.
Researches on the variations of oceanographic conditions in the region of "Ocean Weather Station "Extra" in the North Pacific Ocean. III. The variation of hydrographic conditions discussed from the heat balance point of view and the heat exchanges between sea and atmosphere.
Pap. Meteorol. & Geophys., Tokyo, 6(3/4):273-284.

evaporation, effect of
Konaga, S., 1960
[Water temperature at the sea surface (III). An influence of the weather condition.]
J. Oceanogr. Soc., Japan, 16(3): 128-133.

evaporation
Kondo, Junsei, 1969.
Summary on problems in the evaporation from the water surface. (In Japanese)
Umi to Sora 45 (2/3): 49-75.

evaporation
Kondo, Junsei, 1964.
Evaporation from the Japan Sea in winter.
Sci. Repts., Tohoku Univ., Geophys., (5), 15(2):67-75.

evaporation
Kondo, Junsei, 1962
Evaporation from extensive surfaces of water.
Sci. Repts., Tohoku Univ., (5. Geophys.), 14(3):107-119.

evaporation
Korzh, V.D., 1971.
Computation of relationships between chemical components of the sea water transported from the ocean to the atmosphere through evaporation. (In Russian; English abstract). Okeanologiia 11(5): 881-888.

evaporation
Kraus, E.B., 1959.
The evaporation-precipitation cycle of the trades.
Tellus, 11(2):147-158.

L

evaporation
Laevastu, T., 1960
Factors affecting the temperature of the surface layer of the sea.
Merent. Julk. (Havsforskningsinst. Skr.), No. 195: 136 pp.

evaporation, effect of
LaFond, E.C., 1954.
Environmental factors effecting the vertical temperature structure of the upper layers of the sea. Andhra Univ, Ocean. Mem. 1:94-101, 25 text-figs.

evaporation
Lapworth, C. F., 1947.
Evaporation from a water surface. Water 50:288-291.

evaporation
Latham, J., and C.D. Stow, 1965.
Electrification associated with the evaporation of ice.
J. Atmos. Sci., 22(3):320-324.

evaporation, effect of
Lloyd, R.M. 1966.
Oxygen isotope enrichment of sea water by evaporation.
Geochim. cosmochim. Acta 30(8): 801-814.

M

evaporation
Machta, L., 1969.
Evaporation rates based on tritium measurements for Hurricane Betsy.
Tellus 21(3): 404-408

evaporation
Magin, G.B., Jr., and L.E. Randall, 1960.
Review of literature on evaporation suppression.
Geol. Surv. Prof. Pap., 272-C:53-69.

evaporation
Malkus, Joanne S., 1962
Large-scale interactions. Ch. 4, Sect. II. Interchange of properties between sea and air. In: The Sea, Interscience Publishers, Vol. 1, Physical Oceanography, 88-294.

evaporation(data)
Margalef, Ramón, 1961
Hidrografía y fitoplancton de un área marina de la costa meridional de Puerto Rico.
Inv. Pesq., Barcelona, 18:38-96.

evaporation
Markgraf, H., 1962
Zum Jahresgang der relative Feuchte uber dem Mittelmeer.
Geofisica Pura e Applicata, 52(2):229-230.

English and German summaries.

Mc

evaporation
Miyazaki, M., 1952.
The heat budget of the Japan Sea. Bull. Hokkaido Regional Fish. Res. Lab., Fish. Agency, No. 4: 1-54, 45 charts

evaporation
Mosby, Hakon, 1962
Water, salt and heat balance of the North Polar Sea and of the Norwegian Sea.
Geofys. Publik., Geophysica Norvegica, 24: 289-313.

Mosby, Hakon, 1960 evaporation
The evaporation from the oceans. Introduction to symposium on the water balance of the earth. Proc., AIOP, UGGI, Helsinki, 1960: 81-101.

evaporation
Myers, D.M., and C.W. Bonythorn, 1958.
The theory of recovering salt from sea water by solar evaporation. J. Appl. Chem., 8(4):207-218.

N

evaporation
Neumann, J., 1952.
Evaporation from the Red Sea. Israel Expl. J. 2(3):153-162.

evaporation

Neumann, J., and N. Rosenan, 1954.
The Black Sea: energy balance and evaporation.
Trans. Amer. Geophys. Union, 35(5):767-774,
1 textfig.

evaporation

Norris, R., 1948.
Evaporation from extensive surfaces of water
roughened by waves. Q. J. R. Met. Soc., 74:1-12.

evaporation

Norris, R., 1948
Evaporation from extensive surfaces of
water roughened by waves. Q.J. Roy. Met.
Soc. 74(319):1-12.

O

evaporation

Okuda, Setsuo, and Shoitiro Hayami, 1959

Experiments on evaporation from wavy water surface.
Rec. Ocean. Wks., Japan, n.s., 5(1):6-13.

evaporation

Olson, Boyd E., 1962
Variations in radiant energy and related
ocean temperatures.
J. Geophys. Res., 67(12):4705-4712.

evaporation

Oren, O.H , 1962
A note on the hydrography of the Gulf of
Eylath Contributions to the knowledge of
the Red Sea No. 21.
Sea Fish. Res. Sta., Haifa, Israel, Bull. No.
30: 3-14.

evaporation

Osborne, M.F.M., 1964.
The interpretation of infrared radiation from the
sea in terms of its boundary layer.
Deutsche Hydrogr. Zeits., 17(3):115-136.

evaporation

*Östlund, H. Göte, 1970.
Hurricane tritium III: Evaporation of sea water
in hurricane Faith 1966. J. geophys. Res.,
75(12): 2303-2309.

P

evaporation

Palmén E. and D. Söderman 1966
Computation of the evaporation from
the Baltic Sea from the flux of water
vapor in the atmosphere.
Geophysica, Helsinki 8 (4): 261-279.

evaporation

Panara, R., 1959.
Annotated bibliography on evaporation measurements.
Meteorol. Abstr. & Bibliogr., 10(8):1234-1262.

evaporation

Ponomarenko, G.P., 1959.
[On the question of evaporation from the Caspian
Sea.] Trudy Inst. Okeanol., 3:74-89.

evaporation

Privett, D. W., 1960.
The exchange of energy between the atmosphere and
the oceans of the southern hemisphere. Meteorol.
Off., London, Geophys. Mem., 104(M.O. 631d) 61 pp.

evaporation

Privett, D.W., 1959

Monthly charts of evaporation from the N.
Indian Ocean (including the Red Sea and the
Persion Gulf).
Q.J.R. Meteorol. Soc., 85(366): 424-428.

evaporation

Prusenkov, A.S., 1962
Evaporation from the surface of the Black
Sea.
Okeanologiia, Akad. Nauk, SSSR, 2(1):51-58.

Abstracted in: Soviet Bloc Res., Geophys.
Astron. and Space, 1962(35):
16-17.
(OTS61-11147-35 JPRS13739)

Q

R

evaporation

Raczmarek, Z., 1960

The diurnal course of evaporation from a water
surface.
Przeglad Geofiz., 5(4): 265-274.

evaporation

Ramanadham R., and A.V.S. Murty 1970
Studies of evaporation from the sea at Waltair.
Pure appl. Geophys. 79(2):95-102.

evaporation

Ramdas, L. A., and P. K. Raman, 1946.
A method of estimation of the thickness of the
laminar layer above an evaporating water surface.
Publ. Indian Acad. Sci., A, 23:127-133.

evaporation

Reshetova, O.V., 1969.
On the heat transfer and evaporation over the
ocean. (In Russian). Fizika Atmosfer. Okean.,
Izv. Akad Nauk, SSSR, 5(12): 1318-1323.

evaporation

Roden, G.I., 1958.
Oceanographic and meteorological aspects of the
Gulf of California. Pacific Science 12(1):21-45.

evaporation

Roden, G.I., and G.W. Groves, 1959
Recent oceanographic investigations in the
Gulf of California. J. Mar. Res., 18(1):
10-35.

evaporation

Roll, U., 1949.
Über eine scheinbar Anomalie der Temperatur-
differenz Luft-Wasser. Deut. Hydro. Zeit. 2(4):
134-137, 4 textfigs.

S

evaporation

Saiz, Fernando, Manuel López-Benito y Emilio
Anadón, 1957
Estudio hidrográfico de la Ria de Vigo.
Inv. Pesq., Barcelona, 8:29-88.

evaporation

Sakane, N., 1950.
[Evaporation from the sea water in the vessel.]
Bull. Maizuru Mar. Obs., (1):12-13.

evaporation

Schmitz, H.P., 1963.
Erweiterte Grungleichungen zur Bestimmung der
Verdunstung auf dem Meer aus dem klein- und
grossräumigen Feuchteaustausch.
Deutsche Hydrogr. Zeits., 16(3):105-136.

evaporation

Schmuck, Adam, 1960

[Evaporation from free water surface on the
lower Silesian lowland and in the Sudety
Mountains.]
Przeglad Geofiz., Polski Towarzystwo Meteorol
I Hydrol., 13(3): 183-197.

evaporation

Schooley Allen H. 1969
Evaporation in the laboratory and at
sea.
J. mar. Res. 27 (3): 335-338.

Evaporation

Seelkopf, Carl, and Luis Boscan F., 1960

Hydrochemische Untersuchungen im Maracaibo-See
Deutsche Hydrogr. Zeits., 13(4): 174-180.

evaporation

Sheremetevskaya, O.I., 1964.
Evaporation from the ocean surface in the zone between
40° and 60° S. Lat. (In Russian).
Materiali Vtoroi Konferentsii, Vzaimod. Atmosfer. i
Gidrosfer. v Severn. Atlant. Okean., Mezhd. Geofiz. God,
Leningrad. Gidrometeorol. Inst., 211-221.

evaporation

Siedler, G., 1969.
General circulation of water masses in the Red
Sea.
In: Hot brines and Recent heavy metal deposits in
the Red Sea, E.T.Degens and D.A.Ross, editors,
Springer-Verlag, New York, Inc., 131-137.

evaporation

Simojoki, H., 1949.
Niederschlag und Verdunstung auf dem Baltischen
Meer. Fennia 71(1):1-25.

evaporation

Simojoki, Heikki, 1948.
On the evaporation from the northern Baltic.
Geophys., Geophys. Soc., Finland, 3:123-126.

evaporation

Slaucitajs, L., 1947
Ozeanographie des Rigaischen Meerbusens. Teil 1, Statik. Contrib. Baltic
Univ. No.45, Pinneberg, 110 pp., 69 text
figs.

evaporation
Snopkov, V.G., 1964.
Contact heat exchange and rate of evaporation as determined in the North Indian Ocean during the winter monsoon period. Investigations in the Indian Ocean (33rd voyage of E/S "Vitiaz"). (In Russian).
Trudy Inst. Okeanol., Akad. Nauk, SSSR, 64:5-10.

evaporation
Soliankin, E.B., 1962
The vapor turnover in the Black Sea.
Okeanologiia, Akad. Nauk, SSSR, 2(2):238-250.

evaporation, effect of
Spangenberg, W.G., & W.R. Rowland, 1961
Convective circulation in water induced by evaporative cooling.
Physics of Fluids, 4(6): 743-750.

evaporation
Stefansson, Unnstein, Baldur Lindal, Johan Jakobsson and Isleifur Jonsson, 1961.
The salinity at the shores of southwest Iceland.
Rit Fiskideildar, Atvinnudeild Haskolans-Fiskideild, 2(9):1-26.

evaporation
Strokina, L.A., 1956.
Turbulent heat exchange with the atmosphere and evaporation from the surface of the Baltic Sea.
Meteorol. i Gidrol., 5:56-60.
Abstr. in: Meteorol. Abstr. & Bibl., 9(1):22.

evaporation
Sugden, W., 1963.
The hydrology of the Persian Gulf and its significance in respect to evaporite deposition.
Amer. J. Sci., 261(8):741-755.

evaporation
Sverdrup, H.U., 1943
On the ratio between heat conduction from the sea surface and heat used for evaporation. Ann. N.Y. Acad. Sci., 44: 81-88.

evaporation
Sverdrup, H.U., 1940
On the annual and diurnal variation of the evaporation from the ocean. J. Mar. Res.3: 93-104.

evaporation
Sverdrup, H.U., 1937
On the evaporation from the oceans. J. Mar. Res. 1:3-14.

T

evaporation (data)
Takahashi, T., 1958.
Micro-meteorological observations and studies over the sea. Mem. Fac. Fish., Kagoshima Univ., 6:1-46.

evaporation
Tchernia, Paul, 1960.
Hydrologie d'hiver en Mediterranée Occidentale.
Cahiers Oceanogr. C.C.O.E.C., 12(3):184-198.

evaporation
Timofeyev, M.P., 1958.
The heat balance of water bodies and methods for determining evaporation (from them).
Sovremennyye Problemy Meteorologii Prizemnogo Sloya Vozdukha:43-60.
(A symposium of articles edited by M.I. Budkoy Leningrad)

LC and SLA, mi $2.70, pl $4.50

evaporation
Timonov, V.V., A.I. Smirnova and K.I. Nepop, 1970.
On the centers of air-sea interaction in the north Atlantic. Okeanologiia, 10(5): 745-749.
(In Russian; English abstract)

evaporation
Tixeront, Jean 1970.
Le bilan hydrologique de la mer Noire et de la Méditerranée.
Cah. oceanogr. 22(3): 227-237

evaporation
Tsuchiya, Kiyoshi, and Tetsuya Fujita, 1967.
A satellite study of evaporation and cloud formation over the western Pacific under the influence of the winter monsoon.
J. Met. Soc., Japan, 45(3):232-250.

evaporation
Tsuji, M., 1950.
On the rate of evaporation and condensation of falling drops. Geophys. Mag., Japan, 22(1):11-14, 1 textfig.

evaporation
Tully, John P., 1964.
Oceanographic regions and assessment of temperature structure in the seasonal zone of the North Pacific.
J. Fish. Res. Bd., Canada, 21(5):941-970.

evaporation
Tully, J.P., 1964.
Oceanographic regions and processes in the seasonal zone of the North Pacific Ocean.
In: Studies on Oceanography dedicated to Professor Hidaka in commemoration of this sixtieth birthday, 68-84.

evaporation
Turbovich, L.T., 1947.
Turbulent diffusion and its function in the evaporation from the surface of the ocean.
Zh. Eks. Teor. Fiz. 17(4):6 pp.

evaporation
Turner, J.S., 1965.
Laboratory models of evaporation and condensation
Weather, Roy. Meteorol. Soc., 20(4):124-128, Figs on pp. 120-121.

U

evaporation
Uda, Michitake, 1963.
Oceanography of the subarctic Pacific Ocean.
J. Fish. Res. Bd., Canada, 20(1):119-197.

evaporation
Ulanov, Kh. K., 1960
Loss of water by evaporation from the surface of the Caspian Sea.
Doklady, Akad. Nauk, SSSR, 135(3): 584-586.

V

evaporation
Valyashko, M.G., 1951.
Volume relations of the liquid and solid phases in the evaporation process of ocean water as a determining factor in the formation of layers of potassium salts. Doklady Akad. Nauk, SSSR, 77(6): 1055.

evaporation
Vaux, D., 1953.
Hydrographical conditions in the southern North Sea during the cold winter of 1946-1947.
J. du Cons. 19(2):127-149, 18 textfigs.

evaporation
Venkateswaran, S.V., 1956.
On evaporation from the Indian Ocean.
Indian J. Meteorol., & Geophys., 7(3):551-573.

evaporation
Vives, Francisco, y Manuel López-Benito, 1957
El fitoplancton de la Ría de Vigo desde julio de 1955 a junio de 1956.
Inv. Pesq., Barcelona, 10:45-146.

evaporation
Vladimirov, O.A., 1960.
Computation of increase in the area of the ocean under the influence of wave action.
Meteorol. i Gidrol., (10):36-37.

evaporation
Vowinckel, E., and Bea Taylor, 1966.
Energy balance of the Arctic. IV Evaporation and sensible heat flux over the arctic Ocean.
Arch. Meteorol., Geophys O Bioklimatol. (B), 14:36-52.

W

evaporation
Webb, E.K., 1960
Evaporation from Lake Eucumbene.
C.S.I.R.O. Australia, Div. Meteor. Phys., Tech. Pap. 10: 75 pp.

evaporation
Webb, E.K., 1960.
On estimating evaporation with fluctuating Bowen ratio.
J. Geophys. Res., 65(10):3415-3418.

evaporation
Wüst, G., 1954.
Gesetzmässige Wechselbeziehungen zwischen Ozean und Atmosphäre in der zonalen Verteilung von Oberflächen - Salzgehalt, Verdunstung und Niederslag. Arch. Met., Geophys., u. Bioklim., A, 7: 305-328, 5 textfigs.

evaporation
Wüst, G., 1951.
Der Wasserhaushalt des mittelländischen Meeres und der Ostsee in vergleichender Betrachtung. Atti Convegno Internazionale Meteorol. Marittima, Genova, 20-22 - IX -1951 (Rivista Geofisica Pura e Applicada 21(1952):14 pp., 5 textfigs.

evaporation
Wüst, G., 1936.
Surface salinity, evaporation and rainfall over the oceans.
Länderk. Forsch., Festschr. Norbert Krebs:347-359.
Transl. cited:
USFWS Spec. Sci. Rept., Fish., 227.

evaporation
Wüst, G., W. Brogmus and E. Noodt, 1954.
Die zonale Verteilung von Salzgehalt, Niederschlag, Verdunstung, Temperatur und Dichte an der Oberfläche der Ozeans. Kieler Meeresf. 10(2): 137-161, 10 textfigs.

evaporation
Wüst, Georg (with Arnold Gordon), 1964
Stratification and circulation in the Antillean-Caribbean basins. 1. Spreading and mixing of the water types with an oceanographic atlas. Vema Research Series, Columbia Univ. Press, New York and London, No. 2:201 pp.

evaporation
Wyrtki, Klaus, 1958
Precipitation, evaporation and energy exchange at the surface of the southeast Asian waters. (Abstract).
Proc. Ninth Pacific Sci. Congr., Pacific Sci. Assoc., 1957, 16(Oceanogr.):179.
Published in:
Penjelidikan Laut Di Indonesia (Mar. Res., Indonesia),No. 3:7-40, Djakarta, 1957.

evaporation
Wyrtki, K., 1957.
Precipitation, evaporation and energy exchange at the surface of the southeast Asian waters.
Penjelidikan Laut, Indonesia (Mar. Res.), No. 3: 7-40.

XYZ

evaporation
Zubov, N.N., 1955.
[Index of atmospheric turbulence; the interaction of sea and atmosphere; some features of evaporation in the sea.]
Izbrannye Trudy po Okeanologii, Moscow, Ch. 8, 12, 28, 171-176, 204-214, 443-456.

evaporation
Zurick, S.A., 1960.
Note on evaporation.
J. Appl. Physics, 31(10):1735-1741.

evaporation, effect of

evaporation, effect of
Hsü, K. Jinghwa, and Christoph Siegenthaler, 1969.
Preliminary experiments on hydrodynamic movement induced by evaporation and their bearing on the dolomite problem.
Sedimentology 12 (1/2):11-25

evaporation, effect of
Ledenev, V.G., 1965.
Influence of evaporation on the formation of cold Antarctic water. (In Russian).
Sovetsk. Antarkt. Eksped., Inform. Biull., 44:35

Translation:
Scripta Tecnica, Inc., for AGU, 5(1):50-52. 1965

evaporation, effect of
Paulson, C.A. and T.W. Parker, 1972.
The cooling of a water surface by evaporation, radiation, and heat transfer. J. geophys. Res. 77(3): 491-495.

evaporites

evaporites
Bonatti, E., M. Ball and Carl Schubert 1970
Evaporites and continental drift.
Naturwissenschaften 57(3):107-108.

evaporites
Brongersma-Sanders Margaretha 1971.
Origin of major cyclicity of evaporites and bituminous rocks: an actualistic model.
Mar. Geol. 11 (2): 123-144

evaporites
Brunstrom, R.G.W. and Peter J. Walmsley, 1969
Permian evaporites in North Sea Basin.
Bull. Am. Ass. Petr. Geol., 53(4): 870-883

evaporites
Cramer, Harvey Ross 1969.
Evaporites, a selected bibliography.
Bull. Am. Ass. Petr. Geol. 53(4): 982-1011.

evaporites
Hassan, Fekri, and Sayed El-Dashlouty, 1970.
Miocene evaporites of Gulf of Suez region and their significance.
Bull. Am. Ass. Petrol. Geol. 54(9): 1686-1696

evaporites
Irwin, M.L., 1965.
General theory of epeiric clear water sedimentation.
Bull. Amer. Assoc. Petr. Geol., 49(4):445-459.

evaporites
Kinsman David J.J., 1968.
Modes of formation, sedimentary association and diagnostic features of shallow-water and supratidal evaporites.
Bull. Am. Ass. Petr. Geol., 53(4): 830-840.

evaporites
Pannekoek, A.J., 1965.
Shallow-water and deep-water evaporite deposition.
Amer. J. Sci., 263(3):284-285.

evaporites
Phleger, Fred B, 1969.
A modern evaporite deposit in Mexico.
Bull. Am. Ass. Petr. Geol. 53(4): 824-829.

evaporites
Schmalz, Robert F., 1969.
Deep-water evaporite deposition: a genetic model.
Bull. Am. Ass. Petr. Geol., 53(4): 798-823.

evaporites
Smith Denys B., 1970.
Permian evaporites in North Sea Basin: discussion.
Bull. Am. Ass. Petrol. Geol., 54(4): 662-664

evaporites
Sugden, W., 1963.
The hydrology of the Persian Gulf and its significance in respect to evaporite deposition.
Amer. J. Sci., 261:741-755.

evaporite formation
Raup, Omer B., 1970.
Brine mixing: an additional mechanism for formation of basin evaporites.
Bull. Am. Ass. Petrol. Geol. 54(12): 2246-2259.

Evolution

evolution, geological
Antoine, John, 1968.
The structure, sediments and possible evolution of the Gulf of Mexico. Trans. Fifth Carib. Geol. Conf., St. Thomas, V.I., 1-5 July, 1968, Peter H. Mattson, editor. Geol. Bull. Queens Coll., Flushing 5: 8. (Abstract only).

evolution
Berggren, William A., 1969.
Rates of evolution in some Cenozoic planktonic Foraminifera.
Micropaleontology, 15(3): 351-365

evolution
Chapman, D.M., 1966.
Evolution of the scyphistoma.
Symposia, Zool. Soc., London, Academic Press, 16: 51-75.

evolution
Ma, Ting Ying H., 1958.
The relation of growth rate of reef corals to surface temperature of sea water as basis for study of causes of diastrophisms instigating evolution of life.
Res. on Past Climate and Continental Drift 14: 60 pp.

evolution
May, Raoul-Michel, 1965.
La contribution de l'embryologie des animaux marins à la théorie de l'évolution au XIXe siècle
Colloque Internat., Hist. Biol. Mar., Banyuls-sur-Mer, 2-6 sept., 1963, Suppl., Vie et Milieu, No. 19:239-257.

evolution
Millar, R.H., 1966.
Evolution in ascidians.
In: Some contemporary studies in marine science, H. Barnes, editor, George Allen & Unwin, Ltd., 519-534.

evolution
Rees, W.J., 1966.
The evolution of the Hydrozoa.
Symposia, Zool. Soc., London, Academic Press, 16: 199-222.

evolution, arthropod
Sharov, A.G., 1965.
Origins and main stages of the arthropod evolution. 1. From annelids to arthropods. (In Russian; English abstract).
Zool. Zhurn., Akad. Nauk, SSSR, 44(6):803-817.

evolution
Thiel, Hjalmar, 1966.
The evolution of the Scyphozoa: A review.
Symposia, Zool. Soc. London, Academic Press, 16: 77-117.

evolution
Webb, Michael, 1969.
An evolutionary concept of some sessile and tubiculous animals.
Sarsia 38: 1-8.

excavation nuclear

excavation, nuclear
Gilead, Aviva E., and Wilson K. Talley, 1969.
Nuclear excavation of an Elat-Dead Sea waterway.
J. Waterways Harb. Div. Am. Soc. civ. Engrs. 95(WW3): 329-335

exchange

exchange, chemical

exchange, chemical
Pirogova, M.V., 1953.
[Chemical exchange between the bottom of the Black Sea and the overlying water]
Gidrokhim. Materialy 21:10-18.

Abstr. Chem. Abstr. 11126f, 1954.

exchange, data

Exchange of data (resolution)
India, Indian National Committee on Oceanic Research, Council of Scientific and Industrial Research, New Delhi, 1964.
International Indian Ocean Expedition, Newsletter India, 2(4):48 pp.

exchange, energy
Malkus, Joanne S., 1962
Large-scale interactions. Ch. 4, Sect. II. Interchange of properties between sea and air. IN: The Sea, Interscience Publishers, Vol. 1, Physical Oceanography, 88-294.

exchange rate

exchange rate
Foster, R.F., 1959
Radioactive tracing of the movement of an essential element through an aquatic community with specific reference to radiophosphorus.
Pubb. Sta. Zool. Nap., 31(Suppl): 34-69.

exchange, water

exchange, water
Fukuo, Yoshiaki, 1960
On the exchange of water and the productivity of a bay with special reference to Tanabe Bay (LII). Rec. Oceanogr. Wks., Japan, Spec. No. 4: 29-38.

excretion

excretion

excretion
Bahl, K.N., 1947
Excretion in the Oligochaeta. Biol. Rev., 22(2):109-147, 28 text figs.

excretion
Butler, E.I., E.D.S. Corner and S.M. Marshall, 1969.
On the nutrition and metabolism of zooplankton. VI. Feeding efficiency of Calanus in terms of nitrogen and phosphorus. J. mar. biol. Ass. U.K., 49(4): 977-1001.

excretion
Chapman, G. and A.C. Rae, 1969.
Excretion of photosynthate by a benthic diatom.
Marine Biol., 3(4): 341-351.

excretion
Martin, John H., 1968.
Phytoplankton-zooplankton relationships in Narragansett Bay. III. Seasonal changes in zooplankton excretion rates in relation to phytoplankton abundance.
Limnol. Oceanogr., 13(1):63-71.

excretion
*Tokuda, Hiroshi, 1969.
Excretion of carbohydrate by a marine pennate diatom, Nitzschia closterium.
Rec. oceanogr. Wks. Japan, 10(1):109-122.

excretion, nutrients

excretion, nitrogen
Conover, R.J., and E.D.S. Corner, 1968.
Respiration and nitrogen excretion by some marine zooplankton in relation to their life cycles.
J. mar. biol. Ass., U.K., 48(1):49-75.

Excretion (nitrogen)
Corner, E.D.S., C.B. Cowey, and S.M. Marshall, 1965.
On the nutrition and metabolism of zooplankton. III. Nitrogen excretion by Calanus.
Jour. Mar. Biol. Assoc., U.K., 45(2):429-442.

excretion, phytoplankton
Hellebust, J.A., 1965.
Excretion of some organic compounds by marine phytoplankton.
Limnology and Oceanography, 10(2):192-206.

excretions
Jawed, Mohammad, 1969.
Body nitrogen and nitrogenous excretion in Neomysis rayii Murdoch and Euphausia pacifica Hansen. Limnol. Oceanogr., 14(5): 748-754.

excretion
Kuenzler, Edward J. 1970.
Dissolved organic phosphorus excretion by marine phytoplankton.
J. Phycol. 6(1): 7-13.

excretion, nutrients
Martin, John H., 1965.
Phytoplankton-zooplankton relationships in Narragansett Bay.
Limnology and Oceanography, 10(2):185-191.

excretion
Oguri, Mikio, Naoko Takada and Ryushi Ichikawa, 1965.
Metabolism of radionuclides in fish. IV. Strontium-Calcium discrimination in the renal excretion of fish.
Bull. Jap. Soc. Sci. Fish., 31(6):435-438.

excretion, phosphorus
Satomi, Masako, and Lawrence R. Pomeroy, 1965.
Respiration and phosphorus excretion in some marine populations.
Ecology, 46(6):877-881.

excretion, nutrient
Whitledge, Terry E. and Theodore T. Packard, 1971.
Nutrient excretion by anchovies and zooplankton in Pacific upwelling regions. Investigación pesq. 35(1): 243-250.

excretion, amino acid

excretion, amino acid
Webb, K.L., and R.E. Johannes, 1965.
Dissolved amino acid excretion by zooplankton. (abstract).
Ocean Sci. and Ocean Eng., Mar. Techn. Soc.,-Amer. Soc. Limnol. Oceanogr., 2:1113.

excretion, organic matter

excretion
Hellebust, Johan A., 1967.
Excretion of organic compounds by cultured and natural populations of marine phytoplankton.
In: Estuaries, G.H. Lauff, editor, Publs Am. Ass. Advmt Sci., 83:361-355.

excretion, organic matter
*Khailov, K.M., and Z.P. Burlakova, 1969.
Release of dissolved organic matter by marine seaweeds and distribution of their total organic production to inshore communities.
Limnol. Oceanogr. 14(4):521-527.

excretion of salt

excretion of salt.
Russell, F.S., 1958.
Salt excretion in marine birds.
Nature, 182:1755.

excretion, salt
Schmidt-Nielsen, Knut, C. Barker Jörgensen, and Humio Osaki, 1958.
Extrarenal salt excretion in birds.
Amer. J. Physiol., 193(1):101-108.

exhalations (hydrothermal)

exhalations (hydrothermal)
Bostrom, Kurt, and M.N.A. Peterson, 1966.
Precipitates from hydrothermal exhalations on the East Pacific Rise.
Econ. Geol., 61:1258-1265.

exosmosis
*Gessner, F., and L. Hammer, 1968.
Exosmosis and "free space" in marine benthic algae.
Marine Biol., 2(1):88-91.

expanding earth

expanding earth
Barnett, C.H., 1969.
Oceanic rise in relation to the expanding earth hypothesis.
Nature, Lond., 221(5185):1043-1044.

expatriation

expatriation
Nafpaktitis, Basil G., 1968.
Taxonomy and distribution of the lanternfishes, genera Lobianchia and Diaphus, in the North Atlantic.
Dana Report No. 73: 131 pp.

experimental

experimental
Bainbridge, R., 1949
Movement of zooplankton in diatom gradients
Nature 163(4154):910-911, 2 text figs.

experimental
Barnaby, C. F., 1949.
Effect of rain in calming the sea. Nature 164 (4179):968.

experimental

Francis, J.R.D., 1949.
Laboratory experiments of wind-generated waves.
J. Mar. Res. 8(2):120-131, 4 textfigs.

experimental

Hidaka, K., and M., Koizumi, 1949.
Vertical circulation due to winds as inferred from the buckling experiments of elastic plates.
Ocean. Mag., Tokyo, 1(2):89-98, 12 textfigs.

experimentation (geology)

Kuenen, Ph. H., 1965.
Value of experiments in geology.
Geol. en Mijnbouw, 44(1):22-36.

Experiments

Mosby, H., 1948.
Experiments on turbulence and friction near the bottom of the sea. Bergens Mus. Arb. 1946 of 1947, Naturv. rekke: 6 pp.

exploitation

Exploitation

Collier, J.A., 1964.
The regime of the seas: exploitation and conservation.
In: Report of Conference on Law and Science, held at Niblett Hall, July 11-12, 1964, David Davies Memorial Institute of International Studies and British Institute of Comparative and International Law, 54-60.

Exploitation

Currie, R.I., 1964.
Conservation and exploitation of the sea.
In: Report of Conference on Law and Science, hel at Niblett Hall, July 11-12, 1964, David Davies Memorial Institute of International Studies and British Institute of Comparative and International Law, 49-53.

exploitation

Deacon, G.E.R., 1970.
Exploiting the oceans. Adv. Sci. 26: 313-317.

exploitation

Genty, R., 1968.
Vue prospective de l'exploitation des océans.
Rev. Marit. 259: 1325-1341 (not seen)

exploitation

Sorger, M., 1968.
Exploitation minière au fonds des mers.
Sci. Progrès, Nature (3404): 463-466

exploration

exploration

Anon., 1957.
Spontaneous potential will play big part in future sea exploration.
Ocean Indust. 2(8): 55-63

exploration

Lloyd, Robert E., 1965.
Problems affecting North Sea explorations.
GeoMar. Techn., 1(8):24-27.

exploration, geophysical

Moody, Alton B., 1970.
Satellite systems for geophysical exploration at sea.
Navigation, R.geogr. Soc. 23(4):458-475.

explosions

explosions

Carder, D.S., 1948
Seismic investigations of large explosions.
J. Coast and Geodetic Survey 1:71-73, 5 figs.

explosions, nuclear

Carder, Dean S., and Leslie F. Bailey, 1958.
Seismic wave travel times from nuclear explosions
Bull. Seismol. Soc., Amer., 48(4):377-398.

explosives

Jakosky, J., and J. Jakosky, Jr., 1956.
Explosives for marine seismic exploration.
Geophysics, 21(4):969-991.

explosions, effect of

explosions, effect of

Chan, B.C., M. Holt and R.L. Welsh 1968
Explosions due to pressurized spheres at the ocean surface.
Physics Fluids 11 (4): 714-722.

explosions, effect of

Collins, Richard, and Maurice Holt 1968.
Intense explosions at the ocean surface.
Physics Fluids 11 (4): 701-713.

explosions, effect of

Whalin, Robert W., 1965.
Water waves produced by underwater explosions: propagation theory for regions near the explosion.
J. geophys. Res., 70(22):5541-5549.

explosives, effect on marine life

explosives, effect of, marine life

Aplin, J.A., 1947
The effect of explosives on marine life.
Calif. Fish & Game, 33:23-30.

explosions, effect of

Coker, Coit M., and Edgar H. Hollis 1950
Fish mortality caused by a series of heavy explosions in Chesapeake Bay.
J. wildl. Mgt 14 (4): 435-445 (not seen)

explosions, effect of

Jeffreys, H., 1962.
Travel time of Pacific explosions.
Geophys. J., 7(2):212-219.

explosions, effect of

Kearns, Roger K., and Forbes C. Boyd, 1965.
The effect of a marine seismic exploration on fish populations in British Columbia coastal waters.
Canadian Fish Culturist, No. 34:3-26.

explosions, effect of

Koyama, T., 1954.
Effect of dynamite explosion on fish. Bull Tokai Regional Fish. Res. Lab. No. 8:23-29, 9 textfigs.

explosives, effect of

Ramirez, H., Ernesto, Ilhuicamina Mayes A., y Felipe Brizuela A., 1960
Pruebas experimentales para dedeterminar los efectos de explosiones subacuaticas sobre algunas especies de peces.
Sec. de Indust. y Comercio, Direccion General de Pesca e Industrias Conexas, Oficina Estud. Biol., Ser. Trabajos de Divulgacion, Mexico, 2(14): 46 pp.

explosives, effect of

Schwartz, Frank J., 1961

A bibliography: effects of external forces on aquatic organisms. Chesapeake Biol. Lab., Solomons, Maryland, Contrib. No. 168: 85 pp. (multilithed).

explosions, effect of

Sen, A.R., 1963.
Surface waves due to blasts on and above liquids.
J. Fluid Mechanics, 16(1):65-81.

explosions, effect of

Woodburn, Kenneth D. 1965.
A discussion and selected annotated references of subjective or controversial marine matters.
Techn. Ser. Fla Bd Conserv. 46:50pp.

explosive signals

Wille, P., and R. Thiele 1971.
Transverse horizontal coherence of explosive signals in shallow water.
J. acoust. Soc. Am. 50 (4-2): 348-353.

exposure

exposure

Dommasnes, Are, 1968.
Variations in the meiofauna of Corallina officinalis L. with wave exposure.
Sarsia, 34:117-124.

exposure

Mokyevsky, O.B., 1969.
The biogeocoenotic system of the marine littoral zone. (In Russian; English abstract).
Okeanologiia 9(2): 211-222.

"exposure"

*Muus, Bent J., 1968.
A field method for measuring "exposure" by means of plaster balls: a preliminary account.
Sarsia, 34:61-68.

exposuee, effect of

Preece, G.S., 1971.
The ecophysiological complex of Bathyporeia pilosa and B. pelagica (Crustacea: Amphipoda). II. Effect of exposure. Mar. Biol. 11(1): 28-34.

extension of range

extension of range

Smith, R.I., 1963.
On the occurrence of Nereis (Neanthes) succinea at the Kristineberg Zoological Station, Sweden, and its recent northward spread.
Arkiv. För Zoologi, 15(5):437-441.

extinction coefficient

extinction coefficient (freshwater)

*Smith, Raymond C., 1968.
The optical characterization of natural waters by means of an "extinction coefficient".
Limnol. Oceanogr., 13(3):423-429.

extinction coefficient
Zege, E.P., A.P. Ivanov, B.A. Kagin, and I.L. Katzev 1971.
Determination of the extinction and absorption coefficients in water and atmosphere by distribution in time of the reflected pulse signal. (In Russian; English abstract).
Fizika Atmosfer. Okean. Izv. Akad. Nauk SSSR 7(7): 750-757

extracellular products

extracellular products
Fogg, G.E.,1966.
The extracellular products of algae.
In: Oceanography and marine biology, H. Barnes, editor, George Allen & Unwin, Ltd., 4:195-212.

extracellular material
Guillard, R.R.L. and Johan A. Hellebust 1971
Growth and the production of extracellular substances by two strains of Phaeocystis pouchetii.
J. Phycol. 7 (4): 330-338.

extraterrestrial matter

extraterrestrial matter
Arrhenius, Gustaf, 1967.
Deep-sea sedimentation: a critical review of U. S. work.
Trans. Am. geophys. Un., 48(2):604-631.

extraterrestrial material
Laevastu, T., and O. Mellis, 1955.
Extraterrestrial material in deep-sea deposits.
Trans. Amer. Geophys. Union 36(3):385-388.

extracts, effects of

Extracts, effect of
Kabanova, Iu. G., 1959.
(The influence of extract from "cystozir" and "phyllophora" on some microphytes.)
Trudy Inst. Okeanol. 30:156 -

extracts, effects of
Nordli, Erling, 1957.
Algal flour extract as a stimulating agent for marine dinoflagellate cultures.
Nytt Mag. Botan., 5:13-16.

extrusives
Watkins, N.D., and A. Richardson 1971.
Intrusives, extrusives and linear magnetic anomalies.
Geophys. Jl r. astr. Soc. 23(1): 1-13

exudates,algal
*Hoyt,J.W.,1970.
High molecular weight algal substances in the sea.
Marine Biol., 7(2):93-99.

Eyes, structure of

eyes, cpmpound
Bernhard,C.G., editor,1966.
The functional organization of the compound eye.
Pergamon Press,591 pp.

eyes, structure of
Denton, E.J., 1960.
The 'design' of fish and cephalopod eyes in relation to their environment.
Symposium Zool. Soc., London, 3:53-56.

eyes
Fisher, L.R. (posthumous), 1967.
Pigments of euphausiid eyes.
Proc. Symp. Crustacea, Ernakulam, Jan. 12-15, 1965, 3: 1074-1080

eye movements
Harris, A.J., 1965.
Eye movements of the dogfish Squalus acanthias L.
J. Exp. Biol., 43(1):107-130.

eyes,vertebrate
Munk,Ole,1965.
Ocular degeneration in deep-sea fishes.
Galathea Rep., 8:21-31.

eyes (mollusc).
Newell, G.E., 1965.
The eye of Littorina littorea.
Proc. Zool. Soc., London, 144(1):75-86.

eyes
*Vader,Winn,1968.
A speciman of Hippomedon denticulatus with crystalline eye-lenses: with notes on the developement of the eyes in other Hippomedon species (Amphipoda, Lysianassidae).
Sarsia, 33: 65-72.

facilities

facilities
Hachey, H. B., 1949.
V. Facilities for studying the oceans. Canada's oceans - known and unknown. Symposium. Proc. Roy Soc., Canada, ser. 3, 43:187-190, 1 textfig.

faecal pellets

faecal pellets
Arakawa, Kohman Y. 1970
Scatological studies of the Bivalvia (Mollusca).
Adv. mar. Biol. F.S. Russell and Maurice Yonge, editors, Academic Press 8: 307-436.

faecal pellets
Cowey, C.B., and E.D.S. Corner, 1966.
The amino-acid composition certain unicellular algae and of the faecal pellets produced by Calanus finmarchicus when feeding on them.
In: Some contemporary studies in marine biology, H. Barnes, editor, George Allen & Unwin, Ltd., 225-231.

faeces
Currie, R.I., 1962.
Pigments in zooplankton faecies.
Nature, 193(4819):956-957.

fecal pellets
Dillon, William P., 1964
Flotation technique for separating fecal pellets and small marine organisms from sand.
Limnology and Oceanography, 9(4):601-602.

Faecal pellets
Emery, K. O., and J. Hulsemann, 1961 (1962).
The relationships of sediments,life and water in a marine basin.
Deep-Sea Res., 8(3/4):165-180.

faecal pellets
Frankenberg,Dirk,Stephen Lee Coles, and R.E. Johannes, 1967.
The potential trophic significance of Callianassa major fecal pellets.
Limnol. Oceanogr., 12(1):113-120.

faecal pellets
*Haven,Dexter S., and Reinaldo Morales-Alamo, 1968.
Occurrence and transport of faecal pellets in suspension in a tidal estuary.
Sediment.Geol.,2(2):141-151.

faecal pellets
Johannes, R.E., and Masako Satomi, 1966.
Composition and nutritive value of faecal pellets of a marine crustacean.
Limnol. Oceanogr., 11(2):191-197.

fecal pellets
Johannes, R.E., and Masako Satomi, 1965.
Experimental investigation of the nutritive role of Crustacean fecal pellets in the marine ecosystem. (abstract).
Ocean Sci. and Ocean Eng., Mar. Techn. Soc.,-Amer. Soc. Limnol. Oceanogr., 2:1109.

faecal pellets
Kornicker, L.S., and E.G. Purdy, 1957.
A Bahamian faecal-pellet sediment.
J. Sediment. Petrol., 27(2):126-128.

faecal pellets
Kraeuter, John, and Dexter S. Haven, 1970.
Fecal pellets of common invertebrates of lower York River and lower Chesapeake Bay, Virginia.
Chesapeake Sci. 11 (3): 159-173

faecal pellets
Marr, J.W.S., 1962.
The natural history and geography of the Antarctic krill (Euphausia superba Dana).
Discovery Repts., 32:33-464, Pl. 3.

faecal pellets
Moore, H.B., 1931
The systematic Value of a Study of Molluscan faeces. Proc. Malacological Soc., 19(6):281-290, 22 figs.

faecal pellets
Norris, Robert M., 1964.
Sediments of Chatham Rise.
New Zealand Dept. Sci. Ind. Res., Bull., No. 159;
New Zealand Oceanogr. Inst., Memoir, No. 16:39 pp

faecal pellets
Reineck, Hans-Erich and Indra Bir Singh, 1971.
Der Golf von Gaeta (Tyrrhenisches Meer). III.
Die Gefüge von Vorstrand- und Schelfsedimenten.
Senckenbergiana maritima 3(1): 185-201.

faecal pellets
Shelbourne, J.E., 1962.
A predator-prey size relationship for plaice larvae feeding on Oikopleura.
J. Mar. Biol. Assoc., U.K., 42(2):243-252.

faecal pellets
Smayda, Theodore J., 1969.
Some measurements of the sinking rate of faecal pellets.
Limnol. Oceanogr. 14(4):621-625.

faecal pellets
Wieser, Wolfgang, 1962
Adaptations of two intertidal isopods. 1. Respiration and feeding in Naesa bidentata (Adams) (Sphaeromatidae).
J. Mar. Biol. Assoc., U.K., 42(3):665-682.

fecal pellets, effect of
Schrader, Hans-Joachim, 1971.
Fecal pellets: role in sedimentation of pelagic diatoms. Science 174(4004): 55-57.

Fallout; see: Radioactivity, fallout

fallout
See chiefly under: Chemistry, radioactive fallout

fallout
Dmitrieva, G.V., Ju. V. Krasnopevtsev and S.G. Malakhov, 1970.
Some peculiarities of fission product concentration distributions over oceans in the tropical zone and their connection with atmospheric processes. J. geophys. Res., 75(18): 3675-3685.

fallout
Fabian, P., W.F. Libby and C.E. Palmer, 1968.
Stratospheric residence time and interhemispheric mixing of strontium 90 from fallout in rain. J. geophys. Res., 73(12):3611-3616.

Fallout
Malakhov, S.G. and I.B. Pudovkina, 1970.
Strontium 90 fallout distribution at middle latitudes of the northern and southern hemispheres and its relation to precipitation. J. geophys. Res., 75(18): 3623-3628.

false echoes

false echoes
Mannevy, P., 1943.
Echos anormaux observés au cours de sondage ultra-sonores sur les côtes de Prevence.
Bull. d'Info., C.C.O.E.C. 5(8):341-344, 1 pl.

FALSE echoes
Roumegoux, L., 1953.
Anomalies d'enregistrement observés au cours des sondages sonores sur les côtes de Tunisie.
Bull. d'Info., C.C.O.E.C 5(8):333-340, 6 pls.

fans, deep-sea

fans
See also: sea-fans

fans
Curray, Joseph R., and David G. Moore, 1971.
Growth of the Bengal Deep-Sea Fan and denudation in the Himalayas.
Bull. geol. Soc. Am. 82(3): 563-572

fans
Haner, Barbara E. 1971.
Morphology and sediments of Redondo Submarine Fan, southern California.
Bull. geol. Soc. Am. 82(9): 2413-2432.

fans, deep-sea
Heezen, Bruce C., and A.S. Laughton, 1963.
14. Abyssal plains.
In: The Sea, M.N. Hill, Editor, Interscience Publishers, 3:312-364.

fan, deep-sea
Heezen, Bruce C., and H.W. Menard, 1963.
12. Topography of the deep-sea floor.
In: The Sea, M.N. Hill, Editor, Interscience Publishers, 3:233-280.

fans, deep-sea
Menard, H.W., S.M. Smith and R.M. Pratt, 1965.
The Rhône Deep-Sea Fan.
In: Submarine geology and geophysics, Colston Papers, W.F. Whittard and R. Bradshaw, editors, Butterworth's, London, 17:271-284.

fans (washover)
Pierce, J.W., 1970.
Tidal inlets and washover fans.
J. Geol., 78(2): 230-234

fans
Walker, J.R., and J.V. Massingill 1970.
Slump features on the Mississippi Fan northeastern Gulf of Mexico.
Bull. geol. Soc. Am. 81(10): 3101-3108.

fan valleys

fan valleys
Curray, Joseph R., and David G. Moore, 1971.
Growth of the Bengal Deep-Sea Fan and denudation in the Himalayas.
Bull. geol. Soc. Am. 82(3): 563-572

fan-valleys
Normark, W.R., and D.J.W. Piper, 1969.
Deep-sea fan-valleys past and present.
Bull. geol. Soc. Am. 80(9): 1859-1866

fan valley
Piper, J.W., and William R. Normark 1971.
Re-examination of a Miocene deep-sea fan and fan-valley, southern California.
Bull. geol. Soc. Am. 82(7): 1823-1830.

fan-valleys
Shepard, Francis P., and Edwin C. Buffington, 1968.
La Jolla Submarine Fan-Valley.
Mar. Geol. 6(2): 107-143.

fans, washover
Frey, Robert W. and Taylor V. Mayou, 1971.
Decapod burrows in Holocene barrier island beaches and washover fans, Georgia. Senckenbergiana maritima 3(1): 53-77.

Farming the seas

farming
See also: cultures

farming the sea
Cole, H.A., 1965.
Farming the sea.
Geogr. Mag., 38(5): -

farming
Eaton, Bernard, Editor, 1963.
The undersea challenge.
British SubAqua Club, 182 pp., illus., $3.50.

"sea farms"
*Hanks, Robert W., 1968.
Benthic community formation in a "new" marine environment.
Chesapeake Sci., 9(3):163-172.

farming in the sea
Hardy, A., 1962.
Studying the ever-changing sea.
Nature, 196(4851):207-210.

farming
Holt, S.J., 1969.
The food resources of the ocean.
Scient. Am., 221(3):178-182,187-194. (popular).

farming
Iversen, E.S., 1967.
Farming the sea.
Oceanology Int., May/June 1967:28-30.

farming, oysters
*Krakatitsa, T.F., 1968.
Experience of Ostrea taurica Kryn. breeding in Yagorlytsky Bay of the Black Sea. (In Russian; English abstract).
Gidrobiol. Zh., 4(5):34-38.

farming
McKee, A., 1969.
Farming the sea.
Crowell, N.Y. 198 pp. $6.95 (not seen)

farming
Nash, Colin E. 1969.
Thermal aquaculture.
Sea Frontiers 15(5): 268-276 (popular)

farming, shellfish
*Pierantoni, Angiolo, 1968.
La mitilicoltura nel golfo di Napoli.
Boll. Soc. Nat. Napoli, 76(1-1967):219-228.

farming
Pinchot, Gifford B., 1970.
Marine farming.
Scient. Am., 223(6):14-21. (popular).

fish farming
Pinder, A.C., 1966.
Aquafarming, a brief review of current developments - a report from Tokyo.
GeoMar. Techn., 2(10):16-17.

"farming"
Smith, Thomas E., 1965.
Operation STEELHEAD. (abstract).
Ocean Sci. and Ocean Eng., Mar. Techn. Soc.,-Amer. Soc. Limnol. Oceanogr., 2:1229-1239.

sea farming
*Trevallion, Ann, and Alan D. Ansell, 1967.
Studies on Tellina tenuis Da Costa. II.
Preliminary experiments in enriched sea water.
J. exp. mar. Biol. Ecol., 1(2):257-270.

sea farms
Webber, Harold H., 1968.
Mariculture.
Bio Science 18(10):940-945. (non technical).

farming
Yonge, C.M., 1966.
Farming the sea.
Discovery, 27(7):8-12.

farming, crustaceans

farming the sea
*Coelho, Petrônio Alves, 1965/1966.
Estudio ecológico da lagoa do ôlho d'agua, Pernambuco, com especial referência aos crustáceos decápodos.
Trabhs Inst. Oceanogr., Univ. Fed. Pernambuco, Recife, (7/8):51-69.

farming
Forster J.R.M., 1970.
Further studies on the culture of the prawn, Palaemon serratus Pennant, with emphasis on the post-larval stages.
Fish. Invest. Min. Agric. Fish. Food, London (2) 26 (1): 40 pp

farming
Fujinaga (Hudinaga) Motosaku, 1969.
Kuruma shrimp (Penaeus japonicus) cultivation in Japan.
FAO Fish. Repts. 3 (57) (FRm/57.3 (Trm.)) 811-832. (mimeographed).

farming (shrimp)
Hudinaga, Motosaku, and Mitsutake Miyamura, 1962
Breeding of the "Kuruma" prawn (Penaeus japonicus Bate). (In Japanese; English abstract).
J. Oceanogr. Soc., Japan, 20th Ann. Vol., 694-706.

farming
Ingle, Robert M., and Ross Witham 1968.
Biological considerations in spiny lobster culture.
Proc. Gulf Caribbean Fish Inst. 21st. ann. Sess., 158-162.

farming, shrimp
Lee, Byung Don, and Taek Yuil Lee, 1968
Experiments on the rearing of Metapenaeus joyneri (Miers).
Publ. mar. Lab. Pusan Fish. Coll. 1: 39-42.

farming
Ling S.W., 1969.
Methods of rearing and culturing Macrobrachium rosenbergii (DeMan).
FAO Fish. Repts 3 (57) (FRm/R57.3 (Tri)): 607-619 (mimeographed)

FISH FARMING
Raman, K., and M. Krishna Menon, 1963.
A preliminary note on an experiment in paddy field prawn fishing.
Indian J. Fish. (A) 10(1): 33-38.

farming, crustaceans
Seki, Humitake, and Michael Hardon 1970.
Microbiological studies relevant to a lobster introduction into Fatty Basin, British Columbia.
J. Oceanogr. Soc. Japan 26 (1): 38-51.

farms
Teinsongrusmee, Banchong, 1970.
A present status of shrimp farming in Thailand.
Contrib. Mar. Fish. Lab. Bangkok, 15: 34pp.

farming, shrimp
Webber, H.H., 1970.
The development of a maricultural technology for the penaeid shrimps of the Gulf and Caribbean region. Helgoländer wiss. Meeresunters, 20(1/4) 455-463.

farming, fish

farming, fish
Amanieu, Michel 1967.
Introduction à l'étude écologique des réservoirs à poissons de la région d'Arcachon.
Vie Milieu (B) 18 (2B): 381-452.

farms, fish
Anon, 1968.
Japanese fish farms.
Ocean Industry 3(4): 27-30
(secondary sources)

sea farming
Berton, Robert, 1965.
Britain's first sea fish farm.
Fisheries Newsletter, 24(11):p. 29,31.

farming the sea
*Berry, Frederick, and E.S. Iversen, 1967.
Pompano: biology, fisheries and farming potential.
Proc. Gulf Caribb. Fish. Inst., 19th. Sess., 116-128

sea farming
Blaxter, J.H.S., 1968.
Rearing herring larvae to metamorphosis and beyond.
J. mar. biol. Ass., U.K., 48(1):17-28.

farming, fish
Bowers, Allen Benton, 1966.
Farming marine fish.
Science Journal, 2(6):46-51.

farming
Bowers, 1966.
Farming marine fish. Science J., June 1966:2-7.

farming, fish
Bowers, A.B., 1966.
Marine fish culture in Britain, VI. The effect of the acclimatization of adult plaice to pond condition on the viability of eggs and larvae.
J. Cons. perm. int. Expl. Mer. 30(2):196-203.

fish farming
Bowers, A.B. and Jong Wha Lee, 1971.
The growth of plaice in Laxey Bay (Isle of Man). J. Cons. int. Explor. Mer 34(1): 43-50.

farming, fish
El-Zarka, Salah El-Din, 1963.
Acclimatization of Solea vulgaris (Linn.) in Lake Quarun, Egypt.
J. du Conseil, 28(1):126-136.

farming, fish
Furukawa, Atsushi, Takeshi Umezu, Hiroko Tsukahara, Katsumi Funae and Isamu Iwata, 1966.
Studies on feed for fish. 5. Results of the small floating net culture test to establish the artificial diet as complete yellow-tail foods (1964). (In Japanese; English abstract).
Bull. Naikai reg. Fish. Res. Lab., 23:45-56.

farming, fish
*Harada, T., 1970.
The present status of marine fish cultivation research in Japan. Helgoländer wiss. Meeresunters, 20(1/4): 594-601.

farming
Hempel, G., 1970.
Fish-farming, including farming of other organisms of economic importance. Convener's report of an informal session, held on September 11, during the International Helgoland Symposium 1969. Helgoländer wiss. Meeresunters, 21(4): 445-465.

farms
Hickling, C.F., 1970.
Estuarine fish farming.
Adv. mar. Biol., F.S. Russell and Maurice Yonge, editors, Academic Press, 8: 119-213

farming
Hisaoka, Minoru, Kazuhiko Nogami, Osamu Takeuchi, Masaya Suzuki and Hitomi Sugimoto, 1966.
Studies on sea water exchange in fish farm. 2. exchange of sea water in floating net.
Bull. Naikai reg. Fish. Res. Lab., No. 23:21-43.
(In Japanese; English abstract)

farming, fish
Huet, Marcel, 1970.
Traité de pisciculture. 4th edit. Editions, Ch. de Wyngaert, Bruxelles XIV + 718 pp.
$19.00

farming
Inoue, Hiroo, Yoshiaki Tanaka and Kiyoshi Fukuda, 1970.
On water exchange in a shallow marine fish farm. II Hamachi fish farm at Tanoura.
Bull. Jap. Soc. scient. Fish. 36 (8): 776-782.
(In Japanese; English abstract)

farming, fish

Inoue, Motoo, 1970.
Possibility of tuna cultivation and propagation.
Bull. Japan Soc. Fish. Oceanogr. 16. Also in:
Coll. Repr. Coll. mar. Sci. Techn. Tokai Univ.
4: 197-210. (In Japanese; English abstract).

fish-farming

Inoue ,Motoo,Ryohei Amano, Yukinobu Iwasaki,
and Mitsuyoshi Aoki,1967.
Ecology of various tunas under captivity.I.
Preliminary rearing experiments. (In Japanese;
English abstract).
J. Coll.mar.Sci.Techn.,Tokai Univ.,(2):197-209.

farming, fish

Lloyd, Robert E., 1965.
An approach to aquaculture -- United Kingdon
White Fish Authority.
GeoMar. Techn., 1(7):14-16.

fish culture

Malaysia,Tropical Fish Culture Research
Institute,1966.
Report for 1966, 50 pp.

farming, fish

Martino, P.A. and J.M. Marchello 1968.
Using waste heat for fish farming.
Ocean Industry 3(4): 36-39.

"fish farming"

McIntyre, A.D., and A. Eleftheriou, 1968.
The bottom fauna of a flatfish nursery ground.
J. mar. biol. Ass., U.K., 48(1):113-142.

farming

Milne,P.H., 1970.
Fish farming: A guide to the design and
construction of net enclosures.
Mar.Res., Dept.Agric.Fish., Scotland, 1970(1):
31 pp.

farming the seas

Miyazawa, Hiroshi, Manabu Kitamikado, Takashi
Takahashi and Shinko Tachino, 1960.
Fundamental studies on fish farming food (1).
Rept. Fac. Fish., Pref. Univ. Mie, 3(3):641-648.

farming

Moe, Martin A. Jr., Raymond H. Lewis
and Robert M. Ingle 1968.
Pompano mariculture: preliminary
data and basic considerations.
Tech. Ser. Fla Bd Conserv. 55: 65pp.

fish farming

Page-Jones, R.M., 1971.
A salinity indicator for use in marine fish
farming. J. Cons. int. Explor. Mer 34(1):
121-123.

farming, fish

*Purdom, C.E. and A.E. Howard, 1971.
Ciliate infestations: a problem in marine fish
farming. J. Cons. int. Explor. Mer, 33(3):
511-514.

fish farming

Richardson,I.D.,1967.
Which fish to farm?
Hydrospace,1(1):72-73, 75-77.

farming, fish

Riley, John S., 1966.
Marine fish culture in Britain. VII. Plaice
(Pleuronectes platessa L.) post-larval feeding on
Artemia salina L. nauplii and the effects of varying
feeding levels.
J. Cons. perm. int. Expl. Mer., 30(2):204-221.

farming, fish

Shelbourne, J.E., 1963
Marine fish culture in Britain. II. A plaice
rearing experiment at Port Erin, Isle of Man,
during 1960, in open sea water circulation.
J. du Conseil, 28(1):70-79.

farming, fish

Shelbourne, J.E., 1959
Could fish be farmed. The New Scientist,
19 Feb. 1959: 3 pp.

farming, fish

Shelbourne, J.E., J.D. Riley and G.T. Thacker
1963
Marine fish culture in Britain. I. Plaice
rearing in closed circulation at Lowestoft,
1957-1960.
J. du Conseil, 28(1):50-69.

farming, fish

Sugimoto, Hitomi, Minoru Hisaoka, Kazuhiko Nogami
Osamu Takeuchi and Masaya Suzuki,
1966.
Studies on sea water exchange in fish farm. 1.
Exchange of sea water in fish farm surrounded
by net. (In Japanese; English abstract).
Bull.Naikai reg. Fish. Res. Lab., No. 23:1-20.

farming, fish

Swift, R., 1969.
Fish farming. Oceanol. int. 69, 3: 5 pp.

farming, fish

Thomson, J.M., 19??
Brackish water fish farming.
Fisheries Newsletter, 19(11): reprint with
unnumbered pp.

Also:
Australian Fisheries Leaflet, No. 2.

farming

Vaas, K.F. 1970
Studies on the fish fauna of the newly
created lake near Veere, with special
emphasis on the plaice (Pleuronectes
platessa)
Nethed. J. sea Res. 5(1): 50-95

farming, fish

Varma, P.Udaya, P.R.S. Tampi, and K.V. George,
1963.

Hydrological factors and the primary production
in marine fish ponds. Ind. J. Fish. (A)10(1):
197-208.

seafarms, kelp

North,Wheeler J., and Carl L. Hubbs,editors,
1968.
Utilization of kelp-bed resources in southern
California.
Fish Bull., Dept.Fish Game,Calif.,139-264.

farming, pearls

farms, pearl

*Fujita, Yuji, Tadataka Taniguti and Buhei
Zenitani, 1967.
Microbiological studies on shallow marine areas.
III. On relation of the heterotrphic bacteria
to the changes in carbon dioxide, oxygen
consumption, organic acid and sulfides in the
mud sediment. (In Japanese;English abstract).
Bull. Fac. Fish., Nagasaki Univ., 23: 187-196.

farms, pearl

*Iizuka,Shoji,Yuji Fujita, Buhei Zenitani and
Haruhiko Irie,1968.
Contamination of pearl farm in Imari Bay. (In
Japanese;English abstract).
Bull.Fac.Fish.,Nagasaki Univ.,25:67-78.

pearl farms

Iizuka, Syoji, and Haruhiko Irie, 1967.
Studies on the oceanographic characteristics
of Haikai Channel and the adjacent waters
and of the effects of closing of the channel
on pearl farms. II Movement of sea water
and amount of dissolved oxygen in the farm.
(In Japanese; English abstract).
Bull. Fac. Fish. Nagasaki Univ., 22: 1-14.

farms, pearl

Irie, Haruhiko, and Syozi Iizuka, 1966.
Studies of the oceanographic characteristics of Haiki
Channel and the adjacent waters and of effects of
closing of the channel on pearl farms. 1. Present
status of plankton-biota and its presumptive changes.
(In Japanese; English abstract).
Bull. Fac. Fish., Nagasaki Univ., No. 20:14-21.

farms
pearls

Kawakami, Izumi, 1962
Studies on Mollusca with special reference to
some improvements in pearl cultivation.
Rec. Oceanogr. Wks., Japan, Spec. No. 6:133-1
134.

farms - pearls

Kischi, J., 1955.
Imitation pearl. 3. Pearl essence. 4. Pearly sub
-stance prepared from 9-Phenyluric acid.
Repts. Ind. Res. Inst., Osaka, 8(1):43-45.

farming, pearl oysters

Kuwatani, Yukimasa and Tamotsu
Nishii, 1969.
Effects of pH of culture water on the
growth of the Japanese pearl oyster.
(In Japanese; English abstract).
Bull. Jap. Soc. scient. Fish., 35(4): 342-
350.

farming
pearl oysters

Miyauchi, Tetsuo, and Haruhiko Irie, 1966.
Supplemental report on the relation between pearl-
oysters (Pteria martensii) and the current velocity
of environmental waters on the motion of shell-opening
and-shutting and shell-regeneration. (In Japanese;
English abstract).
Bull. Fac. Fish., Nagasaki Univ., No. 20:22-28.

pearl farming

Miyauti, Tetuo, 1968.
Studies on the effect of shell cleaning in pearl culture III. The influence of fouling organisms upon the oxygen consumption in the Japanese pearl oysters. (In Japanese; English summary)
Japan J. Ecol. 17(1):40-43

farming pearls

Takemura, Y., 1958
Bottom character of pearling bed in the Arafura Sea. 1. Size distribution and mud contents of the Thursday Island and east regions.
Bull. Tokai Reg. Fish. Res. Lab., No. 22: 17-26.

Farming pearls

Uyeno, Fukuzo, 1969.
Relationship between sea water and bottom condition and mechanism of fisheries production in estuarine areas,.....with special reference to the deterioration of pearl oyster farm.
(In Japanese; English abstract). Bull. Jap. Soc. fish. Oceanogr. Spec. No. (Prof. Uda Commem Pap.): 193-196.

farms-pearl

Uyeno, Fukuzo, 1964.
Relationships between production of foods and oceanographical condition of sea water in pearl farms. II. On the seasonal changes of sea water constituents and of bottom condition, and the effect of bottom cultivation. (In Japanese; J. Fac. Fish., Pref. Univ. Mie, 6(2):145-169.
English abstrac

farming, shellfish

farming

Anon.,1969.
Bigger profits for oyster farming.
Ocean Industry,Gulf Publishing Co., 4(1):33-34. (popular).

farming, shellfish

Anon, 1968.
U.K.: University research ship: salvage tug for U.S.Navy.
Hydrospace 1(3): 38-40.

farming

Antunes, S.A., and Yasuzo Ito, 1968.
Chemical composition of oysters from São Paulo and Paraná, Brazil.
Bolm Inst. oceanogr. S.Paulo 17(1): 71-88

farming

Arnaud, Patrick M., 1971.
Les moulières à Mytilus et Aulacomya des îles Kerguelen (Sud de l'Océan Indien). Les "moulières de seuil" et leur intérêt possible pour l'aquaculture des Pelecypodes.
C. r. hebd. Séanc. Acad. Sci. Paris, (D) 272(10): 1423-1425

farming

Azouz, Abderrazak,1966.
Etude des peuplements et des possibilités d'ostreiculture du lac de Bizerte.
Inst.nat.Sci.tech.Océanogr.Pêche,Salammbo,Ann. 15: 69 pp.

farming, oysters

Bonami, Jean Robert, Henri Grizel, Constantin Vago, et Jean-Louis Duthoit 1971.
Recherches sur une maladie épizootique de l'huitre plate, Ostrea edulis Linné.
Revue Trav. Inst. Pêches marit. 35(4): 415-418.

sea farming

Collignont, J.,1967.
La croissance des huitres dans les lagunes marocaines.
Bull.Inst.Pêches marit.,Maroc,15:49-57.

seafarming

Curtin,L.,1968.
Cultivated New Zealand rock oysters.
Fish.Techn.Rept.,N.Z.mar.Dept.,25: 47 pp. (multilithed).

"beginners guide for those intending to engage in farming".

farming oysters

Dardignac-Corbeil, Marie-José 1971.
Étude d'un milieu ostréicole d'après des observations réalisées en Vendée dans le bassin des Chasses des Sables d'Olonne.
Revue Trav. Inst. Pêches marit. 35(4): 419-434.

farming

Dunathan, Jay P., R.M. Ingle and W.K. Havens, Jr. 1969.
Effects of artificial foods upon oyster fattening with potential commercial applications.
Techn. Ser. Fla Bd Conserv., 58: 39 pp.

farming, oysters

Feuillet, Michelle 1971.
Étude du phosphore dans les sédiments ostréicoles du bassin des Chasses des Sables d'Olonne.
Revue Trav. Inst. Pêches marit. 35(4): 443-453.

farming, oysters

Feuillet, Michelle, 1971.
Relations entre les eaux interstitielles des fonds sédimentaires ostréicoles et le milieu hydrobiologique, le bassin des Chasses des Sables d'Olonne.
Revue Trav. Inst. Pêches marit. 35(4): 435-442

farming, oysters

Figueras, A., 1970.
Flat oyster cultivation in Galicia. Helgoländer wiss. Meeresunters, 20(1/4): 480-485.

seafarming

Fontaine,M.,1968.
L'aquiculture marine.
Bull. Cent.Etud.Rech.sci.Biarritz, 7(1):7-25.

farming, oysters

Fujiya, M., 1970.
Oyster farming in Japan. Helgoländer wiss. Meeresunters, 20(1/4): 464-479.

farming, shellfish

Gaarder, T., and R. Spärck, 1932.
Hydrographisch-biochemische Untersuchungen in norwegischen Austern-Pollen. Bergens Mus. Aarbok, Naturvidensk.-rekke, No. 1:5-144, 75 textfigs.

sea farming

*Hrs-Brenko,Mirjana,1967.
Index of conditions in cultured mussels on the Adriatic coast.
Thalassia jugosl., 3(1/6):173-181.

sea-farming

Hughes,John T., 1968.
Grow your own lobsters commerically.
Ocean Industry, 3(12):46-49.

sea farming

LeDantec,Jean,1968.
Ecologie et reproduction de l'huitre portugaise (Crassostrea angulata Lamarck) dans le bassin d'Arcachon et sur la rive gauche de la Gironde.
Revue Trav.Inst.(scient.Tech.) Peches marit. 32(3):241-362.

farming

Lambert, F., et Ph. Polk 1971
Observations sur l'ostréiculture à Cuba. 1. Composition du plancton.
Hydrobiologia 38(1): 9-14.

farming, shellfish

LeDantec, Jean, 1963.
L'ostreiculture dans le bassin d'Arcachon et ses rapports avec les variations du milieu.
Rev. Trav. Inst. Pêches Marit., 27(2):203-210.

mariculture

Medcoff, J.C., 1961.
Oyster farming in the Maritimes.
Fish. Res. Bd., Canada, Bull., No. 131:158 pp.

farming

Meixner, R. 1971.
Wachstum und Ertrag von Mytilus edulis bei Flosskultur in der Flensburger Förde.
Arch. Fischereiwiss. 22(1):41-50.

farming

Millar, R.H., 1968.
Changes in the population of oysters in Loch Ryan between 1957 and 1967.
Mar. Res. Dept. Agric. Fish., Scotland (1): 8 pp.

farming

Morović, Dinko et Ante Šimunović, 1969
Contribution à la connaissance de la croissance de l'huitre (Ostrea edulis L.) et de la moule (Mytilus galloprovincialis Lk.) dans la baie de Mali Ston. (In Jugoslavian; French and Italian abstracts)
Thalassia Jugoslavica 5:237-247

sea farming

*Nokolic, Miroslav, y Santiago Alfonso Melendez, 1968.
El ostion del mangle, Crassostrea rhizophorae Guilding, 1828 (experimentos iniciales en el cultivo).
Nota Invest.Centro Invest.pesq.,Cuba,7:1-30.

"farming"
*Padilla, Miguel, y Julio Orrego, 1967.
La fijación larval de ostras sobre colectores experimentales en Quetalmahue, 1966-67.
Bol. cient., Inst. Fomento Pesquero, Chile, 5:15 pp.

farming
Paul, L. J., 1966.
Observations on past and present distribution of mollusc beds in Ohiwa Harbour, Bay of Plenty.
New Zealand J. Sci., 9(1):30-40.

farming
Quayle, D.B., 1969.
Pacific oyster culture in British Columbia.
Bull. Fish. Res. Bd. Can. 169: 192 pp.

farming, shellfish
Ryther, John H., 1969.
The potential of the estuary for shellfish production.
Proc. nat. shellfish. Ass., 59: 15-22

sea farming
*Saenz, Braulio A., 1965.
El ostion antillano, Crassostrea rhizophorae Guilding y su cultivo experimental en Cuba.
Nota Invest., Centro Invest. Pesq., Cuba, 6:1-34.

farming, shellfish
*Shaw, William N., 1967.
Seasonal fouling and oyster setting on asbestos plates in Broad Creek, Talbot County, Maryland, 1963-65.
Chesapeake Sci., 8(4):228-236.

farming
Šimunović, Ante, 1969.
Contribution à la connaissance de la faune bentique de la baie de Pirovac et la possibilité d'élevage de certains coquillages (Communication préliminaire). (In Jugoslavian; French and Italian abstracts).
Thalassia Jugoslavica 5: 309-314.

farming
Walne, P.R., 1970.
Studies on the food value of nineteen genera of algae to juvenile bivalves of the genera Ostrea, Crassostrea, Mercenaria and Mytilus.
Fish. Invest. Min. Agric. Fish. Food, London (2) 26(5): 62 pp.

farming - shellfish
Walne, P.R., 1969.
Recent developments in shellfish rearing.
Oceanol. int. 69, 3: 7 pp.

"farming"
Walne, P.R., 1966.
Experiments in the large-scale culture of the larvae of Ostrea edulis L.
Minist. Agric., Fish., Food, Fish Invest., Great Britain, (2)25(4):1-53.

fat

fat
Bogorov, B.G., 1960
[Geographic variations in the fatness of plankton in the ocean.] Doklady Akad. Nauk, SSSR, 134(6): 1441-1442.

fats
Brockerhoff, H., and R.J. Hoyle, 1963.
On the structure of the depot of fats of marine fish and mammals.
Arch. Biochem. Biophys., 102(3):452-455.

fat
Collyer, Dorothy M., 1962
Method for the determination of fat percentage in unicellular algae.
J. Mar. Biol. Assoc., U.K., 42(3):485-492.

fat
Fisher, L.R., and Z.D. Hosking, 1962.
Vitamin A and fat in the herring (Clupea harengus) and its food.
Scotland, Dept. Afric. Fish., Marine Research, (4):35 pp.

fats
Igarashi, Hisano, Koichi Zama and Kozo Takama, 1961.
Fatty oil from shellfish. (In Japanese; English abstract).
Bull. Fac. Fish., Hokkaido Univ., 12(3):196-200.

fat content
Jensen, A.J.C., 1953.
Plankton. Danish observations.
Ann. Biol., Cons. Perm. Int. Expl. Mer, 9:160.

fat
Meara, M.L., 1955.
The chemical constitution of some marine animal fats.
Chem. & Ind. 10:247-248.

fat
Petipa, T.S., 1964.
Daily day rhythm of fat accumulation and expenditure in Calanus helgolandicus (Claus) in the Black Sea. (In Russian).
Doklady, Akad. Nauk, SSSR, 156(6):1440-1443.

fat
Sushkina, A.P., 1962.
Rate of fat consumption at various temperatures and life cycle of Calanus finmarchicus (Gunn.) and C. glacialis Jaschn.
Zool Zhurn., Akad. Nauk, SSSR, 41(7):1004-1012.

fat content

fat content
Corlett, J., 1953.
Dry weight and fat content of plankton near Bear Island, 1949-1952.
Ann. Biol., Cons. Perm. Int. Expl. Mer, 9:8-9, 2 textfigs.

fat content
Farkas, Tibor, and Sandor Herodek, 1960
Seasonal changes in the fat contents of the Crustacea plankton in Lake Balaton.
Ann. Inst. Biol. (Tihany) Hungaricae Academiae Sci., 27: 1-7.

fat content
Gopalakrishnan, V., 1953.
Season fluctuations in the fat content of the prawn, Penaeus indicus M.Ed. J. Madras Univ., B, 23(2):193-202, 3 textfigs.

fat content
Gundersen, K.R., 1953.
Zooplankton investigations in some fjords in western Norway during 1950-1951.
Rept. Norwegian Fish. Mar. Invest. 10(6):1-54.

fat content
Hansen, Vagn Kr., 1960.
Investigations on the quantitative and qualitative distribution of zooplankton in the southern part of the Norwegian Sea.
Medd. Danmarks Fiskeri- og Havundersøgelser, n.s., 2(23):1-53.

fat content
Hansen, Vagn, 1955
The food of herring on the Bløden Ground (North Sea) in 1953. J. du Cons., 21(1): 61-64.

fat content
Jensen, A.J.C., 1957.
Observations at Bornholm. Ann. Biol., Cons. Perm. Int. Expl. Mer, 12:113.

fat content
Jensen, Aa. J.C., 1956.
Danish observations on the fat content.
Ann. Biol., Cons. Perm. Int. Expl. Mer, 11:78.

fat content
Jensen, A.J.C., 1954.
Danish observations (plankton).
Ann. Biol., Cons. Perm. Int. Expl. Mer, 10:138.

fat content
Jensen, A.J.C., 1952.
Plankton. Danish observations. Cons. Perm. Int. Expl. Mer, Ann. Biol. 8:132.

fat content
Jensen, A.J.C., 1951.
Plankton (Baltic). Cons. Perm. Int. Expl. Mer, Ann. Biol. 7:114.

"The fat content of the plankton round Bornholm in August varied considerably from place to place ranging between 1.6-20.7%. The average of 19 samples was 11% which was lower than the mean for the previous nine years". (Entire comment).

fat content
Jensen, A.J.C., 1949.
Plankton. Conditions at Bornholm. Ann. Biol. 5: 134.

fat content
Jensen, A.J.C., 1949.
Plankton conditions at Bornholm. Ann. Biol., Int. Cons., 4:143.

fat content
Kwei, Eric A., 1969.
The fat cycle in the sardine (Sardinella aurita, Cuv. and Val.) in Ghanaian waters
Actes Symp. Oceanogr. Ressources halieut. Atlant. trop., Abidjan, 20-28 Oct. 1966, UNESCO, 269-275.

fat content
Nakai, Z., 1955.
The chemical composition, volume weight and size of the important marine plankton.
Tokai Reg. Fish. Res. Lab., Spec. Publ., 5:12-24.

fat content
Ramaswamy, T.S., 1953.
Carbohydrate and fat contents of fishes.
J. Madras Univ., B, 23(3):232-238.

fat content
Rodegker, Waldtraut, and Judd C. Nevenzel, 1964.
The fatty acid composition of three marine invertebrates.
Comp. Biochem. Physiol., 11(1):53-60.

Pisaster ochraceus signis
Mitella polymerus
Mytilus californianus

fat content
Romanini, M. G., 1949.
Sulla presenze di un fattore diffusore nei Celenterati. Atti Soc. Ital. Sci. Nat. e Mus. Civ. Sto. Nat., Milano, 88(1/2):65-68.

fat content
Subrahmanyan, R., and R.S. Gupta, 1963.
Studies on the plankton of the east coast of India 1. Seasonal variation in the fat content of the plankton and its relationship to phytoplankton and fisheries.
Proc. Indian Acad. Sci., (B), 57(1):1-14.

plankton fat
Venkataraman, R., and S.T. Chari, 1953.
Studies of mackerel fat variations: correlation of plankton fat with fat of fish. Proc. Indian Acad. Sci., B, 37(6):224-227, 1 textfig.

faults
Kanenko, Shiro 1966.
Rising promontories associated with a subsiding coast and sea-floor in south-western Japan.
Trans. R. Soc. N.Z., Geol. 4(11): 211-228.

fat content
Wimpenny, R.S., 1953.
The dry-weight and fat content of plankton with estimates from flagellate counts.
Ann. Biol., Coms. Perm. Int. Expl. Mer, 9:119-122, Textfigs. 25-27.

fat content
Wimpenny, R. S., 1949.
The dry weight and fat content of plankton.
Ann. Biol. 5:89.

Coscinodiscus concinnus = 30-40%

fat content
Wimpenny, R.S., 1950.
The dry weight and fat content of plankton.
Ann. Biol. 6:124-126, Textfigs. 6-8.

fat content
Wimpenny, R. S., 1949.
The dry weight and fat content of plankton.
Ann. Biol., Int. Cons., 4:89-90, Textfig. 1.

fat formation

fat formation
Hogg, C.E., 1956.
Photosynthesis and formation of fats in a diatom
Ann. Botany, 20(78):265-286.

fathograms

fathograms
Carstens, R.H., 1953.
Problems in fathogram interpretation. J. CGS No. 5:33-40.

fathometers
Whipp, D.M., 1953.
"808" fathometer errors. J. CGS No. 5:27-32.

faults

faults
Belderson, R.H. , N.H. Kenyon and A.H. Stride, 1970.
10-km wide views of Mediterranean deep-sea floor
Deep-Sea Res., 17(2): 267-270.

Faults
Bott, M.H.P. and A.B. Watts 1970.
The Great Glen Fault in the Shetland area.
Nature, Lond., 227(5255): 268-269.

faults
Dietz, R.S., 1954.
Marine geology of northwestern Pacific: description of Japanese Bathymetric Chart 6901.
Bull. G.S.A. 65(1):1199-1224, 6 textfigs.

faults
Drake, Charles L., 1966.
Recent investigations of the continental margin of eastern United States.
In: Continental margins and island arcs, W.H. Poole, editor, Geol. Surv. Pap., Can., 66-15:33-47.

Fault
Flinn, Derek 1970.
The Great Glen Fault in the Shetland area.
Nature, Lond., 227(5255): 268.

faults (Mongol-Okhotsk)
Gorzhevsky, D.I., and Ye. M. Lazko, 1961.
The Mongol-Okhotsk deep fault. (In Russian).
Doklady, Akad. Nauk, SSSR, 137(5):1177-1180.

Translation:
Consultants Bureau for Amer. Geol. Inst., 137 (1-6):361-363. (1962).

faults
Heezen, Bruce C., and Marie Tharp, 1964.
Oceanic ridges, transcurrent faults and continental displacements. (Abstract).
Geol. Soc., Amer., Special Paper, No. 76:68.

faults
Kishinouye, Fujihiko, and Heihachiro Kobayashi, 1965.
A submarine fault line found near Awashima after the Japan Sea earthquake on June 16, 1964.
Bull. Earthquake Res. Inst., 43(1):205-208.

faults
Krause, Dale C., 1965.
Tectonics, bathymetry and geomagnetism of the southern continental borderland west of Baja California, Mexico.
Geol. Soc., Amer., Bull., 76(6):645-647.

faults
Krishnan, M.S., 1960
The mid-ocean ridges.
Proc. Nat. Inst. Sci., India, (A), 26(Suppl.1) 195-218.

faults
*Lidz, L. (posthumous), Ball, M., and W. Charm, 1968.
Geophysical measurements bearing the problem of the El Pilar Fault in the northern Venezuelan offshore.
Bull. mar.Sci., Miami, 18(3):545-560.

faults
*Morgan, W. Jason, 1968.
Rises, trenches, great faults, and crustal blocks.
J. geophys. Res., 73(6):1959-1982.

faults
Sheridan, R.E., R.M. Berman and D.B. Corman, 1971
Faulted limestone block dredged from Blake Escarpment.
Bull. geol. Soc. Am. 82(1): 199-206

faults
Tobin, Don G., and Lynn R. Sykes 1968.
Seismicity and tectonics of the northeast Pacific Ocean.
J. geophys. Res. 73(12): 3821-3845.

faults
Vacquier, Victor, 1959.
Measurement of horizontal displacement along faults in the ocean floor.
Nature, 183:452-453.

Faults
Vacquier, Victor, Arthur D. Raff and Robert E. Warren, 1961
Horizontal displacements in the floor of the northeastern Pacific Ocean.
Bull. Geol. Soc., Amer., 72(8):1251-1258.

fault-blocks

fault-blocks
Emery, K.O., 1968.
Shallow structure of continental shelves and slopes.
SEast.Geol., Duke Univ., 9(4):173-194.

Fault, block
Fahlquist, Davis A. and David K. Davies, 1971
Fault-block origin of the western Cayman Ridge, Caribbean Sea. Deep-Sea Res., 18(2): 243-253.

fault blocks
Keller, G.H., and G. Peter, 1968.
East-west profile from Kermadec Trench to Valparaiso, Chile.
J.geophys.Res., 73(22):7154-7157.

faults, transcurrent

faults (transcurrent)
Watkins, Joel S., and Wilbur H. Geddes, 1965.
Magnetic anomaly and possible orogenic significance of geologic structure of the Atlantic shelf.
J. Geophys. Res., 70(6):1357-1361.

faults, transcurrent

Wilson, J. Tuzo, 1970.
III. Concepts. 2. Continental Drift, transcurrent, and transform faulting. In: The sea: ideas and observations on progress in the study of the seas, Arthur E. Maxwell, editor, Wiley-Interscience 4(2): 623-644.

faults, transform

faults, transform

*Girdler, R.W., 1968.
Drifting and rifting of Africa.
Nature, 217(5134):1102-1106.

faulting, transform

Isacks, Bryan, Jack Oliver and Lynn R. Sykes, 1968
Seismology and the new global tectonics.
J. geophys. Res., 73(18):5855-5899.

faulting, transform

Larson, Roger L., H.W. Menard and S.M. Smith, 1968.
Gulf of California: a result of ocean-floor spreading and transform faulting.
Science, 161(3843):781-784.

faulting, transform

Moore, David G., and Edwin C. Buffington, 1968.
Transform faulting and growth of the Gulf of California since the late Pliocene.
Science, 161(3847):1238-1241.

faults, transform

Peter, George, Barrett H. Erickson and Paul J. Grim, 1970.
II. Regional observations. 5. Magnetic structure of the Aleutian Trench and Northeast Pacific Basin. In: The sea: ideas and observations on progress in the study of the seas, Arthur E. Maxwell, editor, Wiley-Interscience 4(2): 191-222.

faults transform

Ranalli G., 1968.
Transform faults and large horizontal displacements of the ocean floor.
Annali Geofis. 21(4): 439-457

Faults, transform

Wilson J. Tuzo 1969.
Aspects of the different mechanics of ocean floors and continents.
Tectonophysics, 8(4/6): 281-284.

faults, transform

Vine, Frederick J., 1968.
Evidence from submarine geology.
Proc. Am. phil. Soc., 112(5):325-334.

faulting, transform

Wilson, J. Tuzo, 1970.
III. Concepts. 2. Continental Drift, transcurrent, and transform faulting. In: The sea: ideas and observations on progress in the study of the seas, Arthur E. Maxwell, editor, Wiley-Interscience 4(2): 623-644.

faults (transverse)

McMaster, Robert L. 1971.
A transverse fault on the continental shelf of Rhode Island.
Bull. geol. Soc. Am. 82 (7): 2001-2004.

faulting

faulting

Allen, Clarence R., 1964.
Circum-Pacific faulting. (Abstract).
Geol. Soc., Amer., Special Paper, No. 76:4-5.

block faulting

*Atwater, Tanya M., and John D. Mudie, 1968.
Block faulting on the Gorda Rise.
Science, 159(3816):729-731.

faulting

Chase R.L. and J.B. Hersey 1968.
Geology of the north slope of the Puerto Rico Trench.
Deep. Sea Res. 15(3): 297-317.

faulting, normal

Ball, M.M., C.G.A. Harrison, P.R. Supko and W.D. Bock, and N.J. Maloney, 1968.
Normal faulting on the southern boundary of the Caribbean Sea, Unare Bay, Northern Venezuela
Trans. Fifth Carib. Geol. Conf., St. Thomas, V.I., 1-5 July 1968, Peter H. Mattson, editor.
Geol. Bull. Queens Coll., Flushing, 5: 17-21.

faulting

Cullen, David J., 1969.
Quaternary volcanism at the Antipodes Islands: its bearing on structural interpretation of the southwest Pacific. J. geophys. Res., 74(17) 4213-4220.

faulting

Hodgson, J.H., 1956.
Direction of faulting in some of the larger earthquakes of the North Pacific, 1950-1953.
Publ. Dominion Obs., Ottawa, 18(10):219-252.

faulting

Hodgson, J.H., 1956.
Direction of faulting in some of the larger earthquakes of the southwest Pacific, 1950-1954.
Publ. Dominion Obs., Ottawa, 18(9):171-216.

faulting

Krause, Dale C., 1965.
Submarine geology north of New Guinea.
Geol. Soc., Amer., Bull., 76(1):27-42.

faulting

Lillie, A.R., 1969.
Evidence for Miocene faulting in southern Hawke's Bay.
N.Z. Jl Geol. Geophys. 12:344-345.

faults

Rusnak, Gene A., Robert L. Fisher and Francis P. Shepard, 1964.
Bathymetry and faults of Gulf of California.
In: Marine geology of the Gulf of California, a symposium, Amer. Assoc. Petr. Geol., Memoir, T. van Andel and G.G. Shor, Jr., editors, 3:59-75.

faulting

Scholl, David W., Edwin C. Buffington and David M. Hopkins 1968.
Geologic history of the continental margin of North America in the Bering Sea.
Mar. Geol. 6(4): 297-330.

faulting

Sykes, Lynn R., 1967.
Mechanism of earthquakes and nature of faulting on the mid-oceanic ridges.
J. geophys. Res., 72(8):2131-2153.

faulting

Vacquier, V., 1965.
Transcurrent faulting in the ocean floor.
Phil. Trans. R. Soc., A, 258(1088):77-81.

faulting

Vine, F.J. and H.H. Hess, 1970.
III Concepts. 1. Sea-floor spreading. In: The sea: ideas and observations on progress in the study of the seas, Arthur E. Maxwell, editor Wiley-Interscience 4(2): 587-622.

fauna

fauna

Tortonese, Enrico 1969.
La fauna del Mediterraneo e i suoi rapporti con quelle dei mari vicini.
Pubbl. Staz. Zool. Napoli, 37 (2 Suppl.) 369-384

fauna abyssal

faunas, abyssal

Bruun, A. Fr., 1956.
The abyssal fauna: its ecology, distribution and origin. Nature 177(4520):1105-1108.

fauna, abyssal

Menzies, Robert J., 1965.
Conditions for the existence of life on the abyssal sea floor.
Oceanogr. Mar. Biol., Ann. Rev., 3:195-210.

fauna, deep-water

Zenkevich, L.A., and J. A. Birstein, 1956.
Studies of the deep-water fauna and related problems. Deep-Sea Res. 4(1):54-64.

fauna, Caspian

Mordukhay-Boltovskoy, F.D., 1970.
On the Caspian fauna in the Black Sea (beyond the limits of freshened regions). Gidrobiol. Zh., 6(6): 20-25. (In Russian; English abstract)

faunas, comparison of

Bedot, W., 1909.
La faune eupelagique (holoplancton) de la Baie d'Amboina et ses relations avec celle des autres oceans. Rev. Suisse Zool. 17:121-142.

faunal competition

Segerstråle Sven G., 1969.
The competition factor and the fauna of the Baltic Sea.
Limnologica, Berlin, 7(1): 99-111

faunas, deep-sea
Page, L., 1956.
On the distribution and origin of the deep-sea fauna. Deep-Sea Res. 3(4):294-296.

faunas, hadal
Wolff, Torben, 1970.
The concept of the hadal or ultra-abyssal fauna. Deep-Sea Res., 17(6): 983-1003.

fauna, interstitial

fauna, interstitial
Ax, Peter, 1966.
Die Bedeutung der interstitiellen Sandfauna für allgemeine Probleme der Systematik, Ökologie and Biologie.
Veröff. Inst. Meeresforsch., Bremerh., Sonderband II:15-62.

Fauna, interstitial
Boaden, P.J.S., 1966.
Interstitial fauna from Northern Ireland.
Veröff. Inst. Meeresforsch., Bremerh., Sonderband II:125-130.

fauna, interstitial
Riemann, Franz., 1966.
Die Verbreitung der interstitiellen Fauna im Elbe-Aestuar.
Veröff. Inst. Meeresforsch., Bremerh., Sonderband II:117-122.

intertidal fauna
Tokioka, Takasi, 1963
Supposed effects of the cold weather of the winter of 1962-63 upon the intertidal fauna in the vicinity of Seto.
Publ. Seto Mar. Biol. Lab., Kyoto Univ., 11(2): 415-424(245-254).

fauna lists
Brunel, Pierre, 1963.
Inventaire taxonomique des invertébrés marins du Golfe Saint-Laurent.
Ann. Rept., 1962, Sta. Biol. Mar., Grande-Rivière, Canada, 81-89.

fauna, list of species
Lindberg, G.U., 1959.
List of fauna inhabiting the seas of southern Sakhalin and the Southern Kurile Islands. (In Russian).
Issledovaniya Dal'nevostochnykh Morei, SSSR, 6: 173-256.

fauna, lists of spp.
Wagner, Frances J.E., 1968.
Faunal study, Hudson Bay and Tyrrell Sea.
In: Earth Science Symposium on Hudson Bay, Ottawa, February, 1968, Peter J. Hood, editor, Geol. Surv., Can., GSC Pap. 68-53:7-48.

fauna, littoral

fauna, littoral
Christiansen, Bengt O., 1965.
Notes of the littoral fauna of Bear Island.
Astarte, No. 26:15 pp.

fauna, meio

fauna, meio-
Muus, Kirsten, 1966.
A quantative 3-year-survey on the Meiofauna of known macrofauna communities in the Øresund.
Veröff. Inst. Meeresforsch., Bremerh., Sonderband. II:289-291.

fauna, Meio -
Thiel, Hjalmar, 1966.
Quantitative Untersuchungen uber die Meiofauna des Tiefseebodens.
Veroff. Inst. Meeresforsch., Bremerh., Sonderband II:131-147.

fauna, micro

fauna, micro -
Canning, Björn, 1966.
Short time fluctuations of the Microfauna in a rockpool in the northern Baltic proper.
Veroff. Inst. Meeresforsch., Bremerh., Sonderband II:149-153.

fauna, micro-
Jansson, Ann-Mari, 1966.
Diatoms and Meiofauna-producers and consumers in the Cladophora belt.
Veroff. Inst. Meeresforsch., Bremerh., Sonderband II:281-288.

fauna, midwater
Ebeling, Alfred W., Richard M. Ibara, Robert J. Lavenberg and F. James Rohlf, 1970.
Ecological groups of deep-sea animals off southern California
Bull Los Angeles County Mus. Nat. Hist. Sci. (6): 43pp.

faunas, ultra-abyssal
Wolff, Torben, 1970.
The concept of the hadal or ultra-abyssal fauna. Deep-Sea Res., 17(6): 983-1003.

faunal diversity

faunal diversity
See also: diversity

faunal diversity
Hessler, Robert R., and Howard L. Sanders, 1967.
Faunal diversity in the deep-sea.
Deep-Sea Research, 14(1):65-78.

faunal diversity
Valentine, J.W., and E.M. Moores, 1970.
Plate-tectonic regulation of faunal diversity and sea-level: a model.
Nature, Lond., 228(5272): 657-659.

fauna, extinction of

Faunal extinctions
Hays, James D. 1971.
Faunal extinctions and reversals of the earth's magnetic field.
Bull. geol. Soc. Am. 82(9): 2433-2447.

faunal extinctions
Kennett, J.P., and N.D. Watkins, 1970.
Geomagnetic polarity change, volcanic maxima and faunal extinction in the South Pacific.
Nature, Lond. 227 (5261): 930-934

faunal extinction
Watkins, N.D., and H.G. Goodell, 1967.
Geomagnetic polarity change and faunal extinction in the Southern Ocean.
Science, 156(3778):1083-1087.

faunistic provinces

faunistic provinces
Tommasi, Luiz Roberto, 1965.
Faunistic provinces of the western South Atlantic littoral region. (Abstract).
Anais Acad. bras. cienc., 37(Supl.):261-262.

faunal origins
Zenkovich, L.A., 1949.
[On the antiquity of origin of cold-water fauna and flora.] Trudy Inst. Okeanol., 3:191-199.

feeding of plankton / fish — see under individual groups or in Misc. Index under 'food of plankton organisms / fish'

feeding
See also: food of ....

feeders, deposit
Newell, Richard, 1965.
The role of detritus in the nutrition of two marine deposit feeders, the prosobranch Hydrobia ulvae and the bivalve, Macoma baltica.
Proc. Zool. Soc., London, 144(1):25-45.

feeding
Edmondson, W.T., editor, 1966.
Ecology of Invertebrates, Marine Biology III, Proceeding of the Third International Interdisciplinary Conference.
New York Acad. Sci., 313 pp. ($7.00).

feeding, copepods
Richman, Sumner and Jane N. Rogers, 1969.
The feeding of Calanus helgolandicus on synchronously growing populations of the marine diatom Ditylum brightwellii. Limnol. Oceanogr. 14(5): 701-709.

feeding of fish
Hotta, Hideyuki, and Junko Nakashima 1968.
Experimental study on the feeding activity of jack-mackerel Trachurus japonicus. (In Japanese; English abstract).
Bull. Seikai reg. Fish. Res. Lab. 36: 75-83

Feeding mechanisms
Steele, J.H., editor, 1970.
Marine food chains.
Univ. Calif. Press, 552 pp. $13.50.

feeding succession
Margalef, Ramón 1967.
The food web in the pelagic environment.
Helgoländer wiss. Meeresunters. 15(1/4): 548-558.

feeding mechanisms
Yoshida, Yoichi, 1967.
On the feeding mechanisms of plankton feeders. (In Japanese; English abstract).
Inf. Bull. Planktol. Japan, Comm. No. Dr. Y. Matsue, 271-278.

feeding rhythm
Petipa, T.S. 1964
Суточный ритм в питании и суточные рационы Calanus helgolandicus (Claus) in the Black Sea.
Daily rhythm in the feeding and the daily ration of Calanus helgolandicus (Claus) in the Black Sea. (In Russian).
Trudy Sevastopol Biol. Stants., Akad. Nauk, 15: 69-98

feeding superfluous
Conover, Robert J., 1966.
Factors affecting the assimilation of organic matter by zooplankton and the question of superfluous feeding.
Limnol. Oceanogr., 11(3):346-354.

feldspar

Feldspar
Calvert, S. E., 1966.
Accumulation of diatomaceons silica in the sediments of the Gulf of California.
Geol. Soc., Am., Bull., 77(6):569-596.

feldspar
Chase, R.L., and J.B. Hersey 1968.
Geology of the north slope of the Puerto Rico Trench.
Deep-Sea Res. 15(3): 297-317.

feldspar
Field, Michael E., and Orrin H. Pilkey, 1969.
Feldspar in Atlantic continental margin sands off the southeastern United States.
Bull. geol. Soc. Am. 80(10): 2097-2102.

feldspar
*Giraud, Bernard, Bruce C. Heezen, Wladimir D. Nesteroff, et Germain Sabatier, 1968.
Les feldspaths dans les sédiments marins polaires
C.r. hebd. Séanc., Acad. Sci., Paris, (D)267(25):2099-2100.

feldspar
Hubert, John F., and William J. Neal, 1967.
Mineral composition and dispersal patterns of deep-sea sands in the western North Atlantic petrologic province.
Bull. Geol. Soc., Am. 78(6): 749-772

feldspar
Neiheisel, James, 1965.
Source and distribution of sediments at Brunswick Harbor and vicinity, Georgia.
U.S. Army Coastal Eng. Res. Center, Techn. Memo. No. 12:49 pp.

feldspars
Turekian, Karl K., 1965.
Some aspects of the geochemistry of marine sediments.
In: Chemical oceanography, J.P. Riley and G. Skirrow, editors, Academic Press, 2:81-126.

fermentation

fermentation
Lewin, Ralph A., Editor, 1962.
Physiology and biochemistry of algae.
Academic Press, New York and London, 929 pp.

fertility

fertility
Brongersma-Sanders, Margaretha, 1966.
The fertility of the sea and its bearing on the origin of oil.
Advmt. Sci., Br., Ass., 23(107):41-46.

fertility
Christy, Francis T., Jr., and Anthony Scott 1965.
The common wealth in ocean fisheries, some problems of growth and economic allocation.
Resources for the Future, Inc. Johns Hopkins Univ. Press, 281 pp.

fertility
Costlow, John D. 1971
The fertility of the sea.
Gordon and Breach Science Publishers,
Vol. 1: 308 pp.
2: 309-622

fertility
Fournier, Robert O., 1966.
North Atlantic deep-sea fertility.
Science, 153(3741):1250-1252.

fertility
Huntsman, A.G., 1948
Fertility and fertilization of streams.
J. Fish. Res. Bd. Can. 7(5):248-253, 1 fig.

fertilizers (land)
Lopez-Benito, Manuel, 1963
Estudio de la composicion quimica del Lithothamnium càlcareum (Aresch) y su aplicacion como corrector de terrenos de cultivo.
Inv. Pesq., Barcelona, 23:53-70.

fertility
Prakash, A., 1971.
Terrigenous organic matter and coastal phytoplankton fertility. (English and Portuguese abstracts). In: Fertility of the Sea, John D. Costlow, editor, Gordon Breach, 2: 351-368.

fertilizer
Raja, B.T. Antony, 1969.
The Indian oil sardine.
Bull. cent. mar. Fish. Res. Inst., India, 16:128 pp. (mimeographed).

fertility
Redfield, A.C., B.H. Ketchum and F.A. Richards 1963
2. The influence of organisms on the composition of sea-water. In: The Sea, M.N. Hill, Edit., Vol. 2. The composition of sea water, Interscience Publishers, New York and London, 26-77.

fertility
Tomo, Aldo P., 1971.
Trophic chains observed in Paradise Harbor (Antarctic Peninsula) related to variations in the fertility of its waters. (English and Spanish abstracts). In: Fertility of the Sea, John D. Costlow, editor, Gordon Breach, 2: 593-602.

fertilization

fertilization
Ansell, A.D., J.E.G. Raymont and K.F. Lander, 1963
Studies on the mass culture of Phaeodactylum. III. Small-scale experiments.
Limnol. and Oceanogr., 8(2):207-213.

fertilization
Ansell, A.D., J.E.G. Raymont, K.F. Lander, E. Crowley and P. Shackle, 1963
Studies on the mass culture of Phaeodactylum II. The growth of Phaeodactylum and other species in outdoor tanks.
Limnol. and Oceanogr., 8(2):184-206.

fertilization
Barlow, John P., Carl J. Lorenzen and Richard T. Myren, 1963
Eutrophication of a tidal estuary.
Limnol. and Oceanogr., 8(2):251-262.

fertilization
Brandhorts, W., 1958
Thermocline topography, zooplankton standing crop, and mechanisms of fertilization in the Eastern Tropical Pacific.
J. du Conseil, 24(1):16-31.

fertilization
Brook, A.J., 1958.
Changes in the phytoplankton of some Scottish hill lochs resulting from their artificial enrichment. Verh. Int. Ver. Limnol., 13:298-305.

fertilization
Brook, A.J., 1957.
Fertilization experiments in Scottish freshwater lochs. 1. Loch Kinardochy.
Scottish Home Dept., Freshwater and Salmon Fish. Res., 17:30 pp.

fertilization
Buljan, M., 1957.
Report on the results obtained by a new method of fertilization experimented in the marine bay "Mljetoka Jezera". Acta Adriatica, 6(6):44 pp.

fertilization, effect of
Buljan, Miljenko, Ivo Kačić, Jožica Karlovac, Dinko Morović, Tereza Pucher-Petković, Ante Šimunović, Ante Špan, & Tamara Vučetić, 1969.
Recherches écologiques relatives à la fertilisation artificielle effectuées dans la Baie de Marina. (In Jugoslavian; French and Italian abstracts)
Thalassia Jugoslavica 5:55-66.

fertilization
De Castro Moreira Da Silva, Admiral Paulo, 1971.
Fertilization of the sea as a by-product of an industrial utilization of deep water. In: Fertility of the Sea, John D. Costlow, editor, Gordon Breach, 2: 463-468.

fertilization
Gauld, D.T., 1950.
A fish cultivation experiment in an arm of a sea-loch. 3. The plankton of Kyle Scotnish.
Proc. R. Soc., Edinburgh, Sect. b, 64(1):36-64.

fertilization(artificial)
Hassan, El Sayed Mohamed, 1964
The use of semipermeable membrane to conserve fertilizer in artificially fertilized bays.
Limnology and Oceanography, 9(2):259-260.

fertilization
Hepher, B., 1962
Primary production in fishponds and its application to fertilization experiments.
Limnology and Oceanography, 7(2):131-136.

fertilization
Hepher, B., 1958.
On the dynamics of phosphorus added to fish ponds in Israel. Limnol. & Oceanogr., 3(1):84-100.

fertilization
Kuzhsov, C.I., 1956.
[The knowledge of the application of green and mineral fertilizer added to fish ponds in the Rostobsk district.] Trudy Vses Gidrobiol. Obsh., Akad. Nauk, SSSR 7:37-51.

fertilization
Ladouce, R., 1953.
Utilisation des engrais en ostreiculture.
Sci. et Peche, Off. Sci. Tech. Pêches Marit 1(9):4-6.

fertilization
Mortimer, C.H., 1954.
Fertilizers in fish ponds. Colonial Off., Fish. Publ., No. 5:155 pp.

artificial fertilization
Mortimer, C.H., and C.F. Hickling, 1954.
Fertilizers in Fish Ponds. A review and bibliography. Colonial Office, Fish. Publ., No. 5:4-155.

fertilization
Nelson, P.R., and W.T. Edmondson, 1955.
Limnological effects of fertilizating Bare Lake, Alaska.
U.S. Fish and Wildlife Service, Fish. Bull., No. 102:415-436.

fertilization
Pucher-Petkovic, T., 1960
Effet de la fertilisation artificielle sur le phytoplancton de la region de Mljet.
Acta Adriatica 6 (8): 24 pp.

fertilization
Raymont, J.E.G., 1950.
A fish culture experiment in an arm of a sea loch. IV. The bottom fauna of Kyle Scotnish.
Proc. Roy. Soc., Edinburgh, Sect. B, 64, Pt. 1, No. 4:65-108, 8 textfigs.

fertilization
Raymont, J.E.G., 1949.
Further observations on changes in the bottom fauna of a fertilized sea loch. J.M.B.A.28: 9-19, 1 textfig.

fertilization
Raymont, J.E.G., and R.S. Miller, 1962.
Production of marine zooplankton with fertilization in an enclosed body of water.
Int. Revue Ges. Hydrobiol., 47(2):169-209.

fertilization
Roels, O.A., R.D. Gerard and A.W.H. Bé, 1971.
Fertilizing the sea by pumping nutrient-rich deep water to the surface. (Portuguese and English abstracts). In: Fertility of the Sea, John D. Costlow, editor, Gordon Breach, 2: 401-415.

fertilization
Weatherley, A., and A.G. Nicholls, 1955.
The effects of artificial enrichment of a lake.
Australian J. Mar. & Fresh Water Res. 6(3):443-468.

fetch

fetch
Hino, Mikio, 1966.
A theory on the fetch graph, the roughness of the sea and the energy transfer between wind and wave
Coast. Engng Japan, 9:12-25.

fetch
Ikonnikova, L.N., 1971.
On diversity of wave heights within short-range fetch. (In Russian; English abstract).
Okeanologiia 11(5): 873-880.

fetch
Sager, Günther 1970.
Windwirklängen in der Nordsee.
Beitr. Meeresk. 27:37-45

fetch, ancient
Tanner, William F., 1971.
Numerical estimates of ancient waves, water depth and fetch.
Sedimentology, 16(1/2): 71-88

f/h maps
Moore, D. R., 1969.
Construction of f/H maps and comparison of measured deep-sea motions with the f/H map for the North Atlantic Ocean.
Techn. Rept. Chesapeake Bay Inst., 51 (Ref. 69-4). 61 pp

fighting
Hazlett, Brian A., 1971
Interspecific fighting in three species of brachyuran crabs from Hawaii.
Crustaceana 20(3): 308-314

films

films
*Craik, Alex D.D., 1968.
Wind-generated waves in contaminated liquid films.
J. fluid Mech., 31(1):141-161.

"films"
Dolgopolskaia, M.A., A.Z. Shapiro and IU. A. Gorbenko, 1961.
[Decay of film-forming elements not overgrown with red sea microorganisms.]
Trudy Sevastopol Biol. Sta., 14:303-308.

films, effect of
Dorrestin, R., 1951.
General linearized theory of the effect of surface films on water ripples. Proc. K. Nederl. Akad. Weten., Ser. B, 54:260-272; 350-356.

film accumulation
*Garrett, W.D., 1967.
Damping of capillary waves at the air-sea interface by oceanic surface-active material.
J. mar. Res., 25(3):279-291.

films, effect of
Baudoin, R. 1971.
Les phénomènes de surface en Ecologie, l'écume marine, les sables alvéolaires.
Vie et Milieu Suppl. 22(2): 753-781

films, monomolecular
Goodrich, F.C., 1962.
On the damping of water waves by monomolecular films.
J. Phys. Chem., 66(10):1858-1863.

films
Jones, T. G., B.A. Pethica and D.A. Walker, 1963.
A new interface balance for studying films spread at the oil-water interface.
J. Colloid Sci., 18(5):485-488.
fresh water

films
Kitchener, J.A., 1963.
Surface forces in thin liquid films.
Endeavour, 22(87):118-122.

related to slicks, etc.

filter feeders

filter feeders
Ali, R.M., 1970.
The influence of suspension density and temperature on the filtration rate of Hiatella arctica. Marine Biol., 6(4): 291-302.

filter feeders
Chesher, Richard H., 1969
Filter feeders
Oceans 2(2): 40-47 (popular)

filter feeders
Chipman, W.A., and J.G. Hopkins, 1954.
Water filtration by the bay scallop, Pecten irradians, as observed with the use of radioactive plankton. Biol. Bull. 107(1):80-91.

filter feeders
Davids, C., 1964.
The influence of suspensions of micro-organisms of different concentrations on the pumping and retention of food by the mussel (Mytilus edulis L.).
Netherlands J. Sea-Research, 2(2):233-249.

filter-feeders
Fox, D.L., C.H. Oppenheimer and J.S. Kittredge, 1953.
Microfiltration in oceanographic research. II. Retention of colloidal micelles by adsorptive filters and by filter-feeding invertebrates; proportions of dispersed organic to dispersed inorganic matter and to organic solutes.
J. Mar. Res. 12(2):233-243, 3 textfigs.

filters feeders
Haven, Dexter S. and Reinaldo Morales-Alamo, 1970.
Filtration of particles from suspension by the American oyster Crassostrea virginia.
Biol.Bull.mar.biol.Lab., Woods Hole, 139(2):248-264.

filter feeders
Haven, Dexter S., and Reinalde Morales-Alamo, 1966.
Aspects of biodeposition by oysters and other invertebrate filter feeders.
Limnol. Oceanogr., 11(4):487-498.

filter feeders
Barker Jørgensen, C., 1960
Effeciency of particle retention and rate of water transport in undisturbed lamellibranchs.
J. du Conseil 26(1): 94-116.

filter feeders
Jørgensen, C. Barker, 1955.
Quantitative aspects of filter feeding in invertebrates. Biol. Rev. 30(4):391-354.

filter feeders
Jørgensen, C.B., and E.D. Goldberg, 1953.
Particle filtration in some ascidian and lamellibranchs. Biol. Bull. 105(3):477-489, 8 textfigs.

filter feeders
Lasker, Reuben, 1960.
Utilization of organic carbon by a marine crustacean: analysis with carbon-14.
Science, 131(3407):1098-2000.

filter feeders
Reeve, M.R., 1963.
The filter-feeding of Artemia. 1. In pure cultures of plant cells.
J. Exp. Biol., 40:195-205.

filter-feeders
Reeve, M.R., 1963.
The filter-feeding of Artemia. II. In suspension of various particles.
J. Exp. Biol., 40:207-214.

filter-feeders
Reeve, M.R., 1963.
The filter-feeding of Artemia. III. Faecal pellets and their associated membranes
J. Exp. Biol., 40:215-221.

filter feeders
Ryther, J.H., 1954.
Inhibitory effects of phytoplankton upon the growth, reproduction and survival. Ecology 35(4):522-533.

filter feeders
Sushchenia, L.M., 1963.
The ecological-physiological peculiarities of filter-feeders among the planktonic Crustacea. (In Russian).
Trudy Sevastopol Biol. Sta., 16:256-276.

filter feeders
Sushenia, L.M., 1958.
Quantitative data on filtration feeding of planktonic crustaceans. (In Russian).
Nauchnye Doklady Vyssheishkoly, Biol. Nauk, (1): 16-

filter feeders
Theede, Hans, 1963
Experimentelle Untersuchungen uber die Filtrationsleistung der Miesmuschel Mytilus edulis L.
Kieler Meeresf., 19(1):20-41.
German and English abstract

filter feeders
Umezu, Takeshi, Kazuo Ueda and Atsushi Furukawa, 1967.
A preliminary note on selection of food particles by mussel Mytilus. (In Japanese; English abstract).
Bull. Naikai reg. Fish. Res. Lab., 24:11-20.

filter-feeders
Van Gansen, P., 1960.
Adaptations structurelles des animaux filtrants.
Ann. Soc. R. Zool. Belgique, 90(2):161-231.

filterfeeding
Walters, Vladimir, 1966.
On the dynamics of filter-feeding by the wavyback stickleback (Euthynnus affinis).
Bull. Mar. Sci., 16(2):209-221.

filter feeders
Winter, J.E., 1969.
Über den Einfluß der Nahrungskonzentration und anderer Faktoren auf Filtrierleistung und Nahrungsausnutzung der Muscheln Arctica islandica und Modiolus modiolus. Marine Biol., 4(2): 87-135.

filter-feeders, effect of
Mazzara, M.R., R. Guidone, M.A. Mazzara et H.F. Roecker, 1970.
Etude de l'activité filtrante de certains animaux marins (Salpa, Ciona, Amphioxus) à l'egard de bactéries marquées comparée à l'activité de certains mollusques comestibles (Mytilus, Tellina). Rev. int. Océanogr. méd. 15-16: 131-143

filtrates, effect of

filtrates, effect of
Jørgensen, Erik G., and E. Steemann Nielsen, 1961.
Effect of filtrates from cultures of unicellular algae on the growth of Staphylococcus aureus.
Physiol. Plantarum, 14:896-908.

filters, effect of
Quinn, James G. and Philip A. Meyers, 1971.
Retention of dissolved organic acids in sea-water by various filters. Limnol. Oceanogr. 16(1): 129-131.

filtering rate
Coughlan, J., 1969.
The estimation of filtering rate from the clearance of suspensions. Marine Biol., 2(4): 356-358.

Filtration, effect of
Marvin, K.T., R.R. Proctor, Jr. and R.A. Neal, 1970.
Some effects of filtration on the determination of copper in freshwater and saltwater. Limnol. Oceanogr., 15(2): 320-325.

financial support

financial support
Waterman, Alan T., 1966.
Federal support of science: what is the role of science in achieving national goals? On what basis is its federal support justified.
Science, 153(3742):1359-1361.

fingering interface
Huppert, Herbert E., 1971.
On the stability of a series of double-diffusive layers. Deep-Sea Res. 18(10): 1005-1021.

fiords

FISH
(1) The first part is subdivided to answer questions about fish: their anatomy, physiology (sense of hearing, smell, vision, etc.), schooling, etc.
(2) The second part — far from complete — is a guide to the fish themselves: merely lists of species, in a particular habitat, in a geographical locality, by taxonomic group (family, genus, etc.), by common names (chiefly commercial fish)

FISH - Part I

fish
Cervigón, Fernando, 1967.
Los peces.
In: Ecología marina, Mongr. Fundación La Salle de Ciencias Naturales, Caracas, 14:308-355.

fish
Hardy, Sir Alister, 1959.
The open sea: its natural history. II. Fish and fisheries.
Collins, London, 322 pp.

fish family
Terry, Richard D., editor, 1966.
Ocean Engineering, 1. Energy sources and energy conversion, waste conversion and disposal. 2. Undersea construction, habitation and vehicles; recreation.
Western Periodicals Co., North Hollywood, Calif. Vol. 3:431pp. (multilithed).

fish, age determination
*Alagarswami, K., Yoshio Hiyama and Yukio Nose, 1969.
Studies on age and growth of the Japanese mackeral.
Rec. oceanogr. Wks. Japan, 10(1):39-63.

Fish. Anat. and/or Physiol.
SEE ALSO: SWIM BLADDER in Miscel. file

fish, anat. physiol.
See also: fish, elasmobranch, etc.

fish, anat.-physiol.
Blaxter, J.H.S., 1963
The behaviour and physiology of herring and other clupeids.
In: Advances in Marine Biology, F.S. Russell, Editor, Academic Press, London and New York, 1:261-393.

fishes, anat. physiol.
Cunningham, J. T. 1897
On the histology of the ovary and the ovarian ova in certain marine fishes.
Quart. Jour. Micr. Sci., 40: 101-164

fishes, anat.-physiol.
Denton, Eric, 1971.
Reflectors in fishes.
Scient. Am., 224(6):65-72

fish - anat. -physiol
Fange, Ragnar, and Kjell Fugelli, 1963.
The rectal salt gland of elasmobranchs, and osmoregulation in chimaeroid fishes.
Sarsia, Univ. i Bergen, 10:27-34.

fish
Fourmanoir, P., 1971.
Notes ichtyologiques (III). Cah. ORSTOM, sér. Océanogr. 9(2): 267-278.

fish, anat. -physiol
*Gordon, Malcolm S., 1968.
Oxygen consumption of red and white muscles from tuna fishes.
Science, 159(3810)87-90.

fishes, anat physiol.
*Graham, Jeffrey B., 1971
Temperature tolerances of some closely related tropical Atlantic and Pacific fish species.
Science 172(3985): 861-863.

fish, sharks (anat.-physiol.).
*Katsuki, Yasuji, Keiji Yanafisawa, Albert L. Tester and James I Kendall, 1969.
Shark pit organs: responses to chemicals.
Science, 163(3865):405-407.

fish, anat.-physiol
Marshall, N.B., 1967.
Sound-producing mechanisms and the biology of deep-sea fishes.
In: Marine bio-acoustics, W.N. Tavolga, editor, Pergamon Press, 2:123-132.

fish, anat.
Nair, M.G.K., 1970.
The anatomy and embryology of the heart and its conducting system of the dogfish, Carcharias sorrah Müll. and Henle.
Zool. Anz. 185 (3/4): 265-274.

fish, anat.-physiol
Schneider, Hans, 1967.
Morphology and physiology of sound-producing mechanisms in teleost fishes.
In: Marine bio-acoustics, W.N. Tavolga, editor, Pergamon Press, 2:135-155.

fishes, anat. physiol.
Wright, P.G., 1958.
An electrical receptor in fishes. Nature 181 (4601):64-65.

Fish, artificial propagation

fish, artificial propagation
Dannevig, Gunnar, 1963.
Artificial propagation of cod - some recent results of the liberation of larvae.
Fiskeridirekt. Skrift., Ser. Havundersøgelser, 13(6):73-79.

fish attacks

fish attacks
See chiefly: fish, sharks
There is a subsection on attacks

fish attacks
Loftin, Horace, 1967.
Fish attacks on humans on the Caribbean coast of Panama.
Carib. J. Sci., 7(3/4): 159.

fish bait

fish bait
Hiyama, Y., S. Yoshizuki and H. Nakai, 1955.
An analysis of the fish attracting effect of "Komashi" or fish attracting bait. Jap. J. Ichthy 4:139-152.

fish behavior

fishes
Fishelson, L., 1964.
Observation on the biology and behaviour of Red Sea coral fishes. Contribution to the knowledge of the Red Sea, No. 30.
Sea Fish. Res. Sta., Israel, Bull., No. 37:11-26.

fishes, behavior of
Hasler, A.D., R.M. Horrall, W.J. Wisby and W. Braemer, 1958.
Sun-orientation and homing in fishes.
Limnol. and Oceanogr., 3(4):353-361.

fish behavior
Hester, Frank J., 1968.
Underwater photography in the study of fish behavior. Proceedings, Seminar on Underwater Photo-Optical Instrumentation Applications, Society of Photo-Optical Instrumentation Engineers, 12: 81-83, 1968. Also in: Collected Repr. Div. Biol. Res. Bur. Comm. Fish., U.S. Fish Wildl. Serv., 1968, 1.

fish behavior
Pfeiffer, W., 1962.
The fright reaction of fish.
Biol. Rev., 37(4):495-511.

fish behaviour
Springer, S., 1957.
Some observations on the behavior of schools of fish in the Gulf of Mexico. Ecology 38(1):166-171

fish, behavior of
Suyehiro, Y., 1934.
Some observations on the unusual behaviour of fishes prior to an earthquake.
Bull. Earthquake Res. Inst., Suppl. Vol. 1:228-231.

fishes, behaviour of
Woodhead, P.M.J., 1966.
The behaviour of fish in relation to light in the sea.
In: Oceanography and marine biology, H. Barnes, editor, George Allen & Unwin, Ltd., 4:337-403.

fish behavior
Woods, J.D., and J.N. Lythgoe, editors, 1971.
Underwater science: an introduction to experiments by divers.
Oxford Univ. Press 330pp.

fishbites

"fish bites"
*Forster, G.R., 1967.
A note on two rays lacking parts of the snout.
J. mar. biol. Ass., U.K., 47(3):499-500.

"fishbite"
Frassetto, R., 1966.
Further evidence of "fishbite" damage on deep-sea mooring lines in the Mediterranean.
Limnol. Oceanogr., 11(3):435-437.

fishbites
Turner, Harry J., Jr., 1965.
Fishbite; more experience and analysis.
GeoMarine Techn., 1(6):20-21.

fish bites
Turner, Harry J., Jr., and Bryce Prindle, 1968
The vertical distribution of fishbites on deep-sea mooring lines in the vicinity of Bermuda.
Deep-Sea Res., 15(3):377-379.

fishbites
Turner, Harry J., Jr., and Bryce Prindle, 1965.
Some characteristics of "fishbite" damage on deep-sea mooring lines.
Limnol. and Oceanogr., Redfield Vol., Suppl. to 10:R259-R264.

fish, blood of
Doolittle, R.F., and D.M. Surgenev, 1962.
Blood coagulation in fish.
Amer. J. Physiol., 203(5):964-970.

fishes, blood
Gunter, Gordon, L.L. Sulya and B.E. Box, 1961.
Some evolutionary patterns in fishes' blood.
Biological Bulletin, 121(2):302-306.

fish, blood of
Hemmingsin, Edvard A., and Everett L. Douglas, 1969.
Studies of respiration in Antarctic hemoglobin-free fishes.
Antarctic J., U.S.A., 4(4):108.

fish - blood
Ruud, J.T., 1954.
Vertebrates without erythrocytes and blood pigment. Nature 173(4410):848-850.

fish, blood
Ruud, Johan T., 1965.
The ice fish.
Scientific American, 213(5):108-114.

fish, blood

Scholander, P.F., 1957.
Oxygen dissociation curves in fish blood.
Acta Physiol. Scandinavica, 41(4):340-344.

fishes, blood

Scholander, P.F., and L. Van Dam, 1957.
The concentration of hemoglobin in some cold water Arctic fishes. J. Cell. & Comp. Physiol., 49(1):1-4.

fishes, blood

Van Dam, L., and P.F. Scholander, 1953.
Concentration of hemoglobin in the blood of deep sea fishes. J. Cell. Comp. Physiol. 41(3):522-524.

fishes, blood serum

Ichikawa, R., 1955.
Seasonal differences in the amount of inorganic phosphorus in blood serum of fishes.
Rec. Ocean. Wks., Japan, 2(1):113-116.

fish, blood serum

*Janssen, Werner A., and Caldwell D. Meyers, 1968.
Fish: serologic evidence of infection with human pathogens.
Science, 159(3814):547-548.

fish-blood types

Suzuki, A., 1962.
On the blood types of yellowfin and bigeye tuna.
American Naturalist, 96(889):239-

Paper presented at 10th Pacific Science Congress.

fish - (blood types)

Suzuki, Akimi, 1961.
On the blood types of yellow-fin and big-eye tuna. (Abstract).
Tenth Pacific Sci. Congr., Honolulu, 21 Aug.-6 Sept., 1961, Abstracts of Symposium Papers, 187-188.

fishes, chemistry of

fish, chemistry of

Barashkov, G.K., 1960.
Change in the activity of proteins of herring in the course of a year.
Nauchno-Techn. Biull., 2(12):34-36.

fishes, chemistry of

Blumer, Max, and David W. Thomas, 1965.
"Zamene", isomeric C19 monoolefins from marine zooplankton, fishes, and mammals.
Science, 148(3669):370-371.

fish, chemistry of

Christomanos, Anast., G. Gitsa and Aphr. Demetriadis, 1960.
On the iron content of the liver and spleen in certain fishes.
Res. Proc. Mar. Lab., Suppl. to Mar. Sci. Pages, Natl. Hellenic Oceanogr. Soc., 1:5 pp.

fish, chemistry of

Establier, Rafael, 1970.
Contenido en cobre, hierro, manganeso y cinc de varios ógonos del atún Thunnus thynnus (L.) del golfo de Cádiz.
Inv. pesq., Barcelona, 34(2): 399-408

fish, chemistry of

Establier, Rafael, 1970.
Contenido en cobre, hierro, manganeso y cinc de los ovarios del atún, Thunnus thynnus (L.); bacoreta, Euthynnus alleteratus (Raf.); bonito, Sarda sarda (Bloch); y melva, Auxis thazard (Lac.).
Inv. pesq. Barcelona 34(2): 171-175.

fish, chemistry of

Establier, R., 1966.
Estudios sobre los carotinoides de animales y plantas marinos. III. Carotinoides contenidos en las pinnulas del atún, Thunnus thynnus (L.).
Inv. Pesq., Barcelona, 30:497-500.

fishes, chemistry of

Imanishi, N., 1961
Studies on the inorganic chemical constituents of marine fishes. XII. On the distribution of manganese in marine fishes. (In Japanese; English abstract).
J. Oceanogr. Soc., Japan, 17(3):161-164.

chemical composition fish

Martin, DeCourcey, and Edward D. Goldberg, 1962
Oceanographic studies during Operation "Wigwam". Uptake and assimilation of radiostrontium by Pacific salmon.
Limnol. and Oceanogr., Suppl., to Vol. 7: lxxvi-lxxxi.

fish, chemistry of

*Pearcy, William G., and Charles L. Osterberg, 1968.
Zinc-65 and manganese-54 in albacore Thunnus alalunga from the west coast of North America.
Limnol. Oceanogr., 13(3):490-498.

fish, chemistry of

Raja, B.T. Antony 1969.
The Indian oil sardine.
Bull. cent. mar. Fish. Res. Inst. India 16: 128 pp. (mimeographed).

fish, chemistry

Renfro, William C., Roger V. Phelps, and Thomas F. Guthrie, 1969.
Influence of preservation and storage on radiozinc and total zinc concentration in fish.
Limnol. Oceanogr., 14(1): 168-170.

fish, chemistry of

Shatunovskii, M.I., 1970.
The composition of fatty acids of lipids of spring- and autumn-breeding Baltic herring roe, fingerlings and adults of the Gulf of Riga (the Baltic Sea).
Dokl. Akad. Nauk, SSSR, 195(4): 962-964

fishes, chemistry

Urist, Marshall R., and Karel A. Van de Putte, 1967.
Comparative biochemistry of the blood of fishes: identification of fishes by the chemical composition of serum. In: Sharks, skates, and rays, Perry W. Gilbert, Robert F. Mathewson and David P. Rall, editors, Johns Hopkins Univ. Press, 271-285.

fish culture

See also: farming, fish

fish culture

See: cultures
farming, fish

fish, data only

fish (data only)

California, Marine Research Committee, 1966.
California Cooperative Oceanic Fisheries Investigations, Data Rept., Nos. 5 and 6: 62 pp; 7 pp.

fish, deep records

fish, deep records

Wolff, T., 1961
The deepest recorded fishes.
Nature, 190(4772): 282.

fish, demand for

fish, demand for

Christy, Francis T. Jr. and Anthony Scott 1965.
The common wealth in ocean fisheries; some problems of growth and economic allocation.
Resources for the Future, Inc. Johns Hopkins Univ. Press: 281 pp.

fish diseases

fish diseases

See: diseases, fish

fish distribution

fish

*Miyake, Makoto Peter, 1968.
Distribución del barrilete en el Oceano, Pacifico, basada en los registros de la pesca japonesa palangrera de atunes, segun las capturas incidentales. (In English and Spanish).
Bol. Com. interamer. Atun trop., 12(7):511-608.

fish, diurnal migrations

Kawaguchi, Kouichi, 1969.
Diurnal vertical migration of micronektonic fish in the western North Pacific. Bull. Plankt. Soc. Japan, 16(1): 63-66.
(In Japanese; English abstract)

fish, ecology

fish ecology

*Blaxter, J.H.S., 1968.
Light intensity, vision, and feeding in young plaice.
J. exp. mar. Biol. Ecol., 2(3):293-307.

fish

*Durand, J.R., 1967.
Étude des poissons benthiques du plateau continental congolais. 3. Étude de la repartition de l'abondance et des variations saisonnières.
Cah. ORSTOM, Sér, Océanogr., 5(2):3-68.

fish, ecology of
*Edwards, R.R.C., D.M. Finlayson and J.H. Steele, 1969.
The ecology of O-group plaice and common dabs in Loch Ewe. II. Experimental studies of metabolism.
J. exp. mar. Biol. Ecol., 3(1):1-17.

fish ecology
*Edwards, R., and J.H. Steele, 1968.
The ecology of o-group plaice and common dabs at Loch Ewe. 1. Population and food.
J. exp. mar. Biol. Ecol., 2(3):215-238.

fish, ecol.
Ginés, Hno., y R. Margalef, editores, 1967.
Ecología marina.
Fundación La Salle de Ciencias Naturales, Caracas
Monografía 14: 711 pp.

fish ears
Moulton, J.E., 1969.
Fish ears.
Oceanus, 15(1):18-19. (popular).

fish, effect of

fish, effect of
Bakus, Gerald J., 1966.
Some relationships of fishes to benthic organisms on coral reefs.
Nature, 210(5033):280-284.

fish, effect of
Cook, David O., 1971.
Depressions in shallow marine sediment made by benthic fish.
J. Sedim. Petrol. 41 (2): 577-578.

fish, effect of
Stanley, Daniel J., 1971
Fish-produced markings on the outer continental margin east of the Middle Atlantic States.
J. Sedim. Petrol., 41(1): 159-170.

fish, effect of
Weston, D.E., 1966.
Fish as a possible cause of low-frequency acoustic attenuation in deep water.
J. acoust. Soc., Am., 40(6):1558.

fish, "electroperception"
See: fish, rays

fish eggs

fish eggs (planktonic)
*Einarsson, Herman (posthumous) and George C. Williams, 1968.
Planktonic fish eggs of Faxafloe, southwest Iceland, 1948-1957.
Rit Fiskideild., 4(5):1-15.

fish (eggs & larvae)
Fish, C.J., and M.W. Johnson, 1937
The biology of the zooplankton population in the Bay of Fundy and the Gulf of Maine with special reference to production and distribution. J. Biol. Bd., Canada 3(3):189-322, 29 tables, 45 text figs.

fish eggs
Kubota, Minoru, Koichi Hagiya and Shigeru Kimura, 1971.
Amino acid composition and solubization of fish egg shell membrane. (In Japanese; English abstract). J. Tokyo Univ. Fish. 57(2-1): 87-94.

fish eggs
*Marinaro, J.-Y., and M. Benkara-Mostefa, 1968.
Contribution à l'etude des oeufs et larvas pélagiques de poissons méditerranéens. III. A la recherche d'une aire de ponte de la sardine (Sardinia pilchardus Walb.) dans les eaux Maltaises.
Pelagos, 9:85-92.

fish eggs
Wiborg, K.F., 1944
The production of plankton in a land locked fjord. The Nordåsvatn near Bergen in 1941-1947. With special reference to copepods. Fiskeridirektoratets Skrifter Serie Havundersøkelser (Rept. Norwegian Fish. and Marine Invest.) 7(7):1-83, Map.

fish eggs
Williams, George C. 1968.
Bathymetric distribution of planktonic fish eggs in Long Island Sound.
Limnol. Oceanogr. 13(2): 382-384.

fish farming
see also under cultures and under: farming

fish farming
See: cultures, fish
farming, fish

fish, fat variability
Shulman, G.E. 1971.
Variability of fat in some species of fish from the Azov-Black Sea Basin. (In Russian)
Gidrol. Zh. 7(3): 86-91

fish fauna
Hubbs, C. L., 1948.
Changes in the fish fauna of western North America correlated with changes in ocean temperature. J. Mar. Res. 7(3):459-482.

fish fluctuations

fish fluctuations
*Muzinic, Radosna, 1967.
On fluctuations and spatial distribution of catches in the Yugoslav Sardine, Sprat, anchovy, mackeral and Spanish mackeral fisheries. (In Jugoslavian; English summary).
Acta adriat. 13(3):1-29.

fish, feeding mechanism

fish feeding
Bakus, Gerald J., 1967.
The feeding habits of fishes and primary production at Eniwetok, Marshall Islands.
Micronesica, J. Coll. Guam, 3(2): 135-149.

fish (feeding mechanisms) (plankton feeders)
Tasuda, Fujio, 1963.
The food selectivity of some plankton feeders with regard to the amount and size of bait.
Rec. Oceanogr. Wks., Japan, n.s., 7(1):57-64.

fish flour / meal
SEE ALSO FISH PRODUCTS

fish flour, fish meal, fish protein concentrate have been filed together here

fish flour
Aldrin, J.F., 1965.
Note sur quelques farines de poissons tropicaux.
Rev. Trav. Inst. Pêches Marit., 29(4):421-430.

fish flour
Allison, J.B., et al., 1957.
Notes on the fish meal and fish flour.
Nutrition Division F.A.D.

fish flour
Anonymous, 1962.
Fish flour: National Academy study disputes the Food and Drug Administration's "filthy" label.
Science, 138(3543):880-881.

fish meal
Anonymous, 1962.
Production specifications for concentrated fish protein.
Fish. News Intern., 1(2):58-59.

Abstr. in:
Comm. Fish. Abstr., 1962, 15(7):15
FAO World Fish. Abstr., 1963, 14(1):41 (FWS6.54)

fish meal
Anon., 1957.
Details of new fish flour process.
South African Fish. Ind. Res. Inst. 12(4):63

Capetown tries fish flour in enriched brown bread. Ibid 11(4):53

fish meal
Anon., 1957.
Herstellungsprozess des neuen Fischmehls.
Inform. über Fischwirtschaft des Auslandes 7(3):40

fish meal
Anon., 1957.
Herstellungsprozess des neuen Fischmehls in der südafrikanischen Union,
Die Fischwaren- und Feinkostindust. 29(4):56

fish meal
Anon., 1957.
How flour is made from pilchards.
Fish. Newsletter Australia 16(4):29

fish meal
Braumariage, Dale S., 1968.
Manufacture of fish meal from Florida's fishery waste and under-exploited fishes. 1. Wet reduction. 2. Liquefaction.
Techn. Ser. Fla Bd Conserv. 54: 56 pp.

fish flour
Bender, A.E., and S. Haizelden, 1957.
Biological value of the proteins of a variety of fish meals.
Brit. J. Nutrition, 11:42-43.

fish meal
Booda, Larry L. 1967.
Fish meal in pilot production.
Undersea Techn. 8(4): 16-17.

fish meal
Contreras, Emilio, y Claudio Romo, 1965.
Estudios preliminares sobre digestibilidad de harina de anchoveta en Chile.
Inst. Fomento Pesquero, Chile, Publ., 12:16 pp. (mimeographed)

fish meal
García Cabrejas José, y Antonio E Malaret, 1969.
La harina de pescado en Argentina
Publ. (Ser. Informes Técn.) Mar del Plata, Argentina, 15: 38pp

fish flour
Greenberg, D.S., 1963.
Fish flour: administration's interest has not been matched by funds for needed research.
Science, 139(3558):891-892.

fish flour
Lahiry, N.L., et al., 1962.
Preparation of edible fish flour from oil-sardine (Clupea longiceps).
Food Science, Central Food Techn. Res. Inst. Mysore, India, 11:37-39.

fish meal
*Liebeschutz, Mario, 1966.
Industria de harina de anchoveta en el norte de Chile. Capacidad, equipos y utilization, 1965.
Publ. Inst. Foment pesq., Chile, 18:45 pp. (mimeographed).

fish flour
Metta, V.C., 1960.
Nutritional value of fish flour supplements.
J. Amer. Diet. Assoc. 37(3):234-240.

fish meal
*Morishita, Tatsuo, Takashi Takahashi, Manabu Kitamikado and Shinko Tachino, 1967.
On the fish meal and fish oil from anchovy Engraulis japonicus T. and S., in Ise Bay. Preparation on laboratory scale. (In Japanese; English abstract).
J. Fac. Fish., pref. Univ. Mie, 7(1):47-55.

fish flour
Nath, R.L., N.K. Ghosh and R. Dutt, 1961.
Preparation of protein-rich biscuit with fish flour from hammerhead shark (Zygoena blochii).
Bull., Calcutta School Trop. Med., 9:12-13.
Abstr. in:
J. Sci. Food and Agric., 12(10):194.

fish flour
Nath, R.L., S.K. Pain and R. Dutt, 1961.
Use of fish-protein hydrolysate in the diet. 1. Preparation of biscuit from protein hydrolysate of fish.
J. and Proc. Inst. Chem., India, 33:64-68.
Abstr. in:
Chem. Abstr., 55:276891

fish meal
Nunn, Robert R., 1969.
Fish protein concentrate production is on the rise: the Viobin process and how it works.
Ocean Industry, Gulf Publishing Co., 4(1):36-40. (popular)

fish protein concentrate
Nunn, Robert R., 1968.
Fish protein concentrate production is on the rise.
Ocean Industry, 3(11):47-50.

fish flour
Olden, J.H., 1960.
Fish flour for human consumption.
Commerc. Fish Rev. 22(1):12-18.

fish meal
Purcell Maschke, Arturo, 1965.
Consumo y comercializacion de la harina de pescado en Chile, presente y futuro.
Inst. Fomento Pesquero, Chile, Publ., No. 13: 20 pp.

fish meal
Raja, B.T. Antony, 1969.
The Indian oil sardine.
Bull. cent. mar. Fish. Res. Inst., India, 16:128 pp. (mimeographed).

fish protein concentrate
Ramsey, James E. 1969.
FPC [Fish protein concentrate]
Oceans 2(2): 76-80 (popular)

fish "flour"
Snyder, Donald G., 1966.
Marine protein concentrate
Exploiting the Ocean, Trans. 2nd Mar. Techn. Soc. Conf., June 27-29, 1966, 530-534.

fish meal
Suzuki, Kosaku and Kuman Saruya, 1964.
Fish oil-solvent-water system examined for the foundation of the preparation of fatless fish meal. (In Japanese; English abstract).
Bull. Jap. Soc. Sci. Fish., Tokyo, 30(1):37-41.

fish meal
Tomes, Eilif, y Paul George, 1970
Algunos aspectos de la producción de harina y aceite de pescado
Informe tecn. Proyecto Invest. Desarrollo Pesq. MAC-PNUD-FAO, 3: 32pp

fish flour
Wiechers, S.G., F. Schweigart and M.K. Rowan, 1960.
Microbiological process for the production of a bland fish meal.
Research Rept., C.S.I.R., S. Africa, No. 179:iv+17 pp.

fish: food of: SEE: food of fish

fish food

See: food of fish

fishes, freezing resistance of
*DeVries, Arthur L., and Donald E. Wohschlag, 1969.
Freezing resistance in some Antarctic fishes.
Science, 163(3871):1073-1075.

fish, growth of

fish, growth of
Hohendorf, Kurt, 1966.
Eine Diskussion der Bertelanffy-Funktionen und ihre Anwendung zur Charakterisierung des Wachstums von Fischen.
Kieler Meeresforsch., 22(1):70-97.

fish, growth of
*Iles, T.D., 1968.
Growth studies of North Sea herring II. O-group growth of East Anglican herring.
J. Cons. perm. int. Explor. Mer, 32(1):98-116.

fish growth
*Byung, D. Lee and Yong Uk Kim, 1969.
Studies on the seedlings production of marine fishes in Korea. II. Rearing and growth of Girella punctata Gray. (In Japanese; English abstract). Publ. mar. Lab., Pusan Fish. Coll., 2: 13-18.

fish growth
*Byung, D. Lee and Yong Uk Kim 1969.
Studies on the seedlings production of marine fishes in Korea. 1. Early egg development, larvae and growth of Fugu rubripes (Temminck et Schlegel). (In Japanese; English abstract). Publ. mar. Lab., Pusan Fish. Coll., 2: 1-11.

fish, growth
Morovic, Dinko, 1964.
Contribution a la connaissance de la croissance annuelle de Mugil cephalus, L. et Mugil chelo, Cuv. dans l'Adriatique. (In Jugoslavian; French resume).
Acta Adriatica, 11(1):195-204.

fish, growth
Watson, John E., 1967.
Age and growth of fishes.
Am. Biol. Teacher, 29(6):435-438.

fishes, hearing in

fishes, hearing in
*Buerkle, Udo, 1967.
An audiogram of the Atlantic cod, Gadus morhua L.
J. Fish. Res. Bd., Can., 24(11):2309-2319.

fishes, hearing
Dijkgraaf, S., 1960
Hearing in bony fishes. Proc. Roy. Soc., London, Ser. B, 152(946): 51-53.

fishes, (hearing)
Tavolga, William N., and Jerome Wodinsky, 1963.
Auditory capacities in fishes. Pure tone thresholds in nine species of marine teleosts.
Bull. Amer. Mus. Nat. Hist., 126(2):179-239.

fishes, hearing in
Van Bergeijk, W.A., 1964.
Directional and non directional hearing in fish.
In: Marine bio-acoustics, W.N. Tavolga, Editor, Macmillan Co., 281-299.

fishes, hearing in
Wisby, W.J., J.D. Richard ~~xx~~ D.R. Nelson and S.H. Gruber, 1964.
Sound perception in elasmobranchs.
In: Marine bio-acoustics, W.N. Tavolga, Editor, Macmillan Co., 255-268.

fishes, hearing in
Wodinsky, J., and W.N. Tavolga, 1964.
Sound detection in teleost fishes.
In: Marine bio-acoustics, W.N. Tavolga, Editor, Macmillan Co., 269-280.

fish, homing of

See also: fish - Anguilla
fish - eels

fish, homing of (eels)
Tasch, F.-W. 1968.
Heimfindevermögen von Aalen in der Nordsee.
Umschau 68 (8): 247-248

fish, hydrodynamic resistance

fish, hydrodynamic resistance
Brégnet, A., 1961.
La résistance hydrodynamique du poisson.
Bull. Centre Études Recherches Sci. Biarritz, 3(4):437-448.

fish, large

fish, large
Anon., 1943.
The largest fish in the world. Nat. Hist. 42(2): 70-71.

fish larvae, see LARVAE, fish

fish larvae

See: larvae, fish in "Organismal" index

fish, lateral line organs
Dijkgraaf, S., 1963
The functioning and significance of the lateral-line organs.
Biol. Rev., 38(1):51-106.

fish, lateral line organs
Harris, G.G., and W.A. van Bergeijk, 1962.
Evidence that the lateral-line organ responds to near-field displacements of sound sources in water.
J. Acoust. Soc., America, 34(12):1831-1841.

fish, light organs
Bertelsen, E., 1958.
A new type of light organ in the deep-sea fish Opisthoproctus. Nature 181:862-863.

fish, (luminous organs)
Haneda, Yata, 1963.
Luminous organs of fish, which emit light indirectly.
Rec. Oceanogr. Wks., Japan, 7(1):83-87.
(n.s.)

fish, lists of spp.
Swingle, Hugh A. 1971.
Biology of Alabama estuarine areas - Cooperative Gulf of Mexico Estuarine Inventory.
Bull. Alabama mar. Resources 5: 123 pp

fish meal

fish meal

See: fish flour, etc.

fish - migration

fish
Angelescu, Victor, y Maria L. Fuster de Plaza, 1965.
Migraciones verticales ritmicas de la merluza del sector Bonaerense (Merluciidae, Merluccius hubbsi) y su significado ecologico.
Anais Acad. bras. Cienc., 37 (Supl.):194-214.

fish, migration of
Baggerman, Bertha, 1962
Some endocrine aspects of fish migration.
Gen. and Comp. Endocrinology, Suppl. 1:188-205.

fish migrations
Baggerman, Bertha, 1961.
Hormonal factors in fish migration. (Abstract).
Tenth Pacific Sci. Congr., Honolulu, 21 Aug.-6 Sept., 1961, Abstracts of Symposium Papers, 165-166.

fish-migrations
Fontaine, M., 1946.
Vues actuelles sur les migrations de poissons. Experientia, II:233-237.

fish-migrations
Fontaine, M., 1946
Du rôle joué par les facteurs internes dans certaines migrations de poissons Étude critique des diverses méthodes d'investigations. Rapp. à la Commission internationale pour l'exploration de la mer. Stockholm.

fish migrations
Hasler, A.D., 1956.
Perception of pathways by fishes in migration.
Q. Rev., Biol., 31(3):200-209.

fish
Pearcy, William G., and R.M. Laurs, 1966.
Vertical migration and distribution of mesopelagic fishes off Oregon.
Deep-Sea Res., 13(2):153-165.

salmon
Scheer, B.T., 1939
"Homing instinct" in salmon. Quart. Rev. Biol. 14:408-430.

fish migrations
Sonina, M.A., 1968.
The migrations of the Barents Sea haddock and reasons for their changes. (In Russian; English abstract). Trudy polyar. nauchno-issled. Inst. morsk. ryb. Khoz. Okeanogr. (PINRO) 23: 383-401.

migrations
Talbot, F.H., and M.J. Penrith, 1968.
The tunas of the genus Thunnus in South African waters. 1. Introduction, systematics, distribution and migrations.
Ann. S. Afr. Mus., 52(1):1-41.

fish migrations
Went, Arthur E.J., and David J. Piggins, 1965.
Long-distance migration of Atlantic salmon.
Nature, 205(4972):723.

fish mortality

fish mortality
Hornell, J. 1917
A new Protozoan Cause of Widespread Mortality among Marine Fishes. Madras Fish Bull. No.11:53-66.

fish mortality
Popovici, Z., 1943.
Ungewöhnliches Fischsterben an der Westküste des Schwarzen Meeres. Ann. Inst. Rech. Pisc., Roumanie, 2:337-348.

fish "nursery"

fish "nursery"
Rickards, William L., 1968.
Ecology and growth of juvenile tarpon, Megalops atlanticus, in a Georgia salt marsh.
Bull. mar. Sci., Miami, 18(1):220-239.

fish oil

fish oil
*Morishita, Tatsuo, Takashi Takashi, Manabu Kitamikado and Shinko Tachino, 1967.
On the fish meal and fish oil from anchovy Engraulis japonicus T. and S., in Ise Bay. Preparation on laboratory scale. (In Japanese; English abstract).
J. Fac. Fish., pref. Univ. Mie, 7(1):47-55.

fish, olfaction
Marshall, N.B., 1967.
The olfactory organs of bathypelagic fishes.
In: Aspects of Marine Zoology, N.B. Marshall, editor, Symp. Zool. Soc. Lond., 19: 57-70.

fish orientation

fish orientation
*Bardach, J.E., J.H. Todd and R. Crickmer, 1967.
Orientation by taste in fish of the genus Ictalurus.
Science, 155(3767):1276-1278.

fish, origin of

fish, origin of
Romer, A.S., 1955.
Fish origins - fresh or salt water.
Pap. Mar. Biol. and Oceanogr., Deep-Sea Res., Suppl. to Vol. 3:261-280.

fish otoliths
Dannevig, Alf, 1956.
The influence of temperature on the formation of zones in scales and otoliths of young cod.
Fiskeridirekt. Skrift., Ser. Havundersøgelser, 11(7):16 pp.

fish cod otoliths
Mankevich, E.M., 1962
Biological peculiarities of northwest Atlantic cod differing in otolith structure. (In Russian; English summary).
Sovetskie Ribochoz. Issledov. Severo-Zapad. Atlant. Okeana, VNIRO-PINRO, 331-343.

fish, physiol. of

Bodrova, N.V., and B.V. Krayukhin, 1960.
Role of the receptors of the body surface in the mechanism of the reaction of fish to electric currents. (In Russian).
Trudy Inst. Biol. Vodokhranilishch, 3(6):266-272.

Translation:
OTS 63 - 11111 (1963)

fish, physiol. of.

Brown, M.E., Editor, 1957.
The physiology of fishes. 1. Metabolism.
Academic Press, Inc., N.Y., 447 pp.

fish-physiology

Busnel, R.G., and A. Drilhon, 1946
Recherches sur la physiologie des Salmonidés. II Sur l'oeuf et l'alevin de Salmo salar. Bull. Inst. Océanogr., Monaco, No.893.

fish-physiology

Busnel, R.G., Drilhon,A., and Raffy, A., 1946
Recherches sur la physiologie des Salmonidés. 1. Sur l'adaptation aux eaux saumâtres de Salmo iridaeus à ses differents stades. Bull. Inst. Océanogr., Moncao, No.893

fish-physiology

Fontaine, M., 1946
Du rôle joué par les facteurs internes dans certaines migrations de poissons. Etude critique des diverses méthodes d'investigations. Rapp. à la Commission internationale pour l'exploration de la mer. Stockholm.

fish-physiology

Fontaine, M., and O. Callamand, 1946
Sur les modifications du milieu intérieur des poissons au cours des changements de salinité et leur interprétation. C.R. Acad. Sci. 222:198-200.

fish, physiol.

Gelineo, Stefan, 1964.
Activité spontanée de certains poissons de mer, leur consommation d'oxygène et la concentration d'hémoglobine. (In Jugoslavian; French resume).
Acta Adriatica, 11(1):97-102.

fish, physiology

Saunders, R.L., 1963.
Respiration of the Atlantic cod.
J. Fish. Res. Bd., Canada, 20(2):373-386.

fishes, physiology

Scholander, L. Van Dam, J.W. Kanwisher, H.T. Hammond and M.S. Gordon, 1957.
Supercooling and osmoregulation in Arctic fish.
J. Cell. Com. Physiol., 49(1):5-24.

fish, physiology

Timet, Dubravko, 1963.
Studies on heat resistance in marine fishes. 1. Upper lethal limits in different species of the Adriatic littoral.
Thalassia Jugoslavica, 2(3):5-21.

fishes, physiology of

*Wittenberger,C., 1968.
Biologie du chinchard de la Mer Noire (Trachurus mediterraneus ponticus) XV. Recherches sur le metabolisme d'effort chez Trachurus et Gobius.
Marine Biol., 2(1):1-4.

fish, physiol.

Zeid, Hajdar, 1964.
La premiere maturite sexuelle chez Mullus barbatus dans la partie moyenne de l'Adriatique orientale. (In Jugoslavian; French resume)
Acta Adriatica, 11(1):127-133.

fish, plankton feeders

Margineanu, Carmen et Marius Iliescu, 1965
L'influence de la biomasse trophique du zooplancton sur les poissons planctonophages du littoral roumain de la mer Noire.
Rapp. Proc.-verb. Réun., Comm. int. Explor. scient. Mer Méditerranée 19(3): 423-425.

fish, poisonous

Halstead, Bruce W. 1970
Poisonous and venomous marine animals of the world.
U.S. Govt Printing Office, 1006 pp

fish populations

Andréu, Buenaventura, 1969.
Las branquispinas en la caracterización de las poblaciones de Sardina pilchardus (Walb.).
Inv. pesq. 33(0): 425-607.

fish populations

Tyler, A.V. 1971.
Periodic and resident components in communities of Atlantic fishes.
J. Fish. Res. Bd. Can. 28(7): 935-946

fish, predacious

fish, predacious

Kuroki, Toshiro, and Kazumi Kumanda, 1961.
The effects of underwater explosions for the purpose of killing predacious fishes.
(In Japanese; English abstract).
Bull., Fac. Fish., Hokkaido Univ., 12(1):16-32.

fish products: SEE ALSO fish flour

fish products

Seumenicht, K., 1963.
Fischindustrie, Fischmehl- und Fischölindustrie sowie Garnelendarren.
Jahres Deutsche Fischwirts, 1962/1963, 273-284.

fish production

Cushing, D.H., 1971.
Upwelling and the production of fish. Adv. mar. Biol. 9: 255-334.

fish propulsion

fish propulsion

Gero, D.R., 1952
The hydrodynamic aspects of fish propulsion. Amer. Mus. Novitates, No.1601: 32 pp.

fish propulsion

Gray, James, 1953
The locomotion of fishes. In: Essays in Marine Biology being the Richard Elmhirst Memorial lectures. Oliver + Boyd Edinburgh: 1-16

fish propulsion

Gray, James, 1953
Undulatory propulsion. Q. J. Microscopical Society 94 (4): 551-578.

fish propulsion

Silberberg, George C., 1966.
Sea animal locomotion and fish propulsion studies.
In: Whales, dolphins and porpoises, K.S. Norris, editor, Univ. Calif. Press, 477-481.

fish protein concentrate

See: fish flour, etc.

fish

Bruun, A. Fr., 1932.
A review of W. Schnakenbeck: Zum Rasenproblen bei den Fischen. J. du, Cons. 7(1):100-103.

fish, radioactivity of (bone)

fish, radioactivity of

Boroughs, Howard, 1958
The metabolism of radiostrontium by marine fishes.
Proc. Ninth Pacific Sci. Congr., Pacific Sci. Assoc., 1957, 16(Oceanogr.):146-151.

fish, radioactivity of bone

Hiyama, Yoshio, and Ryushi Ichikawa, 1958.
A measure on level of strontium-90 concentration in sea water around Japan at the end of 1956.
Proc. Ninth Pacific Sci. Congr., Pacific Sci. Assoc., 1957, Oceanogr., 16:141-145.

fish, radioactivity of

Tomiyama, Tetuo, and Kunio Kobahashi, 1962
Direct uptake of radioisotopes by fish.
Proc. Ninth Pacific Sci. Congr., Pacific Sci. Assoc., 1957, 16(Oceanogr.):159-166.

fish, rare

fish, rare

Blacker, R.W., 1962.
Rare fishes from the Atlantic Slope fishing grounds.
Ann. Mag. Nat. Hist., 5(53):261-271.

fish recruitment

Cushing, D.H., 1971.
The dependence of recruitment on parent stock in different groups of fishes. J. Cons. int. Explor. Mer, 33(3): 340-362.

fish, research

fish

Barsukov, V.V., and Ju. E. Permitin, 1960
[Ichthyological research.]
Arktich. i Antarktich. Nauchno-Issled. Inst. Sovetsk. Antarkt. Exped., Mezhd., Geofiz. God, Vtoraia Morsk. Exped., "Ob", 1956-1957, 7: 97-104.

fish scales

fish scales

Blair, A. A. 1937
The validity of age determination from the scales of land-locked salmon.
Science, 86:519-520

"scales" (sharks)

Chernyshev, O.B., and V. A. Earts 1971.
Some peculiarities of scale cover in sharks of different rate groups. (In Russian)
Gidrol. Zh. 7(4): 77-81

Chevey, P. 1936 fish scales
La mètode de "lecture des écailles" et les poissons de la zone intertropicale.
Notes Inst. Oceanogr. Indochine, 29:51-65

fish scales
Dannevig, Alf., 1956.
The influence of temperature on the formation of zones in scales and otoliths of young cod.
Fiskeridirekt. Skrift., Ser. Havundersøgelser, 11(7):16 pp.

fish scales
Dannevig, A. and Dannevig, G. 1937
The season in which "winter" zones in the scales of trout from southern Norway are formed.
Jour. du Cons., 12:192-198

fish scales
Griffiths, P.G., 1968.
An electronic fish scale proportioning system.
J.Cons.perm.int.Explor.Mer, 32(2):280-282.

fish scales
Kaganovskaya, S. M. 1937
The validity of estimating length and rate of growth by scales from various sections of the sardine (Sardinops melanosticta).
Bull. Pacific Sci. Inst. Fish. Ocean., Vladivostock, 12:115-124
English summary

fish scales
Taylor, H. F. 1916
The structure and growth of the scales of squeteague and the pigfish as indicative of life history.
Bull. U. S. Bur. Fish., 34:285-330

fish — scales
Walford, L. A. and K. Mosher, 1943
Studies on the Pacific pelchard or sardine (Sardinops caerulea). 3. Determination of age of adults by scales, and effect of environment on first years growth as it bears on age determination. U. S. Fish and Wildlife Service, Spec. Sci. Rept. No.21:29 pp. (mimeographed), 6 figs. (multilith).

fish — scales
Walford, L. A. and K. H. Mosher, 1943
Studies on the Pacific pilchard or sardine (Sardinops caerulea). 2. Determination of the age of juveniles by scales and otoliths. U. S. Fish and Wildlife Service, Spec. Sci. Rept. No. 20:19 pp. (mimeographed), 32 figs., 8 tables (multilith).

fish schooling

fishes, schooling
Breder, C.M., Jr., 1967.
On the survival value of fish schools.
Zoologica, N.Y., 52(2):25-40.

fish schools
Breder, C.M., Jr., 1965.
Vortices and fish schools.
Zoologica, N.Y. Zool. Soc., 50(2):97-114.

Brock, Vernon E., and Robert H. Riffenburgh, 1960 fish schooling
Fish schooling: a possible factor in reducing predation.
J. du Cons., 35(3)307-317.

fish schooling
*Cahn, Phyllis H., Evelyn Shaw and Ethel H. Atz, 1968.
Lateral-line histology as related in the development of schooling in the atherinid fish, Menidia.
Bull.mar.Sci., Miami, 18(3):660-670.

fish schools
Cushing, D.H., and F.R. Harden Jones 1968.
Why do fish school?
Nature, Lond. 218 (5145): 918-920.

fish schools
Flores, Luis, Oscar Guillén y Rogelio Villanieva, 1966.
Informe preliminar del Crucero de invierno 1965 (Mancora-Morro Sama).
Inst. Mar, Peru, Informe, No.11:1-34(multilithed).

fish schools
Fricke, H.W., 1970
Ein mimetisches Kollektiv - Beobachtungen an Fischschwärmen, die Seeigel nachahmen. Marine Biol., 5(4): 307-314.

Fishes, schooling of
Jorné - Safriel, Ora, and Evelyn Shaw, 1966.
The development of schooling in the atherinid fish Atherina mochon (Cuvier).
Pubbl. Staz. zool., Nappli, 35(1): 76-88.

fish schools
Larranets, M.G., 1967.
Sobre la agregación en peces pelágicos.
Investigación pesq., 31(1):125-135.

fish schools
Mejía, Jorge, y Luis Alberto Poma E., 1966.
Informe preliminar del Crucero de otoño 1966. (Cabo Blanco-Ilo).
Inst.Mar,Peru,Informe No.13:31pp. (multilithed).

fish schooling
Milanovskii, Yu. E., and V.A. Rekubratskii, 1960
Methods of studying the schooling behavior of fishes. (In Russian).
Nauchnye Doklady Vysshei Shkoly, Biol. Nauki, No 4:77-81.

Translation:
OTS 63-11116 - $0.50.

fish schooling
Mitsugi, Shinsuke, Takeo Koyama, Shin'ichi Yajima and Chosei Yoshimuta, 1966.
Behavior of fish schools against luminous substances in waters. (In Japanese; English abstract).
Bull. Tokai reg. Fish. Res. Lab., No. 48: 29-36.

fish schools
Moulton, James M, 1960.
Swimming sounds and the schooling of fishes.
Biol. Bull., 119(2):210-223.

fish schools
*Rosenthal, H., 1968.
Beobachtungen über die Entwicklung des Schwarmverhaltens bei den Larven des Herings Clupea harengus.
Marine Biol., 2(1):73-76.

fishes, schooling of
Shaw, Evelyn 1965.
The optomotor response and the schooling of fish.
ICNAF Spec. Publ. No.6:753-756.

fishes, schooling
Shaw, Evelyn, 1961.
The development of schooling in fishes. II.
Physiol. Zool., 34(4):263-272.

fish, schooling of
Symons, Philip E.K. 1971.
Estimating distances between fish schooling in an aquarium.
J.Fish. Res. Bd Can. 28(11): 1805-1806.

fish schools
Valdivia, Julio E., y Oscar Guillén, 1966.
Informe preliminar del Crucero de primavera 1965 (Cabo Blanco-Morro Sama).
Inst. Mar, Peru, Informe, No. 11:35-70.

fish, sense of smell in

fish, sense of smell
Adrian, E.D., + C. Ludwig, 1938.
Nervous discharges from the olfactory organs of fish. J. Physiol. 94:

fish, sense of smell
Allison, A.C., 1953.
The morphology of the olfactory system in vertebrates. Biol. Revs. 28(2):195-244

fish, sense of smell
Baglioni, S., 1910.
Zur Kenntnis der Leistungen einiger Sinnesorgane (Gesichtssinn, Tastsinn und Geruchsinn) und des Zentralnervensystems der Cephalopoden und Fische.
Z. f. Biol 53:255-286

fish - sense of smell
Baglioni, S. 1908.
Zur Physiologie des Geruchsinnes und des Tastsinnes der Seetiere. Versuch an Octopus und einigen Fischen. Ctbl. f. Physiol. 22:719-723

fish, sense of smell
Craigie, E.H., 1926.
A preliminary experiment upon the relation of the olfactory sense to migration of the sockeye salmon (Oncorhynchus nerca Walbaum) Trans. R. Soc. Canada (5)20:

fish, sense of smell
Dijkgraaf, S., 1933.
Untersuchungen über die Funktion der Seitenorgane an Fischen. Z. vergl. Physiol. 20:162-214.

fish, sense of smell
Göz, Hang, 1941.
Ueber den Art- und Individual-geruch bei Fischen. Zeits. vergleich. Physiol. 29(1/2):1-45.

fish, sense of smell
Mattes, E., 1938.
Olfacto e gosto nos peixes [illegible]. Arq. Mus. Bocage 9:19-46.

fish, sense of smell
Moncrieff, R.W., 1946
The chemical senses. John Wiley and Sons, Inc. $7.50.

"extensive subject index. Proper references are given in the text. This book is not for the uncritical reader, but it is a valuable source of well documented information"
Has chapters "on chemical sensitivity of animals"

fish, sense of smell in
Parker, G.H., 1914.
The directive influence of the sense of smell in the dogfish. Bull. U.S. Bur. Fish. Document 798: 63-68.

fish, sense of smell in
Parker, G.H., and R.E. Sheldon, 1913.
The sense of smell in fishes.
Bull. U.S. Bur. Fish. Doc. 775:35-45.

fish, sense of smell
Patton, Harry D., 1950.
Physiology of smell and taste.
Ann. Rev. Physiol. 12:469-484.

fish, sense of smell
Pipping, M., 1926.
Der Geruchsinn der Fische mit besonderer Berücksichtigung seiner Bedeutung für das Aufsuchen des Futters.
Soc. Sci. Fenn. Comm. Biol. (2)4.

fish, sense of smell
Pipping, M., 1927.
Ergänzende Beobachtungen über den Geruchsinn der Fische mit besonderer Berücksichtigung seiner Bedeutung für das Aufsuchen des Futters.
Soc. Sci. Fenn. Comm. Biol. (2)10.

fish, sense of smell
Powers, E.B., + R.T. Clark, 1953.
Further evidence of chemical factors affecting the chemical movements of fishes, especially the salmon. Ecol. 24

fish, sense of smell
Strieck, R., 1924.
Untersuchungen über den Geruchs- und Geschmackssinn der Elritze. Z. vergl. Physiol. 2:122-154.

fish, sense of smell
Von Buddenbrock, W., 1952.
Vergleichende Physiologie. Bd. 1.
Sinnesphysiologie. Birkhäuser, Basel.
504pp.

fish, sense of smell
Von Frisch, K., 1941.
Die Bedeutung des Geruchsinnes im Leben der Fische. Naturwissenschaften
29:321-333.

fish, sense of smell
von Frisch, K., 1941.
Ueber einen Schreckstoff der Fischhaut und seine biologische Bedeutung. Z. vergl. Physiol.
29(1): 46-145, 19 textfigs.

fish signals
Barham, E.G., W.B. Huckabay, R. Gowdy and B. Burns, 1969.
Microvolt electric signals from fishes and the environment.
Science, 164 (3882):965-968.

fish, stranding of

fish, stranding of (incls. lists of spp.)
Araga, Chuichi, and Hideotomo Tanase, 1966.
Fish stranding caused by a typhoon in the vicinity of Seto.
Publs Seto mar. biol. Lab., 14(2):155-160.

fish, stranding of
Tabeta, Osame, and Hiroshi Tsukahara, 1970.
Ecological studies on fishes stranded upon the beach along the Tsushima Current. III Local characteristics of stranded animals.
Bull. Jap. Soc. scient. Fish. 36(1):1-8

fish strandings
Tabeta, Osamu, and Hiroshi Tsukahara 1967
Ecological studies on the fishes stranded upon the beach along the coast of the Tsushima Current. 1. Fishes and other animals recorded during the first half of 1965 in northern Kyushu.
Bull. Jap. Soc. scient. Fish. 33(4):295-302.

fish, swimbladder
Alexander, R., McN., 1959.
The physical properties of the swimbladder in intact Cypriniformes.
J. Exp. Biol., 36(2):315-332.

fish, swimbladder
Dehadrai, P.V., 1962.
Respiratory function of the swim bladder of Notopterus (Lacepede).
Proc. Zool. Soc., London, 139(2):341-357.

fishes, swimbladder
Jones, F.R. Harden, and N.B. Marshall, 1953.
The structure and functions of the teleostean swim bladder. Biol. Rev. 28:16-83, 7 textfigs.

fish swimbladder
Kanwisher, J., and A. Ebeling, 1957.
Composition of the swimbladder gas in bathypelagic fishes. Deep-Sea Res. 4(3):211-217.

fish swimbladder
Kuhn, W., A. Ramel, H.J. Kuhn and E. Marti, 1963.
The filling mechanism of the swimbladder.
Experientia, 19(10):497-511.

fish swimbladder
Qutob, Z., 1962.
The swimbladder of fishes as a pressure receptor.
Arch. Neerl. Zool., 15(1):1-67.

fish, swimbladder
Scholander, P.F., 1954.
Secretion of gases against high pressures in the swimbladder of deep sea fishes.
II. The rete mirabile. Biol. Bull. 107(2): 260-277, 5 textfigs.

fish, swimbladder
Scholander, P.F., and L. van Dam, 1954.
Secretion of gases against high pressures in the swimbladder of deep sea fishes. 1. Oxygen dissociation in blood. Biol. Bull. 107(2):247-259, 4 textfigs.

fishes, swimbladder
Scholander, P.F., and L. Van Dam, 1953.
Composition of the swim bladder gas in deep sea fishes. Biol. Bull. 104(1):75-86, 5 textfigs.

fishes, swimbladder
Scholander, P.F., L. Van Dam and T. Enns, 1956.
Nitrogen secretion in the swimbladder of whitefish. Science 123(3185):59-60.

fish, swimming of

fish swimming

See also: fish propulsion

fish, swimming of
Gray, James, 1957.
How fishes swim.
Scientific American, 197(2):48-65.

fishes, Swimming
Nursall, J.R., 1958.
The caudal fin as a hydrofoil. Evolution 12(1): 116-120.

fish, swimming speed of
*Shibata, Keishi, 1968.
Analysis of fish-finder records. VII. Swimming speed of fish. (In Japanese; English abstract).
Bull. Fac. Fish., Nagasaki Univ., 25:59-65.

fishes, locomotion of
Van der Stelt, Abraham, 1968.
Spiermechanica en myotoombouw bij vissen.
L.J. Veen's Uitgeversmaatschappij N.V., Amsterdam, 94 pp.

fish, Swimming
Wohlschlag, D.E., 1962
Metabolic requirements for the swimming activity of three Antarctic fishes.
Science, 137(3535):1050-1051.

fish tagging

fish tagging

See also: tagging

fish tagging (sharks)

Davies, David H., and Leonie S. Joubert, 1967.
Tag evaluation and shark tagging in South African waters, 1964-1965. In: Sharks, skates, and rays, Perry W. Gilbert, Robert F. Mathewson, and David P. Rall, editors, Johns Hopkins Univ. Press, 111-140.

---

fish, tagging of (sharks)

Kato, Susumu, and Anatolio Hernandes Carvalho, 1967.
Shark tagging in the eastern Pacific Ocean, 1962-65. In: Sharks, skates and rays, Perry W. Gilbert, Robert F. Mathewson and David P. Rall, editors, Johns Hopkins University Press, 93-109.

---

fishes, taste buds

Iwai, Tamotsu, 1964.
A comparative study of the taste buds in gill rakers and gill arches of teleostean fishes.
Bull. Misaki Mar. Biol. Inst., Kyoto Univ., No. 7:19-34.

---

fish, vertebral count

Ben Tuvia, A., 1963
Variations in vertebral number of young Sardinella aurita in relation to temperature during spawning season.
Rapp. Proc. Verb., Reunions, Comm. Int. Expl. Sci. Mer Méditerranée, Monaco, 17(2):313-318.

---

fish, vertebral count

Itazawa, Y., 1959.
Influence of temperature on the number of vertebrae in fish.
Nature, 183(4672): 1408-1409.

---

fish, vertebrae

Ueyanagi, Shoji, and Hisaya Watanabe, 1965.
On certain specific differences in the vertebrae of istiophorids. (In Japanese; English abstract).
Rept. Nankai Reg. Fish. Res. Lab., No. 22:1-8.

---

fish, vision in

Baylor, Edward R. 1967.
Air and water vision of the Atlantic flying fish, Cypselurus heterurus.
Nature Lond. 214 (5085): 307-309.

---

fish, vision in

Baylor, Edward R. 1967.
Vision of Bermuda reef fishes.
Nature, Lond. 214 (5085): 306-307.

---

fish, vision olfaction

Brett, J.R., and C. Groot, 1963.
Some aspects of olfactory and visual responses in Pacific salmon.
J. Fish. Res. Bd., Canada, 20(2):287-303.

---

fishes, vision

Nicol, J.A.C., 1963
Some aspects of photoreception and vision in fishes.
In: Advances in Marine Biology, F.S. Russell, Editor, Academic Press, London and New York, 1: 171-208.

---

fish, warm-bodied

Barrett, Izadore, and Frank J. Hester, 1964.
Body temperature of yellowfin and skipjack tunas in relation to sea surface temperature.
Nature, 203(4940):96-97.

---

fish, warm-bodied

Carey, F.G., 1969.
Warm fish.
Oceanus, 15(1):22. (popular)

---

fish, warm-bodied

Carey, Francis G., John M. Teal, John W. Kanwisher, Kenneth D. Lawson and James S. Beckett, 1971.
Warm-bodied fish.
Am. Zoologist 11 (1): 137-145

---

fish, year-class strength

Corlett, John, 1965.
Winds, currents, plankton and the year-class strength of cod in the western Barents Sea.
ICNAF Spec. Publ., No. 6:373-378.

---

fish, young

Russell, F.S. and Necla Demir, 1971.
On the seasonal abundance of young fish. XII. The years 1967, 1968, 1969 and 1970. J. mar. biol. Ass., U.K., 51(1): 127-130.

FISH - Part II
Locality

Lists of species
Class, order, suborder
Family
Genus
Common names mostly fishes important in the fisheries
A title is, however, filed in only one place.

---

fish - locality

fish, Adriatic

Zei, Miroslav, 1949.
Razikovanje s travlom na nibolovem področju vzhodnega Jadrana (Prispevek k poznavanju biologije in ekologije bentonskih rib vzhodnega Jadrana) [Investigations with trawl in the northeastern part of the Adriatic. (Contribution to the knowledge of the biology and ecology of the benthonic fishes of eastern Adriatic)] (English summary). Razprave, Slovenska Akademije Znanosti in Umetnosti v Ljubljani, Kniga IV: 89-119, 8 textfigs.

---

fish, Africa, west

*Baudin-Laurençin, F., 1967.
La sélectivité des chaluts et les variations nycthémérales des rendements dans la région de Pointe-Noire.
Cah. ORSTOM, Sér. Océanogr., 5(1):85-121.

---

fish, West Africa

Kähsbauer, Paul, 1965.
Über einige Westafrikanische Syngnathiformes.
Atlantide Rept. 10: 263-290

---

fisheries, Africa, west

Pillay, T.V.R., 1967.
Estuarine fisheries of West Africa.
In: Estuaries, G.H. Lauff, editor, Publs Am. Ass. Advmt Sci., 83:639-646.

---

fish - Africa, northwest

Maurin, Claude, et Marc Bonnet, 1970.
Poissons des côtes nord-ouest Africaines (Campagnes de la Thalassa 1962 et 1968).
Rev. Trav. Inst. Pêches marit. 34(2): 125-170

---

fish, Africa, southwest

Penrith, Mary-Louise, 1970.
Report on a small collection of fishes from the Kunene River mouth.
Cimbebasia (A) 1 (7): 165-176.

---

fish, Antarctic

Bernardo Bellisio, Norberto, 1964.
Peces antarticos del sector Argentino. 1. Taxinomia y biologia de Chaenocephalus aceratus y Notothenis neglecta de orcadas del sur.
Rep. Argentina, Sec. Marina, Serv. Hidrogr. Naval, Publ. H. 900:90 pp.

---

fish, Antarctic

DeWitt, Hugh H., 1965.
Antarctic ichthyology.
BioScience, 15(4):290-293.

---

fisheries, Arctic

Lee, A.J., 1949.
The forecasting of climatic fluctuations and its importance to the Arctic fishery. Rapp. Proc. Verb., Cons. Perm. Int. Expl. Mer, 125:40-41.

---

fisheries, Australian

Pownall, Peter C., 1968.
Australia: the potential. 2. Fisheries.
Hydrospace, 1(4):42-44.

---

fish - Baja California

Ramirez Hernandez, Ernesto, y Joquin Arvizu Martinez, 1967.
Investigaciones ictiologicas en las costas de Baja California. I. Lista de peces marinos de Baja California colectados en el periodo 1961-1965.
Anal. Inst.nac.Invest.biol.-pesqueras, Mexico, 1: 297-358.

---

fisheries, Baltic

Antonov, A.E., 1964.
Basic oceanographic prognosis for the fisheries in the southern Baltic. (In Russian)
Atlantich. Nauchno-Issled. Inst. Ribn. Khoz. i Okeanogr. (ATLANTNIRO), Kaliningrad, 118 pp.

---

fisheries, Bering Sea

Ellison, J. J., D. E. Powell, and H. H. Hildebrand, 1950
Exploratory fishing expedition to the northern Bering Sea in June and July 1949.
Fishery Leaflet 369: 56 pp., 23 figs. (multilith).

---

fish, Black Sea

Oven, L.S., and L.P. Salekhova 1967.
On the problems of the Mediterranization of the ichthyofauna of the Black Sea. (In Russian)
Gidrobiol. Zh. (4): 124-127.

---

fisheries, Black Sea

Rass, T.S., 1949.
Icthyofauna of the Black Sea and its utilization
Trudy Inst. Okeanol., 4:103-123.

fish-Black Sea
Tkacheva, K.S., 1964.
The influence of the temperature and salinity regimes on the constitution (composition) and growth of some fish of the Black Sea (In Russian).
Trudy azov.-chernomorsk. nauchno-issled. Inst. morsk. ryb. khoz. okeanogr. 23:119-130

fish fauna-Black Sea
*Vinogradov,K.A.,editor, 1967.
Biology of the northwestern Black Sea. (In Russian).
Naukova Dumka,Kiev, 268 pp.

fisheries Canada, east
Gutsell, B. V., 1949
An introduction to the geography of Newfoundland. Canada, Dept. Mines and Resources, Geogr. Bur., Info. Ser. No.1: 85 pp., 20 text figs., 3 fold-in maps.

fishes Canada east
Leim,A.H., and W.B. Scott,1966.
Fishes of the Atlantic coast of Canada.
Bull.Fish.Res.Bd Can., 155:485pp. $8.50.

fisheries, Canada, east
Templeman, Wilfred, 1966.
Marine resources of Newfoundland.
Bull. Fish. Res. Bd. Can., 154: 170pp.

fishery resources Canada, north
*Dunbar,M.J.,1970.
On the fishery potential of sea waters of the Canadian north.
Arctic, 23(3):150-174.

fish-Caribbean
Bullis, Harvey R., Jr., and Paul J. Struhsaker 1970
Fish fauna of the western Caribbean upper slope.
Q. Jl Fla. Acad. Sci. 33(1):43-76

fisheries, Cuba
Gonzalez J.R., y N.E. Sálnikov 1967
Investigaciones soviético-cubanas de la economía. Investigaciones de las condiciones oceanológicos y los recursos biológicos marinos pesquera realizadas en 1964-1965. (In Russian; Spanish abstract).
Sovetsk.-Kubinsk. Ribokhoz. Issled. Vses. Nauchno-issled. Inst. Morsk. Ribn. Khoz. Okeanogr. (VNIRO), Centr. Ribokhoz. Issled. Nation Inst. Ribolov. Respibl. Kuba (TSRI) 2: 6-18.

fishes, Florida
Gunter, Gordon, and Gordon E. Hall, 1965.
A biological investigation of the Caloosahatchee Estuary of Florida.
Gulf Res. Repts., Ocean Springs, Mississippi, 2(1):71 pp.

fishes - Great Britain
Wheeler, Alwyne, 1969.
The fishes of the British Isles and North-West Europe. Michigan State University Press, 613 pp.

Fisheries, Gulf of Guinea
Longhurst, Alan R., 1965.
A survey of the fish resources of the eastern Gulf of Guinea.
Journal du Conseil, 29(3):302-334.

fish - Gulf of Maine
Bigelow, H.B., and W.C. Schroeder, 1953.
Fishes of the Gulf of Maine. First revision.
Fish. Bull. 74, Fish. Bull., Fish and Wildlife Service, 53:1-577, 288 textfigs.

fishes - Gulf of Maine
Bigelow, H.B., and W.C. Schroeder, 1936.
Supplemental notes on fishes of the Gulf of Maine. Bull. Bur. Fish. 48:319-343.

fisheries, Gulf of Mexico
Gunter, Gordon, 1967.
Some relationships of estuaries to the fisheries of the Gulf of Mexico.
In: Estuaries, G.H. Lauff, editor, Publs Am. Ass. Advmt Sci. 83: 621-638.

fish, Gulf of Thailand
Ritragsa, Samran, Dirak Dhamniyom, and Sutham Sithichaikasem, 1969.
An analysis of demersal fish catches taken from the otterboard trawling survey in the Gulf of Thailand, 1967. (In Thai; English abstract).
Contrib. Mar. Fish. Lab. Bangkok, 15: 70pp.

fisheries resources Gulf of Thailand
Ritragsa, Samran, and Somsak Pramokchutima, 1970.
The analysis of demersal fish catches taken from the otterboard trawling survey in the Gulf of Thailand, 1968 (In Thai; English abstract)
Contrib. Mar. Fish. Lab., Bangkok, 16: 61 pp.

fishes - Gulf of Tonkin
Pertsova-Ostroumova, T.A., 1965.
Flatfish larvae (Heterosomata) from the Gulf of Tonkin. (In Russian; English abstract).
Trudy Inst. Okeanol., Akad. Nauk, SSSR, 80:177-220.

fisheries, India
Chidambaram, K., and Devidas Menon, 1945.
Correlation of West Coast Fisheries with Plankton and certain Oceanographical Factors.
Proc. Indian Acad. Sci., 22:

fish concentrations, India
Venkataraman, G., and K.C. George, 1964.
On the occurrence of large concentrations of file-fish off the erala coast, India.
J. Mar. biol. Ass., India, 6(2):321-323.

fish, Indian Ocean
Fourmanoir, P., et P. Guéze, 1967.
Poissons nouveaux ou peu connus provenant de la Réunion et de Madagascar.
Cah. ORSTOM, Sér. Océanogr., 5(1):47-58.

fish, India
Jones, S., and M. Kumaran, 1964.
New records of fishes from the seas around India. 1.
J. Mar. biol. Ass., India, 6(2):285-308.

fish - Indian Ocean
Kotthaus, Adolf, 1970.
Fische des Indischen Ozeans: Ergebnisse der ichthyologischen Untersuchungen während der Expedition des Forschungsschiffes Meteor in den Indischen Ozean, Oktober 1964 bis Mai 1965. Percomorphi (2).
Meteor Forsch. Engebn. (D) (6): 56-75

fish, Indian Ocean
Kotthaus, Adolf, 1970.
Fische des Indischen Ozeans: Ergebnisse der ichthyologischen Untersuchungen während der Expedition des Forschungsschiffes Meteor in den Indischen Ozean Oktober 1964 bis Mai 1965. Percomorphi (1).
Meteor Forsch. Engebn. (D) (6): 35-55

fish, Indian Ocean
Maugé, L.A., 1967.
Contribution préliminaire à l'inventaire ichtyologique de la région de Tuléar.
Annales Univ. Madagascar, 5: 215-246

FISHERIES Indian Ocean
Prasad, R. Raghu, 1969.
Zooplankton biomass in the Arabian Sea and the Bay of Bengal with a discussion on the fisheries of the regions.
Proc. Nat. Inst. Sci., India (B) 35 (5): 399-437

fisheries, Indian Ocean
Rass, Th. S., 1965.
Commercial ichthyofauna and fisheries resources of the Indian Ocean. (In Russian; English abstract).
Trudy Inst. Okeanol., Akad. Nauk, SSSR, 80:3-31.

fish, Indian Ocean
Silas, E.G., Daniel Selvaraj and A. Regunathan, 1969.
Rare chimaeroid and elasmobranch fishes from the continental slope off the west coast of India.
Current Sci., 38(5): 105-106

fisheries, Indonesia
Westenberg, J., 1953.
Acoustical aspects of some Indonesian fisheries.
J. du Cons. 18(3):311-325, 2 textfigs.

fish, Irish waters
West, Arthur E. J., 1969.
Rare fishes taken in Irish waters in 1968.
Ir. Nat. J. 16(6): 147-150

fisheries Mandated Islands
Fish and Wildlife Service, 1947
Survey of the fisheries of the former Japanese Mandated Islands. Fishery Leaflet 273, 103 pp., 50 text figs.

fisheries, Mexico
Carranza, Jorge, 1963
Los recursos marinos de Mexico y su approvechamiento.
Inst. Mexicano de Recursos Nat. Renovables Contr. No. 4: 67 pp.

fish, Mexico
Castro Aguirre, José Luis, 1965.
Peces sierra, rayas, mantas y especies afines de México.
Anales Inst. nac. Invest. biolog.-pesqueras, México 1: 171-256.

fish, Mexico
Hildebrand, Henry H., Humberto Cahvez y Henry Compton, 1964.
Aporte al conocimiento de los peces del Arrecife Alacranes, Yucatan (Mexico). Ciencia, Mexico, 23(3):107-134.
Also in:
Inst. Tecnol., Veracruz, Estacion de Biol. Mar, Trabajos e Investigaciones, No. 5.

fish, Mexico
Ramírez, Hernández, Ernesto, 1965.
Estudios preliminares sobre los peces marinos de México.
Anales Inst.nac.Invest.biol.-pesqueras, México, 1;259-292.

fishes, New Guinea
Hardenberg, J.D.F., 1941
Fishes of New Guinea. Treubia 18(2):217-231, 4 textfigs.

fish, New Zealand
Abe, Tokiharu and Ryoichi Arai, 1968.
Notes on some fishes of New Zealand and Balleny Islands. J. Tokyo Univ. Fish., 9 (2): 141-146.

fisheries, New Zealand
Chief Inspector of Fisheries, 1946
Depletion of New Zealand Fish Food Resources.
Fisheries Newsletter - Australia 5(5):15, 24

fishes, North Pacific
Nalbant, T.T. 1970 (Teodor T.)
New contributions to the study of the fishes of the North Pacific. (In Roumanian; English abstract).
Bul. Inst. Cerc. pisc. 29(1/2):57-64.

fisheries, North Sea
Baerends, G. P., (1947)1950.
The rational exploitation of the sea fisheries with particular reference to the fish stock of the North Sea. Paper No. 36, Dept. Fish.
Translated by Jan Hahn, Spec. Sci Fish. Rept., No. 13:102 pp. (mimeographed).

fisheries Norway
Løversen, R., 1946
Undersökelser i Oslofjorden 1936-1940.
Fiskeyngelens forekomst i strand regionen Fiskeridirektoratets Skrifter, Ser. Havundersökelser (Report on Norwegian Fishery and Marine Investigations) VIII (8):1-34, 11 text figs.

fisheries, Norway
Ruud, Johan T., 1968.
Changes since the turn of the century in the fish fauna and fisheries of the Oslofjord.
Helgoländer wiss. Meeresunters. 17: 510-517.

fisheries Pacific
Sette, O. E., 1949.
Pacific oceanic fishery investigations. Copeia, 1949(1):84-85.

fisheries, Pacific, north
Allen, E.H., 1953.
Fishery geography of the North Pacific Ocean.
Geogr. Rev. 43(4):558-563.

fisheries, Pacific, south
Sette, O. E., 1939(1940).
The research program of the south Pacific investigations of the United States Bureau of Fisheries
Proc. 6th Pacific Sci. Congr.:409-411.

fishes, Peru
Abbott, J. F. 1899
The marine fishes of Peru
Proc. Acad. Sci., Phila. 324-364

fishes, Peru
Anon.
Ictiología del Peru
Bol. Mus. Hist. Nat. "Javier Prado"
4:69-84
4:257
4:398-403
4:513-523

fish, Peru
*Chirichigno Fonseca, Norma, 1968.
Nuevos registros para la ictiofauna marina del Peru.
Bol.Inst.Mar Peru 1(8):377-504.

fisheries, Peru
Coker, R.E., 1908.
The fisheries and the guano industry of Peru.
Bull. U.S.B.F. 28:333-365.

fisheries, Peru
Coker, R.E., 1908.
La pesca con dinamita. Bol. del Ministerio de Fomento, Año VI, No. 5:48-53.
Condición en que se encuentra la pesca marina desde Paita hasta Bahía de la Independencia.
Ibid.:53-115.
Las ballenas del Peru. Ibid.:115-125.

fisheries, Peru
Coker, R.E., 1908.
La industria del guano. Bol. del Ministerio de Fomento, Año VI, No. 4:23-34.
Condición en que se encuentra la pesca marina, desde Paita hasta Bahía de la Independencia.
Ibid.:62-99.

fisheries, Peru
Coker, R.E., 1908.
Condición en que se encuentra la pesca marina desde Paita hasta Bahía de la Independencia (continuación). La región del Callao, comprendiendo Chorrillos y Ancon. Bol. del Ministerio de Fomento, Año VI, No. 3:54-95.

fisheries, Peru
Coker, R.E., 1908.
Condición en que se encuentra la pesca marina, desde Paita hasta Bahía de la Independencia. Bol del Ministerio de Fomento, Año VI, No. 2:89-117.

fisheries, Peru
Doucet, W.F., and H. Einarsson, 1967.
A brief description of Peruvian fishes.
Rep. Calif. Coop.Oceanic Fish.Invest., 11:82-87.

fish, Peru
Fowler, H.W.,
Los peces del Peru. Catalogo sistematico de los peces que habitan en aguas peruanas.
Bol. Mus. Hist. Nat. "Javier Prado" 5(3):363-391

fish, Peru
Hildebrandm S. F., and O. Barton, 1939.
A collection of fishes from Talara, Peru.
Smithsonian Misc. Coll., 111(10):36 pp., 9 textfigs.

fish - Peru
Regen, C.T., 1913.
Fishes from Peru collected by Dr. H.O. Forbes.
Ann. Mag. Nat. Hist. (8) 12:278-280.

fishes - Red Sea
Whitehead, P.J.P., 1965.
A review of the elopoid and clupeoid fishes of the Red Sea and adjacent regions.
Bull. Brit. Mus. (N.H.), Zool., 12(7):227-281.

fisheries S. Africa
Gilchrist, J.D.F., 1914
An inquiry into fluctuations in fish supply on the South African coast.
Mar. Biol. Rep. No.2:8-35.

fisheries, S. Africa
von Bonde, C., 1949
Nineteenth Annual Report for the year ended December 1947. Division of Fisheries, Dept. Commerce and Industries. Commerce & Industry, S. Africa, Apr. 1949:415-476, 4 charts.

fisheries, S. America, east
Republica Argentina, Secretaria de Marina.
Servicio de Hidrografia Naval, 1965.
Beneficios economicos de la investagacion oceanografica.
Bol. Servicio Hidrografia Naval, 11(2):90-103.

fish, S. America, east
Saraiva da Costa, Raimundo, e Melquíades Pinto Paiva, 1964.
Notas sôbre a pesca da cavala e da serra no Ceará dados de 1963. (English summary).
Arq. Est. Biol. Mar. Univ. Ceará, Brasil, 4(2):

fisheries, S. America northeast
Alveres Perez, P., 1965.
Trabajos de investigacion y busqueda sobre la plataforma continental de Venezuela. (In Russian; Spanish abstract).
Sovetsk.-Cub. Ribokhoz. Issled., VNIRO:Tsentr. Ribokhoz. Issled. Natsional. Inst. Ribolovsta Republ. Cuba, 285-288.

fisheries, S. America, northeast
Gines, Hermano, y Fernando Cervignon, 1968.
Exploración pesquera en las costas de Guayana y Surinam, año 1967.
Memoria. Soc.Cienc.nat.La Salle, 28(79):5-96.

fisheries, S. America, northeast
Pinto Paiva, Melquiades, y Fernando Cervignon, 1969.
Los recursos pesqueros del nordeste de Sudamérica.
Memoria, Soc. Cienc. nat. La Salle, 29(82): 60-71

fish, S. America (northeast)
Venezuela, Sociedad de Ciencias Naturales La Salle, 1957.
El Archipelago de Los Roques y La Orchila, Editorial Sucre, Caracas, 257 pp., color photos.

fisheries
Saraiva da Costa, Raimundo, e Melquiades Pinto Paiva, 1969.
Notas sôbre a pesca da cavala e da serra no Ceará - dados de 1968.
Arq. Ciên. Mar, Fortaleza, Ceará, Brasil, 9(1): 89-95

fisheries, S. America, west
Castillo, L., 1912.
Contribucion al estudia biologico de los peces maritimos comestibles de Chile.
Bol. Bosques, Pesca i Caza 1:90-102.

fisheries, S. America, west
Delfin, F.T., 1899.
Lista metodica de los pesces de la Bahia de Concepcion i sus alrededores.
Rev. Chil. Hist. Nat. 3:176-178.

fish, S. America, west
Fowler, H.W.,
Fishes from the Pacific slope of Colombia, Ecuador and Peru.
Notulae Naturae, Phila., 33:1-7.

fisheries, S. America, west
Kruuse, H.G., 1882.
The fisheries of the west coast of South America. Rept. U.S. Fish. Comm. 1879; 7:515-522.

fisheries, S. America, west
Kruuse, H.G., 1879
Nogle Meddelelser om fiskeriet paa vest kysten af Sudamerika. Nord. Tidsskr. Fiskeri n.s., 5 Jahrg.:286-299.

fisheries, S. America, west
Perez Canto, C., 1912.
La fauna ictiologica de Chile considerada como riqueza nacional.
Anal. Univ. Chile, 1911, 129:1215-1260.

fisheries, S. America, west
Porter, C.E., 1909.
Breve nota de ictiologia; enumeracion metodica y extension geografica de las especies mas importantes en su mayor parte comestibles, communes a las aguas de Chile y del Peru.
Revist. Univ. Lima, 5:138-149.
Revist. Chil. Hist. Nat. 13:280-293.

fish, S. America, west
Starks, E.C., 1906.
On a collection of fishes made by P.O. Simons in Ecuador and Peru. Proc. U.S. Nat. Mus. 30:761-800.

fishes-South Pacific
Mead, Giles W., 1970.
A history of South Pacific fishes.
In: Scientific exploration of the South Pacific, W.S. Wooster, editor, Nat. Acad. Sci., 236-251.

fishes, Tasmania
Scott, E.D.G., 1965.
Observations on Tasmanian fishes. XIII.
Papers and Proc. R.Soc., Tasmania, 99:53-65.

fishing, experimental USA, east
Schroeder, W.C., 1955.
Report on the results of exploratory otter-trawling along the continental shelf and slope between Nova Scotia and Virginia during the summers of 1952 and 1953.
Pap. Mar. Biol. and Oceanogr., Deep-Sea Res., Suppl. to Vol. 3:358-372.

fishes, by type of water layer

fish, bathypelagic
Blache, J., 1964.
Poissons bathypélagiques rares ou peu connus provenant des eaux de l'Atlantique oriental tropical. 1ère note.
Trav. Centre Océanogr., Pointe-Noire, Cahiers O.R.S.T.O.M., Océanogr., Paris, No. 5:89-96.

fish, bathypelagic
*Grandperrin, R., et M. Legand, 1967.
Influence possible du système des courants équatoriaux du Pacifique sur la répartition et la biologie de deux poissons bathypélagiques.
Cah. ORSTOM, Sér. Océanogr., 5(2):69-77.

fish, bathypelagic
Katthans, A., 1969.
Ein seltener bathypelagischer Fisch (Familie Melanostomiatidae) aus dem östlichen Atlantic.
Helgoländer wiss. Meeresunters. 15 (1/2): 61-68

fishes, bathypelagic
Melchikova, L.I. 1969.
The distribution of certain bathypelagic fish, fry, and larvae in the area of the Kuroshio Current in the summer of 1965. (In Russian).
Izv. Tichookean. nauchno-issled Inst. ribn. Khoz. Okeanogr. (TINRO) 68: 193-202.

fishes, bathypelagic
Tholasilingam, T., G. Venkataraman and K.N. Krishna Kartha, 1964.
On some bathypelagic fishes taken from the continental slope off the south west coast of India.
J. Mar. Biol. Ass., India, 6(2):268-284.

fish, benthic
Collignon, J., 1969.
Première note sur le peuplement en poissons benthiques du plateau continental atlantique marocain.
Bull. Inst. Pêches marit. Maroc. 17: 11-44

fishes, benthic
*Day, Donald S., and William G. Pearcy, 1968.
Species associations of benthic fishes on the continental shelf and slope off Oregon.
J. Fish. Res. Bd., Can., 25(12):2665-2675.

fish, benthic
Marshall, N.B., and D.W. Bourne, 1964.
A photographic survey of benthic fishes in the Red Sea and the Gulf of Aden, with observations on their population density, diversity and habits
Bull., Mus. Comp. Zool., Harvard Coll., 132(2): 225-244, 4 pls.

fish, coral reef
Quiguer, Jean-Paul, 1969.
Quelques données sur la répartition des poissons des récifs coralliens. Cah. Pacifique 13: 181-185.

fishes, deep-sea
Andriashev, A.P., 1955.
On a new Liparid fish from a depth over 7000 meters. Trudy Inst. Oceanol., 12:340-344.

fishes, deep-sea
Mead, Giles W., E. Bertelsen, and Daniel M. Cohen, 1964.
Reproduction among deep-sea fishes.
Deep-Sea Res., 11(4):569-596.

fishes, deep-sea
Munk, Ole, 1966.
Ocular anatomy of some deep-sea teleosts.
Dana Rep., No. 70:63 pp.

fish, deep-sea
Pearcy, William G., Samuel L. Meyer and Ole Munk, 1965.
A four-eyed fish from the deep-sea: Bathylychnops exilis Cohen, 1958.
Nature, 207(5003):1260-1262.

deep-water fishes
Rass, T. S., 1955.
Deep-water fishes of the Kurile-Kamchatka Trench
Trudy Inst. Okeanol., 12:328-339.

Translation
by Lisa Lanz and Robert R. Rofen in:
News Bull., George Vanderbilt Found., Nat. Hist. Mus. Stanford Univ., (undated)
recd. 10 Feb. 1960

fishes, deep-sea
Rass, T.S., 1955.
Deep-Sea fishes of the Kurile-Kamchatka Trench.
Trudy Inst. Oceanol., 12:328-339.

fishes, deep sea
Schroeder, W. C., 1940.
Some deepsea fishes from the North Atlantic.
Copeia 1940(4):231-238.

fisheries, demersal (tropical
Longhurst, Alan R., 1969
Species assemblages in tropical demersal fisheries
Actes Symp. Oceanogr. Ressources halieut. Atlant. trop., Abidjan, 20-28 Oct. 1966, UNESCO 147-168.

fisheries, estuarine
Korringa, P. 1967.
Estuarine fisheries in Europe as affected by man's multiple activities.
In: Estuaries, G.H. Lauff, editor, Publs Am. Ass. Advmt Sci. 83: 658-663.

fisheries, estuarine
Pillay, T.V.R. 1967.
Estuarine fisheries of the Indian Ocean coastal zone.
In: Estuaries G.H. Lauff, editor, Publs Am. Ass. Advmt Sci. 83: 647-657.

fishes beneath flotsam
Hunter, John R., 1968.
Fishes beneath flotsam. (Not technical).
Sea Frontiers, 14(5):280-288.

fish - flotsam
Hunter, John R., and Charles T. Mitchell, 1966
Association of fishes with flotsam in the offshore waters of Central America.
Fish. Bull. Fish Wildlife Service, 66(1): 13-29

fish, freshwater
Boeseman, M., 1956.
Fresh-water sawfishes and sharks in Netherlands New Guinea. Science, 123(3189):222-223.

fish, inshore
Gosline, William A., 1965.
Vertical zonation of inshore fishes in the upper water layers of the Hawaiian Islands.
Ecology, 46(6):823-831.

fish, islands
Cadenat, J., and E. Marchal, 1963.
Resultats des campagnes océanographiques de la Reine-Pokou aux îles Sainte-Helene et Ascension. Poissons.
Bull. Inst. Francais d'Afrique Noire, 25(4):1235-1366.

fishes in kelp
Mitchell, Charles T., and John R. Hunter, 1970.
Fishes associated with drifting kelp, Macrocystis pyrifera, off the coast of Southern California and northern Baja California.
Calif. Fish Game 56(4): 288-297.

fish, littoral
Gibson, R.N., 1969.
The biology and behaviour of littoral fish.
Oceanogr. Mar. Biol. Ann. Rev., H. Barnes, editor, George Allen and Unwin, Ltd., 7: 367-410.

fishes, mesopelagic
Backus, R.H., 1969.
Chain-85.
Oceanus15(1):2-5 (popular)

fish, mesopelagic
Backus, R.H., J.E. Craddock, R.L. Haedrich and D.L. Shores, 1970.
The distribution of mesopelagic fishes in the equatorial and western North Atlantic Ocean.
J. mar. Res., 28(2): 179-201.

fishes, mesopelagic
Backus, R.H., J.E. Craddock, R.L. Haedrich and D.L. Shores, 1969.
Mesopelagic fishes and thermal fronts in the western Sargasso Sea. Marine Biol. 3(2): 87-106.

fish, mesopelagic
Kashkin, N.I. and J.G. Tchindonova, 1971.
Mesopelagic fishes as resonant scatterers in the DSLs of the Atlantic Ocean. (In Russian; English abstract). Okeanologiia 11(3): 482-493.

fish mesopelagic
Legand, Michel, 1967.
Cycles biologiques des poissons mésopélagiques dans l'est de l'Océan Indien.
Cah. O.R.S.T.O.M., Ser. Océanogr., 5(4):47-71.

fishes, mesopelagic
Legand, M. et J. Rivaton, 1970.
Cycles biologiques des poissons mésopélagiques dans l'est de l'océan Indien. Quatrième note: Variations de répartition de 7 espèces synthèse des divers cycles décrits. Cah. O.R.S.T.O.M. ser. Océanogr. 8(1): 59-79.

fish, mesopelagic
Legand, M. et J. Rivaton, 1969.
Cycles biologiques des poissons mésopélagiques de l'est de l'Océan Indien. Troisième note: Action prédatrice des poissons micronectoniques.
Cah. ORSTOM, ser. Océanogr., 7(3): 29-46.

fish, mesopelagic
Legand, Michel, et J. Rivaton, 1968.
Cycles biologiques des poissons mésopélagiques dans l'est de l'Océan Indien.
Cah. O.R.S.T.O.M., Ser. Océanogr., 5(4):73-98.

fishes
Legand, Michel, Philippe Bourret and René Grandperrin, 1970.
A preliminary study of some micronektonic fishes in the equatorial and tropical western Pacific.
In: Scientific exploration of the South Pacific, W.S. Wooster, editor, Nat. Acad. Sci., 226-235.

fishes, midwater
Aron W. and R.H. Goodyear, 1969.
Fishes collected during a midwater trawling survey of the Gulf of Elat and the Red Sea
Israel J. Zool. 18(2/3): 237-244

fishes, midwater
Best, E.A., and J. Gary Smith, 1965.
Fishes collected by midwater trawling from California coastal waters, March 1963.
Calif. Fish Game 51(4):248-251.

fishes, midwater
Bussing, William A., 1965.
Studies of the midwater fishes of the Peru-Chile Trench.
In: Biology of seas, II.
Antarctic Res. Ser., Amer. Geophys. Union, 5: 185-227.

fish, mid-water
Forster, G.R., 1971.
Line-fishing on the continental slope III. Mid-water fishing with vertical lines. J. mar. biol. Ass. U.K., 51(1): 73-77.

fish, midwater
Parin, N.V., 1971.
On the distributional pattern of midwater fishes of the Peru Current Zone.
Trudy Inst. Okeanol. P.P. Shirshova Akad. Nauk SSSR 89: 81-95
(In Russian, English abstract)

fishes, midwater
*Taylor, F.H.C., 1967.
Unusual fishes taken by midwater trawl off the Queen Charlotte Islands, British Columbia.
J. Fish. Res. Bd., Can., 24(10):2101-2115.

fish, pelagic
Aboussouan, Alain, 1971.
Contribution à l'Etude des Téléostéens récoltés au chalut pélagique en relation avec la D.S.L. durant la période du 1er novembre 1967 au 31 décembre 1968.
Cah. océanogr. 23(1): 85-99.

fish, pelagic
Elwertowski, J. et T. Boely 1971.
Repartition saisonnière des poissons pélagiques côtiers dans les eaux mauritaniennes et sénégalaises.
ORSTOM, Centre Rech. océanogr. Dakar-Thiaroye, D.S.P. 32: 15pp. (mimeographed)

fish, pelagic
Gamulin, Tomo, 1964.
Signification de l'Adriatique nord en vue de la connaissance des poissons pelagiques. (In Jugoslavian; English resume).
Acta Adriatic, 11 (1):91-96.

fisheries, pelagic
Hattori, Shigemasa, 1969.
Recruitment of pelagic fish of commercial importance. (In Japanese; English abstract).
Bull. Jap. Soc. fish. Oceanogr. Spec. No. (Prof. Uda Commem. Pap.): 275-277.

fish, pelagic
Jensen, Albert C., 1967.
Observations on pelagic fishes off the west coast of Africa.
Bull. mar. Sci. Miami, 17(1):42-51.

fish, pelagic
*Kačić, Ivo, 1968.
Seasonal and annual fluctuations in the quantity of pelagic fish in Kaštela Bay. (In Jugoslavian; English abstract).
Bilješke, Inst. Oceanogr. Ribarst, Split, 23: 6 pp.

fish, pelagic
Marshall, N.B., 1963.
Pelagic fishes.
(Special Issue on Marine Biology), AIBS Bull., 13(5):65-67.

Fisherie, pelagic
Osipov, V.G., 1968.
Biology and fishery of tuna and other pelagic fishes in the Indian Ocean. (In Russian).
Trudy, Vses. Nauchno-Issled. Inst. Morsk. Ribn. Okeanogr (VNIRO) 64, Trudy Azovo-Chernomorsk. Nauchno-Issled. Inst. Morsk. Ribn. Khoz. Okeanogr. (AscherNIRO), 28: 300-322.

fish (pelagic)
Parin, N.V., 1962
Some peculiar features involved with the distribution of mass pelagic fishes in the zone of equatorial currents in the Pacific Ocean (based on the materials of the 34th cruise of R/V "Vityaz"). (In Russian).
Okeanologiia, Akad. Nauk, SSSR, 2(6):1075-1082.

fish, shore
Kendall, W.C., and L. Radcliffe, 1912.
The shore fishes of the "Albatross" expedition, 1904-1905. Mem. M.C.Z. 25:75-171.

fishes, lists of species

fishes, checklists
Banasopit, Thien, and Thosaporn Wongratana, 1967.
A check list of fishes in the reference collection maintained at the Marine Fisheries Laboratory.
Contrib., Mar. Fish. Lab., Bangkok, 7: 73 pp.

fish
Cadenat, J., et A. Stauch, 1965.
Sur la validité des genres Bathysolea (Roule 1916) et Capartella (Chabanaud 1950).
Cahiers O.R.S.T.R.O.M., Océanogr., 3(3):67-

fishes, checklists
Kami, H.T., I.I. Ikehara and F.P. DeLeon, 1968.
Check-List of Guam fishes. Micronesica, J. Coll. Guam, 4(1): 95-130.

fish, lists of spp.
Birdsong, Ray S., and Alan R. Emery, 1968.
New records of fishes from the western Caribbean.
Q. Jl. Fla. Acad. Sci., 30(3):187-196.

fish, lists of spp.
Cervigron, Fernando, 1965(1967).
Distribución general y local de los peces marinos de Venezuela y su relación con las regiones ecologicas.
Memoria Soc.Cienc.nat.La Salle, 25(70/71/72):359-379.

Fish, lists of spp.
Champagnat, C., F. Conand, J.L. Cremoux et J.P. Rebert 1969.
Campagne océanographique du Jean Charcot (Dakar-Cap Blanc-Iles du Cap Vert) du 29-7 au 5-7-68.
ORSTOM, Sénégal, DSP 17: 87 pp. (mimeographed)

fish, lists of spp.
*Chung, Bu. Kwan, Yong Mun Kim and Yong Sool Kim, 1967.
Zoogeographical studies on the bottom fishes of the Korean coast in the Yellow Sea. (In Korean; English abstract).
Repts.Fish.Resources,Fish.Res.Develop.Agency, Korea, 7:5-27.

fishes, lists of spp.
Crosnier, A., 1964.
Fonds de pêche le long des cotes de la Republique Federale du Camaroun.
Cahiers, O.R.S.T.R.O.M. No. special: 133 pp.

fishes, lists of spp.
Demetropoulos, Andreas, and Daphne Neocleous 1969.
The fishes and crustaceans of Cyprus
Fish. Bull., Cyprus, 1: 21 pp.

fishes, lists of spp.
Eskinazi, Aída Maria, 1970.
Lista preliminar dos peixes estuarinos de Pernambuco e Estabos Vizinhos (Brasil). Trabhs oceanogr. Univ. Fed. Pernambuco 9/11: 265-274.

fish, lists of spp.
Fourmanoir, P., 1970.
Notes Ichtyologiques (I). Cah. ORSTOM, sér. Océanogr., 8(2): 19-33.

fishes, lists of spp.
Fourmanoir, P., et A. Crosnier, 1963.
Deuxième liste complémentaire des poissons du Canal de Mozambique: diagnoses préliminaires de 11 espèces nouvelles.
Trav. Centre Océanogr., Nosy-Bé, Cahiers, O.R.S.T.R.O.M., Océanogr., Paris, No. 6:2-32.

fishes, list of spp.
Grandperrin, R., et J. Rivaton, 1966.
"Coriolis": Croisière "Alizé": individualisation de plusieurs ichtyofaunes le long de l'équateur.
Cah. ORSTOM, Sér. Océanogr., 4(4):35-49.

fishes, lists of spp.
Japan, Hokkaido University, Faculty of Fisheries, 1964.
Data record of oceanographic observations and exploratory fishing, No. 8:303 pp.

photographs of each

Fish, lists of spp.
Matsuda, Seiji 1969.
The studies on fish eggs and larvae occurred in the Nansei Regional Water of Japan. - I. Species occurred and their seasonal variation. (In Japanese; English abstract). Bull. Nansei reg. Fish. Res. Lab., 2: 49-84.

fishes, lists of spp.
Maurin, Claude, Fernando Lozano Cabo, et Marc Bonnet, 1970.
Inventaire faunistique des principales espèces ichthyologiques frequentant les côtes nord-ouest africaines.
Rapp. P.v. Réun. Cons. int. Explor. Mer. 159: 15-21

fish, lists of spp.
Medina, Wenceslao, 1965.
Los peces marinos conocidos del Callao con referencia de su districución geografia.
Biota, 5(42):245-287.

fishes, lists of spp.
Naumov, V.M., 1968.
Fish of the Indian shelf waters in the Bay of Bengal. (In Russian).
Trudy, Vses. Nauchno-Issled. Inst. Morsk. Ribn. Okeanogr (VNIRO) 64, Trudy Azovo-Chernomorsk. Nauchno-Issled. Inst. Morsk. Ribn. Khoz. Okeanogr. (AscherNIRO), 28: 407-430.

fishes, lists of spp.
Naumov, V.M., 1968.
Fecundity of the Indian Ocean fishes. (In Russian).
Trudy, Vses. Nauchno-Issled. Inst. Morsk. Ribn. Okeanogr (VNIRO) 64, Trudy Azovo-Chernomorsk. Nauchno-Issled. Inst. Morsk. Ribn. Khoz. Okeanogr. (AscherNIRO), 28: 431-436.

fish, lists of spp
Pillay, T.V.R. 1967.
Estuarine fisheries of the Indian Ocean coastal zone.
In: Estuaries, G.H. Lauff, editor, Publs Am. Ass. Advmt Sci. 83: 647-657.

fishes (lists of spp.)
Salnikov, N.E., 1965.
Investigaciones de economía pesquera en el Golfo de México y del Mar Caribe. (In Russian; Spanish abstract).
Sovetsk.-Cub. Ribokhoz. Issled., VNIRO:Tsentr. Ribokhoz. Issled. Natsional. Inst. Ribolovsta Republ. Cuba, 93-179.

fish
*Seno, Jiro, and Tatsuyoshi Masuda, 1966.
Fish specimens collected by tuna longline. (In Japanese; English abstract).
J. Tokyo Univ. Fish. (Spec.ed.) 8(2):243-250.

fishes, lists of spp.
Smith, J.L.B., 1968.
New and interesting fishes from deepish water off Durban, Natal and southern Mozambique.
Investl Rep. oceanogr.Res.Inst., S.Afr.Ass.mar. biol.Res. 19:1-30.

fish, lists of spp.
Sokolova, L.V., 1965.
Distribución y la caracteristica biológica de los peces principales comerciales en el Banco de Campeche. (In Russian; Spanish abstract).
Sovetsk.-Cub. Ribokhoz. Issled., VNIRO:Tsentr. Ribokhoz. Issled. Natsional. Inst. Ribolovsta Republ. Cuba, 223-239.

fish, lists of spp.
Thailand, Marine Fisheries Laboratory, Bangkok, and Malaysia, Fisheries Research Institute, 1967.
Results of the Joint Thai-Malasian-German trawling survey off the east coast of the Malay Peninsula, 1967, 64 pp.

fishes
Went, A.E.J., 1971.
Interesting fishes from Irish waters in 1970.
Ir. Nat. J., 17(2): 41-45.

fish, lists of spp
Went, A.E.J., 1966.
Rare fishes taken in Irish waters in 1965.
Irish Naturalists' Jour., 15(6):159-163.

fish, lists of species
Yanulov, K.P., 1968.
Specific composition of catches in the western Indian Ocean. (In Russian).
Trudy, Vses. Nauchno-Issled. Inst. Morsk. Ribn. Okeanogr (VNIRO) 64, Trudy Azovo-Chernomorsk. Nauchno-Issled. Inst. Morsk. Ribn. Khoz. Okeanogr. (AscherNIRO), 28: 282-299.

fish, yearclass
fish
Baughman, J.L., 1948.
Sharks, Sawfishes, and Rays: Their folklore.
Am. Mid. Nat. 39(2):373-381.

fish
Bigelow, H. B., I.P. Farfante and W. C. Schroeder, 1949.
Fishes of the western North Atlantic: Lancelets, Cyclostomes, Sharks. Mem. Sears Found. Mar. Res. 576 pp.

Reviewed: Science 111:95 by L.P. Schultz
J. du Cons. 16(3):397-398 by A. Fr. Bruun

fish (batoid)
Bigelow, H.B., and W.C. Schroeder, 1954.
A new family, a new genus and two new species of batoid fishes from the Gulf of Mexico. Breviora No. 24:16 pp., 4 textfigs.

fish
Bigelow, H.B., and W.C. Schroeder, 1953.
Fishes of the western North Atlantic. Ch. 1. Batoidea. Ch. 2. Holocephali. Mem. Sears Found. Mar. Res. 1(2):xv plus 562, index, figs.

fishes (batoid)
Bigelow, H. B. and W. C. Schroeder, 1948
New genera and species of batoid fishes.
J. Mar. Res. 7(3):543-566, 9 text figs.

fish, batoid
Wallace, John H., 1967.
The batoid fishes of the east coast of Southern Africa. 3. Skates and electric rays.
Investl Rep.Oceanogr.Res.Inst.Un.S.Afr. 17:62 pp.

fish - Chondrichthyes
Forster, Roy P., 1967.
Osmoregulatory role of the kidney in cartilaginous fishes (Chondrichthyes). In: Sharks, skates and rays, Perry W. Gilbert, Robert F. Mathewson and David P. Rall, editors, Johns Hopkins Univ. Press, 187-195.

fish
Kähsbauer, Paul, 1968.
Über einige Westafriakische Syngnathiformes.
Atlantide Rept., 10:263-273.

fishes
Kovalevskaja, N.V., 1965.
The eggs and larvae of Synentognathous fishes (Beloniformes, Pisces) of the Gulf of Tonkin. (In Russian ; English abstract).
Trudy Inst. Okeanol., Akad. Nauk, SSSR, 80:124-146.

fish
Krefft, Gerhard, 1968.
Knorpelfische (Chondrichthyes) aus dem tropicchen Ostatlantik.
Atlantide Rept. 10:33-76.

fishes
Punpoka, Supap, 1964.
A review of the flatfishes (Pleuronectiformes-Heterosomata) of the Gulf of Thailand and its tributaries in Thailand.
Kasetsart Univ. Fish. Res. Bull., No. 1:86 pp.

fish, batoid
Wallace, John H., 1967.
The batoid fishes of the east coast of Southern Africa. 2. Manta, eagle, duckbill, cownose, butterfly and sting rays.
Investl Rep. Oceanogr. Res. Inst. Un. S. Afr. 16: 56 pp.

families

fish - Antennariidae
Le Danois, Yseult, 1970
Etude sur des poissons pediculates de la famille des Antennariidae recoltés dans la Mer Rouge et description d'une espèce nouvelle
Israel J. Zool., 19(2): 83-94.

fish, Balistidae
*Moore, Donald, 1967.
Triggerfishes (Balistidae) of the western Atlantic.
Bull. mar. Sci., Miami, 17(3):689-722.

fish, Bregmacerotidae
D'Ancona, Umberto (posthumous) and Geminiano Cavinato, 1965.
The fishes of the family Bregmacerotidae.
Dana Rept., No. 64: 91 pp.

fish, Brotulidae
Nielsen, Jørgen G., 1965.
On the genera Acanthonus and Typhlonus (Pisces, Brotulidae).
Galathea Rep., 8:33-47.

fish - Carcharinidae
d'Aubrey, J.D., 1964.
Sharks of the Family Carcharinidae of the South West Indian Ocean.
South African Assoc. Mar. Biol. Res., Bull., No. 5:20-26.

fish, ceriatoid
Pietsch, Theodore W., 1969
A remarkable new genus and species of deep-sea angler-fish (Family Oneirodidae) from off Guadalupe Island, Mexico.
Copeia, 1969(2): 365-369

fish (Chaetodontidae)
Klausewitz, Wolfgang 1970.
Forcipiger longirostris und Chaetodon leucopleura (Pisces, Perciformes, Chaetodontidae), zwei Neunachweise für das Rote Meer, und einige zoogeographische Probleme der Rotmeer-Fische.
Meteor - Forsch. Ergebn. (D) 5:1-5.

fish - cheilodipterid
Mead, Giles W., and J.E. De Falla, 1965.
New oceanic cheilodipterid fishes from the Indian Ocean.
Bull. Mus. Comp. Zool., Harvard Univ., 134(7):261-274.

fish - Clupeidae
Solovjev, B.S., 1968.
Distribution and behavior of some Clupeidae in the northwestern Indian Ocean. (In Russian).
Trudy, Vses. Nauchno-Issled. Inst. Morsk. Ribn. Okeanogr (VNIRO) 64, Trudy Azovo-Chernomorsk. Nauchno-Issled. Inst. Morsk. Ribn. Khoz. Okeanogr. (AscherNIRO), 28: 377-389.

fish, Engraulidae
Cervigón, Fernando, 1969.
Las especies de los generos Anchovia y Anchoa (Pisces: Engraulidae) de Venezuela y areas adyacentes del mar Caribe y Atlantico hasta 23° S.
Contrib. Estación Invest. mar. Margarita, Fundación La Salle, Venezuela 39: 193-257

fish - Engraulidae
Fage, L., 1920
Engraulidae, Clupeidae. Dan. Ocean. Exped. II. A9 140 pp.

fish, Engraulidae
Hardenberg, J. D. F. 1933
Notes on some genera of Engraulidae.
Natuurk., Tijdschr. Ned.-Ind., 93:230-256

fish - Engraulidae
Hildebrand, S.F., 1943.
A review of the American anchovies (Family Engraulidae). Bull. Bingham Ocean. Coll. 8(2): 1-165.

fish - engraulids
Peterson, C.L., 1956.
Observations on the taxonomy, biology, and ecology of the engraulid and clupeoid fishes in the Gulf of Nicoya, Costa Rica.
Inter-Amer. Tropical Tuna Comm., Bull., 1(5): 280 pp.

fish, Engraulidae
Salaya, Juan Jose, y Luis Salazar, 1969.
Una nueva cita de Engraulidae para las costas de Venezuela, Anchoa filifera.
Contrib. Estación Invest. mar. Margarita, Fundación La Salle, Venezuela, 40: 253-255.

fish - Exocoetidae
Bruun, A. Fr., 1935
Flying-fishes (Exocoetidae) of the Atlantic. Systematic and biological studies. Dana-Report No.6, 108 pp., 30 text figs. 7 pls.

fishes, Exocoetidae
Bruun, A. Fr., 1933.
On the value of the number of vertebrae in the classification of the Exocoetidae. Vidensk. Medd. fra Dansk Naturh. Foren. 94:375-384, 1 textfig.

fish - Gonostomatidae
Bruun, A. Fr., 1931
On some new fishes of the family Gonostomatidae. Vidensk. Medd. fra Dansk Naturh. Foren., 92:285-291, pl.8.

fish (Gonostomatidae)
*Kawaguchi, Kouichi, and Ryuzo Marumo, 1967.
Biology of Gonostoma gracile (Gonostomatidae). 1. Morphology, life history and sex reversal.
Inf. Bull. Planktol. Japan, Comm. No. Dr. Y. Matsue, 53-69.

fish - Heterocongridae
*Casimir, M.J. und H.W. Fricke, 1971.
Zur Funktion, Morphologie und Histochemie der Schwanzdrüse bei Röhrenaalen (Pisces, Apodes, Heterocongridae). Mar. Biol. 9(4): 339-346.

fish - Hexagrammidae
SSSR, Akademia Nauk, SSSR, Instituta Okeanologii, 1962.
Hexagrammidae fish and their possible transplantation between oceans. (In Russian).
Trudy Inst. Okeanol., Akad. Nauk, SSSR, 59:203pp

fishes, gymnotid
Steinbach, A.B. 1970.
Diurnal movements and discharge characteristics of electric gymnotid fishes in the Rio Negro, Brazil.
Biol. Bull. mar. biol. Lab. Woods Hole 138(2): 200-210.

fish Ipnopidae
Andersen, Knud P., 1966.
Classification of Ipnopidae by means of principal components and discriminant functions.
Galathea Rep. 8:77-90.

fish, Ipnopidae
Nielsen, Jørgen G., 1966.
Synopsis of the Ipnopidae (Pisces, Iniomi) with description of two new abyssal species.
Galathea Rep., 8:49-75.

fish - Kasidoroidae
Thorp, C.H., 1969.
A new species of miripinnaform fish (family Kasidoroidae) from the western Indian Ocean.
J. nat. Hist., 3(1): 61-70.

fish - Luvaridae
Blache, J., 1964.
Sur la présence de Luvarus imperialis Raf. 1810 dans l'Atlantique oriental sud rélevée par la découverte de deux larves au stade Hystricinella de L. Roule (1924)(Pisces, Teleostei, Perciformi, Luvaroidei, Luvaridae).
Trav. Centre Océanogr., Pointe-Noire, Cahiers, O.R.S.T.R.O.M., Océanogr., Paris, No. 5:57-59.

fishes, macrourid
Marshall, N.B., 1965.
Systematic and biological studies of the Macrourid fishes (Anacanthini-Teleostii).
Deep-Sea Res., 12(3):299-322.

fish, macrourid

Phleger, Charles F. 1971.
Biology of macrourid fishes.
Am. Zoologist 11(3): 419-423.

fish (Microdesmidae)

Dawson, C.E. 1970.
A new wormfish (Gobioidea: Microdesmidae) from the northern Red Sea.
Proc. Biol. Soc. Wash., 83(25): 267-272.

fish, Moraenidae

Blache, J., 1967.
Contribution à la connaissance des poissons anguilliformes de la côte occidental d'Afrique. 5. Le genre gymnothorax. 6. Les genres Anarchias, Uropteryzius, et Channomuraena (Moraenidae)
Bull. Inst. fondament. Afr. noire, (A) 29(4): 1695-1705.

fish- myctophids

Aron, William, 1962.
The distribution of animals in the eastern North Pacific and its relationship to physical and chemical conditions.
J. Fish. Res. Bd., Canada, 19(2): 271-314.

fish, myctophids

Barham, Eric G., 1966.
Deep scattering layer migration and composition: observations from a diving saucer.
Science, 151 (3716): 1399-1403.

fish, Myctophids

Bekker, V.E., 1964.
On the moderate cold-water Myctophidae Pisces complex. (In Russian).
Okeanologiia, Akad. Nauk, SSSR, 4(3): 469-476.

fish (Myctophidae)

Blache, J., et A. Stauch, 1964.
Contribution à la connaissance des poissons de la famille des Myctophidae dans la partie orientale du Golfe de Guinée (Teleostei, Clupeiformi, Myxtophidea). 1ère note. Les genres Electrona G. et B. 1895, Hygophum (Tan.) Bolin 1939.
Trav. Centre Océanogr., Pointe-Noire, Cahiers O.R.S.T.R.O.M., Océanogr., Paris, No. 5: 97-104.

fish, myctophids

Coleman, Leonard R., and Basil G. Nafpaktitis 1972.
Dorsadena yaquinae, a new genus and species of myctophid fish from the eastern North Pacific Ocean.
Contrib. Sci. Nat. Hist. Mus. LA County 225: 11 pp.

fish- myctophids

Kubota, Tadashi, 1967.
Brief note on the species of myctophid fish incidently caught along with Sergestes lucens in Suruga Bay. (In Japanese).
J. Coll. mar. Sci. Techn., Tokai Univ., (2): 231-232.

fish (myctophids)

Lourie, A. 1969.
Occurrence of two lanternfishes (Myctophidae) in the open sea off Mount Carmel (Israel).
Israel J. Zool. 18(4): 379-380

fish- Myctophids

Nafpaktitis, Basil, 1966.
Two new fishes of the Myctophid genus Diaphus from the Atlantic Ocean.
Bull. Mus. Comp. Zool., Harvard Univ., 133(9): 403-424.

fish - Myctophids

Nafpaktitis, Basil G., and Mary Nafpaktitis 1969.
Lanternfishes (Family Myctophidae) collected during cruises 3 and 6 of the R/V Anton Bruun in the Indian Ocean.
Bull. Los Angeles County Mus. nat. Hist. Sci., 5: 79 pp.

MYCTOPHIDS

Nakai, Zinziro, and Tadashi Kubota, 1966.
Studies on the morphological characters on alimentary canal and feeding habits of myctophid fishes. Preliminary report on the development of caudal light organs and gonads in some myctophids. (In Japanese)
J. Fac. Oceanogr. Tokai Univ. (1): 168, 169

MYCTOPHIDS

Nakai, Zinziro, Tadashi Kubota and Masahiro Ogura, 1966.
Outline of distribution of some myctophids on the Pacific Coast of Japan in summer-autumn, 1963. (In Japanese)
J. Fac. Oceanogr. Tokai Univ. (1): 170.

fish - myctophids

Rass, Th. S., 1960.
Geographical distribution of bathypelagic fish of the family Myctophidae in the Pacific Ocean.
Trudy Inst. Okeanol. 41: 146-152.

fish, rajids

Bigelow, Henry B., and William C. Schroeder, 1965.
Notes on a small collection of rajids from the sub-Antarctic region.
Limnol. and Oceanogr., Redfield Vol., Suppl. to 10: R38-R49.

fish - Scomberesocidae

Parin, Nikolay V., 1968.
Scomberesocidae (Pisces Synentognathii) of the eastern Atlantic Ocean.
Atlantide Rept., 10: 275-290.

fishes (scombroid)

Gorbunova, N.N., 1965.
On spawning of scombroid fishes (Pisces, Scombroidei) in the Gulf of Tonkin. (In Russian; English abstract).
Trudy Inst. Okeanol., Akad. Nauk, SSSR, 80: 167-176.

fishes (scombroid)

Gorbunova, N.N., 1965.
Seasons and conditions of spawning of the scombroid fishes (Pisces Scombroidei). (In Russian; English abstract).
Trudy Inst. Okeanol., Akad. Nauk, SSSR, 80: 36-61.

fish (scombroid)

Gorbunova, N.N., 1965.
On the distribution of scombroid larvae (Pisces, Scombroidei) in the eastern part of the Indian Ocean. (In Russian; English abstract).
Trudy Inst. Okeanol., Akad. Nauk, SSSR, 80: 32-35.

fish, Scyliorhinidae

Springer, Stewart, 1966.
A review of western Atlantic cat sharks Scyliorhinidae, with descriptions of a new genus and five new species.
Fishery Bull., Fish Wildl. Serv., U.S. 65(3): 581-624.

fish- serranid

Postel, E., 1967.
Deux serranides souvent confondus dans la litterature: Epinephelus fuscoguttatus et Epinephelus dispar.
Cah. Pacif. (10): 51.

fish- Sphyrnidae

Gilbert, Carter R., 1967.
A taxonomic synopsis of the hammerhead sharks (family Sphyrnidae). In: Sharks, skates and rays, Perry W. Gilbert, Robert F. Mathewson and David P. Rall, editors, John Hopkins Univ. Press, 69-77.

fish (Sternoptychidae)

Blache, J., 1964.
Poissons bathypélagiques de la famille des Sternoptychidae (Teleostei, Clupeiformi, Stomiatoidei) provenant des campagnes de l'"Ombango", de la "Reine Pokou" et du "Gerard Treca" dans les eaux africaines de l'Atlantique tropical.
Trav. Centre Océanogr., Pointe-Noire, Cahiers O.R.S.T.R.O.M., Océanogr., Paris, No. 5: 71-87.

fish (Stomiatidae)

Blache, J., 1964.
Contribution à la connaissance des Stomiatidae (Pisces, Teleostei, Cluopeiformi, Stomiatoidei) dans l'Atlantique tropical oriental sud. Mise en évidence d'une sous-espèce de Stomias colubrinus Garmen, 1899, caractéristique des formes du Golfe de Guinée.
Trav. Centre Océanogr., Pointe-Noire, Cahiers O.R.S.T.R.O.M., Océanogr., Paris, No. 5: 61-69.

fish- Synaphobranchidae

Bruun, A. Fr., 1937
Contributions to the Life Histories of the Deep Sea Eels: Synaphobranchidae. Dana Rept. No. 9, 31 pp., 7 textfigs., 1 pl.

fishes - Synodontidae

Zvjagina, O.A., 1965.
Data on development of lizard fishes (Synodontidae, Pisces). (In Russian; English abstract).
Trudy Inst. Okeanol., Akad. Nauk, SSSR, 80: 162-166.

fish-Thonidae

Keyvanfar, A., 1962.
Sérologie et immunologie de deux espèces de thonidés (Germo alalunga Gmelin et Thunnus thynnus Linné) de l'Atlantique et de la Méditerranée.
Rev. Trav. Inst. Pêches Marit., 26(4): 407-456.

fish - Triglidae

Richard, William J., 1968.
Eastern Atlantic Triglidae (Pisces, Scorpaeniformes).
Atlantide Rept., 10: 77-114.

Fish-genera-
Also SEE: Fisheries cards filed under common name of fish.

fish-Alepisaurus

Haedrich, R.L., 1969.
Alepisaurus.
Oceanus, 15(1): 8-9. (popular)

fish, Anchoviella

Carvalho, J. de P., 1951.
Engraulideos brasileiros do genero Anchoviella. Bol. Inst. Paulista Ocean. 2(1): 41-66, 2 pls.

Anguilla

eels Anguilla
D'Ancona, U., 1951.
Intorno alle trasformazioni dell'anguilla nella fase conclusiva del suo ciclo vitale. Riv. Biol., Univ. Perugia, 43(2):171-186, 10 textfigs., 3 pls

Anguilla
Jespersen, P., 1942
Indo-pacific leptocephalids of the genus Anguilla. Systematic and Biological studies. Dana Rept. No.22:128 pp.

fish - Anguilla
eels, migration of
Koops, Harald, 1962
Die Wanderung des Aales, Anguilla anguilla L. in Abhängigkeit vom Sommerhochwasser. Kurze Mitteil., Inst. Fischereibiol., Univ. Hamburg, (12):19-26.

Anguilla
Schmidt, J., 1928.
The freshwater eels of Australia. With some remarks on the short-finned species of Anguilla. Rec. Australian Mus. 16(4):179-210, 14 textfigs.

Anguilla
Schmidt, J., 1928.
The freshwater eels of New Zealand. Trans. N.Z. Inst. 58:379-388, 8 textfigs.

Anguilla anguilla
Bruun, A. Fr., A.M. Hemingsen, and E. Møller-Christensen, 1949.
Attempts to induce experimentally maturation of the gonads of the European eel, Anguilla anguilla L. Acta Endocrinol. 2:212-226.

Anguilla anguilla L.
Callamand, O., 1943
L'Anguille Européené (Anguilla anguilla L.) Les Bases physiologiques de sa migration. Ann. de l'Inst. Ocean., Monaco, 21(6):361-438, 1 pl.

fish
Anguilla anguilla
Deelder, C.L. und F.W. Tesch, 1970.
Heimfindevermögen von Aalen (Anguilla anguilla) die über grosse Entfernungen verpflanzt worden waren. Marine Biol., 6(1): 81-92.

Anguilla anguilla L.
Filuk, Jerzy, 1965.
Eelo (Anguilla anguilla L.) in the Vistule furth (In Polish; English and Russian summary). Prace. Morsk. Inst., Ryback, (A) 13:101-113.

fish - Anguilla anguilla
*Miles, S.G., 1968.
Laboratory experiments on the orientation of the adult American eel, Anguilla rostrata. J. Fish. Res. Bd., Can., 25(10):2143-2155.

fish, Anguilla anguilla
Nishi, Genjirou, and Sadahiko Imai, 1969.
On the juvenile of Anguilla marmorata Quoy et Gaimard in Yakushima (Yaku Island). Its ecology and morphology.
Mem. Fac. Fish. Kagoshima Univ.
18:65-76
(In Japanese; English abstract)

fish - Anguilla anguilla
Scheil, H.-G., 1970.
Beiträge zur Temperaturadaptation des Aales Anguilla anguilla IV. Einfluss der Adaptationstemperatur auf den denervierten Seitenmuskel. Marine Biol., 6(2): 158-166.

Anguilla anguilla
Sharratt, Brenda M., B. Bellamy and J. Chester Jones, 1964.
Adaptation of the silver eel (Anguilla anguilla L) to sea water and to artificial media together with observations on the role of the gut. Comp. Biochem. Physiol., 11(1):19-30.

Anguilla anguilla
Sharratt, Brenda M., I. Chester Jones and D. Bellamy, 1964.
Water and electrolyte composition of the body and renal functions of the eel (Anguilla anguilla L.).
Comp. Biochem. Physiol., 11(1):9-18.

fish - Anguilla anguilla
Sinha, V.R.P., 1969.
A note on the feeding of larger eels Anguilla anguilla (L.).
J. Fish. Biol. 1(3): 279-283.

Anguilla anguilla
Sinha, V.R.P., and J.W. Jones, 1966.
On the sex and distribution of the freshwater eel (Anguilla anguilla).
J. Zool., London, 150(3):371-385.

fish, Anguilla anguilla
Skadhauge, Erik 1969.
The mechanics of salt and water absorption in the intestine of the eel (Anguilla anguilla) adapted to waters of various salinities.
J. Physiol. 204: 135-158
Also in: Trav. Stn zool. Villefranche 30 (1969)

fish - Anguilla anguilla
Tesch, F.-W., 1970.
Heimfindevermögen von Aalen Anguilla anguilla nach Beeinträchtigung des Geruchssinnes, nach Adaptation oder nach Verpflanzung in ein Nachbar-Ästuar. Marine Biol., 6(2): 148-157.

Anguilla anguilla
Tesch, F.W., 1965.
Verhalten der Glasaale (Anguilla anguilla) bei ihrer Wanderung in den Ästuarien deutscher Nordseeflüsse.
Helgoländer wiss Meeresunters. 12(4):404-419.

fish, Anguilla, anguilla
Vladykov, Vadim D. 1971.
Homing of the American eel Anguilla rostrata as evidenced by returns of transplanted tagged eels in New Brunswick.
Can. Fld Nat. 85(3): 241-248.

fish - Anguilla anguilla
Wehrmann, L., 1968.
Messungen der Bewegungsaktivität des A les (Anguilla anguilla L.) unter Verwendung einer Markierung mit Magnetmarken. Erprobung einer neuen Methode. Dipl. Arb. Math.-Naturw. Fakultät Univ. Hamburg, 65 pp. - Autoreferat. Also in: Gesammelte Sonderdrucke, Inst. Hydrobiol. Fischereiwiss. Univ. Hamburg, 1967-1968.

Anguilla japonica
Matsui, I., 1957.
On the records of a leptocephalous and catadromous eels of Anguilla japonica in the waters around Japan with a presumption of their spawning places. J. Shimonoseki Coll. Fish., 7(1):151-167

Anguilla japonica
Matsui, I., 1952.
[Studies on the morphology, ecology and pond-culture of the Japanese eel (Anguilla japonica Temminck & Schlegel).] J. Simonoseki Coll. Fish. 2(2):1-245, 3 pls., 85 textfigs. (English summary).

Anguilla nebulosa labiata
Frost, W.E., 1957.
First record of the elver of the African eel Anguilla nebulosa labiata Peters. Nature, 179 (4559):594.

fish - Anguilla rostrata
Boëtius, Inge, and Jan Boëtius, 1967.
Eels, Anguilla rostrata Le Sueur, in Bermuda.
Vidensk. Meddr. dansk naturh. Foren. 130: 63-84

Anguilla rostrata
Jeffries, Harry P., 1960.
Winter occurrences of Anguilla rostrata elvers in New England and middle Atlantic estuaries. Limnol. & Oceanogr., 5(3):338-340.

fish - Anguilla rostrata
Miles, S.G. 1968.
Rheotaxis of elvers of the American eel (Anguilla rostrata) in the laboratory to water from different streams in Nova Scotia.
J. Fish. Res. Bd, Can. 25(8): 1591-1602

Anguilla rostrata
Parker, G.H., 1945
Melanophore activators in the common American eel Anguilla rostrata Le Sueur. J. Exp. Zool., 98(3):211-234, 1 pl.

Anguilla rostrata
Vladykov, Vadim D., 1964.
Quest for the true breeding area of the American eel (Anguilla rostrata LeSueur). J. Fish. Res. Bd., Canada, 21(6):1523-1530.

Anguilla vulgaris
Bruun, A. Fr., and B. Heiberg, 1932
The "Red Disease" of the eel in Danish Waters. Medd. fra Komm for Danmarks Fiskeri og Havundersøgelser, serie: Fiskeri, IX (6):19 pp.

Anguilla vulgaris
Creutzberg, F., 1961.
On the orientation of migrating elvers (Anguilla vulgaris Turt.) in a tidal area. Netherlands J. of Res., 1(3):257-338.

Anguilla vulgaris
Creutzberg, F., 1958.
Use of tidal streams by migrating elvers (Anguilla vulgaris Turt.). Nature 181(4612):857-858.

Anguilla vulgaris
Deelder, C.L., 1952.
On the migration of the elver (Anguilla vulgaris Tust) at sea. J. du Cons. 18(2):187-218, 4 textfigs.

fish - Anguilla vulgaris
Jankowsky, Hans-Dieter 1966.
The effect of adaptation temperature on the metabolic level of the eel Anguilla vulgaris L.
Helgoländer wiss. Meeresunters. 13(4): 402-407.

fish - Anguilla vulgaris
Malessa, P., 1969.
Beiträge zur Temperaturadaptation des Aales (Anguilla vulgaris). III. Höhe und Verteilung der Aktivität von Succinodehydrogenase und Cytochromoxydase im Seitenmuskel juveniler und adulter Tiere. Marine Biol., 3(2): 143-158.

Anguilla vulgaris
Schultze, Dietmar, 1965.
Beiträge zur Temperaturadaptation des Aales (Anguilla vulgaris L). II.
Zeits. Wiss. Zool., 172(1/2):104-133.

Anisotremis

Anisotremis davidsoni (Sargo or China croaker)
Johnson, M.W., 1943
Underwater Sounds of Biological Origin.
UCDWR No.U28, Sec. No.6.1-sr30-412, File No. 01.33, 20 pp. (mimeographed), 1 chart (photostat), 2 figs. (photo), 15 Feb. 1943.

fish, Aphanopus
Bone, Quentin, 1971.
On the scabbard fish Aphanopus carbo. J. mar. biol. Ass., U.K., 51(1): 219-225.

fish, Aphanopus
Holl, A., und W. Meinel. 1968.
Das Geruchsorgan des Tiefsee fisches Aphanopus carbo (Percomorphi, Trichiuridae).
Helgolander wiss. Meeresunters. 18(4): 404-423.

fish, Ariomma
Haedrich, Richard L. 1968.
First record of Ariomma (Pisces, Stromateoidei) from the South Pacific, and comments on the elongate species of the genus.
Bull. mar. Sci. Miami 18(1): 249-260.

fish, Barbourisia rufa
Struhsaker, Paul, 1965.
The whalefish Barbourisia rufa (Cetunculi) from waters off southeastern United Staes.
Copeia, 1965(3):376-377.

fish - Bathymyrus
Castle, P.H.J., 1968.
Description and osteology of a new eel of the genus Bathymyrus from off Mozambique.
Spec. Publ. Dept. Ichthyol. Rhodes Univ., 4:1-12.

fishes, Bathyprion
Marshall, N.B., 1966.
Bathyprion danae, a new genus and new species of Alepocephaliform fishes.
Dana Rep. No. 68:1-10.

Batrachoides

fish - Batrachoides
Cervigon, Fernando, 1964.
Batrachoides manglae nov. sp., una nueva especi de Batrachoididae de las costas de Venezuela
Novedades Cientificas, Contrib., Ocasionales Mus. Hist. Nat. La Salle, Ser. Zool., No. 32:1-4.

Also:
Contrib. Estacion Invest. Mar. Margarita, No.17

fish - Benthosoma glaciale
Halliday, R.G. 1970
Growth and vertical distribution of the glacier lanternfish, Benthosoma glaciale, in the northwestern Atlantic.
J. Fish. Res. Bd. Can. 27(1): 105-116.

Brama

Brama brama
Mead, Giles W., and Richard L. Haedrich, 1965.
The distribution of the oceanic fish Brama brama
Bull. Mus. Comp. Zool., Harvard Univ., 134(2):29-68.

Carcharinus

Carcharinus
Boeseman, M., 1960
A tragedy of errors: The status of Carcharhinus Blainville, 1816; Galeolamna Owen, 1853; Eulamia Gill, 1861; and the identity of Carcharhinus commersonii Blainville, 1825.
Zoologische Meded., 37(6): 81-100.

Carcharinus sp.
Davies, D.H., and J.D. D'Aubrey, 1961.
Shark attack off the east coast of Africa, 6 January 1961.
S. Afr., Ass. Mar. Biol. Res., Ocean. Res. Inst. Invest. Rept., 3:5 pp.

Sharks, Carcharhinus
Garrick, J.A.F. 1967
A broad view of Carcharhinus species, their systematics and distribution. In: Sharks, skates and rays, Perry W. Gilbert, Robert F. Mathewson and David P. Rall, editors, Johns Hopkins Univ. Press 85-91.

fish - Carcharhinus floridanus
Garricks, J.A.F., Richard H. Backus and Robert H. Gibbs, Jr., 1964.
Carcharhinus floridanus, the silky shark, a synonym of C. calciformis.
Copeia, (2):369-375.

fish - Carcharinus leucas
Urist, Marshall, 1962.
Calcium and other ions in blood and skeleton of Nicaraguan fresh-water shark.
Science, 137(3534):984-986.

Carcharinus leucas

Carcharinus longimanus
Hubbs, C.L., 1951.
Record of the shark, Carcharinus longimanus, accompanied by Naucrates and Remora, from the East-Central Pacific. Pacific Science, 5(1):78-81, 1 textfig.

fish - Carcharhinus melamopterus
Tester, Albert L., and Susumu Kato, 1966.
Visual target discrimination in blacktip sharks (Carcharhinus melamopterus) and gray sharks (C. menisorrah).
Pacif. Sci., 20(4):461-471.

Carcharhinus menisorrah
Fellows, David P., and A. Earl Murchison, 1967.
A non injurious attack by a small shark.
Pacif. Sci., 21(1):150-151.

Carcharhinus milberti
Schwartz, Frank J., 1960.
Measurements and the occurrence of young sandbar shark, Carcharhinus milberti, in Chesapeake Bay, Maryland.
Chesapeake Science, 1(3/4):204-206.

fish - Carcharhinus obscurus
Tibbo, S.N., and R.A. McKenzie, 1963.
An occurrence of dusky sharks, Carcharinus obscurus (Lesueur) 1818, in the northwest Atlantic.
J. Fish. Res. Bd., Canada, 20(4):1101-1102.

Carcharius

Carcharias taurus
Campbell, G.D., David H. Davies and Arthur C. Copley, 1960
A case of shark attack with special reference to attempts to identify the causal species from the wounds. Medical Proc., Mediese Bydraes, S. Africa, 6(26): 612-621.

Carcharias kamoharai
D'Aubrey, Jeannette D., 1964.
A carchariid shark new to South African waters.
S. Afr. Assoc. Mar. Biol. Assoc. Oceanogr. Res. Inst. Investigational Rept., No. 9:16 pp.

Also in:
Collected Reprints, Int. Indian Ocean Exped., UNESCO, 3. 1966.

Carcharias taurus
Davies, D.H., and J.D. D'Aubrey, 1961.
Shark attack off the east coast of South Africa, 24 December 1960, with notes on the species of shark responsible for the attack.
S. Afr. Ass., Mar. Biol. Res., Oceano. Res. Inst. Invest. Rept., 2:8 pp.

Carcharodon

Carcharodon carcharias
Forman, T.B., and D.J. Dunstan, 1967.
Report on an attack by a great white shark off Coledale Beach, N.S.W., Australia, in which both victim and attacker were recovered simultaneously.
Calif. Fish Game, 53(3): 219-223

Carcharodon carcharias
Gudger, E.W., 1950.
A boy attacked by a shark, July 25, 1936 in Buzzards Bay, Massachusetts with notes on attacks by another shark along the New Jersey coast in 1816. Am. Mid. Nat. 44(3):714-719, 2 textfigs.

Carcharodon carcharias
Kato, Susumu, 1965.
White shark, Carcharodon carcharias from the Gulf of California with a list of sharks seen in Mazatlen, Mexico.
Copeia, 1965(3):384.

fish - Carcharodon carcharias
Pike, Gordon C., 1962.
First record of the great white shark (Carcharodon carcharias) from British Columbia.
J. Fish. Res. Bd. Canada 19(2): 363

Carcharodon carcharias
Royce, William F., 1963
First record of white shark (Carcharodon carcharias) from southeastern Alaska.
Copeia, (1):179.

Fish Centrophorus
Hulley, P.A. 1971.
Centrophorus squamosus (Bonnaterre) (Chondrichthyes, Squalidae) in the eastern South Atlantic
Ann. SAfr. Mus. 57(11):265-270

Centroscyllium

Centroscyllium ritteri
Iwai, Tamotsu, 1960.
Luminous organs of the deep-sea squaloid shark, Centroscyllium ritteri Jordan and Fowler.
Pacific Science 14(1): 51-54.

fish - Cetengraulis mysticetus
Bayliff, William H., 1964.
Some aspects of the age and growth of the anchoveta Cetengraulis mysticetus in the Gulf of Panama.
Inter-American Tropical Tuna Commission, Bull., 9(1):1-51.

Cetorhinus

fish - Cetorhinus
Siccardi, Elvira M., 1960
"Cetorhinus" en el Atlantico Sur (Elasmobranchii:Cetorhinidae).
Rev. Mus.Argentino Ciencias Nat. "Bernardino Rivadavia", Ciencias Zool., 6(2):61-101, 3 pls

Cetorhinus maximus
Blumer, Max 1967.
Hydrocarbons in digestive tract and liver of a basking shark.
Science 156 (3773): 390-391.

fish - Cetorhinus maximus
Parker, H.W., and F.C. Stott, 1965.
Age, size and vertebral calcification in the basking shark, Cetorhinus maximus (Gunnerus)
Zool. Mededel., 40 (34):305-318.

fish - Chlamydoselachus anguineus
Wheeler, A., 1962.
New records for distribution of the frilled shark.
Nature, 196(4855):689-690.
Chlamydoselachus anguineus

Clupea

Clupea harengus
Alander, H., 1943.
Investigation on the Baltic herring. Ann. Biol. 1:177-182, map.

Clupea harengus
Anderson, K.A., 1943.
The stock of herring in the Skagerak, the Kattegat, and the Sound in 1939. Ann. Biol. 1: 161-163.

Clupea pilchardus
Corbin, P.G., 1947.
The spawning of mackerel, Scomber scombrus L., and pilchard, Clupea pilchardus Walbaum, in the Celtic Sea in 1937-39 with observations on the zooplankton indicator species, Sagitta and Muggiaea. J.M.B.A., n.s., 27:65-132, 21 textfigs.

Clupea pilchardus
Muzinic, R., 1948/1949.
First tagging experiments on the sardine (Clupea pilchardus) in the Adriatic. Acta Adriatica 3(10):26 pp., 8 textfigs., 1 chart.

Conger

Conger oceanicus
Schmidt, J., 1931.
Eels and conger eels of the North Atlantic.
Nature for 10 Oct. 1931, 8 pp., 3 textfigs.

Conger vulgaris
Schmidt, J., 1931.
Eels and conger eels of the North Atlantic.
Nature for 10 Oct. 1931, 8 pp., 3 textfigs.

Culiceps

Culiceps sthenae
Haedrich, Richard L., 1965.
Culiceps sthense, a new nomeid fish from the western North Atlantic Ocean, and its systematic position among stromateoids.
Copeia, 1965(4):501-505.

fish - Decapterus
Nekrasov, V.V., 1968.
On the biology and fishery of horse mackerel (genus Decapterus) in the northwest Indian Ocean. (In Russian).
Trudy, Vses. Nauchno-Issled. Inst. Morsk. Ribn. Okeanogr (VNIRO) 64, Trudy Azovo-Chernomorsk. Nauchno-Issled. Inst. Morsk. Ribn. Khoz. Okeanogr. (AscherNIRO), 28: 390-400.

fish, Dasyatis
Springer, Stewart, and Bruce B. Collette, 1971
The Gulf of Guinea stingray, Dasyatis rudis.
Copeia 1971(2): 338-341

Engraulis

fish - Engraulis
Chaudhuri, B.L. 1916
Fauna of the Chilka Lake. Fish Pt. I.
Mem. Indian Mus. 5:403-
Engraulis annandalei p.419
E. kempi, p.421
E. rambhae, p.423
E. purava
E. mystax

fish - Engraulis
Day, Francis 1878.
Fishes of India, (London) Pt. IV: 553-778
Engraulis synonomy of 12 Indian species

fish - Engraulis
Delsman, H.C. 1929.
Fish eggs and larvae from the Java Sea. The genus Engraulis.
Treubia 11: 275-286
E. grayi

fish - Engraulis
Evermann, B.W., and L. Radcliffe 1917.
The fishes of the west coast of Peru and the Titicaca basin.
Bull. U.S. Nat. Mus. 95:1-166.
E. ringens
E. nasus doubtful

Engraulis spp.
Reid, Joseph L., Jr., 1967.
Oceanic environments of the genus Engraulis around the world.
Rep.Calif.Coop.Oceanic Fish.Invest., 11:29-33.

Engraulis
Schweigger, E., 1943
Pesqueria y Oceanografia del Peru y Proposiciones para su Desarrollo Futuro; Informe Elevada a la Compania Administradora del Guano. 1. Los Peces y la Vida en el Mar. 2. Oceanografia. 3. Proposiciones para el Futuro. 356 pp. 67 charts in text, 8 graphs, 4 tables in appendix.

E. anchoita

Engraulis anchoita
de Ciechomski, Janina Dz., 1967.
Investigations of food and feeding habits of larvae and juveniles of the Argentine anchovy Engraulis anchoita.
Rep. Calif.Coop.Oceanic Fish.Invest.11:72-81.

Engraulis anchoita
de Ciechomsky, Janina Dz., 1967.
Influence of some environmental factors upon the embryonic development of the Argentine anchovy Engraulis anchoita (Hubbs, Marini).
Rep.Calif.Coop.Oceanic Fish.Invest., 11:67-71.

Engraulis anchoita
de Ciechomski, Janina Dz., 1967.
Present state of the investigations on the Argentine anchovy Engraulis anchoita (Hubbs, Marini).
Rep.Calif.Coop.Oceanic Fish.Invest., 11:58-66.

Engraulis anchoita
de Ciechomski, Juana D., 1965.
Observaciones sobre la reproducción desarrollo embrionario y larval de la anchoita argentina (Engraulis anchoita).
Inst. Biol. Mar., Mar del Plata, Bol., No. 9: 29 pp.

fish - Engraulis anchoita
Marini, T.L. 1935.
La anchoita argentina. Su posición sistemática y su porvenir económica
Physis 11: 445-458.

E. australis

Engraulis australis
Blackburn, Maurice, 1967.
Synopsis of biological information on the Australian anchovy Engraulis australis (White)
Rep. Calif. Coop.Oceanic Fish.Invest., 11:34-43.

Engraulis australis
Blackburn, M., 1950
A biological study of the anchovy, Engraulis australis (White), in Australian waters.
Australian J. Mar. and Freshwater Res. 1 (1): 3, 84, 5 pls., 11 textfigs.

fish, Engraulis australis
Blackburn, M. 1941.
The economic biology of some Australian clupeoid fish
Bull. Counc. Sci. Industr. Res. Comm. Australia (Div. Fish. Rept. 6) 138: 1-135

fish - Engraulis australis
Phillipps, N. J. 1929
Note on an anchovy (Engraulis australis).
New Zealand Jour. Sci. Tech. 10: 345

E. carpenterai

fish - Engraulis carpenteri n.sp.
DeVis, C.W. 1883.
Description of three new fishes of Queensland.
Proc. Linn. Soc. N.S. Wales 7: 318-320.

E. encrasicholus

fish - Engraulis encrasicholus.
Aleksandrov, A. 1927.
Anchois de la mer d'Azoff et de la mer Noire, leurs origines et indications taxonomiques. (In Russian, French résumé).
Rep. Sci. Stat. Fish. Kertch 1 (2/3): 34-99.

Engraulis encrassicholus
Andreu, B., 1950.
Sobre la maduración sexual de la anchoa (Engraulis encrassicholus L.) de la costa del norte de España. Datos Biologicos y biometricos.
Publ. Inst. Biol. Apl. 7:7-36, 26 textfigs., 5 photos.

Engraulis encrasicholus
Andreu, B., and J. Rodriguez-Roda, 1951.
Estudio comparativo del ciclo sexual, engrasamiento y replecion estomacal de la sardina, alacha y anchoa del mar Catalan, acompanado de relacion des pescas de huevos planctonicos de estas especies. Publ. Inst. Biol. Aplic. 9:193-232, 14 textfigs.

Engraulis encrassicholus
Arné, P., 1931.
Contribution à l'étude de l'anchois (Engraulis encrassicholus L.) du Golfe de Gascogne.
Rev. Trav. Pêches Marit., France, 4(2):153-181.

Engraulis encrasicholus
Bas, C., and E. Morales, 1954.
Algunos datos para el estudio de la biologia de la anchoa Engraulis encrasicholus L. de la Costa Brava. Publ. Inst. Biol. Aplic. 16:53-70, 4 textfigs.

Engraulis encrassicholus
Bougis, P., 1952.
La croissance des poissons mediterraneens.
Vie et Milieu, Suppl. No. 2, Ocean. Medit., Jour. Etudes, Lab. Arago:118-146

fish - Engraulis encrassicholus
Cunningham, J. T. 1895
The migration of the anchovy
Jour. Mar. Biol. Assoc., (2) 3: 300-303

Engraulis encrassicholus
DeBuen, F., 1932.
Clupéidés et leur peche. Comm. Inst. Expl. Medit., Rapp. Proc. Verb. 7:319-340.

fish - Engraulis encrassicholus
de Buen, F. 1931
La biología de la anchoa o boquerón (Engraulis encrassicholus Linnaeus 1758
Bol. Com. esp. Union. inter. cien. Biol 2:7-56.

Engraulis encrassicholus
DeBuen, F., 1931.
Clupéidés et leur peche. Comm. Int. Expl. Medit., Rapp. Proc. Verb. 6:289-335.

Engraulis encrassicholus
DeRusso, A., 1925.
Studi sulla pesca nel Golfo di Catania.
Boll. Pesca, Piscicolt. Idrobiol. 1(2):3-55, 3 fold-ins

fish - Engraulis encrassicholus
Dulzetto, F. 1938
Sui caratteri biometrici dell'accuiga Engraulis encrasicholus del Golfo di Catania.
Boll. Pesca Piscicolt. Idrobiol., 14:180-209

fish - Engraulis encrasicholus
Elizarova, S. S. 1936
The influence of the active reaction of hydrogen ions and of salinity on the eggs of Engraulis encrassicholus L.
C. R. Acad. Sci., Moscou, N. S.: 255-260

fish, Engraulis encrasicholus
Ehrenbaum, E. 1892
Die Sardelle (Engraulis encrasicholus L).
Aussichten einer deutschen Sardellenfischerei und Kritik der holländischen Arbeiten über die Sardelle.
Mitth. Deutsche Seefischerei Ver. Suppl. 21 pp.

fish, Engraulis encrasicholus
Ewart, P. 1889-90
Phys. Soc., Edinburgh, p. 333
on occurrence in Scottish waters - E.encrasicholus

Engraulis encrassicholus
Fage, L., 1937.
La ponte et les races locales de l'anchois de la Mediterranee (Engraulis encrassicholus L.).
Rapp. Comm. Expl., Mediterranee, 10:67-71.

fish - Engraulis encrassicholus
Fage, L. 1935
L'anchois de la mer du Nord (Engraulis encrasicholus (L))
et l'assèchement du Zuiderzee.
Bull. Inst. Oceanogr., Monaco, 668:1-7

fish - Engraulis encrasicholus
Fage, L. 1930
Engraulis encrasicholus
Faune Ichthyol. Atlant. Nord 4
(Also in Faune Flora Medit.)

fish - Engraulis encrasicholus
Fage, L. 1920
Engraulidae, Clupeidae.
Rep. Dan. Ocean. Exped. 1908-1910 to the Medit. and adjacent seas 2 Biol A9:1-140

Engraulis encrassicholus
Fage, L., 1911.
Recherche sur la biologie de l'anchois (Engraulis encrassicholus Linne). Race, Age, Migration.
Ann. Inst. Ocean., Paris, 2(4):40 pp., 16 textfigs.

fish - Engraulis encrassicholus
Fage, L. 1911
Sur les races locales de l'anchois (Engraulis encrassicholus Linné)
Réponse à M. Pietro Lo Guidace.
Arch. Zool. Exper. Gen. (5) 8:LXXII-LXXX 45

fish - Engraulis encrassicholus
Fulton, T. W. 1905
Ichthyological notes. II. The anchovy (Engraulis encrassicholus).
23 Ann. Rep. Fish. Board, Scotland: 251-254.

fish - Engraulis encrassicholus
Fulton, T. W. 1902
Ichthyological notes.
20th Ann. Rept. Fish. Board, Scotland: 539-541
I. The anchovy (Engraulis encrassicholus)

Engraulis encrassicholus
Furnestin, J., 1943?
Note preliminaire sur l'anchois (Engraulis encrassicholus L.) du Golfe de Gascogne.
Rev. Trav. Pêches Marit., France, 13:197-209, 4 textfigs.

Engraulis encrassicholus
Furnestin, J., and R. Coupe, 1950.
Les caracteristiques morphologiques des anchois (Engraulis encrassicholus Linn.) du Maroc.
J. du Cons. 16(2):182-184.
Measurements only.

Engraulis encrassicholus
Furnestin, L., and M.L. Furnestin, 1959.
La reproduction de la sardine et de l'anchois des côtes atlantiques du maroc (saisons et aires de ponte).
Rev. Trav. Inst. Pêches Marit., 23(1):79-104.

fish - Engraulis encrassicholus
Gruvel, A. 1913
L'anchois (Engraulis encrassicholus L)
sur la cote occidentale d'Afrique.
Paris C. R. Acad. Sci., 157:1468-1470

Engraulis encrasicholus
Hoeck, P.P.C., 1912.
Les Clupeides (le hareng excepte) et leur migrations. 1. L'anchois (Engraulis encrasicholus L.)
Cons. Perm. Int. Expl. Mer, Rapp. Proc. Verb. 14:1-40.

fish - Engraulis encrasicholus
Hoek, P. P. C. 1892
Mededeelingen omtrent de levenswijze en de voortplanting van de ansjovis in de Zuiderzee.
Verslag van den Staat der Nederl. Zeevisscherijen over 1891, s'Gravenhage:326-339
(Contributions regarding the life history and the propagation of the anchovy and the anchovy fisheries in the Zuiderzee).

fish - Engraulis encrasicholus
Hoek, P. P. C. and others 1886
Bijdragen tot de kennis der levenswijze en der voortplanting van de ansjovis.
Vers. Staat. Nederl. Zeevisscherijen, s'Gravenhage, 1886:173-201
(Contributions to the life history and propagation of the anchovy)

fish - Engraulis encrasicholus 1886
Hoffmann, C.K.
Bijdrage tot de Kennis der levenswijze en der voortplanting van der Ansjovis.
Verslag v. d. Staat d. Nederl. Zeevisscherijen over 1885, p.161

fish - Engraulis encrasicholus 1887
Hubrecht, Hoek, Weber, Wenckebach.
Verslag omtrent op de Ansjovis betrekking hebbende onderzoekingen.
Verslag v.d. Staat d. Nederl. Zeevisscherijen over 1888, p175.

Engraulis encrasicholus
Karlovac, J., 1963.
Contribution à la connaissance de la ponte de l'anchois (Engraulis encrasicholus L.) dans la haute Adriatique (note préliminaire).
Rapp. Proc. Verb., Réunions, Comm. Int. Expl. Sci., Mer Méditerranée, Monaco, 17(2):321-326.

fish, Engraulis encrasicholus
Lo Guidice, p. 1911
Ancora sulla diverse razzi locali di acciughe (Engraulis encrassicholus Cuv.) Replica al Dr. Fage.
Rivist. Mens. Pesca, Pavia, 13 (n.s.,6) 226-236

fish, Engraulis encrasicholus
Marion, A. F. 1894
Notes sur le regime du maquereau et de l'anchois sur les cotes de Marseille durant la campagne 1890.
Ann. Mus. Hist. Nat. Marseille (Zool.) 4:108-112

fish, Engraulis encrasicholus
Marion, A.F. 1889
Notes sur l'Anchois.
Ann. Mus. Hist. nat., Marseille, 3:55.

Engraulis encrassichalus
Miranda y Rivera, A. de., 1930.
Investigaciónes métodicos realizades en 1928 en el Laboratorio de Málaga. Notas y Res., Inst. Español Ocean., 2nd ser., No. 37:26 pp., 16 textfigs.

Engraulis encrassicholus
Navarro, F. de P., and J.M. Navaz, 1946.
Apuntes para la biología y biometría de la sardina, anchoa, boga y chicharro de las costas Vascas. Notas y Res., Inst. Español Ocean., 2nd ser., No. 134:25 pp., 6 figs.

Engraulis encrassicholus
Navaz y Sanz, J.M., 1946.
Nuevos datos sobre la substitucion alternativa en la pesca de peces emigrantes en el litoral de Galicia. Notas y Res., Inst. Español Ocean., 2nd ser., No. 132:1-9, 6 figs.

Engraulis encrassicholus
Navaz, J.M., and F. Navarro, 1950.
Anchois (Engraulis encrassicholus L.) à la côte Basque (1948-49). Cons. Perm. Int. Expl. Mer, Ann. Biol. 6:68-69.

Resumé of papers in
Bol. Inst. Esp. Ocean.
Notas y Res., Inst. Esp. Ocean.

Engraulis encrasicholus
Padoan, P., 1963.
Prime osservazioni sulle acciughe (Engraulis encrasicholus L.) catturate al largo delle foci del Po.
Rapp. Proc. Verb., Réunions, Comm. Int. Expl. Sci., Mer Méditerranée, Monaco, 17(2):327-332.

fish. Engraulis encrasicholus
Patterson, A. H. 1906
Anchovy at Yarmouth
Zoologist, London, 10:435
E. encrasicholus

Engraulis encrasicholus
Planes, A., and F. Vives, 1951.
Sobre la puesta de la anchoa Engraulis encrasichalus L. en el Levante espanol.
Publ. Ins. Biol. Aplic. 9:119-136, 3 textfigs.

fish, Engraulis encrasicholus
Poulsen, E. M. 1937
Fluctuations in the regional distribution of certain fish-stocks within the Transition Area during recent years (1923-1935)
Rapp. Cons. Explor. Mer, 102:1-17
E. encrassicholus

Engraulis encrassichohus
Racovitza, E.G., 1904.
Observations sur un banc d'anchois (Engraulis encrassichohus (L.), recontre pres de l'ile Cabrera (Baleare). Bull. Soc. Zool., France, 29:211-218.

Engraulis engrasicholus L.
Vidalis, E., 1949.
Contribution to the biology of the anchovy in Greek waters. Praktika tou Hellenikou Garobiologou Instituutou, Etos 1949, 3(2):41-70 (English summary).

fish - Engraulis encrasicholus
Redeke, H. C., 1916
Zur Naturgeschichte der Sardelle.
Mitt. Deutsch. Seefisch. Ver., Berlin, 32:194-199
Biology of E. encrasicholus

fish - Engraulis encrasicholus
Redeke, H. C. 1914
Bydragen tot de kennis van de teelt der ansjovis in de Zuiderzee. 2e stuk: Over den groei der Zuiderzee-ansjovis.
Rapporten en Verhandelingen Ryksinstituut Onderzoek der Zee, 1: 241-266
E. encrasicholus

fish - Engraulis encrasicholus
Redecke, H. C. 1907
Rapport over onderzoekingen betreffende de visscherij in de Zuiderzee ingesteld in de jaren 1905 en 1906
Gravenhage 1907 - 339 pp.

fish - Engraulis encrasicholus
Reuter, O. O. 1896
Är den klagen befoged eller icke, som i skargården försporjes beträffande strömmingsfiskets fördämring?
Fiskeritidskr. Finl., 5:168-172

fish - Engraulis encrassicholus
Russo, A. 1924
Le ova di Engraulis encrasicholus L. nel plankton del Golfo di Catania e la pesca delle Acciuhge.
Boll. Acc. Gioenea Catania (2) 53:2-8 (delle Sedute)

fish - Engraulis encrasicholus
Shulman, G.E., and L.M. Kokoz 1967.
Adequate food-resources of the Azov and Black Sea varieties of Engraulis encrasicholus L. (In Russian).
Dokl. Akad. Nauk SSSR 172 (6): 1427-1429.

fish - Engraulis encrasicholus
Smirnoff, A. N. 1938
Distribution of anchovy (Engraulis encrasicholus maeoticus) in the Azov Sea and its food.
Publ. Sci. Inst. Fish. Azov., 11: 53-96
English summary

fish, Engraulis encrasicholus
Wenckebach, K. F. 1887
De embryonale ontwikkeling van de ansjovisch (Engraulis encrasicholus).
Natuurk. Verh. Akad., 26:1-11
(Verh. d. K. Acad. v. Wetensch)

Engraulis encrassicholus
Zolezzi, G., 1938.
Contributo alla conoscenza della alimentazione dei pesci, Engraulis encrassicholus L.
Boll. Pesca, Piscicolt. Idrobiol., 14(5):586-595.

Engraulis encrassicholus ponticus
Tkacheva, K.S., 1955.
[On the biology of the young Black Sea anchovy, Engraulis encrassicholus ponticus (Alex.).]
Trudy Karadagsk Biol. Sta., 13:47-48.

E. holodon

fish, Engraulis holodon n. sp.
Boulenger, G. A.
Descriptions of new Fishes from the Cape of Good Hope.
Mar. Invest., S. Africa, 1:10-12
Engraulis holodon n. sp.

E. iquitensis

fish, Engraulis iquitensis sardina
Nakashima, S. 1941
Algunos peces del oriente peruano
Bol. Mus. Hist. Nat. "Javier Prado" 5:61-78
Engraulis iquitensis sardina

E. japonica

Engraulis japonica
Hayashi, Sigeiti, 1967.
A note on the biology and fishery of the Japanese anchovy Engraulis japonica (Houttuyn).
Rep. Calif. Coop. Oceanic Fish. Invest., 11:44-57.

Engraulis japonicus
Inoue, A., 1949.
[Note on the habit of Engraulis japonicus Temminck & Schlegel collected from the coast of Sumoto City and its neighborhood.] Bull. Japan. Soc. Sci. Fish. 15(8):385-390, 5 textfigs. (In Japanese).

Engraulis japonica
Kondo, Keiichi 1967.
Mode of life of the Japanese anchovy Engraulis japonica (Houttuyn). III. Aggregation of immatures and adults of the Pacific fraction along Honshu in the waters extending between the Suruga Bay and off the Japan coast. (In Japanese; English abstract).
Bull. Tokai reg. Fish. Res. Lab. 52: 13-36.

Engraulis japonica
Kondo, Keiichi 1967.
Mode of life of the Japanese anchovy, Engraulis japonica (Houttuyn). II. Aggregation of immatures of the Pacific fraction along Honshu in the Ise and Mikawa bays. (In Japanese; English abstract).
Bull. Tokai reg. Fish. Res. Lab. 51: 1-28.

fish, Engraulis japonica
Kurogane, Kenji 1967.
Studies on the population of Japanese anchovy in Ise Bay. (In Japanese; English abstract).
Bull. Tokai reg. Fish. Res. Lab. 51: 65-80.

Engraulis japonicus
Kuroki, T., and M. Chuman, 1958.
An example of three-dimensional records of fish-school attracted by underwater lamps.
Mem. Fac. Fish., Kagoshima Univ., 6:77-81.

Engraulis japonicus
*Morishita, Tatsuo, Takashi Takahashi, Manabu Kitamikado and Shinko Tachino, 1967.
On the fish meal and fish oil from anchovy Engraulis japonicus T. and S., in Ise Bay. Preparation on laboratory scale. (In Japanese; English abstract).
J. Fac. Fish., pref. Univ. Mie, 7(1):47-55.

Engraulis japonicus
Nishikawa, T. 1901
Reports on investigations on Engraulis japonicus
Suisan Chosa Hokoku, Tokyo, 10: 1-16
in Japanese

Engraulis japonicus
Oya, T., Shinada, K., and Toyoda, Y 1937
On the fat content of Engraulis japonicus
Bull. Jap. Soc. Sci. Fish., Tokyo, 6:147-150

Engraulis japonicus
Tanaka, S., 1956.
Body length distribution of the Japanese anchovy, Engraulis japonica (Houttuyn) in Ise Bay, Mikawa Bay and Ensyu-Nada. 1. Body length in 1951 and 1952.
Bull. Tokai Reg. Fish. Res. Lab., No. 13:35-50.

E. loreanus

Engraulis loreanus n. sp
Kishinouye, K. 1908
Notes on the natural history of the sardine.
Suisan Ch. Ho, Tokyo (Eng. ed.) 14:71-105

E. loreanus, sp. n.

E. mordax

Engraulis mordax
Baxter, John L., 1967.
Summary of biological information on the northern anchovy (Engraulis mordax Girard.
Rep. Calif. Coop. Oceanic Fish. Invest. 11:110-116.

Engraulis mordax
Berner, Leo, Jr., 1959.
The food of the larvae of the northern anchovy Engraulis mordax.
Inter-American Tropical Tuna Comm., Bull., 4(1):22 pp.

fish, Engraulis mordax
Bolin, R. L. 1936
Embryonic and early larval stages of the California anchovy, Engraulis mordax Girard.
Calif. Fish Game., 22:314-321

Engraulis mordax mordax
California, Department of Fish and Game, Marine Research Committee, 1956.
Cooperative Oceanic Fisheries Investigations, Progress Report, 1 April 1955-30 June 1956:44 pp.

Engraulis mordax mordax
Clark, F.N., and J.B. Phillips, 1952.
The northern anchovy (Engraulis mordax mordax) in the California fishery. Calif. Fish and Game 38(2):189-207, 7 textfigs.

Engraulis mordax
Farris, David A., 1960.
Failure of an anchovy to hatch with continued growth of the larva.
Limnol. & Oceanogr. 5(1):107.

fish, Engraulis mordax
Hubbs, C. L. 1925
Racial and seasonal variation in the Pacific herring, California sardine and California anchovy.
Calif. Fish and Game Comm. Fish. Bull., 8:23 pp. Sacremento

Engraulis mordax
Loukashkin, Anatole S., and Norman Grant, 1965.
Behavior and natural reactions of the northern anchovy Engraulis mordax Girard, under the influence of light of different wave lengths and intensities and total darkness.
Proc. California Acad. Sci., (4), 31(24):631-692.

Engraulis mordax
Marr, J. C., and E. H. Ahlstrom, 1948.
Observations on the horizontal distribution and the numbers of eggs and larvae of the northern anchovy (Engraulis mordax) off California in 1940 and 1941. U. S. Fish and Wildlife Service, Spec. Sci. Rept. No. 56:13 pp. (mimeographed).

Engraulis mordax mordax
McHugh, J.L., 1951.
Meristic variations and populations of northern anchovy (Engraulis mordax mordax). Bull. S.I.O. 6(3):123-160, 9 textfigs.

Engraulis mordax
Miller, D.J., 1955.
Studies relating to the validity of the scale method for age determination of the northern anchovy (Engraulis mordax).
Calif. Dept. Fish and Game, Mar. Fish. Br., Fish Bull., No. 101:6-34.

E. ringens

fish, Engraulis ringens
Boerema, L.K., G. Saetersdal, I. Tsukayama, J.E. Valdivia y B. Alegre 1967.
Informe sobre los efectos de la pesca en el recurso peruano de anchoveta.
Bol. Inst. Mar. Peru 1(4): 135-186.

Engraulis ringens
Brandhorst, W., O. Rojas and J.G. Simpson, 1967.
Seasonal variations of the fat content, solids yield end condition factor of the anchoveta (Engraulis ringens, Jenyns) in Chile.
Arch. Fischereiwissenschaft, 18(1):46-66.

Engraulis ringens
Clark, F.N., 1954.
Biologia de la anchoveta. Biol. Cient., C.A.G., 1(2):98-131, figs.

Engraulis ringens
DeBuen, F., 1955.
El estudio de la edad y el crecimiento en peces, viviendo en medios oceánicos diferentes, y especialmente en la anchoveta o chicora (Engraulis ringens). Bol. Cient., C.A.G., 2:41-47.

Engraulis ringens
Einarsson, H., and B. Royas de Mendiola, 1967.
An attempt to estimate annual spawning intensity of the anchovy (Engraulis ringens Jenyns) by means of regional egg and larval surveys during 1961-1964.
Rep. Calif. Coop. Oceanic Fish Invest. 11:96-104.

Engraulis ringens
Einarsson, H., and B. Rojas de Mendiola, 1963
Descripcion de huevos y larvas de enchoveta Peruana (Engraulis ringens J.).
Bol. Inst. Invest. Jos Recursos Marinos, Callao, Peru, 1(1):1-23.

Engraulis ringens
Jordan, R., 1963
Un analisis del numero de vertebras de la anchoveta Peruana (Engraulis ringens J.).
Bol. Inst. Invest. Recursos Marinos, Callao, Peru, 1(2):25-43.

Engraulis ringens
Jordan S., Romulo, 1959.
Observaciones sobre la biologia de la anchoveta (Engraulis ringens) de la zona pesquera de Huacho.
Bol., C.A.G., 35(11):3-22.

Engraulis ringens
Minano M., Jorge, 1958.
Algunas apreciones relacionadas con la anchoveta peruana (Engraulis ringens J.) y su fecundidad.
Bol. C. A. G. 34(2):9-24.

Engraulis ringens
Minano M., Jorge, 1958.
Algunas apreciaciones relacionadas con la anchoveta peruana y su fecundidad.
Bol. C.A.G., 34(3):11-24.

fish, Engraulis ringens
Rojas De Mendiola, Blanca, 1971.
Some observations on the feeding of the Peruvian anchoveta Engraulis ringens J. in two regions of the Peruvian coast. (English and Spanish abstracts). In: Fertility of the Sea, John D. Costlow, editor, Gordon Breach, 2: 417-440.

Engraulis ringens
Saetersdal, G., I. Tsukayama y B. Alegre, 1965.
Fluctuaciones en la abundancia aparente del stock de anchoveta en 1959-1962.
Bol. Inst. del Mar del Peru, 1(2):33-104.

Engraulis ringens
Saetersdal, G., y J.E. Valdivia, 1964.
Un estudio del crecimiento tamano y relutamiento de la anchoveta (Engraulis ringens J.) basado en datos de frequencia de longitud.
Bol. Inst. Invest. Recursos Marinos, 1(4):88-136.

Engraulis ringens
Saetersdal, G., J. Valdivia, I. Tsukayama and B. Alegre, 1967.
Preliminary results of studies on the present status of the Peruvian stock of anchovy (Engraulis ringens Jenyns).
Rep. Calif. Coop. Oceanic Fish. Invest., 11:88-95.

Engraulis ringens
Sanchez, R., Jorge E., 1966.
General aspects of the biology and ecology of the anchovy. (Engraulis ringens).
Proc. Inst. Mar. Sci., Gulf and Caribbean Fish. Inst., 18th Ann. Sess., 84-93.

fish Engraulis ringens.
Schaefer, Milner B. 1967.
Dinámica de la pesquería de la anchoveta Engraulis ringens, en el Peru. (In Spanish and English).
Bol. Inst. Mar, Peru 1(5): 189-303.

Engraulis ringens
Simpson, John G., y Ramon Buzeta, 1967.
El crecimiento y la edad de la anchoveta (Engraulis ringens Jenyns) en Chile basado en estudios de frequencia de longitud.
Bol. cient. Inst. Fomento Pesquero, Chile, No. 3:53 pp. (mimeographed).

E nasutus

fish Engraulis nasutus n.sp
Castelnau, Count F. de 1879.
Notes on the fishes of the Norman River.
Proc. Linn. Soc. N.S. Wales 3: 41-51.

E tapirulus

fish, Engraulis tapirulus n.sp
Cope, E.D. 1877.
Synopsis of the cold blooded Vertebrata procured by Prof. James Orton during his exploration of Peru in 1876-77.
Proc. Am. Phil. Soc. 17: 33-49.

E telara

Engraulis telara
Nair, K.K., 1940.
On some early stages in the development of the Gangetic anchovy Engraulis telara (Ham).
Rec. Indian Mus. 42(2):277-287.

Eulamia

Eulamia milberti
Springer, S., 1960
Natural history of the sandbar shark, Eulamia milberti. U.S Fish & Wildlife Serv., Fish. Bull., 178: 38 pp.

fish - Euprotomicroides
Hulley, P.A., and M.J. Penrith, 1966.
Euprotomicroides zantedeschia, a new genus and species of pigmy dalatiid shark from South Africa.
Bull. Mar. Sci. 16(2):222-229.

fish - Euprotomicrus bispinatus
Hubbs, Carl L, Tamotsu Iwai and Kiyomatsu Matsubara 1967.
External and internal characters, horizontal and vertical distribution, luminescence and food of the dwarf pelagic shark, Euprotomicrus bispinatus
Bull. Scripps Inst. Oceanogr. 10: 1-64.

fish - Euprotomicrus bispinatus
Parin, N.V., 1964.
Data on biology and distribution of pelagic sharks, Euprotomicrus bispinatus and Isistius brasiliensis (Squalidae, Pisces). (In Russian).
Trudy Inst. Okeanol., Akad. Nauk, SSSR, 73:163-184.

fish, Eurypharynx
Owre, H.B. and F.M. Bayer, 1970.
The deep-sea gulper Eurypharynx pelecanoides Vaillant, 1882 (Order Lyomeri) from the Hispanio Basin. Bull. mar. Sci., Miami, 20(1): 186-192.

fish, Euthynnus alletteratus
Demir, M., 1963.
On the juveniles of Euthynnus alletteratus Raf. appeared in Turkish waters in 1959.
Rapp. Proc. Verb., Réunions, Comm. Int. Expl. Sci. Mer Méditerranée, Monaco, 17(2):375-378.

Galeocerdo

Galeocerdo cuvieri-tiger shark
Tester, Albert L., 1960
Fatal shark attack, Oahu, Hawaii, December 13, 1958.
Pacific Science, 14(2):181-184.

Galeus

Galeus arae
Bullis, Harvey R., Jr., 1967.
Depth segregation and distribution of sex-maturity groups in the marbled catshark, Galeus arae. In: Sharks, skates and rays, Perry W. Gilbert, Robert F. Mathewson and David P. Rall, editors, Johns Hopkins Univ. Press, 141-148.

fish, sharks - Galeus
Hubbs, Carl L., and Leighton R. Taylor, Jr., 1969.
Data on life history and characters of Galeus piperatus, a dwarf shark of Golfo de California.
Fisk Dir. Skr. Ser. HavUnders. 15(3): 310-330.

fish - Gemplus
Stauch, A., 1964.
Sur la capture de trois exemplaires de Gemplus serpens C. 1831 dans la région orientale du Golfe de Guinée (Pisces, Teleostei, Perciformi, Trichiuroidei, Gempylidae).
Trav. Centre Océanogr., Pointe-Noire, Cahiers O.R.S.T.R.O.M., Océanogr., Paris, No. 5:105-107.

fish - Hygophum
Bekker (Becker), V.E., 1965.
The lantern fishes of the genus Hygophum (Myctophidae, Pisces). (In Russian; English abstract).
Trudy Inst. Okeanol., Akad. Nauk, SSSR, 80: 62-103.

fish, sharks - Iago
Compagno, Leonard J.V., and Stewart Springer 1971.
Iago, a new genus of carcharhinid sharks with a redescription of I. omanensis.
Fish. Bull. nat. mar. fish. Serv. NOAA 69(3): 615-626

Icichthys

Icichthys
Haedrich, Richard L., 1966.
The stromateoid fish genus Icichthys: notes and a new species.
Vidensk Meddr. dansk naturh. Foren. 129:199-213.

fish - Isistius brasiliensis
Strasberg, D.W., 1963.
The diet and dentition of Isistius brasiliensis with remarks on tooth replacement in other sharks.
Copeia, (1):33-40.

Isurus

Isurus
Garrick, J.A.F., 1967.
Revision of sharks of genus Isurus with description of a new species.
Proc. U.S. nat. Mus., 118(3537):663-690.

fish - Isurus
Guitart Manday, Darío, 1966.
Nuevo nombre para una especie de tiburón del género Isurus (Elasmobranchii: Isuridae) de aguas cubanas.
Poeyana (A), No.15:9 pp.

fish - Isurus oxyrhinchus
Applegate, Shelton P., 1966
A possible record - sized bonito shark, Isurus oxyrhinchus Refinesque, from Southern California.
Calif. Fish Game, 52(3): 204-207.

fish - Katsuwonus pelamis
Sprague, L.M., and J.R. Holloway, 1962.
Studies of the erythrocyte antigens of the skipjack (Katsuwonus pelamis).
American Naturalist, 96(889):233-238.
Paper presented at 10th Pacific Science Congress.

fish, Katsuwonus
Yuen, Heeny S.H. 1970.
Behavior of skipjack tuna, Katsuwonus pelamis, as determined by tracking with ultrasonic devices.
J. Fish. Res. Bd, Can., 27(11): 2071-2079.

Latimeria

Latimeria
Millot, J., 1955.
First observations on a living coelacanth.
Nature 175(4452):362-363.

coelacanth
Millot, J., and J. Anthony, 1956.
Note préliminaire sur le thymus et la glande Thyroïde de Latimeria chalumnea (Crossopterygien coelacanthidé). C.R. Acad. Sci., Paris, 242(4): 560-562.

Latimeria
Millot, J., and J. Anthony, 1956.
L'organe rostral de Latimeria (Crossoptérygien Coelacanthidé).
Ann. Sci. Nat., Zool. et Biol. Animale, 18(3): 381-390.

Latimeria
Millot, J., and N. Carasso, 1955.
Note préliminaire sur l'oeil de Latimeria chalumnae (Crossopterygien coelacanthidé).
C.R. Acad. Sci., Paris, 241(6):576-577.

Latimeria
Schaeffer, B., 1953.
Latimeria and the history of the coelacanth fishes. Trans. N.Y. Acad. Sci., 2nd ser., 15(6): 170-178, 8 textfigs.

coelacanths
Watson, D.S.M., 1955.
Coelacanths. Proc. R. Inst., Gt. Brit., 35(4) (161):782-786.

Loxodon

Loxodon macrorhinus
Wheeler, J. G. F., 1959.
Sharks of the western Indian Ocean. 1. Loxodon macrorhinus M. & G.
East African Agricult. J., 25(2): 106-109.

fish - Lutjanus purpureus

Fonteles Filho, Antônio Adauto, 1969
Estudo preliminar sôbre a pesca do pargo Lutjanus purpureus Poey, no nordeste Brasileiro.
Arq. Ciên. Mar, Fortaleza, Ceará, Brasil 9(1): 83-88

fishes, Macrouroides

Marshall, N.B., and Å. Vedel Tåning (posthumous), 1966.
The bathypelagic macrourid fish Macrouroides inflaticeps Smith and Radcliffe.
Dana Rep., No. 69:1-6.

fish - Makaira albida

Talbot, F.H., and M.J. Penrith, 1963.
The white marlin Makaira albida (Poey) from the seas around South Africa.
South African J. Sci., 59(7):330-332.

Malania

Malania

Smith, J.L.B., 1953.
The second coelacanth. Nature 171:99-101.

fish - Megagaleus

Bessednov, L.N., 1966.
A new shark species from the Tonkin Gulf - Megagaleus longicaudatus Bessednov sp. n. (Pisces Carcharinidae). (In Russian; English abstract).
Zool. Zhurn., Akad. Nauk, SSSR, 45(2):302-306.

fish - Megalops

Wade, R.A., 1962.
The biology of the tarpon Megalops atlanticus and the ox-eye, Megalops cyprinoides, with emphasis on the larval development.
Bull. Mar. Sci., Gulf and Caribbean, 12(4):545-622.

Menippe

Menippe mercenaria

*Karandieva, O., y A. Silva Lee, 1967.
Intensidad de respiración y osmorregulación del cangrejo commercial Menippe mercenaria (Say) de las aguas cubanas.
Estudios Inst. Oceanol., Acad. Ciencias, Cuba, 2(1): 5-19.

Menippe mercenaria

*Suchenia, L.M., y R. Claro Madruga, 1967.
Datos cuantitavos de las alimentacion del cangrejo commercial Menippe mercenaria (Say) y su relación con el balance energetico delmismo.
Estudios Inst. Oceanol., Acad. Ciencias, Cuba, 2(1): 75-97.

Micropogon

Micropogon ectenes (croaker)

Johnson, M.W., 1943
Underwater Sounds of Biological Origin.
UCDWR No. U28, Sec. No. 6.1-sr30-412, File No. 01.33, 20 pp. (mimeographed), 1 chart (photostat), 2 figs. (photo). 15 Feb. 1943.

Myxine

Myxine

Brodal, Alf, and Ragnar Fänge, Editors, 1963.
The biology of Myxine.
Universitetsforlaget, Oslo, 588 pp.

Myxine

Foss, Gunvor, 1968.
Behaviour of Myxine glutinosa L. in natural habitat.
Sarsia 31:1-13

Myxine (hagfish)

Foss, Gunvor, 1962.
Some observations on the ecology of Myxine glutinosa L.
Sarsia, Univ. i Bergen, 7:17-22.

Myxine

Jensen, David, 1966.
The hagfish.
Scientific American, 214(2):82-90.

fish - Myxine

Tambs-Lyche, Hans, 1969.
Notes on the distribution and ecology of Myxine glutinosa L.
Fiskeridir. Skr. Ser. HavUnders. 15(3): 279-284

Myxine eels

Vogel, I., R. Bösel and H. Adam, 1971.
Lichtmikroskopische Klassifizierung und Quantifizierung von granulärem Neurosekret im hypothalamoneurohypophysären system von Myxine glutinosa L. (Cyclostomata). Norw. J. Zool. 19(1): 93-97.

Negaprion (shark)

Negaprion brevirostris

Gilbert, Perry W., and Steven D. Douglas, 1963
Electrocardiographic studies of free-swimming sharks.
Science, 140(3574):1396.

fish - Negaprion brevirostris

Gruber, S.H., 1967.
A behavioral measurement of dark adaptation in the lemon shark, Negaprion brevirostris. In: Sharks, skates, and rays, Perry W. Gilbert, Robert F. Mathewson and David P. Rall, editors, Johns Hopkins Univ. Press, 479-490.

fish - Negaprion brevirostris

Gruber, S.H., D.H. Hamasaki, and C.D.B. Bridges, 1963.
Cones in the retina of the lemon shark (Negaprion brevirostris).
Vision Research, 3:397-399.

fish, Negaprion brevirostris

Moss, Sanford A., 1967.
Tooth replacement in the lemon shark, Negaprion brevirostris. In: Sharks, skates, and rays, Perry W. Gilbert, Robert F. Mathewson and David P. Rall, editors, Johns Hopkins Univ. Press, 319-329.

Negaprion brevirostris

Nelson, Donald R., 1967.
Cardiac responses to sounds in the lemon shark, Negaprion brevirostris. In: Sharks, skates, and rays, Perry W. Gilbert, Robert F. Mathewson and David P. Rall, editors, Johns Hopkins Univ. Press 533-544.

fish - Negaprion brevirostris

*Nelson, Donald R., 1967.
Hearing thresholds, frequency discrimination, and acoustic orientation in the lemon shark, Negaprion brevirostris (Poey).
Bull. mar. Sci., Miami, 17(3):741-768.

Negaprion brevirostris

O'Gower, A.K., and R.F. Mathewson, 1967.
Spectral sensitivity and flicker-fusion frequency of the lemon shark, Negaprion brevirostris. In: Sharks, skates, and rays, Perry W. Gilbert, Robert F. Mathewson and David P. Rall, editors, Johns Hopkins Univ. Press, 433-446.

fish - Neothunnus albacora

Tsuruta, Saburo, 1964.
Morphometric character of yellowfin tuna, Neothunnus albacora (Lowe) in the eastern waters of the Indian Ocean. (In Japanese; English abstract).
J. Shimonoseki Univ. Fish., 13(1):53-59.

fish, Neothunnus albacora

Tsuruta, Saburo, 1964.
Morphometric comparison of yellowfin tuna, Neothunnus albacora (Lowe) from several areas in the Indian Ocean. (In Japanese; English abstract).
J. Shimonoseki Univ. Fish., 13(1):43-51.

fish - Neothunnus albacora

Tsuruta, S., 1961
Morphometric characters of yellowfin tuna, Neothunnus albacora (Lowe), in the southwestern waters of the Indian Ocean (off the southwest of the Madagascar Island).
J. Shimoneseki Coll. Fish., 11(2): 75-94.

fish, Nettodarus brevirostris

Böhlke, James E., and C. Richard Robins, 1968.
Biological investigations of the deepsea. 36. The eel Nettodarus brevirostris in the western Atlantic.
Bull. mar. Sci., Miami 18(2): 477-480.

Paralepis

Paralepis

Ege, V., 1930
Contributions to the knowledge of the North Atlantic and Mediterranean species of the genus Paralepis Cuv. A systematical and biological investigation. Rept. of the Danish Oceanographical Expeditions 1908-1910 to the Mediterranean and Adjacent Seas, Vol. II A.13, 201 pp., 37 textfigs.

fish - Parathunnus sibi

Kume, Susumu, and Toshio Shiohama, 1965.
Ecological studies on bigeye --. II. Distribution and size composition of bigeye tuna Parathunnus sibi in the equatorial Pacific. (In Japanese; English abstract).
Rept. Nankai Reg. Fish. Res. Lab., No. 22:71-83.

Parexocoetus

Parexocoetus

Bruun, A. Fr., 1935.
Parexocoetus, a Red Sea flying fish from the Mediterranean. Nature 136:553.

Petromyzon (lamprey)

Petromyzon

Applegate, Vernon C., and M.L.H. Thomas, 1965.
Sex ratios and sexual dimorphism among recently transformed sea lampreys, Petromyzon marinus Linnaeus.
J. Fish. Res. Bd., Canada, 22(3):695-711.

Petromyzon marinus
(Sea lamprey)

Bigelow, H.B., and W. C. Schroeder, 1953
Fishes of the Gulf of Maine. First revision.
Fish. Bull. 74, Fish. Bull., Fish and Wildlife Service, 53: 1 - 577

fish - Petromyzon marinus

Hardisty, M. W., 1969.
Information on the growth of the ammocoete larva of the anadromous sea lamprey, Petromyzon marinus in British rivers.
J. Zool., Lond., 159(2): 139-144.

fish - Petromyzon

Hardisty, M.W., 1969.
A comparison of gonadal development in the ammocoetes of the landlocked and anadromous forms of the sea lamprey Petromyzon marinus L.
J. Fish. Biol. 1(2): 153-166

lampreys

Wagner, W.C., and T.M. Stauffer, 1962.
The population of sea lamprey larvae in East Bay Alger County, Michigan.
Pap. Michigan Acad. Sci., Art, Letter, 47(1):235-246.

Porichthys

fishes Porichthys

*Gilbert, Carter R., 1968.
Western Atlantic batrachoidid fishes of the genus Porichthys, including three new species.
Bull. mar. Sci., Miami, 18(3):671-730.

Porichthys notatus (singing fish or midshipman)

Johnson, M.W., 1943
Underwater Sounds of Biological Origin.
UCDWR No.U28, Sec. No.6.1-sr30-412, File No. 01.33, 20 pp. (mimeographed), 1 chart (photostat), 2 figs. (photo), 15 Feb. 1943.

Prionace

fish, Prionace glauca

McKenzie, R.A., and S.N. Tibbo, 1964.
A morphometric description of blue shark (Prionace glauca) from Canadian waters.
J. Fish. Res. Bd., Canada, 21(4):865-866.

sharks

Tucker, D.W., and C.T. Newnham, 1957.
The blue shark Prionace glauca (L.) breeds in British seas. Ann. Mag. Nat. Hist., (12)10(117): 673-688.

Psettodes

fish

Stauch, A., et J. Cadenat, 1965.
Révision du genre Psettodes Bennett 1831 (Pisces, Teleostei, Heterosomata).
Océanographie, Cahiers, O.R.S.T.R.O.M., 3(4):19-30.

fish, Pseudotolithus

Poinsard, F. et J.-C. Le Guen, 1970.
Note sur les variations quotidiennes de vulnérabilité au chalutage de trois espèces de poissons du genre Pseudotolithus de plateau continental Congolais. Cah. ORSTOM, sér. Océanogr., 8(2): 57-64.

fish - Pterolamiops longimanus

Backus, R.H., S. Springer and E.L. Arnold, 1956.
A contribution to the natural history of the white-tip shark, Pterolamiops longimanus (Poey).
Deep-Sea Res., 3(3):178-188.

fish - Pterolamiops longimanus

Tibbo, S.N., 1962.
New records for occurrence of the white-tip shark, Pterolamiops longimanus (Poey), and the dolphin, Coryphaena hippurus L., in the northwest Atlantic.
J. Fish. Res. Bd., Canada, 19(3):517-518.

Raja

fish, Raja

Hulley, P.A., 1969.
The relationship between Raja miraletus Linnaeus and Raja ocellifera Regan based on a study of the clasper.
Ann. S Afr. Mus. 52(6): 137-147.

fish, Raja

Templeman, Wilfred 1965.
Some resemblances and differences between Raja erinacea and Raja ocellata, including a method of separating mature and large immature individuals of these two species.
J. Fish. Res. Bd. Can. 22(4): 899-912.

fish, Thunnus

Gibbs, Robert H. Jr. 1966.
Comparative anatomy and systematics of the tunas, genus Thunnus.
Fish. Bull. Fish Wildlife Serv. 66(1): 65-130.

Raja herwigi

Krefft, G., 1965.
Die ichthyologische Ausbeute der ersten West-afrikas-Fahrt des fischereitechnischen Forschungs-schiffes "Walther Herwig". 1. Raja herwigi spec. nov., eine neue Rochenart aus den Kapverden.
Arch. Fischereiwiss., 15(3):209-216.

fish - Raja richardsoni

Forster, G.R., 1965.
Raja richardsoni from the continental slope off south-west England.
J. mar. biol. Ass., U.K., 45(3):773-777.

Raia undulata

Griffith, D. de G., 1965.
Further occurrences of Raia undulata (Lacépède) in Irish waters.
Ir. Nat. J., 16(1): 26.

fish, Raia

Griffith, David de G., 1966.
Raia undulata (Lacepede), a species new to Irish waters.
Irish Naturalists' Jour., 15(6):166-168.

Remora

Remora

Phillipps, W.J., 1964.
The occurrence of Remora in Cook Strait.
Rec. Dominion Mus., New Zealand, 5(10):73-74.

fish - Rhincodon typus

Kaikini, A.S., V. Ramamohana Rao, M.H. Dhulkhed, 1959
A note on the whale shark Rhincodon typus Smith, stranded off Mangalore. J. Mar. Biol. Ass. India, 1(1): 92-93.
Actual measurements length 12.1 meters weight 17 metric tons

fish - Rhincodon typus

Silas, E.G., and M.D. Rajagopalan, 1963.
On a recent capture of a whale shark (Rhincodon typus Smith) at Tuticorin, with a note on information to be obtained on whale sharks in Indian waters.
J. Mar. Biol. Assoc., India, 5(1):153-157.

fish - Rhinocodon typus

Thomas, M.M., and K.R. Kartha, 1964.
On the catch of a juvenile whale shark Rhinocodon typus Smith from Malabar coast.
J. Mar. Biol. Assoc., India, 6(1):174-175.

Roncador

Roncador stearnsi (spotfin croaker)

Johnson, M.W., 1943
Underwater Sounds of Biological Origin.
UCDWR No.U28, Sec. No.6.1-sr30-412, File No. 01.33, 20 pp. (mimeographed), 1 chart (photostat), 2 figs. (photo), 15 Feb. 1943.

Salmo

Salmo irideus

Bruun, A. Fr., 1936.
En tvekønnet Regnbueørred. Ferskvandfiskeriblad-et, 34 Aargang, No. 12:3 pp., 2 textfigs.

Salmo iridaeus

Busnel, R.G., Drilhon, A., and Raffy, A., 1946
Recherches sur la physiologie des Salmonidés. 1. Sur l'adaptation aux eaux saumâtres de Salmo iridaeus à ses differents stades. Bull. Inst. Océanogr. Monaco, No.893.

Salmo salar

Busnel, R.G., and A. Drilhon, 1946
Recherches sur la physiologie des Salmonidés. II Sur l'oeuf et l'alevin de Salmo salar. Bull. Inst. Océanogr. Monaco, No.893.

fish, Sarda

Barrett, Izadore 1971.
Observaciones preliminares sobre la biología y dinámica pesquera del bonito (Sarda chiliensis) en aguas chilenas. (In Spanish and English).
Bol. cient. Inst. Foment. pesq. Chile 15: 55pp (mimeographed)

Sardinia

fish - Sardina caerulea

Clark, F.N. 1934.
Maturity of the California sardine (Sardina caerulea) determined by ova diameter measurements.
Bull. Div. Fish. Game, Calif. 42: 1-49.

fish, Sardina caerulea

Clark, F.N. 1931.
Dominant size-groups and their influence in the fishery for the California sardine (Sardina caerulea).
Bull. Div. Fish. Game, Calif. 31: 7-42

Sardinia melanostricta

Sano, T., 1949.
[The catch of the "maiwashi" Sardinia melanostricta (T. & S.) and its environmental condition in Tokyo Bay.] Contr. Central Fish. Sta., Japan, 1948-1949(1951):371-374, 6 textfigs.

fish - Sardina pilchardus

Demir, M., 1963.
Analysis of vertebral counts of Sardina pilchardus Walb. from the Sea of Marmara and the Aegean Sea.
Rapp. Proc. Verb., Réunions, Comm. Int. Exp. Sci., Mer Méditerranée, Monaco, 17(2):311.

Sardina pichardus

Furnestin, J., and M.L. Furnestin, 1959.
La reproduction de la sardine et de l'anchois des côtes atlantiques du maroc (saisons et aires de ponte).
Rev. Trav. Inst. Pêches Marit., 23(1):79-104.

Sardina pilchardus
*Karlovac, Jozica, 1967.
Etude de l'écologie de la Sardine, Sardina pilchardus Walb., dans la phase planctonique de sa vie en Adriatique moyenne.
Acta adriat., 13(2):1-109.

fish - Sardina pilchardus
Karlovac, Jozica, 1964.
Ponte de la sardine (Sardina pilchardus Walb.) dans l'Adriatique moyenne, au cours de la saison 1956-1957. (In Jugoslavian; French resume).
Acta Adriatica, 11(1):173-174.

fish - Sardina pilchardus
Lopez, J., 1963
Edad de la sardina (Sardina pilchardus, Walb. de Barcelona. (In Spanish: English summary).
Inv. Pesq., Barcelona, 23:133-157.

fish, Sardinia pilchardus
*Marinaro, J.-Y., and M. Benkara-Mostefa, 1968.
Contribution à l'étude des oeufs et larvas pélagiques de poissons méditerraneens. III. A la recherche d'une aire de ponte de la sardine (Sardinia pilchardus Walb.) dans les eaux Maltaises.
Pelagos, 9:85-92.

fish - Sardina pilchardus
Mossi, C., 1963.
Risultati conseguiti nelle ricerche sulla sardina (Sardina pilchardus Walb.) della parte occidentale dell'alto Adriatico.
Rapp. Proc. Verb., Réunions, Comm. Int. Expl. Sci., Mer Méditerranée, Monaco, 17(2):307-309.

fish - Sardina pilchardus
Muzinic, Radosna, 1964.
Some observations on the reactions of sardine (Sardina pilchardus Walb.) to light under experimental conditions. (In Jugoslavian; English resume).
Acta Adriatica, 11(1):219-226.

Sardina pilchardus
Pinto, J. dos S., and I.F. Barraca, 1958.
Aspects biologiques et biométriques de la sardine (Sardinia pilchardus Walb.) des environs de Lisbonne pendant les années 1952-1957.
Naxt Notas e Estudos, Inst. Biol. Mar. Lisbon, No. 19:1-58.

fish - Sardina pilchardus
Vucetic, Tamara, 1964.
Sur le comportement de la sardine (Sardina pilchardus Walb.) envers les facteurs biotiques du milieu (zooplancton). (In Jugosalvian; French resume).
Acta Adriatica, 11(1):269-276.

Sardinella

fish, Sardinella
Botros, G.A., A.M. El-Magraby and I.A.M. Soliman 1970.
Biometric studies on Sardinella maderensis Lowe and Sardinella aurita Cuv. & Val. from the Mediterranean Sea at Alexandria (U.A.R.).
Bull. Inst. Oceanogr. Fish., U.A.R. 1: 83-128.

fish, Sardinella
Soliman, I.A.M., 1970.
Length-weight relationship and coefficient of condition for Sardinella maderensis Lowe and Sardinella aurita Cuv. & Val. from the Mediterranean Sea at Alexandria (U.A.R.).
Bull. Inst. Oceanogr. Fish., U.A.R. 1: 27-45.

fish, Sardinella
El-Maghraby, A.M., G.A. Botros and I.A.M. Soliman, 1970.
Age and growth studies on Sardinella maderensis Lowe and Sardinella aurita Cuv. and Val from the Mediterranean Sea at Alexandria (U.A.R.).
Bull. Inst. Oceanogr. Fish., U.A.R. 1: 47-82.

Sardinella
Jorvedt, Ole J., 1969.
On the catch statistics of Sardinella in Ghana
Actes Symp. Oceanogr. Ressources halieut. Atlant. trop., Abidjan, 20-28 Oct. 1966, UNESCO, 265-268.

fish - Sardinella
Troadec, J.P., 1964.
Prises par unité d'effort des sardiniers de Pointe-Noire (Congo): variations saisonnières de L'abondance des sardinelles (Sardinella eba C.V. et Sardinella aurita C.V.) dans les eaux congolaises (de 2°30's a 5°30's).
Cahiers, O.R.S.T.R.O.M., Océanogr., 11(4):17-25.

fish - Sardinella anchovia
*Heald, Eric J., y Raymond C. Griffiths, 1967.
La determinación, por medio de la lectura de escamas, de la edad de la sardina, Sardinella anchovia, del Golfo de Cariaco, Venezuela oriental.
Invest. Pesqueras, Ser. Recursos Explotación Pesqueros, Venezuela, 1(10):375-446.

fish, Sardinella anchovia
Simpson, John G., y German Gonalez G., 1968.
Algunos aspectos de las primeras etapas de vida y el medio ambiente de la sardina, Sardinella anchovia, en el oriente de Venezuela.
Ser. Recursos Explot. Pesquer., Invest. Pesquer., Venezuela, 1(2):38-93.

Sardinella aurita
Bardan, E., F. de P. Navarro, and D. Rodrigues 1949.
Nuevos datos sobre la sardina del Mar de Alboran (Agosto de 1948 a Marzo 1949). Bol. Inst. Espanol Ocean., No. 17:11 pp.

fisheries, sardine - Sardinella aurita
Boely, T., 1971
La pêche industrielle de Sardinella aurita dans les eaux sénégalaises de 1966 à 1970
OSTOM-FAO, Dakar, DSP 31: 25+pp (mimeographed)

fish, Sardinella aurita
Boely, T., et C. Champagnat, 1970.
Observations préliminaires sur Sardinella aurita (C. et V.) des côtes sénégalaises.
Rapp. P.-v. Réun. Cons. int. Explor. Mer 159: 176-181

fish - Sardinella aurita
dos Santos, E.P., M.N. de Moraes e Y. Schaeffer, 1969.
Dinâmica da população de sardinha Sardinella aurita (Cuv., Val., 1847) na Costa Sul do Brasil.
Boor. Pesquisa, Rio de Janeiro, Brasil 2(1): 17-28

fish - Sardinella aurita
Marchal, E., 1965.
Note sur deux caractères de Sardinella aurita (C et V) de Côte d'Ivoire.
Cahiers, O.R.S.T.R.O.M. - Océanogr., 3(1):95-99.

fish - Sardinella aurita
Vazzoler, A.E.A. de M., and G. Vazzoler, 1965.
Relation between condition factor and sexual development in Sardinella aurita (Cuv. & Val. 1847).
Anais Acad. bras. Cienc., 37(Supl.):353-359.

fish - Sardinella eba
Marchal, E., 1965.
Etude de quelques caractères de Sardinella eba (C et V) de Côte d'Ivoire.
Cahiers, O.R.S.T.R.O.M. - Océanogr., 3(1):87-94.

Fish, Sardinella longiceps
Krakatitsa, V.V., 1968.
Studies on the early stages of Sardinella longiceps Val. in the Gulf of Aden. (In Russian).
Trudy, Vses. Nauchno-Issled. Inst. Morsk. Ribn. Okeanogr (VNIRO) 64, Trudy Azovo-Chernomorsk. Nauchno-Issled. Inst. Morsk. Ribn. Khoz. Okeanogr. (AscherNIRO), 28: 401-406.

Sardinella sirm
*Raja, B.T. Antony, and Yoshio Hiyama, 1969.
On Sardinella sirm (Walbaum) from Okinawa.
Rec. oceanogr. Wks. Japan, 10(1):105-107.

Sardinops

Sardinops caerulea
Felin, F. E., A.E. Daugherty, and L. Pinkas, 1950
Age and length composition of the sardine catch off the Pacific coast of the United States and Canada in 1949-50. Calif. Fish and Game 36(3): 241-249, Textfig. 81, 9 tables.

fisheries - Sardinops caerulea
Hart, J. L., and Wailes, G. H., 1932
The food of the pilchard, Sardinops caerulea (Gizard), off the coast of British Columbia. Contrib. Canadian Biol. and Fish., N. S. VIII: 245-254

fish - Sardinops caerulea
Marr, J.C., 1961.
A hypothesis of the population biology of the sardine, Sardinops caerulea. (Abstract).
Proc. Ninth Pacific Sci. Congr., Pacific Sci. Assoc., 1957, Fish. 10:63.

Sardinops caerulea
Sette, O. E., 1948.
Estimations of abundance of the eggs of the Pacific pilchard (Sardinops caerulea) off Southern California during 1940 and 1941.
J. Mar. Res. 7(3):511-542, 4 textfigs.

Sardinops caerulea
Sette, O. E., 1943
Studies on the Pacific pilchard or sardine (Sardinops caerulea). 1. Structure of a research program to determine how fishing affects the resource. U. S. Fish and Wildlife Service, Spec. Sci. Rept. 19:27 pp. (mimeographed), 3 figs. (multilith).

Sardinops caerulea
Sette, O.E., L. A. Walford, K.H. Mosher, E. H. Ahlstrom, and R.P. Silliman, 1950.
Studies on the Pacific pilchard or sardine (Sardinops caerulea). Spec. Sci. Rept., Fish., No. 15: 200 pp. (multilith).

Reissue of Spec. Sci. Repts. 19-24, 1943.

Sardinops caerulea
Silliman, R.P., 1946
A study of variability in plankton tow net catches of Pacific pilchard (Sardinops caerulea) eggs. J. Mar. Res., 6(1): 74-83, Fig. 22-23.

Sardinops caerulea

Silliman, R. P., 1943
Studies on the Pacific Pilchard or Sardine (Sardinops caerulea). 6. Thermal and diurnal changes in the vertical distribution of eggs and larvae. U. S. Fish and Wildlife Service, Spec. Sci. Rept. 22:17 pp. (mimeographed), 4 figs. (multilith).

Sardinops caerulea

Silliman, R. P., 1943
Studies on the Pacific pilchard or sardine (Sardinops Caerulea). 5. A method for computing mortalities and replacements. Spec. Sci. Rept. No. 24, U. S. Fish and Wildlife Service, 10 pp. (mimeographed), 2 figs. (multilith).

Sardinops caerulea

Silliman, R.P. and F.N. Clark, 1945
Catch per-unit-of-effort in California waters of the sardine (Sardinops caerulea) 1932-42. Calif. Fish Game, Bull. 62:3-76.

Sardinops caerulea

Walford, L.A., 1946
Correlation between fluctuations in abundance of the Pacific Sardine (Sardinops caerulea) and salinity of the sea water. J. Mar. Res. 6(1): 48-53, fig. 13

Sardinops caerulea

Walford, L. A. and K. Mosher, 1943
Studies on the Pacific pelchard or sardine (Sardinops caerulea). 3. Determination of age of adults by scales, and effect of environment on first years growth as it bears on age determination. U. S. Fish and Wildlife Service, Spec. Sci. Rept. No.21:29 pp. (mimeographed), 6 figs. (multilith).

Sardinops caerulea

Walford, L. A. and K. H. Mosher, 1943
Studies on the Pacific pilchard or sardine (Sardinops caerulea). 2. Determination of the age of juveniles by scales and otoliths. U. S. Fish and Wildlife Service, Spec. Sci. Rept. No. 20:19 pp. (mimeographed), 32 figs., 8 tables (multilith).

fishes pilchards (tagging)

Newman, G.G., 1970.
Stock assessment of the pilchard Sardinops ocellata at Walvis Bay, South West Africa.
Investl Rept. Div. Sea Fish. S.Afr. 85: 13pp.

fish - Saurida

Venkata Subba Rao, K., 1964.
On the occurrence of Saurida dendosquamis (Richardson) off Visakpatnam.
J. Mar. Biol. Ass., India, 6(2):265-267.

fishes - Schindleri

Jones, S., and M. Kumaran, 1964.
On the fishes of the genus Schindleria Giltray from the Indian Ocean.
J. Mar. Biol. Assoc., India, 6(2):257-264.

Schedophilus

Haedrich, Richard L., and Fernando Cervigón, 1969.
Distribution of the centrolophid fish Schedophilus pemarco, with notes on its biology.
Breviora (340): 9pp.

Sciaena

Sciaena saturne (black croaker)

Johnson, M.W., 1943
Underwater Sounds of Biological Origin. UCDWR No.U28, Sec. No.6.1-sr30-412, File No. 01.33, 20 pp. (mimeographed), 1 chart (photostat), 2 figs. (photo), 15 Feb. 1943.

Scomber

fish, Scomber japonicus

Usami, Shuzo, 1969.
Mode of life of the adult mackerel (Japanese mackerel, Scomber japonicus Houttuyn), in the sea off Kanto District - I. (In Japanese; English abstract). Bull. Tokai Reg. Fish. Res. Lab., 58: 97-125.

fishes, Scomber japonicus

Watanabe, T., 1970.
Morphology and ecology of early stages of life in Japanese common mackerel, Scomber japonicus Houttuyn, with special reference to fluctuation of population. Bull. Tokai reg. Fish. Res. Lab. 62: 1-283.

Scomber scombrus

Corbin, P.G., 1947.
The spawning of mackerel, Scomber scombrus L., and pilchard, Clupea pilchardus Walbaum, in the Celtic Sea in 1937-39 with observations on the zooplankton indicator species, Sagitta and Muggiaea. J.M.B.A., ns, 27:65-132, 21 textfigs.

Scomber scombrus

Sette, O. E., 1950.
Biology of the Atlantic mackerel (Scomber scombrus) of North America. Pt. II. Migrations and habits. Fish. Bull. 49:251-358, 21 textfigs.

Also, Fish Bull. Fish and Wildlife Service, 51.

Scomber scombrus

Sette, O. E., 1943.
Biology of the Atlantic mackerel (Scomber scombrus) of North America. Pt. 1. Early life history, including the growth, drift, and mortality of the egg and larval populations. Fish. Bull. 38, U. S. Dept. Interior, Fish and Wildlife Serv. Fish. Bull., Fish and Wildlife Serv. 50:149-237, 18 textfigs.

Scomber scombrus

Steven, G. A., 1949.
Contributions to the biology of the mackerel Scomber scombrus L. II. A study of the fishery in the south-west of England, with special reference to spawning, feeding, and 'fishermen's signs.' J.M.B.A. 28(3):555-576, 5 textfigs., 1 pl.

Scomber scombrus

Steven, G.A., 1948.
Contributions to the biology of the mackerel, Scomber scombrus L., mackerel migrations in the English Channel and Celtic Sea. J.M.B.A., n.s., 27:517-539, 6 textfigs.

fish - Scyliorhynchus canicula

Harris, J.E., 1955.
The development of swimming movements in the embryo of the dogfish, Scyliorhynchus canicula. Ann. Acad. Sci. Fennicae, A, 4. Biol., (29):3-11.

fish - Scyliorhynchus caniculus

Harris, J.E., and H.P. Whiting, 1954.
Structure and function in the locomotory system of the dogfish embryo. The myogenic stage of movement. J. Exp. Biol. 31(4):501-524.

Scyliorhynchus caniculus.

fish - Scymnodon plunketi

Garrick, J.A.F., 1959.
Studies on New Zealand elasmobranchii. Part IX. Scymnodon plunketi (Waite, 1910). An abundant deepwater shark of New Zealand waters. Trans. R. Soc., New Zealand, 87(3/4):271-282.

fish - Sebastes mentella

Surkova, E.I., 1962.
Peculiarities in scale structure of redfish (Sebastes mentella Travin) from the northwest Atlantic.
Sovetskie Riboch. Issledov. v Severo-Zapadnoi Atlant. Okeana, VNIRO-PINRO, Moskva, 313-318.

In Russian; English summary

fish - Sebastes Mentella

Surkova, E.I., 1962.
Size and age composition of redfish (Sebastes mentella Travin) in the northwest Atlantic.
Sovetskie Riboch. Issledov. v Severo-Zapadnoi Atlant. Okeana, VNIRO-PINRO, MOSKVA, 297-311.

In Russian; English summary

fish - Sebastes

Tåning, Å. V., 1949.
On the breeding places and abundance of the Red Fish (Sebastes) in the North Atlantic.
J. du Cons. 16(1):85-95, 6 textfigs.

fish - Sebastes mentella

Yanulov, K.P., 1962.
On groups of redfish (Sebastes mentella Travin) in Labrador-Newfoundland area. (In Russian; English summary).
Sovetskie Riboch. Issledov. v Severo-Zapadnoi Atlant. Okeana, VNIRO-PINRO, Moskva, 285-296.

fish - Sebastes marinus

Zakharov, G.P., 1962.
On the biology of redfish (Sebastes marinus L.) in West Greenland. (IN Russian; English summary).
Sovetskie Riboch. Issledov v Severo-Zapadnoi Atlant. Okeana, VNIRO-PINRO, Moskva, 319-330.

fish (Somniosus pacificus)

Gotshall, Daniel W., and Tom Jow, 1965.
Sleeper sharks (Somniosus pacificus) off Trinidad, California with life history notes.
Calif. Fish. Game, 51(4):294-298.

fishes, Sphyrna

Clarke, Thomas A. 1971.
The ecology of the scalloped hammerhead shark, Sphyrna lewini, in Hawaii.
Pacific Sci. 25(2): 133-144.

fish - Sphyrna

Hernandez Carvalho, Anatolio, 1967.
Observations on the hammerhead sharks (Sphyrna) in waters near Mazatlan, Sinaloa, Mexico. In: Sharks, skates, and rays, Perry W. Gilbert, Robert F. Mathewson and David P. Rall, editors, Johns Hopkins Univ. Press, 79-83.

Squalus

fish - Squalus acanthias

Bearden, C.M., 1965.
Occurrence of spint dogfish Squalus acanthias and other elasmobranchs in South Carolina waters. Copeia, 1965(3):378.

Squalus acanthias

Boylan, John W., 1967.
Gill permeability in Squalus acanthias. In: Sharks, skates, and rays, Perry W. Gilbert, Robert F. Mathewson and David P. Rall, editors, Johns Hopkins Univ. Press, 197-206.

fish - Squalus acanthias

Bulger, Ruth E., 1963.
Fine structure of the rectal (salt-secreting) gland of the spiny dogfish, Squalus acanthias.
Anat. Rec., 147(1):95-127.

Squalus acanthias

Burger, J. Wendell, 1967.
Some aspects of liver function in the spiny dogfish, Squalus acanthias. In: Sharks, skates, and rays, Perry W. Gilbert, Robert F. Mathewson and David P. Rall, editors, Johns Hopkins Univ. Press, 293-298.

Squalus acanthias

Burger, J. Wendell, 1967.
Problems in the electrolyte economy of the spiny dogfish, Squalus acanthias. In: Sharks, skates, and rays, Perry W. Gilbert, Robert F. Mathewson and David P. Rall, editors, Johns Hopkins Univ. Press, 177-185.

Squalus acanthias

Fishman, Alfred P., 1967.
Some features of the respiration and circulation in the dogfish, Squalus acanthias. In: Sharks, skates, and rays, Perry W. Gilbert, Robert F. Mathewson and David P. Rall, editors, Johns Hopkins Univ. Press, 215-219.

Squalus acanthias

Harris, A.J., 1965.
Eye movements of the dogfish Squalus acanthias L.
J. Exp. Biol., 43(1):107-138.

Squalus acanthias

Hogben, C. Adrian M., 1967.
Secretion of acid by the dogfish, Squalus acanthias. In: Sharks, skates, and rays, Perry W. Gilbert, Robert E. Mathewson, and David P. Rall, editors, Johns Hopkins Univ. Press, 299-315.

fish - Squalus acanthias

Holden, M.J., and P.S. Meadows, 1964
The fecundity of the spurdog (Squalus acanthias L).
Journal du Conseil, 28(3):418-424.

Squalus acanthias

Jensen, Albert C., 1967.
New uses for an old pest.
Sea Frontiers, 13(5):276-285. (popular).

Squalus acanthias

Jensen, A.C., 1961

Recaptures of tagged spiny dogfish, Squalus acanthias.
Copeia, 2: 228-229.

Squalus acanthias

Murdaugh, H. Victor, Jr., and E.D. Robin, 1967.
Acid-base metabolism in the dogfish shark. In: Sharks, skates, and rays, Perry W. Gilbert, Robert F. Mathewson and David P. Rall, editors, Johns Hopkins Univ. Press, 249-264.

fish - Squalus acanthias

#Rae, Bennet B., 1967.
The food of the dogfish, Squalus acanthias L.
Mar.Res.Dept.Agric.Fish., Scotland.1967(4):1-19.

Squalus acanthias

Robin, Eugene D., and H. Victor Murdaugh, Jr., 1967.
Gill gas exchange in the elasmobranch, Squalus acanthias. In: Sharks, skates, and rays, Perry W. Gilbert, Robert F. Mathewson and David P. Rall, editors, Johns Hopkins Univ. Press, 221-247.

fish Stomias

Gibbs, Robert H., Jr. 1969.
Taxonomy, sexual dimorphism, vertical distribution and evolutionary zoogeography of the bathypelagic fish genus Stomias (Stomiatidae)
Smithson. Contrib. Zool. 31: 1-25

fish, Sudis

Shores, David L., 1969.
Postlarval Sudis (Pisces: Paralepididae) in The Atlantic Ocean.
Breviora (334): 14pp

Synaphobranchus

Synaphobranchus

Bruun, A. Fr., 1936.
Nordatlantens Synaphobranchus-Arten, Biologi og Udbredelse sammenlignet med Aals og Havaals.
Nordiska (19 Skandinaciska) Naturforskarmötet i Helsingfors 1936:3 pp.

Taeniura

fish - Taeniura

Kewalramani, H.G., and B.F. Chhapgar, 1957.
Occurrence of a rare sting ray [Taeniura melanospila (Bleeker)] in Bombay waters.
J. Bombay Nat. Hist. Soc., 54(3):770-773.

fish - Thunnus

Bullis, H.R., jr., and F.J. Mather, 1956.
Tunas of the genus Thunnus of the North Caribbean. Amer. Mus. Novitates, No. 1765:12 pp.

fish, Thunnus alalunga

Yoshida, Howard O., 1965.
New Pacific records of juvenile albacore Thunnus alalunga (Bonnaterre) from stomach contents.
Pacific Science, 19(4):442-450.

fish, Thunnus albacares

Diaz, Enrique L., 1963.
An increment technique for estimating growth parameters of tropical tunas, as applied to yellowfin tuna (Thunnus albacares) (Spanish and English texts).
Inter-Amer. Trop. Tuna Comm., Bull., 8(7):383-416

fish, Thunnus albacares

Hirs, C.H.W., and H.S. Olcott, 1964.
Amino acid composition of the myoglobin of yellowfin tuna (Thunnus albacares).
Biochim. Biophys. Acta, 82(1):178-180.

fish, Thunnus albacares

Joseph, James, 1963
Fecundity of yellowfin tuna (Thunnus albacares) and skipjack (Katsuwonus pelamis) from the Eastern Tropical Pacific Ocean.
Inter-American Tropical Tuna Commission, Bull 7(4):257-277 (English);278-292(Spanish).

fish, tuna Thunnus albacares

Le Guen, J.C., 1968.
Étude du stock d'albacores (Thunnus albacares) exploité par les palangriers japonais dans l'Atlantique tropical américain de 1956 a 1963.
Cah. ORSTOM, Océanogr., 6(3/4): 27-30.

Thunnus albacares

Royce, William F., 1964.
A morphometric study of yellowfin tuna Thunnus albacares (Bonnaterre).
U.S.F.W.S., Fish. Bull., 63(2):398-443.

fish, Thunnus albacares

Tauruta, Saburo, 1964.
Morphometric comparison of yellowfin tuna Thunnus albacares (Bonnaterre) from western waters of the Line Islands and those of the Gilbert Islands and southeast of the Indian Ocean.
J. Shimonoseki Univ. Fish., 13(2):133-140.

fish, tuna (Thunnus atlanticus)

da Cruz, João Francisco, 1967
Análises da pesca da albacora Thunnus atlanticus (Lesson) no estado do Rio Grande do Norte.
Bolm Inst. Biol. mar. Univ. Fed. Rio Grande Norte, Brasil, 4:43-55.

fish, Thunnus atlanticus

Da Cruz, João Francisco, 1965.
Contribuição ao estudo da biologia pesqueira da albacora Thunnus atlanticus (Lesson), no Nordeste do Brasil.
Bol. Inst. Biol. Mar. Univ. R.G. Norte, 2: 33-40

~~fish, tuna (Thunnus atlanticus)~~

Soares, Leonesa Herculano e João Francisco da Cruz, 1967
Sôbre a biometria da albacora Thunnus atlanticus (Lesson), da costa do estado do Rio G. do Norte.
Bolm Inst. Biol. mar. Univ. Fed. Rio Grande Norte, Brasil, 4: 33-42.

fish - Thunnus germo

Alverson, D.L., 1961.
Ocean temperatures and their relation to albacore tuna (Thunnus germo) distribution in waters off the coast of Oregon, Washington and British Columbia.
J. Fish. Res. Bd., Canada, 18(6):1145-1152.

fish, Thunnus maccoyii

Shingu, Chiomi, and Ikuo Warashina, 1965.
Studies on the southern bluefin, Thunnus maccoyii (Castelnau). 1. Morphometric comparison of southern bluefin. 2. On the distribution of the southern bluefin in the south western Pacific Ocean and on the size of fish taken by the longline fishery.
Rept. Nankai Reg. Fish. Res. Lab., No. 22:85-93; 95-105.

fish, Thunnus obesus

Kume, Susumu, and Yasuo Morita, 1966.
Ecological studies on bigeye tuna ---III. On bigeye tuna, Thunnus obesus, caught by "nighttime longline" in the North Pacific Ocean. (In Japanese; English abstract).
Rep. Nankai reg. Fish. Res. Lab., No. 24:21-30.

fish, Thunnus saliens

Chatwin, B.M., and C.J. Orange, 1960.
Recovery of tagged bluefin tuna (Thunnus saliens).
California Fish & Game, 46(1):107-109.

fish, Thunnus thynnus

Furnestin, J., and J. Dardignac, 1962.
Le thon rouge du Maroc atlantique (Thunnus thynnus Linné).
Rev. Trav. Inst. Pêches Marit., 26(4):381-398.

fish, Thunnus thynnus

Gutiérrez, Manuel, 1967.
Estudios hematológicos en el atún, Thunnus thynnus (L.). Inmunología; grupos Sanuíneos, eritrocitos y fitoaglutininas.
Investigación pesq., 31(1):137-143.

fish, Thunnus thynnus

Gutiérrez, Manuel, 1967.
Estudios hematológicos en el atun, Thunnus thynnus (L.), de la costa Sudatlántica de España.
Investigación pesq., 31(1):53-90.

fish, Thunnus thynnus

Hamre, J., 1965.
Observations on the depthrange of tagged bluefin tuna (Thunnus thynnus L.) based on pressure marks on the Lee tag.
Arch. Fischereiwiss., 15(3):204-209.

fish Thunnus thynnus

Nakamura, Izumi, and Yukio Warashina, 1965.
Occurrence of bluefin tuna, Thunnus thynnus (Linnaeus) in the eastern Indian Ocean and the eastern South Pacific Ocean.
Rept. Nankai Reg. Fish. Res. Lab., No. 22:9-12.

fishes, Thunnus thynnus

Potthoff, Thomas and William J. Richards, 1970.
Juvenile bluefin tuna, Thunnus thynnus (Linnaeus) and other scombrids taken by terns in the Dry Tortugas, Florida. Bull. mar. Sci., 20(2): 389-413.

fish, Thunnus thynnus

Rodríguez-Roda, Julio, 1964.
Biologia del atún, Thunnus thynnus (L.) de la costa sudatlántica de España.
Inv. Pesq., Barcelona, 35:33-146.

fish, Thunnus thynnus

Rodríguez-Roda, Julio, 1967.
Fecundidad del atún, Thunnus thynnus (L.), de la costa Sudatlántica de España.
Investigación pesq., 31(1):33-52.

fish, Thunnus thynnus

Rodríguez-Roda, Julio, 1964.
Talla, peso y edad de los atunes, Thunnus thynnus (L.), capturados por la almadraba de Barbate (costa sudatlántica de España) en 1963 y comparación con el período 1956 a 1962.
Inv. Pesq., Barcelona, 26:3-47.

Trachurus

jack-mackerel

Hotta, Hideyuki, and Junko Nakashima 1968.
Experimental study on the feeding activity of jack-mackerel Trachurus japonicus. (In Japanese; English abstract).
Bull. Seikai reg. Fish. Res. Lab. 36:75-83

jack-mackerel (growth)

Muta, Kunisuke, Nobutsugu Ogawa and Seiichi Hamasaki 1968.
Tag testing experiment on the young jack mackerel Trachurus japonicus (Temminck et Schlegel) by the holding method and its growth in west Japanese water. (In Japanese; English abstract).
Bull. Seikai reg. Fish. Res. Lab. 36:85-101

Trachurus japonicus

Suzuki, Tomoyuki, 1965.
Ecological studies on the jack Mackerel, Trachurus japonicus (Temminck et Schlegel). 1. On the feeding habits. (In Japanese; English abstract).
Bull. Jap. Sea Reg. Fish. Res. Lab., (14):19-29.

Trichurus lepturus

Kosaka, Masaya, Masahiro Ogura, Hideki Shirai and Michiyoshi Maeji, 1967.
Ecological study of the ribbon fish, Trichurus lepturus Linné in Suruga Bay. (In Japanese; English abstract).
J. Coll. mar. Sci. Techn., Tokai Univ., (2):131-159.

fish - Trachurus mediterraneus ponticus

Porumb, Ioan I., 1961
Contributions a la biologie de Trachurus mediterraneus ponticus Aleev de la Mer Noire. (In Roumanian: Russian and French resumes).
Ann. Sci. Univ. "Al. I. Cuza", Jassy, Roumania, 7(2):285-304.

fish - Trachurus

Revina, N.I., 1960
Development and nutrition of Trachurus mediterraneus ponticus Aleev. (In Russian)
Trudy azov.-chernomorsk. nauchno.-issled. Inst. morsk. ryb. Khoz. Okeanog. 13:105-114

fish - Trachurus trachurus

Ferreira Barraca, Ivone, 1964.
Quelques aspects de la biologie et de la pêche du chinchard, Trachurus trachurus (L.) de la côte portugaise.
Notas e Estudos, Inst. Biol. Marit., Lisboa, No. 29:45 pp.

fish - Triaenodon

Wheeler, J.F.G., 1960.
Sharks of the western Indian Ocean. II. Triaenodon (Rüppell).
E. Afr. agric. J., 25(3):202-204.

Also in. Int. Indian Ocean Exped., Coll. Rept. UNESCO, 3. 1966.

Umbrina

Umbrina roncador (yellowfin croaker)

Johnson, M.W., 1943
Underwater Sounds of Biological Origin. UCDWR No.U28, Sec. No.6.1-sr30-412, File No. 01.33, 20 pp. (mimeographed), 1 chart (photostat), 2 figs. (photo), 15 Feb. 1943.

Xiphias

fish, Xiphias gladius

Cavaliere, Antonino 1966.
Studio sulla biologia e pesca di Xiphias gladius L. 4. Prodotta di tale pesca durante le stagioni 1965 e 1966.
Boll. Pesca, Pisic. Idrobiol. n.s. 21(2): 299-303.

fish, Xiphias gladius

Cavaliere, Antonino 1964.
Studi sulla biologia e pesca di Xiphias gladius L. III. Prodotto di tale pesca durante le stagioni 1963 e 1964.
Boll. Pesca, Piscicolt. Idrobiol. n.s. 19(2): 287-291.

fish - Xiphias gladus

Cavaliere, Antonino, 1962.
Studi sulla biologia e pesca di Xiphias gladus L.
Boll. Pesca, Piscicolt. e Idrobiol., n.s., 17(2): 123-143.

fish, Xiphias gladius

Scott, W.B., and S.N. Tibbo 1968.
Food and feeding habits of Swordfish, Xiphias gladius, in the western North Atlantic.
J. Fish. Res. Bd Can. 25(5): 903-919.

fish, Xiphias cavirostris

Paulus, Marcel, 1962
Etude ostéographique et ostéometrique sur un Ziphius cavirostris G. Cuvier, 1823 échoué à Marseille-Estaque en 1879.
Bull. Mus. Hist. Nat., Marseille, 22:17-48.

Xiphias

Tåning, Å.V., 1955.
On the breeding areas of the swordfish (Xiphias).
Pap. Mar. Biol. and Oceanogr., Deep-Sea Res., Suppl. to Vol. 3:438-450.

albacore

fish - fisheries

The references filed here may be to the fish themselves or to the fisheries per se.

albacore

Asano, Masahiro, 1964.
Young albacore taken from the northeastern sea area of Japan in August and September, 1963. (In Japanese; English abstract).
Bull. Tohoku Reg. Fish. Res. Lab., No. 24:20-27.

fish - albacore

Craig, William L., and Edward K. Dean, 1968.
Scouting for albacore with surface salinity data.
Undersea Techn., 9(5):60-61, 90.

fisheries, albacore

Grandperrin, R. et M. Legand, 1971.
Aperçu sur la distribution verticale des germons dans les eaux tropicales du Pacifique sud: nouvelle orientation de la pêche Japonaise et de la pêche expérimentale. Cah. ORSTOM, sér. Océanogr, 9(2): 197-202.

fisheries, albacore

*Griffiths, Raymond C., y Takeshi Nemoto, 1967.
Un estudio preliminar de la pesqueria para atun aleta amarilla y albacora en el mar Caribe y el oceano Atlantico occidental por palangreros de Venezuela.
Invest. Pesqueras, Ser. Recursos Explotación pesqueros, 1(6):209-371.

albacore

Inoue, Motoo, 1965.
Movement of fishing grounds for the albacore and their migration with oceanographic conditions in the northwestern Pacific Ocean. (In Japanese; English abstract).
Rep. Fish. Res. Lab., Tokai Univ., 2(1):1-98.

albacores

Ishii, Takeo, 1965.
Morphometric analysis of the Atlantic albacore populations mainly her eastern areas. (In Japanese; English abstract).
Bull. Jap. Soc., Sci. Fish., 31(5):333-339.

fish - albacore

Koga, Shigeyuki, 1966.
Latitudinal variation of surface water temperature optimum to the fishing of the albacore. (In Japanese; English abstract).
J. Shimonoseki Univ., Fish., 14(3):193-197.

albacore

Koto, Tsutomu, and Koichi Hisada, 1966.
Studies on the albacore. XII. Length frequency distribution of albacore caught by the Japanese longline and pole-and-line fisheries in the western North Pacific in 1960-64 seasons. (In Japanese; English abstract)
Rept. Nankai reg. Fish. Res. lab., No.24:15-20.

albacore
Kurogane, Kenji, and Yoshio Hiyama, 1959.
Morphometric comparison of the albacore from the Indian and the Pacific Ocean.
Rec. Oceanogr. Wks., Japan, n.s., 5(1):68-84.

fish, albacore
Laurencin, F. G. Baudin 1968
Croissance et age de l'albacore du Golfe de Guinée.1.
Centre Recherches océanogr. Côte d'Ivoire, Doc. sci. provisoire 021: 11 pp. (mimeographed)

albacores
Lima, Flávio R., and John P. Wise, 1962
Primeiro estudo da abundância e distribuição da albacora de laje e da albacora branca na região ocidental do Oceano Atlântico Tropica (1957-1961).
Sudene, Bol. de Estudos de Pesca, Recife, 2(10):12-17.

fisheries, albacore
Owen, R.W., Jr., 1963.
Northeast Pacific albacore oceanography survey, 1961.
U.S. F.W.S. Spec. Sci. Rept., Fish., No. 444:1-35

albacore
Suarez Caabro, Jose A., and Pedro Pablo Duarte Bello, 1961
Biología pesquera del bonito (Katsuwonus pelamis) y la albacora (Thunnus atlanticus) en Cuba I. Inst. Cubano de Invest. Tecnol., Ser. Estudios sobre Trabajos de Investigación, No. 15: 150 pp.

albacore
Suda, Akira, 1966.
Catch variations in the North Pacific albacore. VI. The speculations about the influences of fisheries on the catch and abundance of the albacore in the North-West Pacific by use of some simplified mathematical models (Cont.1). (In Japanese; English abstract).
Rep. Nankai reg. Fish Res. Lab. No. 24:1-14.

albacores
Suda, A., 1962.
Studies on the albacore. VIII.
Rept. Nankai Reg. Fish. Res. Lab., 16:127-134.

anchovies

anchovies
Ahlstrom, Elbert H., 1967.
Co-occurrences of sardine and anchovy larvae in the California Current region off California and Baja California.
Rep. Calif. Coop. Oceanic Fish. Invest. 11:117-135.

anchovies
Andreu, B., and J. Rodríguez-Roda, 1951.
Estudio comparativo del ciclo sexual, engrasimiento y repleción estomacal de la sardina, alacha y anchoa del Mar Catalán acompañado de relación de pescado de huevos planctónicos de estas especies. Publ. Inst. Biol. Aplic. 9:1930 232, 14 textfigs.

anchovies
Barredo, M., 1950.
Informe sobre las investigaciones efectuadas con relación al desove de la anchoveta en la Bahía de Pisco. Bol. C.A.G. 26(5):55-63, figs.

anchovies
Bayliff, William H., 1963.
Observations on the life history and identity of intraspecific groups of the anchoveta, Cetengraulis mysticetus in Montijo Bay and Chiriqui Province, Panama.
Inter-Amer. Tropical Tuna Comm., Bull., 8(3): 169-188; (Spanish) 189-197.

anchovies
Bayliff, W.H., and E.F. Klimm, 1962.
Live-box experiments with anchovetas Cetengraulis mysticetus in the Gulf of Panama.
InterAmerican Tropical Tuna Comm. Bull., 6(8): 335-446 (English) 405-440 (Spanish)

anchovies
Berdegue A., Julio, 1958.
Biometric comparison of the anchoveta Cetengraulis mysticetus (Günther) from t en localities of the eastern tropical Pacific Ocean.
Bull. Int.-Amer. Trop. Tuna Comm., 3(1):1-76.

anchovies
Brandhorst, Wilhelm, Mario Carreño y Omar Rojas, 1965.
El número de vértebras de la anchoveta (Engraulis ringens Jenyns) y otras especies de la superfamilia Clupeoidae en aguas chilenas.
Inst. Fomento Pesq., Bol. Cient., No. 1(1):1-16.

anchoveta
*Brandhorst, Wilhelm, y José Raúl Cañón, 1967.
Resultados de estudios oceanografico-pesqueros aeros en el norte de Chile.
Pub. Inst. Fomento Pesquero, Chile, 29:44 pp. (mimeographed)

anchoveta
*Brandhorst, Wilhelm, Omar Rojas, 1967.
Investigaciones sobre los recoursos de la anchoveta (Engraulis ringens J.) y sus relaciones con las condiciones oceanográficas en agosto-septiembre de 1963 y marzo-junio de 1964.
Pub. Inst. Fomento Pesquero, Chile, 31:38 pp. (mimeographed).

anchovies
Cadenat, J., 1946.
Une peche miraculeuse d'anchois a Goree. Bull. Info. Con. Inst. Francais Afrique du Nord 32:28

fish-anchovies
Calderwood, W. L. 1892
Experiments on the relative abundance of anchovies of the south coast of England.
Jour. Mar. Biol. Assoc. n.s. 2:268-271

anchovies
Carvalho, J. de Paivo, 1950.
Engraulideos brasileiros do genero Ancho.
Bol. Inst. Paulista Oceanogr. 1(2):43-69, 2 pls (10 figs.).

anchovies
Del Dolar, E., 1952.
Ensayo sobre la ecología de la anchobeta. Bol. C.A.G. 18(1):3-23, figs.

anchovies
de Ciechomski, Juana D., 1964.
Estudios sobre el desove e influencia de algunos factores ambientales sobre el desarrollo embrionario de la anchoita (Engraulis anchoita).
Contrib. Inst. Biol. Mar., Mar del Plata, Argentina, No. 16:1 pp.

anchovies
Farris, David A., 1960.
Failure of an anchovy to hatch with continued growth of the larva.
Limnol. & Oceanogr. 5(1):107.

fisheries-anchovies
Furnestin, J. 1939-1945.
Note preliminaire sur l'Anchois (Engraulis encrasicholus) du Golfe de Gascogne. Rev. des Trav. de l'Office scientifique et technique DES Peches Maritimes, 12:197-209.

anchovies
Fuster de Plaza, M.L., and E.E. Boschi, 1961
Areas de Migración y ecología de la anchoa Lycengraulis olidus (Günther) en las aguas Argentinas (Pisces, fam. Engraulidae).
Contrib., Inst. Biol. Mar., Mar del Plata, Argentina, No. 1:58 pp.

fish anchovies
Inoue, Makoto, 1970.
The size range of anchovy schools from the viewpoint of their swimming speed. (In Japanese; English abstract). J. Tokyo Univ. Fish. 57(1): 17-24.

anchovies
Jordán, Rómulo, y Aurora Chirinos de Vildosa, 1965.
La anchoveta (Engraulis ringens J.); conocimiente actual sobre su biología ecología y pesquería.
Inst. Mar., Peru, Informe, No. 6:52 pp.

fisheries, anchovies
Klima, E.F., I. Barrett and J.E. Kinnear, 1962.
Artificial fertilization of the eggs and rearing and identification of the larvae of the anchoveta, Cetengraulis mysticetus.
Inter-American Tropical Tuna Comm., 6(4):155-178.

anchovies
Marion, A.F., 1893.
Notes sur le régime du maquereau et de l'anchois sur les côtes de Marseille en 1890.
Ann. Mus., Marseille, Trav. Zool. Appl., 4:108-112.

fish anchovies
Marion, A.F., 1889.
Notes sur l'anchois. Ann. Mus. Marseille, Trav. Zool. Appl., 3:58-64.

fish, anchovies
Matsumura, Shinsaku, Shigeki Yasuie and Isamu Mitani, 1971.
Eggs and larvae of the anchovy Engraulis japonica (Houttuyn) in Bisan-Seto, the Set-Inland Sea, 1970. (In Japanese). Bull. Fish. Exp. Sta. Okayama Pref. (1970): 29-31.

anchovies
Murphy, R.C., 1954.
El guano y la pesca de anchoveta. Informe oficial al supremo gobierno, C.A.G.:1-39,

together with:
Documentos oficiales en pro de la conservacion de las especies productores del guano y su base alimenticia:40-147.

anchovies
Muzinic, R., 1956.
Quelques observations sur la sardine, l'anchois, et le maquereau des captures au chalut dans l'Adriatique. Acta Adriatica, 7(13):39 pp.

anchovies
Pavlovskaia, R.M., 1958.
The survival of anchovy larvae in the northwestern part and other regions of the Black Sea, observed in 1954-1955, and the role played in their survival by feeding conditions.
Dokl. Akad. Nauk, SSSR, 120(2):415-418.

fisheries, anchovies
Peru, Instituto del Mar del Peru, 1970.
Panel de Expertos informe sobre los efectos económicos de diferentes medidas regulatorias de la pesquería de la anchoveta peruana.
Informe, Inst. Mar, Peru 34: 83pp.

fish anchovies
Peru, Instituto del Mar del Peru, 1967.
Informe complementario sobre la pesqueria de la anchoveta.
Informe, 15:13 pp. (multilithed).

anchovies
Peru, Institute del Mar del Peru, 1965.
Efectos de la pesca en el stock de Anchoveta.
Informe No. 7:16 pp. (multilithed).

anchovies
Rojas, E., B., 1953.
Estudios preliminares del contenido estomacal de las anchovetas. Bol. Cient., C.A.G., 1(1):33-42p
7 pls.

anchovies
*Silva S., Jose Oswaldo, 1967.
Un estudio de algunos caracteres meristicos de la rabo amarillo, Cetengraulis edentulus (Cuvier, 1829), de la region oriental de Venezuela.
Invest. Pesqueras, Ser. Recurdos Explotacion Pesqueros, Venezuela, 1(9):333-372.

anchoveta,
Simpson, John G., 1959.
Identification of the egg, early life history and spawning areas of the anchoveta, Cetengraulis mysticetus (Günther), in the Gulf of Panama.
Inter-American Trop. Tuna Comm., Bull. 3(10): 441-580.

anchovies
Simpson, John G., y Mario Carreño R., 1967.
Distribución geografica mensuel y trimestral y composición mensuel de frecuencia de longitud por zona de la anchoveta desembarcada en Arica, Iquique, Tocopolla, Antofagasta, Coquimbo y Talcahuano de enero a junio de 1966.
Publ. Inst. Fomento Pesquero, Chile, 27:117-pp. (mimeographed).

anchovies
Simpson, John G., Mario Carreño, Hans Simon Schlotfeldt y Eduardo Gil R., 1967
Distribución geográfica mensuel y trimestral y composición de las frecuencias de longitudes por mes de las caprutas de anchoveta por zona, desembarcadas en Iquique desde noviembre 1963 hasta diciembre 1965.
Pulb. Inst. Fomento Pesquero, Chile, 26:99 pp. (mimeographed).

anchovies
*Simpson, John G., y Eduardo Gil R., 1967.
Meduración y desove de la anchoveta (Engraulis ringens) en Chile.
Bol. cient. Inst. Fomento Pesquero, Chile, 4: 55pp. (mineographed).

anchovies
Vucetic, Tamara, 1964.
Sur la ponte de l'anchois (Engraulis encrasicholus L.) au large de l'Adriatique.
(In Jugoslavian; French resume).
Acta Adriatica, 11(1):277-284.

resume, p. 284.

fish, anchovies
Yasuie, Shigeki and Isamu Mitani, 1969.
Anchovy eggs and larvae in Bisan-Seto, the Seto Inland Sea, 1969. Bull. Fish. exp. Sta. Okayama, 1969: 14-16. (In Japanese).

larvae, fish
anchovies
Japan

barracuda

barracuda attacks
deSylva, Donald P., 1963.
Systematics and life history of the Great Barracuda Sphyraena barracuda (Walbaum).
Stud. trop. Oceanogr., Miami, 1:viii 179 pp., 32 tables, 36 figs. $2.50.

fishes - billfishes
Strasburg, Donald W., 1970.
A report on the billfishes of the central Pacific Ocean. 575-604.
Bull. mar. Sci., 20(3):

bonito

bonito
Chirinos de Vildoso, Aurora, 1960
Estudios sobre la reproduccion del "bonito" Sarda chilensis (C. y V.) en aguas asyacentes a la costa Peruana. Ministerio de Agricultura, Direccion de Pesqueria y Caza, Serie de Divulgacion Cientifica No. 14: 75 pp. (multilithed).

bonito
Dardignac, J., 1962.
La bonite du Maroc atlantique.
Rev. Trav. Inst. Pêches Marit., 26(4):399-406.

bonito
Suarez, Caabro, Jose A., and Pedro Pablo Duarte Bello, 1961
Biología pesquera del bonito (Katsuwonus pelamis) y la albacora (Thunnus atlanticus) en Cuba. I. Inst. Cubano de Invest. Tecnol., Ser. Estudios sobre Trabajos de Investigación, No. 15: 150 pp.

COD

cod
Mankevich, E.M., 1962.
Biological peculiarities of northwest Atlantic cod differing in otolith structure.
Sovetskie Riboch. Issledov. v Severo-Zapadnoi Atlant. Okeana, VNIRO-PINRO, Moskva, 331-343.

In Russian; English summary

cod
Mankevich, E.M., and V.S. Prokhorov, 1962.
On the biology of cod on the southwestern slope of Flemish Cap. (In Russian; English summary)
Sovetskie Riboch. Issledov. v Severo-Zapadnoi Atlant. Okeana, VNIRO-PINRO, Moskva, 355-360.

cod
Postolsky, A.I., 1962.
Some data on biology of cod from Labrador and Newfoundland areas.
Sovetskie Riboch. Issledov. v Severo-Zapadnoi Atlant. Okeana, VNIRO-PINRO, Moskva, 345-354.

In Russian; English summary

cod
Wise, J.P., 1958.
The world's southernmost indiginous cod.
J. du Cons., 23(2):208-212.

Coelacanth

coelacanths, physiol. of.
Brown, G.W., Jr., and Susan G. Brown, 1967.
Urea and its formation in coelacanth liver.
Science, 155(3762):570-572.

coelacanthes
Manaché, M., 1954.
Étude hydrologique sommaire de la région d'Anjouan en rapport avec la pêche des coelacanthes.
Mém. Inst. Sci., Madagascar, A, 9:151-185, 34 textfigs.

fish, coelacanths
Lutz, Peter L. and James D. Robertson, 1971.
Osmotic constituents of the coelacanth Latimeria chalumnae Smith. Biol. Bull. mar. biol. Lab. Woods Hole 141(3): 553-560.

coelacanth
Millot, J., and A. Policard, 1955.
Sur la structure inframicroscopique du tissu conjonctif du coelacanthe.
Bull. Microscop. Appl. (2) 5(7/8):94-95.

coelacanths, physiol. of.
Pickford, Grace E., and F. Blake Grant, 1967.
Serum osmolality in the coelacanth, Latimeria chalumnae: urea retention and ion regulation.
Science, 155(3762):568-570.

coelacanth, chemistry of
Nevenzel, Judd C., Waldtraut Rodegker, James F. Mead and Malcolm S. Gordon, 1966.
Lipids of the living coelacanth, Latimeria chalumnae.
Science, 152 (3730): 1753-1755.

coelacanth
Smith, J.L.B., 1953.
The second coelacanth. Nature 171:99-101.

fish coelacanth
Thomson, Keith Steward, 1970.
Intracranial movement in the Coelacanth Latimeria chalumnae Smith (Osteichthyes, Crossopterygii).
Postilla 149: 12 pp

crab

cyclostomes

cyclostomes
Kleerekoper, H., 1961.
The role of chemical scents in the orientation of cyclostomes. (Abstract).
Tenth Pacific Sci. Congr., Honolulu, 21 Aug.-6 Sept., 1961, Abstracts of Symposium Papers, 177-178.

dolphins

eels

eels
Asano, Hirotoshi, 1962
Studies on the congrid eels of Japan.
Bull. Misaki Mar. Biol. Inst., Kyoto Univ.,
No. 1:1-143.

eels
Bal, D.V., and K.H. Mohmed, 1957.
A systematic account of the eels of Bombay.
J. Bombay Nat. Hist. Soc., 54(3):732-740.

eels
Bauchot, M.L., et J.M. Bassot, 1958.
Sur Heteroconger longissimus Günther (Teléostéen, Anguilliforme) et quelques aspects de sa biologie.
Bull. Mus. Nat. Hist. Nat., Paris, 30(3):258-261.

fish - eels
Bergeron, Julien, 1970.
Travaux sur l'anguille.
Rapp. ann. 1969, Serv. Biol. Québec, 129-142.

eels
Bertin, Leon, 1956.
Eels: a biological study. Cleaver-Hume Press, Ltd., London, 200 pp., 55 illus., 9 pls.

fish - eels
Blache, J., 1968.
Contribution à la connaissance des poissons anguilliformes de la côte occidentale d'Afrique. 8. La famille des Echelidae. 9. Les Heterancheli-dae.
Bull. Inst. Fondament. Afr. noire, A, 30(4):1501-1539; 1540-1581.

fish - eels
*Castle, P.H.J., 1968.
The congrid eels of the western Indian Ocean and the Red Sea.
Ichthyol. Bull., Rhodes Univ., Grahamstown, 33: 685-726.

fish - EELS
Blache, J., 1967.
Contribution à la connaissance des poissons anguilliformes de la côte occidentale d'Afrique. 6. Les genres Anarchias, Uropterygius, et Channomuraena (Muraenidae).
Bull. Inst. fondament. Afr. noire, (A) 29(4): 1706-1731.

fish - EELS
Blache, J., 1967.
Contribution à la connaissance des poissons anguilliformes de la côte occidentale d'Afrique.
Bull. Inst. fondament. Afr. noire, (A)29(4):1695-1705.

eels
*Boëtius, Inge, and Jan Boëtius, 1967.
Studies in the European eel, Anguilla anguilla (L.). Experimental induction of the male sexual cycle, its relation to temperature and other factors.
Meddelelser Danmarks Fiskeri - og Havundersøgelser, N.S., 4(11):339-405.

eels
Böhlke, James E., and David G. Smith, 1967.
A new xenocongrid eel from the Indian and western Atlantic Oceans.
Notulae Naturae, 408: 6 pp.

eels
Bohun, Sheila, and Howard E. Winn, 1966.
Locomotor activity of the American eel. (Anguilla rostrata).
Chesapeake Sci., 7(3):137-147.

eels, slime
Brodal, Alf, and Ragnar Fänge, Editors, 1963.
The biology of Myxine.
Universitetsforlaget, Oslo, 588 pp.

eels
Bruun, Anton Fr., 1963 (posthumous)
The breeding of the North Atlantic fresh-water eels. In: Advances in Marine Biology, F.S. Russell, Editor, Academic Press, London and New York, 1:137-169.

Callamand, O., 1943. eels, physiology
L'Anguille Européene (Anguilla anguilla L.) Les Bases physiologiques de sa migration. Ann. de l'Inst. Ocean., Monaco, 21(6):361-438, 1 pl.

fish - eels
Castle, P.H.J., 1968.
Synaphobranch eels from the Southern Ocean.
Deep-Sea Res., 15(3):393-396.

eels
Castle, P.H.J., 1967.
Taxonomic notes on the eel, Muraenesox cinereus (Forsskål, 1775).
Spec. Publ. Dept. Ichthyol., Rhodes Univ., 2:10pp.

fish - eels
Castle, P.H.J. 1967.
Heterocongrine eels in the southwest Pacific.
Rec. Dominion Mus. N.Z. 6(2):5-12.

eels
*Castle, P.H.J., 1967.
Two remarkable eel-larvae from off southern Africa.
Spec. Pub. Dept. Ichthyol., Rhodes Univ. 1:1-12.

fish, eels
Castle, P.H.J., 1964.
Eels and eel-larvae of the Tiri Oceanographic Cruise, 1962, to the South Fiji Basin.
Trans. R. Soc., New Zealand, Zool., 5(7):71-84.

eels
Castle, P.H.J., 1963.
The systematics, development and distribution of two eels of the genus Gnathophis (Congridae) in Australasian waters.
Zool. Publ., Victoria Univ., Wellington, (34): 15-47

fish - eels
Castle, P.H.J., 1961
Deep-water eels from Cook Strait, New Zealand.
Zool. Publ., Victoria Univ., Wellington, No. 27: 1-30.

eels
Castle, P.H.J., 1960.
Two eels of the genus Pseudoxenomystax from New Zealand waters.
Trans. Roy. Soc., New Zealand, 88(3):463-472.

fish - eels
Chan, W.L., 1967.
A new species of congrid eel from the South China Sea.
J. nat. Hist., 1(1):97-112

eels
D'Ancona, U., 1958.
Comparative biology of eels in the Adriatic and Baltic. Verh. Int. Ver. Limnol., 13:731-735.

eels
D'Ancona, U., and D.W. Tucker, 1959.
Old and new solutions to the eel problem.
Nature, 183(4672):1405-1406.

eels
Deelder, C.L., 1957.
On the growth of eels in the IJsselmeer.
J. du Cons., 23(1):83-88.

eels
*Della Croce, N., and P.H.J. Castle, 1966.
Leptocephali from the Mozambique Channel.
Boll. Musei Ist. biol. Univ. Genova, 34:149-164.

eel fisheries
*Eales, J. Geoffrey, 1968.
The eel fisheries of eastern Canada.
Bull. Fish Res. Bd., Can., 166:79 pp.

eels, diseases of
Egusa, Syuzo, and Tomoko Nishikawa, 1965.
Studies of a primary infectious disease in the so-called fungus disease of eels.
Bull. Jap. Soc. Sci., Fish., 31(10):804-813.

eels
Francés-Boeuf, C., 1947
Les facteurs physico-chimiques des eaux d'estuaire et l'interpretation des Tropismes au cours des migrations de l'anguille. Bull. Inst. Ocean., Monaco, No. 915, 7 pp.

eels (eggs of
H.P., 1964.
Les oeufs de l'Anguille européenne sont depuis peu connus.
Science Progrès, La Nature, No. 3360:151-152.

eels, migrations of
Jones, J.W., and D.W. Tucker, 1959.
Eel migration.
Nature, 184(4695):1281-1283.

eels
Kanazawa, Robert H., 1961.
Paraconger, a new genus with three new species of eels (Family Congridae).
Proc. U.S. Nat. Mus., 113(3450):1-12.

eels
Kubota, Saburoh S., 1961
Studies on the ecology, growth and metamorphosis in Conger eel, Conger myriaster (Brevoort).
J. Fac. Fish., Pref. Univ., Mie, 5(2):190-370. 66 pls.

fish, eels
Matsumura, Shinsaku, Hirotoshi Hattori and Isamu Mitani, 1971.
Larvae of the (sand-eel Ammodytes personatus Girard) in Bisan-seto, the Seto-Inland Sea, 1971. Bull. Fish. Exp. Sta. Oka yama Pref. (1970): 31-41. (In Japanese)

fish - eels

McCosker, John E., 1970.
A review of the eel genera Leptenchelys and Muraenichthys, with a description of a new genus Schismorhynchus and a new species, Muraenichthys chilensis.
Pacific Sci., 24(4): 506-516

fish - eels

Mead, Giles W., and Sylvia A. Earle, 1970.
Notes on the natural history of snipe eels.
Proc. Calif. Acad. Sci. (4) 38 (G.S. Myers Festschrift): 99-104

eels

Murina, V.V., 1956.
[Nutrition of eels in Kursk and Bislinsk Bays.]
Trusy Vses. Gidrobiol. Obsh., Akad. Nauk, SSSR, 7:148-16?.

eels

Nelson, Gareth J., 1967.
Notes on the systematic status of the eels Neenchelys and Myroconger.
Pacif. Sci., 21(4):562-563.

fish - eel populations

Pantelouris, E.M., A. Arnason and F.-W. Tesch, 1971.
Genetic variation in the eel. III. Comparisons of Rhode Island and Icelandic populations. Implications for the Atlantic eel problem.
Mar. Biol. 9(3): 242-249.

eels

Robins, Catherine H., and C. Richard Robins, 1967.
Biological investigations of the deep sea. 29. The xenocongrid eel Chlopsis bicolor in the western North Atlantic.
Bull. mar. Sci., Miami, 17(1):232-248.

eels

Schmidt, J., 1932.
Danish Eel investigations during 25 years 1905-1930. 16 pp., 24 textfigs.

eels

Schmidt, J., 1931.
Eels and conger eels of the North Atlantic.
Nature for 10 Oct. 1931, 8 pp., 3 textfigs.

eels

Schwartz, Frank, 1961.
Lampreys and eels. Maryland Conservationist, 38(2):11 pp.

eels

*Sinha, V.R.P., and J.W. Jones, 1967.
On the age and growth of the freshwater eel. (Anguilla anguilla).
J. Zool., Lond., 153(1):99-117.

eels

*Sinha, V. R.P., and J.W. Jones, 1967.
On the food of the freshwater eels and their feeding relationship with the salmonids.
J. Zool., Lond., 153(1):119-137.

fish - eels

Smith, David G., 1969.
Biological investigations of the deep sea. 44. Xenocongrid eel larvae in the western North Atlantic. Bull. mar. Sci., 19(2): 377-408.

eels

Smith, J.L.B., 1962.
Sand-dwelling eels of the western Indian Ocean and Red Sea.
Ichthyol. Bull., Rhodes Univ., 24:447-461.

eels

Suslowska, Wieslawa, 1962.
Remarques sur un cas de xantorisme chez une anguille - Anguilla anguilla (L.) venant de la Baie de Gdansk.
Bull. Soc. Sci. et Lettres, Lodz, 13(3):1-8.

fish eels

Tesch, F.-W., 1971.
Aufenthalt der Glasaale (Anguilla anguilla) an den südlichen Nordseeküste vor dem Eindringen in das Süsswasser.
Vie Milieu Suppl. 22(1): 381-392.

eels

*Tesch, Friedrich-Wilhelm, 1967.
Aktivität und Verhalten wandernder Lampetra fluviatilis, Lota lota und Anguilla anguilla.
Helgoländer wiss. Meeresunters., 16(½):92-111.

eels

Tucker, D.W., and U. D'Ancona, 1959.
Old and new solutions to the eel problem.
Nature, 183(4672):1405-1406.

fish - eels

Winn, Howard E., and Susan Hammen 1969
Migratory habits of the American eel pose problems for biologists.
Maritimes, Univ. R.I., 13(1): 3-5

elasmobranchs

fish - elasmobranchs

Bone, Q. and B.L. Roberts, 1969.
The density of elasmobranchs. J. mar. biol. Ass., U. K., 49(4): 913-937.

fish elasmobranch anat-physiol.

Chieffi, Giovanni, 1967.
The reproductive system of elasmobranchs: developmental and endocrinological aspects. In: Sharks, skates, and rays, Perry W. Gilbert, Robert F. Mathewson, and David P. Rall, editors, Johns Hopkins Univ. Press, 553-580.

fish - elasmobranch, anat.-physiol.

Clem, L. William, and Francis De Boutaud, 1967.
Studies on elasmobranch immunoglobulins. In: Sharks, skates, and rays, Perry W. Gilbert, Robert F. Mathewson and David P. Rall, editors, Johns Hopkins Univ. Press, 581-591.

fish - elasmobranchs

Goldstein, Leon, 1967.
Urea biosynthesis in elasmobranchs. In: Sharks, skates, and rays, Perry W. Gilbert, Robert F. Mathewson and David P. Rall, editors, Johns Hopkins Univ. Press, 207-214.

fish - elasmobranch, anat.-physiol.

Grundfest, Harry, 1967.
Comparative physiology of electric organs of elasmobranch fishes. In: Sharks, skates, and rays, Perry W. Gilbert, Robert F. Mathewson and David P. Rall, editors, Johns Hopkins Univ. Press, 399-432.

fish - elasmobranchs, anat.-physiol.

Hamasaki, D.I., C.D.B. Bridges and Kathleen A. Meneghini, 1967.
The electroretinogram of three species of elasmobranchs. In: Sharks, skates, and rays, Perry W. Gilbert, Robert F. Mathewson and David P. Rall, editors, Johns Hopkins Univ. Press, 447-463.

fish - elasmobranchs, anat.-physiol.

Rasmussen, Lois E., and Reinhold A. Rasmussen, 1967.
Comparative protein and enzyme profiles of the cerebrospinal fluid, extradural fluid, nervous tissue, and sera of elasmobranchs. In: Sharks, skates, and rays, Perry W. Gilbert, Robert F. Mathewson, and David P. Rall, editors, Johns Hopkins Univ. Press, 361-379.

fish - elasmobranch

Schaeffer, Bobb, 1967.
Comments on elasmobranch evolution. In: Sharks, skates and rays, Perry W. Gilbert, Robert F. Mathewson and David P. Rall, editors, Johns Hopkins Univ. Press, 3-35.

elvers

flat fish including plaice + flounder

flat-fish

Moiseev, P.A., 1946
The Flat-fish Fishery of the Far East (Promislovie Kambali Dalnego Vostoka)
60 pp.

fish, flat fish

Shelbourne, J.E., 1953.
The feeding habits of plaice post-larvae in the Southern Bight. J.M.B.A. 32(1):149-159, 1 pl., 2 textfigs.

flying fishes

flying fishes

Beklemishev, C.W., and F.A. Pasternak, 1960
[Census of flying fishes in the Atlantic and the problem of evaluation of the productivity of tropical waters.]
Voprosy Ichthyologii (14): 71-77.

flying fish

Bruun, A. Fr., 1936.
Flyvefisk of "Flyvefisk". En Oversigt over Flyvevnens Forekomst og Mekanik indenfor Fiskene
Naturs Verden 20:337-356, 17 textfigs.

fish, flying fish

Imai, Sadahiko, 1957.
[The characteristics and keys for the determination of the species of flying-fishes obtained in Japan.] Mem. Fac. Fish., Kagoshima Univ., 5:91-102.

fishery, flying fish

Kojima, Shumpei, 1969.
Exploitation of flying-fish fishery in Japan Sea. (In Japanese; English abstract). Bull. Jap. Soc. fish. Oceanogr. Spec. No. (Prof. Uda Commem. Pap.): 287-289.

flying fish

Lewis, J.B., J.K. Brundritt and A.G. Fish, 1962.
The biology of the flying fish Hirundichthys affinis (Gunther).
Bull. Mar. Sci., Gulf and Caribbean, 12(1):73-94

flying fish
Parin, N. V., 1960.
Distribution of flying fish (Exocoetidae) in the western and central Pacific. Trudy Inst. Okean. 41:153-162.

flying fish
Parin, N.V., and L.N. Besednov, 1965.
Flying fishes (Oxyporhamphidae and Exocoetidae) of the Gulf of Tonkin. (In Russian; English abstract).
Trudy Inst. Okeanol., Akad. Nauk, SSSR, 80:104-117.

fish - flying fish
Plomley, N.J.B. 1968.
Numbers of flying fish observed during three voyages across the Indian Ocean.
J. Zool. Lond. 155 (1): 111-129.

Flying fishes
Shuntov, V.P., 1965.
Distribution of flying fishes in the Gulf of Tonkin in accordance with oceanographic factors. (In Russian ; English abstract)
Trudy Inst. Okeanol., Akad. Nauk, SSSR, 80:118-123.

hagfish

hagfish
Jensen, David, 1966.
The hagfish.
Scientific American, 214(2):82-90.

lampreys
*Tesch, Friedrich-Wilhelm, 1967.
Aktivität und Verhalten wandernder Lampetra fluviatilis, Lota lota und Anguilla anguilla.
Helgoländer wiss. Meeresunters., 16(½):92-111.

haddock broods

haddock broods
Carruthers, J.N., A.L. Lawford, V.F.C. Veley, B.B. Parrish, 1951.
Variations in brood-strength in the North Sea haddock, in the light of relevant wind conditions
Nature 168(4269):317-319.

herring

fisheries, herring
Balls, R., 1951.
Environmental changes in herring behaviour: a theory of light avoidance, as suggested by echo-sounding observations in the North Sea.
J. du Cons. 17(3):274-298, 4 textfigs.

fishery, herring
Ball, R., 1948.
Herring fishing with the echometer. J. du Cons. 15(2):193-206.

herring
Carruthers, J.N. and W.C. Hodgson, 1937
Similar fluctuations in the herrings of the East Anglian autumn fishery and certain physical conditions. Rapp. Cons. Expl. Mer. 105 pt3(4):10-13.

fisheries - herring
Cheng, C., 1941
Ecological relations between the herring and the plankton off the north-east coast of England.
Hull Bull. Mar. Ecol., 1(5):239-254, 8 textfigs.

herring fisheries
*Cushing, D.H., 1968.
The Downs stock of herring during the period 1955-1966.
J.Cons.perm.int.Explor.Mer, 32(2):262-269.

herring
Drughinin, A.D., 1957.
Материалы по биологии анивской сельди.
Izv. Tichookean. N.I. Inst. Rib. Chozhiaistva + Okean, 44:13-35
[Material on the biology of herring]

herring
Einarsson, H., 1949
Calanus and herring at N. E. Iceland.
Anal. Biol., Int. Cons., 4:28; text fig. 11.

herring fishery
Erdmann, W., 1937.
Die Deutsche grosse Herringsfischerei im Jahre 1936. Ber. Deutsch. Wiss. Komm. Meeresforsch., n.f., 8(3):180-200, 4 textfigs., 6 charts with 6 overlays.

herring
Fisher, L.R., and Z.D. Hosking, 1962.
Vitamin A and fat in the herring (Clupea harengus L.) and in its food.
Scotland, Dept. Agric. Fish., Marine Research, (4):35 pp.

fisheries - herring
Hardy, A. C., 1926.
The herring in relation to its animate environment:Pt, II.
Min. Agric. Fish., Fish. Invest., ser. ii, VIII(7).

fisheries - herring
Hardy, A. C., 1924
The herring in relation to its animate environment. Pt. 1. Min. Agric. Fish., Fish. Invest. Ser.II,VII(1).

fisheries - herring
Hardy, A. C., G. T. D. Henderson, C. E. Lucas, and J. H. Fraser, 1936.
The ecological relations between the herring and the plankton investigated with the plankton indicator. JMBA, ns, XXI:147-291

fish, herring
Iles, T.D., 1971.
Growth studies on North Sea herring.III. The growth of East Anglian herring during the adult stage of the life history for the years 1940 to 1967. J. Cons. int. Explor. Mer, 33(3): 386-420.

fisheries - herring
Jespersen, P., 1928.
Investigations on the food of the herring in Danish waters. Medd. Komm. Havund., Ser. Plankton No. 2

fish - herring
Ketels, E., 1955.
Die norwegische Heringsaison und ihre Bedeutung für die Seeschiffahrt. Der. Seewarte 16(3): 79-81.

fisheries, herring
Manteifel, B.P. 1941
Plankton i seld v Barentsovom More (Plankton and herring in the Barents Sea). Poliarnyi N.-I. Institut Morskovo Rybnovo Khoziaistva i Okeanografii Trudy (Repts. Polar Res. Inst. Sea Fish. and Oceanogr. Murmansk) 7(3):125-218, 21 figs.

herring
Parrish, B.B., and A. Saville, 1967.
Changes in the fisheries of North Sea and Atlanto-Scandian herring stocks and their causes.
Oceanogr. Mar. Biol., Ann.Rev., H. Barnes, editor, George Allen and Unwin, Ltd., 5:409-447.

fisheries - herring
Pearcey, F. G., 1885.
Investigations on the movements and food of the herring and additions to the marine fauna of the Shetland Islands. Proc. Roy. Soc., Edinburgh, VIII:389-415

herring
Pokrovskaia, I.S., 1957
Питание личинок сахалинской сельди.
Izv. Tichookean. N.I. Inst. Rib. Choz. + Okean. 44: 39-56
[The feeding of the larvae of herring]
copepods, lists of spp.
plankton

herring
Probatov, A.N., ~~1957~~ + M.A. Darda, 1957
Биологическая характеристика нерестовой сельди острова Кунашир. Izv. Tichookean. N.I. Inst. Ribnogo Chozhiaistva + Okean. 44: 1-11
[Biological characteristics of some herring of the island of Kunashir.]

fisheries - herring
Savage, R. E., 1937
The food of the North Sea herring, 1930-1934.
Min. Agric. Fish., Fish. Invest., ser. ii, XV(5)

FISHERIES HERRING
Savage, R. E., 1931.
The relation between the feeding of the herring off the East Coast of England and the plankton of the surrounding waters. Min. Agric. Fish., Fish. Invest., Ser. ii, XII(4).

fisheries- herring
Savage, R. E., 1930
The influence of Phaeocystis on the migrations of the herring. Min. Agric. Fish., Fish. Invest., Ser. ii, XII(2)

FISHERIES herring
Savage, R. E. and A. C. Hardy, 1935
Phytoplankton and the herring: Pt. 1.
Min. Agric. Fish., Fish. Invest., ser. ii, XIV(2)

FISHERIES HERRING
Savage, R. E., and R. S. Wimpenny, 1936
Phytoplankton and the herring: Pt.
Min. Agric. Fish., Fish. Invest., ser. ii XV(1)

herring
Sindermann, C.J., 1962.
Serology of Atlantic clupeoid fishes.
American Naturalist, 96(889):225-231.

(Paper presented at 10th Pacific Science Congress).

~~fisheries~~ fisheries
Siromiatnikova, M.G., 1957.
Причины омыления слабосоленой сельди и мероприятия по борьбе с ним. Izv. Tichookean. Nauch.- Issle. Inst. Rib. Choz. i Okean. 44: 179-195
The source of saponification of weak herring

fish-fisheries-herring
Steele, J.H., 1961.
The environment of a herring fishery.
Marine Res., Dept. Agric. Fish., Scotland, 6:19 pp.

herring
Sund, O., 1941.
The fat and small herring on the coast of Norway in 1940. Ann. Biol. 1:58-72.

herring
Yudanov, I.G., 1962.
Herring of the northwest Atlantic.
Sovetskie Riboch. Issledov. v Severo-Zapadnoi Atlant. Okeana, VNIRO-PINRO, Moskva, 368-375.

In Russian; English summary

fish, lantern fish
Nafpaktitis, Basil G., 1968.
Taxonomy and distribution of the lanternfishes, genera Lobianchia and Diaphus, in the North Atlantic.
Dana Report No. 73: 131 pp.

Macrourids

mackerel

fisheries, mackerel
Bullen, G.E., 1908
Plankton studies in relation to the western mackerel fishery. JMBA (N.S) 8:269-303.

fish, mackerel
Iizuka, Keiki, 1971.
On the ecology of young mackerel in the northeastern sea of Japan. 1. Studies of the different broods in same year-class. (In Japanese; English abstract). Bull. Tohku reg. Fish. Res. Lab. 31: 97-108.

mackerel fisheries
Kawasaki, Tsuyoshi, 1969.
Ecology and resources of the Pacific population of the Japanese mackerel (Pneumatophorus japonicus). (In Japanese; English abstract). Bull. Jap. Soc. fish. Oceanogr. Spec. No. (Prof. Uda Commem. Pap.): 249-252.

fish, mackerel
Kim, Wan Soo, 1970.
Studies on the Spanish mackerel populations 1. Age determination. (In Korean; English abstract).
J. oceanogr. Soc., Korea, 5(1): 37-40.

fish mackerel
Kramer, D., 1960.
Development of eggs and larvae of Pacific mackerel and distribution and abundance of larvae.
USFWS Fish. Bull., 60(174):383-438.

fish - mackerel
*Sato, Yuji, 1968.
On the distribution and abundance of the mackerel population migrating to the northern waters of north-eastern Sea of Japan during 1963-1967 fishing seasons. (In Japanese: English abstract).
Bull. Tohoku reg. Fish. Res. Lab., 28:73-115.

fish - mackerel
*Sato Yuji, 1968.
On the relationship between the mackerel populations inhabiting the north eastern sea area of Japan and the north Japan Sea area through Tsugau Strait.
Bull. Tohoku reg. Fish. Res. Lab., 28:51-71.

fish - mackerel
*Sato, Yuji, Keiki Iizuka and Katsumi Kotaki, 1968.
Some biological aspects of the mackerel, Pneumatophorus japonicus (Houttuyn), in the northeastern Sea of Japan. (In Japanese; English abstract).
Bull. Tohoku reg. Fish. Res. Lab., 28:1-50.

fish - mackerel
Sette, O. E., and A.W.H. Needler, 1934.
Statistics of the mackerel fishery off the east coast of North America 1804-1930. U.S. Dept. Commerce, Bur. Fish. Invest. Rept. 19:1-48, 6 textfigs.

fish, mackerel
Usami, Shuzo, 1970.
Mode of life of the adult mackerel in the sea off Kanto District - IV. Characters of aggregation and environmental conditions (2). (In Japanese; English abstract). Bull. Tokai reg. Fish. Res. Lab. 63: 29-60.

fish, mackerel
Usami, Shuzo, 1970.
Mode of life of the adult mackerel in the sea off Kanto District III. Variance in body size composition by grade of sexual maturity. (In Japanese; English abstract). Bull. Tokai reg. Fish. Res. Lab. 63: 17-28.

fish, mackerel
Usami, Shuzo, 1969.
Mode of life of the adult mackerel in the sea off Kanto district. II. Characters of aggregation and environmental condition. (In Japanese; English abstract)
Bull. Tokai Reg. Fish. Res. Lab. 60: 9-27

marlin

fish - marlin
Ishii, Takeo, 1967.
Biometrical studies on the variability of population density of tuna and marlin species caught by tuna longline in the South Pacific.
Bull. Ocean Res. Inst. Tokyo, (2): 81 pp.

fish, marlin
Koga, Shigeyuki 1968.
A study of the fishing conditions of the tuna and marlin in the Tasman Sea. (In Japanese; English abstract).
J. Shimonoseki Univ. Fish. 16(2/3): 51-70.

fish, marlin
Koga, Shigeyuki 1967.
Studies on the fishery biology of the tuna and marlin in the Indian Ocean and the South Pacific Ocean. (In Japanese; English abstract)
J. Shimonoseki Univ. Fish. 15(2): 49-256.

fish - marlin
Koga, Shigeyuki, 1966.
The relationship between length and weight of tun-as and striped marlin in the South Pacific Ocean (In Japanese; English abstract).
Bull. Fac. Fish. Nagasaki Univ., No. 21:23-31.

marlin
Nakagome, Jun, 1963
Variation of catch tonnage per one operation for tuna and marlin in equatorial Atlantic Ocean from 1961 to 1962. (In Japanese; English abstract)
Kanagawa Prefect. Fish. Exp. Sta., No. 12: 8 pp.

fish, marlin
Smith, J.L.B. 1956.
Swordfish, marlin and sailfish in South and East Africa.
Ichthyol. Bull. Rhodes Univ. 2: 34 pp., 2 pls.

menhaden

menhaden
Gunter, G., and J.Y. Christmas, 1960.
A review of literature on menhaden with special reference to the Gulf of Mexico menhaden.
U.S.F.W.S. Spec. Sci. Rept., Fish. No. 363:1-31

fish, menhaden
Turner, William R., 1969.
Life history of menhadens in the eastern Gulf of Mexico.
Trans. Am. Fish. Soc. 98(2): 216-224

fish morays, anat.
*Holl, A., E. Schulte und W. Meinel, 1970.
Funktionelle Morphologie des Geruchsorgans und Histologie der Kopfanhänge der Nasenmuräne Rhinomuraena ambonensis (Teleostei, Anguilliformes).
Helgoländer wiss. Meeresunters, 21(1/2):103-123.

pilchard

fish - pilchard
Blackburn, M., 1950.
Age, growth, and life history of the pilchard in N.S.W. C.S.I.R.O. Bull. 242:

fish - pilchard
Sette, O. E., and E.H. Ahlstrom, 1948.
Estimations of abundance of the eggs of the Pacific pilchard (Sardinops caerulea) off Southern California during 1940 and 1941.
J. Mar. Res. 7(3):511-542, 4 textfigs.

fisheries - pilchard
Went, A.E.J., 1946
The Irish Pilchard Fishery. Proc. Roy. Irish Acad., 51, Sect. B, No. 5:81-120, 6 textfigs.

plaice
Yanulov, K.P., 1962.
On age and growth of American plaice in the northwest Atlantic.
Sovetskie Riboch. Issledov. v Severo-Zapadnoi Atlant. Okeana, VNIRO-PINRO, Moskva, 361-366.

In Russian; English summary

rays

fish rays
Chirichigno, F., Norma, 1963
Nuevas especies de "rayas" para la fauna del Perú.
Servicio de Pesquería, Perú, Serie de Divulgación Cientifica, No. 20: 13 pp.

fish rays
Fourmanon, P., 1963.
Raies et requins-scie de la côte ouest de Madagascar.
Trav. Centre. Océanogr., Nosy-Bé, Cahiers, O.R.S.T.R.O.M., Océanogr., Paris, No. 6:33-58.

fish - rays, manta
*Gilbert, H.H., and L.J. Paul, 1969.
First record of manta rays off New Zealand.
N.Z. Jl mar. Freshwat. Res. 3(2):339-342.

fish, rays
Holden, M.J., D.W. Rout and C.N. Humphreys, 1971.
The rate of egg laying by three species of ray.
J. Cons. int. Explor. Mer, 33(3): 335-339.

fish - rays
Kalmijn, A.J., 1966.
Electro-perception in sharks and rays.
Nature, Lond., 212(5067):1232-1233.

fish rays
Krefft, Gerhard, 1968.
Knorpelfische (Chondrichthyes) aus dem tropischen Ostatlantik.
Atlantide Rept. 10: 33-76

fish rays
Scott E.O.G. 1969.
Notes on some fishes collected in Tasmanian waters by the Umitaka maru in January 1965. I. Sharks and rays.
Tasman. Fish. Res. 3(2): 11-16

fish - rays, anat.
von Wahlert, Gerd., 1966.
Biologie und Evolution der Atemwege bei Haien und Rochen.
Veröff. Inst. Meeresforsch., Bremerh., Sonderband II:337-355.

fish - rays
Wallace, J.H., 1964.
Investigations on skates and rays.
South African Assoc., Mar. Biol. Res. Bull., No. 5:30-37.

redfish

redfish
Kashintsev, M.L., 1962.
Some notes on redfish in the Newfoundland area.
Sovetskie Ribock. Issledov. v Severo-Zapadnoi Atlant. Okeana, VNIRO-PINRO, Moskva, 263-271.
In Russian, English summary

fish, sailfish
Smith, J.L.B. 1956.
Swordfish, marlin and sailfish in South and East Africa.
Ichthyol. Bull. Rhodes Univ. 2: 34 pp., 2 pls.

fish sailfish
Williams, F., 1970.
The sport fishery for sailfish at Malindi, Kenya, 1958-1968, with some biological notes.
Bull. mar. Sci., Miami, 20(4): 830-852.

Salmon

salmon fisheries
Hanamura, Nobuhiko, 1969.
Reproduction, growth and maturity of sockeye salmon in the northern North Pacific. (In Japanese; English abstract). Bull. Jap. Soc. fish. Oceanogr. Spec. No. (Prof. Uda Commem. Pap.): 253-261.

fish salmon
Huntsman, A.G., 1965.
The ecotology of Margaree salmon.
Limnol. and Oceanogr., Redfield Vol., Suppl. to 10:R137-R147.

fisheries salmon
Mattson, C.R., 1962.
Chum salmon resources of Alaska from Bristol Bay to Point Hope.
U.S.F.W.S. Spec. Sci. Rept., Fish., No. 425:22 pp

fisheries, salmon
Neave, F., 1961.
Pacific salmon ocean stocks and fishery developments.
Proc. Ninth Pacific Sci. Congr., Pacific Sci. Assoc., 1957, Fish., 10:59-62.

fish - salmon
Royce, William F., Lynwood S. Smith and Allan C. Hartt, 1965.
Models of oceanic migrations of Pacific salmon and comments on guidance mechanisms
Fish. Bull. U.S. Fish Wildl. Serv., 66(3): 441-462

sardine

fish - sardines
Ahlstrom, Elbert H., 1967.
Co-occurrences of sardine and anchovy larvae in the California Current region off California and Baja California.
Rep. Calif. Coop. Oceanic Fish. Invest. 11:117-135.

fish - sardines
Aldebert, Yvonne, et Henri Tournier 1971.
La reproduction de la sardine et de l'anchois dans le Golfe du Lion.
Rev. Trav. Inst. Pêches marit. 35(1): 57-75.

fisheries, sardine
Bardan E., F. de P. Navarro, and D. Rodriguez, 1949.
Nuevos datos sobre la sardina del Mar de Alboran (Agosto 1948 a Marzo 1949). Bol. Inst. Espanol Ocean., No. 17:11 pp.

fish - sardines
Baron, J.C., 1968.
Étude préliminaire sur le sang de deux espèces de sardinelles (Sardinella aurita C.V., Sardinella eba C.V.).
Document scient. provisoire, Centre Océanogr., Côte d'Ivoire, ORSTOM, 029:48 pp. (mimeographed).

fish sardines
Botros, G.A, A.M. El-Maghraby and I.A.M. Soliman, 1970.
Biometric studies on Sardinella maderensis Lowe and Sardinella aurita Cuv. & Val. from the Mediterranean Sea at Alexandria (U.A.R.). Bull. Inst. Oceanogr. Fish., Cairo, 1: 85-128.

fish - sardines
California Academy of Sciences
California Division of Fish and Game
Scripps Institution of Oceanography
U. S. Fish and Wildlife Service
1950
California Cooperative Sardine Research Program.
Progress Rept. 1950:54 pp., 37 text figs.

fish - sardines
dos Santos, E.P., and Finn M. Fratzen, 1965.
Growth of sardines - quantitative aspects.
Anais Acad. bras. Cienc., 37(Supl.):360-362.

fish sardines
El-Maghraby, A.M., G.A. Botros and I.A.M. Soliman, 1970.
Age and growth studies on Sardinella maderensis Lowe and Sardinella aurita Cuv. and Val. from the Mediterranean Sea at Alexandria (U.A.R.).
Bull. Inst. Oceanogr. Fish., Cairo, 1: 49-82.

fish - sardines
Fonteneau, A. et E.G. Marchal, 1970.
Récolte, stockage et traitement des données statistiques relatives à la pêche des sardiniers (filet tournant) en Côte d'Ivoire. Doc. Scient. Centre Rech. océanogr. Abidjan, 1(1): 21-30.

fish - sardines
Furnestin, J., et M.-L. Furnestin, 1970.
La sardine marocaine et sa pêche. Migrations trophiques et génétiques en relation avec l'hydrologie et le plancton.
Rapp. P.-v. Réun. Cons. int. Explor. Mer 159: 165-175.

fish - sardines
Ghéno, Y. et J.C. LeGuen, 1968.
Détermination de l'age et croissance de Sardinella eba (Val.) dans la région de Pointe-Noire. Cah. ORSTOM, sér. Océanogr., 6(2): 69-82.

fish - sardines
Ghéno, Y. et F. Poinsard, 1968.
Observations sur les jeunes sardinelles de la Baie de Pointe-Noire (Congo). Cah. ORSTOM, sér. Océanogr., 6(2): 53-67.

fish sardines
*Grubišić, Fabjan, 1968.
Sardines (Sardina pilchardus Walb) fishing localities along the Yugoslave coast, their distribution and characteristics. (In Jugoslavian; English abstract).
Posebna Izdanja, Inst. Oceanogr. Ribarst., Split, 24 pp.

fish sardines
Grubisic, Fabjan, 1964.
Sardine fishing grounds along the Yugoslav Adriatic coast, their distribution and characteristics. (In Jugoslavian; English resume).
Acta Adriatica, 11(1):103-110.

fish - sardine fishing
Kusmorskaya, A.P., 1946
Destruction of plankton by fishes grazing upon it and the importance of this phenomenon for estimating the Biomass. CR (Doklady) Acad. Sci. URSS LIII (7):665-

fish sardines (growth of)
Haugen, C.W., 1969
Crecimiento y edad de la sardinia (Sardinella sp.) de las costas nor-orientales de Venezuela.
Memoria Soc. Cienc. nat. La Salle, 29(82): 72-83.

fish - sardines

Kačič, Ivo, 1969.
Some observations on behaviour of sardine in natural habitat. (Jugoslavian and Italian abstracts).
Thalassia Jugoslavica 5:141-147

fish - sardines

Karlovac, Jožica, 1969.
La ponte de la sardine, Sardinia pichardus Walb., en Adriatic moyenne, à l'époque de son maximum, au cours de quatre saisons de recherches (Jugoslavian and Italian abstracts).
Thalassia Jugoslavica, 5:149-157

fish, sardines

L'Herron, Roger 1971.
Étude biologique de la sardine du Golfe de Gascogne et du Plateau Celtique.
Revue Trav. Inst. Pêches marit. 35(4):455-473.

fisheries, sardine

Marchal, E.G., 1967.
La pêche des sardinières ivoriens en 1966.
Document sci. provisoire, Centre Rech. océanogr. Côte d'Ivoire, ORSTOM, No. 019:26 pp. (mimeographed).

fish - sardines

Matta, Francesco, 1964.
Contributo allo studio della morfologia e della biologia della sardina della Manica.
Boll. Pesca, Piscicolt. Idrobiol., n.s., 19(2): 219-248.

fish - sardines

Muzinic, Radosna, 1964.
Fluctuations in the size of the sardine and in the yield of its fishery in the central eastern Adriatic. (In Jugoslavian; English resume).
Acta Adriatica, 11(1):215-218.

fish - sardines

Nakai, Zinziro, 1958.
Fluctuations in the fishery for young sardine off Shichiri-mehama, Kii Peninsula. One of the data relevant to fluctuations of the sardine stock in Japanese waters.
Rec. Oceanogr. Wks., Japan, (Spec. No. 2):128-141.

fish sardines

Oliver, M., and F. Navarro, 1952.
La alacha y la sardina de Baleares. Investigaciones en 1950 y 1951. Bol. Inst. Español Ocean. No. 58:49 pp., 9 textfigs.

fish, sardines

Rafail, S.Z., 1970.
Studies of populations and exploitation status of Egyptian Red Sea abundant sardines. Bull. Inst. Oceanogr. Fish., U.A.R. 1: 129-148.

fish, sardines

Raja, B.T. Antony, 1969.
The Indian oil sardine.
Bull. Cent. mar. Fish. Res. Inst. India. 16: 128 pp.

fish - sardines

*Raja, B.T. Antony, and Yoshio Hiyama, 1969.
Studies on the systematics and biometrics of a few Indo-Pacific sardines.
Rec. oceanogr. Wks. Japan, 10(1):75-103.

fish - sardines

Soliman, I.A.M., G.A. Botros, and A.M. El-Maghraby, 1970.
Length-weight relationship and coefficient of condition for Sardinella maderensis Lowe and Sardinella aurita Cuv. & Vol. from the Mediterranean Sea at Alexandria (U.A.R.). Bull. Inst. Oceanogr. Fish., Cairo, 1: 27-45.

fish sardines

Spain, Instituto Español de Oceanografia, 1962.
Variacion estacional en la composicion de la sardina mediterránea y atlantica.
Bol. Inst. Español Oceanogr., 107:12 pp.

fish - sardines

Zei, M., 1969.
Sardines and related species of the eastern tropical Atlantic.
Actes Symp. Oceanogr. Ressources halieut. Atlant. trop., Abidjan, 20-28 Oct. 1966, UNESCO, 101-108.

Fisheries (Saury)

Han, Hi Soo, and Yeong Gong, 1965.
Relation between oceanographical conditions and catch of the saury in the Eastern Sea of Korea. (In Korean; English abstract).
Bull. Fish. Res. Develop. Agency, Korea, 3:45-56

sea horses

fish - sea horse

Bellomy, Mildred D. 1968.
The sea horse.
Pacific Discovery, Calif. Acad. Sci. 21(3): 18-22.

shark

See also under 'Fish genera' Loxodon, Centroscyllium, Galeocerdo, Negaprion

sharks

Ali, S.A., and L. Rahman, 1962.
A new method for refining shark liver oil on a commercial scale and recovery of vitamin 'A' by partial saponification for production of standard quality as well as higher vitamin 'A' potency oil.
Indo-Pacific Fish. Council, FAO, Proc., 9th Session, (2/3):113-114.

fish - sharks

Anon. 1968
Shark watch in the Gulf Stream.
Ocean Industry 3(9):9.

fish - sharks

Applegate, Shelton P., 1967.
A survey of shark hard parts. In: Sharks, skates and rays, Perry W. Gilbert, Robert F. Mathewson and David P. Rall, editors, Johns Hopkins Univ. Press, 37-67.

sharks

Applegate, Shelton P., 1965.
Tooth terminology and variation in sharks with special reference to the sand shark, Carcharias taurus Rafinesque.
Contrib. in Sci., Los Angeles County Mus., No. 86 :18 pp.

sharks

Aronson, Lester R., 1963.
The central nervous system of sharks and bony fishes with special reference to sensory and integrative systems.
Ch. 6 in: Sharks and Survival, P.W. Gilbert, Editor, D.C. Heath Co., Boston, 165-241.

sharks

Backus, Richard H., 1963.
Hearing in elasmobranchs.
Ch. 7 in: Sharks and Survival, P.W. Gilbert, Editor, D.C. Heath Co., Boston, 243-254.

fish sharks

Baldridge, H. David, 1970.
Sinking factors and average densities of Florida sharks as functions of liver buoyancy.
Copeia, 1970(4):744-754

fish sharks

Baldridge, H. David, Jr., 1969.
Kinetics of onset of responses by sharks to waterborn drugs. Bull. mar. Sci., 19(4): 880-896.

fish, sharks

Bane, Gilbert W. 1968.
The great blue shark
Calif. Currents 1(1):3-4

fish - sharks

*Beck, B., and A.W. Mansfield, 1969.
Observations on the Greenland shark, Somniosus microcephalus, in northern Baffin Island.
J. Fish. Res. Bd., Can., 26(1):143-145.

fish Sharks

Besednov, L.N., 1969.
Fishes of the Tonkin Gulf. 1. Sharks (in Russian).
Izv. Tichookean. nauchno-issled. Inst. ribn. Khoz. okeanogr. (TINRO) 66: 138 pp.

fish - sharks

Bigelow, H.B., W.C. Schroeder, and S. Springer, 1953.
New and little known sharks from the Atlantic and from the Gulf of Mexico. Bull. M.C.Z. 109(3):213-276, 10 textfigs.

fish - sharks

Blanc, Maurice, 1961.
Les poissons des Terres Australes et Antarctiques Francaises.
Mem. Inst. Sci., Madagascar, (7) 4:109-157.

sharks

Boeseman, M., 1956.
Fresh-water sawfishes and sharks in Netherlands New Guinea. Science 123(3189):222-223.

sharks

Buffenbarger, H.D., 1963.
Specifications for a desirable shark repellant.
Ch. 16 in: Sharks and Survival, P.W. Gilbert, Editor, D.C. Heath Co., Boston, 453-454.

sharks
Bullis, Harvey R., Jr., 1961.
Observations on the feeding behavior of white-tip sharks on schooling fishes.
Ecology, 42(1):194-195.

sharks
Cardenat, Jean, 1963
Notes d'ichtyologie ouest-africaine XXXIX. Notes sur les requins de la famille des Carchariidae et formes apparentées de l'Atlantique ouest-africain (avec la description d'une espèce nouvelle: Pseudocarcharias pelagicus, classée dans un sous-genre nouveau.)
Bull. Inst. Français Afrique Noire, (A),25(2) 526-537, 9 figs.

sharks
Casey, John G., 1967.
Sharks - rogue or resource.
Maritimes, Univ. R.I., 11(4):10-13. (popular)

fish, sharks
Castro Aguirre, José Luis, 1965.
Primer registro de los elasmobranzuios en aguas mexicanas.
Anales Inst.nac.Invest.biol-pesqueras, México, 1: 157-167.

fish, sharks
Chan, W.L. 1966.
New sharks from the South China Sea.
J. Zool. 148(1): 218-237.

sharks
de Sylva, Donald P., 1967.
Stingers, biters- and divers.
Sea Frontiers, Miami, 13(6)355-365. (popular).

sharks
Casey, John G., 1964.
Anglers' guide to sharks for the northeastern United States, Maine to Chesapeake Bay.
Bureau of Sport Fisheries, Wildlife Circular, No. 179:132 pp.
32 pp.

sharks
Chirichigno F., Norma, 1963.
Nuevos tiburones para la fauna del Perú.
Servicio de Pesquería, Perú, Serie de Divulgación Cientifica, No. 19:20 pp.

sharks
Clark, Eugenie, 1963.
Massive aggregations of large rays and sharks in and near Sarasota, Florida.
Zoologica, N.Y., 48(2):61-64.

Also in:
Collected Papers, Cape Haze Mar. Lab., Sarasota, 1957-1963, Vol. 1.

sharks
Clark Eugenie, 1963.
The maintenance of sharks in captivity, with a report on their instrumental conditioning.
Ch. 4 in: Sharks and Survival, P.W. Gilbert, Editor, D.C. Heath Co., Boston, 116-149.

sharks
Clark, Eugenie, 1961.
Visual discrimination in lemon sharks. (Abstract).
Tenth Pacific Sci. Congr., Honolulu, 21 Aug.-6 Sept., 1961, Abstracts of Symposium Papers, 175-176.

sharks
Clark, Eugenie, and Katherine von Schmidt, 1965.
Sharks of the central Gulf coast of Florida.
Bull. Mar. Sci., 15(1):13-83.

sharks
D'Aubrey, Jeannette D., 1964.
Preliminary guide to the sharks found off the east coast of South Africa.
S. Afr. Assoc., Mar.Biol. Res., Oceanogr.Res. Inst., Investigational Rept., No. 8:95 pp.

Also in: Collected Reprints, Int.Indian Ocean. Exped., UNESCO, 3. 1966.

sharks
Davies, David H., 1961.
Factors affecting the predacious behavior of sharks in South Africa. (Abstract).
Tenth Pacific Sci. Congr., Honolulu, 21 Aug.-6 Sept., 1961, Abstracts of Symposium Papers, 176.

sharks
Davies, David H., 1961.
Shark research in Natal.
Marine Studies off the Natal Coast, C.S.I.R. Symposium, No. S2:81-88. (Multilithed).

sharks
Davies, David H., Eugenie Clark, Albert L. Tester and Perry W. Gilbert, 1963.
Facilities for the experimental investigation of sharks.
Ch. 5 in: Sharks and Survival, P.W. Gilbert, Editor, D.C. Heath Co., Boston, 151-162.

sharks
Deraniyagala, P.E.P., 1958
A new record of whale sharks from the Eastern Province of Ceylon. Spolia Zeylanica, Ceylon, 28(2): 127-

sharks, behavior of
Essapian, F.S., 1962
Notes on the behavior of sharks in captivity.
Copeia, 2: 457-459.

Sharks
Fitch, John E., 1966.
Fishes and other marine organisms taken during deep trawling off Santa Catalina Island March 3 - 4, 1962.
Calif. Fish Game, 52(3):216-219.

sharks
Fourmanon, P., 1963.
Raies et requins-scie de la côte ouest de Madagascar.
Trav. Centre Océanogr., Nosy-Bé, Cahiers, O.R.S.T.R.O.M., Océanogr., Paris, No. 6:33-58.

sharks
Fourmanoir, P., 1961.
Liste complémentaire des poissons du Canal de Mozambique.
Mém. Inst. Sci., Madagascar, (7) 4:83-107.

sharks
Fourmanoir, P., 1961.
Requins de la côte ouest de Madagascar.
Mém. Inst. Sci., Madagascar, (7)4:1-72.

sharks
Galler, S.R., 1960.
ONR's shark research program.
Naval Research Reviews, Oct., 12-14.

sharks (white)
Garrick, J.A.F., 1964.
Additional information on the morphology of an embryo white shark.
Proc. U.S. Nat. Mus., Smithsonian Inst., 115(3476):1-8.

sharks
Garrick, J.A.F., and Leonard P. Schultz, 1963.
A guide to the kinds of potentially dangerous sharks. Ch. 1 in: Sharks and Survival, P.W. Gilbert, Editor, D.C. Heath Co., Boston, 3-60.

sharks
Gilbert, Perry W., 1968.
The shark: barbarian and benefactor.
Bio Science 18(10):946-950. (non-technical).

sharks
Gilbert, Perry W., Editor, 1963.
Sharks and Survival.
D.C. Heath Co., Boston, 578 pp.

sharks
Gilbert, Perry W., 1963.
The AIBS Shark Research Panel. Ch. 22 in: Sharks and Survival, P.W. Gilbert, Editor, D.C. Heath Co., Boston, 505-507.

sharks
Gilbert, Perry W., 1963.
The visual apparatus of sharks.
Ch. 9 in: Sharks and survival, P.W. Gilbert, Editor, D.C. Heath Co., Boston, 283-326.

sharks
Gilbert, Perry W., 1962
The behavior of sharks.
Scientific American 207(1):60-81.

sharks
Gilbert, Perry W., 1961.
The visual apparatus of sharks and its probable role in predation. (Abstract).
Tenth Pacific Sci. Congr., Honolulu, 21 Aug.-6 Sept., 1961, Abstracts of Symposium Papers, 176-177.

sharks
Gilbert, P.W., 1960.
The SHARK RESEARCH PANEL.
A.I.B.S. Bull., 10(1):19-20.

sharks
Gilbert, Perry W., Edward S. Hodgson and Robert F. Mathewson, 1964.
Electroencephalogram of sharks.
Science, 145(3635):949-951.

sharks
Gilbert, P.W., and H. Kritzler, 1960.
Experimental shark pens at the Lerner Marine Laboratory.
Science, 132(3424):424.

sharks
Gilbert, Perry W., Robert F. Mathewson and David P. Rall, editors, 1967.
Sharks, skates and rays.
Johns Hopkins Univ. Press, 624 pp.

sharks
Gilbert, Perry W., and Stewart Springer, 1963.
Testing shark repellants.
Ch. 19 in: Sharks and survival, P.W. Gilbert, Editor, D.C. Heath Co., Boston, 477-494.

sharks
Gipp, S.K., & A.P. Kuznetzov, 1961
[On the age of the shark teeth (Carcharodon megalodon) found in the recent deposits of the Atlantic Ocean.]
Okeanologia, 2: 305-307.

sharks
Gohar, H.A.F., and F.M. Mazhar, 1964.
The elasmobranchs of the north-western Red Sea.
Publ. Mar. Biol. Sta Ghardaqa, No. 13:1-144, 16 pls.

sharks
Granier, J., 1964.
Les Euselaciens dans le Golfe d'Aigues-Mortes.
Bull. Mus. Hist. Nat., Marseille, 24:33-52.

fish - sharks
Greenberg, Jerry, 1968.
Fishmen fear-sharks.
Seahawk Press, $2.00 (not seen).

sharks
Gudger, E.W., 1952.
Northernmost record of the whale shark. Science 116(3016):432-433.

fish
Sharks
Halstead, Bruce W., 1970.
Poisonous and venomous marine animals of the world. 3. Vertebrates continued.
U.S. Govt. Printing Office, 1006 pp.

sharks
Herre, A.W.C.T., 1955.
Sharks in fresh water. Science 122(3166):417.

sharks
Hobson, Edmund S., 1963
Feeding behavior in three species of sharks.
Pacific Science, 17(2):171-194.

shark industry
Howell Rivero, L., 1950.
La industrie del Tiburon. Bol. Hist. Nat., Soc. "Filipe Poey", 1(2):51-60.

sharks
Jensen, Ad. S., 1914.
The selachians of Greenland. Mindeskrift Japetus Steenstrup 30:32 pp.

dogfish
Jensen, Albert C. 1966.
Life history of the spiny dogfish.
Fishery Bull. Fish Wildl. Serv.U.S., 65(3):527-554.

sharks
Kalmijn, A.J., 1966.
Electro-perception in sharks and rays.
Nature, Lond., 212(5067):1232-1233.

fish
sharks
Krefft, Gerhard, 1968.
Knorpelfische (Chondrichthyes) aus dem tropischen Ostatlantik.
Atlantide Rept. 10: 33-76

fish, sharks
Ledoux, Jean-Claude 1970.
Les dents des squalidés de la Méditerranée occidentale et de l'Atlantique nord-ouest Africain.
Vie Milieu (A) 21(2):309-361

fish - sharks
*Lee, Bo Heng, 1967.
Observation on shark long line fishery and its resources. (In Korea; English abstract).
Repts Fish. Resources, Fish. Res. Develop. Agency, Korea, 7:51-62.

fishes, sharks (chemistry)
Levine, Philip T., Melvin J. Glimcher, Jerome M. Seyer, James I. Huddleston and John W. Hein, 1966.
Noncollagenous nature of the proteins of shark enamel.
Science, 154(3753):1192-1193.

sharks
Limbaugh, Conrad, 1963 (posthumous).
Field notes on sharks. Ch. 2 in: Sharks and Survival, P.W. Gilbert, Editor, D.C. Heath Co., Boston, 63-94.

sharks
Maeda, Hiroshi, 1963.
Distribution pattern of sharks along setline.
Bull. Jap. Soc. Sci. Fish., 29(11):996-999.

fish - sharks
Lineaweaver, Thomas H., III, and Richard H. Backus 1970
The natural history of sharks.
J.P. Lippincott Company Philadelphia and New York and André Deutsch Ltd., England, 256 pp. $6.95, £2.75

sharks
Maul, G.E., 1955.
Five species of rare sharks for Madeira including two new to science.
Notulae Naturae, Acad. Nat. Sci., Phila., No. 279:1-13.

Carcharias ferox
C. noronhai n.sp.
Carcharhinus longimanus
Centrophorus machiquensis n.sp.
Somniosus rostratus

fish - sharks
McLaughlin, Robert H., and A. Kenneth O'Gower 1970.
Underwater tagging of the Port-Jackson shark, Heterodontus portusjacksoni (Meyer).
Bull. Inst. océanogr. Monaco 69 (1410): 11 pp.

sharks
McCormick, Harold W., and Tom Allen, with CAPT William E. Young, 1963.
Shadows in the sea: the sharks, skates and rays.
Chilton Books, Div. Chilton Publishers, 415 pp. $10.00.

sharks (nos.)
McGary, James W., and Joseph J. Graham, 1960.
Biological and oceanographic observations in the central north Pacific, July-September 1950.
U.S.F.W.S. Spec. Sci. Rept., Fish., No. 358: 107 pp.

fish, sharks
McLaughlin, R.H., and A.K. O'Gower 1971
Life history and underwater studies of a heterodont shark.
Ecol. Monogr. 41 (4): 271-289.

sharks
Mihara, Tsunetoshi, y Agustin Brito Leon 1970
Observaciones sobre la pesca del tiburon con palangre de fondo y la del pargo a cordel en el oriente de Venezuela.
Informe tecn., Proyecto Invest. Desarrollo pesq. MAC-PNUD-FAO, 4: 15 pp.

fish, sharks
Mikawa, Masao, 1971.
On the feeding habits of some demersal sharks. (In Japanese; English abstract). Bull. Tohku reg. Fish. Res. Lab., 31: 109-124.

fish, sharks
Nakaya, Kazuhiro, 1971.
Descriptive notes on a porbeagle, Lamna masus, from Argentine waters, compared with the North Pacific salmon shark, Lamna ditropis. Bull. Hokkaido Univ. Fac. Fish. 21(4): 269-279.

sharks
Nelson, Donald R., and Samuel H. Gruber, 1963.
Sharks: attraction by low-frequency sounds.
Science, 142(3594):975-977.

sharks
Nelson, Donald R., and Richard H. Johnson, 1970.
Diel activity rhythms in the nocturnal, bottom-dwelling sharks, Heterodontus francisci and Cephaloscyllium ventriosum.
Copeia, 1970 (4): 732-739

sharks
Olsen, A.M., 1961.
Environmental and behavioral factors which influence the migration of sharks. (Abstract).
Tenth Pacific Sci. Congr., Honolulu, 21 Aug.-6 Sept., 1961, Abstracts of Symposium Papers, 178.

sharks
Orr, R.T., 1959.
Sharks as enemies of sea otters.
J. Mammal., 40(4):617.

sharks
Randall, John E., 1963.
Dangerous sharks of the western Atlantic.
Ch. 11 in: Sharks and Survival, P.W. Gilbert, Editor, D.C. Heath Co., Boston, 339-361.

sharks
Rapson, A.M., 1961.
Food of some tropical predaceous fish including sharks, from net and line fishing records. (Abstract).
Tenth Pacific Sci. Congr., Honolulu, 21 Aug.-6 Sept., 1961, Abstracts of Symposium Papers, 178-180.

sharks
Sadowsky, Victor, 1965.
The hammerhead sharks of the littoral zone of São Paulo, Brazil, with the description of a new species.
Bull. Mar. Sci., 15(1):1-12.

sharks
Schultz, Leopard P., 1961.
Factors affecting the predaceous behavior of sharks in relation to the activities of man. (Abstract).
Tenth Pacific Sci. Cong., Honolulu, 21 Aug.-6 Sept., 1961, Abstracts of Symposium Papers, 180.

fish sharks
Scott, E.O.G. 1969.
Notes on some fishes collected in Tasmanian waters by the Umitaka maru in January 1968. I. Sharks and rays.
Tasman. Fish. Res. 3(2) 11-16

fish - sharks
Shimma, H. and Y. Shimma, 1970.
Comparative studies on shark liver oils from Suruga Bay. (In Japanese; English Abstract).
Bull. Tokai reg. Fish. Res. Lab., 59: 101-110.

sharks
Sivasubrananian, K., 1963.
On the sharks and other undesirable species caught by tuna longline.
Rec. Oceanogr. Wks., Japan, n.s., 7(1):73-81.

fish - sharks
Smith, J.L.B. 1967.
A new squaloid shark from South Africa with notes on the rare Atractophorus armatus Gilchrist.
Occ. Pap., Dept. Ichthyol., Rhodes Univ., Grahamstown, 11:117-136.

sharks
Smith, J.L.B., 1957.
A new shark from South Africa.
South African J. Sci., 53(10):261-264.

sharks
South African Association for Marine Biological Research, 1960.
Bulletin, No. 1:26 pp.

sharks
Springer, Stewart, 1963.
Field observations on large sharks of the Florida-Caribbean region.
Ch. 3 in: Sharks and Survival, P.W. Gilbert, Editor, D.C. Heath Co., Boston, 95-113.

sharks
Springer, Stewart, 1961.
Some environmental factors affecting the feeding behavior of sharks. (Abstract).
Tenth Pacific Sci. Congr., Honolulu, 21 Aug.-6 Sept., 1961, Abstracts of Symposium Papers, 180-181.

sharks
Springer, Stewart, and Perry W. Gilbert, 1963.
Anti-shark measures.
Ch. 18 in: Sharks and Survival, P.W. Gilbert, Editor, D.C. Heath Co., Boston, 465-476.

fish-sharks
* Springer, Stewart, and Richard A. Waller, 1969.
Hexanchus vitulus, a new sixgill shark from the Bahamas. Bull. Mar. Sci., 19(1): 159-174.

sharks
Springer, Victor G., and J.A.F. Garrick, 1964.
A survey of vertebral numbers in sharks.
Proc. U.S. Nat. Mus., 116(3496):73-96.

sharks
Steffens, F.E., and J.D. D'Aubrey, 1967.
Regression analysis as an aid to shark taxonomy.
Investl. Rep. Oceanogr. Res. Inst. S. Afr. 18:16 pp.

sharks
Templeman, Wilfred, 1963.
Distribution of sharks in the Canadian Atlantic (with special reference to Newfoundland waters).
Bull. Fish. Res. Bd., Canada, (140):77 pp.

sharks
Tester, Albert L., 1963.
Olfaction, gustation and the chemical sense organs in sharks.
Ch. 8 in: Sharks and Survival, P.W. Gilbert, Editor, D.C. Heath Co., Boston, 255-282.

sharks
Tester, Albert L., 1963
The role of olfaction in shark predation.
Pacific Science, 17(2):145-170.

fish - shark
Tester, Albert L., 1961.
The role of olfaction in shark predation. (Abstract).
Tenth Pacific Sci. Congr., Honolulu, 21 Aug.-6 Sept., 1961, Abstracts of Symposium Papers, 181-182.

sharks
Tuve, Richard L., 1963.
Development of the U.S. Navy "Shark Chaser" chemical shark repellant.
Ch. 17 in: Sharks and Survival, P.W. Gilbert, Editor, D.C. Heath Co., Boston, 455-464.

sharks
Wheeler, J.F.G., 1962.
Notes on three common species of sharks in the Mauritius-Seychelles area.
Proc. R. Soc., Arts & Sci., Mauritius, 2(2):146-160.

Also in: Int. Indian Ocean Exped., Coll. Rept., UNESCO, 3. 1966.

sharks
Wiles, James, 1962.
Wanted: shark hunters.
Sea Frontiers, 8(2):84-93.

sharks, anat.-physiol.

fish, sharks (anat.-physiol)
Aronson, Lester R., Frederick R. Aronson and Eugenie Clark 1967.
Instrumental conditioning and light-dark discrimination in young nurse sharks.
Bull. mar. Sci., Miami 18(2): 249-256.

fish - sharks, anat-physiol
Best, A.C.G., and J.A.C. Nicol, 1967.
Reflecting cells of the elasmobranch tapetum lucidum.
Contrib. mar. Sci., Port Aransas, 12: 172-201.

fish sharks, anat.-physiol.
Bering, Edgar A., Jr., David P. Rall and W. Walter Oppelt, 1967.
Hydrodynamics of the cerebral ventricular fluid in the shark. In: Sharks, skates, and rays Perry W. Gilbert, Robert F. Mathewson and David P. Rall, editors, Johns Hopkins Univ. Press, 389-398.

sharks, physiology-anatomy
Denton, E.J., 1962
Some recently discovered buoyancy mechanisms in marine animals.
Proc. R. Soc., London, (A), 265(1322):366-370.

fish - sharks, anat.-physiol.
Dixon, Robert L., Richard H. Adamson and David P. Rall, 1967.
Metabolims of drugs by elasmobranch fishes.
In: Sharks, skates, and rays, Perry W. Gilbert, Robert F. Mathewson and David P. Rall, editors, Johns Hopkins Univ. Press, 547-552.

fish sharks (anat.)
Ebbeson, S.O.E. and J.S. Ramsey, 1968.
The optic tracts of two species of sharks (Galeocerdo cuvier and Ginglymostoma cirratum.
Brain Research, 8: 36-53. Also In: Contrib Univ. Puerto Rico, Dept. mar. Sci. 8 (1968-1969).

fish sharks
Gilbert, Perry W., 1963.
Advice to those who frequent or find themselves in, shark-infested waters.
Ch. 21 in: Sharks and survival, P.W. Gilbert, Editor, D.C. Heath Co., Boston, 501-503.

sharks, physiol.
Gilbert, Perry W., and Steven D. Douglas, 1963
Electrocardiographic studies of free-swimming sharks.
Science, 140(3574):1396.

sharks, anat.
Gohar, H.A.F., and F.M. Mazhar, 1964.
The internal anatomy of Selachii of the north western Red Sea.
Publ. Mar. Biol. Sta Ghardaqa, No. 13:145-240.

fish - sharks anat.-physiol.
Hodgson, E.S., R.F. Mathewson and P.W. Gilbert, 1967.
Electroencephalographic studies of chemoreception in sharks. In: Sharks, skates, and rays, Perry W. Gilbert, Robert F. Mathewson and David P. Rall, editors, Johns Hopkins Univ. Press, 491-501.

sharks, anat.-phys.
James, W.W., 1952.
The mechanism of the succession of teeth in the elasmobranchs. Proc. Zool. Soc., London, 122:540.

This was a film demonstration.
The title is merely a paragraph describing situation in a general way.

fish - sharks, anat. physiol.
Katsuki, Yasuji and Toru Hashimoto, 1969.
Shark pit organs: enhancement of mechanosensitivy by potassium ions. Science, 166(3910): 1287-1289.

fish - sharks, anat.-physiol.
Klatzo, Igor, 1967.
Cellular morphology of the lemon shark brain.
In: Sharks, skates, and rays, Perry W. Gilbert, Robert F. Mathewson and David P. Rall, editors, Johns Hopkins Univ. Press, 341-359.

fish - sharks, anat. physiol.
Kuchnow, Karl P., and Perry W. Gilbert, 1967. Preliminary in vivo studies on pupillary and tapetal pigment responses in the lemon shark, Negaprion brevirostris. IN: Sharks, skates, and rays, Perry W. Gilbert, Robert F. Mathewson and David P. Rall, editors, Johns Hopkins Univ. Press 465-477.

fish - sharks, anat.-physiol.
Maren, Thomas H., 1967. Special body fluids of the elasmobranch. In: Sharks, skates, and rays, Perry W. Gilbert, Robert F. Mathewson and David P. Rall, editors, Johns Hopkins Univ. Press, 287-292.

fish, sharks (anat.-physiol.)
Meurling, Patrick 1967. The vascularization of the pituitary in elasmobranchs. Sarsia 28:1-104.

fish- sharks, anat.-physiol.
Oppelt, W.W., 1967. The control of ventricular-fluid production in the shark. In: Sharks, skates, and rays, Perry W. Gilbert, Robert F. Mathewson and David P. Rall, editors, Johns Hopkins Univ. Press, 381-388.

fish. sharks. anat.- physiol.
Sigel, M. Michael, and Annie R. Beasley, 1967. Biological aspects of immunization in sharks. In: Sharks, skates, and rays, Perry W. Gilbert, Robert F. Mathewson, and David P. Rall, editors, Johns Hopkins Univ. Press, 5930610.

fish - sharks, anat-physiol.
* Stockem, W., H. Komnick und K.E. Wohlfarth-Bottermann, 1968. Zur Cytologie der Rectaldrüsen von Knorpelfischen. II Die Mikromorphologie des zentralen Sammel kanels. Helgoländer wiss. Meeresunters. 18(4): 424-452.

fish, sharks (anat.physiol.)
Tester, Albert L., and Gareth J. Nelson 1967. Free neuromasts (pit organs) in sharks. In: Sharks Skates and Rays, Perry W. Gilbert, Robert F. Mathewson and David P. Rall, editors, Johns Hopkins Univ. Press, 503-531.

fish - sharks, anat.-physiol.
Thorson, Thomas B., 1967. Osmoregulation in fresh-water elasmobranchs. In: Sharks, skates and rays, Perry W. Gilbert, Robert F. Mathewson and David P. Rall, editors, Johns Hopkins Univ. Press, 265-270.

sharks, physiol.
Thorson, Thomas B., 1962 Partitioning of body fluids in the Lake Nicaragua shark and three marine sharks. Science, 138(3541):688-690.

sharks, anat.
von Wahlert, Gerd, 1966. Biologie und Evolution der Atemwege bei Haien und Rochen. Veröff. Inst. Meeresforsch., Bremerh., Sonderband II:337-355.

sharks, physiol.
Wisby, W.J., J.D. Richard, D.R. Nelson and S.H. Gruber, 1964 Sound perception in elasmobranchs. Marine Bio-Acoustics. Proc. Symp., Bimini, Apr. 1963, Pergamon Press, 255-267.

shark attacks

sharks, culture of

fish- sharks, culture of
Agalides, Eugene, 1966. Synthetic seawater for sharks. GeoMar. Techn., 2(7):11-15.

fish sharks, freshwater
Thorson, Thomas B., 1971. Movement of bull sharks Carcharhinus leucas, between Caribbean Sea and Lake Nicaragua demonstrated by tagging. Copeia 1971(2): 336-338

shark attack
Anon., 1966. A defense against shark attack. Lightweight screen developed to protect downed pilots. Naval Res.Rev., 19(11):12-17.

fish shark attacks and Joy Williams
Baldridge, H. David, Jr. 1969. Shark attack: feeding or fighting. Military Medicine 134(2): 130-133

fish shark attacks
Baldridge, H. David, and C. Scott Johnson 1969. Antishark measures Naval Res. Rev., April 1969: 15-22.

shark attacks
Bini, Giorgio, 1960 Attacco documentato di pescecane (Charcharoden charcharias). Boll. Pesca, Piscicolt. e Idrobiol., n.s., 15(1):136-139.

shark bites
Caldwell, Melba C., David K. Caldwell and J.B. Sibenaler, 1965. Observations on captive and wild Atlantic bottle-nosed dolphins, Tursiops truncatus in the northeastern Gulf of Mexico. Contrib. in Sci., Los Angeles County Mus., No. 91:10 pp.

shark attacks
Campbell, G.D., David H. Davies and Arthur C. Copley, 1960 A case of shark attack with special reference to attempts to identify the causal species from the wounds. Medical Proc., Mediese Bydraes, S. Africa, 6(26): 612-621.

fish, shark attacks
Clark, Eugenie, 1960. Four shark attacks on the west coast of Florida, summer 1958. Copeia, 1960(1):63-67.
Also in: Collected Papers, Cape Haze Mar. Lab., Sarasota, 1957-1963, Vol. 1.

shark attack
Collier, Ralph S., 1964. Report on a recent shark attack off San Francisco, California. California Fish & Game, 50(4):261-264.

sharks, attacks of
Coppleson, V.M., 1963. Patterns of shark attacks of the world. Ch. 14 in: Sharks and Survival, P.W. Gilbert, Editor, D.C. Heath Co., Boston, 389-421.

fish shark attacks
Coppleson, V.M., 1959. Shark attack. Angus and Roberts, London.

fish shark attacks
Cropp, Ben 1969. Australia's meal of death. Oceans 2(5/6): 15-21 (popular)

shark attacks
D'Aubrey, J.D., and D.H. Davies, 1961. Shark attack off the east coast of South Africa, 1st February 1961. S. Afr., Ass. Mar. Biol. Res., Ocean. Res. Inst. Invest. Rept., 5:3 pp.

shark attacks
Davies, D.H., 1963. Shark attack and its relationship to temperature, beach patronage and the seasonal abundance of dangerous sharks. South African Assoc. Mar. Biol. Res., Oceanogr. Res. Inst. Invest. Rept., No. 6:43 pp.

shark attacks
Davies, David H., 1961. Shark attack on fishing boat in South Africa. S. Afr. Ass., Mar. Biol. Res., Ocean. Res. Inst. Invest. Rept., 1:2 pp.

shark attacks
Davies, D.H., 1961. Shark attack off the east coast of South Africa, 22 January 1961. S. Afr. Ass., Mar. Biol. Res. Ocean. Res. Inst. Invest. Rept., 4:4 pp.

shark attack
Davies, D.H., and J.D. D'Aubrey, 1961. Shark attack off the east coast of South Africa, 24 December 1960, with notes on the species of shark responsible for the attack. S. Afr. Ass., Mar. Biol. Res., Ocean. Res. Inst. Invest. Rept., 2:8 pp.

shark attacks
Davies, D.H., and J.D. D'Aubrey, 1961. Shark attack off the east coast of Africa, 6 January 1961. S. Afr., Ass. Mar. Biol. Res. Oceano. Res. Inst. Invest. Rept., 3:5 pp.

fish shark attacks
Davies, D.H., and J.D. D'Aubrey, 1961 Shark attacks off the east coast of South Africa. S. Africa Assoc. Mar. Biol. Res., Ocean. Res. Inst., Inv. Repts., 1-5: 21 pp.

shark attack
DeWitt, J.W., 1955. A record of an attack by a leopard shark (Triakis semifasciata Girard). Calif. Fish & Game 41(4):348-351.

shark attacks
Eibl-Eibesfeldt, I., and H. Hass, 1959 Erfahrungen mit Haien. Zeits. Tierpsych., 16(6):733-746, 13 pls.

shark attack
Fast, T.N., 1955. Second known shark attack on a swimmer in Monterey Bay. Calif. Fish. & Game 41(4):348-351.

shark attacks
Fellows, David P., and A. Earl Murchison, 1967.
A non injurious attack by a small shark.
Pacif. Sci., 21(1):150-151.

fish shark attacks
Gilbert, Claire K., and Perry W. Gilbert, 1968.
Rare and common sharks off the New York coast.
Conservationist, N.Y. 22(4): 21-25

Gilbert, P.W., L.P. Schultz and S. Springer, 1960 — shark attacks
Shark attacks during 1959. The conditions under which sharks attack man suggest what measures may be taken to reduce risk. Science 132 (3423):323-327.

fish - shark atttack
Forman, T.B., and D.J. Durst an, 1967.
Report on an attack by a great white shark off Coledale Beach, N.S.W., Australia, in which both victim and attacker were recovered simultaneously.
Calif. Fish Game, 53(3): 219-223

shark attack
Gudger, E.W., 1950.
A boy attacked by a shark, July 25, 1936 in Buzzards Bay, Massachusetts with notes on attacks by another shark along the New Jersey coast in 1916. Am. Mid. Nat. 44(3):714-719, 2 textfigs.

sharks, attacks
Halstead, Bruce W., David H. Davies and George D. Campbell, 1963.
First aid treatment for shark bites. Ch. 20 in: Sharks and survival, P.W. Gilbert, Editor, D.C. Heath Co., Boston, 495-503.

fish, shark attacks
Herald, Earl S. 1968.
Size and aggressiveness of the sevengill shark (Notorhynchus maculatus).
Copeia 1968(2): 412-414.

shark attacks
Hess, Paul W., 1961.
Observations on sharks during oceanographic cruises.
Estuarine Bull., Univ. Delaware, 6(3/4):6-11.

fish - shark attacks
Hobson, E.S., F. Mouton and E.S. Reese, 1961.
Two shark incidents at Eniwetok Atoll, Marshall Islands.
Pacific Science, 15(4):605-609.

shark attack
Kean, B.H., 1944.
Death following attack by shark, Carcharodon carcharias. J. Amer. Med. Assoc., July 22, 125(12):845-846.

shark attacks
Lineaweaver, Thomas H. III and Richard H. Backus, 1969
Sharks and the discouragement thereof.
Sea Frontiers, 15(4): 194-206

sharks, attacks of
Llano, George A., 1963.
Open-ocean shark attacks.
Ch. 13 in: Sharks and Survival, P.W. Gilbert, Editor, D. C. Heath Co., Boston, 369-386.

fish - shark, attraction of
Myrberg, A.A., Jr., A. Banner and J.D. Richard, 1969.
Shark attraction using a video-acoustic system.
Marine Biol., 2(3): 264-276.

fish sharks, behavior of
Nelson, Donald R., Richard H. Johnson and Larry G. Waldrop, 1969.
Responses in Bahamian sharks and groupers to low frequency pulsed sounds.
Bull. S. Calif. Acad. Sci. 68(3): 131-137.

sharks, death caused by
Phelan, J.M., 1929.
Fatality from bite of shark. Mil. Surgeon 64:383.

sharks, deaths caused by
Radcliffe, L., 1930.
Youth killed by huge shark. Copeia 1930(3):89-90.

sharks, attacks by
Randall, John E., 1964.
A fatal attack by the shark Carcharhinus galapagensis at St. Thomas, Virgin Islands. (Abstract).
Rept., Assoc. Island Mar. Labs., Caribbean, Fifth Meet., Lerner Mar. Lab., Bimini, Nov. 608, 1963: 2.

shark attacks
Randall, John E., 1963.
A fatal shark attack by the shark Carcharhinus galapagensis at St. Thomas, Virgin Islands.
Caribbean J. Sci., 3(4):201-205.

fish, shark attacks
Schultz, Leonard 1969.
Predation of sharks on man.
Chesapeake Sci. 8(1): 52-62.

sharks, attacks
Schultz, Leonard P., 1963.
Attacks by sharks as related to the activities of man.
Ch. 15 in: Sharks and Survival, P.W. Gilbert, Editor, D.C. Heath Co., Boston, 425-452.

shark attacks
Schultz, L.P., P.W. Gilbert and S. Springer, 1961
Shark attacks.
Science, 134 (3472):87-88.

sharks, attacks
Schultz, Leonard P., and Marilyn H. Malin, 1963.
A list of shark attacks for the world. Appendix in: Sharks and Survival, P.W. Gilbert, Editor, D.C. Heath Co., Boston, 509-567.

shark attacks
Smith, D.E., 1966.
Electrical shark-barrier research.
GeoMar. Techn., 2(10):10-14.

sharks, attacks of
Smith, J.L.B., 1963.
Shark attacks in the South African seas.
Ch. 12 in: Sharks and Survival, P.W. Gilbert, Editor, B.C. Heath Co., Boston, 363-368.

shark attacks
Smith, J.L.B., 1958.
Shark attacks in South Africa.
S. African J. Sci., 54(6):150-152, p. 93.

shark attacks
Tester, Albert L., 1960
Fatal shark attack, Oahu, Hawaii, December 13, 1958.
Pacific Science, 14(2):181-184.

sharks, attacks of
Whitley, Gilbert P., 1963.
Shark attacks in Australia.
Ch. 10 in: Sharks and Survival, P.W. Gilbert, Editor, D.C. Heath Co., Boston, 329-338.

shark control
Davies, D.H., 1964.
The use of anti-coagulants as a means of controlling the shark populations.
South African Assoc., Mar. Biol. Res., Bull., No. 5:51-52.

fish - sharks, protection against
Baldridge, H. David, and C. Scott Johnson, 1969
Antishark measures
Navy Res. Rev. 23(3): 15-22 (popular)

SHARKS, lists of spp

sharks, lists of spp
Union of South Africa, Division of Fisheries, Department of Commerce and Industries, 1961.
Fisheries Research in Natal Waters.
Marine Studies off the Natal Coast, C.S.I.R. No. S2:89-117 (Multilithed).

Shark population

fish - shark populations
Springer, Stewart, 1967.
Social organization of shark populations. In: Sharks, skates and rays, Perry W. Gilbert, Robert F. Mathewson, and David P. Rall, editors, Johns Hopkins Univ. Press, 149-174.

Shark repellants

fish - shark repellants
Anon., 1965.
Shark repellers that don't.
GeoMarine Technology.

Sharks, shocking of

fish, sharks (shocking of)
Smith, E.D. 1967.
Shocking the sharks.
Sea Frontiers 13(2): 83-87.

Sharks, swimming of

fish - sharks, swimming of
Klausewitz, W., 1962.
Wie schwimmen Haifische?
Natur und Museum, 92(6):275-284.

Sharks (tagging)

fish, sharks (tagging)
Davies, David H., and Leonie S. Joubert (undated)
Tag evaluation and shark tagging in South African waters.
Investl Rep. Ocean. Res. Inst. Ass. Mar. Res. Un. SAfr. 12:36 pp.

fish, sharks (teeth)
Belyaev, G.M., and L.S. Glikman 1970.
On the geological age of teeth of the shark Megaselachus megalodon (Ag.). (In Russian; English abstract).
Trudy Inst. Okeanol. Akad. Nauk SSSR 88:277-280.

fish, sharks (teeth)
Belyaev, G.M., and L.S. Glikman 1970
Sharks teeth on the Pacific floor.
(In Russian; English abstract)
Trudy Inst. Okeanol. Akad. Nauk SSSR 88: 252-276.

Skates

fish Skates
Fitch, John E., 1966.
Fishes and other marine organisms taken during deep trawling off Santa Catalina Island March 3 - 4, 1962.
Calif. Fish Game, 52(3): 216-219.

skates
Templeman, Wilfred, 1965.
Rare skates of the Newfoundland and neighboring areas.
J. Fish. Res. Bd., Canada, 22(2):259-279.

fish - skates
Wallace, J.H., 1964.
Investigations on skates and rays.
South African Assoc., Mar. Biol. Res., Bull., No. 5:30-37.

SKIPJACK

fish, skipjack
Asano, Masahiro and Tomatsu Tanaka, 1971.
Studies on the maturation of the skipjack in the western Pacific Ocean - 1. (In Japanese; English abstract). Bull. Tohku reg. Fish. Res. Lab. 31: 153-161.

fish, skipjack
De Campos Rosado, J.M., 1971.
Long-term changes in abundance of yellowfin and skipjack off the Coast of Angola. J. Cons. int. Explor. Mer 34(1): 65-75.

fish - skipjack
*Kasahara, Kohei, and Tamotsu Tanaka, 1969.
On the fishing and the catches of skipjack tuna (Katsuwonus pelamis) in the southern waters.
Bull. Tohoku reg. Fish. Res. Lab., 28:117-136.

fish, skipjack
Kasahara, Kohei, Masakazu Yao, Akira Naganuma, Moriya Anraku and Masahiro Asano, 1971.
Studies on the movement of the skipjack in the Japanese waters by tagging experiments. (In Japanese; English abstract). Bull. Tohku reg. Fish. Res. Lab. 31: 141-152.

fish, skipjack
Kawasaki, Tsuyoshi, 1964.
Population structure and dynamics of skipjack in the North Pacific and its adjacent waters. (In Japanese; English abstract).
Bull. Tohoku Reg. Fish. Res. Lab., No. 24:28-47.

fish, skipjack fishery
Seckel, Gunter H., and Kenneth D. Waldron, 1960
Oceanography and the Hawaiian skipjack fishery.
Pacific Fisherman, February 1960: 3 pp.

fish - skipjack
Uno, Michio, 1965.
On the migration of skipjack shoals in the Pacific coastal region of Japan. (In Japanese; English summary).
J. Fac. Fish., prefect. Univ. Mie, 6(3):351-368.

Stingray

stingrays
Halstead, B.W., R.R. Ocampa and F.R. Modglin, 1955.
A study of the comparative anatomy of the venom apparatus of certain North American stingrays.
J. Morph. 97(1):1-22.

fish, swordfish
Smith, J.L.B., 1956.
Swordfish, marlin and sailfish in South and East Africa.
Rhodes Univ., Ichthyol. Bull., No. 2:34 pp., 2 pls.

fish - sting rays
Struhsaker, Paul, 1969.
Observations on the biology and distribution of the thorny stingray, Dasyatis centroura (Pisces: Dasyatidae). Bull. mar. Sci., 19(2): 456-481.

stingrays
Wilson, Peter C., and James S. Beckett, 1970.
Atlantic Ocean distribution of the pelagic stingray, Dasyatis violacea.
Copeia 1970(4):696-707.

fish swordfish
Tibbo, S.N., and L.M. Lauzier, 1969.
On the origin and distribution of larval swordfish Xiphius gladius L. in the western Atlantic.
Techn. Rept. Fish. Res. Bd., Can., 136: 20 pp.

tunas, anat.-physiol.

Tunas vertical distribution

fish - tunas, vertical distribution
*Osipov, V.G., 1968.
On the vertical distribution of the yellow fin (Neothunnus albacora) and bigeye (Parathunnus obesus) tunas. (In Russian; English abstract).
Zool. Zh. 47(8):1192-1196.

"fisheries" for various invertebrates and vertebrates other than fish.
See also: mammals, seals, etc.
cephalopods in "Organismal" index, etc.

"fisheries" shellfish (molluscs), clams, mussels, etc.
See also: "organismal" index

fishery - scallops
Bullis, Harvey R., Jr., & Robert M. Ingle, 1958.
A new fishery for scallops in western Florida.
Proc. Gulf and Caribbean Fish. Inst. 11th Ann. Sess., Nov., 1958:75-78.

shell fisheries
Corbeil, H. -E., 1949.
Travail sur les mollusques, Mya arenaria L. et Ostrea virginica (Gmelin). Sta. Biol. Saint-Laurent, 8th Rapport, App. 2:45-54.

fisheries scallop
Dow, Robert L., and F.T. Baird, Jr., 1960.
Scallop resource of the United States Passamoquoddy area.
U.S.F.W.S. Spec. Sci. Rept., Fish., No. 367: 1-9.

shellfish, chemical anal.
Etorma, S.T., 1928.
Proximate chemical analysis of some Philippine shellfish. Phil. Agric. 17:125-133.

shellfisheries
Jensen, E.T., 1963.
The shellfish research program of the Public Health Service.
Proc., 15th Ann. Sess., Gulf and Caribbean Fish. Inst., Univ. Miami, 59-65.

fishery - oysters
Preston, Alan, 1965.
The control of radioactive pollution in a North Sea oyster fishery.
Helgoländer wiss. Meeresunters. 17: 269-279.

fisheries - mussel
Scattergood, L.W., and C.C. Taylor, 1950.
The mussel resources of the North Atlantic region. Pt. 1. The survey to discover the locations and areas of the North Atlantic mussel-producing beds. Pt. 2. Observations on the biology and the methods of collecting and processing the mussel. Pt. 3. Development of the fishery and the possible need for conservation measures. Fishery Leaflet 364:33 pp., 7 figs. (multilith).

shellfish
Taylor, Donald M., 1967.
Billion dollar scallop find?
Ocean Industry, 2(12):20-24.

shellfisheries
Turner, H.J., jr., 1952.
Fifth report on investigations of the shellfisheries of Massachusetts. Comm. Mass., Dept. Conserv., Div. Mar. Fish.:1-32, 3 figs., 8 maps.

shellfish
Turner, H.J., jr., 1952.
Fourth report on investigations of the shellfisheries of Massachusetts. Comm. Mass., Dept. Conserv., Div. Mar. Fish., 21 pp., 2 textfigs.

shellfisheries
Turner, H.J., jr., and J.A. Posgay, 1953?
Sixth report on investigations of the shellfisheries of Massachusetts. Comm. Mass. Dept. Nat Res., Div. Mar. Fish:74 pp.

fisheries, molluscs
Wiborg, Kr. Fr., 1963.
Some observations on the Island scallop, Chlamys islandica (Müller) in Norwegian waters.
Fiskeridirect. Skrift., Ser. Havundersøgelser, 13(6):38-53.

"Fisheries" crustaceans (crabs, lobsters, shrimp, prawns, etc.)
See also: "organismal" index

fisheries, shrimp
Anderson, W.W., 1956.
January to April distribution of the common shrimp on the south Atlantic continental shelf. U.S.F.W.S., Spec. Sci. Rept., Fish., No. 171:14 pp.

fisheries, shrimp
Broad, C., 1951.
Results of shrimp research in North Carolina.
Proc. Gulf and Caribbean Fish. Inst., 3rd session,:27-35, 5 textfigs.

fisheries, shrimp
Burkenroad, M. D., 1949.
Occurrence and life histories of commercial shrimp. Science 110:688-689.

shrimp fishery
Butler, T.H., 1967.
Shrimp exploration and fishing in the Gulf of Alaska and Bering Sea.
Tech.Rep.,Fish.Res.Bd.,Can, 18:49 pp. (multilithed)

Carbonneau, Jean, 1966.
Pêche expérimentale au crabe avec le chalut à panneaux aux Iles-de-la-Madeleine en 1965.
Rapp. Stn. Biol. Mar., Grande-Rivière, 1965:137-139.

lobsters
Carbonneau, Jean, 1965.
Pêche expérimentale au homard à l'Île d'Anticosti en 1965.
Rapp. Stn.Biol. Mar., Grande-Rivière, 1965:129-

fisheries, decapod
Christmas, J.Y., and Gordon Gunter, 1967.
A summary of knowledge of shrimps of the genus Penaeus and the shrimp fishery in Mississippi waters.
Proc. Symp. Crustacea, Ernakulam, Jan. 12-15, 1965, 4: 1442-1447

fisheries, prawn
de Zylva, E.R.A., 1958
The prawn fisheries of Ceylon. Proc. Indo-Pacific Fish. Counc., Tokyo, 30 Sept.-14 Oct. 1955, Sect. 2-3 (6th Sess.): 324-327.

fisheries, prawn
Djajadiredja, R. Roestami and M. Sachlan, 1958
Shrimp and prawn fisheries in Indonesia with special reference to the Kroya District.
Proc. Indo-Pacific Fish. Counc., 30 Sept. - 14 Oct. 1955, Sect. 2-3 (6th Sess.): 366-377.

prawn fisheries
Domantay, J.S., 1958.
Prawn fisheries of the Philippines.
Proc. Indo-Pacific Fish. Counc., Tokyo, 30 Sept.-14 Oct. 1955, Sect. 2-3 (6th Session):362-366.

fisheries, decapods
Filewood, L.W.C., and D.A.S.F. Konedobu, 1964.
Commercial sea prawns in New Guinea.
Fisheries Newsletter, Australia, 23(1):24 pp.

fisheries, crab
Fischler, K.J., & C.H. Walburg, 1962
Blue crab movement in coastal South Carolina, 1958-59.
Trans. Amer. Fish. Soc., 91(3):275-278.

fisheries, decapod
George, M.J., 1961(1962).
Studies on the prawn fishery of Cochin and Alleppy coast.
Indian J. Fish., 8(1):75-95.

fisheries, decapod
George, P.C., and K. Ramesh Nayak, 1961(1962).
Observations on the crab fishery of Mangalore coast.
Indian J. Fish., 8(1):44-53.

fisheries, shrimp
Gurney, R., 1927.
Zoological results of the Cambridge Expedition to the Suez Canal. Appendix I to the report on the Crustacea Decapoda (Natantia and Anomura). Trans. Zool. Soc., London, 22:228-229.

prawn fishery
Hallgrimsson, Ingvar, 1961.
Um raekjumidaleit "Asbjarnar". (In Icelandic; English summary).
Atvinnudeild Haskolans Fiskideild, Reprint No. 3 pp.
(from AEGIR).

fisheries, shrimp
Hudinaga, Motosaku, and Mitsutake Miyamura, 1962
Breeding of the "Kuruma" prawn (Penaeus japonicus Bate). (In Japanese; English abstract).
J. Oceanogr. Soc., Japan, 20th Ann. Vol., 694-706.

fisheries, crab
Instituto de Pesca del Pacifico, Guaymas, Sonora, Mexico, 1949.
Informe de las actividades llevadas a cabo por el Instituto de Pesca del Pacifico de Septiembre de 1946 a Marzo de 1949. Rep. Biol. 1(1):48 pp., fig

fisheries, crustacea
Jones, S., 1967.
The crustacean fishery resources of India.
Proc. Symp. Crustacea, Ernakulam, Jan. 12-15, 1965, 4: 1328-1340

FISHERIES, shrimp
Krishna Menon, M., 1955.
Notes on the bionomics and fishery of the prawn Metapenaeus dobsoni Miers on the south-west coats of India. Indian J. Fish. 2(1):41-56.

fisheries, prawn (decapod)
Krishna Menon, M., and K. Raman, 1961(1962).
Observations on the Prawn fishery of the Cochin backwaters with special reference to the stake net catches.
Indian J. Fish., 8(1):1-23.

shrimp
Kutkuhn, Joseph H., 1966.
Dynamics of a penaeid shrimp population and management implications.
Fishery Bull., Fish Wildl. Serv., U.S., 65(2):313-338.

fisheries, decapod
Kutkuhn, J.H., 1962.
Gulf of Mexico commercial shrimp populations - trends and characteristics, 1956-59.
Fish. Bull., 212, U.S.F.W.S. Fish. Bull., 62:343-402.

Shrimp fishery
Lindner, Milton J., 1966.
What we know about shrimp size and the Tortugas fishery.
Proc. Inst. Mar. Sci., Gulf and Caribbean Fish. Inst., 18th Ann. Sess., 18-26.

fisheries, decapod
Miyamoto, H., and A.T. Shariff, 1961(1962).
Lobster fishery off the south-west coast of India. Anchor hook and trap fisheries.
Indian J. Fisheries, 8(1):252-268.

fisheries, shrimp
Neiva, G. de S., 1969.
Observações sôbre a pesca de camarões do litoral Centro-sul do Brasil.
Proc. Pesquisa, Rio de Janeiro, Brasil 2(1):1-16

fisheries, decapod
Radhakrishnan, N., 1969.
On the prawn resources of Karwar region.
Proc. Symp. Crustacea, Ernakulam, Jan. 12-15, 1965, 4: 1421-1436

fisheries, decapod
Ramamurthy, S., 1963.
A note on the prawn fishery of Kutch.
J. Mar. Biol. Assoc., India, 5(1):146-148.

fisheries, decapod
Robinson, Richard K., and Dolores E. Dimitriou, 1963.
The status of the Florida spiny lobster fishery, 1962-1963.
State of Florida, Bd. Conservation, Techn. Ser., (42):30 pp.

lobsters
Scattergood, L. W., 1949.
Translations of foreign literature concerning lobster culture and the early life history of the lobster. Spec. Sci. Repts., Fishery, No. 6: 173 pp. (Mimeographed).

fisheries, spiny lobster
Smith, F.G. Walton, 1958.
The spiny lobster industry of Florida.
Florida, Bd. Conserv., Educ. Ser., No. 11:1-34.

spiny lobster

Smith, F.G.W., 1951.
Caribbean spiny lobster investigations. Proc. Gulf and Caribbean Fish. Inst. 3rd session: 128-134, 4 textfigs.

---

shrimp fishery

Springer, S., and H.R. Bullis, 1952.
Exploratory shrimp fishing in the Gulf of Mexico, 1950-51. Fishery Leaflet 406:34 pp., 9 textfigs. (multilithed).

---

fisheries, lobster

Soares-Rebelo, D.J., 1964.
Spiny lobster industry in southern Africa. S. African J. Sci., 60(3):71-87.

---

fisheries, crab

*Tanoue, Toyotaka, Yukito Suito Takaya Ando, and Yasuhiro Yamagiri, 1967.
On the distribution and the fishing method of Portunus pelagicus in Kagoshima Bay. (In Japanese; English abstract).
Mem. Fac. Fish., Kagoshima Univ., 16:85-92.

---

lobster fisheries

Thomas, H.J., 1958.
Lobster and crab fisheries in Scotland.
Scottish Home Dept. Mar. Res., 1958 (8):107 pp.

---

fishery, shrimp

Tiews, K., 1954.
Einfluss der Gezeiten und der Wassertemperatur auf die Garnelenfischerei.
Ber. Deutschen Wiss. Komm. f. Meeresf., n.f., 13(3):270-282, 6 textfigs.

---

shrimp fisheries

U.S. Department of the Interior, Fish & Wildlife Service, 1958.
Survey of the U.S. shrimp industry, Vol. 1.
Spec. Sci. Rept., Fish., No. 277:1-311.

---

fisheries, lobster

Vincent-Cuaz, Louis, 1964.
Biometry and ecology of the spiny lobster of the Bight of Benin.
CSA Specialist Meeting on Crustaceans, Zanzibar, 19-26, Apr., 1964, Crustaceans (64) 7:37 pp. (mimeographed).

---

fishery, seal

Colman, J. S., 1949.
The Newfoundland seal fishery and the Second World War. J. Animal Ecology 18(1):40-46, 7 textfigs.

---

tuna

fish tunas

Alverson, Franklin G., 1963
The food of yellowfin and skipjack tunas in the eastern tropical Pacific Ocean. (In Spanish and English).
Inter-American Tropical Tuna Commission, Bull. 7(5):295-367 (English); 368-396 (Spanish).

---

fish tuna

Alverson, Franklin G., 1959.

Geographical Distribution of Yellowfin Tuna and Skipjack Catches from the eastern Tropical Pacific Ocean, by quarters of the year 1952-1955.
Inter-American Tropical Tuna Commission.
Bul. Vol. 3, No. 4, p. 167-213.

---

fisheries, tuna

Anonymous, 1963.
Deep sea long-lining is uneconomical at present.
Fish. Newsletter, Australia, 22(12):14.

---

fish - tuna

Barrett, Izadore, and Susumu Kume, 1965.
Observations on bigeye tuna caught in the surface tuna fishery in the eastern Pacific Ocean, 1951-1964.
Calif. Fish Game, 51(4):252-258.

---

fish - tuna

*Baudin Laurencin, F.G. et Marchal, (E.G.), 1968.
Contribution à l'étude biométrique de l'Albacore du Golfe de Guinée. Doc. Centre Rech. océanogr. Abidjan, 024: 22 pp.

---

fish, tuna

Baudin-Laurencin, F. et J.P. Rebert, 1970.
La pêche thonière à Abidjan de 1966 à 1969.
Doc. scient. Centre Rech. océanogr. Abidjan, 1(1): 37-61.

---

fish tuna

Beardsley, Grant L., Jr., 1969.
New Atlantic tuna fishery.
Sea Frontiers, 15(3): 152-159 (popular)

---

fish - tunas

Blackburn, M., 1965.
Oceanography and the ecology of tunas.
In: Oceanography and Marine Biology, H. Barnes, editor, George Allen and Unwin, 3:299-322.

---

fish, tuna

Blackburn, M., et al., 1962.
Tuna oceanography in the eastern Tropical Pacific
U.S.F.W.S. Spec. Sci. Rept., Fish., No. 400:48 pp

---

fisheries, tuna

Borodatov, V.A., 1968.
Tuna fishery in the western Indian Ocean. (In Russian).
Trudy, Vses. Nauchno-Issled. Inst. Morsk. Ribn. Okeanogr (VNIRO) 64, Trudy Azovo-Chernomorsk. Nauchno-Issled. Inst. Morsk. Ribn. Khoz. Okeanogr. (AscherNIRO), 28: 323-343.

---

fish, tuna

Broadhead, Gordon C., and Izadore Barrett, 1964.
Some factors affecting the distribution and apparent abundance of yellowfin and skipjack tuna in the eastern Pacific Ocean.
Inter-American Tropical Tuna Comm., Bull., 8(8): 419-473.

English and Spanish texts

---

fish, tuna

Brock, V.E., 1963.
Will tuna research change direction?
Proc. 15th Ann. Sess., Gulf and Caribbean Fish. Inst., Univ. Miami, 50-52.

---

fish, tuna

Calkins, Thomas P., 1963.
An examination of fluctuations in the "concentration index" of purse-seiners and baitboats in the fishery for tropical tunas in the eastern Pacific, 1951-1961.
Inter-Amer. Trop. Tuna Comm., 8(5):257-316.

(In Spanish and English).

---

fish tuna (migration)

Clemens, Harold B. and Glenn A. Flittner 1969.
Bluefin tuna migrate across the Pacific Ocean.
Calif. Fish Game 55(2): 132-135.

---

fish tuna

Davidoff, Edwin B., 1963.
Size and year class composition of catch age and growth of yellowfin tuna in the eastern tropical Pacific Ocean, 1951-1961. (Spanish and English).
Inter-American Trop. Tuna Comm., 8(4):201-251.

---

tuna

De Angelis, Costanzo M., 1959

Nota su alcuni campioni di plancton raccolti nel 1955 nello stagno di Mariut (Egitto). Boll. Pesca., Piscicult. e Idrobiol., n.s., 14(2): 206-229.

---

fish, yellowfin tuna

De Campos Rosado, J.M., 1971.
Long-term changes in abundance of yellowfin and skipjack off the Coast of Angola. J. Cons. int. Explor. Mer 34(1): 65-75.

---

fish tuna

de Jager, B. van D., C.S. de V. Nepgen and R.J. van Wyk, 1963.
A preliminary report on South African west coast tuna.
Rept. S. Africa, Dept. Comm. & Industr., Div. Sea Fish., Invest. Rept., No. 47:40 pp.

Also in:
Commerce & Industr., Nov., 1963.

---

fish - tuna

Della Croce, N., 1969.
Nota su un avvistamento di tonni nel mar Ligure.
Progr. Ricerca Risorse mar. Fondo mar. Comm. ital. Oceanogr. Consig. naz. Ricerche (B)(30): 1-11

---

fish Tunas

Fink, Bernard D., 1965.
A technique, and the equipment used, for tagging tunas caught by the pole and line method
Journal du Conseil, 29(3):335-339.

---

fish, tuna

Fink, Bernard D., 1965.
Estimacion de las tasas de mortalidad y otros parametros del atun aleta amarilla y del barrilete mediante experimentos de marcacion. (In Spanish and English).
Inter-American Tropical Tuna Comm. Bull., 10(1): 1-82.

---

tuna

Fink, Bernard D., and William H. Bayliff 1970.
Migrations of yellowfin and skipjack tuna in the eastern tropical Pacific Ocean as determined by tagging experiments, 1952-1964.
Bull. inter-Am. trop. tuna Comm. 15(1): 1-227

---

fish tunas

Fonseca, Jose Bonifacio G., e Silvio B. Moraes, 1963.
Conteúdo estomacal e evolução sexual dos atuns.
Sudene, Bol. Estudos, Pesca, Recife, Brasil, 3(9-10):3-6.

---

fish - tuna

Fujiishi, Akio, Satoru Tawara and Makoto Hirose, 1969.
Echo survey of tuna fishing ground in the Indian Ocean. (In Japanese; English abstract.
J. Shimonoseki Univ. Fish. 18(1): 18-25.

fisheries, tuna
*Griffiths, Raymond C., y Takeshi Nemoto, 1967.
Un estudio preliminar de la pesqueria para atun aleta amarilla y albacora en el mar Caribe y el oceano Atlantico occidental por palangreros de Venezuela.
Invest. Pesqueras, Ser. Recursos Explotación pesqueros, 1(6):209-371.

fisheries, tuna
Henmi, Tomio, 1964.
On the relation between the "horizontal tuna-long-line form" and the spot foretelling of the fishing condition in the Celebes. (In Japanese; English abstract).
Mem., Fac. Fish., Kagoshima Univ., 13:93-103.

tuna
Hickling, C.F., 1962.
Tuna.
Proc. R. Soc., London, (A), 265(1322):346-350.

fish - tunas
Higgins, Bruce E., 1970
Juvenile tunas collected by midwater trawling in Hawaiian waters, July - September 1967.
Trans. Am. Fish. Soc., 99(1): 60-69.

fish tunas
Hui-chong, Tan, Yukio Nose and Yoshio Hiyama, 1965.
Age determination and growth of yellowfin tuna, Thunnus albacares Bonnaterre by vertebrae.
Bull. Jap. Soc. Sci. Fish., 31(6):414-422.

fish, tuna
Hynd, J.S., 1965.
Southern bluefin tuna populations in south-west Australia.
Australian J. Mar. Freshwater Res., 16(1):25-32.

fish, tuna
Hynd, J.S., and D. Vaux, 1963.
Report of a survey for tuna in western Australian waters.
C.S.I.R.O., Div. Fish. Oceanogr. Rept., 37:105pp (multilithed).

fish tunas
*Inoue, Motoo, Ryohei Amano, Yukinobu Iwasaki, and Mitsuyoshi Aoki, 1967.
Ecology of various tunas under captivity. I. Preliminary rearing experiments. (In Japanese; English abstract).
J. Coll. mar. Sci. Techn., Tokai Univ., (2):197-209.

fish - tuna
Inoue, Motoo, and Yukinobu Iwasaki, 1969.
Movement of the thermal equator and the fishing grounds mainly for yellowfin tuna in the Indian Ocean. (In Japanese; English abstract).
Bull. Jap. Soc. scient. Fish. 35(10): 957-963

fish tuna
Inoue, Motoo, Yukinobu Iwasaki, Ryohei Amano, Mitsuyoshi Aoki and Minoru Yamauti, 1970
Ecology of various tunas under captivity - II. Behaviour of tuna shown against light and darkness. (In Japanese; English abstract).
J. Coll. mar. Sci. Technol., Tokai Univ., 4:53-58

fish - tuna
Ishii, Takeo, 1967.
Biometrical studies on the variability of population density of tuna and marlin species caught by tuna longline in the South Pacific.
Bull. Ocean Res. Inst. Tokyo, (2): 76 pp

fish - tunas, anatomy
Iwai, Tamotsu, and Izumi Nakamura, 1964.
Olfactory organs of tunas with special reference to their systematic significance.
Bull. Misaki Mar. Biol. Inst., Kyoto Univ., No. 7:1-8.

fish - tunas, anatomy
Iwai, Tamotsu, and Izumi Nakamura, 1964.
Branchial skeleton of the bluefin tuna with special reference to the gill rays.
Bull. Misaki Mar. Biol. Inst., Kyoto Univ., No. 6 :21-25.

fish - tunas
Iwai, Tamotsu, Izumi Nakamura and Kiyomatsu Matsubara, 1965.
Taxonomic study of the tunas. (In Japanese; English abstract).
Misaki Mar. Biol. Inst., Kyoto Univ., Spec. Rept. No. 2:51 pp.

fish tuna
Japan, Kanagawa Prefectural Fisheries Experiment Station, 1963
Tuna fishing survey of M/V Sagamimaru (No. 10 11,12 voyage). (In Japanese).
Kanagawa, Prefect. Fish. Exp. Sta., No. 9: 28 pp.

fish, tunas
Japan, Science Council, National Committee for IIOE, 1966.
General report of the participation of Japan in the International Indian Ocean Expedition.
Rec. Oceanogr. Wks., Japan, n.s. 8(2): 133 pp.

fish tuna
Joseph, J., and I. Barrett, 1963.
The schooling behaviour of Pacific yellowfin and skipjack tuna held in a bait well.
California Gish and Game, 49(1):55.

fish - tuna
Kawasaki, T., 1959.
On the structure of "fish school" of tunas.
Japan. J. Ecol., 9(1):52-54.

fish tuna
Kakagome, Jun, and Shigemichi Suzuki, 1962
Seasonal and annual variation of the hooking-rate and annual variation of the catch-quantity of tuna and marlin in the tropical Atlantic ocean. (In Japanese; English abstract)
Kanagawa Prefect. Fish. Exper. Sta., Jul. 1962, No. 8: 14 pp.

(On two other papers the names was spelled NAKAGOME.)

fish, tuna
Klawe, W.L., 1960.
Larval tunas from the Florida current.
Bull. Mar. Sci., Gulf and Caribbean, 10(2):227-233.

fish, tuna
Koga, Shigeyuki, 1968.
A study of the fishing condition of the tuna and marlin in the Tasman Sea. (In Japanese; English abstract).
J. Shimonoseki Univ. Fish., 16(2/3):51-70.

fish, tuna
Koga, Shigeyuki 1967.
Studies on the fishery biology of the tuna and marlin in the Indian Ocean and the South Pacific Ocean. (In Japanese; English abstract).
J. Shimonoseki Univ. Fish. 15(2):49-256.

fish, tuna
Koga, Shigeyuki, 1966.
The relationship between length and weight of tun-as and striped marlin in the South Pacific Ocean (In Japanese; English abstract).
Bull. Fac. Fish. Nagasaki Univ., No. 21:23-31.

fish - tuna
Konagaya, Shiro, Kazuoki Yamabe and Keishi Amano, 1969.
On body temperature of tunas at the time of haulage.
Bull. Jap. Soc. scient. Fish., 35(4): 410-416

fish tuna
Kume, Susumu, 1967.
Distribution and migration of bigeye tuna in the Pacific Ocean.
Rep. Nankai Reg. Fish. Res. Lab., 25: 75-80

fish - tuna
*Le Guen, J.C., F. Baudin-Laurencin et C. Champagnat, 1969.
Croissance de l'albacore (Thunnus albacares) dans les régions de Pointe-Noire et de Dakar.
Cah. O.R.S.T.O.M., ser. Océanogr., 7(1): 19-40.

fish tuna
Leyandekkers, Jean V., 1964.
Why do tuna migrate?
Fish. Newsletter, Dept. Primary Ind., Fish. Br., Australia, 24(12):23-29.

fish, tuna
Lozano, Fernando, 1964.
Aportaciones españolas durante 1963 a la oceanografía pura y aplicada, nacional e internacional.
Bol. Real Soc. Española, Hist. Nat., 62(1):99-113

tuna
Mather, Frank J., III, 1962.
Transatlantic migration of two large bluefin tuna.
J. du Conseil, 27(3):325-327.

fish - tuna migrations
Mather, Frank J., III, Martin R. Bartlett and James S. Beckett, 1967.
Transatlantic migrations of young bluefin tuna.
J. Fish. Res. Bd., Can., 24(9):1991-1997.

fish tuna
Murphy, Garth I., and Richard S. Shomura, 1958
Variations in yellowfin abundance in the Central Equatorial Pacific.
Proc. Ninth Pacific Sci. Congr., Pacific Sci. Assoc., 1957, 16(Oceanogr.):108-113.

fish tuna

Nakagome, Jun, 1967.
On the cause of annual variation of catch of yellowfin tuna in the tropical western Indian Ocean. IV. Relation between abundance of year class and surface water temperature in spawning and nursery area. (In Japanese; English abstract).
Bull. Jap. Soc. scient. Fish., 33(3):156-160

fisheries, tuna

Nakagome, Jun, 1964.
Relation between montnly variation of hooking xx rate of yellowfin tuna and montnly variation of surface water temperature and between annual variation of hooking-rate of the fish and annual variation of surface water temperature in the tropical Atlantic Ocean. (In Japanese; English abstract).
Bull. Jap. Soc. Sci. Fisn., 30(2):122-126.

fish tuna

Nakagome, Jun, 1963
Relation between annual variation of hooking-rate and age groups of yellowfin tuna in the tropical western Pacific.
Kanagawa Prefect. Fish. Exp. Sta., No. 14: 6 pp.
In Japanese; English abstract

fish tuna

Nakagome, Jun, 1963
Variation of catch tonnage per one operation for tuna and marlin in equatorial Atlantic Ocean from 1961 to 1962. (In Japanese; English abstract).
Kanagawa Prefect. Fish. Exp. Sta., No. 12: 8 pp.

fish - tuna

Nakagome, J., 1962.
Relation between annual variations of hooking-rate and age groups of yellow-fin tuna in the tropical western Pacific Ocean. 1. Annual variation of hooking-rate. (In Japanese; English summary).
Bull. Jap. Soc., Sci. Fish., 28(12):1164-1167.

fish - tuna

Nakagome, Jun, and Eiji Hanamoto, 1967.
On the cause of annual variation of catch of yellowfin tuna in the tropical western Indian Ocean. III Relation between annual variation of hook rate and age group (In Japanese; English abstract)
Bull. Jap. Soc. scient. Fish., 33(3): 151-155.

fisheries, tuna

Nakagome, Jun, and Hirotaka Suzuki, 1964.
Annual variation of lonline catch-rate of big-eyed tuna in the eastern Pacific tropical waters. (In Japanese; English summary).
Bull. Jap. Soc., Sci., Fish., 30(4):331-334.

fish - tunas

Nakagome, Jun, Hisao Tsuchiya, Shigemichi Suzuki, Satoshi Tanaka, Tetsuo Sakakibara and Hideo Honda, 1965.
Age composition of Atlantic tunas related with distribution of water temperature and distance from land. 1. Yellow-fin tuna. II. Albacore. (In Japanese; English abstract).
Bull. Japan. Soc. Sci. Fish., 31(2):97-100;101-104.

fish, tuna

Nakamura, Eugene L., and Walter M. Matsumoto, 1966
Distribution of larval tunas in Marquesan waters.
Fish. Bull. Fish Wildlife Service, 66(1):1-12

fish - tuna distribution

Nakamura, H., and H. Yamanaka, 1959
Relation between the distribution of tunas and ocean structure. J. Oceanogr. Soc., Japan, 15(3): 143-149.

fish - tuna

Nemoto, Takeshi, 1968.
La pesca de atun por palangre.
Informe Tecnico, Invest. Pesquer., Venezuela, 1: 31 pp

fish - tuna

Nepgen, C.S. de V. 1970
Exploratory fishing for tuna off the South African west coast.
Invest. Rep. Div. Sea Fish. S.Afr. 87: 26 pp

fish tunas

Osipov, V.G., 1960.
The distribution, biology and fisheries of the Pacific tunas. (In Russian).
Trudy Soveschanii Ikhtiol. Komissii, Akad. Nauk, SSSR, No. 10:188-194.

translations:
OTS 63-11118 - $0.50.

fish - tuna

Ozawa, Keijiro, and Kiyoshi Inoue, 1966.
Experimental tuna long-line fishing in the eastern Indian Ocean, November, 1963 and January 1964. (In Japanese; English abstract).
J. Tokyo Univ. Fish. (Spec. ed.) 8(2):237-241.

fish - tunas

Pinto Paivo, M., 1962.
Actual status of the knowledge on the biology of tunas in offshore waters of the Brazilian coast.
Bol. Estação Biol. Mar. Univers. Ceará, 5:10 pp.

tuna

Pinto Paivo, M., 1961.
Cartas de posca para os atuns e afins do Atlantico Tropical.
Arqu. Estac. Biol. Mar., Univ. Ceara, 1(2):1-110.

fish tuna

Pinto Paiva, M., 1961
Sôbre a pesca dos atuns e afins nas áreas em exploracão no Atlântico tropical.
Arqu. Estac. Biol. Mar. Univ. Ceará, 1(1): 1-20.

fisheries, tuna

Piskunov, I.A., and A.M. Kharchenko, 1968.
Results of fishery investigations of tuna in the Indian Ocean. (In Russian).
Trudy, Vses. Nauchno-Issled. Inst. Morsk. Ribn. Okeanogr (VNIRO) 64, Trudy Azovo-Chernomorsk. Nauchno-Issled. Inst. Morsk. Ribn. Khoz. Okeanogr. (AscherNIRO), 28: 344-373.

tunas

Postel, E., 1969
Répartition et abondance des thons dans l'Atlantique tropical
Actes Symp. Oceanogr. Ressources halieut. Atlant. trop., Abidjan, 20-28 Oct. 1966, UNESCO, 109-138.

tuna, migrations

*Richards, Williams J., 1969.
An hypothesis on yellowfin tuna migrations in the eastern Gulf of Guinea. Cah. ORSTOM, ser. Océanogr., 7(3): 3-7.

fish tunas

Richards, William J. 1969.
Distribution and relative apparent abundance of larval tunas collected in the tropical Atlantic during Equalant surveys I and II.
Actes Symp. Oceanogr. Ressources halieut. Atlant. trop., Abidjan, 20-28 Oct. 1966, UNESCO, 289-315.

fish - tuna

Ridgeway, George J., 1963.
Distinguishing tuna species by immunochemical methods.
U.S. Fish and Wildlife Service, Fish. Bull., 63(1):205-211.

fish tunas

Rossignol, M., and R. Repelin, 1962
Note sur Neothunnus albacora Lowe et Parathunnus obesus. Differenciation des jeunes-Presence d'un Trematode parasite des sacs nasaux chez N. albacora Lowe.
Trav. Centre Oceanogr., Pointe-Noire, Cahiers O.R.S.T.R.O.M. Oceanogr., 2:175-178.

fish - tuna

Sakamoto, Hisao 1967.
Distribution of bigeye tuna in the Atlantic Ocean (In Japanese; English abstract).
Rep. Nankai Reg. Fish. Res. Lab., 25: 67-73.

fish - tuna

Sakamoto, Hisao, 1967.
Regional and seasonal changes in the distribution of bigeye in the Indian Ocean. (In Japanese; English abstract).
Rep. Nankai Reg. Fish. Res. Lab., 25:49-57.

fish tuna

Sakamoto, Hisao, 1966.
Annual changes in the abundance and age composition of bigeye tuna in the Indian Ocean for the years 1955 through 1963. (In Japanese: English abstract).
Rep. Nankai reg. Fish. Res. Lab., No. 24:31-40.

fish tunas

Sato, T., S. Yamamoto and M. Ueno, 1962.
Studies on the tuna long-line fishery in the sea area off the West Caroline islands. (In Japanese; English abstract).
Bull. Fac. Fish., Hokkaido Univ., 13(2):53-62.

fish tunas

Scaccini, A., 1963.
Quelques notes oecologiques sur les thons génétiques de la Sardaigne.
Rapp. Proc. Verb., Réunions, Comm. Int. Expl. Sci., Mer Méditerranée, Monaco, 17(2):367-369.

fish tuna

Schweigger, E., 1949.
El atun a la costa Peruana. Bol. C.A.G. 25(8): 27 pp., 4 figs.

fish, tuna

Shabotinets, E.I., 1968.
On the age determination of tuna in the Indian Ocean. (In Russian).
Trudy, Vses. Nauchno-Issled. Inst. Morsk. Ribn. Okeanogr (VNIRO) 64, Trudy Azovo-Chernomorsk. Nauchno-Issled. Inst. Morsk. Ribn. Khoz. Okeanogr. (AscherNIRO), 28: 374-376.

fish tunas
Shimada, B.M., 1951.
An annotated bibliography on the biology of Pacific tunas. U.S. Fish Wildlife Service, Fish. Bull. 58:58 pp., (30 cents. U.S. Supt. Documents

fish tuna
Shimazaki, K., K. Otani and S. Mishima, 1962.
Study on the tuna long-line fishing ground of the south off the Island of Java.
Bull. Fac. Fish., Hokkaido Univ., 13(2):98-106.

fish - tuna
Shingu, Chiomi 1967.
Distribution and migration of the southern bluefin tuna.
Rep. Naikai Reg. Fish. Res. Lab. 25:19-36

fish tuna - Atlantic
Shiohama, Toshio, Masako Myojin and Hisao Sakamoto, 1965.
The catch statistic data for the Japanese tuna long-line fishery in the Atlantic Ocean and some simple considerations on it. (In Japanese; English abstract).
Rept. Nankai Reg. Fish. Res. Lab., No. 21:1-131.

fish tuna
Sivasubramaniam, K., 1963.
Efficiency of the tuna longline gear, considering the hooked rate in relation to the hook number and length of mainline of a basket.
Rec. Oceanogr. Wks., Japan, n.s., 7(1):65-72.

fish - tunas
Sokolov, V.A., 1965.
Investigaciones de los atunes del Golfo de México y del Mar Caribe en 1964-1965. (In Russian;Spanish abstract).
Sovetsk.-Cub. Ribokhoz. Issled., VNIRO:Tsentr. Ribokhoz. Issled. Natsional Inst. Ribolovsta Republ. Cuba, 209-221.

fish tunas
Sokolov, V.A., y H. Ramis Ramos, 1965.
Distribución de los atunes en el Atlántico occidental y central. (In Russian;Spanish abstract).
Sovetsk.-Cub. Ribokhoz. Issled., VNIRO:Tsentr. Ribokhoz. Issled. Natsional. Inst. Ribolovsta Republ. Cuba, 189-207.

fish tuna
Springer, S., 1957.
Tuna resources of the tropical and subtropical Atlantic.
Trans. Amer. Fish. Soc., 1955:13-17.

fish, tuna
Squire, J.L., Jr., and F.J. Mather, III, 1963.
Observations on the commercial potential of tuna in the oceanic Northwest Atlantic.
Proc. 15th Ann. Sess., Gulf and Caribbean Fish. Inst., 124-133.
Univ. Miami.

fish
Tunas
Suda, Akiro, y Milner B. Schaefer, 1965.
Composicion de tamanos del atun aleta amarilla capturado en la pesca palangrera Japonesa en el Pacifico oriental tropical al este de los 130°W. (In Spanish and English).
Comision Interamericana del Atun Tropical, Bol., 10(4):267-331.

fish, tuna fishing
Suda, Akira, y Milner B. Schaefer, 1965.
General review of the Japanese tuna long-line fishing in the eastern tropical Pacific Ocean, 1956-1962. (In Spanish and English).
Inter-American Tropical Tuna Comm. Bull., 9(6): 307-462.

fish - tuna
*Sund, Paul N., and William J. Richards, 1967.
Preliminary report on the feeding habits of tunas in the Gulf of Guinea.
Spec. scient. Rep. U.S. Fish. Wildl. Serv., Fish., 551: 6 pp (multilithed).

fish - tunas
Talbot, F.H., and M.J. Penrith, 1968.
The tunas of the genus Thunnus in South African waters. 1. Introduction, systematics, distribution and migrations.
Ann. S. Afr. Mus. 52(1):1-41.

fish, tuna
Tsuruta, S., 1955.
Morphometric comparison of yellowfin tuna of southwestern Great Sunda Islands and of the Pacific waters. J. Shimonoseki Coll. Fish. 4(2):311-319.

fish, tunas
Uno, Michio, 1965.
On the periodicity in tuna migration in the South Pacific. (In Japanese; English summary).
J. Fac. Fish., prefect. Univ. Mie, 6(3):341-349.

fisheries, tuna
U.S. Fish and Wildlife Service, 1948
The Japanese tuna fisheries. Fishery Leaflet 297:60 pp., 22 figs.

fish - tuna fisheries
Van Campen, W., and E.E. Hoven, 1956.
Tunas and tuna fisheries of the world.
An annotated bibliography, 1930-1953.
U.S.F.W.S. Bull. (Fish. Bull. 111) 57:173-249.

fish, tuna
Welsh, J.G. 1968.
A new approach to research on tuna distribution in South African waters.
Fish. Bull. Misc. Contrib. Oceanogr. Fish. Biol. S. Afr. 5:32-34.

fish - tuna
Wilson, Peter C., and Martin R. Barlett, 1967.
Inventory of U.S. exploratory longline fishing effort and catch rates for tunas and swordfish in the northwestern Atlantic, 1957-65.
Spec. scient. Rep. U.S. Fish Wildl. Serv., Fish No. 543:52 pp.

fish tunas
Yamnaka, Hajime, 1962
Tunas and oceanic conditions. (In Japanese; English abstract).
J. Oceanogr. Soc., Japan, 20th Ann. Vol., 663-678.

fish - tuna
Yukinawa, Mori, and Yoichi Yabuta 1967.
Age and growth of the bluefin tuna, Thunnus thynnus (Linnaeus) in the North Pacific Ocean. (In Japanese; English abstract).
Rep. Naikai reg. Fish. Res. Lab. 25:1-18.

fish, tuna
Zharov, V.L., 1966.
Relationship between the distribution of concentrations of tuna and the oceanographic structure in some tropical Atlantic waters. (In Russian).
Trudy vses. nauchno-issled. Inst. morsk. ryb. Khoz. Okeanogr. (VNIRO), 0:135-142.

fish - tuna
Zharov, V.A., 1965.
Resultados de la IV expedición de exploración y búsqueda de atún al Atlántico tropical. (In Russian; Spanish abstract).
Sovetsk.-Cub. Ribokhoz. Issled., VNIRO:Tsentr. Ribokhoz. Issled. Natsional. Inst. Ribolovsta Republ. Cuba, 181-188.

fisheries

fisheries
Armstrong, Terence, 1966.
Soviet sea fisheries since the Second World War.
Polar Record, 13(83):155-186.

fisheries.
Bassindale, R., 1955.
Ch. 5: In: Bristol and its adjoining counties, J. W. Arrowsmith Ltd., Bristol.
Fauna.

fisheries (general)
Berenbeim, D. Ya., and A.N. Probatov, 1963
The oceanologic foundations for the most prospective areas for fishery based on the data produced by oceanic research. (In Russian).
Okeanologiia, Akad. Nauk, SSSR, 3(2):308-312.

fisheries
Bilio, Martin 1969.
La variabilità fra pescate effettuate con rete a strascico nel Mar Ligure.
Pubbl. Staz. Zool. Napoli 37 (2 Suppl.): 115-131

fishing, exploratory
Bogdanov, D.V., 1965.
Investigaciones oceanográficos en la parte tropical del Oceano Atlántico en relación con el desarrollo de la industria pesquera. (In Russian; Spanish abstract).
Sovetsk. -Cub. Ribokhoz. Issled., VNIRO:Tsentr. Ribokhoz. Issled. Natsional. Inst. Ribolovsta Republ. Cuba, 17-21.

fisheries
Burkenroad, M.D., 1952.
Applications of ecology and economics to fisheries. Science 115(2983):251-252.

fisheries
Carruthers, J.N., 1954.
Some inter-relationships of oceanography and fisheries. Arch. Met., Geophys., Bioklim., B, 6(1/2):167-189, 4 textfigs.

fisheries
Carruthers, J.N., A.L. Lawford and V.F.C. Veley, 1952.
Winds and fish fortunes. J. du Cons. 18(3):354-358.

fisheries
Castillo, L., 1912.
Las estaciones de zoologia maritima i reconocimiento hidrografico de los fondos de pesca.
Bol. Bosques, Pesca i Caza 1:466-470.

fishing industry
Chapman, Wilbert MacLeod, 1968.
The U.S. fishing industry- putting the record straight.
Hydrospace, 1(5): 31-33.

fisheries
Christy, Francis T. Jr., and Anthony Scott 1965
The common wealth in ocean fisheries; some problems of growth and economic allocation.
Resources for the Future, Inc. Johns Hopkins Univ. Press 281 pp.

fisheries
Cooper, L.H.N., 1948
Phosphate and fisheries. JMBA, 27(2): 326-336, 3 text figs.

fisheries
F.A.O., Fisheries Division, 1953.
Improving the fisheries contribution to world food supplies. F.A.O. Fish. Bull. 6(5):159-192, 2 textfigs.

fisheries
Felando, August, 1966.
The kind of oceanographic information of direct use to the fisheries.
Exploiting the Ocean, Trans. 2nd Mar. Techn. Soc. Conf., June 27-29, 1966, 336-341.

fisheries
Fishery Board of Scotland, 1932-1939.
Annual reports for 1931-1938. H.M.S.O., Edinburgh

fishing
Fonteneau, A. 1970.
La pêche au chalut sur le plateau continental ivoirien: equilibre maximal des captures.
Doc. scient. Centre Rech. océanogr. Abidjan 1 (1): 31-35.

fisheries
Graham, M., 1953.
English fishery research in northern waters.
Arctic 6(4):252-259, 6 textfigs.

fisheries
*Gulland, J.A., 1968.
The concept of the marginal yield from exploited fish stocks.
J.Cons.perm.int.Explor Mer, 32(2):256-261.

fisheries
Hachey, H.B., 1955.
Water replacements and their significance to a fishery.
Pap. Mar. Biol. and Oceanogr., Deep-Sea Res., Suppl. to Vol. 3:68-73.

fisheries
Hardy, Sir Alister, 1959.
The open sea: its natural history. II. Fish and fisheries.
Collins, London, 322 pp.

fisheries
Herrington, W.C., 1955.
U.S. participation in conservation of international fishery resources.
Pap. Mar. Biol. and Oceanogr., Deep-Sea Res., Suppl. to Vol. 3:398-405.

fisheries
Herrington, W.C., 1954.
50 years of progress in solving fishery problems
Proc. Gulf and Caribbean Fish. Inst., 6th Ann. Session:81-90.

fisheries
Hester, Frank J., 1966.
Man in the sea and fisheries of the future.
Exploiting the Ocean, Trans. 2nd Mar. Techn. Soc. Conf., June 27-29, 1966, 524-549.

fisheries
Hodgson, W.C., 1950.
Echo-sounding and the pelagic fisheries. Fish. Inv., Min. Agric. Fish., Gt. Brit., ser. 2, 17(4):24 pp., 24 figs.

fisheries
Holt, S.J., 1969.
The food resources of the ocean.
Scient.Am., 221(3):178-182,187-194. (popular)

fishing
Hood, Donald W. and C. Peter McRoy, 1971.
Uses of the ocean. (2) In: Impingement of man on the oceans, D.W. Hood, editor, Wiley Interscience: 667-698.

fisheries
Humphrey, G.F., 1969.
Perspectives of fisheries and oceanographic research in Australia. Bull. Jap. Soc. fish. Oceanogr. Spec. No. (Prof Uda Commem. Pap.): 4 47.

fisheries
Iselin, C.O'D., 1955.
Coastal currents and fisheries.
Pap. Mar. Biol. and Oceanogr., Deep-Sea Res., Suppl. to Vol. 3:474-478.

fisheries
Kesteven, G., 1953.
Fisheries and weather. F.A.O. Bull. 6(4):109-118.

fisheries
Kesteven, G.L., and G.E.R. Deacon, 1955.
The contribution of oceanographic research to fisheries science. FAO Fish. Bull. 8(2):67-76.

fisheries
LaFond, E.C., 1955.
On upwelling and fisheries. Current Sci. 24(8): 258-259, 1 textfig.

fisheries
Lonsdale, A.L., 1968.
Soviet fishing- a mass production.
Sea Frontiers 14(2):116-120.

fisheries
Lucas, C.E., 1969.
Present trends in the organisation of world fisheries research. Bull. Jap. Soc. fish. Oceanogr. Spec. No. (Prof Uda Commem. Pap.): 23-29.

fisheries
Lucas, C. E., 1964.
Fisheries research and international collaboration.
Vortrag zur Einweihung des Instituts für Hydrobiologie und Fischereiwessenschaft der Universität Hamburg am 8 Juni 1964:1-7.
also in:
Gesammelte Sonderdrucke, Inst. Hydrobiol. u. Fischereiwiss., Univ. Hamburg, 1963(1964).

fisheries
Lucas, C.E., 1961.
Marine fisheries: their conservation and their potential.
Advance. Sci., 17(70):499-504.

fisheries
Martinsen, G.V., 1969.
Problems of the world sea fisheries. (In Russian; English abstract). Okeanologiia, 9(6): 1049-1055.

fisheries
Martinsen, G. V., 1964.
The recent status of world fisheries and prospects for its development. (In Russian).
Okeanologiia, Akad. Nauk., SSSR, 4(6):939-953.

fisheries
McIlhenny, W.F., 1966.
The oceans: technology's new challenge.
Chem. Engng., (Nov. 7):247-254.

fisheries
Moiseev, P.A., 1958.
Requirements of the fisheries in oceanography.
Trudy Okeanogr. Komissii, Akad. Nauk, SSSR, 3: 109-117.

fisheries
Morgan, R., 1956.
World sea fisheries. Methuen & Co., Ltd., 307 pp.

fisheries
Newcombe, C.L., 1952.
Applications of ecology and economics to fisheries. Science 115(2983):252-253.

fisheries
Office Scientifique et Technique des Pêches Maritimes, 1948.
Bibliographie analytique des publications de l'Office Scientifique et Technique des Pêches Maritimes. Notes et Rapports, n.s., No. 3:72 pp.

fisheries
Otsuka, Kazuyuki, 1969.
The network of scientific studies connected with fisheries oceanography. (In Japanese; English abstract). Bull. Jap. Soc. fish. Oceanogr. Spec. No. (Prof. Uda Commem. Pap.): 203-206.

fisheries, forecasting
Ottestad, P., 1960.
Forecasting the annual yield in sea fisheries.
Nature, 185(4706):183.

fisheries
Popovici, Z., 1949.
Problemas de Hidrobiologia y su Vinculacion con la Explotacion Pesquera. (With Prologue by A.E. Riggi). Inst. Nac. Invest. Ciencias Nat. anexo al Mus. Argentino Cien. Nat. "Bernardino Rivadavia", Misc. 2:50 pp.

Review of needs for research to promote fisheries resources.

2

fisheries

Rass, T.S., 1963
On the possibilities for exploiting the fishery resources of the World Ocean. (In Russian).
Okeanologiia, Akad. Nauk, SSSR, 3(3):495-499.

fisheries

Redeke, H.C., 1927.
River pollution and fisheries.
Rapp. Proc. Verb., Cons. Perm. Int. Expl. Mer, 43:50 pp.

fisheries biology

Rounsefell, G.A., and W.H. Everhart, 1953.
Fishery Science. xii 444 pp., John Wiley and Sons, Inc.

Reviewed:
J. du Cons. 20(1):102-105 by C.E. Lucas.

fisheries

Schüler, Fr., and G. Krefft, 1951.
Versuche zur Beeinflussung von Meerfischen durch Schalldruckwellen und künstliches Licht.
Fischereiwelt und die Fischindustrie 3(1):8-10, 7 figs.

fisheries

Schweigger, E., 1949.
El Empleo de la Bonde Electrica en la Pesca Peruana. Bol. C.A.G. 24(10):185-190, 3 textfigs.

fisheries

Scripps Institution of Oceanography, 1949.
Marine life research program. Progress report, 1 May to 31 July 1949. 24 pp. (mimeographed), 16 figs. (ozalid).

fisheries

Sette, Oscar E., 1966.
Ocean environment and fish distribution and abundance.
Exploiting the Ocean, Trans. 2nd Mar. Techn. Soc. Conf., June 27-29, 1966, 309-318.

fisheries

Shapiro, Sidney, 1964
Fisheries of the world.
In: Farmer's World, the Yearbook of Agriculture, U.S. Dept. Agriculture, 1964:161-177.

fisheries

Skogsberg, T., 1936.
Hydrography of Monterey Bay, California. Thermal conditions, 1929-1933. Trans. Am. Phil. Soc. 29(1):1-152, 45 textfigs.

fisheries

State of Maryland Board of Natural Resources, Department of Research and Education, 1948.
Effects of Underwater explosions on oysters, crabs and fish: a preliminary report. Chesapeake Biol. Lab., Publ. No. 70:43 pp., 13 textfigs.

fisheries

Swain, L.A., 1944.
The Pacific coast dogfish and shark liver oil industry, Prog. Rept. Pacific Coast Stas., Fish. Res. Bd., Canada, 58:3-7.

fisheries

Taning, A.V., 1953.
Pt. 5. Long term changes in hydrography and fluctuations in fish stocks.
Int. Comm. Northwest Atlantic Fish., Ann. Proc. 3:3-11, 10 textfigs.

fisheries, forecasting of

Terada, K., 1962.
Utilization of sea surface temperature data for fisheries and the special forecast for fishing operations in Japan.
The Geophys. Mag., Tokyo, 31(1):105-126.

fisheries

United States, National Academy of Sciences, Committee on Oceanography, 1964.
Economic benefits from oceanographic research.
Nat. Acad. Sci.-Nat. Res. Counc., Publ. No. 1228:50 pp. $2.00.

fisheries

U.S.A., University of California, Institute of Marine Resources, La Jolla, 1965.
California and use of the ocean: a planning study of marine resources.
IMR Ref. 65-21: app. 350 pp. (multilithed, paged by chapters)

fisheries, exploitation

Wakefield, Lowell, 1966.
Role of government in ocean fisheries exploitation—an industrial view.
Exploiting the Ocean, Trans. 2nd Mar. Techn. Soc. Conf., June 27-29, 1966, 173-176.

fisheries

Walford, L.A., 1955.
New directions in fishery research.
Pap. Mar. Biol. and Oceanogr., Deep-Sea Res., Suppl. to Vol. 3:471-473.

fisheries

Watase, Sadao, 1969.
Fisheries resources exploitation in the central and south America. (In Japanese; English abstract). Bull. Jap. Soc. fish. Oceanogr. Spec. No. (Prof. Uda Commem. Pap.): 339-342.

fishing

Wood, H., and B.B. Parrish, 1950.
Echo-sounding experiments on fishing gear in action. J. du Cons. 17(1):25-36, 7 textfigs.

fisheries biology

Cushing, D.H., 1968.
Fisheries biology, a study in population dynamics.
Univ. Wisconsin Press, 200 pp. $7.50

fisheries biology

Graham, M., 1954.
Half a century of fishery biology in Europe.
Proc. Gulf and Caribbean Fish. Inst., 6th Ann. Session:70-81.

fisheries biology

Korringa, Pieter 1969
Triumphs and frustrations of a fishery biologist.
FiskDir. Skr. Ser. HavUnders. 15(3): 114-127

fisheries biology

Kutty, M.Krishman, and S.Z.Qasim, 1968.
The estimation of optimum age of exploitation and potential yield in fish populations.
J.Cons.perm.int.Explor.Mer, 32(2):249-255.

fisheries biology

Marr, J.C., 1954.
Biologia Pesquera marina. Primer Centro Latino-americano de Capacitacion Pesquera, Valparaiso, Chile (Oficina Regional de le FAO), 127 pp.

fisheries oceanography

Uda, Michitaka, 1970.
Recent advances in fisheries oceanography of Japan from 1967 to 1969. Adv. Fish. Oceanogr. (3). Also in: Coll. Repr. Coll. mar. Sci. Techn. Tokai Univ. 4: 171-185.

fisheries biology

Walford, L. A., 1948.
The case for studying normal patterns in fishery biology. J. Mar. Res. 7(3):506-510.

fishery catastrophies

Ishino, Makoto, 1969.
Catastrophic oceanographic conditions influenced the fisheries. (In Japanese; English abstract). Bull. Jap. Soc. fish. Oceanogr. Spec. No. (Prof. Uda Commem. Pap.): 331-334.

fisheries conventions

fisheries conventions

Herrington, William C., 1967.
The convention of fisheries and conservation of living resources: accomplishments of the 1958 Geneva Conference.
In: The law of the sea, L.M.Alexander, editor, Ohio State Univ.Press, 26-35.

fishery dynamics

fishery dynamics

Schaefer, M.B. and R.J.H. Beverton, 1963
21. Fishery dynamics - their analysis and interpretation. In: The Sea, M.N. Hill, Edit. Vol. 2 (IV). Biological Oceanography, Interscience Publishers, New York and London, 464-483.

fisheries, effect of

fishing, effect of

Boerema, L.K., G. Saetersdal, I. Tsukayama, J.E. Valdivia y B. Alegre 1967.
Informe sobre los efectos de la pesca en el recurso peruano de anchoveta.
Bol. Inst. Mar, Peru 1(4): 135-186.

fishing, effects of

Schaefer, Milner B 1968.
Methods of estimating effect of fishing on fish populations.
Trans. Am. Fish. Soc. 97(3): 231-241.

fishing, experimental

Inoue, Kiyoshi, Ryoichi Arai and Tokiharu Abe, 1968.
Experimental fishing during the voyage of the Umitaka-Maru. J. Tokyo Univ. Fish., (2): 135-140.

fisheries failure

Rollefsen, Gunnar 1966.
Norwegian fisheries research.
FiskDir. Skr. Ser. Havunders. 14(1):1-36.

fisheries, international
Gulland, John, 1968.
Emerging patterns of international fisheries development.
Proc. Gulf Carib. Fish. Inst., 20th Ann. Sess. 116-122.

fisheries, international
Kasahara, Hiroshi, 1968.
International aspects of fisheries.
Proc. Gulf Carib. Fish Inst. 20th Ann. Sess. 112-115

fishing, limitations of
Morris, Ian, 1970.
Restraints on the big fish-in.
New Scientist, 48(727): 373-375

Fishery management

fishery management
Beverton, R.J.H., 1953.
Some observations on the principles of fishery regulation. J. du Cons. 19(1):56-68, 6 textfigs.

fisheries
Burkenroad, M.D., 1953.
Theory and practice of marine fishery management J. du Cons. 18(3):300-310.

fisheries management
Crutchfield, James A., editor, 1965.
The fisheries problems in resource management.
Univ. Washington Press, Seattle, 136 pp.

fishery management
Huntsman, A.G., 1953.
Fishery management and research. J. du Cons. 19(1):44-55.

fishery management
Gulland, J.A., 1971.
Science and fishery management. J. Cons. int. Explor. Mer, 33(3): 471-477.

fishery resources
Gulland, J.A. and J.E. Carroz, 1968.
Management of fishery resources.
Advances in Marine Biology 6:1-71.

fisheries, overfishing
Graham, M., 1952.
Overfishing and optimim fishing. Rapp. Proc. Verb., Cons. Perm. Int. Expl. Mer, 132:72-78, 3 textfigs.

fisheries - overfishing
Russell, E. S. 1942
The overfishing problem
Macmillan, 130 pp.

fisheries, research

fisheries research
Boods, Larry, 1967.
Region autonomy aids fisheries research.
Undersea Techn., 8(2):18-19.

fisheries research
Fridriksson, A., 1952.
Marking of fish in Europe during 1927-1951. Rapp. Proc. Verb., Cons. Perm. Int. Expl. Mer, 132:55-64, 4 textfigs.

fisheries research
Jensen, A.J.C., 1952.
Applications to industry and policy. Rapp. Proc. Verb., Cons. Perm. Int. Expl. Mer, 132:79-85.

fisheries research
Tåning, Å.V., 1952.
The transplantation of fish. Rapp. Proc. Verb., Cons. Perm. Int. Expl. Mer, 132:47-54, 5 textfig

fisheries, special

fisheries, year classes
Dragesund, Olav, 1971.
Comparative analysis of year-class strength among fish stocks in the North Atlantic
Fiskeridirektoratets Skr. Ser. Havundersøkelser 16(2): 49-64.

fisheries mollusc
Fiske, John D., John R. Curley and Robert P. Lawton, 1968.
A study of the marine resources of the Westport River.
Monogr. Ser. Comm. Mass. Div. Mar. Fish. 7:51 pp.

fisheries, molluscs
Jerome, William C., Jr., Arthur P. Chesmore, and Charles O. Anderson, Jr. 1969.
A study of the marine resources of the Annisquam River - Gloucester Harbor Coastal system.
Monogr. Ser. Comm. Mass. Div. Mar. Fish. 8: 62 pp.

fisheries, mollusc
Jerome, William C., Jr., Arthur P. Chesmore, and Charles O. Anderson, Jr. 1968.
A study of the marine resources of the Parker-River-Plum Island Sound Estuary.
Monogr. Ser. Comm. Mass. Div. Mar. Fish. 6: 79 pp.

fisheries, shrimp
Von Borstel, J. (Mendoza), 1969.
Posibilidades para incrementar la producción camaronera de alta mar en las aguas del oceano Pacifico frente a México.
FAO Fish. Rept. 3(57) (FRm/57.3 (Trm)): 1141-1147 (mimeographed)

fisheries, shrimp
Raitt, D.F.S. and D.R. Niven, 1969.
Exploratory prawn trawling in the waters off the Niger delta.
Actes Symp. Oceanogr. Ressources halieut. Atlant. trop., Abidjan, 20-28 Oct. 1966, UNESCO, 403-414.

fishery, shrimp
Thomas, D., 1969
Prawn fishing in Nigerian waters.
Actes Symp. Oceanogr. Ressources halieut. Atlant. trop., Abidjan, 20-28 Oct. 1966, UNESCO, 415-417.

tuna fisheries
*Barkley, Richard A., 1969.
Salinity maxima and the skipjack tuna, Katsuwonus pelamis. Bull. Jap. Soc. fish. Oceanogr. Spec. No. (Prof. Uda Commem. Pap.): 243-248.

tuna fisheries
Blackburn, Maurice, 1969.
Outlook for tuna oceanography. Bull. Jap. Soc. fish. Oceanogr. Spec. No. (Prof. Uda Commem. Pap.): 221-225.

tuna fisheries
*Le Guen, J.C., 1968.
Etude du stock d'albacores (Thunnus albacares) exploité par les palangrieres japonais dans l'Atlantique tropical Américain de 1956 a 1963. Cah. ORSTOM, sér. Océanogr., 6(3/4): 27-30.

tuna fisheries
*Inoue, Motoo, 1969.
Perspective on exploitation of fishing grounds for skipjack and young yellowfin tuna in the western tropical Pacific. (In Japanese; English abstract). Bull. Jap. Soc. fish. Oceanogr. Spec. No. (Prof. Uda Commem. Pap.): 235-242

tuna fisheries
*Nakagome, Jun, 1969.
The study of relation between tuna and oceanography - a scope of tuna oceanography. (In Japanese; English abstract). Bull. Jap. Soc. fish. Oceanogr. Spec. No. (Prof. Uda Commem. Pap.): 231-234.

tuna fisheries
*Yamanaka, Hajime, 1969.
Relation between the fishing grounds of tunas and the equatorial current system. (In Japanese English abstract). Bull. Jap. Soc. fish. Oceanogr. Spec. No. (Prof. Uda Commem. Pap.): 227-230.

fission products, action of

fission products
See chiefly: chemistry, chemicals concentrated within organisms
chemistry, radioactivity, etc.

Fission Products, uptake of

Berner, Leo, Jr., Robert Bieri, Edward D. Goldberg, Decourcey Martin and Robert L. Wisner, 1962. Oceanographic studies during Operation "Wigwam". Field studies of uptake of fission products by marine organisms. Limnol. and Oceanogr., Suppl. to Vol. 7:lxxxii-xci.

fission products, action of

Chipman, W.A., 1956. Passage of fission products through the skin of tuna. USFWS Spec. Sci. Rept., Fish., No. 167:6 p

fission products

*Dmitrieva, G.V., Ju. V. Krasnopevtsev and S.G. Malakhov, 1970. Some peculiarities of fission product concentration distributions over oceans in the tropical zone and their connection with atmospheric processes. J. geophys. Res., 75(18): 3675-3685.

fission products, uptake of

Martin, DeCourcey, and Edward D. Goldberg, 1962 Oceanographic studies during Operation "Wigwam". Uptake and assimilation of radiostrontium by Pacific salmon. Limnol. and Oceanogr., Suppl., to Vol. 7: lxxvi-lxxxi.

fission products, accumulation of

Polikarpov, G.G., 1961 Ability of some Black Sea organisms to accumulate fission products. Science, 133(3459): 1127-1128.

Fission products, uptake of

Thomas, William H., Donald W. Lear, Jr., and Francis T. Haxo, 1962. Oceanographic studies during Operation "Wigwam". Uptake of the marine dinoflagellate, Gonyaulax polyedra, of radioactivity formed during an underwater nuclear test. Limnol. and Oceanogr., Suppl. to Vol. 7:lxvilxxi.

fission track

Aumento, F., 1969. The Mid Atlantic Ridge near 45°N. V. Fission track and ferro-manganese chronology. Can. J. Earth Sci. 6(6): 1431-1440

Fjords
See: fiords

fjords anoxic

Adams, Donald D., and Francis A. Richards 1968. Dissolved organic matter in an anoxic fjord, with special reference to the presence of mercaptans. Deep-Sea Res. 15(4): 471-481.

fjords

Beyer, Fredrik, 1968. Zooplankton, zoobenthos, and bottom sediments as related to pollution and water exchange in the Oslofjord. Helgoländer wiss. Meeresunters. 17: 496-509.

fjords

Crary, A.P. 1966. Mechanisms for fiord formation indicated by studies of an ice-covered inlet. Bull. geol. Soc. Am. 77(9): 911-930.

fjords

Føyn, Ernst, 1968. Biochemical and dynamic circulation of nutrients in the Oslofjord. Helgoländer wiss. Meeresunters. 17: 489-495

fjords

Gade, Herman G., 1968. Horizontal and vertical exchanges and diffusion in the water masses of the Oslo Fjord. Helgoländer wiss. Meeresunters., 17: 462-475.

fjords

Gross, M.Grant, 1967. Concentration of minor elements in diatomaceous sediments of a stagnant fjord. In: Estuaries, G.H. Lauff, Publs Am. Ass. Advmt Sci., 83:273-282.

fjords

Hartmann, Jürgen, und Dietrich Schnack, 1969. Verteilung von Heringslarven und Plankton am 28.5.1969 in der Schlei: Absetzvolumen. Ber. dt. wiss. Kommn. Meeresforsch. 20(3/4): 268-296.

fjords

Kändler, Rudolf 1971. Untersuchungen über die Abwasserbelastung der Untertrave. Kieler Meeresforsch. 27(1): 20-29.

fjords

Kaplin, P.A., 1959 [Shore-lines evolution of fiord regions. Questions of studies of marine coasts. Questions of studies of marine coasts.] Trudy Okeanogr., Komissii Akad. Nauk, SSSR, 4:54-65.

fjords

Knight, R.J., 1971. Distributional trends in the Recent marine sediments of Tasiujaq Cove of Exalugad Fiord, Baffin Island, N.W.T. Marit. Sed. 7(1): 1-18.

fjords

Knull, James R. and Francis A. Richards, 1969. A note on the sources of excess alkalinity in anoxic waters. Deep-Sea Res., 16(2): 205-212.

fjords

Munthe-Kaas, Hans, 1968. Surface pollution and light extinction in the Oslofjord. Helgoländer wiss. Meeresunters. 17: 476-488.

fjords

Nicholson, R., 1963. A note on the relation of rock fracture and fjord direction. Geografisker Annaler, 45(4):303-304.

fjords

Pickard, G.L., 1971. Some physical oceanographic features of inlets of Chile. J. Fish. Res. Bd Can. 28(8): 1077-1106

fjords

Pickard, G.L., 1967. Some oceanographic characteristics of the larger inlets of southeast Alaska. J. Fish. Res. Bd. Can., 24(7) 1475-1506

fjords

*Rasmussen, Erik, 1968. Stavns Fjord - et østdansk tidevandsområde. Geografisk Tidsskrift, København, 67(1):70-93.

fjords

Rattray, Maurice, Jr., 1967. Some aspects of the dynamics of circulation in fjords. In: Estuaries, J.H. Lauff, Editor, Publs Am. Ass. Advmt Sci., 83:52-62.

Ruud, Johan T., 1968. Introduction to the studies of pollution in the Oslofjord. Helgoländer wiss. Meeresunters. 17: 455-461.

Ruud, Johan T., 1968. Changes since the turn of the century in the fish fauna and fisheries of the Oslofjord. Helgoländer wiss. Meeresunters. 17: 510-517.

fjords

Saelen, Odd H., 1967. Some features of the hydrography of Norwegian fjords. In: Estuaries, G.H. Lauff, editor, Publs Am. Ass. Advmt Sci., 83:63-70.

fjords

Svansson, Artur 1967. Hydrographical observations on Swedish lightships and fjord stations in 1965. Fish. Bd., Sweden, Ser. Hydrogr. Rep., 14: 95pp.

Flagellae

flagellae

Lewin, Ralph A., Editor, 1962. Physiology and biochemistry of algae. Academic Press, New York and London, 929 pp.

flats, tidal

flats, tidal

Klein, George deVries, 1967. Comparison of Recent and ancient tidal flat and estuarine sediments. In: Estuaries, G.H. Lauff, editor, Publs Am. Ass. Advmt Sci., 83:207-218.

flats, tidal

Kulm, L.D., and John V. Byrne, 1967.
Sediments of Yaquina Bay, Oregon.
In: Estuaries, G.H. Lauff, editor, Publs Am. Ass. Advmt Sci., 83:226-238.

flats, tidal

Reineck, H.E., 1967.
Layered sediments of the tidal flats, beaches, and shelf bottoms of the North Sea.
In: Estuaries, G.H. Lauff, editor, Publs Am. Ass. Advmt Sci., 83:191-206.

flats, tidal

Schou, Axel, 1967.
Estuarine research in the Danish moraine archipelago.
In: Estuaries, G.H. Lauff, editor, Publs Am. Ass. Advmt Sci., 83:129-145.

flint

flint

Emery, K.O., C.A. Kaye, D.H. Loring, and D.J.G. Nota 1968.
European Cretaceous flints on the coast of North America.
Science 160(3833): 1224-1228.

Flip

Bronson, Earl D. 1971.
Deep anchoring "flip".
J. mar. techn. Soc. 5(3): 42-44

Flip

Spiess, F.N., 1968.
Oceanographic and experimental platforms.
In: Ocean engineering: goals, environment, technology, John F. Brahtz, editor, John Wiley & Sons, pp. 553-587.

floats

Floats.

Kazakov, S.P., 1963.
Motion of a hydrometric float. (In Russian).
Trudy Morsk. Gidrofiz. Inst., Akad. Nauk, Ukrain. SSR, 28:67-71.

Translation:
Soviet Oceanography, Issue 2 - 1963 series.
Scripta Tecnica, Inc. for AGU, 45-48.

floats

Nan-niti, Tosio, Hideo Akamatsu and Toshisuke Nakai, 1964.
A further observation of a deep current in the east-north-east sea of Torishima.
Oceanogr. Mag., Tokyo, 16(1/2):11-19.

floats

Tasai, F., T. Suhara and A. Mitsuyasu, 1969.
Methods of studies on floating structures in rough seas. Oceanol. int. 69, 2: 6 pp., 6 figs.

floating islands

floating islands ice

Crary, A.P., 1954.
Seismic studies on Fletcher's/island, T-3.
Trans. Amer. Geophys. Union, 35(2):292-300, 9 textfigs.

floating islands

Fletcher, J.O., 1951.
Floating islands in the Arctic Ocean. Proc. Alaskan Sci. Conf., Bull. Nat. Res. Coun. 122: 122.

floating objects, effect of

floating objects, effect of

Hunter, John R., and Charles T. Mitchell, 1968.
Field experiments on the attraction of pelagic fish to floating objects.
J. Cons. perm. int. Explor. Mer 31(3): 427-434.

flocculation

flocculation

Postma, H., 1967.
Sediment transport and sedimentation in the estuarine environment.
In: Estuaries, G.H. Lauff, editor, Publs Am. Ass. Advmt Sci., 83:158-179.

floods

flood control

Aki, K., 1955.
Some problems on flood control in Japan.
Proc. UNESCO Symp., Typhoons, Nov. 1954:17-21.

floods

Balay, M.A., 1959.
Causes and periodicity of large floods in Rio de la Plata.
Int. Hydrogr. Rev., 36(1):123-151.

floods

Becken, F.W., 1949.
Sturmfluten und Regenmengen in Cuxhaven. Ann. Met. 2(1/2):51-53.

floods

Harrison, A.J.M., 1961

The 1960 Exmouth floods. An investigation of the flood discharges of the Withycombe Brook.
The Surveyor, 4 Feb., 1961: 1-6.

floods

Hayami, Shoitiro, 1957.
On the great flood of 1954 of the Yangtzekiang (In Japanese; English abstract)
Disaster Prevention Res. Inst., Annuals, No. 1: 79-91.

Also in:
Papers on Oceanogr. and Hydrol., (1949-1962), Geophys. Inst., Kyoto Univ., Contrib. No. 16. (1963).

floods

Hayami, Shoitiro, 1953.
Analysis of flood flow. (In Japanese).
Recent Progress in Hydraulic Engineering: 24 pp.

Also in:
Papers on Oceanogr. and Hydrol., (1949-1962), Geophys. Inst., Kyoto Univ., Contrib. No. 3. (1963).

floods

Hayami, Shoitiro, 1951.
On the propagation of flood waves.
Disaster Prevention Res. Inst., Kyoto Univ. Bull., No. 1: 6 pp.

Also in:
Papers on Oceanogr. and Hydrol., (1949-1962), Geophys. Inst., Kyoto Univ., Contrib. No. 2. (1963).

floods

Korringa, P., 1956.
Hydrographical, biological and osteological observations in the Knysna Lagoon with notes on conditions in other South African waters.
S. African Dept. Comm. & Fish., Fish. Div., Invest. Rept., No. 20:63 pp., 23 pls.

floods

Reineck, H.-E., 1962.
Die Oskanflut vom 16 Februar 1962.
Natur und Museum, 92(5):151-172.

floods, effect of

Stephenson W., 1969.
The effects of a flood upon salinities in the southern portion of Moreton Bay.
Proc. R. Soc. Qd., 80(3): 19-34.

floods, prediction of

Titov, L.F., 1960.
Contemporary methods of predicting the Leningrad flood and the course of its development.
Trudy Gosud. Okean. Inst., Leningrad, 56:3-8.

In Russian

flora, micro-

flora, micro

Zebrowski, George, 1965.
Micro-flora of the turbulent eulittoral zone. (Abstract).
Ocean Sci. and Ocean Eng., Mar. Techn. Soc., Amer. Soc. Limnol. Oceanogr., 1:96.

flotation

flotation

Smayda, Theodore J., and Brenda J. Boleyn, 1966
Experimental observations on the flotation of marine diatoms. III. Bacteriastrum hyalinum and Chaetoceros lauderi.
Limnol. and Oceanogr., 11(1):35-43.

flotation

Smayda, Theodore J., and Brenda J. Boleyn, 1966
Experimental observations on the flotation of marine diatoms. II. Skeletonema costatum and Rhizosolenia setigera.
Limnol. and Oceanogr., 11(1):18-34.

flotation

Whitehouse, John W., and Brian G. Lewis, 1966.
The separation of benthos from stream samples by flotation with carbon tetrachloride.
Limnol. and Oceanogr., 11(1):124-126.

flotsam

flotsam, fauna of

David, Peter M., 1965.
The surface fauna of the ocean.
Endeavour, 24(92):95-100.

flotsam

Hunter, John R., and Charles T. Mitchell, 1966.
Association of fishes with flotsam in the offshore waters of Central America.
Fish. Bull. Fish Wildlife Service, 66(1): 13-29

flows

flow

O'Brien, Morrough P., 1967.
Equilibrium flow areas of tidal inlets on sandy coasts.
Proc. 10th Conf. Coast. Engng. Tokyo, 1966, 1:376-386.

flows

Phillips, O.M., 1970.
On flows induced by diffusion in a stably stratified fluid. Deep-Sea Res., 17(3): 435-443.

flow, baroclinic

McIntyre, Michael E., 1970.
On the non-separable baroclinic parallel flow instability problem.
J. fluid Mech. 40 (2): 273-306

flows

Shtokman, V.B., 1946.
[The field of flows in a sea as described in terms of the equation for a fastened plate under bending stress.] Comptes Rendus (Doklady) Akad. Sci. U.S.S.R. 54:885.

flow, equatorial

Hidaka, Koji, 1961
Equatorial flow and inertia terms.
Rec. Oceanogr. Wks., Japan, 6(1):29-35.

flow, horizontal shear

Sasaki, Ken 1971.
Rotational instability of a horizontal shear flow in a stratified rotating fluid.
J. oceanogr. Soc. Japan, 27 (4): 137-141

flow, open channel

*Matsunashi, Junzaburo, 1967.
On a coexistence system of flow and waves.
Proc. 10th Conf. Coast. Engng. Tokyo, 1966, 1:418-433.

flow patterns

flow patterns

*Carpenter, J.H., and H.H. Seliger, 1968.
Studies at Oyster Bay in Jamaica, West Indies.
2. Effects of flow patterns and exchange on bioluminescent distributions.
J.mar. Res., 26(3):256-272.

flow, patterns of

Holland, William R. 1971.
Ocean tracer distributions. 1. A preliminary numerical experiment.
Tellus 23 (4/5): 371-392

flow rate

Shtokman, V.B., 1950.
[Determination of the flow rate and density distribution in the cross-section of an infinite channel in relation to the effect of wind and lateral friction in the field of Coriolis force.] Dok. Akad. Nauk, USSR, 71:41-44.

flow rate

Shtokman, V.B., 1944.
[The stationary distribution of tongue-shaped isotherms in the sea for changes in turbulent diffusion and speed of flow.] Izvest. Akad. Nauk, USSR, Geogr. Geofiz., 8:176-182.

flows, rotating

flows, rotating

Snyder, H. A. 1968.
Experiments on rotating flows between noncircular cylinders.
Physics Fluids 11 (8): 1606-1611.

flow, stratified

Jacobs, S.J., 1964.
On stratified flow over bottom topography.
J. Mar. Res., 22(3):223-235.

flow, stratified

Mandelbrot, L., 1965.
Le nombre de Richardson et les criteres de stabilite des ecoulements stratifies.
La Houille Blanche, 20(1):24-28.

flow theorem

Čadež, M., und H. Ertel, 1970.
Ein Strömungstheorem der Meteorologie und Hydrographie.
Beitr. Geophys. 79 (6): 465-467.

flow, transverse

flow, transverse

Hide, R. and A. Ibbetson with an appendix by M.J. Lighthill 1968.
On slow transverse flow past obstacles in a rapidly rotating fluid.
J. fluid Mech. 32 (2): 251-272.

flow, vertical

Hidaka, Koji, 1968.
Relationship between the meridional and vertical flows at the equator.
J. oceanogr. Soc., Japan 25 (6): 273-280.

flow, vertical

Saito, Y., 1951.
On the velocity of the vertical flow in the ocean. J. Inst. Polytech, Osaka City Univ., B, 2:1-4 (Not seen)

flowerings

flowerings

Bigelow, H.B., L.C. Lillick, and M. Sears, 1940.
Phytoplankton and planktonic protozoa of the offshore waters of the Gulf of Maine. Pt. 1. Numerical distribution. Trans. Am. Phil. Soc., n.s., 31(3):149-191, 10 textfigs.

flowering

Conover, S.A.M., 1956.
Oceanography of Long Island Sound, 1952-1954.
4. Phytoplankton. Bull. Bingham Oceanogr. Coll., 15:62-112.

flowering

Kierstead, H., and L.B. Slobodkin, 1953.
The size of water masses containing plankton blooms. J. Mar. Res. 12(1):141-147.

flowering of diatoms

Sverdrup, H.U., 1953.
On conditions for the vernal blooming of phytoplankton. J. du Cons. 18(3):287-295, 2 textfigs.

Fluctuations in abundance

fluctuations

Cannon, Glenn A., 1971.
Statistical characteristics of velocity fluctuations at intermediate scales in a coastal plain estuary. J. geophys. Res. 76 (24): 5852-5858.

fluctuations

Coe, W.R., 1957.
Fluctuations in littoral populations. Ch. 28 in: Treatise on Marine Ecology and Paleoecology, Vol. 1. Mem. Geol. Soc., Amer., 67:935-940.

annual fluctuation

Harvey, H.W., 1934
Annual Variation of Planktonic Vegetation, 1933. J.M.B.A.XIX:775-392.

fluctuations

Havinga, B., 1950.
De wisselvaltige vangsten bij de garnalenvisserij
Visserij-Nieuws 2(12):137-142.

fluctuations

Kemp, S. 1938
Oceanography and the fluctuations of marine animals. Rep. Brit. Assoc. Adv. Sci., London, p. 85

FLUCTUATIONS

Konaga, Shunji, 1970.
On the short period fluctuation of Kuroshio. (In Japanese; English abstract)
Bull. Kobe mar. Obs. 183: 83-95

fluctuations

Margalef, Ramón, 1967.
Ritmos, fluctuaciones y sucesión.
In: Ecología marina, Monogr. Fundación La Salle de Ciencias Naturales, Caracas, 14:454-592.

fluctuations

Taning, A.V., 1953.
Pt. 5. Long term changes in hydrography and fluctuations in fish stocks.
Int. Comm. Northwest Atlantic Fish., Ann. Proc., 3:3-11, 10 textfigs.

fluctuations, diurnal

fluctuations diurnal

Rudnick, P., and J.D. Cochrane, 1951.
Diurnal fluctuations in bathythermograms.
J. Mar. Res. 10(3):257-262, 2 textfigs.

fluctuations, fisheries

fluctuations, fisheries

Uda, M., 1957.
A consideration on the long years trend of the fisheries fluctuation in relation to sea conditions. Bull. Jap. Soc. Sci. Fish., 23(7/8):368-372.

fluctuations, geomagnetic

Hermance, J.F., 1969.
Resolution of ocean floor magnetotelluric data.
J. geophys. Res., 74(23): 5527-5532.

fluctuations, oceanic

fluctuations, oceanic

Fukuoka, J., 1959.
On the variation of oceanic conditions in the sea adjacent to Japan. Ocean. Mag., Tokyo, 9(1):95-106.

fluctuations, seasonal

Redfield, A. C. and A. Beale, 1940.
Factors determining the distribution of populations of chaetognaths in the Gulf of Maine. Biol Bull., 79(3):459-487, 11 text figs.

fluid dynamics

fluid dynamics

Batchelor, G.K., 1967.
An introduction to fluid dynamics.
Cambridge Univ. Press., 615 pp.

fluid dynamics
Bhatragar, P.L., and G.K. Rajeswari, 1962.
The secondary flows in a non-Newtonian fluid between two parallel infinite oscillating plates. J. Indian Inst. Sci., 44(4):219-238.

fluid dynamics
Roy, M., and R. Legendre, 1963.
Dynamique des fluides - sur le definition et quelques applications d'un tenseur de dynalpie. Comptes Rendus, Acad. Sci., Paris, 256(12):2490-2494.

flume

flume
Kelling, Gilbert, and Peter F. Williams 1967.
Flume studies of the reorientation of pebbles and shells.
J. Geol. 75(3): 243-267.

fluorescence

fluorescence
Duursma, E.K., and J.W. Rommets, 1961.
Interprétation mathématique de la fluorescence des eaux douces, saumâtres et marines.
Netherlands J. Sea Res., 1(3):391-405.

fluorescence
Flemer, David A. 1969.
Continuous measurement of in vivo chlorophyll of a dinoflagellate bloom in Chesapeake Bay.
Chesapeake Sci., 10(2): 99-103.

fluorescence
Ivanoff, A., 1962
Au sujet de la fluorescence des eaux de mer.
Comptes Rendus Acad. Sci., Paris, 254(24): 4190-4192.

fluorescence
Ivanoff, A., M. Baues, J. Bortler, C. Canet, G. Copin-Montegut, M. LeRoy et A. Morel, 1969.
Résultats des observations effectuées en mer Tyrrhénienne à bord du navire océanographique Calypso en juillet 1964.
Cah. océanogr. 21 (Suppl. 2): 193-202.

fluorescent matter
Jerlov, N.G., 1963.
Optical oceanography.
In: Oceanography and Marine Biology, H. Barnes, Editor, George Allen and Unwin, 1:89-114.

fluorescence (blue)
Kalle, Kurt, 1963.
Uber das Verhalten und die Herkunft der in den Gewässern und in der Atmosphäre vorhandenen himmelblauen Fluorescenz.
Deutsche Hydrogr. Zeits., 16(4):153-166.

"fluorescent substance"
Kalle, Kurt, 1962
Über die gelösten organischen Komponenten im Meerwasser.
Kieler Meeresf., 18(3) (Sonderheft): 128-131.

fluorescence
Kalle, K., 1949.
Fluroeszenz und Gelbstoff im Bottnischen und Finnischen Meerbusen. Deut. Hydro. Zeit. 2(4): 117-124, 5 textfigs.

fluorescence
Momzikoff, André 1969.
Recherches sur les composés fluorescents de l'eau de mer. Identification de l'isoxanthoptérine, de la riboflavine et du lumichrome.
Cah. Biol. mar. 10(3): 221-230.

fluorescence
Momzikoff, André, 1969.
Substances fluorescentes des eaux de mer (Méditerranée).
Rapp. P.-v. Réun. Commn int. Explor. scient. Mer Mediterr., 19(4): 779-780.

fluorescence
Momzikoff, André, 1969.
Étude de quelques substances fluorescentes présentes dans deux échantillons de plancton marin.
Cah. Biol. mar., 10(4): 429-437.

flushing

flushing
Arons, A.B., and H. Stommel, 1951.
A mixing-length theory of tidal flushing. Trans. Am. Geophys. Union 32(3):419-421, 1 textfig.

flushing time
Ayers, J.C., 1956.
Population dynamics of the marine clam, Mya arenaria. Limnol., Oceanogr., 1(1):26-34.

flushing time
Craig, R.E., 1954.
A first study of the detailed hydrography of some Scottish West Highland sea lochs (Lochs Inchard, Kanarid and the Cairnbawn group).
Ann. Biol., Cons. Perm. Int. Expl. Mer, 10:16-19, Textfigs. 4-5.

flushing
Dorrestein, R., and L. Otto, 1960.
On the mixing and flushing of the water in the Ems-Estuary.
Verh. Kon. Ned. Geol. Mijnb. k. Gen., Geol. Ser. 19:83-102.

flushing
Garland, C.F., 1952.
A study of water quality in Baltimore Harbor.
Chesapeake Biol. Lab. Publ. 96:1-132, figs.

flushing
Garner, D.M., 1964.
The hydrology of Milford Sound.
New Zealand Dept. Sci. Ind. Res., Bull., No. 157: New Zealand Oceanogr. Inst. Memoir, No. 17:25-33.

flushing
Gibson, Blair W., 1959
A method for estimating the flushing time of estuaries and embayments. HO Tech. Rept. TR-62: 19 pp.

flushing, effect of
Hayami, Shoitiro, and Yoshiaki Fukuo, 1959.
A study on the productivity of Tanabe Bay. VI. On the exchange of water and the productivity of a bay with special reference to Tanabe Bay.
Rec. Oceanogr. Wks., Japan. Spec. No. 3:61-68.

flushing
Hela, I., C.A. Carpenter, Jr., and J.K. McNulty, 1957.
Hydrography of a positive, shallow, tidal, bar-built estuary (report on the hydrography of the polluted area of Biscayne Bay).
Bull. Mar. Sci., Gulf and Caribbean, 7(1):47-99.

flushing
Herlinveaux, R.H., 1962
Oceanography of Saanich Inlet in Vancouver Island, British Columbia.
J. Fish. Res. Bd., Canada, 19(1):1-37.

flushing
Jeffries, Harry P., 1962
The atypical phosphate cycle of estuaries in relation to benthic metabolism.
The Environmental Chemistry of Marine Sediments, Proc. Symp., Univ. R.I., Jan. 13, 1962 Occ. Papers, Narragansett Mar. Lab., No. 1: 58-67.

flushing
Jensen, A.J.C., 1944.
The hydrography of the Praestø Fjord.
Folia Geogr. Danica 3(2):47-55, 2 textfigs.

flushing
Kandeler, R., 1953.
Hydrographische Untersuchungen zum Abwasserproblem in den Buchten und Förden der Ostseeküste Schleswig-Holstein. Kieler Meeresf. 9(2):176-200, Pls. 9-14(18 figs.).

flushing
Ketchum, B.K., 1951.
The flushing of tidal estuaries. Sewage Ind. Wastes 23:198-208.

flushing
Ketchum, B.H., 1951.
The exchange of fresh and salt waters in tidal estuaries. J. Mar. Res. 10(1):18-38, 5 textfigs.

flushing
Ketchum, B. H., 1950.
Hydrographic factors involved in the dispersion of pollutants introduced into tidal waters.
J. Boston Soc., Civil Engineers, 37(3):296-314, 8 textfigs.

flushing
Ketchum, B.H., J.C.Ayers, and R.F. Vaccaro, 1952.
Processes contributing to the decrease of coliform bacteria in a tidal estuary. Ecology 33(2): 247-258, 4 textfigs.

flushing
Ketchum, B.H., and D.J. Keen, 1953.
The exchanges of fresh and salt waters in the Bay of Fundy and in Passamoquoddy Bay.
J. Fish. Res. Bd., Canada, 10(3):97-124, 11 textfigs.

flushing
Kohout, F.A., 1960.
Cyclic flow of salt water in the Biscayne aquifer of southeastern Florida.
J. Geophys. Res., 65(&0:2133-2142.

flushing
Krug, Joachim, 1963
Erneuerung des Wassers in der Kieler Bucht im Verlaufe eines Jahres am Beispiel 1960/61.
Kieler Meeresforsch., 19(2):158-174.

flushing

Marshall, Nelson, 1960.
Studies of the Niantic River, Connecticut with special reference to the bay scallop. Aequipecten irradians. Limnol & Oceanogr. 5(1): 86-105.

flushing

*Neal, Victor T., 1966.
Predicted flushing times and pollution distribution in the Columbia River estuary. Proc. 10th Conf. Coast Engng, 2: 1463-1480. Also in: Coll. Repr., Dep.Oceanogr., Oregon State Univ., 5.

FLUSHING

Newell, B.S., 1969.
Total transport and flushing times in the lower Tamar River.
Rep. Div. Fish. Oceanogr., CSIRO Aust. 45: 1-22 (mimeographed)

flushing

Ohtani, Kiyotaka and Yoshio Aiciba, 1970.
Studies on the change of hydrographic conditions in the Funka Bay. The annual change of the water in the bay. (In Japanese, English abstract) Bull. Fac. Fish. Hokkaido Univ., 20(4): 303-312.

flushing

Powers, Charles F., 1963.
Some aspects of the oceanography of Little Port Walter Estuary, Baranof Island, Alaska. U.S.F.W.S. Fish. Bull., 63(1):143-164.

flushing

Redfield, A.C., 1951.
The flushing of harbors and other hydrodynamic problems in coastal waters. In: Hydrodynamics in Modern Technology, M.I.T. 127-135, 8 textfigs.

Flushing

Saiz, F., M. Lopez-Benito, y E. Anadon, 1961.
Estudio hidrografico de la Ria de Vigo. II.
Inv. Pesq., Barcelona, 18:97-133.

flushing

Saiz, Fernando, Manuel López-Benito y Emilio Anadón, 1957
Estudio hidrográfico de la Ria de Vigo.
Inv. Pesq., Barcelona, 8:29-88.

flushing, tidal

Scrimger, J.A., 1960.
Temperature variations in Esquimalt Lagoon - a small landlocked body of water subject to tidal flushing.
Limnol. & Oceanogr., 5(4):414-424.

flushing

Shaw, Jack T., and G. R. Garrison, 1959.
Formation of thermal microstructure in a narrow embayment during flushing.
J. Geophys. Res., 64(5):533-540.

flushing

Sholkovitz, Edward R. and Joris M. Gieskes, 1971.
A physical-chemical study of the flushing of the Santa Barbara Basin. Limnol. Oceanogr. 16(3): 479-489.

flushing

Yamazi, Isamu, 1959.
A study on the productivity of Tanabe Bay. II. On some plankton indicating the water exchange in Tanabe Bay in August, 1957. Rec. Oceanogr. Wks., Japan, Spec. No. 3:23-30.

flysch

flysch

Esin, N.V., and M.T. Savin, 1970.
Abrasion of the flysch shore of the Black Sea. (In Russian; English abstract). Okeanologiia, 10(1): 126-131.

flysch

Hollister, Charles D., and Bruce C. Heezen, 1964. Primary structure of Atlantic deep-sea turbidite (Abstract).
Geol. Soc., Amer., Special Paper, No. 76:81-82.

flysch

Hsü, K.J., and S.O. Schlanger, 1971.
Ultrahelvetic flysch sedimentation and deformation related to plate tectonics.
Bull. geol. Soc. Am., 82(5): 1207-1218.

flysch

Kuenen, Ph. H. 1967.
Emplacement of flysch-type sand beds.
Sedimentology 9(3): 203-243

flysch

Kuenen, Ph.H., 1966
Light thrown on general problems by the Rolmanian results.
Sedimentology, 7(4): 323-324

flysch

Kuenen, Ph. H., 1964
Deep-sea sands and ancient turbidites.
In: Turbidites, A.H. Bouma and A. Brouwer, Editors, Developments in Sedimentology, Elsevier Publishing Co., 3:3-33.

flysch

Kuenen, Ph. H., 1959.
Turbidity currents a major factor in flysch deposition. Eclogae Geol. Helv. 1958, 58(3): 1009-1021.

flysch sediments

Kuenen, Ph. H., 1958.
Problems concerning source and transportation of flysch sediments.
Geol. en Mijnbouw, n.s., 20:329-339.

flysch

Nesteroff, W.D., and B.C. Heezen, 1963.
Essais de comparaison entre les turbidites modernes et le flysch.
Rev. Géogr. Phys. et Géol. Dyn., (2), 5(2):115-127.

Also in:
Trav. Lab. Géol. Dyn. et Centres de Recherches Géodynamiques, Fac. Sci., Univ. Paris, No. 2:

flysch

Nesteroff, Wladimir D., and Bruce C. Heezen, 1960.
Les dépôts de courantes de , le flysch et leur signification tectonique.
C.R. Acad. Sci., Paris, 250:3690-3692.

Also in:
Trav. Lab. Géol. Sous-Marine, 10(1960).

flysch

Oulianoff, Nicolas, 1960
Problème du flysch et geophysique.
Eclogae Geologicae Helvetiae, 53(1):155-160.

flysch

Stanley, Daniel J., Claude E. Gehin and Carlo Bartolini, 1970.
Flysch-type sedimentation in the Alboran Sea, western Mediterranean.
Nature, Lond., 228(5275): 979-983

flysch

Van der Lingen, G.J. 1968.
Preliminary sedimentological evaluation of some flysch-like deposits from the Makara Basin, central Hawke's Bay, New Zealand.
N.Z. Jl Geol. Geophys. 11(2): 455-477.

flysch

*von Rad, Ulrich, 1968.
Comparison of sedimentation in the Bavarian flysch (Cretaceous) and Recent San Diego Trough (California).
J. sedim. Petr., 38(4):1120-1154.

foaming

foam

Abe, Tomosaburo, 1963
In situ formation of stable foam in sea water to cause salty wind damage.
Papers in Meteorol. and Geophys., Tokyo, 14(2):93-108.

foam

Abe, Tomosaburo, 1962
On the stable foam of sea water in seas. (Preliminary report).
J. Oceanogr. Soc., Japan, 20th Ann. Vol., 242-250.

foam

Abe, T., 1958.
Physical oceanographic significance of the foaming of sea water. Appl. Statistics in Met., 8(3/4):
Also cited but not included in: Collected Reprints, Oceanogr. Lab., Meteorol. Res. Inst., Tokyo, June 1960.

foaming

Abe, T., 1957.
A supplementary note on the foaming of sea water. Rec. Oceanogr. Wks., Japan, n.s., 4(1):1-6.

foaming

Abe, T., 1956.
A study on the foaming of sea water - a tentative analysis of wind wave data in view of the foaming of sea water (II). Pap. Meteorol. Geophys., Tokyo, 7(2):136-143.

foam
Abe, T., 1956.
[A study on the foaming of sea water (12th report) A tentative analysis of wind wave data in view of the foaming of sea water Pt. 2.]
J. Ocean. Soc., Japan, 12(2):39-44.

foaming
Abe, T., 1956.
A study on the foaming of sea water. 11th Rept. A tentative analysis of wind wave data in view of the foaming of sea water.
J. Meteorol. Soc., Japan, (2) 34(3):169-175.

foam
Abe, T., 1955.
A study on the foaming of sea water. On the mechanism of the decay of bubbles and their size distribution in foam layer of sea water. Rec. Ocean. Wks., Japan, 2(1):1-6, 2 textfigs.

foam
Abe, Tomosaburo, 1955.
A study on the foaming of sea water.
Proc. Verb., Assoc. Int. Oceanogr. Phys., UGGI, (6):260-261.

foaming
Abe, T., 1955.
A study on the foaming of sea water - a tentative analysis of wind wave data in view point of the foaming of sea water. Pap. Met., Geophys., Tokyo, 6(2):164-171.

foam
Abe, T., 1955.
A study on the foaming of sea water. A note on the analogy between the coagulation process of colloidal particles and that of bubbles in foam layer of sea.
Pap. Meteorl. Geophys., Tokyo, 6(1):57-62.

foam
Abe, T., 1955.
[A study on the foaming of sea water (10th report) On the analogy between the coagulation process of colloidal particles and that of the bubbles in foam layer of sea water.] J. Ocean. Soc., Japan, 11(1):13-18.

foam
Abe, T., 1955.
A study on the foaming of sea water. (1). On the mechanism of decay of bubbles and their size distribution in foam layer of sea water.
Pap. Met. Geophys., Tokyo, 5(3/4):240-247, 6 textfigs.

foaming
Abe, T., 1954.
A study of the foaming of sea water.
Proc.-Verb., Assoc. Océanogr. Phys., Rome, 6:260-261.

foaming
Abe, T., 1954.
A study on the foaming of sea water (8th report) - on the mechanism of the decay of foam layer of sea water. Part 3. J. Ocean. Soc., Japan, 10(1):15-21, 4 textfigs.

foam
Abe, T., 1953.
A study on the foaming of sea water. On the mechanism of the decay of foam layer of sea water. Rec. Ocean. Wks., Japan, n.s., 1(2):18-24, 5 textfigs., 1 photo.

foam
Abe, T., 1953.
[A study on the foaming of sea water (7th report) - on the mechanism of decay of the foam layer of sea water (Part 3).] J. Ocean. Soc., Japan, 9(2):85-93, 5 textfigs.

foam
Abe, Tomosaburo, Teruaki Ono and Motoaki Kishino, 1963
A fundamental study on the prevention of the salty wind damages due to the foaming of sea water. (Preliminary report). (In Japanese; English abstract).
J. Oceanogr. Soc., Japan, 18(4):185-192.

foam
Abe, Tomosaburo, and Akira Watanabe, 1964
On the stable foam of sea water in situ - the electrical specific conductivity of the stable foam of sea water. (In Japanese; English abstract)
IN: Studies in oceanography dedicated to Professor Hidaka in commemoration of his sixtieth birthday, 254-259.

foam
Akcetin, D., and F.H. Contable 1965.
The cause and the dynamics of the sea foam at Tuzla on the Mediterranean coast of Turkey.
Istanb. Univ. Fen Fak. Mecm. (C) 30(3/4): 125-136

foam
DeVries, S.J., 1958.
Foam stability. 5. Mechanisms of film rupture.
Recueil Trav. Chim. Pays-Bas, 77(6):441-446.

foam
Droppleman, J.D., 1970.
Apparent microwave emissivity of sea foam.
J. geophys. Res., 75(3): 696-698.

foam
Hidaka, Toshiro, 1965.
Les Fucus et la formation de l'écume marine.
C. R. Acad. Sci., Paris, 260(22):5861-5864.

foam
LaFond, E.C., and P.V. Bhavanarayana, 1959.
Foam on the sea.
J. Mar. Biol. Assoc., India, 1(2):228-232.

foam
Maynard, Nancy G. 1968.
Aquatic foams as an ecological habitat.
Zeits. allg. Mikrobiol. 8(2): 119-126

foaming
Miyake, Y., 1943.
Foaming properties of sea water. Science, Japan, 13:358-359.
Chem. Abstr., 1951, 45:9935.

foaming
Miyake, Y., and T. Abe, 1948
A study on the foaming of sea water.
J. Mar. Res. 7(2):67-73, figs.1-3.

foaming
Seilkopf, H., 1956.
Remarks on the effect of plankton-factor of the foaming of sea water in determining the equivalent velocities for Beaufort estimation of the wind force at sea. Ann. Meteorol. 7(10/12):404-409.

FOAM
Southward, A.J., 1953.
Sea foam. Nature 172(4388):1059-1060.

foam
Tsyban, A.V. 1971
Sea foam as an ecological niche for bacteria. (In Russian; English abstract)
Gidrol. Zh. 7(3): 14-24

foam, effect of

foam, effect of
Allen, J.R.L., 1967.
A beach structure due to wind-driven foam.
J. Sed. Petr. 37(2): 691-692

foam, effect of
Nordberg, W., J. Conaway, Duncan B. Ross, and T. Wilheit, 1971.
Measurements of microwave emission from a foam-covered, wind-driven sea.
J. atmos. Sci., 28(3): 429-435

focal mechanisms
Molnar, Peter and Lynn Sykes, 1968.
Focal mechanisms in the Caribbean, middle America, and northern South America. Trans. Fifth Carib. Geol. Conf., St. Thomas, V.I., 1-5 July, 1968, Peter H. Mattson, editor. Geol. Bull. Queens Coll., Flushing 5: 7. (Abstract only).

fog

fog

See also: steam fog

fog
Andreev, V., 1938
Kuril'skie Ostrova. (Kurile Islands) Morsk. sborn. 1938 (11):75-87, (12):87-99.

fog
Anon. 1914
Ice observation, meteorology and oceanography in the North Atlantic Ocean. Rept. on the work carried out by the S.S. "Scotia", 1913, 139 pp.

fog
Balabanova, V.N., 1963.
On the capability of certain substances to crystallize supercooled fog.
Izv., Akad. Nauk, SSSR, Ser. Geofiz. (6):978-984.
Translation for AGU.
(6):598-603.

fog, effect of
Barteneva, O.D., and E.A. Polyakova, 1965.
A study of attenuation and scattering of light in a natural fog due to its microphysical properties. (In Russian; English abstract).
Fisika Atmosferi i Okeana, Izv. Akad. Nauk SSSR 1(2):193-207.

fog
Chakravortty, K.C., 1948
Fog at Calcutta. Sci. Notes; India Meteorol. Dept. X(124):133-140.

fog
Central Meteorological Observatioy, 1949.
Report on sea and weather observations on Antarctic whaling ground (1948-1949). Ocean. Mag., Tokyo, 1(3):142-173, 11 textfigs.

fog
Central Meteorological Observatory, 1949
Report on sea and weather observation on Antarctic Whaling Ground (1947-48). Ocean. Mag., Japan, 1(1):49-88, 17 text figs.

fogs
Chavalier, S., 1893.
Fogs along the northern coasts of China. Ann. Rep Shanghai Met. Soc. 1:18.

fog
Dietrich, G., 1963.
Die Meere.
Die Grosse Illustrierte Länderkunde, C. Bertelsmann Verlag, 2:1523-1606.

fog
Dietrich, G., 1951.
Influences of tidal streams on oceanographic and climatic conditions in the sea as exemplified by the English Channel. Nature 168(4262):8-11, 3 textfigs.

fog
Eggvin, J., and P. Spinnangr., 1944
Fog. forecasting and sea-surface temperature.
Fiskeridirektoratets Skrifter, Serie Havundersøkelse, 7(9):3-20, 12 textfigs.

fog
Eldridge, R.G., & J.C. Johnson, 1962
Distribution of irradiance in haze and fog.
J. Optical Soc. Amer. 52(7): 787-796.

fogs
Evans, F., 1967.
Local sea fogs on the Northumberland Coast.
Proc. Univ. Newcastle-upon-Tyne Phil. Soc.
1 (13):162-169. Also in: Contrib. Dove mar. Lab. 11 (98).

fog.
Fleagle, R. G., 1953.
A theory of fog formation. J. Mar. Res. 12(1): 43-50, 2 textfigs.

fog
Grunow, J., and W. Leistner, 1958
Registrierungen der Verdunstungsgrösse in Wyk/ Föhr und in Hohenpeissenberg. Berichte Deut. Wetterdienstes 7(49): 1-24.

fog
Gutsell, B. V., 1949
An introduction to the geography of Newfoundland. Canada, Dept. Mines and Resources, Geogr. Bur., Info. Ser. No.1: 85 pp., 20 text figs., 3 fold-in maps.

fog
Harami Keiji, 1969.
On the 'steam fog in the Japan Sea.
(In Japanese; English abstract).
Umi to Sora, 44(2/3): 47-53

fog
Hatzikakidis, Athan. D., 1963.
Relations between air and sea temperatures. (In Greek).
Praktika, Hellénic Hydrobiol. Inst., Athens, 8(2) 1:109 pp.

fog
Japan, Hakodate Marine Observatory, 1970.
Report of the marine meteorological observations on sea fog and heat exchange at the ocean-air interface, in June, 1969. Supplement: on the drift-ice east of South Saghalien in the Okhotsk Sea in June 1969. (In Japanese)
Mar. met. Rept., Hakodate mar. Obs. 30: 67 pp.

fog
Japan, Hakodate Marine Observatory, 1969.
Report of the marine meteorological observations on sea fog and heat exchange at the ocean-air interface in June, 1968. (In Japanese)
Mar. Met. Rept. Hakodate mar. Obs., 28:1-40

fog
Japan, Hakodate Marine Observatory, 1968.
Report of the sea fog observations in the adjacent sea of Hokkaido (1965, 1966). (In Japanese)
Mar. Met. Rept., 25:1-19.

fog
Japan, Hakodate Marine Observatory, 1963.
Report of the sea fog observations in the adjacent sea of Hokkaido, 1962. (In Japanese).
Mar. Meteorol. Rept., Hakodate Mar. Obs., 7(1): 241-281.

fog
Japan, Hakodate Marine Observatory, 1962.
Report of the sea fog observations in the adjacent sea off Hokkaido (1961). (In Japanese).
Mar. Meteorol. Rept., Hakodate Mar. Obs., 6(2): 47-100.

fog (data)
Japan, Hakodate Marine Observatory, 1960.
[Report of the sea fog observations in the adjacent sea off Hokkaido (1960).]
Mar. Meteorol. Rept, Hakodate Mar. Obs., 4(2): 1-45.

fog
Japan, Hakodate Marine Observatory, 1959.
[Report of the sea fog observations in the adjacent sea off Hokkaido (1959).]
Mar. Meteorol. Rept., 3(1):249-265.

fog
Japan, Kobe Marine Observatory, 1951
Meteorology and oceanography in the Seto Inland Sea.
Bull. Kobe Mar. Obs., No. 161: 211 pp.

fog
Kawamura, S., F. Tsubata, R. Taguchi, H. Kitamura and J. Sugiura, 1950.
Report of the sea and weather conditions on Antarctic whaling ground (1949-1950). Ocean. Mag. Tokyo, 2(4):149-180, 11 textfigs.

fog
Koninklijk Nederlands Meteorologisch Instituut, 1949.
Rode Zee en Golf van Aden. Oceanographic and meteorological data. No. 129:26 pp. (charts).

fog
Koninklijk Nederlands Meteorogisch Instituut, 1949.
Sea areas round Australia. Oceanographic and meteorological data. No. 124(Atlas):46 charts.

fog
Kopanev, I.E., 1959.
Fog and "snow" haze in Antarctica. (In Russian)
Inform. Biull. Sovetsk. Antarkt. Exped., 1( ):

Translation:
Scripta Tecnica, Inc., Elsevier Publ. Co., 385-386.

fog
Leipper, Dale F. 1968.
The sharp smog bank and California fog development.
Bull. Am. Met. Soc. 49 (4): 354-358.

fog
Leipper, D. F., 1948.
Fog development at San Diego, California.
J. Mar. Res. 7(3):337-346, 3 textfigs.

fog
Lumb, Frank E., 1965.
Meteorology of the North Sea.
Ser. Atlas, Mar. Environment, Folio 9:9 pls.

fogs
Matui, H., 1948.
[Chemical studies of condensation and sublimation nuclei. II. Chemical studies on the fogs at Hitoyosi, Kumamoto Prefecture.] (Japanese)
J. Met. Soc., Japan, 26(6), 2nd ser.:169-172, with English summary, p. 14.

fog
Miyake, Y., and K. Saruhashi, 1950.
The estimation of the height of fog by observing the ultra-violet solar radiation. Geophys. Mag., Tokyo, 21(2):122-125, 4 textfigs.

fog
Nagel, J.F., 1962.
Fog precipitation on Africa's southwest coast.
Notos, 11(1/4):51-60.

fog
Netherlands Hydrographer, 1965.
Some oceanographic and meteorological data of the southern North Sea.
Hydrographic Newsletter, Spec. Issue No. to Vol. I: numerous pp. not sequentially numbered.

fog
Nurminen, A., 1954.
Precipitation of ice needles as a factor causing disappearance of fog. Geophys., Helsinki, 4(4): 226-230, 2 textfigs.

fog
Petterssen, S., 1939
Some aspects of formation and dissipation of fog. Geophys. Publ.12(10).

fog
Ogata, T., 1948.
[The sea fog at the South Kurile islands.] (In Japanese). J. Met. Soc., Japan, 2nd ser., 26(4):107-118, with English summary, p. 9.

fog
Ogiwara, S., 1949.
[Experimental studies on the evaporation of fog particles.] J. Met. Soc., Japan, 27(1):10-15. [In Japanese, with English summary, p. 36.]

fog
Ogiwara, S., and T. Nakanisi, 1949.
[Report on the observation of sea fog at Kusiro.]
J. Met. Soc., Japan, 27(1):23-27, textfigs.
[In Japanese, with English summary, p. 37.]

fog
Ogiwara, S., T. Nakanisi, and M. Siobara, 1948.
[Evaporation of fog particles and the relation between the visibility and the fog particles.]
(In Japanese). J. Met. Soc., Japan, 2nd ser., 26(7):6-12 (184-190), with English summary, p. 23.

fogs
Patton, C.P., 1956.
Climatology of summer fogs in the San Francisco Bay area.
Univ. California Publ., Geogr., 10(3):113-200.

fog
Pedersen, K., and M. Todsen, 1960
Some measurements of the micro-structure of fog and stratus-clouds in the Oslo-area.
Geofys. Publik., Geophysica Norvegica, 21(7): 1-15.

fog
Robinson, E., W.C. Thuman and E.J. Wiggins, 1957.
Ice fog as a problem of air pollution in the Arctic. Arctic, 10(2):88-104.

fog
Rodhe, Bertil, 1962.
The effect of turbulence on fog formation.
Tellus, 14(1):49-86.

Also: Meddelande Meteorol. Inst., Uppsala, No. 83

fog
Roll, U., 1949.
Über eine scheinbar Anomalie der Temperatur-differenz Luft-Wasser. Deutsch. Hydro. Zeit. 2(4):134-137, 4 textfigs.

fog
Roy, A.K., 1951.
On the incidence of fog during winter in Colombo and neighborhood. Indian J. Met. Geophys 2(4):305-306, 1 textfig.

fog
Rubin, M.J., 1958.
An occurrence of steam fog in Antarctic waters.
Weather 13(7):235-238.

fog
Saunders, Peter M., 1964.
Sea smoke and steam fog.
Quart. Jour. R. Meteorol. Soc., 90(384):156-165.

fog
Saunders, W.E., 1950.
Method of forecasting the temperature of fog formation. Met. Mag. 79(938), M.O. 533:213-219, 4 textfigs.

fog
Sawada, T., 1964.
Statistical investigation on the relation between the occurrence of sea fog and oceanic conditions. (In Japanese; English abstract).
Bull. Hakodate Mar. Obs., No. 11:24-30(112-118).

fog
Sawada, Teruo, 1963
A study of the sea fog in the Tsugaru Straits. (In Japanese; English abstract).
Umi to Sora, 39(3):105-113.

fog
Sawada, T., 1963.
13. A study of the sea fog in the north-western Pacific. (1). (In Japanese; English abstract).
Bull. Hakodate Mar. Obs., 10:1-8.

fog
Spinnangr, G., 1950.
Fog and fog forecasting in northern Norway. Met. Annaler 3(4):75-136, 24 textfigs.

fogs
Stevenson, Robert E., 1963.
The summer fogs along the Yorkshire coast, England.
In: Essays in Marine Geology in honor of K.O. Emery, Thomas Clements, Editor, Univ. Southern California Press, 145-170.

fog
Tannehill, I.R., 1938.
Weather around the world. Princeton Univ. Press, 200 pp., 55 textfigs.

fog
Ultrasonic Corporation, 1948.
Contributions to the study of natural fog. Final Rept., Pt. II on (Navy Fog Dispersal Project) June 30, 1948, 1 v. incl. illus. diagrs 279 refs. Unclassified. (Listed as item U4951 in Technical Information Pilot of 31 Aug. 1949).

fog.
Wiener, F.M., J.H. Ball, & C.M. Gogos, 1961
Some micrometeorological measurements in ocean fog.
J. Geophys. Res., 66(11): 3974-3978.

Fog, absence of
Desai, B.N., 1969.
Discussion of the observations of the Oceanographer Discovery off Somalia during August 1964 to explain the absence of fog.
Indian J. Met. Geophys. 20(4):369-372

fog formation

fog formation
Nurminen, A., 1957.
Variations in the difference between the dew point of the air and the water temperature before sea fog. Mitt. Meteorol. Zentralanstalt, No. 42:20 pp.

folding

folding
Scholl, David W., Edwin C. Buffington and David M. Hopkins 1968.
Geologic history of the continental margin of North America in the Bering Sea.
Mar. Geol. 6(4):297-330.

FOOD

food for birds

food of birds
*Ashmole, N. Philip, and Myrtle J. Ashmole, 1967.
Comparative feeding ecology of sea birds of a tropical oceanic island.
Bull. Peabody Mus. N.H., Yale Univ. 24:1-131.

food of birds
*Ashmole, Myrtle J., and N. Philip Ashmole, 1968.
The use of food samples from sea birds in the study of seasonal variation in the surface fauna of tropical oceanic areas.
Pacif. Sci., 22(1):1-10.

food of birds
Bahamonde, N., N., 1955.
XI. Alimentacion de cormoranes o cuevos marinos.
Invest. Zool. Chilenas 2(8):132-133.

food of birds
Beck, J.R., 1969.
Food, moult and age of first breeding in the cape pigeon, Daption capensis Linnaeus.
Br. Antarct. Surv. 21:33-44
Bull.

food of birds
Duncan, K.W., 1969.
The food of the black shag (Phalacrocorax carbo novaehollandiae) in Otago inland waters.
Trans. R. Soc. N.Z., Biol. Sci. 11 (2): 9-23.

food of oceanic birds
Løvenskiold, Herman L., 1963.
Avifauna Svalbardensis.
Norsk Polarinstitutt Skrifter, No. 129:460 pp.

food of birds
MacNae, Wm., 1960
Greater flamingoes eating crabs. Ibis 102: 325-326.

food of birds
Nilsson, Leif, 1969.
Food consumption of diving ducks wintering at the coast of south Sweden in relation to food resources.
Oikos, 20(1): 128-135

food of seabirds
Pearson, T.H., 1968.
The feeding biology of sea-bird species breeding on the Farne Islands, Northumberland.
J. Anim. Ecol., 37(3): 521-552.

food of birds
Rae, Bennet B., 1968.
The food of cormorants and shags in Scottish estuaries and coastal waters.
Marine Res., Dept. Agric. Fish., Scotland, 1: 1-16

food of birds
Threlfall, William, 1968.
The food of three species of gulls in Newfoundland.
Can. Fld. Nat., 82(3):176-180.

food of birds
Tomo, Aldo P., 1971.
Trophic chains observed in Paradise Harbor (Antarctic Peninsula) related to variations in the fertility of its waters. (English and Spanish abstracts). In: Fertility of the Sea, John D. Costlow, editor, Gordon Breach, 2: 593-602.

food of birds

Vauk, Gottfried und Konrad Löhmer, 1969
Ein weiterer Beitrag zur Ernährung der Silbermöwe (*Larus argentatus*) in der Deutschen Bucht.
Veröff. Inst. Meeresforsch. Bremerh. 12(2): 157-160.

food of chaetognaths

Rakusa-Suszczewski, S. 1969.
The food and feeding habits of Chaetognatha in the seas around the British isles.
Pol. Arch. Hydrobiol. 16(29)(2):213-232

food

Borgstrom, Georg, Editor, 1962
Fish as food, Vol. 2. Nutrition, sanitation and utilization.
Academic Press, 778 pp., $25.00.

food

Imai, T., and M. Hatanaka, 1950.
Studies on marine non-colored flagellates, Monas sp., favorite food of larvae of various marine animals. 1. Preliminary research on cultural requirements. Sci. Repts., Tohoku Univ., 4th ser 18(3):304-315.

food chain

food chains

Bernhard, M. 1969.
Studies on the radioactive contamination of the sea: annual report for 1967.
CNEN-EURATOM Rept RT/Bio (68) 60:66 pp. (mimeographed).

food chains

Bernhard, M., preparator, 1965.
Studies on the radioactive contamination of the sea, annual report 1964.
Com. Naz. Energ. Nucleare, La Spezia, Rept., No. RT/BIO (65) 18:35 pp.

food chain

Bieri, Robert, 1966.
Feeding preferences and rates of the snail, *Ianthina prolongata*, the barnacle, *Lepas anserifera*, the nudibranchs, *Glaucus Atlanticus* and *Fiona pinnata*, and the food web in the marine neuston.
Publs Seto mar. biol. Lab., 14(2):161-169.

food chains

Bilio, Martin 1969.
La variabilità fra pescate effettuate con rete a strascico nel Mar Ligure.
Pubbl. Staz. Zool. Napoli 37 (2 Suppl.): 115-131

food chains

Blumer, M., M.M. Mullin and R.R.L. Guillard 1970
A polyunsaturated hydrocarbon (3,6,9,12,15,18-heneicosahexane) in the marine food web.
Mar. Biol. 6(3):226-235.

food chain

Bougis, Paul, 1964.
Note préliminaire sur l'estimation du phytoplancton utilisable par le zooplancton (phytoplancton efficace) à l'aide d'un test biologique.
Vie et Milieu, Lab. Arago, Suppl. No. 17:165-167.

FOOD CHAINS

Chipman, Walter A., 1966.
Food chains in the sea.
In: Radioactivity and Human Diet, Pergamon Press, 421-453.

food chains

Clemens, W.A., 1963.
Pastures of the sea.
Occ. Pap., California Acad. Sci., 41:8 pp.

food chains

Davis, J.J., and R.F. Foster, 1958.
Bioaccumulation of radioisotopes through aquatic food chains.
Ecology, 39(3):530-535.

food chains

Doty, Maxwell S., and Gertrudes Aguilar-Santos 1970.
Transfer of toxic algal substances in marine food chains.
Pacific Sci. 24(3): 351-355.

food chain

Gibor, A., 1957.
Conversion of phytoplankton to zooplankton.
Nature 179(4573):1304.

food chains

Furnestin, Marie Louise, Claude Maurin, Jean Y. Lee, et René Raimbault 1966.
Éléments de planctonologie appliquée.
Rev. Trav. Inst. Pêches marit. 30(2/3): 278 pp.

food chains

Hanaoka, Tasuku, 1968.
International symposium on marine food chains held in Denmark, July 1968. (In Japanese; English abstract). Bull. Plankton Soc., Japan, 15(2): 38-41.

food chain

Isaacs, John D., 1969.
The nature of oceanic life.
Scient. Am., 221(3):146-160,162. (popular).

food chain

Kayama, M., Y. Tsuchiya and J.F. Mead, 1963.
A model experiment of aquatic food chain with special significance in fatty acid conversion.
Bull. Jap. Soc. Sci. Fish., 29(5):452-458.

food chains

Khailov, K.M., T.S. Petipa, Z.Z. Finenko and L. Tsareva, 1971.
Studies of food nets in the marine plankton conducted at the USA Institute of Marine Resources. (In Russian). Okeanologiia 11(3): 549-556.

food chain

Lewin, Ralph A., 1970
Toxin secretion and tail autotomy by irritated *Oxynoe panamensis* (Opisthobranchia, Sacoglossa).
Pacific Sci., 24(3): 356-358.

food chains

Margalef, Ramón 1967.
The food web in the pelagic environment.
Helgoländer wiss. Meeresunters. 15(1/4): 548-558.

food chains

Marples, T.G., 1966.
A radionuclide tracer study of arthropod food chains in a *Spartina* salt marsh ecosystem.
Ecology, 47(2):270-277.

food cycles

McHugh, J.L., 1967.
Estuarine nekton.
In: Estuaries, G.H. Lauff, editor, Publs Am. Ass. Advmt. Sci., 83:581-620.

food chain

Nesis, K.N., 1965.
Some problems involved with the food structure of marine biocoenosis. (In Russian).
Okeanologiia, Akad. Nauk, SSSR, 5(4):701-714.

food chains

Osterberg, Charles, W.G. Pearcy and Herbert Curl, Jr., 1964
Radioactivity and its relationship to oceanic food chains.
J. Mar. Res., 22(1):2-12.

food chains

Riley, G.A., 1963
20. Theory of food-chain relations in the ocean. In: The Sea, M.N. Hill, Edit., Vol. 2. (IV) Biological Oceanography, Interscience Publishers, New York and London, 438-463.

food cycle

Seki, Humitake, 1966.
Studies on microbiol participation to food cycle in the sea. III. Trial cultivation of brine shrimp to adult in a chemostat.
J. Oceanogr. Soc. Japan, 22(3):105-110.

food relationships

Sokolova, M.N., 1957.
[Feeding of some carnivorous benthic deep-sea invertebrates of the Far-Eastern Seas.]
Trudy Inst. Okeanol., 20:279-301.

Food chains

Steele, J.H., editor, 1970.
Marine food chains.
Univ. Calif. Press, 552 pp. $13.50.

food chain

Steele, J. H., 1965.
Some problems in the study of marine resources.
ICNAF. Spec. Publ., No. 6:463-476.

food relationships

Turpaeva, E.P., 1957.
[Food relationships between the dominant species of marine bottom biocenoses.] Trudy Inst. Okeanol. 20:171-185.

food chain

Vinogradov, M.E., and N.M. Voronina, 1965.
Some peculiarities of plankton distribution in the Pacific and Indian Equatorial areas. (In Russian; English abstract).
Okeanolog. Issled. Rezult. Issled. Programme Mezhd. Geofiz. Goda, Mezhd. Geofiz. Komitet Presidiume, Akad. Nauk, SSSR, No. 13:128-136.

food, chemistry of

food, chemistry of

Ogiino, C., 1963.
Studies on the chemical composition of some natural foods of aquatic animals. (In Japanese; English summary).
Bull. Jap. Soc. Sci. Fish., 29(5):459-462.

food of amphipods

food of amphipods

Greze, I.I., 1968.
Feeding habits and food requirements of some amphipods in the Black Sea.
Marine Biol., 1(4):316-321.

food of amphipods

Greze, I.I., 1968.
ПИЩЕВЫЕ СПЕКТРЫ amphipods of the Black Sea. (In Russian)
Revue Roumaine Biol. (Zool.), 13(6): 393-402

foods of amphipods

Keith, Donald E., 1969.
Aspects of feeding in Caprella californica Stimpson and Caprella equilibra Say (Amphipoda).
Crustaceana 16(2): 119-124.

food of amphipods

Saunders, Carolyn Gardella 1966.
Dietary analysis of caprellids (Amphipoda).
Crustaceana 10(3): 314-316.

food of annelids

Hamond, Richard 1969.
On the preferred foods of some autolytoids (Polychaeta, Syllidae).
Cah. Biol. mar. 10(4): 439-445.

food of annelids

Röder, Heinrich, 1971.
Gangsysteme von Paraonis fulgens Levinsen 1883 (Polychaeta) in ökologischer, ethologischer und aktuopaläontologischer Sicht. Senckenbergiana maritima 3(1): 3-51.

food of cirripeds

Moyse, John, 1963
A comparison of the value of various flagellates and diatoms as food for barnacle larvae.
J. du Cons., 28(2):175-187.

food of cladocerans

Inoue, Motoo and Mitsuyoshi Aoki, 1971.
Reproduction of Cladocera, Diaphanosoma sp. cultured with seawater-acclimatized Chlorella as basic diet under different chlorinity.
J. Tokai Univ., Coll. mar. Sci. Technol. 5: 1-8.

Food of Coelenterates

Bieri, Robert, 1970.
The food of Porpita and niche separation in three neuston coelenterates. Publ. Seto mar. biol. Lab., 17(5): 305-307.

food of coelenterates

Fraser, J.H., 1969.
Experimental feeding of some Medusae and Chaetognatha.
J. Fish. Res. Bd., Can., 26(7): 1743-1762.

food of coelenterates

Johannes, R.E., Stephen Lee Coles and Nancy T. Kuenzel, 1970.
The role of zooplankton in the nutrition of some scleractinian corals. Limnol. Oceanogr., 15(4): 579-586.

food of copepods

food of copepods

Anraku, Masateru, and Makoto Omori, 1963
Preliminary survey of the relationship between the feeding habit and structure of the mouthparts of marine copepods.
Limnology and Oceanography, 8(1):116-126.

food of copepods

Arashkevich, E.G., 1969.
The character of feeding of copepods in the northwestern Pacific. (In Russian; English abstract). Okeanologiia, 9(5): 857-873.

Food of Copepods

Arashkevic, E.G. and A.G. Timonin, 1970.
The nutrition of the Copepoda of the tropical Pacific. (In Russian).
Dokl. Akad. Nauk SSSR, 191(4): 935-938.

food, copepods

Gilat, Eliezer, 1967.
On the feeding of a benthonic copepod, Tigropius brevicornis O.F. Müller.
Bull. Sea Fish Res. Stn Israel, 45:79-95.

food of copepods

Itoh, Katsuhiko, 1970.
A consideration on feeding habits of planktonic copepods in relation to the structure of their oral parts. Bull. Plankt. Soc., Japan, 17(1): 1-10. In Japanese; English abstract

food of copepods

Lillelund, Kurt, and Reuben Lasker 1971.
Laboratory studies of predation by marine copepods on fish larvae
Fish Bull. nat. mar. fish. Serv. NOAA. 69(3): 655-667.

food of copepods

Paffenhöfer, G.A. and J.D.H. Strickland, 1970.
A note on the feeding of Calanus helgolandicus on detritus. Mar. Biol., 5(2): 97-99.

food of copepods

Perueva, E.G., 1971.
Some quantitative regularities in the feeding of copepods Calanus. (In Russian; English abstract). Okeanologiia 11(2): 283-292.

food of copepods

Perueva, E.G., and B.J. Vilenkin, 1970.
Nutrition of Calanus glacialis (Jashnov) under different concentrations of algae.
(In Russian)
Dokl. Akad. Nauk SSSR, 194(4): 943-945

food of copepods

Samyshev, E.Z., 1970.
On the nutrition of Nannocalanus minor (Claus) and Eucalanus subtenuis (Giesbrecht) in Guinea Bay.
(In Russian; English abstract).
Gidrobiol. Zh. 6(4): 44-59

food copepods

Urry, D.L., 1965.
Observations on the relationship between the food and survival of Pseudocalanus elongatus in the laboratory.
Jour. Mar. Biol. Assoc., U.K., 45(1):49-58.

food of crustacea

Soldatova, I.N., E.A. Tsikhon-Lukanina, G.G. Nikolaeva, and T.A. Lukasheva, 1969.
On the transformation of the energy of food in marine crustaceans. (In Russian; English abstract). Okeanologiia, 9(6): 1087-1094.

food of shrimp

Belogrudov, E.A., 1971.
About feeding of commerce shrimp in different regions of the Far Eastern seas. (In Russian).
Izv. Tikhookean. nauchno-issled. Inst. ribn. Khoz. Okeanogr. 75: 117-120.

food of decapods

Burukovsky, R.N. 1969.
The feeding habits of the sea shrimp (Penaeus duorarum) off Mauritania. (In Russian).
Trudy vses. nauchno-issled. Inst. morsk. rybn. Okean (VNIRO) 65: 417-423.

food for decapods

Forss, Carl Albert, and Harold G. Coffin, 1960
The use of the brine shrimp nauplii Artemia salina, as food for the laboratory culture of decapods. Walla Walla Coll. Publ. Dept. Biol. Sci. and Biol. Sta. No. 26: 17 pp.

food of decapods

Goldberg, Walter M. 1971.
A note on the feeding behavior of the snapping shrimp Synalpheus fritzmuelleri Coutière (Decapoda, Alpheidae)
Crustaceana 21(3): 318-320.

food of decapods

Hollmann, R., 1969.
Die Entstehung fossilisationsfähiger Schalen-Frassreste dargestellt am Nahrungserwerb von Homarus gammarus (Crustacea, Decapoda)
Helgoländ. wiss. Meeresunters. 19(3): 401-419.

food of shrimp

Judkins, David C., and Abraham Fleminger 1972.
Comparison of foregut contents of Sergestes similis obtained from net collections and albacore stomachs.
Fish. Bull. U.S. Nat. Fish. Serv. 70(1): 217-223

food of decapods

Kanazawa, Akio, Makoto Shimaya, Mitsuyasu Kawasaki and Ken-ichi Kashiwada, 1970.
Nutritional requirements of prawn. 1. Feeding on artificial diet.
Bull. Jap. Soc. scient. Fish. 36(9): 949-954

food, decapods

McLaughlin, P.A., and J.F. Hebard, 1959.
Stomach contents of the Bering Sea king crab.
U.S.F.W.S. Spec. Sci. Rept., Fish., No. 291:1-5.

food of echinoderms

Brun, Einar, 1972.
Food and feeding habits of Luidia ciliaris Echinodermata: Asteroidea. J. mar. biol. Ass. U.K. 52(1): 225-236.

food of echinoderms

Ferguson, John Carruthers, 1969.
Feeding activity in Echinaster and its induction with dissolved nutrients. Biol. Bull. mar. biol. Lab., Woods Hole, 136(3): 374-384.

food of echinoderms

Himmelman, J.H. and D.H. Steele, 1971.
Foods and predators of the green sea urchin Strongylocentrotus droebachiensis in Newfoundland waters. Mar. Biol. 9(4): 315-322.

food of echinoderms

Litvinova, N.M. and M.N. Sokolova, 1971.
On the feeding of deep-sea ophiuroids Amphiophiura. (In Russian; English abstract).
Okeanologiia 11(2): 293-301.

food echinoderms

Roushdy, H.M., and V.K. Hansen, 1960.
Ophiuroids feeding on phytoplankton.
Nature 188(4749):517-518.

food for echinoderms

Rutman, J. and L. Fishelson, 1969.
Food composition and feeding behavior of shallow-water crinoids at Eilat (Red Sea).
Marine Biol., 3(1): 46-57.

food of echinoderms

Schoener, Amy 1971
On the importance of pelagic species of copepods to the diet of some benthic invertebrates (Ophiuroidea).
Crustaceana 21(2): 153-160.

food, effect of

food of fishes, effect of

Crozier, George F., 1969.
Effects of controlled diet on the morphological color change of a marine teleost. J. Exp. mar. Biol. Ecol., 4(1): 1-8.

food, effect of

Dunathan, Jay P. R.M. Ingle and W.K. Havens Jr. 1969.
Effects of artificial foods upon oyster fattening with potential commercial applications.
Techn. Ser. Fla Bd Conserv., 58: 39 pp.

food, effect of

Edmondson, W.T., 1964.
The rate of egg production by rotifers and copepods in natural populations as controlled by food and temperature.
Verh. Internat. Verein. Limnol., 15:673-675.

food, effect of

Edmondson, W.T., 1965.
Reproductive rate of planktonic rotifers as related to food and temperature in nature.
Ecological Monographs, 35:61-111.

food, effect of

Hallam, A., 1965.
Environmental causes of stunting in living and fossil marine benthonic invertebrates.
Paleontology, 8(1):132-155.

food, effect of

*Singarajah, K.V., John Moyse and E.W. Knight-Jones, 1967.
The effect of feeding upon the phototactic behaviour of cirripede nauplii.
J. exp. mar. Biol. Ecol., 1(2):144-153.

food energy

Tsikhon-Lukanina, E.A. and T.A. Lukasheva, 1970.
On food energy transformation by some marine isopod juveniles. (In Russian; English abstrac
Okeanologiia, 10(4): 709-713.

food of euphausiids

food of euphausiids

Lasker, Reuben 1966.
Feeding, growth, respiration and carbon utilization of a euphausiid crustacean.
J. Fish. Res. Bd Can. 23 (9): 1291-1317.

food of euphausiids

Marr, J.W.S., 1962.
The natural history and geography of the Antarctic krill (Euphausia superba Dana).
Discovery Repts., 32:33-464, Pl. 3.

food of euphausiids

Ponomareva, L.A., 1954.
(Feeding of euphausiids of Japan Sea on amphipods)
Dokl. Akad. Nauk, SSSR, 98(1):153.

food of filter feeders

food of filter feeders

Jørgensen, C. Barker, 1962.
The food of filter feeding organisms. Contributions to symposium on zooplankton production, 1961.
Rapp. Proc. Verb., Cons. Perm. Int. Expl. Mer, 153:99-107.

food for fish

food of fish

Alverson, Franklin G., 1963
The food of yellowfin and skipjack tunas in the eastern tropical Pacific Ocean. (In Spanish and English).
Inter-American Tropical Tuna Commission, Bull 7(5):295-367 (English); 368-396) (Spanish).

food of fishes

Anraku, Masateru, and Masanori Azeta, 1965.
The feeding habits of larvae and juveniles of the yellowtail Seriola quinqueradiata Temminck et Schlegel, associated with floating seaweeds. (In Japanese; English abstract)
Bull. Seikai Reg. Fish. Res. Lab., Nagasaki, 33: 13-45.

food for fish

Assman, A.V., 1957.
(On the role of microorganisms as a source of food for young fish.) Zool. Zhurnal, 36(6):900-908

freshwater fish
laboratory experiments.

food of fishes (Sebastes larvae)

Bainbridge, V., 1965.
A preliminary study of Sebastes larvae in relation to the planktonic environment of the Irminger Sea.
ICNAF Spec. Publ., No. 6:303-308.

food of fishes

Bakus, Gerald J., 1967.
The feeding habits of fishes and primary production at Eniwetok, Marshall Islands.
Micronesica, J. Coll. Guam, 3(2):135-149.

food for fishes

Berner, Leo, Jr., 1959.
The food of the larvae of the northern anchovy Engraulis mordax.
Inter-American Tropical Tuna Comm., Bull., 4(1):22 pp.

food of fishes.

Bogorov, B.G., 1934.
(Investigation of nutrition of plankton consuming fishes.)
Biull. Vses. Nauchno-Issled. Inst. Morskogo Chosia. i Okeanogr., No. 1:19-32.

food of fishes

Bright, Thomas J., 1970.
Food of deep-sea bottom fishes. Oceanogr. Stud Texas A & M Univ. 1: 245-252.

food of fishes

Brooks, W. K. 1893
The origin of the food of marine animals.
Bull. U. S. Fish. Comm., 13:87-92

food for fish

Bruun, A. Fr., 1949.
The use of nematodes as food for larval fish.
J. du Cons. 16(1):96-99, 2 textfigs.

food of fish

Brunel, P., 1963.
Variation journalières et saisonnières de l'alimentation de la morue au large de Grande Rivière.
Rapp. Ann., 1962, Sta. Biol. Mar., Grande Rivière 101-117.

food of fish

Carlisle, John G., Jr. 1971
Food of the jack mackerel, Trachurus symmetricus.
Calif. Fish Game 57(3): 205-208

food of fish

Cowey, C.B., J.A. Pope, J.W. Adron and A. Blair, 1971.
Studies on the nutrition of marine flatfish. Growth of the plaice Pleuronectes platessa on diets containing proteins derived from plants and other sources. Mar. Biol. 10(2): 145-153.

food of fishes

Davies, D. H., 1949
Preliminary investigations on the foods of South African fishes (with notes on the general fauna of the area surveyed). Fish & Mar. Biol. Survey Div., Dept. Comm. & Industries, S. Africa Invest. Rept. No.11, Commerce & Industry, Jan. 1949:1-36, 4 pls.

food of fishes

Day, F. 1882
On the food of sea fishes.
Zoologist (3) 6:235-236

food of fish

de Ciechomski, Janina Dz., 1967.
Investigations of food and feeding habits of larvae and juveniles of the Argentine anchovy Engraulis anchoita.
Rep. Calif.Coop.Oceanic Fish.Invest.11:72-81.

food of pilchards

DeSousa e Silva, E., 1954-55.
Some notes on the food of the pilchard Sardinia pilchardus (Walb.) of the Portuguese coasts.
Rev. Fac. Ciencias Univ., Lisboa, (2a)(C) 4(2):28 281-293, 3 pls., tables.

food of tunas

Dragovich, Alexander 1971.
The food of skipjack and yellowfin tunas in the Atlantic Ocean
Fish. Bull. U.S. Nat. Ocean. Atmos. Adm. 68(3):445-460

food of fish

*Dragovich, Alexander, 1970.
The food of bluefin tuna (Thunnus thynnus) in the Western North Atlantic.
Trans.Am. Fish.Soc., 99(4):726-731.

food of fishes

Dunn, M. 1885
Food of mackerel, pilchards, and herring.
Bull. U. S. Fish Comm. 1885:308

food of fish

Dunnington, E., and R. Mansueti, 1955.
School of harvestfish feeds on sea walnuts.
No records of harvestfish feeding on sea walnuts exist. Maryland Tidewater News, 12(5):1, 4.

food of fish

Enomoto, Yoshimasa, 1964.
Studies on the food base in the Yellow and the East China Seas, V. On the main crustaceans as the food item of fish resources.
Bull. Jap. Soc. Sci. Fish., 30(3):216-220.

food of fishes

Ercegovic, Ante, 1939.
The food of sardines (Clupea pilchardus Walb.)
Radovi Inst. za Okeanogr. i Ribarst., Split, 2:26-44.

Translations from Serbo-Croatian, 1962.
OTS-60-21658, pp. 7-18.

food of fishes

Ferreira de Menezes, Mariana 1969.
Alimentação da cavala, Scomberomorus cavalla (Cuvier), em águas costeiras do Estado do Ceará.
Arq. Ciên. Mar, Fortaleza, Ceará, Brasil, 9(1): 15-20

food of fish

Fonseca, Jose Bonifacio G., e Silvio B. Moraes, 1963.
Contéudo estomacal e evolucão sexual dos atuns.
Sudene, Bol. Estudos, Pesca, Recife, Brasil, 3(9-10):3-6.

food of fish

Fourmanoir, P., 1971.
Liste des espèces de poissons contenus dans les estomacs de thons jaunes, Thunnus albacares (Bonnaterre) 1788 et de thons blancs, Thunnus alalunga (Bonnaterre) 1788. Cah. ORSTOM, sér. Océanogr. 9(2): 109-118.

food of fish

Fourmanoir, P., 1969.
Contenus stomacaux d'Alepisaurus (Poissons) dans le sud-ouest Pacifique. Cah. O.R.S.T.O.M. sér Océanogr., 7(4): 51-60.

food of fishes

Franke, J. 1906
Die Ernährung der Fische und die Bedeutung des Planktons.
Stenogr. Ptotok. Verh. Internat. Fisch. Kongr., Wien, 1905: 115-123; 124-127.

food of fishes

Fulton, T. W. 1889
Inquiries into the nature of the food, the spawning habits, etc., of marine food fishes.
7th Ann. Rep. Fish. Board. Scotland, 1889: 182-185

food of fishes

Gapishko, A.I., 1968.
On the feeding of pelagic fishes in the Gulf of Aden. (In Russian).
Trudy, Vses. Nauchno-Issled. Inst. Morsk. Ribn. Okeanogr (VNIRO) 64, Trudy Azovo-Chernomorsk. Nauchno-Issled. Inst. Morsk. Ribn. Khoz. Okeanogr. (AscherNIRO), 28: 278-281.

food for fish (plankton)

Glover, R.S., (1958), 1960.
Fish behaviour and the planktonic environment: the role of field observations.
Proc. Indo-Pacific Fish. Counc., 8th Session, Sec. 3:72-79.

food of fish

Godfriaux, Bruce L., 1969.
Food of predatory demersal fish in Hauraki Gulf.
1. Food and feeding habits of snapper.
N.Z. Jl mar. freshwat. Res. 3(4): 518-544.

food of fish

Haedrich, Richard L., and Jørgen G. Nielsen, 1966.
Fishes eaten by Alopisaurus (Pisces, Iniomi) in the southeastern Pacific Ocean.
Deep-Sea Res., 13(5):909-919.

fish food

Hansen, Vagn, 1955
The food of herring on the Bløden Ground (North Sea) in 1953. J. du Cons., 21(1): 61-64.

Food of myctophids

Hartmann, Jürgen, und Horst Weikert, 1970.
Tagesgang eines Myctophiden (Pisces) und zweier von ihm gefressener Mollusken des Neustons
Kieler Meeresforsch. 25(2): 328-330

food for fishes

Hiyama, Y., and F. Yasuda, 1957.
The method of utilization of plankton by fishes.
Rec. Ocean. Wks., Japan (Spec. No.):67-70.

food of fishes

Hiyama, Yoshio, Fujio Yasuda and Makoto Shimizu, 1963.
An experimental analysis of the rate of utilization of natural animal food by some fishes traced by radioisotope 32P (preliminary report).
Rec. Oceanogr. Wks., Japan, n.s., 7(1):47-50.

food of fishes

Hofer, B. 1896
Die Bedeutung du Planktonstudien für die Fischerei in Seen.
Allgem. Fishherei Zeitg. 21 Jahrg., 355-359

food of fishes

Holton, Arthur Allen, 1969.
Feeding behavior of a vertically migrating lanternfish.
Pacific Sci., 23(3): 325-331

food of fishes

Ivlev, V.S., 1960

On the utilization of food by planktophage fishes. Bull. Math. Biophysics., 22(4): 371-39[illegible]

food of fishes

Keltenkova, M.V., 1961.
Utilization of food organisms by fishes of Azov Sea and the size of catches.
Trudy Vses. Gidrol. Obsh., Akad. Nauk, SSSR, 11: 309-322.

fpod of fish

June, Fred C., and Frank T. Carlson, 1971.
Food of young Atlantic menhaden, Brevoortia tyrannus, in relation to metamorphosis
Fish. Bull. U.S. Nat. Ocean. Atmos. Adm. 68(3): 493-512

food of fish

Kanno, Yasuji and Ikuzô Hamai, 1971.
Food of salmonid fish in the Bering Sea in summer of 1966. Bull. Hokkaido Univ. Fac. Fish. 22(2): 107-128.

food of fishes

Karlovac, J., 1962
Analyse du contenu du tractus digestif du stade planctonique de maquereau (Scomber scombrus L.) en Adriatique. (In Jugoslavian: French summary).
Izvjesca, Inst. Oceanogr. i Ribarstvo, Split, 14(4A): 3-15.

food of fish

King, J.E., and I.I. Ikehara, 1956.
Comparative study of food of bigeye and yellowfin tuna in the Central Pacific.
Fish. Bull., U.S.F.W.S. (Fish. Bull. 108) 57:61-85.

zooplankton as food
King, J.E., and J. Demond, 1953.
Zooplankton abundance in the Central Pacific.
U.S.F.W.S. Bull. 54(82):111-144, 11 textfigs.

food of fish
Kojima, S., 1963.
Studies on fishing conditions of the dolphin Coryphaena hippurus L. in the western region of the Sea of Japan. VII. Relationship between the stomach contents and the pelagic fauna of juveniles. (In Japanese; English abstract).
Bull. Jap. Soc. Sci. Fish., 29(5):407-414.

food of fish
Konstantinov, A.S., 1960.
Method of estimating the production of organisms serving as food for fishes. (In Russian).
Nauchnye Doklady Vysshei Shkoly, Biol. Nauki. (4) 59-62.

Translation:
OTS 63-11115 $0,50.

food of fish
Kotliar, L.K., 1970.
Some regularities of development and quantitative distribution of zooplankton as food supply of herring in the North-East Okhotsk Sea. (In Russian). Izv. Tikhookean. nauchno-issled. Inst. ribn. Khoz. Okeanogr. 71: 59-73.

fish food
Kriss, A.E., and A.V. Assman, 1955.
Microorganisms as fish food.
Dokl. Akad. Nauk, SSSR, 105(3):606-609.

Transl. Mon. 3(9):NRC-C2275.

food of fishes
Kudo, Sinji, Masahiro Toriyama, Osamu OKamura, and Syoshi Morita 1969.
Food studies of bottom fishes in continental slop of Tosa-Wan. (In Japanese; English abstract). Bull. Nansei reg. Fish. Res. Lab., 2: 85-103.

food of fish
*LeBrasseur, R.J., W.E. Barraclough, O.D. Kennedy and T.R. Parsons. 1969.
Production studies in the Strait of Georgia. Part III. Observations on the food of larval and juvenile fish in the Fraser River plume, February to May, 1967.
J. exp. mar. Biol. Ecol., 3(1):51-61.

food of fishes (tuna)
Legand, Michel, 1961
Aspects des recherches oceanographiques dans le Pacifique Sud. Quelques resultats biologiques.
Cahiers du Pacifique, Fondation Singer-Polignac, Mus. Nat. Hist. Nat., No. 3:1-52.

food of fishes
Lewis, John B. and Fritz Axelsen 1967.
Food of the dolphin Coryphaena hippurus Linnaeus, and of the yellowfin tuna Thunnus albacares (Lowe), from Barbados, West Indies.
J. Fish. Res. Bd Can. 24 (3): 683-686.

food of fish
Loukashkin, Anatole S., 1970.
On the diet and feeding behavior of the northern anchovy, Engraulis mordax (Girard).
Proc. Calif. Acad. Sci., 37 (13): 419-458.

fish food
*Maeda, Tatsuaki, Takeji Fujii and Kiyoshi Masuda, 1967.
Studies on the trawl fishing grounds of the eastern Bering Sea. I. On the oceanographical conditions and the distribution of the fish shoals in 1963. (In Japanese; English abstract)
Bull. Jap. Soc. scient. Fish., 33(8):713-720.

food of fishes
Magnuson, John J., 1969.
Digestion and food consumption by skipjack tuna (Katsuwonus pelamis).
Trans. Am. Fish. Soc. 98(3): 379-392

food of fish
Maksimov, V.P. 1970.
Some data on feeding habits of the marlins (Makaira ampla and M. albida) in the tropical Atlantic. (In Russian; English abstract).
Zool. Zh., 49(2): 262-267.

food of fish
Malikova, E.M., 1956.
The nutritional value of some invertebrates as food for fish. Biokhim., Akad. Nauk, SSSR, 21(2): 173-181.

(in freshwater reservoirs)

Translation of journal by Consultants Bureau Inc. in MBL Library.

food of fish
Marak, Robert R., 1960.
Food habits of larval cod, haddock and coalfish in the Gulf of Maine and Georges Bank area.
J. du Cons., 25(2):147-157.

food of fish
Matsui, Isao, Toru Takai and Akiyoshi Kataoka 1970
Leptocephali of the eel Anguilla obscura found in the stomachs of skipjack tuna Katsuwonus pelamis caught near New Guinea.
J. Shimonoseki Univ. Fish. 19 (1): 25-28.

food of fishes
McIntosh, W. C. 1887
On the pelagic fauna of our shores in its relation to the nourishment of the young food fishes.
Ann. Mag. Nat. Hist. (5) 19:137-145

food of fishes
Mehl, John A.P., 1969.
Food of barracouta (Teleostei: Gempylidae) in eastern Cook Strait.
N.Z. Jl mar. freshw. Res. 3(3): 389-396

food of fishes
Mendiola, B. Rojas de, 1969.
The food of the Peruvian anchovy. J. Cons. perm. int. Explor. Mer, 32(3): 433-434.

food of fishes
Merceron, Michel, 1969.
Etude des contenus stomacaux de quelques poissons carnivores du grand récif de Tuléar (Madagascar) et des environs.
Rec. Trav. Sta. mar. Endoume, hors série, Suppl. 9: 3-23

food of fishes
Mobius, K. A. 1882
The food of marine animals
Rept. U. S. Fish. Comm., 7:485-489

food of fishes
Mobius, K. A 1881
Die Nahrung der Seethiere
Zool. Garten 22:208-212

food of fishes
*Müller, Alajos, 1968.
Die Nahrung junger Plattfische in Nord- und Ostsee.
Kieler Meeresforsch., 24(2):124-143.

food of fishes
*Nakai, Zingiro, Masaya Kosaka, Masahiro Ogura, Gosuke Hayashida and Hideaki Shimozono, 1969.
Feeding habit and depth of body and diameter of digestive tract of shirasu, in relation with nutritous condition. (In Japanese; English abstract).
J. Coll. mar. Sci. Technol. Tokai Univ. (3):23-34.

food of fish
Neyman, A.A., 1963
Quantitative distribution of benthos and food supply of demersal fish in the eastern part of the Bering Sea. (In Russian).
Sovetsk. Ribokh. Issled. B Severo-Vostokh. Chasti Tikhogo Okeana, VNIRO 48, TINRO 50(1): 145-206.

food of fish
O'Connell, C.P. and L.P. Raymond, 1970.
The effect of food density on survival and growth of early post yolk-sac larvae of the northern anchovy (Engraulis mordax Girard) in the laboratory. J. exp. mar. Biol. Ecol., 5(2): 187-197.

fish food
*Odum, William E., 1968.
Mullet grazing on a dinoflagellate bloom.
Chesapeake, Sci., 9(3):202-204.

fish, food of
Okamura, K., 1912.
Plankton organisms from bonito.
Rept. Imper. Fish., 1:4-38.

food, fish
Pavlovskaia, R.M., 1958.
The survival of anchovy larvae in the northwestern part and other regions of the Black Sea, observed in 1954-1955, and the role played in their survival by feeding conditions.
Dokl. Akad. Nauk, SSSR, 102(2):415-418.

food of fishes
Peck, J. I. 1896
The sources of marine food.
Bull. U. S. Fish. Comm. 1895, 15:351-368

food of fish
Ponomarenko, I. Ya., 1964.
Effects of the hydrological regime and feeding conditions on the formation of the abundance of the year-class of cod during its first year in the near bottom layers of the Barents Sea. (In Russian).
Trudy, Poliarn. Nauchno-Issled. i Proektn. Inst. Morsk. Ribn. Choz. i Okeanogr. im N.M. Knipovicha, 16:235-249.

euphausids (as food for fish)
Ponomareva, L.A., 1963.
To the problem on availability of food organisms for fishes. Biological investigations of the Ocean (plankton). (In Russian; English abstract).
Trudy Inst. Okeanol., Akad. Nauk, SSSR, 71:72-80.

food of fish
Ponomareva, L.A., 1954.
[The significance of individual species of euphausids as components in the food of fish and whales] Trudy Inst. Okeanol., 8:200-205.

food of fishes, lists of spp.
Popova, O.A., 1962.
Some data on feeding of cod in the Newfoundland area of the northwest Atlantic.
Sovetskie Riboch. Issledov. v Severo-Zapadnoi Atlant. Okeana, VNIRO-PINRO, Moskva, 235-253.

In Russian; English summary

fish, food of, lists of spp. (cod)
Popova, O.A., 1962
Some data on feeding of cod in the Newfoundland area of the northwest Atlantic. (In Russian; English summary).
Sovetskie Ribochoz. Issledov. Severo-Zapad. Atlant. Okeana, VNIRO-PINRO, 235-253.

food of fishes
Rancurel, P., 1970.
Les contenus stomacaux d'Alepisaurus ferox dans le sud-ouest Pacifique (Céphalopodes).
Cah. ORSTOM, sér. Océanogr. 8(4): 3-87.

fish food
Reintjes, J.W., and J.E. King, 1953.
Food of yellowfin tuna in the central Pacific.
U.S.F.W.S. Bull. 54(81):91-110, 10 textfigs.

food of fish
Rice, A.L., 1963
The food of the Irish Sea herring in 1961 and 1962.
J. du Conseil, 28(2):188-200.

food
Rodina, A.G., 1949.
[Bacteria as food of aquatic animals.] Priroda 10:23.

food of anchoveta
Rojas De Mendiola, Blanca, 1971.
Some observations on the feeding of the Peruvian anchoveta Engraulis ringens J. in two regions of the Peruvian coast. (English and Spanish abstracts). In: Fertility of the Sea, John D. Costlow, editor, Gordon Breach, 2: 417-440.

food of fishes
Rojas de Mendiola Blanca, Noemí Ochoa, Ruth Cálienes y Olga Gómez, 1969.
Contenido estomacal de anchoveta en cuatro areas de la costa peruana.
Inform. Inst. Mar Peru, 27: 30pp.

food of fishes
Rosenthal, Harald, 1969.
Verdauungsgeschwindigkeit, Nahrungswahl und Nahrungsbedarf bei den Larven des Herings, Clupea harengus L.
Ber. dt. wiss. Komm. Meeresforsch. 20(1):60-69

food for fishes
Rudakova, V.A., 1956.
Data on the food of the Atlantic herring.
Polyarnyy Nauch.-Issle. Inst. Morsk. Rybn. Khozyay Okean., Vypusk, 9:

Translation: U.S.F.W.S. Spec. Sci. Rept., Fish., 327.

Murmansk,

food of fishes
Sauvage, H. E. 1888
Sur la nourriture de quelques poissons de mer.
Rev. Sci. Nat. Appliq. 38:32-38
Ann. Sta Aquicole de Boulogne, 1:39-51

food of fishes
Shelbourne, J.E., 1962.
A predator-prey size relationship for plaice larvae feeding on Oikopleura.
J. Mar. Biol. Assoc., U.K., 42(2):243-252.

food of fish (tuna) (lists of groups)
Shibata, Keishi, 1964.
Analysis of fish finder records. IV. Report on the deep scattering layers and tuna food (1). (In Japanese; English abstract).
Bull. Fac. Fish., Nagasaki Univ., 15:59-64.

food of fishes
Sicard, A. 1877
Etudes sur la nourriture des poissons de mer.
Bull. Soc. Acclim., Paris, (3)4: 396-400

food of fish, lists of spp.
Sidorenko, I.N., 1962
Feeding of cod in West Greenland. (In Russian; English summary).
Sovetskie Ribochoz. Issledov. Severo-Zapad. Atlant. Okeana, VNIRO-PINRO, 255-261.

food of fishes, lists of spp.
Sidorenko, I.N., 1962.
Feeding of cod of west Greenland waters.
Sovetskie Riboch. Issledov. v Severo-Zapadnoi Atlant. Okeana, VNIRO-PINRO, Moskva, 255-261.

In Russian; English summary

food of fish
Sinha V.R.P. 1969.
A note on the feeding of larger eels Anguilla anguilla (L.).
J. Fish. Biol. 1(3): 279-283.

food of eels
*Sinha, V.R.P., and J. W. Jones, 1967.
On the food of the freshwater eels and their feeding relationship with the salmonids.
J. Zool., Lond., 153(1):119-137.

food of fishes
Steuer, A. 1905
Die Ernahrung der Fische und die Bedeutung des planktons.
Stenogr. Protok. Verh. Intern. Fisch.-Kongr. Wien, :127-131

food of fishes
Steuer, A. 1905
Ueber das Kiemenfilter und die Nahrung adriatische.
Fische. Verh. Zool.-Bot. Ges., Wien, 55:275-299

food of fish
*Sund, Paul N., and William J. Richards, 1967.
Preliminary report on the feeding habits of tunas in the Gulf of Guinea.
Spec.scient.Rep.U.S.Fish.Wildl.Serv.,Fish.,551: 6 pp (multilithed).

food of fish
Suzuki, Tomoyuki, 1965.
Ecological studies on the jack mackerel. Trachurus japonicus (Temminck et Schoegel). 1. On the feeding habits. (In Japanese; English abstract).
Bull.Jap. Sea Reg. Fish. Res. Lab.,(14):19-29.

food of fish
Takano, H., 1954.
The food of the mackerel taken near Oshima Island in 1953. Bull. Jap. Soc. Sci. Fish., 20(8):694-697.

food of fish
*Tello, Felicitas, y Roberto Lam C., 1968.
Estudio sobre la variación del contenido de grasa en el anchoveta Peruanc (Engraulis ringens J.).
Informe Inst.Mar.Peru 24:29 pp.

food for fish
Theilacker, G.H. and M.F. McMaster, 1971.
Mass culture of the rotifer Brachionus plicatilis and its evaluation as a food for larval anchovies. Mar. Biol. 10(2): 183-188.

food of fishes
Troadec, J.P., 1968.
Le régime alimentaire de deux espèces de sciaenidae ouest africains (Pseudotolithus senegalensis V. et Pseudotolithus typus Blkr.).
Doc. Centre Rech. océanogr. Abidjan, 030: 24 pp.

food of fishes
Verrill, A. E. 1871
On the food and habits of some of our marine fishes.
Amer. Nat., 5:397-400
Canad. Nat., 6:107-111

fish food
Vijayaraghavan, P., 1954.
Food of sardines of Madras coast.
J. Madras Univ., B, 23(1):29-39, 1 textfig.

fish food
Vijayaraghavan, P., 1953.
Food of the Indian herring. J. Madras Univ., B, 23(3):239-247.

food of fish
* White, Ray J., 1968.
Importance of appendicularians as food of larval plaice (Pleuronectes platessa L.) off Helgoland.
Ber. dt. wiss. Komm. Meeresforsch. 19(4): 288-291.

food of fishes

Wright, R. R. 1907
The plankton of eastern Nova Scotia waters. An account of floating organisms upon which young food-fishes mainly subsist.
Ann. Rept. Dept. Mar. Fish., Ottawa, 39:1-20

food of tunas

Yamaguchi, Yuichiro, 1969.
Studies on the feeding habits of tunas. (In Japanese; English abstract).
J. Fac. Fish. Pref. Univ. Mie, 8(1):1-16.

food for fishes

Yashda, F., and Y. Hiyama, 1957.
Mechanism of utilization of plankton by some fishes. Rec. Oceanogr. Wks., Japan, n.s., 3(1): 85-91

food for fish (larvae)

food of fish larvae

Detwyler, R. and E.D. Houde, 1970.
Food selection by laboratory-reared larvae of the scaled sardine Harengula pensacolae (Pisces, Clupeidae) and the bay anchovy Anchoa mitchilli (Pisces, Engraulidae). Marine Biol., 7(3): 214-222.

fish larvae, food of

Duka, L.A., and V.I. Sinjukova, 1965.
Some data on the larva nutrition of khamsa in the Black and Mediterranean seas. (In Russian).
Okeanologiia, Akad. Nauk, SSSR, 5 (3):528-533.

food of lancelets

Webb, J.E., 1969.
On the feeding and behaviour of the larva of Branchiostoma lanceolatum. Marine Biol., 3(1): 58-72.

food for fish (sharks)

food of sharks

Alexander, Anne J., 1966
Shark feeding behaviour.
S. Afr. Assoc. Mar. Biol. Res. Bull., No. 6: 49-55.

food of fishes

Bullis, Harvey R., Jr., 1961.
Observations on the feeding behavior of white-tip sharks on schooling fishes.
Ecology, 42(1):194-195.

food of fishes

LeBrasseur, R.J., 1964.
Stomach contents of blue shark (Prionace glauca L.) taken in the Gulf of Alaska.
J. Fish. Res. Bd., Canada, 21(4):861-862.

food of sharks

Porumb, Ioan I., 1961
Contributions a la biologie de Trachurus mediterraneus ponticus Aleev de la Mer Noire. (In Roumanian:Russian and French resumes).
Ann. Sci., Univ. "Al. I. Cuza", Jassy, Roumania 7(2):285-304.

food of sharks

Williamson, G.R., 1963.
Common porpoise from the stomach of a Greenland shark.
J. Fish. Res. Bd., Canada, 20(4):1085-1086.

food of invertebrates

food of isopods

Novotny, Anthony J., and Conrad V.W. Mahnken 1971
Predation on juvenile Pacific salmon by a marine isopod Rocinela belliceps pugettensis (Crustacea Isopoda)
Fish. Bull. nat. mar. fish. Serv. NOAA 69(3): 699-701

Food of isopods

Tsikhon-Lukanina, E.A., 1970.
Assimilation of some kind of food by Idothea baltica (Pallas). Gidrobiol. Zh., 6(3): 87-90.

(In Russian)

food of mammals

food of molluscs

Bubnova, N.P., 1971.
Ration and food assimilation by a detritus feeding mollusc Portlandia arctica. (In Russian; English abstract). Okeanologiia 11(2): 302-305.

food of molluscs

Imeliova, N.N., y J.Sang, 1969.
Respiración y algunas particularidades de la alimentación del ostión Crassostrea rhizophorae Guilding.
Serie Oceanologia, Inst. Oceanol. Acad. Cienc., Cuba, 3: 1-20

food for molluscs

Okiyama, Muheo, 1965.
On the feeding habit of the common squid, Todaroles pacificus Steenstrup, in the off-shore region of the Japan Sea. (In Japanese; English abstract).
Bull. Jap. Sea Reg. Fish. Res. Lab., (14):31-41.

food of molluscs

Okutani, Takashi, 1962
Diet of the common squid, Ommastrephes sloani pacificus landed around Ito port, Shizuoka Prefecture. (In English; Japanese abstract).
Bull. Tokai Reg. Fish. Res. Lab., Tokyo, No. 32: 41-47.

food of oysters

Paulmier, Gérard, 1971.
Cycle des matières organiques dissoutes, du plancton et du micro-phytoplancton dans l'estuaire du Belon: leur importance dans l'alimentation des huîtres.
Rev. Trav. Inst. Pêches marit. 35(2): 157-200.

food of mammals

Tomo, Aldo P., 1971.
Trophic chains observed in Paradise Harbor (Antarctic Peninsula) related to variations in the fertility of its waters. (English and Spanish abstracts). In: Fertility of the Sea, John D. Costlow, editor, Gordon Breach, 2: 593-602.

food of molluscs

*Ukeles, Ravenna, and Beatrice M. Sweeney, 1969
Influence of dinoflagellate trichocysts and other factors on the feeding of Crassostrea virginica larvae on Monochrysis lutheri.
Limnol. Oceanogr., 14(3): 403-410.

food of ostracods

Neale, John W., 1964.
Some factors influencing the distribution of Recent British Ostracoda.
Pubbl. Staz. Zool., Napoli, 33(Suppl.):247-307.

food of plankton organisms

food relationships

food relationships

Edmondson, W.T., editor, 1966.
Ecology of Invertebrates, Marine Biology III, Proceeding of the Third International Interdisciplinary Conference.
New York Acad. Sci., 313pp. ($7.00).

food

Banner, A.H., 1952.
Preliminary report on marine biology study of Onoatoa Atoll, Gilbert Islands, Part 1.
Atoll Res. Bull. 13:1-42.

food from the sea

Brundrett, F., 1963.
The neglected sea.
J. Inst. Navigation, 16(3):332-342.

foods

Davis, Harry C. and Robert R. Guillard 1958.
Relative Value of Ten Genera of Micro-organisms as Foods for Oyster and Clam Larvae.
U.S.F.W.S. Fish. Bull. 136. From Fishery Bul. of F&W S Vol. 58., pps. 293-304.

food

Miller, C.D., and F. Pen, 1959.
Composition and nutritive value of Palolo (Palolo siciliensis Grube).
Pacific Science, 13(2):191-194.

food resources
see: plankton as food

food from the sea

Alverson, D.L., A.R. Longhurst and J.A. Gulland 1970.
How much food from the sea?
Science 168(3930): 503-505.

food resources

Currie, R., 1953.
Food resources of the sea. Times Science Review, Winter 1953.

food from the sea

Dietrich, G., 1968.
Herausforderung des Meeres.
Math. naturw. Unterr., 21(7): 227-236.

food resources

Emery, K.O., and C. O'D. Iselin, 1967.
Human food from the ocean.
Science, 157(3794):1279-1281.

food

Jones, Albert C., 1968.
Food and drugs from the sea: sleeping giant or deceptive illusion. Transactions, Conference on Industry's Future in the Ocean...the Challenge and the Reality (March 1968), Miami, Fla., Florida Commission on Marine Sciences and Technology,: 114-121, 1968. Also in: Collected Repr. Div. Biol. Res., Bur. Comm. Fish., U.S. Fish Wildl. Serv., 1968, 1.

food from the sea

Ryther, John H., 1970.
How much food from the sea? Science, 168(3930): 505.

foods

Schaeffer, Milner B. 1968.
Economic and social needs for marine resources.
In: Ocean engineering: goals, environment, technology, John F. Brahtz, editor, John Wiley + Sons, 6-37.

food of seals

*Calhaem, I., and D.A. Christoffel, 1969.
Some observations of the feeding habits of a Weddell seal and the measurements of its prey, Dissostichus mawsoni, at McMurdo Sound, Antarctica.
N.Z. Jl mar. Freshwat. Res. 3(2):181-190.

food of seals

Rae Bennet B., 1968.
The food of seals in Scottish waters.
Mar. Res. Dept. Agric. Fish., Scotland, (2):23 pp

food of seals

Tikomirov, E.A., 1970.
Cases of eating birds by harbour seals. (In Russian). Izv. Tichookean. nauchno-issled. Inst. Ribn. choz. Okeanogr. 70:249-250.

food of sea snakes

food of sea snakes

Voris, Harold K., 1966.
Fish eggs as the apparent sole food item for a genus of sea snake Emydocephalus (Krefft).
Ecology, 47 (1):153-154.

food of turtles

food of turtles

Bleakney, J. Sherman, 1967.
Food items in two loggerhead sea turtles, Caretta caretta (L.), from Nova Scotia.
Can. Fld. Nat., 81(4):269-272.

food, turtles

Bleakney, J. Sherman, 1965.
Reports of marine turtles from New England and eastern Canada.
Canadian Field-Nat., 79(2):120-128.

food values

food values

*Patton, Stuart, P.T. Chandler, E.B. Kalan, A.R. Loeblich III, G. Fuller and A.A. Benson, 1967.
Food value of red tide (Gonyaulax polyedra).
Science, 158(3802):789-790.

food for whales and other mammals

food of whales

Akimushkin, I.I., 1954.
The basic food-substance of the bottlenose whale (Hyperoodon rostratus Müller).
Dokl. Akad. Nauk, SSSR, 95(2):419.

food of whales

Backus, Richard H., 1966.
A large shark in the stomach of a sperm whale.
J. Mammal., 47(1):142.

food of whales

Bannister, J.L., and A. de C. Baker, 1966.
Observations on food and feeding of baleen whales at Durban.
Norsk. Hvalfangsttid., 56:78-82.

Beklemishev, Constantin W., 1960 food of whales

Southern atmospheric cyclones and the whale feeding grounds in the Antarctic. Nature, 187 (4736):530-531.

food for whales

Betesheva, E.I., 1955.
Food of the whalebone whales in the Kurile Islands region. 18:78-85.
Trudy Inst. Oceanol.,

food for whales

Betesheva, E.I., and I.I. Akimushkin, 1955.
Food of the spermwhale (Physeter catodon) in the Kurile Islands region. Trudy Inst. Oceanol., 18: 86-94.

food of whales.

Brown, S.G., 1968.
Feeding of sei whales at South Georgia.
Norsk Hvalfangsttid., 57(6): 118-125.

whales, food of

Clarke, R., 1955.
A giant squid swallowed by a sperm whale.
Norsk Hval Tid. 10:589-593.

food, whales

Clarke, Malcolm R., 1962
Stomach contents of a sperm whale caught off Madeira in 1959. (In Norwegian and English).
Norsk Hvalfangsttid. (5):173-191.

Also in:
Collected Reprints, Nat. Inst. Oceanogr., Wormley, 10.

food of whales

Dall, W., and D. Dunstan, 1957.
Euphausia superba Dana from a humpback whale, Megaptera nodosa (Bonnaterre) caught off southern Queensland. Norsk Hvalfangst-Tid., 46(1):6-9.

with comments on this article by Åge Jonsgård on pp. 10-12.

food of whales

Gaskin, D.E., and M.W. Cawthorn 1967.
Squid mandibles from the stomachs of spermwhales (Physeter catodon L.) captured in the Cook Strait region of New Zealand.
N.Z. Jl mar. Freshwat. Res. 1(1):59-70

whales, food of

*Gaskin, D.E., and M.W. Cawthorn, 1967.
Diet and feeding habits of the sperm whale (Physeter catodon) in the Cook Strait region. of New Zealand.
N.Z. Jl. mar. Freshwat. Res. 1(2):156-179.

food of whales

Gill, Charles D., and Steven E. Hughes 1971.
A sei whale, Balaenoptera borealis feeding on Pacific saury, Cololabis saira.
Calif. Fish Game 57(3): 218-219.

food of whales

Hosokawa, Hiroshi and Toshiro Kamiya, 1971.
Some observations on the cetacean stomachs, with special considerations on the feeding habits of whales. Scient. Repts. Whales Res. Inst. 23: 91-101.

food of mammals

Hotta, Hideyuki, Hiroshi Mako, Keisuke Okada and Umeyoshi Yamada, 1969.
On the stomach contents of dolphins and porpoises off Kyushu. (In Japanese; English abstract).
Bull. Seikai reg. Fish. Res. Lab., 37: 71-85.

food of whales

Kawamura, Akito, 1971.
Influence of chasing time to stomach contents of baleen and sperm whales. Scient. Repts. Whales Res. Inst. 23: 27-36.

food of whales

Kawamura, Akito, 1970.
Food of sei whale taken by Japanese whaling expeditions in the Antarctic season 1967/68.
Scient. Repts Whales Res. Inst., 22: 127-152.

food of whales

Klumov, S.K., 1961

Plankton and the feeding of the whalebone whales (Mystacoceti).
Trudy Inst. Okean., 51: 142-156.

food of whales

Marr, J.W.S., 1955.
Krill, the whale's food. Zoo Life, 10:56-58.

food for whales

Nakai, Z., 1954.
Species of food plankton for baleen whales in the sea adjacent to Japan.
Tokai Reg. Fish. Res. Lab., Spec. Publ. 4:1-6.

food of whales

Nemoto, Takahisa, 1959

Food of baleen whales with reference to whale movements. Sci. Repts., Whale Research Inst., Tokyo, No. 14:149-290.

food, whales

Nemoto, T., 1957.
Foods of baleen whales in the northern Pacific.
Sci. Rept. Whales Res. Inst., 12:33-90.

food for whales

Nemoto, Takahisa, Koharu Ishikawa and Kimie Kamada, 1969.
A short note on the importance of the krill as a marine biological resource in relation to stock of Baleen whales. (In Japanese; English abstract). Bull. Jap. Soc. fish. Oceanogr. Spec. No. (Prof. Uda Commem. Pap.): 217-220.

food of whales

Nemoto, Takahisa, and Toshio Kasuya, 1965.
Foods of baleen whales in the Gulf of Alaska of the North Pacific.
Sci. Repts., Whales Res. Inst., Tokyo, No. 19:45-51.

Food of whales

Nemoto, Takahisa and Kwang Il Yoo, 1970.
An amphipod, (Parathemisto gaudichaudii) as a food of the Antarctic sei whale. Scient. Repts Whales Res. Inst., 22: 153-158.

food of whales

Okutani, Takashi, and Takahisa Nemoto, 1964.
Squids as the food of sperm whales in the Bering Sea and Alaskan Gulf.
Sci. Repts., Whales Res. Inst., (18):111-122.

food of whales

Ponomareva, L.A., 1954.
[The significance of indicidual species of euphausids as components in the food of fish and whales.] Trudy Inst. Okeanol., 8:200-205.

food of porpoises

Rae, Bennet B., 1965.
The food of the common porpoise (Phocaena phocaena).
J. Zool., Proc. Zool. Soc., London, 146(1):114-122

FOOD OF WHALES

Rice, Dale W., 1968.
Stomach contents and feeding behavior of killer whales in the eastern North Pacific
Norsk Hvalfangsttid. 57(0): 35-38

food of whales

Tarasevich, M.N. 1968.
Food connections of sperm whales in the northern Pacific. (In Russian; English abstract).
Zool. Zh. 47(4):595-601.

food for whales

Tarasevich, M.N., 1963.
Data on feeding of sperm whales in the northern area of Kurile waters (Paramushiro and Onekotan-Shiashkotan regions). Biological investigations of the ocean (plankton). (In Russian). English abstract
Trudy Inst. Okeanol., Akad. Nauk, SSSR, 71:195-206.

food of zooplankton

Barth, Rudolf, 1963
Estudos sôbre o conteúdo intestinal de alguns macroplanctontes.
Ministério da Marinha, Inst. Pesquisas de Marinha, Nota Técnica, 7/63: 23 pp. (multilithed).

food of zooplankton

Beklemishev, C.W., 1962.
17. Superfluous feeding of marine herbivorous zooplankton. Contributions to symposium on zooplankton production, 1961.
Rapp. Proc. Verb., Cons. Perm. Int. Expl. Mer, 153:108-113.

food of zooplankton

Chindonova, Iu. G., 1959.
[Nutrition of some groups of the deep-water macroplankton of the northwest part of the Pacific Ocean.]
Trudy Inst. Okeanol. 30:166-189.

food of zooplankton

Corner, E.D.S., and C.B. Cowey, 1964
Some nitrogenous constituents of the plankton.
In: Oceanography and Marine Biology, Harold Barnes, Editor, George Allen & Unwin, Ltd., 2: 147-167.

food (zooplankton)

Gibor, A., 1957.
Conversion of phytoplankton to zooplankton.
Nature 179(4573):1304.

food of plankton org.

Lebour, M. V., 1923.
The food of plankton organisms. II.
JMBA, XIII:70

food of plankton organ.

Lebour, M. V., 1922.
The food of plankton organisms.
JMBA, XII:644

food of plankton

Pavlovskaja, T.V., and G. A. Pechen', 1971.
Infusoria as a food for some mass species of marine planktonic animals. (In Russian; English abstract).
Zool. Zh. 50(5): 633-640

food for plankton

Seki, Humitake, 1966.
Role of bacteria as food for plankton (review). (In Japanese; English abstract).
Inf. Bull. Planktol. Japan, No. 13:54-62.

food of zooplankton

Vinogradov, M.E., 1962.
18. The feeding of deep-sea zooplankton. Contributions to symposium on zooplankton production, 1961.
Rapp. Proc. Verb., Cons. Perm. Int. Expl. Mer, 153:114-120.

forces on submerged body

forces on submerged body

Havelock, T.H., 1954.
The forces on a submerged body moving under waves
Q. Trans. Inst. Naval Arch., 96(2):77-83.

forcing

Rao, V. Subba, and G.V. Prabhakara Rao 1971
On waves generated in rotating stratified liquids by travelling forcing effects.
J. fluid Mech. 46(3): 447-464

forecasting: see also WAVES, forecasting
STORMS, forecasting

forecasting

Anon., 1965.
Ocean environment prediction.
Travelers Research Center, Inc., 1965:29-33.

forecasting, long-range

Akado, K., 1957.
[On the long-range forecasting in the light of marine data.] Umi to Sora, 33(4/5):54-60.

forecasting

Antonov, A.E., 1962.
Background for forecasting some hydrometeorological phenomena in the basins of the Baltic Sea from 1961-1980. (In Russian).
Trudy, Baltiisk. Nauchno-Issled. Inst. Morsk. Ribn. Khoz. i Okeanogr., (BALTNIRO), 8:62-87.

forecasting

Antonov, A.E., 1964
An application of the periodogram analysis of the forecasting of some hydrometeorological elements (on the example of the Baltic Sea). (In Russian).
Materiali Vtoroi Konferentsii, Vzaimod. Atmosfer. i Gidrosfer. v Severn. Atlant. Okean., Mezhd. Geofiz. God, Leningrad. Gidrometeorol. Inst., 32-40.

forecasting

Baer, Ledolph, Louis C. Adamo and S.I. Adelfang, 1968.
Experiments in oceanic forecasting for the advective region by numerical modeling. 2. Gulf of Mexico.
J. geophys. Res., 73(16):5091-5104.

forecasting

Belinskii, N.A., 1957
[Maritime hydrometeorological forecasting during the past 40 years.] Met. i Gidrol. (11): 68.

forecasting

Belinskiy, N.A., 1957.
The development of methods of marine hydrometeorological forecasts in the Central Institute of Forecasts (TsIP). (In Russian).
Trudy Tsentr. In-ta Prognozov, (55):48-56.

RZhGeofiz 3/58-1877.

forecasting

Belinskiy, N.A., 1956.
Morskiye gidrometeorologicheskiye informatsii i prognozy. (Marine hydrometeorological information and forecasts) (In Russian). [Manual for Hydrometeorological Technicums], 2nd Edit. Revised, L. Gidrometeoizdat, 254 pp.

RZHGeofiz 9?57-8393 K

forecasting

Bulgakov, N.P., 1961
[Significance of different factors and one of the forms of the forecasting equation.]
Trudy Inst. Okean., 52: 166-170.

forecasting

Dmitrieva, A.A., 1968.
On the use of linear extrapolation method for the prognosis of oceanographical elements. (In Russian).
Mater. Ribokhoz. Issled. Severn. Basseina, 12: 139-143.

forecasting

Dmitrieva, A.A., 1967.
On a calculation of probable characteristics of oceanographic processes. (In Russian).
Mater. Rybokhoz. Issled. Severn. Basseina, (PINRO), 10:83-89.

forecasting, oceanograph

Eggvin, Jens, 1968.
The possibility of forecasting oceanographic conditions in northwest European waters and their significance for fisheries.
ICNAF, Spec. Publ., No. 6:903-907.

forecasting

Fukuoka, J., 1961
Variations of oceanographic condition and possibility of forecasting it. (In Japanese)
Weather Service Bull., 28(8):305-313.

Also in: Collected Reprints, Oceanogr. Lab., Meteorol. Res. Inst., Tokyo, Vol. 7, (117).

forecasting
Fukuoka, J., 1961
Variations of oceanographic condition and possibility of forecasting it. Part 2. (In Japanese).
Weather Service Bull., 28(9):345-349.

Also in:
Collected Reprints, Oceanogr. Lab., Meteorol. Res. Inst., Tokyo, Vol. 7(118).

forecasting
Glagoleva, M.G., 1964
Method of computing and prognosticating nonperiodical currents and water temperature. (In Russian).
Materiali Vtoroi Konferentsii, Vzaimod. Atmosfer. i Gidrosfer. v Severn Atlant. Okean., Mezhd. Geofiz. God, Leningrad. Gidrometeorol. Inst., 187-188.

Hamm, D.P., and R.M. Lesser, 1968.
Experiments in oceanic forecasting for the advective region by numerical modeling. 1. The model.
J. geophys. Res., 73(16):5081-5089.

forecasting
*Ishtoshin, Yu.V., and A.I. Karakash, 1967.
Marine hydrographic forecasts. (In Russian; English abstract).
Okeanologiia, Akad. Nauk, SSSR, 7(6):947-956.

forecasting, hydrographic conditions.
Japan, Kobe Marine Observatory, 1953.
[On the hydrographical conditions of the Kuroshio region in 1952 and their forecasting in 1953.] J. Ocean. (2)4(1):31-32.

forecasting
Kaczmarek, Zdzislaw, 1960
[Forecasting random phenomena.]
Przeglad Geofiz., Polskie Towarzystwo Meteorol i Hydrol., 13(3): 172-182.

forecasting, oceanographic
Laevastu, T., 1965.
Is oceanographic forecasting (hydrosis) feasible for fisheries?
ICNAF Spec. Publ., No. 6:881-884.

Laevastu, T., et P.M. Wolff, 1965.
Quelques principes d'analyse et de prévision océanographique synoptique.
Météorologie, 1965: 305-319

forecasting
Laykhtman, D.L., and B.A. Kagan, 1965.
The scheme of the hydrological characteristics precomputation on the sea surface. (In Russian).
Meteorol. i Gidrol., (5):7-13.

forecasting
Marchuk, G.I., 1965.
On the role of research on physics of the atmosphere and ocean for weather forecast. (In Russian).
Fizika Atmosfer. i Okeana, Izv., Akad. Nauk, SSSR, 1(1):3-7.

prediction
Nan-niti, Tosio, 1967.
Some remarks on scale of time and length of phenomena for oceanic prediction.
J. oceanogr. Soc., Japan, 23(2):86-87.

forecasting
Pierson, W.J., Jr., 1960
On the use of time series concepts and spectral and cross-spectral analyses in the study of long-range forecasting problems. J. Mar. Res., 18(2): 112-132.

forecasting
Sauskan, E.M., 1960
[In the Section of Marine Prognoses and Calculations.]
Biull. Oceanogr. Komm. (5): 19-20.

forecasting, oceanographic
Schule, J.J., Jr., 1965.
Some aspects of oceanographic prediction.
ICNAF, Spec. Publ., No. 6:909-912.

forecasting
Soskin, I.M., 1960.
Oscillations in solar activity as the basis for ultra-long-term forecasting of hydrological conditions of the sea (using the Barents Sea as an example). (In Russian).
Unidentified Russian article.

Translation:
LC or SLA, mi $1.80, ph $1.80.

forecasting, oceanographic
Tait, J. B., 1965.
Forecasting environmental conditions in the Faroe-Shetland Channel region.
ICNAF, Spec. Publ., No. 6: 901-902.

forecasting chlorinity
Watanabe, Nobuo, 1959.
On a trial of forecasting of the chlorinity variation during short period in the Kiso-Sansen estuary. (In Japanese; English abstract). Bull. Jap. Soc. fish. Oceanogr. Spec. No. (Prof. Uda Commem. Pap.): 157-171.

forecasting, currents

forecasting, currents
Kan, S.I., 1960.
[Improvement of methods for current forecasting in the Kerchensky Strait.]
Meteorol. i Gidrol., (12):25-27.

forecasting, environment
Wolff, Paul M., 1970.
Environmental forecasting, largest of marine information systems.
J. mar. techn. Soc. 4(6): 7-18.

forecasting, extended range
Sgmagorinsky, Joseph, 1969.
Problems and promises of deterministic extended range forecasting.
Bull. Am. met. Soc., 50(5): 286-311.

forecasting, fisheries

forecasting, fisheries
Adrov, M.M., 1966.
About the so-called systemic basis for forecasting the oceanological conditions and reproduction of commercial fishes. (In Russian).
Mater. Ribokhoz. Issled. severn. Bassaina, Poliarn. Nauchno-Issled. Proektn. Inst. Morsk. Ribn. Khoz. Okeanogr. (PINRO), 7:192-197.

forecasting fisheries
Hirano, Toshiyuki, 1969.
Scope on the forecasting and its service of fishing and oceanographic conditions. (In Japanese; English abstract). Bull. Jap. Soc. fish. Oceanogr. Spec. No. (Prof Uda Commem. Pap.): 149-151.

forecasting, fisheries
Laevastu, Taivo, 1969.
Numerical oceanographic forecasting for fisheries. Bull. Jap. Soc. fish. Oceanogr. Spec. No. (Prof. Uda Commem. Pap.): 139-147.

forecasting, fisheries
Shimomura, Toshimasa, 1969.
The recent and future perspectives on the fisheries oceanography in the Tsushima Current regions. (In Japanese; English abstract). Bull. Jap. Soc. fish. Oceanogr. Spec. No. (Prof Uda Commem. Pap.): 153-156.

forecasting, flushing

forecasting, flushing
*Neal, Victor T., 1966.
Predicted flushing times and pollution distribution in the Columbia River estuary.
Proc. 10th Conf. Coast Engng, 2:1463-1480.
Also in: Coll. Repr., Dep. Oceanogr., Oregon State Univ., 5.

forecasting, fog
Miyazono, S., 1971.
A forecasting technique of fog in Kyushu's west coast and its adjacent sea.
J. Met. Res. Japan met. Agency 21(3):99-103.
Also in: Oceanogr. Met., Nagasaki mar. Obs. 18.

(In Japanese; English abstract)

forecasting, hurricanes

forecasting, hurricanes
*Dunn, Gordon E., R. Cecil Gentry and Billy M. Lewis, 1968.
An eight-year experiment in improving forecasts of hurricane motion.
Mon. Weath. Rev. U.S. Dep. Comm. 96(10):708-713.

forecasting hurricanes
Miller, Banner I., Elbert C. Hill and Peter P. Chase, 1968.
A revised technique for forecasting hurricane movement by statistical methods.
Mon. Weath. Rev. U.S. Dep. Comm. 96(8): 540-548

forecasting, hurricanes
Moore, Paul L., 1966.
Forecasting and warning systems.
Hurricane Symposium, Oct. 10-11, 1966, Houston, Publ. Am. Soc. Oceanogr., 1:102-113.

forcasting, hurricanes
Renard, Robert J., and William H. Levings III, 1969.
The Navy's numerical hurricane and typhoon forecast scheme: application to 1967 Atlantic storm data.
J. appl. Met., 8(5): 717-725

forecasting, hydrography

forecasting, hydrology

Adrov, M.M., 1966.
About the so-called systemic basis for forecasting the oceanological conditions and reproduction of commercial fishes. (In Russian).
Mater. Ribokhoz. Issled. severn. Basseina, Poliarn. Nauchno-Issled. Proektn. Inst. Morsk. Ribn. Khoz. Okeanogr. (PINRO), 7:192-197.

forecasting, hydrology
forecasting, fisheries

forecasting hydrographic conditions

Hirano, Toshiyuki, 1969.
Scope on the forecasting and its service of fishing and oceanographic conditions. (In Japanese; English abstract). Bull. Jap. Soc. fish. Oceanogr. Spec. No. (Prof Uda Commem. Pap.): 149-151.

forecasting, hydrological variations

Smirnov, N.P., E.I. Sarukhanyan and Yu. A. Bochkov, 1967.
Long-term variations in the hydrological regime of the Barents and Norwegian seas and possibilities of forecasting them. (In Russian).
Mater. Sess. Uchen. Sovet. PINRO Rezult. Issled. 1965, Poliarn. Nauchno-Issled. Proektn. Inst. Morsk. Ribn. Khoz. Okeanogr. (PINRO), 8:111-121.

forecasting, hydrography

Stefánsson, Unnsteinn and Gudmundur Gudmundsson 1969.
Hydrographic conditions off the northeast coast of Iceland in relation to meteorological factors.
Tellus, 21(2):245-255.

forecasting, hydrological

Yakolev, V.N., 1966.
On the role of hydrological forecasts in fisheries investigations. (In Russian).
Mater. Ribokhoz. Issled. severn. Basseina, Poliarn. Nauchno-Issled. Proektn. Inst. Morsk. Ribn. Khoz. Okeanogr. (PINRO), 7:198-202.

forecasting, ice

forecasting, ice

*Akagawa, M., 1967.
Practical method for long range sea ice forecasting. (In Japanese-English abstract).
Bull. Hakodate mar. Obs., 13: 85-118.

forecasting, ice

*Kahn, S.I., 1967.
The present state of the methods for ice forecasts in the seas of the USSR. (In Russian; English abstract).
Okeanologiia, Akad.Nauk,SSSR, 7(5):786-791.

forecasting, ice

Khaminov, N.A. 1967.
Many-year variations in ice coverage of the Baltic Sea, their possible causes and ways of forecasting. (In Russian; English abstract)
Okeanologiia, 7(2):269-278.

forecasting, ice

Kryndin, A.N., 1965.
Year-to-year variations in the ice of the Baltic and the possibility of long-term forecasting on the basis of the autumn-prewinter heat content of the ocean and atmosphere. (In Russian).
Trudy, Gosudarst. Okeanogr. Inst., No. 86:3-35.

forecasting, ice

Lebedev, A.A., 1966.
Problems of forecasting ice conditions in the northwest Atlantic. (In Russian).
Materiali, Sess.Uchen.Soveta PINRO Rez. Issled., 1964, Minist.Ribn.Khoz.,SSSR, Murmansk, 108-117.

forecasting, ice

*Volkov, N.A., and Z.M. Gudkovich, 1967.
The state and prospects of the methods for long-term ice forecasts in Arctic seas. (In Russian; English abstract).
Okeanologiia, Akad. Nauk, SSSR, 7(5):792-800.

forecasting, ice

Zorina, V.A., 1965.
A method for forecasting the spring ice in the Kursk and Visla gulfs. (In Russian).
Trudy, Gosudarst. Okeanogr. Inst., No. 86:36-43.

forecasting, ice

Lebedev, A.A., 1964.
Method of forecasting the ice conditions in the North Atlantic and Greenland Sea. (In Russian).
Materialy, Ribochoz. Issled. Severn. Basseina, Gosudarst. Kom. Rib. Choz., SNCH, SSSR, Poliarn. Nauchno-Issled. i Proektn. Inst. Morsk. Rib. Choz i Okeanogr., N.M. Knipovich, (PINRO), 2:135-137.

forecasting

Maksimov, I.V., N.P. Smirnov and V.N. Vorobjev, 1964.
Long-range predictions of the long-term changes in the general ice conditions of the Barents Sea worked out by the harmonic-component method. (In Russian).
Materiali, Ribochoz. Issled. Severn. Basseina, Poliarn. Nauchno-Issled. i Proektn. Inst. Morsk. Rib. Choz. i Okeanogr., im. N.M. Knipovicha, PINRO, Gosud. Proizvodst. Komm. Ribn. Choz. SSSR, 4:73-85.

forecasting, sea level

forecasting, sea level

Groen, P., 1963.
Notes on theory and practice of predicting wind induced variations of sea level.
Atti, Simposio Int. sul Tema "Influenze Meteorol. e Oceanogr. sulle Variazioni del Livello Marino", Ist. Veneto, Sci. Lett. ed Arti, 59-69.

forecasting, SOFAR propagation

forecasting, SOFAR propagation

Hirsch, Peter, and Ashley H. Carter, 1965.
Mathematical models for the prediction of SOFAR propagation effects.
J. Acoust. Soc., Amer., 37(1):90-94.

forecasting, storms

forecasting, storms

*Stas, I.I., 1967.
The possibility of storm forecasting based on underwater wave pressure. (In Russian; English abstract).
Okeanologiia, Akad.Nauk, SSSR, 7(4):683-691.

forecasting, storm surges

forecasting, storm surges

Akedo, K. 1967.
Forecasting diagrams of the storm surges in Osaka Bay. (In Japanese; English abstract)
Bull. Kobe mar. Obs. 178: 759-769.

forecasting, storm surges

Jensen, Hans Erik, and Steen Weywadt, 1966.
Forecasting of storm surges in the North Sea. Part 1.
NATO Subcom., Oceanogr. Res., Techn. Rept., numerous pp. (multilithed).

forecasting, storm surges

Leska, Mieczyslaw, 1966.
The prediction problem of storm surges in the southern Baltic based on numerical calculations. (In Polish; English and Russian abstracts).
Archwm. Hydrotech., 13(2):335-366.

forecasting, swell

forecasting, swell

Fons, Claude, 1966.
Prévision de la houle; la méthode des densités spectro-angulaires No. 5 (D.S.A. 5).
Cahiers Océanogr., C.C.O.E.C., 18(1):15-33.

forecasting, sea state

Graffe, J.L., 1965.
Étude et prévision de l'état de la mer.
Navigation, Inst. Français Navig., 13(52):418.

forecasting, swell

Jalladeau, Philippe, et Alain Gravel 1971.
Prévision de la houle maximale en un point et son application à la region côtière Lorient - Ile de Groix.
Cah. océanogr. 23(10):921-934.

forecasting, waves

Kent, Richard, and R.R. Strange, 1966.
Some aspects of wave forecasting on the Pacific coast.
Exploiting the Ocean, Trans. 2nd Mar. Techn. Soc., Conf., June 27-29, 1966, 211-229.

forecasting

Schule, J.J., K. Terada, H. Walden and G. Verploegh, 1962.
Methods of forecasting the state of sea on the basis of meteorological data.
W.M.O. Techn. Note, No. 46:18 pp.

forecasting temperature

forecasting, temperature

*Adam, Julian, and Warren J. Jacob, 1968.
One-year experiment in numerical prediction of monthly mean temperature in the atmosphere-ocean-continent system.
Mon. Weath. Rev. U.S.Dep.Comm., 96(10):714-719.

forecasting, temperature

Azernikova, S.A., 1967.
Long-term forecast of water temperature in winter in the Barents Sea. (In Russian; English abstract)
Trudy polyar. nauchno-issled. Inst. morsk. ryb. Khoz. Okeanogr. (PINRO), 20:316-322.

forecasting, water temperature

Bochkov, Yu.A., 1966.
The forecast of water temperature in the Barents Sea for 1965/1970. (In Russian).
Materiali, Sess.Uchen.Soveta PINRO Rez. Issled., 1964, Minist.Ribn.Khoz.SSSR, Murmansk, 64-75.

forecasting, temperature gradients

Iselin, C. O'D., 1965.
Notes on the problem of predicting near surface temperature gradients in the open ocean.
ICNAF, Spec. Publ., No. 6:897-899.

forecasting, temperature

Jacob, Warren J., 1967.
Numerical semiprediction of monthly mean sea surface temperature.
J. geophys. Res., 72(6):1681-1689.

forecasting, temperature
Lauzier, L.M., 1965.
Foreshadowing of surface water temperatures at St. Andrews, N.B.
ICNAF Spec. Publ., No. 6:859-867.

forecasting temperature
Maximov, I.V., and N.P. Smirnov 1967.
Genetic method for forecasting long-term fluctuations of climatic characteristics in the ocean as demonstrated by the prognosis of surface water temperature in the Faroe-Shetland Channel. (In Russian; English abstract).
Trudy polyar. nauchno-issled. Inst. morsk. ryb. Khoz. Okeanogr. (PINRO) 20:323-335

forecasting (temperature)
Rossov, V.V., 1964.
On the possible accuracy of the near bottom temperature forecasting in the southern part of the Barents Sea. (In Russian).
Materialy Ribochozh. Issled. Severn. Basseina, Gosudarst. Kom. Ribn. Choz. SNCH, SSSR, Poliarn. Nauchno-Issled. i Proektn. Inst. Morsk. Ribn. Choz. i Okeanogr., N.M. Knipovich, (PINRO), 2:67-74.

forecasting
Seryakov, E. I., 1964.
Forecast of the thermal condition of the unfrozen part of the Barents Sea for 1964-1965. (In Russian).
Material. Sess. Uchen. Sov. PINRO, Rez. Issled., 1962-1963, Murmansk, 41-44.

forecasting, temperature, surface
Wolff, P.M., L.P. Carstensen and T. Laevastu, 1965.
Analyses and forecasting of sea surface temperature (SST).
Fleet Numerical Weather Facility Monterey California, Techn. Note, No. 8:30 pp.

forecasting
Zhukov, L.A., 1961.
Approximate calculation of the temperature and salinity fluctuation in the surface layers of the sea and their relationship to currents. Forecasting and computation of physical phenomena in the Oceans. (In Russian).
Trudy Okeanogr. Komissii, Akad. Nauk, SSSR, 11: 158-163.

forecasting, thermocline

forecasting, thermocline
Cairns, James L., 1966.
Forecasting thermocline behavior in shallow coastal waters. (Abstract only)
Trans. Am. Geophys. Un., 47(3):476.

forecasting, tides

prediction tides
Cirieni, Tite A., and Carlo Stocchino, 1967.
Tide and tidal current predictions by digital computer IBM 7090.
Int. hydrogr. Rev., 44(1):117-128.

forecasting, tides
*Ciriani, T.A., e C. Stocchino, 1966.
La previsione delle altezze di marea e della velocita della corrente di marea medianite l' impiego del calcolatore elettronico IBM*7094.
Atti XV Convegno ann. Ass. Geofis. ital., 175-187.

forecasting, tides
Isozaki, Ichiro 1969.
An investigation on the variations of sea level due to meteorological disturbances on the coasts of the Japanese Islands. 1. On the accuracy of tide predictions.
Pap. Met. Geophys. 19(3):401-426.

forecasting, tsunami

forecasting, tsunamis
Talandier, Jacques 1966.
Contribution à la prévision des tsunamis.
C. r. hebd. Séanc. Acad. Sci. Paris 263(16): 940-942

forecasting typhoon path

forecasting, typhoon movement
*Geraldson, E. L., 1968.
A comparison of the accuracy of objective techniques for forecasting typhoon movement during 1967.
Mon. Weath. Rev. U.S. Dep. Comm., 96(9):649-653.

forecasting, typhoon paths
Hashimoto, M., 1964?
On the forecasting of the movement of typhoon No. 6420. (In Japanese).
Tenki, 12(2):45-50.

Also in:
Bull. Kobe Mar. Obs., No. 175. 1965.

forecasting
Shibayama, Takeshi, 1960.
Forecasting the movements of typhoons by the 500-1000 mb thickness chart.
Kenkyu Jiho (J. Meteorol. Res., Japan), 12(6): 509-517.

forecasting, waves
See also: waves, forecasting

forecasting, waves
*Barnett, T.P., 1968.
On the generation, dissipation, and prediction of ocean wind waves.
J. geophys. Res., 73(2):513-529.

forecasting, waves
Bretschneider, Charles L., 1967.
Wave forecasting.
Ocean Industr., 2(11):38-46.

forecasting, waves
Bretschneider, Charles L., 1967.
Wave forecasting.
Ocean Industry, 2(10):53-58-60.

forecasting, waves
Bunting, D.C., 1970.
Evaluating forecasts of ocean-wave spectra. J. geophys. Res., 75(21): 4131-4143.

forecasting, waves
*Iwagaki, Yuichi, and Tadao Kakinuma, 1966.
Estimated value of bottom friction factors of some Japanese coasts.
Spec. Contr., geophys. Inst., Kyoto Univ., 6:101-106.

forecasting, waves
Rao, N.S. Bhaskara, and S. Mazumdar, 1966.
A technique for forecasting storm waves.
Indian J. Met. Geophys., 17(3):333-346.

forecasting, waves
Rattray, Maurice, Jr., and Wayne V. Burt 1967.
A comparison of methods for forecasting wave generation. (In Japanese)
Umi to Sora 42 (2):89-92.

Forcasting, waves
Savina, A., et C. Fons, 1966.
Analyse et prévision de l'état de la mer. Méthode D. S. A. 5.
La Houille blanche (3):321-337.

forecasting, weather

Forecasting, weather (long range)
Ratcliffe, R.A.S., 1970.
See temperature anomalies and long-range forecasting.
Q. Jl R. met. Soc. 96(408): 337-338.

forecasting (weather)
Republica Argentina, Secretaria de Marina. Servicio de Hidrografia Naval, 1965.
Beneficios economicos de la investagacion oceanografica.
Bol. Servicio Hidrografia Naval, 11(2):90-103.

weather forecasting
United States, National Academy of Scineces, Committee on Oceanography, 1964.
Economic benefits from oceanographic research.
Nat. Acad. Sco.-Nat. Res. Counc., Publ. No. 1228:50 pp. $2.00

foreshocks
Nagumo, Shozaburo, Heihachiro Kobayashi and Sadayuki Koresawa, 1968.
Foreshock phenomena of the 1968 Tokachi-Oki earthquake observed by ocean-bottom seismographs off Sanriku. (In Japanese; English abstract)
Bull. Earthq. Res. Inst. 46:1355-1368.

foreshore changes
Harrison W., 1969
Empirical equations for foreshore changes over a tidal cycle.
Marine Geol., 7(6):529-551

form variation

form variability
Braarud, T., 1948.
On variations in form of Skeletonema costatum and their bearing on the supply of silica in cultures of diatoms. Nytt Mag. f. Naturvidensk. 86:31-44.

form variations
Hasle, G.R., and E. Nordli, 1952.
Form variations in Ceratium fusus and tripos populations in cultures and from the sea. Abhandl. Norske Videnskaps-Akad., Oslo, 1. Math.-Naturv. Kl. 1951(4):1-25, 8 textfigs.

Fortran

Fortran II
Sweers, H.E., 1967.
Some Fortran II programs for computer processing of oceanographic observations.
NATO Subcomm. Oceanogr. Res., Tech. Rept. 37 (Irminger Sea Project):104 pp. (mimeographed).

fossils

fossils
Filarova, Z.A., M.N. Sokolova and R. Ya. Levenstein, 1968.
Mollusc of the Cambro-Devonian class Monoplacophora found in the northern Pacific.
Nature, Lond., 220(5172):1114-1115.

fossils
Fleming, C.A., 1963.
A moa-bone from the sea-floor in Cook Strait.
Rec. Dom. Mus., New Zealand, 4(16):231-233.

fossil shell fragments
Hollmann, R., 1969.
Die Entstehung fossilisationsfähiger Schalen-Frassreste dargestellt am Nahrungserwerb von Homarus gammarus (Crustacea, Decapoda).
Helgoländ. wiss. Meeresunters. 19(3):401-419.

fossil spirals
Hülsemann, Jobst, 1966.
Spiralfährten und "geführte Mäander" auf dem Meeresboden.
Natur. Mus., Frankf., 96(11):449-456.

fossils
Longinelli, A., and S. Nuti 1968.
Oxygen isotope ratios in phosphate from fossil marine organisms.
Science 160 (3830): 879-882.

fossils
*Matthews, Barry, 1967.
Late Quaternary marine fossils from Frobischer Bay (Baffin Island, N.W.T., Canada).
Palaeogr. Palaeoclimatol. Palaeoecol., 3(2):243-263.

fossil fuels, effect of
Joensuu, Oiva I., 1971.
Fossil fuels as a source of mercury pollution.
Science 172(3987): 1027-1028.

fossils
Stuermer, Wilhelm, 1970.
Soft parts of cephalopods and trilobites: some surprising results of x-ray examinations of Devonian slates. Science 170(3964): 1300-1302.

fossiligation

fossiligation
Bajard, Jacques 1966.
Figures et structures sédimentaires dans la gone intertidale de la partie orientale de la Baie du Mont-Saint-Michel.
Rev. Geogr. Phys. Géol. Dyn. (2) 8(1):39-111.

fossiligation
Mathieu, R. 1966.
Structures sédimentaires des dépôts de la gone intertidale dans la partie occidentale de la Baie du Mont-Saint-Michel.
Rev. Geogr. Phys. Géol. Dyn. (2) 8(1):113-122.

fossils
Miller, A.K., and W.M. Furnish, 1956.
Tertiary nautiloids dredged near Cape of Good Hope.
Annals, South African Mus., 42(4)(15):327-328.

fossils
Niino, H., 1955.
Sand pipe from the sea floor off California.
J. Sed. Petr. 25(1):41-44.

fossils, living

living fossils
Marwick, J., 1953.
A Pliocene fossil found living by the Galathea Expedition. N.Z. J. Sci. Tech., B, 35(1):109-112, 6 textfigs.

living fossils
Smith, J.L.B., 1953.
The second coelacanth. Nature 171:99-101.

fossil fuel combustion
Koide, Minoru and Edward D. Goldberg, 1971.
Atmospheric sulfur and fossil fuel combustion.
J. geophys. Res. 76(27): 6589-6596.

FOSSIL MEMBRANES
Degens, Egon T., Stanley W. Watson and Charles C. Remsen, 1970.
Fossil membranes and cell wall fragments from a 7000-year-old Black Sea sediment. Science, 168(3936): 1207-1208.

FOULING

fouling
Abe, Shigeo, Shigeaki Yada, Shoroku Inoue and Yusho Akishige, 1966.
On the test of sectional bottom painting in Nagasaki Harbor. (In Japanese; English abstract).
Bull. Fac. Fish., Nagasaki Univ., No. 20:58-65

fouling
Acker, Robert F., 1967.
The why of marine microbiology.
Naval Res. Rev., (March): 10-17.

fouling
Aksel'band, A.M., 1962.
Ultrasonic protection of ships against fouling.
Protection from marine fouling. (In Russian).
Trudy, Okeanogr. Komissii, Akad. Nauk, SSSR, 13: 7-9.

fouling
Aleem, A.A., 1957.
Succession of marine fouling organisms on test panels immersed in deep water at La Jolla, California. Hydrobiologia, Den Haag, 11(1):40-58.

FOULING
Aljakrinskaya, I.O. 1967.
Distribution of Mytilus and some data about their chemical composition depending on pollution in the region of Novorossiysk. (In Russian)
Trudy Inst. Okeanol. 85:66-76

fouling
Allen, F. E., 1950.
Investigations on underwater fouling. III. Note on the fouling organisms attached to naval mines in North Queensland waters. Australian J. Mar. & Freshwater Res. 1(1):196-109, 1 pl.

fouling
Allen, F.E., and E.J. Ferguson Wood, 1950.
Investigations on underwater fouling. II. The biology of fouling in Australia: results of a year's research. Australia J. Mar. & Freshwater Res. 1(1):92-105, 3 pls. 5 textfigs.

fouling
Almieda, L.J., 1965.
Study of micro-organisms incident on timber submerged in sea at Bombay prior to fouling.
Proc. Symposium on Marine Paints, New Delhi, 1964. 108-109.

fouling
Aragno, Federico Jose, 1964.
Incrustaciones producidas por organismos marinos.
Contrib. Inst. Biol. Mar., Mar del Plata, Argentina, No. 14: 12 pp.

fouling
Aravio-Terre, J., 1947.
Contribucion al estudeo de las peintures submarines para barcos de acero. Inst. Español. Ocean. Notas y Res., Ser. 2, No. 139:15 pp.

fouling
Alvariño, A., 1951.
Incrustaciones marinas. Bol. Inst. Español Ocean. No. 45:12 pp., 14 figs.

fouling
Arbuzova, K.S., 1963.
Fouling in the south-western part of the Baltic Sea. (In Russian; English summary).
Trudy Inst. Okeanol., Akad. Nauk, SSSR, 70:41-51.

fouling
Arias, E., S. Feliu y M.A. Guillén, 1970.
Algunas consideraciones y resultados de la aplicación de compuestos órgano-metálicos de plomo en la formulación de pinturas "antifouling".
Inv. pesquera, Barcelona, 34(2):319-354.

fouling
Aria, E. y E. Morales, 1969.
Ecología del puerto de Barcelona y desarrollo adherencias orgánicas sobre placas sumergidas durante los años 1964 a 1966. Investigación pesq. 33(1): 179-200.

fouling
Arias, E., and E. Morales, 1963
Ecologia del puerto de Barcelona y desarrollo de adherencias organicas sobre embarcaciones.
Inv. Pesq., Barcelona, 24:139-163.

fouling
Bagirov, R.M. 1967.
Fouling of buoys and hydrotechnical installations in Krasnovodsk Bay. (In Russian).
Trudy Inst. Okeanol. 85: 38-42.

FOULING
Bagirov, R.M., 1967.
Fouling in Baku Bay. (In Russian).
Trudy Inst. Okeanol., 85: 34-37.

fouling
Balakrishnan, Nair N., 1962.
Ecology of marine fouling and wood-boring organisms of western Norway.
Sarsia, 8:88 pp., 9 pls.

fouling
Barashkov, G.K., and M. V. Fediakina, 1965.
On stand fouling caused by littoral algae in the area of the Dalne-Zelenetskaya Guba in the Barents Sea. (In Russian).
Okeanologiia, Akad. Nauk, SSSR, 5(5):897-902.

fouling
Barnard, J. L., 1959.
Amphipod crustaceans as fouling organisms in Los Angeles - Long Beach Harbours, with reference to the influence of sea water turbidity.
Calif. Fish & Game, 44: 161-170. (2)
Abs. in Pub. Health Eng. Abs., 39(7): 31

fouling
Barnes, H., and H.T. Powell, 1950.
Some observations on the effect of fibrous glass surfaces upon the settlement of certain sedentary marine organisms. J.M.B.A. 29(2):299-302, 1 pl. 1 textfig.

fouling
Barroso Fernandes, Liane Magilla e Ayston Fernandes da Costa, 1967.
Notas sôbre organismos marinhos incrustantes e perfurantes das embarcações.
Bl. Estud. Pesca, SUDENE, Recife, 7(3): 9-26

Fouling
Bastida, Ricardo 1969.
Las incrustaciónes biologicas en el Puerto de Mar del Plata, periodo 1966/67. (2a. parte).
Lab. Ensayo Materiales Invest. tecnol., Prov. Buenos Aires, 45 pp

fouling
Bellan-Santini, Denise, 1970.
Salissures biologiques de substrats vierges artificiels immergés en eau pure, durant 26 mois, dans la région de Marseille (Méditerranée nord-occidentale). 1. Étude qualitative.
Téthys 2(2): 335-356.

fouling
Bellan-Santini, Denise, Françoise Arnaud, Patrick Arnaud, Gérard Bellan, Jean-Georges Harmelin, Thérèse Le Campion-Alsumard, Leug Tak Kit, Jacques Picard, Loic Pouliquen, Helmut Zibrowius, 1969.
Étude qualitative et quantitative des salissures biologiques de plaques expérimentales immergées en pleine eau.
Téthys 1(3): 709-714

fouling
Bengough, C.D., 1943.
Hull corrosion and fouling. Trans. N.E. Cst. Inst. Engineers Shipb. 59:183-206.

fouling
Berner, L., 1944.
Le peuplement des coques de bateaux à Marseille.
Bull. Inst. Océan., Monaco, 858:1-44.

fouling
Bishop, M.W.H., K.A. Pyefinch, and M.F. Spooner, 1949.
The examination and interpretation of fouling samples from ships. J. Iron Steel Inst. 161:35-40.

fouling
Brattström, H., 1946.
On the epifauna of an anti-submarine net in the northern part of the Sound. K. Fysiogr. Sällsk., Lund Förhandl 16(6):38-44, Fig. 1.

fouling
*Calder, Dale R., and Morris L. Brehmer, 1967.
Seasonal occurrence of epifauna on test panels in Hampton Roads, Virginia.
Int. J. Oceanol. Limnol. 1(3):149-164.

Fouling
Callame, Bernard, 1965.
Action de la lumiere sur la fixation des larves cirripedes.
Trav. Centre de Recherches et d'Etudes Oceanogr. n.s., 6(1/4):413-417.

fouling
Callame, B., 1950.
Missions du navire cablier "Émile-Baudot" sur le plateau continental Breton (Juillet-Août-Septembre, 1949). II. À propos des conditions de fixation des organismes sur les cables sous-marins. Cahiers, C.R.E.O., No. 3:9-11.

fouling
Chadwick, W.L., F.S. Clark, and D.L. Fox, 1950.
Thermal control of marine fouling at Redondo Steam Station of the Southern California Edison Company. Trans. ASME for Feb. 1950:127-131, 3 textfigs.

fouling
Chimenz, Carla, 1965/1966.
Sugli organismi incrostanti del considdetto "fouling" (rivista Sintetica).
Annuario Ist. Mus. Zool., Univ. Napoli, 17(1):1-33.

fouling
Chimenz, Carla, 1965.
Sugli organismi incrostanti del cosiddetto "fouling" (Riviste sintetica).
Ann. Ist. Mus. Zool. Univ. Napoli, 17(1):1-33.

Fouling
Chimenz Gusso, Carla e Ester Taramelli Rivosecchi, 1970
Nuove ricerche sul fouling del porto di Civitavecchia. II. Osservazioni sulle comunità incrostanti piastre metalliche verniciate immerse a varia profondità.
Rc. Accad. naz. XL (4) 20:1-20.

fouling
Christee, A.O., and G.D. Floodgate, 1966.
Formation of microtrees on surfaces submerged by the sea.
Nature, 212 (5059):308-310.

fouling
Corlett, J., 1948.
Rates of settlement and growth of the "pile" fauna of the Mersey estuary. Proc. & Trans. Liverpool Biol. Soc. 61:2-28.

fouling
Cory, Robert L., 1965.
Installation and operation of a water quality data collection system in the Patuxent River Estuary, Maryland.
Ocean Sci. and Ocean Eng., Mar. Techn. Soc., Amer. Soc., Limnol. Oceanogr., 2:728-736.

fouling
Crippen, Robert W., and Donald J. Reish, 1969
An ecological study of the polychaetous annelids associated with fouling material in Los Angeles harbor with special reference to pollution
Bull. S Calif. Acad. Sci. 68(3): 169-187

fouling
Crisp, D.J., and J.S. Ryland, 1960.
Influence of filming and of surface texture on the settling of marine organisms.
Nature, 185(4706):119.

fouling
Daniel, A., 1963.
Factors influencing the settlement of marine foulers and borers in tropical seas.
Recent Advances in Zoology, Zoological Survey of India, 363-382.

fouling
Daniel, A., 1955.
The primary film as a factor in settlement of marine foulers. J. Madras Univ., B, 25(2):189-200

fouling
Daniel, A., 1954.
The seasonal variations and the succession of the fouling communities in the Madras Harbour waters. J. Madras Univ., B, 24(2):189-212, 1 pl., 5 textfigs.

fouling
Dawson, C.E., 1957.
Balanus fouling of shrimp. Science 126(3282):1068

fouling
DePalma, John R., 1963.
Marine fouling and boring organisms off Fort Lauderdale, Florida.
Naval Oceanogr. Off., IMR No. 0-70-62:28 pp.

fouling
Deschamps, P., 1954.
Compte rendu de l'enquete sur les dommages causes par les tarets dans les port de France et d'outre-mer. Trav. C.R.E.O., 1(9):1-8.

Fouling
De Wolf, P., 1965.
The distribution of barnacles on aged anti-fouling paints; consequences and a new hypothesis
Trav. Centre de Recherches et d'Etudes Oceanogr. n.s., 6(1/4):381-387.

fouling
Dolgopolskaia, M.A., A.Z. Shapiro and Iu. A. Gorbenko, 1961.
Decay of film-forming elements not overgrown with Red Sea microorganisms.
Trudy Sevastopol Biol. Sta., 14:303-308.

fouling
Doochin, H., and F.G.W. Smith, 1951.
Marine boring and fouling in relation to velocity of water currents. Bull. Mar. Sci., Gulf and Caribbean 1(3):196-208, 4 textfigs.

fouling
Domskii, V.S., and P.F. Degtiarev, 1962.
On the prevention of the fouling of ships.
Protection from marine fouling. (In Russian).
Trudy, Okeanogr. Komissii, Akad. Nauk, SSSR, 13: 35-39.

fouling
Ealey, E.H.M., and R.G. Chittleborough, 1956.
Plankton, hydrology and marine fouling at Heard Island. Austral. Nat. Antarctic Res. Exped. Interim Rept., 15:81 pp.

Abstr: Meteor. Abstr. & Bibl., 9(2):146.

fouling
Edmondson, C. H., 1944
Incidence of fouling in Pearl Harbor. Occas. Papers, Bernice P. Bishop Mus., 18(1):1-34, 16 figs.

fouling
Eldred, B., 1962.
The attachment of the barnacle Balanus amphitrite niveus Darwin, and other fouling organisms to the rock shrimp, Sicyonia dorsalis Kingsley.
Crustaceana, 3(3):203-206.

fouling
Elliott, F.E., W.H. Myers and W.L. Tressler, 1955.
A comparison of the environmental characteristics of some shelf areas of eastern United States.
J. Washington Acad. Sci. 45(8):248-259, 4 figs.

fouling
Elroi, D., and B. Komarovsky, 1961
On the possible use of the fouling ascidian Ciona intestinalis as a source of vanadium, cellulose and other products.
Proc. Gen. Fis. Counc., Medit., Proc. and Techn. Papers, 6:261-267.

Technical Paper No. 37.

"fouling"
Fancutt, F., and J.C. Hudson, 1947.
The formulation of anti-corrosive compositions for ships' bottoms and under-water service on steel. Pt. II. J. Iron and Steel Inst., No. II, for 1946:273-296, 8 figs., Pl. XLI-XLII.

fouling
Ferguson Wood, E. J., 1950
Investigations on underwater fouling. 1. The role of bacteria in the early stages of fouling.
Australian J. Mar. and Freshwater Res. 1 (1): 85-91, 1 pl.

Fouling
Fink, F. W., R.G. Fuller, L.J. Nowacki, B.G. Brand and W.K. Boyd, 1965.
Navigational buoy corrosion and deterioration.
Trav., Centre de Recherches et d'Etudes Oceanogr n.s., 6(1/4):77-82.

fouling
Frame, David W., and James A. McCann, 1971.
Growth of Molgula complanata Alder and Hancock 1870 attached to test panels in the Cape Cod Canal.
Chesapeake Sci. 12(2): 62-66.

fouling
Fuller, J. L., 1946
Season of attachment and growth of sedentary marine organisms at Lamoine Maine. Ecol., 27(2): 150-158

fouling
Ganapati, P.N., M.V. Lakshmana Rao and R. Nagabhushanam, 1958

Biology of fouling in the Visakhapatnam Harbour
Andhra Univ. Mem., Ocean., 2: 193-209.

fouling
George, R. Yesudian, 1965.
Significance of fungi in the destruction of timber by Sphaeromids.
Proc. Symposium on Marine Paints, New Delhi, 1964: 110-114.

fouling, films
Gorbenko Yu. A. and E. S. Kucherova, 1964
ВЗАИМООТНОШЕНИЯ ДИАТОМОВЫХ ВОДОРОСЛЕЙ И ПАЛОЧКОВИДНЫХ БАКТЕРИЙ В ПЕРВИЧНОЙ ПЛЕНКЕ
Correlation of diatoms and rod-shaped bacteria in primary films. (In Russian).
Trudy Sevastopol. Biol. Stants., Akad. Nauk, 15 485-492.

fouling
Graham, H. W., and H. Gay, 1945.
Season of attachment and growth of sedentary marine organisms at Oakland, California. Ecol., 26(4):375-386, 2 figs.

fouling
Grinbart, S.B., 1937.
[Fouling of sunken vessels in the Black Sea.]
Priroda 26(12):100-102.

RT 3049 Bibl. Trans. Rus. Sci. Tech. Lit. 24.

fouling
Grishin, E.I., and N.A. Krasil'nikov, 1962.
Contemporary ways for protecting the underwater part of ships from fouling. Protection of Marine fouling. (In Russian)
Trudy Okeanogr. Komissii, Akad. Nauk, SSSR, 12:40-41.

fouling organisms
Gunter, G., and R.A. Geyer, 1955.
Studies on fouling organisms of the northwest Gulf of Mexico. Publ. Inst. Mar. Sci., Texas, 4(1):37-67, 9 textfigs.

fouling
Gurevich E.S. and M.A. Dolgopol'skaia 1964
ИСПОЛЬЗОВАНИЕ ОРГАНИЧЕСКИХ ЯДОВ В НЕОБРАСТАЮЩИХ КРАСКАХ
Utilization of organic poisons in paints against new growths. (In Russian).
Trudy Sevastopol. Biol. Stants., Akad. Nauk, 472-484.

fouling
Harris, J.E., 1947.
Report on anti-fouling research, 1942-1944.
J. Iron and Steel Inst., No. 2 for 1946:297-333 7 figs., Pl. XLIII.

fouling
Hendey, N.I., 1951.
Littoral diatoms of Chichester Harbour with special reference to fouling. J.R. Microsc. Soc. 71(1):1-86, figs.

fouling
Heinemann, G., 1951.
Problems of industries using sea water. Texas J. Sci. 3(2):311-321, 4 textfigs.

fouling
Henry, D. P., 1941
Notes on some sessile barnacles from Lower California and the West Coast of Mexico. Proc New England Zool. Club, 18:99-107, 1 pl.

fouling
Hentschel, E., 1924
Das Werden und Vergehen des Bewuches an Schiffen. Mitt. Zool. Mus., Hamburg 41:1-51

fouling
Herrera, Juan, y Ramón Margaleff, 1966.
Estimación de la actividad total añadida y de la autoabsorción en las determinaciones de producción del fitoplancton con 14-C.
Inv. Pesq., Barcelona, 30:37-44.

fouling
Hiraga, Y., 1934
Experimental investigations on the resistance of long planks and Ships.
Zosen Kiokai (J. Soc. Naval Arch. Japan), 55:159-199.

fouling
Horbund, Harold M., and Arnold Freiberger, 1970.
Slime films and their role in marine fouling: a review.
Ocean Engng 1(6): 631-634

fouling
Hori, Katsushige, and Mikio Tonoyama, 1962.
Investigations on the animal fouling organisms in Tsukumo Bay. 1. A preliminary study on the settlement of Spirorbis foraminosus Moore et Bush on the glass slide test panels. (In Japanese; English abstract).
Ann. Rept., Noto Mar. Lab., Univ. Kanazawa, 2:9-14.

fouling
Igić, Ljubimka 1969.
Seasonal aspect of the settling of principal animal groups in the fouling in northern Adriatic. (In Jugoslavian; English and Italian abstracts).
Thalassia Jugoslavica, 5: 127-132.

fouling
Inoue, Shoroku, Shigeo Abe, Shigeaki Yade and Yusho Akishige 1967
A comparative test on the antifouling effect of the ultrasonic wave on hull-fouling life. (In Japanese; English abstract)
Bull. Fac. Fish. Nagasaki Univ. 22: 121-127.

fouling
Iskra, E.V., 1962.
Contemporary methods for preventing fouling.
Protection from marine fouling. (In Russian).
Trudy, Okeanogr. Komissii, Akad. Nauk, SSSR, 13: 3-6.

fouling
Iskra, E.V., E.P. Turpaeva, I.N. Soldatova, and R.G. Simkina, 1963.
Action of some toxic substances on the major fouling organisms in the Gulf of Taganrog. (In Russian).
Trudy Inst. Okeanol., Akad. Nauk, SSSR, 70:259-269.

fouling
Ivanova, K.L., 1961.
Materials on the composition of species of algae in the marine fouling on vessels in the Sea of Azov and the Black Sea. (In Russian).
Trudy Inst. Okeanol., Akad. Nauk, SSSR, 49:137-146.

USN-HO-TRANS 177
M. Slessers 1963
P.O. 32672

fouling
Joint Technical Panel of the Leaching-rate Test of Marine Corrosion Sub-Committee, 1947.
Interim descriptive statement on the leaching-rate test for ships' anti-fouling composition.
J. Iron and Steel Inst., No. II for 1946: 335-339, 2 figs., Pl. XLIV.

fouling
Kalinenko, W.O., and N.A. Mefedova, 1956.
[Bacterial fouling of the submerged parts of a ship.] Mikrobiologiia, Akad. Nauk, SSSR, 25(2): 191-194. (English abstract)

fouling
Kan, S., H. Shiba K. Tsuchida and K. Yokoo, 1956
Effects of fouling of ships hull and propeller upon propulsive performance of ship.
Rept. Transport. Tech. Res. Inst., Tokyo, 22: 115 pp.

fouling
Karajeva, N.I., 1963.
Materials on diatom flora of fouling on the eastern shores of the Caspian Sea. (Preliminary report) (In Russian; English summary).
Trudy Inst. Okeanol., Akad. Nauk, SSSR, 70:29-40.

fouling
Karande Ashok A. 1967.
On cirriped crustaceans (barnacles), an important fouling group in Bombay waters.
Proc. Symp. Crustacea Ernakulam, Jan. 12-15, 1965, 4: 1245-1253

fouling
Kawahara, Tatuo, 1965.
Studies on the marine fouling communities. III. Seasonal changes in the initial development of test block communities.
Rep. Fac.Fish.prefect. Univ.Mie, 5(2):319-364.

fouling
Kawahara, Tatuo, 1963.
Studies on the marine fouling communities. II. Differences in the development of the test block communities with reference to the chronological differences of their initiation.
Rept. Faculty of Fish., Pref. Univ. of Mie, 4(3):391-418,8 pls.

fouling
Kawahara, Tatuo, 1962.
Studies on the marine fouling communities. 1. Development of a fouling community.
Rept. Fac. Fish., Pref. Univ. of Mie, 4(2):27-41.

fouling
Kawahara, Tatuo, 1961
Regional differences in the composition of fouling communities in Ago Bay.
Rept., Fac. Fish., Pref. Univ., Mie, 4(1): 65-80.

fouling
Kawahara, Tatuo, and Hirosi Iizima, 1961.
On the constitution of marine fouling communities at various depths in Ago Bay.
Rept. Fac. Fish., Pref. Univ. Mie, 3(3):582-594.

fouling
Kazihara, Takesi, 1964
Ecological studies on marine fouling animals. (In Japanese; English abstract)
Bull., Fac. Fish., Nagasaki Univ., 16:1-138.

fouling
Koops, H., 1971.
Beitrag zur Bewuchsverhütung bei Netzgehegen in der westlichen Ostsee.
Arch. Fischereiwiss. 22(1): 65-67

fouling
Kucherova Z.S., 1967.
Crecimientos marinos indeseables sobre superficies neutrales y toxicas en la región de la Habana.
Estudios, Inst. Oceanol., Acad. Cienc., Cuba, 2(2): 45-61

fouling
Kuriyan, G.K., 1953.
Biology of fouling in the Gulf of Manaar.
Ecology 34(4):689-692, 4 textfigs.

fouling
Kuriyan, G. K., 1950
The fouling organisms of pearl oyster cages.
J. Bombay Nat. Hist. Soc. 49 (1): 9092., 1 textfig.

fouling
Kuznetsova, I.A. 1967
Fouling in an inlet of Dalne-Zelenoxej Bay and testing of antifouling paints. (In Russian).
Trudy Inst. Okeanol. 85: 49-53.

fouling
Kuznetsova, I.A., and G.B. Zevina 1967.
Fouling in the regions building-up flow electric power station in the Barents and White seas. (In Russian).
Trudy Inst. Okeanol. 85: 18-28.

fouling
Lacombe, Dyrce, 1965.
Observacoes sobre cerrosae biologica em placas de aco na Baie de Guanabara.
Minist. Marinha, Inst. Pesquisas Mar., Notas Tecnicas, Rio de Janeiro, NT 22/1965:20 pp. (mimeographed).

fouling
Laidlaw, F.B., 1952.
The history of the prevention of fouling.
Proc. U.S. Naval Inst. 78(7):769-779.

fouling
LaQue, F.L., 1968.
Materials selection for ocean engineering.
In: Ocean engineering: goals, environment, technology, John F. Brahtz, editor, John Wiley & Sons, pp. 588-632.

fouling
Lebedev, E.M., 1961.
Marine fouling on vessels cruising the Sea of Azov and Kerchenskiy proliv [Kerch Strait]. (In Russian).
Trudy Inst. Okeanol., Akad. Nauk, SSSR, 49:118-136.

Translation:
USN Oceanogr. Off., TRANS 176, M. Slessers, 1963, P.O. 32672.

fouling
Lebedev, E.M., Y.E. Permitin and N.I. Karajeva, 1963.
On the question of fouling of plates in the Black Sea. (In Russian; English summary).
Trudy Inst. Okeanol., Akad. Nauk, SSSR, 70:270-275.

fouling
Lignau, N.A., 1923-1924; 1925-1926.
[The process of marine fouling.]
Russische Hydrobiologische Zeitschrift, 2-3(11-12):280-290
4-5(1-2): 1-10.

fouling
Marine Corrosion Sub-Committee, 1944.
Fouling of ships bottoms: identification of marine growths. J. Iron Steel Inst. 150(2):143p-156p.

fouling
Maruyama, T., and M. Hirai, 1966.
Development of bottom paints for wooden boats.
Indo-Pacific Fish. Counc., Proc. 11th Sess., 147-155.

Fouling
McCoy-Hill, M., 1965.
Marine wood preservation.
Trav. Centre de Recherches et d'Etudes Oceanogr. n.s., 6(1/4):389-396.

fouling
McIlhenny, W.F., 1966.
The oceans: technology's new challenge.
Chem. Engng. (10 Oct.) 191-198.

fouling
McNulty, J.K., 1961
Ecological effects of sewage pollution in Biscayne Bay, Florida; sediments and their distribution of benthic and fouling micro-organisms.
Bull. Mar. Sci., Gulf & Caribbean, 11(3): 394-447.

fouling
Meadows, P.S., and G.B. Williams, 1963.
Settlement of Spirorbis borealis Daudin larvae on surfaces bearing films of micro-organisms.
Nature, 198(4880):610-611.

fouling
Mel'nikov, N.N., and I.L. Vladimirov, 1962.
Synthesis of new toxic admixtures for antifouling. Protection from marine fouling. (In Russian).
Trudy, Okeanogr. Komissii, Akad. Nauk, SSSR, 13: 10-18.

fouling
Menon, N.R. and N.B. Nair, 1971.
Ecology of fouling bryozoans in Cochin waters.
Mar. Biol., 8(4): 280-307.

fouling
Merrill, Arthur S., 1965.
The benefits of systematic biological collecting from navigational buoys.
The ASB Bull. (Assoc. Southeastern Biologists), 12(1):3-8.

fouling
Millard, N. A. H., 1959.
Hydrozoa from ships' hulls and experimental plates in Cape Town docks.
Annals So. Afr. Mus., 44(1):239-256.

fouling
Millard, N., 1952.
Observations and experiments on fouling organisms in Table Bay, South Africa. Trans. R. Soc., S. Africa 33(4):415-445, Pls. 32-33, 7 textfigs.

fouling
Miller, M.A., J.C. Rapean, and W.F. Whedon, 1948
The role of slime film in the attachment of fouling organisms. Biol. Bull. 94(2):143-157, 4 figs.

fouling
Miyadi, D., and S. Mori, 1946.
[Studies pertaining to the control of biofauna attaching to the bottom of concrete vessels.]
Physiol. & Ecol. Contr. Otsu Hydrobiol Exp. Sta., Kyoto Univ., 64:1-4.

fouling
Miyauti, Tetuo, 1968.
Studies on the effect of shell cleaning in pearl culture III. The influence of fouling organisms upon the oxygen consumption in the Japanese pearl oysters. (In Japanese; English summary)
Japan J. Ecol. 17(1):40-43

fouling organisms
Monod, Th., and M. Nickles, 1952.
Notes sur quelques Xylophages et Pétricoles marins de la côte ouest africaine.
I.F.A.N. Catalogues 8:7-68, 151 textfigs.

fouling
Morales, E., y E. Arias, 1965.
Ecologia del puerto de Barcelona y desarrollo de adherencias organicas sobre placas mergidas.
Inv. Pesq., Barcelona, 28:49-79.

fouling
Nair, N. Balakrishnan, 1967.
Seasonal settlement of marine fouling and wood boring crustaceans at Cochin Harbour, south-west coast of India.
Proc. Symp. Crustacea, Ernakulam, Jan. 12-15, 1965, 4:1254-1268

fouling
Nair, N.B., 1964.
Ecology of marine fouling and wood boring organisms of western Norway.
Sarsia, 8:1-88.

fouling
Narasako, Yoshikazu, 1953.
On the effect of some treatment against to the foul of the wooden materials of ship-building in Kagoshima Harbor. (In Japanese).
Mem. Fac. Fish., Kagoshima Univ., 3(1):71-86.

fouling
Nazirov, R.K., I.P. Kuliev, A.M. Ibragimov and L.S. Alimamedov, 1962.
Fouling of marine petroleum installations and its prevention. Protection from Marine fouling. (In Russian).
Trudy, Okeanogr. Komissii, Akad. Nauk, SSSR, 13: 19-25.

fouling
Netherlands Hydrographer, 1965.
Some oceanographic and meteorological data of the southern North Sea.
Hydrographic Newsletter, Spec. Issue No. to Vol. I:numerous pp. not sequentially numbered.

fouling
Neu, W., 1933
Biologishes Arbeiten über den Schiffsbewuchs. Int. Rev. Hydrobiol. Hydrogr., 29:455-458.

fouling
Neu, W., 1932
Untersuchungen über den Schiffsbewuchs.
Int. Rev. Hydrobiol. Hydrogr. 27:105-119.

fouling
Nümann, W., and K. Beth, 1955.
Die Anseidlungszeiten der wichtigsten Bewuchs-organismen in der nordlinchen Adria.
Publ. Hydrobiol. Res. Inst., Istanbul, B, 3(1): 3-34.

fouling
Nilsson-Cantell, C. A., 1939.
Sjöfartens tysta fiender. Naturen och Vi, 10: 32-34, 2 figs.

fouling
O'Neill, Thomas B., and Gary L. Wilcox, 1971.
The formation of a "primary film" on materials submerged in the sea at Port Hueneme, California.
Pacific Sci. 25(1):1-12

fouling
Organization for European Economic Cooperation, European Productivity Agency, 1961.
Hydrological and biological conditions in testing stations in Europe, numerous pp.

fouling
Palikar, V.C., and D.V. Bal, 1955.
Marine organisms injurious to submerged timber in the Bombay Harbour.
J. Bombay Nat. Hist. Soc., 53(2):201-204.

FOULING
Patrikeev, V.V., K.S. Arbuzova and K.I. Orlova, 1967.
The purification of a water piping system and pipes of condensers from rust and fouling. (In Russian)
Trudy Inst. Okeanol. 85: 207-209.

fouling
Persoone, Guido 1971.
Ecology of fouling on submerged surfaces in a polluted harbour.
Vie Milieu Suppl. 22(2): 613-636

fouling
Persoone, Guido, 1968.
Écologie des infusoires dans les salissures de substrats immerges dans un port de mer. 1. Le film primaire et le recouvrement primaire.
Protistologica, 4(2):187-194.

fouling
Persoone, G., 1965.
The importance of fouling in the harbour of Ostend in 1964.
Helgoländer wiss Meeresunters., 12(4):449-450.

fouling
Petukhova, T.A., 1963.
Settlement of larvae of fouling organisms and marine borers in the region of Gelendjik and Novorossysk (Black Sea). (In Russian; English summary).
Trudy Inst. Okeanol., Akad. Nauk, SSSR, 70:151-156.

fouling
Plotner, Robert, 1968.
Sensor fouling on deep submergence vehicles.
Mar. Sci. Instrument. 4: 267-295.

fouling
Pomerat, C. M. and C. M. Weiss, 1946
The influence of texture and composition of surface on the attachment of sedentary marine organisms. Biol. Bull., 91(1);57-65

fouling
Pomerat, C. M., and E. R. Reiner, 1942.
The influence of surface angle and light on the attachment of barnacles and other sedentary organisms. Biol. Bull., 82(1):14

fouling
Pyefinch, K.A., 1950.
Notes on the ecology of ship-fouling organisms.
J. Animal Ecol. 19(1):29-35.

fouling
Pyefinch, K.A., 1947.
The biology of ship fouling. New Biology, London & N.Y., 3:128-148 textfigs.

fouling
Ralph, P.M., and D.E. Hurley, 1952.
The settling and growth of wharf-pile fauna in Port Nicholson, Wellington, New Zealand. Zool. Publ., Victoria Univ. Coll. No. 19:22 pp., 4 textfigs.

fouling
Rathsack, H.A., and E. Rautenberg, 1960.
Untersuchung an Schiffsanstrichen. II.
Abhand. Deutsch. Akad. Wissen., Berlin, Kl. Math. Phy. Tech., 1960(6):1-53.

fouling organisms
Ray, Dixie Lee, Ed., 1960.
Marine boring and fouling organisms.
Univ. Washington Press., 480, pp. $8.50

fouling
Redfield, A.C. and C.M. Weiss, 1948
The resistance of metallic silver to fouling
Biol. Bull. 94(1):25-28.

fouling
Reish, Donald J. 1971.
Seasonal settlement of polychaetous annelids on test panels in Los Angeles-Long Beach harbors 1950-1951.
J. Fish. Res. Bd. Can. 28(10): 1459-1467.

fouling
Relini, Giulio 1969.
Attuali conoscenze sul "fouling" della Liguria.
Pubbl. Staz. Zool. Napoli 37 (2 Suppl.):311-316

fouling
Relini, Giulio, 1968.
Variazioni quantitative stagionali del fouling nel Porto di Genova in relazione alla durata di immersione e alla profondità.
Bol. Mus. Ist. Biol. Univ. Genova, 36(236): 23-40.

fouling
Relini, Guilio, 1966.
Ricerche sul "fouling" nel porto di Genova.
Bolletino Zoologia, 33:178-179.

fouling
Relini, G., G. Dabini Oliva et L. Fernetti, 1970.
Possibilité d'étudier les effets de la pollution sur les organismes benthiques en employant des panneaux immergés.
Rev. intern. Océanogr. méd. 17: 189-199.

fouling
Relini, Giulio e Lidia Relini Orsi 1969.
Alcuni aspetti dell' accrescimento dei Balani nel Porto di Genova.
Pubbl. Staz. Zool. Napol. 37 (2 Suppl.): 327-337.

FOULING
Reznichenko O.G., 1967.
On the interspecific relationship of epibionts (Coelenterata, Bryozoa, Mollusca, Crustacea). (In Russian).
Trudy Inst. Okeanol. 85: 178-184.

fouling
Richards, B. R. and W. F. Clapp, 1944
A preliminary report on the fouling characteristics of Ponce de Leon Tidal Inlet, Daytona Beach, Fla. J. Mar. Res., 5(3): 189-195.

fouling
Richards, Francis A., 1968.
Chemical and biological factors in the marine environment. Ch. 8 in: Ocean engineering: goals, environment, technology, J.F. Brahtz, editor, John Wiley & Sons, 259-303.

fouling
Rivosecchi, Ester Taramelli e Carla Chimenz Gusso 1968.
Nuove ricerche sul fouling del Porto di Civitavecchia. 1. Successione ecologica e progressione stagionale di organismi piastre metalliche verniciate immerse.
Rc. Accad. naz. XL (4) 18 (89): 3-19.

Fouling
Robinson, T.W., and B.W. Sparrow, 1965.
Prevention of fouling of sea water culverts by antifouling paints.
Trav. Centre de Recherches et d'Etudes Oceanogr. n.s., 6(1/4):447-456.

Fouling
Romanovsky, V., Editor, 1965.
Congres International de la Corrosion Marine et des Salissures, 8-12 juin 1964, Cannes.
Trav. Centre de Recherches et d'Etudes Oceanogr., n.s., 6(1/4):509 pp.

fouling
Rosenberg, L.A., 1963.
Quantitative and qualitative characteristic of bacterial fouling of metallic plates. (In Russian; English summary).
Trudy Inst. Okeanol., Akad. Nauk, SSSR, 70:225-230.

Fouling
Ross, F. Fraser, 1965.
The control of mussels in sea water cooling systems.
Trav. Centre de Recherches et d'Etudes Oceanogr. n.s., 6(1/4):437-439.

fouling
Rudwick, M.J.S., 1951.
Notes on some Crustacea (Amphipoda) from Aden.
Ann. Mag. Nat. Hist., ser. 12, 12(38):149-156, 3 textfigs.

FOULING
Rudyakova, N.A., 1967.
Settlement and distribution on sailing vessels. (In Russian).
Trudy Inst. Okeanol., 85: 77-90

FOULING
Rudyakova N.A., 1967.
Vessel fouling in the course of cruises in the Far Eastern area north of the Japan Sea.
Trudy Inst. Okeanol., 85: 1-17.
(In Russian)

fouling
Ryland, J.S., 1967.
Polyzoa.
Oceanogr. Mar. Biol., Ann. Rev., H. Barnes, editor, George Allen and Unwin, Ltd., 5:343-369.

fouling
Rzepischevsky, I.K., G.B. Zevina and I.A. Kuznetsova 1967.
The effect of the velocity of the current on the attachment of the larvae of Balanus balanoides. (In Russian).
Trudy Inst. Okeanol. 85: 94-97.

organisms
Saito, T., 1931
Researches in fouling organisms of the Ships' bottom. Zosen Kiokai (T. Soc. Naval Arch.) 47pp:13-64, 51 figs., 9 graphs.

fouling
Scheer, Bradley T.,
The development of marine fouling communities. Biol. Bull., 89(1):103-121.

fouling
Schütz Liselotte, 1969.
Ökologische Untersuchungen über die Benthosfauna im Nordostseekanal.
III Autökologie der vagilen und hemisessilen Arten im Bewuchs der Pfähle: Makrofauna.
Int. Revue ges. Hydrobiol. 54 (4): 553-592

fouling
Seagren, G. W., M. H. Smith, and G. H. Young, 1945
The comparative antifouling efficacy of DDT.
Science, 102(2652):425-426.

fouling
Sentz-Braconnot, Eveline, 1966.
Données écologiques sur la fixation d'invertébrés sur des plaques immergées dans la rade de Villefranche-sur-mer.
Int. Rev. ges. Hydrobiol., 51(3):461-484.

fouling
*Shaw, William N., 1967.
Seasonal fouling and oyster setting on asbestos plates in Broad Creek, Talbot County, Maryland, 1963-65.
Chesapeake Sci., 8(4):228-236.

fouling
Sieburth, John McN., and John T. Conover, 1965.
Sargassum tannin, an antibiotic which retards fouling.
Nature, Lond., 208(5005):52-53.

fouling
Skarman, T.M., 1956.
The nature and development of primary films on surface submerged in the sea. N.Z.J. Sci. Tech., 38(1):44-57.

fouling
Skerman, T.M., 1959.
Marine fouling at the Port of Aukland.
New Zealand J. Sci., 2(1):57-94.

fouling
Smith, F.G.W., 1948
Surface illumination and barnacle attachment. Biol. Bull. 94(1)133-39, 1 text fig.

fouling
Smith, F.G.W., R.H. Williams, and C.C. Davis, 1950.
An ecological survey of the subtropical inshore waters adjacent to Miami. Ecol. 31(1):119-146, 7 textfigs.

fouling
Smith, Stephen V., and Eugene C. Haderlie, 1969.
Growth and longevity of some calcareous fouling organisms, Monterey Bay, California.
Pacific Sci., 23(4): 447-451

fouling
Snoke, L.R., 1957.
Resistance of organic materials and cable structures to marine biological attack.
Bell System Tech. J., 36(5):1095-1128.

fouling
Soule, Dorothy F., and John D. Soule, 1968.
Bryozoan fouling organisims from Oahu, Hawaii, with a nre species of Watersipora.
Bull. S. Calif. Acad. Sci., 67(4):203-218.

fouling
Sreenivasa Rao, B., 1964.
Studies on marine fouling organisms tolerant to low salinity and copper at Bombay Harbour.
Indian J. Techn. 2(4): 142-146 (not seen).

fouling
Starostin, I.V., 1963.
The fouling of technical water piping on our southern seas and some methods of its controlling. (In Russian; English summary).
Trudy Inst. Okeanol., Akad. Nauk, SSSR, 70:101-123.

fouling
Starostin, I.V., 1961.
[The section of fouling control (antiborer measures).]
Okeanologiia, Akad. Nauk, SSSR, 1(5):931.

fouling
Starostin, I.V., and J.E. Permitin, 1963.
Specific and quantitative characteristics of macrofouling in a water piping system of a metallurgical plant on the Azov Sea. (In Russian; English summary).
Trudy Inst. Okeanol., Akad. Nauk, SSSR, 70:124-141.

fouling
Starostin, I.V., and E.P. Turpaeva, 1963.
Settlement of larvae of fouling organisms at the water works of a metallurgical plant (Sea of Azov). (In Russian; English summary).
Trudy Inst. Okeanol., Akad. Nauk, SSSR, 70:142-150.

FOULING
Starostin I.V. and S.A. Umanskaya 1967.
The experiment of exploitation and hydrobiological characteristic of drainage water of the Novorossiyskaya Heat-Power Electric Station. (In Russian).
Trudy Inst. Okeanol. 85: 210-214

fouling
Straughan Dale 1965.
Intertidal fouling in the Brisbane River Queensland.
Proc. R. Soc. Qd. 79(4): 25-40

fouling
Suzuki, Hiroshi, and Kenjiro Konno 1970.
Basic studies on the antifouling by ultrasonic waves for ship's bottom fouling organisms 1. Influences of ultrasonic waves on the larvae of barnacles, Balanus amphitrite hawaiiensis, and mussels, Mytilus edulis. (In Japanese; English abstract).
J. Tokyo Univ. Fish. 56 (1/2): 31-48.

fouling
Taramelli Rivosecchi, Ester, e Carla Chimenz Gusso 1969.
Ricerche sugli organismi incrostanti del Porto di Civitavecchia.
Pubbl. Staz. Zool. Napoli 37 (2 Suppl.): 359-363.

fouling
Taramelli, Ester, e Carla Chimenz, 1965.
Studi sperimentali e sistematici sul "fouling" nel porto di Civitavecchia.
Rendiconti Accad. Naz. dei XL, (4), 16:1-37.

fouling
Tarasov, N.I., 1962.
On tests for methods for the prevention of fouling Protection from marine fouling. (In Russian).
Trudy Okeanogr. Komissii, Akad. Nauk, SSSR, 13: 26-34.

fouling
Tarasov, N.I., 1959.
Fouling of ships and maritime installations along the shores of the U.S.S.R., (In Russian).
Zool. Zhurn., Akad. Nauk, SSSR, 38(12):1886-1887.

Abstr. in:
Techn. Transl., 3(11):714.

fouling
Tarasov, N.I., 1949.
Corrosion and fouling. Priroda (11):32-

fouling
Tarasov, N.I., and N.A. Rudyakova, 1961.
On the methods of testing fouling of sea-going ships and hydrotechnical structures. (In Russian).
Trudy Inst. Okeanol., Akad. Nauk, SSSR, 49:60-64.

USN-HO-TRANS 175
M. Slessers 1963
P.O. 32672

fouling
Thomas, M.L.H. 1970.
Fouling organisms and their periodicity of settlement at Bideford, Prince Edward Island.
Techn. Rept., Fish. Res. Bd, Can., 158: 49 pp (multilithed).

fouling
Tighe-Ford, David 1971.
Barnacles and hormones.
Sea Frontiers 17(4): 243-250 (popular)

fouling
Turner, Harry J., 1967.
A practical approach to marine fouling.
Geo-Marine Tech., 3(3):28-32.

fouling
Turpaeva, E.P., 1969.
Symphysiological links in the oligomized biocenosis of marine animals fouling underwater construction. (In Russian).
Dokl. Akad. Nauk SSSR 159(2):415-417

fouling
Turpaeva, E.P. 1967.
On the question of the interrelationship of species in marine fouling biocenoses. (In Russian).
Trudy Inst. Okeanol 85:43-48.

fouling
Turpaeva E.P., and B.N. Maximov 1971.
Relative influence of some factors on population of Balanus improvisus Darwin in biocoenosis of marine fouling. (In Russian).
Dokl. Akad. Sci. SSSR. 199(1): 212-215.

fouling
*Tuthill, A.H., and C.M. Schillmoller, 1967.
Guidelines for selection of marine materials.
J. Ocean Tech., 2(1):6-36.

fouling'
United States Navy Hydrographic Office, 1959.
Climatological and oceanographic atlas for mariners. Vol. 1.
North Atlantic Ocean. 182 charts

fouling
Visscher, J.P., 1930
Fouling of ships bottoms. 2. Factors causing fouling. 3. Methods preventing fouling. Paint and Varn. Prod. Mgr. 35,36.

fouling
Visscher, J.P., 1928
Nature and extent of fouling of ships' bottoms. Bull. Bur. Fish, 43(2): 193-252.

fouling
Weiss, C.M., 1948
The seasonal occurrence of sedentary marine organisms in Biscayne Bay, Florida.
Ecol. 29(2):153-172, 17 figs.

fouling
Weiss, C.M., 1948
Observations on the abnormal development and growth of barnacles as related to surface toxicity. Ecol. 29(1):116-119.

fouling
Williams, G.B., 1964.
The effects of extracts of Fucus serratus in promoting the settlement of larvae of Spirorbis borealis (Polychaeta).
J. Mar. Biol. Asso., U.K., 44(2):397-414.

fouling organisms
Wisely, B., 1959.
Factors influencing the settling of the principal marine fouling organisms in Sydney Harbour.
Australian J. Mar. & Freshwater Research, 10(1): 30-44.

fouling
Wood, E.J.F., 1955.
Effect of temperature and rate of flow on some marine fouling organisms. Austral. J. Sci. 18(1): 24-27.

fouling
Wood, E.J.F., 1950.
Investigations on underwater fouling. Pt. 1. The role of bacteria in the early stages of fouling.
Australian J. Mar. and Freshwater Res. 1:85-91.

fouling
Wood, E. J. Ferguson, and F. E. Allen, 1958.
Common marine fouling organisms of Australian waters. Department of the Navy, Navy office, Melborne, 23 pp., 43 pls.

fouling
Woods Hole Oceanographic Institution, 1952.
Marine fouling and its prevention. U.S. Naval Inst. 388 pp., textfigs.

fouling
Yasuda, Toru, 1970.
Ecological studies on the marine fouling organisms occurring along the coast of Fukui Prefecture - observations on the ecology of four species of acorn barnacles found at Otomi Bay. (In Japanese; English abstract)
Bull. Jap. Soc. Scient. Fish. 36(10): 1007-1016

fouling
Yasuda, Toru, 1965.
Studies on the fouling organisms in Nyuura Bay, Fukui III observations on the ecology of a cirriped, Balanus amphitrite amphitrite Darwin. (In Japanese; English abstract)
Japan. J. Ecol., 15(1): 27-32

fouling
Zavodnik, Dušan and Ljubimka Igić 1965.
Observations on fouling in the region of Rovinj (northern Adriatic). (In Jugoslavian; English abstract)
Thalassia Jugoslavica 1: 53-68.

fouling
Zeime, G.B., 1962
[The fouling of vessels docked in Kola Bay (Barents Sea).]
Okeanologiia, Akad. Nauk. SSSR, 2(1):126-133.

Abstracted in: Soviet Bloc Res., Geophys., Astron., and Space, 1962(35):23. (OTS61-11147-35 JPRS13739)

fouling
Zevina, G.B., 1963.
Cirripedia in fouling in the Black Sea. (In Russian; English summary).
Trudy Inst. Okeanol., Akad. Nauk, SSSR, 70:72-75.

fouling
Zevina, G.B., 1963.
Fouling in the White Sea. (In Russian; English summary).
Trudy Inst. Okeanol., Akad. Nauk, SSSR, 70:52-71.

fouling
Zevina, G.B., 1962
Fouling in the Caspian Sea and its variation in the course of the last decade (1951-1961). (In Russian).
Okeanologiia, Akad. Nauk, SSSR, 2(4):715-726.

fouling
Zevina, G.B., 1957.
[Balanus improvisus Darwin and Balanus eburneus Gould observed in overgrowths on ship-bottoms and hydrotechnical constructions of the Caspian Sea.]
Doklady Akad. Nauk, SSSR, 113(2):450-453.

fouling
Zevina, G.B., 1957.
[Biology of bottom organims. On the question of fouling of ships in the Caspian Sea.]
Trudy Vses. Gidrobiol. Obsh., 8:305-320.

fouling
Zevina, G.B., I.A. Zuznetsova, and I.V. Starostin, 1963.
The composition of fouling in the Caspian Sea. (In Russian; English summary).
Trudy Inst. Okeanol., Akad. Nauk, SSSR, 70:3-25.

~~fouling~~
Zibrowius, Helmut et Gérard Bellan, 1969
Sur un nouveau cas de salissures biologiques favorisées par le Clore.
Téthys 1(2): 375-352

fouling
ZoBell, C.E., 1963.
The occurrence, effects and fate of oil polluting the sea.
Int. J. Air Water Pollution, 7(2/3):173-199.

Paper presented at International Conference on Water Pollution Research, London, 3-7 September, 1962.

fouling
Zobell, C.E., 1939
Primary film formation by bacteria and fouling. Collecting Net, 14:2 pp.

Fouling, effect of

fouling, effect of
Arbuzova, K.S., 1961.
The effect of (ship) fouling on steel corrosion in the Black Sea. (In Russian).
Trudy Inst. Okeanol., Akad. Nauk, SSSR, 49:266-273.

USN-HO-TRANS 183
M. Slessers 1963

fouling, effect of
Bartha, S., and S. Henriksson 1971
The growth of sea-organisms and the effect on the corrosion-resistance of stainless steel and titanium.
Trav. Cent. Rech. Etudes océanogr. n.s. 10(4): 20 pp.

fouling, effect of
Mockel, W., 1961.
[Losses in ships voyage speed due to the fouling effect and weather conditions illustrated on the example of three far-range cargo ships.]
Prace Inst. Morskiego, Gdansk (2. Techn. Expl. Floty), No. 7:27-56 (mimeographed).
English summary, pp. 97-106.

Genera

fouling organisms
Drummond, Ainslie H. Jr. 1968.
The foul barnacle.
Sea Frontiers 14(3):158-165 (popular)

fouling organisms (potential)
Gee, J.M., 1964.
On the taxonomy and distribution in South Wales of Filograna, Hydroides and Mercierella (Polychaeta: Serpulidae).
Ann. Mag. Nat. Hist., (13), 6(72):705-715.

fouling organisms
Howie, D.I.D., 1961.
The spawning of Arenicola marina (L.). II. Spawning under experimental conditions.
J.M.B.A., U.K., 41(1):127-144.

fouling
Knight-Jones, E.W., and D.J. Crisp, 1953.
Gregariousness in barnacles in relation to the fouling of ships and to anti-fouling research.
Nature 171(4364):1109-1110.

fouling organisms
L'Hardy, J.P., et C. Quiévreux, 1964.
Observations sur Spirorbis (Laeospira) inornatus (Polychète Serpulidae) et sur la systématique des Spirorbinae.
Cahiers de Biol. Mar., Roscoff, 5(3):287-294.

fouling organisms
Zobell, C.E., 1938
The sequence of events in the fouling of submerged surfaces. Official Digest, Fed'n Paint and Varnish Prod. Clubs, No.178: 379-385.

fouling organisms Mytilus edulis
Clarke, G.L., 1947
Poisoning and recovery in barnacles and mussels Biol. Bull. 92(1):73-91.

fouling organisms Mytilus edulis
Fischer-Piette, E., 1941
Croissance, taille maxima et longévité possible de quelques animaux intercotidaux en function du milieus. Ann. de l'Inst. Océan. Monaco, 21(1):1-1128.

fouling organisms Pomatoceros triqueter
Fischer-Piette, E., 1941
Croissance, taille maxima et longévité possible de quelques animaux intercotidaux en function du milieus. Ann. de l'Inst. Océan. Monaco, 21(1):1-1128.

fouling organisms Pomatocerus triqueter L.
Holm, Th., 1889
Om de paa Fyllas Togt i 1884 foretagne zoologiske Undersögelser i Grönland.
Medd. Grönland. 8(5):151-171.

fouling organisms Spirorbis borealis Daud.
Holm, Th., 1889
Om de paa Fyllas Togt i 1884 foretagne zoologiske Undersögelser i Grönland.
Medd. Grönland. 8(5):151-171.

fouling organisms Spirorbis cancellatus Fabr.
Holm, Th., 1889
Om de paa Fyllas Togt i 1884 foretagne zoologiske Undersögelser i Grönland.
Medd. Grönland. 8(5):151-171.

fouling, lists of spp.
Skerman, T.M., 1960.
Ship-fouling in New Zealand waters: a survey of marine fouling organisms from vessels of the coastal and overseas trades.
New Zealand J. Sci., 3(4):620-648.

fouling organisms
Zibrowius, Helmut 1970 (1971).
Les espèces méditerranéennes du genre Hydroides (Polychaeta Serpulidae): remarques sur le prétendu polymorphisme de Hydroides uncinata.
Téthys 2(3): 691-746

fouling lists of species

fouling, lists of spp.
Bastida, Ricardo 1971.
Las incrustaciones biologicas en el puerto de Mar del Plata, periodo 1966/67.
Revista Mus. argent. Cienc. nat. Bernardino Rivadavia Inst. nac. Invest. Cienc. nat. Hidrobiol. 3(2): 203-285.

fouling, lists of species
Dexter, R.W., 1954.
Fouling organisms attached to the American lobster in Connecticut waters. Ecology 36(1): 159-160.

fouling, lists of spp.
Kawahara, Tatuo 1969.
Studies on the marine fouling communities.
IV Differences in the constitution of fouling communities according to localities. 1. Nagasaki Harbor.
Rept. Fac. Fish. Pref. Univ. Mie, 6(3): 109-126

fouling, lists of spp.
Komarovsky, B., and T. Edelstein, 1961
Preliminary survey of the microflora and microfauna occurring on experimental anti-fouling panels in the Kishon Harbour.
Proc. Gen. Fish. Counc., Medit., Proc. and Tech. Papers, 6:285-287.

Technical Paper No. 42.

fouling (annelids)
Bellan, Gérard, 1970.
Étude du peuplement annelidien d'une structure sous-marine artificielle immergée dans le golfe de Marseille.
Téthys 2(2):365-372

fouling organisms
Wisely, B., 1958.
The development and settling of a serpulid worm, Hydroides norvegica Gunnerus (Polychaeta).
Australian J. Mar. Freshwater Res., 9(3):351-361.

Fouling organisms
Yamaguchi, Masaoki, 1970.
Spawning periodicity and settling time in ascidians, Ciona intestinalis and Styela plicata.
Rec. oceanogr. Wks, Japan, 10(2):147-155

fouling (quantitative)
Bellan-Santini, Denise, 1970.
Salissures biologiques de substrats vierges artificiels immergées en eau pure, durant 26 mois, dans la région de Marseille. II. Résultats quantitatifs.
Téthys 2(2): 357-364.

fouling, prevention

fouling, protection against
Chshun-dao, Dai, Tszn Min-tan and Gu Tszin-chen, 1966.
The significance for the protective coating of calcium-magnesium film for cathodic protection in sea water. (In Chinese; Russian abstract).
Oceanologia et Limnologia Sinica, 8(1):51-59.

fouling, control of
Muller, F.M., 1940.
On the sensitivity of barnacles in different stages of development toward some poisons.
Arch. Néerland. Zool., 4(2/3):113-132.

fouling preventatives
Turner, H.J., jr., D.M. Reynolds and A.C. Redfield 1948.
Chlorine and sodium pentachlorophenate as fouling preventatives in sea water conduits.
Ind. Eng. Chem. 40:450-453.

fouling prevention
Waugh, C.D., F.B. Hawes, and F. Williams, 1952.
Insecticides for preventing barnacle settlement.
Ann. Appl. Biol. 39:407-415.

fouling, protection against
Whiteneck, L.L., C.M. Wake, & H.E. Storer, 1962
Plastic barriers preserve wood piling.
Dock & Harb. Auth., 43(500): 49-51.

foundered land
Heezen, B.C., C. Gray, A.G. Segre and F.F.K. Zarudzki, 1971
Evidence of foundered continental crust beneath the central Tyrrhenian Sea.
Nature, Lond., 229(5283): 327-329

Fourier analysis

Fourier analysis
Miyazaki, Masamori, 1967.
A method of Fourier analysis of tides based on the hourly data of 355 days.
Oceanogrl Mag., 19(1):7-12.

Fourier analysis
Noda, Edward K. 1971.
Fourier analysis of transient wave system.
J. WatWays Harb Div. Am. Soc. civ. Engrs 97 (WW4): 663-670

fracture pattern

fracture pattern
Peter, George, 1966.
Magentic anomalies and fracture pattern in the northeast Pacific Ocean.
J. geophys. Res., 71(22):5365-5374.

fracture zone

fracture zones
Andrews, James E., 1971.
Abyssal hills as evidence of transcurrent faulting on North Pacific fracture zones.
Bull. geol. Soc. Am., 82(2): 463-470.

fracture zone
Auzende, Jean-Marie, Jean-Louis Olivet et Jean Bonnin, 1970.
La marge du Grand Banc et la fracture de Terre-Neuve.
C.r. hebd. Séanc. Acad. Sci. Paris, (D), 271(13): 1063-1066

fracture zones
Bullard, Sir Edward, 1969.
The origin of the oceans.
Scient. Am., 221(3):66-75. (popular).

fracture zones
Crossin, Robert S., and Nikolas I. Christensen, 1969.
Transverse isotropy of the upper mantle in the vicinity of Pacific fracture zones.
Bull. seismol. Soc. Am., 59(1): 59-72.

fracture zones
Christoffel, D.A. and D.I. Ross 1970.
A fracture zone in the South West Pacific Basin south of New Zealand and its implications for sea floor spreading.
Earth Planet. Sci. Letts, 8(2): 125-130.

fracture zones
Cifelli, Richard, Walter H. Blow and William G. Melson 1968.
Paleogene sediment from a fracture zone of the Mid-Atlantic Ridge.
J. mar. Res. 26(2): 105-109.

fracture zones
Cullen, David J., 1969.
Quaternary volcanism at the Antipodes Islands: its bearing on structural interpretation of the southwest Pacific. J. geophys. Res., 74(17) 4213-4220.

fracture zones
Cullen, David J. 1967.
The Antipodes Fracture Zone, a major structure of the south-west Pacific.
N.Z. Jl mar. Freshwat. Res. 1(1): 16-25.

fracture zones
Dehlinger, Peter, R.W. Couch, D.A. McManus and Michael Gemperle, 1970.
II. Regional observations. 4. Northeast Pacific Structure. In: The sea: ideas and observations on progress in the study of the seas, Arthur E. Maxwell, editor, Wiley-Interscience 4(2): 133-189.

fracture zones
Erickson, Barrett H., Frederic P. Naugler, and William H. Lucas, 1970.
Emperor Fracture Zone: a newly discovered feature in the central North Pacific.
Nature, Lond., 225(5227): 53-54

fracture zone
Fail, J.P., L. Montadert, J.R. Delteil, P. Valery, Ph. Patriat et R. Schlich, 1970.
Prolongation des zones de fractures de l'océan Atlantique dans le Golfe de Guinée.
Earth Planet. Sci. Letts 7(5): 413-419.

FRACTURE ZONES
Francheteau, Jean, J.G. Sclater and H.W. Menard, 1970
Pattern of relative motion from fracture zone and spreading rate data in the north-eastern Pacific.
Nature, Lond., 226(5247): 746-748

fracture zones
*Grim, Paul J., and Barrett H. Erickson, 1969.
Fracture Zones and magnetic anomalies south of the Aleutian Trench.
J. Geophys. Res., 74(6):1488-1494.

fracture zones
Heezen, B.C. 1966.
Physiography of the Indian Ocean.
Phil. Trans. R. Soc. (A) 259 (1099): 137-149.

fracture zones
Heezen, B.C., E.T. Bunce, J.B. Hersey and M. Tharp, 1964.
Chain and Romanche Fracture zones.
Deep-Sea Research, 11(1):11-33.

fracture zone
Heezen, Bruce C., R. D. Gerard and Marie Tharp, 1964.
The Vema Fracture Zone in the Equatorial Atlantic
J. Geophys. Res., 69(4):733-739.

fracture zones
Heezen, Bruce C., and H.W. Menard, 1963.
In: The Sea, M.N. Hill, Editor, Interscience Publishers, 3:233-280.

fracture zones
Isacks, Bryan, Jack Oliver and Lynn R. Sykes, 1968
Seismology and the new global tectonics.
J. geophys. Res., 73(18):5855-5899.

fracture zone
*Johnson, G. Leonard, 1967.
North Atlantic fracture zones near 53°.
Earth Planet Sci. Letters 2(5):445-448.

fracture zones
*Johnson, G. Leonard, and Bruce C. Heezen, 1967.
Morphology and evolution of the Norwegian-Greenland Sea.
Deep-Sea Res., 14(6):755-771.

fracture zones
Keen, M.J. 1970.
Fracture zones on the Mid-Atlantic Ridge between 43°N and 44°N.
Can. J. Earth Sci. 7(5): 1352-1355.

fracture zones

Krause, Dale C. 1965.
East and west Azores fracture-zones in The North Atlantic.
In: Submarine geology and geophysics, Colston Papers, W.F. Whittard and R. Bradshaw editors, Butterworths, London 17: 163-172.

fracture zones

Krause, Dale, C, 1964.
Guinea fracture zone in the Equatorial Atlantic.
Science, 146(3640):57-59.

fracture zone

Krause, Dale C., 1964.
Guinea fracture zone in the equatorial Atlantic.
Geol. Soc., Amer., Special Paper, No. 76:310-311.

Abstract only

fracture zone

Krause, Dale C., 1964
Guinea fracture zone in the equatorial Atlantic. (Abstract).
Trans. Amer. Geophys. Union, 45(1): 71

fracture zones

Le Pichon, X., S. Eittreim and J. Ewing, 1971.
A sedimentary channel along Gibbs Fracture Zone.
J. geophys. Res., 76(12): 2891-2896.

fracture zones

Le Pichon, Xavier and Dennis E. Hayes, 1971.
Marginal offsets, fracture zones, and the early opening of the south Atlantic. J. geophys Res. 76(26): 6283-6293.

fracture zones

Malahoff, Alexander, W.E. Strange and G.P. Woollard, 1966.
Molokai Fracture Zone: continuation west of the Hawaiian Ridge.
Science, 153(3735):521-522.

fracture zones

Matthews, D.H. 1965.
The Owen Fracture Zone and the northern end of The Carlsberg Ridge.
Phil. Trans. R. Soc. (A) 259 (1099): 172-186.

fracture zones

McManus, Dean A., 1965.
Blanco Fracture Zone, northeast Pacific Ocean.
Marine Geol., 3(6): 429-455

fracture zones

Menard, H.W., 1969.
The deep-ocean floor.
Scient. Am., 221(3):126-132, 134, 136, 138, 140, 142. (popular).

fracture zone

Menard, H.W., 1967.
Extension of Northeastern-Pacific fracture zones.
Science, 155(3758):72-74.

fracture zones

Menard, H.W., 1966.
Fracture zones and offsets of the East Pacific Rise.
J. geophys. Res., 71(2):682-685.

fracture zones

Menard, H.W. and T.E. Chase, 1970.
1. General observations. 11. Fracture zones (July 1968). In: The sea: ideas and observations on progress in the study of the seas, Arthur E. Maxwell, editor, Wiley-Interscience 4(1): 421-443.

fracture zones

Menard, H.W., T.E Chase and S.M. Smith, 1964
Galapagos Rise in the southeastern Pacific.
Deep-Sea Res., 11(2):233-242.

fracture zones

Metcalf, W.G., B.C. Heezen and M.C. Stalcup, 1964.
The sill depth of the Mid-Atlantic Ridge in the equatorial region.
Deep-Sea Research, 11(1):1-10.

Fracture zones

Moore, David G., and Edwin C. Buffington, 1968.
Transform faulting and growth of the Gulf of California since the late Pliocene.
Science, 161(3847):1238-1241.

fracture zones

Naugler, Frederic P., and Barrett H. Erickson 1968.
Murray Fracture Zone: westward extension.
Science 161 (3846): 1142-1145.

fracture patterns

Peter, George, 1967.
Magnetic anomalies and fracture pattern in the northeast Pacific Ocean.
J.geophys.Res., 71(22):5365.

fouling

*Poore, Gary C.B., 1968.
Succession of a wharf-pile fauna at Lyttelton, New Zealand.
N.Z.Jl.mar.Freshwat.Res. 2(4):577-590.

fracture zones

Sykes, Lynn R., 1964.
Large oceanic fracture zones delineated by earthquake epicenters. (Abstract).
Geol. Soc., Amer., Special Paper, No. 76:162.

fracture zones

Talwani, Manik, Xavier Le Pichon and James R. Heirtzler, 1965.
East Pacific Rise: the magnetic pattern and the fracture zones.
Science, 150(3700):1109-1115.

fractural zones

Wilson, J. Tuzo, 1965.
Submarine fractural zones, aseismic ridges and the International Council of Scientific Unions lines: proposed western margin of the East Pacific Ridge.
Nature, 207(5000):907-911.

frazil ice

Nybrand, G.L., 1943.
Bildning av Bottenis och sörpa i ett rinnende Vattendrag.
Teknisk Tidekr, Väg- och Vattenbyggnadskonst samt Husbyggnadsteknik 1(4):12 pp., 8 figs.

free vehicles

Phleger, Charles F., and Andrew Soutar 1971.
Free vehicles and deep-sea biology
Am. Zoologist 11 (3): 409-418.

freezing

Breitner, H.J., 1953.
Entmischung beim gefrieren wässiger Lösungen. Deutsche Hydrogr. Zeits. 6(2):80-86, 2 textfigs.

freezing

DeVries, A.L., 1969.
Freezing resistance in fishes of the Antarctic Peninsula.
Antarctic J., U.S.A., 4(4):104-105.

freezing

*Foster, Theodore D., 1968.
Haline convection induced by the freezing of sea water.
J. geophys.Res., 73(6):1933-1938.

freezing point

Hansen, H.J., 1904.
Experimental determination of the relation between the freezing point of sea-water and its specific gravity at 0°C. Medd. Komm. Havundersøgelser, Ser. Hydrogr. 1(2):10 pp., 1 textfigs

freezing

Kanwisher, J.W., 1955.
Freezing in intertidal animals. Biol. Bull. 109(1):56-63, 3 figs.

freezing

Lebedev, V.L., 1968.
Maximum width of a wind-generated lead during sea freezing. (In Russian; English abstract).
Okeanologiia, Akad. Nauk, SSSR, 8(3):391-396.

freezing of organisms

Migita, Seiji, 1964.
Freeze-preservation of Porphyra thalli in viable state. 1. Viability of Porphyra tenera preserved at low temperature after freezing in the sea water and freezing under half-dried conditions. (In Japanese; English abstract).
Bull., Fac. Fish., Nagasaki Univ., No. 17:44-54.

freezing

Ouchi, K., 1954.
Freezing mechanism of supercooled water.
Sci. Repts., Tohoku Univ., 5th Ser., Geophys., 6(1):43-61, 7 textfigs.

freezing

Shesterikov, N. P., 1960.
[Water heat supply influence on the freezing terms of the Davis Sea and the adjacent part of the Indian Ocean.]
Inform. Bull., Soviet Antarctic Exped., 19:35-38.

freezing

Simojoki, H., 1951.
On the cooling of sea water and the influence of the water heat content on the time of freezing in the archipelago and north Baltic. J. du Cons. 17(2):121-123, 1 textfig.

freezing
Sömme, Lauritz, 1965/1966.
Seasonal changes in the freezing-tolerance of some intertidal animals.
Nytt Mag. Zool., 13:56-82.

freezing
*Weeks, W.F., and Gary Lofgren, 1967.
The effective solute distribution coefficient during the freezing of NaCL solutions.
In: Physics of snow and ice, H. Oura, editor, Inst. Low Temp. Sci., Hokkaido Univ., 579-597.

freezing, effect of
Jankowsky, H.D., H. Laudien und H. Precht, 1969
Ist eine intrazelluläre Eisbildung bei Tieren tödlich? Versuche mit Polypen der Gattung Laomedea
Marine Biol., 3(1): 73-77.

freezing, effect of
Kanwisher, John, 1959
Histology and metabolism of frozen intertidal animals. Biol. Bull., 116(2): 258-264.

freezing, effect of
Migita, Seiji, 1966.
Freeze-preservation of Porphyra Thalli in viable state - II Effect of cooling velocity and water content of thalli on the frost resistance. (In Japanese :English abstract).
Bull. Fac. Fish. Nagasaki Univ. 21:131-138.

freezing, effect of
Terumoto, Isao, 1965.
Freezing and drying in a red marine alga Porphyra yezoensus Veda. (In Japanese; English Summary).
Low Temp. Sci., Hokkaido Univ., (B) 23:11-20.

freezing, effect of
Thompson, T.G., and K.H. Nelson, 1954.
Desalting sea water by freezing. Refrig. Engineer July 1954.

freezing of organisms
Terumoto, Isao, 1964.
Frost-resistance in some marine algae from the winter intertidal zone.
Low. Temp. Sci., Biol. Sci., (B), 22:19-28.

freezing resistance
*DeVries, Arthur L., and Donald E. Wohlschlag, 1969.
Freezing resistance in some Antarctic fishes.
Science, 163(3871):1073-1075.

freezing water

freezing water
Bain, G.W., 1956.
Concentration of brines and deposition of salts from sea water under frigid conditions: discussion. Amer. J. Sci., 254(12):758-760.

freezing point
Cox, R.A., 1965.
The physical properties of sea water.
In: Chemical oceanography, J.P. Riley and G. Skirrow, editors, Academic Press, 1: 73-120.

freezing
Theede, Hans, 1965.
Vergleichende experimentelle Untersuchungen über die zelluläre Gefrierresistenz mariner Muscheln.
Kieler Meeresforsch., 21(2):154-166.

fresh water

fresh water
Amoureux, L., 1959.
Quelques enclaves d'eau douce en plein milieu marin.
Comptes Rendus, Acad. Sci., Paris, 249(15):1406-1408.

freshwater
Bachmann, Roger W., and Charles R. Goldman, 1964
The determination of microgram quantities of molybdenum in natural waters.
Limnology and Oceanography, 9(1):143-146.

fresh water
Carlston, C.W., 1963.
An early American statement of the Badon Ghyben-Herzberg principle of static fresh-water-salt-water balance.
Amer. J. Sci., 261(1):88-91.

fresh water
Copeland, W.R., 1955.
Freshwater from sea water for 8.4 per 1000 gallons? Water Wks. Eng. 108:422.

Abstr: Publ. Health Eng. Abstr. 35(11):28.

freshwater
Fogg, G.E. and W.D. Watt 1965.
The kinetics of release of extra-cellular products of photosynthesis by phytoplankton.
Mem. Ist. Ital. Idrobiol. 18 (Suppl):165-174.

fresh water
Gilliland, E.R., 1955.
Fresh water for the future. Ind. & Chem. Eng., 47(12):2410-2422.

freshwater
*Manheim, Frank T., 1967.
Evidence for submarine discharge of water on the Atlantic continental slope of the southern United States, and suggestions for further search.
Trans. N.Y. Acad. Sci., (2)29(7):839-853.

freshwater
Mero, John L., 1966.
Reveiw of mineral values on and under the ocean.
Exploiting the Ocean, Trans. 2nd Ann. Mar. Techn. Soc. Conf., June 27-29, 1966, 61-78.

fresh water effluent
Mosby, Hakon, 1962
Water, salt and heat balance of the North Polar Sea and of the Norwegian Sea.
Geofys. Publik., Geophysica Norvegica, 24: 289-313.

fresh water economy
Netherlands, Delta Committee, 1962
Final report delivered by the Advisory Committee to provide an answer to the question of what waterways-technical provisions must be made for the area devastated by the storm flood of February 1, 1953, (Delta Committee) instituted by decree of the Minister of Transport and Waterways of February 18, 1953.
100 pp.

freshwater
Schaeffer, Milner B. 1968.
Economic and social needs for marine resources.
In: Ocean engineering: goals environment, technology, John F. Brahtz, editor, John Wiley and Sons, 6-37.

freshwater contributions
Nutt, D.C., and L.K. Coachman, 1956.
The oceanography of Hebron Fjord, Labrador.
J. Fish. Res. Bd., Canada, 13(5):709-758.

fresh water runoff
Pickard, G.L., 1963.
Oceanographic characteristics of inlets of Vancouver Island, British Columbia.
J. Fish. Res. Bd., Canada, 20(5):1109-1144.

fresh water effluent
Waldichuk, M., 1956.
Oceanography of the Strait of Georgia. VI. Fresh-water budget.
Fish. Res. Bd., Canada, Pacific Coast Stas., Prog. Repts., No. 107:24-27.

freshwater effluent
see also: river effluent

fresh water effluent
See also: river effluent, etc.

freshwater effluent
*Accerboni, E., e F. Mosetti, 1967.
Localizzazione dei deflussi d'acqua dolce in mare mediante un conduttometro elettrico superficiale a registrazione continua.
Boll. Geofis teor. appl., 9(36):255-268.

fresh water effluent
Barber, F.G., 1967.
A contribution to the oceanography of Hudson Bay.
Manuscript Rep. Ser., Dept. Energy, Mines, Resources, Can., No. 4:69 pp. (multilithed).

freshwater effluent
Beyer, Fredrik 1972
Om Vannutvekslingen i Oslofjorden og dens Betydning for faunaen.
Rapp. Inst. mar. Biol. Avd. A+C, 3: 25 pp. (multilithed)

freshwater effluent
Wear, R.G., 1965.
Zooplankton of Wellington Harbour, New Zealand.
Zoology Publl., Victoria Univ., Wellington, No. 38:31 pp.

fresh water reservoir
Emilsson, Ingvar, 1968.
Investigaciones sobre la hidrologia en la ensen-a de La Broa con vista a su posible transformación en un embalse de agua dulce.
Ser. Transformación Naturaleza, Acad. Cienc., Cuba, Inst. Oceanogr., 5:1-45.

fresh water springs in the sea

freshwater springs

Calvino, Floriano, and Antonio Stefanon, 1969.
The submarine springs of fresh water and the problems of their capture

Rapp. P.-v. Réun. Commn int. Explor. scient. Mer Mediterr., 19(4): 609-610.

fresh water springs in the sea

Palausi, Guy, 1968.
Les résurgences sous-marines entre l'Estérel et le Cap d'Antibes (A.M.).
Recl Trav. Stn mar. Endoume 43(59):397-404.

freshwater, source

freshwater

Alfirević, Slobodan 1966.
Les sources sous-marines de la baie de Kaštela: morphologie, structure hydrologique, relations géotectoniques.
Acta adriat. 10 (12): 1-38.

freshwater source

*Gerard, Robert D., and J. Lamar Worzel, 1967.
Condensation of atmospheric moisture from tropical maritime air masses as a freshwater source.
Science, 157(3794):1300-1302.

FRESH WATER from salt
see also: DESALTING

fresh water from salt

Anon., 1963.
Incinerator's waste heat to be used in sea water conversion.
Chem. Engin. News, 42(2):46.

fresh water from salt

Anonymous, 1963.
Sea-water desalting method uses two-solvent extraction. Bench-scale unit shows good separations, high solvent recovery.
Chemical and Engineering News, 41(2):48.

fresh water from salt

Dodge, B.F., 1963.
Fresh water from the oceans.
Discovery, 24(July):35-41.

fresh water from salt

Dodge, B. F., 1960

Fresh water from saline waters--an engineering research problem. Amer. Scient., 48(4): 476-513.

fresh water from salt

Ellis, C.B., 1954.
Fresh water from the ocean. For cities, industry, and irrigation. Ronald, 240 pp., figs. About $5.00

fresh water from salt

Gillam, W.S., and J.W. McCutchan, 1961
Demineralization of saline waters. Current desalination processes and research hope for solution of our impending water crisis.
Science, 134(3485):1041-1048.

fresh water from salt

Gould, G.F., Editor, 1963.
Saline water conversion. II. Based on symposia sponsored by the American Chemical Society. American Chemical Society, Washington, D.C., x+199 pp., illus., Paper, $6.00.

fresh water from salt

Hammond, R.P., 1962.
Large reactors may distill sea water economically.
Nucleonics, 20(12):45-49.

fresh water from salt

Nebbia, G., 1958.
La trasformazione delle acque salmastra in acqua dolce. Boll. Sci. Fac. Chim. Ind., Bologna, 16(2):44-66.

fresh water

Neville-Jones, D.J., 1955.
Fresh water from salt. Research 8(11):423-429.

fresh water from salt

Spiegler, K.S., 1962.
Salt water purification.
John Wiley & Sons, New York, London, 167 pp.

Reviewed:
Kalle, K., 1963. Deutsche Hydrogr. Zeits., 16(2):92-93.

fresh water

Streatfield, E.L., 1961.
Fresh water from the sea.
Chem. & Indus., 19:624-626.

fresh water

Telkes, M., 1953.
Fresh water from sea water by solar distillation.
Ind. Eng. Chem. 45(5):1108.

fresh water from salt

Udall, Stewart L., 1964
Fresh water - valuable product of the sea.
Undersea Technology, 5(10):16-19.

fresh water from salt

U.S. Department of the Interior, Office of Saline Water, 1958.
Proceedings: Symposium on saline water conversion 1957.
NAS-NRC Publ. 568:459 pp.

fresh water from salt

Wilson, J.R., 1961

Demineralization of sea water.
Res. Appl. In Industry, 14(7): 278-284.

fresh water from salt

Yeager, P.B., 1961.
The new age of the sea.
U. S. Naval Inst. Proc., 87(6):38-49.

fresh-salt water interface

fresh-salt water interface

Edgerton, Harold E., 1966.
Sonic detection of a fresh water-salt water interface.
Science, 154(3756):1555.

fresh water from salt

See also: desalination, etc.

freshets, effect of

freshets, effect of

Huntsman, A.G., 1955.
Effect of freshets on Passamoquoddy plankton.
Pap. Mar. Biol., and Oceanogr., Deep-Sea Res. Suppl. to Vol. 3:321-330.

fresh water, effect of

Stone, Alfred N., and Donald Reish, 1965.
The effect of fresh-water run-off on a population of estuarine polychaetous annelids.
Bull. S. Calif. Acad. Sci., 64(3):111-119.

friction

FRICTION

Johnson, J.A., 1970.
Oceanic boundary layers. Deep-Sea Res., 17(3): 455-465.

friction

*Jonsson, Ivar G., 1967.
Wave boundary layers and friction factors.
Proc. 10th Conf. Coast. Engng, Tokyo, 1966, 1:126-148.

friction

Krauss, W., 1965
Theorie des Triftstromes und der virtuellen Reibung im Meer.
Deutsche Hydrogr. Zeits, 18(5):193-210

friction

Selivanov, L.V., 1966.
Determination of friction and energy dissipation in waves using sea wind wave measurements.
(In Russian).
Fisika Atmosferi i Okeana, 2(5):545-547.

friction

Shtokman, V.B., 1950.
Determination of the flow rate and density distribution in the cross-section of an infinite channel in relation to the effect of wind and lateral friction in the field of Coriolis force.
Dok. Akad. Nauk, USSR, 71:41-44.

FRICTION: bottom

friction, bottom

*Iwagaki, Yuichi, and Tadao Kakinuma, 1966.
Estimated values of bottom friction factors of some Japanese coasts.
Spec. Contr., geophys. Inst., Kyoto Univ., 6:101-106.

friction, bottom

Iwagaki, Yuichi, and Tadao Kakinumawa, 1965.
On the transformation of ocean wave spectra in shallow water and the estimation of the bottom friction factor.
Annuals, Disaster Prevention Res. Inst. Kyoto, Univ., No. 8:379-396. Abstract in: Bull. Disaster Prevention Res. Inst., Kyoto, Univ., 15(4):79.

friction, bottom

Iwagaki, Yuichi, Tadae Kakinuma and Hiroshi Miyai, 1965.
On the bottom friction factors of some Japanese coasts.
Proc. 12th Cong. Coastal Eng., Japan, 35-40.
Abstract in: Bull. Disaster Prevention Res., Inst., Kyoto, Univ., 15(4):80.

friction, bottom
Mosby, H., 1950.
Experiments on bottom friction. Univ. i Bergen Årbok 1949, Naturvitenskapelig rekke, No. 10: 12 pp., 8 figs.

friction
Mosby, H., 1948.
Experiments on turbulence and friction near the bottom of the sea. Bergens Mus. Årb. 1946 og 1947, Naturv rekke: 6 pp.

friction, bottom
Mosby, H., 1947.
Experiments on turbulence and friction near the bottom of the sea. Bergens Museums Aarbok 1946/47, Naturvitenskap. rekke, 6 pp., 3 textfigs.

friction coefficient

frictional coefficient
Manabe, Daikaku, and Ichiro Ishida, 1967.
On wave steepness and oceanographic frictional coefficient. (In Japanese; English abstract).
Umi to Sora, 43(1): 11-21.

frictional coefficient
Manabe, Daikaku, and Ichiro Ishida, 1967.
On the oceanographical friction coefficient estimated from the change of wind velocity, ship's rolling angle, and the number of continued wave crest. (In Japanese; English abstract).
Umi to Sora 43(1): 1-13.

frictional coefficient
Manabe, Daikaku, and Ichiro Ishida, 1966.
On the oceanographical frictional coefficient, obtained through statistical treatment from daily observed data. (In Japanese: English abstract).
Umi to Sora, 41(3/4):155-164.

friction (eddy)
Bye, J.A.T., 1970.
Eddy friction in the ocean. J. mar. Res., 28(2) 124-134.

FRICTION: effect of

friction, effect of
Il'in, A.M., and V.M. Kamenkovich, 1963.
The influence of friction on ocean currents. (In Russian).
Doklady, Akad. Nauk, SSSR, 150(6):1274-1277.

friction, effect of
Ivanov, R.N., 1961
Use of a storm basin to study wind currents. Physics of the Sea. (In Russian).
Trudy Morsk. Gidrofiz. Inst., 23: 94-121.

Translation: Scripta Technica, Inc. for Amer. Geophys. Union, pp. 76-96.

friction, effect of
Iwagaki, Yuichi, Yoshito Tsuchiya and Huoxiong Chen, 1967.
On the mechanism of laminar damping of oscillatory waves due to bottom friction.
Bull. Disast. Prev. Res. Inst., Kyoto Univ., 16(3):49-75

friction, bottom, effect of
Iwagaki, Yuichi, Yoshito Tsuchiya and Masayuki Sakai, 1965.
Basic studies on the wave damping due to bottom friction.
Coastal Eng., Japan, 8:37-79.
Abstracted in: Bull. Disaster Prevention Res. Inst., Kyoto, Univ., 15(4):84.

friction, effect of
Kajiura, K., 1964.
On the bottom friction in an oscillatory current.
Bull. Earthquake Res. Inst., Tokyo, 42(1):147-174

friction, effect of
LeBlond, Paul H., 1965.
Über den Einfluss von Reibung und Vermischung auf interne Wellen.
Kieler Meeresforsch., 21(2):127-131.

friction, effect of
Shtokman, V.B., 1950.
[Determination of current-velocities and density distribution on a cross-section of an infinitely long channel, as related to the action of the wind and of side-friction in the Coriolis-force field.] Dok. Akad. Sci., SSSR, 71(1):41-44.

T60R

friction, effect of
Röber, Klaus, 1970
Analytische und numerische Lösungen für Mitschwingungsgezeiten in einem Rechteckbecken konstanter Tiefe unter Berücksichtigung von Bodenreibung, Corioliskraft und horizontalem Austausch.
Mitt. Inst. Meeresk. Univ. Hamburg, 16: 119 pp. (multilithed)

friction, effect of
Stewart, R.W., 1969.
The atmosphere and the ocean.
Scient. Am., 221(3):76-86. (popular).

friction, effect of
Stewart, R.W., 1964.
The influence of friction on inertial models of oceanic circulation.
In: Studies on Oceanography dedicated to Professor Hidaka in commemoration of his sixtieth birthday, 3-9.

friction, effect of
Vapniar, D.U., 1962.
Methods and some results of the calculation of characteristics of friction from observations on tidal currents. (In Russian).
Izv., Akad. Nauk, SSSR, Ser. Geofiz. (12):1804-1814.

friction, effect of
Weigand, J.G., H.G. Farmer, S.J. Prinsenberg, and Maurice Rattray, Jr., 1969.
Effects of friction and surface tide angle of incidence on the coastal generation of internal tides. J. mar. Res., 27(2): 241-259.

friction horizontal-turbulent

friction (horizontal-turbulent)
Kagan, B.A., A.V. Nekrasov and R.E. Tamsalu, 1966.
On the influence of horizontal turbulent friction on the tidal sea level oscillations. (In Russian ; English abstract).
Fizik. Atmosfer. i Okeana, 2(2):174-183.

friction, lateral

friction, lateral
Stockmann, W.B., and V.M. Kamenkovich, 1964.
On the effect of the wind and lateral friction on circulation around islands. (In Russian; English abstract).
In: Studies on Oceanography dedicated to Professor Hidaka in commemoration of his sixtieth birthday, 30-37.

FRICTION, surface

friction, surface
Durst, D. S., 1949.
On surface friction and turbulence in the ocean.
Mon. Not., R. Astron. Soc., Geophys. Suppl. 5(9): 369-373.

FRICTION, terms

friction terms
Ichie, T., 1950.
A note on the friction terms in the equation of ocean currents. Ocean. Mag., Tokyo, 2(2):49-52, 2 textfigs.

FRICTION: tidal

friction, tidal
Groves, Gordon and Walter Munk, 1958
A note on tidal friction. J. Mar. Res., 17:199-214.

friction, tidal, effect of
Schlichter, Louis B., 1963
Secular effects of tidal friction upon the Earth's rotation.
J. Geophys. Res., 68(14):4281-4288.

fronts, oceanic

fronts
Baranov, Ye. I. 1971.
Dynamics and structure of waters of the Gulf Stream frontal zone. (In Russian)
Okeanol. Issled. Rezult. Issled. Mezhd. Geofiz. Proekt. 22: 94-153.

fronts
Baranov, E.I., 1967.
On the study of vortices in the Gulf Stream frontal zone. (In Russian; English abstract).
Okeanologiia, Akad. Nauk, SSSR, 7(1):78-83.

fronts
Baranov, E.I., and M.A. Shmatko, 1966.
Studies of the oceanic thermal structure in the Gulf Stream frontal zone. (In Russian; English abstract).
Okeanologiia, Akad. Nauk, SSSR, 6(5):770-775.

fronts
Barkley, Richard A.
The Kuroshio-Oyashio front as a compound vortex street.
J. mar. Res., 26(2):83-104.

fronts
* Beardsley, Grant L., Jr., 1969.
Distribution and apparent relative abundance of yellowfin tuna (Thunnus albacares) in the eastern tropical Altantic in relation to oceanographic features. Bull. Mar. Sci., 19(1): 48-56.

fronts
Bebnov, V.A., 1960.
On the dynamics of the water in the frontal zone of the Kuroshio and Oyashio currents. Physics of the Sea. Hydrology. (In Russian).
Trudy Morsk. Gidrofiz. Inst., 22:15-25.

Translation: Scripta Technica, Inc. for Amer. Geophys. Union, 11-20.

front
Berger, Wolfgang H. 1971.
Planktonic foraminifera: sediment production in an oceanic front.
J. foram. Res. 1(3): 95-118.

fronts
Berrit, G.R., 1962.
Contribution à la connaissance des variations saisonnières dans le Golfe de Guinée. Observations de surface le long des lignes de navigation Deuxième Partie. Etude regionale.
Cahiers Océanogr., C.C.O.E.C., 14(9):633-643.

fronts
Bogdanov, M.A., S.G. Oradovsky, E.V. Solyankin and N.V. Khvatsky, 1969.
On the frontal zone in the Scotia Sea. (In Russian; English abstract). Okeanologiia, 9(6): 966-974.

fronts
Bubnov, V.A., 1960.
Water dynamics in the frontal zone of the Kuroshio and Oyashio currents. Physics of the sea. Hydrology. (In Russian).
Trudy Morsk. Gidrofiz. Inst., 22:15-25.

Translation:
Scripta Technica, Inc., for Amer. Geophys. Union, pp. 11-20.

fronts
*Bulgakov, N.P., 1967.
The main features of the structure and position of the subarctic fron the in northwestern Pacific. (In Russian;English abstract).
Okeanologiia, Akad.Nauk,SSSR, 7(5):879-888.

fronts
Burkov, V.A., V.S. Aresntyev, and I.M. Ovchinnikov, 1960
[On the notion of northern and southern tropical fronts in the ocean.]
Trudy Inst. Okeanol., 40: 108-120.

fronts
Chernjavskiy, V.I., 1970.
Hydrological front of the North Okhotsk Sea. (In Russian). Izv. Tikhookean. nauchno-issled. Inst. ribn. Khoz. Okeanogr. 71: 3-11.

fronts
Clarke, L.C., and T. Laevastu, 1967.
Numerical methods for synoptic computation of oceanic fronts and water type boundaries.
Int. J. Oceanol. Limnol., 1(1):28-45.

fronts, oceanic
Clarke, L.C., and T. Laevastu, 1966.
Numerical methods for synoptic computation of oceanic fronts and water type boundaries and their significance in applied oceanography.
Fleet Numerical Weather Facility, Techn. Note, Monterey, Calif., No. 20:10 pp. (mimeographed).

Fronts, (in sea)
Cochrane, John D., 1962.
The cool surface water and front on the western flank of the Yucatan Current. (Abstract).
J. Geophys. Res., 67(4):1632.

fronts
Collins, C.A., C.N.K. Mooers, M.R. Stevenson, R.L. Smith and J.G. Pattullo 1969.
Direct current measurements in the frontal zone of a coastal upwelling region.
Jl oceanogr. Soc. Japan, 24(6): 295-306

fronts, oceanic
Cromwell, T., and J.L.Reid, jr., 1956.
A study of oceanic fronts. Tellus 8(1):94-101.

fronts, oceanic
Dietrich, G. and J.M. Gieskes 1968.
The oceanic polar front in the waters off East Greenland in August 1966.
Annls biol. Copenh. (1966) 23: 20-22.

fronts
Ewing, G.C., 1964.
Slithering isotherms and thermal fronts on the ocean surface.
In: Techniques for infrared survey of sea temperature, John Clark, Editor, U.S.F.W.S. Bur. Sport Fish. and Wildlife, Bur. Circ., No. 202:92-93.

fronts
Forsbergh, Eric D., 1969.
On the climatology, oceanography and fisheries of the Panama Bight. (In Spanish and English).
Inter-Am. Trop. Tuna Comm. 14(2): 385 pp.
Bull.

fronts
Gamutilov, A. Ye., and V.M. Grusinov, 1966.
Zonality in the distribution of hydrological characteristics in the Atlantic Ocean. (In Russian).
Trudy, Morsk. Gidrofiz. Inst., 19:93-102.

Translation:
Scripta Technica, for AGU, 1963. Soviet Oceanogr. Transactions, Issue No. 3:69-76.

fronts
Gong, Yeoung, 1971.
A study on the South Korean coastal front.
J. oceanogr. Soc. Korea 6(1): 25-36. (In Korean; English abstract).

fronts
Griffiths, Raymond C., 1965.
A study of ocean fronts off Cape San Lucas, Lower California.
U.S.F.W.S. Spec. Sci. Rept., Fish., No. 499:54 pp

fronts
Gruzinov, V.M., 1965.
Hydrological front as a natural boundary of inherent zones in the ocean. (In Russian).
Trudy, Gosudarst. Oceanogr. Inst., No. 84:252-262.

fronts
Gruzinov, V.M., 1964.
The vertical circulation and the status of frontal zones in the central part of the North Atlantic. (In Russian).
Okeanologiia, Akad. Nauk, SSSR, 4(3):408-411.

fronts, oceanic
Istapoff, Feodor, 1962
The salinity distribution at 200 meters and the Antarctic frontal zones.
Deutsche Hydrographische Zeits., 15(4):133-142.

fronts, oceanic
Ivanov, Yu A., 1961
[On frontal zones in the Antarctic waters.]
Mezhd. Kom. Mezhd. Geofiz. Goda, Presidiume Akad. Nauk, SSSR, Okeanol. Issled., (3) 30-51.

fronts
Ivanov-Frantskevich, G.N., 1961
[Some peculiarities of hydrography and water masses of the Indian Ocean.] (English abstract)
Okeanolog. Issledov., Mezhd. Komitet Proved. Mezhd. Geofiz. Goda, Prezidiume Akad. Nauk, SSSR (4):7-18.

fronts
Katz, E.J., 1969.
Physical oceanography on a biological cruise.
Oceanus, 15(1):6-7. (popular)

fronts
Katz, Eli Joel, 1969.
Further study of a front in the Sargasso Sea.
Tellus 21 (2): 259-269

fronts (oceanic)
Kawai, H., & M. Sasaki. 1962
On the hydrographic condition accelerating the skipjack's northward movement across the Kuroshio Front.
Bull Tohoku Reg. Fish. Res. Lab., 20: 1-27.

fronts, oceanic
Knauss, John A., 1963
10. Equatorial current systems. In: The Sea, M.N. Hill, Edit., Vol. 2. (III) Current Interscience Publishers, New York and London, 235-252.

fronts
Kort, V.G., 1968. (1967).
Frontogenesis in the Southern Ocean. (Translation)
Info.Bull.Soviet Antarct.Exped.65:500-503. (81-89. (Scripta Techica for AGU).

fronts
Krügler, Friedrich, 1966.
Über eine optisch markante Stromgrenze an der Polarfront in der Dänemark Strasse.
Dt. hydrogr. Z., 19(4):159-170.

fronts
Levine, Edward R. and Warren B. White, 1972.
Thermal frontal zones in the eastern Mediterranean Sea. J. geophys. Res. 77(6): 1081-1086.

Fronts
Moroz, I.F. 1969.
General characteristics of the subarctic front of the Pacific Ocean. (In Russian).
Izv. Tichookean. nauchno-issled. Inst. ribn. Khoz. Okeanogr. 68: 15-32. (TINRO)

fronts (oceanic)
Nan'niti, Tosio, 1959.
Relation between the velocity and the location of the front of the Kuroshio off the Tohoku district. Oceanogr. Mag., Tokyo, 10 (2): 185-192. Also in: Collected Reprints, Oceanogr. Lab., Meteorol. Res. Inst., Tokyo, June 1960.
(In English)

fronts

Nan'niti, T., 1958

Relation between the velocity and the location of the front of the Kuroshio off the Tohoku district.
Ocean. Mag., 10(2): 185-192.

frontogenesis (sea)

Pattullo, J. G., and W. Bruce McAllister, 1962.
Evidence for oceanic frontogenesis off Oregon.
Science, 135(3498):106-107.

Fronts

Rochford, D. J., 1966.

Source regions of oxygen maxima in intermediate depths of the Arabian Sea.

Australian J. Mar. freshw. Res., 17(1):1-30.

fronts

Saint-Guily, B., 1968.
Ondes de frontière dans un bassin tournant dont le fond est incliné, C.R. Acad. Sci. Paris, 266(A): 1291-1293.
(hebd. Séanc.

fronts

Schwartzlose, Richard A., and John D. Isaacs, 1969.
Transient circulation event near the deep ocean floor.
Science, 165(3896):889-891.

fronts

Shcherbinin, A.D., 1969.
Water structure of the equatorial area of the Indian Ocean. Okeanologiia, 9(4): 597-607.
(In Russian; English abstract)

fronts

Shpaikher, A.O., and V.N. Moretsky, 1964
The polar hydrologic front in the Greenland and Norwegian seas. (In Russian).
Okeanologiia, Akad. Nauk, SSSR, 4(2):267-276.

fronts

Stewart, R.W., 1969.
The atmosphere and the ocean.
Scient.Am., 221(3):76-86. (popular).

fronts (oceanic)

Uda, Michitaka, 1962
Subarctic oceanography in relation to whaling and salmon fisheries.
Sci. Repts. Whales Res. Inst., No. 16:105-119.

fronts

Uda, Michitaka, and Makoto Ishino, 1960
Researches on the currents of Kuroshio.
Rec. Oceanogr. Wks., Japan, Spec. No. 4: 59-72.

fronts

Vladimirov, O.A., 1965.
Construction and appraisal of charts of isotherms and isohalines. (In Russian).
Trudy, Gosudarst. Okeanogr. Inst., No. 87:105-114.

fronts

*Voorhis, Arthur D., 1969.
The horizontal extent and persistence of thermal fronts in the Sargasso Sea.
Deep-Sea Res., Suppl. 16: 331-337.

fronts

Voorhis, A.D., and J.B. Hersey, 1964.
Oceanic thermal fronts in the Sargasso Sea.
J. Geophys. Res., 69(18):3809-3814.

fronts (oceanic)

Welander, P., 1963.
Steady plane fronts in a rotating fluid.
Tellus, 15(1):33-43.

fronts

*Wooster, Warren S., 1969.
Equatorial front between Peru and Galapagos.
Deep-Sea Res., Suppl. 16: 407-419.

fronts

Wyrtki, Klaus, 1966.
Oceanography of the eastern equatorial Pacific Ocean.
In: Oceanography and marine biology, H. Barnes, editor, George Allen & Unwin, Ltd., 4:33-68.

fronts

Zaneveld, J. Ronald V., Marcello Andrade, and George F. Beardsley, Jr., 1969.
Measurements of optical properties at an oceanic front observed near the Galapagos islands. J. geophys. Res., 74(23): 5540-5541.

fronts, effect of

Timmerman, H. 1971.
On the connection between cold fronts and gust bumps.
Dt. Hydrogr. Z. 24 (4): 159-172.

fronts, polar

front

Alekseyev, A. P., 1959.

Polar front in the Norwegian and Greenland Seas.
Trudy Pol. Nauch. Issle. Inst. Moskogo, 11:106-134.

fronts, polar

Cocho, F. and N. Grijalva 1968.
The influence of a polar front in the Gulf of Mexico.
Mitt. Inst. Meeresk. Univ. Hamburg 10: 184-191.

front, polar

Gorbanev, V.A., 1971.
On the interaction of the Pacific polar front with the atmospheric circulation. (In Russian; English abstract). Okeanologiia 11(4): 745-748.

front, polar

Japan, National Committee for the International Geophysical Year 1960.
The results of the Japanese oceanographic project for the International Geophysical Year, 1957/8: 198 pp.

front, Arctic

Kozlov, V.F., 1971.
The experience of computing currents in the sub-Arctic front zone in the northwestern Pacific. (In Russian; English abstract). Okeanologiia 11(4): 568-577.

fronts, polar

Uda, Michitaka, 1962
Cyclic, correlated occurrence of world-wide anomalous oceanographic phenomena and fisheries conditions.
J. Oceanogr. Soc., Japan, 20th Ann. Vol. 368-376.

fronts

Wyrtki, Klaus, 1960

The Antarctic Circumpolar Current and the Antarctic Polar Front. Deutsche Hydrogr. Zeits., 13(4): 153-174.

front, sub-Arctic

Bulgakov, N.P., 1971.
Thermohaline structure of the sub-Arctic front in the northwestern Pacific. (In Russian; English abstract). Okeanologiia 11(3): 380-389.

frontal organs

frontal organs

Elofsson, Rolf, 1966.
The nauplius eye and frontal organs of the non-Malacostraca (Crustacea).
Sarsia, No. 25: 1-125

frontal organs

Elofsson, Rolf, 1965.
The nauplius eye and frontal organs in Malacostraca (Crustacea).
Sarsia, 19:54 pp.

frozen ground

frozen ground

Werenskiold, W., 1953.
The extent of frozen ground under the sea bottom and glacier beds. J. Glaciol. 2(13): 197-200

fuel cells

fuel cells

Loughman, R., and G. Butenkoff, 1965.
Fuel cells for an underwater research vehicle.
Undersea Techn., 6(9):45-46.

fuel cells

Warszawski, Bernard, Bernard Verger and Jean-Claude Dumas, 1971
Alsthom fuel cells for marine and submarine applications.
J. mar. techn. Soc. 5 (1): 28-40

Fujiwhara effect

Fujiwhara effect

Jager, Gilbert, 1968.
An example of the "Fujiwhara effect" in the West Pacific Ocean.
Mon. Weath. Rev., Dep. Comm., 96(2):125-126.

furca

Bowman, Thomas E. 1971
The case of the nonubiquitous telson and the fraudulent furca.
Crustaceana 21(2): 165-175

furrows

fungi SEE in file after COCCOLITHINAE

Furrows

Dyer, K.R. 1970
Linear erosional furrows in Southampton Water.
Nature, Lond. 225(5227): 56-58.

furrows

*McLean, Roger F., 1967.
Origin and development of ridge-furrow systems in beachrock in Barbados, West Indies.
Mar. Geol., 5(3):181-193.

gabbro

Bonatti, E., J. Honnorez and G. Ferrara 1971
1. Ultramafic rocks. Peridotite-gabbro-basalt complex from the equatorial Mid-Atlantic Ridge.
Phil. Trans. R. Soc. Lond. (A) 268 (1192): 385-402

gabbro

Engel, C.G. and R.L. Fisher, 1969.
Lherzolite, anorthosite, gabbro and basalt dredged from the Mid-Indian Ocean Ridge.
Science, 166(3909): 1136-114 .

gabbro

Green, D.H. 1970.
Peridotite-gabbro complexes as keys to petrology of mid-oceanic ridges: discussion.
Bull. geol. Soc. Am. 81(7): 2161-2166.

gabbro

Hekinian, Roger 1970.
Gabbro and pyroxenite from a deep-sea core in the Indian Ocean.
Marine Geol., 9(4): 287-294.

gabbro

Honnorez, José and Enrico Bonatti, 1970.
Nepheline gabbro from the Mid-Atlantic Ridge.
Nature, Lond., 228(5274): 850-852

gabbros

Miyashiro, Akiho, Fumiko Shido and Maurice Ewing, 1970
Crystallization and differentiation in abyssal tholeiites and gabbros from mid-oceanic ridges.
Earth Planet. Sci. Letters 7(4): 361-365.

gabbro

Chernysheva, V.I., 1969.
Ultrabasite and gabro from the rift zones of the Arabian-Indian and Western Indian bottom ridges. Okeanologiia, 9(4): 637-648.
(In Russian; English abstract)

GALES

gales

Nania, Abele, 1969.
Il vento e l'attrito esterno nelle burrasche Adriatiche - analisi statistica - (Relazione n. 4 del progetto SABA)
Riv. Met. aeronaut. 29(4): 3-13

gales

*Nania, Abele, e Ugo Saponaro, 1969.
Contributo allo studio delle tempeste sul basso Adriatico.
Revta Met. Aeronaut., 29(1):3-16.

gales, effect of

Titov, L.F., N.O. Solnsteva and V.D. Pisarevskaia, 1962.
Calculations of the level in the sea with the period of gales in the western part of the Finnish Gulf. (In Russian).
Trudy Gosud. Okeanogr. Inst., 69:28-45.

gales, effect of

Shuisky, Yu.D., 1969.
On the influence of heavy gales on the shallow sand coasts of the eastern Baltic Sea. Okeanologiia 9(3): 475-478.
(In Russian; English abstract)

galvanic effects

galvanic effects

LaQue, F.L. 1968.
Materials selection for ocean engineering.
In: Ocean Engineering: goals, environment, technology, John F. Brahtz, editor, John Wiley and Sons, pp. 588-632.

garnet

Switzer, George, William G. Melson and Geoffrey Thompson 1970.
Garnet from the Mid-Atlantic Ridge near 43°N. latitude.
Bull. geol. Soc. Am. 81(3): 895-898.

gas

gas

Barrow, Thomas D., 1968.
Oil and gas extraction technology.
Occ. Publ. Narragansett Mar. Lab. Univ. R.I. 4:56-65.

gas

Emery, K.O., 1966
Geological methods for locating mineral deposits on the ocean floor.
Exploiting the Ocean, Trans. 2nd Ann. Mar. Techn. Soc. Conf., June 27-29, 1966, 24-43.

gas

*Gaskell, T.F., 1967.
North Sea gas.
Endeavour, 26:140-143.

gas

Kent, P.E. 1967.
Progress of exploration in North Sea.
Bull. Am. Ass. Petr. Geol. 51 (5): 731-741.

gas

Kent, P.E., 1967.
North Sea exploration- a case history.
Georg. J. 133(3):289-301.

gas

Polgnyi, George, 1965.
North Sea gas - what will it mean?
Manchester, Lit. Phil. Soc., Mem. Proc., 107:58-69.

gas

Weeks, Lewis G. 1968.
The gas, oil and sulfur potentials of the sea.
Ocean Industry, 3(6): 43-51

gastrotrichs

gastrotrichs

Clausen, Claus, 1965.
Tetranchyroderma tribolosum sp. n., a marine gastrotrich with triancres.
Sarsia, 20:9-13.

Gastrotrichs

Schrom, Heinrich, 1966.
Verteilung einiger Gastrotrichen im oberen Eulitoral eines nordadriatischen Sandstrandes.
Veroff. Inst. Meeresforsch., Bremerh., Sonderband II:95-101.

Gaussian variables

gaussian variables.

Longuet-Higgins, M.S., 1964.
Modified gaussian distributions for slightly non-linear variables.
J. Res., Natn. Bur. Stand., 68D, 1049-1062.
Also in:
Collected Reprints, Nat. Inst. Oceanogr., 13.

gazeteers

United States Board on Geographic Names 1969.
Undersea features: official standard names approved by the United States Board on Geographic Names.
Gazeteer 11: 142 pp. (multilitted)

gear

GEK measurements ALSO SEE: Instruments Currents GEK

GEK

Akamatsu, Hideo, and Tsutomu Sawara, 1966
Cruise report of CSK survey by JMA in 1965.
Oceanogr. Mag., Jap. Met. Soc., 18(1/2): 53-56

GEK

Barenov, E.I., 1967.
On the study of vortices in the Gulf Stream frontal zone. (In Russian; English abstract).
Okeanologiia, Akad. Nauk, SSSR, 7(1):78-83.

GEK

Cox, C.S., J.H. Filloux, and J.C. Larsen, 1970.
1. General observations. 17. Electromagnetic studies of ocean currents and electrical conductivity below the ocean-floor. In: The sea: ideas and observations on progress in the study of the seas, Arthur E. Maxwell, editor, Wiley-Interscience 4(1): 637-693.

G.E.K.

Crepon, Michel, 1964.
Participation française à l'Expédition Internationale de l'Océan Indien. Présentation d'observations faites au G.E.K.
Cahiers Océanogr., C.C.O.E.C., 16(10):869-874.

GEK

Darbyshire, J., 1964
A hydrological investigation of the Agulhas Current.
Deep-Sea Res., 11(5):781-815.

GEK

Fressette, R., 1965.
A study of the turbulent flow and character of the water masses over the Sicilian Ridge in both summer and winter. Summary.
Rapp. P.-v. Réun., Comm. int. Explor scient., Mer Mediterr., 18(3):811-815.

GEK

Gorodnicheva, O.P., 1966.
Current measurements with the aid of GEK in the Gulf Stream system. (In Russian; English abstract).
Okeanologiia, Akad. Nauk, SSSR, 6(3):513-519.

GEK

Highley, E., 1967.
Oceanic circulation patterns off the east coast of Australia.
Comm. Sci. Ind. Res. Org., Australia Div. Fish. Oceanogr. Tech. Pap. 23. 1-19.

GEK

Inoue, Motoo, 1965.
Movement of fishing grounds for the albacore and their migration with Oceanographic conditions in the northwestern Pacific Ocean. (In Japanese; English abstract).
Rep. Fish. Res. Lab., Tokai Univ., 2(1):1-98.

GEK

Iwata, E., S. Ohta and T. Suguro, 1968.
Some experiment in geomagnetic electrokinetograph (GEK) - I. Bull. Tokai reg. Fish. Res. Lab., 56: 57-65.
(In Japanese - English abstract).

GEK

Japan, Hakodate Marine Observatory, 1967.
Report of the oceanographic observations in the sea east of Hokkaido and in the southern part of the Okhotsk Sea from May to June, 1964. (In Japanese).
Bull. Hakodate mar. Obs., 13:10-17.

GEK

Japan, Hakodate Marine Observatory, 1967.
Report of the oceanographic observations in the Tsugaru Straits in December, 1964. (In Japanese).
Bull. Hakodate mar. Obs., 13:29-30.

GEK

Japan, Hakodate Marine Observatory, 1967.
Report of the oceanographic observations in the Okhotsk Sea and east of the Kurile islands and Hokkaido from October to November 1964. (In Japanese).
Bull. Hakodate mar. Obs., 13:20-28.

GEK

Japan, Hakodate Marine Observatory, 1967.
Report of the oceanographic observations in the sea southeast of Hokkaido, from June to July, 1964. (In Japanese).
Bull. Hakodate mar. Obs., 13:3-6.

GEK

Japan, Hakodate Marine Observatory, 1967.
Report of the oceanographic observations in the Okhotsk Sea, east of the Kurile islands and Hokkaido and east of the Tohoku district from August to September, 1964. (In Japanese).
Bull. Hakodate mar. Obs., 13:7-19.

GEK

Japan, Hakodate Marine Observatory, 1967.
Report of the oceanographic observations in the Sea southeast of Hokkaido in February, 1964. (In Japanese).
Bull. Hakodate mar. Obs., 13:3-9.

G E K

Japan, Hakodate Marine Observatory, 1966.
Report of the oceanographic observations in the sea east of Honshu and Hokkaido in March 1963.
. . in the sea east of Tohoku district in May 1963
. . in the northern part of the Japan Sea in June
. . in the sea southeast of Hokkaido in July 1963.
. . in the sea southeast of the Kurile Islands and Hokkaido from August to September 1963.
. . in the Okhotsk Sea in November 1963. (In Japanese)
Bull. Hakodate Mar. Obs. No.12:3-8;9-14;15-18; 3-4; 5-11; 12-17.

GEK

Japan, Hakodate Marine Observatory, 1964.
Report of the oceanographic observations in the sea east of the Tohoku District from February to March 1962. Report of the oceanographic observations in the Tsugaru Straits in April 1962.---in May 1962.---in June 1962. from August to September, 1962. Report of the oceanographic observations in the sea south of Hokkaido in June 1962. Report of the oceanographic observations in the sea west of Tsugaru Straits, the Tsugaru Straits and South of Hokkaido in May, 1962. (In Japanese).
Bull. Hakodate Mar. Obs., No. 11: misnumbered pp.

GEK

Japan, Japan Meteorological Agency, 1964.
The results of marine meteorological and oceanographical observations, January-June 1963, No. 33:289 pp.

GEK

Japan Kobe Marine Observatory, 1967.
Report of the oceanographic observations in the sea south of Honshu and in the Seto-Naikai from May to June 1964. (In Japanese)
Bull. Kobe Mar. Obs. No.178: 37-

GEK

Japan Kobe Marine Observatory, 1967.
Report of the oceanographic observations in the sea south of Honshu from February to March 1964. (In Japanese)
Bull. Kobe Mar. Obs. No.178: 27-

GEK

Japan, Kobe Marine Observatory, 1967.
Report of the oceanographic observations in the sea south of Honshu from July to August 1963. (In Japanese).
Bull. Kobe Mar. Obs. No.178: 31-40

GEK

Japan Kobe Marine Observatory 1967
Report of the oceanographic observations in the sea south of Honshu from October to November, 1963.
Bull. Kobe Mar. Obs. No. 178: 41-49
(In Japanese

GEK

Japan, Kobe Marine Observatory, Oceanographical Section, 1964.
Report of the oceanographic observations in the Kuroshio and region east of Kyushu from October to November 1962. (In Japanese)
Res. Mar. Meteorol. and Oceanogr., Japan. Meteorol. Agency, 32: 41-50.

Also in: Bull. Kobe Mar. Obs., 175. 1965.

GEK

Japan, Kobe Marine Observatory, Oceanographical Section, 1964.
Report of the oceanographic observations in the sea south of Honshu from February to March, 1963. Res. Mar. Meteorol. and Oceanogr., Japan Meteorol. Agency, 33: 27-32.

Also in: Bull. Kobe Mar. Obs., 175. 1965.

GEK

Japan, Kobe Marine Observatory, Oceanographical Section, 1964.
Report of the oceanographic observations in the sea south of Honshu from July to August, 1962. Res. Mar. Meteorol. and Oceanogr., Japan. Meteorol. Agency, 32: 32-40. (In Japanese).

Also in: Bull. Kobe Mar. Obs. 175.

GEK

Japan, Maizuru Marine Observatory, 1965.
Report of the oceanographic observations in the central part of the Japan Sea from February to March, 1962.---in the Japan Sea from June to July, 1962.---in the western part of Wakasa Bay from January to April, 1962.---in the central part of the Japan Sea from September to October, 1962.---in the western part of Wakasa Bay from May to November, 1962.---in the central part of the Japan Sea in March, 1963.---in the Japan Sea in June, 1963.---in Wakasa Bay in July, 1963.--- in the central part of the Japan Sea in October, 1963. (In Japanese).
Bull, Maizuru Mar. Obs., No.9:67-73;74-88;89-95; 71-80;81-87;59-65;66-77;80-84;85-91.

GEK (data only)

Japan, Maritime Safety Agency, 1967.
Results of oceanographic observations in 1965.
Data Rept., Hydrogr. Obs., Ser. Oceanogr. (Publ. 792) Oct. 1967, 5:115 pp.

GEK (data only)

Japan, Maritime Safety Agency, 1965.
Results of oceanographic observations in 1962.
Data Rept. Hydrogr. Obs., Ser. Oceanogr. (Publ. 792) Nov. 1965, 1:65 pp.

GEK (data only)

Japan, Maritime Safety Agency 1967.
Results of oceanographic observations in 1964.
Data Rept. Hydrogr. Obs. Ser. Oceanogr. Publ. 792 (4): 88pp.

GEK (data only)

Japan, Maritime Safety Agency 1966.
Results of oceanographic observations in 1963.
Data Rept. Hydrogr. Obs. Ser. Oceanogr. Publ. 792 (3): 74 pp.

GEK

Japan, Nagasaki Marine Observatory 1971.
Report of the oceanographic observations in the sea west of Japan from January to February 1967. (In Japanese)
Oceanogr. Met. Nagasaki mar. Obs. 18:1-10

GEK

Japan, Nagasaki Marine Observatory 1971.
Report of the oceanographic observations in the sea west of Japan from June to August, 1967. (In Japanese)
Oceanogr. Met. Nagasaki mar. Obs. 15: 1-12.

GEK
Japan Nagasaki Marine Observatory 1971.
Report of the oceanographic observations in the sea west of Japan from January to February 1968 (In Japanese)
Oceanogr. Met. Nagasaki mar. Obs. 15: 1-10.

GEK
Japan, Nagasaki Marine Observatory, Oceanographic section, 1965.
Report of the oceanographic observations in the sea west of Japan from February to March 1963,-- from July to August 1963.
Res. Mar. Meteorol. Oceanogr., Japan, Meteorol Agency 33:39-58;34:53-80.
Also in: Oceanogr. Meteorol., Nagasaki, 15 (227-228).

GEK (data only)
Japan, Maritime Safety Agency, 1964.
Tables of results from oceanographic observations in 1961.
Hydrogr. Bull. Tokyo (Publ. 981) 77: 82 pp.

GEK(data only)
Japan, Nagasaki University Research Party for CSK, 1966.
The results of the CSK - NU65S.
Bull. Fac. Fish. Nagasaki Univ., No. 21:273-292.

GEK
Kollmeyer, Ronald C., Robert M. O'Hagan and Richard M. Morse, 1965.
Oceanography of the Grand Banks region and the Labrador Sea in 1964.
U.S.C.G. Oceanogr. Rept., No. 10(CG 373-10):1-24; 34-285.

GEK
*Konaga, Shunji, 1968.
Variations of the oceanographic condition south of Japan in relation to the mean sea level at Kushimoto and Uragami tidal station, Japan No.1- Oceanographic condition at station G 1 and the mean sea level at Kushimoto. (In Japanese; English abstract).
Umi to Sora 43(4):125-134.

GEK
Konaga, Shunji, 1964
On the current velocity measured with geomagnetic electrokinetography. (In Japanese; English abstract).
J. Oceanogr. Soc., Japan, 20(1):1-6.

GEK
Lee, Chang-Ki and Jong-Hon Bong, 1968.
On the current of the Korean Eastern Sea (West of the Japan Sea). (In Korean; English abstract).
Bull. Fish. Res. Develop Agency, Korea, 3: 7-26

GEK
Lipparelli, M.A. and G.F. Beardsley, Jr., 1971.
The GEK signature of internal waves. J. geophys. Res. 76(21): 5043-5047.

GEK
Martin, Jean, Pierre Guibout, Michel Crepon et Jean-Claude Lizeray, 1965.
Circulation superficielle dans l'Océan Indien: résultats de mesures faites à l'aide du courantomètre à électrodes remorquées GEK entre 1955 et 1963.
Cahiers Océanogr., C.C.O.E.C., 17(Suppl.3):222-240, 89 pls.

GEK
Masuzawa, Jotaro, 1968.
Cruise report on multi-ship study of short-term fluctuations of the Kuroshio in October to November 1967.
Oceanogrl Mag., 20(2):91-104.

GEK
Novysh, V.V., 1965.
On the precision of estimating the vertical components of the geomagnetic field in the geomagnetic current meter. (In Russian).
Okeanologiia, Akad. Nauk, SSSR, 5(4):718-725.

GEK
Popov, I.K. 1968.
Use of the GEK in oceanological investigations by the Thirteenth Soviet Antarctic Expedition. (In Russian).
Inform. Biull. Sovetsk. Antarkt. Exped. 71: 65-67.
Translation: Scripta Tecnica Inc. for AGU 7(3): 216-218

G E K
Shoji, Daitaro, 1965.
Description of the Kuroshio (Physical aspect).
Proc. Symp., Kuroshio, Tokyo, Oct. 29, 1963, Oceanogr. Soc., Japan and UNESCO, 1-10.

GEK measurements
Shoji, S., 1957.
Kuroshio during 1955. Rec. Ocean. Wks., Japan, (Spec. No.):21-26.

gelatinous mat

gelatinous mat
*Bathurst, R.G.C., 1967.
Subtidal gelatinous mat, sand stabilizer and food, Great Bahama Bank.
J. Geol., 75(6):736-738.

gelbstoff

GELBSTOFF
*Hickel, W., 1969.
Planktologische und hydrographisch-chemische Untersuchungen in der Eckernförder Bucht (westliche Ostsee) während und nach der Vereisung im extrem kalten Winter 1962/1963. Helgoländer wiss. Meeresunters. 19(2): 318-331.

gelbstoff
Kalle, K., 1966.
The problem of gelbstoff in the sea.
In: Oceanography and marine biology, H. Barnes, editor, George Allen & Unwin, Ltd., 4:91-104.

gelbstoff
Kalle, Kurt, 1962
Über die gelösten organischen Komponenten im Meerwasser.
Kieler Meeresf., 18(3) (Sonderheft): 128-131.

gelbstoff
*Lenz, Jürgen, Heinz Schöne und Bernt Zeitschel, 1967.
Planktonologische Beobachtungen auf einem Schnitt durch die Nordsee von Cuxhaven nach Edinburgh.
Kieler Meeresforsch., 23(2):92-98.

gelbstoff
Lüneberg, H., 1939.
Hydrochemische Untersuchungen in der Elbmündung mittels Elektrokolorimeter.
Arch. Deutschen Seewarte 59(1):1-27, 8 pls.

Gelbstoff
*Sieburth, John McN., 1969.
Studies on algal substances in the sea. III. The production of extracellular organic matter by littoral marine algae.
J. exp. mar. Biol. Ecol., 3(3):290-309.

Gelbstoff
*Sieburth, John McN. and Arne Jensen, 1969.
Studies on algal substances in the sea. II. The formation of Gelbstoff (humic material) by exudates of Phaeophyta.
J. exp. mar. Biol. Ecol., 3(3):275-289.

Gelbstoff
Sieburth John McN. and Arne Jensen, 1968.
Studies on algal substances in the sea. I. Gelbstoff (humic material) in terrestrial and marine waters.
J. exp. mar. Biol. Ecol., 2(2):174-189.

genera
Tortonese, Enrico, 1968.
La valeur biologique du genre comme unité taxonomique.
Revue Roumaine Biol. (Zool.) 13 (6): 467-471

General

general
Boehnecke, G., and G. Neumann, 1948.
Allgemeine Ozeanographie. Geophys., Dieterich'sche Verlagsbuchhandlung, Wiesbaden, Pt. 2, Vol. 18: 76-107.

Oceanography
Fish, C. J. 1946
Japanese Oceanographic and Marine biological stations. Trans Am. Geophys. Un., Vol. 27, No. 4, pp. 501-516.

Oceanography
Fish, C. J. 1946
Oceanography in Japan. Trans. Am. Geophys. Un., Vol. 27, No. 4, pp. 521-522.

general
Fraser, C. M., 1939(1940).
Oceanography in British Columbia. Proc. Sixth Pacific Sci. Congr., 3:20-33.

general
Harvey, H. W., 1939
Biological oceanography. In: Symposium on recent marine sediments. Am. Assoc. Petroleum Geologists:142-152, Tulsa, Okla.

Oceanography
Hidaka, Koji 1946
Cooperation of Japanese oceanographers with army and navy during the Pacific War. Trans. Am. Geophys. Un., Vol. 27, No. 4, pp. 517-520

general
Huntsman, A. G., 1924.
Oceanography. Handbook of the Brit. Assoc. for the Adv. Sci., Toronto, pp. 274-290.

general

Krogh, A., 1934
Conditions of life at great depths in the ocean. Ecol. Monogr., 4:430-439.

general

Proceedings of the Pacific Science Conference of the National Research Council. June 6-8, 1946. Bull. Nat. Res. Counc., No. 114, Sept. 1946

general

Stewart, J. Q. 1945
Coast, Waves, and Weather for Navigators. vii and 348 pp, maps, diagr., ill. bibliogr., index. Ginn and Co. $3.75

general zoology

genetics

genetics

Battaglia, Bruno 1967.
Genetic aspects of benthic ecology in brackish waters.
In: Estuaries, G.H. Lauff, editor, Publs Am. Ass. Advmt Sci. 83: 574-577.

genetics

Battaglia, B. 1960.
Il polimorfismo adattativo e i fattori della selegione nel copepode Tisbe reticulata Bocquet.
Archo Oceanogr. Limnol. 11 (3): 305-356.

genetics

Bocquet, Charles, 1960
Sur un phénotype "semi-pigmenté" réalisé chez de rares Sphaeroma serratum (Fabricius).
Bull. Soc . Linnéene de Normandie, (10), 1: 210-212.

genetics

Bocquet, Charles, 1960
Définition et analyse génétique du phénotype "ornatum noir" de l'"isopode Sphaeroma serratum (F.).
Bull Soc. Lineene de Normandie, (10) 1:204-209.

genetics

Bocquet, Charles, Robert Lejuez, Georges Teissier, 1969
Génétique des populations de Sphaeroma serratum (F.). IX Étude des populations des îles anglo-normandes de Jersey et de Guernesey.
Cah. Biol. mar, 10 (4): 405-427.

genetics

Boquet, C., R. Lejuiz et G. Teissier, 1966.
Génétique des population de Sphaeroma Serratum (F.). VII. Donnees complémentaires sur la pan [illegible] mixie.
Cahiers Biol Mar. 7(1):23-30.

genetics

Bocquet, Charles, Robert Lejuez et Georges Teissier, 1960
Génétique des populations de Sphaeroma serratum (F.). III. Comparaison des populations mères et des populations filles pour les Spheromes du Contentin.
Cahiers de Biol. Marine, 1:279-294.

genetics

Bocquet, Charles, et Georges Teissier, 1960
Génétique des populations de Sphaeroma serratum (F.) 1. Stabilité du polychromatisme local.
Cahiers de Biol. Marine, 1: 103-111.

genetics

Lejuez, R., 1961.
Génétique des populations de Sphaeroma serratum (F.). IV. Étude des populations de la côte septentrionale du Cotentin.
Cahiers Biol. Mar., Roscoff, 2(3):327-342.

genetics

Lejuez, Robert, 1960
Hybridation experimentale et naturelle entre Sphaeroma hookeri Leach et Sphaeroma rugicauda Leach.
C.R. Acad. Sci., Paris, 250: 597-599.

genetics

Lejuez, Robert, 1960
Sur le polychromatisme de Sphaeroma serratum (Fabricius) le long du littoral septentrional du Cotentin.
C.R. Acad. Sci., Paris, 251:1244-1246.

genetics

Kerambrun, Pierre, 1966.
Contribution à l'étude génétique et écologique du polychronatisme de l'isopode Sphaeroma hookeri dans les eaux Saumâtres Méditerranéenes.
Bull. Inst. Océanogr., Monaco, 66(1369):52 pp.

genetics

Miller, R.B., 1957.
Have the genetic patterns of fishes been altered by introductions or by selective breeding.
J. Fish. Res. Bd., Canada, 14(6):797-806.

genetics

Tinturier-Hamelin, Emmanuelle, 1959.
La determination du sexe chez Idotea baltica basteri Audouin (Isopode Valvifère). Premiers resultats d'une étude genetique.
C. R. Adad. Sci., Paris, 248:2660-2662.

geobiochemical interactions

Carritt, Dayton E., 1971.
6. Oceanic residence time and geobiochemical interactions. In: Impingement of man on the oceans, D.W. Hood, editor, Wiley Interscience: 191-199.

geochemistry

Geochemistry

See also: Chemistry

geochemistry

Barth, T.F.W., 1961
Abundance of the elements, areal averages and geochemical cycles.
Geochim et Cosmochim. Acta, 23(1/2): 1-8.

geochemistry

Bender, Michael, Wallace Broecker, Vivian Gornitz, Ursula Middel, Robert Kay, Shine-Soon Sun, and Pierre Biscaye 1971
Geochemistry of three cores from the East Pacific Rise.
Earth planet. Sci. Letts. 12(4): 425-433.

geochemistry, organic

*Blumer, Max, 1967.
Equilibria and nonequilibria in organic geochemistry.
In: Equilibrium concepts in natural water systems, Werner Stumm, editor, Adv. Chem. Ser. 67: 312-318.

geochemistry

Broecker, Wallace, 1960
Geochemistry of ocean water. Trans. Amer. Geophys. Un., 41(2): 259-260.

geochemistry

Carritt, Dayton E., 1965.
Marine geochemistry: some guesses and gadgets.
Narragansett Mar. Lab., Univ. Rhode Island, Occ. Publ., No. 3:203-211.

geochemistry

Deacon, G.E.R., 1965.
Chemistry of the sea. 1. Inorganic.
Chemistry in Britain, 1:48-53.
Also in:
Collected Reprints, Nat. Inst. Oceanogr., 13.

geochemistry(marine)

Goldberg, E.D., 1960
Marine geochemistry.
Ann. Rev. Phys. Chem., 11:29-48.

geochemistry

Goldberg, E.D., 1954.
Marine geochemistry. 1. Chemical scavengers of the sea. J. Geol. 62(3):249-255, 6 textfigs.

geochemistry

Goldschmidt, V.M., 1933.
Grundlagen der quantitaven Geochimie.
Fortschr. Mineral. Krist. Petrog. 17:112-157.

geochemistry

Hoering, T.C., & P.L. Parker, 1961
The geochemistry of the stable isotopes of chlorine.
Geochim. et Cosmochim. Acta, 23(3/4): 186-199.

geochemistry

Koczy, F.F., 1956.
Geochemistry of the radioactive elements of the ocean. Deep-Sea Res. 3(2):93-103.

geochemistry

Koczy, F.F., 1950.
Zur Sedimentation und Geochemie im aequatorischen Atlantischen Ozean. Medd. Oceanografiska Inst Göteborgs 18 (Göteborgs Kungl. Vetenskaps- och Vitterhets- Samhälles Handlingar, Sjätte Fjölden Ser. B, 6(1)):44 pp., 17 textfigs.

geochemistry
Kuenen, Ph. H., 1941. Geochemical calculations concerning the total mass of sediments in the earth. Am. J. Sci. 239:161-190.

geochemistry
Landergran, S., 1954.
On the geochemistry of the North Atlantic sediment core No. 238. Repts. Swedish Deep-Sea Exped. 1947-48, Sediment cores, N. Atlantic Ocean, 7(2):125-148.

geochemistry
Landergren, S., 1948
6. On the geochemistry of Mediterranean sediments. Preliminary report on the distribution of beryllium, boron, and the ferrides in three cores from the Tyrrhenian Sea. Medd. Ocean. Inst., Göteborg 15 (Göteborgs Kungl. Vetenskaps-och Vitterhets -Samhälles Handlingar, Sjätte Följden, Ser. B 5(13)):34-46, figs.6-7.

geology, chemistry of
Lowenstam, H.A., 1962.
Goethite in radular teeth of Recent marine gastropods.
Science, 137(3526):279-280.

geochemistry
Lowenstam, H.A., 1961.
Mineralogy, O18/O16 ratios, and strontium and magnesium contents of Recent and fossil brachiopods and their bearing on the history of the oceans.
J. Geology, 69(3):241-260.

geochemistry
Martin, Dean F., 1968.
Marine geochemistry. 1. Analytical methods. Marcel Dekker, Inc., 280 pp. $5.75

geochemistry
Mason, B., 1952.
Principles of geochemistry. New York, London, 276 pp.

geochemistry
Nakai, Nobuyuki, and Mead LeRoy Jensen, 1960
Biochemistry of sulfur isotopes.
J. Earth Sciences, Nagoya Univ., 8(2): 181-196.

geochemistry
Rankama, K., and Th. G. Sahama, 1950.
Geochemistry. Univ. Chicago Press, 912 pp., 44 textfigs.

geochemistry
Rubey, W.W., 1951.
Geologic history of sea water, an attempt to state the problem. Bull. G.S.A. 62(9):1111-1148, 4 textfigs.

geochemistry
Sayles, Fred L., 1970.
Preliminary geochemistry. Initial Reports of the Deep Sea Drilling Project, Glomar Challenger 4: 645-655.

geochemistry
Teodorovich, G.I., 1949.
The meteorite, geochemical phase of seas and salt water in general as a producer of petroleum. Dok. Akad. Nauk, USSR, 69:227-230.

geochemistry
Tikhomireva, Ye. S., 1961.
The geochemistry of shale-bearing deposits of the Baltic basin. (In Russian).
Doklady Akad. Nauk, SSSR, 136(5):1209-1212.
English Edit., 1962, 136(1-6):17-19.

geochemistry
Tomkeieff, S.I., 1956.
Geochemistry in the U.S.S.R. (1948-1953). In: Physics and Chemistry of the Earth, A.H. Ahrens, K. Rankama and S.K. Runcorn, Eds., Pergamon Press, Ltd., 1(8):235-290.

geochemistry
Turekian, K.K., and K.H. Wedepohl, 1961.
Distribution of the elements in some major units of the earht's crust.
Bull. Geol. Soc., Amer., 72(2):175-192.

geochemistry
Vinogradov, A.P. 1968
Geochemical problems in the evolution of the ocean.
Lithos 1(2): 169-178.

geochronology

geochronology
Baker, B.H., and J.A. Miller, 1963.
Geology and geochronology of the Seychelles islands and structures of the floor of the Arabian Sea.
Nature, 199(4891):346-348.

geochronology
*Baker, I., N.H. Gale and J. Simons.
Geochronology of the St. Helena Volcanoes.
Nature, 215(5109):1451-1456.

geochronology
Goldberg, E.D., 1961
Chemical and mineralogical aspects of deep-sea sediments.
Physics and Chemistry of the Earth, Pergamon Press, 4(8): 281-302.

geochronology
Holmes, C.W., J.K. Osmond and H.G. Goodell, 1966.
The geochronology of Eltanin cores from the South Pacific Ocean.
Antarctic J., United States, 1(5):203-204.

geochronology
Pettersson, H., 1949
Geochronology of the Deep Ocean Bed. Tellus, Quart. J. Geophys. 1(1):1-5, 1 text fig.

GEOCHRONOMETRY

geochronometry
Wells, John W., 1963.
Coral growth and geochronometry.
Nature, 197(4871):948-950.

geodesy

geodetic systems
Heisakanen, W.S., 1955.
Intercontinental connection of geodetic systems.
Int. Hydrogr. Rev. 23(1):141-156, 7 textfigs.

geodesy
Schmidt, A., 1947
Radio aids and geodesy. Int. Hydr. Rev., 24:14-47, 8 text figs.

geodesy
*Thomas, Paul D., 1967.
Geodesy and earth studies.
Int. J. Oceanol. Limnol., 1(4):254-276.

geodesy, marine
Tison, James C., Jr., 1969.
Marine geodesy.
Int. hydrogr. Rev., 46(1): 109-113.
(Reprinted from Bull. Géodés., 88(1)).

geodesy
von Arx, William S., 1970.
1. General observations. 16. Marine geodesy (April 1968). In: The sea: ideas and observations on progress in the study of the seas, Arthur E. Maxwell, editor, Wiley-Interscience 4(1): 597-633.

geodetic positioning

GEODETIC POSITIONING
Campbell, Andrew C., 1968.
Geodetic positioning at sea.
Navigation, U.S.A., 15(6): 22-26

geographical distribution

geographic nomenclature

geographic nomenclature
Beal, M.A., F. Edvalson, K. Hunkins, A. Molloy and N. Ostenso, 1966.
The floor of the Arctic Ocean: geographic names.
Arctic, 19(3):215-219.

Geography

geography
Gutsell, B. V., 1949
An introduction to the geography of Newfoundland. Canada, Dept. Mines and Resources, Geogr. Bur., Info. Ser. No.1: 85 pp., 20 text figs., 3 fold-in maps.

geographic belts
Gvetova, I.A., 1970.
The areas of geographic belts of the Earth, continents and oceans. (In Russian)
Dokl. Akad. Nauk, SSSR, 192(1): 193-195

geoids

geoids
Von Arx, William S., 1966.
Level-surface profiles across the Puerto Rico Trench.
Science, 154(3757):1651-1654.

geoid
Woollard, George P., and Mohammad Asadullah Khan, 1970.
A review of satellite-derived figures of the geoid and their geophysical significance.
Pacific Sci., 24(1):1-28

geological history of
Curry, D., J.W. Murray and W.F. Whittard, 1965.
The geology of the western approaches of the English Channel. III. The Globigerina silts and associated rocks.
In: Submarine geology and geophysics, Colston Papers, W.F. Whittard and R. Bradshaw, editors, Butterworth's, London, 17:239-261.

Geological history

Fischer, Alfred G., Bruce C. Heezen, Robert E. Boyce, David Bukry, Robert G. Douglas, Robert E. Garrison, Stanley A. Kling, V. Krasheninnikov, A.P. Lisitzin and Anthony C. Pimm, 1970.
Geological history of the western North Pacific.
Science, 168(3936): 1210-1214.

geological history of the oceans
Lowenstam, H.A., 1961.
Mineralogy, 018/016 ratios and strontium and magnesium contents of Recent and fossil brachiopods and their bearing on the history of the oceans.
J. Geology, 69(3):241-260.

geological history
Malaise, R., 1956.
Sjunket land i Atlanten. Ymer, 1956(2):121-132.

geology history
Moore, H.B., 1942
The Geological History of the Bermudas.
Science 95 (No.2474):551-552.

geology history
Moore, H.B., and D.M. Moore, 1946
Preglacial history of Bermuda Bull. G.S.A. 57, pp. 207-222, 2 pls.

Geologic time and subdivisions arranged alphabetically

geologic time
Eicher, Don L. 1968.
Geologic time.
Prentice-Hall, Inc. 149 pp. $ 5.95.

geologic history
*Rainwater, E.H., 1967.
Resume of Jurassic to Recent sedimentation history of the Gulf of Mexico.
Trans. Gulf Coast Assoc. Geol. Soc., 17th Ann. Meet.: 179-210.

geological history
Rusnak, Gene A., and Robert L. Fisher, 1964.
Structural history and evolution of Gulf of California.
In: Marine geology of the Gulf of California, a symposium, Amer. Assoc. Petr. Geol., Memoir, T. Van Andel and G.G. Shor, Jr., editors, 3:144-156.

Bathonian

Bathonian
Larsonneur, Claude, et Michel Rioult, 1969.
Le Bathonien et le Jurassique supérieur en Manche centrale.
C. r. hebd. Séanc. Acad. Sci., Paris (D): 268(22): 2645-2648

Cambrian

geologic time, Cambrian
Filatova Z.A. and L.A. Zenkevich 1969.
On the contemporary distribution of the ancient primitive mollusks Monoplacophora in the World Ocean and on Pogonophora fossils in the sediment of Cambrian seas. (In Russian; English abstract).
Okeanologiia 9(1):162-171.

Carboniferous

Carboniferous
Eden R.A., and T.E. Smith 1966.
A submarine outcrop of Lower Carboniferous rocks off south-east Scotland.
Nature, 211 (5056): 1285-1286

Cenozoic

geological ages Cenozoic
Atwater, Tanya, 1970.
Implications of plate tectonics for the Cenozoic tectonic evolution of western North America.
Bull. Geol. Soc. Am. 81 (12): 3513-3536.

geological age Cenozoic
Bada, Jeffrey L. and Bruce P. Luyendyk 1971.
A route to late Cenozoic temperature history?
Science 172 (3982): 503

geological ages Cenozoic
Bandy, Orville L., and Richard E. Casey, 1969.
Major late Cenozoic planktonic datum planes, Antarctica to the tropics.
Antarctic J., U.S., 4(5): 170-171.

Cenozoic
Bartlett, Grant A., and Leigh Smith 1971.
Mesozoic and Cenozoic history of the Grand Banks of Newfoundland: a discussion.
Can. J. Earth Sci. 8 (12): 1608-1610.

geological ages Cenozoic
Bartlett, Grant A., and Leigh Smith, 1971.
Mesozoic and Cenozoic History of the Grand Banks of Newfoundland.
Can. J. Earth Sci., 8(1): 65-84

Cenozoic
Berggren, William A., 1969.
Rates of evolution in some Cenozoic planktonic Foraminifera.
Micropaleontology, 15(3): 351-365

Cenozoic
Clark, David L. 1971
Arctic Ocean ice cover and its late Cenozoic history.
Bull. Geol. Soc. Am. 82 (12): 3313-3324

geological ages Cenozoic
Friend, Jennifer K., and William R. Riedel 1967.
Cenozoic orosphaerid radiolarians from tropical Pacific sediments.
Micropaleontology 13(2): 217-232.

geological ages CENOZOIC
Gibson Thomas G., 1970
Late Mesozoic-Cenozoic tectonic aspects of the Atlantic coastal margin.
Bull. geol. Soc. Am., 81(6): 1813-1822.

Geological ages - Cenozoic
Goll, R.M., 1968.
Classification and phylogeny of Cenozoic Trissocyclidae (Radiolaria) in the Pacific and Caribbean basins.
J. Paleont., 42(6):1409-1432.

geological ages Cenozoic
Heath, G. Ross, 1969.
Mineralogy of Cenozoic deep-sea sediments from the Equatorial Pacific Ocean.
Bull. geol. Soc. Am. 80(10): 1997-2018

geologic ages Cenozoic
Herman, Yvonne, 1970.
Arctic paleo-oceanography in Late Cenozoic time.
Science, 169(3944): 474-477.

geologic ages Caenozoic
Jones, J.G., 1971.
Australia's Caenozoic drift.
Nature, Lond. 230 (5291): 237-239

geologic ages Cenozoic
Krause, Dale C., Mark A. Chramiec, George M. Walsh and Serge Wisotsky, 1966.
Seismic profile showing Cenozoic development of the New England continental margin.
J. geophys. Res., 71(18):4327-4332.

Cenozoic
McIver, N.L. 1972
Cenozoic and Mesozoic stratigraphy of the Nova Scotia Shelf.
Can. J. Earth. Sci., 9(1): 54-70.

geological age Cenozoic
McKenna, Malcolm C., 1971.
A route to late Cenozoic temperature history?
Science 172 (3982): 503.

geological age Cenozoic
Riedel, W.R., 1971.
Cenozoic radiolaria from the western tropical Pacific, Leg 7. Initial Repts. Deep Sea Drill. Proj., 7(2): 1529-1672.

geologic age - Cenozoic
Rowland, Robert W., and David M. Hopkins, 1971.
Comments on the use of Hiatella arctica for determining Cenozoic sea temperatures.
Palaeogr. Palaeoclimatol. Palaeoecol. 9(1): 59-64

geologic ages Cenozoic
Scholl, David W., and David M. Hopkins 1969
Newly discovered Cenozoic basins, Bering Sea shelf, Alaska.
Bull. Am. Ass. Petrol. Geol. 53 (10): 2067-2078.

geological ages - Cenozoic
Silver, Eli A., 1971.
Transitional tectonics and late Cenozoic structure of the continental margin off northernmost California.
Bull. geol. Soc. Am., 82(1): 1-22

geologic ages Cenozoic

Silver, Eli A., 1969.
Late Cenozoic underthrusting of the Continental Margin off northernmost California. Science, 166(3910): 1265-1266.

geologic ages - Cenozoic

Starauch, Friedrich, 1968.
Determination of Cenozoic sea-temperatures using Hiatella arctica (Linné).
Palaeogr. Palaeoclimatol. Palaeoecol., 5(2):213-233. 213-233.

Cenozoic (Late)

Watkins, N.D. and J.P. Kennett, 1971.
Antarctic Bottom Water: major change in velocity during the Late Cenozoic between Australia and Antarctica. Science 173(3999): 813-818.

geologic ages Cenozoic

Yermakov, Yu. G., 1969.
The influence of tectonics on the distribution of the upper Mesozoic and Cenozoic deposits in the northern coastal region of the Black Sea. (In Russian)
Geotekton. Akad Nauk SSSR 1969(4):119-121. Translation: Geotectonics, Scripta Tecnica for AGU (1970): 281-283

Cenozoic

Yorath, C.J., and Eric R. Parker 1971.
Mesozoic and Cenozoic history of the Grand Banks of Newfoundland: discussion.
Can. J. Earth Sci., 8(12): 1606-1608

Cretaceous

Cretaceous

Bandy, Orville L., 1967.
Cretaceous planktonic foraminiferal zonation.
Micropaleontology, 13(1):1-31.

Crétaceaous

Barthe, André, Gilbert Boillot et Raoul Deloffre 1967.
Anticlinaux affectant le Crétacé à l'entrée de la Manche occidentale.
C. r. hebd. Séanc., Acad. Sci., Paris, (D) 264 (24): 2725-2727

Cretaceous

Bartlett, Grant A., 1969.
Cretaceous biostratigraphy of the Grand Banks of Newfoundland.
Marit. Sediments 5(1): 4-14.

Cretaceous

*Burckle, Lloyd H., Tsunemasa Saito and Maurice Ewing, 1967.
A Cretaceous (Turonian) core from the Naturaliste Plateau, southeast Indian Ocean.
Deep-Sea Research, 14(4):421-426.

Cretaceous

Čepek, Pavel and William W. Hay, 1969.
Calcareous nannoplankton and biostratigraphic subdivision of the Upper Cretaceous. Trans. Gulf Coast Ass. geol. Socs, 19 : 323-336.

Cretaceous

Douglas, Robert G., 1971.
Cretaceous Foraminifera from the northwestern Pacific Ocean: leg 6, deep sea drilling project.
Initial Repts. Deep Sea Drilling Project, Glomar Challenger 6: 1027-1053.

Cretaceous

Douglas, Robert G., and Clay Rankin, 1969.
Cretaceous planktonic foraminifera from Bornholm and their zoogeographic significance.
Lethaia, 2(3): 185-217.

geologic age Cretaceous

Dymond, Jack, and Herbert L. Windom 1968.
Cretaceous K-Ar ages from Pacific Ocean seamounts.
Earth Planet. Sci. Letts 4(1): 47-52.

Cretaceous

Emery, K.O., C.A. Kaye, D.H. Loring and D.J.G. Nota, 1968.
European Cretaceous flints on the coast of North America.
Science, 160(3833):1224-1228.

Cretaceous

Ewing, Maurice, Tsunemasa Saito, John I. Ewing and Lloyd H. Burckle, 1966.
Lower Cretaceous sediments from the northwest Pacific.
Science, 152(3723):751-755.

geologic ages, Cretaceous (Upper)

Ewing, John, J.L. Worzel, Maurice Ewing and Charles Windisch, 1966.
Ages of Horizon A and the oldest Atlantic sediments: coring at an outcrop of Horizon A establishes it as a buried abyssal plain of Upper Cretaceous age.
Science, 154(3753):1125-1132.

Cretaceous

Fell, H. Barraclough, 1967.
Cretaceous and Tertiary surface currents in the oceans.
Oceanogr. Mar. Biol., Ann. Rev., H. Barnes, editor George Allen and Unwin, Ltd., 5: 317-341.

Cretaceous

Foreman, Helen P., 1971.
Cretaceous radiolaria, Leg 7, DSDP. Initial Repts Deep Sea Drill. Proj. 7(2): 1673-1693.

Cretaceous (middle)

*Habib, Daniel, 1969.
Middle Cretaceous palynomorphs in a deep-sea core from the seismic reflector Horizon A outcrop area.
Micropaleontology, 15(1):85-101.

Cretaceous

*Habib, Daniel, 1968.
Spores, pollen and mikroplankton from the Horizon Beta outcrop.
Science, 162(3861):1480-1481.

Cretaceous (Lower)

Heezen, Bruce C., Robert E. Sheridan, 1966.
Lower Cretaceous rocks (Neocomian-Albian) dredged from Blake Escarpment.
Science, 154(3757):1644-1647.

Cretaceous

Hopkins, David M., David W. Scholl, Warren O. Addicott, Richard L. Pierce, Patsy B. Smith, Jack A. Wolfe, David Gershanovich, Boris Kotenev, Kenneth E. Lohman, Jere H. Lipps, and John Obradovich, 1969.
Cretaceous, Tertiary, and early Pleistocene rocks from the continental margin in the Bering Sea.
Bull. geol. Soc. Am., 80(8): 1471-1480

Cretaceous

Jacobs, Marian B., 1970.
Clay mineral investigations of Cretaceous and Quaternary deep-sea sediments of the North American Basin.
J. sedim. Petrol. 40(3): 864-868.

Cretaceous (Upper)

*Jones, E.J.W., and B.M. Funnell, 1968.
Association of a seismic reflector and upper Cretaceous sediment in the Bay of Biscay.
Deep-Sea Res., 15(6):701-709.

Cretaceous

McKenzie, Dan, and John G. Sclater 1971.
The evolution of the Indian Ocean since the Late Cretaceous.
Geophys. J. R. astr. Soc. 25(5): 437-528.

Cretaceous

Moore, Clyde H., Jr., 1969.
Factors controlling carbonate sand distribution in the shallow shelf environment: illustrated by the Texas Cretaceous. Trans. Gulf Coast Ass. geol. Socs, 19 : 507.

geologic age, Cretaceous

Saito, Tsunemasa, Lloyd H. Burckle and Maurice Ewing, 1966.
Lithology and paleontology of the reflective layer Horizon A.
Science, 154(3753):1173-1176.

Cretaceous

*Windisch, Charles C., R.J. Leyden, J.L. Worzel, T. Saito and J. Ewing, 1968.
Investigation of Horizon Beta.
Science, 162(3861):1473-1479.

Cretaceous

Worsley, Thomas R., 1971
Terminal Cretaceous events.
Nature, Lond., 230(5292): 318-320

Eocene

Eocene

Burckle, Lloyd H., and Tsunemasa Saito, 1966.
An Eocene dredge haul from the Tuamotu Ridge.
Deep-Sea Res., 13(6):1207-1208.

Eocene

Dott, R. H., Jr., 1966.
Eocene deltaic sedimentation at Coos Bay, Oregon.
J. Geol., 74(4):373-420.

Eocene

Fisher, Robert L., John G. Sclater and Dan P. McKenzie, 1971.
Evolution of the Central Indian Ridge, west Indian Ocean.
Bull. geol. Soc. Am., 82(3): 553-562.

Eocene

Fox, Paul J., Edward Schreiber and Bruce C. Heezen 1971.
The geology of the Caribbean crust. Tertiary sediments, granitic and basic rocks from the Aves Ridge.
Tectonophysics 12 (2): 89-109.

geologic time - Eocene
Geitzenauer, Kurt R., Stanley V. Margolis and Dennis S. Edwards 1968
Evidence consistent with Eocene glaciation in a South Pacific deep-sea sedimentary core.
Earth Planet. Sci. Letts 4(2): 173-177.

Middle Eocene
Lynts, Georges W., and Charles F. Stehman 1971
Factor-vector models of Middle Eocene planktonic foraminiferal fauna of Core 6282 Northeast Providence Channel.
Revta esp. Micropaleontol. 3(2): 205-213.

Eocene
Mattson, P.H. and E.A. Pessagno, Jr., 1971.
Caribbean Eocene volcanism and the extent of horizon A. Science 174(4005): 138-139.

post-Eocene
*Riedel, W.R., 1967.
Radiolarian evidence consistent with spreading of the Pacific floor.
Science, 157(3788):540-542.

geologic age - Eocene
Schlee, John, and Alan H. Cheetham 1967.
Rocks of Eocene age on Fippennies Ledge, Gulf of Maine.
Bull. geol. Soc. Am. 78(5): 681-684.

geological history, Eocene
Schoeffler, J., 1965.
Le "Gouf" de Capbreton, de l'Eocène inférieur à nos jours.
In: Submarine geology and geophysics, Colston Papers, W.F. Whittard and R. Bradshaw, editors, Butterworth's, London, 17:265-268.

Eocene
Stehman, Charles F., 1970.
Eocene deep water sediment from the northeast Providence Channel, Bahamas.
Marit. Sediments, 6(2): 65-67.

geologic time - Flandrian
Olausson, Eric, and Ulf C. Jonasson 1969.
The Arctic Ocean during the Würm and Early Flandrian.
Geol. För. Stockh. Förh. 91: 185-200

Holocene

Holocene
Belderson, R.H., N.H. Kenyon and A.H. Stride, 1970.
Holocene sediments of the continental shelf west of the British Isles. In: The geology of the East Atlantic continental margin. 2. Europe, ICSU/SCOR Working Party 31 Symposium, Cambridge 1970, Rept. 70/14: 157-170.

Holocene
Bloom, A.L., 1970.
Holocene submergence in Micronesia as the standard for Eustatic sea-level changes.
Quaternaria 12: 145-154.

Holocene
Carver, Robert E., 1971.
Holocene and late Pleistocene sediment sources, continental shelf off Brunswick, Georgia.
J. sedim. Petrol. 41(2): 517-525.

geologic time - Holocene
Conolly, J.R., H.D. Needham and Bruce C. Heezen 1967
Late Pleistocene and Holocene sedimentation in the Laurentian Channel.
J. Geol. 75(2): 131-147.

geologic time - Holocene
Cotecchia, V., G. Dai Pra e G. Magri 1969.
Oscillazioni Tirreniane e Oloceniche del livello mare nel Golfo di Taranto, corredate da datazioni col metodo del radiocarbonio.
Geol. appl. Idrogeol., Bari, 4: 93-147.

Holocene
Duncan, John R., L.D. Kulm and G.B. Griggs 1970.
Clay mineral composition of late Pleistocene and Holocene sediments of Cascadia Basin northeastern Pacific Ocean.
J. Geol. 78(2): 213-221.

Holocene
Edgerton, H., E. Seibold, K. Vollbrecht und F. Werner, 1966.
Morphologische Untersuchungen am Mittelgrund (Eckernförder Bucht, westliche Ostsee).
Meyniana, 16:37-50.

Holocene
Fabricius, Frank H., Dietrich Berdau and Karl-Otto Münnich, 1970.
Early Holocene oöids in modern littoral sands reworked from a coastal terrace, southern Tunisia. Science 169(3947): 757-760.

Holocene
Flemming, N.C., 1968.
Holocene earth movements and eustatic sea level change in the Peloponnese.
Nature, Lond., 217(5133):1031-1032.

Holocene
Frey, Robert W. and Taylor V. Mayou, 1971.
Decapod burrows in Holocene barrier island beaches and washover fans, Georgia. Senckenbergiana maritima 3(1): 53-77.

Holocene
*Golik, Abraham, 1968.
History of Holocene Transgression in the Gulf of Panama.
J. Geol. 76(5):497-507.

Holocene
Goodell, H.G., and R.K. Garman, 1969.
Carbonate geochemistry of Superior deep test well, Andros Island, Bahamas.
Bull. Am. Ass. Petrol. Geol., 53(3): 513-536.

Holocene
Gorsline, Donn S., David E. Drake and Peter W. Barnes, 1968.
Holocene sedimentation in Tanner Basin, California continental borderland.
Bull. geol. Soc. Am., 79(6): 659-674

Holocene
Govberg, L.I., 1970.
Distribution of molluscs in Holocene sediments from the White Sea. Oceanologiia, 10(5): 837-846.
(In Russian; English abstract)

Holocene
Govberg, L.I., 1968.
The distribution of molluscs in Holocene sediments of Onega Bay. (In Russian; English abstract).
Okeanologiia, Akad. Nauk, SSSR, 8(4):666-679.

geologic time - Holocene
Greiner, Gary O.G. 1970.
Distribution of major benthonic foraminifera groups on the Gulf of Mexico continental shelf.
Micropaleontology 16(1): 83-101.

Holocene
Griggs, Gary B. and Gerald A. Fowler, 1971.
Foraminiferal trends in a Holocene turbidite.
Deep-Sea Res, 18(6): 645-648.

Holocene
Guilcher, André, 1969.
Pleistocene and Holocene sea level changes.
Earth-Sci. Rev. 5(2): 69-97.

Holocene
Hesse, R., U. von Rad and F.H. Fabricius 1971.
Holocene sedimentation in the Strait of Otranto between the Adriatic and Ionian seas (Mediterranean).
Mar. Geol. 10(5): 293-355.

Holocene
Kaplin, P.A., E.N. Nevessky, Yu. A. Pavlidis, and F.A. Shcherbakov, 1968.
Peculiarities of structure and history of development of the upper shelf and the nearshore zone during the Holocene. (In Russian: English abstract).
Okeanologiia, Akad. Nauk, SSSR, 8(1):3-13.

Holocene
Kendall, Christopher G. St.C. and Patrick A. D'E. Skipwith, 1969.
Holocene shallow-water carbonate and evaporite sediments of Khor al Bazam, Abu Dhabi, Southwest Persian Gulf.
Bull. Am. Ass. Petr. Geol., 53(4): 841-869.

Holocene
McMaster, Robert L., Thomas P. LaChance and Asaf Ashraf, 1970.
Continental shelf geomorphic features off Portuguese Guinea, Guinea, and Sierra Leone (West Africa).
Marine Geol. 9(3): 203-213.

geologic time - Holocene
Medvedev, V.S., E.N. Nevessky, Yu. A. Pavlidis and F.A. Shcherbakov 1968.
Relief and history of the south coast of the Kola Peninsula during the Holocene (In Russian, English abstract)
Okeanologiia 8(2): 257-269.

Holocene
Mörner, Nils-Axel 1971.
The Holocene eustatic sea level problem
Geol. Mijnb. 50(5): 699-702.

Holocene
Müller, German, 1969.
Diagenetic changes in interstitial waters of Holocene Lake Constance sediments.
Nature, Lond., 224(5216): 258-259.

Holocene
Nelson, Campbell S. and K.A. Rodgers, 1969.
Algal stabilisation of Holocene conglomerates by micritic high-magnesium calcite, southern New Caledonia.
N.Z. Jl mar. freshw. Res. 3(3): 395-408.

Holocene
Nevessky, E.N., 1970.
Holocene history of the coastal shelf zone of the USSR in relation with processes of sedimentation and condition of concentration of useful minerals. Quaternaria 12: 79-88.

Holocene
Newman, Walter S., and Craig A. Munsart, 1968.
Holocene geology of the Wachapreague Lagoon, Eastern Shore Peninsula, Virginia.
Mar. Geol., 6(2): 81-105.

Holocene
Piper, D.J.W., 1970.
Transport and deposition of Holocene sediment on La Jolla Deep Sea Fan, California.
Mar. Geol. 8(3/4): 211-227.

Holocene
Piper, David J.W., and Neil F. Marshall, 1969.
Bioturbation of Holocene sediments on La Jolla Deep Sea Fan, California.
J. sedim. Petrol., 39(2): 601-606.

Holocene
Regrain, R., 1971.
L'altitude des marais maritimes. Etude statistique (Le cas du marais de Brouage).
Rev. Géogr. phys. Géol. dyn. (2) 8(2): 123-142

Holocene
Sanders, John E., and Gerald M. Friedman, 1969.
Position of regional carbonate/noncarbonate boundary in nearshore sediments along a coast: possible climatic indicator.
Bull. geol. Soc. Am., 80(9): 1789-1796

Holocene
*Shepard, F.P., J.R. Curray, W.A. Newman, A.L. Bloom, N.D. Newell, J.I. Tracey, Jr., and H.H. Veeh, 1967.
Holocene changes in sea level: evidence in Micronesia.
Science. 157(3788): 542-544.

Holocene
Shinn, Eugene A., 1969.
Submarine lithification of Holocene carbonate sediments in the Persian Gulf.
Sedimentology, 12(1/2): 109-144

Holocene
Taylor, J.C.M., and L.V. Illing, 1969.
Holocene intertidal calcium carbonate cementation, Qatar, Persian Gulf.
Sedimentology, 12(1/2): 69-107.

Holocene
Wladyslaw, Karaszewski 1970.
Holocene changes in level of Black Sea, observed south of Neseber (SE Bulgaria).
Przegl. geogr. 42(3): 517 - 527.

Jurassic

Jurassic
Brookfield, M.E., 1970.
Eustatic changes of sea-level and orogeny in the Jurassic.
Tectonophysics, 9(4): 347-365.

Jurassic
Fox, Paul J., Bruce C. Heezen and G. Leonard Johnson, 1970.
Jurassic sandstone from the tropical Atlantic.
Science, 170(3965): 1402-1404.

Jurassic
Kirkland, D.W., and J.E. Gerhard, 1971.
Jurassic salt, Central Gulf of Mexico, and its temporal relation to circum-gulf evaporites.
Bull. Am. Ass. Petrol. Geol. 55(6): 680-686

Jurassic
Larsonneur, Claude, et Michel Rioult, 1969.
Le Bathonien et le Jurassique supérieur en Manche centrale.
C. r. hebd. Séanc. Acad. Sci., Paris (D): 268(22): 2645-2648

Jurassic
May, Paul R., 1971.
Pattern of Triassic-Jurassic diabase dikes around the North Atlantic in the context of predrift position of the continents.
Bull. geol. Soc. Am. 82(5): 1285-1292.

Jurassic
Taylor, Donald M., 1968.
The Challenger's adventure begins.
Ocean Indust. 3(10): 35-50.

Mesozoic

Mesozoic
Bartlett, Grant A., and Leigh Smith 1971.
Mesozoic and Cenozoic history of the Grand Banks of Newfoundland: a discussion.
Can. J. Earth Sci. 8(12): 1608-1610.

Mesozoic
Bartlett, Grant A., and Leigh Smith, 1971.
Mesozoic and Cenozoic history of the Grand Banks of Newfoundland.
Can. J. Earth Sci., 8(1): 65-84

Pre-Mesozoic
Dietz, Robert S., and John C. Holden, 1971
Pre-Mesozoic oceanic crust in the eastern Indian Ocean (Wharton Basin)?
Nature, Lond., 229(5283): 309-312

Mesozoic
Hallam, A. 1971.
Mesozoic geology and the opening of the North Atlantic.
J. Geol. 79(2): 129-157.

Mesozoic
Hamilton, Warren, 1969.
Mesozoic California and the underflow of the Pacific mantle.
Bull. geol. Soc. Am., 80(12): 2409-2430

Mesozoic
McIver, N.L. 1972
Cenozoic and Mesozoic stratigraphy of the Nova Scotia Shelf.
Can. J. Earth. Sci., 9(1): 54-70.

Mesozoic
Scholl, David W., Edwin C. Buffington and David M. Hopkins, 1966.
Exposure of basement rock on the continental slope of the Bering Sea.
Science, 153(3739): 992-994.

Mesozoic
Yermakov, Yu. G., 1969.
The influence of tectonics on the distribution of the upper Mesozoic and Cenozoic deposits in the northern coastal region of the Black Sea. (In Russian)
Geotekton. AKad Nauk SSSR 1969(4): 119-121. Translation: Geotectonics, Scripta Technica for AGU (1970): 281-283

Mesozoic
Yorath, C.J., and Eric R. Parker 1971.
Mesozoic and Cenozoic history of the Grand Banks of Newfoundland: discussion.
Can. J. Earth Sci., 8(12): 1606-1608

Miocene

Miocene
Foster, John H. and Neil D. Opdyke, 1970.
Upper Miocene to Recent magnetic stratigraphy in deep-sea sediments. J. geophys. Res., 75(23): 4465-4473.

Miocene
*Grekoff, Nicolas, Claude Guernet et Jacqueline Lorenz, 1967.
Existence du Miocène marin, au centre de la mer Egée, dans l'Île de Skyros (Grèce).
C.r.hebd. Séanc., Acad Sci., Paris, (D)265(18): 1276-1277.

Miocene
Hassan, Fekri, and Sayed El-Dashlouty, 1970.
Miocene evaporites of Gulf of Suez region and their significance.
Bull. Am. Ass. Petrol. Geol. 54(9): 1686-1696

Miocene
Hecht, Alan D., 1969.
Miocene distribution of molluscan provinces along the east coast of the United States.
Bull. geol. Soc. Am., 80(8): 1617-1620

geologic time - Miocene
Klingebiel, André, Francis Lapierre, Janine Larroude et Michel Vigneaux 1968.
Présence d'affleurements de roches d'âge miocène sur le plateau continental du Golfe de Gascogne
C.r. hebd. Séanc. Acad Sci. Paris (D) 266 (11): 1102-1104.

Miocene
Ladd, Harry S., Joshua I. Tracey, Jr., and M. Grant Gross, 1967.
Drilling on Midway Atoll, Hawaii.
Science, 156(2778):1088-1094.

Miocene
Lillie, A.R., 1969.
Evidence for Miocene faulting in southern Hawke's Bay.
N.Z. Jl Geol. Geophys. 12:344-345.

Miocene
*Riedel, W.R., 1967.
Radiolarian evidence consistent with spreading of the Pacific floor.
Science, 157(3788):540-542.

Miocene
Scott, G.H., 1968
Comparison of lower Miocene Globigerinoides from the Caribbean and from New Zealand
N.Z. Jl Geol. Geophys. 11(2):376-390

Miocene
Souaya, Fernand Joseph, 1966.
Miocene Foraminifera of the Gulf of Suez, region. U.A.R.
Micropaleontology, 12(4):493-504.

Miocene
Vasiliev, B.I. and N.P. Vasilkovskii 1971.
Discovery of marine Miocene deposits on continental slope of Peter the Great Gulf (Sea of Japan). (In Russian)
Dokl. Akad. Nauk SSSR 198(5): 1195-1198.

Miocene
*Windisch, Charles C., R.J. Leyden, J.L. Worzel, T. Saito and J. Ewing, 1968.
Investigation of Horizon Beta.
Science, 162(3861):1473-1479.

Mio-Pliocene
*Berthois, Léopold, 1968.
Sur la présence d'affleurements calcaires organogènes mio-pliocènes au large de Gibraltar.
C.r.hebd. Séanc., Acad. Sci., Paris, (D)267(15):1186-1189.

Miocene-Pliocene boundary

Miocene-Pliocene boundary
*Bandy, Orville L., James C. Ingle, Jr., and William E. Frerichs, 1967.
Isomorphism in "Sphaeroidinella" and "Sphaeroidinellopsis".
Micropaleontology, 13(4):483-488.

Miocene-Pliocene boundaries
Bandy, Orville L., and Mary E. Wade, 1967.
Miocene-Pliocene boundaries in deep-water environments.
Progress in Oceanography, 4:51-66.

Neogene
Berggren, W.A. 1971.
Multiple phylogenetic zonations of the Cenozoic based on planktonic Foraminifera.
Proc. II Plankton. Conf. Roma 1970, A. Farinacci, editor, 41-56

Neogene
Bronnimann, P., E. Martini, J. Resig, W.R. Riedel, A. Sanfilippo and T. Worsley, 1971.
Biostratigraphic synthesis: Late Oligocene and Neogene of the western tropical Pacific. Initial Repts Deep Sea Drill. Proj. 7(2): 1723-1745.

Neogene
Bronnimann, Paul and Johanna Resig, 1971.
A Neogene globigerinacean biochronologic time-scale of the southwestern Pacific. Initial Repts Deep Sea Drill. Proj. 7(2): 1235-1469.

Neogene
Elsik, William C., 1969.
Late Neogene palynomorph diagrams, northern Gulf of Mexico. Trans. Gulf Coast Ass. geol. Socs, 19 : 509-528.

Neogene
Frerichs, William E. 1971.
Paleobathymetric trends of Neogene foraminiferal assemblages and sea floor tectonism in the Andaman Sea area.
Mar. Geol. 11(3): 159-173

Neogene
Lamb, James L., 1969.
Planktonic foraminiferal datums and late Neogene epoch boundaries in the Mediterranean, Caribbean and Gulf of Mexico. Trans. Gulf Coast Ass. geol Socs, 19 : 559-578.

Neogene
Martini, Erlend, 1971.
Neogene silicoflagellates from the equatorial Pacific. Initial Repts Deep Sea Drill. Proj. 7(2): 1695-1708.

Oligocene

Oligocene
Berggren, W.A. 1971.
Multiple phylogenetic zonations of the Cenozoic based on planktonic Foraminifera.
Proc. II Plankton. Conf. Roma 1970, A. Farinacci, editor, 41-56

Oligocene
Boillot, Gilbert, Alain Cressard et Jean-Pierre Lepretre 1972.
Rôle des mouvements oligocènes sur le plateau continental nord-espagnol entre 3°30 et 5° de longitude ouest.
C.r. hebd. Séanc. Acad. Sci. Paris (D) 274 (1): 27-30.

Oligocene (Late)
Bronnimann, P., E. Martini, J. Resig, W.R. Riedel, A. Sanfilippo and T. Worsley, 1971.
Biostratigraphic synthesis: Late Oligocene and Neogene of the western tropical Pacific. Initial Repts Deep Sea Drill. Proj. 7(2): 1723-1745.

Oligocene
Le Pichon, Xavier, Guy Pautot, Jean-Marie Auzende et Jean-Louis Olivet 1971.
La Méditerranée occidentale depuis l'Oligocène: schéma d'évolution.
Earth planet. Sci. Letts 13 (1): 145-152.

Oligocene
*Riedel, W.R., 1967.
Radiolarian evidence consistent with spreading of the Pacific floor.
Science, 157(3788):540-542.

Ordovician
Strong, D.F. 1972.
Sheeted diabases of central Newfoundland: new evidence for Ordovician seafloor spreading.
Nature, Lond. 235 (5333): 102-104

Palaeocene

Paleogene
Cifelli, Richard, Walter H. Blow and William G. Melson, 1968.
Paleogene sediment from a fracture zone of the Mid-Atlantic Ridge.
J. mar. Res., 26(2):105-109.

Palaeocene
Saito, T., L.H. Burckle and D.R. Horn, 1967.
Palaeocene core from the Norwegian Basin.
Nature, Lond., 216(5113):357-359.

Palaeogene
Stanley, Daniel J., and Emiliano Mutti, 1968
Sedimentological evidence for an emerged land mass in the Ligurian Sea during the Palaeogene.
Nature, Lond., 218(5136): 32-36.

Paleogene

Berggren, W.A. 1971. Paleogene
Paleogene planktonic foraminiferal faunas on Legs I-IV (Atlantic Ocean), JOIDES Deep-Sea Drilling Program – a synthesis.
Proc. II Plankton. Conf., Roma 1970, A. Farinacci, editor, 57-77.

Palaeozoic

Paleozoic
*Bretsky, Peter W., 1968.
Evolution of Paleozoic marine invertebrate communities.
Science, 159(3820):1231-1233.

Late Palaeozoic
Frakes, Lawrence A., and John C. Crowell, 1968.
Late Palaeozoic glacial facies and the origin of the South Atlantic Basin.
Nature, Lond., 217(5131):837-838.

Paleozoic
Lefort, J.P. et J. Deunff, 1971.
Découverte de Paléozoique à microplancton au sud de la Manche occidentale. C.r. hebd. Seanc. Acad. Sci. Paris: 270; 271-274. Also in: Recueil. Trav. Groupe d'Etude Marge Continental, Lab. Géol. sous-marine Univ. Rennes 1.

Paleozoic
Churkin, Michael, Jr. 1969.
Paleozoic tectonic history of the Arctic Basin north of Alaska.
Science, 165 (3893): 549-555

Palaeozoic
McElhinny, M.W., and J.C. Briden, 1971.
Continental drift during the Paleozoic.
Earth planet. Sci. Letts 10 (4): 407-416

Paleozoic
Pitrat, Charles W., 1970.
Phytoplankton and the late Paleozoic wave of extinction
Palaeogeogr., Palaeoclimatol., Palaeoecol. 8(1): 49-55.

Paleozoic
Tappan, Helen 1970.
Phytoplankton abundance and late Paleozoic extinctions: a reply.
Palaeogeogr. Palaeoclimatol. Palaeoecol. 8(1): 56-66.

Permian

Permian
Brunstrom, R.G.W. and Peter J. Walmsley, 1969.
Permian evaporites in North Sea Basin.
Bull. Am. Ass. Petr. Geol., 53(4): 870-883

Phanerozoic
Peterman, Zell E., Carl E. Hedge and Henry A. Tourtelot, 1970.
Isotopic composition of strontium in sea water throughout Phanerozoic time.
Geochim. cosmochim. Acta, 34(1): 105-120.

Permian
Smith, Denys B., 1970.
Permian evaporites in North Sea Basin: discussion.
Bull. Am. Ass. Petrol. Geol., 54(4): 662-664

Pleistocene

Pleistocene
Beard, John H., 1969.
Pleistocene paleotemperature record based on planktonic foraminifers, Gulf of Mexico. Trans. Gulf Coast Ass. geol. Socs, 19 : 535-553.

Pleistocene
Berner, Robert A., 1970.
Pleistocene sea levels possibly indicated by buried black sediments in the Black Sea
Nature, Lond., 227 (5259): 700

Pleistocene
Blackman, Abner, and B.L.K. Somayajulu, 1966.
Pacific Pleistocene cores: faunal analyses and geochronology.
Science, 154(3751):886-889.

Pleistocene
Burckle, Lloyd H., Jessie H. Donahue, James D. Hays and Bruce C. Heezen, 1966.
Radiolaria and diatoms in sediments of the southern oceans.
Antarctic J., United States, 1(5):204.

Pleistocene
Caralp, Michelle 1970.
Essai de stratigraphie du Pleistocène marin terminal d'après les paléoclimats observés dans des carottes du golfe de Gascogne.
Bull. Soc. géol. France (7) 12 (3): 403-412

Pleistocene
Caralp, Michelle, et Michel Vigneaux, 1970
Le Pleistocène marin sur un dôme sous-marin du Golfe de Gascogne.
C.r. hebd. Séanc. Acad. Sci., Paris, (D) 271 (22): 1949-1952.

geologic time - Pleistocene
Chen, Chin 1968.
Pleistocene pteropods in pelagic sediments.
Nature, Lond. 219 (5159): 1145-1149.

geologic time, Pleistocene (late)
Conolly, J.R., H.D. Needham and Bruce C. Heezen 1967.
Late Pleistocene and Holocene sedimentation in the Laurentian Channel.
J. geol. 75 (2): 131-147.

Pleistocene
Costello, W.R., 1970.
River channel and tidal bar deposits Gaspe Nord, Quebec.
Marit. Sediments, 6 (2): 68-71

Pleistocene
Creager, Joe S., and Dean A. McManus, 1965.
Pleistocene drainage patterns on the floor of the Chukchi Sea.
Marine Geology, Elsevier Publ. Co., 3(4):279-290.

PLEISTOCENE
Dansgaard, W., and Henrik Tauber, 1969.
Glacier oxygen-18 content and Pleistocene ocean temperatures.
Science, 166 (3906): 499-502.

Pleistocene
Davvillier, Alexandre, 1970.
La glaciation Pleistocène et les Cañons sous-marins.
C.r. hebd. Séanc. Acad. Sci., Paris (D): 270 (12): 1555-1558

Pleistocene
Deuser, W.G., 1968.
Postdepositional changes in the oxygen isotope ratios of Pleistocene foraminifer tests in the Red Sea.
J. geophys. Res., 73(10):3311-3314.

geologic time - Pleistocene (climatic fluctuations)
Donahue, Jessie G., 1967.
Diatoms as indicators of Pleistocene climatic fluctuations in the Pacific sector of the Southern Ocean.
Progress in Oceanography 4: 133-140.

geologic time, Pleistocene effect of
Dunbar, M.J. 1968.
Ecological development in polar regions: a study in evolution.
Prentice-Hall, Inc., 119 pp. $4.95.

Pleistocene
Duncan, John R., L.D. Kulm and G.B. Griggs 1970.
Clay mineral composition of late Pleistocene and Holocene sediments of Cascadia Basin northeastern Pacific Ocean.
J. Geol. 78 (2): 213-221.

PLEISTOCENE
Emiliani, C., 1970.
Pleistocene paleotemperatures. Science, 168 (3933): 822-824.

Pleistocene
Emiliani, Cesare, 1967.
The Pleistocene record of the Atlantic and Pacific oceanic sediments; correlations with the Alaskan stages by absolute dating; and the age of the last reversal of the geomagnetic field.
Progress in Oceanography, 4:219-224.

Pleistocene
Ericson, Eric B., and Goesta Wollin, 1968.
Pleistocene climates and chronology in deep-sea sediments: magnetic reversals give a time scale of 2 million years for a complete Pleistocene with flour glaciations.
Science, 162(3859):1227-1233.

Pleistocene

Flemming, N.C., 1965
Derivation of Pleistocene marine chronology from morphometry of erosion profiles.
J. Geol. 76: 280-296
Also in: Coll. Repr. Nat. Inst. Oceanogr. Wormley, 16(656)

Pleistocene

Geitzenauer, Kurt R., 1969.
The Pleistocene Coccolithophoridae of the Southern Oceans.
Antarctic J., U.S., 4(5): 176-177.

Pleistocene

Goldberg, Edward D., 1965.
An observation on marine sedimentation rates during the Pleistocene.
Limnol. and Oceanogr., Redfield Vol., Suppl. to 10:R125-R128.

Pleistocene

Guilcher, André, 1969.
Pleistocene and Holocene sea level changes.
Earth-Sci. Rev. 5(2): 69-97.

geologic age, Pleistocene

Hathaway, John C., and Egon T. Degens 1969.
Methane-derived marine carbonates of Pleistocene age.
Science 165 (3894): 690-692.

Pleistocene (climatic fluctuations)

Hays, James D., 1967.
Quaternary sediments of the Antarctic Ocean.
Progress in Oceanography, 4:117-131.

Pleistocene

*Hazel, Joseph E., 1968.
Pleistocene ostracode Zoogeography in Atlantic coast submarine canyons.
J. Paleontol., 42(5):1264-1271.

Pleistocene

Herm, Dietrich 1969
Marines Pliozän und Pleistozän in Nord- und Mittel-Chile unter besonderer Berücksichtigung der Entwicklung der Mollusken-Faunen.
Zitteliana 2: 159pp.

Pleistocen

Hopkins, David M., David W. Scholl, Warren O. Addicott, Richard L. Pierce, Patsy B. Smith, Jack A. Wolfe, David Gershanovich, Boris Kotenev, Kenneth E. Lohman, Jere H. Lipps, and John Obradovich, 1969.
Cretaceous, Tertiary, and early Pleistocene rocks from the continental margin in the Bering Sea.
Bull. geol. Soc. Am., 80(8): 1471-1480

geologic time - Pleistocene (late)

Hoyt, John H. 1967.
Intercontinental correlation of Late Pleistocene sea levels.
Nature, Lond. 215 (5101): 612-614.

Pleistocene

Hoyt, John H., and John R. Hails, 1967.
Pleistocene shoreline sediments in coastal Georgia: deposition and modification.
Science, 155(3769):1541-1543.

Pleistocene

Jafar, S.A., 1970.
A new species of holothurian sclerite from the Pleistocene of the Arabian Sea.
Micropaleontology 16(2): 233-234

Pleistocene

Jousé, A.P., 1971.
Diatoms in Pleistocene sediments from the northern Pacific Ocean. In: Micropalaeontology of oceans, B.M. Funnell and W.R. Riedel, editors, Cambridge Univ. Press, 407-421.

Pleistocene

Kennett, James P., 1970.
Pleistocene paleoclimates and foraminiferal biostratigraphy in subantarctic deep-sea cores.
Deep-Sea Res., 17(1): 125-140.

Pleistocene

Kennett, J.P., N.D. Watkins and P. Vella, 1970.
Paleomagnetic chronology of Pliocene- early Pleistocene climates and the Plio-Pleistocene boundary in New Zealand. Science, 171(3968): 276-279.

Pleistocene

Knott, S.T., and H. Hoskins, 1968.
Evidence of Pleistocene events in the structure of the continental shelf off the northeastern United States.
Marine Geol., 6(1):5-43.

Pleistocene

Land, Lynton S., Fred T. Mackenzie and Stephen J. Gould, 1967.
Pleistocene history of Bermuda.
Bull. geol. Soc. Am., 78(8):993-1006.

Pleistocene

Lehner, Peter, 1969.
Salt tectonics and Pleistocene stratigraphy on continental slope of northern Gulf of Mexico.
Bull. Am. Ass. Petrol. Geol., 53(12): 2431-2479

Pleistocene

Lidz, Louis, 1966.
Deep-sea Pleistocene biostratigraphy.
Science, 154(3755):1448-1451.

Pleistocene

* Lidz, Louis, Walter B. Charm, Mahlon M. Ball, and Sylvia Valdes, 1969.
Marine basins off the coast of Venezuela.
Bull. Mar. Sci., 19(1): 1-17.

Pleistocene

*McIntyre, Andrew, 1967.
Coccoliths as paleoclimatic indicators of Pleistocene glaciation.
Science, 158(3806):1314-1317.

Pleistocene

Lynts, George W. and James B. Judd, 1971.
Late Pleistocene paleotemperatures at Tongue of the Ocean, Bahamas. Science, 171(3976): 1143-1144.

Pleistocene

Mesolella, Kenneth J., 1967.
Zonation of uplifted Pleistocene coral reefs on Barbados, West Indies.
Science, 156 (3775):638-640.

Pleistocene

*Milliman, John D., 1967.
The geomorphology and history of Hogsty Reef, a Bahamian atoll.
Bull. mar. Sci., Miami, 17(3):519-543.

Pleistocene

Montfrans, H.M. van, 1971.
Palaeomagnetic dating in the North Sea Basin.
Earth planet. Sci. Letts 11(3): 226-235

Pleistocene

Moore, George T., 1969.
Interaction of rivers and oceans - Pleistocene petroleum potential.
Bull. Am. Ass. Petrol. Geol., 53(12): 2421-2430

Pleistocene

Morin, Ronald W., Fritz Theyer and Edith Vincent 1970.
Pleistocene climates in the Atlantic and Pacific oceans: a reevaluated comparison based on deep-sea sediments. Science, 169(3943): 365-366.

Pleistocene

Neumann, A. Conrad, 1968.
Elevated Pleistocene sea-level features in the Bahamas. (Abstract only). Trans. Fifth Carib. Geol. Conf., St. Thomas, V.I., 1-5 July 1968, Peter H. Mattson, editor. Geol. Bull. Queens Coll., Flushing, 5: 223.

geologic time - Pleistocene

Ninkovich, Dragoslav 1968.
Pleistocene volcanic eruptions in New Zealand recorded in deep-sea sediments.
Earth Planet. Sci. Letts 4(2): 89-102.

Pleistocene

Olausson, Eric, 1969.
Le Climat au Pleistocène et la Circulation des Océans. Revue Géogr. phys. Géol. dyn. 11(3): 251-264.

Pleistocene

Phipps, Charles V.G., 1966.
Evidence of Pleistocene warping of the New South Wales continental shelf.
In: Continental margins and island arcs, W.H. Poole, editor, Geol. Surv. Pap., Can., 66-15:280-293.

pre-Pleistocene

Ramsay, A.T.S., 1970.
The pre-Pleistocene stratigraphy and palaeontology of the Palmer Ridge area (northeast Atlantic)
Marine Geol., 9(4): 261-285.

Pleistocene

Ruddiman, William F., 1971.
Pleistocene sedimentation in the equatorial Atlantic: stratigraphy and faunal paleoclimatology.
Bull. geol. Soc. Am., 82(2):283-302

Pleistocene

*Shackleton, N.J., and C. Turner, 1967.
Correlation between marine and terrestrial Pleistocene successions.
Nature, Lond., 216(5120):1079-1082.

Pleistocene (late)

Stearns, Charles E., and David L. Thurber, 1967.
Th^{230}/U^{234} dates of Late Pleistocene marine fossils from the Mediterranean and Moroccan littorals.
Progress in Oceanography, 4:293-305.

Pleistocene

Valentine, James W., and H. Herbert Veeh, 1969.
Radiometric ages of Pleistocene terraces from San Nicolas Island, California.
Bull. geol. Soc. Am., 80(7): 1415-1418.

Pleistocene

*Van Andel, Tjeerd H., G. Ross Heath, T.C. Moore and David F.R. McGeary, 1967.
Late Quaternary history, climate and oceanography of the Timor Sea, northwestern Australia.
Am. J. Sci., 265(9):737-758.

Pleistocene

Van den Heuvel, E.P.J., 1966.
On the precession as a cause of Pleistocene variations of the Atlantic Ocean water temperatures.
Geophys. J., R. Astr. Soc., 11(3):323-336.

Pleistocene

Veeh, H. Herbert, 1966.
Th 230/U238 and U234/U238 ages of Pleistocene high sea level stand.
J. geophys. Res., 71(14):3379-3386.

Geologic time - Pleistocene

Webb, S.D., and N. Tessman 1967.
Vertebrate evidence of a low sea level in the middle Pleistocene.
Science 156 (3773): 379.

Pleistocene

Wiles, William W., 1967.
Pleistocene changes in the pore concentration of a planktonic foraminiferal species from the Pacific Ocean.
Progress in Oceanography, 4:153-160.

Pleistocene

Woodcock, Alfred H., Meyer Rubin and R.A. Duce, 1966.
Deep layer of sediments in alpine lake in the tropical mid-Pacific.
Science, 154(3749):647-648.

Pleistocene

Zubakov, V.A., 1966.
A comparison of the radiometric scale of the continental Pleistocene with the chronological diagrams of bathypelagic bottom sediments and the curve of solar radiation. (In Russian).
Dokl., Akad. Nauk, SSSR, 171(5):1153-1155.

Pliocene

Pliocene

Herm, Dietrich 1969
Marines Pliozän und Pleistozän in Nord- und Mittel-Chile unter besonderer Berücksichtigung der Entwicklung der Mollusken-Faunen.
Zitteliana 2: 159pp.

Pliocene

Hoshino, Michihei, 1968.
On sedimentation of marine Pliocene series. (In Japanese; English abstract).
J. geol. Soc. Japan 74(7).
Also in: Coll. Reprs. Coll. mar. Sci., Techn., Tokai Univ., 1967-68, 3: 387-394

Pliocene (late)

Moore, David G., and Edwin C. Buffington, 1968.
Transform faulting and growth of the Gulf of California since the late Pliocene.
Science, 161(3847):1238-1241.

Pliocene

Ostrovskii, A.B., 1966.
On the occurrence of the Kuyalnitskii marine terrace and Upper-Pliocene red-brown clays on the Black Sea shore of the Caucasus, to the south of Anapa. (In Russian).
Dokl., Acad. Nauk, SSSR, 171(5):1160-1163.

Pliocene

*Riedel, W.R., 1967.
Radiolarian evidence consistent with spreading of the Pacific floor.
Science, 157(3788):540-542.

Pliocene-Pleistocene boundary

Pliocene-Pleistocene boundary

Bandy, Orville L., 1967.
Foraminiferal definition of the boundaries of the Pleistocene in southern California, U.S.A.
Progress in Oceanography, 4:27-49.

Pliocence/Pleistocene boundary

*Berggren, W.A., 1968.
Micropaleontology and the Pliocene/Pleistocene boundary in a deep-sea core from the south-central North Atlantic.
G. Geol., (2)35(2):291-312.

Plio-Pleistocene boundary

Bock, W.D., 1971.
Hyalinea baltica and the Plio-Pleistocene boundary in the Caribbean Sea. Science 170 (3960): 847-848.

Pliocene-Pleistocene boundary

Emiliani, Cesare 1971.
Isotopic paleotemperatures and shell morphology of Globigerinoides rubra in the type section for the Pliocene-Pleistocene boundary.
Micropaleontology 17(2): 233-238.

Plio-Pleistocene boundary

Emiliani, Cesare, 1970.
Paleotemperature variations across the Plio-Pleistocene Boundary. Science, 171(3966): 60-62.

Pliscene-Pleistocene boundary

Hays, J.D. and W.A. Berggren, 1971.
Quaternary boundaries and correlations. In: Micropaleontology of oceans, B.M. Funnell and W.R. Riedel, editors, Cambridge Univ. Press, 669-691.

Piocene-Pleistocene boundary

Kennett, J.P., and K.R. Geitzenauer, 1969
Pliocene-Pleistocene boundary in a South Pacific deep-sea core.
Nature, Lond., 224 (5222): 899-

Plio-Pleistocene boundary

Kennett, J.P., N.D. Watkins and P. Vella, 1970.
Paleomagnetic chronology of Pliocene- early Pleistocene climates and the Plio-Pleistocene boundary in New Zealand. Science, 171(3968): 276-279.

Pliocene-Pleistocene boundary

McIntyre, Andrew, Allan W.H. Bé and Resa Preikstas, 1967.
Coccoliths and the Pliocene-Pleistocene boundary.
Progress in Oceanography, 4:3-25.

Pliocene-Pleistocene boundary

Montfrans, H.M. van, 1971.
Palaeomagnetic dating in the North Sea Basin.
Earth planet. Sci. Letts 11 (3): 226-235

Pliocene-Pleistocene boundary

Selli, Raimondo, 1967.
The Pliocene-Pleistocene boundary in Italian marine sections and its relationship to continental stratigraphies.
Progress in Oceanography, 4:67-86.

Precambrian

Precambrian

Lebedev, V.I., 1965.
Calcium content and some other compositional characteristics of the Precambrian Seas.
Geokhimiya, (9):1154-1164.
Translation: Geochemistry Int., Ann Arbor, 2(5):843-852.

Precambrian

Milton, Daniel J., 1966.
Drifting organisms in the Precambrian sea.
Science, 153(3733):293-294.

Precambrian

Weyl, Peter K., 1968.
Precambrian marine environment and the development of life.
Science, 161(3837):158-160.

Quaternary

Quaternary

Amor, I. Asensio, 1970.
Remaniements de formations quaternaires fluviatiles par action marine dans la région littorale Galicio-Asturienne (NW de l'Espagne).
Quaternaria 12: 219-221.

Quarternary
Burckle, Lloyd H., Jessie H. Donahue, James D. Hays and Bruce C. Heezen, 1966.
Radiolaria and diatoms in sediments of the southern oceans.
Antarctic J., United States, 1(5):204.

Quaternary
Ballard, Robert D. and Elazar Uchupi, 1970.
Morphology and Quaternary history of the continental shelf of the Gulf Coast of the United States. Bull. mar. Sci., 20(3): 547-559.

Quaternary
Barash, M.S., 1971.
Paleoclimatic reconstructions based on Quaternary planktonic Foraminifera of the Atlantic Ocean. (In Russian; English abstract). Okeanologiia 11(6): 1049-1056.

Quaternary
*Chamley, Hervé, 1968.
Sur le rôle de la fraction sédimentaire issue du continent comme indicateur climatique durant le Quaternaire.
C.r.hebd.Séanc., Acad.Sci., Paris, 267(16):1262-1265

Quaternary
Chappell, J., 1970.
Quaternary geology of the south-west Auckland coastal region.
Trans. R.Soc. N.Z., Earth Sci., 8(10): 133-153.

Quaternary
Cullen, David J., 1969.
Quaternary volcanism at the Antipodes Islands: bearing on structural interpretation of the southwest Pacific. J. geophys. Res., 74(17
4213-4220.

Quaternary
Curray, Joseph R., 1970.
Quaternary influence, coast and continental shelf of western U.S.A. and Mexico. Quaternaria 12: 19-34.

Quaternary
Curray, Joseph R., Francis P. Shepard and H. Herbert Veeh, 1970.
Late Quaternary sea-level studies in Micronesia: CARMARSEL Expedition.
Bull. geol. Soc. Am., 81(7): 1865-1880.

Quaternary
Deuser, W.G., 1970.
Extreme $^{13}C/^{12}C$ variations in Quaternary dolomites from the continental shelf.
Earth Planet. Sci. Letts., 8(2): 118-124

Quaternary
Dibner, V.D., V.A. Basov, A.A. Gerke, M.F. Soloviyeva, G.P. Sosipatrova and N.I. Shul'gina, 1970.
Age of the pre-Quaternary deposits from the sediments of the Barents Sea shelf. (In Russian; English abstract). Okeanologiia, 10(4): 670-680.

Quaternary
*Dingle, R.V., 1970.
Quaternary sediments and erosional features off the north Yorkshire coast, Western North Sea.
Marine Geol., 9(3):M17-M22.

Quarternary
Emery, K.O. and John D. Milliman, 1970.
Quaternary sediments of the Atlantic Continental Shelf of the United States. Quaternaria 12: 3-18.

Quaternary
Fairbridge, R.W. 1969.
On the absolute fall of sea-level during the Quaternary: a discussion.
Palaeogeogr. Palaeoclimatol. Palaeoecol. 6(3): 241-242.

geologic time, Quaternary
Fevret, Maurice, Jacques Picard et Paul Santaville 1967.
Sur la possibilité de datation de niveaux marins quaternaires par les vermets.
C. r. hebd. Séanc. Acad. Sci. Paris (D) 264(11): 1407-1409.

Quaternary
Frazier, David E., 1969.
Depositional episodes: their relationship to Quaternary sea-level fluctuations in the Gulf Coast region. Trans. Gulf Coast Ass. geol. Socs, 19 :611-612.

Quaternary
Froget, Claude, 1966.
Découverte de formations quaternaires sous-marin au banc du Veyron (baie de Marseille).
C. r. hebd. Séanc., Acad. Sci., Paris, 263 (19):1352-1354.

Quaternary
Fujii, S. and A. Mogi, 1970.
On coasts and shelves in their mutual relations in Japan during the Quaternary.
Quaternaria 12: 155-164.

Quaternary
Groot, J.J., C.R. Groot, M. Ewing, L. Burckle and J.R. Conolly, 1967.
Spores, pollen, diatoms and provenance of the Argentine Basin sediments.
Progress in Oceanography, 4:179-217.

Quaternary
Guilcher, Andre, 1970.
Les variations relatives du niveau de la mer au Quaternaire en Mélanésie et en Polynésie.
Quaternaria 12: 137-143.

Quaternary
Guilcher, A., 1970
Quaternary events on the continental shelves of the world. In: The geology of the East Atlantic continental margin, 1. General and economic papers, ICSU/SCOR Working Party 31 Symposium, Cambridge 1970, Rept. No. 70/13: 31-46.

Quaternary
Hays, James D., 1967.
Quaternary sediments of the Antarctic Ocean.
Progress in Oceanography, 4:117-131.

Quaternary
Holmes, J.W., 1969.
On the absolute fall of sea-level during the Quaternary: a discussion.
Palaeogeogr. Palaeoclimatol. Palaeoecol. 6(3): 242.

Quaternary
Holmes, J.W., 1969.
On the absolute fall of sea-level during the Quaternary.
Palaeogeogr. Palaeoclimatol. Palaeoecol. 6(3): 237-239

Quaternary
Hunkins, Kenneth, and Henry Kutschale, 1967.
Quaternary sedimentation in the Arctic Ocean.
Progress in Oceanography, 4:89-94.

Quaternery
Jacobs, Marian B., 1970.
Clay mineral investigations of Cretaceous and Quaternary deep-sea sediments of the North American Basin.
J. sedim. Petrol. 40(3): 864-868.

Quaternary
Khrustalev, Yu.P. and V.A. Vronsky, 1971.
On biostratigraphy of Late Quaternary sediments of the Sea of Azov. (In Russian; English abstract). Okeanologiia, 11(1): 78-84.

Quaternary
Koh, Ryuji, 1969.
Beach erosion and Quaternary sea level.
Coast. Engng, Japan, 12: 121-128.

Quaternary
Kulakova, L.S., 1970.
Clay minerals in the Upper Quaternary sediments of the middle and southern Caspian Sea. (In Russian; English abstract). Okeanologiia, 10(6) 1034-1043.

Quaternary
*Mars, P., 1967.
Réflexions sur l'étude du Quaternaire Méditerranéen: difficultés, incertitudes et progrès.
Revue Géogr.phys. Géol.dyn., (2)9(5):385-390.

Quarternary
Martin, Louis, 1969.
Datation de deux tourbes quaternaires du plateau continental ivoirien.
C. r. hebd. Séanc. Acad. Sci., Paris, 269 (20) (D): 1925-1927.

Quaternary
Moore, T.C., Jr., and G.R. Heath, 1966.
Manganese nodules, topography and thickness of Quaternary sediments in the Central Pacific.
Nature, Lond., 212(5066):983-985.

Quaternary
Morgan, James P. 1967.
Ephemeral estuaries of the deltaic environment. In: Estuaries, G.H. Lauff, editor, Publs Am. Ass. Advmt Sci., 83:115-120.

Quaternary
Morin, Ronald W. 1971.
Late Quaternary biostratigraphy of cores from beneath the California Current.
Micropaleontology 17(4):475-491.

Quaternary
Mukhina
Mukhina, V.V., 1971.
Problems of diatom and silicoflagellate Quaternary stratigraphy in the equatorial Pacific Ocean. In: Micropaleontology of oceans, B.M. Funnell and W.R. Riedel, editors, Cambridge Univ. Press. 423-431.

geologic ages, Quaternary
Mukhina, V.V., 1966.
On the problem of the boundary surface between the sediments referred to the Quaternary and Tertiary periods in the Pacific Ocean (from the analysis of diatom data). (In Russian).
Okeanologiia, Akad. Nauk, SSSR, 6(1):122-135.

geologic time - Quaternary
Newman, Walter S., and Stanley March 1968.
Littoral of the northeastern United States: late Quaternary warping.
Science 160(3832): 1110-1112.

Quaternary
Oele, E., 1971.
The Quaternary geology of the southern area of the Dutch part of the North Sea.
Geol. Mijnb. 50(3): 461-474.

Quaternary
Oele, E., 1969.
The Quaternary geology of the Dutch part of the North Sea, north of the Frisian isles.
Geol. Mijnbouw, 48(5): 467-480

geologic time - Quaternary
Price, W.A., 1954.
Nonmarine nature of Quaternary Atlantic and Gulf Coastal Plain of southeastern North America. (Abstract). Bull. G.S.A. 65(2):1296-1297.

Quaternary
Quinn, W.H., 1971.
Late Quaternary meteorological and oceanographic developments in the equatorial Pacific.
Nature, Lond., 229(5283): 330-331

Quaternary
Rogers, M.A. and C.B. Koons, 1969.
Organic carbon C^{13} values from Quarternary marine sequences in the Gulf of Mexico: a reflection of paleotemperature changes. Trans. Gulf Coast Ass. geol. Socs, 19 : 529-534.

Quaternary
Romankevich, E.A., P.L. Bezrukov, V.I. Baranov, and L. A. Khristianova, 1966.
Stratigraphy and absolute age of deep-sea sediments in the western Pacific. (In Russian; Rez. Issled. Mezhd. Geofiz. Proekt., Okeanol., Mezhd. Geofiz. Komitet, Prosid.Akad. Nauk, SSSR, No. 14:165 pp.
English abstract

Quarternary
Saidova, Kh. M., 1967.
Distribution of benthonic Foraminifera and depths of the Pacific Ocean during Holocene-Wisconsin. (In Russian; English abstract).
Okeanologiia, Akad. Nauk, SSSR, 7(3):483-489.

Quaternary
Saidova, H.M., 1967.
Sediment stratigraphy and Paleogeography of the Pacific Ocean by benthonic Foraminifera during the Quaternary.
Progress in Oceanography, 4:143-151.

Quaternery
Schell, I.I., 1970.
On the absolute fall of sea-level during the Quaternary: a discussion.
Palaeogeogr. Palaeoclimatol., Palaeoecol. 8(1): 67-68.

Quaternary
Selariu, Octavien, Jeanne Mares-Marinescu, et Mariette Pauliuc, 1969.
Contribution à l'étude des dépôts marins quaternaires de la plateforme continentale de la mer Noire dans le secteur roumain
Rapp. P.-v. Réun. Commn int. Explor. scient. Mer Mediterr., 19(4): 629-631.

Quaternary
Swift, Donald J.P., Gerald L. Shideler, Nicholas F. Avignone, Barry W. Holliday, and Charles E. Dill, Jr., 1970.
Quaternary sedimentation on the inner Atlantic shelf between Cape Henry and Cape Hatteras.
Marit. Sed., 6(1): 5-11

Quaternary
*Van Andel, Tjeerd H., G. Ross Heath, T.C. Moore and David F.R. McGeary, 1967.
Late Quaternary history, climate and oceanography of the Timor Sea, northwestern Australia.
Am. J. Sci., 265(9):737-758.

Quaternary
Veenstra, H.J., 1970.
Quaternary North Sea coasts. Quaternaria 12: 169-184.

Quaternary
Veenstra, H.J., 1965.
Geology of the Dogger Bank area, North Sea. Marine Geology, Elsevier Publ. Co., 3(4):245-262

Quaternary
Verger, F., 1970.
Les rivages quaternaires de la plate-forme continentale de l'Ouest de la France.
Quaternaria 12: 197-206.

Quaternary
Wall, David and Barrie Dale, 1968.
Quaternary calcareous dinoflagellates Calciodinellideae) and their natural affinities.
J. Paleont. 42(6):1395-1408.

Recent
Recent
Mangelsdorf, P.C., Jr., T.R.S. Wilson and Ellen Daniell, 1969.
Potassium enrichments in interstitial waters of Recent marine sediments.
Science, 165(3889):171-174.

Recent-Pleistocene boundary
Recent-Pleistocene boundary
Bandy, Orville L., 1967.
Foraminiferal definition of the boundaries of the Pleistocene in southern California, U.S.A.
Progress in Oceanography, 4:27-49.

Wisconsin
*Frerichs, William E., 1968.
Pleistocene-Recent boundary and Wisconsin glacial stratigraphy in the northern Indian Ocean.
Science, 159(3822):1456-1458.

Tertiary
Tertiary
Addicott, W.O., 1969.
Tertiary climatic change in the marginal northeastern Pacific Ocean.
Science, 165(3893): 583-586.

Tertiary
Barlett, Grant A., 1968.
Mid-Tertiary stratigraphy of the continental slope off Nova Scotia.
Marit. Sed., 4(1):22-31.

Tertiary
Berggren, W.A., 1971.
Tertiary boundaries and correlations. In: Micropaleontology of oceans, B.M. Funnell and W.R. Riedel, editors, Cambridge Univ. Press, 693-809.

Tertiary
Budinger, Thomas F., and Betty J. Enbysk, 1967.
Late Tertiary date from the East Pacific Rise.
J. geophys. Res., 72(3):2271-2274.

Tertiary
*Burckle, Lloyd, H., John Ewing, Tsunemasa Saito and Robert Leyden, 1967.
Tertiary sediment from the East Pacific Rise.
Science, 157(3788):537-540.

Tertiary
Doviess, S.N., 1971.
Barbados: a major submarine gravity slide.
Bull. geol. Soc. Am. 82(9): 2593-2602.

Tertiary
Emiliani, E., and G. Edwards, 1953.
Tertiary ocean bottom temperatures. Nature 171 (4359):887-888.

Tertiary
Fell, H. Barraclough, 1967.
Cretaceous and Tertiary surface currents in the oceans.
Oceanogr. Mar.Biol., Ann. Rev., H. Barnes, editor, George Allen and Unwin, Ltd., 5:317-341.

Tertiary
Fowler, Gerald A., 1966.
Notes on Late Tertiary Foraminifera from off the central coast of Oregon.
Ore Bin, 28(3):53-60.
Also In: Coll. Repr., Dep. Oceanogr., Oregon State Univ., 5.

Tertiary

Fox, Paul J., Edward Schreiber and Bruce C. Heezen 1971.
The geology of the Caribbean crust. Tertiary sediments, granitic and basic rocks from the Aves Ridge.
Tectonophysics 12 (2):89-109.

geologic time - Triassic

Geiger, M.E. and C.B. Hopping 1968.
Triassic stratigraphy of the southern North Sea basin.
Phil. Trans. R. Soc. (B) 254 (790):1-36.

geologic time - Tertiary

Glenie, R.C., J.C. Schofield and W.T. Ward 1968
Tertiary sea levels in Australia and New Zealand.
Palaeogr. Palaeoclimatol. Palaeoecol. 5(1): 141-163.

Tertiary

Hopkins, David M., David W. Scholl, Warren O. Addicott, Richard L. Pierce, Patsy B. Smith, Jack A. Wolfe, David Gershanovich, Boris Kotenev, Kenneth E. Lohman, Jere H. Lipps, and John Obradovich, 1969.
Cretaceous, Tertiary, and early Pleistocene rocks from the continental margin in the Bering Sea.
Bull. geol. Soc. Am., 80 (8): 1471-1480

Tertiary

Le Pichon, Xavier, Guy Pautot, Jean-Marie Auzende et Jean-Louis Olivet 1971.
La Méditerranée occidentale depuis l'Oligocène: schéma d'évolution.
Earth planet. Sci. Letts 13 (1):145-152.

Tertiary

Lipps, Jere H., 1969.
Tertiary plankton from the Clipperton Fracture Zone, equatorial East Pacific.
Bull. geol. Soc. Am., 80 (9): 1801-1808.

Tertiary

Mandra, York T., 1969.
Silicoflagellates: a new tool for the study of Antarctic Tertiary climates.
Antarctic J., U.S. 4 (5):172-174

Tertiary

Margolis, Stanley V. and James P. Kennett, 1970
Antarctic glaciation during the Tertiary recorded in sub-antarctic deep-sea cores.
Science, 170(3962): 1085-1087.

Tertiary

Marlowe, James I., 1969.
A succession of Tertiary strata off Nova Scotia as determined by dredging.
Can. J. Earth Sci., 6(5):1077-1094.

Tertiary

Marlowe, J.I., 1965.
Probable Tertiary sediments from a submarine canyon off Nova Scotia.
Marine Geology, Elsevier Publ. Co., 3(4):263-268.

Tertiary

Martini, E. and T. Worsley, 1971.
Tertiary calcareous nannoplankton from the western equatorial Pacific. Initial Repts Deep Sea Drill. Proj. 7(2): 1471-1507.

Tertiary

Mattson, Peter H., 1966.
Geological characteristics of Puerto Rico.
In; Continental margins and island arcs, W.H. Poole, editor, Geol. Surv. Pap., Can., 66-15:124-138.

geologic time - Tertiary

Mercer, John H. 1968.
The discontinuous glacio-eustatic fall in Tertiary sea level.
Palaeogr. Palaeoclimatol. Palaeoecol. 5(1): 77-85.

geologic ages, Tertiary

Mukhina, V.V., 1966.
On the problem of the boundary surface between the sediments referred to the Quaternary and Tertiary periods in the Pacific Ocean (from the analysis of diatom data). (In Russian).
Okeanologiia, Akad. Nauk, SSSR, 6(1):122-135.

Tertiary

Niini, Aarno, 1969.
The Indian Ocean and ancient Gondwanaland. Appendix: The Asian continent and the Suboceanic ridges.
Annales Acad. scient. Fennicae, (A) III (103): 36 pp.

Tertiary

Park, I., D.B. Clarke, J. Johnson and M.J. Keen 1971.
Seaward extension of the West Greenland Tertiary volcanic province.
Earth Planet. Sci. Letts, 10 (2): 235-238.

Tertiary (Late)

*Pyle, Thomas E., 1968.
Late Tertiary history of Gulf of Mexico based on a core from Sigsbee Knolls.
Bull. Am. Ass. Petr. Geol., 52(11):2242-2246.

Tertiary

*Ramsay, A.T.S., and B.M. Funnell, 1969.
Upper Tertiary microfossils from the Alula Fartak Trench, Gulf of Aden.
Deep-Sea Res. 16(1):24-43.

tertiary

Reshetnjak, V.V., 1971.
Occurrence of Phaeodarian radiolaria in Recent sediments and Tertiary deposits. In: Micropalaeontology of oceans, B.M. Funnell and W.R. Riedel, editors, Cambridge Univ. Press, 343-349.

Tertiary

Roberts, D.G., 1969
New Tertiary volcanic centre on the Rockall Bank, eastern North Atlantic Ocean.
Nature, Lond., 223 (5208): 819-820.

geologic time - Tertiary

Rutford, Robert H., Campbell Craddock and Thomas W. Bastien 1968.
Late Tertiary glaciation and sea-level changes in Antarctica.
Palaeogr. Palaeoclimatol. Palaeoecol. 5 (1): 15-39.

Tertiary

Saito, Tsunemasa, Maurice Ewing and Lloyd H. Burckle, 1966.
Tertiary sediment from the Mid-Atlantic Ridge.
Science, 151(3714):1075-1079.

geological time - Tertiary

Tanner, William F. 1968.
Multiple influences on sea-level changes in the Tertiary.
Palaeogr. Palaeoclimatol. Palaeoecol. 5(1): 165-171.

Tertiary

Von der Borch, C.C. 1971.
Glassy objects in Tertiary deep-sea clays cored by the Deep Sea Drilling Project.
Marine Geol., 10(6): 5-14.

geological time Tertiary

Wilson, John A. 1968
Tertiary shorelines Texas coastal plain.
Palaeogr. Palaeoclimatol. Palaeoecol. 5(1): 135-140

Triassic

Triassic

Blumer, M., and W.D. Snyder, 1967.
Porphyrins of high molecular weight in a Triassic oil shale: evidence by gel permeation chromatography.
Chem. Geol., 2 (1): 35-45.

Triassic

May, Paul R., 1971.
Pattern of Triassic-Jurassic diabase dikes around the North Atlantic in the context of predrift position of the continents.
Bull. geol. Soc. Am. 82(5): 1285-1292.

Triassic

Tagg, A. R., and E. Uchupi, 1966.
Distribution and geologic structure of Triassic rocks in the Bay of Fundy and the northeastern part of the Gulf of Maine.
Geol. Surv. Res., 1966 (B), Geol. Surv. Prof. Paper. 550-B:B95-B98.

Turmian

Olausson, Eric and Ulf C. Jonasson, 1969.
The Arctic Ocean during the Würm and Early Flandrian. Geol. För. Stockh. Förh., 91: 185-200.

Geology

geology, coastal
SEE ALSO: COASTAL GEOLOGY

coastal geology

Bigarella, J.J., 1954.
Nota sôbre os depósitos arenosos recentes do litoral sul-brasileiro.
Bol. Inst. Ocean., Sao Paulo, 5(1/2):233-236.

geology, coastal
Gainanov, A.G., V.P. Petelin, 1963
On physical characteristics of some bottom and coastal rocks in the western part of the Pacific. (In Russian).
Mezhd. Geofiz. Komitet, Prezidiume, Akad. Nauk SSSR, Rezult. Issled. Programme Mezhd. Geofiz. Goda, Okeanol. Issled., (8):112-124.

geology, coastal
Komukai, R., 1959
On the marine geology of beach erosion in O-mori-hama vicinity, Hakodate City.
Publ. 943, Bull. Hydrogr. Office, Japan, 13(2): 217 - 582.
(In English translation as well as Jap.)

coastal geology
Badowsky, V., 1954.
Novas contribuições ao estudo da entrada da Barra de Cananéia.
Bol. Inst. Ocean., Sao Paulo, 5(1/2):151-178.

Geomagnetic anomalies

geomagnetic anomalies
See also: magnetic anomalies

geomagnetic anomalies
Heirtzler, J.R., 1965.
Marine geomagnitic anomalies. (abstract).
J. Geomagnet. Geoelectr., 17(3/4):227-236.

geomagnetic anomalies
Wageman, John M., Thomas W.C. Hilde and K.O. Emery, 1970.
Structural framework of East China Sea and Yellow Sea.
Bull. Am. Ass. Petrol. Geol. 54(9):1611-1643.

geomagnetic anomalies
Veldkamp, J. and H.J.A. Vesseur, 1971.
Geomagnetic anomalies in the continental shelf of Surinam. Hydrogr. Newsletter, R. Netherlands Navy, Spec. Publ. 6: 13-15.

geomagnetic anomalies
Yasui, Masashi, Koichi Nagasaka, Yuichi Hashimoto and Kei Anma 1968.
3. Geomagnetic and bathymetric study of the Okhotsk Sea.
Oceanogrl Mag. Tokyo 20(1):65-72.
Also in: Bull. Maizuru mar. Obs. 11 (1969).

Geomagnetism

geomagnetism
Allan, T.D., 1969
A review of marine geomagnetism.
Earth-Sci. Rev. 5(4): 217-254.

geomagnetic variations
Bullard, E. C., 1966.
Effect of the oceans on geomagnetic variations.
Geophys. J. R. astr. Soc., 10-553.

geomagnetism
Edwards, R.N., L.K. Law and A. White 1971
Geomagnetic variations in the British Isles and their relation to electrical currents in the ocean and shallow seas.
Phil. Trans. R. Soc. (A) 270 (1204): 289-323.

geomagnetism
Fleming, J. A., 1948
Oceanography and geomagnetism. J. Mar. Res. 7(7):147-153.

geomagnetism
Ivanov, M.M., V.V. Novysh and D.L. Finger, 1968.
Results and objectives of the Soviet investigations of the geomagnetic field in the World Oceans (In Russian; English abstract).
Okeanologiia, Akad. Nauk, SSSR, 8(3):363-373.

geomagnetism
Japan, Science Council, National Committee for IIOE, 1966.
General report of the participation of Japan in the International Indian Ocean Expedition.
Rec. Oceanogr. Wks., Japan, n.s., 8(2):133 pp.

geomagnetism
Krause, Dale C., 1968.
Bathymetry, geomagnetism, and tectonics of the Caribbean Sea north of Colombia. (Abstract only). Trans. Fifth Carib. Geol. Conf., St. Thomas, V.I., 1-5 July, 1968, Peter H. Mattson, editor. Geol. Bull. Queens Coll., Flushing 5: 7.

geomagnetism
Krause, Dale C., 1961
Geology of the sea floor east of Guadalupe Island.
Deep-Sea Res., 8(1): 28-38.

GEOMAGNETISM
Malkus, W.V.R., 1968.
Precession of the earth as the cause of geomagnetism.
Science, 160(3825):259-264.

geomagnetism
Miller, E.T., and M. Ewing, 1956.
Geomagnetic measurements in the Gulf of Mexico and in the vicinity of Caryn Peak.
Geophysics, 21(2):406-432, 12 textfigs.

geomagnetic field
Novysh, V.V., 1965.
On the precision of estimating the vertical components of the geomagnetic field in the geomagnetic current meter. (In Russian).
Okeanologiia, Akad. Nauk, SSSR, 5(4):718-725.

geomagnetism
Oguti, T., 1964.
Geomagnetic anomaly around the continental shelf margin southern offshore of Africa.
J. Geophys. Geoelectricity, 16(1):65-67.
Also in:
Geophys. Notes, Tokyo, 18(1)(6). 1965.

geomagnetism
Raspopov, O.M., and V.S. Shneer, 1964.
Observations of short-period pulsations of the geomagnetic field at the North Pole-VI drifting station. (In Russian English abstract).
Geomagnet. Issled., Rez. Issled., Programme Mezhd. Geofiz. Goda, Mezhd. Geofiz. Komitet, Prezidiume, Akad. Nauk, SSSR, No. 6:27-37.

geomagnetism
Slaucitajs, L., 1956.
Mediciones geomagneticas en la region de la peninsula Antarctica, islas adyacentes y mar de Weddell, en 1951-56. Geofis. Pura e Applicata 35: 40-48.

geomagnetism
Vestine, E.H., 1955.
Relations between fluctuations in the earth's rotation, the variation of latitude, and geomagnetism.
(Ann. Geophys., 11(1)), I.A.G.A. Bull., :1pp.

geomagnetism
Whitmarsh, R.B., and M.T. Jones 1969
Daily variation and secular variation of the geomagnetic field from shipboard observations in the Gulf of Aden.
Geophys. J. R. astr. Soc., 18 (5): 477-488.

geomagnetism
*Yasui, Masashi, Yuichi Hashimoto and Seiya Uyeda, 1967.
Geomagnetic and bathymetric study of the Okhotsk Sea - (1).
Oceanogrl Mag., 19(1):73-85.

GEOMAGNETIC ANOMALY
Yasui, Masashi, Yuichi Hashimoto and Seiya Uyeda, 1967.
Geomagnetic studies of the Japan Sea. 1. Anomaly pattern in the Japan Sea.
Oceanogr. Mag., Jap. Met. Agency, 19(2): 221-231

geomagnetism (data only)
United States U.S. Department of Commerce Environmental Science Services Administration, 1969.
Bathymetry, geomagnetic and gravity data, USC&GS Ship Pioneer - 1964.
Int. Indian Ocean Exped., 3: 250 pp.

Geomagnetism, effect

geomagnetism, effect of
Raja Rao, K.S., and C.T. Thomas, 1962.
Lunar and solar geomagnetic tides in the geomagnetic equatorial region. III. Geomagnetic tidal variations at Apia.
Indian J. Meteorol. and Geophys., 13(2):237-240.

geomagnetism, effect of
Shuleykin, V.V., and L.A. Korneva, 1951 1956.
[Geomagnetic effects of oceanic and monsoon-path electric circulation.]
Doklady, Akad. Nauk, SSSR, 76(1):49-52; 57-60.
80(6):879-880
107(5):679-682.

LC and SLA, mi $2.70, ph $4.80

Geomagnetic field

geomagnetic field
*Anma, Kei, 1968.
Geomagnetic and bathymetric study of the Okhotsk Sea. (2).
Oceanogrl Mag., 20(1):65-72.

geomagnetic field
Espersen, J., P. Andreasen, J. Egedal and J. Olsen, 1956.
Measurements at sea of the vertical gradient of the main geomagnetic field during the Galathea Expedition. J. Geophys. Res. 61(4):593-624.

geomagnetic field
Korneva, L.A., 1956.
[On the geomagnetic field part unsymmetric with respect to the earth's axis in the Arctic and the World Ocean.] Dokl. Akad. Nauk, SSSR, 107(5): 679-682.

geomagnetic field
Kornewa, L.A., 1951.
[Anomalous geomagnetic field and its equivalent in a system of currents in the world ocean.] Dok. Akad. Nauk, SSSR, 76:49-52.

geomagnetic field
Runcorn, S.K., 1955.
Core motions and reversals of the geomagnetic field. (Ann. Géophys., 11(1)), I.A.G.A. Bull. 15a :7 pp.

geomagnetic field
Shuleikin, V.V., 1971
Drift of isolines of the western component of geomagnetic field strength across the Atlantic Ocean over 130 years. (In Russian).
Dokl. Akad. Nauk, SSSR 196(3): 575-578

geomagnetic induction
Hyndman, R.D., and N.A. Cochrane 1971.
Electrical conductivity structure by geomagnetic induction at the continental margin of Atlantic Canada.
Geophys. J.R. astron. Soc. 25 (5): 425-446.

Geomagnetic polarity

geomagnetic polarity changes
Kennett J.P., and N.D. Watkins, 1970.
Geomagnetic polarity change, volcanic maxima and faunal extinction in the South Pacific.
Nature, Lond. 227 (5261): 930-934

geomagnetic polarity time scale
Watkins, Norman D. 1972.
Review of the development of the geomagnetic polarity time scale and discussion of prospects for its finer definition.
Bull. geol. Soc. Am. 83 (3): 551-574.

geomagnetic polarity
Watkins, N.D., and H.G. Goodell, 1967.
Geomagnetic polarity change and faunal extinction in the Southern Ocean.
Science, 156(3778):1083-1087.

Geomagnetic reversals

geomagnetic reversals
*Glass, B., D.B. Ericson, B.C. Heezen, N.D. Opdyke and J.A. Glass.
Geomagnetic reversals and Pleistocene chronology.
Nature, Lond., 216(5114):437-442.

geomagnetic field, reversal of
Emiliani, Cesare, 1967.
The Pleistocene record of the Atlantic and Pacific oceanic sediments; correlations with the Alaskan stages by absolute dating; and the age of the last reversal of the geomagnetic field.
Progress in Oceanography, 4:219-224.

geomagnetic reversals
Glass, Bill, and Bruce C. Heezen 1967.
Tektites and geomagnetic reversals.
Nature, Lond. 214 (5086): 372.

geomagnetic field, reversals of
*Heirtzler, J.R., G.O. Dickson, E.M. Herron, W.C. Pitman, III, and X. Le Pichon, 1968.
Marine magnetic anomalies, geomagnetic field reversals and motions of the ocean floor and continents.
J. geophys. Res., 73(6):2119-2136.

geomagnetic reversals
Montfrans, H.M. van, 1971.
Palaeomagnetic dating in the North Sea Basin.
Earth planet. Sci. Letts 11 (3): 226-235

geomagnetic reversal
*Smith, Jerry D. and John H. Foster, 1969.
Geomagnetic reversal in Brunhes normal polarity epoch.
Science 163(3867):565-567.

geomagnetic reversals
Vogt, P.R., C.N. Anderson and D.R. Bracey, 1971.
Mesozoic magnetic anomalies, sea-floor spreading and geomagnetic reversals in the southwestern North Atlantic. J. geophys. Res., 76(20): 4796-4823.

geomagnetic variations

geomagnetic variation
*Backus, G.E., 1968.
Kinematics of geomagnetic secular variation in a perfectly conducting core.
Phil. Trans. R. Soc., (A)263(1141):239-266.

geomagnetic variations
Vanian, L.L., B.E. Marderfeld, and A.V. Rodionov, 1967.
The regional and local coast effect in the geomagnetic variations at the festoon islands of the Far East. (In Russian).
Dokl. Akad. Nauk. SSSR, 176(4):820-821.

Geomicrobial activities
Oppenheimer, Carl H., 1968.
Geomicrobial activities of microorganisms.
Bull. Misaki mar. biol. Inst., Kyoto Univ. 12: 105-109

Geophysics

geophysics
See chiefly: submarine geophysics

geophysics
Defant, A., 1950.
Die Geophysik und ihre Stellung im Rahmen der übrigen Naturwissenschaften. Antrittsrede von Dr. Albert Defant o. Prof. der Meteorologie und Geophysik gehalten anlässlich der Inauguration zum Rector magnificus des Studien Jahres 1950/51 am 18 November 1950 in der Aula der Leopold-Franzens-Universität zu Innsbruck:19 pp.

Geophysics (Soviet-Beloussov)
Maldonado-Koerdell, M., 1961
Actualidades Geological y Geofisicas, 1.
Bol. Asoc. Mexicana de Geol. Petrol, 13(7/8): 283-288.

geophysics
Runcorn, S. K., 1960
Methods and techniques in geophysics. Vol. I. Interscience Publishers, Inc., New York: 374 pp

geophysics
Sverdrup, H. U., 1949.
Theoretical tools in geophysics. Geografiska Annaler, Årg 31(1/4):365-368.

geophysics
Vercelli, F., 1949.
Il superamento delle altezza e delle profundita nelle esplorazioni geographische. Atti Ist. Ven. Sc. Lett. Arti 107(1):41-57.

Pubbl. Ist. Talassografico, Trieste, No. 244.

For geophysics, underwater SEE SUBMARINE GEOPHYSICS

geophysics
Model, Fritz, 1964.
Geophysikalische Bibliographie von Nord- und Ostsee. 1. Chronologische Titelaufzahlung.

Deutsches Hydrogr. Inst., Hydrogr. Dokumentation 1961 pp. (mimeographed) (Unpublished manuscript).

geopotential anomaly

geopotential anomaly
Masuzawa, Jotaro 1967.
An oceanographic section from Japan to New Guinea at 137°E in January, 1967.
Oceanogrl Mag. 19 (2): 95-118.

geopotential anomaly
NORPAC Committee, 1960.
The NORPAC Atlas. Oceanic observations of the Pacific, 1955.

geopotential anomaly
Reid, J.L., 1961
On the geostrophic flow at the surface of the Pacific Ocean with respect to the 1,000-decibar surface.
Tellus, 13(4):489-502.

geopotential anomalies
Wyrtki, K., 1962.
Geopotential topographies and associated circulation in the western South Pacific Ocean.
Australian J. Mar. Freshwater Res., 13(2):89-105.

Geopotential equation

geopotential equation
Katkov, V.L., 1966.
Some exact solutions of the geopotential equation. (In Russian).
Fizika Atmosferi i Okeana, Izv., Akad. Nauk, SSSR, 2(11):1193-1197.

geopotential topography

geopotential topography
*Ingraham, W. James, Jr., 1968.
The geostrophic circulation and distribution of water properties off the coasts of Vancouver Island and Washington spring and fall 1963.
Fishery Bull. Fish Wildl. Serv. U.S. 66(2):223-250.

Geopotential topography
Lee, Chang-Ki and Jong-Hon Bong, 1968.
On the current of the Korean Eastern Sea (West of the Japan Sea). (In Korean; English abstract).
Bull. Fish. Res. Develop Agency, Korea, 3: 7-26

Geopotential topography
Mazeika, P.A., 1968.
Eastward flow within the South Equatorial Current in the eastern South Atlantic.
J. geophys. Res., 73(18):5819-5828.

geopotential topography
Uda, Michitake, 1963.
Oceanography of the subarctic Pacific Ocean.
J. Fish. Res. Bd., Canada, 20(1):119-179.

Geostrophic flow

geostrophic currents
Fukuoka, Jiro, 1967.
Masas de agua y dinámica de los océanos.
In: Ecología marina, Monogr. Fundación La Salle de Ciencias Naturales, Caracas, 14:130-183.

geostrophic flux
Masuzawa, Jotaro 1967.
An oceanographic section from Japan to New Guinea at 137°E in January 1967.
Oceanogrl Mag. 19(2): 95-118.

Geostrophic flow
Murty, C. B., and M. Rattray, Jr., 1962.
On the depth distribution of geostrophic flow in the eastern North Pacific. (Abstract.)
J. Geophys. Res., 67(4):1649.

geostrophic flow
Reid, Joseph L., Jr., 1965.
Intermediate waters of the Pacific Ocean.
The Johns Hopkins Oceanogr. Studies, No. 2L 1-85.

geostrophic circulation
Rotschi, H., and L. Lemasson, 1967.
Oceanography of the Coral and Tasman seas.
Oceanogr. Mar. Biol. Ann. Rev., Harold Barnes, editor, George Allen and Unwin, Ltd., 5:49-97.

geostrophic flow
Seckel, Gunter R., 1968.
A time-sequence oceanographic investigation in the North Pacific trade-wind zone.
Trans. Am. geophys. Un., 49(1):377-387.

Geostrophic flow
Tsuchiya, Mizuki, 1968.
Upper waters of the intertropical Pacific Ocean.
Johns Hopkins Univ. Studies, 4:50 pp.

geostrophic speed
Wooster, W.S., and Malvern Gilmartin, 1961
The Peru-Chile Undercurrent.
J. Mar. Res., 19(3):97-122.

geostrophic flow
Wyllie, John G., 1966.
Geostrophic flow of the California Current at the surface and at 200 meters.
Calif. Coop. Ocean. Fish. Invest., Atlas, 4:288 pp.

geostrophic fluxes

geostrophic flux
Masuzawa, Jotaro, 1965.
Water characteristics of the Kuroshio.
Oceanogr. Mag., Tokyo, 17(1/2):37-47.

geostrophic fluxes
Masuzawa, Jotaro, 1964.
A typical hydrographic section of the Kuroshio extension.
Oceanogr. Mag., Tokyo, 16(1/2):21-30.

geostrophic vertical velocity

geostrophic vertical velocity
Ivanov-Frantskevich, G.N., 1969.
On considering the convergence of meridians in the computation of the geostrophic vertical velocity. (In Russian; English abstract).
Okeanologiia, 9(1):34-37.

geostrophic vortex
Stern, Melvin E., 1965.
Interaction of a uniform wind stress with a geostrophic vortex.
Deep-Sea Res., 12(3):355-367.

geosynclines

geosynclines
Bogdanov, N.A. 1969.
Thalassogeosynclines of the circumpacific ring. (In Russian)
Geotecton. Akad. Nauk, SSSR (3): 3-16.
Translation, Scripta Technica for AGU 1970: 141-148

geosynclines
Crook Keith A.W., 1969.
Contrasts between Atlantic and Pacific geosynclines.
Earth Planet Sci. Letters, 5(7): 429-438.

geosynclines
Dietz, Robert S. 1972.
Geosynclines, mountains and continent-building.
Scient. Am. 226(3): 30-38

geosynclines
Hamilton, Warren, 1966.
Origin of the volcanic rocks of eugeosynclines and island arcs.
In: Continental margins and island arcs, W.H. Poole, editor, Geol. Surv. Pap., Can., 66-15:348-355.

geosynclines
Kraft, John C., Robert E. Sheridan and Marilyn Maisano 1971.
Time-stratigraphic units and petroleum entrapment models in Baltimore Canyon Basin of Atlantic continental margin geosynclines.
Bull. Am. Ass. Petrol. Geol., 55(5): 658-679.

geosynclines
Kuenen, Ph. H., 1966.
Geosynclinal sedimentation.
Geol. Rdsch, 56:1-19.

geosyncline
MacGillavry, H.J., 1968.
Turbidite detritus and geosyncline history.
Trans. Fifth Carib. Geol. Conf., St. Thomas, V.I., 1-5 July 1968, Peter H. Mattson, editor.
Geol. Bull. Queens Coll., Flushing 5: 39.
(Abstract only).

geosynclines
Mitchell, Andrew H., and Harold G. Reading, 1969.
Continental margins, geosynclines, and ocean floor spreading.
J. Geol., 77(6): 629-646.

geosynclines
Peive, A.V., N.A. Shtreis, A.L. Knipper, M.S. Markov, N.A. Bogdanov, A.S. Perfiliev and S.V. Ruzhentsev, 1971.
Oceans and the geosynclinal process. (In Russian).
Dokl. Akad. Nauk, SSSR, 196(3): 657-659

geosynclines
Rodnikov, A.G. 1971
Geotraverses of the Earth's crust in the Pacific mobile belt. (In Russian; English abstract)
Geofiz. Biull. Mezhd. Geofiz. Kom. Presid. Akad. Nauk SSSR 22: 44-51
Mohorovicic discontinuity

geosynclines
Wezel, Forese Carlo 1970.
Interpretazione dinamica della "eugeosinclinale meso-mediterranea"
Riv. Miner. Siciliana 21 (124-126): 187-198.

geothermal flux
Bulatov, R.P., 1971.
On the structure and circulation of the bottom layer in the Atlantic Ocean. (In Russian; English abstract). Okeanol. Issled. Rezult. Issled. Mezhd. Geofiz. Proekt. 21: 43-59.

geothermal gradients
Harper, M.L., 1971.
Approximate geothermal gradients in the North Sea Basin.
Nature, Lond. 230 (5291): 235-236

germs, parasitic
Guelin, A., I. Bychovskaja, P. Lepine et D. Lamblin, 1970.
Distribution des germes parasites des bactéries pathogènes dans les eaux mondiales.
Rev. int. Océanogr. méd. 18-19: 77-83

gibbsite, see clay minerals

Gibbs phase rule

Gibbs phase rule
*Sillén, Lars Gunnar, 1967.
Gibbs phase rule and marine sediments.
In: Equilibrium concepts in natural water systems, Werner Stumm, editor, Adv. Chem. Ser. 67: 57-69.

glacial drift

glacial drift
*Kaye, C.A., 1967.
Fossiliferous bauxite in glacial drift, Martha's Vineyard, Massachusetts.
Science, 157(3792):1035-1037.

glaciation

glaciation
Bauer, Albert, 1967.
Nouvelle estimation du bilan de masse de l'Indlandsis du Greenland.
Deep-Sea Research, 14(1):13-17.

glaciation
*Caralp, Michelle, 1967.
Les foraminifères planctoniques d'une carotte atlantique (golfe de Gascogne) dans la mise en évidence d'une glaciation.
C.r. hebd. Séanc. Acad. Sci., Paris, (D)265(21):1588-1591.

glaciation
Carsola, A.J., 1954.
Extent of glaciation on the continental shelf in the Beaufort Sea. Amer. J. Sci. 252(■):366-371.

glaciation
Chizhov, O.P., 1966.
The problem of interrelations between continental and sea glaciation in The Northern Hemisphere. (In Russian).
Geofiz. Biull., Mezhd. Geofiz. Kom., Prezid. Akad. Nauk. SSSR, No. 17:20-24.

glaciers
Dansgaard, W., 1961
The isotopic composition of natural waters with special reference to the Greenland ice cap.
Medd. om Grønland, 165(2): 1-120.

GLACIATION
Dansgaard, W., and Henrik Tauber, 1969.
Glacier oxygen-18 content and Pleistocene ocean
Science, 166(3904): 499-502.

glaciation
Emiliani, Cesare, 1966.
Isotopic paleotemperatures: Urey's method of paleotemperature analysis has greatly contributed to our knowledge of past climates.
Science, 154'3751):851-857.

glaciation
Frakes, Lawrence A., and John C. Crowell, 1968.
Late Palaeozoic glacial facies and the origin of the South Atlantic Basin.
Nature, Lond., 217(5131):837-838.

glaciation
Goodell, H.G., N.D. Watkins, T.T. Mather and S. Koster 1968.
The Antarctic glacial history recorded in sediments of the Southern Ocean.
Palaeogr. Palaeoclimatol. Palaeoecol. 5(1): 41-62

glaciation
Gram, Ralph, 1969.
Grain surface features in Eltanin cores and Antarctic glaciation.
Antarctic J., U.S., 4(5): 174-175

glaciation
Keen, M.J., B.D. Loncarevic and G.N. Ewing, 1970.
II. Regional observations. 7. Continental Margin of Eastern Canada: Georges Bank to Kane Basin. In: The sea: ideas and observations on the progress in the study of the seas, Arthur E. Maxwell, editor, Wiley-Interscience 4(2): 251-291.

glaciation
Margolis, Stanley V. and James P. Kennett, 1970
Antarctic glaciation during the Tertiary recorded in sub-antarctic deep-sea cores.
Science, 170(3962): 1085-1087.

glaciers
*Page, R.A., 1968.
Sub-surface melt pools in the McMurdo Ice Shelf, Antarctica.
J. Glaciol., 7(51)511-516.

glaciation
Pratt, Richard M., and John Schlee, 1969.
Glaciation on the continental margin off New England.
Bull. geol. Soc. Am., 80(11): 2335-2342.

glaciation
Romanovsky, V., 1945
Résultats des sondages effectués dans la baie du Roi. Bull. Inst. Ocean., Monaco, No.877, 9 pp., 4 figs.

glaciation
Romanovsky, V., 1943
Oscillations de rivage et bathymétre dans la région sud de la Baie du Roi (Spitzberg). Bull. Soc. Géol. de France, No.1-2-3, :81-90, 4 figs., 1 pl.

glaciation
Rutford, Robert H., Campbell Craddock and Thomas W. Bastien 1968.
Late Tertiary glaciation and sea-level changes in Antarctica.
Palaeogr. Palaeoclimatol. Palaeoecol. 5(1): 15-39.

glaciers

glaciers
Reeh, Niels, 1968.
On the calving of ice from floating glaciers and ice shelves.
J. Glaciol., 7(50) 215-232

glaciers, effect of
Dingle, R.V. 1971.
Buried tunnel valleys of the Northumberland Coast, western North Sea.
Geol. Mijnb. 50(5): 679-686

glands.

glands, salt-secreting
Ellis, Richard A., and John H. Abel, Jr., 1964
Intercellular channels in the salt-secreting glands of marine turtles.
Science, 144(3624):1340-1342.

glass

glass
Craig, H., 1969.
Density and refractive index hysteresis in compressed silicate glass.
J. geophys. Res. 74(20):4910-4920.

glass
Fisher, David E., 1969.
Fission track ages of deep sea glasses.
Nature, Lond., 221(5180):549-550.

glass
Pollock, Kenneth J., 1969.
Glass in hydrospace.
Oceanology intl, 4(2): 39-41.

glass
Von der Borch, C.C. 1971.
Glassy objects in Tertiary deep-sea clays cored by the Deep Sea Drilling Project.
Marine Geol., 10(1): 5-14.

glass shards
MacDougall, D., 1971.
Fission track dating of volcanic glass shards in marine sediments.
Earth planet. Sci. Letts 10(4): 403-406

glauconite

glitter

glitter
Griesseier, H., 1953.
Zur Reflexion der direkten Sonnenstrahlung an einen bewegten Wasseroberfläche.
Acta Hydrophysica 1(3):107-133.

glitter
Ljalikov, K.S., and Yu. D. Sharikov, 1956.
[Examination of the method of analyzing by diffraction aerial photos of the agitated surface of the sea.] Trudy Lab. Aerometod., 5:72-82.

glitter
Steleanu, A., 1961
Über die Reflexion der Sonnenstrahlung an Wasserflächern und ihre Bedeutung für das Seenfern.
Gerlands Beitr. z. Geoph., 70(2):90-124.

global network
Oliver, J. and L. Murphy, 1971.
WWNSS: Seismology's global network of observing stations. Science 174(4006):254-261.

GLORIA
Somers, M.L. 1970.
Signal processing in Project Gloria: a long range side-scan sonar.
Electronic Engineering in Ocean Technology, Swansea, 21-24 Sept. 1970, I.E.R.E. Conf. Proc. 19:109-120

glossaries

Gloria
Whitmarsh, R.B., 1971.
Interpretation of long range sonar records obtained near the Azores. Deep-Sea Res., 18(4): 433-440.

glossaries
Anon., 1969.
A navigational glossary.
J. Inst. Navig., 22(2): 255-267.

glossaries
Armstrong, T.E., 1961.
Soviet Terms for the north of the USSR.
Polar Record, 10(69):609-613.

glossaries (ice)
Armstrong, Terence, 1959
Russian-English glossary and Soviet classification of ice found at sea. Compiled by Boris N. Mandrovsky, Washington: Ref. Dept. Library of Congress, 1959. 10 1/4 x 8 inches; vi plus 30 pp; mimeographed $0.30, obtainable from Card Division, Libr. of Congr. Wash. 25, D.C.
Arctic 12(4): 237-238.

glossary, ice
Armstrong, T., and B. Roberts, 1956.
Illustrated ice glossary. Polar Rec. 8(52):4-12, 40 figs.

glossaries, sea ice
Armstrong, Terence, Brian Roberts and Charles Swithinbank, 1965.
Comments on Canadian proposal for changes in WMO sea ice terminology.
Polar Record, 12(81):723.

glossaries
Association d'Océanographie Physique, Union Géodésique et Géophysique, 1940.
Report of the Committee on the Criteria and Nomenclature of the Major Divisions of the Ocean Bottom. Publ. Sci. No. 8:124 pp., 3 fold-in charts.

glossaries
Baker, B.B., Jr., W.R. Deebel and R.D. Geisenderfer, editors, 1966.
Glossary of oceanographic terms (2nd edition)
U.S. Naval Oceanogr. Off., Spec. Publ., SP-35: 204 pp. $2.25.

glossaries
Barth, Rudolf, 1968.
Glossário preliminar de têrmos técnicos usados em biologia marinha.
Publ. Inst. Pesquis. Marinha, Brasil, 027: 62 pp. (multilithed).

glossaries
Beach Erosion Board, 1953.
Shore protection planning and design. (Preliminary, subject to revision). B.E.B. Bull. Special Issue, No. 2:230 pp., 149 figs., plus app.

glossaries
Black, W.A., 1962
Geographical Branch program of ice surveys of the Gulf of St. Lawrence, 1956 to 1962.
Cahiers de Geographie de Quebec, 6(11):65-74.

glossaries
Boltovskoy, E., 1957.
Diccionario foraminiferologbico plurilingüe, en cinco idiomas: Inglés Español, Alemán, Francés y Ruso con indices alphabéticos.
S.H. Pub. Misc., Ministerio de Marina, Direccion General de Navegacion e Hidrografia, Republica Argentina, No. 1001:196 pp.

glossaries
Botrio, A., 1960.
[The problem of investigation and determination of morphological terms along the Jugoslav coast.]
Hidrografski Godisnjak, Split, 1959(30):119-127.

glossaries, ice
Bradford, J.D., and S.M. Smirle, 1970.
Bibliography on northern sea ice and related subjects.
Marine Operations, Ministry of Transport and Marine Sciences Branch, Department of Energy, Mines and Resources, Canada, 158 pp.

glossaries
Brendão, J.M., 1964.
Glossário de nomes dos peixes; português, inglês, sistemático.
Sudene, Bol. Estudos Pesca, Recife, Brasil, 4(4): 40 pp. 4(5): 59 pp. 4(6):59 pp.

glossaries (gives common English, Peruvian, Ecuadorian names)
Chirichigno Fonseca, Norma, 1970
Lista de crustáceos del Perú (Decapoda y Stomatopoda) con datos de su distribución geográfica.
Informe Inst. Mar, Perú, 32: 95 pp.

glossaries
Clausade, Mireille, Nicole Gravier, Jacques Picard, Michel Pichon, Marie-Louise Roman, Bernard Thomassin, Pierre Vasseur, Mireille Vivien et Pierre Weydert 1971.
Morphologie des récifs coralliens de la région de Tuléar (Madagascar): éléments de terminologie récifale.
Téthys Suppl. 2: 1-74

glossaries (ice)
Czekanska, Maria, 1959
Materialy do polskiej terminologii lodow morskich. Acta Geophys. Polonica 7(3/4): 322-342.

glossary (definitions of terms)
Davis, C.C., 1963.
On questions of production and productivity in ecology.
Arch. Hydrobiol., 59(2):145-161.

glossaries, fish (local names)
De Holanda Lima, Hermínia, 1969.
Primeira contribuição ao conhecimento dos nomes vulgares de peixes marinhos do nordeste Brasileiro.
Bol. Ciên. Mar, Fortaleza, Ceará, Brasil, 21: 1-20.

glossaries, ice
Dunbar, Moira, 1967.
International ice nomenclature (in connection with V.L. Tsurikov's article). (In Russian).
Okeanologiia, Akad. Nauk, SSSR, 7(6):1128-1131.

glossaries, sea ice
Dunbar, Moira, 1965.
Canadian proposal for changes in WMO sea ice terminology.
Polar Record, 12(81):717-722.

glossary of terms-English-Polish
Dziembowski, Lech, Jan Kozlowski and Antoni Nowalinski, 1960
Guiding principles for the construction of the laminate boat hulls. Prace Instytutu Morskiego II. Techniczna Eksploatacja Floty, Gdansk, No. 3: 166 pp. (mimeographed).

glossaries
Fairbridge, Rhodes, W., 1958.
What is a consanguineous association.
J. Geol., 66(3):319-324.

glossaries
Fischer, E., and F.E. Elliott, 1950.
A German and English glossary of geographical terms. Amer. Geogr. Soc., Library ser., No. 5: 111 pp.

glossaries
Gonzalez Bonorino, F., and M.E. Teruggi, 1952.
Lexico sedimentologicos. Publ. de Extension Cultural y Didactica, Inst. Nacional Invest. Ciencias Nat., y Mus. Argentino Ciencias Nat. "Bernardino Rivadavia" No. 6:164 pp.

glossaries, geologic features
Gougenheim, André 1970.
Nomenclatures des formes du relief océanique.
Cah. océanogr. 22(8): 769-774.

glossaries
Hoel, A., and Werner Werenskiold, 1962.
Glaciers and snowfields in Norway.
Norsk Polarinst., Skrifter, No. 114:291 pp.

glossaries
Hunt, L., and Groves, 1963.
UST Glossary.
Undersea Technology, 4(1):51-52.

Glossaries
International Hydrographic Bureau, 1962.
French translation of nomenclature of ocean bottom features.
Int. Hydrogr. Rev., 39(1):179-183.

glossaries
International Hydrographic Bureau, 1951.
Hydrographic dictionary. Spec. Publ. No. 32: 89 pp., index.

glossaries
Japan, Maritime Safety Board, 1955.
[On the terminology of sea ice. Pt. 3.]
Publ. 981, Hydrogr. Bull. No. 50:232-236.

glossaries
Japan, Maritime Safety Board, 1955.
[On the terminologies of sea ice. (Part 2.).]
Pub. 981, Hydrogr. Bull., No. 48:136-139.

glossaries
Jerlov, N. G., Chairman, 1964.
Standard terminology on optics of the sea.
Union Géodes. et Géophys. Int., Chronique de l'U.G.G.I., No. 57:246-251.

~~glossaries, ice~~
Koslowski, G. 1969.
Die WMO-Eisnomenklatur.
Dt. hydrogr. Z. 22(6): 256-267

glossaries
Legendre, Vianney, W.B. Scott et Julien Bergeron 1964.
Noms français et anglais des poissons de l'Atlantique canadien.
Cah. inf. Stn. biol. mar. Grande Rivière 23: 178 pp. (multilithed).

glossaries
Leipper, Dale F., 1961.
Oceanography - a definition for academic use.
Trans. Amer. Geophys. Union, 42(4):429-431.

glossaries
Leontiev, O.K., 1956.
[On the terminology in the science of marine coasts. Studies on sea coasts and reservoirs.]
Trudy Okeanogr. Komissii, Akad. Nauk, SSSR, 1: 141-149.

glossary
Matsudaira, Yasuo?
On "Uotsukerin" (oceanographic meaning).
Bull. Kobe Mar. Obs., 158:1-19.

glossaries
Maurstad, A., 1935.
Atlas of sea ice. Geofys. Publ. 10(11):17 pp.

glossary (coccolithophorids)
*McIntyre, Andrew, and Allan W.H. Be, 1967.
Modern Coccolithophoridae of the Atlantic Ocean.
I. Placoliths and cyrtoliths.
Deep-Sea Res., 14(5):561-597.

geochemistry
Miller, R.L., and E.D. Goldberg, 1955.
The normal distribution of geochemistry.
Geochimica and Cosmochimica Acta 8(1/2):53-62.

glossaries
Nusser, Franz, 1964.
Neue Eisbezichnungen für Unterseebootfahrten unter dem Eis.
Deutsche Hydrogr. Zeits., 17(5):236-237.

glossaries
Nusser, F., 1956.
Eine neue internationale Eisnomenklatur.
Deutsche Hydrogr. Zeits., 9(4):174-182.

glossaries
Palosuo, E., 1953.
A treatise on severe ice conditions in the central Baltic. Fennia 77(1):1-130, 55 textfigs.

glossaries
PANZARINI, Rodolfo N. 1963
Nomenclatura del Hielo en el Mar.
Buenos Aires, Instituto Antártico Argentino, 1963.
103 p., fot., gráf., 30 cm. (Publicación del Instituto Antártico Argentino Nº 10).

glossaries
Russell, Richard J. 1969.
Glossary of terms used in fluvial, deltaic and coastal morphology and processes.
Coastal Studies Ser., Louisiana State Univ. Press, 23:97 pp.

glossaries
Schurman, P., 1941.
Tide and current glossary. U.S. Govt. Print, 40 pp.

glossaries
Society of Naval Architects and Marine Engineers, 1950.
Nomenclature for treating the motion of a submerged body through a fluid. Tech. Res. Rept. 1-5:15 pp., 3 textfigs.

glossary
Stamp, L. Dudley,
A glossary of geographical terms.
Wiley, 440 Park Avenue South, New York 16.

glossary
Steidinger, Karen A., Joanne T. Davis and Jean Williams, 1967.
A key to the marine Dinoflagellata genera of the West Coast of Florida.
Bd. Conerv. Fla. Techn. Ser., 52: 44 pp.

glossary, dinoflagellate
Steidinger, Karen A., and Jean Williams 1970
Dinoflagellates.
Mem. Hourglass Cruises, Mar. Res. Lab., Fla. Dept. Nat. Res. 2:1-251.

glossaries
Stocks, Theodor, 1958.
Untermeerische Bodenformen.
Geogr. Taschenbuch, 1958-59, Verlag Steiner, Wiesbaden, 509-516.
In:
Oceanographie, Deutsches Hydrographisches Institut, 1958 (i.e., "Collected Reprints")
terminology in French, German and English

glossary
Strickland, J.D.H., 1965.
Production of organic matter in the primary stages of the marine food chain.
In: Chemical oceanography, J.P. Riley and G. Skirrow, editors, Academic Press, 1:477-610.

Glossaries
USSR Gosudarstvennyi Okeanograficheskii Institut Lingradskoe Otdelenie, 1960.
Album of serial photographs of ice formation on seas, 1-224 pp.
H.O. TRANS.-124 (1961). (Figures omitted)

glossaries
Uusitalo, S., 1964.
About units and dimensions.
Geophysica, Helsinki, 9(1):73-86.

glossaries, sediments
Washburn, A.L., J.E. Sanders and R.F. Flint, 1963
A convenient nomenclature for poorly sorted sediments.
J. Sed. Petrol., 33(2):478-480.

glossaries
Wass, M.L., 1955.
The decapod crustaceans of Alligator Harbor and adjacent inshore areas of northwestern Florida.
Q.J. Florida Acad. Sci., 18(3):129-176.

glossaries
Wiegel, R.L., 1953.
Waves, tides, currents and beaches: glossary of terms and list of standard symbols.
Council on Wave Research, Berkeley, California, 113 pp.

decapods, glossaries
Williams, Austin B., 1965.
Marine decapod crustaceans of the Carolinas.
U.S.F.W.S. Fish. Bull., 65(1): 1-298.

glossaries
Wiseman, J.D.H., and C.D. Ovey, 1953.
Definitions of features on the deep-sea floor.
Deep-sea Res. 1(1):11-16.

Gondwana

Government, effect of

government
Burstyn, Harold L., 1968.
Science and government in the nineteenth century: The Challenger expedition and its report.
Bull. Inst. oceanogr., Monaco, No. special 2: 603-613.

government, effect of
Wakefield, Lowell, 1966.
Role of government in ocean fisheries exploitation - an industrial view.
Exploiting the Ocean, Trans. 2nd Mar. Techn. Soc. Conf., June 27-29, 1966, 173-176.

graben

grabens
Ball, M.M., C.G.A. Harrison, P.R. Supko and W.D. Bock, and N.J. Maloney, 1968.
Normal faulting on the southern boundary of the Caribbean Sea, Unare Bay, Northern Venezuela.
Trans. Fifth Carib. Geol. Conf., St. Thomas, V.I., 1-5 July 1968, Peter H. Mattson, editor.
Geol. Bull. Queens Coll., Flushing, 5: 17-21.

graben
Kaneko Shiro, 1966.
Rising promontories associated with a subsiding coast and sea-floor in north-western Japan.
Trans. R. Soc., N.Z., Geol., 4 (4): 211-228

graben
Knott, S.T., E.T. Bunce and R.L. Chase, 1966.
Red Sea seismic reflection studies.
In: The world rift system, T.N. Irvine, editor, Dept. Mines Techn. Surveys, Geol. Survey, Can., Paper, 66-14:33-61.

grabens
Ma, Ting-Ying H., 1970.
Development of grabens and age of the contained oil-bearing sediments as a basis for re-assessment of past geological deductions.
Acta Geol. Taiwanica, 13: 59-82

graben
Moore, David G., and Joseph R. Curray, 1963
Structural framework of the continental terraces northwest Gulf of Mexico.
J. Geophys. Res., 68(6):1725-1747.

graben
Nesteroff, Wladimir D., 1959
Age des derniers mouvements du graben de la mer Rouge déterminé par la méthode du C14 appliquée aux récifs fossiles.
Bull. Soc. Geol., France (7): 415-418.

graben
Nesteroff, W.D., 1955.
Les récifs coralliens du Banc Farsan Nord (Mer Rouge). Rés. Sci. Camp. Calypso, 1. Camp. 1951-1952 en Mer Rouge. Ann. Inst. Océan., Monaco, 30:8-53, 21 pls., 11 textfigs.

graben
Picard, Leo, 1966.
Thoughts on the graben system in the Levant.
In: The world rift system, T.N. Irvine, editor, Dept. Mines Techn. Surveys, Geol. Surv. Can., Paper, 66-14:22-32.

graded bedding

graded bedding
Kopstein, F.P.H.W., 1954.
Graded bedding of the Harlech Dome.
Geol. Inst., Gronningen, Publ. No. 81:1-97, 3 fold-ins, 47 textfigs.

graded bedding
Kuenen, PH. H., 1953.
Graded bedding with observations on lower Paleozoic rocks of Britain.
Verhandel. K. Nederl. Akad. Wetensch., Afd. Natuurkunde, Eerste reeks, 20(3):1-47, 2 pls., 19 textfigs.

graded bedding
Kuenen, Ph. H., and C.I. Migliorini, 1950.
Turbidity currents as a cause of graded bedding.
J. Geol. 58(2):91-127, 7 textfigs.

gradients, environmental

gradients, environmental
*Ichimura, Shun-ei, 1967.
Environmental gradient and its relation to primary productivity in Tokyo Bay.
Rec. oceanogr. Wks. Japan, 9(1):115-128.

grain size
see also under: bottom sediments, mechanical analysis

grampus

granites

granite
Pratt, Richard M. 1970.
Granitic rocks from the Blake Plateau.
Bull. geol. Soc. Am. 81(10): 3117-3122

granites
Smith, A.J., A. H. Stride and W.F. Whittard, 1965
The geology of the western approaches of the English Channel. IV. A recently discovered Variscan granite west-north-west of the Scilly Islands.
In: Submarine geology and geophysics, Colston Papers, W.F. Whittard and R. Bradshaw, editors, Butterworth's, London, 17:287-298.

granite-free

granitic layer (absent)
Balavadze, B.K., and P. Sh. Mindeli, 1966.
Main results of geophysical investigations of th structure of the earth's crust of the Black Sea Basin. (In Russian; English abstract).
Stroenie Chernomorskoi Vladini, Rez. Issled. Mezhd., Geofiz. Proekt., Mezhd. Geofiz. Komitet, Presid., Akad. Nauk, SSSR, 17-21.

granite - free
Mindeli, P.Sh., Yu. V. Neprochnov and Ye. I. Partaraya, 1966.
Granite-free area in Black Sea Trough, from seismic data.
Int. Geol. Rev., 8(1):36-43. (not seen)

granitic layer, absence of
Rezanov, I. A. and S.S. Chamo, 1969.
Reasons for absence of a 'granitic' layer in basins of the South Caspian and Black Sea type.
Can. J. Earth Sci. 6(4): 671-675

granules

granules
Laughton, A.S., 1963.
18. Microtopography.
In: The Sea, M.N. Hill, Editor, Interscience Publishers, 3:437-472.

grape stones

grape stones
Roberts, Harry H. 1969.
Recently cemented aggregates (grapestones), Grand Cayman Island, B.W.I.
Bull. Coast. Stud. Coast. Stud. Inst., Louisiana State Univ. 3:17-21 (multilith)

graphic aids

graphic aids
Neshyba, Steve, and David E. Amstutz, 1965.
Graphic aid for geostrophic computations from vertical sections.
Deep-Sea Res., 12(3):369-371.

grasses

grasses, sea
Maurer, Larry G., and P.L. Parker, 1967.
Fatty acids in sea grasses and marsh plants.
Contrib. Mar. Sci., Port Aransas, 12: 113-119

grass
Halodule
*Phillips, Ronald C., 1967.
On species of the seagrass, Halodule, in Florida.
Bull. mar. Sci., Miami, 17(3):672-676.

grasses, marine
Taylor, J.D., and M.S. Lewis, 1970.
The flora, fauna and sediments of the marine grass beds of Mahé, Seychelles.
J. nat Hist., 4(2): 199-220.

gravels

Gravels
Cullen, David J., 1966.
Fluviatile run-off as a factor in the primary dispersal of submarine gravels in Foveaux Strait, New Zealand.
Sedimentology, 7(3): 191-201

gravel
Davenport, Joan M. 1971.
Incentives for ocean mining: a case study of sand and gravel.
J. mar. techn. Soc. 5(6): 35-40.

gravel
Dunham, K.C., 1970.
Gravel, sand, metallic placer and other mineral deposits on the East Atlantic continental margin
In: The geology of the East Atlantic continental margin, 1. General and economic papers, ICSU/ SCOR Working Party 31 Symposium, Cambridge 1970, Rept. No. 70/13: 79-85.

gravel
Emery, K.O., 1966.
Geological methods for locating mineral deposits on the ocean floor.
Exploiting the Ocean, Trans. 2nd Ann. Mar. Techn. Soc. Conf., June 27-29, 1966, 24-43.

gravels
Griggs, G.B., L.D. Kulm, A.C. Waters and G.A. Fowler, 1970.
Deep-sea gravel from Cascadia Channel.
J. Geol. 78(5): 611-619.

gravel
Schlee, John, and Richard M. Pratt, 1970
Atlantic continental shelf and slope of the United States - gravels of the northeastern part.
Prof. Pap., U.S. Geol. Surv. 529-H: H-1 - H-39.

gravel
Taney, Norman E., 1971.
Comments on "Incentives for ocean mining - a case study of sand and gravel.
J. mar. techn. Soc. 5(6): 41-43

gravity

A

gravity
Allan, T.D., 1970.
Magnetic and gravity fields over the Red Sea.
Phil. Trans. R. Soc. (A) 267 (1181): 153-180.

gravity
Allan, T.D. and C. Morelli, 1970.
II. Regional observations. 13. The Red Sea (May 1969). In: The sea: ideas and observations on progress in the study of the seas, Arthur E. Maxwell, editor, Wiley-Interscience 4(2): 493-542.

gravity
Allan, T.D., and M. Pisani, 1966.
Gravity and magnetic measurements in the Red Sea. (Summary only).
In: The world rift system, T.N. Irvine, editor, Dept. Mines, Techn. Surveys, Geol. Survey, Can., Paper, 66-14:62-63.

gravity
Allan, T.D., and M. Pisani, 1965.
Gravity, magnetic and depth measurements in the Ligurian Sea.
Rapp. P.-v. Réun., Comm. int. Explor. scient., Mer Méditerr., 18(3):907-909.

specific gravity
Anon., 1951.
Bulletin of the Marine Biological Station of Asamushi 4(3/4): 15 pp.

B

gravity
Bacon, M., and F. Gray, 1970.
A gravity survey in the eastern part of the Bay of Biscay.
Earth planet. Sci. Letts 10(1): 101-105

gravity
Bacon, M., F. Gray and D. N. Matthews, 1969
Crustal structure studies in the Bay of Biscay.
Earth Planet. Sci. Letters, 6(5): 377-385

gravity
Bakkelid, Sivert, 1959.
Gravity observations in a submarine along the Norwegian coast.
Geodetiske Arbeider, Oslo, No. 11:28 pp.

Gravity
Balavadze, B.K., and P. Sh. Mindeli, 1966.
Main results of geophysical investigations of th structure of the earth's crust of the Black Sea Basin. (In Russian; English abstract).
Stroenie Chernomorskoi Vladini, Rez. Issled. Mezhd., Geofiz. Proekt., Mezhd. Geofiz. Komitet, Presid., Akad. Nauk, SSSR, 17-21.

gravity
Ball, Mahlon M., and Woodson R. Oglesby /1969.
A gravity survey of Florida Bay and the lower Keys. Trans. Gulf Coast Ass. geol. Socs, 19: 266.

gravity
Behrendt, John C., and CLETUS S. WOTORSON 1970.
Aeromagnetic and gravity investigations of the coastal area and continental shelf of Liberia, West Africa, and their relation to continental drift.
Bull. Geol. Soc. Am. 81 (12): 3563-3574.

gravity
Beyer, L.A., R.E. von Huene, T.H. McCulloh and J.R. Lovett, 1966.
Measuring gravity on the sea floor in deep water.
J. geophys. Res., 71(8):2091-2100.

Gravity
Bosshard, E. and D.J. Macfarlane, 1970.
Crustal structure of the western Canary Islands from seismic refraction and gravity data. J. geophys. Res., 75(26): 4901-4918.

gravity
Bott, M.H.P. 1965.
The deep structure of the northern Irish Sea, a problem of crustal dynamics. In: Submarine geology and geophysics, Colston Papers, W.F. Whittard and R. Bradshaw, editors, Butterworth's, London 17: 179-201.

gravity
Bott, H.P., 1964.
Gravity measurements in the north-east part of the Irish Sea.
Proc. Geol. Soc., London, No.1613:16-18.

gravity
Bott, Martin Harold Phillips, and David George Garrow Young, 1971.
Gravity measurements in the north Irish Sea.
Q. Jl geol. Soc. Lond., 126(4):413-434.

gravity
Bott, M.H.P., and A.P. Stacy, 1967.
Geophysical evidence on the origin of the Faroe Bank Channel. II. A gravity and magnetic profile.
Deep-Sea Research, 14(1):7-11.

gravity
Bott, M.H.P. and A.B. Watts 1970.
Deep sedimentary basins proved in the Shetland-Hebridean continental shelf and margin.
Nature, Lond. 225(5229): 265-268.

gravity
Bower, Donald R., 1966.
The determination of cross-coupling errors in the measurement of gravity at sea.
J. geophys. Res., 71(2):487-493.

gravity
Bower, Donald R., and Peter A. Watt, 1963
The second-order errors of sea-surface gravity measurements.
J. Geophys. Res., 68(1):245-250.

gravity
Bower, D.R., and P.A. Watt, 1962
Some theoretical comments on the measurements of gravity at sea. (Abstract).
J. Geophys. Res., 67(9):3544.

gravity
Bowin, Carl, 1968.
Some aspects of the gravity field and tectonics of the northern Caribbean region. Trans. Fifth Carib. Geol. Conf., St. Thomas, V.I., 1-5 July, 1968, Peter H. Mattson, editor. Geol. Bull. Queens Coll., Flushing, 5: 1-6.

gravity
Bowin, Carl, 1966.
Gravity over trenches and rifts.
In: Continental margins and island arcs, W.H. Poole, editor, Geol. Surv. Pap. Can., 66-15:430-439.

gravity
Bromery, R.W., and Andrew Griscom, 1964.
A gravity survey in the Mayaguez area of southwest Puerto Rico.
Nat. Acad. Sci.-Nat. Res. Counc., Publ., No. 61-74.

gravity
Browne, B.C., 1954.
Gravity measurements and oceanic structure.
Proc. R. Soc., London, A., 222(1150):398-400.

gravity
Browne, B.C., and R.I.B., Cooper, 1952.
Gravity measurements in the English Channel.
Proc. R. Soc., London, B, 139:426-447, 7 maps.

gravity
Browne, B.D., and B.D. Loncarevic, 1962
Gravity measurement in surface ships. (Abstract).
J. Geophys. Res., 67(9):3545.

gravity
Bunce, Elizabeth T., Joseph D. Phillips, Richard L. Chase, and Carl O. Bowin, 1970.
II. Regional observations. 11. The Lesser Antilles Arc and the eastern margin of the Caribbean Sea. In: The sea: ideas and observations on progress in the study of the seas, Arthur E. Maxwell, editor, Wiley-Interscience 4(2): 359-385.

gravity
Burakovskiy, V.E., 1966.
Some results of the deep structure studies of the earth's crust of the Black Sea depression and of adjoining territories. (In Russian; English abstract).
Stroenie Chermomorskoi Vladini, Rez. Issled. Mezhd. Geofiz. Proekt., Mezhd. Geodiz. Komitet, Presid., Akad. Nauk, SSSR, 22-26.

gravity
Bush, Sam A. and Patricia A. Bush, 1969.
Isostatic gravity map of the eastern Caribbean region. Trans. Gulf Coast Ass. geol. Socs, 19 281-285.

C

gravity
Collette, B.J., J.A. Schouten, K.W. Rutten, D.J. Doornbos and W.H. Staverman, 1971.
Geophysical investigations off the Surinam Coast. Hydrogr. Newsletter, R. Netherlands Navy, Spec. Publ. 6: 17-24.

gravity
Cooper, R.I.B., J.C. Harrison, and P.L. Willmore, 1952.
Gravity measurements in the eastern Mediterranean
Phil. Trans. R. Soc., London, A, 244(889):533-559 3 maps.

gravity
Collette, B.J., R.A. Lagaay, A.R. Ritsema and J.A. Schouten 1967.
Seismic investigations in the North Sea.
Geophys. J., R. astron. Soc. 12(4): 363-373.

gravity
Cordell, Lindrith, and Patrick T. Taylor 1971.
Investigation of magnetization and density of a North Atlantic seamount using Poisson's Theorem.
Geophysics 36(5): 919-937.

gravity
Coster, H. P., 1945.
The gravity field of the western and central Mediterranean.
Proefschrift. Bij J. B. Wolters' Uitgevers-maatschappij, n. v., Groningen-Batavia, 1945: 57 pp.

gravity
Coster, H.P., 1945.
The gravity field of the western and central Mediterranean. Thesis, Utrecht, Groningen.

gravity
Crary, A.P., R.D. Cotell, and J. Oliver, 1952.
Geophysical studies in the Beaufort Sea, 1951.
Trans. Amer. Geophys. Union 33(2):211-216, 3 textfigs.

gravity
Crary, A. P., and Norman Goldstein, 1959.
Geophysical studies in the Arctic Ocean.
Sci. Studies at Fletcher's Ice Island, T-3, Vol., 1, Air Force Cambridge Research Center, Geophys. Res. Pap. 63:7-30.

D

gravity
Dash, B.P., and E. Bosshard, 1969.
Seismic and gravity investigations around the western Canary islands.
Earth Planet. Sci. Letters, 7(2): 169-177.

gravity
Davey, F.J., 1970.
Bouguer anomaly map of the north Celtic Sea and entrance to the Bristol Channel.
Geophys. J., R. astron. Soc. 22(3):277-282.

gravity
Day, G.A., and C.A. Williams 1970.
Gravity compilation in the N.E. Atlantic and interpretation of gravity in the Celtic Sea.
Earth Planet. Sci. Letters, 8(3):205-213.

gravity
Dehlinger, Peter, 1964.
Reliability at sea of gimbal-suspended gravity meters with 0.7 critically damped accelerometers
J. Geophys. Res., 69(24):5383-5394.

gravity
*Dehlinger, P., R.W. Conch, and M. Gemperle, 1968.
Continental and oceanic structure from the Oregon coast westward across the Juan de Fuca Ridge.
Can. J. Earth Sci., 5(4-2):1079-1090.

gravity
Dehlinger, Peter, R.W. Couch and Michael Gemperle, 1967.
Gravity and structure of the eastern part of the Mendocino Escarpment.
J. geophys. Res., 72(4):1233-1247.

gravity

Dehlinger, Peter, R.W. Couch, D.A. McManus and Michael Gemperle, 1970.
II. Regional observations. 4. Northeast Pacific Structure. In: The sea: ideas and observations on progress in the study of the seas, Arthur E. Maxwell, editor, Wiley-Interscience 4(2): 133-189.

gravity

Drake, C.L., J. Heirtzler and J. Hirshman, 1963
Magnetic anomalies off eastern North America
J. Geophys. Res., 68(18):5259-5275.

gravity

Duran, S., Luis G., 1964.
Interpretación geofísica de la plataforma continental del Caribe.
Caldasia, Bol., Inst. Ciencias Nat., Univ. Nac., Colombia, Bogota, 9(42):137-150.

E

gravity

Ewing, J.I., N.T. Edgar and J.W. Antoine, 1970.
II. Regional observations. 10. Structure of the Gulf of Mexico and Caribbean Sea. (Dec. 1968)
In: The sea: ideas and observations on progress in the study of the seas, Arthur E. Maxwell, editor, Wiley-Interscience 4(2): 321-358.

F

GRAVITY

Fleischer, Ulrich 1969
Investigations of rifts by shipboard magnetic and gravity surveys: Gulf of Aqaba - Red Sea and Reykjanes Ridge.
Dt. hydrogr. Z., 22(5): 205-208

gravity

Fleischer, U., 1964.
Schwerestörungen im östlichen Mittelmeer nach Messungen mit einem Askania-Seegravimeter.
Deutsche Hydrogr. Zeits., 17(4):153-164.

gravity

Fleischer, U., 1963.
Surface-ship gravity measurements, in the North Sea.
Geophysical Prospecting, 11(4):535-549.

Also in: Ozeanographie, 1964, Deutsches Hydrogr. Inst.

gravity

Frolov, A.I., 1963.
The problem of taking into account disturbing accelerations when determining gravity at sea aboard surface vessels. (In Russian).
Soobshcheniya Gosudarst. Astron. Inst. imeni Shternberga, Moscow, (128):3-7.

Abstracted in:
Soviet Bloc Res., Geophys., Astron., & Space, No. 93:40.

G

gravity, Eötvös correction

Gantar, C., 1962.
Tables of Eötvo"s correction to continuous sea gravity measurements for Mediterranean sea latitudes.
Boll. Geofis. Teor. Appl., 4(13):37-46.

gravity

Gantar, C., and C. Morelli, 1963
Provisional adjustment 1962 of the southern part of the E.C.L. in the Italian standard for comparison purposes.
Boll. Geofis. Teorica ed Applic., 5(19): 260-282.

gravity

Gantar, C., and C. Morelli, 1962
Misure di gravità sul fondo dei mari intorno alla Sardegna con gravimetro telecomandato.
Boll. Geofis. Teorica ed Applicata, 4(15):286-298.

Pubbl. No. 143, Osservatorio Geofisico Sperimentale di Trieste.

gravity

Gantar, C., C. Morelli e M. Pisani, 1968.
Information report on surface gravity and magnetic measurements with the ship Bannock in the Mediterranean Sea, 1965-1968.5.
Boll. Geofis. teor. appl., 10(38):134-157.

gravity

Gantar, C., C. Morelli, and M. Pisani, 1962
Experimental study of the Graf-Askania Gss2, No. 13, sea gravity meter. (Abstract).
J. Geophys. Res., 67(9):3560.

gravity

Gantar, C., C. Morelli, A.G. Segre, and L. Zampieri, 1961.
Studio gravimetrico e considerazioni geologiche sull'isola de Pantelleria.
Boll. Geofis. Teorica ed Applicata, Trieste, 3(12):267-272.

gravity

Girdler, R.W., and J.C. Harrison, 1957.
Submarine gravity measurements in the Atlantic Ocean, Indian Ocean, Red Sea and Mediterranean Sea. Proc. R. Soc., London, A, 239(1217):202-213.

gravity

Gladun, V.A., 1962.
Marine gravimetric measurements during the seventh voyage of the "Ob". (In Russian).
Inform. Biull., Sovetsk. Antarkt. Exped., 41:39-42.

Translation:
Scripta Tecnica for AGU, 4(6):336-337. 1964

gravity

Glennie, E. A., 1936.
A report on the values of gravity in the Maldive and Laccadive Islands. John Murray Exped., 1933-34, Sci. Repts. 1(4):95-107, 1 pl., 4 charts.

gravity

Goodacre, A.K., 1964.
A shipborne gravimeter testing range near Halifax, Nova Scotia.
J. Geophys. Res., 69(24):5373-5381.

gravity

Goodacre, A.K., and E. Nyland, 1966.
Underwater gravity measurements in the Gulf of St. Lawrence.
In: Continental drift, G.D. Garland, editor, Spec. Publs. R. Soc. Can., No. 9:114-128.

gravity

Grant, A.C. and K.S. Manchester 1970
Geophysical investigations in the Ungava Bay - Hudson Strait region of northern Canada.
Can. J. Earth Sci. 7(4): 1062-1076.

gravity

Gray, F. and A.P. Stacey, 1970.
Gravity and magnetic interpretation of Porcupine Bank and Porcupine Bight. Deep-Sea Res., 17(3): 467-475.

gravity(data)

Great Britain, Admiralty Hydrographic Department, 1963.
Gravity measurements from H.M.S. "Shackleton", August 1961.
Admiralty Mar. Sci. Publ., (3):7 pp. (multilithed).

gravity (data)

Great Britain, Intergovernmental Oceanographic Commission, 1963.
Bathymetric, magnetic and gravity investigations, H.M.S. "Owen", 1961-1962.
Admiralty Mar. Sci. Publ., (4)(Pt. 2):unnumbered pp. (multilithed).

gravity

Griscom, Andrew, 1968.
Tectonic implications of gravity and aeromagnetic surveys on the north and south coasts of Puerto Rico. (Abstract only). Trans. Fifth Carib. Geol Conf., St. Thomas, V.I., 1-5 July 1968, Peter H. Mattson, editor. Geol. Bull. Queens Coll., Flushing, 5: 23-24.

gravity

Grover, J.C., 1966.
Gravity surveys in the British Solomon islands - a narrative.
In: Continental margins and island arcs, W.H. Poole, editor, Geol. Surv. Pap., Can., 66-15:257-277.

H

gravity

Harrison, J.C., 1963
A note on the paper 'Earth-tide observations made during the International Geophysical Year'.
J. Geophys. Res., 68(5):1517-1518.

gravity

Harrison, J.C., 1960
Gravity at sea. Trans. Amer. Geophys. Un. 41 (2): 271-273.

gravity

Harrison, J. C., 1959.
Tests of the LaCoste-Romberg surface ship gravity meter I.
J. Geophys. Res., 64(11):1875-1882.

gravity

Harrison, J.C., G.L. Brown and F.N. Spiess, 1957.
Gravity measurements in the northeastern Pacific Ocean. Trans. Amer. Geophys. Union, 38(6):835-840

gravity

*Harrison, J.C., and Lucien J.B. LaCoste, 1968.
Performance of LaCoste-Romberg shipboard gravity meters in 1963 and 1964.
J. geophys. Res., 73(6):2163-2174.

Gravity

Harrison, J. C., and P. Dehlinger, 1962.
Gravity measurements made in the Mediterranean Sea with a LaCoste-Romberg Surface-ship gravity meter during July and August 1961. (Abstract).
J. Geophys. Res., 67(4):1639-1640.

gravity

Harrison, J.C., M.D. Helfer, and A.T. Edgerton, 1962
Gravity measurements on the R.V. Argo, August 1960-April 1961. (Abstract).
J. Geophys. Res., 67(9):3563.

gravity
Harrison, J.C., N.F. Ness, I M. Longman, R.F.S Forbes, E.A. Kraut and L.B. Slichter, 1963
Earth-tide observations made during the International Geophysical Year.
J. Geophys. Res., 68(5):1497-1516.

Gravity
Harrison, J. C., and F. N. Spiess, 1962.
Tests of the La Coste-Romberg surface-ship gravity meter II. (Abstract).
J. Geophys. Res., 67(4):1640.

gravity
Harrison, J.C., and R.E. von Huene, 1965.
The surface-ship gravity meter as a tool for exploring the geological structure of continental shelves.
Ocean Sci. and Ocean Eng., Mar. Techn. Soc., Amer. Soc. Limnol. Oceanogr., 1:414-431.

gravity
Harrison, J.C., and C.H. Wilcox, 1962
Hughes research program in marine geophysics and oceanography.
Proc. 2nd Interindustrial Oceanogr. Symposium, Lockheed Aircraft Corp., 40-44.

gravity
Hayes, Dennis E., 1966.
A geophysical investigation of the Peru-Chile Trench.
Marine Geol., 4(5):309-351.

gravity
*Hayes, Dennis E., and William J. Ludwig, 1967.
The Manila Trench and West Luzon Trough. II. Gravity and magnetic measurements.
Deep-Sea Res., 14(5):545-560.

gravity
Heezen, B.C., E.T. Bunce, J.B. Hersey and M. Tharp, 1964.
Chain and Romanche Fracture zones.
Deep-Sea Research, 11(1):11-33.

gravity
Heezen, Bruce C., and A.S. Laughton, 1963.
14. Abyssal plains.
In: The Sea, M.N. Hill, Editor, Interscience Publishers, 3:312-364.

gravity
Helfer, Marvin D., Roland E. Von Huene and Michele Caputo, 1962
Gravity measurements in the Santa Barbara Channel with the U.S.S. Rexburg. (Abstract).
J. Geophys. Res., 67(9):3564.

gravity
Henderson, Garry C., 1963.
Preliminary study of the crustal structure across the Campeche escarpment from gravity data.
Geophysics, 28(5)(1):736-744.

gravity
Hill, M.N., 1957.
Recent geophysical exploration of the ocean floor
Physics and chemistry of the earth, Pergamon Press :129-163.

gravity
Hofman, B.J., 1952.
The gravity field of the west Mediterranean area
Geol. en Mijnb., n.s., 8:297-305, 4 figs., 1 map

gravity
Hunkins, Kenneth, 1966.
The Arctic continental shelf north of Alaska.
In: Continental margins and island arcs, W.H. Poole, editor, Geol. Surv. Pap., Can., 66-15:197-205.

gravity
Hunt, Trevor M., and Derek J. Woodward 1971.
Gravity and magnetic measurements in the South Taranaki Bight, New Zealand.
N.Z. Jl Geol. Geophys. 14(1):46-55.

gravity
Hurley, Robert J., 1966.
Geological studies of the West Indies.
In: Continental margins and island arcs, W.H. Poole, editor, Geol. Surv. Pap., Can., 66-15:139-150.

I

gravity
*Innes, M.J.S., A.K. Goodacre, J.R. Weber and R.K. McConnell, 1967.
Structural implications of the gravity field in Hudson Bay and vicinity.
Can. J. Earth Sci., 4(5):977-993.

J

K

gravity
Kanamori, Hiroo, 1962.
A review of gravity measurements at sea. (In Japanese). Zisin(2), 15(4):325-340.
Also in:
Geophysical Notes, Tokyo, 16(1). 1963. (Contrib. No. 3)

gravity
Keen, Charlotte, and B.D. Loncarevic, 1966.
Crustal structure on the eastern seaboard of Canada: studies on the continental margin.
Canadian J. Earth Sci., 3(1):65-76.

gravity
Keen, M.J., B.D. Loncarevic and G.N. Ewing, 1970.
II. Regional observations. 7. Continental Margin of Eastern Canada: Georges Bank to Kane Basin. In: The sea: ideas and observations on the progress in the study of the seas, Arthur E. Maxwell, editor, Wiley-Interscience 4(2): 251-291.

gravity
Kennett, P., 1966.
Reconnaissance gravity and magnetic surveys of part of the Larsen Ice Shelf and adjacent mainland.
Brit. Antarct. Surv., Bull. (8):49-62.

gravity
Kienle, Jürgen, 1971.
Gravity and magnetic measurements over Bowers Ridge and Shirshov Ridge, Bering Sea. J. geophys. Res. 76(29): 7138-7153.

gravity
Kogan, M.G. and E.A. Boyarsky, 1970.
Gravity measurements in the southeast Atlantic on R.V. Academician Kurchatov Cruise 3, March 1968 to July 1968. J. geophys. Res., 75(11): 2137-2140.

gravity
Koryakin, Ye. D., 1963.
The gravity field of the Atlantic Ocean and its relationship to the deep structure of the earth's crust. (In Russian). (Not seen).
Morsk. Gravimetrich. Issled., Moscow, (2):35-50.
Abstracted in:
Soviet Bloc Res., Geophys., Astron., Space, No. 93:38-39.

gravity
Kutschale, Henry, 1966
Arctic Ocean geophysical studies: The southern half of the Siberia Basin.
Geophysics, 31(4):683-710

gravity
Kuzivanov, V.A., 1958.
Determination of gravity over the sea with a gravimeter. (In Russian).
Izv., Akad. Nauk, SSSR, Ser. Geofiz., (5):55- (under checking)

L

gravity
Lacoste, L.J.B., and J.C. Harrison, 1961.
Some theoretical considerations in the measurements of gravity at sea.
Geophys. J., R. Astron. Soc., 5(2):89-103.

gravity
Leenhardt, Olivier, 1971
Über die Struktur des nördlichen Teiles des westlichen Mittelmeeres.
Meteor-Forsch.-Ergebnisse (C) 6:1-13

gravity
Le Pichon, X., and M. Talwani, 1964.
Gravity survey of a seamount near 35° N 46° W in the North Atlantic.
Marine Geol., 2(4):262-277.

gravity
Leyden, R., M. Ewing and E.S.W. Simpson 1971.
Geophysical reconnaissance on African shelf: 1. Cape Town to East London.
Bull. Am. Ass. Petrol. Geol. 55(5):651-657

gravity
*Lidz, L. (posthumous), Ball, M., and W. Charm, 1968.
Geophysical measurements bearing on the problem of the El Pilar Fault in the northern Venezuelan offshore.
Bull. mar. Sci., Miami, 18(3):545-560.

gravity
Loncarevic, B.D., 1964
Geophysical studies in the Indian Ocean.
Endeavor 23(88):43-47.

gravity
Lozinskaya, A.M., I.O. Tsimel'zon and V.V. Laskina, 1956.
Results of a regional survey in the Caspian Sea with underwater gravimeters.
Prikladnaya Geofizika, SSSR, (14):115-129.
LC and SLA, mi $2.40, ph $3.30.

M

gravity
MacDonald, G.J.F., 1964.
The deep structure of continents. Heat-flow and gravity observations and satellite data shed light on the origin of continents and oceans.
Science, 143(3609):921-929.

gravity

MacFarlane, D. J., and W. I. Ridley, 1969.
An interpretation of gravity data for Lanzarote, Canary islands.
Earth planet. Sci. Letts, 6(6): 431-436

gravity

Malahoff, Alexander, 1970.
Some possible mechanisms for gravity and thrust faults under oceanic trenches. J. geophys. Res., 75(11): 1992-200 .

gravity

Malahoff, Alexander and G.P. Woollard, 1970.
II. Regional observations. 3. Geophysical studies of the Hawaiian Ridge and Murray Fracture Zone. In: The sea: ideas and observations on progress in the study of the seas, Arthur E. Maxwell, editor, Wiley-Interscience 4(2): 73-131.

gravity

Malovitskiy, Ya. P., and Yu. P. Neprochnov, 1966
Comparison of seismic and gravimetric data on earth's crust structure of the Black Sea depression. (In Russian; English abstract).
Stroenie Chernomorskoi Vpadini, Rez. Issled. Mezhd. Geofiz. Proekt., Mezhd. Geofiz. Komitet, Presid., Akad. Nauk, SSSR, 5-16.

gravity

Matthews D.H. 1965.
The Owen Fracture Zone and the northern end of the Carlsberg Ridge.
Phil. Trans. R. Soc. (A) 259 (1099): 172-186.

gravity

Matthews, D.H., and D. Davies 1966.
Geophysical studies of the Seychelles Bank.
Phil. Trans. R. Soc. (A) 259 (1099): 227-239.

Mc

gravity

McCahan, A.L., 1962
Surface-ship gravity measurements and evaluation by the U.S. Navy Hydrographic Office. (Abstract).
J. Geophys. Res., 67(9):3576.

gravity

Menard, H.W., 1967.
Transitional types of crust under small ocean basins.
J. geophys. Res., 72(12):3061-3073.

gravity

Morelli, C., 1970.
Physiography, gravity and magnetism of the Tyrrhenian Sea. Boll. Geofis. teor. appl. 13(48): 275-309.

gravity

Morelli, C., 1966.
The geophysical situation in Italian waters (1966.0).
Int. hydrogr. Rev., 43(2):133-147.

gravity

Morelli, C., 1963
The first order and absolute world gravity nets.
Boll. Geofis. Teorica ed Applic., 5(19):195-216.

gravity

Morelli, C., 1961.
Gravity survey of the continental shelf around Italy (with an appendix on the magnetic survey of southern Italy.
Rapp. Proc. Verb., Réunions, Comm. Int. Expl. Sci., Mer Méditerranée, Monaco, 16(3):727-734.

GRAVITY

Morelli, C., 1955.
Rilievo gravimetrico nel mare Adriatico, 1954.
La Ricerca Scientifica 25(10):2846-2872.

Pubbl., Osserv. Geofis., Trieste, n.s., No. 65.

gravity

Morelli, C., 1955.
Rilievo gravimetrico regionale nelle fascia costiera Adriatica. Metano 9(7):429-443.

Pubbl., Osserv. Geofis., Trieste, n.s., No. 66.

gravity

Morelli, C., and F. Mosetti, 1955.
Rilievo gravimetrico e sismico sperimentale nel golfo di Trieste. Metano 9(9):1-13.

Pubbl., Osserv. Geofis., Trieste, n.s., No. 68.

N

gravity

Nakagawa, Ichiro, 1963
On tidal variation of gravity.
Geophysical Papers dedicated to Professor Kenzo Sassa, 327-339.

gravity

Nettleton, L.L., 1957.
Gravity survey over a Gulf coast continental shelf mound. Geophysics. 22(3):630-642.

O

gravity

Orlin, H., 1962
Coast and Geodetic Survey gravity observations in the North Pacific. (Abstract).
J. Geophys. Res., 67(9):3585.

gravity

Orlin, Hyman, B.G. Bassinger and C.H. Gray 1955.
Cape Charles - Wallops Island, Virginia, offshore gravity range.
J. Geophys. Res. 70(24): 6265-6267.

gravity

Ostenso, Ned A., 1968.
A gravity survey of the Chukchi Sea region and its bearing on westward extension of structures in northern Alaska.
Bull. geol. Soc. Am., 79(2): 241-254.

gravity

*Ostenso, N.A., 1968.
Geophysical studies in the Greenland Sea.
Bull. Geol. Soc. Am., 79(1):107-132.

P

gravity

Plouff, Donald, 1964.
Gravity measurements in the Beaufort Sea area.
Arctic, 17(3):151-161.

gravity

Ostenso, N.A., and G.P. Woollard, 1962
Magnetic and gravity studies of the Arctic Ocean Basin. (Abstract).
J. Geophys. Res., 67(9):3585.

gravity

Popov, E.I., 1962
The methods of processing the recordings of marine gravity meters and their accuracy.
Izvestiya Acad. Sci., USSR, Geoph. S. (Eng. Ed.), 2: 203-209.

gravity

Pytkowicz, Ricardo Marcos, 1963
Gravity and the properties of sea water.
Limnol. and Oceanogr., 8(2): 286-287.

Q

R

gravity

Robson, Richard M., and Kenneth L. Cook, 1965.
Some results from a study of the motion problems involved with gravity measurements at sea.
J. Geophys. Res., 70(2):512-514.

gravity

Rose, John C., and John C. Belshé, 1965.
Gravity and magnetic fields over the proposed Moho Hole site north of Maui.
Pacific Science, 19(3):374-380.

gravity

Rose, J.C., G.P. Woollard and Alexander Malahoff, 1968.
Marine gravity and magnetic studies of the Solomon islands.
In: The crust and upper mantle of the Pacific area. Geophys. Monogr. Am. geophys. Un. 12: 379-410.

gravity

Romanouk, V.A., 1958.
[Determination of the intensity of gravity at sea by the pendulum method. III] Izv. Akad. Nauk, SSSR, (1):54-64.

gravity

Romanouk, V.A., 1957.
[Determination of the force of gravity at sea by means of a pendulum.] Izv. Akad. Nauk, SSSR, Ser. Geofys. (4):458-470.

gravity

Ryan, William B.F., Daniel J. Stanley, J.B. Hersey, Davis A. Fahlquist and Thomas D. Allan, 1970.
II. Regional observations. 12. The tectonics and geology of the Mediterranean Sea (June 1969). In: The sea: ideas and observations on progress in the study of the seas, Arthur E. Maxwell, editor, Wiley-Interscience 4(2): 387-492.

S

gravity

Sagitov, M.U., and G.D. Marchuk, 1962
[The Eötvös correction for currents during gravity determinations at sea.]
Geofiz. Biull., Mezhduved. Geofiz. Komitet, Prezid., Akad. Nauk, SSSR, No. 11:40-42.

gravity

Saito, T., 1963
Statistical analysis of the pendulum observations.
Boll. Geofis. Teorica ed Applic., 5(19): 217-234.

gravity
Schoeffler, J., 1965.
Le "Gouf" de Capbreton, de l'Eocène inférieur à nos jours.
In: Submarine geology and geophysics, Colston Papers, W.F. Whittard and R. Bradshaw, editors, Butterworth's, London, 17:265-268.

gravity
Scrutton, R.A., A.P. Stacey and F. Gray 1971.
Evidence for the mode of formation of Porcupine Seabight.
Earth planet. Sci. Letts 11(2): 140-146

gravity
Segre, A.G., 1961.
2. Considerazioni geologiche, commento alle anomalie gravometriche e ai dati geomagnetici sull'Isola di Pantelleria.
Boll. Geofis. Theorica ed Applicata, Trieste, 3(12):273-287.

gravity
Shaver, Ralph, and Kenneth Hunkins, 1964.
Arctic Ocean geophysical studies: Chukchi Cap and Chukchi Abyssal Plain.
Deep-Sea Res., 11(6):905-916.

gravity
Shurbet, G.L., and J.L. Worzel, 1957.
Gravity observations at sea in USS Conger, Cruise II. Trans. Amer. Geophys. Union, 38(1):1-7.

gravity
Shurbet, G.L., J.L. Worzel and M. Ewing, 1956.
Gravity measurements in the Virgin Islands.
Bull., G.S.A., 67(11):1529-1536.

gravity
*Simpson, E.S.W., and A. du Plessis, 1968.
Bathymetric, magnetic, and gravity data from the continental margin of southwestern Africa.
Can. J. Earth Sci., 5(4-2):1119-1123.

gravity
*Sleep, Norman H., 1969.
Sensitivity of heat flow and gravity to the mechanism of sea-floor spreading.
J. geophys. Res., 74(2):542-549.

gravity
Solaini, L., G. Inghilleri and G. Togliatti, 1963
Results of some adjustments of pendulum and gravimeter data on the European calibration line.
Boll. Geofis. Teorica ed Applic., 5(19):235-250.

gravity
Srivastava, S.P., D.L. Barrett, C.E. Keen, K.S. Manchester, K.G. Shih, D.L. Tiffin, R.L. Chase, A.G. Tomlinson, E.E. Davis and C.R.B. Lister 1971.
Preliminary analysis of geophysical measurements north of Juan de Fuca Ridge.
Can. J. Earth Sci. 8(10): 1265-1281.

gravity
Stone, D.B., 1968.
Geophysics in the Bering Sea and surroundings: a review.
Tectonophysics 6(6):433-460

gravity
Sukhodol'skiy, V.V., 1959
[An instrument for recording accelerations and inclinations in determinations of the force of gravity at sea.] Izv. Akad. Nauk, SSSR, Ser. Geofiz., 11:1570-1578
OTS/ $0.50

gravity
Szabo, B., 1963.
World calibration standard, first order gravity net and absolute gravity system.
Boll. Geofis. Teorica ed Applic., 5(19):251-259.

T

gravity
Talwani, Manik, 1970.
1. General observations. 8. Gravity (Nov. 1968)
In: The sea: ideas and observations on progress in the study of the seas, Arthur E. Maxwell, editor, Wiley-Interscience 4(1): 251-297.

gravity
Talwani, M., 1964.
A review of marine geophysics.
Marine Geology, 2(1/2):29-80.

gravity
Talwani, M., 1962.
Gravity measurements on H.M.S. Acheron in South Atlantic and Indian Oceans.
Bull. Geol. Soc., Amer., 73(9):1171-1181.

gravity
Talwani, Manik, and Maurice Ewing, 1966.
A continuous gravity profile over the Sigsbee knolls.
J. geophys. Res., 71(18):4434-4438.

gravity
Talwani, Manik, Xavier Le Pichon and Maurice Ewing, 1965.
Crustal structure of the mid-ocean ridges. 2. Computed model from gravity and seismic refraction data.
J. Geophys. Res., 70(2):341-352.

gravity
Talwani, M., and J.L. Worzel, 1962
Continuous gravity profile across the Romanche Trench. (Abstract).
J. Geophys. Res., 67(9):3602.

gravity
Talwani, M., J.L. Worzel and M. Ewing, 1962
Gravity measurements and crustal structure in the southeast Pacific. (By title only).
J. Geophys. Res., 67(9):3602.

gravity
*Taylor, Patrick T., Isidore Zietz, and Leonard S. Dennis, 1968.
Geologic implications of aeromagnetic data for the eastern Continental Margin of the United States.
Geophysics, 33(5):755-780.

gravity
Thiruvathukal, John V., and Joseph W. Berg, Jr., 1966.
Gravity measurement program in Oregon.
Ore Bin, 28(4):69-75.
Also in: Coll. Repr., Dep. Oceanogr., Oregon. State Univ., 5.

gravity
Thyssen-Bornemisza, Stephen, 1966.
Correlating sea-surface and aerial gravity measurements.
Geophyics, 31(1):264-266.

gravity
Thyssen-Bornemisza, Stephen, 1966.
Possible application of the anomalous free-air vertical gradient to Marine exploration.
Geophysics, 31(1):260-263.

gravity
Tomoda, Yoshibumi, 1968.
Geophysical works of the Umitaka Maru in the Southern Sea 1964-65. J. Tokyo Univ. Fish., 9(1): 5-12.

gravity
Tomoda, Yoshibumi, Keijiro Ozawa and Jiro Segawa, 1968.
Measurement of gravity and magnetic field on board a cruising vessel.
Bull. Ocean Res. Inst. Univ. Tokyo, 3: 169pp.

gravity
Tomoda, Yoshibumi, and Jiro Segawa, 1967.
Measurement of gravity and total magnetic force in the sea near and around Japan (1966).
J. geod. Soc. Japan, 12:157-164
Also in: Coll. Repr. Ocean. Res. Inst. Tokyo 6(1967)

gravity
Tomoda, Yoshibumi, and Jiro Segawa, 1966.
Gravity measurement at sea in the regions: eastern part of the Indian Ocean, off Java and Sumatra, Strait of Malacca, South China Sea, Bashi Channel, East China Sea, off southern coast of Kyushu.
J. Tokyo Univ. Fish., (Spec.ed.)8(2):107-131.

gravity
Tsuboi, C., 1931
Gravity measurements on board a submarine boat.
Kagaku (Science), 1:292-294.

gravity
Tsuboi, C., Y. Tomoda and H. Kanamori, 1961.
Continuous measurements of gravity on board a moving surface ship.
Geophys. Notes, Tokyo Univ., 14(2)(19): reprint
Originally published in:
Proc. Japan Acad., 37(9):571-576.

U

gravity
Uchupi, Elazar, J.D. Milliman, Bruce P. Luyendyk, C.O. Bowin and K.O. Emery, 1971
Structure and origin of southeastern Bahamas
Bull. Am. Ass. Petrol. Geol. 55(5): 687-704.

Gravity
United States, Department of Commerce, Environmental Sciences Services Administration, 1965
International Indian Ocean Expedition, USC&GS Ship Pioneer - 1964.
Vol. 1. Cruise Narrative and scientific results 139 pp.
Vol. 2. Data report: oceanographic stations, BT observations, and bottom samples, 183 pp.

gravity

Ushakov, S.A., and G.E. Lasarev, 1959
[Some conclusions on the data of seismic and gravity measurements on the Little America-Berd profile.]
Info. Biull. Sovetsk. Antark. Exped., (9): 17-20.
Byrd?

gravity

Worzel, J. Lamar 1965.
Deep structure of coastal margins and mid-oceanic ridges.
In: Submarine geology and geophysics, Colston Papers, W.F. Whittard and R. Bradshaw, editors, Butterworths, London 17: 335-359.

gravity charts

Vyskocil, Vincenc, 1963.
Evaluation of the accuracy of gravimetric charts. (In Russian).
Studia Geophysica et Geodaetica, 7: 134-145.

Translation: USN Oceanogr. Off. TRANS. 361. (M. Slessers). 1967.

V

gravity

Van Bemmelen, R.W., 1952.
Gravity field and orogenesis in the West Mediterranean region. Geol. en Mijnb., n.s., 8: 306-313, 1 fig.

gravity

Worzel, J.L., and M. Ewing, 1950.
Gravity measurements at sea, 1947. Trans. Am. Geophys. Union 31(6):917-923, 9 textfigs.

gravity anomalies

Airinei, Stefan, 1969
Cartes d'anomalies gravimétriques-magnétiques ouvertes sur le littoral roumain de la mer Noire.
Studii si cercetari de geologie, geofizica si geografie 5 (1): 147-154
Abstract in: Rapp. P.-v. Reun. Commn int. Explor. scient. Mer Méditerr., 19(4): 659-660.

gravity

van Hees, G.L. Strang, 1971.
Gravity measurements on the continental shelf of Surinam. Hydrogr. Newsletter, R. Netherlands Navy, Spec. Publ. 6: 11-12.

gravity (at sea)

Worsley, B.H., 1952.
On the second-order correction terms to values of gravity measured at sea. Proc. Cambridge Phil. Soc., 48(4):718-732, 4 textfigs.

gravity anomalies

Artem'ev, M.E., 1963.
Gravitational anomalies and seismicity of western Europe and the Mediterranean. (In Russian).
Izv., Akad. Nauk, SSSR, Ser. Geofiz., (2):309-317.

Translation:
Amer. Geophys. Union, NAS-NRC, (2):189-194.

Gravity

Vila, Fernando, 1965.
Conocimiento actual de la plataforma continental Argentina.
Republica Argentina, Servicio Hidrograf. Naval Publ., H. 644:25 pp.

gravity

Worzel, J.L., and G.L. Shurbet, 1957.
Gravity observations at sea in USS Corsair.
Trans. A.G.U., 38(3):292-296.

gravity

Worzel, J.L., and M. Ewing, 1952.
Gravity measurements at sea, 1948 and 1949.
Trans. Amer. Geophys. Union, 33(3):453-460, 7 textfigs.

gravity anomaly

Balavadze, B.K., and P.Sh. Mindeli, 1965.
Earth's crustal structure in the Black Sea Basin based on geophysical data. (In Russian; English abstract).
Seism. Issled. Rezult. Issled. Mezhd. Geofiz. Proekt. ,Mezhd. Geofiz. Komitet, Presidiume Akad. Nauk, SSSR, No. 6:66-76.

Gravity

Vogt, P.R. and N.A. Ostenso, 1970.
Magnetic and gravity profiles across the Alpha cordillera and their relation to Arctic sea-floor spreading. J. geophys. Res., 75(26): 4925-4937.

gravity

Worzel, J. Lamar, and J.C. Harrison, 1963.
9. Gravity at sea.
In: The Sea, M.N. Hill, Editor, Interscience Publishers, 3:134-174.

gravity anomalies

Bott, M.H.P., 1971.
The mantle transition zone as possible source of global gravity anomalies.
Earth Planet. Sci. Letts. 11 (1): 28.

W

gravity

Wall, Robert E., Manik Talwani and J. Lamar Worzel, 1966.
Cross- coupling and off-leveling errors in gravity measurements at sea.
J. geophys. Res., 71(2):465-485.

gravity

Worzel, J.L., and G.L. Shurbet, 1955.
Gravity interpretations from standard oceanic and continental crustal sections.
Spec. Pap., G.S.A., 62:87-100, 5 textfigs.

gravity anomalies

Bott, M.H.P., C.W.A. Browitt and A.P. Stacey 1971.
The deep structure of the Iceland-Faroe Ridge.
Mar. geophys. Researches 1 (3): 328-351

gravity

Wold, Richard J. and Ned A. Ostenso, 1971.
Gravity and bathymetry survey of the Arctic and its geodetic implications. J. geophys. Res. 76(26): 6253-6264.

gravity

Worzel, J. Lamar, 1959.
Continuous gravity measurements on a surface ship with the Graf sea gravimeter.
J. Geophys. Res. 64(9):1299-1316.

gravity anomalies

Bott, M.H.P. and A.B. Watts, 1970.
Deep structure of the continental margin adjacent to the British Isles. In: The geology of the East Atlantic continental margin. 2. Europe, ICSU/SCOR Working Party 31 Symposium, Cambridge 1970, Rept. 70/14: 89-109.

gravity

Worzel, J.L , G.L. Shurbet, M. Ewing, 1955.
Gravity measurements at sea, 1950 and 1951.
Trans. Amer. Geophys. Union, 36(2):335-338, 4 textfigs.

gravity

Wold, R.J., T.L. Woodzick and N.A. Ostenso, 1970.
Structure of the Beaufort Sea continental margin.
Geophysics 35 (5): 849-861.

gravity

Worzel, J.L., G.L. Shurbet and M. Ewing, 1955.
Gravity measurements at sea, 1952 and 1953.
Trans. Amer. Geophys. Union, 36(2):326-334.

gravity anomalies

Browne, B.C., and R.I.B. Cooper, 1950.
The British submarine gravity surveys of 1938 and 1946. Phil. Trans. Roy. Soc., London, Ser. A, No. 847, 242:243-310, 6 charts, 13 textfigs.

gravity

Woodward, Derek J., and Trevor M. Hunt, 1971.
Crustal structure across the Tasman Sea.
N.Z. Jl Geol. Geophys. 14 (1): 39-45.

XYZ

gravity

Zetler, B.D., 1959
The effect of instrumental drift on the harmonic analysis of gravity at Washington, D.C. Boll. Geofis., Trieste, 2(5):234-237.

gravity anomalies

Bowie, W., 1935.
Significance of gravity anomalies at stations in the West Indies. Bull. G.S.A. 46:869-878, 1 fig.

gravity

Woollard, G.P., 1956.
Standardization of the world's gravity data.
Trans. Amer. Geophys. Union, 37(6):669-675.

gravity, accuracy of

Loncarevic, B.D., 1965.
Accuracy of sea gravity surveys: comparison of shipboard and submarine gravity values.
Nature, 205(4966):32-34.

gravity anomalies

Bunce, Elizabeth T., 1966.
The Puerto Rico Trench.
In: Continental margins and island arcs, W.H. Poole, editor, Geol. Surv. Pap., Can., 66-15:165-175.

gravity

Woollard, G.P., and J. Monges Caldera, 1956.
Gravedad, geologia regional y estructura cortical en Mexico. Anales Inst. Geofis., Univ. Nac. Aut., Mexico, 2:60-112.

Gravity anomalies
Bunce, Elizabeth T., M.G. Langseth, R.L. Chase and M. Ewing, 1967.
Structure of the western Somali Basin.
J. geophys. Res., 72(10):2547-2555.

gravity anomalies
Dehlinger, P., and Chiburis, E.F., 1972.
Gravity at sea.
Mar. Geol. 12(1):1-41

gravity anomaly
Dehlinger, Peter, R.W. Couch and Michael Gemperle, 1967.
Gravity and structure of the eastern part of the Mendocino Escarpment.
J. geophys. Res., 72(4):1233-1247.

gravity anomalies
Eldholm, Olav and John Ewing, 1971.
Marine geophysical survey in the southwestern Barents Sea. J. geophys. Res. 76(17): 3832-3841.

gravity anomalies
Ewing, M., and J.L. Worzel, 1954.
Gravity anomalies and stucture of the West Indies. Part 1. Bull. G.S.A. 65(2):165-174, 2 textfigs.

gravity anomalies
Fleischer, Ulrich 1971.
Gravity surveys over the Reykjanes Ridge and between Iceland and the Faeroe Islands.
Mar. geophys. Researches 1 (3): 314-327.

gravity anomalies
Gainanov, A.G., Yu. V. Tulina, I.P. Kosminskaya, S.M. Zverev and P.S. Veitsman, 1965.
A complex interpretation of the materials of geophysical observations carried out in the Sea of Okhotsk and Kurilo-Kamchatskaya zone of the Pacific Ocean. (In Russian; English abstract).
Seism. Issled., Rezult. Issled. Mezhd. Geofiz. Proekt. Mezhd. Geofiz. Komitet, Presidiume Akad. Nauk, SSSR, No. 6:60-65.

gravity anomalies
Harrison, J.C., 1955.
An interpretation of gravity anomalies in the eastern Mediterranean.
Phil. Trans. R. Soc., London, A, 248(947):283-325.

gravity anomalies
Harrison, J. C and W. C Brisbin, 1959
Gravity anomalies off the west coast of North America. 1: Seamount Jasper
Bull. Geol. Soc. Amer 70(7): 929-933.

gravity anomalies
Harrison, J.C., R.I.B. Cooper and R.W. Hey, 1954
An interpretation of the gravity anomalies in the eastern Mediterranean.
Internat. Geol. Congr., Algier, 19th Ser., C.R. Sec. 9(9):39-42.
Reviewed in Geophys. Abstr. 162:162-25

gravity anomalies
Harrison, J.C., and S.P. Mathur, 1964.
Gravity anomalies in Gulf of California.
In: Marine geology of the Gulf of California, a symposium, Amer. Assoc. Petr. Geol., Memoir, T. Van Andel and G.G. Shor, Jr., editors, 3:76-89.

gravity anomalies
Harrison, J.C., R.E. Von Huene and C.E. Corbato, 1966.
Bouguer gravity anomalies and magnetic anomalies off the coast of southern California.
J. Geophys. Res., 71(20):4921-4941.

gravity anomalies
Haworth, R.T., 1971.
A geophysical profile in the south-east Pacific.
Earth planet. Sci. Letts 11(2):83-89.

gravity anomalies
Hayes, Dennis E., 1966.
The Peru-Chile Trench. Summary only).
In: Continental margins and island arcs, W.H. Poole, editor, Geol. Surv. Pap., Can., 66-15:238-242.

gravity anomalies
Hunt, John M., Earl E. Hays, Egon T. Degens and David A. Ross, 1967.
Red Sea: detailed survey of hot-brine areas.
Science, 156(3774):514-516.

gravity anomalies
India, Survey of India, 1953.
Pt. III. Geodetic work. Tech. Rept. 1952: 90 pp., figs.

gravity anomaly
Japan, Science Council, National Committee for IIOE, 1966.
General report of the participation of Japan in the International Indian Ocean Expedition.
Rec. Oceanogr. Wks., Japan, n.s., 8(2):133 pp.

gravity anomalies
Komarov, A.G., 1965.
Oceanic ridges and rift structure; geological nature of the magnetic and gravity anomalies over the rift valley. (In Russian).
Priroda, No. 7:95-98.
Translation:

gravity anomalies
Kumagai, N., 1953.
Relation of the isostatic anomalies of gravity to the geological and seismological facts in north-east Honsyu and over the central part of the Nippon Trench, Japan.
Proc. Seventh Pacific Sci. Congr., 2(Geol.):507-529, 5 textfigs.

gravity anomalies
LaGaay, R.A., 1969.
Geophysical investigations of the Netherlands Leeward Antilles.
Verh. K. ned. Akad. Wet. (1) 25(2):1-86.

gravity anomalies
Lattimore, R.K., L. Austin Weeks and L.W. Mordock 1971.
Marine geophysical reconnaissance of continental margin north of Paria Peninsula, Venezuela.
Bull. Am. Ass. Petrol. Geol. 55(10): 1719-1729.

gravity anomalies
Loncarevic, B. D., C. S. Mason and D. H. Matthews, 1966.
Mid-Atlantic Ridge near 45° north. 1. The median valley.
Canadian J. Earth Sci., 3(3):327-349.

gravity anomalies
*McKenzie, Dan P., 1967.
Some remarks on heat flow and gravity anomalies.
J. geophys. Res., 72(24):6261-6273.

gravity anomalies
Morelli, C., M.T. Carrozzo, P. Ceccherini, I. Finetti, C. Gantar, M. Pisani, P. Schmidt di Friedberg, 1969.
Regional geophysical study of the Adriatic Sea.
Boll. Geofis. teor. appl. 11 (41-42):3-48

gravity anomalies
Morgan, W. Jason, 1966.
A reply.
J. geophys. Res., 71(14):3607-3607.

Morgan, W. Jason, 1965. gravity anomalies
Gravity anomalies and convection currents.
2. The Puerto Rico Trench and the Mid-Atla Rise.
J. geophys. Res., 70(24):6189-6204.

Morgan, W. Jason, 1965. gravity anomalies
Gravity anomalies and convection currents.
1. A sphere and cylinder sinking beneath the surface of a viscous fluid.
J. geophys. Res., 70(24):6175-6187.

gravity anomalies
Peter, George, 1966.
Preliminary results of a systematic geophysical survey south of the Alaska peninsula.
In: Continental margins and island arcs, W.H. Poole, editor, Geol. Surv. Pap., Can., 66-15:223-237.

gravity anomalies
Peter, G., L.A. Weeks, and R.E. Burns, 1965.
A reconnaissance geophysical survey in the Andaman Sea and across the Andaman-Nicobar Island Arc.
In: Int. Indian Ocean Exped., USC&GS Pioneer, 1964, U.S. Dept. Commerce, Environmental Sci. Services Administration, 1:91-107.

gravity anomalies
Peter, G., L.A. Weeks, and R.E. Burns, 1966.
A reconnaissance geophysical survey in the Andaman Sea and across the Andaman-Nicobar Arc.
J. geophys. Res., 71(2):495-509.

gravity-anomalies
Pfannenstiel, Max, 1960
Erläuterungen zu den bathymetrischen Karten des östlichen Mittelmeeres.
Bull. Inst. Oceanogr., Monaco, No. 1192: 60pp 9 charts.
Summary in French, German and English

gravity anomalies
Phillips, J.D., J. Woodside and C.O. Bowin, 1969.
Magnetic and gravity anomalies in the central Red Sea.
In: Hot brines and Recent heavy metal deposits in the Red Sea, E.T. Degens and D.A. Ross, editors, Springer-Verlag, New York, Inc., 98-113.

gravity anomalies
Rabinowitz Philip D., 1971.
Gravity anomalies across the East African Continental Margin. J. geophys. Res. 76(29): 7107-7117.

gravity anomalies
Rabinowitz, Philip D., and William B.F. Ryan, 1970.
Gravity anomalies and crustal shortening in the eastern Mediterranean.
Tectonophysics, 10(5/6): 585-608.

gravity anomalies
*Rikitake, T., S. Miyamura, I. Tsubokawa, S. Murauchi, S. Uyeda, H. Kuno and M. Gorai, 1968.
Gepphysical and geological data in and around the Japan Arc.
Can. J. earth Sci., 5(4-2): 1101-1118.

gravity anomalies
Shurbet, G.L., and J.L. Worzel, 1957.
Gravity anomalies and structures of the West Indies. Bull. G.S.A., 68(2):263-266.
III.

gravity anomalies
Shurbet, G.L., and J.L. Worzel, 1955.
Gravity anomalies associated with seamounts. Bull. GSA 66(6):777-782.

gravity anomalies
Solomon, Sean and Shawn Biehler, 1969.
Crystal structure from gravity anomalies in the southwest Pacific. J. geophys. Res., 74(27): 6696-6701.

gravity anomalies
Stroev P.A. 1971.
Gravity force anomalies of the Sea of Japan aquatory. (In Russian)
Dokl. Akad. Nauk SSSR 198(4): 818-821.

gravity anomalies
Talwani, Manik, 1966.
Gravity anomaly belts in the Caribbean. (Summary only).
In: Continental margins and island arcs, W.H. Poole, editor, Geol. Surv. Pap., Can., 66-15:177.

gravity anomalies
Talwani, Manik, Xavier Le Pichon, Maurice Ewing, George H. Sutton and J. Lamar Worzel, 1966.
Comments on paper by W. Jason Morgan, 'Gravity anomalies and convection currents. 2. The Puerto Rico Trench and the Mid-Atlantic Rise'.
J. Geophys. Res., 71(14):3602-3606.

gravity anomalies
Talwani, Manik, J. Lamar Worzel and Maurice Ewing, 1961
Gravity anomalies and crustal section across the Tonga Trench.
J. Geophys. Res., 66(4): 1265-1278.

gravity anomalies
Vajk, Raoul, 1964.
Correction of gravity anomalies at sea for submarine topography.
J. Geophys. Res., 69(18):3837-3844.

gravity anomalies
Vening-Meinesz, F.A., 1954.
Indonesian archipelago: a geophysical study.
Bull. G.S.A. 65(2):143-164, 4 textfigs., 2 fold-ins.

gravity anomalies
Von Arx, William S., 1966.
Level-surface profiles across the Puerto Rico Trench.
Science, 154(3757):1651-1654.

gravity anomalies
Von Huene, Roland, and J.B. Ridlon, 1966.
Offshore gravity anomalies in the Santa Barbara Channel, California.
J. geophys. Res., 71(2):457-463.

gravity anomalies
Weeks, L.A., R.N. Harbison and G. Peter, 1965.
The island arc system in the Andaman Sea.
In: Int. Indian Ocean, Exped., USC&GS Pioneer, 1964, U.S. Dept. Commerce, Environmental Sci. Services Administration, 1:109-118.

gravity anomalies
Wong, H.K., E.F.K. Zarudzki, J.D. Phillips and G.K.F. Giermann, 1971.
Some geophysical profiles in the eastern Mediterranean.
Bull. Geol. Soc. Am., 82(1): 91-100

gravity anomalies
Woollard, G.P., 1966.
Crust and mantle relations in the Hawaiian area.
In: Continental margins and island arcs, W.H. Poole, editor, Geol. Surv. Pap., Can., 66-15:294-310.

gravity anomalies
Woollard, G. P., 1949.
Gravity anomalies and the nature of the earth's crust. Trans. Am. Geophys. Union 30(2):189-201, 6 textfigs.

gravity anomalies
Yungul, S.H., and Peter Dehlinger, 1962
Preliminary free-air gravity anomaly map of the Gulf of Mexico from surface ship measurements and its tectonic implications.
J. Geophys. Res., 67(12):4721-4728.

Gravity, data only

gravity anomalies (data only)
France, Service Hydrographique de la Marine, 1968.
Anomalies de la pesanteur en mer de Norvège résultats provisoires de mesures effectuées à bord du Paul Goffeny en 1965-1967. Marée à Tuamotu. Hauteurs du niveau moyen à Brest.
Cah. océanogr., 20(10):921-933.

gravity (data only)
Great Britain, Ministry of Defence, Hydrographic Department, 1966.
Bathymetric, magnetic and gravity investigations. H.M.S. Owen, 1962-1963.
Admiralty Marine Sci. Publ., (9)(H.D. 567): 19 pp. numerous charts. (in 2 parts).

gravity (data only)
United States U.S. Department of Commerce Environmental Science Services Administration, 1969.
Bathymetry, geomagnetic and gravity data, USC&GS Ship Pioneer - 1964.
Int. Indian Ocean Exped., 3: 250 pp.

Gravity, effect of

gravity, effect of
Braemer, Wolfgang, and Helga Braemer, 1958.
Orientation of fish to gravity.
Limnol. Oceanogr., 4(3):362-372.

gravity, effect of
Braunss, G., 1965.
Some remarks on gravitational interaction and gravitational waves.
Zeits. f. Naturforschung, 200(4):495-497.

gravity, effect of
Maximov, I.V. 1967.
On the gravitational nature of main peculiarities of the general circulation of the oceans and of the atmosphere in high latitudes of the earth. (In Russian; English abstract).
Trudy polyar. nauchno-issled. Inst. morsk. ryb. Khoz. Okeanogr. (PINRO) 20: 336-342.

gravity, effect of
Nishimura, E., T. Ichinohe and I. Nakagawa, 1957.
A consideration of earth tidal change of gravity
Tellus, 9(1):118-126.

gravity, effect of
Pytkowicz, Ricardo Marcos, 1962
Effect of gravity on distribution of salts in sea water.
Limnol. and Oceanogr., 7(3):434-435.

gravity, measurements

gravity measurements
Browne, B. C., 1950.
Submarine gravity survey. A British Expedition to the eastern Mediterranean. Nature 165(4202):750.

gravity measurements
Gajnanov, A.G., 1955.
[Pendulum gravity measurements in the Sea of Okhotsk and the north-western Pacific.]
Trudy Inst. Oceanol., 12:145-154.

gravity measurements
Graf, A., 1957.
Über die bisherigen Erfahrungen und Messergebnisse mit dem Seegravimeter. Zeitsch. f. Geophys. 23(1):4-25.

gravity measurements
Graf, A., 1954.
Das Problem der Scheremessung auf See mit Gravimetern. Zeits. Geophysik, Deutschen Geophys. Gesellsch. 20(4):208-212.

gravity, measurement of
*La Coste, Lucien J.B., 1967.
Measurement of gravity at sea and in the air.
Reviews Geophys., 5(4):447-526.

gravity measurements
Shurbet, G.L., M.Ewing, 1956.
Gravity reconnaissance survey of Puerto Rico.
Bull., G.S.A., 67(4):511-534, 4 figs., 1 pl.

gravity measurements
Vening Meinesz, F.A., 1953.
The second order corrections for pendulum observations at sea.
Proc. K. Nederl. Akad. Wetens., B, 56(3):218-227, 1 textfig.

gravity sliding

gravity, sliding
Chase, R.L., and J.B. Hersey 1968.
Geology of the north slope of the Puerto Rico Trench.
Deep-Sea Res. 15(3):297-317.

gravity slide
Daviess, S.N., 1971.
Barbados: a major submarine gravity slide.
Bull. geol. Soc. Am. 82(9): 2593-2602.

gravity, tidal

gravity, tidal
France, Service Hydrographique de la Marine and Compafnie générale de Géophysique, 1966.
Tidal gravity correctíons for 1967.
geophys. Prospecting, 14(Suppl. 2):1-53.

gravity stations

gravity stations
Robertson, E.I., 1965.
Gravity base stations in the south-west Pacific.
New Zealand J. Geol. Geophys., 8(3):424-435.

graywackes

graywacke
Hawkins, James W., Jr., and John T. Whetten, 1969
Graywacke matrix minerals: hydrothermal reactions with Columbia River sediments.
Science, 166(3907): 868-870.

graywackes, shallow
Kuenen, Ph. H., 1964
Deep-sea sands and ancient turbidites.
In: Turbidites, A.H. Bouma and A. Brouwer, Editors, Developments in Sedimentology, Elsevier Publishing Co., 3:3-33.

greywackes
Pettijohn, F. J., 1950.
Turbidity currents and greywackes -- a discussion
J. Geol. 58:169-171.

greywacke
Whetten, John T., 1966.
Sediments from the lower Columbia River and origin of graywacke.
Science, 152(3725):1057-1058.

Grazing

grazing
Allen, W.E., 1941
Ocean pasturage in California waters.
Sci. Monthly 52:261-264.

grazing
Anraku, Masateru, 1964.
Some technical problems encountered in quantitative studies of grazing and predation by marine planktonic copepods.
Jour. Oceanogr. Soc. Japan, 20(5):221-132.

grazing
Bainbridge, V., 1960
The plankton of inshore waters off Freetown, Sierra Leone. Colonial Off., Fish. Publ., London, No. 13: 48 pp.

grazing (fish), effect of
Bakus, Gerald J., 1964.
The effects of fish-grazing on invertebrate evolution in shallow tropical waters.
Allan Hancock Found. Publ. Occ. Pap. No. 27:1-29

grazing, effect of
Beklemishev, C.W., 1955.
The influence of the consumption of diatoms by copepods on the numerical fluctuations of the former, based on Far Eastern seas. (In Russian).
Trudy Inst. Okeanol., Akad. Nauk, SSSR, 13:77-82

Transl. cited in:
Techn. Transl., 10(2):182 (1963).
OTS-SLA 62-32834

grazing
Boney, A. D., 1965.
Aspects of the biology of the seaweeds of economic importance.
In: Advances in Marine Biology, Sir Frederick S. Russell, editor, Academic Press, 3:105-253.

grazing
Bursa, Adam S., 1961
The annual oceanographic cycle at Igloolik in the Canadian Arctic. II. The phytoplankton
J. Fish. Res. Bd., Canada, 18(4):563-615.

grazing
Castenholz, Richard W., 1961
The effect of grazing on marine littoral diatom populations.
Ecology, 42(4):783-794.

grazing
*Cushing D.H., 1968.
Grazing by herbivorous copepods in the sea.
J. Cons. perm.int.Explor.Mer, 32(1):70-82.

grazing, effect of
Cushing, D.H., 1959.
On the nature of production in the sea. Ser. 2
Ministry Agric. Fish. Food, Fish. Invest. 22(6):1-40.

grazing
Cushing, D.H., and T. Vucetic, 1963
Studies on a Calanus patch. III. The quantity of food eaten by Calanus finmarchicus
J. Mar. Biol. Assoc., U.K., 43(2):349-371.

grazing
Digby, P.S.B., 1953.
Plankton production in Scoresby Sound, East Greenland. J. Animal Ecology 22(2):289-322, 5 textfigs.

grazing
Ganapati, P.N., and D. Venkata Rama Sarma, 1958.
Hydrography in relation to the production of plankton off Waltair coast.
Andhra Univ. Mem., Oceanogr., 2:168-192.

grazing
Gauld, D.T., 1953.
Diurnal variations in the grazing of planktonic copepods. J.M.B.A. 31(3):461-474, 3 textfigs.

grazing
Gauld, D.T., 1951.
The grazing rate of planktonic copepods. J.M.B.A. 29(3):695-706, 4 textfigs.

grazing
Grall, Jean-René, et Guy Jacques, 1964.
Etude dynamique et variations saisonnières du plancton de la region de Roscoff. B. Phytoplancton.
Cahiers Biol. mar., Roscoff, 5:432-455.

grazing
Halldal, P., 1953.
Phytoplankton investigations from Weather Ship M in the Norwegian Sea, 1948-49 (including observations during the "Armauer Hansen" cruise, July 1949). Hvalrådets Skrifter No. 38:91 pp., 91 pp., 20 tables, 21 textfigs.

Grazing
Hargrave, Barry T., and Glen H. Geen, 1970
Effect of copepod grazing on two natural phytoplankton populations.
J. Fish. Res. Bd, Can., 27(8):1395-1403.

grazing
*Holmes, R.W., P.M. Williams and R.W. Eppley, 1967.
Red water in La Jolla Bay, 1964-1965.
Limnol. Oceanogr., 12(3):503-512.

grazing
Kusmorskaya, A.P., 1946
Destruction of plankton by fishes grazing upon it and the importance of this phenomenon for estimating the Biomass CR (Doklady) Acad. Sci. URSS LIII(7):665-

grazing
Lasker, Reuben 1966.
Feeding, growth, respiration and carbon utilization of a euphausid crustacean.
J. Fish. Res. Bd., Can. 23(9): 1291-1317.

grazing
Longhurst, Alan R., Carl J. Lorenzen and William H. Thomas, 1967.
The role of pelagic crabs in the grazing of phytoplankton off Baja California.
Ecology, 48(2):190-200.

grazing
Martin, John H., 1970.
Phytoplankton-zooplankton relationships in Narragansett Bay. IV. The seasonal importance of grazing. Limnol. Oceanogr., 15(3): 413-418.

grazing
Martin, John H., 1965.
Phytoplankton-zooplankton relationships in Narragansett Bay.
Limnology and Oceanography, 10(2):185-191.

grazing
McAllister, C.D., 1971.
Some aspects of nocturnal and continuous grazing by planktonic herbivores in relation to production studies.
Techn. Rept. Fish. Res. Bd, Canada, 248: 281 pp. (multilithed)

grazing
Paffenhöfer, G.-A., 1971.
Grazing and ingestion rates of nauplii, copepodids and adults of the marine planktonic copepod Calanus helgolandicus. Mar. Biol. 11 (3): 286-298.

grazing

Parsons, T.R., R.J. LeBrasseur and J.D. Fulton, 1967.
Some observations on the dependence of zooplankton grazing on the cell size and concentration of phytoplankton blooms.
J. oceanogr. Soc., Japan, 23(1): 10-17.

grazing

Petipa, T.S., 1965.
Diurnal grazing of phytoplankton in the Black Sea by the copepod Calanus helgolandicus (Claus) (In Russian; English abstract).
Zool. Zhurn., Akad. Nauk, SSSR, 44(6):844-854.

grazing

Platt, T. and R.J. Conover, 1971.
Variability and its effect on the 24th chlorophyll budget of a small marine basin. Mar. Biol. 10(1): 52-65.

grazing

Ramamurthy, S., 1965.
Studies on the plankton of the North Canara coast in relation to the pelagic fishery.
J.mar.biol. Ass.India, 7(1):127-149.

grazing

Randall, J.E., 1961
Overgrazing of algae by herbivorous marine fishes.
Ecology, 42(4): 812.

grazing

Riley, G. A., H. Stommel, and D. F. Bumpus, 1949
Quantitative ecology of the plankton of the western North Atlantic. Bull. Bingham Ocean. Coll. 12(3):169 pp., 39 text figs.

grazing

Steemann Nielsen, E., and E.A. Jensen, 1957.
Primary oceanic production. The autotrophic production of organic matter in the oceans.
Galathea Repts., No. 1:49-136.

grazing

Uyeno, Fukuzo, Kyoichi Kawaguchi, Nagao Terada and Tadashi Okada 1970.
Decomposition, effluent and deposition of phytoplankton in an estuarine pearl oyster area.
Rept. Fac. Fish. Prefect. Univ. Mie 7(1): 7-41

grazing, effect of

Paine, Robert T. and Robert L. Vadas, 1969.
The effects of grazing by sea urchins, Strongylocentrotus spp., on benthic algal populations.
Limnol. Oceanogr., 14(5): 710-719.

grazing, effect of

Randall, John E., 1965.
Grazing effect on sea grasses by herbivorous reef fishes in the West Indies.
Ecology, 46(3):255-260.

"green cells"

Green cells

*Hamilton, R.D., O. Holm-Hansen and J.D.H. Strickland, 1968.
Notes on the occurrence of living microscopic organisms in deep water.
Deep-Sea Res., 15(6):651-656.

green flash

green flash

O'Connell, D.J.K., 1961
The green flash and kindred phenomena.
Endeavour, 20(79): 131-137.

"greenhouse" effect

"Greenhouse" effect

Donn, William L., and David M. Shaw, 1967.
The maintenance of an ice-free Arctic Ocean.
Progress in Oceanography, 4:105-113.

GREENSCHIST

Christensen, Nikolas I., 1970.
Possible greenschist facies metamorphism of the oceanic crust.
Bull. geol. Soc. Am, 81(3): 905-908

greenstones

greenstones

Melson, William G., Vaughn T. Bowen, Tjeerd H. van Andel and Raymond Siever, 1966.
Greenstones from the central valley of the Mid-Atlantic Ridge.
Nature, 209 (5023):604-605.

greenstones

Melson, William G., and Tjeerd H. van Andel, 1966.
Metamorphism in the Mid-Atlantic Ridge, 22° N. latitude.
Marine Geol., 4(3):165-186.

grids

Turikov, V.G. 1969.
On the selection of a grid at the numerical integration of spherical equations of atmospheric and ocean dynamics. (In Russian; English abstract).
Gidrol 1969(5): 34-41

groins

groins

Beach Erosion Board, 1953.
Shore protection planning and design (preliminary subject to revision). B.E.B. Bull., Special Issue No. 2:230 pp., 149 figs., plus app.

groins

Horikawa, K., and C. Sonu, 1958.
An experimental study on the effect of coastal groins.
Coastal Engineering in Japan, 1:59-74.

grottos

grottos

Laborel, J., and J. Vacelet, 1959.
Les grottes sous-marines obscures en Méditerranée.
C. R. Acad. Sci., Paris, 248(18):2619-2621.

ground motion, effect of

Slatkin, Montgomery W., 1971.
Long waves generated by ground motion.
J. fluid Mech. 48(1): 81-90

Groundwater effect of

ground water

*Husmann, Siegfried, 1967.
Klassifizierung mariner, brackiger und limnischer Grundwasserbioyope.
Helgoländer wiss. Meeresunters. 16(3):271-278.

groundwater, effect of

Kohout, F.A., and M.C. Kolipinski, 1967.
Biological zonation related to groundwater discharge along the shore of Biscayne Bay, Miami, Florida.
In: Estuaries, G.H. Lauff, editor, Publs Am. Ass. Advmt Sci., 83:488-499.

growth

Growth

Fischer-Piette, E., 1941
Croissance, taille maxima et longévité possible de quelques animaux intercotidaux en fonction du milieus. Ann. de l'Inst. Océan. Monaco, 21(1):1-II28.

growth

*Hepper, B.T., 1967.
On the growth at moulting of lobsters (Homarus vulgaris) in Cornwall and Yorkshire.
J. mar. biol. Ass., U.K., 47(3):629-643.

Growth

San Feliu, J.M., 1966.
Observaciones sobre la muda y el crecimiento del langostino Penaeus kerathurus (Forskål, 1775) en acuario.
Inv. Pesq., Barcelona, 30:685-705.

growth, effect of

Tanaka, S., 1958
A mathematical consideration of the effect of mortality and growth on a fish population.
Bull. Tokai Reg. Fish. Res. Lab., No. 20: 1-12.

growth efficiency

Reeve, Michael R., 1963
Growth efficiency in Artemia under laboratory conditions.
Biol. Bull., 125(1):133-145.

Growth factor

*Paster, Zvi and Bernard C. Abbott, 1970.
Gibberellic acid: a growth factor in the unicellular alga Gymnodinium breve. Science, 169(3945): 600-601.

growth factors

Provasoli, L., 1963
8. Organic regulation of phytoplankton fertility. In: The Sea, M.N. Hill, Edit., Vol. 2. (II) Fertility of the oceans, Interscience Publishers, New York and London, 165-219.

growth, microbial

Button, D.K., 1970.
Some factors influencing kinetic constants for microbial growth in dilute solution. Occ. Pap. Inst. mar. Sci., Alaska, 1: 537-547.

growth rate

Moore, H. B. and J. A. Kitching, 1939
The biology of Chthamalus stellatus (Poli).
JMBA 23:521-541.

"growth rings"

growth rings

Clarke, Malcolm R., 1965.
"Growth rings" in the beaks of the squid Moroteuthis ingens (Oegopsida: Onychoteuthidae).
Malacologia, 3(2):287-307.

groynes

groynes
Great Britain, Department of Scientific and Industrial Research, 1960.
Waves and sea defences.
Hydraulics Research, 1960:62-69.

groynes, effect of
Kinmont, A., 1961.
The nearshore movement of sand at Durban.
Marine Studies off the Natal Coast, C.S.I.R. Symposium, No. S2:46-58. (Multilithed).

guano

guano
Andersen, 1856
Leone-islands guano
J. agric & Trans Highl. soc., N.S. 55:496-
Z. Ges naturw. 7:492- 1857

guano
Coker, R.E., 1908.
The fisheries and the guano industry of Peru.
Bull. U.S.B.F. 28:333-365.

guano
Coker, R.E., 1921.
An illustration of practical results from the protection of natural resources. Science 53: 295-298.

guano
Coker, R.E., 1908.
Regarding the future of the guano industry and the guano-producing birds. Science 28:58-64.

guano
Coker, R.E., 1908.
La industria del guano. Bol. del Ministerio de Fomento, Año VI, No. 4:25-34.
Condición en que se encuentra la pesca marina, desde Paita hasta Bahía de la Independencia.
Ibid.:62-99.

guano
Hutchinson, G. E., 1950.
Survey of contemporary knowledge of biogeochemistry. 3. The biogeochemistry of vertebrate excretion. Bull. Am. Mus. Nat. Hist. 96:554 pp., 16 pls., 103 textfigs.

guano
Jordán S., Rúmulo, 1961
Les aves guaneras, la cadena alimentaria y la produccion de guano.
Bol., C.A.G., 37(3):19-20.

guano
Meise, Wilhelm 1938
Guano und anderer Vogeldung.
Die Rohstoffe des tierreichs. 1:2113-2172

guano production
Murphy, R.C., 1954.
El guano y la pesca de anchoveta. Informe oficial al supremo gobienno, C.A.G.:1-39,

together with:
Documentos oficiales en pro de la conservacion de las especies productores del guano y su base alimenticoe:44-147.

Murphy, R. C. 1920
The guano industry of modern Peru
Brooklyn Mus. Quart., 7:244-269

guano, effect of
Sieburth, John McNeill, 1963.
Bacterial habitats in the Antarctic environment Ch. 49 in: Symposium on Marine Microbiology, C.H. Oppenheimer, Editor, C.C. Thomas, Springfield, Illinois, 533-548.

guano deposits
Spillmann, F., 1952.
Contribucion al estudio de la genesis de fosforitas. Bol. Soc. Geol., Peru, 25:21 pp., 1 pl., 2 textfigs.

guano birds

guano birds
Ahrens, T.G., 1923.
Einiges über die Guanovögel an den Küsten Perus.
Naturschutz. 4:51-54.

guano birds
Coker, R.E., 1908.
Regarding the future of the guano industry and the guano-producing birds. Science 28:58-64.

guano birds
Forbes, H. O. 1914
Puntos principales del informe presentado al supremo gobierno por el ornitologo Dr. H. O. Forbes.
Mem. Comp. Adminis. Guano, 5:57-105

guano birds
Gamarra Dulanto, L., 1955.
Ensayo sobre la zoonomía de las aves guaneras del Perú. Bol. Cient., C.A.G., 2:73-121.

guanays
Jordán S. Rúmulo, 1961
Les aves guaneras, la cadena alimentaria y la produccion de guano.
Bol., C.A.G., 37(3):19-20.

guanays
Rand, R. W., 1960
The biology of guano-producing sea-birds.
3. The distribution, abundance and feeding habits of the cormorants Phalacrocoracidae off the southwest coast of the Cape Province. Union of S. Africa, Dept. of Commerce and Ind., Div. Fish., Invest. Rept. No. 42:32 pp.

gulfs

gulfs
Cotton, C.A., 1951.
Atlantic gulfs, estuaries and cliffs. Geol. Mag. 88:113-128.

gullies

gullies
Buffington, E.C., and D.G. Moore, 1963.
Geophysical evidence on the origin of gullied submarine slopes, San Clemente, California.
J. Geology, 71(3):356-370.

"gust bumps"

gust bumps
Timmerman, H. 1971.
On the connection between cold fronts and gust bumps.
Dt. Hydrogr. Z. 24 (4):159-172.

"gust bumps"
Wemelsfelder, P.J., 1960
Bui-oscillaties en buistoten tijdens stormvloeden.
Rapp. Deltacommissie, Bijdragen (4):115-130.

guyots

guyots
Anon., 1951.
Travaux bathymetriques. Bull. d'Info., C.O.E.C., 3(2):60-63, 4 figs.

guyots
Bandy, Orville L., 1963
Aquitanian planktonic Foraminifera from Erben Guyot.
Science, 140 (3574):1402.

guyots
Boyce, R.E., and E.L. Smith, 1965
Geomorphology of Erben Guyot.
Mar. Geol. 6(2):179-153

guyots
Christensen, M.N., and C.M. Gilbert, 1964
Basaltic cone suggests constructional origin of some guyots.
Science, 143(3603):240-242.

guyots
Emery, K.O., J.I. Tracy, jr., and H.S. Ladd, 1954.
Geology of Bikini and nearby atolls. Bikini and nearby atolls: Part 1, Geology.
Geol. Survey Prof. Pap., No. 260-A:1-265, 64 pls.

guyots
Erimesco, P., 1961.
Geophysics of ocean basins.
Bull. Inst. Pêches Mar. Maroc, 6:3-56.

guyots
Hamilton, E.L., 1956.
Sunken islands of the mid-Pacific mountains.
Mem. G.S.A. 64:97 pp.

guyots
Hamilton, Edwin L., and Robert W. Rex, 1961.
Lower Eocene phosphatized Globigerina ooze from Sylvania guyot, Marshall Islands. (Abstract).
Proc. Ninth Pacific Sci. Congr., Pacific Sci. Assoc., 1957, 12(Geol.-Geophys.):280.

guyots
Hess, H.H., 1948.
Drowned ancient islands of the Pacific Basin.
Smithsonian Rept, 1947:281-300.

guyots
Lonsdale, Peter, William R. Normark, W.A. Newman, 1972.
Sedimentation and erosion on Horizon Guyot.
Bull. geol. Soc. Am. 83 (2): 289-316.

guyots
Menard, H.W., 1969.
The deep-ocean floor.
Scient. Am., 221(3):126-132, 134, 136, 138, 140, 142. (popular).

guyots
Menard, H.W., and H.S. Ladd, 1963.
15. Oceanic islands, seamounts, guyots, and atolls.
In: The Sea, M.N. Hill, Editor, Interscience Publishers, 3:365-387.

guyots
Menard, H.W., S.M. Smith and T.E. Chase, 1964.
Guyots in the southwestern Pacific basin.
Geol. Soc., Amer., Bull., 75(2):145-148.

guyots
Milliman, John D., 1966.
Submarine lithification of carbonate sediments.
Science, 153(3739):994-996.

guyot
Mogi, A., 1953.
On the flat topped sea mount (guyot) in the North Pacific Ocean. Publ. 981, Hydrogr. Bull., Tokyo, Spec. Number, No. 12:58-61, 1 fig.

guyots
Nayudu, Y.R., 1962
A new hypothesis for origin of guyots and seamount terraces.
Crust of the Pacific Basin, Geophys. Monogr., No. 6:171-180.

guyots
Nayudu, Y. Rammohanroy, 1961.
Origin of seamount terraces and guyots as suggested by the petrographic evidences from Cobb and Bowie seamounts.
Tenth Pacific Sci. Congr., Honolulu, 21 Aug.-6 Sept., 1961, Abstracts of Symposium Papers, 382.

Guyot
Sato, Takahiro, and Shunji Ao, 1961.
A guyot at the north margin of the Philippine basin.
Japan. J. Geol. & Geogr., Trans., 32(2):153-158.

guyots
Sato, Takahiro, and Akio Mogi, 1965.
Guyots found from the Marshall and East Caroline Ridges. Researches on the GEBCO, No. 1.
J. Oceanogr. Soc., Japan, 21(4):139-147.

guyots
Shibata, Keishi, 1966.
An echo survey near the Ojin Seamount.
Bull. Fac. Fish., Nagasaki Univ., No. 20:55-57.

guyots
Silvester, R., 1965.
Coral reefs, atolls and guyots.
Nature, 207(4998):681-688.

guyots
Standard, J.C., 1961.
Submarine geology in the Tasman Sea.
Bull. Geol. Soc., Amer., 72(12):1777-1788.

GUYOTS
Takeuchi, Hitoshi, and Shoki Sakata, 1967.
Guyots and the theory of mantle convection.
Geophys. Notes, Univ. Tokyo, 20(1/2):128-137.
(In Japanese; English abstract)

guyots
Vogt, Peter R., and John R. Conolly
1971
Tasmantid guyots, the age of the Tasman Basin and motion between the Australian Plate and the mantle.
Bull. geol. Soc. Am. 82 (9): 2577-2584

Gyres

gyres
Baker, D. James, Jr., 1971.
A source-sink laboratory model of the ocean circulation.
Geophys. fluid Dyn. 2(1): 17-29

gyres
Chekotillo, K.A., 1961.
Peculiarities of the circulation of intermediate waters in the northern Pacific.
Trudy Inst. Okeanol., Akad. Nauk, SSSR, 45:113-122.

habitat, effect of

habitat, effect of
Southward, A.J., 1955.
On the behaviour of barnacles. II. The influence of habitat and tide-level on cirral activity.
J.M.B.A. 34:423-433.

hadal community

hadal faunas
Bruun, A. Fr., 1961.
New light on the abyssal and hadal faunas of the Pacific. (Abstract).
Proc. Ninth Pacific Sci. Congr., Pacific Sci. Assoc., 1957, Fish., 10:14.

hadal community
Wolff, Torben, 1960.
The hadal community, an introduction.
Deep-Sea Res., 6(2):95-124.

hadal
*Menzies, Robert J., and Robert Y. George, 1967.
A re-evaluation of the concept of hadal or ultra-abyssal fauna.
Deep-Sea Res., 14(6):803-723.

Hadley cell
*Hastenrath, Stefan L., 1968.
On mean meridional circulation in the tropics.
J. atmos. Sci., 25(6):979-983.

halocline

halocline
Barber, F.G., 1958.
Currents and water structure in Queen Charlotte Sound, British Columbia.
Proc. Ninth Pacific Sci. Congr., Pacific Sci. Assoc., 1957, Oceanogr., 16:196-199.

halocline
Dodimead, A.J., F. Favorite and T. Hirano, 1964.
Review of oceanography of the subarctic Pacific Region. Salmon of the North Pacific Ocean. II.
Collected Reprints, Tokai Reg. Fish. Res. Lab., No. 2:187 pp.
(In Japanese).

halocline
*Dodimead, A.J., and G.L. Pickard, 1967.
Annual changes in the oceanic-coastal waters of the eastern subarctic Pacific.
J. Fish. Res. Bd., Can., 24(11):2207-2227.

halocline
Fleming, R. H., 1958
Notes concerning the halocline in the north-eastern Pacific Ocean. J. Mar. Res., 17: 158-173.

halocline
*Meincke, Jens, 1967.
Die Tiefe der jahreszeitlichen Dichteschwankungen im Nordatlantischen Ozean.
Kieler Meeresforsch., 23(1):1-15.

halocline
Tully, J.P., and F.G. Barber, 1961
An estuarine model of the sub-Arctic Pacific Ocean.
In: Oceanography, Mary Sears, Edit., Amer. Assoc. Adv. Sci., Publ. No. 67:425-454.

Previously published under title "An estuarine analogy in the sub-Arctic Pacific Ocean", J. Fish. Res. Bd., Canada, 17(1): 91-112 (1960).

halocline
Tully, J.P., and F.G. Barber, 1960
An estuarine analogy in the Sub-Arctic Pacific Ocean. J. Fish. Res. Bd., Canada, 17(1): 91-112.

halocline, effect of
Biggs, Robert B., and Carolyn D. Wetzel 1968.
Concentration of particulate hydrocarbate at the halocline in Chesapeake Bay.
Limnol. Oceanogr. 13(1):169-171.

Hamilton's principle
Bretherton, Francis P. 1970.
A note on Hamilton's principle for perfect fluids.
J. fluid Mech. 44(1): 19-31.

handbooks

handbooks
Compass Publications, Inc., 1968.
UnderSea Technology Handbook Directory 1968
pages not sequentially numbered. $20.00

haptonema
Leadbeater, B.S.C., 1971.
Observations by means of ciné photography on the behaviour of the haptonema in plankton flagellates of the class Haptophyceae. J. mar. biol. Ass., U.K., 51(1): 207-217.

harbors

harbors
Allen, F. H., and W. A. Price, 1959.
Density currents and siltation in docks and tidal basins.
Dock & Harb. Auth., 40(465):72-76.

harbors
de Rouville, A., 1941/42/43.
Le regime des cotes. Elements hydrographiques des acces des ports. Ann. Ponts et Chaussees:

1941	1942	1943
I-II, p. 5	I-II:5	I-II:1
III-IV:155	III-IV:147	
V-VI:297	V-VI:283	
IX-X:155	VII-VIII:355	
XI-XII:363	IX-X:444	
	XI-XII:555	

harbors
Garrett, C.J.R., 1970.
Bottomless harbors.
J. fluid Mech. 43(3):433-449.

harbors
Iribarren Cavanilles, Ramon, 1962
Ensenanzas adquiridas en los puertos del grupo puerto de Motrico.
Centro de Estudios y Experimentacion de Obras Publicas, Laboratorio de Puertos, February, 1962: unnumbered pp.

harbors
Irribarren Cavanillos, R., and C. Nogales y Olano 1952.
Aportacion española a la technica de puertos.
Rev. Obras Publ., Dec. 1952:7 pp.

harbors
Karo, H.A., 1958.
Basic survey for beach and harbor studies.
Int. Hydrogr. Rev., 35(2):21-32.

harbors
Mackay, J.R., 1961
Notes on small boat harbors of the Yukon coast.
Geogr. Bull., 15: 19-30.

harbors
Miles, John W., 1971.
Resonant response of harbours: an equivalent-circuit analysis
J. fluid Mech. 46(2): 241-265

harbours
Russell, R.C.H., 1961.
The motion of ships moored in waves. To improve harbours by reducing the waves inside them is very expensive. In future, more attention is likely to be paid to the better design of moorings as a means of limiting the movements of ships in harbour. Experiments which have show the effectiveness of this method are here described.
New Scientist, 10:256-259.

harbors
Tanita, S., K. Kato, and T. Okuda, 1951.
Studies on the environmental conditions of shell-fish-fields. In the case of Muroram Harbour (2).
Bull. Fac. Fish., Hokkaido Univ., 2(3):220-230.

harbors
Tayler, J.N., 1840.
Plans for the formation of harbours of refuge improvement of rivers and seaports. London-Plymouth, 66 pp.

harbor protection
Carr, J.H., 1952.
Wave protection aspects of harbor design.
Calif. Inst. Tech., Hydrodynamics Lab. Rept. No. E-11:99 pp.

harmonic analysis

harmonic analysis
Carrozzo, Maria Teresa, 1959
A programme for Lecolazet's method of harmonic analysis of the earth tides in the Bell Interretative System for an electronic computer I.B.M. 650.
Proc. Third Int. Symposium on Earth Tides. Trieste, 1959.
Also:
Ist. Geodesia e Geofisico dell'Universita, Bari (Italia), Pubbl., No. 6.

harmonic analysis
Ichie, Takashi, 1951
[On the hydrography of the Kii-Suido(1951).]
Bull. Kobe Mar. Obs., No. 164:253-278(top of page); 35-60(bottom of page).

harmonic analysis
Japan, Kobe Marine Observatory, 1954.
[The results of the harmonic analysis to use the regular monthly oceanographic observation on board the R.M.S. "Syunpu-maru" in Osaka Wan,1953.]
J. Ocean. (2)5(2):1-15, 11 figs.

harmonic analysis
Japan, Maizuru Marine Observatory, Oceanographical Section, 1961
[Report of the oceanographic observations in the western part of Wakasa Bay from January to March, 1958.]
Bull. Maizuru Mar. Obs., No. 7:50-54.

harmonic analysis
Japan, Maizuru Marine Observatory, 1961
Tidal currents in Maizuru Bay.
Bull. Maizuru Mar. Obs., No. 7:23-29,(111-117).

harmonic analysis
Sweers, H.E., 1967.
Some Fortran II programs for computer processing of oceanographic observations.
NATO Subcomm. Oceanogr. Res., Tech. Rept. 37 (Irminger Sea Project):104 pp. (mimeographed).

harmonic analysis
Tanoue, T., 1955.
[The harmonic analysis of annual variation of temperature in Kagoshima Bay.]
Mem. Fac. Fish., Kagoshima Univ., 4:12-19.

harmonic analysis
Van Isacker, 1961.
Generalized harmonic analysis.
Adv. in Geophysics, 7:189-214.

harvest

see also: resources

harvesting
Bullis, Harvey R. Jr., and John R. Thompson 1968.
Harvesting the ocean in the decade ahead.
Ocean Industry 3(6): 52-60.

harvest of the sea
Fye, Paul M., Arthur E. Maxwell, Kenneth O. Emery and Bostwick H. Ketchum, 1968.
Ocean science and marine resources, In: Uses of the sea, Edmund A. Gullion, editor, Prentice-Hall, Inc., 17-69.

harvest of the sea
Rounsefell, George A. 1971.
Potential food from the sea.
J. mar. Sci., Alabama 1(3):1-82.

hatching

hatching
Davis, Charles C., 1965.
A study of the hatching process in aquatic invertebrates. XIV. An examination of hatching in Palaemonetes vulgaris (Say).
Crustaceana, 8(3):233-238.

hazards

hazards
Busby, Roswell F., Lee M. Hunt and William O. Rainnie 1968.
Hazards of the deep. 3
Ocean Industry 3(9): 53-58.

haze

haze
Srivastava, R.C., and C. Ronne, 1966.
Salt particles and haze in the Indian monsoon air.
Indian J. Met. Geophys., 17(4):587-590.

h cosec θ
Gill, A.E. and R.L. Parker, 1970.
Contours of "h cosec θ" for the world's oceans.
Deep-Sea Res., 17(4): 823-824.

hearing

hearing, fishes
Buerkle, Udo, 1968.
Relation of pure tone thresholds to background noise level in the Atlantic cod (Gadus morhua).
J. Fish. Res. Bd. Can., 25(6): 1155-1160.

hearing
*Buerkle, Udo, 1967.
An audiogram of the Atlantic cod, Gadus morhua L.
J. Fish. Res. Bd., Can., 24(11):2309-2319.

hearing
Dijkgraaf, S., 1960
Hearing in bony fishes. Proc. Roy. Soc., London, Ser. B., 152(946): 51-53.

hearing
Edinger, T., 1955.
Hearing and smell in cetacean history.
Monatsskr. Psychiat. u. Neurol., Basle, 129(1/3):37-58.

hearing
Johnson, C. Scott, 1967.
Sound detection thresholds in marine mammals.
In: Marine bio-acoustics, W.N. Tavolga, editor, Pergamon Press, 2:247-255.

hearing
Mohl, B., 1967.
Frequency discrimination in the common seal and a discussion of the concept of upper hearing limit.
In: Underwater Acoustics, V.M. Albers, editor, Plenum Press, 2: 43-54.

hearing
*Nelson, Donald R., 1967.
Hearing thresholds, frequency discrimination, and acoustic orientation in the lemon shark, Negaprion brevirostris (Poey).
Bull. mar. Sci., Miami, 17(3):741-768.

hearing
Ray, B., 1969.
Directional hearing by divers.
Rept. Underwat. Ass. 4: 49-52

hearing (underwater)
Reysenbach de Haan, F.W., 1960.
Some aspects of mammalian hearing under water.
Proc. R. Soc., London, B, 152(946):54-62.

hearing
Tavolga, W.N., 1967.
Masked auditory thresholds in eleost fishes.
In: Marine bio-acoustics, W.N. Tavolga, editor, Pergamon Press, 2:233-243.

HEAT

heat
Duncan, A.R., and N.M. Pantin, 1969.
Evidence for submarine geothermal activity in the Bay of Plenty.
N.Z. Jl mar. freshwat. Res. 3(4). 602-606

HEAT, advection

Seryakov, E.I., 1960 heat advection
On the advection of heat by currents in the Barents Sea.
Soviet Fish. Invest., North European Seas, VNIRO, PNIRO, Moscow, 1960:89-88.

HEAT, balance

heat balance
Alexander, R.C. 1971.
On the advective and diffusive heat balance in the interior of a subtropical ocean.
Tellus. 23 (4/5): 393-403

heat balance
Arcenieva, N. Ia., 1965.
A method for calculating water temperature in the active layer of the White Sea. (In Russian).
Trudy, Gosudarst. Okeanogr. Inst., No. 86:75-94.

heat balance
Archipova, E.G., 1962.
Peculiarities of the heat balance in the North Atlantic during the IGY. (In Russian).
Trudy Gosud. Okeanogr. Inst., 67:74-85.

heat balance
Archipova, E.G., 1960.
Year to year changes in the heat balance in the northern part of the Atlantic Ocean, based on the last ten years.
Trudy Gosud. Okeanogr. Inst., 54:35-60.

In Russian

heat balance
Ashburn, Edward V., 1966.
Comments on paper by Klaus Wyrtki, "The average annual heat balance of the North Pacific Ocean and its relation to ocean circulation".
J. geophys. Res., 71(6):1758-1759.

heat balance
Batalin, A.M., 1959.
Thermal balance of Far Eastern Seas.
(In Russian).
Izv., Akad. Nauk, SSSR, Ser. Geofiz., (7):1003-1010.

English translation, (7):710-715.

heat balance
Boguslavsky, S.G., 1962.
Heat balance in the North Atlantic in spring.
(In Russian).
Trudy, Morsk. Gidrofiz. Inst., 20:36-43.

Translation:
Scripta Tecnica for AGU (4):27-33. 1964.

heat balance
Boguslavskiy, S.G., 1961.
Latitudinal change in the heat balance of the Atlantic Ocean. Physics of the Sea. (In Russian).
Trudy Morsk. Gidrophys. Inst., 23:139-147.

Translation:
Scripta Technica, Inc. for Amer. Geophys. Union, 111-117.

heat balance
Boguslavskiy, S.G., 1960.
Heat balance in the North Atlantic in spring.
(In Russian).
Trudy, Morsk. Gidrofiz. Inst., Akad. Nauk, SSSR, 20:36-43.

Translation:
Scripta Tecnica for AGU, 27-33.

heat balance
Boguslavsky, S.G., 1960
[Heat balance in the North Atlantic during the first cruise of "Mikhail Lomonosov".]
Biull. Oceanograph. Komm., (5): 60-67.

heat balance
Boudreault, F.-Robert, 1969.
Moyennes climatiques du bilan thermique de Grande-Rivière (Baie des Chaleurs) Québec.
Cahier Info. Sta. Biol. mar. Grand Rivières, 50: 1-52 (multilitted)

heat balance
Boudreault, F. Robert, 1966.
Océanographie physique de la Baie-des-Chaleurs, 1965.
Rapp. Stn. Biol. Mar., Grande-Rivière, 1965:17-30.

heat balance
*Bryan, Kirk, and Michael D. Cox, 1968.
A nonlinear model of an ocean driven by wind and differential heating: I. Description of the three-dimensional velocity and density fields. 2. An analysis of the heat, vorticity and energy balance.
J. atmos. Sci., 25(6):945-967;968-983.

heat balance
Budiko, M.I., Editor, 1963.
Atlas of heat balance of the world. (In Russian)
Mezhd. Geofiz. Komitet, Prezidiume Akad. Nauk, SSSR, Glabnaia Geofiz. Observ. im. A.I. Borikova
69 colored charts.

heat balance
Budyko, M.I., 1961
Radiation balance and heat balance of oceans.
Symposium on the radiant energy in the sea, Helsinki, 4-5 Aug. 1960.
U.G.G.I., Monogr., No. 10:112.

heat balance
Budyko, M.I., R.L. Kagan and L.A. Stokina, 1966.
Concerning anomalies of heat balance of the ocean. Meteorol. i Gidrol., 1966 (1):24-28.

heat balance
Colacino, Michele, et Enrico Rossi, 1971.
Exemples de calcul du bilan thermique dans la mer Tyrrhénienne (mai 1969)
Cah. océanogr. 23 (3): 267-281

heat balance
Colón, José A., 1963
Seasonal variation in heat flux from the sea surface to the atmosphere over the Caribbean Sea.
J. Geophys. Res., 68(5): 1421-1430.

heat balance
Degtyarev, G.M., Yu. A. Men'shev, and B. Ye. Alemasov, 1962.
Characteristics of the heat balance of the northeastern part of the Atlantic Ocean in the summer of 1960. (In Russian)
Izv. Akad. Nauk, SSSR, Ser. Geofiz., (7):965-970.
Eng. Ed. (7): 614-618

English abstract in:
Soviet Bloc Res. Jeophys. Astron., Space, 43:3-4

JPRS 15,253
OTS 61-11147-43.

heat balance
Donn, William L., and David M. Shaw, 1966.
The heat budget of an ice-free and an ice-covered Arctic Ocean.
J. geophys. Res., 71(4):1087-1093.

heat budget
Gade, Herman G., 1961.
Further hydrographic observations in the Gulf of Cariaco, Venezuela. The circulation and water exchange.
Bol. Inst. Oceanogr., Univ. de Oriente, Venezuela, 1(2):359-395.

heat balance
Herbert, Wally, 1970.
The first surface crossing of the Arctic Ocean.
Geographical Jl 136(4): 511-533.

heat balance
Kirillova, T., and T. Ogneva, 1956.
[Characteristics of the heat balance of water surface.] Meteorol. i Gidrol., 4:41-46.

CSIRO-3263 in Transl. Mon. 3(9):304.

heat balance
Kort, V.G., 1963.
Heat exchange in the Southern Ocean.
In Russian; English abstract).
Rez. Issled. Programme Mezhd. Geofiz. Goda, Okeanolog. Issled., Akad. Nauk, SSSR, No. 8:17-23.

heat balance
Lee, Arthur, 1963.
The hydrography of the European Arctic and subarctic seas.
In: Oceanography and Marine Biology, H. Barnes, Editor, George Allen and Unwin, 1:47-76.

heat balance
Makerov, Iu. V., 1961.
The heat balance of the Black Sea.
Trudy Gosud. Okean. Inst., Leningrad, 61:169-198.

In Russian

heat balance
Mosby, Hakon, 1962
Water, salt and heat balance of the North Polar Sea and of the Norwegian Sea.
Geofys. Publik., Geophysica Norvegica, 24: 289-313.

heat balance
Popov, S.M., and S.A. Oiazanov, 1961.
[The significance of the effective radiation in the thermal balance of the ocean.]
Izv. Akad. Nauk, SSSR, Ser. Geofiz., (2):281-293.

heat balance

Privett, D. W., 1960.
The exchange of energy between the atmosphere and the oceans of the southern hemisphere. Meteorol. Off. London, Geophys. Mem., 104(MO. 631d) 61 pp.

heat balance

Roden, Gunnar I., 1959
On the heat and salt balance of the California Current region.
J. Mar. Res., 18(1):36-61.

heat balance

Rossov, V.V., 1961
[On the water and heat balance in the Norwegian and Greenland seas.]
Okeanologiia, Akad. Nauk, SSSR, 1(5):944-947.

heat balance

Samoylenko, W.S., 1963.
The methods of advective terms evaluating in the seas and oceans heat balance. Questions of physical oceanology. (In Russian: English abstract).
Trudy Inst. Okeanol., Akad. Nauk, SSSR., 66:46-58.

heat balance, effect of

Selitskaia, E.S., 1959.
Variability of surface water temperature for separate regions of the North Sea. (In Russian).
Trudy Gosud. Okeanogr. Inst., 48:104-111.

heat balance

Seryakov, E.I., 1968.
Long term fluctuations of the heat balance components in the Barents and Norwegian seas. (In Russian).
Mater. Ribokhoz. Issled. Severn. Basseina, 11:121-133.

heat balance

and

Seryakov, E.I., M.V. Kutseva, 1967.
Long-term fluctuations of the equation components of the heat balance in the Norwegian Sea (in the area of the Weather Ship M). (In Russian).
Mater. Rybokhoz. Issled. Severn. Basseina (PINRO), 10:95-111.

heat balance

Seryakov, E.I., and V.M. Radikevich, 1961

[On the changeability of the heat balance components in the southern part of the Barents Sea during the period of the IGY.]
Issled. Severnoi Chasti Atlanticheskogo Okeana, Mezhd. Geofiz. God, Leningradskii Gidrometeorol Inst., 1: 43-51.

heat balance

Seryakov, E.I., and A.I. Bychkova, 1964.
Peculiarities of heat balance in the part of the Barents Sea not subjected to freezing in the period of the IGY and IGC. (In Russian).
Materiali Vtoroi Konferentsii, Vzaimod. Atmosfer. i Gidrosfer. v Severn. Atlant. Okean. Mezhd. Geofiz. God, Leningrad. Gidrometeorol. Inst., 199-210.

heat balance

Sheremetievskaya, O.I., 1961.
[On the problem of the heat balance in the superficial waters of the Antarctic.]
Okeanologiia, Akad. Nauk, SSSR, 1(5):835-836.

heat balance

Skriptunova, L.I., 1957

[The role of the surface heat balance in the water temperature regime of the northern part of the Atlantic ocean.] Meteorol. i Gidrol. (7): 40-45.

LC/SLA/ mi $2.40 pl $3.30

heat balance

Strokina, L.A., 1964.
Heat balance of the North Atlantic compared with that of other regions in the World Ocean. (In Russian).
Materiali Vtoroi Konferentsii, Vzaimod. Atmosfer. i Gidrosfer. v Severn. Atlant. Okean., Mezhd. Geofiz. God, Leningrad, Gidrometeorol. Inst., 189-198.

heat balance

*Sturm, M., 1968.
Untersuchungen der Wärmebilanz der südlichen Ostsee im Bereich des Feuerscheffes "Fehmarnbelt"
Tellus, 20(3):485-494.

heat balance

Sturm, Manfred, 1963.
Über Methoden zur empirischen Berechnung der Hauptkomponenten des Wärmehaushaltes der Meeresoberfläche aus mittleren hydrometeorologischen Daten.
Beiträge z. Meereskunde, (9):36-66.

heat balance

Timofeev, V.T., 1961.
The effect of the deep layers of Atlantic waters on the hydrological regime of the Kara Sea. (In Russian).
Problemy Severa, 4:46-58.

Translation:
Problems of the North, 1962, 4:45-56.

heat balance

Timofeiv, V.T., 1961.
[The inflow of the Atlantic waters and heat into the Arctic basin.]
Okeanologiia, (3):407-411.

heat balance

Timofeyev, M.P., 1958.
[The heat balance of water bodies and methods for determining evaporation (from them).]
Sovremennyye Problemy Meteorologii Prizemnogo Sloya Vozdukha, 43-60.

(A symposium of articles edited by M.I. Budkoy, Leningrad).

LC and SLA, mi $2.70, ph $4.80

heat balance

Tiutnev, Ia. A., 1961.
Simplified method for computing the heat balance of the sea surface. Forecasting and computing physical phenomena of the oceans. (In Russian).
Trudy Okeanogr. Komissii, Akad. Nauk, SSSR, 11: 142-149.

heat balance

Tyutnev, Yu. A., 1961.
A simplified method for estimation of heat balance. (In Russian).
Meteorol. i Gidrol., (2):36-40.

Translation:
OTS or SLA, $1.10

heat balance

VALERIANOVA, M.A. and E.I. SERYAKOV 1970.
On the many-year variations in the ocean-atmosphere system. Okeanologiia, 10(5): 750-756.
(In Russian; English abstract)

heat balance

Vasyukova, N.G., 1964
The heat balance of the surface waters in some fishing areas of the Bering Sea. (In Russian)
Gosudarst. Kom. Sov. Ministr. SSSR, Ribn. Choz., Trudy VNIRO, 49; Izv. TINRO, 51:77-92.

heat balance

Vives, Francisco, y Manuel López-Benito, 1957
El fitoplancton de la Ria de Vigo desde julio de 1955 a junio de 1956.
Inv. Pesq., Barcelona, 10: 45-146.

heat balance

Wyrtki, Klaus, 1966.
A reply (to E.V. Ashburn, J. geophys. Res., 71 (6):1758-1769).
J. geophys. Res., 71(6):1760.

heat balance

Wyrtki, Klaus, 1965.
The average annual heat balance of the North Pacific Ocean and its relation to ocean circulation.
J. geophys. Res., 70(18):4547-4559.

heat balance

Yakovlev, G.N., 1958.
Solar radiation as the chief component of the heat balance of the Arctic Sea. 4. Sea ice formation, growth and disintegration.
NAS-NRC, Publ., No. 598:181-184.

Zaitsev, G.N., 1960 heat balance

The heat balance of the Norwegian and Greenland seas and factors governing its formation.
Soviet Fish Invest., North European Seas, VNIRO, PNIRO, Moscow, 1960:67-80.

HEAT bottom

heat, bottom

Mosetti, F., 1960
Procedimento per il calcolo della conducibilita termica del terreno dai dati di osservazione delle temperature a piccole profondita.
Atti dell'Ist Veneto di Sci. Lett. ed Arti. Anno acc. 1959-60, Cl. Sci. Mat. e Nat., 118 239-249.
Also: Ist Sperimentale Talassogr., Trieste, Pubbl., No. 369.

HEAT budget

heat budget

Albrecht, Fritz, 1960.
Jahreskarten des Wärme- und Wasserhaushaltes der Ozeane. Berichte des Deutsch. Wetter. 9(66):1-19.

heat budget

Ashburn, Edward V., 1963
The radiative heat budget at the ocean-atmosphere interface.
Deep-Sea Res., 10(5):597-606.

heat budget

Ayers, John C., 1965.
A late summer heat budget of Barnstable Harbor, Massachusetts.
Limnol. and Oceanogr., Redfield Vol., Suppl. to 10: 9-R14.

heat budget

Banks, R.E., 1966.
The cold layer in the Gulf of St. Lawrence.
J. Geophys. Res., 71(6):1603-1610.

Heat Budget

Batalin, A. M., 1960.
An attempt to calculate the heat budget of the Bering Sea. Marine Hydrometeorological Forecasts and Computations.
Trudy Okeanograf. Komissii, 7:23-36.

H. O. TRANS-144 (1962).

heat budget

Batalin, A.M., and N.G. Vasyukova, 1960.
An attempt to calculate the heat budget of the Sea of Okhotsk. (In Russian).
Morskiye Gidrometeorol. Prognozy i Raschety, 7: 37-51.

Translation:
U.S.N. Oceanogr. Off. TRANS-145.

Heat Budget

Bowden, K.F., M.R. Howe and R.I. Tait, 1970.
A study of the heat budget over a seven-day period at an oceanic station. Deep-Sea Res., 17(3): 401-411.

heat budget

Clarke, David B., 1963
Radiation measurements with an airborne radiometer over the ocean east of Trinidad.
J. Geophys. Res., 68(1):235-245.

heat budget

Degtyarev, G.M., Yu. A. Men'shov, B.E. Alemasov, 1962.
Characteristics of the heat budget of the north-western sector of the Atlantic in the summer of 1960. (In Russian).
Izv., Akad. Nauk, SSSR, Ser. Geofiz., (7):965-970.

English Edit., (7):614-618.

heat budget

Dodimead, A.J., F. Favorite and T.Hirano, 1964.
Review of oceanography of the subarctic Pacific Region. Salmon of the North Pacific Ocean. II. (In Japanese).
Collected Reprints, Tokai Reg. Fish. Res. Lab., No. 2:187 pp.

heat budget

Donn, William L., and David M. Shaw, 1967.
The maintenance of an ice-free Arctic Ocean.
Progress in Oceanography, 4:105-113.

heat budget

Donn, William L., and David M. Shaw, 1966.
The heat budget of an ice-free and an ice-covered Arctic Ocean.
J. geophys. Res., 71(4):1087-1093.

heat budget

Dubinskii, G.P., 1957.
On the question of the study of the heat and water regimes of the earth's surface. (In Russian).
Izv., Akad. Nauk, SSSR, Ser. Geogr., (2):94-

Garcia Occhipinti, A., 1959 heat budget
Radiacao solar global e insolacao em Cananeia. Univ. Sao Paulo, Contrib. Avulsas, Inst. Ocean., (1):40 pp.

heat budget

Hastenrath, Stefan, and Marx Steinberg 1971
On the daily variations of temperature and heat budget in the tropical atmosphere.
Arch. Met. Geoph. Biokl. (A) 20 (3): 189-210.

heat budget

Hulburt, Edward M., 1970.
Relation of heat budget to circulation in Casco Bay, Maine.
J. Fish. Res. Bd, Can., 27 (12): 2255-2260

heat budget

Jung, G.H., and R.A. Gilcrest, 1955.
Heat budget of a water column, autumn, North Atlantic Ocean. J. Met. 12(2):152-159.

heat budget (Baltic)

Jurva, R., 1952.
Seas. Fennia 72:136-160.

heat budget

Ketchum, Bostwick H., and Nathaniel Corwin, 1964
The persistence of "winter" water on the continental shelf south of Long Island, New York.
Limnology and Oceanography, 9(4):467-475.

heat budget

Kolesnikov, A.G., and A.A. Pivovarov, 1956.
Sur la possibilité de calculer le bilan thermique à la surface d'un bassin en fonction de la température de l'air.
Izvest. Akad. Nauk, SSSR, Ser. Geophys., 1956(5):

heat budget

Komova, O.N., 1958(1960).
Heat advection in winter in the center of the Arctic basin.
Problemy Severa, (1):

Translation in:
Problems of the North, (1):334-341.

heat budget

Konaga, Shunji, 1961
The water temperature at the sea surface. IV. The influence of weather 2.
J. Oceanogr. Soc., Japan, 17(2):68-73.

English abstract

heat budget

Laevastu, T., 1960

Factors affecting the temperature of the surface layer of the sea.
Merent. Julk. (Havsforskningsinst. Skr.), No. 195: 136 pp.

heat budget

Lyman, John, 1961.
1. Heat and water budget of Antarctica. In: Chapter 3, Antarctic oceanology.
NAS-NRC, Publ., No. 878:25-35.

heat budget

Malkus, Joanne S., 1962
Large-scale interactions. Ch. 4, Sect. II. Interchange of properties between sea and air. In: The Sea, Interscience Publishers, Vol. 1 Physical Oceanography, 88-294.

heat budget

Manabe, Kyohei, Yoshio Watanabe and Akira Wada, 1966.
Study of recirculation of cooling water of Tsuruga nuclear power station sited on Urazoko Bay.
Coast. Engng Japan, 9:157-171.

heat budget

McLaren, I.A., 1961
The hydrography and zooplankton of Ogac Lake, a landlocked fjord on Baffin Island.
Fish. Res. Bd., Canada, MSS Rept. Ser. (Biol.), No. 709: 167 pp. (multilithed).

heat budget

*Meincke, Jens, 1967.
Die Tiefe der jahreszeitlichen Dichteschwankungen im Nordatlantischen Ozean.
Kieler Meesresforsch., 23(1):1-15.

heat budget

Miyazaki, M., 1952.
The heat budget of the Japan Sea. Bull. Hokkaido Regional Fish. Res. Lab., Fish. Agency, No. 4: 1-54, 45 charts

heat budget

Nagayama, M., 1957.
The heat budget of the Eastern China Sea.
Oceanogr. & Meteorol., Nagasaki, 8(1/2):67-75.

heat budget

Neumann, J., 1953.
On a relation between the annual heat budgets of seas and lakes and other terms of the energy equation. Q.J.R. Met. Soc. 79:532-534.

heat budget

Ninomiya, K , 1968.
Heat and water budget over the Japan Sea and the Japan islands in winter season, with special emphasis on the relation among the supply from sea surface, the convective transfer and the heavy snowfall. J. met. Soc. Japan, 46(5): 343-372.

heat budget

Otsuka,K., 1965.
On the unusual low temperature in the East China Sea in winter of 1963 discussed from the heat balance. (In Japanese;English abstract)
J. Meteorol. Res., Japan. Meteorol. Agency, 16(6): 326-333.

Also in: Oceanogr. Meteorol., Nagasaki Mar. Obs. 15 (222).

heat budget (mixed layer)

*Perry, J.D., 1968.
Sea temperature at OWS "I".
Meteorol. Mag. 97(1147):33-43.

heat budget

Peterson, Clifford L., 1960.
The physical oceanography of the Gulf of Nicoya, Costa Rica, a tropical estuary.
Inter-American Tropical Tuna Comm., Bull., 4(4): 139-216.

heat budget

Pickard, G. L., and R. W. Trites, 1957.

Fresh water transport determination from the heat budget with application to British Columbia Inlets.
J. Fish. Res. Bd., Canada, 14(4):605-616.

heat budget

Pickard, G.L., and R.W. Trites, 1957.
Fresh water transport determination from the heat budget with applications to British Columbia inlets. J. Fish. Res. Bd., Canada, 14(4): 605-616.

heat budget

Radok, Uwe, 1956.
The heat economy of the southern ocean. A sampling study. Annales Géophys., 12(4):299-306.

heat budget

Revelle, R., 1958.
Sun, sea and air: IGY studies of the heat and water budget of the earth. Geophysics and the IGY Geophys. Monogr., 2:147-153.

heat budget

Riehl, H., and J.S. Malkus, 1961.
Some aspects of Hurricane Daisy, 1958.
Tellus, 13(2):181-213.

heat budget

Saur, J.F.T., and E.R. Anderson, 1956.
The heat budget of a body of water of varying volume. Limnol. & Oceanogr., 1(4):247-251.

heat budget

Seckel, Gunter R., 1969.
The Hawaiian oceanographic climate, July 1963-June 1965. Bull. Jap. Soc. fish. Oceanogr. Spec. No. (Prof. Uda Commem. Pap.): 105-114.

heat budget

Shesterikov, N. P., 1960

The influence of the heat reserve in water on the time of freezing of Davis Sea and the adjoining part of the Indian Ocean. Info. Biull. Soviet Antark. Exped. (19): 35-38.

heat balance
Shishko, A.F., 1948.
[New calculations of elements in the heat budget balance of the While Sea.] Met. Gidrol. 5:67-75.

heat budget
Stepanov, V.N., 1960.
[The heat budget (balance) in the circulation of the world ocean.]
Trudy Okeanograf. Komissii, Akad. Nauk, SSSR, 10(1):79-81.

heat budget
Stommel, H., and G. Veronis, 1956.
Comments on "Heat budget of a water column, North Atlantic Ocean". J. Meteorol., 13(2):222.

heat budget
Sturm, Manfred 1971.
Extremsituationen im Wärmehaushaltsregime der südlichen Ostsee (Fehmarnbelt) in ihrer Beziehung zur Grosswetterlage.
Beitr. Meeresk. 28:91-110.

heat budget
Sychev, K.A., 1960.
The heat content of Atlantic waters and the expenditure of heat in the Arctic basin.
Problemy Arktiki i Antarkt., 3:5-15.

Listed in Techn. Transl., 8(5):464.

heat budget
Tabata, S., 1958.
Heat budget of the water in the vicinity of Triple Island, British Columbia.
J. Fish. Res. Bd., Canada, 15(3):429-451.

heat budget
Tully, J.P., 1964.
Oceanographic regions and processes in the seasonal zone of the North Pacific Ocean.
In: Studies on Oceanography dedicated to Professor Hidaka in commemoration of his sixtieth birthday, 68-84.

heat budget(ice)
Untersteiner, N., 1964
Calculations of temperature regime and heat budget of sea ice in the Central Arctic.
J. Geophys. Res., 69(22):4755-4766.

heat budget
Untersteiner, N., 1961
On the mass and heat budget of Arctic sea ice.
Arch. f. Meteor., Geoph. u. Bioklim., 12(2): 151-182.

heat budget
Untersteiner, N., and F.I. Badgley, 1958.
Preliminary results of thermal budget studies on Arctic pack ice during summer and autumn. Sec. 3. Physics and mechanics of sea ice.
NAS-NRC, Publ., No. 598:85-92.

heat budget
Vowinckel, E., and S. Orvig, 1966.
Energy balance of the Arctic. V. The heat budget over the Arctic Ocean.
Arch. Meteorol. Geophys. v. Bioklimatol.(A). 14 (3/4):303-325.

heat budget
Vowinkel, E., and S. Orvig, 1962.
Water balance and heat flow of the Arctic Ocean.
Arctic, 15(3):205-223.

heat budget
Waldichuk, Michael, 1957.
Physical oceanography of the Strait of Georgia.
J. Fish. Res. Bd., Canada, 14(3):321-486.

heat budget
*Weller, G.E., 1968.
The heat budget and heat transfer processes in Antarctic Plateau ice and sea ice.
Publ.ANARE sci.Repts.(A) 102:155 pp.

heat budget
Wüst, Georg (with Arnold Gordon), 1964
Stratification and circulation in the Antillean-Caribbean basins. 1. Spreading and mixing of the water types with an oceanographic atlas.
Vema Research Series, Columbia Univ. Press, New York and London No. 2: 201 pp.

HEAT capacity

heat capacity
Bromley, Leroy A., Valeria A. Desaussure, James C. Clipp and James S. Wright, 1967.
Heat capacities of sea water solutions at salinities of 1 to 12% and temperatures of 2° to 80°C.
J.Chem.Engng Data, 12(2):202-206.

heat conservation

heat conservation
*Sudo, Hideo, 1969.
An attempt to estimate the vertical component of the current velocity in the south off the main island of Japan on the basis of heat conservation.
Rec.Oceanogr.Wks.Japan, 10(1):1-11.

heat content

Heat content
Pattullo, June G., Wayne V. Burt, and Sally A. Kulm, 1969.
Oceanic heat content off Oregon: its variations and their causes. Limnol. Oceanogr., 14(2): 279-287.

heat content
Smetanina, N.S., 1961.
[The heat content of the Antarctic waters.]
Okeanologiia, (3):412-417.

heat content
Strokina, L.A. and A.I. Smirnova, 1969.
Heat content of the active layer of the North Atlantic and the change of the surface temperature. Okeanologiia 9(3): 395-403.
(In Russian, English abstract)

heat content
Sturm, Manfred 1970.
Zum Wärmehaushalt der Ostsee im Bereich der südlichen Beltsee (Fehmarnbelt).
Beitr. Meeresk. 27:47-61

heat content
Uusitalo, S, 1965.
The heat content of the Gulf of Bothnia.
Geophysica, Geophys. Soc., Finland, 9(2):149-165

heat distribution
Wyrtki, Klaus and Karl Haberland, 1968.
On the redistribution of heat in the North Pacific Ocean.
J. oceanogr., Soc., Japan, 24(5):220-233.

HEAT exchange

heat exchange
Albrecht, F.H.W., 1958.
Untersuchungen über den Wärmeumsatz an der Meeresoberfläche und die Meeresströmungen im Indischen Ozean. Geofis. Pura e Applic., 39(1958/1):194-215.

heat exchange
Anisimova, E.P., and A.A. Pivovarov, 1966.
The calculation of coefficients of vertical turbulent heat exchange in seas and reservoirs. (In Russian).
Meteorol. i Gidrol., 1966 (2):33-38.

heat exchange
Ball, F.K., 1954.
Sea surface temperatures. Australian J. Physics, 7(4):649-652.

heat exchange
Bespalov, D.P., 1959.
[The heat exchange between the atmosphere and the ocean in the central Arctic.]
Arkticheskiy i Antarkticheskiy Nauchno-Issledova. Inst., Trudy, 226:30-41.

LC n SLA mi $2.40 ph $8.30

heat(sensible) exchange
Bøyum, Gunnvald, 1966.
The energy exchange between sea and atmosphere at Ocean Weather Stations M, I and A.
Geofys. Publr, 26(7):19 pp.

heat exchange
Bøyum, Gunnvald, 1962.
A study of evaporation and heat exchange between the sea surface and the atmosphere.
Geofys. Publikasjoner, Geophysica Norvegica, 22(7):15 pp.

heat exchange
Bogorodsky, M.M., 1964.
A study of tangential friction, vertical turbulent heat exchange and evaporation in the open ocean. (In Russian).
Okeanologiia, Akad. Nauk, SSSR, 4(1):19-25.

heat exchange
Boguslavskii, S.G., 1961.
Latitudinal change in the heat balance of the Atlantic Ocean. Physics of the sea. (In Russian)
Trudy Morsk. Gidrofiz. Inst., 23:139-147.

Translation:
Scripta Technica, Inc. for Amer. Geophys. Union, pp. 111-117.

heat exchange
Boguslavsky, S.G., 1956
[Absorption of sun radiation energy by sea and its effect on the changes of sea temperature.]
Trudy Morskoi Gidrofiz. Inst. 8: 80.

heat exchange
Bryan, K., and E. Schroeder, 1960
Seasonal heat storage in the North Atlantic Ocean. J. Meteorol., 17(6): 670-674. WHOI Contrib. No. 1107.

heat exchange
Budiko, M.I., editor, 1963.
Atlas of heat balance of the world. (In Russian).
Mezhd. Geofiz. Komitet, Prezidiume Akad. Nauk, SSSR, Glabnaia Geofiz. Observ. im. A.I. Borikova, 69 colored charts.

heat exchange
Budyko, M.I., R.L. Kagan and L.A. Stokina, 1966.
Concerning anomalies of heat balance of the ocean. Meteorol. i Gidrol., 1966 (1):24-28.

heat exchange
Charnock, H., 1951.
Energy transfer between the atmosphere and the ocean. Sci. Prog. No. 153:80-95, 4 textfigs.

heat exchange
Dietrich, G., 1951.
Influences of tidal streams on oceanographic and climatic conditions in the sea as exemplified by the English Channel. Nature 168(4262):8-11, 3 textfigs.

heat exchange
Dobroklonsky, S.V., 1944.
[On the diurnal variations of temperature in the surface layer of the sea and on heat currents at the sea-air interface.]
C.R. Acad. Sci. URSS 45(9):371-374.

heat exchange (water-ice)
Doronin, Yu. P., 1961.
[The turbulent heat exchange between water and the ice cover.]
Okeanologiia, Akad. Nauk, SSSR, 1(5):846-850.

heat exchange
Doronin, Yu. P., 1959.
[Turbulent heat exchange between an ice cover and the atmosphere.]
Arktícheskiy i Antarktícheskiy Nauchno-Issled. Inst. Trudy., 226:19-29.

LC and SLA mi, $2.40; ph, $3.30

heat exchange
Eyzhkov, Yu. G., 1963.
Advective heat transfer and formation of a zone of intensified exchange in wind "tide" circulation in the deep sea. (In Russian).
Izv., Akad. Nauk, SSSR, Ser. Geofiz., (5):825-827

Translation:
Amer. Geophys.Union, (5):505-507.

heat exchange
Filatova, T.N., 1959
[Quantitative calculation on heat exchange in certain lakes] Vestnik, Leningrad. Univ., Ser. Geol. i Geogr. 14(6): 107-

heat exchange
Fjeldstad, J.E., 1933.
Wärmeleitung im Meere. Geofys. Publ. 10(7):20pp., 3 textfigs.

heat exchange
Gehlinsch, E., 1948.
[On the process of exchange of solar energy in sea water.] No 68.

Unpublished manuscript in German with English summary on file in HO.

heat exchange
Godbole, Ramesh V., 1963.
A preliminary investigation of radiative heat exchange between the ocean and the atmosphere.
J. Appl. Meteorol., 2(5):674-681.

heat exchange
Gruza, G.V., 1960
Inter-latitude heat exchange in the northern hemisphere. Izv. Akad.Nauk, USSR, Geophys. 2: 226-228. (Volume and pagination that of the English edition).

heat exchange
Haltiner, G.J., 1967.
The effects of sensible heat exchange on the dynamics of baroclinic waves.
Tellus, 19(2): 153-198.

heat exchange
Hasse, Lutz, 1963.
On the cooling of the sea surface by evaporation and heat exchange.
Tellus, 15(4):363-366.

heat exchange
Hishida, Kozo, 1969.
Summary on problems in the synoptic heat exchange. (In Japanese)
Umi to Sora 45(2/3): 77-85

heat exchange
Hishida, K., 1951.
[Second report: Annual variation of sea surface temperature in the oceanic current region.] Bull. Maizuru Mar. Obs., No. 2:9-17, 4 textfigs.

heat exchange
Hishida, Kozo, and Katsunobu Nishiyama, 1969.
On the variation of heat exchange and evaporation at the sea surface in the western North Pacific Ocean.
J. oceanogr. Soc. Japan, 25 (1):1-9.

heat exchange
Hishida, Kozo, and Katsunobu Nishiyama, 1969.
On the diagram for evaluating the amount of heat exchange at the sea surface.
(In Japanese; English abstract).
Umi to Sora 45(1):1-10

heat exchange
Ichie, T., 1950.
On the heat exchange between bay waters and the offshore waters. Mem., Kobe Mar. Obs. 8:26-30, 4 textfigs.

heat exchange
Ivanov, Yu. A., 1967.
On the connection between the processes of heat exchange on the ocean surface and the precipitation-evaporation balance. (In Russian; English abstract).
Fisika Atmosfer. Okean., Izv. Akad. Nauk, SSSR, 3(7):757-763.

heat exchange
Ivanova, Z.S., 1958.
Coefficient of turbulent heat exchange in the surface layers of the Black Sea. (In Russian).
Trudy Morsk. Gidrofiz. Inst., 13:54-64.

heat exchange
Ivanova, Z.S., 1958.
Calculation of the coefficient of vertical turbu-lent heat exchange of different seas. Heat of the sea. Marine Chemistry. (In Russian).
Trudy Morsk. Gidrofiz. Inst., 13:43-53.

heat exchange
Ivanova, Z.S., 1958.
Determination of the differential equation for heat transfer during different systems of variation of turbulent heat exchange. Heat of the sea, sea chemistry. (In Russian).
Trudy Morsk. Gidrofiz. Inst., 13:21-33.

heat exchange
Jacobs, W.C., 1949.
The energy acquired by the atmosphere over the oceans through condensation and through heating from the sea surface. J. Met. 6(4):266-272, 5 textfigs.

heat exchange
Japan, Hakodate Marine Observatory, 1970.
Report of the marine meteorological observations on sea fog and heat exchange at the ocean-air interface, in June, 1969. Supplement: on the drift-ice east of South Saghalien in the Okhotsk Sea in June 1969. (In Japanese)
Mar. met. Rept., Hakodate mar. Obs. 30: 67pp.

heat exchange
Japan Hakodate Marine Observatory, 1969.
Report of the marine meteorological observations on sea fog and heat exchange at the ocean-air interface in June, 1968. (In Japanese)
Mar. Met. Rept. Hakodate mar. Obs., 28:1-40

heat exchange
Johnson, N.G., 1940.
Östersjöns Värmeekonomi.
Svenska Hydrogr.-Biol. Komm. Skr., n.s., Hydrogr 15: 18pp.

heat exchange
Kawabata, Y., H. Kudo, and K. Isiyama, 1949.
[Reflection of heat radiation by sea surface.]
J. Met. Soc., Japan, 27(1):7-9, 2 textfigs.
[In Japanese, with English summary, pp. 34-36.]

heat exchange
Kitaigorodsky, S.A., and Yu. A. Volkov, 1965.
On the calculation of turbulent heat and humidity fluxes in the near-water layer of the atmosphere. (In Russian; English abstract).
Fizika Atmosferi i Okeana, Izv., Akad. Nauk, SSSR 1(12):1319-1336.

heat exchange
Knagos, J. D., 1960
A preliminary investigation of the heat flux from the ocean to the atmosphere in Antarctic regions. J. Geophys. Res., 65(12): 4007-4012.

heat exchange
Koizumi, M., 1956.
Researches on the variations of oceanographic conditions in the region of the Ocean Weather Station "Extra" in the North Pacific Ocean. III. The variations of hydrographic conditions discussed from the heat balance point of view and the exchanges between sea and atmosphere.
Pap. Meteorol. & Geophys., Tokyo, 6(3/4):273-284.

heat exchange
Kolesnikvo, A.G., and A.A. Pivivarov, 1958.
On the correlation between the coefficients of turbulence and heat exchange on the near-water layer of the atmosphere. Heat of the sea, sea chemistry. (In Russian).
Trudy Morsk. Gidrofiz. Inst., 13:65-72.

heat exchange
*Kowalik, Zygmunt, 1966.
Vertical heat exchange in the waters of the Baltic in the years 1950-1962. (In Polish; English abstract).
Przezlad Geofiz., 11(19-3):183-192.

heat exchange
Laevastu, Taivo, 1966.
Daily heat exchange in the North Pacific, its relations to weather and its oceanographic consequences.
Commentat. physico-met., 31(2):1-53.

heat exchange
Laevastu, Taivo, 1965.
Daily heat exchange in the North Pacific, its relation to weather and its oceanographic consequences.
Commentat. physico-math., 31(2):5-53.

heat exchange
Laevastu, T., 1965.
Daily heat exchange in the North Pacific; its effect on the ocean and its relations to weather.
ICNAF, Spec. Publ., No. 6:885-889.

heat exchange
Laevastu, T., 1965.
Synoptic scale heat exchange and its relations to weather.
Fleet Numerical Weather Facility, Monterey, California, Technical Note, No. 7:16 pp.

heat exchange
Laevastu, T., 1963.
Energy exchange in the North Pacific and its relation to weather and its oceanographic consequences. Part 1. Formulas and nomographs for computation of heat exchange components over the sea.
Hawaii Inst. Geophys. Rept., No. 29:15 pp. (mimeographed).

heat exchange
Laevastu, T., 1963/
Energy exchange in the North Pacific; its relations to weather and its oceanographic consequences. Part II. Procedure of computation of heat exchange components and the accuracy of the daily computations.
Hawaii Inst. Geophys., Rept. No. 30:11 pp. (mimeographed)

heat exchange
Laevastu, T., 1963.
Energy exchange in the North Pacific; its relations to weather and its oceanographic consequences. Part III. Daily heat exchange in the North Pacific in different seasons; oceanographic consequences andrelations to weather.
Hawaii Inst. Geophys., Rept. (Rev.)No. 31: 20 pp. (mimeographed).

heat exchange
Laevastu, T., 1960
Factors affecting the temperature of the surface layer of the sea.
Merent. Julk. (Havsforskningsinst. Skr.), No. 195: 136 pp.

heat exchange
Laevastu, T., and E. Ayres, 1966.
Numerical synoptic analyses of heat exchange and their use in ocean thermal structure prediction.
Fleet Numerical Weather Facility, Techn. Note, Monterey, Calif., No. 26:6 pp. (mimeographed)

heat exchange
*Leke,R.A., 1967.
Heat exchange between water and ice in the Arctic Ocean.
Arch.Met.Geophys.Bioklim. (A)16(2/3):242-

heat exchange
Lauscher, F., 1947.
Radiation theory of the hydrophere. Sitzber. Akad. Wiss., Wien, 155, Pt. 2a, Nos. 7-8:281-308.

Abstr.: Phys. Abstr. No. 969(1949).

heat exchange (surface
Lauzier, L.M., et A. Marcotte, 1965.
Comparaison du climat marin de Grande-Rivière (baie des Chaleurs) avec celui d'autres stations de la côte atlantique.
J. Fish. Res. Bd., Canada, 22(6):1321-1334.

heat exchange
Laykhtman, D.L., and Yu. P. Doronin, 1959.
The coefficient of turbulent exchange in the sea, and an estimate of the heat flow from ocean waters. (In Russian).
Arktich. i Antarktich. Nauchno-Issled. Inst., Trudy, SSSR, 226:61-65.

Translation:
LC or SLA, mi $1.80, ph $1.80.

heat exchange
Leahey,D.M., 1966.
Heat exchange and sea ice growth in Arctic Canada.
Mar. Sci. Centre, McGill Univ., Manuscript Rept. No. 1:48 pp. (multilithed).

heat exchanges
Lecompte Pierre et Jacqueline Lenoble, 1966.
Etude theorique des echanges radiatifs mer-atmosphere en grandes longueurs d'onde.
Cahiers Oceanogr., C.C O.E.C., 18(6):497-506.

heat exchange
Manier, G., 1962.
Zur Berechnung der latenten und Fühlbaren Wärmeströme von der Meeresoberfläche an die Luft.
Geofis. Pura e Appl., 52(2):189-213.

heat exchange
Masuzawa, J., 1952.
On the heat exchange between sea and atmosphere in the southern Sea of Japan. Ocean. Mag. 4(2): 49-55, 3 textfigs.

heat exchange
McAlister, E.D., 1962.
Using infrared to measure heat flow from the sea.
Ocean Industry, 2(6):35-39

heat exchange
Miyazaki, M., 1952.
The heat budget of the Japan Sea. Bull. Hokkaido Regional Fish. Res. Lab., Fish. Agency, No. 4: 1-54, 45 charts.

heat exchange
Model, Fr., 1956.
Wärmeumsatz zwischen Meer und Atmosphäre in Atlantischen Südpolarmeer und in Nordpolarmeer.
Ann. Meteorol., 7(1/2):64-66.

heat exchange
Model, F., 1955.
Wärmeumsatz zwischen Meer und Atmosphäre in atlantischen Südpolarmeer und im Nordpolarmeer.
Ann. Met. 7(1/2):64-66.

heat exchange
Newton, Chester W., 1961
Estimates of vertical motions and meridional heat exchange in Gulf Stream eddies and a comparison with atmospheric disturbances.
J. Geophys. Res., 66(3): 853-870.

heat exchange
Nizery, M. A., 1948.
Etudes sur les possibilities d'utilisation de l'énergie thermique des mers et de l'énergie solaire. Bull. Inst. Océan., Monaco, No. 906: 47 pp.

heat exchange
Olevinskaya, S.K., 1961.
[Contribution to theory of thermal interactions between sea, atmosphere and continent.]
Izv. Akad. Nauk, SSSR, Ser. Geofiz., (6):926-932.

English Edit. (6):609-612.

Heat exchange
Pastors, A. A., 1960.
Heat exchange between the Baltic Sea and the Gulf of Riga.
Marine Hydrometeorological Forecasts and Computations.
Trudy Okeanograf. Komissii, 7(98-115).

heat exchange
Ponomarenko, G. P., 1948.
[On a formula for negative heat exchange between air and water.] Trudy Inst. Okean. 2:14-20.

heat exchange
Popov, S.M., and S.A. Piazanov, 1961.
[The significance of the effective radiation in the thermal balance of the ocean.]
Izv. Akad. Nauk, SSSR, Ser. Geofiz., (2):281-293

heat exchange
Privalova, I.V. 1971
Meridional and vertical circulation of waters in the northern part of the Atlantic Ocean. (In Russian)
Okeanol. Issled. Rezult. Issled. Meghd. Geofiz. Proekt. 22:154-219.

heat exchange
Privett, D. W., 1958.
The exchange of heat across the sea surface.
Mar. Obs., 28(179):23-28.

Heat exchange
Samoylenko, V. S. 1965
Experimental and theoretical studies of radiation heat exchange over seas and oceans. Studies of atmospheric circulation and boundary-layer of the air in the Pacific and Indian oceans. (In Russian; English abstract).
Trudy Inst. Okeanol., Akad. Nauk, SSSR, 78:128-153.

heat exchange
Seckel, Gunter R., 1970.
The trade wind zone oceanography pilot study. VIII. Sea level meteorological properties and heat exchange processes, July 1963 - June 1965.
U.S. Fish.Wildl. Serv. spec. scient. Rept. 612: 129 pp. (multilithed).

heat exchange
Shuleikin, V.V., 1952.
[The temperature field above the sea and the land in winter when the coefficient of exchange is varying.]
Doklady Akad. Nauk, SSSR, 83(2):389-

heat exchange
Smetannikova, A.V., 1960.
[Heat exchange between sea and air in the Arctic in winter.]
Trudy Okeanograf. Komissii, Akad. Nauk, SSSR. 10(1):82-89.

Heat exchange
Snopkov, V. G. 1965
On the turbulent exchange of heat and moisture over the ocean. Studies of air in the Pacific and Indian oceans. (In Russian; English abstract).
Trudy Inst. Okeanol., Akad. Nauk, SSSR, 78: 154-178:

heat exchange
Snopkov, V.G., 1964.
Contact heat exchange and rate of evaporation as determined in the North Indian Ocean during the winter monsoon. Investigations in the Indian Ocean (33rd voyage of E/S "Vitiaz"). (In Russian).
Trudy Inst. Okeanol., Akad. Nauk, SSSR, 64:5-10.
period

Heat exchange
Snopkov, V. G., and Yu A. Romanov, 1965
On the turbulent exchange in the surface layer of the ocean. Studies of atmospheric circulation and boundary-layer of the air in the Pacific and Indian oceans. (In Russian; English abstract).
Trudy Inst. Okeanol., Akad. Nauk, SSSR, 78: 203-226.

heat exchange
Stimpson, Robert H., 1968.
Understanding ocean weather.
Oceanol.Intl., 3(7):42-45.

heat exchange
Strokina, L.A., 1956.
[Turbulent heat exchange with the atmosphere and evaporation from the surface of the altic Sea.]
Meteor. i Gidrol., 5:56-60.
Abstr. in: Meteorol. Abstr. & Bibl., 9(1):22.

heat exchange
Tabata, S., 1957.
Heat exchange between sea and atmosphere along the northern British Columbia coast.
Prog. Repts., Pacific Coast Stas., Fish. Res. Bd., Canada, No. 108:18-20.

heat exchange
*Takahashi, Tadas, 1969.
A note on the annual heat exchange across the air-sea boundary.
Rec. oceanogr.Wks.Japan 10(1):13-22.

heat exchange
Takahashi, T., 1958.
[A physical treatment on the accumulated materials of the annual temperature variation in lakes and oceans.]
Mem. Fac. Fish., Kagoshima Univ., 6:47-76.

heat exchange
Terada, K., and K. Osawa, 1953.
On the energy exchange between sea and atmosphere in the adjacent seas. Geophys. Mag. 24(3):155-170, 6 textfigs.

heat exchange
Ulanov, H.K., 1960
[On the problem of heat amount of continental run-off] Meteorol. i Gidrol. (8): 43.

heat exchange
Viebrock, Herbert, 1962
The transfer of energy between the ocean and the atmosphere in the Antarctic region.
J. Geophys. Res., 67(11):4293-4302.

heat exchange
Voskanyan, A.G., A.A. Pivovarov, G.G. Khundzhua, 1970.
Experimental studies of temperature and turbulent heat exchange in the surface layer of the sea. Okeanologiia, 10(4): 588-595.
(In Russian; English abstract)

heat exchange
Watanabe, N., 1955.
Hydrographic conditions of the north-western Pacific. 1. On the temperature change in the upper layer in summer. J. Ocean. Soc., Japan, 11(3):111-122.
Also: Bull. Tokai Reg. Fish. Res. Lab., 11(B-224)

heat exchange
Wooster, Warren S. and Hellmuth A. Sievers, 1970.
Seasonal variations of temperature, drift, and heat exchange in surface waters off the west coast of South America. Limnol. Oceanogr., 15(4): 595-605.

heat exchange
Wyrtki, Klaus, 1966.
Oceanography of the eastern equatorial Pacific Ocean.
In: Oceanography and marine biology, H. Barnes, editor, George Allen & Unwin, Ltd., 4:33-68.

heat exchange
Yakovlev, G.N., 1957.
The turbulent heat exchange between the ice cover and the air in the central Arctic. (In Russian).
Problemy Arktiki, SSSR, (2):193-204.
Translation:
LC or SLA mi $2.70, ph $4.80.

heat exchange
Zakharov, V.F., 1966.
The role of leads off the edge of shore ice in the hydrological and ice regime of the Laptev Sea. (In Russian; English abstract).
Okeanologiia, Akad. Nauk, SSSR, 6(6):1014-1022.

heat exchange
Zhukov, L.A., 1965 (Also: Joukov)
Calculation of the transport and changing of the water temperature of the upper layer of the North Atlantic. (In Russian; English abstract).
Okeanolog. Issled., Rezult. Issled. Programme Mezhd. Geofiz. Goda, Mezhd. Geofiz. Komitet Presidiume, Akad. Nauk, SSSR, No. 13:82-89.

HEAT flow

heat flow
Beloussov, V.V., 1968.
The earth's crust and upper mantle of the oceans. (In Russian; English abstract).
Roz. Issled. Mezhdunarod. Geofiz. Proekt., Lozhduvedomst. Geofiz. Kom., Akad. Nauk, SSSR, 253 pp.

heat flow
Bernal, J.D., R.S. Dietz and J.T. Wilson, 1961.
Continental and oceanic differentiation.
Nature 192(4798):123-128.

heat flow
Birch, Francis, 1967.
Low values of oceanic heat flow.
J. geophys. Res., 72(8):2261-2262.

heat flow
Birch, F.S., 1965.
Heat flow near the New England seamounts.
J. geophys. Res., 70(20):5223-5226.

heat flow
Birch, F., 1956.
Heat flow at Eniwetok Atoll. Bull. G.S.A., 67: 941-942.

heat flow
Boström, K., and M.N.A. Peterson, 1969.
The origin of aluminum-poor ferromanganoan sediments in areas of high heat flow on the East Pacific Rise.
Marine Geol., 7(5): 427-447.

heat flow (bottom)
Bullard, E.C., 1963.
11. The flow of heat through the floor of the ocean.
In: The Sea, M.N. Hill, Editor, Interscience Publishers, 3:218-232.

heat flow
Bullard, E.C., 1962.
The deeper structure of the ocean floor.
Proc. R. Soc., London, (A) 265(1322):386-395.

heat flow
Bullard, E.C., 1954.
The flow of heat through the floor of the Atlantic Ocean. Proc. R. Soc., London, A, 222 (L150):408-420, 17 textfigs.

heat flow
Bullard, E.C., & A. Day, 1961
The flow of heat through the floor of the Atlantic Ocean.
Geophys. J., 4: 282-292.

heat flow (bottom)
Bullard, E.C., R. Revelle and A.E. Maxwell, 1956.
Heat-flow through the deep-sea floor. In: "Advances in Physics", Academic Press, 3:153-181.

heat flow
Burns, Robert E., 1970.
Heat-flow operations at holes 35.0 and 35.1.
Initial Repts. Deep Sea Drilling Project, Glomar Challenger 5: 551-554.

heat flow
Burns, Robert E., 1965.
Sea bottom heat-flow measurements in the Andaman Sea.
In: Int. Indian Ocean Exped., USC&GS Pioneer, 1964, U.S. Dept. Commerce, Environmental Sci. Services Administration, 1:119-120.

heat flow
Burns, Robert E., 1964
Sea bottom heat-flow measurements in the Andaman Sea.
J. Geophys. Res., 69(22):4918-4919.

heat flow
*Burns, Robert E., and Paul J. Grim, 1967.
Heat flow in the Pacific Ocean off central California.
J. geophys. Res., 72(24):6239-6247.

heat flow
Bullett, B.J., R.A. Lagaay, A.P. van Lennep, J.A. Schouten and R.D. Schuiling, 1968.
Some heat-flow measurements in the North Atlantic Ocean.
Proc. K. ned. Akad. Wet. (B) 71(3): 203-208

heat flow
Dehlinger, Peter, R.W. Couch, D.A. McManus and Michael Gemperle, 1970.
II. Regional observations. 4. Northeast Pacific Structure. In: The sea: ideas and observations on progress in the study of the seas, Arthur E. Maxwell, editor, Wiley-Interscience 4(2): 133-189.

heatflow
Diment, William H., and John D. Weaver, 1964.
Subsurface temperatures and heat flow in the AMSOC core hole near Mayaguez, Puerto Rico.
Nat. Acad. Sci.-Nat. Res. Counc., Publ., No. 1188:75-91.

heat flow
*Efimov, A.V., 1967.
Measurements of the heat conductivity coefficient of the bottom sediments with the aid of a thin cylindrical sound in the Indian Ocean. (In Russian; English abstract).
Okeanologiia, Akad. Nauk, SSSR, 7(5):903-907.

heat flow
Epp, David, Paul J. Grim and Marcus G. Langseth, Jr., 1970.
Heat flow in the Caribbean and Gulf of Mexico.
J. geophys. Res., 75(29): 5655-5669.

heat flow
Epp, David, Paul J. Grim and Marcus G. Langseth, Jr., 1968.
Heat flow in the Caribbean and Gulf of Mexico. (Abstract only). Trans. Fifth Carib. Geol. Conf., St. Thomas, V.I., 1-5 July 1968, Peter H. Mattson, editor. Geol. Bull. Queens Coll., Flushing 5: 24.

heat flow
Girdler, R.W., 1970.
A review of Red Sea heat flow.
Phil. Trans. R. Soc. (A) 267 (1181): 191-203

heat flow
Grim, Paul J., 1969.
Heat flow measurements in the Tasman Sea.
J. geophys. Res., 74(15): 3933-3934.

heat flow
*Halunen, A. John, 1968.
Terrestrial heat flow in the Okhotsk Sea. (2).
Oceanogrl Mag., 20(1):73-86.

heat flow
Hanks, Thomas C., 1970.
Model relating heat-flow values near and vertical velocities of mass transport beneath oceanic rises.
J. geophys. Res., 76(2): 537-544.

heat flow
Harper, M.L., 1971.
Approximate geothermal gradients in the North Sea Basin.
Nature, Lond. 230 (5291): 235-236

heat flow, terrestrial
Hart, S. R., and J. S. Steinhart, 1965.
Terrestrial heat flows: measurement in lake bottoms.
Science, 149(3691):1499-1501.

heat flow
Hasebe, K., N. Fujii, and S. Uyeda, 1970
Thermal processes under island arcs.
Tectonophysics 10 (1/3): 335-355.

heat flow
Hastenrath, Stefan L. 1968.
Estimates of the latent and sensible heat flow for the Caribbean Sea and the Gulf of Mexico.
Limnol. Oceanogr. 13 (2): 322-331.

heat flow
Heezen, Bruce C., and A.S. Laughton, 1963.
14. Abyssal plains.
In: The Sea, M.N. Hill, Editor, Interscience Publishers, 3:312-364.

heat flow
Hill, M.N., 1957.
Recent geophysical exploration of the ocean floor
Physics and chemistry of the earth, Pergamon Press:129-163.

heat flow
Horai, Ki-iti, Mary Chessman and Gene Simmons, 1970.
Heat flow measurements on the Reykjanes Ridge.
Nature, Lond., 225 (5229): 264-265.

heat flow
Horai, Ki-iti and Gene Simmons, 1969.
Spherical harmonic analysis of terrestrial heat flow.
Earth Planet. Sci. Letters 6 (5): 386-394

heat flow, effect of
*Horai, Ki-iti, and Gene Simmons, 1968.
Seismic travel time anomaly due to anomalous heat flow and density.
J. geophys. Res., 73(24):7577-7588.

heat flow
Hunt, John M., Earl E. Hays, Egon T. Degens and David A. Ross, 1967.
Red Sea: detailed survey of hot-brine areas.
Science, 156(3774):514-516.

heat flow
Hyndman, R.D., and A.M. Jessop, 1971.
The Mid-Atlantic Ridge near 45°N. XV.
Thermal conductivity of dredge and drill core samples.
Can. J. Earth Sci. 8 (3): 391-393

heat flow
Jones, E.J.W., A.S. Laughton, M.N. Hill and D. Davies, 1966.
A geophysical study of part of the western boundary of the Madeira-Cape Verde Abyssal Plain
Deep-Sea Res., 13(5):889-907.

heat flow
Korgen, Ben J., Gunnar Bodvarsson and Rod S. Mesecar 1971.
Heat flow through the floor of Cascadia Basin.
J. geophys. Res. 76 (20): 4758-4774.

heat flow
Krause, Dale C., 1965.
Tectonics, bathymetry and geomagnetism of the southern continental borderland west of Baja California, Mexico.
Geol. Soc., Amer., Bull., 76(6):645-647.

heat-flow
Lachenbruch, Arthur H., 1970.
Crustal temperature and heat production: implications of the linear heat-flow relation.
J. geophys. Res., 75(17): 3291-3300.

heat flow
Lachenbruch, Arthur H., and B. Vaughn Marshall 1968.
Heat flow and water temperature fluctuations in the Denmark Strait.
J. geophys. Res., 73(18):5829-5842.

heat flow
Lachenbruch, Arthur H., and B. Vaughn Marshall, 1966.
Heat flow through the Arctic Ocean floor: the Canada Basin-Alpha Rise boundary.
J. geophys. Res., 71(4):1223-1248.

heat flow
Lachenbruch, Arthur H., and B. Vaughn Marshall 1964
Heat flux from the Arctic Ocean Basin, preliminary results. (Abstract).
Trans. Amer. Geophys. Union, 45(1):123.

heat flow
*LaFond, Katherine G., 1968.
Messungen des Warmestromes am Meeresboden.
Umschau, 68(1):26.

heat flow
Langseth, Marcus G., Jr., 1966.
Review of heat flow measurements along the Mid-Oceanic Ridge System.
In: World rift system, T.N. Irvine, editor, Dept. Mines Techn. Surveys, Geol. Survey, Can., Paper 66-14:349-362.

heat flow
Langseth, Marcus G., and Paul J. Grim, 1964
New heat-flow measurements in the Caribbean and western Atlantic.
J. Geophys. Res., 69(22):4916-4917.

heat-flow
Langseth, Marcus G., Paul J. Grim and Maurice Ewing, 1965.
Heat-flow measurements in the East Pacific Ocean
J. Geophys. Res., 70(2):367-380.

heat flow
Langseth, Marcus G., Jr., Xavier Le Pichon and Maurice Ewing, 1966.
Crustal structure of the mid-ocean ridges. 5. Heat flow through the Atlantic Ocean floor and convection currents.
J. geophys. Res., 71(22):5321-5355.

heat flow
*Langseth, Marcus G., Jr., and P.T. Taylor, 1967.
Recent heat flow measurements in the Indian Ocean.
J. geophys. Res., 72(24):6249-6260.

heat flow
Langseth, Marcus G., Jr. and Richard P. Von Herzen, 1970.
1. General observations. 9. Heat flow through the floor of the world ocean. In: The sea; ideas and observations on progress in the study of the seas, Arthur E. Maxwell, editor, Wiley-Interscience 4(1): 299-352.

heat flow
Laughton, A.S. 1966.
The Gulf of Aden
Phil. Trans. R. Soc. (A) 259 (1099): 150-171.

heat flow
*Lavenia, A., 1967.
Heat flow measurements through bottom sediments in the southern Adriatic Sea.
Boll. Geofis. teor. appl., 9(36):323-332.

heat flow

Lavenia, A., 1966.
Prime misure sperimentali di flusso termico in mare. (Banneck, 1965. Basso Adriatico).
Boll. Geofis. teor. appl. 8(29):40-55.

heat flow

Law, L. K., W.S.B. Paterson and K. Whitham, 1965.
Heat flow determinations in the Canadian Arctic Archipelago.
Canadian J. Earth Sciences, 2(2):59-71.

heat-flow

Lee, W.H.K., 1970.
On the global variations of terrestrial heat-flow.
Phys. Earth Planet. Interiors, 2(5): 332-341

heat flow

Lee, W.H.K., and C.S. Cox, 1966.
Time variation of ocean temperatures and its relation to internal waves and oceanic heat flow measurements.
J. geophys. Res., 71(8):2101-2111.

heat flow

Lee, W.H.K., S. Uyeda and P.T. Taylor, 1966.
Geothermal studies of continental margins and island arcs.
In: Continental margins and island arcs, W.H. Poole, editor, Geol. Surv. Pap., Can., 66-15:398-414.

Heat

LePichon, X., and M.G. Langseth, J. 1969.
Heat flow from the mid-ocean ridges and sea-floor spreading.
Tectonophysics 8(4/6): 319-344.

heat flow

*Lewis, E.L., 1967.
Heat flow through winter ice.
In: Physics of snow and ice, H. Oura, editor, Inst. Low Temp. Sci., Hokkaido Univ., 611-631.

Heat Flow

Lister, C.R.B., 1970.
Heat flow west of the Juan de Fuca Ridge.
J. geophys. Res., 75(14): 2648-2654.

heat flow

Lister, C.R.B., 1963.
A close group of heat-flow stations.
J. Geophys. Res., 68(19):5569-5573.

heat flow

Lister, C.R.B., and J.S. Reitzel, 1964
Some measurements of heat flow through the floor of the North Atlantic.
J. Geophys. Res., 69(10):2151-2154.

Lubimova, E.A.
+ alternate spellings

heat flow

Lubimova, E.A., 1966.
Marine geothermal investigations and some data on heat flow in the Black Sea basin. (In Russian; English abstract).
Stroenie Chermomorskoi Vladini, Rez. Issled. Mezhd. Geofiz. Proekt., Mezhd. Geofiz. Komitet, Presid., Akad. Nauk, SSSR, 88-93.

heat flow

Liubimora, E.A., and G.A. Tomara and A.L. Alexandrov, 1969.
The heat flow across the bottom of the Arctic Basin in the region of the Lomonossov Ridge. (In Russian)
Dokl. Akad. Nauk. SSSR, 184(2): 403-405.

heat flow

Lubimova, E.A., and G.B. Udintsev, 1965.
Measurements of heat flow through the ocean floor. (In Russian; English abstract).
Okeanolog. Issled., Rez. Issled. po Programme, Mezhd. Geofiz. Goda, Mezhd. Geofiz. Komitet, Prezidiume Akad. Nauk, SSSR, No. 13:236-257.

heat flow(bottom)

Lubimova, E.A., and G.B. Udintsev, 1965.
Measurements of heat flow through the ocean floor. (In Russian; English abstract).
Okeanolog. Issled., Rezult. Issled. Programme Mezhd. Geofiz. Goda, Meshd. Geofiz. Komitet Presidiume, Akad. Nauk, SSSR, No. 13:236-257.

heat flow

MacDonald, G.J.F., 1964.
The deep structure of continents. Heat-flow and gravity observations and satellite data shed light on the origin of continents and oceans.
Science, 143(3609):921-929.

heat flow

Magnitskii, V.A., Iu.P. Neprochnov, and L.N. Rykunov, 1970.
Velocity gradients of elastic waves and temperature gradients beneath the Mohorovicic boundary (Black Sea, Indian Ocean). (In Russian)
Dokl. Akad. Nauk SSSR 195(1): 85-88

heat flow (bottom)

Maxwell, A.E., and R. Revelle, 1956.
Heat flow through the Pacific Ocean basin.
Publ. Bur. Central Seismol., Trav. Sci., (A) 19:394-405.

heat flow

McBirney, Alexander R., and Ian G. Gass, 1967.
Relations of oceanic volcanic rocks to mid-oceanic rises and heat flow.
Earth Planet. Sci. Letters, 2(4): 265-276.

heat flow

McKenzie, D.P., 1970.
Temperature and potential temperature beneath island arcs.
Tectonophysics 10(1/3): 357-366.

heat flow

*McKenzie, Dan P., 1967.
Some remarks on heat flow and gravity anomalies.
J. geophys. Res., 72(24):6261-6273.

heat flow

McKenzie, Dan, and John G. Sclater 1971.
The evolution of the Indian Ocean since the Late Cretaceous.
Geophys. J. R. astr. Soc. 25(5): 437-528.

heat flow

McKenzie, Dan P., and John G. Sclater, 1968.
Heat flow inside the island arcs of the northwestern Pacific. J. geophys. Res., 73(10):3173-3179.

heat flow

Menard, H.W., 1967.
Transitional types of crust under small ocean basins.
J. geophys. Res., 72(12):3061-3073.

heat flow

Mesecar, Roderick S., Gunnar Bodvarsson and Wayne V. Burt, 1969
Time dependent vertical temperature measurements across the water-sediment interface on the continental shelf west of Oregon.
Nature, Lond., 224(5222): 901-902.

heat-flow

Minear, John W., and M. Nafi Toksöz, 1970.
Thermal regime of a downgoing slab.
Tectonophysics, 10(1/3): 367-390.

heat flow

Morelli, C., 1966.
The geophysical situation in Italian waters (1966.0).
Int. hydrogr. Rev., 43(2):133-147.

heat flow

Murray, C.G. 1970.
Magma genesis and heat flow: differences between mid-oceanic ridges and African rift valleys.
Earth Planet. Sci. Letts 9(1): 34-38.

HEAT FLOW

Nagasaka, Koichi, Jean Francheteau and Toshio Kishii, 1970.
Terrestrial heat flow in the Celebes and Sulu seas.
Mar. geophys. Res. 1(1): 99-103.

heat flow

Naidu, Prabhakar S., 1969.
Analysis of scatter in the oceanic heat flow data.
Geophys. J. R. astron. Soc. 17(4): 353-365

heat flow

Naidu, Prabhakar, 1968.
Statistical analysis of East Pacific Rise heat flow data.
Earth Planet Sci. Letters, 4(2): 131-134

heat-flow

Nason, Robert D., and W.H.K. Lee, 1964
Heat-flow measurements in the North Atlantic, Caribbean and Mediterranean.
J. Geophys. Res., 69(22):4875-4883.

heat flow

Nason, R.D., and E.H.K. Lee, 1962.
Preliminary heat-flow profile across the Atlantic.
Nature, 196(4858):975.

heatflow

Oxburgh, E.R., and D.L. Turcotte, 1969
Increased estimate for heat flow at oceanic ridges.
Nature, Lond., 223(5213): 1354-1355

heat flow

Pálmason, Gudmundur, 1967.
On heat flow in Iceland in relation to the Mid-Atlantic Ridge.
Iceland and mid-ocean ridges, Reykjavik 111-127.

heat flow

Paterson, W.S.B., and L.K. law, 1966.
Additional heat flow determinations in the area of Mould Bay, Arctic Canada.
Canadian J. Earth Sci., 3(2):237.

heat flow
Phillips, J.D., G. Thompson, R.P. von Herzen and V.T. Bowen 1969.
Mid-Atlantic Ridge near 43° N. Latitude.
J. geophys. Res. 74 (12): 3069-3081.

heat flow
Popova, A.K., Ya.B. Smirnov, and G.B. Udintsev, 1969.
Deep-water heat flux and its relation to the tectonic structure of the Pacific. Okeanologiia 9(3): 452-461.
(In Russian; English abstract)

heat flow
Rancin, Douglas S., and Roy D. Hyndman, 1971.
Shallow water heat flow measurements in Bras D'or Lake, Nova Scotia.
Can. J. Earth Sci., 8(1): 96-101.

heat flow
Redfield, Alfred C., 1965.
Terrestrial heat flow through salt-marsh peat.
Science, 148(3674):1219-1220.

heat flow
Reitzel, John, 1963
A region of uniform flow in the North Atlantic.
J. Geophys. Res., 68(18):5191-5196.

heat flow, bottom
Revelle, R., A.E. Maxwell, and E.C. Bullard, 1952.
Heat flow through the floor of the Eastern North Pacific Ocean. Nature 170(4318):199-200.

heat flow
Rhea, K., J. Northrop and R.P. Von Herzen, 1964.
Heat-flow measurements between North America and the Hawaiian islands.
Marine Geology, 1(3):220-224.

heat flow
Rikitake, T., and K. Horai, 1960
Studies of the thermal state of the earth. 4th paper. Terrestrial heat flows related to possible geophysical events. Bull. Earthquake Res., Inst., 38(3): 403-420.

heat flow
*Rikitake, T., S. Miyamura, I. Tsubokawa, S. Murauchi, S. Uyeda, H. Kuno and M. Gorai,1968.
Geophysical and geological data in and around the Japan Arc.
Can. J. Earth Sci., 5(4-2):1101-1118.

heat flow
Romanovsky, V., et S. Roobaert, 1967.
Mesure du gradient de température dans les sédiments à grande profondeur dans le golfe de Gascogne.
Trav. Cent. Rech. Étud. océanogr., 7(1):13-18.

heat flow
Ryan, William B.F., Daniel J. Stanley, J.B. Hersey, Davis A. Fahlquist and Thomas D. Allan, 1970.
II. Regional observations. 12. The tectonics and geology of the Mediterranean Sea (June 1969) In: The sea: ideas and observations on progress in the study of the seas, Arthur E. Maxwell, editor, Wiley-Interscience 4(2): 387-492.

Heat flow
Sass, J.H. and Robert J. Munroe, 1970.
Heat flow from deep boreholes on two island arcs.
J. geophys. Res., 75(23): 4387-4395.

heat flow
Schuiling, R.D., 1964.
Serpentinization as a possible cause of high heat-flow values in and near the oceanic ridges.
Nature, 201(4921):807-808.

ice (heat flow)
Schwerdtfeger, Peter, 1963.
Measurement of conducted flow of heat in a sea ice cover.
Nature, 200(4908):769.

heat flow
Sclater, J.G., 1966.
Heat flow in the northwest Indian Ocean and Red Sea.
Phil. Trans. R. Soc., London, A, 259 (1099): 271-278.

heat flow
Sclator, John G., and Charles E. Corey, 1967.
Heat flow, Hawaiian area.
J. geophys. Res., 72(14):3711-3715.

heat flow
Sclater, John G., James W. Hawkins, Jacqueline Mammerickx and Clement G. Chase 1972.
Crustal extension between the Tonga and Lau ridges: petrologic and geophysical evidence.
Bull. geol. Soc. Am. 83 (2): 505-518.

heat flow
Sclater, J.G., E.J.W. Jones and S.P. Miller, 1970.
The relationship of heat flow, bottom topography and basement relief in Peake and Freen deeps, northeast Atlantic.
Tectonophysics 10(1/3): 283-300

heat flow
Sclater, J.G. and J.D. Mudie, 1970.
Detailed geophysical studies on the Hawaiian Arch near 24°25'N, 157°40'W: a closely spaced suite of heat-flow stations. J. geophys. Res., 75(2): 333-348.

heat flow
Sclater, J.G., V. Vacquier and J.H. Rohrhirsch 1970.
Terrestrial heat flow measurements on Lake Titicaca, Peru.
Earth Planet. Sci. Letts 8(1): 45-54.

heat flow
*Shor, G.G., Jr., P. Dehlinger, H.K. Kirk, and W. S. French, 1968.
Seismic refraction studies off Oregon and northern California.
J. geophys. Res., 73(6):2175-2194.

heat flow
Shuiling, R. D., 1966.
Continental drift and oceanic heat-flow.
Nature, 210(5040):1027-1028.

heat flow
*Sleep, Norman H., 1969.
Sensitivity of heat flow and gravity to the mechanism of sea-floor spreading.
J. geophys. Res., 74(2):542-549.

heat flow
Smirnov, J. B., 1966.
The heat-flow at the bottom of aquatories. (In Russian).
Doklady, Akad. Nauk, SSSR, 168(2):428-431.

heat flow
Srivastava, S.P., D.L. Barrett, C.E. Keen, K.S. Manchester, K.G. Shih, D.L. Tiffin, R.L. Chase, A.G. Tomlinson, E.E. Davis and C.R.B. Lister 1971.
Preliminary analysis of geophysical measurements north of Juan de Fuca Ridge.
Can. J. Earth Sci. 8 (10): 1265-1281.

heat flow
Talwani, M., 1964.
A review of marine geophysics.
Marine Geology, 2(1/2):29-80.

heat flow
Yasui, M., K. Horai, S. Uyeda and H. Akamatsu, 1963.
Heat flow measurements in the western Pacific during the JEDS-5 and other cruises in 1962 aboard M/s Ryofu-Maru.
Oceanogr. Mag., Japan Meteorol. Agency, 14(2): 147-156.
JEDS Contrib. No. 47.

heat flow
Taylor, Patrick T., 1966.
Geothermal and magnetic survey off the coast of Sumatra. 2. Interpretation and discussion of results.
Bull. Earthquake Res. Inst., 44(2):541-550.

heat flow
Udintsev, G.B., J.B. Smirnov, A.K. Popova, B.V. Shekhvatov and E.V. Suvilov 1971.
New data on abyssal heat flow through the bottom of the Indian and Pacific Ocean. (In Russian)
Dokl. Akad. Nauk SSSR 2000 (2): 453-456.

Heat - flow
United States, Department of Commerce, Environmental Sciences Services Administration, 1965
International Indian Ocean Expedition, USC&GS Ship Pioneer - 1964.
Vol. 1. Cruise Narrative and scientific results 139 pp.
Vol. 2. Data report: oceanographic stations, BT observations, and bottom samples, 183 pp.

heat flow
Uyeda, S. M. Yasui and K. Horai, 1968.
Heat flow and the magnetic anomalies in oceanic areas around Japan and a review of the present state of marine geophysics. (In Japanese).
J. oceanogr., Soc. Japan, 24(3):137-144.

heat flow
Uyeda, Seiya, Ki-iti Horai, Masashi Yasui and Hideo Akamatsu, 1962.
IX. Heat flow measurements over the Japan Trench during the JEDS-4.
Oceanogr. Mag., Japan Meteorol. Agency, 13(2): 185-189.
JEDS Contrib. No. 37.

heat flow
Uyeda, Seiya, Masashi Yasui, Takahiro Sato, Hideo Akamatsu and Kaoru Kawada, 1964.
Heat flow measurements during the JEDS-6 and JEDS-7 cruises in 1963.
Oceanogr. Mag., Tokyo, 16(1/2):7-10.

heat flow
Vacquier, Victor, Seiya Uyeda, Masashi Yasui, John Sclater, Charles Corry and Teruhiko Watanabe, 1966.
Studies of the thermal state of the earth. 19: Heat-flow measurements in the northwestern Pacific.
Bull. Earthq. Res. Inst., Univ. Tokyo, 44(4) 1519-1535.

Van Andel, Tjeerd H., Richard P. Von Herzen and J.D. Phillips 1971. heat flow
The Vema Fracture Zone and the tectonics of transverse shear zones in oceanic crustal plates.
Mar. geophys. Researches 1 (3): 261-283.

heat flow
*Vaquier, Victor, John G. Sclater and Charles E. Corry, 1967.
Studies of the thermal state of the earth. 21. Heat-flow, eastern Pacific.
Bull. Earthq. Res. Inst., Tokyo Univ., 45(2):375-393.

heat flow
Vaquier, Victor, and P.T. Taylor, 1966.
Geothermal and magnetic survey off the coast of Sumatra. (Summary only).
In: Continental margins and island arcs, W.H. Poole, editor, Geol. Surv. Pap., Can., 66-15:254-255.

heat flow
Vaquier, V., and P.T. Taylor, 1966.
Geothermal and magnetic surveys off the coast of Sumatra. 1. Presentation of data.
Bull. Earthquake Res. Inst., 44(2):531-540.

heat flow
Vaquier, Victor, and R.P. von Herzen, 1964.
Evidence for connection between heat flow and the Mid-Atlantic Ridge magnetic anomaly.
J. Geophys. Res., 69(6):1093-1101.

heat flow
Von Herzen, R.P., 1964.
Ocean-floor heat flow measurements west of the United States and Baja California.
Marine Geology, 1(3):225-239.

heat flow (geothermal)
Von Herzen, R.P., 1963
Geothermal heat flow in the Gulfs of California and Aden.
Science, 140(3572):1207-1208.

heat flow
Von Herzen, R.P., 1963
Heat flow and bottom water temperatures in the eastern Pacific. (Abstract).
Trans. Amer. Geophys. Union, 45(1):73-74.

heat flow
Von Herzen, R.P., 1963
Oceanic heat flow data. In: Nuclear Geophysics.
Nat. Acad. Sci.-Nat. Res. Council, Publ. No. 1075:1-3.

heat flow
Von Herzen, R.P., 1963.
Oceanic heat flow data.
Symposium on the earth's interior. In: Nuclear Geophysics, Proc. Conf., Woods Hole, Massachusetts June 7-9, 1962.
NAS-NRC, Publ., No. 1075:1-3.

heat flow
Von Herzen, Richard P., and M.G. Langseth, 1966
Present status of oceanic heat-flow measurement
In: Physics and Chemistry of the Earth, 6:365-407.

heat flow
Von Herzen, R.P., M.G. Langseth, Jr., 1964
Present status of oceanic heat flow. (Abstract).
Trans. Amer. Geophys. Union, 45(1):74.

heat flow
Von Herzen, R.P., and A.E. Maxwell, 1964.
Measurements of heat flow at the preliminary Mohole site off Mexico.
J. Geophys. Res., 69(4):741-748.

heat flow
Von Herzen, R.P., G. Simmons and A. Folinsbee, 1970.
Heat flow between the Caribbean Sea and Mid-Atlantic Ridge. J. geophys. Res., 75(11): 1973-1984.

heat flow
Von Herzen, R.P., and S. Uyeda, 1963
Heat flow through the eastern Pacific ocean floor.
J. Geophys. Res., 68(14):4219-4250.

heat flow, terrestrial
Von Herzen, R.P., and V. Vacquier, 1967.
Terrestrial heat flow in Lake Malawi, Africa.
J. geophys. Res., 72(16):4221-4226.

heat flow
Von Herzen R.P., and V. Vacquier 1966
Heat flow and magnetic profiles on the Mid-Indian Ocean Ridge.
Phil. Trans. R. Soc. (A) 259 (1099): 262-270.

heat-flow
Watanabe, Teruhiko, David Epp, Seiya Uyeda, Marcus Langseth, and Masashi Yasui, 1970.
Heat flow in the Philippine Sea.
Tectonophysics 10(1/3): 205-224

heat flow
Yasui, Masashi, David Epp, Kiochi Nagasaka and Toshio Kishii 1970.
Terrestrial heat flow in the seas round the Nansei Shoto.
Tectonophysics 10 (1/3): 225-234.

heat flow
Yasui, Masashi, and Toshio Kishii, 1967.
Lowering of heat flow probes in the horizontal attitude.
Oceanogrl Mag., 19(2):193-196.

HEAT FLOW
Yasui, Masashi, Toshio Kishii, Teruhiko Watanabe and Seiya Uyeda, 1966.
Studies of the thermal state of the earth. 18. Terrestrial heat flow in the Japan Sea (2).
Bull. Earthq. Res. Inst., Tokyo Univ., 44:1501-1518.
Also in: Bull. Maizuru mar. Obs. 10 (1967).

heat flow
Yasui, Masashi, Koichi Nagasaka, Toshio Kishii and A. John Halunen 1968.
Terrestrial heat flow in the Okhotsk Sea. (2).
Oceanogrl Mag. 20(1): 73-86
Also in: Bull Maizuru mar. Obs. 11 (1969).

heat flow
Yasui, Masashi, and Teruhiko Watanabe, 1965.
Studies of the thermal state of the earth. 16th paper. Terrestrial heat flow in the Japan Sea (1).
Bull. Maizuru Mar. Obs., No. 9:549-563.

Reprinted: Bull. Earthquake Res. Inst., 43:549-563.

heat flow
*Yasui, Masashi, Toshio Kishii and Ken Sudo, 1967.
Terrestrial heat flow in the Okhotsk Sea - (1).
Oceanogrl Mag., 19(1):87-94.

HEAT FLUX

heat flux
Bortkovsky, R.S., 1962.
The methods of computing the heat and salt fluxes in the ocean, an example of computations in selected regions of the Atlantic. (In Russian)
Trudy Inst. Okeanol., Akad. Nauk, SSSR, 55:130-228.

heat flux
Clauss, E., H. Hinzpeter und J. Müller-Glewe, 1970.
Ergebnisse von Messungen des Temperaturfeldes der Atmosphäre nahe der Grenzfläche Ozean-Atmosphäre.
Meteor Forsch.-Ergebnisse (B) 5: 85-89

heat flux
Colón, José A., 1963
Seasonal variation in hear flux from the sea surface to the atmosphere over the Caribbean Sea.
J. Geophys. Res., 68(5):1421-1430.

heat flux
Dutton, John A., and Reid A. Bryson, 1962.
Heat flux in Lake Mendota.
Limnol. & Oceanogr., 7(1):80-97.

heat flux
Holland, William R. 1971.
Ocean tracer distributions. 1. A preliminary numerical experiment.
Tellus 23 (4/5): 371-392

heat flux
Kraus, E.B., and C. Rooth, 1961
Temperature and steady state vertical heat flux in the ocean surface layers.
Tellus, 13(2):231-237.

heat flux
Lavorko, V.S., 1970.
On the turbulent exchange and heat fluxes in the water near the sea surface. (In Russian).
Fizika Atmosfer. Okean., Izv. Akad. Nauk SSSR, 6(9): 970-972.

heat flux
McAlister, E.D., William McLeish and Ernst A. Corduan, 1971.
Airborne measurements of the total heat flux from the sea during Bomex. J. geophys. Res., 76(18): 4172-4180.

heat flux
O'Brien, Edward E. and Thore Omholt, 1969.
Heat flux and temperature variation at a wavy water-air interface. J. geophys. Res., 74(13): 3384-3385.

heat flux

*Perry, A., 1968.
Turbulent heat flux patterns over the North Atlantic during some recent winter months.
Met. Mag., 97(1153):247-254.

heat flux

Pivovarov, A.A., and Ye. P. Anisimova, 1967.
Turbulent exchange and heat fluxes in the surface layer of the sea. (In Russian)
Vestnik Moskovsk. Univ. 1967(1): 16-23.

heat flux

Radikevich, V.M., 1970.
On computing heat, moisture and momentum fluxes.
Oceanologiia, 10(5): 878-882. (In Russian; English abstract)

heat flux

Roden, Gunnar I., 1971.
Aspects of the transition zone in the northeastern Pacific. J. geophys. Res. 76(15): 3462-3475.

heat flux (oceanic)

Shuleikin, V.V., 1961.
The winter heat flux from the ocean and the effective radiation for a complex continental surface structure. (In Russian).
Doklady, Akad. Nauk, SSSR, 138(2):351-354.

Translation: Consultants Bureau for Amer. Geol. Inst. (Earth Sci. Section only), 138(1-6):618-620. 1962.

heat flux, sensible

Vowinckel, E., and Bea Taylor, 1966.
Energy balance of the Arctic. IV Evaporation and sensible heat flux over the Arctic Ocean.
Arch. Meteorol. Geophys. O Bioklimatol. (B), 14:36-52.

heat flux (surface)

Walmsley, J.L., 1966.
Ice cover and surface heat fluxes in Baffin Bay.
Mar. Sci. Center, McGill Univ., Manuscript Rept., No. 2:94 pp. (multilithed).

heat flux

Wimbush, Mark and John G. Sclater, 1970.
Geothermal heat flux evaluated from turbulent fluctuations above the sea floor. J. geophys. Res., 76(2): 529-536.

heat fluxes

*Zillman, J.W., and J.A. Bell, 1968.
Sea-to-air heat fluxes in the southwest Indian Ocean in summer.
J. geophys. Res., 73(22):7057-7064.

"heat inertia"

"heat inertia"

Kryndin, A.N., and G.P. Isaeva, 1964.
About "heat inertia" of the ocean. (In Russian).
Meteorol. i Gidrol., (11):37-41.

HEAT latent

latent heat, release of

Woodcock, A.H., 1958
The release of latent heat in tropical storms due to the fallout of sea-salt particles.
Tellus, Vol. 10, No. 3, pp. 355-371.

HEATING

heating

Bowden, K. F., 1948.
The process of heating and cooling in a section of the Irish Sea. Mon. Not. Roy. Astron. Soc., 5:270-281.

heating

Kagan, B.A., 1961.
Concerning the method of computing the depth of the sea layer heated during the summer period.
Issled. Severnoi Chasti Atlanticheskogo Okeana, Mezhd. Geofiz. God, Leningradskii Gidrometeorol. Inst., 1:94-97.

heating

Stommel, H., and A.H. Woodcock, 1951.
Diurnal heating of the surface of the Gulf of Mexico in the spring of 1942. Trans. Am. Geophys. Union 32(4):565-571, 5 textfigs.

heating

Takenouti, Y., K. Hata and M. Tori, 1959
On the forecast of surface water temperature for the frontal zone of western North Pacific.
Bull. Hakodate Mar. Obs., (6):96-100.

heating

Tully, John P., 1964.
Oceanographic regions and assessment of temperature structure in the seasonal zone of the North Pacific.
J. Fish. Res. Bd., Canada, 21(5):941-970.

heating

Tully, J.P., 1964.
Oceanographic regions and processes in the seasonal zone of the North Pacific Ocean.
In: Studies on Oceanography dedicated to Professor Hidaka in commemoration of his sixtieth birthday, 68-84.

HEAT loss

heat loss

Anderson, D.V., 1968.
Nocturnal heat loss of a lake and seasonal variation in its vertical thermal structure.
Bull Ass. int. Hydrol. scient. 13(3):33-40

heat loss

Harami, K., 1965.
Some relations between the precipitation in Hokuriku District and the heat energy from sea surface under the monsoonal pressure pattern in winter. (In Japanese; English abstract).
Bull. Maizuru Mar. Obs., No. 9:176-178.

heat loss

Ino, Hideo, and Hiroo Nishida, 1965.
On the relation between the precipitation in Hokuriku District and the heat loss of "Tsushima" warm current during the winter.
(In Japanese; English abstract).
Bull. Maizuru Mar. Obs., No. 9:8-11.

heat loss

Ino, Hideo, and Hiroo Nishida, 1964
On the relation between precipitation in Hokuriku District and the heat loss of "Tsushima" warm current during winter. (In Japanese; English abstract).
Umi to Sora, 40(1):8-11.

heat loss

Leipper, Dale F. 1967.
Observed ocean conditions and Hurricane Hilde 1964.
J. atmos. Sci. 24 (2): 182-196.

heat loss (surface)

Maeda, Akio, 1965.
On the variation of the vertical thermal structure.
J. oceanogr. Soc., Japan, 20(6):255-263.

Also in: Geophys. Notes, Tokyo, 19(1):1966.

heat loss

Samochkin, V.M., 1956.
Use of the heat-balance equation for making more precise the formulas for the calculation of heat loss from water surface. Meteorol. i Gidrol., (2):35-37.

RT4408 in Bibl. Transl. Russ. Sci. Tech. Lit. 38.

heat loss

Timonov, V.V., A.I. Smirnova and K.I. Nepop, 1970.
On the centers of air-sea interaction in the north Atlantic. Okeanologiia, 10(5): 745-749.
(In Russian; English abstract)

HEAT measurements

heat measurements, bottom

Sisoev, N.N., 1961.
The geothermic measurements in the bottom sediments of oceans and seas.
Okeanologiia, Akad. Nauk, SSSR, 1(5):886-887.

HEAT of mixing

heat of mixing

Okubo, A., 1951.
On the heat of mixing of sea water. English abstract.

no reference on reprint in MBL library

heat storage

heat storage

Barber, F.G., 1967.
A contribution to the oceanography of Hudson Bay.
Manuscript Rep. Ser., Dept. Energy, Mines, Resourses, Can., No. 4:69 pp. (multilithed).

heat storage

Bathen, Karl H., 1971.
Heat storage and advection in the North Pacific Ocean. J. geophys. Res., 76(3): 676-687.

HEAT Transfer

heat transfer

Bjørgum, O., 1952.
On the analogy between turbulent transfer of heat and momentum. Univ. Arbok, 1951, Bergen, Naturvit. rekke, No. 10:1-8.

Reviewed: J. du Cons. 19(1):97 by IP.

heat transfer

Bulgakov, N. P., 1961

On the role of convection in the machanism of heat transmission in the abyssal waters of the Atlantic.
Okeanologiya, 1: 45-52.

heat transfer

Burbridge, F.E., 1951.
The modification of continental polar air over Hudson Bay. Q.J. Roy. Soc. Met. 77(333):365-374, 2 textfigs.

heat transfer

Craddock, J.M., 1951.
The warming of arctic air masses over the eastern North Atlantic. Q.J. Roy. Met. Soc. 77(333): 355-364, 4 textfigs.

heat transfer
Crean, P.B. 1967.
Physical oceanography of Dixon Entrance, British Columbia.
Bull. Fish. Res. Bd. Can. 156:1-66.

transfer, heat
Deacon, E.L., and E.K. Webb, 1962
Small-scale interactions. Ch. 3, Sect. II. Interchange of properties between sea and air. In: The Sea, Interscience Publishers, Vol. 1, Physical Oceanography. pp. 43-87.
(Mss received July 1960)

heat transfer
Deardorff, James W., 1968.
Dependence of air-sea transfer coefficients on bulk stability.
J. geophys. Res., 73(8):2549-2557.

heat transfer
Dietrich, G., 1950.
Über systematische Fehler in den beobachteten Wasser- und Lufttemperaturen auf dem Meere und ihre Auswirkung auf die Bestimmung des Warmeumsatzes zwischen Ozean und Atmosphäre. Deutsch Hydro. Zeits. 3(5/6 :314-324, 9 textfigs.

heat transger
Dobrolonsky, S.V., 1944.
[On the diurnal variations of temperature in the surface layer of the sea and on heat currents at the sea-air interface.] Dok. Akad. Nauk, USSR, 45: 371-374.

heat transfer
Drogaytsev, A.A., 1958(1960).
[Winter cooling of Arctic seas.]
Problemy Severa (1):

Translation in:
Problems of the north, 1:44-54.

heat transfer,
Ellison, T.H., 1957.
Turbulent transport of heat and momentum.
J. Fluid Mech., 2(5):456-466.

heat transfer
Ewing, G., and E.D. McAlister, 1960
On the thermal boundary layer of the ocean.
Science, 131(3410): 1374-1376.

heat transfer
Gall, Robert L., and Donald R. Johnson 1971.
The generation of available potential energy by sensible heating: a case study.
Tellus 23(6). 465-482

heat transfer(sea-air)
Gambo, K., 1963
The role of sensible and latent heats in the baroclinic atmosphere.
J. Meteor. Soc., Japan, Ser. II, 41(4):233-246

heat transfer
Gezentsvei, A.N., 1954.
[Divergence of drift currents and heat transmission by the currents in the North Pacific and North Atlantic.] Trudy Okeanol. Inst., 9:54-118.

heat transfer
Hanzawa, M., and N. Takeda, 1950.
Preliminary report on the transformation of the Siberian Pc airmass travelling over sea in winter. Ocean. Mag., Tokyo, 2(2):69-75, 7 textfigs.

heat transfer
Ianes, A.V., 1959
[On the calculation of the heat currents in the ice cover.] Problemi Arktiki i Antarktiki (1): 49-58.

heat transfer
Ichiye, Takashi, and Edward J. Zipser, 1967.
An example of heat transfer at the air-sea boundary over the Gulf Stream during a cold air outbreak.
J. Met. Soc., Japan, 45(3):261-270.

heat transfer
Jensen, A.J.C., 1940
The influence of the currents in the Danish waters on the surface temperature in winter, and on the winter temperature of the air. Medd. Komm. Danmarks Fiskeri-og Havundersø, Ser. Hydrografi, 3(2):52 pp., 14 text figs.

Heat Transfer
Kagan, B. A., 1961. ref.
[The turbulent thermal exchange between the sea surface and the atmosphere and the loss of heat by evaporation in the Arctic seas.]
Probl. Arkt. i Antarkt., (8):78-84.

heat transfer
Kolesnikov, A.G., 1954.
Calculations of the diurnal heat variation of sea temperature from the heat budget at the surface.
Izvest. Akad. Nauk, SSSR, Ser. Geofiz, 1954(2):1 190-194.

heat transfer
Krishna Murthy, B.V., and H.A. Havemann, 1957.
Experiments on free convection heat transfer to water in rectangular vertical gaps open at the upper side. J. Indian Inst. Sci., 39(2):83-88.

heat transfer
Labeish, V.G., 1959
On the dynamics of coastal currents.
Vestnik, Ser. Geol. Geogr., 14(6)91):139-143.

Techn. Transl., 1962,7(4): 205.

heat transfer
Laykhtman, D.L., 1959.
[Several regularities of the heat regime in the Central Arctic.]
Arkticheskiy i Antarkticheskiy, Nauchno-Issled., Inst., Trudy, USSR, 226:42-47.

LC and SLA mi and ph $1.80.

warming-heating
LeGrand, J., 1940.
Nouvelles preuves du rôle de l'échauffement progressif de la mer dans l'affaisement apparent des côtes septentrionales de l'Europe. C. R. Acad Sci., Paris, 210:540-542.

heat transfer
*Leontieva, V.V., 1968.
Some properties of the lower convective layer in the Pacific trenches. (In Russian; English abstract).
Okeanologiia, Akad. Nauk, SSSR, 8(5):807-813.

heat transfer
Leovy, C.B., 1969.
Bulk transfer coefficient heat transfer.
J. geophys. Res., 74(13): 3313-3321.

heat transfer
Liljequist, G.H., 1957.
Long-wave radiation and turbulent heat transfer in the Antarctic winter and the development of surface inversions. J. Atmosph. Terr. Phys., Spec. Suppl., 167-181.

heat transport
Malkus, W.V.R., 1954.
The heat transport and spectrum of thermal turbulence. Proc. R. Soc., London, A, 225(1161): 196-212, 2 textfigs.

heat transport, ocean currents
Manabe, Syukuro 1969
Climate and the ocean circulation 1. The atmospheric circulation and the hydrology of the earth's surface. 2. The atmospheric circulation and the effect of heat transfer by ocean currents.
Mon. Wea. Rev. 97 (11): 739-774; 775-805.

heat-transfer
McAlister, E.D. and William McLeish, 1969.
Heat transfer in the top millimeter of the ocea
J. geophys. Res., 74(13): 3408-3414.

heat transfer
Menshov, Yu. A., and G.M. Degyjarjov, 1963.
About ocean surface effective emission.
(In Russian).
Meteorol. i Gidrol., (7):44-46.

heat transfer
Miyazaki, M., 1949.
The incoming and outgoing heat at the sea surface along the "Tusima" warm current. Ocean. Mag., Tokyo, 1(2):103-111, 5 textfigs.

heat transfer
Model, F., 1950.
Warmwasserheizung Europas. Ber. Deutschen Wetterdienstes in der U.S. Zone, No. 12:51-60, 7 textfigs.

heat transfer
Model, F., 1949.
Berechnungsmethode für die Übertragung von Wärme durch Wassermessen (Stromungen) unter verschieden physikalischen Bedingungen. (dated 25 Oct. 1949). Manuscript, 95 pp. (multilith).

heat transfer
Montgomery, R.B., 1948
Vertical eddy flux of heat in the atmosphere. J. Met. 5(6):265-274.

Heat Transfer
Nazintsev, J. L., 1961. ref.
[The transmission of heat across icebergs in the central Arctic.]
Probl. Arkt. Antarkt., (8):37-45.

heat transfer
Negi, Janardan G., and Rishi Narain Singh, 1967.
On heat transfer in layered oceanic sediments
Earth Planet. Sci. Letters 2(4): 335-336

heat transfer
Privett, D.W., 1958.
The exchange of heat across the sea surface.
Mar. Obs., 28(179):23-28.

heat transfer

Radok, Uwe, 1956.
The heat economy of the Southern Ocean, a sampling study.
Rapp. Sci. Exped. Pol. Franc., n.s., 4(4):299-305.

heat transfer

Reshetova, O.V., 1969.
On the heat transfer and evaporation over the ocean. (In Russian). Fizika Atmosfer. Okean., Izv. Akad Nauk, SSSR, 5(12): 1318-1323.

heat transfer

Savel'ev, B.A., 1958(1960)

[Peculiarities of the ice-thawing processes of the ice cover and in frozen ground.]
Problemy Severa, (1):

Translation in:
Problems of the North, (1): 160-167.

heat transfer

Seryakov, E.I., and Ya. S. Stavissky, 1967.
Heat turnover in the Barents and Norwegian seas. (In Russian).
Mater. Rybokhoz. Issled. Severn. Basseina, (PINRO), 10:126-133.

heat transfer

Shishkov, Yu. A., 1961.
Ice concentrations in the southwest part of the Kara Sea and meriodional heat transfer in the atmosphere. (In Russian).
Problemy Severa, 4:131-137.

Translation, 1962:
Problems of the North, 4:133-139.

heat transfer

Shonting, David H., 1964
Some observations of short-term heat transfer through the surface layers of the ocean.
Limnology and Oceanography, 9(4):576-588.

heat transfer

Shtokman, V.B., 1946.
Vertical distribution of thermal waves in the sea and indirect method of determining the coefficient of turbulent heat conductivity. Trudy Inst. Okean. 1: 44 pp.

heat transfer

Shtokman, V. B., 1946.
[On the relationship between coefficients of horizontal and vertical turbulent heat exchange]
Trudy Inst. Okean. 1: 10 pp.

heat transfer, effect of

Shuleikin, V.V. 1968.
Connection between the climate of Europe and heat transfer in the Atlantic. (In Russian; English abstract).
Fizika Atmosfer. Okean. Izv. Akad Nauk SSSR 4(3): 243-261.

heat transfer

Shuleykin, V.V., 1961.
[Winter transfer of heat from oceans and effective radiation for a continent with a complex surface.]
Doklady, Akad. Nauk, SSSR, 138(2):351-354.

Engl. Abstr., The flow of warm air currents from the Atlantic Ocean into the Eurasian continent.
OTS-61-11147-17 JPRS:8710:2.

heat transfer

Shuleikin, V.V., 1961.
[Winter transfer of heat from the ocean and effective radiation when the surface of the continent is a complex structure.]
Doklady, Akad. Nauk, SSSR, 138(2):351-354.

heat, transfer

Smirnova, A.I., 1970.
Heat transfer by currents in the North Atlantic (In Russian; English Abstract). Okeanologiia, 10(1): 30-37.

heat transfer

Sparger, C.R., 1955.
A relation between wind and heat change in a vertical column of the sea.
Trans. Amer. Geophys. Union 36(5):775-778.

transfer, heat

Stommel, H., 1961.
Thermohaline convection with two stable regimes of flow.
Tellus, 13(2):224-230.

Heat transfer

Tchaplyguine, E. I., 1961. ref.
[The importance of the heat of the currents in the Kara Sea.]
Probl. Ark. i Antarkt., (8):19-28.

heat transfer

Thompson, A.H., and M. Neiburger, 1953.
The radiational temperature changes over the eastern north Pacific Ocean in Jult 1949.
J. Met. 10(3):167-174, 4 textfigs.

heat transfer

Tozer, D.C., 1965.
Heat transfer and convection currents.
Phil. Trans. R. Soc., A. 258(1088):252-271.

heat transfer

Tsikounof, A., 1956.
[Méthode de détermination du coefficient de conductibilité calorifique des couches supérieures de la mer] Inst. Nauchnoj. Informacii, Akad. Nauk, SSSR, Trav. Inst. Océan., d'Etat, No. 33 (45):

Bull. d'Info. 8(10).

heat transfer

*Weller, G.E., 1968.
The heat budget and heat transfer processes in Antarctic Plateau ice and sea ice.
Publ. ANARE sci Repts (A) 102:155 pp.

heat transfer

*Weller, G., 1968.
Heat-energy transfer through a four-layer system: air, snow, sea ice, sea water.
J. geophys. Res., 73(4):1209-1220.

heat transfer

Wickham, J.B., 1951.
Flux of latent and sensible heat at the sea surface off Point Barrow, Alaska. Proc. Alaskan Sci. Conf., Bull. Nat. Res. Coun. 122:81-82.

heat transfer

Zhukov, L.A., 1961

[On the advection of heat by the currents in the upper water layer of the Atlantic Ocean.]
Issled. Severnoi Chasti Atlanticheskogo Okeana, Mezhd. Geofiz. God, Leningradskii Gidrometeorol. Inst., 1:38-42.

heat transfer, effect of

Paulson, C.A. and T.W. Parker, 1972.
The cooling of a water surface by evaporation, radiation, and heat transfer. J. geophys. Res. 77(3): 491-495.

HEAT Transport

transport, heat

Bryan, Kirk, 1962
Measurements of meridional heat transport by ocean currents. (Abstract).
J. Geophys. Res., 67(9):3546.

transport, heat

Bryan, Kirk, 1962
Measurements of meridional heat transport by ocean currents.
J. Geophys. Res., 67(9):3403-3414.

heat transport

*Bryan, Kirk, and Michael D. Cox, 1968.
A nonlinear model of an ocean driven by wind and differential heating: I. Description of the three-dimensional velocity and density fields. 2. An analysis of the heat, vorticity and energy balance.
J. atmos. Sci., 25(6):945-967; 968-983.

heat transport

Gilmour, A.E., 1963.
Hydrological heat and mass transport across the boundary on the ice shelf in McMurdo Sound, Antarctica.
N.Z. J. Geol. Geophys., 6(3):402-422.

heat transport

Emig, Marlies, 1967.
Heat transport by ocean currents.
J. geophys. Res., 72(10):2519-2529.

transport, heat

Holland, William R. 1971.
Ocean tracer distributions. 1. A preliminary numerical experiment.
Tellus 23 (4/5): 371-392

transport, heat

Ivanov, V.M., 1961.
On the relationship of hydrometeorological processes in the Atlantic part of the Arctic and the thermal and dynamic state of the Gulf Stream. (In Russian).
Problemy Severa, 4:27-45.

Translations, 1962:
Problems of the North, 4:25-43.

heat transport

Kort, V.G., 1963.
Heat exchange in the South Ocean. (In Russian).
Mezhd. Geofiz. Kom., Prezidiume, Akad. Nauk, SSSR Rezult. Issled. Programme Mezhd. Geofiz. Goda, Okeanol. Issled., (8):17-23.

transport, heat

Laevastu, T., 1960

Factors affecting the temperature of the surface layer of the sea.
Merent. Julk. (H avsforskningsinst. Skr.), No. 195:136 pp.

transport, heat

Ledniev, A., 1956.
[Le régime thermique des courants de l'Atlantique]
Inst. Nauchnoj Informacii Akad. Nauk SSSR, Trav. Inst. Océan. d'Etat, No. 32(44):

heat transport
Manier, G., 1962
Zur Berechnung der latenten und fühlbaren Wärmeströme von der Meeresoberfläche an die Luft.
Geofisica Pura e Applicata, 52(2):189-213.
English and German summaries

heat transport
Oort, Abraham H., 1964.
Computations of the eddy heat and density transport across the Gulf Stream.
Tellus, 16(1):55-63.

transport, eddy heat
Oort, Abraham H., 1963
Computations of the eddy heat and density transports across the Gulf Stream.
Planetary Circulations Project, Dept. Meteorol Mass. Inst. Techn., Sci. Rept., No. 1:1-30.

heat transport
Penin, V.V. 1966.
On the heat transport by the branches of the Norwegian Current. (In Russian).
Mater. Ribokhoz Issled. Severn. Basseina Poliarn. Nauchno-Issled. Proektn. Inst. Morsk. Ribn. Khoz. Okeanogr. (PINRO) 7: 168-176.

heat transport
Saelen, O.H., 1948.
Temperature and heat transport in the Nordfjord.
Bergens Mus. Årb. 1946 og 1947, Naturv. rekke: 28 pp.

heat transport
Saint-Guily, B., 1963.
On vertical heat convection and diffusion in the South Atlantic.
Deutsche Hydrogr. Zeits., 16(6):263-268.

heat transport
Shuleikin, V.V., 1964.
The transport of heat by currents in the closed cycle of the North Atlantic. (In Russian)
Izv., Akad. Nauk, SSSR, Ser. Geofiz., (2):264-278
Translation:
(AGU) (2):153-160.

heat turnover
Soliankin, E.V., and A.S. Osadchy, 1962
The heat turnover in the Black Sea. (In Russian).
Okeanologiia, Akad. Nauk, SSSR, 2(4):602-613.

heat transport (data)
Soule, F.M., A.P. Franceschetti and R.M. O'Hagen, 1963.
Physical oceanography of the Grand Banks region and the Labrador Sea in 1961.
U.S. Coast Guard Bull., No. 47:19-82.

transport, heat
Soule, Floyd M., and R.M. Morse, 1960
Physical oceanography of the Grand Banks Region and the Labrador Sea in 1958. International Ice Observation and Ice Patrol Service in the North Atlantic.
U.S. Coast Guard Bull., No. 44: 29-99.

transport, heat
Sverdrup, H.U., 1956.
Transport of heat by the currents of the North Atlantic and North Pacific Oceans. Festskrift til Professor Bjørne Helland-Hansen fra Venner og Kolleger ved Chr. Michelsens Institutt, Bergen, 1956:226-236.

Transport, heat
Turner, J.S., 1965.
The coupled turbulent transports of salt and heat across a sharp density interface.
Int. Jour. Heat and Mass Transfer, 8:759-767.

heat wave

heat wave
Howe, M.R., and R.I. Tait, 1969.
Some observations of the diurnal heat wave in the ocean. Limnol. Oceanogr. 14(1): 16-22.

heaving, effect of
Gotheott, J.E.L., and N.H. Benning 1970.
Heave correction in echo sounding.
Electronic Engineering in Ocean Technology, Swansea 21-24 Sept. 1970, I.E.R.E. Conf. Proc. 19: 159-174.

helicopter

helicopter, use of
Boudreault, Y., 1963.
Essai d'un hélicoptère pour les observations océanographiques et la détection du poisson.
Rapp. Ann., 1962, Sta. Biol. Mar., Grande Rivière, 173-181.

Helmholtz equation
Neubert, Jerome A., 1970
Derivation of the stochastic Helmholtz equation for sound propagation in a turbulent fluid.
J. acoust. Soc. Am., 48(5-2):1212-1218

Helmholtz equation
Neubert, Jerome A., 1970.
Asymptotic solution of the stochastic Helmholtz equation for turbulent water.
J. acoust. Soc. Am., 48(5-2):1203-1211

hermaphroditism

hermaphroditism
Baptist, J.P., 1953.
Record of a hermaphroditic horseshoe crab, Limulus polyphemus L. Breviora 14:4 pp., 2 pls.

hermaphroditism
Fréchette, J., G.W. Corrivault et R. Couture, 1970.
Hermaphrodisme protérandrique chez une crevette de la famille des crangonidés, Argis dentata Rathbun.
Naturaliste can. 97 (6):805-822.

heterotrophic activity
Hamilton, R.D. and Janet E. Preslan, 1970.
Observations on heterotrophic activity in the eastern tropical Pacific. Limnol. Oceanogr., 15(3): 395-401.

hermaphroditism
Ulomskii, S.N., 1961.
[A rare case of hermaphroditism in Eudiaptomus graciloides (Lill.)]
Doklady Akad. Nauk, SSSR, 137(3):732-734.

heterotrophic organisms
Williams, P.J. Le B., 1970.
Heterotrophic utilization of dissolved organic compounds in the sea. 1. Size distribution of population and relationship between respiration and incorporation of growth substrates. J. mar. biol. Ass., U.K., 50(4): 859-870.

heterogeneity

Heterogeneity
Priimak, G. I., 1961.
[Certain results of the studies of the statistical micro-heterogeneity of a sea medium.]
Izv. Akad. Nauk, SSSR, Ser. Geofiz., (8):1224-1232.
English Edit., (8):805-810.

heterotrophic activity

heterotrophy
Pintner, I.J. and L. Provasoli 1968.
Heterotrophy in subdued light of 3 Chrysochromulina species.
Bull. Misaki mar. biol. Inst. Kyoto Univ. 12:25-31.

heterotrophy
Sloan, P.R., and J.D.H. Strickland, 1966.
Heterotrophy of four marine phytoplankters at low substrate concentrations.
J. Phycol., 2:29-32.

heterotrophic
Vaccaro, Ralph F., and Holger W. Jannasch, 1966.
Studies on heterotrophic activity in seawater based on glucose assimilation.
Limnol. Oceanogr., 11(4):596-607.

heterothrophic organisms
*Williams, P.J. Le B. and R.W. Gray, 1970.
Heterotrophic utilization of dissolved organic compounds in the sea. 11. Observations on the responses of heterotropic marine populations to abrupt increases in amino acid concentration.
J. mar. biol. Ass., U.K., 50(4):871-881.

high water

high water
Polli, Silvio, 1961
Sul fenemeno dell'acqua alta nell'Adriatico settentrionale.
Rapp. Prelim., Comm. di Studio per la Conservazione della Laguna e della Citta di Venezia, Ist. Veneto di Scienze, Lettere ed Arti, Venezia, 1:1-15.
Also: Ist. Sperimentale Talassografico, Publ. No. 384.

hill, abyssal

hill, abyssal
Matthews, D.H., 1961
Lavas from an abyssal hill on the floor of the North Atlantic Ocean.
Nature, 190(4771):158-159.

hindcasting

see also: forecasting

hindcasting
Huber, K., 1968.
Wind-induced motion in lakes and adjacent seas.
Mitt. Inst. Meeresk. Univ. Hamburg, 10:179-183.

history

History of OCEANOGRAPHY
(and various aspects thereof)
For history of the oceans, see: geological history

history
*Aleem, Anwar Abdel, 1968.
Concepts of marine biology among Arab writers in the Middle Ages.
Bull. Inst. océanogr., Monaco, No. Special 2: 359-367.

history
Brinkmann, August, Jr. 1967.
The Biological Station at Herdla. The Brinkmann reign.
Sarsia 29:99-96.

history
Brattström, Hans 1967.
The biological stations of the Bergens Museum and the University of Bergen 1892-1967.
Sarsia 29:7-80.

history
*Buljan, Miljenko, et Mira Zore-Amanda, 1968.
Apercu historique sur les recherches hydrographiques en mer Adriatique en particulier des navires Ciclope, Najade et Vila Velebita.
Bull. Inst. océanogr., Monaco, No. special 2: 337-349.

history
*Burkhanov, Vasily F., 1968.
Russian researches on the Arctic Ocean.
Bull. Inst. océanogr., Monaco, No. spécial 2: 259-268.

history
Burstyn, Harold L., 1968.
The historian of science and oceanography.
Bull. Inst. océanogr. Monaco, No. special 2: 665-675.

history
*Carpine, Christian, 1968.
Les navires océanographiques dont les noms ont été choisis par S.A. S. le Prince Albert 1er pour figurer sur la facade du Musée océanographique de Monaco.
Bull. Inst. océanogr., Monaco, No. spécial 2: 627-638.

history
*Carpine-Lancre, Jacqueline, 1968.
Les expéditions océanographiques et la publication de leurs résultats (étude bibliographique).
Bull. Inst. océanogr., Monaco, No. special 2: 651-664.

history
*Carruthers, James N., 1968.
Some marine-geological speculations by a British contemporary of Prince Albert I: Professor Percy Fry Kendall F.R.S. (1856-1936).
Bull. Inst. océanogr., Monaco, No. spécial 2: 175-187.

history
*Charlier, Roger Henri, and Eugène Leloup, 1968.
Brief summary of some oceanographic contributions in Belgium until 1922.
Bull. Inst. océanogr., Monaco, No. special 2: 293-310.

history
*Crovetto, Arthur, 1968.
La commission internationale pour l'exploration scientique de la mer Méditerranée: origines, difficultes initiales.
Bull. Inst. océanogr., Monaco, No. spécial 2: 327-335.

history
*Dadic, Zarko, 1968.
The history of the theories of the tide introduced by Jugoslav scientists until the XVIIIth century.
Bull. Inst. océanogr., Monaco, No. spécial 2: 49-54.

history
*Deacon, George E.R., 1968.
Early scientific studies of the Antarctic Ocean.
Bull. Inst. océanogr., Monaco, No. spécial 2: 269-279.

historical
Deacon, G.E.R., 1954.
Exploration of the deep sea. J. Inst. Navig. 7(2):165-174, 5 figs.

history
*Deacon, Margaret, 1968.
Some early investigations of the currents in the Strait of Gibraltar.
Bull. Inst. océanogr., Monaco, No. spécial 2: 63-75.

history
*Destombes, Marcel, 1968.
Les plus anciens sondages portés sur les cartes nautiques aux XVI et XVII Siècles: contribution à l'histone de l'océanographie.
Bull. Inst. océanogr., Monaco, No. spécial 2: 199-222.

history
*Di Paola, Luigi, 1968.
Oceanographic researches by the Hydrographic Institute of the Italian Navy from 1880-192 2.
Bull. Inst. océanogr., Monaco, No. spécial 2: 133-145.

history
Emiliani, C., and R.F. Flint, 1963.
34. The Pleistocene record.
In: The Sea, M.N. Hill, Editor, Interscience Publishers, 3:888-927.

history
*Fedoseyev, Ivan A., 1968.
The advance of knowledge of the quantity of water in the ocean.
Bull. Inst. océanogr., Monaco, No. spécial 2: 99-107.

history
*Friis, Herman R., 1968.
Highlights of the history of the use of conventionalized symbols and signs on large-scale nautical harts of the United States government.
Bull. Inst. océanogr., Monaco, No. spécial 2: 223-241.

history
*Gaskell, Thomas F., 1968.
The history of the Gulf Stream.
Bull. Inst. océanogr., Monaco, No. spécial 2: 77-86.

history
Gigot, J.G., 1965.
Ce que les archives du Roussillon et des Pyrénées-Orientales et en particulier celles de l'Amirauté de Collioure, peuvent apporter à l'étude scientifique de la Méditerranée Roussillonaise.
Colloque Internat., Hist. Biol. Mar., Banyuls-sur-Mer, 2-6 sept., 1963, Suppl., Vie et Milieu, No. 19:123-129.

history
*Gougenheim, Andre, 1968.
Deux ingénieurs hydrographes du XIXth Siècle precurseurs en matière de dynamique des mers.
Bull. Inst. océanogr., Monaco, No. special 2: 87-97.

history
*Grmek, Mirko Drazen, 1968.
Les origines d'une maladie d'autrefois: le scorbut des marins.
Bull. Inst. océanogr., Monaco, No. special 2: 505-523.

history (Canadian oceanography
Hachey, H.B., 1961.
Oceanography and Canadian Atlantic waters.
Fish. Res. Bd., Canada, Bull. No. 134:120 pp.

history
Høisaeter, Tore 1967.
Publications from the biological stations of the Bergens Museum and the University of Bergen 1892-1967.
Sarsia 29:97-136.

history
*Hoppe, Brigitte, 1968.
Influence de la biologie marine sur l'évolution de la pensée écologique au XIX Siècle.
Bull. Inst. océanogr., Monaco, No. spécial 2: 407-416.

history
*Huard, Pierre, 1968.
Hygiène et pathologie navales au XIX Siècle.
Bull. Inst. océanogr., Monaco, No. special 2: 545-554.

history
Huard, Pierre, et Ming Wond, 1965.
Bio-bibliographie de quelques médecins naturalistes voyageurs de la marine au début du XIX e siècle.
Colloque Internat., Hist. Biol. Mar., Banyuls-sur-Mer, 2-6 sept., 1963, Suppl. Vie et Milieu, No. 19:163-217.

history
*Kasumovic, Marijan, 1968.
Histoire du développement de la théorie des marées dans la mer Adriatique.
Bull. Inst. océanogr., Monaco, No. spécial 2: 55-62.

history
*Le Danois, Edouard, 1968.
Présentacion de cartes de pêche à l'Académie des Sciences par S.A.S. le Prince Albert 1er de Monaco (14 février 1921).
Bull. Inst. océanogr., Monaco, No. spécial 2: 255-258.

history
*Le Grand, Yves, 1968.
Les précurseurs de l'optique sous-marine.
Bull. Inst. océanogr., Monaco, No. spécial 2: 163-168.

history
*Lorch, Jacob, 1968.
The history of the sexuality of marine algae.
Bull. Inst. oceanogr., Monaco, No. special 2: 397-406.

history
*Matthäus, Wolfgang, 1968.
The historical development of methods and instruments for the determination of depth-temperatures in the dea in Situ.
Bull. Inst. océanogr. Monaco, No. spécial 2: 35-47.

history
Matthäus, Wolfgang, 1969.
Zur Entdeckungsgeschichte des Äquatorialen Unterstroms im Atlantischen Ozean.
Beitr. Meereskunde, 23:37-70.

history
*Matthäus, Wolfgang, 1968.
Water-level measurements of antiquity.
Bull. Inst. océanogr., Monaco, No. spécial 2: 1-6.

history
*Merriman, Daniel, 1968.
Speculations on life at the depths: a XIXth century prelude.
Bull. Inst. océanogr., Monaco, No. special 2: 377-385.

history
Merriman, D., 1948.
A posse ad esse. J. Mar. Res. 7(3):139-146.

history
*Morovic, Dinko, 1968.
Aperçu historique sur les recherches biologiques effectuées avec le Vila Velebita en Adriatique (1913-1914).
Bull. Inst. océanogr., Monaco, No. special 2: 351-357.

history, chemical
Nicholls, G.D., 1965.
The geochemical history of the oceans.
In: Chemical oceanography, J.P. Riley and G. Skirrow, editors, Academic Press, 2:277-294.

history
*Oren, Oton Haim, 1968.
Jews in cartography and navigation (from the XIth to the beginning of the XVth century).
Bull. Inst. océanogr., Monaco, No. spécial 2: 189-197.

history
*Pérès, Jean-Marie, 1968.
Un précurseur de l'etude du benthos de la Mediterranée: Louis Ferdinand, comte de Marsilli.
Bull. Inst. océanogr., Monaco, No. special 2: 369-376.

history
Peterson, Mendel, 1965.
History under the sea: a handbook for underwater exploration.
Smithsonian Institution, Washington, D.D., 108 pp. 56 pls. $3.00.

history
*Pezzi, Giuseppe, 1968.
L'oeuvre des navigateurs au cours des XVII et XVIII siecles et l'évolution de la médecine Navale.
Bull. Inst. océanogr., Monaco, No. spécial 2: 537-544.

history
*Picotti, Mario, 1968.
L' oceanographie mediterraneene aux XVIII et XIX Siecles.
Bull. Inst. oceanogr., Monaco, No. special 2: 317-326.

history
*Ritchie, George S., 1968.
The Royal Navy's contribution to oceanography in the XIXth century.
Bull. Inst. océanogr., Monaco, No. spécial 2: 121-131.

historical source material
*Ronan, Colin A., 1968.
Some illustrations in the history of oceanography
Bull. Inst. oceanogr., Monaco, No. special 2: 639-649.

history of oceans
Ronov, A.B., 1964.
Common tendencies in the chemical evolution of the earth's crust, ocean and atmosphere.
Geokhimiya, No. 8:715-743.

Translation:
Geochemistry International, No. 4:713-737.

history
Rudwick, M.J.S., 1965.
L'histoire de la paléo-biologie marine.
Colloque Internat., Hist. Biol. Mar., Banyuls-sur-Mer, 2-6 Sept. 1963, Suppl., Vie et Milieu, No. 19:315-320.

history
Runnström, John 1967.
Recollections of the Biological Station at Bergen.
Sarsia 29:81-86.

history
*Sager, Günther, 1968.
The rôle of the tides in Caesar's invasion of Britain.
Bull. Inst. océanogr., Monaco, No. spécial 2: 7-11.

history
*Sager, Günther, 1968.
The tides as an oceanographic factor in the historical development of the North-Central Europe.
Bull. Inst. Océanogr., Monaco, No. spécial 2: 13-23.

history
*Schadewaldt, Hans, 1968.
Idées pathogéniques et thérapeutiques du mal de mer au cours des Siècles.
Bull. Inst. océanogr., Monaco, No. special 2, 525-536.

history
*Schiller, Joseph, 1968.
Controverses autour de certaines structures chez les tuniciers au XIX Siècle.
Bull. Inst. océanogr., Monaco, No. special 2: 387-396.

history
Schopf, Thomas J.M., 1968.
Atlantic Continental Shelf and Slope of the United States - Nineteenth Century Exploration Geol. Surv. Prof. Paper 529-F: F1-F12

history
*Tait, John B., 1968.
Oceanography in Scotland and during the XIXth and early XXth centuries.
Bull. Inst. oceanogr., Monaco, No. special 2: 281-292.

history
*Trégouboff, Grégoire, 1968.
Les précurseurs dans la domaine de la biologie marine dans les eaux des baies de Nice et de Ville franche-Sev-Mer.
Bull. Inst. Océanogr., Monaco, No. spécial 2: 467-480.

history
*Viglieri, Alfredo, 1968.
La carte générale bathymétrique des océans établie par S.A.S. le Prince Albert 1er.
Bull. Inst. océanogr., Monaco, No. spécial 2: 243-253.

history
*Welander, Pierre, 1968.
Theoretical oceanography in Sweden, 1900-1910.
Bull. Inst. océanogr., Monaco, No. spécial 2: 169-174.

history
*Wolff, Torben, 1968.
The Danish Expedition to "Arabia Felix" (1761-1767).
Bull Inst. océanogr., Monaco, No. special 2: 581-601.

history
*Wong, Ming, 1968.
Les navigateurs chinois à la découverte de l' Occident (aspects historiques, techniques et biologiques).
Bull. Inst. océanogr., Monaco, No. special 2: 555-654.

history
Wong, Ming, 1965.
Contribution à l'histoire de la biologie marine chinoise.
Colloque Internat., Hist. Biol. Mar., Banyuls-sur-Mer, 2-6 sept., 1963, Suppl., Vie et Milieu, No. 19:21-34.

history
*Zenkevitch, Lev A., 1968.
Histoires des recherches biologiques quantitatives dans les mers et les océans avant la seconde guerre mondiale.
Bull. Inst. océanogr., Monaco, No. special 2: 491-504.

history
*Wüst, Georg, 1968.
History of investigations of the longitudinal deep-sea circulation (1800-1922).
Bull. Inst. océanogr., Monaco, No. spécial 2: 109-120.

history
Wüst, Georg, 1964.
The major deep-sea expeditions and research vessels 1873-1960 - A contribution to the history of Oceanograpgy.
Progress in Oceanography, Pergamon Press, 2:1-52.

holoplankton

holoplankton
Jeffries, H. Perry, 1967.
Saturation of estuarine zooplankton by congeneric associates.
In: Estuaries, G.H. Lauff, editor, Publs Am. Ass Advmt Sci., 83:500-508.

homing instincts

homing
Cook, Susan Blackford, 1971.
A study of homing behavior in the limpet Siphonaria alternata. Biol. Bull. mar. biol. Lab. Woods Hole, 141(3): 449-457.

Homing
Deelder, C.L. und F.W. Tesch, 1970.
Heimfindevermögen von Aalen (Anguilla anguilla), die über grosse Entfernungen verpflanzt worden waren. Marine Biol., 6(1): 81-92.

homing instincts
Goldsmith, T.H., and D.R. Griffin, 1956.
Further observations of homing terns.
Biol. Bull., 111(2):235-239.

homing (fishes)
Hasler, A.D., R.M. Horrall, W.J. Wisby and W. Braemer, 1958.
Sun-orientation and homing in fishes.
Limnol. and Oceanogr., 3(4):353-361.

homing
Nichols, Paul R., 1960.
Homing tendency of American shad, Alosa sapidissima, in the York River, Virginia.
Chesapeake Science, 1(3/4):200-201.

"homing"
Royce, William F., Lynwood S. Smith and Allan C. Hartt, 1968.
Models of oceanic migrations of Pacific salmon and comments on guidance mechanisms
Fish. Bull. U.S. Fish Wildl. Serv., 66(3):441-462

homing
Tesch, F.-W., 1970.
Heimfindevermögen von Aalen Anguilla anguilla nach Beeinträchtigung des Geruchssinnes, nach Adaptation oder nach Verpflanzung in ein Nachbar-Ästuar. Marine Biol., 6(2): 148-157.

homing

Vladykov, Vadim D. 1971.
Homing of the American eel Anguilla rostrata, as evidenced by returns of transplanted tagged eels in New Brunswick.
Can. Fld Nat. 85(3): 241-248.

homotherms

homotherms

Dunbar, M.J., 1968.
Ecological development in polar regions: a study in evolution.
Prentice-Hall, Inc., 119 pp. $4.95.

horizon

horizon (between sky and sea)

horizon

Saunders, Peter M., 1967.
Shadowing on the ocean and the existence of the horizon.
J. geophys. Res., 72(18):4643-4649.

horizons

"horizons" in crust of earth

Horizon A

Ewing, J., C. Windisch, and M. Ewing, 1970.
Correlation of horizon A with JOIDES bore-hole results. J. geophys. Res., 75(29): 5645-5653.

Horizon A

Ewing, John, J.L. Worzel, Maurice Ewing and Charles Windisch, 1966.
Ages of Horizon A and the oldest Atlantic sediments: coring at an outcrop of Horizon A establishes it as a buried abyssal plain of Upper Cretaceous age.
Science, 154 (3753):1125-1132.

horizon A

Mattson, P.H. and E.A. Pessagno, Jr., 1971.
Caribbean Eocene volcanism and the extent of horizon A. Science 174(4005): 138-139.

Horizon A

Saito, Tsunemasa, Lloyd H. Burckle and Maurice Ewing, 1966.
Lithology and paleontology of the reflective layer Horizon A.
Science, 154(3753):1173-1176.

Horizon Beta

*Habib, Daniel, 1968.
Spores, pollen and mikroplankton from the Horizon Beta outcrop.
Science, 162(3861):1480-1481.

Horizon Beta

*Windisch, Charles C., R.J. Leydebm J.L. Worzel, T. Saito and J. Ewing, 1968.
Investigation of Horizon Beta.
Science, 162(3861):1473-1479.

HORIZONTAL MOVEMENTS

horizontal movements

Anderson, George C., and Karl Banse, 1963
Hydrography and phytoplankton production.
Proc. Conf., Primary Productivity Measurements, Marine and Freshwater, Univ. Hawaii, Aug. 21-Sept. 6, 1961, U.S. Atomic Energy Comm., Div. Techn. Info., TID-7633:61-90.

hormones

hormones

Aubert, M. et C. Margat 1965.
Les facteurs probiotiques du milieu marin.
Cah. C.E.R.B.O.M. 18: 43-48.

hormones

Baggerman, Bertha, 1961.
Hormonal factors in fish migration. (Abstract).
Tenth Pacific Sci. Congr., Honolulu, 21 Aug.-6 Sept., 1961, Abstracts of Symposium Papers, 165-166.

hormones, plant

Bentley, J.A., 1960

Plant hormones in marine phytoplankton, zooplankton and sea water. J.M.B.A., U.K., 39(3): 433-444.

hormones

Briseno, C., B., 1952.
Secreciones internas de los Invertebrados.
Rev. Soc. Mexicana Hist. Nat. 13:1-21.

hormones

Burnett, Allison, L., Richard Davidson and Peter Wiernik, 1963.
On the presence of a feeding hormone in the nematocysts of Hydra pirardi.
Biol. Bull., 125(2):226-233.

hormones

*Fernlund, P., 1968.
Chromactivating hormones of Pandalus borealis.
Bioassay of the red-pigment-concentrating hormone.
Marine Biol., 2(1):13-18.

hormones

*Fernlund, P., and L. Josefson, 1968.
Chromactivating hormones of Pandalus borealis.
On the bioassay of the distal retinal pigment hormone.
Marine Biol., 2(1):19-22.

hormones

Hagerman, D.D., F.M. Wellington and C.A. Villee, 1957.
Estrogens in marine invertebrates. Biol. Bull., 112(2):180-183.

hormones

Jenkin, Penelope M., and John E. Harris, 1962.
Animal hormones: a comparative Survey. Part 1. Kinetic and Metabolic Hormones.
International Series of Monographs on Pure and Applied Biology, Pergamon Press, Vol. 6:310 pp.

hormones, plant

Johnston, R., 1963
Effects of gibberellins on marine algae in mixed cultures.
Limnol. and Oceanogr., 8(2):270-275.

hormones, effect of

Kittredge, James S., Michelle Terry and Francis T. Takahashi 1971.
Sex pheromone activity of the molting hormone, crustecdysone, on male crabs (Pachygrapsus crassipes, Cancer antennarius and C. anthonyi)
Fish. Bull. nat. mar. fish. Serv. 69(2):337-343.

hormones, Crustacea

Knowles, F.G.W., and D.B. Carlisle, 1956.
Endocrine control in the Crustacea.
Biol. Rev., 31(4):396-473.

hormones

Provasoli, L., 1959.

Effect of plant hormones on Ulva.
Biol. Bull., 114(3):375-384.

hormones (pheromone)

Ryan, Edward Parsons, 1966.
Pheromone: evidence in a decapod crustacean.
Science, 151(3708):340-341.

hormones, invertebrate

Sandeen, Muriel I., and John D. Costlow, Jr., 1961

The presence of decapod-pigment-activating substances in the central nervous system of representative Cirripedia.
Biol. Bull., 120(2): 192-205.

hormones (invertebrate)

Scheer, B.T., 1960.
Aspects of the intermoult cycle in Natantians.
Comp. Biochem. Physiol., 1960:3-18.

Also in:
Trav. Sta. Zool., Villefranche-sur-Mer, 19(1960).

hormone

Schneiderman, Howard A., and Lawrence I. Gilbert, 1958

Substances with juvenile hormone activity in Crustacea and other invertebrates.
Biol. Bull., 115(3):530-535.

hornblende

hornblende

Neiheisel, James, 1965.
Source and distribution of sediments at Brunswick Harbor and vicinity, Georgia.
U.S.Army Coastal Eng. Res. Center, Techn. Memo., No. 12:49 pp.

hornblende

Norris, Robert M., 1964.
Sediments of Chatham Rise.
New Zealand Dept. Sci. Ind. Res. Bull., No. 159: New Zealand Oceanogr. Inst., Memoir, No. 16:39pp.

horseshoe crabs

horseshoe crab
Limulus polyphemus

Barber, Saul B., and Wilbur F. Hayes, 1964.
A tendon receptor organ in Limulus.
Comp. Biochem. Physiol., 11:193-198.

Limulus horseshoe crab

Lockwood, S., 1870.
The horsefoot crab.
American Naturalist, 4:257-274.

Limulus horseshoe crab

Pax, Ralph A., and Richard C. Sanborn, 1964.
Cardioregulation in Limulus. 1. Physiology of inhibitor nerves.
Biological Bulletin, 126(1):133-141.

horseshoe crabs

Shuster, Carl N., Jr., 1962.
Serological correspondance among horseshoe "crabs" (Limulidae).
Zoological, N.Y. Zool. Soc., 47(1):1-7.

horseshoe crabs
Shuster, Carl N., Jr., 1960.
Horseshoe crabs. In former years, during the month of May these animals dominated Delaware Bay shores.
Estuarine Bull., 5(2):3-9.

Limulus horseshoe crab
Shuster, Carl, 1953.
Odyssey of the Horseshoe crab.
Audubon Magazine, 55(4):162-163.

horst

horsts
Ball, M.M., C.G.A. Harrison, P.R. Supko and W.D. Bock, and N.J. Maloney, 1968.
Normal faulting on the southern boundary of the Caribbean Sea, Unare Bay, Northern Venezuela
Trans. Fifth Carib. Geol. Conf., St. Thomas, V.I., 1-5 July 1968, Peter H. Mattson, editor.
Geol. Bull. Queens Coll., Flushing, 5: 17-21.

horst
Knott, S.T., E.T. Bunce and R.L. Chase, 1966.
Red Sea seismic reflection studies.
In: The world rift system, T.N. Irvine, editor, Dept. Mines Techn. Surveys, Geol. Survey Can., Paper, 66-14:33-61.

hot brine. See also: salinity, high

hot brines

See also: salinity, high

hot brines
Baturin, G.N., A.V. Kochenov, and E.S. Trimonis, 1969.
On the composition and origin of iron-ore sediments and hot brines in the Red Sea. Okeanologiia 9(3): 442-451.
(In Russian; English abstract)

hot brines
Grice, C. Fitzhugh, 1968.
The Red Sea's hot brines and heavy metals.
Ocean Industry, 3(3):52-53, 55, 57.

hot brine
Riley, J.P., 1967.
The hot saline waters of the Red Sea bottom and their related sediments.
Oceanogr. Mar. Biol. Ann. Rev., H. Barnes, editor, George Allen and Unwin, Ltd., 5:141-157.

hovercraft

hovercraft
Anon., 1964
Hovercraft unveiled.
Undersea Techn., 5(9):24-25.

hovercraft SRN2
Fishlock, David, 1965.
Afloat on a cushion of air.
Sea Frontiers, 11(2):81-89.

hulls, pressure

hulls, pressure
Cornish, R.H., 1969
Comments on "Composite materials for pressure hull structures" by K. Hom.
Ocean Engng 1(3): 325-327

hulls, pressure
* Hom, Kenneth, 1969.
Composite materials for pressure hull structures.
Ocean Engng 1(3): 315-324.

human engineering

human engineering
Miller, James W., 1966.
The measurement of human performance, SEALAB II.
Man's extension into the sea, Trans. Symp., 11-12 Jan. 1966, Mar. Techn. Soc., 156-169.

human engineering
Wenzel, J.G., and W.M. Helvey, 1966.
Manned aspects of deep submersibles.
Man's extension into the sea, Trans. Symp., 11-12 Jan. 1966, Mar. Techn. Soc., 111-133.

human physiology
Mazzone, Walter F., 1966.
Human physiology aspects of SEALAB II.
Man's extension into the sea, Trans. Symp., 11-12 Jan. 1966, Mar. Techn. Soc., 102-110.

humidity gradient

humidity, effect of
Ruppersberg, Gerhard H., 1971.
Die Änderung des maritimen Dunst-Streukoeffizienten mit dem relativen Feuchte.
Meteor Forsch.-Ergebn. (B) 6: 37-60

humidity gradient
Sverdrup, H. U., 1946.
The humidity gradient over the sea surface.
J. Met. 3:1-8.

humus

humidity
Yershinskii, N.V., 1970.
Spectra of humidity pulsations over the Mediterranean Sea. (In Russian).
Dokl. Akad. Nauk SSSR 193(5): 1035-1037.

humus

See: chemistry, humus

hurricanes

hurricanes
Abbot, C.G., 1948
Solar variation attending West Indian hurricanes. Smithsonian Misc. Coll. 110(1):7 pp., 1 text fig.

Hurricanes
Abdullah, Abdul Jabbar, 1966.
The spiral bands of a hurricane: a possible dynamic explanation.
J. Atmospheric Sci., 23(4):367-375.

hurricanes
Anthes, Richard A. 1970.
The role of large-scale asymmetries and internal mixing in computing meridional circulations associated with the steady state hurricane.
Mon. Wea. Rev. U.S. Dept. Comm. 98(7): 521-528.

hurricanes
Anthes, Richard A., James W. Trout and Stellan S. Ostlund 1971.
Three dimensional particle trajectories in a model hurricane.
Weatherwise 24(4): 174-178

hurricanes
Beckhausen, F., 1955.
Zusammentreffen zweier deutscher Schiffe mit Hurrikan "Edna". Der Seewart 16(1):12-15, 2 textfigs.

hurricanes
Bandeen, W.R., V. Kunde, W. Nordberg and H.P. Thompson, 1964.
TIROS III meteorological satellite radiation observations of a tropical hurricane.
Tellus, 16(4):481-502.

hurricanes
Barrientos, Celso S., 1964.
Computation of transverse circulation in a steady state, symmetric hurricane.
J. Appl. Meteorol., 3(6):685-692.

hurricanes
Baum, Robert A., 1970.
The eastern Pacific hurricane season of 1969.
Mon. Wea. Rev. U.S.A. Dept. Comm. 98(4): 280-292.

hurricanes
Bodine, B.R. 1969.
Hurricane surge frequency estimated for the Gulf Coast of Texas.
Techn. Memo. Coast. Engng Res. Cent. U.S. Army Engrs 26: 31 pp. (multilithed)

hurricanes
Bradley, Donald A., 1965.
Tidal components of hurricane development.
Nature, 206(4989):1145.

hurricanes
Bradley, Donald A., 1964.
Tidal components in hurricane development.
Nature, 204(4955):136-138.

hurricanes
Bretschneider, C.L., 1963
Appendix: wave spectra from hurricane Donna 1959.
Ocean wave spectra. Proceedings of a conference. Easton, Maryland, May 1-4, 1961. Prentice-Hall, Inc., viii and 357 pp.

hurricanes
Brooks, C.F., 1939.
Hurricanes in New England. Geogr. Rev. 29:119-127

hurricanes
Brooks, C.F., and C. Chapman, 1945.
The New England hurricane of September 1944.
Geogr. Rev. 35(1):132-136, 1 fig.

hurricanes
Broughner, C.C., 1955.
Hurricane Hazel. Weather 10(6):200-205, 2 textfi

hurricanes
Bunker, A.F., 1957
Turbulence measurements in a young cyclone over the ocean .
Bull. Amer. Meteorol. Soc., Vol. 38, No. 1, pp. 13-16.

hurricanes
Burt, W.V., 1958.
Can mid-latitude storms be as severe as tropical typhoons and hurricanes?
U.S. Naval Inst. Proc., 84(6):127-129.

hurricanes
Carlson, Toby N., 1969.
Synoptic histories of three African disturbances that developed into Atlantic hurricanes.
Mon. Wea. Rev., 97(3): 256-276.

hurricanes
Carrier, G.F., 1971
The intensification of hurricanes.
J. fluid Mech. 49(4): 145-158.

hurricanes
Carrier, G.F., A.L. Hammond and O.D. George, 1971.
A model of the mature hurricane.
J. fluid Mech. 47 (1): 145-170

hurricanes
Colón, José A., 1966.
Some aspects of hurricane Carla (1961).
Hurricane Symposium, Oct. 10-11, 1966, Houston, Publ. Am. Soc. Oceanogr., 1:1-33.

Hurricanes
Cressman, George P., 1969.
Killer storms.
Bull. Am. Met. Soc. 50 (11): 850-855.

hurricanes
Cry, George W., 1965.
Tropical cyclones of the North Atlantic Ocean: tracks and frequencies of hurricanes and tropical storms, 1871-1963.
U.S. Dept. Commerce, Weather Bur., Techn. Paper, No. 55:148 pp.

hurricanes
Davis, W.R., 1954.
Hurricanes of 1954. Mon. Weather Rev. 82(12):370-373.

DeBremaecker, J. Cl., 1965. hurricanes
Microseisms from Hurricane "Hilda".
Science, 148(3678):1725-1727.

hurricanes
Denney, William J., 1971.
Eastern Pacific hurricane season of 1970.
Mon. Weath. Rev., U.S. Dept. Comm. 99 (4): 286-301

hurricanes
Denney, William J., 1969.
The eastern Pacific hurricane season of 1968.
Mon. Wea. Rev., 97(3): 207-224.

hurricanes
Dunn, Gordon E., R. Cecil Gentry and Billy M. Lewis, 1968.
An eight-year experiment in improving forecasts of hurricane motion.
Mon. Weath. Rev. U.S. Dep. Comm. 96(10):708-713.

hurricane
Donn, W.L., 1952.
An investigation of swell and microseisms from the hurricane of September 13-16, 1946.
Trans. Amer. Geophys. Union 33(3):341-344, 3 textfigs.

hurricanes, detection of
Donn, W.R., and M. Blaik, 1953.
A study and evaluation of the tripartite seismic method of locating hurricanes. Bull. Seismol. Soc., Amer., 43(4):311-329, 7 textfigs.

hurricanes
Dunn, Gordon E. 1958.
Hurricane and hurricane tides, Engineering Prog. at U. of Florida Vol XII, no. 12 p. 3-13.

Selected Paper from proceedings of 6th Conf in. Coastal Engineering.

hurricanes
Dunn, G.E., 1956.
Areas of hurricane development. Mon. Weather Rev. 84(2):47-51.

hurricanes
Dunn, G.E., and B.I. Miller, 1960.
Atlantic hurricanes.
Louisiana State Univ. Press, Baton Rouge, xx 326 pp. $10.00

Reviewed by: Alkire, H.L., 1962.
Chesapeake Science, 3(1):50.

hurricanes
Ehhalt, Dieter H. and H. Göte Östlund, 1970.
Deuterium in hurricane Faith 1966: preliminary results. J. geophys. Res., 75(12): 2323-2327.

hurricanes
Erickson, Carl O., 1967.
Some aspects of the development of hurricane Dorothy.
Mon. Weath. Rev. U.S. Dep. Agric., 95(3):121-130.

hurricanes
Fisher, E.L., 1958.
Hurricanes and the sea-surface temperature field.
J. Meteorol., 15(3):328-333.

hurricanes
Fletcher, Robert D., and Karl R. Johannessen, 1965.
Maximum hurricane surface wind computation with typhoon-environment data measured by aircraft.
J. Appl. Meteorol., 4(4):457-462.

hurricanes
Freeman, John C., 1967.
Two theories on the origin of hurricanes.
Ocean Indust. 2(8): 20-22. (popular)

hurricanes
Gangopadhyaya, M., and H. Riehl, 1959
Exchange of heat, moisture and momentum between Hurricane Ella (1958) and its environment. Nat. Hurricane Res. Proj., USWB, No. 29: 1-12.

hurricanes
Gentry, R. Cecil 1967.
Structure of the upper troposphere and lower stratosphere in the vicinity of hurricane Isbell, 1964.
Pap. Met. Geophys. Tokyo 18(4): 293-310.

hurricanes
Gentry, R. Cecil, Tetsuya T. Fujita and Robert C. Sheets, 1970.
Aircraft, spacecraft, satellite and radar observations of hurricane Gladys, 1968.
J. appl. Met. 9(6): 837-850

hurricanes
Gilman, C.S., & V. A. Myers, 1961
Hurricane winds for design along the New England coast.
J. Waterways & Harb. Div., Proc. Amer. Soc. Civil Eng., 87(WW2): 45-66.

hurricanes
Goldman, Joseph L., 1967.
Wind behavior in a hurricane.
Ocean Indust. 2(8): 23-27. (popular)

hurricanes
Gordon, A.H., 1950.
Hurricanes at Barbadoes. Mar. Obs. 20(147)(M.O. 528):37-39, 1 textfig.

hurricanes
Graham, David M., 1968.
'Stormfury' and the hurricane hunt.
Ocean Industry, 3(9): 66-68

hurricanes
Gray, William M., 1966.
On the scales of motion and internal stress characteristics of a hurricane.
J. Atmosph. Sci., 3(3):278-288.

hurricanes
Gray, William M., 1965. vertical
Calculations of cumulus/draft velocities in hurricanes from aircraft observations.
J. Appl. Meteorol., 4(4):463-474.

hurricanes
Gray, W.M., 1962.
On the balance of forces and radial accelerations in hurricanes.
Q.J. R. Meteorol. Soc., 88(378):430-458.

hurricanes
Green, Raymond A., 1964.
The weather and circulation of October 1963 - abnormal warmth and severe drought in the United States and two unusual hurricanes offshore.
Monthly Weather Review, 92(1):37-42.

hurricanes
Groening, H.U., 1968.
Tropische Zyklonen im Südostlichen Nordatlantik.
Der Seewart, 29(5):177-188.

hurricanes
Hadlock, Ronald K., and Seymour L. Hess, 1968.
A laboratory hurricane model incorporating an analog to release of latent heat.
J. atmosph. Sci., 25(2):161-177.

hurricanes
Haggard, William H., 1958
The birthplace of North Atlantic tropical storms. Mo. Weath. Rev., 86(10): 397-404.

hurricanes
Harris, D.L., 1963.
Characteristics of the hurricane storm surge.
U.S. Dept. Commerce, Weather Bureau, Techn. Paper, No. 48:139 pp.

hurricanes
Harris, D. Lee, 1961.
Discussion of Technical Memorandum No. 120 "The prediction of hurricane storm-tides in New York Bay" (and closure by author) [Basil W. Wilson.]
Beach Erosion Bd., Techn. Mem., No. 120-A:29 pp.

hurricane
Harris, D. Lee, 1958
The Hurricane Surge Engineering Prog. at U. of Florida. Vol 12, No. 12 p. 14-32.

Selected paper from Proceedings of 6th Congress on Coastal Engineering

hurricanes
Hawkins, Harry F. 1971.
Comparison of results of the hurricane Debbie (1969) modification experiments with those from Rosenthal's numerical model simulation experiments
Mon. weath. Rev. U.S. Dept. Comm. 99(5): 427-434

hurricanes
Hawkins, Harry F., and Daryl T. Rubsam, 1968.
Hurricane Hilda, 1964 I. Genesis as revealed by satellite photographs, conventional and aircraft data
Mon. Weath. Rev. U.S. Dep. Comm 96 (7): 428-452.

hurricanes
Hawkins, Harry F., and Daryl T. Rubsam, 1968.
Hurricane Hilda, 1964. III. Degradation of the hurricane.
Mon. Weath. Rev. U.S. Dep. Comm., 96(10):701-707.

hurricanes
*Hawkins, Harvey F., and Daryl T. Rubsam, 1968.
Hurricane Hilda, 1964: II. Structure and budgets of the hurricane on October 1, 1964.
Mon. Weath. Rev., U.S. Dep. Comm., 96(9):617-636.

hurricanes
Hilder, B., 1958.
On the avoidance of hurricanes. J. Inst. Navig., 11(2):194-206.

hurricanes
Holliday, Charles R., and Allen F. Flanders, 1966
Redefinition of Hurricane Dora over the Gulf Stream.
Mon. Wea. Rev. 94(10):616-618.

hurricanes
Hoose, Harry M., and José A. Colón, 1970.
Some aspects of the radar structure of hurricane Beulah on September 9, 1967.
Mon. Wea. Rev. U.S. Dept. Comm. 98(7): 529-533

hurricanes
Hoover, E.W., 1961
The effect of differential friction between land and water on the movement of Donna in the vicinity of eastern North Carolina.
Monthly Weather Rev., 89(91):340.

hurricanes
Hoover, R.A., 1957.
Empirical relationships of the central pressures in hurricanes to the maximum surge and storm tide
Mon. Wea. Rev., 85(5):167-174.

hurricanes
Hope, John R., and Charles J. Neumann, 1970.
An operational technique for relating the movement of existing tropical cyclones to past tracks.
Mon. weath. Rev. U.S. Dept. Comm. 98(12): 925-933.

hurricanes
Hubert, L. F., 1969.
Comments on "The eastern Pacific hurricane season of 1968".
Mon. weath. Rev. 97 (2): 521-522

hurricanes
Hubert, L.F., 1955.
Frictional filling of hurricanes.
Bull. Amer. Met. Soc., 36(9):440-445.

hurricanes
Hubert, L.F., 1955.
A case of hurricane formation. J. Met. 12(5): 486-492.

hurricanes
Hubert, Lester F. and Andrew Timchalk, 1969.
Estimating hurricane wind speeds from satellite pictures
Mon. Weath. Rev. 97 (5): 382-383.

hurricanes
Jordan, C.L., 1967.
What are hurricanes and when do they occur.
Ocean Indust. 2(8): 16-20. (popular)

hurricanes
*Jordan, C.L., 1966.
Climatological features of the formation and tracks of hurricanes.
Hurricane Symposium, Oct. 10-11, 1966, Houston, Publ. Am. Soc. Oceanogr., 1:82-101.

hurricanes
Jordan, C.L., 1955.
Tidal forces and the formation of hurricanes.
Nature 175(4444):38-39.

hurricanes
Kasahara, A., 1957.
The numerical prediction of hurricane movement with the barotropic model.
J. Meteorol., 14(5):386-402.

hurricanes
Kessler, E., III., 1958.
Eye region of hurricane Edna, 1954.
J. Meteorol., 15(3):264-270.

hurricanes
Kessler, E., III, 1957
Outer precipitation bands of hurricane Edna and Ione.
Bull. Amer. Meteorol. Soc., 38(6):335-346.

hurricanes
Krishnamurti, T.N., 1962
Some numerical calculations of the vertical velocity field in hurricanes.
Tellus, 14(2): 195-211.

hurricanes
Krishnamurti, T.N., 1961
On the vertical velocity field in a steady symmetric hurricane.
Tellus, 13(2):171-180.

hurricanes
Krueger, A.F., 1954.
The weather and circulation of October 1954 - including a discussion of Hurricane Carol in relation to the larger-scale circulation.
Mon. Weather Rev. 82(10):296-300, 6 textfigs.

hurricanes
*La Seur, N.E., 1966.
On the description and understanding of hurricane structure.
Hurricane Symposium, Oct. 10-11, 1966, Houston, Publ. Am. Soc., Oceanogr., 1:71-81.

hurricanes
Leipper, Dale F. 1967.
Observed ocean conditions and Hurricane Hilda 1964.
J. atmos. Sci. 24 (2): 182-196.

hurricanes
Macdonald, Norman J., 1968.
The evidence for the existence of Rossby-like waves in the hurricane vortex.
Tellus, 20(1): 138-150

hurricanes
Machta, L., 1969.
Evaporation rates based on tritium measurements for Hurricane Betsy.
Tellus 21(3): 404-408

hurricanes
Malkin, W., and G.C. Holzworth, 1954.
Hurricane Edna, 1954. Mon. Wea. Rev. 82(9):267-279.

hurricanes
Malkus, Joanne S., 1962
Large-scale interactions. Ch. 4, Sect. II. Interchange of properties between sea and air. In: The Sea, Interscience Publishers, Vol. 1, Physical Oceanography, 88-294.

hurricanes
Malkus, Joanne S., 1960
Recent developments in studies of penetrative convection and an application to hurricane cumulonimbus towers.
In: Cumulus Dynamics, Pergamon Press, London and New York, pp. 65-84.

hurricanes
Malkus, Joanne S., 1958
On the structure and maintenance of the mature hurricane eye.
Jour. Meteorol., Vol. 15, No. 4, pp. 337-349.

hurricanes
Malkus, J.S., 1958.
Tropical weather disturbances -- why do so few become hurricanes? Weather 13(5):75-89.

hurricanes
Malkus, J.S., 1957.
The origin of hurricanes.
Scientific American, 197(2):33-39.

hurricanes
Malkus, J.S. an H Riehl 1968
On the dynamics and energy transformations in steady state hurricanes.
Tellus 12(1): 1-20.

hurricanes
Malkus, J.S., C. Ronne, & M. Chaffee, 1961
Cloud patterns in hurricane Daisy, 1958.
Tellus, 13(1): 8-30.
WHOI Contrib. No. 1145.

hurricanes
Matano, H., 1958.
On the synoptic structure of Hurricane Hazel 1954, over the eastern United States.
J. Meteorol. Soc., Japan, (2) 36(1):23-31.

hurricanes
*Mayencon, Rene, 1968.
Le cyclone tropical.
Cah. océanogr., 22(8):695-710.

hurricanes
Merino Coronado, J., F. Grivel and H. Cepeda, 1957.
Notas sobre el "tsunami" del 9 de Marzo de 1957.
Rev. Tec. Obr. Marit., 15:37-39.

hurricanes
Miller, Banner I., Elbert C. Hill and Peter P. Chase, 1968.
A revised technique for forecasting hurricane movement by statistical methods.
Mon. Weath. Rev. U.S. Dep. Comm. 96(5): 540-548

hurricanes
Minina, L.S., 1967.
The hurricane "Ines". (In Russian).
Meteorologiya Gidrol., 1967(8):18-31.

hurricanes
Miyazaki, Masamori, 1965.
A numerical computation of the storm surge of Hurricane Carla 1961 in the Gulf of Mexico.
Oceanogr. Mag., Tokyo, 17(1/2):109-140.

hurricanes
Mook, C.P., E.W. Hoover and R.A. Hoover, 1957.
An analysis of the movement of a hurricane off the east coast of the United States, October 12-14, 1947. Mon. Wea. Rev., 85(7):243-250.

hurricanes
Moore, P.L., and staff, 1957.
The hurricane season of 1957.
Mon. Wea. Rev., 85(12):401-408.

hurricane
Myers, V. A., 1959.
Surface friction in a hurricane.
Mon. Wea. Rev., 87(8):307-311

hurricanes
Myers, V.A., and E.S., Jordan, 1956.
Winds and pressures over the sea in the hurricane of September 1938. Mon. Wea. Rev., 84(7): 261-270.

hurricanes
Namias, J., 1955.
Long range factors affecting the genesis and paths of tropical cyclones.
Proc. UNESCO Symp., Typhoons, Nov. 1954, 213-219, 5 textfigs.

hurricanes
Neumann, A. Conrad, 1963.
Processes of recent carbonate sedimentation in Harrington Sound, Bermuda.
Mar. Sci. Center, Lehigh Univ., 130 pp. (Unpublished manuscript).

hurricanes
Nupen, W., and M. Rigby, 1956.
An annotated bibliography of tropical cyclones, hurricanes and typhoons.
Meteorol. Abstr. & Bibl., 7(9):1113-1163.

hurricanes
Östlund, H. Göte, 1970.
Hurricane tritium III: Evaporation of sea water in hurricane Faith 1966. J. geophys. Res., 75(12): 2303-2309.

hurricanes
Östlund, H. Göte, 1967.
Hurricane tritium I: preliminary results on Hilda 1964 and Betsy 1965.
Geophys. Monogr., Am. Geophys. Un., 11:58-60.

hurricanes
Palmen, E., 1958.
Vertical circulation and release of energy during the development of Hurricane Hazel into an extratropical storm. Tellus, 10(1):1-23.

hurricanes
Pardue, L.G., 1971.
The hurricane season of 1970.
Weatherwise, 24(1): 24-33; 50-51

hurricanes
Penn, Samuel, 1965.
Ozone and temperature structure in a hurricane.
J. Appl. Meteorol., 4(2):212-216.

hurricanes
Perlroth, Irving, 1969.
Effects of oceanographic media on equatorial Atlantic hurricanes.
Tellus 21(2): 230-244

hurricanes
Perlroth, Irving 1967.
Hurricane behavior as related to environmental conditions.
Tellus 19(2): 258-268.

hurricanes
Perlroth, I., 1962.
Relationship of central pressure of hurricane Esther (1961) and the sea surface temperature field.
Tellus, 14(4):403-408.

hurricanes
Pfeffer, R.I., 1958.
Concerning the mechanics of hurricanes.
J. Meteorol., 15(1):113-120.

hurricanes
Poley, Wolfgang, 1970.
Hurrikan im Radarschirmbild.
Der Seewart 31(2):60-64

hurricanes
Polushkin, V.A., 1964.
On the practical use of some results of the investigation of the interaction between atmosphere and hydrosphere in the North Atlantic. (In Russian).
Materiali Vtoroi Konferentsii, Vzaimod. Atmosfer i Gidrosfer. v Severn. Atlant. Okean. Mezhd. Geofiz. God, Leningrad. Gidrometeorol. Inst., 27-31.

hurricanes
Price, W.A., 1956.
Hurricanes affecting the coast of Texas from Galveston to Rio Grande.
B.E.B. Tech. Memo. 78:50 pp.
(Beach Erosion Board)

Hurricanes
Rao Gandikota V., 1969.
Role of differential friction and asymmetry of the total flow on hurricane movement.
Mon. weath. Rev. 97(7): 502-509

hurricanes
Redfield, Alfred C. and Arthur R. Miller, 1957
Water levels accompanying Atlantic Coast hurricanes. In: Interaction of sea and atmosphere.
Meteorol. Monogr. Vol. 2, No. 10, pp. 1-23.

hurricanes
Reid, R.O., 1957.
On the classification of hurricanes by storm tide and wave energy indices. Interaction of Sea and Atmosphere, a group of contributions.
Meteorol. Monogr., 2(10):58-66.

hurricanes
Renard, Robert J., and William H. Levings III, 1969.
The Navy's numerical hurricane and typhoon forecast scheme: application to 1967 Atlantic storm data.
J. appl. Met., 8(5): 717-725

hurricanes
Rieck, H., 1962.
Begegnung mit den Hurrikanen Betsy, Debbie und Esther. Der Seewarte, 23(2):73-76.

hurricanes
Riehl, H., 1963
On the origin and possible modification of hurricanes.
Science, 141(3585):1001-1010.

hurricanes
Riehl, Herbert, 1963.
Some relations between wind and thermal structure of steady state hurricanes.
J. Atmos. Sci., 20(4):276-287.

hurricanes
Riehl, H., 1956.
Sea surface temperature anomalies and hurricanes.
Bull. A.M.S., 37(8):413-417.

hurricanes
Riehl, H., and W.H. Haggard, 1955.
A quantitative method for the prediction of tropical storm motion.
Proc. UNESCO Symp., Typhoons, Nov. 1954:247-248.

hurricanes
Riehl, H., W.H. Haggard and R.W. Sanborn, 1956.
On the prediction of 24-hour hurricane motion.
J. Meteorol., 13(5):415-420.

hurricanes
Riehl, H., and J.S. Malkus, 1961.
Some aspects of Hurricane Daisy, 1958.
Tellus, 13(2):181-213.

hurricanes
Riehl, H., and R.W. Sanborn, 1958.
Climatology of three-day hurricane motion.
Bull. A.M.S., 39(2):69-72.

hurricanes
Rodewald, M., 1955.
Hurrikan Hazel (Oktober 1954) in Lichte deutscher Schiffsbeobachtungen. Der Seewart 16(2):43-52, 2 textfigs.

hurricanes
Rodin, G.I., 1958.
Oceanographic and meteorological aspects of the Gulf of California. Pacific Science 12(1):21-45.

hurricanes, effect of
Rosenblatt, M., 1957.
A random model of sea surface generated by a hurricane. J. Math. & Mech., 6(2):235-246.

hurricanes
Ross, R.B., and M.D. Blum, 1957.
Hurricane Audrey, 1957.
Mon. Weather Rev., 85(6):221-227.

hurricanes
Rudoloff, W., 1965.
Tidal components of hurricane development.
Nature, 206(4989):1144-1145.

hurricanes
Sadowski, Alexander, 1959.
Atlantic coastal radar tracking of 1958 hurricanes.
J. Geophys. Res., 64(9):1277-1282.

hurricanes
Senn, H.V., and J.A. Stevens, 1965.
A summary of empirical studies of the horizontal motion of small Radar precipitation echoes in Hurricane Donna and other tropical storms.
Nat. Hurricane Res. Lab., No. 74:55 pp.

hurricanes
Serra C., Sergio, 1971.
Hurricanes and tropical storms of the West coast of Mexico.
Mon. Weath. Rev. U.S. Dept. Comm. 99(4):302-308.

hurricanes
Shabbar, M., 1969.
Vorticity in nature: the mechanics of hurricane development.
Naturaliste can. 96(4):651-665.

hurricanes
Sheets, Robert C. 1969.
Some mean hurricane soundings.
J. appl. Met. 8(1):134-146.

hurricanes
Sherman, L., 1956.
On the wind asymmetry of hurricanes. J. Met., 13(5):500-503.

hurricanes
Sherman, L., 1955.
Reconnaissance of the "complete" hurricane.
Proc. UNESCO Symp., Typhoons, Nov. 1954:109-119, 13 textfigs.

hurricanes
Sherman, L., 1955.
A proposal for the use of SOFAR signals for the reporting of hurricane positions.
Proc. UNESCO Symp., Typhoons, Nov. 1954:245-246.

hurricanes
Shuleikin, V.V. 1971.
Schematized model of nonstationary tropical hurricane. (In Russian).
Dokl. Akad. Nauk SSSR 200(6):1336-1339.

hurricanes
Shuleikin, V.V., 1970.
The dependence of the power of tropical hurricane from the temperature of ocean surface. (In Russian; English abstract).
Fizika Atmosfer. Okean. Izv. Akad. Nauk, SSSR 6(12):1219-1237

hurricanes
Shuleikin, V.V. 1969.
The structure of the tropical hurricane field. (In Russian)
Dokl. Akad. Nauk SSSR, 186(3):578-582

hurricanes
Shuleikin V.V., 1969
Thermal phenomena in the field of tropical hurricanes. (In Russian)
Dokl. Akad. Nauk SSSR, 189(6):1242-1245

hurricanes
Shuleikin, V.V., 1960
Calculations of dimensions of waves possible during Atlantic hurricanes. Izv. Akad. Nauk, SSSR, Ser. Geophys. (7): 1013-1021.
English Ed., (7): 677-682

hurricanes
Shumeyko, G.K., 1962.
Sailing in the tropical hurricane zones.
Morskoi Transport.
Reviewed (1963):
Istoshin, Yu. V., and V.M. Lifshits, Meteorol. i Gidrol., (4):52-53. (In Russian).

hurricanes
Silkin, B.I., 1968.
Aerophotographic survey of a tropical hurricane.
Navigational air to oceanographic work. Studies of the ocean-atmosphere interaction. (In Russian).
Okeanologiia, Akad. Nauk, SSSR, 8(5):933-934.

hurricanes
Simpson, Joanne 1967.
An experimental approach to cumulus clouds and hurricanes.
Weather 22(3):95-114.

hurricanes
*Simpson, Joanne, 1966.
Hurricane modification experiments.
Hurricane Symposium, Oct. 10-11, 1966, Houston, Publ. Am. Soc. Oceanogr., 1:255-292.

hurricanes
Simpson, Joanne, 1967.
Project Stormfury.
Oceanology Int., 2(1):24-26.

hurricanes
Simpson, Robert H., 1963.
Hurricanes and tropical meteorology.
Science, 141(3584):935-938.

hurricanes
Simpson, R.H., 1955.
On the structure of tropical cyclones as studied by aircraft reconnaissance.
Proc. UNESCO Symp., Typhoon, Nov. 1954:129-150, 19 textfigs., 11 pls.

hurricanes
Simpson, Robert H., and James Giraytys, 1966.
The tracking and observation of hurricanes.
Hurricane Symposium, Oct. 10-11, 1966, Houston, Publ. Am. Soc. Oceanogr. 1:34-70.

hurricanes
Simpson, R.H., and Joanne S. Malkus, 1964.
Experiments on hurricane modification.
Scientific American, 211(6):27-37.

hurricanes
Simpson, R.H., and J.S. Malkus, 1963
An experiment in hurricane modification.
Science, 142(3591):498.

hurricanes
Simpson, R.H., and Joseph M. Pelissier, 1971.
Atlantic hurricane season of 1970.
Mon. Weath. Rev., U.S. Dept. Comm. 99(4):269-277

hurricane
Simpson, Robert H., and Joanne Simpson, 1966.
Why experiment on tropical hurricanes?
Trans. N.Y. Acad. Sci., (2) 28 (8):1045-1062.

hurricanes
Simpson, R.H. Arnold L. Sugg et al. 1970.
The Atlantic hurricane season of 1969.
Mon. Wea. Rev., U.S. Dept. Comm. 98(4):293-306.

hurricanes
Smiley, C.H., 1956.
Hurricane vectors. Bull. A.M.S., 37(8):403-405.

hurricanes
Smiley, C.H., 1956.
Hurricanes: a new approach. Navigation 5(3):130-133.

hurricanes
Smiley, C.H., 1955.
Tidal forces and the formation of hurricanes.
Nature 175(4444):39.

hurricanes
Smiley, C.H., 1954.
Tidal forces and the formation of hurricanes.
Nature 173(4400):397.

hurricanes
*Smith,R.K.,1968.
The surface boundary layer of a hurricane.
Tellus,20(3):473-484.

hurricanes
*Stearns,R.D.,1967.
New approaches in hurricane research.
Sea Frontiers 13(5):258-268. (popular).

hurricanes
Steinborn, E., 1965.
Hurrikan ISBELL.
Der Seewart, 26(1):16-18.

hurricanes
Steinborn, E., 1964.
Abby bis Hilda: tropische Wirbelstürme im atlantischen Bereich von Juli bis Anfang Oktober 1964.
Der Seewart, 25(6):241-249.

hurricanes
Stommel, H., 1956.
Electrical data from cable may aid hurricane prediction. Western Union Tech. Rev. 10(1):15-19.

hurricanes
Sugg, Arnold L., 1967.
The hurricane season of 1966.
Monthly Weather Review, 95(3):131-142.

hurricanes
Sugg, Arnold L., and Paul J. Hebert, 1969.
The Atlantic hurricane season of 1968.
Mon. Wea. Rev., 97(3): 225-239.

HURRICANES
Sugg, Arnold L., and Leonard G. Pardue, 1970.
The hurricane season of 1969.
Weatherwise 23(1): 13-17; 31.

hurricanes
Sugg, Arnold, and Joseph M. Pelissier 1968.
The hurricane season of 1967.
Mon. Weath. Rev. US. Dept. Comm. 96(4): 242-250

hurricanes
Tannehill, I.R., 1956.
Hurricanes.
Princeton Univ. Press, 308 pp.

hurricanes
Terauchi, E., 1963
A theoretical treatment of the deformation problem in typhoons, hurricanes or circular vortices.
J. Meteorol. Soc., Japan, (2), 41(3):158-171.

hurricanes
Tisdale, C.F., and P.F. Clapp, 1963.
Origin and paths of hurricanes and tropical storms related to certain physical parameters at the air-sea interface.
J. Applied Meteorol., 2(3):358-367.

hurricanes
Truitt, Reginald V. 1968
High winds ... high tides: a chronicle of Maryland's coastal hurricanes.
Educat. Ser., Nat. Resources Inst., Univ. Maryland, 77: 35pp.

hurricanes
Vanderman, L.W., 1961
Verification of JNWP-Unit hurricane and typhoon forecasts for 1959.
Bull. Amer. Meteor. Soc., 42(4): 239-248.

hurricanes
Wares, David E., 1970.
Temperatures and turbulence at tropopause levels over hurricane Beulah (1967)
Mon. Weath. Rev., U.S. Dept. Comm. 98(10): 749-755.

hurricanes
Whittingham, H.E., 1958
The Bathhurst Bay hurricane and associated storm surges.
Australian Met. Mag., 23: 14-36.

hurricanes
Wilson, Basil W., 1961.
Ch. 30. Hurricane tide prediction for New York Bay.
Proc. 7th Coastal Eng. Conf. (Scheveningen, Holland, Aug. 1960), Counc. Wave Res., Berkeley, 548-584.

hurricanes
Woodcock, A.H., 1958
The release of latent heat in tropical storms due to the fallout of sea-salt particles.
Tellus, Vol. 10, No. 3 pp. 355-371.

hurricanes
Yani, Michio 1968.
Evolution of a tropical disturbance in the Caribbean Sea region.
J. Met. Soc. Japan, 46(2): 86-109.

hurricanes, effect of
hurricanes,effect of
*Ball,Mahlon M., Eugene A. Shinn and Kenneth W. Stockman,1967.
The geological effects of Hurricane Donna in South Florida.
J. Geol.,75(5):583-597.

hurricanes, effect of
Breder, C.M., Jr., 1962
Effects of a hurricane on the small fishes of a shallow bay.
Copeia, 2: 459-462.

hurricanes, effect of
Collins, J. Ian, 1967.
Wave statistics from Hurricane Dora
J. WatWays Harb. Div. Am. Soc. civ. Engrs, 93(WW2): 59-77.

hurricanes, effect of
Conner, W.C., R.H. Kraft and D.L. Harris, 1957.
Empirical methods for forecasting the maximum storm tide due to hurricanes and other tropical storms. Mon. Wea. Rev. 85(4):113-116.

hurricanes, effect of
Cooper, M. J., 1966.
Destruction of marine flora and fauna in Fiji caused by the hurricane of February 1965.
(no abstract).
Pacific Science, 20(1):137-141.

hurricanes, effect of
*Croker, Robert A., 1968.
Distribution and abundance of some intertidal sand beach amphipods accompanying the passage of two hurricanes.
Chesapeake Sci., 9(3):157-162.

hurricanes, effect of
Donn, W.L., and V.F. Jennemann, 1953.
A critical review of the microseisms reported from hurricanes "Easy" and "How" of 1951.
Earthquake Notes 24:1-4.

hurricanes, effect of
Dunn, G.E., 1957.
Some features of the hurricane problem.
1956 Proc. Shellfish Assoc., 47:104-108.

hurricanes
Dunn, G.E., and B.I. Miller, 1960.
Atlantic hurricanes.
Louisiana University Press, 326 pp., $10.00

Reviewed by J. Eaton, 1961.
J. Atmos. Terres. Physics, 20(2/3): 220.

hurricanes, effect of
El-Ashry, Mohamed T., and Harold R. Wanless, 1964
Photo interpretation of coast-line changes due to hurricanes. (Abstract).
Geol. Soc., Amer., Special Paper, No. 76:53.

hurricanes, effect of
Evans, D.J. 1970.
Analysis of wave force data.
J. Petr. Technol. March 1970: 347-358.

Hurricanes, effect of
Franceschini, Guy A., 1962.
A synoptic study of water vapor divergence over the Gulf of Mexico during passage of Hurricane Carla. (Abstract).
J. Geophys. Res., 67(4):1637.

hurricanes, effect of
Geisler, J.E., 1970.
Linear theory of the response of a two layer ocean to a moving hurricane.
Geophys. fluid Dyn. 1(1/2): 249-272.

hurricanes, effect of
Germany, Deutsches Hydrographisches Institut, 1958
[Westindien-Handbuch. 1. Die Nordküste Süd- und Mittelamerikas] 3rd Edit., No. 2048: 118-153.
Translation:
USN-HO TRANS-66
translator: M. Slessers.
M.O. 16086
P.O.:20412

hurricanes, effect of
Glynn, Peter W., Luis R. Almodovar and Jean G. González, 1964.
Effects of Hurricane Edith on marine life in La Parguera, Puerto Rico.
Caribbean J. Sci., 4(2/3):335-346.

hurricanes, effect of
Gordon, Arnold L., Paul J. Grim and Marcus Langseth, 1966.
Layer of abnormally cold bottom water over southern Aves Ridge.
Science, 151(3717):1525-1526.

hurricanes, effect of
Goreau, T.F., 1964
Mass expulsion of zooxanthellae from Jamaican reef communities after Hurricane Flora.
Science, 145(3630):383-386.

hurricanes,effect of
*Goudeau,D.A., and W.C. Conner,1968.
Storm surge over the Mississippi River delta accompanying hurricane Betsy,1965.
Mon.Weath.Rev., U.S. Dep.Comm., 96(2):118-124.

hurricane,effect of
*Harris,D.Lee., 1966
Hurricane storm surges.
Hurricane Symposium,Oct.10-11,1966,Houston,
Publ.Am.Soc.Oceanogr.,1:200-228.

hurricanes, effect of
Hazelworth, John B., 1968.
Water temperature variations resulting from hurricanes.
J. geophys. Res., 73(16):5105-5123.

hurricanes, effect of
Ignatiades, Lydia, and Theodore J. Smayda, 1970.
Autecological studies on the marine diatom Rhizosolenia fragilissima Bergon. II Enrichment and dark variability experiments.
J. Phycol. 6(4):357-364

hurricanes, effect of
Kammer, E.W., 1952.
Directional properties of microseisms during hurricane "Easy" 1951. Earthquake Notes 23:24-26, (Abstr. 23:19).

hurricanes, effect of
Keith, Don C., and Neil C. Hulings, 1965.
A quantitative study of selected nearshore infauna between Sabina Pass and Bolivar Point, Texas.
Publ. Inst. Mar. Sci., Port Aransas, 10:33-40.

HURRICANES, effect of
Landis, Robert C., and Dale F. Leipper 1968.
Effects of Hurricane Betsy upon Atlantic Ocean temperature, based upon radio-transmitted data.
J. appl. Met., 7(4):554-562.

hurricanes, effect of
*Latham, Gary V., Rockne S. Anderson and Maurice Ewing,1967.
Pressure variations produced at the ocean bottom by hurricanes.
J. geophys. Res., 72(22):5693-5704.

hurricanes,effect of
*Leipper,Dale F., 1966
Influence of the hurricane on the structure of the thermocline. 2.
Hurricane Symposium,Oct.10-11,1966,Houston,Publ. Am.Soc.Oceanogr.,1:173-188.

hurricanes effect of
Marinos, George, and Jerry W. Woodward 1968.
Estimation of hurricane surge hydrographs.
J. WatWays Harb. Div. Am. Soc. civ. Engrs. 94 (WW2): 189-216.

hurricanes, effect of
Munk, W., F. Snodgrass and G. Carrier, 1956.
Edge waves on the continental shelf. Science 123(3187):127-132.

Hurricanes, effect of
Murray, Stephen P. , 1970.
Bottom currents near the coast during hurricane Camille. J. geophys. Res., 75(24): 4579-4582.

hurricane, effect of
O'Brien, James J., 1966.
The non-linear response of a two-layer, baroclinic ocean to a stationary, axially-symmetric hurricane: 2. Upwelling and mixing induced by momentum transfer.
J. Atmospl. Sci., 24(2):208-215.

hurricanes, effect of
O'Brien, James J., and Robert O. Reid, 1967.
The non-linear response of a two-layer, baroclinic ocean to a stationary, axially-symmetric hurricane: 1. Upwelling induced by momentum transfer.
J. Atmosph. Sci., 24(2): 197-207

hurricanes, effect of
Orton, Robert, 1970
Tornadoes associated with hurricane Beulah on September 19-23, 1967.
Mon. Wea. Rev., U.S. Dept. Comm. 98(7): 541-547

hurricanes, effect of
*Östlund,H. Göte,1968.
Hurricane tritium II: air-sea exchange of water om Betsy 1965.
Tellus,20(4):577-594.

hurricane, effect of
Paskausky, D.F., 1971.
Numerically predicted changes in the circulation of the Gulf of Mexico accompanying a simulated hurricane passage. J. mar. Res. 29(3): 214-225.

hurricane,effect of
*Pearson,Allen D.,1966.
Hurricane-induced tornadoes- a review.
Hurricane Symposium,Oct.10-11,1966,Houston
Publ.Am.Soc.Oceanogr.,1:114-131.

hurricanes, effect of
*Perkins,R.D., and Paul Enos,1968.
Hurricane Betsy in the Florida-Bahama area-geologic effects and comparison with Hurricane Donna.
J. Geol. 76(6):710-717.

hurricanes, effect of
Pore, A., 1960
Chesapeake Bay hurricane surges.
Chesapeake Science, 1(3-4): 178-186.

hurricanes, effect of
Pore, A., 1957.
Ocean surface waves produced by some recent hurricanes. Mon. Wea. Rev., 85(12):385-392.

hurricanes, effect of
Redfield, A.C., and A.R. Miller, 1957.
Water levels accompanying Atlantic coast hurricanes. Interaction of Sea and Atmosphere, a group of contributions. Meteorol. Monogr., 2(10):1-23.

hurricanes,effect of
Reid,Robert O., and Bernie R.Bodine,1968.
Numerical model for storm surges in Galveston Bay.
J. WatWays Harb.Div.Am.Soc.civ.Engrs.,94(WW1): 33-57.

hurricanes, effect of
Roden, Gunnar I., 1964.
Oceanographic aspects of Gulf of California.
In: Marine geology of the Gulf of California: a symposium, Amer. Assoc. Petr. Geol., Memoir, T. van Andel and G.G. Shor, Jr., Editors, 3:30-58.

hurricanes, effect of
Rodolfo, Kelvin S., Barbara A. Buss and Orrin H. Pilkey 1971.
Suspended sediment increase due to hurricane Gerda in continental shelf waters off Cape Lookout, North Carolina.
J. sedim. Petrol. 41(4): 1121-1125.

hurricanes, eff ect of
Rosenblatt, M., 1957.
A random model of the sea surface generated by a hurricane. J. Math. & Mech., 6(2):235-246.

hurricanes, effect of
Russell, Larry A., 1971
Probability distributions for hurricane effects.
J. WatWays Harb. Div. Am. Soc. civ. Engrs, 97 (WW1): 139-154

hurricanes, effect of
*Schwarz,Francis K., 1970.
The unprecedented rains in Virginia associated with the remnants of hurricane Camille.
Mon.Weath.Rev. U.S. Dept.Comm. 98(11):851-859.

hurricanes, effect of
Shuleikin, V. V., 1960.
Calculation of height of waves, which might be generated by Atlantic hurricanes.
Izv., Akad. Nauk, SSSR, Ser. Geofiz., (7):1013-1021.

hurricanes, effect of
Sirkin, Alan N., 1970.
Hurricane surge computations by computer.
J. WatWays Harb. Div. Am. Soc. civ. Engrs 96 (WW2): 467-482.

hurricanes, effect of
Stelzenmuller, William B., 1965.
Tidal characteristics of two estuaries in Florida
J. Waterways and Harbors Div., Proc. Amer. Soc., Civil Engineers, 91(WW3):25-36.

hurricane,effect of
*Stevenson,Robert E.,1966.
Influence of the hurricane on the structure of the thermocline.Part 1,Part 3.
Hurricane Symposium,Oct.10-11,1966,Houston,Publ. Am.Soc.Oceanogr.,1:158-172;189-199.

hurricanes, effect of
Stoddart, D.R. 1969.
Post-hurricane changes on the British Honduras reefs and cays: resurvey of 1965.
Atoll Res. Bull. 131:1-35.

hurricanes, effects of
Thom, H.C.S., and R.D. Marshall, 1971.
Wind and surge damage due to Hurricane Camille.
J. WatWays Harb. Coast. Engng Div. Am. Soc. Civ. Engrs, 97 (WW2): 355-363.

hurricanes, effect of
Thomas, L.P., D.R. Moore, & R.C. Work, 1961
Effects of hurricane Donna on the turtle grass beds of Biscayne Bay, Florida.
Bull. Mar. Sci. Gulf & Carib., 11(2): 191-197

hurricanes, effect of

Tuck, Leslie M., 1968.
Laughing gulls (Larus atricilla) and black skimmers (Rynchops nigra) brought to Newfoundland by hurricane.
Bird Band. 39(3):200-208.

hurricanes, effect of

Walden, H., 1962.
Betrachtung über Dünung aus der tropische Zyklone "Debbie" an Hand von Beobachtungen des T.S. "Gertrud Fritzen".
Der Seewart, 23(2):45-52.

hurricanes, effect of

Warnke, Detlef A., 1969.
Beach changes at the location of landfall of Hurricane Alma.
SEast Geol. 10(4):189-200.

hurricanes, effect of

Warnke, D A., V. Goldsmith, P. Grose and J.J. Holt, 1966.
Drastic beach changes in a low-energy environment caused by Hurricane Betsy.
J. geophys. Res., 71(8):2013-2016.

hurricanes, effect of

Wilson, Basil W., 1960
The prediction of hurricane storm tides in New York Bay.
B.E.B. Tech. Memo. No. 120: 107 pp.

hurricanes, effect of

Zetler, B.D., 1957.
Hurricane effect on sea level at Charleston.
J. Hydraul. Div., Proc. Amer. Soc. Civ. Eng., 83(Paper 1330):19 pp.

hurricanes, modification of

Gentry, R. Cecil, 1970.
Hurricane Debbie modification experiments, August 1969. Science, 168(3930): 473-475.

hyaloclastites

hybridization

hybridization

Lejuez, Robert, 1959.
Premières recherches sur l'hybridation interspécifique à l'intérieur du genre Sphaeroma.
C. R. Acad. Sci., Paris, 249:1389-1391.

hydraulics

hydraulics

Simmons, H. B., 1950.
Applicability of hydraulic model studies to tidal problems. Evaluation of present state of knowledge of factors affecting tidal hydraulics and related phenomena. Comm. on Hydraulics, Corps Eng., U.S.A., Rept. No. 1:127-145, 6 figs.

hydraulics, tidal

U.S. Waterways Experiment Station, 1954.
Bibliography on tidal hydraulics.
Committee on Tidal Hydraulics, Vicksburg, Miss., 201 pp.

hydraulics

Wicker, C. F., and O. Rosenzweig, 1950.
Theories of tidal hydraulics. Evaluation of present state of knowledge of factors affecting tidal hydraulics and related phenomena. Comm. on Hydraulics, Corps Eng., U.S.A., Rept. No. 1: 101-125.

hydrodynamics

hydrodynamics

Aleyev, Yu. G., 1966.
Buoyancy and hydrodynamic function of the body of nektonic animals. (In Russian; English abstract).
Zool. Zhurn., 45(4): 575-589

hydrodynamics

*Barcilon, V., and J. Pedlosky, 1967.
A unified linear theory of homogeneous and stratified rotating fluids.
J. fluid Mech. 29(3):609-621.

hydrodynamics

*Barcilon, V., and J. Pedlosky, 1967.
On the steady motions produced by a stable stratification in a rapidly rotating fluid.
J. fluid Mech. 29(4):673-690.

hydrodynamics

Bretherton, F.P. and J.S. Turner 1968.
On the mixing of angular momementum in a stirred rotating fluid.
J. fluid Mech. 32(3): 449-464.

hydrodynamics

Beleev, N.T., and V.S. Petrischev, 1966.
A numerical method of solving hydrodynamic equations in the case of a two-dimensional flow. (In Russian).
Doklady, Akad. Nauk SSSR 169(6): 1296-1289.

hydrodynamics

Bye, John A.T., 1966.
Numerical solutions of the steady-state vorticity equation in rectangular basins.
J. fluid Mech., 26(3):577-598.

hydrodynamics

Carter H.H., 1969.
A preliminary report on characteristics of a heated jet discharged horizontally into a transverse current. 1. Constant depth.
Chesapeake Bay Inst. Ref. 69-14: 38 pp. (multi-lithed).

hydrodynamics

Cermak, J.E., 1963
Lagrangian similarity hypothesis applied to diffusion in turbulent shear flow.
J. Fluid Mech., 15(1): 49-64.

hydrodynamics

Crepon, Michel, 1967.
Hydrodynamique marine en régime impulsionnel.
Cah. Océanogr. 19(8):627-655.

hydrodynamics

Curle, N., 1957.
On hydrodynamic stability in unlimited fields of viscous flow.
Proc. R. Soc., London, (A), 238:489-501.

hydrodynamics

Dragos, L., 1962.
La théorie de l'aile mince en magnétohydrodynamique.
Comptes Rendus, Acad. Sci., Paris, 255(8):1251-1253. 255(9):1289-1290.

Hydrodynamics

Drazin, P. G., and L. N. Howard, 1961.
Stability in a continuously stratified fluid.
J. Eng. Mech. Div., Proc. Amer. Soc. Civil. Eng. 87(EM6):101-116.

hydrodynamics

Eckart, Carl, 1963.
Some transformations of the hydrodynamic equations.
Physics of Fluids, 6(8):1037-1041.

hydrodynamics

Eckart, Carl, 1962
The equations of motion of sea-water. Ch. 3, Sect. 1. Fundamentals. In: The Sea, Interscience Publishers, Vol. 1, Physical Oceanography, 31-41.

hydrodynamics

Felsenbaum, A.I., 1968.
A hydrodynamic unstationary model of an inhomogeneous ocean. (In Russian).
Dokl. Akad. Nauk, SSSR, 183(3):580-583.

hydrodynamics

Fischer, G., 1965.
On a finite difference scheme for solving the non-linear primitive equations for a barotropic fluid with application to the boundary current problem.
Tellus, 17(4):405-412.

hydrodynamics

Fofonoff, N.P., 1961.
Energy transformations in the sea.
Fish. Res. Bd., Canada, MSS. Rept. Ser. (Ocean. & Limnol.), No. 109:82 pp.

hydrodynamics

Fofonoff, N.P., 1954.
Steady flow in a frictionless homogeneous ocean.
J. Mar. Res., 13(3):254-262.

Reprinted in:
Wind-driven ocean circulation, A.R. Robinson, Editor, Blaisdell Publ. Co., 69-79.

hydrodynamics

Fortus, M.I., 1971.
Statistical properties of hydrodynamical fields on the sphere at Gaussian stationary distributions. (In Russian). Fizika Atmosfer. Okean. Izv. Akad. Nauk SSSR 7(4): 461-464.

hydrodynamics

*Gates, W.L., 1969.
The Ekman vertical velocity in an enclosed B-plane ocean.
J. mar. Res., 27(1):99-120.

hydrodynamics

Glauert, M.B., 1963.
Magnetohydrodynamic wakes.
J. Fluid Mech., 15(1):1-12.

hydrodynamics

Goldshtik, M.A., 1962.
A mathematical model of discontinuous incompressible fluid flows. (In Russian).
Doklady, Akad. Nauk, SSSR, 147(6):1310-1313.

hydrodynamics

Gormatjuk, Yu. K., and A.S. Sarkisjan, 1965.
The results of calculation of currents in the North Atlantic by means of a four-level model. (In Russian; English abstract).
Fisika Atmos. i Okeana, 1(3):313-326.

hydrodynamics
Gorshkov, A.S., 1969.
Generalization of A.N. Krylov's formulae for computing strain and shape of a flexible thread in a flow. (In Russian; English abstract).
Okeanologiia, 9(6): 953-958.

hydrodynamics
Greenspan, H.P., 1962.
A criterion for the existence of inertial boundary layers in oceanic circulation.
Proc. Nat. Acad. Sci., U.S.A., 48(12):2034-2039.

hydrodynamics
Greenspan, H.P., and D.J. Benney, 1963
On shearlayer instability, breakdown and transition.
J. Fluid Mech., 15(1):133-153. Res. p. 133.

hydrodynamics
Harper, J.F., 1963.
On boundary layers in two-dimensional flow with vorticity.
J. Fluid Mechanics, 17(1):141-153.

hydrodynamics
Hassan, El Sayed Mohamed, 1964
A three dimensional model of the wind driven horizontal velocities in the North Atlantic Ocean.
In: Studies in Oceanography dedicated to Professor Hidaka in commemoration of his sixtieth birthday, 10-14.

hydrodynamics
Hidaka, Koji, 1963
A hydrodynamical computation of an equatorial flow.
Rec. Oceanogr. Wks., Japan, 7(1): 1-7.

hydrodynamics
Hide, R., 1965.
The viscous boundary layer at the free surface of a rotating baroclinic fluid:effects due to the temperature dependence of surface tension.
Tellus, 17(4):440-442.

hydrodynamics
Hide, R., 1964.
The viscous boundary layer at the free surface of a rotating baroclinic fluid.
Tellus, 16(4):523-529.

Hydrodynamics
Holmboe, Jørgen, 1962.
On the behavior of symmetric waves in stratified shear layers.
Geofys. Publik., Geophysica Norvegica, 24:67-114.

hydrodynamics
Honda, M., 1961.
Theory of shear flow through a cascade.
Proc. R. Soc., London, (A), 265(1320):46-70.

Hydrodynamics
Hudimac, Albert A., 1961.
Ship waves in a stratified ocean.
J. Fluid Mechanics, 11(2):229-243.

hydrodynamics
Kamenkovich.V.M., 1966.
On the theory of the inertial-viscous boundary layer in a two-dimensional model of ocean currents. (In Russian;English abstract).
Fisika Atmosfer.Okean.,Izv.Akad. Nauk,SSSR, 2 (12):1274-1295.

hydrodynamics
Kelly, D.L., B.W. Martin and E.S. Taylor, 1964.
A further note on the bathtub vortex.
J. Fluid Mechanics, 19(4):539-542.

hydrodynamics
Kochin, N.E., 1937.
[On the wave-making resistance and lift of bodies submerged in water. Trans. Conf. on Theory of Wave Resistance, Central Aerodynamical Inst., Moscow] 65-134.

Translation:
Tech. Res. Bull., Soc. Naval Architects and Mar. Eng. No. 1-8.

hydrodynamics
Kotchin, N.E., 1940.
[The theory of waves generated by oscillations of a body under the free surface of a heavy incompressible fluid.] Uchenye Zapiski Moskovskogo Gosudarstvennogo Universititeta Mekhanika 46: 85-106.

Translated by J.V. Wehausen in:
Tech. Res. Bull., Soc. Naval Arch. and Mech. Eng. 1-10:38 pp., 1952.

hydrodynamics
Kreiss, Heinz-Otto, 1962
Über die Formulierung der Ranbedingungen und über die Differenzapproximation bei Anfangs-randwertaufgaben.
Proc. Symposium on Mathematical-Hydrodynamical Methods of Phys. Oceanogr., Sept. 1961, Inst. Meeresk., Hamburg: 1-12.

HYDRODYNAMICS
Krivolovich, L.M., 1968.
Non-linear model of the flow of inhomogeneous fluid at the equator. (In Russian; English abstract).
Fisika Atmosfer. Okean., Izv. Akad. Nauk, SSSR, 4(2):186-198.

hydrodynamics
Kuo, H.L., 1969.
Motions of vortices and circulating cylinder in shear flow with friction.
J. atmospher. Sci., 26(3): 390-398

hydrodynamics
Ladikov, Iu. P., 1961.
[Some exact solutions of unsteady motion equations in magnetic hydrodynamics.]
Doklady Akad. Nauk, SSSR, 137(2):303-306.

hydrodynamics
Lakshmana Rao, S.K., 1961.
Some special solutions in viscous fluid motion.
Proc. R. Irish Acad., (A) 62(2):11-16.

hydrodynamics
Lang, Thomas G., 1966.
Hydrodynamic analysis of dolphin fin profiles.
Nature, 209 (5028):1110-1111.

hydrodynamics
Lebedkina, I.G., 1959.
Motion of a viscous fluid on a rotating sphere under the effect of tangential stresses applied to its outer surface. (In Russian).
Trudy Morsk. Gidrofiz. Inst., 18:110-

Translation:
Scripta Tecnica, Inc., for AGU, Issue 4, 1963 ser, 110-129. 1965.

Hydrodynamics
Lineykin, P.S., 1969.
On the theory of currents in an ocean of finite depth. (In Russian; English abstract).
Okeanologiia, 9(1):58-62.

hydrodynamics
Lineikin, P.S., 1967.
Baroclinic ocean current theory. (In Russian).
Meteorologiya Gidrol. (2):3-8.

hydrodynamics,
*Long, Robert R., 1968.
Sources and sinks on a B-earth.
Tellus, 20(3):524-532.

hydrodynamics
Long, R.R., 1961
A turbulent equatorial jet.
J. Fluid Mech., 11(3): 465-477.

hydrodynamics
Longuet-Higgins, M.S., 1965.
The response of a stratified ocean to stationary or moving wind-systems.
Deep-Sea Res., 12(6):923-973.

hydrodynamics
Mack, Lawrence, R., 1962.
Periodic, finite-amplitude, axisymmetric gravity waves.
J. Geophys. Res., 67(2):829-844.

hydrodynamics
Magaard, Lorenz, 1971.
Zur Berechnung von luftdruck- und windbedingten Bewegungen eines stetig geschichteten seitlich unbegrenzten Meeres.
Dt. hydrogr. Z. 24 (4): 145-158.

hydrodynamics
Mamaev, O.I. 1968.
Some aspects of the equation of sea water state. (In Russian)
Fizika Atmosfer Okean. Izv. Akad. Nauk SSSR 4(6): 686-689.

hydrodynamics
Marchuk, G.I. 1967.
Dynamics equations for a baroclinic ocean. (In Russian).
Dokl. Akad. Nauk SSSR 173 (6): 1317-1320.

hydrodynamics
Marchuk, G.I. and V.P. Kochergin 1968.
On vertical current structure in a baroclinic ocean. (In Russian; English abstract).
Met. Gidrol. (1): 3-10.

hydrodynamics
Mihaljan, John M., 1963.
The exact solution of the Rossby adjustment problem.
Tellus, 15(2):150-154.

hydrodynamics
Mikhaylov, V.N., 1960.
Forms of confluence boundaries between a stream and a water basin. Physics of the Sea. Hydrology. (In Russian).
Trudy Morsk. Gidrofiz. Inst., 22:5-14.

Translation:
Scripta Technica, Inc., for Amer. Geophys. Union, 3-10.

hydrodynamics
Miles, John W., 1963.
The Cauchy-Poisson problem for a rotating liquid.
J. Fluid Mechanics, 17(1):75-88.

hydrodynamics
Montgomery, Raymond B., 1964.
Two theorems relating scalar fields.
In: Studies in Oceanography dedicated to Professor Hidaka in commemoration of his sixtieth birthday, 1-2.

hydrodynamics
Namikawa, Tomikazu, 1961.
The occurrence of over-stability of a layer of fluid heated below and subject to the simultaneous action of magnetic field and rotation.
J. Inst. Polytechnics, Osaka City Univ., (G)5(1): 7-11

hydrodynamics
Niiler, P.P., A.R. Robinson and S.L. Spiegel, 1965.
On thermally maintained circulation in a closed ocean basin.
J. Mar. Res., 23(3):222-230.

hydrodynamics
Nowlin, W.D., Jr., 1967.
A steady, wind-driven, frictional model of two moving layers in a rectangular ocean basin.
Deep-Sea Research, 14(1):89-110.

hydrodynamics
Ozmidov, R.V., 1965.
On the energy distribution between oceanic motions of different scales. (In Russian; English abstract).
Fisika Atmosferi i Okeana, 1(4):439-448.

hydrodynamics
Ozmidov, P.V., 1964.
Some data on large scale characteristics of a field in the ocean with a horizontal component of velocity. (In Russian).
Izv., Akad. Nauk, SSSR, Ser. Geofiz., (11):1708-1719.

hydrodynamics
Porter, Gene H., And Maurice Rattray, Jr., 1965.
The influence of variable depth on steady zonal barotropic flow.
Deutsche Hydrogr. Zeits., 17(4):164-174.

hydrodynamics
Radach, Günther 1971.
Ermittlung zufallsangeregter Bewegungsvorgänge für zwei Modellmeere mittels des hydrodynamisch-numerischen Verfahrens.
Mitt. Inst. Meereskunde, Univ. Hamburg, 20: 71pp.

hydrodynamics
Rattray, Maurice, 1964.
Time-dependent motion in an ocean; a unified two-layer, beta-plane approximation.
In: Studies on Oceanography dedicated to Professor Hidaka in commemoration of his sixtieth birthday, 19-29.

hydrodynamics
Richards, J.M., 1961
Experiments on the penetration of an interface by buoyant thermals.
J. Fluid Mech., 11(3): 369-384.

hydrodynamics
Robinson, A.R., 1960
On two-dimensional inertial flow in a rotating stratified fluid. J. Fluid Mech., 9(3): 321-332.

hydrodynamics
Rossby, H.T., 1965.
On thermal convection driven by non-uniform heating from below: an experimental study.
Deep-Sea Res., 12(1):9-16.

hydrodynamics
Saint-Guily, B. 1967.
Études théoriques de dynamique des mers effectuées depuis 1963.
Rapp. Nat. Trav. français 1963-1966, Com. Nat. français Géodés. Géophys, 264-265.

hydrodynamics
Saint-Guily, B., 1964.
Problème d'Ekman dans un océan inhomogène - effet d'une thermocline.
In: Studies on Oceanography dedicated to Professor Hidaka in commemoration of his sixtieth birthday.
15-18.

hydrodynamics
Sarkisian, A.S., 1961.
[On the role of purely drifting advection density in a dynamic wind current of a baroclinic ocean.]
Izv. Akad. Nauk, SSSR, Ser. Geofiz., (9):1396-1407.

English Edit. (9): 911-917

hydrodynamics, theoretical
Schmitz, Hans Pieter, 1961.
Existenzbedingungen für stationäre und Beschleunigungsfreie Strömungsfelder auf der rotierenden Erde, inbesondere am Äquator.
Deutsche Hydrogr. Zeits., 14(4):135-152.

hydrodynamics
Schmitz, H.P., 1955.
Über die vertikale Geschwindigkeitsverteilung in Kanälen und Flüssen und ihre Beeinflussung durch Wind. Acta Hydrophysica 3(1):24-48.

hydrodynamics
Schmitz, H.P., 1955.
Zur Ermittlung der Durchflussmengenänderung in Kanalen und Flüssen infolge zeitlich variabler Windeinwirkung. Acta Hydrophysica 3(1):5-23.

hydrodynamics
Schooley, Allen H., 1962.
Experiments with a self propelled body submerged in a fluid with a vertical density gradient. (Abstract only).
J. Geophys. Res., 67(9):3597.

hydrodynamics
Sekerzh-Zen'kovich, Ya. I., 1961
Free finite oscillations of the boundary surface between two unlimited heavy fluids of differing densities. Physics of the sea. (In Russian).
Trudy Morsk. Gidrofiz. Inst., 23:3-43.

Scripta Technica Inc. Translation for AGU, pp. 1-34.

hydrodynamics
Sekerzh-Zenkovich, Ia. I., 1959.
Concerning the problem of the Cauchy-Poisson drag for liquid spheres and for a layer of fluid covering a solid globe. Swell and currents in the sea. Los of salt from the sea into the atmosphere (In Russian).
Trudy Morsk. Gidrofiz. Inst., 15:3-16.

hydrodynamics
Sells, C.C.L., 1965.
The effect of a sudden change of shape of the bottom of a slightly compressible ocean.
Phil. Trans. R. Soc., London, (A) 258(1092): 495-528.

hydrodynamics
Shuleykin, V.V., 1965.
Analysis of complicated thermal conditions in the region of closed cyclic Atlantic currents. (In Russian, English abstract).
Fisika Atmosferi i Okeana, 1(4):413-425.

hydrodynamics
Snyder, H.A. 1966.
Experiments on rotating flows between noncircular cylinders.
Physics Fluids 11(8): 1606-1611.

hydrodynamics
Snyder, H.A. 1968.
Stability of rotating Couette flow. II. Comparison with numerical results.
Physics Fluids 11(8):1599-1605.

hydrodynamics
Society of Naval Architects and Marine Engineers 1950.
Nomenclature for treating the motion of a submerged body through a fluid. Tech. Res. Rept. 1-5:15 pp., 3 textfigs.

hydrodynamics
Sretenskii, L.N., 1937.
[A theoretical investigation of wave resistance.]
Trudy Tsentral. Aero-Gidrodinam. Inst. No. 319:

hydrodynamics
Sretenskii, L.N., 1939.
[On the damping of the vertical oscillations of the center of gravity of floating bodies.]
Trudy Tsentral. Aero-Gidrodinam. Inst. No. 330:

hydrodynamics
Starr, V.P., 1959.
Hydrodynamical analogy to $E = mc^2$
Tellus, 11(1):135-138.

hydrodynamics
Stern, Melvin E., 1966.
Interaction of a uniform wind stress with hydrostatic eddies.
Deep-Sea Res., 13(2):193-203.

hydrodynamics
Stommel, H., and H.G. Farmer, 1952.
Abrupt change in width in two-layer open channel flow. J. Mar. Res. 11(2): 205-214, 3 textfigs.

hydrodynamics
Stommel, Henry and Robert Frazel, 1968.
Hidaka's onions (Tamanegi).
Rec. oceanogr. Wks, Japan, n.s., 9(2):279-281.

hydrodynamics
Stommel, Henry, and Claes Rooth, 1968.
On the interaction of gravitational and dynamic forcing in simple circulation models.
Deep-Sea Res., 15(2):165-170.

hydrodynamics
Sündermann, Jürgen 1971.
Die hydrodynamisch-numerische Berechnung der Vertikalstruktur von Bewegungsvorgängen in Kanälen und Becken.
Mitteilungen des Instituts für Meereskunde Univ. Hamburg 19: 101 pp.

hydrodynamics
Sundqvist, H., 1963.
A numerical forecast of fluid motion in a rotating tank and a study of how finite-difference approximations affect non-linear interactions.
Tellus, 15(1):44-58.

hydrodynamics
Szesztay, K., 1961.
[Hydraulic characteristics distribution along the rivers and channels.]
Meteorol. i Gidrol., (5):17-23.

hydrodynamics
Takano, Kenzo, 1965.
Courants marins induits par le vent et la non-uniformité de la densité de l'eau superficielle dans un océan.
La Mer: Bull. Soc. franco-japon. Océanogr., 2(2): 81-86.

Also in:
Geophys. Notes, 18(1)(2). 1965.
Tokyo.

hydrodynamics
*Tareev, B.A., 1968.
Nongeostrophic disturbances and baroclinic instability of the two-layer current in the ocean. (In Russian; English abstract).
Fisika Atmosfer. Okean., Izv. Akad. Nauk, SSSR, 4(12): 1275-1284.

hydrodynamics
Tareev, B.A., 1965.
Unstable Rossby waves and nonstationarity of the oceanic currents. (In Russian, English abstract).
Fisika Atmosferi i Okeana, 1(4):426-438.

hydrodynamics
Taylor, J.B., 1963.
The magneto-hydrodynamics of a rotating fluid and the earth's dynamo problem.
Proc. R. Soc., London, (A), 274(1357):274-283.

hydrodynamics
Turner, J.S., 1964.
The dynamics of sphaeroidal masses of buoyant fluid.
J. Fluid Mechanics, 19(4):481-490.

hydrodynamics
Turner, J.S., 1964.
The flow into an expanding spherical vortex.
J. Fluid Mechanics, 18(2):195-208.

hydrodynamics
*Veronis, George, 1967.
Analogous behavior of rotating and stratified fluids.
Tellus, 19(4):620-634.

hydrodynamics
Veronis, George 1967.
Analogous behavior of homogeneous rotating fluids and stratified, non-rotating fluids.
Tellus 19(2): 326-336.

hydrodynamics
Veronis, George, 1963
On the approximations involved in transforming the equations of motion from a spherical to the β-plane. II. Baroclinic systems.
J. Mar. Res., 21(3):199-204.

hydrodynamics,
Veronis, George, 1963.
On the approximations involved in transforming the equations of motion from a spherical surface to the β-plane. I. Barotropic systems.
J. Mar. Res., 21(2):110-124.

hydrodynamics
Veronis, G., 1963.
On inertially-controlled flow patterns in a β-plane ocean.
Tellus, 15(1):59-66.

hydrodynamics
Volkova, L.A., 1959.
Flow in a canal encircling the terrestrial sphere. Physics of the Sea. (In Russian).
Trudy Morsk. Gidrofiz. Inst., 17:41-47.

hydrodynamic equations

hydrodynamic equations
Andreev, A.I., and R.V. Krasochkin, 1964
About one exact solution of complete system of hydrodynamical equations. (In Russian).
Materiali Vtoroi Konferentsii, Vzaimod. Atmosfer. i Gidrosfer. v Severn. Atlant. Okean., Mezhd. Geofiz. God., Leningrad. Gidrometeorol. Inst., 105-113.

hydrodynamic equations
Ertel, H., 1970.
Transformation der Weberschen hydrodynamischen Gleichungen unter Berücksichtigung der Erdrotation.
Beitr. Geophys. 79(6):420-424

hydrodynamic equation
Ovsiannikov, L.V., 1956.
A new solution of the hydrodynamic equations.
Dokl. Akad. Nauk, SSSR, 111(1):47-49.

hydrodynamics of earth's core

hydrodynamics of earth's core
Hide, R., 1956.
The hydrodynamics of the earth's core. In: Physic and Chemistry of the Earth, A.H. Ahrens, K. Rankama, and S.K. Runcorn, Eds., Pergamon Press, Ltd., 1(5):94-137.

hydrofoils

hydrofoil boats
Great Britain, The Dock & Harbour Authority, 1964.
The hydrofoil in ports and harbours. The latest developments.
Dock and Harbour Authority, 44(520):313.

hydrofoils
Kolyshkin, N.N., 1958.
Some data on the hydrofoil motor-vessel "Raketa" and results in operation.
Sudostroyeniye (Shipbuilding) 1958 No. 4 (186): 1-4
T337R by E.R. Hope.
DSIS

hydrofoil ships
Miller, R.T., 1962.
Hydrofoil ships.
Trans. N.Y. Acad. Sci., (2), 24(8):855-878.

hydrographic tables

hydrographic tables
Knudsen, M., 1904.
Tabelle, anhang zu den 1901 herausgegeben hydrographischen Tabellen.
Publ. de Circ., Cons. Perm. Int. Expl. Mer, No. 11:1-23.

hydrographic tables
Knudsen, M., 1903.
Gefrierpunkttabelle fuer Meerwasser.
Publ. de Circ., Cons. Perm. Int. Expl. Mer, No. 5:11-13.

Hydrography

hydrography
Were all references to hydrography to have been inserted here, it became clear very early that they would have been "lost". Hence, in later years they were merely filed under the geographical area.

hydrography
Adrov, M. M., 1959.
Hydrological observations northwards of Iceland in summer 1950.
Trudy Pol. Nauch. Issle. Inst. Moskogo, 11:74-93.

hydrography
Akadmeiia Nauk, SSSR, 1953.
Ocherki po gidregrafii rek SSSR. 234 pp., figs., maps. $5.75.

A. Buschke, 80 East 11th Street, N.Y. 3.

hydrography
Anon., 1951.
Synopsis of hydrographical investigations.
Semi-Ann. Rept., Ocean. Invest., Tokyo, Jan. 1943 - Dec. 1944, No. 72:1-3.

hydrography
Antonov, L.V., 1931
Materialy k gidrologii Amurskogo Lemana. (Material on hydrology of the Amur Liman.) Zap. gidrogr. 65:53-56.

not in USA

hydrography
Asano, H., 1927.
Results of hydrographical observations in the adjacent waters of Japan, Dec. 1925-Nov. 1926. (In Japanese). Suisan Kai 530:48.

hydrography
Asano, H., 1919.
Hydrographic investigations of the Korea Strait and Tokara Islands. (In Japanese). Rep. Imp. Fish. Inst. Sci. Invest. 7(2):1-10.

hydrography
Asano, H., 1918
Report on the oceanographic investigations of Kinkwazan. (In Japanese). Rep. Imp. Fish. Inst. Sci. Invest. 5:27-56.

hydrography
Asano, H., 1915
Hydrographic investigations in the Hokkaido and adjacent seas. (In Japanese) Rep. Imp. Fish. Inst. Sci. Invest. 4:65-75.

hydrography
Asano, H., 1915
Hydrographic investigations of the sea between Izu-Oshima and Shiomisaki of Kii Province. (In Japanese). Rep. Imp Fish. Inst. Sci. Invest. 4:75-97.

Hydrography
Asano, H., and T. Ninagawa, 1918.
On the section within 60 miles off Takobana of the Shimane Prefecture. Rept. Imp. Fish. Inst. Sci. Inv. 6:39-41. (In Japanese.)

hydrography
Asano, H., and T. Ninagawa, 1918.
On the section off Tsukiyama, Kinkesan, and Shioyasaki. (In Japanese). Rept. Imp. Fish. Inst. Sci. Inv. 7:31-46.

hydrography
Asano, H., and T. Ninagawa, 1918.
On the section off Kinkasan and Shioyasaki. (In Japanese). Rept. Imp. Fish. Inst. Sci. Fish. 6:30-38.

hydrography
Barnes, C.A., and T.G. Thompson, 1938.
Physical and chemical investigations in Bering Sea and portions of the North Pacific Ocean. Univ. Washington Publ. Ocean. 3:35-79 and Append.

hydrography
Bassindale, R.F., F.J. Ebling, J. A. Kitching, R.D. Purchon, 1948.
The ecology of the Lough Ine rapids with special reference to water currents. I. Introduction and hydrography. J. Ecol. 36(2):305-322, 1 pl., 9 figs.

hydrography
Bigelow, H. B., 1926
Physical Oceanography of the Gulf of Maine. Bull. Bur. Fish., 40 (Pt.II):511-1027

hydrography
Bigelow, H.B., and C. Iselin, 1927.
Oceanographic reconnaissance of the northern sector of the Labrador Current. Science 65 (1691):551-552.

hydrography
Bigelow, H.B., and M. Leslie, 1930
Reconnaissance of the waters and plankton of Monterey Bay, July 1928. Bull. M.C.Z., 70(5):429-481, 43 text figs.

hydrography
Boerema, J., 1939(1940).
Hydrographic survey in the Netherlands East Indian Archipelago. Proc. Sixth Pacific Sci. Congr., 3:84-89.

hydrography
Bowden, K.F., 1953.
Physical oceanography of the Irish Sea. Brit. Assoc. Sci. Survey, Merseyside:69-80.

Hydrography
Bowers, G.M., 1906.
Dredging and Hydrographic Records of the U.S. Fisheries Steamer "Albatross" for 1904 and 1905. Rep. Comm. Fish. Fiscal Yea 1905 and Special Papers, Bur. Fish. Document No.604:52-66.

hydrography
Boyd, L. A., 1948.
The coast of northeast Greenland with hydrographic studies in the Greenland Sea. Am. Geogr. Soc., Spec. Pub. 30:340 pp., illus., maps.

hydrography
Braarud, T., 1945
A phytoplankton survey of the polluted waters of inner Oslo Fjord. Hvalrådets Skrifter, No.28, 142 pp., 19 text figs., 17 tables.

hydrography
Braarud, T., K. Ringdal Gaarder & O. Nordli, 1958
Seasonal changes in the phytoplankton at various points off the Norwegian west coast. (Observations at the permanent oceanographic stations 1945-46). Fisheridirek. Skrifter 12(3): 5-77.

hydrography
Braarud, T., and A. Klem, 1931.
Hydrographical and chemical investigations in coastal waters off Norway. Hvalrådets Skrifter, No. 1:

hydrography
Braarud, T. and J. T. Ruud, 1937
The Hydrographic conditions and aeration of the Oslo Fjord, 1933-1934. Hvalrådets Skr. No. 15:56 pp., 24 figs.

Reviewed: J. du Cons. XIV(3):406-408. J. N. Carruthers.

hydrography
Braarud, T. and J. T. Ruud, 1932
The "Øst" Expedition to the Denmark Strait 1929. I. Hydrography. Hvalrådets Skr., No. 4:44 pp., 19 text figs.

hydrography
Collier, A., and J.W. Hedgpeth, 1950.
An introduction to the hydrography of tidal waters of Texas. Publ. Inst. Mar. Sci. 1(2):125-194.

hydrography
Dannevig, A., 1945.
Undersøkelser i Oslofjorden 1936-1940. Rept. Norwegian Fish. Mar. Invest. 8(4):91 pp.

hydrography
Deacon, G.E.R. 1934
Die Nordgrenzen antarktischen und subantarktischen Wassers im Weltmeer. Annal. Hydr. u. Marit. Meteorol., 129-136, 3 text figs., 1 pl.

hydrography
Deacon, G. E. R., 1933
A general account of the hydrology of the South Atlantic Ocean. Discovery Repts. 7:173-238, pls.8-10.

hydrography
Deacon, G.E.R., 1933
A general account of the hydrology of the South Atlantic Ocean. Discovery Repts., 7:171-238, pls.8-10

hydrography
Dickerson, R.E., 1925.
Geologic aspects of Philippine hydrography. Handl. Nederl. -Ind. Natuurwet. Congr. 3:381-408.

hydrography
Douguet, M., 1950.
Rapport sur les observations faites par le "Commandant Charcot" au cours de sa croisière dans l'Antarctique durant l'été austral 1948-1949. II. Océanographie. Service Centrale Hydrogr Paris:69-103.

hydrography
Ealey, E.H.M., and R.G. Chittleborough, 1956.
Plankton, hydrology and marine fouling at Heard Island. Austral. Nat. Antarctic Res. Exped., Interim Rept., 15:81 pp.

Abstr. Meteorol. Abstr. & Bibl., 9(2):146.

hydrography
Fleming, J.A., C.C. Ennis, H.U. Sverdrup, S.L. Seaton and W.C. Hendrix 1945.
Scientific results of cruise VII of the Carnegie during 1928-1929 under command of Captain J.P. Ault. Oceanography. 1-B. Observations and results in physical oceanography.
Publ. Carnegie Inst. Washington 545: 1-315, 49 maps.

hydrography
Fleming, R.H., 1940-41
A contribution to the oceanography of the Central American region. Sixth Pac. Sci. Congr., Calif. 1939, Proc. 3:167-175.

hydrography
Fleming, R. H., 1939(1940).
A contribution to the oceanography of the Central American region. Proc. Sixth Pacific Sci. Congr., 3:167-175.

hydrography
Fleming, R.H. 1939
A contribution to the Oceanography of the Central American Region. Proc. Sixth Pacific Sci. Congr.:167-175, 10 text figs.

hydrography
Fleming, R.H., and R. Revelle, 1939
Physical processes in the ocean. Recent Marine Sediments, A. symposium: 48-141 (A.A.P.G., Tulsa, Oklahoma)

hydrography
France, Institut Scientifique et Technique des Pêches Maritimes, 1954.
Rapport d'ensemble sur les travaux de l'Office Scientifique et Technique des Pêches Maritimes de 1945 à 1953. Rev. Trav. Inst. Sci. Tech. Pêches Marit. 18(2/4):144 pp.

Hydrography
Frank, H., 1949.
Meteorologisch-hydrographische Beobachtungen auf der westlichen Ostsee. Ann. Met., 2 Jahrg., (1949), Heft 3/4:118-120, 2 textfigs.

hydrography
Furnestin, J., 1950.
Hydrologie cotière du Maroc (Année 1948). Ann. Biol. 6:46.

hydrography
Furnestin, J., 1949.
L'hydrologie côtière du Maroc. Bull. Sci. Com. Océan. Études Côtes Maroc 4:7-27 and Annex No. 5 31 pp.

hydrography
Gehrke, J. 1910
Beiträge zur Hydrographie des Ostsee bassin
Publ. de. Circ., No. 52.

hydrography
Gomoyunov, K. A., 1945.
Gidrologicheskie issledovaniva v sovetskoy arktiki za 25 let (1920-45). [Hydrological investigations in the Soviet Arctic during 25 years (1920-45).] Izvest. Vses. Geogr. Obshch. 77(6): 328-340.

hydrography
Goodman, J. R., J. H. Lincoln, T.G. Thompson, and F. A. Zeusler, 1942.
Physical and chemical investigations: Bering S a, Bering Strait, Chukchi Sea during the summers of 1937 and 1938. Univ. Washington Publ. Ocean., 3(4):105-169, 37 maps

hydrography
Grainger, E. H., 1959.
The annual oceanographic cycle at Igloolik in the Canadian Arctic. I. The zooplankton and physical and chemical observations. J. Fish. Res. Bd. Canada, 16(4): 453-501.

hydrography
Gran, H. H., 1902.
Das Plankton des Norwegischen Nordmeeres von Biologischen und Hydrographischen Gesichtspunkten behandelt. Rep. Norw. Fish. Mar. Invest., Vol. II (5):36-39, 173-175, pl. 1, figs. 1-9

hydrography
Gran, H.H., and T. Braarud, 1935
A quantitative study of the phytoplankton in the Bay of Fundy and the Gulf of Maine (including observations on hydrography, chemistry, and turbidity). J. Biol. Bd., Canada, 1(5):279-467, 69 text figs.

hydrography
Granqvist, G., 1948.
Den Finländsak Skägårdens hydrografi. Skägårdens boken:104-133.

hydrography
Günther, E. R., 1936.
A report on Oceanographical investigations in the Peru coastal current. Discovery Reports, XIII:107-276, Pls. XIV-XVI.

physics
Harvey, H.W., 1928
Biological chemistry and physics of sea water. 194 pp., 65 textfigs. University Press, Cambridge.

hydrography
Helland-Hansen, B., 1941.
Oceanografiske undersökelser i Norskehavet og Atlanterhavet. Ymer 6(3):161-173.

hydrography
Helland-Hansen, B., 1910
Physical Oceanography and Meteorology. Report on the Scientific Results of the "Michael Sars" North Atlantic Deep-Sea Expedition, 1910, I:1-115.

hydrography
Hicks. Steacy D., 1959.
The physical oceanography of Narragansett Bay. Limnol. & Oceanogr. 4(3):316-327.

hydrography
Holmquist, C., 1963.
Some notes on Mysis relicta and its relatives in northern Alaska.
Arctic, 16(2):109-128.

hydrography
Iselin, C. O'D., 1939
Some physical factors which may influence the productivity of New England coastal waters.
J. Mar. Res., 2:74-85

hydrography
Iselin, C., 1930
A report on the coastal waters of Labrador based on explorations of the "Chance" during the summer of 1926. Proc. Am. Acad. Arts Sci., 66(1):1-37, 14 text figs.

owned by MS

hydrography
Jacobsen, J. P., 1929
Contributions to the Hydrography of the North Atlantic; the "Dana" Expedition 1921-22.
The Danish "Dana" Expeditions 1920-22 in the North Atlantic and Gulf of Panama, Oceanographical Reports edited by the "Dana" Committee, No. 3; 98 pp.

hydrography
Jakhelin, A., 1936
The water transport of gradient currents.
Geofysiske Publ. XI (11):14 pp., 6 text figs

hydrography
Japan, Hydrographic Bureau, 1948.
[Oceanographic phenomena in Izu Shoto and approaches in 1947.] Hydrogr. Bull. 8:31-32, 2 pls.

hydrography
Kiilerich, A., 1945
On the Hydrography of the Greenland Sea. Medd. om Grønland 144(2):1-63.

hydrography
Kiilerich, A.B., 1939.
The Godthaab Expedition, 1928, a theoretical treatment of the hydrographical observation material. Medd. om Grønland 78(5):1-149.

hydrography
Knipovich, N.M., 1938
Hydrography of seas and brack waters (in application to the fisheries): Sci. Inst. Mar. Fish. and Oceanogr. USSR. (In Russian)

hydrography
Kupetski, V. N., 1959.
On causes of anomaly in hydrological conditions of the Olaf Prudce bay (Antarctic)
Izvest. Vsesoy. Geograf. Obshchestva, 91(4):356-358.

hydrography
LaFond, E.C., J.F.T. Saur, and J.P. Tully, 1951.
Physical oceanography of the Bering and Chukchi seas. Proc. Alaskan Sci. Conf., Bull. Nat. Res. Coun. 122:80-81.

hydrography
Makaroff, S., 1894.
Observations hydrologiques faites par les officiers de la corvette "Vitiez" pendant un voyage autour du monde exécuté de 1886 à 1889. St. Petersburg.

hydrography
Marti, P., 1928.
Mission hydrographique sur les côtes d'Algerie et de Tunisie en 1923, 1924, et 1925. Ann. Hydrogr. (3) 8:213-306, 20 figs.

hydrography
McEwen, G.F., 1929
A mathematical theory of the vertical distribution of temperature and salinity in water under the action of radiation, conduction, evaporation, and mixing due to the resulting convection. Bull. Scripps Inst. Oceanogr., Tech. Ser., II: 197-306.

hydrography
McEwen, G.F., 1927
Dynamical oceanography and certain elements of hydrodynamics upon which it is based. Bull. 14 S.I.O.:3-20.

hydrography
McEwen, G.F., 1916
Summary and interpretations of the hydrographic observations made by the Scripps Institution for Biological Research of the University of California, 1908-1915. Univ. Calif. Publ. Zool., XV:255-356.

Hydrography
Merz, A., and L. Möller, 1928
Hydrographische Untersuchungen in Bosphorus und Dardanellen. Veröff. d. Inst. f. Meereskunde, A VIII

hydrography
Mullaney, E. J., 1959
Hydrographic surveys.
J. Boston Soc. Civil Engrs., 46(4):315-325.

hydrography
Navarro, F. de P., 1947.
Exploración oceanográfica del Africa occidental desde el Cabo Ghir al Cabo Juby. Trab. Inst. Españ. Ocean. No. 20:40 pp., folding chart.

hydrography
Nielsen, J.N., 1912.
Hydrography of the Mediterranean and adjacent waters. Rept. Danish Oceanogr. Exped., 1908-1910, 1:76-191, 19 figs., 10 pls.

hydrography
Nutman, S.R., 1950.
A fish cultivation experiment in an arm of a sea loch. 2. Observations on some hydrographic factors in Kyle Scotnish. Proc. R. Soc., Edinburgh, Sect. B, 64(1):5-35.

hydrography
Ovsiannkov, A. N., 1959.
[Experimental study of the hydrometeorological regime of the sea.]
Meteorol. i. Gidiol. (5):48-49.

hydrography
Palmen, E. and E. Laurila, 1938
Über die Einwirkung eines Sturmes auf den hydrografischen Zustand in nördlichen Ostsee gebiet. Soc. Scient. Fennica, Comm. Phys.-Math., X:1

hydrography
Parr, A. E., 1938.
Further observations on the Hydrography of the Eastern Caribbean and adjacent Atlantic waters. Bull. Bingham Ocean. Coll., 6(4):1-29, 22 figs.

hydrography
Parr, A. E., 1937
Report on hydrographic observations at a series of anchor stations across the Straits of Florida. Bull. Bingham Ocean. Coll. VI (3):1-62, 36 text figs.

hydrography
Parr, A. E., 1937
Report on hydrographic observations at a series of anchor stations across the Straits of Florida Bull. Bingham Ocean. Coll., 6(3):62 pp.

hydrography
Parr, A. E., 1937.
A contribution to the hydrography of the Caribbean and Cayman Seas, based upon the observations made by the Research Ship "Atlantis", 1933-1934. Bull. Bingham Ocean. Coll., V(4):1-110, 82 textfigs.

hydrography
Parr, A.E., 1936
On the relationship between dynamic topography and direction of current under the influence of external (climatic) factors. J. Cons. 11(3):299-307, 3 text figs.

hydrography
Parr, A. E., 1936
On the probable relationship between vertical stability and lateral mixing processes. J. Cons., 11(3):299-313.

hydrography
Parr, A. E., 1935
Report on hydrographic observations in the Gulf of Mexico and the adjacent straits made during the Yale Oceanographic Expedition on the "Mabel Taylor" in 1932. Bull. Bingham. Oceanogr. Coll., 5(1):93 pp.

hydrography
Riis-Carstensen, E., 1936.
The Godthaab Expedition 1928. The hydrographical work and material. Medd. om Grønland 78(3):1-101, 25 figs., 12 pls.

hydrography
Robertson, A.J., 1904.
Scottish hydrographic research during 1903. Publ. de Circ., Cons. Perm. Int. Expl. Mer, No. 17:1-6.

hydrographie
Rouch, M.J., 1946.
Océanographie et climatologie des îles Atlantides Contrib. Étude Peuplement Iles Atlant., Mem. Soc. Biogeographie 8:41-57.

hydrography
Rouch, J., 1941.
Echantillons d'eau de mer recueillis dans l'Ocean Indien Austral par l'aviso Bougainville.
C.R. Acad. Sci., Paris, 213:402-404.

hydrography
Rouch, J., 1940.
Observations océanographiques de surface dans l'ocean Atlantique et dans la Mediterranée.
Ann. Inst. Océan. 22(2):51-73.

hydrography
Ruud, J. T., 1932.
On the biology of southern Euphausiidae.
Hvalrådets Skrifter No. 2:1-105, 37 text figs.

hydrography
Saur, J.T.F., jr., 1950.
Oceanographic features of Nodales Channel (British Columbia) with application to underwater sound. USNEL Rept. 188(IV):21 pp.

hydrography
Schmidt, J., 1929
Introduction to the oceanographical reports including list of the stations and hydrographical observations. The Danish "Dana" Expeditions 1920-22 in the North Atlantic and the Gulf of Panama, Oceanographical Reports edited by the "Dana" Committee, No. 1, 9 pp.

hydrography
Schulz, B., 1938
Die hydrographischen Arbeiten der Deutschen wissenschaftlichen Kommission für Meeresforschung. Ber. deutschen wiss. Komm. f. Meeresforschung. n.s. 9(2):186-197, 6 text figs.

hydrography
Seiwell, H. R., 1939.
Die Verwendung der Verteilung von Sauerstoff auf die physische Ozeanographie des Karibischen Meeresgebietes. Gerlands Beiträge z. Geophys. 54(4):1-7, 3 textfigs.

hydrography
Seiwell, H. R., 1938.
Application of the distribution of oxygen to th physical oceanography of the Caribbean Sea Region. P.P.O.M. 6(1):60 pp., 42 figs.

Reviewed: J. du Cons. 14(3):410-411, G. Dietrich

hydrography
Shvisov, P. P., 1938.
[Oceanographical observations.] Compt. Rend. Acad Sci. 19(8):569-580.

hydrography
Skogsberg, T. 1936
Hydrography of Monterey Bay, California. Thermal Conditions, 1929-1933, Trans. Am. Phil. Soc. 29(1):1-152.

hydrography
Skogsberg, T. and A. Phelps, 1946
Hydrography of Monterey Bay, California. Thermal conditions, Part II (1934-1937) Proc. Am. Phil. Soc. 90(5):350-386, 16 text figs., tables.

hydrography
Slaucitajs, L., 1941.
[Hydrologisches Untersuchungen im Rigaischen Meerbusen im Mai 1940.] Acta Univ. Latviendes III3H 35:373-376, with German summary, 376.

hydrography
Slaucitajs, L., 1931.
Latvājas juras udenu hidrafijas iss vesturiska apskets. Kugn. gad. gram. 1930/31. (Riga).

hydrography
Smidt, E.L.B., 1944
Biological Studies of the Invertebrate Fauna of the Harbor of Copehagen. Vidensk. Medd. fra Dansk naturh. Foren. 107:235-316, 23 text figs.

hydrography
Smith, E. H., 1941.
U.S. Coast Guard cutter "Northland"'s ice and oceanographic observation cruise, Baffin Bay and Davis Strait, autumn of 1940. Trans. Am. Geophys Union 3: 788-792.

hydrography
Smith, E.H., 1931.
The Marion Expedition to Davis Strait and Baffin Bay, 1928. U.S.C.G. Bull. 19(3):1-221.

hydrography
Soule, F.M., 1940
Consideration of the depth of the motionless surface near to Grand Banks of Newfoundland Jour. Mar. Research, 2(3): 169-180, figs. 53-58.

hydrography
Soule, F. M., 1940.
Oceanography. Season of 1938. Bull. No. 28, International Ice Observation and Ice Patrol Service in the North Atlantic:113-173, textfigs. 45-56, tables. Ocean

hydrography
Stockman, W.B., 1946
A thoery of J-S curves as a method for studying the mixing of water masses in the Sea. J. Mar. Res., 6(1):1-24, 9 text figs.

hydrography
Sumner, F. B., G.D. Louderback, W. L. Schmitt, and E.C. Johnston, 1914.
A report upon the physical conditions in San Francisco Bay, based upon the operations of the United States Fisheries Steamer "Albatross" during the years 1913 and 1914. Univ. Calif. Publ. Zool. 14(1):1-198, 20 textfigs., 13 pls.

hydrography
Sverdrup, H.U., 1942
Oceanography for meteorologists. New York, Prentice-Hall. 246 pp.

hydrography
Sverdrup, H.U., 1940
Hydrology, Section 2 Discussion. B.A.N.Z. Antartic Research Expedition 1921-1931, Repts., Ser.A, Vol.3. Oceanography, Part 2, Hydrology, Sect.2, 88 and 126

hydrography
Sverdrup, H.U., 1931
Some Oceanographic Results of the "Carnegie's" Work in the Pacific. The Peruvian Current. Hydrographic Rev., VIII: 240-244

hydrography
Sverdrup, H.U., 1930
Some oceanographic results of the "Camegies" work in the Pacific - the Peruvian current. Trans. Am. Geophys. Union, 10th and 11th annual meetings: 257-264, 11 figs.

hydrography
Sverdrup, H.U. 1930
The origin of the Deep-Water of the Pacific Ocean as indicated by the Oceanographic work of the "Carnegie". Internat. Geodet, Geophys. Union, Stockholm Assembly, Section of Oceanography

hydrography
Sverdrup, H.U., J.A. Fleming, F.M. Soule, Scientific results of cruise VII of the Carnegie during 1928-1929 under command of Captain J.P. Ault. Oceanography. 1-A. Observations and results in physical oceanography. Publ. Carnegie Inst. Washington 545: 1-X, 1-156, 17 maps, 5 figs

hydrography
Sverdrup, H.U., and R.H. Fleming, 1941
The waters off the coast of Southern California. March to July 1937. SIO Bull. 4(10):261-378.

hydrography(general)
Tait, J.B., 1952.
Travaux de l'Ecosse. Cons. Perm. Int. Expl. Mer, Ann. Biol. 8:62-63, Textfig. 3.

hydrography
Tait, J.B., 1950.
N.North Sea. Hydrography. Ann. Biol. 6:83-87.

hydrography
Tait, J. B., 1936
Salient features of the hydrography of the Northern North Sea and the Faeroe-Shetland Channel in 1935. J. Cons., XI:164-168

hydrography
Tcherneff, S., 1938.
Observations hydrologiques dans la zone cotiere bulgare. (In Bulgarian, with French resume) Trav. (Trudovie) Stat. Ichtyol. Sozopol-Bulg., 1936, V:35-44, Bourgas.

hydrography
Theisen, E. 1946
Tanafjorden. Enfinmarksfjords oceanografi. Fiskeridirektoratets skrifter ser. Havundersøkelser. (Repts on Norwegian Fishery and Marine Investigations) VIII (8): 1-77, 23 text figs., 8 pls.

hydrography
Thompson, E. F., 1939.
Chemical and physical investigations. The general hydrography of the Red Sea. John Murray Exped., 1933-34, Sci. Repts. 2(3):83-103, 12 textfigs.

hydrography
Thompson, T.G., and C.A. Barnes, 1951.
Physical and chemical oceanography of the Gulf of Alaska and the Aleutian Islands. Proc. Alaska Sci. Conf., Bull. Nat. Res. Coun. 122:82.

hydrography
Thomsen, H., 1952.
The milieu. Rapp. Proc. Verb., Cons. Perm. Int. Expl. Mer, 132:21-27.

hydrography
Thomsen, H., 1948.
The hydrographical conditions in Faxa Bay. Cons. Int. Expl. Mer, Rapp. et Proc. Verb. 120:30-31.

hydrography
Thomsen, H., 1938.
Hydrography of Icelandic waters. Zool. of Iceland 1(4):1-36.

hydrography
Thomsen, H. 1935
Entstehung und Verbreitung einiger characteristischer Wassermassen in dem Indischen und südlichen Pazifischen Ozean. Ann. der Hydrog. und Mar. Met., 63:293-305.

Hydrography
Tibby, R.B., 1941
The water masses off the west coast of North America. J. Mar. Res.4:112-121.

hydrography
Timofeyev, V.T., 1946.
Gidrometeorologicheskiv ocherk zaliva Varangerfiord. [Hydrological and meteorological outline of the Varangerfjorden.] Glav. Uprav. Gidromet. Sluzhby Sov. Ministr., Trudy Nauch. -Issled. Uchrezhd., Ser. V, Gidrol. Morya, 16:14-70.

hydrography
Tully, J.P., 1949.
Oceanography and prediction of pulp mill pollution in Alberni Inlet. Bull. Fish. Res. Bd., Canada, 83:169 pp.

hydrography
Uda, M., 1939(1940).
A sketch of the recent development of hydrographical researches in the sea adjacent to Japan. Proc. Sixth Pacific Sci. Congr., 3:44-82.

hydrography
Ullyott, P. and Orhan Ilgaz, 1946
The Hydrography of the Bosphorus: an Introduction. Geol. Rev., 36 (1), pp. 44-66

hydrography
Van Weel, A. M., 1923.
Meteorological and hydrographical observations made in the western part of the Netherlands East Indian Archipelago. Treubia, 4:559 pp.

hydrography
Wallbrecher, G. O., 1948.
Estudios y trabajos oceanográficos en la Antartida. 47 pp. La Plata, Argentina.

hydrography
Watson, E.E., 1936.
Mixing and residual currents in tidal waters as illustrated in the Bay of Fundy. J. Biol. Bd., Canada, 2(2):141-208, tables, 26 textfigs.

hydrography
Wattenberg, H., 1940.
Der hydrographisch-chemische Zustand der Ostsee im Sommer 1939. Ergebnisse der 'Triton'-Fahrt von 27 Juli bis 10 August, 1939. Ann. d. Hydrogr. usw. 68:185-194.

hydrography
Wendicke, F., 1916.
Hydrographische Untersuchungen des Golfes von Neapel. Mitt. Zool. Sta. Neapel, 22:329.
im Sommer 1913.

hydrography
Wiborg, K.F., 1944
The production of zooplankton in a landlocked fjord, the Nordåsvatn near Bergen, in 1941-42, with special reference to the copepods... (Repts. Norwegian Fish. and Mar. Invest.) 7(7):83 pp., 40 text figs.

hydrography
Wulf, A., 1934
Über Hydrographie und Oberflachenplankton nebst Verbreitung von Phaeocystis in der Deutschen Buch im Mai 1933. Ber. d. Deutsch. Wiss. Komm. f. Meeresforsch., N. F., VII(3):343-350

hydrography
Wüst, G., 1949.
Über die Zweiteilung der Hydrosphäre. Deut. Hydr Zeit. 2(5):218-225, 4 textfigs.

hydrography
Zubov, N.N., 1932
Hydrological investigations in the southwestern part of the Barents Sea during the summer 1928. Trans. oceanogr. Inst. Moscow, 2(4):1-83.

hydrography, effect of

hydrography, effect of
Dementieva, T.F., 1963.
The variation of the commercial fish stocks in the Baltic Sea area under the influence of oceanologic factors. (In Russian). Okeanologiia, Akad.Nauk.SSSR,3(5):876-885.

hydrography, effect of
Fleming, R.H., and T. Laevastu (compilers), 1956
The influence of hydrographic conditions on the behavior of fish (a preliminary literature survey). F.A.O. Fish. Bull., 9(4):181-196.

hydrography, effect of
Tait, J.B., 1952.
Hydrography in relation to fisheries. (Buckland lectures fpr 1938). Edward Arnold & Co., London: lxii & 106 pp., frontispiece and 19 figs.

hydrographic conditions, effect of
Templeman, Wilfred, 1966.
Marine resources of Newfoundland.
Bull. Fish. Res. Bd., Can., 154: 170 pp.

Hydrographic surveys

hydrographic surveys
Anon., 1968.
Hydrographic surveys- at 30 knots.
Hydrospace, 1(5): 34-35.

Hydrographic surveys
Great Britain, Hydrographic Department, 1961.
Report of the Hydrographer of the Navy for the Year 1960. (HW. 733/60) HD519:33 pp.

hydrology

hydrology
Hendricks, E.L., 1962.
Hydrology. An understanding of water in relation to earth processes requires collaboration of many disciplines.
Science, 135(3505):699-705.

hydrology
Stepanov, V.N., 1964
The contemporary status of hydrological investigations of the World Ocean. (In Russian). Okeanologiia, Akad. Nauk. SSSR, 4(5):745-748.

hydrometer size analysis
Buchan, S., A. Jones and P. Simpkin, 1970
A computer program for the calculation of hydrometer size analysis.
Marine Geol., 9(4): M23-M29.

hydrosis

hydrosis
Laevastu, T., 1965.
Is oceanographic forecasting (hydrosis) feasible for fisheries?
ICNAF Spec. Publ., No. 6:881-884.

hydrothermal activity
Bonatti, Enrico, 1968.
Rocks and sediments from the Barracuda Fracture Zone. (Abstract only). Trans. Fifth Carib. Geol. Conf., St. Thomas, V.I., 1-5 July 1968, Peter H. Mattson, editor. Geol. Bull. Queens Coll., Flushing, 5: 45.

hydrothermal activity
Rozanova, T.V. and G.N. Baturin, 1971.
On ore hydrothermal manifestations on the Indian Ocean floor. (In Russian; English abstract). Okeanologiia 11(6): 1057-1064.

hypersalinity

hypersalinity
Carpelan, Lars H., 1967.
Invertebrates in relation to hypersaline habitats
Contrib. mar Sci., Port Aransas, 12: 219-229

hypersalinity
Gunter, Gordon 1967.
Vertebrates in hypersaline waters.
Contrib. mar. Sci., Port Aransas, 12: 230-241

hydrospace

see also: man-in-the-sea

hydrospace simulation
Stachiw, J.D., 1968.
Hydrospace-environment simulation. Ch. 17 in: Ocean engineering: goals, environment, technology, J.F. Brahtzmeditor, John Wiley & Sons, 633-711.

hygiene

hygiene
*Huard, Pierre, 1968.
Hygiène et pathologie navales au XIX Seècle.
Bull. Inst. océanogr., Monaco, No. special 2: 545-554.

hyperplasia
Powell, N.A., C.S. Sayce and D.F. Tufts, 1970.
Hyperplasia in an estuarine bryozoan attributable to coal tar derivatives.
J. Fish. Res. Bd, Can., 27(11): 2095-2096.

hypersthene

hypersthene
Norris, Robert M., 1964.
Sediments of Chatham Rise.
New Zealand Dept. Sci. Ind. Res., Bull., No. 159.
New Zealand Oceanogr. Inst., Memoir, No. 16:39pp.

hyponeuston

hyponeuston, lists of spp.
Champalbert, Gisèle, 1969.
L'hyponeuston dans le Golfe de Marseille.
Téthys 1(3): 585-666.

hyponeuston
David, Peter M., 1965.
The surface fauna of the ocean.
Endeavour, 24(92):95-100.

Hyponeuston
Specchi, Mario, 1968.
Observations préliminaires sur l'hyponeuston du golfe de Trieste.
Rapp. Proc.-verb. Réun., Comm. int. Explor. scient. Mer Méditerranée 19(3): 491-494.

hypothermia (immersion)

hypothermia (immersion)
United States, Weather Bureau and Hydrographic Office, 1961
Climatological and oceanographic atlas for Mariners. Vol. II. North Pacific Ocean. Unnumbered pages.

hypsographic curve

hypsographic curve
Meinardus, W., 1942.
Die bathygraphische Kurve des Tiefseebödens und die hypsographische Kurve der Erdkruste.
Ann. Hydrogr., usw., 70:225-244.

hypsographic curve
Reilly, W.I., 1965.
The mean height and the Hypsographic curve for New Zealand.
New Zealand J. Geol. and Geophys., 8(1):128-139.

hypsometry

hypsometry
Menard, H.W., and Stuart M. Smith, 1966
Hypsometry of ocean basin provinces.
J. geophys. Res., 71 (18): 4305-4325.

hystrichospheres

hystrichospheres
Wall, David, 1971.
Biological problems concerning fossilizable dinoflagellates.
In: Science and Man, Louisiana State Univ, 3: 1-15.

hystrichospheres
*Wall, David, and Barrie Dale, 1967.
The resting cysts of modern marine dinoflagellates and their palaeontological significance.
Palaeobot. Palynol., 2(1/4):349-354.

hystricospheres
Wall, David, and Barrie Dale, 1966.
"Living fossils" in western Atlantic plankton.
Nature 211 (5053):1025-1026.

Ice

ice
Adams, C.M., Jr., D.N. French and W.D. Krugery, 1960.
Solidification of sea ice.
J. Glaciol., 3(28):745-762.

ice
Agureev, S.P., 1958.
Nomograms for reducing observations on the drift of ice. (In Russian).
Meteorol. i Gidrol., (1):78-

ice
Akagawa, M., 1970
Sea ice in the Okhotsk Sea. (In Japanese)
Bull. Hakodate mar. Obs. 15(9):449-479
(reprint from another journal)

ice
Akagawa Masaomi, 1970.
On the drift ice east of South Sakhalin in the Okhotsk Sea in June 1969 and some oceanographic phenomena related to the sea ice situation. (In Japanese; English abstract).
Bull. Hakodate mar. Obs. 15(12):133-144.

Ice
Akagawa, Masaomi, 1970.
On the drift-ice east of South Sakhalin in the Okhotsk Sea in June 1969 and some oceanographic phenomena related to the sea ice situation. (In Japanese; English abstract).
Umi to Sora 45(4):133-144

ice
Akagawa, M., 1964.
On the drifting of pack-ice in the Pacific Ocean off Hokkaido. (In Japanese; English abstract).
Bull. Hakodate Mar. Obs., No. 11:50-35.

ice
Akagawa, M., 1963.
Sea ice conditions in the northern Okhotsk Sea in thawing time. (In Japanese; English abstract)
Bull. Hakodate Mar. Obs., 10(3):54-64.

ice
Akagawa, M., 1960.
[On the melting of pack-ice field.]
Bull. Hakodate Mar. Obs., 7(5):921-932.

ice
Aldoshina, E.I., 1960.
Changes in amount of ice and the location of the ice edge in the Sea of Japan and the Okhotsk Sea in the spring-summer season. (In Russian).
Trudy Gosud. Okeanograf. Inst., Leningrad, 54: 22-34.

ice
Alexeev, N.N., 1935.
O reise ledoreza "Fedor Litke" v Okhotskom More zimoi 1931-1932 g.[Über die Fahrt des Eisbrechers "Fedor Litke" in Ochotoskischen Meer im Winter 1931-1932.] Issled Mor. SSSR, 22:36-40.
(German Summary).

ice
Allen, L.P., 1951.
The geography and morphology of sea ice. Proc. Alaskan Sci. Conf., Bull. Nat. Res. Coun. 122: 121.

ice
Andreev, V., 1938
Kuril'skie Ostrova. (Kurile Islands) Morsk. sborn. 1938 (11):75-87, (12):87-99.

ice
Angino, Ernest E., Kenneth B. Armitage and Jerry C. Tash, 1965.
Ionic content of Antarctic ice samples.
Polar Record, 12(79):407-409.

ice
Anon., 1966.
Ice winter 1964/65 along the Finnish coast.
(In Finnish and English)
Merentutkimuslait. Julk., No. 220-20pp.

ice
Anon., 1965
Ice winter 1963/64 along the Finnish coast.
Merentutkimuslait. Julk., No. 217:24pp.

ice
Anon. (G.A.T.), 1965.
Notes on ice conditions in areas adjacent to the North Atlantic Ocean from January to March 1965.
Mar. Obs. (M.O. 764), 35(209):135-140.

ice
Anon., 1951.
Southern ice reports, April, May, and June 1950.
Mar. Obs. 21(152)(M.O.546):98.

ice
Anon., 1951.
Southern ice reports, January, February and March. Mar. Obs. 121(151)(M.O.546):62-67.

ice
Anon., 1950.
Isforholdene i de Arktiske Have. 1948. Publ. Danske Met. Inst., Årbøger, 16 pp., 5 charts.

ice
Anon., 1949.
Arctic ice and its drift into the North Atlantic Ocean. Pilot Chart of N.A. Ocean, Ed. 9, Suppl.

ice
Anonymous, 1944.
Sea ice: terminology, formation, and movement.
Polar Rec. 4(27):126-133.

ice
Anon., 1942.
Atlas der Eisverhältnisse im deutschen und benachbarten Ost- und Nordseegebiet.
Bearb. Deutscher Seew., Oberkomm. Kriegsmar.
(German marking No. 2198).

ice
Anon. 1942.
Atlas der Vereisungsverhältnisse Russlands und Finnlands, ihrer Küstengewässer sowie wirtschaftlich und militärisch wichtigen Binnenwasserstrassen mit textlichen Vorbemerkungen und Tabellen. Bearb. Deutsche Seew., Oberkomm. Kriegsmar.
(German marking 2197).

ice
Anon., 1940
Eis um Island und Ost-Grönland, mit 10 Jahrgangen Eiskarten für die Monate April bis August.
Neubearb. Deutscher Seew., Oberkomm. Kriegsmar.: 1-69 (German Marking: NR 2011A).

ICE
Anon, 1914
Ice observation, meteorology and oceanography in the North Atlantic Ocean. Rept. on the work carried out by the S.S. "Scotia", 1913, 139 pp.

ice
Anon., 1950.
Southern ice reports, October, November and December, 1948 and 1949. Mar. Obs. 20(150)(M.O. 528) :241-243.

ice
Anon., 1950.
Southern ice reports, July August and September 1948 and 1949. Mar. Obs. 20(149)(M.O.528):180.

ice
Anon., 1950.
Southern ice reports, January 1948, January, February and March 1949. Mar. Obs. 20(147):52-55.

ice
Anon., 1949.
Southern ice reports during the year 1948, October, November, December. Mar. Obs. 19(146)(M.O. 512):238-239.

ice
Anon., 1949.
Southern ice reports, April, May and June 1948.
Mar. Obs. 19(144)(M.O.512):120.

ice
Antonov, A.E., 1964
An application of the periodogram analysis of the forecasting of some hydrometeorological elements (on the example of the Baltic Sea). (In Russian).
Materiali Vtoroi Konferentsii, Vzaimod. Atmosfe i Gidrosfer. v Severn. Atlant. Okean., Mezhd. Geofiz. God, Leningrad. Gidrometeorol. Inst., 32-40.

ice
Antonov, V.S., 1961.
[The variation of the ice drift speed in the Arctic ocean.]
Okeanologiia, (3):418-425.

ice
Antonov, V.S., 1960.
[Causes of a peculiar state of ice cover in one of the basins of the Arctic ocean (1956 observation).] Izv. Vses. Geograf. Obshch. 92:353-356.

ice
Antonov, A.E., and O.S. Roud neva, 1965.
Peculiarities of hydrological and hydrochemical conditions in the south Baltic during the IGY. (In Russian; English abstract).
Okeanolog. Issled. Rezult. Issled. Programme Mezhd. Geofiz. Goda, Mezhd. Geofiz. Komitet Presidiume, Akad. Nauk, SSSR No. 13:90-95.

ice
Apollonio, Spencer, 1965.
Chlorophyll in Arctic sea ice.
Arctic, J. Arctic Inst., N. Amer., 18(2):118-122.

ice
Apollonio, S., 1961.
The chlorophyll a content of Arctic sea-ice.
Arctic, 14(3):197-199.

ice
Arakawa, K., 1955.
The growth of ice crystals in water. J. Glaciol. 2(17)463-468.

ice
Ardus, Dennis A., 1965.
Surface deformation, absolute movement and mass balance of the Brunt Ice Shelf near Halley Bay, 1961.
British Antarct. Survey Bull., No. 6:21-41.

ice
Armstrong, Terence, 1964.
ice atlases.
Polar Record, 12(77):161-163.

ice
Armstrong, Terence, 1959
Russian-English glossary and Soviet classification of ice found at sea. Compiled by Boris N. Mandrovsky, Washington: Ref. Dept. Library of Congress, 1959. 10 1/4 x 8 inches vi + 30 pp; mimeographed $0.30, obtainable from Card Division, Libr. of Congr., Wash. 25, D.C.
Arctic 12(4): 237-238.

ice
Armstrong, T., 1957.
Insurance against ice risks at sea. Polar Rec. 8(51):421-428.

ice
Armstrong, T., 1957.
The ice of the central polar basin. J. Glaciol., 3(22):105-110.

ice
Armstrong, T., 1955.
Sea ice studies. Arctic 7(3/4):201-205.

ice
Armstrong, T., and B. Roberts, 1956.
Illustrated ice glossary. Polar Rec. 8(52):4-12, 140 figs.

ice
Arzybychew, S., and V. Jushakow, 1939.
Das Wärmeleitvermögen der natürlichen Eises.
Ann. Hydr., Jahrg. 67(7):213-215.

ice
Assur, A., 1951.
Über das Dickenwachstum des Eises. Deutsch. Hydro. Zeits. 4(1/2):72-74.

ice
Aufderheide, Arthur C., and Gerald Pitzl, 1970
Observations on ice regions of the Arctic Ocean.
Arctic, 23(2): 133-136.

ice
Augureyev, S.P., 1958.
Nomograms for processing observations for drift ice. (In Russian).
Meteorol. i Gidrol., (1):54-55.

RZh Geofiz8/58-5594.

ice
Avhevich, V.I., 1963.
Some peculiarities of sea ice interpreted from aerial photography. (In Russian).
Voprosy Geogr., No. 62:155-165.

Not seen
Abstract in:
Soviet Bloc Res., Geophys., Astron., Space, No. 95:60.

ice
Avsyuk, G.A., 1958.
Modern glaciation of the Soviet Arctic. Sec. 1. Distribution and character of sea ice.
NAS-NRC, Publ., No. 598:15-21.

ice
Ayers, John C., 1965.
The climatology of Lake Michigan, 1965.
Univ. Michigan, Great Lakes Res. Div., Inst. Sci. and Techn. Publ., No. 12:73 pp.

ice
*Azernikova, O.A., 1967.
On the interrelation between the changes of some behaviour elements of the White and Barents seas. (In Russian; English abstract).
Okeanologiia, Akad. Nauk, SSSR, 7(4):607-616.

B

ice
Badigan, K.S., 1950.
[In the ice of the Arctic.] Vokrug Sveta 1:8-14.

ice
Badigin, K.S., 1940.
[Mit dem Schiff "Georgij Sedov" über das Eismeer. Tagebuch vom Kapitan, redigiert durch N.N. Subow (besonders S.546-606: Einige vorläufige Ergebnisse der vom Eisbrecher "Sedow" ausgeführten wissenschaftlichen Arbeiten.)] Moskau.

ice
Bakaev, V.G., editor, 1966.
Atlas Antarktiki, Sovet skaia Antarktich-eskaia Ekspeditsiia. 1.
Glabnoe Upravlenia Geodesii i Kartografii MG SSSR, Moskva-Leningred, 225 charts.

ice (pack)
Balligand, P., 1955
Les sous-marins sous la banquise.
La Revue Maritime, Ser. 110:737-752.
USN-HO-Transl., No. 321
T. Frontenac

ice, effect of
Banke, Erik G. and Stuart D. Smith, 1971.
Wind stress over ice and over water in the Beaufort Sea. J. geophys. Res. 76(30): 7368-7374.

ice
Barber, F.G., 1970.
Oil spills in ice: some cleanup options.
Arctic, 23(4): 285-286.

ice
*Barber, F.G., 1968.
On the water of Tuktoyaktuk Harbour.
Manuscript Rept. Ser., Mar. Sci. Br. Dept. Energy, Mines, Resources, Ottawa, 9:32 pp.

ice
Barber, F.G., 1967.
A contribution to the oceanography of Hudson Bay.
Manuscript Rep. Ser., Dept. Energy, Mines, Resources, Can., No. 4:69 pp. (multilithed).

ice
Barber, F.G., and C.J. Glennie, 1964.
On the oceanography of Hudson Bay, an atlas presentation of data obtained in 1961.
Manuscr. Rep. Ser., mar. Sci. Br., Dept. Mines tech Surv., Ottawa, 1: numerous pp. (unnumered) (multilithed).

ice
Bardin, V.I., and V.I. Shil'nikov, 1962.
Ice flow from the shores of Antarctica. (In Russian).
Akad. Nauk, SSSR, Antarktika, Doklady, Rom., 1961: 41-57.

Abstract:
Vitaliano, Dorothy B., 1964. Geophys. Abstr., (204):38 (Abstract #204-171).

sea ice
Barnes, C.A., undated (1954?).
Sea-ice problems. In: Oceanographic Instrumentation. NRC Publ. No. 309:85-100.

With discussion by W.K. Lyon
J.F. Holmes
W.G. Metcalf

ice
Barnes, C. A., E.R. Challender, F.M. Soule, and G.H. Read, 1947.
International ice observation and ice patrol service in the North Atlantic Ocean. C.G. Bull. No. 32:188 pp., 56 figs.

ice
Barnes, C.A., E.R. Challender, F.M. Soule, and G.H. Read, 1947
International ice observation and ice patrol service in the North Atlantic Ocean, season of 1946. C.G. Bull. No.32:188 pp., 56 figs.

ice
Barnes, James C., David T. Chang and James H. Willand, 1972.
Image enhancement techniques for improving sea-ice depiction in satellite infrared data.
J. geophys. Res. 77(3): 453-462.

ice
Baschin, O., 1927.
Die Polflucht des Meerwassers. Naturwiss. 15(27): 559-561.

ice
Bates, C.C., 1958.
Sea ice and its relation to surface supply problems in the American Arctic. Mar. Obs., 28(180):82-89.

ice
Bates, C.C., 1957.
Current status of sea ice reconnaissance and forecasting for the American Arctic.
J. Atmosph. Terr. Phys., Spec. Suppl., 285-322.

ice
Bates, C.C., and G.G. Lill, 1950.
Current Naval research in land and sea ice.
Trans. Am. Geophys. Union 31(2):278-281, 1 fig.

ice age
Behrmann, W., 1949.
Golfstrom und Eiszeit. Petermanns Geogr. Mitt. 92 Jahrg., 3/4:154-158, 1 textfig.

ice
Behrmann, W., 1948.
Golfstrom und Eiszeit. Petermanns Geogr. Mitt. 92:154-158.

ice
Bellair, Pierre, 1960.
La Bordure côtière de la Terre Adélie.
Territoire des Terres Australes et Antarctiques Françaises. Terre Adélie, 1960.
Exped. Polaires Françaises, Publ., No. 218:7-40.

ice
Bennington, Kenneth O., 1963.
Some crystal growth features of sea ice.
J. Glaciology, 4(36):669-688.

ice

Berg, V.A., 1961

[Some improvements in the method of computing the ice depth in salt and saltish water bodies.] Issled. Severnoi Chasti Atlanticheskogo Okeana, Mezhd. Geofiz. God, Leningradradskii Gidrometeorol. Inst., 1: 107-111.

ice

Betin, V.V., 1957.
Ice conditions in the region of the Baltic Sea and on the approaches to it according to variations for many years. (In Russian).
Trudy Gosud. Okeanogr. Inst., 41:54-125.

ice

Betin, V.V., and V.V. Preobrazhensky, 1961
Ice research in the Baltic Sea during the International Geophysical Year. Contribution to Special IGY Meeting, 1959.
Cons. Perm. Int. Expl. Mer. Rapp. Proc. Verb., 149:145-146.

ice

Betin, V.V., and Iu. V. Preobrazhenskii, 1959.
Ice investigations in the Baltic during the International Geophysical Year. (In Russian).
Trudy Inst. Okeanol., Akad. Nauk, SSSR, 46: 115-119.

ice

Betin, V.V., and Iu. V. Preobrazhenskii, 1959
[Fluctuations of iciness in the Baltic Sea and the Danish Straits.]
Trudy Gosud Okeanogr. Inst., Leningrad, 37: 1-13.
USN HO TRANS-102(No. 1)
M. Slessers
M.O. 16104
P.O. 13377

ice

Betin, V.V., and Iu. V., Preobrazhenskii, 1959.
Ice observations in the Baltic Sea during the period of the International Geophysical Year. (In Russian).
Trudy Gosud. Okeanogr. Inst., 46:115-119.

ice

Betin, V.V., and K.P. Shirokov, 1961.
Determination of the elements of drifting ice in the sea with a plane.
Trudy Gosud. Okeanogr. Inst., 63:64-77.

ice

Betin, V.V., and A.P. Zaitsev, 1961.
Peculiarities of the freezing of the Baltic Sea in the winter of 1959-60. (In Russian).
Trudy Gosud. Okeanogr. Inst., 63:78-89.

ice

Betin, V.V., S.V. Zhadrinskii and N.S. Uralov, 1959
New methods of aerial observations on sea ice.
Trudy Gosud. Okean. Inst., 37:205-230.

USN-HO-Trans-102(No. 6 of 6)
M. Slessers, P.O. 45449, 1962

ice

Betin, V. V., S. V. Zhadrinskii and N. S. Uralov, 1959.

[Method of instrumental air observations on the continental ice in the sea.]
Meteorol. i Gidiol. (5):51-54.

ice

Betin, V.V., S.V. Zhadrinskii, N.S. Uralov, 1957
On the method of observations on sea ice from aircraft.
Trudy Gosud. Okeanogr. Inst., Leningrad, 40: 147-155.

USN-HO-TRANS-123
M. Slessers
Misc. No. 15868-10
P.O. 41246
1961

ice

Biays, P., 1960.
Le courant du Labrador et quelques-unes de ses conséquences géographiques.
Cahiers Géogr., Quebec, 4(8):237-302.

ice

Bibikov, D.N., 1956.
[On the number of ice crystals in supercooled water.] Dokl. Akad. Nauk, SSSR, 109(6):1123-1125.

ice

Bikina, L.A., 1962.
On the speed of the growth of ice crystals within water. (In Russian).
Izv., Akad. Nauk, SSSR, Ser. Geofiz., (12):1852-1857.

ice

Bilello, M.A., 1961.
Formation, growth and decay of sea-ice in the Canadian Arctic Archipelago.
Arctic, 14(1):3-24.

ice

Bjerrum, N., 1952.
Structure and properties of ice. Science 115(2989):385-390, 6 textfigs.

ice

Bjerrum, N., 1951.
Structure and properties of ice. K. Danske Videnskabernes Selskab., Mat. Fys., Medd. 27(1): 56 pp.

Black, W.A.

ice

Black, W.A., 1962
Geographical Branch program of ice surveys of the Gulf of St.Lawrence, 1956 to 1962.
Cahiers de Geographie de Quebec, 6(11):65-74.

ice

Black, W.A., 1962.
Gulf of St. Lawrence ice survey, winter 1961.
Dept. Mines and Techn. Surveys, Ottawa, Canada, Geogr. Paper, No. 32:52 pp., 46 photos.

ice

Black, W.A., 1961.
A report on sea ice conditions in the eastern Arctic, summer 1960.
Dept. Mines & Tech. Surv., Canada, Geograph. Pap., No. 27: 29 pp.

ice

Black, W.A., (1960 or 1961)

Gulf of St. Lawrence, ice survey, winter of 1960. Canada, Dept. Mines & Techn. Surveys, Geograph. Br., Geograph. Pap., No. 25: 64 pp.

ice

Black, W.A., 1957.
Ice conditions: Gulf of St. Lawrence, 1956.
Geogr. Bull. 10:77-84.

ice

Bloom, G.L., 1956.
Current, temperatures, tide and ice growth measurements, eastern Bering Strait-Cape Prince of Wales, 1953-55.
U.S.N. Electronics Lab. Res. Rept., No. 739:25 pp

ice

Bluthgen, J., 1952.
Wettersregelfälle und Ostseevereisung.
Ber. Deut. Wetterd. 35:75.

ice

Blüthgen, J., 1948.
Der Eiswinter 1946/47 in den deutschen Gewässern der Nord- und Ostsee. Zeitschr. Meteorol., Hft. 12, Jahrg. 2:353-360, 4 figs.

ice

Blüthgen, J., 1938.
Die Eisverhältnisse des Finnischen und Rigaischen Meerbusens. Arch. Deutschen Seewarte 58(3):1-122, 64 textfigs., 9 pls., 5 charts.

ice

Blüthgen, J., 1936.
Die Eisverhältnisse des Bottnischen Meerbusens.
Arch. Deutschen Seewarte 55(3):1-63, 7 pls.

ice

Bogorodskiy, V.V., 1958.
Ultrasonic method of determining ice thickness. (In Russian).
Problemy Arktiki, (4):65-77.

ice

Bokanenko, L.I., and Yu. M. Avsyuk, 1963.
Subglacial relief and thickness of the Lazarev Ice Shelf. (In Russian).
Sovetsk. Antarkt. Eksped., Inform. Biull., 44: 43-

Translation: Scripta Tecnica for AGU, 5(1):55-58

Borovaya, L.I.

ice

Borovaya, L.I., 1967.
Hydrological and ice conditions in the Barents Sea in 1965. (In Russian).
Mater.Sess Uchen. Sovet. PINRO Rezult. Issled., 1965, Poliarn. Nauchno-Issled. Proektn. Inst. Morsk. Ribn. Khoz. Okeanogr. (PINRO), 8:104-111.

ice

Boroveya,L.I.,1966.
Hydrological and ice conditions of the Barents Sea in 1964. (In Russian).
Materiali, Sess.Uchen.Soveta PINRO Rez.Issled., 1964,Minist.Ribn.Khoz.SSSR,Murmansk,58-64.

Borovaya, L.I., 1964.
Hydrological and ice conditions of the Barents Sea in 1964. (In Russian).
Material. Sess. Uchen. Soveta PINRO Result. Issledovan. (6):58-64.

ice

Borovaya, L. I., 1964.
The hydrological conditions in the Barents Sea in 1963. (In Russian).
Material. Sess. Uchen. Sov. PINRO, Rez. Issled., 1962-1963, Murmansk, 35-40.

ice

Braarud, T. and J. T. Ruud, 1932
The "Øst" Expedition to the Denmark Strait 1929. I. Hydrography. Hvalrådets Skr., No. 4:44 pp., 19 text figs.

ice

Bradley, R.S., 1957.
The electrical conductivity of ice.
Trans. Faraday Soc., 53(5):687-691.

ice

Breitner, H.J., 1953.
Über einen besonderen Fall von Grundeisbildung.
Deutsche Hydrogr. Zeits. 6(1):39, 1 textfig.

Lake in Finnland

ice

Breitner, H.J., 1953.
Über die Geschwindigkeit des Dickenwachstums von Eisdecken. Deutsche Hydrogr. Zeits. 6(1):34-39, 5 textfigs.

ice
Breslau, L.R., J.D. Johnson, J.A. McIntosh and L.D. Farmer, 1970.
Environmental research relevant to the development of Arctic Sea transportation.
J. mar. techn. Soc., 4(5):19-43

ice
Brier, G. W., & D. B. Kline, 1959.
Ocean water as a source of ice nuclei.
Science, 130(3377):717.

ice
Brill, R., 1962
The structure of ice.
Angew. Chemie Int. Edit., 1(11):563-567.

ice
Brodsky, A.E., and N.P. Radschneko, 1940.
[The isotope content of Arctic Seas and ices.]
Acta Physicochemica URSS 13:145.

ice
Brodskii, A.I., N.P. Radchenko and B.L. Smolenskaya, 1940.
[Die Isotopenzusammensetzung der arktischen Meere und Eise.] Acta Physicochim., SSSR, 13(1):145-156.

ice
Brodskii, A.I., N.P. Radchenko and B.L. Smolenskaya, 1939.
Isotopic composition of Arctic waters, ices and glaciers. J. Phys. Chem., SSSR, 13:1494-1501.

ice
Brown, A.L., 1954.
An analytical method of ice potential calculatio
Hydrogr. Off. Tech. Rept. 5:13 pp.

ice
Brown, PR., 1951.
Ice in the Newfoundland region during February 1950. M.O. 546, Mar. Obs. 21(152):111-114, 5 figs

ice
Browne, Irene M., and A. P. Crary, 1959.
The movement of ice in the Arctic Ocean.
Sci. Studies at Fletchers Ice Island, T-3, Vol. 1
Air Force Cambridge Res. Center, Geophys. Res. Pap., No. 63:36-49.

ice
Browne, I.M., and A.P. Crary, 1958.
The movement of ice in the Arctic Ocean. 5. Drift and deformation of ice.
NAS-NRC, Publ., No. 598:191-207.

ice
Bryazgin, N.N., 1959
[To the problem of the albedo of the drifting ice surface.] Probl. Artiki i Antarktiki (1): 33-40.

ice
Buck, Beaumont, M., 1965.
Ice drilling in Fletcher's ice island (T-3) with a portable mechanical drill.
Arctic, 18(1):51-54.

ice
Buedel, J., 1947.
Der Eiswinter 1945-1946 an der deutschen Küsten (Brit. Zone) im Vergleich zu den Eiswintern 1903-1904 bis 1942-1943. Rept. Sect. 2, German Hydrogr. Inst. No. 34:
MSS on file at U.S. Navy Hydrographic Office.

ice
Büdel, J., 1943.
Das Eis im Kaspisea. Ann. Hydr., usw., 71(4/6): 118-121, figs.

ice
Buinitsky, V. Kh., 1971.
Most important problems of the sea ice research. (In Russian; English abstract). Okeanologiia 11(5): 827-834.

ice
*Buinitsky, V. Kh., 1968.
The influence of microscopic organisms on the structure and strength of Antarctic sea ice. (In Russian; English abstract).
Okeanologiia, Akad. Nauk, SSSR, 8(6):971-979.

ice
Bulgakov, N.P., 1965.
The winter extreme limit of sea ice in the seas of the North Pacific. (In Russian; English abstract).
Okeanlog. Issled., Rezult. Issled. Programme Mezhd. Geofiz. Goda, Mezhd. Geofiz. Komitet Presidiume, Akad. Nauk, SSSR, No. 13:66-76.

ice
Bulgakov, N.P., 1964
On the distribution of ice in the Bering Sea. (In Russian).
Okeanologiia, Akad. Nauk, SSSR, 4(5):831-841.

ice
Bunt, J.S. and C.C. Lee, 1970.
Seasonal primary production in Antarctic sea ice at McMurdo Sound in 1967. J. mar. Res., 28(3): 304-320.

ice
Bunt, J.S., and E.J.F. Wood, 1963.
Microalgae and Antarctic sea-ice.
Nature, 199:1254-1255.
64-11168 in: Bull. Inst. Pasteur, 62(10:2846.

ice
Burbridge, F.E., 1951.
The modification of continental polar air over Hudson Bay. Q.J. Roy. [Soc.] Met. 77(333):365-374, 2 textfigs.

ice
Burgess, C.R., 1948
Climate and weather in modern naval warfare. Geogr. J. CXI (4/6):235-250.

ice
Burkholder, Paul R., and Enrique F. Mandelli, 1965.
Productivity of microalgae in Antarctic ice.
Science, 149(3686):872-874.

ice
Bursa, Adam S., 1961
The annual oceanographic cycle at Igloolik in the Canadian Arctic. II. The phytoplankton.
J. Fish. Res. Bd., Canada, 18(4):563-615.

ice
Burton, J.M.C., 1960.
The ice-shelf in the neighbourhood of Halley Bay.
Proc. R. Soc., London, (A), 256(1285):197-200.

ice
Buynitskiy, V. Kh., 1968.(1967).
Structure, principal properties, and strength of Antarctic sea ice. (translation).
Info.Bull.Soviet Antarct.Exped., 65:504-510. (90-104).

C

ice
Cairns, Alan A., 1967.
The zooplankton of Tanquary Fjord, Ellesmere Island, with special reference to calanoid copepods.
J. Fish. Res. Bd., Can., 24(3):555-568.

ice
Callaway, E.B., 1954.
An analysis of environmental factors affecting ice growth. H.O. Tech. Rept. 7:31 pp.

ice
Campbell, N.J., and A.E. Collin, 1958
The discoloration of Foxe Basin ice.
J. Fish. Res. Bd., Canada, 15(6):1175-1188.

ice
Canada, Department of Transport, Ottawa, 1950.
Navigation conditions on the Hudson Bay route from the Atlantic seaboard to the Port of Churchill. 82 pp., figs., map. 15 cents.

ice
Capurro, L.R.A., 1955.
Expedicion Argentina al Mar de Weddell (Diciembre 1954 a Enero de 1955). Ministerio de Marina, Argentina, Direccion Gen. de Navegaccion e Hidrografia, 184 pp.

ice
Central Meteorological Observatory, 1949.
Report on sea and weather observations on Antarctic whaling ground (1948-1949). Ocean. Mag., Tokyo, 1(3):142-173, 11 textfigs.

"ice measurer"
Chaplygin, Ye. I., 1958.
[New models of oceanographic apparatus.]
Problemy Arktiki (3):106-108.
T327R ERHope

ice
Chaplygin, Ye. I., and Yu. K. Alekseyev, 1957.
Guide to observations on currents. Chap 6. Study of marine currents and ice drifts with the aid of automatic drifting radio beacons. (In Russian).
Rudovodstvo po Nablyudeniyam nad Techeniyami, Moscow, 152-176.
OTS $0.75

ice
Charcot, J.B., 1934.
Rapport préliminaire sur la campagne du "Pourquoi Pas?" en 1933. Ann. hydrogr., Paris, (3)13:1-85.

ice
Charcot, J.B., 1933.
Rapport préliminaire sur la campagne du "Pourquoi Pas?" en 1932. Ann. Hydrogr., Paris, (3)12:1-29.

ice
Charcot, J.B., 1931-1932.
Rapport préliminaire sur la campagne du "Pourquoi Pas?" en 1931. Ann. Hydrogr., Paris, (3)11:57-139.

ice
Charcot, J.B., 1929.
Rapport préliminaire sur la campagne du "Pourquoi Pas?" en 1928. Ann. Hydrogr., Paris, (3)9:15-84.

ice
Charcot, J.B., 1925-1926.
Rapport préliminaire sur la campagne du "Pourquoi Pas?" en 1925. Ann. Hydrogr., Paris, (3)7:191-389, figs.

ice
Charcot, J.B., 1922.
Rapport préliminaire sur la campagne du "Pourquoi-Pas?" en 1922. Ann. Hydrogr., Paris, (3) 5:99-156, figs.

ice
Cheney, L.A., and F.M. Soule, 1951.
International ice observation and Ice Patrol Service in the North Atlantic Ocean. U.S.C.G. Bull. 35:116 pp., 33 textfigs.

ice
Chernigovskiy, N.T., 1943.
Rol' radiatsionnogo faktora v protsessakh tayaniya i rosta l'da arkticheskikh morey. [Role of the radiation factor in the processes of thawing and growth of ice in Arctic seas.] Probl. Arkt. 1:150-155.

ice
Clark, David L. 1971
Arctic Ocean ice cover and its late Cenozoic history.
Bull. Geol. Soc. Am. 82 (12): 3313-3324

Ice
Clark, P.C., 1970
Arctic sea-ice in summer.
Weather 25(5): 215-218.

ice
Clowes, A.J., 1934.
Hydrology of the Bransfield Strait. Discovery Rept. 9:1-64, 68 textfigs.

ice
Codispoti, L.A., and F.A. Richards 1968.
Micronutrient distributions in the East Siberian and Laptev seas during summer 1963.
Arctic 21 (2): 67-83.

ice
Collin, A.E., and M.J. Dunbar, 1964
Physical oceanography in Arctic Canada.
In: Oceanography and Marine Biology, Harold Barnes, Editor, George Allen & Unwin, Ltd., 2:45-75.

ice
Conseil Permanent International pour l'Exploration de la Mer, Service Hydrographique, 1963.
ICES oceanographic data lists, 1958(1):1-259.

ice
Conseil Permanent International pour l'Exploration de la Mer, 1963.
ICES oceanographic data lists, 1957 (1):277 pp.

ice
Conseil permanent international pour l'Exploration de la Mer, 1927.
Bulletin Hydrographique Trimestrial, No. 3, 1927 (Juillet-Septembre):16 pp. (multilith), charts.

ice
Cook, J.C., 1963.
Seismic reconnaissance on an ice-covered Antarctic Sea.
J. Glaciology, 4(35):559-568.

ice
Cook, J. H., 1946.
Ice contacts and the melting of ice below a water level. Am. J. Sci. 244(7):502-512.

on land in past geologic time.

ice
Cordini, I.R., 1955.
Contribucion al conocimiento del sector Antartico Argentino. Inst. Antartico Argentino, Publ., 1: 277 pp., 82 textfigs., 56 pls.

ice
Corton, E.L., 1955.
Climatology of the ice potential as applied to the Beaufort Sea and adjacent waters. H.O. Tech. Rept. 30:16 pp.

ice
Corton, E.L., 1954.
A study of the Arctic ice pack.
Trans. Amer. Geophys. Union 35(2):375.

ice
Corton, E.L., 1954.
The ice budget on the Arctic pack and its application to ice forecasting. U.S.H.O. Tech. Rept. No. 6:13 pp.

ice
Crary, A.P., 1962.
The Antarctic.
Scientific American, 207(3):60-73.

ice
Crary, A.P., 1960
Seismic studies. Air Force Cambridge Research Center, Geophys. Res. Pap., No. 63 (Sci. Studies at Fletcher's Ice Island, T-3, 1952-1955, Vol. III): 38-44. Taken from Trans. Amer. Geophys. Un., 35: 293-300.

ice
Crary, A.P., 1958.
Arctic ice island and ice shelf studies. 1.
Arctic 11(1):1-42.

ice
Crary, A.P., 1955.
A brief study of ice tremors.
Bull. Seismol. Soc., Amer., 45(1):1-9, 5 textfigs

ice
Crary, A.P., 1955.
Seismic sounding in polar ice. Geogr. Rev. 45(3):428-430.

ice
Czekanska, Maria, 1959.
Materialy do polskiej terminologii lodow morskich. Acta Geophys. Polonica 7(3/4): 322-342.

D

ice
Danielson, Eric W., Jr., 1971.
Hudson Bay ice conditions.
Arctic 24 (2): 90-107

ice
Dayton, Paul K. and Seelye Martin, 1971.
Observations of ice stalactites in McMurdo Sound, Antarctica. J. geophys. Res., 76(6): 1595-1599.

ice
de la Canal, Luis Maria, 1965.
Vuelo de reconocrimierto del estado de los hielos de mer.
Bol. Serv. Hidrogr. Naval, Argentina, (Publ. H.106) 2(3):145-150.

ice
Delneri, Arnaldo Carlos, 1966.
Tareas de investigación en la campaña antartica 1965/66.
Bol. Serv. Hidrogr. Naval, Argentina, 3(3)204-211.

ice
*Denisov, A.S., 1968.
On the effect of the ice cover on sea surface temperature. (In Russian; English abstract)
Okeanologiia, Akad. Nauk, SSSR, 8(4):592-596.

Denmark

ice
Denmark, Det Danske Meteorologiske Institut, 1971.
Isforholdene i de Grønlandske Farvande.
Publ. danske Met. Inst. Årbøg. 25pp. + 32 charts

Ice
Denmark, Det Danske Meteorologiske Institut, 1970.
Isforholdene i de Grønlandske farvande, 1963.
Publikat. (Årbøg.) danske met. Inst. 45 charts

ice
Denmark, Det Danske Meteorologiske Institat, 1968.
Isforholdene i de grønlandsk farvande, 1959.
Publ. danske met. Inst. Årbøger 31 pp., 56 pls.

ice
Denmark, Det Danske Meteorologiske Institut, 1968
Isforholdene i de Grønlandske farvande.
Publikationer, Danske Met. Inst. Arbøg.

ice
Denmark, Det Meteorologiske Institut, 1967.
Isforholdene i de Grønlandske farvande, 1957.
Publikationer fra det Danske Meteorologiske Institut Årbøger, 35 charts
(29 pp.)

ice
Denmark, Danske Meteorologiske Institut, 1967.
Isforhildene i de Grønlandske Farvande, 1958.
Arbøger, Publ. Danske Met. Inst., 1967:37 pp; 47 charts.

ice
Denmark, Det Danske Meteorologiske Institut, 1965.
Isforholdene i de grønlandske Farvande, 1961.
Publ. Dansk. Meteorol. Insy. Årbøger. 33 pp. 63 charts.

ice
Denmark, Det Danske Meteorologiske Institut, 1964.
Isforholdene i de Grønlandske farvande; 1960.
Publikationer, Danske Meteorol. Inst., Arboger.

ice
Denmark, Dankke Meteorologiske Institut, 1955.
Isforholdene i de Arktiske Have, 1952. Tillaeg til Nautisk Meteorol Arbog, 1952:1-22, charts.

ice
Denmark, Danske Meteorolgiske Institut, 1954.
Isforholdene i de Arktiske Have 1951. Tillaeg til Nautisk-Meteorolog. Arbog 1951:26 pp., 5 charts.

ice
Deane, W.F., 1948
Muir Inlet ice factories. J. Coast and Geodetic Survey 1:62-66, 8 figs.

ice
Deckart, M., 1954.
Seltsame Eisbildungen (Pfannkuchen-Eis).
Natur u. Volk 84(5):160-161.

ice
Defant, A., 1949.
Konvektion und Eisbereitschaft in Polaren Schelfmeeren. Geografiska Annaler, Årg. 31(1/4):25-35, 4 textfigs.

ice
Defant, A., 1942.
Der Einfluss des Reflexionsvermogens von Wasser und Eis auf den Wärmeumsatz der Polargebiete. Veröffentlichungen des Deutschen Wissenschaftlichen Instituts zu Kopenhagen, Reihe I:Arktis, 5:7 pp., 4 textfigs.

ice
Denmark, Dansk Meteorologiske Institut, 1959.
Isforholdene i de Arktiske Have, 1956.
App. Nautical Meteorol. Annual, 1956:34 pp., pls.

ice
Denmark, Dansk Meteorologisk Institut, 1958.
Isforholdene i de Arktiske Have 1955.
Tillaeg til Nautisk Meteorologisk Årbog, 1955: 31 pp., charts.

ice
Denmark, Danske Meteorologiske Institut, 1957.
Isforholdene i de arktiske have, 1954.
Tillaeg til Nautisk-Meteorol. Aarbog, 1954:33 pp. charts.

ice
Deryugin, K.K., 1964
The first voyage of the complex training and research oceanological expedition on the "Bataisk" ship designed for practical and research work. (In Russian).
Materiali Vtoroi Konferentsii, Vzaimod. Atmosfer. i Gidrosfer. v Severn. Atlant. Okean., Mezhd. Geofiz. God, Leningrad. Gidrometeorol. Inst., 269-278.

ice observation
Deryugin, K.K., and D.B. Karelin, 1954.
[Sea ice observation.] Ed. Ya. Ya. Gakkel', Leningrad, Gidrometeorologicheskoya Izdatel'stvo, 168 pp., maps.

ice
Deutsches Hydrographisches Institut, 1950.
Atlas der Eisverhältnisse des Nordatlantischen Ozeans und Übersichtskarten der Eisverhältniss des Nord und Südpolargebietes. Deutsches Hydrogr. Inst. No. 2335:24 pp., 27 charts.

ice
Deutsche Seewarte, Hamburg, 1911.
Reise zwischen Wladiwostok und den Hafen Kamchatkas von 14 juni bis 15 august 1908. Ann. Hydro. 39:323-326.

ice(freshwater)
Devik, O., 1942.
Supercooling and ice formation in open waters (Ice studies 1.). Geofys. Publ. 13(8):10 pp., 6 textfigs.

ice
Diaz, E.L., 1956.
La campana Antarctica 1955-1956.
Anales Soc. Cient., Argentina, 162:63-88.

ice
Dichtel, W.J., and G.A. Lundquist, 1951.
An investigation into the physical and electrical characteristics of sea ice. Proc. Alaskan Sci. Conf., Bull. Nat. Res. Coun. 122:122.

ice
Dietrich, G., 1963.
Die Meere.
Die Grosse Illustrierte Länderkunde, 2:1523-1606. Bertelsmann Verlag

ice
Dietrich, Günter, H. Aurich, and A. Kotthaus, 1961
On the relationship between the distribution of redfish and redfish larvae and the hydrographical conditions in the Irminger Sea.
Rapp. Proc. Verb., Cons. Perm. Int. Expl. Mer, 150:124-139.

ice
Digby, P.S.B., 1953.
Plankton production in Scoresby Sound, East Greenland. J. Animal Ecology 22(2):289-322, 5 textfigs.

ice
Dinsmore, R.P., R.M. Morse, Floyd M. Soule, 1960
International ice observations and Ice Patrol Service in the North Atlantic Ocean. Season of 1959.
U.S. Coast Guard Bull. No. 44:1-99.

ice, effect of
Dionne, Jean-Claude 1970.
Exotic pebbles in Quaternary deposits from the South coast of the St. Lawrence estuary, Quebec.
Marit. Sed., 6(3): 110-112.
Halifax

ice
Dionne, Jean-Claude, 1968.
Action of shore ice on the tidal flats of the St. Lawrence Estuary.
Marit. Sed., 4(3): 113-115.

ice
Dobrovolsky, A.D., Editor, 1968.
Hydrology of the Pacific Ocean. (In Russian).
P.P. Shirshor Inst. Okeanol., Akad. Nauk, Isdatel "Nauka, Moskva, 524 pp.

ice
Doronin, Yu. P., 1961.
[The turbulent heat exchange between water and the ice cover at sea.]
Okeanologiia, Akad. Nauk, SSSR, 1(5):846-850.

ice
Doronin, V.P., 1959
[On the problem of sea ice accretion] Problemi Arktiki i Antarktiki (1): 73-80.

ice
Doronin, Yu. P., 1959.
[Turbulent heat exchange between an ice cover and the atmosphere.]
Arktichesky i Antarkticheskiy Nauchno-Issled. Inst., Trudy, SSSR, 226:19-29.
LC or SLA mi $2.40, ph $3.30

ice
Drach, P., and J. Monod, 1935-1936.
Rapport préliminaire sur les observations d'histoire naturelle et de géographie physique.
Ann. Hydrogr., Paris, Ser. 3, 14:29-37.

ice
Drogaytsev, A.A., 1958(1960).
[Winter cooling of Arctic seas.]
Problemy Severa (1):
Translation in:
Problems of the North: 1:44-54.

ice
Drogaicev, D.A., 1956.
[Zones of compression and rarefaction of ice in the atmospheric pressure field.]
Izv. Akad. Nauk, SSSR, Geophys. Ser., (11):1332-1337.
Translation by E.R. Hope - in reference room

ice
Dubrovin, L.I., 1962.
Development of shore ice on the Lazarev Station region. (In Russian).
Sovetsk. Antarktich. Expedits., Inform. Biull., (33):35-41.
Translation:
(Scripta Tecnica, Inc., for AGU) 4(2):81-85.

ice
Dubrovski, A.N., 1944.
Ptitsy - indikatory ledovogo rezhima arkticheskikh morey. [Birds as indicators of the ice regime in Arctic seas.] Priroda 2:67-68.

Dunbar, Max J.

ice
Dunbar, M.J., 1966.
The sea waters surrounding the Quebec-Labrador peninsula.
Cahiers Geogr., Quebec, 10(19):13-35.

Dunbar, Moira

ice
*Dunbar, Moira, 1967.
The monthly and extreme limits of ice in the Bering Sea.
In: Physics of snow and ice, H. Oura, editor, Inst. Low Temp. Sci., Hokkaido Univ., 687-703.

ice
Dunbar, Moira, 1962
Note on the formation process of thrust structures in young sea ice.
J. Glaciol., 4(32):147-150.

ice
Dunbar, M., 1960.
Thrust structures in young sea ice.
J. Glaciol., 3(28):727-732.

ice
Dunbar, Moira, 1958.
Cruious open water feature in the ice at the head of Cambridge Fiord.
C.R. et Rapp., Assembl. Gén., UGGI, Toronto, 4: 514-519.

ice
Dunbar, Moira, 1954.
The pattern of ice distribution in Canadian Arctic seas. Trans. R. Soc., Canada, (3), 48:9-18.

ice
Dyer, J.G., 1951.
Polar ice reconnaissance. Proc. Alaskan Sci. Conf., Bull. Nat. Res. Coun. 122:123-125.

E

ice
Eggvin, Jens, 1963
Tilstanden i havet under den unormale vinter 1963.
Fiskets Gang, (15): 8 pp.

ice
Eigenson, M.S., 1959
[Antarctic ice, climate variations and solar activity.] Info. Biull. Sovetsk Antarkt. Exped. (8): 8-11.

ice
Elliott, F.E., 1956.
Some factors affecting the extent of ice in the Barents Sea area. Arctic 9(4):249-257.

ice (limits)

Ellison, J. J., D. E. Powell, and H. H. Hildebrand, 1950
Exploratory fishing expedition to the northern Bering Sea in June and July 1949.
Fishery Leaflet 369: 56 pp., 23 figs. (multilith).

ice

Emery, K. O., 1949.
Topography and sediments of the Arctic basin.
J. Geol. 57(5):512-521, 1 map.

ice

Emiliani, C., 1969.
Interglacial high sea levels and the control of Greenland ice by the precession of the equinoxes. Science, 166(3912): 1503-1504.

ice

English, T. Saunders, 1961.
Biological oceanography in the North Polar Sea from IGY Drifting Station Alpha, 1957-58.
Trans. Amer. Geophys. Union, 42(4):518-525.

Reprinted from:
"Some biological oceanographic observations in the Central North Polar Sea, Drift Station Alpha, 1957-1958". Arctic Inst. North Amer., Res. Paper, No. 13.
Also:
Air Force, Cambridge Res. Lab. Sci. Rept. 15 (AFCRL-652)

ice

Epstein, S., and R.P. Sharp, 1959

Oxygen isotope studies. Trans. Amer. Geo. U., 40(1): 81-84. I.G.Y. Bull., No. 21: 81-84

ice

Ericson, David B., Maurice Ewing and Goesta Wollin, 1964
Sediment cores from the Arctic and Subarctic seas.
Science, 144(3623):1183-1192.

ice

Evans, R.J. and N. Untersteiner, 1971.
Thermal cracks in floating ice sheets.
J. geophys. Res., 76(3): 694-703.

ice

Evgenov, N.I., 1955.
[Album of ice formation in the sea.] Leningrad, Gidrometeorol. Izd-vo, 1955, 104 pp., 2 pls., 107 photos, 4 tables, colored map. IU. V. Presbrazhenski, Editor.

Latest revision of ice classification and terminology.

ice

Evgenov, N.I., 1935.
[On ice buoys cast in Chukchi Sea and the eastern part of the East-Sibirian Sea from the ice-breaker Krasin in 1934.] Buill. Arktich. Inst., Leningrad, 5(3/4):93

ice

Evteev, S. A., 1960.
Rough estimates of the movement velocity of ice streams according to the width of new-born icebergs. Inform. Bull. Soviet Antarctic Exped., 19:15-17.

ice

Ewing, M., and W.L. Donn, 1958.
A theory of ice ages. II. Science, 127(3307): 1159-1162.

F

ice

Fabricius, Jens S., 1965.
Danish ice-observations in Greenland: history and organization.
Geogr. Tidsskr., 64(2):206-219.

ice

Fabricius, Otto, 1955.
"On the floating ice in the northern waters and particularly in the Straight of Davis". Geogr. Rev. 45(3):405-415.

Introduction by R.H. Dillon.

ice

Farengolts, I.V., 1966.
Ice conditions of Greenland waters. (In Russian).
Okeanologiia, Akad. Nauk, SSSR, 6(3):560.

ice

Federov, Viktor, 1955.
Podl'damiarktiki. Vodnyy Transport (1 Sept. '55) 4.

atomic sub to operate under ice

ice

Fedotov, V.I. 1969.
Influence of surf on the formation of the ice barrier in Atashyev Bight. (In Russian)
Inform. antarct. Biull. Sovetsk. Antarct. Exped. 76: 45-48.
Translation: Am. geophys. Un. 76: 460-462.

ice

Fel'zenbaum, A.I., 1961

[The theory of steady drift of ice and the calculation of the long period mean drift in the central part of the Arctic Basin.] Problems of the North (Transl. of Problemy Severa), 1958, 2: 13-44.

ice

Fel'zenbaum, A.I., 1958
The theory of a steady ice-flow drift and calculations for the drift over a period of many years in the central part of the Arctic basin.
Problemi Severa, 2: 16-46.
Abstr. in: Appl. Mech. Rev., 15(6): #3700.

ice

Felzenbaum, A.I., 1957.
[Compression and rarefaction of ice in the Arctic Basin.] Doklady Akad. Nauk, SSSR, 116(2):217-220.

T271R

ice

Felsenbaum, A.I., 1957.
[Theoretical foundations for calculating the ice drift in the Central Arctic basin.]
Doklady Akad. Nauk, SSSR, 113(2):307-310.

ice

Finland, Havsforskningsinstitutet, 1964
Ice winter 1962-63 along the Finnish coast. (In Finnish and English).
Merent. Julk., (Havsforskningsinst. Skrift), No. 213: 28 pp.

ice

Finland.
Havsforskningsinstitutet, 1960.
Jäätalvi 1960/61 suomen merialueilla (Ice winter 1960/61 along the Finnish coast.) Hakemisto- Index No. 1-200.
Merent. Julk, No. 200:28 pp.

ice

Fleet, M., 1965.
The occurrence of rifts in the Larsen Ice Shelf near Cape Disappointment.
Brit. Antarctic Survey, Bull., No. 6:63-66.

ice

Fletcher, J.O. 1971.
Probing the secrets of Arctic ice
Naval Res. Rev. Mar. 1971: 9-24.

ice

Fletcher, J.O., 1951.
Floating islands in the Arctic Ocean. Proc. Alaskan Sci. Conf., Bull. Nat. Res. Coun. 122: 122.

ice

Flinn, Derek 1967.
Ice front in the North Sea.
Nature, Lond., 215(5106): 1151-1154

ice

Forward, C.N., 1956.
Sea ice conditions along the Hudson Bay route.
Geogr. Bull., Canada, (8):22-50.

ice

Forward, C.N., 1954.
Ice distribution in the Gulf of St. Lawrence during the break-up season. Geogr. Bull., Canada, No. 6:45-84, 18 textfigs.

ice

Fredriksson, K., and L.R. Martin, 1963.
The origin of black spherules found in Pacific islands, deep-sea sediments and Antarctic ice.
Geochim. et Cosmochim. Acta, 27(3):245-248.

ice

Friedman, Irving, Beatrice Schoen and Joseph Harris, 1961.
The deuterium concentration in Arctic Sea ice.
J. Geophys. Res., 66(6):1861-1865.

ice

Fritz, S., 1951.
The growth of ice thickness in Arctic regions.
Proc. Alaskan Sci. Conf., Bull. Nat. Res. Coun. 122:122-123.

ice

Frommeyer, M., 1928.
Die Eisverhältnisse um Spitzbergen und ihre Beziehungen zu klimatischen Faktoren. Ann. Hydrogr. usw., Jahrg. 56 (Heft VII):209-214.

ice

Fujino, Kazuo, 1966.
OCEanographic observations on the Drifting Station ARLIS - II June - November 1964
Low. Temp. Sci., (A)24: 269-284. (In Japanese; English abstract).

ice

Fukuoka, J., 1959.
A note on the ice and ocean conditions in the sea near Syowa Base In Antarctica.
Rec. Oceanogr. Wks., Japan, 5(1):1-5.
Also in: Collected Reprints, Oceanogr. Lab., Meteorol. Res. Inst., Tokyo, June 1960.
(In English).

Fukutomi, T.

ice

Fukutomi, T., 1958.
A theory on the steady drift of sea ice due to wind on the frozen sea. 5. Drift and deformation of sea ice.
NAS-NRC, Publ., No. 598:223-236.

ice
Fukutomi, T., 1953.
Study of sea ice. 20. A study on the formation of sea ice at the surface of the deep sea.
Inst. Low Temp. Sci. (A) Phys. Sci., 95-106.

ice
Fukutomi, T., 1953.
[Study of sea ice. 19. Relation between the beginning date of freezing and air temperature at the coasts of the Okhotsk Sea, the Japan Sea and the Arctic Sea. 20. A study on the formation of sea ice at the surface of the deep sea.]
Low Temp. Sci., Ser. A, Phys. Sci., 87-94; 95-106.

ice
Fukutomi, T., 1953.
[Study of sea ice. 16. On the structure of ice rind, especially on pure-ice percentage, thin ice sheet and ice-sheet block.]
Inst. Low Temp. Sci. 9:113-123.
[17. On the maximum thickness of the winter ice at northern sea coasts.] Ibid.: 125-136.
[18. Drift of sea ice due to wind in the sea of Okhotsk especially in its southern part.] Ibid.: 137-144.

ice
Fukutomi, T., K. Kusunoki and T. Tabata, 1954.
[A report on the survey of sea ice on the Okhotsk Sea coast of Hokkaido.] Low. Temp. Sci. (A), No. 13:59-103.

ice
Fukutomi, T., and K. Kusunoki, 1947.
Temperature distribution in fresh-water ice-plate with cyclic variation of atmospheric temperature. Sea Ice Studies, No. 7, Hokkaido University, Institute of Low Temperature Science. Teion Kagak

Translation by E.R. Hope T 55(7)J

ice
Fukutomi, T., T. Nagashima, and K. Kusunoki, 1949.
[Study of sea ice. (2nd rept.). On the formation of ice crystal in sea water and the texture of sea ice.] "Teion-kagaku"- Low Temp. Sci. 2:73-76, 13 figs.

ice
Fukutomi, T., and K. Kusunoki, 1950.
[Study of sea ice (5). An approximate prediction method for the last drift-ice on the southern Okhotsk Sea coast of Hokkaido.] "Teion-Kagaku" - Low Temp. Sci. 3:159-169, 5 textfigs.

ice
Fukutomi, T., K. Kusunoki, and T. Tabata, 1954.
Study of ice (21st rept.); a report on the survey of sea ice on the Okhotsk Sea east of Hokkaido. Low Temp. Sci., A, 13:59-104.

ice
Fukutomi, T., K. Kusunoki, and T. Tabata, 195
[Study of sea ice. (11th rept.). On chlorinity of the coastal land ice observed at Abashiri and Monbetsu.] "Teion-Kagaku" - Low Temp. Sci. 6:71-83, 10 textfigs.

ice
Fukotomi, T., K. Kusunoki, and T. Tabata, 1950.
[Study of sea-ice. (The 12th Report). Some measurements of the wind-pressure drift of sea ice in open sea.] J. Ocean. Soc., Japan 6(1):18-27, 4 textfigs. (In Japanese with English abstract).

ice
Fukutomi, T., K. Kusunoki, and T. Tabata, 1950.
[Study of sea ice (6th rept.) On the increase of the thickness of sea-ice.] "Teion-Kagaku" - Low Temp. Sci. 3:171-186, 10 textfigs.

Incomplete reference on reprint, but this volume number arrived at from pagination known to be in this volume.

ice
Fukutomi, T., and K. Kusunoki, 1950.
[Study of sea ice (7th rept.) on the temperature distribution in the fresh water ice plate when the surface temperature changes periodically Teion-Kagaku" - Low Temp. Sci. 3:187-192, 3 textfigs.

Incomplete reference on reprints.

ice
Fukutomi, T., K. Kusunoki, and T. Tabata, 1950.
[Study of sea ice (8th rept.). On the vertical distribution of temperature and salinity of sea-water under the coastal land-ice at Abashiri and Monbetsu.] "Teion-Kagaku" - Low Temp. Sci. 3: 193-206, 6 textfigs.

Reference incomplete on reprint.

ice
Fukutomi, T., and K. Kusunoki, 1951.
[On the form and the formation of hummocky ice ranges. 1951. Study of sea ice (15th report).] Low Temp. Sci. 8:59-88, 9 textfigs. (English summary).

ice
Fukutomi, T., K. Kusunoki, and T. Tabata, 1951.
[Study of sea ice. (12th rept.). Some measurement of the wind-pressure drift of sea ice.] "Teion-Kagaku" - Low Temp. Sci. 6:85-93, 5 textfigs.

G

ice
Germany, Oberkommando der Kriegsmarine, 1940.
Die Naturverhaltnisse des sibirischen Seeweges. 169 pp.

ice
Giovinetto, Mario B., and John C. Behrendt, 1964.
The area of the ice shelves in Antarctica.
Polar Record, 12(77):171-173.

ice
Gläde, P., H. Gernandt and V.A. Shamont'yev, 1969.
On the possibility of using meteorological satellite observations for the study of Antarctic ice. (In Russian).
Inform. Biull. Sovetsk. Antarkt. Exped. 73: 27-33
Translation: Scripta Tecnica, Inc. for A.G.U. 7(4): 296-299.

ice
Glebovsky, Ju. S., 1959
Sub-ice ridge in the region of Pionerskaya Station. Info. Biull. Sovetsk. Antarkt. Exped. (7): 5-9.

ice
Goedecke, E., 1965.
Über die schweren Eis- und Schiffahrtsverhältnisse auf dem Hudson und in der Chesapeake-Bucht während des Winters 1960/61.
Der Seewarte, 26(1):21-26.

ice
Goedecke, E., 1958.
Die Aussergewöhnliche Nordatlantische Eissaison 1957. Der Seewart 19(1):26-31.

ice
Goedecke, E., 1958.
Hydrographische Betrachtungen über die regionale Verteilung der mittleren Eisvorbereitungs- und Eisabschmelzzeit in der Deutschen Bucht und der westlichen Ostsee mit Karten.
Deutsche Hydrogr. Zeits., 11(1):1-22.

ice
Goedecke, E., 1957.
Das merkwürdige Verhalten der Eiswinter 1955/54 biss 1955/56. II. Der Seewarte 18(2):50-65.

ice
Goedecke, E., 1957.
Neuere Untersuchungen über die Borbereitungs- und Abschmelzzeit des Eises in der Deutschen Buckt und westlichen Ostsee.
Ann. Meteorol., Hamburg, 8(3/4):80-92.

ice
Goedecke, E., 1956.
Ueber die Bedeutung der Ost-Weterlagen Mitteleuropas für die Vereisung der Deutschen Nord- und Ostseeküsten. Ann. Meteorol., 7(10/12):386-403.

ice
Goedecke, E., 1954.
Über den Zusammenhang zwischen Eis- und Kältesumme sowie Oberflächentemperatur in der Elbe- und Wesermündung während der Winter 1903-1950.
Ann. Met. 6(5/6):151-163, 5 textfigs.

ice
Goedecke, E., 1953/54.
Über Intensität und Reihenfolge der Eiswinter an der deutschen Nordseeküste für die Periode 1872-1950.
Ann. Meteorol. 6(7/8):202-212, 6 textfigs.

ice
Golovkov, M.P., 1951.
On the structure of natural ice-masses of various origins. Dok. Akad. Nauk, SSSR, 78(3):573-575.

T111R

ice
Gonin, G.B., 1957.
Possibilities of deciphering ice according to aerial photographs made with a slit camera. (In Russian).
Problemy Arktiki, No. 2:219-231.

ice
Gordienko, P.A., 1961
The Arctic Ocean.
Scientific American, 204(5): 88-102.

ice
Gordenko, P., 1958.
Arctic ice drift. 5. Drift and deformation of sea ice.
NAS-NRC, Publ., No. 598:210-220.

ice
Gordienko, P.A., 1958 (1960).
[Ice drift in the central part of the Arctic Ocean.]
Problemy Severa, (1):

Translation in:
Problems of the North, (1):1-30.

ice
Gordryenko, P.A., 1958
[Ice drift in the central part of the Arctic Ocean] Problemy Severa, USSR, 2: 5-29. Rev. in Tech. Transl. 4(9): 511

ice
Gordienko, P., 1955.
[Study of ice conditions in Arctic seas and Northern Ice Ocean.] Morskoi Flot. 3:25.

ice
*Gordunov, Yu.A., and L.A. Timokhov, 1968.
On the study of ice dynamics. (In Russian; English abstract).
Fisika Atmosfer.Okean., Izv.Akad.Nauk,SSSR,4(10): 1086-1091.

ice
Gorham, E., 1958.
The salt content of some ice samples from Nordasutlandet (North East Land), Svalbard.
J. Glaciol., 3(23):181-186.

Ice
Goudovitch, Z. M., 1961. ref.
[The relationship between glacial derivatives in the Arctic basin and ice conditions in Arctic seas of the SSSR.]
Trudy Kom. Okeanogr., Akad. Nauk, SSSR, 11:13-20.

ice

Gow, Anthony J., 1971.
Relaxation of ice in deep drill cores from Antarctica. J. geophys. Res., 76(11): 2533-2541.

ice

Gudkovich, Z.M., and Ye. G. Nikiforov, 1963.
Steady state of a single floe. (In Russian).
Trudy Arktich. i Antarktich. Nauchno-Issled. Inst., (253):197-209.

Abstracted in:
Soviet Bloc Res., Geophys., Astron. Space, 93:39

ice

Great Britain, Marine Division, Meteorological Office, 1964.
Notes on ice conditions in areas adjacent to the North Atlantic Ocean, January-March, 1964.
Mar. Obs., 34(205):138-141.

ice

Great Britain, Meteorological Office, Marine Division, 1964.
Notes on ice conditions in areas adjacent to the North Atlantic Ocean.
Mar. Obs., 34(204):91-96.

ice

Great Britain, Meteorological Office, Marine Division, 1963.
Notes on ice conditions in areas adjacent to the North Atlantic Ocean from April to June, 1963.
Marine Observer, 33(202):209-212.

ice

Great Britain, Meteorological Office, Marine Division, 1962.
Notes on ice conditions in areas adjacent to the North Atlantic Ocean from July to September 1961.
Marine Observer, 32(195):40-44.

ice (data)

Great Britain, Meteorological Office, 1961.
Notes on ice conditions in area adjacent to the North Atlantic Ocean from April to June, 1961.
Marine Observer, 31(194):200-205.

ice

Great Britain, Meteorological Office, 1961

Notes on ice conditions in areas adjacent to the North Atlantic Ocean from July to September 1960.
Marine Observer, 31(191): 41-43.

ice

Groen, P., 1967.
The waters of the sea.
D. Van Nostrand Co., Ltd. 328 pp. $9.00

ice

Groissmayr, F.B., 1939.
Schwere und leichte Eisjahre bei Neufundland und das Vorwetter. Ann. Hydrogr. 67:26-30.

ice

Gudkovich, Zh. M., 1961.
Connection of the drift of the ice in the Arctic basin with the ice conditions in Soviet Arctic seas. Forecasting and computation of physical phenomena in the sea. (In Russian).
Trudy Okeanogr. Komissii, Akad. Nauk, SSSR, 11: 13-20.

ice

Gudkovich, Z.M., and Z.G. Nikiforov, 1965.
On some significant features related to the formation of water density anomalies and their effect upon ice and hydrological conditions in the Arctic basin and bordering seas. (In Russian).
Okeanologiia, Akad. Nauk, SSSR, 5(2):250-260.

ice

Gudkovich, Z.M., E.I. Sarukhanyan and N.P. Smirnov 1970.
Pressure "polar tide" and its influence on the ice conditions on the Arctic seas. (In Russian; English abstract).
Okeanologiia 10(3): 426-437.

ice

Guilcher, Andre, 1962.
Chronique océanographique.
Norois, 33(9):65-98.

ice

Gul' K.K., A.N. Kosarev, and V.N. Shiryaev, 1971.
The All-Union Meeting on the ice of the south seas of USSR. (In Russian). Okeanologiia 11(2) 343-345.

ice

Gutsell, B. V., 1949
An introduction to the geography of Newfoundland. Canada, Dept. Mines and Resources, Geogr. Bur., Info. Ser. No.1: 85 pp., 20 text figs., 3 fold-in maps.

H

ice

Hachey, H.B., 1961.
Oceanography and Canadian Atlantic waters.
Fish. Res. Bd., Canada, Bull., No. 134:120pp.

ice

Hachey, H.B., F. Hermann and W.B. Bailey, 1954.
The waters of the ICNAF Convention area.
Int. Comm. Northwest Atlantic Fish., Ann. Proc. 4:67-102, 29 textfigs.

ice

Hagemeister, 1955.
[The ice age and Atlantis.] Priroda 1955(7):92-96.

Translation by E.R. Hope T196R

ice

Hakkel', Ya. Ya., 1959 (see Gakkel in MS file)
[Tectonic deformations modeled naturally in sea ice.]
Izv. Vses. Geograf. Obshch., 91(1):27-41.

Translations, E.R. Hope - T328R

ice

Hansen, P.M., and F. Hermann, 1953.
Fisken of Havet ved Grønland.
Skr. Danmarks Fisk- og Havundersøgelser, No. 15: 1-128, 61 textfigs.

ice

Hanson, Arnold M., 1965.
Studies of the mass budget of Arctic pack-ice floes.
J. Glaciol., 5(41):701-709.

ice

Hanson, K.J., 1961.
The albedo of sea-ice and ice islands in the Arctic Ocean Basin.
Arctic, 14(3):188-196.

ice

Hare, F. K., and M. R. Montgomery, 1949.
Ice, open water, and winter climate in the eastern Arctic of North America. Arctic 2(3): 149-164, figs.

ice

Hay, R.F.M., 1956.
Ice accumulation upon trawlers in northern waters
Meteorol. Mag., 85(1010)(M.O. 613):225-229.

ice

Heap, John A., 1965.
Antarctic pack ice.
In: Antarctica, Trevor Hatherton, editor, Methuen & Co., Ltd., 187-196.

ice

Heap, John A., 1964.
Variations in Antarctic sea-ice distribution. (abstract).
Geol. Soc., Amer., Special Paper, No. 76:309-310.

ice

Heinsheimer, Jorge J., 1967.
Que cantidad de hielo flota en los mares?
Bol. Servicio Hidrografia Naval, Armada Argentina, 4(2):215.

ice

Hela, I., 1958.
The Baltic Sea an an object of ice studies. Sec. 1. Distribution and character of sea ice.
NAS-NRC, Publ., No. 598:29-35.

ice

Hellbardt, G., 1955.
Seismische Versuche auf einer Eisplatte.
Zeits. f. Geophysik 21(1):41-47.

ice

Herbert, Wally, 1970.
The first surface crossing of the Arctic Ocean.
Geographical Jl 136(4): 511-533.

ice

Herdman, H.F.P., 1959

Some notes on sea ice observed by Captain James Cook, R.N., during his circumnavigation of Antarctica, 1772-75.
J. Glaciol. 3: 534-541.

ice

Herdman, H.F.P., 1955.
The Antarctic pack ice. The Trident 17:57-60.

ice

Herdman, H.F.P., 1953.
The Antarctic pack ice in winter. J. Glaciol. (13)2:184-193, 9 textfigs.

ice

Herdman, H.F.P., 1948
Antarctic pack ice. J. Glaciology, London 1(4):156-166.

ice

Herdman, H.F.P., 1948
The Antarctic Pack-Ice. Mar. Obs. M0493, 18(142):205-214, 5 figs.

ice

Hirsch, F.W.P., 1954.
Pfannkuchen-Eis auf der Elbe. Natur u. Volk. 84(2)45-47.

ice

Hodge, Paul W., Frances W. Wright and Chester C. Langway, Jr., 1964.
Studies of particles for extraterrestrial origin. 3. Analyses of dust particles from polar ice deposits.
J. Geophys. Res., 69(14):2919-2931.

ice

Hoel, A., and J. Norvik, 1962.
Glaciological bibliography of Norway.
Norsk Polarinst., Skrifter, No. 126:242 pp.

INcludes papers on sea as well as land.

ice
Hoel, A., and Werner Werenskiold, 1962.
Glaciers and snowfall in Norway.
Norsk Polarinst., Skrifter, No. 114:291 pp.

ice
Hood, A.D., 1958(1961)
An analysis of radar ice reports submitted by Hudson Bay shipping (1953-1957). Nat. Res Counc. Canada, Radio and Elect. Eng. Div. In: Thirty-Third Ann. Rept., Dept. of Transport, Nautical and Pilotage Div., Canada, Season of Navigation, 1961:41-58.

ice
Howard, A.D., 1948
Further observations in the Ross Shelf ice. Bull. G.S.A. 59:919-926.

ice
Holtsmark, B.E., 1955.
Insulating effect of a snow cover on the growth of young sea ice. Arctic 8(1):60-65, 6 textfigs.

ice
Hume, James D., and Marshall Schalk, 1964.
The effects of ice-push on Arctic beaches.
Amer. J. Sci., 262(2):267-273.

ice
Hunkins, K., 1960.
Seismic studies of sea ice.
J. Geophys. Res., 65(10):3469-3472.

ice
Hureau, J.C., 1962.
Observations hydrologiques en Terre Adélie de janvier 1961 à janvier 1962 (Expéditions Polaires Françaises, Missions P.E. Victor).
Bull. Mus. Nat. Hist. Nat., (2), 34(5):412-426.

I

ice
Iakovlev, G.N., 1955.
[Visual observations of the state of the drifting ice cover.] Material. Nabluid. Nauch.-Issledov. Dreifuius. 1950/51, Morskoi Transport 2:6-51.

David Krauss translator in AMS-Astia Doc. 117136.

ice
Iakovlev, G.N., 1955.
[The thermal regime of the ice cover.]
Material Nabluid. Nauch.-Issledov. Dreifuius., 1950/51, Morskoi Transport, 2(7):

David Kraus translator in AMS Astia Doc. 117138.

ice
Iakovlev, G.N., 1955.
[Study of the morphology of the ice cover by surveying.] Material. Nabluid. Nauch.-Issledov. Dreifuius., 1950/51, Morskoi Transport, 2:52-70; 72-102.

David Kraus translator in AMS-Astia Doc. 117136.

ice
Ianes, A.V., 1959
[On the calculation of the heat currents in the ice cover.] Problemi Arktiki i Antarktiki (1): 49-58.

ice
Ingram, R.G., O.M. Johannessen, and E.R. Pounder 1969.
Pilot study of ice drift in the Gulf of St. Lawrence. J. geophys. Res., 74(23): 5453-5459.

ice
Ishida, Tamotsu 1966.
Vibrations of a sea ice sheet on the occasion of its breaking. (In Japanese; English abstract).
Low Temp. Sci. (A) 24: 239-248.

ice
Ishino, Makoto, 1963.
Studies on the oceanography of the Antarctic Circumpolar waters.
J. Tokyo Univ., Fish., 49(2):73-181.

ice
Istoshin, U.V., 1960.
[On the American researches of the sea ice.]
Meteorol. i Gidrol., (11):46-48.

ice
Ivanov, V.M., 1964
Investigation of the regularities of the epoch development of the hydrometeorological processes in the Atlantic zone. (In Russian). Materiali Vtoroi Konferentsii, Vzaimod. Atmosfer. i Gidrosfer. v Severn. Atlant. Okean., Mezhd. Geofiz. God, Leningrad. Gidrometeorol. Inst., 15-26.

ice
Ivanov, V.M., 1961.
On the relationship of hydrometeorological processes in the Atlantic part of the Arctic and the thermal and dynamic state of the Gulf Stream. (In Russian).
Problemy Severa, 4:27-45.

Translation, 1962:
Problems of the North, 4:25-43.

J

ice
Jakovlev, G.N., 1959
[A thermal stream of evaporation from the surface of ice cover in Central Arctic] Problemi Arktiki i Antarktiki (1): 59-64.

Japan, Hakodate Marine Observatory

ice
Japan Hakodate Marine Observatory 1970.
Report of the marine meteorological observations on sea fog and heat exchange at the ocean-air interface in June, 1969. Supplement: on the drift-ice east of South Saghalien in the Okhotsk Sea in June 1969. (In Japanese).
Mar. met. Rept. Hakodate mar. Obs. 30: 67 pp.

ice
Japan, Hakodate Marine Observatory, 1969.
Report of the sea ice observations in the Okhotsk Sea and adjacent sea of Hokkaido (1969). (In Japanese)
Mar. Met. Rept., Hakodate mar. Obs. 29: 1-30

ice
Japan Hakodate Marine Observatory 1968.
Report of the sea ice observations in the Okhotsk Sea and adjacent sea of Hokkaido. (In Japanese).
Mar. Met. Rept. 26: 76 pp.

ice
*Japan, Hakodate Marine Observatory, 1967.
Local ice conditions based on the coastal observations. (In Japanese: English abstract).
Bull. Hakodate mar. Obs., 13:52-63.

ice
*Japan, Hakodate Marine Observatory, 1967.
Sea ice conditions of the Okhotsk.
Sea based on the broadcasting of U.S.S.R. in 1963/1964 and 1964/1965. (In Japanese: English abstract).
Bull. Hakodate mar. Obs., 13:64-79.

ice
*Japan, Hakodate Marine Observatory, 1967.
Sea ice observation along the Okhotsk coast of Hokkaido from January to April 1964, 1965.
Sea ice observations in the sea off Hokkaido from December to May in 1963/1964 and 1964/1965. (In Japanese: English abstract).
Bull. Hakodate mar. Obs., 13:1-51.

ice
Japan, Hakodate Marine Observatory, 1967.
Report of the sea ice observations in the Okhotsk Sea and adjacent sea of Hakkaido 1967. (In Japanese).
Mar. met. Rep., Hakodate mar. Obs., 23: 58 pp.

Ice
Japan, Hakodate Marine Observatory, 1966.
Sea ice observation along the Okhotsk coast of Hokkaido from January to April 1963.
Sea ice conditions in the sea off Hokkaido from December 1962 to April 1963.
Local sea ice conditions based on coastal observations of the Okhotsk Sea based on the broadcasting of U.S.S.R., 1962-1963.
(In Japanese; English abstracts).
Bull. Hakodate Mar. Obs., No 12: 1, 1-35; 38-52; 53-67; abstracts p. 1-2, p.36, p.53.

Ice
Japan, Hakodate Marine Observatory, 1966.

Report of the sea ice observation of the Okhotsk Sea and adjacent sea of Hokkaido (1966).

Mar. Meterol. Rept., Hakodate Mar. Obs. 10(2): 68pp.

ice
Japan, Hakodate Marine Observatory, 1965.
Report of the sea ice observations in the adjacent sea of Hokkaido (1965). (In Japanese).
Mar. Meteorol. Rept., Hakodate Mar. Obs., 9(2): 70 pp.

ice
Japan, Hakodate Marine Observatory, 1964.
Sea ice observations along the Okhotsk coast of Hokkaido from January to April 1962. Sea ice conditions in the sea off Hokkaido from December to April 1962. (In Japanese; English abstract).
Bull. Hakodate Mar. Obs., No. 11:1-34.

ice
Japan, Hakodate Marine Observatory, 1964.
Report of the sea ice observations in the Okhotsk Sea and adjacent sea of Hokkaido, 1964. (In Japanese).
Mar. Meteorol. Rept., Hakodate Mar. Obs., 8(2): 1-64. (multilithed).

ice
Japan, Hakodate Marine Observatory, 1963.
Sea ice observations along the Okhotsk coast of Hokkaido from January to April 1961.
Sea ice conditions observed by aircraft. (In Japanese; English summary), 1-21
Sea ice conditions based on coastal observations. (In Japanese; English abstract), 23-42.
Sea ice conditions based on the information from ships. (In Japanese; English abstract), 43-49
Bull. Hakodate Mar. Obs., 10:49 pp.

ice
Japan, Hakodate Marine Observatory, 1963.
Report of the sea ice observations in the Okhotsk Sea and the adjacent sea of Hokkaido, 1963. (In Japanese).
Mar. Meteorol. Rept., Hakodate Mar. Obs., 7(2):1-54.

ice
Japan, Hakodate Marine Observatory, 1962.
Report of the sea ice observations, 1962.
(In Japanese).
Mar. Meteorol. Rept., Hakodate Mar. Obs., 6(2):1-46.

ice
Japan, Hakodate Marine Observatory, 1961.
[Report of sea ice observations, 1961.]
Mar. Meteorol. Rept., Hakodate Mar. Obs., 5(2): 1-52.

ice (data)
Japan, Hakodate Marine Observatory, 1961.
[Sea ice conditions in the Okhotsk Sea west off Kamchatka Peninsula reported by the crabfishing ships, 1959.] + 1960
Mar. Meteorol. Rept., Hakodate Mar. Obs., 5(2): 53-56, 57-60.

ice
Japan, Hakodate Marine Observatory, 1961.
[Sea ice conditions of 1959 based on air reconnaissance.] (English abstract)
Bull. Hakodate Mar. Obs., (8):1-15.

ice
Japan, Hakodate Marine Observatory, 1961.
[Sea ice conditions of 1959 based on the coastal observations.](English abstract).
Bull. Hakodate Mar. Obs., (8):16-28.

ice
Japan, Hakodate Marine Observatory, 1958.
[On the drift ice in the Okhotsk Sea off Kamchatka Peninsula in April, 1958.]
Mar. Rep Meteorol. Rept., 2(2):

[Sea ice conditions on the basis of coastal observations in 1958.]
Ibid., 2(2):

[Aerial observations of sea ice in the south-western region of the Okhotsk Sea (1958).]
Ibid., 2(2):15-

ice
Japan, Hakodate Marine Observatory, 1957.
[Report of the oceanographic observations of the Okhotsk Sea from April to May 1956.]
Bull. Hakodate Mar. Obs., No. 4:105-112.
1-8.

ice
Japan, Hakodate Marine Observatory, 1955.
[Sea-ice distribution maps in the southern Okhotsk Sea, 1937-1944]:43 pp.

ice
Japan, Hydrographic Bureau, 1948.
[Distribution of sea ice in the northern Japan Sea.] Hydrogr. Bull. No. 8:33-42, 11 pls.

ice
Japan, Maritime Safety Board, 1958.
[Sea ice on the coast of Hokkaido, December 1955-April 1956.] Hydrogr. Bull. (Publ. 981), No. 55: 18-28, 1 pl., 1 fold-in.

ice
Japan, Maritime Safety Board, 1957.
States of sea ice on the coast of Hokkaido in 1955. Hydrogr. Bull. (Publ. 981) 53:6-10.

ice
Japan, Maritime Safety Board, 1955.
[On the terminology of sea ice. Pt. 3.]
Publ. No. 981, Hydrogr. Bull. No. 50:232-236.

ice
Japan, Maritime Safety Board, 1955.
[Sea-ice on the coast of Hokkaido, December 1953-April 1954.] Publ. 981, Hydrogr. Bull., Spec. No. 17:5-12, 17 figs.

ice
Japan, Maritime Safety Board, 1955.
[On the sea-ice in Ryoto Kaiwan.] Publ. No. 981, Hydrogr. Bull., Spec. No. 17:80-88.

ice
Japan, Maritime Safety Board, 1955.
[On the terminologies of sea ice. Part 2.]
Pub. 981, Hydrogr. Bull., No. 48:136-139.

ice
Japan, Maritime Safety Agency, 1953.
[On sea-ice on the coast of Hokkaido, December, 1953-April 1953.] Hydro. Bull. (Publ. No. 981), No. 38:219-224, 2 figs.

ice
Japan, Maritime Safety Agency, 1952.
On the sea-ice along the coast of Hokkaido in the winter, 1951. Publ. 981, Hydro. Bull. No. 32:???

Jelly, K.C.P.

ice
Jelly, K.C.P., and N.B.Marshall, 1967.
Incidence of ice in the approaches to North America during the decade 1956-1965.
Marine Obs., 37(218):186-192.

ice
Jenness, J.L., 1953.
The physical geography of the waters of the western Canadian Arctic. Geogr. Bull., Canada, No. 4: 33-64, 4 text figs.

ice
Jensen, A.J.C., 1940
The influence of the currents in the Danish waters on the surface temperature in winter, and on the winter temperature of the air. Medd. Komm. Danmarks Fiskeri-og Havundersø, Ser. Hydrografi, 3(2):52 pp., 14 text figs.

ice
*John, B.S., 1968.
Directions of ice movement in the southern Irish Sea Basin during the last major glaciation: anhypo thesis.
J. Glaciol., 7(51):507-510.

ice
Johnson, N.G., 1943.
Studier av Isen i Gullmarfjorden.
Svenska Hydro.-Biol. Skr., n.s., Hydrogr., 18:1-21, Fig. 18.
Komm.

ice
Jónsdóttir, S., 1964.
Hydrographic conditions in Icelandic waters in May June 1962.
Ann. Biol., Cons. Perm. Int. Expl. Mer, 1962, 19: 16-17.

ice (Baltic)
Jurva, R., 1952.
Seas. Fennia 72:136-160.

ice
Jurva, R., 1950.
On the cartographical method for studies of the ice off the Finnish coast. J. du Cons. 16(3): 293-306, 11 figs.

ice
Jurva, R., 1950.
Sur les conditions de glace de la Baltique le long des cotes de Finlande. Int. Assoc. Hydr., U.I.G.G. Proc. Verb., 1948, 2:33-37.

ice
Jurva, R., 1947.
Ueber der grösste Gescheindigkeit der Vereisung des Meeres. Sitzber. Finn. Akad. Wiss. 1946: 223-230.

ice
Jurva, R., 1944.
Ueber den allgemeinen Verlauf des Eiswinters in den MeerenFinlands and Über die Schwankungen der grössten Vereisung. Sitzber. Finn. Akad. Wiss. 1941:670112.

ICE
Jurva, R., 1937
Über die Eisverhältnisse des Baltischen Meeres an den Küsten Finnlands. Havsforskningsinst. Hfors., No. 114

ice
Jurva, Risto, and Erkki Palosuo, 1959
Die Eisverhältnisse in den Finnland umgebenden Meeren in den Wintern 1938-1945 und die Baltischen Eiswochen in den Wintern 1938-1939.
Merent Julk., No. 188: 69 pp.

ice
Jurva, R., and E. Palosuo, 1955.
Die Eiswinter 1880-1949 im Licht der Zeitanalyse.
Merent Julk. 169:12 pp.

ice
Just, Jean, 1970.
Marine biological investigations of Jørgen Brønlund Fjord, North Greenland: physiographical and bathygraphical survey, methods and lists of stations.
Medd. Grønland, 184(5):42 pp.

K

ice
Kachurin, L.G., 1960
Calculation of the supercooling of water beneath ice and the rate of growth of ice in waters with allowance for the actual temperature of the crystallization front.
Eng. Ed., Bull. (Izv.) Acad. Sci., USSR, Geoph. S., 10: 1011-1014.

ice
*Kahn, S.I., 1967.
The present state of the methods for ice forecasts in the seas of the USSR. (In Russian; English abstract).
Okeanologiia, Akad. Nauk, SSSR, 7(5):786-791.

ice
Kaminski, H.S., 1955.
Distribution of ice in Baffin Bay and Davis Strait. H.O. Tech. Rept. TR-13:1-32, 14 figs.

ice
Kanavins, E., 1947.
General elements of the long range ice forecast of the Baltic Sea as a result of the large-scale weather situation frequency. Contrib. Baltic Univ. No. 85: 83 pp. (In German with English Summary)

ice
Kapitza, A.P., 1959.
[On the dependence of the East Antarctic ice cap form upon the underice bottom topography and the character of ice spreading.]
Inform. Biull. Sovetsk. Antarkt. Exped., (1):41-46.

ice
Karakash, E.S., 1962.
Forecast of the ice conditions in the northern Caspian. (In Russian).
Trudy Gosud. Okeanogr. Inst., 67:118-131.

ice
Karakash, E.S., 1960.
On the changeability of the amount of ice and the location of the ice edge in the northern Caspian. (In Russian).
Trudy Gosud. Okeanograf. Inst., Leningrad, 54:5-21

ice
Kawakami, K., 1961.
[Sea ice on the coast of Hokkaido, December 1958-April 1959.]
Hydrogr. Bull., Maritime Safety Bd., Tokyo, (Publ. No. 981), No. 65:48-59.

ice
Kawakami, K., 1959.
Sea ice on the coast of Hokkaido December 1956-April 1957.
Hydrogr. Bull., Tokyo, (Publ. 981) (no. 61):50-64.

ice
Kawakami, Kiyoshi, 1958.
[Sea ice on the coast of Hokkaido, Dec. 1955-Apr. 1956.]
Hydrogr. Bull., Tokyo, (Publ. 981); 18-28.

ice
Kehle, Ralph O., 1964.
Deformation of the Ross Ice Shelf, Antarctica.
Geol. Soc. Amer., Bull., 75(4):259-286.

ice
Keller, J.B., and E. Goldstein, 1953.
Water wave reflection due to surface tension and floating ice. Trans. Amer. Geophys. Union 34(1): 43-48, 4 textfigs.

ice
Keller, J.B., and M. Weitz, 1953.
Reflection and transmission coefficients for waves entering and leaving an ice field.
Comm. Pure and Appl. Math. 6(3:415-417.

ice
Kerry, J.G.G., 1947.
Ice blockade of Canadian ports; the winter temperature cycle of the St. Lawrence waters; a plea for more data. Dock and Harbor Auth. 27:273-276, 301-304, 318.

ice
Khaminov, N.A., 1967.
Many-year variations in ice coverage of the Baltic Sea, their possible causes and ways of forecasting. (In Russian; English abstract).
Okeanologiia, Akad. Nauk, SSSR, 7(2):269-278.

ice
Khrol, V.P., 1965.
A method for calculating ice volume in the sea with estimation of hummocks (i.e., Baffin Bay). (In Russian).
Trudy, Gosudarst. Okeanogr. Inst., No. 86:44-74.

ice
Khrol, V.P., 1961.
[Hummock ice as a factor in determining the volume of sea ice with respect to Baffin Sea.]
Issled. Severnoi Chasti Atlanticheskogo Okeana, Mezhd. Geofiz. God, Leningradskii Gidrometeorol. Inst., 1:56-61.

ice
Kingery, W.D., Editor, 1962.
Summary report - Project Ice Way.
Air Force Surveys in Geophysics, No. 145:210 pp.

properties of ice, etc.

ice
Kirillov, A.A., 1958.
Classification of Arctic ice and its distribution in the Soviet sector of the Arctic.
NAS-NRC Publ. No. 598:11-14.

ice
Kirk, T.H., 1948
Ice conditions in the Baltic during the winter 1946-47. Mar. Obs. MO 493, 18(140):80-92, 6 text figs.

ice
Kislyakov, A.G., and L.I. Borovaya, 1963.
Hydrological conditions in the Barents Sea in 1961.
Ann. Biol., Cons. Perm. Int. Expl. Mer, 1961, 18: 29-33.

ice
Knapp, W.W., 1969.
A satellite study of the ice in Antarctic coastal waters.
Antarctic J. US 4(5): 222-223

ice
Knight, C.A., 1962
Polygonization of aged sea ice.
J. Geol., 70(2):240-246.

ice
Koblents, Ya. P., 1960.
[Influence of the Antarctic shelf relief on the development of discharge glaciers.]
Inform. Bull., Soviet Antarctic Exped., 21:10-15

ice
Koch, L., 1945.
The East Greenland ice. Medd. om Grønland 130(3): 1-374.

ice
Koerner, R.M., 1970.
Weather and ice observations of the British Trans-Arctic Expedition, 1968-9.
Weather 25(5):218-228

ice
Kolesnikov, A.G., 1958.
On the growth rate of sea ice. 4. Sea ice formation, growth and disintegration.
NAS-NRC, Publ., No. 598:157-161.

ice
Konovalov, G.V., 1962.
Deflation hollows in the Lazarev ice shelf. (In Russian).
Inform. Biull. Sovetsk. Antarkt. Exped., 39:5-8.

Translation:
Scripta Tecnica for AGU, 4(5):261-263. 1964

ice
Kornilov, N.A., 1968
Disintegration of the Ice in the Sea of the Cosmonauts in the summer of 1967/68.
Inform. Biull. Sovetsk. Antarct. Exped. 71: 50-52.
Translation Scripta Tecnica for A.G.U. 7(3): 206-208.

ice
Korotkevich, E. S., 1959.
The ice regime of the Davis Sea.
Izv. Vse. Geograph. Obshch., (2):152-156.

ice
Korotkevich, E.S., 1958.
[The distribution of icebergs in the Davis Sea region.]
Inform. Biull. Sovetsk. Antarkt. Exped., (1): 65-72.

ice
Kozlovskiy, A.M. 1969
Melting of continental ice in seawater. (In Russian)
Inform. Antarct. biull. Sovetsk. Antarct. Exped. 75:38-42.
Translation: Am. Geophys Un. 75:409-411

ice
Kozlovskiy, A.M., 1964.
Ice conditions in the Davis Sea in January-April 1963. (In Russian).
Sovetsk. Antarkt. Ekspred., Inform. Biull., 46:31-

Translation:
Scripta Tecnica for AGU, 5(2):131-133. 1965.

ice
Kruchinin, Yu. A., 1962.
Ice situation in the Lazarev Station region in 19 1959. (In Russian).
Sovetsk. Antarktich. Expedits., Inform. Biull., (31):26-30.

Translation:
(Scripta Tecnica, Ind., for AGU), 4(1):13-15.

ice
Kruchinin, Yu. A., and Ta. P. Koblents, 1963.
Dynamics of the Trolltung Ice Shelf. (In Russian)
Sovetsk. Antarkt. Eksped., Inform. Biull., 44:49-

Translation:
Scripta Tecnica for AGU, 5(1):58-60. 1965.

ice
Krugler, F., 1952.
Ueber extrem starke örtliche Schwankungen der Wasseroberflächen-temperatur in Eisnähe am Rande des Ostgrönlandstroms. Ann. Met. 5:185-186.

ice
Krutskikh, B.A., 1961.
[The melting of ice as a result of the thermal effects of water.]
Okeanologiia, Akad. Nauk, SSSR, 1(4):642-645.

Kryndin, A.N.

ice
Kryndin, A.N., 1965.
Year-to-year variations in the ice of the Baltic and the possibility of long-term forecasting on the basis of the autumn-prewinter heat content of the ocean and atmosphere. (In Russian).
Trudy, Gosudarst. Okeanogr. Inst., No. 86:3-35.

ice
Kryndin, A.M., 1965.
On the interdependence of ice conditions in some non-Arctic seas. (In Russian).
Okeanologiia, Akad. Nauk, SSSR, 5(3):444-447.

ice
Kryndin, A.M., 1957.
[On the relation between the total amount of ice and the depth of ice on the Baltic.]
Meteorol. i Gidrol., (2):19-

ice
Kudryavaya, K.I., 1961
[Some results obtained from an experimental study of the ice drift coefficient.]
Issled. Severnoi Chasti Atlanticheskogo Okeana, Mezhd. Geofiz. God, Leningradskii Gidrometeorol. Inst., 1:112-120.

ice
Kühnel, Ingrid, 1970.
Eisvorbereitungszeiten für die Ostsee.
Umschau 70(1):13.

ice
Kühnel, Ingrid 1967.
Die Eisvorbereitungszeiten für die Ostsee östlich der Linie Trelleborg-Arkona und für den Finnischen und Rigaischen Meerbusen sowie für die südlichen Randbegirke der Bottensee.
Dt. hydrogr. Z. 20(1):1-6.

ice
Kuksa, V.I., 1959
Hydrological features of the North Kuril waters.
Trudy Inst. Okeanol., 36: 191-214.

Kupetski, V.N.

ice
Kupetski, V.N., 1967.
Luminescence of the sea ice. (In Russian).
Izv.Akad.Nauk,SSSR.Ser.Geograf,99(1):67-70.

ice
Kupetskii, V.N., 1962.
Stationary polynia in Baffin Bay. Northern water. (In Russian). Trudy Gosud. Okeanogr., Inst., 70: 47-60.

Ice
Kupetsky, V. N., 1961. ref.
Open water between the ice in the White sea.
Trudy Inst. Okeanolog. Akad. Nauk. SSSR, 64:78-92.

ice
Kupetski, V.N., 1958.
Hydrobiological peculiarities of stationary polynia. Izv. Vses. Geograph. Obshsh., 90(4):315-323.

ice
Kurasina, S., 1958.
On sea ince in adjacent Sea Japan. (Mainly on the relation between navigation and sea-ice).
Hydrogr. Bull. Maritime Safety Bd., Tokyo, (Publ. 981), No. 56:23-31.

ice
Kurasina, S., 1959.
Sea ice observation and drift-ice movement prediction.
Hydrogr. Bull., Tokyo (Publ. 981) (no. 61):45-49.

ice
*Kurashine, Syoji, Koji Nishida and Syuji Nakabayashi, 1967.
On the open water in the southeastern part of the frozen Okhotsk Sea and the currents through the Kurile islands. (In Japanese; English abstract).
J. oceanogr. Soc. Japan, 23(2):57-62.

ice
Kusunoki, K., 1958.
The present situation of sea ice observations in Japan. Sec. 2. Sea ice observing and reporting techniques.
NAS-NRC, Publ., No. 598:39-47.

ice
Kusunoki, K., 1955.
Observations on the horizontal and vertical distribution of chlorinity of sea-ice.
Low Temp. Sci., A, 14:149-150.

ice
Kusunoki, K., 1955.
Recent advances in the study of sea ice.
Low Temp. Sci., A, 14:155-184.

ice
Kusunoki, K., 1954.
A brief review of studies on sea ice in Japan.
Low Temp. Sci., A, 12:145-159.

ice
Kusunoki, K., 1952.
Bibliography of sea ice in Japan (for the years 1892-1950) J. Jap. Soc. Snow & Ice 13(4):125-128.

ice
Kusunoki, K., and T. Kashima, 1951.
Ocean current variations off the western coast of Hokkaido in the Japan Sea. J. Ocean. Soc., Japan, (Nippon Kaiyo Gakkaisi) 6(3):133-142, 5 textfigs.

ice
Kusunoki, K., and T. Tabata, 1954.
On the method of sampling of sea-ice.
Low Temp. Sci., A, 12:87-94.

ice
Kuznetsov, O.A., 1959(1961).
Aleutian low as a factor in the ice regime of the Chukchi Sea.
Problemy Severa, 3:10-15.

Translation: Problems of the North, 3:9-14.

ice
Kvinge, Tor, 1963.
The "Conrad Holmboe" Expedition to East Greenland waters in 1923.
Årbok, Univ. i Bergen, Mat.-Naturv. Ser., (15): 44 pp.

L

ice
Laevastu, T., 1960
Factors affecting the temperature of the surface layer of the sea.
Merent. Julk. (Havsforskningsinst. Skr.), No. 195:136 pp.

ice
LaFond, E.C., 1954.
Physical oceanography and submarine geology of the seas to the west and north of Alaska. Arctic 7(2):93-101, 11 textfigs.

ice
La Grange, J.J., 1961.
Sea-ice observations in the South Atlantic Ocean during summer 1960/61.
Notos, 10(1/4):119-121.

ice
*Lake, R.A., 1967.
Heat exchange between water and ice in the Arctic Ocean.
Arch.Met.Geophys.Bioklim. (A)16(2/3):242-

ice
Laktionov, A.K., 1958.
Ice observation methods. Sec. 2. Sea ice observing and reporting techniques.
NAS-NRC, Publ. No. 598:48-56.

ice
Laktionov, A.F., V.A. Shamontyev and A.V. Yanes, 1960
Oceanographic characteristics of the North Greenland Sea.
Soviet Fish. Invest., North European Seas, VNIRO, PNIRO, Moscow, 1960:51-65.

ice
Lamb, H.H., 1948
Notes on "Balaena's" Observations of Southern ice. Mar. Obs. M.O. 493, 18(139):34-41, 6 figs.

ice
Lamb, H.H., 1948
Topography and weather in the Antarctic.
Geogr. Jour. CXI (1/3):48-66.

ice
Lambor, J., 1948
Geneza lodu pradowego [La genese de la glace flottante]. Wiadomosci Sluzby Hydrologicznej i Meteorologicznej (Bull. du Service Hydrologique et Meteorologique) 1(3):214-244, 14 text figs. & graphs.

ice
Lamont, A. H., 1949.
Ice conditions over Hudson Bay and related weather phenomena. Bull. Am. Met. Soc. 30: 288-289.

ice
Langleben, M.P., 1971.
Albedo of melting sea ice in the southern Beaufort Sea.
J. Glaciol., 10(58): 101-104.

ice
Langleben, M.P., 1969.
Albedo and degree of puddling of a melting cover of sea ice.
J. Glaciol. 8(54): 407-419.

ice
Langleben, M.P., 1968.
Albedo measurements of an Arctic ice cover from high towers.
J. Glaciol., 7(50): 289-297

ice
Langleben, M.P., 1966.
On factors affecting the rate of ablation of sea ice.
Can. J. Earth Sci., 3(4):431-439.

Ice
Langleben, M. P., 1962.
Young's modulus for sea ice.
Canadian Journal of Physics, 40(1):1-8.

ice
Langway, C.C., Jr., 1958.
400 meter deep ice core in Greenland. Preliminary report. J. Glaciol., 3(23):217.

ice
Lappo, S.S., 1962
A study of empirical formulae dealing with the growth of ice thickness.
Okeanologiia, Akad. Nauk, SSSR, 2(1):59-66.

Abstracted in: Soviet Bloc Res., Geophys., Astron. and Space, 1962(35): 17. (OTS61-11147-35 JPRS13739)

ice
Lappo, S.D., 1958 (1960).
The rotation of drifting ice floes.
Problemy Severa, (1):

Translations in:
Problems of the North, (1):31-43.

ice
Lappo, S.D., 1938.
Lets do a better job of ice forecasting.
Sovetskaya Arktika 2:48-57.

T15R

ice
Latham, J., and C.D. Stow, 1965.
Electrification associated with the evaporation of ice.
J. Atmos. Sci., 22(3):320-324.

ice
Laws, R.M., and R.J.F. Taylor, 1957.
A mass dying of crabeater seals, Lobodon carcinophagus (Gray). Proc. Zool. Soc., London, 129(3):315-324, 1 pl.

ice
Leehey, D.M., 1966.
Heat exchange and sea ice growth in Arctic Canada.
Mar. Sci. Centre, McGill Univ., Manuscript Rept., No. 1:48pp. (multilithed).

Lebedev, A.A.

ice
Lebedev, A.A., 1968.
Probability of the ice occurrence in the Barents and Greenland Seas and North Atlantic. (In Russian; English abstract). Trudy polyar. nauchno-issled. Inst. morsk. ryb. Khoz. Okeanogr. (PINRO) 23: 138-142.

ice
Lebedev, A.A., 1965.
Variability of ice conditions in the northwest Atlantic. (In Russian).
Trudy, Gosudarst. Okeanogr. Inst., No. 87:32-50.

ice
Lebedev, A.A., 1966.
Problems of forecasting ice conditions in the northwest Atlantic. (In Russian).
Materiali, Sess. Uchen. Soveta PINRO Rez. Issled., 1964, Minist. Ribn. Khoz., SSSR, Murmansk, 108-117.

Lebedev, A.A., 1964. ice
Changeability in ice conditions of the Labrador Sea and Davis Strait. (In Russian).
Trudy, Poliarn. Nauchno-Issled. i Proektn. Inst. Morsk. Ribn. Choz. i Okeanogr. im N.M. Knipovicha, 16: 269-233.

ice
Lebedev, A.A., 1964.
Method of forecasting the ice conditions in the North Atlantic and Greenland Sea. (In Russian).
Materialy, Ribochoz. Issled. Severn. Basseina, Gosudarst. Kom. Rib. Choz., SNCH, SSSR, Poliarn. Nauchno-Issled. i Proektn. Inst. Morsk. Rib. Choz. i Okeanogr., N.M. Knipovich, (PINRO), 2:135-137.

ice
Lebedev, A.A., 1964.
To the problem of year-to-year variability in the ice and thermal conditions in the North Atlantic and Greenland Sea. (In Russian).
Materialy, Ribochoz. Issled. Severn. Basseina, Gosudarst. Kom. Rib. Choz., SNCH, SSSR, Poliarn. Nauchno-Issled. i Proektn. Inst. Morsk. Rib. Choz. i Okeanogr., N.M. Knipovich, (PINRO), 2:129-135.

ice
Lebedev, A. A., 1964.
Variability of the ice conditions in the northwest Atlantic. (InRussian).
Material. Sess. Uchen. Sov. PINRO, Rez. Issled., 1962-1963, Murmansk, 123-129.

ice
Lebedev, V. L., 1960.
[On the influence of temperature and swells on the ice field disintegration.]
Inform. Bull., Soviet Antarctic Exped., 21: 38-41

ice
Lebedev, V.L., 1957.
Antarktika. Geografgiz, Moscow.
Translation by E.R. Hope of pages 77-78 "Downwind drift of iceberg fragments"

ice
Lee, A., 1958.
Ice accumulation on trawlers in the Barents Sea.
Mar. Obs., 28(181):138-142.

ice
Lee, O.S., 1955.
Local environmental factors affecting ice formation in Terrington Basin, Labrador.
H.O. Tech. Rept. TR-24:29 pp.

ice
Lee, O.S., and L.S. Simpson, 1958.
A practical method of predicting sea ice formation and growth. H.O. Tech. Rept. TR-4:1-27, 4 figs.

ice
Legen'kov, A.P., and V.D. Uglev, 1970.
Results of measurements of the multi-annual ice field by invar wires. (In Russian).
Fizika Atmosfer. Okean., Izv. Akad. Nauk SSSR, 6(5): 537-540

ice
Lenczyk, R.E., 1965.
Report of the International Ice Patrol Service.
Bull. U.S. Cst. Guard.

ice
LePage, L.S., and A.L.P. Milwright, 1953.
Radar and ice. J. Inst. Navigation 6(2):113-127, 8 textfigs.

ice
Le Shack, Leonard A., 1964.
Long-period oscillations of the ice recorded by continuous gravimeter measurements from drift station T-3.
Arctic, 17(4):272-279.

ice
Lewis, E.L. and R.A. Lake, 1971.
Sea ice and supercooled water. J. geophys. Res. 76(24): 5836-5841.

ice
Limbert, D.W.S., 1964.
The absolute and relative movement and regime of the Brunt Ice Shelf near Halley Bay.
Brit. Antarctic Surv. Bull., (3):1-11.

ice
Linkov, E.M., 1958.
[The study of the elastic characteristics of the ice covers in the Arctic zone.]
Vestnik, Univ. Leningrad, Ser. Fiz. i Khimii, (1):17.

ice
Liser, I. Ya., 1958
[Initial growth of ice and underice sludge.]
Meteorolo. i Gidrol., No. 11: 30-31.
LC mi/ ph $1.80
SLA

Lisignoli, Cesar A., 1964. ice
Movimiento de la barrera de Filchner, Antartida.
Contrib. Inst. Antartico Argentino, No. 81:17 pp.

Lisitzin, A.P.

ice
Lisitzin, A.P., 1966.
Processes of Recent sedimentation in the Bering Sea. (In Russian).
Inst. Okeanol., Kom. Osad. Otdel Nauk o Zemle, Isdatel, Nauka, Moskva, 574 pp.

ice
Lisitzin, A.P., 1959
[Peculiarities of sea geological investigations in the Antarctic.]
Trudy Inst. Okeanol., 35: 121-152.

ice
Lisitzin, E., 1957.
On the reducing influence of sea ice on the piling up of water due to wind stress.
Comm. Phys.-Math., Soc. Sci. Fenn., 20(7):12 pp.

ice
Little, E.M., 1959.
Fluid permeability of sea ice as a good index of its condition.
Amer. Geophys. Union, SW Regional Meeting.
Abstr. in J. Geophys. Res., 64(6):692.

ice
Liubomirova, K.S., 1962.
[Some peculiarities of the attenuation of solar radiation on ice thickness.]
Izv., Akad. Nauk, SSSR, Ser. Geofiz., (5):693-699.

ice
Loewe, F., 1961.
On melting of fresh-water ice in sea-water.
J. Glaciol., 3(30):1051-1052.

ice
Lomniewski, K., 1960.
Hydrographic problems at the Polish Baltic coast.
Przeglad Geograficzny, Suppl. to Vol. 32:78-87.

ice
Lomniewski, Kazimierz, 1958
[The Firth of Vistula] Polska Akad. Nauk, Inst. Geografii, Prace Geograficzne, No. 15: 106 pp.

ice
Loshchilov, V.S., 1959.
[Use of air photography in ice reconnaissance for the determination of the cover mean thickness.]
Problemi Arktiki i Antarktiki, (1):81-86.

ice
Lorck, M. V. L., 1959.
The state of the ice in the Arctic seas. 1956.
Nautisk-Meteor. Aarbog, Charlottenlund: 34 pp.

ice
Lunde, Torbjørn, 1965.
Ice conditions at Svalbard, 1946-1963.
Norsk Polarinstitutt Årbok, 1963:61-80.

ice
Lunde, Torbjørn, 1963.
Sea ice in the Svalbard region, 1957-62.
Norsk Polarinst., Årbok, 1962:24-34.

ice
Lutkovsky, S. V., 1956
The problem of the formation of the temperature field of the sea in conditions of ebb and icing.
Trudi Mor. Gidrofiz. In-ta, Akad. Nauk SSSR, 7:135-152.
Abs. in Appl. Mech. Rev., 12(11): #5831.

ice
Lyon, W.K., 1960
Experiments in the use of explosives in sea ice.
Polar Record, 10(66): 237-247.

ice
Lyons, J.B., S.M. Savin and A.J. Tamburi, 1971.
Basement ice, Ward Hunt Ice Shelf, Ellesmere Island, Canada.
J. Glaciol. 10(58):93-100.

M

ice
MacKay, D.K., 1963
Ice conditions in the Gulf of St. Lawrence and Cabot Strait (with particular reference to the Sydney Bight area).
Cah. Geog. Quebec, 7(14):211-228.

ice
MacKay, G.A., 1952.
The effect of protracted spring thaws on ice conditions in Hudson Bay. Bull. Amer. Met. Soc., 33(3):101-106, 5 textfigs.

ice
Mackintosh, N.A., and H.F.P. Herdman, 1940.
Distribution of the pack-ice in the Southern Ocean. Discovery Repts., 19:287-206, Pls. 49-95.

ice
Malinina, T. I., 1959.
The Yakkimvar Bay (Lagoda Lake) seiches under the ice cover.
Izv. Vse. Geograph. Obshch., (2):156-159.

ice
Mamaev, O.I., 1964
On the theory of the growth and decay of the ice cover at sea. (In Russian).
Okeanologiia, Akad. Nauk, SSSR, 4(5):749-755.

ice
Marr, J.W.S., 1962.
The natural history and geography of the Antarctic krill (Euphausia superba Dana).
Discovery Repts., 32:33-464, Pl. 3.

Ice
Masuzawa, N., 1966.
On the drifting of pack-ice in the adjacent sea of Hokkaido in the early winter. (In Japanese; English abstract).
Bull. Hakodate Mar. Obs., No. 12:72-78.

ice
Maurstad, A., 1935.
Atlas of sea ice. Geofys. Publ. 10(11):17 pp.

Ice
Maximov, I. V., 1961.
[The Antarctic convergence front and the northern limit of icebergs in the Antarctic Ocean over a period of years.]
Probl. Arkt. Antarkt., (8):47-52.

ice
Maksimov, I. V., 1960.
[Level variation of the world with the variation of the continental ice of Antarctica.]
Information Bull., Soviet Antarctic Exped., No. 19:5-7.

ice
Maksimov, I. V., 1959.
[Astronomical reasons of the drift of ice and icebergs.]
Biull. Sovetsk. Antarkt. Exped., No. 5:39-42.

ice
Maximov, I.V., 1954.
Secular fluctuations in the ice-cover of the northern part of the Atlantic Ocean. (In Russian).
Trudy Inst. Okeanol., Akad. Nauk, SSSR, 8:41-91.

ice
Maksimov. I.V., N.P. Smirnov and V.N. Vorobjev, 1964.
Long-range predictions of the long-term changes in the general ice conditions of the Barents Sea worked out by the harmonic-component method. (In Russian).
Materiali, Ribochoz. Issled. Severn. Basseina, Poliarn. Nauchno-Issled. i Proektn. Inst. Morsk. Rib. Choz. i Okeanogr., im. N.M. Knipovicha, PINRO, Gosud. Proizvodst. Komm. Ribn. Choz. SSSR, 4:73-85.

ice
Maykut, Gary A. and Norbert Untersteiner, 1971.
Some results from a time-dependent thermodynamic model of sea ice. J. geophys. Res., 76(6): 1550-1575.

ice
McGough, R.J., 1956.
Local environmental factors affecting ice formation in North Sta Bugt, Greenland.
H.O. Tech. Rept. T-R 23:40 pp.

ice
Mecking, L., 1932.
Die antarktische Treibeisgrenze und ihre Beziehung zur Zyklonenwanderung. Ann. Hydrogr., usw., Jarhg. 60(Heft VI):225-229.

ice
*Meguro, Hiroshi, Kuniyuki Ito, and Hiroshi Fukushima, 1967.
Ice flora (bottom type): a mechanism of primary production in polar seas and the growth of diatoms in sea ice.
Arctic, 20(2):114-113.

ice
Mellor, M., 1963.
Promoting the decay of sea-ice.
Arctic, 16(2):142.

ice
Meshcherskaya, A.V., and L.R. Dmitrieva-Arrago, 1968.
An expansion of annual trend of ice-cover percentage of northern seas by natural orthozonal time functions. (In Russian; English abstract).
Meteorologiya Gidrol. (10):56-64.

ice
Meyer, Arno, 1964
Zusammenhang zwischen Eisdrift, atmosphärischer Zirkulation und Fischerei im Bereich der Fangplätze vor der südostgrönländischen Küste während der ersten Jahreshälfte.
Arch. Fischersiwiss., 15(1):1-16.

ICE
Milne, A.R. 1970.
The transition from moving to fast ice in western Viscount Melville Sound.
Arctic 23(1): 45-46

ice
Milne, A.R., and J.H. Ganton, 1964.
Ambient noise under Arctic sea ice.
J. Acoust. Soc., Amer., 36(5):855-863.

ice
Minchin, C., 1950.
Weather and pack-ice in the Antarctic. Mar. Obs. 20(147)(M.O. 528):50.

ice
Model, F., 1948.
Eisdicke auf der Alster im Winter 1946-47.
Deut. Hydro. Zeitschr. 1(2/3):104-146, 1 fig.

ice
Morecki, V.N., 1965.
Underwater sea ice. (In Russian).
Problemy Arktiki Antarktiki, SSSR 19:32-38.
Translation: E.R. Hope, Rept. T497-R.1968.

ice
Moroshkin, K. V., N. D. Kravtsov, V. S. Nagarov, G. V. Rzheplinskii and Iu. G. Rizhkov, 1958.
[V. Hydrological work.] Opisanie Exped. D/E "Ob" 1955-1956, M.G.G., Trudy Kompaeksnoi Antarkt. Exped. Akad. Nauk, SSSR:48-90.

ice
Morozova, T.P., 1959
[Change of the southern limit of the Arctic pack in the Laptev Sea.]
Problemy Arktiki i Antarktiki (1): 5-10.

ice
Mortimer, C.H., and F.J.H., Mackereth, 1958.
Convection and its consequence in ice-covered lakes. Verh. Int. Ver. Limnol., 13:923-932.

ice
Mosby, Hakon, 1962
Water, salt and heat balance of the North Polar Sea and of the Norwegian Sea.
Geofys. Publik., Geophysica Norvegica, 24: 289-313.

ice
Mosby, O., 1932.
Isforholdene i den nordvestlige del av Atlanterhavet. Naturen 56:208-229.

ice
Moss, R., and A.R. Gleb, 1947.
Investigations on ice during the war.
J. Glaciol. 1(1):8-11.

ice
Murray, J.E., 1969.
Report of the International Ice Patrol Service in the North Atlantic Ocean.
U.S.C.G. Bull. 54(CG-188-23): 105pp

ice
Murray, J.E., 1969.
Report of the International Ice Patrol Service in the North Atlantic Ocean.
Bull. U.S. Coast Guard, 53(CG-188-22):1-52.

ice
Murray, J.E., 1966.
Report of the International Ice Patrol Service in the North Atlantic Ocean. Season of 1966.
Bull. U.S. Cst Guard, 52: 1-27.

N

ice
Naruse, Renji, Tamotsu Ishida, Yasoichi Endo and Yutaka Ageta, 1971.
On the relation between sea ice growth and freezing index at Syowa Station, Antarctica. (In English; Japanese Abstract). Antarct. Rec., Repts. Japan Antarct. Res. Exped, 41: 62-66.

ice
Nazarov, V.S., 1963
Ice in the waters of the World Ocean and the variation of its rate. (In Russian).
Okeanologiia, Akad. Nauk. SSSR, 3(2):243-249.

ice
Nazarov, V.S., 1962
Ice of Antarctic waters. Oceanology, X. section of IGY Program. (In Russian: English summary). Rezult. Issled. Programme Mezhd. Geofis. Gode. Mezhd. Geofiz. Komitet, Prez. A Akad. Nauk, SSSR, 72 pp.

ice
Nazarov, V.S., 1960
[The ice of Antarctica in 1956-1958.]
Okean. Issle., IGY Com., SSSR:91-95.

ice
Nagarov, V.S., 1948.
[Fluctuations of iciness in the North Atlantic regions.] Met. Gidrol. 3:60-65.

ice
Nazarov, V.S., 1947.
[Historical variations of ice conditions in the Kara Sea.] Bull. All-Union Geogr. Soc., USSR, 6: 653-655.

ice
Nazarov, V.S., 1947.
Historical variation of ice conditions in the Kara Sea. Bull. All-Union Geogr. Soc., SSSR 1947 (6):653-655.
T16R

ice
Nazarov, V.S., 1947.
[Historical course of ice in the Kara Sea.]
Izvest. Vses. Geogr. Obshch. 79:654-655.

ice
Nazarov, V.S., 1947.
[Variations in ice conditions in northern seas.]
Nak. Zhizn. No. 3:5-7 (in Russian).

ice
Nazarov, V.S., 1938.
[Contributions to the study of the properties of sea ice.] Trudy Arktischeskogo Inst. 110(1):101-105.

RT-740 Bibl. Trans. Rus. Sci. Tech. Lit. 5.

Nazarov, V. S., 1938 ice
K Izucheniiu Svoistv Morskova Lda. Raboty po Ledovedeniiu, Vypusk 1:101-108.
(Contribution to the study of the properties of ice.
Papers on Cryology, Part 1:101-108.)

Ice
Nazintsev, J. L., 1961.
[The transmission of heat across icebergs in the central Arctic.]
Probl. Arkt. Antarkt., (8):37-45.

ice
Nikolaeva, L.M., 1958 (1960).
[The Ross ice shelf, its regime and the conditions of its formation.]
Problemy Severa, (1):

Translation in:
Problems of the North, (1):303-320.

ice
Normann, C.O.E., 1894
Forslag til en fra Sösiden foretaget undersogelse af Grønlands Østkyst.
Medd. Grønlands 6 (2):35-56.

ice
Novikova, E.M., 1963
On the use of radiolocation technique for the observation of ice drift in the Gorlo (strait) of the White Sea. (In Russian).
Okeanologiia, Akad. Nauk, SSSR, 3(4):730-738.

ice
Nusser, Franz, 1959
Distribution and character of sea ice in the European Arctic. In: "Arctic Sea Ice", NAS-NRC Publ. 598: 1-10. Also in: Deutsches Hydrogr. Inst., Ozeanogr., 1959(1960), No. 1.

ice
Nusser, Franz, 1958.
Zusammenhang zwischen Grosswetterlage und Eisvorkommen an den deutschen Küsten.
Deutschen Hydrogr. Zeits., 11(5):185-194.

ice
Nusser, F., 1956.
Eine neuer internationale Eisnomenklatur.
Deutsche Hydrogr. Zeits., 9(4):174- 182.

ice
Nusser, F., 1956.
Problems of sea ice research. J. Glaciol. 2(19): 619-623.

ice
Nusser, F., 1955.
Probleme der Meereisforschung.
Assoc. Int. Hydrol. (Assemblée Général de Rome 4) Publ. 39:215-219.

ice
Nusser, F., 1955.
Die Eisverhältnisse des Winters 1952/53 an den deutschen Küsten zwischen Ems und Trave.
Deutsche Hydrogr. Zeits. 8(2):65-72, 4 textfigs.

ice
Nusser, F., 1954.
Die Eisverhältnisse des südlichen Atlantischen Ozeans. Handb. Atlant. Ozeans, 2:42-47.

ice
Nusser, F., 1953.
Die Eisverhältnisse des Winters 1951/52 an den deutschen Küsten zwischen Ems und Trave.
Deutsche Hydrogr. Zeits. 6(2):77-80, Pl. 11, 1 textfig.

ice
Nusser, F., 1952.
Die Eisverhältnisse des Winters 1950/51 an den deutschen Küsten zwischen Ems und Trave.
Deutsche Hydrogr. Zeits. 5(4):196-199, 1 textfig., Chart 7 (in back pocket).

ice
Nusser, F., 1950.
Die Vereisung des Lilliehöökfjordes (Spitzbergen) in den Jahren 1941/42 und 1942/43. Deutsch. Hydr. Zeits. 3(5/6):286-293, 1 textfig.

ice
Nusser, F., 1950.
Die Eisverhaltnisse des Winters 1949/50 an den deutschen Küsten zwischen Ems und Oder. Deutsch. Hydro. Zeits. 3(5/6):335-341, 2 textfigs.

ice
Nusser, F., 1950.
Gebiete gleicher Eisvorbereitungszeit an den deutschen Küsten. Deutsch. Hydro. Zeits. 3(3/4): 220-227.

ice
Nusser, F., 1949.
Die Eisverhältnisse des Winters 1948/49 an der deutschen Küsten. Deutsch. Hydro. Zeit. 2(6): 251-255, 1 textfig.

ice
Nusser, Franz, 1948.
[Ice conditions in Lilliehookfiord (Svalbard) in 1941-1942 and 1942-1943.] No. 61.

Unpublished manuscript in German with English Summary, on file in HO.

ice
Nusser, F., 1947.
[The ice winter 1946-1947 on the German coasts.] No. 55.

Unpublished manuscript in German with English summary on file in HO.

ice
Nutt, David C., 1959.
Recent studies of gases in glacier ice.
Polar Notes, Occ. Publ. Stefansson Coll., Dartmouth Coll. Lib., 1:57-65.

ice
Nutt, D.C., and L.K. Coachman, 1956.
The oceanography of Hebron Fjord, Labrador.
J. Fish. Res. Bd., Canada, 13(5):709-758.

ice
Oberkommando der Kriegsmarine, 1940
Die Naturverhältnisse des Sibirischen Seeweges. Beilage zum Handbuch des Sibirischen Seeweges. 169 pp., 64 text figs., 2 bathymetric charts.

ice
Obruchev, V.A., 1955.
[Comments on Hagemeister's paper.] Priroda 1955(7): 96.

Translation by E.R. Hope T196R at WHOI.

ice
Ogata, T., and K. Nakagawa, 1948.
[On the sea ice in the Kurile Islands.]
(In Japanese). J. Met. Soc., Japan, 2nd ser., 26(5):138-146, with English summary, p. 12.

ice
Ogura, Y., 1952.
A supplementary note on the problem of ice formation. Coll. Met. Papers 4(1):231-239, 6 textfigs. (1953)
J. Met. Soc., Jap., 30(7):231-239, 6 textfigs.
and also in:
Also in Geophys. Notes 5(2), No. 9.

Ice
Ohotani, Kiyotaka, 1969.
On the oceanographic structure and the ice formation on the continental shelf in the eastern Bering Sea. (In Japanese; English abstract).
Bull. Fac. Fish. Hokkaido Univ. 20(2):94-117

ice

Olausson, Eric and Ulf C. Jonasson, 1969.
The Arctic Ocean during the Würm and Early Flandrian. Geol. För. Stockh. Förh., 91: 185-200.

ice

Oliver, J., A.P. Crary and R. Cotell, 1954.
Elastic waves in Arctic pack ice.
Trans. Amer. Geophys. Union 35(2):282-292, 9 textfigs.

ice

Omdal, K., 1953.
Drivisen ved Svalbard, 1924-1939.
Norsk Polarinst. Medd. No. 72:1-21.

ice

Ono, Nobuo 1968.
Thermal properties of sea ice.
IV. Thermal constants of sea ice. (In Japanese; English abstract).
Inst. Low Temp. Sci. (A) 26: 329-349.

ice (ships)

Ono, Nobuo, 1964.
Studies on the ice accumulation on ships. II.
On the conditions for the formation of ice and the rate of icing. (In Japanese; English abstract)
Low Temp. Sci., Hokkaido Univ., (A), 22:171-181.

ice

Östman, C.J., 1950.
Om sambandet mellan köldsommor, isläggning och istjocklek. Medd. Sveriges Met. och Hydrol. Inst. Ser. A., No. 1:20 pp., 8 textfigs. (English summary).

P

ice

Paccagnini, Ruben N., and Alberto O. Casellas, 1961
Estudios y resultados preliminares sobre trabajos oceanograficos en el area del Mar de Weddell. (In Spanish; Spanish, English, French German and Italian resumes).
Contrib., Inst. Antartico Argentino, No. 64: 12 pp.

ICE

Paige, Russell A. and Claude W. Lee, 1967.
Preliminary studies on sea ice in McMurdo Sound, Antarctica, during "Deep Freeze 65".
J. Glaciol. 6(46): 515-528

ice

Palosuo, Erkki, 1966.
Ice in the Baltic.
In: Oceanography and marine biology, H. Barnes, editor, George Allen & Unwin, Ltd., 4:79-90.

ice

Palosuo, Erkki, 1965.
Duration of the ice along the Finnish coast 1931-1960. (In Finnish and English).
Merentutkimuslait. Julk., No. 219:49pp.

ice

Palosuo, Erkki, 1963
The Gulf of Bothnia in winter. II. Freezing and ice forms. (In Finnish and English).
Merent. Julk. (Havsforskningsinst. Skrift), No. 209: 64 pp.

ice

Palosuo, Erkki, 1962
Jaatalvi 1961/62 Suomen Merialueilla. Ice winter 1961/62 along the Finnish coast. (In Finnish and English).
Havsforskningsinst. Skrift (Merent.Julk.), No. 206: 24 pp.

ice

Palosuo, Erkki, 1961

Ice winter 1959/60 along the Finnish coast.
Merent. Julk. (Havforskningsinst. Skr.), No. 196: 22 pp.

ice

Palosuo, Erkki, 1960

Ice winter 1958/59 along the Finnish coast.
Merent. Julk., No. 191:18 pp.

ice

Palosuo, E., 1957.
Die Eisverhaltnisse in den Finland umgebenden Meeren im Winter 1955-56.
Havforsk. Skr., 174:14 pp.

ice

Palosuo, E., 1956.
"Kiintojää silloista" suomen ja Ruotsin välilla.
Eripainos Terrasta No. 31:86-96.

ice

Palosuo, E., 1956.
Über verschiedene Vereisungstypen im Finnischen Meerbusen. Deutsche Hydrogr. Zeits., 9(5):222-225.

ice

Palusuo, E., 1953.
A treatise on severe ice conditions in the central Baltic. Merent. Julk. No. 156:1-130, 46 textfigs., 9 charts.

ice

Palosuo, E., 1953.
A treatise on severe ice conditions in the central Baltic. Fennia 77(1):1-130, 55 textfigs.

ice

Palosuo, E., 1951.
Ice in the Baltic and the meteorological factors
J. du Cons. 18(2):124-132, 2 textfigs.

Ice

Panzarini, Rodolfo N.H., 1965.
Vuelos de reconocimiento del estado del hielo en el Antartico, ano 1964.
Bol. Servicio Hidrogr. Naval, Argentina, 11(1): 9-15.

ice

Panzarini, Rodolfo N., 1963
Nomenclature del nielo en el mar.
Inst. Antarctico Argentino, Publ., No. 10: 103 pp.

(English, Spanish, French, German & Russian)

ice

Papov, V. V., 1961.
[The role of the Atlantic in the hydrologic and ice regimes of the Arctic.]
Probl. Arkt. i Antarkt. (8):75-77.

ice

Paterson, M.P. and K.T. Spillane 1968.
The oceans as a source of ice nuclei.
Nature, Lond. 218 (5144): 864.

ice

Pearce, D.C., 1951.
A bibliography on snow and ice. Canada, Nat. Res Counc., 69 pp. (75 cents).

ice

Peschanskiy, I.S., 1957.
Some problems in the Arctic ice studies. (In Russian).
Problemy Arktiki, SSSR, (2):161-170.

Translation
LC or SLA, mi $2.70, ph $4.80.

ice

Petersson, H. v., 1956.
Die Eisverhältnisse der südlichen und mittleren Ostsee im Winter 1951/52. An. f. Hydrogr., (4): 71-78.

ice

Petersson, H.V., 1956.
Die Eisverhältnisse an der Küste der Deutschen Demokratischen Republik im Winter 1952/53.
Ann. f. Hydrogr., 5/6:55-62, 3 pls.

ice

Petrov, I.G., 1955.
[Physical-mechanical properties and thickness of the ice cover.] Material. Nabluid. Nauch.-Issledo 1950/1951, Morskoi Transport, 2:103-165.

David Kraus in AMS-Astia Doc. 117137.

ice

Peyton, Harold R. 1968.
Ice and marine structures. 2. Sea ice properties.
Ocean Industry 3(9): 59-65.

ice

Pickard, G.L., 1967.
Some oceanographic characteristics of the larger inlets of southeast Alaska.
J. Fish. Res. Bd., Can., 24(7) 1475-1506

ice

Poland, Fred, 1968.
Melting the Arctic ice.
New Scientist, 39 (608): 244

ice (rivers)

Popov, E.G., 1957.
Short term forecasts of ice formation and breakup in rivers. (In Russian).
Gidrologicheskiye Prognozy, Ch. 11:391-426.

Translation:
U.S. Naval Oceanogr. Off., by M. Slessers, TRANS-160:44 pp.

ice

Pounder, E.R., 1962
The physics of sea-ice. Sect. VII. In: The Sea, Vol. 1, Physical Oceanography, Interscience Publishers, 826-838.

ice

Pounder, E. R., and E. M. Little, 1959.
Some physical properties of sea ice. 1.
Canadian J. Physics, 37:443-473.

ice

Prahm, G., 1951.
Die Abschmelzzeit des Eises an den deutschen Küsten zwischen Ems und Oder. Deutsch. Hydro. Zeits. 4(1/2):17-28.

ice
Predoehl, Martin C., 1966.
Antarctic pack ice: boundaries established from Nimbus I pictures.
Science, 153(3738):361-363.

ice
Predoehl, Martin C., 1964.
Aerial measurements of albedo of sea ice in the Antarctic. (Abstract).
Trans. Amer. Geophys. Union, 45(1):57.

Q

R

ice
Ragle, R.H., R.G. Blair and L.E. Persson, 1964
Ice core studies of Ward Hunt Ice Shelf, 1960.
J. Glaciol., 5(37):39-59.

Ice
Ragotzkie, Robert A., and Evon P. Ruzecki, 1967.
An ice record of standing waves.
Limnol. Oceanogr., 12(2):326-327.

ice
Rasmussen, B., 1950.
Notes on the ice conditions in Greenland waters in 1949. Ann. Biol. 6:36.

ice
Rekhtzamer, G.R., 1964.
Results of experimental study of ice by aerial methods and a project of international investigations of East Greenland ice transport. (In Russian.)
Materiali Vtoroi Konferentsii, Vzaimod. Atmosfer i Gidrosfer. v Severn. Atlant. Okean., Mezhd. Geofiz. God, Leningrad. Gidrometeorol. Inst., 145-150.

ice
Rekhtzamer, G.R., 1961
[Concerning the methods of airphotography and deciphering the aerial photographs of sea ice.]
Issled. Severnoi Chasti Atlanticheskogo Okeana, Mezhd. Geofiz. God, Leningradskii Gidrometeorol Inst., 1: 121-141.

ice
Rekhtzamer, G.R., 1961
[Determination of ice by aerial methods.]
Issled. Severnoi Chasti Atlanticheskogo Okeana, Mezhd. Geofiz. God, Leningradskii Gidrometeorol Inst., 1: 142-156.

Reece, A., 1949. ice
Snow cover and sea ice. Comments on the Antarctic Research Discussion. J. Glaciol. 1(5):226-227.

ice
Reed, Richard J., and William J. Campbell, 1962.
The equilibrium drift of Ice Station Alpha.
J. Geophys. Res., 67(1):281-298.

ice
Reichelt, W., 1941.
Die ozeanographischen Verhältnisse bis zur warmen Zwischenschicht an der antarktischen Eisgrenze im Südsommer 1936/37. Nach Beobachtungen auf dem Walfangmutterschiff "Jan Wellem" im Weddell-Meer. Arch. Deutschen Seewarte 61(5):54 pp., 11 pls.

ice
Reineck, H.E., 1956.
Abschmelzreste von Treibeis an den Ufersäumen des Gezeiten-Meeres. Senckenberg. Lethaea 37(3/4):288-304.

ice
Richter, J., 1933.
Der Vereisung der Beltsee und südlichen Ostsee im Winter 1928-29. Arch. Deutschen Seewarte 52(5) 1-67, 7 pls.

ice
Ridgeway, N.M., 1962.
Echo sounding through sea ice.
Polar Record, 11(72):298-299.

ice(data)
Rivolier, J., and J. Duhamel, 1956
Terre Adélie 1952. Éléments d'étude de la glace de mer dans l'Archipel de Pointe/Géologie. Expéd. Polaires Franc. Missions Paul-Emile Victor Expéd. Antarct., Res. Sci., No. S. (2) 4: 1-59.

ice
Robin, G. de Q., 1962.
The ice of the Antarctic.
Scientific American, 207(3):132-146.

ice
Robin, G. de Q., 1956.
Determination of the thickness of ice shelves by seismic shooting methods. Nature 177(4508):584-586.

ice
Robin, G. de Q., 1953.
Measurements of ice thickness in Dronning Maud Land, Antarctica. Nature 171(4341):55-58, 4 textfigs.

ice
*Robinson, Edwin S., 1968.
Seismic wave propagation on a heterogeneous polar ice sheet.
J. geophys. Res., 73(2):739-753.

ice
Rodewald, M., 1953.
Der Grönlandkabeljau und die Schwankungen der Eisdrift. Fischwirtschaft 5(10):241-242.

ice
Rodewald, M., 1953.
Zunahme der Eisdrift im Ostgrönlandstrom.
Wetterlotse, Hamburg, (60):161-166.

ice
Rodhe, B., 1958.
Sea ice observing and reporting technique in the Baltic. Sec. 2. Sea ice observing and reporting techniques.
NAS-NRC, Publ., No. 598:57-68.

ice
Rodhe, B., 1956.
The Baltic ice code. Mar. Obs., 26(173):151-153.

ice
Rodhe, B., 1955.
A study of the correlation between the ice extent, the course of air temperature and the sea surface temperature in the Åland Archipelago.
Sver. Meteorol. och Hydrol. Inst. Medd., SB, 12: 143-163.

ice
Rodhe, B., 1955.
A study on the correlation between ice extent, the course of air temperature and the sea surface temperature in the Åland Archipelago.
Geografiska Ann., 37(3/4):141-163.

ice
Rodhe, B., 1952.
On the relation between air temperature and ice formation in the Baltic. Geograf. Ann., 1952 (3/4):175-202, 10 textfigs.

Medd. Sveriges Met. och Hydrologiske Inst., B, No. 10.

ice
Romanov, A.A., 1969.
Average thickness of drift ice in the eastern Antarctic. (In Russian)
Inform. Biull. Sovetsk. Antarct. Exped. 73: 22-26.
Translation: Scripta Tecnica for AGU 7(4):294-296

ice
Romanovsky, V., 1943
Oscillations de rivage et bathymétre dans la région sud de la Baie du Roi (Spitzberg). Bull. Soc. Géol. de France, No.1-2-3,:81-90, 4 figs., 1 pl.

ice
Romanovsky, V., 1945
Résultats des sondages effectués dans la baie du Roi. Bull. Inst. Ocean., Monaco, No.877, 9 pp., 4 figs.

ice
Rouch, J., 1946
Traité d'Océanographie physique. L'eau de mer. Payot, Paris, 349 pp., 150 text figs.

ice
Ryder, C.H., 1889
Undersøgelse af Grønlands Vestkyst fra 72° til 74° 35 NB.
Medd. Grønland, 8, 203-270.

ice
Rzheplinskii, G.V., 1959.
Oceanic swell in the Antarctic. (In Russian).
Trudy Gosud. Okeanogr. Inst., 48:5-85.

S

ice
Sabinin, K.D., 1963
On the problem of the influence of the snow cover and water heat upon the accumulation of ice. (In Russian).
Okeanologiia, Akad. Nauk, SSSR, 3(1):23-29.

ice
Sala, I., 1957.
Einige Messungen der Eisfestigkeit.
Geophys., Helsinki, 5(3):150-152.

ice
Samoilenko, V. S., 1964.
Is there a possibility for the restroation of the natural ice cover in the Arctic Basin in case of its destruction? (In Russian).
Okeanologiia, Akad. Nauk, SSSR, 4(6):997-1007.

ice
Samoilovich, R.L., 1934.
Vo l'dakh Arktiki. [On the ices of the Arctic region; the expedition "Krasin" in the summer of 1928] 3rd ed., 339 pp., illus., maps.

LC G700 1928.S3 1934 50-47624.

ice
Sandford, K.S., 1955.
Tabular icebergs between Spitzbergen and Franz Josef Land. Geogr. J., 121(3):164-170.

ice

Sandford, K.S., 1955.
Tabular icebergs between Spitsbergen and Frans Joseph Land. Geogr. J. 121(2):164-170.

Ice (rivers)

Santema, P., and J. N. Svasek, 1961.
De invloed van de afdamming van de zeegaten op het ijsbezwaar op de Zeeuwse en Zuidhollandse stromen.
Rapp. Deltacommissie, Bijdragen, (5):295-326.

ice

Savelyev, B. A., 1960.
[On some peculiarities of the ice cover structure of the marginal zone in the Mirnyy area.]
Inform. Bull., Soviet Antarctic Exped., 18:10-11

ice

Savel'ev, B.A., 1958 (1960).
[Peculiarities of the ice-thawing processes of the ice cover and in frozen ground.]
Problemy Severa, (1):

Translation in:
Problems of the North, (1):160-167.

ice

Savel'ev, B.A., 1958
[Study of ice in the area of the drifting station N.P.-4 during the period of melting and ice destruction in 1955.]
Problemy Severa, Akad. Nauk, SSSR, 2:47-49.

USN-HO-TRANS- 9
M. Slessers
H.O. 15868
P.O. 61153. 1960

ice

Savel'yev, B. A., 1954.
Procedural notes on determination of content of solid, liquid and gaseous phases in saline ice. Materialy po Laboratornym Issle. Merzlykh Grontov, Sbornik, No. 2:176-192.

ice

Sawada, T., 1960.
[A study of the relationships between the fishery and the drif-ice (sic!) on the Okhotsk Sea.]
Bull. Hakodate Mar. Obs., 7(7):1005-1009.

ice

Sawada, T., 1960.
On the transition of ice-limit and ice thickness for the early drift-ice season on the Okhotsk Sea.
Bull. Hakodate Mar. Obs., 7(6):250-258.

Reprinted from:
J. Meteorol. Soc., Japan, 1960, (2) 38(5):250-258.

ice

Sawada, T., 1958.
A practical method for forecasting ice-area movement. J. Meteorol. Soc., Japan, (2)36(1):32-39.

ice

Sawada, T., 1957.
A study on the statistic forecast of summer temperature over the northern Japan using the amount of sea-ice over Okhotsk Sea. (1).
J. Meteorol. Soc., Japan (2)35(1):60-66.

ice

Sawada, T., 1956.
[Sea ice on the coast of Hokkaido in 1953.]
Bull. Hakodate Mar. Obs., No. 3:107-115.
39-47.

ice

Schaefer, I., 1951.
Bemerkungen zur Nomenklatur der Eiszeitforschung.
Petermans Geogr. Mitt. 95(1):26-31.

ice

Schaeffer, V.J., 1950.
The formation of frazil and anchor ice in cold water. Trans. Am. Geophys. Union 31(6):885-893, 7 textfigs.

ice

*Schell, I.I., 1970.
Arctic ice and sea temperature anomalies in the northeastern North Atlantic and their significance for seasonal foreshadowing locally and to the eastward.
Mon.Weath.Rev.U.S.Dept Comm. 98(11):833-850.

ice

Schell, I.I., 1964
Interrelations of the ice off northern Japan and the weather.
J. Meteorol. Soc., Japan, (2), 42(3):174-185.

ice

Schell, Irving I., 1962.
On the iceberg severity off Newfoundland and its prediction.
J. Glaciol., 4(32):161-172.

ice

Schell, I.I., 1962.
The ice off Iceland and the climate during the last 1200 years, approximately.
Geografiska Annaler, Stockholm, 43(3/4):254-362.

ice

Schell, I.I., 1956.
Interrelationships of Arctic ice with the atmosphere and the ocean in the North Atlantic-Arctic and adjacent areas. J. Met. 13(1):46-58.

ice

Schell, I.I., 1955.
Arctic ice and weather relationships.
Q.J.R. Met. Soc. 81(347):96-97.

ice

Schell, I.I., 1952.
The problem of the iceberg population in Baffin Bay and Davis Strait and advance estimate of the Berg count off Newfoundland. J. Glaciol. 2(11): 58-59.

ice

Schell, I.I., 1952.
On the role of ice off Iceland in the decadal air temperatures of Iceland and other areas.
J. du Cons. 18:11-36, 1 textfig.

ice

Schell, I.I., 1952.
Stability and mutual compensation of relationships with the iceberg severity off Newfoundland
Trans. Amer. Geophys. Union 33(1):27-31.

ice

Schell, I.I., 1950.
Further on foreshadowing the severity of the iceberg season south of Newfoundland. (Abstract).
Trans. Am. Geophys. Union 31(2):332.

ice

Schell, I.I., 1940
Foreshadowing the severity of the iceberg season south of Newfoundland. Bull. Am. Meteorol. Soc. 1940 Publ. Blue Hill Obs. No.1, 4 pp.

ice

Scherhag, R., 1939.
Die Erwärmung des Polargebiets. Ann. Hydr., Jahrg. 67(2):57-67.

ice

Schneider, Otto, 1969.
La barrera de hielo de Filchner en el cincuentenario de su descubrimiento.
Contrib. Inst. Antart. argentino, 63: 44pp.

ice

Schostakowitsch, W.B., 1929.
Über den Aufgang der Flüsse in Ost-Europa.
Ann. Hydrogr., usw., Jahrg. 57(Heft V):129-134.

ice

Schule, J.J., Jr., and W. I. Wittman, 1958.
Comparative ice conditions in North American Arctic, 1953-1955, inclusive.
Trans. Amer. Geophys. Union, 39(3):409-419.

ice

Schulz, B., 1934.
Die Ergebnisse der Polarexpedition auf dem U-Boat "Nautilus". Ann. Hydr., usw., 62:147-152, Pl. 16 with 2 figs.

ice

Schwarzacher, W., 1959.
Pack-ice studies in the Arctic Ocean.
J. Geophys., 64(12):2357-2368.

ice

Schwerdtfeger, Peter, 1964.
The effect of finite heat content and thermal diffusion on the growth of a sea ice cover.
J. Glaciol., 5(39):315-324.

ice

Schwerdtfeger, Peter, 1963.
The thermal properties of sea ice.
J. Glaciology, 4(36):789-807.

ice

Schwerdtfeger, P., 1960.
Observations on estuary ice.
Canadian J. Physics, 38(10):1391-1394.

ice

Seliakov, H.I., 1950.
[Some observations on ice and the processes connected with its formation.] Dok. Akad. Nauk, USSR, 70:821-824.

ice

Serikov, M.I., 1962.
Structure of Antarctic sea ice. (In Russian).
Inform. Biull., Sovetsk. Antarkt. Exped., 39:9-12.

Translation:
Scripta Tecnica for AGU, 4(5):263-264. 1964.

Ice

Serpalov, S. T., 1961.
[Some tests on the ice in the course of work in the central Arctic.]
Probl. Arkt. i Antarkt., (8):90-91.

ice

Shamontiev, V.A., 1967.
On the influence of winter hydrological conditions on some elements of the hydrological regime in the Chuokchi Sea during its navigation (In Russian; English abstract).
Okeanologiia, Akad. Nauk, SSSR, 7(3): 450-456.

ice

Shamont'yev, V.A., 1962.
Brief description of ice conditions in Alasheyev Bight. (In Russian).
Inform Biull. Sovets. Antarkt. Exped., 37:20.

Translation
(AGU) p. 210 1964

ice

Shamont'yev, V.A., 1962.
Local displacements and breakups of shore ice in Alasheyev Bight. (In Russian).
Inform. Biull., Sovetsk. Antarkt. Exped., 42:23-26.

Translation:
Scripta Tecnica for AGU, 4(6):365-366. 1964.

ice

Sharp, R.P., 1947
Suitability of ice for aircraft landings.
Trans. Am. Geophys. Un. 28(1):111-119, 2 textfigs.

ice

Shcherbakov, D.I., 1954.
[First results of the High-Latitude Arctic Expedition of 1954.] Vestnik Akad. Nauk, SSSR (9):10-16.

ice

Sheremetevskaia, O.I., 1961.
Calculation of the distribution of water temperature and the determination of the conditions at the edge of the ice. Forecasting and computation of physical phenomena in the oceans. (In Russian) Trudy Okeanogr. Komissii, Akad. Nauk, SSSR, 11: 150-157.

ice

Shesterikov, N.P., 1959.
Brief description of shore ice in the Davis Sea. (In Russian).
Inform. Biull., Sovetsk. Antarkt. Exped., 1(5): 43-45.

Translation:
Scripta Tecnica, Inc., Elsevier Publ. Co., 239-240.

ice

Shesterikov, N.P., and V.I. Shilnikov, 1959
[Safety measurements during cargo transportation on fast ice in the region of Mirnyy.]
Info. Biull. Sovetsk. Antarkt. Exped., (7): 26-30.

ice

Shil'nikov, V.I., 1961.
[Ice concentration in the Balleny Islands area.]
Biull. Sovetsky Antarkt. Eksped., (28):27-30.

Abstr. in: Soviet Bloc Res. Geophys., Astron., Space, No. 34(1962):24

ice

Shilnikov, V. I., 1960.
[Volume and amount of icebergs in the Antarctic.]
Inform. Bull., Soviet Antarctic Exped., 21:34-37

ice

Shilnikov, V. I., 1959.
[Methodics of iceberg observations.]
Biull. Sovetsk. Antarkt. Exped., No. 5:11-14.

ice

Shishkov, Yu. A., 1961.
Ice concentrations in the southwest part of the Kara Sea and meriodional heat transfer in the atmosphere. (In Russian).
Problemy Severa, 4:131-137.

Translation, 1962:
Problems of the Arctic, 4:133-139.

ice

Shreve, R.L., 1967.
Migration of air bubbles, vapor figures, and brine pockets in ice under a temperature gradient.
J. geophys. Res., 72(16):4093-4100.

ice

Shuleikin, V.V., 1950.
[The present status of the theory of ice field drift.] Pamiati Iuliia Mikhailoviche Shokal'skogo (2):63-82.

Transl. cited in:
USFWS Spec. Sci. Rept., Fish., 227(174)

ice

Shumskiy, P.A., 1955.
Principles of structural studies of ice: petrography of fresh water ice as a method of glaciological investigation. Publ. House (Izdatel(stvo) Akad. Nauk, SSSR, 492 pp., 119 figs.

ice

Shumskij, P.A., 1955.
[Study of ices of Arctic Ocean.]
Vestnik, Akad. Nauk, SSSR (2):33.

ice

Shustov, B.S., 1934.
Ice cover of the Black Sea. (In Russian).
Zemlevedeniye, 36(4):376-395.

Translation:
U.S.N. Oceanogr. Off., TRANS-184.

ice

Shustov, B.S., 1930.
Ice cover of the Sea of Azov. (In Russian).
Izv., Assotsiatsii Nauchno-Issled., Inst. pri Fiz.-Mat. Fakul'tete MGU, 3(1-A):103-128.

Translation:
USN Oceanogr. Off., TRANS-186.

ice

Simijoki, H., 1956.
Die Eisverhältnisse in den Finnland umgebenden Meeren in den Wintern 1951-55. Havsf. Skr., 171: 23 pp.

ice

Simojoki, H., 1953.
The time for freezing and ice break-up and the number of ice days on the Finnish coast in the winters of 1934-53. Merent. Julk. No. 160:1-18, 1 textfig.

ice

Simojoki, H., 1952.
Die Eisverhältnisse in den Finnland umgebenden Meeren in den Wintern 1946-50. Merent. Julk. No. 154:1-88, 80 textfigs.

ice

Simojoki, H., 1950.
Zur Kenntnis der Abkühlung des Meerwassers und die Bedeutung des Wärmevorrates des Wassers in Bezug auf den Zeitpunkt der Vereisubg im Schärenmeer sowie am Rande der Nördlichen Ostsee.
Fennia 73(2):16 pp., 6 textfigs.

ice

Simpson, L.S., 1958.
Estimation of sea ice formation and growth. 4. Sea ice formation, growth and disintegration.
NAS-NRC, Publ., No. 598:162-166.

ice

Skov, Niels Aage, 1970.
The ice cover of the Greenland Sea: an evaluation of oceanographic and meteorological causes for year-to-year variations.
Medd. om Grønland 188(2):54pp.

ice

Skriptunov, N.A., 1960.
Some characteristics of the ice regime on the beaches at the mouth of the Volga. (In Russian).
Trudy Gosud. Okeanogr. Inst., 49:86-97.

ice

Slaucitajs, L., 1947
Ozeanographie des Rigaischen Meerbusens. Teil 1, Statik. Contrib. Baltic Univ. No.45, Pinneberg, 110 pp., 69 text figs.

ice

Smith, E.H., 1941.
U.S. Coast Guard cutter "Northland"'s ice and oceanographic observation cruise, Baffin Bay and Davis Strait, autumn of 1940. Trans. Am. Geophys. Union 3:788-792.

ice

Smith, E.H., 1940.
Recent movements of North Atlantic ice and a proposed Coast Guard experiment in the West Greenland glacier. Trans. Amer. Geophys. Union, 21(2):668-672.

ice

Sokolov, I.N., 1960.
[On the problem of resistant coefficient at the lower surface of ice cover.]
Meteorol. i Gidrol., (4):34-35.

ice

Sokolovskaya, L.J., 1962
The probability of ice formation and parameters responsible for this process. (In Russian)
Okeanologiia, Akad. Nauk, SSSR, 2(4):631-639.

ice

Somma, A., 1952.
Elementi di Meteorologia ed Oceanografia. Pt. II. Oceanografia. Casa Editrice Dott. Antonio Milani, Padova, Italia, xviii 758 pp., 322 textfigs., maps, tables.

(4500 lire)

Ice (BARENTS)

Soskin, I. M., 1960.
Annual fluctuations of hydrological characteristics of the Baltic, Barents and Caspian seas and solar activity.
Marine Hydrometeorological Forecasts and Computations.
Trudy Okeanograf. Komissii, 7:3-22.

H. O. TRANS-142 (1961).

ice

Soule, F. M., 1950.
Arctic ice drift and the Humboldt Current.
Science 112(2898):61-62.

A critique of Lalla R. Boone's "A prediction regarding the Humboldt Current" Science 110:642.

ice

Soule, F. M. and E. R. Challender, 1949.
Discussion of some of the effects of winds on ice distribution in the vicinity of the Grand Bandks and the Labrador Shelf. U.S.C.G. Bull. No. 33:59-61, Figs. 13-14.

ice

Soule, F. M., and E.R. Challender, 1949.
International ice observation and ice patrol service in the North Atlantic Ocean. Season of 1947. U.S.C.G. Bull. 33:1-61, i-v pp., 14 figs.

ice

Soule, Floyd M., Alfred P. Franceschetti, R.M. O'Hagan and V.W. Driggers, 1963.
Physical oceanography of the Grand Banks region, the Labrador Sea and Davis Strait in 1962.
U.S.C.G. Bull., No. 48:29-78;95-153.

ice

SSSR, Akademiia Nauk, 1964
Ice drift observations in the Tartary Strait. (In Russian).
Okeanologiia, Akad. Nauk, SSSR, 4(5):924-925.

ice

SSSR, Gidrologicheskoye Upravleniye, 1926.
Data on ice conditions in the seas of the USSR. (The Sea of Azov and the Black Sea). (In Russian).

Translation: (in part see "author" card)

USN Oceanogr. Off. TRANS-174 (1964).

ice

Speerschneider, C.I.H., 1931.
The state of the ice in Davis Strait, 1820-1930.
Publ. Danske Met. Inst. Medd. No. 8:53 pp., 1 fold-in.

ice
Speerschneider, C.I.H., 1927.
Om isforholdene i Danske farvande, aarene 1861-1906 (fortsaettelse af Meddelser No. 2).
Publ. Danske Met. Inst. Medd. No. 6:83 pp., 1 fig

ice
Speerschneider, C.I.H., 1915.
Om isforholdene i Danske Farvande i aeldre og nyere tid, aarene 690-1860.
Publ. Danske Met. Inst. Medd. No. 2:141 pp., 1 fold-in.

ice
Stefansson, Unstein, 1963.
Hydrographic conditions in North Icelandic waters in June 1961.
Ann. Biol., Cons. Perm. Int. Expl. Mer, 1961, 18: 17-18.

ice
Stefansson, Unnstein, 1962.
North Icelandic waters.
Rit Fiskideildar, Reykjavik, 3:269 pp.

ice
Stefansson, U., 1956.
Astand sjavar a sildveidisvaedinu nordanlands sumareid 1955. [Hydrographic conditions during the summerof 1955 on the North Icelandic herring grounds.] Fjðbrit Fiskideildar, 6:23 pp.

ice
*Strübing, Klaus, 1967.
Über Zusammenhänge zurischen der Eisführung des Ostgronlandstroms und der atmosphärischen Zirkulation über dem Nord polarmeer.
Dt. hydrogr. Z. 20(6):257-265.

ice
Strübing, Klaus, 1970.
Satellitenbild und Meereiserkundung: ein methodischer Versuch für das Baltische Meer.
Dt. hydrogr. Z. 23(5): 193-213

ice
Suda, K., 1948.
On the distribution of sea ice on the North Pacific Ocean.
Hydrogr. Bull., n.s., 8:33-42.

ice
Susuki, Y., 1955.
Observations of ice crystals formed on the sea surface. Low Temp. Sci., A, 14:151-154. (not seen)

ice
Sverdrup, H.U., 1957.
The stress of the wind on the ice of the Polar Sea. Norsk Polarinst. Skr., 111:11 pp.

ice
Swithinbank, C.W.M., 1958.
An ice atlas of the North American Arctic. Sec. 1. Distribution and character of sea ice.
NAS-NRC, Publ., No. 598:22-28.

ice
Swithinbank, C., 1955.
Ice shelves. Geogr. J. 121(1):64-76.

T

ice
Tabata, T., 1958.
On the formation and growth of sea ice, especially on the okhotsk Sea (and discussion, p. 180).
4. Sea ice formation, growth and disintegration.
NAS-NRC, Publ., No. 598:169-179.

ice
Tabata, T., 1955.
A measurement of visco-elastic constants of sea ice. J. Ocean. Soc., Japan, 11(4):185-189.

ice
Tabata, T., 1955.
[A measurement of the visco-elastic constants of sea-ice.] Low Temp. Sci., A, 14:25-32.

ice
Tabata, T., 1953.
[On the prediction as to the last date of disappearance of drift ice at Abashiri in the southern Okhotsk coast of Hokkaido.]
Inst. Low Temp. Sci. 9:149-157.

ice
Tabata, Tadashi, Masaaki Aota, Mosayuki Oi and Masao Ishikawa, 1969.
The distribution of the pack ice field off Hokkaido Island with the sea ice radar net work, February and March, 1969. (In Japanese)
Low Temp. Sci. (A) 27 (44): 23-38.

ice
Tabata, Tadashi, and Kazuo Fujino, 1964.
Studies on mechanical properties of sea ice Vii.
Measurement of flexural strength in situ. (In Japanese; English summary).
Low temperature Sci., Hokkaido Univ., (a), 22: 147-154.

ice
Tabuteau, F., 1956.
Terre Adélie 1950-1951. Observations sur la glace de mer.
Expéd. Polaires Franç., Missions Paul-Emile Victor Expéd. Antarctiques, Res. Tech., n.s., 2(4):1-40, (multilithed).

ice
Takahashi, Tsutomu, 1969.
Electric potential of liquid water on an ice surface. J. atmos. Sci., 26(6): 1253-1258.
Also in: Coll. Pap. Sci. Atmos. Hydrosph., Water Res. Lab., Nagoya Univ., 7(1969).

ice
Tamura, T., 1951.
Observation of the plankton in drift ice. Bull. Fac. Fish., Hokkaido Univ., 1(3/4):134-138.

ice
Tabuteau, Francois, 1956.
Terre Adelie 1950-1960, Observations sur la glace de mer.
Res. Tech., Exped. Antarct., Expeds. Pol. Franc., II(4):1-40, multilithed.

ice
Tarbeev, Iu. V., 1965.
Determination of the stress and strength of shore ice to bending caused by variations in water level. (In Russian).
Trudy, Gosudarst. Okeanogr. Inst., No. 86:124-143

ice
Täubert, H., 1955. (translator)
Sowjetgeographie. Wissenschaftliches Arktis-Forschungsinstitut der UdSSR: Über neue sowjetische Forschungen und Entdeckung in der Zentralarktis. Petermanns Geogr. Mitt., 99(1):70-77, 5 textfigs.

from: Nachr. Akad. Wiss., UdSSR, Geogr. Ser., 1954(5):

ice
Taylor, R.J.F., 1957.
An unusual record of three species of whale being restricted to pools in Antarctic sea-ice.
Proc. Zool. Soc., London, 129(3):325-330, 7 pls.

Ice
Tchikovsky, S. S., 1961.
[The influence of temperature on the resistance of marine ice.]
Probl. Arkt. i Antarkt., (8):93-96.

ice
Thiel, Edward, 1961.
The amount of ice on Planet Earth. (Abstract).
Tenth Pacific Sci. Congr., Honolulu, 21 Aug.-6 Sept., 1961, Abstracts of Symposium Papers, 319-320.

ice
Thiel, Edward, and Ned A. Ostenso, 1961.
Seismic studies on Antarctic ice shelves.
Geophysics, 26(6):706-715.

Thomas, Charles W.

ice
*Thomas, Charles W., 1968.
Antarctic ocean-floor fossils: Their environments and possible significance as indicators of ice conditions.
Pacif. Sci., 22(1):45-51.

ice
Thomas, Charles W., 1960
Late Pleistocene and Recent limits of the Ross Ice Shelf.
J. Geophys. Res., 65(6):1789-1792.

ice
Thomas, C.W., 1958.
The economics of surface transportation of sea ice
7. Sea ice operations.
NAS-NRC, Publ., No. 598:267-269.

Thomsen, Helge

ice
Thomsen, H., 1956.
Hydrography of the Kattegat, Danish Waters.
Ann. Biol., Cons. Perm. Int. Expl. Mer, 11:51.

ice
Thomsen, H., 1950.
Isforholdene i de Arktiske Have. 1947. Publ. Danske Met. Inst., Årbøger:10 pp., 4 charts.

ice
Thomsen, H., 1948.
Sea ice reports: Danish Meteorological Institute.
J. Glaciol. 1(3):140-141.

Ice
Thomsen, H., 1948
Is forholdene i de Arktiske Have. Tillaeg til Nautisk-Meteorologisk Aarbog 1946, 12 pp., 5 charts.

sea-ice
Thomsen, H., 1948
The annual reports on the Arctic sea-ice issued by the Danihs Meteorological Institute.
J. Glacrology 1:140-141.

ice
Thomsen, H., 1946.
The state of the ice in the Arctic seas. Publ. Danske Met. Inst., 1946:1-12, Pl. 5.

ice
Thomsen, H., 1934.
Danish hydrographical investigations in the Denmark Strait and the Irminger Sea during the years 1931, 1932, 1933. Rapp. Proc. Verb., Cons. Perm. Int. Expl. Mer 86:1-14, 15 textfigs.

ice
Thomsen, H., and M.V.L. Lorck, 1956.
Isforholdene i de Arktiske Have, 1953.
Nautisk-Meteorologisk Årbog, 1953, Tillaeg, 27 Pl 5 charts

ice
Thomsen, H., and M.V.L. Lorck, 1953.
Isforholdene i de arktiske have, 1950. Publ., Danske Met. Inst., Aarbøger, Tillaeg til Nautisk-Meteorolog. Aarbog, 1950:24 pp., 5 charts.

ice

Thoroddsen, Th., 1885
[Den Grönländska Drifisen vid Island.]
Ymer Tidskrift, 147-156.
USN HO Transl. S-38
M.O. 16104SER
P.O.:13377, 1961
Translator: M. Slessers

ice

Timofeiv, B.T., 1962
The influence of the deep Atlantic waters upon the formative processes and ice melting in the Kara and Laptev seas.
Okeanologiia, Akad. Nauk, SSSR, 2(2):219-225.

ice

Tomashunas, B.Y., 1959
[Ice observations.]
Arktich. i Antarktich. Nauchno-Issled. Inst., Mezhd. Geofiz. God, Sovetsk. Antarkt. Eksped., 5: 154-158.

ice

Toporkov, L.G., 1961.
Can the Arctic ice cap be removed? (In Russian).
Priroda, 50(11):93-97.

Translation:
OTS or SLA, $1.10.

ice

Treshnikov, A.F., 1958
Some features of the ice regime of the Davis Sea.
Inform. Biull. Sovetsk. Antarkt. Exped., (1): 61-64.

ice

Tressler, Willis L., 1960.
Oceanographic observations at IGY Wilkes station Antarctica.
Trans., Amer. Geophys. Union, 41(1) :98-104.

ice

Tsukernik, V.B., 1963.
Magnetometer survey of the west ice shelf. (In Russian).
Sovetsk. Antarkt. Eksped., Inform. Biull., 43:45-

Translation:
Scripta Tecnica, Inc. for AGU, 5(1):22-23. 1965.

ICE

Tsurikov, V.L., 1963
The analysis of the accumulation of sea-born ice beneath the snow cover. (In Russian).
Okeanologiia, Akad. Nauk, SSSR, 3(3):459-469

ice

Tunnell, G.A., 1961.
Incidence of ice in the approaches to North America during the decade 1946-1955.
Marine Observer, 31(192):78-85.

ice

Tuori, H., 1952.
Sounding through sea and lake ice in Finland.
Polar Record 6:336-339.

U

ice

Untersteiner, N., 1964.
A nomograph for determining heat s torage in sea ice.
J. Glaciol., 5(39):352.

ice

Untersteiner, N., 1964
Calculations of temperature regime and heat budget of sea ice in the Central Arctic.
J. Geophys. Res., 69(22):4755-4766.

ice

Untersteiner, N., 1961
On the mass and heat budget of Arctic sea ice.
Arch. f. Meteor., Geoph. u. Bioklim., 12(2): 151-182.

ice

Untersteiner, N., and F.I. Badgley, 1958.
Preliminary results of thermal budget studies on Arctic pack ice during summer and autumn. Sec. 3. Physics and mechanics of sea ice.
NAS-NRC, Publ., No. 598:85-92.

ice

Uralov, N.S., 1961.
Some peculiarities of the seasonal changes and those of many years standing at the outer boundaries of the ice in the Barents Sea. (In Russian).
Trudy Gosud. Okeanogr. Inst., 64:39-77.

ice

Uralov, N.S., 1959
[The influence of the Nordkapp (North Cape) Current on the iciness of the Barents Sea.]
Trudy Gosud. Okeanogr. Inst., Leningrad, 37: 14-33.

USN-HO-TRANS-102(No. 2)
M. Slessers
M.O. 16104
P.O. 13377. 1961

ice

United States Coast Guard 1971
Report of the International Ice Patrol Service in the North Atlantic Ocean. Season of 1970.
Bull. No. 56: 72 pp.

ice

United States, U.S. Coast Guard, 1965.
Oceanographic cruise USCGC Northwind: Chukchi, East Siberian and Laptev seas, August-September 1963.
U.S. C.G. Oceanogr. Rept., No. 6(CG373-6):69 pp.

ice

U.S.A. U.S. Coast Guard, 1964.
Oceanographic cruise, USCGC Northwind, Bering and Chukchi Sea, July-Sept. 1962.
U.S.C.G. Oceanogr. Rept., No. 1(CG373-1):104 pp.

ice

U.S. National Academy of Sciences - National Research Council, 1958.
Proceedings of the conference conducted by the Division of Earth Sciences, supported by the Office of Naval Research. Arctic Sea Ice, Conference at Easton Maryland, February 24-27, 1958.
NAS-NRC, Publ., No. 598:271 pp.

ice

U.S. Navy Hydrographic Office, 1959.
Climatological and oceanographic atlas for mariners. Vol. 1.
North Atlantic Ocean., 182 charts

ice

U.S. Navy Hydrographic Office, 1959
Long-range ice outlook, eastern Arctic(1959).
H.O. Misc. 15,869-18:24 pp., 15 figs.

ice

U.S.N. Hydrographic Office, 1958
Oceanographic atlas of the Polar seas. Pt. II. Arctic. H.O. Publ. No. 705: 139 pp., charts.

ice

USN Hydrographic Office, 1957.
Oceanographic atlas of the Polar seas. 1. Antarctic. H.O. Pub., No. 705:70 pp.

ice

U.S. Navy Hydrographic Office, 1957.
Operation Deep Freeze II, 1956-1957. Oceanographic survey results. H.O. Tech. Rept., 29: 155 pp. (multilithed).

ice

U. S. Navy Hydrographic Office, 1946.
Ice atlas of the Northern Hemisphere. H. O. 550

ice

U.S. Navy Hydrographic Office, 1945.
Bibliography on ice of the northern hemisphere.
H.O. Pub. No. 240:1-179, i-xii.

ice

U.S. Naval Oceanographic Office, Oceanographic Prediction Division, Forecasting Branch, 1965.
Report of the Arctic ice observing and forecasting program - 1962.
Spec. Publ., SP-70(62):172 pp.

ice

United States, Naval Oceanographic Office, 1963.
Long-range ice outlook, eastern Arctic, 19 pp.

ice

United States, Weather Bureau and Hydrographic Office, 1961
Climatological and oceanographic atlas for Mariners. Vol. II. North Pacific Ocean.
Unnumbered pages.

ice

*Untersteiner, N., 1967.
Natural desalination and equilibrium salinity profile of old ice.
In: Physics of snow and ice, H. Oura, editor, Inst. Low Temp.Sci., Hokkaido Univ., 569-577.

ice

Untersteiner, N., and F. I. Badgley, 1965.
The roughness parameters of sea ice.
J. geophys. Res., 70(18):4573-4577.

ice

Ussachev, P.I., 1949.
[Microflora of polar ice.] Trudy Inst. Okeanol., 3:216-259.

ice

Ussacheu, P., 1946.
Biological indicators of the origin of ice-floes in the Kara Sea and Sea of Brothers Leptev and the Straits of Frans-Josef Land Archipelago.
Trudy Okean. Inst. 1:149

indicators

Ussacheu, P., 1946.
Biological indicators of the origin of ice-flows in the Kara Sea and Sea of Brothers Leptev and the Straits of Frans-Josef Land Archipelago.
Trudy Okean.Inst. 1:149.

Ice

USSR Gosudarstvennyi Okeanograficheskii Institut Lingradskoe Otdelenie, 1960.
Album of aerial photographs of ice formation on seas, 1-224 pp.

H.O. TRANS.-124 (1961). (Figures omitted).

ice

Uusitalo, S., 1957.
Beobachtungen mit Bezug auf das Meereseis.
Geophys., Helsinki, 5(3):139-148.

V

ice

Valdes, Alberto J., y Rolando Nawratil 1961.
Fenomeno glaciologico en el Mar de Bellingshausen durante la campaña Antartica.
Contrnes Inst. Antartic. Argent. 59: 26 pp.

ice
Valdez, Alberto T., and Rolando Nawratil, 1961.
Ice phenomenon in the Bellingshausen Sea in the 1959-1960 season. (Abstract).
Tenth Pacific Sci. Congr., Honolulu, 21 Aug.-6 Sept., 1961, Abstracts of Symposium Papers, 320.

ice
Valeur, Hans H., 1965.
Short-term variations of Polar-ice: Selected examples off south and south east Greenland.
Geogr. Tidsskr., 64(2):220-233.

ice
Varlet, F., 1946
Température de congélation de l'eau de la Méditerranée. Bull. de l'Inst. Oceanogr., Monaco, No.905

ice
Vaux, D., 1953.
Hydrographical conditions in the southern North Sea during the cold winter of 1946-1947.
J. du Cons. 19(2):127-149, 18 textfigs.

ice
Veynbert, B.P., V. Ya. Al'tberg and others, 1940.
[Ice; properties, origin and disappearance of ice (excerpts).]
Led; Svoystva, Vozniknoveniye i Ischeznoveniye L'da, Moscow, 41-45; 177-195; 405-408.

LC and SLA, mi $3.30, ph $7.80

ice
Vinje, Torgny E., 1970.
Sea ice observations.
Årbok, Norsk Polarinst., 1969: 132-138

ice
Vinje, Torgny E., 1970.
Some observations of the ice drift in the East Greenland Current.
Årbok, Norsk Polarinst., 1968: 75-78

ice
Vinje, Torgny E., 1970.
Sea ice observations in 1968.
Årbok Norsk Polarinst. 1968: 95-100.

ice
Vinje, Torgny E., 1969.
The ice condition in Svalbard in 1967.
Årbok, Norsk Polarinst., 1967: 194-196.

ice
Vize, V. Iu.,
[Ice in the Barents Sea and the air temperature in Europe.] (German summary).

pp. 1-30.

Reprint in MBL library.

ice
Vize, V. Iu., ?
[Observations on our knowledge of ice movement in the White Sea.] (German summary).

pp. 113-134.

In MBL library

ice
Vize, V. Iu.,
[On the ice drift in the north Arctic basin.] (German summary).

pp. 327-336.

Reprint in MBL library

ice
Vize, V. Iu.,
[The distribution of ice in the Kara Sea.] (German summary).

pp. 235-242.

Reprint in the MBL library

ice
Vize, V. Iu., ?
[The voyage of the steamer "Stavropol" to the mouth of the Kolyma, 1924-1925.] (German summary)

pp. 135-142.

Reprint in MBL library

ice
Vize, V. Iu., ?
[Über die langfristige Vorhersage der Zeit der Enteisung des Schlundes des Weissen Meeres.]

pp 1-5.

reprint in MBL library

ice
Vize, V. Iu., ?
[Ueber die Möglichkeit einer Vorhersage des Eiszustandes im Barents-Meer.]
Russuskogo Gidrol. Inst., Kazanskaia, gl. No. 35 :1-45.

reprint in MBL library

ice
Vize, V. Iu., 1930.
[Icebergs on the north European coast in 1929.] (German summary).
Izv. Gosud. Gidrol. Inst., No. 29:77-84.

ice
*Volkov, N.A., and Z.M. Gudkovich, 1967.
The state and prospects of the methods for long-term ice forecasts in Arctic seas. (In Russian; English abstract).
Okeanologiia, Akad. Nauk, SSSR, 7(5):792-800.

ice
Vowinckel, Eberhard, 1964.
Ice transport in the East Greenland Current and its causes.
Arctic, 17(2):111-119.

ice
Voznesenskiy, P.I., 1925.
Organization of icebreaker service in the Black and Azov seas. (In Russian).
Ledokol' Noye Delo, No. 5, Ch. II:27-51.

Translation:
U.S. Naval Oceanogr. Off., by M. Slessers, TRANS - 170:25 pp.

ice
Vtjurin, B. I., 1959.

Structure of the one-year fast ice in East Antarctica.
Inform. Biull. Sovetsk. Antarkt. Exped., (4):55-60.

ice
Vtyurin, B.I., 1958.
The study of ice shores and icebergs in Antarctica. (In Russian).
Priroda, (4):124-

W

ice
Wade, F.A., 1945.
The physical aspects of the Ross Shelf ice.
Proc. Am. Phil. Soc. 89(1):160-173.

ice
Walmsley, J.L., 1966.
Ice cover and surface heat fluxes in Baffin Bay.
Mar. Sci. Center, McGill Univ., Manuscript Rept., No. 2:94 pp. (multilithed).

ice
Wandel, C.F., 1894
En Fremstilling af vort Kjendskab til Grønlands. Østkystsamt de med Skonnerten "Ingolf" i 1879 foretagne Undersøgelser i Danmarksstraedet Medd. om Grønland 6, (1):1-32.

ice
Watanabe, Kantaro 1971.
On an understanding of the multiple structure in radiative temperature patterns over current-rips and of the low skin temperature of slicks.
Umi to Sora 46(2): 57-65

ice
Watanabe, Kantaro 1967.
Summary of drift ice in the Okhotsk Sea.
Phys. Snow Ice Inst. Low Temp. Sci. 667-686.
Also in: Bull. Kobe mar. Obs. 180(6)(1968).

ice
Watanabe, Kantaro, 1964.
Summary of ice states in the Okhotsk Sea.
In: Studies in Oceanography dedicated to Professor Hidaka in commemoration of his sixtieth birthday, 265-273.

ice
Watanabe, K., 1963
On an estimation of the origin and drifting speed of ice appearing off the coast of Hokkaido - Study on sea ice in the Okhotsk Sea III.
Oceanograph. Mag., Japan Meteorol. Agency, 14(2):101-116.

ice
Watanabe, Kantaro, 1963.
On the reinforcement of the East Sakhalin Current preceding to the sea ice season off the coast of Hokkaido. Study on sea ice in the Okhotsk Sea. (IV)
Oceanogr. Mag., (Tokyo) 14(2):117-130.

Also in:
Bull. Hakodate Mar. Obs., (10).

ice
Watanabe, Kantaro, 1962
Summary of sea ice off the coast of Hokkaido.
Oceanogr. Mag., Japan Meteorol. Agency, 14(1): 15-28.

ice
Watanabe, Kantaro, 1961
On the theory and technique of an easy method of wide range photogrammetry for the observations of sea ice distribution.
Oceanogr. Mag., Tokyo, 12(2): 77-122.

ice
Watanabe, K., 1960.
[Summary of the aerial sea ice reconnaissance and pack-ice conditions in the Okhotsk Sea.]
Bull. Hakodate Mar. Obs., 7(1):14-21.

ice
Watanabe, K., 1959.
On the first appearance of drift-ice in every winter and the seasonal variation of oceanographical conditions in the southwestern region of the Okhotsk Sea.
Bull. Hakodate Mar. Obs., 6(7):12 pp.

Ice
Watanabe, K., S. Kajihara and M. Akagawa
1966.

Report of sea ice investigation in the Okhotsk Sea by the patrol ship "Soya" in March 1963. (In Japanese; English abstract).

Bull. Hakodate Mar. Obs. No. 12:79-92.

ice
Wattenberg, H., 1949.
Die Salzgehaltsverteilung in der Kieler Bucht und ihre Abhängigkeit von Strom- und Wetterlage. Kieler Meeresforschungen 6:17-30, 10 textfigs.

ice
Weeks, W.F., 1959
United States sea ice physics project. 1954-59.
Polar Record, 9(63):553-555.

ice
Weeks, W.F., and D.L. Anderson, 1958.
Sea ice thrust structure. J. Glaciol., 3(23): 173-175.

ice
Weeks, Wilford F., and Owen S. Lee, 1958.
Observations on the physical properties of sea ice at Hopedale, Labrador.
Arctic, 11(3):135-156.

ice
*Weeks, W.F., and Gary Lofgren, 1967.
The effective solute distribution coefficient during the freezing of NaCL solution.
In: Physics of snow and ice, H. Oura, editor, Inst. Low Temp. Sci., Hokkaido Univ., 579-597.

ice
Weertman, J., 1957.
Deformation of floating ice shelves. J. Glaciol. 3:38-42.

ice
Weitz, M., and J.B. Keller, 1950.
Reflection of water waves from floating ice in water of finite depth. Comm. Pure Applied Math. 3:305-318.

ice
*Weller, G.E., 1968.
The heat budget and heat transfer processes in Antarctic Plateau ice and sea ice.
Publ. ANARE sci. Repts (A) 102:155 pp.

ice
Wellman, H.W., and A.T. Wilson, 1963.
Salts on sea ice in McMurdo Sound, Antarctica.
Nature, 200(4905):462-463.

ice
Wennink, C.J., 1965.
Ice and its effect on navigation in the Baltic Sea.
J. Inst. Navig., 18(3):336-351.

ice
Wexler, H., 1961
Ice budgets for Antarctica and changes in sea-level.
J. Glaciol., 3(29): 867-872.

ice
Wilson, A.T., 1964.
Origin of ice ages: an ice shelf theory for Pleistocene glaciation.
Nature, 201(4915):147-149.

ice
Winchester, J.W., 1954.
A study of the movement of Arctic sea ice in the Canadian Arctic in relation to meteorological, geographical and oceanographic parameters.
Bull. AMS 35(9):417-427.

ice
Winchester, J.W., and C.C. Bates, 1957.
Meteorological conditions and the associated sea ice distribution in the Chukchi Sea during the summer of 1955. J. Atmosp. Terr. Phys., Spec. Sup 323-334.

ice
Woodcock, Alfred H., 1965.
Melt patterns in ice over shallow waters.
Limnol. and Oceanogr., Redfield Vol., Suppl. to 10:R290-R297.

ice
Wordie, J.M., 1957.
Ice in the Weddell Sea. Mar. Obs. (M.O. 624) 27 (175):31-33.

ice
Wurtman, J., 1957.
Deformation of floating ice shelves. J. Glaciol., 3(21):38-41.

XYZ

ice
Yakovlev, G.N., 1958.
Thermal balance of an ice cover of the central Arctic. (In Russian).
Problemy Arktiki, (5):33-44.

Translation:
OTS or SLA, $1.60.

ice
Yakovlev, G.N., 1957.
The turbulent heat exchange between the ice cover and the air in the central Arctic. (In Russian).
Problemy Arktiki, SSSR, (2):193-204.

Translation:
LC or SLA mi $2.70, ph $4.80.

ice
Yeskin, L.I., 1964.
Ice conditions in the Lazarev-Mirny stations area during the 1962/63 navigation season. (In Russian)
Sovetsk. Antarkt. Eksped., Inform. Biull., 46:22-

Translation:
Scripta Tecnica, Inc., for AGU, 5(2):127-129. 1965.

ice
Zakharov, V.F., 1966.
The role of leads off the edge of shore ice in the hydrological and ice regime of the Laptev Sea. (In Russian; English abstract).
Okeanologiia, Akad. Nauk, SSSR, 6(6):1014-1022.

ice
Zakiev, H. Ja., and M. G. Burlachenko, 1959.
[On the dynamics of the margin of continental glaciation in the Davis Sea.]
Biull. Sovetsk. Antarkt. Exped., No. 5:23-25.

ice
Zhukovskii, G.R., 1953.
[Movement of ice, pp. 207-209 in Okeanografia, Moscow, 390 pp.]
RT 2752 in Bibl. Trans. Rus. Sci. Tech. Lit. 21.

ice
Zorina, V.A., 1965.
A method for forecasting the spring ice in the Kursk and Visla gulfs. (In Russian).
Trudy, Gosudarst. Okeanogr. Inst., No. 86:36-43.

ice
Zubenok, L.I., 1963.
The effect of the temperature anomalies in the Arctic ice cover. (In Russian).
Meteorol. i Gidrol., (6):25-30.

ice
Zubov, N.N., 1963?
Arctic ice.
U.S. Navy Electronics Laboratory, San Diego, vi 481 pp.
Translation of l'dy Arktiki. Moscow, 1945.

ice
Zubov, N. N., 1959.
Meteorol. i. Gidiol. (2):22-27.
[On the limiting thickness of the ice in the sea and on the contents.]

ice
Zubov, N.N., 1948.
[Arctic ice and the warming of the Arctic]
In: TSantre Arktiki Ch. 6-7.

Transl. cited:
USFWS Spec. Sci. Rept., Fish., 227.

ice
Zubov, N.N., 1948.
[In the center of the Arctic.] Northern Sea Route Directorate Press, Moscow, Leningrad.

VI Arctic ice
VII The warming of the Arctic.

T14R

ice
Zubov, N.N., 1945.
L'dy Arktiki. Glavsevmorput, 359 pp., Moscow.

ice
Zubov, N.N., 1938
Morskie Vody i Idy. Hydrometeorological Publishing Office, Moscow, 1938: Parts #45-60 (complete) Translation made by US Weather Information Service.

ice
Zubov, M.N., 1938.
[On the maximum thickness of sea ice of many years growth.] Meteorol. i Gidrologiia 4(4):123-131.

RT-981 Bibl. Trans. Rus. Sci. Tech. Lit. 6.

ice
Zubov, N. N., 1933.
The circumnavigation of Franz Josef Land.
Geogr. Rev. 23 (3):394-401, 2 textfigs.

ice
Zubov, N. N., and M. M. Somov, 1940.
The ice drift of the central part of the Arctic basin.
Problemy Arktiki, No. 2:51-68.

LC or SLA. translation $3.00, $6.80.
microfilm photostat

ICE ACCUMULATION (SHIPS)
Tabata, Tadashi, 1969.
Studies on the ice accumulation on ships. III Relation between the rate of ice accumulation and air, sea conditions. (In Japanese; English abstract).
Low Temp. Sci. (A):27(4):339-349

ice ages

ice ages
Donn, W.L., and M. Ewing, 1966.
A theory of ice ages, III. The theory involving polar wandering and an open polar sea is modified and given a quantitative basis.
Science, 152(3730):1706-1712.

ice ages
Ewing, M., and W.L. Donn, 1956.
A theory of ice ages. Science 123(3207):1061-1066

ice age
Hagemeister, E.F., and V.A. Obrucev, 1955.
[Eiszeit und Atlantis] Priroda 7:92-96.

ice ages
Tanner, William F., 1965.
Cause and development of an ice age.
J. Geol., 73(3):413-430.

ice ages
Weyl, Peter K., 1968.
The role of the oceans in climatic change: a theory of the ice ages. Met. Monogr. 8(30): 37-62. Also in: Coll. Reprints, Dept. Oceanogr. Univ. Oregon, 7(1968).

Icebergs

icebergs
Anon., 1953.
Un nouveau système de détection des icebergs.
Navigation, Paris, 1(3):99.

icebergs
Brodie, J.W., and E.W. Dawson 1971.
Antarctic icebergs near New Zealand.
N.Z. Jl mar. Freshwat. Res. 5(1): 80-85.

icebergs
Corkum, D.A., 1971.
Performance of formula for predicting the iceberg count off Newfoundland.
J. appl. Met. 10 (3): 605-607.

icebergs
Evteev, S.A., 1960.
Unusual iceberg. (In Russian).
Inform. Biull., Sovetsk. Antarkt. Ekspedt., 2(16)

Translation:
Scripta Tecnica, Inc., Elsevier Publ. Co., 185-186.

Russian original not in MBL

icebergs, calving of
Holdsworth, G., 1971.
Calving from Ward-Hunt Ice Shelf, 1961-1962.
Can. J. Earth Sci. 8(2): 299-305

icebergs
Holdsworth, G., 1969.
Flexure of a floating ice tongue.
J. Glaciol., 8(54): 385-397

icebergs
Karelin, D. B., 1945.
Aysbergi v polyarnom basseyne. [Icebergs in the Polar basin.] Izvest. Vses. Geogr. Obshch. 77(3): 169-171.

icebergs
Kollmeyer, Ronald C., Robert M. O'Hagan and Richard M. Morse, 1965.
Oceanography of the Grand Banks region and the Labrador Sea in 1964.
U.S.C.G. Rept., No. 10(373-10):1-24, 34-285.

Oceanogr.

icebergs
Kollmeyer, Ronald C., Thomas C. Wolford and Richard M. Morse 1966.
Oceanography of the Grand Banks region of Newfoundland in 1965.
U.S. Cst Gd Oceanogr. Rept. 11 (CG 373-11): 157 pp.

icebergs
Kulikov, N.N., 1962.
Discovery of morainic material on a fragment of an overturned iceberg. (In Russian).
Sovetsk. Antarktich. Expedits., Inform. Biull., (33):15-19.

Translation:
Scripta Tecnica, Inc., for AGU), 4(2):70-73.

ice bergs
Lapina, I. Ia., 1958.
Giant iceberg in the Antarctic. (In Russian).
Priroda, 47(8):117-

icebergs
Ledenev, V.G., 1962.
Study of iceberg discharge from the coast of Antarctica. (In Russian).
Sovetsk. Antarktich. Expedits., Inform. Biull., (35):31-34.

Translation:
(Scripta Tecnica, Inc., for AGU), 4(3):146-151.

icebergs
Loewe, F., 1949.
The formation of dome-shaped icebergs. J. Glaciol. 1(5):283.

icebergs
Maryin, L., 1946.
Eights' pioneer observation and interpretation of glacial erratics in Antarctic icebergs. Bull. G.S.A. 57:1216.

ice bergs, effect of
Moign, Annik, 1965.
Contribution à l'étude littorale et sous-marine de la Baie du Roi (Spitsberg - 79° N).
Cahiers Océanogr., C.C.O.E.C., 17(8):543-563.

icebergs
Murray, J.E., 1966.
Report of the International Ice Patrol Service in the North Atlantic Ocean. Season of 1966.
Bull. U.S. Cst Guard, 52: 1-27.

icebergs
Nazarov, V.S., 1959.
Scale for observations of icebergs. (In Russian).
Meteorol. i Gidrol., (2):54-55.

icebergs
Ozawa, Keijiro and Tsutomu Isouchi, 1968.
Observations of ice bergs and pack-ice. J. Tokyo Univ. Fish., 9(1): 49-56.

icebergs
Post, L.A., 1956.
The role of the Gulf Stream in the prediction of ice-berg distribution in the North Atlantic.
Tellus, 8(1):102-111.

icebergs
Schell, I.I., 1950.
On the distribution of icebergs in the northern hemisphere with special reference to south of Newfoundland. Int. Assoc. Hydrogr., I.U.G.G., Proc. Verb., 2:19-21.

icebergs
Scholander, P.F., W. Dansgaard, D.C. Nutt, H. de Vries, L.K. Coachman and E. Hemmingsen, 1962.
Radio-carbon age and oxygen-18 content of Greenland icebergs.
Medd. om Grønland, 165(1):1-26.

icebergs, chemistry of
Scholander, P.F., E.A. Hemmingsen, L.K. Coachman, & D.C. Nutt, 1961

Composition of gas bubbles in Greenland icebergs.
J. Glaciol., 3(29): 813-822.

icebergs
Scholander, P.F., J.W. Kanwisher and D.C. Nutt, 1956.
Gases in icebergs. Science 123(3186):104-105.

icebergs
Scholander, P.F., and D.C. Nutt, 1960.
Bubble pressure in Greenland icebergs.
J. Glaciol., 3(28):671-678.

icebergs, height of
Shil'nikov, V.I., 1968.(1967).
Accuracy of iceberg height measurements. (translation).
Info.Bull.Soviet Antarct.Exped.,66:579-581. (54-67).

icebergs
Shil'nikov, V.I., and V.I. Bardin, 1962.
Iceberg distribution at the coast of East Antarctica. (In Russian).
Sovetsk. Antarktich. Expedits., Inform. Biull., (36):19-22.

Translation:
(Scripta Tecnica, Inc., for AGU), 4(3):182-184.

icebergs
Shwede, E.E., 1966.
Icebergs of the northwest Atlantic. (In Russian English abstract).
Okeanologiia, Akad. Nauk, SSSR, 6(4):608-614.

icebergs
Atlantic, northwest

icebergs
Sissala, John
Observations of an Antarctic Ocean tabular iceberg from the Nimbus II satellite.
Nature, Lond., 224 (5226): 1285-1287.

icebergs
Zubov, N. N. 1933.
The circumnavigation of Franz Josef Land.
Geogr. Rev. 23(3):394-401, 2 textfigs.

Icebreakers

icebreakers
Fahrenholz, I.V., 1965.
A powerful new Canadian icebreaker. (In Russian).
Okeanologiia, Akad. Nauk, SSSR, 5(5):925-927.

icebreakers
Morley, J.P., 1963.
Polar ships and navigation in the Antarctic.
Brit. Antarctic Survey Bull., (2):1-25.

icebreakers
Morley, J.P., 1962.
Icebreakers, their construction and use.
Polar Record, 11(70):6-12.

icebreaker
Pinkhenson, D.,., 1962.
Makarov's project for an icebreaker for Antarctic expeditions. (In Russian).
Sovetsk. Antarktich. Expedits., Inform. Biull., (36):41-44.

Translation:
(Scripta Tecnica, Inc., for AGU) 4(3):194-195.

Ice breakout

ice breakout
Gribble, M.M., 1968.
Ice breakout, McMurdo Sound, Antarctica.
N.Z. Jl. Geol. Geophys. 11(4): 908-921

ice, brine in

ice, brine in
Richardson, Charles, and E.E. Keller, 1966.
The brine content of sea ice measures with a nuclear magnetic resonans spectrometer.
J. Glaciol., 6(43):89-100.

ice, buoyancy of

ice, buoyancy of
Nazarov, V.S., 1939.
[Buoyancy of ice.] Severnyi Morskoi Put'. 11: 62-63.

RT2760 in Bibl. Trans. Rus. Sci. Tech. Lit. 21.

ice caps

ice cap, effect of
Gough, D.I., 1970.
Did an ice cap break Gondwanaland? J. geophys. Res., 75(23): 4475-4477.

ice caps
Lamb, H.H., and M.R. Bloch 1968.
Volcanic dust, melting of ice caps and sea levels.
Palaeogr. Palaeoclimatol. Palaeoecol. 4(3): 219-226.

ice, characteristics of

ice, characteristics of
*Smirnov, V.I., 1967.
On the possibility of calculating the strength limits of sea ice cover with short-period loads. (In Russian; English abstract).
Okeanologiia, Akad. Nauk, SSSR, 7(3):428-436.

ice, chemistry of

ice, chemistry
Blinov, L.K., 1965.
Salt composition of seawater and ice. (In Russian).
Trudy, Gosudarst Okeanogr. Inst., No. 83:5-55.

ice, chemistry of
Tsurikov, V.L., 1965.
On the formation of ionic composition and salinity of sea ice. (In Russian).
Okeanologiia, Akad. Nauk, SSSR, 5(3):463-472.

ice cracks
Schmidt Ernst 1962.
Über die Ursache von Rissen in der Eiskappe des nördlichen Polarmeeres.
Arch. Met. Geophys. Bioklim. (A) 15 (3/4): 391-394

ice, abnormity distribution in
Kusunoki, K., 1955.
Observations on the horizontal and vertical distribution of sea ice. J. Ocean. Soc., Japan, 11(4):179-183.

ice crystals

ice crystals
DeVries, A.L., 1969.
Freezing resitance in fishes of the Antarctic Peninsula.
Antarctic J., U.S.A.,4(4):104-105.

ice circulation

Ice Circulation
Campbell, William J., 1965.
The wind-driven circulation of ice and water in a polar ocean.
J. Geophys. Res., 70(14):3279-3301

ice (data only)

ICE (data only)
Conseil, Permanent International pour l'Exploration de la Mer, 1968.
ICES Oceanographic Data Lists 1962(3): 27 pp.

ice (data only)
Conseil Permanent International pour l'Exploration de la Mer, 1967.
ICES Oceanographic Data Lists 1960, 10:212 pp.

ice (data only)
Conseil Permanent International pour l'Exploration de la Mer, 1967.
ICES Oceanographic data lists, 1960, No.8:227 pp. (multilithed).

ice (data only)
Conseil Permanent International pour l'Exploration de la Mer 1967.
ICES oceanographic data lists 1959 No. 9: 235pp.

ice (data only)
Conseil Permanent International pour l'Exploration de la Mer 1967.
ICES oceanographic data lists 1957: No. 9: 199pp.

ice (data only)
Conseil Permanent International pour l'Exploration de la Mer 1965.
ICES oceanographic data lists 1958, No. 7: 192pp No. 8: 286 pp.

ice (data only)
Conseil Permanent International pour l'Exploration de la Mer, 1967.
ICES Oceanographic Data lists, 1959, No. 8: 215 pp. (multilithed)

ice (data only)
Conseil Permanent International pour l'Exploration de la Mer, 1967.
ICES oceanographic data lists, 1958, No. 11: 167 pp. (multilithed).

ice (data only)
Conseil Permanent International pour l'Exploration de la Mer, 1964.
ICES oceanographic data lists, 1957, No. 3:167pp. No. 4:178pp. No. 5:255pp. No. 6:160pp. 1958, No. 2:157pp. No. 3:174pp. No. 4:241pp. 1959, No. 1:201pp.

ice (data only)
Conseil Permanent International pour l'Exploration de la Mer, 1963.
ICES Oceanogr. Data Lists, 1957(2):353 pp. (multilithed).

ice (data only)
Japan, Japan Meteorological Agency 1970?
The results of marine meteorological and oceanographical observations, July-December 1966, 40:336 pp.

Ice (data only)
Japan, Japan Meteorological Agency, 1970
The results of marine meteorological and oceanographical observations. (The results of the Japanese Expedition of Deep Sea (JEDS-11); January-June 1967 41: 332 pp.

ice deposits

ice deposits
Hodge, Paul W., Frances W. Wright and Chester C. Langway, Jr., 1967.
Studies of particles for extraterrestrial origin 5. Compositions of the interiors of spherules from Arctic and Antarctic ice deposits.
J. geophys. Res., 72(4):1404-1406.

ice detection

ice, detection of
Anon., 1954.
The detection of ice by radar. Mar. Obs. 24(165) (M.O. 579):150-162, 13 textfigs.

ice drift

ice drift
Bailey, W.B., 1957.
Oceanographic features of the Canadian Archipelago. J. Fish. Res. Bd., Canada, 14(5):731-769.

ice drift
Belyakov, L.N., 1969.
Summer desalination of the sea surface and dynamics of the drift current and ice movement in the Arctic Basin.
Okeanologiia 9(3): 424-429.
(In Russian; English abstract)

Ice drift
Browne, Irene M., 1959.
Ice drift in the Arctic Ocean.
Trans. A.G.U., 40(2):195-200.

ice drift
Burkhanov, V.F., 1957.
[Soviet Arctic research.] Priroda (5):21-30.
T265R

ice drift
Coachman, L.K., and C.A. Barnes, 1961
The contribution of Bering Sea water to the Arctic Ocean.
Arctic 14(3):146-161.

ice drift
Crary, A.P., 1960
Arctic Ice Island and ice shelf studies. Air Force Cambridge Research Center, Geophys. Res. Pap., No. 63 (Sci. Studies at Fletcher's Ice Island, T-3, 1952-1955, Vol. III): 1-37.
Taken from Arctic 11(1): and 13(1) by permission.

ice-drift
Crary, A.P., R.D. Cotell, and J. Oliver, 1952.
Geophysical studies in the Beaufort Sea, 1951.
Trans. Amer. Geophys. Union 33(2):211-216, 3 textfigs.

Ice Drift
Farmer, D., O.M. Johannessen, J.E. Keys, E.R. Pounder, and H. Serson, 1970.
Observations of ice drift from a manned drifting station in the Gulf of St. Lawrence. J. geophys. Res., 75(15): 2863-2867.

ice drift
Felsenbaum, A.I., 1957.
[Theoretical foundations for calculating the ice drift in the Central Arctic basin.]
Doklady Akad. Nauk, SSSR, 113(2):307-310.

ice drift

Gordienko, P.A., 1961

The Arctic Ocean.
Scientific American, 204(5): 88-102.

ice drift

Ishida, Tamotsu, and Nobuo Ono, 1969.
On the correction of sea ice construction to radar pattern. (In Japanese, English abstract).
Low Temp. Sci. (A) 27: 317-325.

ice, tidal drift of

*Kagan, B.A., 1967.
The tidal drift of ice. (In Russian; English abstract).
Fisika Atmosfer. Okean., Izv. Akad. Nauk, SSSR, 3(8): 881-889.

ice, drift of

Laikhtman, D.L., 1968.
Nonlinear theory of wind drift of ice. (In Russian).
Fisika Atmosfer. Okean. Izv. Akad. Nauk, SSSR, 4(4): 1220-1223.

ice drift

Mikhailova, E.N., A.I. Felsenbaum and N.B. Shapiro, 1967.
On the calculation of ice-field drifts and currents in the Arctic basin. (In Russian).
Dokl., Akad. Nauk, SSSR, 175(6): 1273-1276.

ice, drift of

Shleneva, M.V., 1964.
Wind ice drift at different outer conditions. (In Russian).
Materiali Vtoroi Konferentsii, Vzaimod. Atmosfer i Gidrosfer. v Severn. Atlant. Okean., Mezhd. Geofiz. God, Leningrad, Gidrometeorol. Inst., 136-144.

ice drift

*Strübing, Klaus, 1968.
Eisdrift im Nordpolarmeer.
Umschau, 68(21): 662-663.

ice drift

Takeda, Tadashi, Masaaki Aota, Masayuki Oi and Masao Ishikawa, 1969.
Observations on the movement of drift ice with the sea ice radar. (In Japanese; English abstract)
Low Temp. Sci. (A): 27 (44): 295-315

ice, drift of

Vorobjev, V.N., 1966.
The importance of long-period tides for studying the circulation of water and drift of ice in the outlying areas of the Arctic. (In Russian).
Materiali, Sess. Uchen. Soveta PINRO Rez. Issled., 1964, Minist. Ribn. Khoz., SSSR, Murmensk, 75-80.

ice drift

*Watanabe, Kantaro, 1967.
Summary of drift ice in the Okhotsk Sea.
In: Physics of snow and ice, H. Oura, Editor, Inst. Low Temp. Sci., Hokkaido Univ., 667-686.

ice, drift of

Watanabe, Kantaro, 1962
Drift velocities of ice measured from air and separately computed values of their wind-induced and current-induced components. Study on sea ice in the Okhotsk Sea (II).
Oceanogr. Mag., Japan Meteorol. Agency, 14(1): 29-41.

ice drift

Wüst, G., 1942.
Die morphologischen und ozeanographischen Verhältnisse des Nordpolarbeckens. Veröffentlichungen des Deutschen Wissenschaftlichen Instituts zu Kopenhagen. Reihe 1: Arktis. Herausgegeben von Prof. Dr. Phil. Hans Freibold, No. 6: 21 pp., 7 textfigs., 1 pl. (Gebrüder Borntrager, Berlin-Zehlendorf, 1942).

ice drift (data only)

Kusunoki, K., T. Minoda, K. Fujino and A. Kawamura 1967.
Data from oceanographic observations at Drift Station ARLIS II in 1964-1965
Arctic Inst. N.A. unnumbered pp. (duplicated).

ice, drift & properties

ice (drift and properties)

Burkhanov, V.F., 1956.
Across the ocean on drifting ice. National Publ. House for Geographic Lit., Moscow.

A collection of papers edited by Burkhanov.

"Achievements of Soviet geographic exploration and research in the Arctic." Translation by E.R. Hope, T253R.

ice, effect of

ice, effect of

*Bajorunas, L., and D.B. Duane, 1967.
Shifting offshore bars and harbor shoaling.
J. geophys. Res., 72(24): 6195-6205.

ice, effect of

Bianchi, V.V., and V.N. Karpovitsch, 1969.
The influence of abnormal ice-cover of the White Sea and Murman in 1966 upon birds and mammals. (In Russian; English abstract).
Zool. Zh., 48(6): 871-875

ice cover, effect of

Braarud, T., and B. Hope, 1952.
The annual phytoplankton cycle of a landlocked fjord near Bergen (Nordåsvatn). Rep. Norwegian Fish. Mar. Invest. 9(16): 26 pp., 4 textfigs.

ice, effect of

Brown, J.R., 1964.
Reverberation under Arctic ice.
J. Acoust. Soc., Amer., 36(3): 601-603.

ice, effect of

Brown, J.R., and D.W. Brown, 1966.
Reverberation under Arctic sea-ice.
J. acoust. Soc., Am., 40(2): 399-404.

ice, effect of

Brown, J.R., and A.R. Milne, 1967.
Reverberation under Arctic sea-ice.
J. acoust. Soc. Am., 42(1): 78-82

ice, effect of

Chapman, R. P. and H. D. Scott, 1966.
Backscattering strengths of sea ice.

J. acoust. Soc., Amer., 39(6): 1191-1193.

ice, effect of

Dionne, Jean-Claude, 1969.
Tidal flat erosion by ice at La Pocatière, St. Lawrence estuary.
J. sedim. Petrol., 39(3): 1174-1181.

ice, effect of

Donn, William L., and David M. Shaw, 1966.
The heat budget of an ice-free and an ice-covered Arctic Ocean.
J. geophys. Res., 71(4): 1087-1093.

ice effect of

Eaton, R.M., 1963
Airborne hydrographic surveys in the Canadian Arctic.
Int. Hydr. Rev., 40(2): 45-51.

ice, effect of

Ermachenko, I.A., 1965.
Conditions for the formation and decay of the organic matter in the ice areas of the Greenland and Barents seas. (In Russian).
Trudy vses. nauchno-issled. Inst. morsk. ryb. Khoz. Okeanogr. (VNIRO), 57: 161-171.

ice, effect of

Evteev, S.A., 1959.
Relief-forming activity of ice on the eastern shore of Antarctic. (In Russian).
Inform. Biull., Sovetsk. Antarkt. Eksped., 2(12) 17-19.

Translation:
Scripta Tecnica, Inc., Elsevier Publ. Co., 44-45.

ice, effect of

Fedosov, M.V., 1962
Investigations formation conditions of primary productivity in north-west Atlantic. (In Russian; English summary).
Sovetskie Ribochoz. Issledov. Severo-Zapad. Atlant. Okeana, VNIRO-PINRO, 125-135.

ice, effect of

Felsenbaum, A.I., 1965.
The dependence of the wind coefficients in the Arctic basin on wind velocity and ice thickness. (In Russian).
Doklady, Akad. Nauk, SSSR, 164(3): 556-558.

ice, effect of

Finkel'stein, M.I., V.G. Glooshner, A.N. Petrov and V. Ya. Ivashchenko 1970.
On the anistrophy of radio wave attenuation in sea ice. (In Russian).
Fizika Atmosfer. Okean. Izv. Akad. Nauk SSSR 6(3): 311-313.

ice, effect of

Ganjkov, A.A., and O.N. Kiselev, 1964.
The use of echo sounder when fishing under ice. (In Russian).
Materialy, Ribochoz. Issled. Severn. Basseina, Gosudarst. Kom. Rib. Choz., SNCH, SSSR, Poliarn. Nauchno-Issled. i Proektn. Inst. Morsk. Rib. Choz i Okeanogr., N.M. Knipovich, (PINRO), 2: 148-151.

ice, effect of

Ganton, J.H., and A.R. Milne, 1965.
Temperature- and wind-dependent ambient noise under midwinter pack ice.
J. acoust. Soc., Am., 38(3): 406-411.

ice, effect of

Green, Robert E., Jr., 1964.
Reflection of an underwater shock wave from an overlying ice layer.
J. Acoust. Soc., Amer., 36(3): 603.

ice, effect of

Grigoriev, N.F., 1963
The role of the cryogenic factors in the dynamics of the shore area of Yakutia. (In Russian).
Okeanologiia, Akad. Nauk, SSSR, 3(3): 477-481.

ice, effect of
Hashimoto, Tomiju, Yoshinobu Maniwa, Osamu Omoto and Hidekuni Noda, 1963
Echo sounding of frozen lake from surface of ice.
Int. Hydrog. Rev., 40(2):31-40.

ice, effect of
Johannessen, Ola M., 1970.
Note on some vertical current profiles below ice floes in the Gulf of St. Lawrence and near the North Pole. J. geophys. Res., 75(15): 2857-2861.

ice, effect of
Kagan, B.A., 1967.
A vertical structure model of the tidal flow in a homogeneous ice-bound sea. (In Russian).
Dokl. Akad. Nauk, SSSR, 175(2): 338-341.

ice effect of
Kagan, B.A., 1967.
three-dimensional model of tidal flow in ice covered sea. (In Russian; English abstract)
Fisika Atmosfer. Okean., Izv. Akad. Nauk SSSR 3(5): 526-537

ice,effect of
Kagan, B.A., 1967.
On the vertical profile of the tidal current velocity in an ice-covered sea. (In Russian; English abstract).
Fisika Atmosfer. Okean., Akad. Nauk,SSSR, 3(1):69-77.

ice, effect of
Kaleri, E.Yu., A.M. Klouga, A.N. Petrov, and M.I. Finkelstein, 1971.
On the anisotropy of retardation of radiowaves in the sea ice. (In Russian). Fiz. Atmosfer. Okean. Izv. Akad. Nauk SSSR 7(10): 1115-1116.

ice, effect of
Keller, J.B., and M. Weitz, 1953.
Reflection and transmission coefficients for waves entering or leaving an ice field.
Comm. Pure Appl. Math. 6:415-417.

ice, effect of
Kelley, J.J., 1968.
Carbon dioxide in the seawater under the Arctic ice
Nature 215(5144):562-564

ice, effect of
Kelley, J.J., Jr., and D.W. Hood, 1971.
Carbon dioxide in the surface water of the ice-covered Bering Sea.
Nature, Lond., 229(5279): 37-39

ice, effect of
Keys, J., and S.D. Smith, 1970.
Frictional resistance to a ship's passage through converging ice.
Arctic, 23(4): 284-285.

ice, effect of
Kheisin, D.E., 1969.
On the propagation of the bend-gravity waves in the ice covered sea. (In Russian). Fizika Atmosfer. Okean., Izv. Akad Nauk, SSSR, 5(12): 1334.
ice, effect of
waves, gravity

ice, effect of
Kheysin, D. Ye., 1963.
Propagation of long gravitational waves in a floe-covered sea. (In Russian).
Problemy Arktiki i Antarktiki, (14):63-66.

Abstracted in:
Soviet Bloc Res., Geophys., Astron., Space, 93: 40-41.

ice, effect of
Knight, R.J., 1971.
Distributional trends in the Recent marine sediments of Tasiujaq Cove of Ekalugad Fiord, Baffin Island, N.W.T.
Marit. Sed. 7(1): 1-18.

ice, effect of
Kolesnikov, A.G., M.A. Panteleev and V.N. Ivanov, 1965.
Experimental investigations of turbulent layer under drifting ice. (In Russian;English abstract)
Fizika Atmosferi i Okeana, Izv., Akad. Nauk,SSSR, 1(12):1310-1318.

ice, effect of
Laktionov, A.F., 1960.
The problem of the effect of ice on tidal phenomena. (In Russian).
Problemy Arktiki i Antarktiki, (5):53-58.

Translation:
U.S. Naval Oceanogr. Off., by M. Slessers, TRANS -189:11 pp.

ice, effect of
Langleben, M.P., 1969.
Attenuation of sound in sea ice, 10-500 kHz.
J. Glaciol. 8(54): 399-406

ice, effect of
Lazarenko, N.N. and S.M. Losev, 1971.
An experience of measuring near-ice wind from the aircraft during ice drift aerophotography. (In Russian; English abstract). Okeanologiia 11(3): 508-516.

ice, effect of
Legenov, A.P., 1965.
The reflection of Sverdrup waves from the ice edge. (In Russian; English abstract).
Fisika Atmos. i Okeana, 1(3):327-334.

ice, effect of
LeSchack, Leonard A., and Ricahrd A. Haubrich, 1964.
Observations of waves on an ice-covered ocean.
J. Geophys. Res., 69(18):3815-3821.

ice, effect of
Lewis, E.L. and E.R. Walker, 1970.
The water structure under a growing sea ice sheet. J. geophys. Res., 75(33): 6836-6845.

ice, effect of
Likens, Gene E., and Arthur D. Hasler, 1962.
Movements of radiosodium (Na 24) within an ice-covered lake.
Limnol. & Oceanogr., 7(1):48-56.

ice, effect of
Liken, Gene E., and Robert A. Ragotzkie, 1965.
Vertical water motions in a small ice-covered lake.
Jour. Geophys. Res.,70(10):2333-2344.

ice, effect of
Lisitsyn, A.P., 1958.
On the types of marine deposits connected with ice activity. Doklady Akad. Nauk, SSSR, 118(2): 373-376.

ice, effect of
Littlepage, Jack L., 1965.
Oceanographic investigations in McMurdo Sound, Antarctica.
In: Biology of Antarctic seas, II.
Antarctic Res. Ser., Am. Geophys. Union, 5:1-37.

ice, effect of
Marlowe, J.I., 1966.
Mineralogy as an indicator of long-term current fluctuations in Baffin Bay.
Canadian J. Earth Sci., 3(2):191-202.

ice, effect of
Meade, Robert H. 1971.
The coastal environment of New England
New England River Basins Commission, 47 pp.

ice, effect of
Meguro, Hiroshi, Kuniyuki Ito and Hiroshi Fukushima, 1966.
Diatoms and the ecological conditions of their growth in sea ice in the Arctic Ocean.
Science, 152(3725):1089-1090.

ice, effect of
Meixner, R. 1971.
Wachstum und Ertrag von Mytilus edulis bei Flosskultur in der Flensburger Förde.
Arch. Fischereiwiss. 22(1): 41-50.

ice,effect of
Mellen,R.H., 1966.
Underwater acoustic scattering from Arctic ice.
J. acoust.Soc., Am., 40(5):1200-1202.

ice, effect of
Mertins, O., and H. Kruhl, 1953.
Windabschwächung und Nebelbildung am Eisrand.
Der Seewart 14(5):4-7, 4 textfigs.

ice, effect of
Meyerhoff, A.A. and Curt Teichert, 1971.
Discussion of paper by D.I. Gough, 'Did an ice cap break Gonwanaland?'. J. geophys. Res. 76(17): 4038-4044.

ice, effect of
Milne, A. R., 1966.
Statistical description of noise under shore-fast sea ice in winter.
J. acoust. Soc., Amer., 39(6):1174-1182.

ice, effect of
Milne, A.R.,1967.
Sound propagation and ambient noise under sea ice.
In: Underwater Acoustics, V.M. Albers,editor, Plenum Press, 2: 103-138.

ice effect of
Milne, A.R., 1964.
Underwater backscattering strengths of Arctic pack ice.
J. Acoust. Soc., Amer., 36(8):1551-1556.

ice, effect of
Milne, A.R., and S.R. Clark, 1964.
Resonances in seismic noise under Arctic sea ice.
Bull. Seismol. Soc., Amer., 54(6A):1797-1809.

ice, effect of

Milne, A.R., and J.H. Ganton, 1962
Spectrum levels and distribution functions of ambient noise under Arctic Sea ice. (Abstract)
J. Acoust. Soc., Amer., 34(12):1986.

ice, effect of

Milne, A.R., J.H. Ganton and D.J. McMillin 1967
Ambient noise under sea ice and further measurements of wind and temperature dependence.
J. acoust. Soc. Am. 41 (2): 525-528.

ice, effect of

Miyake, Y., and S. Matsuo, 1963
A role of sea ice and sea water in the Antarctic on the carbon dioxide cycle in the atomsphere.
Papers in Meteorol. and Geophys., Tokyo, 14(2)
120-125.

ice, effect of

Miyake, Y., and S. Matsuo, 1963.
A role of sea ice and sea water in the Antarctic on the carbon dioxide cycle in the atmosphere.
Papers in Meteorol. and Geophys., Tokyo, 14(2):
120-125.

ice, effect of

Netherlands, Delta Committee, 1962
Final report delivered by the Advisory Committee to provide an answer to the question of what waterways-technical provisions must be made for the area devastated by the storm flood of February 1, 1953, (Delta Committee) instituted by decree of the Minister of Transport and Waterways of February 18, 1953.
100 pp.

ice, effect of

Nota, D.J.G., and D.H. Loring, 1964.
Recent depositional conditions in the St. Lawrence River and Gulf - a reconnaissance survey.
Marine Geology, 2(3):198-235.

ice, effect of

Ofuya, A.O., and A.J. Reynolds, 1967.
Laboratory simulation of waves in an ice floe.
J. geophys. Res., 72(14):3567-3583.

ice, effect of

Østman, C.J., 1950.
Om sambandet mellan köldsummer isläggning och istjocklek. Med. Sverige Met. Hydrol. Inst., A, 1:20 pp.

ice, effect

Owens E.H. and S.B. McCann 1970.
The role of ice in the Arctic beach environment with special reference to Cape Ricketts, southwest Devon Island, Northwest Territories, Canada.
Am. J. Sci. 268 (5):397-414

ice, effect of

Payne, F.A., 1967
Further measurements on the effect of ice cover on shallow-water ambient noise.
J. acoust. Soc. Am. 41 (5): 1374-1376.

ice, effect of

Payne, F.A., 1964.
Effect of ice cover on shallow-water ambient sea noise.
J. Acoust. Soc., Amer., 36(10):1943-1947.

ice, effect of

Peyton, Harold R., 1968.
Ice and marine structures. 3. The importance of design alternatives.
Ocean Industry, 3(12):51-60.

ice, effect of

Peyton, Harold R., 1968.
Ice and marine structures. 1. The magnitude of ice forces involved in design.
Ocean Industry, 3(3):40-44.

ice, effects of

Popov, E.A., 1959
[Some considerations on the "ice-crust" and "ice-fringe" (prepay effects on the seashore dynamics. Questions of studies of marine coasts.]
Trudy Okeanogr., Komissii Akad. Nauk, SSSR, 4: 109-112.

ice, effect of

Reimnitz, Erk, and Karl Fritz Bruder 1972.
River discharge into an ice-covered ocean and related sediment dispersal, Beaufort Sea, Coast of Alaska.
Bull. geol. Soc. Am. 83 (3): 861-866

ice, effect of

Rex, Robert W., 1964.
Arctic beaches, Barrow, Alaska.
In: Papers in Submarine Geology, R.L. Miller, Editor, Macmillan Co., N.Y., 384-400.

ice, effect of

Rex, R.W., 1955.
Microrelief produced by sea ice grounding in the Chukchi Sea near Barrow, Alaska. Arctic 8:
177-186.

ice, effect of

Robin, G De Q., 1963
Ocean waves and pack ice.
The Polar Record, 11(73):389-393.

ice, effect of

Robin, G. De Q., 1963
Wave propagation through fields of pack ice.
Phil. Trans., R Soc., London, (A), 255(1057):
313-339.

ice, effect of

Robinson, E., W.C. Thuman and E.J. Wiggins, 1957.
Ice fog as a problem of air ollution in the Arctic. Arctic 10(2):88-104.

ice, effect of

Romanov, A.A. 1969.
Ability of the R/V Ob' to operate in Antarctic sea ice. (In Russian)
Inform. antarkt. Biull. Sovetsk Antarkt. Exped. 75: 48-52
Translation: Am. geophys. Un. 75: 414-416.

ice, effect of

Rusanova, M.N., and V.V. Khlebovich, 1967.
On the effects of the anomalous conditions of 1965-1966 on the White Sea fauna. (In Russian; English abstract).
Okeanologiia, Akad. Nauk, SSSR, 7(1):164-167.

ice, effect of

Sawada, T., 1957.
[On the damages brought about by sea-ice in 1956.]
Bull. Hakodate Mar. Obs., No. 4:12-22.
95-104.

ice, effect of

Sawada, T., 1957.
[On the damages brought about by sea-ice in 1956 (II).] Bull. Hakodate Mar. Obs., No. 4:70-74.
6-10.

ice, effect of

Shapiro, A., and L.S. Simpson, 1953.
The effect of broken icefield on water waves.
Trans. Amer. Geophys. Union 34(1):36-42, 4 text-figs.

ice, effect of

Shirshov, P., 1937.
Seasonal changes of the phytoplankton of the Polar seas in connection with the ice regime. (In Russian).
Trudy Arktich. Antarktich. Nauchno-Issledov. 82: 42-112.

ice, effect of

Shvayshteyn, Z.I., 1957.
Laboratory for studying ice and testing models of icebreakers and ships of the icebreaker class V. Sb. (In Russian).
Problemy Arktiki, No. 3:171-178.

ice, effect of

Smith, S.D., E.G. Banke and O.M. Johannessen 1970.
Wind stress and turbulence over ice in the Gulf of St. Lawrence. J. geophys. Res., 75(15):
2803-2812.

ice, effect of

Thomas, C.W., 1966.
Vertical circulation off the Ross Ice Shelf.
Pacific Science, 20(2):239-245.

ice, effect of

Thomas, C.W., 1963
On the transfer of visible radiation through sea ice and snow.
J. Glaciol., 4(34):481-484.

ice, effects of

U.S. Hydrographic Office, 1959
Handling a vessel in ice.
J. Inst. Nav., 12(2): 141-152.

ice, effect of

*Untersteiner, Nobert, 1968.
Natural desalination and equilibrium salinity profile of perennial sea ice.
J. geophys. Res., 73(4):1251-1257.

icebergs, effect of

Urick, R.J., 1971.
The noise of melting ice bergs.
J. acoust. Soc. Am. 50 (1-2): 337-340

ice, effect of

Voronov, P.S., 1964.
Waves in the Antarctic fast ice. (In Russian).
Sovetsk. Antarkt. Eksped., Inform. Biull., 47:76

Translation:
Scripta Tecnica, Inc., for AGU, 5(3):189-190.
1965.

ice, effect of

*Weller, G., 1968.
Heat-energy transfer through a four-layer system: air, snow, sea ice, sea water.
J. geophys. Res., 73(4): 1209-1220.

ice, electrical characteristics
Bogorodskii, V.V., and G.P. Khokhlov, 1969.
Electrical characteristics of drifting Arctic ice within 100 cps — 1 mcps frequency range. (In Russian).
Dokl. Akad. Nauk SSSR, 189(6): 1230-1232

ice fields
*Legen'kov, A.P., 1970.
On the problem of thermal stresses and deformations of ice fields. (In Russian; English abstract).
Fisika Atmosfer. Okean., Izv., Akad. Nauk, SSSR, 6(8):832-839.

ice, flexural strength
Tabata, Tadashi, 1966
Studies on the mechanical properties of sea ice. IX Measurement of flexural strength in situ (3).
Low Temp. Sci., (A) 24: 259-265
(In Japanese; English abstract)

ice floes

ice floes
*Suzuki, Yoshio, 1967.
Wind - and water-drag of an ice floe.
In: Physics of snow and ice, H. Oura, editor, Inst. Low Temp. Sci., Hokkaido Univ., 661-666.

ice fluctuations
Gudkovich, Z.M., E.I. Sarukhanian and N.P. Smirnov, 1970.
A "polar tide" in the atmosphere of high latitudes and fluctuations in the ice of Arctic seas. (In Russian)
Dokl. Akad. Nauk SSSR 190(4): 954-955.

ice foot

ice foot
Rex, Robert W., 1964.
Arctic beaches, Barrow, Alaska.
In: Papers in Marine Geology, R.L. Miller, Editor, Macmillan Co., N.Y., 384-400.

ice forecasting

See also: forecasting, ice

ice, forecasting of
Antipova, E.G., 1960.
[On the methods of forecasting of ice appearance in the rivers of the Severnaya Dvina Basin.]
Meteorol. i Gidrol., (11):31-32.

ice forecasting
Armstrong, T., 1955.
Soviet work on sea-ice forecasting. Polar Record 7:302-311.

ice forecasting
Bates, C.C., H. Kaminski and A.R. Mooney, 1954.
Development of the U.S. Navy's Ice Forecasting Service, 1947-1953 and its geological implication.
Trans. N.Y. Acad. Sci., Ser. 2, 16(4):162-175, 8 textfigs.

ice forecasting
Ginzberg, B.M., 1967.
The questions of technique of long-range ice forecasting. (In Russian).
Meteorologiya Gidrol., 1967(8):82-87.

Ice Forecasting
Gordienko, P. A., and A. A. Kirillov, 1961.
[The state of the practical application of long term ice forecasts in the access to the North Sea and the status of these problems with new techniques.]
Trudy Kom. Oceanogr., Akad. Nauk, SSSR, 11:21-28.

ice, forecasting of
Karakash, E.S., 1961.
On the forecasting of ice on the western shore of the central Caspian.
Trudy Gosud. Okean. Inst., Leningrad, 61:142-152.
In Russian.

ice forecasting
Laktionov, A. F., 1945.
Itogi issledovaniy ledyanogo pokrova morey sovetskoy arktiki i ledovye prognozy. [Results of investigation of the ice cover of the seas of the Soviet Arctic and ice forecasting.] Izvest. Vses. Geogr. Obshch. 77(6):341-350.

ice forecasting
Lappo, S.D., 1938.
Let's do a better job of ice-forecasting.
Sovetskaya Arktika 1938(2):48-57.
T15R

ice, forecasting of
Palosuo, E., 1958.
Methods use in the ice service in the Baltic. 6. Sea ice prediction technique.
NAS-NRC, Publ., No. 598:239-243.

ice, forecasting
Sawada, T., 1960.
[On the sea-ice forecasting and its economic effect.]
Bull. Hakodate Mar. Obs., 7(8):85-90.

ice, forecasting
Sawada, T., 1959.
A method of forecasting the first date of the appearance of drift-ice area along the Okhotsk sea-side coast.
Bull. Hakodate Mar. Obs., 6(6):7 pp.

ice, forecasting of
Shliamin, B.A., 1959.
On forecasting ice conditions in the Caspian Sea and the possibility of its regulation. Problems of the Caspian Sea. (In Russian).
Trudy Okeanogr. Komiss., Akad. Nauk, SSSR, 5: 214-219.

ice, forecasting
Tait, A.J., 1958.
The operational concept for a sea ice reconnaissance and forecasting program conducted during Arctic operations. 7. Sea ice operations.
NAS-NRC, Publ., No. 598:265-266.

ice formation forecasting
Tapager, J.R.D., 1955.
Local environmental factors affecting ice formation in Søndre Strømfjord, Greenland.
U.S.H.O. Tech. Rept., TR-22:26 pp.

ice forecasts
Teleki, G., 1958.
The utilization of aerial photographs in sea ice forecasts. Sec. 2. Sea ice observing and reporting techniques.
NAS-NRC, Publ., No. 598:76-79.

forecasting, ice
U.S. Naval Oceanographic Office, Oceanographic Prediction Division, Forecasting Branch, 1965.
Report of the Arctic ice observing and forecasting program - 1962.
Spec. Publ., SP-70(62):172 pp.

ice, forecasting of
Veslova, L.E., 1959.
Forecast of the ice occurrence in the Caspian Sea and its significance for the different branches of the national economy. Problems of the Caspian Sea. (In Russian).
Trudy Okeanogr. Komiss., Akad. Nauk, SSSR, 5: 111-117.

ice, forecasting
Wittman, W.I., 1962.
The sea ice forecasting program.
Navigation, 8(4):267-272.

ice, forecasting of
Wittmann, W.I., 1958.
Continuity aids in short range ice forecasting. 6. Sea ice prediction technique.
NAS-NRC, Publ., No. 598:244-255.

ice formation

ice formation
Carstens, Torkild, 1966.
Experiments with supercooling and ice formation in flowing water.
Geofys. Publr, 26(9):18 pp.

ice formation
Kolesnikov, A.G., and V.I. Beliaev, 1956.
[On the calculation of the process of the crystallization of supercooled water by turbulent mixing.] Izv. Akad. Nauk, SSSR, Ser. Geofiz., 1956(11):1322-

ice formation
Koslowski, Gerhard, 1964.
Über den Zusammenhang zwischen Groswetterlagen und Eisbildung im Finnischen Meerbusen.
Deutsche Hydrogr. Zeits., 17(6):273-285.

ice formation
Lavrov, V.V., 1956.
[On the question of the formation of ice within the water.] Meteorol. i Gidrol., (5):43-

Lebedev, V.L., 1968. ice formation
Maximum width of a wind-generated lead during sea freezing. (In Russian; English abstract).
Okeanologiia, Akad. Nauk, SSSR, 8(3):391-396.

ice formation
Nelson, K.H., and T.G. Thompson, 1954.
Deposition of salts from sea water by frigid concentration. J. Mar. Res. 13(2):166-182, 11 textfigs.

ice formation
Pruppacher, H.R., E.H. Steinberger and T.L. Wang, 1968.
On the electrical effects that accompany the spontaneous growth of ice in supercooled aqueous solutions.
J. geophys. Res., 73(2):571-584.

ice formation
Shesteperov, I.A., 1969.
Verification of Tsurikov's formula by the measurements in the sea. Okeanologiia, 9(4): 616-618.
(In Russian; English abstract)

ice formation
Sokolovskaya, L. Ya., 1962
The oceanological characteristics of waters with reduced salinity throughout the period of ice formation. (In Russian).
Okeanologiia, Akad. Nauk, SSSR, 2(6):981-987.

ice formation
Suzuki, Y., 1955.
Observations on ice crystals formed on sea surface. J. Ocean. Soc., Japan, 11(3):123-126.

ice formation
Varlet, F., 1946.
Température de congélation de l'eau de la Méditerranée. Bull. Inst. Océan., Monaco, No. 905: 4 pp., 1 textfig.

ice, fossil

Woodcock, A.H., A.S. Furumoto and G.P. Woollard, 1970.
Fossil ice in Hawaii?
Nature, Lond., 226(5248): 873.

ice, frazil

ice, frazil
Bukina, L.A., 1967.
Frazil ice crystals size distribution in turbulent flow. (In Russian; English abstract).
Fisika Atmosfer. Okean., Izv. Akad. Nauk, SSSR, 3(1):58-68.

ice, gas in

ice, gas in
Matsue, Sadao, and Yasue Miyake, 1966.
Gas composition in ice samples from Antarctica.
J. geophys. Res., 71(22):5235-5241.

ice growth
Lake, R.A. and E.L. Lewis, 1970.
Salt rejection by sea ice during growth.
J. geophys. Res., 75(3): 583-597.

ice, heat flow

ice, heat flow
*Lewis, E.L., 1967.
Heat flow through winter ice.
In: Physics of snow and ice, H. Oura, editor, Inst. Low Temp. Sci., Hokkaido Univ., 611-631.

ice, heat of fusion

ice, heat of fusion
*Ono, Nobuo, 1967.
Specific heat and heat of fusion of sea ice.
In: Physics of snow and ice, H. Oura, editor, Inst. Low Temp. Sci., Hokkaido Univ., 599-610.

ice islands

ice islands
Anon., 1965.
Ice Island Arlis II nears end fo four year "Cruise".
Undersea Technology, 6(4):41.

ice islands
Anon., 1957.
Soviet drifting stations in the Arctic Ocean, 1956-1957. Polar Record, 8(57):519-520.

Vodnyy Transport, 28 Mar. 1957.
Moscow Radio, 16 Mar. & 22 Apr. 1957.

ice islands
Biays, P., 1953.
Les îles de glace artiques. Ann. Géogr., Sept-Oct., 62 année, No. 333:377-380.

T-3 ice island
Coachman, L.C., and C.A. Barnes, 1961
The contribution of Bering Sea water to the Arctic Ocean.
Arctic, 14(3):147-161.

ice islands
Crary, A.P., 1958.
Arctic ice island and ice shelf studies. 1.
Arctic 11(1):1-42.

ice ISLANDS
Crary, A.P., 1956.
Arctic ice island research. Adv. Geophysics, 3: 1-41.

ice islands
Crary, A.P., J.L. Kulp and E.W. Marshall, 1955.
Evidences of climatic change from ice island studies. Science 122(3181):1171-1173.

ice islands
Debenham, F., 1954.
The ice islands of the Arctic: a hypothesis.
Geogr. Rev. 44:495-507.

ice islands
Dibner, V.D., 1955.
[The origin of the floating ice-islands.] Priroda 1955(3):89-92.

Translated E.R. Hope, Defence Res. Bd., Canada, T176R

ice islands
Fletcher, J.O., 1950.
Floating ice islands in the Arctic Ocean. Tellus 2:323-324.

ice islands
Fujino, Kazuo 1966.
Oceanographic observations on the Drifting Station ARLIS II, June-November 1964. (In Japanese; English abstract).
Low Temp. Sci. (A) 24: 269-284.

ice islands
*Heirtzler, J.R., 1967.
Measurements of the vertical geomagnetic field gradient beneath the surface of the Arctic Ocean.
Geophys. Prospect. 15(2):194-203.

ice islands
Helk, J.V., and Moira Dunbar, 1953.
Ice islands: evidence from North Greenland.
Arctic 6(4):263-271, 9 textfigs.

ice islands
Hopkins, Thomas L., 1969.

Zooplankton standing crop in the Arctic Basin
Limnol. Oceanogr., 14(1): 80-85.

ice islands
Ivanov, M.I., 1958.
New data on the origin of floating ice islands. (In Russian).
Priroda, 47(7):94-

ice islands
Koenig, L.S., K.R. Greenaway, M. Dunbar, and G. Hattersley-Smith, 1952.
Arctic ice islands. Arctic 5:67-103.

ice islands, drift of
Kusunoki, Kou, 1962
Hydrography of the Arctic Ocean with special reference to the Beaufort Sea.
Contrib. Inst., Low Temp. Sci., Sec. A, (17): 1-74.

ice islands
Lauritzen, K., (Ltr. to G. Hattersley-Smith), 1955 dated 28 Sept. 1955.
Ice islands off East Greenland. Polar Rec. 8(52): 46.

ice island
Mathieu, Guy, 1967.
Fletcher's Ice Island.
Geo-Mar. Tech., 3(2):14-18.

ice islands
Montgomery, M.R., 1952.
Further notes on the ice islands in the Canadian Arctic. Arctic 5:183-187.

ice islands
Muguruma, Jiro, and Keji Higuchi, 1963.
Glaciological studies on ice island T-3.
J. Glaciology, 4(36):709-730.

Ice Islands
Nikitin, M.M., 1961.
[Soviet drifting research stations.]
Mezhd. Geofiz. God, 1957-1958-1959, Akad. Nauk, SSSR, 42 pp. (English summary)

ice islands
Nutt, David C., 1966.
The drift of ice island WH-5.
Arctic, 19(3):244-262.

ice islands
Nutt, D.C., and L.K. Coachman, 1963
A note on Ice Island WH-5.
Arctic, 16(3):205-206.

ice island (drift of)
*Ostenso, N.A., 1968.
Geophysical studies in the Greenland Sea.
Bull. Geol. Soc. Am., 79(1):107-132.

ice island
Polunin, N., 1955.
Attempted dendrochronological dating of ice island T-3. Science 122(3181):1184-1186.

ice islands
Savel-ev, B.A., 1958(1960)

Determination of the age of ice islands in the Arctic (based on foreign research data). Problemy Severa, (1):

Translation in:
Problems of the North, (1): 346-350

ice islands (T-3)
Shaver, Ralph, and Kenneth Hunkins, 1964.
Arctic Ocean geophysical studies: Chukchi Cap and Chukchi Abyssal Plain.
Deep-Sea Res., 11(6):905-916.

Ice Islands
Sisko, R. K., and G. L. Routilevsky, 1961.
[How do the ice islands disappear.]
Probl. Arkt. i Antarkt., (8):103-107.

ice islands
Smith, David D., 1964.
Ice lithologies and structure of ice island Arlis II.
J. Glaciol., 5(37):17-38.

ice islands
Smith, David D., 1964.
Structural glaciology of ice island Arlis II. (Abstract).
Geol. Soc., Amer., Special Paper, No. 76:153-154.

ice islands
Syschev, K.A., 1959.
[Three years of drift of the floating ice island North Pole-6.]
Morskoy Flot, USSR, 19(4):21-23.

ice islands
Untersteiner, N., and R. Sommerfeld, 1963
Supercooled water and the bottom topography of floating ice.
J. Geophys. Res., 69(6):1057-1062.

ice islands
Usachev, P.I., 1961.
[Phytoplankton of the North Pole (based on collections of P.P. Shirshov, under the caommand of I.D. Papanin.]
Trudy Vses. Gidrobiol. Obshch., Akad. Nauk, SSSR, 11:189-208.

ice islands
Zubov, N.N., 1955.
Arctic ice islands and how they drift. Priroda, 1955(2):27-45.

T176R

ice islands
Zubov, N.N., and V.D. Dibner, 1955.
[Arctic ice-islands and how they drift -- The origin of the floating ice islands.] Priroda (2): 37-45; 89-92.

Translation cited in USFWS Spec. Sci. Rept., Fish No. 227.

ice, loading of

ice, loading of effect
*Ross, Bernard, 1967.
Penetration and fracture of sea ice due to impact loading.
In: Physics of snow and ice, H. Oura, editor, Inst. Low Temp.Sci., Hokkaido Univ., 499-521.

ice melting
*Kühnel, Ingrid, 1968.
Die Eisabschmelzzeiten für die Ostsee östlich der Linie Trelleborg-Arkona und für den Finnischen und Rigaischen Meerbusen soure für die Südlichen Randbezirke der Bottensee.
Dt. hydrogr. Z., 21(1):15-20.

ice, microbiology of
Bunt, John 1969.
Microbiology of sea ice.
Antarctic Jl, U.S. 4(5):193.

Ice, navigation under

ice, navigation, under
Redanski, V.G., 1968.
First navigations of submarines under ice in the Far East. (In Russian).
Izv. vses. geogr. Obshch., 100(4):368-370.

ice nomenclature

ice nomenclature
Tsurikov, V.L. 1966.
On the problem of international ice nomenclature. (In Russian; English abstract).
Okeanologiia 6(2):372-378.

ice nuclei

ice nuclei
Sano, I., N. Fukute, Y. Kojima and T. Murai, 1963
Adsorption studies on the mechanism of ice nucleation. I. Oxide nuclei.
J. Meteorol. Soc. Japan, Ser. II, 41(4):189-196.

ice (pack)
Ketchum, Robert D., Jr. 1971.
Airborne laser profiling of the Arctic pack ice
Remote Sensing Environm. 2(1):41-52.

ice (pack)
Ozawa, Keijiro and Tsutomu Isouchi, 1968.
Observations of ice bergs and pack-ice. J. Tokyo Univ. Fish., 9(1): 49-56.

Ice particles
Mossop, S.C., A. Ono, and E.R. Wishart, 1970.
Ice particles in maritime clouds near Tasmania.
Q. Jl R. Met. Soc. 96(409): 487-508.

Ice prevention

ice prevention
Williams, G.P., 1961
Winter water temperatures and ice prevention by air bubbling.
Dock & Harbour Authority, 42(490):111-115.

Ice properties

ice, properties of (electrical)
*Addison, J.R., and E.R. Pounder, 1967.
The electrical properties of saline ice.
In: Physics of snow and ice, H. Oura, editor, Inst. Low Temp. Sci., Hokkaido Univ., 649-660.

ice, properties of
Anderson, D.L., 1958.
A model for determining sea ice properties. 3. Physics and mechanics of sea ice.
NAS-NRC, Publ. No. 598:148-153.

ice, properties of
Anderson, D.L., 1958.
Preliminary results and review of sea ice elasticity and r elated studies.
Trans. Eng. Inst., Canada, 2(3):2-8.

ice, properties of
Anderson, D.L., and W.F. Weeks, 1958.
A theoretical analysis of sea-ice strength.
Trans. Amer. Geophys. Union, 39(4):632-640.

ice, properties of
*Assur, Andrew, 1967.
Flexural and other properties of sea ice sheets (condensed version).
In: Physics of snow and ice, H. Oura, editor, Inst. Low Temp. Sci., Hokkaido Univ., 557-567.

ice, properties of
Assur, A., 1958.
Composition of sea ice and its tensile strength. Sec. 3. Physics and mechanics of sea ice.
NAS-NRC, Publ., No. 598:106-138.

ice, properties of
Bezverkhny, Sh.A., M.A. Bramson and E.V. Moiseeva, 1970.
Emissivity and reflectivity of ice in the infrared region. (In Russian).
Fizika Atmosfer. Okean., Akad. Nauk, SSSR 6(3): 314-317.

ice, properties (tensile)
*Dykins, J.E., 1967.
Tensile properties of sea ice grown in a confined system.
In: Physics of snow and ice, H. Oura, editor, Inst. Low Temp. Sci., Hokkaido Univ., 524-537.

ice, properties (electrical)
*Fujino, Kazuo, 1967.
Electrical properties of sea ice.
In: Physics of snow and ice, H. Oura, editor, Inst. Low Temp. Sci., Hokkaido Univ., 633-648.

ice, properties of
Gordienko, P.A., V.I. Fedorov and V.I. Shil'nikov, 1960.
[1. Ice cover of Antarctic water during summer-autumn period. 2. Determination of physical and mechanical properties of ice and snow.]
Arktich. i Antarkt. Nauchno-Issled. Inst., Mezhd. Geofiz. God, Sovetsk. Antarkt. Ekaped., 11:1-118.

ice, properties of
Hoekstra, Pieter and Patrick Cappillino, 1971.
Dielectric properties of sea and sodium chloride ice at UHF and microwave frequencies. J. geophys. Res., 76(20): 4922-4931.

ice, properties of
*Ishida, Tamotsu, 1967.
Rupture and vibrations of sea ice sheets.
In: Physics of snow and ice, H. Oura, editor, Inst. Low Temp. Sci., Hokkaido Univ., 551-556.

ice, properties of
Langleben, M.P., and E.R. Pounder, 1964.
Arctic sea ice of various ages. 1. Ultimate strength.
J. Glaciol., 5(37):93-98.

ice, properties of
Larin, B.V., 1965.
The influence of the height of the above-water parts of floes on the accuracy of determining the ice cohesion. (In Russian).
Okeanologiia, Akad. Nauk, SSSR, 5(4):727-733.

ice, properties of
Lavrov, V.V., 1965.
Difference in the properties of ice as regards its resistance to compressions and expansion. (In Russian).
Doklady, Akad. Nauk, SSSR, 162(1):54-56.

ice, properties of
Lavrov, V.V., 1958.
The electrical conductivity of ice. (In Russian).
Problemy Arktiki i Antarktiki, SSSR, No. 3:79-82.

Translation:
SLA Translations Center
The John Crerar Library
35 West 33rd Street
Chicago, Ill.

ice, properties of
Nagintsev, U.L., 1959.
[Experimental determination of the heat capacity and thermal conductivity of sea ice.]
Problemi Arktiki i Antarktiki (1):65-72.

ice, specific heat of
*Ono, Nobuo, 1967.
Specific heat and heatof fusion of sea ice.
In: Physics of snow and ice, H. Oura, editor, Inst. Low Temp. Sci., Hokkaido Univ., 599-610.

ice, properties of
Ono, Nobuo, 1965.
Thermal properties of sea ice. 1. Measurements of the thermal conductivity of young winter ice. 2. A method for determining the k/ρP-value of a non-homogeneous ice sheet. (In Japanese; English abstract).
Low. Temp. Sci., Hokkaido Univ., (A)23:167-176; 177-183.

ice, properties of
Oura, Hirobumi, editors, 1967.
Physics of snow and ice.
Inst. Low Temp. Sci., Hokkaido Univ., 711 pp.

ice, properties of
Peschansky, I.S., 1958.
Physical and mechanical properties of Arctic ice and methods of research. Sec. 3. Physics and mechanics of sea ice.
NAS-NRC, Publ., No. 598:100-104.

ice, properties of
Pounder, E.R., and M.P. Langleben, 1964.
Arctic sea ice of various ages. II. Elastic properties.
J. Glaciol., 5(37):99-105.

ice properties
Powell, R.W., 1958.
Thermal conductivities and expansion coefficients of water and ice.
Adv. in Physics, 7(26):276-297.

ice, properties of
Serikov, M.I., 1962.
Bending strength of Antarctic sea ice. (In Russian).
Sovetsk. Antarktich. Expedits., Inform. Biull., (36):30-35.

Translation:
(Scripta Tecnica, Inc., for AGU), 4(3):188-191.

ice, properties of
*Tabata, Tadashi, 1967.
Studies of the mechanical properties of sea ice. X. The flexural strength of small sea ice beams.
In: Physics of snow and ice, H. Oura, editor, Inst. Low Temp. Sci., Hokkaido Univ., 481-497.

ice, properties of
Tabata, T., 1958.
Studies on visco-elastic properties of sea ice. 3. Physics and mechanics of sea ice.
NAS-NRC, Publ. No. 598:139-147.

ice, properties of
Tabata, Tadashi, and Kazuo Fujino, 1965.
Studies of the mechanical properties of sea ice. VIII. Measurement of flexural strength in situ (2). (In Japanese; English abstract).
Low Temp. Sci., Hokkaido Univ., (A), 23:157-166.

ice, properties
*Tabata, Tadashi, Kazuo Fujino and Masaaki Aota, 1967.
Studies of the mechanical properties of sea ice. 11. The flexural strength of sea ice in situ.
In: Physics of snow and ice, H. Oura, editor, Inst. Low Temp. Sci., Hokkaido Univ., 540-550.

ice, properties of
Takahashi, Tsutomu, 1962.
Electric charge generation by breaking of ice piece.
J. Meteorol. Mag., Japan, (2), 40(5):277-286.

Also in:
Collected Papers on Science of Atmosphere and Hydrosphere, 1958-1963, Water Res. Lab., Nagoya Univ., 1(17).

ice, properties
Weeks, W.F., and D.L. Anderson, 1958.
An experimental study of strength of young sea ice.
Trans. Amer. Geophys. Union, 39(4):641-647.

ice, properties of.
Weeks, W.W., 1958.
The structure of sea ice: a progress report. Sec. 3. Physics and mechanics of sea ice.
NAS-NRC, Publ., No. 598:96-98.

Ice, protection against

ice, protection against
Freiberger, A., & H. Lacks, 1961
Ice-phobic coatings for de-icing naval vessels.
Naval Research Reviews, May: 1-7.

Ice rafting

ice-rafting, effect of
Blee, John J., Seymour R. Baker, and Gerald M. Friedman, 1968.
Sedimentological survey of Baffin Bay.
Marit. Sed., 4(1):4-6.

ice rafting
Conolly, John R., and Maurice Ewing, 1965.
Ice-rafted detritus as a climatic indicator in Antarctic deep-sea cores.
Science, 150(3705):1822-1824.

ice rafting
Ericson, David B., Maurice Ewing and Goesta Wollin, 1964
Sediment cores from the Arctic and Subarctic seas.
Science, 144(3623):1183-1192.

ice rafting
Ferguson, Laing, 1970.
"Armored snowballs" and the introduction of coarse terrigenous material into sea-ice.
J. sedim. Petrol. 40(3):1057-1060

ice rafting
Goodell, H.G., N.D. Watkins, T.T. Mather and S. Koster 1968.
The Antarctic glacial history recorded in sediments of the Southern Ocean.
Palaeogr. Palaeoclimatol. Palaeoecol. 5(1): 41-62.

ice rafting
Mullen, Ruth E., Dennis A. Darby and David L. Clark 1971.
Significance of atmospheric dust and ice rafting for Arctic Ocean sediment.
Bull. Geol. Soc. Am. 83(1): 205-212

ICE RAFTING
Ovenshine, A. Thomas, 1970.
Observations of iceberg rafting in Glacier Bay, Alaska, and the identification of ancient ice-rafted deposits.
Bull. geol. Soc. Am. 81(3): 891-894.

ice-rafting
Pelletier, B.R., 1968.
Submarine physiography, bottom sediments, and modern sediment transport in Hudson Bay.
In: Earth Science Symposium on Hudson Bay, Ottawa February, 1968, Peter J. Hood, editor, GSC Pap. Geol. Surv. Can., 68-53:100-136

ice rafting
Perry, P.B., 1961
A study of the marine sediments of the Canadian Eastern Arctic Archipelago.
Fish. Res. Bd., Canada, Manuscript Rept. Ser., (Oceanogr. and Limnol.), No. 89: 80 pp. (multilithed).

ice-rafting
Warnke, Detlef A., 1970.
Glacial erosion, ice rafting, and glacial-marine sediments: Antarctica and the Southern Ocean.
Am. J. Sci., 269(3): 276-290.

ice-rafting
Warnke, Detlef A., and Joseph Richter, 1970.
Sedimentary petrography of till from a floating iceberg in Arthur Harbor, Antarctic Peninsula.
Revue Géogr. phys. Géol. dyn. (2) 12(5): 441-448

ice-rafted rocks
Watkins, N.D., and B.M. Gunn, 1971.
Petrology, geochemistry and magnetic properties of some rocks dredged from the Macquarie Ridge.
N.Z. Jl Geol. Geophys. 14(1): 153-168

ice, salinity of

ice, salinity of
Tsurikov, V.L., 1967.
The changes of sea-ice salinity due to migration of brine pockets. (In Russian; English abstract).
Okeanologiia, Akad. Nauk, SSSR, 7(5):894-902.

ice shelves

ice sheets
Denton, George H., Richard L. Armstrong et Minze Stuiver, 1969.
Histoire glaciaire et chronologie de la région du détroit de McMurdo, sud de la Terre Victoria, Antarctide: note préliminaire.
Revue Géogr. phys. Géol. dynam. (2) 11(3): 265-278

ice shelves
Dort, Wakefield, Jr., 1971.
Shallow subsurface structures of east Antarctic ice shelves. (In English; Japanese abstract).
Antarct. Rec., Repts. Japan Antarct. Res. Exped. 41: 67-80.

ice sheet
Mercer, J.H., 1970.
A former ice sheet in the Arctic Ocean?
Palaeogeogr., Palaeoclimatol., Palaeoecol. 8(1): 19-27

ice shelves
Debenham, Frank, 1965.
The genesis of the McMurdo Ice Shelf, Antarctica.
J. Glaciol., 5(42):829-832.

ice shelf
Dubrovin, L.I., 1962.
Brine in ice shelves. (In Russian).
Sovetsk. Antarktich. Exped., Inform. Biull., (35):35-38.

Translation:
(Scripta Tecnica, Inc., for AGU) 4(3):151-153.

ice shelf
Gilmour, A.E., 1963.
Hydrological heat and mass transport across the boundary on the ice shelf in McMurdo Sound, Antarctica.
N.Z. J. Geol. Geophys., 6(3):402-422.

ice shelf
Gow, A.J., W.F. Weeks, G. Hendrickson and R. Rowland, 1965.
New light on the mode of uplift of the fish and fossiliferous moraines of the McMurdo Ice Shelf, Antarctica.
J. Glaciol., 5(42):813-828.

ice shelf
Heine, A.J. 1968.
Brine in the McMurdo Ice Shelf, Antarctica.
N.Z. Jl Geol. Geophys. 11(4): 829-839.

ice shelf
Heine, A.J. 1967.
The McMurdo ice shelf, Antarctica: a preliminary report.
N.Z. Jl. Geol. Geophys., 10(2): 474-475

ice shelf
Hochstein, M.P., 1967.
Pressure ridges of the McMurdo ice shelf near Scott Base, Antarctica.
N.Z. Jl.Geol.Geophys., 10(4):1165-1168.

ice shelves
Konecry, G., and W. Faig, 1966.
Studies of ice movements on the WARD HUNT ICE shelf by means of triangulation-trilateration.
Arctic, 19(4):337-342.

Ice Shelves
Reeh, Niels, 1968.
On the calving of ice from floating glaciers and ice shelves.
J. Glaciol., 7(50) 215-232

ice shelf
Bisx, G.F., and M.P. Hochstein, 1967.
Subsurface measurements on the McMurdo ice shelf Antarctica.
N.Z. Jl. Geol. Geophys., 10(2):454-497

ice shelves
Thomas, R.H., and P.H. Coslett 1970.
Bottom melting of ice shelves and the mass balance of Antarctica.
Nature, Lond. 228 (5266):47-49.

ice shelves
Zumberge, James H., and Charles Swithinback, 1961.
The dynamics of ice shelves. (Abstract).
Tenth Pacific Sci. Congr., Honolulu, 21 Aug.-6 Sept., 1961, Abstracts of Symposium Papers, 321-322.

ice, specific heat of
Ono, Nobuo, 1966.
Thermal properties of sea ice III. On the specific heat of sea ice. (In Japanese; English abstract)
Low Temp. Sci, (A) 24: 249-258

ice structure

ice, structure of
Kamb, Barclay, 1965
Structure of ice.
Science, 150(3693):205-209.

ice, undersurface of

ice, under surface of
Lyon, Waldo, 1967.
Under surface profiles of sea ice observed by submarine.
In: Physics of snow and ice, H. Oura, editor, Inst. Low Temp.Sci., Hokkaido Univ., 707-711.

ice volumes
Broecker, Wallace S, and Jan Van Donk 1970.
Insolation changes, ice volumes, and the O^{18} record in deep sea cores.
Rev. Geophys. Space Phys. 8(1):169-198

ice winters

ice winters
Goedecke, E., 1957.
Das merkwürdige Verhalten der Eiswinter, 1953/54 bis 1955/56. 1. Teil. Der Seewart 18(1):9-22.

icing

icing
Lebedev, A.A., 1968.
Zone of possible vessel's icing in the Barents and Greenland seas and in the northern Atlantic. (In Russian).
Meter. Ribokhoz. Issled. Severn. Basseina, 12: 154-157.

ice on ships
*Morimura, Shinji, 1969
Studies on the ice accumulation on ships (II). (In Japanese/ English abstract). Bull. Fac. Fish. Hokkaido Univ., 20(3): 185-192.

icing
Stallabrass, J.R. 1971.
Meteorological and oceanographic aspects of trawler icing of the Canadian east coast.
Mar. Obs. (Met. O. 839) 41 (233): 107-121.

icing
Tabata, Tadashi, Shuichi Iwata and Nobuo Ono, 1963.
Studies of ice accumulation on ships. I. (In Japanese; English abstract).
Low Temperature Sci., Hokkaido Univ., (A), 21:173-221.

Translation, E.R.Hope, T93J. Dec. 1967.

icing of ships
Vasil'yeva, G.V., 1966.
Hydrometeorological conditions causing ice accretion on ships. (In Russian).
Ryb. Khoz., 1966(12):43-45.

Translation: E.R. Hope T486R (1967) in WHOI Ref. Rm.

ichthyoplankton
De Ciechomski, Janina D., 1971.
Considerations on the ichthyoplankton in the shelf waters of the southwestern Atlantic, in front of Argentina, Uruguay and southern part of Brazil. (Abstracts in English and Spanish).
In: Fertility of the Sea, John D. Costlow, editor, Gordon Breach, 1: 89-98.

illite
Sea minerals

illumination
Clarke, G.L., and G.K. Wertheim, 1956.
Measurements of illumination at great depths and at night in the Atlantic Ocean by means of a new bathyphotometer. Deep-Sea Res. 3(3):189-205.

illumination
Kampa, E.M., 1955.
A discrepancy between calculation and measurement of submarine illumination.
Proc. Nat. Acad. Sci., 41(11):938-939.

illumination, rate of
Steemann Nielsen, E., and M. Willemoës 1971
How to measure the illumination rate when investigating the rate of photosynthesis of unicellular algae under various light conditions.
Int. Revue ges. Hydrobiol. 56(4): 541-556.

ilmenite
Sea minerals

image contrast
Olszewski, Jerzy 1971.
Utilization of the effect of light polarization for the improvement of image contrast in sea water.
Acta geophys. polon. 19 (2): 103-110

immigrants

immigration
Ivanov, A.I., 1969.
Immigration of Mya arenaria (L.) to the Black Sea, its distribution and quantity. (In Russian; English abstract). Okeanologiia, 9(2): 341-347.

immigrants
See also: migrations (migrants)

immigrants
Steinitz, H, 1967.
A tentative list of immigrants via the Suez Canal.
Israel J. Zool. 16(3): 166-169

immunology

immunology
Cushing, J.E., 1962.
Introduction: Symposium on immunogenetic concepts in marine population research.
Amer. Naturalist, 96(889):193-194.

immunology
Hildemann, W.H., 1962.
Immunogenetic studies of poikilo-thermic animals.
American Naturalist, 96(889):195-204.
Paper presented at 10th Pacific Science Congress.

immunology
Ridgway, G.J., 1962.
The application of some special immunological methods to marine population studies.
American Naturalist, 96(889):219-224.

impedance, sound

impedance, sound
Sabin, Gerald A., 1966.
Acoustic-impedance measurements at high hydrostatic pressure.
J. acoust. Soc., Am., 40(6):1345-1353.

inclination error

inclination error, palaeomagnetic.
Irving, E., 1967.
Evidence for palaeomagnetic inclination error in sediment.
Nature, Lond., 213(5075):483-484.

index

indices, nannoplankton
Loeblich, Alfred R., Jr. and Helen Tappen, 1966.
Annotated index and bibliography of the calcareous nannoplankton.
Phycologia, 5(2/3):81-216.

index of refraction

index of refraction
Gamez, J.C., 1942.
El indice de refraccion en el estudio de la densidad del agua del mar.
Anales Fis. y Quim., Madrid, 38:148-174.

index of refraction
Saint-Guily, B., 1955.
Sur l'indice de réfraction des liquides purs et des solutions, en particulier de l'eau de mer, comme fonction de la température et de la concentration. Bull. Inst. Océan., Monaco, No. 1052:1-24.

indicators, chemical

indicators, climatic

indicators, climatic
Conolly, John R., and Maurice Ewing, 1965.
Ice-rafted detritus as a climatic indicator in Antarctic deep-sea cores.
Science, 150(3705):1822-1824.

indicator, climate
Pflaumann, Uwe 1972.
Porositäten von Plankton-Foraminiferen als Klimaanzeiger?
Meteor-Forsch.—Ergebnisse (C) 7: 4-14.

indicator fossils

indicator fossils
*Thomas, Charles W., 1968.
Antarctic ocean-floor fossils: Their environments and possible significance as indicators of ice conditions.
Pacif. Sci., 22(1):45-51.

indicators, sea level
Lonsdale, Peter, William R. Normark, W.A. Newman, 1972.
Sedimentation and erosion on Horizon Guyot.
Bull. geol. Soc. Am. 83 (2): 289-316.

indicator species

A

indicator species
Abe, Susumu, and Tatsuo Yasui, 1964.
On the zooplankton community in the eastern entrance of Tsugaru Strait in the spring of 1961, especially with regard to the role as an indicator of water mass. (In Japanese; English abstract
Bull. Tohoku Reg. Fish. Res. Lab., No. 24:8-19.

indicator species
Abramova, V.D., 1956.
[Plankton as an indicator of water of different origins in the seas of the North Atlantic.]
Trudy Polar Sci. Inst., Sea-Fish. Econ. & Ocean. No. 9:69-92.

indicator species.
Allen, M.B., 1963.
Our knowledge of the kinds of organisms in Pacific phytoplankton.
Proc. Conf. Primary Productivity Measurements, Marine and Freshwater, Univ. Hawaii, Aug. 21-Sept. 6, 1961, U.S. Atomic Energy Comm., Div. Techn. Info., TID-7633:58-60.

indicator species
Allen, W. E., 1939(1940)
Indicator value of phytoplankton. Proc. Sixth Pacific Sci. Congr., 3:529-531.

indicator species
Aron, William, 1962
The distribution of animals in the eastern North Pacific and its relationship to physical and chemical conditions.
J. Fish. Res. Bd., Canada, 19(2):271-314.

B

indicator species
Bainbridge, V., 1963.
Continuous plankton records: contribution toward a plankton atlas of the North Atlantic and the North Sea, VIII. Chaetognatha.
Bull. Mar. Ecol., 6(2):40-51.

indicator species
Barth, Rudolf, 1970.
Caracterização biológica de diferentes corpos d'água em uma estação de fundeio.
Publ. Inst. Pesquisas Marinha, Brasil, 48: 9pp.

indicator species
Barth, Rudolf, 1965.
Observations on biological indicators in the Brazil Current. (Abstract).
Anais Acad. bras. Cienc., 37 (Supl.):156.

indicator species
Bary, B. McK., 1964
Temperature, salinity and plankton in the eastern North Atlantic and coastal waters of Britain, 1957. IV. The species' relationship to the water body; its role in distribution and in selecting indicator species.
J. Fish. Res. Bd., Canada, 21(1):183-202.

indicator species
Bary, B. McK., 1963.
Temperature, salinity and plankton in the eastern North Atlantic and coastal waters of Britain 1957. II. The relationships between species and water bodies.
J. Fish. Res. Bd., Canada, 20(4):1031-1065.

indicator species
Bogorov, B.G., and M.E. Vinogradov, 1960
[The distribution of the zooplankton in the Kuril-Kamchatka region of the Pacific.] Trudy Inst. Okeanol., 34: 60-84.

indicator species
Boltovskoy, E. 1970.
Masas de agua (caracteristica, distribución, movimientos) en la superficie del Atlantico sudoeste, segun indicadores biologicos - foraminiferos.
Publ. Serv. Hidrograf. Naval, Argentina, H. 643: 99pp.

indicator species
Boltovskoy, Esteban, 1969.
Distribution of planctonic Foraminifera as indicators of water masses in the western part of the tropical Atlantic
Actes Symp. Oceanogr. Ressources halieut. Atlant. trop., Abidjan, 20-28 Oct. 1966, UNESC 45-55.

indicator species
Boltovskoy, Esteban, 1969.
Foraminifera as hydrological indicators.
Proc. First int. Conf. Plankton. Microfossils, Geneve 1967, E.J. Brill, Leiden, 2: 1-14. (reprint).

indicator species
*Boltovskoy, Esteban, 1967.
Indicadores biologicos en la oceanografia.
Cienc. Invest. 23(2):66-75.

indicator species
Boltovskoy, Esteban, 1966.
La zona de convergencia subtropical/subantartica en el Oceano Atlantico (partie occidental). (Un estudio en base a la investigación de Foraminiferos-indicadores).
Republico Argentina, Sec. Marina, Serv. Hidrogr. Naval, Publ., H. 640:69 pp.

indicator species
Boltovskoy, E., 1962.
Plankton foraminifera as indicators of different water masses in the South Atlantic.
Micropaleontology, 8(3):403-408.

indicator species
Boltovskoy, Estaban, 1959
Foraminifera as biological indicators in the study of ocean currents. Micropaleontology, 5(4): 473-481.

indicator species
Bousfield, E.L., 1951.
Pelagic amphipods of the Belle Isle Strait region. J. Fish. Res. Bd., Canada, 8(3):134-163, 14 textfigs.

C

indicator species
Cannicci, Gabriella, 1959
Considerazioni sulla possibilità di stabilire "indicatori ecologici" nel plancton del Mediterraneo. Boll. Pesca. Piscicolt. e Idrobiol. n.s., 14(2): 164-188.

indicator species
Casanova, Jean-Paul, 1968.
Penilia avirostris Dana en Méditerranée occidentale; sa valeur d'indicateur écologique.
Annls Fac. Sci., Marseille, 41: 95-119

indicator species
Casanova, Jean-Paul, 1965.
Penilia avirostris Dana, indicateur d'eaux diluées.
Rev. Trav. Inst. Pêches Marit., 29(2):197-204.

indicator species
Casey, R.E. 1971.
Radiolarians as indicators of past and present water-masses.
In: Micropalaeontology of oceans, B.M. Funnell and W.R. Riedel, editors, Cambridge Univ. Press, 331-341.

Indicator species
Chen, Chin and Norman S. Hillman, 1970.
Shell-bearing pteropods as indicators of water masses off Cape Hatteras, North Carolina.
Bull. mar. Sci., 20(2): 350-367.

indicator species
Citarella, Georges, 1965.
Sur une espèce indicatrice de pollution des eaux marins.
Rev. Trav. Inst. Pêches Marit., 29(2):169-172.

indicators
Chmysnikova, V.L., 1937
Distribution of the biologic indicators in Shanalsky and Vilkitsky Straits. Trans. Arctic Inst., 82:145-157 (Russ); 158-159 (Eng)

indicator species
Clarke, A.H., Jr., 1963.
Littorina littorea as an indicator of Norse settlements.
Science, 142(3595):1022.

indicator species
Clarke, G. L., E.L. Pierce and D.F. Bumpus, 1943
The distribution and reproduction of Sagitta elegans on Georges Bank in relation to the hydrographical conditions. Biol. Bull., 85(3): 201-226, 10 textfigs.

indicator species
Cooper, L.H.N., 1952.
The boar fish, Capros aper (L.), as a possible biological indicator of water movement. J.M.B.A. 31(2):351-362, 1 textfig.

indicator species
Corbin, P.G., 1947.
The spawning of mackerel, Scomber scombrus L., and pilchard, Clupea pilchardus Walbaum, in the Celtic Sea in 1937-39 with observations on the zooplankton indicator species, Sagitta and Muggiaea. J.M.B.A., n.s., 27:65-132, 21 textfigs.

indicator species
Cross, Ford A., and Lawrence F. Small, 1967.
Copepod indicators of surface water movements off the Oregon coast.
Limnol. Oceanogr., 12(1):60-72.

D

indicator species
Della Croce, Norberto, 1964.
The marine cladoceran Penilia avirostris Dana in the American waters of the Pacific Ocean.
Nature, 201(4921):842.

indicator species
Della Croce, Norberto, 1960.
Nuovi ritrovamenti del cladocero marino Penilia avirostris Dana nel basso Tireno.
Boll. Musei e degli Ist. Biol. Univ. Genova, 30(177):5-14.

indicator species
de Silva, N.N., 1963.
Marine bacteria as indicators of upwelling in the sea.
Bull. Fish Res. Sta., Ceylon, 16(2):1-10)

indicator species
Frost, N., 1938.
The genus Ceratium and its use as an indicator of hydrographic conditions in Newfoundland waters. Res. Bull., No. 5, Dept. Nat. Res., St. John's Newfoundland.

indicators
Dubrovski, A.N., 1944.
Ptitsy - indikatori ledovogo rezhima arktiches-kikh morey. [Birds as indicators of the ice regime in Arctic seas.] Priroda 2:67-68.

E

indicator species
*Edwards,C., 1968.
Water movements and the distribution of Hydromedusae in British and adjacent waters.
Sarsia,34:331-346.

indicator species
Ehrhardt, Jean-Paul, Félix Baudin-Laurencin, et Gérard Seguin, 1964
Contribution à l'étude du plancton dans le Canal Corse-Provence.
Cahiers Océanogr., C.C.O.E.C., 16(8):623-636.

F

indicators
Faure, M.-L., 1950.
Le zooplancton des côtes marocaines. Bull. Sci. Com. Local d'Océan et d'Etude des Côtes du Maroc, No. 6:9-17, 1 fold-in.

indicator species
Ferguson Wood, E.J., 1951.
Phytoplankton studies in eastern Australia. Proc Indo-Pacific Fish. Comm., 17-28 Apr. 1950, Cronulla, N.S.W., Australia, Sects. II-III:60-63.

indicator species
(Biddulphia sinensis)
Ferrando, Hugo J., 1959
Estudio del plancton en la zona de pesca de la merluza. Anales, Facultad de Veterinaria, Montevideo 8(6): 89-99.

indicator species
Fincham, A.A., 1969.
Amphipods of the shallow-water sand community in the northern Irish Sea. J. mar. biol. Ass. U.K., 49(4): 1003-1024.

indicator species
Fraser,J.H., 1967.
Scottish plankton investigations in the near northern seas, 1965. Indicator species.
Annls. biol. Copenh. (1965)22:63-65.

indicator species
Fraser, James H., 1965.
Zooplankton indicator species in the North Sea.
Ser. Atlas, Mar. Environment, Folio 8: 2 pp. 3 pls.

Indicator Species
Fraser, J. H., 1962.
Plankton.
Proc. R. Soc., London, (A) 265:335-341.

indicator species
Fraser, J.H., 1961
The survival of larval fish in the northern North Sea according to the quality of the water.
J. Mar. Biol. Assoc., U.K., 41:305-312.

indicator species
Fraser, J.H., 1954.
Zooplankton collections made by Scottish Research Vessels during 1953.
Ann. Biol., Cons. Perm. Int. Expl. Mer, 10:99-101.

indicator species
Fraser, J.H., 1954.
Warm-water species in the plankton off the English Channel entrance. J.M.B.A., U.K., 33: 345-346.

indicator species
Fraser, J.H., 1952.
The Chaetognatha and other zooplankton of the Scottish area and their value as biological indicators of hydrographical conditions.
Scottish Home Dept., Mar. Res., 1952(2):1-52, 3 pls., 21 charts.

indicator species
Fraser, J. H., 1949
Plankton investigations from the Scottish Research Vessel. Ann. Biol., Int. Cons., 4: 66-67.

indicator species
Fraser, J. H. 1937
The distribution of Chaetognatha in Scottish waters during 1936, with notes on the Scottish indicator species. J. Cons., XII(3): 311-320

indicator species
*Furnestin,Marie-Louise,1968.
Le zooplancton de la Mediterranee (bassin occidental). Essai de synthese.
J. Cons.perm.int.Explor.Mer., 32(1):25-69.

indicator species
Furnestin, M.L., 1964.
Les indicateurs planctoniques dans la baie ibero-marocaine.
Rev. Trav. Inst. Pêches Marit., 28(3):257-264.

indicator species
Furnestin, M. L., 1957.
Chaetognathes et zooplancton du secteur atlantique marocain. Rev. Trav. Inst. Pêches Marit. 21(1/2): 9-356.

indicator species
Furnestin, M.-L., 1957.
Chaetognathes et zooplankton du secteur Atlantique marocain. L'Ann. Biol. (3)33(7/8):345-366.

indicator species
Furnestin, Marie-Louise, Claude Maurin, Jean Y. Lee et René Raimbault, 1966.
Éléments de planctonologie appliquée.
Rev. Trav. Inst. Pêches marit. 30(2/3): 278pp.

indicator species
Furuhashi, Kenzo, 1959.
[On the pelagic chaetognatha collected from the Kuroshio warm current region south of Honshu. 1. Notes on some chaetognaths as indicator of "Kuroshio" area and cold water region.]
Umi to Sora 35(4), 34(12):81-84.

G

indicator species
Ganapati, P.N., and P.V. Bhavanarayana, 1958.
Pelagic tunicates as indicators of water movements off Waltair coast. Current Science, 27(2): 57-58.

indicator species
Gaudy, R., 1963.
Sur la présence à Marseille d'espèces planctoniques indicatrices d'eaux d'origine atlantique.
Rapp. Proc. Verb., Réunions, Comm. Int. Expl. Sci., Mer Méditerranée, Monaco, 17(2):539-543.

indicator species
Giacometti-Cannicci, G., 1961.
Considerations sur la possibilité d'établir des "indicateurs écologiques" dans le plancton de la Méditerranée. Note II. Sur les copépodes pelagiques du bassin septentrional de la Mer Tyrrhénienne.
Rapp. Proc. Verb., Réunions, Comm. Int. Expl. Sci. Mer Méditerranée, Monaco, 16(2):207-214.

indicator species
*Golikov,A.N.,1968.
Distribution and variablility of long-lived benthic animals as indicators of currents and hydrological conditions.
Sarsia,34:199-208.

indicator species
Grabert, Brunhilde 1971.
Zur Eignung von Foraminiferen als Indikatoren für Sandwanderung.
Dt. hydrogr. Z. 24(1): 1-14

indicator species
Grainger, E.H., 1963.
Copepods of the genus Calanus as indicators of eastern Canadian waters.
In: Marine Distributions, R. Soc., Canada, Spec. Publ. No. 5:67-94.

H

indicator species
Hada, Yoshine, 1957
[The Tintinnoinea, useful microplankton for judging oceanographic conditions.]
Info. Bull. Plankton, Japan, (5):10-12.

indicator species
Halim, Youssef 1967.
Dinoflagellates of the south-east Caribbean Sea (east Venezuela)
Int. Revue ges. Hydrobiol. 52(5):701-755.

indicator species
Hardy, A. C., 1923.
Notes on the Atlantic plankton taken off the East Coast of England in 1921 and 1922. Publ. de Circ., No. 78.

indicator species
Hart, T.J., 1937.
Rhizosolenia curvata Zacharias, an indicator species in the Southern Ocean. Discovery Repts. 16:415-446, Pl. 14.

indicator species
Hendey, N.I., 1951
Littoral diatoms of Chicester Harbour with special reference to fouling. J.Roy. Microscop. Soc. 71(1): 1-86, 18 pls.

indicator species
Herman, Yvonne and P.E. Rosenberg, 1969.
Pteropods as bathymetric indicators.
Marine Geol., 7(2):169-173.

indicator species
Hida, T.S., 1957.
Chaetognaths and pteropods as biological indicators in the North Pacific. USFWS Spec. Sci. Rept. Fish., No. 215:13 pp.

indicator species
Holtedahl, Hans, 1965.
Recent turbidites in the Hardangerfjord, Norway. In: Submarine geology and geophysics, Colston Papers, W.F. Whittard and R. Bradshaw, editors, Butterworth's, London, 107-140.

plankton indicators
Huntsman, A.G., W.B. Bailey and H.B. Hachey, 1954.
The general oceanography of the Strait of Belle Isle. J. Fish. Res. Bd., Canada, 11(3):198-260, 35 textfigs.

I

indicator species
Inter-American Tropical Tuna Commission, 1961
Annual report for the year 1960: 183 pp.

indicator species
Italy, Stazione Zoologica di Napoli, 1964.
Ostracods as ecological and palaeoecological indicators.
Pubbl., Staz. Zool., Napoli, 33(Suppl.):612 pp.

J

indicator species
Japan, Japan Meteorological Agency, Oceanographical Section, 1962
Report of the oceanographic observations in the sea east of Honshu from February to March 1961. (In Japanese).
Res. Mar. Meteorol. and Oceanogr. Obs., Jan.-June, 1961, No. 29: 13-21.

indicator species
Japan, Kobe Marine Observatory, Oceanographical Section, 1963
Report of the oceanographic observations in the sea south of Honshu from February to March, 1962. (In Japanese).
Res. Mar. Met. & Ocean., J.M.A., 31:37-44.

Also in:
Bull. Kobe Mar. Obs., No. 173(4):1964.

indicator species
Japan Meteorological Agency, 1962
Report of the Oceanographic observations in the sea east of Honshu from August to September, 1960. (In Japanese).
Res. Mar. Meteorol. and Oceanogr., July-Dec., 1960, Japan Meteorol. Agency, No. 28:21-29.

indicator species
Jarrige, Francois, 1968.
On the eastward flow of water in the western Pacific south of the equator.
J. mar. Res., 26(3):286-289.

indicator species
Jashnov, W.A., 1970.
Hydromedusae Mitrocomella polydiademata and M. cruciata as indicators of boreal and arctic waters. (In Russian; English abstract).
Zool. Zh. 49 (12): 1780-1789.

indicator species
Jaschnov, W.A., 1966.
Water masses and plankton. 4. Calanus finmarchicus and Dimophyes arctica as indicators of Atlantic waters in the Polar Basin. (In Russian; English abstract).
Okeanologiia, Akad. Nauk, SSSR, 6(3):493-503.

indicator species
Jaschnov, W.A., 1963.
Water masses and plankton. 2. Calanus glacialis and Calanus pacificus as indicators of definite water masses in the Pacific. (In Russian; English abstract).
Zool. Zhurn., 42(7):1005-1021.

indicator species
Jaschnov, W.A., 1961.
Water masses and plankton. 1. Species of Calanus finmarchicus s.l. as indicators of definite waters masses.
Zool. Zhurn., Akad. Nauk, SSSR, 40(9):1314-1334.

indicator species
Jeffries, Harry P., 1962
Copepod indicator species in estuaries.
Ecology, 43(4):730-733.

indicator species
Jones, James I., 1969.
Planktonic Foraminifera as indicator organisms in the eastern Atlantic Equatorial Current System.
Actes Symp. Oceanogr. Ressources halieut. Atlant. trop., Abidjan, 20-28 Oct. 1966, UNESCO 213-230.

K

Indicator species
Kändler, R., 1961.
Über das Vorkommen von Fischbrut, Decapodenlarver und Medusen in der Kieler Förde.
Kieler Meeresf., 17(1):48-64.

indicator species
Kokubo, S., 1952
Results of the observations on the plankton and oceanography of Mutsu Bay during 1950, reference being made also to the period 1946-1950. Bull Mar. Biol. Sta., Asamushi 5(1/4): 1-54, 3 tables, (fold-in), 1 fold-in.

indicator species
*Kennett, James P., 1968.
Globorotalis truncatulinoides as a paleo-oceanographic index.
Science, 159(3822):1461-1463.

indicator species
Kriss, A. E., 1960.
Micro-organisms as indicators of hydrological phenomena in seas and oceans. 1. Methods
Deep-Sea Res. 6(2):88-94.

indicator species
Kriss, A.E., S.S. Abyzov and I.N. Mitzekevitch, 1960.
Micro-organisms as indicators of hydrological phenomena in seas and oceans. III. Distribution of water masses in the central part of the Pacific Ocean (according to microbiological data.
Deep-Sea Res., 6(4):335-345.

indicator species
Kriss, A.E., M.N. Lebedeva and I.N. Mitzkevich, 1960.
Micro-organisms as indicators of hydrological phenomena in seas and oceans. II. Investigation of the deep circulation of the Indian Ocean using microbiological methods.
Deep-Sea Res., 6(3):173-183.

indicator species
Künne, C., 1937.
Über als "Fremdlinge" zu bezeichnende Grossplanktonen in der Ostsee. Cons. Perm. Int. Expl. Mer, Rapp. Proc. Verb. 102(2):7 pp.

indicator species
Kuzmina, A.I., 1962.
Phytoplankton of the Kuril Straits as an indicator of different water bodies. (In Russian).
Issled. Dalinevostochnich Moreia, SSSR, Zool. Inst., Akad. Nauk, SSSR, 8:6-90.

L

indicator species
LeBrasseur, R. J., 1959.
Sagitta lyra, a biological indicator species in the subarctic waters of the eastern Pacific Ocean.
J. Fish. Res. Bd., Canada, 16(6):795-805.

indicator species
Leloup, E., 1946.
Contributions à l'étude de la faune belge. XV. Margelopsis haeckeli Hartlaub 1897, forme indicatrice du plancton au large de la côte belge.
Bull. R. Mus. Hist. Nat. Belg. 22(15):1-3.

indicator species.
Lubny-Gertsyk, E.A., 1956.
Plankton current indicators.
Trudy Inst. Okeanol. Akad. Nauk, SSSR, 13:67-70.

indicator species
Lucas, C. E., 1949
Notes on continuous plankton records at 10 m depth in the North Sea and Northeastern Atlantic during 1946-1947. Ann. Biol., Int. Cons., 4:63-66, text fig. 4.

M

indicator species
Maeda, R., 1956.
Plankton copepods in the Tsugaru Straits, northern Japan as investigated by underway samplings.
Bull. Fac. Fish., Hokkaido Univ., 7(3):225-232.

indicator species
Mankowski, W., 1963.
The role of plankton and benthos in the hydrological characteristic of the seas. (In Polish; English summary).
Przeglad Zoologiczny, 7(2):125-135.

indicator species (value)
Margaleff, Ramón, 1966.
Análisis y valor indicador de las comunidades de fitoplancton mediterráneo.
Inv. Pesq., Barcelona, 30:429-482.

indicator species
Marshall, N. B., 1948
Continuous plankton records: Zooplankton (other than Copepoda and young fish) in the North Sea 1938-1939. Hull Bull. Mar. Ecol. 2(13):173-213, Pls. 89-108.

indicator species
Marumo, R., 1957.
Plankton as the indicator of water masses and ocean currents. Ocean. Mag. Tokyo, 9(1):55-63.

indicator species
Marumo, R., 1955.
Analysis of water masses by distribution of the microplankton (1). Distribution of the microplankton and their relation to water masses in the North Pacific Ocean in the summer of 1954.
J. Ocean. Soc., Japan, 11(3):133-137.

indicator species
Marumo, Ryuzo, Osamu Asaoka and Kohei Karoji, 1961
On the distribution of Eucampia zoodiacus Ehrenberg with reference to hydrographic conditions.
J. Oceanogr. Soc., Japan, 17(1):45-47.

Mc

indicator species

McGowan, John A., 1960

The relationship of the distribution of the planktonic worm, Poeobius meseres Heath, to the masses of the North Pacific. Deep-Sea Res., 6(2): 125-139.

Indicator species

*McIntyre, Andrew,1967.
Coccoliths as paleoclimatic indicators of Pleistocene glasiation.
Science, 158(3806):1314-1317.

indicator species

Moore, H.B., 1953.
Plankton of the Florida Current. II. Siphonophora.
Bull. Mar. Sci., Gulf and Caribbean 2(4):559-573, 9 textfigs.

indicator species

Müller Melchers F.C., 1954.
Observaciones sobre Biddulphia sinensis Grev.
Rev. Biol. Mar., Valparaiso, 4(1/2/3):203-210, 16 figs.

indicator species

Melchers, F.C.M., 1952.
Biddulphia sinensis Grev., as indicator of ocean currents. Comm. Bot. Mus. Hist. Nat., Montevideo 2(26):1-14.

indicator species

Müller Melchers, F.C., 1952.
Biddulphia chinensis Grev., as an indicator of ocean currents. Comm. Bot. Mus. Hist. Nat., 2(26):1-14, pls.

indicator species

Murakami, Akio, 1959
Marine biological study on the planktonic Chaetognaths in the Seto Inland Sea.
Bull. Naikai Reg. Fish. Res. Lab., Fish. Agency No. 12: 1-186.

indicator species

Murakami, Akio, 1957
[Value of chaetognaths preferring low salinity as indicator forms of water masses.]
Info. Bull., Plankton., Japan, (5):8-10.

N

indicator species
Nair, R. Velappan, and R. Subrahmanyan, 1955.
The diatom, Fragilaria oceanica Cleve, an indicator of abundance of the Indian oil sardine, Sardinella longiceps Cuv. and Val. Current Sci. 24(2):41-42.

Indicator species

Nakai, Zinziro, 1969.
Note on mature female, floating egg and nauplius of Calanus cristatus Kröyer (Crustacea, Copepoda) - A suggestion on biological indicator for tracing of movement of the Oyashio water.
Bull. Soc. Jap. fish. Oceanogr. Spec. No. (Prof. Uda Commem. Pap.): 183-191.

indicator species

Nakajima, F., 1959

Plankton as indicators of sea water pollution.
Bull. Inst. Publ. Health, 7(4): 240-245. Abstr in: Publ. Health Eng. Abstr. 40(10): 29.

indicator species
Nesis, K.N., 1962
Corals and Pennatularia as indicators of the hydrological regime. (In Russian).
Okeanologiia, Akad. Nauk, SSSR, 2(4):705-714.

indicator species
Nesis, K.N., 1960. (benthic)
[The bottom fauna as an indicator of the hydrological regime in the sea.]
Nauchno-Techn. Biull. PNIRO, 3(13):34-36.

indicator species

Nishimura, S., 1959

A short note on the penetration into and the migration in the Japan Sea of some tropical and subtropical aquatic animals.
Ann. Rept., Japan Sea Reg. Fish. Res. Lab., 1(4):113-120

O

indicator species
Odate, Kazuko, 1962
On the properties of zoo-plankton in the northeastern sea region along the Pacific coast of Japan. (Investigations on the wet-weight, the appearance and the composition rates of each species, classified by the sea water types of different properties). (In Japanese; English Summary).
Bull. Tohoku Reg. Fish. Res. Lab., No. 21: 93-103.

P

indicator species

Park, Joo Suck, 1967.
Note sur les chaetognathes indicateurs planctoniques dans la mer coréenne en hiver 1967. (Korean abstract)
J. oceanogr. Soc., Korea, 2(1/2): 34-41.

indicator species

Polo, Francisco Pineda, 1971.
The relationship between chaetognaths, water masses, and standing stock off the Colombia Pacific Coast. (English and Portuguese abstracts)
In: Fertility of the Sea, John D. Costlow, editor, Gordon Breach, 2: 309-335.

indicator species
Prasad, R.R., 1953.
Swarming of Noctiluca in the Palk Bay and its effect on the "Chhodai" fishery, with a note on the possible use of Noctiluca as an indicator species. Proc. Indian Acad. Sci., B, 38(1):40-47, 1 textfig.

indicators
Pytkowicz, R.M., and D.R. Kester, 1966.
Oxygen and phosphate as indicators for the intermediate waters in the northeast Pacific Ocean.
Deep-Sea Research, 13(3):373-379.

Q

R

indicator species

Rakusa-Suszczewski, S. 1967.
The use of chaetognath and copeped population age-structures as an indication of similarity between water masses.
J. Cons. perm. int. Explor. Mer, 31(1):46-55.

indicator species

Rao, T.S. Satyanarayana, 1958

Studies on chaetognatha in the Indian seas. II. The Chaetognata of the Lawson's Bay, Waltair.
Andhra Univ., Mem. Oceanogr., 2:137-146.

indicator species
Redfield, A. C. and A. Beale 1940
Factors determining the distribution of populations of chaetognaths in the Gulf of Maine
Biol. Bull., 79:459-487

indicator species
Reyes Vasquez, Gregorio, 1966.
Ch. 6. Fitoplancton.
Estudios hidrobiologicos en el Estuario de Maracaibo, Inst. Venezolano de Invest. Cient. 122-145.

indicator species

Rottini, Laura, 1969.
I sifonofori quali indicatori idrologici.
Boll. Pesca Piscic. Idrobiol. 24(2):165-169

indicator species
Rowland, Robert W., and David M. Hopkins 1971.
Comments on the use of Hiatella arctica for determining Cenozoic sea temperatures.
Palaeogr. Palaeoclimatol. Palaeoecol. 9(1): 59-64.

indicator species
Russell, F. S. 1939
Hydrological and biological conditions in the North Sea as indicated by plankton organisms.
J. Cons., XIV(2):171-192

indicator species
Russell, F. S., 1935.
On the value of certain plankton animals as indicators of water movements in the English Channel and North Sea. JMBA, ns XX:309-332

S

indicator species
Sherman, Kenneth, and Everett Schaner 1968.
Pontellid copepods as indicators of an oceanic incursion over Georges Bank.
Ecology 49(3): 582-584.

indicator species

Simonsen, Reimer 1969.
Diatoms as indicators in estuarine environments.
Veröff. Inst. Meeresforsch. Bremerhaven, 11(2): 287-291

indicator species

Smayda, Theodore J., 1958

Biogeographical studies of marine phytoplankton
Oikos, 9(2): 158-191.

indicator species
Sohn, I.G., 1964.
The ostracode genus Cytheralloidea, a possible indicator of paleotemperature.
Pubbl., Staz. Zool., Napoli, 33(Suppl.):529-534.

indicator species
Southward, A.J., 1962.
Plankton indicator species and their statistical analysis.
Nature, 193(4822):1245-1246.

indicator species
Spjeldnaes, N., and Kari E. Henningsmoen, 1963.
Comment on Clarke: Littorina littorea as an indicator of Norse settlements.
Science, 142(3595):1022.

indicator species

Strauch, Friedrich, 1971.
Some remarks on Hiatella as an indicator of sea temperatures.
Palaeogr. Palaeoclimatol. Palaeoecol. 9(1): 59-64.

indicator, species, fossil
*Starauch, Friedrich, 1968.
Determination of Cenozoic sea-temperatures using Hiatella arctica (Linné).
Palaeogr. Palaeoclimatol. Palaeoecol. 5(2):213-233.
213-233.

indicator species
Sund, Paul N., 1964.
Los quetognatos en las aguas de la region del Perú
Inter-Amer. Tropical Tuna Comm., Bull., 9(3):115-216.

indicator species
Sund, Paul N., 1961.
Some features of the autecology and distributions of Chaetoghatha in the Eastern Tropical Pacific.
Inter-American Tropical Tuna Comm., Bull., 5(4): 307-340.

indicator species
Sund, P. N., and J. A. Renner, 1959.
The chaetognatha of the Eastropic Expedition, with notes as to their possible value as indicators of hydrographic conditions.
Inter-Amer. Trop. Tuna Comm., Bull., 3(9):395-436

indicator species (lists)
Sushkina, A.P., 1962.
Plankton organism-indicators of currents in the Faroe-Iceland waters and adjacent regions. (In Russian).
Trudy Vses. Nauchno-Issledov. Inst. Morsk. Ribn. Chos. i Okeana, VNIRO, 46:267-287.

T

indicator species
Tabb, Durbin C., 1962
The ivory barnacle, Balanus eburneus, as a biological indicator in brackish waters of South Florida (Abstract).
Proc. Gulf and Caribbean Fish. Inst., Inst. Mar. Sci., Univ. Miami, 14th Ann. Sess.:109-110

indicator species
Thiriot, Alain, 1965.
Campagne du navire océanographique "Zenobe Gramme" (20 mars - 2 avril 1964). Zooplancton du Golfe de Lion. Essai de justification de l'étude du zooplancton dans la détermination des differentes masses d'eau.
Cahiers Oceanogr., C.C.O.E.C., 17(5):331-343.

indicators
Thomson, J. M., 1947
The Chaetognatha of South-eastern Australia.
Counc. Sci. & Ind. Res., Australia, Bull. No. 222, (Div. Fish. Rept. 14), 43 pp., 8 text figs.

indicator species
Truveller, K.A. 1966.
Chaetognatha of the Davis and Denmark straits as indicators of water masses (In Russian).
Mater. Ribokhoz. Issled. severn. Basseina Poliarn. Nauchno-Issled. Proektn. Inst. Morsk. Ribn. Khoz. Okeanogr. (PINRO) 7:114-124.

U

V

indicator species
Vannucci, M., 1963.
On the biology of Brazilian medusae at 25° Lat. S.
Bol. Inst. Oceanogr., Sao Paulo, 13(1):143-184.

indicator species
Vannucci, M., 1963.
On the ecology of Brazilian medusae at 25° Lat. S.
Bol. Inst. Oceanogr., Sao Paulo, 13(1):143-184.

indicator species
Virketis, M.A., 1945.
[Zooplankton as an indicator of hydrological conditions in the Kara Sea.] Problemy Arktiki, 1944(1):67-101, tables, maps.

indicator species
Vives, F., 1971.
L'affleurement d'eau sur la côte catalane et les indicateurs biologiques (Copépodes).
Investigación pesq. 35(1): 161-169.

indicator species
Vučetić, Tamara 1969.
Distribution of Sagitta decipiens and identification of Mediterranean water masses circulation.
Bull. Inst. océanogr. Monaco 69 (1398): 12pp.

indicator species
Vučetić, Tamara, 1969.
Contribution to the knowledge of biologic indicators of water masses in the Mediterranean. (In Jugoslavian; English and Italian abstracts).
Thalassia Jugoslavica, 5: 435-441.

indicator species
Vučetic, T., 1961.
Sur la répartition des chaetognathes en Adriatique et leur utilisation comme indicateurs biologiques des conditions hydrographiques.
Rapp. Proc. Verb., Réunions, Comm. Int. Expl. Sci. Mer Méditerranée, Monaco, 16(2):111-116.

W

indicator species
Wagner, C.W., 1964.
Ostracods as environmental indicators in Recent and Holocene estuarine deposits of the Netherlands
Pubbl., Staz. Zool., Napoli, 33(Suppl):280-295.

indicator species
Williamson, D.I., 1956.
Planktonic evidence for irregular flow through the Irish Sea and North Channel in the autumn of 1954. J.M.B.A., 35(3):461-466.

indicator species
Williamson, D.I., 1952.
Distribution of plankton in the Irish Sea in 1949 and 1950. Proc. & Trans. Liverpool Biol. Soc. 58:1-46.

XYZ

indicator species
Yamamoto, G., 1952.
An ecological note on spawning in the scallop and plankton as an indicator of hydrographic conditions. Ecol. Rev., Mt. Hakkoda Bot. Lab. 13(2): 81-85, 2 textfigs. Tsendai

indicator species
Yamazi, Isamu, 1959.
A study on the productivity of Tanabe Bay. II. On some plankton indicating the water exchange in Tanabe Bay in August, 1957.
Rec. Oceanogr. Wks., Japan, Spec. No. 3:23-30.

indicator species
Yamazi, I., 1958.
A study on the productivity of the Tanabe Bay. (Part 1). IV. On some plankton indicating the water exchange in the Tanabe Bay.
Rec. Oceanogr. Wks., Japan, Spec. No. 2:25-35.

indicator species
Yamazi, I., 1956.
Plankton investigations in inlet waters along the coast of Japan. XIX. Regional characteristics and classification of inlet waters based on the plankton communities.
Publ. Seto Mar. Biol. Lab., 5(2)(9):157-196, Pls. 16-23.

indicator species
Yanulov, K.P., 1962.
Parasites as indicators of local redfish stocks.
Sovetskie Riboch. Issledov. v Severo-Zapadnoi Atlant. Okeana, VNIRO-PINRO, Moskva, 273-283.

In Russian; English summary

indicator species
Yashnov, V.A., 1965.
Water masses and plankton. 3. Halosphaera viridis as an indicator of Mediterranean waters in the North Atlantic. (In Russian).
Okeanologiia, Akad. Nauk, SSSR, 5(5):884-890.

Indicator species
Yonge, C. M., 1962.
On the biology of the mesogastropod Trichotropis cancellata Hinds, a benthic indicator species.
Biol. Bull., 122(1):160-181.

indicator species (fossil)

indicator species
Bandy, Orville L. 1968.
Cycles in Neogene paleoceanography and eustatic changes.
Palaeogr. Palaeoclimatol. Palaeoecol. 5(1):63-75.

indicator species
Cita, Maria Bianca, and Sara d'Onofrio, 1967.
Climatic fluctuations in submarine cores from Adriatic Sea.
Progress in Oceanography, 4:161-178.

indicator species
Donahue, Jessie G. 1967.
Diatoms as indicators of Pleistocene climatic fluctuations in the Pacific sector of the Southern Ocean.
Progress in Oceanography 4:133-140.

indicator species
Hays, James D., 1965.
Radiolaria and Late Tertiary and Quaternary history of Antarctic seas.
In: Biology of Antarctic seas, II.
Antarctic Res. Ser., Amer. Geophys. Union, 5: 125-184.

indicator species
*Kawarada, Yitaka, Masataka Kitou, Kenzo Furuhashi and Akiro Sano, 1968.
Plankton in the western North Pacific in the winter of 1968 (CSK).
Oceanogrl Mag., 20(1):9-29.

indicator species
Riedel, W.R., 1957.
Radiolaria: a preliminary stratigraphy.
Repts., Swedish Deep-Sea Exped., 6(Sediment Cores from the West Pacific, No. 3:61-96, 4 pls.

indicator species
*Seiglie, George A., 1968.
Foraminiferal assemblages as indicators of high organic carbon content in sediments and of polluted waters.
Bull. Am. Ass. Petr. Geol. 52(11):2231-2241.

industrial wastes

industrial wastes
Hancke, Manfred, 1968.
Deichsicherung mit Verhüttungsrückständen.
Helgoländer wiss. Meeresunters., 17: 381-391

industrial waste, effect of

Boëtius, Jan, 1968.
Toxicity of waste from a parathion industry at the Danish North Sea coast.
Helgoländer wiss. Meeresunters. 17: 182-187.

industrial wastes, effect of

Kinne, Otto, und Karl-Heinz Schumann, 1968.
Biologische Konsequenzen schwefelsäure- und eisensulfathaltiger Industrieabwässer. Mortalität junger Gobius pictus und Solea solea (Pisces)
Helgoländer wiss. Meeresunters. 17: 141-155.

industrial waste

Wada, Akira, 1969.
A study of mixing process in the sea caused by outfall of industrial waste water. Coast. Engng. Japan, 12: 147-158.

industry

inertia terms

INERTIA

Johnson, J.A., 1970.
Oceanic boundary layers. Deep-Sea Res., 17(3): 455-465.

inertia terms

Hidaka, Koji, 1961
Equatorial flow and inertia terms.
Rec. Oceanogr. Wks., Japan, 6(1):29-35.

inertia terms

Hidaka, K., and H. Miyoshi, 1950.
On the neglect of the inertia terms in dynamical oceanography. J. Ocean. Soc., Tokyo, 6(1): 68-76, 6 textfigs.

inertia terms

Ichie, T., 1950.
A note on the effect of inertia terms upon the drift currents. Ocean. Mag., Tokyo, 2(2):41-44, 2 textfigs.

inertial flow

inertial flow

Blandford, Robert, 1965.
Inertial flow in the Gulf Stream.
Tellus, 17(1):69-76.

inertial flow

Reed, R.K., 1966.
An observation of inertial flow in the Pacific Ocean.
J. Geophys. Res., 71(6):1764-1765.

inertial oscillations

oscillations, inertial

Belyaev, V.S., and A.G. Kolesnikov, 1966.
On the cause of formation of inertial oscillations in the drift currents. (In Russian). Fizika Atmosferi i Okeana, Izv., Akad. Nauk, SSSR, 2(10):1104.

inertial motions

Kielmann, Jürgen, Wolfgang Krauss und Lorenz Magaard 1970.
Über die Verteilung der kinetischen Energie im Bereich der Trägheits- und Seichesfrequenzen der Ostsee im August 1964 (Internationales Ostseeprogramm).
Kieler Meeresforsch. 25(2): 245-254.

inertial motion

Munk, Walter and Norman Phillips, 1968.
Coherence and band structure of inertial motion in the sea.
Rev. Geophys. 6(4):447-472.

inertial waves

Saint-Guily, Bernard, 1971.
Sur les ondes d'inertie en milieu homogène.
C.r. hebd. Séanc. Acad. Sci. Paris (D) 272 (20): 2531-2532

inertial-period motions

Webster, Ferris, 1968.
Observations of inertial-period motions in the deep sea.
Rev. Geophyics, 6(4):473-490.

inertial rotation

inertial rotation

Reid, Joseph L., Jr., 1962
Observations of inertial rotation and internal waves.
Deep-Sea Res., 9(4):283-289.

infestations

infestations

Herman, Sidney S., and Joseph A. Mihursky, 1964.
Infestation of the copepod Acartia tonsa with the stalked ciliate Zoothamnion.
Science, 146 (3643)543-544.

inflow

Francke, Eberhard, und Dietwart Nehring 1971.
Erste Beobachtungen über einen erneuten Salzwassereinbruch in die Ostsee im Februar 1969.
Beitr. Meeresk. 28: 33-47.

information retrieval

information retrieval

United States, National Oceanographic Data Center, 1964.
Instructions for coding and keypunching the geological information form for core, grab and dredge samples.
Manual Ser., Publ . M-5(Provisional):37 pp.

information retrieval

United States, National Oceanograp hic Data Center, 1965.
Manual for processing current data. 1. Instructions for coding and keypunching drift bottle data.
Manual Ser., Publ. M-6 (Provisional):17 pp.

information

Petersson, H. von 1971
Zur Frage des Einflusses der winterlichen Salzgehaltsschwankungen in der westlichen Ostsee auf die Eisbildung vor der Darss-Küste.
Acta Hydrophysica 16(2):95-104

information theory

information theory

Karbowiak, A.E., 1967.
Elements of information theory.
In: The Collection and processing of field data, E.F. Bradley and O.T. Denmead, editors, Interscience Publishers, 329-372.

Infrared radiation

infrared radiation

Bennett, H.E., and J.M. Bennett, 1962.
Predicting the distribution of infrared radiation from the clear sky.
J. Optical Soc., Amer., 52(11): 1305-1306

infra red

Deschamps, P.Y., 1970.
Température de la mer et radiométrie infrarouge.
Met. Mar. 67: 20-28, 10 figs. Also in: Recl Trav. Lab. Océanogr. phys. Mus. natn. Hist. nat. Paris 8(153).

infrared

*Friedlander, 1967.
World War II: electronics and the U.S. Navy: radar, sonar, loran and infrared techniques.
IEEE Spectrum, 4(11):56-70.

infrared

Kropotkin, M.A., B.P. Kozyrev and V.A. Zaytsev, 1966.
Infrared spectra of reflection from sea and fresh water and from some water solutions. (In Russian)
Fizika Atmosferi i Okeana, Izv. Akad. Nauk, SSSR, 2(4):234-235.

infrared absorption

Moskalenko, N.I. and S.O. Mirumyants, 1969.
The temperature effect on infrared absorption of radiation by H_2O and CO_2 vapour. (In Russian; English abstract). Fizika Atmosfer. Okean., Izv. Akad Nauk, SSSR, 5(12): 1292-1300.

infrared

Osborne, M.F.M., 1965.
The effect of convergent and divergent flow patterns on infrared and optical radiation from the sea.
Deutsche Hydrogr. Zeits., 18(1):1-25.

ingestion

Paffenhöfer, G.-A., 1971.
Grazing and ingestion rates of nauplii, copepodids and adults of the marine planktonic copepod Calanus helgolandicus. Mar. Biol. 11 (3): 286-298.

inhibition (bacterial)

Sieburth, J. McN., 1971.
An instance of bacterial inhibition in oceanic surface water. Mar. Biol. 11(1): 98-100.

inlets

inlets

Battjes, J.A., 1967.
Quantitative research on littoral drift and tidal inlets.
In: Estuaries, G.H. Lauff, editor, Publs Am. Ass Advmt Sci., 83:185-190.

inlets

Brown, P.M., J.A. Battjes, T.Y. Chiu and J.A. Purpura, 1966.
Coastal engineering model studies of three Florida coastal inlets.
Eng. Prog. Univ. Florida, 20 (6) (Bull. No. 122).

inlets

O'Brien, Morrough P., 1969.
Equilibrium flow areas of inlets on sandy coasts.
J. WatWays Harb. Div. Am. Soc. civ. Engrs. 95(WW1): 43-52.

inlets

*O'Brien, Morrough P., 1967.
Equilibrium flow areas of tidal inlets on sandy coasts.
Proc. 10th Conf. Coast. Engng. Tokyo, 1966, 1:376-386.

inlets
Pickard, G.L., 1971.
Some physical oceanographic features of inlets of Chile.
J. Fish. Res. Bd Can. 28(8): 1077-1106

inlets
Pickard, G.L., 1963.
Oceanographic characteristics of inlets of Vancouver Island, British Columbia.
J. Fish. Res. Bd., Canada, 20(5):1109-1144.

inlets
Terwindt, J.H.J. 1971.
Litho-facies of inshore estuarine and tidal-inlet deposits.
Geol. Mijnb. 50 (3): 515-526.

insects

insects (halobates)
Savilov, A.I., 1967.
Oceanic insects Halobates (Hemiptera; Gerridae) in the Pacific Ocean. (In Russian; English abstract).
Okeanologiia, Akad. Nauk, SSSR, 7(2):325-336.

insects (marine)
Scheltema, R.S., 1968.
Ocean insects, 1968.
Oceanus, 14(3):8-12. (non technical).

insecticides

insecticides, effect of
Buchanan, David V., Raymond E. Millemann and Nelson E. Stewart, 1970
Effects of the insecticide Sevin on various stages of the Dungeness crab Cancer magister
J. Fish. Res. Bd. Can., 27(1): 93-104

insecticides, effect of
Fisler, Ronald, 1969.
Acute toxicities of insecticides to marine decapod crustaceans.
Crustaceana, 16(3): 302-310

insecticides, effect of
Eisler, R., 1970.
Latent effects of insecticide intoxication to marine molluscs.
Hydrobiologia, 36 (3/4): 345-352

insecticides, effect of
*Eisler, Ronald, and Melvin P. Weinstein, 1967.
Changes in metal composition of the quahaug clam, Mercenaria mercenaria, after exposure to insecticides.
Chesapeake Sci., 8(4):253-258.

insecticides
Ernst, W., 1969.
Stoffwechsel von Pestiziden in marinen Organismen. 1. Vorläufig Untersuchungen über die Umwandlung und Akkumulation von DDT-14C durch den Polychaeten Nereis diversicolor.
Veröff. Inst. Meeresforsch. Bremerhaven, 11 (2): 327-331.

insecticides, effect of
Koeman, Jan H., Jan Veen, Eduard Brouwer, Leontine Huisman-de Brouwer and Jan L. Koolen, 1968.
Residues of chlorinated hydrocarbon insecticides in the North Sea environment.
Helgoländer wiss. Meeresunters. 17: 375-380.

insecticides, effect
Portmann, John E., 1968.
Progress report on a programme of insecticide analysis and toxicity testing in relation to the marine environment.
Helgoländer wiss. Meeresunters. 17: 247-256

insecticides, effect of
Woodwell, George M., Charles F. Wurster, Jr., and Peter A. Isaacson, 1967.
DDT residues in an east coast estuary: a case of biological concentration of a persistent insecticide.
Science, 156(3776):821-823.

insolation

insolation
Agenov, V.K., 1964.
On the daily fluctuation of radiant energy of the sun in the layers of the ocean in relation with its optical density. Hydrophysical investigations (Results of the investigations of the seventh cruise of the Research Vessel "Mikhail Lomonosov") (In Russian).
Trudy Morsk. Gidrofiz. Inst., Akad. Nauk Ukrain., SSR, 29:76-83.

insolation
Araz, F.E., and O.N. Tolstyakov 1969.
On the penetration of solar radiation into water. (In Russian: English abstract)
Meteorologiya Gidrol. (6): 58-64

insolation
Audouin, Jacques, 1962.
Hydrologie de l'Etang de Thau.
Rev. Trav. Inst. Pêches Marit., 26(1):5-104.

insolation
Boudreault, Yves, 1966.
Mesure du rayonnement solaire incident à Grande-Rivière en 1965.
Rapp. Stn. Biol. Mar., Grande-Rivière, 1965:13-16.

insolation (data)
Boudreault, Yves, 1965.
Mesure du rayonnement solaire incident à Grande Rivière en 1964.
Rapp. Ann., Sta. Biol. Mar., Grande Riviere, 1964:13-16.

insolation
Bourdreault, Y., 1963.
Mesure de la radiation solaire incidente à Grande Rivière en 1962.
Rapp. Ann., 1962, Sta. Biol. Mar., Grande Rivière, 13-17.

insolation
Broecker, Wallace S., and Jan Van Donk 1970.
Insolation changes, ice volumes and the O18 record in deep-sea cores.
Rev. Geophys. Space Phys. 8(1): 169-198.

insolation
Buzovkin, B.A., 1957.
Experience in studying the radiations conditions on the Baltic Sea by expeditions. (In Russian).
Trudy, Gosudarst. Okeanogr. Inst., 41:126-

insolation
Byzovkin, B.A., 1957.
Tests on the expeditionary investigations of radiation in the Baltic Sea. (In Russian).
Trudy Gosud. Okeanogr. Inst., 41:126-141.

insolation
Dodimead, A.J., and J.P. Tully, 1958.
Canadian oceanographic research in the northeast Pacific Ocean.
Proc. Ninth Pacific Sci. Congr., Pacific Sci. Assoc., 1957, Oceanogr., 16:180-195.

insolation
Just, Jean, 1970.
Marine biological investigations of Jørgen Brønlund Fjord, North Greenland: physiographical and bathygraphical survey, methods, and lists of stations.
Meddr Grønland, 154 (5): 42 pp

insolation
*Kido, Takuo, and Isamu Kotake, 1968.
The measurement of the horizontal insolation quantity on the Indian Ocean. (In Japanese; English abstract).
J. Tokyo Univ. Fish., 54(1):21-33.

insolation
Laevastu, T., 1960
Factors affecting the temperature of the surface layer of the sea.
Merent. Julk. (Havsforskningsinst. Skr.), No. 195:136 pp.

insolation, effect of
Lane, R.K., 1962
A review of the temperature and salinity structures in the approaches to Vancouver Island, British Columbia.
J. Fish. Res. Bd., Canada, 19(2):45-91.

insolation
Lumb, F.E., 1964.
The influence of cloud on hourly amounts of total solar radiation at the sea surface.
Q.J.R. Meteorol. Soc., 90(386):493-495.

insolation
Matsudaira, Yasuo, 1964
Cooperative studies on primary productivity in the coastal waters of Japan, 1962-63. (In Japanese; English abstract).
Inform. Bull., Planktol., Japan, No. 11:24-73.

insolation
Mullamaa, Yu. R., 1964.
Penetration of direct radiation in the sea. (In Russian).
Izv., Akad. Nauk, SSSR, Ser. Geofiz., (8):1259-1268.

insolation
Padmanabhamurty, B., and V. P. Subrahmanyam, 1964.
Some studies on radiation at Waltair.
Indian J. Pure and Appl. Physics, 2(9):293-295.

insolation
* Quinn, William H., Wayne V. Bent and Walter M. Pawley, 1969.
A study of several approaches to computing surface insolation over tropical oceans.
J. appl. Met. 8(2): 205-212.

insolation
Ryzhkov, Yu. G., 1961
Actinometric observations during the First Antarctic Voyage of the Diesel-Electric Ship "Ob" (1955-1956). Physics of the Sea. (In Russian).
Trudy Morsk. Gidrofiz. Inst., 23: 131-138.
Translation:
Scripta Technica, Inc., for Amer. Geophys. Union, pp. 104-110.

insolation
*Strokina, L.A., 1968.
The study of the radiation regime of oceans. (In Russian; English abstract).
Meteorologiya Gidrol. (10):77-83.

insolation
Tabata, Susumu, 1964.
Insolation in relation to cloud amount and sun's altitude.
In: Studies on Oceanography dedicated to Professor Hidaka in commemoration of his sixtieth birthday, 202-210.

insolation, effect of
Vives, F., and F. Fraga, 1961.
Pesca y energía solar.
Inv. Pesq., Barcelona, 20:5-16.

insolation
Vives, Francisco, y Manuel López-Benito, 1958
El fitoplancton de la Ría de Vigo y su relación con los factores térmicos y energéticos.
Inv. Pesq., Barcelona, 13:87-124.

instability

instability
Armstrong, J.C., 1947.
The maintenance of instability in the surface waters of the ocean. Ann. N.Y. Acad. Sci. 48(8): 801-809, 6 textfigs.

instability
Armstrong, J.C., 1947
The maintenance of instability in the surface waters of the ocean. Ann. N.Y. Acad. Sci. 48:801-809, with discussion by R.B. Montgomery pp.809-810.

instabilities
Caldwell, D.R., and C.W. van Atta 1970.
Characteristics of Ekman boundary layer instabilities.
J. fluid Mech., 44(1): 79-95.

instability
Fedorov, K.N., 1971.
A case of convection with occurrence of temperature inversion in connection with local instability in oceanic thermal wedge. (In Russian)
Dokl. Akad. Nauk SSSR 198(4): 822-825

instability, hydrodynamic
Frijalva, N., et F. Cocho 1968.
Instabilité hydrodynamique non-linéaire.
Mitt. Inst. Meeresk. Univ. Hamburg 10:30-49.

instability
Kusano, Hiromitsu, 1967.
A growth and vertical structure of instability lines in the sea of Kii straits. (In Japanese; English abstract).
Umi to Sora, 43(2): 54-69

instability
Matsumoto, S., 1962
On some instability properties of multi-level geostrophic models.
J. Meteor. Soc. Japan, Ser. II, 40(2):116-125.

instability, rotational
Sasaki, Ken 1971.
Rotational instability of a horizontal shear flow in a stratified rotating fluid.
J. oceanogn. Soc. Japan, 27(4): 137-141

instability, baroclinic
Schulman, Elliott E. 1967.
The baroclinic instability of a mid-ocean circulation.
Tellus 19(2): 292-305.

instability
Spilhaus, A.F., A. Ehrlich, and A.R. Miller, 1950.
Hydrostatic instability in the ocean. Trans. Am. Geophys. Union 31(2):213-215, 4 textfigs.

instability, baroclinic
Stone, Peter H., 1971.
The symmetric baroclinic instability of an equatorial current.
Geophys. fluid Dynam. 2(2): 147-164

instability
Tatro, P.R., and E.L. Mollo-Christensen 1967.
Experiments on Ekman layer instability
J. fluid Mech. 28(3): 531-544.

instability
Veronis, George, 1965.
On finite amplitude instability in thermohaline convection.
J. Mar. Res., 23(1):1-17.

installations

coastal installations
Ayers, J.R., 1950.
Seawalls and breakwaters. Inst. Coastal Eng., Univ. Ext., Univ. Calif., Long Beach, 11-13 Oct. 1950:18 pp., 18 figs. (multilithed).

coastal installations
Hickson, R.E., and F.W. Rodolf, 1950.
Design and construction of jetties. Inst. Coastal Eng., Univ. Ext., Univ. Calif., Long Beach, 11-13 Oct. 1950:21 pp. (mimeographed), 14 figs. (multilithed).

coastal installations
Hickson, R.E., and F.W. Rodolf, 1950.
Case history of Columbia River jetties. Inst. Coastal Eng., Univ. Ext., Univ. Calif., Long Beach, 11-13 Oct. 1950:16 pp. (mimeographed), 12 pls. (multilithed).

coastal installations
Horton, D.F., 1950.
The design and construction of groins. Inst. Coastal Eng., Univ. Ext., Univ. Calif., Long Beach, 11-13 Oct. 1950:18 pp. (mimeograached), 3 figs. (multilithed).

installations, deep-water
Hromadik, Joseph J. 1968.
Deep ocean installations and fixed structures.
In: Ocean engineering: goals, environment technology, John F. Brahtz, editor, John Wiley and Sons, 310-349.

coastal installations
Hudson, R.Y., and L.F. Moore, 1950.
The hydraulic model as an aid in breakwater design. Inst. Coastal Eng., Univ. Ext., Univ. Calif., Long Beach, 11-13 Oct. 1950:14 pp., 3 figs. (multilithed).

shore construction
Iribarren Cavanilles, R., 1950.
Generalizacion de la formula para el calculo de los diques de escallero y comprobacion de sus coeficientes. Revista de Obras Publicas de Mayo de 1950, Madrid, Spain.

Translated: Bull. B.E.B. 5(1):4-24, 6 figs., tables.

coastal installations
Morrison, J.R., 1950.
Design of piling. Inst. Coastal Eng., Univ. Ext. Univ. Calif., Long Beach, 11-13 Oct. 1950:9 pp. (mimeographed), 3 figs. (multilithed).

coastal installations
Peel, K.P., 1950.
Location of harbours. Inst. Coastal Eng., Univ. Ext., Univ. Calif., Long Beach, 11-13 Oct. 1950: 16 pp. (mimeographed).

installations
Roberts, E.B., 1950.
A seismic sea wave warning system for the Pacific
J.C.G.S. 3:74-79.

coastal installations
Schaufele, H.J., 1950.
Erosion and corrosion on marine structures. Inst. Coastal Eng., Univ. Ext., Univ. Calif., Long Beach, 11-13 Oct. 1950:6 pp., 5 pls. (multilithed).

installations
Snodgrass, F.E., 1950.
Wave recorders and wave data. Inst. Coastal Eng., Univ. Ext., Univ. Calif., Long Beach, 11-13 Oct. 1950:28 pp. (mimeographed), 3 figs. (multilith).

general

institutions
Anonymous, 1962.
A North Atlantic Institute of Science and Technology.
Nature, 196(4860):1127-1129.

institutions
Balech, E., 1955.
Creacio de un organismo coordinator de estudios oceanograficos. Rev. Biol. Mar. 6(1/3):180-182.

institutions
Chapman, William McLeod 1968.
The theory and practice of international fisheries commissions and bodies.
Proc. Gulf Carib. Fish. Inst. 20th Ann. Sess. 7-105.

meetings
Dietrich, G., 1954.
Tagung des Internationalen Rates für Meeresforsching in Paris von 4 bis 12 Oktober 1954.
Deutsche Hydrogr. Zeits. 7(5/6):201-203.

hydrographic offices
Gougenheim, Andre, 1967.
The contribution the hydrographic offices can make to the marine sciences.
Int. hydrogr. Rev., 44(1):61-66.

institutions
Holt, S.J., 1959
[International organisations for fishery problems.]
Biull. Okeanograf. Komissii, Akad. Nauk, SSSR, (3): 20-30.

Institutions
Istoshine, J.V., 1961.
[Organization of contemporary oceanographic research.]
Trudy Kom. Okeanogr., Akad. Nauk, SSSR, 11:29-45.

Institutions
Korganoff, Alexandre, 1963.
Océanographie, science de première urgence.
Bull. Assoc. Française pour l'Etude des Grandes Profondeurs Océaniques, No. 2:9-14.

institutions
Leipper, D.F., 1950.
Developments in oceanography. Science 112(2909): 366.

institutions
Lyman, John, 1963
International Initialography.
Int. Hydrogr. Rev., 40(2):15-19.

Reprint of article presented to Ann. Meet., Inst. Navigation, 18-20 June 1962 at San Diego

Particular

Institutions
Grouped by countries or "international"
See also: committees, meetings, etc.

ALGIERS

institutions
Bernard, Francis, 1963
L'Institut Océanographique d'Alger.
Pelagos (1): 27 pp.

ARCTIC

institutions
Kolman, O.V., 1964
Oceanographic research work of the Arctic Institute of North America in the area of Devon Island, 1961-1964. (In Russian).
Okeanologiia, Akad. Nauk, SSSR, 4(3):542-543.

institutions
U.S.A., Institute of Biological Sciences, 1964.
Centers of Arctic biological research.
Bioscience, 14(5):37-51.

ARGENTINA

institutions
Martinez Fontes, Elena, 1963.
La Estacion Hidrobiologica de Puerto Quequen, pasado, presente y futuro.
Rivista Mus. Argentino Ciencias Nat. "Bernadino Rivadavia" e Inst. Nac. Invest. Ciencias Nat., Hidrobiol., 1(1):18 pp.

institutions
República Argentina, Secretaria de Marine, Servicio de Hidrografia Naval, 1965.
El Servicio de Hidrografia Naval y la cooperacion nacional e internacional.
Bol. Servicio Hidrografia Naval, 11(2):85-89.

Australia

instutitions, Australia
Humphrey, G.F., 1969.
Perspectives of fisheries and oceanographic research in Australia. Bull. Jap. Soc. fish. Oceanogr. Spec. No. (Prof Uda Commem. Pap.): 43-47.

institutions
Takano, Hideaki, 1969.
Research organization of fishery and marine science in Australia. (In Japanese; English abstract) Bull. Plankt. Soc. Japan, 16(1): 67-71

institutions
Bowman, Robert I, 1971.
The Darwin Research Station.
Pacific Discovery 24(1): 18-22. (popular)

institutions
Australia, Commonwealth Scientific and Industrial Research Organization, Division of Fisheries and Oceanography, 1963.
Fifteenth Annual Report, 1962-1963, 162 pp.

institutions
Australia, Parliament of the Commonwealth of Australia, 1959.
Eleventh annual report of the Commonwealth Scientific and Industrial Research Organization for year 1958-1959. 181 pp.

institutions
Harrison, A.J., 1967.
Proposed marine laboratory at Taroona.
Tasman. fish. Res., 1(2):10-12.

Bermuda Biological Station

institutions
Mamaev, O.I., 1961.
The Bermuda Biological Research Station.
Okeanologiia, Akad. Nauk, SSSR, 1(5):939-940.

institutions, Bermuda
Smith F.G. Walton, 1968
Mid-ocean lab.
Sea Frontiers, 14(3): 150-156 (popular)

BRASIL

institutions
Delbouteville, C.D., 1960.
L'Institut de Biologie Marine et d'Océanographie de Récife (Brésil).
Vie et Milieu, 11(2):316-318.

Laboratorio de Biologia Marinha de Sao Sebastiao

institutions- Laboratorio de Biologia Marinha de Sao Sebastiao, Sao Paulo.
Sawaya, P., 1957.
Atividades do Laboratorio de Biologia Marinha de Sao Sebastiao, Sao Paulo.
IV. Reunion del Grupo de Trabajo de Ciencias del Mar, Montevideo, 22-24 Mayo de 1957, Actas de las Sesiones y Trabajo Presentados:143-147.

Canada

institutions - Canada
Loncarevic, B.D., W.L. Ford and R.M. McMullen 1969.
Atlantic Oceanographic Laboratory, Bedford Institute: the first six years.
Polar Rec. 14(93): 807-813.

institutions
Canada, Fisheries Research Board of Canada, 1957
Annual report, 1956-1957 for the fiscal year ending March 31, 1957:195 pp.

Institutions, Canada
Kuznetsov, A.P. and Z.A. Filatova, 1970.
Nanaimo Biological Station (British Columbia, Canada). (In Russian). Oceanologiia, 10(5): 921.

Institutions, Marine Sciences Rearch Lab. Newfoundland
Aldrich, Frederick A., 1967.
Oceanography week in Newfoundland.
Sea Frontiers, Miami, 13(6):332-338.

Institutions
Arsen'ev, V.S., A.P. Kuznetsov and A. Filatova, 1970.
An Oceanographic Institute of the University of British Columbia and its new oceanographic ship. (In Russian). Okeanologiia, 10(6): 1125-1126.

Atlantic Oceanographic Group

institutions
Campbell, N.J., 1958.
Recent oceanographic activities of the Atlantic Oceanographic group in the eastern Arctic.
Prog. Repts., Atlantic Coast Stas., Fish. Res. Bd. Canada, No. 69:18-20.

Station de Biologie Marine, Grande Rivière

institutions
Marcotte, Alexandre, 1964.
Rapport du Directeur de la Station de Biologie Marine pour l'Année 1963.
Rapp. Ann., 1963, Sta. Biol. Mar., Grande Rivière 3-8.

institutions
Brunel, Pierre, 1963.
Recherches sur les invertébrés de fond à la Station de Biologie Marine.
Actualités Marines, 7(3):3-8.

Station Biologique Marine, Grande Rivière
Drainville, G., M. Tiphane and P. Brunel, 1963.
Croisière océanographique dans le fjord de Saguenay, 14-22 juin 1962.
Rapp. Ann., 1962, Sta. Biol. Mar., Grande Rivière 133-146.

institutions
Marcotte, A., 1963.
Rapport du directeur de la Station de Biologie marine pour l'année, 1962.
Rapp. Ann., 1962, Sta. Biol. Mar., Grande Rivière 3-9.

Bedford Institute of Oceanography

institutions
Canada, Bedford Institute of Oceanography, Director, 1963.
Bedford Institute of Oceanography.
Polar Record, 11(24):611.

institutions, Canada - Bedford Institute of Oceanogr.)
Pelletier, B.R., 1965.
Research in marine geology at Bedford Institute of Oceanography.
Maritime Sediments, 1(1):4-7 (mimeographed).

institutions
SSSR, Akademia Nauk, 1963
A new oceanographic institute in Canada. (In Russian).
Okeanologiia, Akad. Nauk, SSSR, 3(3):565.

Fisheries Research Board of Canada

institutions
Anonymous, 1962.
Arctic Unit of Fisheries Research Board of Canada field activities in 1960.
The Polar Record, 11(72):277-278.

Summary of:
Fish. Res. Bd., Canada, Ann. Rept., 1960-61, Ottawa, 1962:117-119.

institutions
Clemens, W.A., 1953.
The Fisheries Research Board of Canada, its organization and programme.
Proc. Seventh Pacific Sci. Congr. 4:408-420.

institutions
Needler, A.W.H., 1958.
Fisheries Research Board of Canada Biological Station, Nanaimo, B.C., 1908-1958.
J. Fish. Res. Bd., Canada, 15(5):759-777.

Institute of Oceanography, University of British Columbia

institutions
Pickard, G.L., and W.M. Cameron, 1951.
The Institute of Oceanography, University of British Columbia. Trans. Am. Geophys. Union 32(1):112-113.

institutions
Scagel, Robert F., 1958.
Report of the Institute of Oceanography, University of British Columbia.
Proc. Ninth Pacific Sci. Congr., Pacific Sci. Assoc., Oceanogr., 16:14-15.
1957,

Pacific Naval Laboratory

institutions
Canada, Pacific Naval Laboratory, 1958.
Report on Physical Oceanography - Pacific Naval Laboratory - 1954-1957.
Proc. Ninth Pacific Sci. Congr., Pacific Sci. Assoc., 1957, Oceanogr., 16:16.

Pacific Oceanographic Group

Institutions
Canada, Pacific Oceanographic Group, 1961.
Annual Report, 97 pp. (restricted).

Canadian Pacific Oceanography
Tully, J. P., 1949.
Review of Canadian Pacific Oceanography since 1938. Trans Am. Geophys. Union 30(6):891-893.

Caribbean

institutions
Association of Island Marine Laboratories of the Caribbean, 1964.
Report, 26 pp.

institution
Zaneveld, J.S., 1958.
The Caribbean Marine Biological Institute, Piscadera Bay, Curacao (N.A.).
Turtox News, 36(12):284-285.

institutions
Zaneveld, J.S., 1956.
Het Caraibisch Marien-Biologisch Instituut to Curacao.
Vakblad voor Biologen, 36(12):6-16.

Chile

institutions
Barros Gonsalez, Guillermo, 1963.
La oceanografia en la Armada de Chile.
Bol. Informativo, Dept. Navegacion e Hidrografia, de la Armada, Republica de Chile, No. 64:1-26.

institutions, (Chile)
DeBuen, Fernando, 1962
La Estacion de Biologia Marina de Montemar.
Ciencia Interamericana, 3(2):7-9.

Central Agency for C14 Determination - Charlottenlund

Columbia, SA

Institutions
Lozano y Lozano, Fabio, 1962.
La Universidad de Bogota y los recursos del mar.
Peces y Conservas, Colombia, No. 12:5-6.

Cuba

institutions
Buesa, R., 1964.
The Cuban center of fishery research.
Okeanologiia, Akad. Nauk, SSSR, 4(2):357-359.

CURACAO

institutions
Zaneveld, J.S., 1957.
El Instituto Marino Biologico del Caribe en Curazo.
Bol. Inform., Fundavec, 2(12):219-222.

institutions
Zaneveld, J.S., 1958.
Instituto de Biologia Maritima das Caraibas.
Contr. Avuls., Inst. Ocean., Univ. Sao Paulo, Biol., 2:1-5.

DENMARK

Danish Biological Sta.
Blegvad, H., 1949.
Report of the Danish Biological Station to the Ministry of Agriculture and Fisheries. 1946, 49:37 pp., 8 figs.

institutions
Menache, M., 1954.
Apercu sur l(organization de le recherche oceanographique au Danemark. Bull. d'Info., C.C.O.E.C. 6(10):455-474.

institutions
Am. Geophysical Union, 1959.
Central Agency for carbon 14 determination of sea water.
Trans. Am. Geophys. Un., 40(1):34.

institutions
Steemann Nielsen, E., 1959.
A central agency for 14C determination (measurements of primary production in the sea) at Charlottenlund Slot, Denmark.
J. du Conseil, 24(2): 372-373.

institutions
U.S.A. American Society of Limnology and Oceanography, 1959.
Communication concerning a Central Agency for C14 Determination at Charlottenlund Slot, Denmark.
Limnol. & Oceanogr., 4(1):106-107.

EAST AFRICA

Institutions
East Africa, High Commission, 1961
East African Marine Fisheries Research Organization, Annual report, 1960: 38 pp.

ECUADOR

institutions
Ecuador, Instituto Nacional de Pesca del Ecuador 1964.
Apuntes e informaciones sobre las pesquerias en las provincias del Guayas y Los Rios.
Bol. Informativo, Guayaquil, Ecuador, 1(4):64pp.

ENGLAND See: Great Britain

institutions, England
See: institutions, Great Britain

FINLAND

institutions
Luther, A., 1957.
Tvärmine Zoologiske Station. Acta Soc. Fauna et Flora Fennica, 73:128 pp.

FISHERIES (when countries not predominant!)

France

institutions, France

institutions
Gougenheim, Andre, 1963
Editorial.
Bull., Assoc. Francaise Étude des Grandes Profondeurs Oceaniques, (1): 1-4.

institutions
Lacombe, H., Secrétaire, 1964.
Section d'Océanographie physique. Rapport de la Section d'Océanographie physique.
Com. Nat. Francais de Géodes. et Géophys., Comptes Rendus - Année 1963:161-176.

Institutions, France
Villat, Philippe, 1970.
L'activité océanographique française et son développement depuis 1961.
Geofis. Met. 19(1/2):34-38.

Bouee Laboratoire

Bouee-Laboratoire
Spiess, F.N., 1968.
Oceanographic and experimental platforms.
In: Ocean engineering: goals, environment, technology, John F. Brahtz, editor, John Wiley & Sons, pp. 555-587.

Centre de Géologie Marine et de Sédimentologie

Centre de Géologie Marine et de Sédimentologie
Rosfelder, A., 1961.
L'activité du Centre de Géologie Marine et de Sédimentologie (C.G.M.S.).
Rapp. Proc. Verb., Réunions, Comm. Int. Expl. Sci., Mer Méditerranée, Monaco, 16(3):719-720.

institutions
Rosfelder, A., P. Mailloux, J.P. Caulet, A. Grovel, L. LeClaire et Ph. Bouysse, 1962.
Les applications de la géologie marine. Réflexions sur l'expérience récente du Centre de Géologie Marine et de Sedimentologie. Éléments de programme.
In: Océanographie Géologique et Géophysique de la Méditerranée Occidentale, Centre National de la Recherche Scientifique, Villefranche sur Mer, 4 au 8 avril 1961, 225-231.

Centre d'Etude et de Recherche de Biologie et d'Océanographie Médicale

institutions - Centre d'Etudes et des Recherches de Biologie et d'Océanographie Medicale, Nice
Aubert, M. 1967.
Les buts et les moyens de l'océanographie médicale.
Scientia, Rivista Scienza (7) 61:29-36.

institutions
France, Bulletin, Municipal de Nice, 1964.
Centre d'Etude et de Recherche de Biologie et d'Océanographie Medicale (C.E.R.B.O.M.).
Cahiers C.E.R.B.O.M., Nice, 15(3):23-28.

institutions
France, Bureau Municipal d'Hygiène, Nice, 1961
Centre d'Études et de Recherches de Biologie et d'Oceanographie appliquées à la Médecine et à l'Hygiène Publique. Les Cahiers du C.E.R.B.O.M. No. 1: 2-4(mimeographed).

Centre National pour l'Exploitation des Oceans (CNEXO)

institutions, Centre National pour l'Exploitation des Oceans (CNEXO).
*Anon., 1967.
France: after the Action Concertée --.
Hydrospace, 1(1):26-28.

Chatou National Hydraulic Laboratory

institutions
Gridel, Henri, undated.
The Chatou National Hydraulic Laboratory.
Travaux (a review), Suppl., No. 247:1-8, 9 figs.

institutions
Laktionov, A.F., 1961
[The French Laboratory of Hydraulics and Electrical Powers.]
Okeanologiia, Akad. Nauk, SSSR, (2):351-354.

institutions
Nizery, A., 1949.
Le Laboratoire National d'Hydraulique. Annales des Ponts et Chaussees, 85 pp., 44 textfigs.

institutions
Nizery, A., 1950.
Le Laboratoire d'Hydraulique au service de l'oceanographie. Conf., C.R.E.O. No. 9:9 pp., 7 textfigs.

Institut de Géologie de la Faculté des Sciences de l'Université de Montpellier

institutions
Avias, Jacques, et Pierre Muraour, 1962.
Programme de recherches d'océanographie géologique et de sédimentologie de l'Institut de Géologie de la Faculté des Sciences de l'Université de Montpellier.
In: Océanographie Géologique et Géophysique de la Méditerranée Occidentale, Centre National de la Recherche Scientifique, Villefranche-sur-Mer, 4 au 8 avril 1961, 103-109.

Institut des Pêches Maritimes à Boulogne-sur-Mer

institutions - Institut Scientifique des Pêches Maritimes
Garcia del Cid, Francisco, 1965.
L'Institut Scientifique des Pêches Maritimes, son histoire.
Colloque Internat., Hist. Biol. Mar., Banyuls-sur-Mer, 2-6 sept., 1963, Suppl., Vie et Milieu, No. 19:361-364.

institutions
Petit, G., 1963.
Rapport sur le fonctionnement du Laboratoire Arago en 1962.
Vie et Milieu, 14(1):183-194.

Laboratoire Arago
Petit, G., 1954.
Rapport sur le fonctionnement du Laboratoire Arago en 1954.
Vie et Milieu, Bull. Lab. Arago, 5(4):593-604.

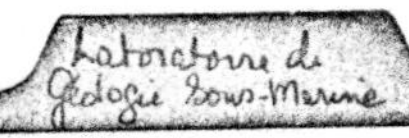

institutions
Gennesseaux, M., 1962.
Travaux du Laboratoire de Géologie Sous-Marine concernant les grands carottages effectués sur le précontinent de la region Nicoise.
In: Océanographie Géologique et Géophysique de la Méditerranée Occidentale, Centre National de la Recherche Scientifique, Villefranche sur Mer, 4 au 8 avril 1961, 177-181.

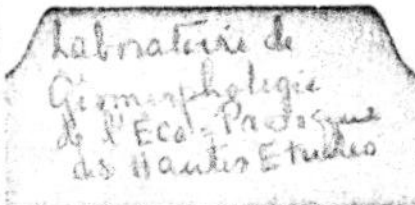

institutions
Ruellan, F., 1955.
Stage, à Dinard, du Laboratoire de Géomorphologie de l'École Pratique des Hautes Études.
Bull. Lab. Marit. Dinard, 41:19-21.

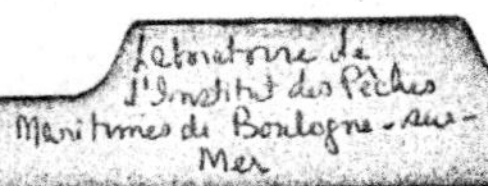

Institutions
Anon., 1965.
Le nouveau laboratoire de l'Institut des Pêches maritimes à Boulogne-sur-Mer.
Science et Pêche, Bull. d'Inform. et Document., Inst. Sci. Techn., Pêches Marit., No. 134:9-12. (not seen).

institutions
I.S.T.M., 1964.
Le nouveau Laboratoire de l'Institut des Pêches Maritimes de Boulogne-sur-Mer.
Rev. Trav. Inst. Pêches Marit., 28(4):317-320.

institutions, France
Salvat, Bernard, 1970.
Les activités du Muséum National d'Histoire Naturelle en Polynésie française (sciences de la mer)
Cah. Pacifique, 14: 255-269.

Rance Tidal Energy Installation

institutions
Gougenheim, André, 1967.
The Rance Tidal Energy Installation.
Int. hydrogr. Rev., 44(2)133-140.
(reprinted from J. Inst. Navig.)

Institutions

institutions
Kiseleva, M.I., 1961.
[The Marine Station in Endoume (France, Marseille)]
Okeanologiia, Akad. Nauk, SSSR, 1(6):1100-1101.

institutions, France (Hydrographic Service)
Gougenheim, André, 1971.
250 years of hydrography in France. Int. hydrogr Rev. 48(2): 31-34.

institutions
Gougenheim A., 1970
Le Corps des Ingénieurs hydrographes et le Service Hydrographique
Navigation, Paris
18(72): 365-377.

Station Océanographique de Villefranche

laboratories, France
*Blacher, Leonidas J., 1968.
La période russe dans l'activité de la Station Zoologique de Villefranche-Sur-Mer.
Bull. Inst. océanogr., Monaco, No. special 2: 481-491.

institutions
Lalou, C., 1960.
Travaux de la Station Océanographique de Villefranche.
Rev. Géogr. Phys., Géol. Syn., (2), 3(1):53-59.

institutions
Lalou, C., 1960
Travaux de la Station Océanographique de Villefranche. Rev. Géogr. Phys. Géol. Dyn. 2(3) (3): 159-166.

Station Zoologique de Villefranche
Petit, G., and G. Tregouboff, 1954.
Rapport sur le fonctionnement de la Station Zoologique de Villefranche en 1954.
Vie et Milieu, Bull. Lab. Arago, 5(4):605-607.

GERMANY

institutions, Germany

institutions, Germany
Dietrich, G., 1969.
Perspectives of oceanography in Germany. Bull. Jap. Soc. fish. Oceanogr. Spec. No. (Prof Uda Commem. Pap.): 37-41.

German program
Dietrich, Gunter, Arwed H. Meyl und Friedrich Schott, 1968.
Deutsche Meeresforschung 1962-73, Fortschritte, Vorhaben und Aufgaben.
Denkschrift II, Franz Steiner Verlag GMBH, Wiesbaden, 78 pp.

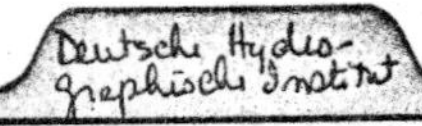

G.H.I.
Böhnecke, G., 1948
The German Hydrographic Institute. Trans. from Deutsche Hydrographische Zeitschrift 1(1): 6 pp.

institutions
Model, Fritz, 1963.
Das Deutsche Hydrographische Institut.
GWF Das Gas- und Wasserfach, 104(22):642-645.

institutions
Schmidt, U., 1963.
Die Arbeiten der Deutschen Wissenschaftlichen Kommission für Meeresforschung.
Jahresber. Deutsch. Fischwirtschaft, 1962/1963: 192-201.

institutions
Schmidt, U., 1962.
Die Arbeiten der Deutschen Wissenschaftlichen Kommission für Meeresforschung.
Jahresbericht Deutsch. Fischwirt., 1961/62:188-197.

institutions, Senckenberg am Meer
*Schäfer, Wilhelm, 1967.
Forschungsanstalt für Meeresgeologie und Meeresbiologie Senckenberg in Wilhelmshaven.
Senckenberg. leth., 48(3/4):191-217.

Institutions
Simonsen, Reimer, 1970.
Der Friedrich-Hustedt-Arbeitsplatz für Diatomeenkunde am Institut für Meeresforschung in Bremerhaven.
Beihefte Nova Hedwigia 31: XVIII-XXIV

Institut für Hydrobiologie und Fischereiwissenschaft der Universität, Hamburg

institutions
Bückmann, A., 1964
Das neue Institut für Hydrobiologie und Fischereiwissenschaft der Universität Hamburg und seine Gerschichte.
Vortrag zur Einweihung des Instituts für Hydrobiologie und Fischereiwissenschaft der UniversitätHamburg am 8 Juni 1964: 10 pp.

Also in:
Gesammelte Sonderdrucke, Inst. Hydrobiol. u. Fischereiwiss., Univ. Hamburg, 1963 (1964).

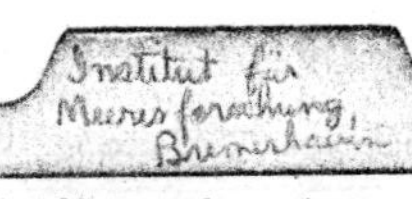

institutions, Germany, Inst. Meeresforschung
Bohnsack, Helmut, 1966.
Der Erweiterungbau des Instituts für Meeresforschung in Bremerhaven.
Veröff.Inst.Meeresforsch.Bremerh.10(1):V-VIII.

institutions, Germany, Inst.für Meeresforschung
Gerlach, Sebastian A., 1966.
Das Institet für Meeresforschung in Bremerhaven.
Veröff. Inst.Meeresforsch., Bremerh., 10(1):

institutions
Pax, F., 1949.
Das Institut für Meeresforschung in Bremerhaven.
Hydrobiologia 2(1):94-96.

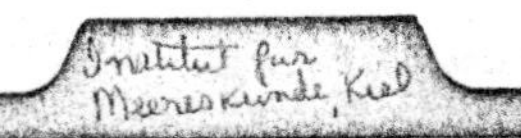

Institut für Meereskunde
Wüst, G., C. Hoffmann, C. Schlieper, R. Kändler, J. Krey and R. Jaeger, 1956.
Das Institut für Meereskunde der Universität Kiel nach seinem Wiederaufbau. Kieler Meeresf., 12(2):127-153.

Institutions, E. Germany, Institut für Meereskunde
Bongelsdorff, Edgar 1971.
Anlagen zur Nachbildung von Tiefsee-Verhältnissen.
Beitr. Meeresk. 28: 9-18.

GREAT BRITAIN

institutions United Kingdom

Ishiguro, Shizuwo, 1969.
Organization of marine science research in the United Kingdom. (In Japanese; English abstract) Bull. Jap. Soc. fish. Oceanogr. Spec. No. (Prof Uda Commem. Pap.): 31-35.

Admiralty Underwater Weapons Establishment

Admiralty Underwater Weapons Establishment
Anon.,1968.
AUEW stresses need for co-operation with industry.
Hydrospace, 1(4):22-24.

Discovery Committee

Discovery Committee
Mackintosh, N. A., 1950.
The work of the Discovery Committee. Proc. Roy. Soc., London, Ser. A. Math. Phys. Sci., 202(1068):1-16, 4 pls., 12 figs.

Discovery Committee
Mackintosh, N. A., 1950.
The work of the Discovery Committee. Proc. Roy. Soc., London, Ser. B, Biol. Sci., 137(887): 137-152, 12 figs., Pls. 8-11.

Hydrographic Department

Institutions
Great Britain, Hydrographic Department, 1961.
Report of the Hydrographer of the Navy for the Year 1960. (HW. 733/60) HD519:33 pp.

Institute of Seaweed Research

Institute of Seaweed Research
Fletcher, H. R., 1947.
The Institute of Seaweed Research. Nature 160 (4072):662.

Liverpool Observatory and Tidal Institute

institutions
Anonymous, 1961.
Liverpool Observatory and Tidal Institute.
Dock & Harbour Authority, 41(486):410-411.

Marine Biological Station, Millport

laboratories
Yonge, C.M., 1946
Jubilee of the Marine Biological Station Millport. Nature 158:506, 2 textfigs.

Institutions
Colman, J.S., and A.B. Bowers, 1968.
The Marine Biological Station of the University of Liverpool at Port Erin: a new wing, and a new research vessel.
J. mar. biol. Ass., U.K., 48(1):259-272.

Marine Sciences Laboratories, Menai Bridge

institutions
Crisp, D.J., 1962.
Marine Sciences Laboratories, Menai Bridge.
Nature, 195(4841):549-551.

National Institute of Oceanography

National Institution of Oceanogr
Bracelin, P.R., 1956.
National Institution of Oceanography.
Mar. Obs., 26(174)(M.O. 608):226-228.

National Institute of Oceanograph
Anon. (P.R.B. - Bracelin?), 1951. -y
National Institute of Oceanography. Mar. Obs. 21(153(M.O. 546):191-193.

institutions, England (National Institute of Oceanography)
Deacon, G.E.R., 1970.
The National Institute of Oceanography.
Mar. Engrs. Jl, Apr. 1970: 4-7.

institutions
Deacon, G.E.R., 1962
The National Institute of Oceanography.
Marine Observer, 32:125-127.

Also in:
Collected Reprints, Nat. Inst. Oceanogr., Wormley, 10.

Insts.- National Institute of Oceanography
Deacon, G.E.R., 1957.
Marine research. The work of the National Institute of Oceanography. Proc. R. Soc., Edinburgh, 64(4):350-368.

institutions
Deacon, G.E.R., 1954.
The National Institute of Oceanography.
J. Inst. Navigation 7(3):252-261.

institutions
Deacon, G.E.R., 1954.
The National Institute of Oceanography. Nature 173(4413):1014-1016, 2 textfigs.

National Institute of Oceanograph
Deacon, G.E.R., 1951.
The National Institute of Oceanography. Polar Record 6(41):88-90.

institutions
McElheny, Victor K., 1964
Britain's National Institute of Oceanography.
Science, 144(3615):160-163.

institutions, National Institute of Oceanography
Stride, A.H., 1965.
Marine geology at the National Institute of Oceanography.
Times Science Review, No. 58:10-11.
Also in:
Collected Reprints, Nat. Inst. Oceanogr., 13.

National Oceanographic Council

institutions
Anon., 1951.
National Oceanographic Council. M.O. 546,
Mar. Obs. 21(152):101-102.

institutions
Anon., 1951.
The National Oceanographical Council. Nature 167(4246):415-416.

National Oceanographic Council (institutions)
Great Britain, National Oceanographic Council, 1960
Annual report of the National Oceanographic Council, 1959-1960.

institutions
Great Britain, National Oceanographic Council, 1957.
Annual report, 1956-57:34 pp.

institutions
Samarin, V.G., 1962.
New Naval Hydrodynamic Laboratory in England. (In Russian).
Trudy Morsk. Gidrofiz. Inst., 26:144-154.

Translation:
Scripta Tecnica, Inc., for Amer. Geophys. Union, 26:110-119.

Plymouth Laboratory

MBL, Plymouth
Cooper, L.H.N., 1948.
Recent work of the Marine Biological Laboratory Plymouth. British Sci. News 1(3):8-10.

Marine Biological Laborator
Cooper, L.H.N., 1947
Recent work at the Marine Biological Laboratory, Plymouth. Brit. Sci. News 1(3): 8-10, 4 text figs.

institutions
Russell, F.S., 1959
Marine biological and oceanographic institutions of the world 1. The Plymouth Laboratory of the Marine Biological Association of the U.K. J. Mar. Biol. Ass. India 1(1): 89-90.

institutions (MBL-UK)
Wilson, D.P., 1960.
The new aquarium and new sea-water circulation systems at the Plymouth laboratory.
J. Mar. Biol. Assoc., U.K., 39(2):391-412.

Port Erin Marine Biological Station

institutions
Colman, John S., 1962
Marine biological and oceanographic stations of the world. V. The Port Erin Marine Biological Station of the University of Liverpool.
J. Mar. Biol. Assoc., India, 3(1-2):249-250.

institutions
Colman, J.S., 1961.
Marine biological and oceanographic institutions of the world. V. The Port Erin Marine Biological Station of the University of Liverpool.
J. Mar. Biol. Assoc., India, 3(1/2):249-250.

institutions
Lucas, C.E.,1956.
The Scottish Home Departments Marine Laboratory in Aberdeen. Proc. R. Soc., Edinburgh (B) 66(2): 222-234, 4 pls.

Royal Society, London

institutions, Royal Society, London
Deacon, Margaret, 1965.
Founders of marine science in Britain: the work of the Fellows of the Royal Society.
Notes Rec. R. Soc. Lond., 20:28-50.
Also in:
Collected Reprints, Nat. Inst. Oceanogr., 13.

INDIA

institutions
Anraku, Masateru, 1969. (In Japanese; English abstract)
The Indian Ocean Biological Centre. Bull. Plankt. Soc. Japan, 16(1): 72-74.

institutions, Indian
Anraku, M., 1968.
Indian Ocean Biological Centre. (In Japanese; English translation). Bull. Plankton Soc., Japan, 15(1): 45-49.

India, Indian Ocean Biol.Center
*Anraku,M.,1967.
Indian Ocean Biological Center.(In Japanese; English abstract).
Inf.Bull.Planktol.Japan, 14:66-68.

institutions
Bruun, A. Fr., 1954.
An oceanographic organization for the Indo-Pacific region. Deep-Sea Res. 2(1):84-86.

Consultative Committee for the Indian Ocean Biological Center
India, Indian National Committee on Oceanic Research, Council of Scientific and Industrial Research, New Delhi, 1964.
International Indian Ocean Expedition, Newsletter India, 2(4):48 pp.

Indian Ocean Biological Center
Hansen, Vagn Kr., 1966.
The Indian Ocean Biological Centre: The centre for sorting plankton samples of the International Indian Ocean Expedition.
Deep-Sea Res., 13(2):229-234.

institutions, India, Indian Ocean Biological Centre
Hempel, G., und J. Krey, 1966.
Das erste internationale Plankton-Institut der Welt.
Umschau, 66:80-84.

institutions, Indian
Motoda, S., 1968.
Training of CSK zooplankton sorters at Regional Marine Biological Centre for Southeast Asia, Singapore. (In Japanese; English translation).
Bull. Plankton Soc., Japan, 15(1): 50-54.

institutions - India
Motoda, Sigeru, 1962
International Biological Center to be established in India. (In Japanese; English abstract).
Info. Bull., Planktology, Japan, No. 8:26-29.

institutions, India
Omori, Makoto, 1970.
The Indian Ocean Biological Centre. (In Japanese English abstract). Bull. Plankt. Soc., Japan, 17(1): 59-62.

institutions, India
Tranter, D.J. 1969.
Guide to the Indian Ocean Biological Center (IOBC), Cochin (India).
UNESCO techn. Pap. mar. Sci. 10:49 pp. (mimeographed).

International Meterorological Center, Bombay
India, Indian National Committee on Oceanic Research, Council of Scientific and Industrial Research, New Delhi, 1964.
International Indian Ocean Expedition, Newsletter India, 2(4):48 pp.

institutions
Motoda, Sigeru, 1963.
Activity at the Indian Ocean Biological Center. (In Japanese; English abstract).
Info. Bull., Plankt., Japan, No. 10:65-67.

At Oceanographic Laboratory of Kerala University Ernakulam-Cochin, South India.

institutions, Univ. of Kerala
India, University of Kerala, 1965.
University Department of Marine Biology and Oceanography. Oceanographic Laboratory, 18 pp.

INDIAN OCEAN area

institutions
Bogorov, B.G., 1961.
[Scientific organizations of marine investigation in the Indian Ocean area.]
Okeanologiia, Akad. Nauk, SSSR, 1(5):937-939.

INDO-CHINA

Institut Oceanographique de l'Indochine
Chevey, P., and R. Serene, circa 1948.
Rapport sur le fonctionnement de l'Institut Océanographique de l'Indochine pendant les années 1938-1947. Nhatrang, Annam, 33 pp., 8 pls.

Oceanographic Institute of Indo-China
Serene, R., 1952.
An outline of the work of the Oceanographic Institute of Indo-China from 1939 to 1948.
Proc. Seventh Pacific Science Congr., Met. Ocean. 3:192-195.

INDONESIA

Laboratory for Investigation of the Sea, Batavia
Hardenberg, J.D.F., 1952.
Report of the work carried out since 1939 in the Java Sea by the Laboratory for Investigation of the Sea, Batavia. Proc. Seventh Pacific Sci. Congr., Met. Ocean., 3:190-192.

INDO-PACIFIC Fisheries Council

Institutions
F.A.O., 1957.
Indo-Pacific Fisheries Council. Directory of Fisheries Institutions (Asia and the Far East).

INTERNATIONAL

institutions
Skolnikoff, Eugene B., 1968.
National and international organization for the seas. In: Uses of the sea, Edmund A. Gullion, editor, Prentice-Hall, Inc., 98-112.

FAO

conferences
United Nations, Food and Agriculture Organization 1955.
International conference on marine biological laboratories. F.A.O. Fish. Bull. 8(3):146-147.

United Nations, Food and Agriculture Organization 1955.
Report on the United Nations Technical Conference on the Conservation of the Living Resources of the Sea. F.A.O. Fish. Bull. 8(3):117-130.

Inter-Governmental Oceanographic IOC

institutions
Currie, R.I., 1961.
The Intergovernmental Oceanographic Commission.
Nature, 192(4807):1015-1016.

International Association for Geodesy

institutions
Kneissl, Mas, Editor, 1963.
Festschrift zur Hundertjahrfeier der Internationalen Assoziation für Geodäsie am 12. und 13. Oktober 1962 in München.
Deutsche Geodätische Komm., 147 pp.

International Union of Biological Sciences

institutions
United Nations, Food and Agriculture Organization 1955.
International Union of Biological Sciences (IUBS). F.A.O. Fish. Bull. 8(3):144-146.

ICNAF International Commission for the Northwest Atlantic Fisheries

institutions
Adrov, M.M., 1964
The main results and problems of fishery-oceanographic investigations in the region of the activity of the International Commission of the Northwest Atlantic Fishery (ICNAF). (In Russian).
Materiali Vtoroi Konferentsii, Vzaimod. Atmosfer. i Gidrosfer. v Severn. Atlant. Okean., Mezhd. Geofiz. God, Leningrad. Gidrometeorol. Inst., 238-243.

institutions
International Commission for the Northwest Atlantic Fisheries, 1963.
Annual proceedings for the year 1962-63., 14:44pp

International Commission for the Northwest Atlantic Fisheries.
United Nations, Food and Agriculture Organization, 1955.
International Commission for the Northwest Fisheries. F.A.O. Fish. Bull. 8(3):148-150.

International Committee on Bottom Nomenclature
Bourcart, J., 1955.
Note sur les définitions des formes du terrain sous-marin, proposées par la comité international de Monaco, 1952. Deep-Sea Res. 2(2):140-144.

International Committee on the Nomenclature of Ocean Bottom Features
Japan, Maritime Safety Board, 1955.
International Committee on the Nomenclature of Ocean Bottom Features. Publ. No. 981, Hydrogr. Bull. No. 49:189-191.

International Committee on Bottom Nomenclature
Wiseman, J.D.H., and C.D. Ovey, 1955.
Proposed names of features on the deep-sea floor, 1. The Pacific Ocean. Deep-Sea Res. 2(2):93-106.

International Committee on the Nomenclature of Ocean Bottom Features
Wiseman, J.D.H., and C.D. Ovey, 1953.
Definitions of features on the deep-sea floor.
Deep-sea Res. 1(1):11-16.

International Commission for the Scientific Exploration of the Mediterranean Sea

Institutions
Anon., 1949.
International Council for the Exploration of the Sea. Nature 164(4182):1072-1073.

institutions, internation
Crovetto, Arthur, 1968.
La commission internationale pour l'exploration scientifique de la mer Méditerranée: Origines, difficultés initiales.
Bull. Inst. océanogr., Monaco, No. spécial 2: 327-335.

Institutions
Petit, G., 1956.
La Commission Internationale pour l'Exploration de la Mer Méditerranée. Son Histoire. Sa raison d'être. Balik ve Balikcilik, Istanbul, 4(9):

International Committee on the Oceanography of the Pacific

International Committee on the Oceanography of the Pacific
Thompson, T.G., 1952.
Report by the Chairman of the International Committee on the Oceanography of the Pacific.
Proc. Seventh Pacific Sci. Congr., Met. Ocean., 3:136-142.

International Council for the Exploration of the Seas

ICES

institutions
Denmark, Conseil Permanent International pour l'Exploration de la Mer, 1961.
Procès-Verbaux de la Réunion, 1960:124 pp.

ICES
Lozano, Gernando, 1964.
Aportaciones españolas durante 1963 a la oceanografía pura y aplicada, nacional e internacional.
Bol. Real Soc. Española, Hist. Nat., 62(1):99-113

institutions - International Council for the Exploration of the Sea
Smed, Jens, 1968.
The Service Hydrographique of the International Council for the Exploration of the Sea.
J.Cons.perm.int.Explor.Mer, 32(2):155-171.

institutions - ICES
UNESCO, 1965.
Report on the intercalibration measurements in Copenhagen, 9-13 June 1965 organized by ICES.
UNESCO Techn. Papers, Mar. Sci., No. 3:14 pp. mimeorgraphed).

International Geophysical Year

A

International Geophysical Year
Absalom, H.W.L., 1956.
International Geophysical Year. Meteorol. Mag. 85(1004)(M.O. 613):33-38.

IGY
Akademiia Nauk SSSR, Okeanologish. Komm., 1961.
[On the World Data Centre of the International Geophysical Year.]
Okeanologiia, Akad. Nauk, SSSR, 1(5):933.

IGY
Alekseyev, A.P., 1957.
Oceanographic investigations of PINRO according to the program of the International Geophysical Year. (In Russian).
Nauchno-Tekhn. Byul. Polyarn. n.-i. In-ta. Morsk. Rybn. zva i Okeanogr. (Sci.-Techn. Bull., Polar Sci. Res. Inst. Mar. Fish. Economy and Oceanogr.) No. 2-3:35-37.

RHZGeofiz. 8/58-5603.

IGY
Alekseiv, A.P., S.I. Potaitchuk, 1960
[The first cruise of R/S "Sevastopol" in the Norwegian Sea in accordance with the IGY program.]
Biull. Okeanogr. Komissii, Akad. Nauk, SSSR, (6): 45-49.

IGY
Alekseev, A.P., G.N. Zaitsev and S.I. Potaichuk, 1961
Results of hydrological and hydrochemical studies in the Norwegian Sea in 1958. Contribution to Special IGY Meeting, 1959.
Cons. Perm. Int. Expl. Mer, Rapp. Proc. Verb. 149:53-55.

IGY
Aliverti, G., 1958.
I lavori italiani di oceanografia nel quadro dell'Anno Geofisico Internazionale.
Comm. Naz. Ital. A.G.I., 1957-1958:9 pp.

IGY
Aliverti, G., 1958.
Ricerche talassografiche Italiane per l'A.G.I., 1957-1958. Comm. Naz. Ital. AGI 1957-1958:8 pp.

IGY
Aliverti, G., 1958.
Travaux italiens de thalassographie dans le cadre de l'année géophysique internationale.
Journées des 24 et 25 février 1958: 9-17.
Publ. by Centre Belge d'Ocean. Recherches Sous-marines.

International Geophysical Year
American Geophysical Union, 1956.
Antarctica in the International Geophysical Year, based on a symposium on the Antarctic.
Geophys. Monogr. (Publ. 462) 1:133 pp.

IGY
Anon., 1957.
International Geophysical Year, 1957-58: Antarctica, 1956. Polar Record 8(55):357-361.

IGY
Antonov, A.E., and O.S.Roudneva, 1965.
Peculiarities of hydrological and hydrochemical conditions in the South Baltic during the IGY. (In Russian; English abstract).
Okeanolog. Issled., Rez. Issled. po Programme Mezhd. Geofiz. Goda, Mezhd. Geofiz. Komitet, Prezidiume Akad. Nauk, SSSR, No.13:90-95.

IGY
Archipova, E.G., 1962.
Peculiarities of the heat balance in the North Atlantic during the IGY. (In Russian).
Trudy Gosud. Okeanogr. Inst., 67:74-85.

IGY
Argentina, Servicio de Hidrografia Naval, 1959.
Operacion Vema-Sanaviron II, Ano Geofisico Internacional 1957-58. unnumbered mimeographed sheets.
Ejemplar No. 33 (27 Jan. 1959 - 1215)

IGY
Argentina, Ministerio de Marina, Servicio de Hidrografico Naval, 1957.
Año Geofisico Internacional, 1957-58. Operacion Atlantico Sur.

I.G.Y.
Arsenjev, V.A., 1960
[Observations on the sea animals and birds of the Antarctic.] Arktich. i Antarkt. Nauchno-Issled. Inst. Sovetsk. Antarkt. Exped., Mezhd. Geofiz. God, Vtoraia Morsk. Exped., "Ob", 1956-1957, 7: 85-96.

IGY
Arsenyev, V.S., A.D. Shcherbinin, 1963
Investigation of currents in the Aleutian waters and in the Bering Sea. (In Russian).
Mezhd. Geofiz. Komitet, Prezidiume, Akad. Nauk SSSR, Rezult. Issled. Programme Mezhd. Geofiz. Goda, Okeanol. Issled., (8):58-66.

IGY
Artanova, A.K., 1960
[Peculiarities of the wind regimen in the southwestern part of the northern Pacific during the 1 st (25th) cruise of the r/v "Vitiaz".]
Trudy Inst. Okeanol., 40: 29-39.

IGY
Aurich, H.J., 1959.
Versuch einer biologischen Gliederung des nördlichen Nordatlantischen Ozeans auf Grund der Verteilung des Grossplanktons während der deutschen Forschungsfahrten im Internationalen Geophysikalischen Jahr 1958.
Deutsch. Hydrogr. Zeits., Ergänzungsheft Reihe B, 4(3):93-98.

I.G.Y.
Australia, Royal Society of Victoria, 1956.
Symposium - Australias Part in the I.G.Y. in Antarctica.
Proceedings Roy. Soc. Victoria New Zer. vol 69(1957) pp 55-68.

IGY
Aver'yanov, A.G., P.S. Veizman, E.I. Halperin, S.M. Zverev, M.A. Zaionchkovski, I.P. Kosminskaya, R.M. Krakshina, G.G. Mikhota, Yu. V. Tulina, 1961.
[Deep seismic sounding in the zone of transition from the Asian continent to the Pacific Ocean during the IGY.]
Izv., Akad. Nauk, SSSR, Ser. Geofiz., (2):169-184.
Engl. Edit., (2):109-117.

B

IGY
Baker, F.W.G., 1969.
Carbon dioxide.
Annls int. geophys. Year, 46: 191-194

IGY
Belyaev, G.M., 1969.
Study of sea floor fauna conducted by the Soviet Union during the IGY and IGC.
Annls int geophys. Year, 46: 210-216

IGY
Bezrukov, P.L. and A.P. Lisitzin, 1969.
Bottom sediments.
Annls int. geophys. Year. 46: 145-184.

IGY
Böhnecke, G., 1957.
Die ozeanographischen Arbeiten im Internationalen Geophysikalischen Jahr (IGJ). Übersicht über die Planungen und Zeile. Deutsche Hydrogr. Zeits., 10(2):33-38.

IGY
Böhnecke, G., and A. Bückmann, 1959.
Die Expeditionen von F.F.S. "Anton Dohrn" und V.F.S. "Gauss" im Internationalen Geophysikalischen Jahr 1957/58.
Deutsch. Hydrogr. Zeits., Ergänzungsheft, Reihe B, 4(3):108 pp.

IGY
Bordovskiy, O.K., 1961
[On chemistry of sediments in the Indian Ocean] (In Russian; English abstract).
Okeanolog. Issled., Mezhd. Komitet Proved. Mezhd. Geofiz. Goda, Prezidiume, Akad. Nauk, SSSR. (4):91-99.

I.G.Y.
Bossolasco, M. & I. Dagnino, 1959
Richerche di fisica marina nell'Anno Geofisico Internazionale 1957-1958: 1. Geofisica e Meteor. 7(5-6): 99-113.

IGY
Bourkov, V.A., 1962.
A survey of works on physical oceanography in the Pacific. (In Russian).
Mezhd. Geofiz. Komitet, Prezidiume, Akad. Nauk, SSSR, Rezult. Issled., Programme Mezhd. Geofiz. Goda, Okeanol. Issled., No. 7:19-31.

IGY
Brasil, Diretoria de Hidrografia e Navegacao, 1957.
Ano Geofisico Internacional. Publicacao, DG-06-V: 3 pp., 18 figs., 3 pp. of data.

IGY
Bregant, D., 1961.
Talassografia adriatica, A.G.I., 1959-1960, elementi oligodinamici: Rame
Rapp. Proc. Verb., Réunions, Comm. Int. Expl. Sci., Mer Méditerranée, Monaco, 16(3):587-589.

IGY
Bregant, Davide, 1961.
Elementi oligodinamici: Rame. Talassografia Adriatica, A.G.I. 1959-1960.
Consiglio Naz. delle Ricerche, Commissione Naz. Ital. per la Cooperazione Geofisica Internaz., Pubbl., No. 3:3-6.
Also:
Ist. Sperimentale Talassogr., Trieste, Pubbl., No. 372. (1961).

I.G.Y.

Bückmann, A., Günter Dietrich and Joachim Joseph, 1959
Die Forschungsfahrten von F.F.S. "Anton Dohrn" und V.F.S. "Gauss" im nördlichen Nordatlantischen Ozean im Rahmen des Polarfront-Programms des Internationalen Geophysikalischen Jahres 1958. Deutschen Hydrogr. Zeits., Ergänzungsheft Reihe B (4°), No. 3: 7-21.

Also in: Deutsches Hydrogr. Inst., Ozeanogr., 1959(1960) No. 3.

IGY

Burkov, V.A., 1969.
Review of IGY and IGC investigations of physical oceanography in the Pacific (1957-9)
Annls int. geophys. Year, 46:78-93.

IGY

Burkov, V.A., V.S. Arsentyev, and I.M. Ovchinnikov, 1960

[On the notion of northern and southern tropical fronts in the ocean.]
Trudy Inst. Okeanol., 40: 108-120.

IGY

Burkov, V.A., and I.M. Ovchinnikov, 1960

[Investigations of equatorial currents to the north of New Guinea.]
Trudy Inst. Okeanol., 40: 121-134.

IGY

Burkov, V.A., and I.M. Ovchinnikov, 1960

[Structure of zonal streams and meridional circulation in the central Pacific during the Northern Hemisphere winter.]
Trudy Inst. Okeanol., 40: 93-107.

IGY

Burkov, V.A., and Yu. V. Pavlova, 1963
On the use of statistical characteristics for investigation of the Northern trade-wind current in the western tropical part of the Pacific. (In Russian).
Mezhd. Geofiz. Komitet, Prezidiume, Akad. Nauk SSSR, Rezult. Issled. Programme Mezhd. Geofiz. Goda, Okeanol. Issled., (8):52-57.

C

IGY

Cadez, M., 1961.
General solution of the wave equation.
Annals, Int. Geophys. Year, 11:28-30.

IGY

Campbell, N.I., 1961
Canadian IGY project "Deep Water Circulation" North Atlantic, 1958. Contribution to Special IGY Meeting, 1959.
Cons. Perm. Int. Expl. Mer, Rapp. Proc. Verb. 149:114.

IGY

Chaen, Masaaki, Tomio Hemmi and Tadao Takahashi, 1960

Report of the Keiten-maru IGY cruise 1958.
Mem. Fac. Fish., Kagoshima Univ., 8: 87-88.

I G Y

Chapman, Sydney, 1959.

The International Geophysical Year.
Trans. A. G. U., 40(2):112-119.

International Geophysical Year

Chapman, S., 1955.
Scientific programme of the International Geophysical Year 1957-58. Int. Hydrogr. Rev. 23(1):157-164.

institutions IGY

Chapman, S., 1955.
The International Geophysical Year, 1957-1958.
Nature 175(4445):55-56.

I.G.Y.

Chile, Departamento de Navegacion e Hidrografia de la Armada, 1959
Trabajos oceanograficos realizados durante el Ano Geofisico Internacional: 53 pp.

IGY

Brazil, Diretoria de Hidrografia e Navegacao, 1957.
Ano Geofisico Internacional. Publ. DG-06-III+IV. (mimeographed).

IGY

Chile, Departamento de Navegacion e Hidrografia de la Armada, 1957.
Expedicion "Downwind" de oceanografia.
Bol. Informativo, 13(50):1-3.

IGY

Chile, Departamento de Navegacion e Hidrografia de la Armada, 1955.
Bol. Informativo, 11(38):1-4.

IGY

Cooper, L.H.N., 1961
Vertical and horizontal movements in the ocean derived from the work during the IGY. Contribution to Special IGY Meeting, 1959.
Cons. Perm. Int. Expl. Mer, Rapp. Proc. Verb. 149:111-112.

IGY

Corlett, J., 1961
Zooplankton observations at Ocean Weather stations I and J in 1957 and 1958. Contribution to Special IGY Meeting, 1959.
Cons. Perm. Int. Expl. Mer, Rapp. Proc. Verb. 149:200-201.

IGY

Cox, R.A., 1961
Results obtained during IGY by the Institute of Oceanography and some provisional interpretations. Contribution to Special IGY Meeting, 1959.
Cons. Perm. Int. Expl. Mer, Rapp. Proc. Verb 149:129.

D

IGY

Dalinger, Rene E., and Otto Freytag, 1960
Observaciones sobre el skua polar del sur en Bahia Margarita.
Contrib. Inst. Antartico Argentino, No. 51: 19 pp.

IGY

d'Ancona, U., 1961.
Les recherches planctoniques italiennes pour l'Année Géophysique Internationale.
Rapp. Proc. Verb., Réunions, Comm. Int. Expl. Sci. Mer Méditerranée, Monaco, 16(2):225-226.

IGY

Deacon, G.E.R., 1964
Sea current measurements.
The Royal Society IGY Expedition, Halley Bay, 1955-59, 4:348-352.

IGY

Deacon, G.E.R., 1961.
Outstanding problems in the Antarctic Ocean. (Summary).
Ann. Int. Geophys. Year, 11:309.

IGY

Deacon, G.E.R., 1961.
The measurement of deep currents. (Summary).
Ann. Int. Geophys. Year, 11:308-309.

IGY

Deacon, G.E.R., 1957.
Oceanography in the International Geophysical Year. Mar. Obs. (M.O. 624) 27(175):33-36.

IGY

Dietrich, Günter, 1970.
Zur Hydrographie des nördlichen Nordatlantischen Ozeans während des Internationalen Geophysikalischen Jahres 1957/58.
Ber. dt. wiss. Komm. Meeresforsch. 21(1/4): 399-402.

IGY

Dietrich, G., 1969.
North Atlantic Polar Front Survey in the IGY 1957/58.
Annls int. geophys. Year. 46:63-77

IGY

Dietrich, G., 1961
Some thoughts on the working up of the observations made during the "Polar Front Survey" in the IGY, 1958. Contribution to Special IGY Meeting, 1959.
Cons. Perm. Int. Expl. Mer, Rapp. Proc. Verb. 149:103-110.

IGY

Dietrich, G., Editor, 1960
Temperatur-, Salzgehalts-und Sauerstoff-Verteilung auf den Schnitten von F.F.S. "Anton Dohrn" und V.F.S. "Gauss" im Internationalen Geophysikalischen Jahr 1957/1958.
Deutsche Hydrogr. Zeits., Ergänzungsheft, Reihe B, 4(3):1-103.

I.G.Y.

Dietrich, Gunter, 1959

Zur Topographie und Morphologie des Meeresboden im nördlichen Nordatlantischen Ozean.
Deutschen Hydrogr. Zeits., Erganzungsheft Reihe B (4): No. 3; 26-34. Also in: Deutsches Hydrogr. Inst., Ozeanogr., 1959(1960), No. 8.

IGY

Dietrich, Günter, 1958.

Die Meereskunde im Internationalen Geophysikalischen Jarh, 1957/58.
Orion, Zeits. F. Natur u. Technik, (4):4 pp.

IGY

Dietrich, Günter, 1958.

Die Meereskunde im Internationalen Geophysikalen Jahr 1957/58 und der Deutsche Beitrag.
In: Ozeanographie, 1958, Deutsches Hydrogr. Inst. 366-379. (Original journal not obvious from this reprint!)

IGY

Dietrich, G., 1957.
Ozeanographische Probleme der deutschen Forschungsfahrten im Internationalen Geophysikalischen Jahr, 1957/58.
Deutsche Hydrogr. Zeits. 10(2):39-61.

IGY

Dobrovolskii, A. D., 1958.

[First trip with the Vitiaz for the IGY programme.]
Info. Biull. Mezhd. Geofiz. God., No. 5: 74-76.

IGY

Dobrovolsky, A.D., V.V. Leontyeva, and V.I. Kuksa, 1960

[Observations of trade wind inversion during the 26th cruise of the r/v "Vitiaz".]
Trudy Inst. Okeanol., 40:47-57.

[The title on the paper reads from the Russian: "On the characteristic structure and water masses of the western and central Pacific."]

E

IGY

Eggvin, J., 1961
Some results of the Norwegian hydrographical investigations in the Norwegian Sea during the IGY. Contribution to Special IGY Meeting 1959.
Cons. Perm. Int. Expl. Mer. Rapp. Proc. Verb. 149:212-218.

IGY

Epstein, S., and R.P. Sharp, 1959
Oxygen isotope studies. Trans. Amer. Geo. U., 40(1): 81-84. I.G.Y. Bull., No. 21:81-84

IGY

Eroschev-Shak, V.A., 1963
On distribution of argillaceous minerals in deep water sediments of the Atlantic. (In Russian).
Mezhd. Geofiz. Komitet, Prezidiume, Akad. Nauk SSSR, Rezult. Issled. Programme Mezhd. Geofiz. Goda, Okeanol. Issled., (8):125-136.

IGY

Ewing, W.M., 1961.
Voyage to the South Atlantic and the Indian Ocean. (Summary).
Ann. Int. Geophys. Year, 11:311.

F

IGY

Fedorov, C.N., 1960
[Estimation of deep currents in the region under investigation.]
Trudy Inst. Okeanol., 40:162-166.

IGY

Fedosov, M.V., and V. Andreev, 1961
Hydrochemical observations in the Davis Strait in Spring 1958. Contribution to Special IGY Meeting, 1959.
Cons. Perm. Int. Expl. Mer. Rapp. Proc. Verb 149:93-96.

IGY

Filarski, J., 1961
Hydrographic investigations in the North Sea in May and September, 1958. Contribution to Special IGY Meeting, 1959.
Cons. Perm. Int. Expl. Mer. Rapp. Proc. Verb. 149:138-139.

IGY

Fomichev, A.V., 1960
[Characteristics of water masses of the Peru current.]
Trudy Inst. Okeanol., 40:83-92.

(Easter Island-Valparaiso)

IGY

France, Centre de Recherches et d'Études Océanographiques, 1960
Stations hydrologiques effectuées par le "Passeur du Printemps" dans le cadre des travaux de l'Année Géophysique Internationale.
Travaux, C.R.E.O., 3(4):17-22.

International Geophysical Year
France, Comité Central d'Océanographie et d'Étude des Côtes, 1955.
Assemblée générale du Comité special de l'Année Géophysique Internationale, (Bruxelles, 8-14 septembre 1955). Bull. d'Info., C.C.O.E.C. 7(9):397-406.

IGY

France, Comité National Française Géodesique et Géophysique, 1957.
Participation française à l'Année Géophysique International (par P. Lejay)
Bibliography of principal publications on physical oceanography in France, 1957.
C.R. Assemblée Générale du Comité National Français, Un. Géod. et Géophys. Int., Paris, 16-21; 143-159.

I.G.Y.

France, Service Central Hydrographique de la Marine 1959.
Resultats d'observations océanographiques. 1. Observations hydrologiques de la station AGI de Madagascar.
Cahiers Océanogr., C.C.O.E.C., 11(5):323-331.

IGY

France, Service Central Hydrographique, 1957.
Année Géophysique Internationale. Réunion à Goteborg du Groupe de Travail d'Océanographie Physique (15-17 janvier 1957).
Bull. d'Info., C.C.O.E.C., 9(4):204-211.

IGY

France, Service Central Hydrographique, 1957.
Information. Bull. d'Info., C.C.O.E.C., 9(9): 469-471; 473.

See author card

IGY

France, Service Hydrographique de la Marine, 1963
Résultats d'observations.
Cahiers Océanogr., C.C.O.E.C., 15(5):344-355.

IGY

France, Service Hydrographique de la Marine, 1962
Résultats d'observations.
Cahiers Oceanogr., C.C.O.E.C., 14(4):274-283.

IGY

France, Service Hydrographique de la Marine, 1960.
II. Observations du niveau marin.
Cahiers Oceanogr., C.C.O.E.C., 12(3):236-237.

I.G.Y.

France, Service Hydrographique de la Marine, 1959
III. Observations du niveau marin à la station littorale de l'A.G.I. de Dakar, A.O.F. Cahiers Océan. C.C.O.E.C. 11(6): 461-464.

I.G.Y.

France, Service Hydrographique de la Marine, 1959
II. Observations du niveau marin à la station littorale de l'A.G.I. à Casablance (Maroc).
Cahiers Océan. C.C.O.E.C., 11(6): 459-460.

IGY

Fraser, J.H., 1961
The plankton of the Iceland-Faroe Ridge in 1958. Contribution to Special IGY Meeting, 1959.
Cons. Perm. Int. Expl. Mer. Rapp. Proc. Verb 149:179-182.

IGY

Fronert, V.V., 1960
Zone of intertrade western winds in the central Pacific.
Trudy Inst. Okeanol., 40: 40-43.

G

IGY

Gainanov, A.G., V.P. Petelin, 1963
On physical characteristics of some bottom and coastal rocks in the western part of the Pacific. (In Russian).
Mezhd. Geofiz. Komitet, Prezidiume, Akad. Nauk SSSR, Rezult. Issled. Programme Mezhd. Geofiz. Goda, Okeanol. Issled., (8):112-124.

IGY

Galerkin, L. I., 1962.
[On definition of the reading table for calculations of density fluctuations of the sea level.]
Mezhd. Geofiz. Komitet, Prezidiume Akad. Nauk, SSSR, Rezult. Issled. Programme Mezhd. Geofiz. Goda, Okeanol. Issled., No. 5:25-30.

IGY

Galperin, E.I., 1961.
Investigation of the crustal structure in the zone of transition from the Asiatic continent to the Pacific Ocean.
Annals, Int. Geophys. Year, 11:406-411.

IQSY

Germany, Deutsche Hydrographisches Institut, 1965.
Atlantische Expedition 1965 (IQSY) Forschungsschiff "Meteor" im Rahmen der Internationalen Jahre der Ruhigen Sonne, 31 pp.

IGY

Gillbricht, M., 1961
The distribution of microplankton and organic matter in the northern North Atlantic during late winter and late summer, 1958. Contribution to Special IGY Meeting, 1959.
Cons. Perm. Int. Expl. Mer. Rapp. Proc. Verb 149:189-193.

IGY

Glowinska, A., and B. Wojnicz, 1961
Temperature and salinity conditions in the southern Baltic during the period 1957 to 1959.
Contribution to Special IGY Meeting, 1959.
Cons. Perm. Int. Expl. Mer. Rapp. Proc. Verb 149:140-144.

IGY

Gololobov, Ya. K. 1963
Hydrochemical characteristics of the Aegean Sea during autumn of 1959. (In Russian)
Mezhd. Geofiz. Komitet, Prezidiume, Akad. Nauk SSSR, Rezult. Issled. Programme Mezhd. Geofiz. Goda, Okeanol. Issled., (8): 90-96.

IGY

González, G.B., 1962.
La oceanografia en la Armada de Chile.
Chile, Bol. Informativo, No. 58:7 pp.

IGY

Gordejev, Y.A., 1959.
[Hydrographical works.]
Arktich. i Antarktich. Nauchno-Issled. Inst., Mezhd. Geofiz. God, Sovetsk. Antarkt. Eksped., 5:144-153.

IGY

Gordienko, P.A., 1961
The Arctic Ocean.
Scientific American, 204(5): 88-102.

IGY

Gordienko, P.A., V.I. Fedotov, and V.I. Shil'nikov, 1960.
[1. Ice cover of Antarctic water during summer-autumn period. 2. Determination of physical and mechanical properties of ice and snow.]
Arktich. i Antarkt. Nauchno-Issled. Inst., Mezhd. Geofiz. God, Sovetsk Antarkt. Eksped., 11:1-118.

IGY

Gordienko, P.A., and A.F. Laktionov, 1969.
Circulation and physics of the Arctic Basin waters.
Annls. int. geophys. Year. 46: 94-112.

IGY

Gorshkova, T.I., and I.V. Solyankin, 1961
The grounds, water masses and regime of the Norwegian-Greenland Basin and adjacent areas Contribution to Special IGY Meeting, 1959.
Cons. Perm. Int. Expl. Mer. Rapp. Proc. Verb 149:44-45.

IGY

Gougenheim, Andre, 1962.
Introduction générale (1955-1959).
Océanogr. Phys., Année Géophys. Internat., Part. Francais, Centre Nat. Recherche Sci., Ser. 10 (1):1-10.

I G Y

Gougenheim, A., 1959.
Les recherches océanographique de l'année géophysique internationale.
La Nature, 3203: 396-400

IGY

Greece, National Hellenic Oceanographic Society 1960
Is the earth contracting or expanding? A summarized review of the last IGY findings.
Marine Scient. Pages, 4(6): 113-126. (In Greek with English summary)

IGY

Greece, National Hellenic Oceanographic Society, 1957
[The National Hellenic Oceanographical Society and the IGY.] Thalassia Phylla, No. 7:1 pp.

IGY

Gripenberg, S., 1961
Alkalinity and boric acid content of Barents Sea water. Contribution to Special IGY Meeting, 1959.
Cons. Perm. Int. Expl. Mer. Rapp. Proc.Verb. 149:31-37.

I.G.Y.

Gusev, A.V., 1960
[Parasitological research.] Arktich. i Antarkt. Nauchno-Issled. Inst., Sovetsk. Antarkt. Exped. Mezhd. Geofiz. God, Vtoraia Morsk. Exped., "Ob" 1956-1957, 7: 105-110.

I.G.Y.

Gutnikov, V.P., 1959
[Synoptic processes in the Southern Hemisphere.] Arktich. i Antarktich. Nauchno-Issled. Inst., Mezhd. Geofiz. God, Sovetsk. Antarkt. Exped. 5: 85-100.

H

IGY

Hachey, H.B., 1956.
Oceanography - its part in the International Geophysical Year. Trade News, Dept. Fish., Canada 8(11):5-8.

I.G.Y.

Hamon, B.V., and F.D. Stacey, 1960.
Sea levels around Australia during the International Geophysical Year.
Australian J. Mar. Freshwater Res., 2(3):269-281

IGY

Hansen, V.K., 1961
Danish investigations on the primary production and the distribution of chlorophyll at the surface of the North Atlantic. Contribution to Special IGY Meeting, 1959.
Cons. Perm. Int. Expl. Mer. Rapp. Proc. Verb 149:160-166.

IGY

Harrison, J.C., N.F. Ness, I.M. Longman, R.F.S. Forbes, E.A. Kraut and L.B. Slichter 1963
Earth-tide observations made during the International Geophysical Year.
J. Geophys. Res. 68(5):1497-1516.

IGY

Hela, I. 1969.
Mean sea level.
Annls int. geophys. Year. 46: 25-45

IGY

Hela, I., 1961
Hydrographic results of the "Aranda" Expedition to the Barents Sea in July 1957. Contribution to Special IGY Meeting, 1959.
Cons. Perm. Int. Expl. Mer. Rapp. Proc. Verb. 149:17-24.

IGY

Hermann, F., 1961
Danish hydrographic observations under the Polar Front Programme. Contribution to Special IGY Meeting, 1959.
Cons. Perm. Int. Expl. Mer. Rapp. Proc. Verb. 149:118-121.

IGY

Höhnk, W., 1959
Ein Beitrag zur ozeanischen Mykologie.
Deutsche Hydrogr. Zeits., Ergänzungsheft, Reihe B, 4(3):81-86.

I

IGY

I.G.Y. Bulletin, No. 6, 1957.
Arctic Ocean submarine ridges.
Trans. Amer. Geophys. Union, 38(6):1016-1018.

IGY

IGY Bulletin, No. 6, 1957.
Oceanographic island observatory in Iceland.
Trans. Amer. Geophys. Union, 38(6):1019-1022.

IGY

Ivanov, Yu. A., 1961
[Horizontal circulation of the Antarctic waters.]
Mezhd. Kom. Mezhd. Geofiz. Goda, Presidiume Akad. Nauk, SSSR, Okeanolog. Issled., (3):5-29.

IGY

Ivanov, Yu. A., 1961
[On frontal zones in the Antarctic waters.]
Mezhd. Kom. Mezhd. Geofiz. Goda, Presidiume Akad. Nauk, SSSR, Okeanol. Issled., (3):30-51.

IGY

Ivanov-Frantskevich, G.N., 1961
[Some peculiarities of hydrography and water masses of the Indian Ocean.] (English abstract
Okeanolog. Issledov., Mezhd. Komitet Proved. Mezhd. Geofiz. Goda, Prezidiume Akad. Nauk, SSSR (4):7-18.

J

I.G.Y.

Japan, Maritime Safety Board 1961.
Tables of results from oceanographic observations in 1958.
Hydrogr. Bull. Tokyo 66 (Publ. 981): 153 pp.

I.G.Y.

Japan, National Committee for the International Geophysical Coordination, Science Council of Japan, 1962
X. Oceanography.
Japan Contrib. Int. Geophys. Year and Int. Geophys. Cooperation, 4:142-150.

IGY

Japan National Committee for the International Geophysical Year 1960.
The results of the Japanese oceanographic project for the International Geophysical Year, 1957/8: 198 pp.

IGY

Jones, Sir Harold Spencer, 1957.
The International Geophysical Year. J. Inst. Navigation 10(1):17-30.

International Geophysical Year

Jones, H.S., 1956.
British contribution to the International Year.
Research 9(1):33-36.

IGY

Joseph, J., 1961.
On the turbidity distribution in the surface layer of the northern North Atlantic Ocean during the I.G.Y. Symposium on the radiant energy in the sea, Helsinki, 4-5 Aug., 1960.
U.G.G.I. Monogr., No. 10:115.

I.G.Y.

Joseph, Joachim, 1959
Über die vertikalen Temperatur-und Trübungsregistrierungen in einer 500 m mächtigen Deckschicht des nördlichen Nordatlantischen Ozeans.
Deutsche Hydrogr. Zeits., Ergänzungsheft Reihe B (4°) No. 2: 48-55. Also in: Deutsches Hydrogr. Inst., Ozeanogr., 1959(1960), No. 2.

IGY

Josse, A.P., G.S. Koroleva and G.A. Nagaeva, 1963.
Stratigraphical and paleogeographical investigations in the Indian Sector of the South Ocean. (In Russian).
Mezhd. Geofiz. Kom., Prezidiume, Akad. Nauk, SSSR Rezult. Issled., Programme Mezhd. Geofiz. Goda, Okeanol. Issled., (8):137-161.

K

IGY

Kabanova, Yu. G., 1961
[Primary production and nutrients in the Indian Ocean.] (In Russian; English abstract).
Okeanolog. Issled., Mezhd. Komitet Proved. Mezhd. Geofiz. Goda, Prezidiume, Akad. Nauk, SSSR, (4):72-75.

IGY

Kalashnikov, A.G., 1961.
On some types of pulsations of the geomagnetic field and the earth's currents occurring simultaneously on the USSR territory.
Annals, Int. Geophys. Year, 11:96-

IGY

Kanaeva, I.P., 1962.
First report of the Soviet plankton investigations in the program of the IGY-IGC in the Atlantic Ocean. (In Russian)
Trudy Vses. Nauchno-Issledov. Inst. Morsk. Ribn. Chos. i Okean., VNIRO, 46:201-214.

IGY

Khromov, S.P., 1959.
[Atmospheric circulation and weather en route during the voyage of the "Ob" in 1956-1957.]
Arktich. i Antarkt. Nauchno-Issled. Inst., Mezhd. Geofiz. God, Sovetsk. Antarkt. Exped., 5:27-84.

IGY

Klepikova, V.V., Edit., 1961
Third Marine Expedition on the D/E "Ob", 1957-1958. Data.
Trudy Sovetskoi Antarktich. Exped., Arktich. i Antarktich. Nauchno-Issled. Inst., Mezhd. Geofiz. God, 22:1-234 pp.

I.G.Y.

Koopman, Georg., 1959

Thermo-haline Schichtung im jahreszeitlichen Wechsel Zwischen Kap Farvel und der Flämischen Kappe. Deutsche Hydrogr. Zeits., Ergänzungsheft Reihe B (4) No. 3: 42-46. Also in: Deutsches Hydrogr. Inst., Ozeanogr., 1959(1960) No. 7.

IGY

Korotkevich, V.S., and K.V. Beklemishev, 1960

[Zooplankton research.]
Arktich. i Antarkt. Nauchno-Issled. Inst. Sovetsk. Antarkt. Exped., Mezhd. Geofiz. God, Vtoraia Morsk. Exped., "Ob", 1956-1957, 7: 111-125.

IGY

Kort, V.G., 1969.
Oceanographic investigations in the Antarctic.
Annls int. geophys Year, 46: 113-127.

IGY

Kort, V.G., 1963.
Heat exchange in the South Ocean (In Russian).
Mezhd. Geofiz. Kom., Prezidiume, Akad. Nauk, SSSR Rezult. Issled. Programme Mezhd. Geofiz. Goda, Okeanol. Issled., (8):17-23.

IGY

Kort, V.G., 1963
Exchange of waters in the South Ocean. (In Russian).
Mezhd. Geofiz. Komitet, Prezidiume, Akad. Nauk SSSR, Rezult. Issled. Programme Mezhd. Geofiz. Goda, Okeanol. Issled., No. 8:5-16.

IGY

Kort, V.G., 1962.
Oceanographic researches in Antarctica. (In Russian).
Mezhd. Geofiz. Komitet, Prezidiume, Akad. Nauk, SSSR, Rezult. Issled., Programme Mezhd. Geofiz. Goda, Okeanol. Issled., No. 7:7-18.

IGY

Kort, V.G., 1960

[I.G.Y. Results. Oceanography.] Inform. Biull. Mezhdun Geofiz, God, No. 8: 26-34.

I.G.Y.

Koslowski, Gerhard, 1959

Über die Verteilung der Oberflächentemperatur im nördlichen Nordatlantischen Ozean im Spätwinter und Spätsommer, 1958. Deutschen Hydrogr. Zeits., Ergänzungsheft Reihe B (4°) No. 3: 34-40. Also in: Deutsches Hydrogr. Inst., Ozeanogr. 1958(1960).

IGY

Kosminskaya, I.P., S.M. Zverev, P.S. Veitsman, Yu. V. Tulina and R.M. Krakshina, 1963.
Basic features of the crustal structure of the Sea of Okhotsk and the Kuril-Kamchatka zone of the Pacific Ocean from deep seismic sounding data.(In Russian).
Izv., Akad. Nauk, SSSR, Ser. Geofiz., (1):20-41.
Translation:
Amer. Geophys. Union, NAS-NRC, (1):11-27.

IGY

Koutyurin, V. M., A. P. Lisitsin, 1961.
[Vegetable pigments in the suspended material and in bottom sediments of the Indian Ocean.]
Mezhd. Kom. Mezhd. Geofiz. Goda, Presidiume Akad. Nauk, SSSR, Okeanol. Issled., (3):90-116.

IGY

Kovylin, V.M., 1961
[Multichannel acoustic network for seismoacoustic measurements in the ocean.] (In Russia English abstract).
Okeanolog. Issled., Mezhd. Komitet Proved. Mezhd. Geofiz. Goda, Prezidiume, Akad. Nauk, SSSR (4):100-109.

IGY

Kozlyaninov, M.V., 1960

[On optical water characteristics in the central Pacific.]
Trudy Inst. Okeanol., 40:167-174.

IGY

Krauss, W., 1961
Meteorologically forced infernal waves in the region south-west of Iceland. Contribution to Special IGY Meeting, 1959.
Cons. Perm. Int. Expl. Mer, Rapp. Proc. Verb 149:89-92.

IGY

Krey, J., 1961
The vertical distribution of seston in the northern North Atlantic, as observed during cruises of "Gauss" and "Anton Dohrn" in late winter and late summer of 1958.
Contribution to Special IGY Meeting, 1959.
Cons. Perm.Int. Expl. Mer, Rapp. Proc. Verb., 149: 194-199.

IGY

Kusmorskaya, A.P., 1961
Distribution of plankton in the North Atlantic in spring and autumn, 1958. Contribution to Special IGY Meeting, 1959.
Cons. Perm. Int. Expl. Mer, Rapp. Proc. Verb., 149: 183-188.

IGY

Kutyurin, E.M., and A.P. Lisitsin, 1962
[Vegetative pigments present in suspensions and bottom sediments in the western part of the Indian Ocean. Communication. 2. Quantitative distributions and qualitative contents of pigments present in suspensions.]
Mezhd. Geofiz. Komitet, Prezidiume Akad. Nauk, SSSR, Rezult. Issled. Programme Mezhd. Geofiz. Goda, Okeanol. Issled., No. 5:112-129.

IGY

Kutyurin, V.M., 1959.
[Determination of chlorophyll content in sea water and the spectral analysis of phytoplankton pigments.]
Arktich. i Antarktich. Nauchno-Issled. Inst., Mezhd. Geofiz. God, Sovetsk. Antarkt. Eksped., 5:173-176.

L

I.G.Y.

Lacombe, Henri, 1961

Mesures de courant à 1000 metres de profondeur à l'ouest de la côte espagnol (Cap Finisterre).
Cahiers Océanogr., C.C.O.E.C. 13(1): 9-13.

IGY

Lacombe, Henri, et Mme C. Richez, 1961.
Contribution à l'étude du régime du Détroit de Gibraltar. II. Étude hydrologique. Année Géophysique Internationale, 1957-1958. Participation française.
Cahiers Océanogr., C.C.O.E.C., 1395):276-291.

I.G.Y.

Lacombe, Henri, and Paul Tchernia, 1960.
Listes des stations M.O.P. Calypso 241 a 297 (campagne 1958) pour servir a l'etude des echanges entre la Mer Mediterranee et l'Ocean Atlantique. Cahiers Oceanogr., 12(6):417-439

IGY

Lacombe, Henri, and Paul Tchernia, 1960

Liste des stations M.O.P. CALYPSO 176à 234 (Campagne 1957) pour servir à l'étude des échanges entre la mer Méditerranée et l'océan Atlantique. Cahiers Oceanogr., C.C.O.E.C., 12 (3): 204-235.

IGY

Lacombe, Henri et Paul Tchernia, 1958.
Participation française aux recherches océanographiques de l'Anée Géophysique Internationale, 1957-1958.
Bull. d'Info., C.C.O.E.C., 10(9):527-530.

IGY

Laktionov, A.F., 1961.
Oceanographic research in the Central Arctic.
Ann. Int. Geophys. Year, 11:316-318.

IGY

Laktionov, A.F., 1959
[Oceanographic researches in Central Arctic. Some problems and results of oceanographic investigations.]
Mezhd. Kom. Proved. Mezhd. Geofiz. Goda, Prez. Akad. Nauk, SSSR, 17-19.
Moscow 5th Meeting CSAGI
To be published in French and English in Ann. IGY.

IGY

Lauzier, L.M., 1961
Canadian serial temperatures and salinities relative to IGY sea level recording, North Atlantic, 1957-1958.
Contribution to Special IGY Meeting, 1959.
Cons. Perm. Int. Expl. Mer, Rapp. Proc. Verb 149:114.

IGY

Law, P., 1959

The IGY in Antarctica.
Australian J. Sci., 21(9): 285-294.

IGY

Lebedeva, M.N., 1960.
[Microbiological research.]
Arktich. i Antarktich. Nauchno-Issled. Inst., Mezhd. Geofiz. God, Sovetsk. Antarkt. Eksped., Vtoraia Morsk. Eksped., "Ob", 1956-1957, 7:153-164.

IGY

Lee, A., 1961
Hydrographic conditions in the Barents Sea and Greenland Sea during the IGY compared with those in previous years.
Contribution to Special Meeting, 1959.
Cons. Perm. Int. Expl. Mer, Rapp. Proc. Verb. 149:40-43.

I.G.Y.

Leipper, D.F., 1961.
I.G.Y. Cruise of the Hidalgo. (Summary).
Ann. Int. Geophys. Year, 11:310.

IGY

LeJay, P. (1958).
Participation française à l'Année Géophysique Internationale.
C.R. Com. Nat. Fran. Géod. et Géophysique, Année, 1957:16-21.

IGY

Leontyeva, V. V., 1962.
Hydrological studies of oceanic trenches of the Pacific Ocean and some problems of further research.
Mezhd. Geofiz. Komitet, Prezidiume Akad. Nauk, SSSR, Rezult. Issled. Programme Mezhd. Geofiz. Goda, Okeanol. Issled., No. 5:31-42.

IGY

Leontyeva, V. V., 1961.
Current and water masses in the western part of the Pacific Ocean in summer 1957.
Mezhd. Kom. Mezhd. Geofiz. Goda, Presidiume Akad. Nauk, SSSR, Okeanol. Issled., (3):137-150.

IGY

Leontyeva, V.V., 1960
Some data on the hydrology of the Tonga and Kermadec Trenches.
Trudy Inst. Okeanol., 40:72-82.

IGY

Linden, N.A., 1961.
Seismicity of the Arctic region.
Annals, Int. Geophys. Year, 11:375-387.

IGY

Linkov, Ye. M., and Ye. F. Savarensky, 1960.
The apparatus for recording the motion trajectory during the microseismic oscillations.
Seism. Issle. Mezhd. Kom. Proved. Mezhd. Geofiz Goda, Praes., Akad. Nauk, SSSR, 133-137.

IGY

Lisitsin, A. P., 1962.
Distribution and contents of suspensions of the Indian Ocean waters. Communication 3. Comparison of granulometric contents of suspensions and bottom sediments.
Mezhd. Geofiz. Komitet, Prezidiume Akad. Nauk, SSSR, Rezult. Issled. Programme Mezhd. Geofiz. Goda, Okeanol. Issled., No. 5:130-139.

IGY

Lisitsin, A. P., 1961.
Distribution and content of suspended material in the waters of the Indian Ocean.
Mezhd. Kom. Mezhd. Geofiz. Goda, Presidiume Akad. Nauk, SSSR, Okeanol. Issled., (3):52-89.

I.G.Y.

Lisitsin, A.P., 1960
Marine geological research. Arktich. i Antarktich. Nauchno-Issled. Inst., Sovetsk. Antarkt. Exped., Mezhd. Geofis. God, Vtoraia Morsk. Exped., "Ob", 1956-1957, 7: 7-43. (In Russian)

IGY

Lisitzin, A.P., and A.V. Zhivago, 1958
VII. Submarine geology.
Opisanie Exped. D/E "Ob" 1955-1956, Mezhd. Geofiz. God, Trudy Kompaeksnoi Antarkt. Exped., Akad. Nauk, SSSR: 103-144.

IGY

Lukyanov, V.V., and L.K. Moiseyev, 1961
Distribution of water temperatures in the northern Indian Ocean. (In Russian; English abstract).
Okeanolog. Issled., Mezhd. Komitet Proved. Mezhd. Geofiz. Goda, Prezidiume, Akad. Nauk, SSSR (4):31-43.

IGY

Lukyanov, V.V., N.P. Nefedyev, Yu. A. Romanov, 1962
On the scheme of surface currents in the Indian Ocean during the winter monsoon.
Mezhd. Geofiz. Komitet, Prezidiume Akad. Nauk, SSSR, Rezult. Issled. Programme Mezhd. Geofiz. Goda, Okeanol. Issled., No. 5:19-24.

IGY

Lumby, J.R., 1961
A note on the IGY cruise data in the Atlantic Ocean. Contribution to Special IGY Meeting, 1959.
Cons. Perm. Int. Expl. Mer, Rapp. Proc. Verb., 149:97-102.

IGY

Lumby, J.R., 1959.
The I.G.Y. world data centers for oceanography.
Texas J. Sci., 11(3):259-269.

Also in: Contributions in Oceanography and Meteorology, A. & M. College of Texas, Vol. 5, Contrib. No. 139.

IGY

Lyakov, M.E., 1960.
Characteristics of the climatic zones in the central Pacific.
Trudy Inst. Okeanol., 40:23-28.

IGY

Lyman, John, 1961.
The U.S. Navy IGY Antarctic Program in oceanography. (Summary).
Annals Int. Geophys. Year, 11:310.

IGY

Lyman, John, 1958.
The U.S. Navy International Geophysical Year program in oceanography.
Int. Hydrogr. Rev., 35(2):111-126.

M

IGY

Man, I.A., 1959.
The second cruise of the diesel-electric ship "Ob".
Arktich. i Antarktich. Nauchno-Issled. Inst., Mezhd. Geofiz. God, Sovetsk. Antarkt. Eksped., 5:19-26.

IGY

Martin, Jean, 1962
Utilisation du courantomètre électrique à électrodes remorquées (1955-1956).
Océanogr. Phys., Année Géophys. Internat., Part. Francais, Centre Nat. Recherche Sci., Ser. 10, (1):11-55.

IGY

Massey, Harrie, 1963
The International Geophysical Year in retrospect.
Endeavour, 22: 70-74.

I.G.Y.

Maximov, I.V., 1959
The Second Antarctic Sea Expedition. Arktich. i Antarktich. Nauchno-Issled. Inst., Mezhd. Geofiz. God, Sovetsk. Antarkt. Exped. 5: 7-18.

Mc

IGY

McGill, David A., 1964.
The distribution of phosphorus and oxygen in the Atlantic Ocean as observed during the I.G.Y., 1957-1958.
Progress in Oceanography, Pergamon Press, 2:127-211.

IGY

McLellan, Hugh J., 1960.
An I.G.Y. cruise from Texas A. & M.
Int. Hydrogr. Rev., Jan. 1960, 139-148.

Also in:
Contrib. Oceanogr. and Meteorol., A. & M. Coll. of Texas, 5(145).

IGY

Menendez, N., 1961
Distribution de la salinité et de la température au méridien de Tarifa en août 1958. Contribution to Special IGY Meeting, 1959.
Cons. Perm. Int. Expl. Mer, Rapp. Proc. Verb 149:133-137.

IGY

Mertins, H.O., and F. Krügler, 1959.
Über die Wetterverhältnisse während der Forschungsfahrten von F.F.S. "Anton Dohrn" und V.F.S. "Gauss" in Internationalen Geophysikalischen Jahr 1958.
Deutsche Hydrogr. Zeits., Ergänzungsheft Reihe B (4) No. 3:21-25.

IGY

Midttun, L., 1961
Norwegian hydrographical investigations in the Barents Sea during the International Geophysical Year. Contributions to Special IGY Meeting, 1959.
Cons. Perm. Int. Expl. Mer, Rapp. Proc. Verb. 149: 25-30.

IGY

Minkina, A. L., 1961.
Organic nitrogen and phosphorus in the water masses of the Barents Sea.
Mezhd. Kom. Mezhd. Geofiz. Goda, Presidium Akad. Nauk, SSSR, Okeanol. Issled., (3):162-171.

IGY

Miyake, Y. 1969.
Artificial radioactivity.
Annls int. geophys. Year, 46: 187-190

IGY

Miyake, Y., 1961.
The distribution of man-made radioactivity in the North Pacific (through the summer of 1955). (Summary).
Ann. Int. Geophys. Year, 11:310-311.

IGY

Mokiyevskaya, V.V., 1961
Some peculiarities of hydrochemistry in the northern Indian Ocean. (In Russian; English abstract).
Okeanolog. Issled., Mezhd. Komitet Proved. Mezhd. Geofiz. Goda, Prezidiume, Akad. Nauk, SSSR, (4):50-61.

IGY

Monakhov, F.I., 1961.
The conditions of the origin and propagation of North Atlantic microseisms.
Annals, Int. Geophys. Year, 11:418-438.

IGY

Monakhov, F.I., 1960
Microseisms in connection with cyclones at seas and in oceans.
Seism. Issle. Mezhd. Kom. Proved., Mezhd. Geofiz. God, Praes., Akad. Nauk, SSSR: 78-104.

IGY

Monakhov, F.I., and O.A. Korchagina, 1962
Conditions of the formation and propagation of microseisms in the northwestern part of the Pacific Ocean.
Geophys. Mag., Japan Meteorol. Agency, 31(2): 257-274.

IGY

Monges Lopez, R., 1957.
El programa de Mexico para el Ano Geophisico Internacional. Anales Inst. Geofis., U.N.A.M., 2: 113-131.

IGY
Monges Lopes, R., 1956.
El programa de Mexico para el Año Geofisico Internacional. Anales Inst. Geofis., Univ. Nac. Aut., Mexico, 2:113-131.

IGY
Morandi, G., 1962
L'anno geofisico internaxionale: l'oceanografia (prima parte).
Scientia, 97(607-608):239-246.

IGY
Moroshkin, K.V., 1959.
Hydrological investigations.
Arktich. i Antarkt. Nauchno-Issled. Inst., Mezhd. Geofiz. God, Sovetsk. Antarkt. Eksped., 5:107-124.

IGY
Moroshkin, K.V., and M.A. Bogdanov, 1959
Results of the use of the geomagnetic electrokinetograph in the Indian Ocean and the southern part of the Pacific Ocean.
Arktich. i Antarktich. Nauchno-Issled. Inst., Mezhd. Geofiz. God, Sovetsk. Antarkt. Eksped., 5: 125-139.

IGY
Morozov, A.P., 1959.
Observations on rough sea.
Arktich. i Antarkt. Nauchno-Issled. Inst., Mezhd. Geofiz. God, Sovetsk. Antarkt. Eksped., 5:140-143.

N

IGY
National Academy of Sciences, 1960.
Seismic crustal studies during the IGY. 1. Marine program. I.G.Y. Bull., No. 33.
Trans. Amer. Geophys. Union, 41(1):107-113.

IGY
National Academy of Sciences, 1959.
Seismic studies in the western Caribbean.
IGY Bull., No. 21.
Trans. Amer. Geophys. Union, 40(1):73-75.
Based on a report made at IGY symposium, Dec. 1958 by John W. Antoine.

IGY
Navrotsky, V.V., 1964.
Some results of research studies on the interaction of theocean and atmosphere in the Gulf Stream during the IGY. (In Russian).
Okeanologiia, Akad. Nauk, SSSR, 4(4):603-611.

IGY
Nekrasov, V.A., 1961.
Preliminary charts of the main diurnal and main semi-diurnal tidal waves in the Norwegian Sea and the Greenland Sea.
Issled. Severnoi Chasti Atlanticheskogo Okeana, Mezhd. Geofiz. God, Leningradskii Gidrometeorol. Inst., 1:81-87.

IGY
Nekrasova, V.A., and V.N. Stepanov, 1963.
Meridional hydrological profiles of the oceans from IGY data. (In Russian).
Mezhd. Geofiz. Kom., Prezidiume, Akad. Nauk, SSSR Rezult. Issled. Programme Mezhd. Geofiz. Goda, Okeanol. Issled., (8):34-51.

Nemchenko, V.I., and V.I. Tishunina, 1963. IGY
Investigation of the thermal structure of the surface layer of the ocean in the northern part of the Atlantic. (In Russian; English abstract).
Rez. Issled. Programme Mezhd. Geofiz. Goda, Okeanolog. Issled., Akad. Nauk, SSSR, No. 8:97-103.

IGY
Neyman, V.G., 1961.
Factors conditioning the oxygen minimum in the subsurface waters of the Arabian Sea.
Okeanol. Issledov., Mezhd. Komit., Mezhd. Geofiz. God, Presidiume, Akad. Nauk, SSSR,:62-65.

IGY
Neyman, V.G., 1961
Dynamic map of Antarctica.
Mezhd. Kom. Mezhd. Geofiz. Goda, Presidiume Akad. Nauk, SSSR, Okeanol. Issled., 3:117-123.

IGY
Newell, H. E., and J. W. Townsend, Jr., 1959.
IGY Conference in Moscow.
Science, 129(3341):79-84.

IGY
Nguyen-Hai and Nguyen-Dinh-Ba, 1961
Quelques observations hydrologiques dans la région Cap St. Jacques-Poulo Condore.
Ann. Fac. Sci., Saigon, 1961:319-324.
Also: Contrib. Inst. Océanogr. Nhatrang No.47.

IGY
Novitsky, V.P., 1963
Hydrological features of the Aegean Sea in October 1959. (In Russian).
Mezhd. Geofiz. Komitet, Prezidiume, Akad. Nauk, SSSR, Rezult. Issled. Programme Mezhd. Geofiz. Goda, Okeanol. Issled., (8):78-89.

O

IGY
Ovchinnikov, I.M., 1963.
Some peculiarities of water circulation in the Alaska Bay. (In Russian).
Mezhd. Geofiz. Kom., Prezidiume, Akad. Nauk SSSR, Rezult. Issled. Programme Mezhd. Geofiz. Goda, Okeanol. Issled., (8):67-75.

IGY
Ovchinnikov, I.M., 1961
Circulation of waters in the northern Indian Ocean during the winter monsoon. (In Russian; English abstract).
Okeanolog. Issledov., Mezhd. Komitet Proved. Mezhd. Geofiz. Goda, Prezidiume Akad. Nauk, SSSR, (4):18-24.

P

I.G.Y.
Pasternak, F.A., and A.V. Gusev, 1960
Benthonic research.
Arktich. i Antarkt. Nauchno-Issled. Inst., Sovetsk. Antarkt. Exped., Mezhd. Geofiz. God, Vtoraia Morsk. Exped., "Ob", 1956-1957, 7: 126-142.

IGY
Pattullo, June G., 1960
Seasonal variations in sea level in the Pacific Ocean during the International Geophysical Year, 1957-1958. J. Mar. Res., 18(2): 168-184.

IGY
Pavlov, V.M., 1961
Optical properties of the main water masses in the northern Indian Ocean. (In Russian; English abstract).
Okeanolog. Issled., Mezhd. Komitet Proved. Mezhd. Geofiz. Goda, Prezidiume, Akad. Nauk, SSSR, (4):44-49.

IGY
Pavshtiks, E.A., and L.N. Grusov, 1961
Distribution of plankton in the Norwegian Sea in 1958. Contribution to Special IGY Meeting 1959.
Cons. Perm. Int. Expl. Mer, Rapp. Proc. Mer, 149:176-178.

IGY
Peluchon, G., 1961
Observations hydrologiques effectuées par "L'Aventure" en 1957 et en 1958 au nord du 48ème parallèle. Contribution to Special IGY Meeting, 1959.
Cons. Perm. Int. Expl. Mer, Rapp. Proc. Verb. 149: 113.

IGY
Petelin, V.P., and N.N. Sisoyev, 1960.
The third and fourth cruises of the VITYAZ IGY programme.
Inform. Biull., Mezhdun. Geofiz. God, No. 8:46-59.

IGY
Plachotnik, A.F., 1962.
Hydrology of the northeastern part of the Pacific Ocean (literature survey). Investigations on the program of the International Geophysical Year.
Trudy Vses. Nauchno-Issled. Inst. Morsk. Ribn. Chos. i Okean, VNIRO, 46:190-201.
In Russian

IGY
Plachotnik, A., 1958.
The importance of the IGY for ocean voyages.
Mor. Flot (11):20-22.

IGY
Potaichuk, M.S., 1962.
Tests on calculation of the amount of heat advection in North Atlantic currents based on the data of the I.G.Y. (In Russian).
Trudy Gosud. Okeanogr. Inst., 67:86-103.

IGY
Proskouryakova, T.A., 1960.
Some results of microseismic observations in Pulkovo and Yalta.
Seismol. Issle., Mezhd. Kom. Proved., Mezhd. Geofiz. Goda, Praes. Akad. Nauk, SSSR, 105-115.

Q

R

I.G.Y.
Rikhter, G.D., 1960
Processes of denudation in Antarctica. Arktich i Antarktich. Nauchno-Issled. Inst. Mezhd. Geofiz. God, Sovetsk. Antarkt. Exped., Vtoraia Morsk. Exped., "Ob", 1956-1957. 7: 73-84. (In Russian)

IGY
Riznichkenko, Ou. V., 1957.
On a study of the formation of the earth's crust in the period of Third International Geophysical Year. Izv. Akad. Nauk, Ser. Geofiz., 1957(2):129-140.

IGY
Romankevich, E.A., 1962
Organic substance in the surface layer of bottom sediments.
Mezhd. Geofiz. Komitet, Prezidiume Akad. Nauk, SSSR, Rezult. Issled. Programme Mezhd. Geofiz. Goda, Okeanol. Issled., No. 5:67-111.

IGY
Romanov, Yu. A., 1961
Dynamic method as applied to the Equatorial Indian Ocean. (In Russian; English abstract).
Okeanolog. Issled., Mezhd. Komitet Proved. Mezhd. Geofiz. Goda, Prezidiume, Akad. Nauk, SSSR (4):25-30.

S

IGY
Sabinin, C.D., and A.E. Gamutilov, 1960
Possible use of the laboratory interpherometer ITR-2 for sea water salinity determinations.
Trudy Inst. Okeanol., 40: 175-183.

IGY

Saelen, O.H., 1961
Preliminary report on current measurements in 1958 on the Galicia Bank west of Cape Finisterre. Contribution to Special IGY Meeting, 1959.
Cons. Perm. Int. Expl. Mer. Rapp. Proc. Verb., 149:130-132.

IGY

Saelen, O.H., 1961
Preliminary report on the hydrographic sections made in the Norwegian Sea 1958 by the Geophysical Institute, Bergen. Contribution to Special IGY Meeting, 1958.
C ons. Perm. Int. Expl. Mer. Rapp. Proc. Verb 148:56-59.

IGY

Savarensky, E.F., L.N. Rykunov, T.A. Proskuryakova and W.M. Prosvirnine, 1961.
The influence of the Scandinavian relief on the propagation of microseisms.
Annals, Int. Geophys. Year, 11:439-444.

IGY

Savarensky, Ye. F., and N.G. Valdner, 1960
[The waves Lg and Rg from the Black Sea basin earthquakes and some comments on their origin.]
Seism. Issle. Mezhd. Kom. Proved. Mezhd. Geofiz. Goda, Praes.Akad. Nauk, SSSR,55-77.

IGY

Schaefer, H., 1961
On the variation in the pattern of free amino acids in the skeletal muscle of freshly caught oceanic fish from different localities. Contribution to Special IGY Meeting, 1959.
Cons. Perm. Int. Expl. Mer. Rapp. Proc. Verb 149:167-174.

IGY

Seryakov, E.I., and V.M. Radikevich, 1961
[On the changeability of the heat balance components in the southern part of the Barents Sea during the period of the IGY.]
Issled. Severnoi Chasti Atlanticheskogo Okeana, Mezhd. Geofiz. God, Leningradskii Gidrometeorol Inst., 1: 43-51.

IGY

Shellard, H.C., 1957.
The International Geophysical Year.
Mar. Obs. (M.O. 624) 27(175):20-26.

IGY

Sirotov, K.M., 1963.
Swell accompanying waves. (In Russian).
Mezhd. Geofiz. Kom., Prezidiume, Akad.Nauk SSSR, Rezult. Issled. Programme Mezhd. Geofiz. Goda, Okeanol. Issled., (8):76-77.

IGY

Sirotov, K.M., 1962.
The study of sea waves. (In Russian).
Mezhd. Geofiz. Komitet, Prezidiume Akad. Nauk, SSSR, Rezult. Issled. Programme Mezhd. Geofiz. Goda, Okeanol. Issled., No. 7:105-113.

IGY

Smetanin, D.A., 1960
[Some features of the chemistry of water in the central Pacific.]
Trudy Inst. Okeanol., 40: 58-71.

I.G.Y.

Sobolev, L.G., 1959
[The works of the aerometeorological party.]
Arktich. i Antarktich. Nauchno-Issled., Inst., Mezhd. Geofiz God, Sovetsk. Antarkt. Exped., 5: 101-106. (In Russian)

IGY

Solovyev, V.S., 1960
[Zero dynamic surface in the western Pacific.]
Trudy Inst. Okeanol., 40: 152-161.

Spar, J., 1962 IGY
Earth, sea and air. A survey of the geophysical sciences. Addison-Wesley Publishing Co., Reading, Mass., 152 pp.
$2.95 bound Hard
$1.75 paper-back

IGY

Spilhaus, A.F., 1957.
The International Geophysical Year.
J. Madras Univ., B, 27(1):211-224.

IGY

Steele, J.H., 1961
Notes on the deep water overflow across the Iceland-Faroe Ridge. Contribution to Special IGY Meeting, 1959.
Cons. Perm. Int. Expl. Mer. Rapp. Proc. Verb 149:84-88.

IGY

Steemann Nielsen, E., and V.K. Hansen, 1961
The primary production in the waters west of Greenland during July 1958. Contribution to Special IGY Meeting, 1959.
Cons. Perm. Int. Expl. Mer. Rapp. Proc. Verb. 149:158-159.

IGY

Sutton, O.G., 1957.
The Meteorological Office and the International Geophysical Year. Mar. Obs. (M.O. 624) 27(175): 19-20.

IGY

Sysoev, N.N., editor, 1961.
[Oceanological investigations of the "Vitiaz" in the northern part of the Pacific Ocean as part of the IGY program.]
Trudy Inst. Okeanol., Akad. Nauk, SSSR, 45:283 pp

IGY

Sysoyev, N.N., 1961.
[The study of deep currents in the Pacific during the IGY period.]
Informats. Biull., Mezhd. Geofiz. God, Akad. Nauk, SSSR, No. 9:17-21.

IGY

Sytinsky, A.D., 1960.
Microseisms in Mirny and their connection with hydrometeorological conditions.]
Seism. Issle. Mezhd. Kom. Proved., Mezhd. Geofiz. Goda, Praes. Akad. Nauk, SSSR, 116-132.

T

IGY

Tait, J.B., and J.H.A. Martin, 1961
The Atlantic Current and water masses in the Faroe-Shetland Channel and over the Iceland-Faroe Ridge during the IGY. Contribution to Special IGY Meeting, 1959.
Cons. Perm. Int. Expl. Mer. Rapp. Proc. Verb. 149: 60-83.

IGY

Takahasi, Ryutaro, Kintaro Hiranom Isamu Aida, Tokutaro Hatori and Shizuko Shimizu, 1961
Observations at Miyagi-Enoshima Tsunami Observatory during the IGY period.
Bull. Earthquake Res. Inst., 39(3):491-521.

IGY

Takahashi, Tadao, Masaaki Chaen and Soichi Ueda, 1960.
Report of the Kagoshima-maru IGY cruise, 1958.
Mem. Fac. Fish., Kagoshima Univ., 8: 82-86.

IGY

Tareyev, B.A., A.V. Fomitchev, 1963
On surface currents of the South Ocean. (In Russian).
Mezhd. Geofiz. Komitet, Prezidiume, Akad. Nauk SSSR, Rezult. Issled. Programme Mezhd. Geofiz. Goda, Okeanol. Issled., (8):24-33.

IGY

Taylor, E.G.R., 1957.
Operacion I.G.Y.,:review. Geogr. J. 123(4):514-516.

IGY

Teisseyre, R., 1961.
The dislocation processes in the Pacific Ocean.
Annals, Int. Geophys. Year, 11:411-413.

IGY

Timonov, V.V., 1961
[A study of variations of the hydrometeorological condition in the northern Atlantic in relation to the IGY and IGC.]
Okeanologia, 2: 220-225.

IGY

Tomczak, G., 1961
Ergebnisse der ozeanographischen Forschung im Internationalen Geophysikalischen Jahr. II. Topographie des Meeresbodens - Die grossen Meereströmungen im Pazifischen Ozean Weiteres Forschungsziel: Indischer Ozean.
Umschau, (18):570-572.

I.G.Y.

Tomashunas, B.Y., 1959
[Ice observations.]
Arktich. i Antarktich. Nauchno-Issled. Inst., Mezhd. Geofiz. God, Sovetsk. Antarkt. Eksped., 5: 154-158.

IGY

Tomczak, G., 1961.
Ergebnisse der ozeanographischen Forschung im Internationalen Geophysikalischen Jahr. I. Erforschung der Zirkulation im Atlantischen Ozean.
Umschau, (16):499-502.

Abstract in:
Oceanogr., 1961, Deutsch. Hydrogr. Inst.,

IGY

Tressler, Willis L., 1960
Oceanographic and hydrographic observations at Wilkes IGY Station, Antarctica.
J. Washington Acad. Sci., 50(5):1-13.

U

IGY

Ulintsev G.B., 1969.
IGY investigations of submarine topography.
Annls int. geophys. Year, 46:191-147

IGY
Udintsev, G.B., 1962.
Studies of the submarine relief. (In Russian; English abstract).
Mezhd. Geofiz. Komitet, Presidiume, Akad. Nauk, SSSR, Rezult. Issled. Programme Mezhd. Geofiz. Goda, Okeanol. Issled., No. 7:33-48.

IGY
Ulrich, J., 1962
Echolotprofile der Forschungsfahrten von F.F.S Anton Dohrn und V.F.S. Gauss im Internationalen Geophysikalischen Jahr, 1957/1958.
Deutsch. Hydrogr. Zeits., Erganzungsheft, 4(6) 15 pp.

IGY
U.S.A., National Academy of Sciences, 1961.
Arctic Basina seismic studies from IGY Drifting Station Alpha - Hunkins, K.
Trans. Amer. Geophys. Union, IGY Bull., No. 46: 1-5.

Original Report prepared by the Lamont Geological Observatory:
Seismic studies of the Arctic Ocean floor, 1960.
Geophys. Res. Directorate, Air Force Cambridge Res. Lab., AFCRC-TN-60,257.

IGY
U.S. National Academy of Sciences-National Research Council, 1961.
Science in Antarctica. The physical sciences in Antarctica. Part II, Report by the Committee of Polar Research.
NAS-NRC, Publ., No. 878:131 pp.

IGY
U.S. National Academy of Science -National Research Council, 1961.
Science in Antarctica. The Life sciences in Antarctica. Part I Report by the Committee on Polar Research.
NAS-NRC, Publ., No. 839:162 pp.

IGY
USSR, Akademia Nauk, 1960.
[Personnel taking part in the three cruises of the e/s "Vitiaz" on the plans of the International Geophysical Year.]
Trudy Inst. Okeanol., 40:189-192.

IGY
USSR, Committee for the International Geophysical Year, 1959.
[Some problems and results of the oceanographic investigations.] 33 pp.

IGY
USSR
SSSR, Mezhduvedomstvennii Geofizicheskii Komitet pri Presidiume Akademii Nauk, SSSR, 1961.
[Bibliographic index to the literature in the Russian language, 1960.]
Mezhd. Geofiz. God, 124 pp.

V

IGY
Van Dorn, W.G., and W.F. Dorn, 1969.
Long waves.
Annls int. geophys. Year, 46: 46-60

IGY
Vasiliev, V.G., P.S. Veitsman, Ye. I. Galperin, V.A. Gladun, A.V. Goryachev, S.M. Zverev, I.P. Kosminskaya, R.M. Krakshina, V.L. Panteleyev, O.N. Solovyov, Ye. A. Starshinova, and S.A. Fedotov, 1960.
[Crustal structure researches in the transition region from the Asiatic continent to the Pacific. The Pacific Geological and Geophysical Expedition of the USSR Akademy of Sciences, 1958.]
(over)

IGY
Vilensky, J.G., B.H. Gluhovsky, J. M. Krilov and A.A. Yuschak, 1959
[Some results and methods of the study of sea wind disturbance. Some problems and results of oceanographic investigations.]
Mezhd. Kom. Proved. Mezhd. Geofiz. Goda, Prez. Akad. Nauk, SSSR, 29-33.
Moscow 5th CSAGI meeting
To beppublished in French and English in Ann. IGY.

IGY
Vinogradov, L.A., 1962
[Qualitative and quantitative distribution of phytoplankton in different water masses of the Norwegian Sea in October 1958.]
MEZHD. Geofiz. Komitet, Prezidiume Akad. Nauk SSSR, Rezult. Issled. Programme Mezhd. Geofiz. Goda, Okeanol. Issled., No. 5:140-154.

I.G.Y.
Vinogradov, M.E. 1969.
The study of plankton during IGY cruises.
Annls int. geophys. Year 46: 177-209.

IGY
Vinogradov, M. Ye., 1962
The study of plankton by the Soviet and foreig expeditions. (In Russian; English summary).
Rez. Issled. Programme Mezhd. Geofiz. Goda, Okean. Issled., No. 7: 84-96.

IGY
Voipio, A., 1961
The iodine content of Barents Sea water. Contribution to Special IGY Meeting, 1959.
Cons. Perm. Int. Expl. Mer, Rapp. Proc. Verb 149:38-39.

W

IGY
Waterman, A.T., 1956.
The International Geophysical Year.
Amer. Scientist 44(2):130-133.

IGY
Weidemann, H., 1961
Results of current measurements by towed electrodes in the northern North Atlantic during the IGY. Contribution to Special IGY Meeting, 1959.
Cons. Perm. Int. Expl. Mer, Rapp. Proc. Verb., 149:115-117.

I.G.Y.
Weidemann, Hartwig, 1959
Strommessungen vom fahrenden Schiff auf F.F.S. "Anton Dohrn" und V.F.S. "Gauss" während der Fahrten im Internationalen Geophysikalischen Jahr 1958. Deutschen Hydrogr. Zeits., Ergänzungsheft Reihe B (4°), No. 3: 59-66. Also in: Deutsches Hydrogr. Inst., Ozeanogr., 1959(1960)

IGY
Wexler, H., 1957.
Meteorology in the International Geophysical Year.
Scientific Monthly, 84(3):141-145.

International Geophysical Year
Wexler, H., 1956.
Antarctic research. Internationa Geophysical Year. Geophysics 21(3):681-690.

IGY
Wiborg, K.F., 1961
Some remarks on the distribution of zooplankton, fish eggs and larvae in the Norwegian Sea in 1958. Contribution to Special IGY Meeting, 1959.
Cons. Perm. Int. Expl. Mer, Rapp. Proc. Verb. 149:175.

IGY
Worthington, L.V., and W.G. Metcalf, 1961
The relationship between potential temperature and salinity in deep Atlantic water. Contribution to Special IGY Meeting, 1959.
Cons. Perm. Int. Expl. Mer, Rapp. Proc. Verb 149:122-128.

XYZ

IGY
Yoshida, S., H. Nitani and N. Suzuki, 1959.
Report of multiple ship survey in the equatorial region (I.G.Y.), Jan.-Feb., 1958.
Publ. 981, Hydrogr. Bull., No. 59:1-30.

IGY
Zaitsev, G.N., 1962.
Oceanographic investigations in the Norwegian Sea according to the IGY-IGS plan. (In Russian)
Trudy Vses. Nauchno-Issledov. Inst. Morsk. Ribn. Chos. i Okean., VNIRO, 46:6-13.

IGY
Zaitsev, G.N., M.V. Fedosov, N.L. Iljina and I.A. Ermachenko, 1961
Components of the water: thermic and chemical budget of the Greenland and Norwegian seas. Contribution to Special IGY Meeting, 1959.
Cons. Perm. Int. Expl. Mer, Rapp. Proc. Verb. 149:46-52.

IGY
Zaitsev, G.N., and S.I. Potaitchuk, 1960
[The second cruise of R/S "Sevastopol" in the Norwegian Sea in accordance with the IGY program.]
Biull. Okeanogr. Komissii, Akad. Nauk, SSSR, (6): 50-54.

IGY
Zalemaee, B.S., 1957.
On the problem and practicability of the advisory committee affiliated with the presidium of the SSSR in the preparation and conduct of the International Geophysical Year.
Izv. Akad. Nauk, SSSR, Ser. Geofiz., (7):965-967

IGY
Zatonskiy, L. K., V. F. Kanayev and G. B. Udintsev, 1961.
[Geomorphology of the submarine part of the Kuril-Kamchatka island arc.]
Mezhd. Kom. Mezhd. Geofiz. Goda, Presidiume Akad. Nauk, SSSR, Okeanol. Issled., (3):124-136.

IGY
Zenkevitch, L.A., 1961.
Certain aspects of ocean depth studies.
Annals. Int. Geophys. Year, 11:303-308.

IGY
Zenkevich, L.A., 1959
[Some questions connected with the study of the ocean depths.
Some problems and results of oceanographic investigations.]
Mezhd. Kom. Proved. Mezhd. Geofiz Goda, Prez, Akad Nauk, SSSR, 7-11.
To be published in French and English in Ann. IGY. Moscow 5th Meeting CSAGI.

IGY
Zhivago, A.V., 1962
[Investigation methods of floor topography in the Antarctic.]
Mezhd. Geofiz. Komitet, Prezidiume Akad. Nauk, SSSR, Rezult. Issled. Programme Mezhd. Geofiz. Goda, Okeanol., Issled., No. 5: 60-66.

I.G.Y.

Zhivago, A.V., 1960
[Geomorphological research.] Arktich. i Antarktich. Nauchno-Issled. Inst. Mezhd. Geofiz. God Sovetsk. Antarkt. Exped., Vtoraia Morsk. Exped "Ob", 1956-1957, 7: 44-72. (In Russian)

IGY

Ziegelmeier, E., 1961
Investigations on the bottom fauna in the shelf region off SW Iceland and SE Greenland during the research cruises with "Anton Dohrn" in the IGY. Contribution to Special IGY Meeting, 1959.
Cons. Perm. Int. Expl. Mer, Rapp. Proc. Verb 149:209.

IGY

Ziegelmeier, E., 1959
Untersuchungen der Bodenfauna auf den Schelfgebeiten vor Südostgrönland und vor SS Island im Spätwinter und Spätsommer 1958.
Deutsch. Hydrogr. Zeits., Ergänzungsheft, Reihe B, 4(3):99-102.

IGY

Zvereva, A.A., Edit., 1959.
[Data, 2nd Marine Expedition of the "Ob",1956-1957] Arktich. i Antarkt. Nauchno-Issled. Inst., Mezhd. Geofiz. God, Sovetsk. Antarkt. Exped., 6: 1-387.

IGY-IGC

Istapoff, Feodor, 1962
The salinity distribution at 200 meters and the Antarctic frontal zones.
Deutsche Hydrographische Zeits., 15(4):133-142.

institutions

Dean, J.R., 1963.
The International Hydrographic Bureau.
Geogr. J., London, 129(4):503-504.

institutions, internationa

Moitoret, Victor A., 1971.
The International Hydrographic Bureau: 50 years of progress. Int. hydrogr. Rev. 48(2): 7-30.

International Indian Ocean Expedition

International Indian Ocean Expedition

India, Indian National Committee on Oceanic Research, Council of Scientific and Industrial Research, New Delhi, 1964.
International Indian Ocean Expedition, Newsletter India, 2(4):48 pp.

International Indian ocean Expedition

Rice, Anthony Leonard, 1970.
Decapod crustacean larvae collected during the International Indian Ocean Expedition. Families Raninidae and Homolidae.
Bull. Brit. Mus. (N.H.), Zool. 21(4): 1-24

Joint Commission on Oceanography

Joint Commission on Oceanography

Ovey, C.D., 1955.
The work of the Joint Commission on Oceanography, 1951-54, of the International Council of Scientific Unions. Deep-Sea Res. 2(2):159.

International Joint Commision on Oceanography.

Wiseman, J.D.H., 1953.
International collaboration in deep-sea research
Deep-sea Res. 1(1):3-10.

Pacific Science Association

conferences

Anon - 1946
Proceedings of the Pacific Science Conference of the National Research Council. Bull. Nat. Res. Council No.114, 79 pp.

SACLANT-La Spezia

SACLANT -La Spezia

Allan, T.D., 1961.
Oceanographic group.
Rapp. Proc. Verb., Réunions, Comm. Int. Expl. Sci., Mer Méditerranée, Monaco, 16(3):717.

Committee on the Criteria and Nomenclature of the Major Subdivisions of the Ocean, Bottom.
Union Géodésique et Géophysique International, 1940.
Report of the Committee on the Criteria and Nomenclature of the Major Subdivisions of the Ocean Bottom.
Publ. Sci., Assoc. Océan. Phys., No. 8:124 pp., charts.

UNESCO

IOC

Lacombe, H., 1967.
La coopération internationale en océanographie.
Revue Géogr. phys. Géol. dyn., 9(3): 191-198.

organizations

United Nations, Food and Agriculture Organization 1955.
UNESCO International Advisory Committee on Marine Sciences. F.A.O. Fish. Bull. 8(3):141-143.

institutions

Union Geodesique et Geophysique Internationale, 1961.
UNESCO Office of Oceanography.
I.U.G.G. Chronicle, 35:67-68.

World Meteorological Organization

World Meteorological Organization WMO

Snodgrass, James M., 1962
An oceanographic data and communications system.
Proc. 2nd Interindustrial Oceanogr. Symposium Lockheed Aircraft Corp., 31-39.

Italy

institutions

Salvatori, B., 1963.
Réorganisation du réseau marégraphique italien.
Rapp. Proc. Verb., Réunions, Comm. Int., Expl. Sci., Mer Méditerranée, Monaco, 17(3):873-874.

institutions

Vercelli, F., 1948.
Centrò di studi talassografi. Ric. Sci. 18:1286-1290.

Laboratorio Centrale di Idrobiologia

Institutions

Maldura, Carlo M., 1958(1959)

Relazione sull'attivita del Laboratorio Centrale di Idrobiologia durante l'anno 1958.
Boll. Pesca, Piscicolt. Idrobiol., n.s., 13(1/2):7-8.

institutions

Maldura, Carlo, 1957.

Relazione sull'attivita del Laboratorio Centrale di Idrobiologia durante l'anno 1957.
Boll. Pesca, Piscicolt. e Idrobiol., n.s., 12(2): 124-125.

institutions

Bernhard, Michael 1968.
Il laboratorio di Fiascherino per lo studio della contaminazione radioattiva del mar
Notiziario, Com. Naz. Energia nucleare (11): 4-19

Istituto Sperimentale Talassografico

Institutions

Picotti, Mario, 1958(1959)

Sull'attivita svolta dall'Istituto Sperimentale Tallasografico di Trieste nel 1958.
Boll. Pesca, Piscicolt., Idrobiol., n.s., 13(1/2):9-11.

institutions

Picotti, Mario, 1957.

Relazione sull attivita dell' Istituto Sperimentale Talassografico di Trieste per l'anno 1957. Boll. Pesca, Piscicolt. Idrobiol., n.s., 12(2):121-123.

institutions

Picotti, M., 1954.
Origini, sviluppo e pubblicazioni (Del. No. 1-300) dell' Istituto Talassografico di Trieste (1841-1953).
Ist. Talassograf. "Francesco Vercelli", Trieste, Pubbl., No. 301:1-23.

Istituto Talassografico di Taranto

Naples Zoological Station

institutions

Cerruti, A., 1943.
L'Istituto Talassografico di Taranto del Consiglio Nazionale della Ricerche. (Memoria CCCVII), Arch. Ocean. Limn., Anno III(1/2)1943 XXI:87-132, 11 pls.

Naples Zoological Station

Dohrn, R., 1954.
The Zoological Station at Naples. Endeavour 13(49):22-26, 10 color figs.

institutions, Italy

Bolognari, Arturo 1964.
La stazione di biologia marina e di idrobiologia e pescicoltura di Ganzirri (Messina).
Atti Soc. Peloritana Sci. fis. mat. nat. 10(4): 503-507

Japan

laboratories

Hidaka, K., 1948
Activities in physical oceanography in Japan since 1939. Internat. Assoc. Ocean., Oslo Assembly, 8 pp.

institutions, Japanese

Hogetsu, K., 1968.
Report of recent activities of International and Japanese IBP/PM Section. (In Japanese; English translation). Bull. Plankton Soc., Japan, 15(1): 35-37.

institutions, Japan

Honda, Nobuo, 1971.
Outline of researches in 1970. (In Japanese).
Bull. Fish. Exp. Sta. Okayama Pref. (1970): 1-5.

institutions

International Hydrographic Bureau, 1962.
Activities of the Hydrographic Office of Japan.
Int. Hydrogr. Rev., 39(2):9-24.

institutions
Japan, Maritime Safety Board, Tokyo, 1961.
Hydrographic Bulletin, Special number for the 90th Anniversary of Founding (Publ. No. 981), No. 67:1-100.

institutions
Motoda, S., 1958.
Plankton associations around Japan.
Proc. Indo-Pacific Fish. Counc. (6th Sess.), Tokyo, 30 Sept.-14 Oct., 1955, Sect. 2-3:247-249.

institutions
Japan, Nansei Regional Fisheries Research Laboratory, 1970, 39pp

Akkeshi Marine Biological Station

institutions
Uchida, Tohru, Mayumi Yamada, Fumio Iwata, Chitaru Ohure and Zen Nagee, 1963.
The zoological environs of the Akkeshi Marine Biological Station.
Publ. Akkeshi Mar. Biol. Sta., No. 13:36 pp.

Amakusa Marine Biological Laboratory

institutions
Kikuchi, Taiji, 1968.
The zoological environment of the Amakusa Marine Biological Laboratory.
Publ. Amakusa mar. biol. Lab., 1(2):117-127.

institution
Kurabara, Masuraro, 1968.
The Amakusa Marine Biological Laboratory of Kyushu University.
Publ. Amakusa mar. biol. Lab., 1(2):107-116.

Institutions -Japanese CSK Sorting Service Center
Yamazi, Isamu, 1969.
Progress of sorting on CSK zooplankton in Japan. Bull. Plankt. Soc. Japan, 16(1): 82-83
(In Japanese; English abstract)

JAPAN Meteorological Agency

Japan Meteorological Agency
Terada, K., and M. Hanzawa, 1957.
Recent works in relation to the activities of the Marine Division of the Japan Meteorological Agency. Geophys. Mag., Tokyo, 28(1):117-133.

Kobe Marine Observatory

institutions
Matsudaira, Y., 1957.
Short history of plankton research programme in Kobe Marine Observatory.
Info. Bull., Plankton., Japan, No. 4:1-2.

Ocean Research Institute, University of Tokyo

institutions
Japan, Tokyo University, 1962
The Ocean Research Institute of the University of Tokyo. 8 pp. (August 1962).

institutions
Komaki, Yuzo, 1962
Establishment of the Ocean Research Institute, University of Tokyo. (In Japanese: English abstract).
Info. Bull., Planktology, Japan, No. 8:22-26.

institutions
Taka, Nobuo, 1963.
Information on Ocean Research Institute, University of Tokyo. (In Japanese; English abstract).
Info. Bull. Plankto., Japan, No. 10:68-70.
28-Sakaechodori 1, Nakano ward, Tokyo.

Matoya Oyster Research Laboratory

institutions, Japan
Sato Tadao (Chuyu), 1968.
Matoya Oyster Research Laboratory: History and present status with a note on oceanographical characteristics of Matoya Bay. (In Japanese; English abstract). Bull. Plankton Soc., Japan, 20-37.

Shimoda Marine Biological Laboratory

institutions, Japan
Ichimura, S., 1968. Kyoiku
Shimoda Marine Biological Station, Faculty of Science, Tokyo University of Education (In Japanese; English translation). Bull. Plankton Soc., Japan, 15(1): 67-69.

Jugoslavia

Institutions, Bioskog Institute u Rovinju
Gamulin, Tomo, Miroslav Nikolic i Dusan Zavodnik 1964.
70 Godina Bioloskog Instituta u Rovinju 1891-1961 (In Jugoslavian; English abstract)
Thalassia Jugoslavica, 2(5/6):26 pp.

institutions
Filjushkin, B.N., 1961
[The Institute of Marine Biology in Rovinj (Jugoslavia).]
Okeanologiia, Akad. Nauk, SSSR, (2):354-355.

institutions
Jugoslavia, Godionjica Instituta za Oceanografiju i Ribarstvo, Split, 1961.
Spomen-Knjiga, 1930-1960: 35 pp.

institutions
Leontiev, N.F., 1962
The Institute of Oceanography and Fishery Trade in Split, Yogoslavia. (In Russian).
Okeanologiia, Akad. Nauk, SSSR, 2(5):945.

Korea

Institutions
Korea, Korean National Commission for UNESCO, 1962.
The Central Fisheries Experimental Station.
Korean Journal, 2(2):16.

MADAGASCAR

Station Océanographique de Nossi-Bé
Paulian, R., 1954.
La station océanographique de Nossi-Bé.
Vie et Milieu, Lab. Arago, Univ. Paris, 5(3): 266-269.

MEXICO

institutions
Campbell, B., 1963.
A new marine research station at Guaymas, Mexico.
Veliger, 5(3):122.
Sea of Cortez Marine Research Center (California)

institutions
Ushakov, P.V. 1968.
Oceanographic Museum in Monaco. (In Russian).
Okeanologiia, Akad. Nauk, SSSR, 8(6):1107-1109.

MOZAMBIQUE

institutions
Gonçalves Sanches, J., 1960.
A Inhaca e a sua Estação de Biologia Marítima
Notas Mimeografadas do Centro de Biologia Piscatória, 14:10 pp.

laboratories
Kalk, M., 1954.
Marine biological research at Inhaca Island, Mozambique. S. African J. Sci. 51(4):107-115, 5 textfigs.

institutions
Moreira Rato, J., 1959
The Marine Biological Station of Inhaca.
So. Afr. J. Sci., 55(7): 161-162.

NETHERLANDS

institutions
Desakorn, V., 1957.
Progress achieved by the Hydrographic Department of the Royal Thai Navy during the last five years (1952-1957). Int. Hydrogr. Rev., 34(2):15-18.

institutions
Vaas, K., 1963.
Annual report of the Delta Division of the Hydrobiological Institute of the Royal Netherlands Academy of Sciences for the years 1960 and 1961.
Netherlands J. Sea Res., 2(1):68-76.

institutions
Verwey, J., 1964.
Annual report of the Netherlands Institute for Sea Research for the year 1962.
Netherlands J. Sea Res., 2(2):293-318.

institutions
Verwey, J., 1963.
Annual report of the Netherlands Institute for Sea Research for the year 1961.
Netherlands J. Sea Research, 2(1):102-122.

institutions
Verwey, J., 1961.
Annual report of the Zoological Station of the Netherlands Zoological Society for the year 1959.
Netherlands J. Sea Res., 1(1/2):241-256.

NEW ZEALAND

institutions
Lowry, C.C., 1957.
Starting up a small hydrographic service.
The story of the New Zealand surveying service.
Int. Hydrogr. Rev., 34(2):9-14.

institutions
Zenkevich, L.A., 1963.
South African Association for Marine Biological Research, Durban, 2 West Street, P.O. Box 736 and its Oceanographic Institute and Oceanarium in Durban. (In Russian).
Okeanologiia, Akad. Nauk, SSSR, 3(6):1128.

NIGERIA

institutions
Longhurst, A.R., 1961.
Report on the Fisheries of Nigeria, 1961.
Federal Fisheries Service, Ministry of Economic Development, Lagos, numerous pp. (mimeographed).

NORWAY

institutions
Alekseyev, A.P., 1958.
Hydrographic laboratory of the Marine Research Institute in Bergen. (In Russian).
Nauchno-Tekhn. Byul. Polyarn. N.-I. In-ta Morsk. Rybn. (Sci. Techn. Bull., Polar Sci. Res. Inst., Mar. Fish. Economy and Oceanogr.), No. 2(6):66-67.
RZHGeofiz 7/59-6762

institutions
Brattström, H., 1961.
Sarsia - an introduction.
Sarsia, 1:1-2.

Norges Sjökartverk

institutions
Sundby, E., 1952.
The tidal work of Norges Sjökartverk.
Int. Hydrogr. Rev. 29(2):60-62.

Norsk Polarinstitutt

institutions
Sverdrup, H.U., 1957.
Tatigkeit des norvegischen Polarinstuts.
Petermanns Geogr. Mitt., 101(2):114.

OCEANIA

institutions
Bugnicourt, F., 1958.
Note on the activities of the Oceanographic Section of the Institut Francais d'Oceanie, 1953-1957.
Proc. Ninth Pacific Sci. Congr., Pacific Sci. Assoc., 1957, Oceanogr., 16:36.

PACIFIC area

Institutions
Elkin, A.P., 1961.
Pacific Science Association - its history and role in international cooperation.
Bernice P. Bishop Mus. Spec. Publ., 48:80 pp.

institutions
Bushnell, O.A., 1948
A list of scientific institutions in the Pacific area. Pacific Science II(4):243-261.

PERU

institutions
Sparre, T., 1963
Prologo: El Instituto de Investigacion de los Recursos Marinos. (In Spanish and English).
Bol. Inst. Invest. Recursos Marinos, 1(1): iii-viii.

POLAND

institutions
Poland, Prace Instytutu Morskiego, Gdansk, 1961
Information on the Maritime Institute. (English summary).
Prace Inst. Morsk., Gdansk, (1) Hydrotech. II. Sesja Naukowa Inst. Morsk., 20-21 wrzesnia 1960: 151-155. (mimeographed).

PORTUGAL

institutions
Martins, Nuno, 1962
O canal hidrodinâmico do Laboratório Nacional de Engenhara Civil.
Lab. Nac. Engenhara Civil, Memoria, Lisbon, No. 185: 10 pp.

Roumania

institutions (Roumania)
Andriescu, Ionel, 1968.
Dix ans d'activité de recherche et d'enseignement à la Station de Biologie Marine "Prof. I. Borcea" d'Agigea (1957-1966).
In: Lucrările Sesiunii Stiintifice a Stațiunii de Cercetări Marine "Prof. Ioan Borcea", Agigea (1-2 Noiembrie 1966), Volum Festiv, Iași, 1968: 1-49.

institutions
Podina, V., 1963.
The Agigea station for marine zoology (In Rumanian).
Stiinta si Tehnica, 15(3):10-11.

Techn. Transl., 10(1):47.
OTS $0.50
63-21721.

institutions
Nicolae, Ionescu, 1964.
Cercetari efectuarte in ultimi 20 de ani la litoralul Rominesc al Mării Negre de catre statiunea maritima de cercetari piscicole Constanta.
Bulet. Inst. Cercetari si Proiectari Piscicole, 23(2/3):135-155.

French resume

SINGAPORE

institutions
*Motoda, S., 1967.
Proposed Marine Biological Centre in Singapore. (In Japanese; English abstract).
Inf. Bull. Planktol. Japan, 14:68-69.

institutions
Tham, Ah Kow, 1958.
Report on the Fisheries Division, Singapore.
Proc. Ninth Pacific Sci. Congr., Pacific Sci. Assoc., 1957, Oceanogr., 16:41.

SOUTH AFRICA

Fishing Industry Research Inst.
Dreosti, G. M., 1947.
Fishing Industry Research Institute. Bull. Fish. Industr. Res. Inst. 1:1-13. (Cape Town).

institutions
Mallory, J.K. 1969.
Current oceanographic research around Southern Africa.
S.Afr. J. Sci. 65(6): 187-191

institutions
South Africa, Department of Commerce and Industries, Division of Sea Fisheries, 1963.
Thirty-first annual report for the period 1 April 1959 to 31 March 1960:242 pp.

institutions (The Oceanographic Research Institute)
South Africa, Association for Marine Biological Research, 1960
Bulletin No. 1: 26 pp.

institutions
South African Association for Marine Biological Research, 1960.
Bulletin, No. 1:26 pp.

SPAIN

institutions
Garcia, R., 1948
Instituto Español de Oceanografia. J. du Cons. 15(3):253-259, 4 text figs.

SWEDEN

institutions
Kirpichnikov, A.A., 1964
The Oceanographic Institute in Goteborg. (In Russian).
Okeanologiia, Akad. Nauk, SSSR, 4(3):540-542.

United States of America

institutions
Abel, Robert, 1964.
Federal organization in oceanography.
Undersea Techn., 5(6):17-19.

institutions (ONR)
Galler, Sidney R., 1963.
National agencies: interest and support. (Special Issue on Marine)Biology, AIBS Bull., 13(5):26-27.

institutions
Malone, Thomas F., 1959.
A national institute for atmospheric research.
Trans. A. G. U., 40(2):95-111.

institutions
Munske, R.E., 1962.
Let's get organized.
Undersea Technology, 3(4):6.

institutions
United States, National Academy of Sciences-National Research Council, 1962
Oceanography 1960 to 1970. 11. A history of oceanography: a brief account of the development of oceanography in the United States, 28 pp.

Allan Hancock Foundation

institutions
Emery, K.O., 1958.
Report of the Allan Hancock Foundation, University of Southern California.
Proc. Ninth Pacific Sci. Congr., Pacific Sci. Assoc., 1957, Oceanogr., 16:45-46.

American Geophysical Union

institutions
Fleming, J.A., 1954.
Origin and development of the American Geophysical Union, 1919-1952. Trans. Amer. Geophys. Un. 35(1):5-46.

Arctic Research Laboratory

institutions
Britton, M.E., 1967.
US Office of Naval Research Arctic Research Laboratory, Point Barrow, Alaska.
Polar Rec., 13(85):421-423.

Bear Bluff Laboratory

Bear Bluff Laboratory
Hull, E.W. Seabrook, 1965.
Vision - ingenuity - sweat.
GeoMar. Techn., 1(8):20-23.

Bingham Oceanographic Laboratory

institutions
Child, A.M., 1960.
Old wine, new vessels.
Sea Frontiers, 6(2):95-99.

Bingham Oceanographic Laboratory

Bureau of Commercial Fisheries

institutions, USA, Bureau of Commercial Fisheries.
Long, E. John, 1965.
B.C.F. (Bureau of Commercial Fisheries).
GeoMarine Techn., 1(6):33-35.

Bureau of Sport Fisheries and Wildlife

Bureau of Sport Fisheries and Wildlife
Long, E. John, 1965.
B.S.F. & W.
GeoMarine Technology, 1(3):38-40.

California

institutions
Pinkas, Leo, 1958.
Report on oceanographic activities of the Marine Fisheries Branch of the Department of Fish and Game of the State of California.
Proc. Ninth Pacific Sci. Congr., Pacific Sci. Assoc., 1957, Oceanogr., 16:49.

institutions
Ahlstrom, Elbert H., 1958.
Research being done by the California Cooperative Oceanographic Fisheries Investigations on oceanographic activities in the Pacific.
Proc. Ninth Pacific Sci. Congr., Pacific Sci. Assoc., 1957, Oceanogr., 16:47-48.

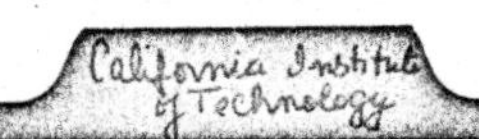

institutions
Miller, Robert C., 1958.
Report of the California Academy of Sciences, 1951-1957.
Proc. Ninth Pacific Sci. Congr., Pacific Sci. Assoc., 1957, Oceanogr., 16:44.

California Institute of Technology

California Institute of Technology
Anon., 1965.
Marine science center will soon take shape. Marine Biology lab will be southern California center's first unit.
Chemical & Engineering News, 43(22):40.

California State College System

California State College System
Anon., 1965.
Marine science center will soon take shape. Marine Biology lab will be southern California center's first unit.
Chemical & Engineering News, 43(22):40.

institutions
Motoda, Sigeru, 1969.
Center for Short-Lived Phenomena, Cambridge, Mass., U.S.A. Bull. Plankt. Soc. Japan, 16(1): 75. (In Japanese; English abstract)

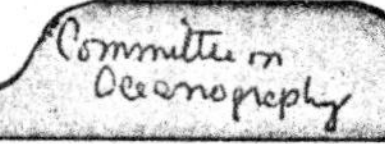

institutions
Gorsky, N.N., 1960
[Committee on Oceanography (U.S.A.).]
Biull. Okeanogr. Komissii, Akad. Nauk, SSSR, (6): 77.

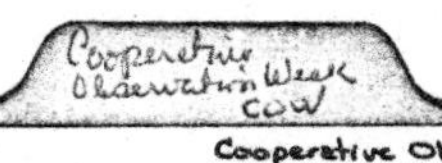

Cooperative Observational Week COW
Frank, Sidney R., 1962
Operation COW, the Cooperative Observational Week.
First Interindustrial Oceanogr. Symposium, Lockheed Aircraft Corp., 2-5.

Cooperative Observational Week
Hovind, E.L., and S.R. Frank, 1962
Narrated film-COW operations.
Proc. 2nd Interindustrial Oceanogr. Symposium Lockheed Aircraft Corp., 45-49.

Cooperative Observational Week COW
Lesser, Robert M., 1962
Temperature structure of the Santa Barbara Channel.
First Interindustrial Oceanogr. Symposium, Lockheed Aircraft Corp., 13-20.

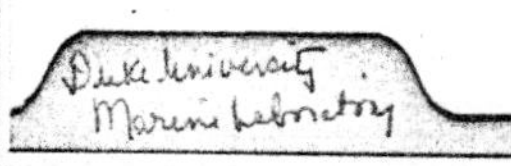

Duke Univ. Marine Lab.
Bookhout, C.G., 1955.
Duke University Marine Laboratory. A.I.B.S. Bull. 5(2):30-31.

Eniwetok Marine Biological Laboratory

Institutions, Marshall Islands
Hiatt, Robert W., 1966. (1967).
Eniwetok Marine Biological Laboratory.
Micronesica, J. Coll. Guam, 2(2): 265-267.

Environmental Science Services Administration

institutions, U.S.A., Environmental Science Services Administration
Boods, Larry L., 1966.
ESSA encompasses sea, earth, air, space.
Undersea Techn., 7(3):31-33.

ESSA
Minevich, A. Ya., 1966.
The new USA Oceanographic Institute. (In Russian).
Okeanologiia, Akad. Nauk, SSSR, 6(6):1121.

(Taken from the New York Times)

institutions, Florida State University
Goodell, H. G., and Dennis S. Cassidy, 1969.
The Antarctic Marine Geology Research Facility.
Antarctic J., U.S. 4(5): 177-178

Gulf Universities Research Corp.

Gulf Universities Research Corporation
Anon., 1967.
GURC (Gulf Universities Research Corporation) will direct extensive Gulf of Mexico research.
Ocean Industry, 2(12):49-50.

Hawaii Institute of Geophysics

institutions
Ramage, Colin S., 1962.
The Hawaii Institute of Geophysics.
Trans. Amer. Geophys. Union, 43(1):14-19.

Also:
Contrib. No. 14, Hawaii Inst. Geophys., Univ. Hawaii, 1960-1962.

Hawaii Marine Laboratory

institutions
Banner, A.H., 1958.
Report of the Hawaii Marine Laboratory.
Proc. Ninth Pacific Sci. Congr., Pacific Sci. Assoc., 1957, Oceanogr., 16:17.

Hawaii Oceanic Institute

institutions-Oceanic Institute Hawaii
Anon., 1966.
The oceanic foundation.
Undersea Techn., 7(9):24-29.

institutions - Oceanic Institute - Hawaii

Hopkins Marine Station

institutions - Hopkins Marine Station
Abbott, Donald P., Lawrence R. Blinks and John H. Phillips, 1964.
An experiment in undergraduate teaching and research in the biological sciences.
The Veliger, 6 (Suppl.):1-6.

Hopkins Marine Lab.
Blinks, L.R., 1955.
Hopkins Marine Laboratory. A.I.B.S. Bull. 5(2): 33-34.

INstitutions
Bolin, Rolf L., 1958.
Report of oceanographic activity of the Hopkins Marine Station, 1953-1957.
Proc. Ninth Pacific Sci. Congr., Pacific Sci. Assoc., 1957, Oceanogr., 16:50.

institutions
Vaas, K.F., 1964.
Annual report of the Delta Division of the Hydrobiological Institute of the Rpyal Netherlands' Academy of Sciences for the year 1962.
Netherlands J. Sea Res., 2(2):284-292.

House Merchant Marine and Fisheries Committee Subcommittee on Oceanography

House Merchant Marine and Fisheries Committee, Subcommittee on Oceanography
Walsh, John, 1965.
Oceanography: House Subcommittee encourages use of merchant ships to gather data on the high seas
Science, 148(3668):349-350.

Hughes Aircraft Co.

Hughes Aircraft Company
Harrison, J.C., and C.H. Wilcox, 1962
Hughes research program in marine geophysics and oceanography.
Proc. 2nd Interindustrial Oceanogr. Symposium Lockheed Aircraft Corp., 40-44.

institutions
Chestnut, A.F., 1960
Institute of Fisheries Research, University of North Carolina. AIBS Bull. 7(1): 5-6.

International Ice Patrol

International Ice Patrol
Graves, G. Van A., 1953.
International Ice Patrol. Mar. Obs. (M.O. 566) 23(160):109-110, 2 photos.

institutions
Kolman, O.V., 1962
International Ice Patrol.
Okeanologiia, Akad. Nauk, SSSR, 2(2):376-377.

Johns Hopkins University

institutions, USA (Johns Hopkins Univ.)
Raup, Susan, 1963.
Chesapeake Bay Institute and the Department of Oceanography, the Johns Hopkins University.
American Zoologist, 3(3):3 pp.

Also in:
Collected Reprints, The Johns Hopkins Univ., Chesapeake Bay Inst. and Dept. Oceanogr., 6. 1965

Lamont Geological Observatory

institutions
Burkholder, Paul R. and Oswald A. Roels, 1968?
Biological oceanology at Lamont Geological Observatory.

Selected Papers from the Governor's Conference on Oceanography, October 11 and 12, 1967, at the Rockefeller University, New York, N.Y : 72-85.

institutions
Ewing, Maurice and J. Lamar Worzel, 1968?
Geophysical oceanographic studies at Lamont Geological Observatory.
Selected Papers from the Governors Conference on Oceanography, October 11 and 12, 1967, at the Rockefeller University, New York, N.Y. 8-35

Institutions
Worzel, J. Lamar, 1967.
Lamont Geological Observatory.
Science, 158(3803):948-949.

Marine Science Research Center Fishermen's Cove, Santa Catalina Island

Marine Science Research Center, Fisherman's Cove, Santa Catalina Island.
Anon., 1965.
Marine science center will soon take shape.
Marine Biology lab will be southern California center's first unit.
Chemical & Engineering News, 43(22):40.

Massachusetts Institute of Technology

institutions 1959
Great Britain, Royal Meteorological Society, 1959
Center for Earth Sciences- Massachusetts Institute of Technology.
Weather, 14(6):210-211

institutions
Shrock, Robert R., 1959.
Earth Science Center at Massachusetts Institute of Technology.
Trans. A. G. U., 40(2):120-122.

Mount Desert Island Biological Station

institutions-Mount Desert Island Biological Station
Smith F. G. Walton, 1967.
Biomedical progress by sea.
Sea Frontiers, 13(2):94-97

Narragansett Marine Laboratory

institutions
Conover, R.J., 1957.
The Narragansett Marine Laboratory.
Sea Frontiers, 3(4):217-223.

institutions
Fish, Charles J., 1964.
The Narragansett Marine Laboratory seawater system.
U.S.F.W.S. Res. Rept., 63:187-190.

National Aquarium

institutions
Feldman, G.E., 1965.
The U.S. National Aquarium. (In Russian).
Okeanologiia, Akad. Nauk, SSSR, 5(2):372.

National Oceanographic Data Center

institutions, USA, NODC
Austin, Thomas S., and William H. Myers, 1969.
NODC takes a new tack.
Undersea Techn., 10)1):58-60. (popular).

institutions
International Hydrographic Bureau, 1962.
The U.S. National Oceanographic Data Center.
Int. Hydrogr. Rev., 39(2):49-51.

institution
Picciolo, Anthony R., 1963.
National Oceanographic Data Center.
(Special Issue on Marine Biology), AIBS Bull., 13(5):38-40.

institutions
Samarin, V.G., 1962
The U.S. National Data Center. (In Russian).
Okeanologiia, Akad. Nauk, SSSR, 2(5):947.

institutions
United States, American Society of Limnology and Oceanography, 1962.
The National Oceanographic Data Center.
Limnol. & Oceanogr., 7(1):108-109.

National Oceanographic Program

National Oceanographic Program
Anon., 1965.
The National Oceanographic Program, Fiscal Year 1966.
Naval Res. Rev., April, 21.

National Science Foundation

institutions
Sprugel, George, Jr., 1963.
National Science Foundation.
(Special Issue on Marine Biology), AIBS Bull., 13(5):33-35.

Naval Ordnance Laboratory Corona

institutions
Anonymous, 1964.
The Naval Ordnance Laboratory, Corona.
Undersea Technology, 5(7):47-48.

institutions
Blenman, C., Jr., 1963.
U.S. Naval Ordnance Test Station.
Undersea Technology, 4(8):19-20.

institutions
Witten, D.E., 1963.
U.S. Naval Ordnance Laboratory.
Undersea Tech., 4(6):24-25.

New York Aquarium

institutions
Nigrelli Ross F., 1968?
The New York Aquarium: a brief history.
Selected Papers from the Governor's Conference on Oceanography, October 11 and 12, 1967, at the Rockefeller University, New York, N.Y.: 86-93.

Institution, New York Aquarium
Ruggieri, G.D., 1967.
New York Aquarium and Osborn Laboratories of Marine Sciences.
Science, 158(3801):675-676.

New York University

Institutions
Neumann, Gerhard, 1968?
Physical oceanographic studies at New York University.
Selected Papers from the Governor's Conference on Oceanography, October 11 and 12, 1967, at the Rockefeller University, New York, N.Y.: 172-181.

Occidental College

Occidental College
Anon., 1965.
Marine science center will soon take shape.
Marine Biology lab will be southern California center's first unit.
Chemical & Engineering News, 43(22):40.

Office of Naval Research

institutions
Greenbaum, R., 1963.
The Office of Naval Research.
Undersea Technolgy, 4(1):30-31.

Oceanographic Institute, Florida State University

Oceanographic Inst., Florida State Univ.
Mayer, S.L., and R.H. Williams, 1955.
Oceanographic Institute of Florida State University. A.I.B.S. Bull. 5(2):28-30.

Oceanographic Laboratories

institutions
Kolman, O.V., 1963
The oceanographic laboratory in Honeywell, Seattle, U.S.A. (In Russian).
Okeanologiia, Akad. Nauk, SSSR, 3(2):362.

institutions
Walton Smith, F. G., 1959.
Proposed Hatteras laboratory
Sea Frontiers, 5 (4):216-220.

Orange County Schools Floating Laboratory

institutions, Orange County Schools Floating Laboratory
Linsky, Ronald B. 1969.
Developing environmental awareness: a new approach to an old problem.
Oceans Mag. 1(5): 52-58. (popular)

Oregon State College

institutions
Burt, Wayne V., 1966.
The Department of Oceanography at Oregon State University.
U.S. Naval Res.Rev., 19(4):12-19.
Also in: Coll. Repr., Dep. Oceanogr., Oregon State Univ., 5.

institutions
Burt, Wayne V., 1958.
Activities in oceanography at Oregon State College.
Proc. Ninth Pacific Sci. Congr., Pacific Sci. Assoc., 1957, Oceanogr., 16:57-58.

Osborn Laboratories for Marine Sciences

Institutions, Osborn Laboratories for Marine Sciences
Ruggieri, G.D., 1967.
New York Aquarium and Osborn Laboratories of Marine Sciences.
Science, 158(3801):675-676.

Pacific Oceanic Fisheries Investigations

institutions
Godfrey, Mary Lynne, 1958.
Review of POFI's oceanographic program, January 1952-June 1957.
Proc. Ninth Pacific Sci. Congr., Pacific Sci. Assoc., 1957, Oceanogr., 16:18-20.

Pacific Oceanic Fisheries Investigations
Stroup, E.D., and T.S. Austin, 1955.
Review of the oceanographic program of the Pacific Oceanic Fisheries Investigations.
Trans. Amer. Geophys. Union, 36(5):881-884.

Pomona College

Pomona College
Anon., 1965.
Marine science center will soon take shape.
Marine Biology lab will be southern California center's first unit.
Chemical & Engineering News, 43(22):40.

President's Scientific Advisory Council

institutions

Greenberg, D.S., 1966.
Oceanography: PSAC Panel calls for setting up new agency.
Science, 153(3734):391-393.

Scripps Institution of Oceanography

institutions

Beklemishev, C.W., 1960
[Scripps Institution of Oceanography.]
Biull. Okeanogr. Komissii, Akad. Nauk, SSSR, (6): 69-74.

institutions, Scripps Institution of Oceanography

Booda, Larry L., 1966.
Scripps Institution of Oceanography: Research - but emphasis on education.
Undersea Techn., 7(5):37-40.

institutions

Hubbs, C., 1956.
Scripps Institution of Oceanography, University of California. A.I.B.S. Bull., 6(1):11-13.

institutions

Johnson, Martin W., 1958.
Report of the Scripps Institution of Oceanography, 1953-1957.
Proc. Ninth Pacific Sci. Congr., Pacific Sci. Assoc., 1957, Oceanogr., 16:51-56.

institutions

Manar, T.A., 1959.
Marine biological and oceanographic institutions of the world. II. Research in marine biology at the Scripps Institution of Oceanography.
J. Mar. Biol. Assoc., India, 1(2):247-249.

University of California
Scripps Instiution of
Oceanography

NORPAC Committee, 1955.
Oceanic observations of the Pacific, 1955.
The NORPAC data, 532 pp.

institutions

Raitt, Helen, and Beatrice Moulton 1968.
The Scripps Institution of Oceanography: first fifty years.
The Ward Ritchie Press: 217 pp.

SIO

Sverdrup, H.U., and Staff, 1947.
Research within physical oceanography and submarine geology at the Scripps Institution of Oceanography during April 1946 to April 1947.
Trans. Am. Geophys. Union, 28(5):801-802.

SIO

Sverdrup, H.U., 1940-41
Activities of the Scripps Institution of Oceanography (1933-1938). Sixth Pac. Sci. Congr., Calif., 1939, 3:114-123.

Scripps Institution
of Oceanography

Sverdrup, H.U., and Staff, 1945
Research within Physical Oceanography and Submarine Geology at the Scripps Institution of Oceanography during April, 1944 to April, 1945. Trans. Am. Geophys. Union 26(1):127-128.

SIO

Sverdrup, H.U. and Staff, 1945
Research within Physical Oceanography and Submarine Geology at the Scripps Institution of Oceanography during April 1943 to April 1944. Am. Geophys. Union Trans. of 1944, Part IV:605

Scripps Institution

Sverdrup, H.U., and Staff, 1943
Oceanographic observations of the Scripps Institution in 1939. SIO Rec. of Obs. 1(2): 65-159.

SIO

Zobell, C.E., 1941
The Scripps Institution of Oceanography.
Turtox News. 19:22-23.

Sea Grant Colleges

institutions

Wildman, Robert D., 1968?
The Sea Grant Concept
Selected Papers from the Governor's Conference on Oceanography, October 11 and 12, 1967, at the Rockefeller University, New York, N.Y. 182-185.

Seismological Society of America

"institutions"

Byerly, Perry, 1968.
History of the Seismological Society of America. (Abstract).
Geol. Soc., Amer., Special Paper, No. 76:193.

Smithsonian

institutions

Kolman, Ov., 1965.
Oceanography in the Smithsonian Institution (U.S.A.)
Meteorol. i Gidrol. (9):43-44.

institutions, Smithsonian

Long, E. John, 1965.
Smithsonian.
GeoMar. Techn., 1(7):25-26.

Texas Agricultural and Mechanical College

institutions Texas A+M

Leipper, D.F., 1950.
The establishment of the Department of Oceanography in the Agricultural and Mechanical College of Texas. Trans. Am. Geophys. Union 31(5):789-791.

U.S. Army Coastal Engineering Research Center

institutions

Anonymous, 1964.
The Coastal Engineering Research Center - summary of capabilities.
U.S. Army, Coastal Engineering Res. Center, Misc. Paper, No. 3-64:20 pp. (multilithed).

U.S. Atomic Energy Commission

institutions

Joseph, Arnold, 1963.
U.S. Atomic Energy Commission.
(Special Issue on Marine Biology), AIBS Bull., 13(5):29-31.

U.S. Coast and Geodetic Survey

U.S. Coast and Geodetic Survey

Colbert, L.O., 1952.
Oceanographic activities of the U.S. Coast and Geodetic Survey, 1939-1949. Proc. Seventh Pacific Sci. Congr., Met. Ocean., 3:154-157.

U.S. Coast and Geodetic Survey

Colbert, L. O., 1949.
Oceanographic activities of the U.S. Coast and Geodetic Survey. Trans. Am. Geophys. Union 30(6):897-899.

USCGS

Colbert, L.O., 1945
Oceanographic activities of the United States Coast and Geodetic Survey. Am. Geophys. Union Trans. of 1944, Part IV: 604-605.

institutions, U.S.A. (USC&GS)

Hicks, Steacy D., 1967.
The tide prediction centenary of the United States Coast and Geodetic Survey.
Int. hydrogr. Rev., 44(2):121-131.

USC&GS

Kolman, O.V., 1962
Oceanography survey works in the Pacific Ocean performed by the U.S. Coast and Geodetic Survey.
Okeanologiia, Akad. Nauk, SSSR, 2(5):946-947.

institutions - U.S. Coast and Geodetic Survey

Long, E. John, 1964.
U.S.C.&.G.S.
GeoMar. Techn., 1(1):25-26.

institutions

Pierce, Charles, 1958.
Report of the U.S. Coast and Geodetic Survey.
Proc. Ninth Pacific Sci. Congr., Pacific Sci. Assoc., 1957, Oceanogr., 16:180-195.

institutions

Shalowitz, Aaron L., 1962 and 1964.
Shore and sea boundaries, with special reference to the interpretation and use of Coast and Geodetic Survey data.
U.S. Dept. Commerce, Coast and Geodetic Survey Publ. 10-1:420 pp. $3.50.

institutions

Studds, R.F.A., 1950.
Oceanographic activities of the U.S. Coast and Geodetic Survey. Trans. Am. Geophys. Union 31(5): 786-788.

U.S. Coast Guard

Coast Guard

Farley, J.F., 1947
Coast Guard plans for oceanographic work
Trans. Am. Geophys. Un. 28(5):800.

U.S. Coast Guard

Hufford, Gary L., and Robert B. Elder, 1969.
U.S. Coast Guard Oceanographic Unit's participation in IWSOE-1969.
Antarctic J., USA 4(4):98-99.

U. S. Coast Guard

Long, E. John, 1965.
U.S.C.G. 175-year history of working at sea.
GeoMar. Techn., 1(2):22-23.

Coast Guard

Waesche, R.R., 1945
Oceanographic Interests and Activities of the Coast Guard. Trans. Am. Geophys. Union 26(1):129-130.

U S Geological Survey

institutions, U.S. Geol. Survey

Corwin, Gilbert and Edward Bradley, 1965.
Geological Survey's marine program.
Undersea Techn., 6(9):35-37.

U.S. Public Health Service

institutions
Bowman, Paul W., 1963.
Public Health Service.
(Special Issue on Marine Biology), AIBS Bull., 13(5):37-38.

U.S. National Museum

institutions
Wallen, I.E., 1963.
U.S. National Museum, Smithsonian Institution.
(Special Issue on Marine Biology), AIBS Bull., 13(5):35-37.

institutions
U.S.A., National Museum (Smithsonian Institution) 1963.
Annual report for the year ended June 30, 1963, 226 pp.

U.S. Navy

Meteorol. Buoy program, U.S. Navy
Mottern, R.E., E.F. Corwin and A.F. Pyle, 1967.
The Meteorological Buoy Programme of the U.S. Navy.
Marine Obs., 37(218):178-185.

institutions, United States, Navy
Waters, Odale D., 1966.
The Navy's new oceanographic program.
Undersea Techn., 7 (11):33-35.

U.S. Navy Hydrographic Office

Hydrographic Office
Bates, C.C., and R.H. Fleming, 1948
Oceanography in the Hydrographic Office.
Int. Hydro. Rev. 25(1):63-72, 5 figs.

Reproduced "Military Engineer" August 1947.

Hydrographic Office
Bates, C.C. and R.H. Fleming, 1947
Oceanography in the Hydrographic Office
The Military Engineer, Aug. 1947; 2-8, 4 textfigs.

HO
Bryan, G.S., 1945
The Oceanographic activities of the Hydrographic Office during the past year.
Am. Geophys. Union. Trans. of 1944, Part IV: 603-604.

institutions
Cohen, Philip M., 1964
U.S. Naval Oceanographic Office.
Int. Hydrogr. Rev., 41(1):23-27.

Reprinted from:
Undersea Technology, May 1963.

institutions
Kolman, O.V., 1963
Oceanographic expeditions sponsored by the Hydrographic Office of the U.S. Navy Department in the Arc tic area in 1961. (In Russian)
Okeanologiia, Akad. Nauk, SSSR, 3(3):564.

institutions
Lyman, J., 1953.
Oceanographic activities of the Hydrographic Office, 1946-1952. Trans. Amer. Geophys. Union 34(1):122-124.

USN Marine Engineering Laboratory

institutions (U.S. Navy Marine Engineering Laboratory)
Johnston, David H., and Martin G. Imbach, 1965.
A deep ocean environmental laboratory. (abstract)
Ocean Sci. and Ocean Eng., Mar. Techn. Soc.,-Amer. Soc. Limnol. Oceanogr., 2:1205-1211.

USN Office of Naval Research

U.S.F.W.S.

institutions
Galtsoff, Paul S., 1962.
The story of the Bureau of Commercial Fisheries Biological Laboratory, Woods Hole, Massachusetts
U.S. Dept. Interior Circular, 145:121 pp.

institutions
Galtsoff, P. S., 1959.
New for old.
Sea Frontiers, 5(2):96-100
New F.W.S. building Woods Hole.

institutions
McKernan, Donald L., and J.L. McHugh, 1963.
Bureau of Commercial Fisheries.
(Special Issue on Marine Biology), AIBS Bull., 13(5):31-33.

institutions
Janzen, Daniel H., 1963.
Bureau of Sport Fisheries and Wildlife.
(Special Issue on Marine Biology), AIBS Bull., 13(5):28.

University of California

University of California, L.A., Riverside and Irvine
Anon., 1965.
Marine science center will soon take shape.
Marine Biology lab will be southern California center's first unit.
Chemical & Engineering News, 43(22):40.

University of North Carolina

University of Rhode Island

Institutions- Univ. R.I.
United States Navy, Office of Naval Research, 1966
The University of Rhode Island's graduate school of Oceanography.
Naval Research Reviews, 19(9):8-15.

University of Southern California

University of Southern California
Anon., 1965.
Marine science center will soon take shape.
Marine Biology lab will be southern California center's first unit.
Chemical & Engineering News, 43(22):40.

University of Washington

University of Washington
Thompson, T.G., 1952.
Report on the Oceanographic Laboratories of the University of Washington. Proc. Seventh Pacific Sci. Congr., Met. Ocean., 3:164-168.

institutions
Van Cleave, R., 1953.
The School of Fisheries, University of Washington
Proc. Seventh Pacific Sci. Congr. 4:429-436.

Woods Hole Oceanographic Institution

institutions
Emery, K.O. 1966.
The Woods Hole Oceanographic Institution-U.S. Geological Survey program for The Atlantic continental margin: status at end of 1965.
Maritime Sediments 2 (2): 55-68

WHOI (pictures)
Engel, F.M., 1953.
Unterseeische Lebenswunden: ein Überblick über ältere und neue Ergebnisse der Meeresbiologischer Forschung. Atlantis, Heft 1:30-41, photos.

Woods Hole Oceanographic Inst.
Clarke, G.L., 1935.
Attacking ocean problems. Nat. Hist. 36(5):405-416.

Woods Hole Oceanographic Institution
Colman, J.S., 1933.
The Woods Hole Oceanographic Institution: Sonic soundings and an account of a winter cruise to the Somers Islands. Geogr. Jour. 82(4):326-336, 2 textfigs.

Pacific Oceanic Biology Project
Fish, C.J., 1952.
The Pacific Oceanic Biology Project.
Proc. Seventh Pacific Sci. Congr., Met. Ocean., 3:161-163.

institutions
The Director (Paul M. Fye), 1963.
Marine biological and oceanographic institutions of the world. VII. The Woods Hole Oceanographic Institution, Woods Hole, Mass., U.S.A.
J. Mar. Biol. Assoc., India, 5(1):133-136.

WHOI
Hahn, J., 1948
Woods Hole Oceanographic Institution.
Turtox News 26(4):7 pp., 5 photos.

institutions
Lill, G.G., 1954.
Office of Naval Research Laboratory of Oceanography and Hydraulics Laboratory, Woods Hole, Massachusetts. Nature 173(4413):1017-1019, 2 textfigs.

institutions (WHOI)
USA, American Meteorological Society, 1963.
Expansion of oceanographic research program.
Bull. Amer. Meteorol. Soc., 44(1):27.

USSR

institutions - USSR
Dobrovol'sky, A.D., 1971.
The anniversary plenum of the Oceanographic Commission of the USSR Academy of Sciences. (In Russian). Okeanologiia 11(4): 752-753.

institutions USSR-Floating Marine Research Institute
Shuleikin, V.V., 1971.
Contemporary branches of marine physics originating from the Floating Marine Research Institute. (In Russian; English abstract).
Okeanologiia 11(5): 784-794.

institutions, Russian
Zhirnov, V.M., 1960
[Activities of the Aral and Caspian Seas Section.]
Biull. Okeanograf. Komissii, Akad. Nauk, SSSR, (6):14.

Akademia Nauk

institutions, USSR Academy of Sciences
Dobrovol'sky, A.D., 1971.
The twentieth year of work of the Oceanographic Commission of the USSR Academy of Sciences. (In Russian). Okeanologiia 11(5):917-919.

meetings
Mamaeva, R.B., 1964.
Scientific conferences sponsored by the Oceanographic Commission of the USSR Academy of Sciences. (In Russian).
Okeanologiia, Akad. Nauk, SSSR, 4(4):727-728.

institutions
Mokievskii, O.B., 1962.
Section on Submarine Investigations of the Oceanographic Commission, Academy of Sciences, SSSR, and its mission. Methods and results of submarine investigations. (In Russian).
Trudy, Okeanogr., Komissii, Akad. Nauk, SSSR, 14: 5-6.

institutions
Zenkovitsch, V.P., 1957.
Studies of sea coasts and the origin of the western coastal subdivisions of the Oceanographic Commission of the Praesidium AN, SSSR. Material subdivisions for the study of sea coasts and reservoirs.
Trudy Okeanogr. Komissii, Akad. Nauk, SSSR, 2:3-9

Komissiia po Problemam Severa pri Presidiume AN, SSSR

Komissiia po Problemam Severa pri Presidiume AN,SSSR
Klenova, M.V., 1960
Geology of the Barents Sea. Izdatel. Akad. Nauk, SSSR, Moscow, 1960, 367 pp.

institutions
Fedosov, M.V., and A.D. Starostin, 1960
All-Union Scientific Institute of Marine Fishery and Oceanography.
Biull. Oceanograph. Komm. (5): 21-25.

Arctic Institute

Arctic Institute
Armstrong, T.E., 1947.
The Arctic Institute, Leningrad. Polar Rec. 5: 33-34, 88-92. (Cambridge, England).

institutions
Roberts, B., and T. Armstrong, 1957.
The Arctic Institute, Leningrad. Polar Record, 8(55):306-316.

institutions, USSR, Arctic-Antarctic Research Institute
Tsurikov, V.L. and E.N. Dvorkin, 1971.
The 50th anniversary of the Arctic and Antarctic Research Institute. (In Russian). Okeanologii 11(5): 919-921.

institutions, Russian biological stations
Blacher, L.J., 1965.
Esquisse de l'histoire des stations biologiques maritimes russes.
Colloque Internat., Hist. Biol. Mar., Banyuls-sur-Mer, 2-6 sept. 1963, Suppl., Vie et Milieu, No. 19:261-263.

Black Sea

institutions
Aibulatov, N.A., and V.P. Nekolaev, 1960.
Jubilee Conference of the Black Sea Experimental Scientific Research Station of the Institute of Okeanology of the USSR Academy of Sciences. (In Russian).
Izv., Akad. Nauk, SSSR, Ser. Geograf., (4):139-142.

Translations:
OTS or SLA, $1.10.

institutions
Takeuchi, I., 1957.
Fisheries institutions in USSR.
Info. Bull., Plankton, Japan, No. 4:3.

Hydrometeorological Service

institutions, Russian
Filippov, B.A., 1960
On the status of the Hydrometeorological Service in providing the ships of the marine transport and fishery fleets with weather information at sea.
Biull. Okeanograf. Komissii, Akad. Nauk, SSSR, (6): 9-13.

institutions
Zobel, R.F., 1961.
The Soviet Hydrometeorological Service.
Nature, 192(4808):1141-1142.

institutions
Zolotuchin, A. A., 1961.
40 years of the Soviet Hydrometeorological Service.
Meteorol. i Gidrol., (6):3-12.

Institute of Oceanology

Institutions, Institute of Okeanology.
Ionin,A.S., and V. (B.) G. Bogorov,1966.
Twenty years of the Institute of Oceanology of the USSR Academy of Sciences. (In Russian).
Okeanologiia,Akad.Nauk,SSSR,6(6):1093-1099.

institutions (Institute of Oceanology)
Ionin, A.S., P.A. Kaplin, and V.S. Medvedev, 1963.
Underwater geomorphological research in the USSR.
Vestnik, Moskovsk. Univ., Ser. V, Geogr., (3):17-23.

English abstract:
Soviet Bloc Res. Geophys., Astron., Space, 1963, (64):12.

Institute of Oceanology of the Arctic Basin
Kort, V.G., 1955.
The work of the Institute of Oceanology of the Arctic Basin. Vestuik Akad. Nauk, SSSR 25(1):41-42.

Translated by J.C. Arnell, Defence Res. Bd., Defence Info. Service, Canada, T178R, Mar. 1955.

institutions
Kort, V.G., and S.V. Suyetov, 1957.
Outline of the history of the organization and development of the Institute of Okeanology of the Akademy of Sciences SSSR, (In Russian).
Izv., Akad. Nauk, SSSR, Ser. Geogr., (5):119-123.

institutions
Zenkevich, L.A., 1958.
Oceanographic Institute of the Academy of Sciences of the U.S.S.R. (In Russian).
Biull., Okeanogr. Komissiia, Akad. Nauk, SSSR, (1):21-

institutions -SSSR
Zenkevitch, L., 1957.
VI. Quelques projets pour les recherches futures. Propositions adressées au colloque.
L'Année Biol., (3)33(5/6):275-277.

lists marine biological stations in SSSR

Institutions, USSR
Deryugin, K.K. and V.E. Dzhus, 1970.
Some results of scientific research carried out by the Laboratory for underwater studies of the Leningrad Hydrometeorological Institute. (In Russian). Oceanologiia, 10(5): 906-910.

Marine Hydrophysical Institute

institutions
Ivanov, A.A., 1960
Marine Hydrophysical Institute of the Academy of Sciences of the USSR.
Biull. Oceanograph. Komm. (5): 26-29.

institutions
Kazakov, S.P., and V.G. Samarin, 1961
Participation of the Marine Hydrophysical Institute of the Academy of Sciences of the USSR in the scientific conference in Kiev on methods and instruments for hydraulic investigations. Physics of the sea. (In Russian).
Trudy Morsk. Gidrofiz. Inst., 23:167-171.

Translation: Scripta Technica, Inc. for Amer. Geophys. Union, pp. 133-136.

institutions
Lednev, V.A., 1964.
On expeditional work of the Marine Hydrophysical Institute (MGI) in the Atlantic Ocean. (In Russian).
Materiali Vtoroi Konferentsii, Vzaimod, Atmosfer. i Gidrosfer. v Severn. Atlant. Okean., Mezhd. Geofiz. God, Leningrad, Gidrometeorol. Inst., 225-259.

institutions
Cwiek, Z., 1955.
Activity of the Marine Institute in the last five years. Technika i Gospodarka Morska, 5(4):81-82.

Oceanographic Commission AN SSSR

Institutions, USSR
Dobrovol'sky, A.D., 1970.
The Oceanographic Commission of the USSR Academy of Sciences in 1969. (In Russian). Oceanologiia, 10(5): 910-912.

Institutions, Russian (Oceanogr. Commission AN SSSR)
Mokievsky, O.B., 1966.
Submarine research section of the Oceanographic Commission, AN SSSR. (In Russian).
Okeanologiia, Akad. Nauk, SSSR, 6(3):548-551.

institutions
SSSR, Akademia Nauk, 1963
On the establishment of a center for the development of oceanographic instruments. (In Russian).
Okeanologiia, Akad. Nauk, SSSR, 3(3):565.

institutions
Vinogradov, K.A., 1963
The Odessa Biological Station attached to the Institute of Hydrobiology, Academy of Sciences of the Ukrainian SSR; its historical background and research work. (In Russian).
Okeanologiia, Akad. Nauk, SSSR, 3(3):554-558.

PINRO

institutions
Adrov, M.M., 1960
Oceanographic observations and investigations of the Polar Research Institute of Marine Fisheries and oceanography in the region of West-Greenland, Labrador, and Newfoundland banks.
Biull. Okeanogr. Komissii, Akad.Nauk, SSSR, (6): 59-61.

institutions
Adrov, M.M., and A.P. Alexeev, 1964
The ultimate results of oceanographic investigations of PINRO during 1958-1960. (In Russian) Materiali Vtori Konferentsii, Vzaimod. Atmosfer i Gidrosfer. v Severn. Atlant. Okean., Mezhd. Geofiz. God, Leningrad. Gidrometeorol. Inst., 264-268.

institutions
Alexeiv, A.P., 1960
[The research vessels of the Polar Research Institute of Marine Fisheries and Oceanography] Biull. Okeanograf. Komissii. Akad. Nauk. SSSR, (6): 62-65.

institutions
Alekseyev, A.P., 1957.
Oceanographic investigations of PINRO according to the program of the International Geophysical Year. (In Russian).
Nauchno-Tekhn. Byul. N.-I. In-ta Morsk. Rybn. x-va i Okeanogr. (Sci.-Techn. Bull., Polar Sci. Res. Inst., Mar. Fish. Economy and Oceanogr, No. 2-3:35-37.

RHZGeofiz 8/58-5603.

Polar Institute of Marine Fisheries and Oceanography, SSSR
Ydanov, I.G., 1961.
[The first cruise of the research exploratory expedition of the Polar Institute of Marine Fisheries and Oceanography into the northwestern part of the Atlantic Ocean (summer 1960) in search of herring.]
Okeanologiia, Akad. Nauk, SSSR, 1(4):756.

institutions (PNIRO)
USSR, Institut Morskogo Ribnogo Chozhiaestva i Okeanografii, 1961.
[The development of scientific investigations in the northern basin.]
Nauchno-Techn. Biull., PNIRO, 1(15):5-21.

Sevastopol Marine Biological Station

institutions
Aronov, M.P., 1962.
Submarine work at the Sevastopol Marine Biological Station. Methods and results of submarine investigations. (In Russian).
Trudy, Okeanogr. Komissii, Akad. Nauk, SSSR, 14: 76-81.

institutions
Vodianitskii, V.A., 1963.
On the ninetieth year of the Sevastopol Biological Station A.O. Kovalevskii. (In Russian).
Trudy Sevastopol Biol. Sta., 16:3-25.

institutions
Nikiforov, Y.D., 1959
[State Oceanographical Institute.]
Biull. Okeanograf. Komissii. Akad. Nauk, SSSR, (3):5-6.

institutions
Akademiia Nauk, SSSR, Okeanologish. Komm., 1961.
[On the World Data Centre of the International Geophysical Year.]
Okeanologiia, Akad. Nauk, SSSR, 1(5):933.

institutions
Belyaev, Yu I., 1967.
The oceanographic Data Center. (In Russian).
Meteorologiya Gidrol., 1967(7):103-104.

institutions
Belyaev, Yu.M., and D.M. Filippov, 1967.
Oceanographic Data Center. (In Russian).
Okeanologiia, Akad. Nauk, SSSR, 7(6):1132-1133.

institutions
Anon., 1958.
Soviet Biological station on the Barents Sea.
Polar Record 9(58):42-43.

Kamshilov, M.M., 1957. [Murmansk Biological station.] Priroda (3):59-63.

VENEZUELA

institutions
Anon., undated (1964?)
Instituto Oceanografico, Universidad de Oriente, Cerro Colorado, Cumana, Venezuela, unnumbered pp.

institutions
Venezuela, Ministerio de Educacion, 1960
Instituto Oceanografico, Universidad de Oriente, Cumana, 33 pp.

instruments

INSTRUMENTATION

In general, an attempt is made to group instruments which serve similar purposes together and/or under the general name of the instrument, if it may serve several purposes. Instrument journals per se have not been scanned nor instruments which might have sea-going applications included.

INSTRUMENTATION

See also: Methods

instruments
Aksakoba, E.A., 1954.
[New apparatus for the investigation of the deep sea.] Priroda 1954(11):108-110.

instrumentation
Alt, Fred, 1966.
NavOceano's instrument test center.
GeoMar.Techn., 2(10):22-26.

instrumentation, at sea
Al'tman, E.N., 1962
Research at sea using new oceanographic apparatus. (In Russian).
Meteorol. i Gidrol., (12):33-38.

Abstr. in:
Soviet Bloc Res. Geophys., Astron. & Space, 52:

instrumentation, misc.
Boods, Larry L., 1966.
Sound is 'king' but non-acoustic systems show future promise.
Undersea Techn., 7(11):43-44, 46, 49-50.

instrumentation,
Branham, D. W., 1965.
Shipboard oceanographic survey system.
Ocean Sci. and Ocean Eng., Mar. Techn. Soc. Amer. Soc. Limnol. Oceanogr., 1:341-362.

instrumentation
Bruns, E.V. 1967
Main problems and directions of development of modern oceanographic equipment. (In Russian; English abstract).
Fizika Atmosfer. Okean Izv. Akad Nauk SSSR 3(5): 544-554.

instrumentation
California Academy of Sciences
California Division of Fish and Game
Scripps Institution of Oceanography
U. S. Fish and Wildlife Service
1950
California Cooperative Sardine Research Program.
Progress Rept. 1950:54 pp., 37 text figs.

instrumentation
Chaplygin, Ye. I., 1958.
[New models of oceanographic apparatus.]
Problemy Arktiki (3):106-108

T327R ERHope

instrumentation
Clayson, C.H., 1970.
Instrumentation in oceanography. Metron, 2 (2): 41-48.

instrumentation
Covey, Charles W., 1965.
Oceanographic instrumentation heads for major expansion.
Undersea Technology, 6(4):20-22.

instrumentation
Covey, Charles W., 1965.
Ocenaographic instrumentation needs coordination. (an editorial).
Undersea Technology, 6(4):4.

instrumentation
Davis, B. W., 1966.
Exploration engineering and instrumentation problems in the marine environment.
Exploiting the Ocean, Trans. 2nd Ann. Mar. Techn. Soc. Conf., June 27-29, 1966, 134-146.

instruments
Defant, A., G. Böhnecke, H. Wattenberg, 1936.
I. Plan und Reiseberichte die Tiefenkarte das Beobachtungsmaterial. Die Ozeanographischen Arbeiten des Vermessungsschiffes "Meteor" in der Dänemarkstrasse und Irmingersee während der Fischereischutzfahrten 1929, 1930, 1933 und 1935. Veroffentlichungen des Instituts für Meereskunde, n.f., A. Geogr.-naturwiss. Reihe, 32:1-152 pp., 7 text figs., 1 plate.

instrumentation
Edgerton, Harold E., and Samuel O. Raymond, 1960
Instrumentation for exploring the oceans.
Electronics Mag., 8 Apr. 1960:62-63.

instrumentation
Frederick, Dolores A., 1966.
Tools of the sea.
Industrial Research, 8(3):115-124.

instrumentation
Gaul, R.D., D.D Ketchum, J.T. Shaw, & J.M. Snodgrass, Editors, 1962.
Marine Sciences instrumentation. Volume I.
Plenum Press, New York: 354 pp.

instrumentation
Gonella, J. et J. Martin, 1968.
Ocean data measuring device. Proc. I.E.R.E. Conf. Electronic Engineering in Oceanography, I.E.R.E. Conf. Proc. No. 8, suppl. vol. contr. no. 47.

instrumentation
Great Britain, Hydraulics Research Board, 1957.
Hydraulics research 1956, 54 pp.

instrumentation
Great Britain, Department of Scientific and Industrial Research, 1956
Hydraulics research, 1955:56 pp.

INSTRUMENTATION,

Greiner, Leonard, 1968.
Ocean survey with instrumented underwater missil
Undersea Techn., 9(5):52-53.

instrumentation (misc.)
Hehn, Jan, 1967.
Oceanographic measuring instruments.
Oceanology Int., 2(1):31-32.

instruments
Hamaker, H.C., 1941.
Ch. I. The oceanographic instruments and the accuracy of the temperature observations. Pt. 1. Methods and instruments. Snellius Exped., 1929-1930, Ocean. Res. 2:45 pp., 22 figs.

instrumentation
Hunley, William H., 1968.
Deep-ocean work systems. Ch. 14 in: Ocean engineering; goals, environment, technology, J.F. Brahtz, editor, John Wiley & Sons, 493-552.

instrumentation
Injutkina, A.I., 1962
The drifting automatic station of the U.S. Navy for the collection of meteorological and oceanographic data. (In Russian).
Okeanologiia, Akad. Nauk, SSSR, 2(4):745.

instrumentation, oceanographic
Isaacs, J.D., undated (1954?).
Considerations of oceanographic instrumentation. In: Oceanographic Instrumentation. NRS Publ. No. 309:1-12.

With discussion by R. Revelle
A.C. Vine
R.G. Paquette

instruments
Isaacs, J.D., R.P. Huffen and L.W. Kidd, 1957.
Instrument stations in the deep sea. Science 125(3243):341.

gear
Iselin, C. O'D., 1949.
Principal instrumentation problems of deep-sea exploration. Instruments 22:898-901.

instruments
Joseph, J., 1948.
Meereskundliche Merrgerate. Geophys., Dieterich'sche Verlagsbuchhandlung, Wiesbaden, Pt. 2, Vol. 18:178-214.

instrumentation, misc.
Kanal, Leveen, 1962.
An analysis of a class of pattern recognition networks.
Marine Sciences Instrumentation, Instr. Soc., Amer., Plenum Press, N.Y., 1:340-346.

instrumentation
Kolesnikov, A.G., N.A. Panteleyev and A.N. Paramonov 1966.
Modern trends in the development of instruments and methods for deep-sea investigations.
Metody i Pribory dlya Issledovaniya Fizicheskikh Protsessov v Okeane, Morsk Gidrofiz. Inst. Kiev, 36:3-14.

*Kreitner, F.J., 1968. instrumentation
Mobile instruments for aiding underwater research.
Instrument. Techn., 15(3):35-40.

instrumentation
Lacombe, Henri, 1965.
Cours d'oceanographie physique. (Theories de la circulation. Houles et vagues).
Gauthier-Villars, Paris, 392 pp.

instruments
Longard, J.R., 1953.
Modern instruments of physical oceanography and their use in waters of interest to Canada.
Eng. J., 6 pp., 8 textfigs.

instrumentation
Margetts, A.R., 1969.
Gear research without disturbing the environment
Oceanol. int. 69, 3: 5 pp.

instruments
Megia, T.G., and A.R. Sebastian, 1953.
Equipment used in oceanographic investigations in the Philippines. Philippines J. Fish. 2(2): 165-171.

instrumentation
Moreira, da Silva, P., 1956.
Um novo abáco para a prévisas da mare. (DG-02).
Anais Hidr., 15:167-172.

instrumentation
Morgan, Charles J., and Julian Josephson, 1964.
Needs for oceanographic instrumentation.
Undersea Technology, 5(7):33-35.

instruments
Mosby, H., 1938
Svalbard waters. Geophys. Publ. 12(4): 85 pp., 34 text figs.

instrumentation,
Murty, A.V.S., 1967.
A simple slide device for the use of vertical profile studies in oceanography.
J.Mar.biol.Ass.India, 9(2):423-425.

instrumentation
Nakano, M., 1957.
Some oceanographical instruments recently devised by the members of the Central Meteorological observatory and its subordinate organs.
Proc. UNESCO Symp., Phys. Ocean., Tokyo, 1955: 30-37.

instrumentation
Nakano, M., 1956.
Some oceanographical instruments recently devised in Japan. Misc. Geofis., Serv. Met. Angola Comem. X Aniv. Serv. Met. Nac., Luanda,:206-222.

instruments
Narasako, Yoshikazu, Sukehiro Oya and Akira Yamabe, 1960.
[Simplified ultrasonic shell-plate-flaw-detector for fishing vessel with the use of the standing wave technique in the thickness measurement.] Mem. Fac. Fish., Kagoshima Univ., 8: 111-120.

instrumenation
Niskin, Shale, 1968.
"Free" and expendable instruments.
Oceanology intl, 3(2):31-32, 34.

instrumentation, misc.
Niskin, Shale J., 1962.
Some new mechanical devices for oceanographic research.
Marine Sciences Instrumentation, Instr. Soc., Amer., Plenum Press, N.Y., 1:246-250.

instrumentation
Pickard, G.L., 1961.
Oceanograhic instrumentation in Canada. (Abstract).
Tenth Pacific Sci. Congr., Honolulu, 21 Aug.-6 Sept., 1961, Abstracts of Symposium Papers, 346.

instrumentation (general)
Pritchard, D.W., 1961
Problems in oceanographic instrumentation.
Proc. Instrument Soc., Amer., Aero-Space Instrument. Symposium, May 1-4, 1961, Dallas, Texas, Preprint I-6: 3 pp.

Also in:
Chesapeake Bay Inst., Collected Reprints, 5 (1963)

instrumentation
Rotschi, H., 1953.
Expédition océanographique "Capricorn" de la "Scripps Institution of Oceanography" de l'Université de Californie, Novembre 1952-Février 1953. Bull. d'Info., C.C.O.E.C. 5(10):439-466, 2 pls.

instruments
Ryther, J. H., C. S. Yentsch and G. H. Lauff, 1959
Sources of Limnological and oceanographic appartus and supplies.
Limnol. & Oceanogr., 4(3):357-365.

instrumentation
Sasaki, T., 1957.
Three instruments constructed and employed in Japan.
Proc. UNESCO Symp., Phys. Oceanogr., Tokyo, 1955 :46-48.

instruments
Schumacher, A., 1942.
Le puiseur d'eau de surface de O. Sund, perfectionné. Rev. Hydrogr. 19(37):137-139.

equipment
Scripps Institution of Oceanography, 1949.
Marine life research program. Progress report, 1 May to 31 July 1949. 24 pp. (mimeographed), 16 figs. (ozalid).

instrumentation
Sheffield Publishing Company, 1963.
Problems in undersea technology.
Undersea Techn., 4(1):48.

instrumentation
Shekhvatov, B.V., 1970.
Development of oceanographic instrumentation at the Institute of Oceanology of Academy of Sciences of the USSR. Okeanologiia, 10(4): 573-587.

(In Russian; English abstract)

Instrumentation
Shouleykin, V.V., 1970.
The floating sea Scientific Institute and the cruises of Persey. (In Russian). Fizika Atmosfer. Okean., Izv. Akad. Nauk, SSSR, 6(4): 334-351.

instrumentation
Sinha, Evelyn, 1966.
State of the art - instrumentation.
Oceanic Abstracts, 1:204 pp.

instrumentation
Sisoev, N.N., 1963
Exposition and seminar exhibiting recent oceanological devices and apparatuses. (In Russian).
Okeanologiia, Akad. Nauk, SSSR, 3(1):179-181.

instrumentation

Sysoev, N.N., editor, 1960
Conference on oceanographic instruments and other articles.
Biull. Okeanogr. Komissii Pri Prezidiume Akad. Nauk, SSSR, (4):68 pp.

H.O. TRANS-124(1962)

instrumentation

Sysoev, N.N., 1960
The design and improvement of oceanographic instruments in the Institute of Oceanology of the Academy of Sciences USSR.
Biull. Okeanogr. Komissii pri Prezidiume Akad. Nauk, SSSR, (4):34-36.

H.O. TRANS-124(1962)

instrumentation

Sysoiev, N.N., 1957.
Developments and future in the oceanographic instrument construction in the USSR.
Proc. UNESCO Symp., Phys. Ocean., Tokyo, 1955: 246-250.

instruments

Sysoev, N.N., 1956.
[On methods and apparatus for deep-water oceanographic measurements.] Trudy Inst. Okeanol., 19: 156-163.

instrumentation

Sysoev, N.N., et al, 1959
[Papers on oceanography and oceanographic instruments.]
Trudy Akad. Nauk SSSR, Inst. Okean., 35: 78-85, 113-117, 190-224, 238-244.
Listed in Techn. Transl., 6(10): 753.

instrumentation

Tessier, M., J.C. Garrec, 1969.
Mechanical apparatus for lowering into the sea and deep sea towing. Oceanol. int., 69, 1: 8 pp 2 figs.

instruments

Tomczak, G., 1952.
Das Bourdonrohr als ozeanographisches Messelement.
Ber. Deutschen Wetterdienstes in der U.S.-Zone, No. 38:347-353, 1 textfig. (DK551.46.018.2).

instrumentation

Tucker, M.J., and R.Bowers, 1967.
Oceanographic equipment; the oceanographer's view.
Hydrospace, 1(1):47-51.

instrumentation

Valembois, J., 1957.
Appareils pour mesures océanographiques.
La Nature (3267):282-285.

instrumentation

Verber, James L., 1965.
Standardizing oceanographic instrumentation is practical.
Undersea Techn., 6(8):20-21.

instrumentation

Vine, Allyn, 1965.
Tools for ocean research: vehicles, platforms, buoys, sensors, instruments are all becoming more effective.
Internat. Sci., Technol., Conover-Mast Publ., Dec. 1965:77-80;82-84.

instrumentation

Vine, A.C., 1957.
Some trends in oceanographic instrumentation.
Proc. UNESCO Symposium on Phys. Oceanogr., Tokyo, 1955:49-52.

instruments

Vine, A.C., 1955.
Oceanographic instrumentation.
Science 122(3173):748-751.

instrumentation

Wüst, G., 1932.
Program, Ausrüstung, Methoden der Serienmessungen
Deutsch. Atlant. Exped., "Meteor", 1925-27, Wiss. Ergebn., 4(1):1-59.

Transl. cited:
USFWS Spec. Sci. Rept., Fish., 227.

instrumentation,

Zolotov, S.V., 1962.
On the principal developments of submarine surveying equipment. Methods and results of submarine investigations. (In Russian).
Trudy, Okeanogr. Komissii, Akad. Nauk, SSSR, 14: 24-29.

Acoustic

instrumentation, acoustic

Some instruments using acoustics are to be found under other groupings such as instrumentation, depth, etc.

instrumentation, acoustic

See also: instrumentation, navigational (sonar)
instrumentation, geophysics (sonar)
instrumentation, echo-sounding, etc., etc.

instrumentation, acoustic

Albers, Vernon M., 1965.
Underwater acoustics handbook II.
The Pennsylvania State University Press, 356 pp. $12.50.

Instrumentation, acoustic

Alden, J. M., and L. A. Farrington, 1962.
High resolution direct graphic recording of underwater sound.
Int. Hydrogr. Rev., 39(1):125-137.

instrumentation, acoustic

Anderson, P.R., 1963.
Cable assembly with integral hydrophones and instrumentation.
Mar. Sci. Instrument., Plenum Press, 2:75-79.

instrumentation, acoustic

Anderson V.C., 1958.
Arrays for the investigation of ambient noise in the ocean. J. Acoust. Soc., Amer., 30(5):470-477.

instrumentation, acoustic

Anderson, Victor C., and John C. Munson, 1962
Directivity of spherical receiving arrays. (Abstract).
J. Acoust. Soc., Amer., 34(12):1971-1972.

instrumentation, acoustic

Anon., 1969.
S*DAR: submerged electrode detection and ranging.
UnderSea Techn., 10(5):43.

instrumentation, acoustic

Baird, D.L., and C.M. McKinney, 1962.
Investigation of the line-and-cone underwater sound transducer.
J. Acoust. Soc., Amer., 34(10):1576-1581.

instrumentation, acoustic

Bauer, B.B., and E.L. Torick, 1966.
Calibration and analysis of underwater earphones by loudness- balance method.
J. Acoust. Soc., Amer., 39(1):35-39.

instrumentation, acoustics

Bauer, B.B., and E.L. Torick, 1966.
Experimental studies in underwater directional communication.
J. Acoust. Soc., Amer., 39(1):25-34.

instrumentation, acoustic

Beckerle, John C., 1963.
Effects of ocean waves on acoustic signals to very deep hydrophones.
J. Acoust. Soc., Amer., 35(3):267-272.

instrumentation, acoustic

Bell, John, and Ilmar G. Raudsep 1967.
Phase comparison acoustic systems: how they work, what they are capable of doing.
Ocean Industry 2(4):39-44.

INSTRUMENTATION, acoustic

Bell, Thaddeus G., 1968
Effect of phase errors on the direction-finding precision of hydrophone arrays.
J. acoust. Soc. Am., 43(4):704-705

instrumentation, acoustic

Bobber, R.J., 1970.
An active transducer as a characteristic impedance of an acoustic transmission line.
J. acoust. Soc. Am., 48(1-2):317-324

instrumentation, acoustic

Bryan, George M., Marek Truchan and John I. Ewing, 1963.
Long-range SOFAR studies in the South Atlantic Ocean.
J. Acoust. Soc., Amer., 35(3):273-278.

instrumentation, acoustic

Buchanan, C.L., 1963.
Wide band transducers for sound velocimeters.
Mar. Sci. Instrument., Plenum Press, 2:157-161.

instrumentation, acoustic

Carey, George F., John E. Chlupsa, and Howard H. Schloemer 1967.
Acoustic turbulent water-flow tunnel.
J. acoust. Soc. Am. 41(2):373-379.

instrumentation, acoustic

Carson, D.L., 1962.
Diagnosis and cure of erratic velocity distributions in sonar projector arrays.
J. Acoust. Soc., Amer., 34(9):1191-1196.

instrumentation, acoustic

Caulfield, David D., 1962.
Predicting sonic pulse shapes of underwater spark discharges.
Deep-Sea Research, 9(4):339-348.

instrumentation, acoustics

Chapman, R.M., and A.D. Mugnier, 1964
Computer processing of acoustic data.
Undersea Techn., 5(10):26-28.

instrumentation, acoustic
Chesterman, W. Deryck, J.M.P. St. Quinton, Y. Chan and H.R. Matthews, 1967.
Acoustic surveys of the sea floor near Hong Kong.
Int. hydrogr. Rev., 44(1):35-54.

instrumentation, acoustic
Cobb, A. Donn, 1962.
The quiet platform key to successful oceanographic acoustic research.
Marine Sciences Instrumentation, Inst. Soc., Amer., Plenum Press, N.Y., 1:6-7.

instrumentation, acoustic
Covey, Charles W., 1965.
Digital systems checks acoustic underwater transducers.
Undersea Techn., 6(12):39-41.

instrumentation, acoustic
Crews, Arthur A., and Robert L. Erath, 1966.
Unique directional reflectors for omnidirectional underwater acoustics transducers.
J. Acoust. Soc., Am., 40(6):1556-1557.

instrumentation, acoustic
Cushing, D.H., 1969.
Acoustic fish counting. Oceanol. int. 69, 4: 4 pp., 4 figs.

instruments, acoustics
Daniels, Charles K., 1969.
Acoustic instruments.
Oceanology Intl 4(4):21-23. (popular)

instrumentation, acoustic
Daniels, D. and R. Henderson, 1969.
An integrated acoustic underwater survey system. Oceanol. int. 69, 4: 11 pp.

instrumentation, acoustic
Degtyarev, A.A., and P.E. Shatunov, 1964.
Small rotary-sliding arrangement for hydro-acoustic devices. (In Russian).
Materiali, Ribochoz. Issled. Severn. Basseina. Poljarn. Nauchno-Issled. i Proektn. Inst. Morsk. Rib. Choz. i Okeanogr. im N.M. Knipovicha, PINRO, Gosud. Proizvodst. Komm. Ribn. Choz. SSSR. 4:123-127.

instrumentation, acoustic
Delaney, R.P., 1962.
Deep transducer design.
Marine Sciences Instrumentation, Instr. Soc., Amer., Plenum Press, N.Y., 1:25-27.

instrumentation, acoustics
Dobroklonskiy, S.V., and V.V. Filippov, 1962.
Absolute calibration of large hydrophones in the infrasonic frequency range by the pistonphone method. (In Russian).
Trudy, Morsk. Gidrofiz. Inst., 20:3-19.

Translation:
Scripta Tecnica for Agu, (4):1-14. 1964.

instrumentation, acoustic
Dobroklonskiy, S.V., and V.V. Filipov, 1960.
Absolute calibration of large hydrophones in the infrasonic frequency range by the pistonphone method. (In Russian).
Trudy, Morsk. Gidrofiz. Inst., Akad. Nauk, SSSR., 20:3-19.

Translation:
Scripta Tecnica, Inc., for AGU, 1-14.

instrumentation, acoustic
Donovan, John E., 1970.
Triboelectric noise generation in some cables commonly used with underwater electroacoustic transducers.
J. acoust. Soc. Am. 48 (3-2): 714-724.

instrumentation, acoustic
Dow, Willard, and Stephen L. Stillman, 1962.
Inverted echo sounder.
Marine Sciences Instrumentation, Instr. Soc., Amer., Plenum Press, N.Y., 1:263-272.

instrumentation, acoustic
Dranetz, Abraham I., and Harvey Rathbun Jr. 1966.
Analysis of electroacoustical transducers by differential immittance techniques.
J. acoust. Soc. Am. 40 (2): 412-416.

instruments, acoustic
Dziedzic, A., 1962.
Étude d'un compteur d'impulsions acoustiques, destinée à des mesures de rhythmes d'activité biologique d'animaux marins.
Cahiers Biol. Mar., Roscoff, 3(4):417-431.

instrumentation, acoustic
Epstein, M., and B. Harris, 1967.
A digital hyperbolic chirp generator for oceanographic applications.
1967 IEEE Int. Conv. Rec. (8):81-83.

instrumentation, acoustic
Folche, M.P., 1963.
Underwater sound calibration stations at Le Brusc Laboratory.
In: Underwater acoustics, V.M. Albers, Edit., Plenum Press, New York, 103-123.

instrumentation, acoustic
Folsom, T.R., 1963.
Transducers for oceanic research, a wide scope.
Mar. Sci. Instr., Plenum Press, 2:1-7.

instrumentation, acoustic
Friedman Bob 1967.
Fish charts ocean floor.
Sea Frontiers 13(2): 98-101.

instrumentation, acoustics
Frosch, R.A., V.C. Anderson and H. Bradner, 1964
Instrumenting the sea floor.
IEEE Spectrum, 1(11):101-114.

instrumentation, acoustic
Gaul, R.D., Editor, 1963.
Proceedings of the symposium on transducers for oceanic research, November 8-9, 1962, San Diego, California.
Marine Sci. Instrument., Plenum Press, 2:195 pp.

instrumentation, acoustics
Gaul, R.D., and David D. Ketchum, 1963.
Preface-proceedings of the symposium on transducers for oceanic research, November 8-9, 1962.
Mar. Sci. Instrument., Plenum Press, 2:v.

instrumentation, acoustic
Gilchrist, Richard B., and Wayne A. Strawderman, 1965.
Experimental hydrophone-size correction factor for boundary-layer pressure fluctuations.
J. Acoust. Soc., Amer., 38(2):298-302.

instrumentation, acoustic
Givan, G.R., and M. Levine, 1968.
An electro-mechanical marine environment
Mar. Sci. Instrument. 4: 381-386

instrumentation, acoustic
Gondet, P., and L. Erdely, 1963
Un nouveau manomètre piezoélectrique pour pressions relatives fonctionnant sous très haute pression ambiente.
Bull., Assoc. Francaise Étude des Grandes Profondeurs Océaniques, (1):10-17.

instrumentation, acoustic
Gray, F., and T.R.E. Owen 1969.
A recording sono-radio buoy for seismic refraction work.
Oceanol. int. '69 (4): 5 pp.

instrumentation, acoustic
Green, P.S., J.L.S. Bellin and G.C. Knollman, 1968.
Acoustic imaging for viewing in turbid water.
Undersea Techn., 9(5):48-51.

instrumentation, acoustic
Greenspan, M., and C.E. Tschiegg, 1963.
A sing-around velocimeter for measuring the speed of sound in the sea.
In: Underwater acoustics, V.M. Albers, Edit., Plenum Press, N.Y., 87-101.

instrumentation, acoustic
Groves, Ivor D., Jr., and Allan C. Tims 1970.
Standard probe hydrophone for acoustic measurements from 10 Hz to 200 kHz.
J. acoust. Soc. Am. 48 (3-2): 725-728.

instrumentation, acoustic
Hart, James A., 1969.
Full power testing of sonar transducers.
UnderSea Techn., 10(5):44-45, 50.

instrumentation, acoustic
Haslett, R.W.G., G. Pearce, A.W. Welsh and K. Hussey, 1966.
The underwater acoustic camera.
Acustica, 17(4):187-203.

instrumentation, acoustics
Heaps, H.S., 1966.
Effect of turbulence nearfield on a shielded transducer.
J. acoust. Soc., Am., 40(6):1331-1336.

instrumentation, acoustic
Henriquez, T.A., 1969.
Calibration at high pressure of piezoelectric elements for deep-submergence hydrophones.
J. acoust. Soc. Am., 46 (5-2): 1251-1253

instrumentation, acoustic
Henriquez, T.A., and L.E. Ivey 1970.
Standard hydrophone for the infrasonic and audio-frequency range at hydrostatic pressure to 10 000 psig.
J. acoust. Soc. Am. 47 (1-2): 276-280.

instrumentation, acoustic
Hersey, J.B., undated (1954?).
Acoustic instrumentation as a tool in oceanography. In: Oceanographic Instrumentation.
NRS. Publ. 309:101-122, 7 textfigs.

With discussion by R.J. Christiensen
H.E. Nash
L.N. Liebermann.

instrumentation, acoustic
Hoffman, D., R. Sadowy, Jr. and B.B. Adams 1968.
Extended aperture deltic correlators.
Mar. Sci. Instrument. 4: 87-93.

instrumentation, acoustics
Horita, R.E., 1967.
Free-flooding unidirectional resonators for deep-ocean transducers.
J. acoust. Soc. Am., 41(1):158-166.

instrumentation, acoustics
Hueter, T.F., & P.H. Morse, 1961
Optimum hydrophone design for low frequencies.
J. Acous. Soc. Amer., 33(11): 1628-1629.

instrumentation acoustics
Hutchins, R.W., 1969.
Broadband hydro-acoustic sources for high resolution sub bottom profiling. Oceanol. int., 69, 1: 7 pp., 7 figs.

instrumentation, acoustics
Hyde, J.L., 1963.
Nuclear digital transducers.
Mar. Sci. Instrument., Plenum Press, 2:173-181.

instrumentation, acoustic
Hyde, James L., and Edwin Joscelyn, 1965.
A temperature-insensetive pressure transducer.
Mar. Sci. Instrument., Plenum Press, 3:197-210.

instrumentation, acoustic
Kendig, Paul M., 1967.
Advanced transducer developments.
In: Underwater acoustics, V.M. Albers, editor, Plenum Press, 2: 7-34.

instrumentation, acoustic
Kendig, Paul M., and Hugh J. Clarke, 1965.
Experimental liquid-filled transducer array for deep-ocean operation.
J. Acoust. Soc., Amer., 37(1):99-104.

instrumentation, acoustic
Ketchum, David D., 1966.
Acoustic command.
GeoMarine Techn., 2(6):39-44.

instrumentation, acoustic
Kolman, O.V., 1968.
New television and hydroacoustic devices for underwater study. (In Russian).
Okeanologiia, Akad. Nauk, SSSR, 8(3):561.

instrumentation, acoustic
Kroebel, Werner, und Johannes Wick, 1963
Registrierungen in situ im Nordatlantik mit der Bathysonde und einem neuen Messgerät für Schallgeschwindigkeit im Meerwasser mit extrem hoher Genauigkeit.
Kieler Meeresforsch., 19(2):133-141.

instrumentation, acoustic
Kronengold, M., R. Dann, W.C. Green and J.M. Loewenstein, 1964
Description of the system.
In: Marine Bio-Acoustics, Pergamon Press, Ltd 11-25.

instrumentation acoustic
Liebich, R.E., 1969.
Hydrophones-undersea microphones.
Oceanology intl, 4(3):50-54. (popular)

instrumentation, acoustic
Lord, N.W., 1962.
Precise calibration of sea-going velocimeters.
J. Acoust. Soc., Amer., 34(11):1790-1791.

instrumentation, acoustic
Lovett, J.R., 1962.
The SVTP [sound-velocity-temperature-pressure] instrument and some applications to oceanography.
Marine Sciences Instrumentation, Instr. Soc., Amer., Plenum Press, N.Y., 1:168-172.

instrumentation, acoustic
Lukman, Frank J., 1962
Electroacoustic sensitivity of pressure equalized spherical and cylindrical hydrophones. (abstract).
J. Acoust. Soc., Amer., 34(12):1971.

instrumentation, acoustic
Lutsch, A., 1958.
An apparatus for measuring and recording the velocity of sound and temperature versus depth in sea waters.
Acustica, 8(9):386-391.

instrumentation, sound (velocimeters)
Mackenzie, Kenneth V. 1970.
A decade of experience with velocimeters.
J. acoust. Soc. Am. 50(5-2): 1321-1333

instrumentation, acoustic
Mackenzie, K.V., 1961.
Sound-speed measurements utilizing the bathyscaph "Trieste".
J. Acoustical Soc., Amer., 33(8):1113-1119.

INSTRUMENTATION, acoustics
Maidanik, G., and W.T. Reader, 1968.
Filtering action of a blanket dome.
J. acoust. Soc. Am., 44(2): 497-502

instrumentation, acoustics
McCartney, B.S., 1966.
Low frequency sound sources: statement of problem and some possible solutions.
Proc. I.E.R.E. Conf. Electronic Eng. Oceanogr., 6 pp.

instrumentation, acoustics
McGehee, Maurice S., and D.E. Boegeman, 1966.
MPL acoustic transponder.
37(11):1450-1455.
Rev. scient. Instrum.

instrumentation, acoustic
McKinney, C.M., and W.R. Owens, 1957.
Wedge-shaped acoustic horns for underwater sound applications. J. Acoust. Soc., Amer., 29(8):940-947.

instrumentation, acoustic
Mohammed, A., 1966.
Expressions for the electromechanical coupling factor in terms of critical frequencies.
J. Acoust. Soc., Amer., 39(2):289-293.

instrumentation, acoustic
Moore, David G., 1964.
Acoustic-reflection reconnaissance of continental shelves: eastern Bering and Chukchi seas.
In: Papers in Marine Geology, R.L. Miller, Editor, Macmillan Co., N.Y., 319-362.

instrumentation, acoustic
Morand, Genevieve, Maurice Meton and Gerard Lejeune, 1963.
Description et construction d'une sonde thermo-électrique pour la mesure de l'intensité acoustique dans l'eau.
Comptes Rendus, Acad. Sci., Paris, 257(5):1018-1020.

instrumentation, acoustic
Morrissey, James H., 1962.
Location of underwater or surface sound sources by means of computer-linked cabled-hydrophone fields.
Marine Sciences Instrumentation, Instr. Soc., Amer., Plenum Press, N.Y., 1:28-30.

instrumentation, acoustic
Nelson, Ralph A., Jr., and Larry H. Royster, 1971.
Development of a mathematical model for the Class I flextensional underwater acoustic transducers.
J. acoust. Soc. Am., 49(5-2): 1609-1620

instrumentation, acoustic
Nickles, J.C., and R.K. Johnson 1971.
A digital system for volume reverberation studies.
J. acoust. Soc. Am. 50(1-2): 314-320.

instrumentation, acoustic
Norton, H.N., 1963
Piezoelectric pressure transducers.
Instr. and Control Systems, 36(2):83-86.

instrumentation acoustic
Oberlin, Richard P., 1962.
The design and installation of the fixed acoustic buoy.
Marine Sciences Instrumentation, Instr. Soc., Amer., Plenum Press, N.Y., 1:1-5.

instrumentation, acoustic
O'Neill, E.T., 1962.
Pressure-balanced high-pressure hydrophone.
J. Acoust. Soc., Amer., 34(10):1661-1662.

instrumentation acoustic
Owens, W.R., and C.M. McKinney, 1957.
Experimental investigation of conical horns used with underwater sound transducers.
J. Acoust. Soc., Amer., 29(6):744-748.

instrumentation, acoustic
Parfitt, G.G., 1963.
The improvement of vibration isolation.
In: Underwater acoustics, V.M. Albers, Edit., Plenum Press, N.Y., 67-86.

instrumentation, acoustic
Perry, Kenneth E., and Paul Ferris Smith, 1965.
Digital methods of handling oceanographic transucer.
Mar. Sci. Instrument., Plenum Press, 3:25-39.

instrumentation, acoustic
Rao, V.N., 1963.
An instrument for measuring small changes in the velocity of sound in the sea.
Indian J. Pure Appl. Physics, 1(2):69-72.

instrumentation, acoustic
Royster, L.H., 1969.
Flextensional underwater acoustics transducer.
J. acoust. Soc. Am. 45(3): 671-682.

instrumentation, acoustic
Rusby, J.S.M., 1965.
The possibility of an interaction anomaly in acoustic arrays and radio arrays.
Radio and Electronic Engineer, 29:113-116.
Also in:
Collected Reprints, Nat. Inst. Oceanogr., 13.

instrumentation, acoustic
Sanders, D.R., 1965.
Underwater sound transducer with a conical-shell radiation pattern.
J. acoust. Soc., Am., 38(3):412-415.

instrumentation, acoustics
Savage, G.H., and J.B. Hersey, 1969.
Project Seaspider: The design, assembly, construction and sea trials of a tri-moored buoyant structure with neutrally buoyant legs to provide a near motionless instrument base for oceanographic research.
J. mar. techn Soc. 3(2):95-112.

instrumentation acoustic
Schaefer, Jan V., 1960.
Remote pre-amplifiers for under-ocean work.
Electronics, (8July, 1960):60-62.

instrumentation, acoustic
Schofield, D., 1963.
Transducers.
In: Underwater acoustics, V.M. Albers, Edit., Plenum Press, N.Y., 5-27.

instrumentation, acoustical
Shatova, O.E., 1960.
On recording fish density with self-writing and electronic marks of hydroacoustic instruments.
Nauchno-Techn. Biull., PNIRO, 1(15):61-64.

instrumentation, acoustic
Shurbet, D.H., 1962.
Note on use of a SOFAR geophone to determine seismicity of regional oceanic areas.
Bull. Seismol. Soc., Amer., 52(3):689-691.

instrumentation, acoustics
Sims, C.C., 1960.
Bubble transducer for radiating high-power low-frequency sound in water.
J. Acoust. Soc., Amer., 32(10):1305-1307.

instrumentation, acoustic
Sims, C.C., and T.A. Henriquez, 1962.
Cavity-loaded piston resonators.
J. Acoust. Soc., Amer., 34(9):1204-1206.

instrumentation, acoustic
Smith, P.F., and E. Fredkin, 1963.
Computers and transducers.
Mar. Sci. Instrument., Plenum Press, 2:81-86.

instrumentation, acoustic
Snyder, A.E., and A. D'Onofrio, 1963.
Digital pressure transducer study.
Mar. Sci. Instrument., Plenum Press, 2:189-194.

instrumentation, acoustic
Solberg, Ruell F., Jr., and Leonardt F. Kreisle, 1968.
Small low-frequency, omnidirectional broad-band hydrophone.
J.acoust.Soc.Am., 44(6):1699-1705.

instrumentation, acoustics
Spiess, Fred N., Michael S. Loughridge, Maurice S. McGehee and D.E. Boegeman, 1966.
An acoustic transponder system.
Navigation, 13(2):154-161.

instrumentation, acoustic
Stahl, Raymond A., and Robert L. Miller, 1965.
The aquasond.
Mar. Sci. Instrument., Plenum Press, 3:133-147.

instrumentation, acoustical
Stas', I.I., 1957.
Remote pressure transducer. (In Russian).
Trudy Morsk. Gidrofiz. Inst., Akad. Nauk, SSSR, (11):42-55.

Instrumentation, acoustic
Stepanishen, P.R. 1971.
The transient response of arrays of transducers.
J. acoust. Soc. Am. 50(3-2):964-974.

instrumentation, acoustic
Stubbs, A.R., and R.G.G. Laurie, 1962.
Asdic as an aid to spawning ground investigations.
J. du Conseil, 27(3):248-260.

instrumentation, acoustic
Stumpf, F.B., and P.M. Kendig, 1963.
Determination of the power radiated by a piston-like underwater sound transducer from near field axial pressure measurements.
J. Acoustic Soc., Amer., 35(2):254-255.

instrumentation, acoustic
Stumpf, F.B., and Y. Y. Lam, 1970.
Radiation resistance of a small transducer at a water surface near plane boundaries.
J. acoust. Soc. Am., 47(1-2):332-338

instrumentation, acoustic
Suellentrop, F.J., A.E. Brown and Eric Rule, 1962.
An instrument for the direct measurement of the speed of sound in the ocean.
Marine Sciences Instrumentation, Instr. Soc., Amer., Plenum Press, N.Y., 1:186-189.

instrumentation, acoustic
Sussman, H., 1957.
On the uses of unicellular rubber for underwater acoustic devices.
J. Acoustical Soc., Amer., 29(1):145.

instrumentation, acoustic
Taira, Keisuke, and Atsushi Takeda, 1969.
On the response characteristics of the sonic wave gauge.
J. oceanogr. Soc. Japan, 25(6):299-306.

instrumentation, acoustics
Teer, K., 1965.
On the optimum configuration for a condenser microphone.
Acustica, 15(5):256-263.

instrumentation, acoustic
Trott, W. James, 1964.
Underwater-sound-transducer calibration from nearfield data.
J. Acoust. Soc., Amer., 36(8):1557-1568.

instrumentation, acoustic
Tschiegg, C.E., & D.D. Hays, 1959
Transistorized velocimeter for measuring the speed of sound in the sea.
J. Acoust. Soc. Amer., 31(7): 1038-1039.

instrumentation
Tucker, D.G., and J.G. Henderson, 1960
Automatic stabilization of underwater acoustic beams without mechanical motion of the transducer.
Int. Hydrogr. Rev., 37(1): 69-78.

instrumentation, acoustic
Ulonska, Arno, und Joachim Jarke, 1966.
Ein Gerät zur in situ- Messung der Schallgeschwindigkeit in marinen Sedimenten.
Dt. hydrogr. Z., 19(3):113-120.

instrumentation, acoustics
United States, Office of Naval Research, 1964.
Seagoing platform for acoustic research.
Naval Res. Rev., Oct., 1964:8-11, 28.

instrumentation, acoustics
Urick, R.J., and C.L. Kelly, 1968.
Underwater sound measurements from an aircraft.
Mar. Sci. Instrument. 4:306-313

instrumentation, acoustic
Vigoureux, P., and J.B. Hersey, 1962
Sound in the sea. Ch. 12, Sect. IV. Transmission of energy within the sea. In: The Sea, Vol. 1, Physical Oceanography, Interscience Publishers, 476-497.

instrumentation, acoustics
Vines, Darrell L., 1968.
Acceleration sensitivity measurements for hydrophones.
Mar. Sci. Instrument., 4: 622-634.

instrumentation, acoustic
Walther, K., 1963.
Logarithmic-periodic underwater sound projector.
J. Acoust. Soc., Amer., 35(1):81-90.

instrumentation, acoustic
Watanabe, Kishimatsu, 1969.
A device with outriggers for increasing sounding width.
Int. hydrogr. Rev., 46(1): 131-138.

instrumentation, acoustic
Watkins, William A., 1963.
Portable underwater recording system.
Undersea Technology, Sheffield Pubs. Co., Inc., 4(9):23-24.

instrumentation, acoustic
Weber, Harry P. 1968.
An experimental near real time sound velocity profile system.
Mar. Sci. Instrument. 4:196-202.

instrumentation, acoustic
Weber, Peter F. 1968.
A low-frequency, deep-bottomed sound source.
Mar. Sci. Instrument. 4:687-693

instrumentation, acoustic
Weinberg, N.L., J.M. Loewenstein and H. Yedid 1968.
A wide range, phase coherent demodulator for oceanographic research.
Mar. Sci. Instrument 4:661-670.

instrumentation, acoustic
Weston, D.E., 1966.
Acoustic interaction in arrays of small spheres.
J. Acoust. Soc., Amer., 39(2):316-322.

instrumentation, acoustic
Weston, D.E., 1963.
Explosive sources.
In: Underwater acoustics, V.M. Albers, Edit., Plenum Press, N.Y., 51-66.

instrumentation, acoustic
Williamson, R.L., G. Hodges and E. Eady, 1968.
A new sound velocity meter.
Int. hydrogr. Rev., 45(1):177-188.

instrumentation, acoustic
Woollett, R.S., 1966.
Effective coupling factor of single-degree- of freedom transducers.
J. acoust. Soc., Am., 40(5):1112-1123.

instrumentation, acoustic
Woolston, D.D., and E.A. Bitz, 1963.
Times Fax, an underwater acoustic range reference.
Naval Ordnance Lab., NOLTR, 62-219:19 pp.

instrumentation, acoustic
Wyber, R., 1967.
Acoustic and related measurements in the ocean.
In: The collection and processing of field data, E.F. Bradley and O.T. Denmead, editors, Interscience Publishers, 205-211.

airborne

instrumentation, airborne
Wilkerson, John W., 1966.
Research in airborne oceanography.
Geo-Marine Techn., 2(8):9-15.

Air-sea interface

instrumentation, air-sea interface
Deardorff, James W., 1962.
An experimental ocean buoy for air-sea transfer studies. Unpublished manuscript.).
Univ. Washington, Dept. Meteorol. and Climatol., Occ. Rept., No. 13:1-79.

instrumentation, air-sea interface
Doe, L.A.E., and John Brooke, 1965.
A moored stable platform for air-sea interaction studies.
Limnol. and Oceanogr., Redfield Vol., Suppl. to 10:R79-R86.

instrumentation, air-sea interface
Kestner, A.P., 1969.
An instrument for recording aerodynamic pressure.
Okeanologiia, 9(4): 718-724. (In Russian; English abstract)

instrumentation, air-sea interface
Kuznetsov, O.A., and I.A. Filipov, 1963.
A gradient device for the measurement of wind velocity in the lower section of the nearearth air layer above the sea surface. (In Russian).
Okeanologiia, Akad. Nauk, SSSR, 5(1):166-169.

instrumentation, air-sea interface
McFadden, James D., and John W. Wilkerson, 1967.
Compatibility of aircraft and shipborne instruments used in air-sea interaction research.
Mon. Wealth. Rev. U.S. Dep. Comm. 95(12):936-941.

instrumentation
Munk, W.H., undated (1954?).
Air-sea boundary processes. In: Oceanographic Instrumentation. NRC Publ. 309:210-224.

With discussion by R.O. Reid
A.H. Woodcock
R.B. Montgomery

instrumentation, air-sea interface
Schooley, Allen H., 1965.
Simple instrumentation for measuring temperature gradients at the sea-air interface. (abstract).
Ocean Sci. and Ocean Eng., Mar. Techn. Soc., - Amer. Soc. Limnol. Oceanogr., 2:989.

anchoring

instrumentation (anchoring)
Sergeev, I.V., 1959
[Experience of placing the ship at anchor at great depths.]
Trudy Inst. Okeanol., 35: 206-224.

instrumentation, anchoring
Mills, A.A., and T.C. Leung, 1964.
A delayed-release link.
Deep-Sea Research, 11(2):257-260.

instrumentation, antennas
Umezono, Shigeru 1966.
Studies on experimentally manufactured television antennas for ships. (In Japanese; English abstract)
Bull. Fac. Fish. Nagasaki Univ. 21:265-272.

Arctic

instrumentation, Arctic
Ganton, J.H. 1968.
Arctic field equipment.
Arctic 21(2): 92-97.

automatic station recorder

instrumentation autonomous
Kudrjavtsev, N.F., 1959.
[On the current influence on the immersion depth of automatic station recorders.]
Probl. Arktiki i Antarktiki, (1):11-24.

McCalmont, Arnold M., 1966.
A system for remote automated oceanographic data handling and telemetry. (Abstract only).
Trans. Am. geophys. Un., 47(3):476-477.

instrumentation, autonomous
Schick, George B., John D. Isaacs and Meredith H. Sessions 1968.
Autonomous instruments in oceanographic research.
Mar. Sci. Instrument. 4:203-230.

instrumentation, bathymetry

See: instrumentation, depth

bathypage

instrumentation, bathypage
Mills, A.A., 1962.
The Bathypage.
Marine Sciences Instrumentation, Instr. Soc. Amer., Plenum Press, N.Y., 1:239-243.

bathysonde

instrumentation, bathysonde
Bonnot, Jean-François 1971.
Howaldt bathysonde: uses, operation and calibration procedures.
Int. hydrogr. Rev. 48(1): 155-195.

instrumentation, temperature salinity (bathysonde)
Kroebel, Werner, und Johannes Wick, 1963
Registrierungen in situ im Nordatlantik mit der Bathysonde und einem neuen Messgerät für Schallgeschwindigkeit im Meerwasser mit extrem hoher Genauigkeit.
Kieler Meeresforsch., 19(2):133-141.

instrumentation, bathysonde (temperature, conductivity, pressure)
Siedler, Gerold, 1963
On the in situ measurement of temperature and electrical conductivity of sea-water.
Deep-Sea Res., 10(3):269-277.

bathythermometer

batteries

instrumentation batteries
Cooper, D.D., 1969.
Alkaline primary batteries. Oceanol. int. 69, 4: 7 pp., 7 figs.

instrumentation, batteries
Cunningham, Wm. A. 1967.
New-design charge retaining batteries.
Trans. 2nd Int. Buoy Techn. Symp. 105-113.

instrumentation, batteries
Gray, J., and F. Twentyman, 1970.
Fuel batteries for marine use.
Electronic Engineering in Ocean Technology, Swansea, 21-24 Sept. 1970, I.E.R.E. Conf. Proc. 19: 447-460.

instrumentation, batteries
Momsen, Donald F., and Joseph C. Clerici, 1971.
First silver zinc batteries used in deep submergence.
J. mar. techn. Soc. 5(2): 31-36

instrumentation, batteries
Ream, Joseph T., 1971
Radioisotope fueled thermoelectric nuclear batteries.
J. mar. techn. Soc 5(2): 16-30

biological

instrumentation, biological
Ahlstrom, E.H., undated (1954?).
Biological instruments. In: Oceanographic Instrumentation. NRC Publ. No. 309:36-54.

With discussion by G.L. Clarke
G.A. Riley.

instrumentation, biological (at sea)
Aron, W., 1962.
5. Some aspects of sampling the macroplankton.
Contributions to symposium on zooplankton production, 1961.
Rapp. Proc. Verb., Cons. Perm. Int. Expl. Mer, 153:29-38.

instrumentation, biological
Bernhard, M., e L. Rampi, 1963
4. Botanica oceanografica.
Rapp. Attività Sci. e Tecn., Lab. Studio della Contaminazione Radioattiva del Mare, Fiascherino, La Spezia (maggio 1959-maggio 1962), Comit. Naz. Energia Nucleare, Roma, RT/BIO(63) 8:57-123. (multilithed) plus Appendice 4.

instrumentation, plankton (at sea)
Nassogne, Armand, e Michael Bernhard, 1969.
Campionamento quantitativo di zooplancton con un campionatore ad alta velocità (Delfino).
Pubbl. Staz. Zool. Napoli 37 (2 Suppl.): 219-336.

instrumentation, fish (sharks)
Snodgrass, James M., and Perry W. Gilbert, 1967.
A shark-bite meter. In: Sharks, skates, and rays Perry W. Gilbert, Robert F. Mathewson and David P. Rall, editors, Johns Hopkins Univ. Press, 331-337.

instrumentation, fisheries
Schuyf, A. and S.J. de Groot, 1971.
An inductive locomotion detector for use in diurnal activity experiments in fish. J. Cons. int. Explor. Mer 34(1): 126-131.

instrumentation, biological (at sea)

ACOUSTIC DETECTION

instrumentation, biological, at sea
Bass, George A., and Mark Rascovich, 1965.
A device for sonic tracking of large fishes.
Zoological, N.Y. Zool. Soc., 50(2):75-81.

instrumentation, biological (at sea)

ALGAE

instrumentation, algae
Anon., 1969.
Automatic harvester may solve pollution problem.
Ocean Industry., 4(3):26.

instrumentation, biological (field)
Williams, Richard B., 1963
Use of netting to collect motile benthic algae.
Limnology and Oceanography, 8(3):360-361.

instrumentation biological bacteria

instrumentation, biological (at sea)

BACTERIA

instrumentation, sterile sampling
Bernhard, M., preparator, 1965.
Studies on the radioactive contamination of the sea, annual report 1964.
Com. Naz. Energ. Nucleare, La Spezia, Rept., No. RT/BIO (65) 18:35 pp.

instrumentation biological (at sea) benthos
Brinkhurst, R.O., K.E. Chua and E. Batoosingh, 1969.
Modifications in sampling procedures as applied to studies on the bacteria and tubificid oligochaetes inhabiting aquatic sediments
J. Fish. Res. Bd. Can. 26(10): 2581-2593.

instrumentation, biological, at sea
Brisou, Jean, 1965.
Le "bathyrophe" appareil destiné a prélever aseptiquement les eaux aux grandes profondeurs.
Cahiers Oceanogr., C.C.O.E.C., 17(1):53-54.

instrumentation, biological, at sea.
Conover, John T., and John McN. Sieburth, 1963.
Leveling device for preparation of petri plates at sea.
J. Bacteriol., 86(5):1129-1130.

gear, bacteriological
Devèze, L., 1950.
Description d'un appareil de prélevements d'eau à differentes profondeurs en vue d'une étude microbiologique. Vie et Milieu, Bull. Lab. Arago, 1(2):178-184, 5 textfigs.

instrumentation, biological - bacteria
Jannasch, H.W., and W.S. Maddux 1967.
A note on bacteriological sampling in sea water.
J. mar. Res. 25(2): 185-189.

instrumentation
Emery, K.O., 1958.
Bacterial bottom sampler. Limnol. & Oceanogr., 3(1):109-111.

instrumentation, biological, at sea [water bottles
Fauvel, Yves, 1961.
Un nouvel appareil pour prélèvement d'eau en vue d'analyse bactériologiques.
Rev. Trav. Inst. Pêches Marit., 25(4):413-415.

instrumentation, biol. at sea
Kriss, A.E., 1962.
On the fitness of Nansen bathometres [water bottles] for collecting water samples from seas and oceans for microbiological research.
(In Russian).
Microbiologiia, 31(6):1067-1075.

instrumentation, microbiology (at sea)
Lewis, William M., O.D. McNail and R.C. Summerfelt, 1963
A device for taking water samples in sterile bottles at various depths.
Ecology, 44(1):171-173.

instrumentation, microbiology
Melchiorri-Santolini, U., 1963
5. Microbiologia oceanografica.
Rapp. Attività Sci. e Tecn., Lab. Studio della Contaminazione Radioattiva del Mare, Fiascherino, La Spezia (maggio 1959-maggio 1962), Comit. Naz. Energia Nucleare, Roma, RT/BIO(63) 8:125-142. (multilithed) plus Appendice 5.

instruments
Mortimer, C.H., 1940.
An apparatus for obtaining water from different depths for bacteriological examination.
J. Hygiene 40(6):641-646, 3 textfigs.

instrumentation, biological, at sea
Niskin, Shale J., 1962
A water sampler for microbiological studies.
Deep-Sea Res., 9(5):501-503.

instrumentation, bacteria
Persoone, Guido, 1964.
Contributions à l'étude des bactéries marines du littoral Belge. 1. Un appareil simple pour detacher et obtenir en suspension les bactéries contaminant des surfaces.
Inst. r. Sci. Nat. Belg., Bull., 41(5): 6 pp.

instrumentation, biological (at sea), bacteria
Schegg, Ernst, 1970.
A new bacteriological sampling bottle. Limnol. Oceanogr., 15(5): 820-822.

instrumentation, biological (at sea)
Sieburth, John McN., 1963.
A simple form of the ZoBell bacteriological sampler for shallow water.
Limnology and Oceanography, 8(4):489-492.

instrumentation, microbiological
Sieburth, John McN., James A. Frey and John T. Conover, 1963
Microbiological sampling with a piggy-back device during routine Nansen bottle casts.
Deep-Sea Research, 10(6):757-758.

instrumentation, bacteria
Sorokin, J.I., 1964
A quantitative study of the microflora in the central Pacific Ocean.
Journal du Conseil, 29(1):25-40.

instrumentation, biological (at sea) - bacteria
Williams, E.D.F., 1969.
A submerged membrane filter apparatus for microbiological sampling. Marine Biol., 3(1): 78-80.

instruments
Young, D.C., D.B. Finn, and R.H. Bedford, 1931.
A deep sea bacteriological water bottle.
Contr. Canadian Biol. Fish., n.s., 6(18):417-422, 1 textfig.

instrumentation, biological (at sea) (bacteria)
ZoBell, C.E., 1941
Apparatus for collecting water samples from different depths for bacteriological analysis. J. Mar. Res. 4:173-188.

instrumentation, biological (at sea)

BIOLUMINESCENCE

instrumentation, bioluminescence
Backus, R.H., C.S. Yentsch and A. Wing, 1961.
Bioluminescence in the surface waters of the sea.
Nature, 192(4802):518.

instrumentation biological (at sea)
Seliger, H.H., W.G. Fastie and W.D. McElroy, 1969.
Towable photometer for rapid area mapping of concentrations of bioluminescent marine dinoflagellates. Limnol. Oceanogr., 14(5): 806-813.

instrumentation, bioluminescence
Seliger, H.H., W. G. Fastie, W.R. Taylor and W.D. McElroy, 1962.
Bioluminescence of marine dinoflagellates. 1. An underwater photometer for day and night measurements.
J. Gen. Physiol., 45(5):1003-1017.

Samplers bottom

instrumentation, biological (benthos)
Horikoshi, M., 1962.
Distribution of benthic organisms and their remains at the entrance of Tokyo Bay, in relation to submarine topography, sediments and hydrography.
Nat. Sci. Rept., Ochanomizu Univ., 13(2):47-122.

instrumentation, biological (at sea)

BENTHOS (miscellaneous devices)

instrumentation, biological (at sea)
Barnett, P.R.O., 1969.
A stabilizing framework for the Knudsen bottom sampler.
Limnol. Oceanogr. 14(4):648-649.

instrumentation, biological (at sea)
Corey, Susan, and J.S. Craib, 1966.
A new quantitative bottom sampler for meiofauna.
J. Cons. perm. int. Expl. Mer., 30(3):346-353.

instrumentation, biological (at sea)
*Emig, Christian C., et Roland Lienhart, 1967.
Un nouveau moyen de récolte pour les substrats meubles infra-littoraux: l'aspirateur sous-marin.
Recl. Trav. Stn. mar. Endoume, 42(58):115-120.

instrumentation, biology (at sea)
Fielder, D.R., and C.L. French, 1970.
An activity recorder for bottom living marine crustaceans.
Crustaceana, 19(2): 208-210

instrumentation, biological (at sea) (benthos)
Finland, Finnish IBP-PM Group, 1969.
Quantitative sampling equipment for the littoral benthos.
Int. Revue ges. Hydrobiol. 54(2): 155-193

instrumentation, biol, sea
Gilat, E., 1965.
Methods of study in marine benthonic ecology.
Méthodes quantitatives d'étude du benthos et echelle dimensionnell des benthontes: Colloque du Comité du Benthos (Marseille Novembre 1963).
Comm. Int. Expl. Sci. Mer. Medit., Monaco, 7-13.

instrumentation
Grøntved, Jul., 1957.
A sampler for underwater macrovegetation in shallow waters. J. du Cons., 22(3):293-297.

instrumentation
Horikoshi, Masuoki, 1962.
Distribution of benthic organisms and their remains at the entrance of Tokyo Bay, in relation to submarine topography, sediments and hydrography.
Nat. Sci. Rept., Ochanomizu Univ., 13(2):47-122.
"submarine climograph"

instrumentation, biological (at sea)
Jacobi, G. Zolton, 1971
A quantitative artificial substrate sampler for benthic macroinvertebrates.
Trans. Am. fish. Soc. 100(1): 136-138

instrumentation, biological (at sea) - benthos
Massé, Henri 1967.
Emploi d'une suceuse hydraulique transformée pour les prélèvements quantitatifs dans les substrats meubles infralittoraux.
Helgoländer wiss. Meeresunters. 15(1/4): 500-504.

instrumentation, biology (at sea)
Menzies, Robert J. and Gilbert T. Rowe.
The LUBS, a large undisturbed bottom sampler.
Limnol. Oceanogr., 13(4):708-714.

instrumentation, biological (at sea)
Milbrink, Göran 1971.
A simplified tube bottom sampler.
Oikos 22(2): 260-263.

instrumentation, biological at sea.
Taylor, John L., 1965.
Bottom samplers for estuarine research.
Chesapeake Science, 6(4):233-234.

instrumentation, biological (at sea)
Tonolli, V., 1962.
Nuovi strumenti per la raccolta e la separazione dei popolamenti bentonici. Problemi ecologici delle zone litorale del Mediterraneo 17-23

instrumentation
True, M.A., 1965.
Dispositif pour récolte total du peuplement sur substrat dur
Méthodes Quantitative d'Étude du Benthos et Echelle Dimensionnelle des Benthontes: Colloque du Comité du Benthos (Marseille, Nov. 1963).
Comm. Int. Expl. Sci. Mer. Médit., Monaco, 25-27.

instrumentation biological (at sea) - benthos
True, M.A., J.-P. Reys and H. Delauze 1968.
Progress in sampling the benthos: the Benthic Suction Sampler.
Deep-Sea Res. 15(2): 239-242.

instrumentation
Usinger, R.L., and P.R. Needham, 1956.
A drag-net riffle-bottom sampler.
Progr. Fish-Culturist 18(1):42-44.

instrumentation, biological (at sea)

BOTTOM SKIMMER

instrumentation, biological, at sea
Carpine, C., G. Fredj. et R. Vaissiere, 1965.
Note préliminaire sur une methode d'utilisation de la "Troika" sous-marine.
Méthodes Quantitatives d'Étude du Benthos et Echelle Dimensionnelle des Benthontes: Colloque du Comité du Benthos (Marseille, Nov. 1963).
Comm. Int. Expl. Sci. Mer. Médit., Monaco, 41-43.

instrumentation, biological at sea
Frolander, H.F., and Ivan Pratt, 1962.
A bottom skimmer.
Limnol. and Oceanogr., 7(1):104-106.

instrumentation, biological (at sea)
Ockelmann, Kurt W., 1964.
An improved detritus-sledge for collecting meiobenthos.
Ophelia, Helsingor, 1(2):217-222.

instrumentation, water sampler
Murray, J.W., 1962
A new bottom-water sampler for ecologists.
J. Mar. Biol. Assoc., U.K., 42(3):499-501.

instrumentation, biological (at sea)
Pequegnat, Willis E., Thomas Bright and Bela M. James, 1970.
The benthic skimmer, a new biological sampler for deep-sea studies. Oceanogr. Stud. Texas A & M Univ. 1: 17-20.

instrumentation, biological (at sea)

CLOSING DEVICES for plankton nets and samplers

instrumentation, biological (at sea)
Clutter, Robert I., 1965.
Self-closing device for sampling plankton near the sea bottom.
Limnology and Oceanography, 10(2):293-296.

instrumentation, biological, at sea
Currie, R.I., 1962.
7. Net closing gear. Contributions to symposium on zooplankton production, 1961.
Rapp. Proc. Verb., Cons. Perm. Int. Expl. Mer, 153:48-54.

instruments
Currie, R.I., and P. Foxton, 1956.
The Nansen closing method with vertical plankton nets. J.M.B.A., 35(3):483-492.

instrumentation, biological (at sea)
Foxton, P., 1963
An automatic opening-closing device for large plankton nets and mid-water trawls.
J. Mar. Biol. Assoc., U.K., 43(2):295-308.

instrumentation, biological (at sea)
Kinzer, Johannes, 1962
Ein einfacher Schliessmechanismus für die Planktonröhre "Hai".
Kurze Mitteilungen, Inst. Fischereibiol., Univ. Hamburg, No. 12:13-17.

instruments
VanCleve, R., 1937
An Electrical Plankton-net closing device.
J. Cons. 12(2):171-173, 1 text fig.

instrumentation, biological, at sea
Yentsch, C.S., G.D. Grice, and A.D. Hart, 1962.
9. Some opening-closing devices for plankton nets operated by pressure, electrical and mechanical action. Contributions to symposium on zooplankton production, 1961.
Rapp. Proc. Verb., Cons. Perm. Int. Expl. Mer, 153:59-65.

instrumentation, biological (at sea)

CORERS

instrumentation, biological (at sea)
Hamilton, Andrew L., William Burton and John F. Flannagan 1970.
A multiple corer for sampling profundal benthos.
J. Fish. Res. Bd., Can. 27(10): 1867-1869.

instrumentation, biology (at sea) corers
McIntyre, A.D., 1971
Deficiency of gravity corers for sampling meiobenthos and sediments.
Nature, Lond., 231 (5300): 260.

instrumentation, biological (at sea)

IN SITU CULTURE CHAMBERS

instrumentation, biology, at sea
Chapman, V.J., 1964.
A submarine algal growth chamber for studies in benthic productivity.
Ecology, 45(4):889-892.

instrumentation, biological (at sea)
Owen, R.W., Jr., 1963
Trailing incubator for studies of C14 uptake by phytoplankton.
Limnol. and Oceanogr., 8(2):297-298.

instrumentation (at sea)
Paka, V.T., M.F. Naumenko and K.I. Cnigrakov, 1964.
An apparatus for leading off distant dosimeters from the board of a ship. (In Russian).
Okeanologiia, Akad. Nauk, SSSR, 4(2):313-314.

instrumentation
Strickland, J.D.H., and L.D.B. Terhune, 1961
The study of in-situ marine photosynthesis using a large plastic bag.
Limnol. & Oceanogr., 6(1): 93-96.

instrumentation, biological
Terhune, L.D.B., 1963
Construction of a large plastic bag for in situ studies of marine photosynthesis.
Fish. Res. Bd., Canada, Manuscript Rept. Ser. (Oceanogr. and Limnol.), No. 173:20 pp. (multilithed).

instrumentation, biological (at sea)
Watt, W.D., 1965.
A convenient apparatus for in situ primary production studies.
Limnology and Oceanography, 10(2):298-300.

instrumentation, biological (at sea)

DEPRESSOR

instrumentation, biological (at sea)
Bercaw, J.S., 1966.
A folding midwater trawl depressor.
Limnol. Oceanogr., 11(4):633-635.

instrumentation, biological, at sea
Graham, Joseph J., and George B. Vaighan, 1965.
A new depressor design for high speed samplers. (abstract).
Ocean Sci. and Ocean Eng., Mar. Techn. Soc.,-Amer. Soc. Limnol. Oceanogr., 2:742.

instrumentation, biological (at sea)-depressor
Shibata, Keishi 1967
Field test of Gyoken depressor.
Bull. Fac. Fish. Nagasaki Univ. 22:105-108.

instrumentation, biological, depressors
Zhuravlev, V.F., V.T. Paka and G. V. Puzyrev, 1969.
A sinker for towed oceanographic instruments. (In Russian; English abstract).
Okeanologiia, 9(1):175-179.

instrumentation, biological (at sea)

DIVING ADJUNCTS

instrumentation, biological (at sea)-diver-operated
Barnet, Peter R.O., and Bernard R.S. Hardy 1967.
A diver-operated quantitative bottom sampler for sand macrofauna.
Helgoländer wiss. Meeresunters 15(1/4): 390-398.

instrumentation, biological (at sea)

DRAG NET

instrumentation, biological, at sea
Higo, Nobio, 1964
Studies on the drag net. 1. An increases of the current velocity inside net. (In Japanese; English abstract).
Mem., Fac. Fish., Kagoshima Univ., 13:78-91.

instrumentation, biological (at sea)
Larsonneur, C., 1964.
Petite drague pour fonds meubles et peu profonds.
Bull. Soc. Linnéenne, Normandie, (10), 5:94-96.

instrumentation, biological (at sea) bottom dredges

instrumentation, biological (at sea)

DREDGES

See also: instrumentation, geological (at sea), dredges

instrumentation, biological (at sea)-dredges
Carey, Andrew G. Jr., and Danil R. Hancock 1965.
An anchor-box dredge for deep-sea sampling.
Deep-Sea Res. 12(6): 983-984.

instrumentation, biological (at sea)-dredges
Castagna, Michael, 1967.
A benthic sampling device for shallow water.
Limnol. Oceanogr. 12(2): 357-359.

instruments
Delbouttevillle, C.D., 1954.
Description d'un appareil pour la capture de la faune des eaux souterraines littorales sous la mer. Premiers resultats.
Vie et Milieu, Bull. Lab. Arago, 4(3):411-421, 1 textfig.

instrumentation, biological (at sea)
MacIntyre, R.J., 1964
A box dredge for quantitative sampling of benthic organisms.
Limnology and Oceanography, 9(3):460-461.

instrumentation. biol. at sea
Merna, James W., 1962
Quantitative sampling with the orange-peel dredge.
Limnol. and Oceanogr., 7(3):432-433.

instrumentation, biological (at sea)
Takeuchi, Isamu, 1970.
On the newly improved dredge. (In Japanese; English abstract)
Bull. Hokkaido reg. Fish. Res. Lab., 36:15-24.

instrumentation, biological (at sea)
Zoutendyk, Peter, 1964.
A new bottom dredge.
Trans. R. Soc., S. Africa, 37(2):109-110.

instrumentation, biological (at sea)

GRABS

instrumentation, biological (at sea) (grab)
Emig, Christian-Charles, et Roland Lienhart 1971.
Principe de l'aspirateur sous-marin automatique pour sédiments meubles.
Vie Milieu Suppl. 22(2): 573-598.

instrumentation, biological (at sea) grabs
Flannagan, John F., 1970.
Efficiencies of various grabs and corers in sampling freshwater benthos.
J. Fish. Res. Bd, Can., 27(10): 1691-1700.

instrumentation, biological (at sea)
Ford, J.B., and R.S. Hall, 1958.
A grab for quantitative sampling in stream muds.
Hydrobiologia 11 (3/4):198-204.

instrumentation, biological (at sea) dredges
Forster, G.R., 1953.
A new dredge for collecting burrowing animals.
J.M.B.A. 32(1):193-198, 2 textfigs.

instrumentation, biological (at sea)
Brett, C. Everett, 1964.
A portable hydraulic diver-operated dredge-sieve, for sampling subtidal macrofauna.
J. Mar. Res., 22(2):205-209.

instruments
Holme, N.A., 1955.
An improved "vacuum" grab for sampling the sea-floor. J.M.B.A. 34:445-555, 1 textfig.

instrumentation, biology (at sea)
Menzies, Robert James, Logan Smith and K.O. Emery, 1963.
A combined underwater camera and bottom grab: a new tool for investigation of deep-sea benthos.
Int. Revue Ges. Hydrobiol., 48(4):529-545.

instruments, biological (at sea)
Paterson, C.G., and C.H. Fernando, 1971.
A comparison of a simple corer and an Ekman grab for sampling shallow-water benthos.
J. Fish. Res. Bd. Can. 28(3): 365-368

instrumentation, biological (at sea) grabs
Wigley, Roland L. 1967.
Comparative efficiencies of Van Veen and Smith-McIntyre grab samplers as revealed by motion pictures.
Ecology 48(1): 168-169.

instrumentation, biological (at sea)

IN SITU METABOLISM MEASUREMENTS

instrumentation,
Pamatmat, Mario M., 1965.
A continuous=flow apparatus for measuring metabolism of benthic communities.
Limnol. Oceanogr., 10(3):486-489.

instrumentation,
Pamatmat, Mario M., and Douglas Fenton, 1968.
An instrument for measuring subtidal benthic metabolism in situ.
Limnol. Oceanogr., 13(3):537-540.

instrumentation, biological (at sea)

NET DEPTH METER

instrumentation, biological (at sea)
Boudreault, Yves, 1964.
Utilisation du bathymètre Furuno avec le chalut méso-pélagique "Isacc-Kidd".
Rapp. Ann., 1963, Sta. Biol. Mar., Grande Rivière, 109-113.

instrumentation, biology at sea
Hester, Frank J., Donald C. Aasted and Robert W. Gilkey, 1963
The bathykymograph, a depth-time recorder.
U.S.F.W.S. Spec. Sci. Rept., Fish., No. 441: 5 pp.

instrumentation
Holmes, R.W., & J.M. Snodgrass, 1961
A multiple-detector irradiance meter and electronic depth-sensing unit for use in biological oceanography.
J. Mar. Res., 19(1): 40-56.

instrumentation, biology (trawls)
Matuda, Kô, Kunio Sato and Tasae Kawakami, 1969
On lateral deviation of midwater trawl net from course of towing in turning. Bull. Jap. Soc. fish. Oceanogr. Spec. No. (Prof. Uda Commem. Pap.): 323-330.

instrumentation, nets (nekton)
Mosely, Frank N., and B.J. Copeland, 1969.
A portable drop-net for representative sampling of nekton.
Contrib. mar. Sci., Port Aransas, 14:37-45

neuston net

instrumentation, biological (at sea)

NEUSTON NET

INSTRUMENTATION,
Bartlett, Martin R., and Richard L. Haedrich, 1968.
Neuston nets and South Atlantic larval blue marlin (Makaira nigricans).
Copeia, (1968) (3): 469-474

instrumentation, biological (at sea)
Ben-Yami, M., A. Herzberg, S. Pisanty and Al Lourie, 1970.
A side-tracking neuston net. Marine Biol., 6(4): 312-316.

instrumentation, biological (at sea)
Bieri, Robert, and Thomas K. Newbury, 1966.
Booby II, a quantitative neuston sampler for use from small boats.
Publ. Seto Mar. Biol. Lab., 13(5):405-410.

Instrumentation, biological, at sea
David, P.M., 1965.
The neuston net; a device for sampling the surface fauna of the ocean.
Jour. Mar. Biol. Assoc., U.K., 45(2):313-320.

instrumentation, biological (at sea)
David, Peter M., 1963.
The Neuston net, a device for sampling the surface fauna of the ocean.
N.I.O. Internal Rept., No. B3:7 pp. (unpublished manuscript).

instrumentation, biological (at sea) - neuston net
Marinaro, J.-Y. et J. Henry 1968.
Sur un nouvel engin collecteur d'hyponeuston et son utilisation dans la baie d'Alger.
Pelagos 8:71-78.

instrumentation, biology (at sea) net neuston)
Sameoto, D.D., and L.O. Jaroszynski 1969.
Otter surface sampler: a new neuston net.
J. Fish. Res. Bd, Can., 26(8): 2240-2244

instrumentation, biological (at sea)

OBSERVATION CHAMBER

instrumentation, biology (at sea).
Fuss, C.M., Jr., and L.H. Ogren, 1965.
A shallow water observations chamber.
Limnology and oceanography, 10(2):290.

instruments
Tregouboff, G., 1955.
Sur l'emploi de la tourelle submersible Galeazzi pour des observations biologiques sous-marines à faibles profondeurs.
Bull. Inst. Océan., Monaco, No. 1070:1-5.

instrumentation, biological (at sea)

PARTICLE COUNTER

instrumentation, biological (at sea)
Maddux, William S., and John W. Kanwisher, 1965.
An in situ particle counter.
Limnol. and Oceanogr., Redfield Vol., Suppl. to 10:R162-R168.

instrumentation, biological (at sea)

PLANKTON

instrumentation
Allen, D.M., and A. Ingles, 1958.
A pushnet for quantitative sampling of shrimp in shallow estuaries. Limnol. & Oceanogr., 3(2): 239-241.

instrumentation (nets)
Anichini, Carlo, 1959.
Risultati delle ricerche planctonologiche effettuate nella stazione fissa di Cagliari.
Boll. Pesca, Pisc Idrobiol. 35(14)(1):59-70.

instrumentation, biological (at sea)
Bernard, Michelle, 1958.
Méthodes et engins pour le prélèvement quantitatif du zooplancton, en usage a la station de zoologie marine de l'Université d'Alger.
Rapp. Proc. Verb., Comm. Int. Expl. Sci., Mer Medit., n.s., 14:139-145.

instrumentation, biol., at sea
Bourdillon, A., 1964.
Quelques aspects du problème de l'échantillonnage du planeton marin.
Le Terre et le Vie (1):77-93.

instrumentation
Cachon, J., 1957.
Sur quelques techniques de pêches planktoniques pour l'étude biologiques.
Bull. Inst. Océan., Monaco, No. 1103:6 pp.

instrumentation, plankton
Cassie, R.M., 1963
Microdistribution of plankton. In: Oceanography and Marine Biology, H. Barnes, Edit., George Allen & Unwin, 1:223-252.

instrumentation
Cassie, R. Morrison, 1958
Apparatus for investigating spatial distribution of plankton. N.Z.J. Sci. 1(3): 436-448.

~~instrumentation~~
Fenaux, Robert, 1968.
Un dispositif pour pêches planctoniques dans les couches superficielles.
Rapp. Proc.-verb. Réun., Comm. int. Explor. scient. Mer Méditerranée 19(3): 379-381.

instrumentation, biological (at sea)
Motoda, Sigeru, 1962
Proposed biological apparatus and methods to be used on Japanese ships in the International Indian Ocean Expedition. (In Japanese; English abstract).
Info. Bull., Planktology, Japan, No. 8:40-53.

instrumentation, biological (at sea)
Omaly, Nicole, 1966.
Moyens de prelevement du zooplancton: essai historique et critique.
Pelagos, 5:169 pp.

instrumentation, biological, at sea
Raymont, J.E.G., and B.G.A. Carrie, 1964
The production of zooplankton in Southampton Water.
Int. Rev. Ges. Hydrobiol., 49(2):185-232.

instrumentation, biological (at sea)
Schlichting, Harold E., Jr., and James E. Hudson, Jr., 1967.
Radio-controlled model boat samples air and plankton.
Science, 156(3772):238-239.

plankton nets

instrumentation, biological (at sea)

PLANKTON NETS

instruemnts
Aikawa, H., 1935.
On the reliability of plankton net.
Bull. Jap. Soc. Sci. Fish. 3:331-345.

instrumentation, biological, at sea.
Aizawa, Yasushi, Ryuzo Marumo and Makoto Omori, 1965.
On the movement of plankton net in oblique haul. (In Japanese; English abstract).
Info. Bull. Planktol., Japan, No. 12:60-66.

nets
Anon., 1949.
Campañas del "Xauen" en 1947 y 1948 en el mar de Alboran y en el estrecho de Gibraltar. Registro de operaciones. Bol. Inst. Español Ocean. 18: 1-53.

plankton nets
Anraku, M., 1956.
Some experiments of the variability of horizontal plankton hauls and on the horizontal distribution of plankton in a limited area.
Bull. Fac. Fish., Hokkaido Univ., 7(1):1-16.

instrumentation, biological (at sea)
Anraku, Masateru, and Masanori Azeta, 1966.
Observations on four models of a fish larve net made of two different fabrics having unequal porosities. (In Japanese; English abstract).
Inf. Bull. Planktol. Japan, No. 13:34-40.

instrumentation, biological (at sea)
Arnaud, Jean. et Jacques Mazza, 1965.
Pêches planctoniques au filet Juday-Bogorov modifié (Matériel-techniques-résultats).
Bull. Inst. Océanogr., Monaco, 65(1343):26 pp.

plankton nets, efficiency of
Asaoka, O., and M. Ohwada, 1960.
[On the influence of the plankton organisms and other factors upon the filtering efficiency of the plankton net.]
K. Oceanogr. Soc., Japan, 16(3):146-149.

instrumentation, biological, sea
Asaoka, O., and M. Ohwada, 1960.
[On the influence of the planktonic organisms and other factors upon the filtering efficiency of the plankton net.]
J. Oceanogr. Soc., Japan, 16(3):36-39.

Reprinted in: Bull. Hakodate Mar. Obs., 1961(8): 36-39.

instrumentation, biological, at sea
Barnes, H., and D.J. Tranter, 1965.
A statistical examination of the catches, numbers and biomass taken by three commonly used plankton nets.
Australian J. Mar. Freshwater Res., 16(3):293-306.

instrumentation, nets
Bourdillon, André 1968
Essais d'un collecteur multiple pour filets à plancton
Rapp. Proc.-verb. Réun., Comm. int. Explor. scient. Mer Mediterranée 19(3): 373-375.

instruments, biological, at sea
Bourdillon, A., 1963
Essais comparés de divers filets à plancton.
Rapp. Proc. Verb., Reunions, Comm. Int. Expl. Sci. Mer Méditerranée, Monaco, 17(2):455-461.

instrumentation, biological (at sea)
Brown, Charles L., Jr., 1968.
The use of multiple sampling plankton nets.
Bio Science, 18(10):962.

instrumentation, biological (at sea)
Currie, Ronald I., 1963
The Indian Ocean standard net.
Deep-Sea Res., 10(1/2):27-32.

instrumentation
Currie, R.I., and P. Foxton, 1957.
A new quantitative plankton net. J.M.B.A., 36(1):17-32.

instrumentation
Ferrando, H. J., 1958.
Red para fitoplancton con copo intercambiable.
Contra. Avulsas Inst. Ocean., Sao Paulo, 1:5 pp.

instruments
Gauld, D.T., and T.B. Bagenel, 1951.
A high speed tow-net. Nature 168(4273):523, 1 textfig.

gear
Gibbons, S.G., 1939
The Hensen net. J. du Cons. 14:242-248.

plankton nets
*Heron, A.C., 1968.
Plankton gauze.
In: Zooplankton sampling, D.J. Tranter, editor, UNESCO, 19-25.

instrumentation
Ihara, J., Y. Yamaguchi and K. Ozawa, 1956.
On the collection-net used in the case of observations of DSL. Journal Tokyo Univ. Fish., 42(2):151-156.

instrumentation, biological (at sea)
Ishida, Teruo,
A note on the efficiency of plankton nets by mesh size. (In Japanese; English abstract).
Bull. Hokkaido Reg. Fish. Res. Lab., No. 30: 25-30.

plankton nets
Ito, S., 1958.
[Variability of catches in the vertical net haul of fish eggs and larvae.]
Ann. Rept., Japan Sea Reg. Fish. Res. Lab., 1(4):33-42.

plankton net
Ito, S., and S. Nishimura, 1958
[Experiments on the filtering rate of the Maru-Toku type plankton net used in Japan for quantitative sampling.]
Ann. Rept., Japan Sea Reg. Fish. Res. Lab., 1(4):57-64.

instrumentation, biological (at sea)
Jashnov, V.A., 1961
[The velocity of planktonic nets=A high-speed plankton net.]
Zool. Zhurnal, 40(1):122-127.

instrumentation, biological (at sea)
Jossi, Jack W., 1966.
The ICITA one-meter plankton net: description and evaluation.
Limnol. Oceanogr., 11(4):640-642.

instrumentation, biological (at sea)
Kato, Akira and Toshimasa Shimomura, 1959.
Some characteristics of the D.S.L. peculiar to the Japan Sea.
Bull. Japan Sea Reg. Fish. Res. Lab., (7):67-84.

plankton net, performance of
Kobayashi, K., and S. Igarashi, 1956.
Mathematical analysis of the filtering rate of plankton net. Bull. Fac. Fish., Hokkaido Univ., 7(1):17-20.

instruments
Kulukova, E.V., 1956.
[Plankton nets on R/V Vityez.]
Trudy Inst. Okeanol., 19:330-333.

instruments
Känne, C., 1933
Weitere Untersuchungen zum vergleich der Fangfähigkeit verschiedener modelle von vertikal Fischer den Plankton-netzen. Rapp. et Proc., verb. Cons. perm. int. pour l'explor. de la mer, 83:19pp.

instrumentation, biology (at sea)
LeBrasseur, R.J., C.D. McAllister, J.D. Fulton and O.D. Kennedy, 1967.
Selection of a zooplankton net for coastal observations.
Tech. Rep., Fish. Res. Bd., Can., 37: 13 pp. (multilithed).

instrumentation, biology (at sea)
Mahadeva, N., 1962
Preliminary report on the experiments with Marutoku-B net and a 45 cm x 90 cm net of No. 0 bolting silk off Trincomalee, Ceylon, during October 1960.
Indo-Pacific Fish. Council, FAO, Proc., 9th Sess., (2/3):17-24.

instrumentation, biological (at sea) - plankton nets
Malmken, C.V.W., and J.W. Jossi 1967
Flume experiments on the hydrodynamics of plankton nets.
J. Cons. perm. int. Explor. Mer, 31(1): 38-45.

instrumentation
Motoda, S., 1957.
[North Pacific standard plankton net.]
Info. Bull., Plankton, Japan, No. 4:13-15.

plankton net, volume of water filtered
Nishizawa, S., and M. Anraku, 1956.
[A note on measuring of the volume of water filtered by plankton net by means of a flow meter in a towing tank.]
Bull. Fac. Fish., Hokkaido Univ., 6(4):298-309.

instrumentation
Metelkin, L.I., 1957.
[On the accuracy of the determination of the average value for the strength of the noose of the net cloth.] Izv. Tichookean. Nauch.-Issle. Inst. Rib. Choz. i Okean., 44:111-117.

instrumentation, biological, at sea
Morioka, Yasuhiro, 1965.
Intercalibration of catch efficiency between bolting silk net and nylon net. (In Japanese: (English abstract).
Info. Bull. Planktol., Japan, No. 12:54-60.

instrumentation biological (at sea) (plankton Nets)
Motoda, Shigeru, 1969.
Devices of simple plankton apparatus IV.
Bull. Fac. Fish. Hokkaido Univ., 20(3): 180-184.

instrumentation, biological (at sea)
Motoda, Sigeru, 1967.
Devices of simple plankton apparatus.
Bull. Fac. Fish., Hokkaido Univ., 18(1):3-8.

instruments
Motoda, S., and M. Anraku, 1954.
Daily change of vertical distribution of plankton animals near western entrance to the Tsugaru Strait, northern Japan.
Bull. Fac. Fish., Hokkaido Univ., 5(1):15-19, 2 textfigs.

instrumentation
Motoda, S., M. Anraku and T. Minoda, 1957.
Experiments on the performance of plankton samplings with net. Info. Bull., Plankton., Japan 8(1):1-22.

instrumentation, biological (at sea)
Motoda, Sigeru, Kenjiro Konno, Akito Kawamura, and Keisuke Osawa, 1963
Zooplankton samplings accomplished on the "Umitaka Maru" and the "Oshoro Maru" in the Indian Ocean, December 1962-January 1963. (In Japanese; English abstract).
Info. Bull. Plankt., Japan, No. 9:37-50.

instrumentation, biological (at sea)
Motoda, Sigeru and Keisuke Osawa, 1964
Filtration, variance of samples and estimated distance of haul in vertical hauls with Indian Ocean standard net. (In Japanese; English abstract).
Inform. Bull., Planktol., Japan, No. 11:11-24.

instrumentation, biology (at sea)
Pathansali, D., 1962
Comparisons of efficiency of plankton nets.
Indo-Pacific Fish. Council, FAO, Proc., 9th Sess. (2/3):25-36.

instrumentation, biological (at sea)
Percier, A., et J. Harambillet, 1964.
Essai de vieillissement de fils de pêche en polyamide (nylon) au contact du fer oxydé.
Bull., C.E.R.S., Biarritz, 5(2):221-251.

INSTRUMENTATION, biological (at sea)
Pullen, E.J., C.R. Mock and R.D. Ringo, 1968.
A net for sampling the intertidal zone of an estuary.
Limnol. Oceanogr., 13(1):200-202.

instrumentation, biological (at sea) (net)
Repelin, R. et J.-A. Gueredrat, 1970.
Efficiences comparées de filets à plancton coniques de mêmes dimensions et de mailles différentes. II. Sélectivité de la maille No 000 pour les amphipodes et les copépodes
J. Cons. int. Explor. Mer, 33(2): 256-280.

instruments
Rochford, D.J., 1947
Oceanographic Research and Fisheries Development. The Collection of plankton.
Fisheries Newsletter, Australia, 6(3): 10-11, 3 text figs.

instrumentation
Rose, M., 1948
Un filet automatique pour la pêche du Plankton en profondeur. Bull. Inst. Océan., Monaco, No.933:8 pp., 4 text figs.

gear
Russell, F. S. and J. S. Colman, 1931
The zooplankton. 1. Gear, methods and station lists. Brit. Mus. (N.H.) Great Barrier Reef Expedition 1928-29, Sci. Repts. II(2):5-35, 1 pl., 9 text figs.

instrumentation
Saville, A., 1958.
Mesh selection in plankton nets.
J. du Cons., 23(2):192-201.

plankton net (high speed)
Scripps Institution of Oceanography, 1949
Marine Life Research Program. Progress Report to 30 April 1949, 25 pp. (mimeographed), numerous figs. (photo.rozalid). 1 May 1949.

instrumentation, biological (at sea) - plankton nets
Gentz-Braconnet, E. 1967.
Sur un filet permettant la récolte des animaux semi-planctoniques des herbiers de posidonies
Vie Milieu (B) 18 (2B): 453-455.

instrumentation, biological (at sea)
*Simonsen, Reimer, 1967.
"Multinetz" ein Mehrfachschliessnetz für Phytoplankton.
Meteor Forschungsergebn. (D)(1)(Biol):85-88.

instruments
Slack, H.D., 1955.
A quantitative plankton net for horizontal sampling. Hydrobiologia 7(3):264-268.

instrumentation
Southward, A.J., 1961.
The distribution of some plankton animals in the English Channel and western approaches. 1. Samples taken with stramin nets in 1955 and 1957.
J.M.B.A., U.K., 41(1):17-35.

instrumentation, biological at sea
Suzuki, Otohiko, and Ko Matuda, 1965.
Drag force of plane net set parallel to stream.
Bull. Jap. Soc. Sci. Fish., 31(8):579-584.

instruments
Sysoev, N.N., 1956.
[Some hydrodynamics of plankton nets.]
Trudy Inst. Okeanol., 19:324-329.

instruments
Takane, H., 1954.
A simple method for frequent sampling of plankton from deep layers. Bull. Jap. Soc. Sci. Fish. 19(12):1197-1199, 2 textfigs.

gear
Tamura, T., 1948.
The method of collecting the plankton while the boat is in motion. J. Fisheries, No. 53:

INSTRUMENTATION, nets
Tanaka, H., Y. Imayama, M. Azeta and M. Anraku, 1965.
A hydrodynamic study of a modified model of the Norpac net.
Marine Biol. 1(3):204-209.

instrumentation, biological (at sea) plankton nets
Tonolli, V., 1954.
Rete da plancton con registrazione di profundita e temperature.
Mem. Ist. Ital. Idrobiol., "Dott. Marco de Marchi" 8:153-162, 3 textfigs.

instrumentation, biological (nets)
Tranter, D.J. 1967.
A formula for the filtration coefficient of a plankton net.
Aust. J. mar. Freshwat. Res., 18(1): 113-121.

instrumentation, biological (nets)
Tranter, D.J., and A.C. Heron, 1967.
Experiments on filtration in plankton nets.
Aust. J. mar. Freshwat. Res., 18(1): 89-111.

plankton net, filtration of
*Tranter, D.J., and P.E. Smith, 1968.
Filtration performance.
In: Zooplankton sampling, D.J.Tranter, editor, UNESCO, 27-56.

gear
Van der Werff, A., 1950.
On a new material for screening plankton.
Amsterdam Naturhist. 1(2):53-54, 2 figs.

plankton net meshes, effect of
*Vannuci, M., 1968.
Loss of organism through the meshes.
In: Zooplankton sampling, D.J.Tranter, editor, UNESCO, 77-86.

instrumentation, biological (at sea)
Vives, Francisco, 1965.
Sur la selectivité des filets à zooplancton.
Rapp. Proc. Verb., Reunions, Comm. Int. Expl. Sci., Mer Méditerranée, Monaco, 18(2):333-334.

nets, drift,
Von Brandt, A., 1955.
The efficiency of drift nets. J. du Cons. 21(1): 8-16.

instrumentation, biological (plankton nets
Wheeler, J. F. G., 1941.
The parachute net. J. Mar. Res., 4(1):92-98, 1 fig.

instrumentation
Wilson, Woodrow L., 1962.
Gear data report from Atlantic plankton cruises for the R.V. Pathfinder, March 1961-March 1962. Virginia Inst., Mar. Sci., Gloucester Pt., Spec. Sci. Rept., No. 32: 3 pp., 3 tables.
(mimeographed)

instrumentation, biological (at sea)

PLANKTON NETS (CLOSING)

instruments
Barnes, H., 1953.
A simple and inexpensive closing net.
Mem. Ist. Ital. Idrobiol., "Dott. Marco De Marchi", Pallanza, 7:189-198, 6 textfigs.

Instrumentation, nets, micronekton
Davies, I.E., and E.G. Barham, 1969.
The Tucker opening-closing micronekton net and its performance in a study of the deep scattering layer.
Marine Biol., 2(2):127-131.

instrumentation, biological (at sea), plankton nets
Grice, George D., and Kuni Hülsemann 1968.
Contamination in Nansen-type vertical plankton nets and a method to prevent it.
Deep-Sea Res. 15(2): 229-233.

instrumentation, biology at sea
Kawamura, Akito, 1968.
Performance of Petersen type vertical closing net. (In Japanese; English abstract). Bull. Plankton Soc., Japan, 15(1): 11-12.

instrumentation, biological (at sea) - plankton closing nets
Kawarada, Yutaka, and Hideo Akamatsu 1966.
Automatic opening-and-closing multiple net.
Oceanogrl Mag. 18(1/2): 25-29.

instrumentation, biological (at sea)
Motoda, Sigeru, 1962
Specifications of zooplankton standard net to be used in the International Indian Ocean Expedition, and a design of a new closing net. (In Japanese; English abstract).
Info. Bull., Planktology, Japan, No. 8:30-40.

instrumentation, biological (at sea)
Omori, Makoto, 1965.
A 160-cm opening-closing plankton net. 1. Description of the gear.
J. Oceanogr. Soc., Japan, 21(5):212-220.

instrumentation - biological (plankton recorder)
Hardy, A. C., 1936
The continuous plankton recorder. Discovery Reports, XI:457-510

instrumentation - biological (plankton recorder)
Hardy, A. C., 1926
A new method of plankton research. Nature, CXVIII:630

sampler (benthos)
Massé, Henri, 1970.
La suceuse hydraulique, bilan de quatre années d'emploi, sa manipulation, ses avantages et inconvénients peuplements benthiques.
Téthys 2(2): 547-556.

samplers

instrumentation, biological (at sea)

PLANKTON SAMPLERS

instrumentation, biological (at sea) (plankton sampler)
Ackefors Hans, 1971.
A quantitative plankton sampler.
Oikos, 22(1): 114-118.

Applegate, Richard L., Alfred C. Fox and Victor J. Starostka 1968.
A water core plankton sampler.
J. Fish. Res. Bd. Can. 25(8): 1741-1742.

instrumentation
Arnold, E.L., jr., 1952.
A high speed plankton sampler (Model Gulf 1-A).
USFWLS Spec. Sci. Rept., Fish. No. 88:1-6, 4 figs. (multilithed).

instrumentation, biological (at sea)
Arnold, Z.M., 1962.
A high-speed plankton sampler for manual operation.
Micropaleontology, 8(4):515-518.

instrumentation, biological (at sea)
Aron, W., E.H. Ahlstrom, B. McK. Bary, A.W.H. Bé and W.D. Clarke, 1965.
Towing characterisitcs of plankton sampling gear.
Limnol. Oceanogr., 10(3):333-340.

instrumentation, biolgical (at sea)
Aron, William, Newell Raxter, Roy Noel and William Andrews, 1964.
Description of a discrete depth plankton sampler with some notes on the towing behavior of a 6-foot Isaacs-Kidd mid-water trawl and a one-meter ring.
Limnology and Oceanography, 9(3):324-333.

instrumentation
Bary, B.M., J.G. de Stefano, M. Forsyth and J. van den Kerkhof, 1958.
A closing, high-speed plankton catcher for use in vertical and horizontal towing. Pacific Science 12(1):46-59.

instrumentation, biol (at sea)
Bary, B. McK. and E.J. Frazer, 1970.
A high-speed, opening-closing plankton sampler (Catcher II) and its electrical accessories.
Deep-Sea Res., 17(4): 825-835.

instrumentation, biological, at sea
Bé, Allan W.H., 1962
Quantitative multiple opening-and-closing plankton samplers.
Deep-Sea Res., 9(2):144-151.

instrumentation, plankton sampler
Be, Allan W.H., 1962.
Quantitative multiple opening-and-closing plankton sampler.
Marine Sciences Instrumentation, Instr. Soc., Amer., Plenum Press, N.Y., 1:156-162.

instrumentation
Bé, Allan W.H., M. Ewing, and Laurence W. Linton, 1959
A quantitative multiple opening-and-closing plankton sampler for vertical towing. J. du Cons. 25(1): 36-46.

instrumentation, biological (at sea)
*Beverton, R.J.H., and D.S. Tungate, 1967.
A multi-purpose plankton sampler.
J. Cons. perm. int. Explor. Mer, 31(2):145-157.

Instrumentation, biol. (at sea)
Bitjukov, E. P., 1961.
[A new model planktonsampler.]
Trudy Vses. Gidrobiol. Obshch., 11:419-423.

instruments
Bossanyi, J., 1951.
An apparatus for the collection of plankton in the immediate vicinity of the sea-bottom.
J.M.B.A. 30(2):265-279, 3 textfigs.

instrumentation
Bridges, J.P., 1958.
On efficiency tests made with a modified Gulf III high-speed tow-net. J. du Cons., 23(3):357-365.

instrumentation
Brouardel, J., 1958.
Appareils de prélèvement (F.N.R.S. III).
Ann. Inst. Océan., 35:255-258.

instrumentation, biological (at sea)
Clarke, William D., 1965.
Preliminary evaluation of the jet net.
Mar. Sci. Instrument., Plenum Press, 3:9-14.

instrumentation, biological (at sea)
Clarke, William D., 1964.
The jet net, a new high-speed plankton sampler.
J. Mar. Res., 22(3):284-287.

samplers, avodance of
*Clutter, Robert I., and Masateru Anraku, 1968.
Avoidance of samplers.
In: Zooplankton sampling, D.J. Tranter, editor, UNESCO, 57-76.

instrumentation
Collier, A., 1957.
Gulf-II semiautomatic plankton sampler for inboard use. Spec. Sci. Rept., Fish., No. 199: 11 pp.

instrumentation
Dorris, Troy C., 1961
A plankton sampler for deep river waters.
Limnol. & Oceanogr., 6(3): 366-367.

gear
Erdmann, W., 1937.
Ein neues Planktongerät für Horizontalfänge in verschiedenen Tiefen und seine Bedeutung für die praktische Fischerei. Ber. Deutsch. Wiss. Komm. Meeresforsch., n.f., 8(3):165-179, 18 textfigs.

instrumentation
Evans, J.H., 1961
A phytoplankton multi-sampler and its use in Lake Victoria.
Nature, 191(4783): 53-55.

instrumentation, biological, at sea
Fish, C.J., and J.M. Snodgrass, 1962.
3.A. The Scripps-Narragansett high-speed multiple plankton sampler. Contribution to symposium on zooplankton production, 1961.
Rapp. Proc. Verb., Cons. Perm. Int. Expl. Mer, 153:23-24.

instrumentation
Gehringer, J.W., 1952.
An all metal plankton sampler (Model Gulf III).
USFWLS Spec. Sci. Rept., Fish., No. 88:7-12, 8 figs. (multilithed).

instrumentation, biological, at sea
Gehringer, J.W., 1962.
The Gulf III and other modern high-speed plankton samplers. Contributions to symposium on zooplankton production, 1961.
Rapp. Proc. Verb., Cons. Perm. Int. Expl. Mer, 153:19-23.

instrumentation, biological, at sea
Glover, R.S., 1961
The multi-depth plankton indicator.
Bulletins, Mar. Ecol., 5(44):151-164.

instrumentation
Griffith, R.E., 1957.
A portable apparatus for collecting horizontal plankton samples. Ecology 38(3):538-540.

instrumentation, plankton samplers
Harding, D. and G.P. Arnold, 1971.
Flume experiments on the hydrodynamics of the Lowestoft high-speed plankton samplers: I.
J. Cons. int. Explor. Mer, 34(1): 24-36.

instrumentation, biology (at sea)
Harding, D., E. Shreeve, D.S. Tungate and D. Mummery, 1971.
A net-changing mechanism for the Lowestoft multi-purpose sampler. J. Cons. int. Explor. Mer, 33(3) 483-491.

instrumentation, biological (at sea)
Hempel, Gotthilf, 1964.
Die Filterleistung der Planktonröhre "Hai" bei verscheidener Schleppgeschwindigkeit. Eine vorläufige Mitteilung.
Helgoländer Wiss. Meeresuntersuchunge, 11(3/4): 161-167.

instruments, plankton, at sea
Ishida, T., 1964.
Quantitative plankton sampler. II. (In Japanese; English abstract).
Bull. Hokkaido Reg. Fish. Res. Lab., (28):61-64.

instrumentation, biological (sea)
Ishida, Teruo, 1963
A new quantitative sampler for large crustacean plankton. (In Japanese; English abstract).
Bull. Hokkaido Reg. Fish. Res. Lab., No. 26: 73-74.

instruments
Langford, R.R., 1953.
Methods of plankton collection and a description of a new sampler. J. Fish. Res. Bd., Canada, 10(5):238-252, 4 textfigs.

instrumentation, biological (at sea) plankton samplers
Macer, C.T. 1967.
A new bottom-plankton sampler.
J. Cons. perm. int. Explor. Mer 31(2): 158-163

instrumentation, biological, at sea
Marples, Timothy G., 1962
An interval plankton sampler for use in ponds.
Ecology, 43(2):323-324.

instrumentation, biological (at sea)
Mohr, J.L., and S.R. Geiger, 1962.
Comparison of results from a new automatic plankton sampler and from conventional methods in the Arctic Basin. Contributions to symposium on zooplankton production, 1961.
Rapp. Proc. Verb., Cons. Perm. Int. Expl. Mer, 153:205-206.

instrumentation, biological (at sea) plankton samplers
Motoda, Sigeru, 1971.
Devices of simple plankton apparatus V.
Bull. Hokkaido Univ. Fac. Fish. 22(2): 101-106.

instrumentation, biological, at sea
Motoda, Sigeru, 1963
Devices of simple plankton apparatus II.
Bull. Fac. Fish., Hokkaido Univ., 14(3): 152-162.

instrumentation, biological, at sea
Motoda, S., 1962.
8. Plankton sampler for collecting uncontaminated materials from several different zones by a single vertical haul. Contributions to symposium on zooplankton production, 1961.
Rapp. Proc. Verb., Cons. Perm. Int. Expl. Mer, 153:55-58.

instrumentation
Motoda, Sigeru, 1959
Devices of simple plankton apparatus
Mem., Fac. Fish., Hokkaido Univ., 7(1/2): 73-94, 8 pls.

instruments
Motoda, S., 1954.
Handy underway plankton catchers.
Bull. Fac. Fish., Hokkaido Univ., 5(2):149-152, 2 textfigs.

instruments
Motoda, S., 1953.
New plankton samplers. Bull Fac. Fish., Hokkaido Univ., 3(3):181-186, 7 textfigs.

instrumentation, at sea
Muller K., 1965.
An automatic stream drift sampler.
Limnol. Oceanogr., 10(3):483-484.

instrumentation, biological (at sea)
Nakai, Zinziro, 1962
Apparatus for collecting macroplankton in the spawning surveys of Iwashi (Sardine, anchovy and round herring) and others. (In English; Japanese summary).
Bull. Tokai Reg. Fish. Res. Lab., Tokyo, No. 9:221-237, 10 pls.

instrumentation, biological (at sea)
Nellen, W., and G. Hempel 1969.
Versuche zur Fängigkeit des "Hai" und des modifizierten Gulf-V-Plankton-Samplers "Nachthai".
Ber. dt. wiss. Komm. Meeresforsch. n.s. 20(2): 141-154.

instrumentation
Niskin, S., and J.J. Jones, 1963.
New collecting andrecording devices for limnological and oceanographic research (Abstract).
Great Lakes Res. Div., Inst. Sci. & Techn., Univ Michigan, Publ., 10:266.

instrumentation
Omori, Makoto, 1961.
[Trial tow with high-speed underway plankton catcher, Models V and VI.]
Info. Bull., Plankton, Japan, No. 7:15-36.

instrumentation, biological (at sea).
Omori, Makoto, Ryuzo Marumo and Yasushi Aizawa, 1965.
A 160-cm opening-closing net. II. Some notes on the towing behaviour of the net.
J. Oceanogr. Soc., Japan, 21(6):245-252.

instruments
Patalas, K., 1954.
Comparative studies on a new type of self acting water sampler for plankton and hydrochemical investigations.
Ekologia Polska, Polska Akad. Nauk, 2(2):231.

instruments, water sampler
Pettersson, O., 1929.
Water bottle with apparatus for current measurement and quantitative catch of plankton.
Svenska Hydrogr.-Biol. Komm. Skr., n.s., Hydrogr. 3:13-15.

instruments (biol.-at sea)
Pettersson, V., 1926.
Improvements in hydrographic technique. 1. A registering photothermograph. 2. A new plancton-catcher.
Svenska Hydrogr.-Biol. Komm. Skr., n.s., Hydrogr. 2:6 pp.

instrumentation, biological (at sea), plankton samplers
Quayle, D.B., and L.D.B. Terhune 1967.
A plankton sampler for oyster larvae.
J. Fish. Res Bd Can. 24 (4): 883-885.

instruments
Roberts, C.H., 1952.
An automatic collector for water and plankton.
J. du Cons. 18(2):107-116, 4 textfigs.

instrumentation, plankton
Schindler, D.W., 1969.
Two useful devices for vertical plankton and water sampling.
J. Fish. Res. Bd., Can., 26(7): 1948-1955.

instrumentation, biol. (at sea)
Smith, Robert E., Donald P. de Sylva and Richard A. Livellara, 1964.
Modification and operation of the Gulf 1-A high speed plankton sampler.
Chesapeake Science, 5(1/2):72-76.

instrumentation
Suyehiro, Yasuo, Yoshiyuki Matsue, Yaichiro Okada Takeharu Kumagori, Sigeo Takayama, Ryuzo Marumo, Osamu Asaoka, Makoto Ishino, Eiji Iwai, Yuzo Komaki, Masuoki Horikoshi, 1960.
Notes on the sampling gears and the animals collected on the second cruise of the Japanese Expedition of Deep Seas.
Repts. of JEDS, 1:187-200.

instrumentation, biological (at sea)
Swanson, George A., 1965.
Automatic plankton sampling system.
Limnology and Oceanography, 10(1):149-152.

instrumentation, biological (at sea)
Tanaka, H., H. Kasai, Y. Imayama, S. Kimura, M. Azeta and M. Anraku, 1969.
Hydrodynamic and towing characteristics of a modified model of the Clarke Jet Net. Marine Biol., 2(4): 297-306.

gear
Tonolli, V., 1951.
Un nuovo apprechio per la cattura di plancton in modo continuo e quantitativo: il "Plancton-Bar". Mem. Ist. Italiano Idrobiol. Dott. Marco Marchi, 6:193-202.

instrumentation, samplers
Troutner, Richard T., 1968.
Samplers and sampling; a brief review.
Bio Science, 18(10):960-961.

instruments
Westley, R.E., 1954.
A multiple-depth running plankton sampler.
Fish. Res. Pap. 1(2):46-49.

instruments
Wickstead, J., 1953.
A new apparatus for the collection of bottom plankton. J.M.B.A. 32(2):347-355, figs.

instrumentation, biological(sea)
Williamson, D.I., 1963
An automatic plankton sampler.
Bull. Mar. Ecology, 6(1):1-16, Pls. 1-5.

instrumentation, biology, at sea
Williamson, D.I., 1962.
2. An automatic sampler for use in surveys of plankton distribution. Contributions to symposium on zooplankton production, 1961.
Rapp. Proc. Verb., Cons. Perm. Int. Expl. Mer, 153:16-18.

instrumentation, biological (at sea)
Yamazi, Isamu, 1960.
Automatic plankton sampler with multiple nets.
Publ. Seto Mar. Biol. Lab., 8(2):(29):451-453.

Clarke-Bumpus sampler

instrumentation, biological (at sea)

CLARKE-BUMPUS SAMPLER

plankton sampler
Clarke, G.L., and D.F. Bumpus, 1940.
The plankton sampler (an instrument for quantitative plankton investigations). Limnol. Soc., Amer Spec. Publ. 5:2-8 pp., 5 figs.

Reviewed: J. du Cons. 17(2):202-203, RS.Wimpenny

instrumentation, biological (plankton sampler)
Clarke, G. L., and D. F. Bumpus, 1939.
Brief account of the plankton sampler. Intern. Revue Gesam. Hydrobiol. u. Hydrogr., 39(1/2):

instrumentation
Comita, G.W., and J.J., 1957.
The internal distribution patterns of a calanoid copepod population and a description of a modified Clarke-Bumpus plankton sampler.
Limnol. & Oceanogr. 2(4):321-332.

instrumentation, biological, at sea
Heinle, Donald R., 1965.
A screen for excluding jellyfish and ctenophores from Clarke-Bumpus plankton samplers.
Chesapeake Science, 6(4):231-232.

instrumentation, biological (at sea
Nival, P., 1965.
Modification du filet à plancton de type Clarke-Bumpus.
Rapp. Proc. Verb., Réunions, Comm. Int. Expl. Sci., Mer Méditerranée, Monaco, 18(2):329-331.

instruments
Paquette, R.G., and H.F. Frolander, 1957.
Improvements in the Clarke-Bumpus plankton sampler. J. du Cons., 22(3):284-288.

instrumentation

Paquette, Robert G., Eugene L. Scott and Paul N. Sund, 1961

An enlarged Clarke-Bumpus plankton sampler. Limnol. & Oceanogr., 6(2): 230-233.

instrumentation, biolofical, at sea

Tranter, D.J., and A.C. Heron, 1965. Filtration characteristics of Clarke-Bumpus Samplers. Australian J. Mar. Freshwater Res., 16(3):281-291.

instrumentation

Yentsch, C.S., and A.C. Duxbury, 1956. Some of the factors affecting the Clarke-Bumpus quantitative plankton sampler. Limnol. & Oceanogr., 1(4):268-273.

Hardy recorder-

instrumentation, biological (at sea)
continuous plankton recorder
Hardy plankton recorder

instrumentation

Butler, P.A., and A.J. Wilson, 1957. A continuous water sampler for estimation of daily changes in plankton. 1956 Proc., Natl. Shellfish. Assoc., 47:109-113.

continuous plankton recorder

Colebrook, J.M., R.S. Glover and G.A. Robinson, 1961

Continuous plankton records: a plankton atlas of the north-eastern Atlantic and the North Sea General Introduction. Bulls. Mar. Ecology, 5(42): 67-80.

continuous plankton recorder

Colebrook, J.M., Dore E. John and W.W. Brown, 1961

Contribution towards a plankton atlas of the north-eastern Atlantic and the North Sea. 2. Copepoda. Bulls. Mar. Ecol., 5(42): 90-97, Pls. 21-27.

continuous plankton recorder

Colebrook, J.M., and G.A. Robinson, 1961. The seasonal cycle of the plankton in the North Sea and the north-eastern Atlantic. J. du Conseil, 26(2):156-165.

Hardy plankton recorder

Colton, John B., Jr., and Robert R. Marak, 1962. Use of the Hardy plankton recorder in a fishery research program. Bull. Mar. Ecology, 5(49):231-246.

instrumentation, biological, at sea

Glover, R.S., 1962. 1. The continuous plankton recorder. Contributions to symposium on zooplankton production, 1961. Rapp. Proc. Verb., Cons. Perm. Int. Expl. Mer, 153:8-15.

instruments

Glover, R.S., 1953. The Hardy plankton indicator and sampler: a description of the various models in use. Bull. Mar. Ecol. 4(26):7-20, 5 textfigs.

Hardy plankton sampler

Hardy, Sir Alister, 1964
Plankton studies and our understanding of major changes in the great oceans.
In: Studies in Oceanography dedicated to Professor Hidaka in commemoration of his sixtieth birthday, 550-560.

instrumentation, biology, at sea

Hardy, A., 1963. Some developments in plankton research. Roy. Soc., New Zealand, Proc., 90(1):73-88.

Hardy plankton recorder

instruments

Hardy, A.C., 1936. The continuous plankton recorder. Rapp. Proc. Verb., Cons. Perm. Int. Expl. Mer, 95:35-48.

apparatus

Hardy, A.C. and W.N. Paton, 1947
Experiments on the vertical migration of plankton animals. JMBA 26(4): 467-526, pls. 8-11, 20 text figs.

Hardy plankton recorder

Henderson, G.T.D., 1961. Continuous plankton records: the distribution of young Sebastes marinus (L.). Bull. Marine Ecology, 5(46):173-193.

Hardy plankton recorder

Henderson, G.T.D., 1961. The distribution of young Sebastes marinus (L.) Int. Comm. Northwest Atlantic Fish., Ann. Proc. 11:103-110.

Hardy plankton sampler

Légaré, J.E.H., 1957. The qualitative and quantitative distribution of plankton in the Strait of Georgia inrelation to certain oceanographic factors. J. Fish. Res. Bd., Canada, 14(4):521-552.

instrumentation, biological at sea

Miller, David, 1961
A modification of the small Hardy plankton sampler for simultaneous high speed plankton hauls.
Bulletins, Mar. Ecol., 5(45):165-172.

continuous plankton recorder

Robinson, G.A., 1961

Contribution towards an atlas of the north-eastern Atlantic and the North Sea. 1. Phytoplankton. Bulls. Mar. Ecology, 5(42): 81-89, Pls. 15-20.

continuous plankton recorder

Vane, F.R., 1961

Contribution towards a plankton atlas of the north-eastern Atlantic and the North Sea. 3. Gastropoda. Bulls. Mar. Ecol., 5(42): 98-101, Pl. 28.

continuous plankton recorder

Vane, F.R., and J.M. Colebrook, 1962
Continuous plankton records: Contributions towards a plankton atlas of the north-eastern Atlantic and North Sea. VI. The seasonal and annual distributions of the Gastropoda.
Bull. Mar. Ecology, 5(50):247-253.

instrumentation biology (pressure)

McDonald, A.G. and I. Gilchrist, 1969.
Life in the ocean depths. The Physiological problems and equipment for the recovery and study of deep sea animals. Oceanol. int., 69, 1: 6 pp., 7 figs.

instrumentation biological (at sea) pumps

instrumentation, biological (at sea)

PUMPS

instrumentation

Aron, W., 1958.

The use of a large capacity portable pump for plankton sampling, with notes on plankton patchiness. J. Mar. Res., 16(2):158-173.

instrumentation, biological (at sea) - pump.

Beers, John R., Gene L. Stewart and John D.H. Strickland 1967.
A pumping system for sampling small plankton.
J. Fish. Res. Bd. Can. 24 (8): 1811-18

instrumentation, biological (at sea)

Bernard, Jean-Guy, and Robert Lagueux, 1970.
Addition d'un antiturbulent à la pompe à plancton.
Nat. can. 97 (4): 421-429

instrumentation, biological, at sea

Cassie, R. Morrison, 1964. Improved filter-changer from a plankton pump. New Zealand J. Sci., 7(3):409-416.

instrumentation, biology (at sea)

Della Croce, N. and A. Chiarabini, 1971.
A suction pipe for sampling mid-water and bottom organisms in the sea. Deep-Sea Res., 18(8): 851-854.

instrumentation, pumps

Freudenthal, Hugo D., and Walter Blogoslowski and Roy Stoecker, 1968.
The use of the pneumatic ejector pump in oceanography.
Limnol. Oceanogr., 13(4):706-708.

instrumentation, biological (at sea)

Harvey, George W., 1966.
A low velocity plankton siphon.
Limnol. Oceanogr., 11(4):646-647.

instrumentation, biological (at sea) - pump

Leong, Roderick 1967.
Evaluation of a pump and reeled hose system for studying the vertical distribution of small plankton.
Spec. scient. Rept., Fish., U.S. Fish Wildl. Serv. 545: 9 pp.

instrumentation, biological (at sea)

O'Connell, Charles P., and Roderick J.H. Leong 1963
A towed pump and shipboard filtering system for sampling small zooplankters.
U.S.F.W.S. Spec. Sci. Rept., Fish., No. 452: 19 pp.

instrumentation, biological, at sea

VanHaagen, Richard H., 1965.
Evolution of a submerged-pump sampling system. (Abstract).
Ocean Sci. and Ocean Eng., Mar. Techn. Soc., Amer. Soc. Limnol. Oceanogr., 1:540/

instrumentation, biology (at sea)

Whaley, R.C., 1958
A submersible sampling pump. Limnol. & Oceanogr., 3:476-477.

instrumentation, biological (at sea). plankton sled.

instrumentation, biol. (at sea).
Dovel, William L., 1964.
An approach to sampling estuarine macroplankton.
Chesapeake Science, 5(1/2):77-90.

instrumentation, biological (at sea)

SHELL FISH HARVESTER

instrumentation biological (at sea) DIGGER
Dickie, L.M., J.S. MacPhail, 1957.
An experimental mechanical shellfish-digger.
Prog. Repts., Atlantic Coast Stations, Fish. Res. Bd., Canada, No. 66:3-9.

instruments, biological
MacPhail, J.S., 1961.
A hydraulic esculator shellfish harvester.
Bull., Fish. Res. Bd., Canada, 128:1-24.

instrumentation, biological (at sea)

TELERECORDING UNIT

instruments
Boden, B.P., E.M. Kampa, J.M. Snodgrass and R.F. Devereux, 1955.
A depth telerecording unit for marine biology. J. Mar. Res. 14(2):205-209, 2 textfigs.

instrumentation, biological (at sea)

TRAPS

instruments
Lishov, S.M., and L.P.Zhidkov, 1953.
The bottom-trap - a sampler for studying transport of bottom organisms in river currents.
Zool. Zhurn. 32(5):1020-1024.

traps, floating fish larval
Scripps Institution of Oceanography, 1949
Marine Life Research Program. Progress Report to 30 April 1949, 25 pp. (mimeographed), numerous figs. (photo.+ozalid). 1 May 1949.

instrumentation, traps (light)
Zismann, Lyka, 1969.
A light-trap for sampling aquatic organisms.
Israel J. Zool. 18(4):343-348

instrumentation, biological (at sea) trawls

instrumentation, biological (at sea)

TRAWLS and their adjuncts

See also: instrumentation, biological (at sea) midwater trawls.

instrumentation, biology (at sea) (shrimp)
Butler, T.H., and R.W. Sheldon, 1965.
Trawl-Board sediment sampler.
J. Fish. Res. Bd, Can. 26(10): 2751-2753

instrumentation, biological (at sea) trawl nets
Kobayashi, K., and T. Deguchi, 1952.
An experiment of a beam-type trawl net for fish larvae, at the various depths of sea water.
Bull. Fac. Fish. Hokkaido Univ., 3(1):104-108.

instrumentation, biological, at sea
Koyama, Takeo, 1965.
On the results of field trials with various big otter trawls. (In Japanese; English abstrac
Bull. Tokai Reg. Fish. Res. Lab., No. 43:13-71.

instrumentation, biological, at sea
Lavrov, V.P., E.A. Liamin, A.N. Paramonov, B.M. Romanov, and B.A. Shmatko, 1963
A device for sight-guided trawling within a range of different depths. (In Russian).
Okeanologiia, Akad. Nauk, SSSR, 3(1):137-143.

instrumentation, biological (at sea)
Lebedev, E.A., 1959
[Behavior of a trawl in water.]
Rybnoe Khozyaistvo (Moscow), 35(2):39-42.

OTS-60-51180

instrumentation, biological (at sea)
Lestev, A.V., and G.E. Grishchenko, 1959
[Some suggestions on the standardization of far-eastern trawls.]
Rybnoe Khozyaistvo, 35(2):33-39.

Translation OTS 60-51129

instrumentation, biological (at sea)
Little, Frank, Jr., and Bill Mullins, 1964.
Diving plate modification of Blake (beam) trawl for deep-sea sampling.
Limnology and Oceanography, 9(1):148-150.

instrumentation, biological (at sea)
Lyapin, N.A., 1959
[Our observations on the behavior of the trawl in water.]
Rybnoe Khozyaistvo (Moscow), 35(2):43-46.

OTS-60-51080

instrumentation - biological - trawls
Menzies, Robert J., 1964
Improved techniques for benthic trawling at depths greater than 2000 meters.
Biology of Antarctic Seas, Antarctic Res. Ser., Amer. Geophys. Union, 1:93-109

instrumentation, biological (at sea) (nets)
Miyazaki, Chihiro, Yozo Tahara and Tokio Yokoi, 1970.
Comparative catching efficiencies between nylon-monofilament net and polyethylene-multifilament net for small otter trawlers. Bull. Tokai reg. Fish. Res. Lab., 63: 107-114.

instrumentation, biological (at sea) (trawling)
Rowe, Gilbert T., and Robert J. Menzies 1967.
Use of sonic techniques and tension recording as improvements in abyssal trawling.
Deep-Sea Res. 14(2): 271-274.

instrumentation, biological (at sea)
Sysoev, N.N., 1961.
[Use of bottom trawls.]
Trudy Inst. Okeanol., Akad. Nauk, SSSR, 47:119-124.

instrumentation, biological at sea
Takayama, S., and T. Koyama, 1961.
Studies on trawl net. VI. Comparative efficiency between two types of the otter board with different ratios as to the length and the height.
Bull. Tokai Reg. Fish. Res. Lab., No. 31:297-310.

instrumentation, biological, at sea.
Takayama, S., and T. Koyama, 1961.
Studies on trawl net. V. A measuring device and a formula to obtain the opening breadth of wing nets.
Bull. Tokai Reg. Fish. Res. Lab., No. 31:289-296.

instrumentation, biological(at sea)
Takayama, S., and T. Koyama, 1958
Studies on trawl-net. II. Determination of the angle of inclination of triangular brackets giving the optimal angle of attack to the otter board.
Bull. Tokai Reg. Fish. Res. Lab. No. 22: 37-46.

instrumentation, biological (at sea)
Trent, W. Lee, 1967.
Attachment of hydrofoils to otter boards for taking surface samples of juvenile fish and shrimp.
Chesapeake Sci., 8(2):130-133.

instrumentation, biological (at sea)

MIDWATER TRAWLS

instrumentation midwater trawl
Aboussouan, Alain, 1971.
Contribution à l'étude des téléostéens recoltés au chalut pélagique en relation avec la D.S.L. durant la période du 1er novembre 1967 au 31 décembre 1968
Cah. océanogr. 23(1): 85-99.

Isaac-Kidd midwater trawl
Aron, William, 1962
The distribution of animals in the eastern North Pacific and its relationship to physical and chemical conditions.
J. Fish. Res. Bd., Canada, 19(2):271-314.

instrumentation, trawling
Backus, Richard H., 1966.
The "Pinger" as an aid in deep trawling.
J. Cons. perm. int. Expl. Mer, 30(2):270-277.

instrumentation biological (at sea) midwater trawl
Bieri, Robert, and Takesi Tokioka 1968.
Dragonet II, an opening-closing quantitative trawl for the study of microvertical distribution of zooplankton and the meio-epibenthos.
Publ. Seto mar. biol. Lab. 15(5):373-390.

instrumentation, biological (at sea)
Clarke, Malcolm R., 1969.
A new midwater trawl for sampling discrete depth horizons. J. Mar. biol. Ass. U.K., 49(4) 945-960.

instrumentation, biological (at sea) (trawls)
Friedl, William A., 1971.
The relative sampling performance of 6- and 10-foot Isaacs-Kidd midwater trawls.
Fish. Bull. nat. mar. Fish. Serv. 69(2): 427-432.

instrumentation, biology (at sea)
Hamuro, C., 1962
Studies on mid-water trawl gears and mid-water telemeters.
Indo-Pacific Fish. Council, FAO, Proc., 9th Sess. (2/3):84-89.

instrumentation, biological, at sea
Imanishi, Hajime, Takeo Taniguchi and Akiyoshi Kataoka, 1965.
Hydrodynamic studies on the Isaacs-Kidd mid-water trawl. II. Field experiments of the 10-foot S-II type larva-net.
Bull. Jap. Soc. Sci. Fish., 31(9):663-668.

instrumentation, biological (at sea)
Iwai, Eiji, 1962.
On a new-type trawl-net for collecting mid-water animals.
Repts. Res. Oceanogr. Studies, Sci. & Techn. Agency, 1960, 42-46. (Not seen)
Abstract in:
Records of Researches, Fac. Agric., Univ. Tokyo, No. 12(1962-1963): 50.

instrumentation, biological (at sea)
Kobayashi, K., 1956.
[An experiment on a mid-water trawl. IV. On its mechanism and trial practice.]
Bull. Fac. Fish., Hokkaido Univ., 7(1):21-30.

instrumentation, midwater trawl
Michel A. et R. Grandperrin, 1970.
Selection du chalut pélagique Isaacs-Kidd 10 pieds. Mar. Biol., 6(3): 200-212.

instrumentation, biological, mid-water trawl
Nakasai, Kei 1967.
Application of chart for estimation of trolling depth.
Bull. Fac. Fish. Nagasaki Univ. 24:101-105.

instrumentation, biological-mid-water trawl
Nakasai, Kei, and Tasae Kawakami, 1968.
Mechanical studies on the mid-water trawl gear in operation.
Bull. Fac. Fish., Nagasaki Univ., 26: 49-61.

instrumentation, biological (at sea)
Nakasai, Kei, and Tasae Kawakami, 1965.
On a simple estimation of working depth of mid-water trawl.
Bull. Jap. Soc., Sci. Fish., 31(4):277-280.

instrumentation, biological, at sea
Pearcy, William G., and Lyle Hubbard, 1964
A modification of the Isaacs-Kidd midwater trawl for sampling at different depth intervals.
Deep-Sea Research, 11(2):263-264.

instrumentation, biology (at sea)
Takayama, S., and T. Koyama, 1962
Application of net gauge to midwater trawling.
Indo-Pacific Fish. Council, FAO, Proc., 9th Sess. (2/3):90-92.

instrumentation, biological (at sea)
Takayama, Shigene, and Takeo Koyama, 1961
Studies on midwater trawling II. Field experiments of a net sonde.
Bull. Tokai Reg. Fish. Res. Lab., No. 27: 47-54.

instrumentation, biological (at sea)
Taniguchi, Takeo, Akiyoshi Kataoka and Hajima Imanishi, 1965.
Hydrodynamic studies on the Isaacs-Kidd mid-water trawl. 1. Field experiments of the 10 foot S-1 type larva net.
Bull. Jap. Soc., Sci. Fish., 31(5):327-332.

instrumentation, biological (at sea)

VOLUME METER

gear
Usachev, P.I., 1939.
[A description of some new devices (volume meter) for determining the volume of plankton under field conditions.] USSR Inst. Mar. Fish., Ocean., Sbornik Knipovich:99-114.

instrumentation, biological (at sea) - water-bottom interface
Milbrink, Göran 1968.
A microstratification sampler for mud and water.
Oikos 19(1): 105-110

instrumentation, biological (at sea)

WATER SAMPLER
for biological purposes

water sampler, biol.
Jensen, E.A., and E. Steemann Nielsen, 1953.
A water sampler for biological purposes.
J. du Cons. 18(3):296-299, 2 textfigs.

instrumentation, biological (at sea)
Jitts, H.R., 1964.
A twin six-liter plastic water sampler.
Limnology and Oceanography, 9(3):452.

instrumentation, biological (at sea)
Marshall, Nelson, and Stanley Rubinsky, 1964.
A microstratification water sampler.
Ecology, 45(1):193-195.

instrumentation, biological (at sea)
Murray, J.W., 1962.
A new bottom-water sampler for ecologists.
J. Mar. Biol. Assoc., U.K., 42(3):499-501.

instruments
Wohlenberg, E., 1950.
Der horizontale Wasserschöpfer. Deutsch. Hydro. Zeitschr. 3(5/6):365-368, 2 textfigs.

instrumentation, biological (at sea)

WIRE ANGLE INDICATOR

instruments
Moore, H.B., 1941
A wire-angle indicator for use when towing plankton nets. JMBA 25:419-422, 4 textfigs.

instruments, biological laboratory

instrumentation, biological (laboratory)

This contains items peculiar to biological oceanography only.

aquaria

instrumentation, biological (laboratory)

AQUARIA

instrumentation, aquaria
Belloy, Maxime, 1969
Regeneration de l'eau de mer dans un petit aquarium en circuit fermé.
Bull. Inst. Pêches marit. Maroc. 17:45-48

instrumentation, biology (lab.)
Breder, C.M., Jr., 1957.
Miniature circulating systems for small laboratory aquaria.
Zoologica, 42(1):1-10.

instrumentation, biological (laboratory)
Flüchter, Jürgen, 1964.
Eine besonders wirksame Aquarienfilterung und die Messung ihrer Leistung.
Helgoländer Wiss. Meeresuntersuchungen 11(3/4): 168-177.

instrumentation, aquaria
Goldizen, V.C., 1970.
Management of closed-system marine aquariums.
Helgoländer wiss. Meeresunters. 20(1/4): 637-641.

instrumentation, aquaria
Hirayama, Kazutsugu, 1970.
Studies on water control by filtration through sand bed in a marine aquarium with closed circulating system. VI. Acidification of aquarium water.
Bull. Jap. Soc. scient. Fish. 36(1): 26-34

instrumentation, aquaria
Scott, K.R. 1971.
Monitoring of fish tank oxygen concentration using a digital recorder.
J. Fish. Res. Bd. Can. 28(8): 1196-1197

instrumentation, biological
Rzhepishevsky, I.K., I.A. Kuznetsova and P.A. Vlasov, 1968.
A device for mixing water in experimental vessels. (In Russian; English abstract).
Okeanologiia, Akad. Nauk, SSSR, 8(2):347-349.

instrumentation, biological (lab)
Stewart, James E., and H.E. Power, 1963.
A sea-water aquarium for marine animal experiments.
J. Fish. Res. Bd., Canada, 20(4):1081-1084.

Instrumentation,

Strand, John A., Joseph T. Cummins, and Burton E. Vaughan, 1969.
A fast-flow sealed disk filter system for marine aquaria. Limnol. Oceanogr., 14(3): 444-448.

instrumentation, biological (lab)
Sundnes, G., 1962.
A pressure aquarium for experimental use.
Fiskeridirek. Skrift., Ser. Havundersg., 13(4): 7 pp.

instrumentation, biological (laboratory)

BENTHOS

instrumentation
Fedikov, N. F., 1960.
Device for washing bottom fauna samples.
Trudy Inst. Okeanol., 41:254-256.

instrumentation
Fremling, Calvin R., 1961.
Screened pail for sifting bottom-fauna samples.
Limnol. & Oceanogr., 6(1):96.

instruments, biological
Lauff, G.H., K.W. Cummins, C.H. Eriksen and M. Parker, 1961.
A methods for sorting bottom fauna samples by elutriation.
Limnol. & Oceanogr., 6(4):462-466.

bioluminescence

instrumentation, bioluminescence
Hardy, A.C., and R.H. Kay, 1964
Experimental studies of plankton luminescence.
Jour. Mar. Biol. Assoc., U.K., 44(2):435-484.

instrumentation, luminescence
Karatashev, G.S., 1970.
On the methods of photoluminescence studies of the sea water. Oceanologiia, 10(5): 883-888. (In Russian; English abstract)

instrumentation, biological

BUOYANCY

instrumentation, biological
Lange, R. and R. Haaland, 1970.
A simple device for continuous recording of buoyancy variations in marine bottom invertebrates
Nytt Mag. Zool., 18(1): 81-83.

instrumentation, biological (laboratory)
Ganning, Björn, and Fredrik Wulff, 1966.
A chamber for offering alternative conditions to small motile aquatic animals.
Ophelia, 3:151-160.

instrumentation, biological (laboratory)
Yentsch, Charles S., 1965.
A perfusion chamber for electrode studies on the physiology of planktonic algae.
J. Mar. Res., 23(1):39-43.

instrumentation, biological (laboratory)

COUNTING "CHAMBER"

instrumentation, biological (lab.)
Dawson, William A., 1960.
Home-made counting chambers for the inverted microscope. Limnol. & Oceanogr., 5(2): 235-236.

instrumentation, biological (counting cells)
Gannon, John E., 1971.
Two counting cells for the enumeration of zooplankton micro-Crustacea.
Trans. Am. microsc. Soc. 90(4): 486-490

instrumentation, biological
Julià, A., 1969
Contador óptico de partículas en suspensión de tamaño mediano y grande (Micro mesoplancton)
Investigación pesq. 33(1): 201-21 .

instrumentation, biological (lab.)
Lund, J.W.G., 1962
Concerning a counting chamber for nannoplankton described previously.
Limnology and Oceanography, 7(2):261-262.

instrumentation
Lund, J.W.G., 1959.
A simple counting chamber for nannoplankton.
Limnol. & Oceanogr., 4(1):57-65.

Instrumentation, biol. (lab.)
Mednikov, B. M., and Ya. I. Starobogatov, 1961.
A random camera for counting small sized biological objects
Trudy, Vses. Gidrobiol. Obshch., 11:426-428.

instrumentation, biology, laboratory
Mitson, R.B., 1963
Marine fish culture in Britain. V. An electronic device for counting the nauplii of Artemia salina L.
J. du Conseil, 28(2):262-269.

instruments
Nielsen, P. H., 1950.
An auxiliary apparatus for plankton studies by means of the sedimentation method. J. du Cons. 16(3):307-309, 2 textfigs.

instrumentation
Palmer, C.M., and T.E. Maloney, 1954.
A new counting slide for nannoplankton.
Amer Soc. Limnol. Oceanogr., Spec. Publ., No. 21:6 pp., 2 textfigs.

instrumentation, biological (lab)
Throndsen, J., 1970.
A small sedimentation chamber for use on the Wild M 40 and the Zeiss plankton microscopes.
J. Cons. int. Explor. Mer, 33(2): 297-298.

instrumentation, biological (laboratory)
*Tungate D.S., 1967.
A new sedimentation chamber.
J. Cons. perm.int.Explor. Mer, 31(2):284-285.

instruments
Ward J., 1955.
A description of a new zooplankton counter.
Q.J. Micr rosc. Soc., 96(3):371-374.

instrumentation
Warren, P.J., 1958.
Description of an improved rotating counter for zooplankton. J. du Cons., 23(3):337-339.

instrumentation, biological (laboratory)

CULTURES

instrumentation, cultures
Bernhard, M., preperator, 1965.
Studies on the radioactive contamination of the sea, annual report 1964.
Com. Naz. Energ. Nucleare, La Spezia, Rept., No. RT/BIO (65) 18:35 pp.

instrumentation, biology (laboratory)
Carpenter, Edward J., 1968.
A simple, enexpensive algal chemostat.
Limnol.Oceanogr., 13(4):720-721.

Instrumentation, biol. (lab)
Davis, W.P., 1970.
Closed systems and the rearing of fish larvae.
Helgoländer wiss. Meeresunters, 20(1/4): 691-696.

instrumentation, biology (laboratory)
Duthie, H.C. and J. Shaw, 1968.
A compressed air motor shaker for use in small incubators.
Limnol.Oceanogr., 13(4):719-720.

instrumentation, biological, laboratory
Fahy, William E., 1964
A temperature-controlled salt-water circulating apparatus for developing fish eggs and larvae.
Journal du Conseil, 28(3):364-384.

INSTRUMENTATION, culturing
Greve, W., 1968.
The 'planktonkreisel' a new device for culturing zooplankton.
Marine Biol. 1(3): 201-203

instrumentation, biol. (lab)
Jatzke, P., 1970.
The trichterkreisel, an in situ device for cultivating marine animals in tidal currents.
Helgoländer wiss. Meeresunters, 20(1/4): 685-690.

instrumentation, cultures
Knaggs, F.W. 1965.
A simplified system for the controlled illumination of algal cultures.
Netherlands J. Sea Res. 4(1): 21-26

instrumentation, biological (laboratory) cultures
Margalef, Ramón 1967.
Laboratory analogues of estuarine plankton systems.
In: Estuaries, G.H. Lauff, editor, Publs Am. Ass. Advmt Sci. 83: 515-521.

instruments
Margalef, R., 1954.
Un aparato para el cultivo de algas en condiciones regulables. Publ. Inst. Biol. Aplic. 17:65-69, 3 textfigs.

instrumentation, cultures
*Miller, R.A., J.P. Shyluk, O.L. Gamborg, and J.W. Kirkpatrick, 1968.
Phytostat for continuous culture and automatic sampling of plant-cell suspensions.
Science, 159(3814):540-542.

instrumentation, biology, at sea
Neushul, M., and J.H. Powell, 1964.
An apparatus for experimental cultivation of benthic marine algae
Ecology, 45(4):983-984.

instrumentation, biology(lab.)
Thomas, William H., Harold L. Scotten and John S. Bradshaw, 1963.
Thermal gradient incubators for small aquatic organisms.
Limnology and Oceanography, 8(3):357-360.

instrumentation, biology (lab)
Von Oertzen, J.A. und V. Motzfeld, 1969.
Eine Apparatur zur kontinuierlichen Respirationsmessung an marinen Organismen.
Marine Biol., 3(4): 336-340.

instrumentation, biological (laboratory)
Whitford, L.A., and Gary E. Dillard, and George J. Schumacher, 1964
An artificial stream apparatus for the study of lotic organisms.
Limnology and Oceanography, 9(4):598-600.

instrumentation, cultures (lab.)
Woods, Langley, 1965.
A controlled conditions system (CCS) for continuously flowing seawater.
Limnol. Oceanogr., 10(3):475-477.

instrumentation, biological (lab)
Yokohama, Yasutsugu, and Shun-ei Ichimura 1969.
A new device of differential gas-volumeter for ecological studies on small aquatic organisms.
J. oceanogr. Soc., Japan, 25(2): 75-80.

instrumentation, biological (laboratory)

ELECTRON MICROSCOPE

electron microscope
Braarud, T., K.R. Gaarder, J. Markali and E. Nordli, 1952.
Coccolithophorids studied in the electron microscope. Observations on Coccolithus huxleyi and Syracosphaera carterae. Nytt Mag. Bot. 1:129-133, 2 pls., 4 figs.

instrumentation, biological (laboratory)

LENGTH MEASUREMENTS

instruments
Cole, H.A., and M.N. Mistakidis, 1953.
A device for the quick and accurate measurement of carapace length in prawns and shrimps.
J. du Cons. 19(1):77-79, 3 textfigs.

instrumentation, biol (lab) plankton
Cooke, R.A., L.D.B. Terhune, J. Stott and W.H. Bell, 1970.
An opto-electronic plankton sizer.
Techn. Rept Fish. Res. Bd, Can. 172: 40pp. (multilithed)

instrumentation, biological, lab.
Griffiths, P.G., 1968.
An electronic fish scale proportioning system.
J. Cons. perm. int. Explor. Mer, 32(2):280-282.

plankton scanner
Scripps Institution of Oceanography, 1949
Marine Life Research Program. Progress Report to 30 April 1949, 25 pp. (mimeographed), numbrous figs. (photo.+ozalid). 1 May 1949.

instrumentation, biological
Watson, J., and P.G. Wells, 1970.
A gauge for carapace measurements of crabs.
J. Fish. Res. Bd, Can., 27(6):1158-1161

instrumentation, biological (laboratory) - microscope condenser
Heron, A.C. 1969.
A dark-field condenser for viewing transparent plankton animals under a low power stereomicroscope.
Mar. Biol. 2 (4): 321-324.

instrumentation, biological (laboratory)

MISCELLANEOUS

instrumentation, biological (laboratory)
Hirata, Hachiro, 1963
An automatic solution adder.
Limnol. and Oceanogr., 8(2):299-300.

instrumentation, biological
Khudadov, G.D., 1961.
[An apparatus for Arthropoda labeling by means of spraying them with radioactive isotopes.]
Biull. M. O.-Va Inst. Prirodi. Otd. Biol., 66(3):40-50.

instrumentation, biological (laboratory)
Macleod, J.C., 1967.
A new apparatus for measuring maximum swimming speeds of small fish.
J. Fish. Res. Bd, Can., 24 (6): 1241-1252

instrumentation, biological (laboratory)
Moore, Johnes K., 1963
Refinement of a method for filtering and preserving marine phytoplankton on a membrane filter.
Limnol. and Oceanogr., 8(2):304-305.

Instrumentation biological (lab)
Soli, Giorgio, and Raymond Vaissière 1968.
An apparatus for studying the response of copepods to simulated dinoflagellate luminescence.
Bull. Inst. océanogr., Monaco, 67(1381): 5pp

instrumentation biological (lab)
Stewart, James E., 1963.
An arrangement for automatically and reproducibly controlling and varying illumination in biological experiments.
J. Fish. Res. Bd., Canada, 20(4):1103-1107.

instrumentation, biological (laboratory)
Wells, J.M., and J.E. Warinner 1968
A recording volumetric respirometer for aquatic animals.
Limnol. Oceanogr. 13(2): 376-378

instrumentation, biological (laboratory)

PLANKTON PICKER

instrumentation, biological (laboratory)
Hamlin, William H., and Allan W.H. Bé, 1963.
A plankton picker.
Deep-Sea Research, 10(4):459-461.

instrumentation
Nicholson, H.F., 1958.
Mechanical pipette for picking out diatoms.
J. du Cons., 23(2):189-191.

instrumentation, Biological (lab)
Throndsen, J., 1969.
A simple micropipette for use on the Wild M 40 and the Zeiss plankton microscopes. J. Cons. perm. int. Explor. Mer, 32(3): 430-432.

plankton splitter

instrumentation, biological (laboratory)

PLANKTON "SPLITTERS"

instruments, biologica, lab.
Cushing, C.E., jr., 1961.
A plankton sub-sampler.
Limnol. & Oceanogr., 6(4):489-490.

instrumentation, biological (lab.)
Hopkins, Thomas L., 1962
A zooplankton subsampler.
Limnol. and Oceanogr., 7(3):424-427.

instruments
Kott, P., 1953.
Modified whirling apparatus for the subsampling of plankton. Australian J. Mar. Freshwater Res. 4(2):387-393, 5 textfigs.

instrumentation, biological (laboratory) (plankton splitter)
Longhurst, Alan R., and Don L.R. Seibert, 1967.
Skill in the use of Folsom's plankton sample splitter.
Limnol. Oceanogr., 12(2):334-335.

instrumentation, biological laboratory
McGowan, John A., and Vernie J. Frandorf, 1964
A modified heavy fraction zooplankton sorter.
Limnology and Oceanography, 9(1):152-155.

instrumentation, biological (laboratory)
McHardy, Robert A., 1964.
A vacuum-assisted subsampler for use with small planktonic organisms.
J. Fish. Res. Bd., Canada, 21(3):639-640

INSTRUMENTATION, biological (lab.)
Scarola, John F., and Anthony J. Novotny, 1968.
Folsom plankton splitter modified for enumeration of Entomostraca.
Limnol. Oceanogr., 13(1):195-196.

plankton sample splitter
Scripps Institution of Oceanography, 1949
Marine Life Research Program. Progress Report to 30 April 1949, 25 pp. (mimeographed), numbrous figs. (photo.+ozalid). 1 May 1949.

instruments
Wiborg, K.F., 1951.
The whirling vessel. An apparatus for the fractioning of plankton samples. Rept. Norwegian Fish Mar. Invest. 9(13):16 pp., 3 textfigs.

instrumentation, biological (laboratory)

PLANKTON VOLUME

instruments
Berardi, G., 1953.
Apparecchio per una precisa valutazione volumetrica di campioni di plancton. Mem. Ist. Ital. Idrobiol., "Dott. Marco De Marchi", Pallanza, 8: 221-228, 1 textfig.

instrumentation
Frolander, H.F., 1957.
A plankton volume indicator. J. du Cons., 22(3): 279-283.

instruments
Gnanamuthu, C.P., 1952.
A simple device for measuring the volume of an aquatic animal.
Nature 170:587-588.

instrumentation, biological (lab)
Jashnov, V.A., 1959.
A new model of a volume meter for rapid and precise plankton evaluation under field conditions. (In Russian; English summary).
Zool. Zhurn., Akad. Nauk, SSSR, 38(11):1741-1744.

instrumentation, biological (laboratory) - volumeter
Persoone, G., 1971.
A simple volumeter for small invertebrates.
Helgoländer wiss. Meeresunters. 22(1): 141-143.

instrumentation, biological (lab)
Robertson, A.A., 1970.
An improved apparatus for determining plankton volume.
Fish. Bull. S.Afr. 6: 23-26

Tranter, D.J., 1960 instrumentation
A method for determining zooplankton volumes.
J. du Cons., 35(3):272-278.

instrumentation
Yentsch, C.S., and J.F. Hebard, 1957.
A gauge for determining plankton volume by the mercury immersion method. J. du Cons., 22(2):184-190.

instrumentation, biological (laboratory)

PLANKTON WEIGHT

instrumentation, biological, laboratory
Reeve, M.R., 1964
A simple torsion microbalance for weighing small animals.
Journal du Conseil, 28(3):285-392.

instrumentation, biological (laboratory)
Reeve, M.R., 1964.
A simple torsion microbalance for weighing small animals.
J. du Conseil, 28(3):385-392.

instrumentation, biological (laboratory)

PRESSURE VESSEL

instrumentation, biological (laboratory)
Lincoln, R.J. and I. Gilchrist, 1970.
An observational pressure vessel for studying the behaviour of planktonic animals. Marine Biol., 6(1): 1-4.

instrumentation, biological (laboratory)
Sundnes, G., 1962.
A pressure aquarium for experimental use.
Fiskeridirek. Skrift., Ser. Havundersø. 13(4): 7 pp.

boats

instrumentation, boats
Driver, E.A., 1964
A simple speedometer for small boats.
Limnology and Oceanography, 9(2):264-265.

instrumentation, boats
Pfanstiehl, Alfred, 1962
An aid for small boat maneuvers.
Limnol. and Oceanogr., 7(3):431-432.

boomerangs

instrumentation, boomerangs
Raymond, Samuel O., 1965.
Glass floats and boomerangs.
Mar. Sci. Instrument. 4: 231-240

buoys

"buoys"
Alldredge, Leroy R., and J.C. Fitz, 1964.
Submerged stabilized platform.
Deep-Sea Res., 11(6):935-942.

instrumentation, buoys
Anon., 1967.
Litton-Ameoom Navy Acre Buoy.
Geo-Mar. Tech., 3(2):19-23.

instrumentation, buoys
Anon., 1965.
U.S. Weather Bureau turns to ocean buoys.
GeoMarine Technology, 1(3):41-45.

instrumentation, buoys
Anon., 1965.
Winch buoys.
GeoMar. Techn., 1(2):26-28.

instrumentation, buoys
Baker, W.F. 1970.
A directional wavebuoy operating up to 15 Hertz.
Electronic Engineering in Ocean Technology, Swansea, 21-24 Sept. 1970, I.E.R.E. Conf. Proc. 19: 307-322.

instrumentation, buoys
Balin, A.A., 1963.
Automatic radio beacon buoy for long-operating automatic stations. (In Russian). (Not seen).
Trudy, Arktich. i Antarktich. Nauchno-Issled. Inst., No. 254:67-68.

Abstracted in:
Soviet Bloc Res., Geophys., Astron., Space, 95: 52-53.

instrumentation, markers
Beccasio, Anthony J., 1964
Dye markers for sonobuoys.
Undersea Techn., 5(8):27-28.

instrumentation, buoys
Bivins, Luther E., and Benjamin R. Swann 1967.
Moored buoy performance measurement.
Trans. 2nd int. buoy Techn. Symp. 245-267.

instrumentation, buoys
Bsm, R.M., D.M. Brown, J.D. Isaacs, R.A. Schwartzlose and M.H. Sessions 1970.
Deep-moored instrument station design and performance, 1967-1970.
SIO Ref. Ser. 70-19: 38 pp. (unpublished manuscript)

instrumentation, buoys
Brainard, Edward C. 1967.
Evaluation of The P.O.E. buoy with conventional buoy designs.
Trans. 2nd int. buoy Techn. Symp. 161-182.

instrumentation buoys
Bronzino, J.D., R.W. Corell, S.L. Heller, F.R. Hess, D.J. Meeker, and D.W. Melvin, 1967.
Full-scale simulation of the IRLS buoy program.
1967 NEREM Record, Inst. Electr. Electron. Engrs, 9:194-195.

instrumentation, buoys
Capart, G., 1967.
Digital transponding oceanographic buoy.
Trans. 2nd Int. Buoy Techn. Symp., 235-243.

instruments
Carruthers, J.N., 1939.
The Lowestoft crossbow float. Hydrogr. Rev. 16(2):149-153, 5 figs.

Also: Nautical Mag., Glasgow, 141:

instrumentation, buoys
Chramiec, M.A., and R.A. Helton, 1967.
A deep-ocean acoustically controlled actuator system.
1967 NEREM Record, Inst. Electr. Electron. Engrs, 9:192-193.

instrumentation, buoys
Crayson, C.H., and N.D. Smith, 1970.
Recent advances in wave buoy techniques at the National Institute of Oceanography.
Electronic Engineering in Ocean Technology, Swansea, 21-24 Sept. 1970, I.E.R.E. Conf. Proc. 19: 289-306.

instrumentation, buoys
Daubin, S.C., 1964.
Deep-sea buoy systems.
Inst. Soc. Amer., Journal, 11(3):49-53.

Dellies, O.J., 1965. instrumentation, buoys
Syntactic foam buoys.
GeoMarine Techn., 1 (6):27-28.

instrumentation, buoys
Favorite, F., D. Fisk and W.J. Ingraham, Jr., 1965.
First transponding oceanographic buoys in the Pacific.
J. Fish. Res., Bd., Canada, 22(3):689-694.

instrumentation, buoys

Ford, James R., 1967.
Direct measurement of two-dimensional wave characteristics from a spar buoy.
Trans. 2nd Int. Buoy Techn. Symp., 201-212

instrumentation, buoys

Francis, T.J. G., 1964.
A long range seismic-recording buoy.
Deep-Sea Res., 11(3):423-425.

instrumentation, buoys

Garstang, Michael, Paul Ferris Smith and Kenneth E. Perry, 1965.
An unattended buoy system for digital recording of air-sea energy exchange parameters.
Trans. 2nd Int. Buoy Techn. Symp., 193-200

instrumentation, buoys

Gandillere, Philippe, 1970.
An experimental oceanographic automatic station.
Electronic Engineering in Ocean Technology, Swansea, 21-24 Sept. 1970, I.E.R.E. Conf. Proc. 19: 481-488

Hodgman, James A., 1969.
Telemetering data buoys.
Oceanology intl, 4(4):29-30. (popular).

instrumentation, buoys

Hodgman, James A., 1968.
U.S. Coast Guard National data buoy systems project.
J. Ocean Techn., 2(4):126-131.

instrumentation, buoys

Huff, Lloyd C., 1967.
Evaluation of the Nomad wind measurement subsystem.
Trans. 2nd Int Buoy Techn System, 395-408

instrumentation buoys

Iwasa, K., 1964.
The study of the radio buoy for oceanographical observations. (In Japanese; English abstract).
Hydrogr. Bull., Tokyo. (Publ. No. 981).
No. 78:27-49.

instrumentation, buoys

Kazakov, S.P., 1963.
On the motion of hydrometric buoys (floats).
Physics of the sea. (In Russian).
Trudy Morsk. Gidrofiz. Inst., Akad. Nauk Ukrain., SSR, 28:67-71.

instrumentation, buoys

Kosic, R.F., K.A. Morgan, L.A. Scott and R.F. Devereux, 1967.
Long-range telemetering from the monster buoy.
Trans 2nd Int. Buoy Techn. Symp, 335-355

"buoys"

Krause, G., and G. Siedler, 1964.
Ein System zur kontinuierlichen Messung physikalischer Grössen im Meere.
Kieler Meeresf., 20(2):130-135.

moored buoys

Krauss, W., 1962
The spectra of turbulence and internal waves in the Baltic Sea.
Proc. Symposium on Mathematical-Hydrodynamical Methods of Phys. Oceanogr., Sept. 1961, Inst. Meeresk., Hamburg, 375-384.

instrumentation, buoys

Kuznetzov, O.A., 1964.
A buoy for undistorted meteorological measurements over the sea. Investigations in the Indian Ocean (33rd voyage of E/S "Vitiaz"). (In Russian)
Trudy Inst. Okeanol., Akad. Nauk, SSSR, 64:154-157.

instrumentation, buoys

Lambert, David R. 1969.
Sonodive dive pattern analysis
J. mar. techn. Soc. 3(5):17-29

instrumentation, buoys

Lampietti, Francois M.J., and Robert M. Snyder, 1965.
Practical approach to estimating buoy scope.
GeoMarine Techn., 1(6):29-32.

instrumentation, buoys

Lebus, U. 1967.
Automatic meteorological oceanographic buoy (AMOB).
Trans. 2nd int. buoy Techn. Symp. 307-309

instrumentation, buoys

Marcus, Sidney O., Jr., 1967.
Evaluation of Nomad-II data in their meteorological and oceanographic applications
Trans. 2nd Int. Buoy Techn. Symp., 379-394.

instrumentation, buoys

McCalmont, Arnold M., and Alfred L. Girard, 1967.
A CSE command telemetry system for meteorological/oceanographic buoys.
Trans. 2nd Int. Buoy Techn. Symp., 359-378.

instrumentation, buoys

Meshcherskiy, V.I., 1957.
Captive buoy with illumination. (In Russian).
Meteorol. i Gidrol., (5):56.

instrumentation, buoys

Meyer, Robert P., Thomas R. Meyer, Lee A. Powell and William L. Unger 1967
A radio controlled seismic recording buoy.
Trans. 2nd Int. Buoy Techn. Symp., 293-306

instrumentation, buoys

Morgan, K.A., L.A. Scott, and R.F. Devereux, 1967.
The monster buoy, its data acquisition and telemetry/command systems.
Trans. 2nd Int. Buoy Techn. Symp, 269-291.

instrumentation, buoys,

Morin, J.O., and R.G. Walden, 1963.
Radio telemetering buoys.
Undersea Technology, 4(8):15-17.

instrumentation buoys

Morrison, J.H., 1969.
Radio-isotope power sources for buoy-type applications. Oceanol. int. 69, 1: 6 pp., 6 fig

instrumentation, buoys

Mottern, R.E., E.F. Corwin and A.F. Pyle, 1967.
The Meteorological Buoy Programme of the U.S. Navy.
Marine Obs., 37(218):178-185.

instrumentation, buoys

Neprochnov, Y.P., 1964.
The structure and thickness of the sedimentary layer of the Arabian Sea, the Bay of Bengal and the Andaman Sea. Investigations in the Indian Ocean (33rd Voyage of E/S "Vitiaz"). (In Russian).
Trudy Inst. Okeanol., Akad. Nauk, SSSR, 64:214-226.

instrumentation, buoys

*Pond, S., 1967.
The effects of buoy motion on measurements of wind speed and stress. (In Russian; English abstract).
Fisika Atmosfer. Okean., Izv. Akad. Nauk, SSSR, 3(12): 1305-1311.

instrumentation, buoys

*Pond, S., 1968.
Some effects of bupy motion on measurements of wind speed and stress.
J. geophys. Res., 73(2):507-512.

instrumentation, buoys

Richardson, William S., 1962.
Current measurements from moored buoys.
Marine Sciences Instrumentation, Instr. Soc., Amer., Plenum Press, N.Y., 1:205-209.

instrumentation, buoys

Rubinstein, M.A., 1967.
State-of-the-art of direct energy conversion power supply for buoys.
Trans. 2nd Int. Buoy Techn. Symp., 213-224

instrumentation, buoys

Ruff, Ronald E., 1967.
Long-term data acquisition electronic buoy.
Trans. 2nd Int. Buoy Techn. Symp., 225-233

instrumentation, buoys

Schick, George B., 1965.
Shoot a line for buoy recovery.
Geo-Marine Techn., 1(10):V-337-.

sono buoys (biol. noise)

Scripps Institution of Oceanography, 1949
Marine Life Research Program, Progress Report to 30 April 1949, 25 pp. (mimeographed), numerous figs. (photo+ozalid). 1 May 1949.

instrumentation, buoys

Seelinger, P.E., R.A. Wallston, B.H. Erickson, J.E. Masterson and W.E. Hoehne, 1967.
An oceanographic data collection system.
Trans. 2nd Int. Buoy Techn. Symp., 311-333.

instrumentation

Sharukov, P., 1959
[Determining the velocity and direction of currents by anchored spar buoys.]
Trudy, Morskoi Gidrofiz. Inst., Akad. Nauk, SSSR, 15: 80-

instrumentation, buoy platforms

Shonting, D.H. and A.H. Barrett, 1971.
A stable spar-buoy platform for mounting instrumentation. J. mar. Res. 29(2): 191-196.

instrumentation, detection

Sosnovsky, I.A., 1961.
[Device for locating autonomous buoy stations.]
Trudy Inst. Okeanol., Akad. Nauk, SSSR, 47:15-28.

instrumentation, buoy stations

Sisoev, N.N., 1961.
[New self-recording apparatuses for anchored buoy stations.]
Okeanologiia, Akad. Nauk, SSSR, 1(6):1079-1082.

Instrumentation - buoys

Sisoev, N. N., 1958.

Autonomous buoy stations Mezhd. Geofiz. God. Inform. Biull. Mezhduvedomstb. Kom. po provedeniiu MGG, pri Prezid. ANCCCR (4): 48-55.

instrumentation, buoys

Snyder, Robert M., and Donald Darms 1967.
Non-linear response of buoy shapes.
Trans. 2nd int. buoy Techn. Symp. 149-160.

buoys

Stas, I. I., 1964.
On some new methods of oceanograppic observations. (In Russian).
Okeanologiia, Akad. Nauk, SSSR, 4(6):1096-1100.

instrumentation, buoys

Sukhovey, A.G., and V.S. Nazerov, 1966.
Transisterized radio buoy. (In Russian).
In: Metody i Pribory dlya Issledovaniya Fizicheskikh Protsessov v Okeane, A.N. Paramonov, editor, Izd-vo Naukova Dumka, Kiev, 168-172.
Translation: JPRS: 39,88, 13 Feb. 1967 (Clearinghouse Fed.Sci. Tech.Info., U.S. Dept. Commerce), 150-155 (multilithed).

instrumentation, buoys

USA, National Academy of Sciences-National Research Council, Committee on Oceanography, 1959.
Oceanography, 1960-1970. 7. Engineering needs for ocean exploration. 22 pp.

instrumentation, buoys

United States, Office of Naval Research, 1964.
The first ocean data station; air-sea data collection on world-wide basis foreseen in network of large sophisticated buoys.
Naval Res. Rev., Oct., 1964:1-3, 7.

instrumentation, buoy stations

Vasil'yev, A.S., and V.S. Nazerov, 1966.
Equipment for finding anchor buoy stations. (In Russian).
In: Metody i Pribory dlya Issledovaniya Fizicheskikh Protsessov v Okeane, A.N. Paramonov, editor, Izd-vo Naukova Dumka, Kiev, 163-167.
Translation: JPRS; 39,881, 13 Feb. 1967 (Clearinghouse Fed.Sci.Tech.Info.U.S.Dept.Commerce), 144-149 (multilithed).

instrumentation, buoys

Veeth, J. Gordon, 1966.
Weather, oceanographic buoys- should they be combined?
Undersea Techn., 7(12):17-30.

instrumentation, buoys (moored)

Vershinsky, N.V., and P.A. Borovikov, 1965.
On the calculation of stations with automatically variable depth. Electronic instruments for oceanographic investigations. (In Russian; English abstract).
Trudy Inst. Okeanol., Akad. Nauk, SSSR, 74:85-89.

drift buoys

Vize, V.I., 1945.
[The drift buoys in arctic seas.] Problemy Arktiki 1944(2):122, tables.

instrumentation, buoys

Volkov, V. G. 1959
Acoustic apparatus for sending the buoy to the surface. (In Russian).
Trudy Inst. Okeanol. 35: 238-244.

instrumentation, buoys

Walden, Robert G. 1968.
A review of oceanographic and meteorological buoy capabilities and effectiveness.
Mar. Sci., Instrument, 4: 99-105

instrumentation, buoys

Zlotky, Richard A., 1962.
A concept for a remotely interrogated synoptic oceanographic data sampling buoy.
Marine Sciences Instrumentation, Inst. Soc., Amer., Plenum Press, N.Y., 1:80-87.

buoy stations

instrumentation, buoy station

Zjikov, I.D., 1963.
About the use of deep water anchor appliances for the buoy stations setting in the ocean. (In Russian).
Meteorol. i Gidrol., (9):51-52.

buoys, telemetering

instrumentation, buoys, telemetering

Devereux, Robert, 1964.
Development of a long-range telemetering buoy.
Undersea Techn., 5(7):28-31.

buoyancy

instrumentation buoyancy

Stechler, B.G., and G.J. Ponerbs 1968.
Parametric analysis of optimum buoyancy module designs with computer applications.
Ocean Engng 1(1): 17-37.

cable

instrumentation, cables

Anon., 1969.
Winches and cable for oceanography.
Undersea Techn. 10(9): 37, 52-54, 63.

INSTRUMENTATION, cables

Burchard, Stewart, 1968.
Deepsea cables.
Oceanology Int., 3(3):31-33.

instrumentation, cables

McLond, Kenneth, 1966.
Cables in the sea.
Oceanology intl, 4(4):23. (popular)

instrumentation, cables

Ovsyannikov, A.N., 1964.
About transference of the distant oceanographical electricity cable through the wave breakers zone. (In Russian).
Meteorol. i Gidrol., (2):54-55.

instruments

Reed, D. G., and T. Stewart, 1949.
A streamline cable depressor. J. Mar. Res. 8(3): 226-236, 9 textfigs.

instrumentation

Sisoev, N.N., 1959
[Cables, cable winches and the extent of cable submergence of the moving ship.]
Trudy Inst. Okeanol., 35: 86-91.

instruments

Sisoev, N.N., 1951.
[Wire cables for hydrological work.]
Trudy Inst. Okeanol., 5:152-168.

Instrumentation, cables

Stimson, P.B., 1965.
Synthetic-fiber deep-sea mooring cables: their life expectancy and susceptibility to biological attack.
Deep-Sea Res., 12(1):1-8.

instrumentation, cables

Thompson, J.C., and R.K. Logan, 1963.
Wire cables for oceanographic operations.
U.S. Navy Electronics Lab., USNEL Rept., 1199: 38 pp.

cable

Watson, E.E., 1953.
An experiment to determine the hydrodynamic forces on a cable inclined to the direction of flow. J. Mar. Res. 12(3):245-248, 1 textfig.

instrumentation

Wood, K.A., 1965.
An ocean platform and submarine cable system for the North Atlantic.
J. Inst. Navig., 18(4):437-445.

calibration

instrumentation, calibration

Stas' I.I., 1961
Experimental pressure chamber for calibration. Physics of the sea. (In Russian).
Trudy Morsk. Gidrofiz. Inst., 23:164-166.

Translation:
Scripta Technica, Inc., for Amer. Geophys. Union, pp. 131-132.

Calorimetry

instrumentation, calorimetry

Birkett, L., 1969.
Calorimetry: modification of a standard bomb for small heat outputs. J. Cons. perm. int. Explor. Mer, 32(3): 437-440.

instrumentation, cameras

Cramer, Ted, 1969.
Selecting an underwater camera.
Oceanology intl. 4(7): 30-32 (popular)

instrumentation, cartography

Boyle, A.R., 1971.
Automatic cartography, special problems of hydrographic charting. Int. hydrogr. Rev. 48(2): 85-92.

Cathodic protection

Instrumentation, cathodic protection

Morgan, J.H., 1965.
Instrumentation and automation in marine cathodic protection.
Trav. Centre de Recherches et d'Etudes Oceanogr. n.s., 6(1/4):251-253.

instrumentation, charting (bottom)

Thiel Hj., 1970.
Ein Fotoschlitten für biologische und geologische Kartierungen des Meeresbodens. Marine Biol., 7(3): 223-229.

Chemical

instrumentation, chemistry

instruments
Anderson, L. J., and R.R. Revelle, 1947.
Apparatus for rapid conductometric titrations.
Analyt. Chem. 19:264-268.

instruments
Baier, C.R., 1949.
Wasserschöpfer zu chemischen und bakteriologischen Probenahme aus tiefe und flachern Gewässern.
Arch. Hydrobiol. 42(3):356-364, 3 figs.

instrumentation
Barnes, H., 1959.
Apparatus and methods of oceanography. Part one: Chemical
G. Allen and Unwin Ltd., London: 341 pp.

instruemtns
Barnes, H., 1953.
A double syringe-pipette for dissolved oxygen estimations. The Analyst 78(929):501-503.

instrumentation, chemistry
Ben-Yaakov, S., and I.R. Kaplan, 1971
An oceanographic instrumentation system for in situ application.
J. mar. techn. Soc. 5(1):41-46

instrumentation, chemistry (gases)
Bieri, Rudolf H. 1965.
Thermetically-sealing seawater sampler.
J. Mar. Res., 23(1):33-38.

instruments
Briggs, R., G. Knowles and L.J. Scragg, 1954.
A continuous recorder for dissolved oxygen.
The Analyst 79(945):744-751.

instrumentation, chemical
Brouardel, Jean, et E. Rinck, 1958.
Appareillage employé en Méditerranée pour l'étude de la production de matière organique.
J. du Conseil, 24(1):10-15.

instrumentation, chemistry
Bruns, E., 1956.
Einige sowjetische Vorschläge neuer ozeanologischer Geräte für Strömungsmessungen, für die Entnahme von Boden- und Wasserproben und für hydrochemische Analysen.
Annalen f. Hydrogr., (5/6):43-54.

instrumentation, chemical
Carritt, D.E., 1963
5. Chemical instrumentation. In: The Sea, M.N. Hill, Edit., Vol. 2. The composition of sea water, Interscience Publishers, New York and London, 109-123.

instrumentation
Carritt, D.E., and J.W. Kanwisher, 1959
An electrode system for measuring dissolved oxygen. Anal. Chem. 31: 5-9.

bottles, plastic
Cox, R.A., 1954.
Water transmission of polythene bottles.
J. du Cons. 19(3):297-300.

instruments
Crinnell, J.T., and W.C.G. Wheeler, 1956.
Zinc anodes for use in sea water.
J. Appl. Chem., 6(10):415-420.

instruments (mass spectrometer)
Dansgaard, W., 1961
The isotopic composition of natural waters with special reference to the Greenland ice cap.
Medd. om Grønland, 165(2): 1-120.

instrumentation, pH
Disteche, Albert, 1964.
Nouvelle cellule à électrode de verre pour la mesure directe du pH aux grandes profondeurs sous-marines. Résultats obtenus au cours d'une plongée du bathyscaphe Archimède en Méditerranée.
Bull. Inst. Océanogr., Monaco, 64(1320):10 pp.

instruments
Feuer, I., 1955.
Simple micropipette. Nucleonics 13(3):68.

instruments
Ford, W.L., 1950.
Seagoing photoelectric colorimeter. Analyt. Chem. 22:1431-1435.

instrumentation, chemistry (oxygen)
Føyn, Ernst 1967.
Density and oxygen recording in water.
In: Chemical environment in the aquatic habitat, H.L. Golterman and R.S. Clymo, editors, Proc. I.B.P. Symp. Amsterdam Oct. 1966, 127-132.

Instrumentation, Chemistry, Oxygen
Føyn, Ernst., 1965.
The oxymeter.
Progress in Oceanography, 3:137-144.

instrumentation (chemistry)
Føyn, Ernst, 1960.
Et apparat for Kontinuerlig maling av oppløst oksyzen i sjøvann.
Tidsskrift for Kjemi, Bergvesen og Metallurgi, 5:109-110.

"oxymeter"

instruments
Føyn, E., 1955.
Continuous oxygen recording in sea water.
Repts. Norwegian Fish. Mar. Invest. 11(3):1-8, 4 textfigs.

instrumentation, chemistry (autoanalyzer)
Grasshoff, K. 1969.
Über ein Gerät zur gleichzeitigen Bestimmung von sechs chemischen Komponenten aus dem Meerwasser mit analoger und digitaler Ausgabe.
Ber. dt. wiss. Kommn Meeresforsch. (n.f.) 20(2):155-166.

instrumentation, chemistry, autoanalyzer
Grasshoff, Klaus, 1966.
Über automatische Methoden zur Bestimmung von Fluorid, gelöstem anorganischem Phosphat und Silikat in Meerwasser.
Kieler Meeresforsch., 22(1):42-46.

instrumentation, chemistry
Grasshoff, Klaus, 1964.
Uber ein neues Gerat zur Herstellung eines absoluten Standards für die Sauerstoffbestimmung nach der Winklermethode.
Kieler Meeresf., 20(2):143-147.

instrumentation, chemical (oxygen)
Grasshof, Klaus, 1963.
Untersuchungen über die Sauerstoffbestimmung im Meerwasser. 3. Über eine Bestimmung von Sauerstoff in Meerwasser mit der schnelltropfenden Quecksilberelektrode.
Kieler Meeresf., 19(1):8-15.

instrumentation, oxygen determination
Grasshoff, Klaus, 1962
Untersuchungen über die Sauerstoffbestimmung im Meerwasser, 2.
Kieler Meeresforschungen, 18(2):151-160.

instrumentation, chemistry
Holmes, R.W., and R.J. Linn, 1961.
A modified Beckman DU spectrophotometer for seagoing use.
U.S.F.W.S. Spec. Sci. Rept. Fish., No. 382:1-4.

instruments
Ishibashi, M., M. Shinagawa, and T. Suzube, 1943
[Chemical research concerning the ocean. XVIII.]
Umi to Sora 23:317-337.

polarograph to detect minute quantities of iron, copper, etc., and oxygen.

instrumentation, pH
Kabanov, V.V., 1962
On the application of the pH-meter LP-57 M for the determination of the active reaction in sea water under vessel conditions aboard R/V "Mikhail Lomonosov". (In Russian).
Okeanologiia, Akad. Nauk, SSSR, 2(6):1085-1092.

instrumentation, chemical
Kanwisher, John W., 1962.
Oxygen and carbon dioxide instrumentation.
Marine Sciences Instrumentation, Instr. Soc. Amer., Plenum Press, N.Y., 1:334-339.

instrumentation
Kanwisher, John, 1959
Polarographic oxygen electrode. Limn. & Ocean. 4(2): 210-217.

apparatus
Korshun, M.A., and N.E. Gel'man, 1946.
[Apparatus for direct microdetermination of oxygen.] Zavofskaia Laboratoriia 12(4-5):500-502.
RT-1020 Bibl. Transl. Rus. Sci. Tech. Lit. 7.

instrumentation, chemistry, trace gases
Kühme, Heinrich 1971.
Eine Apparatur zur Bestimmung atmosphärischer Vertikalprofile von Spurengaskonzentrationen.
Meteor. -Forsch.-Ergebnisse (B) 7: 78-83

instruments
Lincoln, J.H., R.G. Paquette and M. Rattray Jr., 1955.
Microsalinometer for oceanographic model studies
Trans. Amer. Geophys. Union, 36(3):406-412.

instrumentation, oxygen
Linskens, H.F., 1963.
Anwendung des Sauerstoff-Lot nach Tödt im Salzwasser.
Netherlands J. Sea Res., 2(1):77-84.

gear
Lundbak, A., 1949.
A new burette for small volumes. Proc. Verb., Assoc. d'Océan. Phys., Union Géodés. et Géophys. Internat. 4:144-145.

instrumentation, chemistry (oxygen)
McLeod, G.C., F. Bobblis, Jr., and C.S. Yentsch, 1965.
A graphite electrode system for measuring dissolved oxygen.
Limnology and Oceanography, 10(1):146-149.

instrumentation, chemistry
Noble, Vincent E., and John C. Ayers, 1962.
A portable photocell fluorometer for dilution measurements in natural waters.
Limnol. and Oceanogr., 6(4):457-461.

instrumentation chemistry, alkalinity
Park, Kilho, Malcolm Oliphant and Harry Freund, 1963.
Conductometric determination of alkalinity of sea water.
Analytical Chemistry, 35:1549-1550.

instrumentation, chemical
Proctor, Charles M., 1959.
Precise electronic titration instrument (chlorinity titration).
Analyt. Chem., 31:1278.

Also in:
Contrib. Oceanogr. & Meteorol., A. & M. College of Texas, 5:No. 142.

instruments
Roberts, C.H., 1950.
A photo-electric comparator for colorimetric analysis at sea. J. du Cons. 17(1):17-24, 3 textfigs.

instrumentation, chemistry
Rochford, D.J., 1963
SCOR UNESCO chemical intercalibration tests. Results of 2nd series, R.S. Vityaz August 2-9. 1962, Australia.
Int. Indian Ocean Exped., Cronulla, 30 pp. (multilithed).

instruments
Rochford, D.J., 1954.
A chainomatic pycnometer for the determination of chlorine in sea water. Nature 174(4431):641-642.

instruments
Scholander, P.F., L. Van Dam, C. Lloyd Claff and J.W. Kanwisher, 1955.
Micro gasometric determination of dissolved oxygen and nitrogen. Biol. Bull. 109(2):328-334.

instrumentation, chemistry
Scott, P.G.W., and T.A. Strivens, 1962.
An automatic coulometric-titration assembly.
Analyst, 87(1034):356-361.

autopipette
Scripps Institution of Oceanography, 1949
Marine Life Research Program. Progress Report to 30 April 1949, 25 pp. (mimeographed), numerous figs. (photo.r ozalid). 1 May 1949.

instrumentation, chemistry
Sims, M.H., 1968.
Titanium Tubes now fight sea water.
Power, Chicago, 112(9):63-65. (not seen).

instrumentation, chemical
Sinukov, V.V., 1962
[An attempt to use on board the "M. Lomonossov" An electrophotocolorimeter F.E.K.-M. to determine nitrites, silicates and phosphates in sea water.]
Trudy Inst. Gidrofiz. Morsk., 25:130-141.

instrumentation, chemistry at sea
Sinjukov, V.V., and B.A. Skopintsev, 1963
The application of a new photoelectrocolorimeter model for the estimation of biogenic elements [nutrients] in sea water under field work conditions. (In Russian).
Okeanologiia, Akad. Nauk, SSSR, 3(1):127-136.

instrumentation
Snodgrass, J.M., D.E. Carritt, and W.S. Wooster, 1953.
Automatic servo-operated filterphotometer.
Analyt. Chem., 26(1):249-250.

instrumentation, chemistry (oxygen)
*Solovyev, L.G., 1967.
Some examples and peculiarities of oxygen determinations in sea water with the aid of the oxmeter IOAN. (In Russian; English abstract)
Trudy Inst. Okeanol., 83:63-67.

instrumentation, chemistry (oxygen)
*Solovyev, L.G., 1967.
A ship oxymeter of high sensitivity. (In Russian; English abstract).
Trudy Inst. Okeanol., 83(177-181.

instrumentation, chemistry (oxygen)
Solovyiev, L.G., 1964.
An apparatus for the estimation of the oxygen content in sea water. (In Russian).
Okeanologiia, Akad. Nauk, SSSR, 4(1):149-155.

instrumentation, chemical
Stock, J.T., 1962.
A simple automatic potentiometric titrator incorporating end-point anticipation and delayed termination.
Analyst, 87(1040):908-909.

instrumentation, chemical
Trotti, L., and D. Sacks, 1961.
A photometric modification of the Winkler method.
Rapp. Proc. Verb., Réunions, Comm. Int. Expl. Sci., Mer Méditerranée, Monaco, 16(3):673.

instrumentation, chemistry, oxygen
VanLandingham, John W., and Melbone W. Greene 1971.
An in situ molecular oxygen profiler: a quantitative evaluation of performance.
J. mar. techn. Soc. 5 (4): 11-23

instruments
Walker, C.R., 1955.
A modification of the Kemmerer water bottle for sampling shallow waters. Progr. Fish. Cult. 17(1):41.

instrumentation, chemistry
Weichart, Günter, 1963.
Apparatur zur kontinuierlichen quantitativen Analyse von Meerwasser.
Deutsche Hydrogr. Zeits., 16(6):272-281.

instrumentation, chemistry
*Weiss, R.F., 1968.
Piggyback sampler for dissolved gas studies on sealed water samples.
Deep-Sea Res., 15(6):695-699.

instrumentation, chemistry, gases
Werner, A.E., and M. Waldichuk, 1967.
A sampler for gases in bottom sediments.
Limnol. Oceanogr., 12(1):158-161.

instrumentation, chemistry, potentiometric sensor
Whitfield, M., 1971.
A compact potentiometric sensor of novel design. In situ determination of pH, pS^{2-} and Eh. Limnol. Oceanogr. 16(5): 829-837.

instrumenation, Eh
*Whitfield, M., 1969.
Eh as an operational parameter in estuarine studies.
Limnol. Oceanogr., 14(4):547-558.

gear
Wooster, W.S., D.E. Carritt, J.D. Isaacs, 1951.
An automatic reagent dispenser for shipboard use J. Mar. Res. 10(2):194-196 2 textfigs.

instrumentation, chemistry
Yasui, Z., 1938
[A test on the water-tightness and the cleanliness of a Nansen bottle.] Umi to Sora, 18(3): 98-101.

instrumentation, chemical
Zhavoronkina, V.K., 1959.
New polarograph of the Czechoslovakian Akademy of Sciences. (In Russian) Hydrometeorology. Hydrochemistry.
Trudy Morsk. Gidrofiz. Inst., 16:161-166.

Clamps

instrumentation, clamps
Lukin, V.P., 1964
Capler's clamp. (In Russian).
Materiali, Ribochoz. Issled. Severn. Basseina, Poliarn. Nauchono-Issled. i Proektn. Inst. Morsk. Rib. Choz. i Okeanogr. im. N.N. Knipovicha, PINRO, Gosud. Proizvodst. Komm. Ribn. Choz. SSSR. 4:115-117.

instrumentation, clamps
Renshaw, R. Ward, and William G. Pearcy, 1964.
A new swivel cable clamp for towing large plankton nets at different depths.
Deep-Sea Res., 11(6):933-934.

coasts

instrumentation, coasts
Kestner, A.P., 1957
[Utilization of------tensometric---- for the registration of wave parameters for work in the surf zone. Material subdivisions for the study of sea coasts and reservoirs.]
Trudy Okeanogr., Komissii Akad. Nauk, SSSR, 2:126-128.

instrumentation, coasts
Kestner, A.P., 1956.
[Work with electrical devices in the coastal zone. Studies on sea coasts and reservoirs.]
Trudy Okeanogr. Komissii, Akad. Nauk, SSSR, 1: 134-140.

instrumentation, coastal
*Koontz, W.A., and D.L. Inman, 1967.
A multi-purpose data acquisition system for instrumenation of the nearshore environment.
Tech. Memo., U.S. Army Coast Engny Res. Center, 21: 1-38.

Color

instrumentation, color
Wolf, G., 1961
Farbmessungen nach dem Spektralverfahren mit dem Pulfisch-Photometer.
Beitr. Meereskunde, 2/3: 26-32.

Communications

instrumentation
Anonymous, 1960.
New VIF to reach all Polaris subs. Worldwide underwater range. Electronics, 33(29):34-35.

instrumentation, communications
Snodgrass, James M., 1968
Instrumentation and communications
In: Ocean engineering: goals, environment technology, John F. Brahtz, editor, John Wiley and Sons, 393-477.

Compasses

instrumentation, compasses
Christoph, Peter, 1969.
Deutung von Eigenfrequenzen in Zweikreiselkompasssystemen.
Dt. hydrogr. Z. 22(2): 49-56.

instrumentation, compasses (magnetic)

Sasaki, Yukiyasu and Tsutomu Isouchi, 1968. Usable limitation of magnetic compass near south magnetic pole. J. Tokyo Univ. Fish., 9(1): 45-47.

instrumentation compasses

Uhlig, Hans-Rüdiger 1968. Krängungsfehler beim Kompensierten Magnetkompass. Dt. hydrogr. Z. 20(4): 168-176.

computers: SEE ALSO METHODS computers

instrumentation, computers

Abbott, J. Lynn, G. Scott Morris, Jr. and John D. Mudie, 1969. Scripps' seagoing computer centers. Trans. Applications of Sea Going Computers Symposium, Mar. Tech. Soc., 187-200.

instrumentation, computers

Anon, 1968. Marine computer uses diversified. UnderSea Techn., 9(10):30-33.

instrumentation, computers

Antonov, V.I., V.I.Babiy and V.K. Kupriyanov, 1966. System for recording and reproducing information on magnetic tape by the method of frequency modulation. (In Russian). In: Metody i Pribory dlya Issledovaniya Fizicheskikh Protsessov v Okeane, A.N. Paramonov, editor, Izd-vo Naukova Dumka, Kiev, 192-198. Translation: JPRS:39,881, 13 Feb. 1967 (Clearinghouse Fed.Sci.Tech.Info., U.S.Dept Commerce), 177-184 (multilithed).

instrumentation, computers

Belshe, J.C., 1969. Sea-going computers in range instrumentation. Trans. Applications of Sea Going Computers Symposium, Mar. Tech. Soc., 337-343.

instrumentation, computers

Bennett, A.S., 1969. Computer data display. Trans. Applications of Sea Going Computers Symposium, Mar. Tech. Soc., 99-103.

instrumentation, computer

Bennett, A.S., C.S. Mason and E.A. Bendell, 1969. The Bedford Institute shipboard data logging system. Trans. Applications of Sea Going Computers Symposium, Mar. Tech. Soc., 63-69.

instrumentation, computers

Bogdanov, K.T., and V.A. Magarik, 1963. Computation of the S2 tidal component for the Pacific water area by means of the BESM-2 electronic computer (In Russian). Doklady, Akad. Nauk, SSSR, 151(6):1315-1318.

instrumentation, computers

Born, Roger L., 1969. WHEO-701: USCG oceanographic research ship. Rationale and design concept. Trans. Applicatic of Sea Going Computers Symposium, Mar. Tech. Soc 311-335.

Instrumentation computers

Bown, Carl O., 1969. Experience with a sea-going computer system: lessons, recommendations and predictions. Trans. Applications of Sea Going Computers Symposium, Mar. Tech. Soc. 141-157

instrumentation, computers

Boyd, Carl M. and Glen W. Johnson, 1969. Studying zooplankton populations with an electronic zooplankton counting device and the linc-8 computer. Trans. Applications of Sea Going Computers Symposium, Mar. Tech. Soc., 83-90.

instrumentation, computers

Branhan, Don W. 1968. The shipboard computer's place in oceanography. Mar. Sci. Instrument. 4: 600-606.

instruments, computers

Bruch, Walter F., Leslie C. Merrill and Duane E. Maddux 1969. Deep ocean underwater digital data acquisition systems. J. Mar. techn. Soc., 3(2): 81-85

instrumentation, computers

Brock, Fred V., and Daniel J. Provine, 1962. A standard deviation computer. J. Appl. Meteorol., 1(1):81-90.

computers, use of

Burkov, V.A., 1963 On the uniform pattern of deep-sea recordings of hydrologic observations for the systemization and the machine treatment of the latter. (In Russian). Okeanologiia, Akad. Nauk, SSSR, 3(2):338-345.

instrumentation, computers

Capart, Jean-Jacques, 1965. Transmission digitale de mesures oceanographiques NATO Subcommittee Oceanogr. Res., Techn. Rept., ("Ship Buoys Project"), 13:15 pp. (Multilithed).

instrumentation, computers

Churgin, James, 1969. A new computer at NODC-- its impact on data management. Trans. Applications of Sea Going Computers Symposium, Mar. Tech. Soc., 357-361.

instrumentation, computers

Clark E.E., and B. Matthews, 1970. Experience with shipborne computer systems for research and survey applications. Electronic Engineering in Ocean Technology, Swansea, 21-24 Sept. 1970, I.E.R.E. Conf. Proc. 19: 211-238.

instrumentation computers

Coughran, Edward H., 1969. Shipboard computers. Oceanology intl, 4(4):28-29. (popular)

instrumentation, computers

Coughran, Edward H., 1969. Scope of current usage of shipboard computers. Trans. Applications of Sea Going Computers Symposium, Mar. Tech. Soc., 5-9.

instrumentation, computers

Crease, J., 1971. Experience with a computer in oceanographic research at sea. J. Inst. Navig. Lond. 24 (3): 294-299.

instrumentation, computers

de la Morinière, T.C., 1969. A data activities and data system concept. Trans. Applications of Sea Going Computers Symposium, Mar. Tech. Soc., 285-310.

instrumentation, computers

Edwards, M.G., and M.J. McCann, 1970. A portable computer system for oceanographic and acoustic research. Electronic Engineering in Ocean Technology, Swansea, 21-24 Sept. 1970, I.E.R.E. Conf. Proc. 19: 251-258.

instrumentation computers

Ewart, Terry E., 1969. Design philosophy and operational experiences of a seagoing computer and data acquisition system. Trans. Applications of Sea Going Computers Symposium, Mar. Tech. Soc., 159-174.

instrumenation, data acquisition systems

*Ewing, Gifford C., 1969. On the design efficiency of rapid oceanographic data acquisition systems. Deep-Sea Res., Suppl. 16: 35-44.

Instrumentation, computers

Fein, Kenneth J. and David G. William, 1969. Development and use of portable digital computer systems for submarine experiments. Trans. Applications of Sea Going Computers Symposium, Mar. Tech. Soc., 209-222.

instrumentation, computer

Flittner, Glenn A., 1969. Computers are better on the shore. Trans. Applications of Sea Going Computers Symposium, Mar. Tech. Soc., 11-16.

computers, use of

Foster, J.J., 1968. Using computers on the samller research vessel. Hydrospace, 1(4):10-15.

computers

Fredkin, Edward, and Malcolm Bivar, 1964. Oceanographic data display system. (Abstract). Trans. Amer. Geophys. Union, 45(1):68.

instrumentation, computers

Hansen, Palle G., 1969. A sea going UNIVAC 1218 interfaced with NUWC'S thermistor chain. Trans. Applications of Sea Going Computers Symposium, Mar. Tech. Soc., 105-114.

instrumentation computers

Hasse, R.W. and R.L. Martin, 1969. Real-time processing of acoustic and oceanographic data at sea. Trans. Applications of Sea Going Computers Symposium, Mar. Tech. Soc., 115-130.

instrumentation, computer

Holcombe, Troy L., William B.F. Ryan and Edwin E. Westerfield, 1969. An experiment in computer processing of oceanographic data aboard the USNS Kane. Trans. Applications of Sea Going Computers Symposium, Mar. Tech. Soc., 45-61.

instrumentation, computers

Ignat'ev, M.B., F.M. Kulakov, A.M. Pokrovsky and V.S. Yastrebov, 1970.
On the problem of developing a deep-sea automatic manipulator. (In Russian; English abstrac
Okeanologiia, 10(6): 1090-1100.

instrumentation, computers
Ishihara, Tojiro, Shoitiro Hayami and Shigenori Hayashi, 1954
On the elctronic analog computer for flood routing.
Proc. Jap. Acad. 30(9):891-895.

Also in:
Papers on Oceanogr. and Hydrol., Geophys. Inst., Kyoto Univ., (1949_1962).

instrumentation, computers
Jackson, Charles B., 1967.
Computers at sea.
Oceanol. int., 2(5):28-31. (popular)

Instrumentation computers

Jones, Edward E., 1969.
Computer data acquisition systems aboard ESSA survey ships. Trans. Applications of Sea Going Computers Symposium, Mar. Tech. Soc., 201-208.

computers, use of
Kamenskaya, O.A., 1964 (deceased)
About the use of electronic computer at oceanological computation. (In Russian).
Materiali Vtoroi Konferentsii, Vzaimod. Atmosfer. i Gidrosfer. v Severn. Atlant. Okean. Mezhd. Geofiz. God, Leningrad. Gidrometeorol. Inst., 161-165.

instrumentation Computer

Kelley, James C. and C. Stephen Smyth, 1969.
Reduction and presentation of acoustic reflection data with the aid of a shipboard computer. Trans. Applications of Sea Going Computers Symposium, Mar. Tech. Soc., 35-44.

instrumentation, computers
Khokholov, A.V., 1966.
Device for visual readout of binary data. (In Russian).
In: Metody i Pribory dlya Issledovaniya Fizicheskikh Protsessov v Okeane, A.N. Paramonov, editor, Izd-vo Naukova Dumka, Kiev, 181-186.
Translation: JPRS:39,881, 13 Feb. 1967 (Clearinghouse Fed.Sci. Tech.Info., U.S.Dept.Commerce), 164-170 (multilithed).

instrumentation, computors
Knighting, E., 1964.
Handling meteorological data by electronic computor.
Mar. Obs., 34(204):83-86.

Instrumentation, computers

Loncarevic, B.D., 1969.
Buoy plot as a survey aid: a data acquisition application. Trans. Applications of SeaGoing Computers Symposium, Mar. Tech. Soc., 27-33.

instrumentation, computers
Long, Fred S., 1969.
Redundancy gives Researcher computer system high reliability.
UnderSea Technol. 10(8):35-37, 40

instrumentation, computers

Lowenstein, Carl D., 1969.
Navigation and data logging for the MPL deep tow. Trans. Applications of Sea Going Computers Symposium, Mar. Tech. Soc., 91-98.

instrumentation, computers

Mabey, D.J., 1969.

An automatic data logging and computing system for the hydrographer of the Royal Navy. Trans. Applications of Sea Going Computers Symposium, Mar. Tech. Soc., 175-186.

instrumentation, computers
MacDonald, Frank H., 1969.
Bottom loss computer. Trans. Applications of Sea Going Computers Symposium, Mar. Tech. Soc., 229-241.

instrumentation, computers
Mason, C.S. 1966.
A geophysical data logging system for shipboard use.
J. Ocean Techn. 1(1):35-44.

instrumentation, computers
McLennan, Miles W., 1969.
Use of a land-based computer in a real-time underwater acoustic tracking and noise data reduction system. Trans. Applications of Sea Going Computer Symposium, Mar. Tech. Soc., 275-284.

instrumentation, computors
Metzler, A.R., 1963.
Untended digital data acquisition system philosophy.
Mar. Sci. Instrument., Plenum Press, 2:87-90.

instrumentation, computers
Morenoff, Edward, 1971.
Marine information systems: what we have, what we need.
Oceanol. int. 6(9): 34-36.

instrumentation, computers

Neidell, Norman S., 1969.
Reflection seismic data processing at sea? Trans. Applications of Sea Going Computers Symposium, Mar. Tech. Soc., 259-273.

instrumentation, computers
O'Hagan, Robert M., 1968.
Digital computer application in the marine sciences.
UnderSea Techn., 9(10):35-37.

instrumentation, computers

Riffenburgh, Robert H., 1969
Some considerations of real-time data analysis of uncontrolled phenomena. Trans. Applications of Sea Going Computers Symposium, Mar. Tech. Soc., 243-258.

instrumentation, computors
Rogozin, A.A., 1963.
The experience of use of the electronic calculator for the determination of elements of the waves by the stereophotogrammetrical method. (In Russian).
Meteorol. i Gidrol., (9):42-45.

instrumentation, computors
Shekhtman, A.N., 1965.
The use of mechanical computors for ship-board treatment of hydrometeorological data. (In Russian).
Okeanologiia, Akad. Nauk, SSSR, 5(6):1110-1112.

computers
Shilling, Charles W., and Joe W. Tyson, 1963.
Is aquatic biology ready for the computor age. (Special Issue on Marine Biology), AIBS Bull., 13(5):44-45.

instrumentation, computers
Siler, William Jack Lubowsky, Stuart A. Kahan, and Phyllis H. Cahn, 1969.
A hybrid sea-going computer for bioacoustical studies. Trans. Applications of Sea Going Computers Symposium, mar Tech. Soc., 223-227.

instrumentation, computors
Smith, P.F., and E. Fredkin, 1963.
Computors and transducers.
Mar. Sci. Instrument., Plenum Press, 2:81-86.

instrumentation, computers

Stockton, Thomas R., 1969.
A digital computer-controlled dynamic ship positioning system. Trans. Applications of Sea Going Computers Symposium, Mar. Tech. Soc., 71-81.

instrumentation, computers
Sweers, H.E., 1967.
Some Fortran II programs for computer processing of oceanographic observations.
NATO Subcomm. Oceanogr. Res., Tech. Rept. 37 (Irminger Sea Project):104 pp. (mimeographed).

instrumentation, computers

Talwani, Manik, James Dorman and Robert Kittredge, 1969.
Experiences with computers aboard research vessels Vema and Robert D. Conrad. Trans. Applications of Sea Going Computers Symposium, Mar. Tech. Soc., 17-25.

computers

Tsiklauri, I.D., and L.I. Boris, 1961

[Computation of tidal phenomena with the help of the electronic computer (Ural-1).]
Issled. Severnoi Chasti Atlanticheskogo Okeana, Mezhd. Geofiz. God, Leningradskii Gidrometeorol. Inst., 1: 157-172.

instrumentation, computers

Wells, D.E., 1969.
The automatic real-time plotting of ship's tracks. Trans. Applications of Sea Going Computers Symposium, Mar. Tech. Soc., 131-139.

Concrete hulls

instrumentation, concrete hulls
Stachiw, J.D. 1967.
Concrete pressure hulls for ocean floor installations.
J. Ocean Techn. 1(2):19-28.

conductivity

instrumentation, conductivity
Aagaard, E.E., and R.H. van Haagen, 1963.
A probe type induction conductivity cell.
Mar. Sci. Instrument., Plenum Press, 2:11-17.

instrumentation, conductivity

Dauphine, T.M., 1968.
In situ conductivity measurements using low-frequency square-wave a.c.
Mar. Sci. Instrument., 4:555-562.

instrumentation, conductivity
Khristoforov, G.N., and G. Yu. Aretinskiy, 1966.
Device for measuring turbulent fluctuations of conductivity and temperature of sea water. (In Russian).
In: Metody i Pribory dlya Issledovaniya Fizicheskikh Protsessov v Okeane, A.N. Paramonov, editor, Izd-vo Naukova Dumka, Kiev, 208-214.
Translation: JPRSL39,881, 13 Feb.1967 (Clearinghouse Fed.Sci.Tech.Info., U.S.Dept. Commerce), 192-201 (mutlilithed).

instrumentation, conductivity
Skinner, D.D., 1963.
An in situ conductivity meter.
Mar. Sci. Instrument., Plenum Press, 2:25-28.

Convection

instrumentation
Nakagawa, Y., 1957.
Apparatus for studying convection under the simultaneous action of a magnetic field and rotation. Rev. Sci. Instr., 28(8):603-609.

instrumentation, correlator
Wolf, Alfred A., and J.H. Dietz, 1962.
An adaptive correlator for underwater measurements.
Marine Sciences Instrumentation, Instr. Soc., Amer., Plenum Press, N.Y., 1:347-354.

Cosmic rays

instrumentation, cosmic rays
Vavilov, Yu.N., B.A. Nelepo, G.I. Pugacheva, and V.M. Fedorov 1966.
Apparatus for measuring the intensity of cosmic rays at great depths. (In Russian).
Metody i Pribory dlya Issledovaniya Fizicheskikh Protsessov v Okeane, Morsk Gidrofiz. Inst. Kiev, 36: 31-36.

instrumentation, crane
Tolbert, W.H., R.H. Payne and G.G. Salsman, 1964.
An underwater c rane.
Limnology and Oceanography, 9(1):150-151.

Currents

A

instrumentation, currents
Aliverti, G., 1961.
Considérations sur les mesures directes de courants marins par la méthode de Swallow et sur dispositif de recuperation des "pingers".
Rapp. Proc. Verb., Réunions, Comm. Int. Expl. Sci., Mer Méditerranée, Monaco, 16(3):593.
See:
Consiglio Naz. Ricerche, Roma, Comm. Nat. Ital. Geofiz. Int., Publ. No. 1:
for complete paper

instrumentation, currents
Altman, E.N., 1962
The sea research with the application of the new oceanography instruments. (In Russian).
Meteorol. i Gidrol., (12):33-38.

instrumentation, currents
Anon., 1969.
An intercomparison of some current meters: report on an experiment at W.H.O.I. Mooring Site "D", 16-24 July 1967.
UNESCO Tech. Techn. Pap., mar. Sci. 11: 70 pp. (mimeographed)

instrumentation, currents
Anon., 1967.
Measuring waves and currents.
Ocean Industr., 2(11):18-19; 53.

instruments
Anon., 1953.
The ONO self-recording current meter.
Int. Hydrogr. Rev. 30(2):177-178.

instrumentation, currents
Artemenko, N.P., 1963.
Damping of oscillations of the magnetic system of BPV-2 current meters. (In Russian.)(Not seen).
Trudy, Arktich. i Antarktich. Nauchno-Issled. Inst., No. 254:40-42.

Abstract in:
Soviet Bloc Res., Geophys., Astron., Space, No. 94:45.

instrumentation, currents
*Augarde, Jacques, 1967.
Le clydonomètre, appareil de mesure et d'enregistrement des pressions et des courants pour l'étude des conditions hydrodynamiques superficielles en milieu marin.
C.r. hebd. Séanc., Acad. Sci., Paris, (D)265(18).

instrumentation, currents
Avery, Don E. 1968.
An integrating current meter.
Deep-Sea Res. 15(2): 235-236.

B

instrumentation, current
Banchero, Louis A., 1967.
Deep ocean mooring of temperature and current measuring systems.
Trans. 2nd Int. Buoy Techn. Symp. 37-5

instrumentation, currents
Bartoli, F., G. Sandrelli and G. Sorrentino, 1970.
Instrument to measure weak sea currents.
Rev. intern. Océanogr. méd. 17: 109-123

instrumentation, current
Beliakov, L.N., 1961.
The influence of swell on readings of current meters of different types. Methods for hydrological observations for stations aboard ship. Methods for oceanological investigations, a collection of papers. (In Russian).
Trudy Arktich. i Antarktich. Nauchno-Issled. Inst., 210:91-93.

electromagnetic barographs
Benioff, H., and B. Gutenberg, 1939.
Waves and currents recorded by electromagnetic barographs. Bull. Amer. Met. Soc., 20:422-426.

instrumentation, currents
Beyer, F., E. Føyn, J.T. Ruud and E. Totland, 1967.
Stratified currents measured in the Oslofjord by means of a new, continuous depth-current recorder, the bathyrheograph.
J. Cons. perm. int. Explor. Mer, 31(1):5-26.

instruments
Black, R.P., 1954.
The Black current meter with "S" rotor.
Ass. Int. Hydr. Sci., Assem. Gen., Rome, 3:304-310.

instrumentation, currents
Bogdanova, A.K., 1959
[Methodics of near-shore current observations for hydro-meteorological coast stations.]
Biull. Okeanogr. Komissii, Akad. Nauk, SSSR, (3): 57-68.

instruments, currents
Böhl, D., 1961
Untersuchungen zur Strömungsmessung mit Fix-und Schleppelektroden.
Beitr. Meeres., 1:48-55.

instruments
Böhnecke, G., 1937
III. Bericht über die Stromessungen auf der Ankerstation 369. Bericht über die erste Teilfahrt der Deutschen Nordatlantischen Expedition des Forschungs-und Vermessungsschiffes "Meteor" Februar bis Mai 1937. Ann. Hydr. u. mar. Meteorol., 1937, 14-16, 3 text figs.

"water bottles"
Bougis, P., and M. Ruivo, 1953.
Un nouveau type de flotteur en matière plastique pour l'étude des courants de surface.
Vie et Milieu, Bull. Lab. Arago, 4(2):171-176, 3 textfigs.

instrumentation
Boyar, H.C., and F.E. Schueler, 1960
A photoelectric current meter. USFWS Spec. Sci. Rept., Fish. No. 330: 7 pp.

instrumentation, currents
Bruns, E., 1956.
Einige sowjetische Vorschläge neuer ozeanologischer Geräte für Strömungsmessungen, für die Entnahme von Boden- und Wasserproben und für hydrochemische Analysen.
Annalen f. Hydrogr., (5/6):43-54.

instrumentation
Burkov, V.A., and K. M. Gobdanov, 1957
[Short-term floating stations for observations upon the currents.] Met. i Gidrol. (10): 37.

C

instrumentation, currents
Canney, Heyward, Jr., 1966.
A buoy-satellite system for current measurement.
Undersea Technology, 5(3):32-35.

instrumentation, current meters
Cannon, Glenn A., 1969.
Observations of motion at intermediate and large scales in a coastal plain estuary.
Techn. Rept. Chesapeake Bay Inst., 52 (Ref. 69-5): 114 pp. (multilithed) (unpublished manuscript)

instrumentation, currents
Cannon, G.A. and D.W. Pritchard, 1971.
A biaxial propeller current-meter system for fixed-mount applications. J. mar. Res. 29(2): 181-190.

instrumentation, currents
Carrothers, P.J.G., 1966.
New speed calibration curves for the Braincon Type 316 Histogram Current Meter.
J. Fish Res. Bd., Can., 23(11):1805-1806.

instrumentation, currents
Carruthers, James N., 1970.
The plastic seabed "oyster" for measuring bottom currents.
Cah. océanogr. 22 (9): 881-894

instrumentation, bottom currents
Carruthers, James N., 1969.
The plastic seabed "oyster" for measuring bottom currents.
FiskDir. Skr. Ser. HavUnders. 15(3): 163-171

instrumentation, currents
*Carruthers, J.N., 1968.
The Pooh-bah automatic float.
Cah. océanogr., 20(1):13-17.

instrumentation, currents
Carruthers, J.N., 1964
Various desiderata in current-measuring and a new instrument to meet some of them.
In: Studies in Oceanography dedicated to Professor Hidaka in commemoration of his sixtieth birthday, 296-301.

instrumentation, currents (deep)
Carruthers, J.N., 1962
Measurement of ocean bed currents. A new cheap method and a suggested plan.
Nature, 195(4845):976-981.

instrumentation, currents
Carruthers, J.N., 1959.
A simple device for observing bottom currents.
Int. Hydrogr. Rev., 36(1):167-169.

instrumentation (currents)
Carruthers, J. N., 1958.
A leaning-tube current indicator ("Pisa").
Bull. Inst. Oceanogr., Monaco, No. 1126:34 pp.

instrumentation (currents)
Carruthers, J. N., 1958.
Seine net fishermen can easily observe bottom currents for themselves.
Fishing News (2362):6-7.

instrumentation
Carruthers, J.N., 1956.
Ch. 24. Carruthers current meter and wire angle gage. Proc. First Conf. Coastal Eng. Instr., Berkeley, Calif., 31 Oct.-2 Nov., 1955:255-259.

instrumentation
Carruthers, J.N., 1954.
A new automatic current float. Water Sanit. Eng. 5(1):

instruments
Carruthers, J.N., 1954.
A new automatic current float. An improvised self-anchoring, self-signalling and self-timing drifting buoy suggested for use in estuarine and coastwise effluent dispersal investigations, etc. Int. Hydrogr. Rev. 31(2):155-162, 3 textfigs.

Reprinted from Water and Sanitary Eng. 5(1): (1954).

instrumentation
Carruthers, J.N., 1954.
A penetrometer for use on water covered beaches. J.M.B.A., u.k., n.s., 33:637-643, 2 textfigs.

gear
Carruthers, J.N., 1949.
A new method of continuous current-measuring from an unattended buoy. Proc. Verb., Assoc. Océan. Phys., Union Géodés. et Géophys. 4:130.

instruments
Carruthers, J.N., 1947
Realism in current-measuring in the upper layers of the sea. Int. Hydr. Rev. 24:3-13, 4 tables, 5 text figs.

vertical log
Carruthers, J.N., 1939.
First annual report on vertical log observations in the southern North Sea and eastern English Channel. Cons. Perm. Int. Expl. Mer, Rapp. Proc. Verb., 109(3):37-45.

instrumentation
Carruthers, J.N., and A.J. Woods, 1955.
Some simple oceanographical instruments to aid in certain forms of commercial fishing and in various problems of fisheries research.
F.A.O. Fish. Bull. 8(3):130-140, 8 textfigs.

instrumentation, currents
Castelbon, J.C., 1971.
Description d'une bouée assurant automatiquement la descente par paliers d'un courantographe.
Cah. océanogr. 23(2):135-144.

instrumentation, currents
Chalupnik, J.D., and P.S. Green, 1962.
A Doppler-shift ocean-current meter.
Marine Sciences Instrumentation, Instr. Soc. Amer., Plenum Press, N.Y., 1:194-199.

instrumentation (neutral buoyancy floats)
Cooper, L.H.N., 1961
Vertical and horizontal movements in the ocean. Oceanography, Amer. Assoc. Adv. Sci. Publ. No.67:599-621.

instruments
Crestani, G., 1931.
Floats used in Italy for the study of surface currents of the ocean. Hydrogr. Rev. 8:185-189.

D

gear
Dahl, O., and J.E. Fjeldstad, 1949.
A new repeating current-meter. Proc. Verb., Assoc. d'Océan. Phys., Union Géodés. et Géophys. Internat. 4:131-132.

instrumentations
Dahl, Odd, and J.E. Fjeldstad, 1949.
A new repeating current-meter. Assoc. Océanogr. Phys., Proc.-verb. 4:131-132.

instruments
da Silva, P. de C.M., 1952.
Nota sôbre un correntômetro de Pendulo utilizado pelo Chesapeake Bay Institute (Technical Report No. 1). Bol. Inst. Ocean., Univ. Sao Paulo 3(1/2) 119-124, 5 figs.

instruments
Dedow, H.R.A., 1955.
A miniature current meter and a free running torary flowmeter. First Conf. Coastal Eng. Instr. Berkeley, Calif., 31 Oct.-2 Nov., 1955:8 pp.

instrumentation
Dermody, J., 1960
A modification to the T.S. Ekman-Merz current meter.
Deep-Sea Res., 7(3): 218.

instrumentation, currents
Dextraze, Raymond E. 1968.
A three axis ducted impeller current meter system.
Mar. Sci. Instrument. 4: 671-674.

instrumentation, currents
Dietrich, G., and G. Siedler, 1963.
Ein neuer Dauerstrommesser.
Kieler Meeresf., 19(1):3-7.

instrumentation
Dmitrenko, O., 1959
[Use of cadmium non-polarizing electrodes in the study of natural electric currents in the sea.] Trudy Inst. Okeanol., 35: 102-112.

Instrumentation, currents
Dohler, G. C., 1961.
Current survey, St. Lawrence River, Montreal-Quebec, 1960. Canadian Hydrographic Survey, Surveys and Mapping Branch, Department of Mines and Technical Surveys, Ottawa, 58 pp. (multilithed).

instrumentation, currents
Donguy, J.R., 1962.
Quelques problemes d'equipment en océanographie physique.
In: Océanographie Géologique et Géophysique de la Méditerranée Occidentale, Centre National de la Mediterranee Occidentale, Centre National de la Recherche Scientifique, Villefranche sur Mer, 4 au 8 avril 1961, 51-58.

instrument
Doodson, A.T., 1940.
A current meter for measuring turbulence.
Int.Hydrogr. Rev. 17(1):79-100, 8 textfigs.

instrumentation, currents
Dooley, H.D., D.N. MacLennan and M.J.D. Mowat, 1971.
A computer tape translation for the Bergen and Plessey current meters. Deep-Sea Res., 18(5): 543-544.

instrumentation, current
Drever, Robert G., and Thomas Sanford, 1970.
A free-fall electromagnetic current meter - instrumentation.
Electronic Engineering in Ocean Technology, Swansea, 21-24 Sept. 1970, I.E.R.E. Conf. Proc. 19: 353-370.

instrumentation, currents
Düing, Walter, 1965.
Strömungsverhältnisse im Golf von Neapel.
Pubbl. Staz. Zool., Napoli, 34:256-316.

instruments
Dupin, P., 1955.
Un appareil pour la mesure de la vitesse des fluides.
C.R. Acad. Sci., Paris, 240(17):1687-1689.

instruments
Duroche, J., 1953.
Le courantographe BBT-Neypyric.
Bull. d'Info., C.C.O.E.C. 5(2):67-79, 5 photos.

instrumentation
Duroche, J., and J. Serpaud, 1958.
Le courantographe BBT-Neyrpic. Description - fonctionnement - utilisation. Trav. C.R.E.O., n.s., 3(1):9-15.

E

instrument
Ekman, G., 1913.
Zur Technik der Strommessung. Strommessung im offenen Meer in beliebiger Tiefe mittelst automatisch registriender Instrumenten von unter der Oberfläche verankerten Bojen.
Svenska Hydrogr.-Biol. Komm. Skr., 5:3-9.

currents meters
Ekman, V.W., 1953.
Studies on ocean currents. Results of a cruise on board the "Armauer Hansen" in 1930 under the leadership of Bjørn Helland-Hansen. Geograf. Pub. 19(1):106 pp. (text), 122 pp. (92 pls.).

instruments
Ekman, W.K., 1905.
Kurze Beschreibung eines propellstrommessers.
Publ. de Circ., Cons. Perm. Int. Expl. Mer, No. 24:1-4, 1 pl.

instrumentation currents
Ewing, Maurice, Dennis E. Hayes and E.M. Thorndike 1967.
Corehead camera for orientation of currents and core orientation.
Deep-Sea Res. 14(2): 253-258.

F

instrumentation-currents
Filimonov, A.I., 1965.
The use of Alekseev's printing current meter for measuring storm currents in coastal waters. (In Russian).
Okeanologiia, Akad. Nauk, SSSR, 5(6):1095-1099.

instrumentation, currents
Filippov, D.M., S.T. Mihailov, and V.G. Krivosheja, 1961.
[Use of aviation parachutes for deep-sea current measurements.]
Meteorol. i Gidrol., (5):42-43.

INSTRUMENTATION, CURRENTS
Foerster, John W. 1968.
A portable non-electrical current meter.
Chesapeake Sci. 9(1): 52-55.

instrumentation, currents (sea bed drifters)
Folger, D.W., 1971.
Nearshore tracking of seabed drifters.
Limnol. Oceanogr. 16(3): 588-589.

instrumentation, current meters

Forstner, Helmut and Klaus Rützler, 1969.
Two temperature-compensated thermistor current meters for use in marine ecology. J. mar. Res., 27(2): 263-271.

instrumentation, currents

Frassetto, R., 1967.
A neutrally-buoyant, continuously self-recording ocean current meter for use in compact, deep-moored systems.
Deep-Sea Research, 14(2):145-157.

instrumentation, currents

Fukuda, Masaaki, 1965.
The spherical current meter.
J. Oceanogr. Soc., Japan, 21(3):109-111.

G

instrumentation, currents

Gakkel, J.J., and L.P. Samsonia, 1961.
The first drifting radio buoys.
Okeanologiia, Akad. Nauk, SSSR, 1(4):691-700.

instrumentation, currents

Gaul, R.D., J.M. Snodgrass and D.J. Cretzler, 1963.
Some dynamical properties of the Savonius rotor current meter.
Mar. Sci. Instrument., Plenum Press, 2:115-125.

Instrumentation, currents

Gerard, Robert and Mark Salkind, 1965.
A note on the depth stability of deep parachute drogues.
Deep-Sea Res., 12(3):377-379.

instrumentation, currents

Germany, Deutsches Hydrographisches Institut, 1960.
Strombeobachtungen in der Deutschen Bucht in den Jahren 1956-1958.
Meereskundliche Beobachtungen und Ergebnisse, No. 15(2122/15): numerous unnumbered pp.

instrumentation(currents)

Glazunov, V.A., 1959
Velocity and direction registering element for the "Vitjaz-56" self-recorder.
Trudy Inst. Okeanol., 35: 182-189.

instrumentation

Gougenheim, A., 1958.
Le courantometre à électrodes remorquées.
Navigation 6(22):143-148.

instruments, flow meter

Great Britain, Department of Scientific and Industrial Research, 1955.
Hydraulics research, 1953, 50 pp.

H

instrumentation, currents

Hanff, Michel, 1970.
Static electromagnetic currentmeter.
Electronic Engineering in Ocean Technology, Swansea, 21-24 Sept. 1970, I.E.R.E. Conf. Proc. 19: 371-380.

instrumentation, currents

Helm, R., 1961
Drei neue Strömungsmessgeräte des Instituts für Meereskunde.
Beitr. Meereskunde, 2/3: 33-41.

instrumentation, currents

*Hodges, G.F., 1967.
The engineering for production of a recording current meter.
Int. hydrogr. Rev., 44(2):151-167.

instrumentation, currents

Horrer, Paul L., 1967.
Methods and devices for measuring currents.
In: Estuaries, G.H. Lauff, editor, Publs Am. Ass. Advmt Sci., 83:80-89.

instrumentation, currents

Hughes, P., 1962.
Towed electrodes in shallow water.
Geophys. J., R. Astron Soc., 7(1):111-124.

I

instrumentation

Iwasa, K., 1957.
An instrument for measuring directly the velocity, direction of the current and the temperature (the direct-reding current meter, Model CM-3).
Proc. UNESCO Symp., Phys. Ocean., Tokyo, 1955: 27-29.

J

instruments

Japan, Maritime Safety Board, 1955.
On an electro-magnetic induction current meter. (Preliminary report.) Publ. 981, Hydrogr. Bull., Spec. No. 17:22-27.

instrumentation, currents (parachute drogues)

Japan, National Committee for the International Geophysical Year 1960.
The results of the Japanese oceanographic project for the International Geophysical Year, 1957/8, 198pp.

instrumentation, currents

*Joseph, J., 1967.
Current measurements during the International Iceland-Faroe Ridge Expedition, 30 May to 18 June, 1960.
Rapp. P.-V. Réun. Cons. perm. int. Explor. Mer, 157:157-172.

instrumentation, current meters

Joseph, J., 1961.
Meereskundliche Messgeräte. FIAT Review of German Science, 1939-1946. Geophysics II: 192-202

Translation:
USNHO TRANS-82, M. Slessers
P.O. 20113 (1961)

K

instrumentation, currents

Karnaushenko, N.N., 1965.
A buoy with neutral buoyancy for measuring currents from large expedition ships. (In Russian).
Trudy Morsk. Gidrofiz. Inst., Akad. Nauk, SSSR, 31:105-111.

instrumentation, currents

Karpukhov, S., 1958.
On the possibility of using thermistors for measuring the speed of water current. (In Russian)
Byul. Stud. Nauchn. O-va, Leningrad Univ., No. 1: 82-84.

instrumentation, currents

*Karwowski, J., 1967.
On N.F. Kudriaсеv and E.G. Nikiforov's article "about the choice of effective constructions of the measurement instruments and their optimal parameters for the measurement of sea-currents in the undulating water layer".
Int. hydrogr. Rev., 44(2):41-72.

instruments

Kawakami, T., and Y. Iitake, 1955.
On a simple instrument, Siomi-ito, for detecting underwater current. Bull. Jap. Soc. Sci. Fish. 20(11):962-964, 1 textfig.

instrumentation

Kestner, A.P., 1961
Inductional gauge for measuring of hydrodynamical pressure.
Trudy Okeanogr. Komissii, Akad. Nauk, SSSR, 8: 179-185.

instrumentation, currents

Khobta, P.M., 1956.
Integral current meter. (In Russian).
Meteorol. i Gidrol., No. 12:49-50.

instrumentation, currents (surface)

Khundjua, G.G., A.G. Voshanyan, A.A. Pivovarov, and Yu.G. Pyrkin, 1967.
Methods of remote recording of sea current velocity modulus and direction in the sea surface layer. (In Russian; English abstract).
Okeanologiia, Akad. Nauk, SSSR, 7(1):177-181.

instrumentation, currents

Knauss, John A., 1963
14. Drogues and neutral-buoyant floats. In: The Sea, M.N. Hill, Edit., Vol. 2, (III) Currents, Interscience Publishers, New York and London, 303-305.

instrumentation, currents

Koczy, F.F., M. Kronengold and J.M. Loewenstein, 1963.
A doppler current meter.
Mar. Sci. Instrument., Plenum Press, 2:127-134.

instruments

Kolupaila, S., 1949.
Recent developments in current-meter design.
Trans. Am. Geophys. Union 30(6):916-918.

instrumentation, currents

Kudriavtsev, N.F., and E.G. Nikiforov, 1964.
On the selection of rational constructions of instruments and the estimation of their optimum parameters intended for current measurements in a layer enveloped by wave activity. (In Russian).
Okeanologiia, Akad. Nauk, SSSR, 4(2):479-487.

L

instrumentation, currents

Lacroix, Pierre, et Lucien Laubier, 1964.
Description, étalonnages et essais préliminaires d'un nouveau courantomètre enrigistreur.
Cahiers Océanogr., C.C.O.E.C., 16(9):725-753.

instrumentation (currents

Lacroix, P., and L. Laubier, 1961.
Un nouvel appareil enregistreur de courants et de mouvements de turbulence près du fond.
C. R. Acad. Sci., Paris, 252(15):2280-2282.

instrumentation, currents

Laktionov, A.F., 1962
Driftograph designed by M.V. Izvekov.
(In Russian).
Okeanologiia, Akad. Nauk, SSSR, 2(5):901-903.

instrumentation, currents

Ledenev, V.G., 1961.
On the use of electromagnetic current meters in Arctic seas. Methods for hydrological observations for stations aboard ship. Methods for oceanological investigations, a collection of papers. (In Russian).
Trudy Arktich. i Antarktich. Nauchno-Issled. Inst., 210:106-110.

instrumentation, current meters

Lee, Arthur, and John Ramster, 1968.
The hydrography of the North Sea. A review of our knowledge in relation to pollution problems.
Helgoländer wiss. Meeresunters. 17: 44-63

instrumentation, currents

Le Grand, A., et V. Romanovsky 1969.
Courantomètre C.R.E.O. à Rotor horizontal.
Trav. Cent. Rech. Etud. océanogr. ns. 8(3): 9-11.

instrumentation, currents
Lester, R.A., 1962.
High-accuracy, self-calibrating acoustic flow meters.
Marine Sciences Instrumentation, Instr. Soc., Amer., Plenum Press, N.Y., 1:200-204.

instrumentation
Leterrier, G., 1955.
Mesure des vitesse et des directions de courants marins. Communication présentée a l'Association Internationale de Recherches Hydrauliques, La Haye, 31 Aout-6Septembre, 1955 (B18):1-8.
au 6ᵉ Congrès de

instruments
Leterrier, G., 1955.
Current recorder. Int. Hydrogr. Rev. 23(1):221-224, 5 figs.
Reprinted from Bull. d'Info. 6(9):Nov. 1954.

instrumentation
Leterrier, G., 1954.
Courantographe. Bull. d'Info., C.C.O.E.C. 6(9): 381-386, 2 pls.

instruments
Lozano Rey, L., 1945.
Anteproyectos de nuevos tipos de correntimetros marinos. Notas y Res., Inst. Espanol Ocean., 2nd ser., No. 125:1-44, 17 textfigs.

instrumentation, currents
Lupachev, Yu. V., 1968.
New bottom current meter. (In Russian).
Okeanologiia, Akad. Nauk, SSSR, 8(3):560-561.

M

instruments
Malkus, W.V.R., 1953.
A recording bathypitotmeter. J. Mar. Res. 12(1): 51-59, 5 textfigs.

instrumentation, currents
Man'kowskiy, V.I., 1961
Instrument for measuring the speed and direction of slow currents and water temperature. Physics of the sea. (In Russian).
Trudy Morsk. Gidrofiz. Inst., 23:122-130.
Translation:
Scripta Technica, Inc. for Amer. Geophys. Union, pp 97-103.

Instruments Current Meters
Maslennikova, T. K., 1960.
Some of the samples of foreign oceanographic equipment.
Biull. Okeanogr. Komissii Pri Prezidiume Akad. Nauk, SSSR, (4):47-52.
H. O. TRANS.-124 (1962).

Mc

instrumentation, currents
McAllister, Raymond F., 1962.
Deep current measurements near Bermuda.
Marine Sciences Instrumentation, Instr. Soc., Amer., Plenum Press, N.Y., 1:210-222.

instrumentation currents
Meschersvii, V.I., 1959
[Measurement of deep currents.]
Trudy Gosud. Okeanogr. Inst., Leningrad, 37: 79-84.
Translator: M. Slessers
M.O. 15047-62
P.O.:08836
USN-HO-TRANS-98-1960
see author card for remarks on figures

parachute drogue
Metcalf, W.G., and M.C. Stalcup, 1967.
Origin of the Atlantic Equatorial Undercurrent.
J. geophys. Res., 72(20):4959-4975.

instruments
Middleton, F.H., 1955.
An ultrasonic current meter for estuarine research. J. Mar. Res. 14(2):176-186, 5 textfigs.

instruments
Mitchell, H., 1934.
The Mitchell current-meter. Hydrogr. Rev. 11(1): 144.

instrumentation, currents (drogues)
Montgomery, R.B., and E.D. Stroup, 1962.
Equatorial waters and currents at 150°W in July-August 1952.
The Johns Hopkins Oceanogr. Studies, No. 1:68 pp.

instrumentation, currents
Moroshkin, K.V., 1957.
Testing the operation of the electromagnetic current meter in the open sea. (In Russian).
Trudy Inst. Okeanol., Akad. Nauk, SSSR, 25:78-87.
USN-HO-TRANS 198
M. Slessers 1963
P.O. 39008

instrumentation, currents
Murray, Stephen P., 1969.
Current meters in use at the Coastal Studies Institute.
Bull Coast. Stud. Coast Stud. Inst. Louisiana State Univ. 3:1-15. (multilithed)

N

instrumentation, currents
Nagata, Yutaka, 1964.
An electromagnetic current meter.
J. Oceanogr. Soc., Japan, 20(2):71-80.

instrumentation
Nagata, Y., 1960
Applications of electro-magnetic current meter for beach problems. Geophys. Notes. 13(1): No. 10 from Proc. 6th Conf. Coast Eng. Japan, 1959: 45-48. No copy of reprint available for Geophys. Notes.

instrumentation, currents
Nakamura, Shigehisa, 1965.
A study on photoelectric current meters.
Bull. Disaster Prevention Res. Inst., Kyoto Univ. 15(1) (89):63-70.

instrumentation
Nan'niti, T., 1959.
A new simple current profile recorder - Nan'niti-Iwamiya current meter.
J. Oceanogr. Soc., Japan, 15(2):57-60.
Also in: Coll. Reprints, Oceanogr. Lab., Meteorol. Res. Inst., Tokyo, 6.

instrumentation
Nan'niti, T., 1959.
On the structure of ocean currents, III. A new Nan'niti-Iwamiya current meter and current profile observed by it. Pap. Meteorol. Geophys., Tokyo, 10(2):124-134.
Also in: Coll. Reprints, Oceanogr. Lab., Meteorol. Res. Inst., Tokyo, 6

instruments
Nan'niti, T., 1955.
On the structure of ocean currents. (1). Test manufacture of a new current meter and some experiments on the sea.
Pap. Met. Geophys., Tokyo, 5(3/4):248-252.

instruments
Nan'niti, T., 1954.
A current meter using a phototube (III). Experiment on a tidal current.
Pap. Met. Geophys., Tokyo, 5(1):54-63, 13 textfigs.

instrumentation, currents
Nan'niti, T., 1953.
A current meter using a phototube. (II).
Pap. Meteorol. Geophys., Tokyo, 4(2):66-73.

instruments
Nan'niti, T., 1953.
A current meter using a phototube.
Int. Hydrogr. Rev. 30(2):171-176, 8 textfigs.

instruments
Nan'niti, T., 1953.
A current meter using a phototube.
Rec. Ocean. Wks., Japan, n.s., 1(1):44-51, 9 textfigs.

instrumentation
Nan'niti, T., 1953.
A current meter using a phototube. Papers Met. Geophys. 3(4):286-294, 13 textfigs.

instrumentation
Nan'niti, T., 1947.
[A current meter using a photocell.] Umi-to-Sora 25(5):64- (in Japanese).

instrumentation, current
Nan'niti, Tosio and Kimio Ito, 1967.
New Nan'niti-Ito current meter.
J. Oceanogr. Soc., Japan, 23(5):256-257.

instrumentation
Nannichi, T., M. Iwamiya and H. Akamatsu, 1959.
[A new simple current profile recorder NANNICHI-IWAMIYA current meter.] J. Oceanogr. Soc., Japan, 15(2): 57-60.

instrumentation, currents
Netherlands, Hydrographic Office, 1963
Current meter of the Netherlands Hydrographic Office.
Int. Hydrogr. Rev., 40(2):187-189.

Instrumentation, currents
Nikitin, M.M., 1968.
On the possibility of design improvements of the recording meters BPV and the Ekman-Merz current meters. (In Russian; English abstract).
Okeanologiia, Akad. Nauk, SSSR, 8(3):520-533.

instrumentation, currents
Niskin, Shale J., 1965.
A method and arrangement for recording current fields and total volume transport in vertical profile from sea surface to any desired depth.
Mar. Sci. Instrument. 4:15-18

O

Instrumentation, design of
O'Brien, M.P. and R.G. Folsom, 1948
Notes on the design of current meters.
Trans. Am. Geophys. Union 29(2):243-250, 5 text figs.

instruments
Olson, F.C.W., 1952.
A contact indicator for the Price current meter.
Pap. Ocean. Inst., Florida State Univ. Studies No. 7:15-16, 1 textfig.

instrumentation
Ono, K., 1957.
The Ono's self-recording current meter.
Proc. UNESCO Symp., Phys. Ocean., Tokyo, 1955: 26-27.

instrumentation
Ono, K., and K. Iwasa, 1957.
The new current meters, recording and direct reading.
Proc. UNESCO Symp., Phys. Ocean., Tokyo, 1955: 26.

instruments
Ono, 1953.
The Ono self-recording current meter.
Int. Hydrogr. Rev. 30(2):177-178, figs.

instrumentation, currents
Ortolan, Georges, 1966.
Nouveau courantometre pour la mesure des courants pres du fond.
Cahiers Océanogr., 18(9):801-810.

instrumentation, currents
Ozmidov, R.V., 1962
Some problems related to the methods for the measurement of sea currents with the aid of the BPV-2 device. (In Russian).
Okeanologiia, Akad. Nauk, SSSR, 2(5):916-921.

P

instrumentation, currents
Paquette, R.G., 1963.
Practical problems in the direct measurement of ocean currents.
Mar. Sci. Instrument., Plenum Press, 2:135-146.

instruments
Pettersson, O., 1929.
Current meter for determination of the direction and velocity of the movement of the water at the bottom of the ocean.
Svenska Hydrogr.-Biol. Komm. Skr., n.s., Hydrogr. 3:9-11.

instruments
Pettersson, O., 1913.
Photographisch registriender Tiefenstrommesser für Dauerbeobachtungen.
Svenska Hydrogr.-Biol. Komm. Skr., 5:1-4

instruments
Pettersson, O., 1905.
Beschreibung des bifilar-strommessers.
Publ. de Circ., Cons. Perm. Int. Expl. Mer, No. 25:1-6, 1 pl.

instrumentation, currents
Pyrkin, Yu. G., 1962.
Autonomous electrovane for the registration of deep current velocities. (In Russian)
Geofiz. Biull., Mezhduved. Geofiz. Komitet, Presid., Akad. Nauk, SSSR, No. 11:43-46.

Q

R

instruments
Rauschelbach, H., 1929.
Beschreibung eines bifilar aufgehängten an Bord elektrische registrierende Strommessens. Ann. Hydr. 57, March Beih.:71 pp.

instruments
Rauschelbach, H., 1932.
Zur Geschichte des Hochseepegels. Ann. Hydr. 60: 73-76.

instrumentation, currents (drogues)
Reid, Joseph L., Jr., 1963
Measurements of the California Countercurrent off Baja California.
J. Geophys. Res., 68(16):4819-4822.

instrumentation, currents
Richardson, William S., 1962.
Current measurements from moored buoys.
Marine Sciences Instrumentation, Instr. Soc., Amer., Plenum Press, N.Y., 1:205-209.

instruments, currents
*Richardson, W.S., A.R. Carr, and H.J. White, 1969.
Description of a freely dropped instrument for measuring current velocity.
J.mar.Res., 27(1):153-157.

instrumentation, currents
Richardson, W.S., P.B. Stimson, and C.H. Wilkins, 1963.
Current measurements from moored buoys.
Deep-Sea Res., 10(4):369-388.

gear
Renner, O., 1947.
Ein neuer electro-mechanischer Druckmesser und seine Anwendung als Küstenpegel. G.H.I. Rept. No. 49.

gear
Renner, O., 1946.
Ein neues Gerät zum Messen der Meeresströmung. G.H.I. Rept. No. 25.

instruments
Roberts, E.B., 1955.
An improved radio current meter.
Int. Hydrogr. Rev. 32(2):163-166.

instrumentation
(Roberts, E.B.) 1947
The Roberts radio current meter. Int. Hydr. Rev. 24:210-212, 2 figs.

current meter calibration
Robson, A.D., 1954.
The effect of water temperature upon the calibration of a current meter. Trans. Amer. Geophys. Union, 35(4):647-648.

instrumentation, currents
Romanovsky, V., 1963.
Le courantographe américain Richardson. (Résumé). Rapp. Proc. Verb., Réunions, Comm. Int., Expl. Sci., Mer Méditerranée, Monaco, 17(3):887.

instruments
Romanovsky, V., 1949.
Les mesures directes des courants marins.
Ann. Hydr. No. 1357:34 pp., 15 textfigs.

instrumentation, currents
Rossby, T. and D. Webb, 1970.
Observing abyssal motions by tracking Swallow floats in the SOFAR Channel. Deep-Sea Res., 17(2): 359-365.

instrumentation, currents
Ryzhkov, Yu. G., 1961.
Measurement of deep currents in the Black Sea by neutrally buoyant floats.
Doklady Akad. Nauk, SSSR, 141(1):74-75.
(ultrasonic)

instrumentation, currents
Ryzhkov, Iu. G., and N.N. Karnaushenko, 1961.
Measurement of deep currents in the Black Sea with the aid of an ultrasonic buoy of neutral buoyancy.
Doklady Akad. Nauk, SSSR, 141(1):74-76.

S

instrumentation, currents
Sasaki, Tadayoshi, Seiichi Watanabe and Gohachiro Oshiba, 1967.
Current measurements on the bottom in the deep waters of the western Pacific.
Deep-Sea Research, 14(2):159-167.

Sasaki, Tadayoshi, Seiichi Watanabe and Gohachiro Oshiba, 1965.
New current meters for great depths.
Deep-Sea Res., 12(6):815-824.

instruments
Savur, S.R., 1954.
A new ocean current meter.
Andhra Univ. Ocean. Mem. 1:122-124, 2 textfigs.

flowmeters, depth
Scripps Institution of Oceanography, 1949
Marine Life Research Program, Progress Report to 30 April 1949, 25 pp. (mimeographed), numerous figs. (photo.+ozalid). 1 May 1949.

instrumentation
Sisoev, N.N., 1959
Buoy stations for measuring currents in the ocean. Trudy Inst. Okeanol., 35: 78-85.

instrumentation
Sisoev, N.N., 1959
Sea electro-contact current meter.
Trudy Inst. Okeanol., 35: 71-77.

instruments
Sisoev, N.N., 1951.
Shortcomings of construction and manufacture of propeller-type current meters and methods of increasing their accuracy. Trudy Inst. Okeanol., 5:132-135.

instrumentation, currents
Smith, Paul Ferris, 1969.
The significance of direct current measurements an evaluation. Oceanol. int., 69, 1: 7 pp., 6 figs.

instrumentation, flowmeters
Squier, E.D., 1968.
A Doppler shift flowmeter.
Mar. Sci. Instrument. 4: 585-593.

instrumentation, currents
Snodgrass, Frank E., 1968.
Deep sea instrument capsule.
Science, 162(3849):78-87.

instrumentation
Solovyev, L.G., 1959
Electrodes for an electromagnetic current meter
Trudy Inst. Okeanol., 35: 92-101

instrumentation, currents
Stas, I.I., 1962
Automatic photoelectric deep-water instrument for recording ocean current elements. (In Russian).
Trudy Morsk. Gidrofiz. Inst., 26:128-131.

Translation:
Scripta Tecnica, Inc., for Amer. Geophys. Union, 26:97-99. 1964.

instrumentation, currents
Stas, I.I., 1961.
An autonomous instrument for recording temperature and currents in the sea.
Biull. Izobreteniy, No. 20:54.

Complete translation in:
Soviet Bloc Res. Geophys., Astron., and Space, No. 34(1962),

instrumentation, currents
Stas, J. J., 1959
Device for element recording of currents.
Biull. Okeanogr. Komissii, Akad. Nauk, SSSR, (3):53-56.

instrumentation, currents
Stas, I.I., and A.G. Ishutin, 1962
An abyssal autonomous photo-selfrecorder of currents. (In Russian).
Okeanologiia, Akad. Nauk, SSSR, 2(5):898-901.

instrumentation, currents
Stasy, I.I., 1961.
Automatic recorder of ocean currents. Sea physics (In Russian).
Trudy Morsk. Gidrofiz. Inst., 23:160-163.

Translation:
Scripta Technica, Inc. for Amer. Geophys. Union, pp. 128-130.

instrumentation, currents
(Swallow floats)
Steele, J.H., J.R. Barrett and L.V. Worthington, 1962
Deep currents south of Iceland.
Deep-Sea Res., 9(5):465-474.

instrumentation, currents, bottom
Sternberg, R.W., 1969.
Camera and dye-pulser system to measure bottom boundary-layer flow in the deep-sea. Deep-Sea Res., 16(5): 549-554.

instruments, currents
Suellentrop, F.J., A.E. Brown and Eric Rule, 1962.
An acoustic ocean-current meter.
Marine Sciences Instrumentation, Instr. Soc., Amer., Plenum Press, N.Y., 1:190-193.

instruments
Sverdrup, H.U., and O. Dahl, 1927.
Two new recording current meters. Hydrogr. Rev. 4(2):208-209.

Extracted from: Two oceanographic current recorders designed and used on the "Maud" Expedition. J. Opt. Soc., Amer., and Rev. Sci Inst. 12(5):
and J. du Cons. 2(1):79-80.

instrumentation
Swallow, J.C., 1955.
A neutral-buoyancy float for measuring deep currents. Deep-Sea Res. 3(1):74-81.

instrumentation, currents
Sweers, H.E., 1967.
Some Fortran II programs for computer processing of oceanographic observations.
NATO Subcomm. Oceanogr. Res., Tech. Rept. 37 (Irminger Sea Project):104 pp. (mimeographed).

instrumentation
Sysoev, N.N., 1956.
[On the application of the radiodirection finder in the observation of currents at sea.]
Trudy Inst. Okeanol., 19:117-128.

instrumentation, currents
Sysoev, N.N., 1961.
[Self-recording current meter of the type the "Ocean".]
Trudy Inst. Okeanol., Akad. Nauk, SSSR, 47:22-28.

T

instrumentation, currents
Terjune, L.D.B., 1968.
Free-floating current followers.
Techn.Rept.Fish.Res.Bd.,Can. 85: 21 pp. (mimeographed).

instrumentation, currents
Thorndike, E.M., 1963
A suspended-drop current meter.
Deep-Sea Res., 10(3):263-267.

instrumentation, current meters
Titov, V.B. and V.G. Krivosheya, 1971.
On disadvantages of measuring velocity modulus with the BPV current meters. (In Russian; English abstract). Okeanologiia 11(5): 906-913.

U

V

instruments
Van Roosendaal, A.M., and C.H. Wind, 1905.
Pruefung von Strommessern und Strommessungsversuche in der Nordsee.
Publ. de Circ., Cons. Perm. Int. Expl. Mer, No. 26:1-10, 2 pls.

instrumentation, currents
Veed, V.A., 1962.
Pole-integrator for the measurement of mean velocity on the vertical of hydrometric gauge lines. (In Russian).
Meteorol. i Gidrol., (3):55-57.

instrumentation, currents
Volkmann, G.H., 1963
13. Deep-current measurements using neutrally buoyant floats. In: The Sea, M.N. Hill, Edit., Vol. 2. (III) Currents, Interscience Publishers, New York and London, 297-302.

instruments
Volkmann, G., J. Knauss and A. Vine, 1956.
The use of parachute drogues in the measurement of subsurface currents.
Trans. Amer. Geophys. Union, 37(5):573-577.

instrumentation currents
Volkov, V.G., 1957.
Suppression of wave noises during work with electromagnetic current meters. (In Russian).
Trudy Gos. Okeanogr. Inst., No. 40:57-64.

instrumentation, currents
Volkov, V.G., 1956.
A method and apparatus for an electromagnetic current meter. (In Russian).
Trudy Inst. Okeanol., Akad. Nauk, SSSR, No. 19: 98-106.

instrumentation, currents
Von Arx, W.S., undated (1954?).
Measurements of the oceanic circulation in temperate and tropical latitudes. In: Oceanographic Instrumentation. NRS Publ. No. 309:13-33.

With appendix on current measurement
With discussion by C.O'D. Iselin
R.H. Fleming
D.W. Pritchard

gear
von Arx, W.S., 1950.
Some current meters designed for suspension from an anchored ship. J. Mar. Res. 9(2):93-99, 5 textfigs.

W

instrumentation, currents (vertical)
Webb, D.C., D.L. Dorson and A.D. Voorhis, 1970
A new instrument for the measurement of vertical currents in the ocean.
Electronic Engineering in Ocean Technology, Swansea, 21-24 Sept. 1970, I.E.R.E. Conf. Proc. 19: 323-331

current meters
Wiegel, R.L., and J.W. Johnson, 1960
Ocean currents, measurement of currents and analysis of data.
Waste disposal in the marine environment, Pergamon Press:175-245.

instruments
Wilkie, M.J., 1955.
A recording water velocity meter.
J. Sci. Instr., 32:350-353.

instrumentation, currents
*Wiseman, William Joseph, Jr., 1968.
Doppler-shift current meter signals and the drop-out problem.
Tech.Rep.Chesapeake Bay Inst.(Ref.68-9) 41: 11pp. (multilithed).

instruments
Witting, R., 1905.
Kurze Beschreibung eines elektrisch registrierenden Strommessers.
Publ. de Circ., Cons. Perm. Int. Expl. Mer, No. 30:1-8, 1 pl.

instrumentation, current meters
Wolf, Roland, 1960.
Le courantographe type Chausey.
Cahiers Oceanogr., C.C.O.E.C., 12(4):275-278.

instrumentation, currents
Woods, A.J., and D.H. MacMillan, 1959.
A new development in current meters.
Dock and Harbour Authority, 40(469):205-208.

XYZ

instrumentation, currents
Zeigler, J.M. and H.J. Tasha, 1969.
Measurement of coastal currents. Proc. Eleventh Conf. on Coastal Engin., London: 436-445.
Also in: Contrib. Univ. Puerto Rico, Dept. mar. Sci. 8 (1968-1969).

instrumentation, currents
Zvorykin, K.A., 1956
Initial velocity and sensitivity of a hydrometric current meter. (In Russian).
Trudy Leningr. Gidrometeorol. Inst., No. 5-6: 108-115.

instrumentation, data processing
Kolesnikov, A.G., 1971.
Automation of oceanographic research. (In Russian; English abstract). Okeanologiia 11(5): 795-801.

instrumentation dyes

instrumentation, currents

DYES

instrumentation, dyes
Gerard, Robert, and J. Michael Costin, 1967.
Underwater dye pumping system.
Deep-Sea Research, 14(4):479-480.

GEK

instrumentation, currents

GEK = Geomagnetic electrokinetograph

GEK
Bennett, Edward B., and Milner B. Schaefer, 1960
Studies of physical, chemical and biological oceanography in the vicinity of the Revilla Gigedo Islands during "Island Current Survey" of 1957.
Bull. Inter-American Tropical Tuna Comm., 4(5): 219-317. (Also in Spanish).

GEK (data)
Blackburn, M., R.C. Griffiths, R.W. Holmes and W.H. Thomas, 1962.
Physical, chemical and biological observations in the eastern tropical Pacific Ocean: three cruises to the Gulf of Tehuantepec, 1958-1959.
U.S.F.W.S. Spec. Sci. Repts., Fish., No. 420:170p

GEK
Bourkov, V.A., 1963
Some results of oceanographic observations with express methods to the east and south of Japan.
Okeanol. Issled., Rezhult. Issled., Programme Mezhd. Geofiz. Goda, Mezhd. Geofiz. Komitet, Presidiume Akad. Nauk, SSSR, No. 9:32-41.

GEK
Bowden, K.F., 1953.
Measurement of wind currents in the sea by the method of towed electrodes. Nature 171(4356): 735-737, 1 textfig.

GEK
Burkov, V.A., 1963
Some results of oceanographic observations with express methods to the east and south of Japan. (In Russian; English abstract).
Okeanolog. Issled. Rezhult. Issled. po Programme Mezhd. Geofiz. Goda, Mezhd. Geofiz. Kom. pri Presidiume Akad. Nauk, SSSR, No. 9:32-41.

GEK
Capart, Jean-Jacques, and Marc Steyaert, 1963
Mission OTAN en mer d'Alboran, juillet-août 1962. Rapport préliminaire. Températures et courants de surface enregistrés à bord du navire Belge "Euren".
Documents de Travail, Inst. Roy. Sci. Nat. Belg. No. 1: 16 pp. numerous charts. (Mimeographed and multilithed).

GEK
Chew, Frank, 1967.
On the cross-stream variation of the k-factor for geomagnetic electrokinetograph data from the Florida Current off Miami.
Limnol. Oceanogr., 12(1):73-78.

GEK
Chew, Frank, 1958.
An interpretation of the transverse component of the geomagnetic electrokinetograph readings in the Florida Straits off Miami.
Trans. Amer. Geophys. Union, 39(5):875-884.

GEK
Chew, F., and L.P. Wagner, 1954.
Note on correcting G.E.K. observations of Florida Current for tidal currents.
Bull. Mar. Sci., Gulf and Caribbean, 4(4):336-345, 2 textfigs.
of Miami

instrumentation
Colton, John B., Jr., 1959.
The multiplane kite-otter as a depressor for high-speed plankton samplers.
J. du Cons., 25(1):29-35.

GEK
Dmitrenko, O., 1959
[Use of cadmium non-polarizing electrodes in the study of natural electric currents in the sea.] Trudy Inst. Okeanol., 35: 102-112.

instrumentation, GEK
*Fonarev, G.A., 1968.
The effect of sea telluric currents on the operation of the GEK. (In Russian; English abstract).
Okeanologiia, Akad. Nauk, SSSR, 8(4):727-735.

GEK
France, Service Central Hydrographique, 1957.
Peut-on utiliser le courantomètre à électrodes remorquées dans les travaux hydrographiques.
Bull. d'Info., C.C.O.E.C., 9(6):317-324.

GEK
Fukuoka, Jiro, 1961.
An analysis of the mechanism of the cold and warm water masses in the seas adjacent to Japan.
Rec. Oceanogr. Wks., Japan, 6(1):63-100.

instrumentation
Gougenheim, A., 1958.
Le courantomètre à électrodes remorquées.
Navigation 6(22):143-148.

G. E. K.
Guelke, R. W. & C. A. Schoute-Vanneck, 1947.
The measurement of sea-water velocities by electromagnetic induction.
J. Inst. Elect. Engineers, 94(37): 71-74.

GEK
Guibout, P., and J. C. Lizeray, 1959
Mesures effectuées dans l'océan Indien à l'aide du courantometre à electrodes remorquées.
Cahiers Océanogr., C.C.O.E.C., 11(3):155-157.

GEK
Hamon, B.V., 1965.
The East Australian Current, 1960-1964.
Deep-Sea Res., 12(6):899-921.

GEK
Ichiye, T., 1957.
A note on the horizontal eddy viscosity in the Kuroshio. Rec. Oceanogr. Wks., Japan, n.s., 3(1):16-25.

GEK
Japan, Hakodate Marine Observatory, Oceanographical Section, 1962
Report of the oceanographic observations in the sea east of Tohoku District in May, 1961. (In Japanese).
Res. Mar. Meteorol. and Oceanogr. Obs., Jan.-June, 1961, No. 29:9-12.

GEK
Japan, Hakodate Marine Observatory, Oceanographical Section, 1962
Report of the oceanographic observations in the sea east of Tohoku District from February to March, 1961. (In Japanese).
Res. Mar. Meteorol., and Oceanogr. Obs., Jan. June, 1961, No. 29:3-8.

GEK
Japan, Hakodate Marine Observatory, 1962
Report of the oceanographic observations in the sea east of Tohoku District and in the Okhotsk Sea from August to September 1960. (In Japanese).
Results, Mar. Meteorol. and Oceanogr., July-Dec., 1960, Japan Meteorol. Agency, No. 28: 7-16.

GEK
Japan, Hakodate Marine Observatory, 1962
Report of the oceanographic observations in the sea south of Hokkaido in July, 1960. (In Japanese).
Res. Mar. Meteorol. and Oceanogr., July-Dec., 1960, Japan Meteorol. Agency, No. 28; 3-6.

GEK
Japan, Hakodate Marine Observatory, 1962
Report of the oceanographic observations in the sea west of Hiyama (Hokkaido) and in the sea east of Tohoku District in November 1960. (In Japanese).
Res. Mar. Meteorol. and Oceanogr., July-Dec., 1960, Japan Meteorol. Agency, No. 28:17-20.

GEK
Japan, Nagasaki Marine Observatory, 1962
Report of the oceanographic observations in the sea west of Japan from October to November, 1960. (In Japanese).
Res. Mar. Meteorol. and Oceanogr., July-Dec., 1960, Japan Meteorol. Agency, No. 28:52-59.

GEK
Japan, Hakodate Marine Observatory, 1961
[Report of the oceanographic observations in the sea east of Tohoku District from February to March 1959.]
Bull. Hakodate Mar. Obs., (8):3-7.

GEK
Japan, Hakodate Marine Observatory, 1957
[Report of the oceanographic observations in the sea east of the Tohoku district in August 1956.]
Bull. Hakodate Mar. Obs., No. 4: 1-12.

GEK
Japan, Hakodate Marine Observatory, 1957
[Report of the oceanographic observations in the sea east of Tohoku district from October to November 1956.]
Bull. Hakodate Mar. Obs., No. 4: 1-8.

GEK
Japan, Hydrographic Office, Maritime Safety Board, Undated
State of the adjacent seas of Japan, 1955-1959. Vol. 1: Numerous charts.

GEK
Japan, Japan Meteorological Agency, Oceanographical Section, 1962
Report of the oceanographic observations in the sea east of Honshu from February to March, 1961. (In Japanese).
Res. Mar. Meteorol. and Oceanogr. Obs., Jan.-June, 1961, No. 29:13-21.

GEK
Japan, Kobe Marine Observatory, 1963.
Report of the oceanographic observations in the sea south of Honshu from July to August, and from the cold water region south of Enshu Nada October to November 1960. (In Japanese).
Bull. Kobe Mar. Obs., 171(3):36-52.

GEK
Japan, Kobe Marine Observatory, Oceanographical Section, 1963
Report of the oceanographic observations in the sea south of Honshu from May to June, 1962. (In Japanese).
Res. Mar. Met. & Ocean., J.M.A., 31:45-49.
Also in:
Bull. Kobe Mar. Obs., No. 173(4):1964.

GEK
Japan, Kobe Marine Observatory, Oceanographical Section, 1963
Report of the oceanographic observations in the sea south of Honshu from February to March 1962. (In Japanese).
Res. Mar. Met. & Ocean., J.M.A., 31:37-44.
Also in:
Bull. Kobe Mar. Obs., No. 173(4):1964.

GEK
Japan, Kobe Marine Observatory, Oceanographical Section, 1962
Report of the oceanographic observations in the cold water region off Kii Peninsula from October to November 1961. (In Japanese).
Res. Mar. Meteorol. & Oceanogr., No. 30:49-55.
Also in:
Bull. Kobe Mar. Obs., No. 173(3). 1964

GEK
Japan, Kobe Marine Observatory, Oceanographical Section, 1962
Report of the oceanographical observations in the sea south of Honshu from July to August 1961. (In Japanese).
Res. Mar. Meteorol. & Oceanogr., 30:39-48.
Also in:
Bull. Kobe Mar. Obs., No. 173(3) 1964.

GEK
Japan, Maizuru Marine Observatory, 1963
Report of the oceanographic observation s in the Japan Sea in June, 1961. (In Japanese).
Bull. Maizuru Mar. Obs., No. 8:59-79.

GEK
Japan, Maizuru Marine Observatory and Hakodate Marine Observatory, Oceanographical Sections, 1962
Report of the oceanographic observations in the Japan Sea in June, 1961. (In Japanese).
Res. Mar. Meteorol. and Oceanogr. Obs., Jan.-June, 1961, No. 29:59-79.

GEK (data only)
Japan, Maritime Safety Agency, Tokyo, 1964.
Tables of results from oceanographic observations in 1960.
Hydrogr. Bull., (Publ. No. 981), No. 75: 86 pp.

GEK

Japan, Japanese Meteorological Agency, 1966.
The results of the Japanese Expedition of Deep Sea (JEDS-8).
Results mar.met.oceanogr.Obsns.Tokyo, 35:328 pp.

GEK

Japan Meteorological Agency, 1962
Report of the Oceanographic observations in the east of Honshu from August to September, 1960. (In Japanese).
Res. Mar. Meteorol. and Oceanogr., July-Dec., 1960, Japan Meteorol. Agency, No. 28:21-29.

GEK (data only)

Japan, Japan Meteorological Agency, 1958
The results of marine meteorological and oceanographical observations, July-December 1957, No. 22: 183 pp.

GEK

Japan, Kobe Marine Observatory, Oceanographical Observatory. 1962
Report of the oceanographic observations in the cold water region off Enshu Nada in May 1961. (In Japanese).
Res. Mar. Meteorol. and Oceanogr. Obs., Jan.-June, 1961, No. 29:28-35.

GEK

Japan, Kobe Marine Observatory, 1962
Report of the oceanographic observations in the cold water region south of Enshu Nada from October to November, 1960. (In Japanese
Res. Mar. Meteorol. and Oceanogr., July-Dec., 1960, Japan Meteorol Agency, No. 28: 43-51.

GEK

Japan, Kobe Marine Observatory, 1961
[Report of the oceanographic observations in the sea south of Honshu in March 1958.]
Bull. Kobe Mar. Obs., No. 167(21-22):30-36.

--from May to June, 1958(21-22):37-42
--from July to September, 1958(23-24):34-40
--from October to December, 1958(23-24):41-4
--from February to March, 1959(25-26):33-47.

GEK

Japan, Kobe Marine Observatory, 1958.
[Report of the oceanographic observations in the sea south of Honshu from November to December, 1957.]
J. Oceanogr., Kobe Mar. Obs., (2) 10(1):21-28.

in May 1957.] Ibid., 9(2):69-78.

in August, 1957.] Ibid., 9(2):79-86.

GEK

Japan, Meteorological Agency, 1962
Report of the Oceanographic Observations in the sea south and west of Japan from October to November, 1960. (In Japanese).
Res. Mar. Meteorol. and Oceanogr., July-Dec., 1960, Japan Meteorol. Agency, No. 28:30-35.

GEK

Japan, Japan Meteorological Agency, 1960.
The results of marine meteorological and oceanographical observations, July-December, 1959, No. 26:256 pp.

GEK (data only)

Japan, Japan Meteorological Agency, 1958.
The results of marine meteorological and oceanographical observations, January-July, 1957, No. 21:168 pp.

currents (GEK) (Data only)

Japan, Japanese Meteorological Agency, 1957
The results of marine meteorological and oceanographical observations, Jan.-June, 1956: 184 pp.
July-December, No. 20: 191 pp.

GEK

Japan, Maizuru Marine Observatory, Oceanographical Section, 1961
[Report of the oceanographic observations in the Japan Sea from June to July, 1959.]
Bull. Maizuru Mar. Obs., No. 7:57-64.

GEK

Japan, Maizuru Marine Observatory, Oceanographical Section, 1961
Report of the oceanographic observations in the Japan Sea from June to July, 1958.
Bull. Maizuru Mar. Obs., No. 7:60-67.

GEK

Japan, Maritime Safety Board, 1961
Tables of results from oceanographic observations in 1958.
Hydrogr. Bull., Tokyo, No. 66 (Publ. No. 981): 153 pp.

currents (GEK) (data only)

Japan, Maritime Safety Board, 1961
Tables of results from oceanographic observations in 1958.
Hydrogr. Bull., Tokyo, No. 65 (Publ. No. 981): 153 pp.

GEK

Japan, Maritime Safety Board, 1955.
[Measurement of the currents in the sea with geomagnetic electro-kinetograph.] Publ. No. 981, Hydrogr. Bull., Spec. No. 17:28-32.

GEK

Japan, Maritime Safety Board, 1955.
[On Droop correction of G.E.K.] Publ. No. 981, Hydrogr. Bull. Spec. No. 17:33-34.

GEK

Japan, Nagasaki Marine Observatory, Oceanographical Section, 1962
Report of the oceanographic observations in the sea west of Japan from April to May, 1961. (In Japanese).
Res. Mar. Meteorol. and Oceanogr. Obs., Jan.-June, 1961, No. 29: 45-53.

GEK

Japan, Nagasaki Marine Observatory, Oceanographical Section, 1962
Report of the oceanographic observations in the sea west of Japan from February to March, 1961. (In Japanese).
Res. Mar. Meteorol. and Oceanogr. Obs., Jan.-June, 1961, No. 29:36-44.

GEK

Japan, Nagasaki Marine Observatory, Oceanographical Section 1960.
Report of the oceanographic observation in the sea west of Japan from January to February, 1960. Report of the Oceanographic observation in the sea north-west of Kyushu from April to May, 1960.
Results Mar. Meteorol. & Oceanogr., J.M.A., 27:42-50; 51-67.

Also in:
Oceanogr. & Meteorol., Nagasaki Mar. Obs., (1961) 11 (202).

GEK

Japan, Nagasaki Marine Observatory, Oceanographical Section, 1960
[Report of the oceanographic observation in the sea west of Japan from January to February 1959.]
Res. Mar. Meteorol. & Ocean., J.M.A., 25: 48-56.
Also in:
Oceanogr. & Meteorol., Nagasaki Mar. Obs., (1961) 11(199).

GEK

Japan, Nagasaki Marine Observatory, Oceanographical Section, 1960
[Report of the oceanographic observations in the sea west of Japan from June to July 1959.]
Res. Mar. Meteorol. & Oceanogr., J.M.A., 26: 51-57.
Also in:
Oceanogr. & Meteorol., Nagasaki Mar. Obs., (1961), 11(200).

GEK

Japan, Nagasaki Marine Observatory, 1959.
Report of the oceanographic observations in the sea west of Japan from January to February, 1958
Results Mar. Meteorol. & Oceanogr., J.M.A., 23: 43-49.

Also in: Oceanogr. & Meteorol., Nagasaki Mar. Obs., 10(194):1960.

GEK

Japan, Nagasaki Marine Observatory, 1959.
Report of the oceanographic observations in the sea west of Japan from June to July 1958 and in the sea north west of Kyushu in October 1958.
Results Mar. Meteorol. & Oceanogr., J.M.A., 24: 47-60.

Also in:
Oceanogr. & Meteorol., Nagasaki Mar. Obs., 10(195): 1960.

instrumentation GEK

Knauss, J.A., and J.L. Reid, 1957.
The effects of cable design on the accuracy of th the GEK. Trans. AGU 38(8):320-325.

GEK

Kontorovitch, V.M., 1961.
[Magnetohydrodynamic effects in the ocean.]
Doklady Akad. Nauk, USSR, 137(3):576-579.

GEK

Kormyshev, N.S., 1961.
Experiment of synchronous survey of currents by means of electromagnetic measuring instruments. (In Russian).
Meteorol. i Gidrol., (8):49-51.

GEK

Korneva, L., 1956?
[Force electromotrice induite dans les conducteurs lors de la mesure des courants électriques dans l'eau de mer.]
Akad. Nauk, SSSR., Trav. Inst. Marit. Hydrophys., Moscou, 7:

Cited in Bull d'Info., 8(10):

G.E.K.

Kubota, T., and K. Iwasa, 1961
[On the currents in Tugaru Strait.]
Hydrogr. Bull., Maritime Safety Bd., Tokyo, (Publ. No. 981), No. 65:19-26.

G.E.K.

Lacombe, Henri, 1961
Année Geophysique Internationale 1957-1958, participation française. Contribution à l'étude du regime du detroit de Gibraltar. 1. Etude dynamique. Cahier Océanogr., C.C.O.E.C., 13 (2): 73-106.

G.E.K.,

Lacombe, Henri, Paul Tchernia, Claude Richez et Lucien Gamberoni, 1964.
Deuxieme contribution a l'etude du regime du detroit de Gibraltar (Travaux de 1960).
Cahiers Oceanogr., C.C.O.E.C., 16(4):283-327.

geomagnetic electrokinetograph

Lebedev, A. A., 1957.
Re the influence of geomagnetic disturbance on geomagnetic electrokinetograph current measurements in the Barents Sea.
Trudy Gos. Okean. Inst., No. 40: 50-56.

Translation ERHope T335R

GEK

LeFloch, Jean, 1961
Mesures de courants par électrodes remorquées dans le Canal de Corse. Relation avec le relief dynamique. Application au choix d'une surface de référence pour le calcul des vitesses en profondeur et des debits.
Cahiers Océanogr., C.C.O.E.C., 13(9):619-626.

GEK

Lim, Ki Bong and Minoru Fujimoto, 1971.
A study on the results of GEK measurement in Satsunan Sea of Japan. (In Korean; English abstract). J. oceanogr. Soc., Korea 6(2): 99-103.

GEK

Lizeray, J.-C., 1957.
Travaux océanographiques recents de l'"Elie Monnier" en Méditerranée occidentale.
Bull. d'Info., C.C.O.E.C., 9(8):413-415.

GEK

Longuet-Higgins, M.S., M.E. Stern and H. Stommel, 1954.
The electrical field induced by ocean currents and waves, with applications to the method of towed electrodes. P.P.O.M. 13(1):1-37, 28 textfigs.

GEK

Malmberg, Svend-Aage, 1962
Schichtung und Zirkulation in den südisländischen Gewässern.
Kieler Meeresf., 18(1):3-28.

GEK

Martin, Jean, 1962
Utilisation du courantomètre électrique à électrodes remorquées (1955-1956).
Océanogr. Phys. Année Géophys. Internat., Part. Francais, Centre Nat. Recherche Sci., Ser. 10, (1):11-55.

GEK

Martin, J., 1956.
Résultats des premiers travaux français effectuées avec un mesureur électrique de courant à électrodes remorquées. C.R. Acad. Sci., Paris, 243(19): 1432-1434.

GEK

Martin, J., 1956.
Utilisation du courantometre electrique a electrodes remorquees. Bull. d'Info., C.C.O.E.C. 8(9):465-505, pls.

GEK

Martin, J., 1956.
Utilisation du courantometre electrique à electrodes remoquées. Bull. d'Info., C.C.O.E.C., 8(8): 355-398.

GEK

Masuzawa, Jotaro, 1964.
A typical hydrographic section of the Kuroshio extension.
Oceanogr. Mag., Tokyo, 16(1/2):21-30.

GEK

Meshcherskiy, V.I., 1955.
Experience of the work with the geoelectromagnetic current meter in the Baltic Sea in 1954. (In Russian).
Trudy Gos. Okeanogr. Inst., (30):113-118.

instrumentation, currents

Morhange, Jacques, 1964.
Etude et réalisation d'un courantomètre pendule. Premiers résultats experimentaux.
Bull. Inst. Océanogr., Monaco, 61(1283A, 1283B): 28 pp., 3 folded charts.

G.E.K.

Moroshkin, K.V., 1959
[Measurements of currents in the South-west Indian Ocean by the use of electromagnetic method.]
Info. Biull. Sovetsk. Antarkt. Exped., (7): 22-25.

GEK

Moroshkin, K.V., and M.A. Bogdanov, 1959
[Results of the use of the geomagnetic electrokinetograph in the Indian Ocean and the southern part of the Pacific Ocean.]
Arktich. i Antarktich. Nauchno-Issled. Inst., Mezhd. Geofiz. God, Sovetsk. Antarkt. Eksped., 5: 125-139.

GEK

Moroshkin, K.V., 1957.
[The works with the electro-magnetic current meter in the open sea.] Trudy Inst. Okeanol., 25:62-87.

GEK (data only)

Mosby, Hakon, 1962
Current measurements in the Faroe-Shetland Channel, 1960 and1961. Tables.
NATO Subcommittee on Oceanogr. Res., 173 pp. (mimeographed).

GEK

Murray, K.M., 1952.
Short period fluctuations of the Florida Current from geomagnetic-electrokinetograph observations
Bull. Mar. Sci., Gulf and Caribbean 2(1):360-375, 9 textfigs.

GEK

Nitani, H., 1961.
[On the general oceanographic conditions at the western boundary region of the North Pacific Ocean.]
Hydrogr. Bull., Maritime Safety Bd., Tokyo, (Publ. No. 981), No. 65:27-35.

GEK

Nitani, H., K. Iwasa and W. Inada, 1959.
On the oceanic and tidal current observation in the channel by making use of induced electric potential.
Hydrogr. Bull., Tokyo, (Publ. 981) (61):14-24.

instrumentation currents GEK

Novysh, V.V., 1957.
On the depth of towing the cables and electrodes of the electromagnetic currents meter (EMIT). (In Russian).
Trudy N.-I. Inst. Zemn. Magnetizma, Ionosfery i Rasprost. Radiovoln., No. 12(22):225-240.

GEK

Nowlin, W.D., Jr., and H.J.McLellan, 1967.
A characterization of the Gulf of Mexico waters in winter.
J.mar.Res., 25(1):29-59.

GEK

Peluchon, Georges, and Jean Rene Donguy, 1962.
Travaux océanographiques de l'"Origny" dans le détroit de Gibraltar - 15 mai, 15 juin 1961. 2. Courants de surface dans le détroit de Gibraltar.
Cahiers Océanogr., C.C.O.E.C., 14(7):474-483.

GEK

Rattray, M., Jr., 1962.
GEK observations of inertial oscillations. (Abstract).
J. Geophys. Res., 67(4):1654.

GEK

Reid, Joseph L., Jr., 1962
Observations of intertial rotation and internal waves.
Deep-Sea Res., 9(4):283-289.

GEK

Reid, G.L., jr., 1958.
A comparison of drogue and GEK measurements in deep water. Limnol. & Oceanogr., 3(2):160-165.

GEK

Reid, Joseph L., Jr., and Richard A. Schwartzlose, 1962
Direct measurements of the Davidson Current off central California.
J. Geophys. Res., 67(6):2491-2497.

instrumentation, currents

Rymsha, V.A., 1958.
An instrument for measuring the low velocities of a water current. (In Russian).
Meteorol. i Gidrol., (9):44-46.

Translation:
OTS, $0.50.

GEK

Ryzhkov, Yu. G., 1957.
[Measurement of electric current in the ocean.]
Doklady Akad. Nauk, SSSR, 113(4):787-790.

OTS translation $0.50.

instrumentation, G.E.K.

Sasaki, Tadayoshi, 1955
Three instruments constructed and employed in Japan.
Proc. UNESCO Symp. Phys. Oceanogr., Tokyo, 1955: 46-48.

instrumentation, currents

Sasaki, Tadayoshi, Seiichi Watanabe, Gohachiro Oshiba, Masahiro Kajohara, Noboru Okami and Tadashi Takahashi, 1962
Studies on deep-sea floor current (1). (In Japanese; English abstract).
J. Oceanogr. Soc., Japan, 20th Ann. Vol. 213-222.

GEK

Shoji, D., R. Watanabe, N. Suzuki and K. Hasuike, 1958.
On the "shiome" at the boundary zone of the Kuroshio and the coastal waters off Shionomisaki.
Rec. Oceanogr. Wks., Japan, Spec. No. 2):78-84.

GEK

Sisoev, N.N., 1958.
The electromagnetic method for detecting currents on ships underway.
Trudy Okeanogr. Komissii, Akad. Nauk, SSSR, 3:138-146.

In Russian.

GEK

Sitarz, J., 1958.
Mise au point et essais d'un courantomètre électrique à électrodes remorquées (G.E.K.) le long de la côte méditérranéenne d'Antibes à Menton en été 1956. Trav. C.R.E.O., n.s., 3(1):23-50.

GEK

Solovyev, L.G., 1959
[Electrodes for an electromagnetic current meter.]
Trudy Inst. Okeanol., 35: 92-101

GEK

Soule, F.M., P.S. Branson, and R.P. Dinsmore, 1952.
Physical oceanography of the Grand Banks region and the Labrador Sea in 1951. U.S.C.G. Bull. No. 37:17-85, 21 textfigs.

GEK

Suda, K., and D. Syôzi, 1954.
[Geomagnetic Electro Kinetograph.] Publ 981, Hydrogr. Bull., Maritime Safety Tokyo, Spec. Number (Ocean. Repts.), No. 4:1-16, 16 figs.

GEK (data)

Sugiura, J., 1960.
On the currents south off Hokkaido in the western North Pacific. II.
Bull. Hakodate Mar. Obs., 7(4):79-97.

Reprinted from:
Oceanogr. Mag., 1960, 11(2):79-97.

GEK

Tchernia, P., H. Lacombe and P. Guibout, 1958.
Sur quelques nouvelles observations hydrologiques relatives à la région équatoriale de l'océan Indien. Bull. d'Info., C.C.O.E.C., 10(3):115-143.

GEK

Tourlygine, S., and L. Korneva, 1956?
Sur les électrodes de contract appliquées à la mesure des courants électriques et de leur intensité dans l'eau de mer.
Akad. Nauk, SSSR, Trav. Inst. Marit. Hydrophys., Moscou, 7:

Cited from Bull d'Info. 8(10)

GEK
Tourlygine, S., and N. Karelina, 1956?
Sur les électrodes impolarisables utilisées pour la mesure des courants faibles circulant dans un milieu quelconque.
Akad. Nauk, SSSR, Trav. Inst. Marit. Hydrophys., Moscou, 7:

Cited from Bull. d'Info., 8(10):

GEK
Trites, R.W., and D.G. MacGregor, 1962.
Flow of water in the passages of Passamaquoddy Bay measured by the electromagnetic method.
J. Fish. Res. Bd., Canada, 19(5):895-919.

GEK, data
Webster, Ferris, 1961
The effect of meanders on the kinetic energy balance of the Gulf Stream.
Tellus, 13(3):392-401.

GEK
Weidemann, H., 1961
Results of current measurements by towed electrodes in the northern North Atlantic during the IGY. Contribution to Special IGY Meeting, 1959.
Cons. Perm. Int. Expl. Mer, Rapp. Proc. Verb. 149:115-117.

GEK
Wiegel, R.L., and J.W. Johnson, 1960
Ocean currents, measurement of currents and analysis of data.
Waste disposal in the marine environment, Pergamon Press: 175-245.

G.E.K.
Yoshida, S., 1961
[On the short period variation of the Kurosio in the adjacent sea of Izu Islands.]
Hydrogr. Bull., Maritime Safety Bd., Tokyo, (Publ. No. 981), No. 65:1-18.

GEK (data only)
Zvereva, A.A., Edit., 1959.
[Data, 2nd Marine Expedition, "Ob", 1956-1957.]
Arktich. i Antarkt. Nauchno-Issled. Inst., Mezhd. Geofiz. God, Sovetsk. Antarkt. Exped., 6: 1-387.

surface drifters

instrumentation, currents

SEA-BED DRIFTERS

instrumentation currents
Lee, Arthur J., Dean F. Bumpus and Louis M. Lauzier, 1965.
The sea-bed drifter.
Res. Bull. Int. Comm. NW Atlantic Fish., (2) 42-47

"instrumentation", currents (bottom)
Lee, Arthur J., Dean F. Bumpus et Louis M. Lauzier, 1965.
Le "deriveur de fond": nouveau modèle de materiel derivant sous l'influence des courants près du fond.
Chaiers Oceanogr., C.C.O.E.C., 17(2):123-125.

sea drifters
Leloup, Eugène, 1966.
Observations sur la derive des courants au large de la côte belge au moyen de flotteurs de fond.
Bull. Inst. r. Sci. nat. Belg., 42(20):1-20.

instrumentation, currents

SURFACE DRIFTERS

instrumentation, currents, surface drifters
Harvey, J.G., and W.J. Gould, 1966.
A note on the design of sea-surface drifters.
J. Cons. perm. int. Explor. Mer, 30(3):358-360.

instrumentation, data acquisition
Ketchum, David D., and Raymond G. Stevens, 1962.
A data acquisition and reduction system for oceanographic measurements.
Marine Sciences Instrumentation, Inst. Soc., Amer., Plenum Press, N.Y., 1:55-60.

data analyzer processing

instrumentation, data analyzer (processing)

automatic data handling
Boyle, A. Raymond, 1964
Automatic data handling in hydrographic work.
Int. Hydrogr. Rev., 41(2):149-155.

instrumentation, data-processing
Furuhata, Tsuneo, 1962
New automatic data processing machine. MERIAC-1-F.
J. Oceanogr. Soc., Japan, 18(3):117-129.

instrumentation, data collection
Johnson, E.W., 1962.
ASWEPS Shipboard System, a new concept in the automated collection of oceanographic data.
Proc. Inst. Radio Eng., 50(11):2252-2254.

instrumentation, data processing
Laing, Joseph T., 1962.
A data processing and display instrument for oceanographic research.
Marine Sciences Instrumentation, Inst. Soc., Amer., Plenum Press, N.Y., 1:65-76.

automatic data handling
Mattila, Erkki, and Sakari Pajunen, 1964
Automation in hydrographic surveys: digital instruments and their operation on board the Finnish Survey Vessel "Tauvo".
Int. Hydrogr. Rev., 41(2):157-163.

instrumentation, data analyzer
Seleznev, W.M., and G.P. Sobolev, 1961.
[Semi-automatic integral analyzer for the processing of obs ervation data.]
Trudy Inst. Okeanol., Akad. Nauk, SSSR, 47:108-118.

instrumentation, data logging
Sudweeks, R.W. 1970.
An automatic data-logging system at sea in HMS Hecla.
Electronic Engineering in Ocean Technology. Swansea, 21-24 Sept. 1970. I.E.R.E. Conf. Proc. 19: 237-250.

instrumentation, data logging
Whitfield, G.R. 1970.
A data recording and transmitting system.
Electronic Engineering in Ocean Technology, Swansea, 21-24 Sept. 1970, I.E.R.E. Conf. Proc. 19: 477-480.

density

instrumentation, density

instrumentation, density
Føyn, Ernst 1967.
Density and oxygen recording in water.
In: Chemical environment in the aquatic environment, H.L. Golterman and R.S. Clymo, editors, Proc. I.B.P. Symp. Amsterdam, Oct. 1966: 127-132.

instrumentation
Føyn, Ernst, 1958.
Hydrographical and biological investigations in Oslofjord by means of the densigraph and the oxymeter. J. Mar. Res., 17:174-177.

instruments
Føyn, E., 1953.
The densigraph (an apparatus for continuous density recording of sea water). Rept. Norwegian Fish. Mar. Invest. 10(5):1-9, app., 7 textfigs.

instrumentation
Hargens, C.W., 1957.
Portable liquid density instrument employing transistors. Rev. Sci. Instr., 28(11):921-923.

instrumentation, density
Møller, K. Max, and P. Ottolenghi, 1964.
The manufacture of small standardized glass float for calibrating density gradients.
C. R. Trav. Lab. Carlsberg, 34(3):169-185.

instrumentation
Richardson, William S., 1959
A vibrating rod densitometer.
Nat. Acad. Sci.-Nat. Res. Counc., Publ. No. 600:113-117.

instrumentation, density
Vershinsky, N.V., and O.N. Mikheilova, 1965.
On application of some vibratory system for water density determinations. Electronic instruments for oceanographic investigations. (In Russian; English abstract).
Trudy, Inst. Okeanol., Akad. Nauk, SSSR, 74:13-16

depressor

instrumentation, depressor

See: instrumentation, biological (at sea)—DEPRESSOR

depth

instrumentation, depth

See also: instrumentation, echo-sounders

instrumentation, depth
Anon., 1956.
Two new 'fish master' echo sounders.
World Fish., London, 5(11):44-45.

instrumentation, depth
Anon., 1957.
The Simrad master sounder. Norwegian Fish. News, 4(1):19

instrumentation, depth
Abbotts, W.E., 1963
The solartron transdeptor.
Int. Hydrogr. Rev., 40(2):179-185.

instrumentation

Beckmann, W. C., A. C. Roberts & B. Luskin, 1959.

Sub-bottom depth recorder.
Geophysics, 24(4):749-760.

instrumentation, depth

Branham, D. W., 1965.
Shipboard oceanographic survey system.
Ocean Sci. and Ocean Eng., Mar. Techn. Soc., Amer. Soc. Limnol. Oceanogr., 1:341-362.

instrumentation, depth

Cushing, D.H., and I.D. Richardson, 1955.
A triple frequency echo sounder.
Min. Agric., Fish. & Food., Fish. Invest. (2) 20(1):18 pp.

instruments

EDO Corporation, 1953.
EDO Model 185 Depth Sounder. Int. Hydrogr. Rev. 30(2):157, 1 fig.

instrumentation, depth

Haines, R.G., 1966.
A gating system for deep water sounding.
Int. Hydrogr. Rev., 43(1):43-51.

instrumentation, depth

Haines, R.G., 1963
Developments in ultrasonic instruments.
Int. Hydrogr. Rev., 40(1):49-57.

instrumentation, depth

Hamon, B.V., D.J. Tranter, and A.C. Heron, 1963.
A simple integrating depth recorder.
Deep-Sea Res., 10(4):457-458.

instrumentation, bathymetry

Hess, Frederick R., 1967.
Project ADD.
J. Ocean Techn., 1(1): 32-36

instrumentation, depth

Hester, Frank J., Donald C. Aasted and Robert W. Gilkey, 1963
The bathykymograph, a depth-time recorder.
U.S.F.W.S. Spec. Sci. Rept., Fish., No. 441: 5 pp.

instrumentation, depth-sensing

Holmes, Robert W., and James M. Snodgrass, 1961

A multiple-detector irradiance meter and electronic depth-sensing unit for use in biological oceanography.
J. Mar. Res., 19(1):40-56.

instruments

Isaacs, J.D., and A.E. Maxwell, 1952.
The ball-breaker, a deep water bottom signalling device. J. Mar. Res. 11(1):63-68, 3 textfigs.

instruments

Japan, Hydrographic Office, 1953.
Precise echo sounder for shallow water.
Int. Hydrogr. Rev. 30(2):143-144, 6 figs.

instrumentation, depth

Jollymore, P.G., 1971.
A portable digital sounding system for Arctic use. Int. hydrogr. Rev. 48(2): 35-42.

instrumentation, depth

Kelvin, Hughes (Marine) Ltd., 1961
The new shallow water depth indicator.
Int. Hydrogr. Rev., 38(2):67-70.

instrumentation, depth

Key, Charles R., Gerg J. Moss, Jr. and Frank H. MacDonald, 1968
Precision bathymetric detector.
Mar. Sci. Instrument. 4: 78-86

instruments

Kiernan, E. F., 1947.
Telemetering fathometer. Electronics 20(10): 96-98.

instrumentation, depth

MacPhee, S.B., and G.E. Awalt, 1971.
Automatic digital instruments for deep ocean bathymetry.
UnderSea Technol. 12(2): 23-24.

instruments

Millán, L., 1954.
Un nuevo aparato de sondeo. Industria Conservera 20(179):159.

depth-distance recorder

Miller, S.M., H.B. Moore and K.R. Kvammen, 1953.
Plankton of the Florida Current. 1. General conditions. Bull. Mar. Sci. Gulf and Caribbean 2(3): 465-485, 27 textfigs.

instrumentation, bathymetry

Mitson, R.B., P.G. Griffiths and C.R. Hood, 1967.
An underwater acoustic link.
Deep-Sea Research, 14(2):259-270.

instrumentation, depth

Popovici, Richard G., and Thomas K. DeWitt, 1965/
Precision Fathometer recorder.
Ocean Sci. and Ocean Eng., Mar. Techn. Soc., Amer. Soc. Limnol. Oceanogr., 2: 663-686.

instrumentation, depth

Rebikoff, D., 1969.
Underwater mapping telemetry for deep offshore installations. Oceanol. int. 69, 4: 6 pp., 8 figs.

Instrumentation, depth

Sasaki, Tadayoshi, 1955.
Three instruments constructed and employed in Japan.
Proc. UNESCO Symp. Phys. Oceanogr., Tokyo, 1955: 46-48.

instrumentation, depth

Saunders, John M., and Robert W. Havey 1968.
Development of a narrow-beam, bathymetric programmer.
Mar. Sci. Instrument. 4: 70-77.

depth meters, max.

Scripps Institution of Oceanography, 1949
Marine Life Research Program, Progress Report to 30 April 1949, 25 pp. (mimeographed), numerous figs. (photo + ozalid). 1 May 1949.

instrumentation, depth

Single, H.C., 1960.
Portable depth finder for small boats.
Electronics, 33(9):50-51.

instrumentation

Stephens, F.H., jr., and F.J. Shea, 1956.
Underwater telemeter for depth and temperature.
Spec. Sci. Rept., Fish., No. 181:22 pp.

instruments

Sysoev, N.N., 1956.
[Method and apparatus for measuring ocean depths]
Trudy Inst. Okeanol., 19:156-163.

instrumentation
depth determination

Takenouti, Yositada, and Masashi Yasui, 1960.
A note on the ball breaker and ball.
Repts. of JEDS, 1:173-180.

Also in Oceanogr. Mag., 11(2) (same pagination)

instrumentation, depth recorder

Terekhin, Yu.V., and V.F. Shermazan, 1966.
Instrument for determining the range of depths during their measurement using the "Ladoga" recorder. (In Russian).
In: Metody i Pribory dlya Issledovaniya Fizicheskikh Protsessov v Okeane, A.N. Paramonov, editor, Izd-vo Naukova Dumka, Kiev, 202-207.
Translation: JPRS:39,881, 13 Feb. 1967 (Clearinghouse Fed.Sci.Tech.Info., U.S. Dept.Commerce), 185-191 (multilithed).

instrumentation, depth

Tucker, M.J., R. Bowers, F.E. Pierce and B.J. Barrow, 1963.
An acoustically telemetering depth gauge.
Deep-Sea Res., 10(4):471-478.

instrumentation, depth

Udintsev, G.B., G.N. Lunarsky, V.I. Marakuiev, L.G. Barinov and V.N. Sedelnikov, 1962
The use of the phototelegraphic apparatus "Ladoga" for depth recordings measured with echo sounders. (In Russian).
Okeanologiia, Akad. Nauk, SSSR, 2(6):1093-1103

instrumentation, depth

Weeks, C.G. McQ., 1971.
The use of a dual frequency echo sounder in sounding an irregular bottom. Int. hydrogr. Rev. 48(2): 43-49.

depth indicators

Wiegel, R.L., and J.W. Johnson, 1960
Ocean currents, measurement of currents and analysis of data.
Waste disposal in the marine environment, Pergamon Press: 175-245.

instrumentation, depth

Wilson, R.R., 1962
A precision echo-sounder.
Muirhead Technique, Jour. Instr. Engineering, Beckenham, 16(1):6-8.

Also in:
Collected Reprints, Nat. Inst. Oceanogr., Wormley, 10.

depth-determination of thermometer

instruments, depth determination of instruments

Krause, Gunther, and Gerold Siedler, 1962
Zur kontinuierlichen Bestimmung der Tiefenlage von Schleppgeräten im Meere.
Kieler Meeresf., 18(1):29-33.

German and English summary

instrumentation

Nakai, J., 1935
Determination of the unprotected thermometer depth. Umi to Sora, 15(2): 53-55.

instrumentation, desalination
Coit, R.L., and Y.S. Touloukian, 1957.
Flash evaporator for distillation of sea water.
Westinghouse Engineer, 7:58-60.

Abstr. in: Publ. Health Eng. Abstr., 37(9):28.

detection

instrumentation, detection

Instrumentation, detection
Bouigues, A., and P. Michaud, 1962.
Un appareil de detection sous-marine.
Annales de radioelectricite, Paris, 17(67):59-70.

instrumentation (Asdic)-detection
Haines, R.G., 1963.
Locating underwater obstructions - procedure adopted for use in port approaches.
Dock & Harbour Authority, 44(514):121-122.

diffusion

instrumentation, diffusion

instrumentation, diffusion
Kullenberg, Gunnar, 1968.
Measurements of horizontal and vertical diffusion in coastal waters.
Rept. København Univ. Inst. fys. Oceanogr., 3:50 pp. (multilithed).

distance measurement

instrumentation, distance
Iuda, Kanji, and Kanji Hiraoka 1963
Measurement of the nautical distance by a hydrodist. (In Japanese; English abstract).
Bull. Coast. Oceanogr. 2(1):1-9.
Also in: Coll. Repr. Fac. Oceanogr. Tokai Univ. 1965.

diving

instrumentation diving
Achurch, I.C. and M.A. Garnett, 1969.
A comparative study of the economics of four diving systems. Oceanol. int. 69, 1: 10 pp., 8 figs.

Anon., 1965. instrumentation, diving
Dual mask prevents divers jaw ailments, side effects.
Undersea Technology, 6(6):15.

instruments
Bashchieri-Salvadori, F., 1959
The aqualung: an aid in underwater exploration.
J. Mar. Biol. Ass. India 1(1): 50-53.

instrumentation, diving
Breslau, L.R., J.M. Zeigler and D.M. Owen, 1962.
A self-contained portable tape recording system for use by SCUBA divers.
Bull. Inst. Océanogr., Monaco, 59(1235):1-4.

instrumentation, biological (at sea)
Fager, E.W., A.O. Flechsig, R.F. Ford, R.I. Clutter and R.J. Ghelardi, 1966.
Equipment for use in ecological studies using Scuba.
Limnol. Oceanogr., 11(4):503-509.

instrumentation, diving
Flemming, N.C., 1965.
Operational diving with oxy-helium self-contained diving apparatus.
Malta '65:3-12.

instrumentation, diving
Gill, J.S., R.J. Morris and M.G. Edwards 1970.
A helium speech processor operating in the time domain.
Electronic Engineering in Ocean Technology, Swansea, 21-24 Sept. 1970, I.E.R.E. Conf. Proc. 19:523-528.

instrumentation, diving
Greene, Maltone W., 1968.
Oxygen and carbon dioxide partial pressure sensors for hyperbaric applications
Mar. Sci. Instrument 4: 116-126.

instrumentation
Grousson, R., 1960
Report on RANA equipment of the French Navy Hydrographic Office. Suppl. Intern. Hydro. Rev., 1: 45-55.

instrumentation, diving
Haigh, Kenneth R., 1970.
Future electronic instrumentation for submersibles, habitats and divers.
Electronic Engineering in Ocean Technology, Swansea, 21-24 Sept. 1970, I.E.R.E. Conf. Proc. 19:381-402.

instrumentation, diving
Halfon, A. 1968.
Instrumentation for an operational saturation diving system.
Mar. Sci. Instrument 4: 127-134.

diving apparatus, use of
Japan, Hydrographic Office, 1954.
[Underwater photography by using a diving apparatus.] Publ. No. 981, Hydrogr. Bull. 45:263-268, 4 pls.

instrumentation, diving
Krasberg, Alan R., 1968.
An operational closed-circuit constant O_2 diving apparatus.
Mar. Sci. Instrument. 4:135-41

instrumentation, diving
Mellquist, V.G., 1968.
Electric tools for diver operation.
J. Ocean Techn., 2(4):49-50.

instrumentation, diving
Nashimoto, Ichiro, 1969.
A method of adjusting decompression pressure and time for caisson and diving work. Oceanol. int. 69, 4: 2 pp.

instrumentation, diving
Padden, J.B., 1971.
Filtration system for divers' breathing air.
J. mar. techn. Soc. 5 (3):35-41.

instrumentation, diving
Parker, Eugene K., 1962.
Scuba as a tool for scientists.
Marine Sciences Instrumentation, Instr. Soc. Amer., Plenum Press, N.Y., 1:310-320.

instrumentation, diving
Partridge, C.J., 1970.
Electronic aids for exploration divers.
Electronic Engineering in Ocean Technology, Swansea, 21-24 Sept. 1970, I.E.R.E. Conf. Proc. 19: 423-446.

instrumentation, diving
Roworth, D.A.A. 1970.
A practical processor for helium speech.
Electronic Engineering in Ocean Technology, Swansea, 21-24 Sept. 1970, I.E.R.E. Conf. Proc. 19: 529-540.

instrumentation, diving
Sigl, Walter, Ulrich von Rad, Hansjörg Oeltzschner, Karl Braune and Frank Fabricius 1969.
Diving sled: a tool to increase the efficiency of underwater mapping by Scuba divers
Marine Geol., 7(4):357-363

instrumentation, diving
Thompson, Tommy, 1969.
Diving equipment for professionals.
Oceanology intl, 4(2): 42-44.

instrumentation, divers
Williams, J.S., 1968
Underwater "toolsheds" will aid diving industry.
Hydrospace, 1(5): 22-23.

instrumentation, diving
Zhilzov, A.A., and V.P. Nikolejev, 1965.
The underwater talk-back device. Electronic instruments for oceanographic investigations. (In Russian; English abstract).
Trudy Inst. Okeanol., Akad. Nauk, SSSR, 74:82-84.

instrumentation drogues

instrumentation, drogues
See: instrumentation, currents

droplets

instrumentation, droplets
Hanson, A.R., E.G. Domich and H.S. Adams, 1963.
Shock tube investigations of the breakup of drops by air blast.
Physics of Fluids, 6(8):1070-1080.

dropsonde

instrumentation, dropsonde
Smith, Paul Ferris, William S. Richardson and Dennis Coburn 1968.
A digital tape recording dropsonde.
Mar. Sci. Instrument. 4: 578-584.

Echo sounders

instrumentation, geological (at sea), echo-sounders
See also: instrumentation, depth

instrumentation
Ahrens, E., 1960.
Automatic echo-sounding systems.
J. Inst. Navig., 13(2):173-177.

instrumentation, echo sounding
Brookes and Gatehous, Ltd., 1961
A miniature echo sounder for inshore navigation and survey.
Int. Hydrogr. Rev., 38(2): 57-65.

instrumentation, echo-sounding
Cooke, C.H., 1970
Automatic estimation of catch in sea-bed trawling.
Electronic Engineering in Ocean Technology. Swansea, 21-24 Sept. 1970, I.E.R.E. Conf. Proc. 19:81-94.

instrumentation
Eide, R., undated
Locating fish with echo sounder. Simonsen Radio, A/S, Oslo.

instrumentation, echosounders
Engelmann, I., 1967.
Towed echosounders for parallel sounding.
Int. hydrogr. Rew., 44(2):7-10.

echo-sounders
Fahrentholz, S., 1963
Profile and area echograph for surveying and location of obstacles in waterways.
Int. Hydrogr. Rev., 40(1):23-37.

instrumentation - echo-sounding machine
Farquharson, W. I., 1936.
Topography with an appendix on magnetic observations. John Murray Exped., 1933-34, Sci. Repts. 1(2):43-61, 6 pls., 6 charts.

instrumentation, echo-sounding
Forbes, S.T., 1970.
Quantitative fish-counting echosounder
Electronic Engineering in Ocean Technology. Swansea, 21-24 Sept. 1970, I.E.R.E. Conf. Proc. 19:95-108.

instrumentation
Gerhardsen, T.S., Undated
The use of echo ranging equipment in the herring fisheries in the Norwegian Sea. An account of the present status and an indication of the future trend. Sonar Dept. Simonsen Radio AS, Ensjøveisen 20, Oslo. unnumbered pp.

instrumentation, echo-sounding
Hashimoto, Tomiju, Yoshinobu Maniwa and Masuo Kato, 1965.
Research on telesounders (wireless remote control echosounding).
Int. Hydrogr. Rev., 42(2):59-65.

echo sounders, ultrasonic
Hashimoto, T., and M. Nishimura, 1959.
Reliability of bottom topography obtained by ultrasonic echo-sounder.
Int. Hydrogr. Rev., 36(1):43-50.

instrumentation, echo sounding
Hervieu, R., 1961.
Simulateur automatique de sondage continu à la mer suivant une route determinée.
Navigation 9(33):64-67.

instrumentation, echo-sounding
Hickley, Thomas J., 1965.
Some recent systems development by U.S. Coast Geodetic Survey.
Int. Hydrogr. Rev., 42(1):43-56.

instrumentation, echo-sounding
Huestis, Thomas H.W., 1964.
Coast & Geodetic gets narrow beam sounder.
Undersea Techn., 5(6):28-29.

instrumentation, sounding
Kroebel, W., 1961.
Zur Messmethodik von ozeanographischen Sondenmessgeräten.
Kieler Meeresf., 17(1):17-24.

instruments
Meschkat, A., 1951.
Der Entwicklungstand das schreibenden Echolotes.
Fischereiwelt 3(12):191-193.

echo sounders
Miyajima, J., 1963
Miniature recording echo sounder.
Int. Hydrogr. Rev., 40(1):39-45.

instrumentation, echo-sounding
Murray, H.W., 1942.
The 808 fathometer. Ch. 5. Echo Sounding, pp. 437-553 in: Hydrographic Manual (reised 1942 Edit.

instrumentation, echo sounder
Nawar, F.G., W.L. Liang and C.S. Clay, 1966.
A pulse-compression echo sounder for ocean bottom survey.
J. geophys. Res., 71(22):5279-5282.

instrumentation, echo-sounding
Sothcott, J.E.L., and N.H. Benning, 1970.
Heave correction in echo sounding.
Electronic Engineering in Ocean Technology, Swansea, 21-24 Sept. 1970, I.E.R.E. Conf. Proc. 19: 159-174.

instrumentation, echo sounding
Spiess, F.N. and John D. Mudie, 1970.
1. General observations. 7. Small scale topographic and magnetic features. In: The sea: ideas and observations on progress in the study of the seas, Arthur E. Maxwell, editor, Wiley-Interscience 4(1): 205-250.

instruments, echosounding
Truskanov, M.D., and M.N. Shcherbino, 1960.
The zone of functioning of the echo sounder.
Nauchno-Techn. Biull., PNIRO, 1(11):24-28.

instrumentation, echosounding
Truskanov, M.D., and M.N. Shcherbino, 1959.
Determination of the capability of the echo-sounder.
Nauchno-Techn. Biull., PNIRO, 2(10):22-24.

instrumentation echo sounding
Tscherning, H., 1947
New types of ultra-sonic sounders. Int. Hydr. Rev., 24:90-95, 12 figs.

instrumentation echo sounding
Tucker, D.J., 1960
Directional echo sounding, some possible improvements in equipment and technique.
Int. Hydrogr. Rev., 37(2): 43-53.

echo sounder corrections
Udintsev, G.B., 1954.
On the question of correction of depths recorded with echo sounders. Trudy Inst. Okeanol., 8:305-314.

instrumentation, echo-sounding
Whitfield, G.R. 1970.
A data recording and transmitting system.
Electronic Engineering in Ocean Technology. Swansea, 21-24 Sept. 1970, I.E.R.E. Conf. Proc. 19: 477-480.

eddy flux

instrumentation, eddy flux
Högström, Ulf., 1967.
A new sensitive eddy flux instrumentation.
Tellus, 19(2): 230-239

electric field

instrumentation, electric field
Bomanitsky, E.A., and M.G. Strunsky, 1964
The variations of a natural electric field at sea. (In Russian).
Okeanologiia, Akad. Nauk, SSSR, 4(2):325-339.

ELECTRICAL

instrumentation, electrical

instruments
Barnes, H., and A.G. Randall, 1954.
Electrical lead-through plugs. Letter to Editor, Deep-Sea Res. 1(4):281-282, 2 textfigs.

instrumentation, electrodes
Duxbury, Alyn C., 1963.
Calibration and use of a galvanic type oxygen electrode in field work.
Limnology and Oceanography, 8(4):483-485.

instrumentation, at sea
Covey, Charles W., 1965.
Deck handling equipment and electrical systems.
UnderSea Techn., 6(10):24-27; 30-33.

instrumentation, electrical
Grant, H.L., 1965.
Compressed rubber junction box for underwater use
Deep-Sea Research, 12(2):221-223.

instrumentation, electrical
Hess, F.R., and L.V. Slabaugh, 1965.
A shipboard cable-hauling system for large electrical cables.
Deep-Sea Res., 12(4):537-538.

instrumentation, electrical
Holzkamm, Fritz, 1964.
Bodenberührungsscaalter für Geräte mit Einleiter-kabeln.
Kieler Meeresf., 20(2):136-137.

instrumentation
Isaev, B.A., 1965.
Second model of an electrometric ventilator. (In Russian).
Trudy Morsk. Gidrofiz. Inst., Akad. Nauk, SSSR, 31:75-80.

instrumentation, electrical
Lakshmanan, T.K., 1963.
An improved selenium photovoltaic cell.
Mar. Sci. Instrument., Plenum Press, 2:183-187.

instruments
Leukus, D.I., and B.N. Kabanov, 1956.
Non-polarized electrodes for measurement of small differences in potential in the sea.
Trudy Inst. Okeanol., 19:112-116.

instruments
Mironov, , 1950.
[Electrodes for measurements in the sea.] Dok. Akad. Nauk, USSR, 70:825-828.

instrumentation, electrical
Mühlersen, R., 1961.
Electrode effect measurements above the sea.
J. Atmos. Terres. Phys., 20(1):79-81.

instrumentation, potentiometers
Musiakov, L.A., 1965.
On the verification of automatic potentiometers. (In Russian).
Trudy, Gosudarst. Okeanogr. Inst., No. 87:166-172.

instrumentation, electrical
Nagasaka, Koichi, 1967.
Handy charger for oceanographic instruments.
Oceanogrl Mag., 19(2):197-199.

instruments
Tucker, M.J., 1955.
A twin-T R C oscillator. Electron. Eng., 27:346-347.

electrical conductivity

instrumentation, electrical conductivity
Aranson, V.A., Yu.M. Gusev and B.V. Shekhvatov, 1971.
An instrument to measure electrical conductivity of the sea water from aboard the ship. (In Russian; English abstract). Okeanologiia 11(2): 314-319.

instrumentation, electrical conductivity
Khundzhua, G.G., and G.N. Khristoforov, 1965.
The registration of turbulent fluctuations of sea water electrical conductivity. (In Russian).
Okeanologiia, Akad. Nauk, SSSR, 5(4):734-739.

instrumentation, electrical conductivity
Pavlenko, Yu.V., and G.N.Khristoforov, 1966.
New design of a contactless pickup for determining electrical conductivity of sea water. (In Russian).
In: Metody i Pribory dlya Issledovaniya Fizicheskikh Protsessov v Okeane, A.N. Paramonov, editor, Izd-vo Naukova Dumka, Kiev, 96-102.
Translation: JPRS:39,881, 13 Feb.1967 (Clearing-house Fed.Sci.Tech.Info., U.S.Dept.Commerce), 72-79 (multilithed).

ELECTROCHEMISTRY

instrumentation, electrochemical
Collins, J.L., 1963.
Solion electrochemical devices.
Mar. Sci. Instrument., Plenum Press, 2:163-167.

electromagnetic

instrumentation, electromagnetic
English, W.N., D.J. Evans, J.E. Lokken, J.A. Shand and C.S. Wright, 1962.
Equipment for observation of the natural electromagnetic background in the frequency range 0.01-30 cycles per second.
Marine Sciences Instrumentation, Instr. Soc. Amer., Plenum Press, N.Y., 1:321-333.

electronics

instrumentation, electronics
Arkwright, J.H. 1970.
Pressure-balanced electronics - performance testing and application.
Electronic Engineering in Ocean Technology, Swansea, 21-24 Sept. 1970, I.E.R.E. Conf. Proc. 19: 67-79.

instrumentation, electronics
Berman, Alan, 1967.
The role of electronics in underwater search, rescue and recovery.
1967 IEEE Int.Conv.Rec.(5):12-16.

instrumentation, electronic components
Buchanan, Chester L., and Matthew Flato, 1962.
Influence of a high hydrostatic pressure environment on electronic components.
Marine Sciences Instrumentation, Instr. Soc., Amer., Plenum Press, N.Y., 1:119-136.

instrumentation, electronics
Covey, Charles W., 1965.
Shipboard electronics for oceanography and marine science. Undersea Techn., 6(9): 24-34.

instrumentation
Gigas, E., and K. Ebeling, 1957.
Elektrisches Auge. Mitt. Inst. Angew. Geodäsie, No. 26:15 pp.

instrumentation, electronics
Jackson, James M., and George R. Koonce, 1965.
Thin film circuits for deep sea applications.
Undersea Technology, 6(6):31-32,34.

instrumentation, electronic
Kirkley, John, and George Nussear, 1965.
Pressure insensitive components.
Undersea Techn., 6(9):39.

instrumentation,
Lawrence, L. George, 1967.
Electronics in oceanography.
Howard W. Sams & Co., The Bobbs-Merrill Co., Inc. 288 pp. $4.95.

instrumentation, electronics
Ramsay, Selwyn P., 1968.
Electronics at sea.
Oceanol. Intl., 3(7):46-50.

instrumentation, electronic
Tucker, M.J., 1966.
Electronic engineering in oceanography - an introductory survey.
Proc. I.E.R.E. Conf. Electronic Eng. Oceanogr., 2 pp.

instruments
Tucker, M.J., and L. Draper, 1955.
A high-Q R C feedback filter for low audio frequencies. Electron. Eng., 27:451-453.

instrumentation, electrical
Van Haagen, R.H., and E.E. Aagaard, 1963.
Ultrasonically cleaned electrode.
Mar. Sci. Instrument., Plenum Press, 2:29-31.

instrumentation, electronics
Vershinsky, N.V., 1965.
Marine electronics. Electronic Instruments for oceanographic investigations. (In Russian)
Trudy Inst. Okeanol., Akad. Nauk, SSSR, 74:3-12.

instrumentation, electronic
Wisotsky, S., 1965.
Sonic underwater long range aid to navigation.
Undersea Techn., 6(9):47-49.

instrumentation, electrolytic integrator
Kurpakov, Yu.A., and G.V. Gindjuk 1968.
Experience of using electrolytic integrator X-603 in sea conditions. (In Russian)
Met. Gidrol. (1): 96-97.

electron microscope

instrumentation, electron microscope
*Hay, William W., and Philip A. Sandberg, 1967.
The scanning electron microscope, a major break-through for micropaleontology.
Micropaleontology, 13(4):407-418.

instrumentation, electron microscope
*Sandberg, Philip A., 1968.
A new specimen stub for sterophotography with the scanning electron microscope.
Micropaleontology, 14(4):489-498.

instrumentation, electrophoresis
Camps, J.M., E. Arias y C. Martínez, 1971.
Aparato de electroforesis vertical con gel de poliacrilamida. Investigación pesq. 35(2): 521-530.

elutriator

elutriator
Scripps Institution of Oceanography, 1949
Marine Life Research Program. Progress Report to 30 April 1949, 25 pp. (mimeographed), numerous figs. (photo.+ozalid). 1 May 1949.

instrumentation

instrumentation
Faure, M., 1953.
Calcul des pertes d'énergie dans un estuaire à marée (Gironde). Principe et exécution du calcul a l'aide d'une machine mathématique.
La Houille Blanche, No. Spécial B/ 1953:157-169.

engineering

instrumentation, engineering
USA, National Academy of Sciences-National Research Council, Committee on Oceanography, 1959.
Oceanography, 1960 to 1970. 7. Engineering needs for ocean exploration, 22 pp.

evaporation

instrumentation, evaporation
Cheng, Y., 1955.
[An instrument for recording the rate of evaporation from water surface.] Umi to Sora 32(3):44-46.

instrumentation, evaporation
Taylor, R.J., and A.J. Dyer, 1958.
An instrument for measuring evaporation from natural surfaces. Nature 181(4606):408-409.

instrumentation, evaporation
Walker, Donald C., 1963.
New instrument for measuring evaporation: application to evaporation of water through thin films.
Rev. Sci. Instruments, 34(9):1006-1009.

instrumentation expendable

instrumentation, expendable

See also: instrumentation, temperature (bathythermograph, expendable)

instruments expendable
Eden, H. Francis, and C. Michael Mohr, 1969.
Orifice flow and expendable oceanographic instrumentation.
J. mar. techn. Soc., 3(2): 71-72

instrumentation, expendable
Haynes, Alex H., Walter L. Heid and Gerald F. Appell, 1965.
Depth accuracy of expendable oceanographic instruments.
Mar. Sci. Instrument., Plenum Press, 3:149-163.

instrumentation, expendable
Knoll, Denys W., 1965.
Oceanographic sensors and expendable instrumentation.
Mar. Sci. Instrument, Plenum Press, 3:1-5.

instrumentation, expendables
Robertson, Robert M., 1965.
Expendable instrumentation.
Mar. Sci. Instrument., Plenum Press, 3:99-121.

facsimile recorder

instrumentation, facsimile recorder
Umezono, Shigeru 1967.
The sensitivity measurement and maintenance of the facsimile-receiving apparatus. (In Japanese; English abstract).
Bull. Fac. Fish. Nagasaki Univ. 22: 109-114.

filter

filter, pier end
Scripps Institution of Oceanography, 1949
Marine Life Research Program, Progress Report to 30 April 1949, 25 pp. (mimeographed), numerous figs. (photo.+ozalid). 1 May 1949.

filter, subsurface
Scripps Institution of Oceanography, 1949
Marine Life Research Program, Progress Report to 30 April 1949, 25 pp. (mimeographed), numerous figs. (photo.+ozalid). 1 May 1949.

fishing

instrumentation, fisheries
See also: instrumentation, biological (at sea)
trawls
mid-water trawls
instrumentation, echo sounders

instruments
Anon., 1951.
Designed for fishing boats. Fish. Newsletter 10(7):21, 1 fig.

Hughes MS22 echo sounder.

instrumentation, fish
Beckett, J.S., 1968.
A harpoon adapter for tagging large free-swimming fish at the surface.
J. Fish. Res. Bd., Can., 25(1):177-179.

instrumentation
Carruthers, J. N., 1959.
Simple devices for studying the geometry of various gears and for relating some commercial fishing operations to the existing water movements. In: Modern Fishing Gear of the World, Fishing News (Books). Ltd, London, pp. 254-255.

instruments
Carruthers, J.N., 1955.
Some simple oceanographical instruments to aid in certain forms of commercial fishing and various problems of fisheries research.
F.A.O. Fish. Bull., 8:130-140.

instrumentation fisheries
Chaplin, P.D., 1969.
Current developments on trawl gear. Oceanol. int., 69, 3: 6 pp., 2 figs.

instrumentation, fisheries
Cooke, C.H., 1970
Automatic estimation of catch in sea-bed trawling.
Electronic Engineering in Ocean Technology, Swansea, 21-24 Sept. 1970, I.E.R.E. Conf. Proc. 19: 81-94.

instrumentation, fisheries
Crowther, Harold E., 1969.
Hardware research - key to improved fishing.
Undersea Techn., 10(1):54-55, 61. (popular).

instrumentation
Dickson, W., 1961.
Trawl performance. A study relating models to commercial trawls.
Marine Research, Edinburgh, (1):1-48.

instrumentation fisheries
Elliott, F.E., 1969.
Summary on the status of electric fishing.
Oceanol. int. 69, 3: 7 pp.

instruments
Huzita, T., 1955.
[An experiment on the trial manufacture of lead plate depth-finder for fishing tools.]
Mem. Fac. Fish., Kagoshima Univ., 4:42-46

instrumentation, fisheries
Kahl, M. Philip, Jr., 1963
Technique for sampling population density of small shallow water fish.
Limnol. and Oceanogr., 8(2):302-304.

instrumentation, fisheries
Klima, Edward F., 1970.
Development of an advanced high seas fishery and processing system.
J. Mar. Techn. Soc., 4(5): 80-87

instrumentation fisheries
MacLennan, D.N., 1969.
Instrumentation for the on-line measurement of the environment and performance of fishing gears. Oceanol. int. 69, 3: 9 pp., 7 figs.

instruments
Mitson, R.B., and R.J. Wood, 1961
On automatic method of counting fish echoes.
J. du Conseil, 26(3): 281-291.

instrumentation, fisheries
Moore, W.H., 1954.
A new type of electrical fish-catcher.
J. Animal Ecology 23(2):373-375.

gear
Rollefsen, G., 1953.
The selectivity of different fishing gear used in Lofoten. J. du Cons. 19(2):1910194, 1 textfig.

instrumentation, fisheries
Sasakawa, Yasuo, 1965.
Fishing efficiency of King Crab tangle net in relation to gear structures.
Bull. Hokkaido Reg. Fish. Res. Lab., No.30: 31-44.

instrumentation fisheries
Stevenson, W.H. and E.A. Scaefers, 1969.
Ocean engineering in United States fisheries production. Oceanol. int. 69, 3:9 pp., 4 figs.

instrumentation, fisheries
Terentyev, A.V., editor, 1966.
Mechanization and automation of labour-consuming processes in the fishing industry. (In Russian; English summary).
Trudy vses. nauchno-issled. Inst. morsk. ryb. Khoz. Okeanogr. (VNIRO), 59:206 pp.

instrumentation, fisheries
Trefethen, P.S., J.W. Dudley and M.R. Smith, 1957.
Ultrasonic tracer follows tagged fish.
Electronics, 30(4):156-160.

instrumentation, fisheries
Yuen, H.S.H., 1958.
A preliminary report on the sea-scanar, an ultra-sonic fish-finder (Abstract).
Indo-Pacific Fish. Counc., Proc., 7th Session, (2/3):83.

instrumentation
Zupanovic, S., and F. Grubisic, 1958.
Fishing effectiveness of trawl nets as resulting from experiments with wire cable bridles and with manilla covered bridles.
Acta Adriatica, 8(12):27 pp.

floats

instrumentation, floats
*Adler, Cyrus, 1967.
Isobaric buoys- a new class of subsurface floats.
J. Ocean Tech., 2(1):43-49.

instrumentation, floats
Anderson, D.V., and D.H. Matheson, 1967.
An articulated limnological float.
Limnol. Oceanogr., 12(1):162-163.

instrumentation, floats
Carruthers, J.N., 1967.
The Pooh-bah automatic float.
Wat. Waste Treat. J., Sept./Oct. 1967: 2 pp.

instrumentation, floats
Matthews J.B. and D.L. Nebert, 1971
A sub-arctic pooh-bah float.
Cah. océanogr. 23(9): 795-799.

instrumentation, glass floats
Raymond, Samuel O., 1968.
Glass floats and boomerangs.
Mar. Sci. Instrument., 4: 231-240

instrumentation, glass spheres
Whitmarsh, R.B., 1966.
Sea trials of glass spheres.
GeoMar. Techn., 2(7):16-18.

instrumentation flow meter

instrumentation, flow meter
Holmes, John F., 1965.
Wide range flow meter for oceanographic measurements.
Mar. Sci. Instrument., Plenum Press, 3:251-255.

instrumentation, flowmeter
Olson, Jack R., 1972.
Two-component electromagnetic flowmeter.
Jl Mar. Techn. Soc. 6(1): 19-24.

instrumentation, flowmeter
Tungate, D.S., and D. Mummery, 1965.
An inexpensive mechanical digital flowmeter.
J. Cons. perm. int. Explor. Mer, 30(1):86.

instrumentation, flumes
Narasakao, Y., and M. Kanamori, 1957.
[On the Kagoshima University large-sized experiment-tank consisted of twin symmetric elliptical circuits.] Mem. Fac. Fish., Kagoshima Univ., 5: 64-77.

instrumentation, free-fall vehicle
Isaacs, J.D., & G.B. Schick, 1960
Deep-sea free instrument vehicle.
Deep-Sea Res., 7(1): 61-67.

instrumentation, gages
Samarin, V.G., and V.F. Tsyplukhin, 1961
Practical method of designing gas-filled, membrane-type leak pressure gages. Physics of the Sea. (In Russian).
Trudy Morsk. Gidrofiz. Inst., 23: 85-93.

Translation:
Scripta Technica, Inc. for Amer. Geophys. Union, pp. 69-75.

instrumentation, generators
Somlo, P, 1969.
Glandless turbine generators for subsea well heads. Oceanol. int. 69, 4: 4 pp., 3 figs.

Geochemistry

instrumentation, geochemistry
Carritt, Dayton E., 1965.
Marine geochemistry: some guesses and gadgets.
Narragansett Mar. Lab., Univ. Rhode Island, Occ. Publ., No. 3:203-211.

geological

instrumentation, geological
Dietz, R.S., undated(1954?).
Methods of exploring the ocean floor. In: Oceanographic Instrumentation. NRC Publ. 309: 194-209.
With discussion by B.C. "eezen
J.D. Frautschy
R. Revelle.

geological (at sea

instrumentation, geological (at sea)

instrumentation
Lisitzin, A.P., 1959
[Experience of sea geological investigations in the tropics.]
Trudy Inst. Okeanol., 35: 153-174.

instrumentation
Lisitzin, A.P., 1959
[Peculiarities of sea geological investigations in the Antarctic.]
Trudy Inst. Okeanol., 35: 121-152.

instrumentation, geological (at sea)
Lomachenkov, V.S., and K.P. Samsonov 1968
On the application of the hydroacoustic station "Paltus-M" for geologic-geomorphological investigations. (In Russian; English abstract)
Okeanologiia 8(1): 158-160.

instrumentaion, geology
Richards, Adrian F., editor, 1967.
Marine Geotechnique.
Univ. Illinois Press, 327 pp. $8.95.

instrumentation
Rouch, J., 1955.
La composition du fond de la mer. Les instruments modernes pour son étude. Scientia (6)90(73):345-350.

instruments
Waterways Experiment Station, 1949.
Subsurface exploration and sampling of soils for civil engineering purposes. Ed. H.V. Hvorslev, 521 pp.

instrumentation, geological (at sea)

MISCELLANEOUS

Bowers, R., 1963. instrumentation, geological
Oceanography - modern methods of sub-bottom using sonic reflection technique.
Science Progress, 51(201):80-84.

instrumentation, geological
Constalin, Jean-Baptiste, 1969
Appareil simple pour la séparation des fractions fines et grossières d'un sédiment.
Tethys 1(2): 569-572

instrumentation, geology, at sea
Dill, Robert F., and David G. Moore, 1965.
A diver-held vane-shear apparatus.
Marine Geology, 3(5):323-327.

instrumentation, geological
Egorov, E.N., and B.A. Popov, 1958.
[An experimental cable-way for the study of sand bottom-drifting.] Trudy Inst. Okeanol., 28:30-36.

instrumentation, geological
*Folger, D.W., 1968.
New particulate matter sampling devices and effects of technique on marine suspensate recovery.
Deep-Sea Res., 15(6):657-664.

instrumentation
France, Institut Français du Pétrole, 1967.
Techniques marines pour la recherche et l'exploitation du pétrole.
Publ. Inst. Francais Pétrole, Coll. Colloques Semin., 4: 315 pp.
Editions Technip,

instrumentation, geological (at sea
Inderbitzen, A.L., F. Simpson and G. Goss 1971.
A comparison of in situ and laboratory vane shear measurements.
J. mar. techn. Soc. 5(4): 24-34

instrumentation, geological
Jennings, David, Norman Cutshall and Charles Osterberg, 1965.
Radioactivity: detection of gamma-ray emission in sediments in situ.
Science, 148(3672):948-950.

instrumentation, geological
Kravitz, Joseph H., 1970.
Repeatability of three instruments used to determine the undrained shear strength of extremely weak, saturated cohesive sediments.
J. sedim. Petrol. 40(3): 1026-1037.

instruments
LaFond, E. C., R.S. Dietz, and J. A. Knauss, 1950
A sonic device for underwater sediment surveys.
J. Sed. Petr. 20(2):107-110, 2 textfigs.

instruments, geological (magnetic separator)
McDougall, J.C., 1961
Ironsand deposits offshore from the west coast, North Island, New Zealand.
New Zealand J. Geol. Geophys., 4(3):283-300.

instrumentation, geological (at sea)
McNary, J.F. and N. Frohlich 1970
An in situ sediment vane shear testing device.
Mar. Geol., 8(5): 367-370

instrumentation
Okolski, P., 1956.
[Un appareil pour definir la limite des terrains instables.] Inst. Nat. pour les Projets de Ports Marit., Moscow, Trav., 3:

instrumentation, geological (at sea)
Reidy, F.A., 1967.
Vehicle makes ocean bottom surveys.
Ocean Industry, 2(6): HD-44.

instruments, geology
Sato, Magoshichi, 1960
Capped cylinder bucket.
Rec. Oceanogr. Wks., Japan, Spec. No. 4: 207-209.

instrumentation, geological, at sea
Shipek, Carl J., 1965.
A new deep sea oceanographic system.
Ocean Sci. and Ocean Eng., Mar. Techn. Soc.-Amer. Soc. Limnol. Oceanogr., 2:999-1008.

instrumentation, geological
Sternberg, Richard W., and Joe S. Creager, 1965.
An instrument system to measure boundary-layer conditions at the sea floor.
Marine Geol., 3(6): 475-482.

instrumentation, geological
Wunderlich, Friedrich, 1969.
Kombiniertes Gerät zum Farben der Unterwasser-bodenfläche und zum Gewinnen eines Kernes - 100-1-Horizontalwasserschöpfer. Senckenberg. marit. [1] 50: 147-150.

geological (at sea) accessory

instrumentation, geological (at sea) - acoustic
Biriukov, S.N., E.F. Dubrov and J.A. Ivanov 1970.
Some results of the geoacoustic sounding in the Bay of Gdansk. (In Russian; English and German abstracts).
Baltica, Lietuv. TSR Mokslu Akad. Geogr. Skyr. INQUA Taryb. Sek. Vilnius 4: 169-180.

geological, beaches

instrumentation, beaches
Zinn, D.J. 1969.
An inclinometer for measuring beach slopes.
Mar. Biol. 2(2): 132-134.

instrumentation, geology (at sea) bottom topography
Mudie, J.D., W.R. Normark and E.J. Gray Jr. 1970.
Direct mapping of the sea floor using side-scanning sonar and transponder navigation.
Bull. geol. Soc. Am. 81(5): 1547-1554

instrumentation, geological (at sea)

COMPASSES

instrumentation, geological (at sea)

COMPASSES

instruments, geological (at sea)
Bouma, Arnold H., 1964.
Self-locking compass.
Marine Geology, 1(2):181-186.

instrumentation, geological (at sea)—compasses
Harrison, C.G.A., J.C. Belshé, A.S. Dunlap, J.D. Mudie and A.I. Rees 1967.
A photographic compass inclinometer for the orientation of deep-sea sediment samples.
J. Ocean Techn. 1(1): 37-39.

instrumentation, geological (at sea)
Morrison, Douglas R. and Bobb Carson, 1971.
A gyrocompass for measurement of core orientation and core behavior. Deep-Sea Res. 18(9): 935-939.

instruments
Shumway, G., 1955.
Compass-inclinometer for underwater outcrop mapping. Amer. Assoc. Petr. Geol. Bull. 39(7):1403-1404

geological (at sea) corers

instrumentation, geological (at sea)

CORERS (and accessories)

instruments
Bader, R.G., and R.G. Paquette, 1956.
A piston coring device. Deep-Sea Res. 3(4):289.

instrumentation geological (at sea) corer
Blackman Abner, and Rockne S. Anderson, 1969.
Description of a dual corer for obtaining paired piston or gravity cores.
J. mar. techn. Soc. 3(2): 73-74

instrumentation, geological
Bouma, Arnold H., and Neil F. Marshall, 1964.
A method for obtaining and analysing undisturbed oceanic sediment samples.
Marine Geology, Elsevier, Publishing Co., 2(1/2): 81-99

instrumentation, geological (at sea)—corers
Burke, John C. 1968.
Davit for handling piston corers.
J. mar. Res. 26(2): 178-181.

instrumentation, geology (at sea)
Burke, John C., 1968.
A sediment coring device of 21-cm diameter with sphincter core retainer,
Limnol. Oceanogr., 13(4):714-718.

instrumentation, geological (at sea)—corers
Burns, Robert E. 1966.
Free-fall behavior of small, light-weight gravity corers.
Mar. Geol. 4(1):1-9.

instrumentation, geological
Byrne, John V., and L.D. Kulm, 1962.
An inexpensive lightweight piston corer.
Limnol. & Oceanogr., 7(1):106-108.

instrumentation
Cawley, John H. and Donald T. Bray, 1965.
Design for a free fall instrument package.
Undersea Techn., 6(2): 18-21.

instrumentation, geology (at sea).
Craib, J.S., 1965.
A sampler for taking short undisturbed marine cores.
J. Cons. perm. Int. Explor. Mer, 30(1):34-39.

instrumentation, geology
Piston corers
Cubit, John, 1970.
A simple piston corer for sampling sand beaches.
Limnol. Oceanogr., 15(1): 155-156.

instrumentation, geological (at sea) corers
Donovan, D.T., 1967.
Henry Marc Brunel: The first submarine geological survey and the invention of the gravity corer.
Marine Geol., 5(6). 5-14

instruments
Emery, G.R., and D.E. Broussard, 1954.
A modified Kullenberg piston corer. J. Sed. Petr. 24:207-211.

instruments
Emery, K.O., and R.S. Dietz, 1941
Gravity coring instrument and mechanics of sediment coring. Bull. G.S.A. 52:1685-1714.

instrumentation, geological
Erofeev, P.N., 1961.
[Use of a corer with a ship under way.]
Trudy Inst. Okeanol., Akad. Nauk, SSSR, 47:135-138.

instrumentation, geological (at sea)
Farengeltz, I.V., 1966.
A new instrument: a new core sampler without wire, (In Russian).
Okeanologiia, Akad. Nauk, SSSR, 6(4):730-731.

instrumentation, geological (at sea)—corers
Farrell, Thomas P. 1968.
Design and development of a hard-sediment propellant corer.
Mar. Sci. Instrument. 4: 675-681.

instrumentation, geological
Felsher, Murray, 1964
A core aligner designed to recover orientated marine cores.
Limnology and Oceanography, 9(4):603-605.

instrumentation, geological (at sea)
Fowler, Gerald A., and L.D. Kulm, 1966.
A multiple corer.
Limnol. Oceanogr., 11(4):630-633.

instrumentation, (corers)
Frohlich, H., and J.F. McNary, 1969.
A hydrodynamically activated deep sea hard rock corer.
J. mar. Techn. Soc. 3(3): 53-60

instrumentation
Ginsburg, R.N., and R.M. Lloyd, 1956.
A manual piston coring device for use in shallower water. J. Sed. Petr. 26(1):64-66.

instrumentation
Glezen, E.H., 1957.
An external core catcher for oceanographic work.
J. Mar. Res. 16(1):55-59.

instrumentation, geological (at sea) (corer)
Gleason, Gale R., Jr., and Fredrick J. Ohlmacher 1965.
A core sampler for in situ freezing of benthic sediments. (abstract).
Ocean Sci. and Ocean Eng., Mar. Techn. Soc.—Amer. Soc. Limnol. Oceanogr., 2:737-741.

instrumentation, geological (at sea)—corers
Greene, Michael L. 1971.
A line cutter and a coring device powered by elastic bands
J. mar. techn. Soc. 5(3): 48-50.

instruments
Hanna, M.A., 1954.
A simple coring tube for soft sediments.
J. Sed. Petr. 24(4):263-269, 2 pls.

Instrumentation
Hersey, J. B., 1960.
Acoustically monitored bottom coring.
Deep-Sea Res., 6(2):170-172.

instrumentation, geological (at sea)
Houbolt, J.J.H.C., 1971.
Transferable deep-sea coring gear.
Marine Geol. 10(2):121-131

instrumentation, geological (corers)
Hvorslev, M.J., and H.C. Stetson, 1946
Free-fall coring tube: a new type of gravity bottom sampler. Bull. G.S.A. 57:935-950, 4 pls., 4 figs.

instrumentation, corers
Igarashi, Y., J.B. Ridlon, J.R. Campbell and R.L. Allman, 1970
Note on a mode of piston core disturbance.
J. sedim. Petrol. 40(4): 1351-1355

INSTRUMENTATION, corers
Inderbitzen, John D., and Samuel M. Brown, 1968.
"Bootstrap" corer.
J. Sed. Petr. 38(1): 159-162

instrumentation, geology (corers)
Izbekov, M.V., 1958.
Device for taking a soil core with a free-falling soil corer. (In Russian).
Problemy Arktiki, No. 4:91-93.

instruments
Janiszewski, F., 1954.
Sukcesy radzieckiej oceanologii (coring device).
Przeglad Meteor. i Hydrol. 7(3/4):192-195.

instrumentation, geological
Keller, George H., Adrian H. Richards and John H. Recknagel, 1961.
Prevention of water loss through CAB plastic sediment core liners. Instrumental Note.
Deep-Sea Res., 8(2):148-151.

instrumentation, geology, (at sea)
Kermabon, A., P. Blavier, V. Cortis and H. Delauze, 1966.
The "sphincter" corer: a wide diameter corer with watertight core-catcher.
Marine Geology, 4(2): 149-162.

instrumentation geological (at sea) (corers)
Kermabon, A., and V. Cortis 1969.
A new "sphincter" corer with a recoilless piston.
Marine Geol., 7(2): 147-159.

instruments, geological
Klovan, J.E., 1964.
Box-type sediment-coring device.
J. Sed. Petr., 34(1):185-198.

instrumentation, geology (laboratory)
*Kravitz, Joseph H., 1968.
Splitting core liners with a mica undercutter.
J. sedim. Petr., 38(4):1358-1361.

instrumentation
Kudinov, E.I., 1957.
[Vibro-piston core sampler.] Trudy Inst. Okeanol., 25:143-152.

instruments, geological
Kudinov, Ye. I., 1957.
Vibropiston soil corer. (In Russian).
Trudy Inst. Okeanol., Akad. Nauk, SSSR, (25):143-152.

instruments
Kudinov, E.I., 1951.
[Hydraulic expeller of bottom cores from corers.]
Trudy Inst. Okean., 5:11-13.

instrumentation
Kullenberg, B., 1955.
A new core sampler. Göteborgs K. Vetenskaps- och Vitterhets-Samhalles Handl., Sjätte Följden, (B), 6(15):17 pp., 8 textfigs.

Also:
Medd. Oceanogr. Inst., Göteborg, No. 26.

instruments
Kullenberg, B., 1955.
Deep-Sea coring. Repts. Swedish Deep-Sea Exped., 1947-48, Bottom Invest., 4(2):37-96.

Piston Core Sampler
Kullenberg, B., 1948
The Piston Core Sampler. Int. Hydr. Rev. 25(1):53-60, 10 figs., 1 pl.

Extract from Svenska Hydrografisk-Biologiska Kommissionens Skrifter, 3rd Ser. Hydrographi 1(2); 1947.

instruments
Kullenberg, B., 1947
The piston core sampler. Svenska Hydrografisk - Biologiska Kommissionens Skrifter. Ser.3 Hydrografi. 1(2): 46 pp. 21 text figs., 1 plate.

instruments
Kullenberg, B., 1945
(The New Swedish Core Sampler). Dansk Geologisk Forening, Bd 10, Hft.5:557-560. (Dansk Geofysisk Forening, Meddelelse No.3) Translated from the Swedish by H.U. Sverdrup and F.P. Shepard, 4 pp., 1 fig. (mimeographed). 1945.

instrumentation, geology (at sea) - corers
Langford, A., J.C. McDougall and N. Robertson 1969.
A new large-diameter piston corer and core-liner cutter.
N.Z. Jl mar. Freshwat. Res. 3(4): 595-601.

instruments
Lisicyn, A.P., V.P. Petelin and G.B. Udincev, 1954.
[A new achievement of Soviet marine geology. (The Sysoyev-Kudinov Hydrostatic core-sampler.)]
Priroda 1954(6):63-66.

T206R, E.R. Hope.

instrumentation
Lisitzin, A.P., and L.P. Barinov, 1960.
[New core sampler "the Antarctica" having a large diameter.]
Trudy Inst. Okeanol., Akad. Nauk, SSSR, 44:123-133.

instrumentation, geological (at sea) (corers)
Livingstone, D.A., 1967.
The use of filament tape in raising long cores from soft sediment.
Limnol. Oceanogr., 12(2):346-348.

instrumentation, geology, at sea (corer)
Lüneburg, Hans 1969.
Ein Rammlot mit Druckluftverschluss zur Entnahme von Sedimentproben im Litoralbereich.
Veröff. Inst. Meeresforsch, Bremerhaven, 11(2): 165-172.

instrumentation, corers
Mackereth, F.J.H., 1969.

A short core sampler for subaqueous deposits.
Limnol. Oceanogr., 14(1): 145-150.

instrumentation geological (corers)
Mackereth, F.J.H., 1958.
A portable core sampler for lake deposits.
Limnol. & Oceanogr., 3(2):181-191.

instrumentation, corers
Maitland, Peter S., 1969.

A simple corer for sampling sand and finer sediments in shallow water. Limnol. Oceanogr., 14(1): 151-156.

instrumentation, geological (corers)
Marlowe, J.I., 1967.
A piston corer for use through small ice holes.
Deep-Sea Research, 14(1):129-131.

instrumentation, geology (at sea) corers
McCoy, F.W., and R.P. Von Herzen 1971.
Deep-sea corehead camera photography and piston coring.
Deep-Sea Res. 18: 361-373.

instrumentation, geological
McManus, Dean A., 1965.
A large-diameter coring device.
Deep-Sea Research, 12(2):227-232.

instrumentation, geology
Mills, A.A., 1962.
An external core-retainer.
Marine Sciences Instrumentation, Instr. Soc., Amer., 1:244-245.
(Plenum Press, N.Y.)

instrumentation
Mills, A.A., 1961

An external core-retainer.
Deep-Sea Res., 7(4): 294-295.

instrumentation, corers
*Nesteroff, W.D., et Y. Lancelot, 1968.
Deux perfectionnements apportés au carottier à piston.
Cah. océanogr., 20(8):675-682.

instrumentation
Nevessky, E.N., 1958.
[Methods of investigation of littoral deposits by means of the vibtr-piston core sampler.]
Trudy Inst. Okeanol., 28:3-13.

instrumentation, geological, (corer)
*Pautot, Guy, 1967.
Mesures de résistivités électriques sur des carottes de sédiments marins et lacustres.
Bull. Inst. océanogr., Monaco, 67(1376):8 pp.

instrumentation, geology (corers).
Pautot, Guy, et Marceau Flucha, 1969.
Prelèvements d'échantillons par carottages et traitement.
Cah. océanogr., 21(1):47-55.

instruments
Pratje, O., 1952.
Erfarhungen bei der Gewinnung von Grundproben vom fahrenden Schiff. Deutsch. Hydrogr. Zeits. 5(1):28-31, 4 textfigs.

gear
Pratje, O., 1951.
A new core sampler and its first test. Int. Hydro. Rev. 28(1):116-121, 1 textfig.

Extract from Deutsch. Hydro. Zeit. 1950:100-107.

core sampler
Pratje, O., 1950.
Eine neue Lotröhre und ihre erste Erprobung.
Deutsche Hydro. Zeits. 3(1/2):100-107, 2 textfigs.

Instrumentation, geology (at sea)
Prych, Edmund A., and D.W. Hubbell, 1966.

A sampler for coring sediments in rivers and estuaries.

Geol. Soc., Amer., Bull., 77(5):549-556.

instrumentation, geological (at sea)
Raymond, Samuel O., and Peter L. Sacks, 1965.
Development of the boomerang sediment corer.
Mar. Sci. Instrument., Plenum Press, 3: 55-70.

instrumentation, geological (at sea)(corer)
*Reineck, Hans-Erich, 1967.
Ein Kolbenlot mit Plastik-Rohren.
Senckenberg. leth., 48(3/4):285-289.

instrumentation, geological (at sea) (corers)
Riedl, Rupert J. and Jörg A. Ott, 1970.
A suction-corer to yield electric potentials in coastal sediment layers. Senckenberg marit. 2: 67-84.

instrumentation, geology, cores
Rosfelder, Andre M., 1966.
A tubular spring value used as a tight and thin-walled core-retainer.
J. Sedim. Petrol., 36(4):973-976.

instrumentation, corers
*Rosfelder, Andre M., and Neil F. Marshall, 1967.
Obtaining large, undisturbed and orientated samples in deep water.
In: Marine geotechnique, A.F. Richards, editor, Univ. Illinois Press, 243-263.

instrumentation, geology (at sea)
Rosfelder, André M., and Neil F. Marshall, 1966.
Oriented marine cores: a description of new locking compasses and triggering mechanisms.
J. Mar. Res., 24(3):353-364.

instrumentation, geological (at sea
Rouvillois, Armelle, et Marie Rosset-Moulinier, 1969.
Mise au point d'un petit carottier pour le prélèvement sans perturbation de la partie superficielle des sédiments marins
Cah. océanogr., 21(10): 933-941

instrumentation, geological (at sea)
Sachs, Peter L., and Samuel O. Raymond, 1965.
A new unattached sediment sampler.
J. Mar. Res., 23(1):44-53.

instrumentation, geology (corers)
*Sanders, John E., 1968.
Diver-operated simple hand tools for coring nearshore sands.
J. Sedim. Petr. 38(4):1381-1386.

instrumentation, geology (at sea)
Sanders, John E., 1962
A new internal curved-gate core retainer.
J. Mar. Res., 20(3):217-222.

instrumentation, geological (corers)
Shapiro, Joseph, 1958.
The core-freezer - a new sampler for lake sediments.
Ecology, 39(4):758-759.

instrumentation, geological (at sea
Schiemer, E.W., and J.R. Schubel 1971.
A poppet valve for the Benthos gravity corer.
Chesapeake Sci., 12(2):116-118.

instrumentation
Silverman, M., and R.C. Whaley, 1952.
Adaptation of the piston coring device to shallow water sampling. J. Sed. Petr. 22(1):11-16, 4 textfigs.

instruments
Sisoev, N.N., 1951.
[Question of the application and construction of bottom corers.] Trudy Inst. Okeanol., 5:3-10.

instruments- coring device
Sysoev, N.N., 1956.
[Rational construction for percussion coring tube.]
Trudy Inst. Okeanol., 19:238-239.

instruments
Udintsev, G.B., A.P. Lisitzin, V.P. Kanaev, N.L. Zenkevich and F.I. Ganantzerov, 1956.
[Construction of a piston tube with automatic stabilized piston.] Trudy Inst. Okeanol., 19: 232-237.

instruments
Vallentyne, J.R., 1955.
A modification of the Livingstone piston sampler for lake deppsits. Ecology 36(1):139-141, 1 textfig.

instrumentation, geology
Van Den Bussche, H.K.J., and J.J.H.S. Houbolt, 1964.
A corer for sampling shallow-marine sands.
Sedimentology, 3(2):155-159.

instruments
Walker, C.R., 1955.
A core sampler for obtaining samples of bottom muds. Progr. Fish. Cult. 17(3):140.

instrumentation, geological (at sea) - corers

Winterhalter, Boris 1970.
An automatic-release piston for use in piston coring.
Mar. Geol. 8(5): 371-375.

instrumentation
Wiseman, J.D.H., 1947
Piston core-sampler. Nature 160:410

instrumentation geological (corers)
Woodruff, James L., 1970.
A self-deactivating piston for a piston corer.
Ocean Engng. 1(6): 597-599.

instrumentation, geological (piston corers)
*Woodruff, James L., 1967.
A device for releasing a piston corer and deactivating the piston.
Deep-Sea Res., 14(6):809-810.

geological (at sea) Dredges

instrumentation, geological (at sea)

DREDGES

instrumentation, geological (dredges)
Aumento, F., 1970.
Improved positioning of dredges on the sea floor.
Can. J. Earth Sci. 7(2-1): 534-539

instrumentation, geology (at sea) (dredges)
Filippov, L.A., I.A. Kraush, M.S. Barash, V.M. Lavrov, and L.V. Dmitriev, 1971.
A large cylindrical dredge. (In Russian; English abstract). Okeanologiia, 11(1):169-171.

instrumentation, geological (at sea)
Larsonneur, C., 1964.
Petite drague pour fonds meubles et peu profonds.
Bull. Soc. Linnéenne, Normandie, (10), 5:94-96.

gear
MacGinitie, G.E., 1948.
Dredges for use at marine laboratories. Turtox News 26(12):280-281.

instrumentation, dredging
Monroe, Frederick F., 1967.
A feasibility study of a wave-powered device for moving sand.
Misc. Pap. U.S. Army Engng Res. Center, 3-67: 39 pp. (multilithed)

gear
Reilly, G.P., 1950.
Dredging at coastal inlets: The seagoing hopper dredge. Inst. Coastal Eng., Univ. Ext., Univ. Calif., Long Beach, 11-13 Oct. 1950:7 pp. (mimeographed).

instrumentation, geological (at sea)
Sachs, Peter, L. 1964.
A tension recorder for deep-sea dredging and coring.
J. Mar. Res., 22(3):279-283.

geological (at sea) drilling

instrumentation, geological (at sea)

DRILLS

instrumentation, geological
Anonymous, 1961.
Plastic float for underwater drilling.
Dock & Harbour Authority, 41(486):414.

instrumentation, Geological (at sea), drills
Brooke, John, and R.L.G. Gilbert, 1968.
The development of the Bedford Institute Deep-Sea Drill.
Deep-Sea Res., 15(4):483-490.

instrumentation, geology
Gaskell, T.F., 1961.
Drilling of deep bore-holes.
Advance. Sci., 17(70):534-540.

instrumentation, Mohole
Ragland, H.W., 1965.
A dynamic positioning system for the Mohole drilling platform.
Ocean Sci. and Ocean Eng., Mar. Techn. Soc.,-Amer. Soc., Limnol. Oceanogr., 2:1145-1161.

geological (at sea) grabs

instrumentation, geological (at sea)

GRABS

instrumentation, geology
Bandy, Orville L., 1965.
The pinger as a deep-water grab control.
Undersea Techn., 6(3):36.

instrumentation, geological
Brika, Cl., et J.P. Reys, 1968.
Modifications d'une benne "orange-peel" pour des prélèvements quantitatifs du benthos de substrats meubles.
Rec. Trav. Stn. mar. Endoume, 41(57):117-121

instrumentation, bottom water
Carey, Andrew G., Jr., and Roger R. Paul, 1968.
A modification of the Smith-McIntyre grab for simultaneous collection of sediment and bottom water.
Limnol. Oceanogr., 13(3):545-549.

instrumentation, geological - at sea
Ford, J.B., and R.E. Hall, 1958.
A grab for quantitative sampling in stream muds.
Hydrobiologia, 11(3/4):198-204.

bottom sampler
LaFond, E.C., and Dietz, R.S., 1948
New snapper-type sea floor sediment sampler. J. Sed. Petrol. 18(1):34-37, 2 figs.

instrumentation (grabs)
Kudinov, E.I., 1959
[Grab net "D.B.-57" bottom bathometers.] Trudy Inst. Okeanol., 35: 175-177.

Instrumentation, geological
Lassig, J., 1965.
An improvement to the van Veen bottom grab.
Journal du Conseil, 29(3):352-353.

instrumentation, geology (at sea)
Lie, Ulf, and Mario M. Pamatmat, 1965.
Digging characteristics and sampling efficiency of the 0.1 m2 Van Veen grab.
Limnol. Oceanogr., 10(3):379-384.

instrumentation, geology
Menzies, Robert James, Logan Smith and K.O. Emery, 1963.
A combined underwater camera and bottom grab: a new tool for investigation of deep-sea benthos.
Int. Revue Ges. Hydrobiol., 48(4):529-545.

instrumentation, coastal geology
Reineck, Hans-Erich, 1963
Der Kastengreifer.
Natur un Museum, Frankfurt a.M., 93(3):102-108.

instruments
Tamura, T., 1953.
On a new bottom snapper. Bull. Fac. Fish., Hokkaido Univ., 3(4):240-242, 3 textfigs.

geological (at sea) interstitial water

instrumentation, geological (at sea)

INTERSTITIAL WATER

instrumentation, geological (at sea)-interstitial water
Burbanck, W.D., and George P. Burbanck 1967.
Parameters of interstitial water collected by a new sampler from the biotopes of Cyathura polita (Isopoda) in six southeastern states.
Chesapeake Sci. 8(1): 14-27.

instrumentation, geological ("at sea")
Creaser, Edwin P. Jr. 1971.
An interstitial water-sampling receptacle for intertidal mud flats.
J. Fish. Res. Bd Can. 28(7): 1049-1051

instrumentation, geological
Hartmann, Martin, 1965.
An apparatus for the recovery of interstitial water from Recent sediments.
Deep-Sea Research, 12(2):225-226.

instrumentation, interstitial water
*Presley, Bob J., Robert R. Brooks and Henry M. Kappel, 1967.
A simple squeezer for removal of interstitial water from ocean sediments.
J. mar. Res., 25(3):355-357.

instrumentation, geological
Reeburgh, W.S., 1967.
An improved interstitial water sampler.
Limnol. Oceanogr., 12(1):163-165.

geological (at sea) ...

instrumentation, geological (at sea)

IN MUD

instrumentation, geological (at sea)-mud
Bergeron, J., O. Leenhardt and C. Véysseyre, 1963.
De l'utilisation du "mud penetrator" dans les études des sédiments immergés superficiels.
Comptes Rendus, Acad. Sci., Paris, 256(24):5179-5181

instruments
Kondo, M., G. Hasegawa, and Y. Imura, 1953.
A mud-meter. Rept. Transport. Tech. Res. Inst. No. 4:59-65, 5 textfigs.

instrumentation, submarine geology (at sea)
Leenhardt, Olivier, 1966.
Suggestions for probing a soft mud bottom.
Int. Hydrogr. Rev., 43(1):59-68.

instrumentation, geological (at sea)-mud
Leenhardt, Olivier, 1964.
Le mud penetrator.
Bull. Inst. Océanogr., Monaco, 62(1303):44 pp.

instrumentation (mudtrap)
Lüneberg, H., 1952.
Ein neues Gerät zur Messung des Schlickfalles in Küstengewässern und Häfen. Veröff. Inst. Meeresf. Bremerhaven 1(2):129-138.

instrumentation
Somers, M.L., and A.R. Stubbs, 1962.
A mud echo-sounder.
J. Inst. Water Engrs., 16:501-502.

Also in:
Collected Reprints, N.I.O., Vol. 11(457):1963.

instrumentation, geological (at sea)

PROBES

instrumentation, geological (penetrometers)
Anon., 1954.
A new instrument, a penetrometer, that provides a method for measuring the compactness of underwater sediments without disturbing them.
Science, 121(3132):12-13.

instrumentation, geological
Hinrichs, F. Woods, 1961
Recent results of marine sonoprobe surveys. (Abstract).
Proc. Ninth Pacific Sci. Congr., Pacific Sci. Assoc., 1957, 12(Geol--Geophys.):352.

instrumentation, geological (at sea)-probes
Keller, George H. 1965.
Deep-sea nuclear sediment density probe.
Deep-Sea Res. 12(3): 373-376

instrumentation, geological (at sea)
Keller, George H., 1965.
Nuclear density probe for in place measurement in deep-sea sediments.
Ocean Sci. and Ocean Eng., Mar. Techn. Soc.,-Amer. Soc. Limnol. Oceanogr., 1:363-372.

instrumentation
Kermabon, A., C. Gelin and P. Blavier, 1969.
A deep-sea electrical resistivity probe for measuring porosity and density of unconsolidated sediments.
Geophysics 34(4): 554-571

instrumentation, geological (sediments)
Lister, C.R.B., 1970.
Measurement of in situ sediment conductivity by means of a Bullard-type probe.
Geophys. J. R. astr. Soc. 19(5): 521-532

instrumentation, geological (at sea)
Schmidt, Wolfgang, Erhard Eycke and Otto Kolp, 1969.
Beschreibung and Ergebnisse der Erprobung des 9m langen vibrationsstechrohrs 4701/1 mit neuer Kolben zieheinnichtung.
Beitr. Meereskunde, 24-25: 175-185

instrumentation, geological (at sea)-probes
Williams, J. 1956
The sediment probe.
Chesapeake Bay Inst., Techn. Rept. 10: 5pp

geological (at sea) samplers

instrumentation, geological (at sea)

SAMPLERS

instrumentation, geology (at sea)
Andresen, A., S. Sollie and A.F. Richards, 1964.
N.G.I. gas-operated, sea-floor sampler.
NATO Subcommittee on Oceanographic Research, Techn, Rept., No. 15:unnumbered pp. (multilithed)

instrumentation (geological)
Blanc, M., 1952.
Description d'un deuxième appareil pour prélèvement en eau deuce de sédiments submergés.
Bull. Mus. Nat. Hist. Nat., 2 ser., 24(6):591-593.

instrumentation, geological
Dietrich, G., and Helmut Hunger, 1962.
Gezeilte Tiefsee-Beobachtungen: eine neue Tiefsee-Fernsehkamera mit eingebauter Fotokamera und mit gekoppelten Sammelgeräten.
Deutschen Hydrogr. Zeits., 15(6):229-242.

instrumentation, geology
Elgmork, K., 1962
A bottom sampler for soft mud.
Hydrobiologia, Acta Hydrobiol., Hydrogr., et Protist., Den Haag, 20(2):167-172.

gear
Emery, K.O., and A.R. Champion, 1948.
Underway bottom sampler. J. Sed. Petr. 18(1): 30-33, 2 pls.

bottom sampler
Emery, K.O., and A.R. Champion, 1948
The Underway bottom sampler (known as the "Scoopfish") Int. Hydrogr. Rev. 25(1):61-62, 2 pls.

Extract from J. Sed. Petrol.

instrumentation
Fairchild, J.C., 1956.
Development of a suspended sediment sampler for laboratory use under wave action. Bull. B.E.B., 10(1):41-60.

instrumentation, geology samplers
Franklin, W.R., and D.V. Anderson, 1961
A bottom-sediment sampler.
Limnol. & Oceanogr., 6(2): 233-234.

instrumentation, geological
Fukushima, Hisao and Masakazu Kashiwamura, 1961
Some experiments on bamboo samplers.
Coastal Eng., Japan, 4:61-63.

instruments
Furukawa, A., 1955.
Studies of the soil resistence to penetration in the Inland Sea. 1. A new designed penetrometer.
Bull. Jap. Soc. Sci. Fish. 20(12):1071-1075, 2 textfigs.

instrumentation, geological
Göhren, Harold, 1965.
Ein Neuss Schöpfgerät für Schwebstoff-untersuchungen im Wett.
Die Küste, 13:133-139.

instrumentation, geological (samplers)
Gordeiev, V.D., 1945
The prismatic bottom sampler. Bull. Pacific Scientific Institute of Fisheries and Oceanography 19:99-104, 1 textfig.

gear
Holme, N. A., 1949.
A new bottom sampler. J.M.B.A. 28(2):323-332, 1 pl., 4 textfigs.

instrumentation, geological
Hopkins, Thomas L., 1964
A survey of marine bottom samplers.
Progress in Oceanography, Pergamon Press, 2: 213-256.

instrumentation, geological (sea)
Jakobsen, A.B., 1962.
Vadehavets sedimentomsaetning belyst ved kvantitativer målinger.
Medd. Skalling-Lab., 17:87-103.

Reprinted from:
Geograf. Tidsskr., 60:87-103.

instrumentation,geological (at sea)
Jonasson,Axel, and Eric Olausson,1966.
New devices for sediment sampling.
Marine Geol., 4(5):365-372.

gear
Kellen, W.R., 1954.
A new bottom sampler. Amer. Soc. Limnol. Oceanogr., Spec. Publ. No. 22: 3pp., 1 fig.

instrumentation geological (sedimentation)
Krämer, J., and H. Beth, 1955.
Ein integrierendes Sinkstoff-Banggerät.
Die Küste, 4:93-101.

instruments geological (sedimentation)
Kleerekoper, H., 1952.
A new apparatus for the study of sedimentation in lakes. Canadian J. Zool., 30:185-190.

instrumentation, geological (at sea) - samplers
Kutty, M. Krishnan, and B.N. Desai 1968.
A comparison of the efficiency of the bottom samplers used in benthic studies off Cochin.
Mar. Biol. 1(3):168-171.

instruments
Livingstone, D.A., 1955.
A lightweight piston sampler for lake deposits.
Ecology 36(1):137-138, 1 textfig.

instrument
Moore, H.B., and R.G. Neill, 1930
An instrument for sampling marine muds.
JMBA XVI (2):589-594, 5 textfigs.

instruments
Mundie, J.H., 1956.
A bottom sampler for inclined rock surfaces in lakes. J. Animal Ecology, 25(2):429-432.

instrumentation,geology, at sea
Munroe,Frederick F., 1966.
Scuba diving to investigate in-situ behavior of mobile suspended-sediment sampler.
CERC Bull. and Summary Rept. of Res.Prog., 1965-1966:60-61.

instrumentation,geological
Nicholls,Mayard M., and Robert L. Ellison,1966.
Light-weight bottom samplers for shallow water.
Chesapeake Sci., 7(4):215-217.

instrumentation, geological, at sea
* Pestrong, Raymond, 1969.
A multipurpose soft sediment sampler.
J. sed. Petr. 39(1): 327-330.

instruments
Pettersson, O., 1929.
A new apparatus for taking bottom samples.
Svenska Hydrogr.-Biol. Komm. Skr., n.s., Hydrogr 3:6-8.

instruments
Pettersson, O., 1928.
A new apparatus for taking bottom samples.
Svenska Hydrogr.-Biol. Komm. Skr., n.s., Hydrogr 6:6-7.

instrumentation, geology
Prokopovich, Nikola P., 1966.
Ecologic sampler for soft sediments.
Ecology, 47(5):856-858.

instruments
Rowley, J.R., and A.O. Dahl, 1956.
Modifications in design and use of the Livingstone piston sampler. Ecology 37(4):849-851.

instrumentation, geological (at sea)

Schiemer, E.W. and J.R. Schubel, 1970.
A near-bottom suspended sediment sampling system for studies of resuspension. Limnol. Oceanogr., 15(4): 644-646.

instrumentation, geological(at sea)
Schink, David R., Kent A. Fanning and John Piety, 1966.
A sea-bottom sampler that collects both water and sediments simultaneously.
J. mar. Res., 24(3):365-373.

instruemtns
Smith, W., and A.D. McIntyre, 1955.
A spring-loaded bottom sampler.
Int. Hydrogr. Rev. 23(1):213-219, 3 figs.

Reprinted from J.M.B.A. 33:257-264.

instruments
Smith, W., and A.D. MacIntyre, 1954.
A spring-loaded bottom-sampler. J.M.B.A. 33: 257-264, 3 textfigs.

bottom sampler
Tanita, S., 1951.
On a new sampler for marine muds. Bull. Fac. Fish., Hokkaido Univ., 1(2):63-65, 2 figs.

instrumentation, geological at sea
Taylor, John L., 1965.
Bottom samplers for estuarine research.
Chesapeake Science, 6(4):233-234.

instruments
Ursin, E., 1956.
Efficiency of marine bottom samplers with special reference to the Knudsen sampler.
Medd. Danmarks Fiskeri- og Havundersøgelser, n.s. 1(14):6 pp.

instrumentation,geological (at sea)
Walker,Bryan,1967.
A diver-operated pneumatic core sampler.
Limnol. Oceanogr., 12(1):144-146.

instruments
Watts, G.M., 1953.
Development and field tests of a sampler for suspended sediment in wave action.
B.E.B. Tech. Memo. No. 34:41 pp., 20 figs.

instrumentation, geological (at sea)
Werner,A.E., and M. Waldichuk,1967.
A sampler for gases in bottom sediments.
Limnol. Oceanogr., 12(1):158-161.

instrumentation, geological
Whitley, L. Stephen, 1962
New bottom sampler for use in shallow streams
Limnology and Oceanography, 7(2):265-266.

instruments
Worzel, J. L., 1950.
Ocean bottom sampler for ships under way. Hydr. Rev. 27(1):103-106,3 textfigs.

gear
Worzel, J. L., 1948.
Ocean bottom sampler for ships under way.
Geophysics 13(3):452-456.

instrumentation, geological (at sea)

SURF ZONE

instrumentation, geophysical (at sea)
Reinmitz, Erik, and David A. Ross 1971
The sea sled - a device for measuring bottom profiles in the surf zone.
Mar. Geol. 11 (3):M27-M32.

instrumentation,geology
Kolessar,Michael A., and John L. Reynolds,1966.
The Sears sea sled for surveying in the surf zone.
CERC Bull. and Summary Rept. of Res.Prog., 1965-1966:47-53.

instrumentation, geological (at sea)

TRACERS

instrumentation, tracers
*DeVries,M.,1967.
Compteur photométrique pour tracers fluorescents (Photometric counter for fluorescent tracers.
Houille blanche, 22(7):717-722.

instrumentation, luminescence
Karabashev,G.S., and A.N. Solov'ev,1968.
Towed sensor of concentrations of luminescence tracers in the sea. (In Russian).
Fisika Atmosfer. Okean.,Izv.Akad.Nauk,SSSR, 4(12) 1331-1333.

instrumentation, geological
Nachtigall, K.H., 1964.
Ein Gerät zur automatischen Zählung von Luminophoren.
Meyniana, 14:48-51.

Geological (laboratory)

instrumentation, geological (laboratory)

instruments
Appel, D.W., 1953.
An instrument for rapid size-frequency analysis of sediment. Proc. Fifth Hydraulics Conf., June 9-11, 1952, State Univ. Iowa, Studies in Eng., Bull. 34(426):287-300, 9 textfigs.

instrumentation, geological
Bascomb, C.L., and D.T. Pritchard, 1963.
A Cartesian hydrometer and its application to sedimentation analysis.
J. Sci. Instr., 40(1):30-31.

instrumentation, geological
Bezrukov, P.L., and A.P. Lisitsin, 1962.
The study of bottom sediments. (In Russian; English abstract).
Mezhd. Geofiz. Komitet, Prezidiume, Akad. Nauk, SSSR, Rezult. Issled. Programme Mezhd. Geofiz. Goda, Okeanol. Issled., No. 7:49-83.

instrumentation, geological (lab) (transport)
Channon, R.D. 1971.
The Bristol fall column for coarse sediment grading.
J. Sedim. Petrol. 41 (3): 867-870

instrumentation, geological
Craig, G.Y., and J. Hogg, 1962.
A rapid sorting device for microfossils.
Micropaleontology, 8(1):107-108.

bottom sampler
Enequist, P., 1941.
Einneuer Zerteiler-Bodenstecher für Sedimentanalyse zu ökologischen Zwecken. Zool. Bidrag., Uppsala, 20:461-464.

instrumentation, geological

Fabricius, F., and St. Müller, 1970.
A burst cylinder for grain-size analysis of silt and clay (with ALGOL-program)
Sedimentology 14(1/2):39-50

instrumentation, geological (lab.)

*Gadow, Sibylle, 1969.
Verglich von Schlämmanalysen mit der Pipette nach Köhn und dem Atterberg-Zylinder.
Senckenberg. maritima 50:79-84.

instrumentation, geological (lab.)

* Lawson, R.I., and G.F. Day, 1969.
Note on a non-laminated joint-free sample splitter.
J. sed. Petr., 39(1): 373-375.

instrumentation

Lovegrove, T., 1960
An improved form of sedimentation apparatus for use with an inverted microscope. J. du Cons., 35 (3):279-284.

instrumentation, geological

Möckel, Friedrich, 1965.
Bemerkungen zur Funktion des Vibrations-Stechrohr 4700/1.
Beiträge Meeresk., Berlin, (12/14):149-151.

instrumentation,

Ottmann, Francis et Guy Courtois, 1970.
Réalisation et utilisation d'un banc de gammadensimetrie des carottes au Laboratoire de Géologie Marine de Nantes.
Cah. océanogr. 22(8): 815-825.

instrumentation, geological (laboratory)

Prokoptsev, N.G., and O.L. Stelman, 1963
The setting of the Vasiliev pipette for total granulometric analysis of the ground's mechanical composition. (In Russian).
Okeanologiia, Akad. Nauk, SSSR, 3(2):313-315.

instrumentation, geology (lab.)

Sathapathi, N., 1957.
A direct recorder instrument for size analysis of coarse granular material.
Nature, 179(4566):913-914.

Instrumentation, geological (lab.)

Schlee, John, 1966
A modified Woods Hole rapid sediment analyzer.
J. Sed. Petr., 36(2):403-413.

instrumentation, geological

Schmidt, Wolfgang, und Otto Kulp, 1965.
Beschreibung und Ergebnisse der Erprobung eines im Auftrage des Instituts für Meereskunde Warnemünde gebauten Vibrationsstechrohrs 4700/1.
Beiträge Meeresk., Berlin, (12/14):143-148.

instrumentation, geological

Zeigler, John M., Carlyle R. Hayes and Douglas C. Webb, 1964
Direct readout of sediment analyses by settling tube for computer processing.
Science, 145(3627):51.

instrumentation geological (laboratory)

Zeigler, J.M., G.G. Whitney, Jr. and C.R. Hayes 1960
Woods Hole rapid sediment analyzer. J. Sed. Petr. 30(3): 490-495.

geophysical

instrumentation, geophysical

instrumentation, geophysical

Båth, M., and V. Kårnik, 1963.
Investigations of the earth's crust.
Union Géodés. et Géophys. Int., Monogr., No. 22: 50 pp. (multilithed).

Meetings:
Working group on investigations of the earth's crust, March, 19-22, 1962, Paris.

instrumentation geophysical

Bedenbender, J.W., R.C. Johnston and E.B. Neitzel, 1970.
Electroacoustic characteristics of marine seismic streamers.
Geophysics 35(6): 1054-1072.

instrumentation, geophysics

Brede, E.C., R.C. Johnston, L.B. Sullivan and H.L. Viger, 1970.
A pneumatic seismic energy source for shallow water/marsh areas.
Geophys. Prospect. 18(4): 581-599.

instrumentation, geophysical

Buffington, E.C., and D.G. Moore, 1963.
Geophysical evidence on the origin of gullied submarine slopes, San Clemente, California.
J. Geology, 71(3):356-370.

instrumentation, geophys. (Flexotir)

Cassand, J, J.P. Fail and L. Montadert, 1970.
Sismique reflexion en eau profonde (Flexotir)
Geophys. Prospect. 18(4): 600-614

Instrumentation, geophysics

Knott, S.T., 1962.
Use of the precision graphic recorder (PGR) in oceanography.
Marine Sciences Instrumentation, Instr. Soc. Amer., Plenum Press, N.Y., 1:251-262.

instrumentation, geophysical

Leenhardt, Olivier, 1963.
Aperçu sur les méthodes et techniques de sismique marine.
Bull. Inst. Océanogr., Monaco, 60(1269):64 pp.

instrumentation, geophysical

Leenhardt, O., 1962.
Quelques appareils et techniques americains de geophysique sous-marine.
Rev. Geogr. Phys. Geol. Dyn., 5(1):67-71.

Abstract in: Geophys. Abstr., 1963 (200):835.

instrumentation, geophysics at sea

Neidell, Norman S., 1969.
Reflection seismic data processing at sea?
Trans. Applications of Sea-Going Computers Symposium, Mar. Tech. Soc., 259-273.

instrumentation, geophysics

Orlenok, V.V. and N.I. Sviridov, 1971.
The use of the improved spectrometer IFS for operative frequency analysis of seismic waves. (In Russian; English abstract). Okeanologiia 11(3): 524-529.

instrumentation, geophysical

Parasnis, D.S., 1960
The compaction of sediments and its bearing on some geophysical problems.
Geophys. J., R. Astron. Soc., 3(1): 1-28.

instrumentation, geophysical

Raitt, R.W., undated (1954?).
Geophysical measurements. In: Oceanographic Instrumentation. NRC Publ. No. 309:70-84.

With discussion by F. Press
J.L. Worzel

instrumentation, seismic)

Rykunov, L.N., V.V. Khorosheva, and V.V. Sedov, 1961.
A two-dimensional model of a seismic wave guide with soft boundaries.
Izv. Akad. Nauk, SSSR, Ser. Geofiz., 11:1601-1603.

English Ed., 11:1069-1071.

instrumentation, geophysical (seismograph)

Schneider, W.A., P.J. Farrell and R.E. Brannian, 1964.
Collection and analysis of Pacific Ocean-bottom seismic data.
Geophysics, 29(5):745-771.

instrumentation, geophysical

Schoenberger, Michael, 1971.
Optimization and implementation of marine seismic arrays.
Geophysics, 35(6): 1038-1053.

instrumentation, geophysical

Semerchan, A.A., and D.B. Balashov, 1962.
Design and test of containers for geophysical investigations at great sea depths. (In Russian).
Doklady Akad. Nauk, SSSR, 146(3):592-595.

Geophysical "boomer"

instrumentation, geophysical - "boomer"

"boomer"

Horn, Robert, Jean-René Vanney, Gilbert Boillot, Philippe Bouysse et Lucien Leclaire, 1966.
Résultats géologiques d'une prospection sismique par la méthode "boomer" au large du massif armoricain méridional.
C.r. hebd. Séanc., Acad. Sci., Paris, (D)263(21): 1560-1563.

"boomer"

Leenhardt, Olivier, 1965.
Le sondage sismique continu.
Rev. Geogr. Phys. et Géol. Dyn., (2), 7(4):285-294.

boomer

*Ostenso, N.A., 1968.
Geophysical studies in the Greenland Sea.
Bull. Geol. Soc. Am., 79(1):107-132.

geophysical, magnetometer

instrumentation, geophysical, MAGNETOMETERS

instrumentation, magnetism

Belshé, J.C., 1964.
Studies with deep submergence magnetometers. (Abstract).
Trans. Amer. Geophys. Union, 45(1):71.

instrumentation, geophysics (magnetometer)

*Filloux, J.H., 1967.
An ocean bottom, D component magnetometer.
Geophysics, 32(6):978-987.

instrumentation magnetism
Foster, John H., 1966.
A paleomagnetic spinner magretometer using a fluxgate gradiometer.
Earth Planet. Sci. Letters, 1(6):463-466.

instrumentation, magnetism
Hess, Heinrich, 1963.
Aufbau und Theorie eines Gerätes zur Messung der magnetischen Horizontalfeldstärke aud See mit der Förster-Sonde.
Deutsche, Hydrogr. Zeits., 16(1):15-43.

instrumentation
Hill, M.N., 1960
A shipborne nuclear-spin magnetometer. Int. Hydrogr. Rev., 37(2): 113-115. Reproduced from Deep-Sea Res., 5(4): 309-311

instruments
Meyer, O., and D. Voppel, 1954.
Ein Theodolit zur Messung des erdmagnetischen Feldes mit der Förstersonde asl Nullfeldindikator.
Deutsche Hydrogr. Zeits. 7(3/4):73-77.

instrumentation, magnetism
Raff, A.D., 1961.
The magnetism of the ocean floor.
Scientific American, 205(4):146-156.

INSTRUMENTATION, MAGNETISM
Royer G., 1969.
New developments concerning the high sensitivity CSF magnetometer.
Techn. Bull. Econ. Comm. Asia Far East, Comm. Co-Ordin. Joint Prospect. Mineral. Res. Asian Offshore Areas 2: 59-77.

instrumentation, geophysical (at sea)
Sarma D. Gupta G. Varadarajan T. R.M. Prasad and D. Panduranga m 1969
A proton precession magnetometer and recorder for off-shore use.
Bull. natn. geophys. res. Inst. 7(4): 133-139.

instrumentation, magnetism
Till, K., 1961
Der Seemagnetograph, ein neues Schleppgerät zur Messung der horizontalen und vertikalen Komponente des erdmagnetische Feldes auf See.
Beitr., Meereskunde, 2/3: 102-109.

instrumentation, magnetism
Tomoda, Yoshibumi, Keijiro Ozawa and Jiro Segawa, 1968.
Measurement of gravity and magnetic field on board a cruising vessel.
Bull. Ocean Res. Inst. Univ. Tokyo, 3: 169 pp.

instrumentation, magnetism
Verzhbitsky, E.V., E.N. Isaev and A.A. Shreyder, 1969.
Equipment and methods used for the geomagnetic studies on the 40th and 40st cruises of the R/V Vityaz. (In Russian; English abstract).
Okeanologiia, 9(1):187-192.

instrumentation, magnetism
Voppel, D. 1969.
Ein Spulentheodolit zur Messung der erdmagnetischen Komponenten mit dem Protonenmagnetometer.
Z. Geophys. 35: 151-159
Also in: Ozeanogr. 1969, Dt. Hydrogr. Inst. (1970).

Geophysical, Pinger

instrumentation, geophysical-"pingers"

instrumentation, geophysical
Breslau, Lloyd R., J.B. Hersey, Harold E. Edgerton and Francis S. Birch, 1962
A precisely timed submersible pinger for tracking instruments in the sea.
Deep-Sea Res., 9(2):137-144.

instrumentation, geophysics (pinger)
Deacon, G.E.R., 1957.
The pinger.
Oceanus, 4(3/4):19-26.

Instrumentation, geophysics
Hersey, J. B., H. E. Edgerton, S. O. Raymond, and G. Hayward 1961.
Pingers and thumpers advance deep-sea exploration.
J. Instr. Soc., Amer., 8(1):72-77.

instrumentation, geophysical - pinger
Holgkamm, F. 1965.
Der Einsatz des Sonar Pingers während der IOE "Meteor".
Dt. hydrogr. Z. 18(4): 172-178.

instrumentation, pingers
Jones, Ronald E., 1965.
Deep ocean installations.
Ocean Sci. and Ocean Eng., Mar. Techn. Soc., - Amer. Soc. Limnol. Oceanogr., 1: 200-266.

instrumentation, geophysics
Van Reenan, Earl D., 1962.
A complete sonar thumper seismic system.
Marine Sciences Instrumentation, Instr. Soc., Amer., Plenum Press, N.Y., 1:283-288.

instrumentation, geophysical pingers
Volkmann, G.H., 1963
13. Deep-current measurements using neutrally buoyant floats. In: The Sea, M.N. Hill, Edit., Vol. 2. (III) Currents, Interscience Publishers, New York and London, 297-302.

Geophysical profiler

instrumentation, geophysical

PROFILERS

instrumentation, geophysics
Barnes E.J., 1970.
Development of a sound reflection system for profiling sediments in shelf and slope depths.
N.Z. Jl mar. freshwat. Res. 4(1): 70-86.

instrumentation, geophysics
Bowers, R., 1963.
A high-power, low-frequency sonar for sub-bottom profiling.
J. Brit. Inst. Radio Engrs., 25:457-460.
Also in:
Collected Reprints, N.I.O., 11(408):1963.

instrumentation, geophysical
Caulfield, David D., Hartley Hoskins and Richard T. Nowak, 1965.
Improvements in the continuous seismic profiler.
Geophysics, 30(1):133-138.

instrumentation, geophysical
Chramiec, Mark A., and George M. Walsh, 1966.
Sub-bottom profiling with a replica correlation receiver.
Undersea Techn., 7(9):49-51.

instrumentation, geophysical
Clay, C.S., and W.L. Liang, 1964.
Seismic profiling with a hydroacoustic transducer and correlation receiver.
J. Geophys. Res., 69(16):3419-3428.

instrumentation, seismic
Edgerton, Harold E., and Gary G. Haywood, 1964.
The 'boomer' sonar source for seismic profiling.
J. Geophys. Res., 69(14):3033-3042.

instrumentation, geophysical profiler records
Ewing, John, and Maurice Ewing, 1962
Reflection profiling in and around the Puerto Rico Trench.
J. Geophys. Res., 67(12): 4729-4740.

instrumentation, submarine geophysics
Ewing, J., and G.B. Tirey, 1961
Seismic profiler.
J. Geophys. Res., 66(9):2917-2928.

instrumentation, geophysical
Ewing, John and Roger Zaunere, 1964
Seismic profiling with a pneumatic sound source
J. Geophys. Res., 69(22):4913-4915.

instrumentation, geophysical
Fain, Gilbert, F.H. Middleton and Robert S. Haas 1965.
A continuous seismic profiling display system.
Ocean Engng 1(1): 3-7.

instrumentation, seismic profiling
Fain, G., F.H. Middleton and R.S. Haas, 1967.
A photo-electronic-seismic profiling display system.
1967 N EREM Record, Inst. Electr. Electron. Engrs, 0:190-191.

instruments, geophysical
Hoskins, H., and S.T. Knott, 1961.
Geophysical investigation of Cape Cod Bay, Massachusetts, using the continuous seismic profiles.
J. Geol., 69(3):330-340.

instrumentation, seismic profiler
Keen, M.J., B.D. Loncarevic and G.N. Ewing, 1970.
II. Regional observations. 7. Continental Margin of Eastern Canada: Georges Bank to Kane Basin. In: The sea: ideas and observations on the progress in the study of the seas, Arthur E. Maxwell, editor, Wiley-Interscience 4(2): 251-291.

instrumentation seismic profiling
Leenhardt, Olivier, 1971.
Resolution in continuous seismic profiling.
Int. hydrogr. Rev. 48(2): 71-83.

instrumentation (geophysics) (Rayflex Electro-Sonic profiler)
Moore, David G., and Joseph R. Curray, 1963
Structural framework of the continental terraces northwest Gulf of Mexico.
J. Geophys. Res., 68(6):1725-1747.

instrumentation, geophysics
Moore, T.C., Jr. and L.D. Kulm, 1970.
A high-resolution sub-bottom profiling system for use in ocean basins. J. mar. Res., 28(2): 271-280.

instrumentation, geophysics (at sea) profiler
*Rusnak, Gene A., 1967.
High efficiency subbottom profiling.
Prof. Pap., U.S. geol. Surv., 575-C:081-091.

instrumentation, geophysical; seismic profiling
Schiemer, E.W., and J.R. Schubel 1971.
An improved vehicle for towing a displacement type sound source for continuous seismic profiling.
Mar. geophys. Researches 1(3): 352-353.

instrumentation, geophysics (Electro-Sonic Profiler)
Shor, George G., Jr., David G. Moore, and William B. Huckaby, 1963
Deep-sea tests of a new nonexplosive reflection profiler.
J. Geophys. Res., 68(5):1567-1571.

instrumentation seismic profiling
Sieck, Herman, 1967.
Continuous seismic profiling
Ocean Industry, 2(1): 1A-7A.

instrumentation, geophysical
Van Reenan, Earl D., 1967.
Deep seismic profiling becomes major subsurface geological probe.
Oceanology Int., 2(2):39-40.

instrumentation, geophysics
VanReenan, Earl, 1965.
Low-voltage, deep-penetration marine-seismic profiling system.
Mar. Sci. Instrument., Plenum Press, 3:167-182.

instrumentation geophysics - probes
Kermabon, A., C. Gehin, and P. Blavier, 1969.
A deep sea electrical resistivity probing device. Oceanol. int. 69, 1: 8 pp., 12 figs.

instrumentation, geophysical
SEISMOGRAPH

instrumentation, seismic
Benoit, R., and Nguyen-hai, 1958
L'installation de la station séismologique de Nhatrang et l'étude des caractéristiques de l'appareillage. (Station de Nhatrang: 12 12' 6N-109 12' 7E).
Ann. Fac. Sci., Saigon, 1958:11-30.
Also: Contrib., Inst. Océanogr., Nhatrang, No. 32.

instrumentation, geophysical
Bradner, Hugh, 1964
Seismic measurements on the ocean bottom. New instruments are used to study earth's crustal structure and seismic background.
Science, 146(3641):208-216.

instrumentation, seismic
Ewing, John and Maurice Ewing, 1961
A telemetering ocean-bottom seismograph.
J. Geophys. Res., 66(11):3863-3878.

instrumentation, seismograph
Hasegawa, Shiji and Shozaburo Nagumo, 1970.
Construction of a long life magnetic tape recorder and some features of ocean-bottom seismographs. (In Japanese; English abstract).
Bull. Earthq. Res. Inst. Tokyo Univ., 48(5): 967-981.

instrumentation, geophysics (seismograph)
Kishinouye, Fuyuhiko 1966.
The submarine seismograph, the second paper.
Bull. Earthquake Res. Inst. Tokyo Univ. 44(4):1443-1447.

instruments, geophysics
Kishinaye, Fuyuhiko, Yoshio Yamazaki, Heihachiro Kobayashi and Sadayuki Koresawa, 1963.
49. A submarine seismograph: the first paper. (Japanese abstract).
Bull. Earthquake Res. Inst., Univ. Tokyo, 41(4): 819-824.

instrumentation, seismograph
Nagumo, Shozaburo, Heihachiro Kobayashi and Sadayuki Koresawa, 1970.
Pressure vessels for ocean-bottom seismograph. (In Japanese; English abstract)
Bull. Earthq. Res. Inst. Tokyo Univ. 48(5): 955-966

instrumentation - seismic
Shima, E., 1960.
[Note on the magnetic circuit of the moving-coil type seismometer.]
Bull. Earth Res. Inst., Tokyo, 38(4):545-557.

instrumentation, geophysical
Thomson, J.T., and W. Schneider, 1962.
An automatic marine seismic monitoring and recording device.
Proc. Inst. Radio Eng., 50(11):2209-2215.

instrumentation
Tucker, M. J., 1958.
An electronic feedback seismograph.
J. Sci. Instr., 35:167-171.

instruments
Volk, J.A., 1950.
The photoelectric seismograph. Bull. Seismol. Soc. America, 40(3):169-173, 7 textfigs.

instruments
Volk, J. A., and F. Robertson, 1950.
The electronic seismograph. Bull. Seismol. Soc. America, 40(2):81-93, 13 textfigs.

instrumentation, gravity
Wing, Charles G., 1969.
MIT vibrating string surface-ship gravimeter.
J. geophys. Res., 74(25): 5882-5894.

instrumentation, geophysical (at sea)
Whitmarsh, R.B., 1970
An ocean bottom pop-up seismic recorder
Mar. geophys. Res. 1(1): 91-98.

SOFAR
Anderson, E.R., 1950.
Distribution of sound velocity in a section of the eastern North Pacific. Trans. Am. Geophys. Union 31(2):221-228, 10 textfigs.

instrumentation SOFAR
Arase, T., 1959.
Some characteristics of long-range explosive sound propagation.
J. Acoust. Soc., Amer., 31(5):588-595.

instrumentation SOFAR
Aubrat, J., 1963.
Ondes T reflechies dans la mer des Antilles.
Ann. Geophys., Centre Nat. Recherche Sci., 19(4):386-405.

SOFAR
Bose, S.K., 1964.
SOFAR and microseisms from storms over deep sea.
Gerlands Beiträge zur Geophysik, 73(5/6):334-341.

SOFAR
Brekhovskikh, L.M., 1949.
O rasprostranenii zvuka v podvodnom zvukovom kanale. [The propagation of sound in an underwater channel.] Dokl. Akad. Nauk, SSSR 69(2):157-160, 1 textfig.
Translation (mimeographed) at WHOI. Received from HO.

SOFAR
Bryan, George M., Marek Truchan and John I. Ewing, 1963.
Long-range SOFAR studies in the South Atlantic Ocean.
J. Acoust. Soc., Amer., 35(3):273-278.

SOFAR
Dietz, R.S., and M.F. Sheehy, 1953.
TransPacific detection by underwater sound of Myojin volcanic explosions.
J. Ocean. Soc., Japan, 9(2):53-83, 6 textfig
Also published: Bull. G.S.A. 65(10):941-956, 4 pls., 4 textfigs. in 1954

instrumentation - SOFAR
Di Napoli, F.R., and F.H. Middleton 1966.
Surface-wave propagation in a continuously stratified medium.
J. acoust. Soc. Am. 39(5-1): 899-903.

instrumentation SOFAR
Frosch R.A. M. Klerer & L. Tyson 1961
Long-range localization of small underwater charges.
J. Acoust. Soc., Amer., 33(12):1804-1805.

instrumentation - SOFAR.
Garner, D.M. 1967.
Oceanic sound channels around New Zealand.
N.Z. Jl mar. Freshwat. Res. 1(1): 3-15.

SOFAR?
Gorbachevski, O.S., 1951.
[On the superdistance propagation of sound in stretches of deep water.] Priroda 1951(2):54.
Translation: E.R. Hope, Sci. Info. Center, DRB, Canada, 1 June 1951, T38R.

instrumentation SOFAR
Grousson, R., 1956.
La propagation du son dans le chenal sonore.
Ann. Hydrogr., Paris, (4) 6:353-369.

instrumentation SOFAR
Guthrie, Albert N., Ivan Tolstoy and John Shaffer, 1960
Propagation of low-frequency cw sound signals in the deep ocean. J. Acoust. Soc. Amer., 32(6): 645-647.

SOFAR
Hirsch, Peter, and Ashley H. Carter, 1965.
Mathematical models for the prediction of SOFAR propagation effects.
J. Acoust. Soc., Amer., 37(1):90-94.

SOFAR
Ivanov, I.D., 1955.
Distribution of sound in the sea over a very large distance. Priroda (3):85.

instrumentation
SOFAR
Johnson, Rockne H., 1969.
Synthesis of point data and path data in estimating sofar speed. J. geophys. Res., 74(18): 4559-4570.

instrumentation SOFAR
Johnson, Rockne H., 1966.
Routine location of T-phase sources in the Pacific.
Bull. Seismol. Soc., Amer., 56(1):109-118.

instrumentation SOFAR
Johnson, Rockne H., and Roger A. Norris, 1968.
Geographic variation of Sofar speed and axis depth in the Pacific Ocean.
J. geophys. Res., 73(14):16954700.

SOFAR
Kutschale, Henry, 1961.
Long-range sound transmission in the Arctic Ocean
J. Geophys. Res., 66(7):2189-2198.

SOFAR
LaFond, E. C., 1949.
Oceanographic research at the U.S. Navy Electronics Laboratory. Trans. Am. Geophys Union 30(6):894-896.

instrumentation
Sofar
* Lovett, J.R., 1969.
Comments concerning the determination of absolute sound speeds in distilled and seawater and Pacific sofar speeds.
J. acoust. Soc. Am., 45(4): 1051-1053.

instrumentation SOFAR
Milder D. Michael, 1969.
Ray and wave invariants for SOFAR channel propagation.
J. acoust. Soc. Am. 46 (5-2): 1259-1263

instrumentation
SOFAR
Nguyen-Hai and Nguyen-Duc-Khang, 1961
Sur les ondes T des séismes des Philippines enregistrees a Nha-Trang.
Ann. Fac. Sci., Saigon, 1961:343-368.

Also:
Contrib. Hai-Hoc-Vien Nhatrang, Inst. Océanogr., Nhatrang. No. 48.

instrumentation
SOFAR
*Norris, Roger A., and Rockbe H. Johnson, 1969.
Submarine volcanic eruptions recently located in the Pacific by Sofar hydrophones.
J. geophys. Res., 74(2):650-664.

Sofar channel
Northrop, John, L. E. Tyson and J. O. Lee, 1962.
Long-range recordings of bottomed shots in the Gulf of Maine. (Abstract).
J. Geophys. Res., 67(4):1650.

SOFAR
Piip, Ants T., 1964.
Fine structure and stability of the sound channel in the ocean.
J. Acoust. Soc., Amer., 36(10):1948-1953.

instrumentation
SOFAR channel
Rossby, T. and D. Webb, 1970.
Observing abyssal motions by tracking Swallow floats in the SOFAR Channel. Deep-Sea Res., 17(2): 359-365.

SOFAR
Russell, R.D., 1949.
SOFAR: A new oceanographic research tool. Proc. Verb., Assoc. d'Océan. Phys., Union Géodés. et Géophys. Internat. 4:150.

instrumentation SOFAR
Sherman, L., 1955.
A proposal for the use of SOFAR signals for the reporting of hurricane positions. Proc. UNESCO Sy Symp., Typhoons, Nov. 1954:245-246.

SOFAR
Shurbet, D.H., 1962.
Note on use of SOFAR geophone to determine seismicity of regional oceanic areas.
Bull. Seismol. Soc., Amer., 52(3):689-691.

instrumentation - SOFAR
Shurbet, D.H., 1955.
Bermuda T Phases with large continental paths.
Bull. Seismol. Soc., Amer., 45(1):23-35, 7 text-figs.

SOFAR
Stifler, W.W., jr. and W.F. Saars, 1948
SOFAR. A discussion of underwater SOund Fixing And Ranging. Electronics 21:98-101.

instrumentation
Sofar
Tolstoy, Ivan, 1960
Guided waves in a fluid with continuously variable velocity overlying an elastic solid theory and experiment.
J. Acoust. Soc., Amer., 31(1):81-87

instrumentation
SOFAR
Tolstoy, I., and J. May, 1960
A numerical solution for the problem of long-range sound propagation in continuously stratified media with applications to the deep ocean.
J. Acoust. Soc., Amer., 32(6): 655-660.

SOFAR
Williams, A.O., Jr., and William Horne, 1967.
Axial focusing of sound in the SOFAR Channel.
J. acoust. Soc., Am., 41(1):189-198.

SOFAR
USSR, Anonymous, 1951.
On long-range distribution of sound in a deep-water basins. A most important meteorological work
Priroda (2):54.

geophysical, sonar

instrumentation, geophysics (sonar)

See also: instrumentation, navigation (sonar)

instrumentation sonar
Anderson, Victor C., and Carl D. Lowenstein 1968
Improvements in side-looking sonar for deep vehicles.
Mar. Sci. Instrument. 4: 260-266

instrumentation sonar
Anon., 1969.
GLORIA - successful first trials
Hydrospace 2(3): 10-11

instrumentation, sonar
Anon., 1967.
CTFM sonar goes deep: profile - Straza Industries.
Undersea Techn., 8(5):49-51.

sonar, side looking
Bascom, Willard, 1969.
Technology and the ocean.
Scient. Am., 221(3):198-204,206,208,210,213,214, 216-217. (popular).

instrumentation Sonar
Beatty, Louis G., Robert J. Bobber and David L. Phillips, 1966.
Sonar Transducer in a high-pressure tube.
J. Acoust. Soc., Amer., 39(1):48-54.

instrumentation sonar
*Bucker, H.P. and Halcyon E. Morris, 1968.
Normal mode reverberation in channels or ducts.
J. acoust. Soc. Am., 44(3):827-828.

instrumentation - sonar
Buford, W.H., Jr. and J.M. Johnston, 1969.
Development and testing of doppler sonar navigation systems. Oceanol. int. 69, 3: 10 pp. 4 figs.

instrumentation, sonar
Cole, Bernard F., and John J. Hanrahan 1968.
Influence of beamwidth and multiple transmissions on reverberation-limited sonars.
J. acoust. Soc. Am. 43(6): 1373-1377.

instrumentation sonar
Colldeweih I.R. E.L. Walls and R.D. Lee, 1961
Portable sonar for frogmen.
Electronics, 34(52):37-39.

sonar
Craig, Robert E., and Sinclair T. Forbes, 1969.
Design of a sonar for fish counting.
FiskDir. Skr. Ser. HavUnders. 15(3): 210-219.

instrumentation, sonar
Creasey, D.J., 1970.
High resolution sonars for ocean operation.
Electronic Engineering in Ocean Technology Swansea, 21-24 Sept. 1970, I.E.R.E. Conf. Proc., 19: 175-186.

instrumentation - sonar
Creasey, D.J., 1969.
Sonar systems in the micro electronic era.
Oceanol. int. 69, 4: 6 pp., 7 figs.

instrumentation, sonar
Crum, L.A., and F.B. Stumpf 1968.
Effect of compliant-tube arrays on the radiation impedance of small sonar transducers at an air-water interface.
J. acoust. Soc. Am. 43(6): 1378-1382.

sonar, scanning
Cushing, D.H., and F.R. Harden Jones, with an appendix by J.A. Gulland, 1966.
Sea trials with modulation sector scanning sonar.
J. Cons. perm. int. Explor. Mer, 30(3):324-345.

instrumentation, sonar

Dulberger, L., 1961.
Deep-ocean velocimeter aids sonar systems design
Electronics, 34(22):41-43.

instrumentation sonar

Edgerton, Harold E., 1960.
Uses of sonar in oceanography.
Electronics, 24 June 1960:93-95.

instrumentation sonar

*Friedlander, 1967.
World War II: electronics and the U.S. Navy: radar, sonar, loran and infrared techniques.
IEEE Spectrum, 4(11):56-70.

instrumentation, sonar

Glenn, Morris F., 1970.
Introducing an operational multi-beam array sonar.
Int. hydrogr. Rev. 47(1): 35-39.

instrumentation, sonar(side-scan)

Hopkins, J.C. 1970
Cathode ray tube display and correction of side scan sonar signals.
Electronic Engineering in Ocean Technology, Swansea, 21-24 Sept. 1970, I.E.R.E. Conf. Proc. 19: 151-158

instrumentation sonar

Horton, J.W., 1957.
Fundamentals of sonar. U.S. Naval Inst., Annapolis, 387 pp.

instrumentation, sonar

Hudson, J.E. 1970
New types of high resolution sonar displays.
Electronic Engineering in Ocean Technology, Swansea, 21-24 Sept. 1970, I.E.R.E. Conf. Proc. 19:121-128

instrumentation sonar

Jones, F.R. Harden, and B.S. McCartney, 1962.
The use of electronic sector-scanning sonar for following the movements of fish shoals: sea trials of R.R.S. "Discovery II".
J. du Cons., 27(2):141-149.

instrumentation sonar

Keller, A.C., 1947
Submarine detection by sonar. Bell Lab. Rec. 25:55-60.

instrumentation sonar

Kellogg, W.N., 1962.
Sonar system of the blind; new research measures their accuracy in detecting the texture, size and distance of objects "by ear".
Science, 137(3528):399-404.

instrumentation sonar

Kellogg, W.N., 1961

Porpoises and sonar.
Univ. Chicago Press, Chicago 37: 177 pp.

instrumentation, sonar (side-scan)

Klein, Martin, 1967.
Side scan sonar.
Undersea Techn., 8(4):24-26, 38.

instrumentation sonar, F.M.

*Kramer, S.A., 1967.
Doppler and acceleration tolerances of high-fain wideband linear FM correlation sonars.
Proc. Inst. Electr. Electronics Engr, 55(5):627-636.

Sonar

Laevastu, Taivo, 1969
Effects of ocean thermal structure on fish finding with sonar.
Fiskdir. Skr. Ser. HavUnders. 15(3): 202-208

instrumentation, sonar

Maidanik, G. 1968.
System of small-size transducers as elemental unit in a sonar system.
J. acoust. Soc. Am. 44(2): 488-496.

instrumentation, sonar

Maidanik, G. 1968.
Domed sonar system.
J. acoust. Soc. Am. 44(1): 113-124.

instrumentation, geophysical (sonar)

McGrath, Thomas D. 1968.
Deep water sonar transducers.
Undersea Techn. 9(7): 27-30, 31, 36.

instrumentation sonar array

Mellen, R.H., D.G. Browning and W.L. Konrad, 1971.
Parametric sonar transmitting array measurements.
J. acoust. Soc. Am. 49(3-2): 932-935.

instrumentation, sonar

Mitson, R.B., and J.C. Cook, 1970.
Shipboard installation and trials of an electronic sector scanning sonar.
Electronic Engineering in Ocean Technology, Swansea, 21-24 Sept. 1970, I.E.R.E. Conf. Proc. 19: 187-210.

instrumentation, sonar

Mudie, J.D., W.R. Normark and E.J. Cray, Jr. 1970.
Direct mapping of the sea floor using side-scanning sonar and transponder navigation.
Bull. geol. Soc. Am., 81(5): 1547-1554

instrumentation, sonar

Perkins, Paul J., 1966.
Passive Sonar aids deep-sea research on TRIDENT.
Undersea Techn., 7(3):36-37.

instrumentation, sonar

Perry, Darrell E., 1961.
Airborne bathythermograph boosts sonar device effectiveness.
Underwater Engineering, 2(1):36.

instrumentation, sonar

* Poulter, Thomas C., 1969.
Sonar of penguins and fur seals.
Proc. Calif. Acad. Sci., 36(13): 363-380.

instrumentation "Sonar"

Poulter, T.C., 1963
Sonar signals of the sea lion.
Science, 139(3556):753-757.

sonar

Poulter, Thomas C., and Richard A. Jennings, 1969.
Sonar discrimination ability of the California sea lion, Zalophus californicus.
Proc. Calif. Acad. Sci., 36(14): 381-389.

instrumentation, sonar

Rand, G., 1960.
Designing transducers for sonar systems.
Electronics, 33(9):62-65.

instrumentation sonar

Rand, G., 1960.
Determining sonar system capability.
Electronics, 33(8):41-45.

instrumentation, geophysics (sonar)

Rusby, J.S.M., 1970.
A long range side-scan sonar for use in the deep sea: (GLORIA project). Int. hydrogr. Rev. 47(2): 25-39.

instrumentation Sonar, FM

Russo, Donato M., and Charles L. Bartberger, 1965
Ambiguity diagram for linear FM sonar.
J. Acoust. Soc., Amer., 38(2):183-190.

instrumentation,

Rychkov, V.S. and A.N. Kosarev, 1971.
An acoustic current meter. (In Russian).
Okeanologiia 11(4): 758-759.

instrumentation Sonar

Sanders, John E., 1966.
Geological calibration attempt of side-looking Sonar, north shore of Minas Basin, Nova Scotia.
Marit. Sediments, 2(1):23-25. (mimeographed).

instrumentation, geophysics (sonar)

Somers, M.L., 1970.
Signal processing in project GLORIA, a long range side-scan sonar. Electronic Engineering in Ocean Technology, Swansea, 21-24 Sept. 1970, I.E.R.E. Conf. Proc. 19: 109-120.

instrumentation, side-scanning sonar

Spiess, F.N. and John D. Mudie, 1970.
1. General observations. 7. Small scale topographic and magnetic features. In: The sea: ideas and observations on progress in the study of the seas, Arthur E. Maxwell, editor, Wiley-Interscience 4(1): 205-250.

instrumentation sonar

Stewart, J.L., E.C. Westerfield and M.K. Brandon, 1961
Optimum frequencies for active sonar detection.
J. Acoust. Soc., Amer., 33(9):1216-1222.

instrumentation - sonar

Tucker, D.G., 1967.
Sonar in fisheries - a forward look.
Fishing News (Books) Ltd., 136 pp.

instrumentation - Sonar

Tucker, D.G., 1966.
Underwater observation using sonar.
Fishing News (Books) LTD. 144 pp.

instrumentation, sonar

Tucker, D.G., 1963.
Sonar arrays, systems and displays.
In: Underwater acoustics, V.M. Albers, Edit., Plenum Press, N.Y., 29-49.

instrumentation, geophysical (sonar)

Tucker, M.J., 1966.
The use of sideways looking sonar for marine geology.
Geo-Mar. Techn., 2(9):18-23.

instrumentation SONAR

Urick, R.J., 1962.
Generalized form of the Sonar equations.
J. Acoust. Soc., Amer., 34(5):547-550.

instrumentation
sonar
Urick, R.J., 1962
Radar and sonar observations of some abnormal objects in the air and in the sea. (Abstract).
J. Geophys. Res., 67(9):3605-3606.

sonar, ultrasonic (mammals).
Waterman, Talbot H., 1959.
Animal navigation in the sea.
Gunma J. Med. Sci., 8(3):243-262.

instrumentation
SONAR
Welsby, V. G., 1968
Pressure and displacement sensors used simultaneously in underwater sonar.
Nature, 218(5144): 890-891.

instrumentation, sonar(side-scan)
White, W. St., and P. Diederich
A data logger for oceanographic surveys.
Electronic Engineering in Ocean Technology.
Swansea, 21-24 Sept. 1970, I.E.R.E. 19:271-280

instrumentation
Sonar, side-scan
Wong, H.K., W.D. Chesterman and J.D. Bromhall, 1970.
Comparative side-scan sonar and photographic survey of a coralbank.
Int. Hydrogr. Rev. 47(2): 11-23

instrumentation, sonar
Volberg, H.W., 1964.
CTFM sonar for deep submergence.
Undersea Technology, 5(7):38-41.

sonar
Zietz, I., and L.C. Pakiser, 1957.
Note on an application of sonar to the shallow reflection problem. Geophysics, 22(2):345-347

sonar, biological
Hester, Frank J., 1967.
Identification of biological sonar targets from body-motion doppler shifts.
In: Marine bio-acoustics, W.N. Tavolga, editor, Pergamon Press, 2:59-73.

instrumentation
sonar, doppler
Goulet, Thomas A., 1970.
The use of pulsed doppler sonar for navigation of manned deep submergence vehicles.
Navigation, J. Inst. Navig., U.S.A. 17(2): 136-141

sonar, side-scanning
Heaton, M.J.P., 1968.
Profile, plan and section: three developments for sea-bed survey.
Int. hydrogr. Rev., 45(1):73-80.

instrumentation, geophysical (sonoprobe)
McClure, C.D., H.F. Nelson and W.B. Huckabay, 1958.
Marine sonoprobe system, new tool for geologic mapping. Bull. Amer. Assoc. Petr. Geol., 42(4): 701-716.

geophysical "sparker"

"Sparker"
Leenhardt, Olivier, 1965.
Le sondage sismique continu.
Rev. Géogr. Phys. et Geol. Dyn., (2), 7(4):285-294.

Glass spheres

gravity

instrumentation, geophysical

GRAVITY

instrumentation, gravity
Allen, T.D., P. Dehlinger, C. Gantar, C. Morelli M. Pisani and J. C. Harrison, 1964.
Comparison of Graf-Askania and LaCoste-Romberg surface-ship gravity meters.
J. Geophys. Res., 67(13):5157-5162.

instrumentation, gravity
Allan, T.D., P. Dehlinger, C. Gantar, C. Morelli, M. Pisani and J.C. Harrison, 1962
Comparison of Graf-Askania and LaCoste-Romberg surface-ship gravity meters.
J. Geophys. Res., 67(13):5157-5162.

instrumentation gravity
Bowin, Carl, Charles G. Wing, and Thomas C. Aldrich, 1969.
Test of the MIT vibrating string gravimeter, 1967. J. geophys. Res., 74(12): 3278-3280.

instrumentation, gravity
Branham, D. W., 1965.
Shipboard oceanographic survey system.
OceanSci. and Ocean Eng., Mar. Techn. Soc., Amer. Soc. Limnol. Oceanogr., 1:341-362.

instruments
Browne, B.C., and R.I.B. Cooper, 1950.
The British submarine gravity surveys of 1938 and 1946. Phil. Trans. Roy. Soc., London, Ser. A, No. 847, 242:243-310, 6 charts, 13 textfigs.

instrumentation
Clarkson, H.N., and L.J.B. Lacoste, 1957.
Improvements in tidal gravity meters and their simultaneous comparison. Trans. Amer. Geophys. Union, 38(1):8-16.

instrumentation, gravity
Dehlinger, Peter, 1964.
Reliability at sea of gimbal-suspended gravity meters with 0.7 critically damped accelermomters
J. Geophys. Res., 69(24):5383-5394.

instrumentation, gravity
Dehlinger, Peter, and S.H. Yungul, 1962
Experimental determination of the reliability of the LaCoste and Romberg surface-ship gravity meter S-9.
J. Geophys. Res., 67(11):4389-4394.

instrumentation, gravity
Dehlinger, P., and S.H. Yungul, 1962
Reliability of the LaCoste & Romberg surface-ship gravity meter S-9. (Abstract).
J.Geophys. Res., 67(9):3553.

instrumentation, gravity
Fleisher, U., and J. Fritsch, 1962
Three years' experience with an Askania sea gravimeter after Graf. (Abstract).
J. Geophys. Res., 67(9):3558.

instrumentation, gravity
Fritsch, Jürgen, 1962
Erfahrungsbericht über Messungen mit dem Askania-Seegravimeter.
Deutsche Hydrograph. Zeits., 15(4):142-173.

instrumentation, geophysics (gravimeter)
Frowe, E.W., 1947
A diving bell for underwater gravimeter operation. Geophys. 12:1-12.

instrumentation, gravity
Gantar, C., and C. Morelli, 1969
First tests on the new Bell Aerosystems sea gravity meter.
Boll. Geofis. teor. appl. 11 (43-44): 173-190

instrumentation, gravity
Gantar, C., and C. Morelli, 1963.
New experimental data about temperature and pressure effects on Worden gravity-meters.
Boll. Geofis. Teorica ed Applicata, 5(19):175-194.

instrumentation, gravity
Gantar, C., C. Morelli and M. Pisani, 1962
Experimental study of the response of the Graf-Askania Gss2, No. 13, sea gravity meter.
J. Geophys. Res., 67(11):4411-4420.

instrumentation, gravity
Gantar, C., C. Morelli, and M. Pisani, 1962
Experimental study of the Graf-Askania Gss2, No. 13, sea gravity meter. (Abstract).
J. Geophys. Res., 67(9):3560.

instrumentation, gravity
Goodacre, A.K., 1964.
A shipborne gravimeter testing range near Halifax, Nova Scotia.
J. Geophys. Res., 69(24):5373-5381.

instruments
Graf, A., 1956.
Beschreibung eines neuentwickelten Seegravimeters und Ergebnisse der ersten Messfahrt auf dem Starn-berger See an Bord den "Seeshaupt".
Bayerische Akad. Wiss., Abhandl., Mat.-Naturwiss. Kl., n.s., 75:5-16, figs.

instruments, gravity
Graf, A., and R. Schulze, 1961.
Improvements on the sea gravimeter GSS2.
J. Geophys. Res., 66(6):1813-1821.

instrumentation, gravity meters
Grearson, Lloyd S., 1969.
Discussion on "Quantitative evaluation of a stabilized platform shipboard gravity meter" by T.R. LaFehr and L.L. Nettleton (Geophysics, February 1967, p. 110-118).
Geophysics 34(3):479.

instrumentation, gravity
Grushinskii, N. P. 1960.
Conditions for gravimetric measurements on the Diesel Electric Ship "Ob" during its Antarctic voyages. (In Russian).
Inform. Biull. Sovetsk. Antarkt. Exped., 2(17):

Translation:
Scripta Tecnica, Inc., Elsevier Publ. Co., 200-203.

instrumentation, gravity
Harrison, J. C., 1959.
Tests of the LaCoste-Romberg surface ship gravity meter I.
J. Geophys. Res., 64(11):1875-1882.

instrumentation, gravity meters
*Harrison, J.C., and Lucien J.B. LaCoste, 1968.
Performance of LaCoste-Romberg shipboard gravity meters in 1963 and 1964.
J. geophys. Res., 73(6):2163-2174.

instrumentation, gravity
Harrison, J. C., and F N. Spiess, 1963
Tests of the LaCoste and Romberg surface-ship gravity meter II.
J. Geophys. Res., 68(5):1431-1438.

instrumentation, gravity
Haworth, R.T., 1971.
Cross-coupling errors as a function of the orientation of a Graf-Askania sea gravimeter Gss2. J. geophys. Res., 76(11): 2663-2673.

instrumentation, gravity
Kuzivanov, V.A., I.A. Maslov, I.T. Naumenko-Bondarenko, and O.A. Potapov, 1971.
The use of optical filter when making gravimetric traverses at sea. (In Russian)
Dokl. Akad. Nauk SSSR 196(3): 573-574.

instrumentation, geophysics (gravity)
LaCoste, Lucien, Neal Clarkson and George Hamilton 1967.
LaCoste and Romberg stabilized platform shipboard gravity meter.
Geophysics 32(1): 99-109.

instrumentation, geophysics (gravity)
Lafehr, T.R., and L.L. Nettleton 1967.
Quantitative evaluation of a stabilized platform shipboard gravity meter.
Geophysics 32(1): 110-118

instrumentation
Lozinskaya, A.M., 1959.
The string gravimeter for the measurement of the gravity at sea.
Izv. Akad. Nauk, SSSR, Ser. Geofiz., 3:398-409.
Eng. Edit., 3:272-278.

instrumentation, gravity
Nettleton, L.L., L.J.B. LaCoste and Milton Glicken, 1962
Quantitative evaluation of precision of airborne gravity meter.
J. Geophys. Res., 67(11):4395-4410.

instrumentation, gravity
*Orlin, Hyman, 1967.
Marine gravity surveying instruments and practice.
Int. J. Oceanol. Limnol. 1(3):205-221.

instrumentation, gravity
Popov, E.I., and V.V. Sukhodol'ski, 1964.
Test-stand studies of marine gravimetric apparatus. (In Russian).
Izv., Akad. Nauk, SSSR, Ser. Geofiz., (6):801-818
Translation:
(AGU)(6):487-497.

instrumentation, gravity
*Schultz, Oscar T., and Joseph A. Winokur, 1969.
Shipboard or aircraft gravity vector determination by means of a three-channel inertial navigator.
J. geophys. Res., 74(20):4882-4896.

instrumentation, gravity
Schulze, Reinhard, 1962
Automation of the sea gravimeter Gss2.
J. Geophys. Res., 67(9):3397-3401.

instrumentation, gravity
Schulze, R., 1962
The new automatic servo control for the sea gravity meter Gss2. (Abstract).
J. Geophys. Res., 67(9):3597.

instrumentation, gravity
Stacey, A.P., M.J.R. Fasham, D.I. Black and R.A. Scrutton, 1970.
Design and application of digital filters for the Graf-Askania Gss2, No. 11 sea gravity meter.
Mar. geophys. Res. 1(2): 220-232

instrumentation
Sukhodol'skiy, V.V., 1959
An instrument for recording accelerations and inclinations in determinations of the force of gravity at sea. Izv. Akad. Nauk, SSSR, Ser. Geofiz., 11: 1590-1595
OTS/ $0.50

instrumentation, geophysical
Tomoda, Y., 1958.
Self exciting short period bifilar gravity pendulum designed for the purpose of gravity measurement on board a moving vessel. 1.
Geophys. Notes, Tokyo, 11(2):No. 33.
Reprinted from an unidentified journal, 4(4):107-124.

instrumentation, gravity
Tomoda, Y., and H. Kanamori, 1962.
23. Tokyo surface ship gravity meter a-1.
Geophys. Notes, Tokyo, 15(2):116-145.

instrumentation
Tomoda, Y., T. Maruyama and H. Kanamori, 1960.
Self-exciting short-period bifilar gravity pendulum designed for the purpose of gravity measurements on board a moving vessel. (Pt. 3).
Geophysical Notes, 13(2):39-46. In Japanese
Contribution No. 35.

instrumentation, gravity
Tomoda, Yoshibumi, Keijiro Ozawa and Jiro Segawa, 1968.
Measurement of gravity and magnetic field on board a cruising vessel.
Bull. Ocean Res. Inst. Univ. Tokyo, 3: 169pp.

instrumentation, gravity
Veselov, K. Ye., 1956.
On the static method of gravity measurements at sea with the use of an elastic rotating system.
Prikladnaya Geofizika, SSSR, (15):91-102.
LC and SLA, mi $2.40; ph $3.30

instrumentation, gravity
Wing, Charles G., 1967.
An experimental deep-sea-bottom gravimeter.
J. geophys. Res., 72(4):1249-1257.

instrumentation, gravity
Worzel, J.L., 1962
Comparison of Graf and LaCoste-Romberg sea gravimeters. (Abstract).
J. Geophys. Res., 67(9):3611.

instrumentation, gravity
Worzel, J. Lamar, and J.C. Harrison, 1963.
9. Gravity at sea.
In: The Sea, M.N. Hill, Editor, Interscience Publishers, 3:134-174.

instrumentation, gravity
Worzel, J.L., and M. Talwani, 1962
Investigations of the cross-coupling effect of the Graf sea gravimeter. (Abstract).
J. Geophys. Res., 67(9):3611.

instrumentation, gyropendulum
Von Arx, W.S., 1963
16. Applications of the gyropendulum. In: The Sea, M.N. Hill, Edit., Vol. 2. (III) Currents, Interscience Publishers, New York and London, 325-345.

harbor installations

instrumentation, harbor installation
Anonymous, 1961.
Aluminum pontoon for Milford Haven.
Dock and Harbour Authority, 41(486):414.

heat flow

instrumentation, heat flow (bottom)
Bullard, E.C., 1963.
11. The flow of heat through the floor of the ocean.
In: The Sea, M.N. Hill, Editor, Interscience Publishers, 3:218-232.

instrumentation, heat flow
Corry, Charles, Carl Dubois and Victor Vacquier 1968
Instrument for measuring terrestrial heat flow through the ocean floor.
J. mar. Res. 26(2): 165-177.

instrumentation, heat flow
Korgen, Ben 1971.
Extending the versatility of the Bullard heat probe.
Mar. geophys. Res 1(3): 354-357.

instrumentation, heat flow
Lister, C.R.B., 1963.
Geothermal gradient measurement using a deep-sea corer.
Geophysical Journal, 7(5):571-583.

instrumentation, heat flow (sea surface
McAlister, E.D., and W. McLeish 1970.
A radiometric system for airborne measurement of the total heat flow from the sea.
Appl. Optics 9: 2697-2705.
Also in: Coll. Repr. Scripps Inst. Oceanogr. 40: 1231-1239

instrumentation, heat flow
*Romanovsky, V., et S. Roobaert, 1967.
Mesure du gradient de température dans les sédiments à grande profondeur dans le golfe de Gascogne.
Trav. Cent. Rech. Étud. océanogr., 7(1):13-18.

instrumentation, heat flow
Yasui, Masashi, and Toshio Kishii, 1967.
Lowering of heat flow probes in the horizontal attitude.
Oceanogrl Mag., 19(2):193-196.

heat flux

instrumentation, heat flux
McAlister, E.D. 1968.
Airborne measurements of the total heat flux from the sea surface: progress report.
Mar. Sci. Instrument. 4: 301-305.

housing

instrumentation, housings
Raymond, S.O. 1969.
Glass instrument housings for deep ocean use.
Oceanol. int '69, 1: 7pp. 12 figs.

instrumentation, housing (deep sea)
Sullivan, Jere, Jr., 1965.
Filament wound housing for deep sea research equipment.
Undersea Techn., 6(2):26-27.

hydraulics

instrumentation, hydraul-ics
Great Britain, Department of Scientific and Industrial Research, 1962.
Hydraulics Research, 1961: the report of the Hydraulics Research Board, with the report of the Director of Hydraulics Research (Wallingford) 96 pp.

instrumentation, hydrodist
Mallory, J.K., 1960.
Hydrodist.
Suppl. Intern. Hydro. Rev., 1:57-62.

hydrographic

instrumentation (hydrostat
Azhazha, V.G., 1962.
Explorers of the ocean depths. (In Russian).
Priroda, (6):112-113.
Abstracted in:
Soviet-bloc Res. Geophys. Astr., Space, 1962 (42):8
JPRS 15,068 OTS 61-11147-42.

instrumentation
Dooge, J. C. I., 1959.
A general theory of the unit hydrograph.
J. Geophys. Res. 64(2):241-256.

instruments, hydrographic surveys
Fagerholm, P.O., and A. Thunberg, 1962
Electronic positioning in Swedish coastal hydrographic surveys (A report on operational trials with the Hydrodrist MRB-2 in 1961-62).
Int. Hydrogr. Rev., Suppl. 3:43-58.

instrumentation
Hall, D.M., P.H. Barker and N.D. Anderson, 1958.
An improved hydrographic survey radar.
J. Inst. Navig., 11(1):95-97.

instrumentation (moored mast)
Krauss, Wolfgang, 1960
Hydrographische Messungen mit einem Beobachtungsmast in der Ostsee. Kieler Meeresf., 16 (1): 13-27.

instrumentation
Sauzay, A., 1960
A new plotting method for hydrographic sounding.
Int. Hydrogr. Rev., 37(2):37-42.

instruments
Stupishin, A.V., 1955.
[The contribution of Russian researches to the creation of hydrologic instruments.] Priroda 44(1):76-77.
Translation 3 pp. typed.
RT-3227 in Bibl. Transl. Russ. Sci. Tech. Lit. 26

instrumentation, hydrographi
Vicariot, Jean, 1963.
Mission hydrographique de Terre Adelie (1960-1961).
Ann. Hydrogr., Paris, (4), 11:145-156.

hydrometeorological
see also: Marine meteorology

Instrumentation
Efremychev, V. I., 1960.
New models of oceanographic instruments devised for Hydrometeorological Service.
Biull. Okeanogr. Komissii Pri Prezidiume Akad. Nauk, SSSR, (4):28-33.
H. O. TRANS-124 (1962).

instrumentation
Levchenko, S.P., 1961.
[An informative communication on the scientific and technical conference devoted to the construction of hydrometeorological instruments.]
Okeanologiia, Akad. Nauk, SSSR, 1(5):927-928.

instrumentation
Stas, I.I., 1967.
On some variants of marine hydrometeorological stations on mobile platforms. (In Russian; English abstract).
Okeanologiia, Akad. Nauk, SSSR, 7(1):168-176.

instrumentation, hydroscopes
Thorndike, E.M., 1955.
Hydroscopes. Trans. Amer. Geophys. Union, 36(3): 470-472.

instrumentation, hydrostatic pressure
Appell, Gerald F., and Jacob Frankel 1971
Test structure for tensile loading in hydrostatic pressure environment.
J. mar. techn. Soc. 5(6):17-19